LIST OF THE ELEMENTS WITH THEIR SYMBOLS AND ATOMIC WEIGHTS

	Symbol	Atomic number	Atomic weight		Symbol	Atomic number	Atomic weight		Symbol	Atomic number	Atomic weight
Actinium	Ac	89	227.0278	Hahnium[b]	Ha	105	(262)	Protactinium	Pa	91	231.0359
Aluminum	Al	13	26.98154	Helium	He	2	4.00260	Radium	Ra	88	226.0254
Americium	Am	95	(243)[a]	Holmium	Ho	67	164.9304	Radon	Rn	86	(222)
Antimony	Sb	51	121.75	Hydrogen	H	1	1.0079	Rhenium	Re	75	186.207
Argon	Ar	18	39.948	Indium	In	49	114.82	Rhodium	Rh	45	102.9055
Arsenic	As	33	74.9216	Iodine	I	53	126.9045	Rubidium	Rb	37	85.4678
Astatine	At	85	(210)	Iridium	Ir	77	192.22	Ruthenium	Ru	44	101.07
Barium	Ba	56	137.33	Iron	Fe	26	55.847	Rutherfordium[b]	Rf	104	(261)
Berkelium	Bk	97	(247)	Krypton	Kr	36	83.80	Samarium	Sm	62	150.36
Beryllium	Be	4	9.01218	Lanthanum	La	57	138.9055	Scandium	Sc	21	44.9559
Bismuth	Bi	83	208.9804	Lawrencium	Lr	103	(260)	Selenium	Se	34	78.96
Boron	B	5	10.81	Lead	Pb	82	207.2	Silicon	Si	14	28.0855
Bromine	Br	35	79.904	Lithium	Li	3	6.941	Silver	Ag	47	107.8682
Cadmium	Cd	48	112.41	Lutetium	Lu	71	174.967	Sodium	Na	11	22.98977
Calcium	Ca	20	40.078	Magnesium	Mg	12	24.305	Strontium	Sr	38	87.62
Californium	Cf	98	(251)	Manganese	Mn	25	54.9380	Sulfur	S	16	32.06
Carbon	C	6	12.011	Mendelevium	Md	101	(258)	Tantalum	Ta	73	180.9479
Cerium	Ce	58	140.12	Mercury	Hg	80	200.59	Technetium	Tc	43	(98)
Cesium	Cs	55	132.9054	Molybdenum	Mo	42	95.94	Tellurium	Te	52	127.60
Chlorine	Cl	17	35.453	Neodymium	Nd	60	144.24	Terbium	Tb	65	158.9254
Chromium	Cr	24	51.996	Neon	Ne	10	20.179	Thallium	Tl	81	204.383
Cobalt	Co	27	58.9332	Neptunium	Np	93	237.048	Thorium	Th	90	232.0381
Copper	Cu	29	63.546	Nickel	Ni	28	58.69	Thulium	Tm	69	168.9342
Curium	Cm	96	(247)	Niobium	Nb	41	92.9064	Tin	Sn	50	118.69
Dysprosium	Dy	66	162.50	Nitrogen	N	7	14.0067	Titanium	Ti	22	47.88
Einsteinium	Es	99	(254)	Nobelium	No	102	(259)	Tungsten	W	74	183.85
Erbium	Er	68	167.26	Osmium	Os	76	190.2	Uranium	U	92	238.0289
Europium	Eu	63	151.96	Oxygen	O	8	15.9994	Vanadium	V	23	50.9415
Fermium	Fm	100	(257)	Palladium	Pd	46	106.4	Xenon	Xe	54	131.29
Fluorine	F	9	18.998403	Phosphorus	P	15	30.97376	Ytterbium	Yb	70	173.04
Francium	Fr	87	(223)	Platinum	Pt	78	195.08	Yttrium	Y	39	88.9059
Gadolinium	Gd	64	157.25	Plutonium	Pu	94	(244)	Zinc	Zn	30	65.38
Gallium	Ga	31	69.72	Polonium	Po	84	(209)	Zirconium	Zr	40	91.22
Germanium	Ge	32	72.59	Potassium	K	19	39.0983	[106][b]		106	(263)
Gold	Au	79	196.9665	Praseodymium	Pr	59	140.9077	[107]		107	(262)
Hafnium	Hf	72	178.49	Promethium	Pm	61	(145)	[108]		108	(265)
								[109]		109	(266)

[a]Approximate values for radioactive elements are listed in parentheses.
[b]The official name and symbol have not been agreed to.

CHEMISTRY
The Central Science

Fifth Edition

CHEMISTRY
The Central Science

Theodore L. Brown
University of Illinois

H. Eugene LeMay, Jr.
University of Nevada

Bruce E. Bursten
The Ohio State University

Prentice Hall, Englewood Cliffs, NJ 07632

Library of Congress Cataloging-in-Publication Data

Brown, Theodore L.
Chemistry: the central science/Theodore L. Brown, H. Eugene LeMay, Jr., Bruce E. Bursten.—5th ed.
p. cm.
Includes index.
ISBN 0-13-126202-5
1. Chemistry. I. LeMay, H. Eugene (Harold Eugene), 1940– II. Bursten, Bruce E. 1954– III. Title.
QD31.2.B78 1991
540—dc20 90-41267
CIP

Acquisitions editor: Dan Joraanstad
Development editor: Robert J. Weiss
Supplements editor: Alison Munoz
Design director: Florence Dara Silverman
Cover and interior design: Judith A. Matz-Coniglio
Page layout: Diane Koromhas
Prepress buyer: Paula Massenaro
Manufacturing buyer: Lori Bulwin
Photo editor: Lorinda Morris-Nantz
Photo researcher: Yvonne Gerin
Cover photo: © Aaron Jones

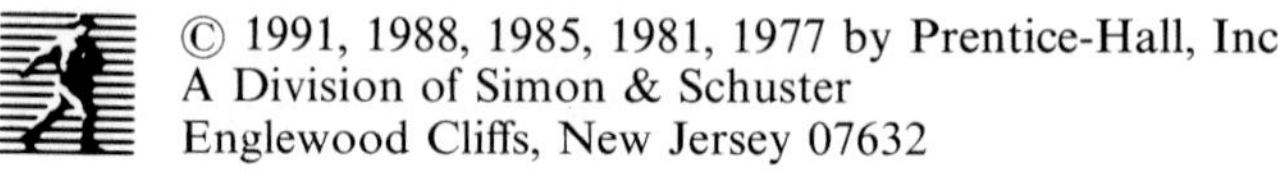

Printed in the United States of America
10 9 8 7 6 5

ISBN 0-13-126202-5

Prentice-Hall International (UK) Limited, *London*
Prentice-Hall of Australia Pty. Limited, *Sydney*
Prentice-Hall Canada Inc., *Toronto*
Prentice-Hall Hispanoamericana, S.A., *Mexico*
Prentice-Hall of India Private Limited, *New Delhi*
Prentice-Hall of Japan, Inc., *Tokyo*
Simon & Schuster Asia Pte. Ltd., *Singapore*
Editora Prentice-Hall do Brasil, Ltda., *Rio de Janeiro*

To our students, whose enthusiasm and
curiosity have often inspired us,
and whose questions and suggestions have
sometimes taught us.

Brief Contents

Gene Aurens, Shostal Associates/Superstock

Contents

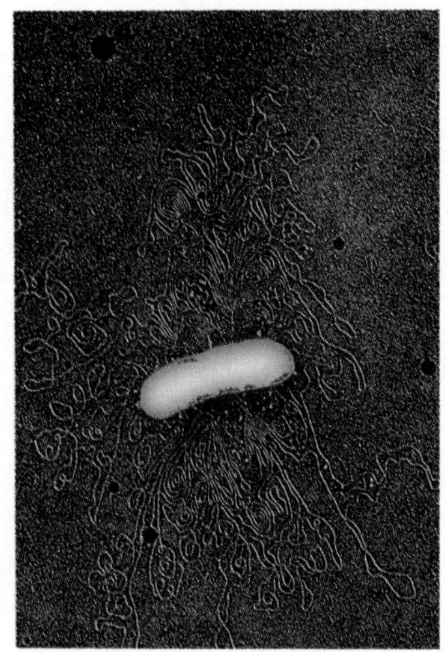

© Dr. Gopal Murti, Science Photo Library, Science Source/Photo Researchers

Dr. E. R. Degginger

6 Electronic Structures of Atoms 167

Four By Five/Superstock

7 Periodic Properties of the Elements 211

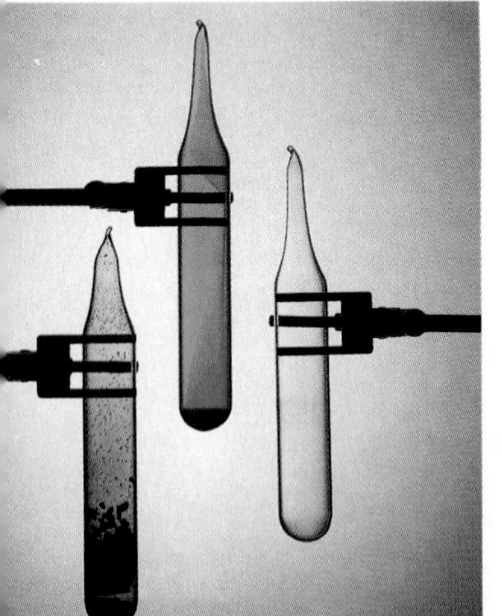

Fundamental Photographs

8 Basic Concepts of Chemical Bonding 246

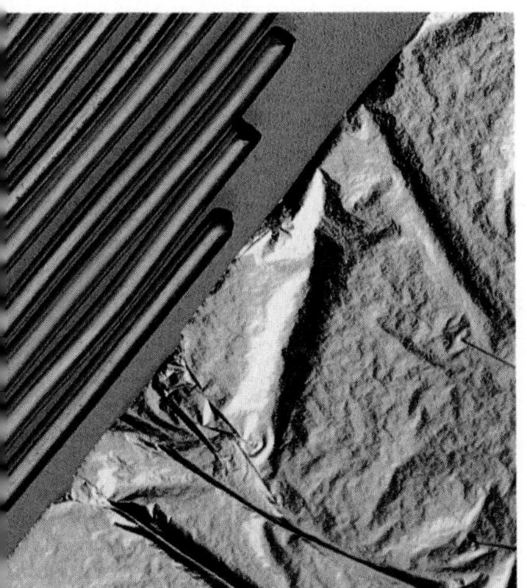

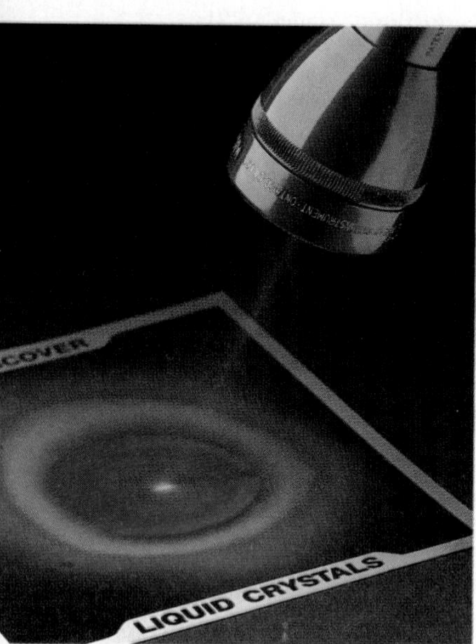

© Richard Megna/Fundamental Photographs

Dr. Ahmed H. Zewail, California Institute of Technology

© Richard Megna/Fundamental Photographs

© Paul Silverman/Fundamental Photographs

Lawrence Berkeley Laboratory, University of California

U.S. Dept. of the Navy, Naval Photographic Center

Gordon Garradd, Science Photo Library, Science Source/Photo Researchers

© Richard Megna/Fundamental Photographs

John Walsh, Science Photo Library/Science Source/Photo Researchers

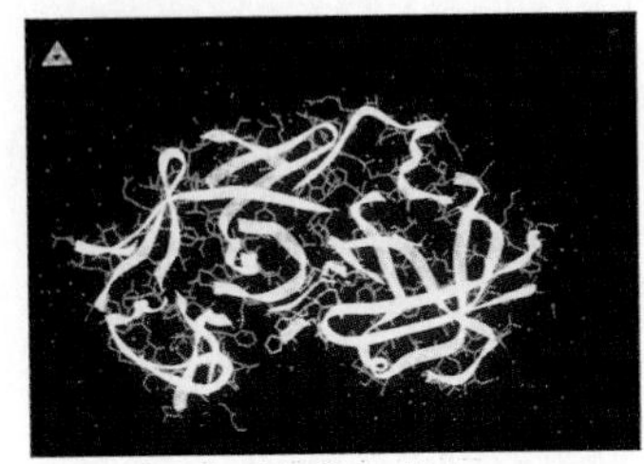

Dr. Christine Humblet/Parke Davis

Tim Fitzharris, Animals for Advertising Collection/ALLSTOCK

Chemical Applications and Essays

CHEMISTRY AT WORK

A CLOSER LOOK

A HISTORICAL PERSPECTIVE

Preface

TO THE STUDENT

Chemistry: The Central Science, Fifth Edition, has been written to introduce you to modern chemistry. During the many years we have been practicing chemists, we have found chemistry to be an exciting intellectual challenge and an extraordinarily rich and varied part of our cultural heritage. We hope that as you advance in your study of chemistry, you will share with us some of that enthusiasm and appreciation. We also hope that you will come to realize the importance of chemistry in your everyday life. As authors, we have, in effect, been engaged by your instructor to help you learn chemistry. Based on the comments of students and instructors who have used this book in its previous editions, we believe that we have done that job well. Of course, we expect the text to continue to evolve through future editions. We invite you to write to us to tell us what you like about the book, so that we will know where we have helped you most. Also, we would like to learn of any shortcomings, so that we might further improve the book in subsequent editions. Our addresses are given at the end of the Preface.

Using this Text

As you page through any chapter of this text, you will see sections entitled *Sample Exercise.* Each exercise illustrates the use of a key concept or skill. The accompanying solution provides the reasoning required to answer the exercise. Paired with every Sample Exercise is a *Practice Exercise*, which addresses the same concept. You can test your understanding by working the Practice Exercise and comparing your answer with the one that is given.

Full-color illustrations, which include photographs, charts, graphs, and diagrams, help you to visualize abstract ideas. Don't neglect to study the figures and their captions as you read the text.

Three kinds of brief *supplemental sections* appear throughout the text as enrichment material. The *Chemistry at Work* sections introduce interesting applications of the concepts under consideration in the text. Those supplements entitled *A Closer Look* enrich the discussion by delving more deeply into a topic. Those entitled *A Historical Perspective* expand your knowledge of what led to a particularly important development or discovery in chemistry.

At the end of each chapter you will find a *summary* that points out the chapter highlights. The list of *key terms* gives the vocabulary that has been introduced in the chapter. The terms listed here are printed in boldface type in the chapter and also appear in the end-of-book *glossary*.

The *Exercises* at the end of each chapter test your understanding of the material. Many exercises are grouped according to topic, and these are also arranged in matched pairs. Each exercise in a pair deals with the same principle or procedure, so if you have difficulty with a particular exercise, its companion will provide you with further practice. The answer to one member of each pair is given in a section following the appendices. *Additional exercises* appear at the end of each chapter's exercise set. The additional exercises test your ability to solve problems that are not identified by topic. These exercises often combine ideas from more than one part of the chapter. The additional exercises and the matched-pair exercises whose answers appear near the end of the book are numbered in color. The more challenging exercises are marked with brackets.

Finally, you should note that there are several *appendices* near the back of the book, as well as useful tables in the front and back inside covers.

Advice for Studying Chemistry

Keep up with your studying day to day. New material builds on the old. It is important not to fall behind; if you do, you will find it much harder to follow the lectures and discussions on current topics. Trying to "cram" just before an exam is generally a very ineffective way to study chemistry.

Focus your study. The amount of information you will receive in your chemistry course can sometimes seem overwhelming. Certainly, there are more facts and details than any student can hope to assimilate in a first course. It is important to focus on the key concepts and skills. Listen intently for what your lecturer and discussion leader emphasize. Pay attention to the skills stressed in the sample exercises and homework assignments. Notice the italicized statements in the text, and study the concepts presented in the chapter summaries.

Keep good lecture notes, so that you have a clear and concise record of the required material. You will find it easier to take useful notes if you *skim topics in the text before they are covered in lecture.* To skim a chapter, first read the introduction and summary. Then quickly read through the chapter, skipping Sample Exercises and supplemental sections. Pay attention to section heads and subheads, which give you a feeling for the scope of topics. Avoid the feeling that you must learn and understand everything right away.

After lecture, *carefully read the topics covered in class.* Remember, though, that you'll need to read assigned material more than once to master it. As you read, pay particular attention to the Sample Exercises. Once you think you understand a Sample Exercise, test your understanding by working the accompanying Practice Exercise.

Finally, *attempt all of the assigned end-of-chapter exercises.* Working out these exercises provides necessary practice in recalling and using the essential ideas of the chapter. You cannot learn merely by observing; you must be a participant. In particular, there is little value in merely copying answers from the Solutions Manual or from another student. If however, you really get stuck on a problem, get help from your instructor, teaching assistant, tutor, or a fellow student. Spending more than 20 minutes on a single exercise is rarely effective unless you know that it is particularly challenging and requires extensive thought and effort.

Philosophy

Throughout the evolution of this text, certain goals have guided our writing efforts. The first is that a text should endeavor to show students the usefulness of chemistry in their major areas of study as well as in the world around them. It has been our experience that as students become aware of the importance of chemistry to their own goals and interests, they become more enthusiastic about learning the subject. With this in mind, we have attempted, as much as space and our imaginations permit, to bring in interesting and significant applications of the subject matter. We attempt to show that chemistry is indeed the *central science.* At the same time, of course, we seek to provide students with the necessary background in modern chemistry for their specialized studies, including more advanced chemistry courses.

Second, we want to show not only that chemistry provides the basis for much of what goes on in our world, but that it is a vital, continually developing science. We have tried to keep the book up-to-date in terms of new concepts and applications and to convey some of the excitement of the field.

Third, we feel that any text should be written to the students and not just to their instructors. We have sought to keep our writing clear and interesting and the book attractive and well illustrated. Furthermore, we have provided numerous in-text study aids for students. A more subtle aspect of this student orientation is the care we have taken to describe problem-solving strategies.

Organization

In the present edition, the first five chapters give a largely macroscopic, phenomenological view of chemistry. They introduce basic concepts (including nomenclature and stoichiometry) that provide the necessary background for many of the laboratory experiments usually performed in general chemistry. Chapter 4, in particular, is an earlier treatment of chemical reactions in aqueous solutions than that found in previous editions.

The next four chapters (Chapters 6–9) deal with electronic structure and bonding. The focus then changes to the next level of the organization of matter: the states of matter (Chapters 10 and 11) and solutions (Chapter 13). Also included in this section of the text is a chapter on the chemistry of modern materials (Chapter 12), which builds on the student's understanding of chemical bonding and intermolecular interactions and their relationships to the properties of matter.

The next several chapters examine the factors that determine the speed and extent of chemical reactions: kinetics (Chapter 14), equilibria (Chapters 15–17), thermodynamics (Chapter 19), and electrochemistry (Chapter 20). Also in this section is an optional chapter on environmental chemistry (Chapter 18), in which the concepts developed in preceding chapters are applied to a discussion of the atmosphere and hydrosphere.

After a discussion of nuclear chemistry (Chapter 21), the final chapters survey the chemistry of nonmetals, metals, organic chemistry, and biochemistry (Chapters 22–27). These chapters are developed in a parallel fashion and can be treated in any order.

Although our chapter sequence provides a fairly standard organization, we recognize that not everyone teaches all of the topics in exactly our order.

We have therefore structured our writing so that instructors can make common changes in teaching sequence with no loss in student comprehension. In particular, many instructors prefer to introduce gases (Chapter 10) after stoichiometry or after thermochemistry rather than with states of matter. The chapter on gases has been written to permit this change with *no* disruption in the flow of material. It is also possible to treat the balancing of redox equations (Sections 20.1 and 20.2) earlier, after the introduction of oxidation numbers in Section 8.10, or even with the introduction to redox reactions in Section 4.6.

We have introduced students to the properties of elements and their compounds (*descriptive chemistry*) in several ways throughout the text. Most obvious are those chapters that directly address these topics (especially Chapters 4, 7, 12, 18, and 22–27). In addition, material on the properties of elements and compounds is woven throughout the text to illustrate principles and applications. Furthermore, such descriptive chemistry is also used in end-of-chapter exercises.

Changes in this Edition

Our major goal in the fifth edition has been to strengthen an already strong textbook while retaining its effective and popular style. The traditional strengths of *Chemistry: The Central Science* include its clarity of writing, its scientific accuracy and currency, its strong end-of-chapter exercises, and its consistency in level of coverage.

The most obvious change is the addition of a *new coauthor* who brings a fresh outlook and new insights to the text. Probably the least obvious changes are the countless small ways that the sentences, paragraphs, and even whole sections have been rewritten to improve clarity.

The text has a *new design.* Sample exercises are now set in a single-column format that reduces the need to break equations and calculations. Key terms, which appear in boldface type in the text and which are listed at the end of each chapter, are now defined in a glossary near the end of the book. Learning goals, which once appeared at the end of each chapter, are now given only in the Student's Guide and in the Annotated Instructor's Edition of the text.

The text also contains an *increase in the number of color photos* that illustrate chemical reactions and that help to convey the beauty and excitement of chemistry. Illustrations have also been added to summarize common problem-solving strategies. Our goal has been to use color in a nondistracting, functional way to help emphasize important points, to focus the student's attention, and to make the text attractive and inviting.

Two new chapters have been added: Chapter 4 (Aqueous Reactions and Solution Stoichiometry) and Chapter 12 (Modern Materials). Chapter 4, which introduces molarity, electrolytes, reactions in aqueous solution, and solution stoichiometry, provides valuable background for laboratory programs, which usually involve a great deal of aqueous chemistry. For those who prefer to postpone treatment of these topics, the chapter can be treated near the middle of the course, when solutions are discussed in Chapter 13.

Chapter 12 is an optional chapter that applies structure and bonding concepts from preceding chapters to a discussion of the structures, properties, and uses of polymers, ceramics, liquid crystals, and films. These materials play important roles in modern society and illustrate the vibrancy and relevance of chemistry.

Many chapters have been extensively revised: For example, the chapters on electronic structure and periodic properties have been rewritten and consolidated. Chapter 6 (Electronic Structures of Atoms) now not only introduces the quantum theory of the atom but also treats many-electron atoms, ending with their electron configurations. Chapter 7 (Periodic Properties of the Elements) now includes not only discussions of atomic radii, ionization energies, and electron affinities but also examines periodic trends in the descriptive chemistry of the active metals and selected nonmetals. Chapter 11 (Intermolecular Forces, Liquids, and Solids) has been restructured so that intermolecular forces are treated first, providing a better bridge to the bonding concepts discussed in earlier chapters. The material on the structures of solids has also been rewritten.

Finally, several features of the text have been enhanced: The new edition has substantially more supplemental Chemistry at Work essays than the previous edition. Greater attention has been paid to problem-solving strategies. The number of multistep problems and the number of more difficult and challenging problems in the end-of-chapter exercises have been increased. About half of the end-of-chapter exercises are new to this edition.

ACKNOWLEDGEMENTS

This book owes its final shape and form to the assistance and hard work of many people. Several colleagues reviewed the manuscript and helped us immensely by sharing their insights and criticizing our initial writing efforts. We would like especially to thank the following:

Wayne Anderson
Bloomsburg State College

Richard Beitzel
Bemidji State University

James Birk
Arizona State University

Alfred Boyd
University of Maryland, College Park

Susan Brennan
University of North Florida

James Davis
Harvard University

Marcia Davies
Creighton University

Walter Deal
University of California, Riverside

Arthur Ellis
University of Wisconsin, Madison

Paul Fleitz
Syracuse University

Charles Fosha
University of Colorado

Wade Freeman
University of Illinois

L. Peter Gold
Pennsylvania State University

Thomas Greenbowe
Iowa State University

Albert Haim
State University of New York—Stony Brook

David Harris
University of California—Santa Barbara

C. Alton Hassell
Baylor University

Stephen Hawkes
Oregon State University

Jack Healey
Las Positas College

Charles Henrickson
Western Kentucky University

Roger Hoburg
University of Nebraska at Omaha

Colin Hubbard
University of New Hampshire

Harold Hunt
Georgia Institute of Technology

Donald Jicha
University of North Carolina, Chapel Hill

Walter Klemperer
University of Illinois—Urbana

Stephen Kozub
University of Pennsylvania

Larry Krannich
University of Alabama–Birmingham

Robert M. Kren
University of Michigan, Flint

Russell D. Larsen
Texas Tech University

Kenneth Magnell
Central Michigan University

Joel Mague
Tulane University

Edward Mellon
Florida State University

Harvey Moody
United States Air Force Academy

Leon Morgan
University of Texas

Lee Pederson
University of North Carolina, Chapel Hill

William Perry
Auburn University

Robert Pfaff
University of Nebraska—Omaha

Helen Place
Washington State University

Guy Rosenthal
University of Vermont

Theodore Sakano
Rockland Community College

William Schulz
Eastern Kentucky University

Curtis T. Sears, Jr.
Georgia State University

Mary Sohn
Florida Institute of Technology

Robert Sprague
Lehigh University

George Stanley
Louisiana State University

Mable Ruth Stephanic
Oklahoma State University

Jack Stocker
University of New Orleans

Wayne Tikkanen
California State University, Los Angeles

Donald Titus
Temple University

Charles Trapp
University of Louisville

Richard Treptow
Chicago State University

Kenneth Watters
University of Wisconsin, Milwaukee

Gregory Williams
California State University—Fullerton

Byron Wilson
Brigham Young University

James Zubrick
Hudson Valley Community College

We would also like to express our sincere appreciation to the people at Prentice Hall who have worked with us to make this edition possible: Dan Joraanstad, our chemistry editor, who pulled the project together with energy and enthusiasm and whose phone calls were often needed to prod us to meet deadlines that always seemed to come too soon; Bob Weiss, our fine developmental editor, whose attention to detail and whose ideas for improving our exposition were invaluable to this revision; John Morgan, our production editor, who so efficiently managed the incredibly complex task of bringing the design, photos, artwork, and writing together; and Diane Goodman, our marketing manager, for her excellent work in publicizing our efforts.

Many others were intimately involved in the complex task of putting together this textbook and deserve special recognition: Richard Megna and Kip Peticolas of Fundamental Photographs, whose photographs brought our crude drawings to life with considerable artistic flair; Roxy Wilson (University of Illinois) for performing the difficult job of working end-of-chapter exercises and checking our calculations; Roy Garvey (North Dakota State University), David Todd (Worcester Polytechnic Institute), and Nicholas Kildahl (Worcester Polytechnic Institute) for their helpful proofreading; and especially Robert Pfaff (University of Nebraska, Omaha) for his fine work in preparing the Annotated Instructor's Edition of the text.

Other coworkers and friends deserve a special thanks: Kenneth Kemp (University of Nevada, Reno) and James Hill (California State University, Sacramento) for their frequent advice; Steve McKee and Kevin Novo-Gradac (The Ohio State University) for providing helpful suggestions for end-of-chapter exercises; Harold Shechter (The Ohio State University) for adding helpful insights on several of the Chemistry at Work essays.

HEL would also like to thank the University of Nevada, Reno, for a sabbatical leave while this edition was in progress and the members of the Chemistry Department at U.C.L.A. for their hospitality and helpful conversations.

Theodore L. Brown
School of Chemical Sciences
University of Illinois
Urbana, IL 61801

H. Eugene LeMay, Jr.
Department of Chemistry
University of Nevada
Reno, NV 89557-0020

Bruce E. Bursten
Department of Chemistry
The Ohio State University
Columbus, OH 43210-1173

About the Authors

Theodore L. Brown

Theodore L. Brown received his Ph.D. from Michigan State University in 1956. Since that time he has been a member of the faculty of the University of Illinois, Urbana, where he is Professor of Chemistry. He has served as Vice-Chancellor for Research and as Dean, The Graduate College. In March 1987 he became Director of the Beckman Institute.

Professor Brown has been an Alfred P. Sloan Research Fellow and has received the American Chemical Society Award for Research in Inorganic Chemistry. He has also been awarded a Guggenheim Fellowship and has held several offices with the American Chemical Society. He is a member of the Board of Governors of Argonne National Laboratory and Chairman of the Scientific and Technical Advisory Committee of the Laboratory.

H. Eugene LeMay, Jr.

H. Eugene LeMay, Jr., received his Ph.D. in 1966 from the University of Illinois. He then joined the faculty of the Chemistry Department at the University of Nevada, Reno, where he has served as Associate Chairman and Acting Chairman. He is presently Professor of Chemistry and Freshman-Chemistry Coordinator. He has enjoyed Visiting Professorships at the University of North Carolina at Chapel Hill, at the University College of Wales in Great Britain, and at U.C.L.A.

Professor LeMay has received several university awards for outstanding teaching at the undergraduate and graduate levels. He is the author of many research publications and review articles, mainly in the areas of solid-phase reactions and coordination chemistry.

Bruce E. Bursten

Bruce E. Bursten received his Ph.D. from the University of Wisconsin in 1978. Following two years as a National Science Foundation Postdoctoral Fellow at Texas A&M University, he joined the faculty of The Ohio State University, where he is currently Professor of Chemistry.

Professor Bursten has been a Camille and Henry Dreyfus Foundation Teacher-Scholar and an Alfred P. Sloan Foundation Research Fellow. At Ohio State, he has received the university distinguished teaching award, the Arts and Sciences Student Council Outstanding Teaching Award, and the University Distinguished Scholar Award. He has been a Visiting Associate at the California Institute of Technology, and is a Laboratory Collaborator at Los Alamos National Laboratory. He is presently a member of the Board of Editors of the American Chemical Society journal, *Inorganic Chemistry*.

The New York Times and Prentice Hall are sponsoring *A Contemporary View*, a program designed to enhance student access to current information of relevance in the classroom.

Through this program, the core subject matter provided in the text is supplemented by a collection of time-sensitive articles from one of the world's most distinguished newspapers, *The New York Times*. These articles demonstrate the vital, ongoing connection between what is learned in the classroom and what is happening in the world around us.

To enjoy the wealth of information of *The New York Times* daily, a reduced subscription rate is available. For information, call toll-free: 1-800-631-1222.

Prentice Hall and *The New York Times* are proud to cosponsor *A Contemporary View*. We hope it will make the reading of both textbooks and newspapers a more dynamic, involving process.

In the fields of observation chance favors only those minds that are prepared.

Louis Pasteur (1822–1895)
Innaugural Lecture as Professor and Dean of the Faculty of Science
University of Lille, Douay, France
December 7, 1854

Genius is one per cent inspiration and ninety-nine per cent perspiration.

Thomas Alva Edison (1847–1931)
Life, 1931, Ch. 24

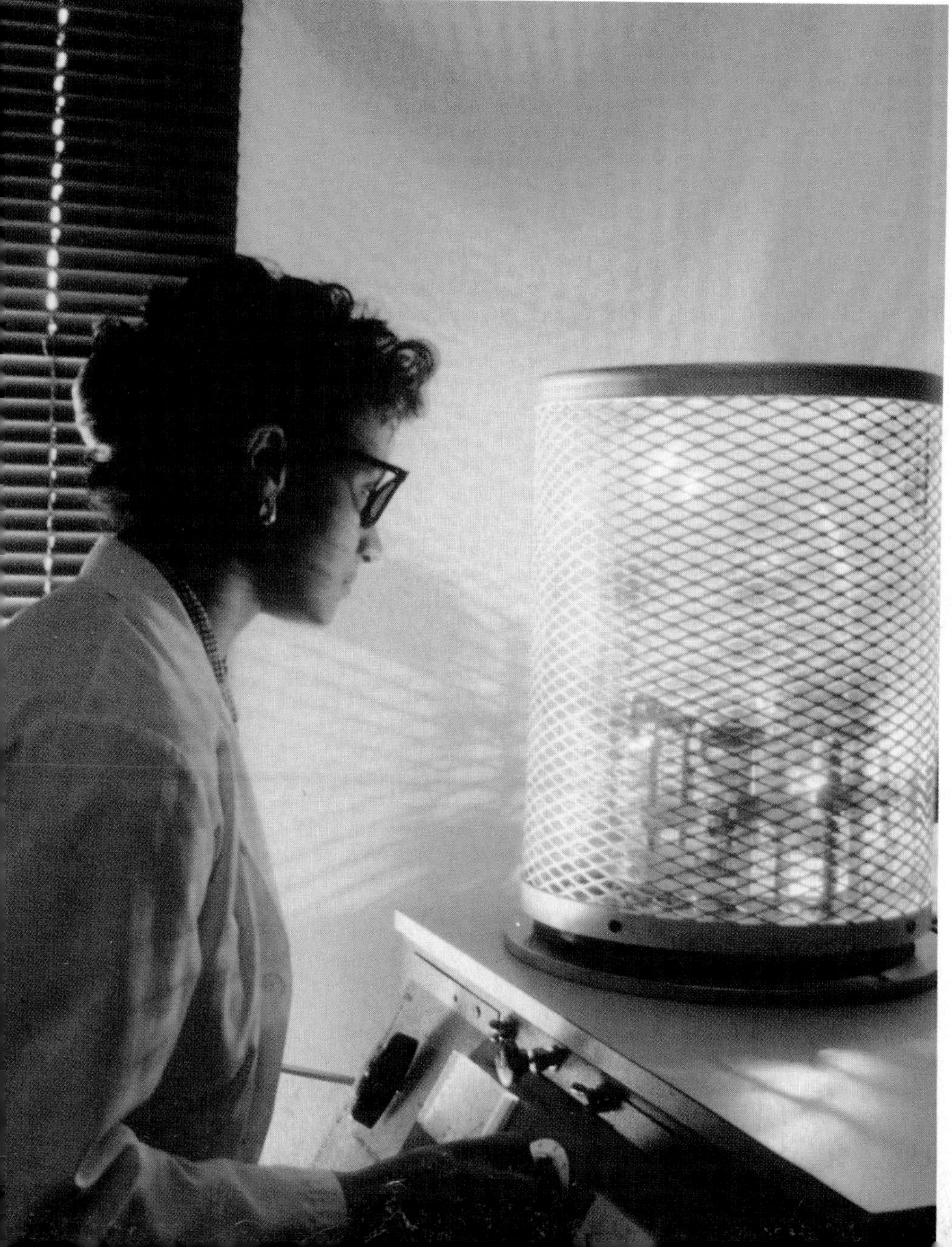

Introduction: Some Basic Concepts 1

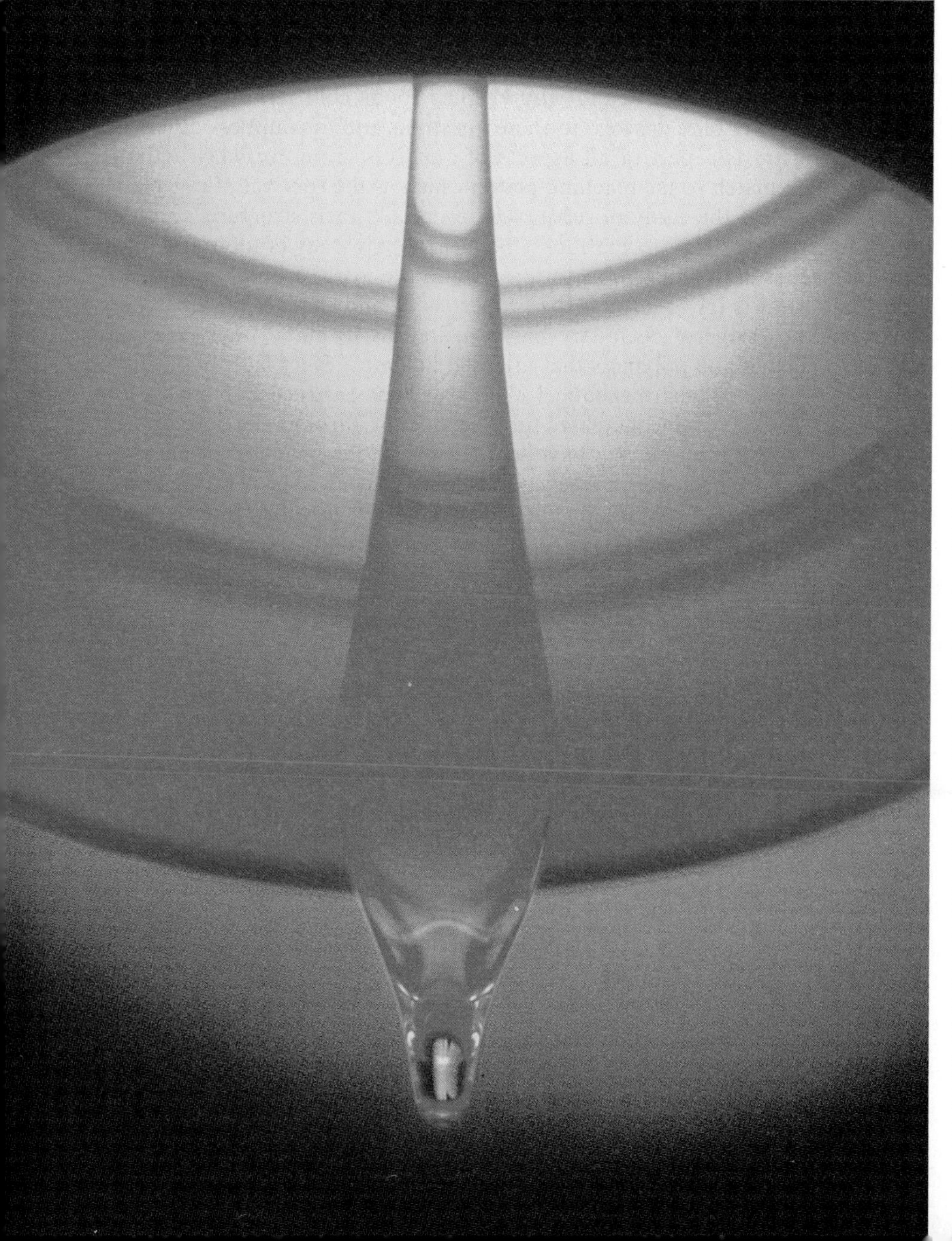

Chemistry has many applications, both in the laboratory and in the world around us. (© by SIU/ Biomedical Communications, Science Source/Photo Researchers)

CONTENTS

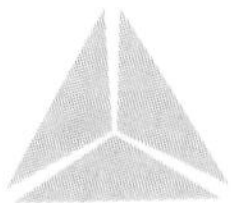

The greatest inventions are those inquiries which tend to increase the power of man over matter.

—Benjamin Franklin (1706–1790)

Have you ever wondered why ice melts and water evaporates? Why leaves turn colors in the fall and why iron rusts when it is left out in the rain? How acid rain is formed and why it is so dangerous to our environment? How a battery generates electricity, and why keeping foods cold retards their spoilage? Chemistry supplies answers to these questions and to countless others like them. Chemistry is a part of all aspects of our lives, from everyday activities like lighting a match to far-reaching matters such as the survival of our planet.

Chemistry is the study of substances, especially their structure, composition, properties, and transformations. Chemists use a variety of tools and scientific approaches in their attempts to understand the world around us. But they do more. Chemists try to *harness* the properties of, and changes in, substances.

Our knowledge of chemistry has benefited us in many ways. We have developed plastics and polymers that are used in almost every facet of our lives. We have discovered pharmaceutical chemicals that enhance our health and prolong our lives. We have increased food production through the development of fertilizers and pesticides. We have learned to use the products of naturally occurring chemical processes, namely petroleum and coal, to provide the energy and raw materials that have made our modern lifestyle possible. It is clear that we are more than passive observers of the chemistry around us; in nearly everything we do, we are active participants in the development and application of chemical processes.

The benefits that chemistry has given us have not come without a price. Increasingly, we find that our reliance on chemistry has led to environmental damage, unexpected health problems, and human tragedies such as the 1984 leak of toxic chemicals at Bhopal, India, which killed or injured over 40,000 people. We produce vast amounts of chemicals (Table 1.1) that touch our lives in a variety of ways. It is in our best interests, then, to understand the profound effects—both positive and negative—that chemistry has on our lives.

Most of you are studying chemistry because it is an essential part of your curriculum. Your major might be chemistry, education, agriculture, engineering, geology, biology, or one of many other related fields. Why do so many diverse subjects share an essential tie to chemistry? The answer is that chemistry, by its very nature, is the *central science.* Our interactions with the material world raise basic questions about the materials around us: What are their compositions and properties? How do they interact with us and their environment? How, why, and when do they undergo change? These questions are important whether the material is a modern polymer involved in the production of high-

Table 1.1 The Top Ten Chemicals Manufactured by the Chemical Industry in 1988[a]

Rank	Chemical	Formula	1988 Production (billions of pounds)
1	Sulfuric acid	H_2SO_4	85.56
2	Nitrogen	N_2	52.10
3	Oxygen	O_2	37.09
4	Ethylene	C_2H_4	36.56
5	Ammonia	NH_3	33.89
6	Lime	CaO	32.34
7	Sodium hydroxide	NaOH	23.97
8	Phosphoric acid	H_3PO_4	23.43
9	Chlorine	Cl_2	22.66
10	Propylene	C_3H_6	19.97

[a] *Source: Chemical and Engineering News*, June 19, 1989, p. 39.

tech computer chips, an aged pigment used by a Renaissance painter, or unknown extraterrestrial material collected during a space mission (Figure 1.1). Chemistry provides us with answers to these questions. For this reason, chemistry will probably play a significant role in your future. You will be better prepared for the future if you increase your understanding of chemical principles.

There is another very good reason to study chemistry: It can be a lot of fun! In the course of your studies, you will find that many of the mysterious

Figure 1.1 (*a*) A microscopic view of a computer logic chip. (*b*) A Renaissance painting, "L'alchemista." (*c*) A "moon rock" collected from the lunar surface. (John Walsh/Photo Researchers; Scala/Art Resource; NASA)

(*a*)

(*b*)

(*c*)

Figure 1.2 Many common supermarket products have very simple chemical compositions. (Donald Clegg and Roxy Wilson)

transformations of matter you may have observed make much more sense within a scientific framework. In the laboratory portion of your course, you will have the opportunity to carry out controlled chemical reactions and to see many beautiful color changes. You will become familiar with the chemical compositions of many common household products (Figure 1.2) and learn how and why they do certain things. Finally, you will have the opportunity to make sense out of a variety of seemingly unconnected observations. To do so is a goal shared by all the sciences, and it is one of the greatest thrills that scientists can experience.

This text introduces you to chemical facts and theories, not as ends in themselves, but as a means to help you understand the world around you. We will begin by discussing some basic ideas about matter and its properties.

1.1 INTRODUCTION TO MATTER

Our present understanding of the changes we see around us—such as the melting of ice and the burning of wood—is intimately tied to our understanding of the nature and composition of matter. **Matter** is the physical material of the universe; it is anything that occupies space and has mass.

Matter can exist in three physical *states*: gas (also known as vapor), liquid, and solid (Figure 1.3). These states differ in some of their simple observable

Figure 1.3 The three physical states of water are common and familiar to us: water vapor, liquid water, and ice. We cannot see water vapor. What we see when we look at steam is tiny droplets of liquid water dispersed through the water vapor. In this photo we see both the liquid and solid states of water. (Robert W. Hernandez)

A CLOSER LOOK: Impurities

No substance in nature is ever totally pure, even when we take great care to separate it from other substances. One substance may be the major component of a given sample, but other substances, called *impurities*, are always present to some extent. A substance may be described as 99.9 percent pure if it contains only 0.1 percent impurities. For many purposes we may describe such a sample as "pure." Sometimes, though, much greater purity may be required. For example, the silicon used to make computer chips has to be 99.99999 percent pure. At times we may need to know not only the percent purity of a substance but also the nature of the impurities. Labels for chemical substances used in laboratories often list the major impurities and their amounts (Figure 1.4).

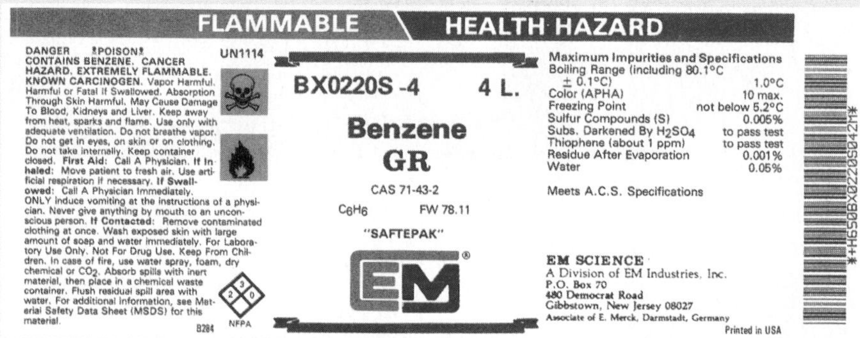

Figure 1.4 Label for bottle of benzene, C_6H_6, indicating physical properties, health hazards, and impurities. The benzene is more than 99.9% pure. The impurities present are listed on the label. (EM Science, A Div. of EM Industries, Inc.)

properties. A **gas** has no fixed volume or shape; rather, it conforms to the volume and shape of its container. A gas can be compressed to occupy a smaller volume, and it will expand to occupy a larger one. A **liquid** has a distinct volume independent of its container but has no specific shape. It assumes the shape of the portion of the container that it occupies. A **solid** has both a definite shape and a definite volume; it is rigid. Neither liquids nor solids can be compressed to any appreciable extent.

Substances

Most forms of matter that we encounter—for example, the air we breathe (a gas), gasoline for cars (a liquid), and a sidewalk on which we walk (a solid)—are not chemically pure. We can, however, resolve, or separate, these kinds of matter into different pure substances. A **pure substance** (from here on referred to simply as a *substance*) is matter that has a fixed composition and distinct properties. For example, seawater can be separated into several different substances, the most abundant of which are water and ordinary table salt (sodium chloride).

Chemical and Physical Properties

Every substance has a unique set of *properties*—characteristics that allow us to recognize it and to distinguish it from other substances. These properties can be grouped into two categories: physical and chemical. **Physical properties** are those that we can measure without changing the basic identity of the substance. They include color, odor, density, melting point, boiling point, and hardness. **Chemical properties** describe the way a substance may change or "react" to form other substances. For example, a common chemical property is *flammability*, the tendency of a substance to burn in the presence of oxygen.

A HISTORICAL PERSPECTIVE: Pasteur and the "Handedness" of Substances

The three-dimensional nature of matter can lead to a remarkable physical property, namely, the "handedness" of substances. Your left and right hands are virtually identical in every respect, and yet they are different: They are mirror images of one another (Figure 1.5). Similarly, two substances can be identical, except that one is "left-handed" while the other is "right-handed."

Figure 1.5 Left and right hands are mirror images of one another. This painting is by the German artist Albrecht Dürer (1471–1528). (Bridgeman Art Library/Art Resource)

Figure 1.6 Micrographs of left- and right-handed crystals of sodium ammonium tartrate. (From "The Resolution of Racemic Acid" by George B. Kauffman and Robin D. Myers, *Journal of Chemical Education,* Vol. 52, p. 777, Dec. 1975. © 1975 Div. of Chemical Education, American Chemical Society)

Louis Pasteur (1822–1895), the French chemist and bacteriologist, was the first to prove that substances can exist in left- and right-handed versions. He demonstrated this for an organic substance, tartaric acid, and showed that the two forms could be separated. What is perhaps most amazing about Pasteur's discovery is the means he used to separate the two forms: Solid crystals of the left- and right-handed forms appear as mirror images of one another under a microscope (Figure 1.6). Pasteur undertook the arduous task of separating the left- and right-handed crystals by hand and was able to demonstrate that they were identical except for their handedness.

Pasteur's discovery led to the development of the field of *stereochemistry,* the study of the spatial arrangements of atoms in molecules. Stereochemistry is an extremely important aspect of organic and inorganic chemistry, as many of you will find in your continued studies of chemistry.

Chemical and Physical Changes

As with the properties of substances, the changes that substances undergo can be classified as either physical or chemical. **Physical changes** are processes during which a substance changes its physical appearance but not its basic identity. The evaporation of water is a physical change. When water evaporates, it changes from the liquid state to the gas state, but it is still water; it has not changed into any other substance. All **changes of state** (for example, from liquid to gas or from liquid to solid) are physical changes. In **chemical changes** (also called **chemical reactions**), a substance is transformed into a chemically different substance. For example, when hydrogen burns in air it undergoes a chemical change in which it is converted to water.

Chemical changes can be dramatic. The following account describes a young man's first experiences with chemical reactions. The writer is Ira Remsen, who later became the author of a popular chemistry text published in 1901. The chemical reaction that he observed is shown in Figure 1.7.

> While reading a textbook of chemistry, I came upon the statement "nitric acid acts upon copper," and I determined to see what this meant. Having located some nitric acid, I had only to learn what the words "act upon" meant. In the interest of knowledge I was even willing to sacrifice one of the few copper cents then in my possession. I put one of them on the table, opened a bottle labeled "nitric acid," poured some of the liquid on the copper, and prepared to make an observation. But what was this wonderful thing which I beheld? The cent was already changed, and it was no small change either. A greenish-blue liquid foamed and fumed over the cent and over the table. The air became colored dark red. How could I stop this? I tried by picking the cent up and throwing it out the window. I learned another fact: nitric acid acts upon fingers. The pain led to another unpremeditated experiment. I drew my fingers across my trousers and discovered nitric acid acts upon trousers. That was the most impressive experiment I have ever performed. I tell of it even now with interest. It was a revelation to me. Plainly the only way to learn about such remarkable kinds of action is to see the results, to experiment, to work in the laboratory.

Figure 1.7 The chemical reaction between a copper penny and nitric acid. The dissolved copper produces the blue-green solution; the reddish brown gas produced is nitrogen dioxide. (Donald Clegg and Roxy Wilson)

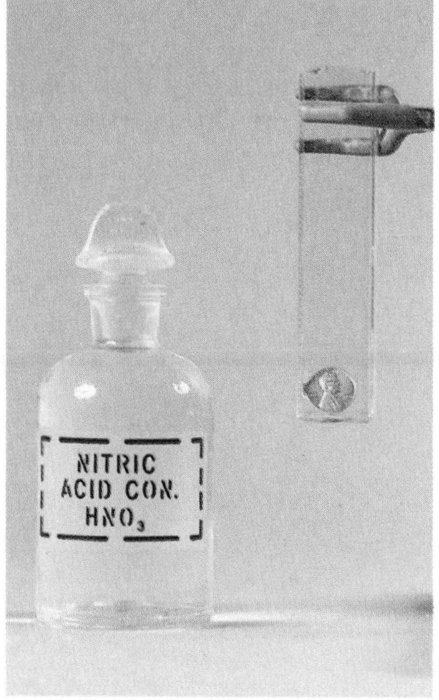

SAMPLE EXERCISE 1.1

The alcoholic beverage vodka can be separated into several substances, the main two of which are the liquids water and ethanol (grain alcohol). Based on our everyday experiences, what differences are there in the physical and chemical properties of these substances?

Solution: We will list only some of the better known properties. *Physical properties:* Water is colorless and odorless. Ethanol is colorless but has a distinctive odor. Ethanol evaporates faster than water. Ethanol remains a liquid at the temperature at which water freezes. *Chemical properties:* Ethanol is flammable; water is not. Also, an excess of ethanol, when taken internally, reacts differently with your body than does an excess of water. We do not recommend that you attempt this experiment yourself! We will learn about many more chemical differences in the course of this text.

PRACTICE EXERCISE

Based on your everyday experiences, what are the differences in the physical and chemical properties of the solid metals iron and gold? ***Answer:*** We will list only a few. *Physical properties:* Gold is yellow; iron is gray. Iron is attracted to magnets; gold is not. *Chemical properties:* Iron reacts with oxygen in the presence of water to form rust. Gold does not react with oxygen under normal conditions.

Mixtures

Most of the matter we encounter consists of combinations of different substances. Chemists use the term **mixtures** to refer to combinations of two or more substances in which each substance retains its own chemical identity. Some mixtures, such as sand, rocks, and wood, do not have the same composition, properties, and appearance throughout. Such mixtures are *heterogeneous* [Figure 1.8(a)]. Mixtures that are uniform throughout are *homogeneous.* Air is a homogeneous mixture of the gaseous substances nitrogen, oxygen, water vapor, and carbon dioxide and smaller amounts of other substances. The nitrogen in air has all the properties that pure nitrogen does. Salt, sugar, and many other substances dissolve in water to form homogeneous mixtures [Figure 1.8(b)]. Homogeneous mixtures are also called **solutions.** Air is a gaseous solution; gasoline is a liquid solution; brass is a solid solution.

The compositions of mixtures can vary widely. Because each component of a mixture retains its own properties, however, we can separate a mixture into its component substances by taking advantage of the differences in their physical properties. For example, a heterogeneous mixture of iron filings and

(*a*)

(*b*)

Figure 1.8 (*a*) Many common materials, including rocks, are heterogeneous. This photo shows a rock containing a copper mineral called malachite. (*b*) Homogeneous mixtures are called solutions. Many substances, including the blue solid shown in this photo (copper sulfate), dissolve in water to form solutions. (© Paul Silverman, 1990/ Fundamental Photographs; Richard Megna/Fundamental Photographs)

Figure 1.9 Separation by filtration. A mixture of a solid and a liquid is poured through a porous medium, in this case filter paper. The liquid passes through the paper while the solid remains on the paper. (Donald Clegg and Roxy Wilson)

gold filings could be sorted individually by color into iron and gold. A more clever approach would be to use a magnet to attract the iron filings and leave the gold ones behind. We can also use differences in chemical properties to separate mixtures into their components. Of course, chemical reactions will change some of the substances into others. In the case of our mixture of iron and gold, we could take advantage of an important chemical difference between these metals: Many acids dissolve iron but not gold. Thus, if we put our mixture into an appropriate acid, the iron will dissolve and the gold will be left behind. The two could then be separated by filtration, a procedure illustrated in Figure 1.9. We would have to use other chemical reactions, which we will learn about later, to transform the dissolved iron back into metal.

We can separate homogeneous mixtures into their components in a similar fashion. For example, water has a much lower boiling point than table salt. If we boil a solution of salt and water, the water will evaporate and the salt will be left behind. We could use a cold object, a *condenser*, to change the water vapor back into liquid (Figure 1.10).

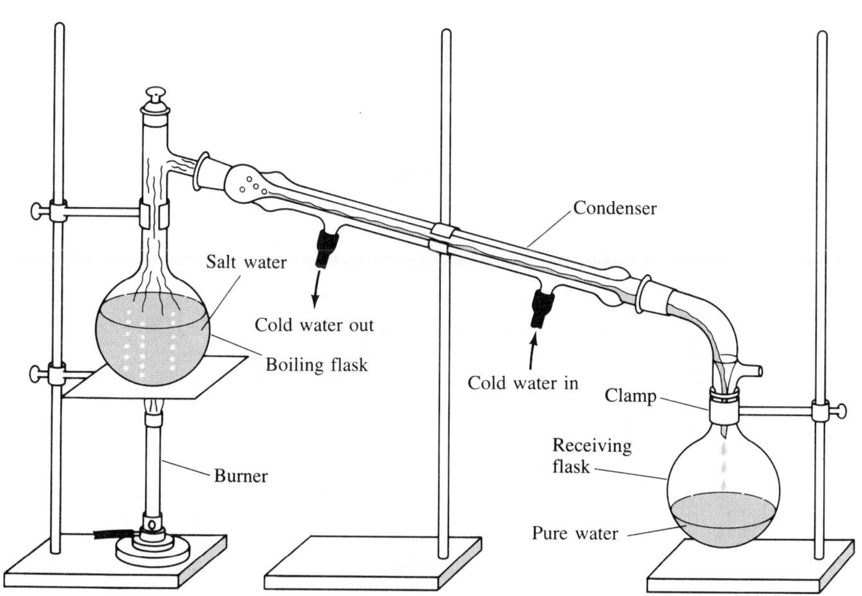

Figure 1.10 A simple apparatus for the separation of a sodium chloride solution (salt water) into its components. Boiling the solution evaporates the water, which is condensed and collected in the receiving flask. After all the water has boiled away, pure sodium chloride remains in the boiling flask.

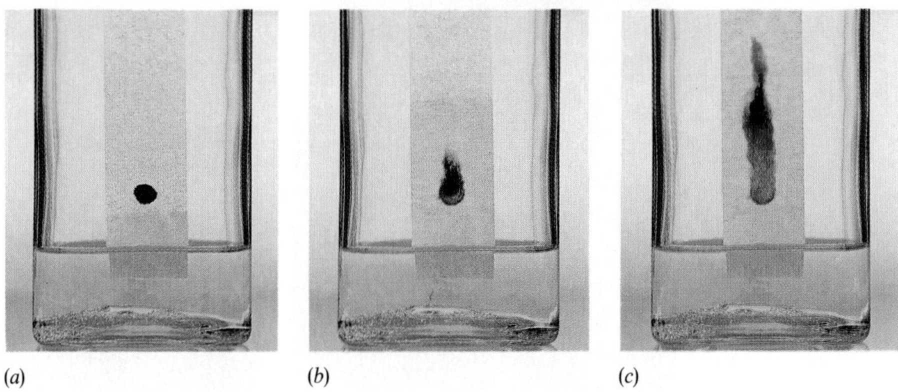

Figure 1.11 Separation of ink into components by paper chromatography. (*a*) Water begins to move up the paper. (*b*) Water moves past the ink spot, lifting different components of the ink at different rates. (*c*) Water has separated the ink into several different components. (© Richard Megna, 1990/ Fundamental Photographs)

The differing abilities of substances to move through various media can also be used to separate mixtures. This is the basis of *chromatography* (literally "the writing of colors"), a technique that can give beautiful and dramatic results. An example of the chromatographic separation of ink is shown in Figure 1.11.

1.2 ELEMENTS AND COMPOUNDS

Pure substances have a constant, invariable composition. We can classify substances as either elements or compounds. **Elements** are substances that cannot be decomposed into simpler substances by chemical means. **Compounds**, in contrast, can be decomposed by chemical means into two or more elements. Figure 1.12 summarizes the classification of matter into mixtures, compounds, and elements.

Elements

Elements are the basic substances out of which all matter is composed. In light of the seemingly endless variety of materials in our world, it is perhaps surprising

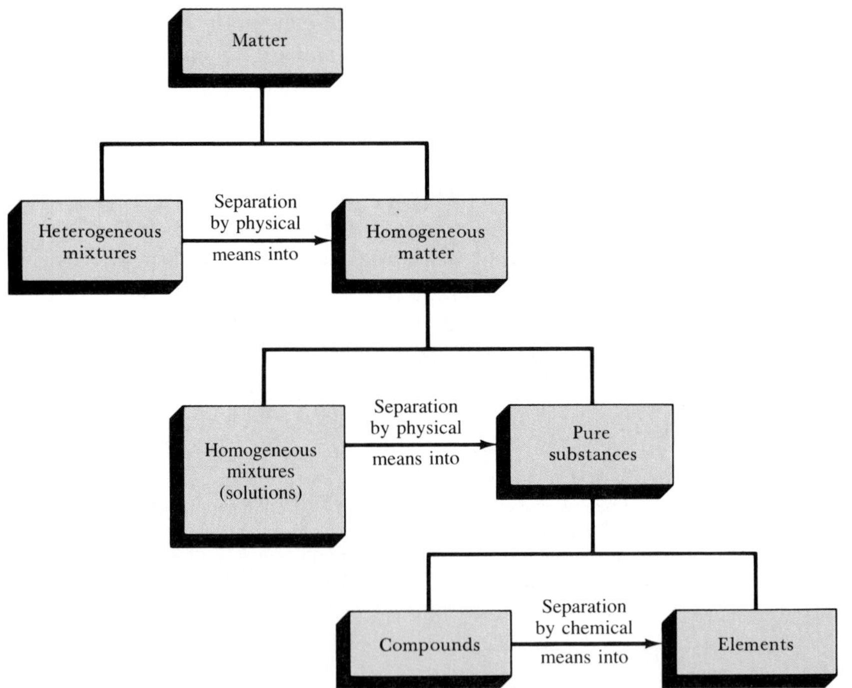

Figure 1.12 A diagram showing the classification of matter. "Separation by physical means" involves using differences in the physical properties of substances to separate them. "Separation by chemical means" involves transforming a substance into one or more other substances.

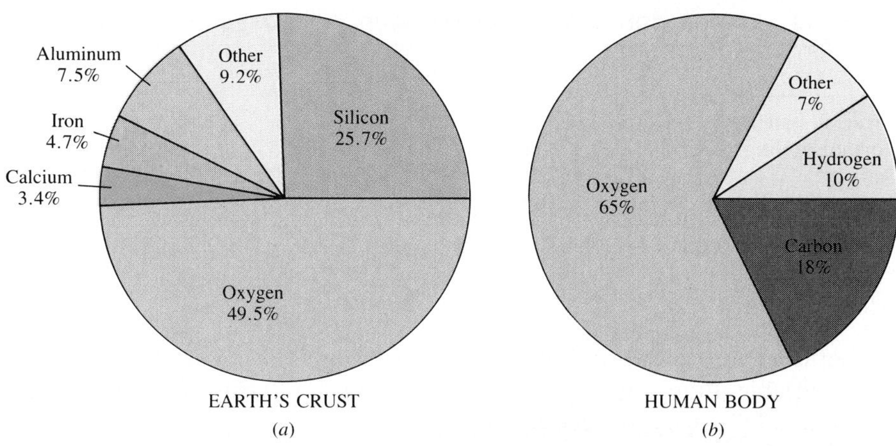

Figure 1.13 Elements, in percent by mass, in (*a*) the earth's crust (including oceans and atmosphere) and (*b*) the human body.

that there are only 109 known elements (as of October 1990). They vary greatly in abundance, as shown in Figure 1.13. For example, over 90 percent of the mass of the earth's crust consists of only five elements: oxygen, silicon, aluminum, iron, and calcium. In contrast, over 90 percent of the mass of the human body is accounted for by just three elements: oxygen, carbon, and hydrogen.

Some of the more familiar elements are listed in Table 1.2 along with the chemical abbreviations—or *symbols*—that we use to denote them. All of the known elements are listed on the front inside cover of this text. The symbol for an element consists of one or two letters, with the first letter capitalized. These symbols are often derived from the English name for the element (first and second columns in Table 1.2), but sometimes they are derived from a foreign name instead (third column). You will need to know these symbols and to learn others as we encounter them in the text.

Figure 1.14 Decomposition of the compound water into the elements hydrogen and oxygen by passing a direct electrical current through it.

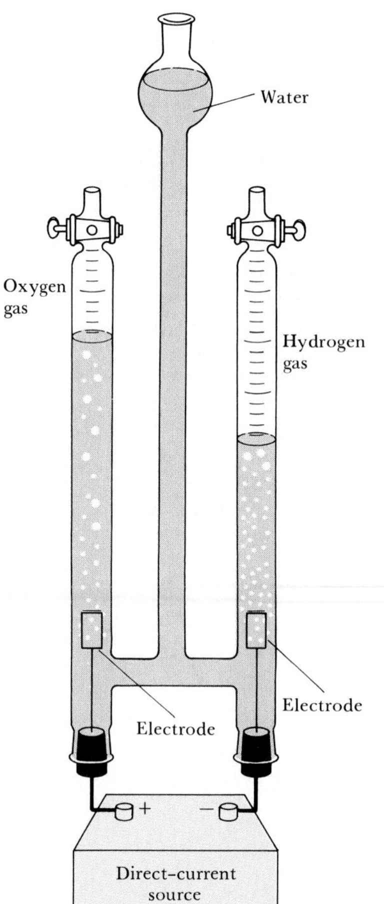

Compounds

Compounds are substances composed of two or more elements united chemically in definite proportions by mass. We can gain a clearer understanding of the distinctions among elements, compounds, and mixtures by examining a common substance, water. With the discovery of methods of generating electricity, chemists found that water could be decomposed into the elements hydrogen and oxygen, as shown in Figure 1.14. This decomposition clearly indicates that water is not an element. However, it is not merely a mixture of hydrogen and oxygen either. The properties of water are much different from those of its con-

Table 1.2 Some Common Elements and Their Symbols

Carbon (C)	Aluminum (Al)	Copper (Cu, from *cuprum*)
Fluorine (F)	Barium (Ba)	Iron (Fe, from *ferrum*)
Hydrogen (H)	Calcium (Ca)	Lead (Pb, from *plumbum*)
Iodine (I)	Chlorine (Cl)	Mercury (Hg, from *hydrargyrum*)
Nitrogen (N)	Helium (He)	Potassium (K, from *kalium*)
Oxygen (O)	Magnesium (Mg)	Silver (Ag, from *argentum*)
Phosphorus (P)	Platinum (Pt)	Sodium (Na, from *natrium*)
Sulfur (S)	Silicon (Si)	Tin (Sn, from *stannum*)

Table 1.3 **Comparison of Water, Hydrogen, and Oxygen**

	Water	Hydrogen	Oxygen
Physical state[a]	Liquid	Gas	Gas
Normal boiling point	100°C	−253°C	−183°C
Density[a]	1.00 g/mL	0.084 g/L	1.33 g/L
Combustible?	No	Yes	No

[a] At room temperature and atmospheric pressure.

stituent elements, as seen in Table 1.3. Furthermore, the composition of water is not variable. Pure water, regardless of its source, consists of 11 percent hydrogen and 89 percent oxygen by mass.

The observation that the elemental composition of a pure compound is always the same is known both as the **law of constant composition** and the **law of definite proportions**. It was first put forth by the French chemist Joseph Louis Proust (1754–1826) in about 1800. Although this law has been known for almost 200 years, the general belief persists among some people that a fundamental difference exists between compounds prepared in the laboratory and the corresponding compounds found in nature. However, a pure compound has the same composition and properties regardless of its source. Both chemists and nature must use the same elements and operate under the same natural laws. Differences in composition and properties between substances indicate that the compounds are not the same or that they differ in purity.

1.3 UNITS OF MEASUREMENT

Many properties of matter are *quantitative*, that is, they are associated with numbers. When a number represents a measured quantity, the units of that quantity must always be specified. To say that the length of a pencil is 17.5 is meaningless. To say that it is 17.5 centimeters (cm) properly specifies the length. The units used for scientific measurements are those of the **metric system**.

The metric system, which was first developed in France during the late eighteenth century, is used as the system of measurement in most countries throughout the world. The United States has traditionally used the English system, although use of the metric system has become more common in recent years. For example, the contents of most canned goods and soft drinks in grocery stores are now given in metric as well as in English units. Figure 1.15 illustrates the application of the metric system to another aspect of our culture.

In order to standardize scientific measurement, an international agreement was reached in 1960 specifying certain basic metric units that all scientists should use. The preferred units are known as **SI units**, after the French *Système International d'Unités*. There are seven fundamental scientific quantities; their SI units are given in Table 1.4. All other quantities (for example, speed and volume) can be derived from these seven. Speed, which is the ratio of distance to elapsed time, has the SI unit of meters per second, or m/s, and volume, which has the units of length cubed, has the SI unit cubic meter, or m^3.

The SI system employs a series of prefixes to indicate decimal fractions or multiples of various units. For example, the prefix *milli-* represents a 10^{-3} fraction of a unit: A milligram (mg) is 10^{-3} gram (g), a millimeter (mm) is 10^{-3} meter, and so forth. Table 1.5 presents the prefixes most commonly encountered

Figure 1.15 Metric measurements are becoming common in sporting activities, such as baseball and track and field. (Courtesy Montreal Expos)

Table 1.4 Basic SI Units

Physical quantity	Name of unit	Abbreviation
Mass	Kilogram	kg
Length	Meter	m
Time	Second	s[a]
Electric current	Ampere	A
Temperature	Kelvin	K
Luminous intensity	Candela	cd
Amount of substance	Mole	mol

[a] The abbreviation sec is frequently used.

Table 1.5 Selected Prefixes Used in the SI System

Prefix	Abbreviation	Meaning	Example
Mega-	M	10^{6}	1 megameter (Mm) = 1×10^{6} m
Kilo-	k	10^{3}	1 kilometer (km) = 1×10^{3} m
Deci-	d	10^{-1}	1 decimeter (dm) = 0.1 m
Centi-	c	10^{-2}	1 centimeter (cm) = 0.01 m
Milli-	m	10^{-3}	1 millimeter (mm) = 0.001 m
Micro-	μ[a]	10^{-6}	1 micrometer (μm) = 1×10^{-6} m
Nano-	n	10^{-9}	1 nanometer (nm) = 1×10^{-9} m
Pico-	p	10^{-12}	1 picometer (pm) = 1×10^{-12} m
Femto-	f	10^{-15}	1 femtometer (fm) = 1×10^{-15} m

[a] This is the Greek letter mu (pronounced "mew").

in chemistry. In using the SI system and in working problems throughout this text, it is important to have a comfortable familiarity with exponential notation. If you are unfamiliar with exponential notation or want to review it, refer to Appendix A.1.

Although non-SI units are being phased out, there are still some that are commonly used by scientists. Whenever we first encounter a non-SI unit in the text, the proper SI unit will also be given.

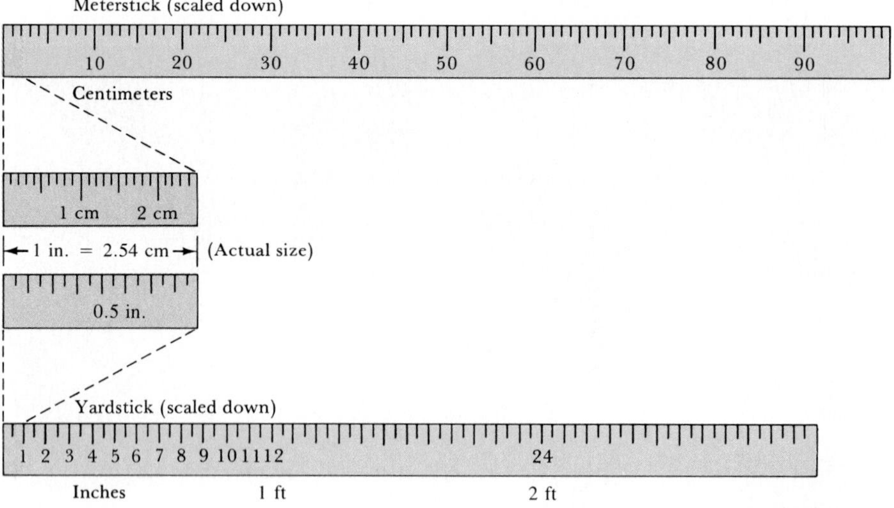

Figure 1.16 Comparison of common metric measures of length.

Length

The basic SI unit of length is the meter (m). From the comparison shown in Figure 1.16, we see that the meter is only slightly longer than a yard. We will discuss how to convert English units into metric units, and vice versa, in Section 1.5. For the moment, it is more important to understand clearly the use of the prefixes given in Table 1.5. (The relations between the English and metric system units that we will use most frequently in this text appear on the back inside cover.)

Volume

Because the units of volume are (length)^3, the basic SI unit of volume is the cubic meter, or m^3. A cubic meter is the volume of a cube that is 1 m on each edge. Because this is a very large volume, for most applications of chemistry some smaller units are commonly used. The cubic centimeter, cm^3 (sometimes written as cc), is one such unit. The cubic decimeter, dm^3, is also used. This

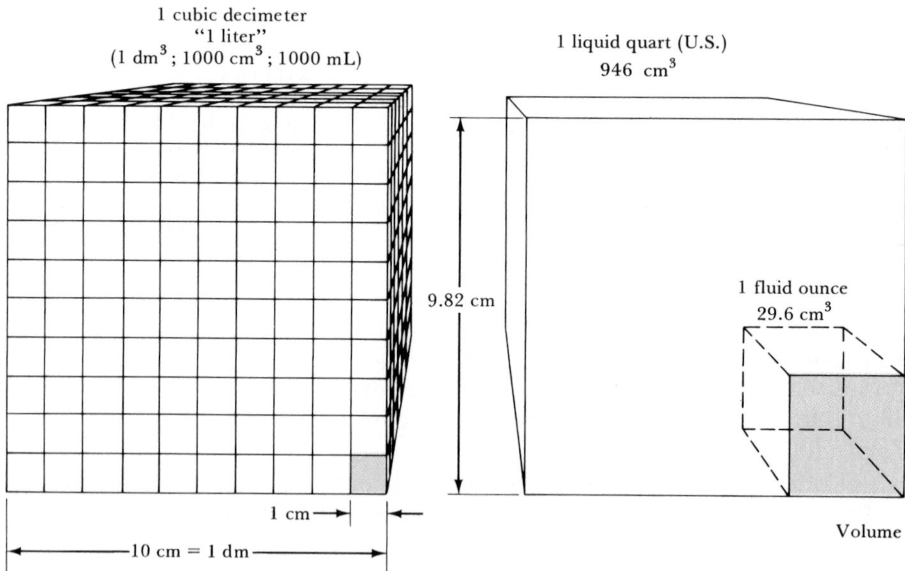

Figure 1.17 Comparison of common measures of volume.

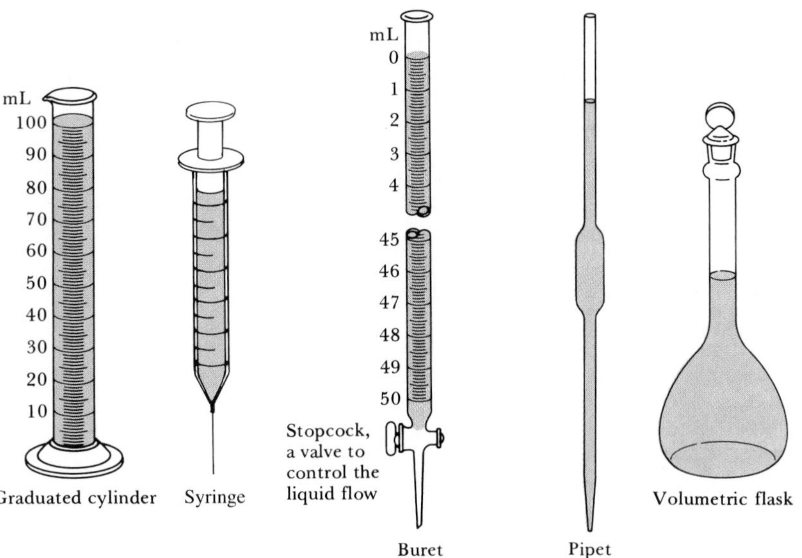

Figure 1.18 Common devices used in chemistry laboratories for the measurement and delivery of volumes of liquid. The cylinder, syringe, and buret are used to measure variable volumes of liquid; the pipet and flask are used to deliver a specific volume of a liquid.

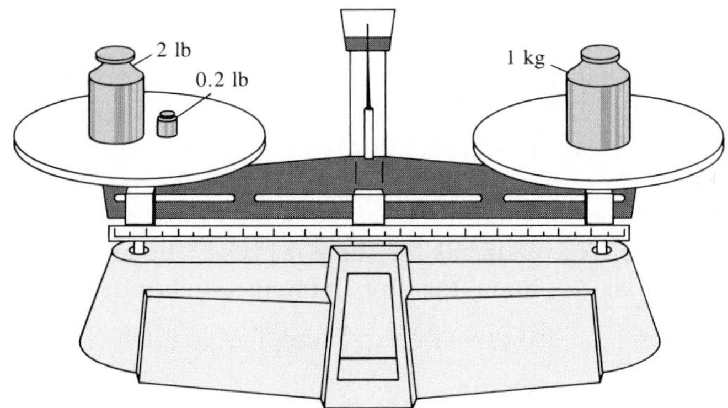

Figure 1.19 A kilogram is about 2.2 lb.

volume is more commonly known as the *liter* (L) and is roughly the size of a quart. The liter is the first metric unit that we have encountered that is *not* an SI unit. There are 1000 milliliters (mL) in a liter (Figure 1.17), and each milliliter is the same volume as a cubic centimeter: $1\ mL = 1\ cm^3$. The terms *milliliter* and *cubic centimeter* are used interchangeably in expressing volume.

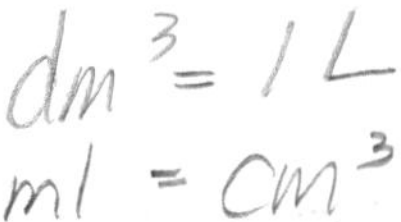

The devices used most frequently in chemistry to measure volume are illustrated in Figure 1.18. Syringes and burets allow the delivery of a variable amount of liquid with more accuracy than do graduated cylinders. Pipets and volumetric flasks are used for the precise delivery of a specific volume of liquid.

Mass

The basic SI unit of **mass*** is the kilogram (kg). This base unit is unusual because it uses a prefix, *kilo-*, instead of the word *gram* alone. As shown in Figure 1.19, a kilogram is equal to about 2.2 lb. We obtain other units for mass by adding prefixes to the word *gram*.

* Mass and weight are often incorrectly thought to be the same. *Mass* is a measure of the amount of material in an object. The *weight* of the object, however, is the force that the mass exerts because of gravity. In space, where gravitational forces are very weak, an astronaut can be *weightless*, but he or she cannot be *massless*. In fact, the astronaut's mass in space is the *same* as it is on Earth. Nevertheless, it is common practice to use the terms *mass* and *weight* interchangeably.

Table 1.6 Densities of Some Selected Substances at 25°C

Substance	Density (g/cm^3)
Air	0.001
Balsa wood	0.16
Water	1.00
Table salt	2.16
Iron	7.9
Gold	19.32

SAMPLE EXERCISE 1.2

What is the name given to the unit that equals **(a)** 10^{-9} gram; **(b)** 10^{-6} liter; **(c)** 10^{-3} meter?

Solution: In each case we can refer to Table 1.5, finding the prefix related to each of the decimal fractions: **(a)** nanogram, ng; **(b)** microliter, μL; **(c)** millimeter, mm.

PRACTICE EXERCISE

What fraction of a second is a picosecond, ps? ***Answer:*** 10^{-12} second

Density

Density is widely used to characterize substances. It is defined as the amount of mass in a unit volume of the substance:

$$\text{Density} = \frac{\text{mass}}{\text{volume}} \qquad [1.1]$$

Although the basic SI unit of density is kg/m^3, density is more commonly expressed in units of grams per cubic centimeter (g/cm^3). The densities of some common substances are listed in Table 1.6. It is no coincidence that the density of water is 1.00 g/cm^3; the gram was originally defined as the mass of 1 cm^3 of water at a specific temperature.

The terms *density* and *weight* are sometimes confused. A person who says that iron weighs more than air generally means that iron has a higher density than air; 1 kg of air has the same mass as 1 kg of iron, but the iron is confined to a smaller volume, thereby giving it a higher density.

SAMPLE EXERCISE 1.3

(a) Calculate the density of mercury if 1.00×10^2 g occupies a volume of 7.36 cm^3. **(b)** Calculate the mass of 65.0 cm^3 of mercury.

Solution

(a) $\text{Density} = \frac{\text{mass}}{\text{volume}} = \frac{1.00 \times 10^2\ \text{g}}{7.36\ \text{cm}^3} = 13.6\ \text{g/cm}^3$

(b) Solving Equation 1.1 for mass gives

$$\text{Mass} = \text{volume} \times \text{density}$$

Using the density of mercury that we calculated in part (a) gives

$$\text{Mass} = (65.0\ \text{cm}^3)(13.6\ \text{g/cm}^3) = 884\ \text{g}$$

PRACTICE EXERCISE

A student needs 15.0 g of ethanol for an experiment. If the density of the alcohol is 0.789 g/mL, how many milliliters of alcohol are needed? ***Answer:*** 19.0 mL

Temperature

We sense temperature as a measure of the hotness or coldness of an object. Indeed, temperature determines the direction of heat flow. Heat always flows spontaneously from a substance at higher temperature to one at lower temperature. Thus, we feel the influx of energy when we touch a hot stove, and we know that the stove is at a higher temperature than our hand.

The temperature scales commonly employed in scientific studies are the Celsius (formerly called *centigrade*) and Kelvin scales. The kelvin is the SI unit of temperature. The **Celsius scale** is based on assignment of 0°C to the freezing point of water and 100°C to its boiling point at sea level. The corresponding temperatures on the *Fahrenheit scale* are 32°F and 212°F. There are 100° between the freezing point and boiling point of water on the Celsius scale, and there are 180° between these points on the Fahrenheit scale (Figure 1.21). Consequently, the Celsius and Fahrenheit scales are related as follows:

$$^{\circ}\mathrm{C} = \frac{100}{180}(^{\circ}\mathrm{F} - 32^{\circ}) = \frac{5}{9}(^{\circ}\mathrm{F} - 32^{\circ}) \qquad [1.2]$$

Figure 1.20 Many countries employ the Celsius temperature scale in everyday use, as is illustrated by this Australian stamp. (Australian Postal Service/National Standards Commission)

or, equivalently,

$$^{\circ}\mathrm{F} = \frac{9}{5}(^{\circ}\mathrm{C}) + 32^{\circ} \qquad [1.3]$$

The **Kelvin scale** is based on the properties of gases, and its origins will be considered more fully in Chapter 10. Zero on this scale corresponds to −273.15°C, and a kelvin is the same size as a degree Celsius. The Kelvin and Celsius scales are therefore related by

$$\mathrm{K} = {}^{\circ}\mathrm{C} + 273.15 \qquad [1.4]$$

For most purposes we will use 273 instead of 273.15 to convert temperature. [According to the SI convention, a degree sign (°) is not used with the Kelvin scale. Thus we write 273 K and not 273°K.] Further comparisons of the Celsius, Kelvin, and Fahrenheit scales are made in Figure 1.20 and Table 1.7.

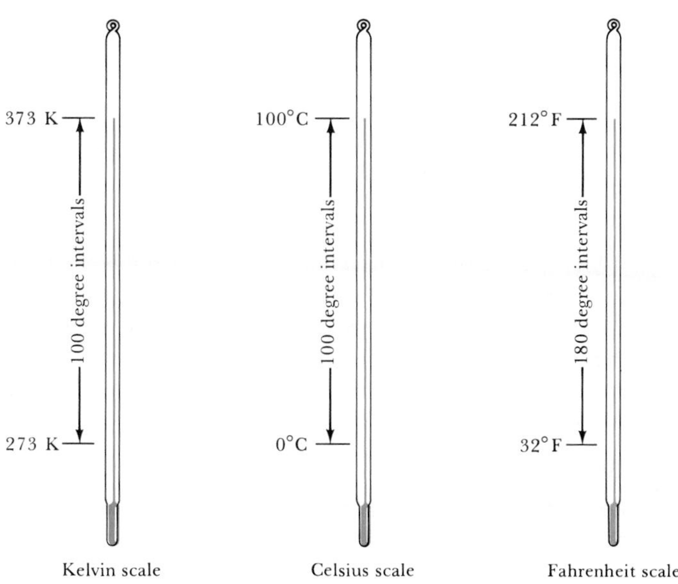

Figure 1.21 Comparison of the Kelvin, Celsius, and Fahrenheit temperature scales.

Table 1.7 Some Comparisons of Fahrenheit, Celsius, and Kelvin Temperatures

Absolute zero	−460°F	−273°C	0 K
Freezing point of water	32°F	0°C	273 K
Average room temperature	68°F	20°C	293 K
Normal body temperature	98.6°F	37.0°C	310 K
Boiling point of water	212°F	100°C	373 K

SAMPLE EXERCISE 1.4

If a weather forecaster predicts that the temperature for the day will reach 30°C, what is the predicted temperature **(a)** in K; **(b)** in °F?

Solution

(a) K = 30 + 273 = 303 K

(b) $°F = \frac{9}{5}(°C) + 32°$

$= 54° + 32°$

$= 86°F$

PRACTICE EXERCISE

Ethylene glycol, the major ingredient in antifreeze, boils at 199°C. What is the boiling point in **(a)** K; **(b)** °F? *Answers:* **(a)** 472 K; **(b)** 390°F

Intensive and Extensive Properties

Temperature is an **intensive property**, meaning that its value does not depend on the amount of material chosen. Two samples of a liquid may have the same temperature, though one has a volume of one cup and the other sample fills a bathtub. Density is another example of an intensive property. The density of mercury, calculated in Sample Exercise 1.3, is 13.6 g/cm^3 whether one has 100 g or 1000 g. In contrast, both volume and mass are **extensive properties**: they depend on the amount of material. Similarly, the *heat content* of a sample (the total amount of heat "held" by a sample) is an extensive property. In some types of solar heating systems for homes, air is heated by passage through solar panels exposed to the sun. The heated air is then passed through a large bed of stones. Because the heat content of the stones is large, the heat stored in them during the day is sufficient to heat the house during the night. Note the important distinction between heat content, an extensive property, and temperature, an intensive property.

SAMPLE EXERCISE 1.5

In the following description, label each property or characteristic as intensive or extensive: The yellow sample is solid at 25°C. It weighs 6.0 g and has a density of 2.3 g/cm^3.

Solution: Mass is an extensive property; color, physical state (that is, solid), temperature, and density are intensive properties.

PRACTICE EXERCISE

Is the heat per gram required to evaporate liquid water an intensive or extensive property? *Answer:* Intensive

CHEMISTRY AT WORK: Chemistry in the News

Because chemistry is so central to our lives, it is not surprising that we can read about it in the news nearly every day. Some of the news is good, such as recent breakthroughs in the development of new pharmaceuticals, materials, and products. Some of the news is more somber, dealing with the negative effects of chemicals on the environment or on society in general. The "chemical headlines" listed below indicate both the advances promised by chemistry and the current chemical problems we must confront. An important aspect of being a student of chemistry is to become more aware of the chemistry that is continually going on around you. We hope that as you proceed through this text, your awareness and understanding of the impact of chemistry on your everyday life is increased.

"New Class of Superconductors Pushing Temperatures Higher"

Since the early 1900s, scientists have been aware that when certain metals and alloys become cooled, they are *superconductors*—materials that conduct electricity with no resistance (and hence no power loss). The ability to utilize such materials could lead to huge advances in, for example, the transmission of energy, biomedical imaging, and solid-state electronic devices. Until the mid-1980s, however, superconductivity was observed only at extremely low temperatures (23 K and below). In 1986, two researchers in Switzerland discovered that certain chemical compounds called *metal oxides* were superconductors at about 30 K, a significant jump in the superconducting temperature. Since that discovery, chemical modification of the metal oxides has led to the production of new materials that exhibit superconductivity at temperatures greater than 100 K. We discuss the story of high-temperature superconductors in more detail in Chapter 12.

"Greenhouse Effect Leads to Global Warming"

Gases such as water and carbon dioxide in the Earth's atmosphere act as a natural "greenhouse," trapping heat near the Earth's surface. This process makes our planet habitable; without it, the surface temperature of the Earth would decline by about 35°C. Unfortunately, the growth in human population and the increase in per-capita energy consumption is having an adverse effect on this natural temperature regulation system. As we discuss in Chapters 3 and 18, the burning of fossil fuels, the use of aerosol propellants, and the deforestation of the Earth all serve to increase the amount of greenhouse gases. Many scientists believe that the resulting "greenhouse effect" is causing a general increase in global temperatures. This issue is discussed in more detail in the "Chemistry at Work" box in Chapter 3.

"Drug Shows Promise for AIDS Therapy"

The search for a cure for acquired immune deficiency syndrome (AIDS) has been one of the most intense scientific research efforts ever undertaken. It has also been expensive and frustrating. In 1984, scientists discovered that human immunodeficiency virus (HIV) causes AIDS. However, because the means by which HIV attacks the human body is not well understood, an effective treatment remains elusive. In 1987, the Food and Drug Administration (FDA) approved the drug azidothymidine (AZT) for AIDS treatment. Although AZT is a comparatively simple molecule, its development and synthesis have been extremely expensive. As a result, a year's supply of AZT can cost nearly $10,000. Furthermore, AZT only slows the progression of AIDS; it does not cure it. In addition, the drug has toxic side effects for many patients. The search for alternative and more effective treatments promises to be a major research focus of the 1990s.

"'Hole' in Ozone Layer Grows Larger"

Ozone is a form of the element oxygen whose molecules contain three atoms of oxygen each, as we discuss in Chapters 2 and 7. It is less common than the "usual" form of oxygen, which contains only two atoms per molecule. The action of sunlight on oxygen produces a layer of ozone in the stratosphere, more than 10 kilometers (km) above the earth's surface. This ozone layer serves as a natural sun screen, filtering out harmful ultraviolet radiation that can cause skin cancer and cataracts. Unfortunately, our widespread use of a class of substances called *chlorofluorocarbons* (CFCs) is believed to be threatening the ozone layer. These substances are widely used as refrigerants, aerosol propellants, cleaning solvents, and foaming agents for plastics. Normally, CFC molecules are unreactive. When exposed to sunlight, however, they produce chlorine atoms which serve to decompose ozone into normal oxygen. In the mid-1980s, scientists detected a hole in the ozone layer over Antarctica, an ominous warning of the effects of continued use of CFCs on our environment. We will discuss this topic further in Chapter 18.

"Is There a Methanol Car in Your Future?"

The use of clean-burning fuels as alternatives to gasoline could help reduce both air pollution and our dependence on petroleum. The most promising gasoline substitute is the simple chemical methyl alcohol, or *methanol*. Methanol is safer and cleaner and has a higher octane rating than gasoline. It is readily made from natural gas (1988 production: 3.33×10^9 kg) and costs about the same on a per-mile basis as gasoline. In 1989, President George

Bush submitted a clean air plan to Congress that directs automobile makers to start selling cars that run on gasoline substitutes such as methanol, with a goal of 1 million vehicles per year in 1997. We will consider the fuel energy of methanol and other gasoline substitutes in Chapter 5.

"Radon: The Invisible Danger in Your Home"

The invisible gas radon is a rare, very unreactive radioactive element (Chapter 7). Radon is formed from the decay of more common radioactive elements, particularly radium and uranium. During the 1980s, scientists came to realize that the decay of naturally occurring uranium in soil can lead to unusually high concentrations of radon in the basements of homes in many parts of the United States. Because of its radioactivity and the fact that it is inhaled, radon is considered a serious health risk that can induce lung cancer. Several companies now market kits for detecting radon in the home (Figure 1.22), and the Environmental Protection Agency (EPA) is reevaluating the level of exposure to radon gas that it considers safe. We discuss radon further in Chapter 21.

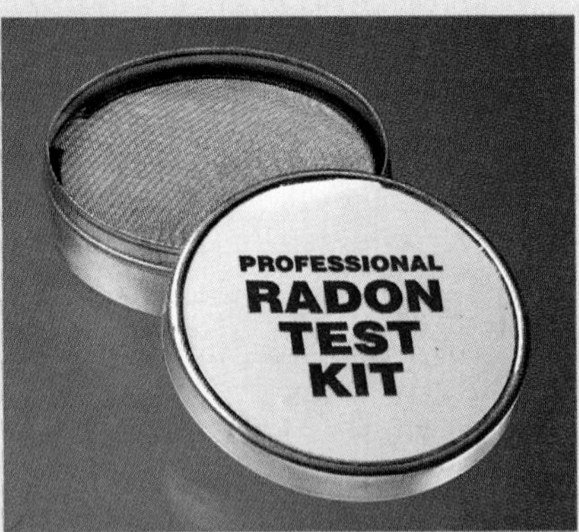

Figure 1.22 Radon home-testing kit. Kits such as these are available for measuring radon levels in the home. (© Richard Megna/Fundamental Photographs)

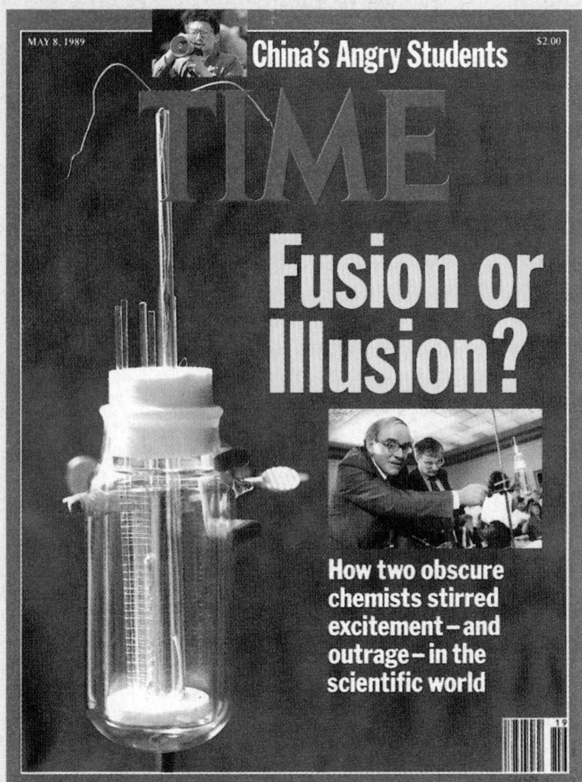

Figure 1.23 News of the "discovery" of cold fusion initially caused a great deal of excitement, both within the scientific community and among the public at large. (Copyright © 1989 The Time Inc. Magazine Company. Reprinted by permission)

"Cold Fusion: Excitement Ebbs as Disbelief Grows"

In March 1989, two chemists at the University of Utah claimed to have achieved the "cold fusion" of hydrogen into helium at room temperature. The fusion process, the same one that accounts for the energy of the sun, was previously thought to occur only at extremely high temperatures. The claim of cold fusion set off a furor in the scientific and popular press (Figure 1.23), particularly because cold fusion offers the possibility of producing nearly limitless energy. The process by which cold fusion was reportedly achieved involves *electrochemistry*, which we will discuss in Chapter 20. By late 1989, hundreds of attempts had been made to verify independently the occurrence of cold fusion. Most of them were decidedly negative, and it seems likely that the original claims of achieving cold fusion were in error.

1.4 UNCERTAINTY IN MEASUREMENT

In scientific work, we recognize two kinds of numbers: *exact numbers* (those whose values are known exactly) and *inexact numbers* (those whose values have some uncertainty). Exact numbers are those that have defined values or integers that result from counting. For example, there are exactly 3 ft in a yard, exactly 1000 g in a kilogram, and exactly 12 eggs in a dozen. By definition, there are exactly 2.54 cm in an inch. The number 1 in any conversion factor between units, as in 1 m = 1.0936 yd, is also an exact number.

Numbers obtained by measurement are always *inexact*. There are always built-in errors in the equipment used to measure quantities (equipment errors), and there are differences in how different people make the same measurement (human errors). Suppose that ten students with ten different balances are given the same dime to weigh. The ten measurements will vary slightly. The balances might be calibrated slightly differently, and there might be differences in how each student reads the mass from the balance. Remember: *Uncertainties always exist in measured quantities.*

Precision and Accuracy

Two terms are usually used in discussing the uncertainties in measured values: precision and accuracy. **Precision** is a measure of how closely individual measurements agree with each other. **Accuracy** refers to how closely individual measurements agree with the correct, or "true," value. In general, the more precise a measurement, the more accurate it will be. We gain confidence in the accuracy of a measurement if we obtain nearly the same value in many different experiments. Thus, in the laboratory, you will often perform several different "trials" of the same experiment. It is possible, however, for a precise value to be inaccurate. If a very sensitive balance is poorly calibrated, for example, the masses measured will be precise but inaccurate.

Significant Figures

Suppose you weigh a dime on a balance capable of measuring to the nearest 0.0001 g. You could report the mass as 2.2405 ± 0.0001 g. The $\pm$ notation (read "plus or minus 0.0001") is a useful way to express the uncertainty of a measurement. In much scientific work, we drop the $\pm$ notation with the understanding that an uncertainty of at least one unit exists in the last digit of the measured quantity. That is, measured quantities are generally reported in such a way that only the last digit is uncertain. All the digits, including the uncertain one, are called **significant figures**. The number 2.2405 has five significant figures. The number of significant figures indicates the preciseness of a measurement.

SAMPLE EXERCISE 1.6

What is the difference between 4.0 g and 4.00 g?

Solution: Many people would say there is no difference, but a scientist would note the difference in the number of significant figures in the two measurements. The value 4.0 has two significant figures, while 4.00 has three. This implies that the second measurement is more precise. A mass of 4.0 g indicates that the mass is between 3.9 and 4.1 g; the mass is 4.0 ± 0.1 g. A measurement of 4.00 g implies that the mass is between 3.99 and 4.01 g; the mass is 4.00 ± 0.01 g.

PRACTICE EXERCISE

A balance has a precision of ± 0.001 g. A sample that weighs about 25 g is weighed on this balance. How many significant figures should be reported for this measurement? ***Answer:*** 5

The following rules apply to determining the number of significant figures in a measured quantity:

1. All nonzero digits are significant—457 cm (three significant figures); 0.25 g (two significant figures).
2. Zeros between nonzero digits are significant—1005 kg (four significant figures); 1.03 cm (three significant figures).
3. Zeros to the left of the first nonzero digit in a number are not significant; they merely indicate the position of the decimal point—0.02 g (one significant figure); 0.0026 cm (two significant figures).
4. Zeros that fall both at the end of a number and to the right of the decimal point are significant—0.0200 g (three significant figures); 3.0 cm (two significant figures).
5. When a number ends in zeros that are not to the right of a decimal point, the zeros are not necessarily significant—130 cm (two or three significant figures); 10,300 g (three, four, or five significant figures). We describe how to remove this ambiguity below.

The use of exponential notation (Appendix A) avoids the potential ambiguity of whether the zeros at the end of a number are significant (rule 5). For example, a mass of 10,300 g can be written in exponential notation showing three, four, or five significant figures:

1.03×10^4 g (three significant figures)
1.030×10^4 g (four significant figures)
1.0300×10^4 g (five significant figures)

In these numbers all the zeros to the right of the decimal point are significant (rules 2 and 4). (All significant figures come before the exponent; the exponential term does not add to the number of significant figures.)

Exact numbers can be treated as if they have an infinite number of significant figures. This rule applies to many definitions between units. Thus, when we say, "There are 12 inches in 1 foot," the number 12 is exact, and we need not worry about the number of significant figures in it.

SAMPLE EXERCISE 1.7

How many significant figures are in each of the following numbers (assume that each number is a measured quantity): **(a)** 4.003; **(b)** 6.023×10^{23}; **(c)** 5000?

Solution: **(a)** Four; the zeros are significant figures. **(b)** Four; the exponential term does not add to the number of significant figures. **(c)** One, two, three, or four. In this case, the ambiguity could have been avoided by using standard exponential notation. Thus 5×10^3 has only one significant figure; 5.00×10^3 has three.

PRACTICE EXERCISE

How many significant figures are in each of the following measurements: **(a)** 3.549 g **(b)** 2.3×10^4 cm **(c)** 0.00134 m^3?
Answers: **(a)** four **(b)** two **(c)** three

Significant Figures in Calculations

In carrying measured quantities through calculations, observe this point: The precision of the result is limited by the least precise measurement. *In multiplica-*

tion and division the result must be reported as having no more significant figures than the measurement with the fewest significant figures. When the result contains more than the correct number of significant figures, it must be rounded off.

For example, the area of a rectangle whose edge lengths are 6.221 cm and 5.2 cm should be reported as 32 cm^2:

$$\text{Area} = (6.221\text{ cm})(5.2\text{ cm}) = 32.3492\text{ cm}^2 \longrightarrow \text{round off to } 32\text{ cm}^2$$

We round off to two significant figures because 5.2 cm has only two significant figures.

In rounding off numbers, the following rules are followed (each example is rounded to two digits):

1. If the leftmost digit to be removed is more than 5, the preceding number is increased by 1; 2.376 rounds to 2.4.
2. If the leftmost digit to be removed is less than 5, the preceding number is left unchanged; 7.248 rounds to 7.2.
3. If the leftmost digit to be removed is 5, the preceding number is not changed if it is even and is increased by 1 if it is odd; 2.25 rounds to 2.2; 4.35 rounds to 4.4.

The rule used to determine the number of significant figures in multiplication and division cannot be used for *addition and subtraction.* For these operations, *the result should be reported to the same number of decimal places as that of the term with the least number of decimal places.* In the following example the uncertain digits appear in color:

This number limits	20.4	⟵	one decimal place
the number of significant	1.322	⟵	three decimal places
figures in the result ⟶	83	⟵	zero decimal places
	104.722	⟶	round off to 105 (one uncertain digit)

SAMPLE EXERCISE 1.8

A person's height is measured to be 67.50 in. What is this height in centimeters?

Solution: There are 2.54 cm in an inch; this is an exact number and can be treated as if it had an infinite number of significant figures. The precision of the answer is thus limited by the measurement in inches and should be reported to four significant figures (67.50 has four significant figures). The answer is

$$(67.50\text{ in.})\left(2.54\,\frac{\text{cm}}{\text{in.}}\right) = 171.45\text{ cm, which is rounded to } 171.4\text{ cm}$$

PRACTICE EXERCISE

There are 1609.4 m in a mile. How many meters are in a distance of 1.35 mi?
Answer: 2.17×10^3 m

SAMPLE EXERCISE 1.9

A gas at 25°C exactly fills a container previously determined to have a volume of 1.05×10^3 cm^3. The container plus gas are weighed and found to have a mass of 837.6 g. The container, when emptied of all gas, has a mass of 836.2 g. What is the density of the gas at 25°C?

Solution: The mass of the gas is just the difference in the two masses: (837.6 − 836.2) g = 1.4 g. Notice that 1.4 g has only two significant figures, even though the masses from which it is obtained have four.

From the definition of density we have

$$\text{Density} = \frac{\text{mass}}{\text{volume}} = \frac{1.4\ \text{g}}{1.05 \times 10^3\ \text{cm}^3}$$
$$= 0.0013\ \text{g/cm}^3$$

There are two significant figures in this quantity, corresponding to the smaller number of significant figures in the two numbers that form the ratio.

PRACTICE EXERCISE

To how many significant figures should the container be weighed (with and without the gas) in Sample Exercise 1.9 in order for the density to be calculated to three significant figures? ***Answer:*** 5

It is important to have a feeling for significant figures when you use a calculator, because calculators ordinarily display more digits than are significant. For example, a typical calculator would give 1.3333333×10^{-3} as the answer to the calculation in Sample Exercise 1.9. This result must be rounded off because of the uncertainties in the measured quantities used in the calculation.

When a calculation involves two or more steps, retain at least one additional digit—past the number of significant figures—for intermediate answers. This procedure ensures that small errors from rounding at each step do not combine to affect the final result. In using a calculator, you may enter the numbers one after another, rounding only the final answer. Accumulated round-off errors often account for small differences between results you obtain and answers given in the text for numerical problems.

1.5 PROBLEM SOLVING WITH DIMENSIONAL ANALYSIS

Before we continue, perhaps a word of caution is in order. To have a real knowledge of chemistry, you must understand both the basic concepts and how to use these concepts to solve problems. Your instructor and this text will show you how to solve different types of problems, many of which involve numbers. But is that enough? Is watching someone else solve problems an effective way for you to learn how to solve problems yourself (as you will have to do on exams)? The answer is no.

One of our goals in this text (and, we hope, one of your goals in studying chemistry) is to make you comfortable solving chemical problems. Achieving this goal requires active practice in problem solving through the use of homework problems and the like. Throughout the book, we have provided sample exercises in which the solutions are shown in detail. A practice exercise, for which only the answer is given, accompanies each sample exercise; it is important that you use these exercises as learning aids. End-of-chapter exercises provide additional questions to help you understand the material in the chapter. Colored numbers indicate exercises whose answers are given at the back of the book. A review of basic mathematics is given in Appendix A.

Throughout the text, we use an approach called **dimensional analysis** as an aid in problem solving. In dimensional analysis, we carry units through all calculations. Units are multiplied together, divided into each other, or "canceled." Dimensional analysis will help ensure that the solutions to problems yield the proper units. More important, though, dimensional analysis provides a means by which problems can be solved.

The key to using dimensional analysis is the correct use of **conversion factors** to change one unit into another. A conversion factor is a fraction whose numerator and denominator are different units of the same quantity. Consider the statement, "There are 2.54 centimeters per inch," or the equivalent equation, 2.54 cm = 1 in. This relationship allows us to write two conversion factors:

$$\frac{2.54 \text{ cm}}{1 \text{ in.}} \qquad \frac{1 \text{ in.}}{2.54 \text{ cm}}$$

The first of these factors is used when we want to convert inches to centimeters. For example, the length in centimeters of an object that is 8.50 in. long is given by

$$\text{Number of centimeters} = (8.50 \text{ in.}) \frac{2.54 \text{ cm}}{1 \text{ in.}} = 21.6 \text{ cm}$$

Desired unit

Given unit

Note that the units of inches in the denominator of the conversion factor cancel the units of inches in the data we were given (8.50 *inches*). The centimeters in the numerator of the conversion factor become the units of the final answer. In general, the units multiply and divide as follows:

$$\text{Given unit} \times \frac{\text{desired unit}}{\text{given unit}} = \text{desired unit}$$

SAMPLE EXERCISE 1.10

A man weighs 185 lb. What is his mass in grams?

Solution: From the back inside cover we have 1 lb = 453.6 g. In order to cancel pounds and leave grams, we will use the conversion factor with grams in the numerator and pounds in the denominator:

$$\text{Mass in grams} = (185 \text{ lb})\left(\frac{453.6 \text{ g}}{1 \text{ lb}}\right) = 8.39 \times 10^4 \text{ g}$$

Note that the answer can be given only to three significant figures.

PRACTICE EXERCISE

By using a conversion factor from the back inside cover, determine the length in kilometers of a 500.0-mi automobile race. ***Answer:*** 804.7 km

We can use more than one conversion factor in the solution of a problem. For example, suppose we want to know the length in inches of an 8.00-m rod. The table at the end of the text doesn't give the relationship between meters and inches. It *does* give the relationship between centimeters and inches, though, and we know that 1 m = 100 cm. Thus, we can convert first from meters to centimeters, and then from centimeters to inches:

Length in inches

= (length in m)(factor converting m → cm)(factor converting cm → in.)

The relationship between meters and centimeters gives us the first conversion factor. Because we need to cancel meters, we write meters in the denominator:

$$\frac{100 \text{ cm}}{1 \text{ m}}$$

We then use the relationship 2.54 cm = 1 in. to write the second conversion factor with the desired units, inches, in the numerator:

$$\frac{1 \text{ in.}}{2.54 \text{ cm}}$$

Thus, we have

$$\text{Number of inches} = (8.00 \text{ m})\left(\frac{100 \text{ cm}}{1 \text{ m}}\right)\left(\frac{1 \text{ in.}}{2.54 \text{ cm}}\right) = 315 \text{ in.}$$

The conversion factors above convert from length to length or, more generally, from one unit of a given measure to another unit of the same measure. We also have conversion factors that convert from one measure to a different one. The density of a substance, for example, is a conversion factor between mass and volume. Suppose that we want to know the mass in grams of a cubic inch of gold, which has a density of 19.3 g/cm^3. The density gives us the following conversion factors:

$$\frac{19.3 \text{ g}}{1 \text{ cm}^3} \qquad \frac{1 \text{ cm}^3}{19.3 \text{ g}}$$

Because the answer we want is a mass in grams, we can see that we will use the first of these factors, which has mass in grams in the numerator. To use this factor, however, we must first relate cubic centimeters to cubic inches. We know the conversion factor from inches to centimeters, and the cube of this gives us the desired conversion factor:

$$\left(\frac{2.54 \text{ cm}}{1 \text{ in.}}\right)^3 = \frac{(2.54)^3 \text{ cm}^3}{(1)^3 \text{ in.}^3} = \frac{16.4 \text{ cm}^3}{1 \text{ in.}^3}$$

Notice that both the numbers and the units are cubed. Applying our conversion factors, we can now solve the problem:

$$\text{Mass in grams} = (1 \text{ in.}^3)\left(\frac{16.4 \text{ cm}^3}{1 \text{ in.}^3}\right)\left(\frac{19.3 \text{ g}}{1 \text{ cm}^3}\right) = 317 \text{ g}$$

Summary of Dimensional Analysis

In using dimensional analysis to solve problems, we will always ask three questions:

1. *What data are we given in the problem?*
2. *What quantity do we wish to obtain in the problem?*
3. *What conversion factors do we have available to take us from the given quantity to the desired one?*

When solving numerical problems, always ask yourself whether your answer makes sense. (*Calvin and Hobbes*. Copyright 1986 Universal Press Syndicate. Reprinted with permission. All rights reserved.)

If you carry the units through during your calculations, you will always know whether you are using the correct conversion factors. Finally, whenever you finish a calculation, look at the numerical value of your answer (as well as the units) and ask yourself whether your answer makes any sense.

SAMPLE EXERCISE 1.11

A certain printed page has an average of 25 words per square inch of paper. The average length of the words is 5.3 letters. What is the average number of letters per square centimeter of paper?

Solution: We are given the "print density" in the units of words per square inch. We want to convert this to a print density in the units of letters per square centimeter. We are given the relation 1 word = 5.3 letters and, from the back inside cover, 1 in. = 2.54 cm. The solution is given by

$$\text{Number of letters/cm}^2 = \left(\frac{25\ \cancel{\text{words}}}{1\ \cancel{\text{in.}^2}}\right)\left(\frac{5.3\ \text{letters}}{1\ \cancel{\text{word}}}\right)\left(\frac{1\ \cancel{\text{in.}}}{2.54\ \text{cm}}\right)^2$$
$$= 21\ \text{letters/cm}^2$$

PRACTICE EXERCISE

In a certain part of the country, there is an average of 710 people per square mile and 0.72 telephones per person. What is the average number of telephones in an area of 5.0 km^2? ***Answer:*** 990 telephones

SAMPLE EXERCISE 1.12

What is the mass in grams of 1.00 gal of water? The density of water is 1.00 g/mL.

Solution: Following the procedure summarized above, we note the following:

1. We are given 1.00 gal of water.
2. We wish to obtain the mass in grams.
3. We have the following conversion factors either given, commonly known, or available on the back inside cover of the text:

$$\frac{1.00\ \text{g water}}{1\ \text{mL water}} \qquad \frac{1\ \text{L}}{1000\ \text{mL}} \qquad \frac{1\ \text{L}}{1.057\ \text{qt}} \qquad \frac{1\ \text{gal}}{4\ \text{qt}}$$

We realize that the first of these conversion factors must be used as written (with grams in the numerator) to give the desired result. We also realize that the last conversion factor must be "turned over" in order to cancel gallons. The solution is given by

$$\text{Mass in grams} = (1.00\ \cancel{\text{gal}}\ \text{of water})\left(\frac{4\ \cancel{\text{qt}}}{1\ \cancel{\text{gal}}}\right)\left(\frac{1\ \cancel{\text{L}}}{1.057\ \cancel{\text{qt}}}\right)\left(\frac{1000\ \cancel{\text{mL}}}{1\ \cancel{\text{L}}}\right)\left(\frac{1.00\ \text{g}}{1\ \cancel{\text{mL}}}\right)$$
$$= 378 \times 10^3\ \text{g water}$$

PRACTICE EXERCISE

A car travels 28 mi/gal of gasoline. How many kilometers per liter will it go?
Answer: 12 km/L

FOR REVIEW

SUMMARY

Chemistry is the study of the properties, composition, and changes of matter. Matter exists in three states: gas, liquid, and solid. Most matter consists of a mixture of substances. Mixtures can be either homogeneous or heterogeneous; homogeneous mixtures are called solutions. Mixtures can be separated into two types of pure substances: elements and compounds. Each substance has a unique set of physical and chemical properties that can be used to identify it. Matter can undergo physical changes and chemical changes (chemical reactions).

Measurements in chemistry are made using the metric system. Special emphasis is placed on a particular set of metric units called SI units, which are based on the meter, kilogram, and second as the basic units of length, mass, and time, respectively. The metric system employs a set of prefixes to indicate decimal fractions or multiples of the base units.

All measured quantities are inexact to some extent. The number of significant figures indicates the exactness of the measurement. Certain rules must be followed so that a calculation involving measured quantities is reported to the proper number of significant figures.

In the dimensional analysis approach to problem solving we keep track of units as we carry measurements through calculations. The units are multiplied together, divided into each other, or canceled like algebraic quantities. Obtaining the proper units for the final result is an important means of checking the method of calculation. In converting units, and in several other types of problems, conversion factors can be used. These factors are ratios constructed from valid relations between equivalent quantities.

KEY TERMS

matter (Sec. 1.1)
gas (Sec. 1.1)
liquid (Sec. 1.1)
solid (Sec. 1.1)
pure substance (Sec. 1.1)
physical properties (Sec. 1.1)
chemical properties (Sec. 1.1)
physical changes (Sec. 1.1)
changes of state (Sec. 1.1)
chemical changes (Sec. 1.1)
chemical reactions (Sec. 1.1)
mixtures (Sec. 1.1)
solution (Sec. 1.1)
elements (Sec. 1.2)
compounds (Sec. 1.2)
law of constant composition (Sec. 1.2)
law of definite proportions (Sec. 1.2)
metric system (Sec. 1.3)
SI units (Sec. 1.3)
mass (Sec. 1.3)

density (Sec. 1.3)
Celsius scale (Sec. 1.3)
Kelvin scale (Sec. 1.3)
intensive property (Sec. 1.3)
extensive property (Sec. 1.3)
precision (Sec. 1.4)
accuracy (Sec. 1.4)
significant figures (Sec. 1.4)
dimensional analysis (Sec. 1.5)
conversion factors (Sec. 1.5)

EXERCISES

Introduction to Matter

1.1 You are told that a substance is either a gas or a liquid. What additional information do you need to determine which it is?

1.2 Which tests would you perform on a substance to determine whether it is a liquid or a solid?

1.3 Identify each of the following substances as a gas, a liquid, or a solid under ordinary conditions: **(a)** gold; **(b)** ethanol; **(c)** helium; **(d)** bromine; **(e)** carbon monoxide.

1.4 Give the state of matter (gas, liquid, or solid) for each of the substances in Table 1.1 under ordinary conditions.

1.5 Indicate which of the following are physical processes and which are chemical processes: **(a)** tarnishing of silverware; **(b)** melting of ice; **(c)** cutting a diamond; **(d)** burning of gasoline; **(e)** wine turning into vinegar.

1.6 A match is lit and held under a cold piece of metal. The following observations are made: **(a)** The match burns. **(b)** The metal gets warmer. **(c)** Water condenses on the metal. **(d)** Soot (carbon) is deposited on the metal. Which of these occurrences are due to physical changes and which are due to chemical changes?

1.7 In the process of attempting to characterize a substance, a chemist makes the following observations: The substance is a silvery-white, lustrous metal. It melts at 649°C and boils at 1105°C. Its density at 20°C is 1.738 g/cm^3. The substance burns in air, producing an intense white light. It reacts with chlorine to give a brittle, white solid. The substance can be pounded into thin sheets or drawn into wires. It is a good conductor of electricity. Which of these characteristics are physical properties and which are chemical properties?

1.8 Classify the following observations about a substance as either physical or chemical properties: **(a)** color; **(b)** melting point; **(c)** reactivity with water; **(d)** boiling point; **(e)** state of matter under ordinary conditions; **(f)** flammability; **(g)** density; **(h)** electrical conductivity; **(i)** decomposition products upon heating.

1.9 Classify each of the following as a pure substance or a mixture; if a mixture, indicate whether it is homogeneous or heterogeneous: **(a)** silicon dioxide (quartz); **(b)** gasoline; **(c)** a fruit cake; **(d)** a pure gold coin; **(e)** bronze (an alloy of copper and tin).

1.10 Classify each of the following as an element, compound, or mixture; if a mixture, indicate whether it is homogeneous or heterogeneous: **(a)** diamond; **(b)** ammonia water; **(c)** iodine crystals; **(d)** salad dressing; **(e)** magnesium chloride (the salt used to melt snow and ice).

Elements and Compounds

1.11 Give the chemical symbol for each of the following elements: **(a)** boron; **(b)** lithium; **(c)** chromium; **(d)** phosphorus; **(e)** potassium; **(f)** silver; **(g)** tungsten; **(h)** antimony.

1.12 Identify the chemical elements represented by the following symbols: **(a)** Si; **(b)** Be; **(c)** F; **(d)** Na; **(e)** Hg; **(f)** Au; **(g)** Ar; **(h)** As.

1.13 Classify each of the substances in Table 1.1 as an element or a compound.

1.14 How many total elements are represented by the substances in Table 1.1? What fraction of all of the known elements is represented?

Metric System; SI Units

1.15 What basic SI units are appropriate for expressing the following quantities: **(a)** the diameter of the earth; **(b)** the surface area of a tennis ball; **(c)** the volume of a gasoline tank; **(d)** the mass of a brick; **(e)** the speed of light; **(f)** the temperature of the air?

1.16 What basic SI units are appropriate for expressing the following: **(a)** miles; **(b)** quarts; **(c)** square inches; **(d)** fluid ounces; **(e)** pounds; **(f)** days?

1.17 What decimal power do the following abbreviations represent: **(a)** d; **(b)** c; **(c)** f; **(d)** μ; **(e)** M; **(f)** k; **(g)** n; **(h)** m; **(i)** p?

1.18 Use the appropriate abbreviation to replace the decimal power in each of the following values: **(a)** 3.4×10^{-12} m; **(b)** 4.8×10^{-6} mL; **(c)** 7.23×10^{3} g; **(d)** 2.35×10^{-6} m^3; **(e)** 5.8×10^{-9} s; **(f)** 3.45×10^{-3} mol.

1.19 What type of quantity (for example, length, volume, density) do the following represent: **(a)** cm^3; **(b)** kg/dL; **(c)** mm; **(d)** ML; **(e)** km^2; **(f)** ps; **(g)** μL.

1.20 What basic SI units are appropriate for expressing each of the quantities in Problem 1.19?

1.21 Convert each of the following values into its basic SI unit: **(a)** 1.2 kg/dL; **(b)** 39.7 pm; **(c)** 10.07 μs; **(d)** 83,645 mg; **(e)** 150 km; **(f)** 320 mmol.

1.22 Perform the following conversions: **(a)** 32.2 mm to μm; **(b)** 47.6 g/mL to kg/dL; **(c)** 32.4×10^{-12} m to pm; **(d)** 4.5×10^{8} pm^3 to m^3.

1.23 **(a)** Water has a density of 1.00 g/mL or 1.00 g/cm^3. What is the mass of 338 dL of water? **(b)** A cube of plastic 1.2×10^{-5} km on a side weighs 1.1 g. What is the density of this material? Will it float in water? **(c)** Table salt (sodium chloride) has a density of 2.16 g/cm^3. What volume would 36.2 μg of this salt occupy?

1.24 **(a)** A certain sodium chloride solution has a density of 1.033 g/mL. What is the mass of 120 L of this solution? **(b)** A piece of a pure unknown metal has a mass of 46.5 g and a volume of 5.3 cm^3. What is the density of this metal? Could this metal be gold? (see Table 1.6) **(c)** An iron bar has a density of 7.20 g/cm^3 and a mass of 431 g. What is the volume of this piece of iron?

1.25 Make the following temperature conversions: **(a)** 25.4 K to °C; **(b)** 3.6°C to °F; **(c)** 3000°F to K; **(d)** −50°F to °C; **(e)** 354°C to K; **(f)** 12 K to °F.

1.26 Make the following conversions: **(a)** 9.2 K to °C; **(b)** −36.7°C to °F; **(c)** −102°F to K; **(d)** 120°F to °C; **(e)** −70°C to K; **(f)** 1500 K to °F.

Uncertainty in Measurement

1.27 Indicate which of the following are exact numbers: **(a)** the number of inches in a mile; **(b)** the value of π; **(c)** the mass of a 12-oz bag of potato chips; **(d)** the number of ounces in a pound; **(e)** the number of micrometers in a kilometer; **(f)** the number of inches in a kilometer.

1.28 Indicate which of the following are exact numbers: **(a)** the mass of a penny; **(b)** the height of a building; **(c)** the volume of a bottle; **(d)** the number of picoseconds in a year; **(e)** the temperature of the surface of the sun; **(f)** the number of players on a basketball team.

1.29 What is the number of significant figures in each of the following measured quantities: **(a)** 101.3 g; **(b)** 0.00005 dL; **(c)** 365.4 km; **(d)** 12,000 mg; **(e)** 6.54×10^{-12} m?

1.30 Indicate how many significant figures are in each of the following measured quantities: **(a)** 3.141 cm; **(b)** −1.200°C; **(c)** 0.002004 L; **(d)** 3,490,400 ps; **(e)** 6.000×10^{-3} km.

1.31 Round each of the following numbers to four significant figures: **(a)** 12,345,670; **(b)** 2.35500; **(c)** 456,500; **(d)** 3.218×10^3; **(e)** 0.000657030; **(f)** 100,500.1.

1.32 Round each of the following numbers to three significant figures: **(a)** 10.000; **(b)** 0.05000; **(c)** 23,000; **(d)** 1.565×10^1; **(e)** 9,834.05; **(f)** −1235.

1.33 Carry out the following operations and express the answers with the appropriate number of significant figures: **(a)** 1.23056 + 67.809; **(b)** 23.67 − 500; **(c)** 890.05 × 12.3; **(d)** 88,132/22.500.

1.34 Carry out the following operations and express the answer with the appropriate number of significant figures:
(a) $324.55 - (6104.5/22.3)$;
(b) $[(285.3 \times 10^6) - (12.000 \times 10^3)] \times 22.8954$;
(c) $(0.0045 \times 30{,}000.0) + (283 \times 12)$;
(d) $869 \times [1255 - (3.45 \times 10^3)]$.

Dimensional Analysis

1.35 Perform the following conversions: **(a)** 5.0 ft to m; **(b)** 2.55 gal to mL; **(c)** 3.00 days to s; **(d)** 66.2 ft^3 to cm^3; **(e)** 55 mi/hr to km/hr; **(f)** 25.2 mi/gal to km/L.

1.36 Make the following conversions: **(a)** 3.2×10^3 mi to Mm; **(b)** 100 mi/hr to m/s; **(c)** 2.00 yd^3 to m^3; **(d)** 37.2 in./hr to mm/s; **(e)** \$10.50 per pound to pennies per gram.

1.37 The density of air at ordinary atmospheric pressure and 25°C is 1.19 g/L. What is the mass, in kilograms, of the air in a room that measures 8.2 × 13.5 × 2.75 m?

1.38 The maximum allowable concentration of carbon monoxide in urban air is 10 mg/m^3 over an 8-hr period. At this level, what mass of carbon monoxide is present in a room measuring 8 × 12 × 20 ft?

1.39 Gold is presently selling for \$380 per ounce. How many grams of gold can you buy with \$10,000?

1.40 The Morgan silver dollar has a weight of 26.73 g. By law, it was required to contain 90 percent silver, with the remainder being copper. **(a)** When the coin was minted in the late 1800s, silver was worth \$1.18 per ounce. At this price, what is the value of the silver in the silver dollar? **(b)** Today, silver sells for \$6.10 per ounce. How many silver dollars are required to obtain \$25.00 of pure silver?

1.41 In March 1989, the *Exxon Valdez* ran aground and spilled 240,000 barrels of crude petroleum off the coast of Alaska. One barrel of petroleum is equal to 42 gal. How many milliliters of petroleum were spilled?

1.42 A pound of coffee beans yields 50 cups of coffee. How many milliliters of coffee can be obtained from 1 g of coffee beans?

Additional Exercises

1.43 Which of the following are intensive properties: **(a)** mass; **(b)** density; **(c)** temperature; **(d)** area; **(e)** color; **(f)** volume

1.44 Magnesium is used in automobile wheels because it is "lighter" than steel. What is a more scientifically correct statement of this?

1.45 You aren't feeling well, so you go to the doctor. The nurse takes your temperature, and it is 312 K. Do you have a fever?

1.46 How many grams of sulfuric acid were produced in 1988? (See Table 1.1.)

1.47 The density of sodium hydroxide is 2.130 g/cm^3. How many cubic kilometers of sodium hydroxide were produced in 1987? (See Table 1.1.)

1.48 The distance from the Earth to the moon is approximately 240,000 mi. **(a)** What is this distance in millimeters? **(b)** The Concorde SST has an airspeed of about 2400 km/hr. If the Concorde could fly to the moon, how many megaseconds would it take?

1.49 You are given a bottle that contains 2.36 mL of a yellow liquid. The total mass of the bottle and the liquid is 5.26 g. The empty bottle weighs 3.01 g. What is the density of the liquid?

1.50 A U.S. quarter has a mass of 5.67 g and is approximately 1.55 mm thick. **(a)** The Washington Monument is 575 ft tall. How many quarters would have to be stacked to reach this height? **(b)** How much would this stack weigh? **(c)** How much money would this stack contain? **(d)** At this writing, the national debt stands at \$2.9 trillion. How many stacks like the one described would be necessary to pay off this debt?

1.51 Is the use of significant figures in each of the following statements appropriate? Why or why not? **(a)** The 1976 circulation of *Reader's Digest* was 17,887,299. **(b)** There are more than 1.4 million people in the United States who have the surname Brown. **(c)** The average annual rainfall in San Diego, California, is 20.54 in. **(d)** The population of East Lansing, Michigan, was 51,237 in 1979.

1.52 A cylindrical container of radius r and height h has a volume of $\pi r^2 h$. **(a)** Calculate the volume in cubic centimeters of a cylinder with a radius of 16.5 cm and a height of 22.3 cm. **(b)** Calculate the volume in cubic meters of a cylinder that is 6.3 ft high and 2.0 ft in diameter. **(c)** Calculate the mass in kilograms of a volume of mercury equal to the volume of the cylinder in part (b). The density of mercury is 13.6 g/cm^3 (see Sample Exercise 1.3).

1.53 A cylindrical glass tube 15.0 cm in length is filled with ethanol. The mass of ethanol needed to fill the tube is found to be 9.64 g. Calculate the inner diameter of the tube in centimeters. The density of ethanol is 0.789 g/mL.

[1.54] Chromatography (Figure 1.11) is a simple but reliable method for separating a mixture into its constituent substances. Suppose you are using chromatography to separate a mixture of two substances. How would you know whether the separation is successful? Can you propose a means of quantifying how good or how poor the separation is?

[1.55] Suppose you are given a sample of a homogeneous liquid. What would you do to determine whether it is a solution or a pure substance?

[1.56] Using the *Handbook of Chemistry and Physics* or a similar source of data, determine: **(a)** the solid element with the greatest density; **(b)** the solid element with the highest known melting point; **(c)** the element with the lowest known melting point; **(d)** the only two elements that are liquids at room temperature (about 20°C).

2 Atoms, Molecules, and Ions

Chemical elements are produced in stellar interiors. Supernovae, such as the one shown here, are stellar "explosions" that scatter the elements throughout the universe. (© Roger Ressmeyer/Starlight)

CONTENTS

In our discussion of matter in Chapter 1, we introduced the concept of *elements.* The notion that fundamental elements exist in nature originated with ancient Greek philosophers, such as Anaximander of Miletus (about 610–545 B.C.), Empidocles of Agrigentum (about 495–435 B.C.), Plato (427–347 B.C.), and Aristotle (384–322 B.C.). These philosophers proposed that nature was composed of four elements: fire, earth, air, and water.

We now have a very different view of the elements. The modern use of the term *element* was first proposed by Robert Boyle (1627–1691) in his 1661 book *The Sceptical Chymist.* Boyle defined elements as substances that cannot be decomposed into simpler substances. He demonstrated that there was no sound basis for regarding fire, earth, air, and water as elements. His ideas were the stimulus for many of the advances in chemistry made in the eighteenth century.

Classifying substances as elements and compounds provides us with a system for organizing basic chemical information. However, it also raises a number of questions: What makes one element different from another? Why do some elements have very similar chemical and physical properties? Why is a compound different from a mixture? How and why do elements combine to form compounds? These were among the basic questions that chemists tried to answer in the eighteenth century. The answer that emerged, *the atomic theory of matter*, has become the foundation of modern chemistry. It forms the basis for understanding why elements and compounds react in the ways they do and why they exhibit specific physical properties. In this chapter, we will introduce you to the atomic view of matter. We will examine some basic concepts of atomic structure and briefly discuss the formation of molecules and ions. Finally, we will introduce the systematic procedure used to name compounds.

2.1 THE ATOMIC THEORY

The origin of the atomic theory goes back at least to the time of ancient Greece. Greek philosophers pondered the question, Can matter be divided endlessly into smaller and smaller pieces, or is there a point at which it cannot be divided any further? Most of these philosophers, including Plato and Aristotle, believed that matter was infinitely divisible. One person who disagreed with this view was Democritus (460–370 B.C.). He argued that matter is composed of small, indivisible particles called *atomos*, meaning "indivisible."

Despite Democritus's proposition, however, the erroneous idea that matter could be divided infinitely was widely believed until the early nineteenth century. During this long period, scientists were continually accumulating data on how substances react with one another. Some of these data arose from the practice of *alchemy*, in which people attempted to turn cheap metals, such as lead, into gold. As interest in the basic structure of substances grew, more-quantitative methods for studying chemical reactions were developed, and patterns in chemical reactivity were discovered. These patterns seemed inconsistent with the idea of infinitely divisible matter. As a result, Democritus's notion of atoms reemerged.

A meaningful atomic theory was finally published during the period 1803–1807 by John Dalton, an English schoolteacher (Figure 2.1). Dalton designed his theory to explain several experimental observations. His efforts were so insightful that his theory has remained basically intact up to the present.

The basic postulates of Dalton's theory are as follows:

1. Each element is composed of extremely small particles called atoms.
2. All atoms of a given element are identical.
3. Atoms of different elements have different properties (including different masses).
4. Atoms of an element are not changed into different types of atoms by chemical reactions; atoms are neither created nor destroyed in chemical reactions.
5. Compounds are formed when atoms of more than one element combine.
6. In a given compound, the relative number and kind of atoms are constant.

Dalton's theory provides us with a conceptual picture of matter. As shown in Figure 2.2, we can visualize an element as being composed of tiny particles called atoms (Figure 2.3). **Atoms** are the basic building blocks of matter. They are the smallest units of an element that can combine with other elements in a chemical reaction. In compounds, the atoms of two or more elements combine in definite arrangements [Figure 2.2(*b*)]. Mixtures do not involve the intimate interactions between atoms that are found in compounds [Figure 2.2(*c*)].

Dalton's theory embodies several simple laws of chemical combination that were known at the time. Because atoms are neither created nor destroyed in the course of chemical reactions (postulate 4), it follows that matter is neither created nor destroyed in such reactions. This is the *law of conservation of matter*

Figure 2.1 John Dalton (1766–1844) was the son of a poor English weaver. Dalton began teaching at the age of 12; he spent most of his years in Manchester, where he taught both grammar school and college. His lifelong interest in meteorology led him to study gases and hence to chemistry and eventually to the atomic theory. (Library of Congress)

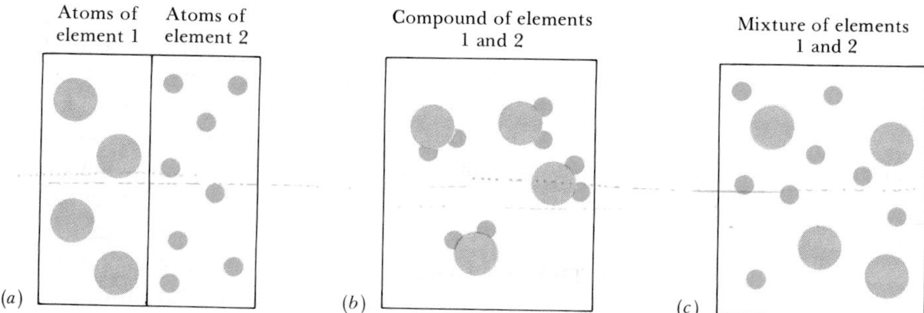

Figure 2.2 Difference between elements, compounds, and mixtures as visualized through Dalton's atomic theory. (*a*) Elements are composed of small particles called atoms. All atoms of a given element are identical; atoms of different elements are different. (*b*) Compounds involve atoms of two or more elements combined in definite arrangements. (*c*) Mixtures have variable compositions. There is no restriction on the relative number of atoms of elements 1 and 2.

(also known as the *law of conservation of mass*): The total mass of materials present after a chemical reaction is the same as the total mass before the reaction. This law is the basis for much of what we will discuss in Chapter 3. Dalton's postulate 6 explains the *law of constant composition* (Section 1.2): In a given compound, the relative number and kind of atoms are constant.

Dalton's atomic theory also explains the **law of multiple proportions:** If two elements A and B combine to form more than one compound, then the masses of B that can combine with a given mass of A are in the ratio of small whole numbers. We can illustrate this law by considering the substances water and hydrogen peroxide, both of which consist of the elements hydrogen and oxygen. We find that, in forming water, 8.0 g of oxygen reacts with 1.0 g of hydrogen. In hydrogen peroxide, there are 16.0 g of oxygen per 1.0 g of hydrogen. In other words, the ratio of the mass of oxygen per gram of hydrogen in the two compounds is 2:1. Using the atomic theory, we can conclude that hydrogen peroxide contains twice as many atoms of oxygen per hydrogen atom as does water.

We can now answer our earlier question about what makes one element different from another: They have different types of atoms. However, this explanation only leads to a more fundamental question: How do the atoms of the various elements differ from each other? To answer this question, we now need to consider the structure of the atom.

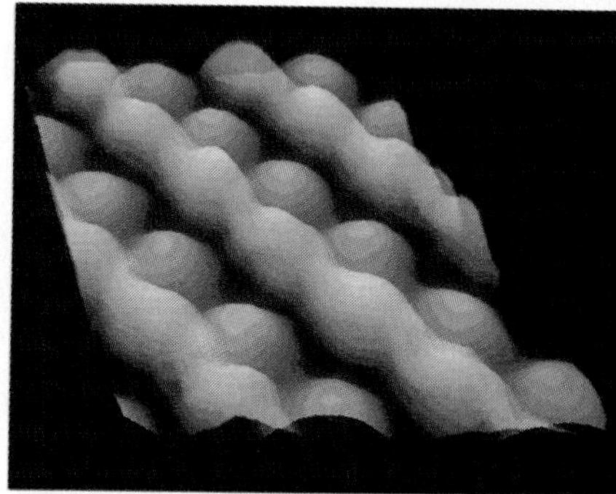

Figure 2.3 The image of the surface of the semiconductor GaAs (gallium arsenide) as obtained by a relatively new technique called tunneling electron microscopy. The blue spheres are gallium atoms, and the red spheres are arsenic atoms. (IBM Research)

2.2 THE DISCOVERY OF ATOMIC STRUCTURE

Scientists presently have a large arsenal of sophisticated equipment with which to measure the properties of individual atoms in great detail. Consequently, we now know a great deal about the structure of atoms. However, only 150 years ago very little was known about atoms beyond what was contained in Dalton's atomic theory. Dalton and his contemporaries viewed the atom as an indivisible object, like a tiny, indestructible and unchangeable ball. By 1850, scientists had begun to accumulate data indicating that the atom is composed of even smaller particles. Before we summarize the current model of atomic structure, we will consider a few of the most important experiments that led to that model. In order to understand these experiments we need to keep in mind a basic rule regarding the behavior of electrically charged particles: *Like charges repel each other; unlike charges attract.*

Cathode Rays and Electrons

In the mid-1800s, scientists began to study electrical discharge through partially evacuated tubes (tubes that had been pumped almost empty of air), such as those shown in Figure 2.4. A high voltage produces radiation within the tube. This radiation became known as **cathode rays** because it originated from the negative electrode, or cathode. Although the rays themselves could not be seen, their movement could be detected because the rays cause certain materials, including glass, to *fluoresce*, or give off light. (Television picture tubes are cathode-ray tubes; a television picture is the result of fluorescence from the television screen.) In the absence of magnetic or electric fields, cathode rays travel in straight lines. However, magnetic and electric fields "bend" the rays in the manner expected for negatively charged particles, as shown in Figure 2.5. Moreover, a metal plate exposed to cathode rays acquired a negative charge. These observations of the properties of cathode rays suggested that the radiation consists of a stream of negatively charged particles, which we now call *electrons*. In addition, it was found that the cathode rays emitted by different cathode materials were

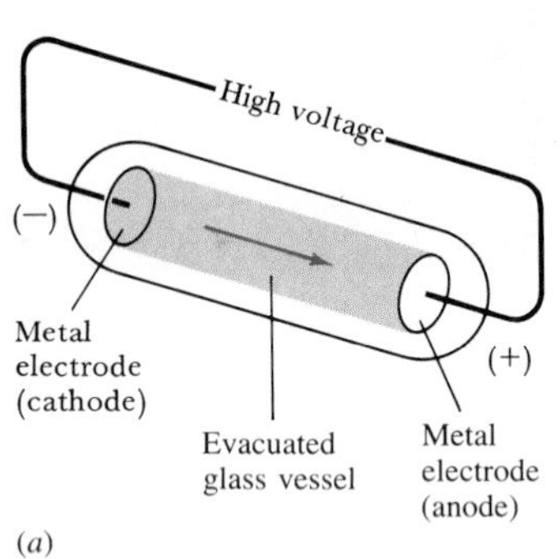

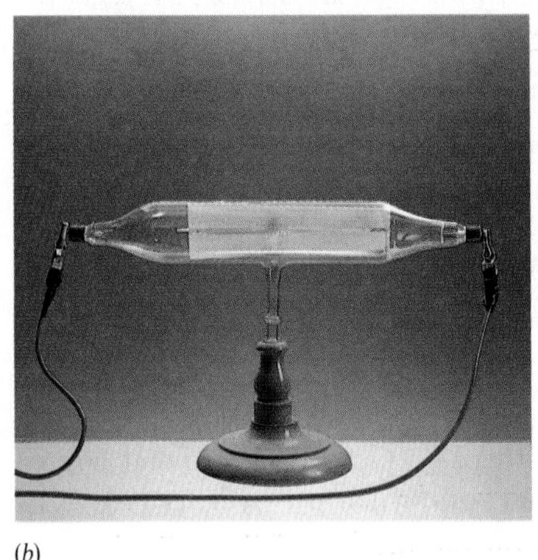

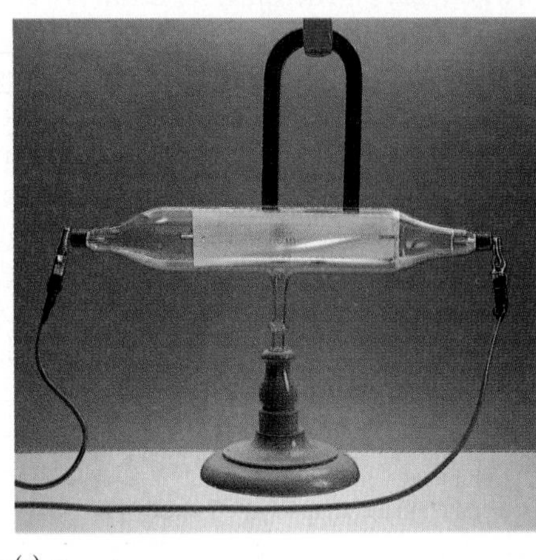

Figure 2.4 (*a*) In a cathode-ray tube, electrons move from the negative electrode (cathode) to the positive electrode (anode). (*b*) A photo of a cathode-ray tube containing a fluorescent screen to show the path of the cathode rays. (*c*) The path of the cathode rays is deflected by the presence of a magnet. (© Richard Megna/ Fundamental Photographs)

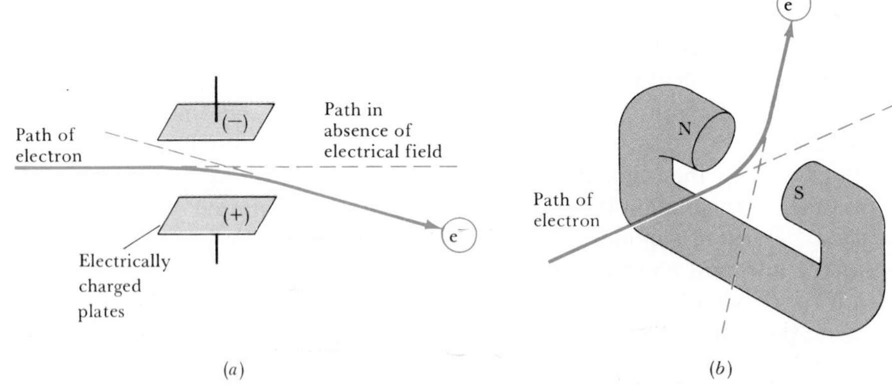

Figure 2.5 The behavior of a negatively charged particle such as an electron moving through an electric field (*a*) and a magnetic field (*b*). The path of the electron in (*a*) is bent because the electron is attracted toward the plate of opposite charge and repelled by the plate carrying the same charge. Magnetic fields, as in (*b*), cause charged particles moving at right angles to the magnetic field to follow a curved path.

Figure 2.6 J. J. Thomson (1856–1940) was appointed professor of physics at Cambridge University when he was not quite 28 years old. He received the Nobel Prize in physics in 1906 for his characterization of the electron. (The Granger Collection)

the same. All of these observations led to the conclusion that electrons are a basic component of matter.

In 1897, the British physicist J. J. Thomson (Figure 2.6) measured the ratio of the electrical charge to the mass of the electron using a cathode-ray tube such as that shown in Figure 2.7. When only the magnetic field is turned on, the electron strikes point A of the tube. When the magnetic field is off and the electric field is on, the electron strikes point C. When both the magnetic and electric fields are off or when they are balanced so as to cancel each other's effects, the electron strikes point B. By carefully and quantitatively determining the effects of magnetic and electric fields on the motion of the cathode rays, Thomson was able to determine the charge-to-mass ratio of 1.76×10^8 coulombs per gram.*

Once the charge-to-mass ratio of the electron was known, a scientist who could measure either the charge or mass of an electron could easily calculate the other quantity. In 1909, Robert Millikan of the University of Chicago devised a way to determine the charge of an electron, an experiment known as

* The coulomb (C) is the SI unit for electrical charge.

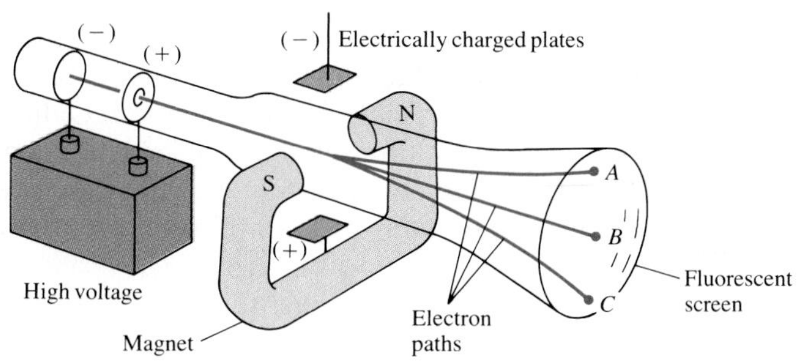

Figure 2.7 Cathode-ray tube with perpendicular magnetic and electric fields. The cathode rays (electrons) originate from the negative plate on the left. The electrons are accelerated toward the positive plate, which has a hole in its center. Many of the electrons pass through this hole and enter the region where their motions are affected by the fields generated by the electrically charged plates and by the magnet.

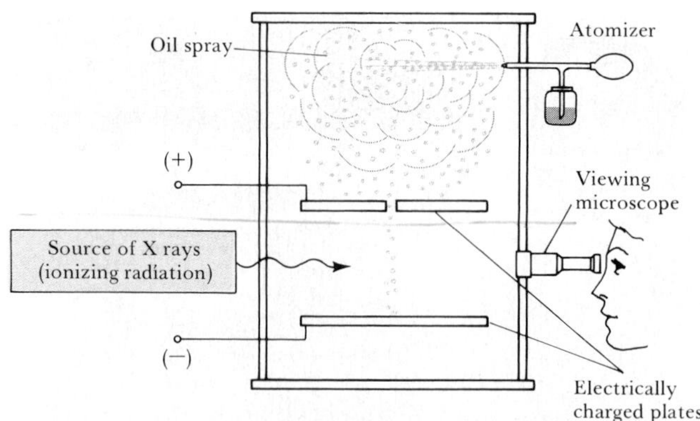

Figure 2.8 Schematic representation of Millikan's apparatus for studying the rate of fall of oil droplets.

"the Millikan oil-drop experiment." The apparatus Millikan used is represented in Figure 2.8. A fine mist of very small oil drops was produced and allowed to fall in air under the influence of both gravity and air resistance. Some of the drops fell through a small hole in the upper of two metal plates. The rate at which a particular drop falls (which Millikan monitored through a microscope) is determined by its size and mass. Thus, by measuring the rate of fall of a drop, Millikan could calculate its mass. The apparatus had a source of high-energy radiation in the lower part of the chamber. This radiation removed electrons from molecules in the air, which in turn became attached to the oil drops. When an electrical potential was applied to the plates, the resultant electric field acted on the charged oil drops. Depending on the charge on the drops and the size and direction of the electric field, the fall of the drops was accelerated, slowed down, or even reversed. Millikan used the difference in the rate of fall of the drops when the electric field was on and off to calculate the charge on the oil drops. After careful measurement on many oil drops, Millikan found that the charge on the drops was always an integral multiple of 1.60×10^{-19} C. He deduced that this must be the charge of the electron.

The mass of the electron was then calculated by combining Thomson's charge-to-mass ratio and Millikan's value of the electron charge:

$$\text{Mass} = \frac{1.60 \times 10^{-19}\ \text{C}}{1.76 \times 10^{8}\ \text{C/g}} = 9.10 \times 10^{-28}\ \text{g}$$

Using slightly more accurate values, we obtain the presently accepted value for the mass of the electron, 9.10939×10^{-28} g. This mass is about 2000 times smaller than that of hydrogen, the lightest atom.

Radioactivity

Throughout the late 1800s, scientists continually discovered new physical phenomena. Many of these discoveries grew out of studies on cathode rays. In 1895, Wilhelm Roentgen (1845–1923) found that when cathode rays struck certain materials, a new type of invisible ray was emitted. Unlike cathode rays, these new rays passed unimpeded through many objects, and they were unaffected by magnetic fields. He also found that they produced an image on photographic plates. Roentgen called this startling new phenomenon *X rays.* Today we know that X rays are a high-energy form of radiation.

When Roentgen announced the discovery of X rays in December 1895, he caused a frenzy within the scientific community. A few months later, the French

Figure 2.9 Marie Sklodowska Curie (1867–1934). When M. Curie presented her doctoral thesis, it was described as the greatest single contribution of any doctoral thesis in the history of science. Among other things, two new elements, polonium and radium, had been discovered. In 1903, Becquerel, M. Curie, and her husband, Pierre, were jointly awarded the Nobel Prize in physics. In 1911, M. Curie won a second Nobel Prize, this time in chemistry. Irene Curie, daughter of Marie and Pierre Curie, was also a scientist. She and her husband, Frederic Joliot, shared the 1935 Nobel Prize in chemistry for their work in artificial production of radioactive substances by bombarding certain elements with particles. (The Bettman Archive)

scientist Henri Becquerel (1852–1908) made an equally startling discovery. Becquerel had been studying substances that become luminous after exposure to sunlight, a phenomenon referred to as *phosphorescence*. After Roentgen's announcement, Becquerel sought to determine whether phosphorescent substances emitted X rays. While working with a phosphorescent uranium mineral, Becquerel accidentally discovered that, even in the dark, the mineral spontaneously produced high-energy radiation. This spontaneous emission of radiation is called **radioactivity**. At Becquerel's suggestion, Marie Sklodowska Curie (Figure 2.9) and her husband, Pierre, began their famous experiments to isolate the radioactive components of the mineral, called pitchblende.

Further study of the nature of radioactivity, principally by the British scientist Ernest Rutherford (1871–1937), revealed three types of radiation: alpha (α), beta (β), and gamma (γ) radiation. Each type differs in its response to an electric field, as shown in Figure 2.10. Both α and β radiation are bent by the electric field, although in opposite directions. In contrast, γ radiation is not affected by the electric field.

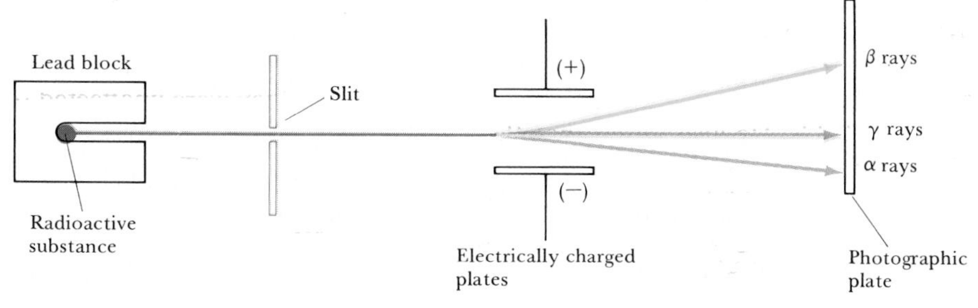

Figure 2.10 Behavior of alpha (α), beta (β), and gamma (γ) rays in an electric field.

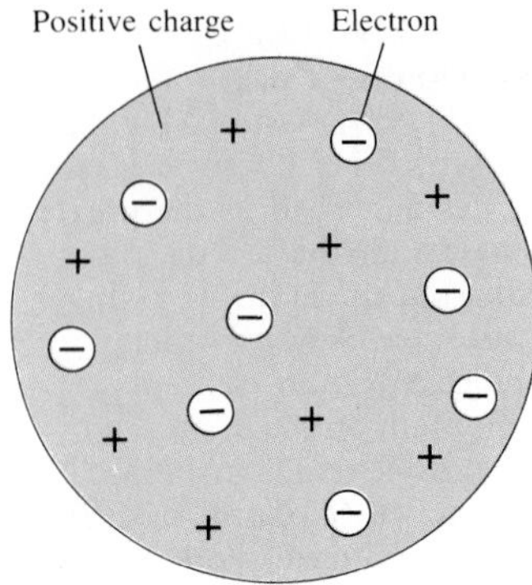

Figure 2.11 J. J. Thomson's "plum pudding" model of the atom, in which the small, negatively charged electrons are imbedded in the large, positively charged portion of the atom. Ernest Rutherford proved this model wrong.

Rutherford showed that both α and β rays consist of fast-moving particles, which were called α and β particles. In fact, β particles are high-speed electrons and can be considered the radioactive equivalent of cathode rays. The α particles are much more massive than the β particles and have a positive rather than negative charge. In units of the charge of the electron, β particles have a charge of $1-$, and α particles a charge of $2+$. Rutherford showed further that α particles combine with electrons to form atoms of helium. He thus concluded that an α particle consists of the positively charged core of the helium atom. He further concluded that γ radiation is high-energy radiation similar to X rays; it does not consist of particles. We will discuss radioactivity in greater detail in Chapter 21.

The Nuclear Atom

Rutherford's studies on radioactivity had profound effects on the model of the atom. After the mass of the electron became known through Millikan's experiments, J. J. Thomson proposed a model for the structure of atoms. He reasoned that because electrons comprised only a very small fraction of the mass of an atom, they probably were responsible for an equally small fraction of the atom's size. He proposed that atoms consisted of a large, positively charged portion in which the very small, negatively charged electrons were imbedded (Figure 2.11). This model, which accounted for the small mass of electrons and the overall electrical neutrality of atoms, became known as the "plum pudding" model because of its resemblance to this popular English dessert. Thomson's atomic model was very short-lived.

In 1910, Rutherford and his co-workers performed an experiment that led to the downfall of Thomson's model. Rutherford was studying the angles at which α particles were scattered as they passed through a thin gold foil. He had found only slight scattering, on the order of 1 degree, which was consistent with Thomson's model. One day Hans Geiger, an associate of Rutherford's, proposed that Ernest Marsden, a 20-year-old undergraduate working in their laboratory, get some experience in conducting such experiments. Rutherford suggested that Marsden see if α particles were scattered through large angles. In Rutherford's own words:

> I may tell you in confidence that I did not believe they would be since we knew that the α particle was a very massive particle with a great deal of energy. . . . Then I remember two or three days later Geiger coming to me in great excitement and saying, "We have been able to get some α particles coming backwards." . . . It was quite the most incredible event that has ever happened to me in my life. It was almost as if you fired a 15-inch shell into a piece of tissue paper and it came back and hit you.

Rutherford and his co-workers observed that almost all of the α particles passed directly through the foil without deflection. However, a few were deflected, some even bouncing back in the direction from which they had come, as shown in Figure 2.12.

By 1911, Rutherford was able to explain these observations; he postulated that most of the mass of the atom, and all of its positive charge, reside in a very small, extremely dense region which he called the **nucleus**. Most of the total volume of the atom is empty space in which electrons move around the nucleus. In the α-scattering experiment, most α particles pass directly through the foil

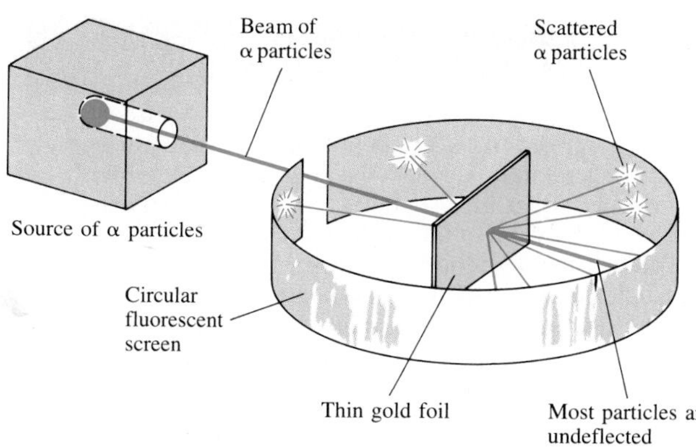

Figure 2.12 Rutherford's experiment on the scattering of α particles.

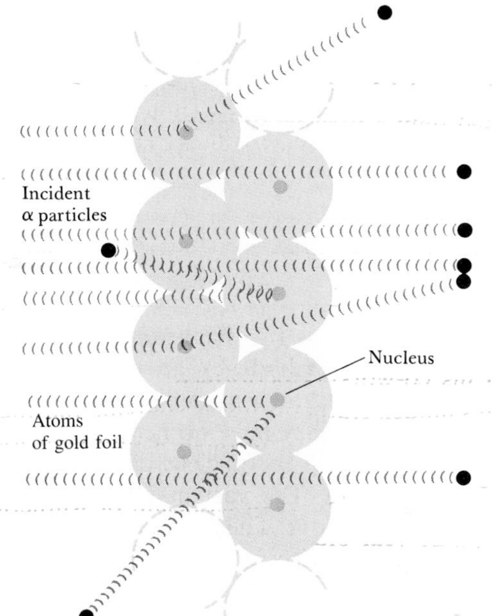

Figure 2.13 Rutherford's model explaining his experiment on the scattering of α particles (Figure 2.12). The gold foil is actually several thousand atoms thick. Both the α particles and the gold nuclei are positively charged; according to Coulomb's law, like charges repel one another. Thus, when an α particle collides with (or passes very close to) a gold nucleus, it is strongly repelled. The less massive α particle is deflected from its path by this repulsive interaction. Because the gold nuclei are extremely small, only a small fraction of the particles is deflected.

because they do not encounter the minute nucleus; they merely pass through the empty space of the atom. Occasionally an α particle comes into the close vicinity of a gold nucleus, however. The repulsion between the highly charged gold nucleus and the α particle is strong enough to deflect the less massive α particle, as depicted in Figure 2.13.

2.3 THE MODERN VIEW OF ATOMIC STRUCTURE

Since the time of Rutherford, physicists have learned much about the detailed composition of atomic nuclei. In the course of these discoveries the list of particles that make up nuclei has grown long and continues to increase. As chemists, we can take a very simple view of the atom, because only three subatomic particles—the **proton**, **neutron**, and **electron**—have a bearing on chemical behavior.

The charge of an electron is -1.602×10^{-19} C, and that of a proton is $+1.602 \times 10^{-19}$ C. The quantity 1.602×10^{-19} C is called the **electronic charge**. For convenience, the charges of atomic and subatomic particles are usually expressed as multiples of this charge rather than in coulombs. Thus, the charge

A CLOSER LOOK: Basic Forces

There are four basic forces, or interactions, known in nature: gravity, electromagnetism, and the strong and the weak nuclear forces. *Gravitational forces* act between all objects in proportion to their masses. Gravitational forces between atoms or subatomic particles are so small that they are of no chemical significance.

Electromagnetic forces act between electrically charged or magnetic objects. Electric and magnetic forces are intimately related. Electric forces are of fundamental importance in understanding the chemical behavior of atoms. The magnitude of the electrical force between two charged particles is given by *Coulomb's law:* $F = kQ_1Q_2/d^2$, where Q_1 and Q_2 are the magnitudes of the charges on the two particles, d is the distance between their centers, and k is a constant determined by the units for Q and d. A negative value for the force indicates attraction, and a positive value indicates repulsion.

All nuclei except those of hydrogen atoms contain two or more protons. Because like charges repel, the electrical repulsion would cause the protons to fly apart if a stronger force did not keep them together in the nucleus. This stronger force is called the *strong nuclear force.* It acts between subatomic particles that are extremely close together, as they are in the nucleus. At this small distance this force is stronger than the electrical force, so the nucleus holds together. The *weak nuclear force* is weaker than the electric force but stronger than gravity. We are aware of its existence only because it shows itself in certain types of radioactivity. The implications of nuclear forces in our daily lives are much less evident than the effects of electromagnetic forces, and we will not consider them further.

All forces that we experience around us are derived from these four basic interactions. For example, the force that your fingers exert on a pencil is the result of electrical repulsion between the outer electrons of the atoms of your fingers and those of the pencil. We will often refer to electrical interactions as we discuss chemical behavior in the coming chapters.

of the electron is 1−, and that of the proton is 1+. Neutrons, which were discovered in 1932 by the British scientist James Chadwick, are uncharged, that is, electrically neutral (which is how they received their name). Because atoms have an equal number of electrons and protons, they have no net electrical charge.

Protons and neutrons reside together in the nucleus of the atom which, as Rutherford proposed, is extremely small. The vast majority of an atom's volume is the space in which the electrons move. The electrons are attracted to the protons in the nucleus by the force that exists between particles of opposite electrical charge. In later chapters, we will see that the strength of the attractive forces between electrons and nuclei can be used to explain many of the differences between different elements.

Atoms have extremely small masses. For example, the mass of the heaviest known atom is on the order of 4×10^{-22} g. Because it would be cumbersome continually to have to express such small masses in grams, we instead use a unit called the *atomic mass unit*, or amu.* An amu equals 1.66053×10^{-24} g. A proton has a mass of 1.0073 amu, a neutron 1.0087 amu, and an electron 5.486×10^{-4} amu. The masses of the proton and neutron are very nearly equal, and both are much greater than that of an electron. In fact, it would take 1836 electrons to equal the mass of 1 proton. Thus the nucleus contains most of the mass of an atom. Table 2.1 summarizes the charges and masses of the subatomic particles. We will have more to say about atomic masses in Section 3.3.

Atoms are extremely small; most of them have diameters between 1×10^{-10} m and 5×10^{-10} m, or 100–500 pm. Another convenient, although non-SI, unit of length used to express atomic dimensions is the **angstrom** (Å). One angstrom equals 10^{-10} m. Thus, atoms have diameters on the order of 1–5 Å.

* The SI abbreviation for the atomic mass unit is merely u. We will use the more common abbreviation amu.

Table 2.1 Comparison of the Proton, Neutron, and Electron

Particle	Charge	Mass (amu)
Proton	Positive (1+)	1.0073
Neutron	None (neutral)	1.0087
Electron	Negative (1−)	5.486×10^{-4}

For example, the diameter of a chlorine atom is 200 pm, or 2.0 Å. Both picometers and angstroms are commonly used to express the dimensions of atoms and molecules, and you should be familiar with both units.

SAMPLE EXERCISE 2.1

The diameter of a U.S. penny is 19 mm. How many chlorine atoms would fit side by side along this diameter?

Solution: As mentioned in the text, the diameter of a single chlorine atom is 2.0 Å. Starting with the diameter of the penny, we can convert its dimensions to angstroms and then, using the diameter of the Cl atom, calculate the number of Cl atoms:

$$\text{Cl atoms} = (19 \text{ mm})\left(\frac{1 \text{ m}}{10^3 \text{ mm}}\right)\left(\frac{10^{10} \text{ Å}}{1 \text{ m}}\right)\left(\frac{1 \text{ Cl atom}}{2.0 \text{ Å}}\right)$$

$$= 9.5 \times 10^7 \text{ Cl atoms}$$

This exercise helps to illustrate how very small atoms are compared to more familiar dimensions.

PRACTICE EXERCISE

The diameter of a carbon atom is 1.5 Å. **(a)** Express this diameter in picometers. **(b)** How many carbon atoms could be aligned side by side in a straight line across the width of a pencil line that is 0.10 mm wide? ***Answers:*** **(a)** 150 pm; **(b)** 6.7×10^5 C atoms

The diameters of atomic nuclei are on the order of 10^{-4} Å, only a small fraction of the diameter of the atom as a whole. If the atom were scaled upward in size so that the nucleus were 2 cm in diameter (about the diameter of a penny), the atom would have a diameter of 200 m (about twice the length of a football field). Because the tiny nucleus carries most of the mass of the atom in such a small volume, it has an incredible density—on the order of 10^{13}–10^{14} g/cm^3. A matchbox full of material of such density would weigh over 2.5 billion tons! Astrophysicists have suggested that the interior of a collapsed star may reach nearly this density.

An illustration of the atom that incorporates the features we have just discussed is shown in Figure 2.14. The electrons, which take up most of the

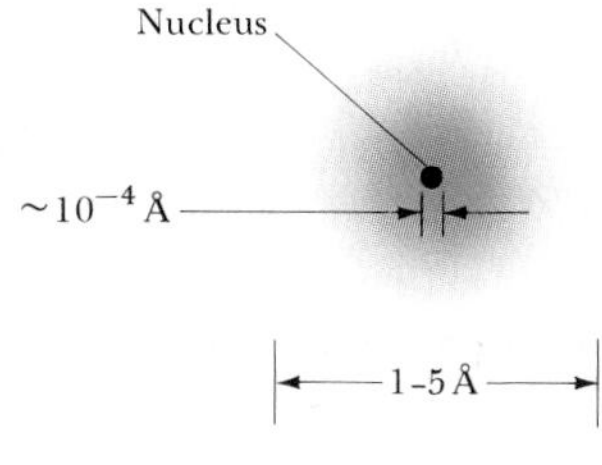

Figure 2.14 Schematic cross-sectional view through the center of an atom. The nucleus, which contains protons and neutrons, is the location of virtually all the mass of the atom. The rest of the atom is the space in which the light, negatively charged electrons move.

volume of the atom, play the major role in chemical reactions. The significance of representing the region containing the electrons as an indistinct cloud will become clear in later chapters when we consider the energies and spatial arrangements of the electrons.

Isotopes, Atomic Numbers, and Mass Numbers

What makes an atom of one element different from an atom of another element? The answer to this question centers on the number of protons in the nucleus of the atom: *All atoms of an element have the same number of protons in the nucleus.* The specific number of protons is different for different elements. Furthermore, because an atom has no net electrical charge, the number of electrons in it must equal its number of protons. For example, all atoms of the element carbon have six protons and six electrons. Most carbon atoms also have six neutrons, although some have more and some have less.

Atoms of a given element that differ in the number of neutrons, and consequently in mass, are called **isotopes**. The symbol $^{12}_{6}C$ or simply ^{12}C (read "carbon twelve," carbon-12) represents the carbon atom with six protons and six neutrons. The number of protons, which is called the **atomic number**, is shown by the subscript. The atomic number of each element is listed with the name and symbol of the element on the front inside cover of the text. Because all atoms of a given element have the same atomic number, the subscript is redundant and hence is often omitted. The superscript is called the **mass number**; it is the total number of protons plus neutrons in the atom. For example, some carbon atoms contain six protons and eight neutrons and are consequently represented as ^{14}C (read "carbon fourteen"). The known isotopes of carbon are listed in Table 2.2.

We will generally use the notation with subscripts and superscripts only when making reference to a particular isotope of an element. An atom of a specific isotope is called a **nuclide**. Thus, an atom of $^{14}_{6}C$ is referred to as a $^{14}_{6}C$ nuclide.* We will have more to say about the isotopic compositions of the elements in Section 3.3, when we examine atomic masses.

* We can now state, more precisely, that Rutherford's α particles are the nuclei of $^{4}_{2}He$ nuclides.

Table 2.2 The Known Isotopes of Carbon[a]

Symbol	Number of protons	Number of electrons	Number of neutrons
^{10}C	6	6	4
^{11}C	6	6	5
^{12}C	6	6	6
^{13}C	6	6	7
^{14}C	6	6	8
^{15}C	6	6	9
^{16}C	6	6	10

[a] Almost 99 percent of the carbon found in nature consists of ^{12}C.

SAMPLE EXERCISE 2.2

How many protons, neutrons, and electrons are in an atom of ^{197}Au?

Solution: According to the list of elements given on the front inside cover of this text, gold has an atomic number of 79. Consequently, an ^{197}Au atom has 79 protons, 79 electrons, and 197 − 79 = 118 neutrons.

PRACTICE EXERCISE

How many protons, neutrons, and electrons are in a ^{39}K atom?
Answer: 19 protons, 19 electrons, and 20 neutrons

SAMPLE EXERCISE 2.3

Hydrogen has three isotopes, with mass numbers 1, 2, and 3. Write the complete chemical symbol for each of these.

Solution: Hydrogen has atomic number 1, so all atoms of hydrogen contain one proton. The three isotopes are therefore represented by $^{1}_{1}H$, $^{2}_{1}H$, and $^{3}_{1}H$.

PRACTICE EXERCISE

Give the complete chemical symbol for the nuclide that contains 18 protons, 18 electrons, and 22 neutrons. *Answer:* $^{40}_{18}Ar$

All atoms are made up of protons, neutrons, and electrons. Because these particles are the same in all atoms, the difference between atoms of distinct elements (gold and oxygen, for example) is due entirely to the difference in the number of subatomic particles in each atom. We can therefore consider an atom to be the smallest sample of an element, because breaking an atom into subatomic particles destroys its identity.

It is enlightening now to reconsider the goal of early alchemists, namely, the conversion of a common metal like lead (atomic number 82) into precious gold (atomic number 79). This transformation would require the removal of three protons from the nucleus of the lead atom, an exceedingly difficult task because of the strong binding forces between nuclear particles. As we will discuss in Chapter 21, the energies necessary to cause changes in the nucleus are much greater than those associated with even the most vigorous chemical changes. Thus, we still agree with Dalton's initial hypothesis that atoms of an element are not changed into different types of atoms by chemical reactions.

2.4 THE PERIODIC TABLE

Dalton's atomic theory set the stage for a vigorous growth in chemical experimentation during the early 1800s. As the body of chemical observations grew and the list of known elements expanded, attempts were made to find regularities in chemical behavior. These efforts culminated in the development of the periodic table in 1869. We will have much to say about the periodic table in later chapters, but it is so important and useful that you should become acquainted with it now.

Many elements show very strong similarities to each other. For example, lithium (Li), sodium (Na), and potassium (K) are all soft, very reactive metals. The elements helium (He), neon (Ne), and argon (Ar) are very nonreactive gases. If the elements are arranged in order of increasing atomic number, their chemical and physical properties are found to show a repeating, or periodic,

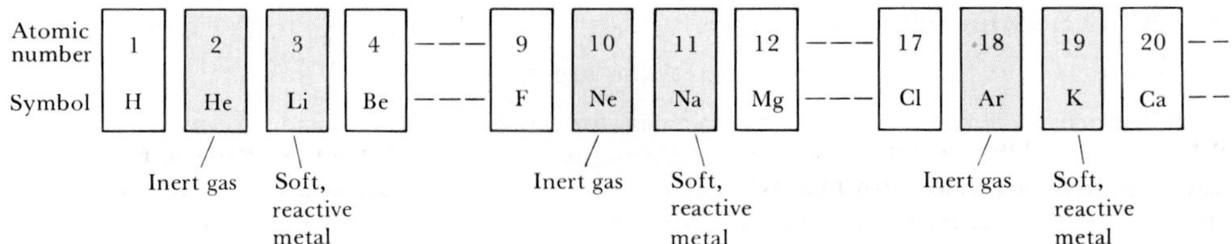

Figure 2.15 Arranging the elements by atomic number illustrates the periodic, or repeating, pattern in properties that is the basis of the periodic table.

pattern. For example, each of the soft, reactive metals—lithium, sodium, and potassium—comes immediately after one of the nonreactive gases—helium, neon, and argon—as shown in Figure 2.15. The arrangement of elements in order of increasing atomic number, with elements having similar properties placed in vertical columns, is known as the **periodic table**. The periodic table is shown in Figure 2.16 and is also given on the front inside cover of the text. You will see large periodic tables hung on the walls of most chemistry classrooms—a testimony to their usefulness. You may notice slight variations in periodic tables from one book to another, or between those in the lecture hall and in the text. These are matters of style or concern the particular information included; there are no fundamental differences.

The elements in a column of the periodic table are known as a **family** or **group**. They are identified as group 1A, 2A, and so forth along the top of the

Figure 2.16 Periodic table of the elements, showing the division of elements into metals, semimetals, and nonmetals.

1A	2A	3B	4B	5B	6B	7B	8B	8B	8B	1B	2B	3A	4A	5A	6A	7A	8A
1 H																	2 He
3 Li	4 Be											5 B	6 C	7 N	8 O	9 F	10 Ne
11 Na	12 Mg											13 Al	14 Si	15 P	16 S	17 Cl	18 Ar
19 K	20 Ca	21 Sc	22 Ti	23 V	24 Cr	25 Mn	26 Fe	27 Co	28 Ni	29 Cu	30 Zn	31 Ga	32 Ge	33 As	34 Se	35 Br	36 Kr
37 Rb	38 Sr	39 Y	40 Zr	41 Nb	42 Mo	43 Tc	44 Ru	45 Rh	46 Pd	47 Ag	48 Cd	49 In	50 Sn	51 Sb	52 Te	53 I	54 Xe
55 Cs	56 Ba	57 La	72 Hf	73 Ta	74 W	75 Re	76 Os	77 Ir	78 Pt	79 Au	80 Hg	81 Tl	82 Pb	83 Bi	84 Po	85 At	86 Rn
87 Fr	88 Ra	89 Ac	104 Rf	105 Ha	[106]	[107]	[108]	[109]									

58 Ce	59 Pr	60 Nd	61 Pm	62 Sm	63 Eu	64 Gd	65 Tb	66 Dy	67 Ho	68 Er	69 Tm	70 Yb	71 Lu
90 Th	91 Pa	92 U	93 Np	94 Pu	95 Am	96 Cm	97 Bk	98 Cf	99 Es	100 Fm	101 Md	102 No	103 Lw

Metals

Semimetals

Nonmetals

Table 2.3 Family Names for Some of the Groups in the Periodic Table

Group	Name	Elements
1A	Alkali metals	Li, Na, K, Rb, Cs, Fr
2A	Alkaline earth metals	Be, Mg, Ca, Sr, Ba, Ra
5A	Pnicogens ("choking-gas formers")	N, P, As, Sb, Bi
6A	Chalcogens ("chalk formers")	O, S, Se, Te, Po
7A	Halogens ("salt formers")	F, Cl, Br, I, At
8A	Noble gases (or inert gases or rare gases)	He, Ne, Ar, Kr, Xe, Rn

table.* Elements that belong to the same group often exhibit some similarities in their physical and chemical properties. You can see, for example, that the "coinage" metals—copper (Cu), silver (Ag), and gold (Au)—all belong to group 1B.

As their name suggests, the coinage metals are used throughout the world to make coins. Many other groups in the periodic table have family names. For example, the members of group 1A—lithium, sodium, potassium, rubidium (Rb), cesium (Cs), and francium (Fr)—are known as the *alkali metals*. The names of several of the groups in the periodic table are given in Table 2.3. You should note the origins of these family names; they arise from some of the chemical characteristics of each group, as we will discuss in later chapters.

We will learn in Chapters 6 and 7 that the elements in a family of the periodic table have similar properties because they have the same type of arrangement of electrons at the periphery of their atoms. However, we need not wait until then to make good use of the periodic table; after all, the table in pretty much its current form was invented by chemists who knew nothing of the electronic structures of atoms! We can use the table, as they intended, to correlate the behaviors of elements and to aid in remembering many facts. You will find it helpful to refer to the periodic table frequently in studying the remainder of this chapter.

Even in our first viewing of the periodic table, we can see patterns in the physical properties of the elements. All the elements on the left side and in the middle of the periodic table (except for hydrogen) are **metallic elements**, or **metals**. The majority of elements are metallic. Metals share many characteristic properties, such as luster and high electrical and heat conductivity. All metals, with the exception of mercury (Hg), are solids at room temperature. The metals are separated from the **nonmetallic elements** by a diagonal steplike line that runs from boron (B) to astatine (At), as shown in Figure 2.16. Hydrogen, although on the left side of the periodic table, is also a nonmetal. Some of the nonmetals are gaseous, some are liquid, and some are solid at room temperature. They generally differ from the metals in appearance (Figure 2.18) and in other physical properties. Many of the elements that lie along the line that separates metals from nonmetals, such as antimony (Sb), have properties that fall between those of metals and nonmetals. These elements are often referred to as **semimetals** or **metalloids**.

* The labeling of the families is basically arbitrary, and three different labeling schemes are presently in use: (1) the scheme shown in Figure 2.16, which is currently most common; (2) the scheme that numbers the columns from 1A through 8A and then from 1B through 8B, thus giving the label 8B to the noble gases; and (3) the scheme recently proposed that numbers the columns from 1 (for the alkali metals) through 18 (for the noble gases), with no A or B designations. When using the first two schemes, Roman numerals, rather than Arabic ones, are sometimes employed. Thus the halogens (group 7A) are often labeled VIIA.

CHEMISTRY AT WORK: Radioactive Isotopes and the Chernobyl Accident

All known isotopes of uranium, plutonium, and many other heavy elements are radioactive and thus intrinsically unstable. As we will see in Chapter 21, these isotopes release radiation as they spontaneously change or "decay" into other elements. Radioactive isotopes also exist for other elements, a fact dramatically demonstrated by a recent tragic event.

On April 26, 1986, the worst nuclear reactor accident in history occurred at Chernobyl, Russia (Figure 2.17). While testing a modification to the reactor, the operators of the power plant inexcusably bypassed the reactor's emergency protection systems. Because the safety system was bypassed, the reactor overheated. As a result, cooling water used to remove heat from the reactor core was converted quickly to steam. The increasing pressure eventually caused the reactor to explode, blowing off its 1000-ton concrete lid. The contents of the core of the reactor were blown into the atmosphere. Dust from the explosion was dispersed through the atmosphere, settling to the earth over a wide area. By July 1986, significant levels of radioactive isotopes from Chernobyl were found in the Canadian arctic and the western United States, as well as in the soil and vegetation of most parts of Europe.

The primary fuel used in the reactor was an isotope of uranium, ^{235}U. This isotope was also the principal component of the atomic bomb dropped on Hiroshima, Japan, in 1945. The ^{235}U produces radioactive nuclides of several elements, including Xe, Kr, I, Te, Cs, and Sr, as byproducts. Of these nuclides, ^{131}I, ^{134}Cs, ^{137}Cs, and ^{90}Sr are particularly noteworthy because of their long-term effects on human health. Iodine, cesium, and strontium move readily into the human food chain. Radioactive ^{131}I, like the nonradioactive isotopes of iodine, concentrates in the thyroid gland. Cesium is chemically similar to the essential nutrient potassium, which is just above it in the alkali metal family. Radioactive cesium can therefore be mistaken for potassium by living organisms and thus gain entry to cells. Strontium, which is just below calcium in the alkaline earth family, can replace calcium in bone tissue. Once inside the body, all these radioactive isotopes can induce cancer. The Chernobyl accident is therefore expected to increase the number of cancer deaths in the coming decades, although exact predictions are impossible to make. By one estimate, the radioactive material released at Chernobyl will increase the fatal cancer rate worldwide by about 28,000 over the normal rate of 600 million during the next 50 years. Time will tell us how accurate these forecasts are.

The amount of radioactive material released at Chernobyl dwarfs that from the 1979 accident at the Three Mile Island nuclear power plant near Harrisburg, Pennsylvania. No detectable radioactive cesium was released at Three Mile Island, and the release of ^{131}I was only $\frac{1}{2000}$ as great as that at Chernobyl.

Want to read more? We recommend: C. H. Atwood, "Chernobyl—What Happened?," *Journal of Chemical Education*, 65, 1037–1041 (1988).

Figure 2.17 The Chernobyl nuclear power plant, site of a major accident in April 1986. (Photo by V. Zufarov, Tass/Sovfoto)

Figure 2.18 Some familiar examples of metals and nonmetals. The metals pictured are in the form of silver dollars, an aluminum whisk, brass keys, copper pennies, and gold jewelry. The nonmetals are in the form of tincture of iodine, graphite and diamonds (both forms of carbon), and sulfur. (Richard Megna/Fundamental Photographs)

SAMPLE EXERCISE 2.4

Which of the following elements would you expect to show the greatest similarity in chemical and physical properties: Li, Be, F, S, Cl?

Solution: The elements F and Cl should be most alike because they are in the same family (group 7A, the halogen family).

PRACTICE EXERCISE

Locate P (phosphorus) and K (potassium) on the periodic table. Give the atomic number of each and label each a metal, semimetal, or nonmetal. ***Answer:*** P, atomic number 15, is a nonmetal; K, atomic number 19, is a metal.

2.5 MOLECULES AND IONS

We have seen that the atom is the smallest representative sample of an element. However, only the noble gas elements are normally found in nature as isolated atoms. Most matter is composed of molecules or ions, both of which are formed from atoms.

Molecules

A **molecule** is an assembly of two or more tightly bound atoms. The resultant "package" of atoms behaves in many ways as a single, distinct object, just as a television set composed of many parts can be recognized as a single object. We will discuss the forces that hold the atoms together (the chemical bonds) in Chapters 8 and 9.

Many elements are found in nature in molecular form; that is, two or more of the same type of atom are bound together. For example, oxygen as it is normally found in air consists of molecules that contain two oxygen atoms. We represent this molecular form of oxygen by the **chemical formula** O_2 (read "oh two"). The subscript in the formula tells us that two oxygen atoms are present in each molecule. Any molecule that is made up of two atoms is said to be a **diatomic molecule**. Oxygen also exists in another molecular form known as *ozone*. Molecules of ozone consist of three oxygen atoms, so its chemical formula is O_3. Even though "normal" oxygen (O_2) and ozone are both composed only

Figure 2.19 Common elements that exist as diatomic molecules at room temperature.

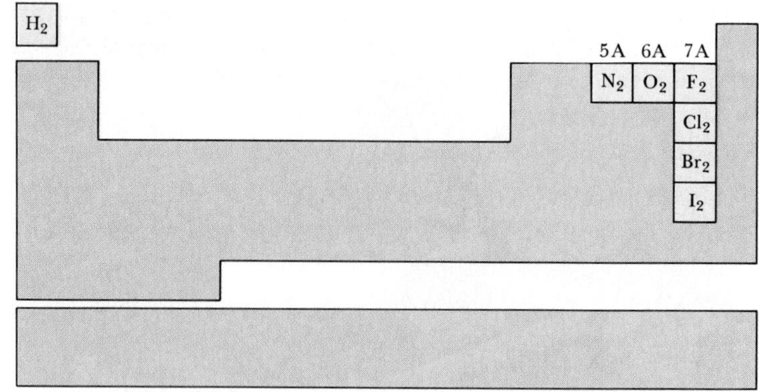

of oxygen atoms, they exhibit very different chemical and physical properties. For example, O_2 is essential for life, but O_3 is toxic; O_2 is odorless, whereas O_3 has a sharp, pungent smell.

The elements that normally occur as diatomic molecules are hydrogen, oxygen, nitrogen, and the halogens. Their locations in the periodic table are shown in Figure 2.19. When we speak of the substance hydrogen, we mean H_2 unless we indicate explicitly otherwise. Likewise, when we speak of oxygen, nitrogen, or any of the halogens, we are referring to O_2, N_2, F_2, Cl_2, Br_2, or I_2. Thus, the properties of oxygen and hydrogen listed in Table 1.3 are those of O_2 and H_2. Other forms of these elements behave much differently.

Molecules of compounds contain more than one type of atom. For example, a molecule of water consists of two hydrogen atoms and one oxygen atom. It is therefore represented by the chemical formula H_2O (read "aitch two oh"). Lack of a subscript on the O implies one atom of O per molecule. Another compound composed of these same elements is hydrogen peroxide, H_2O_2. The properties of these two compounds are very different.

Several common molecules are shown in Figure 2.20. You can see that the

Figure 2.20 Representation of some common simple molecules.

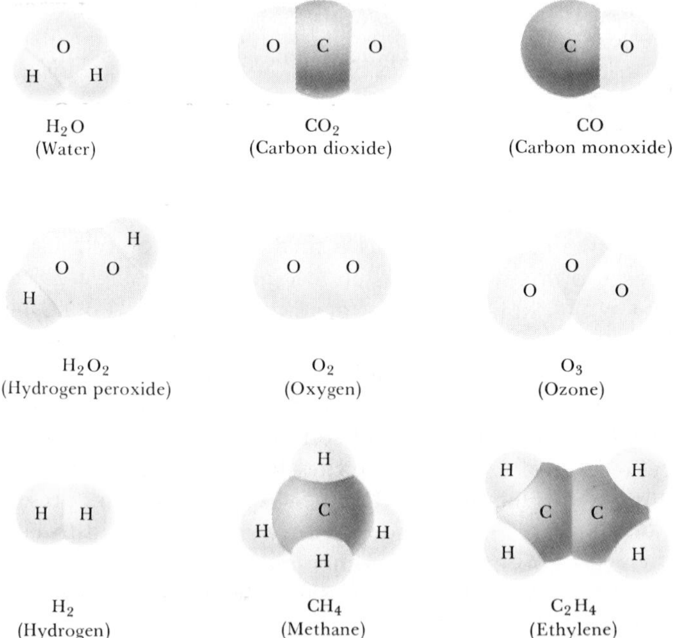

composition of each molecule is given by its chemical formula. Note also that these substances are composed only of nonmetallic elements. Most molecular substances that we will encounter contain only nonmetals.

SAMPLE EXERCISE 2.5

Which of the molecules shown in Figure 2.20 are compounds and which are elements?

Solution: The compounds are H_2O, CO_2, CO, H_2O_2, CH_4, and C_2H_4. The elements are O_2, O_3, and H_2 (one type of atom).

PRACTICE EXERCISE

Which of the following formulas represent molecules of elemental substances: P_4, HCl, NH_3? ***Answer:*** P_4

Chemical formulas that indicate the *actual* numbers and types of atoms in a molecule are called **molecular formulas**. (The formulas in Figure 2.20 are molecular formulas.) Chemical formulas that give only the *relative* number of atoms of each type in a molecule are called **empirical** or **simplest formulas**. The subscripts in an empirical formula are always the smallest whole-number ratios; conversely, the subscripts in a molecular formula are always integral multiples of the subscripts in the empirical formula of the substance. For example, the empirical formula for hydrogen peroxide is HO; its molecular formula is H_2O_2. The empirical formula for ethylene is CH_2; its molecular formula is C_2H_4. For many substances, the empirical formula and molecular formula are identical, as in the case of water, H_2O.

SAMPLE EXERCISE 2.6

Write the empirical formulas for the following molecules: **(a)** glucose, a substance also known as blood sugar and as dextrose, whose molecular formula is $C_6H_{12}O_6$; **(b)** nitrous oxide, a substance used as an anesthetic and commonly called laughing gas, whose molecular formula is N_2O.

Solution: **(a)** The empirical formula has subscripts that are the smallest whole-number ratios. The smallest ratios are obtained by dividing each subscript by the largest common factor, in this case 6. The resultant empirical formula is CH_2O.

(b) Because the subscripts in N_2O are already the lowest integral numbers, the empirical formula for nitrous oxide is the same as its molecular formula, N_2O.

PRACTICE EXERCISE

Give the empirical formula for the substance whose molecular formula is Si_2H_6.
Answer: SiH_3

Molecular formulas are preferred over empirical formulas because they provide more information. Many substances do not exist as discrete molecules, however, and for these we can write only empirical formulas. For example, the element carbon normally exists in extended three-dimensional structures rather than as isolated atoms or molecules. Because there are no distinct molecules of carbon in this structure, we cannot write a molecular formula. The empirical formula for any element is simply the symbol for that element, so carbon is represented by its symbol, C.

Often the formula of a molecule is written to show how its atoms are joined together. For example, the formulas for water and hydrogen peroxide can be written as follows:

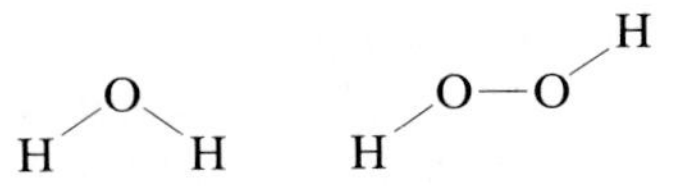

H—O—H (Water) H—O—O—H (Hydrogen peroxide)

Water Hydrogen peroxide

Such formulas are known as **structural formulas**. The lines between the symbols for the elements represent the bonds that hold the respective atoms together. These formulas indicate which atoms are attached to which; however, they do not necessarily tell anything about the shapes of molecules (that is, about the actual angles at which the atoms are joined together).

Ions

The nucleus of an atom is unchanged by ordinary chemical processes, but atoms can readily gain or lose electrons. If electrons are removed or added to a neutral atom, a charged particle called an **ion** is formed. An ion with a positive charge is called a **cation** (pronounced CAT-ion); a negatively charged ion is called an **anion** (AN-ion). For example, the sodium atom, which has 11 protons and 11 electrons, easily loses an electron. The resulting cation has 11 protons and 10 electrons, and hence has a net charge of 1+. The net charge on an ion is represented by a superscript; +, 2+, and 3+ mean a net charge resulting from the loss of one, two, or three electrons, respectively. The superscripts −, 2−, and 3− represent net charges resulting from the gain of one, two, or three electrons, respectively. The formation of the Na^+ ion from a Na atom is shown schematically below:

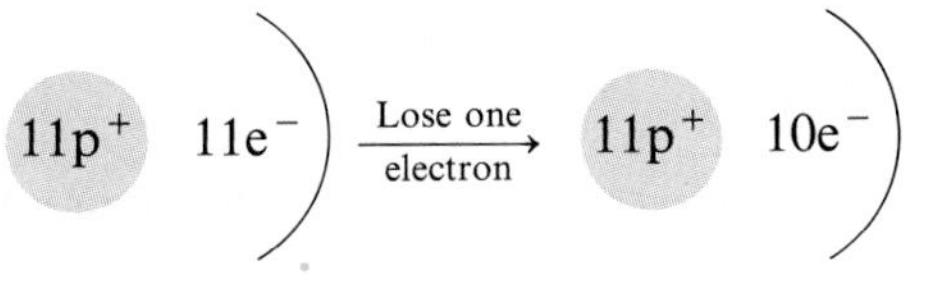

$11p^+$ $11e^-$ $\xrightarrow[\text{electron}]{\text{Lose one}}$ $11p^+$ $10e^-$

Na atom Na^+ ion

Chlorine, with 17 protons and 17 electrons, often gains an electron in chemical reactions, producing the Cl^- ion:

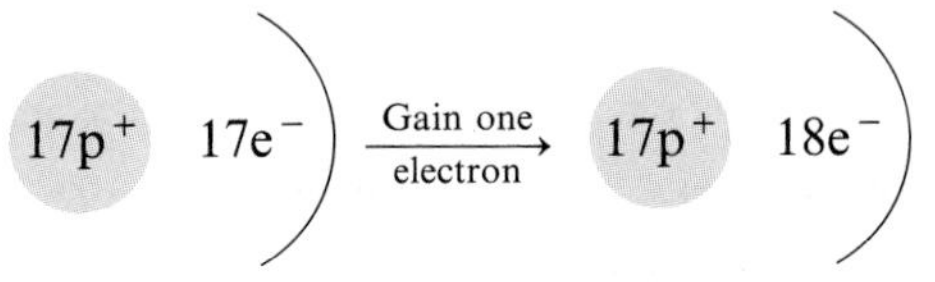

$17p^+$ $17e^-$ $\xrightarrow[\text{electron}]{\text{Gain one}}$ $17p^+$ $18e^-$

Cl atom Cl^- ion

In general, metal atoms lose electrons most readily, and nonmetal atoms tend to gain electrons.

SAMPLE EXERCISE 2.7

Give the chemical symbol for the ion with 26 protons and 24 electrons.

Solution: The element whose atoms have 26 protons (atomic number 26) is Fe (iron). If the ion has two more protons than electrons, it has a net charge of 2+; thus the symbol for the ion is Fe^{2+}.

PRACTICE EXERCISE

How many protons and electrons does the Se^{2-} ion possess? ***Answer:*** 34 protons and 36 electrons

In addition to simple ions such as Na^+ and Cl^-, there are **polyatomic ions** such as NO_3^- (nitrate ion) and SO_4^{2-} (sulfate ion). These ions consist of atoms joined together as in a molecule, but they have a net positive or negative charge. We will consider further examples of polyatomic ions in Section 2.6.

The chemical properties of ions are greatly different from those of the atoms from which they are derived. The change of an atom or molecule to an ion is like that from Dr. Jekyll to Mr. Hyde: Although the body may be essentially the same (plus or minus a few electrons), the behavior is much different.

Many atoms gain or lose electrons so as to end up with the same number of electrons as the noble gas closest to it in the periodic table. The members of the noble gas family are chemically very nonreactive and form very few compounds. We might deduce that this is because their electron arrangements are very stable. Nearby elements can obtain these same stable arrangements by losing or gaining electrons. For example, loss of one electron from an atom of sodium leaves it with the same number of electrons as the neutral neon atom (atomic number 10). Similarly, when chlorine gains an electron, it ends up with 18, the same as argon (atomic number 18). We will content ourselves with this simple observation in explaining the formation of ions until later chapters in which we consider chemical bonding.

SAMPLE EXERCISE 2.8

Predict the charges expected for the most stable ions of barium and oxygen.

Solution: Refer to the periodic table. Barium has atomic number 56. The nearest noble gas is xenon, atomic number 54. Barium can obtain the stable arrangement of 54 electrons by losing two of its electrons, thereby forming the Ba^{2+} cation.

Oxygen has atomic number 8. The nearest noble gas is neon, atomic number 10. Oxygen can obtain this stable electron arrangement by gaining two electrons, thereby forming an anion of 2− charge, O^{2-}.

PRACTICE EXERCISE

Predict the charge of the most stable ion of aluminum. *Answer:* 3+

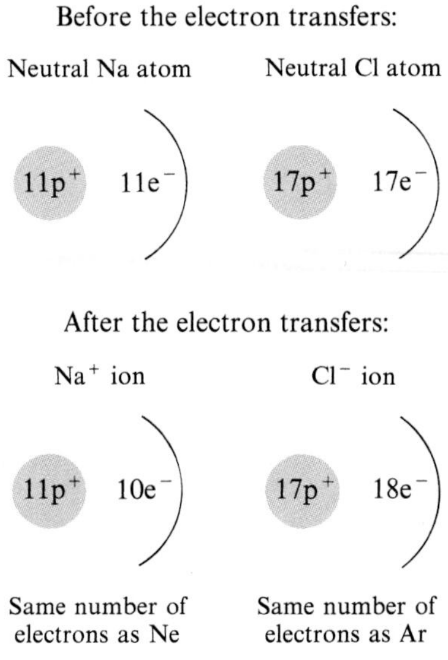

We will see in later chapters that a great deal of chemical activity involves the transfer of electrons between substances. Ions are formed when one or more electrons are transferred from one neutral atom to another. Such a scenario is depicted in Figure 2.21: An electron is transferred from a neutral sodium atom to a neutral chlorine atom. We are left with a Na^+ ion and a Cl^- ion. Note that both ions have the same number of electrons as a noble gas. In general, positively charged ions (cations) are metal ions; negatively charged ions (anions) are nonmetal ions. Consequently, *ionic compounds are generally combinations of metals and nonmetals.* Ordinary table salt, NaCl (which can be made by reacting elemental sodium and elemental chlorine), is an example of an ionic compound; it consists of equal numbers of Na^+ and Cl^- ions that result from the transfer of electrons between the atoms. In contrast, *molecular compounds are generally composed of nonmetals only.*

Figure 2.21 The transfer of an electron from a neutral Na atom to a neutral Cl atom leads to the formation of a Na^+ ion and a Cl^- ion.

SAMPLE EXERCISE 2.9

Which of the following compounds would you expect to be ionic: N_2O, Na_2O, $CaCl_2$, SF_4?

Figure 2.22 Arrangement of ions in solid sodium chloride, NaCl.

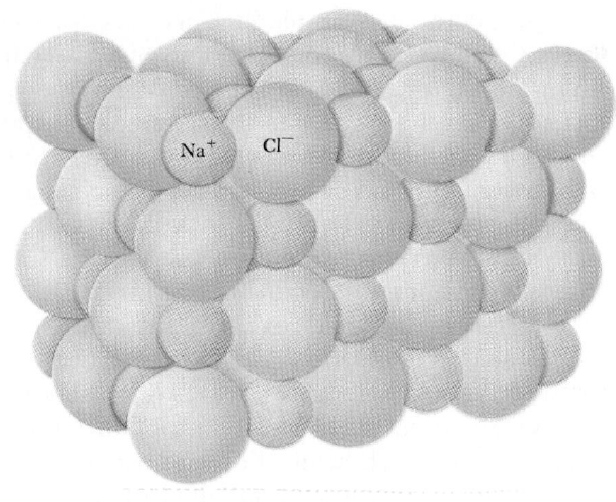

Solution: We would predict that the ionic compounds are Na_2O and $CaCl_2$ because they are composed of a metal combined with a nonmetal. The other two compounds, which are composed entirely of nonmetals, are molecular.

PRACTICE EXERCISE

Which of the following compounds are molecular: CI_4, FeS, P_4O_6, PbF_2?
Answer: CI_4 and P_4O_6

The ions in ionic compounds are arranged in three-dimensional structures. The arrangement of Na^+ and Cl^- ions in NaCl is shown in Figure 2.22. Because there is no discrete molecule of NaCl, we are able to write only an empirical formula for this substance. In fact, only empirical formulas can be written for most ionic compounds.

It is a simple matter to write the empirical formula for an ionic compound if we know the charges of the ions of which it is composed. Chemical compounds are always electrically neutral. Consequently, the ions in an ionic compound always occur in such a ratio that the total positive charge is equal to the total negative charge. Thus, there is one Na^+ to one Cl^- giving NaCl, one Ba^{2+} to two Cl^- giving $BaCl_2$, and so forth.

SAMPLE EXERCISE 2.10

What are the empirical formulas of the compounds formed by **(a)** Al^{3+} and Cl^- ions; **(b)** Al^{3+} and O^{2-} ions; **(c)** Mg^{2+} and NO_3^- ions?

Solution: **(a)** It requires three Cl^- ions to balance the charge of one Al^{3+} ion. Thus the formula is $AlCl_3$.

(b) It requires two Al^{3+} ions to balance the charge of three O^{2-} ions (that is, the total positive charge is 6+ and the total negative charge is 6−). Thus the formula is Al_2O_3. **(c)** Two NO_3^- ions are needed to balance the charge of one Mg^{2+}. Thus the formula is $Mg(NO_3)_2$. In this case the formula for the entire negative ion must be enclosed in parentheses so that it is clear that the subscript 2 applies to all the atoms of that ion.

PRACTICE EXERCISE

Write the empirical formulas for the compounds formed by the following ions: **(a)** Na^+ and PO_4^{3-}; **(b)** Zn^{2+} and SO_4^{2-}; **(c)** Fe^{3+} and CO_3^{2-}. ***Answers:*** **(a)** Na_3PO_4; **(b)** $ZnSO_4$; **(c)** $Fe_2(CO_3)_3$

2.6 NAMING INORGANIC COMPOUNDS

As you proceed in your study of this text and in chemistry laboratory work, you will need to refer to specific chemical substances by name. We present here some of the basic rules for naming simple compounds. You may not have immediate use for some of the rules, but they are gathered here in one place for your convenience to use whenever a question of *nomenclature*, or naming of substances, arises.

There are now about 10 million known chemical substances. Naming them all would be a hopelessly complicated task if each had a special name independent of all the others. Many important substances that have been known for a long time, such as water, H_2O, and ammonia, NH_3, do have individual, traditional names. For most substances, however, we rely upon a set of rules that lead to an informative, systematic name for each substance.

One of the earliest classification schemes in chemistry was the distinction between inorganic and organic compounds. Organic compounds contain carbon, usually in combination with hydrogen, oxygen, nitrogen, or sulfur. Organic compounds were first associated only with plants and animals. However, a great number of organic compounds have now been prepared that do not occur in nature. We will discuss the chemistry and naming of organic compounds in Chapter 26. However, we will have many occasions throughout the text to illustrate chemical principles with organic compounds as examples. In this section, we will consider the basic rules for naming inorganic compounds, those that early chemists associated with the nonliving portion of our world. Let us first consider the naming of ionic compounds.

Ionic Compounds

The names of ionic compounds are based on the names of the ions of which they are composed. For example, NaCl is called sodium chloride after the Na^+ or sodium ion and the Cl^- or chloride ion. The positive ion is always named first and listed first in writing the formula for the compound. The negative ion is named and written last. To see how the names of these ions arise, consider first the naming of positive ions. Ions may be monatomic (composed of a single atom) or polyatomic (formed from two or more atoms). Monatomic cations are most commonly formed from metallic elements. These ions take the name of the element itself:

Na^+ sodium ion $\quad$ Zn^{2+} zinc ion $\quad$ Al^{3+} aluminum ion

If an element can form more than one positive ion, the positive charge of the ion is indicated by a Roman numeral in parentheses following the name of the metal:

Fe^{2+}	iron(II) ion	Cu^+	copper(I) ion
Fe^{3+}	iron(III) ion	Cu^{2+}	copper(II) ion

At this stage you have no way of knowing which elements commonly exist in more than one charge state. This need not be a source of difficulty. If there is any doubt in your mind, use the Roman numeral designation of charge as part of the name. It is never wrong to use this form of charge designation, even though it may sometimes be unnecessary.

An older method still widely used for distinguishing between two differently charged ions of a metal is to apply the endings *-ous* or *-ic*. These endings

Figure 2.23 Compounds of ions of the same element but with different charge can be very different in appearance. Both substances shown are complex salts of iron with K^+ and CN^- ions. The one on the left is potassium ferrocyanide, which contains the Fe(II) bound to CN^- ions. The one on the right is potassium ferricyanide, which contains the Fe(III) bound to CN^- ions. Both substances are used extensively in blueprinting and other dying processes. (Richard Megna/ Fundamental Photographs)

represent the lower and higher charged ions, respectively. They are added to the root of the Latin name of the element:

Fe^{2+}	ferrous ion	Cu^+	cuprous ion
Fe^{3+}	ferric ion	Cu^{2+}	cupric ion

Compounds of differently charged ions of the same element generally exhibit very different properties, including physical appearance (Figure 2.23).

The only common polyatomic cations are those given below:

NH_4^+	ammonium ion	Hg_2^{2+}	mercury(I) or mercurous ion

The name mercury(I) ion is given to Hg_2^{2+} because it can be considered to consist of two Hg^+ ions. Mercury also occurs as the monatomic Hg^{2+} ion, which is known as the mercury(II) or mercuric ion.

Monatomic anions (those derived from a single atom) are most commonly formed from atoms of the nonmetallic elements. They are named by dropping the ending of the name of the element and adding the ending *-ide:*

H^-	hydride ion	O^{2-}	oxide ion	N^{3-}	nitride ion
F^-	fluoride ion	S^{2-}	sulfide ion	P^{3-}	phosphide ion

Only a few common polyatomic ions end in *-ide:*

OH^-	hydroxide ion	CN^-	cyanide ion
O_2^{2-}	peroxide ion	N_3^-	azide ion

Table 2.4 lists the most common cations and anions. Notice that many polyatomic anions contain oxygen. Anions of this kind are referred to as **oxyanions.** A particular element such as sulfur may form more than one oxyanion. When this occurs, there are rules for indicating the relative numbers of oxygen atoms in the anion. When an element forms only two oxyanions, the name of the one that contains more oxygen ends in *-ate;* the name of the one with less oxygen ends in *-ite:*

NO_2^-	nitrite ion	SO_3^{2-}	sulfite ion
NO_3^-	nitrate ion	SO_4^{2-}	sulfate ion

When the series of anions of a given element extends to three or four members, as with the oxyanions of the halogens, prefixes are also employed. The prefix *hypo-* indicates less oxygen, and the prefix *per-* indicates more oxygen:

ClO^-	hypochlorite ion (one less oxygen than chlorite)
ClO_2^-	chlorite ion (one less oxygen than chlorate)
ClO_3^-	chlorate ion
ClO_4^-	perchlorate ion (one more oxygen than chlorate)

Notice that if you memorize the rules just indicated, you need know only the name for one oxyanion in a series to deduce the names for the other members.

Table 2.4 Common Ions

Positive ions (cations)	Negative ions (anions)
1+	1−
Ammonium (NH_4^+)	Acetate ($C_2H_3O_2^-$)
Cesium (Cs^+)	Azide (N_3^-)
Copper(I) or cuprous (Cu^+)	Bromide (Br^-)
Hydrogen (H^+)	Chlorate (ClO_3^-)
Lithium (Li^+)	Chloride (Cl^-)
Potassium (K^+)	Cyanide (CN^-)
Silver (Ag^+)	Dihydrogen phosphate ($H_2PO_4^-$)
Sodium (Na^+)	Fluoride (F^-)
2+	Hydride (H^-)
Barium (Ba^{2+})	Hydrogen carbonate or bicarbonate (HCO_3^-)
Cadmium (Cd^{2+})	Hydrogen sulfate or bisulfate (HSO_4^-)
Calcium (Ca^{2+})	Hydroxide (OH^-)
Cobalt(II) or cobaltous (Co^{2+})	Iodide (I^-)
Copper(II) or cupric (Cu^{2+})	Nitrate (NO_3^-)
Iron(II) or ferrous (Fe^{2+})	Nitrite (NO_2^-)
Lead(II) or plumbous (Pb^{2+})	Perchlorate (ClO_4^-)
Magnesium (Mg^{2+})	Permanganate (MnO_4^-)
Manganese(II) or manganous (Mn^{2+})	Thiocyanate (SCN^-)
Mercury(I) or mercurous (Hg_2^{2+})	2−
Mercury(II) or mercuric (Hg^{2+})	Carbonate (CO_3^{2-})
Nickel (Ni^{2+})	Chromate (CrO_4^{2-})
Strontium (Sr^{2+})	Dichromate ($Cr_2O_7^{2-}$)
Tin(II) or stannous (Sn^{2+})	Hydrogen phosphate (HPO_4^{2-})
Zinc (Zn^{2+})	Oxide (O^{2-})
3+	Peroxide (O_2^{2-})
Aluminum (Al^{3+})	Sulfate (SO_4^{2-})
Chromium(III) or chromic (Cr^{3+})	Sulfide (S^{2-})
Iron(III) or ferric (Fe^{3+})	Sulfite (SO_3^{2-})
	3−
	Arsenate (AsO_4^{3-})
	Nitride (N^{3-})
	Phosphate (PO_4^{3-})
	Phosphide (P^{3-})

SAMPLE EXERCISE 2.11

The formula for the selenate ion is SeO_4^{2-}. Write the formula for the selenite ion.

Solution: The selenite ion should have one less oxygen than the selenate ion; hence, SeO_3^{2-}.

PRACTICE EXERCISE

The formula for the bromate ion is BrO_3^-. Write the formula for the hypobromite ion. *Answer:* BrO^-

Because many names of ions predate the establishment of systematic rules, there are many exceptions to these rules. For example, the permanganate ion is MnO_4^-; we thus expect that the manganate ion should be MnO_3^-, but this ion is unknown. The name manganate is given to the species MnO_4^{2-}.

Many polyatomic anions that have high charges readily add one or more hydrogen ions (H^+) to form anions of lower charge. These ions are named by prefixing the word *hydrogen* or *dihydrogen*, as appropriate, to the name of the hydrogen-free anion. An older method, which is still used, is to use the prefix *bi-:*

HCO_3^-	hydrogen carbonate (or bicarbonate) ion
HSO_4^-	hydrogen sulfate (or bisulfate) ion
$H_2PO_4^-$	dihydrogen phosphate ion

We are now in a position to combine the names of cations and anions to name and write the formulas for ionic compounds. The following examples illustrate the relationship between formula and name:

barium bromide	$BaBr_2$	mercurous chloride	Hg_2Cl_2
copper(II) nitrate	$Cu(NO_3)_2$	aluminum oxide	Al_2O_3

SAMPLE EXERCISE 2.12

Name the following compounds: **(a)** K_2SO_4; **(b)** $Ba(OH)_2$; **(c)** $FeCl_3$.

Solution: **(a)** This compound is composed of K^+ and SO_4^{2-} ions. Because K^+ is called the potassium ion and SO_4^{2-} is called the sulfate ion, the name of the compound is potassium sulfate. **(b)** This compound is composed of Ba^{2+} and OH^- ions. Ba^{2+} is the barium ion; OH^- is the hydroxide ion. Thus, the compound is called barium hydroxide. **(c)** This compound is composed of Fe^{3+}, which is called iron(III) or ferric ion, and chloride ions, Cl^-. The compound is iron(III) chloride or ferric chloride.

PRACTICE EXERCISE

Name the following compounds: **(a)** NH_4Cl; **(b)** Cr_2O_3; **(c)** $Co(NO_3)_2$.
Answers: **(a)** ammonium chloride; **(b)** chromium(III) oxide; **(c)** cobalt(II) nitrate

SAMPLE EXERCISE 2.13

Write the chemical formulas for the following compounds: **(a)** calcium carbonate; **(b)** stannous fluoride; **(c)** iron(II) perchlorate.

Solution: **(a)** The calcium ion is Ca^{2+}; the carbonate ion is CO_3^{2-}. Because of the charges of each ion, there will be one Ca^{2+} ion for each CO_3^{2-} in the compound, giving the empirical formula $CaCO_3$. **(b)** The stannous ion, also known as the tin(II) ion, is Sn^{2+}. The fluoride ion is F^-. Two F^- ions are needed to balance the positive charge of Sn^{2+}, giving the formula SnF_2. This compound is a common tooth-decay preventative found in many toothpastes. **(c)** The iron(II) ion is Fe^{2+}; the perchlorate ion is ClO_4^-. Two ClO_4^- ions are required to balance the charge on one Fe^{2+}, giving $Fe(ClO_4)_2$.

These examples should indicate the importance of remembering the charges of the common ions. Indeed, you may find it necessary to memorize many of the entries in Table 2.4.

PRACTICE EXERCISE

Give the chemical formula for **(a)** magnesium sulfate; **(b)** silver sulfide; **(c)** lead(II) nitrate. ***Answers:*** **(a)** $MgSO_4$; **(b)** Ag_2S; **(c)** $Pb(NO_3)_2$

Acids

The important class of compounds known as acids are named in a special way. These compounds will be discussed further in Chapter 4 and then extensively considered in Chapter 16. An acid may be defined as a substance whose molecules each yield one or more hydrogen ions (H^+) when dissolved in water. The formula for an acid is formed by adding sufficient H^+ ions to balance the anion's charge, as seen in the examples in the table below. The name of the acid is related to the name of the anion. Anions whose names end in *-ide* have associated acids that have the *hydro-* prefix and an *-ic* ending, as in these examples:

Anion	Corresponding acid
Cl^- (chloride)	HCl (hydrochloric acid)
S^{2-} (sulfide)	H_2S (hydrosulfuric acid)

Many of the most important acids are derived from oxyanions. If the anion has an *-ate* ending, the corresponding acid is given an *-ic* ending. Anions whose names end in *-ite* have associated acids whose names end in *-ous*. Prefixes in the name of the anion are retained in the name of the acid. These rules are illustrated by the oxyacids of chlorine:

Anion	Corresponding acid
ClO^- (hypochlorite)	$HClO$ (hypochlorous acid)
ClO_2^- (chlorite)	$HClO_2$ (chlorous acid)
ClO_3^- (chlorate)	$HClO_3$ (chloric acid
ClO_4^- (perchlorate)	$HClO_4$ (perchloric acid)

A summary of the rules for naming the acid corresponding to a given anion is shown in Figure 2.24. Although these rules apply to most acids, you should be aware that, as is often the case in nomenclature, there are exceptions. We will encounter some of these later in the text.

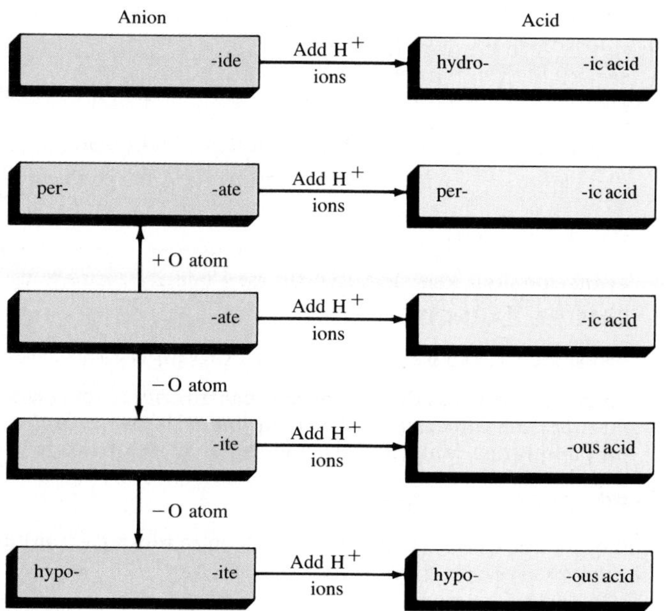

Figure 2.24 A summary of the general way in which the name of an acid is derived from that of its corresponding anion.

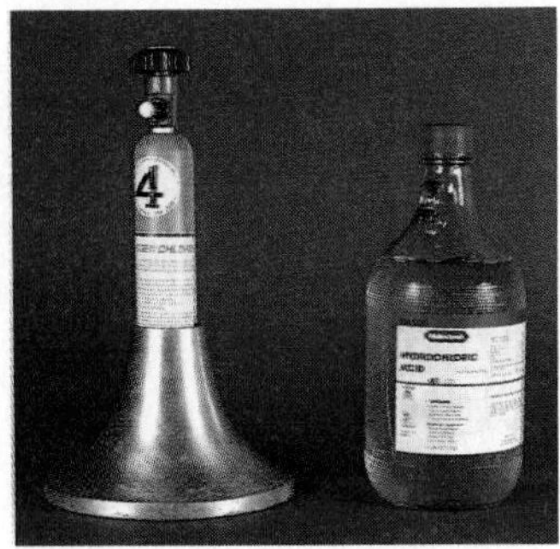

Figure 2.25 Hydrogen chloride, which is a gas at room temperature, is shown in a compressed-gas cylinder (left). Hydrochloric acid, which is a water solution of hydrogen chloride, is shown in a glass bottle (right). (Donald Clegg and Roxy Wilson)

SAMPLE EXERCISE 2.14

Name the following acids: **(a)** HCN; **(b)** HNO_3; **(c)** H_2SO_4; **(d)** H_2SO_3.

Solution: **(a)** The anion from which this acid is derived is CN^-, the cyanide ion. Because this ion has an *-ide* ending, the acid is given a *hydro-* prefix and an *-ic* ending: hydrocyanic acid. Incidentally, only water solutions of HCN are referred to as hydrocyanic acid; the pure compound is called hydrogen cyanide. **(b)** Because NO_3^- is the nitrate ion, HNO_3 is called nitric acid (the *-ate* ending of the anion is replaced with an *-ic* ending in naming the acid). **(c)** Because SO_4^{2-} is the sulfate ion, H_2SO_4 is called sulfuric acid. **(d)** Because SO_3^{2-} is the sulfite ion, H_2SO_3 is sulfurous acid (the *-ite* ending of the anion is replaced with an *-ous* ending).

PRACTICE EXERCISE

Give the chemical formulas for **(a)** hydrobromic acid; **(b)** phosphoric acid.
Answers: **(a)** HBr; **(b)** H_3PO_4

Molecular Compounds

The procedures for naming binary (two-element) molecular compounds are similar to those for naming ionic compounds. In these molecular compounds it is possible to associate a more positive nature with one element in the molecule and a more negative nature with the other element. (In Chapter 8, we will consider how the more positive atom is selected.) The element with the more positive nature is named first and also appears first in the chemical formula. The second element is named with an *-ide* ending. For example, the name for HCl is hydrogen chloride. (This is the name used when referring to the pure compound; water solutions of HCl are referred to as hydrochloric acid; see Figure 2.25).

Often a pair of elements can form several different molecular compounds. For example, carbon and oxygen form CO and CO_2. To distinguish these compounds from one another, the prefixes given in Table 2.5 are used to denote the numbers of atoms of each element present. Thus, CO is called carbon *mon*oxide, and CO_2 is called carbon *di*oxide. When the prefix ends in *a* or *o* and the name of the anion begins with a vowel (such as *oxide*), the *a* or *o* is often dropped. A few examples follow:

NF_3	nitrogen trifluoride
N_2O_4	dinitrogen tetroxide
SO_3	sulfur trioxide

Table 2.5 Prefixes Used in Naming Binary Compounds Formed Between Nonmetals

Prefix	Meaning
Mono-	1
Di-	2
Tri-	3
Tetra-	4
Penta-	5
Hexa-	6
Hepta-	7
Octa-	8

SAMPLE EXERCISE 2.15

Name the following compounds: **(a)** PCl_5; **(b)** N_2O_3.

Solution: Because the compounds consist entirely of nonmetals, you should expect that they are molecular rather than ionic. Using the prefixes in Table 2.5, we have **(a)** phosphorus pentachloride, and **(b)** dinitrogen trioxide.

PRACTICE EXERCISE

Give the chemical formula for **(a)** silicon tetrabromide; **(b)** disulfur dichloride.
Answers: **(a)** $SiBr_4$; **(b)** S_2Cl_2

FOR REVIEW

SUMMARY

Atoms are the basic building blocks of matter; they are the smallest units of an element that can combine with other elements. Atoms are composed of a nucleus (containing protons and neutrons) and electrons that move around the nucleus. We considered some of the historically significant experiments that led to this model of the atom: Thomson's experiments on the behavior of cathode rays (a stream of electrons) in magnetic and electric fields; Millikan's oil-drop experiment; Becquerel's and Rutherford's studies on radioactivity, and Rutherford's studies on the scattering of α particles by thin metal foils.

Elements can be classified by atomic number, the number of protons in the nucleus of an atom. All atoms of a given element have the same atomic number. The mass number of an atom is the sum of the number of protons and neutrons. Atoms of the same element that differ in mass number are known as isotopes.

The periodic table is an arrangement of the elements in order of increasing atomic number. Elements with similar properties are placed in vertical columns. The elements in a column are known as a periodic family or group. The metallic elements, which comprise the majority of the elements, dominate the left side and middle of the table; the nonmetallic elements are located on the upper right side.

Atoms can combine to form molecules. Atoms can also either gain or lose electrons, thereby forming charged particles called ions. Metals tend to lose electrons, becoming positively charged ions (cations). Nonmetals tend to gain electrons, forming negatively charged ions (anions). Because ionic compounds are electrically neutral, containing both cations and anions, they usually contain both metallic and nonmetallic elements. Compounds composed of molecules (molecular compounds) usually contain only nonmetallic elements.

The chemical formulas used for ionic compounds are empirical formulas. The empirical formula of an ionic compound can be written readily if the charges of the ions are known. Although empirical formulas can also be written for molecular substances, the molecular formula is preferred because it gives the actual number of each type of atom in a molecule of the substance. Structural formulas show the order in which the atoms in a molecule are connected.

A set of systematic rules has been developed for naming inorganic compounds. We considered ionic compounds, acids, and binary molecular compounds.

KEY TERMS

atoms
law of multiple proportions (Sec. 2.1)
cathode rays (Sec. 2.2)
radioactivity (Sec. 2.2)
nucleus (Sec. 2.2)
proton (Sec. 2.3)
neutron (Sec. 2.2)
electron (Sec. 2.3)
electronic charge (Sec. 2.3)
angstrom (Sec. 2.3)
isotopes (Sec. 2.3)
atomic number (Sec. 2.3)
mass number (Sec. 2.3)
nuclide (Sec. 2.3)
periodic table (Sec. 2.4)
family (group) (Sec. 2.4)
metallic elements (metals) (Sec. 2.4)
nonmetallic elements (nonmetals) (Sec. 2.4)
semimetals (metalloids) (Sec. 2.4)
molecule (Sec. 2.5)
chemical formula (Sec. 2.5)
diatomic molecule (Sec. 2.5)
molecular formula (Sec. 2.5)
empirical formula (Sec. 2.5)
structural formula (Sec. 2.5)
ion (Sec. 2.5)
cation (Sec. 2.5)
anion (Sec. 2.5)
polyatomic ions (Sec. 2.5)
oxyanion (Sec. 2.6)

EXERCISES

Atomic Theory and Atomic Structure

2.1 Using the atomic theory, explain the difference in the following two statements: **(a)** Nitrogen dioxide is a *compound* of nitrogen and oxygen. **(b)** Air is a *mixture* composed mostly of nitrogen and oxygen.

2.2 Classify each of the following as an element, a compound, or a mixture: **(a)** hydrogen; **(b)** water; **(c)** oxygen; **(d)** a flask that contains some hydrogen gas and some oxygen gas; **(e)** hydrogen peroxide.

2.3 A chemist prepared a series of compounds containing only sulfur and fluorine and determined the amount of each element in each compound:

Compound	Mass of sulfur (g)	Mass of fluorine (g)
A	23.2	55.0
B	16.6	9.8
C	19.3	68.6

(a) Calculate the mass of fluorine per gram of sulfur in each compound. **(b)** Do the numbers in part (a) follow the law of multiple proportions?

2.4 A chemistry student finds that 15.20 g of nitrogen will react with either 17.37 g, 34.74 g, or 43.43 g of oxygen to form three different compounds. **(a)** Calculate the mass of oxygen per gram of nitrogen in each compound. **(b)** Do the numbers in part (a) conform to the law of multiple proportions?

2.5 What was the evidence used to conclude that cathode rays consist of negatively charged particles?

2.6 A negatively charged particle is directed between two electrically charged plates, as shown in Figure 2.5(*a*). **(a)** Why does the path of the charged particle bend? **(b)** As the charge on the plates is increased, would you expect the bending to increase, decrease, or stay the same? **(c)** As the mass of the particle is increased, would you expect the bending to increase, decrease, or stay the same?

2.7 Natural gas (methane) burns in the presence of oxygen to produce water, carbon dioxide, and heat. We know that 4.0 g of natural gas requires 16.0 g of oxygen for complete combustion and that 9.0 g of water is produced in this reaction. Can the mass of carbon dioxide produced be determined from this information? Explain.

2.8 When the magnesium foil within a flashbulb burns, the flashbulb undergoes no change in mass. When magnesium ribbon burns in open air, the magnesium gains mass. What is the difference between these two experiments? Does the second observation violate the law of conservation of mass? Explain.

2.9 Static electricity, such as that given to a piece of amber by rubbing it with wool, is due to a buildup of electrons. A sample of amber is measured to have a static charge of 3.24×10^{-16} C. How many excess electrons are on the piece of amber?

2.10 In a physics laboratory a student carried out the Millikan oil-drop experiment, using several oil droplets for her measurements, and calculated the charges on the drops. She obtained the following data:

Droplet	Calculated charge (C)
A	1.60×10^{-19}
B	3.15×10^{-19}
C	4.81×10^{-19}
D	6.31×10^{-19}

What is the significance of the fact that the droplets carried different charges? What conclusion can the student draw from these data regarding the charge of the electron? What value (and to how many significant figures) should she report for the electronic charge?

2.11 Why is Rutherford's nuclear model of the atom more consistent with the results of the α-particle-scattering experiment than Thomson's "plum pudding" model?

2.12 What differences would you expect if beryllium foil were used instead of gold foil in the α-particle-scattering experiment depicted in Figures 2.10 and 2.12?

2.13 The diameter of the cesium (Cs) atom is about 4.7 Å. **(a)** Express this distance in nanometers (nm); in picometers (pm). **(b)** How many cesium atoms would have to be lined up to span 1.0 cm?

[2.14] One cubic centimeter of gold contains about 5.9×10^{22} gold atoms. From this, estimate the diameter of a single gold atom. What assumptions did you have to make in order to arrive at your estimate?

2.15 How many protons, neutrons, and electrons are in the following atoms: **(a)** ^{51}V; **(b)** ^{119}Sn; **(c)** ^{127}Te; **(d)** ^{165}Ho **(e)** ^{16}O; **(f)** ^{234}Th; **(g)** ^{112}Cd; **(h)** ^{113}Cd?

2.16 Each of the following nuclides is used in medicine. Indicate the number of protons and neutrons in each nuclide: **(a)** cobalt-60; **(b)** iodine-131; **(c)** technetium-99; **(d)** phosphorus-32; **(e)** chromium-51; **(f)** iron-59.

2.17 The uranium isotope used to generate nuclear power has 143 neutrons in its nucleus. The most common isotope of uranium has 146 neutrons in its nucleus. What are the full chemical symbols, with both superscript and subscript, for these isotopes of uranium?

2.18 Write the correct symbol, with both superscript and subscript, for each of the following (use the list of elements on the front inside cover): **(a)** the isotope of sodium with mass 23; **(b)** the nuclide of vanadium that contains 28 neutrons; **(c)** an α particle; **(d)** the isotope of chlorine with mass 37; **(e)** the nuclide of magnesium that has an equal number of protons and neutrons.

2.19 Fill in the gaps in the following table:

Symbol	$^{17}_{8}O$	$^{232}_{90}Th$			
Protons			27		
Neutrons				56	53
Electrons					42
Mass no.			60	100	

2.20 Fill in the gaps in the following table:

Symbol	$^{11}_{5}B$	$^{137}_{56}Ba$			
Protons			40		
Neutrons			51	29	126
Electrons					83
Mass no.				53	

The Periodic Table; Molecules and Ions

2.21 For each of the following elements, write its chemical symbol, locate it in the periodic table, and indicate whether it is a metal, semimetal, or nonmetal: **(a)** sulfur; **(b)** silver; **(c)** strontium; **(d)** arsenic; **(e)** xenon; **(f)** aluminum; **(g)** yttrium; **(h)** iodine.

2.22 Locate each of the following elements in the periodic table, indicate whether it is a metal, semimetal, or nonmetal, and give the name of the element: **(a)** Se; **(b)** Be; **(c)** Hg; **(d)** Kr; **(e)** Te; **(f)** Ga; **(g)** Br; **(h)** Ba.

2.23 For each of the following elements, write its chemical symbol, determine the name of the group to which it belongs (Table 2.3), and indicate whether it is a metal, semimetal, or nonmetal: **(a)** radon; **(b)** polonium; **(c)** rubidium; **(d)** phosphorus; **(e)** magnesium; **(f)** chlorine.

2.24 Classify each of the group 5A elements (the pnicogens) as metallic, semimetallic, or nonmetallic.

2.25 Which conveys more information, the empirical formula of a molecule or its molecular formula? Explain.

2.26 Two molecules have the same empirical formula. Does it necessarily follow that they have the same molecular formula?

2.27 Write the empirical formula corresponding to each of the following molecular formulas: **(a)** C_4H_6; **(b)** H_2O_2; **(c)** P_2O_5; **(d)** $C_2H_4O_2$; **(e)** N_2H_4; **(f)** B_2H_6.

2.28 From the following list, find the groups of compounds that have the same empirical formula: C_2H_2, N_2O_4, C_2H_4, C_6H_6, NO_2, C_3H_6, C_4H_8.

2.29 Each of the following elements is capable of forming an ion in chemical reactions. By referring to the periodic table, predict the charge found on the most stable ion formed by each: **(a)** Li; **(b)** Br; **(c)** Mg; **(d)** O; **(e)** Al; **(f)** Y.

2.30 Which of the following ions would you *not* expect to form: **(a)** F^+; **(b)** S^{2-}; **(c)** Be^-; **(d)** P^{3-}; **(e)** Br^-; **(f)** K^{2+}?

2.31 Predict the empirical formula for the ionic compound formed from each of the following pairs of ions: **(a)** Cu^{2+} and $C_2H_3O_2^-$; **(b)** NH_4^+ and HCO_3^-; **(c)** Al^{3+} and Br^-; **(d)** Ca^{2+} and ClO_4^-; **(e)** K^+ and SO_4^{2-}; **(f)** NH_4^+ and PO_4^{3-}.

2.32 Predict the empirical formula for the ionic compound formed from each of the following pairs of elements: **(a)** Na, S; **(b)** Ca, F; **(c)** Mg, O; **(d)** Al, O; **(e)** Be, S; **(f)** Li, N.

2.33 Predict whether each of the following compounds is molecular or ionic: **(a)** NO_2; **(b)** BF_3; **(c)** Li_2O; **(d)** Sc_2O_3; **(e)** CsBr; **(f)** PF_5; **(g)** NF_3; **(h)** LaP.

2.34 Which of the following are ionic substances and which are molecular: **(a)** Fe_2O_3; **(b)** NOCl; **(c)** $Ca_3(PO_4)_2$; **(d)** $AgNO_3$; **(e)** B_2H_6; **(f)** $SiCl_4$?

Naming Inorganic Compounds

2.35 Provide names for the following ionic compounds: **(a)** RbI; **(b)** $NaNO_3$; **(c)** Ag_2S; **(d)** $KMnO_4$; **(e)** Hg_2Cl_2; **(f)** MnO_2; **(g)** CaH_2; **(h)** $ScCl_3$; **(i)** $Ba(OH)_2$; **(j)** $(NH_4)_2SO_4$.

2.36 Name the following ionic compounds: **(a)** $ZnCl_2$; **(b)** $PbCrO_4$; **(c)** $Hg(NO_3)_2$; **(d)** $Ca(CN)_2$; **(e)** FeF_3; **(f)** Na_2CO_3; **(g)** $KClO_4$; **(h)** $Cu(OH)_2$; **(i)** Cr_2O_3; **(j)** Ag_3PO_4.

2.37 Give the chemical formula for each of the following ionic compounds: **(a)** copper(II) acetate; **(b)** ferric chloride; **(c)** ammonium sulfide; **(d)** aluminum oxide; **(e)** calcium nitride; **(f)** magnesium phosphate.

2.38 Give the chemical formula for each of the following ionic compounds: **(a)** chromium(III) nitrate; **(b)** manganese(II) fluoride; **(c)** calcium bicarbonate; **(d)** mercurous chloride; **(e)** potassium hydride; **(f)** sodium peroxide.

2.39 Give the name or chemical formula, as appropriate, for each of the following molecular compounds: **(a)** CS_2; **(b)** sulfur hexafluoride; **(c)** SO_3; **(d)** diphosphorus pentoxide; **(e)** CBr_4; **(f)** sulfur dioxide.

2.40 Give the name or chemical formula, as appropriate, for the following molecular compounds: **(a)** SF_4; **(b)** nitrogen dioxide; **(c)** IF_7; **(d)** hydrogen selenide; **(e)** N_2O_5; **(f)** boron trifluoride.

2.41 Provide chemical formulas or names for the following acids: **(a)** H_3PO_4; **(b)** sulfurous acid; **(c)** HBr; **(d)** nitric acid; **(e)** $HClO_3$; **(f)** hypochlorous acid.

2.42 Provide chemical formulas or names for the following acids: **(a)** H_2CO_3; **(b)** nitrous acid; **(c)** $HClO_2$; **(d)** hydrocyanic acid; **(e)** H_2CrO_4; **(f)** periodic acid.

2.43 Write the chemical formula for each substance mentioned in the following word descriptions (use the front inside cover to find the symbols for the elements you don't know). **(a)** Zinc carbonate can be heated to form zinc oxide and carbon dioxide. **(b)** On treatment with hydrofluoric acid, silicon dioxide forms silicon tetrafluoride and water. **(c)** Sulfur dioxide reacts with water to form sulfurous acid. **(d)** The substance hydrogen phosphide is commonly called phosphine. **(e)** Perchloric acid reacts with cadmium to form cadmium(II) perchlorate. **(f)** Vanadium(III) bromide is a colored solid.

2.44 Assume that you encounter the following phrases in your reading. What is the chemical formula for each substance mentioned? **(a)** Potassium chlorate is used as a laboratory source of oxygen. **(b)** Sodium hypochlorite is used as a household bleach. **(c)** Ammonia is important in the synthesis of fertilizers such as ammonium nitrate. **(d)** Hydrofluoric acid is used to etch glass. **(e)** The smell of rotten eggs is due to hydrogen sulfide. **(f)** When hydrochloric acid is added to sodium bicarbonate (baking soda), carbon dioxide gas forms.

Additional Exercises

2.45 Describe the contributions to atomic theory made by the following scientists: **(a)** Dalton; **(b)** Thomson; **(c)** Millikan; **(d)** Rutherford.

2.46 Deuterium and tritium are the names given to the isotopes of hydrogen that have one and two neutrons in the nucleus, respectively. **(a)** Write the full chemical symbols for deuterium and tritium. **(b)** Describe the similarities and differences between an atom of deuterium and one of tritium.

2.47 Fill in the gaps in the following table:

Symbol	$^{19}_{9}F$	$^{201}_{80}Hg^{+}$			
Protons	9		11		17
Neutrons			12	30	20
Electrons		79		23	18
Net charge	0	1+	0	2+	

2.48 Fill in the gaps in the following table:

Symbol	$^{16}_{8}O$	$^{185}_{75}Re^{3+}$			
Protons			16		78
Neutrons			17	41	117
Electrons				34	74
Net charge			2−	2−	

2.49 Consider the substances listed in Table 1.1. **(a)** How many of them contain a halogen atom? **(b)** How many contain an alkali metal atom? **(c)** How many contain a chalcogen atom? **(d)** How many contain a pnicogen atom? **(e)** How many are acids? **(f)** Classify each substance as molecular or ionic.

2.50 From the following list of elements—Ar, H, Ga, Al, Ca, Br, Ge, K, O—pick the one that best fits each description; use each element only once: **(a)** an alkali metal; **(b)** an alkaline earth metal; **(c)** a noble gas; **(d)** a halogen; **(e)** a semi-metal; **(f)** a nonmetal listed in group 1A; **(g)** a metal that forms a 3+ ion; **(h)** a nonmetal that forms a 2− ion; **(i)** an element that resembles aluminum.

[2.51] Hydrocarbons are molecules that contain only carbon and hydrogen. A hydrocarbon with empirical formula CH contains 0.923 g of carbon per gram of hydrocarbon; the remaining mass is hydrogen. Listed below is the carbon content per gram of substance for several common hydrocarbons:

Hydrocarbon	**Grams of carbon per gram of hydrocarbon**
Methane	0.750
Ethylene	0.857
Ethane	0.800
Propane	0.818
Butane	0.828

(a) For the hydrocarbon with empirical formula CH, calculate the mass of H that combines with 1 g of C. **(b)** For the hydrocarbons listed in the table, calculate the mass of H that combines with 1 g of C. **(c)** By comparing the results for part (b) to those for part (a), determine small-number ratios of hydrogen atoms per carbon atom for the hydrocarbons in the table. **(d)** Write empirical formulas for the hydrocarbons in the table.

2.52 The following are pairs of anions and cations *without their charges given.* For each pair, indicate the charge on the cation and the anion, provide an empirical formula for the ionic compound that results from these, and provide the proper name for the compounds: **(a)** NH_4, SO_4; **(b)** Ca, HCO_3; **(c)** Al, HSO_4; **(d)** K, CrO_4; **(e)** Na, $C_2H_3O_2$; **(f)** Ag, S; **(g)** Be, Br; **(h)** K, Te.

2.53 Give the chemical names of each of the following familiar compounds: **(a)** NaCl (table salt); **(b)** $NaHCO_3$ (baking soda); **(c)** NaOCl (in many bleaches); **(d)** NaOH (caustic soda); **(e)** $(NH_4)_2CO_3$ (smelling salts); **(f)** $CaSO_4$ (plaster of paris).

2.54 Many familiar substances have common, unsystematic names. For each of the following, give the correct systematic name: **(a)** saltpeter (KNO_3); **(b)** soda ash (Na_2CO_3); **(c)** lime (CaO); **(d)** muriatic acid (HCl); **(e)** Epsom salt ($MgSO_4$); **(f)** milk of magnesia ($Mg(OH)_2$).

2.55 Many ions and compounds have very similar names, and there is great potential for confusing them. Write the correct chemical formulas to distinguish between **(a)** calcium sulfide and calcium hydrogen sulfide; **(b)** hydrobromic acid and bromic acid; **(c)** aluminum nitride and aluminum nitrite; **(d)** iron(II) oxide and iron(III) oxide; **(e)** ammonia and ammonium ion; **(f)** potassium sulfite and potassium bisulfite; **(g)** mercurous chloride and mercuric chloride; **(h)** chloric acid and perchloric acid.

2.56 Iodic acid has the molecular formula HIO_3. Write the formulas for the following: **(a)** the iodate anion; **(b)** the periodate anion; **(c)** the hypoiodite anion; **(d)** hypoiodous acid; **(e)** periodic acid.

[2.57] Using a suitable reference such as the *Handbook of Chemistry and Physics*, look up the following information for sulfur: **(a)** the number of known isotopes; **(b)** the natural abundance of the four most abundant isotopes.

[2.58] Using the *Handbook of Chemistry and Physics*, find the density, melting point, and boiling point for PF_3.

Stoichiometry: Calculations with Chemical Formulas and Equations

3

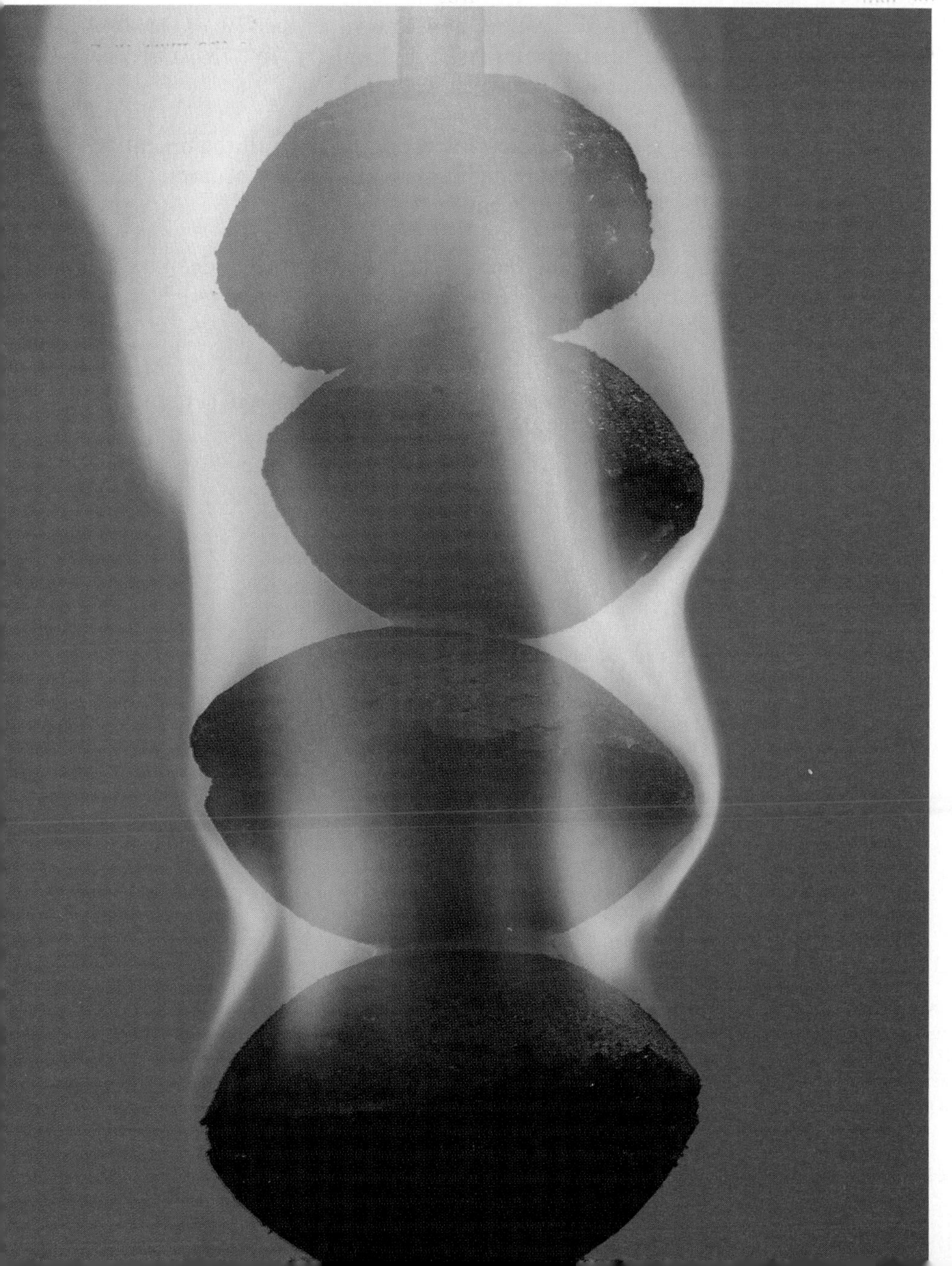

Burning charcoal briquettes. Combustion reactions were among the first systematically studied chemical reactions. (Will & Deni McIntyre/Photo Researchers)

CONTENTS

In the late 1700s, a French nobleman named Antoine Lavoisier (Figure 3.1) undertook a series of careful, quantitative studies of chemical reactions. Perhaps his most important observation was that the total mass of all substances present after a chemical reaction is the same as the total mass before the reaction. This important observation, now known as the **law of conservation of mass**, is one of the fundamental laws of chemistry. In 1789, Lavoisier published a textbook on chemistry in which he stated:

> We may lay it down as an incontestable axiom that, in all the operations of art and nature, nothing is created; an equal quantity of matter exists both before and after the experiment.*

With the advent of the atomic theory, chemists came to understand the basis for this law: *Atoms are neither created nor destroyed during any chemical reaction.* Thus, the same collection of atoms is present after a reaction as when

* A. L. Lavoisier, *Traite élémentaire de chimie* (Paris: 1789); English translation in *Great Books*, vol. 45 (Chicago: Encyclopedia Britannica, 1952), pp. 9–133.

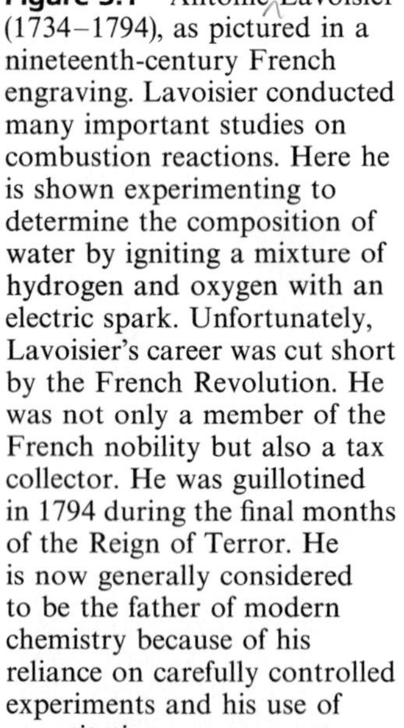

Figure 3.1 Antoine Lavoisier (1734–1794), as pictured in a nineteenth-century French engraving. Lavoisier conducted many important studies on combustion reactions. Here he is shown experimenting to determine the composition of water by igniting a mixture of hydrogen and oxygen with an electric spark. Unfortunately, Lavoisier's career was cut short by the French Revolution. He was not only a member of the French nobility but also a tax collector. He was guillotined in 1794 during the final months of the Reign of Terror. He is now generally considered to be the father of modern chemistry because of his reliance on carefully controlled experiments and his use of quantitative measurements. (The Granger Collection)

the reaction began. The changes that occur during any reaction merely involve the rearranging of atoms.

In this chapter, we consider the quantities of substances consumed and produced in chemical reactions. This area of study is known as **stoichiometry** (pronounced stoy-key-OM-uh-tree), a name derived from the Greek words *stoicheion* ("element") and *metron* ("measure"). The law of conservation of mass serves as the guiding principle in our discussions.

Stoichiometry is an essential tool in chemistry. Such diverse problems, for example, as measuring the concentration of ozone in the atmosphere, determining the potential yield of gold from an ore, and assessing the suitability of different processes for converting coal into gaseous fuels all involve aspects of stoichiometry.

3.1 CHEMICAL EQUATIONS

Chemical reactions are represented in a concise way by **chemical equations.** For example, when hydrogen, H_2, burns, it reacts with oxygen, O_2, in the air to form water, H_2O. We write the chemical equation for this reaction as follows:

$$2H_2 + O_2 \longrightarrow 2H_2O \qquad [3.1]$$

We read the + sign to mean "reacts with" and the arrow as "produces." The chemical formulas on the left of the arrow represent the starting substances, called **reactants.** The substances produced in the reaction, called **products**, are shown to the right of the arrow. The numbers in front of the formulas are called *coefficients.* (As in algebraic equations, the numeral 1 is usually not written.)

Because atoms are neither created nor destroyed in any reaction, an equation must have an equal number of atoms of each element on each side of the arrow. When this condition is met, the equation is said to be *balanced.* For example, on the right side of Equation 3.1 there are two molecules of H_2O, each containing two atoms of hydrogen and one atom of oxygen. Thus, 2 H_2O contains $2 \times 2 = 4$ H atoms and $2 \times 1 = 2$ O atoms. Because there are also 4 H atoms and 2 O atoms on the left side of the equation, the equation is balanced.

Before we can write the chemical equation for a reaction, we must determine by experiment those substances that are reactants and those that are products. Once we know the chemical formulas of the reactants and products in a reaction, we can write the unbalanced chemical equation. We then balance the equation by determining the coefficients that provide equal numbers of each type of atom on each side of the equation.

Consider the reaction that occurs when methane, CH_4, the principal component of natural gas, burns in air to produce carbon dioxide gas, CO_2, and liquid water, H_2O. Both of these products contain oxygen atoms that come from O_2 in the air. We say that combustion in air is "supported by oxygen," meaning that oxygen is a reactant. The unbalanced equation is

$$CH_4 + O_2 \longrightarrow CO_2 + H_2O \quad \text{(unbalanced)} \qquad [3.2]$$

It is usually best to balance first those elements that occur in only one substance on each side of the equation. In our example, both C and H appear in only one reactant and, separately, in one product each, so we begin by focusing attention on CH_4. We consider first carbon and then hydrogen.

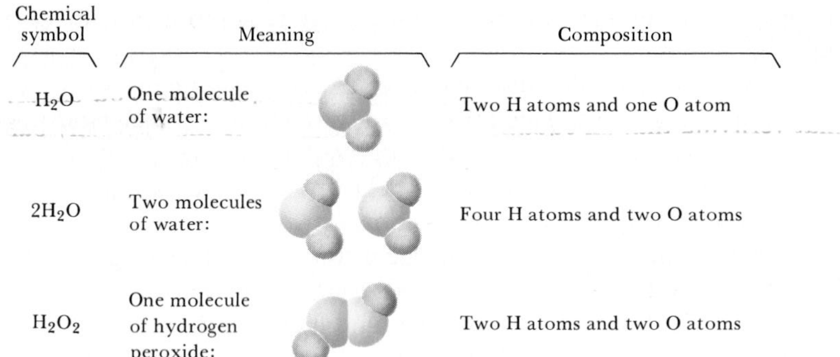

Chemical symbol
Meaning
Composition
H₂O
One molecule of water:
Two H atoms and one O atom
2H₂O
Two molecules of water:
Four H atoms and two O atoms
H₂O₂
One molecule of hydrogen peroxide:
Two H atoms and two O atoms

Figure 3.2 Illustration of the difference between a subscript in a chemical formula and a coefficient in front of the formula. Notice that the number of atoms of each type (listed under composition) is obtained by multiplying the coefficient and the subscript associated with each element in the formula.

Notice that one molecule of CH_4 contains the same number of C atoms (one) as does one molecule of CO_2. Therefore, the coefficients for these substances *must* be the same, and we choose them both to be 1 as we start the balancing process. However, the reactant CH_4 contains more H atoms (four) than does the product H_2O (two). If we place a coefficient 2 in front of H_2O, there will be four H atoms on each side of the equation:

$$CH_4 + O_2 \longrightarrow CO_2 + 2H_2O \quad \text{(unbalanced)} \qquad [3.3]$$

Before we continue to balance this equation, let's make sure that we clearly understand the distinct difference between a coefficient in front of a formula and a subscript in a formula. Refer to Figure 3.2. Notice that changing a subscript in a formula—from H_2O to H_2O_2, for example—changes the identity of the chemical. The substance H_2O_2, hydrogen peroxide, is quite different from water. *Subscripts should never be changed in balancing an equation.* Placing a coefficient in front of a formula changes only the *amount* and not the *identity* of the substance; $2H_2O$ means two molecules of water, $3H_2O$ means three molecules of water, and so forth.

Now let's continue balancing the example equation. Equation 3.3 has equal numbers of C atoms and H atoms on both sides. However, the products have more total O atoms (four, two from CO_2 and two from $2H_2O$) than the reactants have (two). If we place a coefficient 2 in front of O_2, we will make the number of O atoms equal on both sides of the equation:

$$CH_4 + 2O_2 \longrightarrow CO_2 + 2H_2O \quad \text{(balanced)} \qquad [3.4]$$

The balanced equation is shown schematically in Figure 3.3. For most purposes, a balanced equation should contain the smallest whole-number coefficients, as in this example.

One methane molecule	+	Two oxygen molecules	→	One carbon dioxide molecule	+	Two water molecules
CH_4	+	$2O_2$	→	CO_2	+	$2H_2O$
(1 C, 4 H)		(4 O)		(1 C, 2 O)		(2 O, 4 H)

Figure 3.3 Balanced chemical equation for the combustion of CH_4. The drawings of the molecules involved call attention to the conservation of atoms through the reaction.

The approach we have taken to balancing Equation 3.4 is largely trial and error. We balance each kind of atom in succession, adjusting coefficients as necessary. This approach works for most chemical equations. You should be aware that verifying that an equation is balanced is easier than actually balancing one, so practice balancing equations until you are comfortable with this process.

The physical state of each chemical in a chemical equation is often indicated parenthetically. We use the symbols (*g*), (*l*), (*s*), and (*aq*) for gas, liquid, solid, and aqueous (water) solution, respectively. Thus, the balanced equation above can be written

$$CH_4(g) + 2O_2(g) \longrightarrow CO_2(g) + 2H_2O(l) \qquad [3.5]$$

Often the conditions under which the reaction proceeds appear above or below the arrow between the two sides of the equation. For example, the temperature or pressure at which the reaction occurs could be so indicated. The symbol Δ is often placed above the arrow to indicate the addition of heat.

SAMPLE EXERCISE 3.1

Balance the following equation:

$$Na(s) + H_2O(l) \longrightarrow NaOH(aq) + H_2(g)$$

Solution: A quick count of the atoms reveals an equal number of Na and O atoms on both sides of the equation. Furthermore, we see that the coefficients on both Na and H_2O must equal the coefficient on NaOH in order for both the Na and O atoms to be balanced. A problem arises, however, with the H atoms: There are two H atoms among the reactants and three H atoms among the products. To increase the number of H atoms among reactants, we can place a coefficient 2 in front of H_2O:

$$Na(s) + 2H_2O(l) \longrightarrow NaOH(aq) + H_2(g)$$

In order to regain the balance in O atoms, we must put a coefficient 2 in front of NaOH. This also brings the H atoms into balance, but it requires a coefficient 2 in front of Na to rebalance the Na atoms:

$$2Na(s) + 2H_2O(l) \longrightarrow 2NaOH(aq) + H_2(g)$$

Finally, we check the number of atoms of each element and find that we have two Na atoms, four H atoms, and two O atoms on each side of the equation. The equation is therefore balanced.

PRACTICE EXERCISE

Balance the following equations by providing the missing coefficients:
(a) $\_\_C_2H_4 + \_\_O_2 \longrightarrow \_\_CO_2 + \_\_H_2O$
(b) $\_\_Al + \_\_HCl \longrightarrow \_\_AlCl_3 + \_\_H_2$
Answers: **(a)** 1, 3, 2, 2; **(b)** 2, 6, 2, 3

3.2 PATTERNS OF CHEMICAL REACTIVITY

Our discussion in Section 3.1 focused on how to balance chemical equations given the reactants and products for the reactions. You were not asked to say anything about the type of reaction or to predict the products. Discussing this matter is a bit of a diversion from the basic theme of this chapter. However, an introductory discussion now will help you feel more at ease with the reactions used to illustrate concepts in the chapter. It will also help you in the lab. We will further examine this topic in Chapter 4 when we discuss reactions in aqueous solutions.

(a)

(b)

Figure 3.4 The reaction of sodium with water (*a*) and of potassium with water (*b*). An acid-base indicator (phenolphthalein) has been added to both solutions; the pink color indicates the presence of hydroxide ions in the solution. Although the overall reactions of the two metals with water are very similar, potassium reacts more vigorously. (Donald Clegg and Roxy Wilson)

Using the Periodic Table

We may identify reactants and products experimentally by their properties. However, we can often predict what will happen in a reaction if we have seen a similar reaction before. Naturally, recognizing a general pattern of reactivity for a class of substances gives you a broader understanding than merely memorizing a large number of unrelated reactions. The periodic table can be a powerful ally in this regard. For example, knowing that sodium (Na) reacts with water, H_2O, to form NaOH and H_2 (see Sample Exercise 3.1), we can predict what happens when potassium (K) is placed in water. Both sodium and potassium are in the same family of the periodic table (the alkali metal family, family 1A). We would expect them to behave similarly, producing the same kinds of products. Indeed, this prediction is correct:

$$2K(s) + 2H_2O(l) \longrightarrow 2KOH(aq) + H_2(g) \qquad [3.6]$$

In fact, all alkali metals behave in this general fashion. If we let M represent any alkali metal, we can write the general reaction as follows:

$$2M(s) + 2H_2O(l) \longrightarrow 2MOH(aq) + H_2(g) \qquad [3.7]$$

Figure 3.4 shows photographs of the reactions of sodium and of potassium with water. In the chapters ahead, we will sometimes encounter reactions that are characteristic of a particular family of elements. We will also see some reactions that are characteristic of particular classes of compounds. For example, different combustible carbon compounds almost always form the same products when they burn in air.

Combustion in Air

Combustion reactions are rapid reactions that produce a flame. Most of the combustion reactions we observe involve O_2 from air as a reactant. Equation 3.5 and Practice Exercise 3.1(a) illustrate a general class of reactions involving the burning or combustion of hydrocarbon compounds (compounds that contain only carbon and hydrogen, such as CH_4 and C_2H_6).

When hydrocarbons are combusted, they react with O_2 to form CO_2 and H_2O.* The number of molecules of O_2 required in the reaction and the number

* When there is an insufficient quantity of O_2 present, carbon monoxide, CO, will be produced. Even more severe restriction of O_2 will cause the production of fine particles of carbon that we call soot. *Complete* combustion produces CO_2.

Figure 3.5 Propane, C_3H_8, burns in air, producing a blue flame. The liquid propane vaporizes and mixes with air as it escapes from the bottle through the nozzle. The combustion reaction between C_3H_8 and O_2 produces CO_2 and H_2O along with a high temperature. (Richard Megna/ Fundamental Photographs)

of molecules of CO_2 and H_2O formed depend on the composition of the hydrocarbon. For example, the combustion of propane, C_3H_8, a gas used for cooking and home heating, is described by the following equation:

$$C_3H_8(g) + 5O_2(g) \longrightarrow 3CO_2(g) + 4H_2O(l) \qquad [3.8]$$

The blue flame produced when propane burns is shown in Figure 3.5.

Combustion of compounds containing oxygen atoms as well as carbon and hydrogen (for example, CH_3OH and $C_6H_{12}O_6$) also produce CO_2 and H_2O. The chemical equation for one of these reactions is considered in the following sample exercise.

SAMPLE EXERCISE 3.2

Write the balanced chemical equation for the reaction that occurs when methanol, $CH_3OH(l)$, is burned in air.

Solution: The C atoms in CH_3OH end up in $CO_2(g)$ in the products; the H atoms end up in $H_2O(l)$. The additional O atoms that are needed come from $O_2(g)$ in air, which is a reactant. Thus the unbalanced equation is

$$CH_3OH(l) + O_2(g) \longrightarrow CO_2(g) + H_2O(l)$$

Because CH_3OH has only one C atom, we can start with the coefficient 1 for CO_2. Because CH_3OH has four H atoms, we place a coefficient 2 in front of H_2O to balance the H atoms:

$$CH_3OH(l) + O_2(g) \longrightarrow CO_2(g) + 2H_2O(l)$$

This gives four O atoms among the products and three among the reactants (one in CH_3OH and two in O_2). We can use the fractional coefficient $\frac{3}{2}$ in front of O_2 to provide four O atoms among the reactants (three O atoms in $\frac{3}{2}O_2$):

$$CH_3OH(l) + \tfrac{3}{2}O_2(g) \longrightarrow CO_2(g) + 2H_2O(l)$$

Although the equation is now balanced, it is not in its most conventional form because it contains a fractional coefficient. If we multiply each side of the equation by 2, we will remove the fraction and achieve the following balanced equation:

$$2CH_3OH(l) + 3O_2(g) \longrightarrow 2CO_2(g) + 4H_2O(l)$$

PRACTICE EXERCISE

Write the balanced chemical equation for the reaction that occurs when ethanol, $C_2H_5OH(l)$, is burned in air.
Answer: $C_2H_5OH(l) + 3O_2(g) \longrightarrow 2CO_2(g) + 3H_2O(l)$

CHEMISTRY AT WORK: CO_2 and the Greenhouse Effect

Coal and petroleum provide the fuels that we use to move our vehicles, heat our homes and factories, and power our industrial machinery. These fuels are composed primarily of hydrocarbons and other carbon-containing compounds (for a more detailed discussion, see Section 5.8). Combustion of these fuels annually releases about 20 billion tons of CO_2 into the atmosphere.

Chemists have systematically monitored atmospheric CO_2 concentrations since 1958. Analysis of air trapped in ice cores taken from Antarctica and Greenland permit us to determine the atmospheric levels of CO_2 during the past 160,000 years. These measurements reveal that the level of CO_2 remained fairly constant from the last ice age some 10,000 years ago until roughly the beginning of the Industrial Revolution, about 300 years ago. Since that time, the concentration of CO_2 has increased by about 25 percent (Figure 3.6). It is now higher than at any time during the past 160,000 years.

Although CO_2 is a minor component of the atmosphere, it plays a significant role in trapping heat near the surface of the planet. By absorbing radiant heat, it acts much like the glass over a greenhouse. For this reason, we often refer to CO_2 and other heat-trapping gases as greenhouse gases and call the warming caused by these gases the *greenhouse effect*. Some climatologists believe that the accumulation of CO_2 and other heat-trapping gases such as CH_4 has begun to change the climate of our planet (Figure 3.7). They cite as evidence the fact that the six warmest years on record were all in the 1980s.

One long-term outcome of global warming is the melting of the polar icecaps, causing a rise in the sea level, which could lead to the flooding of low-lying coastal areas. Other effects are shifts in rainfall distribution over the continental land masses. Eventually, heat and drought could become commonplace over the vast region of the United States that now provides most of our agricultural products.

If we are to reverse the present trends, we must develop new energy sources, intensify energy conservation efforts, and halt deforestation throughout the globe. We will examine the characteristics of our planet's atmosphere in Chapter 18.

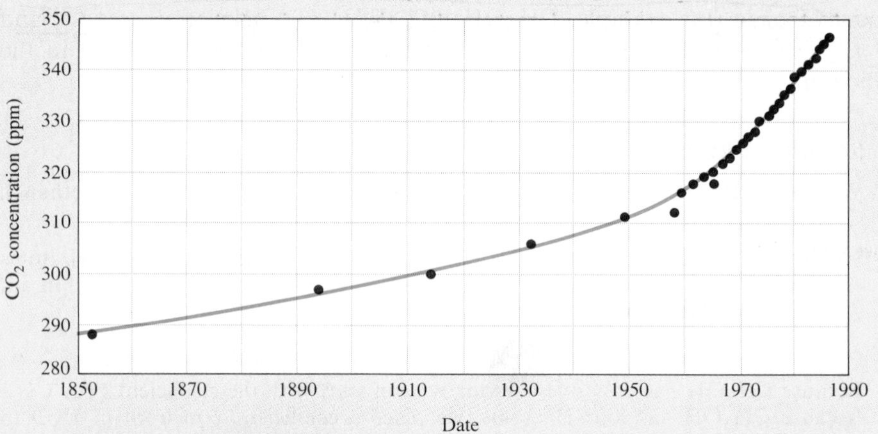

Figure 3.6 The concentration of atmospheric CO_2 over the past 140 years. Data before 1958 come from analyses of air trapped in bubbles of glacial ice. The concentration in ppm (vertical scale) is the number of molecules of CO_2 per million (10^6) molecules of air. CO_2 is a relatively minor component of air, but its presence has significant consequences.

(*a*)

(*b*)

Figure 3.7 Rhône Glacier in Switzerland is shown in a lithograph after an 1848 watercolor by Henri Hogard (*a*). Four sets of moraines are clearly visible. The outermost set has been dated to 1602, the second to 1818, the third to 1826 and the fourth to 1848; the pattern indicates that the glacier had been retreating up the valley for at least 250 years. A photograph from 1970 (*b*) shows that the glacier has retreated still farther up the valley. The retreat is additional evidence for recent global warming. (Markus Aellen, Chief of Section, Swiss Federal Institute of Technology, Zurich, Switzerland)

Other Common Reaction Types

Table 3.1 summarizes four other common types of reactions. Although not all reactions fall into these categories, there are a large number of reactions in which atoms or groups of atoms rearrange in these ways. However, merely knowing that there are such patterns of reactivity doesn't allow us to predict what will happen when we heat a substance or when we mix two or more reactants. To make such predictions, we need to look more closely at chemical behavior, as we have done in considering the combustion of hydrocarbons. In the chapters that follow, we will take a closer look at some important reactions that fit some of these patterns. For the time being, we will consider only the first two types of reactions.

In **combination reactions**, two or more substances react to form one product. There are many examples of such reactions, especially those in which different elements combine to form compounds. For example, magnesium metal burns in air with a dazzling brillance to produce magnesium oxide, as shown in Figure 3.8:

$$2Mg(s) + O_2(g) \longrightarrow 2MgO(s) \quad [3.9]$$

In a **decomposition reaction**, one substance undergoes a reaction to produce two or more other substances. Many compounds undergo decomposition reactions when they are heated. For example, many metal carbonates decompose to form metal oxides and carbon dioxide when heated:

$$CaCO_3(s) \longrightarrow CaO(s) + CO_2(g) \quad [3.10]$$

$$PbCO_3(s) \longrightarrow PbO(s) + CO_2(g) \quad [3.11]$$

Table 3.1 Some Common Types of Reactions

Reaction	Description
Combination reactions	
$A + B \longrightarrow C$	Two reactants combine to form a single product. Many elements react with one another in this fashion to form compounds.
$C(s) + O_2(g) \longrightarrow CO_2(g)$	
$N_2(g) + 3H_2(g) \longrightarrow 2NH_3(g)$	
$CaO(s) + H_2O(l) \longrightarrow Ca(OH)_2(s)$	
Decomposition reactions	
$C \longrightarrow A + B$	A single reactant breaks apart to form two or more substances. Many compounds behave in this fashion when heated.
$2KClO_3(s) \longrightarrow 2KCl(s) + 3O_2(g)$	
$CaCO_3(s) \longrightarrow CaO(s) + CO_2(g)$	
Single displacement reactions	
$A + BX \longrightarrow AX + B$	One element replaces another in a compound. (The elements are often hydrogen or a metal.)
$Fe(s) + 2HCl(aq) \longrightarrow FeCl_2(aq) + H_2(g)$	
$Zn(s) + CuSO_4(aq) \longrightarrow ZnSO_4(aq) + Cu(s)$	
$2Na(s) + 2H_2O(l) \longrightarrow 2NaOH(aq) + H_2(g)$	
Double displacement (metathesis) reactions	
$AX + BY \longrightarrow AY + BX$	Atoms or ions exchange partners.
$BaBr_2(aq) + K_2SO_4(aq) \longrightarrow 2KBr(aq) + BaSO_4(s)$	
$Ca(OH)_2(aq) + 2HCl(aq) \longrightarrow CaCl_2(aq) + 2H_2O(l)$	

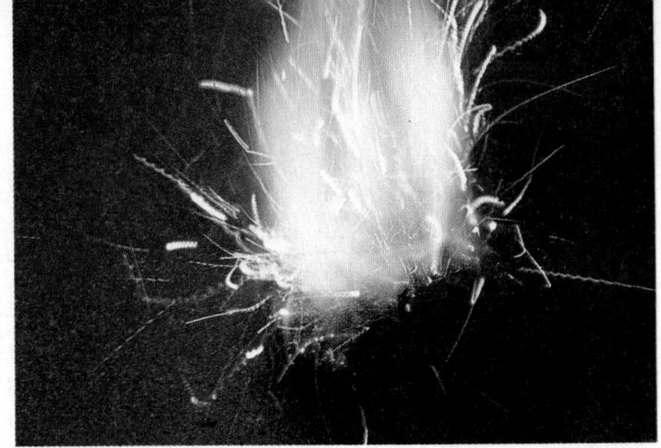

Figure 3.8 Magnesium metal burns with a brilliant flame. (Dr. E. R. Degginger)

Single displacement reactions and double displacement reactions are also very common types of reactions. They occur readily in aqueous solutions, as we will see in Chapter 4.

3.3 ATOMIC AND MOLECULAR WEIGHTS

Chemical formulas and chemical equations have a *quantitative* significance; that is, the subscripts in formulas and the coefficients in equations represent precise quantities. The formula H_2O is not merely an abbreviation for the word *water;* it indicates that a molecule of this substance contains exactly two atoms of hydrogen and one atom of oxygen. Similarly, the chemical equation for the combustion of propane—$C_3H_8(g) + 5O_2(g) \longrightarrow 3CO_2(g) + 4H_2O(l)$, shown in Equation 3.8—indicates more than the qualitative idea that propane reacts with O_2 to form CO_2 and H_2O. The equation is a quantitative statement that the combustion of one molecule of C_3H_8 requires five molecules of O_2 and produces exactly three molecules of CO_2 and four of H_2O. Although we cannot directly count atoms or molecules, we can indirectly determine their numbers if we know the masses of atoms. Therefore, before we can pursue the quantitative aspects of chemical formulas or equations further, we must explore the concept of atomic and molecular masses.

The Atomic Mass Scale

Although scientists of the nineteenth century knew nothing about subatomic particles, they were aware that atoms of different elements have different masses. They found, for example, that each 100 g of water contains 11.1 g of hydrogen and 88.9 g of oxygen. Thus, water contains $88.9/11.1 = 8$ times as much oxygen, by mass, as hydrogen. Once scientists understood that water contains two hydrogen atoms for each oxygen, they concluded that an oxygen atom must weigh $2 \times 8 = 16$ times as much as a hydrogen atom. Hydrogen, the lightest atom, was arbitrarily assigned a relative mass of 1 (no units), and atomic masses of other elements were determined relative to this value. Thus, oxygen was assigned an atomic mass of 16.

Today we can measure the masses of individual atoms with a high degree of accuracy. For example, we know that the hydrogen-1 atom has a mass of 1.6736×10^{-24} g and the oxygen-16 atom has a mass of 2.6561×10^{-23} g. As we saw in Section 2.3, it is convenient to use a unit called the **atomic mass unit** (amu):

$$1 \text{ amu} = 1.66056 \times 10^{-24} \text{ g}$$

The amu is defined by assigning a mass of exactly 12 amu to the ^{12}C isotope of carbon. In these units, the mass of the hydrogen-1 atom is 1.0080 amu and that of the oxygen-16 atom is 15.995 amu.

HISTORICAL PERSPECTIVE: Gay-Lussac, Avogadro, and Berzelius

At the time of Dalton, the problem of establishing even a relative atomic weight scale for the elements was formidable. Because atoms and molecules cannot be seen, scientists of the nineteenth century had no simple way to be sure about the relative numbers of atoms in any compound. Dalton thought that the formula for water was HO. However, in 1808 the French scientist Joseph Louis Gay-Lussac demonstrated that two volumes of hydrogen gas were required to react with one volume of oxygen to form two volumes of water vapor. This observation was inconsistent with Dalton's formula for water. Furthermore, if one assumed that oxygen was a monatomic gas, as Dalton did, one could obtain two volumes of water vapor only by splitting the oxygen atoms in half, which of course violates the concept of the atom as indivisible in chemical reactions.

In 1811, the Italian physicist Amedeo Avogadro resolved many of these problems when he suggested that Gay-Lussac's results could be explained if it were assumed that both hydrogen and oxygen exist in the gas phase as diatomic molecules, H_2 and O_2. Avogadro also proposed that equal volumes of gases, if measured at the same temperature and pressure, contain equal numbers of molecules. (When we talk about gases in Chapter 10 we will see the basis for Avogadro's hypothesis, which is entirely correct.) The way in which Avogadro's hypothesis explains Gay-Lussac's observations is illustrated in Figure 3.9. The key here is in assuming that the reacting gases are diatomic. Thus it is possible to obtain two volumes of water vapor from one volume of oxygen gas. Because the combining volume of hydrogen is twice that of oxygen, the formula for water must be H_2O, not HO as Dalton had suggested.

Not until about 1860 did Avogadro's ideas gain acceptance. In the meantime chemists such as Berzelius (Figure 3.10) had been making painstaking measurements of the masses of the elements that combined with one another and had established the atomic weights of many elements with quite good accuracy.

Figure 3.10 Jons Jakob Berzelius (1779–1848), Swedish chemist. Berzelius is one of the truly great figures in nineteenth-century science. He developed the modern system of symbols and formulas in chemistry and discovered several elements. Berzelius was an exceptionally gifted experimentalist. His determinations of the atomic weights of several elements were the most accurate known for many years. (The Granger Collection)

Figure 3.9 Gay-Lussac's experimental observation of combining volumes shown together with Avogadro's explanation of this phenomenon.

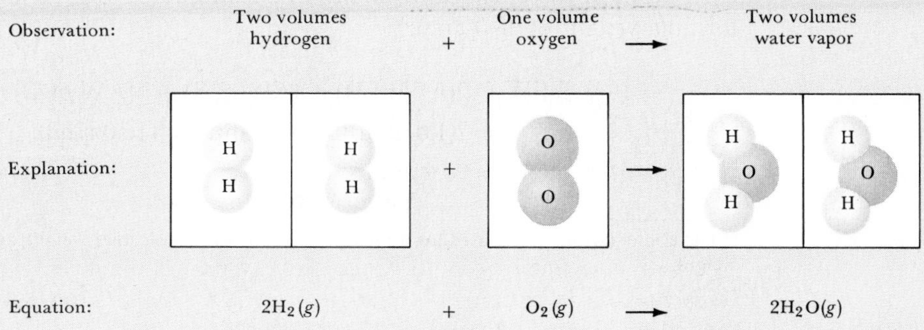

Average Atomic Masses

We can determine the average atomic mass of each element by using the masses of the various isotopes of an element and their relative abundances. For example, naturally occurring carbon contains three isotopes: ^{12}C (98.892 percent abundance), ^{13}C (1.108 percent abundance), and ^{14}C (2×10^{-10} percent abundance). The masses of these nuclides are 12 amu (by definition), 13.00335 amu, and 14.00317 amu, respectively. The computation of the average atomic mass of carbon follows:

$$(0.98892)(12 \text{ amu}) + (0.01108)(13.00335 \text{ amu}) + (2 \times 10^{-12})(14.00317 \text{ amu}) = 12.011 \text{ amu}$$

The average atomic mass of each element (expressed in amu) is also known as its **atomic weight**. Although the term *average atomic mass* is more proper, the term *atomic weight* has become so commonly used that there is little point in attempting to change. The atomic weights of the elements are listed below the symbol for the element in the periodic table found on the front inside cover of this text. They are also listed in the table of the elements on the front inside cover.

SAMPLE EXERCISE 3.3

Naturally occurring chlorine is 75.53 percent ^{35}Cl, which has an atomic mass of 34.969 amu, and 24.47 percent ^{37}Cl, which has a mass of 36.966 amu. Calculate the average atomic mass (that is, the atomic weight) of chlorine.

Solution

$$\begin{aligned}\text{average atomic mass} &= (75.53\%)(34.969 \text{ amu}) + (24.47\%)(36.966 \text{ amu}) \\ &= (0.7553)(34.969 \text{ amu}) + (0.2447)(36.966 \text{ amu}) \\ &= 26.41 \text{ amu} + 9.05 \text{ amu} \\ &= 35.46 \text{ amu}\end{aligned}$$

PRACTICE EXERCISE

Three isotopes of silicon occur in nature: ^{28}Si (92.21 percent), which has a mass of 27.97693 amu; ^{29}Si (4.70 percent), which has a mass of 28.97649 amu; and ^{30}Si (3.09 percent), which has a mass of 29.97376 amu. Calculate the atomic weight of silicon.
Answer: 28.09 amu

Formula and Molecular Weights

The **formula weight** of a substance is merely the sum of the atomic weights of each atom in the chemical formula. For example, H_2SO_4, sulfuric acid, has a formula weight of 98.0 amu:*

$$\begin{aligned}\text{FW} &= 2(\text{AW of H}) + (\text{AW of S}) + 4(\text{AW of O}) \\ &= 2(1.0 \text{ amu}) + 32.0 \text{ amu} + 4(16.0 \text{ amu}) \\ &= 98.0 \text{ amu}\end{aligned}$$

* The abbreviation AW is used for atomic weight, MW for molecular weight, and FW for formula weight.

Here we have rounded off the atomic weights to one place beyond the decimal point. We will round off the atomic weights in this way for most problems.

If the chemical formula of a substance is its molecular formula, then the formula weight is often called the **molecular weight**. For example, the molecular formula for glucose (the sugar transported by the blood to body tissues in order to provide energy) is $C_6H_{12}O_6$. The molecular weight of glucose is therefore

$$\begin{aligned} \text{MW} &= 6(12.0 \text{ amu}) + 12(1.0 \text{ amu}) + 6(16.0 \text{ amu}) \\ &= 180.0 \text{ amu} \end{aligned}$$

With ionic substances such as NaCl that exist as three-dimensional arrays of ions (Figure 2.22), it is inappropriate to speak of molecules. Thus, we cannot write molecular formulas and molecular weights for such substances. The formula weight of NaCl is

$$\begin{aligned} \text{FW} &= 23.0 \text{ amu} + 35.5 \text{ amu} \\ &= 58.5 \text{ amu} \end{aligned}$$

SAMPLE EXERCISE 3.4

Calculate the formula weight of **(a)** sucrose, $C_{12}H_{22}O_{11}$ (table sugar); **(b)** calcium nitrate, $Ca(NO_3)_2$.

Solution: **(a)** By adding the weights of the atoms in sucrose we find it to have a formula weight of 342.0 amu:

$$\begin{aligned} 12 \text{ C atoms} &= 12(12.0 \text{ amu}) = 144.0 \text{ amu} \\ 22 \text{ H atoms} &= 22(1.0 \text{ amu}) = 22.0 \text{ amu} \\ 11 \text{ O atoms} &= 11(16.0 \text{ amu}) = \underline{176.0 \text{ amu}} \\ & \qquad\qquad\qquad\qquad 342.0 \text{ amu} \end{aligned}$$

(b) If a chemical formula has parentheses, the subscript outside the parentheses is a multiplier for all atoms inside. Thus for $Ca(NO_3)_2$ we have

$$\begin{aligned} 1 \text{ Ca atom} &= 1(40.1 \text{ amu}) = 40.1 \text{ amu} \\ 2 \text{ N atoms} &= 2(14.0 \text{ amu}) = 28.0 \text{ amu} \\ 6 \text{ O atoms} &= 6(16.0 \text{ amu}) = \underline{96.0 \text{ amu}} \\ & \qquad\qquad\qquad\qquad 164.1 \text{ amu} \end{aligned}$$

PRACTICE EXERCISE

Calculate the formula weight of **(a)** $Al(OH)_3$; **(b)** CH_3OH.
Answers: **(a)** 78.0 amu; **(b)** 32.0 amu

Percentage Composition from Formulas

Occasionally we must calculate the *percentage composition* of a compound (that is, the percentage by mass contributed by each element in the substance). For example, we may wish to compare the calculated composition of a substance with that found experimentally in order to verify the purity of the compound. Calculating percentage composition is a straightforward matter if the chemical formula is known. The calculation of such percentages is illustrated in Sample Exercise 3.5.

SAMPLE EXERCISE 3.5

Calculate the percentage composition of $C_{12}H_{22}O_{11}$.

Solution: In general, the percentage of a given element in a compound is given by

$$\frac{\text{(Atoms of element)(AW)}}{\text{FW of compound}} \times 100$$

The formula weight of $C_{12}H_{22}O_{11}$ is 342 amu (Sample Exercise 3.4). Therefore, the percentage composition is

$$\%C = \frac{12(12.0 \text{ amu})}{342 \text{ amu}} \times 100 = 42.1\%$$

$$\%H = \frac{22(1.0 \text{ amu})}{342 \text{ amu}} \times 100 = 6.4\%$$

$$\%O = \frac{11(16.0 \text{ amu})}{342 \text{ amu}} \times 100 = 51.5\%$$

The same elemental composition is obtained from the empirical formula of a substance as from its molecular formula. In Section 3.7 we shall see how the experimentally determined percentage composition of a compound can be used to calculate the empirical formula of the compound.

PRACTICE EXERCISE

Calculate the percentage of nitrogen, by mass, in $Ca(NO_3)_2$.
Answer: 17.1 percent

3.4 THE MASS SPECTROMETER

The most direct and accurate means for determining atomic and molecular weights is provided by the **mass spectrometer**. This instrument is illustrated schematically in Figure 3.11. A gaseous sample is introduced into the instrument at *A* and then bombarded by a stream of high-energy electrons at *B*. Collisions between the electrons and the atoms or molecules of the gas produce positive ions. These ions (of mass *m* and charge *e*) are accelerated toward a negatively charged wire grid (*C*). After they pass through the grid, they encounter two slits that produce a narrow beam of the ions. This beam is then passed between

Figure 3.11 (*a*) Schematic diagram of a mass spectrometer. (*b*) A modern mass spectrometer. (Courtesy of Finnigan MAT. Photo by Dave Pierce)

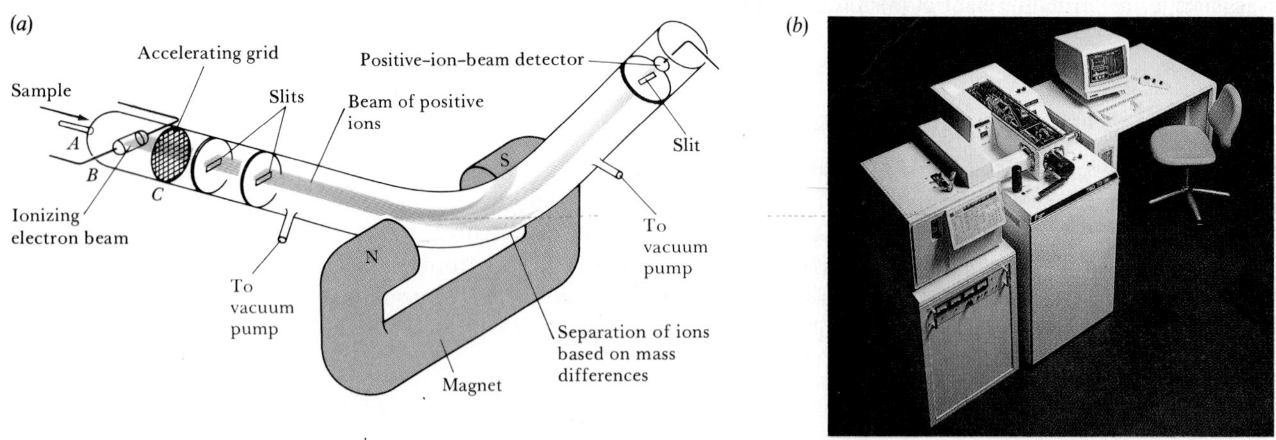

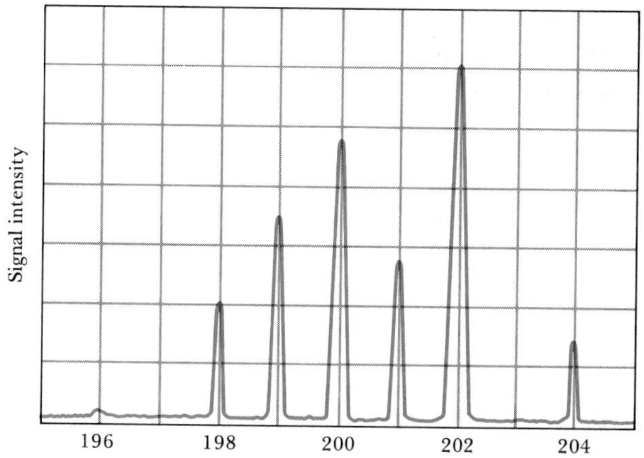

Figure 3.12 Mass spectrum of Hg^+ ions in mercury vapor.

the poles of a magnet, which forces the ions into a circular path whose radius depends on the mass-to-charge ratio of the ions, m/e. Ions of larger m/e ratio follow a curved path of larger radius than those having a smaller m/e ratio. Consequently, ions with equal charges but different masses are separated from each other. By continuously changing either the strength of the magnetic field or the accelerating voltage on the negatively charged grid, ions of varying m/e ratios can be selected to enter the detector at the end of the instrument. Because most of the ions carry a 1+ charge, the separation is primarily on the basis of mass. A graph of the intensity of the signal from the detector versus the mass of the ion is called a *mass spectrum.*

The mass spectrometer provided the first unambiguous evidence for the existence of isotopes. Suppose we allow a stream of mercury vapor to enter the mass spectrometer. If all the atoms of mercury were in fact identical, all of them would have the same m/e ratio upon ionization and would therefore appear at the same place in the mass spectrum. But the mass spectrum actually observed for mercury is as shown in Figure 3.12. It is clear that there are several kinds of atoms of mercury, differing slightly in their masses.

Using the mass spectrometer, it is possible to measure relative values of the ratio m/e for ions with great accuracy. It is also possible to measure with high accuracy the relative amounts of the different isotopes of an element. As we have noted previously, the atomic weight is the weighted average of the masses of all the isotopes of that element. For example, from the mass spectrum of mercury (Figure 3.12), we obtain the data shown in Table 3.2, from which the atomic weight of mercury, 200.59 amu, can be calculated.

Table 3.2 Mass Spectrum of Mercury

Mass number	Atomic weight (amu)	Fractional abundance
196	195.965	0.0014
198	197.967	0.10039
199	198.967	0.1683
200	199.968	0.2312
201	200.970	0.1323
202	201.970	0.2979
204	203.973	0.0685

3.5 THE MOLE

Even the smallest samples that we deal with in the laboratory contain enormous numbers of atoms, ions, or molecules. For example, a teaspoon of water (about 5 mL) contains 2×10^{23} water molecules! It is convenient to have a special unit for describing such large numbers of objects.

In everyday life, we use counting units like dozen (12 objects) and gross (144 objects) to deal with large quantities. In chemistry, the unit we use for dealing with atoms, ions, and molecules is the **mole**, abbreviated mol.* A **mole** is defined as the amount of matter that contains as many objects (atoms, molecules, or whatever objects we are considering) as the number of atoms in exactly 12 g of ^{12}C. From numerous experiments, scientists have determined the number of atoms in this quantity of ^{12}C to be 6.0221367×10^{23}. This number is given a special name: **Avogadro's number**. For most purposes, we will use 6.02×10^{23} for Avogadro's number throughout the text. You should commit this number to memory.

A mole of ions, molecules, or anything else contains Avogadro's number of these objects:

$$1 \text{ mol } ^{12}\text{C atoms} = 6.02 \times 10^{23}\ ^{12}\text{C atoms}$$

$$1 \text{ mol } H_2O \text{ molecules} = 6.02 \times 10^{23}\ H_2O \text{ molecules}$$

$$1 \text{ mol } NO_3^- \text{ ions} = 6.02 \times 10^{23}\ NO_3^- \text{ ions}$$

Avogadro's number is so large that it is difficult to imagine. Spreading 6.02×10^{23} marbles over the surface of the entire earth would produce a layer about 3 mi thick!

SAMPLE EXERCISE 3.6

How many C atoms are in 0.350 mol of $C_6H_{12}O_6$?

Solution: There are 6.02×10^{23} $C_6H_{12}O_6$ molecules in 1 mol. Each molecule contains 6 C atoms:

$$\text{C atoms} = (0.350 \text{ mol } C_6H_{12}O_6)\left(\frac{6.02 \times 10^{23} \text{ molecules}}{1 \text{ mol}}\right)\left(\frac{6 \text{ C atoms}}{1 \text{ molecule}}\right)$$

$$= 1.26 \times 10^{24} \text{ C atoms}$$

PRACTICE EXERCISE

How many nitrogen atoms are in 0.25 mol of $Ca(NO_3)_2$? ***Answer:*** 3.0×10^{23}

Molar Mass

A single ^{12}C atom has a mass of 12 amu, whereas a single ^{24}Mg atom is twice as massive, 24 amu. Because a mole always has the same number of particles, a mole of ^{24}Mg must be twice as massive as a mole of ^{12}C atoms. Because a mole of ^{12}C weighs 12 g (by definition), then a mole of ^{24}Mg must weigh 24 g. Notice that the mass of a single atom of an element (in amu) is numerically equal to the mass (in grams) of 1 mol of atoms of that element. This fact is true regardless of the element:

One ^{12}C atom weighs 12 amu; 1 mol ^{12}C weighs 12 g.
One ^{24}Mg atom weighs 24 amu; 1 mol ^{24}Mg weighs 24 g.
One ^{197}Au atom weighs 197 amu; 1 mol ^{197}Au weighs 197 g.

* The term *mole* comes from the Latin word *moles*, meaning "a mass." The term *molecule* is the diminutive form of this word and means "a small mass."

Table 3.3 Mole Relationships

Name	Formula	Formula weight (amu)	Mass of 1 mol of formula units (g)	Number and kind of particles in 1 mol
Atomic nitrogen	N	14.0	14.0	6.02×10^{23} N atoms
Molecular nitrogen	N_2	28.0	28.0	6.02×10^{23} N_2 molecules; $2(6.02 \times 10^{23})$ N atoms
Silver	Ag	107.9	107.9	6.02×10^{23} Ag atoms
Silver ions	Ag^+	107.9[a]	107.9	6.02×10^{23} Ag^+ ions
Barium chloride	$BaCl_2$	208.2	208.2	6.02×10^{23} $BaCl_2$ units; 6.02×10^{23} Ba^{2+} ions; $2(6.02 \times 10^{23})$ Cl^- ions

[a] Recall that the electron has negligible mass; thus ions and atoms have essentially the same mass.

The mass in grams of 1 mol of a substance is called its **molar mass**. *The molar mass (in grams) of any substance is always numerically equal to its formula weight (in amu):*

One H_2O molecule weighs 18.0 amu; 1 mol H_2O weighs 18.0 g.

One NO_3^- ion weighs 62.0 amu; 1 mol NO_3^- weighs 62.0 g.

One NaCl unit weighs 58.5 amu; 1 mol NaCl weighs 58.5 g.

Further examples of mole relationships are shown in Table 3.3. Figure 3.13 shows 1-mol quantities of several substances.

The first entries in Table 3.3, those for N and N_2, point out the importance of stating the chemical form of a substance exactly when we use the mole concept. Suppose you read that 1 mol of nitrogen is produced in a particular reaction. You might interpret this statement to mean 1 mol of nitrogen atoms (14.0 g). Unless otherwise stated, what was probably meant is 1 mol of nitrogen molecules, N_2 (28.0 g), because N_2 is the usual chemical form of the element. However, to avoid ambiguity it is always best to state explicitly the chemical form being discussed. Using the chemical formula N_2 or referring to the substance as "dinitrogen" avoids ambiguity.

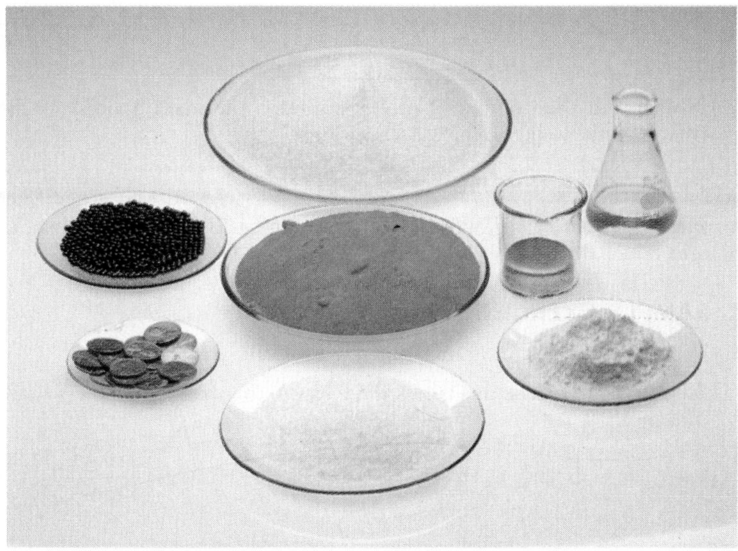

Figure 3.13 One mole each of several substances. Clockwise from top: $C_{12}H_{22}O_{11}$ (sucrose), H_2O, Hg, S, NaCl, Cu, Pb. The colored compound in the center is $K_2Cr_2O_7$. (Richard Megna/Fundamental Photographs)

SAMPLE EXERCISE 3.7

What is the mass of 1 mol of glucose, $C_6H_{12}O_6$?

Solution: By adding the weights of the atoms in glucose, we find it to have a formula weight of 180 amu:

$$
\begin{aligned}
6\text{ C atoms} &= 6(12.0\text{ amu}) = 72.0\text{ amu} \\
12\text{ H atoms} &= 12(1.0\text{ amu}) = 12.0\text{ amu} \\
6\text{ O atoms} &= 6(16.0\text{ amu}) = \underline{96.0\text{ amu}} \\
& \qquad\qquad\qquad\quad 180.0\text{ amu}
\end{aligned}
$$

Hence 1 mol of $C_6H_{12}O_6$ weighs 180 g; that is, $C_6H_{12}O_6$ has a molar mass of 180 g. Glucose is sometimes called dextrose. Also known as blood sugar, glucose is found widely in nature, occurring, for example, in honey and fruits. Other types of sugars used as food must be converted into glucose in the stomach or liver before they are used by the body as energy sources. Because glucose requires no conversion, it is often given intravenously to patients who need immediate nourishment.

PRACTICE EXERCISE

Calculate the molar mass of $Ca(NO_3)_2$. ***Answer:*** 164.1 g

Interconverting Masses, Moles, and Numbers of Particles

The conversions of mass to moles and of moles to mass are frequently encountered in calculations using the mole concept. These calculations are made easy through dimensional analysis, as illustrated in the next two sample exercises.

SAMPLE EXERCISE 3.8

How many moles of glucose, $C_6H_{12}O_6$, are in **(a)** 538 g and **(b)** 1.00 g of this substance?

Solution: **(a)** One mole of $C_6H_{12}O_6$ weighs 180 g (Sample Exercise 3.7). Therefore, there must be more than 1 mol in 538 g.

$$
\text{Moles } C_6H_{12}O_6 = (538\text{ g } C_6H_{12}O_6)\left(\frac{1\text{ mol } C_6H_{12}O_6}{180\text{ g } C_6H_{12}O_6}\right) = 2.99\text{ mol}
$$

(b) In this case there must be less than 1 mol.

$$
\text{Moles } C_6H_{12}O_6 = (1.00\text{ g } C_6H_{12}O_6)\left(\frac{1\text{ mol } C_6H_{12}O_6}{180\text{ g } C_6H_{12}O_6}\right) = 5.56 \times 10^{-3}\text{ mol}
$$

Notice that the number of moles is always the mass divided by the mass of 1 mol (the formula weight expressed in grams).

PRACTICE EXERCISE

How many moles of $NaHCO_3$ are present in 5.08 g of this substance?
Answer: 0.0605 mol

SAMPLE EXERCISE 3.9

What is the mass, in grams, of 0.433 mol of $C_6H_{12}O_6$?

Solution: Because this is less than 1 mol, the mass will be less than 180 g, the mass of 1 mol.

$$
\text{Grams } C_6H_{12}O_6 = (0.433\text{ mol } C_6H_{12}O_6)\left(\frac{180\text{ g } C_6H_{12}O_6}{1\text{ mol } C_6H_{12}O_6}\right) = 77.9\text{ g}
$$

The mass of a certain number of moles of a substance is always the number of moles times the mass of 1 mol.

PRACTICE EXERCISE

What is the mass, in grams, of 6.33 mol of $NaHCO_3$? ***Answer:*** 532 g

In the introduction to Section 3.3, we noted that the concept of atomic weights is important because it permits us to count atoms indirectly by weighing samples. The mole concept provides the bridge between masses and numbers of particles. To illustrate how we can interconvert masses and numbers of particles, let's calculate the number of copper atoms in a traditional copper penny. Such a penny weighs 3 g, and we'll assume that it is 100 percent copper:

$$\text{Cu atoms} = (3 \text{ g Cu})\left(\frac{1 \text{ mol Cu}}{63.5 \text{ g Cu}}\right)\left(\frac{6.02 \times 10^{23} \text{ Cu atoms}}{1 \text{ mol Cu}}\right)$$
$$= 3 \times 10^{22} \text{ Cu atoms}$$

Notice how we were able to use dimensional analysis (Section 1.5) in a straightforward manner to go from grams to numbers of atoms. The molar mass and Avogadro's number are used as conversion factors to convert grams $\longrightarrow$ moles $\longrightarrow$ atoms. Notice also that our answer is a very large number. Any time you calculate the number of atoms, molecules, or ions in an ordinary sample of matter, you can expect the answer to be very large. On the other hand, the number of moles in a sample will usually be much smaller, often less than 1.

SAMPLE EXERCISE 3.10

How many glucose molecules are in 5.23 g of $C_6H_{12}O_6$?

Solution: In this case we need to carry out unit conversions in the order grams $\longrightarrow$ moles $\longrightarrow$ molecules. Because the mass we begin with corresponds to less than a mole, there should be fewer than 6.02×10^{23} molecules.

Molecules $C_6H_{12}O_6$

$$= (5.23 \text{ g } C_6H_{12}O_6)\left(\frac{1 \text{ mol } C_6H_{12}O_6}{180 \text{ g } C_6H_{12}O_6}\right)\left(\frac{6.02 \times 10^{23}\ C_6H_{12}O_6 \text{ molecules}}{1 \text{ mol } C_6H_{12}O_6}\right)$$
$$= 1.75 \times 10^{22} \text{ molecules}$$

PRACTICE EXERCISE

How many oxygen atoms are present in 4.20 g of $NaHCO_3$?
Answer: 9.03×10^{22}

3.6 EMPIRICAL FORMULAS FROM ANALYSES

The empirical formula for a substance tells us the relative number of atoms of each element it contains. Thus the formula H_2O indicates that water contains two H atoms for each O atom. This ratio also applies on the molar level; thus, 1 mol of H_2O contains 2 mol of H atoms and 1 mol of O atoms. Conversely, the ratio of the number of moles of each element in a compound gives the subscripts in a compound's empirical formula. Thus, the mole concept provides a way of calculating the empirical formulas of chemical substances, as shown in the following examples.

Mercury forms a compound with chlorine that is 73.9 percent mercury and 26.1 percent chlorine by mass. This means that if we had a 100-g sample

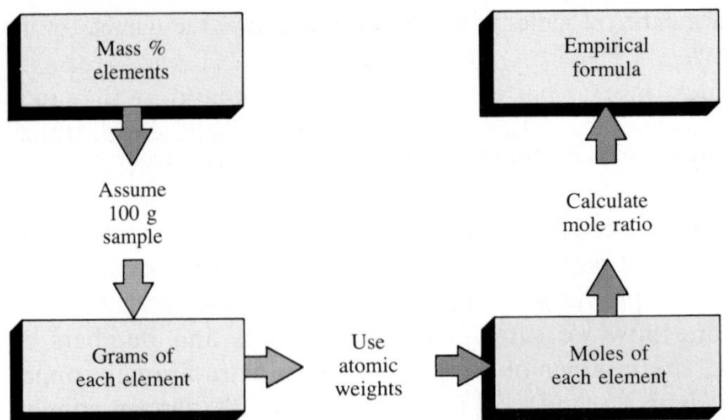

Figure 3.14 Outline of the procedure used to calculate the empirical formula of a substance from its percentage composition.

of the solid, it would contain 73.9 g of mercury, Hg, and 26.1 g of chlorine, Cl. We divide each of these masses by the appropriate atomic weight to obtain the number of moles of each element in 100 g:

$$73.9 \text{ g Hg}\left(\frac{1 \text{ mol Hg}}{200.6 \text{ g Hg}}\right) = 0.368 \text{ mol Hg}$$

$$26.1 \text{ g Cl}\left(\frac{1 \text{ mol Cl}}{35.5 \text{ g Cl}}\right) = 0.735 \text{ mol Cl}$$

We then divide the larger number of moles by the smaller to obtain the ratio 1.99 mol of Cl/mole of Hg. Because of experimental errors, the results of an analysis may not lead to exact integers for the ratios of moles. The ratio obtained in this case is very close to 2, and we can confidently conclude that the formula for the compound is $HgCl_2$. This is the simplest, or empirical, formula because it uses as subscripts the smallest set of integers that express the correct ratios of atoms present (Section 2.5). The general procedure is outlined in Figure 3.14.

SAMPLE EXERCISE 3.11

Phosgene, a poison gas used during World War I, contains 12.1 percent C, 16.2 percent O, and 71.7 percent Cl by mass. What is the empirical formula of phosgene?

Solution: For simplicity, we may assume that we have 100 g of material (although any number can be used). The number of moles of each element is then

$$\text{Moles C} = (12.1 \text{ g})\left(\frac{1 \text{ mol C}}{12.0 \text{ g}}\right) = 1.01 \text{ mol C}$$

$$\text{Moles O} = (16.2 \text{ g})\left(\frac{1 \text{ mol O}}{16.0 \text{ g}}\right) = 1.01 \text{ mol O}$$

$$\text{Moles Cl} = (71.7 \text{ g})\left(\frac{1 \text{ mol Cl}}{35.5 \text{ g}}\right) = 2.02 \text{ mol Cl}$$

The simplest ratio, found by dividing each number by the smallest, 1.01, is C:O:Cl = 1:1:2, and the empirical formula is $COCl_2$.

PRACTICE EXERCISE

A 5.325-g sample of methyl benzoate, a compound used in the manufacture of perfumes, is found to contain 3.758 g of carbon, 0.316 g of hydrogen, and 1.251 g of oxygen. What is the empirical formula of this substance? ***Answer:*** C_4H_4O

Remember that the formula determined from percentage compositions is always the empirical formula. To obtain the molecular formula from the empirical formula we must know the molecular weight of the compound. In the practice exercise above, the empirical formula of methyl benzoate is C_4H_4O, which has a formula weight of 68.0 amu. The experimentally determined molecular weight of the compound is 136.0 amu. Because the molecular weight is twice the formula weight, the subscripts in the molecular formula must be twice those in the empirical formula: $C_8H_8O_2$.

Combustion Analysis

When a compound containing carbon and hydrogen is combusted in an apparatus such as that shown in Figure 3.15, all the carbon of the compound is converted to CO_2, and all the hydrogen to H_2O. The amount of CO_2 produced can be measured by determining the mass increase in the CO_2 absorber. Similarly, the amount of H_2O formed is determined from the increase in mass of the water absorption tube. As an example, let's consider an analysis of a sample of ascorbic acid (vitamin C), a compound that contains only C, H, and O. Combustion of 1.000 g of ascorbic acid produces 1.500 g of CO_2 and 0.405 g of H_2O. From these two bits of experimental information we can calculate the quantities of C and H in the 1.000-g sample of ascorbic acid:

$$(1.500 \text{ g } CO_2)\left(\frac{1 \text{ mol } CO_2}{44.00 \text{ g } CO_2}\right)\left(\frac{1 \text{ mol C}}{1 \text{ mol } CO_2}\right)\left(\frac{12.01 \text{ g C}}{1 \text{ mol C}}\right) = 0.409 \text{ g C}$$

$$(0.405 \text{ g } H_2O)\left(\frac{1 \text{ mol } H_2O}{18.0 \text{ g } H_2O}\right)\left(\frac{2 \text{ mol H}}{1 \text{ mol } H_2O}\right)\left(\frac{1.01 \text{ g H}}{1 \text{ mol H}}\right) = 0.045 \text{ g H}$$

Because ascorbic acid contains only C, H, and O, the amount of oxygen in the compound must be

$$1.000 \text{ g} - (0.409 \text{ g} + 0.045 \text{ g}) = 0.546 \text{ g}$$

From these data, we can now calculate the number of moles of each element present in 1.000 g of ascorbic acid:

$$\text{Moles C} = (0.409 \text{ g C})\left(\frac{1 \text{ mol C}}{12.0 \text{ g C}}\right) = 0.0341 \text{ mol C}$$

$$\text{Moles H} = (0.045 \text{ g H})\left(\frac{1 \text{ mol H}}{1.01 \text{ g H}}\right) = 0.045 \text{ mol H}$$

$$\text{Moles O} = (0.546 \text{ g O})\left(\frac{1 \text{ mol O}}{16.0 \text{ g O}}\right) = 0.0341 \text{ mol O}$$

Figure 3.15 Apparatus for determining the percentages of carbon and hydrogen in a compound. Copper oxide serves to oxidize traces of carbon and carbon monoxide to carbon dioxide, and to oxidize hydrogen to water. Magnesium perchlorate, $Mg(ClO_4)_2$, is used to absorb water, whereas sodium hydroxide, NaOH, absorbs carbon dioxide.

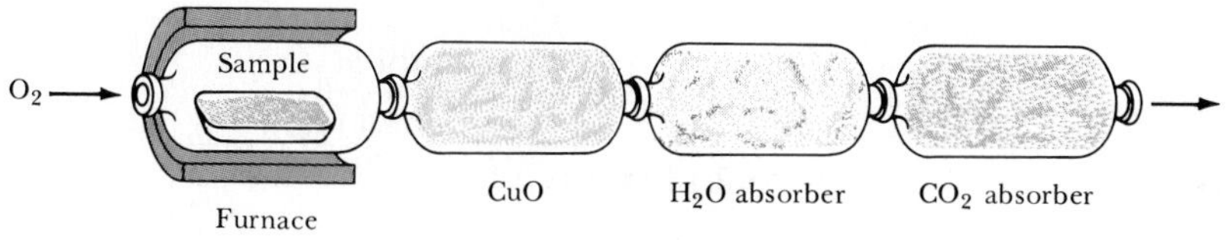

The relative number of moles of each element can be found by dividing each number by the smallest number, 0.0341. The ratio of C:H:O so obtained is 1:1.32:1. The second number is too far from 1 to attribute the difference to experimental error; in fact, it is quite close to $1\frac{1}{3}$. This suggests that if we multiply the ratio by 3, we will get whole numbers: 3(1:1.32:1) = 3:3.96:3, which we may take to be 3:4:3. Thus, the empirical formula is $C_3H_4O_3$.

The experimentally determined molecular weight of ascorbic acid is 176 amu. Using the molecular weight and the empirical formula, we can determine the molecular formula. The subscripts in the molecular formula of a substance are always a whole-number multiple of those in its empirical formula (Section 2.5). The formula weight associated with the empirical formula of ascorbic acid, $C_3H_4O_3$, is 3(12.0 amu) + 4(1.0 amu) + 3(16.0 amu) = 88.0 amu. The molecular weight is twice this value: 176/88.0 = 2.00. Therefore, the subscripts in the empirical formula must be multiplied by 2 to obtain the molecular formula, $C_6H_8O_6$.

3.7 QUANTITATIVE INFORMATION FROM BALANCED EQUATIONS

The mole concept allows us to use the quantitative information available in a balanced chemical equation on a practical, macroscopic level. Consider the following balanced equation:

$$2H_2(g) + O_2(g) \longrightarrow 2H_2O(l) \qquad [3.12]$$

The coefficients tell us that two molecules of H_2 react with each molecule of O_2 to form two molecules of H_2O. It follows that the relative numbers of moles are identical to the relative numbers of molecules:

$2H_2(g)$	+	$O_2(g)$	$\longrightarrow$	$2H_2O(l)$
2 molecules		1 molecule		2 molecules
2(6.02 × 10^{23} molecules)		6.02 × 10^{23} molecules		2(6.02 × 10^{23} molecules)
2 mol		1 mol		2 mol

The coefficients in a balanced chemical equation can be interpreted both as the relative numbers of molecules (or formula units) involved in the reaction and as the relative numbers of moles.

The quantities 2 mol H_2, 1 mol O_2, and 2 mol H_2O, which are given by the coefficients in Equation 3.12, are called *stoichiometrically equivalent quantities.* The relationship between these quantities can be represented as follows:

$$2 \text{ mol } H_2 \simeq 1 \text{ mol } O_2 \simeq 2 \text{ mol } H_2O$$

where the symbol $\simeq$ is taken to mean "stoichiometrically equivalent to." In other words, Equation 3.12 shows 2 mol of H_2 and 1 mol of O_2 forming 2 mol of H_2O. These stoichiometric relations can be used to give conversion factors for relating quantities of reactants and products in a chemical reaction. For example, the number of moles of H_2O produced from 1.57 mol of O_2 can be calculated as follows:

$$\text{Moles } H_2O = (1.57 \text{ mol } O_2)\left(\frac{2 \text{ mol } H_2O}{1 \text{ mol } O_2}\right)$$
$$= 3.14 \text{ mol } H_2O$$

As an additional example, consider the following reaction:

$$2CuFeS_2(s) + 5O_2(g) \longrightarrow 2Cu(s) + 2FeO(s) + 4SO_2(g) \qquad [3.13]$$

This equation describes a process in the smelting of copper using chalcopyrite, $CuFeS_2$, as the mineral source of the copper. Suppose we want to calculate the mass of Cu that can be produced from 1.00 g of $CuFeS_2$. The coefficients in Equation 3.13 tell us how the number of moles of $CuFeS_2$ consumed is related to the number of moles of Cu produced: 2 mol $CuFeS_2 \simeq$ 2 mol Cu. In order to use this relationship, however, the quantity of $CuFeS_2$ must be converted to moles. Because 1 mol $CuFeS_2$ = 183 g $CuFeS_2$, we have

$$\text{Moles } CuFeS_2 = (1.00 \text{ g } CuFeS_2)\left(\frac{1 \text{ mol } CuFeS_2}{183 \text{ g } CuFeS_2}\right)$$
$$= 5.46 \times 10^{-3} \text{ mol } CuFeS_2$$

The stoichiometric factor from the balanced equation, 2 mol $CuFeS_2 \simeq$ 2 mol Cu, can then be used to calculate moles of Cu:

$$\text{Moles Cu} = (5.46 \times 10^{-3} \text{ mol } CuFeS_2)\left(\frac{2 \text{ mol Cu}}{2 \text{ mol } CuFeS_2}\right)$$
$$= 5.46 \times 10^{-3} \text{ mol Cu}$$

Finally, the mass of the Cu, in grams, can be calculated using the molar mass of Cu: 1 mol Cu = 63.5 g Cu:

$$\text{Grams Cu} = (5.46 \times 10^{-3} \text{ mol Cu})\left(\frac{63.5 \text{ g Cu}}{1 \text{ mol Cu}}\right)$$
$$= 0.347 \text{ g Cu}$$

Thus, the conversion sequence is

grams reactant ⟶ moles reactant ⟶ moles product ⟶ grams product

These steps, of course, can be combined in a single sequence of factors:

$$\text{Grams Cu} = (1.00 \text{ g } CuFeS_2)\left(\frac{1 \text{ mol } CuFeS_2}{183 \text{ g } CuFeS_2}\right)\left(\frac{2 \text{ mol Cu}}{2 \text{ mol } CuFeS_2}\right)\left(\frac{63.5 \text{ g Cu}}{1 \text{ mol Cu}}\right)$$
$$= 0.347 \text{ g Cu}$$

We can similarly calculate the amount of SO_2 produced in the production of this quantity of copper using the relationship 2 mol $CuFeS_2 \simeq$ 4 mol SO_2, which also comes from the coefficients in Equation 3.13.

$$\text{Grams } SO_2 = (1.00 \text{ g } CuFeS_2)\left(\frac{1 \text{ mol } CuFeS_2}{183 \text{ g } CuFeS_2}\right)\left(\frac{4 \text{ mol } SO_2}{2 \text{ mol } CuFeS_2}\right)\left(\frac{64.1 \text{ g } SO_2}{1 \text{ mol } SO_2}\right)$$
$$= 0.701 \text{ g } SO_2$$

Figure 3.16 Outline of the procedure used to calculate the number of grams of a reactant consumed or of a product formed in a reaction, starting with the number of grams of one of the other reactants or products.

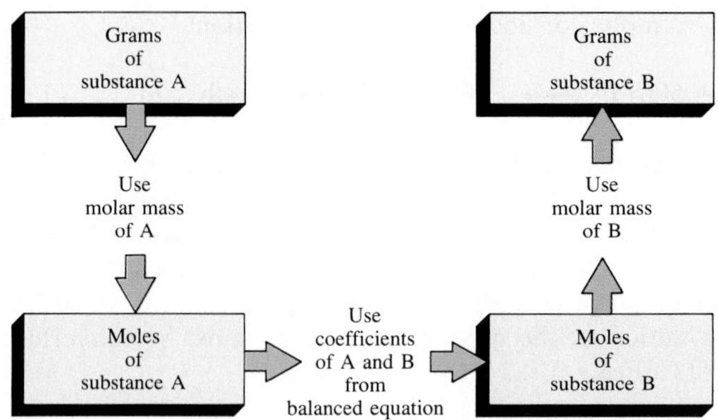

It is interesting to note that the mass of SO_2 produced in this reaction is approximately twice the mass of the copper. Sulfur dioxide, which is a major air pollutant (Chapter 18), is generated in substantial quantities by copper smelters.

Figure 3.16 summarizes the general approach used to calculate the quantities of substances consumed or produced in chemical reactions. As we have seen, the balanced chemical equation tells us the relative numbers of moles of reactants and products involved in the reaction. To use this information, we must first convert the quantities of the substances into moles as shown in the diagram and illustrated in the preceding examples.

SAMPLE EXERCISE 3.12

How much water is produced in the combustion of 1.00 g of glucose, $C_6H_{12}O_6$:

$$C_6H_{12}O_6(s) + 6O_2(g) \longrightarrow 6CO_2(g) + 6H_2O(l)$$

Solution: The chemical equation indicates how much H_2O is produced from $C_6H_{12}O_6$: 1 mol $C_6H_{12}O_6 \simeq 6$ mol H_2O. To use this information, we must convert the amount of $C_6H_{12}O_6$ from grams to moles. Thus the problem can be solved by a stepwise conversion of grams of $C_6H_{12}O_6$ to moles of $C_6H_{12}O_6$ to moles of H_2O to grams of H_2O:

$$\text{Grams } H_2O = (1.00 \text{ g } C_6H_{12}O_6)\left(\frac{1 \text{ mol } C_6H_{12}O_6}{180 \text{ g } C_6H_{12}O_6}\right)\left(\frac{6 \text{ mol } H_2O}{1 \text{ mol } C_6H_{12}O_6}\right)\left(\frac{18.0 \text{ g } H_2O}{1 \text{ mol } H_2O}\right)$$

$$= 0.600 \text{ g } H_2O$$

We may note that an average person ingests 2 L of water daily and eliminates 2.4 L. The difference is produced in the metabolism of foodstuffs. (*Metabolism* is a general term used to describe all the chemical processes of a living animal or plant.) The desert rat (kangaroo rat) is able to take great advantage of its metabolic water to help it survive in the dry desert. In fact, it apparently never drinks water.

PRACTICE EXERCISE

A common laboratory method for preparing small amounts of O_2 involves the decomposition of $KClO_3$: $2KClO_3(s) \longrightarrow 2KCl(s) + 3O_2(g)$. How many grams of O_2 can be prepared from 4.50 g of $KClO_3$? ***Answer:*** 1.77 g

SAMPLE EXERCISE 3.13

Solid lithium hydroxide is used in space vehicles to remove exhaled carbon dioxide. The lithium hydroxide reacts with gaseous carbon dioxide to form solid lithium carbonate and liquid water. How many grams of carbon dioxide can be absorbed by each 1.00 g of lithium hydroxide?

Solution: Using the verbal description of the reaction we can write the unbalanced chemical equation:

$$LiOH(s) + CO_2(g) \longrightarrow Li_2CO_3(s) + H_2O(l)$$

The balanced equation is

$$2LiOH(s) + CO_2(g) \longrightarrow Li_2CO_3(s) + H_2O(l)$$

The problem can be solved by a stepwise conversion of grams of LiOH to moles of LiOH to moles of CO_2 to grams of CO_2. The conversion from grams of LiOH to moles of LiOH requires this substance's formula weight ($6.94 + 16.00 + 1.01 = 23.95$). The conversion of moles of LiOH to moles of CO_2 is based on the balanced chemical equation: 2 mol LiOH $\simeq$ 1 mol CO_2. To convert the number of moles of CO_2 to grams, we must use the formula weight of CO_2: $12.01 + 2(16.00) = 44.01$:

$$(1.00 \text{ g LiOH})\left(\frac{1 \text{ mol LiOH}}{23.95 \text{ g LiOH}}\right)\left(\frac{1 \text{ mol } CO_2}{2 \text{ mol LiOH}}\right)\left(\frac{44.01 \text{ g } CO_2}{1 \text{ mol } CO_2}\right) = 0.919 \text{ g } CO_2$$

PRACTICE EXERCISE

Propane, C_3H_8, is a common fuel used for cooking and home heating. What mass of O_2 is consumed in the combustion of 1.00 g of propane? ***Answer:*** 3.64 g

3.8 LIMITING REACTANT

In many situations, an excess of one or more substances is available for chemical reaction. Some of these excess substances will therefore be left over when the reaction is complete; the reaction stops as soon as one of the reactants is totally consumed. Before we examine a chemical example, consider a simple analogy: Suppose that you have ten slices of bread and seven pieces of bologna. You wish to make as many sandwiches as possible using only one slice of bologna and two slices of bread per sandwich. You will be able to make only five sandwiches, at which point you run out of bread; at that point you have two slices of bologna left over.

Now consider the combustion of hydrogen:

$$2H_2(g) + O_2(g) \longrightarrow 2H_2O(l)$$

Suppose this reaction is initiated by passing a spark through a reaction vessel containing 10 mol H_2 and 7 mol O_2. Because 2 mol $H_2 \simeq$ 1 mol O_2, the number of moles of O_2 needed to react with all the H_2 is

$$\text{Moles } O_2 = (10 \text{ mol } H_2)\left(\frac{1 \text{ mol } O_2}{2 \text{ mol } H_2}\right)$$
$$= 5 \text{ mol } O_2$$

Because 7 mol O_2 was available at the start of the reaction, 7 mol O_2 − 5 mol O_2 = 2 mol O_2 will be present when the H_2 is all consumed. This situation is shown on a molecular scale in Figure 3.17.

The substance that is completely consumed in a reaction is called the **limiting reactant** or **limiting reagent** because it determines, or limits, the amount of product formed. The other reactants are sometimes called *excess reactants* or *excess reagents.* In our example above, H_2 is the limiting reactant, and O_2 is the excess reactant. This means that once all the H_2 has been consumed, the reaction stops.

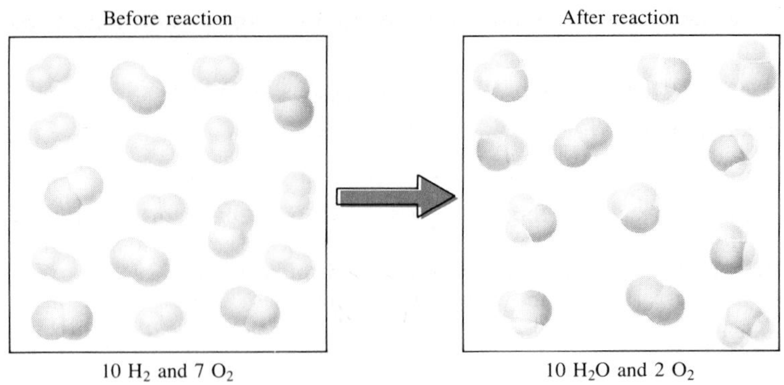

Figure 3.17 Diagram showing the complete utilization of a limiting reagent in a reaction. Because the H_2 is completely consumed, it is the limiting reagent in this case. Because there is a stoichiometric excess of O_2, some is left over at the end of the reaction.

SAMPLE EXERCISE 3.14

Part of the SO_2 that is introduced into the atmosphere ends up being converted to sulfuric acid, H_2SO_4. The net reaction is

$$2SO_2(g) + O_2(g) + 2H_2O(l) \longrightarrow 2H_2SO_4(l)$$

How much H_2SO_4 can be formed from 5.0 mol of SO_2, 1.0 mol of O_2, and an unlimited quantity of H_2O?

Solution: The number of moles of O_2 needed for complete consumption of 5.0 mol of SO_2 is

$$\text{Moles } O_2 = (5.0 \text{ mol } SO_2)\left(\frac{1 \text{ mol } O_2}{2 \text{ mol } SO_2}\right)$$
$$= 2.5 \text{ mol } O_2$$

This quantity of O_2 is not available; therefore, all of the SO_2 cannot be consumed. O_2 is the limiting reactant. We use the quantity of the limiting reactant, O_2, to calculate the quantity of H_2SO_4 produced.

$$\text{Moles } H_2SO_4 = (1.0 \text{ mol } O_2)\left(\frac{2 \text{ mol } H_2SO_4}{1 \text{ mol } O_2}\right)$$
$$= 2.0 \text{ mol } H_2SO_4$$

We might note that in forming 2.0 mol of H_2SO_4, 2.0 mol of SO_2 is consumed. Therefore, 3.0 mol of SO_2 is left over.

PRACTICE EXERCISE

Consider the reaction $2Al(s) + 3Cl_2(g) \longrightarrow 2AlCl_3(s)$. A mixture of 1.5 mol of Al and 3.0 mol of Cl_2 are allowed to react. **(a)** What is the limiting reactant? **(b)** How many moles of $AlCl_3$ are formed? ***Answers:*** **(a)** Al; **(b)** 1.5 mol

Another approach to the problem of the limiting reactant is to calculate the amount of product that could be formed from each of the given amounts of reactants, assuming they were all completely consumed. The reagent that leads to the smallest amount of product is the limiting reactant.

SAMPLE EXERCISE 3.15

Consider the following reaction:

$$2Na_3PO_4(aq) + 3Ba(NO_3)_2(aq) \longrightarrow Ba_3(PO_4)_2(s) + 6NaNO_3(aq)$$

Suppose that a solution containing 3.50 g of Na_3PO_4 is mixed with a solution containing 6.40 g of $Ba(NO_3)_2$. How many grams of $Ba_3(PO_4)_2$ can be formed?

Solution: From the balanced equation we have the following stoichiometric relations:

$$2 \text{ mol } Na_3PO_4 \simeq 3 \text{ mol } Ba(NO_3)_2 \simeq 1 \text{ mol } Ba_3(PO_4)_2$$

The mass of 1 mol of each substance can be found by determining the formula weight for each substance. The results are as follows:

$$1 \text{ mol } Na_3PO_4 = 164 \text{ g } Na_3PO_4$$
$$1 \text{ mol } Ba(NO_3)_2 = 261 \text{ g } Ba(NO_3)_2$$
$$1 \text{ mol } Ba_3(PO_4)_2 = 602 \text{ g } Ba_3(PO_4)_2$$

Let us now calculate the amount of product that could be formed from each of our given amounts of reactants, assuming that each in turn is the limiting reactant. Assuming first that Na_3PO_4 is completely consumed gives:

$$\text{Grams } Ba_3(PO_4)_2 = (3.50 \text{ g } Na_3PO_4)\left(\frac{1 \text{ mol } Na_3PO_4}{164 \text{ g } Na_3PO_4}\right) \times \left(\frac{1 \text{ mol } Ba_3(PO_4)_2}{2 \text{ mol } Na_3PO_4}\right)\left(\frac{602 \text{ g } Ba_3(PO_4)_2}{1 \text{ mol } Ba_3(PO_4)_2}\right) = 6.42 \text{ g } Ba_3(PO_4)_2$$

Assuming that $Ba(NO_3)_2$ is completely consumed gives:

$$\text{Grams } Ba_3(PO_4)_2 = (6.40 \text{ g } Ba(NO_3)_2)\left(\frac{1 \text{ mol } Ba(NO_3)_2}{261 \text{ g } Ba(NO_3)_2}\right) \times \left(\frac{1 \text{ mol } Ba_3(PO_4)_2}{3 \text{ mol } Ba(NO_3)_2}\right)\left(\frac{602 \text{ g } Ba_3(PO_4)_2}{1 \text{ mol } Ba_3(PO_4)_2}\right) = 4.92 \text{ g } Ba_3(PO_4)_2$$

The lesser quantity, 4.92 g, is the amount of $Ba_3(PO_4)_2$ that will form, indicating that $Ba(NO_3)_2$ is the limiting reactant.

PRACTICE EXERCISE

A strip of zinc metal weighing 2.00 g is placed in an aqueous solution containing 2.50 g of silver nitrate, causing the following reaction to occur:

$$Zn(s) + 2AgNO_3(aq) \longrightarrow 2Ag(s) + Zn(NO_3)_2(aq)$$

How many grams of Ag will form? ***Answer:*** 1.59 g

Theoretical Yields

The quantity of product that is calculated to form when all of the limiting reagent reacts is called the **theoretical yield**. The amount of product actually obtained in a reaction is called the *actual yield*. The actual yield is almost always less than (and can never be greater than) the theoretical yield. There are many reasons for this difference. For example, part of the reactants may not react, or they may react in a way different from that desired (side reactions). In addition, it is not always possible to recover all of the reaction product from the reaction mixture. The **percent yield** of a reaction relates the actual yield to the theoretical (calculated) yield:

$$\text{Percent yield} = \frac{\text{actual yield}}{\text{theoretical yield}} \times 100 \qquad [3.14]$$

For example, in the experiment described in Sample Exercise 3.15, we calculate that 4.92 g of $Ba_3(PO_4)_2$ should form when 3.50 g of Na_3PO_4 is mixed with

6.40 g of $Ba(NO_3)_2$. This is the theoretical yield of $Ba_3(PO_4)_2$ in the reaction. If the actual yield turned out to be 4.70 g, the percent yield would be

$$\frac{4.70\text{ g}}{4.92\text{ g}} \times 100 = 95.5\%$$

SAMPLE EXERCISE 3.16

Adipic acid, $H_2C_6H_8O_4$, is a raw material used for the production of nylon. It is made commercially by oxidation of cyclohexane, C_6H_{12}:

$$2C_6H_{12} + 5O_2 \longrightarrow 2H_2C_6H_8O_4 + 2H_2O$$

(a) Assume that you carry out this reaction starting with 25.0 g of cyclohexane, and that cyclohexane is the limiting reactant. What is the theoretical yield of adipic acid?

(b) If you obtain 33.5 g of adipic acid from your reaction, what is the percent yield of adipic acid?

Solution: **(a)** The theoretical yield is merely the calculated quantity of adipic acid formed in the reaction:

$$\text{Grams } H_2C_6H_8O_4 = (25.0\text{ g } C_6H_{12})\left(\frac{1\text{ mol } C_6H_{12}}{84.0\text{ g } C_6H_{12}}\right)$$
$$\times \left(\frac{2\text{ mol } H_2C_6H_8O_4}{2\text{ mol } C_6H_{12}}\right)\left(\frac{146\text{ g } H_2C_6H_8O_4}{1\text{ mol } H_2C_6H_8O_4}\right)$$
$$= 43.5\text{ g } H_2C_6H_8O_4$$

(b) Using the definition of percent yield, Equation 3.14, we have

$$\text{Percent yield} = \frac{\text{actual yield}}{\text{theoretical yield}} \times 100 = \frac{33.5\text{ g}}{43.5\text{ g}} \times \; = 77.0\%$$

PRACTICE EXERCISE

Imagine that you are working on ways to improve the process by which iron ore containing Fe_2O_3 is converted into iron. In your tests you carry out the following reaction on a small scale:

$$Fe_2O_3(s) + 3CO(g) \longrightarrow 2Fe(s) + 3CO_2(g)$$

(a) If you start with 150 g of Fe_2O_3 as the limiting reagent, what is the theoretical yield of Fe? **(b)** If the actual yield of Fe in your test was 87.9 g, what was the percent yield? ***Answers:*** **(a)** 105 g Fe; **(b)** 83.7%

FOR REVIEW

SUMMARY

The law of conservation of mass states that the total mass of the products of a chemical reaction is the same as the total mass of the reactants. The same number of atoms of each type are present before and after a chemical reaction. A balanced equation shows equal numbers of atoms of each element on each side of the equation, consistent with the law of conservation of mass. Equations are balanced by placing coefficients in front of the chemical formulas for the reactants and products of a reaction.

You can sometimes predict the products of a reaction from a knowledge of similar reactions and by using the periodic table. Among the reaction types seen in this chapter are (1) combustion reactions in oxygen, in which a hydrocarbon reacts with O_2 to form CO_2 and H_2O, (2) combination reactions, in which two reactants combine to form one product, and (3) decomposition reactions, in which a single reactant forms two or more products.

Much quantitative information can be determined from chemical formulas and equations by using atomic weights. The atomic weight (average atomic mass) of an element can be calculated from the relative abundances and masses of that element's isotopes. Atomic weights are conveniently expressed in atomic mass units (amu). The amu is defined by assigning a mass of exactly 12 amu to the ^{12}C atom. The formula weight of a compound equals the sum of the atomic weights of the atoms in its formula. If the formula is a molecular formula, the formula weight is also called the molecular weight. The mass spectrometer provides the most direct and accurate means of experimentally determining atomic and molecular weights.

A mole of any substance is Avogadro's number (6.022×10^{23}) of formula units of that substance. The mass of a mole of atoms, molecules, or ions is the formula weight of that material expressed in grams (the molar mass). For example, a single molecule of H_2O weighs 18 amu; a mole of H_2O weighs 18 g. The mole concept is very useful in chemistry, and knowing how to interconvert grams and moles for any substance is important.

The empirical formula of any substance can be determined from its percent composition by calculating the relative number of moles of each atom in 100 g of the substance. If the substance is molecular in nature, its molecular formula can be determined from the empirical formula if the molecular weight is known.

The mole concept can be used to calculate the relative quantities of reactants and products involved in chemical reactions. The coefficients in a balanced equation give the relative numbers of moles of these substances. A limiting reactant is the reactant that is completely consumed in a reaction. When it is gone, the reaction stops, thus limiting the quantities of products formed.

KEY TERMS

law of conservation of mass
stoichiometry
chemical equation (Sec. 3.1)
reactants (Sec. 3.1)
products (Sec. 3.1)
combustion reaction (Sec. 3.2)
combination reaction (Sec. 3.2)
decomposition reaction (Sec. 3.2)
atomic mass unit (Sec. 3.3)
atomic weight (Sec. 3.3)
formula weight (Sec. 3.3)
molecular weight (Sec. 3.3)
mass spectrometer (Sec. 3.4)
mole (Sec. 3.5)
Avogadro's number (Sec. 3.5)
molar mass (Sec. 3.5)
limiting reactant (reagent) (Sec. 3.8)
theoretical yield (Sec. 3.8)
percent yield (Sec. 3.8)

EXERCISES

Balancing Chemical Equations

3.1 **(a)** What scientific principle or law is used in the process of balancing chemical equations? **(b)** What are the symbols used to represent gases, liquids, solids, and aqueous solutions in chemical equations? **(c)** What is the difference between P_4 and 4P in a chemical equation?

3.2 **(a)** What is the difference between a reactant and a product in a chemical equation? **(b)** In balancing equations, why shouldn't subscripts in chemical formulas be changed? **(c)** Is the following chemical equation, as written, consistent with the law of conservation of mass?

$$H_2SO_4(aq) + Ca(OH)_2(aq) \longrightarrow H_2O(l) + CaSO_4(s)$$

Why or why not?

3.3 Balance the following equations by providing the missing coefficients:

(a) $\_La_2O_3(s) + \_H_2O(l) \longrightarrow \_La(OH)_3(s)$

(b) $\_CH_4(g) + \_Cl_2(g) \longrightarrow \_CCl_4(g) + \_HCl(g)$

(c) $\_PCl_5(l) + \_H_2O(l) \longrightarrow \_H_3PO_4(aq) + \_HCl(aq)$

(d) $\_Al(OH)_3(s) + \_HCl(aq) \longrightarrow \_AlCl_3(aq) + \_H_2O(l)$

(e) $\_Mg_3N_2(s) + \_HCl(aq) \longrightarrow \_MgCl_2(aq) + \_NH_4Cl(aq)$

(f) $\_Cu(s) + \_AgNO_3(aq) \longrightarrow \_Cu(NO_3)_2(aq) + \_Ag(s)$

(g) $\_C_6H_6(l) + \_O_2(g) \longrightarrow \_CO_2(g) + \_H_2O(l)$

(h) $\_C_3H_5NO(g) + \_O_2(g) \longrightarrow \_CO_2(g) + \_NO_2(g) + \_H_2O(l)$

3.4 Balance the following equations:
(a) $N_2O_5(g) + H_2O(l) \longrightarrow HNO_3(aq)$
(b) $Na_2O_2(s) + H_2O(l) \longrightarrow NaOH(aq) + H_2O_2(aq)$
(c) $BF_3(g) + H_2O(l) \longrightarrow HF(aq) + H_3BO_3(aq)$
(d) $HClO_4(aq) + Ca(OH)_2(aq) \longrightarrow Ca(ClO_4)_2(aq) + H_2O(l)$
(e) $Au_2S_3(s) + H_2(g) \longrightarrow H_2S(g) + Au(s)$
(f) $C_4H_{10}(g) + O_2(g) \longrightarrow CO_2(g) + H_2O(g)$
(g) $Pb(NO_3)_2(aq) + H_3AsO_4(aq) \longrightarrow PbHAsO_4(s) + HNO_3(aq)$
(h) $NO_2(g) + H_2O(l) \longrightarrow HNO_3(aq) + NO(g)$

3.5 Write a balanced chemical equation to correspond to each of the following descriptions: **(a)** When sulfur trioxide gas reacts with water, a solution of sulfuric acid forms. **(b)** Boron sulfide, $B_2S_3(s)$, reacts violently with water to form dissolved boric acid, H_3BO_3, and hydrogen sulfide gas. **(c)** Phosphine, $PH_3(g)$, combusts in oxygen gas to form gaseous water and solid tetraphosphorus decaoxide. **(d)** When solid mercury(II) nitrate is heated, it decomposes to form solid mercury(II) oxide, gaseous nitrogen dioxide, and oxygen. **(e)** When hydrogen sulfide gas is passed over solid hot iron(III) hydroxide, the resultant reaction produces solid iron(III) sulfide and gaseous water.

3.6 Write balanced chemical equations to correspond to each of the following descriptions: **(a)** When ammonia gas, $NH_3(g)$, is passed over hot liquid sodium metal, hydrogen gas is released and sodium amide, $NaNH_2$, is formed as a solid product. **(b)** Solid zinc metal reacts with sulfuric acid to form hydrogen gas and an aqueous solution of zinc sulfate. **(c)** When solid potassium nitrate is heated, it decomposes to form solid potassium nitrite and oxygen gas. **(d)** When liquid phosphorus trichloride is added to water, it reacts violently to form aqueous phosphorous acid, $H_3PO_3(aq)$, and aqueous hydrochloric acid. **(e)** Copper metal reacts with hot concentrated sulfuric acid to form aqueous copper(II) sulfate, sulfur dioxide gas, and water.

Patterns of Chemical Reactivity

3.7 Write a balanced chemical equation for the reaction that occurs when: **(a)** $C_6H_{12}(l)$ is burned in air; **(b)** $CH_3CO_2C_2H_5(l)$ is combusted in air; **(c)** $Li(s)$ is added to water; **(d)** $PbCO_3(s)$ is heated.

3.8 Write a balanced chemical equation for the reaction that occurs when: **(a)** $C_8H_{18}(l)$ is burned in air; **(b)** $CH_3OC_2H_5(l)$ is combusted in an atmosphere of pure O_2; **(c)** $Mg(s)$ reacts with $S(s)$.

3.9 Balance the following equations and classify each reaction into one of the four categories given in Table 3.1:
(a) $H_2O_2(aq) \longrightarrow H_2O(l) + O_2(g)$
(b) $Fe(s) + Cl_2(g) \longrightarrow FeCl_3(s)$
(c) $Na_2CO_3(aq) + Ca(OH)_2(aq) \longrightarrow CaCO_3(s) + NaOH(aq)$
(d) $Cu(NO_3)_2(aq) + Fe(s) \longrightarrow Cu(s) + Fe(NO_3)_2(aq)$

3.10 Balance the following equations and classify each reaction into one of the four categories given in Table 3.1:
(a) $KClO_3(s) \longrightarrow KCl(s) + O_2(g)$
(b) $CO(g) + O_2(g) \longrightarrow CO_2(g)$
(c) $CuSO_4(aq) + Al(s) \longrightarrow Al_2(SO_4)_3(aq) + Cu(s)$
(d) $Pb(NO_3)_2(aq) + H_2SO_4(aq) \longrightarrow PbSO_4(s) + HNO_3(aq)$

3.11 From the hints provided, write complete balanced equations for the following reactions: **(a)** Reaction of potassium with liquid ammonia is very much like reaction of this metal with water. **(b)** Combustion of nitromethane, $CH_3NO_2(g)$, leads to $NO_2(g)$ as one of the products. **(c)** Fluorine, like oxygen, can support combustion. For example, methane, $CH_4(g)$, can "burn" in an atmosphere of $F_2(g)$.

3.12 Using a periodic table or the hints provided, write complete balanced equations for the following reactions: **(a)** Combustion of $Si_2H_6(g)$. **(b)** Combustion of $CH_3SH(g)$ leads to $SO_2(g)$ as one of the products. **(c)** Magnesium metal undergoes a vigorous combination reaction when exposed to chlorine gas.

Atomic Weights and the Mass Spectrometer

3.13 **(a)** What isotope is used as the standard in establishing the atomic mass scale? **(b)** What is an atomic mass unit?

3.14 **(a)** What is the mass in amu of a carbon-12 atom? **(b)** If the mass of a carbon-12 atom were redefined to be exactly 20 new mass units (nmu), what would be the mass of an average hydrogen atom in new mass units?

3.15 The element neon consists of three isotopes with masses 19.99, 20.99, and 21.99 amu. The relative abundances of these three isotopes are 90.92, 0.25, and 8.83 percent, respectively. From these data, calculate the average atomic mass of neon.

3.16 The element magnesium consists of three isotopes: ^{24}Mg (mass = 23.9924 amu; abundance = 78.70 percent), ^{25}Mg (mass = 24.9938 amu; abundance = 10.13 percent), and ^{26}Mg (mass = 25.9898 amu; abundance = 11.17 percent). Calculate the atomic weight (average atomic mass) of magnesium.

3.17 The mass spectrum of a sample of lead oxide contains ions of the formula PbO^+. The lead oxide has been prepared from isotopically pure ^{16}O. The ion masses seen and their relative intensities are listed as follows:

PbO^+ ion mass	Fractional intensity
220.002	0.0137
222.056	0.2630
223.050	0.2080
224.055	0.5153

The mass of ^{16}O is 15.9948. Calculate the average atomic mass of lead in this sample.

[3.18] The element bromine consists of two isotopes. The mass spectrum of molecular bromine, Br_2, consists of three peaks:

Mass (amu)	Relative size
157.84	0.2534
159.84	0.5000
161.84	0.2466

(a) What is the origin of each peak (that is, of what isotopes does each consist)? **(b)** What is the mass of each isotope?

(c) Determine the average molecular mass of a Br_2 molecule. **(d)** Determine the average atomic mass of a bromine atom. **(e)** Calculate the abundances of the two isotopes.

3.19 In his determination of the atomic weight of the element zinc, Berzelius determined in 1818 that the weight ratio of Zn to O in the oxide of zinc was 4.032. He thought that the formula of the compound was ZnO_2. Assuming an atomic weight of 16.00 for oxygen, what value does this give for the atomic weight of zinc? By comparing this with the presently accepted value, what can you say about Berzelius's assumption? If it was in error, what should it have been?

[3.20] Stas reported in 1865 that he had reacted a weighed amount of pure silver with nitric acid and had recovered all the silver as pure silver nitrate, $AgNO_3$. The weight ratio of Ag to $AgNO_3$ was found to be 0.634985. Using only this ratio and the presently accepted values for the atomic weights of silver and oxygen, calculate the atomic weight of nitrogen. Compare this calculated atomic weight with the currently accepted value.

Formula Weights; Percent Composition

3.21 Calculate the formula weights of the following: **(a)** CO_2; **(b)** Si_3H_8; **(c)** K_3PO_4; **(d)** $(NH_4)_2CO_3$; **(e)** $Mg(OH)_2$; **(f)** $Ca(C_2H_3O_2)_2$; **(g)** $CH_3OC_2H_5$; **(h)** $Cr_2(SO_4)_3$.

3.22 Determine the formula weight of each of the following compounds: **(a)** nitrous oxide, N_2O, a compound known as laughing gas; **(b)** acetylene, C_2H_2, a gas used in welding; **(c)** barium sulfate, $BaSO_4$, a compound used in making X-ray pictures of the gastrointestinal tract; **(d)** ascorbic acid, $C_6H_8O_6$, also known as vitamin C; **(e)** xenon tetrafluoride, XeF_4, one of the first noble gas compounds prepared by chemists; **(f)** $PtCl_2(NH_3)_2$, a chemotherapy agent called cisplatin.

3.23 Calculate the percentage by mass of each element in the following compounds: **(a)** CO_2; **(b)** SF_4; **(c)** NH_4Br; **(d)** C_2H_5OH; **(e)** $Mg(HCO_3)_2$; **(f)** $Zn(NO_3)_2$.

3.24 Calculate the percentage by mass of the indicated element in each of the following compounds: **(a)** hydrogen in hydrogen peroxide, H_2O_2; **(b)** oxygen in lactic acid, $HC_3H_5O_3$; **(c)** carbon in glycerol, $C_3H_8O_3$; **(d)** sulfur in cobalt(II) thiocyanate, $Co(SCN)_2$; **(e)** phosphorus in calcium phosphate, $Ca_3(PO_4)_2$; **(f)** nitrogen in morphine, $C_{17}H_{19}NO_3$.

The Mole

3.25 The mole is a key concept in chemistry. **(a)** Define the term *mole*. **(b)** What is Avogadro's number? **(c)** What is the relationship between the formula weight of a substance and its molar mass?

3.26 **(a)** What is the mass, in grams, of a mole of ^{12}C? **(b)** What is the mass, in grams, of a single ^{12}C atom?

3.27 Answer the following questions: **(a)** What is the molar mass of magnesium nitrate, $Mg(NO_3)_2$? **(b)** What is the mass, in grams, of 0.0885 mol $Mg(NO_3)_2$? **(c)** How many moles are there in 5.20 g of $Mg(NO_3)_2$? **(d)** How many nitrogen atoms are there in 3.75 mg of $Mg(NO_3)_2$?

3.28 Aspartame, the artificial sweetener marketed by G. D. Searle as NutraSweet, has a molecular formula $C_{14}H_{18}N_2O_5$. **(a)** What is the mass of 1.00 mol of aspartame? **(b)** How many moles are present in 75.8 g of aspartame? **(c)** What is the mass, in grams, of 0.736 mol of aspartame? **(d)** How many hydrogen atoms are present in 8.22 mg of aspartame?

3.29 Calculate the mass, in grams, of each of the following: **(a)** 0.00850 mol SO_2; **(b)** 3.58×10^{22} atoms of Ar; **(c)** 1.50×10^{20} molecules of caffeine, $C_8H_{10}N_4O_2$.

3.30 Calculate the mass, in grams, of each of the following: **(a)** 0.0590 mol of aspirin, $C_9H_8O_4$; **(b)** 8.25×10^{20} molecules of ozone, O_3; **(c)** 1.83×10^{24} molecules of cholesterol, $C_{27}H_{46}O$.

3.31 Calculate the number of molecules present in each of the following samples: **(a)** 0.150 mol acetylene, C_2H_2, a fuel used in welding; **(b)** a 500-mg tablet of vitamin C, $C_6H_8O_6$; **(c)** an average snowflake containing 5.0×10^{-5} g of H_2O.

3.32 Calculate the number of molecules in **(a)** 0.0350 mol propane, C_3H_8, a hydrocarbon fuel; **(b)** a 100-mg tablet of paracetamol, $C_8H_9O_2N$, an analgesic sold under the name Tylenol; **(c)** a tablespoon of table sugar, $C_{12}H_{22}O_{11}$, weighing 12.6 g.

3.33 The allowable concentration level of vinyl chloride, C_2H_3Cl, in the atmosphere in a chemical plant is 2.05×10^{-6} g/L. How many moles of vinyl chloride in each liter does this represent? How many molecules per liter?

3.34 About 25 μg minimum of tetrahydrocannabinol (THC), the active ingredient in marijuana, is required to produce intoxication. The molecular formula of THC is $C_{21}H_{30}O_2$. How many moles of THC does this 25 μg represent? How many molecules?

Empirical Formulas

3.35 What is the difference between an empirical formula and a molecular formula?

3.36 Explain why percentage compositions can be used to give empirical formulas but not necessarily molecular formulas.

3.37 Determine the empirical formula of each of the following compounds if a sample contains: **(a)** 0.36 mol H and 0.090 mol C; **(b)** 11.66 g Fe and 5.01 g O; **(c)** 87.5 percent N and 12.5 percent H by mass.

3.38 Give the empirical formula for each of the following compounds if a sample contains: **(a)** 0.014 mol S and 0.042 mol O; **(b)** 5.28 g tin and 3.37 g fluorine; **(c)** 29.1 percent sodium, 40.6 percent sulfur, and 30.3 percent oxygen by mass.

3.39 Determine the empirical formulas of the compounds with the following percentage compositions: **(a)** 1.6 percent H, 22.2 percent N, and 76.2 percent O; **(b)** 62.1 percent C, 5.21 percent H, 12.1 percent N, and 20.7 percent O.

3.40 Determine the empirical formulas of the compounds with the following percentage compositions: **(a)** 10.4 percent C, 27.8 percent S, and 61.7 percent Cl; **(b)** 32.79 percent Na, 13.02 percent Al, and 54.19 percent F.

3.41 Determine the empirical and molecular formulas of each of the following substances: **(a)** epinephrine (adrenaline), a hormone secreted in the bloodstream in times of danger or stress: 59.0 percent C, 7.1 percent H, 26.2 percent

O, and 7.7 percent N by mass; MW about 180 amu; **(b)** nicotine, a component of tobacco: 74.1 percent C, 8.6 percent H, and 17.3 percent N by mass; molar mass = 160 ± 5 g.

3.42 Determine the empirical and molecular formulas of each of the following substances: **(a)** ethylene glycol, the substance used as the primary component of most antifreeze solutions: 38.7 percent C, 9.7 percent H, and 51.6 percent O by mass; MW = 62.1 amu; **(b)** caffeine, a stimulant found in coffee: 49.5 percent C, 5.15 percent H, 28.9 percent N, and 16.5 percent O by mass; molar mass about 195 g.

3.43 Cyclopropane, a substance used with oxygen as a general anesthetic, contains only two elements, carbon and hydrogen. When 1.00 g of this substance is completely combusted, 3.14 g of CO_2 and 1.29 g of H_2O are produced. What is the empirical formula of cyclopropane?

3.44 The characteristic odor of pineapple is due to ethyl butyrate, a compound containing carbon, hydrogen, and oxygen. Combustion of 2.78 mg of ethyl butyrate produces 6.32 mg of CO_2 and 2.58 mg of H_2O. What is the empirical formula of the compound?

3.45 Epsom salts, a strong laxative used in veterinary medicine, is a hydrate, which means that a certain number of water molecules are included in the solid structure. The formula for epsom salts can be written as $MgSO_4 \cdot xH_2O$, where x indicates the number of moles of water per mole of $MgSO_4$. When 5.061 g of this hydrate is heated to 250°C, all the water of hydration is lost, leaving 2.472 g of $MgSO_4$. What is the value of x?

3.46 Washing soda, a compound used to prepare hard water for washing laundry, is a hydrate. Its formula can be written as $Na_2CO_3 \cdot xH_2O$, where x is the number of moles of H_2O per mole of Na_2CO_3. When a 2.558-g sample of washing soda is heated at 125°C, all the water of hydration is lost, leaving 0.948 g of Na_2CO_3. What is the value of x?

[3.47] Fungal laccase, a blue protein found in wood-rotting fungi, is approximately 0.39 percent copper by mass. If a laccase molecule contains four copper atoms, what is its approximate molecular weight?

[3.48] Hemoglobin, the oxygen-carrying protein in red blood cells, has four iron atoms per molecule and contains 0.340 percent iron by mass. Calculate the molecular mass of hemoglobin.

Calculations Based on Chemical Equations

3.49 Why is it essential to use balanced chemical equations in solving stoichiometry problems?

3.50 What parts of chemical equations give information about the relative numbers of moles of reactants and products involved in a reaction?

3.51 The complete combustion of butane, C_4H_{10} (lighter fluid), proceeds as follows:

$$2C_4H_{10}(l) + 13O_2(g) \longrightarrow 8CO_2(g) + 10H_2O(l)$$

(a) How many moles of O_2 are needed to burn 10.0 mol of butane in this way? **(b)** When 10.0 g of butane are burned, how many grams of O_2 are needed?

3.52 The alcohol in "gasohol" burns according to the following equation:

$$C_2H_5OH(l) + 3O_2(g) \longrightarrow 2CO_2(g) + 3H_2O(l)$$

(a) How many moles of CO_2 are produced when 5.00 mol of C_2H_5OH is burned in this way? **(b)** How many grams of CO_2 are produced when 5.00 g of C_2H_5OH is burned in this way?

3.53 Hydrofluoric acid, HF(*aq*), cannot be stored in glass bottles because compounds called silicates in the glass are attacked by the HF(*aq*). For example, sodium silicate, Na_2SiO_3, reacts in the following way:

$$Na_2SiO_3(s) + 8HF(aq) \longrightarrow H_2SiF_6(aq) + 2NaF(aq) + 3H_2O(l)$$

(a) How many moles of HF are required to dissolve 2.50 mol of Na_2SiO_3 in this reaction? **(b)** How many grams of NaF form when 5.00 mol of HF reacts in this way? **(c)** How many grams of Na_2SiO_3 can be dissolved by 5.00 g of HF?

3.54 The fermentation of glucose, $C_6H_{12}O_6$, produces ethyl alcohol, C_2H_5OH, and CO_2:

$$C_6H_{12}O_6(aq) \longrightarrow 2C_2H_5OH(aq) + 2CO_2(g)$$

(a) How many moles of CO_2 are produced when 0.350 mol of $C_6H_{12}O_6$ reacts in this fashion? **(b)** How many grams of $C_6H_{12}O_6$ are needed to form 10.0 mol of C_2H_5OH? **(c)** How many grams of CO_2 form when 10.0 g of C_2H_5OH is produced?

3.55 The reusable booster rockets of the U.S. space shuttle use a mixture of aluminum, Al, and ammonium perchlorate, NH_4ClO_4, for fuel. The reaction between these substances is as follows:

$$3Al(s) + 3NH_4ClO_4(s) \longrightarrow Al_2O_3(s) + AlCl_3(s) + 3NO(g) + 6H_2O(g)$$

What mass of ammonium perchlorate should be used in the fuel mixture for each kilogram of aluminum?

3.56 The fizz produced when an Alka-Seltzer tablet is dissolved in water is due to the reaction between sodium bicarbonate, $NaHCO_3$, and citric acid, $H_3C_6H_5O_7$:

$$3NaHCO_3(aq) + H_3C_6H_5O_7(aq) \longrightarrow 3CO_2(g) + 3H_2O(l) + Na_3C_6H_5O_7(aq)$$

How many grams of citric acid should be used for each 1.00 g of sodium bicarbonate?

3.57 Many antacids contain aluminum hydroxide, $Al(OH)_3$, as their active ingredient. **(a)** Write the balanced chemical equation for the reaction of HCl in stomach acid with solid $Al(OH)_3$ to form water and aqueous $AlCl_3$. **(b)** How many grams of HCl react with 5.00 g of $Al(OH)_3$?

3.58 Automotive airbags inflate when sodium azide, NaN_3, rapidly decomposes to the elements. **(a)** Write a balanced chemical equation for this reaction. **(b)** How many grams of NaN_3 are required to form 1.00 g of N_2? **(c)** How many grams of NaN_3 are required to produce 12.0 ft^3 of $N_2(g)$ if the gas has a density of 1.25 g/L?

Limiting Reagents; Theoretical Yields

3.59 **(a)** Define the terms *limiting reagent* and *excess reagent*. **(b)** Why is the amount of products formed in a reaction determined only by the amount of the limiting reagent?

3.60 **(a)** Define the terms *theoretical yield, actual yield*, and *percent yield*. **(b)** Why is the actual yield in a reaction almost always less than the theoretical yield?

3.61 Silicon carbide, SiC, is commonly known as carborundum. This hard substance, which is used commercially as an abrasive, is made by heating SiO_2 and C to high temperatures:

$$SiO_2(s) + 3C(s) \longrightarrow SiC(s) + 2CO(g)$$

(a) How many grams of SiC are formed by complete reaction of 5.00 g of SiO_2? **(b)** How many grams of C are required to react with 5.00 g of SiO_2? **(c)** How many grams of SiC can form when 2.50 g of SiO_2 and 2.50 g of C are allowed to react? **(d)** In part (c), which reactant is the limiting reactant and which is the excess reactant? **(e)** In part (c), how much of the excess reactant remains after the limiting reactant is completely consumed?

3.62 One of the steps in the commercial process for converting ammonia to nitric acid involves the catalytic oxidation of NH_3 to NO:

$$4NH_3(g) + 5O_2(g) \longrightarrow 4NO(g) + 6H_2O(g)$$

(a) How many grams of NO are formed by complete reaction of 2.50 g of NH_3? **(b)** How many grams of O_2 are required to react with 2.50 g of NH_3? **(c)** How many grams of NO form when 1.50 g of NH_3 reacts with 1.00 g of O_2? **(d)** In part (c), which reactant is the limiting reactant and which is the excess reactant? **(e)** In part (c), how much of the excess reactant remains after the limiting reactant is completely consumed?

3.63 Consider the following reaction:

$$H_2S(g) + 2NaOH(aq) \longrightarrow Na_2S(aq) + 2H_2O(l)$$

How many grams of Na_2S are formed if 3.05 g of H_2S is bubbled into a solution containing 1.84 g of NaOH, assuming that the limiting reagent is completely consumed?

3.64 Ethylene, C_2H_4, burns in air:

$$C_2H_4(g) + 3O_2(g) \longrightarrow 2CO_2(g) + 2H_2O(l)$$

How many grams of CO_2 can form when a mixture of 2.93 g of C_2H_4 and 4.29 g of O_2 is ignited, assuming only the reaction above occurs?

3.65 A student reacts benzene, C_6H_6, with bromine, Br_2, in an attempt to prepare bromobenzene, C_6H_5Br:

$$C_6H_6 + Br_2 \longrightarrow C_6H_5Br + HBr$$

(a) What is the theoretical yield of bromobenzene in this reaction when 30.0 g of benzene reacts with 65.0 g Br_2? **(b)** If the actual yield of bromobenzene was 56.7 g, what was the percentage yield?

3.66 Azobenzene, $C_{12}H_{10}N_2$, is an important intermediate in the manufacture of dyes. It can be prepared by the reaction between nitrobenzene, $C_6H_5NO_2$, and triethylene glycol, $C_6H_{14}O_6$, in the presence of zinc and potassium hydroxide:

$$2C_6H_5NO_2 + 4C_6H_{14}O_4 \xrightarrow[KOH]{Zn} C_{12}H_{10}N_2 + 4C_6H_{12}O_4 + 4H_2O$$

(a) What is the theoretical yield of azobenzene when 115 g of nitrobenzene and 327 g of triethylene glycol are allowed to react? **(b)** If the reaction yields 55 g of azobenzene, what is the percent yield of azobenzene?

Additional Exercises

3.67 Balance the following equations:

(a) $Li_3N(s) + H_2O(l) \longrightarrow NH_3(g) + LiOH(aq)$;
(b) $PBr_5(s) + H_2O(l) \longrightarrow H_3PO_4(aq) + HBr(aq)$;
(c) $La(NO_3)_3(aq) + Ba(OH)_2(aq) \longrightarrow La(OH)_3(s) + Ba(NO_3)_2(aq)$;
(d) $C_3H_7OH(l) + O_2(g) \longrightarrow CO_2(g) + H_2O(l)$.

3.68 Write the balanced chemical equation for the complete combustion of butyric acid, $C_4H_8O_2$, a compound produced when butter becomes rancid.

[3.69] Which would be larger: an atomic mass unit based on the current standard, or one based on the mass of a beryllium-9 atom set at exactly 9 amu? Explain briefly. (Beryllium-9 is the only naturally occurring isotope of beryllium.)

[3.70] Gallium (Ga) consists of two naturally occurring isotopes with masses of 68.926 and 70.926 amu. **(a)** How many protons and neutrons are in the nucleus of each isotope? Write the complete atomic symbol for each, showing the atomic number and mass number. **(b)** The average atomic mass of Ga is 69.72 amu. Calculate the abundance of each isotope.

3.71 Calculate the percent by mass of the indicated element in each of the following compounds: **(a)** the percent samarium (Sm) in Co_5Sm, an alloy used to form permanent magnets in very lightweight headsets such as the Sony Walkman; **(b)** the percentage of N in N_2O, a gas used as an anesthetic by dentists (known as laughing gas); **(c)** the percentage of C in dopamine, $C_8H_{11}O_2N$, a neurotransmitter; **(d)** the percentage of platinum (Pt) in cisplatin, $Pt(NH_3)_2Cl_2$, a compound used as an antitumor agent.

3.72 Aluminum oxide, Al_2O_3, occurs in nature as the mineral corundum. This mineral, which is noted for its hardness, has a density of 3.97 g/cm^3. Calculate the number of Al atoms in 10.0 cm^3 of Al_2O_3.

3.73 The molecule pyridine, C_5H_5N, adsorbs on the surfaces of certain metal oxides. A 5.0-g sample of finely divided zinc oxide, ZnO, was found to adsorb 0.068 g of pyridine. How many pyridine molecules are adsorbed? What is the ratio of pyridine molecules to formula units of zinc oxide? If the surface area of the oxide is 48 m^2/g, what is the average area of surface per adsorbed pyridine molecule?

[3.74] An element X forms an iodide XI_3 and a chloride XCl_3. The iodide is quantitatively converted to the chloride when it is heated in a stream of chlorine:

$$2XI_3 + 3Cl_2 \longrightarrow 2XCl_3 + 3I_2$$

If 0.5000 g of XI_3 is treated, 0.2360 g of XCl_3 is obtained. **(a)** Calculate the atomic weight of the element X. **(b)** Identify the element X.

[3.75] One of the earliest accurate formula weight measurements involved measurement of the weight ratio of $KClO_3$ to KCl, based on decomposition of $KClO_3$:

$$2KClO_3(s) \longrightarrow 2KCl(s) + 3O_2(g)$$

In 1911 Stähler and Meyer measured this ratio and found it to be 1.64382. Using only this ratio and the presently accepted atomic weight of oxygen, calculate the formula weight for KCl. Compare this calculated formula weight with the presently accepted value.

3.76 The koala bear dines exclusively on eucalyptus leaves. Its digestive system detoxifies the eucalyptus oil, a poison to other animals. The chief constituent in eucalyptus oil is a substance called eucalyptol, which contains 77.87 percent C, 11.76 percent H, and the remainder O. **(a)** What is the empirical formula for this substance? **(b)** What is its molecular formula if the substance has a molecular mass of 154.2 amu?

3.77 Vanillin, the dominant flavoring in vanilla, contains three elements: C, H, and O. When 1.05 g of this substance is completely combusted, 2.43 g of CO_2 and 0.50 g of H_2O are produced. What is the empirical formula of vanillin?

[3.78] An oxybromate compound, $KBrO_x$, where x is unknown, is analyzed and found to contain 52.92 percent Br. What is the value of x?

3.79 A mixture of $N_2(g)$ and $H_2(g)$ is caused to react in a closed container to form ammonia, $NH_3(g)$. The reaction ceases before either reactant has been totally consumed. At this stage, 2.0 mol N_2, 2.0 mol H_2, and 2.0 mol NH_3 are present. How many moles of N_2 and H_2 were present originally?

[3.80] A mixture containing $KClO_3$, $KHCO_3$, K_2CO_3, and KCl was heated, producing CO_2, O_2, and H_2O gases according to the following equations:

$$2KClO_3(s) \longrightarrow 2KCl(s) + 3O_2(g)$$

$$2KHCO_3(s) \longrightarrow K_2O(s) + H_2O(g) + 2CO_2(g)$$

$$K_2CO_3(s) \longrightarrow K_2O(s) + CO_2(g)$$

The KCl does not react under the conditions of the reaction. If 100.0 g of the mixture produces 1.80 g of H_2O, 13.20 g of CO_2, and 4.00 g of O_2, what was the composition of the original mixture? (Assume complete decomposition of the mixture.)

3.81 Chloromycetin is an antibiotic with the formula $C_{11}H_{12}O_5N_2Cl_2$. A 1.03-g sample of an ophthalmic ointment containing chloromycetin was chemically treated to convert its chlorine into Cl^- ions. The Cl^- was precipitated as AgCl. If the AgCl weighed 0.0129 g, calculate the weight percentage of chloromycetin in the sample.

3.82 A particular coal contains 2.8 percent sulfur by mass. When this coal is burned, the sulfur is converted into $SO_2(g)$. This SO_2 is reacted with CaO to form $CaSO_3(s)$. If the coal is burned in a power plant that uses 2000 tons of coal per day, what is the daily production of $CaSO_3$?

3.83 Aspirin, $C_9H_8O_4$, is produced from salicylic acid, $C_7H_6O_3$, and acetic anhydride, $C_4H_6O_3$:

$$C_7H_6O_3 + C_4H_6O_3 \longrightarrow C_9H_8O_4 + HC_2H_3O_2$$

(a) How much salicylic acid is required to produce 1.5×10^2 kg of aspirin, assuming that all of the salicylic acid is converted to aspirin? **(b)** How much salicylic acid would be required if only 80 percent of the salicylic acid is converted to aspirin? **(c)** What is the theoretical yield of aspirin if 185 kg of salicylic acid is allowed to react with 125 kg of acetic anhydride? **(d)** If the situation described in part (c) produces 182 kg of aspirin, what is the percentage yield?

[3.84] A mixture of pure AgCl and pure AgBr is found to contain 60.94% Ag by mass. What are the mass percentages of Cl and Br in the mixture?

Aqueous Reactions and Solution Stoichiometry

4

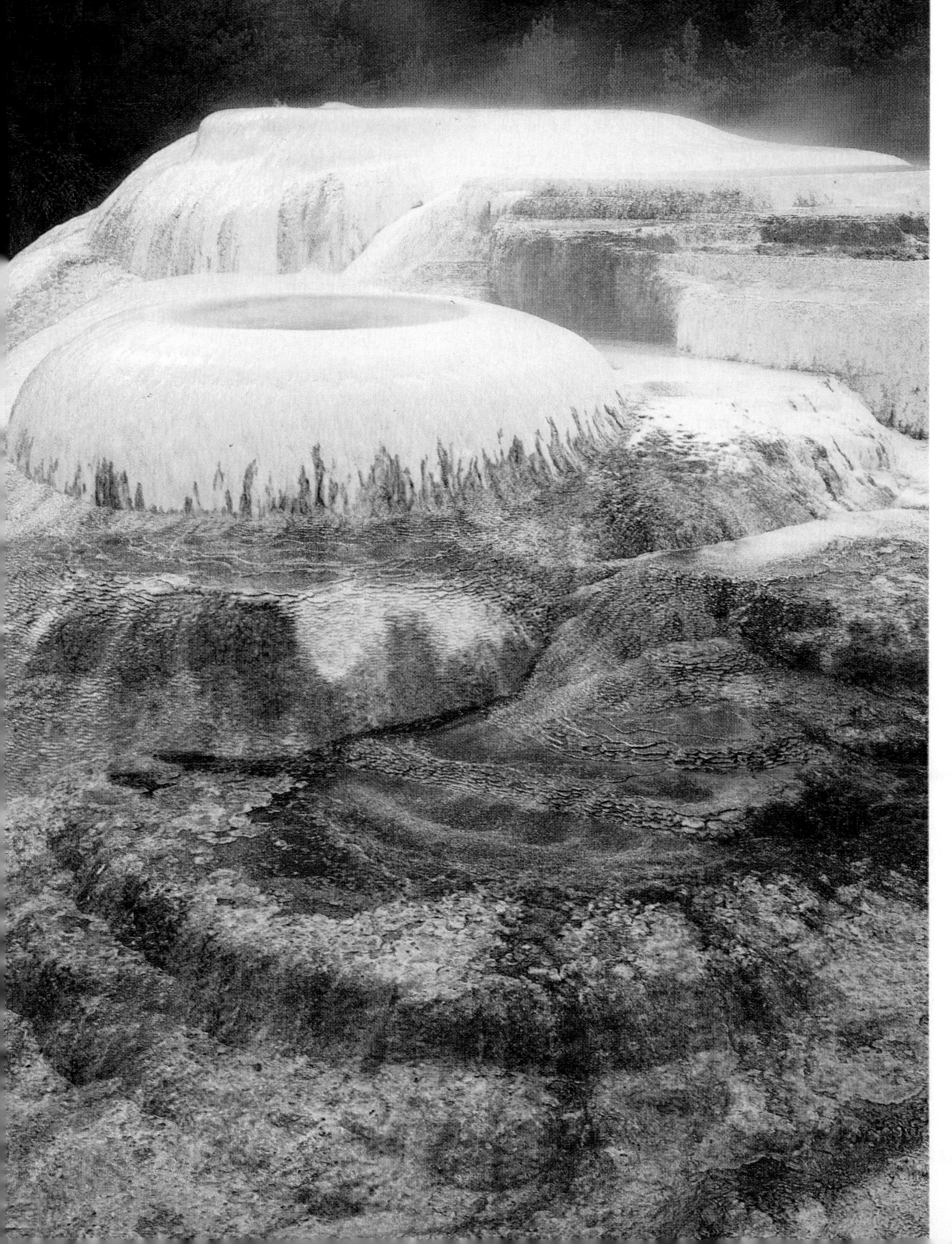

"Narrow-Gauge" Spring, part of Mammoth Hot Springs in Yellowstone National Park. The water in the hot springs is rich in dissolved minerals, which are deposited as the solution reaches the surface and cools. (Larry Ulrich/DRK Photo)

CONTENTS

Water is a remarkable substance. It is not only one of the most abundant compounds on earth, but it also possesses many unusual properties. For example, it melts and boils at higher temperatures than other substances of comparable molar mass, and it is one of the few substances that expands upon freezing. One of its most important properties is its ability to dissolve a wide variety of substances. Solutions in which water is the dissolving medium are called **aqueous solutions**.

Many of the chemical reactions that take place around us and affect our lives involve substances dissolved in water. Nutrients dissolved in blood are carried to our cells, where they enter into reactions that help keep us alive. Automobile parts rust when they come into frequent contact with aqueous solutions that contain various dissolved substances. Spectacular limestone caves, such as Mammoth Cave in Kentucky (Figure 4.1), were formed by the dissolving action of underground water containing carbon dioxide:

$$CaCO_3(s) + H_2O(l) + CO_2(aq) \longrightarrow Ca(HCO_3)_2(aq) \quad [4.1]$$

In this chapter we will consider some common reactions in aqueous solutions. Much of the chemistry we will discuss involves the reactions of three categories of substances: acids, bases, and salts. Before we consider these sub-

Figure 4.1 Mammoth Cave in Kentucky. The great vaults in these and similar caves were formed by the action of water containing dissolved CO_2 on limestone formations. (National Park Service, Mammoth Cave National Park)

stances and their reactions, however, we need to address two background questions: How do we express solution composition? What are the chemical forms in which substances occur in aqueous solutions?

4.1 SOLUTION COMPOSITION

A *solution* is a homogeneous mixture of two or more substances (Section 1.1). One of these substances is called the **solvent**; it is usually the component that is present in the greater quantity. The other substances in the solution are known as **solutes**; they are said to be dissolved in the solvent. For example, when a small amount of sodium chloride, NaCl, is mixed with a large quantity of water, we refer to water as the solvent and sodium chloride as the solute.

Molarity

Scientists use the term **concentration** to designate the amount of solute dissolved in a given quantity of solvent or solution. The concept of concentration is a common one. We may describe a laundry detergent as concentrated, meaning that it contains a significant amount of active ingredients in a given quantity of the detergent. Physicians monitor the concentrations of substances in body fluids in diagnosing illnesses. In the chapters ahead, we will encounter many properties of solutions that depend on their concentrations.

One of the most widely used ways of expressing concentration is **molarity**. The molarity (symbol M) of a solution is defined as the number of moles of solute in a liter of solution (soln):

$$\text{Molarity} = \frac{\text{moles solute}}{\text{volume of soln in liters}} \qquad [4.2]$$

A 1.00 molar solution (written 1.00 M) contains 1.00 mol of solute in every liter of solution. Figure 4.2 shows the preparation of 250 mL of a 1.00 M solution

Figure 4.2 Procedure for preparation of 0.250 L of 1.00 M solution of $CuSO_4$. (*a*) Weigh out 0.250 mol (39.9 g) of $CuSO_4$ (formula weight = 159.6 amu). (*b*) Put the $CuSO_4$ (solute) into a 250-mL volumetric flask and add a small quantity of water. (*c*) Dissolve the solute by swirling the flask. (*d*) Add more water until the solution just reaches the calibration mark etched on the neck of the flask. Shake the stoppered flask to ensure complete mixing (Donald Clegg and Roxy Wilson)

(*a*)

(*b*)

(*c*)

(*d*)

of $CuSO_4$ in a 250-mL volumetric flask. First 0.250 mol of $CuSO_4$ (39.9 g) is weighed out and placed in the volumetric flask. Water is added to dissolve the salt, and the resultant solution is diluted to a total volume of 250 mL.

SAMPLE EXERCISE 4.1

Calculate the molarity of a solution made by dissolving 23.4 g of sodium sulfate, Na_2SO_4, in enough water to form 125 mL of solution.

Solution

$$\text{Molarity} = \frac{\text{moles } Na_2SO_4}{\text{liters soln}}$$

$$\text{Moles } Na_2SO_4 = (23.4 \text{ g } Na_2SO_4)\left(\frac{1 \text{ mol } Na_2SO_4}{142 \text{ g } Na_2SO_4}\right)$$

$$= 0.165 \text{ mol } Na_2SO_4$$

$$\text{Liters soln} = (125 \text{ mL})\left(\frac{1 \text{ L}}{1000 \text{ mL}}\right) = 0.125 \text{ L}$$

$$\text{Molarity} = \frac{0.165 \text{ mol } Na_2SO_4}{0.125 \text{ L soln}}$$

$$= 1.32 \frac{\text{mol } Na_2SO_4}{\text{L soln}}$$

$$= 1.32\ M$$

PRACTICE EXERCISE

Calculate the molarity of a solution made by dissolving 5.00 g of $C_6H_{12}O_6$ (MW = 180 amu) in sufficient water to form 100 mL of solution. ***Answer:*** 0.278 M

If we know the molarity of a solution, we can easily calculate the number of moles of solute in a given volume. Thus, molarity can be used as a conversion factor between volume of solution and moles of solute. Calculation of the number of moles of HNO_3 in 2.0 L of 0.200 M HNO_3 solution illustrates the conversion of volume to moles:

$$\text{Moles } HNO_3 = (2.0 \text{ L soln})\left(\frac{0.200 \text{ mol } HNO_3}{1 \text{ L soln}}\right)$$

$$= 0.40 \text{ mol } HNO_3$$

Notice how dimensional analysis can be used in this conversion if we express molarity as moles/liter soln. Notice also that to obtain moles we multiplied liters and molarity: moles = liters × molarity. To illustrate the conversion of moles to volume, consider the calculation of the volume of 0.30 M HNO_3 solution required to supply 2.0 mol of HNO_3:

$$\text{Liters soln} = (2.0 \text{ mol } HNO_3)\left(\frac{1 \text{ L soln}}{0.30 \text{ mol } HNO_3}\right)$$

$$= 6.7 \text{ L soln}$$

In this case we needed to apply the reciprocal of molarity to convert moles to volume: liters = moles × $1/M$.

SAMPLE EXERCISE 4.2

How many grams of Na_2SO_4 are required to make 0.350 L of 0.500 *M* Na_2SO_4?

Solution

$$M\ Na_2SO_4 = \frac{\text{moles } Na_2SO_4}{\text{liters soln}}$$

Thus,

$$\text{Moles } Na_2SO_4 = \text{liters soln} \times M\ Na_2SO_4$$
$$= (0.350 \text{ L soln})\left(\frac{0.500 \text{ mol } Na_2SO_4}{\text{L soln}}\right)$$
$$= 0.175 \text{ mol } Na_2SO_4$$

Because each mole of Na_2SO_4 weighs 142 g, the required number of grams of Na_2SO_4 is

$$(0.175 \text{ mol } Na_2SO_4)\left(\frac{142 \text{ g } Na_2SO_4}{1 \text{ mol } Na_2SO_4}\right) = 24.8 \text{ g } Na_2SO_4$$

PRACTICE EXERCISE

(a) How many grams of Na_2SO_4 are there in 15 mL of 0.50 *M* Na_2SO_4? **(b)** How many milliliters of 0.50 *M* Na_2SO_4 solution are required to supply 0.035 mol of this salt? ***Answers:*** **(a)** 1.1 g; **(b)** 70 mL

Dilution

Solutions that are used routinely in the laboratory are often purchased or prepared in concentrated form (called *stock solutions*). For example, sulfuric acid, H_2SO_4, is purchased as an 18 *M* solution (concentrated sulfuric acid). Solutions of lower concentrations can then be obtained by adding water,* a process called **dilution**.

When solvent is added to dilute a solution, the number of moles of solute remains unchanged:

$$\text{Moles solute before dilution} = \text{moles solute after dilution} \qquad [4.3]$$

Because number of moles = $M \times$ liters, we can write

$$\text{(Initial molarity)(initial volume)} = \text{(final molarity)(final volume)}$$
$$M_{\text{initial}}V_{\text{initial}} = M_{\text{final}}V_{\text{final}} \qquad [4.4]$$

Suppose you had to prepare 250 mL of 0.10 *M* $CuSO_4$ solution by diluting 1.0 *M* $CuSO_4$. You can calculate the volume of the more concentrated solution that must be diluted:

$$(1.0\ M)(V_{\text{initial}}) = (0.10\ M)(250 \text{ mL})$$
$$V_{\text{initial}} = \frac{(0.10\ M)(250 \text{ mL})}{1.0\ M} = 25 \text{ mL}$$

* In diluting an acid or base, the acid or base should be added to water and then further diluted by addition of more water. Adding water directly to concentrated acid or base can cause spattering because of the intense heat generated.

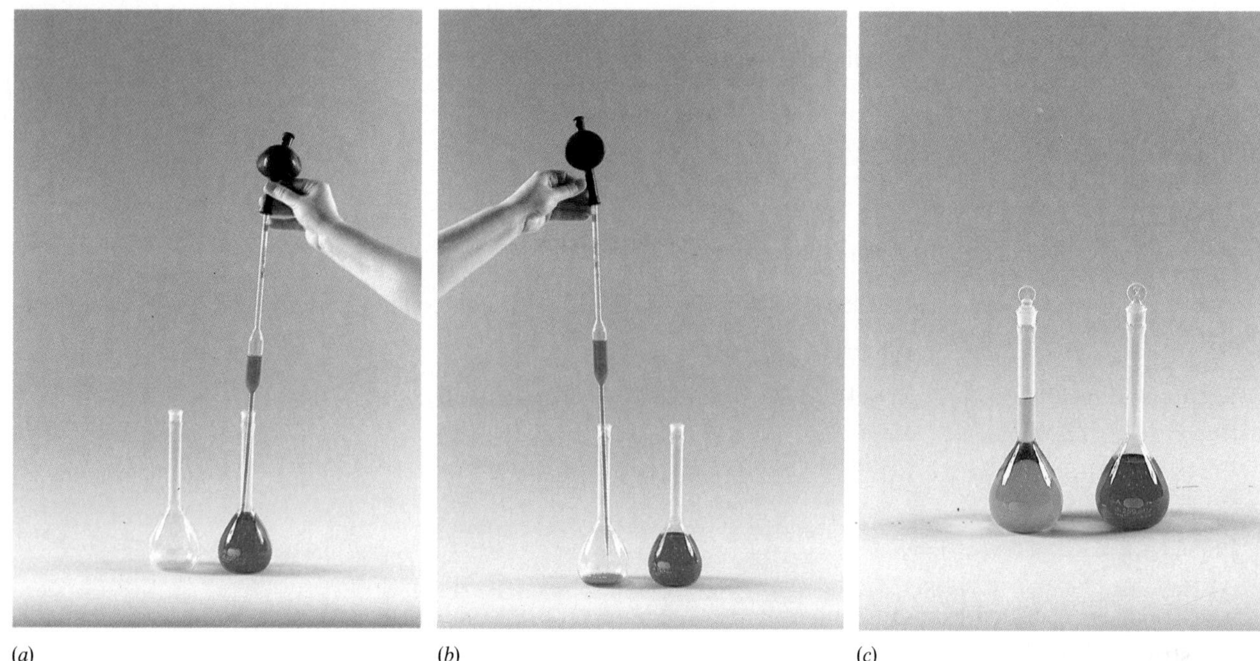

(*a*) (*b*) (*c*)

Figure 4.3 Procedure for preparation of 250 mL of 0.10 *M* $CuSO_4$ by dilution of 1.0 *M* $CuSO_4$. (*a*) Draw 25 mL of the 1.0 *M* solution into a pipet. (*b*) Add this amount to a 250-mL volumetric flask. (*c*) Add water to dilute the solution to a total volume of 250 mL (Donald Clegg and Roxy Wilson)

Thus, this dilution is achieved by withdrawing 25 mL of the 1.0 *M* solution using a pipet, adding it to a 250-mL volumetric flask, and then diluting it to 250 mL, as shown in Figure 4.3.

SAMPLE EXERCISE 4.3

How much 3.0 *M* H_2SO_4 would be required to make 500 mL of 0.10 *M* H_2SO_4?

Solution: Using Equation 4.4, $M_{initial}V_{initial} = M_{final}V_{final}$, we can write

$$V_{initial} = \frac{M_{final}V_{final}}{M_{initial}}$$

$$= \frac{(0.10\ M)(500\ \text{mL})}{3.0\ M} = 17\ \text{mL}$$

We see that if we start with 17 mL of 3.0 *M* H_2SO_4 and dilute it to a total volume of 500 mL, the desired 0.10 *M* solution will be obtained.

PRACTICE EXERCISE

How many milliliters of 5.0 *M* $K_2Cr_2O_7$ solution must be diluted in order to prepare 250 mL of 0.10 *M* solution? ***Answer:*** 5.0 mL

4.2 ELECTROLYTES

A solution prepared by dissolving 1.0 mol of HCl(*g*) in enough water to make 1.0 L of solution will be labeled 1.0 *M* HCl. Because HCl dissociates into H^+ and Cl^- ions in water, however, the solution does not contain 1.0 mol of HCl molecules per liter of solution. Rather, it contains 1.0 mol of H^+ ions and 1.0 mol of Cl^- ions.

Many substances dissociate to form ions (or *ionize*) when they dissolve in water. Although water itself is a poor conductor of electricity, the presence of ions causes aqueous solutions to become good conductors. For this reason, solutes that exist as ions in solution are called **electrolytes**. Not surprisingly, soluble ionic compounds are electrolytes. For example, when NaCl dissolves in water, the Na^+ and Cl^- ions are dispersed throughout the solution.

However, many molecular substances do not form ions when they dissolve in water. We call these nonionizing substances **nonelectrolytes** because they form nonconducting solutions. Sucrose, or table sugar, $C_{12}H_{22}O_{11}$, is a nonelectrolyte. An aqueous solution of sucrose contains neutral sucrose molecules dispersed throughout the water.

A device such as that shown in Figure 4.4 can be used to test whether ions are present in a solution, provided the solution is not too dilute. If ions are present, they permit electrical charges to move between the electrodes immersed in the solution. The ions complete the electrical circuit, causing the light bulb to glow. The larger the concentration of ions in solution, the greater the electrical current, and the brighter the bulb glows.

Strong and Weak Electrolytes

Substances such as HCl that exist in solution almost completely as ions are called **strong electrolytes.** Most ionic compounds are strong electrolytes. We can usually infer the nature of the ions present in a solution of an ionic compound from the chemical name of the substance. For example, barium iodide, BaI_2, dissociates in aqueous solution into barium ions, Ba^{2+}, and iodide ions, I^-. Sodium sulfate, Na_2SO_4, dissociates to form sodium ions, Na^+, and sulfate ions, SO_4^{2-}. It is important for you to remember the formulas and charges of common ions (Section 2.6) in order to understand the forms in which ionic compounds exist in solution.

Figure 4.4 A device for distinguishing solutions containing strong electrolytes, weak electrolytes, and nonelectrolytes. In (*a*) the bulb glows brightly because the ions present in the strong electrolytic solution provide a large number of electrical current carriers. In (*b*) the bulb glows weakly because the solution of a weak electrolyte has comparatively few ions to serve as current carriers. In (*c*) the bulb does not glow at all because the nonelectrolytic solution has no charged species to serve as current carriers.

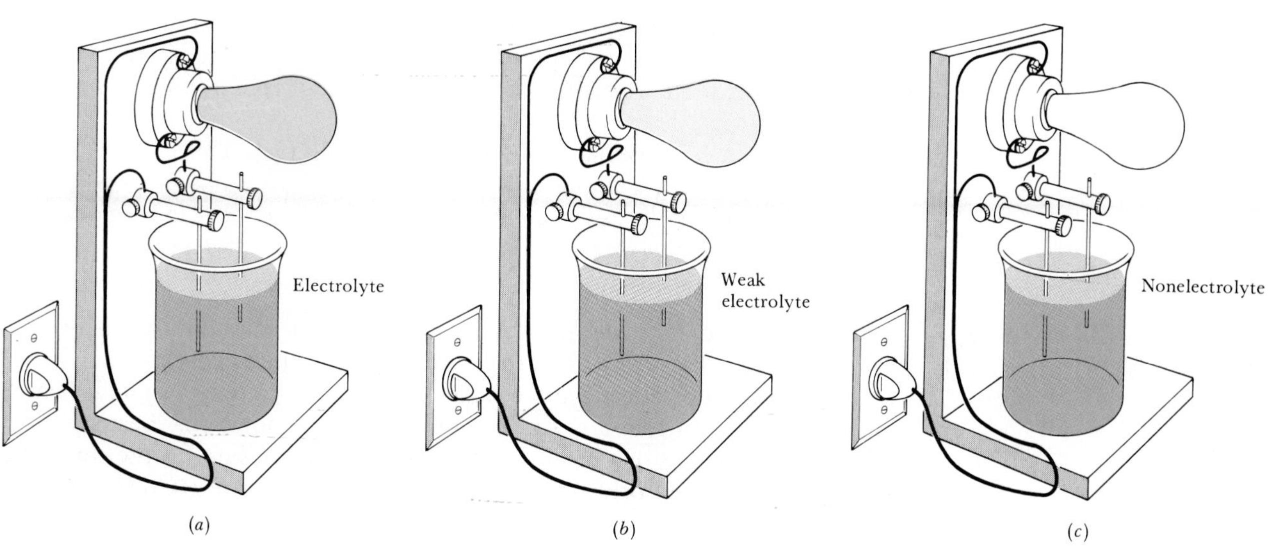

SAMPLE EXERCISE 4.4

What are the molar concentrations of all ions present in a 0.025 *M* aqueous solution of calcium nitrate?

Solution: Calcium nitrate is an ionic compound composed of calcium ions, Ca^{2+}, and nitrate ions, NO_3^-. Because most soluble ionic compounds are strong electrolytes, we may correctly predict that each mole of $Ca(NO_3)_2$ that dissolves dissociates into 1 mol of Ca^{2+} and 2 mol of NO_3^-. Thus a solution that is 0.025 *M* in $Ca(NO_3)_2$ is 0.025 *M* in Ca^{2+} and $2 \times 0.025\ M = 0.050\ M$ in NO_3

PRACTICE EXERCISE

How many moles of Cl^- ions are present in 0.25 L of 0.015 *M* $NiCl_2$ solution? ***Answer:*** 0.0075 mol

Some compounds ionize incompletely upon dissolving in water. For example, in a 1.0 *M* solution of acetic acid, $HC_2H_3O_2$, only a small fraction (about 1 percent) of the $HC_2H_3O_2$ is present as H^+ and $C_2H_3O_2^-$ ions. Compounds that only partly dissociate into ions in a solution are called **weak electrolytes.**

We must be careful not to confuse the extent to which an electrolyte dissolves with whether it is strong or weak. For example, $HC_2H_3O_2$ is extremely soluble in water but is a weak electrolyte. In contrast, $Ba(OH)_2$ is not very soluble, but the amount of the substance that does dissolve dissociates almost completely. Therefore, $Ba(OH)_2$ is a strong electrolyte.

When a weak electrolyte such as acetic acid ionizes in solution, we write the reaction in the following manner:

$$HC_2H_3O_2(aq) \rightleftharpoons H^+(aq) + C_2H_3O_2^-(aq) \qquad [4.5]$$

The double arrow means that the reaction is significant in both directions. At any given moment, some $HC_2H_3O_2$ molecules are dissociating to form H^+ and $C_2H_3O_2^-$. As the same time, H^+ and $C_2H_3O_2^-$ ions are recombining to form $HC_2H_3O_2$. The balance between these opposing processes determines the relative concentrations of neutral molecules and ions. This balance, which produces a state of **chemical equilibrium**, varies from one weak electrolyte to another. Chemical equilibria are extremely important, and we will devote several later chapters to examining them.

Chemists use a double arrow to represent the ionization of weak electrolytes and a single arrow to represent the ionization of strong electrolytes. For example, we write the equation for the dissociation of HCl into ions as follows:

$$HCl(aq) \longrightarrow H^+(aq) + Cl^-(aq) \qquad [4.6]$$

The single arrow indicates that the H^+ and Cl^- ions have no tendency to recombine in water to form HCl molecules.

4.3 ACIDS, BASES, AND SALTS

Acids, bases, and salts are among the most familiar compounds that we encounter (Figure 4.5). They are also the most common electrolytes.

Acids are substances that are able to donate a hydrogen ion and thereby increase the concentration of $H^+(aq)$ ions in aqueous solutions. Because a hydrogen atom consists of a proton and an electron, H^+ is simply a proton. Thus, acids are often called proton donors. Examples of acids are **HCl**, $\mathbf{HNO_3}$, and $\mathbf{HC_2H_3O_2}$.

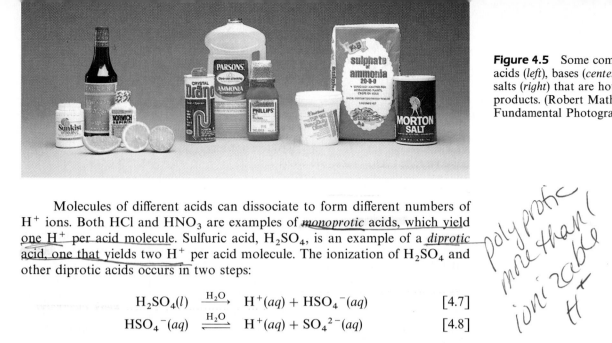

Figure 4.5 Some common acids (*left*), bases (*center*), and salts (*right*) that are household products. (Robert Mathena/Fundamental Photographs)

Molecules of different acids can dissociate to form different numbers of H^+ ions. Both HCl and HNO_3 are examples of *monoprotic* acids, which yield one H^+ per acid molecule. Sulfuric acid, H_2SO_4, is an example of a *diprotic* acid, one that yields two H^+ per acid molecule. The ionization of H_2SO_4 and other diprotic acids occurs in two steps:

$$H_2SO_4(l) \xrightarrow{H_2O} H^+(aq) + HSO_4^-(aq) \qquad [4.7]$$

$$HSO_4^-(aq) \xrightleftharpoons{H_2O} H^+(aq) + SO_4^{2-}(aq) \qquad [4.8]$$

Although H_2SO_4 is a strong electrolyte, only the first ionization is complete. Thus, aqueous solutions of sulfuric acid contain a mixture of $H^+(aq)$, $HSO_4^-(aq)$, and $SO_4^{2-}(aq)$.

Bases

Bases are substances that can react with or accept H^+ ions. Hydroxide ions, OH^-, readily react with H^+ ions to form water:

$$H^+(aq) + OH^-(aq) \longrightarrow H_2O(l) \qquad [4.9]$$

Thus, we can also define a base as any substance that increases the concentration of $OH^-(aq)$ when added to water. Ionic hydroxide compounds such as NaOH, KOH, and $Ca(OH)_2$ are among the most common bases. When dissolved in water, they dissociate, introducing OH^- ions into the solution. Compounds that do not contain OH^- ions can also be bases. For example, ammonia, NH_3, is a common base. When added to water, it accepts a H^+ ion from the water molecule and thereby increases the concentration of OH^- ions in water (Figure 4.6):

$$NH_3(aq) + H_2O(l) \rightleftharpoons NH_4^+(aq) + OH^-(aq) \qquad [4.10]$$

Because only a small fraction of the NH_3 forms NH_4^+ and OH^- ions, ammonia is a weak electrolyte.

Acids and bases that are strong electrolytes (completely ionized in solution) are referred to as **strong acids** and **strong bases**. Those that are weak electrolytes (partly ionized) are referred to as **weak acids** and **weak bases**. Strong acids are more reactive than weak acids when the reactivity depends only on the concentration of $H^+(aq)$. The reactivity of an acid, however, can depend on the anion as well as on $H^+(aq)$. For example, hydrofluoric acid, HF, is a weak acid; a 0.1 *M* solution of HF is only 8 percent ionized. However, HF is very reactive and vigorously attacks many substances, including glass. This reactivity is due to the combined action of $H^+(aq)$ and $F^-(aq)$.

Figure 4.6 An H_2O molecule acts as a proton donor (acid) and NH_3 as a proton acceptor (base). Only a fraction of the NH_3 reacts with H_2O; NH_3 is a weak electrolyte.

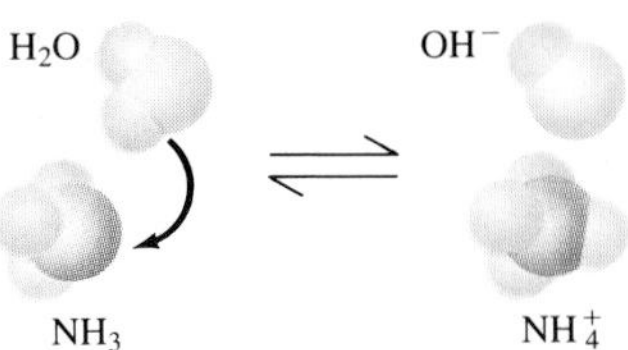

Table 4.1 Common Strong Acids and Bases

Strong acids	Strong bases
Hydrobromic, HBr Hydrochloric, HCl Hydroiodic, HI Nitric, HNO_3 Perchloric, $HClO_4$ Sulfuric, H_2SO_4	Group 1A metal hydroxides (LiOH, NaOH, KOH, RbOH, CsOH) Heavy group 2A metal hydroxides [$Ca(OH)_2$, $Sr(OH)_2$, $Ba(OH)_2$]

Table 4.1 lists the common strong acids and bases. You should commit these to memory. We can make several observations about the acids and bases listed in this table. First, some of the most common and familiar acids, such as HCl, HNO_3, and H_2SO_4, are strong. Second, three of the strong acids result from combining a hydrogen atom and a halogen atom. (HF is a weak acid.) Third, the list of strong acids is very short. Most acids are weak. Fourth, the only common strong bases are the hydroxides of Li^+, Na^+, K^+, Rb^+, and Cs^+ (the alkali metals, group 1A) and the hydroxides of Ca^{2+}, Sr^{2+}, and Ba^{2+} (the heavy alkaline earths, group 2A). The most common weak base is NH_3.

Salts

Salts are ionic compounds that can be formed by replacing one or more of the hydrogen ions of an acid by a different positive ion. For example, replacing the H^+ ion of HCl with Na^+ gives the salt NaCl. Similarly, replacing two H^+ ions of H_2SO_4 with Ni^{2+} gives the salt $NiSO_4$. Almost all salts are strong electrolytes. The only exceptions are some salts of the heavy metals such as mercury and lead. For example, $HgCl_2$ and $Pb(C_2H_3O_2)_2$ are weak electrolytes.

Acids, bases, and salts are among the most important compounds in industry and in the chemical laboratory. Table 4.2 lists several acids, bases, and

Table 4.2 U.S. Production of Some Acids, Bases, and Salts in 1988

Compound	Formula	Annual production (kg)	Principal uses and end products
Acids			
Sulfuric	H_2SO_4	3.9×10^{10}	Fertilizers (70%), metallurgy (10%), petroleum (5%)
Phosphoric	H_3PO_4	1.1×10^{10}	Fertilizers (90%), detergents (5%)
Nitric	HNO_3	7.2×10^9	Fertilizers (80%), plastics (10%), explosives (5%)
Bases			
Ammonia	NH_3	1.5×10^{10}	Fertilizers (80%), plastics and fibers (10%)
Calcium hydroxide (lime)	$Ca(OH)_2$	1.5×10^{10}	Metallurgy (40%), water treatment (25%), chemicals (10%)
Sodium hydroxide (caustic soda)	NaOH	1.1×10^{10}	Chemicals (50%), pulp and paper (20%)
Salts			
Sodium carbonate (soda ash)	Na_2CO_3	8.7×10^9	Glass (50%), chemicals (25%)
Ammonium nitrate	NH_4NO_3	6.5×10^9	Fertilizers (90%)

salts and the amount of each compound produced annually in the United States. You can see that these substances are produced in enormous quantities.

Identifying Strong and Weak Electrolytes

The following generalizations are useful in recognizing which substances are strong electrolytes and which are weak:

1. Most *salts* are strong electrolytes.
2. Most *acids* are weak electrolytes. However, HCl, HBr, HI, HNO_3, H_2SO_4, and $HClO_4$ are strong acids.
3. The common strong *bases* are the hydroxides of the alkali metals and the heavy alkaline earths. Ammonia, NH_3, is a weak electrolyte.
4. Most other substances are nonelectrolytes.

SAMPLE EXERCISE 4.5

Classify each of the following substances as nonelectrolyte, weak electrolyte, or strong electrolyte: $CaCl_2$, HNO_3, CH_3OH (methanol), $HCHO_2$ (formic acid), KOH.

Solution: One approach is to identify those substances that are acids, bases, or salts. Salts are ionic compounds and as such tend to be compounds of metals with nonmetals. Only one of the substances, $CaCl_2$, is a salt. Like most salts, $CaCl_2$ is a strong electrolyte.

Two of the substances, HNO_3 and $HCHO_2$, are acids. We should recognize HNO_3 as nitric acid, a common strong acid (a strong electrolyte). The name of $HCHO_2$ given in the question identifies it as an acid. (Recall that the chemical formulas of acids are generally written with H listed first.) Because most acids are weak acids, our best guess would be that $HCHO_2$ is a weak acid (weak electrolyte). This is correct.

The bases that we have discussed so far are NH_3 and metal hydroxides. There is one base, KOH, in the list. It is one of the common strong bases (a strong electrolyte) because it is a hydroxide of an alkali metal.

The remaining compound, CH_3OH, is not an acid, base, or salt. Although CH_3OH has an OH group, it is not a *metal* hydroxide and thus not a base; CH_3OH is a nonelectrolyte.

PRACTICE EXERCISE

Predict which one of the following 0.1 *M* solutions in water will cause the light bulb in the apparatus in Figure 4.4 to glow most brightly: C_2H_5OH (ethanol), $HC_2H_3O_2$ (acetic acid), $NaC_2H_3O_2$ (sodium acetate). ***Answer:*** $NaC_2H_3O_2$

Neutralization Reactions

Solutions of acids and bases have very different properties. Acids have a sour taste, whereas bases have a bitter taste.* Acids can change the colors of certain dyes in a specific way that differs from the effect of a base. For example, the dye known as litmus is changed from blue to red by an acid, and from red to blue by a base (Figure 4.7). In addition, acidic and basic solutions differ in chemical properties in several important ways that we will explore in this chapter and in later chapters.

---

* Tasting chemical solutions is, of course, not a good practice. However, we have all had acids such as ascorbic acid (vitamin C), acetylsalicylic acid (aspirin), and citric acid (in citrus fruits) in our mouths, and we are familiar with the characteristic sour taste. It differs from the taste of soaps, which are mostly basic.

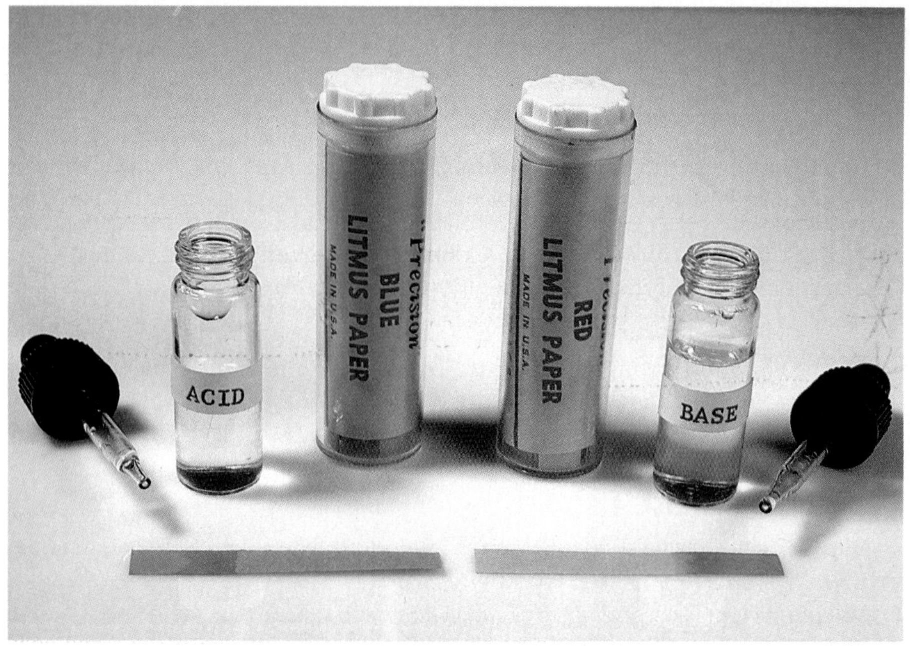

Figure 4.7 The acid-base indicator litmus changes from blue to red upon addition of an acid (left), and from red to blue upon addition of a base (right). (Dr. E. R. Degginger)

When a solution of an acid is mixed with a solution of a base, a **neutralization reaction** occurs. The products of the reaction have none of the characteristic properties of either acidic or basic solutions. For example, when 1 mol of hydrochloric acid is mixed with a solution containing 1 mol of sodium hydroxide, the products of the reaction are water and the salt sodium chloride:

$$\underset{\text{(acid)}}{HCl(aq)} + \underset{\text{(base)}}{NaOH(aq)} \longrightarrow \underset{\text{(water)}}{H_2O(l)} + \underset{\text{(salt)}}{NaCl(aq)} \qquad [4.11]$$

In general, *a neutralization reaction between an acid and a metal hydroxide produces water and a salt.*

SAMPLE EXERCISE 4.6

Write a balanced equation for the reaction of hydrobromic acid, HBr, with barium hydroxide, $Ba(OH)_2$ in aqueous solution.

Solution: The products of the acid-base reaction are a salt and water. The salt is that formed from the cation of the base, $Ba(OH)_2$, and the anion of the acid, HBr. The charge on the barium ion is 2+ (see Table 2.3), and that on the bromide ion is 1−. Therefore, to maintain electrical neutrality, the formula for the salt must be $BaBr_2$. The unbalanced equation for the neutralization reaction is therefore

$$HBr(aq) + Ba(OH)_2(aq) \longrightarrow H_2O(l) + BaBr_2(aq)$$

To balance the equation we must provide two molecules of HBr to furnish the two Br^- ions and to supply the two H^+ ions needed to combine with the two OH^- ions of the base. The balanced equation is thus

$$2HBr(aq) + Ba(OH)_2(aq) \longrightarrow 2H_2O(l) + BaBr_2(aq)$$

PRACTICE EXERCISE

Write a balanced equation for the reaction between phosphoric acid, H_3PO_4, and potassium hydroxide, KOH. ***Answer:*** $H_3PO_4 + 3KOH \longrightarrow 3H_2O + K_3PO_4$

4.4 IONIC EQUATIONS

In writing chemical equations for reactions in solution, it is often useful to indicate explicitly whether the dissolved substances are present *predominantly* as ions or molecules. Consider again the neutralization reaction between HCl and NaOH:

$$HCl(aq) + NaOH(aq) \longrightarrow H_2O(l) + NaCl(aq)$$

Equations written in this fashion, showing the complete chemical formulas of the reactants and products, are called **molecular equations**. This term is a bit of a misnomer in this case, because all of the substances in the reaction except H_2O are present predominantly as ions. HCl, NaOH, and NaCl are strong electrolytes, and we can write the chemical equation to indicate that they are completely ionized in solution:

$$H^+(aq) + Cl^-(aq) + Na^+(aq) + OH^-(aq) \longrightarrow H_2O(l) + Na^+(aq) + Cl^-(aq) \quad [4.12]$$

An equation written in this form—with all soluble strong electrolytes shown as ions—is known as a *complete ionic equation.*

Notice that $Na^+(aq)$ and $Cl^-(aq)$ appear in identical forms on both sides of Equation 4.12. Ions that appear in identical forms and with the same coefficent among both the reactants and products of a complete ionic equation are called **spectator ions**. When spectator ions are omitted from the equation (they cancel out like algebraic quantities), we are left with the **net ionic equation**. In the present example, omitting the spectator ions $Na^+(aq)$ and $Cl^-(aq)$ gives

$$H^+(aq) + OH^-(aq) \longrightarrow H_2O(l) \quad [4.13]$$

The net ionic equation includes only those ions and molecules that are directly involved in the reaction. Because charge is conserved in reactions, the sum of the charges of the ions must be the same on both sides of a balanced net ionic equation.

Net ionic equations are widely used because they can illustrate the similarities between large numbers of reactions involving electrolytes. For example, Equation 4.13 expresses the essential feature of the neutralization reaction between *any* strong acid and *any* strong base: $H^+(aq)$ and $OH^-(aq)$ ions combine to form H_2O. Thus, a net ionic equation helps us appreciate that more than one set of reactants can lead to the same net reaction. The molecular equation, on the other hand, identifies the actual reactants that participate in a reaction.

In order to write an ionic equation, you need to ask yourself two questions. First, is the substance soluble? [An (*aq*) following a chemical formula in a molecular equation indicates that the substance is soluble.] Second, if it is soluble, is it a strong electrolyte? Only if the answers to both these questions are yes should the substance be written in ionic form. Only soluble strong electrolytes are written in ionic form. Soluble weak electrolytes, soluble nonelectrolytes, and insoluble substances (whether solid, liquid, or gas) are written in "molecular" form. Finally, spectator ions are omitted from net ionic equations.

SAMPLE EXERCISE 4.7

Write the net ionic equation for neutralization of *two* of the acidic hydrogens of phosphoric acid by sodium hydroxide in aqueous solution.

Solution: The chemical formula for phosphoric acid is H_3PO_4. The formula for sodium hydroxide is NaOH. Sodium hydrogen phosphate, Na_2HPO_4, is the salt formed when two H^+ ions from H_3PO_4 are replaced by two Na^+. The other product is H_2O. The unbalanced molecular equation for the reaction is

$$H_3PO_4(aq) + NaOH(aq) \longrightarrow H_2O(l) + Na_2HPO_4(aq)$$

Balancing this equation gives

$$H_3PO_4(aq) + 2NaOH(aq) \longrightarrow 2H_2O(l) + Na_2HPO_4(aq)$$

Because H_3PO_4 is not on the list of common strong acids, we would predict that it is a weak acid. This prediction is indeed correct. Both NaOH and Na_2HPO_4 are strong electrolytes; Na_2HPO_4 ionizes to form $2Na^+(aq)$ and $HPO_4^{2-}(aq)$ ions. Thus, the complete ionic equation for the reaction is

$$H_3PO_4(aq) + 2Na^+(aq) + 2OH^-(aq) \longrightarrow 2H_2O(l) + 2Na^+(aq) + HPO_4^{2-}(aq)$$

Elimination of the spectator ions, $2Na^+(aq)$, gives the net ionic equation:

$$H_3PO_4(aq) + 2OH^-(aq) \longrightarrow 2H_2O(l) + HPO_4^{2-}(aq)$$

In sum, the weak acid H_3PO_4 reacts with the basic hydroxide ion to form water and the aqueous dihydrogen phosphate ion. Notice that each side of this balanced net ionic equation has a total charge of 2−.

PRACTICE EXERCISE

Write the net ionic equation for the reaction between aqueous solutions of HF and $Ba(OH)_2$. ***Answer:*** $HF(aq) + OH^-(aq) \longrightarrow H_2O(l) + F^-(aq)$

4.5 METATHESIS REACTIONS

In the molecular equations for many aqueous reactions, positive ions (cations) and negative ions (anions) appear to exchange partners. These reactions conform to the following general equation:

$$AX + BY \longrightarrow AY + BX \quad [4.14]$$

Example: $AgNO_3(aq) + KCl(aq) \longrightarrow AgCl(s) + KNO_3(aq)$

Such reactions are known as **double displacement** or **metathesis reactions** (meh-TATH-eh-sis, which is the Greek word for "to transpose"). Acid-base neutralization reactions involving ionic hydroxides are a type of metathesis reaction. The H^+ from the acid combines with the OH^- from the base to form H_2O; the anion of the acid and the cation of the base form the salt.

We can predict whether a particular metathesis reaction can occur by examining its net ionic equation. Consider, for example, the possible metathesis between $CaCl_2(aq)$ and $HNO_3(aq)$:

$$CaCl_2(aq) + 2HNO_3(aq) \longrightarrow Ca(NO_3)_2(aq) + 2HCl(aq)$$

Because each reactant and product in our equation is a soluble strong electrolyte, the complete ionic equation is

$$Ca^{2+}(aq) + 2Cl^-(aq) + 2H^+(aq) + 2NO_3^-(aq) \longrightarrow Ca^{2+}(aq) + 2NO_3^-(aq) + 2H^+(aq) + 2Cl^-(aq)$$

We see that no net reaction occurs in this case; all the ions are in the same form on both sides of the equation. Thus, all are spectator ions, and all cancel, leaving no net equation.

Our example illustrates an important principle: Metathesis occurs only when some form of *driving force* impels the reaction to proceed. The driving force can be one of the following:

- The formation of an insoluble solid (a precipitate)
- The formation of a gas that escapes from solution
- The formation of either a weak electrolyte or nonelectrolyte.

In the following sections, we further examine these aspects of metathesis.

Precipitation Reactions

Metathesis reactions that result in the formation of an insoluble solid are known as **precipitation reactions**. A **precipitate** is an insoluble solid formed by a reaction in solution. As an example, consider the reaction between a solution of potassium iodide, KI, and a solution of lead nitrate, $Pb(NO_3)_2$. When the two solutions are mixed, a yellow precipitate forms, as shown in Figure 4.8. This precipitate is lead iodide, PbI_2, a salt that has a very low solubility in water.

The **solubility** of a substance is the amount of that substance that can be dissolved in a given quantity of solvent. Only 1.2×10^{-3} mol of PbI_2 dissolves in a liter of water at 25°C. In our discussions, any substance whose solubility is less than 0.01 mol/L will be referred to as *insoluble.*

Solubility Rules

In order to predict whether a precipitate will form when solutions are mixed, we must have some knowledge of the solubilities of different compounds. Unfortunately, there are no rules based on simple physical properties such as ionic charge to guide us. Experimental observations, however, have led to a set of empirical solubility rules for ionic compounds. The rules are given in Table 4.3, where they are organized according to the anion in the compound. If you look closely, you will also see an important generalization that applies to cations: *All common ionic compounds of the alkali metal ions (group 1A) and of the ammonium ion, NH_4^+, are water-soluble.*

Figure 4.8 The addition of a solution of the colorless potassium iodide, (KI), to one of the colorless lead nitrate, $Pb(NO_3)_2$, produces a yellow precipitate of lead iodide, PbI_2. (Lawrence Migdale/Photo Researchers)

Table 4.3 Solubility Rules for Common Ionic Compounds in Water

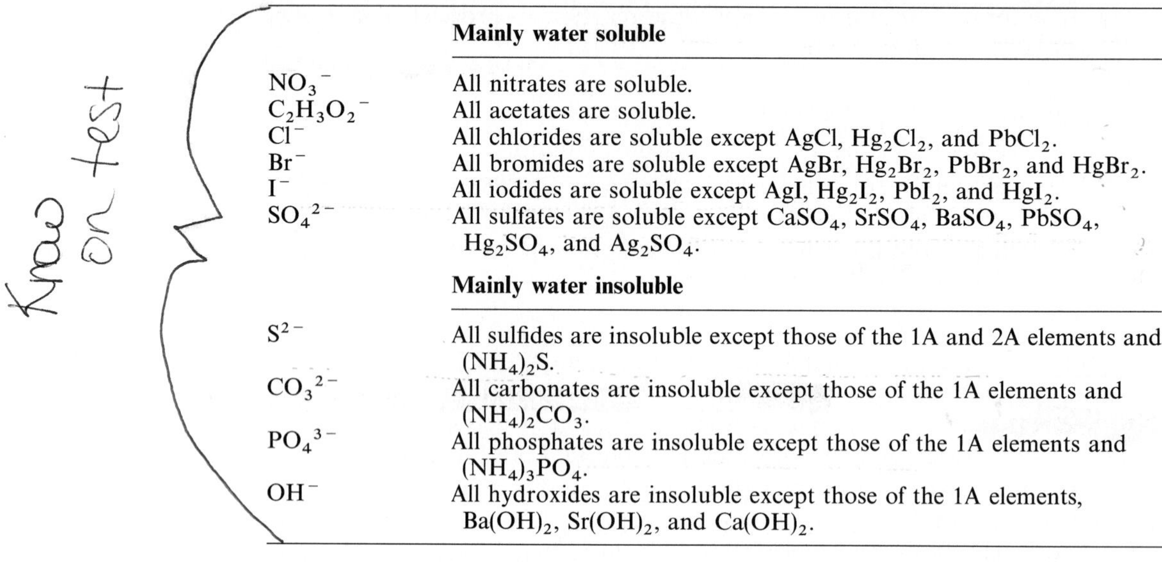

	Mainly water soluble
NO_3^-	All nitrates are soluble.
$C_2H_3O_2^-$	All acetates are soluble.
Cl^-	All chlorides are soluble except AgCl, Hg_2Cl_2, and $PbCl_2$.
Br^-	All bromides are soluble except AgBr, Hg_2Br_2, $PbBr_2$, and $HgBr_2$.
I^-	All iodides are soluble except AgI, Hg_2I_2, PbI_2, and HgI_2.
SO_4^{2-}	All sulfates are soluble except $CaSO_4$, $SrSO_4$, $BaSO_4$, $PbSO_4$, Hg_2SO_4, and Ag_2SO_4.
	Mainly water insoluble
S^{2-}	All sulfides are insoluble except those of the 1A and 2A elements and $(NH_4)_2S$.
CO_3^{2-}	All carbonates are insoluble except those of the 1A elements and $(NH_4)_2CO_3$.
PO_4^{3-}	All phosphates are insoluble except those of the 1A elements and $(NH_4)_3PO_4$.
OH^-	All hydroxides are insoluble except those of the 1A elements, $Ba(OH)_2$, $Sr(OH)_2$, and $Ca(OH)_2$.

Table 4.3 informs us, for example, that all common ionic compounds containing NO_3^- are soluble in water. Thus, if $Fe(NO_3)_2$ is one of the products in a metathesis reaction in aqueous solution, you will know that it is soluble and does not precipitate. In contrast, Table 4.3 indicates that most ionic compounds containing OH^- are insoluble in water. Therefore, we would expect $Mg(OH)_2$ to precipitate from solution if Mg^{2+} and OH^- ions are present. The precipitation of $Mg(OH)_2$ is the first step in the reactions used to extract magnesium metal from seawater.

SAMPLE EXERCISE 4.8

Write balanced molecular, ionic, and net ionic equations for the precipitation reactions (if any) that occur when solutions of the following compounds are mixed: **(a)** $BaCl_2$ and Na_2SO_4; **(b)** KCl and Na_2SO_4.

Solution: **(a)** Both $BaCl_2$ and Na_2SO_4 are soluble and ionize to give Ba^{2+}, Cl^-, Na^+, and SO_4^{2-} ions. The possible precipitation products are $BaSO_4$ and NaCl. The $BaSO_4$ is insoluble according to the rule given in Table 4.3 for SO_4^{2-} compounds. Therefore, the molecular equation is

$$BaCl_2(aq) + Na_2SO_4(aq) \longrightarrow BaSO_4(s) + 2NaCl(aq)$$

The ionic equation is

$$Ba^{2+}(aq) + 2Cl^-(aq) + 2Na^+(aq) + SO_4^{2-}(aq) \longrightarrow BaSO_4(s) + 2Na^+(aq) + 2Cl^-(aq)$$

Both $Na^+(aq)$ and $Cl^-(aq)$ are spectator ions. The net ionic equation is

$$Ba^{2+}(aq) + SO_4^{2-}(aq) \longrightarrow BaSO_4(s)$$

(b) Both reactants are soluble and ionize in solution. There are no possible insoluble salts resulting from reaction. Both NaCl and K_2SO_4 are soluble. Therefore, there is no reaction; the solutes merely mix in the solution.

PRACTICE EXERCISE

Write net ionic equations for the reactions that occur when solutions of the following compounds are mixed: **(a)** NaOH and $Co(NO_3)_2$; **(b)** $Ca(C_2H_3O_2)_2$ and H_2SO_4 (acetic acid, $HC_2H_3O_2$, is a weak acid).

Answers:
(a) $2OH^-(aq) + Co^{2+}(aq) \longrightarrow Co(OH)_2(s)$
(b) $Ca^{2+}(aq) + 2C_2H_3O_3^-(aq) + 2H^+(aq) + SO_4^{2-}(aq) \longrightarrow 2HC_2H_3O_2(aq) + CaSO_4(s)$

Reactions in Which a Gas Forms

Sometimes the product of a metathesis reaction is a gas that has a low solubility in water. For example, hydrogen sulfide, H_2S, the substance that gives rotten eggs their foul odor, forms when a strong acid such as $HCl(aq)$ reacts with a metal sulfide such as Na_2S:

Molecular equation:

$$2HCl(aq) + Na_2S(aq) \longrightarrow H_2S(g) + 2NaCl(aq) \quad [4.15]$$

Net ionic equation:

$$2H^+(aq) + S^{2-}(aq) \longrightarrow H_2S(g) \quad [4.16]$$

Carbonates and sulfites also react with acids to form gases. In these cases, metathesis reactions form carbonic acid, H_2CO_3, and sulfurous acid, H_2SO_3, respectively. These acids, however, are too unstable to be isolated as pure compounds; if present in solution in sufficient concentrations, they decompose:

$$H_2CO_3(aq) \longrightarrow H_2O(l) + CO_2(g) \quad [4.17]$$

$$H_2SO_3(aq) \longrightarrow H_2O(l) + SO_2(g) \quad [4.18]$$

For example, when hydrochloric acid is added to sodium bicarbonate, the following metathesis occurs:

$$HCl(aq) + NaHCO_3(aq) \longrightarrow NaCl(aq) + H_2CO_3(aq) \quad [4.19]$$

The resultant H_2CO_3 decomposes readily into H_2O and CO_2, causing bubbles to form, as shown in Figure 4.9. The overall reaction is

Molecular equation:

$$HCl(aq) + NaHCO_3(aq) \longrightarrow NaCl(aq) + H_2O(l) + CO_2(g) \quad [4.20]$$

Net ionic equation:

$$H^+(aq) + HCO_3^-(aq) \longrightarrow H_2O(l) + CO_2(g) \quad [4.21]$$

Figure 4.9 Carbonates react with acids to form carbon dioxide gas. Here $NaHCO_3$ (white solid) reacts with hydrochloric acid; the bubbles contain CO_2. (Richard Megna/Fundamental Photographs)

CHEMISTRY AT WORK: Antacids

The stomach secretes acids (digestive acids or stomach acids) to help digest foods. These acids, which include hydrochloric acid, are about 0.1 M in H^+. The stomach and digestive tract are protected from the corrosive effects of the acids by a mucosal lining. In some cases, holes develop in this lining, allowing the acid to attack the underlying tissue. These holes are known as *ulcers*. Ulcers can be caused either by the secretion of excess acid or by the inability of the digestive lining to resist the attack of the acid. Between 10 and 20 percent of Americans develop ulcers at some point in their lives, and many others suffer occasional indigestion or heartburn caused by high levels of digestive acids.

Treatment of these problems often focuses on neutralizing the digestive acids using substances called *antacids*. Antacids (Figure 4.10) are merely simple bases. Their ability to neutralize acids is due to the hydroxide, carbonate, or bicarbonate ions they contain. Table 4.4 lists the active ingredients in some popular antacids.

There is no ideal formulation for antacids. Preparations containing $Al(OH)_3$ tend to cause constipation, whereas those containing $Mg(OH)_2$ act as laxatives. (This is one reason for formulations that contain both ingredients.) Antacids containing calcium ions have been advertised as calcium supplements useful in combating osteoporosis (see the Chemistry at Work box in Section 7.6). Calcium, however, stimulates acid production after it is absorbed, a phenomenon known as acid rebound. Thus, calcium-containing antacids are somewhat out of favor with physicians.

Table 4.4 Some Common Antacids

Commercial name	Acid-neutralizing agents
Alka-Seltzer	$NaHCO_3$
Amphojel	$Al(OH)_3$
Di-Gel	$Mg(OH)_2$ and $CaCO_3$
Milk of magnesia	$Mg(OH)_2$
Maalox	$Mg(OH)_2$ and $Al(OH)_3$
Mylanta	$Mg(OH)_2$ and $Al(OH)_3$
Rolaids	$NaAl(OH)_2CO_3$
Tums	$CaCO_3$

Figure 4.10 Some common antacids. (Coco McCoy/Rainbow)

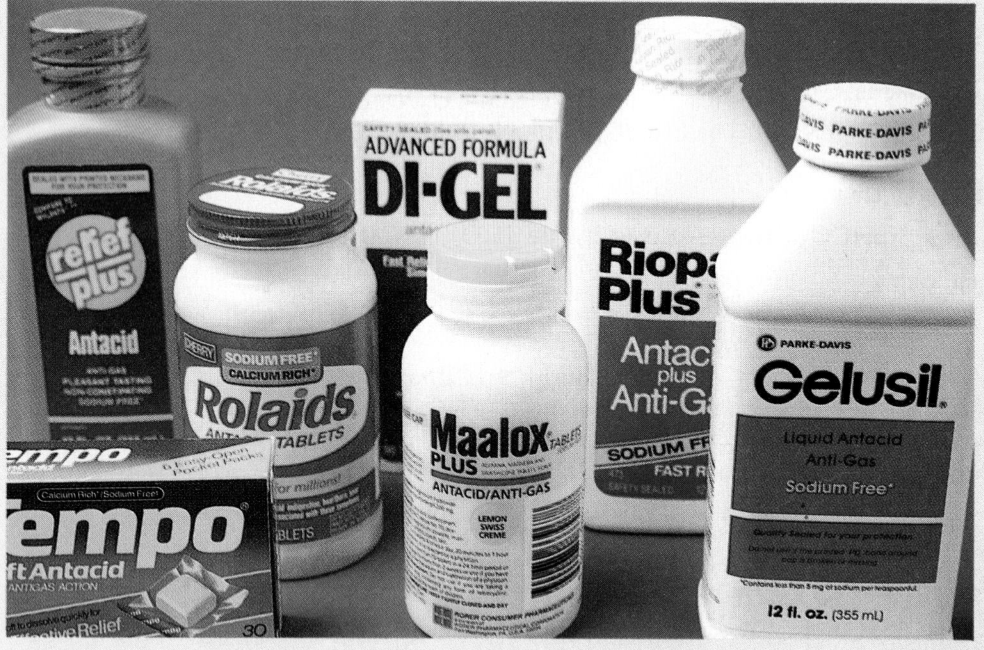

Both $NaHCO_3$ and Na_2CO_3 are used as acid neutralizers in acid spills. The bicarbonate or carbonate salt is added until the fizzing due to the formation of $CO_2(g)$ stops. Sometimes sodium bicarbonate is used as an antacid to soothe an upset stomach. In that case, the HCO_3^- reacts with stomach acid to form $CO_2(g)$. The fizz of Alka-Seltzer tablets is due to the reaction of sodium bicarbonate and citric acid in the tablets.

Sulfite compounds react with acids in a similar fashion to form H_2SO_3, which quickly decomposes into H_2O and SO_2:

$$2H^+(aq) + SO_3^{2-}(aq) \longrightarrow H_2SO_3(aq) \longrightarrow H_2O(l) + SO_2(g) \quad [4.22]$$

SAMPLE EXERCISE 4.9

Write balanced complete ionic and net ionic equations for any reactions that occur when the following compounds are mixed: **(a)** $FeCO_3(s)$ and $HCl(aq)$; **(b)** $NiS(s)$ and $HCl(aq)$.

Solution: **(a)** This reaction is similar to Equation 4.21, but the reactant carbonate is a solid, insoluble in water.

Complete ionic:

$$FeCO_3(s) + 2H^+(aq) + 2Cl^-(aq) \longrightarrow Fe^{2+}(aq) + 2Cl^-(aq) + H_2O(l) + CO_2(g)$$

Net ionic:

$$FeCO_3(s) + 2H^+(aq) \longrightarrow Fe^{2+}(aq) + H_2O(l) + CO_2(g)$$

(b) This reaction is of the type of Equation 4.16. Most sulfides will react with acids even though they are insoluble. Similarly, solid carbonates and hydroxides will generally react with acids, as exemplified in part (a).

Complete ionic:

$$NiS(s) + 2H^+(aq) + 2Cl^-(aq) \longrightarrow Ni^{2+}(aq) + 2Cl^-(aq) + H_2S(g)$$

Net ionic:

$$NiS(s) + 2H^+(aq) \longrightarrow Ni^{2+}(aq) + H_2S(g)$$

Notice that in writing the net ionic equation, in both (a) and (b), both solids and gases are included. The only species that goes through the reaction unchanged is the Cl^- ion.

PRACTICE EXERCISE

Write the net ionic equation for reaction of solid barium sulfite with aqueous sulfuric acid.

Answer: $BaSO_3(s) + H^+(aq) + HSO_4^-(aq) \longrightarrow BaSO_4(s) + SO_2(g) + H_2O(l)$

Reactions in Which H_2O or a Weak Electrolyte Forms

We have seen that reactions occur when ions are removed from solution as solids (precipitates) or gases. Reactions also occur when ions interact to form water or some other nonelectrolyte or weak electrolyte that remains dissolved in the solution.

The formation of water accompanies acid-base neutralization reactions involving metal hydroxides. Even water-insoluble hydroxides react with acids. For example, $Mg(OH)_2(s)$, which is sold as a milky-white mixture called milk of magnesia, dissolves when it reacts with $HCl(aq)$ in this fashion:

Molecular equation:

$$Mg(OH)_2(s) + 2HCl(aq) \longrightarrow MgCl_2(aq) + 2H_2O(l) \quad [4.23]$$

Net ionic equation:

$$Mg(OH)_2(s) + 2H^+(aq) \longrightarrow Mg^{2+}(aq) + 2H_2O(l) \qquad [4.24]$$

Insoluble metal oxides also react with acids because the oxide ion can combine with two H^+ ions to give water. For instance, NiO(s) reacts with HNO_3 (Figure 4.11):

Molecular equation:

$$NiO(s) + 2HNO_3(aq) \longrightarrow Ni(NO_3)_2(aq) + H_2O(l) \qquad [4.25]$$

Net ionic equation:

$$NiO(s) + 2H^+(aq) \longrightarrow Ni^{2+}(aq) + 2H_2O(l) \qquad [4.26]$$

Metals are often washed with acid before electroplating in order to remove any oxide coating that might be on the surface of the metal.

A net reaction will also occur when ions are removed from solution by the formation of a weak electrolyte such as a weak acid. For example, reaction of HCl with $NaC_2H_3O_2$ results in the formation of acetic acid, $HC_2H_3O_2$, a weak acid:

Molecular equation:

$$HCl(aq) + NaC_2H_3O_2(aq) \longrightarrow HC_2H_3O_2(aq) + NaCl(aq) \qquad [4.27]$$

Net ionic equation:

$$H^+(aq) + C_2H_3O_2^-(aq) \longrightarrow HC_2H_3O_2(aq) \qquad [4.28]$$

(*a*)

(*b*)

Figure 4.11 (*a*) Nickel oxide (NiO), nitric acid (HNO_3), and water. (*b*) The NiO is insoluble in water but reacts with HNO_3 to give a green solution of $Ni(NO_3)_2$. (Richard Megna/Fundamental Photographs)

A CLOSER LOOK: Chemical Synthesis

Formation of new substances is an important aspect of chemistry. The design and execution of a plan for developing a new drug or for preparing in the laboratory a complex substance found in nature often require great ingenuity and lengthy research. In contrast, the preparation of simple salts from aqueous solution reactions can often be accomplished quite easily by taking advantage of one or another of the driving forces that impel reactions.

Suppose that we need a sample of solid lead sulfate, $PbSO_4$. How might it be prepared? We note from the solubility rules (Table 4.3) that $PbSO_4$ is insoluble in water. Thus, we can prepare the solid material by mixing a soluble lead salt with a soluble sulfate salt:

$$Na_2SO_4(aq) + Pb(NO_3)_2(aq) \longrightarrow PbSO_4(s) + 2NaNO_3(aq) \quad [4.29]$$

In choosing the reagents for precipitation reactions, we normally employ the chloride salt or nitrate salt of the desired cation, because these are generally the least expensive soluble salts of the cations. The common sources of anions are the acids, sodium salts, or potassium salts. To make the most cost-effective use of chemicals, the amounts of solution and their concentrations should be chosen so that stoichiometrically equivalent quantities of the two soluble salts are employed: for example, one mole of Na_2SO_4 for each mole of $Pb(NO_3)_2$. If, however, there were an excess of one or the other substance, solid $PbSO_4$ would still be formed in a quantity determined by the amount of limiting reagent (Section 3.8). After formation, the solid would be filtered to remove the solution, washed with pure water to remove traces of the solution, and then dried. The net ionic equation for the reaction is

$$Pb^{2+}(aq) + SO_4^{2-}(aq) \longrightarrow PbSO_4(s) \quad [4.30]$$

We can see from this equation that formation of the desired product is independent of the particular soluble lead salt or sulfate salt we choose, as long as the accompanying ions do not themselves react.

In the preparation of $PbSO_4$ just outlined, we took advantage of the formation of a precipitate as the driving force for reaction. This driving force can be combined with another force to obtain $PbSO_4$ by a different route. We react the metal oxide PbO, which is soluble in acidic solution, with H_2SO_4 (see Figure 4.12):

$$PbO(s) + H_2SO_4(aq) \longrightarrow PbSO_4(s) + H_2O(l) \quad [4.31]$$

In this case there remains no soluble salt in solution; as a result, if we employ just a tiny excess of H_2SO_4 to ensure that all the PbO reacts, the products are solid $PbSO_4$ and a solution containing only the slight excess of H_2SO_4. The net ionic equation for this route is

$$PbO(s) + 2H^+(aq) + SO_4^{2-}(aq) \longrightarrow PbSO_4(s) + H_2O(l) \quad [4.32]$$

Now suppose we wish to prepare a salt that is water-soluble. In this event we need to find a reaction in which the soluble compound is left as the only solute. We could then use the solution directly in a subsequent reaction or evaporate the water to recover the desired material as a solid. For example, suppose that we wish to prepare copper(II) bromide, which is very water-soluble. One approach would be to mix solutions of $CuSO_4$ and $BaBr_2$. The driving force for reaction is the formation of insoluble $BaSO_4$:

$$CuSO_4(aq) + BaBr_2(aq) \longrightarrow BaSO_4(s) + CuBr_2(aq) \quad [4.33]$$

(Note that the substance in which we are most interested, $CuBr_2$, would not appear in the net ionic equation for this reaction because the constituent ions remain in solution. Thus, we find that sometimes the complete ionic equation, or the molecular equation, as written above, is the most useful description of the reaction.) The $BaSO_4$ can be removed from the reaction mixture by filtration, and the $CuBr_2$ is obtained by evaporating the remaining solution to dryness.

Alternatively, we might prepare $CuBr_2$ by taking advantage of the formation of a gas as a driving force; for example,

$$CuCO_3(s) + 2HBr(aq) \longrightarrow CuBr_2(aq) + H_2O(l) + CO_2(g) \quad [4.34]$$

Figure 4.12 Action of sulfuric acid on lead(II) oxide. Notice that the colored oxide is converted to a white solid, $PbSO_4$. (Donald Clegg and Roxy Wilson)

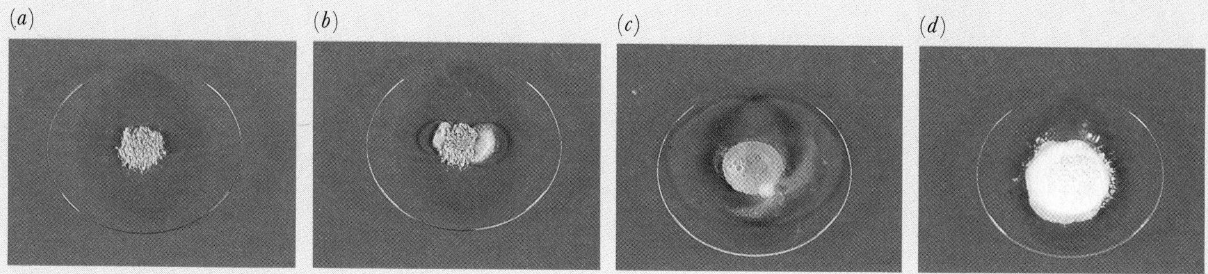

4.6 REACTIONS OF ACIDS AND METAL SALTS WITH METALS

Have you ever noticed corrosion at the terminals of your automobile battery? What we term *corrosion* is merely the conversion of a metal into a metal compound by a reaction between the metal and some substance in its environment. The most common corrosion reactions involve oxygen, water, acids, or salts as reactants. The corrosion shown in Figure 4.13 results from the reaction of battery acid, H_2SO_4, with the metal clamp.

Oxidation and Reduction

When a metal undergoes corrosion, it loses electrons. For example, calcium is vigorously attacked by acids to form calcium, Ca^{2+}, ions:

$$Ca(s) + 2H^+(aq) \longrightarrow Ca^{2+}(aq) + H_2(g) \qquad [4.35]$$

When an atom, ion, or molecule has become more positively charged (that is, when it has lost electrons), we say that it has been *oxidized.* Loss of electrons by a substance is called **oxidation.** Thus, Ca, which has no net charge, is oxidized (undergoes oxidation) in Equation 4.35; that is, it reacts to form the Ca^{2+} ion.

The term *oxidation* is used because the first reactions of this sort to be studied thoroughly were reactions with oxygen. Many metals react directly with O_2 in air to form metal oxides. In these reactions, the metal loses electrons to oxygen, forming an ionic compound of the metal ion and oxide ion. For example, when calcium metal is exposed to air, the bright metallic surface of the metal tarnishes as CaO forms (Figure 4.14):

$$2Ca(s) + O_2(g) \longrightarrow 2CaO(s) \qquad [4.36]$$

The analogous reaction of iron with oxygen in the presence of water is responsible for the formation of rust, an oxide of iron.

As Ca is oxidized in Equation 4.36, oxygen is transformed from O_2 (no net charge) to the O^{2-} ion. When an atom, ion, or molecule has become more negatively charged (gained electrons), we say that it is *reduced;* the gain of electrons by a substance is called **reduction.** When one reactant loses electrons, another reactant must gain them; the oxidation of one substance is always accompanied by the reduction of another as electrons are transferred between them, as shown in Figure 4.15. We will examine oxidation and reduction more closely in Chapter 20.

Figure 4.13 Corrosion at the terminals of a battery caused by attack of the metal by sulfuric acid from the battery. (Dr. E. R. Degginger)

Figure 4.14 Calcium metal oxidizes readily in air. The surface of the metal on the left is already dull due to the formation of oxide. The sample on the right is appreciably oxidized. (Richard Megna/Fundamental Photographs)

Oxidation of Metals by Acids

Many metals react with acids to form salts and hydrogen gas. For example, magnesium metal reacts with hydrochloric acid to form magnesium chloride and hydrogen gas:

$$\text{Metal + acid} \longrightarrow \text{salt + hydrogen}$$

$$Mg(s) + 2HCl(aq) \longrightarrow MgCl_2(aq) + H_2(g) \qquad [4.37]$$

Similarly, iron reacts with sulfuric acid (see Figure 4.16):

$$Fe(s) + H_2SO_4(aq) \longrightarrow FeSO_4(aq) + H_2(g) \qquad [4.38]$$

In each instance, the metal is oxidized by the acid to form the metal cation; the H^+ ion of the acid is reduced to form H_2.

The oxidation of iron by acids occurs when acidic foods are cooked in cast-iron cookware, providing a useful dietary source of iron. Half a cup of spaghetti sauce contains only 3 mg of iron. However, spaghetti sauce simmered several hours in an iron pot increases in iron content to at least 50 mg.

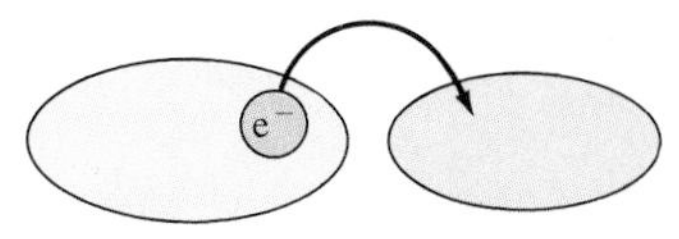

Figure 4.15 Oxidation is the loss of electrons by a substance; reduction is the gain of electrons by a substance. Oxidation of one substance is always accompanied by reduction of another.

SAMPLE EXERCISE 4.10

Write the balanced molecular and net ionic equations for the reaction of aluminum with hydrobromic acid.

Solution: Aluminum reacts with hydrobromic acid, HBr, to form H_2 and the salt of Al^{3+} and Br^-:

$$2Al(s) + 6HBr(aq) \longrightarrow 2AlBr_3(aq) + 3H_2(g)$$

Both HBr and $AlBr_3$ are strong electrolytes. When the complete ionic equation is written, we see that Br^- is a spectator ion:

$$2Al(s) + 6H^+(aq) + 6Br^-(aq) \longrightarrow 2Al^{3+}(aq) + 6Br^-(aq) + 3H_2(g)$$

Thus, the net ionic equation is

$$2Al(s) + 6H^+(aq) \longrightarrow 2Al^{3+}(aq) + 3H_2(g)$$

PRACTICE EXERCISE

Write the balanced molecular and net ionic equations for the reaction between hydrochloric acid and lead.

Answer: $Pb(s) + 2HCl(aq) \longrightarrow PbCl_2(s) + H_2(g)$

$Pb(s) + 2H^+(aq) + 2Cl^-(aq) \longrightarrow PbCl_2(s) + H_2(g)$

Figure 4.16 Many metals, such as the iron in the nail shown here, react with acids such as sulfuric acid to form hydrogen gas. The bubbles are due to the hydrogen gas. (Dr. E. R. Degginer)

Single Displacement Reactions

The reaction of a metal with an acid to form hydrogen gas is a member of a class of reactions called **single displacement reactions.** Single displacement reactions are ones in which element A reacts with compound BX to form compound AX and the element B:

$$A + BX \longrightarrow AX + B \qquad [4.39]$$

In the reactions of present interest to us, A is a metal and BX is either an acid or a metal salt, as in the following examples:

Molecular equation:

$$Fe(s) + Ni(NO_3)_2(aq) \longrightarrow Fe(NO_3)_2(aq) + Ni(s) \qquad [4.40]$$

Net ionic equation:

$$Fe(s) + Ni^{2+}(aq) \longrightarrow Fe^{2+}(aq) + Ni(s) \qquad [4.41]$$

Molecular equation:

$$Zn(s) + 2HCl(aq) \longrightarrow ZnCl_2(aq) + H_2(g) \qquad [4.42]$$

Net ionic equation:

$$Zn(s) + 2H^{+}(aq) \longrightarrow Zn^{2+}(aq) + H_2(g) \qquad [4.43]$$

In each case the single displacement reaction results in oxidation of a metal.

The Activity Series

One of the most important properties of metals is the ease with which they oxidize. When a metal is oxidized, it forms compounds that lack metallic properties. The metal appears to be eaten away as it reacts to form various compounds. This is important because metals are widely used as structural materials and as parts of machinery. Extensive oxidation can therefore lead to failure of metal parts or deterioration of metal structures.

Comparing single displacement reactions between metals allows us to rank metals according to the ease with which they can undergo oxidation in solution. For example, although zinc reacts with aqueous solutions of copper(II) sulfate, silver does not. We conclude that zinc loses electrons more readily than does silver.

A list of metals arranged in order of decreasing ease of oxidation is called an **activity series**. Table 4.5 gives the activity series for many of the most com-

Table 4.5 Activity Series of Metals

Metal	Oxidation reaction		
Lithium	Li	$\longrightarrow$	$Li^{+} + e^{-}$
Potassium	K	$\longrightarrow$	$K^{+} + e^{-}$
Barium	Ba	$\longrightarrow$	$Ba^{2+} + 2e^{-}$
Calcium	Ca	$\longrightarrow$	$Ca^{2+} + 2e^{-}$
Sodium	Na	$\longrightarrow$	$Na^{+} + e^{-}$
Magnesium	Mg	$\longrightarrow$	$Mg^{2+} + 2e^{-}$
Aluminum	Al	$\longrightarrow$	$Al^{3+} + 3e^{-}$
Manganese	Mn	$\longrightarrow$	$Mn^{2+} + 2e^{-}$
Zinc	Zn	$\longrightarrow$	$Zn^{2+} + 2e^{-}$
Chromium	Cr	$\longrightarrow$	$Cr^{3+} + 3e^{-}$
Iron	Fe	$\longrightarrow$	$Fe^{2+} + 2e^{-}$
Cobalt	Co	$\longrightarrow$	$Co^{2+} + 2e^{-}$
Nickel	Ni	$\longrightarrow$	$Ni^{2+} + 2e^{-}$
Tin	Sn	$\longrightarrow$	$Sn^{2+} + 2e^{-}$
Lead	Pb	$\longrightarrow$	$Pb^{2+} + 2e^{-}$
Hydrogen	H_2	$\longrightarrow$	$2H^{+} + 2e^{-}$
Copper	Cu	$\longrightarrow$	$Cu^{2+} + 2e^{-}$
Silver	Ag	$\longrightarrow$	$Ag^{+} + e^{-}$
Mercury	Hg	$\longrightarrow$	$Hg^{2+} + 2e^{-}$
Platinum	Pt	$\longrightarrow$	$Pt^{2+} + 2e^{-}$
Gold	Au	$\longrightarrow$	$Au^{3+} + 3e^{-}$

Ease of oxidation decreases (downward arrow)

mon metals. Hydrogen is also included in the table. The metals at the top of the table are most easily oxidized; that is, they most readily react to form compounds. Notice that the alkali metals and alkaline earth metals are at the top. They are called the *active metals*. The metals at the bottom of the activity series are very stable and form compounds less readily. Notice also that the transition elements from groups 8B and 1B are at the bottom of the list. These metals, which are used in making coins and jewelry, are called noble metals because of their low reactivity.

The activity series can be used to predict the outcome of single displacement reactions between metals and either metal salts or acids. *Any metal on the list is able to displace the elements below it from their compounds.* For example, copper is above silver in the series. Thus, copper metal will displace silver from aqueous solutions of its compounds, as pictured in Figure 4.17.

$$Cu(s) + 2Ag^{+}(aq) \longrightarrow Cu^{2+}(aq) + 2Ag(s) \qquad [4.44]$$

Similarly, because aluminum is above iron in the series, aluminum metal will displace iron from aqueous solutions of its compounds:

$$2Al(s) + 3Fe^{2+}(aq) \longrightarrow 2Al^{3+}(aq) + 3Fe(s) \qquad [4.45]$$

However, because copper is below iron, copper metal will not displace iron. Therefore, no reaction occurs when copper metal is added to a solution of iron(II) sulfate.

Figure 4.17 The displacement reaction between copper metal and a solution of silver nitrate. The product, silver metal, is evident on the surface of the copper wires. The other product, copper(II) nitrate, produces the blue color in the solution. (Fundamental Photographs)

SAMPLE EXERCISE 4.11

Will an aqueous solution of iron(II) chloride oxidize magnesium metal? If so, write the balanced molecular and net ionic equations for the reaction.

Solution: We see in Table 4.5 that Mg is above Fe in the activity series, which indicates that a displacement reaction will occur. To write the equations for displacement reactions, we must remember the charges of common ions. Magnesium is always present in compounds as the Mg^{2+} ion; the chloride ion is Cl^{-}. Thus, the compound formed is $MgCl_2$. The balanced molecular equation is

$$Mg(s) + FeCl_2(aq) \longrightarrow MgCl_2(aq) + Fe(s)$$

Both $FeCl_2$ and $MgCl_2$ are soluble salts (see Table 4.3). Thus, they are strong electrolytes and can be written in ionic form. When we do this, we find that Cl^{-} is a spectator ion. The net ionic equation is

$$Mg(s) + Fe^{2+}(aq) \longrightarrow Mg^{2+} + Fe(s)$$

As the net ionic equation so clearly shows, Mg is oxidized and Fe^{2+} is reduced in this reaction.

PRACTICE EXERCISE

Which of the following metals will displace lead from $Pb(NO_3)_2$: Zn, Cu, Fe?
Answer: Zn and Fe

Only those metals above hydrogen in the activity series are able to react with acids to form H_2. For example, Mn reacts with HCl to form H_2:

$$Mn(s) + 2HCl(aq) \longrightarrow MnCl_2(aq) + H_2(g) \qquad [4.46]$$

Because elements below hydrogen in the activity series do not displace hydrogen from acids, Cu does not react with HCl. Interestingly, copper does react with nitric acid, as shown earlier in Figure 1.7. However, this reaction is not a single displacement reaction. Instead, the nitrate ion of the acid, rather than the hydrogen ion, removes electrons from the copper. The reaction between copper and concentrated nitric acid proceeds as follows:

$$Cu(s) + 4HNO_3(aq) \longrightarrow Cu(NO_3)_2(aq) + 2H_2O(l) + 2NO_2(g) \quad [4.47]$$

4.7 SOLUTION STOICHIOMETRY

In Chapter 3, we examined the procedures for calculating the quantities of reactants and products involved in a chemical reaction. Recall that the coefficients in a balanced equation give the relative number of moles of reactants and products (Section 3.7). To use this information, we have to convert the amounts of substances involved in a reaction into moles. When we are dealing with grams of a substance, as we were in Chapter 3, we use the molar mass to achieve this conversion. When we are working with solutions of known molarity, however, we use the molarity and volume to determine the number of moles. As we saw in Section 4.1, moles solute = $M \times L$. This approach is illustrated in the following sample exercise.

SAMPLE EXERCISE 4.12

How many moles of H_2O form when 25.0 mL of 0.100 M HNO_3 solution is completely neutralized by NaOH?

Solution: The product of the molar concentration of a solution and its volume in liters gives the number of moles of solute (Section 4.1):

$$\text{Moles } HNO_3 = M \times L = \left(0.100 \frac{\text{mol } HNO_3}{L}\right)(0.0250 \text{ L})$$
$$= 2.50 \times 10^{-3} \text{ mol } HNO_3$$

The relationship between the number of moles of HNO_3 and the number of moles of H_2O is given by the balanced chemical equation for the reaction. Because this is an acid-base neutralization reaction, we know that HNO_3 and NaOH react to form H_2O and the salt containing Na^+ (the cation of the base) and NO_3^- (the anion of the acid), $NaNO_3$, as shown in the following molecular equation:

$$HNO_3(aq) + NaOH(aq) \longrightarrow H_2O(l) + NaNO_3(aq)$$

Thus, 1 mol $HNO_3 \simeq$ 1 mol H_2O. Therefore,

$$\text{Moles } H_2O = 2.50 \times 10^{-3} \text{ mol } HNO_3 \left(\frac{1 \text{ mol } H_2O}{1 \text{ mol } HNO_3}\right)$$
$$= 2.50 \times 10^{-3} \text{ mol } H_2O$$

Had the question asked for the number of grams of H_2O instead of the number of moles, we would proceed in the same way and then convert moles of H_2O into grams using its molar mass (18.0 g/mol); 0.0450 g of H_2O is produced.

PRACTICE EXERCISE

What volume of 0.500 M HCl(aq) is required to react completely with 0.100 mol of $Pb(NO_3)_2(aq)$, forming a precipitate of $PbCl_2(s)$? ***Answer:*** 0.400 L

CH [illegible] ed Silver

Silv[illegible] tact [illegible] exar[illegible] for [illegible] Like[illegible] may [illegible] and [illegible] silve[illegible] of t[illegible]

that [illegible] ben[illegible] pow[illegible] seri[illegible] silv[illegible]

$$2Al(s) + [illegible] \longrightarrow 6Ag(s) + 2Al(OH)_3(s) + 3H_2S(aq)$$

This same process, incidentally, occurs if the tarnished silver is in contact with aluminum foil in warm water containing some sodium bicarbonate, $NaHCO_3$.

Figure 4.18 Silver reacts with hydrogen sulfide in air to form a thin coating of silver sulfide, commonly referred to as "tarnish." (Robert Mathena/Fundamental Photographs)

Titrations

We will end this chapter by addressing an important question: How can we determine the concentration of a solution? One common way is to use a second solution of known concentration, called a **standard solution**, that undergoes a specific chemical reaction with the solution of unknown concentration. This procedure is known as a **titration**. As an example, suppose we have an HCl solution of unknown concentration and an NaOH solution we know to be 0.100 *M*. To determine the concentration of the HCl solution, we take a specific volume of that solution, say 20.00 mL. We then slowly add the standard NaOH solution to it until the neutralization reaction between the HCl and NaOH is complete. The point at which stoichiometrically equivalent quantities are brought together is known as the **equivalence point** of the titration.

In order to titrate an unknown with a standard solution, there must be some way to determine when the equivalence point of the titration has been reached. In acid-base titrations, dyes known as acid-base **indicators** are used for this purpose. For example, the dye known as phenolphthalein is colorless in acidic solution but is red in basic solution. If we add phenolphthalein to an unknown solution of acid, the solution will be colorless. We can then add standard base from a buret until the solution barely turns from colorless to red. This color change indicates that the acid has been neutralized and the drop of base that caused the solution to become colored has no acid to react with. The solution therefore becomes basic, and the dye turns red. The color change signals the *end point* of the titration, which coincides with the equivalence point. Care must be taken to choose indicators whose end points correspond to the equivalence point of the titration. We will consider this matter further in Chapter 17. The titration procedure is summarized in Figure 4.19.

Figure 4.19 Procedure for titrating an acid against a standardized solution of NaOH. (*a*) A known quantity of acid is added to a flask. (*b*) An acid-base indicator is added, and standardized NaOH is added from a buret. (*c*) Equivalence point is signaled by a color change in the indicator.

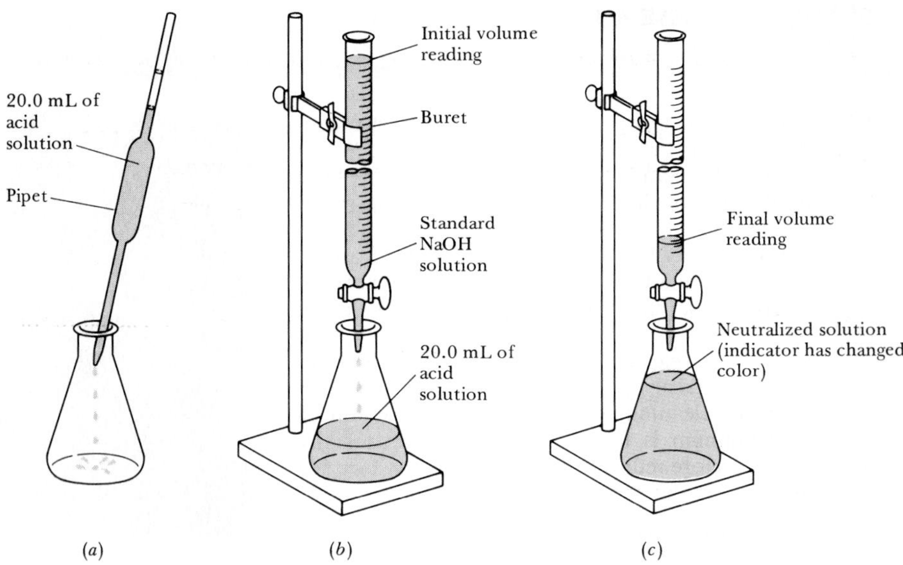

SAMPLE EXERCISE 4.13

One method used commercially to peel potatoes is to soak them in a solution of NaOH for a short time, remove them from the NaOH, and spray off the peel. The concentration of NaOH is normally in the range 3 to 6 *M*. The NaOH is analyzed periodically. In one such analysis, 45.7 mL of 0.500 *M* H_2SO_4 is required to react completely with a 20.0-mL sample of NaOH solution:

$$H_2SO_4(aq) + 2NaOH(aq) \longrightarrow 2H_2O(l) + Na_2SO_4(aq)$$

What is the concentration of the NaOH solution?

Solution

$$\text{Moles } H_2SO_4 = (45.7 \text{ mL soln})\left(\frac{1 \text{ L soln}}{1000 \text{ mL soln}}\right)\left(0.500\,\frac{\text{mol } H_2SO_4}{\text{L soln}}\right)$$
$$= 2.28 \times 10^{-2} \text{ mol } H_2SO_4$$

According to the balanced equation, 1 mol $H_2SO_4 \simeq 2$ mol NaOH. Therefore,

$$\text{Moles NaOH} = (2.28 \times 10^{-2} \text{ mol } H_2SO_4)\left(\frac{2 \text{ mol NaOH}}{1 \text{ mol } H_2SO_4}\right)$$
$$= 4.56 \times 10^{-2} \text{ mol NaOH}$$

Knowing the number of moles of NaOH present in 20.0 mL of solution allows us to calculate the concentration of this solution:

$$\text{Molarity NaOH} = \frac{\text{mol NaOH}}{\text{L soln}}$$
$$= \left(\frac{4.56 \times 10^{-2} \text{ mol NaOH}}{20.0 \text{ mL soln}}\right)\left(\frac{1000 \text{ mL soln}}{1 \text{ L soln}}\right)$$
$$= 2.28\,\frac{\text{mol NaOH}}{\text{L soln}} = 2.28\ M$$

Thus, the solution is too dilute and needs to be replaced.

PRACTICE EXERCISE

What is the molarity of a NaOH solution if 48.0 mL is needed to neutralize 35.0 mL of 0.144 *M* H_2SO_4? The equation for the chemical reaction is given in the above Sample Exercise. ***Answer:*** 0.210 *M*

SAMPLE EXERCISE 4.14

The quantity of Cl^- in a water supply is determined by titrating the sample with Ag^+:

$$Ag^+(aq) + Cl^-(aq) \longrightarrow AgCl(s)$$

What mass of chloride ion is present in a 10.0-g sample of the water if 20.2 mL of 0.100 *M* Ag^+ is required to react with all the chloride in the sample?

Solution: We must first determine the number of moles of Ag^+ used in the titration:

$$(20.2 \text{ mL soln})\left(\frac{1 \text{ L soln}}{1000 \text{ mL soln}}\right)\left(0.100 \frac{\text{mol Ag}^+}{\text{L soln}}\right) = 2.02 \times 10^{-3} \text{ mol Ag}^+$$

From the balanced equation, we see that 1 mol $Ag^+ \simeq$ 1 mol Cl^-. Therefore, the sample must contain 2.02×10^{-3} mol of Cl^-:

$$\text{Moles Cl}^- = (2.02 \times 10^{-3} \text{ mol Ag}^+)\left(\frac{1 \text{ mol Cl}^-}{1 \text{ mol Ag}^+}\right)$$

$$= 2.02 \times 10^{-3} \text{ mol Cl}^-$$

The number of moles of Cl^- can then be converted to grams:

$$\text{Grams Cl}^- = (2.02 \times 10^{-3} \text{ mol Cl}^-)\left(\frac{35.5 \text{ g Cl}^-}{1 \text{ mol Cl}^-}\right)$$

$$= 7.17 \times 10^{-2} \text{ g Cl}^-$$

We might note that the weight percentage of Cl^- in the water is

$$\% \text{ Cl}^- = \frac{7.17 \times 10^{-2} \text{ g Cl}^-}{10.0 \text{ g soln}} \times 100 = 0.717\%$$

Chloride ion is one of the major ions in water and sewage. Ocean water contains 1.92 percent Cl^-. Whether water containing Cl^- exhibits a salty taste depends on the other ions present. If the accompanying ions are Na^+ ions, a salty taste may be detected with as little as 0.03 percent Cl^-.

PRACTICE EXERCISE

What mass of chloride ion is present in a sample of water if 15.7 mL of 0.108 *M* $AgNO_3$ is required to titrate the sample? Refer to Sample Exercise 4.14 for the reaction. *Answer:* 0.0602 g

FOR REVIEW

SUMMARY

The composition of a solution expresses the relative quantities of solvent and solutes that it contains. The molarity is the number of moles of solute per liter of solution. This concentration unit allows us to interconvert solution volume and number of moles of solute. Solutions of known molarity can be formed either by weighing out the solute and diluting it to a known volume or by diluting a more concentrated solution of known concentration.

Substances that exist in solution as ions are called electrolytes. Substances that are completely ionized in solution are called strong electrolytes, whereas those that are partially ionized are called weak electrolytes. The most important electrolytes are acids, bases, and salts.

Acids are proton donors; they increase the concentration of $H^+(aq)$ in aqueous solutions to which they are added. Bases are proton acceptors; they increase the concentration of $OH^-(aq)$ in aqueous solutions. When solutions of acids and bases are mixed, a neutralization reaction results. The neutralization reaction between an acid and a metal hydroxide produces water and a salt.

Chemical equations can be written to show whether dissolved substances are present in solution predominantly as ions or molecules. In a net ionic equation, those ions that go through the reaction unchanged (spectator ions) are omitted.

One significant class of reactions involving ions is metathesis, which includes both neutralization and precipitation reactions. Metathesis requires a driving force such as the formation of a precipitate, a gas, a weak electrolyte, or a nonelectrolyte. Solubility rules help in determining whether a precipitate will form. The gases commonly formed in metathesis are H_2S, CO_2, and SO_2. The most common nonelectrolyte produced in metathesis is H_2O.

Many metals are oxidized by O_2 and by acids and salts. The reaction of a metal and an acid to form a salt and H_2 belongs to a class of reactions called single displacement reactions. This class also includes the reactions of metals with metal salts. Comparing single displacement reactions allows us to rank metals according to their ease of oxidation. A list of metals arranged in order of decreasing ease of oxidation is called an activity series. Any metal on the list is able to displace the metals below it from their compounds.

In the process called titration, we bring a solution of known concentration (a standard solution) into reaction with a solution of unknown concentration in order to determine the unknown concentration or the quantity of solute in the unknown.

KEY TERMS

aqueous solution
solvent (Sec. 4.1)
solutes (Sec. 4.1)
concentration (Sec. 4.1)
molarity (Sec. 4.1)
dilution (Sec. 4.1)
electrolyte (Sec. 4.2)
nonelectrolyte (Sec. 4.2)
strong electrolyte (Sec. 4.2)
weak electrolyte (Sec. 4.2)
chemical equilibrium (Sec. 4.2)
acids (Sec. 4.3)
bases (Sec. 4.3)
strong acids (Sec. 4.3)
weak acids (Sec. 4.3)
strong bases (Sec. 4.3)
weak bases (Sec. 4.3)
salts (Sec. 4.3)
neutralization reaction (Sec. 4.3)
molecular equation (Sec. 4.4)
spectator ions (Sec. 4.4)
net ionic equation (Sec. 4.4)
metathesis reaction (Sec. 4.5)
double displacement reaction (Sec. 4.5)
precipitation reaction (Sec. 4.5)
precipitate (Sec. 4.5)
solubility (Sec. 4.5)
oxidation (Sec. 4.6)
reduction (Sec. 4.6)
single displacement reaction (Sec. 4.6)
activity series (Sec. 4.6)
standard solution (Sec. 4.7)
titration (Sec. 4.7)
equivalence point (Sec. 4.7)
indicators (Sec. 4.7)

EXERCISES

Solution Composition; Molarity

4.1 Describe the steps involved in preparing a solution of known molar concentration starting with a pure substance and using a volumetric flask.

4.2 Suppose you prepare 500 mL of a 0.10 *M* solution of some salt and then accidently spill some of it. What happens to the concentration of the solution left in the container?

4.3 What is the difference between 0.50 mol HCl and 0.50 *M* HCl?

4.4 Consider a solution identified by its label as 0.50 *M* HCl. **(a)** Does the label tell us how much HCl is in the bottle? **(b)** What does the label tell us about the solution?

4.5 **(a)** Calculate the molarity of a solution containing 0.0335 mol Na_2CrO_4 in 200 mL. **(b)** How many moles of HCl are present in 25.0 mL of a 12.0 *M* solution of hydrochloric acid? **(c)** How many milliliters of 2.00 *M* NaOH solution are needed to obtain 0.100 mol of NaOH?

4.6 **(a)** Calculate the molarity of a solution made by dissolving 0.0670 mol $NaHCO_3$ in enough water to form

250 mL of solution. **(b)** How many moles of $K_2Cr_2O_7$ are present in 50.0 mL of a 0.105 *M* solution? **(c)** How many milliliters of 9.0 *M* H_2SO_4 solution are required to obtain 0.050 mol of H_2SO_4?

4.7 Calculate the number of grams of solute present in each of the following solutions:
(a) 0.200 L of 0.125 *M* KBr
(b) 100 mL of 0.150 *M* Na_2SO_4
(c) 250 mL of 0.0500 *M* $KBrO_3$
(d) 50.0 mL of 1.70 *M* $C_6H_{12}O_6$

4.8 Calculate the molar concentration of solute in each of the following solutions:
(a) 0.250 L containing 5.75 g of $NaNO_3$
(b) 100.0 mL containing 22.57 g of H_2SO_4
(c) 50.0 mL containing 1.48 g of $AgNO_3$
(d) 2.00 L containing 138 g of $NiCl_2 \cdot 6H_2O$

4.9 Describe how you would prepare 100.0 mL of 0.1000 *M* glucose solution starting with solid glucose, $C_6H_{12}O_6$.

4.10 Describe how you would prepare 250 mL of 0.100 *M* $K_2Cr_2O_7$ solution starting with the pure solute.

4.11 Describe how you would prepare 250 mL of 0.100 *M* $C_6H_{12}O_6$ solution starting with 1.00 L of 2.00 *M* $C_6H_{12}O_6$.

4.12 An experiment calls for you to use 200 mL of 1.0 *M* HNO_3 solution. All you have available is a liter bottle of 6.0 *M* HNO_3. How would you prepare the desired solution?

Electrolytes

4.13 Although pure water is a nonconductor of electricity, we are often cautioned not to operate electrical appliances around water. Why?

4.14 Describe a simple way to compare qualitatively the electrical conductivities of solutions. What kinds of particles move through aqueous solutions to conduct electrical current (atoms, ions, molecules, or electrons)?

4.15 Write equations to show how the following electrolytes dissociate into ions in aqueous solution:
(a) HI **(b)** K_2SO_4 **(c)** NH_4NO_3 **(d)** $CaCl_2$

4.16 Write equations to show how the following electrolytes dissociate into ions in aqueous solutions:
(a) $FeBr_2$ **(b)** $HClO_4$ **(c)** $Al(NO_3)_3$ **(d)** K_2CrO_4

4.17 Classify each of the following aqueous solutions as a nonelectrolyte, weak electrolyte, or strong electrolyte: **(a)** HBrO; **(b)** HNO_3; **(c)** KOH; **(d)** $CoSO_4$; **(e)** sucrose, $C_{12}H_{22}O_{11}$; **(f)** O_2.

4.18 Classify each of the following substances as a nonelectrolyte, weak electrolyte, or strong electrolyte in water: **(a)** HF; **(b)** ethanol, C_2H_5OH; **(c)** NH_3; **(d)** $KClO_3$; **(e)** $Cu(NO_3)_2$.

4.19 Indicate the total concentration of all solute species present in each of the following solutions: **(a)** 0.14 *M* NaOH; **(b)** 0.25 *M* $CaBr_2$; **(c)** 0.25 *M* CH_3OH; **(d)** a mixture of 50.0 mL of 0.20 *M* $KClO_3$ and 25.0 mL of 0.20 *M* Na_2SO_4.

4.20 Indicate the total concentration of each ion present in the solution formed by mixing: **(a)** 20.0 mL of 0.100 *M* HCl and 10.0 mL of 0.220 *M* HCl; **(b)** 15.0 mL of 0.300 *M* Na_2SO_4 and 10.0 mL of 0.100 *M* NaCl; **(c)** 3.50 g of KCl in 60.0 mL of 0.500 *M* $CaCl_2$ solution (assume no volume change).

Acids, Bases, and Salts

4.21 An aqueous solution is tested with litmus paper and found to be acidic. The solution is weakly conducting compared with a solution of NaCl of the same concentration. Which of the following substances could the unknown be: KOH, NH_3, HNO_3, $KClO_2$, H_3PO_3, CH_3COCH_3?

4.22 Label each of the following substances as an acid, base, salt, or none of the above. Indicate whether the substance exists in aqueous solution entirely in molecular form, entirely as ions, or as a mixture of molecules and ions. **(a)** HF; **(b)** acetonitrile, CH_3CN; **(c)** $NaClO_4$; **(d)** $Ba(OH)_2$.

4.23 Why do we use a single arrow in the chemical equation for the ionization of HNO_3 but a double arrow for the ionization of HCN?

4.24 Although sulfuric acid is a strong electrolyte, a 0.1 *M* aqueous solution of H_2SO_4 is not 0.2 *M* in $H^+(aq)$. Explain.

4.25 What is the difference between: **(a)** a monoprotic acid and a diprotic acid; **(b)** a weak acid and a strong acid; **(c)** an acid and a base?

4.26 Explain the following observations: **(a)** NH_3 contains no OH^- ions and yet its aqueous solutions are basic; **(b)** HF is called a weak acid, and yet it is very reactive.

4.27 Complete and balance the following equations:
(a) $Fe(OH)_2(s) + HClO_3(aq) \longrightarrow$
(b) $HI(aq) + Ca(OH)_2(aq) \longrightarrow$
(c) $Al(OH)_3(s) + H_2SO_4(aq) \longrightarrow$

4.28 Write balanced chemical equations for the following neutralization reactions: **(a)** $HC_2H_3O_2(aq)$ is neutralized by KOH(*aq*); **(b)** hypochlorous acid solution is neutralized by sodium hydroxide solution; **(c)** $NH_3(aq)$ is neutralized by HCl(*aq*).

Ionic Equations; Metathesis Reactions

4.29 Write balanced net ionic equations for the following reactions, and identify the spectator ion or ions present in each:
(a) $Pb(NO_3)_2(aq) + Na_2SO_4(aq) \longrightarrow PbSO_4(s) + 2NaNO_3(aq)$
(b) $Zn(s) + 2HCl(aq) \longrightarrow ZnCl_2(aq) + H_2(g)$
(c) $FeO(s) + 2HClO_4(aq) \longrightarrow H_2O(l) + Fe(ClO_4)_2(aq)$

4.30 Balance the following equations, and then write their balanced net ionic equations:
(a) $Cr(OH)_3(s) + HNO_3(aq) \longrightarrow H_2O(l) + Cr(NO_3)_3(aq)$
(b) $Na_2CO_3(aq) + HCl(aq) \longrightarrow H_2O(l) + CO_2(g) + NaCl(aq)$
(c) $CuBr_2(aq) + NaOH(aq) \longrightarrow Cu(OH)_2(s) + NaBr(aq)$

4.31 Explain what is meant by the term *driving force* when referring to metathesis reactions. What are the driving forces in each of the reactions in Exercise 4.30?

4.32 Give a specific example of a metathesis that occurs because of each of the following driving forces: **(a)** formation of an insoluble solid; **(b)** formation of a gas; **(c)** formation of a weak electrolyte or nonelectrolyte.

4.33 Using solubility rules or reasonable extensions of them, predict whether each of the following compounds is soluble in water: **(a)** $PbCl_2$; **(b)** CsBr; **(c)** $Co(OH)_2$; **(d)** $SrSeO_4$.

4.34 Using solubility rules or reasonable extensions of them, predict whether each of the following compounds is soluble in water: **(a)** $NiCl_2$; **(b)** $Mn(C_2H_3O_2)_2$; **(c)** $Cr_2(SO_4)_3$; **(d)** $CaCO_3$.

4.35 Separate samples of a solution of an unknown salt are treated with dilute solutions of HBr, H_2SO_4, and NaOH. A precipitate forms only with H_2SO_4. Which of the following cations could the solution contain: K^+; Pb^{2+}; Ba^{2+}?

4.36 Separate samples of a solution of an unknown salt are treated with dilute $AgNO_3$, $Pb(NO_3)_2$, and $BaCl_2$. Precipitates form in all three cases. Which of the following anions could be the anion of the unknown salt: Br^-; SO_4^{2-}; NO_3^-?

4.37 Write balanced net ionic equations for the reactions, if any, that occur between
(a) $ZnS(s)$ and $HCl(aq)$;
(b) $Na_2CO_3(aq)$ and $BaCl_2(aq)$;
(c) $Na_3PO_4(aq)$ and $HBr(aq)$;
(d) $Ba(OH)_2(aq)$ and $HCl(aq)$;
(e) $Sr(C_2H_3O_2)_2(aq)$ and $NiSO_4(aq)$;
(f) $ZnSO_3(aq)$ and $HCl(aq)$;
(g) $Pb(NO_3)_2(aq)$ and $H_2S(aq)$;
(h) $Fe(OH)_3(s)$ and $HClO_4(aq)$.

4.38 Write balanced net ionic equations for the reactions, if any, that occur when each of the following pairs is mixed:
(a) $H_2SO_4(aq)$ and $BaCl_2(aq)$;
(b) $NaCl(aq)$ and $(NH_4)_2SO_4(aq)$;
(c) $AgNO_3(aq)$ and $Na_2CO_3(aq)$;
(d) $KOH(aq)$ and $HNO_3(aq)$;
(e) $Ca(OH)_2(aq)$ and $HC_2H_3O_2(aq)$;
(f) $K_2SO_3(s)$ and $H_2SO_4(aq)$;
(g) $Pb(NO_3)_2(aq)$ and $MgSO_4(aq)$.

[4.39] Suggest a method for synthesis of each of the following substances, and write a balanced molecular equation or equations for the process: **(a)** $Cu(NO_3)_2$; **(b)** MnS; **(c)** $Fe(OH)_2$.

[4.40] Suggest a method for synthesis of each of the following substances, and write a balanced molecular equation or equations for the process: **(a)** $ZnBr_2$; **(b)** HgI_2; **(c)** $CoCO_3$.

Oxidation of Metals

4.41 Where, in general, do the most easily oxidized metals occur in the periodic table? Where do the least easily oxidized metals occur in the periodic table?

4.42 Why are platinum and gold called noble metals? Why are the alkali metals and alkaline earth metals called the active metals?

4.43 Write balanced molecular and net ionic equations for the reactions of **(a)** hydrochloric acid with nickel; **(b)** sulfuric acid with iron; **(c)** hydrobromic acid with zinc; **(d)** acetic acid, $HC_2H_3O_2$, with magnesium.

4.44 Write balanced molecular and net ionic equations for the reactions of **(a)** manganese with sulfuric acid; **(b)** chromium with hydrobromic acid; **(c)** tin with hydrochloric acid; **(d)** aluminum with formic acid, $HCHO_2$.

4.45 Based on the activity series (Table 4.5), what is the outcome of each of the following reactions?
(a) $Al(s) + NiCl_2(aq) \longrightarrow$
(b) $Ag(s) + Pb(NO_3)_2(aq) \longrightarrow$
(c) $Cr(s) + NiSO_4(aq) \longrightarrow$
(d) $Mn(s) + HBr(aq) \longrightarrow$
(e) $H_2(g) + CuCl_2(aq) \longrightarrow$
(f) $Ba(s) + H_2O(l) \longrightarrow$

4.46 Using the activity series (Table 4.5), write balanced chemical equations for the following reactions. If no reaction occurs, simply write NR. **(a)** Zinc metal is added to a solution of silver nitrate; **(b)** iron metal is added to a solution of aluminum sulfate; **(c)** hydrochloric acid is added to cobalt metal; **(d)** hydrogen gas is bubbled through an aqueous solution of $FeCl_2$; **(e)** lithium metal is added to water.

4.47 Use the following equations to prepare an activity series for the hypothetical elements A, B, C, and D: $A + DX \longrightarrow AX + D$; $B + DX$ gives no reaction; $C + AX \longrightarrow CX + A$.

4.48 **(a)** Use the following reactions to prepare an activity series for the halogens: $Br_2(aq) + 2NaI(aq) \longrightarrow 2NaBr(aq) + I_2(aq)$; $Cl_2(aq) + 2NaBr(aq) \longrightarrow 2NaCl(aq) + Br_2(aq)$. **(b)** Relate the positions of the halogens in the periodic table with their locations in this activity series. **(c)** Predict whether reaction occurs when the following reagents are mixed: $Cl_2(aq)$ and $KI(aq)$; $Br_2(aq)$ and $LiCl(aq)$.

4.49 Why is HCl called a nonoxidizing acid and HNO_3 an oxidizing acid?

4.50 Nitric acid dissolves silver, but hydrochloric acid does not. Explain.

Solution Stoichiometry; Titrations

4.51 **(a)** What volume of 0.105 *M* $HClO_4$ solution is required to neutralize 50.0 mL of 0.0875 *M* NaOH? **(b)** What volume of 0.158 *M* HCl is required to neutralize 2.87 g of $Mg(OH)_2$? **(c)** If 25.8 mL of $AgNO_3$ is required to precipitate all the Cl^- ion in a 895-mg sample of KCl (forming AgCl), what is the molarity of the $AgNO_3$ solution? **(d)** If 35.8 mL of 0.117 *M* HCl solution is required to neutralize a solution of KOH, how many grams of KOH must be present in the solution?

4.52 **(a)** How many milliliters of 0.210 *M* HCl are needed to neutralize completely 35.0 mL of 0.101 *M* $Ba(OH)_2$ solution? **(b)** How many milliliters of 3.50 *M* H_2SO_4 are needed to neutralize 75.0 g of NaOH? **(c)** If 45.2 mL of $BaCl_2$ solution is needed to precipitate all of the sulfate in a 544-mg sample of Na_2SO_4 (forming $BaSO_4$), what is the molarity of the solution? **(d)** If 42.7 mL of 0.250 *M* HCl solution is needed to neutralize a solution of $Ca(OH)_2$, how many grams of $Ca(OH)_2$ must be present in the solution?

4.53 Some sulfuric acid is spilled on a lab bench. It can be neutralized by sprinkling sodium bicarbonate on it and then mopping up the resultant solution. The sodium bicarbonate reacts with sulfuric acid in the following way:

$$2NaHCO_3(s) + H_2SO_4(aq) \longrightarrow Na_2SO_4(aq) + 2CO_2(g) + 2H_2O(l)$$

Sodium bicarbonate is added until the fizzing due to the formation of $CO_2(g)$ stops. If 25 mL of 6.0 M H_2SO_4 was spilled, what is the minimum quantity of $NaHCO_3$ that must be added to the spill to neutralize the acid?

4.54 The distinctive odor of vinegar is due to acetic acid, $HC_2H_3O_2$. Acetic acid reacts with sodium hydroxide in the following fashion:

$$HC_2H_3O_2(aq) + NaOH(aq) \longrightarrow H_2O(l) + NaC_2H_3O_2(aq)$$

If 2.50 mL of vinegar requires 34.9 mL of 0.0960 M NaOH to reach the equivalence point in a titration, how many grams of acetic acid are in a 1.00-qt sample of this vinegar?

4.55 A sample of solid $Ca(OH)_2$ is allowed to stand in contact with water at 30°C for a long time, until the solution contains as much dissolved $Ca(OH)_2$ as it can hold. A 100-mL sample of this solution is withdrawn and titrated with 5.00×10^{-2} M HBr. It requires 48.8 mL of the acid solution for neutralization. What is the molarity of the $Ca(OH)_2$ solution? What is the solubility of $Ca(OH)_2$ in water at 30°C, in grams of $Ca(OH)_2$ per 100 mL of solution?

4.56 In the laboratory, 6.67 g of $Sr(NO_3)_2$ is dissolved in enough water to form 0.750 L. A 0.100-L sample is withdrawn from this stock solution and titrated with a 0.0460 M solution of Na_2CrO_4. What volume of Na_2CrO_4 solution is required to precipitate all the $Sr^{2+}(aq)$ as $SrCrO_4$?

Additional Exercises

4.57 The following equation is not balanced:

$$NO_2(aq) + H_2O(l) \longrightarrow NO_3^-(aq) + 2H^+(aq)$$

What is wrong with this equation?

4.58 Commercial concentrated hydrochloric acid is 12 M HCl. What volume of concentrated hydrochloric acid is required to prepare 1.00 L of 1.5 M HCl solution?

4.59 Calculate the molarity of the solution produced by mixing **(a)** 50.0 mL of 0.200 M NaCl and 100.0 mL of 0.100 M NaCl; **(b)** 24.5 mL of 1.50 M NaOH and 20.5 mL of 0.850 M NaOH. (Assume that the volumes are additive.)

4.60 Using modern analytical techniques it is possible to detect sodium ions in concentrations as low as 50 pg/mL. What is this detection limit expressed in **(a)** molarity of Na^+; **(b)** Na^+ ions per cubic centimeter?

4.61 Antacids are often used to relieve pain and promote healing in the treatment of mild ulcers. Write balanced net ionic equations for the reactions that occur between the $HCl(aq)$ in the stomach and each of the following substances used in various antacids: **(a)** $Al(OH)_3(s)$; **(b)** $Mg(OH)_2(s)$; **(c)** $MgCO_3(s)$; **(d)** $NaAl(CO_3)(OH)_2(s)$; **(e)** $CaCO_3(s)$.

4.62 Suppose you have a solution that might contain any or all of the following cations: Ni^{2+}, Ag^+, Sr^{2+}, and Mn^{2+}. Addition of HCl solution causes a precipitate to form. After filtering off the precipitate, H_2SO_4 solution is added to the resultant solution and another precipitate forms. This is filtered off, and a solution of NaOH is added to the resulting solution. No precipitate is observed. Which ions are present in each of the precipitates? Which of the four ions listed above must be absent from the original solution?

4.63 When we say that magnesium metal is more reactive than iron, what property of these metals are we concerned with?

4.64 Use Table 4.5 to predict which of the following ions can be reduced to their metal forms by reacting with zinc: **(a)** $Na^+(aq)$; **(b)** $Pb^{2+}(aq)$; **(c)** $Mg^{2+}(aq)$; **(d)** $Fe^{2+}(aq)$; **(e)** $Cu^{2+}(aq)$; **(f)** $Al^{3+}(aq)$. Write the balanced net ionic equation for each reaction that occurs.

4.65 Tartaric acid, $H_2C_4H_4O_6$, has two acidic hydrogens. The acid is often present in wines and precipitates from solution as the wine ages. A solution containing an unknown concentration of the acid is titrated with NaOH. It requires 22.62 mL of 0.2000 M NaOH solution to titrate both acidic protons in 40.00 mL of the tartaric acid solution. Write a balanced net ionic equation for the neutralization reaction and calculate the molarity of the tartaric acid solution.

[4.66] The arsenic in a 1.22-g sample of a pesticide was converted to AsO_4^{3-} by suitable chemical treatment. It was then titrated using Ag^+ to form Ag_3AsO_4 as a precipitate. If it took 25.0 mL of 0.102 M Ag^+ to reach the equivalence point in this titration, what is the percentage of arsenic in the pesticide?

[4.67] A solid sample of $Zn(OH)_2$ is added to 0.400 L of a 0.550 M solution of HBr. The solution that remains is still acidic. It is then titrated with 0.500 M NaOH solution, and 165 mL of the NaOH solution is required to reach the equivalence point. What was the mass of $Zn(OH)_2$ added to the HBr solution?

[4.68] Federal regulations set an upper limit of 50 parts per million (ppm) of NH_3 in the air in a work environment (that is, 50 mL NH_3 per 10^6 mL of air). The density of $NH_3(g)$ at room temperature is 0.771 g/L. Air from a manufacturing operation was drawn through a solution containing 100 mL of 0.0105 M HCl. The NH_3 reacts with HCl as follows:

$$NH_3(aq) + HCl(aq) \longrightarrow NH_4Cl(aq)$$

After drawing air through the acid solution for 10.0 min at a rate of 10.0 L/min, the acid was titrated. The remaining acid required 13.1 mL of 0.0588 M NaOH to reach the equivalence point. **(a)** How many grams of NH_3 were drawn into the acid solution? **(b)** How many ppm of NH_3 were in the air? **(c)** Is this manufacturer in compliance with regulations?

5 Energy Relationships in Chemistry: Thermochemistry

The thermite reaction, in which powdered aluminum reacts with Fe_2O_3, is a highly exothermic reaction. (Dr. E. R. Degginger)

When you eat an orange, the sugar it contains reacts in your body with oxygen to form CO_2 and H_2O. During this chemical process, one other important change occurs—energy is released. The food you eat is the fuel that your body uses to operate your muscles and to maintain proper body temperature. This example illustrates a general point: Chemical reactions involve changes in energy. Some reactions, like the oxidation of sugar, produce energy. Others, like the splitting of water into hydrogen and oxygen, use energy. At the present time, over 90 percent of the energy produced in our society comes from chemical reactions, principally from the combustion of coal, petroleum products, and natural gas.

The study of energy and its transformations is known as **thermodynamics** (Greek: *therme*, "heat"; *dynamis*, "power"). This area of study began during the Industrial Revolution as the relationships among heat, work, and the energy content of fuels were studied in an effort to maximize the performance of steam engines. Thermodynamics is important not only to chemistry but to other areas of science and to engineering as well. It touches our daily lives as we use energy for manufacturing, travel, and communications. Thermodynamics relates to such diverse topics as the metabolism of foods, the operation of batteries, and the design of engines.

In this chapter, we examine the relationships between chemical reactions and energy changes. This aspect of thermodynamics is called **thermochemistry**. We will discuss additional concepts of thermodynamics in Chapter 19.

CONTENTS

5.1 THE NATURE OF ENERGY

What do we mean when we talk about energy? Because energy is not tangible, as are material objects, it is somewhat difficult to define. To get a better grasp of this important concept, let's consider a simple process that requires us to expend energy, such as lifting a barbell off the ground (Figure 5.1). Gravity attracts the barbell to the ground, and in order to lift it we must overcome this attraction. What is the relationship between this attraction and the amount of energy we must use?

The gravitational attraction of the barbell is an example of a force. We can define **force** as any kind of *push* or *pull* exerted on an object. In our example, the force of gravity "pulls" the barbell to the ground. In chemistry, we are more concerned with forces other than gravity. For example, the gas molecules in a balloon exert a "push" on the skin of the balloon. Electrically charged particles also generate forces. Thus, because of electrostatic attraction, positively charged nuclei exert a "pull" on negatively charged electrons.

When we lift a barbell, we perform **work** to overcome the force of gravity. Similarly, when we increase the distance between a proton and an electron, we perform work to overcome the attractive electrostatic force between them.

Figure 5.1 Lifting a barbell requires energy in the form of mechanical work to overcome the force of gravity. If the force of gravity acting on the barbell is denoted as F, then the work, w, required to lift the barbell a distance, d, off the ground is $w = F \times d$. (Paul J. Sutton/ Duomo)

These examples lead us to our first definition of energy: **Energy** is the capacity to do work. The most familiar kind of work, **mechanical work**, w, is defined as the amount of energy required to move an object a certain distance, d, when it is subject to a force, F. The work is the product of the force and the distance:

$$w = F \times d \tag{5.1}$$

Energy, in the form of work, must be used to overcome the effects of a force. It requires more work to lift a heavy barbell than a light one, because the gravitational force exerted on the heavy barbell is greater.

After spending time lifting barbells, we also notice that we feel warmer. Our bodies generate heat as we perform this arduous task. Chemists define **heat** as energy that is transferred from one object to another because of a difference in temperature. This observation leads us to a second definition of energy: Energy is the capacity to transfer heat. As we get hotter while performing work, we transfer heat from our bodies to our surroundings. When we lift barbells, we expend energy in the forms of mechanical work and heat.

In our studies of thermochemistry, we will be concerned with the performance of work by chemical systems and the flow of heat between them. For example, we must perform work to compress a gas or to separate particles of opposite electrical charge, such as an electron and a nucleus. We also know, of course, that chemical processes such as the burning of gasoline release energy in the form of heat. In order to understand how chemical systems perform work and transfer heat, we must look more closely at the two ways in which molecules can possess energy: kinetic energy and potential energy.

Kinetic and Potential Energy

Kinetic energy is the energy of motion. The magnitude of the kinetic energy of an object depends on its mass, m, and velocity, v:

$$E_k = \tfrac{1}{2}mv^2 \tag{5.2}$$

This equation expresses quantitatively what our experience teaches—that both the mass and speed of an object determine how much work it can accomplish. For example, the more massive a hammer is and the faster it moves, the more work it can accomplish with each blow in pounding a nail.

An object can also possess energy by virtue of its position relative to other objects. This stored energy is called **potential energy**. It is a result of the attractions and repulsions an object experiences in relation to other objects. A brick held high in the air has potential energy because of the force of gravity acting on it. As it falls, its potential energy is converted into kinetic energy, and the brick does work on whatever object it strikes. Likewise, an electron has potential energy when located near a proton because of the attractive electrostatic force between them.

Besides the kinetic and potential energies of objects, other forms of energy exist. These forms include electrical energy, nuclear energy, radiant energy (light), and chemical energy. However, with the advent of atomic theory, these other forms of energy have come to be considered kinetic or potential energy at the atomic or molecular level. For example, the chemical energy of gasoline can be regarded as potential energy stored in the arrangements of the electrons and nuclei within the molecules that comprise the gasoline.

Energy Units

The SI unit of energy is the **joule** (J); 1 J = 1 kg-m^2/s^2. A mass of 2 kg moving at a velocity of 1 m/s possesses a kinetic energy of 1 J:

$$E = \tfrac{1}{2}(2\text{ kg})(1\text{ m/s})^2 = 1\text{ kg-m}^2/\text{s}^2 = 1\text{ J}$$

Many other units with which we are familiar are based on the joule. For example, the watt is a unit of **power** (a measure of energy per time) equal to 1 joule per second. Thus, a 100-watt (W) light bulb produces 100 J of energy each second it operates.

Traditionally, energy changes accompanying chemical reactions have been expressed in units of calories. A **calorie** (cal) is the amount of energy required to raise the temperature of 1 g of water by 1°C, from 14.5°C to 15.5°C.* A kilocalorie (1000 cal) is the same as the Calorie (capitalized), used in expressing the energy values of foods. One calorie is the same amount of energy as 4.184 J:

$$1\text{ cal} = 4.184\text{ J}$$

In much of engineering work, it is common to use the British thermal unit (Btu); 1 Btu = 1.05 kJ. A Btu is the amount of heat required to raise the temperature of 1 lb of water by 1°F. In most places in this text, the kilojoule will be used as the unit of energy.

SAMPLE EXERCISE 5.1

(a) A 145-g baseball is thrown with a speed of 25 m/s. Calculate the kinetic energy of the ball in joules. **(b)** What is the kinetic energy of the ball in calories?

Solution: **(a)** To calculate the kinetic energy, we use Equation 5.2. Recall that a joule is 1 kg-m^2/s^2. Thus, mass needs to be expressed in kilograms: 145 g = 0.145 kg.

$$E_k = \tfrac{1}{2}mv^2 = \tfrac{1}{2}(0.145\text{ kg})(25\text{ m/s})^2 = 45\text{ J}$$

(b) To obtain the kinetic energy in calories, we use the conversion factor 1 cal = 4.184 J:

$$(45\text{ J})\left(\frac{1\text{ cal}}{4.184\text{ J}}\right) = 11\text{ cal}$$

PRACTICE EXERCISE

(a) Calculate the kinetic energy, in joules, of a 6.0-kg object moving at the speed of 5.0 m/s. **(b)** What is its kinetic energy in calories? ***Answers:*** **(a)** 75 J; **(b)** 18 cal

Systems and Surroundings

When we study energy changes, we focus our attention on a limited and well-defined part of the universe. The portion we single out for study is called the **system**; everything else is called the **surroundings**. When we observe a chemical reaction in the laboratory, the chemicals usually constitute the system. The container and everything beyond it are considered the surroundings. The systems

* The temperature interval 14.5°C to 15.5°C is specified because the energy required to raise the temperature of water by 1°C is slightly different for different temperatures. However, it is constant to three significant figures, 1.00 cal, over the entire range of liquid water, from 0°C to 100°C.

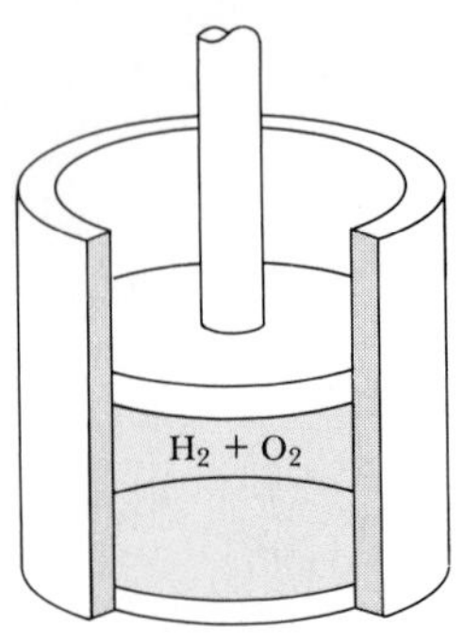

Figure 5.2 Hydrogen and oxygen gases in a cylinder. If we are interested in studying the properties of the gases only, the gases are the system and the cylinder and piston are considered part of the surroundings.

we can most readily study are those that exchange energy but not matter with their surroundings. As an example, consider a mixture of hydrogen gas, H_2, and oxygen gas, O_2, in a cylinder, as illustrated in Figure 5.2. The system in this case is just the hydrogen and oxygen; the cylinder, piston, and everything beyond them (including us) are the surroundings. If the hydrogen and oxygen react to form water, energy is produced:

$$2H_2(g) + O_2(g) \longrightarrow 2H_2O(l) + \text{energy}$$

Although the chemical form of the hydrogen and oxygen atoms in the system is changed by this reaction, the system has not lost or gained mass; it undergoes no exchange of matter with its surroundings. However, it does exchange energy with its surroundings in the form of heat and work. These are quantities that we can measure.

5.2 THE FIRST LAW OF THERMODYNAMICS

Thus far we have discussed the *transfer* of energy. One basic fact concerning energy is that although it can be converted from one form to another, it can be neither created nor destroyed: *Energy is conserved.* The energy lost by a system equals the energy gained by its surroundings. Likewise, the energy gained by a system equals that lost by its surroundings. This important observation is known as the **law of conservation of energy**. Because it is the most elementary of thermodynamic concepts, it is also known as the **first law of thermodynamics**.

Internal Energy Changes

The total energy of a system is the sum of all the kinetic and potential energies of its component parts. For the system shown in Figure 5.2, total energy includes not only the motions and interactions of the H_2 and O_2 molecules themselves but also of their component nuclei and electrons. This total energy is called the **internal energy** of the system. Because there are so many types of motions and interactions, we cannot determine the exact energy of any system of practical interest. We can, however, measure the *changes* in internal energy that accompany chemical and physical processes.

We define the change in internal energy, which we represent as ΔE (read "delta E"),* as the difference between the internal energy of the system at the completion of a process and that at the beginning:

$$\Delta E = E_{\text{final}} - E_{\text{initial}} \qquad [5.3]$$

Thermodynamic quantities such as ΔE have two parts: a number giving the magnitude of the change, and a sign giving the direction. A *positive* ΔE results when $E_{\text{final}} > E_{\text{initial}}$, indicating that the system has gained energy from its surroundings. A negative ΔE is obtained when $E_{\text{final}} < E_{\text{initial}}$, indicating that the system has lost energy to its surroundings. These two situations are compared in Figure 5.3.

* The symbol Δ is commonly used to denote *change*. For example, a change in volume can be represented by ΔV.

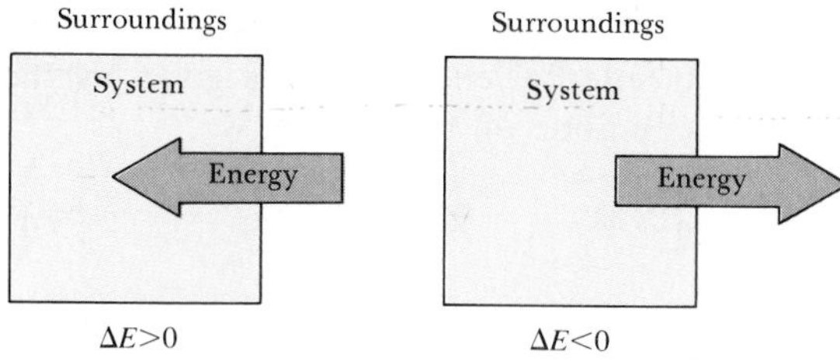

Figure 5.3 If the system gains energy (left), ΔE will be positive. If the system loses energy (right), ΔE will be negative.

In a chemical reaction, of course, the initial state of the system refers to the reactants, and the final state refers to the products. When hydrogen and oxygen react to form water, the system loses energy; ΔE for this process is negative. This means that the internal energy of the hydrogen and oxygen is greater than that of the water, as shown in the energy diagram in Figure 5.4.

A basic fact of nature, and one that governs much of chemistry, is that *systems tend to attain as low an energy as possible.* For example, consider a bicyclist at the top of a steep hill, as shown in Figure 5.5. Because of the force of gravity, the potential energy of the bicyclist and her bicycle (which comprise the system) is much greater at the top of the hill than it would be at the bottom. As a result, the bicycle easily proceeds down the hill with no effort on the part of the rider. As it does so, the stored potential energy of the system is converted into kinetic energy, which can do work on, and transfer heat to, the surroundings.

We will see throughout this text that chemical reactions follow this fundamental idea as well. Systems with a high potential energy are less stable and more likely to undergo change than systems with a low potential energy. Like the bicycle in Figure 5.5, chemical reactants move spontaneously toward a lower potential energy when possible.

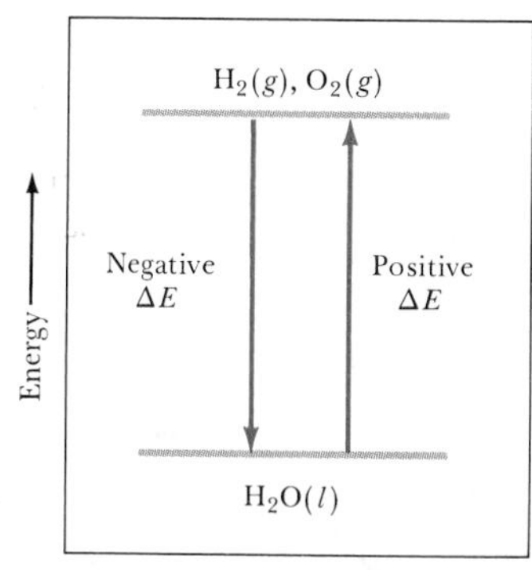

Figure 5.4 A system composed of $H_2(g)$ and $O_2(g)$ has a greater energy content than one composed of $H_2O(l)$. The system loses energy (negative ΔE) when H_2 and O_2 are converted to H_2O. It gains energy (positive ΔE) when H_2O is decomposed into H_2 and O_2.

Relating ΔE to Heat and Work

Any system can exchange energy with its surroundings in two general ways: as heat or as work. The internal energy of a system changes in magnitude as heat is added to or removed from the system, or as work is done on it or by it. We can use these ideas to write a very useful algebraic expression of the first

Figure 5.5 A bicycle at the top of a hill (left) has a high potential energy. As the bicycle proceeds down the hill (right), the potential energy is converted into kinetic energy; the potential energy is lower at the bottom than at the top. Energy is transferred from the system (the bicycle) to the surroundings by friction and air resistance.

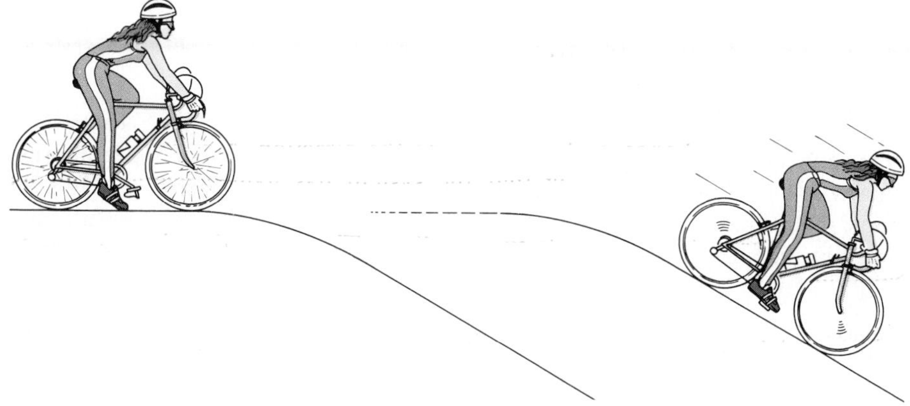

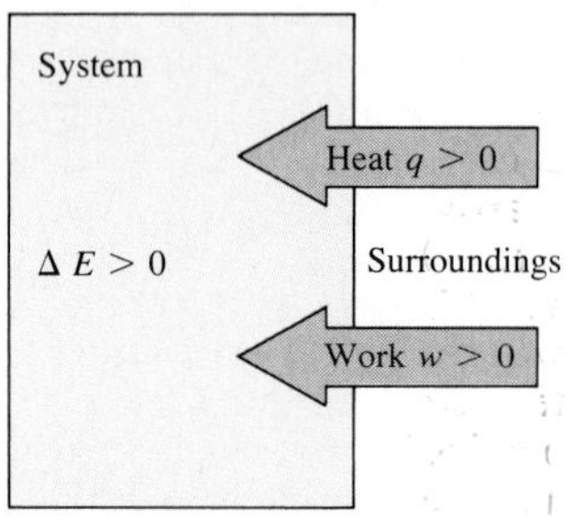

Figure 5.6 Heat, q, absorbed by the system and work, w, done on the system are both positive quantities. Both serve to increase the internal energy, E, of the system: $\Delta E = q + w$.

law of thermodynamics. When a system undergoes any chemical or physical change, the accompanying change in its internal energy, ΔE, is given by the heat added to the system, q, plus the work done on the system, w:

$$\Delta E = q + w \qquad [5.4]$$

Heat added to the system is assigned a positive sign. Likewise, *work done to the system is positive* (Figure 5.6). Both the heat added to the system and the work done on it increase the internal energy. On the other hand, both the heat lost by the system and the work done by the system on its surroundings are negative; they reduce the internal energy. For example, if the system absorbs 50 J of heat and does 10 J of work on its surroundings, $q = 50$ J and $w = -10$ J. Thus, $\Delta E = 50 \text{ J} + (-10 \text{ J}) = 40$ J.*

A reaction that results in the evolution of heat is **exothermic** (*exo-* is a prefix meaning "out of"); that is, heat flows out of the system and into its surroundings. Consider an exothermic process occurring in a container that is initially at room temperature. If we, as part of the surroundings, touch the container, it feels hot to us, because heat has passed from the container to our hands. Combustion reactions are examples of exothermic reactions. Reactions that absorb heat from their surroundings are **endothermic** (*endo-* is a prefix meaning "into"). If we touch a container that is initially at room temperature in which an endothermic process is occurring, it feels cold; what we perceive as cooling is in fact heat being absorbed by the system from the surroundings. The melting of ice is an example of an endothermic process.

SAMPLE EXERCISE 5.2

The hydrogen and oxygen gases in the cylinder illustrated in Figure 5.2 are ignited. As the reaction occurs, the system loses 550 J of heat to its surroundings (the reaction is exothermic). The reaction also causes the piston to move upward as the gases expand. It is determined that the expanding gas does 240 J of work on the surroundings. What is the change in the internal energy of the system?

Solution: From the sign conventions for q and w, we have $q = -550$ J and $w = -240$ J (that is, energy flows from the system both as heat and as work). Using Equation 5.4, we have

$$\Delta E = q + w = (-550 \text{ J}) + (-240 \text{ J}) = -790 \text{ J}$$

Thus, 790 J of energy has been transferred from the system to the surroundings.

PRACTICE EXERCISE

Calculate ΔE for an endothermic process in which the system absorbs 65 J of heat and also receives 12 J of work from its surroundings. ***Answer:*** 77 J

State Functions

Although we usually have no way of knowing the precise value of the internal energy of a system, we do know that it has a fixed value for a given set of conditions. The conditions that influence this energy include the temperature and pressure. We also know that the total internal energy of a system is propor-

* Equation 5.4 is sometimes written $\Delta E = q - w$. When written this way, work done *by* the system is defined as positive. This convention has merit in the many engineering applications that focus interest on a machine that does work on its surroundings.

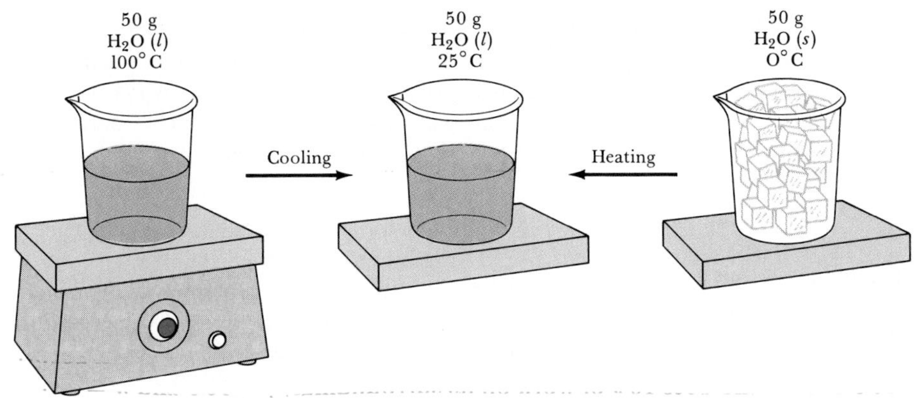

Figure 5.7 Internal energy, a state function, depends only on the present state of the system and not on the path by which it arrived at that state. The internal energy of 50 g of water at 25°C is the same whether the water is cooled from a higher temperature to 25°C or is obtained by melting 50 g of ice and then warming to 25°C.

tional to the total quantity of matter in the system; energy is an extensive property (Section 1.3).

Suppose we define our system as 50 g of water at 25°C, as in Figure 5.7. Our system could have arrived at that state by our cooling 50 g of water from 100°C or by our melting 50 g of ice and subsequently warming the water to 25°C. The internal energy of the water is the same in either case. The internal energy of a system is a **state function**, a property of a system that is determined by specifying its condition, or its state (in terms of temperature, pressure, location, and so forth). *The value of a state function does not depend on the particular previous history of the sample, only on its present condition.* Because E is a state function, ΔE depends only on the initial and final states of the system and not on how the change occurs.

You may ask: What is an example of something that is *not* a state function? The work done by a system in a given process is not a state function. Rather, it depends on the manner in which the process is carried out. As an example, let's consider a flashlight battery as our system, and let the change in the system be the complete discharge of the battery at constant temperature. If the battery is discharged in a flashlight [Figure 5.8(*a*)], no mechanical work is accomplished. All the energy lost from the battery appears as radiant energy and heat. If the battery is used in a mechanical toy [Figure 5.8(*b*)], the same change in state of the battery produces mechanical work and heat. The change

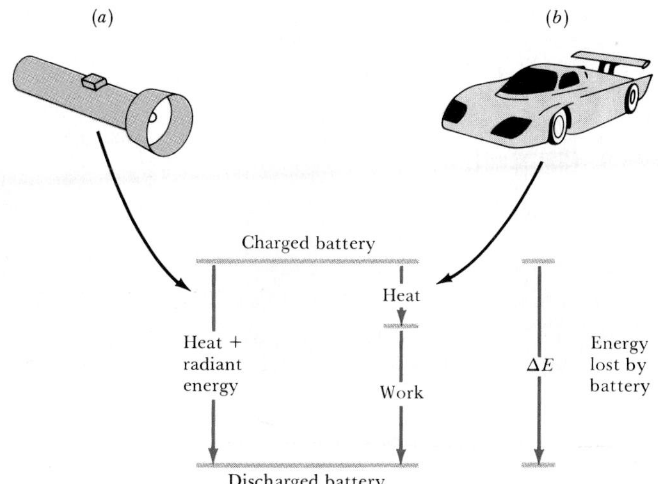

Figure 5.8 When a battery is discharged in lighting a flashlight, all the energy of the battery appears as radiant energy and heat; no work is done. When the battery is used in the toy car, work is done in moving the car from place to place. Thus, the work done by the system (the battery) is not a state function, because its magnitude depends on the particular path by which the system gets from its initial to its final state.

in state of the system, and thus the change in ΔE, is the same in both cases. However, the amount of work done in the two cases is different, as is the amount of heat released. We see from this example that although ΔE is always the same for a given change in the system, the way in which the change is performed will determine the relative contributions of q and w, the means by which energy is transferred.

5.3 *P-V* WORK AND ENTHALPY

There are many varieties of work, but in chemistry we are interested mainly in electrical work and in the mechanical work done by expanding gases. In this chapter, we shall restrict our discussion to the latter type of work. Expansion work, which results from a change in the volume of the system, is usually called pressure–volume or *P-V* work. Expanding gases in the cylinder of an automobile engine do *P-V* work on the piston, and this work eventually turns the wheels. Expanding gases from an open reaction vessel do *P-V* work in pushing back the atmosphere, but this work is lost to us; it accomplishes nothing in a practical sense. However, all kinds of work, whether useful or not, must be considered if we desire to keep track of all the energy changes of a system.

To understand *P-V* work better, consider a gas confined to a cylinder with a movable piston, as shown in Figure 5.9. A downward force, F, acts on the piston. The pressure on the gas is defined as the ratio of this force to the area of the piston, A. That is, $P = F/A$. We will assume that the piston itself is weightless and that the only force acting on it, F, is the weight of the earth's atmosphere. This pressure is called the *atmospheric pressure*. We will have more to say about atmospheric pressure and about the properties of gases in Chapter 10; at this point we need only be aware that the atmosphere exerts a pressure. Now assume that the gas in the cylinder expands against this constant external pressure. This expansion could be caused by heating the gas or by a reaction that increases the number of moles of gas in the system. The magnitude of the work accomplished by the expanding gas is given by the product of the pressure and the volume change of the piston, ΔV; because the expanding gas does work on its surroundings in pushing back the atmosphere, the work has a negative sign:

$$w = -P\Delta V \qquad [5.5]$$

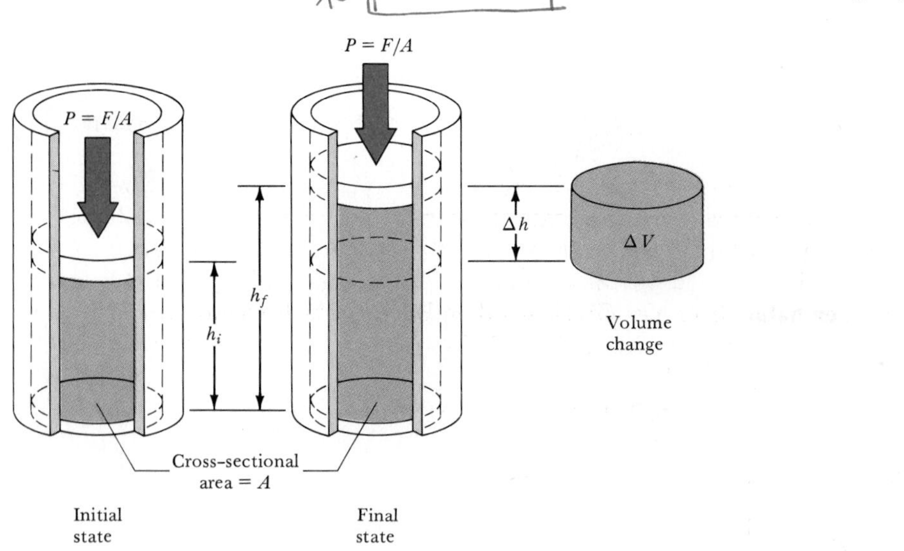

Figure 5.9 A piston moving a distance, Δh, against a pressure, P, does work on the surroundings. The amount of work done is given by the product of the pressure and the change in volume of the gas: $w = -P\Delta V$.

A CLOSER LOOK: Derivation of $w = -P\Delta V$

Equation 5.5 is readily derived from the definitions of work and pressure. The magnitude of the work accomplished by a moving piston, such as that shown in Figure 5.9, is given by the product of the force, F, and the distance the piston moves, Δh:

$$\text{Work} = \text{force} \times \text{distance} = F \times \Delta h$$

Rearranging the equation that defines pressure, $P = F/A$, gives $F = P \times A$. Thus,

$$\text{Work} = F \times \Delta h = P \times A \times \Delta h$$

The product $A \times \Delta h$ is the volume change, ΔV, resulting from the piston moving. Thus,

$$\text{Work} = P \times A \times \Delta h = P \times \Delta V$$

But in our derivation we have ignored the sign of this work. Because an expanding gas must do work on its surroundings, the sign of the work must be negative. Thus, we have Equation 5.5: $w = -P\Delta V$. This equation applies equally well to the case of a contracting gas. When a gas is compressed, ΔV is negative. In that case, the product $-P\Delta V$ will be positive, as it should be when work is done *on* the system by the surroundings.

For the case where only P-V work is done, Equation 5.5 can be substituted into Equation 5.4 to give

$$\begin{aligned}\Delta E &= q + w \\ &= q - P\Delta V\end{aligned} \qquad [5.6]$$

When a reaction is carried out in a constant-volume container ($\Delta V = 0$), the heat flow equals the change in internal energy:

$$\Delta E = q_V \qquad \text{(constant volume)} \qquad [5.7]$$

(We put a subscript V on q to remind ourselves that we are considering a special case where volume is constant). This equation tells us that if the volume of the system is constant, any heat added to or lost from the system equals the change in internal energy. In contrast, if the volume is allowed to change and the pressure is constant, we have

$$\Delta E = q_P - P\Delta V$$

hence

$$q_P = \Delta E + P\Delta V \qquad \text{(constant pressure)} \qquad [5.8]$$

Thus, the heat needed to bring about any change at constant pressure is the sum of the internal energy change plus the P-V work.

The condition of constant pressure is especially important in chemistry. Most chemical reactions, including those in living systems, take place under the essentially constant pressure of the earth's atmosphere. For this reason, we find it convenient to give the quantity $E + PV$ a special symbol, H, and a special name, **enthalpy** (from the Greek word *enthalpein*, which means "to warm"). At constant pressure, the change in enthalpy for a process is given by

$$\Delta H = H_{\text{final}} - H_{\text{initial}} = \Delta E + P\Delta V \qquad [5.9]$$

From Equations 5.8 and 5.9 we see that

$$\Delta H = q_P \qquad [5.10]$$

Equation 5.10 tells us that the enthalpy change, ΔH, is the heat added to (or lost by) a system at constant pressure. Therefore, if we study a reaction in a container open to the atmosphere, such as a beaker in the laboratory, ΔH is equal to the energy flow as heat. For this reason, the enthalpy change for a reaction is commonly called the *heat of reaction*. Like internal energy, enthalpy is a state function.

Enthalpy changes are extremely important in thermochemistry. In fact, chemists tend to think in terms of ΔH rather than ΔE. The volume change accompanying many reactions is close to zero, so ΔH is close to ΔE. Even when the volume change is significant, the difference in magnitude between ΔH and ΔE is generally small.

5.4 ENTHALPIES OF REACTION

The enthalpy change for a chemical reaction is given by the enthalpy of the products minus that of the reactants:

$$\Delta H = H(\text{products}) - H(\text{reactants}) \quad [5.11]$$

If the products of the reaction have a greater enthalpy than the reactants, ΔH will be *positive*. In this case, the system absorbs heat, and the reaction is *endothermic*. An example of an endothermic process is the formation of nitric oxide, NO, from nitrogen, N_2, and oxygen, O_2:

$$\text{Heat} + N_2(g) + O_2(g) \longrightarrow 2NO(g)$$

This reaction accompanies high-temperature combustion reactions in air and contributes greatly to smog (Section 18.4). The enthalpy content of the $2NO(g)$ is greater than that of the $N_2(g) + O_2(g)$, as represented graphically in Figure 5.10.

Figure 5.10 The enthalpy change for the combustion of $N_2(g) + O_2(g)$ to form $2NO(g)$. The system absorbs heat from its surroundings, increasing in enthalpy. The process is endothermic, and ΔH is positive.

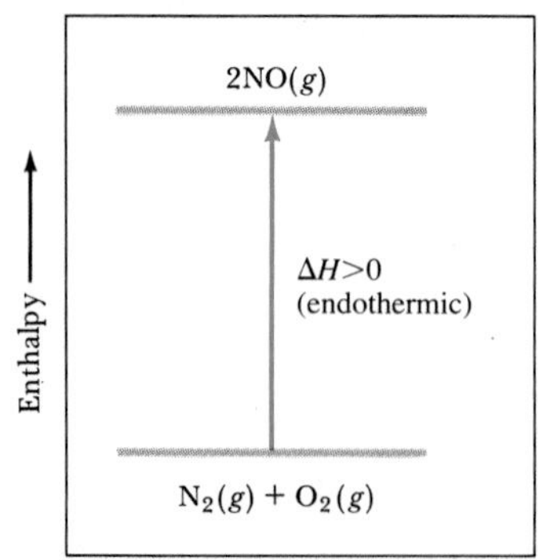

If the enthalpy content of the products of a reaction is less than that of the reactants (an *exothermic* reaction), ΔH will be *negative*. Using the example of the bicyclist from Figure 5.5, we can think of exothermic reactions as going "downhill" in enthalpy. The higher potential energy of the reactants is released as heat as the reaction proceeds to give products of lower potential energy. This situation arises, for example, when hydrogen is combusted in the presence of oxygen:

$$2H_2(g) + O_2(g) \longrightarrow 2H_2O(g) + \text{heat}$$

This reaction produced the disastrous explosions of the German airship *Hindenburg* in 1937 and the space shuttle *Challenger* in 1986 (Figure 5.11). Figure 5.12 shows the explosion of a balloon filled with H_2 as it is ignited by a burning candle. The change in the enthalpy of the system during this reaction is diagramed in Figure 5.12.

Before exploring further how we might use the concept of enthalpy, let's consider three of its important characteristics.

1. *Enthalpy is an extensive property.* This fact means that the magnitude of ΔH is directly proportional to the amount of reactant consumed in the process. Consider the combustion of methane to form carbon dioxide and water. It is found experimentally that 802 kJ of heat is produced when

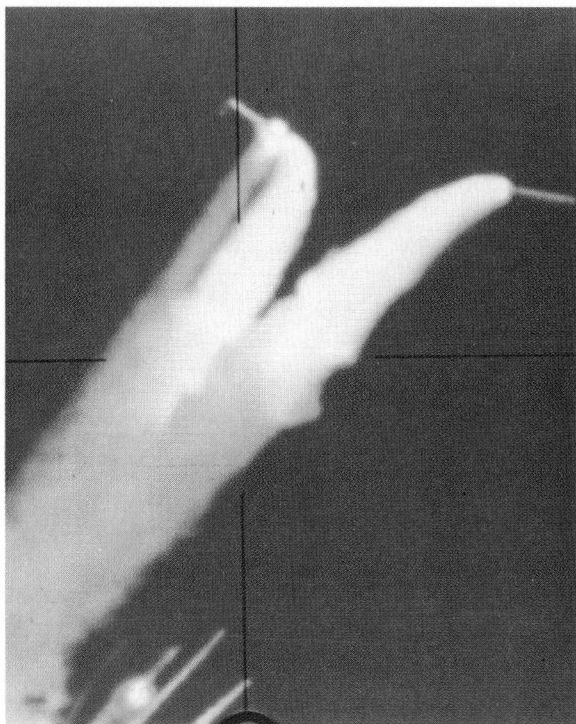

Figure 5.11 The tragic explosion of the space shuttle *Challenger* was a result of the uncontrolled exothermic reaction between hydrogen and oxygen. Hydrogen, stored as a liquid, is the fuel used to propel space shuttles out of the Earth's gravitational pull. (NASA)

1 mol of CH_4 is burned in a constant-pressure system. We can express this fact as follows:

$$CH_4(g) + 2O_2(g) \longrightarrow CO_2(g) + 2H_2O(g) \qquad \Delta H = -802 \text{ kJ} \quad [5.12]$$

The negative sign for ΔH tells us that this reaction is exothermic. Notice that ΔH is reported at the end of a balanced equation, with no explicit mention of the amounts of chemicals involved. In such cases, it is understood that the coefficients in the balanced equation represent the number of moles of reactants producing the associated enthalpy change. Thus,

Figure 5.12 The combustion of $H_2(g)$ in $O_2(g)$ to form $H_2O(g)$. The system gives off heat to the surroundings, decreasing in enthalpy. The reaction is exothermic, and ΔH is negative, as shown in (*c*). (Donald Clegg and Roxy Wilson)

(*a*)

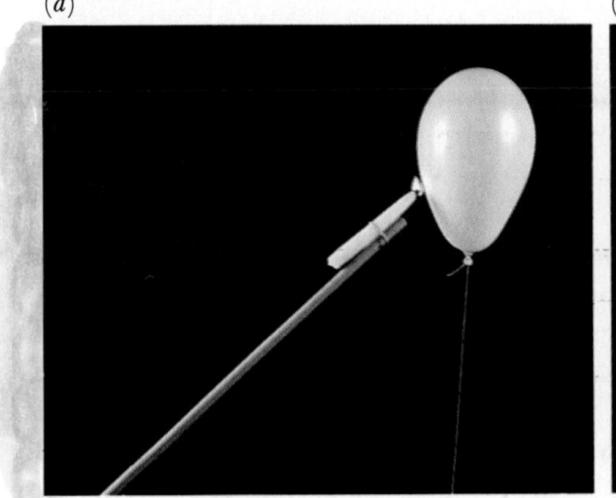

(*b*)

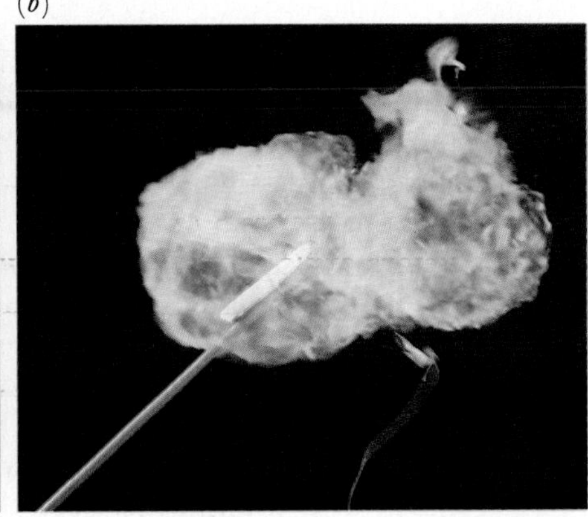

(*c*)

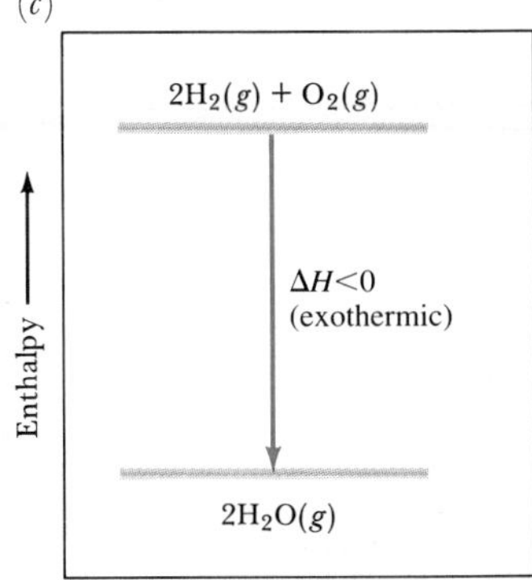

combustion of 1 mol of CH_4 with 2 mol of O_2 produces 802 kJ of heat. The combustion of 2 mol of CH_4 with 4 mol of O_2 produces 1604 kJ.

SAMPLE EXERCISE 5.3

How much heat is produced when 4.50 g of methane gas is burned in a constant-pressure system? (Use the information given in Equation 5.12).

Solution: According to Equation 5.12, 802 kJ is produced when a mole of CH_4 is burned ($\Delta H = -802$ kJ/mol). A mole of CH_4 has a mass of 16.0 g.

$$\text{Heat} = (4.50 \text{ g } CH_4)\left(\frac{1 \text{ mol } CH_4}{16.0 \text{ g } CH_4}\right)\left(-802 \frac{\text{kJ}}{\text{mol } CH_4}\right) = -226 \text{ kJ}$$

PRACTICE EXERCISE

Ammonium nitrate can decompose explosively by the following reaction:

$$NH_4NO_3(s) \longrightarrow N_2O(g) + 2H_2O(g) \qquad \Delta H = -37.0 \text{ kJ}$$

Calculate the quantity of heat produced when 2.50 g of NH_4NO_3 decomposes at constant pressure. ***Answer:*** -1.16 kJ

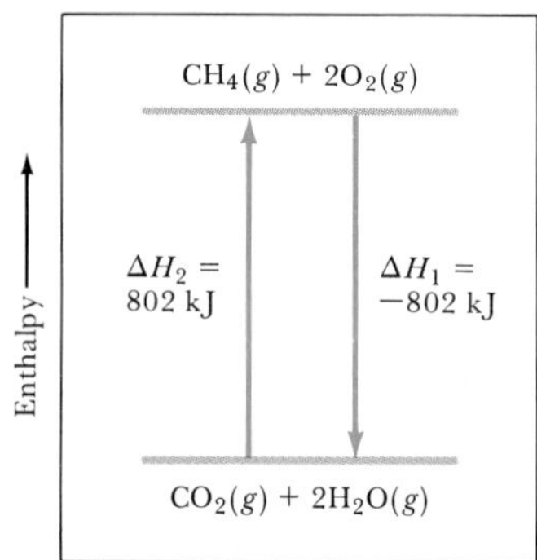

Figure 5.13 Reversing a reaction changes the sign but not the magnitude of the enthalpy change: $\Delta H_2 = -\Delta H_1$.

2. *The enthalpy change for a reaction is equal in magnitude but opposite in sign to ΔH for the reverse reaction.* For example, when Equation 5.12 is reversed, ΔH for the process is $+802$ kJ:

$$CO_2(g) + 2H_2O(g) \longrightarrow CH_4(g) + 2O_2(g) \qquad \Delta H = 802 \text{ kJ} \qquad [5.13]$$

When we reverse a reaction, we reverse the roles of the products and the reactants; the reactants in a reaction become the products of the reverse reaction, and so forth. From Equation 5.11, we can see that reversing the products and reactants leads to the same magnitude but a change in sign for ΔH. This relationship is diagramed for Equations 5.12 and 5.13 in Figure 5.13.

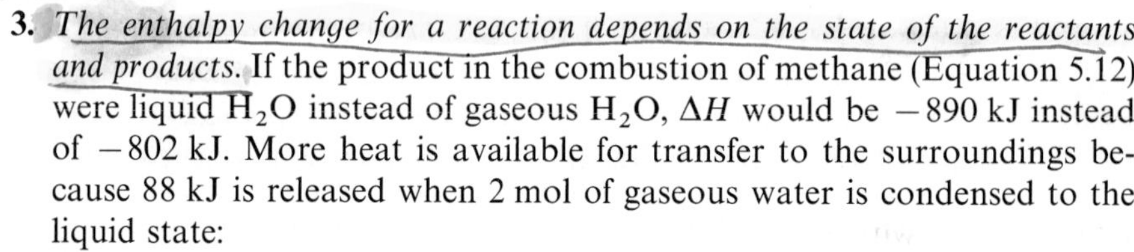

3. *The enthalpy change for a reaction depends on the state of the reactants and products.* If the product in the combustion of methane (Equation 5.12) were liquid H_2O instead of gaseous H_2O, ΔH would be -890 kJ instead of -802 kJ. More heat is available for transfer to the surroundings because 88 kJ is released when 2 mol of gaseous water is condensed to the liquid state:

$$2H_2O(g) \longrightarrow 2H_2O(l) \qquad \Delta H = -88 \text{ kJ} \qquad [5.14]$$

Therefore, the states of the reactants and products must be specified. In addition, we will generally assume that the reactants and products are both at the same temperature, 25°C, unless otherwise indicated.

The enthalpy change associated with a given chemical process is often of great significance. In the sections that follow, we will consider some ways in which we can evaluate this important quantity. As we shall see, ΔH for a reaction can be either directly determined by experiment or calculated from a knowledge of the enthalpy changes associated with other reactions by invoking the first law of thermodynamics. Let's begin by discussing how we make use of the concept of a state function to carry out calculations of enthalpy changes.

Because enthalpy is a state function, the enthalpy change, ΔH, associated with any chemical process depends only on the amount of matter that undergoes change and on the nature of the initial state of the reactants and the final state of the products. This means that if a particular reaction can be carried out in one step or in a series of steps, the sum of the enthalpy changes associated with the individual steps must be the same as the enthalpy change associated with the one-step process. For example, the combustion of methane gas to form carbon dioxide and liquid water can occur in two distinct steps. First, methane is combusted to $CO_2(g)$ and gaseous water (Equation 5.12). Second, the gaseous water is condensed to liquid water (Equation 5.14). The overall enthalpy change is simply the sum of the enthalpy changes for these two steps:

$$\begin{array}{llll} & CH_4(g) + 2O_2(g) \longrightarrow & CO_2(g) + 2H_2O(g) & \Delta H = -802\ \text{kJ} \\ \text{(Add)} & 2H_2O(g) \longrightarrow & 2H_2O(l) & \Delta H = -88\ \text{kJ} \\ \hline & CH_4(g) + 2O_2(g) + 2H_2O(g) \longrightarrow & CO_2(g) + 2H_2O(l) + 2H_2O(g) & \Delta H = -890\ \text{kJ} \end{array}$$

Net equation:

$$CH_4(g) + 2O_2(g) \longrightarrow CO_2(g) + 2H_2O(l) \qquad \Delta H = -890\ \text{kJ}$$

To obtain the net equation, the sum of the reactants of the two equations is placed on one side of the arrow, and the sum of the products on the other side. Because $2H_2O(g)$ occurs on both sides of the arrow, it can be canceled like an algebraic quantity that appears on both sides of an equal sign.

Hess's law states that *if a reaction is carried out in a series of steps, ΔH for the reaction will be equal to the sum of the enthalpy changes for the individual steps.* The overall enthalpy change for the process is independent of the number of steps or the particular nature of the path by which the reaction is carried out. We can therefore calculate ΔH for any process, as long as we find a route for which ΔH is known for each step. This important fact permits us to use a relatively small number of experimental measurements to calculate ΔH for a vast number of different reactions.

Hess's law provides a useful means of calculating energy changes that are difficult to measure directly. For instance, it is not possible to measure directly the heat of combustion of carbon to form carbon monoxide. Combustion of 1 mol of carbon with $\frac{1}{2}$ mol of O_2 produces not only CO but also CO_2, leaving some carbon unreacted. However, both solid carbon and carbon monoxide can be completely burned in O_2 to produce CO_2. We can use the enthalpy changes of these reactions to calculate the heat of combustion of C to CO, as shown in Sample Exercise 5.4.

SAMPLE EXERCISE 5.4

The heat of combustion of C to CO_2 is -393.5 kJ/mol C, and the heat of combustion of CO to CO_2 is -283.0 kJ/mol CO:

$$(1) \quad C(s) + O_2(g) \longrightarrow CO_2(g) \qquad \Delta H = -393.5\ \text{kJ}$$

$$(2) \quad CO(g) + \tfrac{1}{2}O_2(g) \longrightarrow CO_2(g) \qquad \Delta H = -283.0\ \text{kJ}$$

Using these data, calculate the heat of combustion of C to CO:

$$(3) \quad C(s) + \tfrac{1}{2}O_2(g) \longrightarrow CO(g)$$

Solution: In order to use equations (1) and (2), we need to arrange them so that C(*s*) is on the reactant side and CO(*g*) is on the product side of the arrow, as we see in the target reaction, equation (3). Because equation (1) has C(*s*) as a reactant, we can use that equation just as it is. Notice that the target reaction has CO as a product. Thus, we need to turn equation (2) around so that CO(*g*) is a product. Remember that when reactions are turned around, the sign of ΔH is reversed. We arrange the two equations so that they can be added to give the target equation:

$$\begin{array}{rcll} C(s) + O_2(g) & \longrightarrow & CO_2(g) & \Delta H = -393.5 \text{ kJ} \\ CO_2(g) & \longrightarrow & CO(g) + \frac{1}{2}O_2(g) & \Delta H = 283.0 \text{ kJ} \\ \hline C(s) + \frac{1}{2}O_2(g) & \longrightarrow & CO(g) & \Delta H = -110.5 \text{ kJ} \end{array}$$

When we add the two equations, $CO_2(g)$ appears on both sides of the arrow and therefore cancels out. Likewise, $\frac{1}{2}O_2(g)$ is eliminated from each side when the two equations are summed.

PRACTICE EXERCISE

Carbon occurs in two forms, graphite and diamond. The heat of combustion of graphite is -393.5 kJ/mol, and that of diamond is -395.4 kJ/mol:

$$\text{C(graphite)} + O_2(g) \longrightarrow CO_2(g) \qquad \Delta H = -393.5 \text{ kJ}$$

$$\text{C(diamond)} + O_2(g) \longrightarrow CO_2(g) \qquad \Delta H = -395.4 \text{ kJ}$$

Calculate ΔH for the conversion of graphite to diamond:

$$\text{C(graphite)} \longrightarrow \text{C(diamond)}$$

Answer: $+1.9$ kJ

SAMPLE EXERCISE 5.5

Calculate ΔH for the reaction

$$2C(s) + H_2(g) \longrightarrow C_2H_2(g)$$

given the following reactions and their respective enthalpy changes:

$$C_2H_2(g) + \tfrac{5}{2}O_2(g) \longrightarrow 2CO_2(g) + H_2O(l) \qquad \Delta H = -1299.6 \text{ kJ}$$

$$C(s) + O_2(g) \longrightarrow CO_2(g) \qquad \Delta H = -393.5 \text{ kJ}$$

$$H_2(g) + \tfrac{1}{2}O_2(g) \longrightarrow H_2O(l) \qquad \Delta H = -285.9 \text{ kJ}$$

Solution: Because the target equation has C_2H_2 as a product, we turn the first equation around; the sign of ΔH is therefore changed. Because the target equation has 2C(*s*) as a reactant, we multiply the second equation and its ΔH by 2. Because the target equation has H_2 as a reactant, we keep the third equation as it is. We then add the three equations and their enthalpy changes in accordance with Hess's law:

$$\begin{array}{rcll} 2CO_2(g) + H_2O(l) & \longrightarrow & C_2H_2(g) + \frac{5}{2}O_2(g) & \Delta H = 1299.6 \text{ kJ} \\ 2C(s) + 2O_2(g) & \longrightarrow & 2CO_2(g) & \Delta H = -787.0 \text{ kJ} \\ H_2(g) + \frac{1}{2}O_2(g) & \longrightarrow & H_2O(l) & \Delta H = -285.9 \text{ kJ} \\ \hline 2C(s) + H_2(g) & \longrightarrow & C_2H_2(g) & \Delta H = 226.7 \text{ kJ} \end{array}$$

When the equations are added, there are $2CO_2$, $\frac{5}{2}O_2$, and H_2O on both sides of the arrow. These are canceled in writing the net equation.

PRACTICE EXERCISE

Calculate ΔH for the reaction

$$NO(g) + O(g) \longrightarrow NO_2(g)$$

given the following information:

$$NO(g) + O_3(g) \longrightarrow NO_2(g) + O_2(g) \qquad \Delta H = -198.9 \text{ kJ}$$

$$O_3(g) \longrightarrow \tfrac{3}{2}O_2(g) \qquad \Delta H = -142.3 \text{ kJ}$$

$$O_2(g) \longrightarrow 2O(g) \qquad \Delta H = 495.0 \text{ kJ}$$

Answer: -304.1 kJ

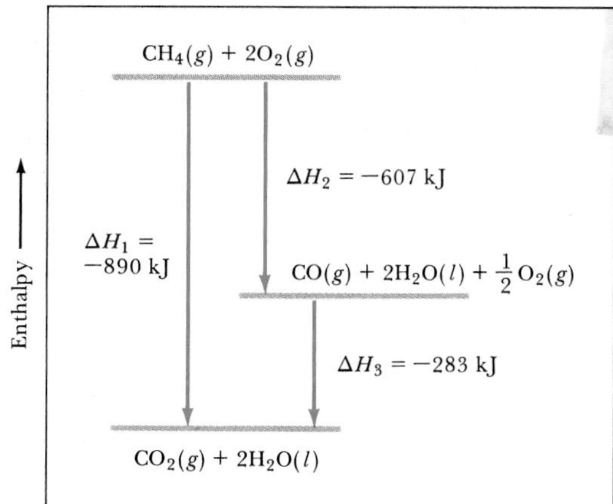

Figure 5.14 The quantity of heat generated by combustion of CH_4 is independent of whether the reaction takes place in one or more steps ($\Delta H_1 = \Delta H_2 + \Delta H_3$).

The first law of thermodynamics, in the form of Hess's law, teaches us that we can never expect to obtain more (or less) energy from a chemical reaction by changing the method of carrying out the reaction. For example, for the reaction of methane, CH_4, and oxygen, O_2, to form CO_2 and H_2O, we may envision the reaction to occur either directly or with the initial formation of CO, which is then subsequently combusted. This set of choices is illustrated in Figure 5.14. Because ΔH is a state function, either path produces the same change in the enthalpy content of the system. That is, $\Delta H_1 = \Delta H_2 + \Delta H_3$. Can we now obtain a series of ΔH values that will allow us to use Hess's law to calculate enthalpy changes for most common chemical reactions? The answer is yes, as we discuss in the next section.

5.6 HEATS OF FORMATION

By using the methods we have just discussed, we can calculate the enthalpy changes for a great many reactions from a few tabulated values. Many experimental data are tabulated according to the type of process. For example, extensive tables exist of heats of vaporization (ΔH for converting liquids to gas), heats of fusion (ΔH for melting solids), heats of combustion (ΔH for combusting a substance in oxygen), and so forth. A particularly important process used for tabulating thermochemical data is the formation of a compound from its constituent elements. The enthalpy change associated with this process is called the **heat of formation**, labeled ΔH_f.

The magnitude of any enthalpy change depends on the conditions of temperature, pressure, and state (gas, liquid, solid, crystalline form) of the reactants and products. A standard enthalpy change, ΔH°, is one that takes place with all reactants and products in their **standard states.** That is, all substances are in the forms most stable at the particular temperature of interest and at standard atmospheric pressure (1 atm; see Section 10.2). The temperature usually chosen for purposes of tabulating data is 298 K (25°C). Thus, the **standard heat of formation** of a compound, ΔH_f°, is the change in enthalpy that accompanies the formation of 1 mol of that substance from its elements, with all substances in their standard states. For example, the standard heat of formation for ethanol, C_2H_5OH, is the enthalpy change for the following reaction:

$$2C(\text{graphite}) + 3H_2(g) + \tfrac{1}{2}O_2(g) \longrightarrow C_2H_5OH(l) \qquad \Delta H_f^\circ = -277.7 \text{ kJ} \quad [5.15]$$

Table 5.1 Standard Heats of Formation, ΔH_f°, at 25°C

Substance	Formula	ΔH_f° (kJ/mol)	Substance	Formula	ΔH_f° (kJ/mol)
Acetylene	$C_2H_2(g)$	226.7	Hydrogen chloride	$HCl(g)$	−92.30
Ammonia	$NH_3(g)$	−46.19	Hydrogen fluoride	$HF(g)$	−268.6
Benzene	$C_6H_6(l)$	49.04	Hydrogen iodide	$HI(g)$	25.9
Calcium carbonate	$CaCO_3(s)$	−1207.1	Methane	$CH_4(g)$	−74.85
Calcium oxide	$CaO(s)$	−635.5	Methanol	$CH_3OH(l)$	−238.6
Carbon dioxide	$CO_2(g)$	−393.5	Propane	$C_3H_8(g)$	−103.85
Carbon monoxide	$CO(g)$	−110.5	Silver chloride	$AgCl(s)$	−127.0
Diamond	$C(s)$	1.88	Sodium bicarbonate	$NaHCO_3(s)$	−947.7
Ethane	$C_2H_6(g)$	−84.68	Sodium carbonate	$Na_2CO_3(s)$	−1130.9
Ethanol	$C_2H_5OH(l)$	−277.7	Sodium chloride	$NaCl(s)$	−411.0
Ethylene	$C_2H_4(g)$	52.30	Sucrose	$C_{12}H_{22}O_{11}(s)$	−2221
Glucose	$C_6H_{12}O_6(s)$	−1273	Water	$H_2O(l)$	−285.8
Hydrogen bromide	$HBr(g)$	−36.23	Water vapor	$H_2O(g)$	−241.8

The elemental source of oxygen is O_2, not O or O_3, because O_2 is the stable form of oxygen at 25°C and standard atmospheric pressure. Similarly, the elemental source of carbon is graphite and not diamond, because the former is the stable (lower-energy) form at 25°C and standard atmospheric pressure. The conversion of graphite to diamond requires the addition of energy:

$$\text{C(graphite)} \longrightarrow \text{C(diamond)} \qquad \Delta H^\circ = 1.88 \text{ kJ} \qquad [5.16]$$

The most stable form of hydrogen under standard conditions is $H_2(g)$, so this is used as the source of hydrogen in Equation 5.15.

The stoichiometry of formation reactions is always such that 1 mol of the desired product is produced, as in Equation 5.15. As a result, heats of formation are reported in kJ/mol, where we understand that it is heat per mole of product. Several standard heats of formation are given in Table 5.1. A more complete table is provided in Appendix C. *By definition, the standard heat of formation of the most stable form of any element is zero.* Thus, ΔH_f° for C(graphite), $H_2(g)$, and $O_2(g)$ are zero by definition.

Using Heats of Formation to Calculate Heats of Reaction

We can determine the standard enthalpy change for any reaction, ΔH_{rxn}°, by using standard heats of formation and Hess's law. We sum the heats of formation of all reaction products, taking care to multiply each molar heat of formation by the coefficient of that substance in the balanced equation. From this, we subtract a similar sum of the heats of formation for the reactants:

Simplification

$$\Delta H_{rxn}^\circ = \sum n\,\Delta H_f^\circ(\text{products}) - \sum m\,\Delta H_f^\circ(\text{reactants}) \qquad [5.17]$$

The symbol $\sum$ (sigma) means "the sum of," and n and m are the stoichiometric coefficients of the chemical reaction. For example, we can calculate ΔH° for the combustion of propane from the ΔH_f° values of the products and reactants:

$$C_3H_8(g) + 5O_2(g) \longrightarrow 3CO_2(g) + 4H_2O(l) \qquad [5.18]$$

$$\Delta H_{rxn}^\circ = [3\,\Delta H_f^\circ(CO_2) + 4\,\Delta H_f^\circ(H_2O)] - [\Delta H_f^\circ(C_3H_8) + 5\,\Delta H_f^\circ(O_2)] \qquad [5.19]$$

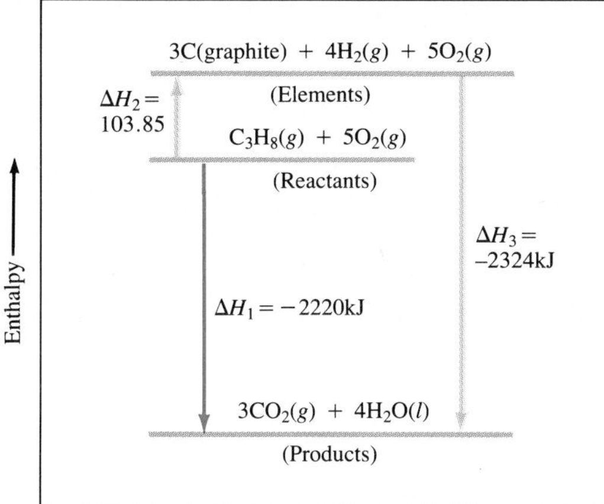

Figure 5.15 Because ΔH is a state function, $\Delta H_1 = \Delta H_2 + \Delta H_3$. Note that ΔH_2 is $-\Delta H_f^\circ(C_3H_8)$, whereas ΔH_3 is $4\Delta H_f^\circ(H_2O) + 3\Delta H_f^\circ(CO_2)$. That is, the enthalpy change for combustion of C_3H_8 equals the sum of the enthalpy changes for converting the reactants into the standard states of the elements and then forming the products from the elements. This same result is given by Equation 5.19, because $\Delta H_f^\circ(O_2)$ is zero.

Because $O_2(g)$ is the most stable form of the element, $\Delta H_f^\circ(O_2) = 0$. The other heats of formation can be obtained from Table 5.1, giving

$$\begin{aligned}\Delta H^\circ_{\text{rxn}} &= \left[(3\text{ mol CO}_2)\left(-393.5\,\frac{\text{kJ}}{\text{mol CO}_2}\right) + (4\text{ mol H}_2\text{O})\left(-285.8\,\frac{\text{kJ}}{\text{mol H}_2\text{O}}\right)\right]\\ &\quad - \left[(1\text{ mol C}_3\text{H}_8)\left(-103.85\,\frac{\text{kJ}}{\text{mol C}_3\text{H}_8}\right) + (5\text{ mol O}_2)\left(0\,\frac{\text{kJ}}{\text{mol O}_2}\right)\right]\\ &= (-2324\text{ kJ}) - (-103.85\text{ kJ})\\ &= -2220\text{ kJ}\end{aligned}$$

The general relationship expressed in Equation 5.17 is just an application of Hess's law. It follows directly from the fact that H is a state function. This fact allows calculation of the enthalpy change for any reaction from the energies required to convert the initial reactants into elements and then combine the elements into the desired products. This reaction pathway is shown in Figure 5.15 for the combustion of propane.

SAMPLE EXERCISE 5.6

Compare the quantity of heat produced by combustion of 1.00 g of propane, C_3H_8, with that produced by 1.00 g of benzene, C_6H_6.

Solution: The example worked in the text gave $\Delta H^\circ = -2220$ kJ for the combustion of a mole of propane. The molecular weight of propane is 44 amu. Therefore, the heat produced per gram is

$$(-2220\text{ kJ/mol})\left(\frac{1\text{ mol}}{44.0\text{ g}}\right) = -50.5\text{ kJ/g}$$

For 1 mol of benzene:

$$C_6H_6(l) + \tfrac{15}{2}O_2(g) \longrightarrow 6CO_2(g) + 3H_2O(l)$$

$$\begin{aligned}\Delta H^\circ_{\text{rxn}} &= [6\,\Delta H_f^\circ(CO_2) + 3\,\Delta H_f^\circ(H_2O)] - [\Delta H_f^\circ(C_6H_6) + \tfrac{15}{2}\Delta H_f^\circ(O_2)]\\ &= 6(-393.5\text{ kJ}) + 3(-285.8\text{ kJ}) - (49.04\text{ kJ})\\ &= (-2361 - 857.4 - 49.04)\text{ kJ}\\ &= -3267\text{ kJ}\end{aligned}$$

The molecular weight of benzene is 78 amu. Therefore, the heat produced per gram of benzene is

$$(-3267 \text{ kJ/mol})\left(\frac{1 \text{ mol}}{78.0 \text{ g}}\right) = -41.9 \text{ kJ/g}$$

Both propane and benzene are hydrocarbons. As a rule of thumb, the energy obtained from the combustion of a gram of hydrocarbon is between 40 and 50 kJ.

PRACTICE EXERCISE

Using the standard heats of formation listed in Table 5.1, calculate the enthalpy change for the combustion of 1 mol of ethanol:

$$C_2H_5OH(l) + 3O_2(g) \longrightarrow 2CO_2(g) + 3H_2O(l)$$

Answer: -1366.7 kJ

SAMPLE EXERCISE 5.7

The standard enthalpy change for the reaction

$$CaCO_3(s) \longrightarrow CaO(s) + CO_2(g)$$

is 178.1 kJ. From the values for the standard heats of formation given in Table 5.1, calculate the standard heat of formation of $CaCO_3(s)$.

Solution: The standard enthalpy change in the reaction is

$$\Delta H^\circ_{rxn} = [\Delta H^\circ_f(CaO) + \Delta H^\circ_f(CO_2)] - \Delta H^\circ_f(CaCO_3)$$

Inserting the known values, we have

$$178.1 \text{ kJ} = -635.5 \text{ kJ} - 393.5 \text{ kJ} - \Delta H^\circ_f(CaCO_3)$$

Solving for $\Delta H^\circ_f(CaCO_3)$, we have

$$\Delta H^\circ_f(CaCO_3) = -1207.1 \text{ kJ/mol}$$

PRACTICE EXERCISE

Given the following standard heat of reaction, use the standard heats of formation in Table 5.1 to calculate the standard heat of formation of $CuO(s)$:

$$CuO(s) + H_2(g) \longrightarrow Cu(s) + H_2O(l) \qquad \Delta H^\circ = -130.6 \text{ kJ}$$

Answer: -155.2 kJ/mol

5.7 CALORIMETRY

The value of ΔH can be determined experimentally by measuring the heat flow accompanying a reaction at constant pressure. When heat flows into or out of a substance, the temperature of the substance changes. Experimentally, we can determine the heat flow associated with a chemical reaction by measuring the temperature change it produces. The measurement of heat flow is called **calorimetry;** an apparatus that measures heat flow is called a **calorimeter**.

The temperature change experienced by a body when it absorbs a certain amount of heat is determined by its **heat capacity**—the energy required to raise the temperature of the body by 1°C. An object of large heat capacity requires more heat to undergo a certain temperature change than does an object of smaller heat capacity. For example, considerably more heat is required to raise the temperature of water in an outdoor swimming pool from 15°C to 25°C than to produce the same temperature change in a fish tank. The heat capacity, C, of an object can be expressed as the ratio $q/\Delta T$, where q is the total heat

Table 5.2 Specific Heats of Some Selected Substances

Compound	Temperature (°C)	Specific heat (J/g-°C)
$H_2O(l)$	15	4.184
$H_2O(s)$	−11	2.03
$Al(s)$	20	0.89
$C(s)$	20	0.71
$Fe(s)$	20	0.45
$Hg(l)$	20	0.14
$CaCO_3(s)$	0	0.85
$MgO(s)$	0	0.87
$HgS(s)$	0	0.21

flow into or out of an object and ΔT is the temperature change produced. Thus, the heat flow equals the heat capacity times the temperature change:

$$q = C \times \Delta T \qquad [5.20]$$

The heat capacity of an object depends on its mass and its composition. For pure substances, the heat capacity is usually given for a specified amount of the substance. The heat capacity of 1 mol of a substance is called the **molar heat capacity,** which is expressed in units of joules per mole per degree (J/mol-°C). The heat capacity for 1 g of a substance is called the **specific heat capacity,** or more commonly the **specific heat.*** Specific heat has the units J/g-°C. If we know the number of moles or the mass of a substance, we can use the molar heat capacity or specific heat to determine the quantity of heat transferred when the temperature change is ΔT:

$$\begin{aligned} q &= (\text{moles of substance}) \times (\text{molar heat capacity}) \times \Delta T \\ &= (\text{grams of substance}) \times (\text{specific heat}) \times \Delta T \end{aligned} \qquad [5.21]$$

The specific heats of several substances are listed in Table 5.2.† As we saw in Section 5.1, the calorie is defined as the amount of heat needed to raise the temperature of 1 g of liquid water by 1°C. Thus, the specific heat of $H_2O(l)$ is 1 cal/g-°C = 4.184 J/g-°C. The specific heat of liquid water is one of the highest known; it is, for example, about five times as great as that of aluminum metal. Thus, 4.184 J of heat will increase the temperature of 1 g of aluminum by nearly 5°C, but it will raise the temperature of water by only 1°C. The high specific heat of water makes our body's important task of maintaining a constant body temperature, 37°C, much easier. The adult body is about 60 percent water by mass and consequently has the ability to absorb or release considerable energy with little effect on its temperature.

* Specific heat was originally defined as a ratio: the quantity of heat needed to raise the temperature of 1 g of a substance by 1°C divided by the quantity of heat needed to raise the temperature of 1 g of water by 1°C. This definition would make specific heat a dimensionless quantity. The definition given above (the heat capacity per gram of substance) is more commonly used now, however.

† The specific heat for a given form (solid, liquid, gas) of a substance can, in principle, vary with the temperature at which the measurement is made. However, these variations are small for most substances, and we will usually ignore them.

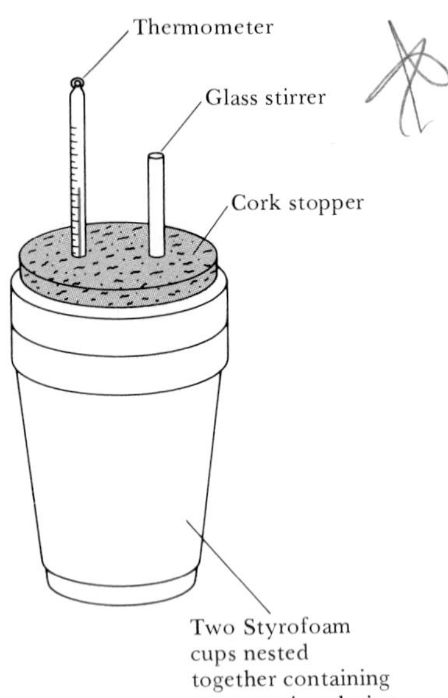

Figure 5.16 "Coffee cup" calorimeter, in which reactions occur at constant pressure.

SAMPLE EXERCISE 5.8

The specific heat of iron(III) oxide, Fe_2O_3, is 0.75 J/g-°C. **(a)** What is the heat capacity of a 2.00-kg brick of Fe_2O_3? **(b)** What quantity of heat is required to increase the temperature of 1.75 g of Fe_2O_3 from 25°C to 380°C?

Solution: **(a)** To obtain the total heat capacity of the brick we simply multiply its mass by its specific heat. Because 2.00 kg = 2.00×10^3 g, we have

$$C = (2.00 \times 10^3 \text{ g})(0.75 \text{ J/g-°C}) = 1.5 \times 10^3 \text{ J/°C}$$

(b) Using Equation 5.21, we have

$$\begin{aligned} q &= (\text{mass of } Fe_2O_3) \times (\text{specific heat of } Fe_2O_3) \times \Delta T \\ &= (1.75 \text{ g})(0.75 \text{ J/g-°C})(355\text{°C}) \\ &= 4.7 \times 10^2 \text{ J} \end{aligned}$$

PRACTICE EXERCISE

How much heat, in kilojoules, is required to increase the temperature of 150 g of water from 25°C to 42°C? The specific heat of water is 4.18 J/g-°C.
Answer: 11 kJ

Constant-Pressure Calorimetry

The techniques and equipment employed in calorimetry depend on the nature of the process being studied. For many reactions, such as those occurring in solution, it is a simple matter to control pressure so that ΔH can be measured directly. (Recall that $\Delta H = q_P$, Equation 5.10.) Although the calorimeters used for highly accurate work are precision instruments, a very simple "coffee cup" calorimeter, such as that shown in Figure 5.16, is often used in general chemistry labs to illustrate the principles of calorimetry. Because the calorimeter is not sealed, the reaction occurs under the essentially constant pressure of the atmosphere. The heat of a reaction is determined from the temperature change of a known quantity of solution in the calorimeter, as shown in Sample Exercise 5.9.

SAMPLE EXERCISE 5.9

When a student mixes 50 mL of 1.0 *M* HCl and 50 mL of 1.0 *M* NaOH in a "coffee cup" calorimeter, the temperature of the resultant solution increases from 21.0°C to 27.5°C. Assuming that the calorimeter absorbs only a negligible quantity of heat, that the total volume of the solution is 100 mL, that its density is 1.0 g/mL, and that its specific heat is 4.18 J/g-°C, calculate the enthalpy change for the reaction

$$HCl(aq) + NaOH(aq) \longrightarrow H_2O(l) + NaCl(aq)$$

Solution: Because the total volume of the solution is 100 mL, the mass of the solution is

$$(100 \text{ mL})(1.0 \text{ g/mL}) = 100 \text{ g}$$

The temperature change is 27.5°C − 21.0°C = 6.5°C. Because the temperature increases, the reaction must be exothermic. The heat is given by the product of the mass, specific heat, and temperature change (Equation 5.21):

$$\begin{aligned} q_P = -\Delta H &= (100 \text{ g})(4.18 \text{ J/g-°C})(6.5\text{°C}) \\ &= 2700 \text{ J} = 2.7 \text{ kJ} \end{aligned}$$

This is the heat gained by the water. Thus, the heat lost by the system (the reacting chemicals) is $q = -2.7$ kJ. To put the enthalpy change on a molar basis, we use the fact that the quantity of HCl and of NaOH is given by the product of the respective solution volumes (50 mL = 0.050 L) and concentrations:

$$(0.050 \text{ L})(1.0 \text{ mol/L}) = 0.050 \text{ mol}$$

Thus, the heat evolved per mole is

$$-2.7 \text{ kJ}/0.050 \text{ mol} = -54 \text{ kJ/mol}$$

PRACTICE EXERCISE

When 50.0 mL of 0.100 *M* $AgNO_3$ and 50.0 mL of 0.100 *M* HCl are mixed in a constant-pressure calorimeter, the temperature of the mixture increases from 22.30°C to 23.11°C. The temperature increase is caused by the following reaction:

$$AgNO_3(aq) + HCl(aq) \longrightarrow AgCl(s) + HNO_3(aq)$$

Calculate ΔH for this reaction, assuming that the combined solution has a mass of 100.0 g and a specific heat of 4.18 J/g-°C. ***Answer:*** −68000 J/mol = −68 kJ/mol

Bomb Calorimetry

One of the most important types of reaction studied by means of calorimetry is combustion (Section 3.2). A compound, usually an organic compound, is allowed to react completely with excess oxygen. Equation 5.12 is an example of such a reaction. Combustion reactions are most conveniently studied by means of a **bomb calorimeter**, a device shown schematically in Figure 5.17. The substance to be studied is placed in a small cup within a sealed vessel called a *bomb*. The bomb, which is designed to withstand high pressures, has an inlet valve for adding oxygen and also has electrical contacts to initiate the combustion reaction. After the sample has been placed in the bomb, the bomb is sealed and pressurized with oxygen. It is then placed in the calorimeter, which is essentially an insulated container, and covered with an accurately measured quantity of water. When all the components within the calorimeter have come to the same temperature, the combustion reaction is initiated by passing an electrical current through a fine wire that is in contact with the sample. When the wire gets sufficiently hot, the sample ignites.

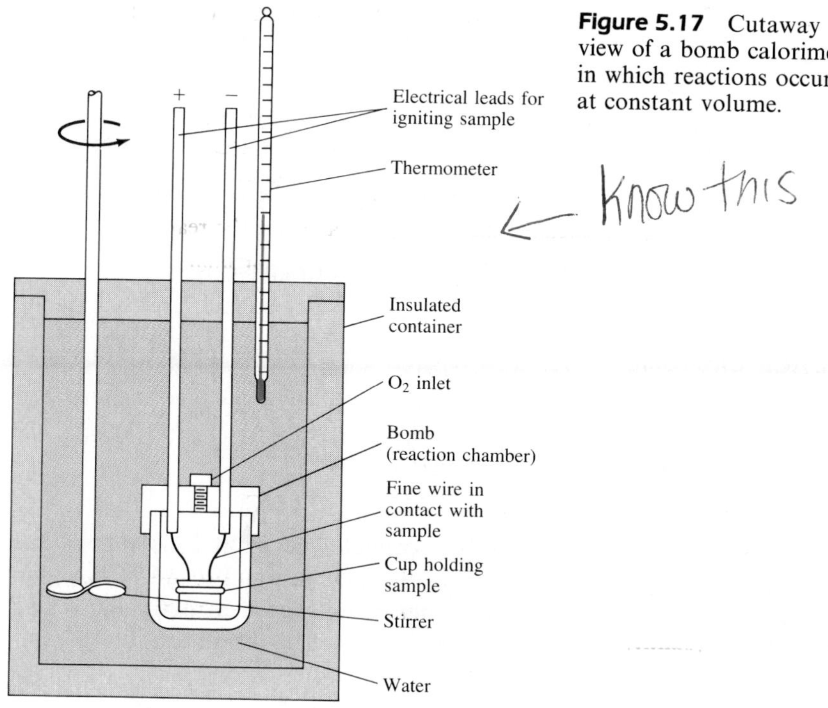

Figure 5.17 Cutaway view of a bomb calorimeter, in which reactions occur at constant volume.

← Know this

Heat is evolved (released) when combustion occurs. This heat is absorbed by the calorimeter contents, causing a rise in the temperature of the water. The temperature of the water is very carefully measured before reaction and then after reaction when the contents of the calorimeter have again arrived at a common temperature. The heat evolved in the combustion of the sample is absorbed by its surroundings, namely the calorimeter contents.

To calculate the heat of combustion from the measured temperature increase in the calorimeter, it is necessary to know the heat capacity of the calorimeter. This is normally ascertained by combusting a sample that gives off a known quantity of heat. For example, it is known that combustion of exactly 1 g of benzoic acid, $C_7H_6O_2$, in a bomb calorimeter produces 26.38 kJ of heat. Suppose that 1 g of benzoic acid is combusted in our calorimeter, and it causes a temperature increase of 5.022°C. The heat capacity of the calorimeter is then given by 26.38 kJ/5.022°C = 5.253 kJ/°C. Once we know the value of the heat capacity of the calorimeter, we can measure temperature changes produced by other reactions, and from these we can calculate the heat, q, evolved in the reaction:

$$q_{\text{evolved}} = -C_{\text{calorimeter}} \times \Delta T \qquad [5.22]$$

SAMPLE EXERCISE 5.10

Hydrazine, N_2H_4, and its derivatives are widely used as rocket fuels. The combustion of hydrazine with oxygen produces $N_2(g)$ and $H_2O(g)$:

$$N_2H_4(l) + O_2(g) \longrightarrow N_2(g) + 2H_2O(g)$$

When 1.00 g of hydrazine is burned in a bomb calorimeter, the temperature of the calorimeter increases by 3.51°C. If the calorimeter has a heat capacity of 5.510 kJ/°C, what is the quantity of heat evolved? What is the heat evolved upon combustion of a mole of N_2H_4?

Solution: The quantity of heat evolved is simply the heat capacity times the temperature change (Equation 5.22).

$$-\left(\frac{5.510 \text{ kJ}}{1°\text{C}}\right)3.51°\text{C} = -19.3 \text{ kJ}$$

Because this is the amount of heat that results from combustion of 1.00 g of hydrazine, the amount released by combustion of 1 mol of N_2H_4 is

$$\left(\frac{-19.3 \text{ kJ}}{1 \text{ g } N_2H_4}\right)\left(\frac{32.0 \text{ g } N_2H_4}{1 \text{ mol } N_2H_4}\right) = -618 \text{ kJ/mol } N_2H_4$$

PRACTICE EXERCISE

A 0.5865-g sample of lactic acid, $HC_3H_5O_3$, is burned in a calorimeter whose heat capacity is 4.812 kJ/°C. The temperature increases from 23.10°C to 24.95°C. Calculate the heat of combustion of lactic acid per gram and then per mole.
Answer: −15.2 kJ/g; −1370 kJ/mol

Because the reactions in a bomb calorimeter occur at constant volume rather than constant pressure, the heats evolved correspond to ΔE rather than to ΔH. The measured ΔE values can readily be converted to ΔH values using Equation 5.9, $\Delta H = \Delta E + P\Delta V$. However, even when the formation or consumption of gases in a reaction leads to large pressure changes, the correction factor $P \Delta V$ is quite small.* For example, the combustion of 1 mol of octane, C_8H_{18}, gives $\Delta E = -5461$ kJ and $\Delta H = -5450$ kJ. Thus, we will not concern ourselves with the correction procedure. The important point is that ΔH can be calculated from measurements made at constant volume.

5.8 FOODS AND FUELS

Most chemical reactions used to produce heat are combustion reactions. The energy released when 1 g of a material is combusted is often called its **fuel value**. Because all heats of combustion are exothermic, it is common to report fuel values without their associated negative sign. The fuel value of any food or fuel can be measured by calorimetry.

Foods

Most of the energy our bodies need comes from carbohydrates and fats. Carbohydrates are decomposed in the intestines into glucose, $C_6H_{12}O_6$. Glucose is soluble in blood and is known as blood sugar. It is transported by the blood to cells, where it reacts with O_2 in a series of steps, eventually producing $CO_2(g)$, $H_2O(l)$, and energy:

$$C_6H_{12}O_6(s) + 6O_2(g) \longrightarrow 6CO_2(g) + 6H_2O(l) \qquad \Delta H^\circ = -2803 \text{ kJ}$$

The breakdown of carbohydrates is rapid, so their energy is quickly supplied to the body. However, the body stores a very small amount of carbohydrates. The average fuel value of carbohydrates is 17 kJ/g (4 kcal/g).

Like carbohydrates, fats produce CO_2 and H_2O in both their metabolism and their combustion in a bomb calorimeter. The reaction of tristearin, $C_{57}H_{110}O_6$, a typical fat, is as follows:

$$2C_{57}H_{110}O_6(s) + 163O_2(g) \longrightarrow 114CO_2(g) + 110H_2O(l) \qquad \Delta H^\circ = -75{,}520 \text{ kJ}$$

The body puts the chemical energy from foods to different uses: to maintain body temperature, to drive muscles, and to construct and repair tissues. Any excess energy is stored as fats. Fats are well suited to serve as the body's energy reserve for at least two reasons: (1) They are insoluble in water, which permits their storage in the body; and (2) they produce more energy per gram than either proteins or carbohydrates, which makes them efficient energy sources on a mass basis. The average fuel value of fats is 38 kJ/g (9 kcal/g).

In the case of proteins, metabolism in the body produces less energy than combustion in a calorimeter does because the products are different. Proteins contain nitrogen, which is released in the bomb calorimeter as N_2. In the body, this nitrogen ends up mainly as urea, $(NH_2)_2CO$. Proteins are used by the body mainly as building materials for organ walls, skin, hair, muscle, and so forth. On the average, the metabolism of proteins produces 17 kJ/g (4 kcal/g), the same as for carbohydrates.

---

* To calculate $P \Delta V$, you need to know about the pressure–volume properties of gases, a subject discussed in Chapter 10. It works out that the $P \Delta V$ term is equal to $RT \Delta n$, where Δn is the total number of moles of gas in the products minus the total number of moles of gas in the reactants, R is the molar gas constant (8.314 J/K-mol), and T is the temperature in kelvins.

Table 5.3 Fuel Values and Compositions of Some Common Foods

	Approximate composition (%)			Fuel value	
	Protein	Fat	Carbohydrate	kJ/g	kcal/g
Apples (raw)	0.4	0.5	13	2.5	0.59
Beer[a]	0.3	0	1.2	1.8	0.42
Bread (white, enriched)	9	3	52	12	2.8
Cheese (cheddar)	28	37	4	20	4.7
Eggs	13	10	0.7	6	1.4
Fudge	2	11	81	18	4.4
Green beans (frozen)	1.9	—	7.0	1.5	0.38
Hamburger	22	30	—	15	3.6
Milk	3.3	4.0	5.0	3.0	0.74
Peanuts	26	39	22	23	5.5

[a] Beers typically contain 3.5 percent ethanol, which has fuel value.

The fuel values for a variety of common foods are shown in Table 5.3. The amount of energy the body requires varies considerably depending on such factors as body weight, age, and muscular activity. When a person is doing average work, his or her daily energy requirement is about 10,000 to 13,000 kJ (2500 to 3000 kcal). This is about the same amount of energy as a 100-W light bulb consumes in operating for a 24-hr period.

SAMPLE EXERCISE 5.11

Scientists estimate that a person of average weight requires the consumption of 100 Cal/mi when running or jogging. What weight of hamburger provides the fuel value requirements for running 3 mi?

Solution: Recall that the dietary Calorie is equivalent to 1 kcal. The running requires 300 Cal, or 300 kcal. Using data from Table 5.3:

$$\text{Hamburger required} = 300 \text{ kcal} \times \left(\frac{1 \text{ g hamburger}}{3.6 \text{ kcal}}\right)$$
$$= 83 \text{ g hamburger}$$

Thus, the calories consumed in running 3 mi are replaced by less hamburger than a "quarter-pounder" offers.

PRACTICE EXERCISE

Dry red beans contain 62 percent carbohydrate, 22 percent protein, 1.5 percent fat, and the remainder water. Estimate the fuel value of these beans. ***Answer:*** 15 kJ/g

Fuels

Several common fuels are compared in Table 5.4. Notice that an increase in the percentage of carbon or hydrogen in the fuel increases the fuel value. For example, the fuel value of bituminous coal is greater than that of wood because of its greater carbon content.

Coal, oil, and natural gas, which are presently our major sources of energy, are known as **fossil fuels**. All are thought to have formed over millions of years from the decomposition of plants and animals. All are presently being depleted far more rapidly than they are being formed. **Natural gas** consists of gaseous

CHEMISTRY AT WORK: Hydrogen as a Fuel

> I believe that water will one day be employed as a fuel, that hydrogen and oxygen which constitutes it will furnish an inexhaustible source of heat and light. / Jules Verne, *The Mysterious Island* (1870).

As the above quotation indicates, hydrogen, H_2, has long been recognized as a very attractive fuel. Hydrogen has a very high fuel value (Table 5.4), and its combustion produces water, a "clean" byproduct with no negative environmental effects. However, hydrogen cannot be used as a primary energy source because there is so little H_2 in nature. Most hydrogen is produced by decomposing water or hydrocarbons. This decomposition requires energy; in fact, because of heat losses, more energy must be used to generate hydrogen than can be reclaimed when the hydrogen is subsequently used as a fuel. However, should large, cheap sources of energy become available (as might be the case in solar-power generation), a portion of this energy could be used to generate hydrogen, which is both portable and storable.

Hydrogen can also be made from the treatment of coal with superheated steam:

$$\text{Coal} + H_2O(g) \longrightarrow CO(g) + H_2(g)$$

The conversion of high-sulfur coal to hydrogen is especially important because the sulfur is removed in this process: the direct burning of high-sulfur coal generates emissions that are particularly harmful to the environment (Chapter 18). Hydrogen can also be generated from methane, the principal component of natural gas, through "steam reforming":

$$CH_4(g) + H_2O(g) \longrightarrow CO(g) + 3H_2(g)$$

Once generated, hydrogen could serve as a convenient energy carrier. It would be cheaper to transport hydrogen using existing natural-gas pipelines than to transport electrical energy. Because present industrial technology is based on combustible fuels, hydrogen could replace petroleum and natural gas as these fuels become scarcer and more expensive. Hydrogen-fueled internal combustion engines suitable for automobiles have already been developed; they are 20 to 50 percent more efficient than gasoline engines, as based on mileage-per-fuel value.

Of course, there are drawbacks to the use of hydrogen as a fuel. Most notable among these is the explosive flammability of hydrogen gas (Figures 5.11 and 5.12). The most promising solution to this problem is the use of metal alloys to store hydrogen. For example, an alloy of iron and titanium acts as a "sponge" in which the hydrogen gas combines with the alloy to form a fine silvery powder. Heating the powder safely releases the hydrogen gas for use as a fuel. Such storage systems are much safer than the storage of hydrogen as a gas or liquid.

hydrocarbons, compounds of hydrogen and carbon. It varies in composition, but it contains primarily methane, CH_4, with small amounts of ethane, C_2H_6, propane, C_3H_8, and butane, C_4H_{10}. We determined the fuel value of propane and benzene (another hydrocarbon) in Sample Exercise 5.6. **Oil**, which is also known as petroleum, is a liquid composed of hundreds of compounds. Most of these compounds are hydrocarbons, with the remainder being mainly organic compounds containing sulfur, nitrogen, or oxygen. **Coal**, which is solid, contains

Table 5.4 Fuel Values and Compositions of Some Common Fuels

	Approximate elemental composition (%)			
	C	**H**	**O**	**Fuel value (kJ/g)**
Wood (pine)	50	6	44	18
Anthracite coal (Pennsylvania)	82	1	2	31
Bituminous coal (Pennsylvania)	77	5	7	32
Charcoal	100	0	0	34
Crude oil (Texas)	85	12	0	45
Gasoline	85	15	0	48
Natural gas	70	23	0	49
Hydrogen	0	100	0	142

hydrocarbons of high molecular weight as well as compounds containing sulfur, oxygen, and nitrogen. The sulfur in oil and coal is important from the standpoint of air pollution, as we shall discuss in Chapter 18.

5.9 ENERGY USAGE: TRENDS AND PROSPECTS

The average energy consumption per person in the United States each day amounts to about 8.8×10^5 kJ. This amount of energy is about 100 times greater than the individual food-energy requirement. We are a very energy-intensive society.

During the past century, the fossil fuels—first coal, later oil and gas—have become the major energy sources in our society. Together they account for more than 80 percent of national energy consumption. Eventually, such fossil fuel energy sources must decline in importance as we exhaust the readily available supplies. Current projections are that 80 percent of all the known and estimated oil reserves will have been consumed before the children born during the 1980s have lived out their lives. Clearly we must develop alternative energy sources to replace fossil fuels if the 8 billion people expected to be living in the year 2025 are to enjoy a decent standard of living.

Of the various sources of energy that could serve as alternatives to fossil fuels, only nuclear and solar energies are potentially capable of furnishing sufficient quantities of energy to satisfy the world's needs. Nuclear power supplied about 4 percent of the gross energy consumed in the United States in 1985. Aspects of the use of nuclear reactors for energy production are discussed in Chapter 21. In the paragraphs that follow, we shall consider some aspects of coal and solar radiation as energy sources.

Coal

Coal is the most abundant fossil fuel; it constitutes 80 percent of the fossil fuel reserves of the United States and 90 percent of those of the world. However, the use of coal presents a number of problems. Coal is a complex mixture of substances, and it contains components that produce air pollution. Because it is a solid, recovery from its underground deposits is expensive and often dangerous. Furthermore, coal deposits are not always close to the locations where energy use is high, so there are often substantial shipping costs as well.

Some experts feel that coal could be used most effectively if it were converted to a gaseous form, often called **syngas** (for "*syn*thesis *gas*"). In such a

Figure 5.18 Basic processes involved in the gasification of coal to form synthesis gas (syngas). A catalyst is a substance that increases the speed of a reaction without itself being consumed in the reaction.

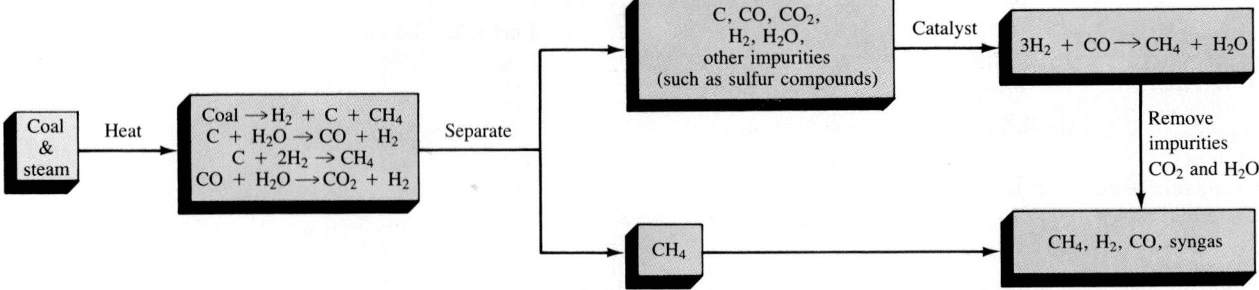

Figure 5.19 View of a large, new plant for conversion of coal to acetic anhydride, $(CH_3CO)_2O$. This plant is designed to process 900 tons of high-sulfur-content Appalachian coal daily, for both feedstock and to furnish plant power. Acetic anhydride is used in the manufacture of photographic film, synthetic fibers, cigarette filters, and coatings. (Eastman Chemical Company)

conversion the sulfur is removed, thereby decreasing air pollution when the syngas is burned. Syngas could be easily transported in pipelines and could supplement our diminishing supplies of natural gas. Gasification of coal requires the addition of hydrogen to coal. Typically, the coal is pulverized and treated with superheated steam. The product contains a mixture of CO, H_2, and CH_4, all of which can be used as fuels. However, conditions are maintained to maximize production of CH_4. A simplified schematic showing some of the reactions that occur is given in Figure 5.18. The syngas produced in coal gasification can also be used to make chemicals. Methanol, CH_3OH, and acetic anhydride, $(CH_3CO)_2O$, are examples of chemicals produced in large quantities from syngas (Figure 5.19).

Solar Energy

Solar energy is the world's largest energy source. The solar energy that falls on only 0.1 percent of the land area of the United States is equivalent to all the energy that this nation currently uses. The problem with the use of solar energy is that it is dilute (because it is distributed over a wide area) and fluctuates with time and weather conditions. The effective use of solar energy will depend on the development of some means of storing the energy collected for use at a later time. Any practical means for doing this will almost certainly involve use of an endothermic chemical process that can be reversed at a later time to release heat. One such reaction is the following.

$$CH_4(g) + H_2O(g) + \text{heat} \rightleftharpoons CO(g) + 3H_2(g) \qquad [5.23]$$

This reaction can be made to proceed in the forward direction at high temperatures, which can be obtained directly in a solar furnace. The CO and H_2 formed in the reaction could then be reacted later, with the resulting evolution of heat put to useful work.

Solar energy can be converted directly into electricity by use of photovoltaic devices, sometimes called *solar cells*. The efficiencies of solar energy conversion by use of such devices have increased dramatically during the past few years as a result of intensive research efforts. Photovoltaics are vital to the generation of power of satellites. However, for large-scale generation of useful energy at the earth's surface, they are not yet practical because of high unit cost. Even if the costs are reduced, some means must be found to store the energy produced by the solar cells, because the sun shines only intermittently and during only part of the day at any place. Once again, the solution to this problem will almost certainly be to use the energy to run a chemical reaction in the direction in which it is endothermic.

FOR REVIEW

SUMMARY

This chapter has focused on energy and the first law of thermodynamics. Thermodynamics is the study of heat, work, and energy, and the rules that govern their interconversions. To study the thermodynamic properties of matter, we define some specific amount of matter as our system and study the interactions between this system and its surroundings. The system may exchange heat with its surroundings, have work done on it by the surroundings, or do work on the surroundings. Work comes in many forms. Mechanical work, w, is measured by the product of a force, F, acting through a distance, d: $w = F \times d$. The term heat refers to the flow of energy between two bodies at different temperatures when they are placed in thermal contact.

The joule (J) is the SI unit of energy: $1\ \text{J} = 1\ \text{kg-m}^2/\text{s}^2$. Another common energy unit is the calorie (cal). A calorie is the quantity of energy necessary to increase the temperature of 1 g of water by 1°C: 1 cal = 4.184 J.

The internal energy, E, of a system is all the energy possessed by the atomic, molecular, or ionic units of which the system is composed. We do not in general know the magnitude of this internal energy; we are concerned primarily with *changes* in the internal energy, ΔE, that accompany changes in the system. The first law of thermodynamics tells us that the change in internal energy is given by the heat added to the system during the change, q, plus the work done on the system by the surroundings: $\Delta E = q + w$. The internal energy is a state function; that is, the value of the internal energy for the system depends only on the condition of the system, as described by variables such as temperature, pressure, and so forth. It does not depend on the details of how the system came to be in that state.

The enthalpy, H, of a system, is a state function defined as $E + PV$, where P is the pressure acting on the system and V is its volume. Enthalpy is a most useful function for dealing with changes that occur at constant pressure; then $\Delta H = \Delta E + P\Delta V$. The enthalpy change for any process is the heat flow into or out of a system maintained at constant pressure, so $q_P = \Delta H$. For endothermic processes (those that absorb heat), ΔH is positive. For exothermic processes (those that transfer heat from the system to the surroundings), ΔH is negative.

Enthalpy is an extensive property; the magnitude of the enthalpy change for any process is proportional to the quantity of reactants. The magnitude of ΔH also depends on the physical states of the reactants and products. If a reaction is reversed, the sign of ΔH is also reversed.

Because enthalpy is a state function, the enthalpy change for a given chemical process is the same whether the process is carried out in one step or in a series of steps. Hess's law states this fact in a very useful form: If a reaction is carried out in a series of steps, ΔH for the reaction will be equal to the sum of the enthalpy changes for the steps. This means that we can calculate the enthalpy changes for any reaction as long as we can find a route for which ΔH is known for each step. Tabulated values for the standard heat of formation, ΔH_f°, are particularly helpful for such calculations. The standard heat of formation of a substance is defined as the enthalpy change for the formation of a mole of that substance from the elements, with all reactants and products in their standard states. The standard state refers to the conditions of 1 atm pressure, with each reactant and product in its stable form (gas, liquid, or solid) at the temperature in question, usually 298 K (25°C). By convention, the enthalpies of formation of elements in their standard states are zero. The enthalpy change in any reaction can be calculated from the heats of formation of the reactants and products in the reaction:

$$\Delta H^\circ_{rxn} = \sum n\,\Delta H^\circ_f(\text{products}) - \sum m\,\Delta H^\circ_f(\text{reactants})$$

The quantity of heat absorbed or evolved in chemical and physical processes is measured experimentally by calorimetry. A calorimeter measures the temperature changes accompanying the process being investigated. The magnitude of the temperature change that an object undergoes when it absorbs a certain quantity of heat depends on its heat capacity. The magnitude of the heat flow is given by the product of the heat capacity and the temperature change: $q = C\,\Delta T$. The molar heat capacity of a substance is the heat required to raise the temperature of 1 mol of the substance by 1°C. The specific heat of a substance is the heat required to raise the temperature of 1 g of the substance by 1°C. Thus, the quantity of heat transferred to a substance is given by the product of its mass, specific heat, and temperature change, or by the product of its mole amount, molar heat capacity, and temperature change. If the system is at constant pressure, the heat flow equals ΔH for the process; if the system is at constant volume, the heat flow equals ΔE. A bomb calorimeter is a constant-volume device used to measure the heat evolved when a substance is combusted in oxygen.

The fuel value of any substance or mixture is defined as the heat energy released when a gram of that material is combusted. We considered briefly the fuel values of some common foods and fuels, including various forms of fossil fuels. Finally, we have considered current and projected rates of fuel consumption and the prospects for various sources of energy in the future.

KEY TERMS

thermodynamics
thermochemistry
force (Sec. 5.1)
work (Sec. 5.1)
energy (Sec. 5.1)
mechanical work (Sec. 5.1)
heat (Sec. 5.1)
kinetic energy (Sec. 5.1)
potential energy (Sec. 5.1)
joule (Sec. 5.1)
power (Sec. 5.1)
calorie (Sec. 5.1)
system (Sec. 5.1)
surroundings (Sec. 5.1)
law of conservation of energy (Sec. 5.2)
first law of thermodynamics (Sec. 5.2)
internal energy (Sec. 5.2)
exothermic (Sec. 5.2)
endothermic (Sec. 5.2)
state function (Sec. 5.2)
enthalpy (Sec. 5.3)
Hess's law (Sec. 5.5)
heat of formation (Sec. 5.6)
standard states (Sec. 5.6)
standard heat of formation (Sec. 5.6)
calorimetry (Sec. 5.7)
calorimeter (Sec. 5.7)
heat capacity (Sec. 5.7)
molar heat capacity (Sec. 5.7)
specific heat capacity (Sec. 5.7)
specific heat (Sec. 5.7)
bomb calorimeter (Sec. 5.7)
fuel value (Sec. 5.8)
fossil fuels (Sec. 5.8)
natural gas (Sec. 5.8)
oil (Sec. 5.8)
coal (Sec. 5.8)
syngas (Sec. 5.9)

EXERCISES

Nature of Energy; the First Law

5.1 What is the difference between force and energy? How do we determine the amount of work done given the magnitude of the associated force?

5.2 For each of the following, identify the force present and explain whether work is being performed: **(a)** A spring is stretched to twice its normal length. **(b)** Two positively charged particles are held a fixed distance from one another. **(c)** A weight lifter lifts a barbell off a weight rack. **(d)** Two magnets are pulled apart.

5.3 **(a)** Calculate the kinetic energy, in joules, of a 58-g tennis ball traveling at 55 m/s. **(b)** Convert this energy to calories. **(c)** What happens to this energy when the ball is struck by a tennis racquet?

5.4 **(a)** Calculate the kinetic energy, in kilojoules, of a car weighing 4.8×10^3 lb if it is traveling at a speed of 45 mi/hr. (Use the necessary conversion factors from the back inside cover of the text.) **(b)** Convert this energy to calories. **(c)** What happens to this energy when the car stops for a red light?

5.5 **(a)** Verbally state the first law of thermodynamics. **(b)** Define the term *system.* **(c)** How is the energy change of a system related to that of its surroundings?

5.6 **(a)** Define the term *surroundings.* **(b)** Will the energy change of a system depend on what is defined to be the system? Explain. **(c)** Write an equation that expresses the first law of thermodynamics.

5.7 Calculate the change in internal energy of the system in each of the following processes. **(a)** A gas expands very rapidly, so that there is no heat exchange with the surroundings; in the expansion it does 450 J of work on the surroundings. **(b)** A chemist heats 200 g of water from 30°C to 40°C, a process that requires approximately 8360 J of heat. **(c)** A gas contracts as it is cooled; it has 300 J of work done on it and loses 146 J of heat to its surroundings.

5.8 Calculate ΔE for the following cases: **(a)** A system absorbs 137 J of heat and does 185 J of work. **(b)** $q = -2.5$ kJ and $w = -850$ J. **(c)** A system releases 128 J of heat while the surroundings does 243 J of work on it. **(d)** $q = 650$ J and $w = -320$ J. In each case, indicate whether the internal energy of the system has increased or decreased.

5.9 **(a)** What is meant by the term *state function?* **(b)** Is the internal energy of a system a state function? **(c)** Is the work done on the system a state function? **(d)** If you know the value of ΔE for a process, can you determine the value of q?

5.10 Indicate which of the following is independent of the path by which a change occurs: **(a)** the change in potential energy when a book is transferred from table to shelf; **(b)** the heat evolved when a cube of sugar is oxidized to $CO_2(g)$ and $H_2O(g)$; **(c)** the work accomplished in burning a gallon of gasoline.

Enthalpy

5.11 Why are constant-pressure processes so important in chemistry? For a system under a constant pressure, what is the relationship between the change in enthalpy and the amount of heat exchanged with the surroundings?

5.12 During a process, the volume of a system remains constant. What is the value of the P-V work for the process? Will the heat exchanged between the system and its surroundings be equal to ΔH or ΔE?

5.13 Classify the following processes as endothermic or exothermic: **(a)** A match burns. **(b)** Ice melts. **(c)** Molten metal solidifies. **(d)** Sodium metal reacts with water. **(e)** Rubbing alcohol evaporates.

5.14 Predict the sign of ΔH for each of the following processes: **(a)** $H_2O(l) \longrightarrow H_2O(g)$; **(b)** $C_2H_4(g) + 3O_2(g) \longrightarrow 2CO_2(g) + 2H_2O(g)$; **(c)** $NaOH(aq) + HCl(aq) \longrightarrow H_2O(l) + NaCl(aq)$ (the temperature of the solution increases); **(d)** $2H_2O(g) \longrightarrow 2H_2(g) + O_2(g)$.

5.15 A gas is confined to a cylinder under constant atmospheric pressure, as illustrated in Figure 5.2. When 600 J of heat is added to the gas, it expands and does 140 J of work on the surroundings. What are the values of ΔH and ΔE for this process?

5.16 A gas is confined to a cylinder under constant atmospheric pressure, as illustrated in Figure 5.2. When the gas undergoes a particular chemical reaction, it releases 135 kJ of heat to its surroundings and does 63 kJ of P-V work on its surroundings. What are the values of ΔH and ΔE for this process?

5.17 Consider the following reaction:

$$2N_2(g) + O_2(g) \longrightarrow 2N_2O(g) \qquad \Delta H = +163.2 \text{ kJ}$$

(a) Is the reaction exothermic or endothermic? **(b)** Draw a diagram similar to Figure 5.10 for this reaction. **(c)** Calculate the amount of heat transferred when 10.0 g of $N_2O(g)$ forms by this reaction at constant pressure. **(d)** How many grams of nitrogen gas must react to produce an enthalpy change of 1.00 kJ? **(e)** How many kilojoules of heat are produced when 15.0 g of $N_2O(g)$ is decomposed into $N_2(g)$ and $O_2(g)$ at constant pressure?

5.18 Consider the following reaction:

$$2Na(s) + Cl_2(g) \longrightarrow 2NaCl(s) \qquad \Delta H = -821.8 \text{ kJ}$$

(a) Is the reaction exothermic or endothermic? **(b)** Draw a diagram similar to Figure 5.10 for this reaction. **(c)** Calculate the amount of heat transferred when 8.0 g of $Na(s)$ reacts according to this reaction at constant pressure. **(d)** How many grams of NaCl are produced during an enthalpy change of 10.0 kJ? **(e)** How many kilojoules of heat are absorbed when 25.0 g of $NaCl(s)$ is decomposed into $Na(s)$ and $Cl_2(g)$ at constant pressure?

5.19 A common means of forming small quantities of oxygen gas in the laboratory used to be to heat $KClO_3$:

$$2KClO_3(s) \longrightarrow 2KCl(s) + 3O_2(g) \qquad \Delta H = -89.4 \text{ kJ}$$

For this reaction, calculate ΔH for the formation of **(a)** 0.345 mol of O_2; **(b)** 7.85 g of KCl; **(c)** 9.22 g of $KClO_3$ from KCl and O_2.

5.20 When solutions containing silver ions and chloride ions are mixed, silver chloride precipitates:

$$Ag^+(aq) + Cl^-(aq) \longrightarrow AgCl(s) \qquad \Delta H = -65.5 \text{ kJ}$$

(a) Calculate ΔH for formation of 0.200 mol of AgCl by this reaction. **(b)** Calculate ΔH for formation of 2.50 g of AgCl. **(c)** Calculate ΔH when 0.350 mol of AgCl dissolves in water.

5.21 Which of the following has the highest enthalpy at a given temperature and pressure: $H_2O(s)$, $H_2O(l)$, or $H_2O(g)$? Which has the lowest enthalpy?

5.22 Consider the following reaction:

$$3O_2(g) \longrightarrow 2O_3(g) \qquad \Delta H = +284.6 \text{ kJ}$$

(a) Under the conditions of this reaction, does $O_2(g)$ or $O_3(g)$ have the higher enthalpy? **(b)** Would you expect this reaction to proceed readily? Explain.

Hess's Law

5.23 State Hess's law. Why is it important to thermochemistry?

5.24 What is the connection between the fact that H is a state function and Hess's law?

5.25 From the following heats of reaction:

$$2SO_2(g) + O_2(g) \longrightarrow 2SO_3(g) \qquad \Delta H = -196 \text{ kJ}$$

$$2S(s) + 3O_2(g) \longrightarrow 2SO_3(g) \qquad \Delta H = -790 \text{ kJ}$$

calculate the heat of the reaction

$$S(s) + O_2(g) \longrightarrow SO_2(g)$$

5.26 From the following heats of reaction:

$$N_2(g) + 2O_2(g) \longrightarrow 2NO_2(g) \qquad \Delta H = +67.6 \text{ kJ}$$

$$NO(g) + \tfrac{1}{2}O_2(g) \longrightarrow NO_2(g) \qquad \Delta H = -56.6 \text{ kJ}$$

calculate the heat of the reaction

$$N_2(g) + O_2(g) \longrightarrow 2NO(g)$$

5.27 From the following heats of reaction:

$$H_2(g) + F_2(g) \longrightarrow 2HF(g) \qquad \Delta H = -537 \text{ kJ}$$

$$C(s) + 2F_2(g) \longrightarrow CF_4(g) \qquad \Delta H = -680 \text{ kJ}$$

$$2C(s) + 2H_2(g) \longrightarrow C_2H_4(g) \qquad \Delta H = +52.3 \text{ kJ}$$

calculate the heat of reaction of ethylene with F_2:

$$C_2H_4(g) + 6F_2(g) \longrightarrow 2CF_4(g) + 4HF(g)$$

5.28 Given the following data:

$$2C_2H_6(g) + 7O_2(g) \longrightarrow 4CO_2(g) + 6H_2O(l) \qquad \Delta H = -3120 \text{ kJ}$$

$$C(s) + O_2(g) \longrightarrow CO_2(g) \qquad \Delta H = -394 \text{ kJ}$$

$$2H_2(g) + O_2(g) \longrightarrow 2H_2O(l) \qquad \Delta H = -572 \text{ kJ}$$

use Hess's law to calculate ΔH for the reaction

$$2C(s) + 3H_2(g) \longrightarrow C_2H_6(g)$$

Heats of Formation

5.29 Write the balanced equations that describe the formation of the following compounds from their elements in their standard states (as was done for C_2H_5OH in Equation 5.15) and use Appendix C to obtain the heat of formation of each: **(a)** $FeCl_3(s)$; **(b)** $BaCO_3(s)$; **(c)** $NOCl(g)$; **(d)** $NH_3(g)$.

5.30 Write the balanced equations that describe the formation of the following compounds from their elements in their standard states and use Appendix C to obtain the heat of formation of each: **(a)** $NaHCO_3(s)$; **(b)** $Fe_3O_4(s)$; **(c)** $CH_3COOH(l)$; **(d)** $HI(g)$.

5.31 Nitric acid is obtained when nitrogen dioxide dissolves in water:

$$3NO_2(g) + H_2O(l) \longrightarrow 2HNO_3(aq) + NO(g)$$

Using heats of formation from Appendix C, calculate $\Delta H°$ for this reaction.

5.32 The following reaction is known as the thermite reaction:

$$2Al(s) + Fe_2O_3(s) \longrightarrow Al_2O_3(s) + 2Fe(s)$$

This highly exothermic reaction is used for welding massive units such as propellers for large ships. Using heats of formation from Appendix C, calculate $\Delta H°$ for this reaction.

5.33 Nitric oxide reacts with oxygen to form nitrogen dioxide:

$$2NO(g) + O_2(g) \longrightarrow 2NO_2(g) \qquad \Delta H° = -113.1 \text{ kJ}$$

$\Delta H_f°$ for $NO_2(g)$ is 33.8 kJ/mol. Calculate $\Delta H_f°$ for $NO(g)$.

5.34 Ferrous chloride reacts with chlorine to form ferric chloride:

$$2FeCl_2(s) + Cl_2(g) \longrightarrow 2FeCl_3(s) \qquad \Delta H° = -115.4 \text{ kJ}$$

$\Delta H_f°$ for $FeCl_3(s)$ is -400 kJ/mol. Calculate $\Delta H_f°$ for $FeCl_2(s)$.

5.35 Using values from Appendix C, calculate the standard enthalpy change for each of the following reactions:
(a) $CO(g) + 2NH_3(g) \longrightarrow NH_4CN(s) + H_2O(g)$;
(b) $NH_4NO_3(s) \longrightarrow N_2O(g) + 2H_2O(g)$;
(c) $P_4O_6(s) + 2O_2(g) \longrightarrow P_4O_{10}(s)$;
(d) $KClO_3(s) + 3PCl_3(l) \longrightarrow 3POCl_3(l) + KCl(s)$.

5.36 Using values from Appendix C, calculate the standard enthalpy change for each of the following reactions:
(a) $2NOCl(g) \longrightarrow 2NO(g) + Cl_2(g)$;
(b) $N_2(g) + 3H_2(g) \longrightarrow 2NH_3(g)$;
(c) $CH_3COOH(l) \longrightarrow CH_4(g) + CO_2(g)$;
(d) $4NH_3(g) + 5O_2(g) \longrightarrow 4NO(g) + 6H_2O(g)$.

5.37 Calcium carbide, CaC_2, reacts with water to form acetylene, C_2H_2, and $Ca(OH)_2$. From the following heat of reaction data and the data in Appendix C, calculate $\Delta H_f°$ for $CaC_2(s)$:

$$CaC_2(s) + 2H_2O(l) \longrightarrow Ca(OH)_2(s) + C_2H_2(g) \qquad \Delta H° = -127.2 \text{ kJ}$$

5.38 Complete combustion of acetone, C_3H_6O, results in the liberation of 1790 kJ:

$$C_3H_6O(l) + 4O_2(g) \longrightarrow 3CO_2(g) + 3H_2O(l) \qquad \Delta H° = -1790 \text{ kJ}$$

Using this information together with the data in Appendix C, calculate the heat of formation of acetone.

Calorimetry

5.39 **(a)** What is the specific heat of water? **(b)** What is the heat capacity of 348 g of water? **(c)** How many kilojoules of heat are needed to raise the temperature of 2.06 kg of water from 35.14°C to 76.37°C?

5.40 **(a)** What is the molar heat capacity of water? **(b)** What is the heat capacity of 6.35 mol of water? **(c)** How many kilojoules of heat are needed to raise the temperature of 165 mol of water from 10.55°C to 47.32°C?

5.41 The specific heat of ethanol is 2.46 J/g-°C. How many joules of heat are required to heat 193 g of ethanol from 19.00°C to 35.00°C?

5.42 The specific heat of lead is 0.129 J/g-°C. How many joules of heat are required to raise the temperature of 382 g of lead from 22.50°C to 37.20°C?

5.43 When a 6.50-g sample of solid sodium hydroxide dissolves in 100.0 g of water in a "coffee cup" calorimeter (Figure 5.16), the temperature rises from 21.6°C to 37.8°C. Calculate ΔH (in kJ/mol NaOH) for the solution process

$$NaOH(s) \longrightarrow Na^+(aq) + OH^-(aq)$$

Assume that the specific heat of the solution is the same as that of pure water.

5.44 When a 4.25-g sample of solid ammonium nitrate dissolves in 60.0 g of water in a "coffee cup" calorimeter (Figure 5.16), the temperature drops from 22.0°C to 16.9°C. Calculate ΔH (in kJ/mol NH_4NO_3) for the solution process

$$NH_4NO_3(s) \longrightarrow NH_4^+(aq) + NO_3^-(aq)$$

Assume that the specific heat of the solution is the same as that of pure water.

5.45 A 1.80-g sample of octane, C_8H_{18}, was burned in a bomb calorimeter whose total heat capacity is 11.66 kJ/°C. The temperature of the calorimeter and its contents increased from 21.36°C to 28.78°C. What is the heat of combustion per gram of octane? Per mole of octane?

5.46 A 2.20-g sample of quinone, $C_6H_4O_2$, is burned in a bomb calorimeter whose total heat capacity is 7.854 kJ/°C. The temperature of the calorimeter increases from 23.44°C to 30.57°C. What is the heat of combustion per gram of quinone? Per mole of quinone?

5.47 Under constant-volume conditions, the heat of combustion of benzoic acid, $HC_7H_5O_2$, is 26.38 kJ/g. A 1.200-g sample of benzoic acid is burned in a bomb calorimeter. The temperature of the calorimeter increased from 22.45°C to 26.10°C. **(a)** What is the total heat capacity of the calorimeter? **(b)** If the calorimeter contained 1.500 kg of water, what is the heat capacity of the calorimeter when it contains no water? **(c)** What temperature increase would be expected in this calorimeter if the 1.200-g sample of benzoic acid were combusted when the calorimeter contained 1.000 kg of water?

5.48 Under constant-volume conditions, the heat of combustion of glucose is 15.57 kJ/g. A 2.500-g sample of glucose is burned in a bomb calorimeter. The temperature of the calorimeter increased from 20.55°C to 23.25°C. **(a)** What is the total heat capacity of the calorimeter? **(b)** If the calorimeter contained 2.700 kg of water, what is the heat capacity of the dry calorimeter? **(c)** What temperature increase would be expected in this calorimeter if the glucose sample had been combusted when the calorimeter contained 2.000 kg of water?

Foods, Fuels, and Sources of Energy

5.49 The heat of combustion of ethanol, $C_2H_5OH(l)$, is −1371 kJ/mol. A 12-oz (355-mL) bottle of beer contains 3.7 percent ethanol by mass. Assuming the density of the beer to be 1.0 g/mL, what caloric content does the alcohol in a bottle of beer have?

5.50 The heat of combustion of fructose, $C_6H_{12}O_6$, is −2812 kJ/mol. If a freshly harvested Golden Delicious apple weighing 4.23 oz (120 g) contains 16.0 g of fructose, what caloric content does the fructose contribute to the apple?

5.51 A pound of peanut brittle contains 214 g of carbohydrate, 146 g of fat, and 79 g of protein. What is the fuel value in kilojoules in a 50-g bar of peanut brittle? How many Calories does it provide?

5.52 A 1-oz serving of a popular "high-protein" breakfast cereal contains 6 g of protein, 19 g of carbohydrate, and 1 g of fat. What is the fuel value in kilojoules and in Calories for this cereal?

5.53 From the following data for three prospective fuels, calculate which could provide the most energy per unit volume.

Fuel	Density (g/cm^3) at 20°C	Molar heat of combustion (kJ/mol)
Nitroethane, $C_2H_5NO_2(l)$	1.052	−1348
Ethanol, $C_2H_5OH(l)$	0.789	−1371
Diethyl ether, $(C_2H_5)_2O(l)$	0.714	−2727

5.54 The standard enthalpies of formation of gaseous propyne, C_3H_4, propylene, C_3H_6, and propane, C_3H_8, are +185.4, +20.4, and −103.8 kJ/mol, respectively. **(a)** Calculate the heat evolved per mole on combustion of each substance to yield $CO_2(g)$ and $H_2O(g)$. **(b)** Also calculate the heat evolved on combustion of 1 kg of each substance. **(c)** Which is the most efficient fuel in terms of heat evolved per unit mass?

5.55 In 1977, the "proved" U.S. reserves of natural gas were estimated to be 2.09×10^{11} ft^3. Assuming the fuel value of the gas to be 980 kJ/ft^3, calculate the total fuel value of this quantity of natural gas. What weight of anthracite coal has an equivalent total fuel value?

5.56 The United States uses about 2.0×10^{16} kJ of natural gas annually. Using the fuel value of the natural gas given in Table 5.4, and assuming its density to be 0.70 g/L, estimate the number of cubic meters of natural gas burned each year in the United States.

Additional Exercises

5.57 When a mole of dry ice, $CO_2(s)$, is converted to $CO_2(g)$ at atmospheric pressure and −78°C, the heat absorbed by the system exceeds the increase in internal energy of the CO_2. Why is this so? What happens to the remaining energy?

5.58 An expanding gas absorbs 1.55 kJ of heat. If its internal energy increases by 1.32 kJ, does the system do work on its surroundings or have work done on it? What quantity of work is involved?

5.59 Limestone stalactites are formed in caves by the following reaction:

$$Ca^{2+}(aq) + 2HCO_3^-(aq) \longrightarrow CaCO_3(s) + CO_2(g) + H_2O(l)$$

If 1 mol of $CaCO_3$ forms at 298 K under 1 atm pressure, the reaction performs 2.47 kJ of expansion work, pushing back the atmosphere as the gaseous CO_2 forms. At the same time, 38.95 kJ of heat are absorbed from the environment. What are the values of ΔH and of ΔE for this reaction?

5.60 When fruits and grains are fermented, glucose is converted to ethyl alcohol. The following data are given:

$$C_6H_{12}O_6(s) \longrightarrow 2C_2H_5OH(l) + 2CO_2(g) \qquad \Delta H = -69.4 \text{ kJ}$$

(a) Is this reaction exothermic or endothermic? **(b)** Which has higher enthalpy, the reactants or products of this reaction? **(c)** Calculate ΔH for the formation of 5.00 g of C_2H_5OH. **(d)** What quantity of heat is liberated when 95.0 g of C_2H_5OH is formed at constant pressure?

5.61 Compare the ΔH_f° values for $CH_3OH(g)$ and $CH_3OH(l)$ in Appendix C. Which phase has the lower enthalpy? Explain why this must be the case.

5.62 Titanium tetrachloride reacts with water to form titanium dioxide and hydrogen chloride:

$$TiCl_4(l) + 2H_2O(l) \longrightarrow TiO_2(s) + 4HCl(g) \qquad \Delta H^\circ = 67.0 \text{ kJ}$$

ΔH_f° for $TiCl_4(l)$ is -804.2 kJ/mol. Using additional information from Appendix C, determine ΔH_f° for $TiO_2(s)$.

5.63 Calculate the standard enthalpy changes for the following reactions. Indicate which are endothermic and which are exothermic. **(a)** $6H_2(g) + P_4(s) \longrightarrow 4PH_3(g)$ [$P_4(s)$ is white phosphorus]; **(b)** $2HF(g) + Br_2(l) \longrightarrow 2HBr(g) + F_2(g)$; **(c)** $CaO(s) + CO_2(g) \longrightarrow CaCO_3(s)$; **(d)** $3NO(g) \longrightarrow NO_2(g) + N_2O(g)$.

5.64 The following reaction, known as the *water-gas shift reaction*, is important in coal gasification technology:

$$CO(g) + H_2O(g) \longrightarrow CO_2(g) + H_2(g)$$

Use data in Appendix C to determine whether this reaction is endothermic or exothermic under standard conditions. Does the answer change if $H_2O(l)$ is used instead of $H_2O(g)$?

5.65 The two common sugars, glucose, $C_6H_{12}O_6$, and sucrose, $C_{12}H_{22}O_{11}$, are both carbohydrates. Their standard heats of formation are given in Table 5.1. Using these data: **(a)** Calculate the molar heat of combustion to $CO_2(g)$ and $H_2O(l)$ for the two sugars. **(b)** Calculate the heat of combustion per gram of each sugar. **(c)** How do your answers to part (b) compare to the average fuel value of carbohydrates discussed in Section 5.8?

[5.66] A "coffee cup" calorimeter of the type shown in Figure 5.16 contains 150 g of water at 24.6°C. A 110-g block of molybdenum metal is heated to 100°C and then placed in the water in the calorimeter. The temperature of the water rises to 28.0°C, at which point it stops rising. What is the specific heat of molybdenum metal? Ignore the heat capacity of the plastic cups.

5.67 A house is being designed to have passive solar energy features. Brickwork is to be incorporated into the interior of the house to act as a heat absorber. Each brick weighs approximately 1.8 kg. The specific heat of the brick is 0.85 J/g-°C. How many bricks will need to be incorporated into the interior of the house to provide the same total heat capacity as 1000 gal of water?

5.68 When 50.0 mL of 1.00 *M* $CuSO_4$ and 50.0 mL of 2.00 *M* KOH are mixed in a constant-pressure calorimeter, the temperature of the mixture rises from 21.5°C to 27.7°C. From these data, calculate ΔH for the process

$$CuSO_4(aq; 1\ M) + 2KOH(aq; 2\ M) \longrightarrow Cu(OH)_2(s) + K_2SO_4(aq; 0.5\ M)$$

Assume that the calorimeter absorbs only a negligible quantity of heat, that the total volume of the solution is 100.0 mL, and that the specific heat and density of the solution following mixing is the same as that of pure water.

5.69 Burning methane in oxygen can produce three different carbon-containing products: soot (very fine particles of graphite), $CO(g)$, and $CO_2(g)$. **(a)** Write three balanced equations for the reaction of methane gas with oxygen to produce these three products. In each case, assume that $H_2O(l)$ is the only other product. **(b)** Determine the standard enthalpies for the reactions in part (a). **(c)** Why, if the oxygen supply is adequate, is $CO_2(g)$ the predominant carbon-containing product of the combustion of methane?

5.70 A 1-oz serving of a popular oat cereal provides 21 g of carbohydrates, 2 g of fat, and 4 g of protein. The fuel value of this serving is listed as 100 Cal. Is this value higher or lower than the average fuel value of a food with the above composition?

5.71 Acetylene, C_2H_2, and benzene, C_6H_6, are hydrocarbons that have the same empirical formula. Benzene is an example of an "aromatic" hydrocarbon, one that is unusually stable because of its structure. **(a)** By using the data in Appendix C, determine the standard enthalpy change for the reaction $3C_2H_2(g) \longrightarrow C_6H_6(l)$, **(b)** Which has greater enthalpy, 3 mol of acetylene gas or 1 mol of liquid benzene? **(c)** Determine the fuel value in kJ/g for acetylene and benzene.

[5.72] Ammonia, NH_3, boils at -33°C; at this temperature, it has a density of 0.81 g/cm^3. The heat of formation of $NH_3(g)$ is -46.2 kJ/mol, and the heat of vaporization of $NH_3(l)$ is 4.6 kJ/mol. Calculate the heat evolved when 1 L of liquid NH_3 is burned in air to give $N_2(g)$ and $H_2O(g)$. How does this compare with the heat evolved upon complete combustion of a liter of liquid methanol, CH_3OH [density at 25°C = 0.792 g/cm^3, and heat of formation $\Delta H_f^\circ(CH_3OH(l)) = -239$ kJ/mol]?

[5.73] **(a)** When a 0.235-g sample of benzoic acid is combusted in a bomb calorimeter, a 1.642°C rise in temperature is observed. When a 0.265-g sample of caffeine, $C_8H_{10}O_2N_4$, is burned, a 1.525°C rise in temperature is measured. Using the value 26.38 kJ/g for the heat of combustion of benzoic acid, calculate the heat of combustion per mole of caffeine at constant volume. **(b)** Assuming that there is an uncertainty of 0.002°C in each temperature reading and that the masses of samples are measured to 0.001 g, what is the estimated uncertainty in the value calculated for the heat of combustion per mole of caffeine?

[5.74] Aspirin is produced commercially from salicylic acid, $C_7O_3H_6$. A large shipment of salicylic acid is contaminated with boric oxide, which, like salicylic acid, is a white powder. The heat of combustion of salicylic acid at constant volume is known to be -3.00×10^3 kJ/mol. Boric oxide, because it is fully oxidized, does not burn. When a 3.556-g sample of the contaminated salicylic acid is burned in a bomb calorimeter, the temperature increases 2.556°C. From previous measurements, the heat capacity of the calorimeter is known to be 13.62 kJ/°C. What is the amount of boric oxide in the sample, in terms of percent by weight?

[5.75] Three common hydrocarbons that contain four carbons are listed below, along with their standard heats of formation:

Hydrocarbon	Formula	ΔH_f° (kJ/mol)
1,3-Butadiene	$C_4H_6(g)$	111.9
1-Butene	$C_4H_8(g)$	1.2
n-Butane	$C_4H_{10}(g)$	−124.7

(a) For each of these, calculate the molar heat of combustion to $CO_2(g)$ and $H_2O(l)$. **(b)** Calculate the fuel value in kJ/g for each of these compounds. **(c)** For each hydrocarbon, determine the percentage of hydrogen by mass. **(d)** By comparing your answers for parts (b) and (c), propose a relationship between hydrogen content and fuel value in hydrocarbons.

Electronic Structures of Atoms

6

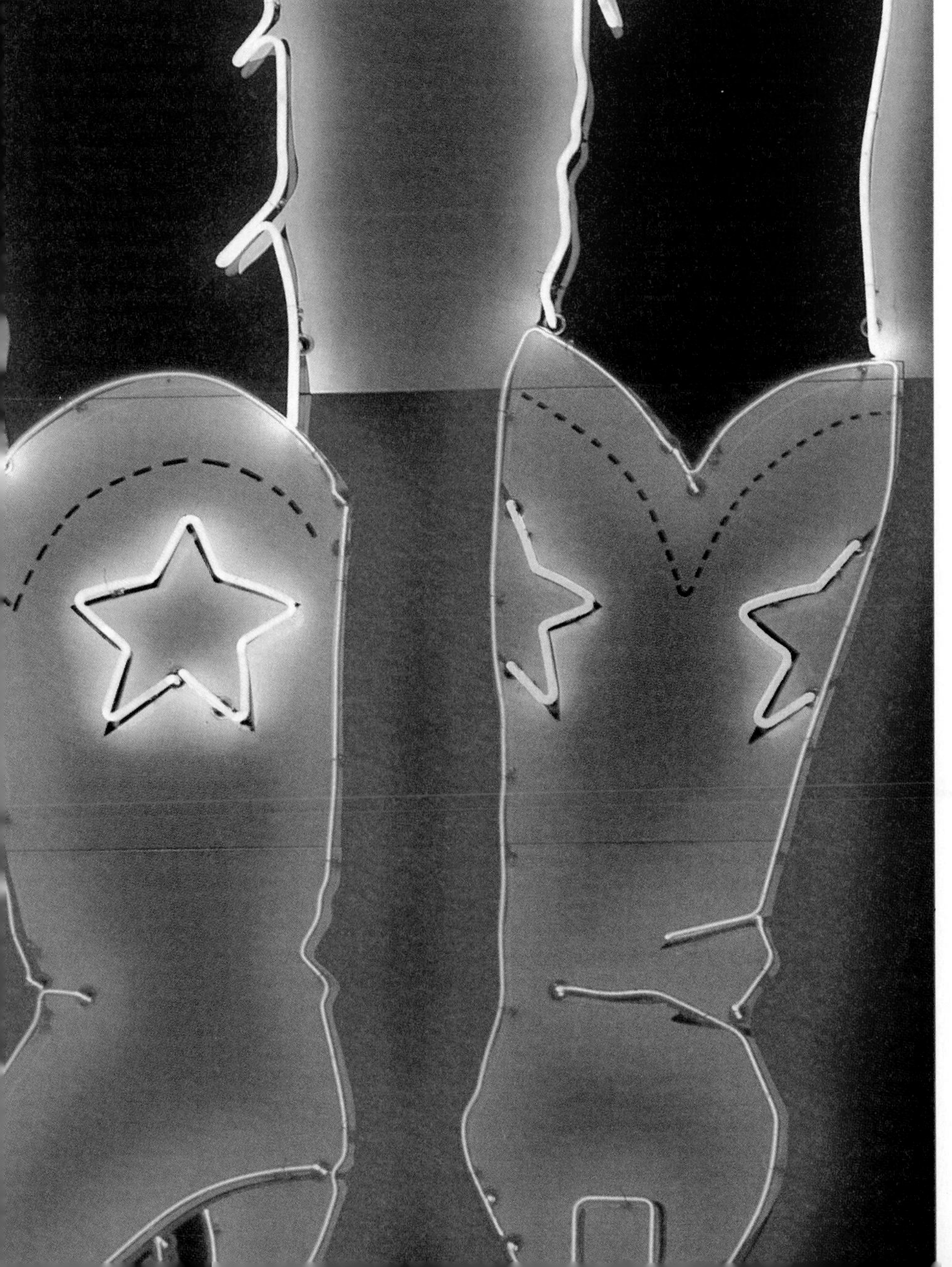

In "neon" signs, light is produced when the electrons in an electrically-excited gas atom relax to a lower-energy state. Different gases produce light of different color. (© Raymond Genreau/ Allstock)

CONTENTS

The periodic table, discussed in Chapter 2, evolved largely as a result of experimental observation. Elements that exhibit similar properties were placed together in the same column of the table. But what is the fundamental reason for these similarities? Why, for example, are sodium and potassium both soft, reactive metals? And why are helium and neon both unreactive gases? Why do the halogens all react with hydrogen to form compounds that contain exactly one hydrogen atom? To answer these questions and many others like them, we must take a closer look at atoms, especially at how electrons are arranged in atoms.

In this chapter, you will be introduced to the arrangements of electrons in atoms, which we refer to as their **electronic structure**. In later chapters, we will see how electronic structure is used to explain trends in the periodic table and to understand the bonding between atoms. Our knowledge of electronic structure is the result of one of the major developments of twentieth-century science, the **quantum theory**. We shall see how the quantum theory was developed and how it relates to the behavior of electrons in atoms. The quantum theory arose from the need for a better description of the properties of radiant energy (light). We will therefore begin this chapter by considering the characteristics of radiant energy.

6.1 RADIANT ENERGY

Different kinds of radiant energy—such as the warmth from a glowing fireplace, the light reflected off snow in the mountains, and the X rays used by a dentist—*seem* very different from one another, yet they share certain fundamental characteristics. All types of radiant energy, also called **electromagnetic radiation**, move through a vacuum at a speed of 2.9979250×10^8 m/s, the "speed of light." The speed of light is one of the most accurately known physical constants. For our purposes it will be sufficient to use 3.00×10^8 m/s.

All radiant energy has wavelike characteristics similar to those of waves that move through water. Water waves are the result of energy imparted to the water, perhaps by the dropping of a stone, the movement of a boat, or the force of wind on the water surface. This energy is expressed as the up-and-down movements of the water. A cork bobbing on water as waves pass by is not swept along with the wave. Rather, it moves up and down with the wave motion.

If we look at a cross section of a water wave (Figure 6.1), we see that it is *periodic:* The pattern of peaks and troughs repeats itself at regular intervals. The distance between successive peaks (or troughs) is called the **wavelength**. The number of complete wavelengths that pass a given point in 1 s is the **frequency** of the wave. We can measure the frequency of the water wave by counting

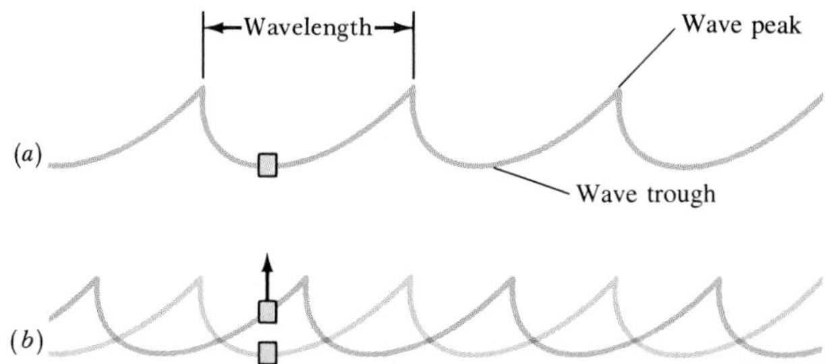

Figure 6.1 Characteristics of water waves. (*a*) The distance between corresponding points on each wave is called the *wavelength*. (*b*) The number of times per second that the cork bobs up and down is called the *frequency*.

the number of times per second that the cork moves through a complete cycle of upward and downward motion.

In a similar way, we can assign a frequency and a wavelength to radiant energy. These are illustrated in Figure 6.2. Just as for the water wave, the frequency of the radiation is the number of complete wavelengths, or **cycles**, that pass a given point in 1 s. Because all radiant energy moves at the speed of light, wavelength and frequency are related. If the wavelength is long, there will be fewer cycles of the wave passing a point per second, and the frequency will be low. Conversely, in order for a wave to have a high frequency, the distance between the peaks of the wave must be small (the wavelength is short). This inverse relationship between the frequency and the wavelength of radiant energy can be expressed as a simple equation: The product of the frequency, ν (nu), and the wavelength, λ (lambda), equals the speed of light, c:

$$\nu\lambda = c \quad [6.1]$$

Why do different forms of radiant energy have very different properties? We characterize the different types of radiant energy by their wavelengths, as shown in Figure 6.3. We can see that these span a tremendous range. The

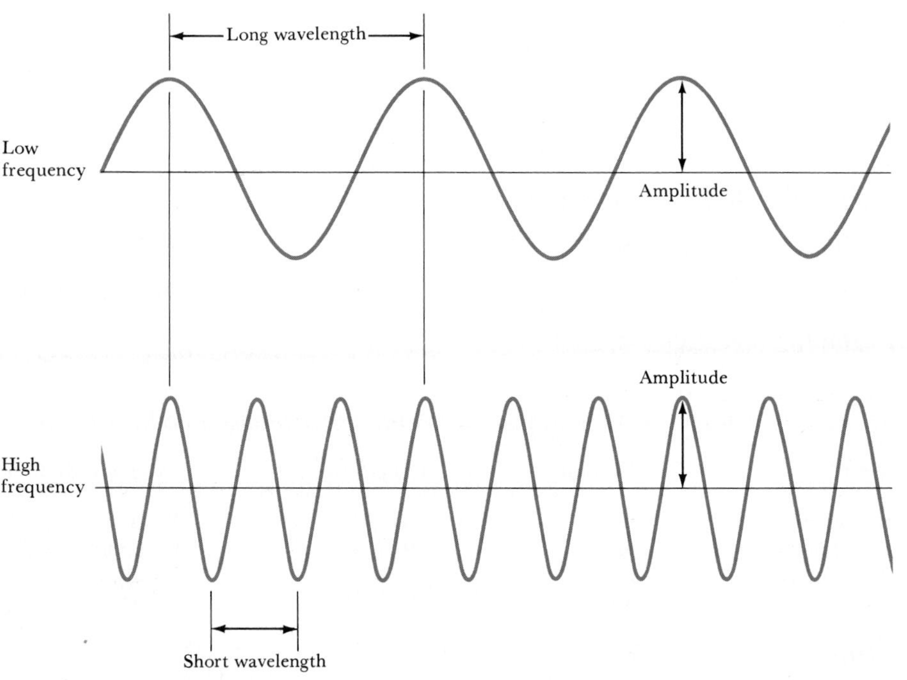

Figure 6.2 Wave characteristics of radiant energy. Radiant energy is associated with periodic changes in the electric and magnetic fields. The frequency of the high-frequency radiation in this illustration is three times that of the low-frequency radiation. The *amplitude* of the wave relates to the intensity of the radiation. Here both waves have the same amplitude.

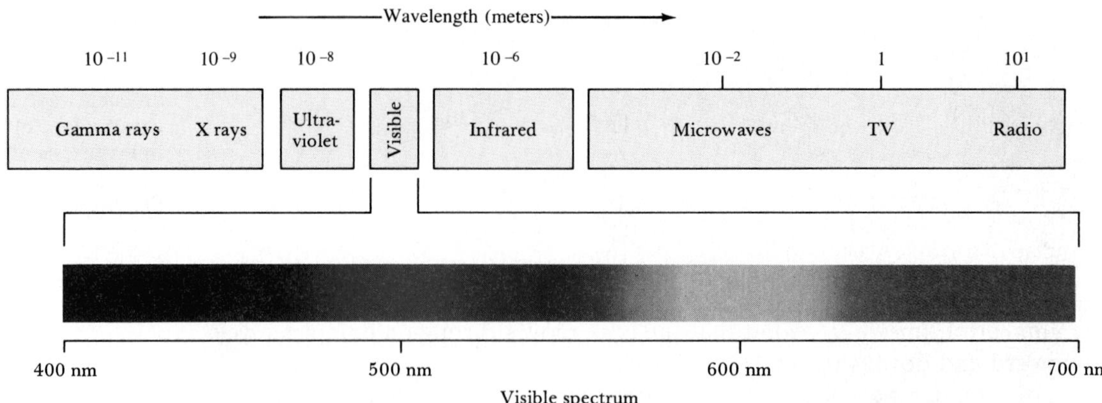

Figure 6.3 Wavelengths of electromagnetic radiation characteristic of various regions of the electromagnetic spectrum. Notice that color can be expressed quantitatively by wavelength.

wavelengths of cosmic rays are similar to the diameters of atomic nuclei, whereas those of radio waves can be longer than a football field. The unit of length normally chosen to express wavelength depends on the type of radiation, as shown in Table 6.1. Frequency is expressed in cycles per second, a unit also called a hertz (Hz). Because it is understood that cycles are involved, the units of frequency are normally given simply as "per second," which is denoted by s^{-1} or /s. For example, a frequency of 820 kilohertz (kHz), a typical frequency for an AM radio station, could also be written as 820,000 s^{-1}.

SAMPLE EXERCISE 6.1

The yellow light given off by a sodium lamp has a wavelength of 589 nm. What is the frequency of this radiation?

Solution: We can rearrange Equation 6.1 to give $\nu = c/\lambda$. We insert the values for c and λ and then convert nanometers to meters. This gives

$$\nu = \left(\frac{3.00 \times 10^8 \text{ m/s}}{589 \text{ nm}}\right)\left(\frac{10^9 \text{ nm}}{1 \text{ m}}\right)$$
$$= 5.09 \times 10^{14} \text{ s}^{-1}$$

Note that frequency has units of "per second" or s^{-1}.

PRACTICE EXERCISE

A laser used to weld detached retinas produces radiation with a frequency of 4.69×10^{14} s^{-1}. What is the wavelength of this radiation?
Answer: 6.40×10^{-7} m = 640 nm

Table 6.1 Common Wavelength Units for Electromagnetic Radiation

Unit	Symbol	Length (m)	Type of radiation
Angstrom	Å	10^{-10}	X ray
Nanometer	nm	10^{-9}	Ultraviolet, visible
Micrometer	μm	10^{-6}	Infrared
Millimeter	mm	10^{-3}	Infrared
Centimeter	cm	10^{-2}	Microwaves
Meter	m	1	TV, radio

Figure 6.4 Max Planck (1858–1947), physicist. Born in Kiel, Germany, Planck was the son of a law professor at the University of Kiel. When he announced his intention to study physics, Planck was warned that all the major discoveries had already been made in this field. Nevertheless, Planck became a physicist and in 1892 was named professor of physics at the University of Berlin. In 1900, he presented a paper before the Berlin Physical Society that launched one of the greatest intellectual revolutions in the history of science. (Library of Congress)

Figure 6.3 illustrates that visible light is an extremely small portion of the electromagnetic spectrum. Visible light can be seen because of chemical reactions that it triggers in your eyes. We are also familiar with the effects of other forms of radiation. For example, overexposure of a part of your body to infrared radiation (heat) can cause a burn, overexposure to ultraviolet light can cause suntan and sunburn, and overexposure to X rays can cause tissue damage or even cancer. These diverse effects are due to differences in the energy of the radiation. Radiation of high frequency (and thus short wavelength) is more energetic than radiation of lower frequency (and thus longer wavelength). The quantitative relation between frequency and energy was developed at the turn of the century in the revolutionary *quantum theory* of the German physicist Max Planck (Figure 6.4).

6.2 THE QUANTUM THEORY

The quantum theory is concerned with the rules that govern the energy changes of an object. In our everyday experience, we are used to energy changes that are smooth and continuous. Consider, for example, a ball rolling down a ramp (Figure 6.5). The laws of physics set forth by Sir Isaac Newton (1642–1727) state that the potential energy of the ball is proportional to its height above the ground. We will call the potential energy at the top of the ramp E_{top}. As the ball rolls down the ramp, its potential energy decreases until it comes to the

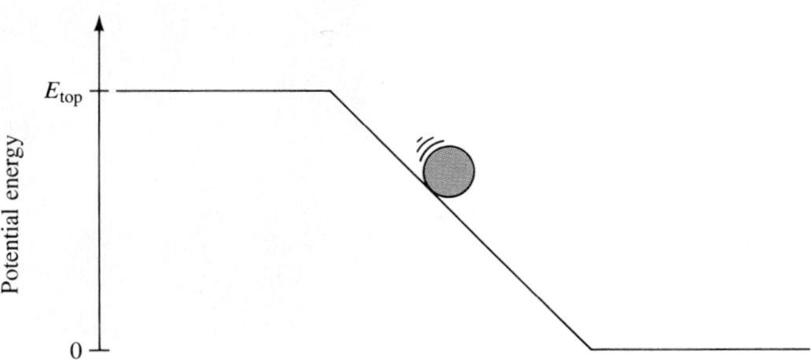

Figure 6.5 A ball rolling down a frictionless ramp is an example of the continuous conversion of one form of energy into another. In this case, potential energy decreases continuously from E_{top} to 0 as the ball rolls down; the potential energy is converted into kinetic energy.

bottom of the ramp, where the potential energy is zero. During its journey down the ramp, the ball's potential energy has had every possible value between E_{top} and zero; the potential energy has decreased *continuously*. Where has the energy gone? Because energy is conserved, the lost potential energy must have been converted into kinetic energy (Section 5.1), which increases continuously as the ball rolls down the ramp. We can view the ball rolling down the ramp as a processs in which potential energy is continuously converted into kinetic energy. Throughout the nineteenth century, scientists thought that all energy changes were continuous, that there were no restrictions on the amount of one form of energy that could be converted into another. However, this view soon changed.

Planck's Theory

In the late 1800s, a number of physicists were studying a phenomenon known as *blackbody radiation*, the radiation emitted by a hot object (Figure 6.6). The prevailing laws of physics could not account for the spectrum of light emitted from a blackbody. In 1900, Planck was able to explain blackbody radiation by making a radical assumption: The energy changes associated with blackbody radiation are *not* continuous. Planck proposed that energy can be released only in "chunks" of some minimum size.

Planck termed the smallest possible increment of energy gained or lost a **quantum**. The gain or loss of radiant energy, such as that absorbed or emitted by a blackbody, involves discrete energy changes that must be whole-number multiples of a constant times the frequency, ν, of the radiant energy. Thus, if ΔE is the amount of radiant energy gained or lost, then according to Planck's theory,

$$\Delta E = h\nu, 2h\nu, 3h\nu, \ldots \qquad [6.2]$$

The constant h, known as *Planck's constant*, has a value of 6.63×10^{-34} joule-seconds (J-s). The gain or loss of energy by a fractional multiple of $h\nu$, for example $0.8h\nu$ or $2.3h\nu$, is forbidden in Planck's theory; we say that the allowed energies are *quantized*. The smallest increment of energy at a given frequency, $h\nu$, is called a *quantum* of energy.

Figure 6.6 A hot object, such as this electric stove burner, emits visible radiation. (Paul Silverman/Fundamental Photographs)

SAMPLE EXERCISE 6.2

Calculate the smallest increment of energy (that is, the quantum of energy) that an object can absorb from yellow light whose wavelength is 589 nm.

Solution: We obtain the magnitude of a quantum of energy from Equation 6.2, $\Delta E = h\nu$. The value of Planck's constant is given both in the text above and in the table of physical constants on the back inside cover of the text: $h = 6.63 \times 10^{-34}$ J-s. The frequency, ν, is calculated from the given wavelength, as shown in Sample Exercise 6.1: $\nu = c/\lambda = 5.09 \times 10^{14}\ \text{s}^{-1}$. Thus, we have

$$\begin{aligned}\Delta E &= (6.63 \times 10^{-34}\ \text{J-s})(5.09 \times 10^{14}\ \text{s}^{-1}) \\ &= 3.37 \times 10^{-19}\ \text{J}\end{aligned}$$

Planck's theory tells us that an atom or molecule emitting or absorbing radiation whose wavelength is 589 nm cannot lose or gain energy by radiation except in multiples of 3.37×10^{-19} J. It cannot, for example, gain 5.00×10^{-19} J from this radiation because this amount is not a multiple of 3.37×10^{-19} J.

If one quantum of radiant energy supplies 3.37×10^{-19} J, then one mole of these quanta will supply (6.02×10^{23} quanta)(3.37×10^{-19} J/quantum), which equals 2.03×10^{5} J. This is the magnitude of heats of reactions (Section 5.4). Indeed, radiation can cause chemical bonds to break, producing what are called *photochemical reactions.*

PRACTICE EXERCISE

A laser that emits light energy in pulses of short duration has a frequency of $4.69 \times 10^{14}\ \text{s}^{-1}$ and deposits 1.3×10^{-2} J of energy during each pulse. How many quanta of energy does each pulse deposit? (Hint: First calculate the energy of one quantum of this frequency.) ***Answer:*** 4.2×10^{16} quanta

If the notion of quantized rather than continuous energies seems confusing, it might be helpful to draw an analogy with some musical instruments, specifically a violin and a piano (Figure 6.7). A violinist can play every pitch between two notes (say B and C) by changing the position of the fingers on the strings; the pitch of a violin can be varied continuously. In contrast, a pianist can change notes only by a set amount. It is not possible to play any notes between B and C. We could say, therefore, that the notes on a piano are quantized.

Figure 6.7 The pitch of a violin can be varied continuously, but that of a piano can be varied only in steps; the notes on the piano are quantized. (Four By Five/Superstock; © Christian Steiner)

If Planck's quantum theory is correct, why can we still use Newton's laws successfully to predict, for example, the motion of the planets or the trajectory of a projectile? You will note that Planck's constant is an extremely small number. Thus, a quantum of energy, $h\nu$, will be an extremely small amount of energy. Planck's rules regarding the gain or loss of energy are always the same, whether we are concerned with objects on the size scale of our ordinary experience or with microscopic objects. For macroscopic objects, such as humans, the gain or loss of a quantum of energy is completely unnoticed. When dealing with matter at the atomic level, however, the impact of quantized energies is far more significant.

At this point you may be wondering about the practical applications of Planck's quantum theory. A few years after Planck presented his theory, scientists began to see its applicability to a great many experimental observations. It soon became apparent that Planck's theory had within it the seeds of a revolution in the way the physical world is viewed. Let's consider a few applications that are of special importance for chemistry.

The Photoelectric Effect

Light shining on a clean metallic surface can cause the surface to emit electrons. This phenomenon, known as the *photoelectric effect*, is illustrated in Figure 6.8. For each metal, there is a minimum frequency of light below which no electrons are emitted, regardless of how intense the beam of light. The photoelectric effect was counter to the predictions of pre-1900 physics. In 1905, Albert Einstein (1879–1955) used the quantum theory to explain the photoelectric effect. He assumed that the radiant energy striking the metal surface is a stream of tiny energy packets. Each energy packet, called a **photon**, is a quantum of energy, $h\nu$. Thus, radiant energy itself is considered to be quantized. *Photons of higher-*

Figure 6.8 The photoelectric effect. When photons of sufficiently high energy strike a metal surface, electrons are emitted from the metal, as in (*a*). The photoelectric effect is the basis of the photocell shown in (*b*). The emitted electrons are drawn toward the other electrode, which is a positive terminal. As a result, current flows in the circuit. Photocells in automatic doors use this current to control the door-opening circuits.

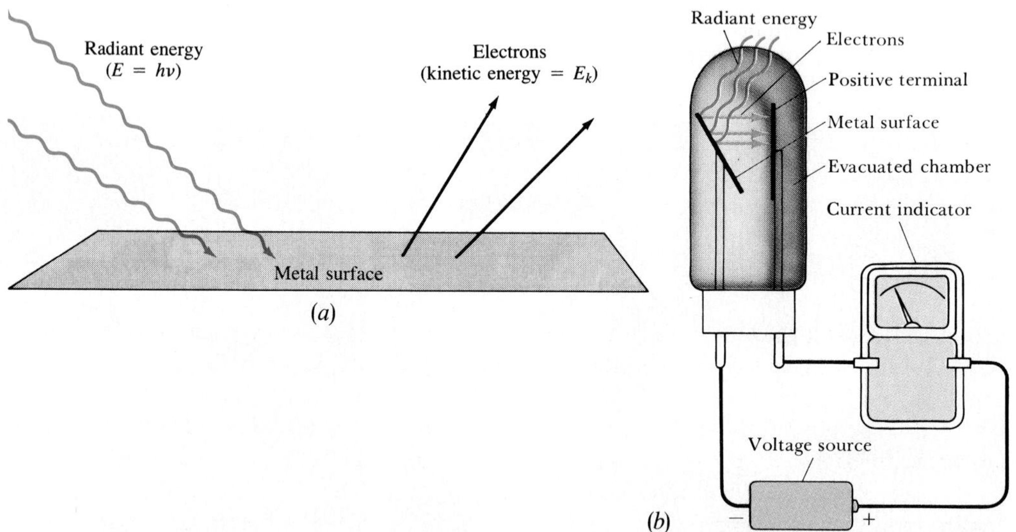

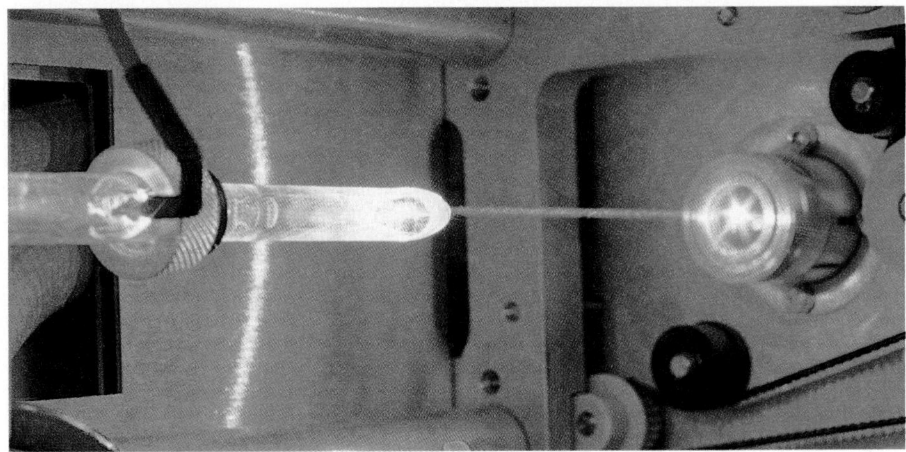

Figure 6.9 Monochromatic light being emitted by a krypton gas laser. (Dan McCoy/Rainbow)

frequency radiation have higher energies, and photons of lower-frequency radiation have lower energy:

$$E_{photon} = h\nu \quad [6.3]$$

When a photon is absorbed by the metal, its energy is transferred to an electron in the metal. A certain amount of energy is required for the electron to overcome the attractive forces that hold it within the metal. If the photons of the radiation have less energy than this energy threshold, the electron cannot escape from the metal surface, even if the light beam is intense. If a photon does have sufficient energy, the electron is emitted. If a photon has more than the minimum energy required to free an electron, the excess appears as the kinetic energy of the emitted electron. Thus, the kinetic energy of the emitted electron, E_k, equals the energy supplied by the photon, $h\nu$, minus the binding energy holding the electron in the metal, E_b:

$$E_k = h\nu - E_b \quad [6.4]$$

Continuous and Line Spectra

A particular source of radiant energy may emit a single wavelength, as in the light from a laser (Figure 6.9). Radiation composed of a single wavelength is said to be **monochromatic**. However, most common radiation sources, including light bulbs and stars, produce radiation containing many different wavelengths. When radiation from such sources is separated into its different wavelength components, a **spectrum** is produced. Figure 6.10 shows how "white" light from a light bulb can be dispersed by a prism. The spectrum so produced consists of a continuous range of colors: Violet merges into blue, blue into green, and so forth, with no blank spots. This rainbow of colors, containing light of all wavelengths, is called a **continuous spectrum**. The most familiar example of a continuous spectrum is the rainbow, produced by the dispersal of sunlight by raindrops or mist.

Not all radiation sources produce a continuous spectrum. For example, when hydrogen is placed under reduced pressure in a tube such as that depicted in Figure 6.11 and a high voltage is applied, light is emitted. Different gases emit different colors of light (Figure 6.12). The light emitted by neon gas is the

defraction grating

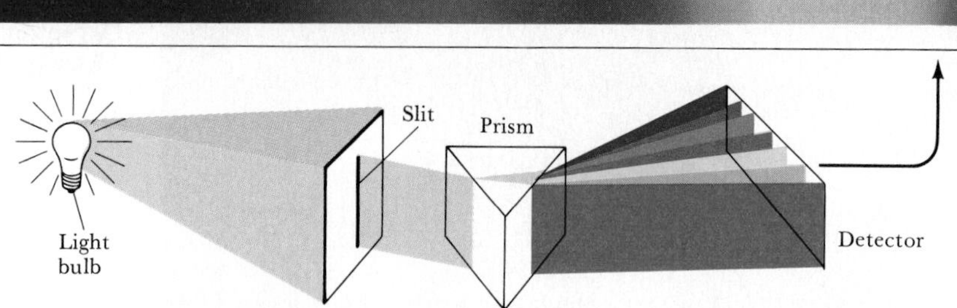

Figure 6.10 A continuous visible spectrum is produced when a narrow beam of white light is passed through a prism. The white light could be sunlight or light from an incandescent lamp.

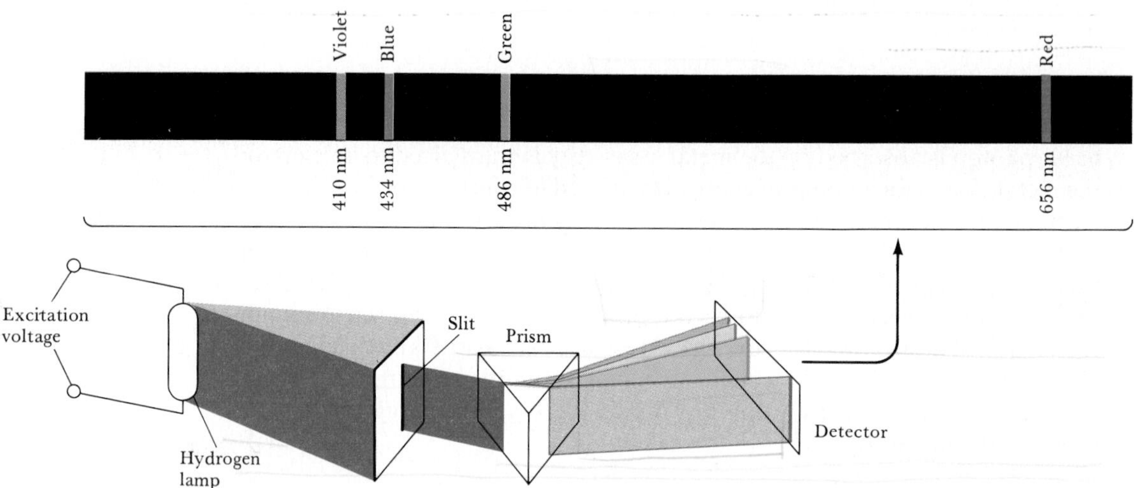

Figure 6.11 Part of the emission line spectrum of hydrogen showing the four most intense lines. This spectrum is obtained by passing a narrow beam of the light emitted by a hydrogen lamp through a prism. The lamp emits light when hydrogen atoms are excited in an electrical discharge.

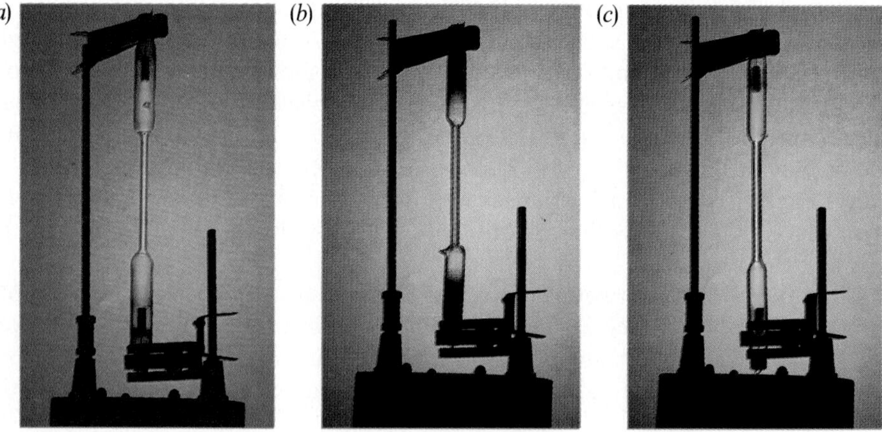

Figure 6.12 Different gases emit light of different characteristic colors upon excitation in an electrical discharge: (*a*) Hg; (*b*) H; (*c*) N. (Richard Megna/Fundamental Photographs)

A CLOSER LOOK: Discovery of Helium

The spectrum of the radiation emitted by a substance is called an *emission spectrum.* The line spectrum of hydrogen shown in Figure 6.11 is an emission spectrum. Substances also exhibit *absorption spectra.* When continuous electromagnetic radiation, such as that from a light bulb, passes through a substance, certain wavelengths may be absorbed. The spectrum of the radiation that passes through is called an absorption spectrum. The absorption spectrum of hydrogen consists of what looks like a continuous spectrum interrupted by black lines at 656.3 nm, 486.3 nm, and the other wavelengths found in the emission spectrum. The absorption spectrum of hydrogen is a line spectrum that is complementary to its emission spectrum.

Each element has a characteristic line spectrum that can be used to identify it. Figure 6.13 shows the spectrum of the light emitted by the sun. Notice that there are several dark bands in the spectrum. Because of its high temperature, the core of the sun emits a continuous spectrum of radiation; however, elements that are present in the outer regions of the sun, where the temperatures are not so high, absorb radiation at characteristic wavelengths. Absorptions due to hydrogen, helium, iron, and other atoms give rise to the dark bands evident in Figure 6.13.

Helium was first discovered in 1868 from a similar spectrum of our sun. Some of the absorption lines of the sun's spectrum could not be matched with those of any element then known. It was concluded that the sun contained an element previously unknown on earth. This element was named helium after *helios*, the Greek word for "sun." Helium was subsequently isolated and characterized in the laboratory in 1895.

Figure 6.13 Spectrum of the sun's radiation. Notice the presence of dark lines, called *Fraunhofer lines,* due to absorption of radiation by hydrogen and other atoms at the outer limits of the sun. The lower spectrum is that of an incandescent source of the same temperature as the sun's surface. (Courtesy Bausch & Lomb)

familiar red-orange glow of many "neon" lights (many "neon" lights actually contain other gases); sodium vapor emits the yellow light characteristic of many modern streetlights. When the light coming from such tubes is dispersed by a prism, only certain wavelengths are found to be present. A spectrum containing radiation of only specific wavelengths is called a **line spectrum**. A line spectrum of the visible light emitted by hydrogen is shown in Figure 6.11.

When scientists first detected the line spectrum of hydrogen in the mid-1800s, they were fascinated by its simplicity. In 1885, a Swiss schoolteacher named Johann Balmer observed that the frequencies of the four lines shown in Figure 6.11 fit an intriguingly simple formula:

$$\nu = C\left(\frac{1}{2^2} - \frac{1}{n^2}\right) \qquad n = 3, 4, 5, 6 \qquad [6.5]$$

In this formula, C is a constant equal to $3.29 \times 10^{15}\ s^{-1}$. How could the remarkable simplicity of this equation be explained? It took nearly 30 more years to answer this question, as we will see in the next section.

6.3 BOHR'S MODEL OF HYDROGEN

In 1914, the Danish physicist Niels Bohr (Figure 6.14) used Planck's quantum theory to explain the line spectrum of hydrogen. Bohr introduced the fundamental idea that the absorptions and emissions of light by hydrogen atoms corresponded to energy changes of the electrons within the atoms. The fact that only certain frequencies are absorbed or emitted by an atom tells us that only certain energy changes are possible: The energy changes within an atom are quantized.

Bohr's model of the hydrogen atom included two important developments that were relatively new at that time. The first was Rutherford's establishment of the nuclear nature of the atom (Section 2.2). The other was Einstein's discovery that radiant energy could be thought of as a stream of discrete bundles of energy called photons. Bohr's model, although not strictly correct, was the first to propose the quantization of the energy of electrons in atoms.

After Rutherford discovered the nuclear nature of the atom, scientists were tempted to think of atoms as "microscopic solar systems" in which the electrons orbited the nucleus. There was a problem with this description, however. For this model, classical physics predicted that continuous radiation would be emitted and that as the electron lost energy it would spiral into the nucleus. Bohr solved this problem in much the same way that Planck had solved the problem of blackbody radiation: He assumed that classical physics was inadequate to describe atoms. Bohr's model consists of a series of postulates that we can summarize as follows:

1. The electron of a hydrogen atom moves about the central proton in a circular orbit. Only orbits of certain radii, corresponding to certain definite energies, are permitted. An electron in a permitted orbit is said to be in

Figure 6.14 Niels Bohr (1885–1962), a Danish physicist, born and educated in Copenhagen. From 1911 to 1913, he studied in England, working first with J. J. Thomson at Cambridge University and then with Ernest Rutherford at the University of Manchester. He published his quantum theory of the atom in 1914 and was awarded the Nobel Prize in physics in 1922. (Alan W. Richards/AP/Wide World)

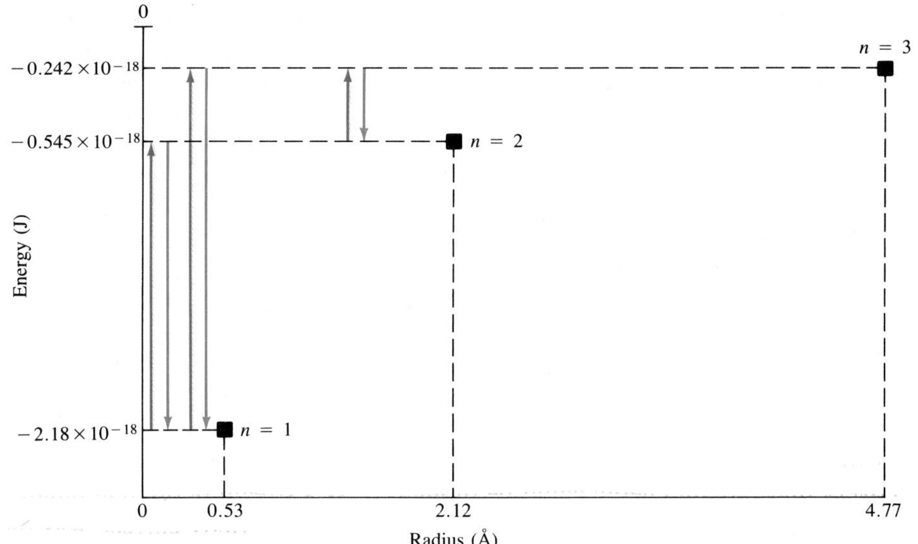

Figure 6.15 Radii and energies of the three lowest-energy orbits in the Bohr model of hydrogen. The arrows show transitions of the electron from one allowed energy state to another. When absorption occurs (blue arrows), the transition takes the electron from a lower- to a higher-energy state. When emission occurs (red arrows), the transition is from a higher- to a lower-energy state.

an "allowed" energy state. An electron in an allowed energy state will not spiral into the nucleus.

2. The electron can change from one allowed energy state to another one by absorbing or emitting radiant energy. The frequency, ν, of the radiant energy corresponds exactly to the energy difference, ΔE, between the two allowed energy states: $\Delta E = h\nu$.

We need not concern ourselves with the details of how Bohr used quantum theory to calculate the energies of the electron. The main point is that he was able to calculate a set of allowed energies. Each of these allowed energies corresponds to a circular path of different radius. In Bohr's model, each allowed orbit was assigned an integer, n, known as the **principal quantum number**, that may have values from 1 to infinity. The radius of the electron orbit in a particular energy state varies as n^2:

$$\text{Radius} = n^2(5.30 \times 10^{-11}\text{ m}) \qquad n = 1, 2, \ldots \qquad [6.6]$$

Thus, the larger the value of n, the farther the electron is from the nucleus.

The energy of the electron depends on the orbit it occupies:

$$E_n = -R_H\left(\frac{1}{n^2}\right) \qquad [6.7]$$

The constant R_H in Equation 6.7 is called the *Rydberg constant;* it has the value of 2.18×10^{-18} J. From Equation 6.7, we see that the energy of the electron is -2.18×10^{-18} J when the electron is in the orbit closest to the nucleus, $n = 1$. When it is in the second orbit, $n = 2$, its energy is

$$E_2 = (-2.18 \times 10^{-18}\text{ J})\left(\frac{1}{2^2}\right) = -5.45 \times 10^{-19}\text{ J}$$

Figure 6.15 illustrates the orbit radii and energies for $n = 1$, $n = 2$, and $n = 3$. A more common way of representing the allowed energies is shown in Figure 6.16.

Figure 6.16 Energy levels in the hydrogen atom from the Bohr model. The arrows refer to transitions of the electron from one allowed energy state to another, as described in Figure 6.15. Only the lowest six energy levels are shown.

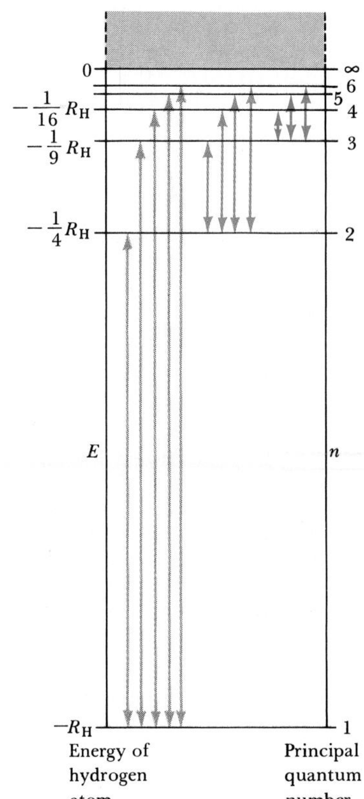

All the energies given by Equation 6.7 are negative. The lower (more negative) the energy is, the more stable the atom will be. (One way to remember this convention is to think of a ball rolling down a stairway. The ball will naturally proceed to the bottom stair; it is most stable when its potential energy is lowest.) The lowest-energy state (analogous to the bottom stair) is that for which $n = 1$. This is called the **ground state** of the atom. When the electron is in a higher (less negative) energy orbit—namely, $n = 2$ or higher—the atom is said to be in an **excited state**.

What happens to the orbit radius and the energy as n becomes infinitely large? The radius increases as n^2; we reach a point at which the electron is completely separated from the nucleus. The energy for $n = \infty$ becomes

$$E_\infty = (-2.18 \times 10^{-18}\ \text{J})\left(\frac{1}{\infty^2}\right) = 0$$

Thus, the state in which the electron is removed from the nucleus is the reference, or zero-energy, state of the hydrogen atom. It is important to remember that this zero-energy state is *higher* in energy than the states with negative energies.

Bohr's model of the hydrogen atom quantitatively explained the observed line spectra of that substance. He proposed that the absorptions or emissions in the line spectra correspond to transitions of the electron from one orbit to another. Radiant energy must be absorbed for the electron to move from one orbit to another one having a larger radius; energy is required to pull the electron away from the nucleus. Conversely, energy is emitted when the electron moves from a larger orbit to one having a smaller radius. The changes in energy, ΔE, are given by the difference between the energy of the final state of the electron, E_f, and the initial state, E_i:

$$\Delta E = E_f - E_i$$

Substituting the expression for the energy of the electron, Equation 6.7, gives

$$\Delta E = \left(\frac{-R_H}{n_f^2}\right) - \left(\frac{-R_H}{n_i^2}\right) = R_H\left(\frac{1}{n_i^2} - \frac{1}{n_f^2}\right)$$

Because $\Delta E = h\nu$, we have

$$\Delta E = h\nu = R_H\left(\frac{1}{n_i^2} - \frac{1}{n_f^2}\right) \qquad [6.8]$$

In this expression, n_i and n_f represent the quantum numbers for the initial and final states, respectively. Notice that when the final-state quantum number, n_f, is larger than the initial-state quantum number, n_i, the term in parentheses is positive, and ΔE is positive. This means that the system has absorbed a photon and thus increased in energy (an endothermic process). When n_i is larger than n_f, as happens in emission, energy is given off; ΔE is negative.

Bohr's model provided an explanation for Balmer's empirical observations about the line spectra of hydrogen. The relationship between Balmer's formula (Equation 6.5) and Bohr's formula (Equation 6.8) is made clear in Sample Exercise 6.3.

SAMPLE EXERCISE 6.3

Calculate the wavelength of light that corresponds to the transition of the electron from the $n = 4$ to the $n = 2$ state of the hydrogen atom. Is the light absorbed or emitted by the atom?

Solution: We use Equation 6.8, substituting $n_i = 4$ and $n_f = 2$ because these are the quantum numbers for the initial and final orbits, respectively:

$$\Delta E = h\nu = R_H\left(\frac{1}{n_i^2} - \frac{1}{n_f^2}\right)$$

$$\nu = \frac{R_H}{h}\left(\frac{1}{n_i^2} - \frac{1}{n_f^2}\right)$$

$$= \frac{2.18 \times 10^{-18}\ \text{J}}{6.63 \times 10^{-34}\ \text{J-s}}\left(\frac{1}{4^2} - \frac{1}{2^2}\right)$$

$$= \frac{2.18 \times 10^{-18}\ \text{J}}{6.63 \times 10^{-34}\ \text{J-s}}\left(\frac{1}{16} - \frac{1}{4}\right)$$

$$= \frac{2.18 \times 10^{-18}\ \text{J}}{6.63 \times 10^{-34}\ \text{J-s}}\left(-\frac{3}{16}\right)$$

$$= -6.17 \times 10^{14}\ \text{s}^{-1}$$

work out 6.7 twice & solving for ν

The negative sign indicates that the light is emitted by the atom. We expect this; the electron is moving from a higher to a lower energy state, so energy is released. A negative frequency is physically meaningless, so we ignore the negative sign in calculating the wavelength of the radiation:

$$\lambda = \frac{c}{\nu} = \frac{3.00 \times 10^8\ \text{m/s}}{6.17 \times 10^{14}\ \text{s}^{-1}}$$

$$= 4.86 \times 10^{-7}\ \text{m} = 486\ \text{nm}$$

We see in Figure 6.11 that this is the wavelength of the green emission line in the spectrum of hydrogen. All of the emission lines shown in Figure 6.11 correspond to transitions of the electron from higher orbits to the $n = 2$ orbit.

We can now also see the relationship between Balmer's formula (Equation 6.5) and Equation 6.8. Only those lines for which $n_i = 2$ appear in the visible portion of the electromagnetic spectrum. Thus, Balmer had a factor of $1/2^2$ in his formula. We also see from this problem that the constant C in Equation 6.5 is related to R_H:

$$C = \frac{R_H}{h}$$

PRACTICE EXERCISE

Calculate the wavelength of the hydrogen line that corresponds to the transition of the electron from the $n = 4$ to the $n = 1$ state. ***Answer:*** 97.3 nm, in the ultraviolet portion of the electromagnetic spectrum

Complete removal of the electron from a hydrogen atom, corresponding to a transition from the $n = 1$ (ground) state to the $n = \infty$ state, is known as **ionization**. This is represented as

$$\text{H}(g) \longrightarrow \text{H}^+(g) + \text{e}^-$$

The energy required for ionization from the ground state is called the **ionization energy**.

SAMPLE EXERCISE 6.4

Calculate the energy required for ionization of an electron from the ground state of the hydrogen atom.

Solution: The ionization energy may be written as the difference between the final and initial state energies. We have $n_f = \infty$, $n_i = 1$.

$$\Delta E = E_f - E_i = R_H\left(\frac{1}{1^2} - \frac{1}{\infty^2}\right)$$
$$= R_H(1 - 0)$$

This is just equal to R_H, 2.18×10^{-18} J.

It is often useful to express this energy on a molar basis. To do this we simply multiply by Avogadro's number:

$$\left(2.18 \times 10^{-18} \frac{\text{J}}{\text{atom}}\right)\left(6.02 \times 10^{23} \frac{\text{atoms}}{\text{mol}}\right)\left(\frac{1 \text{ kJ}}{1000 \text{ J}}\right) = 1.31 \times 10^3 \text{ kJ/mol}$$

PRACTICE EXERCISE

Calculate the energy, in kJ/mol, required for ionization of an electron from the $n = 2$ state of the hydrogen atom. ***Answer:*** 328 kJ/mol

Bohr's model was very important because it introduced the idea of quantized energy states for electrons in atoms. This feature is incorporated into our current model of the atom. However, Bohr's model was adequate for explaining only atoms and ions with one electron, such as H, He^+, or Li^{2+}. It could not explain the atomic spectra of other atoms or ions, except in a rather crude way. Consequently, Bohr's model was eventually replaced by a new way of viewing atoms that is called **quantum mechanics** or **wave mechanics**. Although this newer model maintains the concept of quantized energy states, it adds further applications of Planck's quantum theory.

6.4 MATTER WAVES

In the years following Bohr's development of a model for the hydrogen atom, the dual nature of radiant energy became a familiar concept. Depending on the experimental circumstances, radiation appears to have either a wavelike or a particlelike (photon) character. Louis de Broglie (1892–1987), who was working on his Ph.D. thesis in physics at the Sorbonne in Paris, made a daring extension of this idea. If radiant energy could, under appropriate conditions, behave as though it were a stream of particles, could not matter under appropriate conditions possibly show the properties of a wave? Suppose that the electron in orbit around the nucleus of a hydrogen atom could be thought of as a wave, with a characteristic wavelength. De Broglie suggested that the electron in its circular path about the nucleus has associated with it a particular wavelength. He went on to propose that the characteristic wavelength of the electron or of any other particle depends on its mass, m, and velocity, v:

$$\lambda = \frac{h}{mv} \qquad [6.9]$$

(h is Planck's constant). The quantity mv for any object is called its **momentum**. De Broglie used the term **matter waves** to describe the wave characteristics of material particles.

Because de Broglie's hypothesis is applicable to all matter, any object of mass m and velocity v would give rise to a characteristic matter wave. However, Equation 6.9 indicates that the wavelength associated with an object of ordinary size, such as a golf ball, is so tiny as to be completely out of the range of any possible observation. This is not so for electrons because their mass is so small.

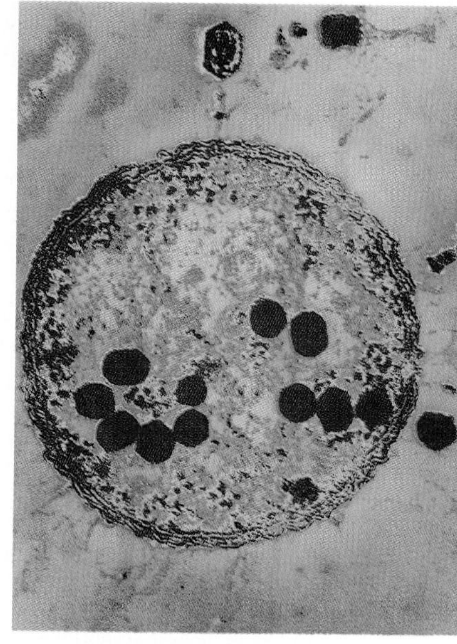

Figure 6.17 Electron micrograph of a newly synthesized T2 bacteriophage. In an electron microscope, the wave behavior of a stream of electrons is utilized in the same way that a conventional microscope uses the wave behavior of a beam of light. The bacteriophage has been magnified about 50,000 times. (Lee Simon/Science Source/ Photo Researchers)

SAMPLE EXERCISE 6.5

What is the characteristic wavelength of an electron with a velocity of 5.97×10^6 m/s? (The mass of the electron is 9.11×10^{-28} g.)

Solution: The value of Planck's constant, h, is 6.63×10^{-34} J-s (1 J = 1 kg-m²/s²).

$$\lambda = \frac{h}{mv}$$

$$= \frac{6.63 \times 10^{-34}\ \text{J-s}}{(9.11 \times 10^{-28}\ \text{g})(5.97 \times 10^6\ \text{m/s})}\left(\frac{1\ \text{kg-m}^2/\text{s}^2}{1\ \text{J}}\right)\left(\frac{10^3\ \text{g}}{1\ \text{kg}}\right)$$

$$= 1.22 \times 10^{-10}\ \text{m} = 0.122\ \text{nm}$$

By comparing this value with the wavelengths of electromagnetic radiations shown in Figure 6.3, we see that the characteristic wavelength is about the same as that of X rays.

PRACTICE EXERCISE

At what velocity must a neutron be moving in order for it to exhibit a wavelength of 500 pm? The mass of a neutron is given in the table on the back inside cover of the text. ***Answer:*** 7.92×10^2 m/s

Within a few years after de Broglie published his theory, the wave properties of the electron were demonstrated experimentally. Electrons were diffracted by crystals, just as X rays, which are definitely radiant energy, are diffracted. (We shall have more to say about the X-ray diffraction experiment in Chapter 11.)

The technique of electron diffraction has been highly developed. In the electron microscope, the wave characteristics of electrons are used to obtain pictures of tiny objects. The electron microscope is an important tool for studying surface phenomena at the very highest magnifications. Figure 6.17 is an example of an electron microscope picture. Such pictures are powerful demonstrations that tiny particles of matter can indeed behave as waves.

The Uncertainty Principle

The discovery of the wave properties of matter raised some new and interesting questions about classical physics. Consider, for example, the ball rolling down the ramp in Figure 6.5. By using Newtonian physics, we can calculate *exactly* its position, its direction of motion, and its speed of motion at any time. Can we do the same for an electron that exhibits wave properties? A wave extends in space, and its location is not precisely defined. We might therefore anticipate that it is not possible to determine exactly where an electron is located at a specific time. We will have to account for this difficulty in our description of the distribution of electrons in atoms.

The German physicist Werner Heisenberg (1901–1976) concluded that there is a fundamental limitation on how precisely we can know both the loca-

A CLOSER LOOK: Measurement and the Uncertainty Principle

Whenever any measurement is made, some uncertainty exists. Our experience with objects of ordinary dimensions, like balls or trains or laboratory equipment, indicates that the uncertainty of a measurement can be decreased by using more precise instruments. In fact, we might expect that the uncertainty in a measurement can be made indefinitely small. However, the uncertainty principle states that there is an actual limit to the accuracy of measurements. This limit is not a restriction on how well instruments can be made; rather, it is inherent in nature. This limit has no practical consequences when we are dealing with ordinary-sized objects, but its implications are enormous when we are dealing with subatomic particles, such as electrons.

To measure an object, we must disturb it, at least a little, with our measuring device. Imagine that you use a flashlight to locate a large rubber ball in a dark room. You see the ball when the light from the flashlight bounces off the ball and strikes your eyes. When a beam of photons strikes an object of this size, it does not alter its position or momentum to any practical extent. Imagine, however, that you wish to locate an electron by similarly bouncing light off it into some detector. Objects can be located to an accuracy no greater than the wavelength of the radiation used. Thus, if we want an accurate position measurement for an electron, we must use a short wavelength. This means that photons of high energy must be employed. The more energy the photons have, the more momentum they impart to the electron when they strike it, which changes the electron's motion in an unpredictable way. The attempt to measure accurately the electron's position introduces considerable uncertainty in its momentum; the act of measuring the electron's position at one moment makes our knowledge of its future position inaccurate.

Suppose, then, that we use photons of longer wavelength. Because these photons have lower energy, the momentum of the electron is not so appreciably changed during measurement, but its position will be correspondingly less accurately known. This is the essence of the uncertainty principle: *There is an uncertainty in either the position or the momentum of the electron that cannot be reduced beyond a certain minimum level.* The more accurately one is known, the less accurately the other is known. Although we can never know with certainty the exact position and motion of the electron, we can talk about the probability of the electron being at certain locations in space. In the next section, we introduce a model of the atom that provides the probability of finding electrons of specific energies at certain positions in atoms.

tion and the momentum of any object. The limitation becomes important only when we deal with matter at the subatomic level, that is, with masses as small as that of an electron. Heisenberg's principle is called the **uncertainty principle**. When applied to the electrons in an atom, this principle states that it is inherently impossible for us to know simultaneously both the exact momentum of the electron and its exact location in space. Thus, it is not appropriate to imagine the electrons as moving in well-defined circular orbits about the nucleus.

De Broglie's hypothesis and Heisenberg's uncertainty principle set the stage for a new and more broadly applicable theory of atomic structure. In this new approach, any attempt to define precisely the instantaneous location and momentum of the electrons is abandoned. The wave nature of the electron is recognized, and the electron's behavior is described in terms appropriate to waves.

6.5 QUANTUM MECHANICS AND THE HYDROGEN ATOM

Quantum mechanics can be used to determine the allowed electron energy states in the hydrogen atom. Even for such a simple system, however, the mathematics used is quite advanced, and we will not present the details here. We will instead present a qualitative description that will give you the essence of the full quantum-mechanical treatment.

The energy of the electron is a combination of its potential energy (due to the attraction of the electron to the nucleus) and its kinetic energy. In quantum mechanics, we also impose wave properties on the electron. The Austrian physicist Erwin Schrödinger (1887–1961) proposed an equation, now known as **Schrödinger's wave equation**, that properly incorporated the wavelike behavior of the electron. Schrödinger was awarded the Nobel Prize in physics in 1933. Schrödinger's equation leads to a series of solutions that describe the allowed energy states of the electron. These solutions, usually represented by the symbol ψ (the Greek lowercase letter psi), are called **wave functions.***

A wave function provides information about an electron's location in space when it is in a certain allowed energy state. The allowed energies are the same as those predicted by the Bohr model. However, the Bohr model assumes that the electron is in a circular orbit of some particular radius about the nucleus. In the quantum-mechanical model for hydrogen, it is not so simple to describe the electron's location. The uncertainty principle suggests that if we know the momentum of the electron with high accuracy, our knowledge of its location is very uncertain. Thus, we cannot hope to specify the location of an individual electron around the nucleus. Rather, we must be content with a kind of statistical knowledge. In the quantum-mechanical model, we therefore speak of the *probability* that the electron will be in a certain region of space at a given instant. As it turns out, the square of the wave function, ψ^2, at a given point in space represents the probability that the electron will be found at that location. For this reason, ψ^2 is called the **probability density**.

One way of representing the probability of finding the electron in various regions of an atom is shown in Figure 6.18. In this figure, the density of the dots represents the probability of finding the electron. The regions with a high density of dots correspond to relatively large values for ψ^2. **Electron density** is another way of expressing probability: Regions where there is a high probability of finding the electron are said to be regions of high electron density. In Section 6.6, we will say more about the ways in which we can represent electron density.

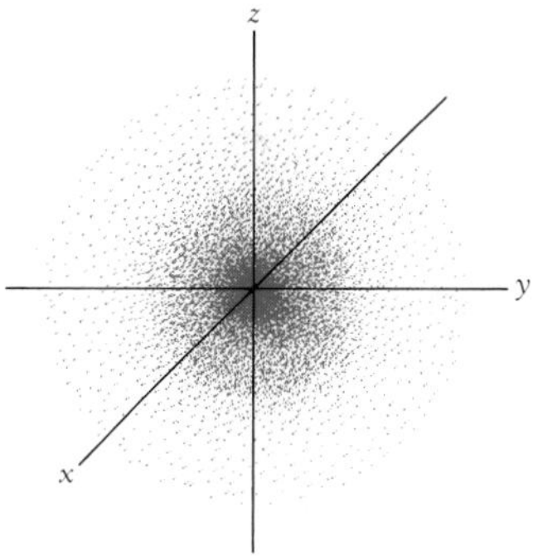

Figure 6.18 Electron-density distribution in the ground state of the hydrogen atom.

Orbitals and Quantum Numbers

The complete quantum-mechanical solution to the hydrogen atom problem yields a set of wave functions and a corresponding set of energies. Each wave function satisfies Schrödinger's equation. These allowed wave functions for the hydrogen atom are called **orbitals**. Each orbital describes a specific distribution of electron density in space, as given by its probability density. An orbital therefore has both a characteristic energy and a characteristic shape. For example, the lowest energy orbital in the hydrogen atom has an energy of -2.18×10^{-18} J and the shape illustrated in Figure 6.18. Note that an *orbital* (quantum-mechanical model) is not the same as an *orbit* (Bohr model).

The Bohr model introduced a single quantum number, n, to describe an orbit. The quantum-mechanical model uses three quantum numbers, n, l, and m_l, to describe an orbital. We will consider what information we obtain from each of these and how they are interrelated.

1. The *principal quantum number*, n, can have integral values of 1, 2, 3, and so forth. As n increases, the orbital becomes larger, and the electron spends more time farther from the nucleus. An increase in n also means that the

* In mathematics, a *function* has a specific value that depends on one or more variables. A wave function of an electron has a specific value for each location (x, y, z) in space: ψ is a function of x, y, and z.

Table 6.2 Letters Used to Label Atomic Orbitals

Value of l	0	1	2	3
Letter used	s	p	d	f

Table 6.3 Relationship Among Values of n, l, and m_l through $n = 4$

n	l	Subshell designation	m_l	Number of orbitals in subshell
1	0	1s	0	1
2	0	2s	0	1
	1	2p	1, 0, −1	3
3	0	3s	0	1
	1	3p	1, 0, −1	3
	2	3d	2, 1, 0, −1, −2	5
4	0	4s	0	1
	1	4p	1, 0, −1	3
	2	4d	2, 1, 0, −1, −2	5
	3	4f	3, 2, 1, 0, −1, −2, −3	7

electron has a higher energy and is therefore less tightly bound to the nucleus. For hydrogen, $E_n = -(2.18 \times 10^{-18}\text{ J})(1/n^2)$, as in the Bohr model.

2. The second quantum number—the *azimuthal quantum number*, l—can have integral values from 0 to $n - 1$ for each value of n. This quantum number defines the shape of the orbital. (We will consider these shapes in Section 6.6.) The value of l for a particular orbital is generally designated by the letters s, p, d, and f,* corresponding to l values of 0, 1, 2, and 3, respectively, as summarized in Table 6.2.

3. The *magnetic quantum number*, m_l, can have integral values between l and $-l$, including zero. This quantum number describes the orientation of the orbital in space. (The orientations will be considered in Section 6.6.)

A collection of orbitals with the same value of n is called an **electron shell**. For example, all the orbitals with $n = 3$ are said to be in the third shell. One or more orbitals with the same set of n and l values is called a **subshell**. Each subshell is designated by a number (the value of n) and a letter (s, p, d, or f, corresponding to the value of l). For example, all of the orbitals of an atom with $n = 3$ and $l = 1$ are collectively referred to as 3p orbitals and are said to be in the 3p subshell. The possible values of the three quantum numbers through $n = 4$ (the fourth shell) are summarized in Table 6.3.

* The letters s, p, d, and f come from the words *sharp*, *principal*, *diffuse*, and *fundamental*, which were used to describe certain features of spectra before quantum mechanics was developed.

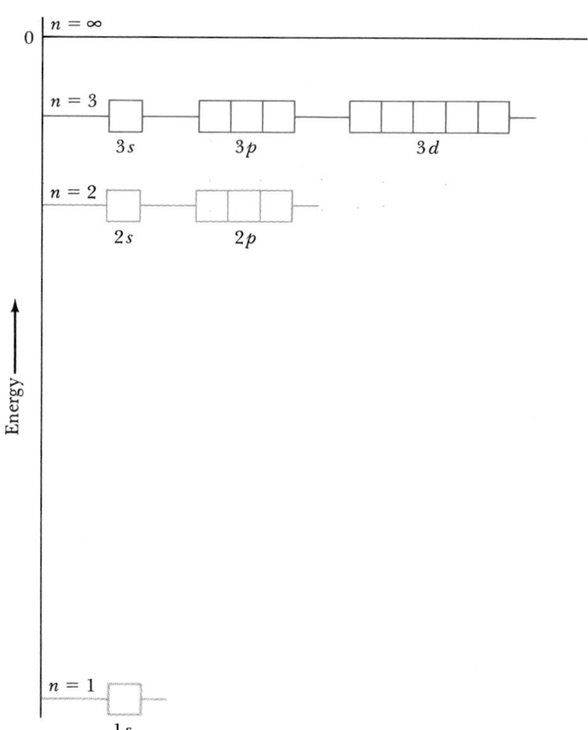

Figure 6.19 Orbital energy levels in the hydrogen atom and in hydrogenlike ions (those containing just one electron). Note that all orbitals with the same value for the principal quantum number, n, have the same energy. This is true only in one-electron systems.

The restrictions on the possible values of the quantum numbers give rise to a pattern that is very important:

1. Each shell is divided into the number of subshells equal to the principal quantum number, n, for that shell. The first shell consists of only the 1*s* subshell; the second shell consists of two subshells, 2*s* and 2*p*; the third shell consists of three subshells, 3*s*, 3*p*, and 3*d*. Thus, every shell has an *s* subshell; every shell beginning with the second has a *p* subshell; every shell beginning with the third has a *d* subshell, and so forth.
2. Each subshell is divided into orbitals. Each *s* subshell consists of one orbital; each *p* subshell consists of three orbitals; each *d* subshell consists of five orbitals; each *f* subshell consists of seven orbitals. (Notice that these are the odd numbers, 1, 3, 5, 7.)

Figure 6.19 shows the number and relative energies of all hydrogen atom orbitals through $n = 3$. Each box represents an orbital; orbitals of the same subshell, such as the 2*p*, are grouped together. When the electron is in the lowest energy orbital (the 1*s* orbital), the hydrogen atom is said to be in its *ground state*. When the electron is in any other orbital, the atom is in an *excited state*. At ordinary temperatures, essentially all hydrogen atoms are in their ground states. The electron may be promoted to an excited-state orbital by absorption of a photon of appropriate energy.

SAMPLE EXERCISE 6.6

(a) Without referring to Table 6.3, predict the number of subshells in the fourth shell (that is, for $n = 4$). **(b)** Give the label for each of these subshells. **(c)** How many orbitals are in each of these subshells?

Solution: **(a)** There are four subshells in the fourth shell, corresponding to the four possible values of l (0, 1, 2, and 3).

(b) These subshells are labeled 4*s*, 4*p*, 4*d*, and 4*f*. The number given in the designation of a subshell is the principal quantum number, *n*; the following letter designates the value of the azimuthal quantum number, *l*, as given in Table 6.2.

(c) There is one 4*s* orbital (when $l = 0$, there is only one possible value of m_l: 0). There are three 4*p* orbitals (when $l = 1$, there are three possible values of m_l: 1, 0, and -1). There are five 4*d* orbitals (when $l = 2$, there are five allowed values of m_l: 2, 1, 0, -1, -2). There are seven 4*f* orbitals (when $l = 3$, there are seven permitted values of m_l: 3, 2, 1, 0, -1, -2, -3).

PRACTICE EXERCISE

(a) What is the designation for the subshell with $n = 5$ and $l = 1$? **(b)** How many orbitals are in this subshell? **(c)** Indicate the values of m_l for each of these orbitals.
Answers: **(a)** 5*p*; **(b)** 3; **(c)** 1, 0, -1

6.6 REPRESENTATIONS OF ORBITALS

In our discussion of orbitals, we have so far emphasized their energies. But the wave function also provides information about the electron's location in space when it is in a particular allowed energy state. We need to examine the ways that we can picture the orbitals.

The *s* Orbitals

The lowest-energy (most stable) orbital, the 1*s* orbital, is spherically symmetric, as shown in Figure 6.18. Figures of this type, showing electron density, are one of the several ways we have to help us visualize orbitals. This figure indicates that the probability of finding the electron around the nucleus decreases as we move away from the nucleus in any direction. When the probability function, ψ^2, for the 1*s* orbital is graphed as a function of the distance from the nucleus, *r*, it rapidly approaches zero, as shown in Figure 6.20(*a*). This effect indicates

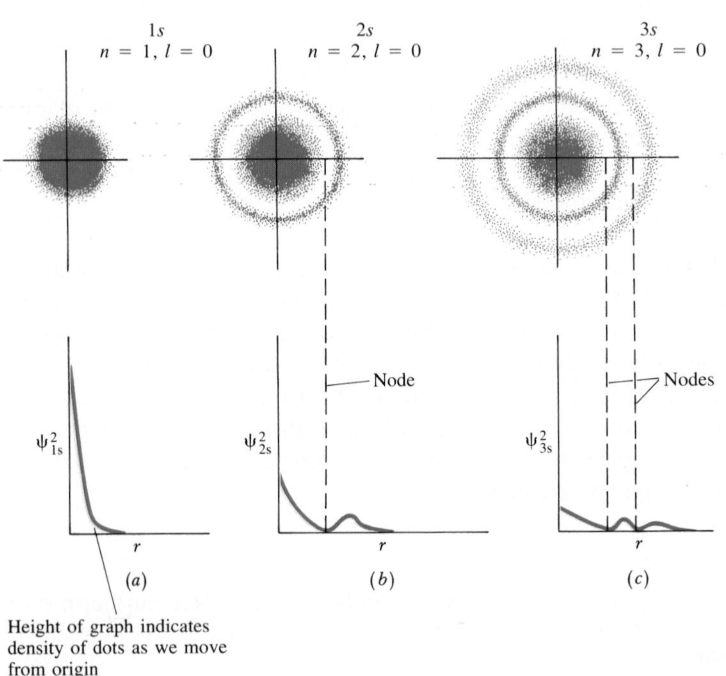

Figure 6.20 Electron-density distribution in 1*s*, 2*s*, and 3*s* orbitals. The lower part of the figure shows how the electron density, represented by ψ^2, varies as a function of distance from the nucleus. In the 2*s* and 3*s* orbitals, the electron-density function drops to zero at certain distances from the nucleus. The spherical surfaces around the nucleus at which ψ^2 is zero are called *nodes*.

that the electron, which is drawn toward the nucleus by electrostatic attraction, is unlikely ever to get very far from the nucleus.

If we similarly consider the 2*s* and 3*s* orbitals of hydrogen, we find that they are also spherically symmetric. Indeed, *all s* orbitals are spherically symmetric. The manner in which the probability function, ψ^2, varies with *r* for the 2*s* and 3*s* orbitals is shown in Figure 6.20(*b*) and (*c*). Notice that for the 2*s* orbital, ψ^2 goes to zero and then increases again in value before finally approaching zero at a larger value of *r*. The intermediate regions where ψ^2 goes to zero are called **nodal surfaces**, or simply **nodes**. The number of nodes increases with increasing value for the principal quantum number, *n*. The 3*s* orbital possesses two nodes, as illustrated in Figure 6.20(*c*). Notice also that as *n* increases, the electron is more and more likely to be located farther from the nucleus. That is, the size of the orbital increases as *n* increases.

The most widely used method of representing orbitals is to display a boundary surface that encloses some substantial fraction, say 90 percent, of the total electron density for the orbital. For the *s* orbitals, these contour representations are merely spheres. The contour or boundary surface representations of the 1*s*, 2*s*, and 3*s* orbitals are shown in Figure 6.21. They have the same shape, but they differ in size. The fact that there are nodes within the 2*s* and 3*s* surfaces is lost in these representations. This is not a serious disadvantage; it turns out that for more qualitative discussions the most important features of orbitals are their size and shape. These features are adequately represented by the contour diagrams.

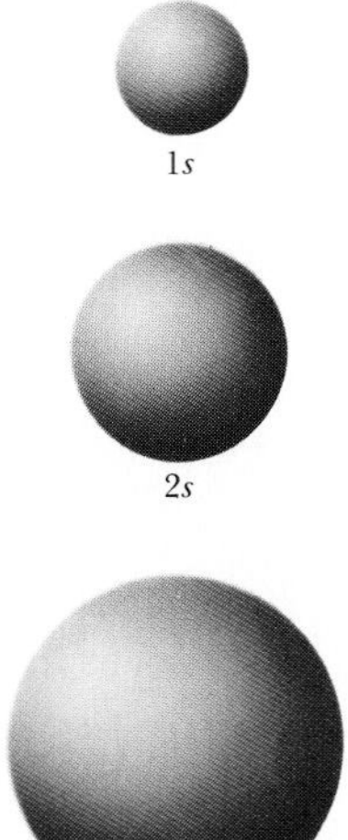

Figure 6.21 Contour representations of the 1*s*, 2*s*, and 3*s* orbitals. The relative radii of the spheres correspond to a 90 percent probability of finding the electron within each sphere.

The *p* Orbitals

The distribution of electron density for a 2*p* orbital is shown in Figure 6.22. As we can see from this figure, the electron density is not distributed in a spherically symmetric fashion as in an *s* orbital. Instead, the electron density is concentrated on two sides of the nucleus, separated by a node at the nucleus; we often say that this orbital has two lobes. It is useful to recall that we are making no statement of how the electron is moving within the orbital; Figure 6.22 portrays the *averaged* distribution of the 2*p* electron in space.

Figure 6.22 Electron-density distribution in a 2*p* orbital.

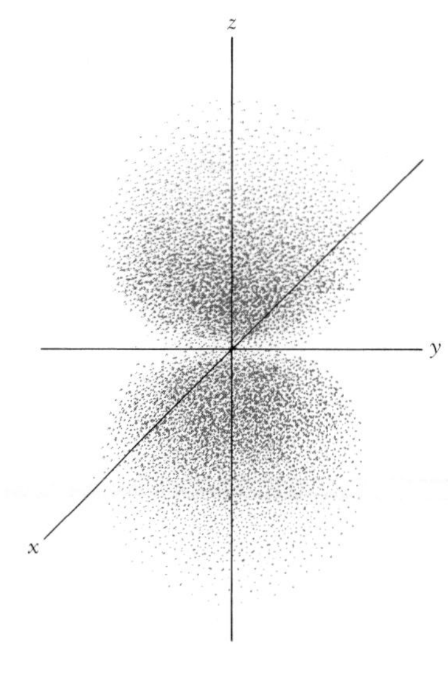

Each shell beginning with $n = 2$ has three *p* orbitals. For example, there are three 2*p* orbitals, three 3*p* orbitals, and so forth. The orbitals of a given principal quantum number have the same size and shape but differ from each other in orientation. The contour surfaces of the three 2*p* orbitals are shown in Figure 6.23. It is convenient to label these as the $2p_x$, $2p_y$, and $2p_z$ orbitals. The letter subscript indicates the axis along which the orbital is oriented. As it turns out, there is no necessary connection between one of these subscripts

Figure 6.23 Contour representations of the three 2*p* orbitals. Note that the subscript on the orbital label indicates the axis along which the orbital lies.

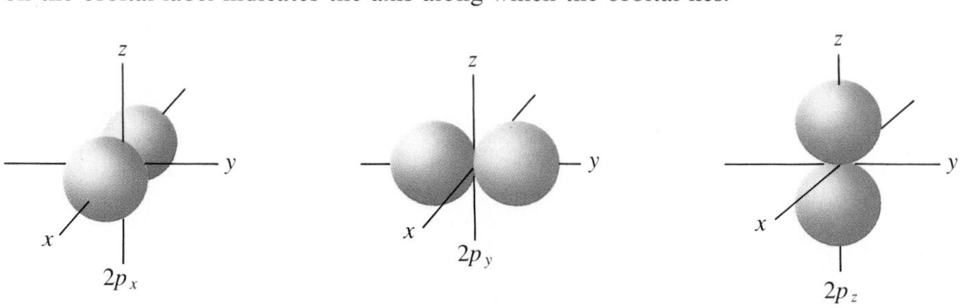

and a particular value of m_l. To explain why this is so would require discussion of material beyond the scope of an introductory text.

The distance from the nucleus to the center of the electron density moves outward as we go from the 2*p* to 3*p* to 4*p* orbitals. In other words, orbital size increases with increase in the principal quantum number, *n*. In accurate contour representations of the 3*p* and higher *p* orbitals, small regions of electron density separate the major lobes. These details of the shapes of the *p* orbitals are not of major chemical importance. The general overall shape of the *p* orbitals is more significant. We shall therefore always represent the *p* orbitals as shown in Figure 6.23, regardless of the value for the principal quantum number, *n*.

The *d* and *f* Orbitals

When $l = 2$, the orbital is a *d* orbital. No *d* orbitals can exist with *n* lower than 3 because *l* can never be larger than $n - 1$. There are five equivalent 3*d* orbitals corresponding to the five possible values for m_l: 2, 1, 0, -1, and -2. Similarly, there are five 4*d* orbitals, and so forth. Just as in the case of the *p* orbitals, the differing values of m_l correspond to different orientations of orbitals in space. The most useful representations of the 3*d* orbitals are shown in Figure 6.24. Notice that four of these orbitals have lobes centered in the plane indicated by the subscript on the orbital label. The lobes of both the d_{xy} and $d_{x^2-y^2}$ orbitals lie in the *xy* plane. However, the lobes of the $d_{x^2-y^2}$ orbital lie along the *x* and *y* axes, whereas the lobes of the d_{xy} orbital lie between the axes. The lobes of the d_{xz} and d_{yz} orbitals also lie between the respective axes. The one orbital of unique shape is the d_{z^2}, which has two lobes along the *z* axis and a "doughnut" in the *xy* plane. Although the d_{z^2} orbital looks different, it has the same energy as the other four *d* orbitals.

The representations of higher *d* orbitals are very much like those for the 3*d*. The contour representations shown in Figure 6.24 are commonly employed for all *d* orbitals, regardless of principal quantum number.

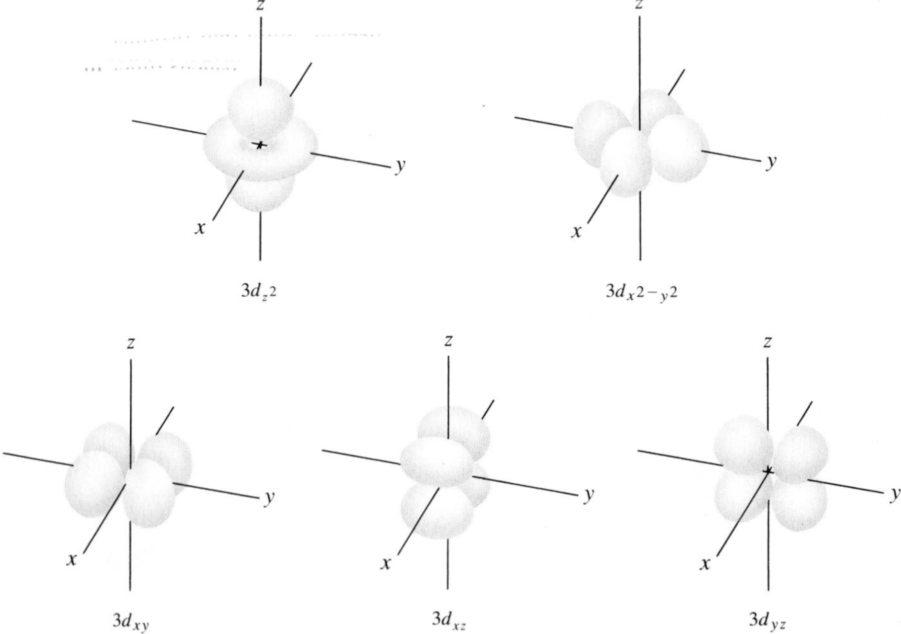

Figure 6.24 Contour representations of the five 3*d* orbitals.

There are seven equivalent f orbitals (for which $l = 3$) for each value of n of 4 or greater. The f orbitals are difficult to represent in three-dimensional contour diagrams. We shall have no need to concern ourselves with orbitals having values for l greater than 3.

As we shall see in Chapter 9, an understanding of the number and shapes of atomic orbitals is important to a proper understanding of the molecules formed by combining atoms. *You should commit to memory the orbital representations shown in Figures 6.21, 6.23, and 6.24.*

6.7 ORBITALS IN MANY-ELECTRON ATOMS

We have seen that quantum mechanics leads to a very elegant description of the hydrogen atom. The hydrogen atom is the simplest atom, of course, because it has only one electron. How does our description of atomic electronic structure have to be changed when we discuss atoms with two or more electrons? Fortunately, we can describe an atom with more than one electron (a *many-electron* atom) in terms of orbitals like those for hydrogen. Thus, we can continue to designate orbitals as 1s, $2p_x$, and so forth. Furthermore, these orbitals have the same shapes as the corresponding hydrogen orbitals.

Although the shapes of the orbitals for many-electron atoms are the same as those for hydrogen, the presence of more than one electron greatly changes the energies of the orbitals. In hydrogen, the energy of an orbital depends only on its principal quantum number, n (Figure 6.19); the 3s, 3p, and 3d subshells all have the same energy, for instance. In a many-electron atom, however, the electron-electron repulsions cause different subshells to be at different energies, as shown in Figure 6.25. For example, the 2s subshell is lower in energy than

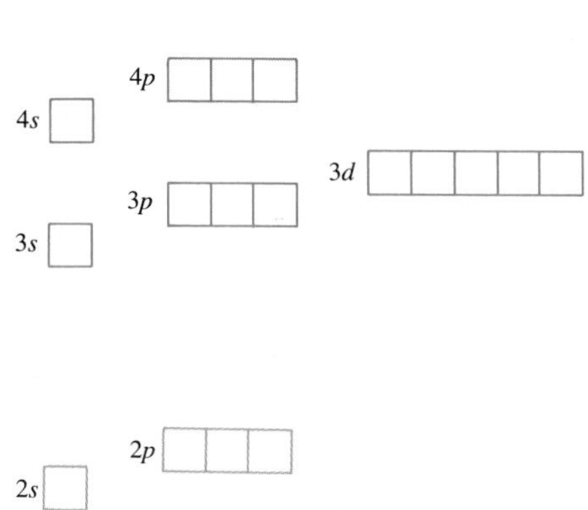

Figure 6.25 Ordering of orbital energy levels in many-electron atoms, through the 4p orbitals. As in Figure 6.19, which shows the orbital energy levels for the hydrogen atom, each box represents an orbital. Note that orbitals in different subshells differ in energy. In some atoms, the 4s orbital is lower in energy than the 3d orbitals.

the $2p$ subshell. To understand why this is so, we need to consider the forces between the electrons and how these forces are affected by the shapes of the orbitals.

Effective Nuclear Charge

If an atom or ion has only one electron (H, He^+, Li^{2+}, and so forth), the energy of that electron can be calculated exactly. That energy depends on the principal quantum number, n, and the charge of the nucleus, Z:

$$E_n = -R_H\left(\frac{Z^2}{n^2}\right) = -(1312\ \text{kJ/mol})\left(\frac{Z^2}{n^2}\right) \qquad [6.10]$$

Notice that Equation 6.6, which applies to hydrogen, is just a special case of this equation in which $Z = 1$.

Now let us consider a many-electron atom. As well as being attracted to the nucleus, each electron is repelled by the electron density due to the other electrons in the atom. In general, there are so many electron-electron repulsions that we cannot analyze the situation exactly. We can, however, estimate the energy of each electron by considering how it interacts with the *average* environment created by the nucleus and all the other electrons in the atom.

Any electron density between the nucleus and the electron of interest will reduce the nuclear charge acting on that electron. The net positive charge attracting the electron is called the **effective nuclear charge**. The effective nuclear charge, Z_{eff}, equals the number of protons in the nucleus, Z, minus the average number of electrons, S, that are between the nucleus and the electron in question:

$$Z_{eff} = Z - S \qquad [6.11]$$

Thus, the positive charge experienced by outer-shell electrons is always less than the full nuclear charge because the inner-shell electrons partly offset the positive charge of the nucleus. The inner electrons are said to shield or screen the outer electron from the full charge of the nucleus. This effect, which is called the **screening effect**, is illustrated in Figure 6.26.

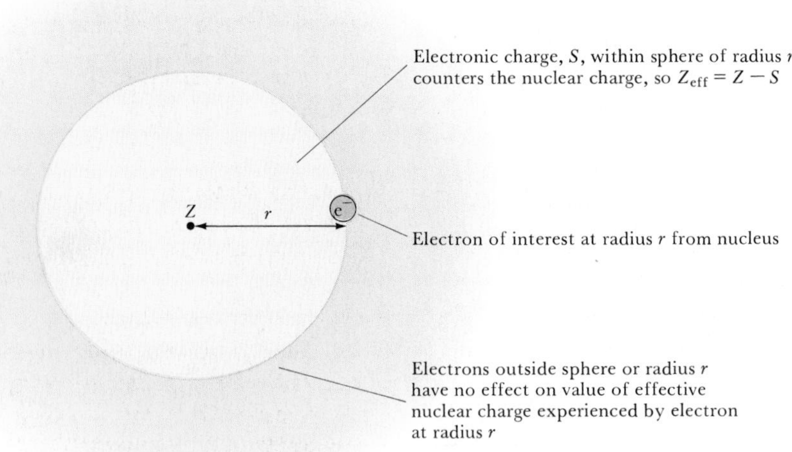
Electronic charge, S, within sphere of radius r counters the nuclear charge, so $Z_{eff} = Z - S$
Z
r
e⁻
Electron of interest at radius r from nucleus
Electrons outside sphere or radius r have no effect on value of effective nuclear charge experienced by electron at radius r

Figure 6.26 Shielding of the nuclear charge from an electron by other electrons in an atom. As an example, if the nuclear charge were 5 and the sphere of radius r contained three electrons, the effective nuclear charge at the radius of the sphere would be $5 - 3 = 2$.

Energies of Orbitals

The extent to which an electron will be screened by the other electrons depends on its electron distribution as we move outward from the nucleus. For a given value of n, this distribution differs for each value of the azimuthal quantum number l. Consider the orbitals for which $n = 3$. An electron in the 3*s* orbital is more likely to be close to the nucleus than an electron in the 3*p* orbital; an electron in the 3*p* orbital, in turn, has a greater probability of being close to the nucleus than one in the 3*d* orbital. As a result, the other electrons in the atom (that is, the 1*s*, 2*s*, and 2*p*) shield the 3*s* electrons least effectively and the 3*d* electrons most effectively. Thus, the 3*s* electrons experience a larger Z_{eff} than do the 3*p* electrons, which in turn experience a larger Z_{eff} than do the 3*d* electrons. We can generalize these observations: *In a many-electron atom, for a given value of n,* Z_{eff} *decreases with increasing value of l.*

The energy of an electron depends on the effective nuclear charge, Z_{eff}. Because Z_{eff} is larger for the 3*s* electrons, they have a lower energy (that is, they are more stable) than the 3*p*, which in turn are lower in energy than the 3*d*. Again, we can generalize: *In a many-electron atom, for a given value of n, the energy of an orbital increases with increasing value of l.* The relative energies of orbitals in a many-electron atom are shown in Figure 6.25. Figure 6.25 is a *qualitative* energy-level diagram; the exact energies and their spacings differ from one atom to another. Notice that all orbitals of a given subshell (such as the 3*d* orbitals) still have the same energy, just as they do in the hydrogen atom. Orbitals that have the same energy are said to be **degenerate**.

SAMPLE EXERCISE 6.7

Based on the energy-level diagram in Figure 6.25, would you expect the average distance from the nucleus of a 3*d* electron to be greater or less than that of a 2*p* electron? Explain.

Solution: The energy of the 2*p* orbitals is considerably lower than for a 3*d*. This indicates that Z_{eff} for the 2*p* electron is much greater than that for an electron in the 3*d* orbital. The increased attractive interaction is due to a smaller average distance of the 2*p* electron from the nucleus.

PRACTICE EXERCISE

The sodium atom has 11 electrons. Two of these occupy a 1*s* orbital, two occupy a 2*s* orbital, and one occupies a 3*s* orbital. Which of these electrons experience the lowest effective nuclear charge? ***Answer:*** The electron in the 3*s* orbital

6.8 ELECTRON SPIN AND THE PAULI EXCLUSION PRINCIPLE

We have seen that we can use orbitals to describe many-electron atoms. What determines the orbitals in which electrons reside? We will address this question in the next section. First, however, we must introduce a new property of the electron.

When scientists studied the line spectra of many-electron atoms in great detail, they noticed a very puzzling feature: Lines that were originally thought to be single were actually closely spaced pairs. This meant, in essence, that there were twice as many energy levels as there were "supposed" to be. In 1925, the Dutch physicists George Uhlenbeck and Samuel Goudsmit proposed a solution to this dilemma. They postulated that electrons have an intrinsic property,

A CLOSER LOOK: Experimental Evidence for Electron Spin

Even before electron spin had been proposed, there was experimental evidence that electrons had an additional property that needed explanation. In 1921, Otto Stern and Walter Gerlach succeeded in separating a beam of neutral atoms into two groups by passing them through an inhomogeneous magnetic field. Their experiment is diagramed in Figure 6.27. Let us assume that they used a beam of hydrogen atoms (in actuality, they used silver atoms). We would normally expect that neutral atoms would not be affected by a magnetic field. However, the magnetic field arising from the electron's spin interacts with the magnet's field, deflecting the atom from its straight-line path. As shown in Figure 6.27, the magnetic field splits the beam in two, suggesting that there are two (and only two) equivalent values for the electron's own magnetic field. The Stern-Gerlach experiment could be readily interpreted once it was realized that there are exactly two values for the spin of the electron. These values will produce equal magnetic fields that are opposite in direction.

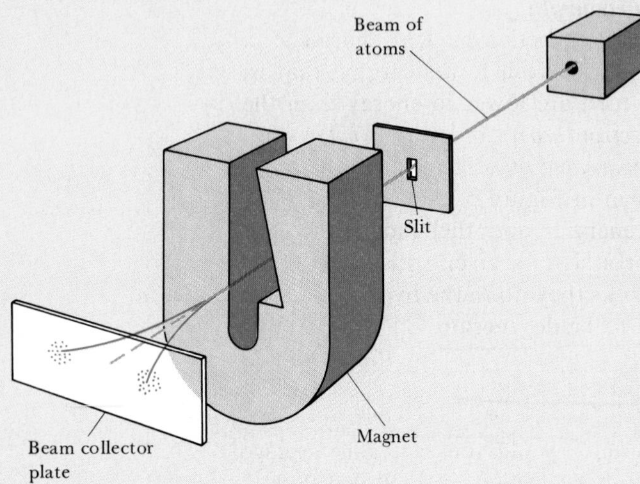

Figure 6.27 Diagrammatic illustration of the Stern-Gerlach experiment. A beam of atoms is allowed to pass through an inhomogeneous magnetic field. Atoms in which the electron-spin quantum number m_s is $+\frac{1}{2}$ are deflected in one direction, whereas those in which m_s is $-\frac{1}{2}$ are deflected in the other.

called **electron spin**. If we view the electron as a sphere, we can envision it as spinning on its own axis.*

By now it probably does not surprise you to learn that electron spin is quantized. This observation led to the assignment of a new quantum number for the electron, in addition to n, l, and m_l that we have already discussed. This new quantum number, the **electron spin quantum number**, is denoted m_s (the subscript s stands for *spin*). Only two possible values are allowed for m_s, $+\frac{1}{2}$ or $-\frac{1}{2}$, which we interpret as indicating the two opposite directions in which the electron can spin. A spinning charge produces a magnetic field. The two opposite directions of spin produce oppositely directed magnetic fields, as shown in Figure 6.28. These two opposite magnetic fields lead to the splitting of spectral lines into closely spaced pairs.

Electron spin is crucial for understanding the electronic structures of atoms. In 1925, the Austrian-born physicist Wolfgang Pauli (Figure 6.29) dis-

* As we discussed earlier, the electron has both particlelike and wavelike properties. Thus, the picture of an electron as a spinning charged sphere is, strictly speaking, not correct. It is a useful conceptual picture of electron spin, however.

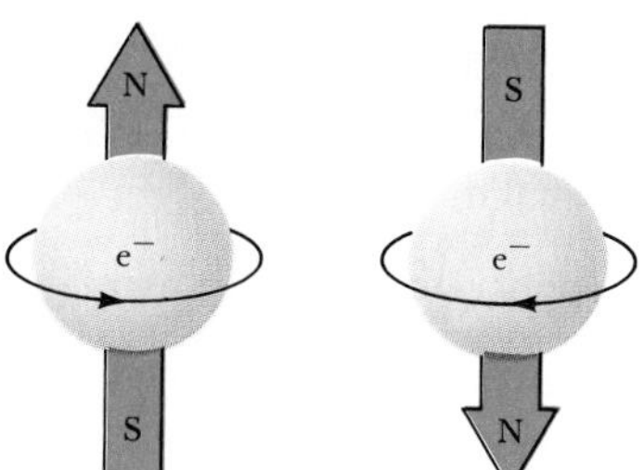

Figure 6.28 The electron behaves as if it were spinning about an axis through its center. The two directions of spin correspond to the two possible values for the spin quantum number, m_s.

covered the principle that governs the arrangements of electrons in many-electron atoms. The **Pauli exclusion principle** states that *no two electrons in an atom can have the same set of four quantum numbers n, l, m_l, and m_s*. For a given orbital (1*s*, 2p_z, and so forth), the values of *n*, *l*, and m_l are fixed. Thus, if we want to put more than one electron in an orbital *and* satisfy the Pauli exclusion principle, our only choice is to assign different m_s values to the electrons. Because there are only two such values, we conclude that *an orbital can hold a maximum of two electrons, and they must have opposite spins.* This restriction provides the key to one of the great problems in chemistry—understanding the structure of the periodic table of the elements. We shall address this issue in the next section.

Figure 6.29 Wolfgang Pauli (1900–1958), who formulated the exclusion principle. Born in Vienna, Pauli received his Ph.D. in 1921 from the University of Munich. Following further study with Niels Bohr, he became a professor at the University of Zurich. Pauli received the Nobel Prize in physics in 1945. (Austrian Press Information Service)

CHEMISTRY AT WORK: Nuclear Spin and Magnetic Resonance Imaging

A major challenge facing medical diagnosis is to see inside the human body from the outside. Until recently, this was accomplished primarily by using X rays to yield images of human bones, muscles, and organs. However, there are several drawbacks to using X rays for medical imaging. First, X rays do not give well-resolved images of overlapping physiological structures. Moreover, because damaged or diseased tissue often yields the same image as healthy tissue, X rays frequently fail to detect illness or injuries. Finally, X rays are high-energy radiation that can cause physiological harm, even in low doses.

In the 1980s, a new technique called *magnetic resonance imaging* (MRI) moved to the forefront of medical imaging technology. The foundation of MRI is a phenomenon called nuclear magnetic resonance (NMR), which was discovered in the mid-1940s. Today, NMR has become one of the most important spectroscopic methods used in chemistry. It is based on the observation that, like electrons, the nuclei of many elements possess an intrinsic spin. Like electron spin, nuclear spin is quantized. For example, the nucleus of ^1H (a proton) has two possible nuclear-spin quantum numbers, $+\frac{1}{2}$ and $-\frac{1}{2}$. The hydrogen nucleus is the most common one studied by NMR.

A spinning hydrogen nucleus acts like a tiny magnet. In the absence of external effects, the two spin states have the same energy. However, when the nuclei are placed in an external magnetic field, they can align either parallel or opposed (antiparallel) to the field, depending on their spin. The parallel alignment is lower in energy than the antiparallel one (Figure 6.30) by a certain amount, ΔE. If the nuclei are irradiated with photons with energy equal to ΔE, the spin of the nuclei can be "flipped," that is, excited from the parallel to the antiparallel alignment. Detection of the flipping of nuclei between the two spin states leads to an NMR spectrum. The radiation used in an NMR experiment is in the radiofrequency range, typically 100 to 500 MHz.

Because hydrogen is a major constituent of aqueous body fluids and fatty tissue, the hydrogen nucleus is the most convenient one for study by MRI. In MRI, a person's body is placed in a strong magnetic field. By irradiating the body with pulses of radio-frequency radiation and using sophisticated detection techniques, tissue can be imaged at specific depths within the body, giving pictures with spectacular detail (Figure 6.31). The ability to sample at different depths allows medical technicians to construct a three-dimensional picture of the body.

Figure 6.30 Like electron spin, nuclear spin generates a small magnetic field and has two allowed values. In the absence of an external magnetic field (left), the two spin states have the same energy. If an external magnetic field is applied (right), the parallel alignment of the nuclear magnetic field is lower in energy than the antiparallel alignment. The energy difference, ΔE, is in the radio-frequency portion of the electromagnetic spectrum.

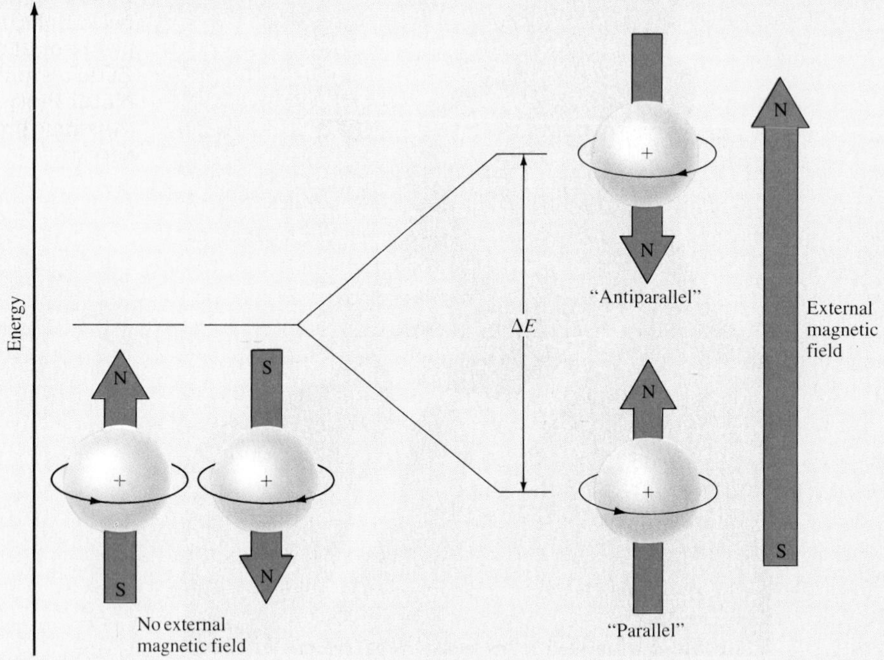

MRI has none of the disadvantages of X rays. Diseased tissue appears very different from healthy tissue, resolving overlapping structures at different depths in the body is much easier, and the radio-frequency radiation is not harmful to humans in the doses used. The major drawback of MRI is expense: The current cost of a new MRI instrument for clinical applications is over $1,000,000.

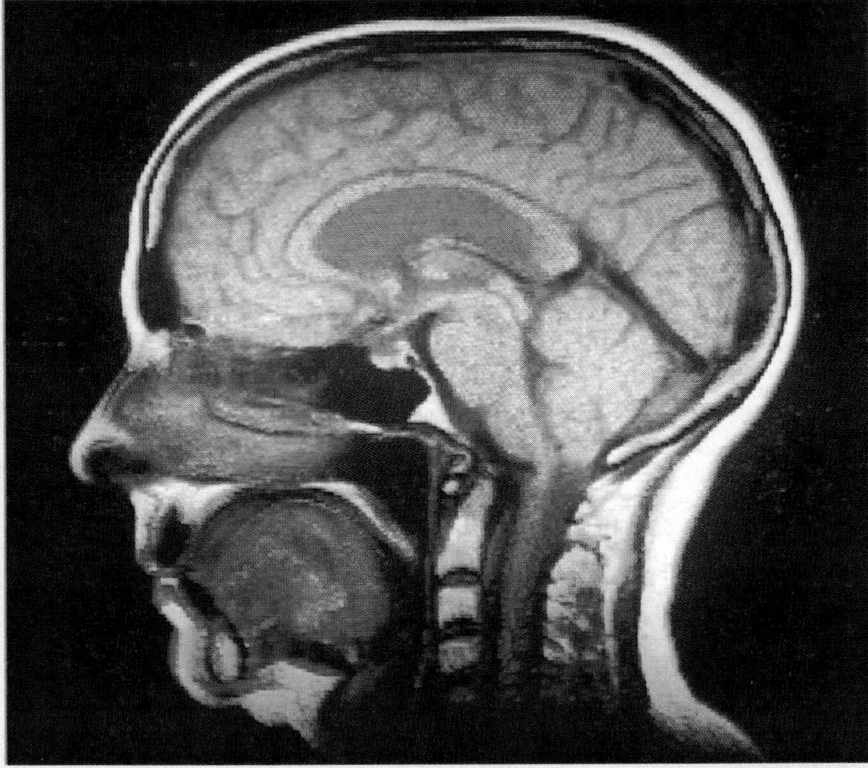

Figure 6.31 An MRI image of a human head. (Dan McCoy/Rainbow).

6.9 ELECTRON CONFIGURATIONS

Elements are arranged in the periodic table according to atomic number. The atomic number corresponds not only to the number of protons in the nucleus of the neutral atom, but also to the number of electrons in that atom. With Rutherford's discovery of the nuclear atom in 1911, scientists began to seek relationships between atomic structure and chemical behavior. The American chemist Gilbert N. Lewis (1875–1946), reasoned that if the chemical properties of the elements are repeated when the elements are arranged by atomic number, then their electron configurations must be repeating in some way. He suggested that electrons in atoms are arranged in shells and that once a shell of electrons is filled, additional electrons must go into a new shell. He further reasoned that the noble gases, which are chemically very unreactive, have closed or filled shells of electrons. Chemical behavior might then be understood in terms of the tendency of an atom to achieve a closed shell by gaining, losing, or sharing electrons. For example, we saw in Section 2.5 that sodium, atomic number 11, loses one electron in chemical reactions to form the sodium ion, Na^+. This ion has the stable, closed-shell electron arrangement of the noble gas neon, atomic number 10.

These ideas of Lewis generate additional questions. Why do certain electron configurations occur periodically? Why are electrons arranged in shells?

To answer these questions and to account in more detail for the ways in which certain atomic properties vary with atomic number, we must apply the ideas about orbitals that we have learned earlier in this chapter.

The Electron Configurations of the Elements

The arrangement of electrons in an atom is called the **electron configuration**. The most stable, or ground, electron configuration of an atom is that in which the electrons are in the lowest possible energy states. If there were no restrictions on the possible values for the quantum numbers of the electrons, all the electrons would crowd into the 1*s* orbital, because this is the lowest in energy (Figure 6.25). The Pauli exclusion principle, however, tells us that there can be at most two electrons in any single orbital. Thus, the orbitals are filled in order of increasing energy, with no more than two electrons per orbital. For example, consider the lithium atom, which has three electrons. (Recall that the number of electrons in a neutral atom is equal to its atomic number, *Z*.) The 1*s* orbital can accommodate two of the electrons. The third one goes into the next lowest energy orbital, the 2*s*.

We can summarize any electron configuration by writing the symbol for the occupied subshell and adding a superscript to indicate the number of electrons in that subshell. For example, for lithium we write $1s^2 2s^1$ (read "1*s* two, 2*s* one"). We can also show the arrangement of the electrons in the following way:

Li	[↿⇂]	[↿]
	1*s*	2*s*

In this kind of representation, which we shall call an **orbital diagram**, each orbital is represented by a box and each electron by a half-arrow. A half-arrow pointing upward (↿) represents an electron with a positive spin quantum number ($m_s = +\frac{1}{2}$), and a downward half-arrow (⇂) represents an electron with a negative spin quantum number ($m_s = -\frac{1}{2}$). This pictorial representation of electron spin is quite convenient. In fact, chemists and physicists often refer to electrons as "spin-up" and "spin-down" rather than specifying the value for m_s.

Electrons having opposite spins are said to be *paired* when they are in the same orbital. An *unpaired electron* is not accompanied by a partner of opposite spin. In the lithium atom, the two electrons in the 1*s* orbital are paired, and the electron in the 2*s* orbital is unpaired.

Writing Electron Configurations

It is informative to consider how the electron configurations of the elements change as we move from element to element across the periodic table. Hydrogen has one electron, which occupies the 1*s* orbital in its ground state:

H	[↿]	: $1s^1$
	1*s*	

The choice of a spin-up electron here is arbitrary; we could equally well show the ground state with one spin-down electron in the 1*s* orbital.

The next element, helium, has two electrons. Because two electrons with opposite spins can occupy an orbital, both of helium's electrons are in the 1*s* orbital:

$$\text{He} \quad \underset{1s}{\boxed{\uparrow\downarrow}} : \quad 1s^2$$

The two electrons present in helium complete the filling of the first shell. This arrangement represents a very stable configuration, as is evidenced by the chemical inertness of helium.

The electron configurations of lithium and several elements that follow it in the periodic table are shown in Table 6.4. For the third electron of lithium, the change in principal quantum number represents a large jump in energy and a corresponding jump in the average distance of the electron from the nucleus. It represents the start of a new shell of electrons. As you can see by examining the periodic table, lithium represents the start of a new row of the periodic table. It is the first member of the alkali metals family (group 1A).

The element that follows lithium is beryllium; its electron configuration is $1s^22s^2$ (Table 6.4). Boron, atomic number 5, has the electron configuration $1s^22s^22p^1$. The fifth electron must be placed in a 2*p* orbital because the 2*s* orbital is filled. Because the three 2*p* orbitals are of equal energy, it doesn't matter which 2*p* orbital is occupied.

With the next element, carbon, we come to a new situation. We know that the sixth electron must go into a 2*p* orbital. However, does this new electron go into the 2*p* orbital that already has one electron, or into one of the others? This question is answered by **Hund's rule**, which states that *for degenerate orbitals, the lowest energy is attained when the number of electrons with the same spin is maximized.* This means that electrons will occupy orbitals singly to the maximum extent possible, with their spins parallel. Thus, for a carbon atom to

Table 6.4 Electron Configurations of Several Lighter Elements

Element	Total electrons	Orbital diagram 1s	2s	2p			3s	Electron configuration
Li	3	↑↓	↑					$1s^22s^1$
Be	4	↑↓	↑↓					$1s^22s^2$
B	5	↑↓	↑↓	↑				$1s^22s^22p^1$
C	6	↑↓	↑↓	↑	↑			$1s^22s^22p^2$
N	7	↑↓	↑↓	↑	↑	↑		$1s^22s^22p^3$
Ne	10	↑↓	↑↓	↑↓	↑↓	↓↑		$1s^22s^22p^6$
Na	11	↑↓	↑↓	↑↓	↑↓	↑↓	↑	$1s^22s^22p^63s^1$

achieve its lowest energy, the two $2p$ electrons will have the same spin. In order for this to happen, the electrons must be in different $2p$ orbitals, as shown in Table 6.4. We see that a carbon atom in its ground state has two unpaired electrons. Similarly, for nitrogen in its ground state, Hund's rule requires that the three $2p$ electrons singly occupy each of the three $2p$ orbitals. This is the only way that all three electrons can have the same spin. For oxygen and fluorine, we need to place four and five electrons, respectively, in the $2p$ orbitals. In order to achieve this, we must pair up electrons in the $2p$ orbitals, as we will see in Sample Exercise 6.8.

Hund's rule is based in part on the fact that electrons repel one another. By occupying different orbitals, the electrons remain as far as possible from one another, thus minimizing electron-electron repulsions.

Returning to our survey of electron configurations, we now have neon, the last member of the second row. Neon has ten electrons. Two electrons fill the $1s$ orbital, two electrons fill the $2s$ orbital, and the remaining six electrons fill the $2p$ orbitals. The electron configuration is thus $1s^2 2s^2 2p^6$ (Table 6.4). In neon, all of the orbitals with $n = 2$ are filled. The filling of the $2s$ and $2p$ orbitals by the eight electrons that they can hold represents a very stable configuration. As a result, neon is chemically quite inert. Lewis, in his model for the electron configurations of elements, noted that the octet of electrons (eight) in the outermost shell of an atom or ion represents an especially stable arrangement.

Sodium, atomic number 11, marks the beginning of a new row of the periodic table. Sodium has a single $3s$ electron beyond the stable configuration of neon. We can abbreviate the electron configuration of sodium as follows:

$$\text{Na} \qquad [\text{Ne}]3s^1$$

The symbol [Ne] represents the electron configuration of the ten electrons of neon, $1s^2 2s^2 2p^6$. Writing the electron configuration in this manner helps us focus attention on the outermost electrons of the atom. The outer electrons are the ones largely responsible for the chemical behavior of an element. For example, we can write the electron configuration of lithium as follows:

$$\text{Li} \qquad [\text{He}]2s^1$$

By comparing this with the electron configuration for sodium, it is easy to appreciate why lithium and sodium are so similar chemically: They have the same type of outer-shell electron configuration. All the members of the alkali metal family (group 1A) have a single s electron beyond a noble-gas configuration. The outer-shell electrons are often referred to as **valence electrons**. The electrons in the inner shells are called the **core electrons**.

Kernel

SAMPLE EXERCISE 6.8

Draw the orbital diagram representation for the electron configuration of oxygen, atomic number 8.

Solution: The ordering of orbitals is shown in Figure 6.25. Two electrons each go into the $1s$ and $2s$ orbitals. This leaves four electrons for the three $2p$ orbitals. Following Hund's rule, we put one electron into each $2p$ orbital until all three have one each. The fourth electron must then be paired up with one of the three electrons already in a $2p$ orbital, so that the correct representation is

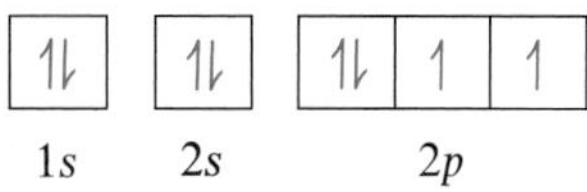

The corresponding electron configuration is written $1s^22s^22p^4$ or $[He]2s^22p^4$. The $1s^2$ or [He] electrons are the inner-shell or core electrons of the oxygen atom. The $2s^22p^4$ electrons are the outer-shell or valence electrons.

PRACTICE EXERCISE

Write the electron configuration of phosphorus, element 15.
Answer: $1s^22s^22p^63s^23p^3 = [Ne]3s^23p^3$

The noble-gas element argon marks the end of the row started by sodium. The configuration for argon is $1s^22s^22p^63s^23p^6$. The element following argon in the periodic table is potassium (K), atomic number 19. In all its chemical properties, potassium is clearly a member of the alkali metal family. The experimental facts about the properties of potassium leave no doubt that the outermost electron of this element occupies an *s* orbital. But this means that the highest-energy electron has *not* gone into a 3*d* orbital, which we might naively have expected it to do. In this case the ordering of energy levels is such that the 4*s* orbital is lower in energy than the 3*d* (see the caption for Figure 6.25).

Following complete filling of the 4*s* orbital (this occurs in the calcium atom), the next set of equivalent orbitals to be filled is the 3*d*. (You will find it helpful as we go along to refer often to the periodic table on the front inside cover.) Beginning with scandium and extending through zinc, electrons are added to the five 3*d* orbitals until they are completely filled. Thus, the fourth row of the periodic table is ten elements wider than the previous rows. These ten elements are known as the **transition elements** or **transition metals**. Note the position of these elements in the periodic table.

In accordance with Hund's rule, electrons are added to the 3*d* orbitals singly until all five orbitals have one electron each. Additional electrons are then placed in the 3*d* orbitals with spin pairing until the shell is completely filled. The orbital diagram representations and electron configurations of two transition elements are as follows:

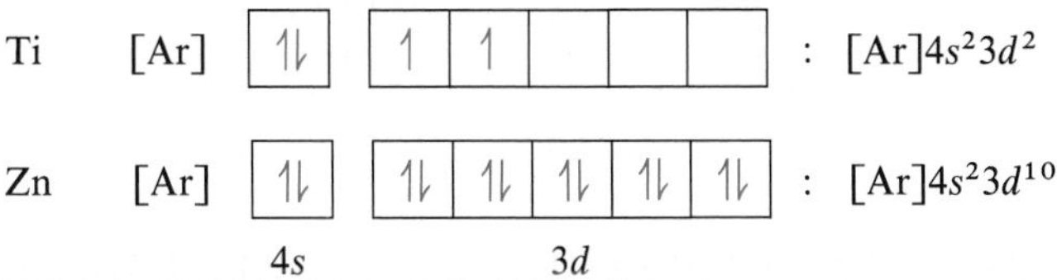

: $[Ar]4s^23d^2$

: $[Ar]4s^23d^{10}$

Upon completion of the 3*d* transition series, the 4*p* orbitals begin to be occupied, until the completed octet of outer electrons ($4s^24p^6$) is reached with krypton (Kr), atomic number 36. Krypton is another of the noble gases. Rubidium (Rb) marks the beginning of the fifth row of the periodic table. This row is in every respect like the preceding one, except that the value for *n* is 1 greater. The sixth row of the table begins similarly to the preceding one: one electron in the 6*s* orbital of cesium (Cs) and two electrons in the 6*s* orbital of barium (Ba). The next element, lanthanum (La), represents the start of the third series of transition elements. But with cerium (Ce), element

58, a new set of orbitals, the $4f$, enter the picture. The energies of the $5d$ and $4f$ orbitals are very close. For lanthanum itself, the $5d$ orbital energy is just a little lower than the $4f$. However, for the elements immediately following lanthanum, the $4f$ orbital energies are a little lower, so that the $4f$ orbitals fill before the $5d$ orbitals.

There are seven equivalent $4f$ orbitals, corresponding to the seven allowed values of m_l, ranging from 3 to -3. Thus it requires 14 electrons to fill the $4f$ orbitals completely. The 14 elements corresponding to the filling of the $4f$ orbitals are elements 58 to 71, known as the **rare-earth**, or **lanthanide, elements**. In order not to make the periodic table unduly wide, the rare-earth elements are set together below the other elements. The properties of the rare-earth elements are all quite similar, and they occur together in nature. For many years it was virtually impossible to separate them from one another.

After the rare-earth series, the third transition element series is completed, followed by the filling of the $6p$ orbitals. This brings us to radon (Rn), heaviest of the noble-gas elements. The final row of the periodic table begins as the one before it. The **actinide elements**, of which uranium (U, element 92) and plutonium (Pu, element 94) are the best known, are built up by completion of the $5f$ orbitals. Most of the actinides are not found in nature. In Chapter 21, we will discuss how these elements are made.

6.10 USING THE PERIODIC TABLE TO WRITE ELECTRON CONFIGURATIONS

Our rather brief survey of electron configurations of the elements has taken us through the periodic table. We have seen that the electron configurations of elements are related to their location in the periodic table. The periodic table is structured so that elements with the same type of outer-shell electron configuration are arranged in columns. For example, for groups 2A and 3A we have

Group 2A		Group 3A	
Be	$[He]2s^2$	B	$[He]2s^22p^1$
Mg	$[Ne]3s^2$	Al	$[Ne]3s^23p^1$
Ca	$[Ar]4s^2$	Ga	$[Ar]3d^{10}4s^24p^1$
Sr	$[Kr]5s^2$	In	$[Kr]4d^{10}5s^25p^1$
Ba	$[Xe]6s^2$	Tl	$[Xe]4f^{14}5d^{10}6s^26p^1$
Ra	$[Rn]7s^2$		

If you understand how the periodic table is organized, it is not necessary to memorize the order in which orbitals fill. You can write the electron configuration of an element based on its location in the periodic table. The pattern is summarized in Figure 6.32. Notice that the elements can be grouped in terms of the *type* of orbital into which the electrons are placed. On the left are *two* columns of elements. These elements, known as the **active metals**, are those in which the outer-shell *s* orbitals are being filled. On the right is a block of *six* columns. These are the elements in which the outermost *p* orbitals are being filled. The *s* block and the *p* block of the periodic table contain the **representative** or **main-group elements**. In the middle of the table is a block of *ten* columns that contains the transition metals. These are the elements in which the *d* orbitals are being filled. Below the main portion of the table are two rows

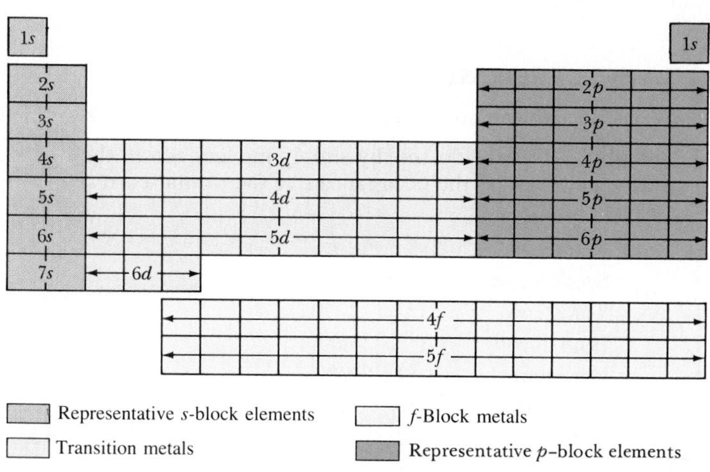

Figure 6.32 Block diagram of the periodic table showing the groupings of the elements according to the type of orbital being filled with electrons.

that contain *fourteen* columns. These elements, are often referred to as the ***f*-block metals** because they are the ones in which the *f* orbitals are being filled. Recall that the numbers 2, 6, 10, and 14 are precisely the number of electrons that can fill the *s*, *p*, *d*, and *f* subshells, respectively. Recall also that the 1*s* subshell is the first *s* subshell, the 2*p* is the first *p* subshell, the 3*d* is the first *d* subshell, and the 4*f* is the first *f* subshell.

SAMPLE EXERCISE 6.9

What is the characteristic outer-shell electron configuration of the group 7A elements, the halogens?

Solution: The first member of the halogen family is fluorine, atomic number 9. The abbreviated form of the electron configuration for fluorine is

$$\text{F} \quad [\text{He}]2s^2 2p^5$$

Similarly, the abbreviated form of the electron configuration for chlorine, the second halogen, is

$$\text{Cl} \quad [\text{Ne}]3s^2 3p^5$$

From these two examples we see that the characteristic outer-shell electron configuration of a halogen is ns^2np^5, where n ranges from 2 in the case of fluorine to 5 in the case of iodine.

PRACTICE EXERCISE

What family of elements is characterized by having an ns^2 electron configuration?
Answer: the alkaline earth metals, group 2A

In some cases, electrons shift from one orbital to another in ways that appear to violate the rules we have just discussed. For example, the ground electron configuration of chromium is $[\text{Ar}]4s^1 3d^5$ rather than $[\text{Ar}]4s^2 3d^4$, as we might have expected. Similarly, the configuration of copper is $[\text{Ar}]4s^1 3d^{10}$ instead of the expected $[\text{Ar}]4s^2 3d^9$. This anomalous behavior is largely a consequence of the closeness of the 3*d* and 4*s* orbital energies. It usually occurs when there are enough electrons to lead to precisely half-filled sets of degenerate orbitals (as in chromium) or to a completely filled *d* subshell (as in copper). There are a few similar cases among the heavier transition metals (those with partially filled 4*d* or 5*d* orbitals) and among the *f*-block metals. Although these minor departures from the expected are interesting, they are not of great chemical significance.

SAMPLE EXERCISE 6.10

Write the electron configuration for the element bismuth, atomic number 83.

Solution: We can do this by simply moving across the periodic table one row at a time and writing the occupancies of the orbital corresponding to each row (refer to Figure 6.32),

First row	$1s^2$
Second row	$2s^22p^6$
Third row	$3s^23p^6$
Fourth row	$4s^23d^{10}4p^6$
Fifth row	$5s^24d^{10}5p^6$
Sixth row	$6s^24f^{14}5d^{10}6p^3$
Total:	$1s^22s^22p^63s^23p^63d^{10}4s^24p^64d^{10}4f^{14}5s^25p^65d^{10}6s^26p^3$

Note that 3 is the lowest possible value that n may have for a d orbital, and that 4 is the lowest possible value of n for an f orbital.

The total of the superscripted numbers should equal the atomic number of bismuth, 83. It does not matter a great deal precisely in which order the orbitals are listed. They may be listed, as shown above, in the order of increasing major quantum number. However, it is also possible to list them in the sequence read from the periodic table: $1s^22s^22p^63s^23p^64s^23d^{10}4p^65s^24d^{10}5p^66s^24f^{14}5d^{10}6p^3$.

It is a simple matter to write the abbreviated electron configuration of an element using the periodic table. First locate the element of interest (in this case element 83) and then move backward until the first noble gas is encountered (in this case Xe, element 54). Thus the inner core is [Xe]. The outer electrons are then read from the periodic table as before. Moving from Xe to Cs, element 55, we find ourselves in the sixth row. Moving across this row to Bi gives us the outer electrons. The complete electron configuration is thus $[Xe]6s^24f^{14}5d^{10}6p^3$.

PRACTICE EXERCISE

Use the periodic table to write the electron configurations for the following atoms by giving the appropriate noble-gas inner core plus the electrons beyond it: **(a)** Co (atomic number 27); **(b)** Te (atomic number 52).
Answers: **(a)** $[Ar]4s^23d^7$; **(b)** $[Kr]5s^24d^{10}5p^4$

SAMPLE EXERCISE 6.11

Draw the orbital diagram representation for zirconium, atomic number 40; show only those electrons beyond the krypton inner core.

Solution: Zirconium has four electrons beyond the nearest noble gas, krypton, atomic number 36. Examining the periodic table, we see that zirconium is a transition element from the fifth row of the table. This means that its outermost electrons are in $5s$ and $4d$ orbitals. Two electrons occupy the $5s$ orbital; two must be placed in the five $4d$ orbitals. As indicated by Hund's rule, the $4d$ electrons occupy separate orbitals. Thus, we have

		$5s$	$4d$				
Zr	[Kr]	↑↓	↑	↑			

PRACTICE EXERCISE

Using orbital diagrams for the outer electrons, determine the number of unpaired electrons in **(a)** Ni (atomic number 28); **(b)** Br (atomic number 35)
Answers: **(a)** 2; **(b)** 1

Table 6.5 gives a complete list of the ground electron configurations of the elements. You should use this table to check your answers as you practice writing electron configurations. We have written these configurations as they

Table 6.5 The Electron Configurations of the Elements

Atomic number	Symbol	Electron configuration	Atomic number	Symbol	Electron configuration	Atomic number	Symbol	Electron configuration
1	H	$1s^1$	37	Rb	[Kr]$5s^1$	73	Ta	[Xe]$6s^24f^{14}5d^3$
2	He	$1s^2$	38	Sr	[Kr]$5s^2$	74	W	[Xe]$6s^24f^{14}5d^4$
3	Li	[He]$2s^1$	39	Y	[Kr]$5s^24d^1$	75	Re	[Xe]$6s^24f^{14}5d^5$
4	Be	[He]$2s^2$	40	Zr	[Kr]$5s^24d^2$	76	Os	[Xe]$6s^24f^{14}5d^6$
5	B	[He]$2s^22p^1$	41	Nb	[Kr]$5s^14d^4$	77	Ir	[Xe]$6s^24f^{14}5d^7$
6	C	[He]$2s^22p^2$	42	Mo	[Kr]$5s^14d^5$	78	Pt	[Xe]$6s^14f^{14}5d^9$
7	N	[He]$2s^22p^3$	43	Tc	[Kr]$5s^24d^5$	79	Au	[Xe]$6s^14f^{14}5d^{10}$
8	O	[He]$2s^22p^4$	44	Ru	[Kr]$5s^14d^7$	80	Hg	[Xe]$6s^24f^{14}5d^{10}$
9	F	[He]$2s^22p^5$	45	Rh	[Kr]$5s^14d^8$	81	Tl	[Xe]$6s^24f^{14}5d^{10}6p^1$
10	Ne	[He]$2s^22p^6$	46	Pd	[Kr]$4d^{10}$	82	Pb	[Xe]$6s^24f^{14}5d^{10}6p^2$
11	Na	[Ne]$3s^1$	47	Ag	[Kr]$5s^14d^{10}$	83	Bi	[Xe]$6s^24f^{14}5d^{10}6p^3$
12	Mg	[Ne]$3s^2$	48	Cd	[Kr]$5s^24d^{10}$	84	Po	[Xe]$6s^24f^{14}5d^{10}6p^4$
13	Al	[Ne]$3s^23p^1$	49	In	[Kr]$5s^24d^{10}5p^1$	85	At	[Xe]$6s^24f^{14}5d^{10}6p^5$
14	Si	[Ne]$3s^23p^2$	50	Sn	[Kr]$5s^24d^{10}5p^2$	86	Rn	[Xe]$6s^24f^{14}5d^{10}6p^6$
15	P	[Ne]$3s^23p^3$	51	Sb	[Kr]$5s^24d^{10}5p^3$	87	Fr	[Rn]$7s^1$
16	S	[Ne]$3s^23p^4$	52	Te	[Kr]$5s^24d^{10}5p^4$	88	Ra	[Rn]$7s^2$
17	Cl	[Ne]$3s^23p^5$	53	I	[Kr]$5s^24d^{10}5p^5$	89	Ac	[Rn]$7s^26d^1$
18	Ar	[Ne]$3s^23p^6$	54	Xe	[Kr]$5s^24d^{10}5p^6$	90	Th	[Rn]$7s^26d^2$
19	K	[Ar]$4s^1$	55	Cs	[Xe]$6s^1$	91	Pa	[Rn]$7s^25f^26d^1$
20	Ca	[Ar]$4s^2$	56	Ba	[Xe]$6s^2$	92	U	[Rn]$7s^25f^36d^1$
21	Sc	[Ar]$4s^23d^1$	57	La	[Xe]$6s^25d^1$	93	Np	[Rn]$7s^25f^46d^1$
22	Ti	[Ar]$4s^23d^2$	58	Ce	[Xe]$6s^24f^15d^1$	94	Pu	[Rn]$7s^25f^6$
23	V	[Ar]$4s^23d^3$	59	Pr	[Xe]$6s^24f^3$	95	Am	[Rn]$7s^25f^7$
24	Cr	[Ar]$4s^13d^5$	60	Nd	[Xe]$6s^24f^4$	96	Cm	[Rn]$7s^25f^76d^1$
25	Mn	[Ar]$4s^23d^5$	61	Pm	[Xe]$6s^24f^5$	97	Bk	[Rn]$7s^25f^9$
26	Fe	[Ar]$4s^23d^6$	62	Sm	[Xe]$6s^24f^6$	98	Cf	[Rn]$7s^25f^{10}$
27	Co	[Ar]$4s^23d^7$	63	Eu	[Xe]$6s^24f^7$	99	Es	[Rn]$7s^25f^{11}$
28	Ni	[Ar]$4s^23d^8$	64	Gd	[Xe]$6s^24f^75d^1$	100	Fm	[Rn]$7s^25f^{12}$
29	Cu	[Ar]$4s^13d^{10}$	65	Tb	[Xe]$6s^24f^9$	101	Md	[Rn]$7s^25f^{13}$
30	Zn	[Ar]$4s^23d^{10}$	66	Dy	[Xe]$6s^24f^{10}$	102	No	[Rn]$7s^25f^{14}$
31	Ga	[Ar]$4s^23d^{10}4p^1$	67	Ho	[Xe]$6s^24f^{11}$	103	Lr	[Rn]$7s^25f^{14}6d^1$
32	Ge	[Ar]$4s^23d^{10}4p^2$	68	Er	[Xe]$6s^24f^{12}$	104	Rf	[Rn]$7s^25f^{14}6d^2$
33	As	[Ar]$4s^23d^{10}4p^3$	69	Tm	[Xe]$6s^24f^{13}$	105	Ha	[Rn]$7s^25f^{14}6d^3$
34	Se	[Ar]$4s^23d^{10}4p^4$	70	Yb	[Xe]$6s^24f^{14}$	106	[106]	[Rn]$7s^25f^{14}6d^4$
35	Br	[Ar]$4s^23d^{10}4p^5$	71	Lu	[Xe]$6s^24f^{14}5d^1$	107	[107]	[Rn]$7s^25f^{14}6d^5$
36	Kr	[Ar]$4s^23d^{10}4p^6$	72	Hf	[Xe]$6s^24f^{14}5d^2$	108	[108]	[Rn]$7s^25f^{14}6d^6$
						109	[109]	[Rn]$7s^25f^{14}6d^7$

would be read off the periodic table. They are sometimes written with orbitals of a given principal quantum number grouped together. Thus, you could write the electron configuration of arsenic (atomic number 33) as [Ar]$3d^{10}4s^24p^3$ rather than [Ar]$4s^23d^{10}4p^3$ as shown in Table 6.5.

We have covered a lot of ground in this chapter: We started with the origins of quantum mechanics and eventually developed a systematic way of using the periodic table to write the electron configurations of atoms. In the next chapter, we shall examine the periodic trends of atomic properties more closely and discuss the chemistry of several families of representative elements.

FOR REVIEW

SUMMARY

Radiant energy moves through a vacuum at the "speed of light," $c = 3.00 \times 10^8$ m/s. It has wavelike characteristics that allow it to be described in terms of wavlength, λ, and frequency, ν, which are interrelated: $c = \lambda\nu$. The dispersion of radiation into its component wavelengths produces a spectrum. If all wavelengths are present, the spectrum is said to be continuous; if only certain wavelengths are present, it is called a line spectrum.

The quantum theory describes the minimum amount of radiant energy that an object can gain or lose, $E = h\nu$; this smallest quantity is called a quantum. A quantum of radiant energy is called a photon. The quantum theory was used to explain the photoelectric effect and the line spectrum of the hydrogen atom. The absorptions or emissions of light by an atom, which produce its line spectrum, correspond to energy changes of electrons within the atom; the energy of the electron in an atom is quantized.

Electrons exhibit wave properties and can be described by a wavelength, $\lambda = h/mv$. Discovery of the wave properties of the electron led to the uncertainty principle, which indicates that the position and momentum of an electron can be determined simultaneously with only limited accuracy.

In the quantum-mechanical model of the hydrogen atom we speak of the probability of the electron being found at a particular point in space. Although the positions of the electrons are defined in this averaged sense, their energies are precisely known. Each allowed state of an electron in the atom corresponds to a particular set of values for three quantum numbers. Each such allowed energy state is termed an orbital. An orbital is described by a combination of an integer and letters, corresponding to the three values for the quantum numbers. The principal quantum number, n, is indicated by the integers 1, 2, 3, ... This quantum number relates most directly to the size and energy of an orbital. The azimuthal quantum number, l, is indicated by the letters s, p, d, f, and so on, corresponding to values of l of 0, 1, 2, 3, ... The l quantum number defines the shape of the orbital. The magnetic quantum number, m_l, describes the orientation of the orbital in space. For example, the three $3p$ orbitals are designated $3p_x$, $3p_y$, and $3p_z$, the subscript letters indicating the axis along which the orbital is oriented.

Restrictions on the values of the three quantum numbers give rise to the following allowed subshells:

$1s$
$2s, 2p$
$3s, 3p, 3d$
$4s, 4p, 4d, 4f$
⋮

There is one orbital in an s subshell, three in a p subshell, five in a d subshell, and seven in an f subshell. The contour representations are the most generally useful way to visualize the spatial characteristics of the orbitals.

In many-electron atoms, we saw that the electron configurations can be written by placing electrons into orbitals in the following order:

$$1s, 2s, 2p, 3s, 3p, 4s, 3d, 4p, \ldots$$

Subshells with a given principal quantum number, such as the $3s$, $3p$, and $3d$ subshells, do not have the same energies. This fact can be understood in terms of the effective nuclear charge and the average distance of an electron from the nucleus in each of these subshells.

The Pauli exclusion principle places a limit of two on the number of electrons that can occupy any one atomic orbital. These two electrons differ in their electron-spin quantum number, m_s. As the electrons populate orbitals of equal energy, they do not pair up until each orbital contains one electron; this observation is called Hund's rule. Using the relative energies of the orbitals, the Pauli exclusion principle, and Hund's rule, it is possible to write the electron configuration of any atom. When we do so, we see that the elements in any given family in the periodic table have the same type of electron arrangements in their outermost shells. For example, the electron configurations of the halogens fluorine and chlorine are $[He]2s^22p^5$ and $[Ne]3s^23p^5$, respectively. This periodicity in electron configurations, summarized in Figure 6.32, allows us to write the electron configurations of an element from its position in the periodic table.

KEY TERMS

electronic structure
quantum theory
electromagnetic radiation (Sec. 6.1)
wavelength (Sec. 6.1)
frequency (Sec. 6.1)
cycles (Sec. 6.1)
quantum (Sec. 6.2)
photon (Sec. 6.2)
monochromatic (Sec. 6.2)
spectrum (Sec. 6.2)
continuous spectrum (Sec. 6.2)
line spectrum (Sec. 6.2)
principal quantum number (Sec. 6.2)
ground state (Sec. 6.3)
excited state (Sec. 6.3)
ionization (Sec. 6.3)
ionization energy (Sec. 6.3)
quantum (wave) mechanics (Sec. 6.3)
momentum (Sec. 6.4)
matter waves (Sec. 6.4)
uncertainty principle (Sec. 6.4)
Schrödinger's wave equation (Sec. 6.5)
wave functions (Sec. 6.5)
probability density (Sec. 6.5)
electron density (Sec. 6.5)
orbitals (Sec. 6.5)
electron shell (Sec. 6.5)
subshell (Sec. 6.5)
nodal surfaces (nodes) (Sec. 6.6)
effective nuclear charge (Sec. 6.7)
screening effect (Sec. 6.7)
degenerate (Sec. 6.7)
electron spin (Sec. 6.8)
electron spin quantum number (Sec. 6.8)
Pauli exclusion principle (Sec. 6.8)
electron configuration (Sec. 6.9)
orbital diagram (Sec. 6.9)
Hund's rule (Sec. 6.9)
valence electrons (Sec. 6.9)
core electrons (Sec. 6.9)
transition elements (Sec. 6.9)
transition metals (Sec. 6.9)
rare-earth (lanthanide) elements (Sec. 6.9)
actinide elements (Sec. 6.9)
active metals (Sec. 6.10)
representative (main-group) elements (Sec. 6.10)
f-block metals (Sec. 6.10)

EXERCISES

Radiant Energy

6.1 List the following types of electromagnetic radiation in order of increasing wavelength: **(a)** radiation from a microwave oven; **(b)** the red light emitted by a hot electric stove burner; **(c)** the infrared radiation emitted by a hot electric stove burner; **(d)** the ultraviolet light from a sun lamp; **(e)** cosmic radiation from outer space.

6.2 List the following types of electromagnetic radiation in order of increasing wavelength: **(a)** radiation from an FM radio station at 89.7 on the dial; **(b)** radiation from an AM radio station at 1460 on the dial; **(c)** X rays used in medical diagnosis; **(d)** the red light of a light-emitting diode, such as in a calculator display; **(e)** the gamma rays produced in the radioactive decay of particular ^{99}Tc nuclei, which are used for medical imaging.

6.3 **(a)** What is the wavelength of radiation whose frequency is $4.62 \times 10^{14}\ s^{-1}$? **(b)** What is the frequency of radiation whose wavelength is 10.5 cm? **(c)** Would you be able to see either of the radiations specified in parts (a) and (b)? **(d)** What distance does light travel in 3.00 min?

6.4 **(a)** What is the wavelength of radiation whose frequency is $9.23 \times 10^{7}\ s^{-1}$? **(b)** What is the frequency of radiation whose wavelength is 180 nm? **(c)** Would you be able to detect either of the radiations specified in parts (a) and (b) with an ultraviolet detector? **(d)** What distance does light travel in 0.50 ps?

6.5 A neon light emits radiation of 616-nm wavelength. What is the frequency of this radiation? Using Figure 6.3, predict the color associated with this wavelength.

6.6 Excited barium atoms emit visible light whose frequency is $6.59 \times 10^{14}\ s^{-1}$. What is the wavelength of this light? Use Figure 6.3 to predict its color.

Quantum Theory

6.7 The quantum theory assumes that energy changes are not continuous. Why do we not notice this effect in our everyday activities?

6.8 Could a xylophone be considered a "quantized" musical instrument? Explain.

6.9 **(a)** Calculate the smallest increment of energy (a quantum) that can be emitted or absorbed at a wavelength of 381 nm. **(b)** Calculate the energy of a photon of frequency $3.6 \times 10^{13}\ s^{-1}$. **(c)** What wavelength of radiation has photons of energy 8.23×10^{-19} J?

6.10 **(a)** Calculate the smallest increment of energy (a quantum) that can be emitted or absorbed at a wavelength of 703 nm. **(b)** Calculate the energy of a photon of frequency

$8.9 \times 10^{14}\ s^{-1}$. **(c)** What frequency of radiation has photons of energy 6.18×10^{-18} J?

6.11 Calculate and compare the energy of an infrared photon of wavelength 5.4 μm with that of an ultraviolet-light photon of wavelength 156 nm.

6.12 Under appropriate conditions, copper emits X rays that have a characteristic wavelength of 1.54 Å. Calculate and compare the energy of photons of these X rays to those emitted by a microwave source that radiates at a frequency of $5.87 \times 10^{10}\ s^{-1}$.

6.13 A high-powered laser is pulsed for a period of 100 ns. During that time, it emits a signal with a total energy of 8300 J. If the wavelength of the signal is 351 nm, how many photons have been emitted?

6.14 If the human eye receives a 1.45×10^{-17} J signal from photons whose wavelength is 550 nm, how many photons have hit the eye?

6.15 The energy from radiation can be used to cause the rupture of chemical bonds. A minimum energy of 495 kJ/mol is required to break the oxygen-oxygen bond in O_2. What is the longest wavelength of radiation that possesses the necessary energy to break the bond? What type of electromagnetic radiation is this?

6.16 A certain photographic film requires a minimum radiation energy of 80 kJ/mol to cause exposure. What is the longest wavelength of radiation that possesses the necessary energy to expose the film? Could this film be used for infrared photography?

6.17 Molybdenum metal must absorb radiation with a minimum frequency of $1.09 \times 10^{15}\ s^{-1}$ before it can emit an electron from its surface via the photoelectric effect. **(a)** What is the minimum energy required to produce this effect? **(b)** What wavelength radiation will provide a photon of this energy? **(c)** If molybdenum is irradiated with light whose wavelength is 120 nm, what is the maximum possible kinetic energy of the emitted electrons?

6.18 It requires 222 kJ/mol to eject electrons from potassium metal. **(a)** What is the minimum frequency of light necessary to emit electrons from potassium via the photoelectric effect? **(b)** What is the wavelength of this light? **(c)** If potassium is irradiated with light of wavelength 400 nm, what is the maximum possible kinetic energy of the emitted electrons?

Bohr's Model; Matter Waves

6.19 Is energy emitted or absorbed when the following electronic transitions occur in hydrogen: **(a)** from $n = 3$ to $n = 6$; **(b)** from an orbit with radius 4.77 Å to one with radius 2.12 Å; **(c)** ionization of an electron from the ground state?

6.20 Indicate whether energy is emitted or absorbed when the following electronic transitions occur in hydrogen: **(a)** from $n = 5$ to $n = 2$; **(b)** from an orbit with radius 0.53 Å to one with radius 8.48 Å; **(c)** ionization of an electron from the $n = 3$ state.

6.21 For each of the following electronic transitions in the hydrogen atom, calculate the energy, frequency, and wavelength of the associated radiation: **(a)** from $n = 1$ to $n = 3$; **(b)** from $n = 2$ to $n = 5$; **(c)** from $n = 6$ to $n = 7$. Will the radiation be absorbed or emitted during these transitions?

6.22 For each of the following electronic transitions in the hydrogen atom, calculate the energy, frequency, and wavelength of the associated radiation: **(a)** from $n = 4$ to $n = 1$; **(b)** from $n = 5$ to $n = 3$; **(c)** from $n = 6$ to $n = 2$. Will the radiation be absorbed or emitted during these transitions?

6.23 What wavelength of light is emitted when an electron moves from the $n = 5$ to the $n = 2$ state in hydrogen? Using Figure 6.11, identify the line in the emission spectrum that corresponds to this transition.

6.24 What wavelength of light is necessary to excite a hydrogen atom from its ground state to its lowest-energy excited state ($n = 2$). What type of radiation is necessary to make this transition occur?

6.25 Use the de Broglie relation to determine the wavelengths of the following: **(a)** a 58-g tennis ball traveling at 130 mi/hr; **(b)** an 85-kg person skiing at 60 km/hr; **(c)** a helium atom that has a speed of 1.5×10^5 m/s.

6.26 Calculate the wavelengths of the following objects: **(a)** a 50-g golf ball traveling at 400 m/s; **(b)** a 3000-lb automobile moving at 55 mi/hr; **(c)** an argon atom that has a speed of 2.5×10^4 m/s.

6.27 Neutron diffraction is an important technique for determining the structures of molecules. Calculate the velocity of a neutron that has a characteristic wavelength of 0.88 Å. (Refer to the back inside cover for the mass of the neutron.)

6.28 The electron microscope has been widely used to obtain highly magnified images of biological and other types of materials. When an electron is accelerated through 100 V, it attains a speed of 5.93×10^6 m/s. What is the characteristic wavelength of this electron? Is the wavelength comparable to the size of atoms?

Wave Functions; Orbitals; Quantum Numbers

6.29 **(a)** For $n = 5$, what are the possible values of l? **(b)** For $l = 3$, what are the possible values of m_l?

6.30 **(a)** For $n = 4$, what are the possible values of l? **(b)** For $l = 3$, what are the possible values of m_l?

6.31 Give the values for n, l, and m_l **(a)** for each orbital in the 4*d* subshell; **(b)** for each orbital in the $n = 3$ shell.

6.32 Give the values for n, l, and m_l **(a)** for each orbital in the 4*f* subshell; **(b)** for each orbital in the $n = 2$ shell.

6.33 Which of the following are permissible sets of quantum numbers for an electron in a hydrogen atom: **(a)** $n = 2$, $l = 1$, $m_l = 1$; **(b)** $n = 1$, $l = 0$, $m_l = -1$; **(c)** $n = 4$, $l = 2$, $m_l = -2$; **(d)** $n = 3$, $l = 3$, $m_l = 0$? For those combinations that are permissible, write the appropriate designation for the subshell to which the orbital belongs (that is, 1*s*, and so on).

6.34 Which of the following sets of quantum numbers are allowed for an electron in a hydrogen atom: **(a)** $n = 1$, $l = 1$, $m_l = 0$; **(b)** $n = 3$, $l = 0$, $m_l = 0$; **(c)** $n = 4$, $l = 1$, $m_l = -1$; **(d)** $n = 2$, $l = 1$, $m_l = 2$? For those combinations that are allowed, write the designation for the subshell to which the orbital belongs.

6.35 In the quantum-mechanical description of the hydrogen atom, what is the physical significance of the square of the wave function, ψ^2?

6.36 In the Bohr model of the ground state of hydrogen, the electron orbits the nucleus with a radius of 0.53 Å. Is this also true in the quantum-mechanical description of the hydrogen atom? Explain.

6.37 Sketch the contour representation for the following types of orbitals: **(a)** s; **(b)** p_x; **(c)** $d_{x^2-y^2}$.

6.38 Sketch the contour representation for the following types of orbitals: **(a)** p_y; **(b)** d_{xz}; **(c)** d_{z^2}.

6.39 What are the similarities and differences between a 2*s* and a 3*s* orbital? In the hydrogen atom, which is lower in energy?

6.40 What are the similarities and differences between a 2*s* and a $2p_x$ orbital? In the hydrogen atom, which is lower in energy?

Many-Electron Atoms; Electron Spin

6.41 Which quantum numbers must be the same in order that orbitals be degenerate (have the same energy) **(a)** in a hydrogen atom, and **(b)** in a many-electron atom?

6.42 Within a given shell, how do the energies of the *s*, *p*, *d*, and *f* subshells compare for a many-electron atom? How do the energies of the orbitals of a given subshell compare?

6.43 Explain why the effective nuclear charge experienced by a 3*s* electron in iron is greater than that for a 3*d* electron.

6.44 Compare the average distance from the nucleus of a 2*s* electron in neon with that for a 2*p* electron.

6.45 Explain why the effective nuclear charge experienced by a 3*s* electron in magnesium is larger than that experienced by a 3*s* electron in sodium.

6.46 Which should experience a greater effective nuclear charge, a 4*s* electron in titanium or a 4*s* electron in copper?

6.47 What is the maximum number of electrons that can occupy each of the following subshells: **(a)** 3*d*; **(b)** 4*s*; **(c)** 2*p*; **(d)** 5*f*?

6.48 What is the maximum number of electrons in an atom that can have the following quantum numbers: **(a)** $n = 3$; **(b)** $n = 4$, $l = 2$; **(c)** $n = 4$, $l = 3$, $m_l = 2$; **(d)** $n = 2$, $l = 1$, $m_l = 0$, $m_s = -\frac{1}{2}$?

6.49 List the possible values of the four quantum numbers for each electron in the lithium atom.

6.50 List the possible values of the four quantum numbers for a 2*p* electron in boron.

Electron Configurations of the Elements

6.51 Write the electron configurations for the following atoms using the appropriate noble-gas inner core for abbreviation: **(a)** K; **(b)** Ga; **(c)** Se; **(d)** Fe; **(e)** Sm; **(f)** Sc.

6.52 Write the electron configurations for the following atoms using the appropriate noble-gas inner core for abbreviation: **(a)** Mn; **(b)** Rb; **(c)** Zn; **(d)** V; **(e)** Tl; **(f)** Yb.

6.53 Draw the orbital diagrams for the valence electrons of each of the following elements: **(a)** As; **(b)** Mn; **(c)** Sn; **(d)** Lu; **(e)** Te. How many unpaired electrons would you expect in each of these?

6.54 Using orbital diagrams, determine the number of unpaired electrons in each of the following atoms: **(a)** Ge; **(b)** Ni; **(c)** B; **(d)** Kr; **(e)** In.

6.55 Arrange the following subshells in the order in which they fill with electrons: 3*d*, 4*s*, 5*p*, 3*s*, 4*f*, 3*p*, 5*s*.

6.56 Referring only to Figure 6.32, write the order in which the occupied orbitals of radon fill with electrons.

6.57 Identify the specific element that has the following electron configuration: **(a)** $1s^22s^22p^4$; **(b)** $[Ne]3s^23p^5$; **(c)** $[Ar]4s^1$; **(d)** $[Ar]4s^13d^5$

6.58 Identify the group of elements that corresponds to the following electron configurations: **(a)** [noble gas]ns^1; **(b)** [noble gas]ns^2np^3; **(c)** [noble gas]$ns^2(n-1)d^{10}np^2$; **(d)** [noble gas]$ns^2(n-1)d^6$

Additional Exercises

6.59 Consider the two electromagnetic waves shown below:

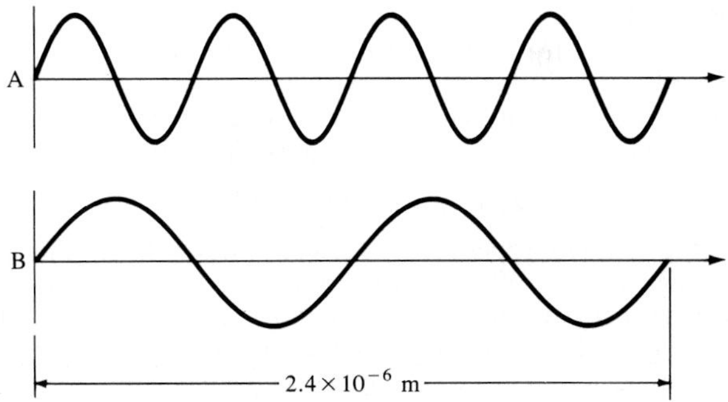

(a) What is the wavelength of wave A? Of wave B? **(b)** What is the frequency of wave A? Of wave B? **(c)** Identify the regions of the electromagnetic spectrum to which waves A and B belong.

6.60 Certain elements emit light of a specific color when they are burned. Historically, chemists used such "flame tests" to determine whether specific elements were present in a sample. Some characteristic wavelengths for some of the elements are:

Ag	328.1 nm	Fe	372.0 nm
Au	267.6 nm	K	404.7 nm
Ba	455.4 nm	Mg	285.2 nm
Ca	422.7 nm	Na	589.6 nm
Cu	324.8 nm	Ni	341.5 nm

(a) For each of these elements, determine the color associated with its flame. **(b)** Which element emits the most energetic light? The least energetic light? **(c)** When burned, a sample of an unknown substance is found to emit light of frequency $6.59 \times 10^{14}\ s^{-1}$. Which of the above elements is probably in the sample?

6.61 The planetary space probe *Voyager 2* has sent pictures from Neptune to Earth, a distance of 2.82×10^9 mi. How long did it take for the pictures transmitted from *Voyager 2* to reach Earth?

6.62 The laser in an audio compact disc player uses light whose wavelength is 780 nm. **(a)** What is the frequency of this radiation? **(b)** What is the energy (in joules) of a single photon of this wavelength?

6.63 Chlorophyll absorbs blue light, $\lambda = 460$ nm, and emits red light, $\lambda = 660$ nm. Calculate the net energy change in the chlorophyll system (in kilojoules) when a mole of 460-nm photons is absorbed and a mole of 660-nm photons is emitted.

[6.64] A phototube, such as that illustrated in Figure 6.8(*b*), is a device used to measure the intensity of light. In a certain experiment, when light of wavelength 550 nm is shined on the phototube, electrons are emitted at the rate of 8.6×10^{-13} C/s. Assume that each photon that impinges on the phototube emits one electron. How many photons per second are striking the phototube? How much energy per second is the phototube absorbing?

6.65 When the light from a neon light is dispersed in a prism to form a spectrum, the spectrum is not continuous; rather, it consists of several sharp lines, each of a specific frequency. Explain in general terms why a line spectrum is produced.

[6.66] The stratospheric ozone (O_3) layer helps to protect us from harmful ultraviolet radiation. It does so by absorbing ultraviolet light and falling apart into an O_2 molecule and an oxygen atom, a process known as *photodissociation:*

$$O_3(g) \longrightarrow O_2(g) + O(g)$$

Use the data in Appendix C to calculate the enthalpy change for this reaction. What is the maximum wavelength a photon can have if it is to possess sufficient energy to cause this dissociation? In what portion of the spectrum does this wavelength occur?

[6.67] Hydrogen displays an emission line in the infrared region of the spectrum, at 1875.6 nm. Determine the values of n for the two energy levels that are responsible for this emission.

6.68 What is the maximum wavelength of light capable of removing an electron from a hydrogen atom in the energy state characterized by **(a)** $n = 1$; **(b)** $n = 2$; **(c)** $n = 3$?

6.69 Use Equation 6.10 to calculate the ionization energy (in kJ/mol) for Li^{2+}. Explain why this value is much greater than that for a mole of hydrogen atoms.

[6.70] Heisenberg's uncertainty principle can be expressed mathematically as $\Delta x \cdot \Delta p \geq h/2\pi$, where Δx and Δp denote the uncertainty in position and momentum, respectively, and h is Planck's constant. **(a)** An electron has a speed of 3.0×10^6 m/s. If the accuracy with which this can be measured is 1.0 percent, what is the minimum uncertainty in the position of the electron? **(b)** Repeat the calculation in part **(a)**, but for a 12-g bullet whose speed is 200 m/s. **(c)** Compare the results for parts (a) and (b): How does each compare with the size of the object itself? (Remember that momentum is mass times velocity: $p = mv$.)

6.71 What is the subshell designation for each of the following cases: **(a)** $n = 2$, $l = 0$; **(b)** $n = 4$, $l = 2$; **(c)** $n = 5$, $l = 1$; **(d)** $n = 3$, $l = 2$; **(e)** $n = 4$, $l = 3$?

6.72 Indicate the number of orbitals that can have each of the following designation: **(a)** $n = 5$; **(b)** $3p$; **(c)** $4f$; **(d)** $5d_{xy}$; **(e)** $6s$; **(f)** $4d$.

6.73 In a chlorine atom, which electrons experience the largest effective nuclear charge? Which experience the smallest?

6.74 The quantum numbers listed below are for four different electrons in the same atom. Arrange them in order of increasing energy. Indicate whether any two have the same energy.
(a) $n = 4$, $l = 0$, $m_l = 0$, $m_s = \frac{1}{2}$;
(b) $n = 3$, $l = 2$, $m_l = 1$, $m_s = \frac{1}{2}$;
(c) $n = 3$, $l = 2$, $m_l = -2$, $m_s = -\frac{1}{2}$;
(d) $n = 3$, $l = 1$, $m_l = 1$, $m_s = -\frac{1}{2}$.

6.75 Why can the $2p$ subshell of an atom hold more electrons than the $2s$ subshell?

6.76 Using only a periodic table as a guide, write the complete electron configurations for the following atoms: **(a)** Cd; **(b)** As; **(c)** La; **(d)** Pd; **(e)** S.

6.77 The Stern-Gerlach experiment illustrated in Figure 6.31 was carried out using a beam of silver atoms. Draw the orbital-diagram representation for the valence electron configuration of silver. How many unpaired electrons does this atom have?

6.78 Consider the ground electron configurations for the elements of the fourth row of the periodic table. **(a)** Which of these will have no unpaired electrons? **(b)** Which will have one unpaired electron? **(c)** Which will have the most unpaired electrons?

6.79 For each of the following electron configurations, determine the element to which it corresponds, and determine whether it is a ground- or excited-state electron configuration: **(a)** $[He]2s^12p^5$; **(b)** $[Ar]4s^23d^{10}4p^5$; **(c)** $[Ne]3s^23p^24s^1$; **(d)** $[Kr]5s^24d^{10}5p^1$.

[6.80] Cite evidence to support or refute the following claims: **(a)** The energy levels of a hydrogen atom are quantized. **(b)** The electron in a hydrogen atom can be located exactly. **(c)** Electrons possess spin. **(d)** The effective nuclear charge experienced by a $3p$ electron in a Cl^- ion is less than that experienced by a $3p$ electron in a neutral Cl atom. **(e)** The actinide elements are the only ones that have electrons residing in f orbitals.

Periodic Properties of the Elements

7

Orange cup corals off the coast of California. Corals are composed mainly of calcium carbonate, $CaCO_3$. They absorb and release carbon dioxide gas and serve as regulators of the amount of dissolved CO_2 in the oceans. (Jeff Rotman)

CONTENTS

Scientists constantly seek ways to organize facts so they can identify similarities, differences, and trends among these facts. The most significant tool for organizing and remembering chemical facts is the periodic table. As we saw in Chapter 6, the periodic table arises from the periodic nature of electron configurations. Elements in the same column contain the same number of electrons in their outer-shell orbitals, or **valence orbitals**. For example, N ($[He]2s^22p^3$) and As ($[Ar]4s^23d^{10}4p^3$) are both members of group 5A; the similarity in their electronic structures leads to many similarities in their properties.

When we compare N and As, however, it is apparent that they exhibit differences as well as similarities (Figure 7.1). Some clues as to the distinctions of these elements are evident in the *differences* in their configurations. The outermost electrons of N have $n = 2$, whereas those of As have $n = 4$. Also, As has a completely filled $3d$ subshell, whereas N has no electrons in d orbitals. We shall see that variations in electron configurations can be used to explain differences in the properties of elements.

In this chapter, we shall use the electronic structure of atoms to gain a deeper understanding of how the properties of elements change as we move across a row or down a column of the periodic table. We shall see that for many properties, the trends within a row or column form patterns. Our discussion will begin with a brief history of the periodic table, followed by a con-

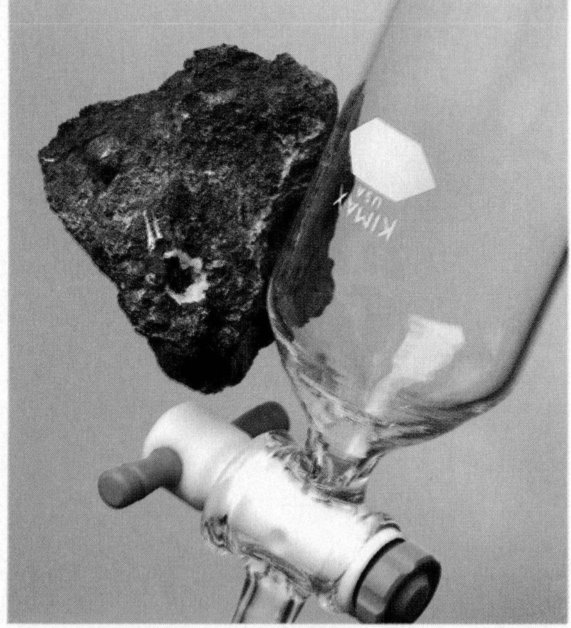

Figure 7.1 Nitrogen (right) and arsenic (left) are both group 5A elements. As such, they have many chemical similarities. However, they also have many differences including, as shown here, the forms they take as elements. Nitrogen is a colorless diatomic gas, whereas arsenic is a gray, solid semimetal. (Richard Megna/Fundamental Photographs)

sideration of the periodic trends in several properties of individual atoms. We shall then discuss trends in the behavior of collections of atoms of the elements, especially whether a given element will be a metal, a semimetal, or a nonmetal. Finally, we shall examine the properties of the elements within several families of the periodic table.

7.1 DEVELOPMENT OF THE PERIODIC TABLE

The discovery of new chemical elements has been an ongoing process since ancient times (Figure 7.2). Certain elements, such as gold and silver, appear in nature in elemental form and were thus discovered thousands of years ago. In contrast, some elements are radioactive and intrinsically unstable. We know about them only because of twentieth-century technological developments, as we shall see in Chapter 21.

The majority of elements, although stable, are found in nature only combined with other elements. Thus, for centuries, scientists were unaware of the existence of most elements. Finally, in the early nineteenth century, advances in chemistry made it far easier for scientists to isolate these elements. As a result, the number of known elements more than doubled from 31 in 1800 to 63 by 1865.

As the number of known elements increased, scientists began to investigate the possibilities of classifying them in useful ways. In 1869, Dmitri Mendeleev in Russia (Figure 7.3) and Lothar Meyer in Germany published nearly identical schemes for classifying the elements. Both scientists noted that similar

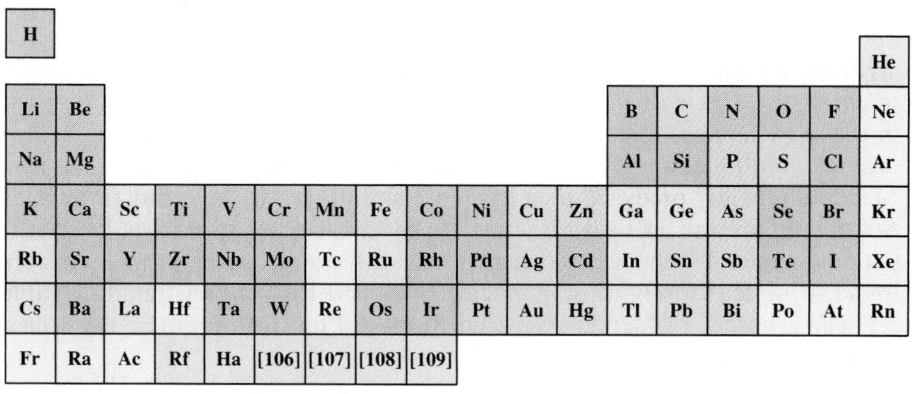

Figure 7.2 Periodic table showing the dates of discovery of the elements. (V. Ringnes, *J. Chem. Educ*, **1989**, *66*, 731)

Figure 7.3 Dmitri Mendeleev. Mendeleev rose from very poor beginnings to a position of great eminence in nineteenth-century science. He was born in Siberia, the youngest child in a family of at least 14. His mother endured great personal sacrifice to make it possible for him to enroll in a university in Saint Petersburg. Mendeleev proved to be a brilliant student in science and mathematics and eventually was able to study in France and Germany. He spent most of his career as a professor of chemistry at the University of Saint Petersburg. Despite his eminence as a scientist, he was often in trouble because of his liberal, unorthodox opinions. (Library of Congress)

chemical and physical properties recur periodically when the elements are arranged in order of increasing atomic weight. Scientists at that time had no knowledge of atomic numbers. Atomic weights, however, generally increase with increasing atomic number, so both Mendeleev and Meyer unwittingly arranged the elements in proper sequence. The tables of elements advanced by Mendeleev and Meyer were the forerunners of the modern periodic table.

Although Mendeleev and Meyer came to essentially the same conclusion about the periodicity of the properties of the elements, Mendeleev is given credit for advancing his ideas more vigorously and stimulating much new work in chemistry. His insistence that elements with similar characteristics be listed in the same families forced him to leave several blank spaces in his table. For example, both gallium (Ga) and germanium (Ge) were at that time unknown. Mendeleev boldly predicted their existence and properties, referring to them as eka-aluminum and eka-silicon, after the elements they appear under in the periodic table. When these elements were discovered, their properties were found to match closely those predicted by Mendeleev, as illustrated in Table 7.1. The accuracy of Mendeleev's predictions did much to promote the acceptance of the periodic table.

Table 7.1 Comparison of the Properties of Eka-Silicon Predicted by Mendeleev with the Observed Properties of Germanium

Property	Mendeleev's predictions for eka-silicon (made in 1871)	Observed properties of germanium (discovered in 1886)
Atomic weight	72	72.59
Density (g/cm^3)	5.5	5.35
Specific heat (J/g-K)	0.305	0.309
Melting point (°C)	High	947
Color	Dark gray	Grayish white
Formula of oxide	XO_2	GeO_2
Density of oxide (g/cm^3)	4.7	4.70
Formula of chloride	XCl_4	$GeCl_4$
Boiling point of chloride (°C)	A little under 100	84

In 1913, just 2 years after Rutherford had proposed the nuclear model of the atom, Henry Moseley (1887–1915) discovered the concept of atomic numbers. Moseley bombarded different elements with energetic electrons and studied the resultant X rays. He observed that the frequencies of the X rays were different for each element. He was able to arrange these frequencies in order by assigning each element a unique whole number, which he called the *atomic number*. He correctly proposed that the atomic number is the charge on the nucleus of the atom. We know today that the atomic number equals not only the number of protons in the nucleus of an atom, but also the number of electrons in that atom. In the sections that follow, we shall examine how the arrangement of electrons helps determine an element's chemical properties.

7.2 ELECTRON SHELLS IN ATOMS

When we move down a column of the periodic table, we change the principal quantum number, n, of the valence orbitals of the atoms. In Section 6.5, we referred to all of the orbitals with the same value of n as a *shell*. The origin of this term actually predates the quantum-mechanical model of the atom. Even before Bohr had proposed his theory of the hydrogen atom, Gilbert N. Lewis had suggested that electrons in atoms are arranged in spherical shells around the nucleus. How does the quantum-mechanical description of electron configurations correspond to Lewis's idea of electron shells? Consider the noble gases helium, neon, and argon, whose electron configurations follow:

He	$1s^2$
Ne	$1s^22s^22p^6$
Ar	$1s^22s^22p^63s^23p^6$

The total electronic charge distribution in these atoms can be accurately calculated using large computers. These distributions are shown in Figure 7.4. The quantity plotted on the vertical axis is called the *radial electron density*. It corresponds to the probability of finding the electron at a particular distance from the nucleus. As Figure 7.4 shows, the radial electron density does not fall off continuously as we move away from the nucleus. Rather, it shows maxima

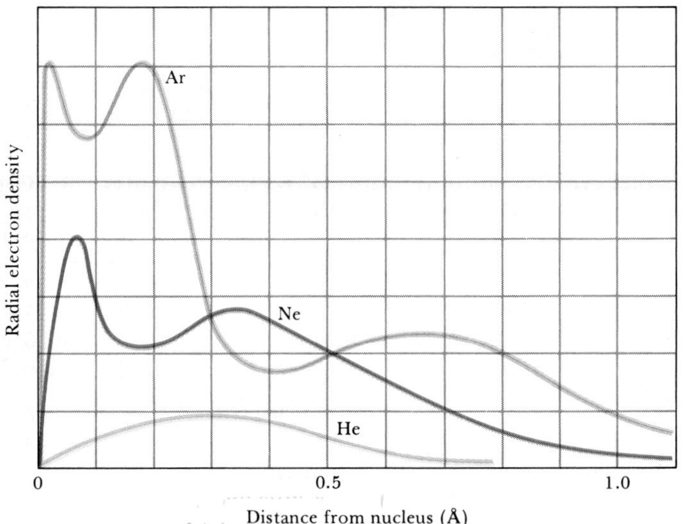

Figure 7.4 Radial electron-density graphs for the first three rare-gas elements, He, Ne, and Ar. The maxima that occur in the radial electron density correspond to electrons with the same value of the principal quantum number, n.

corresponding to distances at which there are higher probabilities of finding electrons. These maxima correspond to Lewis's idea of shells of electrons; however, these shells are diffuse and overlap considerably.

Helium shows a single shell, neon two, and argon three. Each of these maxima is due mainly to electrons that have the same principal quantum number, n. Thus, for helium the $1s$ electrons show a maximum in radial electron density at about 0.3 Å. In argon, the maximum in the $1s$ radial electron density occurs at only 0.05 Å. The second maximum, which occurs at larger radial distance, is due to both the $2s$ and $2p$ electrons. The third maximum is due to $3s$ and $3p$ electrons.

Why is the $1s$ shell in argon so much closer to the nucleus than the $1s$ shell in helium? The reason is clear when we recall that the nuclear charge of helium is only $2+$, whereas that for argon is $18+$. Because the $1s$ electrons are the innermost electrons of the atom, they are not shielded from the nucleus very effectively by other electrons. Thus, as the nuclear charge increases, the $1s$ electrons are "pulled" closer and closer to the nucleus. Similarly, the $n = 2$ shell of argon is closer to the nucleus than that of neon: Even though the $n = 3$ shell weakly shields the $n = 2$ shell in argon, this is more than offset by the attraction due to the greater nuclear charge of argon.

Now that we have some understanding of the electronic structure of atoms, we can examine some properties of atoms that depend largely on their electron configurations. We shall consider three properties that provide important insights into chemical behavior: atomic size, ionization energy, and electron affinity.

7.3 SIZES OF ATOMS

We often think of atoms as spherical objects with well-defined boundaries. However, one conclusion we can draw from the quantum-mechanical model is that an atom does not have a sharply defined boundary that determines its size. This is evident in Figure 7.4. The electron-density distributions illustrated do not end sharply; rather, they slowly drop off with increasing distance from the nucleus, approaching zero at large distance. Given this state of affairs, we might ask whether it makes sense to speak of a well-defined radius for an atom. Such a concept would be difficult to define for an isolated atom. Suppose, however, that two atoms form a chemical bond between them, as in Br_2. We can define the radius of an atom (the **atomic radius**) as the radius of a sphere that leads to observed bond lengths when the spheres are just touching one another. For example, the distance between the centers of the two bromine atoms in Br_2 can be thought of as the sum of two bromine atomic radii. The Br—Br distance in Br_2 is 2.286 Å (228.6 pm)*; we might then say that, to the nearest 0.01 Å, the radius of the bromine atom is 1.14 Å. Similarly, in compounds containing carbon-carbon bonds, the C—C distance is found to be very close to 1.54 Å. We can thus assign an atomic radius of 0.77 Å to carbon. If our concept of an atomic radius is going to be very useful, these atomic radii should remain nearly constant when the atom is bound to another element. Thus, the distance between carbon and bromine in a C—Br bond should be 1.14 + 0.77 Å, or 1.91 Å. It turns out that carbon-bromine bonds in various compounds are about this length. Atomic radii are not so easy to evaluate for metallic ele-

* Remember: The angstrom ($1 \text{ Å} = 10^{-10}$ m), although a convenient metric unit for atomic measurements of length, is not an SI unit. The most commonly used SI unit for such measurements is the picometer ($1 \text{ pm} = 10^{-12}$ m). **1 Å = 100 pm.**

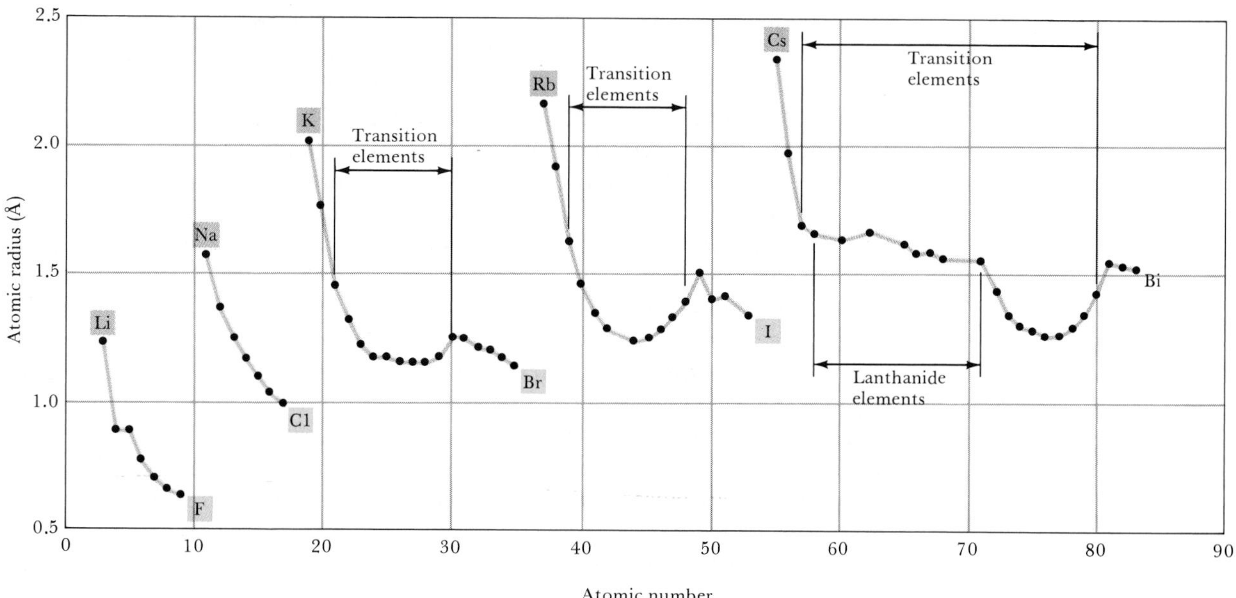

Figure 7.5 Atomic radius versus atomic number. The noble gases are not included in this graph because there is no simple way of relating their radii to those of the other elements on the basis of solid-state structure determinations. Gaps in the graph are due to lack of experimental data.

ments, but a list of atomic radii has been assembled for many elements, based on a large body of experimental data. These radii are graphed in Figure 7.5 as a function of atomic number.

Provided we keep in mind that there are uncertainties in these values because of the means by which they are obtained, we can discern some interesting trends in the data:

1. Within each group (column), the atomic radius tends to increase going from top to bottom.
2. Within each period (row), the atomic radius tends to decrease moving left to right.

These general trends, which are summarized in Figure 7.6, are the result of two factors that determine the size of the outermost orbital: its principal quantum number and the effective nuclear charge acting on its electrons. Increasing the principal quantum number increases the size of the orbital; increasing the effective nuclear charge reduces the size.

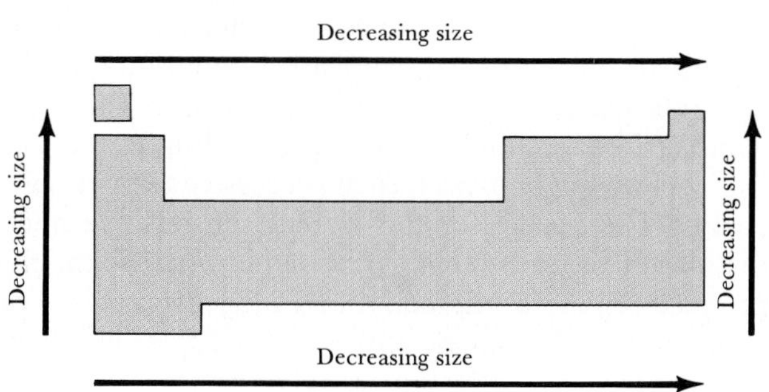

Figure 7.6 General periodic trends in atomic size.

Proceeding across any row of the table, the number of core electrons remains the same, while the nuclear charge increases. The electrons that are added to counterbalance the increasing nuclear charge are very ineffective in shielding each other. The effective nuclear charge therefore increases steadily, while the principal quantum number remains constant. For example, the inner $1s^2$ electrons of lithium ($1s^22s^1$) shield the outer $2s$ electron from the 3+ charged nucleus. Consequently, the outer electron experiences an effective nuclear charge of about 1+. For beryllium ($1s^22s^2$), the effective nuclear charge experienced by each outer $2s$ electron is larger; in this case the inner $1s^2$ electrons are shielding a 4+ nucleus, and each $2s$ electron only partially shields the other from the nucleus. As the effective nuclear charge increases, the electrons are drawn closer to the nucleus. Thus, the radius of the atom decreases as we proceed from left to right.

In going down a column of elements, the effective nuclear charge remains relatively constant, while the principal quantum number increases. Thus, the size of the orbital and consequently of the atomic radius increases.

SAMPLE EXERCISE 7.1

Referring to a periodic table, arrange the following atoms in order of increasing size: O, S, F.

Solution: Notice that F is to the right of O in the same period; thus we expect that F is smaller than O. Now notice that S is below O in group 6A: hence S is larger than O. The resultant order of increasing radii is F < O < S.

PRACTICE EXERCISE

Arrange the following atoms in order of increasing atomic radius: Na, Be, Mg.
Answer: Be < Mg < Na

7.4 IONIZATION ENERGY

As we saw in Section 6.3, the ionization energy, I, is the energy required to remove an electron from a gaseous atom or ion. The **first ionization energy** for an element, I_1, is the energy required for the process shown in Equation 7.1:

$$M(g) \longrightarrow M(g)^+ + e^- \qquad [7.1]$$

where M is a gaseous, neutral atom. The **second ionization energy**, I_2, is then the energy for the removal of the second electron, Equation 7.2:

$$M(g)^+ \longrightarrow M(g)^{2+} + e^- \qquad [7.2]$$

Successive ionization energies are defined in a similar manner. The values of successive ionization energies are known for many elements. Values for the elements sodium through argon are listed in Table 7.2. Remember that a larger value of I corresponds to stronger binding of the electron to the atom or ion.

As we might expect, each successive removal of an electron requires more energy. The reason for this is that the positive nuclear charge that provides the attractive force remains the same, whereas the number of electrons, which produce repulsive interaction, steadily decreases. For example, the electronic configuration for silicon (Si) is $1s^22s^22p^63s^23p^2$. If we look at the successive

Table 7.2 Successive Values of Ionization Energies, *I*, for the Elements Sodium through Argon (kJ/mol)[a]

Element	I_1	I_2	I_3	I_4	I_5	I_6	I_7
Na	496	4560		(Inner-shell electrons)			
Mg	738	1450	7730				
Al	577	1816	2744	11,600			
Si	786	1577	3228	4354	16,100		
P	1060	1890	2905	4950	6270	21,200	
S	999	2260	3375	4565	6950	8490	27,000
Cl	1256	2295	3850	5160	6560	9360	11,000
Ar	1520	2665	3945	5770	7230	8780	12,000

[a] Although the ionization energies are given here in units of kJ/mol, they are also often given in units of electron volts; 1 electron volt is equal to 96.49 kJ/mol.

ionization energies for silicon given in Table 7.2, we see a steady increase from 786 kJ/mol to 4354 kJ/mol for the four values of I that correspond to loss of the four valence-shell electrons with principal quantum number $n = 3$. The fifth electron, however, requires considerably more energy for removal: 16,100 kJ/mol. This sharp increase in ionization energy occurs because the fifth electron is an inner-shell electron. This electron is in a $2p$ orbital and consequently penetrates closer to the nucleus than do the $3s$ and $3p$ electrons. The $2p$ electron in silicon not only has a smaller average distance from the nucleus, but it also experiences a larger effective nuclear charge because it penetrates the charge distribution of the other electrons.

The above observation about ionization energies is general: For every element, we see a huge increase in ionization energy after we have removed enough electrons to achieve a noble-gas configuration. The next electron to be ionized comes from the shell with the next lower value of n and will require considerably more energy for removal.

These ionization energy data support the idea that only the outermost electrons, those beyond the noble-gas core, are involved in the sharing and transfer of electrons that give rise to chemical change. The inner electrons are too tightly bound to the nucleus to be lost from the atom or even shared with another atom.

SAMPLE EXERCISE 7.2

As can be seen in Table 7.2, the energy required to remove an electron from P^{4+} is 6270 kJ/mol, as compared with 16,100 kJ/mol for removal of an electron from Si^{4+}. Account for the large difference.

Solution: The outer electron configuration of phosphorus is $3s^2 3p^3$. After removal of four of these electrons, the highest-energy electron remaining is a $3s$. The outer electron configuration of silicon is $3s^2 3p^2$. After removal of these four electrons the highest-energy electron remaining is a $2p$. It requires considerably less energy to remove the $3s$ electron, which lies largely outside the $1s^2 2s^2 2p^6$ core of electrons, than to remove an electron from the $2p$ level.

PRACTICE EXERCISE

Which atom should have the larger second ionization energy, lithium or beryllium?
Answer: lithium

Periodic Trends in Ionization Energies

We have seen that, for a given element, the ionization energy increases as we remove successive electrons. What trends do we observe in the ionization energies as we move from one element to another in the periodic table? Figure 7.7 shows a graph of the first ionization energies, I_1, versus atomic number. An overall periodicity in I_1 is evident, as noted in the following trends:

1. Within each period, I_1 generally increases with increasing atomic number. The alkali metals show the lowest ionization energy in each row, and the noble gases the highest. There are slight irregularities in this trend that we will overlook for the moment.
2. Within each group, ionization energy generally decreases with increasing atomic number. For example, less energy is required to remove an electron from a potassium atom than from a lithium atom.

A few simple considerations help to explain these trends. The energy needed to remove an electron from the outer shell depends on both the effective nuclear charge and the average distance of the electron from the nucleus. Either increasing the effective nuclear charge or decreasing the distance from the nucleus increases the attraction between the electron and the nucleus. As this attraction increases, it becomes harder to remove the electron, and thus the ionization energy increases. As we move across a period, there is both an increase in effective nuclear charge and a decrease in atomic radius, causing the ionization energy to increase. However, as we move down a column, the atomic radius increases, while the effective nuclear charge remains essentially constant. Thus, the attraction between the nucleus and the electron decreases in this direction, causing the ionization energy to decrease.

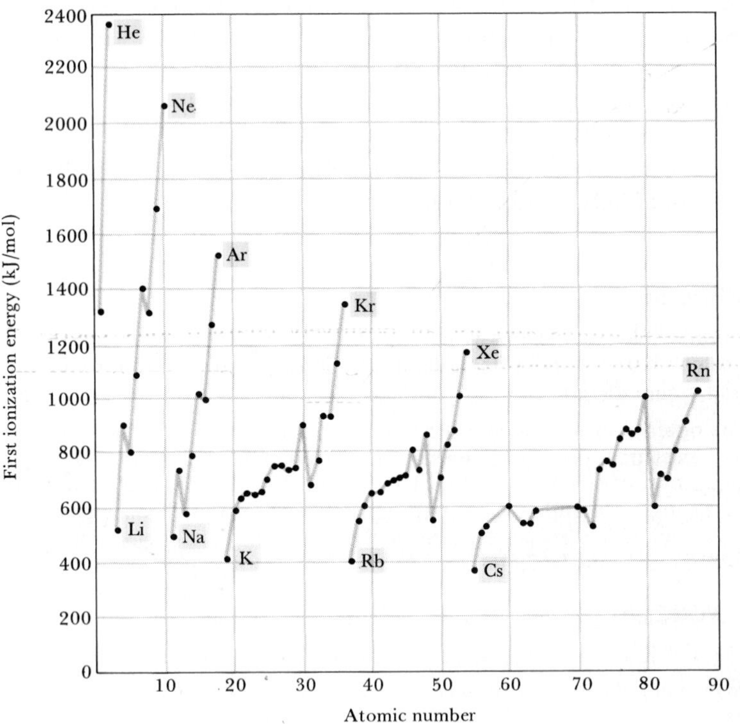

Figure 7.7 First ionization energy versus atomic number.

The irregularities within a given period are somewhat more subtle but are readily explained. For example, the decrease in ionization in going from beryllium ([He]$2s^2$) to boron ([He]$2s^22p^1$) arises because the electrons in the filled $2s$ orbital are more effective at shielding the electrons in the $2p$ subshell than they are at shielding each other. This is essentially the same reason that in many-electron atoms, the $2p$ orbital is at a higher energy than the $2s$ (Figure 6.25). The decrease in ionization energy on going from nitrogen ([He]$2s^22p^3$) to oxygen ([He]$2s^22p^4$) is due to repulsion of paired electrons in the p^4 configuration. (Remember that, according to Hund's rule, each electron in the p^3 configuration resides in a different p orbital.)

For the third, fourth, and fifth periods, we see several sequences of elements that have similar values of I_1, with only a slight trend toward increasing I_1 with increasing atomic number. From the periodic table, we see that these sequences correspond to the transition metal and f-block elements. In general, periodic trends are not as dramatic in the properties of these elements as they are for the representative elements.

SAMPLE EXERCISE 7.3

Referring to the periodic table, select the atom from the following list that has the greatest first ionization energy: S, Cl, Se, Br.

Solution: Because ionization energy tends to increase as we move from the bottom of any family to the top, S and Cl should have greater ionization energies than Se and Br. Because the ionization energy increases as we move left to right in any period, Cl should have a greater ionization energy than S. Thus, Cl has the largest ionization energy of these four elements.

PRACTICE EXERCISE

Which of the following atoms—B, Al, C, and Si—has the lowest ionization energy?
Answer: Al

7.5 ELECTRON AFFINITIES

Atoms not only lose electrons to form positively charged ions, they also gain them to form negatively charged ones. The ionization energy measures the energy changes associated with removing electrons from gaseous atoms. The energy change that occurs when an electron is added to a gaseous atom or ion is called the **electron affinity**, E. The process may be represented for a neutral atom as

$$M(g) + e^- \longrightarrow M^-(g) \qquad [7.3]$$

For most neutral atoms and for all positively charged ions, energy is evolved when an electron is added; E is thus negative in sign.* The greater the attraction between the species and the added electron, the more negative the electron affinity. On the other hand, the electron affinities of anions and some neutral atoms are positive, meaning that work must be done to force the electron onto the species.

Figure 7.8 shows the variation of electron affinities for the first 20 elements in the periodic table. Notice that the electron affinities of the elements

* We have defined electron affinity in such a way that a negative E is associated with an exothermic process. Thus, the more negative the value for E, the greater the attraction for electrons. However, E is sometimes defined as the energy given off when an electron is added to a gaseous atom or ion. In references that define E in this second way, the more positive the E value, the greater the attraction for electrons.

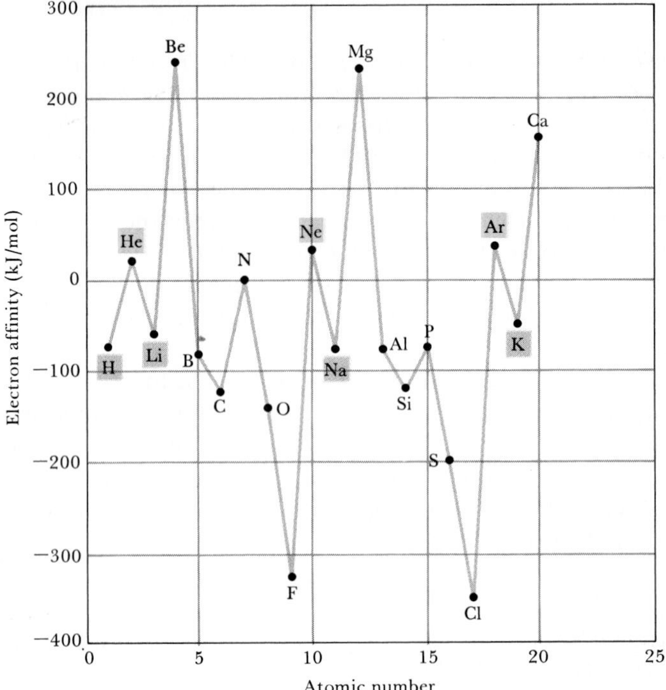

Figure 7.8 Electron affinity versus atomic number.

in group 2A (Be, Mg, and Ca) and group 8A (He, Ne, and Ar) are positive. These elements have filled subshells, so adding an electron requires energy. The outer *s* subshells are filled for the 2A elements; both the *s* and *p* subshells are filled for the noble gases. Notice also that the halogens (F and Cl) have the most negative electron affinities. By picking up an electron, they form very stable negative ions with noble-gas electron configurations.

Notice also how the electron affinities change as we move from Li to F and from Na to Cl. In general, electron affinities become more negative as we move across a period. The group 5A elements (N and P) have electron affinities that are less negative than those of the preceding group 4A elements (C and Si) because of half-filled subshells. For example, nitrogen has a half-filled $2p$ subshell ($1s^22s^22p^3$); thus an added electron must be placed into an orbital that is already occupied by an electron. The resultant electron-electron repulsion causes the nitrogen to have less attraction for an extra electron than does carbon. In general, atoms with filled or half-filled subshells have more positive electron affinities than do elements on either side of them in the periodic table.

As we go down a group, the electron affinities do not change greatly. For example, consider the electron affinities of halogens, which are listed in Table 7.3. As we proceed from fluorine to iodine, the added electron is going into a *p* orbital of increasing principal quantum number. The average distance of the

Table 7.3 Electron Affinities of Hydrogen and the Halogens

Element	Ion formed	*E* (kJ/mol)
H	H^-	−73
F	F^-	−332
Cl	Cl^-	−349
Br	Br^-	−325
I	I^-	−295

electron from the nucleus steadily increases, and electron-nucleus attraction should steadily decrease. If this were all that is involved, fluorine would have the highest electron affinity. But the orbitals that hold the outermost electrons of the halogen are increasingly spread out as we proceed from fluorine to iodine. The electron-electron repulsions between these electrons and the added one therefore decrease with increasing atomic number of the halogen. A lower electron-nucleus attraction is thus counterbalanced by lower electron-electron repulsion. The overall result is that the electron affinities differ very little among the halogens. Other families likewise show little variation in electron affinities among their members.

7.6 METALS, NONMETALS, AND SEMIMETALS

The periodic properties that we have discussed thus far are properties of individual atoms. With the exception of the noble gases, however, none of the elements exists in nature as isolated, individual atoms. In fact, the properties of elements that we usually observe are those of large collections of the atoms. We saw in Section 2.4 that elements can be divided into three broad categories: *metals*, *nonmetals*, and *semimetals*, also called *metalloids*. Further, we saw that each of these categories dominates a certain portion of the periodic table. Why is this so? As you might expect, a close relationship exists between the electron configurations of atoms and the forms in which the elements are found in nature. We shall discuss this in greater detail in later chapters on chemical bonding. For now, we shall reexamine the properties of metals, nonmetals, and semimetals, using the periodic table as a guide.

The periodic table in Figure 7.9 shows the classification of the elements as metals, nonmetals, and semimetals. Roughly three-quarters of the elements

Figure 7.9 The periodic table showing metals, semimetals, and nonmetals.

1A	2A	3B	4B	5B	6B	7B	8B	8B	8B	1B	2B	3A	4A	5A	6A	7A	8A
1 H																	2 He
3 Li	4 Be											5 B	6 C	7 N	8 O	9 F	10 Ne
11 Na	12 Mg											13 Al	14 Si	15 P	16 S	17 Cl	18 Ar
19 K	20 Ca	21 Sc	22 Ti	23 V	24 Cr	25 Mn	26 Fe	27 Co	28 Ni	29 Cu	30 Zn	31 Ga	32 Ge	33 As	34 Se	35 Br	36 Kr
37 Rb	38 Sr	39 Y	40 Zr	41 Nb	42 Mo	43 Tc	44 Ru	45 Rh	46 Pd	47 Ag	48 Cd	49 In	50 Sn	51 Sb	52 Te	53 I	54 Xe
55 Cs	56 Ba	57 La	72 Hf	73 Ta	74 W	75 Re	76 Os	77 Ir	78 Pt	79 Au	80 Hg	81 Tl	82 Pb	83 Bi	84 Po	85 At	86 Rn
87 Fr	88 Ra	89 Ac	104 Rf	105 Ha	[106]	[107]	[108]	[109]									

58 Ce	59 Pr	60 Nd	61 Pm	62 Sm	63 Eu	64 Gd	65 Tb	66 Dy	67 Ho	68 Er	69 Tm	70 Yb	71 Lu
90 Th	91 Pa	92 U	93 Np	94 Pu	95 Am	96 Cm	97 Bk	98 Cf	99 Es	100 Fm	101 Md	102 No	103 Lw

Metals

Semimetals

Nonmetals

Table 7.4 Characteristic Properties of Metallic and Nonmetallic Elements

Metallic elements	Nonmetallic elements
Distinguishing luster	Nonlustrous; various colors
Malleable and ductile as solids	Solids are usually brittle, may be hard or soft
Good thermal and electrical conductivity	Poor conductors of heat and electricity
Most metallic oxides are basic, ionic solids	Most nonmetallic oxides are molecular, acidic compounds
Exist in aqueous solution mainly as cations	Exist in aqueous solution mainly as anions or oxyanions

are metals, and these are situated in the left and middle portions of the table. The nonmetals are located in the upper right-hand corner of the table, and the semimetals lie between the metals and nonmetals. Notice that hydrogen, which is located at the top left corner, is a nonmetal. It is for this reason that we set off hydrogen from the remaining group 1A elements. Some of the distinguishing properties of metals and nonmetals are summarized in Table 7.4.

Metals

Most metals look much alike, exhibiting a characteristic metallic luster (Figure 7.10). Metals exhibit good electrical and thermal conductivity. They are malleable (can be pounded into thin sheets), ductile (can be drawn into wire), and can be formed into various shapes (Figure 7.11). All are solids at room temperature except mercury (melting point = $-39°C$), which is a liquid. Two metals melt at slightly above room temperature: cesium at 28.4°C and gallium at 29.8°C. At the other extreme, many metals melt at very high temperatures. For example, chromium melts at 1900°C. Figure 7.12 shows how the melting points of the metals vary through the fourth, fifth, and sixth periods of the periodic table. Notice that the melting points are periodic in character. In each period, they tend to increase with increasing atomic number up to group 6B, near the center of the transition metals.

Figure 7.10 Metals are readily recognized by their characteristic luster. (Dr. E. R. Degginger)

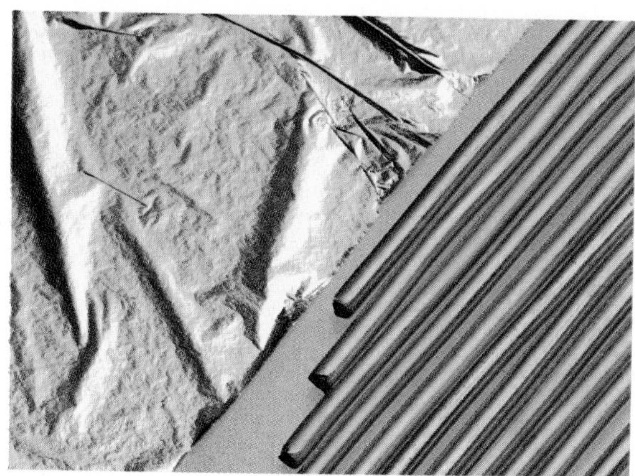

Figure 7.11 The gold leaf (left) and copper wire (right) shown here demonstrate the characteristic malleability and ductility of metals. (Richard Megna/Fundamental Photographs)

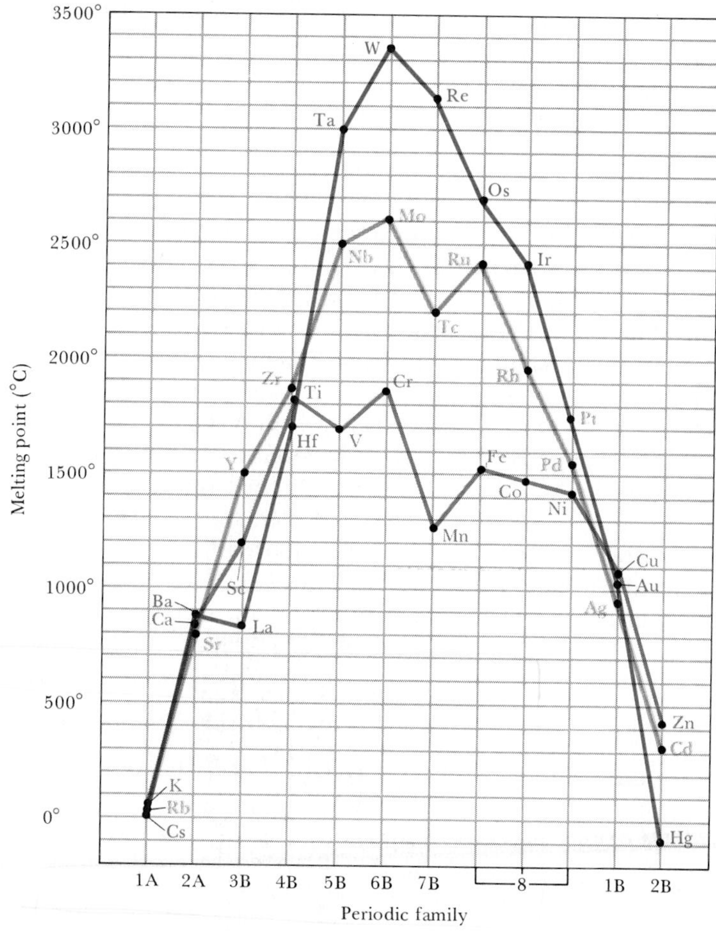

Figure 7.12 Melting points of some metals as a function of their location in the periodic table.

Metals tend to lose electrons when they undergo chemical reactions; that is, they transfer electrons to other substances and become cations (positively charged ions) (Section 2.5). For example, the reaction between nickel metal and oxygen produces nickel oxide, an ionic compound containing Ni^{2+} and O^{2-} ions:

$$2Ni(s) + O_2(g) \longrightarrow 2NiO(s) \qquad [7.4]$$

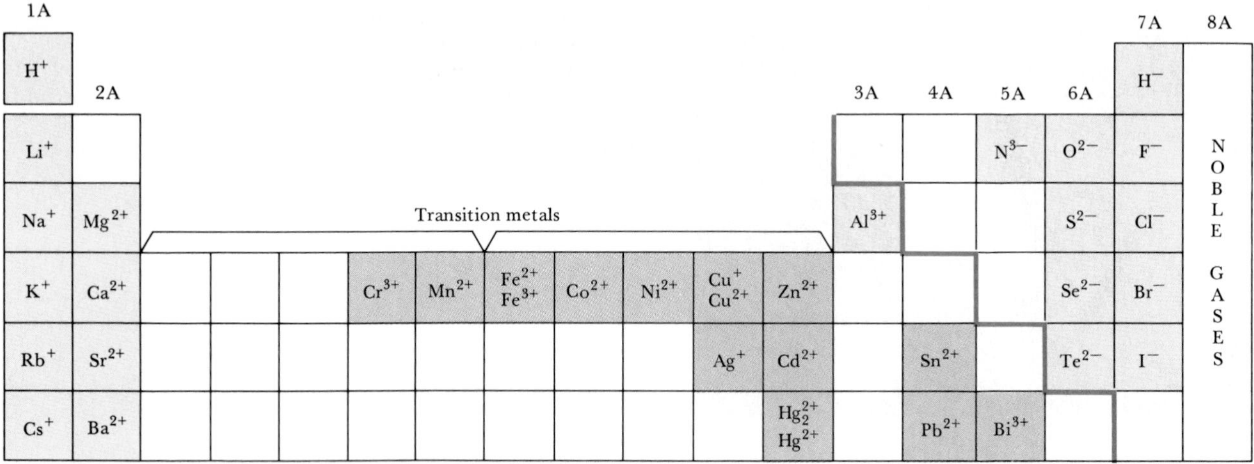

Figure 7.13 Charges of some common ions found in ionic compounds. Notice that the steplike line that divides metals from nonmetals also separates cations from anions. Ions shown in tan have the same number of electrons as the nearest noble-gas atom.

Figure 7.13 shows the charges of some common ions. Notice that the charges of alkali metals are always 1+ and that the charges of the alkaline earth metals are always 2+ in their compounds. For each of these families, the outer *s* electrons are easily lost, yielding a noble-gas configuration. The charges on the transition metal ions do not follow an obvious pattern. Many transition metal ions have 2+ charges, but 1+ and 3+ are also encountered. One of the characteristic features of the transition metals is their ability to form more than one positive ion. For example, iron may be 2+ in some compounds and 3+ in others. Figure 7.13 shows other examples.

Compounds of metals with nonmetals tend to be ionic substances with relatively high melting points. For example, most oxides, halides, and hydrides of metals are ionic solids. The oxides are particularly important because of the great abundance of oxygen in our environment.

Most metal oxides are *basic oxides;* those that dissolve in water react to form metal hydroxides, as in the following examples:

$$\text{Metal oxide + water} \longrightarrow \text{metal hydroxide}$$

$$Na_2O(s) + H_2O(l) \longrightarrow 2NaOH(aq) \quad [7.5]$$

$$CaO(s) + H_2O(l) \longrightarrow Ca(OH)_2(aq) \quad [7.6]$$

Metal oxides also react with acids to form salts and water (Section 4.5):

$$\text{Metal oxide + acid} \longrightarrow \text{salt + water}$$

$$MgO(s) + 2HCl(aq) \longrightarrow MgCl_2(aq) + H_2O(l) \quad [7.7]$$

$$NiO(s) + H_2SO_4(aq) \longrightarrow NiSO_4(aq) + H_2O(l) \quad [7.8]$$

SAMPLE EXERCISE 7.4

(a) Write the chemical formula for aluminum oxide. **(b)** Would you expect this substance to be a solid, liquid, or gas at room temperature? **(c)** Write the balanced chemical equation for the reaction of aluminum oxide with nitric acid.

Solution: **(a)** In all its compounds, aluminum has a 3+ charge, Al^{3+}; the oxide ion is O^{2-}. Consequently, the chemical formula of aluminum oxide is Al_2O_3.

(b) Because aluminum oxide is the oxide of a metal, we would expect it to be a solid. Indeed it is, and it happens to have a very high melting point, 2072°C.

(c) Metal oxides generally react with acids to form salts and water. In this case the salt is aluminum nitrate, $Al(NO_3)_3$. The balanced equation is

$$Al_2O_3(s) + 6HNO_3(aq) \longrightarrow 2Al(NO_3)_3(aq) + 3H_2O(l)$$

PRACTICE EXERCISE

Write the balanced chemical equation for the reaction between copper(II) oxide and sulfuric acid. *Answer:* $CuO(s) + H_2SO_4(aq) \longrightarrow CuSO_4(aq) + H_2O(l)$

Nonmetals

Nonmetals vary greatly in appearance (Figure 7.14). They are not lustrous and are generally poor conductors of heat and electricity. Although diamond, a form of carbon, has a high melting point (3570°C), the melting points of nonmetals are generally lower than those of metals. Seven of the nonmetals exist under ordinary conditions as diatomic molecules. Included in this list are gases (H_2, N_2, O_2, F_2, and Cl_2), one liquid (Br_2), and one volatile solid (I_2). The remaining nonmetals are solids that can be hard like diamond or soft like sulfur.

Nonmetals, in reacting with metals, tend to gain electrons and become anions (negatively charged ions) (Section 2.5). For example, the reaction of aluminum with bromine produces aluminum bromide, an ionic compound containing the bromide ion, Br^-, and the aluminum ion, Al^{3+}:

$$\text{Metal} + \text{nonmetal} \longrightarrow \text{salt}$$

$$2Al(s) + 3Br_2(l) \longrightarrow 2AlBr_3(s) \quad [7.9]$$

Notice in Figure 7.13 that the nonmetals commonly gain enough electrons to fill their outer *p* subshell completely, giving a noble-gas electron configuration.

Figure 7.14 Nonmetals are very diverse in their appearances. Shown here are, (left to right) sulfur, white phosphorus (stored under water), bromine, and carbon. (Dr. E. R. Degginger)

SAMPLE EXERCISE 7.5

Predict the formulas of compounds formed between **(a)** Ba and Te; **(b)** Ga and Br.

Solution: **(a)** Notice the location of these elements in the periodic table. The charge on a barium ion is 2+; the charge on a tellurium ion (telluride) is 2−, like that of oxides and sulfides. Thus, the formula of the compound is BaTe. **(b)** Gallium is in group 3A. Like the more familiar aluminum ion, gallium forms a 3+ ion. Bromine forms a 1− ion (bromide). Thus, the formula is $GaBr_3$.

PRACTICE EXERCISE

Predict the formula of the compound formed by Rb and Se. ***Answer:*** Rb_2Se

Compounds composed entirely of nonmetals are molecular substances. For example, the oxides, halides, and hydrides of the nonmetals are molecular substances that tend to be gases, liquids, or low-melting solids.

Most nonmetal oxides are *acidic oxides;* those that dissolve in water react to form acids, as in the following examples:

$$\text{Nonmetal oxide} + \text{water} \longrightarrow \text{acid}$$

$$CO_2(g) + H_2O(l) \longrightarrow H_2CO_3(aq) \quad [7.10]$$

$$P_4O_{10}(s) + 6H_2O(l) \longrightarrow 4H_3PO_4(aq) \quad [7.11]$$

The reaction of carbon dioxide with water (Figure 7.15) accounts for the acidity of carbonated water, and, to some extent, rainwater. Nonmetal oxides also dissolve in basic solutions to form salts, as in the following examples:

$$\text{Nonmetal oxide} + \text{base} \longrightarrow \text{salt} + \text{water}$$

$$CO_2(g) + 2NaOH(aq) \longrightarrow Na_2CO_3(aq) + H_2O(l) \quad [7.12]$$

$$SO_3(g) + 2KOH(aq) \longrightarrow K_2SO_4(aq) + H_2O(l) \quad [7.13]$$

Figure 7.15 When dry ice, CO_2, dissolves in water, the solution becomes acidic. Here we see the color of the acid-base indicator change from blue to yellow, indicating the increase in the acidity of the water. (Dr. E. R. Degginger)

(*a*)

(*b*)

SAMPLE EXERCISE 7.6

Write the balanced chemical equations for the reactions of solid selenium dioxide with **(a)** water; **(b)** sodium hydroxide.

Solution: **(a)** Selenium dioxide is SeO_2. Its reaction with water is like that of carbon dioxide (Equation 7.10):

$$SeO_2(s) + H_2O(l) \longrightarrow H_2SeO_3(aq)$$

(It doesn't matter that SeO_2 is a solid and CO_2 is a gas; the point is that both are nonmetal oxides.)

(b) The reaction with sodium hydroxide is like the reaction summarized by Equation 7.12:

$$SeO_2(s) + 2NaOH(aq) \longrightarrow Na_2SeO_3(aq) + H_2O(l)$$

PRACTICE EXERCISE

Write the balanced chemical equation for the reaction of tetraphosphorus hexoxide with water. ***Answer:*** $P_4O_6(s) + 6H_2O(l) \longrightarrow 4H_3PO_3(aq)$

Figure 7.16 A large crystal of elemental silicon, which is a semimetal. Although it looks metallic, silicon is brittle and is a poor thermal and electrical conductor as compared to metals. Large crystals of ultrapure silicon are sliced into very thin wafers for use in integrated circuits. (Richard Megna/Fundamental Photographs)

Semimetals

Semimetals have properties intermediate between those of metals and nonmetals. They may have *some* characteristic metallic properties but lack others. For example, silicon *looks* like a metal (Figure 7.16), but it is brittle rather than malleable and is a much poorer conductor of heat and electricity than metals. Several of the semimetals, most notably silicon, are electrical semiconductors and are the principal elements used in the manufacture of integrated circuits and computer chips.

Trends in Metallic Character

The more fully an element exhibits the physical and chemical properties characteristic of metals, the greater its **metallic character**. Similarly, we can speak of the *nonmetallic character* of an element. Metallic and nonmetallic character exhibit some important periodic trends, summarized in Figure 7.17 and in the following list:

Figure 7.17 Periodic trends in metallic and nonmetallic character. Notice the positions of the most active metals and the most active nonmetal.

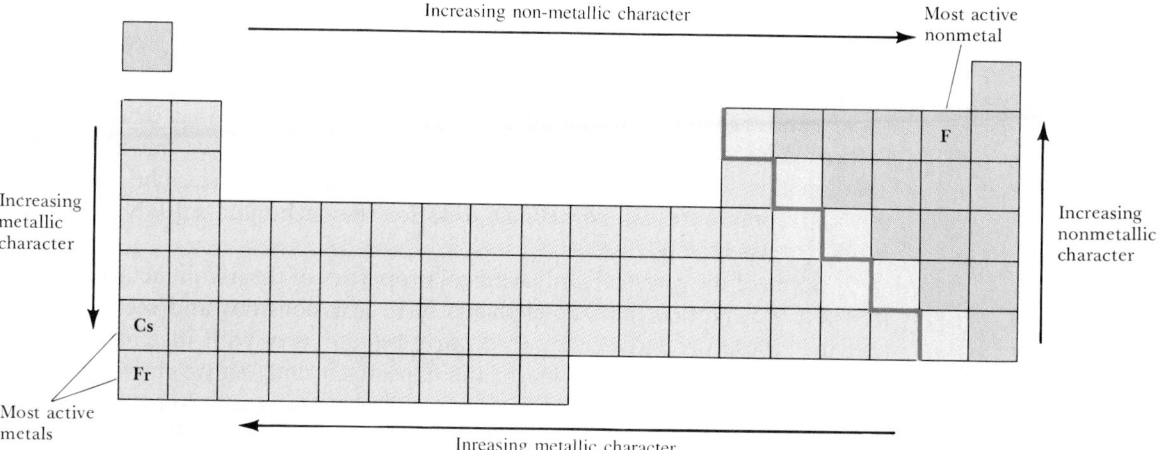

1. *Metallic character is strongest for the elements in the leftmost part of the periodic table and tends to decrease as we move to the right in any period.* For example, as we move across the fourth period from K to Kr, the ease with which each element loses electrons decreases. This is, of course, the same trend that we observed in the I_1 values for the elements (Section 7.4). For the fourth period, K displays the greatest reactivity toward substances that readily accept electrons, such as oxygen.
2. *Within any group of representative elements, the metallic character increases progressively from top to bottom.* For example, Ba loses electrons more readily than Mg. This rule is generally not followed among the transition metals. The first-row transition metals are generally more active than those in the second or third rows. For example, Fe is a more active metal than Ru or Os.
3. *Nonmetallic character is greatest for the elements in the upper right of the table and increases moving from left to right across a period.* For example, Br gains electrons more readily than As and thus displays greater chemical reactivity toward metals and other substances that readily lose electrons.
4. *Nonmetallic character decreases moving down a family.* The general shift from nonmetallic to metallic character as we go down a family of representative elements is most evident in groups 3A, 4A, and 5A. For example, carbon, the first element in group 4A, is clearly a nonmetal. Below carbon are the semimetals silicon and germanium. Below them are the metals tin and lead.

7.7 GROUP TRENDS: THE ACTIVE METALS

Our discussion of atomic size, ionization energy, electron affinity, and metallic character gives some indication of the way the periodic table can be used to organize and remember facts. Not only do elements in a family possess general similarities, but there are also trends as we move through a family or from one family to another. In this section, we shall use the periodic table and our knowledge of electron configurations to examine the chemistry of the **alkali metals** (group 1A) and the **alkaline earth metals** (group 2A). We shall compare these elements briefly to those of groups 1B and 2B. In the following section, we shall look at the group trends in selected families of nonmetals.

1A
3 Li
11 Na
19 K
37 Rb
55 Cs
87 Fr

Group 1A: The Alkali Metals

The alkali metals are soft, metallic solids (Figure 7.18). All have characteristic metallic properties such as a silvery, metallic luster and high thermal and electrical conductivities. The name *alkali* comes from an Arabic word meaning "ashes." Many compounds of sodium and potassium, the two most abundant alkali metals, were isolated from wood ashes by early chemists. The names soda ash and potash are still sometimes used for the carbonate salts Na_2CO_3 and K_2CO_3, respectively.

Some of the physical and chemical properties of the alkali metals are given in Table 7.5. Notice that the elements have low densities and melting points and that these properties vary in a fairly regular way with increasing atomic number. We can also see some of the expected trends as we move down the family, such as increasing atomic radius and decreasing first ionization energy. The alkali metals have the lowest I_1 values of the elements (Figure 7.7), which

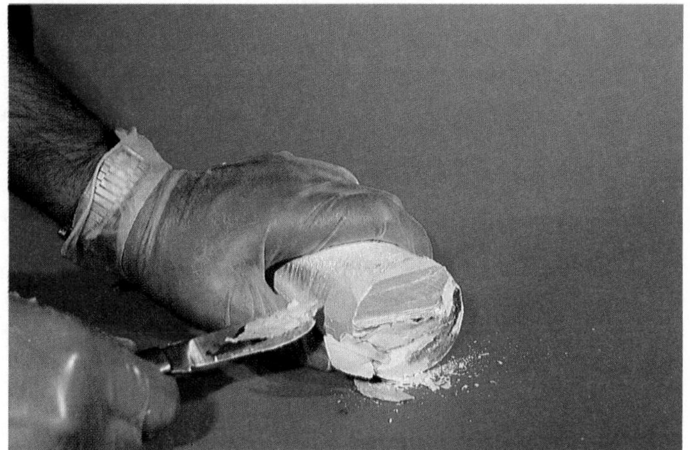

Figure 7.18 Sodium and the other alkali metals are soft. Here we see sodium being cut with a knife. The shiny metallic surface quickly tarnishes as the metal reacts with oxygen in the air. (H. E. LeMay, Jr.)

reflects the ease with which their outer *s* electron can be removed. As a result, the alkali metals are all very reactive, readily losing one electron to form ions with a 1+ charge:

$$M \longrightarrow M^+ + e^- \qquad [7.14]$$

(The symbol M in this equation and others represents any one of the alkali metals.) The alkali metals are the most active metals (Section 4.6) and thus exist in nature only as compounds. The metals can be obtained by passing an electric current through a molten salt, a process known as *electrolysis.* For example, sodium is prepared commercially by the electrolysis of molten NaCl. The electrical energy is used to remove electrons from Cl^- ions and to force them onto Na^+ ions:

$$2Cl^- \longrightarrow Cl_2 + 2e^- \qquad [7.15]$$

$$2Na^+ + 2e^- \longrightarrow 2Na \qquad [7.16]$$

We will discuss electrolysis in detail in Chapter 20.

Not surprisingly, the chemistry of the alkali metals is dominated by their tendency to lose an electron, thus forming a 1+ cation (Equation 7.14). The metals combine directly with most nonmetals. For example, they react with hydrogen to form solid hydrides, with sulfur to form solid sulfides, and with chlorine to form solid chlorides:

$$2M(s) + H_2(g) \longrightarrow 2MH(s) \qquad [7.17]$$

$$2M(s) + S(s) \longrightarrow M_2S(s) \qquad [7.18]$$

$$2M(s) + Cl_2(g) \longrightarrow 2MCl(s) \qquad [7.19]$$

Table 7.5 Some Properties of the Alkali Metals

Element	Electron configuration	Melting point (°C)	Density (g/cm^3)	Atomic radius (Å)	I_1 (kJ/mol)
Lithium	$[He]2s^1$	181	0.53	1.52	520
Sodium	$[Ne]3s^1$	98	0.97	1.86	496
Potassium	$[Ar]4s^1$	63	0.86	2.27	419
Rubidium	$[Kr]5s^1$	39	1.53	2.48	403
Cesium	$[Xe]6s^1$	29	1.90	2.65	376

In the hydrides of the alkali metals (LiH, NaH, and so forth), hydrogen is present as H^-, called the **hydride ion**. The hydride ion is to be distinguished from the hydrogen ion, H^+, formed when a hydrogen atom loses its electron.

The alkali metals react vigorously with water, producing hydrogen gas and solutions of alkali metal hydroxides:

$$2M(s) + 2H_2O(l) \longrightarrow 2MOH(aq) + H_2(g) \qquad [7.20]$$

These reactions are very exothermic. In many cases, enough heat is generated to ignite the H_2, producing a fire or explosion (Figure 3.4). This reaction is most violent in the case of the heavier members of the family, in keeping with their weaker hold on the single outer-shell electron.

The reactions between the alkali metals and oxygen are more complex. When oxygen reacts with metals, metal oxides, which contain the O^{2-} ion, are usually formed. Indeed, lithium shows this reactivity:

$$4Li(s) + O_2(g) \longrightarrow \underset{\text{lithium oxide}}{2Li_2O(s)} \qquad [7.21]$$

In contrast, the other alkali metals all form metal peroxides, which contain the $O_2{}^{2-}$ ion. For example,

$$2Na(s) + O_2(g) \longrightarrow \underset{\text{sodium peroxide}}{Na_2O_2(s)} \qquad [7.22]$$

Potassium, rubidium, and cesium also form compounds that contain the $O_2{}^-$ ion, called superoxides:

$$K(s) + O_2(g) \longrightarrow \underset{\text{potassium superoxide}}{KO_2(s)} \qquad [7.23]$$

You should realize that the reactions shown in Equations 7.22 and 7.23 are somewhat surprising; in most cases, the reaction of oxygen with a metal forms the metal oxide.

As is evident from Equations 7.20 through 7.23, the alkali metals are extremely reactive toward water and oxygen. Because of this, the metals are usually stored under a hydrocarbon, such as kerosene or mineral oil.

The small size of lithium confers some special properties on the metal and its compounds. For example, Li is the only alkali metal that reacts directly with N_2 to form a nitride, Li_3N. Thus, when lithium is burned in air, it forms not only lithium oxide but lithium nitride as well.

Alkali metal salts and their aqueous solutions are colorless unless they contain a colored anion like the yellow $CrO_4{}^{2-}$. Color is produced when an electron in an atom is excited from one energy level to another by visible radiation. Alkali metal ions, having lost their outermost electrons, have no electrons that can be excited by visible radiation.

When alkali metal compounds are placed in a flame, they emit characteristic colors, as shown in Figure 7.19. The alkali metal ions are reduced to gaseous metal atoms in the lower, central region of the flame. The atoms are electronically excited by the high temperature of the flame; they then emit

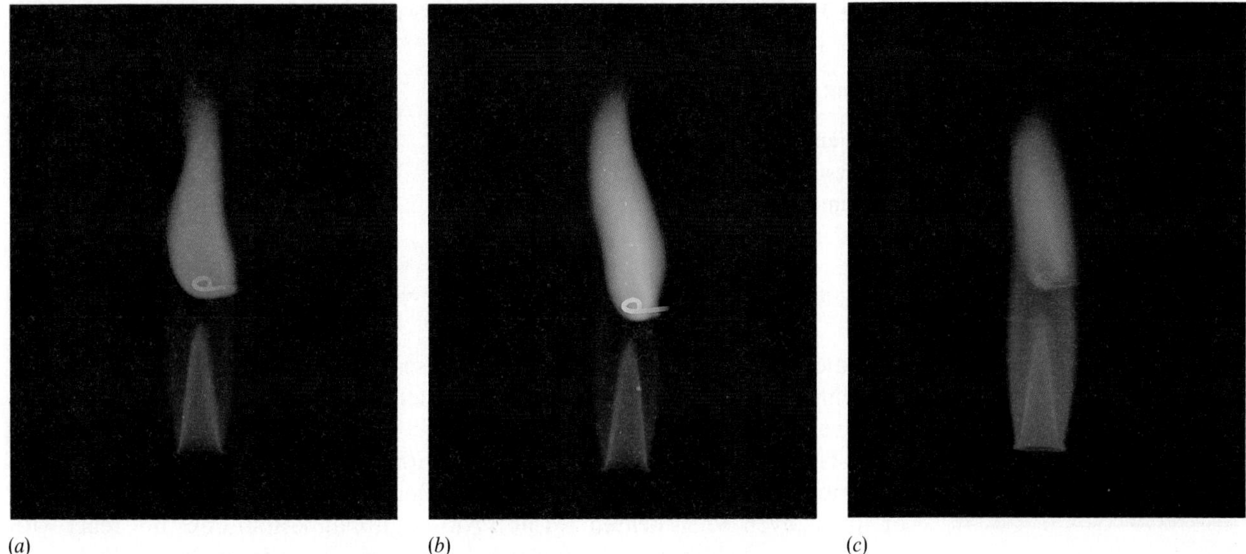

Figure 7.19 Flame test for (*a*) Li (crimson-red), (*b*) Na (yellow), and (*c*) K (lilac). (H. E. LeMay, Jr.)

energy in the form of visible light as they return to the ground state. For example, sodium gives a yellow flame due to emission at 589.2 nm. This wavelength is produced by transition of the excited valence electron from the 3*p* to the 3*s* subshell.

SAMPLE EXERCISE 7.7

Write balanced chemical equations for the reactions of cesium with **(a)** $H_2(g)$; **(b)** $Cl_2(g)$; **(c)** $H_2O(l)$.

Solution: By analogy to the equations above, we have

$$2Cs(s) + H_2(g) \longrightarrow 2CsH(s)$$

$$2Cs(s) + Cl_2(g) \longrightarrow 2CsCl(s)$$

$$2Cs(s) + 2H_2O(l) \longrightarrow 2CsOH(aq) + H_2(g)$$

In each case, cesium forms a 1+ ion in its compounds, Cs^+. The hydride ion, chloride ion, and hydroxide ion all have 1− charges: H^-, Cl^-, and OH^-

PRACTICE EXERCISE

Solid sodium hydride reacts with water to form an aqueous solution of sodium hydroxide and hydrogen gas. Write a balanced chemical equation for this reaction.
Answer: $NaH(s) + H_2O(l) \longrightarrow NaOH(aq) + H_2(g)$

2A
4 Be
12 Mg
20 Ca
38 Sr
56 Ba
88 Ra

Group 2A: The Alkaline Earth Metals

Like the alkali metals, the group 2A elements are all solids with typical metallic properties, some of which are listed in Table 7.6. Compared with the alkali metals, the alkaline earth metals are harder, are more dense, and melt at higher temperatures.

The first ionization energies of the alkaline earth elements are low, but not as low as those of the alkali metals. Consequently, the alkaline earths are less reactive than their alkali metal neighbors. As we have noted in Section 7.4, the ease with which the elements lose electrons decreases as we move across the

Figure 7.20 Calcium metal reacts with water to form hydrogen gas and calcium hydroxide, $Ca(OH)_2$. (Dr. E. R. Degginger)

Table 7.6 Some Properties of the Alkaline Earth Metals

Element	Electron configuration	Melting point (°C)	Density (g/cm^3)	Atomic radius (Å)	I_1 (kJ/mol)
Beryllium	$[He]2s^2$	1287	1.85	1.12	899
Magnesium	$[Ne]3s^2$	649	1.74	1.60	738
Calcium	$[Ar]4s^2$	839	1.55	1.97	590
Strontium	$[Kr]5s^2$	768	2.63	2.15	549
Barium	$[Xe]6s^2$	727	3.62	2.22	503

periodic table from left to right and increases as we move down a family. Thus, beryllium and magnesium, the lightest members of the family, are the least reactive.

The trend of increasing reactivity within the family is illustrated by the behavior of the elements toward water. Beryllium does not react with water or steam, even when heated red hot. Although magnesium does not react with liquid water, it does react with steam to form magnesium oxide and hydrogen:

$$Mg(s) + H_2O(g) \longrightarrow MgO(s) + H_2(g) \quad [7.24]$$

Calcium and the elements below it react readily with water at room temperature (although more slowly than the alkali metals adjacent to them in the periodic table), as shown in Figure 7.20:

$$Ca(s) + 2H_2O(l) \longrightarrow Ca(OH)_2(aq) + H_2(g) \quad [7.25]$$

Figure 7.21 The red color of the fireworks and flares is produced by strontium salts. (Peter Sahula/Photo Researchers)

The above two reactions are illustrative of the dominant pattern in the reactivity of the alkaline earth elements—the tendency to lose their two outer *s* electrons and form 2+ ions. For example, magnesium reacts with chlorine at room temperature to form $MgCl_2$, and it burns with dazzling brilliance in air to give MgO (Figure 3.8):

$$Mg(s) + Cl_2(g) \longrightarrow MgCl_2(s) \quad [7.26]$$

$$2Mg(s) + O_2(g) \longrightarrow 2MgO(s) \quad [7.27]$$

In the presence of O_2, magnesium metal is protected from many chemicals by a thin surface coating of water-insoluble MgO. Thus, even though it is high in the activity series (Section 4.6), Mg can be incorporated into lightweight structural alloys used in, for example, automobile wheels (so-called mag wheels). The heavier alkaline earth metals (Ca, Sr, and Ba) are even more reactive toward nonmetals than magnesium and must be stored in such a way as to protect them from oxidation by O_2 and H_2O.

Like the 1+ ions of the alkali metals, the 2+ ions of the alkaline earth elements have a noble-gas electron configuration. As such, they form colorless or white compounds unless they are combined with a colored anion. The heavier alkaline earths have characteristic flames. The calcium flame is brick red, strontium is crimson red, and barium is green. The strontium flame produces the familiar red color of flares and fireworks (Figure 7.21).

CHEMISTRY AT WORK: Osteoporosis and Calcium Intake

The calcium ion is an essential component of our diet, required for clotting blood and maintaining heartbeat rhythm. However, most calcium is used in forming and maintaining bones and teeth. Bone is composed of protein fibers called collagen, water, and inorganic salts (or "minerals"). The primary mineral is known as hydroxyapatite, $Ca_5(PO_4)_3OH$.

The recommended daily allowance for calcium for adults is 1000 mg a day. Most of our dietary calcium is provided by dairy products, such as milk and cheese. However, many weight-conscious people shy away from these foods because they are high in calories. Thus, many people fail to meet the daily requirement of calcium in their diet.

One of the major health problems associated with calcium metabolism is osteoporosis (Greek *osteon*, meaning "bone," and *poros*, meaning "a passage"), the increased porosity or softening of bone (see Figure 7.22). This condition can lead to severe curvature of the spine and to increased likelihood of hip fractures. Osteoporosis is more a problem for women than for men because calcium metabolism is affected by levels of the female sex hormone, estrogen. Reduced estrogen production, which occurs after menopause, changes the normal reactions by which the removal of calcium from bones is balanced by new deposits. Calcium leaves faster than it can be replaced, and the bones become thin. Another problem for older women is reduced vitamin D production, which in turn decreases the efficiency with which dietary calcium is absorbed from the intestine.

Many experts now recommend the use of calcium supplements, especially for women with a family history of osteoporosis. Supplements must be taken with caution, though, because an excess of calcium can lead to kidney stones and other medical problems.

(*a*)

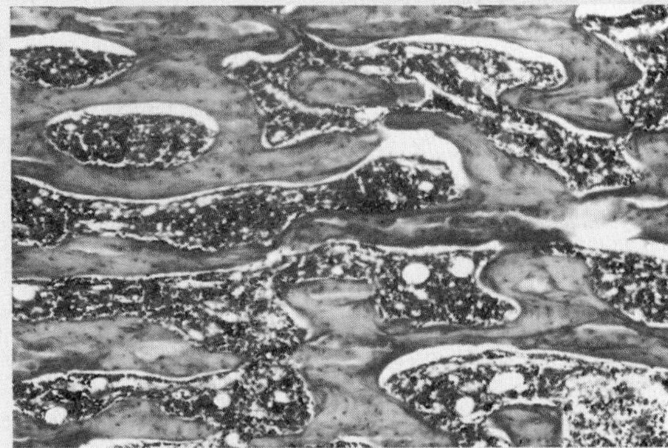

(*b*)

Figure 7.22 (*a*) Normal, cancellous (spongy) bone, composed of a network of bony material separated by a labyrinth of interconnecting spaces containing bone marrow. (*b*) Decalcified, cancellous bone, which shows a reduction in the white areas composed of bony material. (© Biophoto Associates/Photo Researchers; Patrick Lynch/Science Source Photo Researchers)

Comparison of A and B Groups

The label attached to each family in the periodic table has either an A or B designation.* Elements of the same group number but different letter labels may share certain characteristics, but in general there is not a close connection. Let's consider the distinction between A and B elements in terms of electron configuration. As an example, locate the group 1A and 1B elements in the periodic table. In both families of group 1 elements the outer electron configuration has a single *s* electron. In the case of the alkali metals (group 1A), this *s* electron is outside a stable noble-gas core of electrons. For example, the single 4*s* electron in potassium is outside a set of eight electrons associated with

* The designation of groups as A or B is purely arbitrary. In fact, some periodic tables use a different system from that used in this text; however, the system we use is presently the most common one for designating groups. (See the footnote in Section 2.4, p. 47).

Table 7.7 A Comparison of the Ionization Energies (in kJ/mol) for Group 1A, 1B, 2A, and 2B Elements

1A	I_1	1B	I_1	2A	I_1	I_2	2B	I_1	I_2
K	419	Cu	745	Ca	590	1145	Zn	906	1733
Rb	403	Ag	731	Sr	549	1064	Cd	877	1631
Cs	376	Au	890	Ba	503	965	Hg	1007	1809

filled 3*s* and 3*p* subshells, $3s^2 3p^6$. In the corresponding group 1B element, copper, the single electron is outside filled 3*s*, 3*p*, and 3*d* subshells, $3s^2 3p^6 3d^{10}$. The presence of the additional ten *d* electrons in copper, and of course the increase of ten in nuclear charge that goes along with them, has a profound effect on the chemical behavior of the single *s* electron. Because these *d* electrons only partially shield the nucleus, the *s* electron of the 1B group element experiences a larger effective nuclear charge than does the *s* electron of the 1A group. As a result, the 1B elements have higher ionization energies and are therefore much less reactive than the corresponding 1A elements. Table 7.7 lists the ionization energies for the group 1 and 2 elements. Compare the values for the A and B groups.

7.8 GROUP TRENDS: SELECTED NONMETALS

Hydrogen

1
H

The first element in the periodic table, hydrogen, has a $1s^1$ electron configuration and is usually placed above the alkali metals. However, it is a unique element and does not truly belong to any family. Hydrogen is a nonmetal that occurs as a colorless diatomic gas, $H_2(g)$, under most conditions. Whereas the chlorides and oxides of metals tend to be solids at room temperature, HCl is a gas and H_2O is a liquid.

Owing to the complete absence of nuclear shielding of its sole electron, the ionization energy of hydrogen, 1312 kJ/mol, is markedly higher than that of the active metals. In fact, it is comparable to I_1 of other nonmetals, such as oxygen and chlorine. Hydrogen generally reacts with other nonmetals to form molecular compounds. These reactions can be quite exothermic, as evidenced by the combustion reaction between hydrogen and oxygen to form water:

$$2H_2(g) + O_2(g) \longrightarrow 2H_2O(g) \qquad \Delta H^\circ = -571.7 \text{ kJ} \qquad [7.28]$$

We have also seen that hydrogen reacts with active metals to form solid metal hydrides, which contain the H^- ion. For example,

$$2Na(s) + H_2(g) \longrightarrow 2NaH(s) \qquad [7.29]$$

$$Ca(s) + H_2(g) \longrightarrow CaH_2(s) \qquad [7.30]$$

The aqueous chemistry of hydrogen, which was introduced in Chapter 4, is dominated by the $H^+(aq)$ ion. We shall take up its study in detail in Chapter 16.

Group 6A: The Oxygen Family

As we proceed down group 6A, the increase in metallic character is clearly evident. Oxygen is a colorless gas at room temperature; all of the others are solids.

Table 7.8 Some Properties of the Group 6A Elements

Element	Electron configuration	Melting point (°C)	Density	Atomic radius (Å)	I_1 (kJ/mol)
Oxygen	$[He]2s^2 2p^4$	−219	1.43 g/L	0.73	1314
Sulfur	$[Ne]3s^2 3p^4$	112	2.07 g/cm³	1.03	999
Selenium	$[Ar]4s^2 3d^{10} 4p^4$	217	4.79 g/cm³	1.40	941
Tellurium	$[Kr]5s^2 4d^{10} 5p^4$	452	6.24 g/cm³	1.60	869
Polonium	$[Xe]6s^2 4f^{14} 5d^{10} 6p^4$	254	9.32 g/cm³	1.64	813

6A
8 O
16 S
34 Se
52 Te
84 Po

Oxygen, sulfur, and selenium are typical nonmetals. Tellurium has some metallic properties and is classified as a semimetal. Polonium, which is radioactive and quite rare, is a metal. Some of the physical properties of the group 6A elements are given in Table 7.8.

As we saw in Section 2.5, oxygen is encountered in two molecular forms, O_2 and O_3. The O_2 form is the common one. People generally mean O_2 when they say "oxygen," although the name *dioxygen* is more descriptive. O_3 is called **ozone**. The two forms of oxygen are examples of **allotropes**. Allotropes are different forms of the same element in the same state (in this case, both forms are gases). About 21 percent of dry air consists of O_2 molecules. Ozone, which is toxic and has a pungent odor, is present in very small amounts in the upper atmosphere and in polluted air. It is also formed in electrical discharges, such as in lightning storms:

$$3O_2(g) \longrightarrow 2O_3(g) \qquad \Delta H^\circ = 264.6 \text{ kJ} \qquad [7.31]$$

As indicated by the endothermic nature of this reaction, ozone is less stable than O_2.

Oxygen has a great tendency to attract electrons from other elements (in other words, to *oxidize* them). As we have discussed, oxygen in combination with metals is almost always present as the oxide, O^{2-}, ion. This ion has a noble-gas configuration and is particularly stable. As shown in Equation 7.28, the formation of nonmetal oxides is also often very exothermic and thus energetically favorable.

In our discussion of the alkali metals, we noted two less common oxygen anions, namely, the peroxide, $O_2^{\ 2-}$, and superoxide, $O_2^{\ -}$, ions. Compounds of these ions often react with themselves to produce an oxide and O_2. For example, aqueous hydrogen peroxide, H_2O_2, slowly decomposes into water and O_2 at room temperature:

$$2H_2O_2(aq) \longrightarrow 2H_2O(l) + O_2(g) \qquad \Delta H^\circ = -196.1 \text{ kJ} \qquad [7.32]$$

For this reason, bottles of aqueous hydrogen peroxide are topped with caps that are able to release the $O_2(g)$ produced before the pressure inside becomes too great (Figure 7.23).

After oxygen, the most important member of group 6A is sulfur. Sulfur also exists in several allotropic forms, the most common and stable of which is the yellow solid with molecular formula S_8. This molecule consists of an eight-membered ring of sulfur atoms, as shown in Figure 7.24.

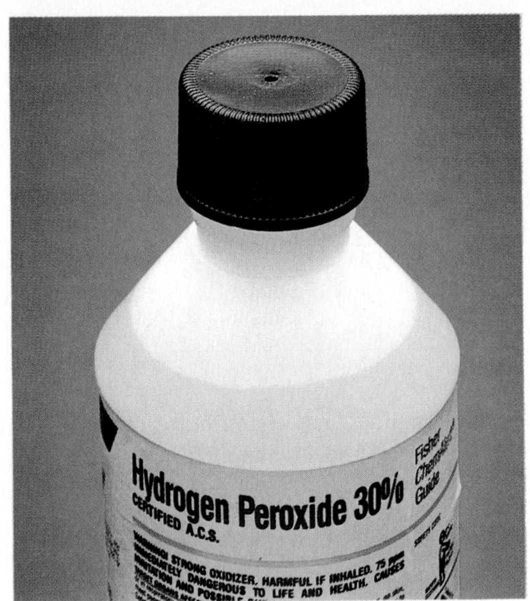

Figure 7.23 Hydrogen peroxide bottles are topped with caps that "don't fit right." The cap allows any excess pressure of $O_2(g)$ to be released from the bottle. Hydrogen peroxide is often stored in dark-colored or opaque bottles to minimize exposure to light, which accelerates its decomposition. (Richard Megna/Fundamental Photographs)

Like oxygen, sulfur has a tendency to gain electrons from other elements to form sulfides, which contain the S^{2-} ion. This is particularly true for the active metals, as in the example of sodium metal reacting with $S_8(s)$:

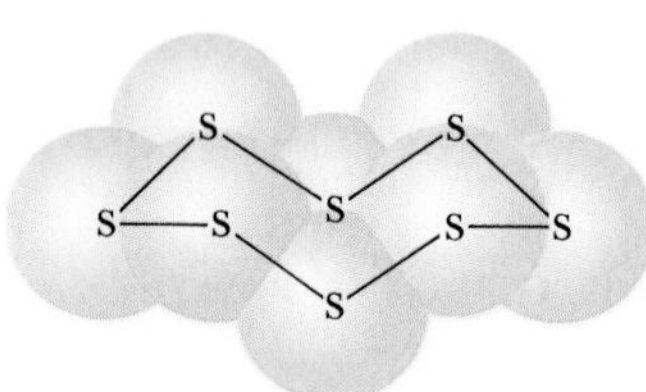

Figure 7.24 Structure of S_8 molecules as found in the most common allotropic form of sulfur at room temperature.

$$16\text{Na}(s) + \text{S}_8(s) \longrightarrow 8\text{Na}_2\text{S}(s) \qquad [7.33]$$

Most sulfur in nature is present as metal sulfide compounds. Because sulfur is below oxygen in the periodic table, its tendency to form sulfide anions is not as great as oxygen's ability to form oxide ions. As a result, the chemistry of sulfur is more complex and varied than that of oxygen. In fact, sulfur or sulfur-containing compounds (including those in coal and petroleum) can be burned in oxygen, yielding mainly sulfur dioxide:

$$\text{S}_8(s) + 8\text{O}_2(g) \longrightarrow 8\text{SO}_2(g) \qquad [7.34]$$

As we shall discuss further in Chapter 18, SO_2 is one of the principal air pollutants and is the major cause of acid rain.

Group 7A: The Halogens

7A
9 F
17 Cl
35 Br
53 I
85 At

The group 7A elements are known as the **halogens**, after the Greek words *halos* and *gennao*, meaning "salt formers." Some of the properties of these elements are given in Table 7.9. Astatine, which is both extremely rare and radioactive, is omitted because many of its properties are not yet known.

Table 7.9 Some Properties of the Halogens

Element	Electron configuration	Melting point (°C)	Density	Atomic radius (Å)	I_1 (kJ/mol)
Fluorine	$[\text{He}]2s^2 2p^5$	−219	1.81 g/L	0.72	1681
Chlorine	$[\text{Ne}]3s^2 3p^5$	−101	3.21 g/L	0.99	1256
Bromine	$[\text{Ar}]4s^2 3d^{10} 4p^5$	−7.3	3.19 g/cm^3	1.14	1143
Iodine	$[\text{Kr}]5s^2 4d^{10} 5p^5$	114	4.94 g/cm^3	1.33	1009

As we move from group 6A to group 7A, the nonmetallic behavior of the elements increases, as we would expect. Unlike the group 6A elements, all the halogens are typical nonmetals. Their melting and boiling points increase with increasing atomic number. Fluorine and chlorine are gases at room temperature, whereas bromine is a liquid and iodine is a solid. Each element consists of diatomic molecules: F_2, Cl_2, Br_2, and I_2. Fluorine gas is pale yellow; chlorine gas has a yellow-green color; bromine liquid is reddish brown and readily forms a reddish-brown vapor; and solid iodine is grayish black and readily forms a violet vapor (Figure 7.25).

The electron affinities of the halogens are among the greatest (most negative) of the elements (Table 7.3). Thus, it is not surprising that the chemistry of the halogens is dominated by their tendency to gain electrons from other elements to form halide ions:

$$X_2 + 2e^- \longrightarrow 2X^- \qquad [7.35]$$

(In this and subsequent equations, X indicates any one of the halogen elements.) Because electron affinity generally decreases as we progress down the periodic table, fluorine and chlorine are more reactive than bromine and iodine. In fact, fluorine removes electrons from almost any substance with which it comes into contact, including water, and usually does so very exothermically, as in the following examples:

$$2Na(s) + F_2(g) \longrightarrow 2NaF(s) \qquad \Delta H^\circ = -1147\ \text{kJ} \qquad [7.36]$$

$$2H_2O(l) + 2F_2(g) \longrightarrow 4HF(aq) + O_2(g) \qquad \Delta H^\circ = -758.7\ \text{kJ} \qquad [7.37]$$

As a result, fluorine gas is difficult and dangerous to use, requiring special apparatus.

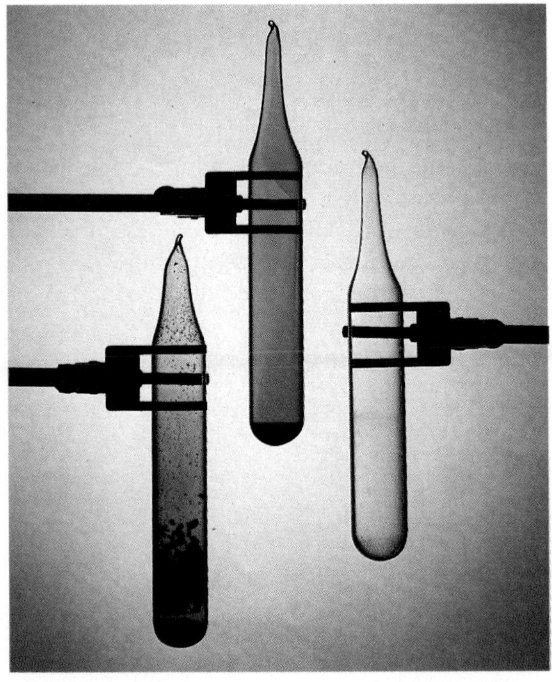

Figure 7.25 Chlorine, Cl_2, is a greenish-yellow gas at room temperature and at ordinary pressure; bromine, Br_2, is a reddish-brown liquid that vaporizes readily to form a vapor of the same color; iodine, I_2, is a grayish-black solid that readily vaporizes (sublimes) to form a violet vapor. (Richard Megna/ Fundamental Photographs)

Chlorine is the most industrially useful of the halogens. In 1988, total production was 22.66 billion pounds, making it the ninth most produced chemical in the United States. Chlorine is usually produced by the electrolysis of molten NaCl (Equations 7.15 and 7.16) or by the electrolysis of *brine*, a concentrated aqueous solution of NaCl:

$$2NaCl(aq) + 2H_2O(l) \xrightarrow{\text{electricity}} 2NaOH(aq) + H_2(g) + Cl_2(aq) \quad [7.38]$$

Unlike fluorine, Cl_2 reacts slowly with water to form relatively stable aqueous solutions of HCl and HOCl (hypochlorous acid):

$$Cl_2(g) + H_2O(l) \longrightarrow HCl(aq) + HOCl(aq) \quad [7.39]$$

Chlorine is often added to drinking water and swimming pools, where the HOCl(*aq*) that is generated serves as a disinfectant.

Inasmuch as the halogens readily gain electrons and active metals easily lose electrons, it is not surprising that the halogens react with metals to form salts. We have already seen several such reactions in the preceding section. The halogens also react with hydrogen to form gaseous hydrogen halide compounds:

$$H_2(g) + X_2 \longrightarrow 2HX(g) \quad [7.40]$$

These compounds are all very soluble in water and dissolve to form the hydrohalic acids. As we discussed in Section 4.3, HCl(*aq*), HBr(*aq*), and HI(*aq*) are strong acids, whereas HF(*aq*) is a weak acid.

Group 8A: The Noble Gases

8A
2 He
10 Ne
18 Ar
36 Kr
54 Xe
86 Rn

The group 8A elements, known as the **noble gases**, are all nonmetals that are gases at room temperature. They are all *monoatomic* (that is, they consist of single atoms rather than molecules). Some physical properties of the noble-gas elements are given in Table 7.10. The high radioactivity of Rn has inhibited the study of its chemistry.

The noble gases are characterized by completely filled *s* and *p* subshells. All elements of group 8A have large first ionization energies, and we see the expected decrease as we move down the column. Because the noble gases possess such stable electron configurations, they are exceptionally unreactive. In fact, until the early 1960s the elements were called the *inert gases* because they were thought to be incapable of forming chemical compounds. In 1962, Neil Bartlett at the University of British Columbia reasoned that the ionization

Table 7.10 Some Properties of the Noble Gases

Element	Electron configuration	Boiling point (K)	Density (g/L)	Atomic radius (Å)	I_1 (kJ/mol)
Helium	$1s^2$	4.2	0.18	0.53	2372
Neon	$[He]2s^22p^6$	27.1	0.90	0.71	2080
Argon	$[Ne]3s^23p^6$	87.3	1.78	0.98	1520
Krypton	$[Ar]4s^23d^{10}4p^6$	120	3.75	1.12	1351
Xenon	$[Kr]5s^24d^{10}5p^6$	165	5.90	1.31	1170
Radon	$[Xe]6s^24f^{14}5d^{10}6p^6$	211	9.73	—	1037

A HISTORICAL PERSPECTIVE: Discovery of the Noble Gases

None of the noble gases was known when Mendeleev proposed his periodic table. Their discovery created considerable scientific controversy but proved invaluable in understanding chemical behavior. The first significant hint of their existence was obtained during the early 1890s in the laboratories of the British physicist Lord Rayleigh. Rayleigh was involved in a program of making exact measurements of the densities of simple gases. During his studies he discovered a discrepancy between the densities of nitrogen obtained from air and nitrogen obtained from chemical reactions, as in the decomposition of ammonia:

$$2NH_3(g) \longrightarrow N_2(g) + 3H_2(g)$$

When all the other gases then known were removed from air, the remaining nitrogen was found to have a density of 1.2572 g/L at 0°C and standard atmospheric pressure. However, the density of nitrogen from chemical sources was found to be 1.2506 g/L under the same conditions. The difference was slight but reproducible. In April 1894, Rayleigh published a paper devoted entirely to the matter of the densities of nitrogen from the two sources. This report excited the interest of the British chemist Sir William Ramsay (Figure 7.26), who began to study the atmospheric nitrogen.

Figure 7.26 Sir William Ramsay (1852–1916), from the *Vanity Fair* series of caricatures. He was one of the principals in the discovery and study of the noble gases. He was awarded the Nobel Prize in chemistry in 1904 for this work. (The Granger Collection)

Rayleigh and Ramsay deduced that air must contain a previously unidentified component. They were able to isolate small quantities of this substance and measure some of its properties. The gas was composed of a new element with an atomic weight of 39.95 amu, which presented some problems in placing the element in the periodic table. Based on its atomic weight, the element would fall between potassium (K) and calcium (Ca), but this placement made no chemical sense. The scientists met with a second surprise: The element had no chemical reactivity. Try as they may, they were unable to form any compounds of the element. They consequently named the element *argon*, which means "the lazy one" in Greek.

Rayleigh and Ramsay publicly announced the discovery of argon in August 1894. Chemists found the existence of this curious element hard to believe, and many ingenious but erroneous alternative proposals were put forth for its identity. In January 1895, Ramsay presented a paper on the discovery of argon before the Royal Society. The largest crowd in that scientific society's history (over 800 people) assembled to hear him.

Later that same year, Ramsay isolated helium (He), the lightest of the noble gases, from uranium ores. The helium in these ores forms when alpha particles produced in radioactive decay (Section 2.2) pick up electrons. During the spring and summer of 1898, Ramsay and his co-workers isolated three additional noble gases from air: neon (Ne), krypton (Kr), and xenon (Xe). The discovery of these elements contributed to a partial resolution of the problem of argon's location in the periodic table. Clearly, there is an entire family of elements whose most obvious characteristic is their lack of chemical reactivity. The atomic weights of family members other than argon placed them after the halogens, a position that made chemical sense. The apparent problem of argon's position in the periodic table was finally resolved by Moseley's work (Section 7.1), which led to arranging elements by their atomic numbers.

energy of Xe might be low enough to allow it to form compounds. In order for this to happen, Xe would have to react with a substance with an extremely high ability to remove electrons from other substances, such as fluorine. Bartlett synthesized the first noble-gas compound by reacting Xe with the fluorine-containing compound PtF_6. Xenon also reacts directly with $F_2(g)$ to form the molecular compounds XeF_2, XeF_4, and XeF_6. Krypton has a higher I_1 value than xenon and is therefore less reactive. In fact, only a single stable compound of krypton is known, KrF_2. No compounds of He, Ne, or Ar are yet known; they still warrant the label "inert."

FOR REVIEW

SUMMARY

Many properties of atoms exhibit periodic character. Among the most important of these are atomic radii, ionization energy, and electron affinity. Electron configurations and the periodic table help us understand the trends in these properties. In general, atomic radii increase as we go down a column and decrease as we proceed left to right in a row. Ionization energies show exactly the opposite behavior: They tend to decrease as we move down a column and increase as we proceed from left to right in a row.

There is also periodicity in the classification of the elements as metals, nonmetals, or semimetals (also called metalloids). Most elements are metals; they occupy the left side and the middle of the periodic table. Nonmetals appear in the upper-right section of the table. Semimetals occupy a narrow band between the metals and nonmetals.

Metals have a characteristic luster. They are good electrical and thermal conductors. Metals tend to transfer electrons to other substances to form cations when they react. The compounds of metals with nonmetals, such as the oxides, hydrides, and chlorides, are mostly ionic solids. Metal oxides are basic; they react with acids to form salts and water.

Nonmetals lack metallic luster. Several are gases at room temperature. When nonmetals react with metals, the nonmetal generally gains electrons from the metal to become an anion. Compounds composed of nonmetals are molecular; thus, the oxides, hydrides, and chlorides of the nonmetals are usually gases, liquids, or solids that melt at low temperatures. Nonmetal oxides are acidic; they tend to react with bases to form salts and water.

Metallic character increases as we move down a family of representative elements; it also increases moving to the left in any period. Metallic character is strongest for elements at the bottom left of the periodic table; nonmetallic character is greatest for elements in the upper right.

We use electron configurations to help understand many of the chemical and physical properties of families of the representative elements. The alkali metals (group 1A) are soft metals of low density that melt at low temperatures. They have the lowest ionization energies of the elements. As a result, they are very reactive, easily losing their outer *s* electron to give 1+ ions. The metals in group 2A, the alkaline earth metals, are harder, more dense, and melt at higher temperatures than the alkali metals. They are also very reactive, readily losing their two outer *s* electrons to yield 2+ ions.

Hydrogen is a nonmetal and exists as a diatomic molecule, H_2. Hydrogen tends to form molecular compounds with other nonmetals, such as oxygen or the halogens. Hydrogen also reacts with active metals to give metal hydrides, ionic compounds that contain the H^- ion. Oxygen and sulfur are the most important of the group 6A elements. Oxygen is usually found as a diatomic molecule, O_2. Ozone, O_3, is an important allotrope of oxygen. Oxygen has a strong tendency to take electrons from other substances, hence oxidizing them. In combination with metals, oxygen is usually found as the oxide ion, O^{2-}. Alkali metal peroxides and superoxides are formed on reaction of the heavier alkali metals with O_2. Sulfur is most commonly found as S_8 molecules.

The group 7A elements are the halogens. They are all nonmetals that exist as diatomic molecules. The halogens have the highest electron affinities of the elements. As a result, their chemistry is dominated by their tendency to gain an electron to form 1− ions. The members of group 8A are called the noble gases. They are characterized by a completely filled shell of electrons, which is the most stable of electron configurations. As a result, the noble gases exhibit very low chemical reactivity, and only the heavier members of the family—Kr, Xe, and Rn—form compounds.

KEY TERMS

valence orbitals
atomic radius (Section 7.3)
first ionization energy (Section 7.4)
second ionization energy (Section 7.4)
electron affinity (Section 7.5)
metallic character (Section 7.6)

alkali metals (Section 7.7)
alkaline earth metals (Section 7.7)
hydride ion (Section 7.7)
ozone (Section 7.8)
allotropes (Section 7.8)
halogens (Section 7.8)
noble gases (Section 7.8)

EXERCISES

Sizes of Atoms

7.1 What is the correspondence between Lewis's idea of electron shells in atoms and the quantum-mechanical description of many-electron atoms?

7.2 How can a plot of the radial electron density of an atom be used to determine the number of electron shells in the atom?

7.3 Why is the $n = 2$ electron shell in neon closer to the nucleus than the same shell in carbon?

7.4 Arrange the following atoms in order of increasing distance of the $n = 3$ shell from the nucleus: Ti, Rh, P, K, and Mg.

7.5 Why does the quantum-mechanical description of many-electron atoms make it difficult to define the term *atomic radius*?

7.6 The distance between the fluorine nuclei in the F_2 molecule is 1.43 Å. What is the atomic radius of fluorine?

7.7 By referring to a periodic table, arrange the following atoms in order of increasing atomic size: Mg, F, P, O, and Ca.

7.8 Arrange the following atoms in order of increasing atomic radius: In, Ar, Rb, Ge, and He.

Ionization Energies; Electron Affinities

7.9 Write equations that show the processes corresponding to the second and third ionization energies of a scandium atom.

7.10 The fourth ionization energy of sulfur is 4565 kJ/mol. Write the chemical equation that corresponds to this energy.

7.11 Why is the second ionization energy of lithium much greater than that of beryllium?

7.12 Why is the second ionization energy of K much greater than its first ionization energy?

7.13 Based on their positions in the periodic table, select the atom with the larger first ionization energy from each of the following pairs: **(a)** B, Cl; **(b)** N, P; **(c)** Hf, Cs; **(d)** O, N; **(e)** Ga, Ge.

7.14 For each of the following pairs, indicate which element has the larger first ionization energy: **(a)** P, Cl; **(b)** Al, Ga; **(c)** Cs, La; **(d)** La, Hf. In each case provide an explanation in terms of electron configuration and effective nuclear charge.

7.15 What is the trend in first ionization energies as you proceed across the fourth period from K to Kr? How does this compare with the trend in atomic sizes?

7.16 What is the trend in first ionization energies as you proceed down the group 8A elements? How does this compare with the trend in atomic sizes?

7.17 Is the effective nuclear charge operating on the 3*s* electron greater in Al^{2+} or in Si^{3+}? How do the data in Table 7.2 provide support for your answer?

7.18 Although SiO_2—which can be thought of as containing Si^{4+} ions—is common, AlO_2 is not a known compound. Explain this fact in terms of the data in Table 7.2.

7.19 The electron affinity of chlorine is strongly negative; that is, addition of an electron to Cl is an exothermic process. In contrast, addition of an electron to Ar is an endothermic process. Account for this difference in terms of the electron configurations of the two elements.

7.20 Addition of an electron to Na(*g*) is a slightly exothermic process, whereas addition of an electron to Mg(*g*) is strongly endothermic. Explain this difference in terms of the electron configurations of the two elements.

Properties of Metals and Nonmetals

7.21 Use the periodic table to arrange the following pure, solid elements in order of increasing electrical conductivity at room temperature: Ge, Ca, S, and Si. Explain the reason for the order you chose.

7.22 Use the periodic table to arrange the following elements in order of increasing melting point: Al, F, Na, and Mg. Explain the reason for the order you chose.

7.23 For each of the following pairs, which element will have the greater metallic character: **(a)** Li and Be; **(b)** Li and Na; **(c)** Sn and P; **(d)** B and Al?

7.24 **(a)** Arrange the following elements in order of increasing metallic character: As, P, Bi, Sb, N. **(b)** Arrange the following elements in order of increasing nonmetallic character: S, Hg, Ge, F, In.

7.25 Which of the following oxides are ionic and which are molecular: N_2O, Na_2O, CaO, CO, P_2O_5, Cl_2O_7, Fe_2O_3? Explain the reason for your choices.

7.26 Which of the following compounds are solids at room temperature and which are gases: NO, Na_2S, BaO, CO_2, $PbCl_2$, OF_2, MgF_2? Explain the reason for your choices.

7.27 What is meant by the terms *acidic oxide* and *basic oxide?* Give an example of each.

7.28 Arrange the following oxides in order of increasing acidity: CO_2, MgO, Al_2O_3, SO_3, BaO, SiO_2, and P_2O_5.

7.29 Write balanced chemical equations for the following reactions: **(a)** sodium oxide with water; **(b)** copper(II) oxide with nitric acid; **(c)** sulfur trioxide with water; **(d)** selenium dioxide with aqueous sodium hydroxide.

7.30 Write balanced chemical equations for the following reactions: **(a)** dichlorine heptoxide with water; **(b)** iron(II) oxide with hydrochloric acid; **(c)** barium oxide with water; **(d)** carbon dioxide with aqueous potassium hydroxide.

Group Trends in Metals and Nonmetals

7.31 Compare the elements sodium and magnesium with respect to the following properties: **(a)** electron configuration; **(b)** most common ionic charge; **(c)** first ionization energy; **(d)** reactivity toward water; **(e)** melting point; **(f)** atomic radius. Account for the differences between the two elements.

7.32 Compare the elements rubidium and silver with respect to the following properties: **(a)** electron configuration; **(b)** most common ionic charge; **(c)** first ionization energy; **(d)** reactivity toward water; **(e)** melting point (see Figure 7.12); **(f)** atomic radius (see Figure 7.5). Account for the differences between the two elements.

7.33 **(a)** Why is strontium more reactive than beryllium toward water? **(b)** Why is the product of reacting sodium with oxygen somewhat surprising? **(c)** Why is barium more reactive than mercury toward hydrochloric acid?

7.34 **(a)** Why is calcium generally less reactive than potassium? **(b)** Why do salts of lithium often show very different properties than those of sodium? **(c)** Why is gold metal found in nature but cesium metal not?

7.35 Write a balanced chemical equation for the reaction that occurs in each of the following cases: **(a)** Potassium is added to water. **(b)** Barium is added to water. **(c)** Lithium is heated in nitrogen. **(d)** Magnesium burns in oxygen.

7.36 Write a balanced chemical equation for the reaction that occurs in each of the following cases: **(a)** Sodium vapor reacts with bromine vapor. **(b)** Hydrogen gas is bubbled through molten sodium. **(c)** Lithium is burned in oxygen. **(d)** Strontium oxide is added to water.

7.37 Explain, in terms of electron configurations, why hydrogen exhibits properties similar to those of both Li and F.

7.38 **(a)** Write balanced equations for the reaction of sodium with chlorine and for the reaction of hydrogen with chlorine. What are the similarities and differences among the products of these two reactions? **(b)** Write balanced equations for the reaction of fluorine with calcium and for the reaction of hydrogen with calcium. What are the similarities among the products of these two reactions?

7.39 Compare the elements oxygen and fluorine with respect to the following properties: **(a)** electron configuration; **(b)** most common ionic charge; **(c)** first ionization energy; **(d)** reactivity toward water; **(e)** reactivity toward hydrogen; **(f)** atomic radius. Account for the differences between the two elements.

7.40 Compare the elements fluorine and chlorine with respect to the following properties: **(a)** electron configuration; **(b)** most common ionic charge; **(c)** first ionization energy; **(d)** reactivity toward water; **(e)** electron affinity; **(f)** atomic radius. Account for the differences between the two elements.

7.41 Until the early 1960s, the group 8A elements were called inert gases. Why is this name no longer used?

7.42 Why does xenon react with fluorine whereas neon does not?

7.43 Write a balanced chemical equation for the reaction that occurs in each of the following cases: **(a)** Sulfur reacts with lithium. **(b)** Ozone decomposes to dioxygen. **(c)** Aqueous potassium bromide is electrolyzed. **(d)** Chlorine reacts with calcium.

7.44 Write a balanced chemical equation for the reaction that occurs in each of the following cases: **(a)** Iodine reacts with lithium. **(b)** Sulfur is burned in air. **(c)** Krypton reacts with fluorine. **(d)** Oxygen reacts with potassium.

7.45 **(a)** Why does xenon react with fluorine but not with iodine? **(b)** When we talk of "molecular oxygen," why do we mean O_2 rather than O_3? **(c)** Why must special apparatus be used to carry out reactions with fluorine gas?

7.46 **(a)** Which would you expect to be a better conductor of electricity, tellurium or iodine? **(b)** How does a molecule of sulfur (in its most common room-temperature form) differ from a molecule of oxygen? **(c)** Why is chlorine generally more reactive than bromine?

Additional Exercises

7.47 By examining the modern periodic table, find as many examples as you can of the violations of Mendeleev's periodic law that the chemical and physical properties of the elements are periodic functions of their atomic weight.

7.48 Explain why the radii of atoms do not increase uniformly as the atomic number of the atom increases.

7.49 Why is the second ionization energy, I_2, of an atom always larger than the first, I_1?

7.50 Arrange the following elements in terms of **(a)** increasing ionization energy and **(b)** increasing atomic radius: O, Se, C, Si, F.

7.51 The successive ionization energies for boron are 801, 2430, 3670, 25,000, and 32,800 kJ/mol. **(a)** Why do the values increase with successive ionization? **(b)** Why is there such a large change in ionization energy in moving from the third to the fourth ionization energy? **(c)** Why are the values for the first four ionization energies greater than the corresponding values for aluminum (Table 7.2)?

7.52 Atomic radii normally increase going down a group in the periodic table. Suggest a reason why hafnium breaks this rule, as shown in the following data.

Atomic radii (Å)			
Sc	1.62	Ti	1.47
Y	1.80	Zr	1.60
La	1.87	Hf	1.59

[7.53] By examining the distances between ions in a salt it is possible to define *ionic radii* in a manner similar to the way that atomic radii were defined. Listed below are the atomic and ionic (2+) radii for calcium and zinc:

		Radii (Å)	
Ca	1.97	Ca^{2+}	1.00
Zn	1.34	Zn^{2+}	0.74

(a) Explain why the atomic radius of calcium is larger than that of zinc. **(b)** Suggest a reason why the difference in the ionic radii is much less than the difference in the atomic radii.

7.54 As you go down a group in the periodic table, the change in electron affinity is not nearly as great as the change in first ionization energy. Explain.

7.55 The electron affinities of F and of the O^- ion are given below:

$$F(g) + e^- \longrightarrow F^-(g) \qquad \Delta H = -332 \text{ kJ/mol}$$

$$O^-(g) + e^- \longrightarrow O^{2-}(g) \qquad \Delta H = +710 \text{ kJ/mol}$$

(a) What can be said about the electron configurations of F and O^-? **(b)** What is the essential difference in these two processes? **(c)** What prediction can you make concerning the electron affinity of gaseous N^{2-}?

[7.56] On the basis of electron configurations, explain the following observations: **(a)** The first ionization energy of phosphorus is greater than that of sulfur. **(b)** The electron affinity of nitrogen is lower (less negative) than those of both carbon and oxygen. **(c)** The second ionization energy of oxygen is greater than that of fluorine. **(d)** The third ionization energy of manganese is greater than those of both chromium and iron.

[7.57] What is the electron affinity of the $Na^+(g)$ ion? Explain your reasoning.

[7.58] Make a graph of all of the second ionization energies, I_2, listed in Table 7.2, as a function of atomic number. Account for the general trend in the series from Mg through Ar. Account for the exceptionally large value for Na. Suggest why the value for Al is larger than one might expect from the general trend.

[7.59] There are certain similarities in properties that exist between the first member of any periodic family and the element located below it and to the right in the periodic table. For example, in some ways Li resembles Mg, Be resembles Al, and so forth. This observation is called the *diagonal relationship.* In terms of what we have learned in this chapter, offer a possible explanation for this relationship.

[7.60] The ionization energy of the oxygen *molecule* is the energy required for the following process:

$$O_2(g) \longrightarrow O_2^+ + e^-$$

The energy needed for this process is 1175 kJ/mol, very similar to the first ionization energy of Xe. Would you expect O_2 to react with F_2? If so, suggest a product or products of this reaction.

[7.61] The elements at the bottom of groups 1A, 2A, 6A, 7A, and 8A—Fr, Ra, Po, At, and Rn—are all radioactive. As a result, much less is known about their physical and chemical properties than for the elements above them. Based upon what we have learned in this chapter, which of these five elements would you expect **(a)** to have the most metallic character; **(b)** to have the most nonmetallic character; **(c)** to have the largest ionization energy; **(d)** to have the smallest ionization energy; **(e)** to have the greatest (most negative) electron affinity; **(f)** to have the largest atomic radius; **(g)** to resemble least in appearance the element immediately above it; **(h)** to have the highest melting point; **(i)** to react most readily with water?

[7.62] Using the *Handbook of Chemistry and Physics*, determine whether the physical properties of tellurium (density, melting point, boiling point, and so on) most closely resemble those of its neighbor to the left, to the right, or above it in the periodic table. Which would you expect the chemical properties of tellurium to most closely resemble?

8 Basic Concepts of Chemical Bonding

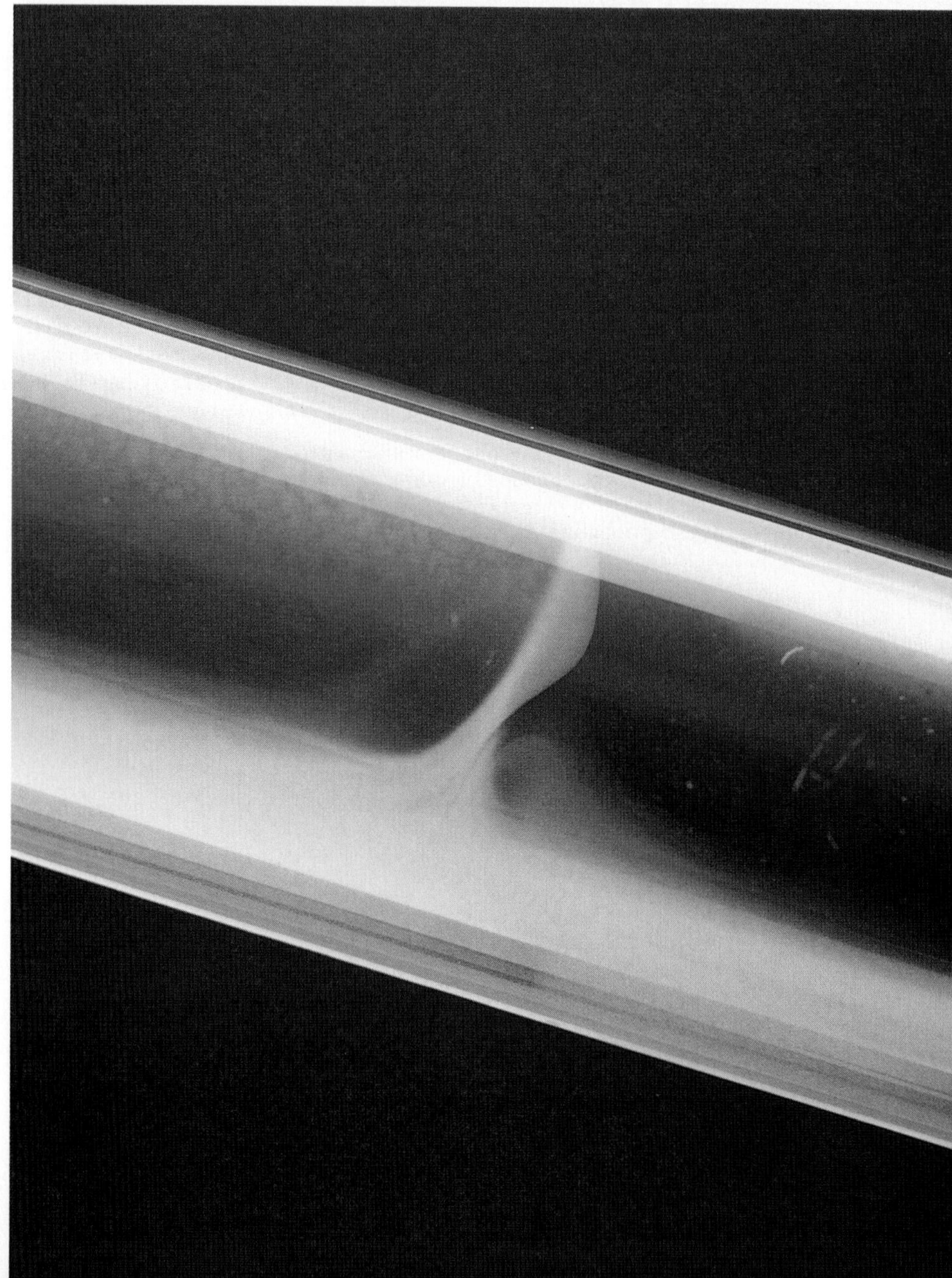

Two gases reacting to form an ionic solid. Gaseous HCl enters the glass tubing from the left, while gaseous NH_3 enters from the right. When the gases meet, they react to form the white ionic solid, NH_4Cl. (© Richard Megna/Fundamental Photographs)

Deep within the earth, below the city of Detroit and beneath the rolling plains of Kansas, lie enormous deposits of the white mineral halite (see Figure 8.1). This substance, also known as sodium chloride, NaCl, was deposited in these and other places millions of years ago, when extensive primordial seas dried upon the changing surface of the earth. Sodium chloride is the most abundant dissolved substance present in seawater and is found in human body tissues in large quantities. However, it is most familiar to us as ordinary table salt. This substance consists of sodium ions and chloride ions, Na^+ and Cl^-.

Water, H_2O, is another very abundant substance. We drink it, swim in it, and use it as a cooling agent. It is essential to life as we know it. This substance is composed of molecules.

Why are some substances composed of ions and others are composed of molecules? The key to this question is found in the electronic structures of the atoms involved, which we studied in Chapters 6 and 7, and in the nature of the chemical forces within the compounds. In this chapter and the next, we shall examine the relationships between electronic structure, chemical bonding forces, and the properties of substances. As we do this, we shall find it useful

CONTENTS

Figure 8.1 One million tons of salt awaiting shipment near a saltworks. (© Georg Gerster/Comstock)

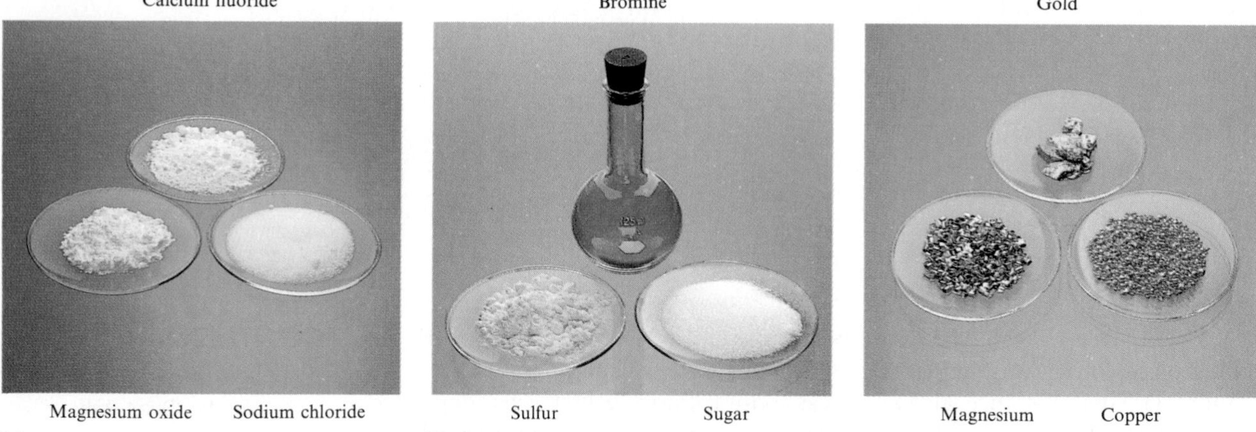

Figure 8.2 Examples of substances in which (*a*) ionic, (*b*) covalent, and (*c*) metallic bonds are found. (Richard Megna/Fundamental Photographs)

to classify chemical forces into three broad groups: (1) ionic bonds, (2) covalent bonds, and (3) metallic bonds. Figure 8.2 shows examples of substances in which we find these types of bonds.

The term **ionic bond** refers to the electrostatic forces that exist between particles of opposite charge. As we shall see, ions may be formed from atoms by the transfer of one or more electrons from one atom to another. Ionic substances generally result from the interaction of metals from the far left side of the periodic table with nonmetallic elements from the far right side (excluding the noble gases, group 8A).

A **covalent bond** results from the sharing of electrons between two atoms. The most familiar examples of covalent bonding are seen in the interactions of nonmetallic elements with one another.

Metallic bonds are found in solid metals such as copper, iron, and aluminum. In the metals, each metal atom is bonded to several neighboring atoms. The bonding electrons are relatively free to move throughout the three-dimensional structure. Metallic bonds give rise to such typical metallic properties as high electrical conductivity and luster. We will postpone further discussion of metallic bonding until Chapter 24. We shall start our discussion by examining the preferred arrangements of electrons in atoms when they form chemical compounds.

8.1 LEWIS SYMBOLS AND THE OCTET RULE

The term **valence** (from the Latin *valere*, "to be strong") is commonly used in discussions of both ionic and covalent bonding. The valence of an element is a measure of its capacity to form chemical bonds. Originally, it was determined by the number of hydrogen atoms with which an element combined. Thus, the valence of oxygen is 2 in H_2O; in CH_4 the valence of carbon is 4.

We speak of **valence electrons** when referring to the electrons that take part in chemical bonding. These electrons are the ones residing in the outermost electron shell of the atom, the valence shell. **Electron-dot symbols** (also known as **Lewis symbols**, after G. N. Lewis) are a simple and convenient way of showing the valence electrons of atoms and keeping track of them in the course of bond formation. The electron-dot symbol for an element consists of the chemical symbol for the element plus a dot for each valence electron. For example, sulfur has the electron configuration [Ne]$3s^23p^4$; its electron-dot symbol therefore shows six valence electrons:

$$\cdot\underset{\cdot\cdot}{\ddot{\mathrm{S}}}\cdot$$

Notice that the dots are placed in four "regions" around the atomic symbol: the top, the bottom, and the left and right sides. We envision each of these regions as being able to accommodate a pair of electrons. Two of the regions around sulfur contain a pair of electrons each, while the other two contain single electrons. Other examples are shown in Table 8.1.

The number of valence electrons of any representative element is the same as the column number of the element in the periodic table. For example, the electron-dot symbols for both oxygen and sulfur, members of family 6A, show six dots.

Atoms often gain, lose, or share electrons to achieve the same number of electrons as the noble gas closest to them in the periodic table. The noble gases, you will recall, have very stable electron arrangements, as evidenced by their high ionization energies, low affinity for additional electrons, and general lack of chemical reactivity. Because all noble gases (except He) have eight valence electrons, many atoms undergoing reactions also end up with eight valence electrons. This observation has led to what is known as the **octet rule**: Atoms tend to gain, lose, or share electrons until they are surrounded by eight valence electrons. An octet of electrons can be thought of as four pairs of valence electrons arranged around the atom. An example is the configuration for Ne in Table 8.1. Of course, because He has only two electrons, atoms near it in the periodic table, such as H, will generally tend to obtain an arrangement of two electrons. As we shall see, there are many exceptions to the octet rule. Nevertheless, it provides a useful framework for introducing many important concepts of bonding.

Table 8.1 Electron-Dot Symbols

Element	Electron configuration	Electron-dot symbol
Li	[He]$2s^1$	$\mathrm{Li}\cdot$
Be	[He]$2s^2$	$\cdot\mathrm{Be}\cdot$
B	[He]$2s^22p^1$	$\cdot\dot{\mathrm{B}}\cdot$
C	[He]$2s^22p^2$	$\cdot\underset{\cdot}{\dot{\mathrm{C}}}\cdot$
N	[He]$2s^22p^3$	$\cdot\underset{\cdot}{\dot{\mathrm{N}}}:$
O	[He]$2s^22p^4$	$:\underset{\cdot}{\dot{\mathrm{O}}}:$
F	[He]$2s^22p^5$	$\cdot\underset{\cdot\cdot}{\ddot{\mathrm{F}}}:$
Ne	[He]$2s^22p^6$	$:\underset{\cdot\cdot}{\ddot{\mathrm{Ne}}}:$

8.2 IONIC BONDING

When sodium metal is brought into contact with chlorine gas, Cl_2, a violent reaction ensues (see Figure 8.3). The product of this reaction is sodium chloride, NaCl, a substance composed of Na^+ and Cl^- ions:

$$2Na(s) + Cl_2(g) \longrightarrow 2NaCl(s)$$

These ions are arranged throughout the solid NaCl in a regular three-dimensional array, as shown in Figure 8.4.

The formation of Na^+ from Na and of Cl^- from Cl_2 indicates that an electron has been lost by a sodium atom and gained by a chlorine atom. Such electron transfer to form oppositely charged ions occurs when the atoms involved differ greatly in their attraction for electrons. Our example of NaCl is rather typical for ionic compounds; it involves a metal of low ionization energy

Figure 8.3 The reaction between sodium metal and chlorine gas to form sodium chloride. (*a*) A container of chlorine gas (left) and sodium metal (right). (*b*) Formation of NaCl begins as sodium is added to the chlorine. (*c*) The reaction a few minutes later. (Donald Clegg and Roxy Wilson)

Figure 8.4 (*a*) The crystal structure of sodium chloride. Each of the Na^+ ions is surrounded by six Cl^- ions, and each Cl^- ion is surrounded by six Na^+ ions. This is made clearer in (*b*), in which the cubic arrangement of the lattice is emphasized.

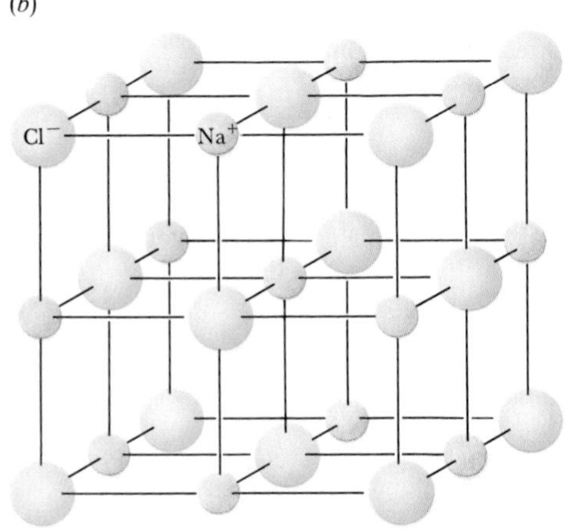

and a nonmetal with a high affinity for electrons. Using electron-dot symbols (and showing a chlorine atom rather than the Cl_2 molecule), we can represent this reaction as follows:

$$\text{Na}\cdot + \cdot\ddot{\underset{\cdot\cdot}{\text{Cl}}}: \longrightarrow \text{Na}^+ + \left[:\ddot{\underset{\cdot\cdot}{\text{Cl}}}:\right]^-$$

The arrow indicates the transfer of an electron from the Na atom to the Cl atom. Each ion has an octet of electrons, the octet on Na^+ being the $2s^22p^6$ electrons that lie below the single $3s$ valence electron of the Na atom.

Energetics of Ionic Bond Formation

As is evident in Figure 8.3, the reaction between Na(*s*) and $Cl_2(g)$ to form NaCl(*s*) is very exothermic. Indeed, we see in Appendix C that ΔH_f° for NaCl(*s*) is very negative:

$$\text{Na}(s) + \tfrac{1}{2}\text{Cl}_2(g) \longrightarrow \text{NaCl}(s) \qquad \Delta H_f^\circ = -410.9 \text{ kJ/mol} \qquad [8.1]$$

In fact, the enthalpies of formation of all ionic substances are quite negative. Why is this so? What are the energetic factors that make the formation of ionic compounds so exothermic? We will address this question for sodium chloride by envisioning the formation of NaCl(*s*) as occurring in a series of well-defined steps. We will then use Hess's law (Section 5.5) to put these steps together in a way that gives us ΔH_f° for NaCl(*s*). By so doing, we will construct a **Born-Haber cycle**, a thermochemical cycle named after the German scientists Max Born (1882–1970) and Fritz Haber (1868–1934), who introduced it to analyze the factors leading to the stability of ionic compounds.

In the Born-Haber cycle, we envision the first step in the formation of NaCl(*s*) to be the generation of gaseous atoms of sodium and chlorine. We then remove an electron from Na(*g*) to form $Na^+(g)$, and we add the electron to Cl(*g*) to form $Cl^-(g)$. Finally, we combine the gaseous sodium and chloride ions to form solid sodium chloride.

The energies involved in the first of these steps are available to us as enthalpies of formation (Appendix C):

$$\text{Na}(s) \longrightarrow \text{Na}(g) \qquad \Delta H_f^\circ[\text{Na}(g)] = 107.7 \text{ kJ/mol} \qquad [8.2]$$

$$\tfrac{1}{2}\text{Cl}_2(g) \longrightarrow \text{Cl}(g) \qquad \Delta H_f^\circ[\text{Cl}(g)] = 121.7 \text{ kJ/mol} \qquad [8.3]$$

Notice that both of these processes are endothermic; energy is required to generate gaseous sodium and chlorine atoms.

We now remove an electron from Na(*g*) and add it to Cl(*g*). The energies required for these steps are the ionization energy of Na and the electron affinity of Cl, respectively (Sections 7.4 and 7.5):

$$\text{Na}(g) \longrightarrow \text{Na}^+(g) + e^- \qquad I_1(\text{Na}) = 496 \text{ kJ/mol} \qquad [8.4]$$

$$\text{Cl}(g) + e^- \longrightarrow \text{Cl}^-(g) \qquad E(\text{Cl}) = -349 \text{ kJ/mol} \qquad [8.5]$$

Finally, we bring the gaseous ions together to form NaCl(*s*):

$$\text{Na}^+(g) + \text{Cl}^-(g) \longrightarrow \text{NaCl}(s) \qquad \Delta H = ? \qquad [8.6]$$

Figure 8.5 A Born-Haber cycle shows the energetic relationships in the formation of ionic solids from the elements. In this example, the enthalpy of formation of NaCl(*s*) from elemental sodium and chlorine (Equation 8.1) is equal to the sum of the energies of several individual steps (Equations 8.2 through 8.6) by Hess's law.

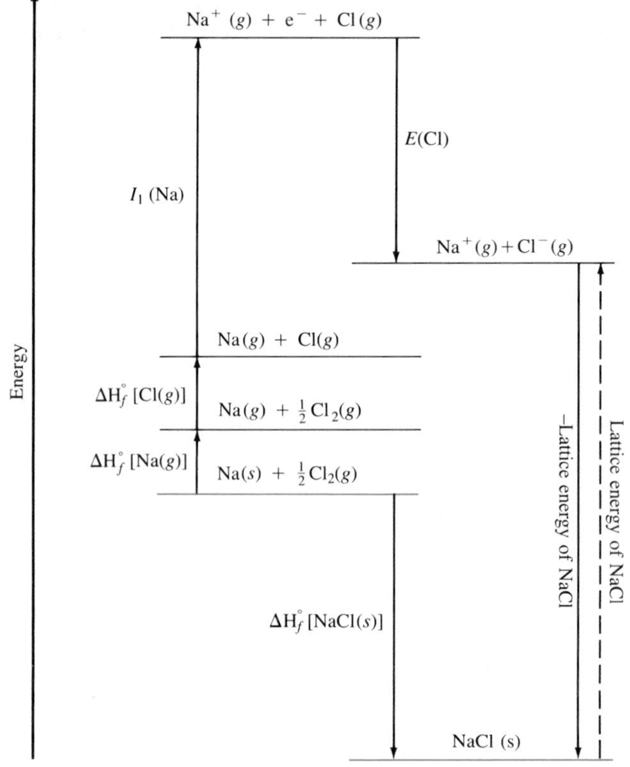

We do not know what the enthalpy of this last step is. However, we can calculate its value by using Hess's law. We see that adding up Equations 8.2 through 8.6 gives us Equation 8.1, a situation that is diagramed in Figure 8.5:

$$\Delta H_f^\circ[\text{NaCl}(s)] = \Delta H_f^\circ[\text{Na}(g)] + \Delta H_f^\circ[\text{Cl}(g)] + I_1(\text{Na}) + E(\text{Cl}) + \Delta H \text{ (Eq. 8.6)}$$

Therefore,

$$\begin{aligned}\Delta H \text{ (Eq. 8.6)} &= \Delta H_f^\circ[\text{NaCl}(s)] - \{\Delta H_f^\circ[\text{Na}(g)] + \Delta H_f^\circ[\text{Cl}(g)] + I_1(\text{Na}) + E(\text{Cl})\} \\ &= -411 - \{108 + 122 + 496 + (-349)\} \qquad \text{(all in kJ/mol)} \\ &= -788 \text{ kJ/mol} \qquad [8.7]\end{aligned}$$

Table 8.2 Lattice Energies for Some Ionic Compounds

Compound	Lattice energy (kJ/mol)
LiF	1024
LiI	744
NaF	911
NaI	693
KF	815
KI	641
MgF_2	2910
$SrCl_2$	2130
MgO	3938

Thus, we see that bringing together gaseous ions to form a solid is a very exothermic process, the main one that releases energy in the Born-Haber cycle.

Because Equation 8.6 is exothermic, the reverse of Equation 8.6 is an endothermic process. The energy required for this reverse process is called the **lattice energy** of NaCl(*s*):

$$\text{NaCl}(s) \longrightarrow \text{Na}^+(g) + \text{Cl}^-(g) \qquad \text{Lattice energy} = +788 \text{ kJ/mol} \qquad [8.8]$$

The lattice energy of an ionic compound is the energy required to separate 1 mol of the solid ionic substance completely into gaseous ions. The large value of the lattice energy for NaCl(*s*) accounts for the fact that sodium chloride is a stable, solid substance with a high melting point. Table 8.2 lists the lattice energies of some other ionic compounds. We see that these lattice energies are all large, positive values, indicating that the ions are strongly attracted to one another in these solids. Because of these strong interactions, most ionic substances are hard, brittle solids with high melting points.

Lattice energies are large and positive because of the attraction between positive and negative ions. The potential energy (Section 5.1) of two interacting charged particles is given by

$$E = k\frac{Q_1 Q_2}{d} \qquad [8.9]$$

In this equation, Q_1 and Q_2 are the charges on the particles in coulombs, and d is the distance between their centers in meters. The constant k has the value 8.99×10^9 J-m/C^2. If Q_1 and Q_2 have the same sign, E is greater than zero, indicating that the potential energy increases as the ions approach one another. This increase represents the repulsion between particles of like charge. Similarly, when Q_1 and Q_2 have the opposite sign, E is negative, indicating that the potential energy decreases as the ions approach one another. This decrease in potential energy represents the attraction between particles of opposite charge. In sodium chloride (Figure 8.4), each Na^+ ion is surrounded by six Cl^- ions, and each Cl^- ion is surrounded by six Na^+ ions. Of course, each Na^+ ion experiences a repulsion from the other Na^+ ions in the lattice, but these ions are farther away than its six Cl^- "nearest neighbors." The Cl^- ions also repel one another. The lattice energy is the result of all the electrostatic attractions and repulsions between the ions in the lattice. The lattice energy is very large because the attractions between ions of opposite charge far outweigh the repulsions between ions of like charge. Lattice energies are positive because they represent the energy that must be supplied in order to separate the ions of a crystal into gaseous ions.

Equation 8.9 indicates that the attractive interaction between two oppositely charged ions increases as the magnitudes of their charges increase and as the distance between their centers decreases. Thus, for a given arrangement of ions, the lattice energy increases as the charges on the ions increase and as their radii decrease. The magnitude of lattice energies depends primarily on the ionic charges because ionic radii do not vary over a very wide range.

SAMPLE EXERCISE 8.1

Arrange the following ionic compounds in order of increasing lattice energy: LiF, KBr, and MgO.

Solution: LiF consists of Li^+ and F^- ions, KBr of K^+ and Br^- ions, and MgO of Mg^{2+} and O^{2-} ions. According to Equation 8.9, the electrostatic attraction between oppositely charged ions increases with the charge on the ions. For this reason, we expect the lattice energy of MgO, which has 2+ and 2− ions, to be the greatest of the three. To compare LiF and KBr, we need to determine in which substance the ions are closer together. Because the radius of a Li atom is less than that of a K atom, we might expect that a Li^+ ion is smaller than a K^+ ion. Similarly, we expect that F^- is a smaller ion than Br^-. (We will discuss the sizes of ions in Section 8.3.) Therefore, the distance between the ions in LiF should be less than the distance between the ions in KBr, and the lattice energy of LiF should be greater than that of KBr. Table 8.2 confirms the order KBr $<$ LiF $<$ MgO.

PRACTICE EXERCISE

Which substance would you expect to have the greater lattice energy, FeO or Fe_2O_3?
Answer: Fe_2O_3

If lattice energy increases as the charges of the ions increase, why doesn't sodium lose two electrons to form Na^{2+}? The second electron would have to come from the inner shell of the sodium atom. We can see from the large value for the second ionization energy of Na (Table 7.2) that too much energy would be required to form the Na^{2+} ion in chemical compounds. Thus, sodium and the other group 1A metals are found in ionic substances only as 1+ ions. Similarly, addition of a second electron to a chloride ion to form a hypothetical Cl^{2-} is never observed. The other group 7A elements (the halogens) are also found in ionic compounds only as the 1− ions F^-, Br^-, or I^-, which possess noble-gas configurations.

Magnesium, an element of group 2A, also forms an ionic compound with chlorine, $MgCl_2$. In this instance, the metal achieves a noble-gas configuration by losing two electrons, forming Mg^{2+}. Much less energy is required to remove two electrons from Mg (Table 7.2) than from Na because both Mg electrons are in its valence shell. Of course, removing two electrons requires more energy than removing just one. This energy is more than recovered, however, in the increased lattice energy of $MgCl_2$, which comes from the higher charge on the metal ion. Similarly, the group 6A elements form O^{2-}, S^{2-}, and so forth, in which the ion possesses a noble-gas configuration. In the formation of MgO, both magnesium and oxygen attain the noble-gas configuration by the transfer of two electrons:

$$\dot{\text{Mg}}\cdot + :\dot{\ddot{\text{O}}}: \longrightarrow \text{Mg}^{2+} + \left[:\ddot{\underset{\cdot\cdot}{\text{O}}}:\right]^{2-}$$

Thus, the balance between ionization energies and lattice energies determines the magnitude of the positive charge on a metal ion. Similarly, the balance between electron affinities and lattice energies determines the magnitude of the negative charge on nonmetal ions.

Transition-Metal Ions

Ionic bond theory correctly predicts the charges found on many simple ions of the representative elements, based on the notion that attainment of a noble-gas configuration leads to maximum stability. For some elements, however, the rule must be modified, and for others it is not applicable at all. For example, metals of group 1B (Cu, Ag, Au) often occur as the 1+ ions (as in CuBr and AgCl). Silver possesses a $5s^14d^{10}$ outer electron configuration. In forming Ag^+, the 5s electron is lost, leaving a completely filled shell of 18 electrons in the $n = 4$ level. Because it is a completed shell, it is somewhat like a noble-gas arrangement. Similarly, the group 2B elements most commonly are seen as the 2+ ions (Zn^{2+}, Cd^{2+}, Hg^{2+}) in ionic compounds. The valence-shell s electrons are lost in forming the ions, leaving an electronic arrangement consisting of 18 electrons in the highest occupied level.

For most of the transition metals, the attainment of a noble-gas configuration by loss of electrons is not feasible; that would require the loss of too many electrons. The outer electron configurations of these elements are either $ns^2(n-1)d^x$ or $ns^1(n-1)d^x$, where n is 4, 5, or 6, and x may vary from 1 to 10. *In forming ions, the transition metals lose the valence-shell s electrons first, then as many d electrons as are required to form an ion of particular charge.* Most transition metals are found in more than one charge state. For example, the element chromium is found in compounds as Cr^{2+} or Cr^{3+}. No simple rules tell which charge state of a transition-metal ion will exist in a particular case.

SAMPLE EXERCISE 8.2

Write the electron configuration for the Co^{2+} ion and for the Co^{3+} ion.

Solution: Cobalt (atomic number 27) has an electron configuration $[Ar]4s^2 3d^7$. To form a 2+ ion, two electrons must be removed. As discussed in the text above, the 4*s* electrons are removed before the 3*d*. Consequently, the Co^{2+} ion has an electron configuration of $[Ar]3d^7$. To form Co^{3+} requires the removal of an additional electron; the electron configuration for this ion is $[Ar]3d^6$.

PRACTICE EXERCISE

Write the electron configuration for the Cr^{3+} ion. ***Answer:*** $[Ar]3d^3$

This is a good point at which to review Table 2.4, which lists common ions. Recall that positively charged ions are called *cations* and negatively charged ones are called *anions*. Notice that some ions are polyatomic. Examples of polyatomic cations are the vanadyl ion, VO^{2+}, and the familiar ammonium ion, NH_4^+. However, most polyatomic ions are anions. Examples are the carbonate ion, CO_3^{2-}, found in many mineral deposits, and the brightly colored yellow chromate ion, CrO_4^{2-}.

In polyatomic ions, two or more atoms are bound together by predominantly covalent bonds. They form a stable grouping that carries a charge, either positive or negative. We will examine the covalent bonding forces in these ions in Chapter 9. For now, you must realize only that the group of atoms as a whole acts as a charged species in forming an ionic compound with an ion of opposite charge.

SAMPLE EXERCISE 8.3

The dichromate ion, $Cr_2O_7^{2-}$, is readily obtained in the form of its ammonium salt. Write the formula for ammonium dichromate.

Solution: The charge on the dichromate ion is 2−; that on the ammonium ion is 1+, NH_4^+. We therefore require two ammonium ions to balance the charge of the dichromate ion. The formula for the salt is thus $(NH_4)_2Cr_2O_7$.

PRACTICE EXERCISE

Write the chemical formula for sodium sulfate. ***Answer:*** Na_2SO_4

8.3 SIZES OF IONS

Ionic size plays a crucial role in determining the structure and stability of ionic solids. For example, the sizes of ions are important in determining both the way in which ions pack in an ionic solid and the lattice energy of the solid. Ionic size is also a major factor governing the properties of ions in solution. For example, a small difference in ionic size is often sufficient for one metal ion to be biologically important and another not to be.

The size of an ion depends on its nuclear charge, the number of electrons it possesses, and the orbitals in which the outer-shell electrons reside. Consider first the relative sizes of an ion and its parent atom. Positive ions are formed by removing one or more electrons from the outermost region of the atom. Thus, the formation of a cation not only vacates the most spatially extended orbitals, it also decreases the total electron-electron repulsions. As a consequence, *cations are smaller than their parent atoms*, as illustrated in Figure 8.6. The opposite is true of negative ions. When electrons are added to form an anion, the increased electron-electron repulsions cause the electrons to spread out more in space. Thus, *anions are larger than their parent atoms.*

Figure 8.6 Relative sizes of atoms and ions. The values in parentheses are radii (Å).

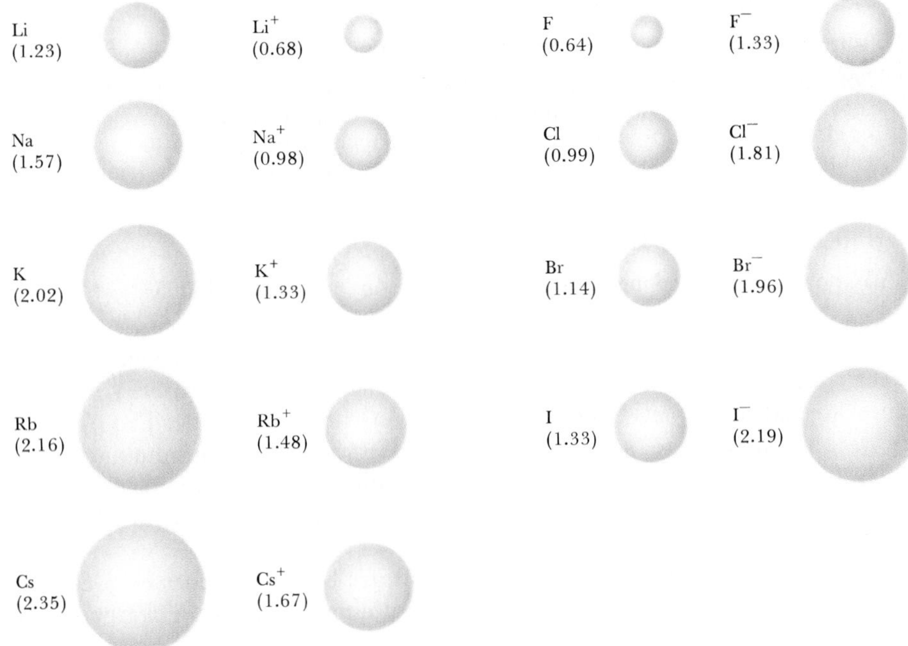

It is also important to note that *for ions of the same charge, size increases as we go down a family in the periodic table.* This trend is seen in Figure 8.6 and also in Table 8.3, which gives the radii for several ions with noble-gas electron configurations. As the principal quantum number of the outer occupied orbital of an ion increases, the size of both the ion and its parent atom increases.

SAMPLE EXERCISE 8.4

Arrange the following atoms and ions in order of decreasing size: Mg^{2+}, Ca^{2+}, and Ca.

Solution: Cations are smaller than their parent atoms; thus Ca^{2+} is smaller than the Ca atom. Because Ca is below Mg in group 2A of the periodic table, Ca^{2+} is larger than Mg^{2+}. These observations lead to the following order: $Ca > Ca^{2+} > Mg^{2+}$.

PRACTICE EXERCISE

Which of the following atoms and ions is largest: S^{2-}, S, O^{2-}. ***Answer:*** S^{2-}

The effect of varying nuclear charge on ionic radii is seen in the variation in radius in an **isoelectronic series** of ions. The term *isoelectronic* means that the ions possess the same number of electrons. For example, each ion in the series O^{2-}, F^-, Na^+, Mg^{2+}, and Al^{3+} has ten electrons arranged in a $1s^22s^22p^6$

Table 8.3 Radii (Å) of Ions with Noble-Gas Electron Configurations

Group 1A		Group 2A		Group 3A, 3B		Group 6A		Group 7A	
Li^+	0.68	Be^{2+}	0.30			O^{2-}	1.45	F^-	1.33
Na^+	0.98	Mg^{2+}	0.65	Al^{3+}	0.45	S^{2-}	1.90	Cl^-	1.81
K^+	1.33	Ca^{2+}	0.94	Sc^{3+}	0.68	Se^{2-}	2.02	Br^-	1.96
Rb^+	1.48	Sr^{2+}	1.10	Y^{3+}	0.90	Te^{2-}	2.22	I^-	2.19
Cs^+	1.67	Ba^{2+}	1.31						

electron configuration, like neon. The nuclear charge in this series increases steadily in the order listed. Because the number of electrons remains constant, the radius of the ion decreases with increasing nuclear charge, as the electrons are more strongly attracted to the nucleus:

Increasing nuclear charge ⟶

O^{2-}	F^-	Na^+	Mg^{2+}	Al^{3+}
1.45 Å	1.33 Å	0.98 Å	0.65 Å	0.45 Å

The charge on the nucleus of an atom or a monatomic ion, of course, is given by the atomic number of the element. Notice the positions of these particular elements in the periodic table and also their atomic numbers. The nonmetal anions precede the noble gas Ne in the table. The metal cations follow Ne. Oxygen, the largest ion in this isoelectronic series, has the lowest atomic number, 8. Aluminum, the smallest of these ions, has the highest atomic number, 13.

SAMPLE EXERCISE 8.5

Arrange the ions S^{2-}, Cl^-, K^+, and Ca^{2+} in order of decreasing size.

Solution: This is an isoelectronic series of ions, with all ions having 18 electrons. In an isoelectronic series, size decreases as the nuclear charge (atomic number) of the ion increases. The atomic numbers of the ions are S (16), Cl (17), K (19), and Ca (20). Thus the ions decrease in size in the following order: $S^{2-} > Cl^- > K^+ > Ca^{2+}$.

PRACTICE EXERCISE

Which of the following ions is largest: Rb^+, Sr^{2+}, Y^{3+}? ***Answer:*** Rb^+

One further comparison worth keeping in mind is the relative sizes of ions from the A and B subgroups. You may recall from the discussion in Section 7.7 that the outermost *s* electron of the group 1B elements experiences a higher effective nuclear charge. This happens because the ten *d* electrons added in going from a group 1A element to a group 1B element in the same row (for example, in going from K to Cu) do not completely shield the valence *s* electron from the nucleus. They also do not completely shield one another from the nucleus. As a result, not only is the atom of the group 1B element smaller, the ion formed by removal of the valence *s* electron is smaller also. For example, the radius of Cu^+, 0.96 Å, is less than that for the corresponding group 1A ion, K^+, radius 1.33 Å.

8.4 COVALENT BONDING

We have seen that ionic substances possess several characteristic properties. They are usually brittle substances with high melting points. They are usually also crystalline, meaning that the solids have flat surfaces that make characteristic angles with one another. Ionic crystals can often be cleaved; that is, they break apart along smooth, flat surfaces. The characteristics of ionic substances result from the electrostatic forces that maintain the ions in a rigid, well-defined, three-dimensional arrangement such as that illustrated in Figure 8.4.

The vast majority of chemical substances do not have the characteristics of ionic materials; water, gasoline, banana peelings, hair, antifreeze, and plastic bags are examples. Most of the substances with which we come in daily contact tend to be gases, liquids, or solids with low melting points. Many vaporize readily—for example, mothball crystals. Many in their solid forms are pliable rather than rigidly crystalline—for example, paraffin and plastic bags.

For the very large class of substances that do not behave like ionic substances, we need a different model for the bonding between atoms. G. N. Lewis reasoned that an atom might acquire a noble-gas electron configuration by sharing electrons with other atoms. A chemical bond formed by sharing a pair of electrons is called a **covalent bond**.

The hydrogen molecule, H_2, furnishes the simplest possible example of a covalent bond. Using Lewis symbols, formation of the H_2 molecule by combination of two hydrogen atoms can be represented as

$$\mathrm{H}\cdot + \cdot\mathrm{H} \longrightarrow \mathrm{H}:\mathrm{H}$$

The shared pair of electrons provides each hydrogen atom with two electrons in its valence shell (the 1*s*) orbital, so that in a sense it has the electron configuration of the rare gas helium (the shared electrons are counted with both atoms). Similarly, when two chlorine atoms combine to form the Cl_2 molecule, we have:

$$:\ddot{\underset{\cdot\cdot}{\mathrm{Cl}}}\cdot + \cdot\ddot{\underset{\cdot\cdot}{\mathrm{Cl}}}: \longrightarrow :\ddot{\underset{\cdot\cdot}{\mathrm{Cl}}}:\ddot{\underset{\cdot\cdot}{\mathrm{Cl}}}:$$

Each chlorine atom, by sharing the bonding electron pair, acquires eight electrons (an octet) in its valence shell. It thus achieves the noble-gas electron configuration of argon. The structures shown above for H_2 and Cl_2 are called **Lewis structures**. In writing Lewis structures, we usually show each electron pair shared between atoms as a line and the unshared electron pairs as dots. The Lewis structures for H_2 and Cl_2 are shown as follows:

$$\mathrm{H}\text{—}\mathrm{H} \qquad :\ddot{\underset{\cdot\cdot}{\mathrm{Cl}}}\text{—}\ddot{\underset{\cdot\cdot}{\mathrm{Cl}}}:$$

Figure 8.7 Electron distribution in the H_2 molecule.

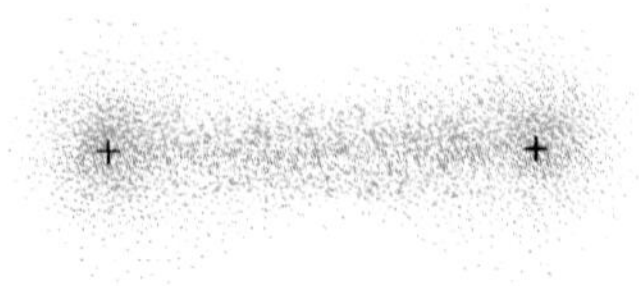

Of course, the shared pairs of electrons are not located in fixed positions between nuclei. Figure 8.7 shows the distribution of electron density in the H_2 molecule. Notice that electron density is concentrated between the nuclei. The two atoms are bound into the H_2 molecule principally because of the electrostatic attractions of the two positive nuclei for the concentration of negative charge between them. In Chapter 9, in which we start with the Lewis symbols of the atoms, we shall look more closely at the use of orbitals to describe the electron density within molecules. Meanwhile, our discussions will rely primarily on Lewis structures.

In the Lewis model, the valence of an element is the number of electron pairs shared to complete the octet of electrons. Because the number of valence electrons is the same as the group number for the nonmetals, one might predict that the 7A elements, such as F, would form one covalent bond to achieve an octet; 6A elements, such as O, would form two covalent bonds; 5A elements, such as N, would form three covalent bonds; and 4A elements, such as C, would form four covalent bonds. These predictions are borne out in many compounds. For example, consider the simple hydrides of the nonmetals of the second row (period) of the periodic table:

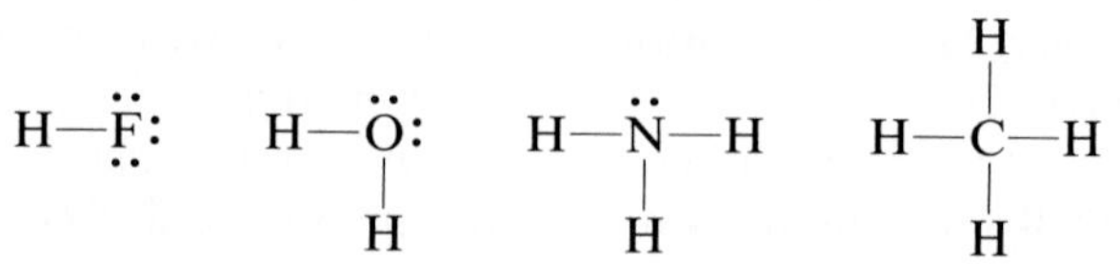

$$\mathrm{H}\text{—}\ddot{\underset{\cdot\cdot}{\mathrm{F}}}: \qquad \mathrm{H}\text{—}\underset{\mathrm{H}}{\underset{|}{\ddot{\mathrm{O}}}}: \qquad \mathrm{H}\text{—}\underset{\mathrm{H}}{\underset{|}{\ddot{\mathrm{N}}}}\text{—}\mathrm{H} \qquad \mathrm{H}\text{—}\underset{\mathrm{H}}{\underset{|}{\overset{\mathrm{H}}{\overset{|}{\mathrm{C}}}}}\text{—}\mathrm{H}$$

Thus the Lewis model succeeds in accounting for the compositions of compounds of the nonmetals, in which covalent bonding predominates.

SAMPLE EXERCISE 8.6

Using the Lewis theory and the theory of ionic bonding described earlier, explain the formulas of the following hydrides: NaH; MgH_2; AlH_3; SiH_4; PH_3; H_2S; HCl.

Solution: The hydrides of the metallic elements are ionic compounds consisting of metallic cations and the hydride ion, H^-. These ionic substances are formed by transfer of one or more electrons from the metal to hydrogen atoms. Each hydrogen atom accepts one electron to form H^-. One hydride ion is required for each electron removed from the metal to form the noble-gas configuration. The first three compounds thus correspond to the compositions Na^+H^-; $Mg^{2+}2H^-$; $Al^{3+}3H^-$. The remaining compounds are best formulated as covalent, in which an electron pair is shared between the central atom and each hydrogen. Silicon, an element of group 4A, requires four electrons to attain the noble-gas configuration of eight valence-shell electrons. Phosphorus requires three, sulfur two, and chlorine one. The formulas of the hydrides are in accord with the number of electrons needed.

PRACTICE EXERCISE

How many H atoms must bond to selenium (atomic number 34) to give the selenium atom an octet of valence-shell electrons? ***Answer:*** Two

Multiple Bonds

The sharing of a pair of electrons constitutes a single covalent bond, generally referred to simply as a **single bond**. In many molecules, atoms attain complete octets by sharing more than one pair of electrons between them. When two electron pairs are shared, two lines are drawn, representing a **double bond**. A **triple bond** corresponds to the sharing of three pairs of electrons. Such **multiple bonding** is found, for example, in the N_2 molecule:

$$:\dot{\underset{\cdot}{N}}\cdot + \cdot\dot{\underset{\cdot}{N}}: \longrightarrow :N:::N: \qquad (\text{or } :N{\equiv}N:)$$

Because each nitrogen atom possesses five electrons in its valence shell, the sharing of three electron pairs is required to achieve the octet configuration. The properties of N_2 are in complete accord with this Lewis structure. Nitrogen gas is a diatomic gas with exceptionally low reactivity that results from the very stable nitrogen-nitrogen bond. Study of the structure of N_2 reveals that the nitrogen atoms are separated by only 1.10 Å. The short N≡N bond distance is a result of the triple bond between the atoms. From structure studies of many different substances in which nitrogen atoms share one or two electron pairs, we have learned that the average distance between bonded nitrogen atoms varies with the number of shared electron pairs:

N—N	N=N	N≡N
1.47 Å	1.24 Å	1.10 Å

As a general rule, the distance between bonded atoms decreases as the number of shared electron pairs increases.

Carbon dioxide, CO_2, provides a further example of a molecule containing multiple bonds:

$$:\dot{\underset{\cdot}{O}}: + \cdot\dot{\underset{\cdot}{C}}\cdot + :\dot{\underset{\cdot}{O}}: \longrightarrow \ddot{\underset{\cdot\cdot}{O}}::C::\ddot{\underset{\cdot\cdot}{O}} \qquad (\text{or } \ddot{\underset{\cdot\cdot}{O}}{=}C{=}\ddot{\underset{\cdot\cdot}{O}})$$

A CLOSER LOOK: Formal Charge and Lewis Structures

The concept of **formal charge** is sometimes used as an aid in deciding between alternative Lewis structures. The formal charge is largely a means of "bookkeeping" for the valence electrons. To establish the formal charge on any atom in a molecule or ion, we assign electrons to the atom as follows:

1. All bonding electrons are divided equally between the atoms that form bonds.
2. All nonbonding electrons are assigned entirely to the atom on which they are found.

The formal charge is defined as *the number of valence electrons in the isolated atom, minus the number of electrons assigned to the atom in the Lewis structure.* That is,

$$\text{Formal charge} = \text{number of valence electrons} - (\tfrac{1}{2}\text{ number of bonding electrons} + \text{number of nonbonding electrons})$$

Let's illustrate these rules by calculating the formal charge on the central atom in the second-row nonmetal hydrides as shown in the table below. Note that the formal charge on the central atom is zero in all cases.

To see how the idea of formal charge can help in making a distinction between alternative Lewis structures, consider the cyanate ion, NCO^-. There are three possible orders for the atoms in this ion. For each, we can write a Lewis structure that yields an octet about each atom. For each structure, we can then calculate the formal charge on each atom. The results are as follows:

	$[\ddot{N}{=}C{=}\ddot{O}]^-$	$[\ddot{C}{=}O{=}\ddot{N}]^-$	$[\ddot{O}{=}N{=}\ddot{C}]^-$
Formal charge	1− 0 0	2− 2+ 1−	0 1+ 2−

Because the ion as a whole has a charge of 1−, the formal charges of all atoms must sum to 1−. As a general rule, the most stable Lewis structure will be that in which the atoms bear the smallest formal charges. The Lewis structure on the left is clearly superior to the other two in producing the smallest variations in formal charge among the atoms. This suggests that the arrangement shown at the left is the preferred structure for the ion; indeed, it is the observed structure.

Although the concept of formal charge is useful in helping to decide between alternative Lewis structures, you must keep in mind that *the formal charges do not represent real charges on the atoms.* Other factors that we will be discussing in the material ahead contribute to determine the actual net charges on atoms in molecules and ions.

	Hydride			
	H—F:	H—O:—H	H—N—H (H)	H—C—H (H, H)
Bonding electrons assigned	1	2	3	4
Nonbonding electrons assigned	6	4	2	0
Total electrons assigned	7	6	5	4
Electrons in isolated atom	7	6	5	4
Formal charge	0	0	0	0

8.5 DRAWING LEWIS STRUCTURES

We have seen in the foregoing section some simple examples of Lewis structures. It is actually fairly easy to draw the Lewis structures for most compounds and ions formed from nonmetallic elements. Doing so is a skill that is important in mastering the material in this chapter and the next.

When drawing Lewis structures, you should follow a regular procedure. We'll first outline the procedure, then go through several examples to show its application.

1. *Sum the valence electrons from all atoms.* (Use the periodic table as necessary to help you determine the number of valence electrons in each atom.) If the species is an ion, add an electron for each negative charge or subtract

an electron for each positive charge. Do not worry about keeping track of which electrons come from which atoms. Only their total number is important.

2. *Write the symbols for the atoms involved so as to show which atoms are connected to which.* Atoms are often written in the order in which they are connected in the molecule or ion, as in HCN. When a central atom has a group of other atoms bonded to it, we usually write the central atom first, as in CO_3^{2-} and CCl_4. In other cases, you may need more information before you can draw the Lewis structure.
3. *Draw a single bond between each pair of atoms bonded together.*
4. *Complete the octets of the atoms bonded to the central atom.* (Remember, however, that hydrogen needs only two electrons.)
5. *Place any leftover electrons on the central atom, even if doing so results in more than an octet.*
6. *If there are not enough electrons to give the central atom an octet, try multiple bonds.* Use one or more of the unshared pairs of electrons on the atoms bonded to the central atom to form double or triple bonds.

These rules are illustrated in the following sample exercise.

SAMPLE EXERCISE 8.7

Draw Lewis structures for the following molecules: **(a)** PCl_3; **(b)** HCN; **(c)** ClO_3^-.

Solution: **(a)** Phosphorus (group 5A) has five valence electrons, and each chlorine (group 7A) has seven. The total number of valence-shell electrons is therefore $5 + (3 \times 7) = 26$. There are various ways the atoms might be arranged. However, in binary (two-element) compounds the first element listed in the chemical formula is generally surrounded by the remaining atoms. Thus, we begin with a skeleton structure that shows single bonds between phosphorus and each chlorine:

```
Cl—P—Cl
   |
   Cl
```

(It is not crucial to place the atoms in exactly this arrangement; Lewis structures are not drawn to show geometry. However, it is important to show correctly which atoms are bonded to which.) The octets around each Cl are completed, accounting for 24 electrons. The remaining 2 electrons are then placed on P, completing the octet around that atom as well:

```
 ..   ..  ..
:Cl—P—Cl:
 ..   |   ..
     :Cl:
      ..
```

(Remember that in achieving an octet, the bonding electrons are counted for both atoms.)

(b) Hydrogen has one valence-shell electron, carbon (group 4A) has four, and nitrogen (group 5A) have five. The total number of valence-shell electrons is therefore $1 + 4 + 5 = 10$. Again there are various ways we might choose to arrange the atoms. Because hydrogen can accommodate only one electron pair, it always has only one single bond associated with it in any compound. This fact causes us to reject C—H—N as a possible arrangement. The remaining two possibilities are H—C—N and H—N—C. The first is the arrangement found experimentally. You might have guessed this to be the atomic arrangement because the formula is written with the atoms in this order. Thus we begin with a skeleton structure that shows single bonds between hydrogen, carbon, and nitrogen:

H—C—N

These two bonds account for four electrons. If we then place the remaining six electrons around N to give it an octet, we do not achieve an octet on C:

$$\text{H—C—}\ddot{\underset{..}{\text{N}}}\text{:}$$

We therefore try a double bond between C and N, using an unshared pair of electrons that we had placed on N. Again, there are fewer than eight electrons on C, so we try a triple bond. This structure gives an octet around both C and N:

$$\text{H—C}\equiv\text{N:}$$

(c) Chlorine has seven valence electrons, and oxygen (group 6A) has six. An extra electron is added to account for the ion having a 1− charge. The total number of valence-shell electrons is therefore $7 + (3 \times 6) + 1 = 26$. After putting in the single bonds and distributing the unshared electron pairs, we have

$$\left[\begin{array}{c}:\ddot{\underset{..}{\text{O}}}\text{—}\ddot{\text{Cl}}\text{—}\ddot{\underset{..}{\text{O}}}: \\ | \\ :\ddot{\underset{..}{\text{O}}}:\end{array}\right]^-$$

For oxyanions—ClO_3^-, SO_4^{2-}, NO_3^-, CO_3^{2-}, and so forth—the oxygen atoms surround the central nonmetal atom.)

PRACTICE EXERCISE

Draw the Lewis electron dot structure for the NO^+ ion. ***Answer:*** $[:\text{N}\equiv\text{O}:]^+$

8.6 RESONANCE FORMS

We sometimes encounter substances in which the known arrangement of atoms is not adequately described by a single Lewis structure. The structural chemistry of nonmetallic elements affords several examples. Consider ozone, O_3, about which we shall have much to say in Chapter 18. This fascinating substance consists of bent molecules with both O—O distances the same, as shown in Figure 8.8. Because each oxygen atom contributes 6 valence-shell electrons, the ozone molecule has 18 valence-shell electrons. In writing the Lewis structure, we find that we must have one double bond to attain an octet of electrons about each atom:

$$:\ddot{\underset{..}{\text{O}}}\text{—}\ddot{\text{O}}\text{=}\ddot{\text{O}}$$

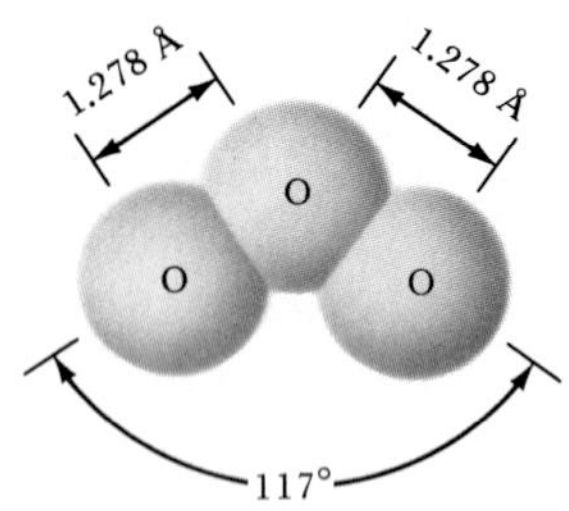

Figure 8.8 Molecular structure of ozone, O_3.

But this structure cannot by itself be correct, because it requires that one O—O bond be different from the other, contrary to the observed structure; we would expect the O=O double bond to be shorter than the O—O single bond. However, in drawing the Lewis structure we could just as easily have put the O=O bond on the left:

$$\ddot{\text{O}}\text{=}\ddot{\text{O}}\text{—}\ddot{\underset{..}{\text{O}}}:$$

The two alternative Lewis structures for ozone are equivalent except for the placement of electrons. Equivalent Lewis structures of this sort are called **resonance forms**. To describe the structure of ozone properly, we write both Lewis structures and indicate that the real molecule is described by an average of the structures suggested by the two resonance forms:

$$\ddot{\text{O}}\text{=}\ddot{\text{O}}\text{—}\ddot{\underset{..}{\text{O}}}: \longleftrightarrow :\ddot{\underset{..}{\text{O}}}\text{—}\ddot{\text{O}}\text{=}\ddot{\text{O}}$$

The double-headed arrow indicates that the structures shown are resonance forms.

The fact that we must write more than one Lewis structure to describe a molecule (or ion) does not imply there is anything especially different about the molecule. You must not suppose that the molecule really exists in two or more different forms and oscillates rapidly between them. There is only one form of the molecule, that which is observed experimentally. We need to write two or more different resonance forms only because Lewis structures are limited in describing the electron distributions in molecules.

One rule that must be followed in writing resonance forms is that the arrangement of the nuclei must be the same in each structure. That is, the same atoms must be bonded to one another in all the structures, so that the only differences are in the arrangements of electrons.

As an additional example of resonance forms, let us draw the Lewis structure for the nitrate ion, NO_3^-, one of the most commonly encountered anions. We find that three equivalent Lewis structures are required in this instance:

$$\left[\begin{array}{c} :\text{O}: \\ \| \\ \text{N} \\ :\ddot{\text{O}}\quad\ddot{\text{O}}: \end{array}\right]^- \longleftrightarrow \left[\begin{array}{c} :\ddot{\text{O}}: \\ | \\ \text{N} \\ \text{O}\quad\ddot{\text{O}}: \end{array}\right]^- \longleftrightarrow \left[\begin{array}{c} :\ddot{\text{O}}: \\ | \\ \text{N} \\ :\ddot{\text{O}}\quad\text{O} \end{array}\right]^-$$

Note that the arrangement of nuclei is the same in each structure; only the placement of electrons differs. All three Lewis structures taken together adequately describe the nitrate ion, which is observed to be planar (all atoms in the same plane), with all three N—O distances equal.

In the examples of resonance structures we have seen so far, all the resonance forms have the same importance in contributing to the overall description of the molecule or ion. In the example in Sample Exercise 8.8, the contributing resonance structures are not equally important.

SAMPLE EXERCISE 8.8

The molecule chlorine dioxide, ClO_2, is bent, with a central chlorine atom bound to two oxygen atoms. The two Cl—O distances are observed to be equal. Describe the ClO_2 molecule in terms of three possible resonance forms.

Solution: The chlorine atom has 7 valence-shell electrons, and oxygen has 6. We therefore have a total of 19 electrons $[(2 \times 6) + 7]$ to place in this molecule. Note that ClO_2 has an odd number of electrons. This is an unusual situation; it means that one electron must remain unpaired.

We draw the arrangement of atoms in accord with the experimental facts and put one single bond between chlorine and each oxygen.

$$\text{O}\diagup\text{Cl}\diagdown\text{O}$$

This leaves us with 15 electrons to place. We put electron pairs on the atoms to achieve an octet on each atom, or to come as close to it as we can get. Because the molecule possesses an odd number of electrons, we can be sure that at least one atom will not possess an octet of electrons. The odd electron must go on one or the other of the atoms, so that three possible structures result:

$$:\dot{\text{O}}\diagup\ddot{\text{Cl}}\diagdown\ddot{\text{O}}: \longleftrightarrow :\ddot{\text{O}}\diagup\ddot{\text{Cl}}\diagdown\dot{\text{O}}: \longleftrightarrow :\ddot{\text{O}}\diagup\dot{\text{Cl}}\diagdown\ddot{\text{O}}:$$

The first two of these are equivalent, but the third is different; the odd electron is on chlorine rather than oxygen. To find out how much weight to attach to this resonance structure in comparison with the first two, we would have to have experimental information on how the odd electron is distributed in the real ClO_2 molecule.

PRACTICE EXERCISE

Draw two equivalent resonance structures for the NO_2^- ion.

Answer: $[\ddot{\underset{\cdot\cdot}{O}}=\ddot{N}-\ddot{\underset{\cdot\cdot}{O}}:]^- \longleftrightarrow [:\ddot{\underset{\cdot\cdot}{O}}-\ddot{N}=\ddot{\underset{\cdot\cdot}{O}}]^-$

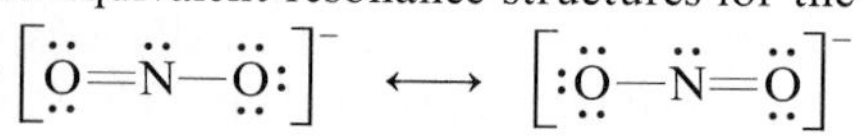

8.7 EXCEPTIONS TO THE OCTET RULE

The octet rule is so simple and useful in introducing the basic concepts of bonding that you might assume it is always obeyed. However, in many situations the octet rule fails. These exceptions are of three main types:

1. Molecules with an odd number of electrons
2. Molecules in which an atom has less than an octet
3. Molecules in which an atom has more than an octet

Odd Number of Electrons

In the vast majority of molecules, the number of electrons is even, and complete pairing of electron spins occurs. However, in a few molecules such as ClO_2, NO, and NO_2, the number of electrons is odd. For example, NO contains $5 + 6 = 11$ valence electrons. Obviously, complete pairing of these electrons is impossible, and an octet around each atom cannot be achieved.

Less Than an Octet

A second type of exception occurs when there are fewer than eight electrons around an atom in a molecule or ion. This is also a relatively rare situation and is most often encountered in compounds of boron and beryllium. For example, let's consider boron trifluoride, BF_3. If we follow the procedure on pages 260–261 for drawing Lewis structures, then through the first five steps we have the following structure:

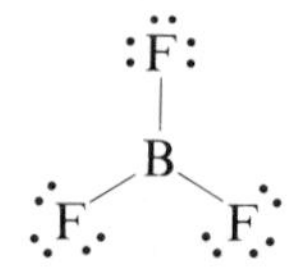

We see that there are only six electrons around the boron atom. We could complete the octet around boron by forming a double bond. In so doing, we see that there are three such equivalent resonance structures:

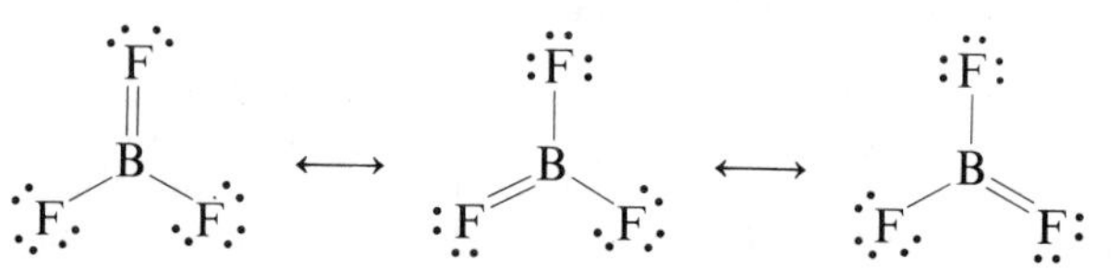

However, in forming these Lewis structures, we have forced a fluorine atom to share additional electrons with the boron atom. This is inconsistent with the high electron affinity of fluorine, though: We don't expect the fluorine atoms to share additional electrons with the boron atom. We conclude that the Lewis structures in which there is a B=F double bond are less important than the one in which there is less than an octet around boron:

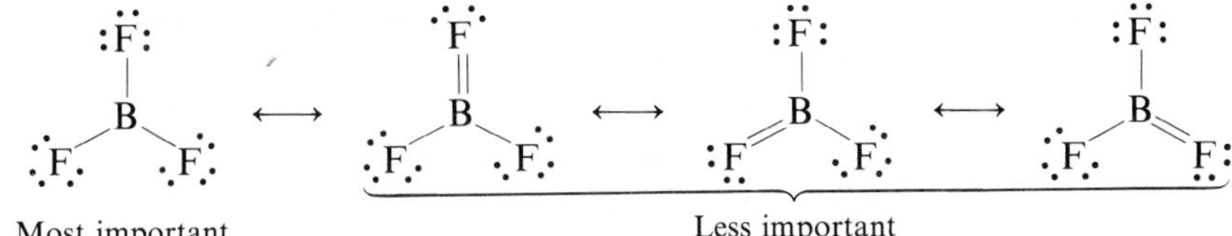

We usually represent BF_3 solely by the leftmost resonance structure in which there are only six electrons around boron. The chemical behavior of BF_3 is consistent with this representation. Thus, BF_3 reacts very energetically with molecules that have an unshared pair of electrons that can be used to form a bond with boron. For example, it reacts with ammonia, NH_3, to form the compound NH_3BF_3:

$$H_3N{:} + BF_3 \longrightarrow H_3N{-}BF_3$$

In this stable compound boron has an octet of electrons.

More Than an Octet

The third and largest class of exceptions consists of molecules or ions in which there are more than eight electrons in the valence shell of an atom. As an example, consider PCl_5. When we draw the Lewis structure for this molecule, we are forced to "expand" the valence shell and place ten electrons around the central phosphorus atom:

$$PCl_5$$

Among the other molecules and ions with "expanded" valence shells are SF_4, AsF_6^-, and ICl_4^-. The corresponding molecules with a second-period atom, such as NCl_5 and OF_4, do *not* exist. Let's take a look at why expanded valence shells are observed only for elements in period 3 and beyond in the periodic table.

The octet rule works as well as it does because the representative elements usually employ only an *ns* and three *np* valence-shell orbitals in bonding, and these orbitals hold eight electrons. Because elements of the second period have only 2*s* and 2*p* orbitals available for bonding, they can never have more than an octet of electrons in their valence shells. However, from the third period on, the elements have unfilled *nd* orbitals that can be used in bonding. For example, the orbital diagram for the valence shell of a phosphorus atom is as follows:

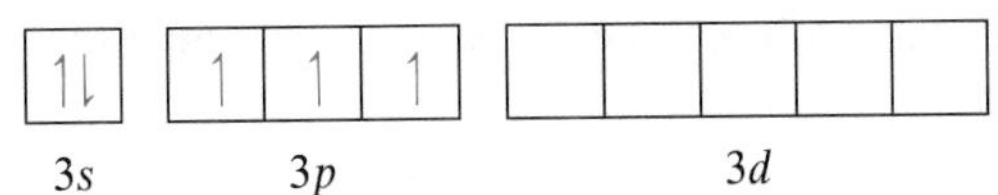

Although third-period elements such as phosphorus often satisfy the octet rule, as in PF_3, they also often exceed it by using their empty *d* orbitals to accommodate electrons.

Size also plays an important role in determining whether an atom can accommodate more than eight electrons. The larger the central atom, the larger the number of atoms that can surround it. The occurrences of expanded valence shells therefore increase with increasing size of the central atom. The size of the surrounding atoms is also important. Expanded valence shells occur most often when the central atom is bonded to the smallest and most strongly electron-attracting atoms, such as F, Cl, and O.

SAMPLE EXERCISE 8.9

Draw the Lewis structure for ICl_4^-.

Solution: Iodine (group 7A) has seven valence electrons; each chlorine (group 7A) also has seven; an extra electron is added to account for the 1− charge of the ion. Therefore, the total number of valence electrons is $7 + 4(7) + 1 = 36$. The I atom is the central atom in the ion. Putting eight electrons around each Cl atom (including a pair of electrons between I and each Cl to represent the single bonds between these atoms) requires $8 \times 4 = 32$ electrons. We are thus left with $36 - 32 = 4$ electrons to be placed on the larger iodine:

$$\left[\begin{matrix} :\ddot{\underset{..}{Cl}} & & \ddot{\underset{..}{Cl}}: \\ & \ddot{\underset{..}{I}} & \\ :\ddot{\underset{..}{Cl}} & & \ddot{\underset{..}{Cl}}: \end{matrix}\right]^-$$

Therefore, iodine has 12 electrons around it, exceeding the common octet of electrons.

PRACTICE EXERCISE

Which of the following atoms is never found with more than an octet of electrons around it: S, C, P, Br? *Answer:* C

8.8 STRENGTHS OF COVALENT BONDS

The stability of a molecule can be related to the strengths of the covalent bonds it contains. The strength of a covalent bond between two atoms is determined by the energy required to break that bond. The **bond-dissociation energy**, also called the **bond energy**, is the enthalpy change, ΔH, required to break a particular bond in a mole of gaseous substance. For example, the dissociation energy for the bond between chlorine atoms in the Cl_2 molecule is the energy required to dissociate a mole of Cl_2 into chlorine atoms:

$$:\ddot{\underset{..}{Cl}}-\ddot{\underset{..}{Cl}}:(g) \longrightarrow 2\,:\ddot{\underset{..}{Cl}}\cdot(g) \qquad \Delta H = D(\text{Cl—Cl}) = 242 \text{ kJ}$$

We use the designation *D*(bond type) in this equation and elsewhere to represent bond-dissociation energies.

It is a relatively easy matter to assign bond energies to bonds in diatomic molecules. As we have seen, the bond energy is just the energy required to break the diatomic molecule into its component atoms. However, for bonds that occur only in polyatomic molecules (such as the C—H bond) we must often utilize average bond energies. For example, the enthalpy change for the process shown below (called "atomization") can be used to define an average bond strength for the C—H bond:

$$\begin{matrix} & H & \\ & | & \\ H- & C & -H(g) \\ & | & \\ & H & \end{matrix} \longrightarrow \cdot\dot{\underset{\cdot}{C}}\cdot(g) + 4\,H\cdot(g) \qquad \Delta H = 1660 \text{ kJ}$$

Because there are four equivalent C—H bonds in methane, the heat of atomization is equal to the total bond energies of the four C—H bonds. Therefore, the average C—H bond energy is $D(\text{C—H}) = (1660/4)$ kJ/mol = 415 kJ/mol.

The bond energy for a given set of atoms, say C—H, depends on the rest of the molecule of which it is a part. However, the variation from one molecule to another is generally small. This supports the idea that the bonding electron pairs are localized between atoms. If we consider C—H bond strengths in many different compounds, we find that the average strength is 413 kJ/mol, which compares closely with the 415 kJ/mol value calculated from CH_4.

Table 8.4 lists several average bond energies. Notice that the bond energy is always a positive quantity; energy is always required to break chemical bonds. Conversely, energy is released when a bond forms between two gaseous atoms or molecular fragments. Of course, the greater the bond energy, the stronger the bond.

A molecule with strong chemical bonds generally has less tendency to undergo chemical change than does one with weak bonds. This relationship between strong bonding and chemical stability helps explain the chemical form in which many elements are found in nature. For example, Si—O bonds are among the strongest ones that silicon forms. It is not surprising therefore that SiO_2 and other substances containing Si—O bonds (silicates) are so common; it is estimated that over 90 percent of the earth's crust is composed of SiO_2 and silicates. We shall discuss these compounds in greater detail in Chapter 23.

Bond Energies and the Enthalpy of Reactions

We can use the average bond energies given in Table 8.4 to estimate the enthalpies of chemical reactions in which bonds are broken and new ones are made. This procedure enables us to estimate quickly whether a given reaction will be

Table 8.4 Average Bond Energies (kJ/mol)

Single bonds							
C—H	413	N—H	391	O—H	463	F—F	155
C—C	348	N—N	163	O—O	146		
C—N	293	N—O	201	O—F	190	Cl—F	253
C—O	358	N—F	272	O—Cl	203	Cl—Cl	242
C—F	485	N—Cl	200	O—I	234		
C—Cl	328	N—Br	243			Br—F	237
C—Br	276			S—H	339	Br—Cl	218
C—I	240	H—H	436	S—F	327	Br—Br	193
C—S	259	H—F	567	S—Cl	253		
		H—Cl	431	S—Br	218	I—Cl	208
Si—H	323	H—Br	366	S—S	266	I—Br	175
Si Si	226	H—I	299			I—I	151
Si—C	301						
Si—O	368						
Multiple bonds							
C=C	614	N=N	418	O_2	495		
C≡C	839	N≡N	941				
C=N	615			S=O	523		
C≡N	891			S=S	418		
C=O	799						
C≡O	1072						

endothermic ($\Delta H > 0$) or exothermic ($\Delta H < 0$), even if we do not know ΔH_f° for all the chemical species involved. Our strategy for estimating reaction enthalpies is straightforward: (1) We determine which bonds are being broken in the reaction and sum the bond energies of these bonds. This sum is the total amount of energy that must be supplied to break bonds during the reaction. (2) We determine which bonds are being formed in the reaction and sum these bond energies. This total is the amount of energy released by bond formation during the reaction. (3) The enthalpy of the reaction is estimated as the energy required to break bonds, minus the total bond energy of the new bonds formed:

$$\Delta H = \sum (\text{bond energies of bonds broken}) - \sum (\text{bond energies of bonds formed}) \quad [8.10]$$

As an example, consider the reaction between 1 mol of chlorine and 1 mol of methane:

$$\text{Cl—Cl}(g) + \text{H—CH}_3(g) \longrightarrow \text{H—Cl}(g) + \text{Cl—CH}_3(g) \quad [8.11]$$

In the course of this reaction, the following bonds are broken and made:

Bonds broken: 1 mol Cl—Cl, 1 mol C—H
Bonds formed: 1 mol H—Cl, 1 mol C—Cl

Using Equation 8.10 and the data in Table 8.4, we estimate the enthalpy of the reaction as

$$\begin{aligned} \Delta H &= [D(\text{Cl—Cl}) + D(\text{C—H})] - [D(\text{H—Cl}) + D(\text{C—Cl})] \\ &= (242 \text{ kJ} + 413 \text{ kJ}) - (431 \text{ kJ} + 328 \text{ kJ}) \\ &= -104 \text{ kJ} \end{aligned}$$

We see that the reaction is exothermic because the bonds in the products (especially the H—Cl bond) are stronger than those in the reactants (especially the Cl—Cl bond).

We usually use bond energies to estimate ΔH for a reaction only if we do not have all the needed ΔH_f° values readily at hand. Thus, in the example above we cannot calculate the enthalpy from ΔH_f° values and Hess's law because ΔH_f° for $CH_3Cl(g)$ is not given in Appendix C (although we could certainly find that ΔH_f° in other sources). It is also important to remember that bond energies are derived for gaseous molecules and that they are often averaged values. Nonetheless, the use of bond energies often gives a result very close to the correct value, as illustrated in Sample Exercise 8.10.

SAMPLE EXERCISE 8.10

Using Table 8.4, estimate ΔH for the following reaction (where we show explicitly the bonds involved in the reactants and products):

$$\text{H—}\underset{\text{H}}{\overset{\text{H}}{\underset{|}{\overset{|}{\text{C}}}}}\text{—}\underset{\text{H}}{\overset{\text{H}}{\underset{|}{\overset{|}{\text{C}}}}}\text{—H}(g) + \tfrac{7}{2}\text{O}_2(g) \longrightarrow 2\text{O=C=O}(g) + 3\text{H—O—H}(g)$$

Solution: Among the reactants, we must break six C—H bonds and a C—C bond in C_2H_6; we also break $\frac{7}{2}O_2$ bonds. Among the products, we form four C=O bonds (two in each CO_2) and six O—H bonds (two in each H_2O). Using Equation 8.10 and data from Table 8.4, we have

$$\begin{aligned}\Delta H &= 6D(\text{C—H}) + D(\text{C—C}) + \tfrac{7}{2}D(\text{O}_2) - 4D(\text{C=O}) - 6D(\text{O—H}) \\ &= 6(413 \text{ kJ}) + 348 \text{ kJ} + \tfrac{7}{2}(495 \text{ kJ}) - 4(799 \text{ kJ}) - 6(463 \text{ kJ}) \\ &= 4558 \text{ kJ} - 5974 \text{ kJ} \\ &= -1416 \text{ kJ}\end{aligned}$$

This estimate can be compared with the value of −1428 kJ calculated from more precise thermochemical data; the agreement is excellent.

PRACTICE EXERCISE

Using Table 8.4, estimate ΔH for the following reaction:

$$\text{H—}\underset{\text{H}}{\underset{|}{\text{N}}}\text{—}\underset{\text{H}}{\underset{|}{\text{N}}}\text{—H}(g) \longrightarrow \text{N}{\equiv}\text{N}(g) + 2\text{H—H}(g)$$

Answer: −86 kJ

Bond Strength and Bond Length

Just as we can define an average bond strength, we can also define an average bond length for a number of common bond types. Some of these are listed in Table 8.5. Of particular interest to us here is the relationship between bond strength, bond length, and the number of bonds between the atoms. For example, we can use data in Tables 8.4 and 8.5 to compare the bond lengths and bond energies of carbon-carbon single, double, and triple bonds:

C—C	C=C	C≡C
1.54 Å	1.34 Å	1.20 Å
348 kJ/mol	614 kJ/mol	839 kJ/mol

As the number of bonds between the carbon atoms increases, the bond energy increases and the bond length decreases; that is, the carbon atoms are held more closely and more tightly together. In general, *as the number of bonds between two atoms increases, the bond grows shorter and stronger.*

Table 8.5 Average Bond Lengths for Some Single, Double, and Triple Bonds

Bond	Bond length (Å)	Bond	Bond length (Å)
C—C	1.54	N—N	1.47
C=C	1.34	N=N	1.24
C≡C	1.20	N≡N	1.10
C—N	1.43	N—O	1.36
C=N	1.38	N=O	1.22
C≡N	1.16	O—O	1.48
C—O	1.43	O=O	1.21
C=O	1.23		
C≡O	1.13		

CHEMISTRY AT WORK: Explosives

Enormous amounts of energy can be stored in chemical bonds. Perhaps the most graphic example of the chemical energy stored in bonds is in the chemistry of explosives. An explosive is a liquid or solid substance that satisfies three principal criteria: (1) It must decompose very exothermically; (2) the products of its decomposition must be gaseous, so that a tremendous gas pressure accompanies the decomposition; and (3) its decomposition must occur extremely rapidly. The combination of these three effects leads to the violent generation of heat and gases that we associate with explosions (Figure 8.9).

What are the most desirable products into which an explosive should decompose? Ideally, to give the most exothermic reaction, an explosive should have weak chemical bonds, and it should decompose into substances with very strong bonds. Looking at bond energies (Table 8.4), we see that the N≡N, C≡O, and C=O bonds are among the strongest. Not surprisingly, explosives are usually designed to produce $N_2(g)$, $CO(g)$, and $CO_2(g)$. Water vapor is nearly always produced as well.

Many common explosives are organic molecules that contain nitro (NO_2) or nitrate (NO_3) groups attached to a carbon backbone. The structures of two of the most familiar explosives, glyceryl trinitrate (more commonly called nitroglycerin) and trinitrotoluene (TNT), are shown below.

Nitroglycerin

TNT

Figure 8.9 Explosives being used in the demolition of buildings. (Hoffman-Spooner/Gamma-Liaison)

Nitroglycerin is a pale-yellow, oily liquid. It is highly *shock-sensitive:* Merely shaking the liquid can cause its explosive decomposition into nitrogen, carbon dioxide, water, and oxygen gases:

$$4C_3H_5N_3O_9(l) \longrightarrow 6N_2(g) + 12CO_2(g) + 10H_2O(g) + O_2(g)$$

The high energies of the bonds in the N_2 molecules (941 kJ/mol), CO_2 molecules (2 × 799 kJ/mol), and water molecules (2 × 463 kJ/mol) make this reaction enormously exothermic. Nitroglycerin is an exceptionally unstable explosive because it is in nearly perfect *explosive balance:* Ideally, with the exception of a small amount of $O_2(g)$ produced, the only products are N_2, CO_2, and H_2O. Note also that, unlike combustion reactions (Section 3.2), explosions are entirely self-contained. No other reagent, such as $O_2(g)$, is needed for the explosive decomposition.

Because nitroglycerin is so unstable, it is difficult to use as a controllable explosive. The Swedish inventor Alfred Nobel (1833–1896) found that mixing nitroglycerin with an absorbent solid material such as diatomaceous earth gives a solid explosive (*dynamite*) that is much safer than liquid nitroglycerin. Nobel's discovery was hugely profitable and led to his endowment of the Nobel Prizes and other philanthropic causes.

8.9 BOND POLARITY AND ELECTRONEGATIVITIES

The electron pairs shared between two different atoms are not necessarily shared equally. We can visualize two extreme cases in the degree to which electron pairs are shared. On the one hand, we have bonding between two identical atoms, as in Cl_2 or N_2, where the electron pairs must be equally shared. At the other extreme, illustrated by NaCl, there will be essentially no sharing of electrons. We know that in this case the compound is best described as composed of Na^+ and Cl^- ions. The 3*s* electron of the Na atom is, in effect, transferred completely to chlorine. The bonds occurring in most covalent substances fall somewhere between these extremes. In practice we describe bonds as either ionic or covalent depending on which extreme the bond more closely resembles.

The concept of **bond polarity** is useful in describing the sharing of electrons between atoms. A **nonpolar bond** is one in which the electrons are shared equally between two atoms. In a **polar covalent bond**, one of the atoms exerts a greater attraction for the electrons than the other. If the difference in relative ability to attract electrons is large enough, an ionic bond is formed.

Electronegativity

We use a quantity called electronegativity to estimate whether a given bond will be nonpolar, polar covalent, or ionic. **Electronegativity** is defined as the ability of an atom *in a molecule* to attract electrons to itself. The greater an atom's electronegativity, the greater its ability to attract electrons to itself. We might expect the electronegativity of an atom in a molecule to be related to its ionization energy and electron affinity, which are properties of isolated atoms. The ionization energy measures how strongly the atom holds on to its electrons. Likewise, the electron affinity is a measure of how strongly the atom attracts additional electrons. An atom with a very negative electron affinity and high ionization energy will both attract electrons from other atoms and resist having its electrons attracted away; it will be highly electronegative.

Numerical estimates of electronegativity can be based on a variety of properties, not just ionization energy and electron affinity. For example, the American chemist Linus Pauling (b. 1901), who first developed the concept of electronegativity, based his scale on bond-energy relationships. We will not be concerned in detail with how the electronegativity values are obtained, but rather with using the concept in discussing chemical bonding.

Figure 8.10 shows Pauling's electronegativity values for many of the elements. Fluorine is the most electronegative element, with an electronegativity of 4.0. The least electronegative element, cesium, has an electronegativity of 0.7. The values for all other elements lie between these extremes.

Note that each horizontal row of the table generally exhibits a steady increase in electronegativity from left to right; that is, from the most metallic to the most nonmetallic elements. Notice also that, with some exceptions (especially within the transition metals), electronegativity decreases with increasing atomic number in any one group. This is what we might expect, because we know that ionization energies tend to decrease with increasing atomic number in a group and electron affinities don't change very much. You do not need to memorize numerical values for electronegativity. However, you should know the periodic trends so you can predict which of two elements is more electronegative.

Keep in mind that electronegativities are *approximate* measures of the *relative* tendencies of these elements to attract electrons to themselves in a chemical bond. The electronegativity varies with the type of chemical environ-

1A	2A	3B	4B	5B	6B	7B	8B			1B	2B	3A	4A	5A	6A	7A
H 2.1																
Li 1.0	Be 1.5											B 2.0	C 2.5	N 3.0	O 3.5	F 4.0
Na 0.9	Mg 1.2											Al 1.5	Si 1.8	P 2.1	S 2.5	Cl 3.0
K 0.8	Ca 1.0	Sc 1.3	Ti 1.5	V 1.6	Cr 1.6	Mn 1.5	Fe 1.8	Co 1.9	Ni 1.9	Cu 1.9	Zn 1.6	Ga 1.6	Ge 1.8	As 2.0	Se 2.4	Br 2.8
Rb 0.8	Sr 1.0	Y 1.2	Zr 1.4	Nb 1.6	Mo 1.8	Tc 1.9	Ru 2.2	Rh 2.2	Pd 2.2	Ag 1.9	Cd 1.7	In 1.7	Sn 1.8	Sb 1.9	Te 2.1	I 2.5
Cs 0.7	Ba 0.9	La 1.0	Hf 1.3	Ta 1.5	W 1.7	Re 1.9	Os 2.2	Ir 2.2	Pt 2.2	Au 2.4	Hg 1.9	Tl 1.8	Pb 1.9	Bi 1.9	Po 2.0	At 2.2

Figure 8.10 Electronegativities of the elements.

ment in which an element is situated. For example, the electronegativity of chlorine in its bonding to phosphorus in PCl_3 is likely to be different from its value in the chlorate ion, ClO_3^-, in which it is in a much different bonding situation. Variations of this sort are not so large as to render the concept of electronegativity useless, but we must avoid placing too much reliance on precise values for electronegativities. As long as we remain aware of their limitations, electronegativity values provide a useful guide to polarities in chemical bonds.

Electronegativity and Bond Polarity

We can use the difference in electronegativity between two atoms to gauge the polarity of the bonding between them. Consider the three fluorine-containing compounds listed below:

Compound	F_2	HF	LiF
Electronegativity difference	0	1.8	3.0
Type of bond	Nonpolar	Polar covalent	Ionic

In F_2, the electrons are shared equally between the fluorine atoms, and the bond is nonpolar. In HF, the fluorine atom has a greater electronegativity than the hydrogen atom. Consequently, the sharing of electrons is unequal (the bond is polar), with the more electronegative fluorine attracting electron density off the hydrogen. We represent this situation in the following two ways:

$$\overset{\delta+}{\mathrm{H}}-\overset{\delta-}{\mathrm{F}} \quad \text{or} \quad \overset{+\!\!\longrightarrow}{\mathrm{H}-\mathrm{F}}$$

The $\delta+$ and $\delta-$ are meant to represent partial positive and negative charges, respectively. (The symbol δ is the Greek lowercase letter "delta.") The arrow represents the pull of electron density off the hydrogen by the fluorine, leaving the hydrogen with a partial positive charge; the head of the arrow points in the direction in which the electrons are attracted, toward the more electronegative atom. In LiF, the far greater electronegativity of fluorine as compared to lithium leads to the complete transfer of the valence electron of Li to F. This transfer results in the formation of Li^+ and F^- ions; the resultant bond is therefore ionic.

We shall consider the molecular consequences of bond polarities further in Chapter 9. For now, remember: *The greater the difference in electronegativity, the more polar the bond.*

SAMPLE EXERCISE 8.11

Which of the following bonds is more polar: **(a)** B—Cl or C—Cl; **(b)** P—F or P—Cl? Indicate in each case which atom has the partial negative charge.

Solution: **(a)** The difference in the electronegativities of boron and chlorine is $3.0 - 2.0 = 1.0$; the difference between carbon and chlorine is $3.0 - 2.5 = 0.5$. Consequently, the B—Cl bond is the more polar; the chlorine atom carries the partial negative charge because it has a higher electronegativity. We should be able to reach this same conclusion without using a table of electronegativities; instead, we can rely on periodic trends. Because boron is to the left of carbon in the periodic table, we would predict that it has a lower attraction for electrons. Chlorine, being on the right side of the table, has a strong attraction for electrons. The most polar bond will be the one between the atoms having the lowest attraction for electrons (boron) and the highest attraction (chlorine).

(b) Because fluorine is above chlorine in the periodic table, we would predict it to be more electronegative. Consequently, the P—F bond will be more polar than the P—Cl bond. You should compare the electronegativity differences for the two bonds to verify this prediction. The fluorine atom carries the partial negative charge.

PRACTICE EXERCISE

Which of the following bonds is most polar: S—Cl, S—Br, Se—Cl, or Se—Br?
Answer: Se—Cl

8.10 OXIDATION NUMBERS

In light of our discussion of polar covalent bonds, it is useful to examine the similarities between reactions in which electrons are completely transferred and those in which only partial shifts of electron density occur. Recall that in Section 4.6 we discussed *oxidation* and *reduction*. In oxidation electrons are lost, and in reduction electrons are gained. Thus, when sodium and chlorine react to form NaCl (Figure 8.3), Na atoms are oxidized to Na^+ ions and Cl atoms are reduced to Cl^- ions.

The definitions of oxidation and reduction in terms of the loss and gain of electrons apply readily to the formation of ionic compounds, because electrons are transferred completely. As we shall see, the concepts of oxidation and reduction are also useful in treating the reactions of molecular compounds.

Consider the reaction of hydrogen atoms and chlorine atoms to form HCl, which is a gaseous substance at room temperature. Because ionic compounds are solids at room temperature, we see that HCl is not ionic. The bonding in HCl is best described as polar covalent. Because chlorine is more electronegative than hydrogen, the electrons in the H—Cl bond are displaced toward chlorine:

$$\mathrm{H\cdot + \cdot\underset{\cdot\cdot}{\ddot{Cl}}: \longrightarrow \overset{+\!\!\longrightarrow}{H:\underset{\cdot\cdot}{\ddot{Cl}}:}}$$

The H atom therefore carries a somewhat positive charge, and the Cl atom a somewhat negative one.

In keeping track of electrons, it is a reasonable simplification to assign the shared electrons to the more electronegative Cl atom:

$$\text{H} \quad [\,:\ddot{\underset{..}{\text{Cl}}}:$$

This procedure gives Cl eight valence-shell electrons, one more than the neutral atom. We have in effect given a 1− charge to the chlorine. Hydrogen, stripped of its electron, is assigned a charge of 1+.

Charges assigned in this fashion are called **oxidation numbers** or **oxidation states**. The oxidation number of an atom is the charge that results when the electrons in a covalent bond are assigned to the more electronegative atom; it is the charge an atom would possess *if* the bonding were ionic. In HCl, the oxidation number of H is +1 and that of Cl is −1. (In writing oxidation numbers, we will write the sign before the number to distinguish them from actual electronic charges, which we write with the number first.)

Although we can determine oxidation numbers for atoms using Lewis structures and electronegativities as we have done for HCl, we seldom use this procedure. It is generally easier to determine oxidation numbers using the following set of rules.

1. *The oxidation number of an element in its elemental form is zero.* For an isolated atom—like an Na atom, where there is no bonding and no net charge—the oxidation state must be 0. It is also zero for any elemental substance in which there is bonding. For example, in Na metal, N_2, Cl_2, and P_4 the bonding electrons are shared equally between identical atoms, giving each atom an oxidation state of 0.
2. *The oxidation number of a monoatomic ion is the same as its charge.* For example, the oxidation number of sodium in Na^+ is +1 and that of sulfur in S^{2-} is −2.
3. *In binary compounds (those with two different elements), the element with greater electronegativity is assigned a negative oxidation number equal to its charge in simple ionic compounds of the element.* For example, consider the oxidation state of Cl in PCl_3. Cl is more electronegative than P. In its simple ionic compounds, chlorine appears as the chloride ion, Cl^-. Thus, in PCl_3, Cl is assigned an oxidation number of −1.
4. *The sum of the oxidation numbers equals zero for an electrically neutral compound and equals the overall charge for an ionic species.* For example, PCl_3 is a neutral molecule. Thus, the sum of the oxidation numbers of the P and Cl atoms must equal zero. Because the oxidation number of each Cl in this compound is −1 (rule 3), the oxidation number of P must be +3. In like manner, the sum of the oxidation numbers of C and O in CO_3^{2-} must equal −2. The oxidation number of O in this ion is −2, because O is more electronegative than C and −2 is the charge on the oxide ion (rule 3). Therefore, the oxidation number on C must be +4, because $+4 + 3(-2) = -2$.

The periodic table provides us with many additional guidelines for assigning oxidation numbers. As shown in Figure 8.11, oxidation numbers exhibit periodic trends. Some observations are particularly helpful. The alkali metals (group 1A) exhibit only the oxidation state of +1 in their compounds. The

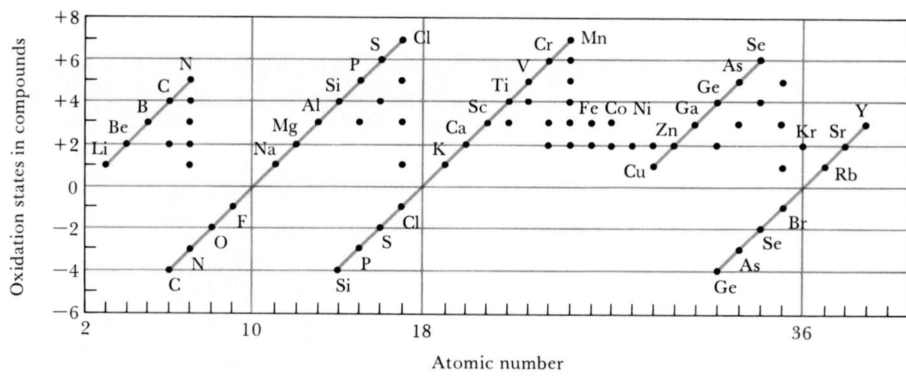

Figure 8.11 Common nonzero oxidation numbers for elements with atomic numbers 3 through 39. Notice that the maximum and minimum oxidation states (through which lines have been drawn for emphasis) are a periodic function of atomic number.

alkaline earth metals (group 2A) are always found in compounds in the +2 oxidation state. The most commonly encountered element in group 3A, Al, is always found in the +3 oxidation state.

The most electronegative element, F, is always found in compounds in the −1 oxidation state. Oxygen in compounds is nearly always in the −2 oxidation state. The only common exception to this general rule occurs in peroxides. In the peroxide ion, O_2^{2-}, and in molecular peroxides, such as H_2O_2, oxygen has an oxidation number of −1. Hydrogen has an oxidation number of +1 when it is bonded to a more electronegative element (most nonmetals) and −1 when bonded to a less electronegative element (most metals).

SAMPLE EXERCISE 8.12

Determine the oxidation state of sulfur in each of the following molecules or ions: **(a)** H_2S; **(b)** S_8; **(c)** SCl_2; **(d)** Na_2SO_3; **(e)** SO_4^{2-}.

Solution: **(a)** This is a binary (two-element) compound. Because S is more electronegative than H, S must have a negative oxidation number equal to its ionic charge (rule 3). Thus, S has an oxidation number of −2. This is consistent with the oxidation number of +1 expected for H.

(b) Because this is an elemental form of sulfur, the oxidation state of S is 0 (rule 1).

(c) Chlorine is more electronegative than sulfur. Therefore, the Cl in this binary compound must have a negative oxidation number equal to its ionic charge (rule 3), which is −1. The sum of the oxidation numbers must equal zero (rule 4). Letting x equal the oxidation number of S, we have $x + 2(-1) = 0$. Consequently, the oxidation state of S, x, must be +2.

(d) Sodium, an alkali metal, is always found in compounds in the +1 oxidation state. Oxygen has a common oxidation state of −2. If we let x equal the oxidation number of S, we have $2(+1) + x + 3(-2) = 0$. Therefore, the oxidation number of S in this compound is +4.

(e) The oxidation state of O is −2. The sum of the oxidation numbers equals −2, the net charge of the SO_4^{2-} ion. Letting x equal the oxidation number of S, we have $x + 4(-2) = -2$. From this relation we conclude that the oxidation state of S is +6.

These examples illustrate the range of oxidation states exhibited by sulfur, from −2 to +6. In general, the most negative oxidation state of a nonmetal corresponds to the number of electrons that must be added to the atom to give it an octet. In this case, S, which belongs to periodic family 6A, has six valence-shell electrons. Thus, two more are needed to give it an octet, as in the S^{2-} ion. The most positive oxidation state, in this case +6, corresponds to loss of all valence-shell electrons.

PRACTICE EXERCISE

What is the oxidation state of the underlined element in each of the following: **(a)** $\underline{P}_2O_5$; **(b)** $Na\underline{H}$; **(c)** $\underline{Cr}_2O_7^{2-}$; **(d)** $\underline{Sn}Br_4$; **(e)** $Ba\underline{O}_2$? ***Answers:*** **(a)** +5; **(b)** −1; **(c)** +6; **(d)** +4; **(e)** −1

A CLOSER LOOK: Binary Oxides

The concepts discussed in this chapter help us organize and understand many important chemical observations. To illustrate this point, let's consider the binary compounds of oxygen, which we treated briefly in Sections 7.6 and 7.8.

With an electronegativity of 3.5, oxygen is more electronegative than all other elements except fluorine. Because metals tend to have low electronegativities, oxides of metals in low oxidation states (especially +1 and +2) are usually ionic solids with large lattice energies (Table 8.2). As a result, these metal oxides typically have high melting points (Table 8.6). Many metal oxides are therefore used as heat-resistant materials, such as firebrick and ceramics. Many other applications of metal oxides, such as the use of Y_2O_3 in fluorescent lights (Figure 8.12), also depend on the high thermal stability of these materials.

Because they contain the O^{2-} ion, ionic metal oxides are basic (Section 7.6). For example, CaO, also known as lime, is an important commerical base: It reacts with water to form $Ca(OH)_2$. Similarly, Na_2O dissolves in water to form NaOH. In keeping with their high lattice energies, most metal oxides are rather insoluble in water.

The electronegativity of most nonmetals is sufficiently high that nonmetal oxides are not ionic. Rather, these elements form molecular oxides that contain polar covalent bonds with oxygen. These substances are typically gases, liquids, or solids with low melting points. Nonmetal oxides, especially those in which the nonmetal is in a high oxidation state, are generally acidic (Section 7.6). For example, chlorine(VII) oxide, Cl_2O_7, dissolves in water to form perchloric acid, $HClO_4$:

$$Cl_2O_7(l) + H_2O(l) \longrightarrow 2HClO_4(aq)$$

Notice that Cl has an oxidation state of +7 in both Cl_2O_7 and $HClO_4$. When two nonmetals are in the same oxidation state, the acidity of the oxide increases with increasing electronegativity of the element. For example, in both SO_2 and SeO_2 the oxidation number of the central atom is +4. Because S is more electronegative than Se, SO_2 is more acidic than SeO_2.

Table 8.6 Melting Points of Some Metal Oxides

Oxide	Melting point (°C)
Na_2O	1132
TiO_2	1857
Al_2O_3	2045
BeO	2530
CaO	2610
MgO	2826

Figure 8.12 Yttrium oxide, Y_2O_3, is a white solid. Here we see the red fluorescence of a sample of yttrium oxide caused by the UV lamp seen in the background. This oxide material is used to coat the walls of fluorescent lamps and TV picture tubes to enhance the red light they produce. This use of Y_2O_3 relies on its thermal stability at high operating temperature. (Courtesy of GTE Products Corporation)

In general, the acidity of oxides increases with increasing oxidation state of the element. A fascinating example of this trend is seen in the oxides of the transition metals. For example, manganese(II) oxide, MnO, is an ionic solid with a high melting point that has the same structure as NaCl. Although insoluble in water, MnO is strongly basic. It dissolves in acid, neutralizing the acid and giving Mn^{2+} ions:

$$MnO(s) + 2H^+(aq) \longrightarrow Mn^{2+}(aq) + H_2O(l)$$

In contrast, manganese(VII) oxide, Mn_2O_7 is a green liquid (melting point 5.9°C) that dissolves readily in water to form a strongly acidic solution.

Finally, there are certain oxides that exhibit *both* acidic and basic properties. These oxides, which are usually insoluble in water, are soluble in both acids and bases. They are called *amphoteric* (from the Greek *amphoteros,* meaning "each of two"). Most of the representative elements that form amphoteric oxides lie near the diagonal line in the periodic table that divides metals from nonmetals. For example, aluminum oxide, Al_2O_3, is amphoteric and reacts with both acids and bases:

$$Al_2O_3(s) + 6HCl(aq) \longrightarrow 2AlCl_3(aq) + 3H_2O(l)$$

$$Al_2O_3(s)) + 6NaOH(aq) \longrightarrow 2Na_3AlO_3(aq) + 3H_2O(l)$$

We shall consider amphoterism again in Chapter 17, when we discuss acids and bases and solubility in more detail.

Remember that the oxidation numbers, or oxidation states, do not correspond to real charges on the atoms, except in the special case of simple ionic substances. Nevertheless, they furnish a useful means of organizing chemical facts, especially in the case of metallic elements. Their most frequent use is in naming compounds, in balancing chemical equations for reactions in which changes in oxidation numbers occur, and in examining trends in chemical properties.

Oxidation Numbers and Nomenclature

The rules for naming simple inorganic compounds were discussed in Chapter 2. This is a good point at which to review those rules. Now that the concept of oxidation number has been explained, we can add another widely used method for naming binary compounds. You learned in Section 2.6 that the name of a binary compound consists of the name of the less electronegative element first, followed by the name of the more electronegative element, modified to have an *-ide* ending. The following examples are illustrative:

BaSe	barium selenide
ZnS	zinc sulfide
MgH_2	magnesium hydride

When one or the other of the elements involved has more than one possible oxidation state, the number of atoms may be included in the name as a Greek prefix (1 = *mono*-, 2 = *di*-, 3 = *tri*-, 4 = *tetra*-, 5 = *penta*-, 6 = *hexa*-, 7 = *hepta*-), *or* the oxidation state is indicated using a roman numeral, as in these examples:

MnO_2	manganese dioxide *or* manganese(IV) oxide
Mn_2O_3	dimaganese trioxide *or* manganese(III) oxide
P_2O_5	diphosphorus pentoxide *or* phosphorus(V) oxide
P_2O_3	diphosphorus trioxide *or* phosphorus(III) oxide
$SnCl_4$	tin tetrachloride *or* tin(IV) chloride
$SnCl_2$	tin dichloride *or* tin(II) chloride

The use of roman numerals to indicate the oxidation state of the element has the disadvantage that it gives only the simplest formula for the compound, and not the molecular formula. For example, the compounds NO_2 and N_2O_4 are both nitrogen(IV) oxide and are not distinguished by this name. However, the names nitrogen dioxide and dinitrogen tetroxide do distinguish them. This shortcoming is often unimportant because many compounds are referred to only by their simplest formula. As an example, phosphorus(V) oxide normally exists as P_4O_{10} molecules, but it is frequently represented by its simplest formula, P_2O_5.

SAMPLE EXERCISE 8.13

(a) Give the name of TiO_2. **(b)** What is the chemical formula of chromium(III) sulfide?

Solution: **(a)** One way to name the compound is to use the Greek prefix *di-* to indicate the presence of two O atoms: titanium dioxide. The compound can also be named using oxidation numbers. Because O has an oxidation number of −2, the oxidation number of Ti must be +4 [that is, when $x + 2(-2) = 0$, x must be +4]. Therefore, the compound can be called titanium(IV) oxide, the roman numeral indicating the oxidation state of Ti.

(b) Chromium(III) indicates that Cr has an oxidation state of +3. The sulfide ion is S^{2-}. It takes two Cr^{3+} ions to balance the charge of three S^{2-} ions to give a neutral compound, Cr_2S_3.

PRACTICE EXERCISE

(a) Name the chemical substance, Cl_2O_7. **(b)** Give the chemical formula for sulfur(IV) fluoride. ***Answers:*** **(a)** dichlorine heptoxide or chlorine(VII) oxide; **(b)** SF_4

FOR REVIEW

SUMMARY

In this chapter, we have dealt with the interactions that lead to the formation of chemical bonds. The tendencies of atoms to gain, lose, or share their valence electrons to form bonds can often be viewed in terms of attempts to achieve a noble-gas electron configuration (the octet rule).

Ionic bonding results from the complete transfer of electrons from one atom to another, with formation of a three-dimensional lattice of charged particles. The stabilities of ionic substances result from the powerful electrostatic attractive forces between an ion and all the surrounding ions of opposite charge. These interactions are measured by the lattice energy. We can calculate lattice energies by using a Born-Haber cycle. The magnitude of the lattice energy depends primarily on the charges and sizes of the ions. In general, lattice energies increase as the charges of the ions increase and as their sizes decrease.

Cations are smaller than their parent atoms; anions are larger than their parent atoms. For ions of the same charge, size increases going down a family. For an isoelectronic series, size decreases with increasing nuclear charge (atomic number). Not all ions have noble-gas configurations. Many transition metal ions do not. In forming transition-metal ions, the atom first loses its outer *s* electrons.

Covalent bonding results from the sharing of electrons. The octet rule is useful in describing this sharing. We can represent shared electron-pair structures of molecules by means of Lewis structures, which show the sharing of electron pairs between atoms. The sharing of one pair of electrons produces a single bond; the sharing of two or three pairs of electrons between atoms produces double and triple bonds, respectively.

Sometimes a single Lewis structure is inadequate to represent a particular molecule, but an average of two or more Lewis structures does form a satisfactory representation. In these cases, the Lewis structures are called resonance forms. Sometimes the octet rule is not obeyed; this situation occurs mainly when a large atom is surrounded by small, electronegative atoms like F, O, or Cl. In such instances, the large atom has unfilled *d* orbitals in its valence shell to accommodate more than an octet of electrons. Thus, expanded octets are observed for atoms in the third period and beyond in the periodic table.

The strength of a covalent bond is measured by its bond energy. The strengths of covalent bonds increase with the number of electron pairs shared between two atoms. We can use bond energies to estimate the enthalpy changes during chemical reactions.

It is important to recognize that even in covalent bonding, electrons may not be shared equally between two atoms. Electronegativity is a measure of the ability of an atom to compete with other atoms for the electrons shared between them. Highly electronegative elements strongly attract electrons. The electronegativities of the elements, which show a regular periodic relationship, are an important guide to chemical behavior. We shall be using the concept of electronegativity often throughout the text. The difference in electronegativities of bonded atoms is used to determine the polarity of a bond.

Another application of electronegativity is in the assignment of oxidation numbers, formal whole-number charges assigned to atoms in molecules and ions. Although the oxidation numbers do not represent the real charges on atoms except in simple ionic substances, they are of great value in helping us to organize chemical facts, to balance equations, and to name compounds.

KEY TERMS

ionic bond
covalent bond
metallic bond
valence (Sec. 8.1)
valence electrons (Sec. 8.1)
electron-dot symbols (Lewis symbols) (Sec. 8.1)
octet rule (Sec. 8.1)
Born-Haber cycle (Sec. 8.2)
lattice energy (Sec. 8.2)
isoelectronic series (Sec. 8.3)
covalent bond (Sec. 8.4)
Lewis structures (Sec. 8.4)
single bond (Sec. 8.4)
double bond (Sec. 8.4)
triple bond (Sec. 8.4)
multiple bonding (Sec. 8.4)
formal charge (Sec. 8.5)
resonance forms (Sec. 8.6)
bond-dissociation energy (Sec. 8.8)
bond energy (Sec. 8.8)
bond polarity (Sec. 8.9)
nonpolar bond (Sec. 8.9)
polar covalent bond (Sec. 8.9)
electronegativity (Sec. 8.9)
oxidation numbers (states) (Sec. 8.10)

EXERCISES

Valence Electrons; Lewis Symbols; Ionic Bonding

8.1 Write the Lewis symbol for each of the following elements: **(a)** phosphorus, P; **(b)** gallium, Ga; **(c)** silicon, Si; **(d)** helium, He.

8.2 What is the Lewis symbol for each of the following atoms or ions: **(a)** S; **(b)** I; **(c)** P^{3-}; **(d)** Ba^{2+}?

8.3 Use Lewis symbols to diagram the reaction that occurs between sodium and hydrogen atoms to give Na^+ and H^- ions.

8.4 Using Lewis symbols, diagram the reaction that occurs between Al and F atoms.

8.5 Predict the chemical formula of the ionic compound formed between the following pairs of elements: **(a)** Ca, Cl; **(b)** Y, O; **(c)** Sr, S; **(d)** Mg, N.

8.6 Indicate whether each of the following formulas is likely to represent a stable compound, and give an explanation for your answer: **(a)** Rb_2O; **(b)** BaF; **(c)** Mg_2O; **(d)** $ScBr_3$; **(e)** Na_3N.

8.7 Write the electron configuration for each of the following ions, and state which possess noble-gas configurations: **(a)** Ba^{2+}; **(b)** I^-; **(c)** Fe^{2+}; **(d)** Zr^{4+}; **(e)** Au^{3+}; **(f)** Se^{2-}.

8.8 Write the electron configuration for each of the following ions, and state which possess noble-gas configurations: **(a)** As^{3-}; **(b)** Ag^+; **(c)** La^{3+}; **(d)** Mn^{2+}; **(e)** Cl^+; **(f)** In^+.

8.9 Energy is required to remove two electrons from Ca to form Ca^{2+} and is also required to add two electrons to O to form O^{2-}. Why, then, is CaO stable relative to the free elements?

8.10 List the individual steps used in constructing a Born-Haber cycle for the formation of $CaBr_2$ from the elements. Which of these steps would you expect to be exothermic?

8.11 Which factors govern the magnitude of the lattice energy of an ionic compound?

8.12 Explain the following trends in lattice energy: **(a)** CaS > KCl; **(b)** LiF > CsBr; **(c)** MgO > MgS; **(d)** MgO > BaO.

8.13 Use data from Appendix C, Table 7.3, and Table 7.5 to calculate the lattice energy of RbCl. Is this value greater than or less than the lattice energy of NaCl? Explain.

8.14 By using data from Appendix C, Table 7.3, and Table 7.7, calculate the lattice energy of CaF_2. Is this value greater than or less than the lattice energy of NaF? Explain.

8.15 The lattice energy of LiH is 858 kJ/mol. Use this value along with data from Appendix C, Table 7.3, and Table 7.5 to calculate ΔH_f° for LiH.

8.16 The lattice energy of MgH_2 is 2790 kJ/mol. Use this value along with data from Appendix C, Table 7.2, and Table 7.3 to calculate ΔH_f° for MgH_2.

Sizes of Ions

8.17 **(a)** Why are monoatomic anions larger than the corresponding neutral atoms? **(b)** Why are monoatomic cations smaller than the corresponding neutral atoms? **(c)** Why does the size of ions increase as we proceed down a family in the periodic table? **(d)** Why does the size of isoelectronic ions decrease with increasing nuclear charge?

8.18 Explain the following variations in atomic or ionic radii: **(a)** $I^- > I > I^+$; **(b)** $Ca^{2+} > Mg^{2+} > Be^{2+}$; **(c)** $Br^- >$ Kr > Rb^+; **(d)** $N^{3-} > O^{2-} > F^-$.

8.19 Arrange the atoms and ions in each of the following sets in order of increasing size: **(a)** Li^+, Rb^+, K^+; **(b)** Br^-, Na^+, Mg^{2+}; **(c)** Ar, Cl^-, S^{2-}, K^+; **(d)** Cl, Cl^-, Ar.

8.20 For each of the following sets of atoms and ions, arrange the members in order of increasing size: **(a)** Se^{2-}, Te^{2-}, Se; **(b)** Co^{3+}, Fe^{2+}, Fe^{3+}; **(c)** Ca, Ti^{4+}, Sc^{3+}; **(d)** Be^{2+}, Na^+, Ne.

8.21 Based on the data in Figure 8.6, how would you compare the effective nuclear charge experienced by a 3*p* electron in K with that of a 3*p* electron in K^+? Explain.

8.22 Compare the effective nuclear charge experienced by a $2p$ electron in Na^+ with that experienced by a $2p$ electron in F^-.

8.23 Which neutral atom is isoelectronic with each of the following ions: **(a)** Cl^-; **(b)** I^+; **(c)** Ba^{2+}; **(d)** Si^{2-}?

8.24 Select the ions or atoms from each of the following sets that are isoelectronic with each other: **(a)** K^+, Rb^+, Ca^{2+}; **(b)** Cu^+, Ca^{2+}, Sc^{3+}; **(c)** S^{2-}, Se^{2-}, Ar; **(d)** Fe^{2+}, Co^{3+}, Mn^{2+}.

Lewis Structures; Resonance Forms

8.25 Draw the Lewis structures for **(a)** SiH_4; **(b)** ClO_2^-; **(c)** CO; **(d)** $HBrO_3$ (H is bonded to O); **(e)** $TeCl_2$.

8.26 Write the Lewis structures for **(a)** H_2O_2 (the O atoms are bonded to each other); **(b)** CN^-; **(c)** BH_4^-; **(d)** HOCl; **(e)** ONCl.

8.27 Draw the Lewis structures for each of the following compounds. Identify those that do not obey the octet rule, and explain why they do not. **(a)** NO_2; **(b)** GeF_4; **(c)** TeF_4; **(d)** BCl_3; **(e)** XeF_4.

8.28 Draw the Lewis structures for each of the following ions. Identify those that do not obey the octet rule, and explain why they do not. **(a)** SO_3^{2-}; **(b)** BH_3; **(c)** I_3^-; **(d)** AsF_6^-; **(e)** O_2^-.

8.29 Draw resonance structures for each of the following: **(a)** SO_3; **(b)** $C_2O_4^{2-}$ (each C atom is bonded to two O atoms and to the other C); **(c)** HNO_3 (H is bonded to O); **(d)** ClO_3.

8.30 Draw the resonance forms for the following: **(a)** NO_2^-; **(b)** CO_3^{2-}; **(c)** SCN^-; **(d)** HCO_2^- (H and both O atoms are bonded to C).

8.31 Based on their Lewis structures, predict the ordering of N—O bond lengths in NO^+, NO, NO_2^- and NO_3^-.

8.32 Predict the ordering of the S—O bond lengths in SO_2, SO_3, SO_3^{2-}, and SO_4^{2-}.

8.33 Calculate the formal charge on the indicated atom in each of the following molecules or ions: **(a)** the central oxygen atom in O_3; **(b)** phosphorus in PF_6^-; **(c)** nitrogen in NO_2; **(d)** iodine in ICl_3; **(e)** chlorine in $HClO_4$ (H is bonded to O).

8.34 Use the concept of formal charge to choose the more likely skeleton structure in each of the following cases: **(a)** NNO or NON; **(b)** HCN or HNC; **(c)** NOBr or ONBr; **(d)** NSC^- or SNC^- or SCN^-.

Bond Energies

8.35 Using the bond energies tabulated in Table 8.4, estimate ΔH for each of the following gas-phase reactions:

(a) $H_3C{-}O{-}H + H{-}I \longrightarrow H_3C{-}I + H{-}O{-}H$ (structural formulas: H—C(H)(H)—O—H + H—I ⟶ H—C(H)(H)—I + H—O—H)

(b) $H{-}CCl_3 + O \longrightarrow Cl{-}C({=}O){-}Cl + H{-}Cl$ (structural formulas: H—C(Cl)(Cl)—Cl + O ⟶ Cl—C(=O)—Cl + H—Cl)

(c) $C{\equiv}O + H{-}O{-}H \longrightarrow H{-}H + O{=}C{=}O$

8.36 Using bond energies, estimate ΔH for the following gas-phase reactions:

(a) $H_2C{=}CH_2 + H{-}O{-}O{-}H \longrightarrow H{-}O{-}CH_2{-}CH_2{-}O{-}H$

(b) $H_2C{=}CH_2 + H{-}C{\equiv}N \longrightarrow H{-}CH_2{-}CH_2{-}C{\equiv}N$

(c) $2Cl{-}N(Cl){-}Cl \longrightarrow N{\equiv}N + 3Cl{-}Cl$

8.37 Using the bond energies in Table 8.4, estimate the enthalpy change for each of the following gas-phase reactions:
(a) $HCN + 3H_2 \longrightarrow CH_4 + NH_3$
(b) $HBr + 2F_2 \longrightarrow BrF_3 + HF$

8.38 Using bond energies (Table 8.4), estimate ΔH for the following gas-phase reactions:
(a) $CO + 2H_2 \longrightarrow CH_3OH$
(b) $CH_2{=}CH_2 + F_2 \longrightarrow CH_2F{-}CH_2F$

8.39 Given the following bond dissociation energies, calculate the average bond energy for the Ti—Cl bond.

Reaction	ΔH (kJ/mol)
$TiCl_4(g) \longrightarrow TiCl_3(g) + Cl(g)$	335
$TiCl_3(g) \longrightarrow TiCl_2(g) + Cl(g)$	423
$TiCl_2(g) \longrightarrow TiCl(g) + Cl(g)$	444
$TiCl(g) \longrightarrow Ti(g) + Cl(g)$	519

[8.40] Using the ΔH_f° values in Appendix C for $C_2H_6(g)$, $C(g)$, and $H(g)$, and the average C—H bond energy from Table 8.4, calculate the C—C bond energy in ethane, C_2H_6. How does this value compare to the average C—C bond energy given in Table 8.4?

Electronegativity; Bond Polarities

8.41 **(a)** What is meant by the term *electronegativity?* **(b)** How does the electronegativity of an atom differ from its electron affinity? **(c)** Does the same element have the greatest electronegativity and the greatest (most negative) electron affinity?

8.42 **(a)** How does the electronegativity of the elements generally vary as you proceed from left to right in a row of the periodic table? **(b)** What is the general trend in electronegativity as you proceed down a family of the periodic table? **(c)** How do the periodic trends in electronegativity relate to those for ionization energy and electron affinity?

8.43 By referring only to a periodic table, arrange the members of each of the following sets in order of increasing electronegativity: **(a)** O, P, S; **(b)** Mg, Al, Si; **(c)** S, Cl, Br; **(d)** C, Si, N.

8.44 Using only a periodic table, select the most electronegative atom in each of the following sets: **(a)** Be, B, C, Si; **(b)** Cl, S, Br, Se; **(c)** Si, Ge, Al, Ga; **(d)** K, Ca, As, Se.

8.45 Which of the following bonds are polar: **(a)** B—Cl; **(b)** Cl—Cl; **(c)** Hg—Sb; **(d)** As—F; **(e)** C—Br? Which is the more electronegative atom in each polar bond?

8.46 Arrange the bonds in each of the following sets in order of increasing polarity: **(a)** C—S, B—F, N—O; **(b)** Pb—Cl, Pb—Pb, Pb—C; **(c)** H—F, O—F, Be—F.

Oxidation Numbers

8.47 Determine the oxidation number of the underlined element in each of the following: **(a)** $K\underline{Mn}O_4$; **(b)** $\underline{Fe}_2O_3$; **(c)** $\underline{Xe}OF_4$; **(d)** $\underline{Sn}Cl_4$; **(e)** $K_2\underline{O}_2$; **(f)** $\underline{N}O_2{}^-$; **(g)** $\underline{U}O_2{}^{2+}$; **(h)** $\underline{Cu}Cl_4{}^{2-}$.

8.48 Determine the oxidation number of the underlined element in each of the following: **(a)** $\underline{S}_8$; **(b)** $\underline{Hg}_2Cl_2$; **(c)** $Mg_2\underline{P}_2O_7$; **(d)** $\underline{Cl}O_4{}^-$; **(e)** $\underline{Cr}_2O_7{}^{2-}$; **(f)** $\underline{O}F_2$; **(g)** $\underline{N}O^+$; **(h)** $H\underline{As}O_4{}^{2-}$.

8.49 What are the maximum and minimum oxidation numbers exhibited by each of the following elements: **(a)** Br; **(b)** As; **(c)** Ba; **(d)** Cr?

8.50 What are the maximum and minimum oxidation numbers exhibited by each of the following elements: **(a)** Se; **(b)** Ge; **(c)** Al; **(d)** Mn?

8.51 Give the name or chemical formula, as appropriate, for each of the following substances: **(a)** iron(III) fluoride; **(b)** molybdenum(VI) oxide; **(c)** arsenic(V) bromide; **(d)** V_2O_3; **(e)** CoF_3; **(f)** MnS.

8.52 Give the name or chemical formula, as appropriate, for each of the following substances: **(a)** manganese(IV) oxide; **(b)** gallium(III) sulfide; **(c)** selenium(VI) fluoride; **(d)** Cu_2O; **(e)** ClF_3; **(f)** TeO_3.

8.53 Indicate the oxidation states of the elements that undergo a change in oxidation state in each of the following reactions:

(a) $Xe(g) + 2F_2(g) \longrightarrow XeF_4(g)$

(b) $2CuSO_4(aq) + 4KI(aq) \longrightarrow 2CuI(s) + 2K_2SO_4(aq) + I_2(s)$

(c) $NH_3(g) + 3Cl_2(g) \longrightarrow NCl_3(g) + 3HCl(g)$

(d) $I_2(aq) + SO_2(aq) + 2H_2O(l) \longrightarrow 2HI(aq) + H_2SO_4(aq)$

(e) $2PbS(s) + 3O_2(g) \longrightarrow 2PbO(s) + 2SO_2(g)$

8.54 Indicate the oxidation states of the elements that undergo a change in oxidation state in each of the following reactions:

(a) $2KI(aq) + F_2(g) \longrightarrow 2KF(aq) + I_2(s)$

(b) $MnO_2(s) + 4HCl(aq) \longrightarrow MnCl_2(aq) + Cl_2(aq) + 2H_2O(l)$

(c) $2H_2O_2(aq) \longrightarrow 2H_2O(l) + O_2(g)$

(d) $2CH_4(g) + O_2(g) \longrightarrow 2CH_3OH(g)$

(e) $5Fe^{2+}(aq) + MnO_4{}^-(aq) + 8H^+(aq) \longrightarrow 5Fe^{3+}(aq) + Mn^{2+}(aq) + 4H_2O(l)$

Additional Exercises

8.55 In each of the following examples of a Lewis symbol, indicate the group in the periodic table in which the element X belongs: **(a)** $\cdot\overset{\cdot}{\underset{\cdot}{X}}\cdot$; **(b)** $\cdot X \cdot$; **(c)** $:\overset{\cdot}{\underset{\cdot}{X}}\cdot$.

8.56 Which of the following contain metal ions that do not have noble-gas electron configurations: **(a)** CuCl; **(b)** CdO; **(c)** TiO_2; **(d)** ScF_3.

8.57 Write the electron configuration for each of the following ions: **(a)** Cu^{2+}; **(b)** Pb^{2+}; **(c)** Co^{3+}; **(d)** Sc^{3+}.

8.58 The $+2$ oxidation state is common for transition-metal ions. Explain why this might be expected.

8.59 Describe electron configurations other than a completed octet in the valence shell that are relatively stable arrangements often found in ions.

8.60 Explain the following trend in lattice energies: LiH, 858 kJ/mol; NaH, 782 kJ/mol; KH, 699 kJ/mol; RbH, 674 kJ/mol.

8.61 From the ionic radii given in Table 8.3, calculate the potential energy of each of the following ion pairs, assuming that they are separated by the sum of their ionic radii (the magnitude of the electronic charge is given on the back inside cover): **(a)** Mg^{2+}, O^{2-}; **(b)** Na^+, Br^-.

[8.62] From the ionic radii given in Figure 8.6, calculate the potential energy of a Na^+ and Cl^- ion pair that are just touching (the magnitude of the electronic charge is given on the back inside cover). Calculate the energy of a mole of such pairs. How does this value compare with the lattice energy of NaCl? Explain the difference.

[8.63] The standard enthalpy of formation for Sr(*g*) is 164.4 kJ/mol. Using this value, along with data from Appendix C, Table 7.3, Table 7.7, and Table 8.2, calculate ΔH_f° for $SrCl_2$.

[8.64] The electron affinity of oxygen is -141 kJ/mol, corresponding to the reaction

$$O(g) + e^- \longrightarrow O^-(g)$$

The lattice energy of SrO is 3369 kJ/mol. Using these data, along with the data referred to in Problem 8.63, calculate the "second electron affinity" of oxygen, corresponding to the reaction

$$O^-(g) + e^- \longrightarrow O^{2-}(g)$$

8.65 Use the octet rule to predict the formula of the simplest compound formed between **(a)** hydrogen and silicon; **(b)** sulfur and fluorine; **(c)** phosphorus and chlorine.

8.66 Although $I_3{}^-$ is known, $F_3{}^-$ is not. Using Lewis structures, explain why $F_3{}^-$ does not form.

8.67 Ammonia is produced directly from nitrogen and hydrogen by using the *Haber process*. The chemical reaction is

$$N_2(g) + 3H_2(g) \longrightarrow 2NH_3(g)$$

(a) Use bond energies to estimate whether this reaction is exothermic or endothermic. **(b)** Compare the enthalpy change you calculate in (a) to the true enthalpy change as obtained using ΔH_f° values.

8.68 An important reaction for the conversion of natural gas to other useful hydrocarbons is the conversion of methane to ethane:

$$2CH_4(g) \longrightarrow C_2H_6(g) + H_2(g)$$

In practice, this reaction is carried out in the presence of oxygen, which converts the hydrogen produced to water:

$$2CH_4(g) + \tfrac{1}{2}O_2(g) \longrightarrow C_2H_6(g) + H_2O(g)$$

Use bond energies to estimate ΔH for these two reactions. Why is the conversion of methane to ethane more favorable when oxygen is used?

8.69 Use bond energies (Table 8.4), electron affinities (Table 7.3), and the ionization energy of hydrogen (1312 kJ/mol) to estimate ΔH for the following reactions:

(a) $HF(g) \longrightarrow H^+(g) + F^-(g)$
(b) $HCl(g) \longrightarrow H^+(g) + Cl^-(g)$
(c) $HBr(g) \longrightarrow H^+(g) + Br^-(g)$

[8.70] The hydrogen molecule can, in principle, break apart into either neutral atoms or ions, as shown in the following reactions:

$$H_2(g) \longrightarrow H(g) + H(g)$$
$$H_2(g) \longrightarrow H^+(g) + H^-(g)$$

The bond energy of an H—H bond is defined by the first reaction. By using the data referred to in Problem 8.69, determine ΔH for the second reaction. Based on a comparison of the enthalpy changes for these reactions, which should be the preferred means by which hydrogen molecules break apart?

[8.71] The enthalpy of formation of $NH(g)$ at 25°C is 360 kJ/mol. From this value, and using the data in Table 8.4, calculate the N—H bond dissociation energy in NH.

8.72 Two compounds are *isomers* if they have the same chemical formula but a different arrangement of atoms. Use bond energies to estimate ΔH for each of the following isomerization reactions and indicate which isomer has the lower enthalpy:

(a) CH_3CH_2OH (H—C(H)(H)—C(H)(H)—O—H) $\longrightarrow$ CH_3OCH_3 (H—C(H)(H)—O—C(H)(H)—H)

Ethanol   Dimethyl ether

(b) Ethylene oxide (three-membered ring: O bonded to both carbons; H—C(H)—C(H)—H) $\longrightarrow$ Acetaldehyde (H—C(H)(H)—C(=O)—H)

Ethylene oxide   Acetaldehyde

(c) Cyclopentene (five-membered ring of carbons, C=C, with H atoms) $\longrightarrow$ Pentadiene (H_2C=C(H)—C(H)=C(H)—C(H)(H)—H)

Cyclopentene   Pentadiene

(d) H—C(H)(H)—N≡C $\longrightarrow$ H—C(H)(H)—C≡N

Methyl isocyanide   Acetonitrile

[8.73] With reference to the Chemistry at Work box on explosives: **(a)** Use bond energies to estimate the enthalpy change for the explosion of 1.00 g of nitroglycerin. **(b)** Write a balanced equation for the decomposition of TNT. Assume that, upon explosion, TNT decomposes into $N_2(g)$, $CO_2(g)$, $H_2O(g)$, and $C(s)$.

8.74 The two sulfur-oxygen bonds in sulfur dioxide, SO_2, are equal in length and short (1.43 Å). **(a)** Write a Lewis structure for SO_2 that satisfies the octet rule. Would you expect more than one resonance form for this Lewis structure? **(b)** Write a Lewis structure in which there is an expanded valence shell for sulfur. Would you expect more than one resonance form for this Lewis structure? **(c)** For the Lewis structures in (a) and (b), determine the formal charge for the atoms in SO_2. **(d)** If the criterion of minimizing the formal charges is used, would the Lewis structure from (a) or (b) be a better one for SO_2? [After you cover Chapter 9, you might want to read more about the Lewis structures of SO_2. See: G. H. Purser, *J. Chem. Ed.* **(1989)**, 66, 710–713.]

8.75 Sample Exercise 8.8 gives several different Lewis structures for ClO_2. On the basis of formal charge, which of these is most important?

8.76 Use the concepts of formal charge and electronegativity to explain why the best Lewis structure of BF_3 is the one with less than an octet around boron.

8.77 Based on the positions of the elements in the periodic table, select the most polar and least polar bond from the following list: P—N, P—O, P—P, P—S.

8.78 Assign oxidation states to all atoms in each of the following compounds: **(a)** N_2O; **(b)** $KBiO_3$; **(c)** ClF_3; **(d)** PBr_3; **(e)** $HAsO_2$; **(f)** N_2H_4; **(g)** $Na_2S_2O_3$; **(h)** $(NH_4)_2SO_4$.

8.79 A white solid, melting at 1115°C and insoluble in water but slightly soluble in aqueous NaOH, is likely to be which of the following: SrO; GeO_2; SeO_2; N_2O_3? Explain.

[8.80] The bond lengths of carbon-carbon, carbon-nitrogen, carbon-oxygen, and nitrogen-nitrogen single, double, and triple bonds are listed in Table 8.5. Plot bond energy (Table 8.4) versus bond length for these bonds. What do you conclude about the relationship between bond length and bond energy? What do you conclude about the relative strengths of C—C, C—N, C—O, and N—N bonds?

[8.81] An alternative definition of the electronegativity of an atom is the first ionization energy minus the electron affinity ($I_1 - E$). Using the data given in Problem 8.69, Table 7.3, and Table 7.9, determine the electronegativities of hydrogen and the halogens according to this definition. Plot these values versus the Pauling electronegativity values given in Figure 8.10. What do you conclude about the correspondence of the two definitions for these elements?

9 Molecular Geometry and Bonding Theories

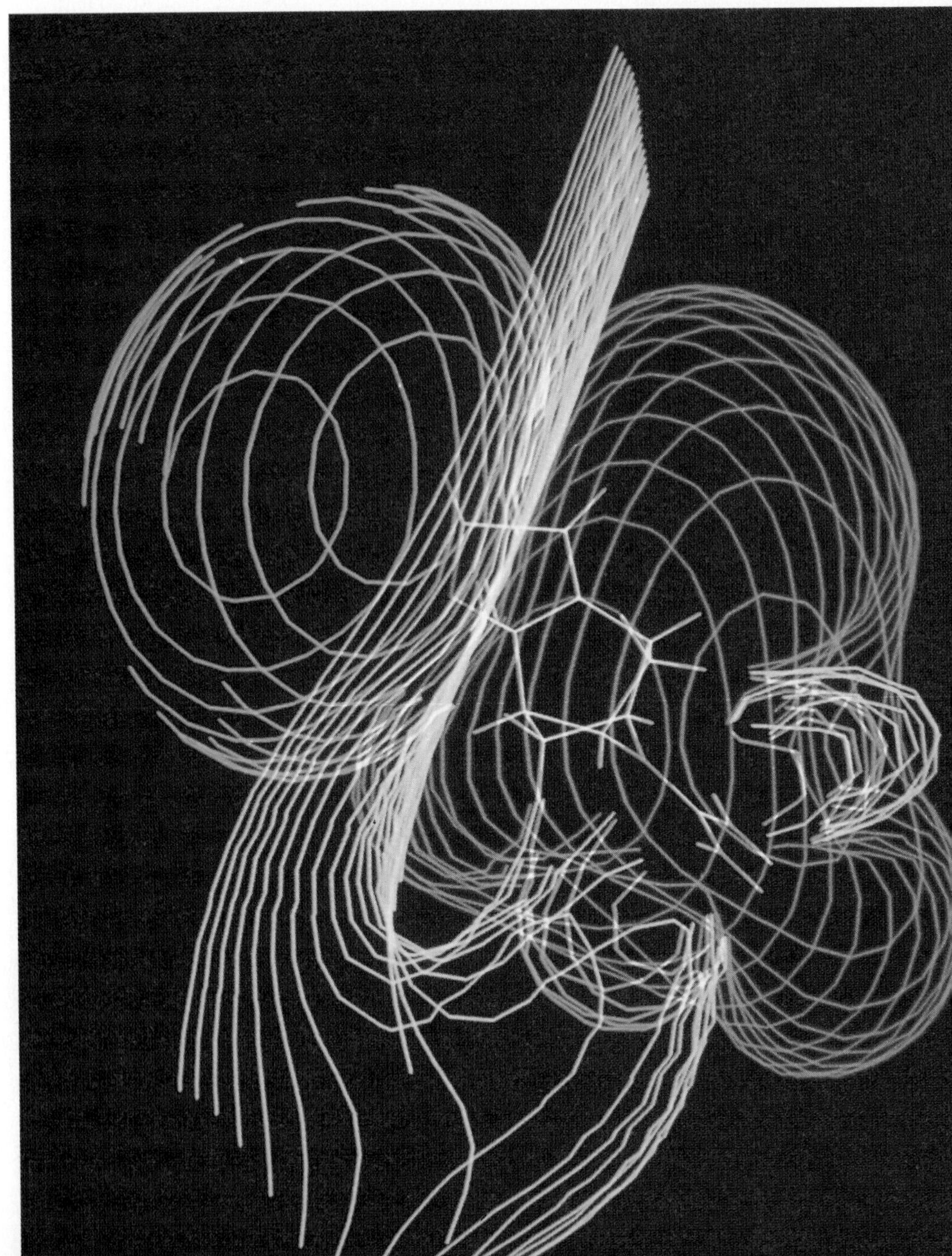

Computer graphical rendition of the charge distribution in an organic molecule. The blue lobes represent negatively charged regions, the red lobes represent positively charged regions, and the yellow surface is electrically neutral. (Courtesy of Tripos Associates, St. Louis, Missouri)

When you smell the aroma of freshly baked bread, molecules from the bread stimulate receptor sites in the oral cavity of your nose. The shapes and sizes of the molecules determine their ability to fit properly on these receptor sites. The interactions between the molecules and the receptors sensitize nerve endings that transmit impulses to the brain. The brain then identifies these impulses as a particular aroma. The nose is so good at molecular recognition that two substances may produce different sensations of odor even when their molecular structures differ as subtly as your right hand differs from your left.

The size and shape of a molecule of a particular substance, together with the strength and polarity of its bonds, largely determine the physical and chemical properties of that substance. Some of the most dramatic examples of the importance of size and shape are seen in biochemical reactions. For example, a small change in the size or shape of a drug molecule may enhance its effectiveness or reduce its side effects.

In this chapter, we shall consider how the shapes of simple molecules can be described and predicted. We shall also examine why molecules have the shapes they do and look more closely at how atoms use their electrons to form bonds.

CONTENTS

9.1 MOLECULAR GEOMETRIES

In Chapter 8, we used Lewis structures to account for the formulas of covalent compounds. Two-dimensional Lewis structures do not indicate the shapes of molecules. Rather, they simply indicate the presence of electron-pair bonds between atoms. For example, the Lewis structure of carbon tetrachloride tells us only that four Cl atoms are bonded to a central C atom by four single bonds:

```
      ..
     :Cl:
  ..  |  ..
 :Cl—C—Cl:
  ..  |  ..
     :Cl:
      ..
```

The Lewis structure is drawn with the atoms in the same plane. However, as shown in Figure 9.1, the actual three-dimensional arrangement of the atoms has the Cl atoms at the vertices of a *tetrahedron*, a solid object with four corners and four faces, each of which is an equilateral triangle.

The overall shape of a molecule is determined by its **bond angles**, the angles made by the lines joining the nuclei of the atoms in the molecule. The bond angles of a molecule, together with the bond lengths (Section 8.8), precisely define the size and shape of the molecule. In CCl_4, all four C—Cl bonds are the same length (1.78 Å), and all six Cl—C—Cl angles have the same value

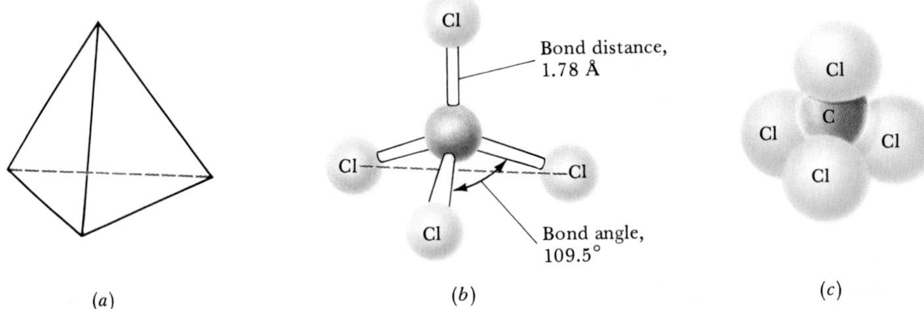

Figure 9.1 (*a*) A tetrahedron is an object with four faces and four vertices. Each face is an equilateral triangle. (*b*) The geometry of the CCl_4 molecule. Each C—Cl bond in the molecule points toward a vertex of a tetrahedron. All the C—Cl bonds are the same length, and all Cl—C—Cl bond angles are the same. This type of drawing of CCl_4 is called a "ball-and-stick" model. (*c*) A representation of CCl_4 called a "space-filling" model. It shows the relative sizes of the atoms, but the geometry is somewhat harder to see.

(109.5°, which is characteristic of a regular tetrahedron). Thus, the size and shape of CCl_4 is completely described by stating that it is tetrahedral with C—Cl bonds of length 1.78 Å.

In describing molecular geometries, we shall first consider molecules that contain two or more B atoms bonded to a central A atom. We denote these as AB_n molecules, where n is the number of B atoms bonded to the central atom. Thus, CCl_4 is an example of an AB_4 molecule. Although an enormous number of different AB_n molecules exist, the common shapes exhibited by them is rather limited. For example, the shape of an AB_2 molecule must be either linear (bond angle = 180°) or bent (bond angle ≠ 180°). Figure 9.2 shows these and several other common molecular geometries. We see that several different geometries can be exhibited by, for example, AB_3 molecules, and it is natural for us to ask

Figure 9.2 The geometries of some simple molecules.

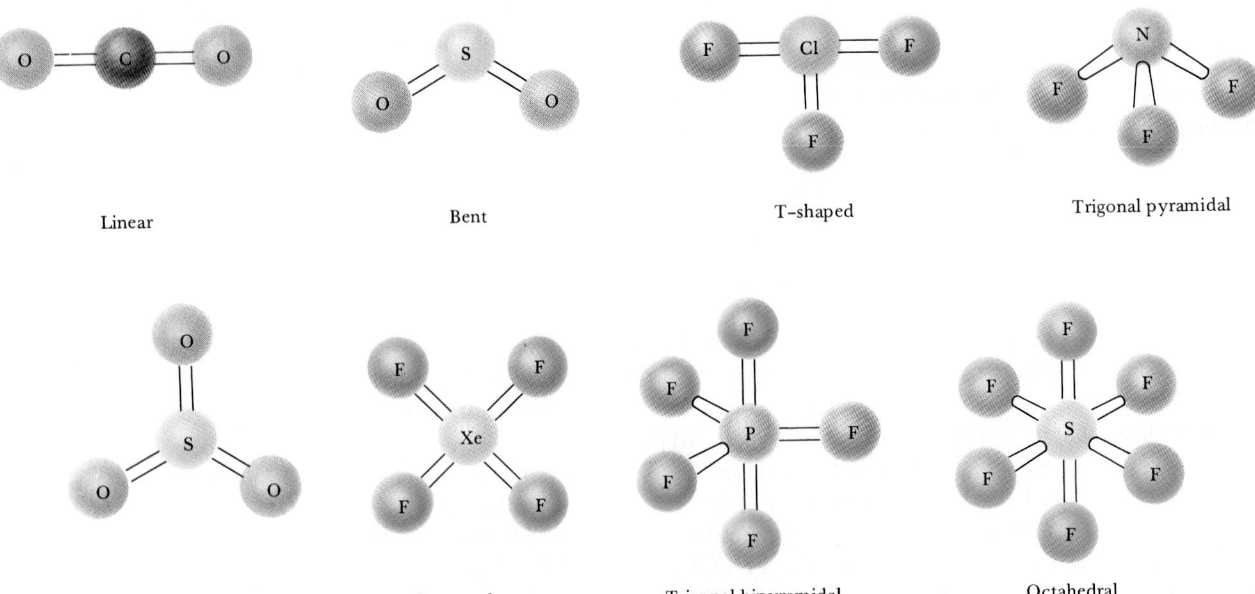

whether the geometry of a given AB_n molecule can be predicted. The answer is that it can be, at least when A is a nonmetal. To do so we use the **valence-shell electron-pair repulsion (VSEPR) model**. Although the name is rather imposing, the model is quite simple.

The Valence-Shell Electron Pair Repulsion Model

Imagine taking two identical balloons and tying them together at their ends. As shown in Figure 9.3(*a*), the balloons naturally orient themselves so that each is pointing away from the other. They do so in order to minimize the interactions between themselves; that is, they try to "get out of each other's way" as much as possible. If we add a third balloon, the balloons orient themselves toward the vertices of an equilateral triangle [Figure 9.3(*b*)]. If we add a fourth balloon, the balloons naturally adopt a tetrahedral shape, as shown in Figure 9.3(*c*). We see that there is an optimum geometry for each number of balloons.

We have seen that atoms are bonded to each other in molecules by the sharing of pairs of valence-shell electrons. Electron pairs repel one another. Therefore, like the balloons in Figure 9.3, electron pairs try to stay out of each other's way. *The best arrangement of a given number of electron pairs is the one that minimizes the repulsions among them.* This simple idea is the basis of the VSEPR model. In fact, the analogy between electron pairs and balloons is so close that the same preferred geometries are found in both cases.

Let's consider the case of two electron pairs around an atom. The repulsion between the electron pairs is at a minimum when they are exactly opposite one another. This leads to an angle of 180° between the electron pairs. For example, the BeF_2 molecule has the following Lewis structure:

$$:\ddot{\underset{\cdot\cdot}{F}}—Be—\ddot{\underset{\cdot\cdot}{F}}:$$

There are two electron pairs around the beryllium atom, which leads us to predict (and indeed observe) a linear arrangement of the atoms. Table 9.1 summarizes the best (minimum-repulsion) arrangements of electron pairs around a central atom for up to six electron pairs. It also gives the angles between the electron pairs in each arrangement and the name associated with the geometry. You should familiarize yourself with these arrangements.

Figure 9.3 Balloons tied together at their ends naturally adopt their lowest-energy arrangement. (*a*) Two balloons adopt a linear arrangement. (*b*) Three balloons adopt a trigonal planar arrangement. (*c*) Four balloons adopt a tetrahedral arrangement. (Kristen Brochmann/Fundamental Photographs)

(*a*)

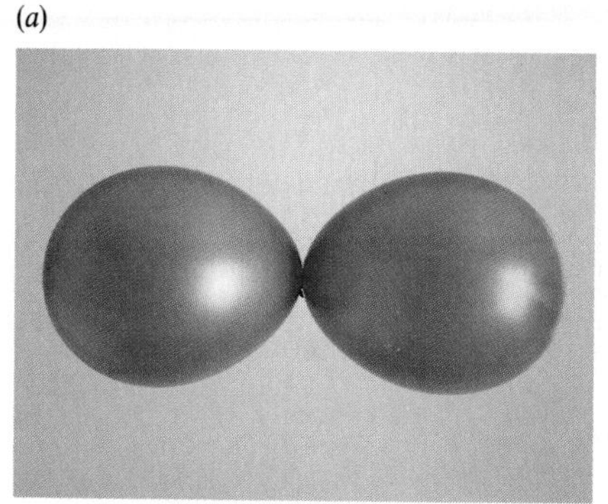

(*b*)

(*c*)

Table 9.1 Electron-Pair Geometries as a Function of the Number of Electron Pairs

Number of electron pairs	Arrangement of electron pairs	Electron-pair geometry	Predicted bond angles
2	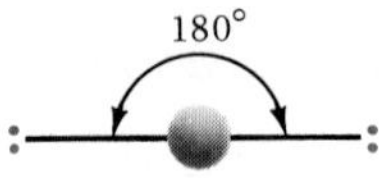 180°	Linear	180°
3	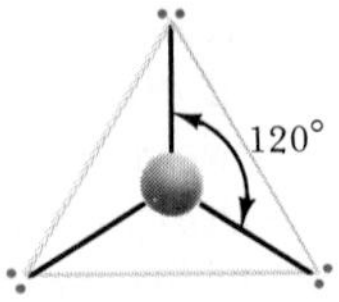 120°	Trigonal planar	120°
4	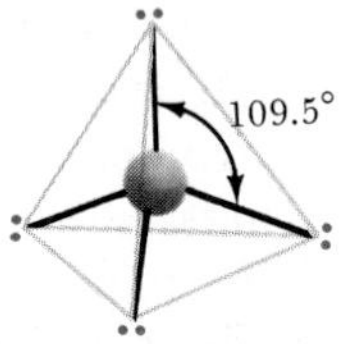 109.5°	Tetrahedral	109.5°
5	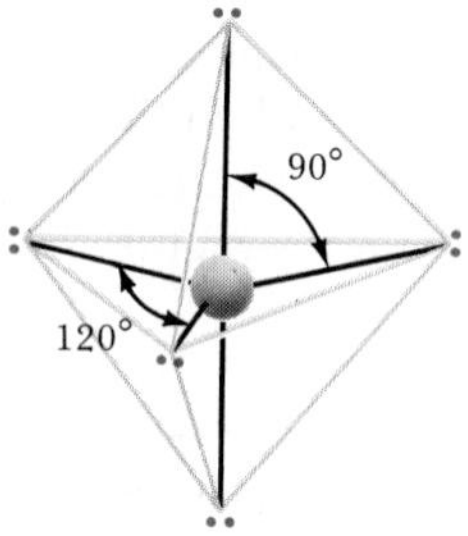 90° 120°	Trigonal bipyramidal	120° 90°
6	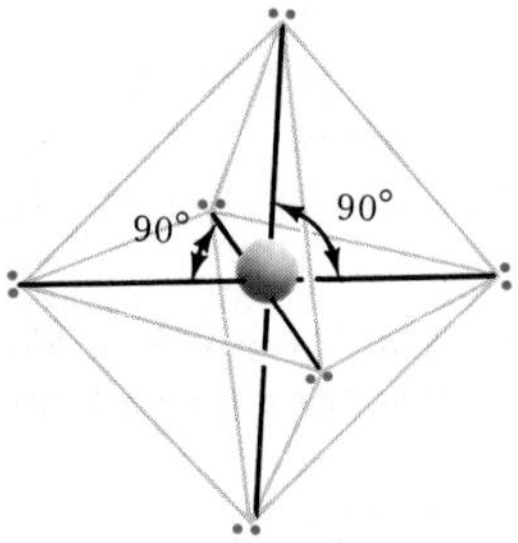 90° 90°	Octahedral	90°

In drawing Lewis structures (Section 8.5), we see that there are two types of valence-shell electron pairs: **bonding pairs**, which are shared by atoms in bonds, and **nonbonding pairs** (also called *lone pairs*). For example, the Lewis structure of ammonia reveals three bonding pairs and one nonbonding pair around the central nitrogen atom:

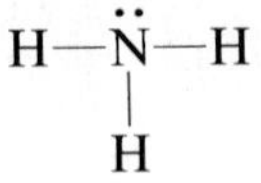

$$\begin{array}{c} \mathrm{H{-}\ddot{N}{-}H} \\ | \\ \mathrm{H} \end{array}$$

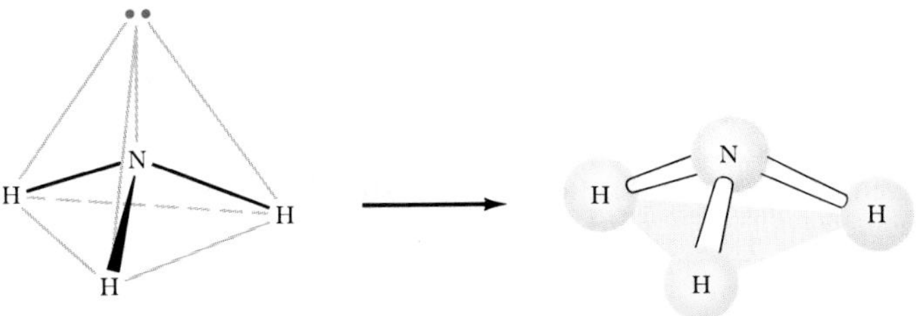

Figure 9.4 The molecular geometry of NH_3.

The electron-pair repulsions about the nitrogen are minimized when these four electron pairs are arranged tetrahedrally (Table 9.1). Hence, the arrangement of electron pairs, known as the **electron-pair geometry**, around the nitrogen atom in NH_3 is tetrahedral.

When we experimentally determine the structure of a molecule, we locate the atoms, not the electron pairs. The **molecular geometry** of a molecule (or ion) is the arrangement of the *atoms* in space. We can determine the molecular geometry of a molecule from its electron-pair geometry. In NH_3, for example, the three bonding pairs point toward three of the vertices of a tetrahedron. The hydrogen atoms are therefore located at three of the vertices of a tetrahedron that has the nitrogen atom at its center. The nonbonding electron pair on nitrogen points toward the fourth vertex (Figure 9.4). Thus, the VSEPR model correctly predicts that the atoms in NH_3 do not all lie in a plane; rather, NH_3 resembles a tetrahedral molecule but with an atom missing.

When describing the shapes of molecules, we always give the *molecular* geometry rather than the *electron-pair* geometry. For example, we describe the shape of NH_3 as *trigonal pyramidal.* Even though the electron-pair geometry of NH_3 is tetrahedral, the arrangement of the atoms is a trigonal pyramid (a pyramid with a triangular base).

The following steps are used in predicting molecular geometries using the VSEPR model:

1. Sketch the Lewis dot structure of the molecule or ion.
2. Count the total number of electron pairs around the central atom and arrange them in the way that minimizes electron-pair repulsions (see Table 9.1).
3. Describe the molecular geometry in terms of the angular arrangement of the *bonding* pairs. (The angular arrangement of the bonding pairs corresponds to the angular arrangement of the bonded atoms.)

In the following discussions, we apply the VSEPR model first to molecules and ions that obey the octet rule and then to those that have expanded octets.

Four or Fewer Valence-Shell Electron Pairs Around a Central Atom

The molecular geometries that can arise when a central atom has four or fewer valence-shell electron pairs are summarized in Table 9.2. These geometries are important because they include all of the commonly occurring structural types found for molecules or ions that obey the octet rule. Application of the VSEPR model to molecules containing double or triple bonds reveals one further rule: *Multiple bonds are considered the same as single bonds in determining molecular geometry.* That is, a double or triple bond has essentially the same effect on bond angles as a single bond and should therefore be counted as one bonding pair when predicting geometry.

Table 9.2 Electron-Pair Geometries and Molecular Shapes for Molecules with Two, Three, and Four Electron Pairs about the Central Atom

Number of electron pairs	Electron-pair geometry	Bonding pairs	Nonbonding pairs	Molecular geometry	Example
2	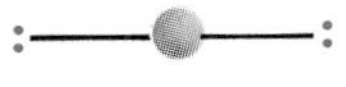 Linear	2	0	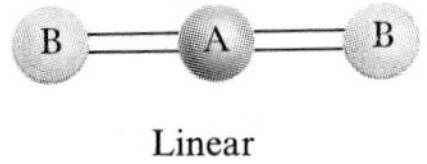 Linear	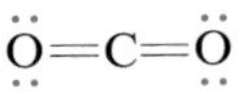 $O=C=O$
3	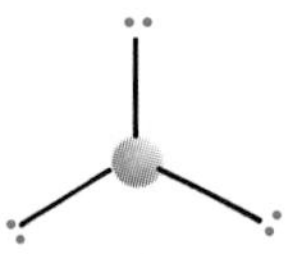 Trigonal planar	3	0	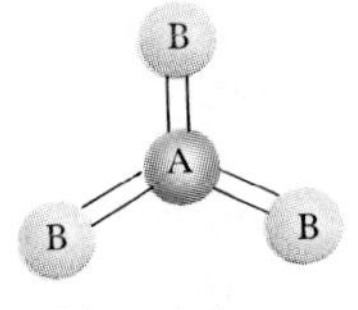 Trigonal planar	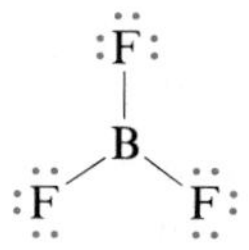 BF_3
		2	1	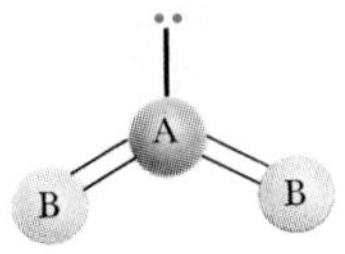 Bent	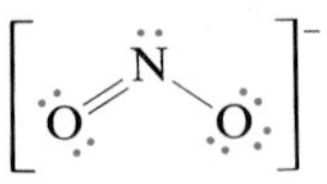 $[NO_2]^-$
4	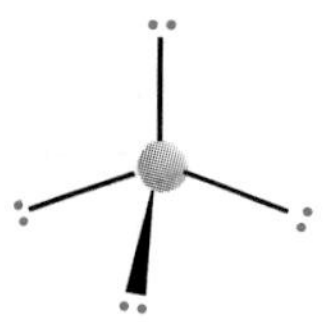 Tetrahedral	4	0	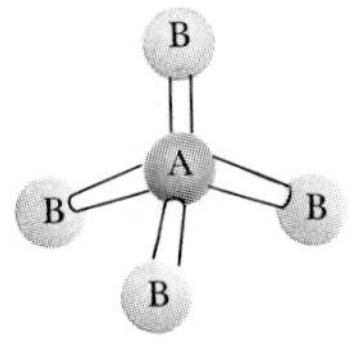 Tetrahedral	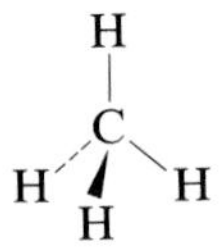 CH_4
		3	1	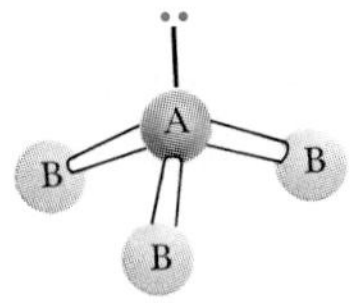 Trigonal pyramidal	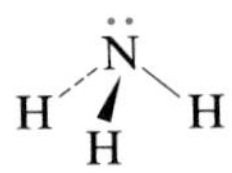 NH_3
		2	2	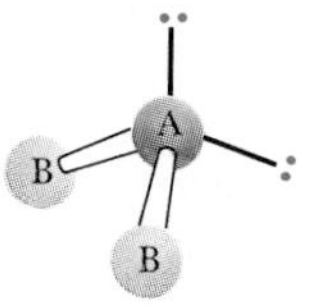 Bent	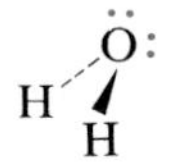 H_2O

SAMPLE EXERCISE 9.1

Using the VSEPR model, predict the molecular geometries of the following: **(a)** $SnCl_3^-$; **(b)** O_3.

Solution: **(a)** The Lewis structure for the $SnCl_3^-$ ion is as follows:

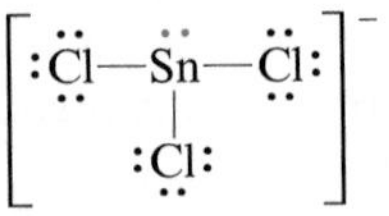

$$\left[:\ddot{Cl}-\ddot{Sn}-\ddot{Cl}: \atop \quad | \atop :\ddot{Cl}: \right]^-$$

The central Sn atom is surrounded by one nonbonding electron pair and three single bonds. Thus, the electron-pair geometry is tetrahedral. That is, the four electron pairs are disposed at the corners of a tetrahedron. Three of the corners are occupied by the bonding pairs of electrons. The molecular geometry is thus pyramidal:

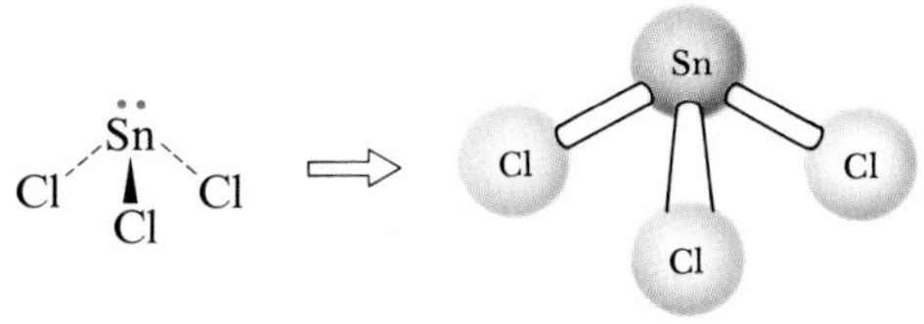

(b) Two resonance structures for the O_3 molecule can be drawn:

$$:\ddot{\underset{..}{O}}-\ddot{O}=\ddot{O} \longleftrightarrow \ddot{O}=\ddot{O}-\ddot{\underset{..}{O}}:$$

Both resonance structures show one nonbonding electron pair, one single bond, and one double bond about the central O atom. When we predict geometry, a double bond is counted as one electron pair. Thus, the arrangement of valence-shell electrons is trigonal planar. Two of these positions are occupied by O atoms, so the molecule has a bent shape:

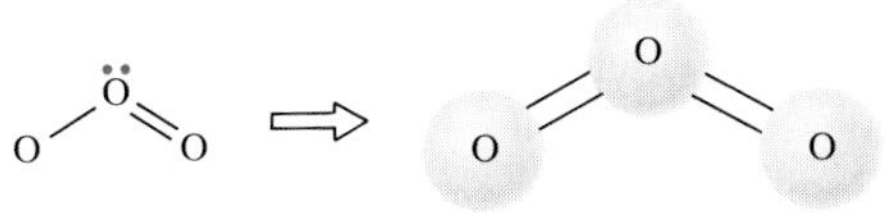

As this example illustrates, when a molecule exhibits resonance, any one of the resonance structures can be used to predict the geometry.

PRACTICE EXERCISE

Predict the electron-pair geometry and the molecular geometry for **(a)** H_2S; **(b)** CO_3^{2-}. ***Answers:*** **(a)** tetrahedral, bent; **(b)** trigonal planar, trigonal planar

The Effect of Nonbonding Electrons and Multiple Bonds on Bond Angles

We can refine the VSEPR model to predict and explain slight distortions of molecules from the ideal geometries summarized in Table 9.2. For example, consider methane, CH_4, ammonia, NH_3, and water, H_2O. All three have tetrahedral electron-pair geometries, but their bond angles differ slightly:

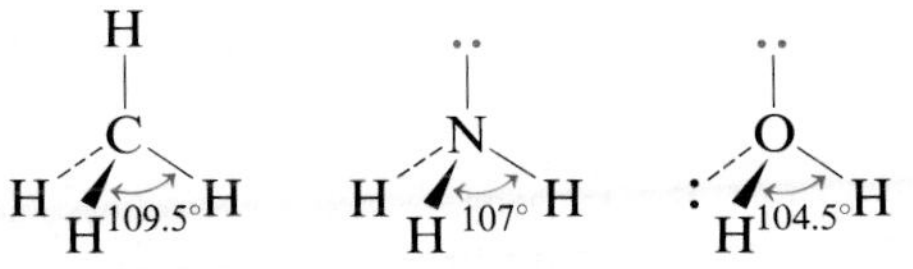

Notice that the bond angles decrease as the number of nonbonding electron pairs increases. Bonding pairs are relatively confined by the attractive influence of the two nuclei of the bonded atoms. By contrast, nonbonding electrons move under the attractive influence of only one nucleus and thus spread out more in space. As a result, *nonbonding electron pairs exert greater repulsive forces on adjacent electron pairs and thus tend to compress the angles between the bonding pairs.* Using the analogy in Figure 9.3, we can envision nonbonding electron pairs as represented by slightly larger and fatter balloons than bonding pairs.

Because multiple bonds contain a higher electronic-charge density than do single bonds, multiple bonds also affect bond angles. Notice how the bond angles in the formaldehyde molecule, H_2CO, differ from the ideal 120° angles:

H
←122°
116° C=O
←122°
H

The double bond seems to act much like a nonbonding pair of electrons, reducing the H—C—H bond angle from 120° to 116°. *Multiple bonds, like nonbonding pairs, exert a greater repulsive force on adjacent electron pairs than do single bonds.*

Geometries of Molecules with Expanded Valence Shells

When the central atom of an AB_n molecule is from the third period of the periodic table and beyond, that atom may have more than four electron pairs around it (Section 8.7). Molecules with five or six electron pairs around the central atom display a variety of molecular geometries, as shown in Table 9.3. We can use the VSEPR model to predict the structures of these expanded-valence-shell molecules.

The most stable electron-pair geometry for five electron pairs is the trigonal bipyramid (two trigonal pyramids sharing a common base). Unlike the electron-pair geometries we have seen to this point, the trigonal bipyramid contains two geometrically distinct types of electron pairs. Two pairs are called *axial pairs*, and the remaining three are called *equatorial pairs*, as shown in Figure 9.5. In an axial position, an electron pair is situated at 90° from the three equatorial pairs. In an equatorial position, an electron pair is situated 120° from the other two equatorial pairs and 90° from the two axial pairs.

Suppose a molecule has five electron pairs, one of which is a nonbonding pair. Will the nonbonding pair occupy an axial or an equatorial position? Because we could place the lone pair at two distinct sites, we must determine which site will minimize the electron-pair repulsion. Electron-pair repulsions are much greater when pairs are situated 90° from each other than when they are at 120°. Because they are 90° from only two other pairs, the equatorial pairs experience less repulsion than do the axial pairs, which are 90° from *three* other pairs. Because nonbonding pairs exert larger repulsions than bonding pairs, they always occupy the equatorial positions, as shown in Table 9.3.

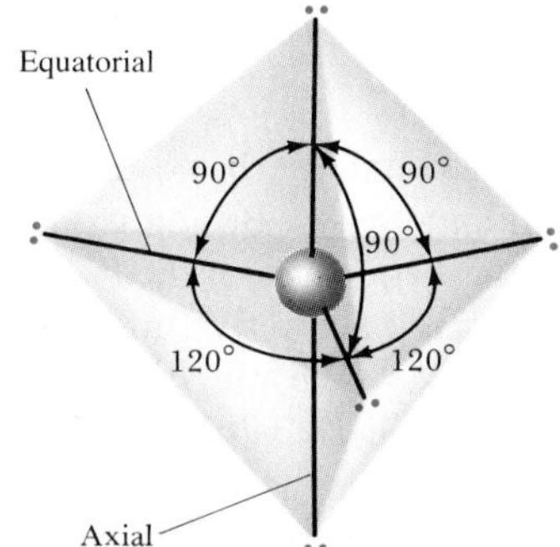

Figure 9.5 Trigonal bipyramidal arrangement of five electron pairs about a central atom. There are two geometrically distinct types of electron pairs, two axial and three equatorial pairs. Unshared electron pairs occupy the equatorial positions.

Table 9.3 Electron-Pair Geometries and Molecular Shapes for Molecules with Five and Six Electron Pairs About the Central Atom

Number of electron pairs	Electron-pair geometry	Bonding pairs	Nonbonding pairs	Molecular geometry	Example
5	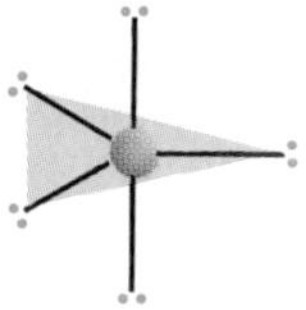 Trigonal bipyramidal	5	0	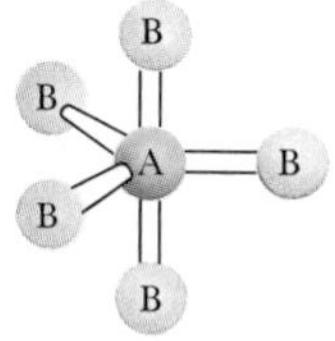 Trigonal bipyramidal	PCl_5
		4	1	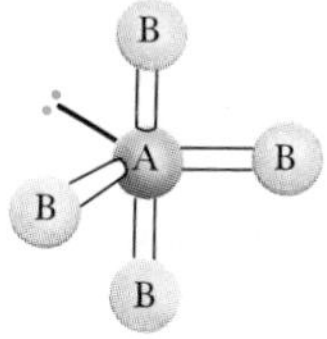 Seesaw	SF_4
		3	2	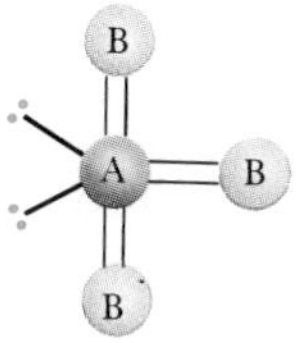 T-shaped	ClF_3
		2	3	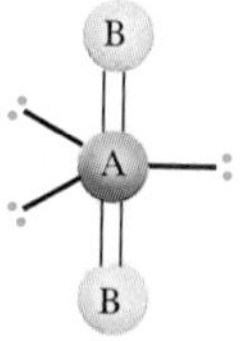 Linear	XeF_2
6	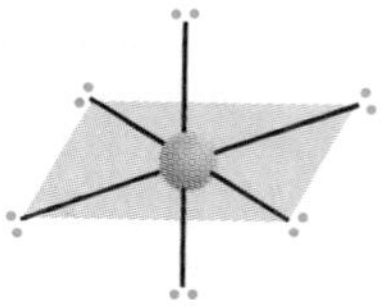 Octahedral	6	0	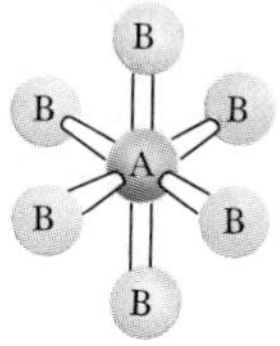 Octahedral	SF_6
		5	1	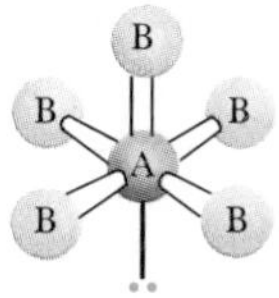 Square pyramidal	BrF_5
		4	2	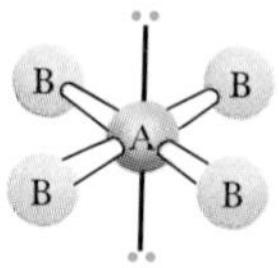 Square planar	XeF_4

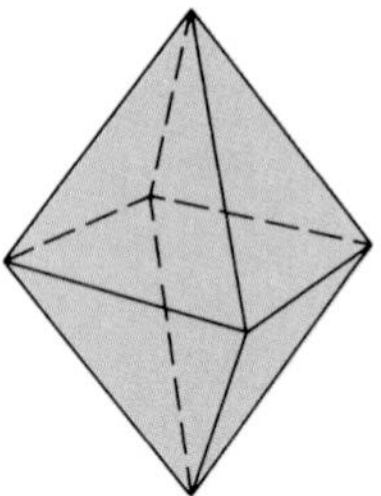

Figure 9.6 An octahedron is an object with eight faces and six vertices. Each face is an equilateral triangle.

The most stable electron-pair geometry for six electron pairs is the *octahedron.* As shown in Figure 9.6, an octahedron is a solid object with six vertices and eight faces, each of which is an equilateral triangle. In an octahedral molecule such as SF_6 (Figure 9.2), the central atom is located in the center and the electron pairs point toward the six vertices. All the angles in an octahedron are 90° or 180°, and all six positions are equivalent. Therefore, if a molecule has five bonding pairs of electrons and one nonbonding pair, it makes no difference where we place them on the octahedron. However, when there are two nonbonding electron pairs, their repulsions are minimized by placing them on opposite sides of the octahedron, as shown in Table 9.3.

SAMPLE EXERCISE 9.2

Using the VSEPR model, predict the molecular geometry of **(a)** SF_4; **(b)** IF_5.

Solution: **(a)** The Lewis dot structure for SF_4 is

The sulfur has five valence-shell electron pairs around it. Each pair points toward a corner of a trigonal bipyramid. The one nonbonding pair occupies an equatorial position. The four bonding electron pairs occupy the remaining four positions:

186° 116°

The experimentally observed structure is shown on the right. From this structure, we can infer that the unshared electron pair resides in an equatorial position, as predicted. Note that the axial S—F bonds are slightly bent back, suggesting that they are pushed by the nonbonding electron pair, with its greater repulsive effect.

(b) The Lewis structure of IF_5 (nonbonding pairs excluded from F) is

The iodine has six pairs of valence-shell electrons around it, five of them bonding pairs. The electron pairs should point toward the corners of an octahedron. The geometrical arrangement of atoms is therefore square pyramidal (Table 9.3).

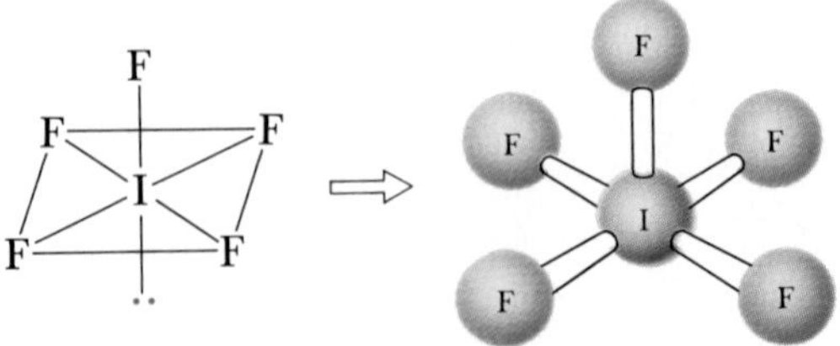

PRACTICE EXERCISE

Predict the molecular geometries of **(a)** ClF_3; **(b)** ICl_4^-. *Answers:* **(a)** T-shaped; **(b)** square planar

Molecules with No Single Central Atom

The molecules and ions whose structures we have thus far considered contain only a single central atom. The VSEPR model, however, can be readily extended to more complex molecules. Consider the acetic acid molecule, whose Lewis structure is shown below:

```
     H :O:
     |  ||  ..
  H—C—C—O—H
     |      ..
     H
```

Using the VSEPR model, we can predict the geometry around the leftmost C atom, the central C atom, and the rightmost O atom.

Notice that the leftmost C has four electron pairs around it, all of them bonding pairs. Thus the geometry around that atom is tetrahedral. The central C has effectively three bonding pairs around it (counting the double bond as if it were one pair). Thus the geometry around that atom is trigonal planar. The O atom has four electron pairs, giving it a tetrahedral electron-pair geometry. However, only two of these pairs are bonding pairs, so the molecular geometry around O is bent. The molecular geometry is shown in Figure 9.7.

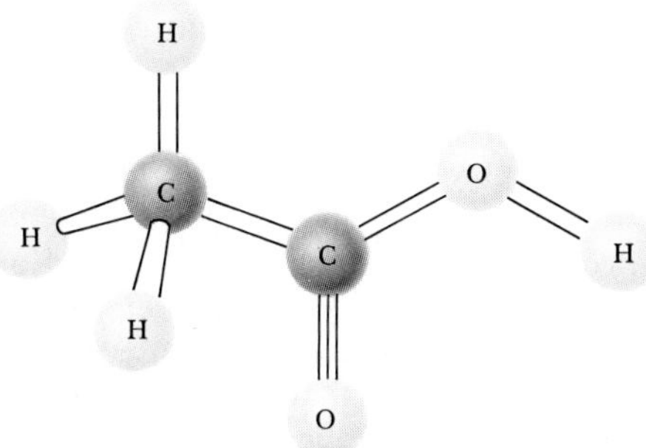

Figure 9.7 The molecular structure of acetic acid, $HC_2H_3O_2$.

SAMPLE EXERCISE 9.3

Predict the approximate values for the H—O—C and O—C—O bond angles in oxalic acid:

```
        :O::O:
   ..    ||  ||   ..
H—O—C—C—O—H
   ..              ..
```

Solution: To predict the H—O—C bond angle, consider the number of electron pairs around the central O of this angle. Because there are four electron pairs (two single bonds and two nonbonding electron pairs), the electron-pair geometry is tetrahedral, and thus the bond angle is approximately 109°.

In the case of the O—C—O bond angle, the central C atom is surrounded by a double bond and two single bonds. To predict their arrangement, we count the double bond as a single bond, so we have three electron pairs. Thus the electron-pair geometry is trigonal planar, and the bond angle is approximately 120°.

PRACTICE EXERCISE

Predict the C—C—C bond angle in the following molecule:

```
     H
     |
  H—C—C≡C—H
     |
     H
```

Answer: 180°

9.2 DIPOLE MOMENTS

The shape of a molecule and the polarity of its bonds together determine the charge distribution in the molecule. A molecule is said to be **polar** if its centers of negative and positive charge do not coincide. One end of a polar molecule has a slight negative charge and the other a slight positive charge.

Any diatomic molecule with a polar bond (Section 8.9) is a polar molecule. For example, the HF molecule is polar, having a concentration of negative charge on the more electronegative F atom, leaving the less electronegative H atom as the positive end:

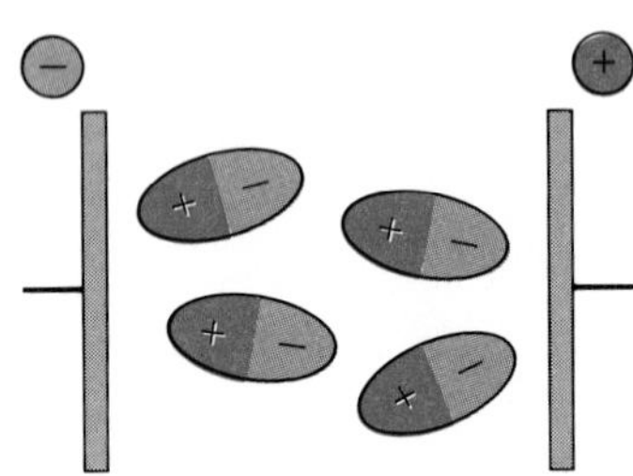

Figure 9.8 Polar molecules align themselves in an electric field, with their negative sides pointing toward the positive plate.

$$\overset{+\!\!\longrightarrow}{\text{H—F}}$$

The arrow denotes the shift in electron density toward the fluorine atom.

Polar molecules align themselves in an electric field, as shown in Figure 9.8. They also align themselves with respect to each other and with respect to ions. The negative end of one polar molecule and the positive end of another attract each other. Polar molecules are likewise attracted to ions. The negative end of a polar molecule is attracted to a positive ion; the positive end is attracted to a negative ion. These interactions are important in explaining the properties of liquids, solids, and solutions, as you will see in Chapters 11, 12, and 13.

The degree of polarity of a molecule is measured by its **dipole moment**. The dipole moment, μ, is defined as the product of the charge at either end of the dipole, Q, times the distance, r, between the charges: $\mu = Qr$. Therefore, the dipole moment increases as the quantity of charge that is separated increases and as the distance between the positive and negative centers increases.

Dipole moments are generally reported in units of debye, D. A debye is 3.33×10^{-30} coulomb-meters (C-m). Let's consider what the dipole moment value tells us about the separation of charge in a polar molecule. As an example, the dipole moment of HCl is 1.03 D. The H—Cl bond length in this molecule is 1.36 Å. From these data, we can calculate the value of Q in the formula for the dipole moment, assuming that the charges are centered on the H and Cl atoms. We have

$$\mu = Qr$$

$$(1.03\text{ D})\left(3.33 \times 10^{-30}\,\frac{\text{C-m}}{\text{D}}\right) = Q \times (1.36\text{ Å})\left(10^{-10}\,\frac{\text{m}}{\text{Å}}\right)$$

$$Q = \frac{3.43 \times 10^{-30}\text{ C-m}}{1.36 \times 10^{-10}\text{ m}}$$

$$= 2.52 \times 10^{-20}\text{ C}$$

Table 9.4 Bond Lengths, Electronegativity Differences, and Dipole Moments of the Hydrogen Halides

Compound	Bond length (Å)	Electronegativity difference	Dipole moment (D)
HF	0.92	1.9	1.91
HCl	1.27	0.9	1.03
HBr	1.41	0.7	0.79
HI	1.61	0.4	0.38

The charge of the electron is 1.60×10^{-19} C. Thus, as a percentage of the electronic charge, Q is $100 \times (2.52 \times 10^{-20})/(1.60 \times 10^{-19}) = 15.8\%$. If the H—Cl bond were ionic, there would be a full 1+ charge on H and a full 1− charge on Cl. In that case Q would equal the electronic charge. In reality, Q is substantially less than this because the H—Cl bond is polar covalent rather than ionic. The dipole moments of the hydrogen halides are listed in Table 9.4. Notice that the dipole moment decreases as the electronegativity difference decreases.

The Polarity of Polyatomic Molecules

The polarity of a molecule containing more than two atoms depends on both the polarities of the bonds and the geometry of the molecule. For each polar bond in a molecule, we can consider the **bond dipole**, that is, the dipole moment due only to the two atoms bonded together. We must then ask what *overall* dipole moment results from adding up the individual bond dipoles. For example, consider the CO_2 molecule, which is linear. As shown in Figure 9.9(*a*), each C—O bond is polar and, because the C—O bonds are identical, the bond dipoles are equal in magnitude.

Does the fact that both C—O bonds are polar mean that the CO_2 molecule is polar? Not necessarily. Bond dipoles and dipole moments are *vector* quantities; that is, they have both a magnitude and a direction. The overall dipole moment of a polyatomic molecule is the sum of its bond dipoles. Both the magnitudes *and* the directions of the bond dipoles must be considered in this sum of vectors. The two bond dipoles in CO_2, although equal in magnitude, are exactly opposite in direction. Adding them together is the same as adding two numbers that are equal in magnitude but opposite in sign, such as 100 + (−100): The bond dipoles, like the numbers, "cancel" each other. Therefore, the overall dipole moment of CO_2 is zero. Note that the oxygen atoms in CO_2 do carry a partial negative charge and that the carbon atom carries a partial positive charge, as we expect for polar bonds. Even though the individual bonds are polar, the geometry of the molecule dictates that the overall dipole moment be zero.

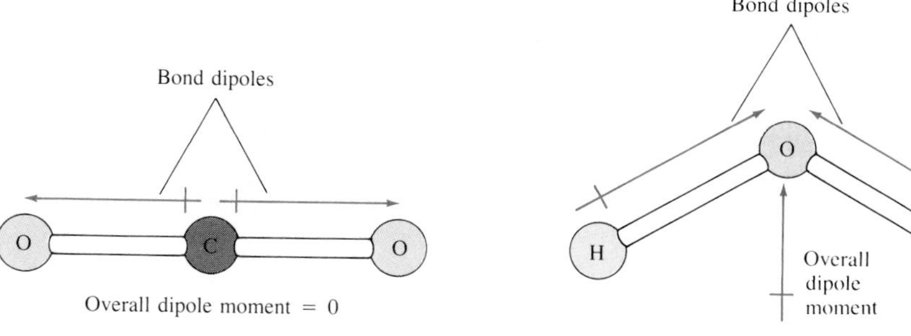

Figure 9.9 The overall dipole moment of a molecule is the sum of its bond dipoles. (*a*) In CO_2, the bond dipoles are equal in magnitude but exactly oppose each other. The overall dipole moment is zero. (*b*) In H_2O, the bond dipoles are also equal in magnitude but do not exactly oppose each other. The molecule has a nonzero overall dipole moment.

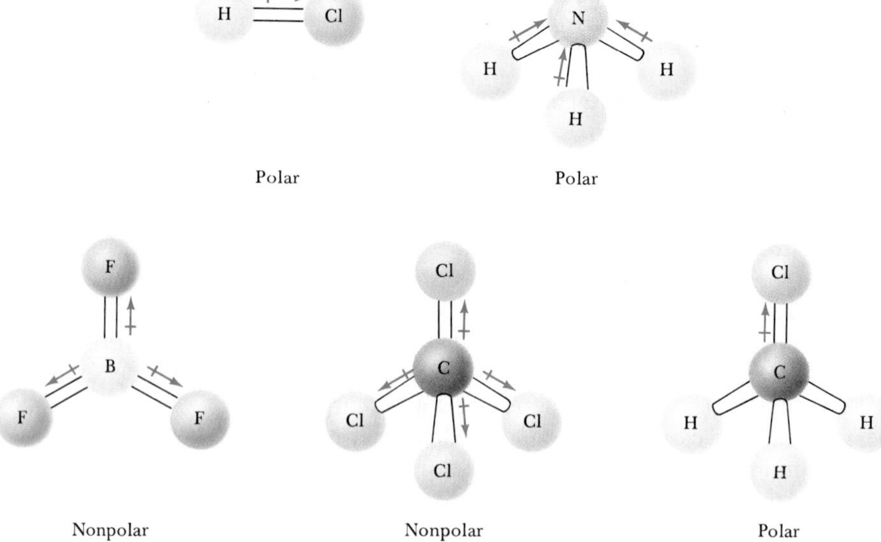

Figure 9.10 Examples of molecules with polar bonds. Some of these molecules are nonpolar because their bond dipoles cancel one another.

Now let's consider H_2O, which is a bent molecule with two polar bonds [Figure 9.9(*b*)]. Again, both the bonds are identical, so the bond dipoles are equal in magnitude. Because the molecule is bent, however, the bond dipoles do not directly oppose each other and therefore do not cancel each other. Hence, the water molecule has an overall dipole moment ($\mu = 1.84$ D). The oxygen atom carries a partial negative charge, and the hydrogen atoms each have a partial positive charge.

Figure 9.10 shows several other examples of polar and nonpolar molecules, all of which have polar bonds. Notice that the molecules in which the central atom is symmetrically surrounded by identical atoms (BF_3 and CCl_4) are nonpolar. For AB_n molecules in which all the B atoms are the same, certain symmetrical geometries—linear (AB_2), trigonal planar (AB_3), tetrahedral and square planar (AB_4), trigonal bipyramidal (AB_5), and octahedral (AB_6)—must lead to nonpolar molecules regardless of how polar the individual bonds are.

SAMPLE EXERCISE 9.4

Predict whether the following molecules are polar or nonpolar: **(a)** BrCl; **(b)** SO_2; **(c)** SF_6.

Solution: **(a)** Chlorine is more electronegative than bromine. Consequently, BrCl will be polar with chlorine carrying the negative charge:

$$\overset{\longleftrightarrow}{\text{Br—Cl}}$$

Experimentally, the dipole moment of the molecule is 0.57 D. All diatomic molecules with polar bonds are polar molecules.

(b) Because oxygen is more electronegative than sulfur, the molecule has polar bonds. Several resonance forms for SO_2 can be written (see Exercise 8.74):

$$:\ddot{\text{O}}=\ddot{\text{S}}=\ddot{\text{O}}: \longleftrightarrow :\ddot{\text{O}}-\ddot{\text{S}}=\ddot{\text{O}}: \longleftrightarrow :\ddot{\text{O}}=\ddot{\text{S}}-\ddot{\text{O}}:$$

For each of these, the VSEPR model predicts a bent geometry. Because the molecule is bent, the bond dipoles do not cancel and the molecule is polar ($\mu = 1.62$ D):

S
O O

(c) Fluorine is more electronegative than sulfur. The bond dipoles therefore point toward fluorine. The six S—F bonds are arranged in an octahedral fashion around the central sulfur:

F
F
F—S—F
F
F

The symmetric octahedral geometry of the molecule leads to cancellation of the bond dipoles, and the molecule is nonpolar ($\mu = 0$).

PRACTICE EXERCISE

Are the following molecules polar or nonpolar: **(a)** NF_3; **(b)** BCl_3?
Answers: **(a)** polar; **(b)** nonpolar

9.3 COVALENT BONDING AND ORBITAL OVERLAP

The VSEPR model provides a simple means for predicting the shapes of molecules. However, it does not explain why bonds exist between atoms. In developing a theory of covalent bonding, chemists have approached the problem from another direction, namely, using quantum mechanics. How can we account for the observed geometries of molecules in terms of the atomic orbitals used by the atoms in forming bonds with one another? In this section, we will first use atomic orbitals to address Lewis's notion that atoms share electrons in making covalent bonds. We will then describe a model in which the orbitals of an atom can mix with one another to give a picture that corresponds nicely to the VSEPR model.

Valence Bond Theory

In the Lewis theory, covalent bonding occurs when atoms share electrons. Such sharing concentrates electron density between the nuclei. The marriage of Lewis's notion of electron-pair bonds to the idea of atomic orbitals leads to a model of chemical bonding called **valence bond theory**. In valence bond theory, the buildup of electron density between two nuclei occurs when a valence atomic orbital of one atom merges with that of another atom. The orbitals are then said to share a region of space, or to **overlap**. The overlap of orbitals allows the two electrons in a bond to share the common space between the nuclei, much as in a Lewis structure the two electrons in a bond are shared by the two atoms.

The overlap of the 1*s* orbitals from two H atoms to form H_2 is depicted in Figure 9.11. Because the overlap region is between the two hydrogen nuclei, the probability of finding the electrons between the nuclei is high. A covalent bond is thus formed from the two electrons. The overlap region is a favorable one for the electrons to reside in, for they are simultaneously attracted to both positively charged nuclei.

Figure 9.11 The overlap of two 1*s* orbitals from two H atoms to form H_2.

Atoms approach each other

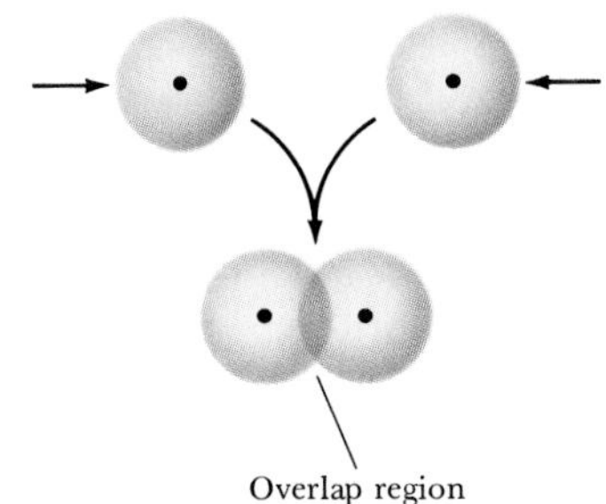

There is always an optimum distance between the two bonded nuclei in any covalent bond. This fact is illustrated in Figure 9.12, which shows two H atoms coming together to form an H_2 molecule. As the atoms approach one another, the overlap between their 1*s* orbitals increases. Because of the resultant increase in electron density between the nuclei, the potential energy of the system decreases. That is, the strength of the bond increases, as shown by the decrease in the energy on the curve. However, the figure also shows that as the atoms

Figure 9.12 The change in potential energy during the formation of the H_2 molecule. The minimum in the energy, at 0.74 Å, represents the equilibrium bond distance. The energy at that point, −436 kJ/mol, corresponds to the energy change for formation of the H—H bond.

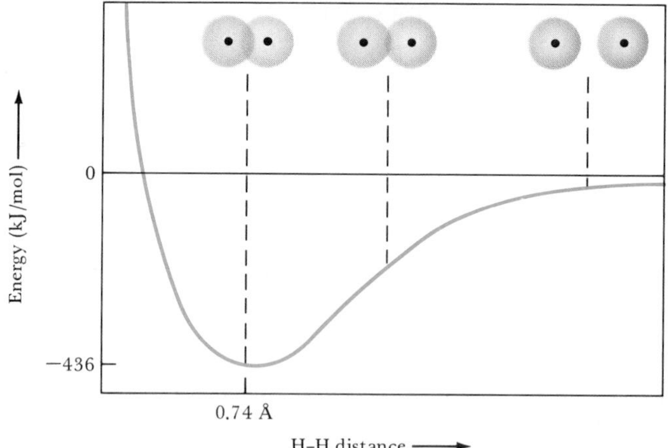

come very close together, the energy increases rapidly. This rapid increase is due to the electrostatic repulsion between the nuclei, which becomes significant at short internuclear distances. The internuclear distance at the minimum on the potential-energy curve corresponds to the observed bond length. Thus, the observed bond length is a compromise between increased overlap of the atomic orbitals, which draws the atoms together, and nuclear-nuclear repulsion, which forces them apart.

The bond that results from the overlap of two *s* orbitals, one from each atom, is called a **sigma (σ) bond**. A σ bond is one in which the electron density is concentrated symmetrically about the line that connects the nuclei (the *internuclear axis*). That is, in a σ bond the line joining the two nuclei passes through the middle of the overlap region. The overlap of an *s* orbital with a *p* orbital and the overlap of two *p* orbitals pointing toward each other can also produce σ bonds, as shown in Figure 9.13. In each of these cases, the overlap of the orbitals leads to the buildup of electron density in the region between the two nuclei.

A different kind of bond results from the overlap between two *p* orbitals oriented perpendicularly to the internuclear axis (Figure 9.14). This sideways overlap of *p* orbitals produces a **pi (π) bond**. A π bond is a covalent bond in which the overlap regions lie above and below the internuclear axis. The internuclear axis itself is a node of both *p* orbitals; unlike a σ bond, there is no probability of finding the electron in a π bond on this axis.* Because the total overlap in a π bond tends to be less than that in a σ bond, π bonds are generally weaker than σ bonds.

In almost all cases, single bonds are σ bonds. A double bond consists of one σ bond and one π bond, and a triple bond consists of one σ bond and two π bonds:

H—H	$H_2C{=}CH_2$	:N≡N:
One σ bond	One σ bond plus one π bond	One σ bond plus two π bonds

* Actually, the node of a π bond is a *plane* that contains the internuclear axis and is perpendicular to the main direction of the *p* orbitals. This plane is called a *nodal plane* of the π bond.

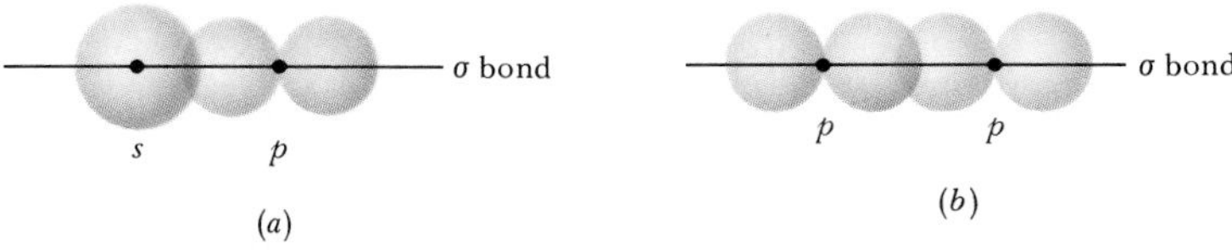

Figure 9.13 Formation of σ bonds by overlap (*a*) of an *s* orbital with a *p* orbital, and (*b*) of two *p* orbitals.

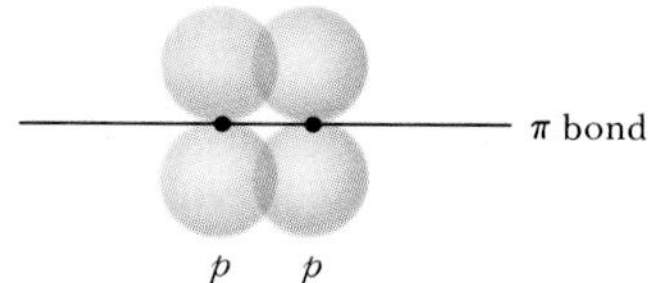

Figure 9.14 Formation of a π bond by overlap of two *p* orbitals. The two regions of overlap constitute *one* π bond.

Double and triple bonds (and hence π bonds) are more common in molecules with small atoms, especially C, N, and O. Larger atoms, such as S, P, and Si, form π bonds less readily. We will take a closer look at π bonds in Section 9.5. Meanwhile, we will consider how σ bonding and molecular geometry are related.

9.4 HYBRID ORBITALS

sp Hybrid Orbitals

The idea of the overlap of atomic orbitals allows us to understand why covalent bonds form. However, we must still reconcile the formation of bonds from atomic orbitals with the observed geometries of molecules. For example, heating the salt BeF_2 to high temperatures generates gaseous molecules of BeF_2, the Lewis structure of which is

$$:\ddot{\underset{..}{F}}-Be-\ddot{\underset{..}{F}}:$$

The VSEPR model predicts that this molecule is linear; indeed, experiments show that the molecule is linear with two identical Be—F bonds. How can we use valence bond theory to describe the bonding in linear BeF_2? We have no problem with the fluorine atoms, because the electron configuration of F ($1s^22s^22p^5$) tells us that there is an unpaired electron in a 2*p* orbital. This 2*p* electron can be paired with one from the Be atom to form a polar covalent bond. However, we are now faced with a more difficult question: Which orbitals on the Be atom overlap with those on the F atoms to form the Be—F bonds?

The orbital diagram for a ground-state Be atom is as follows:

⇅	⇅			
1*s*	2*s*		2*p*	

Because it has no unpaired electrons, the Be atom in its ground state is incapable of forming bonds with the fluorine atoms. We can envision the atom obtaining the ability to form two bonds by "promoting" one of the 2*s* electrons to a 2*p* orbital:

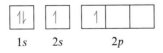

⇅	↑	↑		
1*s*	2*s*		2*p*	

Because the 2*p* orbital is of higher energy than the 2*s*, this promotion requires energy. The Be atom now has two unpaired electrons and can therefore form two polar covalent bonds with the F atoms. The Be 2*s* orbital would be used to form one of the bonds, and a 2*p* orbital would be used for the other. We

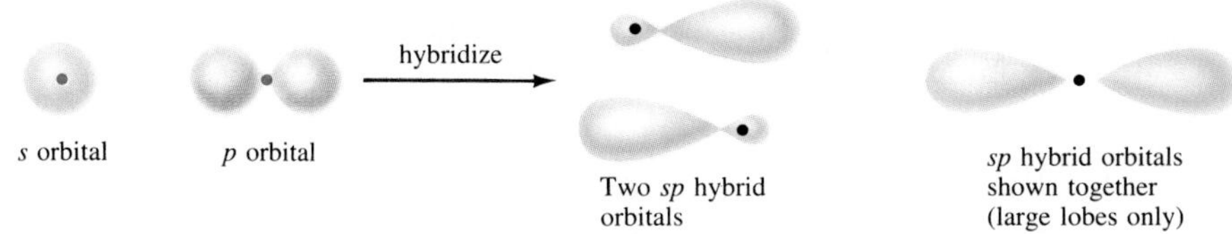

Figure 9.15 One *s* orbital and one *p* orbital can hybridize to form two equivalent *sp* hybrid orbitals. The two hybrid orbitals have their large lobes pointing in opposite directions, 180° apart.

certainly would not expect these two bonds to be identical. Therefore, although the promotion of an electron allows the formation of two Be—F bonds, we still haven't explained the structure of BeF_2.

We can solve our dilemma by envisioning a "mixing" of the 2*s* orbital and one of the 2*p* orbitals to generate two new orbitals, as shown in Figure 9.15. Like *p* orbitals, each of the new orbitals has two lobes. However, unlike *p* orbitals, one lobe is much larger than the other. The two new orbitals are identical in shape, but their large lobes point in opposite directions. We have created two **hybrid orbitals**, orbitals that we form by mixing two or more atomic orbitals on an atom, a procedure called **hybridization**. In this case, we have hybridized one *s* and one *p* orbital, so we call each hybrid an *sp hybrid orbital.*

For the Be atom of BeF_2, we write the orbital diagram for the formation of two *sp* hybrid orbitals as follows:

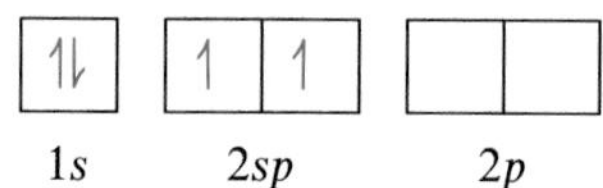

The electrons in the *sp* hybrid orbitals can form shared electron bonds with the two fluorine atoms (Figure 9.16). Because the *sp* hybrid orbitals are equivalent to one another but point in opposite directions, BeF_2 has two identical σ bonds and a linear geometry.

As noted above, the promotion of a 2*s* electron to a 2*p* orbital in Be requires energy. Why, then, is it convenient for us to envision the formation of hybrid orbitals? Hybrid orbitals have one large lobe and can therefore be directed at other atoms more effectively than can atomic orbitals. Hence, they can overlap more strongly with the orbitals of other atoms than can atomic orbitals, and stronger bonds result. The energy released by the formation of chemical bonds more than offsets the energy that must be expended to promote electrons.

sp^2 and sp^3 Hybrid Orbitals

Whenever we mix a certain number of atomic orbitals, we get the same number of hybrid orbitals. Each of these hybrid orbitals is equivalent to the others but points in a different direction. Thus, mixing one 2*s* and one 2*p* orbital yields

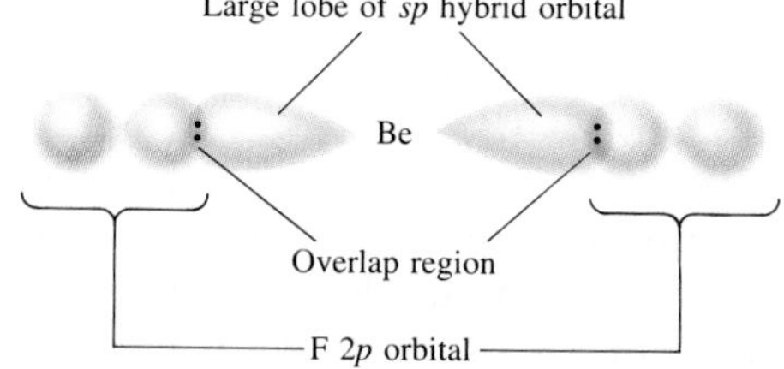

Figure 9.16 The formation of two equivalent Be—F bonds in BeF_2. Each of the *sp* hybrid orbitals on Be overlaps with a 2*p* orbital on F to form an electron-pair bond.

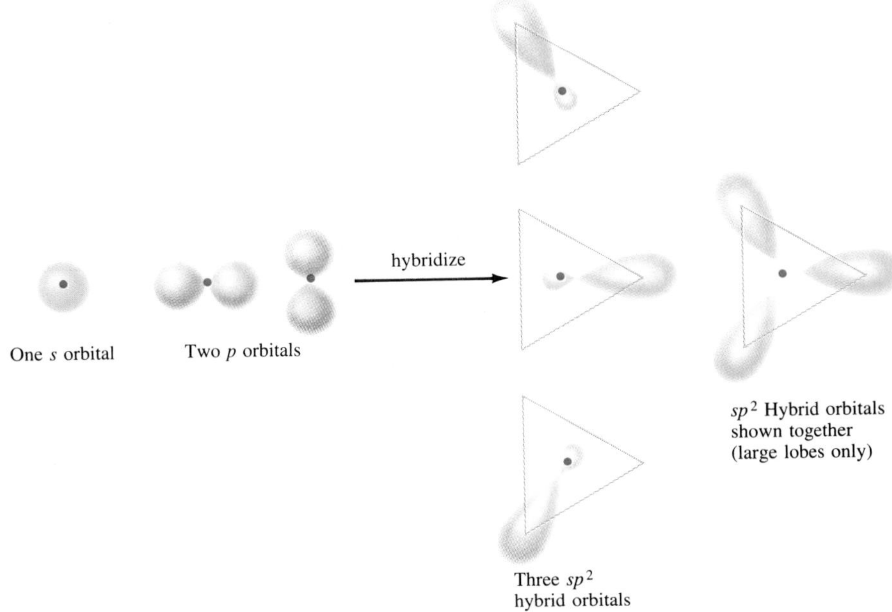

Figure 9.17 One *s* orbital and two *p* orbitals can hybridize to form three equivalent sp^2 hybrid orbitals. The large lobes of the hybrid orbitals point toward the corners of an equilateral triangle.

two equivalent *sp* hybrid orbitals that point in opposite directions (Figure 9.15). Other combinations of atomic orbitals can be hybridized to obtain different geometries of hybrid orbitals about a central atom. For example, in BF_3, a 2*s* electron on the B atom can be promoted to a vacant 2*p* orbital. Mixing the 2*s* and two of the 2*p* orbitals yields three equivalent sp^2 (pronounced "*s*-*p*-two") hybrid orbitals:

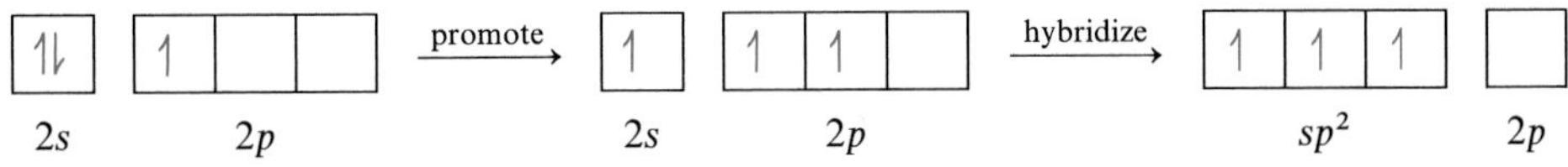

The three sp^2 hybrid orbitals lie in the same plane, 120° apart from one another (Figure 9.17). These are used to make three equivalent σ bonds with the three fluorine atoms, leading to the trigonal planar geometry of BF_3.

A very common type of hybridization occurs when an *s* orbital mixes with all three *p* orbitals in the same subshell. For example, the carbon atom in CH_4 forms four equivalent σ bonds with the four hydrogen atoms. We envision this process as resulting from the mixing of the 2*s* and all three 2*p* atomic orbitals of carbon to create four equivalent sp^3 (pronounced "*s*-*p*-three") hybrid orbitals:

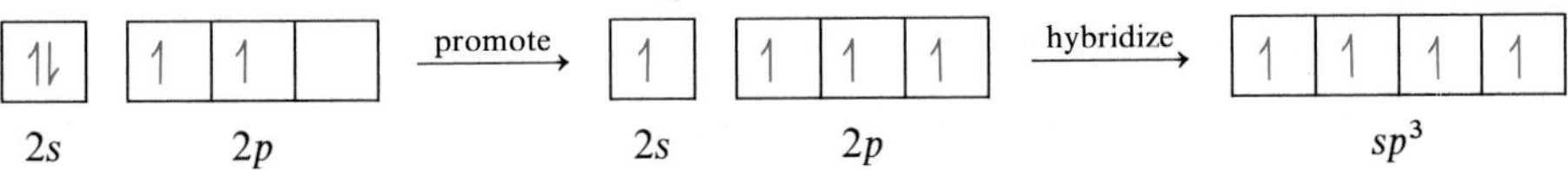

Each of the sp^3 hybrid orbitals has a large lobe that points toward a vertex of a tetrahedron, as shown in Figure 9.18. These hybrid orbitals can be used to form two-electron bonds by overlap with the atomic orbitals of another atom,

Figure 9.18 Formation of four sp^3 hybrid orbitals from a set of one s orbital and three p orbitals.

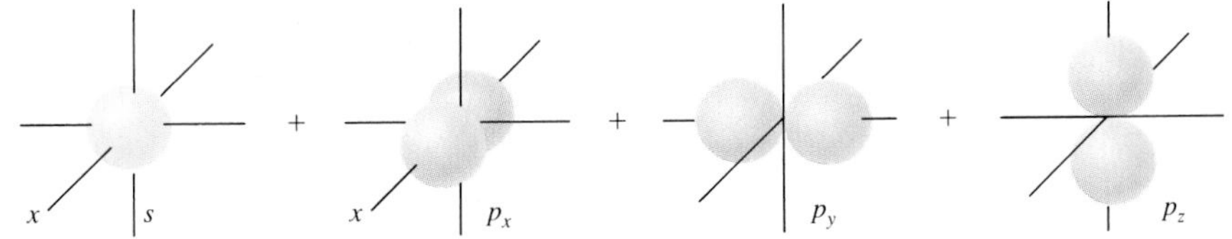

Hybridize to form four sp^3 hybrid orbitals

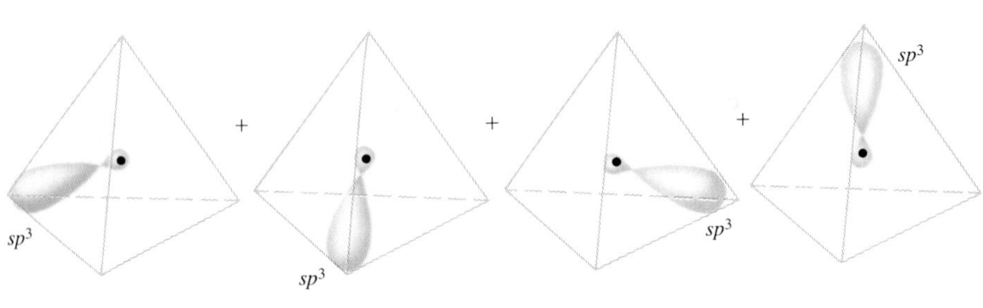

Shown together (large lobes only)

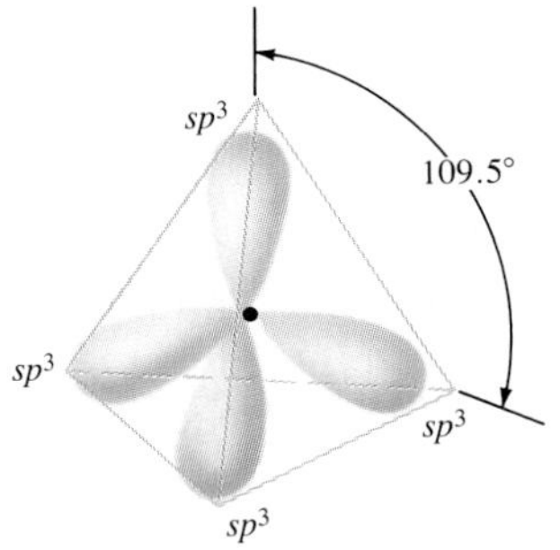

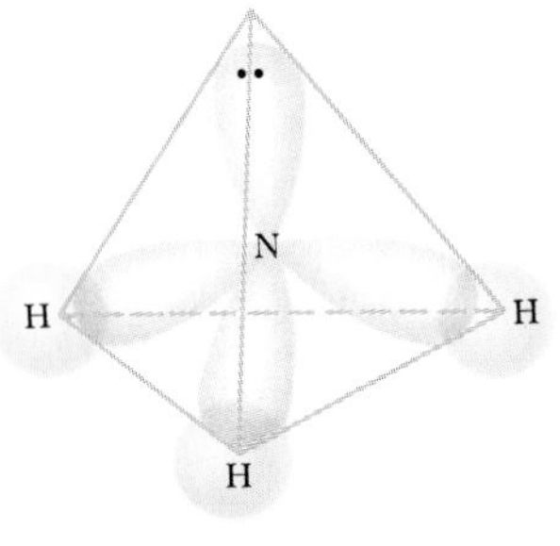

(*a*) NH_3

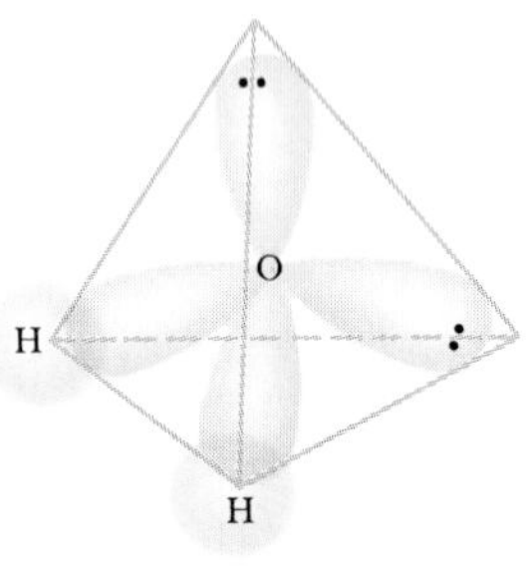

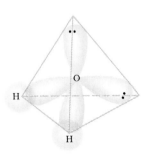

(*b*) H_2O

Figure 9.19 (*a*) Ammonia, NH_3, and (*b*) water. Note overlap of hydrogen 1*s* orbitals with the sp^3 hybrid orbitals of the central atom in both molecules. Ammonia has a lone pair occupying one hybrid orbital. Water has two lone pairs.

for example, H. Thus, within valence bond theory, we can describe the bonding in CH_4 as the overlap of four equivalent sp^3 hybrid orbitals on C with the 1*s* orbitals of the four H atoms.

The idea of sp^3 hybridization can also be used to describe the bonding in other molecules, such as NH_3 and H_2O, where the electron-pair geometry around the central atom is approximately tetrahedral. In NH_3, one of the sp^3 hybrid orbitals contains the nonbonding pair of electrons, and the other three contain the bonding pairs, as shown in Figure 9.19(*a*). In H_2O, two of the hybrid orbitals contain nonbonding pairs of electrons, while the other two are used in forming σ bonds with hydrogen atoms, as shown in Figure 9.19(*b*).

Hybridization Involving *d* Orbitals

Atoms in the third period and beyond can also use *d* orbitals to form hybrid orbitals. Mixing one *s* orbital, three *p* orbitals, and one *d* orbital leads to five sp^3d hybrid orbitals. These hybrid orbitals are directed at the vertices of a trigonal bipyramid. The formation of sp^3d hybrids is exemplified by the phosphorus atom in PF_5:

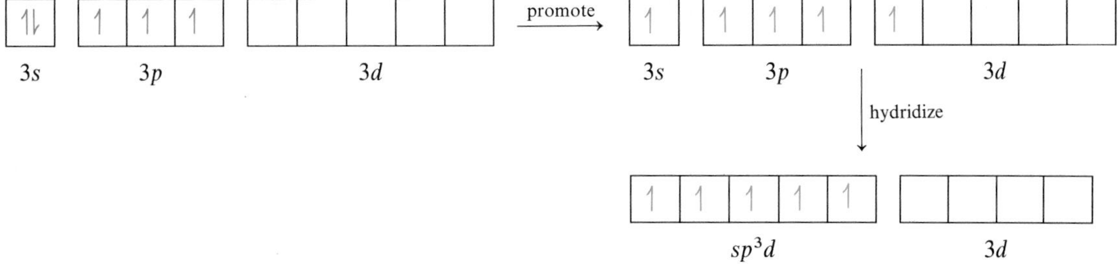

Similarly, mixing one *s* orbital, three *p* orbitals, and two *d* orbitals gives six sp^3d^2 hybrid orbitals, which are directed at the vertices of an octahedron. The use of *d* orbitals in constructing hybrid orbitals nicely corresponds to the notion of an expanded valence shell (Section 8.7). The geometrical arrangements characteristic of hybrid orbitals are summarized in Table 9.5.

Summary

You should realize that the purpose of hybrid orbitals is to provide a convenient model for using valence bond theory to describe covalent bonds in molecules. The picture of hybrid orbitals has limited predictive value; that is, we cannot say in advance that the nitrogen atom in NH_3 uses sp^3 hybrid orbitals. When we are given the molecular geometry, however, we can employ the concept of hybridization to describe the atomic orbitals used by the central atom in bonding. We therefore use the following three steps to predict the hybrid orbitals used by an atom in bonding:

1. Draw the Lewis structure for the molecule or ion.
2. Determine the electron-pair geometry using the VSEPR model.
3. Specify the hybrid orbitals needed to accommodate the electron pairs based on their geometrical arrangement (Table 9.5).

SAMPLE EXERCISE 9.5

Indicate the hybridization of orbitals employed by the central atom in each of the following: **(a)** NH_2^-; **(b)** SF_4 (see Sample Exercise 9.2).

Solution: **(a)** The Lewis structure for NH_2^- is

$$\left[\mathrm{H}:\ddot{\underset{\cdot\cdot}{\mathrm{N}}}:\mathrm{H}\right]^-$$

From the VSEPR model, we conclude that the four electron pairs around N should be arranged in a tetrahedral fashion. Such an arrangement is characteristic of sp^3 hybridization (Table 9.5): Two of the hybrid orbitals contain unshared electron pairs, and the other two contain pairs shared with hydrogen.

(b) As shown in Sample Exercise 9.2, there are ten valence-shell electrons around sulfur in SF_4. With an expanded octet of ten electrons, the use of a *d* orbital on the sulfur is indicated. The trigonal-bipyramidal arrangement of valence-shell electron pairs shown in Sample Exercise 9.2 corresponds to sp^3d hybridization (Table 9.5). One of the hybrid orbitals contains an unshared electron pair; the other four are bonded to fluorine.

PRACTICE EXERCISE

Predict the electron-pair geometry and the hybridization of the central atom in **(a)** SO_3^{2-}; **(b)** SF_6. ***Answers:*** **(a)** tetrahedral, sp^3; **(b)** octahedral, sp^3d^2

Table 9.5 Geometrical Arrangements Characteristic of Hybrid Orbital Sets

Atomic orbital set	Hybrid orbital set	Geometry	Examples
s,p	Two *sp*	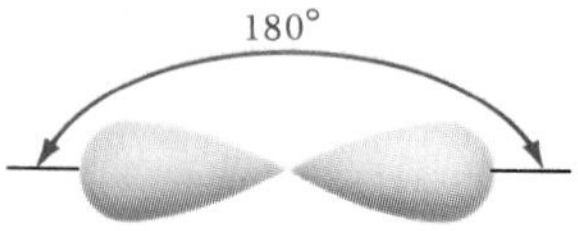 180° Linear	BeF_2, $HgCl_2$
s,p,p	Three sp^2	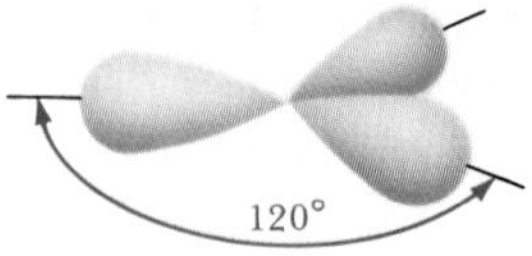 120° Trigonal planar	BF_3, SO_3
s,p,p,p	Four sp^3	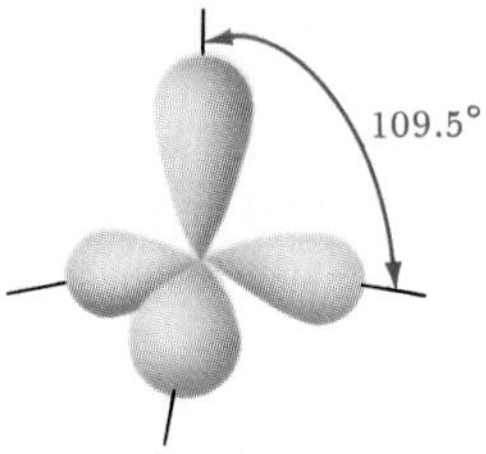 109.5° Tetrahedral	CH_4, NH_3, H_2O
s,p,p,p,d	Five sp^3d	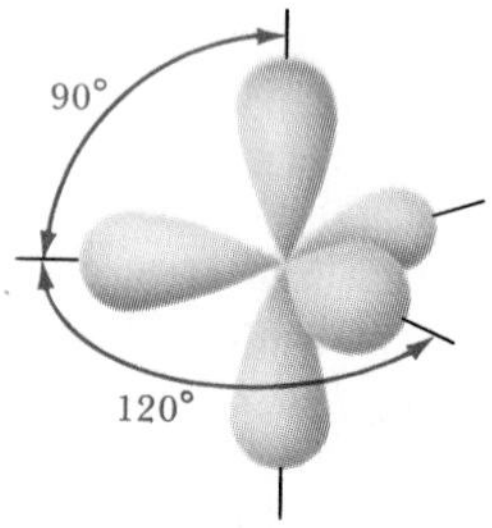 90° 120° Trigonal bipyramidal	PF_5, SF_4, BrF_3
s,p,p,p,d,d	Six sp^3d^2	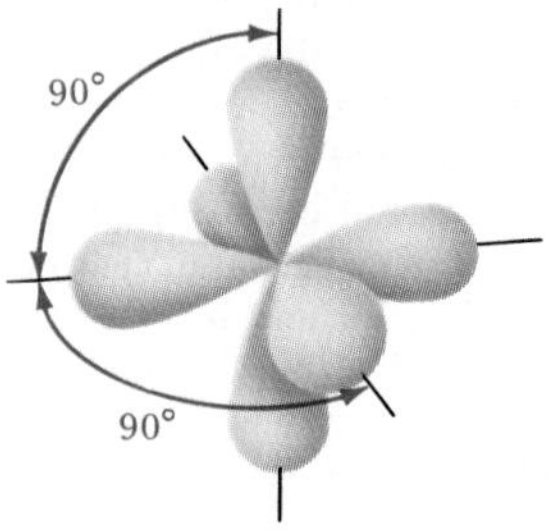 90° 90° Octahedral	SF_6, ClF_5, XeF_4

A CLOSER LOOK: Molecular Symmetry

Like many objects in nature, molecules can be beautifully symmetric. For example, we have seen that carbon tetrachloride, CCl_4, adopts the shape of a regular tetrahedron (Figure 9.1). We all have an intuitive sense of what makes an object highly symmetric, and we can apply this intuition to molecules. Suppose we compare, for example, carbon tetrachloride and chloroform, $CHCl_3$:

Cl — C(Cl)(Cl)(Cl)   H — C(Cl)(Cl)(Cl)

Carbon tetrachloride strikes us as the more highly symmetric molecule because the carbon atom is bonded to four identical chlorine atoms.

Symmetry serves more than an aesthetic purpose in chemistry. As we will see in Chapter 11, the symmetry of a molecule can profoundly affect its physical properties, for example, its melting point. Many of the interactions of molecules with electromagnetic radiation (the *spectroscopic* properties of molecules) are governed by their degree of symmetry. Chemists use a branch of mathematics called *group theory* to analyze the influence of symmetry on molecular properties. An important aspect of using group theory to predict properties of molecules is the quantitative determination of how much symmetry a molecule has.

We quantify the symmetry of a molecule by considering its *symmetry operations*—manipulations of the molecule that leave it looking unchanged. For example, if we rotate $CHCl_3$ by 120° about the C—H bond axis, it will look as if nothing has changed:

H — C(Cl)(Cl)(Cl) $\xrightarrow{\text{rotate } 120°}$ H — C(Cl)(Cl)(Cl)

In carbon tetrachloride, a 120° rotation about *any* of the C—Cl bond axes leaves the molecule unchanged. The higher symmetry of carbon tetrachloride is reflected in its number of symmetry operations: Carbon tetrachloride has 24 symmetry operations, whereas chloroform has only 6.

One of the most beautifully symmetric molecules is *dodecahedrane*, $C_{20}H_{20}$ [Figure 9.20(*a*)]. Dodecahedrane derives its name from the dodecahedron, which consists of 12 regular pentagons joined along their edges [Figure 9.20(*b*)]. Dodecahedrane was first synthesized in 1981 by Leo Paquette and co-workers at Ohio State University. Each of the carbon atoms in this compound sits at the vertex of a dodecahedron. The hybridization at each carbon atom is sp^3. Three of the hybrids are used to make σ bonds with other carbon atoms, and the fourth is used to form a C—H σ bond. Dodecahedrane has extremely high symmetry: It has 120 symmetry operations.

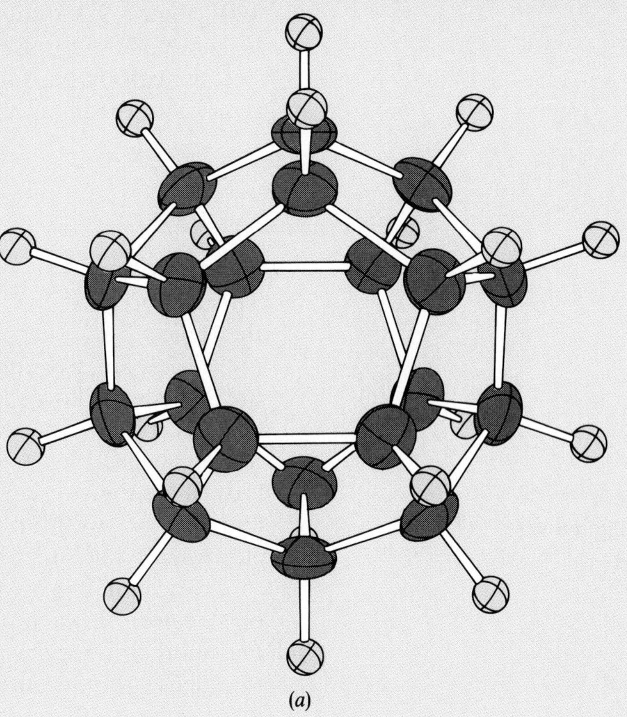
(*a*)

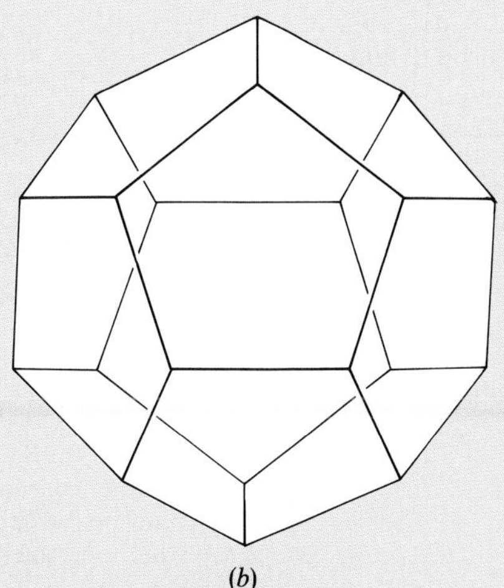
(*b*)

Figure 9.20 (*a*) A ball-and-stick representation of dodecahedrane, $C_{20}H_{20}$. The larger balls represent carbon atoms, and the smaller ones represent hydrogen atoms. (*b*) A dodecahedron, the solid that gives dodecahedrane its name. The drawing in part (a) is courtesy of Prof. L. A. Paquette, Ohio State University.

9.5 HYBRID ORBITALS AND MULTIPLE BONDS

We can also use the concept of hybridization to explain the bonding in molecules containing multiple bonds. For example, consider ethylene, C_2H_4, which possesses a C=C double bond. The bond angles in ethylene are all approximately 120° (Figure 9.21), suggesting that each carbon atom uses sp^2 hybrid orbitals (Figure 9.17) to form σ bonds with the other carbon and with two hydrogens. Because carbon has four valence electrons, after sp^2 hybridization one electron remains in the *unhybridized* $2p$ orbital:

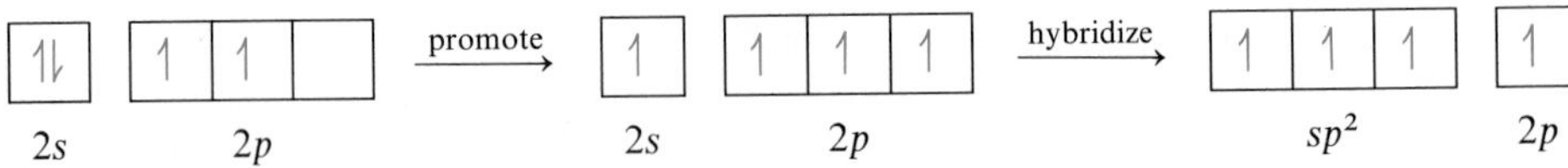

The unhybridized $2p$ orbital is directed perpendicular to the plane that contains the three sp^2 hybrid orbitals.

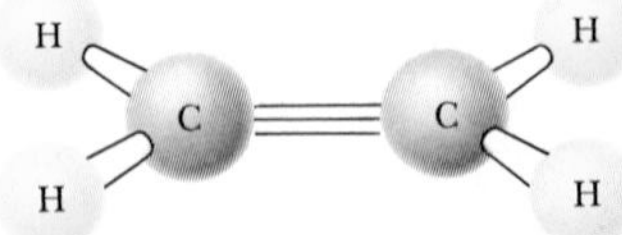

Figure 9.21 The molecular geometry of ethylene, C_2H_4.

Each sp^2 hybrid orbital on a carbon atom contains one electron. Figure 9.22 shows how the four C—H σ bonds are formed by the overlap of sp^2 hybrid orbitals on C with the $1s$ orbitals on each hydrogen atom. We therefore use eight electrons to form four electron-pair bonds. The C—C σ bond is formed by the overlap of two sp^2 hybrid orbitals, one on each carbon atom, and requires two more electrons. The C_2H_4 molecule has a total of 12 valence electrons, 10 of which are used to form one C—C and four C—H σ bonds.

The remaining two valence electrons reside in the unhybridized $2p$ orbitals, one electron on each of the carbon atoms. These $2p$ orbitals can overlap with one another in a sideways fashion, as shown in Figure 9.23. The resultant electron density is concentrated above and below the C—C bond axis; this is therefore a π bond (Figure 9.14). Thus, the C=C double bond in ethylene consists of one σ bond and one π bond.

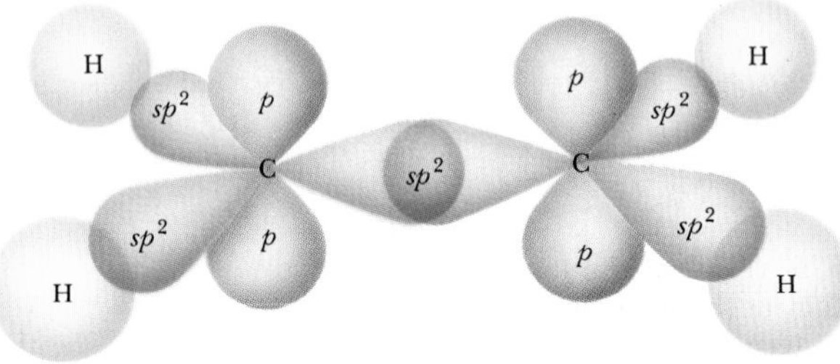

Figure 9.22 Hybridization of carbon orbitals in ethylene. The σ bonding framework, formed from sp^2 hybrid orbitals on the carbon atoms, determines the observed structure of the molecule.

Figure 9.23 Formation of the π bond in ethylene by overlap of the $2p$ orbitals on each carbon atom. Note that the centers of charge density in the π bond are above and below the bond axis, whereas in the σ bonds, the centers of charge density lie on the bond axes. The two lobes constitute *one* π bond.

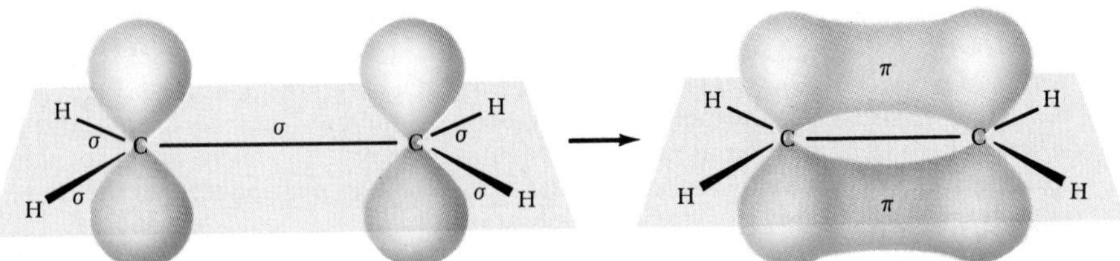

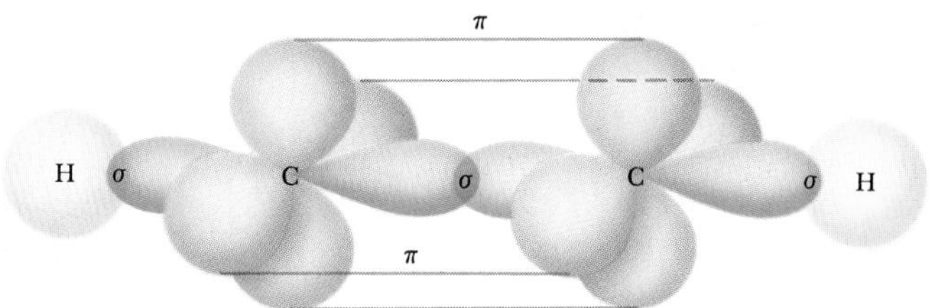

Figure 9.24 Formation of two π bonds in acetylene, C_2H_2, from the overlap of two sets of carbon $2p$ orbitals.

Although we cannot experimentally observe a π bond directly (all we can observe are the positions of the atoms), the structure of ethylene provides strong support for the presence of a π bond. First, the C—C bond length in ethylene (1.34 Å) is much shorter than that in compounds with C—C single bonds (1.54 Å), consistent with the presence of a C—C double bond. Second, all six atoms in C_2H_4 lie in the same plane. Only when the two CH_2 fragments lie in the same plane can the $2p$ orbitals that comprise the π bond achieve a good overlap. If the π bond were not present, there would be no reason to expect the two CH_2 fragments of ethylene to lie in the same plane. The fact that π bonds require that portions of a molecule be planar can introduce rigidity into molecules. This molecular rigidity can strongly affect the properties of substances, as we shall see in Chapter 12.

Although it is possible to make π bonds from d orbitals, the only kind of π bond that we shall consider is that formed by the overlap of p orbitals. This π bond can form only if unhybridized p orbitals are present on the bonded atoms. Therefore, only atoms having sp or sp^2 hybridization can be involved in such π bonding.

Triple bonds can also be explained by using hybrid orbitals. Consider acetylene, C_2H_2, a linear molecule containing a triple bond: H—C≡C—H. The linear geometry suggests that each carbon atom uses sp hybrid orbitals to form σ bonds with the other carbon and one hydrogen. Each carbon atom then has two remaining unhybridized $2p$ orbitals at right angles to each other and to the axis of the sp hybrid set (Figure 9.24). These overlap to form a pair of π bonds. Thus, the triple bond in acetylene consists of one σ bond and two π bonds.

SAMPLE EXERCISE 9.6

Formaldehyde, which is a planar molecule, has the following Lewis structure:

```
H
 \
  C=Ö:
 /
H
```

Describe the bonding in formaldehyde in terms of an appropriate set of hybrid orbitals at the carbon atom.

Solution: Using the VSEPR model, we would predict the bond angles around C to be about 120° (trigonal-planar geometry). The 120° bond angles about the central atom suggest sp^2 hybrid orbitals for the σ bonds (Table 9.5). There remains a $2p$ orbital on carbon, perpendicular to the plane of the three σ bonds.

Using the VSEPR model, we predict that oxygen will have two unshared electron pairs and that the O—C σ bond will be in a trigonal plane, with approximately 120° angles between them. In this case also, there remains a $2p$ orbital on oxygen, perpendicular to the plane of the three σ bonds. This orbital overlaps with the similarly oriented $2p$ orbital on carbon to form a π bond between carbon and oxygen, as illustrated in Figure 9.25.

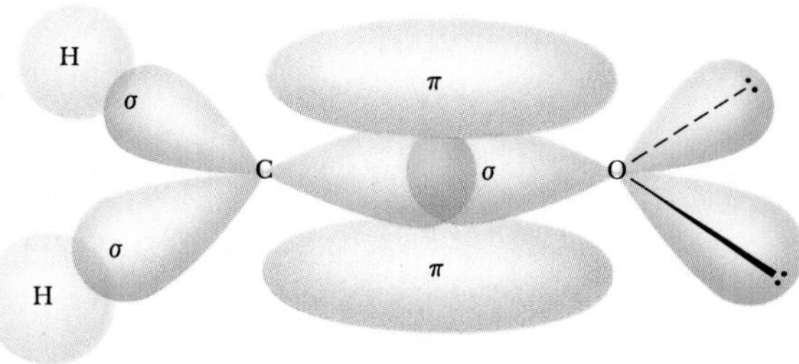

Figure 9.25 Formation of σ and π bonds in formaldehyde, H_2CO.

PRACTICE EXERCISE

Consider the acetonitrile molecule:

```
     H
     |
  H—C—C≡N:
     |
     H
```

(a) Predict the bond angles around each carbon; **(b)** give the hybridizations on both carbon atoms; **(c)** determine the total number of σ and π bonds in the molecule. ***Answers:*** **(a)** 109° around the leftmost C and 180° on the rightmost C; **(b)** sp^3, sp; **(c)** five σ bonds and two π bonds.

Delocalized Bonding

In each of the molecules we have discussed in this chapter, the bonding electrons are *localized*. By this we mean that the σ and π electrons are associated totally with the two atoms forming the bond. In some molecules, particularly those with more than one resonance form, we cannot accurately describe the bonding as localized. Benzene, C_6H_6, is an example of such a molecule. Benzene has two resonance forms:

```
        H                       H
        |                       |
  H    C     H            H    C     H
   \ //  \  /              \ /  \\  /
    C     C                 C     C
    |     ||      ⟷        ||     |
    C     C                 C     C
   /  \\ /  \              /  \  // \
  H     C     H           H     C     H
        |                       |
        H                       H
```

All six C—C bonds in benzene are of equal length—1.40 Å—intermediate between the values for a C—C single bond (1.54 Å) and a C=C double bond (1.34 Å). The molecule is planar, and the bond angles around each carbon are 120°.

To describe the bonding in benzene in terms of hybrid orbitals, we first choose a hybridization scheme consistent with the geometry of the molecule. Because each carbon is surrounded by three atoms at 120° angles, the appropriate hybrid set is sp^2, as it is in ethylene. Six C—C σ bonds and six C—H σ bonds are formed from the sp^2 hybrid orbitals, as shown in Figure 9.26(*a*). This leaves a $2p$ orbital on each carbon that is oriented perpendicularly to the plane of the molecule. The situation is very much like that in ethylene, except that we now have six carbon $2p$ orbitals arranged in a ring [Figure 9.26(*b*)].

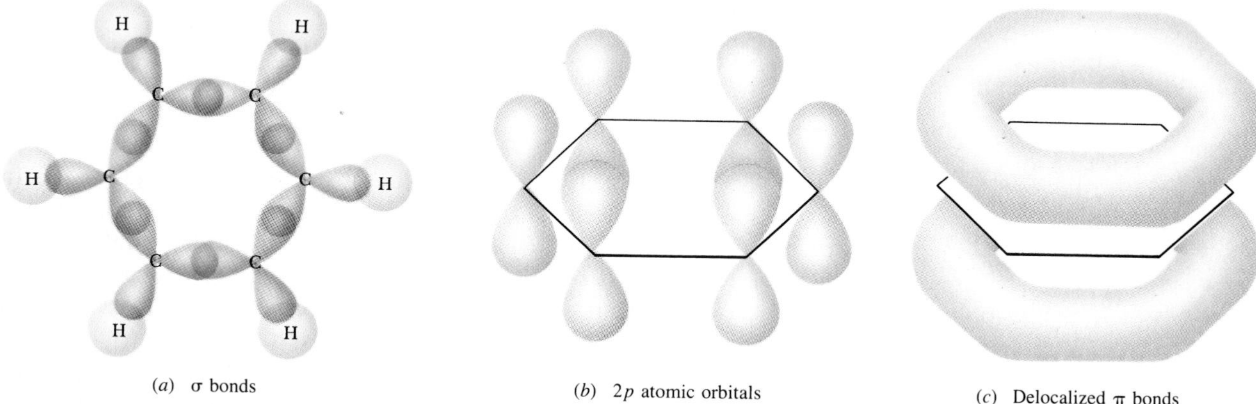

Figure 9.26 The σ and π bond networks in benzene, C_6H_6. (*a*) The C—C and C—H σ bonds all lie in the plane of the molecule and are formed by using carbon sp^2 hybrid orbitals. (*b*) Each carbon atom has an unhybridized $2p$ orbital that lies perpendicular to the molecular plane. (*c*) A representation of the smearing out, or delocalization, of three C—C π bonds among the six carbon atoms.

Each of the $2p$ orbitals in benzene has one electron in it, so three π bonds can be formed. Now, if localized π bonds were formed between pairs of carbon atoms, the molecule would have three short double bonds and three longer single bonds. This corresponds precisely to the situation described by either of the individual Lewis structures above. However, a description that better reflects *both* resonance structures is to envision the three π bonds "smeared out" among all six carbon atoms, as shown in Figure 9.26(*c*). This model leads to the description of each bond between neighboring carbon atoms as an average of a single bond and a double bond, consistent with the observed C—C bond lengths in benzene.

Because we cannot describe the π bonds in benzene in terms of electron-pair bonds between neighboring atoms, we say that the π bonds are **delocalized** among the six carbon atoms. Delocalization of the electrons in its π bonds gives benzene a special stability, called (rather peculiarly) *aromaticity*. For example, bromine readily attacks the π bond of ethylene, but the π-bonding system in benzene is far less reactive.

It is common to represent carbon- and hydrogen-containing molecules such as benzene by omitting the hydrogen atoms attached to carbon and showing only the carbon-carbon framework with the vertices unlabeled. The π bonds are shown either by using one of the Lewis structures or by placing a circle in the center of the carbon ring:

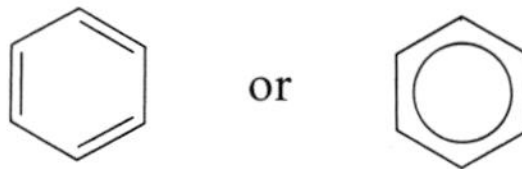

The representation on the left gives a better sense of the number of electrons used in forming C—C σ and π bonds, but it implies localized bonding. The representation on the right correctly shows the delocalization of the π bonds, but it does not explicitly indicate the number of electrons used in forming these bonds. Chemists use both representations of benzene interchangeably.

General Conclusions

On the basis of all the examples we've seen, we can formulate a few general conclusions that are helpful in using the concept of hybrid orbitals to discuss molecular structures.

1. Every pair of bonded atoms shares one or more pairs of electrons. In every bond, at least one pair of electrons is localized in the space between the atoms, in a σ bond. The appropriate set of hybrid orbitals used to form the σ bonds between an atom and its neighbors is determined by the observed geometry of the molecule. The relationship between hybrid orbital set and geometry about an atom is given in Table 9.5.
2. The electrons in σ bonds are localized in the region between two bonded atoms and do not make a significant contribution to the bonding between any other two atoms.
3. When atoms share more than one pair of electrons, the additional pairs are in π bonds. The centers of charge density in a π bond lie above and below the bond axis.
4. Molecules can have π bonds that extend over more than two bonded atoms. Electrons in π bonds that extend over more than two atoms are said to be delocalized.

9.6 MOLECULAR ORBITALS

The models of covalent bonding and molecular geometry discussed in this and the preceding chapter are very useful. Valence bond theory provides a nice way of relating atomic orbitals and Lewis dot structures. Hybrid orbitals allow us to rationalize the observed geometries of molecules. For example, we can understand why methane has the formula CH_4, how the carbon and hydrogen atomic orbitals are used to form electron-pair bonds, and why the arrangement of the C—H bonds about the central carbon is tetrahedral. But in all of this discussion we have only briefly touched on a very important question: Why do atoms combine to form covalent bonds in the first place? The answer to this question has to be expressed in terms of energy.

In Chapters 6 and 7, we saw that electrons in atoms exist in allowed energy states, which we call atomic orbitals. The quantum theory tells us that, in a similar way, electrons in molecules exist in allowed energy states called **molecular orbitals**. The model we use to describe molecular orbitals is called **molecular orbital theory**.

We usually think of a molecular orbital as forming from a combination of atomic orbitals. As atoms approach each other and their atomic orbitals overlap, molecular orbitals are formed. Molecular orbitals have many of the same characteristics as atomic orbitals. For example, they can hold a maximum of two electrons (with opposite spins), and their electron-density distributions can be visualized by using contour representations, as we did when discussing atomic orbitals. Furthermore, whenever two atomic orbitals interact, two molecular orbitals form.

The Hydrogen Molecule

We shall first look at molecular orbitals by examining the hydrogen molecule, H_2. We can combine the $1s$ orbitals on the hydrogen atoms to form two molecular orbitals for the H_2 molecule. In the first molecular orbital, we combine the 1*s* atomic orbitals in a fashion that concentrates electron density in the region between the two hydrogen nuclei, as shown in Figure 9.27. We can envision this sausage-shaped molecular orbital as resulting from the sum of the two atomic orbitals in such a way that the atomic orbitals reinforce each other in the bond region. Because an electron in this molecular orbital is strongly attracted to both nuclei, the electron is more stable (lower energy) than it is in the 1*s* orbital of a hydrogen atom. This molecular orbital is called a **bonding molecular orbital** because the concentration of electron density between the nuclei holds the atoms together in a covalent bond.

What is the other molecular orbital that results from interaction of the two 1*s* atomic orbitals? In the bonding molecular orbital described above, the two 1*s* atomic orbitals add in such a way as to increase the charge density between the nuclei. In the other molecular orbital, precisely the opposite occurs: The 1*s* atomic orbitals combine in a way that leads to very little electron density between the nuclei, as shown in Figure 9.27. Now, instead of reinforcing each other in the region between the nuclei, the atomic orbitals cancel each other in this region. There is a node between nuclei, at which the electron density is zero; in fact, the greatest electron density is on opposite sides of the nuclei. Thus, this molecular orbital excludes electrons from the very region in which a bond must be formed. For this reason, it is called an **antibonding molecular orbital**. An electron in this orbital is actually repelled from the bonding region and is therefore less stable (at higher energy) than it is in the 1*s* orbital of a hydrogen atom.

The electron density in both the bonding and the antibonding molecular orbitals of H_2 is centered about an imaginary line passing through the two nuclei (see Figure 9.27). Molecular orbitals of this type are called **sigma (σ) molecular orbitals**. The bonding sigma molecular orbital of H_2 is labeled σ_{1s}, the subscript indicating that the molecular orbital is formed from two 1*s* orbitals. The antibonding sigma orbital of H_2 is labeled σ^*_{1s}, the asterisk denoting that the orbital is antibonding.

The interaction between two 1*s* orbitals to form σ_{1s} and σ^*_{1s} molecular orbitals can be represented by an **energy-level diagram** (also called a **molecular**

Figure 9.27 Contour representations of the combination of two hydrogen 1*s* atomic orbitals into two molecular orbitals of H_2. In the bonding molecular orbital, σ_{1s}, the atomic orbitals combine constructively, leading to a buildup of electron density between the nuclei. In the antibonding molecular orbital, σ^*_{1s}, the orbitals combine destructively in the bonding region: The two atomic orbitals partially cancel each other, leading to an exclusion of electron density from the region between the nuclei. Note that the σ^*_{1s} orbital has a node between the two nuclei.

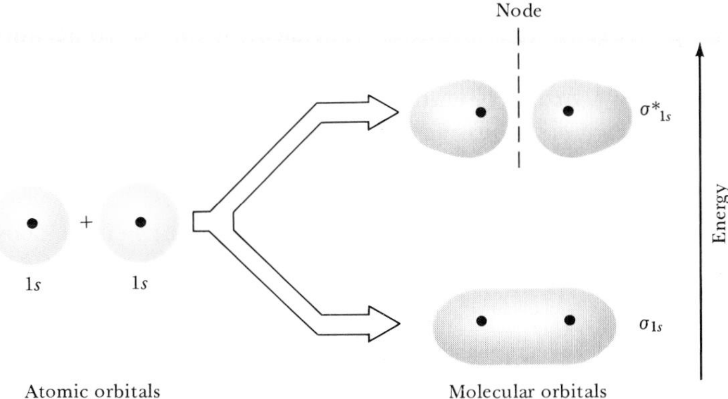

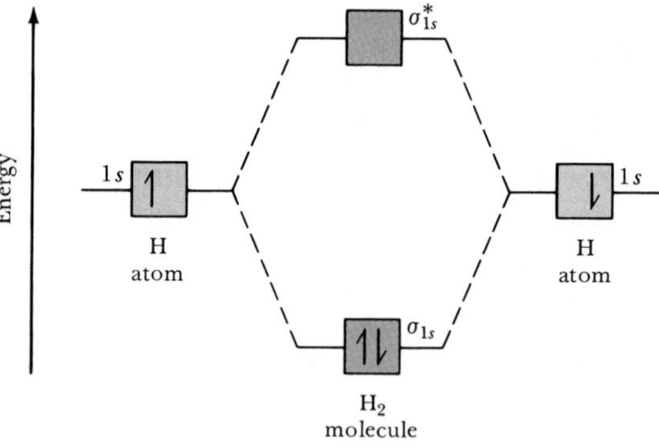

Figure 9.28 Energy-level diagram for the molecular orbitals in the H_2 molecule.

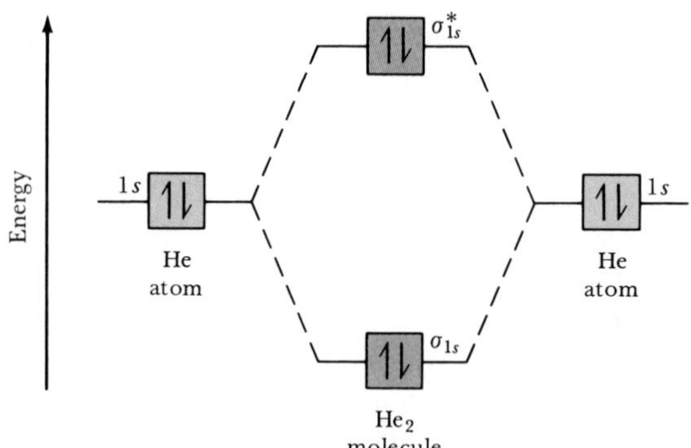

Figure 9.29 Energy-level diagram for the molecular orbitals in the hypothetical He_2 molecule.

orbital diagram), as shown in Figure 9.28. Such a diagram shows the interacting atomic orbitals in the left and right columns and the molecular orbitals in the middle column. Note that the bonding molecular orbital, σ_{1s}, is lower in energy than the hydrogen atom 1*s* orbitals, whereas the antibonding orbital, σ_{1s}^*, is higher in energy than the 1*s* orbitals. Like atomic orbitals, each of the molecular orbitals can accommodate two electrons. Each H atom contains one electron, so there are two electrons in H_2. These electrons occupy the lower-energy bonding (σ_{1s}) molecular orbital, with their spins paired. Electrons occupying a bonding molecular orbital are called *bonding electrons.* Because the σ_{1s} orbital is lower in energy than the isolated 1*s* orbitals, the H_2 molecule is more stable than the two separate H atoms.

In contrast, let us consider formation of the hypothetical He_2 molecule. When two helium atoms come together, each atom has two electrons in a 1*s* orbital. The 1*s* orbitals of He combine to form σ_{1s} and σ_{1s}^* molecular orbitals, just as described for hydrogen. Now, however, four electrons must be accommodated. Because only two electrons can be placed in the σ_{1s} orbital, the other two must be placed in the σ_{1s}^* orbital (Figure 9.29). Thus, there are two bonding electrons and two *antibonding electrons.* The energy decrease resulting from the bonding electrons is offset by the energy increase resulting from the antibonding electrons.† Hence, He_2 is not a stable molecule. Molecular orbital theory correctly predicts that hydrogen forms diatomic molecules but that helium does not.

† In fact, antibonding molecular orbitals are slightly more unfavorable than bonding orbitals are favorable. Whenever there is an equal number of electrons in bonding and antibonding orbitals, the energy is higher than it is for the isolated atoms; no bond is formed.

Bond Order

In molecular orbital theory, the stability of a covalent bond is related to its **bond order**, defined as follows:

$$\text{Bond order} = \tfrac{1}{2}(\text{number of bonding electrons} - \text{number of antibonding electrons})$$

That is, the bond order is half the difference between the number of bonding electrons and the number of antibonding electrons. We take half the difference because we are used to thinking of bonds in terms of pairs of electrons. A bond order of 1 represents a single bond, a bond order of 2 represents a double bond, and a bond order of 3 represents a triple bond. Because molecular orbital theory also treats molecules with an odd number of electrons, bond orders of $\frac{1}{2}$, $\frac{3}{2}$, or $\frac{5}{2}$ are possible.

Because H_2 has two bonding electrons and no antibonding ones (Figure 9.28), it has a bond order of $\frac{1}{2}(2 - 0) = 1$. Because He_2 has two bonding electrons and two antibonding ones (Figure 9.29), it has a bond order of $\frac{1}{2}(2 - 2) = 0$. A bond order of 0 means that no bond exists.

SAMPLE EXERCISE 9.7

What is the bond order of the He_2^+ ion? Would you expect this ion to be stable relative to the separated He atom and He^+ ion?

Solution: The energy-level diagram for this system is shown in Figure 9.30. The He_2^+ ion has a total of three electrons. Two are placed in the bonding orbital, the third in the antibonding orbital. Thus the bond order is

$$\text{Bond order} = \tfrac{1}{2}(2 - 1) = \tfrac{1}{2}$$

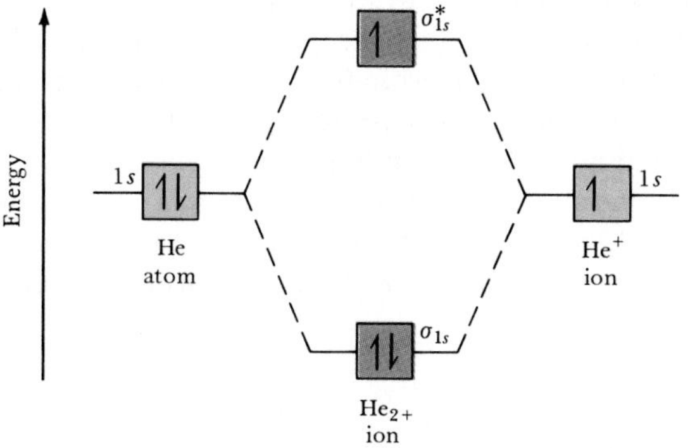

Figure 9.30 Energy-level diagram for the He_2^+ ion.

Because the bond order is greater than 0, the He_2^+ molecular ion is predicted to be stable relative to the separated He and He^+. Formation of He_2^+ in the gas phase has been demonstrated in laboratory experiments.

PRACTICE EXERCISE

Determine the bond order of the H_2^- ion. ***Answer:*** $\frac{1}{2}$

9.7 MOLECULAR ORBITALS FOR SECOND-PERIOD DIATOMIC MOLECULES

Just as we treated the bonding in H_2 by using molecular orbital theory, we can also consider the molecular orbital description of other diatomic molecules. In this section, we shall restrict our discussion to *homonuclear* diatomic molecules (those composed of two identical atoms) of elements in the second period. As we shall see, the procedure for determining the distribution of electrons in these molecules closely follows the one we used for H_2.

Second-period atoms have more than one atomic orbital. We shall therefore consider some rules that will simplify the construction of molecular orbitals from these atomic orbitals. Further, we shall look at some rules that govern the way in which we place electrons in molecular orbitals. Some of the most important of these rules follow:

1. The number of molecular orbitals formed equals the number of atomic orbitals combined.
2. Atomic orbitals combine most effectively with other atomic orbitals of similar energy.
3. The effectiveness with which two atomic orbitals combine is proportional to their overlap with one another; that is, as the overlap increases, the bonding orbital is lowered in energy, and the antibonding orbital is raised in energy.
4. Each molecular orbital can accommodate at most two electrons, with their spins paired (Pauli exclusion principle).
5. When molecular orbitals have the same energy, one electron enters each orbital (with parallel spins) before spin pairing occurs (Hund's rule).

Molecular Orbitals for Li_2 and Be_2

Lithium, the first atom of the second period, has a $1s^2 2s^1$ electron configuration. When it is heated above its boiling point (1347°C), Li_2 molecules are found in the vapor phase. The Lewis structure for Li_2 indicates a Li—Li single bond. We shall now address the description of the bonding in Li_2 in terms of molecular orbitals.

Because the 1*s* and 2*s* orbitals of Li are so different in energy, we may assume that the 1*s* orbital on one Li atom interacts only with the 1*s* orbital on the other atom (rule 2). Likewise, the 2*s* orbitals interact only with each other. The resulting energy-level diagram is shown in Figure 9.31. The 1*s* orbitals combine to form σ_{1s} and σ^*_{1s} bonding and antibonding orbitals, as they did for H_2. The 2*s* orbitals interact with one another in exactly the same way, producing bonding (σ_{2s}) and antibonding (σ^*_{2s}) molecular orbitals. Because the 2*s* orbitals of Li extend farther from the nucleus than do the 1*s*, the 2*s* orbitals overlap

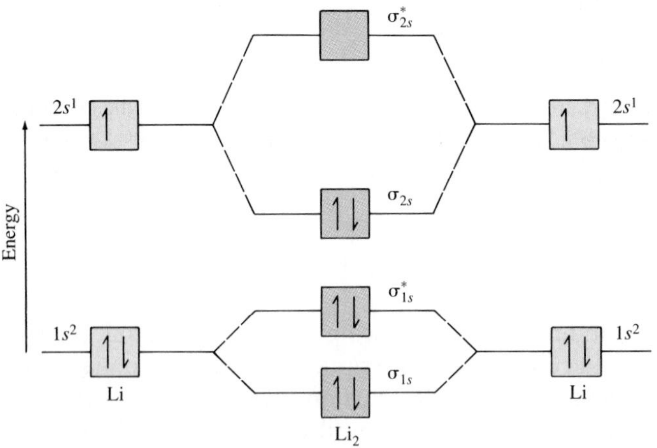

Figure 9.31 Energy-level diagram for the Li_2 molecule.

more effectively. As a result, the energy separation between the σ_{2s} and σ^*_{2s} orbitals is greater than that for the 1*s*-based molecular orbitals. Note that, despite being an antibonding orbital, the σ^*_{1s} molecular orbital is lower in energy than the σ_{2s} bonding orbital. The 1*s* orbitals of Li are so much lower in energy that their antibonding orbital is still well below the 2*s*-based bonding orbital.

Each Li atom has three electrons, so six electrons must be placed in the molecular orbitals of Li_2. As shown in Figure 9.31, these occupy the σ_{1s}, σ^*_{1s}, and σ_{2s} molecular orbitals, each with two electrons. There are four electrons in bonding orbitals and two in antibonding orbitals, so the bond order equals $\frac{1}{2}(4 - 2) = 1$. The molecule has a single bond, in accord with its Lewis structure.

Because both the σ_{1s} and σ^*_{1s} molecular orbitals of Li_2 are completely filled, the 1*s* orbitals contribute almost nothing to the bonding. The single bond in Li_2 is due essentially to the interaction of the valence 2*s* orbitals on the Li atoms. This example illustrates the general rule that *filled atomic subshells usually do not contribute significantly to bonding in molecule formation.* This rule is equivalent to our use of only the valence electrons when drawing Lewis structures. Thus, we need not consider further the 1*s* orbitals while discussing the other second-period diatomic molecules.

The molecular orbital description of Be_2 follows readily from the energy-level diagram for Li_2. Each Be atom has four electrons ($1s^2 2s^2$), so we must place eight electrons in molecular orbitals. Thus, we completely fill the σ_{1s}, σ^*_{1s}, σ_{2s}, and σ^*_{2s} molecular orbitals. We have an equal number of bonding and antibonding electrons, so the bond order equals 0. Be_2 does not exist.

Molecular Orbitals from 2*p* Atomic Orbitals

Before we can consider the next molecule, B_2, we must look at the molecular orbitals that result from combining 2*p* atomic orbitals. The interaction between *p* orbitals is shown in Figure 9.32, where we have arbitrarily chosen the internuclear axis to be the *z* axis. The $2p_z$ orbitals face each other in a "head-to-head" fashion. Just as we did for the *s* orbitals, we can combine the $2p_z$ orbitals in two ways. One combination concentrates electron density between the nuclei and is therefore a bonding molecular orbital. The other combination excludes electron density from the bonding region; it is an antibonding molecular orbital. In each of these molecular orbitals, the electron density lies along the line through the nuclei. Hence, they are σ molecular orbitals: σ_{2p} and σ^*_{2p}.

The other 2*p* orbitals overlap in a sideways fashion and thus concentrate electron density on opposite sides of the line through the nuclei. Molecular orbitals of this type are called **pi (π) molecular orbitals**. We get one π bonding molecular orbital by combining the $2p_x$ atomic orbitals and another one from the $2p_y$ atomic orbitals. These two π_{2p} molecular orbitals have the same energy; they are degenerate. Likewise, we get two degenerate π^*_{2p} antibonding molecular orbitals.

The $2p_z$ orbitals on two atoms point directly at one another. Hence, the overlap of two $2p_z$ orbitals is greater than that for two $2p_x$ or $2p_y$ orbitals. From rule 3, we therefore expect the σ_{2p} molecular orbital to be lower in energy (more stable) than the π_{2p} molecular orbitals. Similarly, the σ^*_{2p} molecular orbital should be higher in energy (less stable) than the π^*_{2p} molecular orbitals. This would indeed be the case were it not for the influence of the 2*s* atomic orbitals, as we shall see shortly.

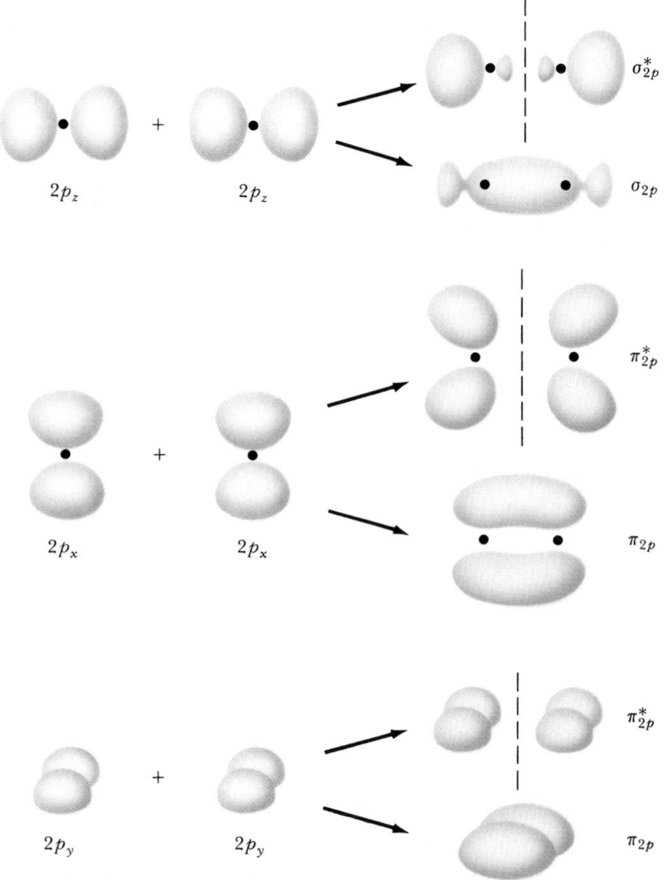

Figure 9.32 Contour representations of the molecular orbitals formed by the $2p$ orbitals on two atoms. Each time we combine two atomic orbitals, we obtain two molecular orbitals, one bonding and one antibonding. The two π_{2p} molecular orbitals have the same energy; they are degenerate. Likewise, the two π^*_{2p} molecular orbitals are degenerate.

Electron Configurations for B_2 Through F_2

The elements boron through fluorine all have valence $2s$ and $2p$ electrons. We can use the energy-level diagram in Figure 9.33 to discuss all the homonuclear diatomic molecules of these elements. This diagram is as we would expect, with one exception: We see that the σ_{2p} molecular orbital is at a higher energy than the π_{2p} molecular orbitals, contrary to what we would expect from rule 3. This reversal in ordering is the result of an unfavorable interaction between the σ_{2p} molecular orbital and the σ_{2s} molecular orbital. Both of these orbitals try to concentrate electron density between the nuclei on the internuclear axis and, because they are not very different in energy, they can interact. The σ_{2p} molecular orbital is literally pushed upward in energy to the point where it is above the π_{2p} orbitals, and the σ_{2s} molecular orbital is pushed down in energy.[†]

Given the energy ordering of the molecular orbitals shown in Figure 9.33 it is a simple matter to place the appropriate number of electrons in them for the second-period diatomic molecules. For example, a boron atom has three valence electrons (remember, we are ignoring the inner-shell $1s$ electrons). Thus, for B_2 we must place six electrons in molecular orbitals. Four of these fully occupy the σ_{2s} and σ^*_{2s} molecular orbitals, leading to no net bonding. The last

[†] The interaction between the σ_{2p} and σ_{2s} molecular orbitals decreases as we move from left to right in the second period. As a result, the σ_{2p} orbital is higher in energy than the π_{2p} orbitals for B_2, C_2, and N_2. For O_2, F_2, and Ne_2, the σ_{2p} orbital is lower in energy than the π_{2p} orbitals.

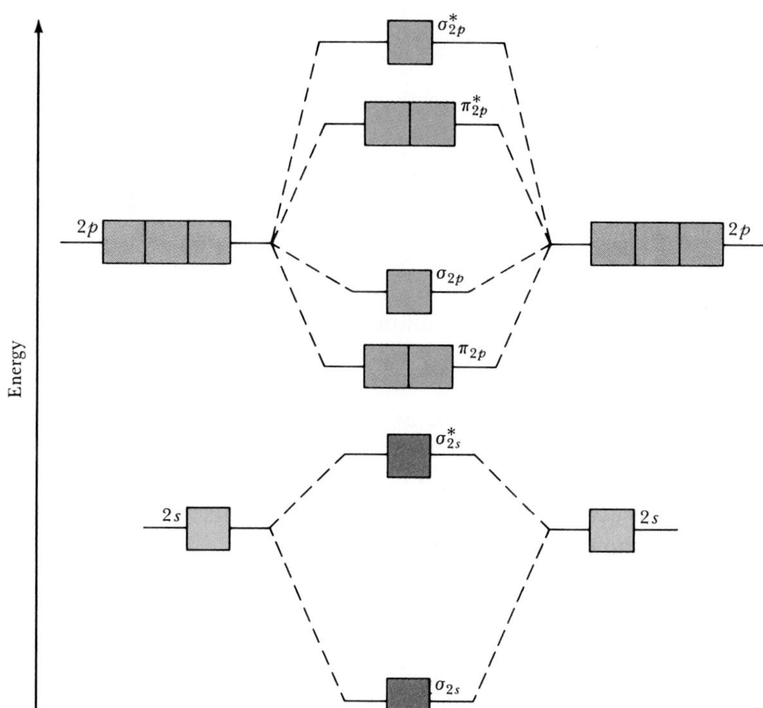

Figure 9.33 General energy-level diagram for molecular orbitals of homonuclear second-row diatomic molecules.

two electrons are put in the π_{2p} bonding molecular orbitals; one electron is put in each π_{2p} orbital with the same spin (rule 5). Therefore, B_2 has a bond order of 1. Each time we move to the right in the second period, two more electrons must be placed in the diagram. For example, on moving to C_2, we have two more electrons than in B_2, and these electrons also are placed in the π_{2p} molecular orbitals, completely filling them. The electron configurations and bond orders for the diatomic molecules B_2 through F_2 are given in Table 9.6.

Table 9.6 Molecular Orbital Electron Configurations and Some Experimental Data for Several Second-Row Diatomic Molecules

	B_2	C_2	N_2	O_2	F_2
σ^*_{2p}	□	□	□	□	□
π^*_{2p}	□ □	□ □	□ □	↑ ↑	↑↓ ↑↓
σ_{2p}	□	□	↑↓	↑↓	↑↓
π_{2p}	↑ ↑	↑↓ ↑↓	↑↓ ↑↓	↑↓ ↑↓	↑↓ ↑↓
σ^*_{2s}	↑↓	↑↓	↑↓	↑↓	↑↓
σ_{2s}	↑↓	↑↓	↑↓	↑↓	↑↓
Bond order	1	2	3	2	1
Bond energy (kJ/mol)	290	620	941	495	155
Bond length (Å)	1.59	1.31	1.10	1.21	1.43
Magnetic behavior	Paramagnetic	Diamagnetic	Diamagnetic	Paramagnetic	Diamagnetic

Electron Configurations and Molecular Properties

It is interesting to compare the electronic structures of these diatomic molecules with their observable properties. The behavior of a substance in a magnetic field provides an important insight into the arrangements of its electrons. Molecules with one or more unpaired electrons are attracted into a magnetic field. The more unpaired electrons in a species, the stronger the force of attraction. This type of magnetic behavior is called **paramagnetism**.

Substances with no unpaired electrons are weakly repelled from a magnetic field. This property is called **diamagnetism**. Diamagnetism is a much weaker effect than paramagnetism. A straightforward method for measuring the magnetic properties of a substance, illustrated in Figure 9.34, involves weighing the substance in the presence and absence of a magnetic field. If the substance is paramagnetic, it will appear to weigh more in the magnetic field; if it is diamagnetic, it will appear to weigh less. The magnetic behaviors observed for the diatomic molecules of the second-period elements agree with the electron configurations shown in Table 9.6.

The electron configurations can also be related to the bond distances and bond-dissociation energies of the molecules. As bond orders increase, bond distances decrease, and bond-dissociation energies increase. Notice the short bond distance and high bond-dissociation energy of N_2, whose bond order is 3. The N_2 molecule does not react readily with other substances to form nitrogen compounds. The high bond order of the molecule helps explain its exceptional stability. We should also note that molecules with the same bond orders do not have the same bond distances and bond-dissociation energies. Bond order is only one factor influencing these properties. Other factors, including the nuclear charges and the extent of orbital overlap, also contribute.

The bonding in the dioxygen molecule, O_2, is especially interesting. The Lewis structure of this molecule shows a double bond and complete pairing of electrons:

$$\ddot{\underset{\cdot\cdot}{O}}=\ddot{\underset{\cdot\cdot}{O}}$$

Figure 9.34 Experiment for determining the magnetic properties of a sample. (*a*) The sample is first weighed in the absence of a magnetic field. (*b*) When a field is applied, a diamagnetic sample tends to move out of the field and thus appears to have a lower mass. (*c*) A paramagnetic sample is drawn into the field and thus appears to gain mass. Paramagnetism is a much stronger effect than is diamagnetism.

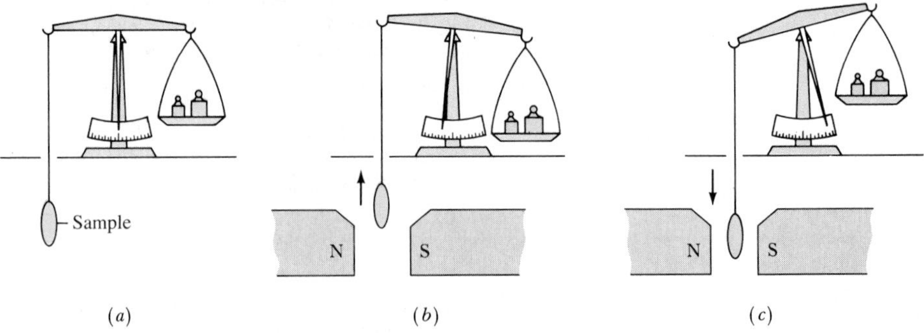

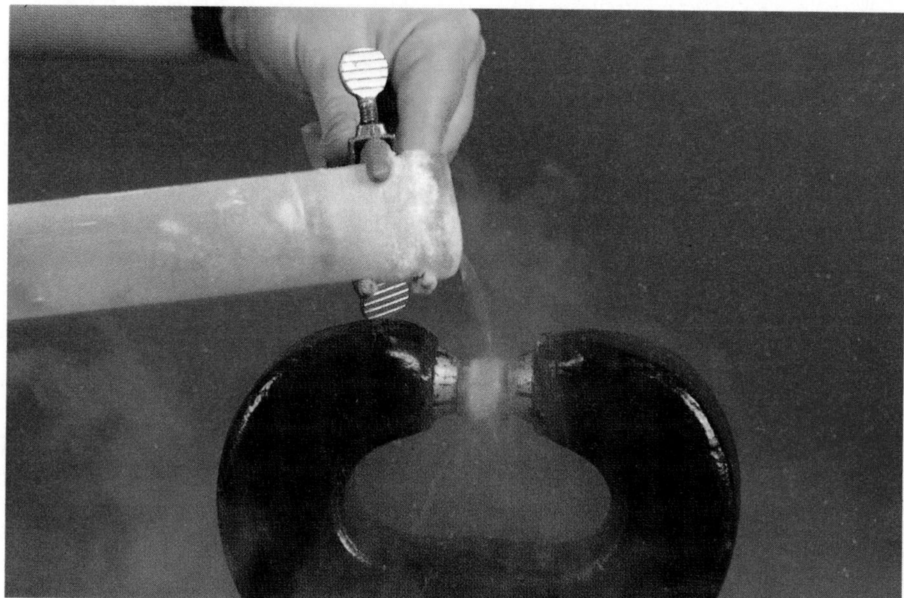

Figure 9.35 Liquid O_2 being poured between the poles of a magnet. Because the O_2 is paramagnetic, it is attracted into the magnetic field and forms a bridge between the magnetic poles. (Donald Clegg and Roxy Wilson)

The short O—O bond distance (1.21 Å) and the relatively high bond-dissociation energy (495 kJ/mol) of the molecule are in agreement with the presence of a double bond. However, the molecule is found to contain two unpaired electrons. The paramagnetism of O_2 is demonstrated in Figure 9.35. Although the Lewis structure fails to account for the paramagnetism of O_2, molecular orbital theory correctly predicts that there are two unpaired electrons in the π_{2p}^* orbital of the molecule (Table 9.6). The molecular orbital description also correctly indicates a bond order of 2.

SAMPLE EXERCISE 9.8

Predict the following properties of O_2^+: **(a)** number of unpaired electrons; **(b)** bond order; **(c)** bond-dissociation energy and bond length.

Solution: **(a)** The O_2^+ ion has one electron less than O_2. The electron removed from O_2 to form O_2^+ is one of the two unpaired π^* electrons (see Table 9.6). Therefore, O_2^+ should have just one unpaired electron left.

(b) The molecule has eight bonding electrons (the same number as O_2) and three antibonding ones (one less than O_2). Thus, its bond order is

$$\text{Bond order} = \tfrac{1}{2}(8 - 3) = 2\tfrac{1}{2}$$

(c) The bond-dissociation energy and bond length should be about midway between that for O_2 and N_2, say 720 kJ/mol and 1.15 Å, respectively. The observed bond-dissociation energy and bond length of the ion are 625 kJ/mol and 1.123 Å, respectively.

PRACTICE EXERCISE

Predict the magnetic properties and bond order of **(a)** the peroxide ion, O_2^{2-}; **(b)** the acetylide ion, C_2^{2-} *Answers:* **(a)** diamagnetic, 1; **(b)** diamagnetic, 3

CHEMISTRY AT WORK: Organic Dyes

The chemistry of color has fascinated people since ancient times. The brilliant colors around you—those of your clothes, the photographs in this book, the foods you eat—are due to the selective absorption of light by chemicals. Light excites electrons in molecules. In a molecular orbital picture, we can envision light exciting an electron from a filled molecular orbital to a higher-energy empty one. Because the molecular orbitals have a definite energy, only light of the proper wavelength can excite electrons. The situation is analogous to the atomic line spectra we discussed in Section 6.2. If the appropriate wavelength for exciting electrons is in the visible portion of the electromagnetic spectrum, the substance will appear colored: Certain wavelengths of white light are absorbed, others are not. For example, a red traffic light appears red because only red light is transmitted through the lens. The other wavelengths of visible light are absorbed by it.

In using molecular orbital theory to discuss the absorptions of light by molecules, it is useful to focus on two molecular orbitals in particular. The *highest occupied molecular orbital* (HOMO) is the molecular orbital of highest energy that has electrons in it. The *lowest unoccupied molecular orbital* (LUMO) is the molecular orbital of lowest energy that does not have electrons in it. In N_2, for example, the HOMO is the π_{2p} molecular orbital and the LUMO is the π^*_{2p} molecular orbital (Table 9.6). The energy difference between the HOMO and the LUMO—known as the HOMO-LUMO gap—is related to the minimum energy needed to excite an electron in the molecule. Colorless or white substances usually have such a large HOMO-LUMO gap that visible light is not energetic enough to excite an electron to the higher level. For example, the minimum energy needed to excite an electron in N_2 corresponds to light with a wavelength of less than 200 nm, which is far into the ultraviolet part of the spectrum (Figure 6.3). As a result, N_2 cannot absorb any visible light and is therefore colorless.

Many of the rich colors with which we are familiar are due to *organic dyes*, organic molecules that strongly absorb selected wavelengths of visible light. The first synthetic dye, called mauveine or aniline purple, was discovered by accident in 1856 as a derivative of coal tar (Figure 9.36). Its discoverer, William Perkin, was 18 years old at the time; he immediately quit school to commercialize his discovery, thus spawning the coal-tar dye industry. The design of new dyes that are "tuned" to absorb very specific wavelengths of light is an important part of today's chemical industry.

Organic dyes contain extensively delocalized π electrons. The molecules contain atoms that are predominantly sp^2 hybridized, like the carbon atoms in benzene (Figure 9.26). This leaves one unhybridized p orbital to form π bonds with neighboring atoms. The p orbitals are arranged so that electrons can be delocalized throughout the entire molecule; we say that the π bonds are *conjugated*. The HOMO-LUMO gap in such molecules

Figure 9.36 Perkin's original sample of mauveine, the first synthetic organic dye ever made. The yarn in the background has been colored by mauveine. Also shown is an old sample of alizarin, another organic dye. (Science Museum, London. Photo © Michael Halford)

gets smaller as the number of conjugated double bonds increases. For example, consider butadiene, C_4H_6, a molecule that has alternating carbon-carbon double and single bonds:

H H H H H H (structural formula of butadiene, $H_2C=CH-CH=CH_2$) or (line structure)

The representation on the right is the shorthand notation that chemists use for organic molecules. There are implicitly carbon atoms at the ends of the straight segments, and there are implicitly enough hydrogen atoms to make a total of four bonds at each carbon. Butadiene is planar, so the unhybridized p orbital on each carbon is pointing in the same direction. The π electrons are delocalized among the four carbon atoms; the double bonds are conjugated.

Because butadiene has only two conjugated double bonds, it still has a fairly large HOMO-LUMO gap. Butadiene absorbs light at 217 nm, still well into the ultraviolet part of the spectrum. It is therefore colorless. If we keep adding new conjugated double bonds, however, the HOMO-LUMO gap keeps shrinking until visible light is absorbed. For example, the molecule shown on the next page is β-carotene, the substance chiefly responsible for the bright orange color of carrots.

Because β-carotene contains 11 conjugated double bonds, its π electrons are very extensively delocalized. It absorbs light of wavelength 500 nm, in the middle of the visible part of the spectrum. The human body converts β-carotene into vitamin A, which in turn is converted into a protein, *rhodopsin,* found in the retinas of your eyes. The absorption of visible light by rhodopsin is a major reason why "visible" light is indeed visible. There would thus seem to be a good basis for the maxim that eating carrots is good for your eyesight.

FOR REVIEW

SUMMARY

In this chapter, we've applied the basic principles of chemical bonding to several important areas of chemical structure and behavior. The three-dimensional structures of molecules are determined by the distances between bonded atoms and by the directions of chemical bonds with respect to one another around a particular atom. The valence-shell electron-pair repulsion (VSEPR) model explains these relative directions in terms of the repulsions that exist between electron pairs. According to this model, electron pairs around an atom orient themselves so as to minimize electrostatic repulsions; that is, they remain as far apart as possible. By recognizing that unshared electron pairs take up more space (exert greater repulsive forces) than shared electron pairs, we can account for the departures of bond angles from idealized values and explain many other aspects of molecular structure. The shape of a molecule and the bond polarities determine whether the molecule will be polar. The degree of polarity of a molecule is measured by its dipole moment.

The Lewis model for covalent bonding introduced in Chapter 8 can be extended to account for the geometric properties of molecules. In valence bond theory, we imagine that the atoms in a molecule are bonded to one another by electron pairs that occupy pairs of overlapping atomic orbitals. The extent to which the atomic orbitals share the same region of space, called overlap, is important in determining the amount of stability that results from bond formation. The bonds directed along the internuclear axes are called σ bonds. It is possible to formulate orbitals on an atom that are directed toward each of the other atoms surrounding it by forming hybrid orbitals. These orbitals are made up of mixtures of the familiar *s*, *p*, and *d* atomic orbitals. Depending on the particular number of other atoms bonded to an atom and their arrangement in space, a particular set of hybrid orbitals can be formulated that has the necessary directional characteristics. For example, sp^3 hybrid orbitals are directed toward the corners of a tetrahedron.

In addition to the σ bonds, which determine the geometry of the bonding around a particular atom, there may be also π bonds constructed from remaining, unhybridized atomic orbitals. Thus a double bond, consisting of a σ and a π bond, or a triple bond, consisting of a σ and two π bonds, may be formed. In some molecules the π bonds may extend, or be delocalized, over several atoms. Delocalization of the π electrons in a cyclic structure, such as in benzene, leads to a special stability.

The coming together of atoms to form molecules may be viewed also as the coming together of atomic orbitals to form molecular orbitals. Atomic orbitals may combine with one another in various ways. The rules for combining atomic orbitals on atoms to form molecular orbitals allow us to account very well for the observed properties of the diatomic molecules formed by the first several elements of the periodic table. The molecular orbital model is particularly impressive in explaining why the O_2 molecule contains two unpaired electrons.

KEY TERMS

bond angle (Sec. 9.1)
valence-shell electron-pair repulsion (VSEPR) model (Sec. 9.1)
bonding pair (Sec. 9.1)
nonbonding pair (Sec. 9.1)
electron-pair geometry (Sec. 9.1)
molecular geometry (Sec. 9.1)
polar (Sec. 9.2)
dipole moment (Sec. 9.2)
bond dipole (Sec. 9.2)
valence bond theory (Section 9.3)
overlap (Sec. 9.3)
sigma (σ) bond (Sec. 9.3)
pi (π) bond (Sec. 9.3)
hybrid orbital (Sec. 9.4)
hybridization (Section 9.4)
delocalized (Sec. 9.5)
molecular orbital (Sec. 9.6)
molecular orbital theory (Section 9.6)
bonding molecular orbital (Sec. 9.6)
antibonding molecular orbital (Sec. 9.6)
sigma (σ) molecular orbital (Sec. 9.6)
energy-level diagram (Sec. 9.6)
molecular orbital diagram (Section 9.6)
bond order (Sec. 9.6)
pi (π) molecular orbital (Sec. 9.7)
paramagnetism (Sec. 9.7)
diamagnetism (Sec. 9.7)

EXERCISES

Molecular Geometry; the VSEPR Model

9.1 Describe the characteristic electron-pair geometry of each of the following numbers of electron pairs about a central atom: **(a)** 3; **(b)** 4; **(c)** 5; **(d)** 6.

9.2 Indicate the number of electron pairs about a central atom, given the following angles between them: **(a)** 120°; **(b)** 180°; **(c)** 109°; **(d)** 90°.

9.3 What is the difference between the electron-pair geometry and the molecular geometry of a molecule? Use the water molecule as an example in your discussion.

9.4 Give examples of the following: **(a)** a molecule that has trigonal planar electron-pair and molecular geometries; **(b)** a molecule that has a tetrahedral electron-pair geometry and a trigonal pyramidal molecular geometry; **(c)** a molecule that has a trigonal bipyramidal electron-pair geometry and a linear molecular geometry.

9.5 Give the electron-pair geometry and the molecular geometry for each of the following molecules and ions: **(a)** ClO_2^-; **(b)** BH_4^-; **(c)** SO_3; **(d)** SO_3^{2-}; **(e)** PCl_3; **(f)** ICl_3.

9.6 Give the electron-pair geometry and the molecular geometry for each of the following molecules and ions: **(a)** ClO_3^-; **(b)** CO_3^{2-}; **(c)** CF_4; **(d)** SCl_2; **(e)** Cl_2SO; **(f)** ICl_2^-.

9.7 The molecules NF_3, BF_3, and ClF_3 all have molecular formulas of the type XF_3, but the molecules have different molecular geometries. Predict the shape of each molecule, and explain the origin of the differing shapes.

9.8 The molecules SiF_4, SF_4, and XeF_4 all have molecular formulas of the type XF_4, but the molecules have different molecular geometries. Predict the shape of each molecule, and explain the origin of the differing shapes.

9.9 The three species NO_2^+, NO_2, and NO_2^- all have a central N atom. The ONO bond angles in the three species are 180°, 134°, and 115°, respectively. Explain this variation in bond angles.

9.10 The three species NH_2^-, NH_3, and NH_4^+ have H—N—H bond angles of 105°, 107°, and 109°, respectively. Explain this variation in bond angles.

9.11 Give approximate values for the indicated bond angles in the following molecules:

(a) H—Ö—N̈=Ö (angles 1, 2)
(c) H—N(H)—Ö—H (angles 5, 6)
(b) H—C(H)(H)—C(H)=Ö (angles 3, 4)
(d) H—C(H)(H)—C≡N: (angles 7, 8)

9.12 Give the approximate values for the indicated bond angles in the following molecules:

(a) H—Ö—C̈l(:Ö:)—Ö: (angles 1, 2)
(c) H—C≡C—H (angle 5)
(b) H—C(H)(H)—Ö—H (angles 3, 4)
(d) H—C(=O)—Ö—C(H)(H)—H (angles 6, 7, 8)

Dipole Moments

9.13 Indicate whether each of the following molecules possesses a dipole moment: **(a)** CCl_4; **(b)** NF_3; **(c)** SO_3; **(d)** CS_2; **(e)** SCl_2; **(f)** N_2O.

9.14 Predict whether the following molecules are polar or nonpolar: **(a)** PH_3; **(b)** $SiCl_4$; **(c)** BF_3; **(d)** IF; **(e)** C_2H_4; **(f)** SO_2.

9.15 Despite the larger electronegativity difference between the bonded atoms, $BeCl_2(g)$ has no dipole moment, whereas $SCl_2(g)$ does possess one. Account for this difference in polarity.

9.16 The PF_3 molecule has a dipole moment of 1.03 D, but SiF_4 has a dipole moment of zero. How can you explain this difference?

9.17 The substance dichloroethylene, $C_2H_2Cl_2$, has the following three isomers:

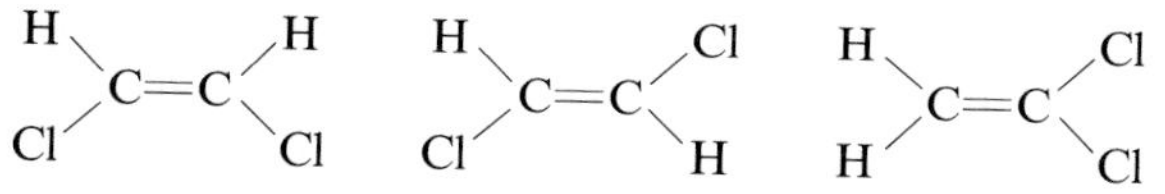

H(Cl)C=C(H)(Cl)   H(Cl)C=C(Cl)(H)   H(H)C=C(Cl)(Cl)

A pure sample of one of the isomers is found experimentally to have a dipole moment of zero. Can we determine which of the three isomers was measured?

9.18 Dichlorobenzene, $C_6H_4Cl_2$, exists in three different isomers, called *ortho*, *meta*, and *para*:

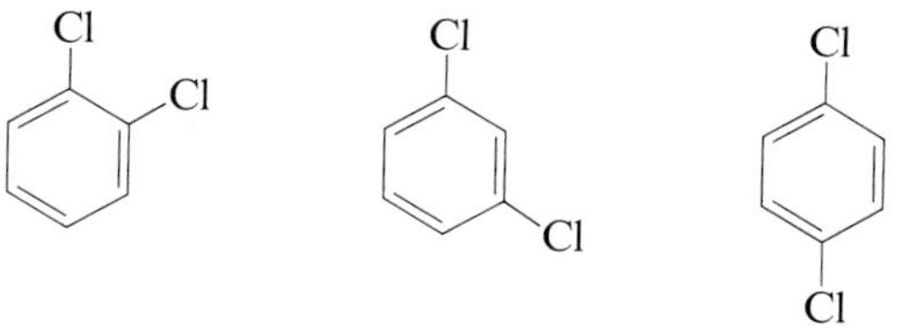

ortho   *meta*   *para*

Which of these would have a nonzero dipole moment? Explain.

9.19 The bond length in HBr is 1.41 Å. From its dipole moment (Table 9.4), calculate the percentage ionic character in the H—Br bond, assuming that the partial charges are centered on H and Br.

9.20 The bond length in HF is 91.7 pm. From its dipole moment (Table 9.4), calculate the percentage ionic character in the H—F bond, assuming that the partial charges are centered on H and F.

Orbital Overlap; Hybrid Orbitals

9.21 **(a)** What is meant by the term *orbital overlap*? **(b)** What is the significance of overlapping orbitals in valence bond theory? **(c)** What two fundamental concepts are incorporated in valence bond theory?

9.22 Draw sketches illustrating the overlap between the following orbitals on two carbon atoms: **(a)** the 2*s* orbital on each; **(b)** the $2p_z$ orbital on each (assume that the atoms are on the *z* axis); **(c)** the 2*s* orbital on one and the $2p_z$ orbital on the other.

9.23 Without referring to tables or figures in the chapter, indicate the designation for the hybrid orbitals formed from each of the following combinations of atomic orbitals: **(a)** one *s* and two *p*; **(b)** one *s*, three *p*, and one *d*; **(c)** one *s*, three *p*, and two *d*. What bond angles are associated with each?

9.24 Without referring to tables or figures in the chapter, indicate the hybridization and bond angles associated with each of the following electron-pair geometries: **(a)** linear; **(b)** tetrahedral; **(c)** trigonal planar; **(d)** octahedral; **(e)** trigonal bipyramidal.

9.25 Draw the Lewis structure of H_2S. What is its electron-pair geometry and molecular shape? Describe the bonding of the molecule, using hybrid orbitals.

9.26 Draw the Lewis structure of BF_3. What is its electron-pair geometry and molecular shape? Describe the bonding of the molecule using hybrid orbitals.

9.27 Indicate the hybrid orbital set used by the central atom in each of the following molecules and ions: **(a)** NF_3; **(b)** PF_5; **(c)** $AlCl_4^-$; **(d)** CS_2; **(e)** PF_6^-; **(f)** SO_2.

9.28 For each of the following molecules and ions, predict the molecular geometry (including approximate bond angles), and indicate the hybrid orbitals on the central atom: **(a)** OF_2; **(b)** SiF_4; **(c)** SeF_6; **(d)** NO_2^-; **(e)** ClF_5; **(f)** XeF_2.

9.29 If an atom uses an sp^2 hybrid orbital set, how many unhybridized *p* orbitals in the same valence shell remain on the atom? How many π bonds can the atom form?

9.30 Indicate the number of unhybridized *p* orbitals on a particular atom available for π bonding when the following hybrid orbital sets are used: **(a)** *sp*; **(b)** sp^3; **(c)** sp^3d^2.

9.31 Consider the Lewis structure for glycine, the simplest amino acid:

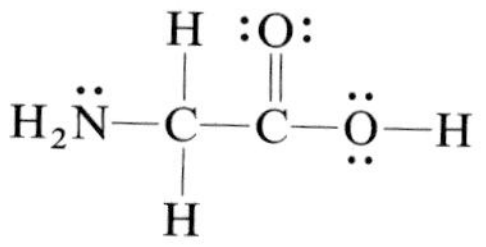

H :O:
$H_2\ddot{N}$—C—C—Ö—H
H

(a) What are the approximate bond angles about each of the two carbon atoms, and what are the hybridizations of the orbitals on each of them? **(b)** What are the hybridizations of the orbitals on the two oxygens and the nitrogen atom, and what are the approximate bond angles at the nitrogen? **(c)** What is the total number of σ bonds in the entire molecule, and what is the total number of π bonds?

9.32 The compound whose Lewis structure is shown below is acetylsalicylic acid, better known as aspirin:

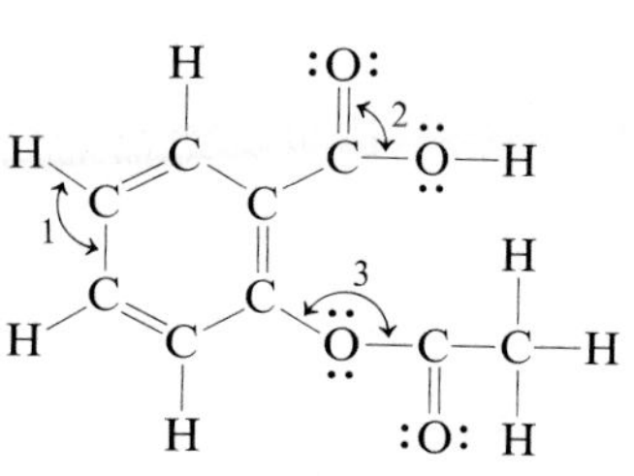

(a) What are the approximate values of the bond angles marked 1, 2, and 3? **(b)** What hybrid orbitals are used about the central atom of each of these angles? **(c)** How many σ bonds are in the molecule?

9.33 The geometrical structure of the nitrate ion, NO_3^-, can be accounted for either in terms of Lewis structures, using resonance forms, or in terms of delocalized π bonding. Explain the structure of NO_3^- in both of these terms.

9.34 Butadiene, C_4H_6, is a planar molecule that exhibits the following carbon-carbon bond lengths:

1.34 Å — 1.48 Å — 1.34 Å

Compare these bond lengths to the average bond lengths listed in Table 8.5. Can you explain any differences?

Molecular Orbitals

9.35 How do bonding and antibonding molecular orbitals differ with respect to **(a)** energies; **(b)** the spatial distribution of electron density?

9.36 How do σ and π molecular orbitals differ with respect to the spatial distribution of electron density? Sketch the shapes of the σ_{2p} and π_{2p} molecular orbitals.

9.37 What is meant by the following terms: **(a)** bond order; **(b)** paramagnetism; **(c)** energy-level diagram?

9.38 How do bond order, bond length, and bond energy correlate for a series of bonds between the same two elements?

9.39 If we assume that the energy-level diagram for homonuclear diatomic molecules (Figure 9.33) can be applied to heteronuclear diatomic molecules and ions, predict the bond order and magnetic behavior of the following: **(a)** CO; **(b)** NO^-; **(c)** CN^-; **(d)** OF.

9.40 Using Figures 9.31 and 9.33 as a guide, give the molecular orbital electron configuration for each of the following cations: **(a)** Li_2^+; **(b)** B_2^+; **(c)** C_2^+; **(d)** Ne_2^{2+}. In each case, indicate whether the addition of an electron would increase or decrease the stability of the species.

9.41 The ions O_2^-, O_2^{2-}, and O_2^+ occur in several compounds. Compare these three ions with O_2 by listing the four in order of increasing bond length.

9.42 List the members of the following series in order of increasing bond length: N_2^+, N_2, N_2^-.

9.43 Provide explanations for the following observations about molecular orbital energy-level diagrams: **(a)** The σ_{1s} molecular orbital in H_2 is lower in energy than the σ_{1s}^* molecular orbital. **(b)** The σ_{1s}^* molecular orbital in H_2 is higher in energy than the atomic H 1*s* orbitals. **(c)** In Li_2, the energy separation between the σ_{1s} and σ_{1s}^* molecular orbitals is less than that between the σ_{2s} and σ_{2s}^* molecular orbitals.

9.44 Explain the following observations: **(a)** In Li_2, the σ_{1s}^* antibonding molecular orbital is lower in energy than the σ_{2s} bonding molecular orbital. **(b)** In B_2, there are two degenerate π_{2p} molecular orbitals. **(c)** In B_2, the π_{2p}^* molecular orbitals are lower in energy than the σ_{2p}^* molecular orbital.

Additional Exercises

9.45 Using the VSEPR model, predict the molecular geometry of each of the following: **(a)** PO_4^{3-}; **(b)** AsF_3; **(c)** OCN^-; **(d)** H_2CO; **(e)** ICl_4^-; **(f)** I_3^-.

9.46 The H—P—H bond angle in PH_3 is 93°; in PH_4^+ it is 109.5°. Account for this difference.

9.47 When applying the VSEPR model, we count double and triple bonds as a single pair of electrons. Present an argument as to why this is justified.

9.48 From their Lewis structures, determine the number of σ and π bonds in each of the following molecules or ions: **(a)** H_2CO; **(b)** CN^-; **(c)** SO_2; **(d)** SO_3^{2-}.

9.49 Which of the following molecules will have a nonzero dipole moment: **(a)** HI; **(b)** SCl_2; **(c)** BCl_3; **(d)** HCN; **(e)** HC≡CH; **(f)** $Cl_2C{=}CH_2$?

[9.50] The water molecule, H_2O, has O—H bonds of length 0.96 Å, and the H—O—H angle is 104.5°. The dipole moment of the water molecule is 1.85 D. **(a)** In what directions do the bond dipoles of the O—H bonds point? In what direction does the dipole moment vector of the water molecule point? **(b)** Calculate the magnitude of the bond dipole for the O—H bonds. (Note: You will need to use vector addition to do this.)

[9.51] By using vector addition, prove that the dipole moment of BF_3 must be zero even though the bond dipoles are not zero.

[9.52] The Lewis structure for allene is

$$H_2C{=}C{=}CH_2$$

Make a sketch of the structure of this molecule that is analogous to Figure 9.24. In addition, answer the following two questions: **(a)** Is the molecule planar? **(b)** Does it have a nonzero dipole moment?

9.53 Consider the unstable molecule called diimine, HN=NH. **(a)** Draw the Lewis structure. (b) Use the VSEPR model to predict the molecular geometry. (c) Indicate the hybrid orbitals used by the nitrogen atoms. **(d)** Sketch the formation of σ and π bonds in the molecule from the hybrid and unhybridized orbitals. (e) It is found experimentally that the dipole moment of diimine is zero. Is this consistent with your molecular geometry? Is there a geometry for the molecule, with the same Lewis structure, that could have a nonzero dipole moment?

9.54 Indicate the hybrid orbital set used by the underlined atom in each of the following molecules and ions: **(a)** SCl_2; **(b)** HC≡<u>C</u>—CH_3; **(c)** $\underline{Br}F_4^+$; **(d)** $\underline{Tl}Cl_3$; **(e)** $\underline{Se}O_3^{2-}$; **(f)** $\underline{Sn}Cl_6^{2-}$.

9.55 Cumene hydroperoxide, for which the structure is

$$CH_3{-}C(O{-}O{-}H)(C_6H_5){-}CH_3$$

is an intermediate in the formation of phenol, C_6H_5OH, an important industrial chemical. Indicate the hybrid orbital set employed by all the carbon and oxygen atoms of cumene hydroperoxide.

9.56 What change in the hybridization of orbitals at the central atom occurs in each of the following reactions?
(a) $GaCl_4^- \longrightarrow GaCl_3 + Cl^-$
(b) $PCl_5 \longrightarrow PCl_3 + Cl_2$
(c) $SF_6 \longrightarrow SF_4 + F_2$

9.57 The nitrogen-nitrogen bond lengths in N_2H_4, N_2F_2, and N_2 are 1.45, 1.25, and 1.10 Å, respectively. How can this trend be explained?

[9.58] The azide ion, N_3^-, is a linear ion with two N—N bonds of equal length, 1.16 Å. **(a)** Draw a Lewis structure for the azide ion. **(b)** With reference to Table 8.5, is the observed N—N bond length consistent with your Lewis structure? **(c)** What hybridization scheme would you expect at each of the nitrogen atoms in N_3^-? **(d)** Show which hybridized and unhybridized orbitals are involved in the formation of σ and π bonds in N_3^-. **(e)** It is often observed that σ bonds that involve an *sp* hybrid orbital are shorter than those that involve only sp^2 or sp^3 hybrid orbitals. Can you propose a reason for this? Is this observation applicable to the observed bond lengths in N_3^-?

9.59 Use average bond energies (Table 8.4) to estimate ΔH for the atomization of benzene, C_6H_6:

$$C_6H_6(g) \longrightarrow 6C(g) + 6H(g)$$

Compare the value to that obtained by using ΔH_f° data given in Appendix C and Hess's law. To what do you attribute the large discrepancy in the two values?

[9.60] The reaction of three molecules of fluorine gas with a Xe atom produces the substance xenon hexafluoride, XeF_6:

$$Xe(g) + 3F_2(g) \longrightarrow XeF_6(g)$$

(a) Draw a Lewis structure for XeF_6. **(b)** If you try to use the VSEPR model to predict the molecular geometry of XeF_6, you run into a problem. What is it? **(c)** What could you do to attempt to resolve the difficulty in part (b)? **(d)** Suggest a hybridization scheme for the Xe atom in XeF_6. **(e)** The molecule IF_7 has a *pentagonal bipyramidal* structure (five equatorial fluorine atoms at the vertices of a regular pentagon and two axial fluorine atoms). Based on the structure of IF_7, suggest a structure for XeF_6.

[9.61] In ozone, O_3, the two oxygen atoms on the ends of the molecule are equivalent to one another. **(a)** What is the best choice of hybridization scheme for the atoms of ozone? **(b)** For one of the resonance forms of ozone, which of the orbitals are used to make bonds and which are used to hold nonbonding pairs of electrons? **(c)** Which of the orbitals can be used to make a delocalized π system? **(d)** How many electrons are delocalized in the π system of ozone?

9.62 What are the similarities and differences in the Lewis structure and molecular orbital descriptions of the O_2 molecule?

[9.63] The nitric oxide molecule, NO, is a *heteronuclear* diatomic molecule (one that consists of two different atoms). We shall assume that the energy-level diagram for homonuclear diatomic molecules (Figure 9.33) can be applied to NO. **(a)** In terms of molecular orbitals, write the electron configuration for the NO molecule. **(b)** Would NO be diamagnetic or paramagnetic? **(c)** In chemical reactions, NO readily loses one electron to form the NO^+ ion. Why is this consistent with the electronic structure of NO? **(d)** Predict the order of the N—O bond strengths in NO, NO^+, and NO^-. **(e)** With what neutral homonuclear diatomic molecules are the NO^+ and NO^- ions isoelectronic (same number of electrons)?

9.64 Consider the methyl cation, CH_3^+. **(a)** Predict the molecular geometry of CH_3^+. **(b)** Would you expect the molecular geometry of the methyl anion, CH_3^-, to be the same as or different from that of CH_3^+? **(c)** The lowest-energy bonding molecular orbital of CH_3^+ results from adding the 1*s* orbitals of the three hydrogen atoms to the 2*s* orbital of the carbon atom. Produce a sketch analogous to Figure 9.27 that shows the formation of this molecular orbital.

9.65 The fact that B_2 is paramagnetic is consistent with the π_{2p} molecular orbitals being lower in energy than the σ_{2p} molecular orbital in Figure 9.33. Why must this be the energy ordering of these molecular orbitals for B_2?

9.66 Would either Ne_2 or Ne_2^+ be expected to exist? Explain.

[9.67] Many compounds of the transition-metal elements contain direct bonds between metal atoms. We shall assume that the *z* axis is defined as the metal-metal bond axis. **(a)** Which of the 3*d* orbitals (Figure 6.24) can be used to make a σ bond between metal atoms? **(b)** Sketch the σ_{3d} bonding and σ_{3d}^* antibonding molecular orbitals. **(c)** Sketch the energy-level diagram for the Sc_2 molecule, assuming that only the 3*d* orbital from part (a) is important. **(d)** What is the bond order in Sc_2?

10 Gases

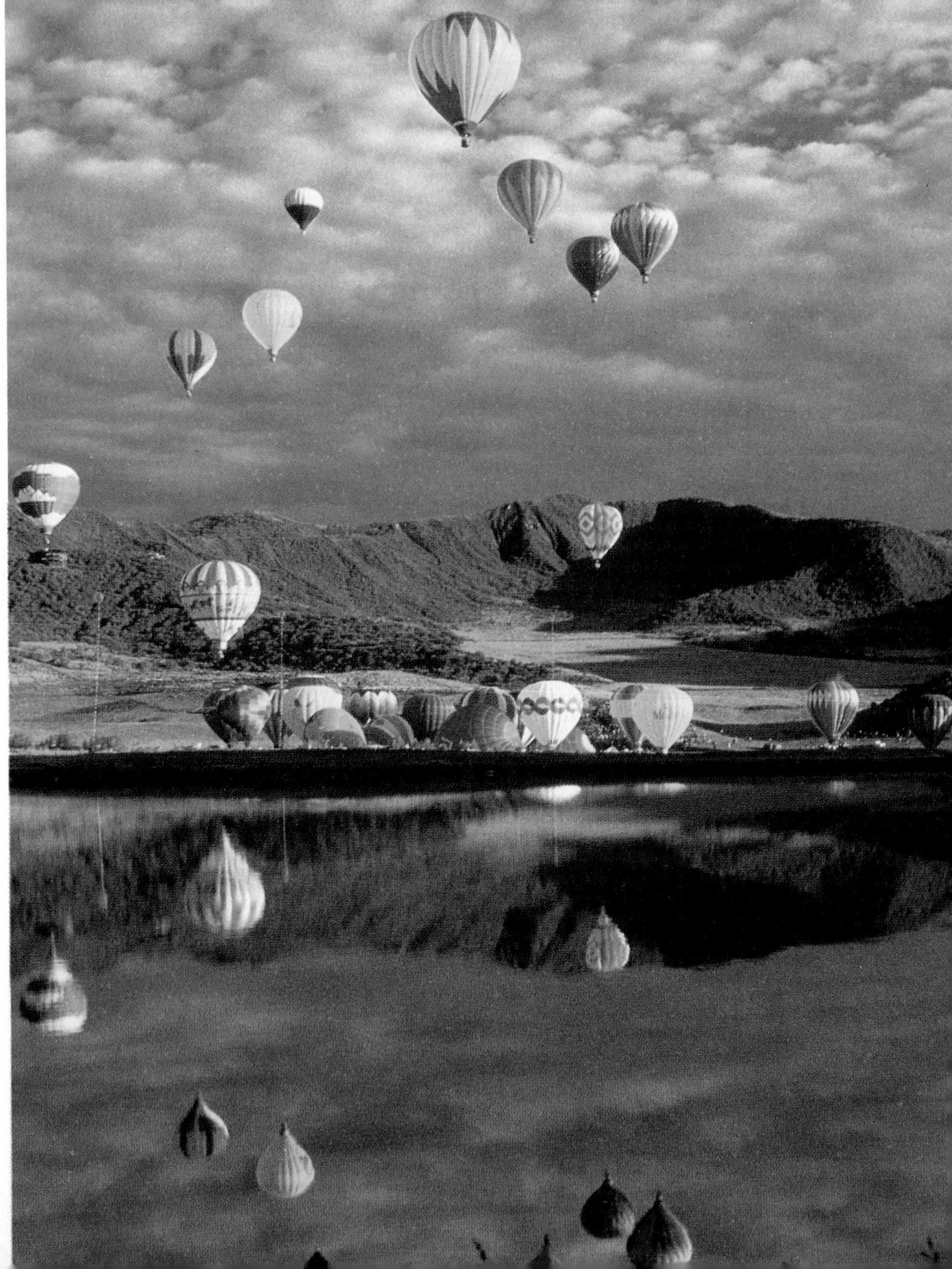

Hot-air balloons. These balloons are buoyant because gases decrease in density with increasing temperature. (Tom Martin/The Stock Market)

In the past several chapters, we have learned about the electronic structures of atoms and about how atoms come together to form molecules or ionic substances. In everyday life, however, we encounter matter not on the atomic or molecular level, but as solids, liquids, and gases. In the next few chapters, we will consider some important characteristics of these states of matter. We will learn why substances are found in a particular state, what forces operate within states, and what some of the characteristic properties of each state are. This chapter will focus on gases; Chapter 11 will discuss liquids and solids.

We are surrounded by an atmosphere composed of a mixture of gases that we refer to as air. The behavior of air determines our weather, and the oxygen, O_2, in air supports life. We also encounter gases in countless other situations. For example, chlorine gas, Cl_2, is used to purify drinking water. Acetylene gas, C_2H_2, is used in welding. Nitrous oxide gas, N_2O, sometimes called laughing gas, is used as an anesthetic in dentistry. Although different gases may vary widely in their chemical properties, they share many physical properties. Our goal in this chapter is to develop a deeper understanding of the physical properties of gases.

CONTENTS

10.1 CHARACTERISTICS OF GASES

Air consists primarily of two substances: oxygen and nitrogen. About 78 percent of the molecules in air are N_2 molecules, and about 21 percent are O_2 molecules. Several other nonmetallic elements also exist as gases under ordinary conditions of temperature and pressure: H_2, F_2, Cl_2, and the noble gases (group 8A in the periodic table: He, Ne, Ar, Kr, and Xe). Many molecular compounds are also gases. Table 10.1 lists a few of the more common gaseous compounds.

Table 10.1 Some Common Compounds that Are Gases

Formula	Name	Characteristics
HCN	Hydrogen cyanide	Very toxic, slight odor of bitter almonds
HCl	Hydrogen chloride	Toxic, corrosive, choking odor
H_2S	Hydrogen sulfide	Very toxic, odor of rotten eggs
CO	Carbon monoxide	Toxic, colorless, odorless
CO_2	Carbon dioxide	Colorless, odorless
CH_4	Methane	Colorless, odorless, flammable
N_2O	Nitrous oxide	Colorless, sweet odor, laughing gas
NO_2	Nitrogen dioxide	Red-brown, irritating odor
NH_3	Ammonia	Colorless, pungent odor
SO_2	Sulfur dioxide	Colorless, irritating odor

Notice that all of these gases are composed entirely of nonmetallic elements. Furthermore, all have simple molecular formulas and, therefore, low molar masses.

Under appropriate conditions, substances that are ordinarily liquids or solids can also exist in the gaseous state, where they are often referred to as **vapors**. The substance H_2O, for example, can exist as liquid water, solid ice, or water vapor. Frequently, a substance exists in all three states of matter, or phases, at the same time. A thermos bottle containing a mixture of ice and water at 0°C has some water vapor in the gas phase over the liquid and solid phases.

Gases differ significantly from solids and liquids in several respects. For example, a gas expands to fill its container. Consequently, the volume of a gas is equal to the volume of the container in which it is held. Volumes of solids and liquids, in contrast, are not determined by the container. A corollary to this is that gases are highly compressible. When pressure is applied to a gas, its volume readily decreases. Liquids and solids, in contrast, are not very compressible; great pressures must be applied to cause the volume of a liquid or solid to diminish by even as little as 5 percent.

Gases also form homogeneous mixtures with each other regardless of the identities or relative proportions of the component gases. For example, when water and gasoline are poured into a bottle, the water vapor and gasoline vapors above the liquids form a homogeneous gas mixture. The two liquids, in contrast, remain largely separate.

The characteristic properties of gases arise because the individual molecules are relatively far apart. For example, in the air we breathe, the molecules take up only about 0.1 percent of the total volume, with the rest being empty space. Thus, each molecule behaves largely as though the others weren't present. As a result, different gases behave similarly, even though they are made up of different molecules. In contrast, the individual molecules in a liquid are close together and occupy perhaps 70 percent of the total space. The attractive forces among the molecules keep the liquid together.

10.2 PRESSURE

Among the most readily measured properties of a gas are its temperature, volume, and pressure. It is not surprising, therefore, that many early studies of gases focused on relationships among these properties. We have already discussed volume and temperature (Section 1.3). Let us now consider the concept of pressure.

In general terms, **pressure** conveys the idea of a force, something that tends to move something else in a given direction. Pressure, P, is, in fact, the force, F, that acts on a given area, A:

$$P = \frac{F}{A} \qquad [10.1]$$

Gases exert a pressure on any surface with which they are in contact. For example, the gas in an inflated balloon exerts a pressure on the inside surface of the balloon.

To understand better the concept of pressure and the units in which it is expressed, consider the aluminum cylinder illustrated in Figure 10.1. Because of gravity, this cylinder exerts a downward force upon the surface on which it rests. The force exerted by an object is the product of its mass, m, times its acceleration, a, that is, $F = ma$. The acceleration due to the gravitational force of the earth is 9.81 m/s^2. The mass of the cylinder is 1.06 kg. Thus, the force with which the earth attracts it is

$$(1.06 \text{ kg})(9.81 \text{ m/s}^2) = 10.4 \text{ kg-m/s}^2 = 10.4 \text{ N}$$

A kg-m/s^2 is the SI unit for force; it is called the *newton*, abbreviated N: $1 \text{ N} = 1 \text{ kg-m/s}^2$. The cylinder has a cross-sectional area of $7.85 \times 10^{-3} \text{ m}^2$; the pressure exerted by the cylinder is therefore

$$P = \frac{F}{A} = \frac{10.4 \text{ N}}{7.85 \times 10^{-3} \text{ m}^2} = 1.32 \times 10^3 \text{ N/m}^2$$

A N/m^2 is the standard unit of pressure in SI units. It is given the name **pascal** (Pa) after Blaise Pascal (1623–1662), a French mathematician and scientist: $1 \text{ Pa} = 1 \text{ N/m}^2$.

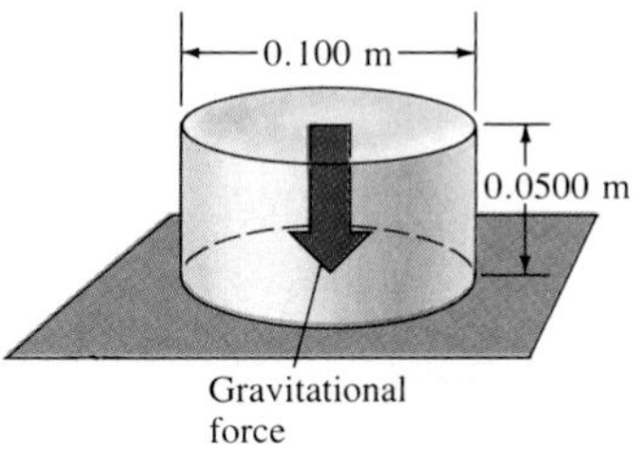

Figure 10.1 Aluminum cylinder resting on a flat surface. The diameter of the cylinder is 0.100 m and its height is 0.0500 m. The area of its base is thus $A = \pi r^2 = \pi(0.0500 \text{ m})^2 = 7.85 \times 10^{-3} \text{ m}^2$. Its volume is $V = \pi r^2 h = \pi(0.0500 \text{ m})^2(0.0500 \text{ m}) = 3.93 \times 10^{-4} \text{ m}^3$. The density of aluminum is $2.70 \text{ g/cm}^3 = 2.70 \times 10^3 \text{ kg/m}^3$. The mass of the cylinder is then $M = d \times V = (2.70 \times 10^3 \text{ kg/m}^3) \times (3.93 \times 10^{-4} \text{ m}^3) = 1.06 \text{ kg}$. Calculation of the pressure exerted by this cylinder on the surface upon which it rests is described in the text.

Atmospheric Pressure and the Barometer

Like the aluminum cylinder in our example above, the earth's atmosphere is also attracted toward earth by gravitational attraction. A column of air 1 m^2 in cross section extending through the atmosphere has a mass of roughly 10,000 kg and produces a resultant pressure of about 100 kPa (Figure 10.2):

$$P = \frac{F}{A} = \frac{(10{,}000 \text{ kg})(9.81 \text{ m/s}^2)}{1 \text{ m}^2} = 1 \times 10^5 \text{ Pa} = 1 \times 10^2 \text{ kPa}$$

Of course, the actual atmospheric pressure at any location depends on altitude and weather conditions.

Atmospheric pressure can be measured by use of a mercury **barometer** like that illustrated in Figure 10.3. Such a barometer can be made from a glass tube more than 76 cm long, which is closed at one end. The tube is filled with mercury and inverted in a dish of mercury. Care must be taken that no air gets into the tube. When the tube is inverted in this manner, some of the mercury runs out, but a column remains.

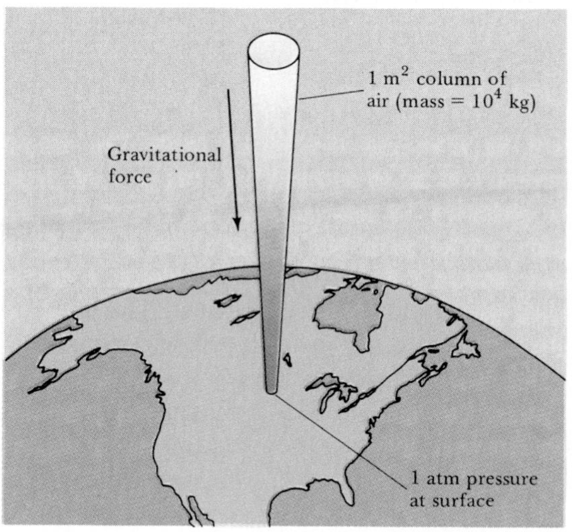

Figure 10.2 Illustration of the manner in which earth's atmosphere exerts pressure at the surface of the planet. The mass of a column of atmosphere 1 m^2 in cross-sectional area and extending to the top of the atmosphere exerts a force of 1.01×10^5 N. Thus, the pressure is 101 kPa, corresponding to 760 mm Hg.

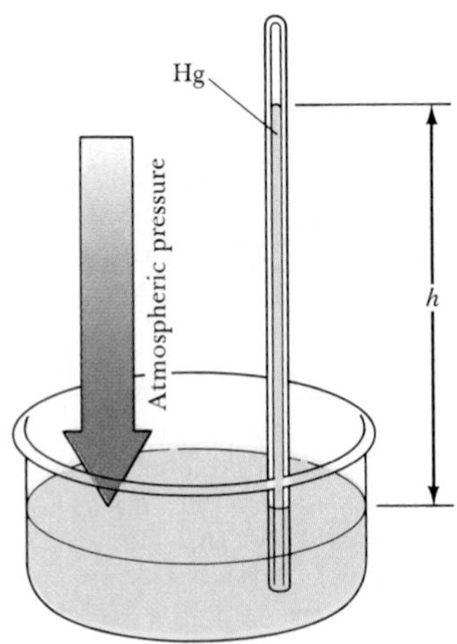

Figure 10.3 Mercury barometer invented by Torricelli. The space in the tube above mercury is nearly a vacuum; a negligible amount of mercury vapor occupies it.

The mercury surface outside the tube experiences the full force of the earth's atmosphere over each unit area. However, the atmosphere is not in contact with the mercury surface within the tube. The atmosphere pushes the mercury up the tube until the pressure due to the mass of the mercury column balances the atmospheric pressure. Thus, the height of the mercury column fluctuates as the atmospheric pressure fluctuates.

Standard atmospheric pressure, which corresponds to the typical pressure at sea level, is the pressure sufficient to support a column of mercury 760 mm in height. This pressure equals 1.01325×10^5 Pa and is used to define another unit in common use, the *atmosphere* (atm):

$$1 \text{ atm} = 760 \text{ mm Hg} = 1.01325 \times 10^5 \text{ Pa} = 101.325 \text{ kPa}$$

One mm Hg pressure is also referred to as a **torr**, after the Italian scientist Evangelista Torricelli (1608–1647), who invented the barometer: 1 mm Hg = 1 torr.

In this text, we will ordinarily express gas pressure in units of atm or mm Hg. However, you should be able to convert gas pressures from one set of units to another.

SAMPLE EXERCISE 10.1

(a) Convert 0.605 atm to mm Hg. **(b)** Convert 3.5×10^{-4} mm Hg to atm.

Solution: **(a)** We convert atm to mm Hg by multiplying the number of atm by the factor 760 mm Hg/1 atm:

$$(0.605 \text{ atm})\left(\frac{760 \text{ mm Hg}}{1 \text{ atm}}\right) = 460 \text{ mm Hg}$$

Notice that the units cancel in the required manner.

(b) To convert from mm Hg to atm, we must multiply by the conversion factor 1 atm/760 mm Hg:

$$(3.5 \times 10^{-4} \text{ mm Hg})\left(\frac{1 \text{ atm}}{760 \text{ mm Hg}}\right) = 4.6 \times 10^{-7} \text{ atm}$$

PRACTICE EXERCISE

In countries that use the metric system—for example, Canada—atmospheric pressure is expressed in weather reports in units of kPa. Convert a pressure of 735 mm Hg to kPa. *Answer:* 98.0 kPa

Pressures of Enclosed Gases and Manometers

We use various devices to measure the pressures of enclosed gases. The simple tire gauges used to measure the pressure of air in automobile tires give the pressure over and above atmospheric pressure. In laboratories, we sometimes use a simple device called a **manometer**, whose principle of operation is similar to that of a barometer. Figure 10.4(*a*) shows a closed-tube manometer, a device normally used to measure pressures below atmospheric pressure. The pressure is just the difference in the heights of the mercury levels in the two arms.

An open-tube manometer, like that pictured in Figure 10.4(*b*) and (*c*), is often employed to measure gas pressures that are near atmospheric pressure. The difference in the heights of the mercury levels in the two arms of the mano-

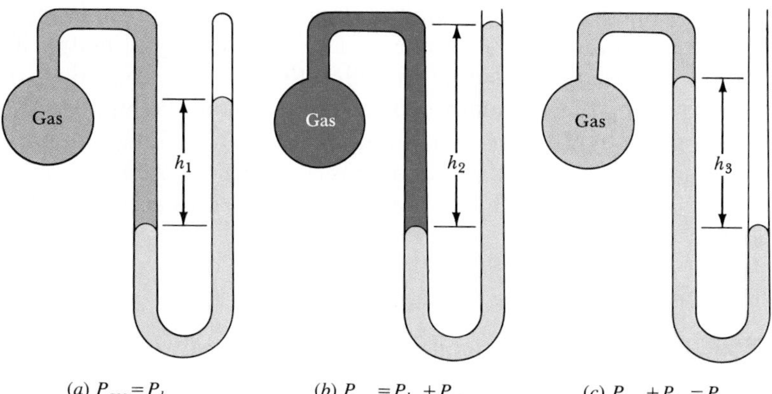

Figure 10.4 Closed-end manometer (*a*) and open-end manometers (*b*) and (*c*). In (*b*), gas pressure exceeds atmospheric pressure; in (*c*), gas pressure is less than atmospheric pressure.

meter relates the gas pressure to atmospheric pressure. If the pressure of the enclosed gas is the same as atmospheric pressure, the levels in the two arms are equal. If the pressure of the enclosed gas is greater than atmospheric pressure, mercury is forced higher in the arm exposed to the atmosphere, as in Figure 10.4(*b*). Conversely, if atmospheric pressure exceeds the gas pressure, the mercury is higher in the arm exposed to the gas, as in Figure 10.4(*c*).

Although mercury is the liquid most often used in a manometer, other liquids can be employed. For a given pressure difference, the difference in heights of the liquid levels in the two arms of the manometer is inversely proportional to the density of the liquid. That is, the greater the density of the liquid, the smaller the difference in column heights. The high density of mercury (13.6 g/mL) allows us to build smaller manometers than we could with less dense liquids, as is shown in the following sample exercise.

SAMPLE EXERCISE 10.2

Consider a container of gas with an attached open-tube manometer. The manometer is not filled with mercury but rather with another nonvolatile liquid, dibutylphthalate. The density of mercury is 13.6 g/mL; that of dibutylphthalate is 1.05 g/mL. If the conditions are as shown in Figure 10.4(*b*) with $h = 12.2$ cm when atmospheric pressure is 0.964 atm, what is the pressure of the enclosed gas in mm Hg?

Solution: Converting atmospheric pressure to mm Hg, we have

$$(0.964 \text{ atm})\left(\frac{760 \text{ mm Hg}}{1 \text{ atm}}\right) = 733 \text{ mm Hg}$$

The pressure associated with a 12.2-cm column of dibutylphthalate is equivalent to a mercury column of

$$(12.2 \text{ cm})\left(\frac{1.05 \text{ g/mL}}{13.6 \text{ g/mL}}\right) = 0.94 \text{ cm} = 9.4 \text{ mm}$$

If the situation is like that in Figure 10.4(*b*), the pressure of the enclosed gas exceeds atmospheric pressure by this amount:

$$P = 733 \text{ mm Hg} + 9 \text{ mm Hg} = 742 \text{ mm Hg}$$

PRACTICE EXERCISE

What height of a dibutylphthalate column would be supported by 1 atm pressure?
Answer: 9.84×10^3 mm (over 32 ft)

10.3 THE GAS LAWS

Experiments with a large number of gases reveal that the four variables temperature, T, pressure, P, volume, V, and the quantity of matter in the gaseous sample are usually sufficient to define the state, or condition, of a gas. The quantity of matter is usually expressed in terms of the number of moles, which we designate by the symbol n. Recall from Section 3.5 that the number of moles in a sample equals the number of grams divided by the molar mass of the substance. The equations that express the relationships among P, T, V, and n are known as the *gas laws.*

The Pressure–Volume Relationship: Boyle's Law

If the pressure on a balloon is decreased, the balloon expands. That is why weather balloons expand as they rise through the atmosphere (Figure 10.5). Conversely, when a volume of gas is compressed, the pressure of the gas increases. The first person to investigate the relationship between the pressure of a gas and its volume was Robert Boyle, who, as we saw in the introduction to Chapter 2, also introduced the basic concept of elements.

To perform his gas experiments, Boyle used a J-shaped tube like that shown in Figure 10.6. A quantity of gas is trapped in the tube behind a column of mercury. Boyle changed the pressure on the gas by adding mercury to the

Figure 10.5 The volume of gas in this weather balloon will increase as it ascends into the high atmosphere, where the atmospheric pressure is lower than on the earth's surface. (NASA/Science Course/Photo Researchers)

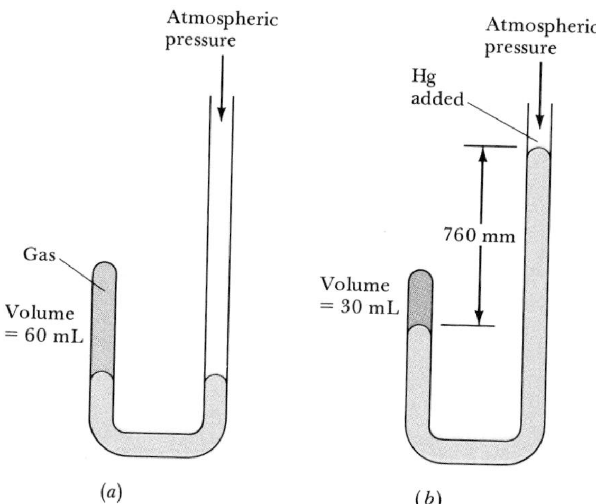

Figure 10.6 An illustration of Boyle's experiment. In (*a*), the volume of the gas trapped in the J-tube is 60 mL when the gas pressure is 760 mm Hg. When additional mercury is added, as shown in (*b*), the trapped gas is compressed. The volume is 30 mL when its total pressure is 1520 mm Hg, corresponding to atmospheric pressure plus the pressure exerted by the 760-mm column of mercury.

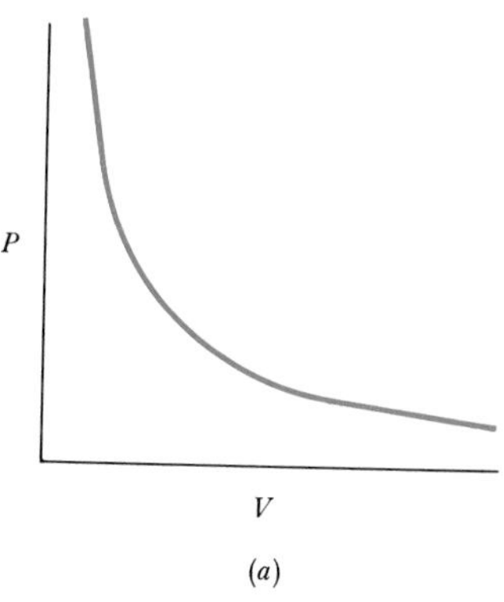

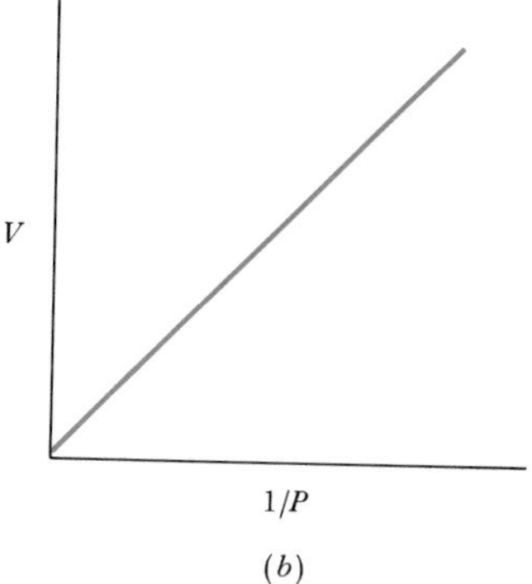

Figure 10.7 Graphs based on Boyle's law: (*a*) pressure versus volume; (*b*) volume versus $1/P$.

tube. He found that the volume of the gas decreased as the pressure increased. For example, doubling the pressure caused the gas volume to decrease to one-half its original value.

Boyle's law, which summarizes these observations, states that *the volume of a fixed quantity of gas maintained at constant temperature is inversely proportional to the pressure.* When two measurements are inversely proportional, one gets smaller as the other gets larger. Boyle's law can be expressed in mathematical terms:

$$V = \text{constant} \times \frac{1}{P} \quad \text{or} \quad PV = \text{constant} \qquad [10.2]$$

The value of the constant depends on the temperature and the amount of gas in the sample. The graph of P versus V in Figure 10.7(*a*) shows the type of curve always obtained for a given quantity of gas at a fixed temperature. A linear relationship is obtained when V is plotted versus $1/P$, as shown in Figure 10.7(*b*).

The Temperature–Volume Relationship: Charles's Law

Hot-air balloons rise because air expands as it is heated. The warm air in the balloon is less dense than the surrounding cool air at the same pressure. The difference in densities causes the balloon to ascend.

The relationship between gas volume and temperature was discovered in 1787 by the French scientist Jacques Charles (1746–1823). Charles found that the volume of a fixed quantity of gas at constant pressure increases linearly with temperature. Some typical data are shown in Figure 10.8. Notice that the extrapolated (extended) line (which is dashed) through the data points passes through $-273.15°C$. Note also that the gas is predicted to have zero volume at this temperature. Of course, this condition is never fulfilled because all gases liquefy or solidify before reaching this temperature.

In 1848, William Thomson (1824–1907), a British physicist whose title was Lord Kelvin, proposed an absolute-temperature scale, now known as the Kelvin scale. On this scale 0 K, which is called *absolute zero*, equals $-273.15°C$. In terms of the Kelvin scale, **Charles's law** can be stated as follows: *The volume of a fixed amount of gas maintained at constant pressure is directly proportional*

Figure 10.8 Volume of an enclosed gas as a function of temperature at constant pressure.

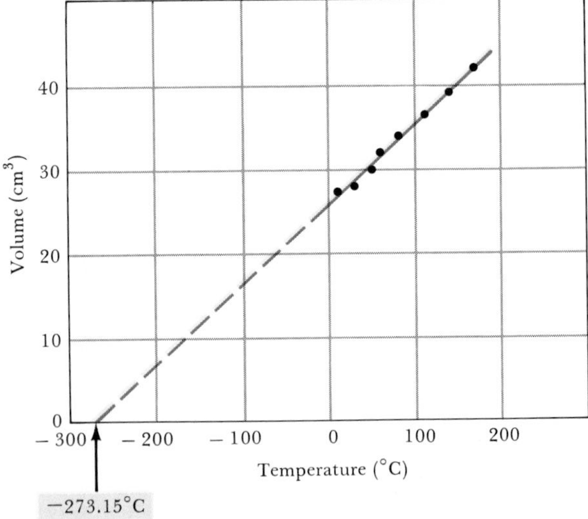

to its absolute temperature. Thus, doubling absolute temperature, say from 200 K to 400 K, causes the gas volume to double. Mathematically, Charles's law takes the following form:

$$V = \text{constant} \times T \qquad \text{or} \qquad \frac{V}{T} = \text{constant} \qquad [10.3]$$

The value of the constant depends on the pressure and amount of gas.

The Quantity–Volume Relationship: Avogadro's Law

As we add gas to a balloon, the balloon expands. The volume of a gas is affected not only by pressure and temperature, but by the amount of gas as well. The relationship between the quantity of a gas and its volume follows from the work of Joseph Louis Gay-Lussac and Amedeo Avogadro (see the Historical Perspective in Section 3.3.).

Gay-Lussac is one of those extraordinary figures in the early history of modern science who can truly be called an adventurer. He was interested in lighter-than-air balloons and in 1804 made an ascent to 23,000 ft. This exploit set the altitude record for several decades, but Gay-Lussac had other reasons for making the flight. He tested the variation of the earth's magnetic field and sampled the composition of the atmosphere as a function of elevation.

To control lighter-than-air balloons properly, Gay-Lussac needed to know more about the properties of gases. He therefore carried out several experiments, the most important of which led to his discovery in 1808 of the law of combining volumes (Section 3.3). Recall that this law states that the volumes of gases that react with one another at the same pressure and temperature are in the ratios of small whole numbers.

Gay-Lussac's work led Avogadro in 1811 to propose what is now known as **Avogadro's hypothesis:** *Equal volumes of gases at the same temperature and pressure contain equal numbers of molecules.* To illustrate, suppose we have three 1-L bulbs containing Ar, N_2, and H_2, respectively (Figure 10.9), and that each gas is at the same pressure and temperature. According to Avogadro's hy-

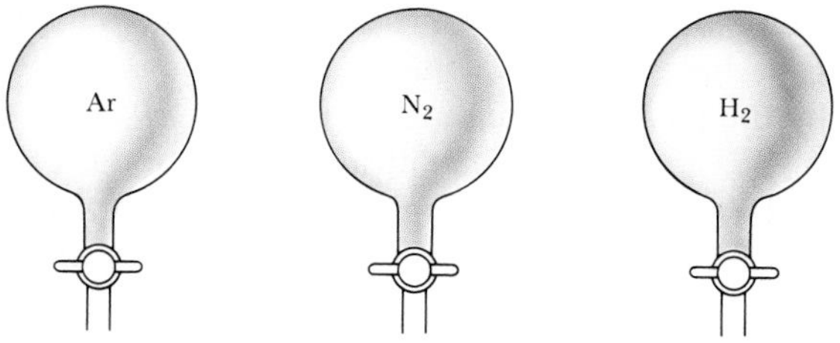

Volume	1 L	1 L	1 L
Pressure	1 atm	1 atm	1 atm
Temperature	0°C	0°C	0°C
Mass of gas	1.783 g	1.250 g	0.0899 g
Number of gas molecules	2.688×10^{22}	2.688×10^{22}	2.688×10^{22}

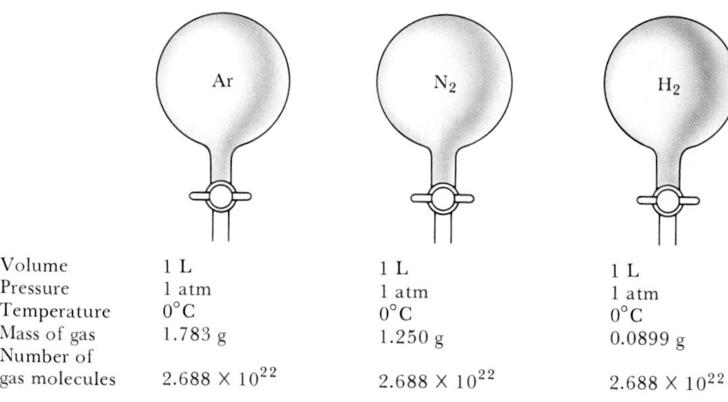

Figure 10.9 Comparison illustrating Avogadro's hypothesis. Note that argon gas consists of argon atoms; we can regard these as one-atom molecules. Each gas has the same volume, temperature, and pressure and thus contains the same number of molecules. Because a molecule of one substance differs in mass from a molecule of another, the masses of gas in the three containers differ.

pothesis, these bulbs contain equal numbers of gaseous particles, although the masses of the substances in the bulbs differ greatly. Experiments show that 1 mol of any gas (that is, 6.02×10^{23} gas molecules) at 1 atm and 0°C occupies approximately 22.4 L.

Avogadro's law follows from Avogadro's hypothesis: *The volume of a gas maintained at constant temperature and pressure is directly proportional to the number of moles of the gas.* That is,

$$V = \text{constant} \times n \qquad [10.4]$$

Thus, doubling the number of moles of gas will cause the volume to double if T and P remain constant.

10.4 THE IDEAL-GAS EQUATION

In the preceding section, we examined three historically important gas laws. Each was obtained by holding two variables constant in order to see how the other variables affect each other. Using the symbol $\propto$, which is read "is proportional to," we have

$$\text{Boyle's law:} \quad V \propto \frac{1}{P} \quad (\text{constant } n, T)$$

$$\text{Charles's law:} \quad V \propto T \quad (\text{constant } n, P)$$

$$\text{Avogadro's law:} \quad V \propto n \quad (\text{constant } P, T)$$

We can combine these relationships to make a more general gas law:

$$V \propto \frac{nT}{P}$$

If we call the proportionality constant R, we have

$$V = R\left(\frac{nT}{P}\right)$$

Rearranging, we have this relationship in its more familiar form:

$$PV = nRT \qquad [10.5]$$

Table 10.2 Numerical Values of the Gas Constant, *R*, in Various Units

Units	Numerical value
L-atm/K-mol	0.08206
cal/K-mol	1.987
J/K-mol[a]	8.314
m^3-Pa/K-mol[a]	8.314
L-torr/K-mol	62.36

[a] SI unit.

This equation is known as the **ideal-gas equation**. An **ideal gas** is a hypothetical gas whose pressure, volume, and temperature behavior is completely described by the ideal-gas equation.

The term R in the ideal-gas equation is called the **gas constant**. The value and units of R depend on the units of P, V, n, and T. Temperature must *always* be expressed on an absolute-temperature scale, normally the Kelvin scale. The quantity of gas, n, is normally expressed in moles. The units chosen for pressure and volume are most often atm and liters, respectively. However, other units can be used. Table 10.2 shows the numerical value for R in various units. Because the product PV gives energy units (Section 5.3), the units of R can include calories or joules. In working problems with the ideal-gas equation, the units of P, V, n, and T must agree with the units in the gas constant. Throughout this chapter, we will use the value $R = 0.08206$ L-atm/K-mol (four significant figures) or 0.0821 L-atm/K-mol (three significant figures) whenever we use the ideal-gas equation.

Now suppose we have 1.000 mol of an ideal gas at 1.000 atm and 0.00°C (273.15 K). Then, from the ideal-gas equation, the volume of the gas is given by

$$V = \frac{nRT}{P} = \frac{(1.000 \text{ mol})(0.08206 \text{ L-atm/K-mol})(273.15 \text{ K})}{1.000 \text{ atm}} = 22.41 \text{ L}$$

The conditions 0°C and 1 atm are referred to as the **standard temperature and pressure (STP)**. Many properties of gases are tabulated for these conditions. The volume occupied by 1 mol of ideal gas at STP, 22.41 L, is known as the *molar volume* of an ideal gas at STP.

The ideal-gas equation does not always accurately describe real gases. For example, the measured volume, V, for given conditions of P, n, and T, might differ from the volume calculated from $PV = nRT$. Ordinarily, the difference between ideal and real behavior is so small that we may ignore it. We will examine deviations from ideal behavior in Section 10.9.

SAMPLE EXERCISE 10.3

A flashbulb of volume 2.6 cm^3 contains O_2 gas at a pressure of 2.3 atm and a temperature of 26°C. How many moles of O_2 does the flashbulb contain?

Solution: Because we know volume, temperature, and pressure, the only unknown quantity in the ideal-gas equation (Equation 10.5) is the number of moles, n. Solving Equation 10.5 for n gives

$$n = \frac{PV}{RT}$$

The values of all quantities involved are tabulated and changed to units consistent with those for R:

$$P = 2.3 \text{ atm}$$
$$V = 2.6 \text{ cm}^3 = 2.6 \times 10^{-3} \text{ L}$$
$$n = ?$$
$$R = 0.0821 \text{ L-atm/K-mol}$$
$$T = (26 + 273) \text{ K} = 299 \text{ K}$$

Thus, we have

$$n = \frac{(2.3 \text{ atm})(2.6 \times 10^{-3} \text{ L})}{(0.0821 \text{ L-atm/K-mol})(299 \text{ K})} = 2.4 \times 10^{-4} \text{ mol}$$

PRACTICE EXERCISE

We can make an exotic—but quite accurate—thermometer by measuring the volume of a known quantity of gas at a known pressure. If 0.200 mol of helium (He) occupies a volume of 64.0 L at a pressure of 0.150 atm, what is the temperature of the gas? ***Answer:*** 585 K

Relationship Between the Ideal-Gas Equation and the Gas Laws

The simple gas laws, such as Boyle's law, that we discussed in Section 10.3 are special cases of the ideal-gas equation. For example, when the temperature and quantity of gas are held constant, n and T have fixed values. Therefore, the product nRT is the product of three constants and must itself be a constant:

$$PV = nRT = \text{constant} \qquad \text{or} \qquad PV = \text{constant} \qquad [10.6]$$

Thus, we have Boyle's law. Equation 10.6 expresses the fact that even though the individual values of P and V can change, their product PV remains constant.

Similarly, when n and P are constant, the ideal-gas equation gives the following relationship:

$$V = \left(\frac{nR}{P}\right)T = \text{constant} \times T \qquad \text{or} \qquad \frac{V}{T} = \text{constant} \qquad [10.7]$$

As we have noted earlier, this relationship was first observed by Charles and is known as Charles's law.

SAMPLE EXERCISE 10.4

The pressure of nitrogen gas in a 12.0-L tank at 27°C is 2300 lb/in.^{2}. What volume would the gas in this tank have at 1 atm pressure (14.7 lb/in.^{2}) if the temperature remains unchanged?

Solution: Let us begin by making a table that lists the initial and final values for the pressure, temperature, and volume of the gas. (Always convert temperature to the Kelvin scale.)

	Temperature	Pressure	Volume
Initial	300 K	2300 lb/in.^{2}	12.0 L
Final	300 K	14.7 lb/in.^{2}	?

If the quantity of gas, n, and the temperature, T, do not change, the product PV must remain constant (see Equation 10.6). Thus, if we have two different sets of conditions for the same quantity of gas at constant temperature, we can write

$$P_1V_1 = P_2V_2$$

In our example, P_1 is 2300 lb/in.^{2}, P_2 is 14.7 lb/in.^{2}, V_1 is 12 L, and V_2 is unknown. Inserting all the known quantities and solving for V_2, we obtain

$$V_2 = \frac{P_1V_1}{P_2} = \frac{(2300\ \text{lb/in.}^2)(12.0\ \text{L})}{14.7\ \text{lb/in.}^2} = 1880\ \text{L}$$

You can see from this expression that we have multiplied the initial volume, V_1, by a ratio of pressures, P_1/P_2. You can check that the result makes sense. If the pressure is to decrease in going from the initial state to the final state, the volume should increase, as it does.

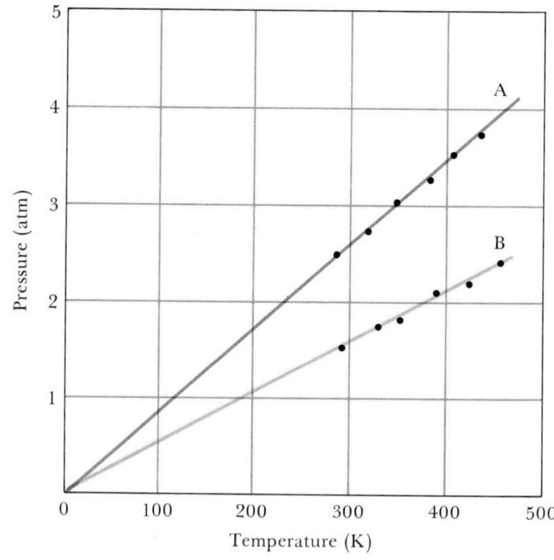

Figure 10.10 Variation of gas pressure with temperature under constant-volume conditions.

PRACTICE EXERCISE

A large natural-gas storage tank is arranged so that the pressure is maintained at 2.20 atm. On a cold day in December when the temperature is $-15°C$ (4°F), the volume of gas in the tank is 28,500 ft^3. What is the volume of the same quantity of gas on a warm July day when the temperature is 31°C (88°F)? ***Answer:*** 33,600 ft^3

We know that when a confined gas is heated at constant volume, the pressure increases. For example, a popcorn kernel bursts open under the pressure of steam that forms within the kernel when it is heated. We could make quantitative measurements of the change in pressure of a confined gas by placing the gas in a steel container fitted with a pressure gauge and then varying the temperature. We would find that the pressure increases linearly with absolute temperature, perhaps as shown by the sample data labeled A in Figure 10.10. If the experiment were repeated with a different-sized sample of the same gas, we might obtain the results labeled B in the figure.

If both n and V in the ideal-gas equation, Equation 10.5, are fixed, the pressure varies with temperature as expressed in the equation

$$P = \left(\frac{nR}{V}\right)T = \text{constant} \times T \qquad \text{or} \qquad \frac{P}{T} = \text{constant} \qquad [10.8]$$

Thus, the ideal-gas equation predicts a linear relationship between pressure and absolute temperature, extrapolating to zero pressure at 0 K. Again, we must remind ourselves that real gases lose their gaseous properties before absolute zero is reached.

SAMPLE EXERCISE 10.5

The gas pressure in an aerosol can is 1.5 atm at 25°C. Assuming that the gas inside obeys the ideal-gas equation, what would the pressure be if the can were heated to 450°C?

Solution: Let us proceed by writing down the initial and final conditions of temperature, pressure, and volume that the problem gives.

	Volume	Pressure	Temperature
Initial	V_1	1.5 atm	298 K
Final	V_1	P_2	723 K

From Equation 10.5 we can see that $P/T = nR/V$. In this problem the quantity of gas, n, and the volume are constant. Thus, P/T must be constant. If we have two different sets of conditions for the same quantity of gas at constant volume, we have

$$\frac{P_1}{T_1} = \frac{P_2}{T_2}$$

Rearranging gives

$$P_2 = P_1\left(\frac{T_2}{T_1}\right)$$

$$P_2 = (1.5 \text{ atm})\left(\frac{723 \text{ K}}{298 \text{ K}}\right) = 3.6 \text{ atm}$$

It is evident from this example why aerosol cans carry the warning not to incinerate.

PRACTICE EXERCISE

An inflatable raft is filled with gas at a pressure of 800 mm Hg at 16°C. When the raft is left in the sun, the gas heats up to 44°C. Assuming no volume change, what is the gas pressure in the raft under these conditions? ***Answer:*** 878 mm Hg

Combined Gas Law

The three variables P, V, and T may all change for a given sample of gas. Under these circumstances we have, from the ideal-gas law,

$$\frac{PV}{T} = nR = \text{constant}$$

Thus, as long as the total quantity of gas, n, is constant, PV/T is a constant. If we represent the initial and final conditions of pressure, temperature, and volume by subscripts 1 and 2, respectively, we can write the following expression:

$$\frac{P_1V_1}{T_1} = \frac{P_2V_2}{T_2} \qquad [10.9]$$

SAMPLE EXERCISE 10.6

A quantity of helium gas occupies a volume of 16.5 L at 78°C and 45.6 atm. What is its volume at STP?

Solution: It is best to begin problems of this sort by writing down all we know of the initial and final values of temperature, pressure, and volume.

	Pressure	Volume	Temperature
Initial	45.6 atm	16.5 L	351 K
Final	1 atm (exactly)	V_2	273 K

Putting the quantities into Equation 10.9, we have

$$\frac{(45.6\text{ atm})(16.5\text{ L})}{351\text{ K}} = \frac{(1\text{ atm})(V_2)}{273\text{ K}}$$

$$V_2 = \left(\frac{45.6\text{ atm}}{1\text{ atm}}\right)\left(\frac{273\text{ K}}{351\text{ K}}\right)(16.5\text{ L}) = 585\text{ L}$$

Does this answer make sense? Notice that the initial volume, 16.5 L, is multiplied by a ratio of temperatures and by a ratio of pressures. The temperature of the gas decreases; this should cause a decrease in volume. Thus, the temperature ratio should be less than 1. The decrease in pressure from 45.6 atm to 1 atm will cause expansion of the gas. Thus the pressure ratio should be greater than 1. Our intuitive sense of what should happen is in accord with the expression we have used. It is always a good idea in checking your work to ask whether the answer makes sense.

PRACTICE EXERCISE

A pocket of gas is discovered in a deep drilling operation. The gas has a temperature of 480°C and is at a pressure of 12.8 atm. Assuming ideal behavior, what volume of the gas is required at the surface to provide 18.0 L at 1.00 atm and 22°C? ***Answer:*** 3.59 L

10.5 MOLAR MASS AND GAS DENSITIES

We can make many applications of the ideal-gas equation in measuring and calculating gas density. Density has the units of mass per unit volume. We can arrange the gas equation to obtain

$$\frac{n}{V} = \frac{P}{RT}$$

Now n/V has the units of moles per liter. Suppose that we multiply both sides of this equation by the molar mass, $\mathcal{M}$, which is the number of grams in 1 mol of a substance:

$$\frac{n\mathcal{M}}{V} = \frac{P\mathcal{M}}{RT} \qquad [10.10]$$

The product of the quantities n/V and $\mathcal{M}$ equals density, because the units multiply as follows:

$$\frac{\text{Moles}}{\text{Liter}} \times \frac{\text{grams}}{\text{mole}} = \frac{\text{grams}}{\text{liter}}$$

Thus, the density of the gas is given by the expression on the right in Equation 10.10:

$$d = \frac{P\mathcal{M}}{RT} \qquad [10.11]$$

This relationship can be rearranged to calculate the molar mass of a gas:

$$\mathcal{M} = \frac{dRT}{P} \qquad [10.12]$$

SAMPLE EXERCISE 10.7

What is the density of carbon dioxide gas at 745 mm Hg and 65°C?

Solution: The molar mass of carbon dioxide, CO_2, is 12.0 + (2)(16.0) = 44.0 g/mol. If we are to use 0.0821 L-atm/K-mol for R, we must convert pressure to atm. Using Equation 10.11 we have

$$d = \frac{(\frac{745}{760}\text{ atm})(44.0\text{ g/mol})}{(0.0821\text{ L-atm/K-mol})(338\text{ K})} = 1.55\text{ g/L}$$

PRACTICE EXERCISE

The mean molar mass of the atmosphere at the surface of Titan, Saturn's largest moon, is 28.6 g/mol. The surface temperature is 95 K, and the pressure is 1.6 earth atm. Assuming ideal behavior, calculate the density of Titan's atmosphere.
Answer: 5.9 g/L

SAMPLE EXERCISE 10.8

A large flask is evacuated and found to weigh 134.567 g. It is then filled to a pressure of 735 mm Hg at 31°C with a gas of unknown molar mass and then reweighed; its mass is 137.456 g. The flask is then filled with water and again weighed; its mass is now 1067.9 g. Assuming that the ideal-gas equation applies, what is the molar mass of the unknown gas? (The density of water at 31°C is 0.997 g/cm^3.)

Solution: First we must determine the volume of the flask. This is given by the difference in weights of the empty flask and the flask filled with water, divided by the density of water at 31°C, which is 0.997 g/cm^3:

$$V = \frac{1067.9\text{ g} - 134.6\text{ g}}{0.997\text{ g/cm}^3} = 936\text{ cm}^3$$

Because the mass of the gas is 137.456 g − 134.567 g = 2.889 g, its density is 2.889 g/0.936 L = 3.087 g/L. Using Equation 10.12, we have

$$\begin{aligned}\mathcal{M} &= \frac{dRT}{P} \\ &= \frac{(3.087\text{ g/L})(0.0821\text{ L-atm/K-mol})(304\text{ K})}{(735/760)\text{ atm}} \\ &= 79.7\text{ g/mol}\end{aligned}$$

PRACTICE EXERCISE

One method for accurately determining the molar mass of a gas is to measure its density as a function of pressure. The graph of the quantity d/P against pressure is extrapolated to zero pressure to obtain a limiting value. In one set of experiments a certain gas was shown to have a limiting value d/P of 2.86 g/L-atm at 0°C. Calculate the molar mass. *Answer:* 64.1 g/mol

10.6 GAS MIXTURES AND PARTIAL PRESSURES

The pressure of a gas under conditions of constant volume and temperature is directly proportional to the number of moles of gas:

$$P = \left(\frac{RT}{V}\right)n = \text{constant} \times n \qquad [10.13]$$

Suppose that the gas with which we are concerned is not made up of a single kind of gas particle but is rather a mixture of two or more different substances. We expect that the total pressure exerted by the gas mixture is the sum of pressures due to the individual components. Each of the individual components, if present alone under the same temperature and volume conditions as the mixture, would exert a pressure that we term the **partial pressure**. John Dalton, who developed the atomic theory (Section 2.1), was the first to observe that the *total pressure of a mixture of gases equals the sum of the pressures that each gas would exert if it were present alone:*

$$P_t = P_1 + P_2 + P_3 + \cdots \qquad [10.14]$$

This statement is known as **Dalton's law of partial pressures.**

If each of the gases obeys the ideal-gas equation, we can write

$$P_1 = n_1\left(\frac{RT}{V}\right), \quad P_2 = n_2\left(\frac{RT}{V}\right), \quad P_3 = n_3\left(\frac{RT}{V}\right), \quad \text{and so forth.}$$

All of the gases experience the same temperature and volume. Therefore, by substituting into Equation 10.14, we obtain

$$P_t = \frac{RT}{V}(n_1 + n_2 + n_3 + \cdots) \qquad [10.15]$$

That is, the total pressure at constant temperature and volume is determined by the total number of moles of gas present, whether that total represents just one substance or a mixture.

SAMPLE EXERCISE 10.9

What pressure, in atm, is exerted by a mixture of 2.00 g of H_2 and 8.00 g of N_2 at 273 K in a 10.0-L vessel?

Solution: The pressure depends on the total moles of gas (Equation 10.15). Calculating the moles of H_2 and N_2, we have

$$n_{H_2} = (2.00\text{ g }H_2)\left(\frac{1\text{ mol }H_2}{2.02\text{ g }H_2}\right) = 0.990\text{ mol }H_2$$

$$n_{N_2} = (8.00\text{ g }N_2)\left(\frac{1\text{ mol }N_2}{28.0\text{ g }N_2}\right) = 0.286\text{ mol }N_2$$

Using Equation 10.15 gives

$$\begin{aligned} P_t &= \frac{RT}{V}(n_{H_2} + n_{N_2}) \\ &= \frac{(0.0821\text{ L-atm/K-mol})(273\text{ K})}{10.0\text{ L}} \times (0.990\text{ mol} + 0.286\text{ mol}) \\ &= 2.86\text{ atm} \end{aligned}$$

This total pressure is the sum of the partial pressures of H_2 and N_2.

PRACTICE EXERCISE

If a 0.30-L sample of O_2 at 0°C and 0.80 atm pressure and a 0.10-L sample of N_2 at 0°C and 1.8 atm pressure are both placed in a 0.40-L container at 0°C, what is the total pressure in the container? ***Answer:*** 1.05 atm

Partial Pressures and Mole Fractions

The compositions of gas mixtures are often described in terms of the fractions or percentages of molecules of each gas present in the mixture. For example, 78 percent of the molecules in air are N_2 molecules. That is, the fraction of N_2 molecules in air is 0.78.

Because the number of moles is proportional to the number of molecules, the fraction of moles of any gas in a mixture must equal the fraction of molecules of that gas. Thus, the mole fraction of N_2 in air is 0.78. The **mole fraction**, X, of any component of a mixture is simply the ratio of the number of moles of that component to the total number of moles in the mixture:

$$\text{Mole fraction of component 1} = X_1 = \frac{\text{moles of component 1}}{\text{total moles in mixture}} \quad [10.16]$$

We can relate mole fractions to partial pressures. Because $P = nRT/V$, we can write

$$\frac{P_1}{P_t} = \frac{n_1RT/V}{n_tRT/V} = \frac{n_1}{n_t} = X_1$$

where n_t equals the total number of moles of all gases in the mixture. Thus, the ratio of the partial pressure of a gas to the total pressure of the mixture gives the mole fraction of that gas. By rearranging this relationship, we see that the partial pressure of a gas in a mixture is equal to the product of its mole fraction and the total pressure:

$$P_1 = X_1P_t \quad [10.17]$$

Therefore, the partial pressure of N_2 in air at a total barometric pressure of 720 mm Hg is

$$P_{N_2} = (0.78)(720 \text{ mm Hg}) = 560 \text{ mm Hg}$$

SAMPLE EXERCISE 10.10

A study of the effects of certain gases on plant growth requires a synthetic atmosphere composed of 1.5 mol percent CO_2, 18.0 mol percent O_2, and 80.5 mol percent Ar. **(a)** Calculate the partial pressure of O_2 in the mixture if the total pressure of the atmosphere is to be 745 mm Hg. **(b)** If this atmosphere is to be held in a 120-L space at 295 K, how many moles of O_2 are needed?

Solution: **(a)** The mole percent is just the mole fraction times 100. Therefore, the mole fraction of O_2 is 0.180. Using Equation 10.17, we have

$$P_{O_2} = (0.180)(745 \text{ mm Hg}) = 134 \text{ mm Hg}$$

(b) Tabulating the given variables and changing them to appropriate units, we have

$$P_{O_2} = (134 \text{ mm Hg})\left(\frac{1 \text{ atm}}{760 \text{ mm Hg}}\right) = 0.176 \text{ atm}$$

$$V = 120 \text{ L}$$

$$n_{O_2} = ?$$

$$R = 0.0821 \frac{\text{L-atm}}{\text{K-mol}}$$

$$T = 295 \text{ K}$$

Solving the ideal-gas equation for n_{O_2}, we have

$$n_{O_2} = P_{O_2}\left(\frac{V}{RT}\right) = (0.176 \text{ atm}) \frac{120 \text{ L}}{(0.0821 \text{ L-atm/K-mol})(295 \text{ K})} = 0.872 \text{ mol}$$

PRACTICE EXERCISE

From data gathered by *Voyager 1*, scientists have estimated the composition of the atmosphere of Titan, Saturn's largest moon. The total pressure on the surface of Titan is 1220 mm Hg. The atmosphere consists of 82 mol percent N_2, 12 mol percent Ar, and 6.0 mol percent CH_4. Calculate the partial pressure of each of these gases in Titan's atmosphere. ***Answer:*** 1.0×10^3 mm Hg N_2, 1.5×10^2 mm Hg Ar, and 73 mm Hg CH_4

10.7 VOLUMES OF GASES IN CHEMICAL REACTIONS

A knowledge of the properties of gases is important for chemists because gases are often reactants or products in chemical reactions. For this reason, we are often faced with calculating the volumes of gases required as reactants or produced as products in reactions. We saw in Chapter 3 that the coefficients in balanced chemical equations tell us the relative amounts (in moles) of reactants and products in a reaction. The number of moles of a gas, of course, is related to P, V, and T.

SAMPLE EXERCISE 10.11

The industrial synthesis of nitric acid involves the reaction of nitrogen dioxide gas with water:

$$3NO_2(g) + H_2O(l) \longrightarrow 2HNO_3(aq) + NO(g)$$

How many moles of nitric acid can be prepared using 450 L of NO_2 at a pressure of 5.00 atm and a temperature of 295 K?

Solution: The balanced chemical equation tells us that 3 mol of NO_2 will produce 2 mol of HNO_3 in this reaction. To use this information, we must first determine the number of moles of NO_2:

$$n = \frac{PV}{RT} = \frac{(5.00 \text{ atm})(450 \text{ L})}{(0.0821 \text{ L-atm/K-mol})(295 \text{ K})} = 92.9 \text{ mol}$$

From here we can use the number of moles of NO_2 and the coefficients in the balanced equation to calculate the number of moles of HNO_3:

$$92.9 \text{ mol } NO_2\left(\frac{2 \text{ mol } HNO_3}{3 \text{ mol } NO_2}\right) = 61.9 \text{ mol } HNO_3$$

PRACTICE EXERCISE

In the first step in the industrial process for making nitric acid, ammonia reacts with oxygen at 850°C and 5.00 atm in the presence of a suitable catalyst. The following reaction occurs:

$$4NH_3(g) + 5O_2(g) \longrightarrow 4NO(g) + 6H_2O(g)$$

How many liters of $NH_3(g)$ at 850°C and 5.00 atm are required to react with 1.00 mol of $O_2(g)$ in this reaction? ***Answer:*** 14.8 L

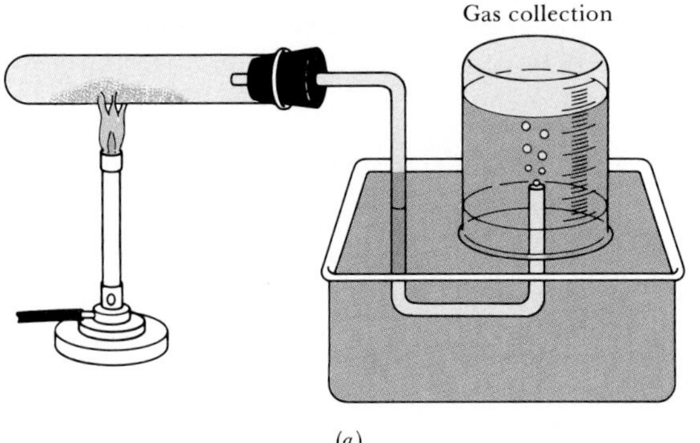

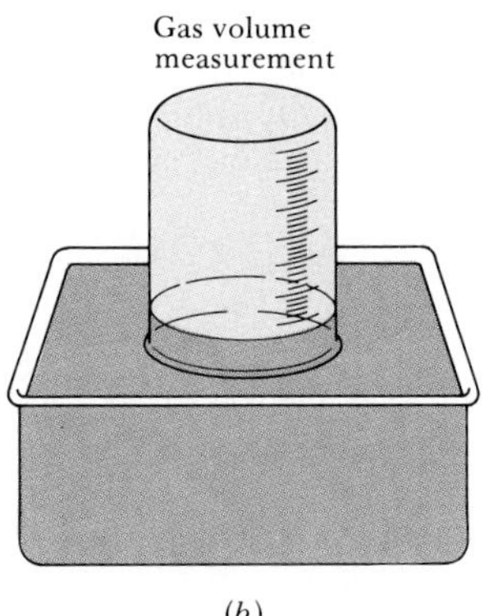

Figure 10.11 (*a*) Collection of gas over water. (*b*) When the gas has been collected, the bottle is raised or lowered so the heights of the water inside and outside the collection vessel are equalized. The total pressure of the gases inside the vessel is then equal to the atmospheric pressure.

Collecting Gases over Water

An experiment that often comes up in the course of laboratory work involves determining the number of moles of gas collected from a chemical reaction. Sometimes this gas is collected over water. For example, solid potassium chlorate, $KClO_3$ can be decomposed by heating it in a test tube in an arrangement shown in Figure 10.11. The balanced equation for the reaction is

$$2KClO_3(s) \longrightarrow 2KCl(s) + 3O_2(g)$$

The oxygen gas is collected in a bottle that is initially filled with water and inverted in a water pan.

The volume of gas collected is measured by raising or lowering the bottle as necessary until the water levels inside and outside the bottle are the same. When this condition is met, the pressure inside the bottle is equal to the atmospheric pressure outside. The total pressure inside is the sum of the pressure of gas collected and the pressure of water vapor in equilibrium with liquid water:

$$P_{\text{total}} = P_{\text{gas}} + P_{H_2O} \qquad [10.18]$$

The pressure exerted by water vapor, P_{H_2O}, at various temperatures is shown in Appendix B.

SAMPLE EXERCISE 10.12

Suppose that 0.200 L of oxygen gas is collected over water, as shown in Figure 10.11. The temperature of the water and gas is 26°C, and the atmospheric pressure is 750 mm Hg. **(a)** How many moles of O_2 are collected? **(b)** What volume would the O_2 gas collected occupy when dry, at the same temperature and pressure?

Solution: **(a)** The pressure of O_2 gas in the vessel is the difference between the total pressure, 750 mm Hg, and the vapor pressure of water at 26°C, 25 mm (Appendix B):

$$P_{O_2} = 750 - 25 = 725 \text{ mm Hg}$$

Solving the ideal-gas equation for n, we have

$$n = \frac{PV}{RT} = \frac{(\frac{725}{760} \text{ atm})(0.200 \text{ L})}{(0.0821 \text{ L-atm/K-mol})(299 \text{ K})} = 7.77 \times 10^{-3} \text{ mol}$$

(b) Suppose that we removed the water from the gas sample while maintaining the same total pressure. The O_2 pressure after drying would be 750 mm Hg. The corrected O_2 volume would thus be less for the dry gas, because its partial pressure is greater than that of the wet gas:

$$V_{O_2} = (0.200 \text{ L})\left(\frac{725 \text{ mm Hg}}{750 \text{ mm Hg}}\right) = 0.193 \text{ L}$$

PRACTICE EXERCISE

Assume that 260 mL of dry nitrogen at 20°C and 760 mm Hg are bubbled through water and stored in an inverted beaker, as shown in Figure 10.11(*b*). What is the volume of the stored gas if its pressure is 740 mm Hg and its temperature is 26°C? ***Answer:*** 282 mL

SAMPLE EXERCISE 10.13

A 2.55-g sample of ammonium nitrite, NH_4NO_2, is heated in a test tube, as shown in Figure 10.11. The ammonium nitrite is expected to decompose according to the equation

$$NH_4NO_2(s) \longrightarrow N_2(g) + 2H_2O(g)$$

If it does decompose in this way, what volume of N_2 will be collected in the flask when the water and gas temperature is 26°C and the barometric pressure is 745 mm Hg?

Solution: We begin by calculating the number of moles of N_2 gas formed:

$$2.55 \text{ g } NH_4NO_2\left(\frac{1 \text{ mol } NH_4NO_2}{64.0 \text{ g } NH_4NO_2}\right)\left(\frac{1 \text{ mol } N_2}{1 \text{ mol } NH_4NO_2}\right) = 0.0398 \text{ mol } N_2$$

To predict the volume of N_2 gas that will be collected, we might be tempted simply to calculate the volume that would be occupied by 0.0398 mol of N_2 at 745 mm Hg. But we must also account for the partial pressure of water vapor at 26°C, because the gas within the bottle is saturated with water vapor. In Appendix B, we can see that the vapor pressure of water at 26°C is 25 mm Hg. The pressure of nitrogen gas in the flask when the water levels inside and out have been equalized is thus 745 − 25 = 720 mm Hg = 0.947 atm. We must calculate the predicted volume of gas using this pressure. Rearranging Equation 10.5, we obtain

$$V = \frac{nRT}{P}$$

Inserting all known quantities in correct units, we find

$$V = \frac{(0.0398 \text{ mol})(0.0821 \text{ L-atm/K-mol})(299 \text{ K})}{0.947 \text{ atm}} = 1.03 \text{ L}$$

PRACTICE EXERCISE

A 1.60-g sample of $KClO_3$ is heated to produce O_2 according to the equation

$$2KClO_3(s) \longrightarrow 2KCl(s) + 3O_2(g)$$

Assume complete decomposition and ideal-gas behavior. What volume of O_2 collects over water at 26°C, 740 mm Hg pressure? ***Answer:*** 511 mL

10.8 KINETIC-MOLECULAR THEORY

The ideal-gas equation describes *how* gases behave, but it doesn't explain *why* they behave as they do. For example, why does a gas expand when heated at constant pressure? Or why does its pressure increase when the gas is compressed at constant temperature? To understand the physical properties of gases, we need a model that helps us picture what happens to gas particles as experimental conditions such as pressure or temperature change. Such a model, known as the **kinetic-molecular theory**, was developed over a period of about 100 years, culminating in 1857 when Rudolf Clausius (1822–1888) published a complete and satisfactory form of the theory.

The kinetic-molecular theory (the theory of moving molecules) is summarized by the following statements:

1. Gases consist of large numbers of molecules that are in continuous, random motion. (The word *molecule* is used here to designate the smallest particle of any gas; some gases, such as the noble gases, consist of uncombined atoms.)
2. The volume of all the molecules of the gas is negligible compared to the total volume in which the gas is contained.
3. Attractive and repulsive forces between gas molecules are negligible.
4. Energy can be transferred between molecules during collisions, but the *average* kinetic energy of the molecules does not change with time, as long as the temperature of the gas remains constant. In other words, the collisions are perfectly elastic.
5. The average kinetic energy of the molecules is proportional to absolute temperature. At any given temperature the molecules of all gases have the same average kinetic energy.

The kinetic-molecular theory gives us an understanding of both pressure and temperature at the molecular level. The pressure of a gas is caused by collisions of the molecules with the walls of the container; it is determined both by the frequency of collisions per unit area and by the impulse imparted per collision (that is, by how often and by how "hard" the molecules strike the walls).

The absolute temperature of a gas is a measure of the *average* kinetic energy of its molecules. If two different gases are at the same temperature, their molecules have the same average kinetic energy. If the temperature of a gas is doubled (say from 200 K to 400 K), the average kinetic energy of its molecules doubles. Thus, molecular motion increases with increasing temperature.

Although the molecules in a sample of gas have an *average* kinetic energy and hence an average speed, the individual molecules move at varying speeds. At one instant, some of them are moving rapidly, others slowly. Figure 10.12 illustrates the distribution of molecular speeds within nitrogen gas at 0°C (blue

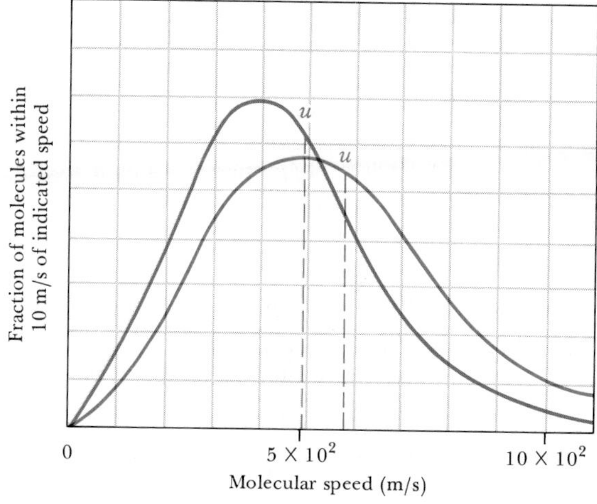

Figure 10.12 Distribution of molecular speeds for nitrogen at 0°C (blue line) and 100°C (red line).

line) and at 100°C (red line). The curve tells us the fraction of molecules moving at each speed. Notice that at higher temperatures a greater fraction of molecules is moving at greater speeds; the distribution curve has shifted toward higher speeds and hence toward higher average kinetic energy.

Figure 10.12 also shows the value of the **root-mean-square (rms) speed**, u, of the molecules at each temperature. This quantity is the speed of a molecule possessing average kinetic energy. The rms speed is not quite the same as the average speed. The difference between the two, however, is small.*

The rms speed is important because the average kinetic energy of the gas molecules, ϵ, is related directly to u^2:

$$\epsilon = \tfrac{1}{2}mu^2 \qquad [10.19]$$

where m is the mass of the molecule. Because mass doesn't change with temperature, the rms speed (and also the average speed) of molecules must increase as temperature increases.

Application to the Gas Laws

The empirical observations of gas properties as expressed in the various gas laws are readily understood in terms of the kinetic-molecular theory. The following examples illustrate this point.

1. *Effect of a volume increase at constant temperature:* The fact that temperature remains constant means that the average kinetic energy of the gas molecules remains unchanged. This in turn means that the rms speed of the molecules, u, is unchanged. However, if the volume is increased, the molecules must move a longer distance between collisions. Consequently, there are fewer collisions per unit time with the container walls, and pressure decreases. Thus the model accounts in a simple way for Boyle's law.
2. *Effect of a temperature increase at constant volume:* An increase in temperature means an increase in the average kinetic energy of the molecules, and thus an increase in u. If there is no change in volume, there will be more collisions with the walls per unit time. Furthermore, the change in momentum in each collision increases (the molecules strike the walls harder). Hence the model explains the observed pressure increase.

SAMPLE EXERCISE 10.14

A sample of O_2 gas initially at STP is transferred from a 2-L container to a 1-L container at constant temperature. What effect does this change have on **(a)** the average kinetic energy of O_2 molecules; **(b)** the average speed of O_2 molecules; **(c)** the total number of collisions of O_2 molecules with the container walls in a unit time; **(d)** the number of collisions of O_2 molecules with a unit area of container wall in a unit time?

---

* To illustrate the difference between rms speed and average speed, suppose that we have four objects with speeds of 4.0, 6.0, 10.0, and 12.0 m/s. Their average speed is $\tfrac{1}{4}(4.0 + 6.0 + 10.0 + 12.0) = 8.0$ m/s. The rms speed, u, however, is the square root of the average squared speeds of the molecules:

$$\sqrt{\tfrac{1}{4}(4.0^2 + 6.0^2 + 10.0^2 + 12.0^2)} = \sqrt{74.0} = 8.6 \text{ m/s}$$

For an ideal gas, the average speed equals $0.921 \times u$. Thus the average speed is directly proportional to the rms speed, and the two are in fact nearly equal.

A CLOSER LOOK: The Ideal-Gas Equation

Beginning with the postulates of the kinetic-molecular theory, it is possible to derive the ideal-gas equation. Rather than proceed through a derivation, let's consider in somewhat qualitative terms how the ideal-gas equation might follow. As we have seen (Section 10.2), pressure is force per unit area. The total force of the molecular collisions on the walls, and hence the pressure produced by these collisions, depend both on how strongly the molecules strike the walls (impulse imparted per collision) and on the rate at which these collisions occur:

$$P \propto \text{impulse imparted per collision} \times \text{rate of collisions} \quad [10.20]$$

For a molecule traveling at the rms speed, u, the impulse imparted by a collision with a wall depends on the momentum of the molecule; that is, it depends on the product of its mass and speed, mu. The rate of collisions is proportional to both the number of molecules per unit volume, n/V, and their speed, u. If there are more molecules in a container, there will be more frequent collisions with the container walls. As the molecular speed increases or the volume of the container decreases, the time required for molecules to traverse the distance from one wall to another is reduced, and the molecules collide more frequently with the walls. Thus, we have

$$P \propto mu \times \frac{n}{V} \times u \propto \frac{nmu^2}{V} \quad [10.21]$$

Because the average kinetic energy, $\frac{1}{2}mu^2$, is proportional to temperature, we have $mu^2 \propto T$. Making this substitution into Equation 10.21 gives

$$P \propto \frac{n(mu^2)}{V} \propto \frac{nT}{V} \quad [10.22]$$

Let us now convert the proportionality sign to an equal sign by expressing n as the number of moles of gas; we then insert a proportionality constant—R, the molar gas constant:

$$P = \frac{nRT}{V} \quad [10.23]$$

This expression, of course, is the familiar ideal-gas equation.

Solution: **(a)** The average kinetic energy of O_2 molecules is determined only by temperature. The average kinetic energy is not changed by the compression of O_2 from 2 L to 1 L at constant temperature. **(b)** If the average kinetic energy of O_2 molecules doesn't change, the average speed remains constant. **(c)** The total number of collisions with the container walls in a unit time must increase, because the molecules are moving within a smaller volume but with the same average speed as before. Under these conditions they must encounter a wall more frequently. **(d)** The number of collisions with a unit area of wall increases because the total number of collisions with the walls is higher and the area of wall is smaller than before.

PRACTICE EXERCISE

How is the rms speed of N_2 molecules in a gas sample changed by **(a)** an increase in temperature; **(b)** an increase in volume of sample; **(c)** mixing with an Ar sample at the same temperature? ***Answers:*** **(a)** increases; **(b)** no effect; **(c)** no effect

10.9 MOLECULAR EFFUSION AND DIFFUSION

The motion of the molecules in a gas is continuous and totally random. Nevertheless, as long as we hold the temperature constant, the *average* kinetic energy and *average* speed of the molecules will remain constant. If two gases are at the same temperature, their molecules must have the same average kinetic energy, $\frac{1}{2}mu^2$, even though their masses may differ. In order to have the same average kinetic energy, the lighter molecules must move with a higher average speed. An equation that expresses this fact quantitatively can be derived from the kinetic-molecular theory:

$$u = \sqrt{\frac{3RT}{\mathcal{M}}} \quad [10.24]$$

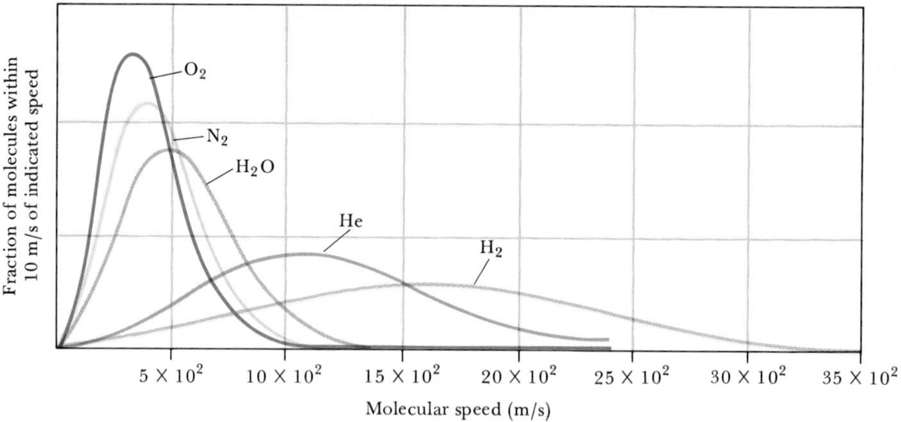

Figure 10.13 Distribution of molecular speeds for different gases at 25°C.

Notice that the molar mass, $\mathcal{M}$, appears in the denominator. Thus, the less massive the gas molecules, the higher the rms speed, u. Figure 10.13 shows the distribution of molecular speeds for several different gases at 25°C. Notice how the distributions are shifted toward higher speeds for gases of lower molar masses.

SAMPLE EXERCISE 10.15

Calculate the rms speed, u, of a N_2 molecule at 25°C.

Solution: In using Equation 10.24, we should convert each quantity to SI units, so all the units are compatible:

$$T = 25 + 273 = 298 \text{ K}$$

$$\mathcal{M} = 28.0 \text{ g/mol} = 28.0 \times 10^{-3} \text{ kg/mol}$$

$$R = 8.314 \text{ J/K-mol} = 8.314 \text{ kg-m}^2/\text{s}^2\text{-K-mol}$$

(These units follow from the fact that $1 \text{ J} = 1 \text{ kg-m}^2/\text{s}^2$)

$$u = \sqrt{\frac{3(8.314 \text{ kg-m}^2/\text{s}^2\text{-K-mol})(298 \text{ K})}{28.0 \times 10^{-3} \text{ kg/mol}}} = 5.15 \times 10^2 \text{ m/s}$$

This corresponds to a speed of 1150 mi/hr.

PRACTICE EXERCISE

What is the rms speed of a He atom at 25°C? ***Answer:*** 1.36×10^3 m/s

The dependence of molecular speeds on mass has several interesting consequences. For example, the rate at which a gas is able to escape through a tiny hole, as when a gas escapes through a hole in a balloon, depends on the molecular mass of the gas. This process is known as **effusion**. Effusion is related to, but is not quite the same as, **diffusion**. The latter term refers to the spread of one substance throughout a space or throughout a second substance. For example, the molecules of a perfume diffuse through a room.

Graham's Law of Effusion

In about 1830, Thomas Graham discovered that the effusion rate of a gas is inversely proportional to the square root of its molar mass. Assume that we have two gases at the same initial pressure contained in identical containers, each with an identical pinhole in one wall. Let the rate of effusion be called r. **Graham's law** states that

$$\frac{r_1}{r_2} = \sqrt{\frac{\mathcal{M}_2}{\mathcal{M}_1}} \qquad [10.25]$$

Equation 10.25 compares the *rates* of effusion of two different gases; the lighter gas effuses more rapidly (see Figure 10.14).

Graham's law follows from our previous discussion if we assume that the rate of effusion is proportional to the rms speed of the molecules. Because R and T are constant, we have from Equation 10.24:

$$\frac{r_1}{r_2} = \frac{u_1}{u_2} = \sqrt{\frac{3RT/\mathcal{M}_1}{3RT/\mathcal{M}_2}} = \sqrt{\frac{\mathcal{M}_2}{\mathcal{M}_1}} \qquad [10.26]$$

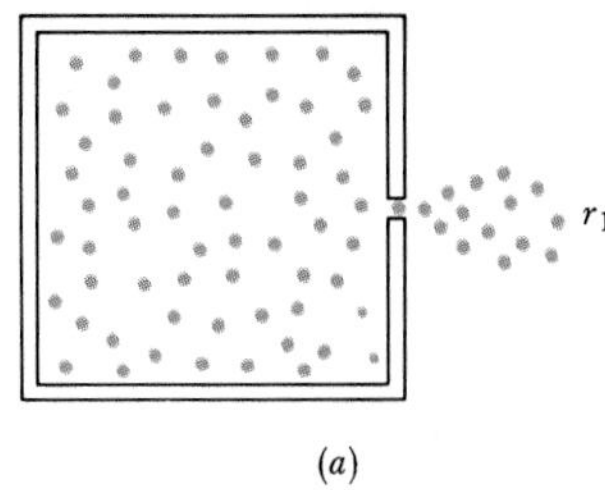

(*a*)

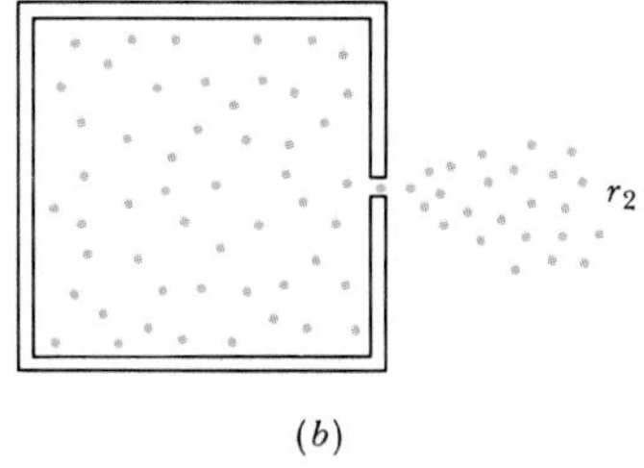

(*b*)

Figure 10.14 Effusion of gases through a pinhole. In this illustration the molar mass of gas molecules in (*a*) is higher than in (*b*). The rate of effusion r_1, from box (*a*), is slower than r_2, from box (*b*), in accordance with Equation 10.25.

SAMPLE EXERCISE 10.16

If an unknown gas effuses at a rate that is only 0.468 times that of O_2 at the same temperature, what is the molar mass of the unknown gas?

Solution: Using Equation 10.25, we have

$$\frac{r_x}{r_{O_2}} = \sqrt{\frac{\mathcal{M}_{O_2}}{\mathcal{M}_x}}$$

Thus

$$\frac{r_x}{r_{O_2}} = 0.468 = \sqrt{\frac{32.0 \text{ g/mol}}{x}}$$

Solving for the unknown molar mass, $\mathcal{M}_x$, yields

$$\frac{32.0 \text{ g/mol}}{\mathcal{M}_x} = (0.468)^2 = 0.219$$

$$\mathcal{M}_x = \frac{32.0 \text{ g/mol}}{0.219} = 146 \text{ g/mol}$$

PRACTICE EXERCISE

Calculate the ratio of the effusion rates of N_2 and O_2, r_{N_2}/r_{O_2}.
Answer: $r_{N_2}/r_{O_2} = 1.07$

Diffusion and Mean Free Path

Diffusion, like effusion, is faster for light molecules than for heavy ones. In fact, the relative rates of diffusion of two gases under identical experimental conditions is also given by Equation 10.25. The ratio of the rates is inversely proportional to the square root of the ratio of the molar masses. Nevertheless, molecular collisions make diffusion more complicated than effusion.

We can see from the horizontal scale in Figure 10.12 that the speeds of molecules are quite high. The average speed of N_2 at room temperature is 515 m/s (1150 mi/hr). Yet we know that if someone opens a vial of perfume at one end of a room, some time elapses, perhaps a few minutes, before the odor is detected at the other end. The diffusion of gases is much slower than molecular speeds because of molecular collisions. These collisions occur quite frequently for a gas at atmospheric pressure—about 10^{10} times per second for each molecule. Because of these collisions, the direction of motion of a gas molecule is constantly changing. Therefore, the diffusion of a molecule from one point to another consists of many short, straight-line segments as collisions buffet it around in random directions, as depicted in Figure 10.15. First the molecule moves in one direction, then in another, at one instant at high speed, next at low speed.

Figure 10.15 Schematic illustration of the diffusion of a gas molecule. For the sake of clarity, no other gas molecules in the container are shown. The path of the molecule of interest begins at the dot. Each short segment of line represents travel between collisions. The red arrow indicates the net distance traveled by the molecule.

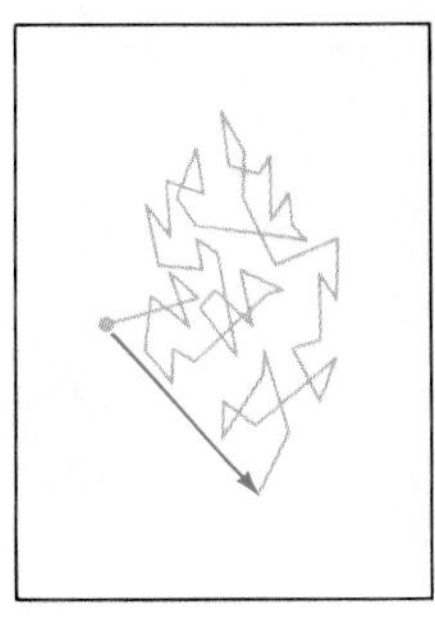

CHEMISTRY AT WORK: Gas Separations

The fact that lighter molecules move at higher average speeds than heavier ones has many interesting consequences and applications. If you buy a helium-filled balloon at a carnival, you will find that it deflates overnight. The lighter He atoms have a higher average speed than the heavier N_2 and O_2 molecules in air. Thus, the helium escapes faster than the N_2 and O_2 molecules can enter. As the number of moles of gas in the balloon decreases, the balloon deflates.

The effort to develop the atomic bomb during World War II required scientists to separate the relatively low-abundance uranium isotope ^{235}U (0.7 percent) from the much more abundant ^{238}U (99.3 percent). This task was accomplished by converting the uranium into a volatile compound, UF_6, which boils at 56°C. The gaseous UF_6 was allowed to pass through porous barriers. The slight difference in molar mass between the compounds of the two isotopes caused the molecules to move at slightly different rates:

$$\frac{r_{235}}{r_{238}} = \sqrt{\frac{352.04}{349.03}} = 1.0043$$

Thus, the gas initially appearing on the opposite side of the barrier would be very slightly enriched in the lighter molecule. The diffusion process was repeated thousands of times, leading to a nearly complete separation of the two nuclides of uranium.

The rate of diffusion of a gas through a porous medium is not always determined solely by the molecular mass of the gas molecules. Even weak interactions between the molecules of gas and the molecules of the porous medium will affect the rate. Attractive intermolecular interactions, which we will discuss in more detail in the next chapter, slow the rate at which a gas molecule passes through the narrow passages of the porous medium.

The medium's molecules may attract one substance in a mixture more than another. Thus, the rates of diffusion of a mixture's components through a medium are not uniform. Dow Chemical Company has recently developed a method of separating gaseous mixtures into their components by using the difference in the rate at which molecules pass through a porous membrane. Figure 10.16 shows Dow's tiny hollow fiber made of polyolefin, an organic polymer. The fiber is smaller than a human hair. Its walls are permeable to the passage of N_2 and O_2. However, O_2 and water vapor diffuse through the fiber walls considerably more readily than does N_2. To take advantage of this property, thousands of the fibers are formed into bundles. Compressed air is introduced around the outside of the bundles, as shown schematically in Figure 10.17. The gas stream taken from inside the fibers is selectively enriched in O_2. The gas stream on the outside of the fibers is correspondingly enriched in N_2.

Figure 10.16 Polyolefin fibers used to carry out gas separations. The solid cylinder is human hair. Gases such as nitrogen and oxygen differ significantly in the rate at which they pass through the wall of the hollow fiber. This difference in permeability provides the basis for gas separations, as shown in Figure 10.17. (Dow Chemical Company)

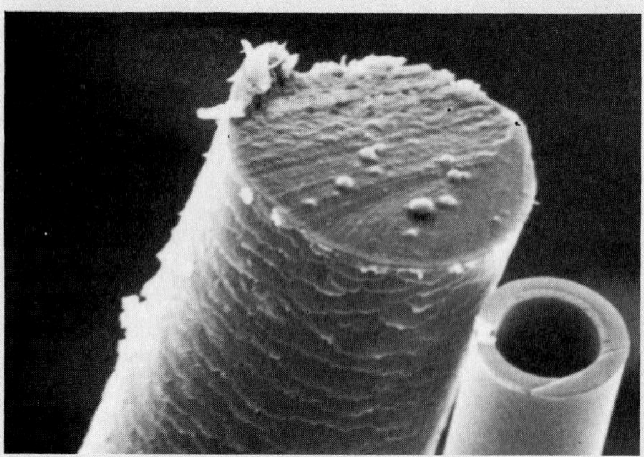

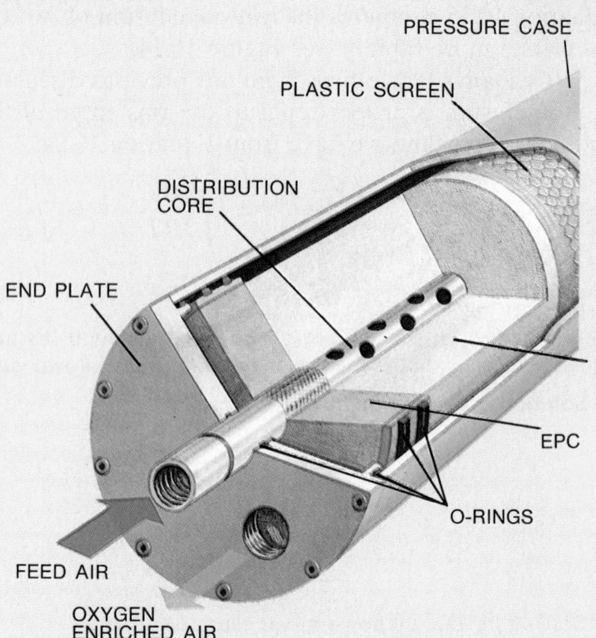

Figure 10.17 A diagrammatic illustration of gas separation that takes advantage of the difference in ease of diffusion of gases through polyolefin fiber walls. The gas that enters under pressure (feed air) passes from the distribution core all around the outsides of the tiny hollow fibers. Oxygen and water vapor preferentially pass through the fiber walls, are collected at the ends of the fibers, and pass out as oxygen-enriched air. The gas that has not passed through the fibers exits the assembly as nitrogen-enriched gas. (Dow Chemical Company)

The average distance traveled by a molecule between collisions is called the **mean free path**. The higher the density of a gas, the smaller the mean free path. That is, the more molecules there are in a given volume, the shorter the average distance traveled between collisions. The mean free path for air molecules at sea level is about 60 nm (6×10^{-6} cm). At about 100 km in altitude, where the air density is much lower, the mean free path is about 10 cm, about *1 million* times longer than at the earth's surface.

10.10 DEVIATIONS FROM IDEAL BEHAVIOR

Although the ideal-gas equation is a very useful description of gases, all real gases fail to obey the relationship to some degree. The extent to which a real gas departs from ideal behavior may be seen by slightly rearranging the ideal-gas equation:

$$\frac{PV}{RT} = n \qquad [10.27]$$

For a mole of ideal gas ($n = 1$), the quantity PV/RT equals 1 at all pressures. In Figure 10.18, PV/RT is plotted as a function of P for 1 mol of several different gases. At high pressures, the deviation from ideal behavior ($PV/RT = 1$) is large and is different for each gas. Clearly, real gases do not behave ideally at high pressure. At lower pressures (usually below 10 atm), however, the deviations from ideal behavior are small, and we can use the ideal-gas equation without generating serious error.

The deviation from ideal behavior also depends on temperature. Figure 10.19 shows graphs of PV/RT for 1 mol of N_2 at three different temperatures. As temperature increases, the properties of the gas more nearly approach that of the ideal gas. In general, gases deviate significantly from ideal behavior at temperatures near their liquefaction points; that is, the deviations increase as temperature decreases, becoming significant near the temperature at which the gas is converted into a liquid.

We can understand these pressure and temperature effects on nonideality by considering two factors that are considered negligible in the kinetic-molecular theory: (1) The molecules of a gas possess finite volumes, and (2) at short distances of approach they exert attractive forces upon one another.

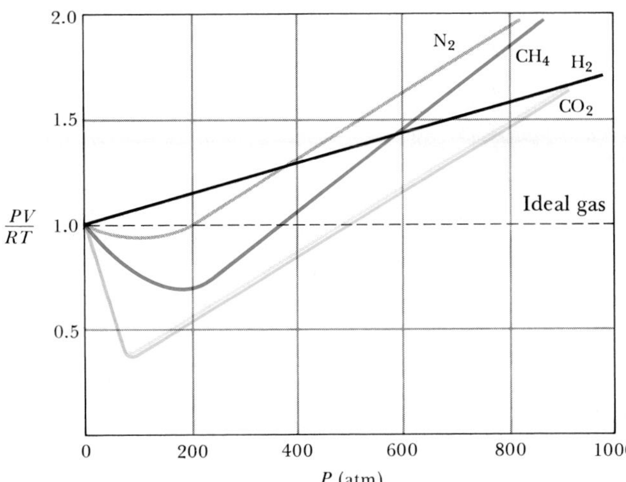

Figure 10.18 PV/RT versus pressure for 1 mol of several gases at 300 K. The data for CO_2 pertain to a temperature of 313 K because CO_2 liquefies under high pressure at 300 K.

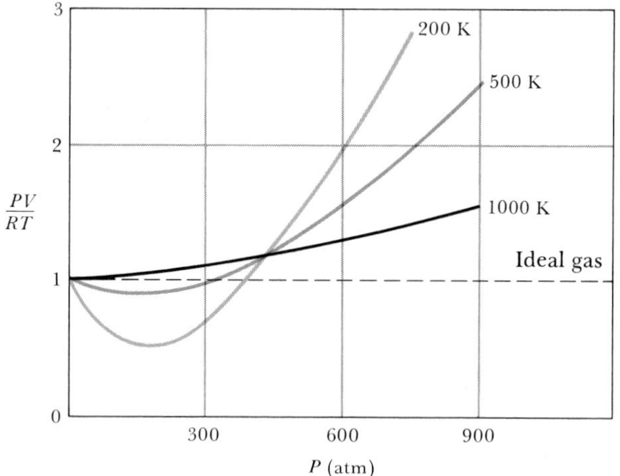

Figure 10.19 PV/RT versus pressure for 1 mol of nitrogen gas at three different temperatures. As temperature increases, the gas more closely approaches ideal behavior.

Real molecules have finite, though small, volumes. Thus, the free, unoccupied space in which the molecules can move is somewhat less than the container volume. At relatively low pressures, the difference between the container volume and the unoccupied volume of the gas is negligible. At much higher pressures, however, the free volume can be considerably smaller than the container volume, as is illustrated in Figure 10.20. In contrast, the volume of the molecules of an ideal gas are negligible (assumed to be zero) under all conditions. Thus, the free volume available to an ideal gas always equals the container volume. Using the container volume instead of the free volume in calculating PV/RT for real gases gives a quantity that is larger than that for an ideal gas. The increase in PV/RT at high pressures for the gases in Figure 10.18 is due to the finite volumes of the gas molecules.

In addition, the attractive forces between molecules come into play at short distances, as when molecules are crowded together at high pressures. Because of these attractive forces, the impact of a given molecule with the wall of the container is lessened. If we could stop the action in a gas, the position of the molecules might resemble the illustration in Figure 10.21. The molecule about to make contact with the wall experiences the attractive forces of nearby molecules. These attractions lessen the force with which the molecule hits the

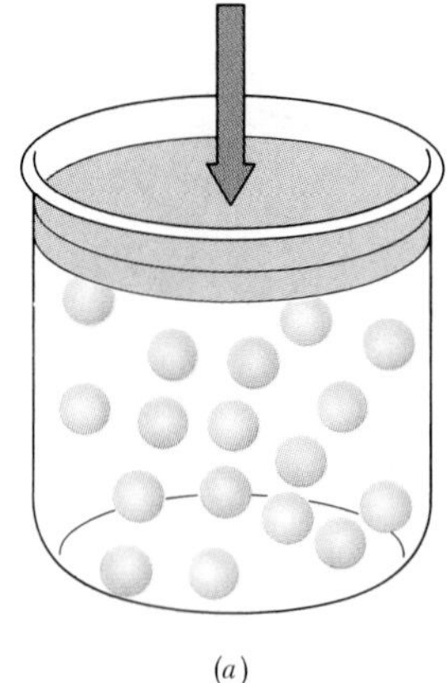

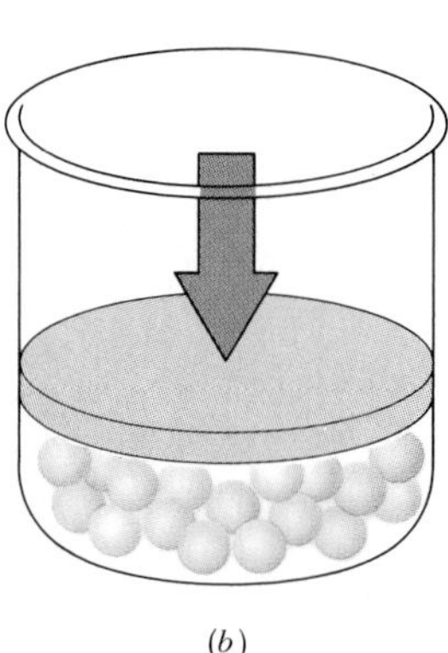

Figure 10.20 Illustration of the effect of the finite volume of gas molecules on the properties of a real gas at high pressure. In (*a*), at low pressure, the volume of the gas molecules is small compared with the container volume. In (*b*), at high pressure, the volume of the gas molecules themselves is a large fraction of the total space available.

wall, and thus the change in momentum when it hits the wall is lessened. As a result, the pressure is less than that of an ideal gas. This effect serves to decrease PV/RT, as seen in Figure 10.19. When the pressure is sufficiently high, the volume effects discussed above dominate, and PV/RT increases.

Temperature determines how effective attractive forces between gas molecules are. As the gas is cooled, the average kinetic energy decreases while intermolecular attractions remain constant. In a sense, cooling a gas deprives molecules of the energy they need to overcome their mutual attractive influence. The effects of temperature shown earlier in Figure 10.19 illustrate this point very well. Notice that as temperature increases, the negative departure of PV/RT from the ideal-gas behavior disappears. The difference that remains at high temperature stems mainly from the effect of volume.

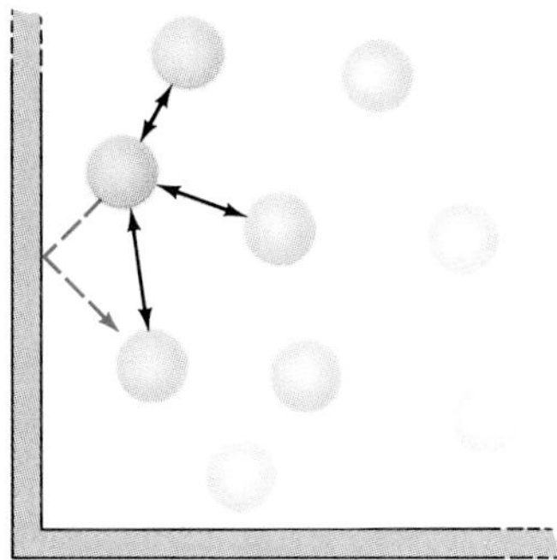

Figure 10.21 Effect of attractive intermolecular forces on the pressure exerted by a gas on its container walls. The molecule that is about to strike the wall in the figure experiences attractive forces from nearby molecules, and its impact on the wall is thereby lessened. The attractive forces become noticeable only under high-pressure conditions, when the average distance between molecules is small.

The van der Waals Equation

Engineers and scientists who work with gases at high pressures often cannot use the ideal-gas equation to predict the pressure–volume properties of gases because departures from the ideal-gas behavior are too large. Various equations have been developed to predict more realistically the pressure–volume behavior of real gases. One of the earliest and most useful of these equations was proposed by the Dutch scientist Johannes van der Waals:

$$\left(P + \frac{n^2 a}{V^2}\right)(V - nb) = nRT \qquad [10.28]$$

The van der Waals equation differs from the ideal-gas equation by the presence of two correction terms; one corrects the volume, the other modifies the pressure. The term nb in the expression $(V - nb)$ is a correction for the finite volume of the gas molecules; the van der Waals constant, b, different for each gas, has units of L/mol. It is a measure of the actual volume occupied by the gas molecules. Values of b for several gases are listed in Table 10.3. Note that b increases with an increase in mass of the molecule or in the complexity of its structure.

Table 10.3 Van der Waals Constants for Gas Molecules

Substance	a (L^2-atm/mol^2)	b (L/mol)
He	0.0341	0.02370
Ne	0.211	0.0171
Ar	1.34	0.0322
Kr	2.32	0.0398
Xe	4.19	0.0510
H_2	0.244	0.0266
N_2	1.39	0.0391
O_2	1.36	0.0318
Cl_2	6.49	0.0562
H_2O	5.46	0.0305
CH_4	2.25	0.0428
CO_2	3.59	0.0427
CCl_4	20.4	0.1383

The correction to the pressure takes account of the intermolecular attractions between molecules. Notice that it consists of the constant a, different for each gas, times the quantity $(n/V)^2$. The units of n/V are mol/L. This quantity is squared because the number of molecular interactions, as illustrated in Figure 10.21, is proportional to the square of the number of molecules per unit volume. Values of the van der Waals constant, a, are listed in Table 10.3 for several gases. The magnitude of a reflects how strongly gas molecules attract each other. Notice that a, like b, increases with an increase in molar mass and with an increase in complexity of molecular structure.

SAMPLE EXERCISE 10.17

If 1.000 mol of an ideal gas were confined to 22.41 L at 0.0°C, it would exert a pressure of 1.000 atm. Use the van der Waals equation and the constants in Table 10.3 to estimate the pressure exerted by 1.000 mol of $Cl_2(g)$ in 22.41 L at 0.0°C.

Solution: If we solve van der Waals equation, Equation 10.28, for pressure, we have

$$P = \frac{nRT}{V - nb} - \frac{n^2a}{V^2}$$

Substituting $n = 1.000$ mol, $R = 0.08206$ L-atm/K-mol, $T = 273.2$ K, $V = 22.41$ L, $a = 6.49$ L^2 atm/mol^2, and $b = 0.0562$ L/mol:

$$P = \frac{(1.000\ \text{mol})(0.08206\ \text{L-atm/K-mol})(273.2\ \text{K})}{22.41\ \text{L} - (1.000\ \text{mol})(0.0562\ \text{L/mol})} - \frac{(1.000\ \text{mol})^2(6.49\ \text{L}^2\text{-atm/mol}^2)}{(22.41\ \text{L})^2}$$

$$= 1.003\ \text{atm} - 0.013\ \text{atm} = 0.990\ \text{atm}$$

Notice that 1.003 atm is the pressure corrected for molecular volume. This value is higher than the ideal value, 1.000 atm, because the volume in which the molecules are free to move is smaller than the container volume, 22.41 L. Thus, the molecules must collide more frequently with the container walls. The second factor, 0.013 atm, corrects for intermolecular forces. The intermolecular attractions between molecules reduce the pressure to 0.990 atm. We can conclude, therefore, that the intermolecular attractions are the main cause of the slight deviation of $Cl_2(g)$ from ideal behavior under the stated experimental conditions.

PRACTICE EXERCISE

Consider a sample of 1.000 mol of $CO_2(g)$ confined to a volume of 3.000 L at 0.0°C. Calculate the pressure of the gas using **(a)** the ideal gas equation, and **(b)** the van der Waals equation. ***Answer:*** **(a)** 7.473 atm; **(b)** 7.182 atm

FOR REVIEW

SUMMARY

Many substances are capable of existing in any one of the three states of matter—solid, liquid, or gas. This chapter has been concerned with the gaseous state. To describe the state or condition of a gas, we must specify four variables: pressure, temperature, volume, and quantity of gas. Volume is usually measured in liters (L), and temperature in the Kelvin scale. Pressure is defined as the force per unit area. It is expressed in SI units as pascals, Pa (1 Pa = 1 N/m^2 = 1 kg/m-s^2), or more commonly in millimeters of mercury (mm Hg) or in atmosphere (atm). One atmosphere pressure equals 101.325 kPa, or 760 mm Hg. A barometer is often used to measure the atmospheric pressure. A manometer can be used to measure the pressure of enclosed gases.

The ideal-gas equation, $PV = nRT$, is the equation of state for an ideal gas. Most gases at pressures of about 1 atm and temperatures of 300 K and above obey the ideal-gas equation reasonably well. We can use the ideal-gas equation to calculate variations in one variable when one or more of the others is changed. For example, for a constant quantity of gas at constant temperature, the volume of the gas is inversely proportional to the pressure (Boyle's law). Similarly, for a constant quantity of gas at constant pressure, the volume of a gas is directly proportional to temperature (Charles's law). Avogadro's law states that at constant temperature and pressure the volume of a gas is directly proportional to the quantity of gas, that is, to the number of gas molecules. In gas mixtures, the total pressure is the sum of the partial pressures that each gas would exert if it were present alone under the same conditions (Dalton's law of partial pressures). In all applications of the ideal-gas equation we must remember to convert temperatures to the absolute-temperature scale, Kelvin.

We use the ideal-gas equation to solve problems involving gases as reactants or products in chemical reactions. From the gas density, d, under given conditions of pressure and temperature, we may calculate the molar mass of the gas: $\mathcal{M} = dRT/P$. In calculating the quantity of gas collected over water, correction must be made for the partial pressure of water vapor in the container.

The kinetic-molecular theory accounts for the properties of an ideal gas in terms of a set of assumptions about the nature of gases. Briefly, these assumptions are that molecules are in continuous chaotic motion; that the volume of gas molecules is negligible in relation to the volume of their container; that the gas molecules have no attractive forces for one another; and that the average kinetic energy of the gas molecules is proportional to absolute temperature.

The molecules of a gas do not all have the same kinetic energy at a given instant. Their speeds are distributed over a wide range; the distribution varies with the molar mass of the gas and with temperature. The root-mean-square (rms) speed, u, varies in proportion to the square root of absolute temperature and inversely with the square root of molar mass: $u = (3RT/\mathcal{M})^{1/2}$. It follows that the rate at which a gas escapes (effuses) through a tiny hole is inversely proportional to the square root of its molar mass (Graham's law). Molecules in a real gas possess finite volume and thus undergo frequent collisions with one another. Because of these collisions, the mean free path—the mean distance traveled between collisions—is short. Collisions between molecules limit the rate at which a gas molecule can diffuse through the space occupied by other gas molecules.

Departures from ideal behavior increase in magnitude as pressure increases and as temperature decreases. The extent of nonideality of a real gas can be seen by examining the quantity PV/RT for 1 mol of the gas as a function of pressure; for an ideal gas this quantity is exactly 1 at all pressures. Real gases depart from the ideal behavior because the molecules possess finite volume (leads to $PV/RT > 1$) or because the molecules experience attractive forces for one another upon collision (leads to $PV/RT < 1$). The van der Waals equation is an equation of state for gases that modifies the ideal-gas equation to represent more faithfully the pressure and volume behavior of real gases.

KEY TERMS

vapors (Sec. 10.1)
pressure (Sec. 10.2)
pascal (Sec. 10.2)
barometer (Sec. 10.2)
standard atmospheric pressure (Sec. 10.2)
torr (Sec. 10.2)
manometer (Sec. 10.2)
Boyle's law (Sec. 10.3)
Charles's law (Sec. 10.3)
Avogadro's hypothesis (Sec. 10.3)
Avogadro's law (Sec. 10.3)
ideal-gas equation (Sec. 10.4)
ideal gas (Sec. 10.4)
gas constant (Sec. 10.4)
standard temperature and pressure (Sec. 10.4)
partial pressure (Sec. 10.6)
Dalton's law of partial pressures (Sec. 10.6)
mole fraction (Sec. 10.6)
kinetic-molecular theory (Sec. 10.8)
root-mean-square speed (Sec. 10.8)
effusion (Sec. 10.9)
diffusion (Sec. 10.9)
Graham's law (Sec. 10.9)
mean free path (Sec. 10.9)

EXERCISES

Introduction; Pressure

10.1 Compressibility, which is the change in volume of a substance in response to a change in pressure, is much greater for gases than for liquids or solids. Explain.

10.2 The density of water vapor at 100°C and 1 atm pressure is 0.59 g/L, whereas liquid water under the same conditions has a density of 0.96 g/mL. Explain on a molecular basis why the vapor and the liquid differ so greatly in their densities.

10.3 Explain why the height of mercury in a mercury barometer is independent of the diameter of the mercury column.

10.4 **(a)** How high must a column of water be to exert a pressure equal to that of a column of mercury that is 760 mm high? **(b)** What is the pressure on the body of a diver if he is 25 ft below the surface of the water?

10.5 Perform the following conversions: **(a)** 0.950 atm to kPa; **(b)** 735 mm Hg to atm; **(c)** 670 mm Hg to torr **(d)** 1.03 atm to mm Hg.

10.6 Perform the following conversions: **(a)** 100.8 kPa to atm; **(b)** 0.430 atm to mm Hg; **(c)** 745 torr to mm Hg; **(d)** 23.0 mm Hg to atm.

10.7 **(a)** If the weather in Chicago is reported to be 89°F with a barometric pressure of 29.32 in. Hg, what is the temperature in degrees Celsius and pressure in kilopascals? **(b)** If the weather report for Montreal states that the temperature is 15°C and the barometric pressure is 99.8 kPa, what is the temperature in degrees Fahrenheit and the pressure in inches of mercury?

10.8 **(a)** On Titan, the largest moon of Saturn, the atmospheric pressure is 1.6 earth atmospheres. What is the atmospheric pressure of Titan in kilopascals? **(b)** On Venus the surface atmospheric pressure is about 90 earth atmospheres. What is the Venusian atmospheric pressure in kilopascals?

10.9 A phonograph stylus resting on a record has an effective mass of 2.5 g. The area of the stylus making contact with the record is 2.7×10^{-4} mm^2. What pressure, in pascals, does the stylus exert on the record?

10.10 Suppose that a woman weighing 130 lb and wearing high-heeled shoes momentarily places all her weight on the heel of one foot. If the area of the heel is 0.50 in.^{2}, calculate the pressure exerted on the underlying surface in kilopascals.

10.11 An open-tube manometer containing mercury is connected to a container of gas. What is the pressure of the enclosed gas in mm Hg in each of the following situations? **(a)** The mercury in the arm attached to the gas is 4.5 cm higher than the one open to the atmosphere; atmospheric pressure is 0.955 atm. **(b)** The mercury in the arm attached to the gas is 3.8 cm lower than the one open to the atmosphere; atmospheric pressure is 1.02 atm.

10.12 Suppose that the manometer illustrated in Figure 10.4(*c*) contains mineral oil (density 0.752 g/cm^3) in place of mercury, which has a density of 13.6 g/cm^3. The oil has no significant partial pressure of its own. When h is 36.5 cm and the atmospheric pressure is 742 mm Hg, what is the pressure of the enclosed gas?

The Gas Laws

10.13 **(a)** What do we mean when we say two quantities are directly proportional? Inversely proportional? **(b)** Is volume of a gas directly or inversely proportional to pressure? To absolute temperature? To the quantity of gas?

10.14 Write an equation or proportionality expression that expresses each of the following statements: **(a)** For a given quantity of gas at constant temperature, the product of pressure times volume is constant. **(b)** For a given temperature and pressure, the volume of a gas is proportional to the number of moles of gas present. **(c)** For a given volume and quantity of gas, the pressure is proportional to the absolute temperature. **(d)** For a given quantity of gas, the product of pressure and volume is proportional to absolute temperature.

10.15 **(a)** A fixed quantity of gas is compressed at constant temperature from a volume of 638 mL to 208 mL. If the initial pressure was 522 mm Hg, what is the final pressure? **(b)** What is the final volume of a gas if a 1.50-L sample is heated from 22°C to 450°C at constant pressure?

10.16 **(a)** A fixed quantity of gas is allowed to expand at constant temperature from 2.45 L to 5.38 L. If the original pressure was 0.950 atm, what is the final pressure? **(b)** A gas originally at 15°C and having a volume of 282 mL is reduced in volume to 82.0 mL, while its pressure is held constant. What is its final temperature?

The Ideal-Gas Equation

10.17 **(a)** What is an ideal gas? **(b)** Write the ideal-gas equation and give the units used for each term in the equation when $R = 0.0821$ L-atm/K-mol.

10.18 **(a)** What conditions are represented by the abbreviation STP? **(b)** What is the molar volume of an ideal gas at STP?

10.19 Starting with the ideal-gas equation, derive Avogadro's law.

10.20 Starting with the ideal-gas equation, show that at constant volume the pressure of a given quantity of gas is proportional to its absolute temperature.

10.21 Calculate each of the following quantities for an ideal gas: **(a)** the pressure, in atm, if 8.25×10^{-2} mol occupies 174 mL at −24°C; **(b)** the quantity of gas, in moles, if 1.50 L at −15°C has a pressure of 2.08 atm; **(c)** the volume of gas, in liters, if 2.38 mol has a pressure of 350 mm Hg at a temperature of 22°C; **(d)** the absolute temperature of the gas at which 9.87×10^{-2} mol occupies 164 mL at 722 mm Hg.

10.22 For an ideal gas, calculate the following quantities: **(a)** the pressure of the gas if 1.34 mol occupies 3.28 L at 28°C; **(b)** the volume occupied by 0.150 mol at −14°C and

60.0 atm; **(c)** the number of moles in 1.50 L at 37°C and 725 mm Hg; **(d)** the temperature at which 0.270 mol occupies 15.0 L at 2.54 atm.

10.23 A deep breath of air has a volume of 1.05 L at a pressure of 740 mm Hg and body temperature, 37°C. Calculate the number of molecules in the breath.

10.24 A neon sign is made of glass tubing whose inside diameter is 2.0 cm and whose length is 4.0 m. If the sign contains neon at a pressure of 1.5 mm Hg at 35°C, how many grams of neon are in the sign? (The volume of a cylinder is $\pi r^2 h$.)

10.25 At 36°C and 1.00 atm pressure, a gas occupies a volume of 0.600 L. How many liters will it occupy **(a)** at 0°C and 0.205 atm; **(b)** at STP?

10.26 Chlorine is widely used to purify municipal water supplies and to treat swimming pool waters. Suppose that the volume of a particular sample of Cl_2 is 6.18 L at 740 torr and 33°C **(a)** What volume will the Cl_2 occupy at 107°C and 680 torr? **(b)** What volume will the Cl_2 occupy at STP? **(c)** At what temperature will the volume be 3.00 L if the pressure is 8.00×10^2 mm Hg? **(d)** At what pressure will the volume be 5.00 L if the temperature is 67°C?

10.27 Many gases are shipped in high-pressure containers. Consider a steel tank whose volume is 42.0 L and which contains O_2 gas at a pressure of 18,000 kPa at 23°C. **(a)** What mass of O_2 does it contain? **(b)** What volume would the gas occupy at STP?

10.28 Fluorine gas, which is dangerously reactive, is shipped in steel cylinders of 30.0-L capacity at a pressure of 160 lb/in.^{2} at 26°C. **(a)** What mass of F_2 is contained in a cylinder? **(b)** What volume would the gas occupy at STP?

10.29 In an experiment recently reported in the scientific literature, male cockroaches were made to run at different speeds on a miniature treadmill while their oxygen consumption was measured. The average cockroach running at 0.08 km/hr consumed 0.8 mL of O_2 at 1 atm pressure and 24°C per gram of insect weight per hour. **(a)** How many moles of O_2 would be consumed in 1 hr by a 5.2-g cockroach moving at this speed? **(b)** This same cockroach is caught by a child and placed in a 1-qt fruit jar with a tight lid. Assuming the same level of continuous activity as in the research, will the cockroach consume more than 20 percent of the available O_2 in a 48-hr period? (Air is 21 mol percent O_2.)

10.30 After the large eruption of Mount St. Helens in 1980, gas samples from the volcano were taken by sampling the downwind gas plume. The unfiltered gas samples were passed over a gold-coated wire coil to absorb mercury (Hg) present in the gas. The mercury was recovered from the coil by heating it, and then analyzed. In one particular set of experiments, scientists found a mercury vapor level of 1800 ng of Hg per cubic meter in the plume, at a gas temperature of 10°C. Calculate **(a)** the partial pressure of Hg vapor in the plume; **(b)** the number of Hg atoms per cubic meter in the gas; **(c)** the total mass of Hg emitted per day by the volcano if the daily plume volume was 1600 km^3.

Density and Molar Mass

10.31 **(a)** Calculate the density of CO_2 gas at 700 mm Hg and 22°C. **(b)** Calculate the molar mass of a gas if 0.835 g occupies 800 mL at 400 mm Hg and 34°C.

10.32 **(a)** Calculate the density of sulfur trioxide gas at 2.50 atm and 25°C. **(b)** Calculate the molar mass of a gas if it has a density of 2.18 g/L at 66°C and 720 mm Hg.

10.33 Cyanogen, a highly toxic gas, is composed of 46.2 percent C and 53.8 percent N by mass. At 25°C and 750 mm Hg, 1.05 g of cyanogen occupies 0.500 L. What is the molecular formula of cyanogen?

10.34 Cyclopropane, a gas used with oxygen as a general anesthetic, is composed of 85.7 percent C and 14.3 percent H by mass. If 1.56 g of cyclopropane has a volume of 1.00 L at 0.984 atm and 50.0°C, what is the molecular formula of cyclopropane?

10.35 In the Dumas-bulb technique for determining the molar mass of an unknown liquid, you vaporize the sample of a liquid that boils below 100°C in a boiling-water bath and determine the mass of vapor required to fill the bulb (see Figure 10.22). From the following data, calculate the molar mass of the unknown liquid; mass of unknown vapor, 1.012 g; volume of bulb, 354 cm^3; pressure, 742 mm Hg; temperature, 99°C.

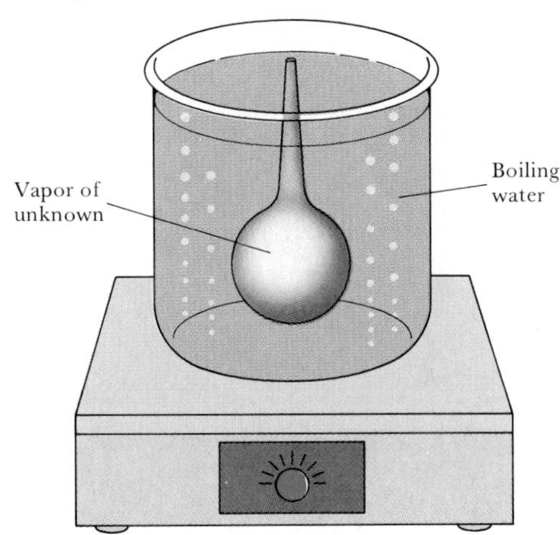

Figure 10.22

10.36 The molar mass of a volatile substance was determined by the Dumas bulb method described in Exercise 10.35. The unknown vapor had a mass of 0.846 g; the volume of the bulb was 354 cm^3, pressure 752 mm Hg, temperature 100°C. Calculate the molar mass of the unknown vapor.

Gas Mixtures

10.37 **(a)** Define the term *partial pressure*. **(b)** State Dalton's law of partial pressures.

10.38 Can the partial pressures of the components of an enclosed mixture of gases be measured with a manometer? If not, how are the partial pressures determined?

10.39 A mixture containing 0.125 mol $CH_4(g)$, 0.250 mol $C_2H_6(g)$, and 0.075 mol $O_2(g)$ is confined in a 2.00-L vessel at 22°C. **(a)** Calculate the partial pressure of CH_4 in the mixture. **(b)** Calculate the total pressure of the mixture.

10.40 A gaseous mixture containing 5.00 g N_2, 2.00 g O_2, and 1.20 g Ar is confined to a volume of 500 mL at 27°C. **(a)** Calculate the partial pressure of O_2 in the mixture. **(b)** Calculate the total pressure of the mixture.

10.41 A mixture of gases contains 0.45 mol N_2, 0.25 mol O_2, and 0.10 mol CO_2. If the total pressure of the mixture is 1.32 atm, what is the partial pressure of each component?

10.42 A mixture of gases contains 3.50 g of N_2, 1.30 g of H_2, and 5.27 g of NH_3. If the total pressure of the mixture is 2.50 atm, what is the partial pressure of each component?

10.43 A quantity of N_2 gas originally held at 4.60 atm pressure in a 1.00-L container at 26°C is transferred to a 10.0-L container at 20°C. A quantity of O_2 gas originally at 3.50 atm and 26°C in a 5.00-L container is transferred to this same container. What is the total pressure in the new container?

10.44 Consider the arrangement of bulbs shown in Figure 10.23. Each of the bulbs contains a gas at the pressure shown. What is the pressure of the system when all the stopcocks are opened, assuming that the temperature remains constant? (We can neglect the volume of the capillary tubing connecting the bulbs.)

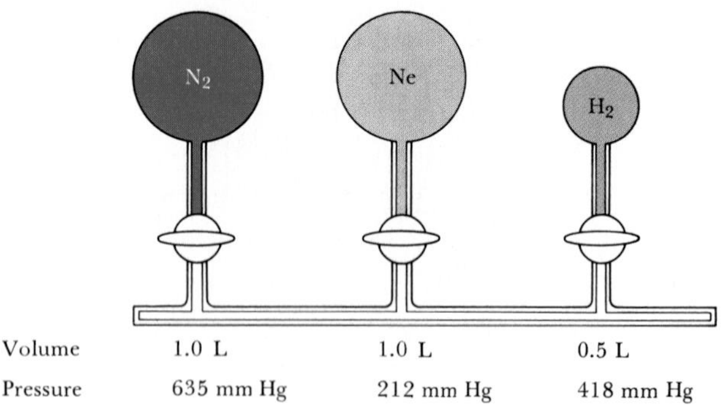

Figure 10.23

Quantities of Gases in Chemical Reactions

10.45 Calcium hydride, CaH_2, reacts with water to form hydrogen gas:

$$CaH_2(s) + 2H_2O(l) \longrightarrow Ca(OH)_2(aq) + 2H_2(g)$$

This reaction is sometimes used to inflate life rafts, weather balloons, and the like, where a simple, compact means of generating H_2 is desired. How many grams of CaH_2 are needed to generate 10.0 L of H_2 gas if the partial pressure of H_2 is 740 mm Hg at 23°C?

10.46 Magnesium can be used as a "getter" in evacuated enclosures, to react with the last traces of oxygen. (The magnesium is usually heated by passing an electric current through a wire or ribbon of the metal.) If an enclosure of 0.382 L has a partial pressure of O_2 of 3.5×10^{-6} mm Hg at 27°C, what mass of magnesium will react according to the following equation?

$$2Mg(s) + O_2(g) \longrightarrow 2MgO(s)$$

10.47 The metabolic breakdown of glucose, $C_6H_{12}O_6$, in our bodies produces CO_2, which is expelled from our lungs as gas:

$$C_6H_{12}O_6(s) + 6O_2(g) \longrightarrow 6CO_2(g) + 6H_2O(l)$$

Calculate the volume of dry CO_2 produced at body temperature (37°C) and 1.00 atm when 5.00 g of glucose is consumed in this reaction.

10.48 Ammonium sulfate, an important fertilizer, can be prepared by the reaction of ammonia with sulfuric acid:

$$2NH_3(g) + H_2SO_4(aq) \longrightarrow (NH_4)_2SO_4(aq)$$

Calculate the volume of $NH_3(g)$ needed at 20°C and 25.0 atm to react with 150 kg of H_2SO_4.

10.49 Hydrogen gas is produced when zinc reacts with sulfuric acid:

$$Zn(s) + H_2SO_4(aq) \longrightarrow ZnSO_4(aq) + H_2(g)$$

If 124 mL of wet H_2 is collected over water at 24°C and a barometric pressure of 725 mm Hg, how many grams of Zn have been consumed? (The vapor pressure of water is recorded in Appendix B.)

10.50 Small quantities of oxygen gas are sometimes generated in the laboratory by heating $KClO_3$ in the presence of MnO_2 as a catalyst:

$$2KClO_3(s) \longrightarrow 2KCl(s) + 3O_2(g)$$

What volume of O_2 is collected over water at 23°C by reaction of 0.2890 g of $KClO_3$ if the barometric pressure is 742 mm Hg. (The vapor pressure of water is recorded in Appendix B.)

Kinetic-Molecular Theory; Graham's Law

10.51 Why doesn't the size or identity of the molecules of a gas enter into the gas laws?

10.52 Use the kinetic molecular theory to explain why **(a)** gases have low density; **(b)** gases exert a pressure on the surfaces they contact; **(c)** gases are highly compressible; **(d)** gases assume the shape of their container; **(e)** gases can be mixed in any proportion to form a homogeneous mixture.

10.53 Suppose you have two 1-L flasks, one containing N_2 at STP, the other containing SF_6 at STP. How do these systems compare with respect to **(a)** number of molecules? **(b)** density? **(c)** average kinetic energy of the molecules? **(d)** rate of effusion through a pinhole leak?

10.54 Vessel A contains H_2 gas at 0°C and 1 atm. Vessel B contains O_2 gas at 20°C and 0.5 atm. The two vessels have the same volume. **(a)** Which vessel contains more molecules? **(b)** Which contains more mass? **(c)** In which vessel is the average kinetic energy of molecules higher? **(d)** In which vessel is the average speed of molecules higher?

10.55 Indicate which of the following statements regarding the kinetic-molecular theory of gases are correct. For those that are false, formulate a correct version of the statement.

(a) The average kinetic energy of a collection of gas molecules at a given temperature is proportional to $\mathcal{M}^{1/2}$. (b) The gas molecules are assumed to exert no forces on each other. (c) All the molecules of a gas at a given temperature have the same kinetic energy. (d) The volume of the gas molecules is negligible in comparison to the total volume in which the gas is contained.

10.56 What change or changes in the state of a gas bring about each of the following effects? **(a)** The number of impacts per unit time on a given container wall increases. **(b)** The average energy of impact of molecules with the wall of the container decreases. **(c)** The average distance between gas molecules increases. **(d)** The average speed of molecules in the gas mixture is increased.

10.57 **(a)** Place the following gases in order of increasing average molecular speed at 25°C: CO, SF_6, H_2S, Cl_2, HI. **(b)** Calculate the rms speed of CO at 25°C.

10.58 **(a)** Place the following gases in order of increasing average molecular speed at 300 K: CO_2, N_2O, HF, F_2, H_2. **(b)** Calculate and compare the rms speeds of H_2 and CO_2 at 300 K.

10.59 A gas of unknown molecular mass is allowed to effuse through a small opening under constant pressure conditions. It required 72 s for 1 L of the gas to effuse. Under identical experimental conditions it required 28 s for 1 L of O_2 gas to effuse. Calculate the molar mass of the unknown gas. (Remember that the faster the rate of effusion, the shorter the time required for effusion of 1 L; that is, rate and time are inversely proportional.)

10.60 Arsenic(III) sulfide sublimes readily, even below its melting point of 320°C. The molecules of the vapor phase are found to effuse through a tiny hole at 0.28 times the rate of effusion of Ar atoms under the same conditions of temperature and pressure. What is the molecular formula of arsenic(III) sulfide in the gas phase?

Nonideal-Gas Behavior

10.61 **(a)** Under what experimental conditions of temperature and pressure do gases usually behave nonideally? **(b)** What two properties or characteristics of gas molecules cause them to behave nonideally?

10.62 The planet Jupiter has a mass 318 times that of earth, and its surface temperature is 140 K. Mercury has a mass 0.05 times that of earth, and its surface temperature is between 600 and 700 K. On which planet is the atmosphere more likely to obey the ideal-gas law? Explain.

10.63 Explain how the function PV/RT can be used to show how gases behave nonideally at high pressures.

10.64 **(a)** What causes real gases to have values of PV/RT that are less than those of an ideal gas? **(b)** What causes PV/RT to be greater for real gases than for an ideal gas? **(c)** Why does the difference in PV/RT values for real gases and ideal gases become smaller as the temperature increases?

10.65 Based on their respective van der Waals constants, is O_2 or Cl_2 expected to behave more nearly like an ideal gas at high pressures? Explain.

10.66 Briefly explain the physical significance of the constants a and b in the van der Waals equation.

10.67 Calculate the pressure that CCl_4 will exert at 40°C if 1.00 mol occupies 28.0 L, assuming that **(a)** CCl_4 obeys the ideal-gas equation; **(b)** CCl_4 obeys the van der Waals equation. (Values for the van der Waals constants are given in Table 10.3).

10.68 It turns out that the van der Waals constant b is equal to four times the total volume actually occupied by the molecules of a mole of gas. Using this figure, calculate the fraction of the volume in a container actually occupied by Ar atoms **(a)** at STP; **(b)** at 100 atm pressure and 0°C. (Assume for simplicity that the ideal-gas equation still holds.)

Additional Exercises

10.69 Suppose the mercury used to make a barometer has a few small droplets of water trapped in it that rise to the top of the mercury in the tube. Will the barometer read the correct atmospheric pressure? Explain.

10.70 The mass of Venus is 0.82 times that of the earth. The atmosphere is 96 percent CO_2, and the surface temperature is over 700 K. Given these facts, explain how it is possible for the atmospheric pressure on Venus to be 90 times greater than that on the earth.

10.71 Suppose that the manometer illustrated in Figure 10.4(*c*) contains mineral oil (density 0.752 g/mL) in place of mercury (density 13.6 g/mL). The oil has no significant partial pressure of its own. When h is 25.6 cm and the atmospheric pressure is 0.975 atm, what is the pressure of the enclosed gas?

10.72 The *Hindenburg* was a famous hydrogen-filled dirigible that exploded in 1937. If the *Hindenburg* held $2.0 \times 10^5\ m^3$ of hydrogen gas at 27°C and 1.0 atm, what mass of hydrogen was present?

10.73 Propane, C_3H_8, liquefies under modest pressure, allowing a large amount to be stored in a container. **(a)** Calculate the number of moles of propane gas in a 1.00-L container at 3.00 atm and 27°C. **(b)** Calculate the number of moles of liquid propane that can be stored in the same volume if the density of the liquid is 0.590 g/mL. **(c)** Explain why more moles are present in the liquid.

10.74 To minimize the rate of evaporation of the tungsten filament, 10^{-5} mol of argon is placed in a 200-cm^3 light bulb. What is the pressure of argon in the light bulb at 23°C?

10.75 Some aerosol spray cans will explode if their internal pressure exceeds 3.0 atm. If an aerosol can has a pressure of 2.2 atm at 24°C, at what temperature will its pressure equal 3.0 atm?

10.76 The atmosphere of Titan, Saturn's largest moon, is believed to consist of 82 mol percent N_2, 12 mol percent Ar, and 6.0 mol percent CH_4 as its principal constituents. The average temperature at the surface is 95 K, and the pressure is 1.6 earth atmospheres. Assuming ideal-gas behavior: **(a)** calculate the average molecular weight of Titan's atmosphere; **(b)** calculate the density of Titan's atmosphere.

10.77 Assume that a single-cylinder of an automobile engine has a volume of 600.0 cm^3. **(a)** If the cylinder is full of air at 80°C and 0.980 atm, how many moles of O_2 are present? (The mole fraction of O_2 in dry air is 0.2095.) **(b)** How many grams of C_8H_{18} could be combusted by this quantity of O_2 assuming complete combustion with formation of CO_2 and H_2O?

10.78 If the partial pressure of ozone, O_3, in the stratosphere is 3.0×10^{-3} atm and the temperature is 250 K, how many ozone molecules are in a liter?

10.79 Nickel carbonyl, $Ni(CO)_4$, is one of the most toxic substances known. The present maximum allowable concentration in laboratory air during an 8-hr workday is 1 part in 10^9. Assume 24°C and 1 atm pressure. What mass of $Ni(CO)_4$ is allowable in a laboratory that is 110 m^2 in area, with a ceiling height of 2.7 m?

10.80 Ammonium nitrate can be decomposed thermally to form nitrous oxide, N_2O:

$$NH_4NO_3(s) \xrightarrow{\Delta} N_2O(g) + 2H_2O(l)$$

In one experiment, 860 cm^3 of N_2O was collected over mineral oil, which has no appreciable vapor pressure, at 22°C and 740 mm Hg pressure, using the experimental arrangement shown in Figure 10.11. **(a)** Assuming complete decomposition, how much NH_4NO_3 was decomposed? **(b)** What reason can you give to explain why mineral oil rather than water would be used in this case?

[10.81] A mixture of methane, CH_4, and acetylene, C_2H_2, occupies a certain volume at a total pressure of 70.5 mm Hg. The sample is burned, forming CO_2 and H_2O. The H_2O is removed, and the remaining CO_2 is found to have a pressure of 96.4 mm Hg at the same volume and temperature as the original mixture. What fraction of the gas was acetylene?

[10.82] A gaseous mixture of O_2 and Kr has a density of 1.104 g/L at 435 mm Hg and 300 K. What is the mole percent O_2 in the mixture?

[10.83] A glass vessel fitted with a stopcock has a mass of 337.428 g when evacuated. When filled with Ar, it has a mass of 339.712 g. When evacuated and refilled with a mixture of Ne and Ar, under the same conditions of temperature and pressure, it weighs 339.218 g. What is the mole percent of Ne in the gas mixture?

10.84 A balloon made of rubber permeable to small molecules is filled with helium. This balloon is then placed in a box that contains pure hydrogen, H_2. Will the balloon expand or contract? Explain.

[10.85] Scandium metal reacts with excess hydrochloric acid to produce H_2 gas. When 2.25 g of scandium is treated in this way, it is found that 2.41 L of H_2 measured at 100°C and 722 mm Hg pressure is liberated. Write the balanced chemical equation for the reaction that occurs.

[10.86] Consider the experiment illustrated in Figure 10.24. A gas is confined in the leftmost cylinder under 1 atm pressure, and the other cylinder is evacuated. When the stopcock is opened, the gas expands to fill both cylinders. Only a very small temperature change is noted when this expansion occurs. Explain how this observation relates to assumption 3 of the kinetic-molecular theory (Section 10.8.)

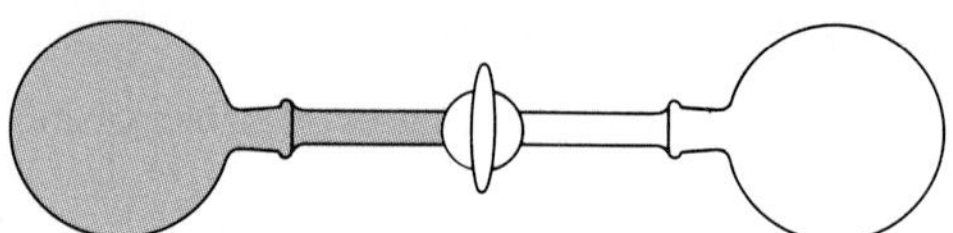

Figure 10.24

[10.87] The density of a gas of unknown molar mass was measured as a function of pressure at 0°C:

Pressure (atm)	1.00	0.666	0.500	0.333	0.250
Density (g/L)	2.3074	1.5263	1.1401	0.7571	0.5660

(a) Determine a precise molar mass for the gas. (Hint: Graph d/P versus P.) **(b)** Why is d/P not a constant as a function of pressure?

[10.88] Steam at high pressure and high temperature is used to drive turbines to generate electricity at utility power plants. A boiler in such a plant has a volume of 2000 m^3 and contains 115 metric tons (1 metric ton = 10^3 kg) of steam at a temperature of 550°C. Use the van der Waals equation to estimate the pressure of the steam in the boiler.

Intermolecular Forces, Liquids, and Solids

11

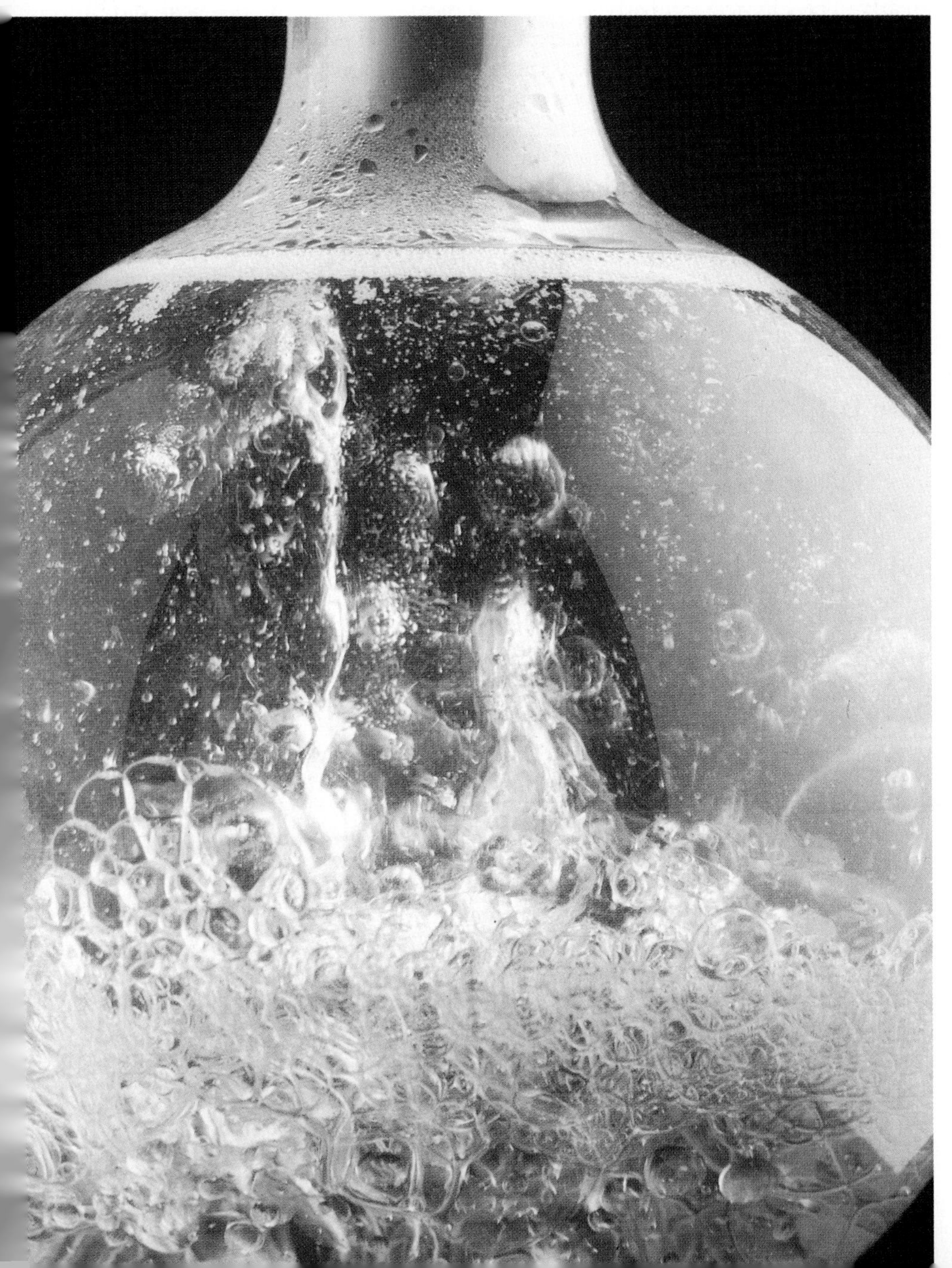

A vigorously boiling aqueous solution. (© Richard Megna/ Fundamental Photographs)

CONTENTS

The water vapor in air (which we recognize as humidity), the water in a lake, and the ice in a glacier are all forms of the same substance, H_2O. All have the same chemical properties; however, their physical properties differ greatly. The physical properties of a substance depend on its physical state. Some of the characteristic properties of each of the states of matter are given in Table 11.1. In Chapter 10, we discussed the gaseous state in some detail. In this chapter, we turn our attention to the physical properties of liquids and solids.

Many of the substances that we will consider are molecular. In fact, virtually all substances that are liquids at room temperature are molecular substances. Covalent bonds, which are forces *within* molecules, influence molecular shape, bond energies, and many aspects of chemical behavior. The physical properties of molecular liquids and solids, however, are due largely to **intermolecular forces**, the forces that exist *between* molecules. By understanding the nature and strength of intermolecular forces, we can begin to relate the composition and structure of molecules to their physical properties.

Table 11.1 Some Characteristic Properties of the States of Matter

Gas	1. Assumes both the volume and shape of container. 2. Is compressible. 3. Diffusion within a gas occurs rapidly. 4. Flows readily.
Liquid	1. Assumes the shape of the portion of the container it occupies. 2. Does not expand to fill container. 3. Is virtually incompressible. 4. Diffusion within a liquid occurs slowly. 5. Flows readily.
Solid	1. Retains its own shape and volume. 2. Is virtually incompressible. 3. Diffusion within a solid occurs extremely slowly. 4. Does not flow.

11.1 THE KINETIC-MOLECULAR DESCRIPTION OF LIQUIDS AND SOLIDS

In Chapter 10, we learned that the physical properties of gases can be understood in terms of the kinetic-molecular theory. Gases consist of a collection of widely separated molecules in constant, chaotic motion. The average kinetic energy of the molecules is much larger than the average energy of the attractions between them. The lack of strong attractive forces between molecules allows a gas to expand to fill its container.

In liquids, the intermolecular attractive forces are strong enough to hold molecules close together. Thus, liquids are much denser and far less compress-

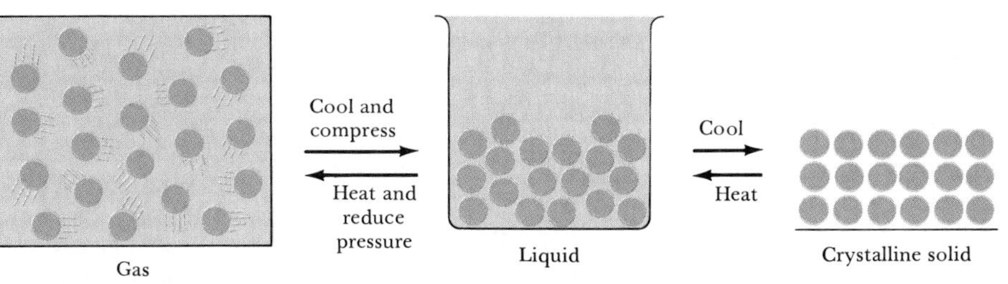
Cool and
compress
Heat and
reduce
pressure
Cool
Heat
Gas
Liquid
Crystalline solid
Total disorder; much
empty space; particles
have complete freedom of
motion owing to separation
and ineffective attractive
forces.
Disorder; particles
or clusters of
particles are free
to move relative to
each other; particles
close together.
Ordered arrangement;
particles can vibrate,
but are in fixed
positions; particles
close together.

Figure 11.1 Molecular-level comparison of gases, liquids, and solids. The particles can be atoms, ions, or molecules. The density of particles in the gas phase is exaggerated as compared with most real situations.

ible than gases. Unlike gases, liquids have a definite volume, independent of the size and shape of their container. The attractive forces in liquids are not strong enough, however, to keep the molecules from moving past one another. Thus, liquids can be poured, and they assume the shapes of their containers.

In solids, the intermolecular attractive forces are strong enough not only to hold molecules close together but virtually to lock them in place. Solids, like liquids, are not very compressible because the molecules have little free space between them. Often the molecules take up positions in a highly regular pattern. Solids that possess highly ordered structures are said to be *crystalline*. (The transition from a liquid to a crystalline solid is rather like the change that occurs on a military parade ground when the troops are called to formation.) Because the particles of a solid are not free to undergo long-range movement, solids are rigid.

Figure 11.1 compares the three states of matter. The particles that compose the substance can be individual atoms, as in Ar; molecules, as in H_2O; or ions, as in NaCl. The state of a substance depends largely on the balance between the kinetic energies of the particles and the interparticle energies of attraction. The kinetic energies, which depend on temperature, tend to keep the particles apart and moving. The interparticle attractions tend to draw the particles together. Those substances that are gases at room temperature have weaker interparticle attractions than those that are liquids; those that are liquids have weaker attractions than those that are solids.

We can change a substance from one state to another by heating or cooling, which changes the average kinetic energy of the particles. For example, NaCl, which is a solid at room temperature, melts at 804°C and boils at 1465°C under 1 atm pressure. Conversely, N_2O, which is a gas at room temperature, liquefies at −88.5°C and solidifies at −102.4°C under 1 atm pressure.

11.2 INTERMOLECULAR FORCES

The strengths of intermolecular forces of different substances vary over a wide range. However, they are generally much weaker than ionic or covalent bonds. For example, only 16 kJ/mol is required to overcome the intermolecular attractions between HCl molecules in liquid HCl in order to vaporize it. In contrast, the energy required to break the covalent bond to dissociate HCl into H and Cl atoms is 431 kJ/mol. Less energy is required to vaporize a liquid or to melt a solid than to break covalent bonds in molecules. Thus, when a molecular substance like HCl changes from solid to liquid to gas, the molecules remain intact.

Many properties of liquids, including their *boiling points*, reflect the strengths of the intermolecular forces. A liquid boils when bubbles of its vapor form within the liquid. The molecules of a liquid must overcome their attractive forces in order to separate and form a vapor. The stronger the attractive forces, the higher the temperature at which the liquid boils. Similarly, the *melting points* of solids increase with an increase in the strengths of the intermolecular forces.

Three types of intermolecular attractive forces are known to exist between neutral molecules: dipole-dipole forces, London dispersion forces, and hydrogen-bonding forces. The first two of these are also called *van der Waals forces* after Johannes van der Waals, who developed the equation for predicting the deviation of gases from ideal behavior (Section 10.10). Another kind of attractive force, the ion-dipole force, is important in solutions. As a group, intermolecular forces tend to be less than 15 percent as strong as covalent or ionic bonds. As we consider these forces, notice that each is electrostatic in nature, involving attractions between positive and negative species.

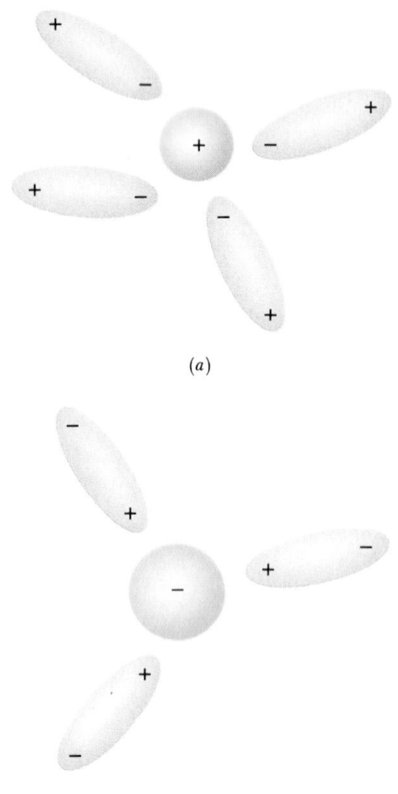

Figure 11.2 Illustration of the preferred orientation of dipolar molecules toward ions. The negative end of the dipolar molecule is oriented toward a cation (*a*), the positive end toward an anion (*b*).

Ion-Dipole Forces

An **ion-dipole force** exists between an ion and the partial charge on the end of a polar molecule. Polar molecules are dipoles; they have a positive end and a negative end (Section 9.2). For example, HCl is a polar molecule because of the difference in the electronegativities of the H and Cl atoms. The extent of the charge separation in a polar molecule is measured by its dipole moment. The dipole moment of HCl is 1.03 Debyes (D).

Positive ions are attracted to the negative end of a dipole, whereas negative ions are attracted to the positive end, as shown in Figure 11.2. The magnitude of the interaction energy depends on the charge on the ion (Q), the dipole moment of the dipole (μ), and the distance from the center of the ion to the midpoint of the dipole (d): $E \propto Q\mu/d^2$.

Ion-dipole forces are especially important in solutions of ionic substances in polar liquids, for example, a solution of NaCl in water. We will have more to say about such solutions later (Section 12.2).

Dipole-Dipole Forces

A **dipole-dipole force** exists between neutral polar molecules. Polar molecules attract each other when the positive end of one molecule is near the negative end of another as in Figure 11.3(*a*) and (*b*). Dipole-dipole forces are effective only when polar molecules are very close together, and they are generally weaker than ion-dipole forces.

Figure 11.3 Variation in the dipole-dipole interaction with orientation. In (*a*) and (*b*), the dipoles are aligned so as to produce an attractive interaction. In (*c*) and (*d*), the interactions are repulsive.

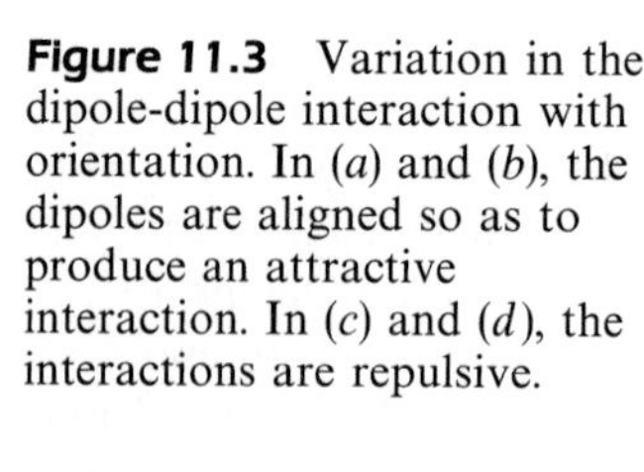

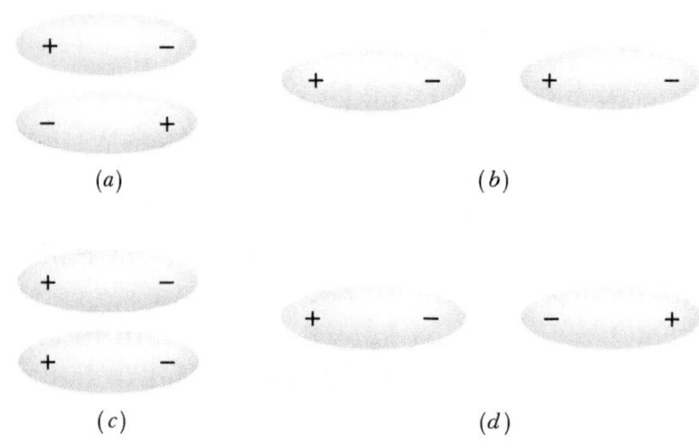

Table 11.2 Molecular Masses, Dipole Moments, and Boiling Points of Several Simple Organic Substances

Substance	Molecular mass (amu)	Dipole moment, μ (D)	Boiling point (K)
Propane, $CH_3CH_2CH_3$	44	0.1	231
Dimethyl ether, CH_3OCH_3	46	1.3	249
Methyl chloride, CH_3Cl	50	2.0	249
Acetaldehyde, CH_3CHO	44	2.7	293
Acetonitrile, CH_3CN	41	3.9	355

In liquids, dipolar molecules are free to move with respect to one another. They will sometimes be in an orientation that is attractive, and sometimes in an orientation that is repulsive. Two molecules that are attracting each other spend more time near each other than do two that are repelling each other. Thus, the overall effect is a net attraction. When we examine various liquids, we find clear evidence that *for molecules of approximately equal mass and size, the strengths of intermolecular attractions increase with increasing polarity.* We can see this trend for the liquids listed in Table 11.2: The boiling points increase with increasing magnitude of the dipole moment.

London Dispersion Forces

What kind of interparticle forces can exist between nonpolar atoms or molecules? Clearly, there can be no dipole-dipole forces when the particles are nonpolar. Yet the fact that nonpolar gases can be liquefied tells us that there must be some kind of attractive interactions between the particles. The origin of this attraction was first proposed in 1930 by Fritz London, a German-American physicist. London recognized that the motion of electrons in an atom or molecule can create an *instantaneous* dipole moment. Let's consider helium atoms as an example.

In a collection of helium atoms, the *average* distribution of the electrons about each nucleus is spherically symmetrical. The atoms are nonpolar and possess no permanent dipole moment. The instantaneous distribution of the electrons, however, can be different from the average distribution. (After all, it is only an average.) For example, if we could freeze the motion of the electrons in a helium atom at any given instant, both electrons could be on one side of the nucleus. At just that instant, then, the atom would have an instantaneous dipole moment. Because electrons repel one another, the motions of electrons on one atom influence the motions of electrons on its near neighbors. Thus, the temporary dipole on one atom can induce a similar dipole on an adjacent atom, causing the atoms to be attracted to each other as shown in Figure 11.4.

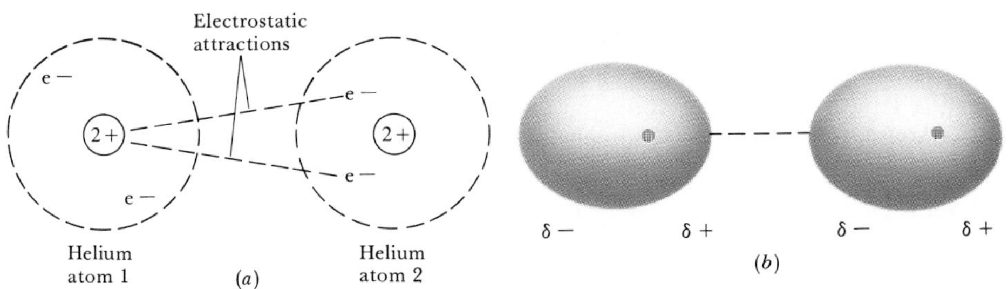

Figure 11.4 Two schematic representations of the instantaneous dipoles on two adjacent helium atoms, showing the electrostatic attraction between them.

Table 11.3 Boiling Points of the Halogens and Noble Gases

Halogen	Boiling point (K)	Noble gas	Boiling point (K)
F_2	85.1	He	4.6
Cl_2	238.6	Ne	27.3
Br_2	332.0	Ar	87.5
I_2	457.6	Kr	120.9
		Xe	166.1

This attractive interaction is called the **London dispersion force** (or merely the dispersion force). This force is significant only when molecules are very close together.

The ease with which the charge distribution in a molecule can be distorted by an external force is called its **polarizability**. The greater the polarizability of a molecule, the more easily its electron cloud can be distorted to give a momentary dipole, and thus the greater the strength of the London dispersion force. In general, larger molecules tend to have greater polarizabilities because their electrons are farther from the nuclei. Therefore, the strength of the London dispersion forces tends to increase with increasing molecular size. Because molecular size and mass generally parallel each other, *dispersion forces tend to increase in strength with increasing molecular weight.* Thus, the boiling points of the substances listed in Table 11.3 increase with increasing molecular weight.

SAMPLE EXERCISE 11.1

Which of the following substances is most likely to exist as a gas at room temperature and normal atmospheric pressure: P_4O_{10}, Cl_2, AgCl, or I_2?

Solution: In essence, the question asks which substance has the weakest intermolecular attractive forces. The weaker these forces, the more likely the substance is to exist as a gas at any given temperature and pressure. We should therefore select Cl_2, because it is a nonpolar molecule and also has the lowest molecular weight. In fact, Cl_2 does exist as a gas at room temperature and normal atmospheric pressure, whereas the others are solids. Of the other substances, AgCl is least likely to be a gas because it exists as Ag^+ and Cl^- ions with very strong ionic bonds holding the ions within the solid.

PRACTICE EXERCISE

Of Br_2, Ne, HCl, and N_2, which is likely to have **(a)** the largest intermolecular dispersion forces; **(b)** the largest dipole-dipole attractive forces? ***Answers:*** **(a)** Br_2; **(b)** HCl

The shapes of molecules can also play a role in the magnitudes of dispersion forces. For example, *n*-pentane* and neopentane, illustrated in Figure 11.5, have the same molecular formula, C_5H_{12}, yet the boiling point of *n*-pentane is 27 K higher than that of neopentane. The difference can be traced to the different shapes of the two molecules. The overall attraction between molecules is greater in the case of *n*-pentane because the molecules can come in contact over the entire length of the long, somewhat cylindrically shaped molecule. Less contact is possible between the more nearly spherical molecules of neopentane.

* The *n* in *n*-pentane is an abbreviation for the word *normal.* A normal hydrocarbon is one whose carbon atoms are arranged in a straight chain.

n-Pentane   Neopentane

Figure 11.5 Illustration of the effect of molecular shape on intermolecular attraction. The boiling point of *n*-pentane is 309.4 K, and that of neopentane is 282.7 K.

Dispersion forces operate between all molecules, whether they are polar or nonpolar. In fact, dispersion forces between polar molecules can contribute more to *overall* attractive forces than do dipole-dipole forces. For example, the fact that the boiling point of HBr (206.2 K) is higher than that of HCl (189.5 K) indicates that the overall attractive forces are stronger for HBr. The stronger attractive forces for HBr cannot be attributed to greater dipole-dipole forces, because HBr is less polar than HCl (0.79 D compared to 1.03 D). Because HBr is more massive and polarizable than HCl, however, the dispersion forces are stronger in HBr, and these lead to stronger overall attractive forces for it.

It is difficult to make generalizations about the relative strengths of intermolecular attractions unless we restrict ourselves to comparing molecules of either similar size and shape or similar polarity and shape. If molecules are of similar size and shape, dispersion forces are approximately equal, and therefore attractive forces increase with increasing polarity. If molecules are of similar polarity and shape, attractive forces tend to increase with increasing molecular mass because dispersion forces are greater.

Hydrogen Bonding

Figure 11.6 shows the boiling points of the simple hydrides of group 4A and 6A elements. In general, the boiling point increases with increasing molecular weight, owing to increased dispersion forces. The notable exception to this trend is H_2O, whose boiling point is much higher than we would expect on the basis of its molecular weight. NH_3 and HF also have abnormally high boiling points. These compounds also have many other characteristics that distinguish them from other substances of similar molecular weight and polarity. For example, water has a high melting point, a high heat capacity per gram, and a high heat of vaporization. Each of these properties indicates that the intermolecular forces between H_2O molecules are abnormally strong.

These strong intermolecular attractions result from hydrogen bonding. **Hydrogen bonding** *is a special type of intermolecular attraction that exists between the hydrogen atom in a polar bond (particularly an H—F, H—O, or H—N bond) and an unshared electron pair on a nearby electronegative atom (usually an F, O, or N atom on another molecule).* For example, a hydrogen bond exists

Figure 11.6 Boiling points of the group 4A (bottom) and 6A (top) hydrides as a function of molecular weight.

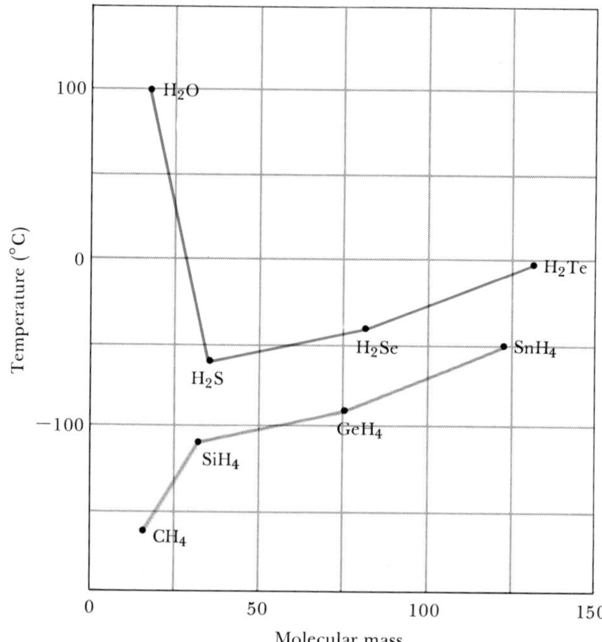

between the H atom in an HF molecule and the F atom of an adjacent HF molecule, F—H· · ·F—H (where the dots represent the hydrogen bond between the molecules). Several additional examples are shown in Figure 11.7.

Hydrogen bonds can be considered unique dipole-dipole attractions. Because F, N, and O are so electronegative, a bond between hydrogen and any of these three elements is quite polar, with hydrogen at the positive end:

$$\overset{\longleftarrow\!\!\!+}{\text{N—H}} \quad \overset{\longleftarrow\!\!\!+}{\text{O—H}} \quad \overset{\longleftarrow\!\!\!+}{\text{F—H}}$$

The hydrogen atom has no inner core of electrons. Thus, the positive side of the bond dipole has the concentrated charge of the partially exposed, nearly bare proton of the hydrogen nucleus. This positive charge is attracted to the negative charge of an electronegative atom in a nearby molecule. Because the electron-poor hydrogen is so small, it can approach an electronegative atom very closely and thus interact strongly with it.

The energies of hydrogen bonds vary from about 4 kJ/mol to 25 kJ/mol or so. Thus, they are much weaker than ordinary chemical bonds (see Table 8.4). Nevertheless, hydrogen bonding is generally much stronger than dipole-dipole or dispersion forces and therefore has important consequences for the properties of many substances, including those in biological systems.

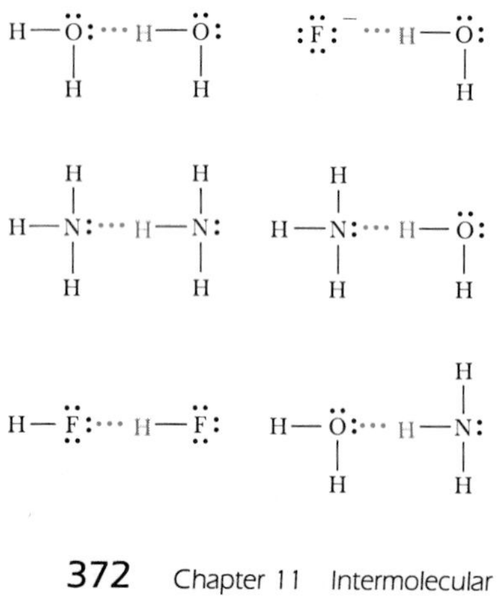

Hydrogen bonding is responsible, for example, for the rather open structure for ice (Figure 11.8), which causes ice to have a *lower* density than liquid water. (By contrast, for most substances the solid phase is *more* dense than the liquid, as illustrated in Figure 11.9.)

The low density of ice compared to water can be understood in terms of hydrogen-bonding interactions between water molecules. The interactions in the liquid are random. However, when water freezes, the molecules assume the ordered arrangement shown in Figure 11.8. This structure, which extends in all directions in space, permits the maximum number of hydrogen-bonding interactions between the H_2O molecules. Because the structure has large hexagonal holes, ice is more open and less dense than the liquid.

Figure 11.7 Examples of hydrogen bonding. The solid lines represent covalent bonds; the red dotted lines represent hydrogen bonds.

H—Ö:···H—Ö: (each O bonded to H below)

:F̤:⁻ ···H—Ö: (O bonded to H below)

H—N:···H—N: (each N bonded to H above and H below)

H—N:···H—Ö: (N bonded to H above and below; O bonded to H below)

H—F̤:···H—F̤:

H—Ö:···H—N: (O bonded to H below; N bonded to H above and below)

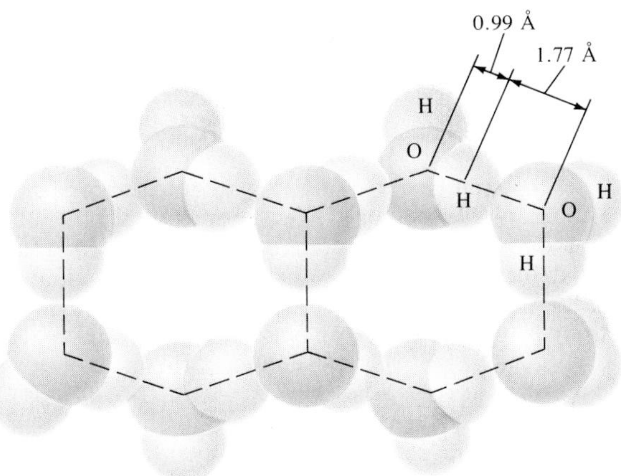

Figure 11.8 Arrangement of water molecules in ice. Each hydrogen atom on one water molecule is oriented toward a nonbonding pair of electrons on an adjacent water molecule. In the full structure each oxygen atom has a hydrogen bond with two O—H groups.

This unusual property of water has some important consequences. It permits ice to float on water. When ice forms in cold weather, it covers the top of the water, thereby insulating the water below. If ice were more dense than water, ice forming at the top of a lake would fall to the bottom, and the lake could freeze solid. Most aquatic life could not survive under these circumstances. The expansion of water upon freezing is also what causes water pipes to break in freezing weather.

SAMPLE EXERCISE 11.2

List the substances $BaCl_2$, H_2, CO, HF, and Ne in order of increasing boiling points.

Solution: The boiling point depends in part on the attractive forces in the liquid. These are stronger for ionic substances than for molecular ones, so $BaCl_2$ has the highest boiling point. The intermolecular forces of the remaining substances depend on molecular weight, polarity, and hydrogen bonding. The other molecular weights are H_2 (2), CO (28), HF (20), and Ne (20). The boiling point of H_2 should be the lowest because it is nonpolar and has the lowest molecular weight. The molecular weights of CO, HF, and Ne are roughly the same. Because HF has hydrogen bonding, it has the highest boiling point of the three. Next is CO, which is slightly polar and has the highest molecular weight. Finally, Ne, which is nonpolar, comes last of these three. The predicted boiling points are therefore

$$H_2 < Ne < CO < HF < BaCl_2$$

The actual normal boiling points are H_2 (20 K), Ne (27 K), CO (83 K), HF (293 K), and $BaCl_2$ (1813 K).

PRACTICE EXERCISE

Which of the following can form hydrogen bonds with H_2O: **(a)** CH_3OCH_3; **(b)** CH_4; **(c)** NH_4^+? *Answer:* CH_3OCH_3 and NH_4^+

Figure 11.9 Like most substances, the solid phase of paraffin is denser than the liquid phase and the solid therefore sinks below the liquid (left). The solid phase of water, ice, is less dense than its liquid phase (right). (Richard Megna/Fundamental Photographs

A CLOSER LOOK: Hydrogen-Bonding Trends

If the hydrogen bond is indeed the result of an electrostatic interaction between the X—H bond dipole and an unshared electron pair on another atom, Y, then the strength of hydrogen bonding should increase as the X—H bond dipole increases. Thus, for the same Y, we would expect hydrogen-bonding strength to increase in the series

$$\mathrm{N{-}H\cdots Y < O{-}H\cdots Y < F{-}H\cdots Y}$$

This is indeed true. But what property of Y is important? The atom Y must possess an unshared electron pair that attracts the positive end of the X—H dipole. This electron pair must not be too diffuse in space; if the electrons occupy too large a volume, the X—H dipole does not experience a strong, directed attraction. For this reason, we find that hydrogen bonding is not very strong unless Y is one of the smaller atoms: N, O, or F, specifically. Among these three elements, we find that hydrogen bonding is stronger when the electron pair is not attracted too strongly to its own nuclear center. The ionization energy of the electrons on Y is a good measure of this aspect. For example, the ionization energy of an unshared-pair electron on nitrogen in a covalent molecule is less than the corresponding value for oxygen. Nitrogen is thus a better donor of the electron pair to the X—H bond. For a given X—H, hydrogen-bond strength increases in the order

$$\mathrm{X{-}H\cdots F < X{-}H\cdots O < X{-}H\cdots N}$$

When X and Y are the same, the energy of hydrogen bonding increases in the order

$$\mathrm{N{-}H\cdots N < O{-}H\cdots O < F{-}H\cdots F}$$

When the Y atom carries a negative charge, the electron pair is able to form especially strong hydrogen bonds. The hydrogen bond in the $\mathrm{F{-}H\cdots F^-}$ ion is among the strongest known; the reaction

$$F^-(g) + HF(g) \longrightarrow FHF^-(g)$$

has a ΔH value of about -155 kJ/mol.

11.3 PROPERTIES OF LIQUIDS: VISCOSITY AND SURFACE TENSION

Now that we are familiar with intermolecular forces, we can consider some of the properties of liquids and solids. In this section, we examine two important properties of liquids: viscosity and surface tension.

Viscosity

Some liquids literally flow like molasses; others flow easily. The resistance of liquids to flow is called their **viscosity**. The greater the viscosity, the more slowly the liquid flows. Liquids such as molasses and motor oils are relatively viscous. Water and organic liquids such as carbon tetrachloride are not. Viscosity can be measured by timing how long it takes a certain amount of the liquid to flow through a thin tube, under gravitational force. More viscous fluids take longer. In another method, the rate of fall of steel spheres through the liquid is measured. The spheres fall more slowly through the more viscous liquids.

In the United States, the Society of Automotive Engineers (SAE) has established numbers to indicate the viscosity of motor oils. The higher the number, the greater the viscosity at any given temperature (Figure 11.10). An oil designated SAE 40 is more viscous than one designated SAE 10. As temperature increases, the viscosity of a liquid decreases. A multigrade oil designated SAE 10W/40 has the viscosity of an SAE 10 oil at $-18°C$ (denoted W for winter) and that of an SAE 40 oil at 99°C.

Viscosity is related to the ease with which individual molecules of the liquid can move with respect to one another. It thus depends on the attractive forces between molecules, and on whether structural features exist that cause the molecules to become entangled. Viscosity decreases with increasing temperature, because at higher temperature the greater average kinetic energy of the molecules more easily overrides the attractive forces between molecules. The viscosities of some common liquids are listed in Table 11.4.

Table 11.4 Viscosities and Surface Tensions of Some Common Liquids at 20°C, and of Water at Several Temperatures

Substance	Formula	Viscosity ($N\text{-}s/m^2$)[a]	Surface tension, (J/m^2)
Benzene	C_6H_6	0.65×10^{-3}	2.89×10^{-2}
Ethanol	C_2H_5OH	1.20×10^{-3}	2.23×10^{-2}
Ether	$C_2H_5OC_2H_5$	0.23×10^{-3}	1.70×10^{-2}
Glycerin	$C_3H_8O_3$	1490×10^{-3}	6.34×10^{-2}
Mercury	Hg	1.55×10^{-3}	46×10^{-2}
Water at:	H_2O		
20°C		1.00×10^{-3}	7.29×10^{-2}
40°C		0.652×10^{-3}	6.99×10^{-2}
60°C		0.466×10^{-3}	6.70×10^{-2}
80°C		0.356×10^{-3}	6.40×10^{-2}

[a] This is the SI unit: $1\ N\text{-}s/m^2 = 1\ kg/m\text{-}s$.

Figure 11.10 The more viscous motor oil (SAE 40, *left*) flows more slowly than the less viscous oil (SAE 10, *right*) (Kristen Brochmann/Fundamental Photographs)

Surface Tension

When water is placed on a waxy surface, it "beads up," forming distorted spheres. This behavior is due to an imbalance of intermolecular forces at the surface of the liquid, as shown in Figure 11.11. Notice that molecules in the interior are attracted equally in all directions, whereas those at the surface experience a net inward force. This inward force pulls molecules from the surface into the interior, thereby reducing the surface area. (Spheres have the smallest surface area for their volume.) The inward force also makes the molecules at the surface pack closely together, causing the liquid to behave almost as if it had a skin. This effect permits a carefully placed needle to float on the surface of water and some insects to "walk" on water (Figure 11.12), even though their densities are greater than that of water.

A measure of the inward forces that must be overcome in order to expand the surface area of a liquid is given by its surface tension. **Surface tension** is the energy required to increase the surface area of a liquid by a unit amount. Table 11.4 gives the surface tensions of several liquids. A surface tension of $7.29 \times 10^{-2}\ J/m^2$ for water at 20°C means that an energy of 7.29×10^{-2} J must be supplied to increase the surface area of a given amount of water by $1\ m^2$. Water has a high surface tension because of its strong hydrogen bonds. Mercury has an even higher surface tension because of even stronger metallic bonds between atoms.

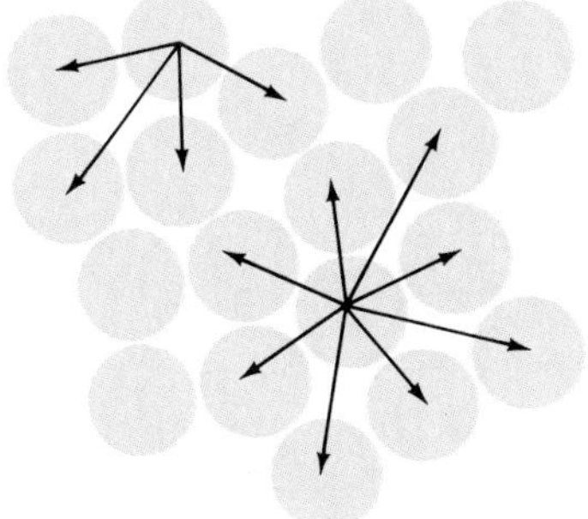

Figure 11.11 Molecular-level view of the intermolecular forces acting on a molecule at the surface of a liquid compared to that in the interior.

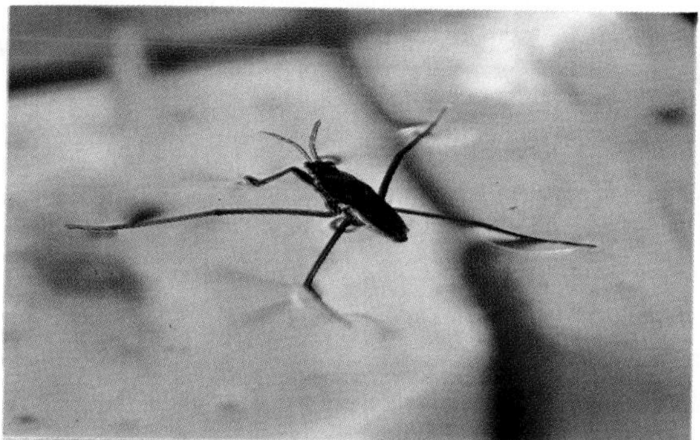

Figure 11.12 Surface tension permits an insect such as the water strider to "walk" on water. (© 1977 Michael P. Godomski, National Audubon Society, Photo Researchers, Inc.)

Figure 11.13 Comparison of the shape of the water meniscus in a glass tube and the shape of the mercury meniscus. (Richard Megna/ Fundamental Photographs)

Forces that bind like molecules to one another are called **cohesive forces**. Forces that bind a substance to a surface are called **adhesive forces**. Water placed in a glass tube adheres to the glass because of the adhesive forces between water and glass. The curved upper surface, or **meniscus**, of the water is therefore U-shaped (Figure 11.13). For mercury, however, the meniscus is curved downward where the mercury contacts the glass. In this case, the cohesive forces between the mercury atoms are much greater than the adhesive forces between the mercury atoms and the glass.

When a small-diameter glass tube, or capillary, is placed in water, water rises in the tube. The rise of liquids up very narrow tubes is called **capillary action**. The adhesive forces between the liquid and the walls of the tube tend to increase the surface area of the liquid. The surface tension of the liquid tends to reduce the area, thereby pulling the liquid up the tube. The liquid climbs until the adhesive and cohesive forces are balanced by the force of gravity on the liquid. Capillary action helps water and dissolved nutrients to move upward through plants.

11.4 CHANGES OF STATE

Many important properties of liquids and solids relate to the ease with which they change from one state to another. Everyone has seen examples of such changes. Water left uncovered in a glass for several days evaporates. An ice cube left in a warm room quickly melts. Solid CO_2 (sold as "Dry Ice") *sublimes* at room temperature; that is, it changes directly from the solid to the vapor state. In general, each state of matter can change into either of the other two states. Figure 11.14 shows the name associated with each of these changes. These changes are called **phase changes**, phase transitions, or merely changes of state.

Energy Changes Accompanying Changes of State

Each change of state is accompanied by a change in the energy of the system. Whenever a change of state involves going to a less ordered state, energy must be supplied to overcome intermolecular forces. Thus, energy is required to melt a solid. The attractive forces that hold particles in fixed positions in the solid

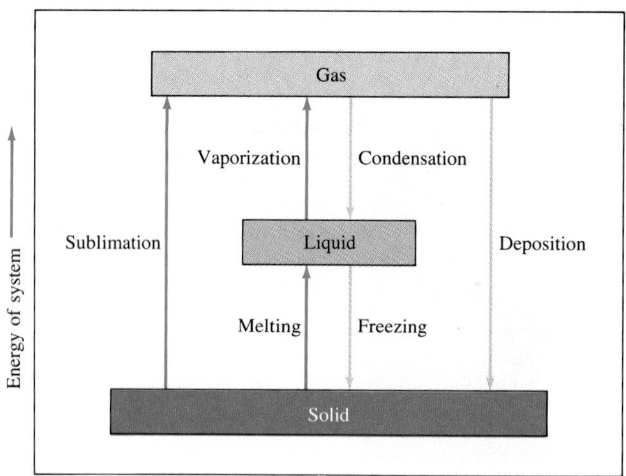

Figure 11.14 Energy changes accompanying phase changes between the three states of matter, and the names associated with them.

must be overcome to form the liquid. Likewise, vaporization requires energy. The attractive forces that hold particles close to each other in the liquid must be overcome to form the gas. Sublimation also requires energy. As the strengths of the intermolecular forces increase, the amounts of energy required to cause a change of state also increase.

The melting process for a solid is also referred to as *fusion*. Thus, the enthalpy change associated with melting a solid is often called the enthalpy of fusion, or **heat of fusion**. The heat of fusion of ice is 6.01 kJ/mol. The heat needed for the vaporization of a liquid is called the **heat of vaporization** (or enthalpy of vaporization). The heat of vaporization of water is 40.67 kJ/mol. Notice that the heat of fusion is smaller than the heat of vaporization. Less energy is needed to allow particles to move past one another than to separate them totally.

The cooling effect accompanying vaporization is evident when we get out of a swimming pool. As the water evaporates from our skin, it removes heat from our bodies. The evaporation of water through perspiring is important in keeping our bodies cool on hot days and during strenuous exercise. A refrigerator also relies on the cooling effect accompanying vaporization. Its mechanism contains an enclosed gas, usually Freon, CCl_2F_2, that can be liquefied under pressure. The Freon absorbs heat as it evaporates, thus cooling the interior of the refrigerator. The Freon vapor is then recycled to a compressor where it is again liquefied.

Because melting, vaporization, and sublimation are endothermic processes, the reverse processes (freezing, condensation, and deposition) are all exothermic. That is why, for example, steam can cause severe burns. When steam comes in contact with skin, it condenses, releasing considerable heat.

Enthalpy and Temperature Changes Accompanying Heating

Figure 11.15 shows how enthalpy and temperature change as H_2O is heated. As we add heat to ice, its temperature increases (line segment *AB*); the slope of this line depends on the heat capacity of ice. At 0°C, the temperature remains constant while the ice melts (segment *BC*). Once melting is complete, the temperature of the liquid increases until it reaches 100°C (segment *CD*). The slope of segment *CD* is different from that of *AB* because the molar heat capacities

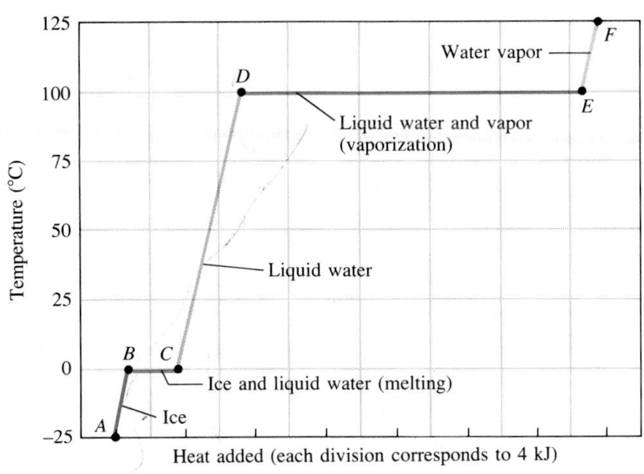

Figure 11.15 Molar enthalpy of water between −25°C and 125°C. When a solid, liquid, or gas is heated, its enthalpy always increases. However, its temperature does not change during phase changes, for example, melting ($B \rightarrow C$) and vaporization ($D \rightarrow E$).

of water and ice are different. The temperature then remains constant while the liquid boils (segment DE). Once the liquid has boiled away, the temperature of the vapor again increases as further heat is absorbed (segment EF).

The horizontal line segment BC (at 0°C) corresponds to the enthalpy of fusion, and the horizontal segment DE (at 100°C) corresponds to the enthalpy of vaporization. Temperature remains constant during phase transitions because the added energy is used to overcome attractive forces between molecules rather than to increase their average kinetic energy.

SAMPLE EXERCISE 11.3

Calculate the enthalpy change associated with converting 1.00 mol of ice at −25°C to water vapor at 125°C and 1 atm pressure. The heat capacities per gram (specific heats) of ice, water, and steam are 2.09 J/g-°C, 4.18 J/g-°C, and 1.84 J/g-°C, respectively. The heat of fusion of ice is 6.01 kJ/mol, and the heat of vaporization of water is 40.67 kJ/mol.

Solution: The heat required to bring the ice from −25°C to 0°C is given by the product of the mass, specific heat, and temperature change (Section 5.7). Thus, for 1.00 mol (18.0 g) of water, we have

$$\Delta H = (18.0 \text{ g})(2.09 \text{ J/g-°C})(25\text{°C}) = 940 \text{ J} = 0.94 \text{ kJ}$$

The heat then required to melt a mole of ice is given by the molar enthalpy of fusion, 6.01 kJ.

The next step in the process, heating the liquid water from 0°C to 100°C requires

$$\Delta H = (18.0 \text{ g})(4.18 \text{ J/g-°C})(100\text{°C}) = 7520 \text{ J} = 7.52 \text{ kJ}$$

The heat then required to vaporize the water at 100°C is given by the molar enthalpy of vaporization, 40.67 kJ.

Finally, to heat the water vapor from 100°C to 125°C requires

$$\Delta H = (18.0 \text{ g})(1.84 \text{ J/-g°C})(25\text{°C}) = 830 \text{ J} = 0.83 \text{ kJ}$$

The total enthalpy change is the sum of the enthalpy changes noted above:

$$\begin{aligned}\Delta H &= 0.94 \text{ kJ} + 6.01 \text{ kJ} + 7.52 \text{ kJ} + 40.67 \text{ kJ} + 0.83 \text{ kJ} \\ &= 55.97 \text{ kJ}\end{aligned}$$

PRACTICE EXERCISE

What is the enthalpy change associated with converting 10.0 g of steam at 115°C and 1 atm pressure to water at 60.0°C? (Use the enthalpies for phase changes and the specific heats given in the sample exercise above.)
Answer: −0.28 kJ − 22.59 kJ − 1.67 kJ = −24.54 kJ

Critical Temperature and Pressure

Gases can be liquefied by compressing them at a suitable temperature. As temperature rises, however, gases become more difficult to liquefy because of the increasing kinetic energies of their molecules. For every substance there exists a temperature above which the gas cannot be liquefied, regardless of the pressure. The highest temperature at which a substance can exist as a liquid is called its **critical temperature**. The **critical pressure** is the pressure required to bring about liquefaction at this critical temperature. The greater the intermolecular attractive forces, the more readily a gas is liquefied, and thus the higher the critical temperature of the substance.

CHEMISTRY AT WORK: Supercritical Fluid Extraction

At ordinary pressures, a substance above its critical temperature behaves as an ordinary gas. However, as pressure increases up to several hundred atmospheres, its character changes. Like a gas, it still expands to fill the confines of its container, but its density approaches that of a liquid. (For example, the critical temperature of water is 374.4°C, and its critical pressure is 217.7 atm. At this temperature and pressure, the density of water is 0.4 g/mL.) It is perhaps more appropriate to speak of a substance at critical temperature and pressure as a *supercritical fluid* rather than as a gas.

Like liquids, supercritical fluids can behave as solvents, dissolving a wide range of substances. This ability forms the basis of a system for separating the components of mixtures, a process known as supercritical fluid extraction. The solvent power of a supercritical fluid increases as its density increases. Conversely, lowering its density (either by decreasing pressure or increasing temperature) causes the supercritical fluid and the dissolved material to separate. With skillful manipulation of temperature and pressure, it is possible to separate the components of very complicated mixtures.

The process of supercritical fluid extraction is now under extensive study in the chemical, food, drug, and energy industries. A process for removing caffeine from green coffee beans by extraction with supercritical carbon dioxide has been in commercial operation for several years. (The critical temperature of CO_2 is 31.1°C, and its critical pressure is 73.0 atm.) At the proper temperature and pressure, the CO_2 removes caffeine from the beans but leaves the flavor and aroma components, producing decaffeinated coffee. Other applications of supercritical CO_2 extraction include removal of nicotine from tobaccos and removal of oil from potato chips, producing a lower-calorie product that is less greasy but has the same flavor and texture.

Table 11.5 Critical Temperatures and Pressures of Selected Substances

Substance	Critical temperature (K)	Critical pressure (atm)
Ammonia, NH_3	405.6	111.5
Argon, Ar	150.9	48
Carbon dioxide, CO_2	304.3	73.0
Dinitrogen, N_2	126.1	33.5
Dioxygen, O_2	154.4	49.7
Water, H_2O	647.6	217.7

The critical temperatures and pressures of substances are often of considerable importance to engineers and other people working with gases because they provide information about the conditions under which gases liquefy. Sometimes we want to liquefy a gas, other times we want to avoid liquefying it. It is useless to try to liquefy a gas by applying pressure if the gas is above its critical temperature. For example, O_2 has a critical temperature of 154.4 K. It must be cooled below this temperature before it can be liquefied by pressure. In contrast, ammonia has a critical temperature of 405.6 K. Thus, it can be liquefied at room temperature (approximately 295 K) by compressing the gas to a sufficient pressure. The critical temperatures and pressures of some selected substances are given in Table 11.5.

11.5 VAPOR PRESSURE

We have seen that molecules can escape from the surface of a liquid into the gas phase by vaporization or evaporation. Suppose we conduct an experiment in which we place a quantity of ethyl alcohol (ethanol) in an evacuated, closed container such as that in Figure 11.16. The ethanol will quickly begin to evaporate. As a result, the pressure exerted by the vapor in the space above the liquid will begin to increase. After a short time, the pressure of the vapor will attain a constant value, which we call the **vapor pressure** of the substance.

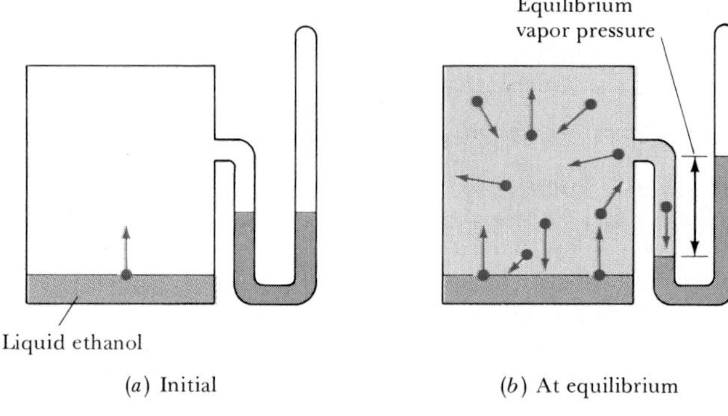

Figure 11.16 Illustration of the equilibrium vapor pressure over liquid ethanol. In (*a*), we imagine that no ethanol molecules exist in the gas phase; there is zero pressure in the cell. In (*b*), the rate at which molecules of ethanol leave the surface equals the rate at which gas molecules pass into the liquid phase. Thus, the rates of condensation and of vaporization are equal. This produces a stable vapor pressure that does not change with time, as long as temperature remains constant.

Explaining Vapor Pressure on the Molecular Level

The molecules of a liquid move at various speeds. Figure 11.17 shows the distribution of kinetic energies of the particles at the surface of a liquid at a particular temperature. The distribution curve is like those shown earlier for gases (Figure 10.13). At any instant, some of the molecules on the surface of the liquid possess sufficient energy to escape from the attractive forces of their neighbors. The weaker the attractive forces, the larger the number of molecules that are able to escape into the gas phase, and hence the higher the vapor pressure.

The movement of molecules from the liquid to the gas phase goes on continuously. However, as the number of gas-phase molecules increases, the probability increases that a molecule in the gas phase will strike the liquid surface and stick there [Figure 11.16(*b*)]. Eventually, the number of molecules returning to the liquid exactly equals the number escaping from it. The number of molecules in the gas phase then reaches a steady value, and the pressure of the vapor at this stage becomes constant.

The condition in which two opposing processes are occurring simultaneously at equal rates is called a **dynamic equilibrium**. A liquid and its vapor are in equilibrium when evaporation and condensation occur at equal rates. The observer may conclude that nothing is occurring during an equilibrium, because there is no net change in the system. In fact, a great deal is happening; molecules continuously pass from the liquid state to the gas state and from the gas state to the liquid state. All equilibria between different states of matter possess this dynamic character. *The vapor pressure of a liquid is the pressure exerted by its vapor when the liquid and vapor states are in dynamic equilibrium.*

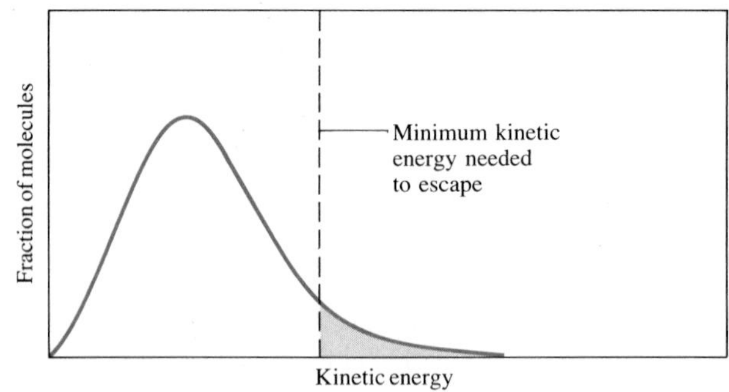

Figure 11.17 Distribution of kinetic energies of surface molecules of a hypothetical liquid compared to the minimum kinetic energy needed to escape from the surface. This minimum energy depends on the magnitude of the attractive forces between molecules. The fraction of molecules having sufficient kinetic energy to escape the liquid is given by the shaded area.

Volatility, Vapor Pressure, and Temperature

When vaporization occurs in an open container, as when water evaporates from a bowl, the vapor spreads away from the liquid. Little, if any, is recaptured at the surface of the liquid. Equilibrium never occurs, and the vapor continues to form until the liquid evaporates to dryness. Substances with high vapor pressure (such as gasoline) evaporate more quickly than substances with low vapor pressure (such as motor oil). Liquids that evaporate readily are said to be **volatile**.

Hot water evaporates more quickly than cold water because vapor pressure increases with temperature. As the temperature of a liquid is increased, the molecules move more energetically and can therefore escape more readily from their neighbors. Figure 11.18 depicts the variation in vapor pressure with temperature for four common substances that differ greatly in volatility. Note that in all cases the vapor pressure increases nonlinearly with increasing temperature.

Vapor Pressure and Boiling Point

A liquid boils when its vapor pressure equals the external pressure acting on the surface of the liquid. At this point, bubbles of vapor are able to form within the interior of the liquid. The temperature of boiling increases with increasing external pressure. The boiling point of a liquid at 1 atm pressure is called its **normal boiling point**. From Figure 11.18, we see that the normal boiling point of water is 100°C.

SAMPLE EXERCISE 11.4

Using Figure 11.18, estimate the boiling point of ethanol at 400 mg Hg.

Solution: From Figure 11.18, we see that the boiling point must be about 64°C.

PRACTICE EXERCISE

If we wanted to establish a boiling point of 40°C for diethyl ether, what vapor pressure would we need to maintain in the container? ***Answer:*** about 960 mm Hg

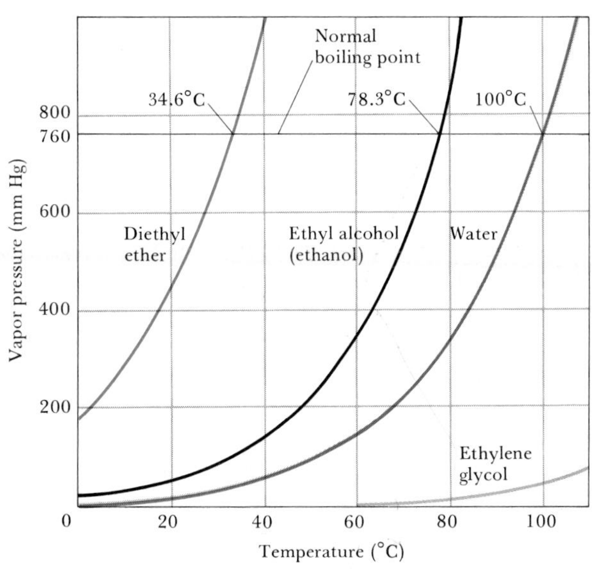

Figure 11.18 Vapor pressure of four common liquids shown as a function of temperature. The temperature at which the vapor pressure is 760 mm Hg is the normal boiling point of each liquid.

A CLOSER LOOK: Relating Vapor Pressure to Temperature

We can represent the dependence of the vapor pressure, P, on temperature as a log function of the form

$$\log P = \frac{-\Delta H_v}{2.303RT} + C \qquad [11.1]$$

In this equation, T is absolute temperature, R is the gas constant (Table 10.2), ΔH_v is the enthalpy of vaporization, and C is a constant. This equation, called the *Clausius-Clapeyron* equation, tells us that a graph of log P versus $1/T$ should be linear. The slope of the line can be used to calculate the enthalpy of vaporization.

As an example of the application of the Clausius-Clapeyron equation, the vapor-pressure data for ethanol shown in Figure 11.18 are graphed in Figure 11.19 as log P versus $1/T$ (temperature is in kelvin). Note that a linear relationship results. Because a straight line is easy to extrapolate, we could extend it to obtain values for the vapor pressure of ethanol at temperatures above and below the temperature range for which we have data.

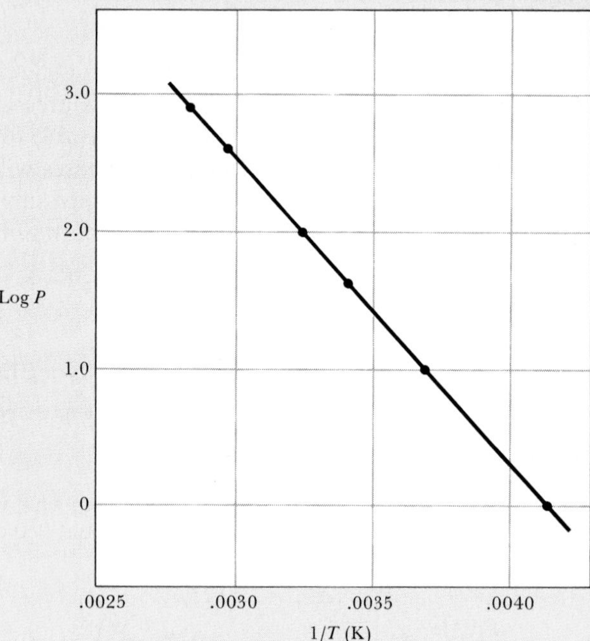

Figure 11.19 Application of the Clausius-Clapeyron equation, Equation 11.1, to the vapor-pressure-versus-temperature data for ethanol. The slope of the line equals $-\Delta H_v/2.3R$.

The boiling point is important to many processes that involve heating liquids, including cooking. The time required to cook food depends on the temperature. As long as water is present, the maximum temperature of the cooking food is the boiling point of water. Pressure cookers work by allowing steam to escape only when it exceeds a predetermined pressure; the pressure above the water can therefore increase above atmospheric pressure. The higher pressure causes water to boil at a higher temperature, thereby allowing the food to get hotter and so cook more rapidly. The effect of pressure on boiling point also explains why it takes longer to cook food at higher elevations than at sea level. At higher altitudes the atmospheric pressure is lower, so water boils at a lower temperature.

11.6 PHASE DIAGRAMS

The equilibrium between a liquid and its vapor is not the only dynamic equilibrium that can exist between states of matter. Under appropriate conditions of temperature and pressure, a solid can be in equilibrium with its liquid state or even with its vapor state. A **phase diagram** is a graphical way to summarize the conditions under which equilibria exist between the different states of matter. It also allows us to predict the phase of a substance that is stable at any given temperature and pressure.

The general form of a phase diagram for a substance that exhibits three phases is shown in Figure 11.20. The diagram contains three important curves, each of which represents the conditions of temperature and pressure at which the various phases can coexist at equilibrium.

1. The line from A to B is the vapor-pressure curve of the liquid. It represents the equilibrium between the liquid and gas phases. The point on this curve where the vapor pressure is 1 atm is the normal boiling point of the substance. The vapor-pressure curve ends at the *critical point* (B), which is the critical temperature and critical pressure of the substance. Beyond the critical point, the liquid and gas phases become indistinguishable.
2. The line AC represents the variation in the vapor pressure of the solid as it sublimes at different temperatures.
3. The line from A through D represents the change in melting point of the solid with increasing pressure. This line normally slopes slightly to the right as pressure increases. For most substances, the solid is denser than the liquid; therefore, an increase in pressure favors the more compact solid phase. Thus, higher temperatures are required to melt the solid at higher pressures. The *melting point* of a substance is identical to its *freezing point*. The two differ only in the temperature direction from which the phase change is approached. The melting point at 1 atm is the **normal melting point**.

Point A, where the three curves intersect, is known as the **triple point**. All three phases are at equilibrium at this temperature and pressure. Any other

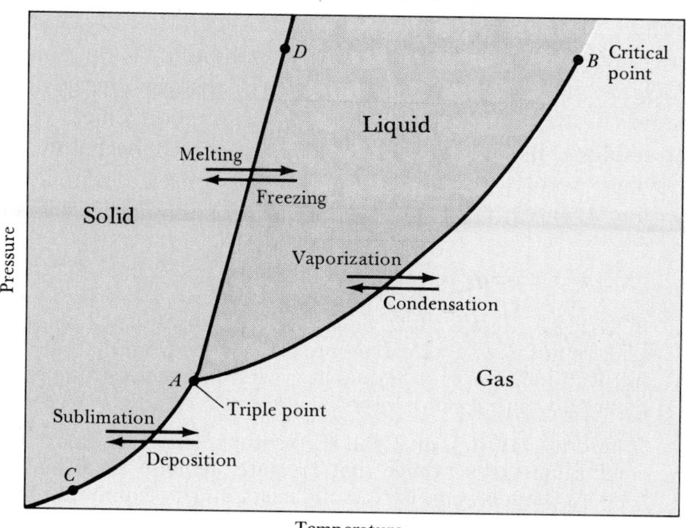

Figure 11.20 General shape for a phase diagram of a system exhibiting three phases: gas, liquid, and solid.

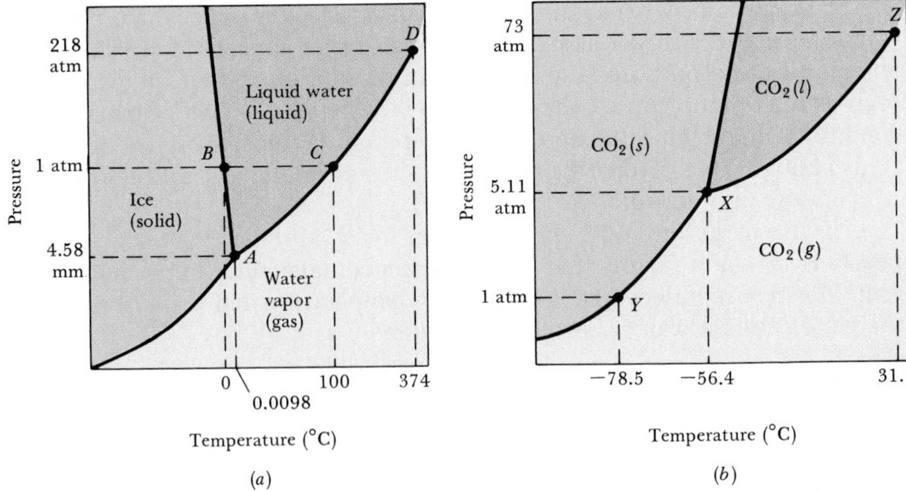

Figure 11.21 Phase diagram of (*a*) H_2O and (*b*) CO_2. The axes are not drawn to scale in either case. In (*a*), for water, note the triple point *A* (0.0098°C, 4.58 mm Hg), the normal melting (or freezing) point *B* (0°C, 1 atm), the normal boiling point *C* (100°C, 1 atm), and the critical point *D* (374.4°C. 217.7 atm). In (*b*), for carbon dioxide, note the triple point *X* (−56.4°C, 5.11 atm), the normal sublimation point *Y* (−78.5°C, 1 atm), and the critical point *Z* (31.1°C, 73.0 atm).

point on the three curves represents an equilibrium between two phases. Any point on the diagram that does not fall on a line corresponds to conditions under which only one phase is present. Notice that the gas phase is the stable phase at low pressures and high temperatures. The conditions under which the solid phase is stable extend to low temperatures and high pressures. The stability range for liquids lies between the other two regions.

The Phase Diagrams of H_2O and CO_2

Figure 11.21 shows the phase diagrams of H_2O and CO_2. Notice that the solid-liquid equilibrium (melting point) line of CO_2 is normal; its melting point increases with increasing pressure. On the other hand, the melting point of H_2O *decreases* with increasing pressure. Water is among the very few substances whose liquid form is more compact than the solid form (Section 11.2).

The triple point of H_2O (0.0098°C and 4.58 mm Hg) is at much lower pressure than that of CO_2 (−56.4°C and 5.11 atm). For CO_2 to exist as a liquid, the pressure must exceed 5.11 atm. Consequently, solid CO_2 does not melt but rather sublimes when heated at 1 atm. Thus, CO_2 does not have a normal melting point; instead, it has a normal sublimation point, −78.5°C. Because CO_2 sublimes rather than melts as it absorbs energy at ordinary pressures, solid CO_2 (commonly called "Dry Ice") is a convenient coolant. For water (ice) to sublime, however, its vapor pressure must be below 4.58 mm Hg. Freeze-drying of food is accomplished by placing frozen food in a low-pressure chamber (below 4.58 mm Hg) so that the ice in it sublimes.

SAMPLE EXERCISE 11.5

Referring to Figure 11.22, describe any changes in the phases present when H_2O is **(a)** kept at 0°C while the pressure is increased from that at point 1 to that at point 5 (vertical line); **(b)** kept at 1.00 atm while the temperature is increased from that at point 6 to that at point 9 (horizontal line).

Solution: **(a)** At point 1, the H_2O exists totally as a vapor. At point 2, a solid-vapor equilibrium exists. Above that pressure, at point 3, all the H_2O is converted to a solid. At point 4, some of the solid melts, and an equilibrium between solid and liquid is achieved. At still higher pressures, all the H_2O melts, so that only the liquid phase is present at point 5.

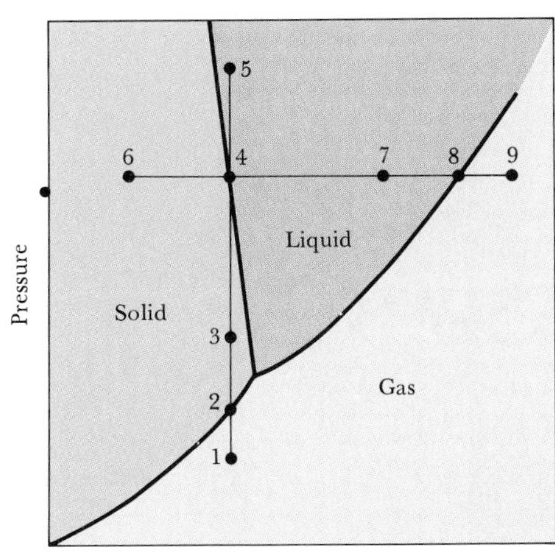

Figure 11.22 Phase diagram of H_2O.

(b) At point 6, the H_2O exists entirely as a solid. When the temperature reaches point 4, the solid begins to melt, and an equilibrium condition occurs between the solid and the liquid phases. At a yet higher temperature, point 7, the solid has been converted entirely to a liquid. When point 8 is encountered, vapor forms and a liquid-vapor equilibrium is achieved. Upon further heating, to point 9, the H_2O is converted entirely to the vapor phase.

PRACTICE EXERCISE

Using Figure 11.21(*b*), describe what happens when the following changes are made in a CO_2 sample initially at 1 atm and $-60°C$. **(a)** Pressure increases at constant temperature to 60 atm. **(b)** Following (a) temperature increases to $-20°C$ at constant 60 atm pressure. ***Answers:*** **(a)** $CO_2(g) \longrightarrow CO_2(s)$; **(b)** $CO_2(s) \longrightarrow CO_2(l)$

11.7 STRUCTURES OF SOLIDS

Throughout the remainder of this chapter we will focus on how the properties of solids relate to their structures and bonding. Solids can be either crystalline or amorphous (noncrystalline). A **crystalline solid** is a solid whose atoms, ions, or molecules are ordered in well-defined arrangements. These solids usually have flat surfaces or faces that make definite angles with one another. The orderly stacks of particles that produce these faces also cause the solids to have highly regular shapes (Figure 11.23). Quartz and diamond are examples of crystalline solids.

Figure 11.23 Crystalline solids come in a variety of forms and colors. (*a*) Calcite, (*b*) fluorite twin crystals, (*c*) quartz. (Thomas Taylor/Photo Researchers; Runk, Schoenberger/Grant Heilman Photography)

(*b*)

(*c*)

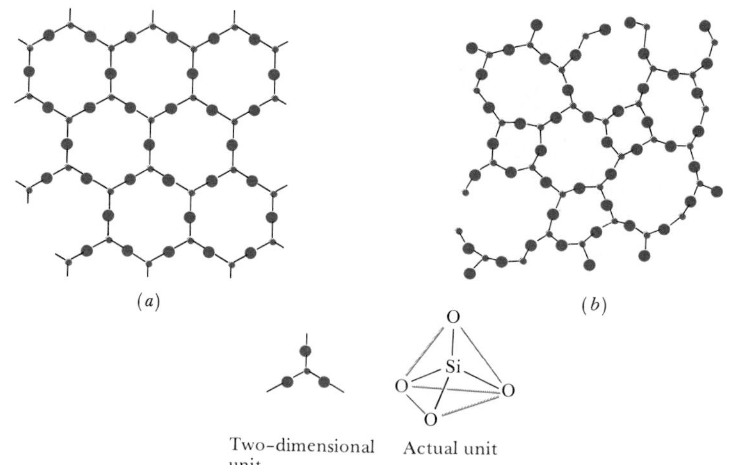

Figure 11.24 Schematic comparisons of (*a*) crystalline SiO_2 (quartz) and (*b*) amorphous SiO_2 (quartz glass). The blue dots represent silicon atoms; the red dots represent oxygen atoms. The structure is actually three-dimensional and not planar as drawn. The unit shown as the basic building block (silicon and three oxygens) actually has four oxygens, the fourth coming out of the plane of the paper and capable of bonding to other silicon atoms.

An **amorphous solid** (from the Greek words for "without form") is a solid whose particles have no orderly structure. These solids lack well-defined faces and shapes. Many amorphous solids are mixtures of molecules that do not stack together well. Most others are composed of large, complicated molecules. Familiar amorphous solids include rubber and glass.

Quartz, SiO_2, is a crystalline solid with a three-dimensional structure like that shown in Figure 11.24(*a*). When quartz melts (at about 1600°C), it becomes a viscous, tacky liquid. Although the silicon-oxygen network remains largely intact, many Si—O bonds are broken, and the rigid order of the quartz is lost. If the melt is rapidly cooled, the atoms are unable to return to an orderly arrangement. An amorphous solid known as quartz glass or silica glass results [Figure 11.24(*b*)].

Because the particles of an amorphous solid lack any long-range order, intermolecular forces vary in strength throughout a sample. Thus, amorphous solids do not melt at specific temperatures. Instead, they soften over a temperature range as intermolecular forces of various strengths are overcome. A crystalline solid, on the other hand, melts at a specific temperature.

Unit Cells

The order characteristic of crystalline solids allows us to convey a picture of an entire crystal by looking at only a small part of it. We can think of the solid as being built up by stacking together identical building blocks, much as a brick wall is formed by stacking individual, identical bricks. The repeating unit of a solid—the crystalline "brick"—is known as the **unit cell**. A simple two-dimensional example appears in the sheet of wallpaper shown in Figure 11.25. There are several ways of choosing the repeat pattern, or unit cell, of the design, but the choice is usually the smallest one that shows clearly the symmetry characteristic of the entire pattern.

Figure 11.25 Wallpaper design showing a characteristic repeat pattern. Each dashed blue square denotes the unit cell of the repeat pattern.

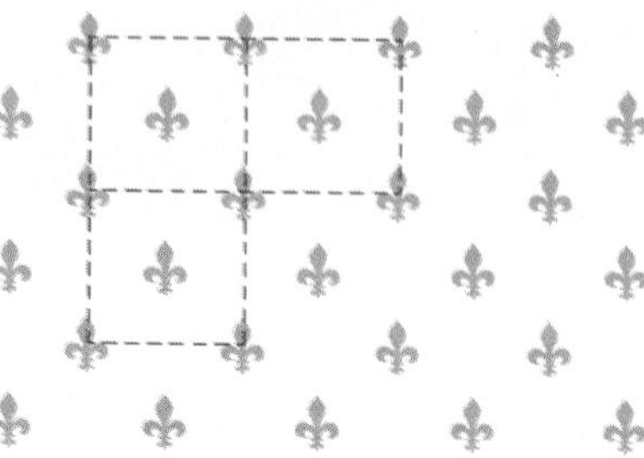

A crystalline solid can be represented by a three-dimensional array of points, each of which represents an identical environment within the crystal. Such an array of points is called a **crystal lattice**. We can imagine forming the entire crystal structure by arranging the contents of the unit cell repeatedly in a network of points.

Figure 11.26 shows a crystal lattice and its associated unit cell. In general, unit cells are parallelepipeds (six-sided figures whose faces are parallelograms).

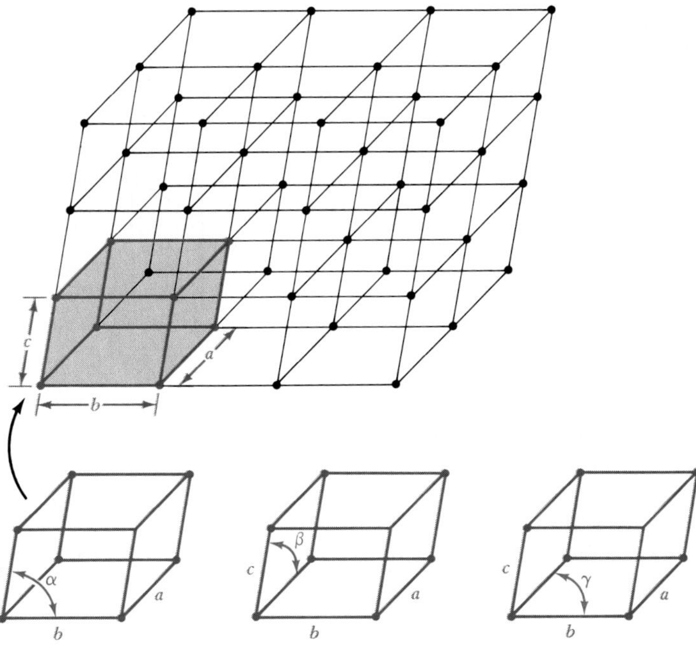

Figure 11.26 Simple crystal lattice and its associated unit cell. Each view of the unit cell shows one of the characteristic angles between unit-cell axes. Angle α is the angle between the b and c axes; β is the angle between the a and c axes; γ is the angle between the a and b axes.

Each unit cell can be described in terms of the lengths of the edges of the cell (a, b, and c) and by the angles between these edges (α, β, and γ), as shown in Figure 11.26. The lattices of all crystalline compounds can be described in terms of seven basic types of unit cells. The simplest of these is the cubic unit cell, in which all the sides are equal in length and all the angles are 90°.

There are three kinds of cubic unit cells, as illustrated in Figure 11.27. When lattice points are at the corners only, the unit cell is described as **primitive cubic**. When a lattice point also occurs at the center of the unit cell, the cell is known as **body-centered cubic**. A third type of cubic cell has lattice points at the center of each face, as well as at each corner, an arrangement known as **face-centered cubic**.

The Crystal Structure of Sodium Chloride

It is instructive to examine the crystal structure of NaCl, Figure 11.28. This figure shows two ways of locating the lattice points so they are in identical environments. In Figure 11.28(a), the points are centered on the Cl^- ions; in Figure 11.28(b), they are centered on Na^+. In both cases, the structure possesses a lattice with a face-centered-cubic unit cell. The cubic character of the unit cell is reflected in the shapes of well-formed crystals of NaCl, Figure 11.29.

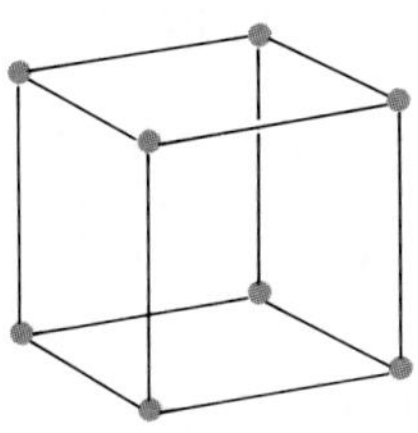
Primitive cubic

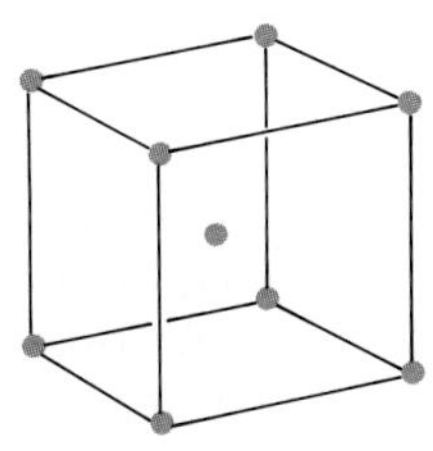
Body-centered cubic

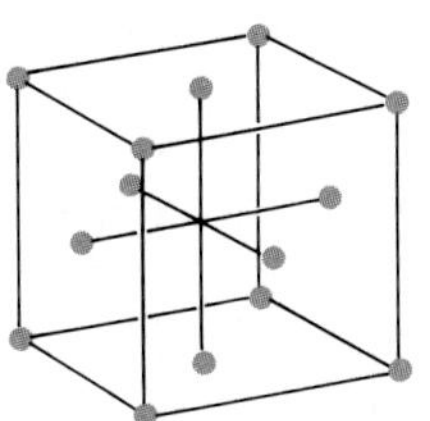
Face-centered cubic

Figure 11.27 The three types of unit cells found in cubic lattices.

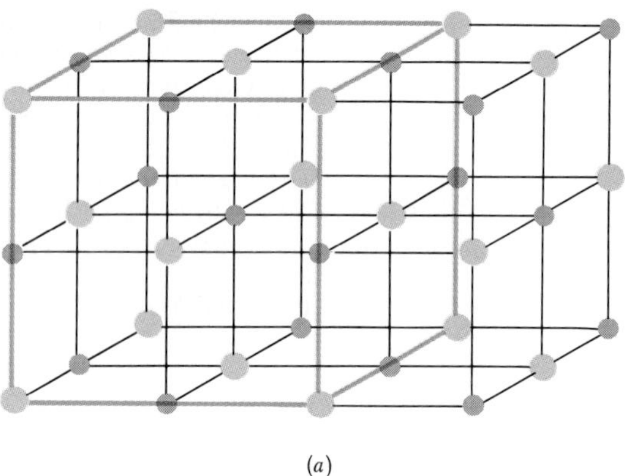

(a)

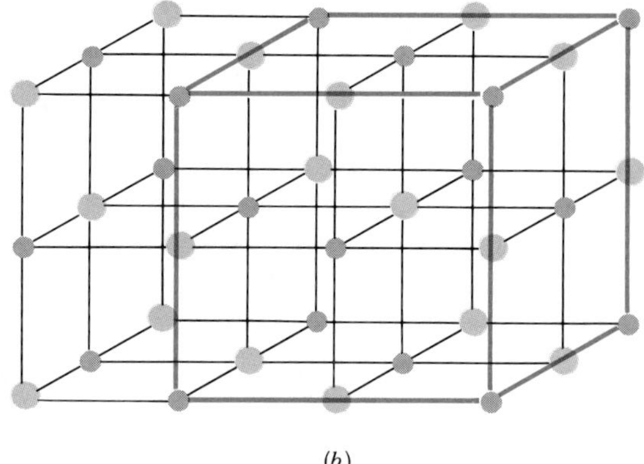

(b)

Figure 11.28 Portion of the crystal structure of NaCl illustrating two ways of positioning its crystal lattice. Gray spheres represent Na^+ ions, and green spheres represent Cl^- ions. Colored lines define the face-centered unit cell of each lattice. In (*a*), the lattice points are centered on Cl^- ions. In (*b*), the lattice points are centered on Na^+ ions. However, the lattice points need not be centered on ions at all. The only restriction on where lattice points are placed is that the points must be in positions that have identical environments throughout the solid. Regardless of where the lattice points are placed in the NaCl structure, the result is a face-centered-cubic lattice.

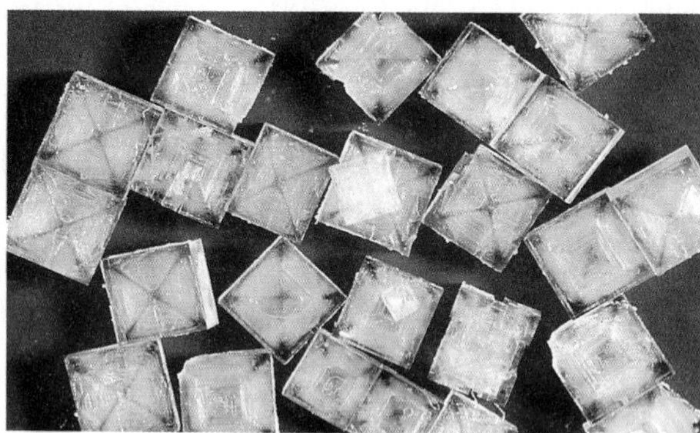

Figure 11.29 Crystals of NaCl, showing well-defined crystal planes based on the underlying cubic structure. (Dr. E. R. Degginger)

In Figure 11.28, the Na^+ and Cl^- ions have been moved apart so the symmetry of the structure can be seen more clearly. In this representation, no attention is paid to the relative sizes of the ions. In contrast, Figure 11.30(*a*) provides a representation that shows the relative sizes of the ions and how they fill the unit cell. Notice that the particles at the corners, edges, and faces do not lie wholly within the unit cell. Instead, these particles are shared by other unit cells. A particle at a corner is shared by eight unit cells [Figure 11.30(*b*)], one at the center of a face is shared by two [Figure 11.30(*c*)], and one at the edge is shared by four.

The total cation-to-anion ratio of a unit cell must be the same as that for the entire crystal. Therefore, within the unit cell of NaCl there must be an equal number of Na^+ and Cl^- ions. Similarly, the unit cell for $CaCl_2$ would have one Ca^{2+} for each two Cl^-, and so forth for other crystals. In Sample Exercises 11.6 and 11.7, the contents of the NaCl unit cell are determined and used to calculate the density of the solid.

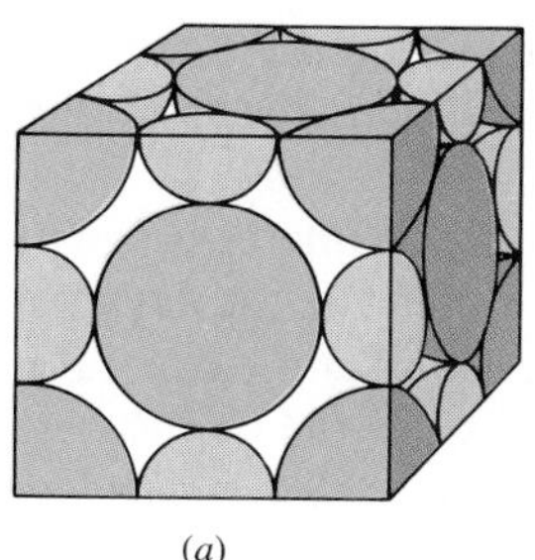

(a)

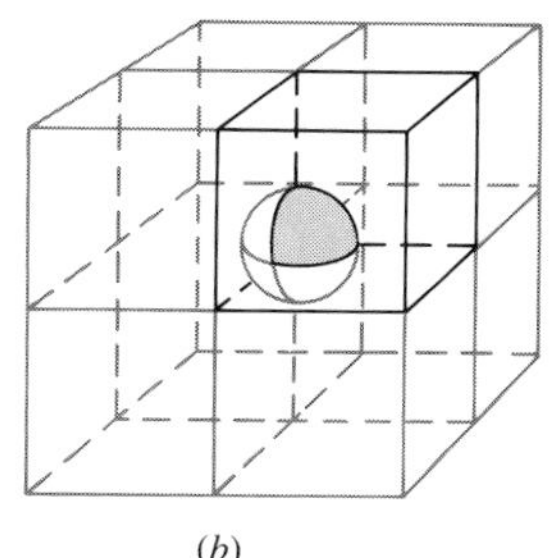

(b)

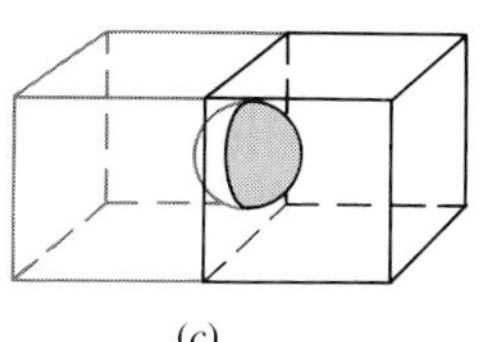

(c)

Figure 11.30 (*a*) The unit cell of NaCl showing the relative sizes of the Na^+ ions (gray color) and Cl^- ions (green color). Notice that only portions of most of the ions lie within the boundaries of the single unit cell. (*b*) The sharing of a corner atom or ion by eight unit cells. (*c*) The sharing of a face-centered atom or ion by two unit cells.

SAMPLE EXERCISE 11.6

Determine the net number of Na^+ and Cl^- ions in the NaCl unit cell (Figure 11.30).

Solution: There is one-fourth of a Na^+ on each edge, a whole Na^+ in the center of the cube (refer also to Figure 11.28), one-eighth of a Cl^- on each corner, and one-half of a Cl^- on each face. Thus, we have the following:

$$
\begin{aligned}
Na^+: \quad & (\tfrac{1}{4}Na^+ \text{ per edge})(12 \text{ edges}) = 3\ Na^+ \\
& (1\ Na^+ \text{ per center})(1 \text{ center}) = 1\ Na^+ \\
Cl^-: \quad & (\tfrac{1}{8}Cl^- \text{ per corner})(8 \text{ corner}) = 1\ Cl^- \\
& (\tfrac{1}{2}Cl^- \text{ per face})(6 \text{ faces}) = 3\ Cl^-
\end{aligned}
$$

Thus, the unit cell contains four Na^+ and four Cl^-. This result agrees with the compound's stoichiometry: one Na^+ for each Cl^-.

PRACTICE EXERCISE

The element iron crystallizes in a form called α-iron, which has a body-centered-cubic unit cell. How many iron atoms are in the unit cell? ***Answer:*** two

SAMPLE EXERCISE 11.7

If the unit cell of NaCl is 5.64 Å on an edge, calculate the density of NaCl.

Solution: The volume of the unit cell is $(5.64\ \text{Å})^3$. Because each unit cell contains $4\,Na^+$ and $4\,Cl^-$ (Sample Exercise 11.6), its mass is

$$4(23.0\ \text{amu}) + 4(35.5\ \text{amu}) = 234.0\ \text{amu}$$

The density is mass/volume:

$$\text{Density} = \frac{234.0\ \text{amu}}{(5.64\ \text{Å})^3}\left(\frac{1\ \text{g}}{6.02 \times 10^{23}\ \text{amu}}\right)\left(\frac{1\ \text{Å}}{10^{-8}\ \text{cm}}\right)^3 = 2.17\ \text{g/cm}^3$$

This value agrees with that found by simple density measurements: 2.165 g/cm³. The size and contents of the unit cell are therefore consistent with the macroscopic density of the substance.

PRACTICE EXERCISE

The body-centered-cubic unit cell of α-iron is 2.8664 Å on each side. Calculate the density of α-iron. ***Answer:*** 7.878 g/cm³

Close Packing of Spheres

The structures adopted by crystalline solids are those that bring particles in closest contact to maximize the attractive forces between them. In many cases, the particles that make up the solids are spherical or approximately so. Such is the case for atoms in metallic solids. Many molecules can also be approximated

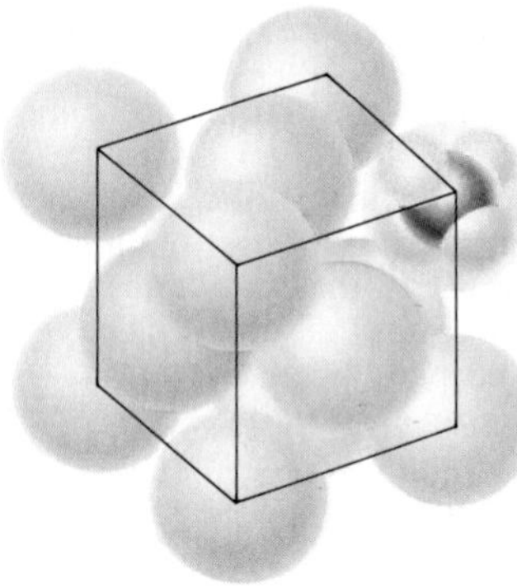

Figure 11.31 Unit cell of solid methane. Each large sphere represents a CH_4 molecule, as shown in the upper right.

as spheres, as seen for CH_4 in Figure 11.31. It is therefore instructive to consider how equal-sized spheres can pack most efficiently (that is, with the minimum amount of empty space).

The most efficient arrangement of a layer of equal-sized spheres is shown in Figure 11.32(*a*). Each sphere is surrounded by six others in the layer. A second layer of spheres will pack efficiently in the depressions of the first layer [Figure 11.32(*b*)]. The spheres of the third layer likewise sit in the depressions of the second layer. However, there are two types of depressions, and they result in different structures.

If the spheres of the third layer are placed immediately above those of the first layer, as shown in Figure 11.32(*c*), the result is a structure known as **hexagonal close packing**. The third layer repeats the first layer, the fourth layer repeats the second layer, and so forth, giving a layer sequence *ABABAB*. . . . The stacking sequence in the hexagonal close-packed structure is viewed from a different perspective in Figure 11.33(*a*).

If the spheres of the third layer are placed in slightly different positions, as shown in Figure 11.32(*d*), the resulting structure is known as **cubic close packing**. In this case the fourth layer repeats the first layer, giving a layer sequence *ABCABC*. . . . The stacking sequence is pictured in Figure 11.33(*b*). Although it cannot be seen in either Figure 11.32(*d*) or Figure 11.33(*b*), the unit cell of the cubic close-packed structure is face-centered cubic.

In both of the close-packed structures, each sphere has 12 equidistant nearest neighbors: six in one plane, three above that plane, and three below, as seen in Figure 11.33. We say that each sphere has a **coordination number** of 12. The coordination number is the number of particles immediately surrounding a particle in the crystal structure. In both types of close packing, 74 percent of the total volume of the structure is occupied by spheres; 26 percent is empty space between the spheres. By comparison, each sphere in the body-centered-cubic structure has a coordination number of 8, and only 68 percent of the space is occupied. In the simple cubic structure, the coordination number is 6, and only 52 percent of the space is occupied.

When unequal-sized spheres are packed in a lattice, the large particles sometimes assume one of the close-packed arrangements with small particles

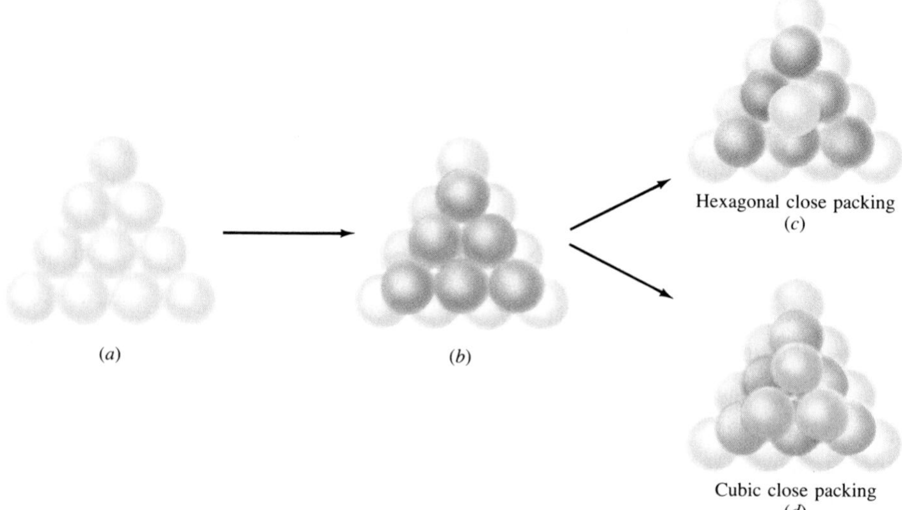

Figure 11.32 Close packing of equal-sized spheres. (*a*) One layer; each sphere is in contact with six others. (*b*) Spheres in the second layer sit in the depressions of the first layer. (*c*) In the hexagonal close-packed structure, each sphere in the third layer sits directly over a sphere in the first layer. (*d*) In the cubic close-packed structure, the spheres in the third layer are not directly over the first-layer spheres.

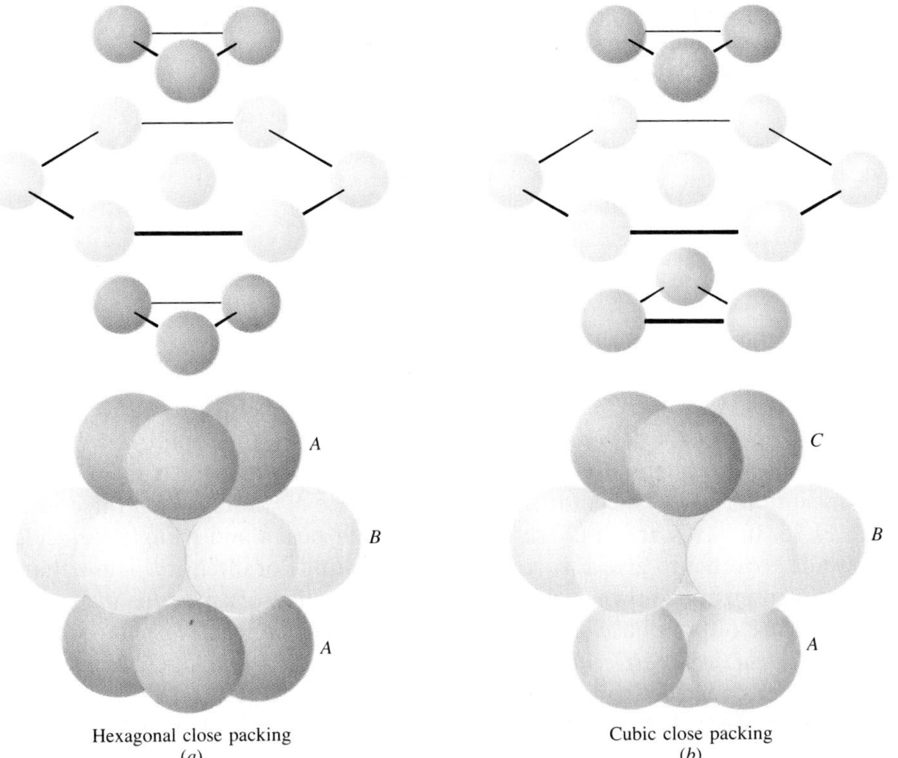

Figure 11.33 (*a*) Hexagonal close packing. In this arrangement, the atoms of the third layer lie directly over those in the first layer. Thus, the order of layers is *ABABAB*. (*b*) Cubic close packing. In this arrangement, the atoms of the third layer are not directly over those in the first, but are offset by a bit. As a result, the layers are arranged vertically as *ABCABC*.

occupying the holes between the large spheres. For example, in Li_2O the oxide ions assume a cubic close-packed structure, and the Li^+ ions occupy small cavities that exist between oxide ions.

X-Ray Diffraction by Crystals

Much of what we know about crystal structures has been obtained by studies of the diffraction of X rays by solids. **X-ray diffraction** results from the scattering of X rays by a regular arrangement of atoms or ions. Figure 11.34 shows the diffraction of a beam of X rays as it passes through a crystal.

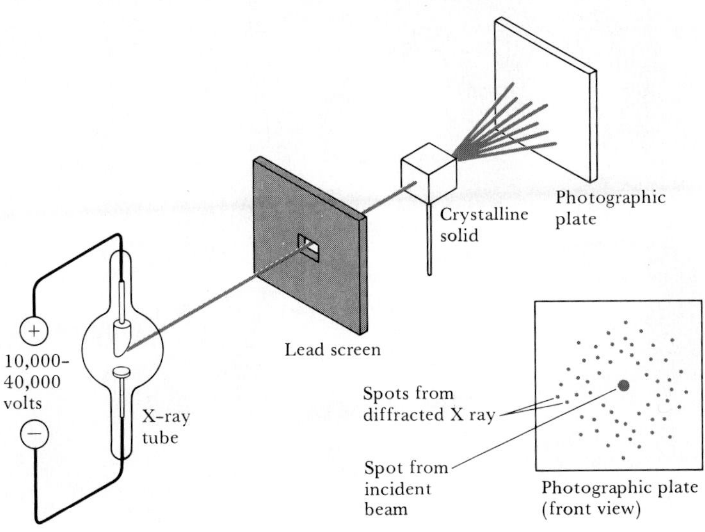

Figure 11.34 X-ray diffraction pattern for a crystal and the experimental method by which it is obtained.

Figure 11.35 Illustration of interference. In (*a*), the waves are in phase, and the interference is constructive; in (*b*), the waves are out of phase, and the interference is destructive.

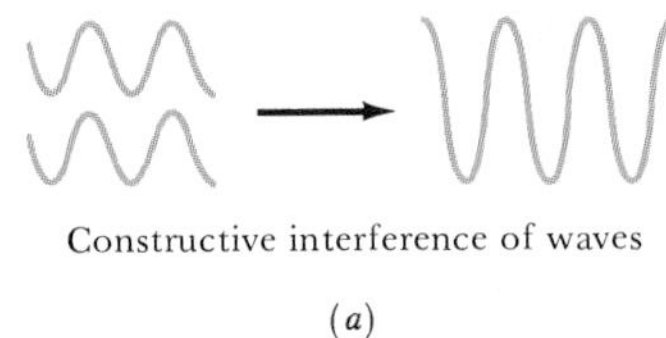

Constructive interference of waves

(*a*)

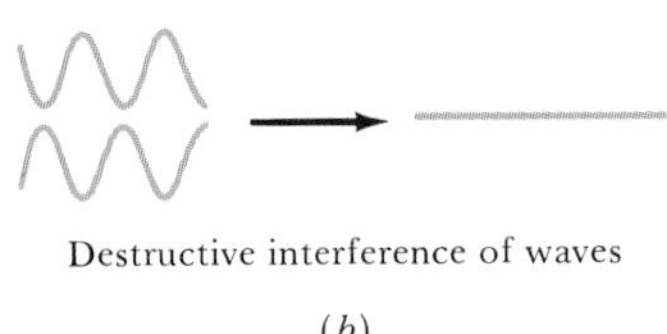

Destructive interference of waves

(*b*)

When a beam of X rays passes through a crystalline solid, each atom in the beam scatters some of the radiation. The waves scattered from different atoms interfere with one another. The interference of waves is illustrated in Figure 11.35. If waves are in phase—that is, if their peaks and troughs coincide—they add together to give a wave of greater amplitude or intensity. This enhancement of intensity, shown in Figure 11.35(*a*), is called *constructive interference.* If waves are out of phase, as in Figure 11.35(*b*), they cancel. This cancellation is called *destructive interference.* The spots on the photgraphic film in Figure 11.34 result from constructive interference.

A modern X-ray diffraction instrument, such as that shown in Figure 11.36, is capable of mapping the diffraction pattern of a crystal very quickly (sometimes in a matter of hours). It is then possible to work backward from the diffraction pattern to deduce the arrangement of atoms that produced the diffraction.

Figure 11.36 A modern X-ray diffractometer. A detector moves under computer control to find the angles at which X rays are constructively scattered from the crystal. The intensity of the scattering at this angle is measured. The data from many points, typically several thousand, are analyzed using a computer associated with the diffractometer to produce a picture of the arrangement of atoms within the crystal. (Courtesy of Nicolet)

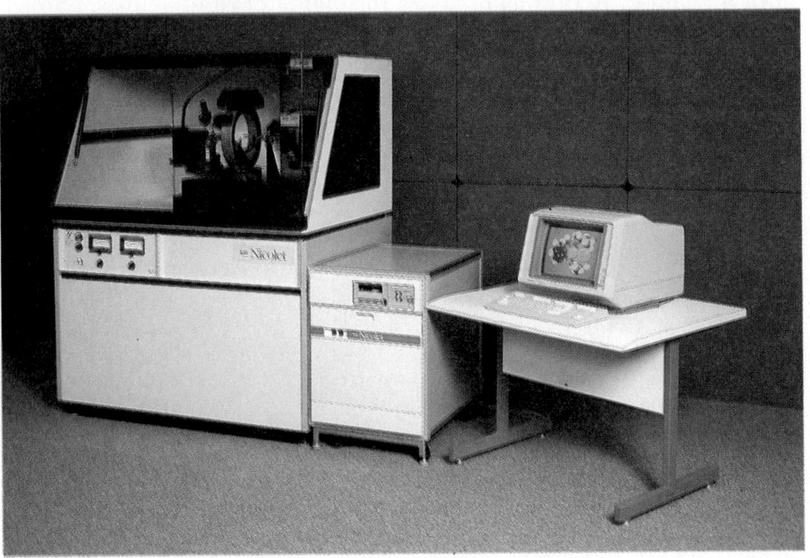

A CLOSER LOOK: The Bragg Equation

In 1913, the English father and son scientists William and Lawrence Bragg found that diffraction patterns can be interpreted by treating layers of atoms in crystals as reflecting planes. Consider a solid with planes of atoms separated by a distance d, as shown in Figure 11.37.

The incoming rays are in phase at AB. Wave ACA' is scattered, or reflected, by an atom in the first layer of the solid. Wave BEB' is reflected by an atom in the second layer. If these two waves are to be in phase at $A'B'$, the extra distance covered by BEB' must be a whole-number multiple of the wavelength, λ. In Figure 11.37, the extra distance, DEF, is 2λ. Now notice that the triangle CDE is a right triangle. Using trigonometry, it can be shown that the distance DE is $d \sin \theta$, where d is the distance between the planes and θ is the angle between the incoming wave and the plane. Because the distance DEF is twice DE, we have

$$DEF = 2\lambda = 2d \sin \theta$$

It can be shown that the general equation for constructive interference is

$$n\lambda = 2d \sin \theta \qquad \text{where } n = 1, 2, 3, 4, \ldots \qquad [11.2]$$

This relationship, known as the *Bragg equation*, allows determination of the spacing between planes from the known wavelength of the light and from experimentally determined values of θ at which constructive interference occurs. The Braggs were awarded the Nobel Prize in physics in 1915 for their pioneering work.

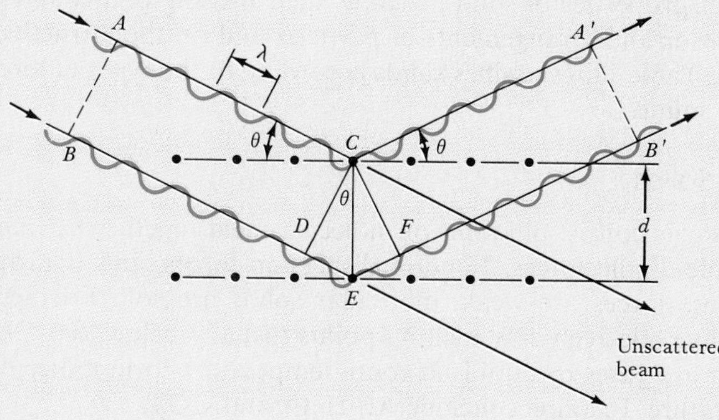

Figure 11.37 Scattering of X rays by atoms in parallel planes. The incoming X rays of wavelength λ are diffracted by atoms, which are represented as dots. The atoms are arranged in planes separated by a distance d. The incoming X rays make an angle θ with the planes and are scattered at an angle 2θ from the unscattered beam.

Crystal Defects

Although we have generally pictured crystalline solids as being composed of perfectly ordered arrays of particles, this image is merely a useful abstraction, like that of an ideal gas. Real crystals contain imperfections whose number and type can play an important role in determining the properties of the solid. For example, a crystal lattice that has many sites where particles are missing (vacancies) can be more readily deformed than can a perfect crystal lattice of the same substance. It is not hard to appreciate that structural imperfections can occur readily. In forming a solid, the lattice is built up very rapidly, and misplacements can readily occur. Defects can also arise through thermal motion of the units that make up the lattice, or through strains that might be externally imposed on the solid.

An edge dislocation is a common defect in many solids. In Figure 11.38, we see that one of the planes within the solid has discontinued, and the adjacent planes have closed up over the edge of the discontinued plane.

Dislocations and other crystal defects are important because they represent sites where the atoms or molecules within the solid are in strained, high-energy environments. Thus, oxidation of graphite would proceed more readily at an edge dislocation than elsewhere on the surface of the crystal.

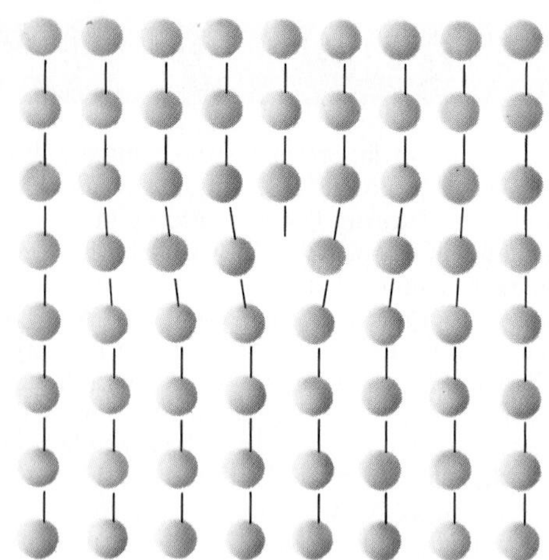

Figure 11.38 Schematic illustration of an edge dislocation. The circles represent repeating units of the solid structure. An edge is created where the plane of units in the center stops. The adjacent planes shift to close up over the gap that is created. In the process, strains are created in the lattice.

11.8 BONDING IN SOLIDS

The physical properties of solids, such as their melting points and hardnesses, depend both on the arrangements of particles and on the attractive forces between them. Table 11.6 classifies solids according to the types of forces between particles in solids.

Molecular Solids

Molecular solids consist of atoms or molecules held together by intermolecular forces (dipole-dipole forces, London dispersion forces, and hydrogen bonds). Because these forces are weak, molecular solids are soft. Furthermore, they normally have relatively low melting points (usually below 200°C). Most substances that are gases or liquids at room temperature form molecular solids at low temperature. Examples include Ar, H_2O, and CO_2.

Table 11.6 Types of Solids

Type of solid	Form of unit particles	Forces between particles	Properties	Examples
Molecular	Atoms or molecules	London dispersion, dipole-dipole forces, hydrogen bonds	Fairly soft, low to moderately high melting point, poor thermal and electrical conduction	Argon, Ar; methane, CH_4; sugar, $C_{12}H_{22}O_{11}$; Dry Ice, CO_2
Covalent (network)	Atoms that are connected in a covalent-bond network	Covalent bonds	Very hard, very high melting point, often poor thermal and electrical conduction	Diamond, C; quartz, SiO_2
Ionic	Positive and negative ions	Electrostatic attractions	Hard and brittle, high melting point, poor thermal and electrical conduction	Typical salts—for example, NaCl, $Ca(NO_3)_2$
Metallic	Atoms	Metallic bonds	Soft to very hard, low to very high melting point, excellent thermal and electrical conduction, malleable and ductile	All metallic elements—for example, Cu, Fe, Al, W

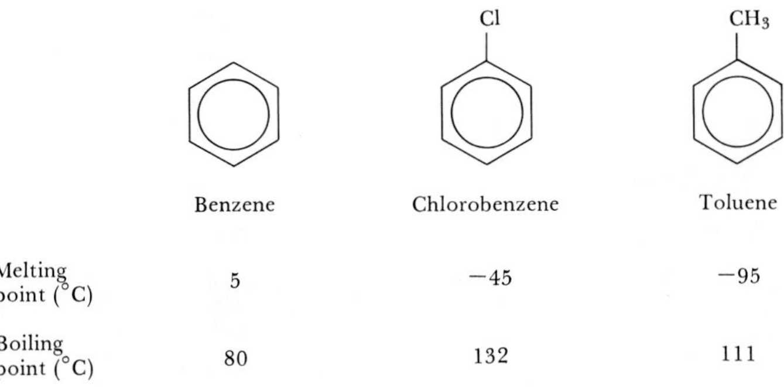

	Benzene	Chlorobenzene	Toluene
Melting point (°C)	5	−45	−95
Boiling point (°C)	80	132	111

Figure 11.39 Comparative melting and boiling points for benzene, chlorobenzene, and toluene.

The properties of molecular solids depend not only on the strengths of the forces that operate between molecules but also on the abilities of the molecular units to pack efficiently in three dimensions. For example, benzene—C_6H_6, a highly symmetric molecule—has a higher melting point than substituted benzene compounds such as chlorobenzene and toluene. The lower symmetry of the substituted molecules, shown in Figure 11.39, prevents them from packing as efficiently as benzene. As a result, the intermolecular forces that depend on close contact do not operate as effectively, and the melting point is lower. In contrast, the boiling points of these compounds are higher than that of benzene, indicating that the attractive forces between these molecules in the liquid state are larger than those between liquid benzene molecules.

Covalent-Network Solids

Covalent-network solids consist of atoms held together in large networks or chains by covalent bonds. Because covalent bonds are much stronger than intermolecular forces, these solids are much harder and have higher melting points than molecular solids. Two of the most familiar examples of covalent-network solids are diamond and graphite, two allotropes of carbon. Other examples include quartz, SiO_2, silicon carbide, SiC, and boron nitride, BN.

In diamond, each carbon atom is bonded to four other carbon atoms as shown in Figure 11.40(*a*). This interconnected three-dimensional array of strong carbon-carbon single bonds contributes to diamond's unusual hardness. Industrial-grade diamonds are employed in the blades of saws for the most demanding cutting jobs. In keeping with its structure and bonding, diamond also has a high melting point, 3550°C.

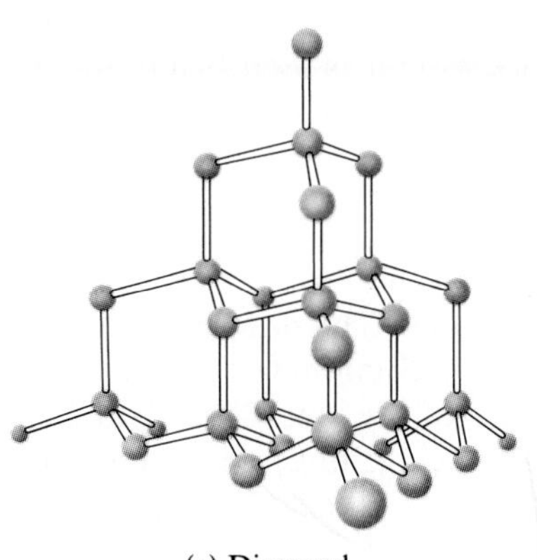

(*a*) Diamond

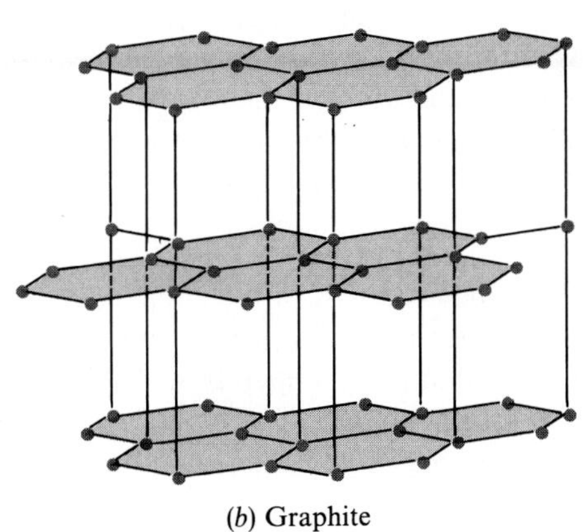

(*b*) Graphite

Figure 11.40 Structures of (*a*) diamond and (*b*) graphite.

Figure 11.41 Piece of graphite photographed with an electron microscope (magnified about 15 million times). The bright bands are layers of carbon atoms that are only 3.41 Å apart. (P. A. Marsch and A. Voet. J. M. Hunder Corp.)

In graphite, the carbon atoms are arranged in layers of interconnected hexagonal rings as shown in Figure 11.40(*b*). Each carbon atom is bonded to three others in the layer. The distance between adjacent carbon atoms in the plane, 1.42 Å, is very close to the C—C distance in benzene, 1.395 Å. In fact, the bonding resembles that of benzene (Section 9.4), with delocalized π bonds extending over the layers. Electrons move freely through the delocalized orbitals, making graphite a good conductor of electricity along the layers. (If you have ever taken apart a flashlight battery, you know that the central electrode in the battery is made of graphite.) The layers, which are separated by 3.41 Å, are held together by weak dispersion forces. The layers readily slide past one another when rubbed, giving the substance a greasy feel. Graphite is used as a lubricant and in making the "lead" in pencils. Figure 11.41 is a photomicrograph of graphite that reveals its layered structure.

Ionic Solids

Ionic solids consist of ions held together by ionic bonds. The principles governing the formation of ionic bonds were introduced in Section 8.2. The strength of an ionic bond depends strongly on the charges of the ions. Thus, NaCl, in which the ions have charges of 1+ and 1−, has a melting point of 804°C, whereas MgO, in which the charges are 2+ and 2−, melts at 2826°C.

CHEMISTRY AT WORK: Diamond Coatings

Besides being one of the hardest known substances, diamond is also highly resistant to corrosion. At the present time, commercial diamonds are widely used to strengthen cutting and grinding tools. These diamonds are embedded in the tools, not intimately and uniformly part of the material.

Scientists have recently developed procedures for applying ultrathin layers of synthetic diamond coatings on materials. Diamond films promise to give diamond's hardness and durability to a variety of materials—glass, paper, plastics, metals, and semiconductor devices, for example. Imagine scratchproof glass; cutting tools that virtually never need sharpening (Figure 11.42); surfaces that are chemical-resistant. Because diamond is a good heat conductor, diamond-coated silicon chips could be packed closer together in electronic devices, allowing them to be more compact and therefore faster. Because diamond is compatible with biological tissue, it also can be used to coat prosthetic materials and biosensors.

One procedure for generating diamond films involves exposing a mixture of methane gas, CH_4, and hydrogen gas, H_2, to intense microwave radiation in the presence of the object to be coated. Under appropriate conditions, the CH_4 decomposes, depositing a thin film of diamond. The H_2 dissociates into atomic hydrogen, which impedes formation of graphite. Atomic hydrogen reacts faster with graphite than with diamond, effectively removing graphite from the growing film. Although the process is not fundamentally expensive or sophisticated, its widespread commercial application is still some years away. Researchers are presently trying to reduce the deposition temperatures, increase the deposition rate, and develop techniques for coating the surfaces of a wider variety of sizes and shapes than is now practical.

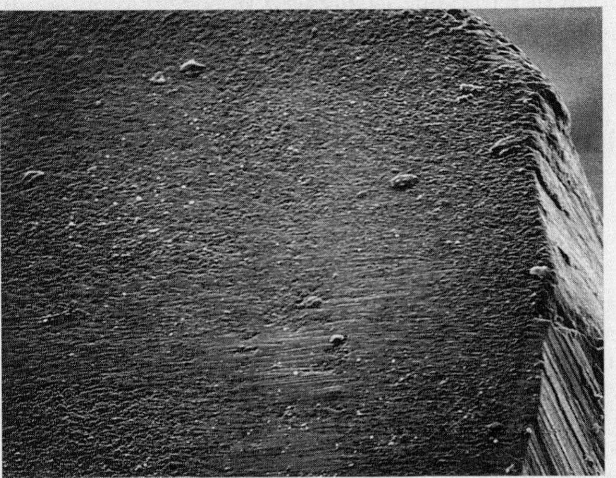

Figure 11.42 Two cutting-tool inserts of the same composition and used under similar conditions. The one on the left has been coated with a diamond film and shows less wear at the cutting edge. (Photomicrographs courtesy of William Drawl, Pennsylvania State University; *Chemical Engineering News* **1989**, May 15, p. 37)

The structures of simple ionic solids can be classified as a few basic types. The NaCl structure is a representative example of one type. Other compounds that possess this same structure include KCl, AgCl, and CaO. Three other common types of crystal structures are shown in Figure 11.43.

The structure adopted by an ionic solid depends largely on the charges and relative sizes of the ions. In the CsCl structure [Figure 11.43(*a*)], each Cs^+ is surrounded by eight Cl^- ions. The fact that NaCl and CsCl adopt different structures is due to the different sizes of the Na^+ and Cs^+ ions. The coordination number of a cation in an ionic solid tends to increase with an increase in

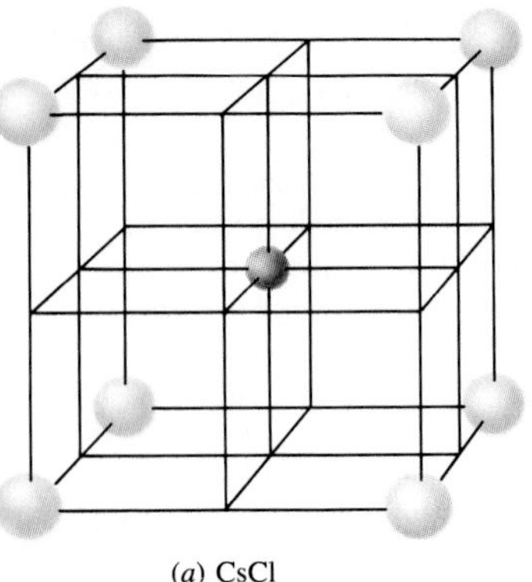

(*a*) CsCl

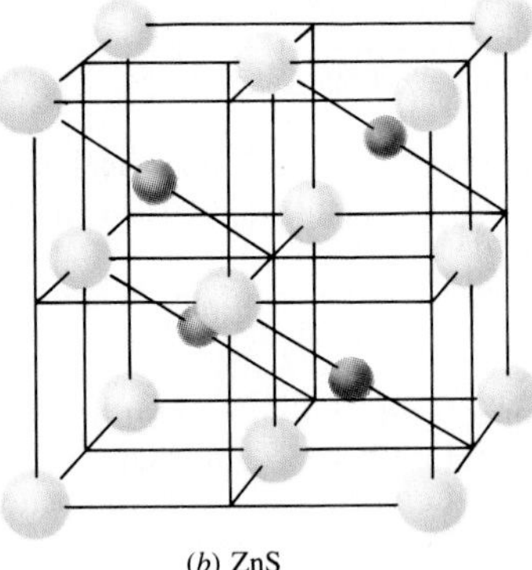

(*b*) ZnS

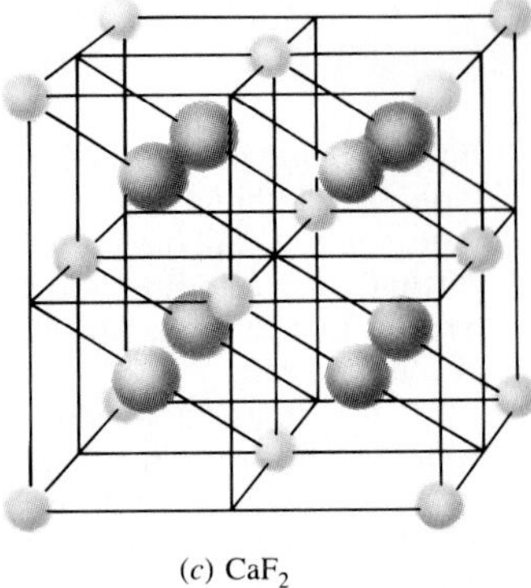

(*c*) CaF_2

Figure 11.43 Unit cells of some common types of crystal structures: (*a*) CsCl; (*b*) ZnS (zinc blende); (*c*) CaF_2 (fluorite).

the ratio of the cation radius to the anion radius, r_c/r_a. Table 11.7 provides some guidelines relating coordination numbers to radius ratios. These rules are based on the assumption of hard spheres engaged in purely ionic bonding. Because ions, particularly anions, are polarizable, and because some covalent character may be present in the bonds, the rules are not always strictly obeyed.

SAMPLE EXERCISE 11.8

The radii of Cs^+, Na^+, and Cu^+ are 1.67 Å, 0.98 Å, and 0.74 Å, respectively. The radius of a Cl^- ion is 1.81 Å. Calculate the cation-to-anion-radius ratio for CsCl, NaCl, and CuCl, and use Table 11.7 to predict the coordination number of each cation.

Solution: For CsCl, we have 1.67 Å/1.81 Å = 0.923. Therefore, a coordination number of 8 is expected for Cs^+. For NaCl, the ratio is 0.98 Å/1.81 Å = 0.54. Thus, we would predict a coordination number of 6 for the Na^+. For CuCl, the ratio is 0.74 Å/1.81 Å = 0.41. We would predict a coordination number of 4 for Cu^+.

PRACTICE EXERCISE

Predict the coordination number of Fe^{3+} in Fe_2O_3; the radius of Fe^{3+} is 0.65 Å, and that of O^{2-} is 1.45 Å. *Answer:* 6

In the zinc blende, ZnS, structure, Figure 11.43(*b*), the S^{2-} ions adopt a face-centered-cubic arrangement, with the smaller Zn^{2+} ions arranged so they are each surrounded tetrahedrally by four S^{2-} ions. CuCl also adopts this structure.

Table 11.7 Radius Ratios and Cation Location in an Anion Lattice

r_c/r_a[a]	Coordination number of cation	Arrangement of anions about the cation
0.225–0.414	4	Tetrahedral
0.414–0.732	6	Octahedral
0.732–1.000	8	Cubic

[a] If $r_c > r_a$, the ratio r_a/r_c gives the coordination number of the anion and the arrangement of cations.

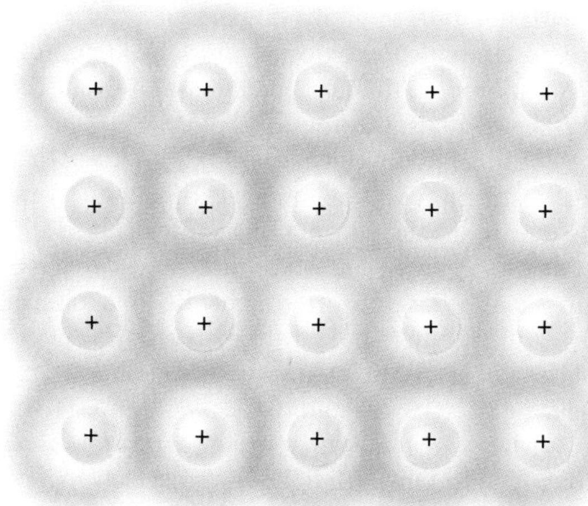

Figure 11.44 A cross section of a metal. Each sphere represents the nucleus and inner-core electrons of a metal atom. The surrounding colored "fog" represents the mobile sea of electrons that binds the atoms together.

In the fluorite, CaF_2, structure, Figure 11.43(*c*), the Ca^{2+} ions are shown in a face-centered-cubic arrangement. As required by the chemical formula of the substance, there are twice as many F^- ions in the unit cell as there are Ca^{2+} ions. Other compounds that have the fluorite structure include $BaCl_2$ and PbF_2.

Metallic Solids

Metallic solids consist entirely of metal atoms. Metallic solids usually have hexagonal close-packed, cubic close-packed (face-centered-cubic), or body-centered-cubic structures. Thus, each atom typically has 8 or 12 adjacent atoms.

The bonding in metals is too strong to be due to London dispersion forces, and yet there are not enough valence electrons for ordinary covalent bonds between atoms. The bonding is due to valence electrons that are delocalized throughout the entire solid. In fact, we can visualize the metal as an array of positive ions immersed in a sea of delocalized valence electrons, as shown in Figure 11.44.

Metals vary greatly in the strength of their bonding, as is evidenced by their wide range of physical properties such as hardness and melting point. In general, however, the strength of the bonding increases as the number of electrons available for bonding increases. Thus, sodium, which has only one valence electron per atom, melts at 97.5°C, whereas chromium, with six electrons beyond the noble-gas core, melts at 1890°C. The mobility of the electrons explains why metals are good conductors of heat and electricity. The bonding and properties of metals will be examined more closely in Chapter 24.

FOR REVIEW

SUMMARY

Substances that are gases or liquids at room temperature are composed of molecules. In gases, the intermolecular attractive forces are negligible compared to the kinetic energies of the molecules; thus, the molecules are widely separated and undergo constant, chaotic motion. In liquids, the intermolecular forces are strong enough to keep the molecules in close proximity; nevertheless, the molecules are free to move with respect to one another. Solids are composed of atoms, molecules, or ions. The interparticle attractive forces are strong enough to restrain motion and to force the particles to occupy specific locations in a three-dimensional arrangement.

Three types of intermolecular forces exist between neutral molecules: dipole-dipole, London dispersion, and hydrogen bonding. Ion-dipole forces are important in solutions. London dispersion forces operate between all molecules. The relative strengths of the dipole-dipole and dispersion forces depend on the polarity, polarizability, size, and shape of the molecule. Dipole-dipole forces increase in strength with increasing polarity. Dispersion forces tend to increase in strength with increasing molecular mass. Hydrogen bonding occurs in compounds containing O—H, N—H, and F—H bonds. Hydrogen bonds are generally stronger than dipole-dipole or dispersion forces.

The stronger the intermolecular forces, the greater the viscosity, or resistance to flow, of a liquid. The surface tension of a liquid also increases as intermolecular forces increase in strength. Surface tension is a measure of the tendency of a liquid to maintain a minimum surface area. The adhesion of a liquid to the walls of a narrow tube and the cohesion of the liquid account for capillary action and for the formation of a meniscus.

A substance may exist in more than one state of matter, or phase. Changes of a solid to liquid (melting), solid to gas (sublimation), and liquid to gas (vaporization) are all endothermic processes. That is, the heats of melting (also called fusion), sublimation, and vaporization are all positive quantities. The reverse processes are thus exothermic. A gas cannot be liquefied by application of pressure if the temperature is above its critical temperature.

The vapor pressure of a liquid measures the tendency of the liquid to evaporate. The vapor pressure is the partial pressure of the vapor when it is in equilibrium with the liquid. The equilibrium is a dynamic process in which the rate of transfer of molecules from the liquid to the vapor equals the rate of transfer from the vapor to the liquid. The higher the vapor pressure of a liquid, the more readily it evaporates and the more volatile it is said to be. Vapor pressure increases nonlinearly with temperature. Boiling occurs when the vapor pressure equals the external pressure. The normal boiling point is the temperature at which the vapor pressure equals 1 atm.

The equilibria between the solid, liquid, and gas phases of a substance as a function of temperature and pressure are displayed on a phase diagram. Equilibria between any two phases are indicated by a line. The point on the diagram at which all three phases coexist in equilibrium is called the triple point.

Solids whose particles are arranged in a regularly repeating pattern are said to be crystalline. Those whose particles show no such order are said to be amorphous. Crystalline solids have imperfections, or defects, that cause them to have less than perfect order.

The essential structural features of a crystalline solid can be represented by its unit cell, the smallest part of the crystal that can, by simple displacement, reproduce the three-dimensional structure. The three-dimensional structures of crystals can also be represented by their crystal lattices. The points in a crystal lattice represent positions in the structure where there are identical environments.

Many solids have a close-packed structure in which spherical particles are arranged so as to leave the minimal amount of empty space. Two closely related forms of close packing, cubic close packing and hexagonal close packing, are possible. In both, each sphere has a coordination number of 12.

The structures of crystalline solids can be determined from their X-ray diffraction patterns. Diffraction patterns result from the constructive interference of X rays scattered by the atoms in the solid.

The properties of solids depend both on the arrangements of particles and on the attractive forces between them. Molecular solids, which consist of atoms or molecules held together by intermolecular forces, are soft and low melting. Covalent-network solids, which consist of atoms held together by covalent bonds that extend throughout the solid, are hard and high melting. Ionic solids are hard and brittle and have high melting points. Metallic solids, which consist of metal cations held together by a sea of electrons, exhibit a wide range of properties.

KEY TERMS

intermolecular forces
ion-dipole force (Sec. 11.2)
dipole-dipole force (Sec. 11.2)
London dispersion force (Sec. 11.2)
polarizability (Sec. 11.2)
hydrogen bonding (Sec. 11.2)

viscosity (Sec. 11.3)
surface tension (Sec. 11.3)
cohesive forces (Sec. 11.3)
adhesive forces (Sec. 11.3)
meniscus (Sec. 11.3)
capillary action (Sec. 11.3)
phase changes (Sec. 11.4)
heat of fusion (Sec. 11.4)
heat of vaporization (Sec. 11.4)
critical temperature (Sec. 11.4)
critical pressure (Sec. 11.4)
vapor pressure (Sec. 11.5)
dynamic equilibrium (Sec. 11.5)
volatile (Sec. 11.5)
normal boiling point (Sec. 11.5)
phase diagram (Sec. 11.6)
normal melting point (Sec. 11.6)
triple point (Sec. 11.6)
crystalline solid (Sec. 11.7)
amorphous solid (Sec. 11.7)
unit cell (Sec. 11.7)
crystal lattice (Sec. 11.7)
primitive cubic (Sec. 11.7)
body-centered cubic (Sec. 11.7)
face-centered cubic (Sec. 11.7)
hexagonal close packing (Sec. 11.7)
cubic close packing (Sec. 11.7)
coordination number (Sec. 11.7)
X-ray diffraction (Sec. 11.7)
molecular solids (Sec. 11.8)
covalent-network solids (Sec. 11.8)
ionic solids (Sec. 11.8)
metallic solids (Sec. 11.8)

EXERCISES

Kinetic-Molecular Theory

11.1 Account for the difference in compressibility of solids, liquids, and gases.

11.2 Compare the degree of order within a solid with that within a liquid. What accounts for the difference?

11.3 Why does lowering the temperature cause a substance to change in succession from the gaseous to the liquid to the solid state?

11.4 Why is it easier to liquefy gases when they are compressed?

Intermolecular Forces

11.5 Define and give an example of each of the following types of intermolecular forces: **(a)** dipole-dipole force; **(b)** ion-dipole force; **(c)** London dispersion force; **(d)** van der Waals force; **(e)** hydrogen bond.

11.6 Which type of intermolecular attractive force operates between: **(a)** all molecules; **(b)** polar molecules; **(c)** the hydrogen atom of a polar bond and a nearby electronegative atom?

11.7 What kind of attractive forces must be overcome to: **(a)** melt ice; **(b)** boil Br_2; **(c)** melt NaCl; **(d)** dissociate F_2 into F atoms?

11.8 What kind of attractive forces must be overcome to: **(a)** boil water; **(b)** melt CCl_4; **(c)** dissociate NaCl into ions; **(d)** sublime I_2?

11.9 What is the nature of the major intermolecular attractive force in each of the following: **(a)** $Xe(l)$; **(b)** $NH_3(l)$; **(c)** $PCl_3(l)$; **(d)** between H_2O and Fe^{3+} in $FeCl_3(aq)$?

11.10 What is the nature of the major intermolecular attractive force in each of the following: **(a)** $CH_3OH(l)$; **(b)** $Br_2(l)$; **(c)** $CH_3F(l)$; **(d)** between H_2O and Cl^- in $CaCl_2(aq)$?

11.11 **(a)** Explain the term *polarizability*. **(b)** Which of the following molecules would you expect to be most polarizable: F_2, Cl_2, Br_2, or I_2? Explain.

11.12 **(a)** How does the polarizability of a molecule relate to the strengths of dispersion forces between molecules? **(b)** How do the strengths of dispersion forces vary with molecular size?

11.13 Which member of each of the following pairs of substances would you expect to have the higher boiling point: **(a)** N_2 or O_2; **(b)** CH_4 or SiH_4; **(c)** NaCl or CH_3Cl; **(d)** C_2H_5Cl or C_2H_5OH? Explain.

11.14 Rationalize the difference in boiling point between the members of the following pairs of substances: **(a)** HF (20°C) and HCl (−85°C); **(b)** $CHCl_3$ (61°C) and $CHBr_3$ (150°C); **(c)** Br_2 (59°C) and ICl (97°C); **(d)** $CH_3CH_2CH_2CH_3$ (0°C) and $(CH_3)_3CH$ (−10°C).

11.15 Cite three properties of water that can be attributed to the existence of hydrogen bonding.

11.16 Explain why water expands when it freezes.

Viscosity and Surface Tension

11.17 How do the viscosities and surface tensions of liquids change as intermolecular forces become stronger?

11.18 How do the viscosities and surface tensions of liquids change as temperature increases?

11.19 Explain the following observations: **(a)** The viscosity of ethanol, CH_3CH_2OH, is greater than that of ether, $CH_3CH_2OCH_2CH_3$. **(b)** In contact with a narrow capillary tube made of polyethylene, water forms a concave-downward meniscus like that of mercury in a glass tube.

11.20 Explain the following observations: **(a)** The surface tension of $CHBr_3$ is greater than that of $CHCl_3$. **(b)** As temperature increases, oil flows faster through a narrow tube. **(c)** Raindrops that collect on a waxed automobile hood take on a nearly spherical shape.

Changes of State

11.21 Name all the possible phase changes that can occur between different states of matter. Which of these are exothermic, and which are endothermic?

11.22 Explain the following observations: **(a)** During the cold winter months, snow often gradually disappears without melting. **(b)** The heat of fusion for any substance is generally lower than its heat of vaporization. **(c)** Ethyl chloride, C_2H_5Cl, boils at 12°C. When liquid C_2H_5Cl under pressure is sprayed on a surface at atmospheric pressure, the surface is cooled considerably. **(d)** When heated above 279°C, carbon disulfide, CS_2, cannot be liquefied regardless of how great the pressure exerted on the gas is.

11.23 Freon 12, CCl_2F_2, is used as a refrigerant. Its heat of vaporization is 289 J/g. What mass of Freon 12 must evaporate in order to freeze 100 g of water initially at 18°C? (The heat of fusion of water is 334 J/g; the heat capacity of water per gram is 4.18 J/g-°C).

11.24 For many years, drinking water has been cooled in hot climates by evaporating it from the surfaces of canvas bags or porous clay pots. How many grams of water can be cooled from 35°C to 22°C by the evaporation of 10 g of water? The heat of vaporization of water in this temperature range is 2.4 kJ/g. The heat capacity of water per gram is 4.18 J/g-°C.

11.25 Ethyl alcohol melts at −114°C and boils at 78°C. The enthalpy of fusion at −114°C is 105 J/g, and the enthalpy of vaporization at 78°C is 870 J/g. If the heat capacity per gram of solid ethyl alcohol is taken as 0.97 J/g-°C and that for the liquid as 2.3 J/g-°C, how much heat is required to convert 16.0 g of ethyl alcohol at −130°C to the vapor phase at 78°C?

11.26 Calculate the heat required to convert 10.0 g of propanol, C_3H_7OH, from a solid at −140°C into a vapor at 110°C. The normal melting point and boiling point of C_3H_7OH are −127°C and 97°C, respectively. The heat of fusion is 5.18 kJ/mol, and the heat of vaporization is 41.7 kJ/mol. The heat capacities of the solid, liquid, and gas states are 142, 170, and 108 J/mol-°C, respectively.

11.27 What does the critical temperature tell us about the conditions required for liquefaction of gases?

11.28 Select those substances in Table 11.5 that can be liquefied by applying pressure at room temperature (about 20°C).

Vapor Pressure and Boiling Point

11.29 Explain why the boiling point of a liquid varies substantially with pressure, whereas the melting point of a solid depends little on pressure.

11.30 Explain how each of the following affects the vapor pressure of a liquid: **(a)** surface area; **(b)** temperature; **(c)** intermolecular attractive forces; **(d)** volume of the liquid; **(e)** pressure of the air above the liquid.

11.31 Which of the following liquids would you expect to be most volatile: CCl_4, CBr_4, or CI_4? Explain.

11.32 Imagine replacing one H atom of a methane molecule, CH_4, with another atom or group of atoms. Account for the order of the boiling points in the following series of such compounds: CH_4 (−161°C), CH_3F (−78°C), CH_3Cl (−24°C), CH_3Br (3.6°C), and CH_3OH (65°C).

11.33 Two pans of water are on different burners of a stove. One pan of water is boiling vigorously, while the other is boiling gently. What can be said about the temperature of the water in the two pans?

11.34 Explain the following observations: **(a)** Water evaporates more quickly on a hot, dry day than on a hot, humid day. **(b)** It takes longer to boil eggs at high altitudes than at lower ones.

11.35 **(a)** Use the vapor pressure curve in Figure 11.18 to estimate the boiling point of ethanol at 500 mm Hg. **(b)** Use the vapor pressure table in Appendix B to determine the temperature at which water boils when the pressure is 30 mm Hg.

11.36 **(a)** Given the vapor pressures of the following substances at 25°C, which is most volatile? **(b)** Which is likely to have the highest boiling point: carbon disulfide, CS_2 (309 mm Hg); carbon tetrachloride, CCl_4 (107 mm Hg); or acetone, CH_3COCH_3 (185 mm Hg)?

11.37 Reno, Nevada, is about 4500 ft above sea level. If the barometric pressure is 680 mm Hg in Reno, at what temperature will water boil? Refer to Appendix B.

11.38 Mt. Kilimanjaro in Tanzania is the tallest peak in Africa (19,340 ft). If the barometric pressure at the top of the mountain is 350 mm Hg, at what temperature will water boil there? Refer to Appendix B.

Phase Diagrams

11.39 Refer to Figure 11.21(*a*), and describe all the phase changes that would occur in each of the following cases. **(a)** Water vapor originally at 1.0×10^{-3} atm and −0.10°C is slowly compressed at constant temperature until the final pressure is 10 atm. **(b)** Water originally at −10°C and 0.30 atm is heated at constant pressure until the temperature is 80.0°C.

11.40 Refer to Figure 11.21(*b*), and describe the phase changes (and the temperatures at which they occur) when CO_2 is heated from −80°C to −20°C at **(a)** a constant pressure of 3 atm; **(b)** a constant pressure of 6 atm.

11.41 The normal melting and boiling points of sulfur dioxide, SO_2, are −72.7°C and −10.0°C, respectively. Its triple point is at −75.5°C and 1.65×10^{-3} atm, and its critical point is at 157°C and 78 atm. Sketch the phase diagram for SO_2, showing the four points given above and indicating the areas in which each phase is stable.

11.42 The normal melting and boiling points of xenon are −112°C and −108°C, respectively. Its triple point is at −121°C, at a pressure of 282 mm Hg. Sketch the phase diagram for xenon, showing the three points given above and indicating the areas in which each phase is stable.

Structures of Solids

11.43 How does an amorphous solid differ from a crystalline one?

11.44 Amorphous silica has a density of about 2.2 g/cm^3, whereas the density of crystalline quartz is 2.65 g/cm^3. Account for this difference in density.

11.45 What is a unit cell? What properties does it have?

11.46 Calculate the net number of spheres in: **(a)** a primitive cubic unit cell; **(b)** a body-centered-cubic unit cell; **(c)** a face-centered-cubic unit cell.

11.47 What is the coordination number of each sphere in: **(a)** a three-dimensional, close-packed array of equal-sized spheres; **(b)** a primitive cubic structure; **(c)** a body-centered-cubic lattice?

11.48 What is the coordination number of: **(a)** Na^+ in the NaCl structure; **(b)** Zn^{2+} in the ZnS unit cell, Figure 11.43(*b*); **(c)** Ca^{2+} in the CaF_2 unit cell, Figure 11.43(*c*)?

11.49 Nickel metal crystallizes in a cubic close-packed structure (face-centered-cubic unit cell). **(a)** How many Ni atoms are in one unit cell? **(b)** What is the coordination number of Ni in this structure? **(c)** If each Ni atom has a radius of 1.24 Å, what is the length of a side of the unit cell? (The atom in the center of each face is in contact with the corner atoms, as shown in Figure 11.45.) **(d)** Calculate the density of nickel metal.

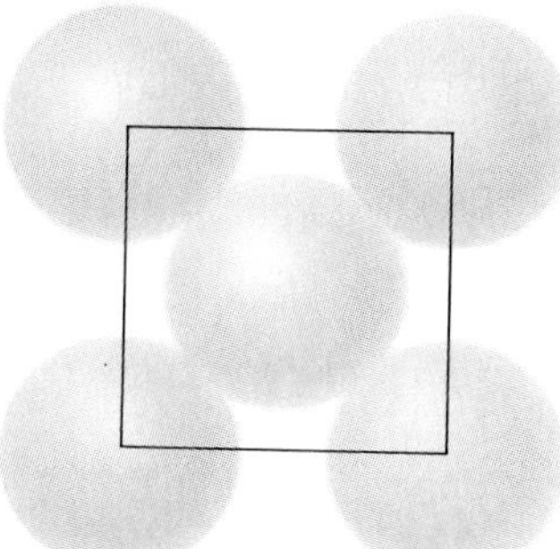

Figure 11.45

11.50 Copper crystallizes in a face-centered-cubic unit cell whose edge length is 3.61 Å. The atom in the center of each face is in contact with the corner atoms, as shown in Figure 11.45. **(a)** Calculate the atomic radius of copper. **(b)** Calculate the density of copper metal.

11.51 Potassium fluoride has the NaCl type of crystal structure (Figure 11.28). The density of KF at 25°C is 2.468 g/cm^3. Calculate the dimensions of the KF unit cell.

11.52 The unit cell of aluminum metal is cubic with an edge length of 4.05 Å. Determine the type of unit cell (primitive, face-centered, or body-centered) if the metal has a density of 2.70 g/cm^3.

11.53 An element crystallizes in a body-centered-cubic lattice. The edge of the unit cell is 2.86 Å, and the density of the crystal is 7.92 g/cm^3. Calculate the atomic weight of the element.

11.54 KCl has the same structure as NaCl. The length of the unit cell is 628 pm. The density of KCl is 1.984 g/cm^3, and its formula mass is 74.55. Using this information, calculate Avogadro's number.

11.55 Describe what is meant by *constructive interference* and *destructive interference* and explain how these terms apply to the diffraction of X rays by a crystal.

11.56 From which of the following materials would you expect to obtain well-defined X-ray diffraction patterns: **(a)** a sugar crystal; **(b)** KBr; **(c)** liquid water; **(d)** pure iron; **(e)** ice; **(f)** a section of a rubber stopper.

11.57 In the diffraction of a crystal using X rays with wavelength of 1.54 Å, a first-order ($n = 1$) reflection was obtained at an angle of 12.5°. What is the distance between the planes of atoms that cause this diffraction?

11.58 The first-order ($n = 1$) diffraction of X rays from crystal planes separated by 2.81 Å occurs at 11.8°. **(a)** What is the wavelength of the X rays? **(b)** Calculate the angle at which the second-order ($n = 2$) diffraction will appear.

Bonding in Solids

11.59 What kinds of attractive forces exist between particles in: **(a)** molecular crystals; **(b)** covalent-network crystals; **(c)** ionic crystals; **(d)** metallic crystals?

11.60 Indicate the type of crystal (molecular, metallic, covalent-network, or ionic) each of the following would form upon solidification: **(a)** HBr; **(b)** Ar; **(c)** Mn; **(d)** $Co(NO_3)_2$; **(e)** C; **(f)** paradichlorobenzene (moth balls).

11.61 Name two substances that form covalent-network solids.

11.62 Covalent bonding occurs in both molecular and covalent-network solids. Why do these two kinds of solids differ so greatly in their hardnesses and melting points?

11.63 For each of the following pairs of substances, predict which will have the higher melting point and indicate why: **(a)** KBr, Br_2; **(b)** SiO_2, CO_2; **(c)** Se, CO; **(d)** NaF, MgF_2.

11.64 For each of the following compounds, predict which will have the higher melting point and indicate why: **(a)** C_6Cl_6, C_6H_6; **(b)** HF, HCl; **(c)** KO_2, SiO_2; **(d)** Ar, Xe.

11.65 You are given a white substance that sublimes at 3000°C; the solid is a nonconductor of electricity and is insoluble in water. Which type of solid (Table 11.6) might this substance be?

11.66 A white substance melts with some decomposition at 730°C. As a solid it is a nonconductor of electricity, but it dissolves in water to form a conducting solution. Which type of solid (Table 11.6) might the substance be?

11.67 Using the cation-to-anion-radius ratio, predict the coordination number of the cation in each of the following compounds: **(a)** CaO; **(b)** MgO; **(c)** BeO. The ionic radii are Ca^{2+}, 1.14 Å; Mg^{2+}, 0.86 Å; Be^{2+}, 0.41 Å; O^{2-}, 1.26 Å.

11.68 Using the cation-to-anion-radius ratio, predict the coordination number of the cation in each of the following compounds: **(a)** CsF; **(b)** KF; **(c)** LiF. The ionic radii are Cs^+, 1.88 Å; K^+, 1.52 Å; Li^+, 0.90 Å; F^-, 1.19 Å.

Additional Exercises

11.69 What evidence can you cite for the existence of intermolecular forces between N_2 molecules?

11.70 Using the thermodynamic data listed in Appendix C, calculate ΔH for the following processes at 25°C:

$$Br_2(l) \longrightarrow Br_2(g) \qquad Br_2(g) \longrightarrow 2Br(g)$$

Discuss the relative magnitude of these enthalpy changes in terms of the forces involved in each case.

11.71 What kinds of intermolecular forces exist between pairs of the given molecules: **(a)** H_2; **(b)** CH_3OH; **(c)** SO_2? Which substance is likely to have the higher boiling point? Explain.

11.72 **(a)** Which of the following substances can exhibit dipole-dipole attractions between its molecules: CO_2, SO_2, H_2, IF, HBr, CCl_4? **(b)** Which of the following substances exhibit hydrogen bonding in their liquid and solid states: CH_3NH_2, CH_3F, PH_3, HCOOH?

11.73 In dichloromethane, CH_2Cl_2 ($\mu = 1.60$ D), the dispersion force contribution to the intermolecular attractive forces is about five times larger than the dipole-dipole contribution. Would you expect the relative importance of the two kinds of intermolecular attractive forces to differ **(a)** in dibromomethane ($\mu = 1.43$ D); **(b)** in difluoromethane ($\mu = 1.93$ D)? Explain.

11.74 What measurable physical property of a liquid relates to each of the following: **(a)** its ability to flow; **(b)** the temperature at which its vapor pressure equals the pressure on the surface of the liquid; **(c)** its tendency to bead up on a surface for which it exhibits no appreciable adhesive forces; **(d)** the amount of heat that must be added to vaporize it.

11.75 Which of the following chlorofluorocarbons would you expect to have the highest normal boiling point: CCl_3F (Freon 11), CCl_2F_2 (Freon 12), $CHClF_2$ (Freon 22), or CCl_2FCClF_2 (Freon 113)? Explain.

11.76 Trimethylamine, $(CH_3)_3N$, and propylamine, $CH_3CH_2CH_2NH_2$, have fishy, ammonialike odors. Explain why propylamine has a lower vapor pressure than trimethylamine.

11.77 What is the main intermolecular force that must be overcome to: **(a)** boil methanol, CH_3OH; **(b)** sublime solid CO_2; **(c)** melt solid $CaSO_4$?

11.78 Ethylene glycol, the major component of antifreeze, is a slightly viscous liquid that is not very volatile at room temperature and boils at 198°C. Its chemical formula is $CH_2(OH)CH_2(OH)$. Pentane, C_5H_{12}, which has about the same molecular weight, is a nonviscous liquid that is highly volatile at room temperature and whose boiling point is 36°C. Explain the differences in the physical properties of the two substances.

11.79 Liquid butane, C_4H_{10}, is stored in cylinders to be used as a fuel. Suppose 3.00 L of butane at 17°C and 735 mm Hg is removed from a cylinder. How much heat must be added to vaporize this much butane if its heat of vaporization is 21.3 kJ/mol?

11.80 What is meant by *dynamic equilibrium?* When ice is in equilibrium with water, what processes are occurring at the same rate?

11.81 The percent relative humidity of air equals 100 times the partial pressure of the water vapor in the air divided by the vapor pressure of water. If the partial pressure of water in air at 18°C is 5.50 mm Hg, use the data in Appendix B to calculate the percent relative humidity of the air.

11.82 What mass of water vapor can you expect to find in a bathroom measuring 4.0 m × 4.0 m × 3.0 m if water has been left in a bathtub at 40°C? See Appendix B.

11.83 A chlorofluorocarbon called dichlorodifluoromethane, CCl_2F_2, has a normal boiling point of −29.8°C, a critical temperature of 112°C, and a critical pressure of 39.6 atm. If CCl_2F_2 gas is compressed to 20.0 atm at a temperature of 19°C, will the substance liquefy? (Hint: Sketch the vapor pressure curve.)

11.84 The triple point of benzene is 5°C, 21 mm Hg. The density of solid benzene is 1.005 g/cm^3, whereas that of the liquid is 0.894 g/cm^3. The normal boiling point of benzene is 80°C; its critical point is 289°C, 48 atm. Sketch the phase diagram.

11.85 Test the applicability of the Clausius-Clapeyron equation (Equation 11.1), using the following vapor pressure-versus-temperature data for mercury:

Temperature (°C)	Vapor pressure of Hg (mm Hg)
50.0	0.01267
60.0	0.02524
70.0	0.04825
80.0	0.0880
90.0	0.1582

By plotting these data, determine whether the Clausius-Clapeyron equation is obeyed. If so, use the slope of the line to calculate the heat of vaporization of mercury.

[11.86] We might guess that the Clausius-Clapeyron equation (Equation 11.1) would be applicable also to the vapor pressure data for a solid. Use this equation to estimate the heat of sublimation of ice from the following data:

Temperature (°C)	Vapor pressure (mm Hg)
−20.0	0.640
−16.0	1.132
−12.0	1.632
−8.0	2.326
−4.0	3.280
0.0	4.579

[11.87] The critical temperature and pressure of Freon 13, $CClF_3$, are 29°C and 39 atm, respectively. For methyl chloride, $CClH_3$, the corresponding values are 143°C and 66 atm. What do these values tell us about the relative intermolecular attractive forces between molecules in the two substances? Are the results surprising? If so, why?

[11.88] Chromium crystallizes in a body-centered-cubic unit cell whose edge length is 2.884 Å. If the atoms touch along the body diagonal of the unit cell, calculate the atomic radius of a Cr atom.

[11.89] It is widely recognized that impurities in solids tend to concentrate at sites of dislocations. On the basis of Figure 11.38, explain why.

[11.90] Assume that 3.000 g of H_2O is introduced into an evacuated flask whose volume is 1.000 L. If the temperature of the water and flask is 30.0°C, what mass of water will evaporate? The vapor pressure of water at 30.0°C is 31.82 mm Hg.

Modern Materials 12

Assembly of *Intelsat IV,* one of a series of six communications satellites built for a 114-nation consortium by Hughes Aircraft Company. A special portion of the mirror-coated outer surface of this spacecraft helps radiate to cold space the excess heat generated by the solar-powered electronic components within. (Photo by Gary Panton, Hughes Aircraft Company)

CONTENTS

Since the beginning of the modern era of chemistry in the nineteenth century, one of the important goals of chemical research has been the discovery and development of materials with useful properties. Chemists have invented both entirely new substances and the means for processing naturally occurring materials to form fibers, films, coatings, adhesives, and substances with special electrical, magnetic, or optical properties. Today, we are entering a new era in which advances in technology will depend more than ever upon the discovery and development of useful new materials. Here are some examples of how these materials will affect all aspects of our lives in the near future:

1. They will make possible electronic-display devices that are thin and light, for example, a television set that can be mounted on the wall like a picture.
2. They will serve as repositories of vast quantities of information. Already the storage of information in a tiny space is possible (Figure 12.1), although not yet ready for widespread use. In the future, this technology will develop as a practical means of accessing information.
3. The expected lifetimes of biological replacement parts such as hip and knee joints will increase from the present 10 yr or less to the lifetime of the recipient.
4. Solar-cell technology for energy conversion with practical efficiencies will lead to more widespread utilization of solar energy.
5. Catalytic converters that reduce automobile pollution more effectively will contribute to improvements in the environment at lower costs.

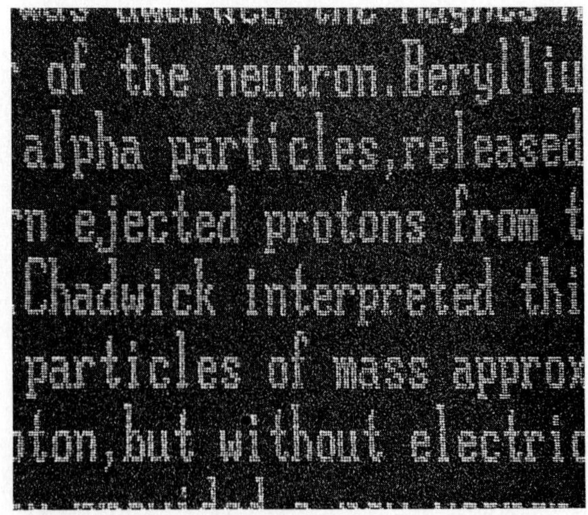

Figure 12.1 A portion of the *Encyclopedia Brittanica* written into an AlF_3 surface using an electron beam. The diameter of the beam is only 0.5 nm. At this scale, the entire contents of the *Encyclopedia Brittanica* could be contained within the area of the period at the end of this sentence. (Courtesy of Professor Colin Humphreys, University of Cambridge)

In this chapter, we will discuss the properties and applications of several kinds of materials. The aim is to show how we can understand many special physical or chemical properties in terms of the principles we have discussed in earlier chapters. We will see that the observable properties of materials are the result of atomic- and molecular-level structures and processes. At this stage of the text, we cannot cover all kinds of materials. Instead, we will examine four types: liquid crystals, polymers, ceramics, and thin films.

12.1 LIQUID CRYSTALS

In 1888, Frederick Reinitzer, an Austrian botanist, discovered that an organic compound he was studying, called cholesteryl benzoate, has interesting and unusual properties. When heated, the substance melts at 145°C to form a milky liquid. At 179°C, the milky liquid suddenly becomes clear. When the substance is cooled, the reverse processes occur: The clear liquid turns milky at 179°C, as shown in Figure 12.2, and the milky liquid solidifies at 145°C. Reinitzer's work represents the first systematic report of what we now call **liquid-crystal** behavior.

As they melt, many substances pass through a temperature region in which they possess properties intermediate between those of the solid and liquid phases. The region in which they exhibit these properties is marked by sharp transition temperatures, as in Reinitzer's example. The intermediate phase or phases are described as liquid-crystalline, in which the molecules retain a partial ordering, as distinct from the random, or *isotropic*, character of the ordinary liquid phase. Because of this partial ordering, a liquid-crystal phase may be

Figure 12.2 (*a*) Molten cholesteryl benzoate at a temperature above 179°C. In this temperature region, the substance is a clear, isotropic liquid. Note that the printing on the surface of the beaker in back of the sample test tube is readable. (*b*) Cholesteryl benzoate at a temperature between 179°C and 145°C, the melting point of cholesteryl benzoate. In this temperature interval, cholesteryl benzoate exhibits a milky liquid-crystalline phase. © Richard Megna/Fundamental Photographs)

(*a*)

(*b*)

very viscous. The liquid-crystalline phase possesses properties, not seen in either the solid or liquid phase, that form the basis of useful applications in computer, calculator, and wristwatch displays and in temperature measurements.

Structures of Liquid Crystals

Liquid-crystalline substances have characteristic molecular shapes and structures. The molecular structures of some organic liquid-crystalline substances are shown in Figure 12.3. You need not concern yourself with the details of

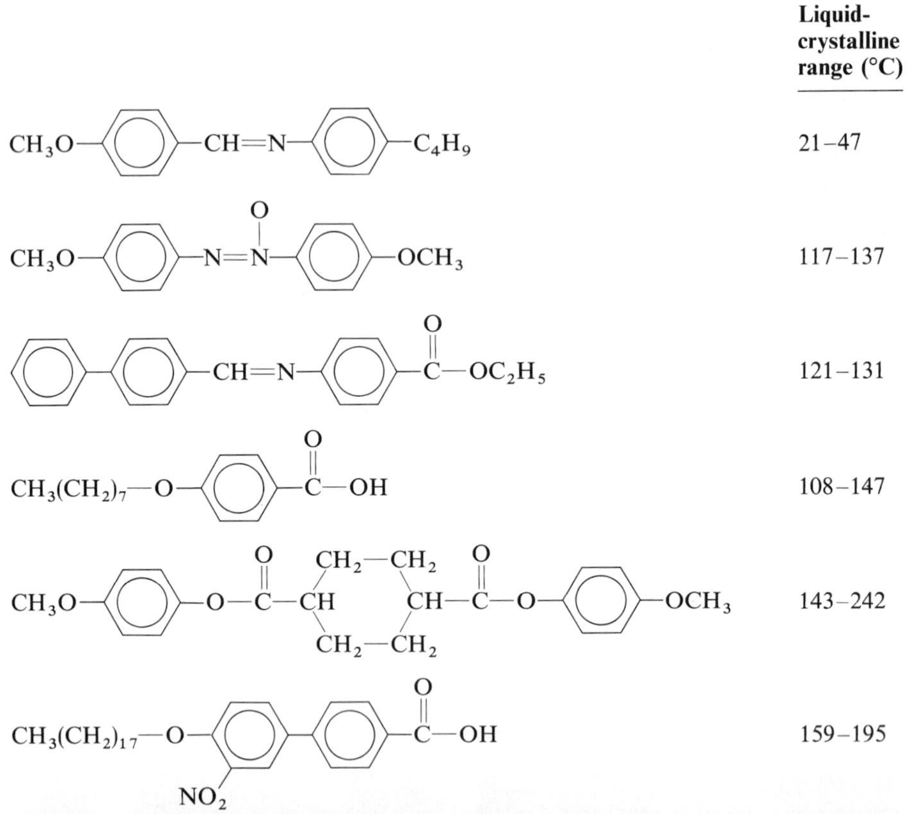

Figure 12.3 Structures of some typical liquid crystals.

Structure	Liquid-crystalline range (°C)
$CH_3O-C_6H_4-CH{=}N-C_6H_4-C_4H_9$	21–47
$CH_3O-C_6H_4-N{=}N(O)-C_6H_4-OCH_3$	117–137
$C_6H_5-C_6H_4-CH{=}N-C_6H_4-C(=O)-OC_2H_5$	121–131
$CH_3(CH_2)_7-O-C_6H_4-C(=O)-OH$	108–147
$CH_3O-C_6H_4-O-C(=O)-CH(CH_2-CH_2)_2CH-C(=O)-O-C_6H_4-OCH_3$	143–242
$CH_3(CH_2)_{17}-O-C_6H_3(NO_2)-C_6H_4-C(=O)-OH$	159–195

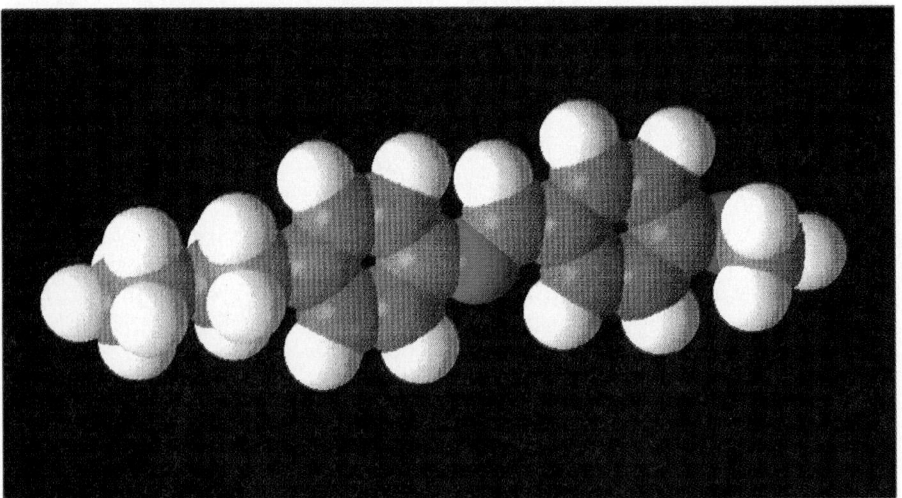

Figure 12.4 Three-dimensional structure of $CH_3OC_6H_4CH{=}NC_6H_4C_4H_9$, which is liquid-crystalline in the range 21°C to 47°C (the first compound listed in Figure 12.3).

(a)

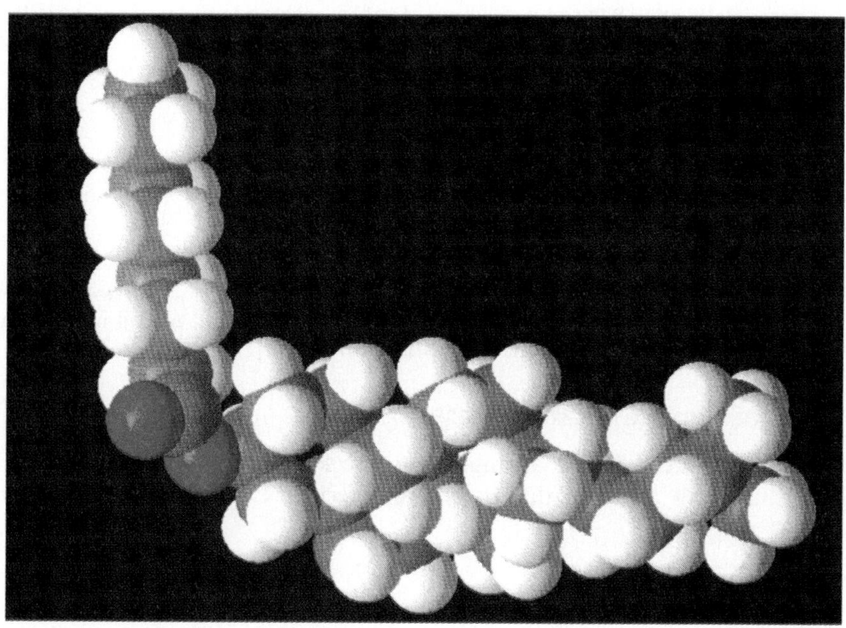

(b)

Figure 12.5 (*a*) Molecular structure of cholesteryl octanoate. Note that the rings in this molecule are not aromatic. Each corner of the rings has a carbon atom and as many attached hydrogen atoms or other bonds as needed to satisfy the carbon atom valency. (*b*) A three-dimensional model of cholesteryl octanoate.

these formulas; the main point to notice is that the molecules generally tend to have a long, rodlike shape and are somewhat rigid. To make this more evident, Figure 12.4 shows the three-dimensional structure of the first compound on the list in Figure 12.3. The characteristic rodlike shape allows for intermolecular interactions that maintain a parallel ordering in the liquid phase. Figure 12.5 shows the molecular and three-dimensional structures of cholesteryl octanoate, a compound similar to the one Reinitzer studied. Even though this is a complex molecule, you can see that it has a rather long, rigid structure that permits alignment and ordering of the molecules with respect to one another.

SAMPLE EXERCISE 12.1

Which of the following substances is most likely to exhibit liquid-crystal behavior?

$$\begin{array}{c} CH_3 \\ | \\ CH_3{-}CH_2{-}CH{-}CH_2{-}CH_3 \\ | \\ CH_3 \end{array}$$

(i)

$$CH_3CH_2-C_6H_4-N{=}N-C_6H_4-\overset{\displaystyle O}{\overset{\|}{C}}-OCH_3$$

(ii)

$$C_6H_5-CH_2-\overset{\displaystyle O}{\overset{\|}{C}}-O^-Na^+$$

(iii)

Solution: Molecule (i) is not likely to be liquid-crystalline because it does not have a long, axial structure. Molecule (iii) is ionic; the generally high melting points of ionic materials (Section 8.2) and the absence of a characteristic long axis make it unlikely that this substance will exhibit liquid-crystalline behavior. Molecule (ii) possesses the characteristic long axis and the kinds of functional groups that are often seen in liquid crystals (Figure 12.3).

PRACTICE EXERCISE

Suggest a reason why the following molecule, *n*-decane, does not exhibit liquid-crystalline behavior:

$$CH_3CH_2CH_2CH_2CH_2CH_2CH_2CH_2CH_2CH_3$$

Answer: Because rotation can occur around carbon-carbon single bonds, the simple carbon-chain molecules are too flexible; the molecules tend to coil in random ways and thus appear to be round rather than rodlike

Types of Liquid-Crystalline Phases

Many different ordering arrangements are known for liquid crystals. Figure 12.6 illustrates a few of the many characteristic types. The simplest is called

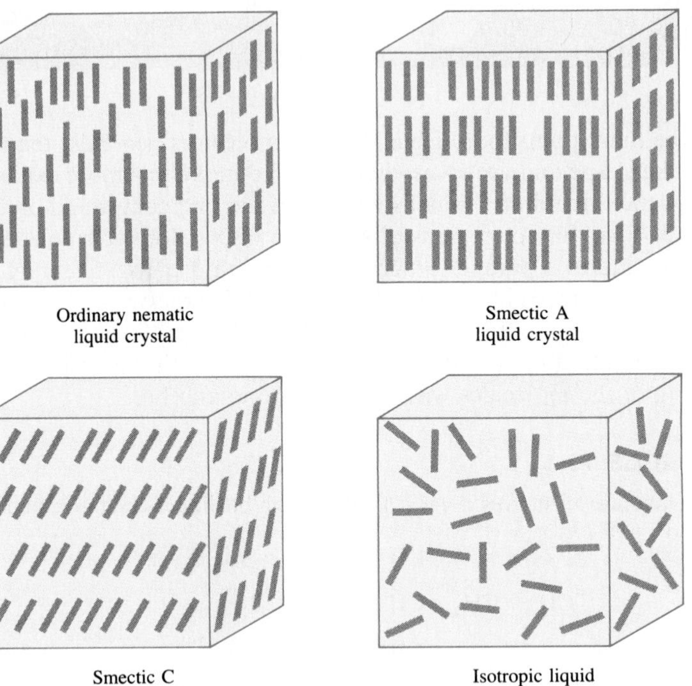

Figure 12.6 Ordering in liquid-crystalline phases, as compared with an isotropic (non-liquid-crystalline) liquid.

nematic liquid crystal. In the nematic phase, the molecules are aligned along their long axes, but there is no ordering with respect to the ends of the molecules. It is as if you picked up a handful of pencils without attempting to line up the ends. The **smectic liquid-crystalline** phases exhibit additional ordering of the molecules. In the smectic A phase, the molecules are arranged in sheets, with their long axes parallel and their ends aligned as well. Other smectic phases display different types of alignments. For example, in the smectic C phase shown in Figure 12.6, the molecules are aligned with their long axes tilted with respect to a line perpendicular to the plane in which the molecules are stacked. In these liquid-crystalline phases, the ordering that persists above the melting temperature of the solid is due to intermolecular forces, mainly London dispersion forces, and in some cases dipole-dipole interactions. The attractive interactions are optimized when the molecules align with their long axes parallel.

The **cholesteric liquid-crystalline** phase, characteristic of derivatives of the cholesterol molecule, is composed of layers or sheets of nematic liquid-crystal material. Cholesteryl benzoate, the substance Reinitzer studied, belongs to this class. In a typical cholesteric liquid crystal, shown in Figure 12.7, the molecules in each layer are ordered with respect to the other molecules in that layer and

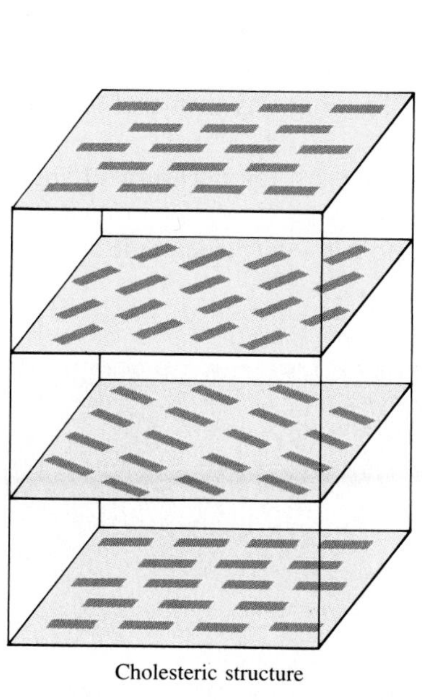

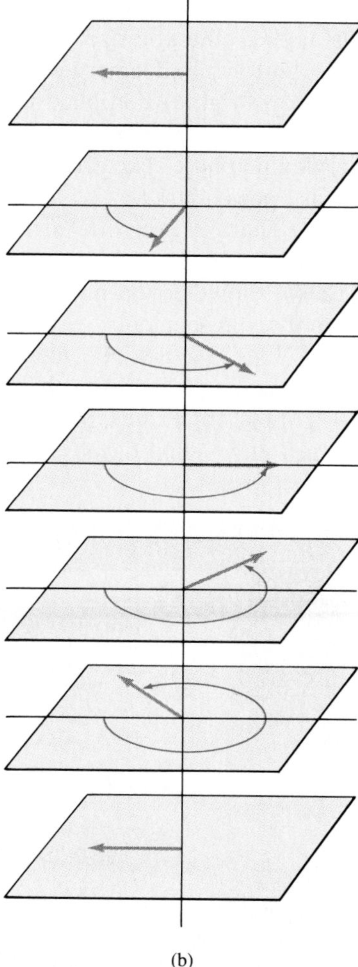

Figure 12.7 (*a*) Ordering in a cholesteric liquid crystal. The molecules in successive layers are oriented at a characteristic angle with respect to those in adjacent layers, to avoid unfavorable interactions. The result is a screwlike axis, as shown in (*b*).

with respect to the layers above and below. The molecules in a given layer are twisted with respect to those in the adjacent layers to avoid a direct repulsive interaction between groups of atoms that protrude out of the plane of the molecule.

Applications of Liquid Crystals

The attractive forces that cause the alignments of molecules in liquid crystals are not large in comparison with the energies of chemical bonds. Nevertheless, these relatively weak interactions lead to physical properties of considerable technological importance. Once viewed as chemical curiosities, liquid crystals are now widely used in electrically controlled liquid-crystal-display (LCD) devices in watches, calculators, and computer screens, as illustrated in Figure 12.8. These applications arise from the ability of an applied electrical field to cause a change in the orientation of liquid-crystal molecules and thus affect the optical properties of a layer of liquid-crystalline material.

Liquid-crystal devices come in a variety of designs, but the structure shown in Figure 12.9 is typical. The liquid-crystalline material is placed between two glass plates coated with transparent, electrically conducting material, as shown in Figure 12.9(*a*). In a typical watch or calculator display, the light that strikes the display surface first passes through a thin sheet called a *polarizer*, which passes only light rays that lie along a certain direction. When no voltage is applied, the polarized light passes through the liquid crystal, is reflected at the bottom, and emerges again through the cover plate. Thus, the surface appears bright. An applied voltage changes the orientation of the liquid-crystal molecules. As a result, the polarized light does not pass through the liquid-crytalline phase. The areas of the display to which the voltage has been applied thus appear dark.

The top electrode surface is divided into tiny areas. Applying a voltage to combinations of these areas forms letters, numerals, or other figures. Figure 12.9(*b*) shows how a numeric display capable of forming any numeral is made up of seven segments.

Liquid-crystalline materials that change color as temperature changes measure temperatures in situations where conventional methods are not feasible (Figure 12.10). For example, they can detect hot spots in microelectronic circuits, which may signal flaws.

Figure 12.8 A laptop computer with a liquid-crystal display (LCD) panel. (Apple Computer, Inc.)

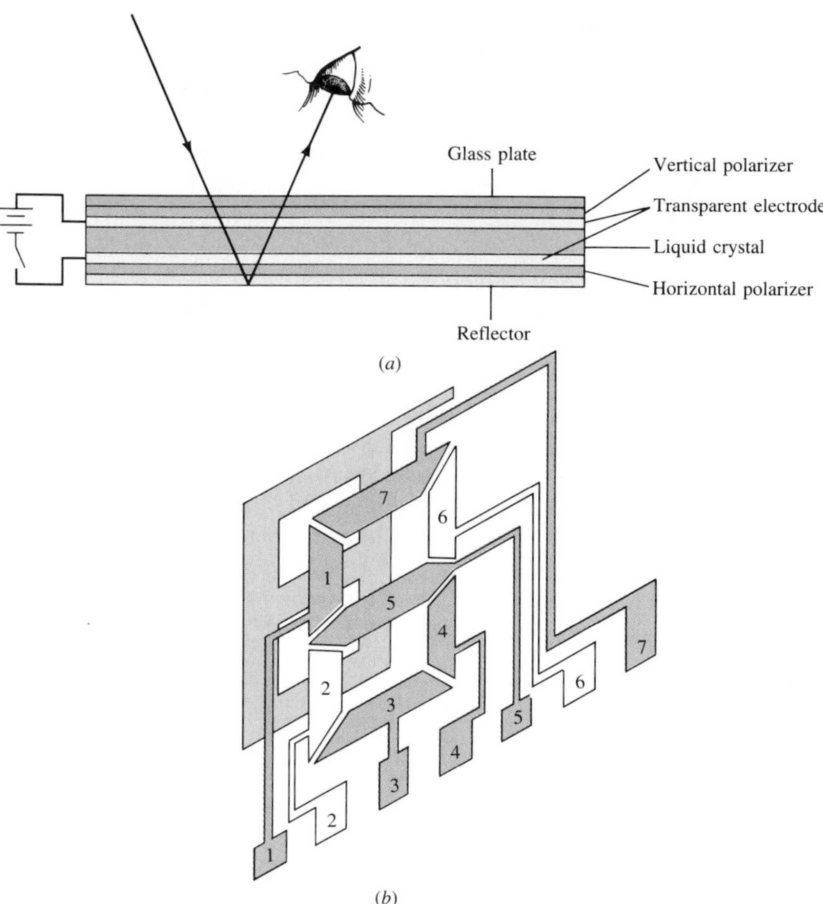

Figure 12.9 (*a*) Cross-section of a liquid-crystal display device. (*b*) Segmenting of the area of a transparent electrode into seven areas to form numerals. By applying voltage to the appropriate segments, we can form any numeral. For example, the numeral 5 would be formed by applying voltage to segments 7, 1, 5, 4, and 3.

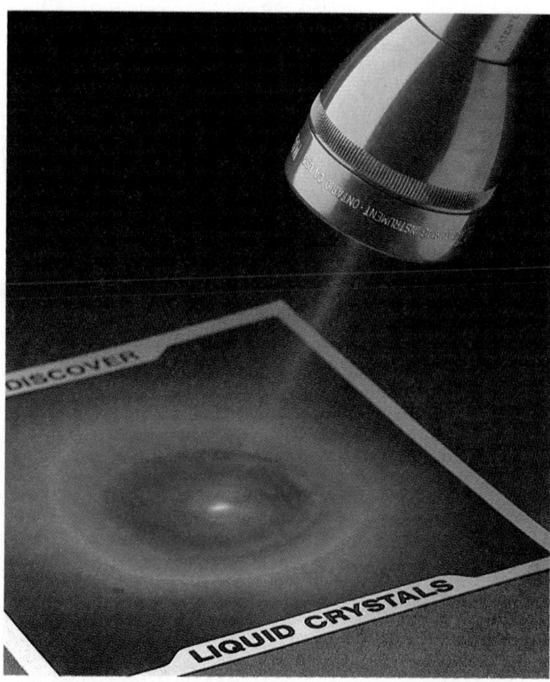

Figure 12.10 Color change in a cholesteric liquid-crystalline material as a function of temperature. (Richard Megna/ Fundamental Photographs)

12.2 POLYMERS

Our discussions of chemistry to this point have focused primarily on molecules of fairly low molecular mass. However, in nature we find many substances of very high molecular mass, running into millions of amu. Starch and cellulose abound in plants; proteins and nucleic acids are found in both plants and animals. In 1827, Jons Jakob Berzelius coined the word **polymer** (from the Greek *polys*, "many," and *meros*, "parts") to denote molecular substances of high molecular mass formed by the polymerization (joining together) of **monomers**, molecules with low molecular masses. Starch and cellulose are formed by polymerization of the monomer glucose, the structure of which is shown in Figure 12.11. Starch and cellulose differ only in the different spatial arrangements of the linkages joining the glucose molecules.

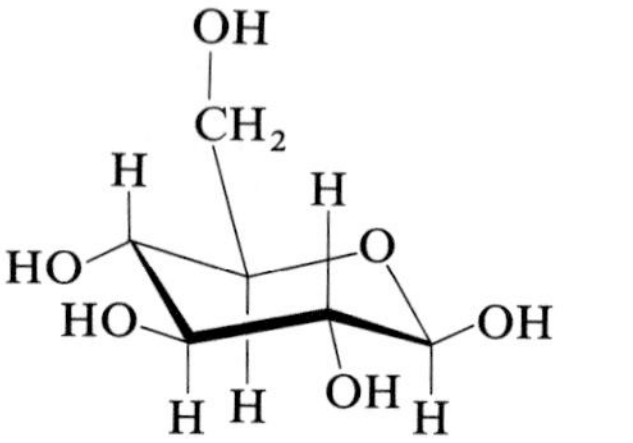

Figure 12.11 The structure of glucose, the monomer from which starch and cellulose are formed.

For a long time, humans made use of naturally occurring polymers to form useful materials. The spinning of wool, the tanning of leather, and the manufacture of natural rubber are all examples of processing natural polymers. During the past 50 years or so, chemists have learned to form synthetic polymers by polymerizing monomers through controlled chemical reactions.

The simplest example of a polymerization reaction is the formation of **polyethylene** from ethylene molecules. In the polymerization reaction, the double bond in each ethylene molecule "opens up" to form new single carbon-carbon bonds with two other ethylene molecules:

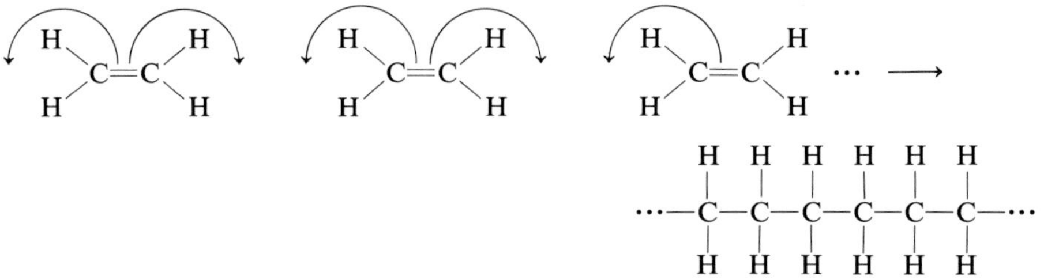

Polymerization that occurs through such coupling of monomers with one another, with no other products formed in the reaction, is termed **addition polymerization**.

We can write the polymerization reaction this way:

$$nCH_2{=}CH_2 \longrightarrow \left[-\overset{\displaystyle H}{\underset{\displaystyle H}{\overset{|}{\underset{|}{C}}}} - \overset{\displaystyle H}{\underset{\displaystyle H}{\overset{|}{\underset{|}{C}}}} - \right]_n \qquad [12.1]$$

Here the letter n is the large number—ranging from hundreds to many thousands—of monomer molecules (ethylene in this case) that react to form one large polymer molecule. Within the polymer, a repeat unit (the unit in brackets) appears along the entire chain. The ends of the chain are capped by carbon-hydrogen bonds or by some other bond that satisfies the valency requirement of the end carbons.

Polyethylene is a very important material; about 10 million tons is produced in the United States each year. Although its formula is the most simple of all the organic polymers, it is not readily formed from ethylene. Only after many years of research were the right conditions identified for manufacturing

Table 12.1 Polymers of Commercial Importance

Polymer	Structure	Uses	Quantity (tons/yr)
Polyethylene	$\left(CH_2-CH_2\right)_n$	Films, packaging, bottles	10,000,000
Polyisoprene	$\left[CH_2-C(CH_3)=CH-CH_2\right]_n$	Sporting goods, caulking compounds, tires	75,000
Polystyrene	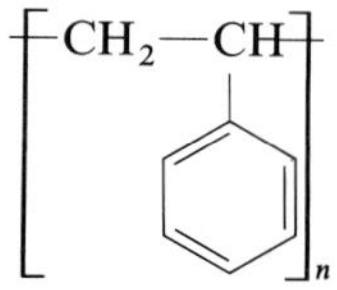 $\left[CH_2-CH(C_6H_5)\right]_n$	Packaging, disposable food containers, insulation	1,900,000
Polyurethane	$\left[C(=O)-NH-R-NH-C(=O)-O-R'-O\right]_n$ R, R′ = for example, $-CH_2-CH_2-$	"Foam" furniture stuffing, spray-on insulation, automotive parts, footwear, water-protective coatings	500,000
Polyvinyl chloride	$\left[CH_2-CH(Cl)\right]_n$	Pipe fittings, phonograph records, clear film for meat packaging	3,000,000
Polyethylene terephthalate (a polyester)	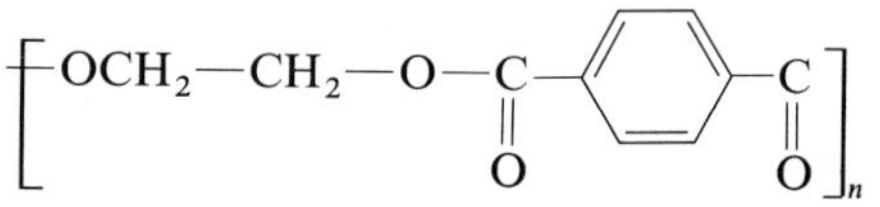 $\left[OCH_2-CH_2-O-C(=O)-C_6H_4-C(=O)\right]_n$	Tire cord, magnetic tape, apparel, soft-drink bottles	2,000,000
Nylon 6,6	$\left[NH\left(CH_2\right)_6NH-C(=O)-(CH_2)_4-C(=O)\right]_n$	Home furnishings, apparel, carpet fibers, fishing line	1,300,000 (all nylons)

a commercially useful product. Today, many different forms of polyethylene, varying widely in physical properties, are known. Polymers of other chemical compositions provide still greater variety in physical and chemical properties. Table 12.1 shows examples of some polymers used in forming structural materials, packaging, films, fibers, and other products.

Types of Polymers

Before we continue this discussion of polymers, we should clarify the meanings of some commonly used terms. The word **plastic** is generally applied to materials that can be formed into various shapes, usually by the application of heat and pressure. **Thermoplastic** materials can be reshaped. For example, plastic milk containers are made from polyethylene of high molecular weight. These containers can be melted down and the polymer recycled for some other use. In contrast, a **thermosetting plastic** is shaped through certain irreversible chemical processes and therefore cannot be reshaped readily. The term **elastomer** is applied to a material that exhibits rubbery or elastic behavior. When subjected to stretching or bending, it regains its original shape upon removal of the distorting force, provided that it has not been distorted beyond some elastic limit. Some polymers can also be formed into **fibers** that, like hair, are very long in relation to cross-sectional area and are not elastic.

Polymer Processing

Polymeric substances can be processed in several ways. The process used helps determine the characteristics of the final product. As an illustration, let's consider formation of fibers of nylon 6,6 (Table 12.1). In this case, polymerization occurs through **condensation polymerization**. This reaction involves the splitting out of water between a diamine (a compound that has six carbon atoms in its chain and an amino, $—NH_2$, group on each end) and adipic acid, which has six carbons in the molecule:

$$\underset{\text{Diamine}}{n\mathrm{H_2N{+}CH_2{+}_6{—}NH_2}} + \underset{\text{Adipic acid}}{n\mathrm{HO\overset{O}{\overset{\|}{C}}{+}CH_2{+}_4\overset{O}{\overset{\|}{C}}{—}OH}} \longrightarrow \underset{\text{Nylon 6,6}}{\left[\mathrm{NH(CH_2)_6NH{—}\overset{O}{\overset{\|}{C}}(CH_2)_4\overset{O}{\overset{\|}{C}}}\right]_n} \qquad [12.2]$$

The reaction occurs at each end of the diamine and the acid to form a long polymer chain, as indicated in Table 12.1. It is brought about by heating the acid and the diamine under pressure until the desired average chain length is reached. The pressure is then reduced, and the polymer is extruded (squeezed through a small hole) in spaghettilike strands. It can be cooled and cut into little chips for transport or conveyed directly to the next stage in manufacture.

To make a useful fiber, the polymer is melted and again extruded, after which it is drawn, or stretched. These operations orient the polymer molecules along the direction of the stretch, as illustrated in Figure 12.12. The fibers gain strength in the direction of their orientation.

In contrast, to make a rigid solid such as an auto panel or bumper, the reactants that form the thermoset polymer are injected into a mold in which the polymerization reaction occurs. This process is called *reaction-injection molding*.

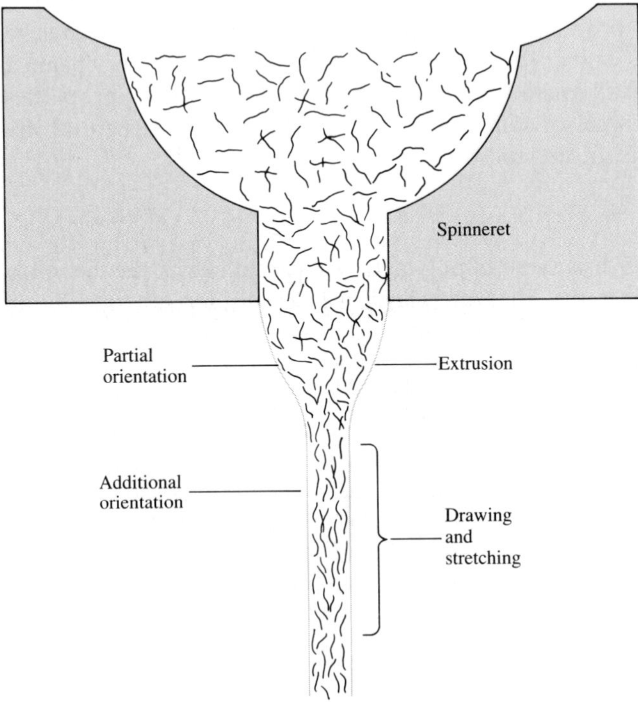

Figure 12.12 Extrusion and drawing of polymeric material to form a fiber. As the extruded material is drawn, the individual polymer molecules align along the direction in which the stretch occurs.

Structures and Physical Properties of Polymers

Polymers are amorphous materials. Rather than exhibiting well-defined crystalline phases with sharp melting points, they soften over a range of temperatures. Although we classify polymers as amorphous materials, they actually possess some amount of ordering, which we refer to as **crystallinity**. Interactions between the polymer molecules can arise from chemical bonds formed between two polymer chains or through intermolecular forces operating between the chains.

Thermoplastic materials consist of independent linear chains of large polymer molecules. The individual chains are flexible and can assume a variety of complex shapes, depending on the rotations that occur about chemicals bonds along the chains. The simple linear chains are often complicated by branching—that is, by side chains that extend off the main chain, like spur lines that branch from a main railway line. Because of intermolecular forces between adjacent chains, there may be regions in which the polymer chains assume a more-or-less ordered arrangement, with other regions remaining more disordered, as illustrated in Figure 12.13. The extent of such ordering depends on the particular chemical groups along the chain and on the average length of the chains.

As an example, we have seen that polyethylene can be a very simple polymer. The simple linear structure of polyethylene is conducive to intermolecular interactions that lead to crystallinity. However, the degree of crystallinity in polyethylene is strongly dependent on the average molecular mass. The so-called low-density polyethylene used in forming films and sheets has average molecular masses in the range of 10^4 amu and has substantial chain branching. High-density polyethylene, used to form bottles, drums, and pipes, has molecular masses in the range of 10^6 amu. These differing forms of polyethylene are illustrated in Figure 12.14. Table 12.2 shows how the properties of polyethylene vary as the degree of crystallinity increases.

The properties of polymeric materials can be extensively modified by adding substances with lower molecular mass. For example, substances may be added to provide protection against degradation of the material in sunlight. In other cases, **plasticizers** are added to reduce the extent of interactions between chains and thus to make the polymer more pliable. As an example, polyvinyl chloride (PVC) (Table 12.1) is a hard, rigid material of high molecular mass that is used to manufacture sewer pipes. When blended with a suitable substance of lower molecular mass, however, it forms a flexible polymer that can be used to make rain boots and doll parts. In some applications, the plasticizer in a plastic object may be lost over time because of evaporation. As this happens, the plastic loses its flexibility and becomes subject to cracking.

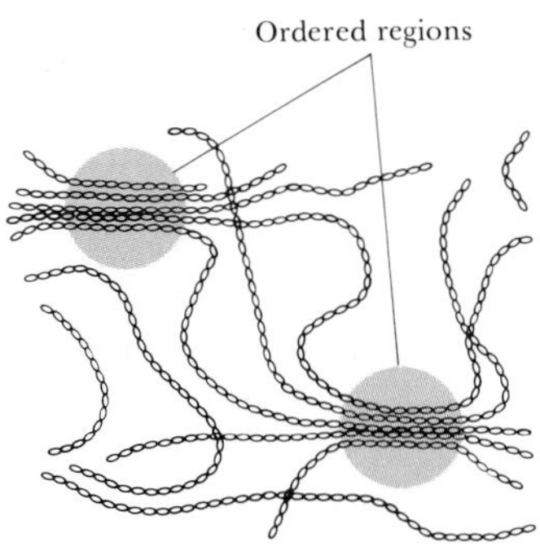

Figure 12.13 Interactions between polymer chains. In the circled regions, the forces that operate between adjacent polymer-chain segments lead to ordering analogous to the ordering in crystals.

Figure 12.14 (*a*) Schematic illustration of the structure of low-density polyethylene (LDPE) and a typical use of LDPE film to form food-storage bags. (*b*) Schematic illustration of the structure of high-density polyethylene (HDPE) and 1-gal containers formed from HDPE. (© Richard Megna/ Fundamental Photographs)

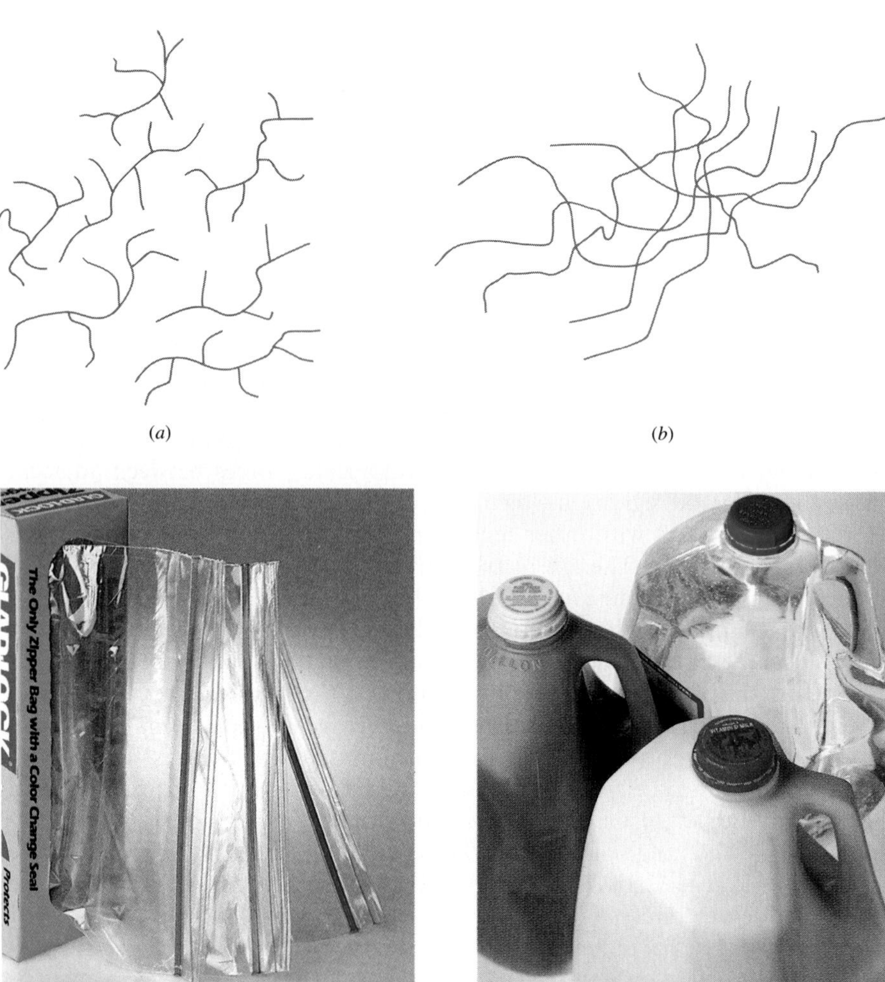

Table 12.2 Properties of Polyethylene as a Function of Crystallinity

	Degree of crystallinity				
	55%	**62%**	**70%**	**77%**	**85%**
Melting point (°C)	109	116	125	130	133
Density (g/cm^3)	0.92	0.93	0.94	0.95	0.96
Stiffness[a]	25	47	75	120	165
Yield stress[a]	1700	2500	3300	4200	5100

[a] These test results show that the mechanical strength of the polymer increases with increased crystallinity. The physical units for the stiffness test are psi × 10^{-3} (psi = pounds per square inch); those for the yield stress test are psi. Discussion of the exact meaning and significance of these tests is beyond the scope of this text.

Crosslinking of Polymers

Thermosetting polymeric materials have chemical bonds that crosslink the polymer chains, as illustrated schematically in Figure 12.15. The greater the density of crosslinks in a unit volume of the material, the more rigid the material. The best-known example of such crosslinking is the **vulcanization** of nat-

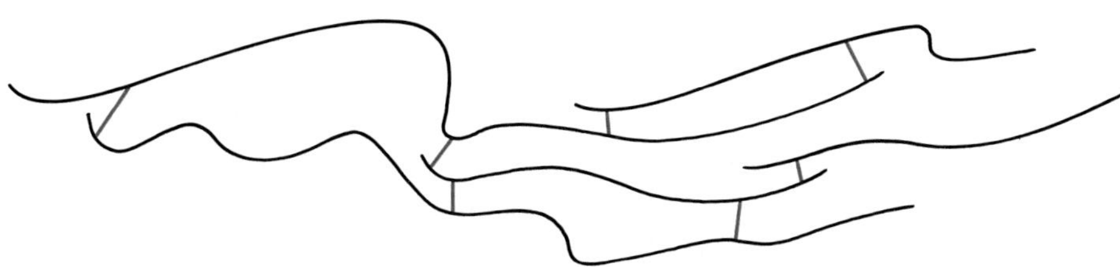

Figure 12.15 Crosslinking of polymer chains. The crosslinking groups constrain the relative motions of the polymer chains, making the material harder and less flexible.

ural rubber, discovered by Charles Goodyear in 1839. **Natural rubber** is formed from a liquid resin derived from the inner bark of the *Hevea brasiliensis* tree. Chemically, it is a polymer of isoprene, C_5H_8:

$$(n+2)\ \underset{\text{Isoprene}}{CH_2{=}C(CH_3){-}CH{=}CH_2} \longrightarrow$$

$$\underset{\text{Rubber}}{\sim\!\sim CH_2{-}C(CH_3){=}CH{-}CH_2{-}\left[CH_2{-}C(CH_3){=}CH{-}CH_2\right]_n{-}CH_2{-}C(CH_3){=}CH{-}CH_2\!\sim\!\sim} \qquad [12.3]$$

Because rotation about the carbon-carbon double bond does not readily occur, the orientation of the groups bound to the carbons is rigid. In naturally occurring rubber, the chain extensions are on the same side of the double bond, as shown in Equation 12.3. This form is called *cis*-polyisoprene; the prefix *cis*- is derived from a Latin phrase meaning "on this side."

Natural rubber is not a useful plastic because it is too soft and too chemically reactive. Goodyear accidentally discovered that adding sulfur to rubber and then heating the mixture makes the rubber harder and reduces its susceptibility to oxidation or other chemical attack. The sulfur changes rubber into a thermosetting polymer by crosslinking the polymer chains through reactions at some of the double bonds, as shown schematically in Figure 12.16. Crosslinking about 5 percent of the double bonds creates a flexible, resilient rubber. When the rubber is stretched, the crosslinks help prevent the chains from slipping, so that the rubber retains its elasticity.

SAMPLE EXERCISE 12.2

What mass of sulfur per gram of isoprene, C_5H_8, is required to establish a crosslink with every isoprene unit, as shown in Figure 12.16?

Solution: We see from the figure that each crosslink involves two sulfur atoms and two isoprene, C_5H_8, units. That is, the ratio of S to C_5H_8 is 1:1. Thus, we have

$$(1.0\ g\ C_5H_8)\left(\frac{1\ mol\ C_5H_8}{68.6\ g\ C_5H_8}\right)\left(\frac{1\ mol\ S}{1\ mol\ C_5H_8}\right)\left(\frac{32.1\ g\ S}{1\ mol\ S}\right) = 0.47\ g\ S$$

PRACTICE EXERCISE

In a particular vulcanization process, 6.7 g S was used per 100 g polyisoprene. Assuming that all the sulfur reacted to form crosslinks, what is the percentage of crosslinking in the product? ***Answer:*** 14 percent

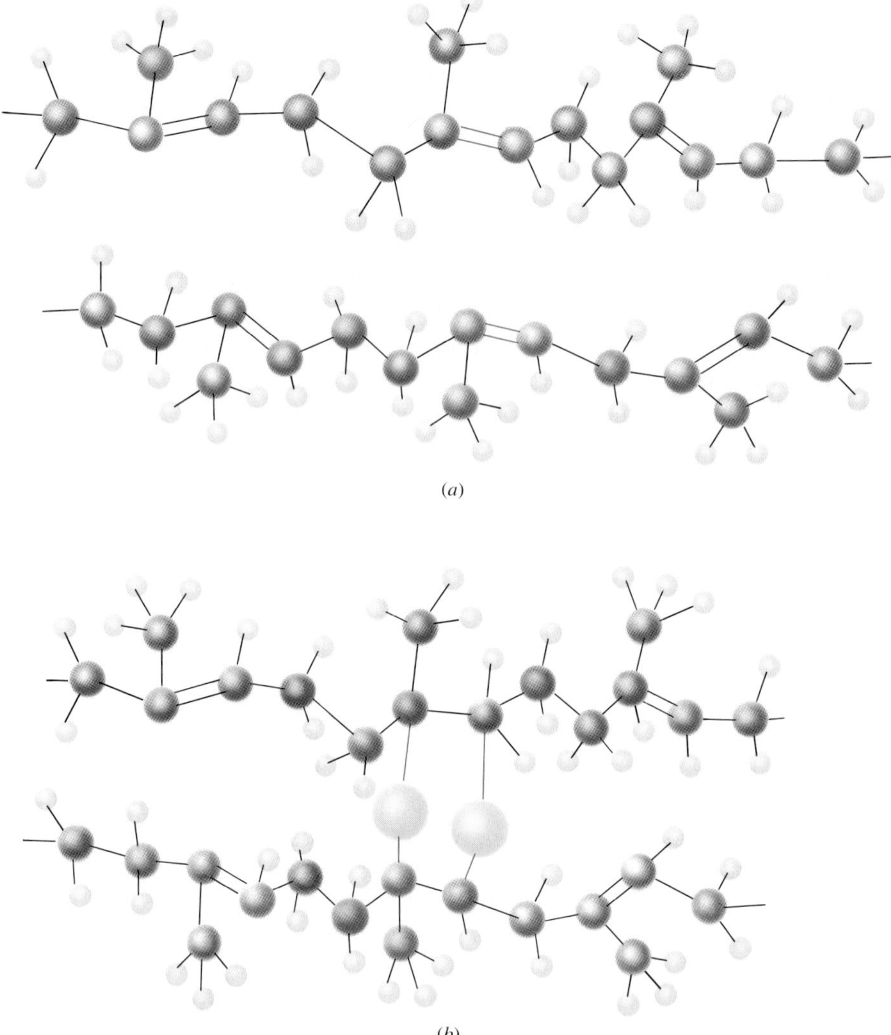

Figure 12.16 The structure of natural rubber is disclosed in (*a*). The carbon-carbon double bonds marked in red are opened as two sulfur atoms are added to form an interchain link. This crosslinking, shown in (*b*), is referred to as *vulcanization*.

The polymerization of melamine and formaldehyde produces a highly crosslinked molecular structure, illustrated in Figure 12.17. Although it is not obvious from this two-dimensional drawing, the crosslinking extends in three dimensions, creating a hard, rigid, chemically stable material. This class of thermosetting polymers is found in products such as dinnerware, coatings, and table tops (Formica).

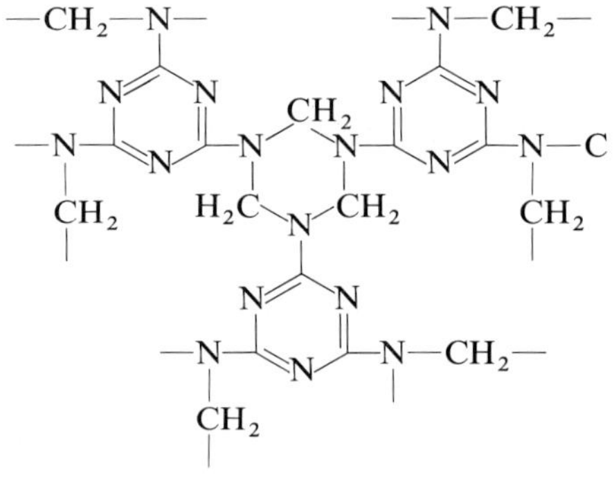

Figure 12.17 The structure of the melamine-formaldehyde polymer, a highly crosslinked material.

CHEMISTRY AT WORK: Kevlar: An Advanced Material

Polymers are used today in many ways that were entirely unanticipated at the time of their discovery. For example, nylon is not used merely as a fiber, which was its initial application. It is also used in the manufacture of bearings, insulators, fishing line, and tire cord. These extended applications for nylon and other polymers have stimulated a strong demand for new "super" fibers with the heat resistance of asbestos, the stiffness of glass, and strengths much greater than steel.

One new substance that has many of these properties is called Kevlar and has the following chemical structure:

$$\left[-\overset{\overset{\large O}{\|}}{C}-C_6H_4-\overset{\overset{\large O}{\|}}{C}-NH-C_6H_4-NH-\right]_n$$

The average molecular mass of each polymer chain is 10^5 amu. Kevlar owes its special properties to the way in which the polymer molecules interact with one another, as illustrated in Figure 12.18(*a*). Notice that the individual chains are hydrogen-bonded to adjacent chains. This hydrogen-bonding network causes the chains to align, forming a sheet structure. The individual polymer fibers contain these sheets arranged in a radial manner around the center, as depicted in Figure 12.18(*b*).

Kevlar has many exceptional properties due to the strong bonding in the individual polymer chains, the strong hydrogen-bonding network in the sheets, and the regular arrangement of the sheets in the fibers. Kevlar ropes have replaced steel ropes and cables in many applications, especially on off-shore oil-drilling platforms. For a given diameter, Kevlar ropes in seawater have 20 times the strength of steel. Because it has very good high-temperature stability, Kevlar is used in protective gloves and clothing worn by firefighters. It is also a major component in very strong and damage-resistant structures such as the hull of the racing craft shown in Figure 12.19.

Figure 12.19 The *KAAMA* offshore racer. The hull of this craft is reinforced with Kevlar. (E.I. Du Pont de Nemours & Co, Textile Fibers Department KEVLAR Special Products)

Hydrogen = bonded sheet

Fiber axis

(*a*)

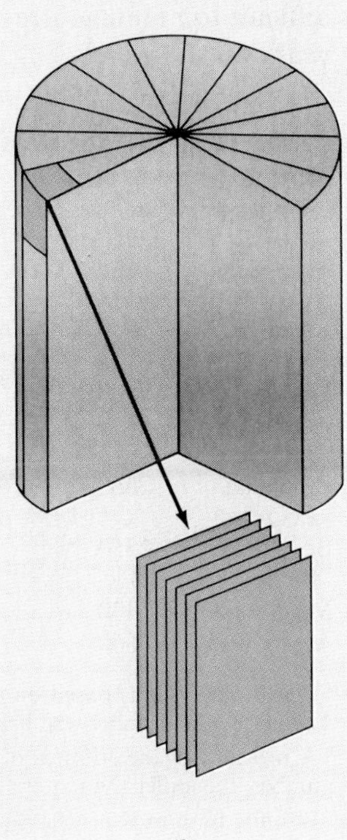

(*b*)

Figure 12.18 (*a*) Hydrogen bonding structure in sheets of Kevlar molecules. (*b*) Individual sheets of molecules, as shown in (*a*), are arranged in fiberlike sheets radiating from a center.

12.3 CERAMICS

Ceramics are inorganic, nonmetallic, solid materials. They can be crystalline or noncrystalline. Noncrystalline ceramics include glass and a few other materials with amorphous structures. Ceramics can possess a covalent network structure (Table 11.6), ionic bonding, or some combination of the two. They are normally hard and brittle and are stable to very high temperatures. Familiar examples of ceramic materials are pottery, china, cement, roof tiles, refractory bricks used in furnaces, and the insulators in spark plugs.

Ceramic materials come in a variety of chemical forms, including *silicates* (silica, SiO_2, with metal oxides), *oxides* (oxygen and metals), *carbides* (carbon and metals), *nitrides* (nitrogen and metals), and *aluminates* (alumina, Al_2O_3, with metal oxides). Although most ceramic materials contain metal ions, some do not. Table 12.3 lists a few ceramic materials and contrasts their properties with those of metals.

Engineering Ceramics

Ceramic materials that might be selected to replace other engineering materials, such as metal, wood, or plastics, are termed **engineering ceramics**. Ceramics are highly resistant to heat, corrosion, and wear, do not readily deform under stress, and are less dense than the metals used for high-temperature applications. Some ceramics used in aircraft, missiles, and spacecraft weigh only 40 percent as much as the metal component they replace (Figure 12.20). In addition, in some cases using the ceramic replacement means the United States does not have to import an expensive metal.

In spite of these many advantages, the use of ceramics as engineering materials has been limited because of their extremely brittle nature. Whereas a metal component might suffer a dent when struck, a ceramic part typically shatters into many pieces. A second problem is that ceramic components are difficult to manufacture free of defects. Attention has therefore focused in recent years on the processing of ceramic materials.

Table 12.3 Properties of Some Ceramic and Selected Nonceramic Materials

Material	Melting point (°C)	Density g/cm^3	Hardness (mohs)[a]	Modulus of elasticity[b]	Coefficient of thermal expansion[c]
Alumina, Al_2O_3	2050	3.8	9	34	8.1
Silicon carbide, SiC	2800	3.2	9	65	4.3
Zirconia, ZrO_2	2660	9.9	8	24	6.6
Beryllia, BeO	2550	3.0	9	40	10.4
Mild steel	1370	7.9	5	17	15
Aluminum	660	2.7	3	7	24

[a] The Mohs scale is based on the relative ability of a material to scratch another, softer material. The larger the number, the harder the material.

[b] A measure of the stiffness of a material when subjected to a load (MPa $\times 10^4$). The larger the number, the stiffer the material.

[c] In units of m/m/K ($\times 10^{-6}$). The larger the number, the greater the size change upon heating or cooling.

Figure 12.20 A variety of ceramic parts made of silicon nitride, Si_3N_4. These ceramic components could replace metal parts in engines and/or be used in other applications where high temperatures and wear are involved. (Photo courtesy GTE Laboratories, Incorporated)

Processing of Ceramics

Ceramic parts often develop random, undetectable microcracks and voids (hollow spaces) during processing. When the parts are used under stress, these defects concentrate the stress, ultimately causing the part to fail. To "toughen" a ceramic—that is, to increase its resistance to fracture—scientists frequently produce very pure uniform particles of the ceramic material that are less than a micrometer ($\frac{1}{1000}$ of a millimeter) in diameter. These are then sintered (heated at high temperature under pressure so that the individual particles bond together) to form the desired object.

The **sol-gel process** is an important method of forming extremely fine particles of uniform size. A typical sol-gel procedure begins with a metal alkoxide. An alkoxide contains organic groups bonded to a metal atom through oxygen atoms. Alkoxides are produced when the metal is reacted with an alcohol, which is an organic compound containing an OH group bonded to carbon. To illustrate this process, we will use titanium as the metal and ethyl alcohol, CH_3CH_2OH, as the alcohol:

$$Ti(s) + 4CH_3CH_2OH(l) \longrightarrow Ti(OCH_2CH_3)_4(s) + 2H_2(g) \quad [12.4]$$

The alkoxide product, titanium tetraethoxide, $Ti(OCH_2CH_3)_4$, is dissolved in an appropriate alcohol solvent. Water is then added, which reacts with the alkoxide to form metal —OH groups and to regenerate the ethyl alcohol:

$$Ti(OCH_2CH_3)_4(soln) + 4H_2O(l) \longrightarrow Ti(OH)_4(s) + 4CH_3CH_2OH(l) \quad [12.5]$$

The $Ti(OH)_4$ is present at this stage as a **sol**, a suspension of extremely small particles. The acidity or basicity of the sol is adjusted to cause water to be split out from between two of the Ti—OH bonds:

$$(HO)_3Ti—O—H(s) + H—O—Ti(OH)_3(s) \longrightarrow (HO)_3Ti—O—Ti(OH)_3(s) + H_2O(l) \quad [12.6]$$

This is an example of a very common reaction called a **condensation reaction**, which involves the splitting out of a small molecule, such as water, from between two reactants. In the above reaction, condensation also occurs at some of the other OH groups bonded to the central titanium atom, producing a three-dimensional network. The resultant material, called a **gel**, is a suspension of extremely small particles with the consistency of Jell-O. When this material is heated carefully at 200°C to 500°C, all the liquid is removed, and the gel is converted to a finely divided metal oxide powder with particles in the range of 0.003 to 0.1 μm in diameter. Figure 12.21 shows SiO_2 particles formed with remarkably uniform, spherical shapes by a precipitation process similar to the sol-gel process.

SAMPLE EXERCISE 12.3

Write balanced chemical equations showing **(a)** the formation of a metal alkoxide by reaction of aluminum with isopropyl alcohol, $(CH_3)_2CHOH$; **(b)** the reaction of dissolved aluminum isopropoxide with water to form a sol of aluminum hydroxide.

Solution: **(a)** Because the normal valence of aluminum is 3, we expect that 3 mol of the alcohol will be required per mole of Al:

$$2Al(s) + 2(CH_3)_2CHOH(l) \longrightarrow 2[(CH_3)_2CHO]_3Al(s) + 3H_2(g) \quad [12.7]$$

(b) The reaction of aluminum isopropoxide with water proceeds in the same manner as the reaction of titanium alkoxide, as shown in Equation 12.5. The solvent would in all likelihood be an alcohol, to ensure that aluminum isopropoxide is in solution at the start of the reaction:

$$[(CH_3)_2CHO]_3Al(soln) + 3H_2O(l) \longrightarrow Al(OH)_3(s) + 3(CH_3)_2CHOH(l) \quad [12.8]$$

PRACTICE EXERCISE

What is the molecular formula of the alkoxide formed from silicon and ethyl alcohol?
Answer: $Si(OCH_2CH_3)_4$

The sol-gel process is particularly useful in producing ceramic coatings and films. The gel is simply applied and then heated to form the ceramic product. It is more difficult to form a ceramic object with a complex three-dimensional shape. One possible approach is to add the gel to a casting and heat it to drive out the liquid component. However, when this is done there is considerable shrinkage, and it is difficult to prevent formation of void spaces and other imperfections. Another approach is to form a useful object from the ceramic powder. The finely divided material, possibly mixed with other powders, is compacted under pressure and then sintered at high temperature. The tempera-

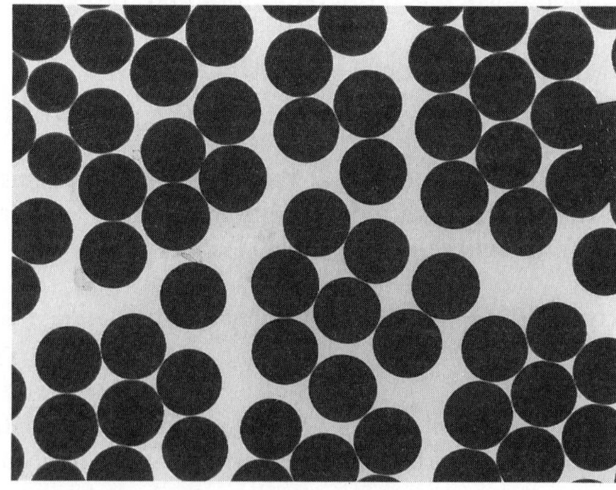

Figure 12.21 Uniformly sized spheres of amorphous silica, SiO_2, formed by precipitation from an alcohol solution of $Si(OCH_3)_4$ upon addition of water and ammonia. The average diameter of these spheres is 550 nm. (Professor C. Zukowki, University of Illinois, Urbana-Champaign)

tures required are about 1650°C for alumina, 1700°C for zirconium oxide, and 2050°C for silicon carbide. During sintering, the ceramic particles partially coalesce without actually melting (compare the sintering temperatures with the melting temperatures listed in Table 12.3).

Ceramic Composites

Ceramic objects are much tougher when they are formed from a complex mixture of two or more materials. Such a mixture is called a **composite**. The most effective composites are formed by addition of **ceramic fibers** to a ceramic material. Thus, the composite consists of a ceramic matrix containing embedded fibers of a ceramic material, which may or may not be of the same chemical composition as the matrix.

By definition, a fiber has a length at least 100 times its diameter. Fibers typically have great strength with respect to loads applied along the long axis. When they are embedded in a matrix, they strengthen it by resisting deformations that exert a stress on the fiber along its long axis.

The formation of ceramic fibers is illustrated by the case of silicon carbide, SiC, or carborundum. The first step in the production of SiC fibers is the synthesis of a polymer, polydimethylsilane:

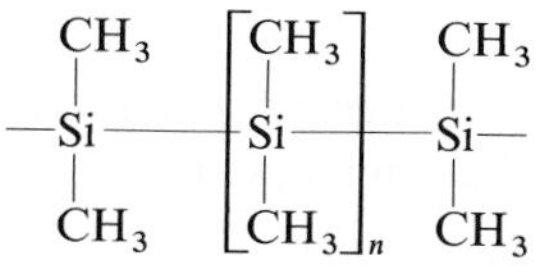

$$-\underset{CH_3}{\overset{CH_3}{|}}\!Si - \left[\underset{CH_3}{\overset{CH_3}{|}}\!Si\right]_n - \underset{CH_3}{\overset{CH_3}{|}}\!Si-$$

When this polymer is heated to about 400°C, it converts to a material that has alternating carbon and silicon atoms along the chain:

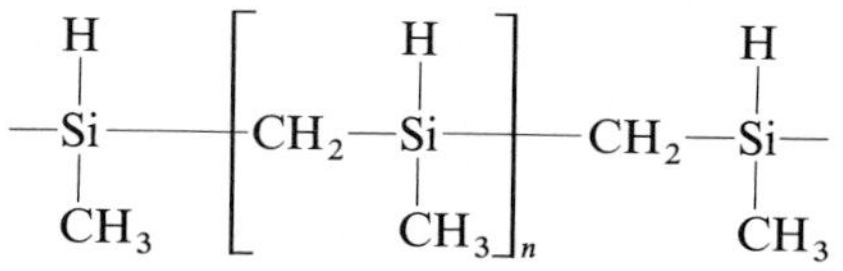

$$-\underset{CH_3}{\overset{H}{|}}\!Si - \left[CH_2-\underset{CH_3}{\overset{H}{|}}\!Si\right]_n - CH_2-\underset{CH_3}{\overset{H}{|}}\!Si-$$

Fibers formed from this polymer are then heated slowly to about 1200°C in a nitrogen atmosphere to drive off all the hydrogen and all carbon atoms other than those that directly link the silicon atoms. The final product is a ceramic material of composition SiC, in the form of fibers ranging in diameter from 10 to 15 μm. By similar procedures, beginning with an appropriate organic polymer, ceramic fibers of other compositions—such as boron nitride, BN—can be fabricated. When the ceramic fibers are added to a ceramic material processed as described above, the resulting product has a much higher resistance to catastrophic crack failure.

Applications of Ceramics

Ceramics, particularly new ceramic composites, are widely used in the cutting-tool industry. For example, alumina reinforced with silicon carbide whiskers (extremely fine fibers) is used to cut and machine-cast iron and harder nickel-based alloys. Ceramic materials are also used in grinding wheels and as abrasives because of their exceptional hardness (Table 12.3). Silicon carbide is the most widely used abrasive.

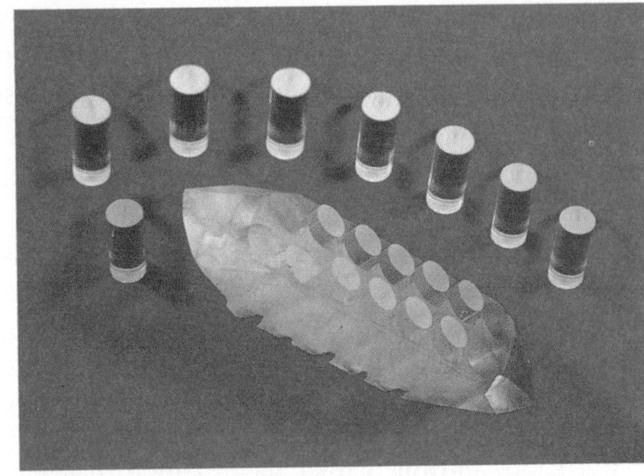

Figure 12.22 Quartz crystals used to control frequency in a quartz watch are cut from these cylinders of quartz, which in turn were cut from the larger block. (Matec Corporation)

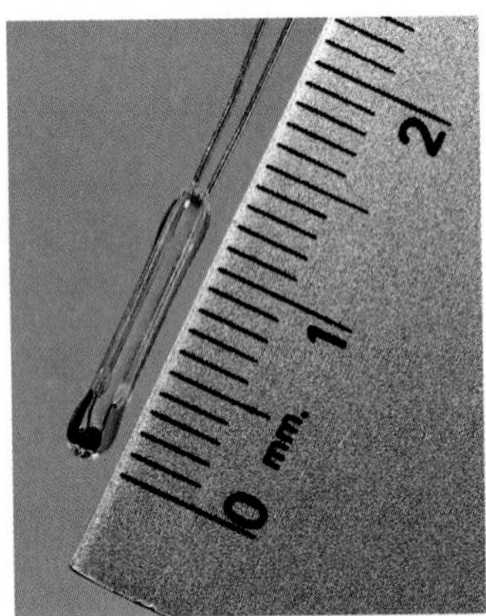

Figure 12.23 A thermistor temperature sensor. The electrical resistance of the ceramic element at the ends of the wires decreases reproducibly with increasing temperature. Therefore, the device can be used to measure temperature after it is calibrated. (© Richard Megna/ Fundamental Photographs)

Ceramic materials play an important role in the electronics industry. Semiconductor integrated circuits are typically mounted on a ceramic substrate, usually alumina. Some ceramics, notably quartz (crystalline SiO_2), are **piezoelectric**, which means that they generate an electrical potential when subjected to mechanical stress. This property enables us to use piezoelectric materials to control frequencies in electronic circuits, as in quartz watches and ultrasonic generators (Figure 12.22).

Thermistors are ceramic materials with limited electrical conductivity that increases with temperature. Thermistors are widely used in devices that measure or control temperature. They are also used as heating elements and electrical switches. Figure 12.23 shows a thermistor used as a temperature sensor.

One of the most highly publicized uses of ceramic materials is in the manufacture of ceramic tiles for the surfaces of space shuttle vehicles to protect against overheating on reentry into the earth's atmosphere (Figure 12.24). The tiles are made of short, high-purity silica fibers reinforced with aluminum borosilicate fibers. The material is formed into blocks, sintered at over 1300°C, and then cut into tiles. The tiles have a density of only 0.2 g/cm^3, yet they are able to keep the shuttle's aluminum skin below 180°C while sustaining a surface temperature of up to 1250°C.

Figure 12.24 Workers apply thermally insulating ceramic tiles to the body of the space shuttle orbiter. (Lyndon B. Johnson Space Center, NASA)

Superconducting Ceramics

In 1911, Dutch physicist H. Kamerlingh Onnes discovered that when mercury is cooled below 4.2 K it loses all resistance to the flow of an electrical current. Since that discovery, scientists have found that many substances exhibit this "frictionless" flow of electrons. This property has become known as **superconductivity**. Substances that exhibit superconductivity do so only when cooled below a particular temperature, called the **superconducting transition temperature, T_c**. The observed values of T_c are generally very low. In fact, before the 1980s, the highest value that had been observed for T_c was about 23 K for a niobium-germanium compound.

In 1986, J. G. Bednorz and K. A. Muller, working at the IBM research laboratories in Zürich, Switzerland, discovered superconductivity above 30 K in a ceramic oxide containing lanthanum, barium, and copper. This material represents the first **superconducting ceramic**. That discovery, for which Bednorz and Muller received the Nobel Prize in 1987, set off a flurry of research activity all over the world. Before the end of 1986, scientists had verified the onset of superconductivity at 95 K in a compound of yttrium-barium-copper oxide of approximate composition $YBa_2Cu_3O_x$, where x varies typically between 6.5 and 7.2. More recently, zero resistance has been observed at temperatures over 125 K. There are reports of the onset of zero resistance at or above room temperature (about 295 K), though these have yet to be confirmed.

The discovery of so-called high-temperature (high-T_c) superconductivity is of great significance. Using materials that carry electrical current with zero resistance could save great amounts of energy in many applications, including electrical generators and large electric motors, and could lead to the production of smaller and faster computer chips. In addition, superconducting materials exhibit a property, called the *Meissner effect*, in which they exclude from their volume all magnetic fields (Figure 12.25). For this reason, scientists might be

Figure 12.25 Demonstration of the Meissner effect. (*a*) The block of material at the base is a superconducting ceramic that has been cooled below its transition temperature. The circular object is a magnet. Because the superconductor excludes magnetic field lines, the small magnet is levitated in space: The repulsion of the magnetic field lines balances the gravitational force. (*b*) An experimental magnetically levitated train near Miyazaki, Japan. (© Lawrence Manning/West Light; © Chuck O'Rear/West Light)

(*a*)

(*b*)

able to use these materials to design magnetically levitated trains. All these applications became feasible only with the discovery of high-temperature superconductivity, because the costs of maintaining extremely low temperatures is excessive. The only readily available safe coolant at temperatures below 77 K is liquid helium, which costs about \$2.60 per liter. However, for materials that undergo the superconducting transition at temperatures well above 77 K, liquid nitrogen, which costs only about \$0.05 per liter, can be used.

One high-T_c superconductor of particular importance is the ceramic $YBa_2Cu_3O_x$. Researchers discovered early that replacing Y or Ba with certain closely related elements has little effect on T_c. Of far greater significance are the copper and oxygen contents. The Y-Ba-Cu-O system takes up or releases oxygen at high temperatures, depending on the partial pressure of O_2 at the time the material is fired.

The structure of $YBa_2Cu_3O_7$ is shown in Figure 12.26; the unit cell is defined by the lines. A few oxygen atoms that lie outside the unit cell are shown to illustrate the arrangement of oxygens about each copper atom. (In counting atoms in this structure, remember from Section 11.7 that one-eighth of each atom situated on the corner and one-fourth of each atom on a unit-cell edge are contained within the unit cell.) The structure shown is idealized; the actual structure is more complex because some oxygen sites may be vacant or oxygen atoms may be situated at other places in the lattice.

Figure 12.26 Unit cell of $YBa_2Cu_3O_7$. A few oxygen atoms that fall outside the unit cell are also shown to illustrate the arrangement of oxygen atoms about each copper atom. The unit cell is defined by the lines that describe a rectangle.

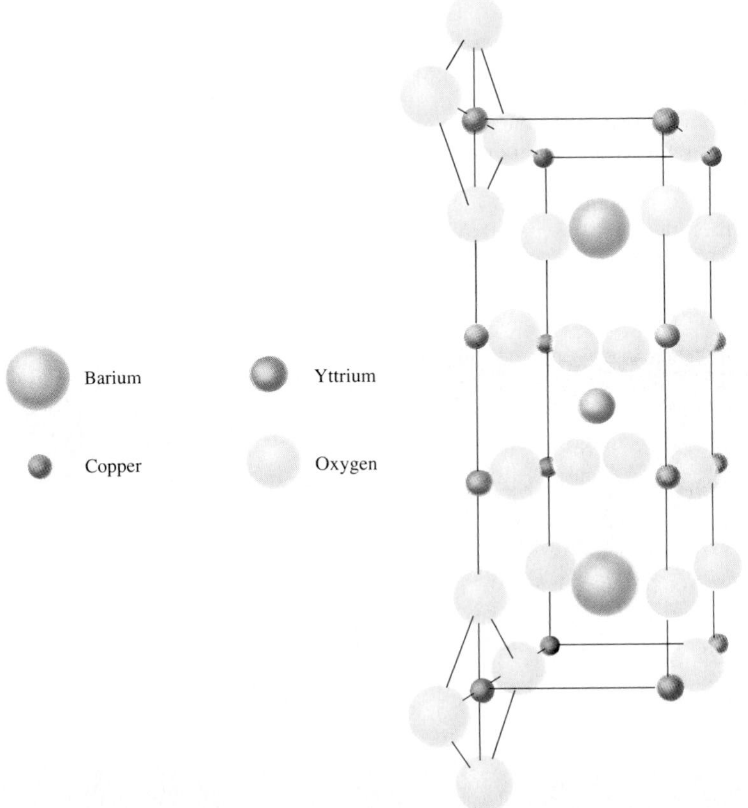

Theories of Superconductivity

During the 1960s, U.S. physicists John Bardeen (Figure 12.27), Leon Cooper, and J. Robert Schrieffer developed a theory that explains the type of superconductivity observed at low temperatures in metals and related substances. The BCS theory, as it is called, earned its creators the Nobel Prize for Physics in 1972. The theory assumes that at low temperatures the electrons move as pairs, which enables them to escape the interactions with atoms in the structure that lead to electrical resistance. It is not clear, however, that the same theory can account for high-temperature superconductivity in ceramic materials such as $YBa_2Cu_3O_x$. Although many variants of the BCS theory, as well as several entirely new approaches, have been proposed, no generally accepted model for high-temperature superconductivity exists, and theoreticians throughout the world continue to work on this problem.

The new superconducting ceramic materials have immense promise, but a great deal of research is needed before they can be applied on a practical basis. At present it is difficult to mold ceramics, which are brittle materials, into useful shapes like wires on a large scale. Although science has made some progress in this area (Figure 12.28), the attainable current densities (that is, the current that can be carried by a wire of a certain cross-sectional area) are not yet high enough for many applications. A related problem is the tendency of ceramics to interact with their environment, particularly with water and carbon dioxide. For example, the reaction of $YBa_2Cu_3O_7$ with atmospheric water liberates O_2 and forms $Ba(OH)_2$, $Y_2BaCu_3O_5$, and CuO. Because these materials are so reactive, they must be protected against long-term exposure to the atmosphere.

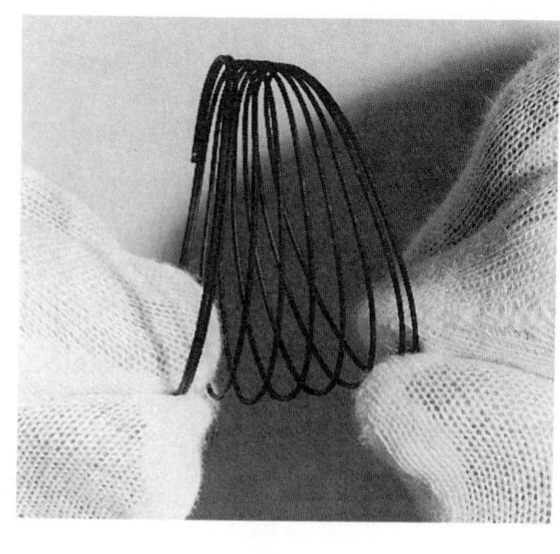

Figure 12.28 A wire coil of the high-temperature superconductor $YBa_2Cu_3O_7$. The wire was formed by extruding a wire from a paste of reactant materials and shaping it into a coil before heating to form the ceramic oxide. (Argonne National Laboratory)

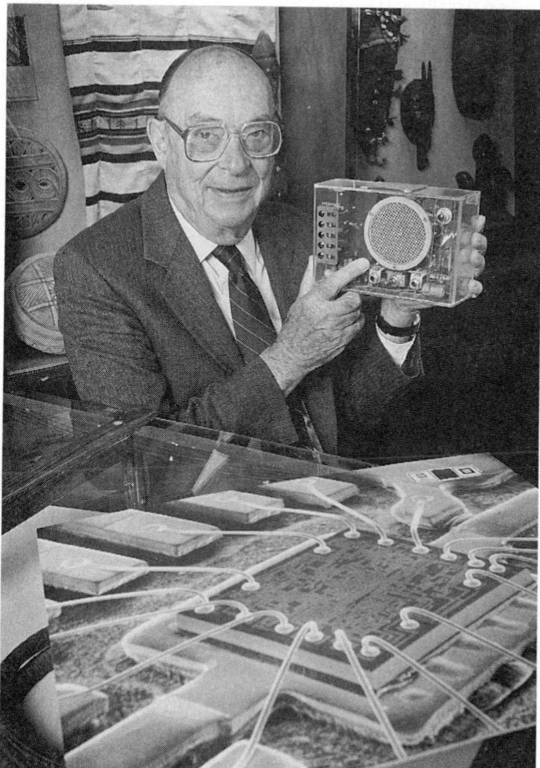

Figure 12.27 John Bardeen, Professor of Physics and Electrical and Computer Engineering at the University of Illinois, Urbana-Champaign. In 1956, Professor Bardeen shared the Nobel Prize with Walter H. Brattain and William Shockley for the invention of the transistor. He also shared the 1972 Nobel Prize in Physics with Leon Cooper and J. Robert Schrieffer for development of a theory of superconductivity, becoming the only person ever to win a Nobel Prize twice in the same field. (University of Illinois News Bureau)

The discovery of new high-temperature superconducting materials with superior properties and the fabrication of useful devices from the known superconducting materials are subjects of very active research. Even so, scientists estimate that the new discoveries will not be translated into important practical applications for several years. In time, however, this new class of ceramic materials may become part of our everyday lives.

12.4 THIN FILMS

Thin films were first used for decorative purposes. In the seventh century, artists learned how to paint a pattern on a ceramic object with a silver salt solution and then heat the painted object to cause decomposition of the salt, leaving a thin film of metallic silver. Thin films are used today for decorative or protective purposes, to form conductors, resistors, and other types of films in microelectronic circuits, to form photovoltaic devices for conversion of solar energy to electricity, and for many other applications (Figure 12.29). A thin film might be made of any kind of material, including metals, metal oxides, or organic substances.

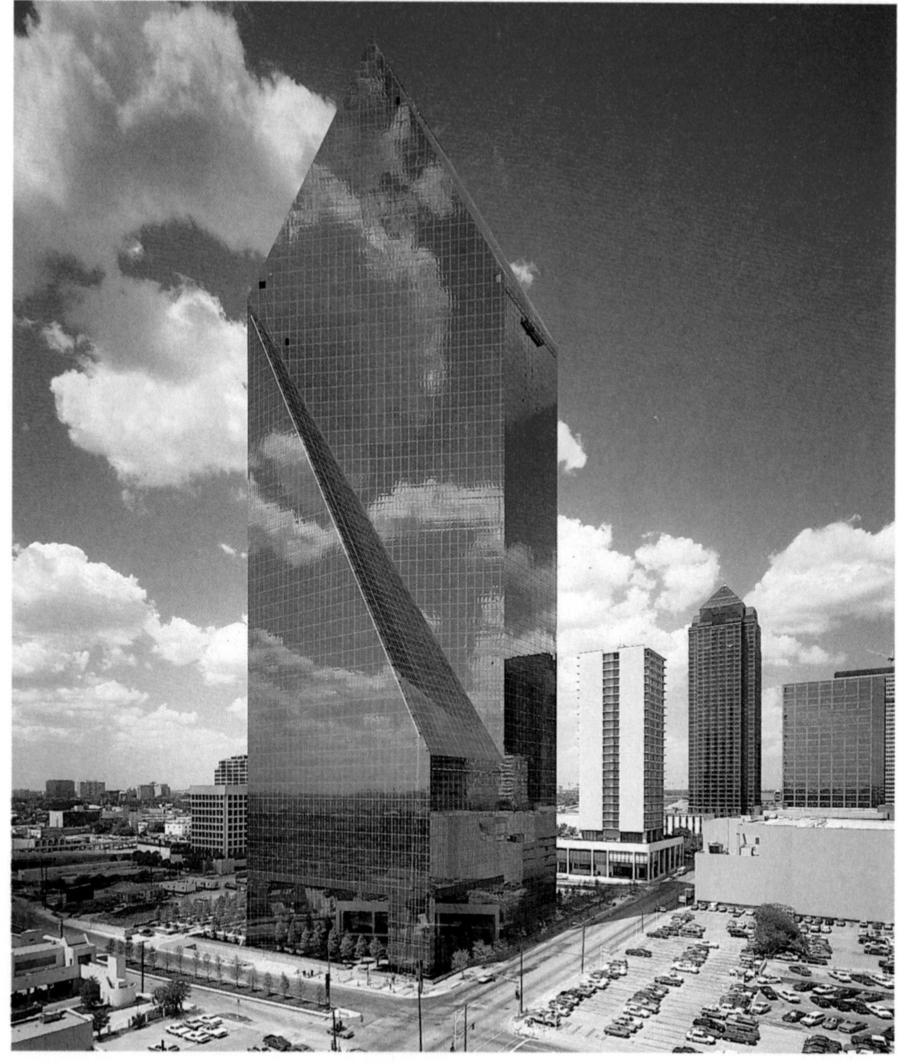

Figure 12.29 The glass panels forming the outer wall of this building have a thin film of metal, which acts to reflect a significant fraction of the outdoor light. The reflective glass provides privacy, reduces interior glare, and reduces the cooling load on the building in hot weather. The building is the Allied Bank, Dallas, designed by I.M. Pei. (© Wes Thompson/ The Stock Market)

The term **thin film** does not have one precise definition. In general, it refers to films with thicknesses ranging from 0.1 μm to about 300 μm. It does not normally refer to coatings such as paint and varnish, which are typically much thicker. For a thin film to be useful, it should possess all or most of the following properties: (a) It should be chemically stable in the environment in which it is to be used; (b) it should adhere well to the substrate it covers; (c) it should have a uniform thickness; (d) it should be chemically pure or of controlled chemical composition; and (e) it should have a low density of dislocations and other imperfections. In addition to these general characteristics, special properties might be required for certain applications. For example, the film might need to be an insulator or a semiconductor, or to possess special optical or magnetic properties.

A thin film must adhere to its underlying substrate if it is to perform usefully. Because the film is inherently fragile, it must depend on the substrate for structural support. To attain that support, the film must be bound to the substrate by strong forces. The bonding forces may be chemical in nature; that is, a chemical reaction at the interface can connect the film to the underlying material. For example, when a metal oxide is deposited on glass, the oxide lattices of the metal oxide and the glass blend at the interface, forming a thin zone of intermediate composition. In these cases, the bonding energies between the film and the substrate are of the same magnitudes as chemical bonds, in the range of 250 to 400 kJ/mol. In some cases, however, the bonding between the film and the substrate is based solely on intermolecular van der Waals and electrostatic forces, as might be the case when an organic polymer film is deposited on a metal surface. The energies that bind the film to the substrate in such cases might be in the range of 50 to 100 kJ/mol. Thus, films in which only physical bonding of this sort is present are not as robust.

Uses of Thin Films

Thin films are very important in microelectronics. They are employed as conductors, resistors, and capacitors. Thin films are widely used as optical coatings on lenses (Figure 12.30) to reduce the amount of light reflected from the lens surface and to protect the lens. Thin metallic films have been used for a long time as protective coatings on metals. They are usually deposited from solutions by the use of electrical currents, as in silver plating and "chrome" plating. (We

Figure 12.30 The lenses of these binoculars are coated with a ceramic thin film to reduce reflectance and protect the softer glass against scratching. (© Kristen Brochmann/Fundamental Photographs)

will defer discussion of electrochemical methods for forming films until Chapter 20.) Metal tool surfaces are coated with ceramic thin films to increase their hardness. For example, a hard steel drill bit may be coated with a thin film of tungsten carbide (Figure 12.31). Diamond thin films are employed to confer hardness and wear resistance on cutting tools and to enhance the high-frequency response of diaphragms in high-fidelity audio speakers (Section 11.8).

Figure 12.31 The tip of this masonry drill bit has been coated with a thin film of tungsten carbide to impart hardness and wear resistance. (© Richard Megna/ Fundamental Photographs)

Formation of Thin Films

Thin films are formed by a variety of techniques. We will discuss three of the most commonly used methods: vacuum deposition, sputtering, and chemical-vapor deposition.

Vacuum deposition is used to form thin films of substances that can be vaporized or evaporated without destroying their chemical identities. These substances include metals, metal alloys, and simple inorganic compounds such as oxides, sulfides, fluorides, and chlorides. For example, optical lenses are coated with inorganic materials such as MgF_2, Al_2O_3, and SiO_2. The material to be deposited as a thin film is heated—either electrically or by electron bombardment—in a high-vacuum chamber with a pressure of 10^{-5} mm Hg or less. The process is shown diagrammatically in Figure 12.32. To obtain a film of uniform thickness, all parts of the surface to be coated must be equally accessible to the vapor phase from which the thin film material is deposited. Sometimes this uniformity is obtained by rotating the piece to be coated.

Sputtering involves the use of a high voltage to remove material from a source, or target. Atoms removed from the target are carried through the ionized gas within the chamber and deposited on the substrate. The target surface is the negative electrode, or cathode, in the circuit, and the substrate may be attached to the positive electrode, or anode. Figure 12.33 depicts this process. The chamber contains an inert gas such as argon that is ionized in the high-voltage field. The positively charged ions are accelerated toward the target surface, which they strike with sufficient energy to dislodge atoms of the target material. Many of these atoms are accelerated toward the substrate surface. On striking it, they form a thin film.

The sputtered atoms have a lot of energy. The initial atoms striking the surface may penetrate several atomic layers into the substrate, which helps to ensure good adhesion of the thin-film layer to the substrate. An additional advantage of sputtering is that it is possible to change the target material from

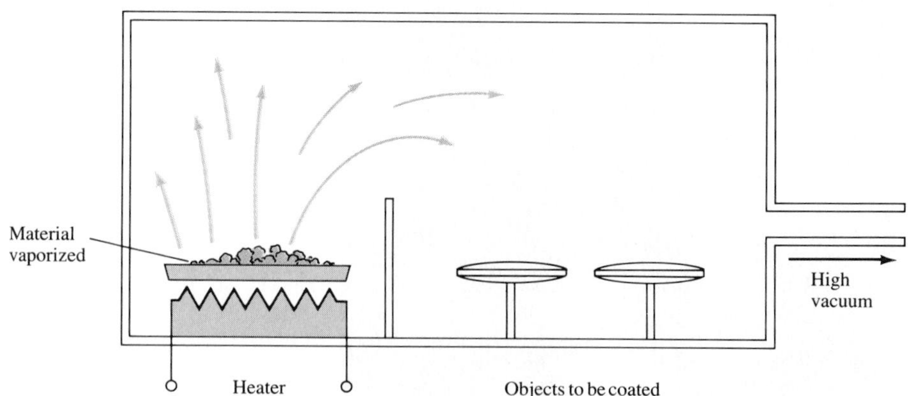

Figure 12.32 A schematic illustration of a vacuum-deposition apparatus.

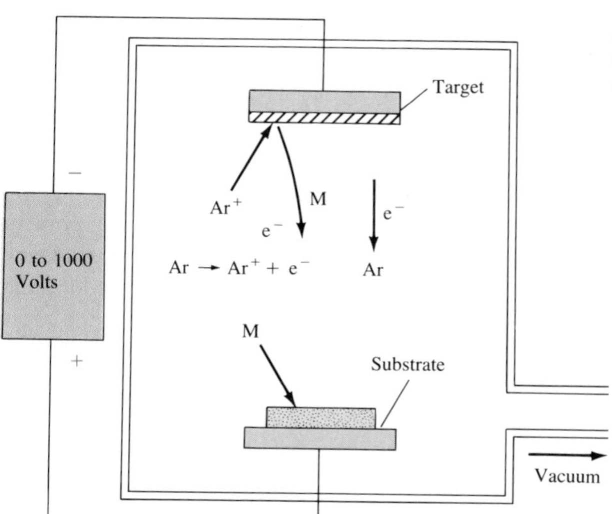

Figure 12.33 A schematic illustration of a sputtering apparatus.

which the sputtered atoms arise without disturbing the system, so that multilayer thin films can be formed.

Sputtering is widely used to form thin films of such elements as silicon, titanium, niobium, tungsten, aluminum, gold, and silver. It is also employed to form thin films of refractory materials such as carbides, borides, and nitrides on metal tool surfaces, to form soft lubricating films such as molybdenum disulfide, and to apply antiglare coatings of metal oxides to optical equipment. It is even possible with appropriate equipment to sputter nonconducting organic polymers.

In **chemical-vapor deposition**, the surface is coated with a volatile, stable chemical compound at a temperature below the melting point of the surface. The compound then undergoes some form of chemical reaction to form a stable, adherent coat. For example, titanium tetrabromide is evaporated, and the gaseous $TiBr_4$ is mixed with hydrogen. The mixture is then passed over a surface heated to about 1300°C. The heated substrate is ordinarily a ceramic such as silica or alumina. The metal halide undergoes reaction with hydrogen to form a coating of titanium metal:

$$TiBr_4(g) + 2H_2(g) \longrightarrow Ti(s) + 4HBr(g) \qquad [12.9]$$

Similarly, it is possible to make silicon films by decomposition of $SiCl_4$ in the presence of H_2 at 1100°C to 1200°C:

$$SiCl_4(g) + 2H_2(g) \longrightarrow Si(s) + 4HCl(g) \qquad [12.10]$$

Films of silica, SiO_2, are formed by decomposition of $SiCl_4$ in the presence of both H_2 and CO_2 at 600°C to 900°C:

$$SiCl_4(g) + 2H_2(g) + 2CO_2(g) \longrightarrow SiO_2(s) + 4HCl(g) + 2CO(g) \qquad [12.11]$$

Films of silicon nitride can be formed by the reaction of silane, SiH_4, with ammonia at 900°C to 1100°C:

$$3SiH_4(g) + 4NH_3(g) \longrightarrow Si_3N_4(s) + 12H_2(g) \qquad [12.12]$$

FOR REVIEW

SUMMARY

In this chapter, we have considered four important classes of modern materials: liquid crystals, polymers, ceramics, and thin films. Liquid crystals are substances that exhibit one or more ordered phases in a temperature range above the melting point of the solid. In a nematic liquid crystal, the molecules are aligned along a common direction, but the ends of the molecules are not lined up. In smectic phases, the ends of the molecules are lined up as well, so that the molecules form sheets. Nematic and smectic phases are generally composed of molecules with fairly rigid, rodlike shapes. The cholesteric liquid-crystalline phase is formed from molecules that form ordered layers of molecules. The most important applications of liquid-crystalline materials are in screens for portable computers and in other digital-display devices such as watches and calculators.

A polymer contains many identical molecules (monomers) joined together to form a molecular chain. Two of the most important kinds of reactions that join monomers into polymers are addition polymerization, exemplified by the formation of polyethylene, and condensation polymerization, exemplified by the formation of nylon 6,6. The properties of a given polymer vary with the average length of the chain and with the degree of branching along the chain.

Plastics are materials that can be formed into particular shapes using heat and pressure. Thermoplastic materials can be reshaped, whereas thermosetting plastics, once formed, are not readily reshaped. An elastomer is a material that, when stretched or bent, regains its original shape upon removal of the distorting force.

To be useful, polymeric materials must be processed, that is, made into film, fiber, or another form. Fibers are produced by extruding polymerized material under pressure through a small hole. The fiber is then formed by further extrusion accompanied by pulling, or drawing. The stretching of the fiber in these steps has the effect of aligning the large molecules along their long axes, adding strength to the fiber.

Polymers are considered amorphous materials, but they may possess regions in which the molecules are aligned. These regions are referred to as being *crystalline;* they impart hardness and raise the temperature at which the materials soften. A plasticizer is a substance with a low molecular mass that is added to a polymer to disrupt partially the alignment of the polymer molecules, thus imparting greater flexibility to the material.

Greater hardness and lower flexibility can be achieved by crosslinking, that is, by forming chemical bonds between polymer chains. An example is the vulcanization of rubber, which occurs when natural rubber is heated with sulfur. Sulfur atoms bridge between adjacent polymer chains, making the rubber harder and less subject to decomposition.

Ceramics are inorganic materials that typically exhibit hardness, rigidity, and high-temperature stability. They are employed as insulators and furnace bricks and in the manufacture of pottery and dinnerware. Engineering ceramics find use in a variety of applications where metals, wood, or plastics were traditionally used. The advantages of engineering ceramics include high thermal stability, rigidity, chemical stability, and resistance to wear. However, the use of ceramics in many applications is limited by their brittle nature, which makes them susceptible to catastrophic failure through crack formation. Ceramics can be toughened (made resistant to cracking) by sintering of extremely small particles. One important method for forming small ceramic particles of uniform size is called the sol-gel process. Ceramics can also be made tougher by forming a composite, which is a solid mixture of two or more component materials.

Ceramics find many applications in modern technology. Ceramic substrates are employed in the manufacture of microelectronic circuits. Certain ceramic materials, including quartz, are piezoelectric; that is, they develop an electrical potential upon deformation. A thermistor is a ceramic whose electrical resistance decreases with increasing temperature. Superconducting ceramics are relatively new materials that exhibit no electrical resistance below a certain transition temperature. The superconducting transition temperatures for ceramic superconductors, such as $YBa_2Cu_3O_7$, are higher than for any nonceramic superconductors. Most importantly, many of them lie above 77 K, the boiling point of liquid nitrogen. Thus, this relatively inexpensive coolant can be employed to maintain the superconducting state.

A thin film is a layer of one substance, ranging in thickness from 0.1 μm to perhaps several hundred micrometers, covering an underlying substrate. Thin films are very important in the production of microelectronic devices, in solar-energy conversion, and in many applications in which the thin film acts as a protective layer for the substrate. Thin films are formed primarily by vacuum deposition, in which a substance is sublimed onto the substrate. A second method, called sputtering, involves use of a high-voltage gaseous discharge. In chemical-vapor deposition, a substance is deposited on a surface, after which a chemical reaction forms the desired film.

KEY TERMS

liquid crystal (Sec. 12.1)
nematic liquid crystal (Sec. 12.1)
smectic liquid crystal (Sec. 12.1)
cholesteric liquid crystal (Sec. 12.1)
polymer (Sec. 12.2)
monomer (Sec. 12.2)
polyethylene (Sec. 12.2)
addition polymerization (Sec. 12.2)
plastic (Sec. 12.2)
thermoplastic (Sec. 12.2)
thermosetting plastic (Sec. 12.2)
elastomer (Sec. 12.2)
fibers (Sec. 12.2)
condensation polymerization (Sec. 12.2)
crystallinity (Sec. 12.2)
plasticizers (Sec. 12.2)
vulcanization (Sec. 12.2)
natural rubber (Sec. 12.2)
ceramics (Sec. 12.3)
engineering ceramics (Sec. 12.3)
sol-gel process (Sec. 12.3)
sol (Sec. 12.3)
condensation reaction (Sec. 12.3)
gel (Sec. 12.3)
composite (Sec. 12.3)
ceramic fibers (Sec. 12.3)
piezoelectric (Sec. 12.3)
thermistors (Sec. 12.3)
superconductivity (Sec. 12.3)
superconducting transition temperature (T_c) (Sec. 12.3)
superconducting ceramic (Sec. 12.3)
thin film (Sec. 12.4)
vacuum deposition (Sec. 12.4)
sputtering (Sec. 12.4)
chemical-vapor deposition (Sec. 12.4)

EXERCISES

Liquid Crystals

12.1 Describe how a liquid-crystalline material behaves differently from one that is not liquid-crystalline.

12.2 What observations made by Reinitzer on cholesteryl benzoate suggested that this substance possesses a liquid-crystalline phase?

12.3 What are the most common characteristics of the shapes of molecules of liquid-crystalline substances?

12.4 Although a liquid-crystalline phase may form upon melting, it disappears on further heating when a certain temperature is reached, to form the normal, isotropic liquid. Using the kinetic-molecular theory, account for formation of the liquid-crystalline phase and for its conversion to the isotropic liquid at higher temperature.

12.5 Name the three major classes of liquid-crystalline phases and indicate the distinctions between them.

12.6 The molecules shown in Figure 12.3 all possess polar groups; that is, groupings of atoms that give rise to fairly large dipole moments within the molecules. In terms of the factors that give rise to intermolecular interactions (Section 11.2), how might the presence of such polar groups enhance the tendency toward liquid-crystal formation?

12.7 List as many required properties as you can think of for a liquid-crystalline substance that is to be employed in a digital watch display.

12.8 When a second substance is dissolved in a substance that has a liquid-crystalline phase, only a small concentration of solute is often sufficient to eliminate formation of the liquid-crystalline phase. Why?

Polymers

12.9 What is a polymer?

12.10 Name six naturally occurring or synthetic materials familiar to you that are polymeric.

12.11 Write a chemical equation that describes the formation of polyacrylonitrile from acrylonitrile,

$$\begin{array}{c} CH_2{=}CH \\ \quad\;\; | \\ \quad\;\; CN \end{array}$$

(Polyacrylonitrile is employed in home furnishings, craft yarns, wearing apparel, and many other uses.)

12.12 Write a chemical equation that represents the formation of polychloroprene from chloroprene,

$$CH_2{=}CH{-}\underset{\underset{Cl}{|}}{C}{=}CH_2$$

(Polychloroprene is employed in highway-pavement seals, expansion joints, conveyor belts, and wire and cable jacketing.)

12.13 Write a chemical equation representing the formation of poly(phenylene oxide) by condensation polymerization of 2,6-dimethyl-*p*-hydroxyphenol,

$$HO{-}\underset{CH_3}{\overset{CH_3}{C_6H_2}}{-}OH$$

[Poly(phenylene oxide) is employed in fabricating automotive parts, appliances, and computer chasses.]

12.14 The nylon Nomex has the following structure:

$$\left[\overset{O}{\overset{\|}{C}}{-}C_6H_4{-}\overset{O}{\overset{\|}{C}}{-}NH{-}C_6H_4{-}NH \right]_n$$

Draw the structures of the two monomeric compounds that yield Nomex as a polymerization product when subjected to a condensation reaction.

12.15 Most polymers are amorphous materials. Why is this so?

12.16 What is meant by the term *crystallinity* in referring to a polymer such as high-density polyethylene?

12.17 Provide a brief description of each of the following: **(a)** elastomer; **(b)** thermoplastic; **(c)** thermosetting plastic.

12.18 Provide a brief description or definition of each of the following: **(a)** plasticizer; **(b)** crosslinking; **(c)** condensation reaction.

Ceramics

12.19 List some of the ways in which ceramics differ from polymers in terms of molecular structure and physical properties.

12.20 List the potential advantages of ceramics as engineering materials over other materials such as plastics and metals.

12.21 Account for each of the following applications of ceramic materials in terms of electronic structure of the substance: **(a)** Zirconia, ZrO_2, is employed in high-temperature engine blocks and in gas turbine rotors. **(b)** The compound $Mg_3Al_2(SiO_4)_3$, a garnet mineral, is employed as the abrasive material in sandpaper.

12.22 Silicon carbide, SiC, has the three-dimensional structure shown in Figure 12.34. Describe how the bonding and structure of SiC lead to its great thermal stability (to 2700°C) and exceptional hardness.

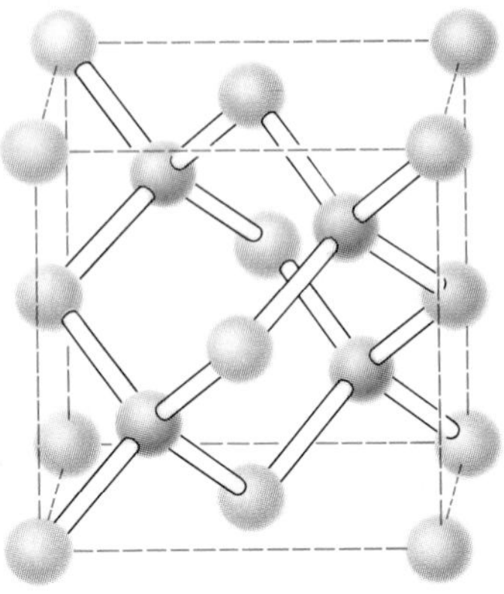

Figure 12.34

12.23 Why are ceramic materials generally brittle?

12.24 Describe the steps that can be taken to increase the resistance of ceramic materials to mechanical failure.

12.25 In a sol-gel process, a metal alkoxide is reacted with water to produce a metal hydroxide, which then undergoes condensation reactions to yield a polymeric network of metal-oxygen-metal linkages. Suggest a reason why the process is not simply begun with the metal hydroxide.

12.26 In a typical insulator composed primarily of Al_2O_3, the solid is formed by sintering grains of Al_2O_3 that are 5 to 10 μm in diameter. What is the reason behind sintering small particles to form a larger solid piece?

12.27 What is a composite? In what respects is a ceramic composite likely to be an improvement over the primary ceramic material from which it is formed?

12.28 Ceramic materials typically fail by catastrophic crack propagation. Offer an explanation of how the addition of ceramic fibers to a ceramic matrix toughens the material against this.

12.29 What general properties of ceramic materials are important in the use of ceramic tiles to protect space shuttle vehicles against overheating during reentry into the earth's atmosphere?

12.30 What property or properties of the ceramic material are most important in each of the following applications of ceramics: **(a)** thermistor temperature sensor; **(b)** silicon carbide-reinforced machine cutting tool; **(c)** silicon dioxide substrate for a semiconductor integrated circuit; **(d)** silicon carbide gas turbine rotor in an aircraft engine.

12.31 How does a superconducting material differ in its electrical properties from ordinary electrical conductors?

12.32 In what ways does the superconducting ceramic $YBa_2Cu_3O_7$ differ from superconducting materials known before 1980?

Thin Films

12.33 What is meant by the term *thin film*? List the characteristics that would be desirable in a thin film of tantalum carbide applied to the surface of a cutting tool to impart hardness.

12.34 A gold film strip is applied to an SiO_2 surface to form a conductor in an integrated circuit. List as many desirable characteristics for such a film as you can think of.

12.35 Describe the methods employed for formation of thin films.

12.36 Indicate the method or methods applicable to formation of a thin film of each of the following materials: **(a)** MgO; **(b)** Teflon, an organic polymer of empirical formula CF_2; **(c)** titanium.

Additional Exercises

12.37 Ceramics are generally brittle, subject to crack failure, and stable to high temperatures. In contrast, plastics are generally deformable under stress and have limited thermal stability. Discuss these differences in terms of the structures and bonding in the two classes of materials.

12.38 Classify each of the following as a ceramic, polymer, or liquid crystal.

(a) $\left[-CH_2-C(CH_3)(COOCH_3)- \right]_n$ **(b)** $LiNbO_3$

(c) SiC **(d)** $\left[-Si(CH_3)_2- \right]_n$

(e) $CH_3O-C_6H_4-N{=}N(O)-C_6H_4-OCH_3$

[12.39] A particular liquid-crystalline substance has the phase diagram shown in Figure 12.35. By analogy with the phase diagram for a non-liquid-crystalline substance (Section 11.6), label the areas as to the phase present.

12.40 In most applications of ceramic composites, the matrix and the added fiber should have closely similar coefficients of thermal expansion. Why is this so?

12.41 Describe the process by which a thin film is formed by the technique known as sputtering. What is the role of the inert gas present in the sputtering chamber?

12.42 List an application of thin films that involves each of the following characteristics or types of material: **(a)** an electrical conductor; **(b)** a metal oxide; **(c)** a hard ceramic material.

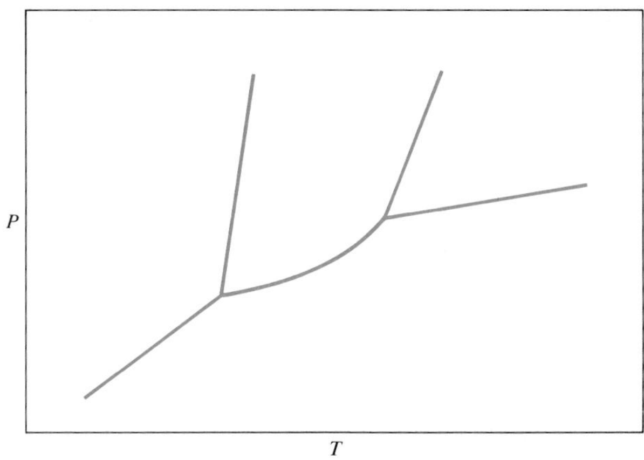

Figure 12.35

[12.43] In ceramic composites, a strong interaction between the embedded fibers and the matrix host is essential. What kinds of interactions might exist in these systems? How would the high-temperature processing of the ceramic be likely to affect the nature of the interactions?

12.44 Write balanced chemical equations to describe: **(a)** formation of a silicon carbide whisker by the two-step thermal decomposition of poly(dimethylsilane) (two equations); **(b)** formation of a thin film of niobium by thermal decomposition of $NbBr_5$ on a hot surface under an H_2 atmosphere; **(c)** formation of $Si(OCH_2CH_3)_4$ by reaction of $SiCl_4$ with ethyl alcohol; **(d)** polymerization of styrene,

$$C_6H_5-CH{=}CH_2,$$

to form polystyrene.

[12.45] The operation of a liquid-crystal display, as illustrated in Figure 12.9, depends upon a change in the orientation of the molecules in an electric field. Suppose the molecules in a nematic substance were aligned as follows in the absence of an applied electrical potential:

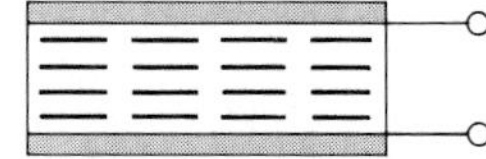

How would these molecules be aligned when an electrical potential is applied by connecting a battery across the contacts? What molecular properties would be likely to cause such a realignment to occur? (Hint: See Section 11.2.)

[12.46] In the superconducting ceramic $YBa_2Cu_3O_7$, what is the average oxidation state of copper, assuming that Y and Ba are in their expected oxidation states? Yttrium can be replaced with a rare-earth element such as Ln, and Ba can be replaced with other similar elements without fundamentally changing the superconducting properties of the material. However, general replacement of copper by any other element leads to a loss of superconductivity. In what respects is the electronic structure of copper different from that of the other two metallic elements in this compound?

[12.47] Indicate the nature of the film formed by thermal decomposition on a heated surface of each of the following: **(a)** SiH_4 with H_2 as carrier gas, in the presence of CO_2 (CO is a product); **(b)** $TiCl_4$ in the presence of water vapor; **(c)** GeH_4 in the presence of H_2 as carrier gas.

[12.48] Whereas thin films of metal oxides can be formed by vacuum deposition, a general rule is that inorganic compounds having anionic components with names ending in *-ite* or *-ate* cannot be successfully vacuum-deposited. Why is this so?

[12.49] Polypropylene is an addition polymer formed from propylene, $CH_3—CH=CH_2$. Depending on the conditions under which it is made, polypropylene can vary greatly in crystallinity, elasticity, and density. It has a much wider range of properties in these respects than does polyethylene. Account for these observations in light of the comparative structures of the two polymers.

Properties of Solutions 13

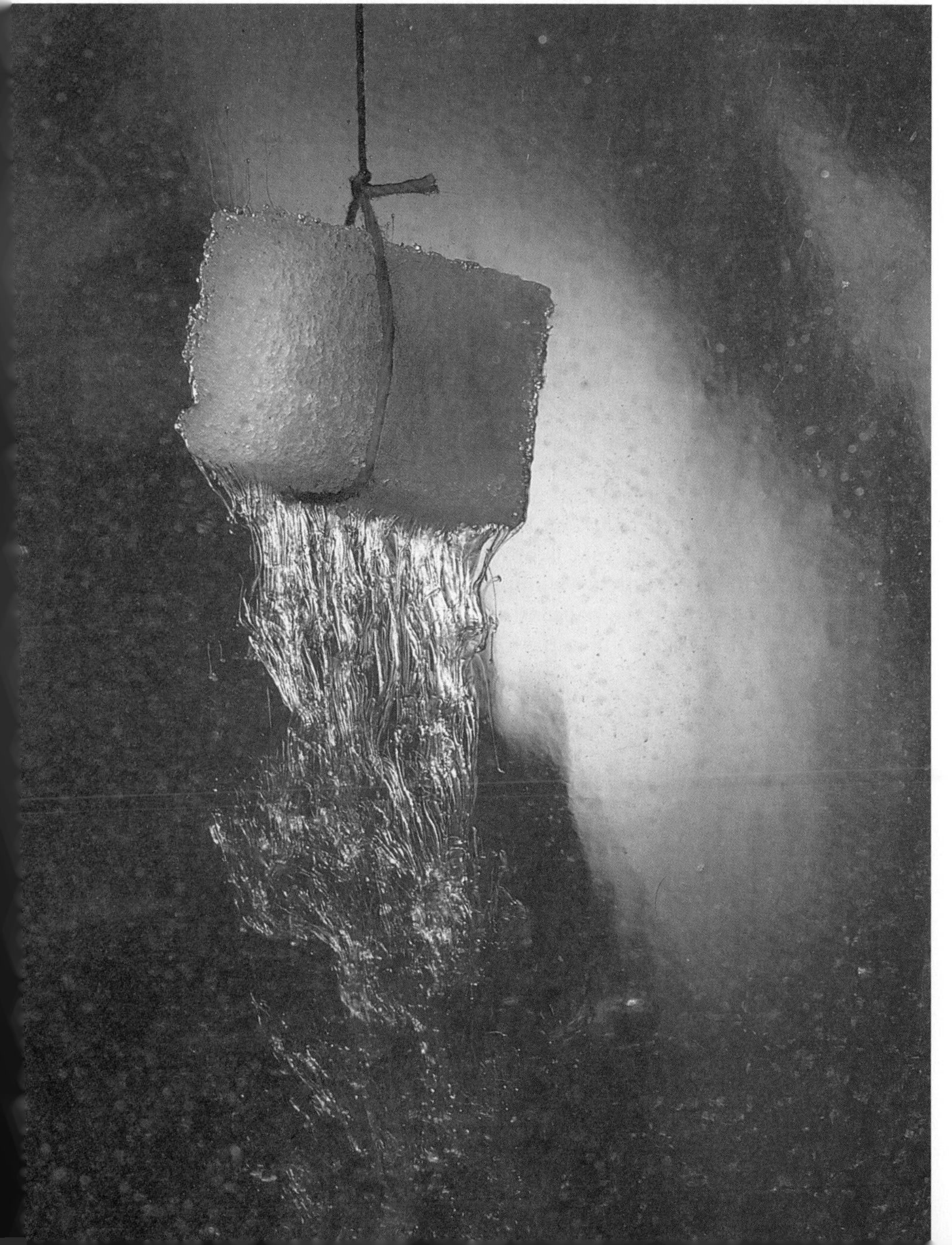

Dissolution of a sugar cube in water As the sugar dissolves, the density of the solution changes, creating the pattern seen below the cube. (© Richard Megna/Fundamental Photographs)

CONTENTS

Very few of the materials that we encounter in everyday life are pure substances; most are mixtures. Many of these mixtures are homogeneous; that is, their components are uniformly intermingled on a molecular level. We have seen that homogeneous mixtures are called *solutions* (Sections 1.2 and 4.1). Examples of solutions abound in the world around us. The air we breathe is a homogeneous mixture of several gaseous substances. The familiar metal brass is a solution of zinc in copper. The oceans are a solution of many dissolved substances in water. The fluids that run through our bodies are solutions, carrying a great variety of essential nutrients, salts, and other materials.

Solutions may be gases, liquids, or solids. Examples of each kind are given in Table 13.1. As we saw in Chapter 4, the *solvent* is normally the component present in greatest amount. Other components are called *solutes*. Because liquid solutions are the most common, we will focus our attention on them in this chapter.

In Chapter 11, we considered the various types of intermolecular forces that exist between molecular and ionic particles. In this chapter, we will see that these forces are also involved in the interactions between solutes and solvents. We will examine the solution process, the factors that determine the amount of solute that can be dissolved in a given quantity of solvent, and some properties of the solutions that result. We will be particularly concerned with aqueous solutions of ionic substances because of their central importance in chemistry and in our daily lives. Near the end of the chapter we will consider a type of mixture, known as a colloid, that is on the borderline between heterogeneous mixtures and solutions. Before we examine these topics, however, it is useful to discuss ways of describing the concentrations of solutions, that is, the amount of solute dissolved in a given quantity of solvent or solution.

Table 13.1 Examples of Solutions

State of solution	State of solvent	State of solute	Example
Gas	Gas	Gas	Air
Liquid	Liquid	Gas	Oxygen in water
Liquid	Liquid	Liquid	Alcohol in water
Liquid	Liquid	Solid	Salt in water
Solid	Solid	Gas	Hydrogen in platinum
Solid	Solid	Liquid	Mercury in silver
Solid	Solid	Solid	Silver in gold (certain alloys)

13.1 WAYS OF EXPRESSING CONCENTRATION

The concentration of a solution can be expressed either qualitatively or quantitatively. The terms *dilute* and *concentrated* are used to describe a solution qualitatively. A solution with a relatively small concentration of solute is said to be dilute; one with a large concentration is said to be concentrated.

Several quantitative expressions of concentration are employed in chemistry. One of the simplest is the **weight percentage**. The weight percentage of a component of a solution is given by

$$\text{Wt \% of component} = \frac{\text{mass of component in soln}}{\text{total mass of soln}} \times 100 \qquad [13.1]$$

where we have abbreviated "solution" as "soln." A *percent* is defined as a part per hundred. Thus, if a solution of hydrochloric acid is 36 percent HCl by weight, it has 36 g of HCl for each 100 g of solution.

For very dilute solutions, concentrations are often expressed in **parts per million (ppm):**

$$\text{ppm of component} = \frac{\text{mass of component in soln}}{\text{total mass of soln}} \times 10^6 \qquad [13.2]$$

A solution whose solute concentration is 1 ppm contains 1 g of solute for each million (10^6) grams of solution, or equivalently, 1 mg of solute per kilogram of solution. Because dilute aqueous solutions have densities of 1 g/mL = 1 kg/L, 1 ppm also corresponds to 1 mg of solute per liter of solution. For example, the maximum allowable concentration of arsenic in drinking water in the United States is 0.05 ppm (that is, 0.05 mg of arsenic per liter of solution). For solutions that are even more dilute, parts per billion (ppb) is used. A concentration of 1 ppb represents 1 g of solute per billion (10^9) grams of solution.

SAMPLE EXERCISE 13.1

(a) A solution is made containing 6.9 g of $NaHCO_3$ per 100 g of water. What is the weight percentage of solute in this solution? **(b)** A 2.5-g sample of ground water was found to contain 5.4 micrograms (μg) of Zn^{2+}. What is the concentration of Zn^{2+} in parts per million?

Solution

(a)
$$\text{Wt \% of solute} = \frac{\text{mass solute}}{\text{mass soln}} \times 100$$
$$= \frac{6.9\ \text{g}}{6.9\ \text{g} + 100\ \text{g}} \times 100 = 6.5\%$$

Notice that the mass of solution is the sum of the mass of solvent and the mass of solute. The weight percentage of solvent in this solution is $(100 - 6.5)\% = 93.5\%$.

(b) Because 1 μg is 1×10^{-6} g, 5.4 μg = 5.4×10^{-6} g. Thus,

$$\text{ppm} = \frac{\text{mass of solute}}{\text{mass of soln}} \times 10^6$$
$$= \frac{5.4 \times 10^{-6}\ \text{g}}{2.5\ \text{g}} \times 10^6 = 2.2\ \text{ppm}$$

PRACTICE EXERCISE

A commercial bleaching solution contains 3.62 weight percent sodium hypochlorite, NaOCl. What is the mass of NaOCl in a bottle containing 2500 g of bleaching solution? *Answer:* 90.5 g of NaOCl

Mole Fraction, Molarity, and Molality

Several concentration expressions are based on the number of moles of one or more components of the solution. Three are commonly used in chemistry: mole fraction, molarity, and molality. We encountered mole fraction earlier, in Section 10.6. The *mole fraction* of a component of a solution is given by

$$\text{Mole fraction of component} = \frac{\text{moles component}}{\text{total moles all components}} \qquad [13.3]$$

The symbol X is commonly used for mole fraction, with a subscript to indicate the component on which attention is being focused. For example, the mole fraction of HCl in a hydrochloric acid solution can be represented as X_{HCl}. The sum of the mole fractions of all components of a solution must equal 1.

SAMPLE EXERCISE 13.2

Calculate the mole fraction of HCl in a solution of hydrochloric acid containing 36 percent HCl by weight.

Solution: Assume that there is 100 g of solution. (You can verify for yourself that assuming any other quantity will not change the result, though it can make the arithmetic more difficult.) The solution therefore contains 36 g of HCl and 64 g of H_2O.

$$\text{Moles HCl} = (36 \text{ g HCl})\left(\frac{1 \text{ mol HCl}}{36.5 \text{ g HCl}}\right) = 0.99 \text{ mol HCl}$$

$$\text{Moles } H_2O = (64 \text{ g } H_2O)\left(\frac{1 \text{ mol } H_2O}{18 \text{ g } H_2O}\right) = 3.6 \text{ mol } H_2O$$

$$X_{HCl} = \frac{\text{moles HCl}}{\text{moles } H_2O + \text{moles HCl}} = \frac{0.99}{3.6 + 0.99} = \frac{0.99}{4.6} = 0.22$$

PRACTICE EXERCISE

Calculate the mole fraction of NaOCl in a commercial bleach solution containing 3.62 weight percent NaOCl. ***Answer:*** 9.00×10^{-3}

Molarity (M), which we discussed in Section 4.1, is defined as the number of moles of solute in a liter of solution:

$$\text{Molarity} = \frac{\text{moles solute}}{\text{liters soln}} \qquad [13.4]$$

SAMPLE EXERCISE 13.3

(a) Calculate the molarity of an ascorbic acid, $C_6H_8O_6$ (vitamin C), solution prepared by dissolving 1.80 g in enough water to make 125 mL of solution. **(b)** How many milliliters of this solution contain 0.0100 mol of ascorbic acid?

Solution: **(a)** We use the molecular weight of $C_6H_8O_6$ (176 amu) to figure the number of moles of the substance:

$$(1.80 \text{ g } C_6H_8O_6)\left(\frac{1 \text{ mol } C_6H_8O_6}{176 \text{ g } C_6H_8O_6}\right) = 0.0102 \text{ mol } C_6H_8O_6$$

Using the number of moles of $C_6H_8O_6$ and the volume in liters (125 mL = 0.125 L), we have

$$\text{Molarity} = \frac{\text{mol } C_6H_8O_6}{\text{L soln}} = \frac{0.0102 \text{ mol } C_6H_8O_6}{0.125 \text{ L soln}} = 0.0818\ M$$

(b) Rearranging Equation 13.4, solving for liters, we have

$$\text{Liters soln} = \text{moles solute} \times \frac{1}{\text{molarity}}$$

$$= (0.0100 \text{ mol } C_6H_8O_6)\left(\frac{1 \text{ L soln}}{0.0818 \text{ mol } C_6H_8O_6}\right) = 0.122 \text{ L}$$

$$\text{Milliliters soln} = (0.122 \text{ L})\left(\frac{10^3 \text{ mL}}{1 \text{ L}}\right) = 122 \text{ mL}$$

PRACTICE EXERCISE

Calculate the molarity of a commercial bleach solution that contains 9.65 g of CaCl(OCl) per liter. ***Answer:*** 0.0760 *M*

The **molality** (*m*) of a solution is defined as the number of moles of solute in a kilogram of solvent:

$$\text{Molality} = \frac{\text{moles solute}}{\text{kilograms of solvent}} \qquad [13.5]$$

Notice the difference between molality and molarity. Because these two ways of expressing concentration are so similar, they can be easily confused. Molality is defined in terms of the mass of *solvent*, but molarity is defined in terms of the volume of *solution*. A 1.50 molal (written 1.50 *m*) solution contains 1.50 mol of solute for every kilogram of solvent. A 1.50 *M* solution contains 1.50 mol of solute for every liter of solution. (When water is the solvent, the molality and molarity of a dilute solution are numerically about the same, because 1 kg of solvent is nearly the same as 1 kg of solution, and 1 kg of the solution has a volume of about 1 L.)

The molality of a given solution does not vary with temperature, because masses do not vary with temperature. Molarity, however, changes with temperature because of the expansion or contraction of the solution.

SAMPLE EXERCISE 13.4

What is the molality of a solution made by dissolving 5.0 g of toluene (C_7H_8) in 225 g of benzene (C_6H_6)?

Solution: We determine the number of moles of solute, C_7H_8, by using its molecular weight (92 amu):

$$(5.0 \text{ g } C_7H_8)\left(\frac{1 \text{ mol } C_7H_8}{92 \text{ g } C_7H_8}\right) = 0.054 \text{ mol } C_7H_8$$

Using the number of moles of C_7H_8 and the number of kilograms of solvent (225 g = 0.225 kg), we have

$$\text{Molality} = \frac{\text{mol } C_7H_8}{\text{kg } C_6H_6} = \frac{0.054 \text{ mol } C_7H_8}{0.225 \text{ kg } C_6H_6} = 0.24\ m$$

PRACTICE EXERCISE

Determine the molality of a solution that contains 36.5 g of naphthalene, $C_{10}H_8$, in 420 g of toluene, C_7H_8. ***Answer:*** 0.678 *m*

For most purposes in ordinary chemical laboratory work, molarity is the most useful expression of concentration. However, the other ways we have just considered find use in special situations, and you should be familiar with them. Sample Exercise 13.5 shows how the various expressions of concentration are related.

SAMPLE EXERCISE 13.5

Given that the density of a solution of 5.0 g of toluene and 225 g of benzene (Sample Exercise 13.4) is 0.876 g/mL, calculate the concentration of the solution in **(a)** molarity; **(b)** mole fraction of solute; **(c)** weight percentage of solute.

Solution: **(a)** The total mass of the solution is equal to the mass of the solvent plus the mass of the solute:

$$\text{Mass soln} = 5.0\text{ g} + 225\text{ g} = 230\text{ g}$$

The density of the solution is used to convert the mass of the solution to its volume:

$$\text{Milliliters soln} = (230\text{ g})\left(\frac{1\text{ mL}}{0.876\text{ g}}\right) = 263\text{ mL}$$

Density must be known in order to interconvert molarity and molality, because one is based on mass and the other is based on volume. The number of moles of solute must be known to calculate either molarity or molality:

$$\text{Moles } C_7H_8 = (5.0\text{ g } C_7H_8)\left(\frac{1\text{ mol } C_7H_8}{92\text{ g } C_7H_8}\right) = 0.054\text{ mol}$$

Molarity is moles of solute per liter of solution:

$$\text{Molarity} = \frac{\text{moles } C_7H_8}{\text{liter soln}} = \left(\frac{0.054\text{ mol } C_7H_8}{263\text{ mL soln}}\right)\left(\frac{1000\text{ mL soln}}{1\text{ L soln}}\right) = 0.21\ M$$

Compare this value with the molality of the same solution calculated in Sample Exercise 13.4, 0.24 *m*.

(b) The mole fraction of solute is expressed as

$$X_{C_7H_8} = \frac{\text{moles } C_7H_8}{\text{moles } C_7H_8 + \text{moles } C_6H_6}$$

We have already calculated the number of moles of C_7H_8.

$$\text{Moles } C_6H_6 = (225\text{ g } C_6H_6)\left(\frac{1\text{ mol } C_6H_6}{78.0\text{ g } C_6H_6}\right) = 2.88\text{ mol}$$

$$X_{C_7H_8} = \frac{0.054\text{ mol}}{0.054\text{ mol} + 2.88\text{ mol}} = \frac{0.054}{2.93} = 0.018$$

(c) The weight percentage of solute is calculated as follows:

$$\text{Wt \% } C_7H_8 = \frac{5.0\text{ g } C_7H_8}{5.0\text{ g } C_7H_8 + 225\text{ g } C_6H_6} \times 100 = \frac{5.0\text{ g}}{230\text{ g}} \times 100 = 2.2\%$$

PRACTICE EXERCISE

A solution containing equal masses of glycerol, $C_3H_8O_3$, and water has a density of 1.10 g/mL. Calculate the **(a)** molarity; **(b)** mole fraction of glycerol; **(c)** molality of the solution. ***Answers:*** **(a)** 5.97 *M*; **(b)** mole fraction $C_3H_8O_3 = 0.163$; **(c)** 10.9 *m*

13.2 THE SOLUTION PROCESS

A solution is formed when one substance disperses uniformly throughout another. With the exception of gas mixtures, all solutions involve substances in a condensed phase. We learned in Chapter 11 that substances in the liquid and solid state experience intermolecular attractive forces that hold the individual particles together. Intermolecular forces also operate between a solute particle and the solvent that surrounds it.

A CLOSER LOOK: Normality

The concentration units that we have just discussed appear throughout the rest of this chapter and elsewhere in this text. However, one further concentration expression, called *normality*, may be encountered in other places. Normality (abbreviated *N*) is defined as the number of *equivalents* of solute per liter of solution:

$$\text{Normality} = \frac{\text{equivalents solute}}{\text{liters soln}} \qquad [13.6]$$

An equivalent is defined according to the type of reaction being examined. For acid-base reactions, an equivalent of an acid is the quantity that supplies 1 mol of H^+; an equivalent of a base is the quantity reacting with 1 mol of H^+. In an oxidation-reduction reaction, an equivalent is the quantity of substance that gains or loses 1 mol of electrons.

An equivalent is always defined in such a way that 1 equivalent of reagent A will react with 1 equivalent of reagent B. For example, in acid-base reactions, 1 mol H^+ (1 equivalent of acid) reacts with 1 mol OH^- (1 equivalent of base). Similarly, in oxidation-reduction reactions, for each mole of electrons lost by one substance (1 equivalent), 1 mol must be gained by another substance (1 equivalent).

Table 13.2 gives the masses of 1 equivalent of several substances in either acid-base or oxidation-reduction reactions. For example, when H_2SO_4 reacts as an acid to form SO_4^{2-}, it loses *two* H^+ ions. Thus, 1 mol of H_2SO_4 (98.0 g) will be 2 equivalents. If 1 mol of H_2SO_4 is dissolved in sufficient water to form 1 L of solution, its concentration can be expressed as either 1 *M* or 2 *N*:

$$\frac{1\ \text{mol}}{1\ \text{L}} = 1\ M \qquad \frac{2\ \text{equivalents}}{1\ \text{L}} = 2\ N$$

Likewise, a 0.255 *M* solution of H_2SO_4 will be $2 \times 0.255 = 0.510\ N$. The normality of the H_2SO_4 solution is twice its molarity. Normality is always a whole-number multiple of molarity. In acid-base reactions, the whole number is the number of H^+ or OH^- available in a formula unit of the substance. In oxidation-reduction reactions, the whole number is the number of electrons gained or lost by one formula unit of the substance.

Table 13.2 Equivalent-Mass Relationships

Reactant	Product	Reaction type	Mass of 1 mol of reactant (g)	Mass of 1 equivalent of reactant (g)
H_2SO_4	SO_4^{2-}	Acid (2 H^+)	98.0	98.0/2 = 49.0
$Al(OH)_3$	Al^{3+}	Base (3 OH^-)	78.0	78.0/3 = 26.0
$KMnO_4$	Mn^{2+}	Reduction (5 e^-)	158.0	158.0/5 = 31.6
$KMnO_4$	MnO_2	Reduction (3 e^-)	158.0	158.0/3 = 52.7
$Na_2C_2O_4$	CO_2	Oxidation (2 e^-)	134.0	134.0/2 = 67.0

Any of the various kinds of intermolecular forces that we discussed in Chapter 11 can operate between solute and solvent particles in a solution. As a general rule, we expect solutions to form when the attractive forces between solute and solvent are comparable in magnitude with those that exist between the solute particles themselves or between the solvent particles themselves. For example, the ionic substance NaCl dissolves readily in water because of the attractive interaction between the ions and the polar H_2O molecules.

When NaCl is added to water, the water molecules orient themselves on the surface of the NaCl crystals, as shown in Figure 13.1. The positive end of the water dipole is oriented toward the Cl^- ions, and the negative end of the water dipole is oriented toward the Na^+ ions. The ion-dipole attractions between Na^+ and Cl^- ions and water molecules are sufficiently strong to pull these ions from their positions in the crystal.

Once separated from the crystal, the Na^+ and Cl^- ions are surrounded by water molecules, as shown in Figure 13.2. Such interactions between solute and solvent molecules are known as **solvation**. When the solvent is water, the interactions are known as **hydration**.

Figure 13.1 Interactions between H_2O molecules and the Na^+ and Cl^- ions of a NaCl crystal.

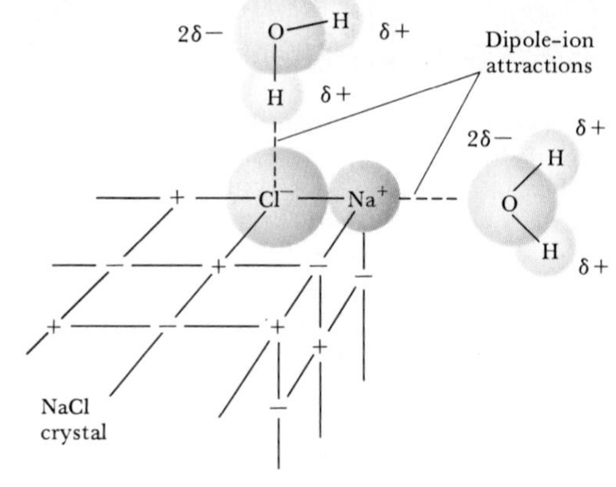

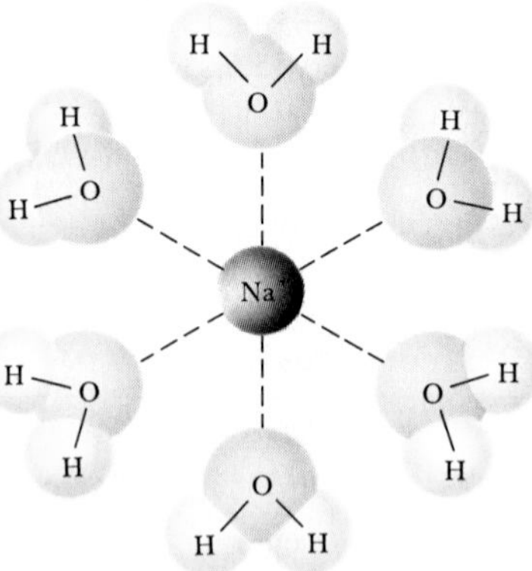

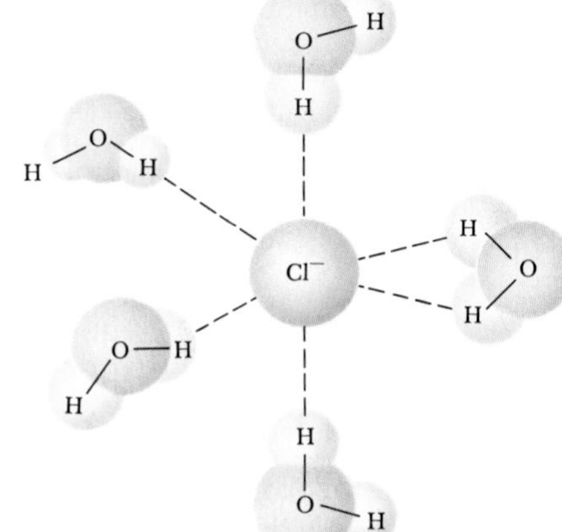

Figure 13.2 Hydrated Na^+ and Cl^- ions. The negative ends of the water dipole point toward the positive ion. The positive ends of the water dipole point toward the negative ion. We do not know whether one or both positive hydrogens are oriented toward the negative ion.

Energy Changes and Solution Formation

Sodium chloride dissolves in water because the water molecules have a sufficient attraction for the Na^+ and Cl^- ions to overcome the attraction of these two ions for one another in the crystal. To form an aqueous solution of NaCl, water molecules must also separate from one another to make room for the solute particles. This example suggests that there are three attractive interactions involved in solution formation:

1. Solute-solute interactions
2. Solvent-solvent interactions
3. Solute-solvent interactions

The net enthalpy change in forming a solution is the sum of the enthalpy changes that correspond to each of these three interactions:

$$\Delta H_{\text{soln}} = \Delta H_1 + \Delta H_2 + \Delta H_3$$

A CLOSER LOOK: Hydrates

Frequently, hydrated ions remain in crystalline salts that are obtained by evaporation of water from aqueous solutions. Common examples include $FeCl_3 \cdot 6H_2O$ [iron(III) chloride hexahydrate] and $CuSO_4 \cdot 5H_2O$ [copper(II) sulfate pentahydrate]. The $FeCl_3 \cdot 6H_2O$ consists of $Fe(H_2O)_6{}^{3+}$ and Cl^- ions; the $CuSO_4 \cdot 5H_2O$ consists of $Cu(H_2O)_4{}^{2+}$ and $SO_4(H_2O)^{2-}$ ions. Water molecules can also occur in positions in the crystal lattice that are not specifically associated with either a cation or anion. $BaCl_2 \cdot 2H_2O$ (barium chloride dihydrate) is an example. Compounds such as $FeCl_3 \cdot 6H_2O$, $CuSO_4 \cdot 5H_2O$, and $BaCl_2 \cdot 2H_2O$, which contain a salt and water combined in definite proportions, are known as *hydrates;* the water associated with them is called *water of hydration.* Figure 13.3 shows an example of a hydrate and the corresponding anhydrous (water-free) substance.

Figure 13.3 Samples of hydrated copper sulfate, $CuSO_4 \cdot 5H_2O$ (left) and anhydrous $CuSO_4$. (Dr. E. R. Degginger)

Figure 13.4 depicts the enthalpy change for each process. Separation of the solute particles from one another requires an input of energy to overcome their attractive interactions. This process is therefore endothermic (positive ΔH_1). The separation of solvent molecules from one another is similarly endothermic (positive ΔH_2). These first two steps require that energy be added to the system. However, when the solute and solvent particles are allowed to interact, we have an exothermic process (negative ΔH_3) that releases energy from the system.

As shown in Figure 13.4, the overall enthalpy change, ΔH_{soln}, can be either exothermic or endothermic. For example, when sodium hydroxide, NaOH, is added to water, the resultant solution gets quite warm; $\Delta H_{soln} = -44.48$ kJ/mol.

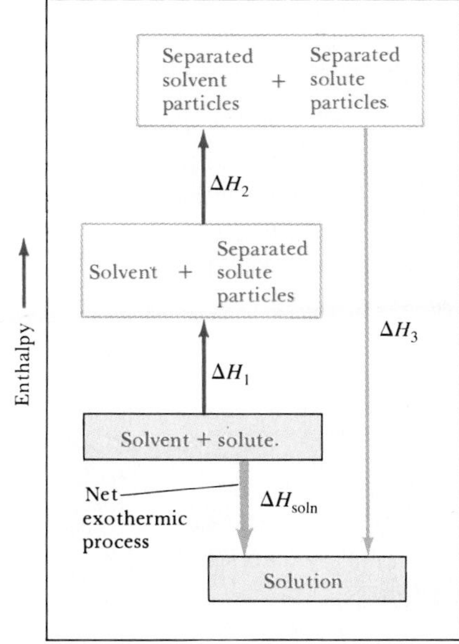

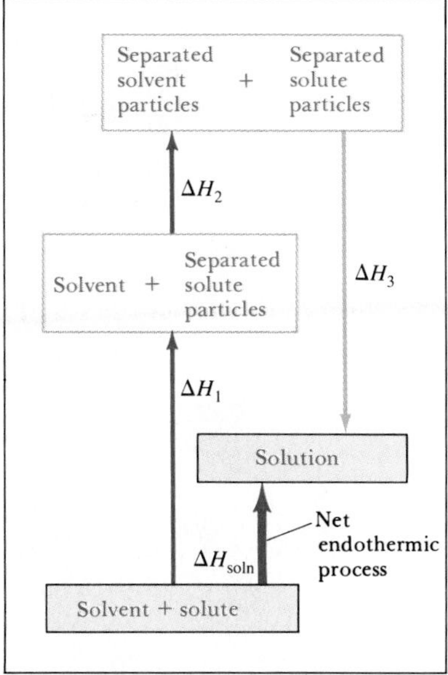

Figure 13.4 Analysis of the enthalpy changes accompanying the solution process: ΔH_1 is the enthalpy required to separate solute particles; ΔH_2 is the enthalpy required to separate solvent particles; ΔH_3 is the enthalpy released when solute and solvent particles interact with each other. The figure on the left shows a net exothermic heat of solution, whereas the one on the right shows a net endothermic heat of solution.

In contrast, the dissolution of ammonium nitrate, NH_4NO_3, is endothermic: $\Delta H_{soln} = 26.4$ kJ/mol. Ammonium nitrate has been used to make instant ice packs, which are used to treat athletic injuries (Figure 13.5).

A solution will not form if ΔH_{soln} is too endothermic. The solvent-solute interaction must be strong enough to make ΔH_3 comparable in quantity to $\Delta H_1 + \Delta H_2$. This is why ionic solutes like NaCl do not dissolve in nonpolar liquids like gasoline. The nonpolar hydrocarbon molecules of the gasoline would experience only weak attractive interactions with the ions, and these interactions would not go very far toward compensating for the energies required to separate the ions from one another.

By similar reasoning, we can understand why polar liquids such as water do not form solutions with nonpolar liquids such as carbon tetrachloride, CCl_4. The water molecules experience strong hydrogen-bonding interactions with one another (Section 11.2); these attractive forces would need to be overcome to disperse the water molecules throughout the nonpolar liquid. The energy required to separate the H_2O molecules is not recovered in the form of attractive interactions between H_2O and CCl_4 molecules.

Solution Formation, Spontaneity, and Disorder

When two nonpolar substances such as CCl_4 and hexane, C_6H_{14}, are mixed, they readily dissolve in one another in all proportions. The attractive forces between molecules in both of these substances are London dispersion forces. The two substances have similar boiling points: CCl_4 boils at 77°C, and C_6H_{14} boils at 69°C. It is therefore reasonable to suppose that the magnitudes of the attractive forces between molecules are comparable in the two substances. When the two are mixed, there is little or no energy change. Yet the dissolving process occurs spontaneously; that is, it occurs to an appreciable extent without any extra input of energy from outside the system. Two distinct factors are involved in processes that occur spontaneously. The most obvious is energy; the other is disorder.

Figure 13.5 Photo of instant ice pack, ammonium nitrate, used to treat athletic injuries. To activate the ice pack, the container is kneaded, breaking the seal separating solid NH_4NO_3 from water. (© Richard Megna/Fundamental Photographs)

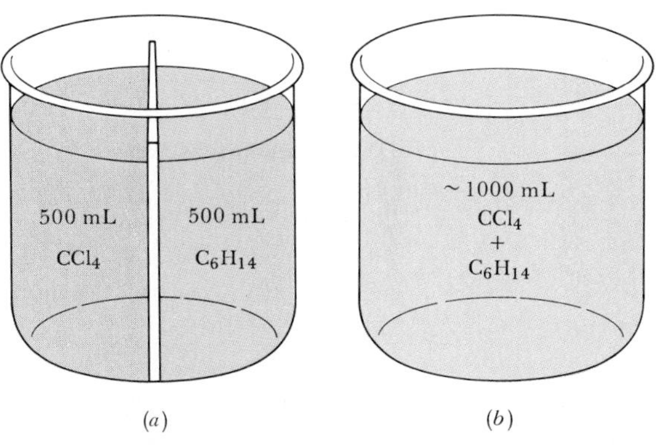

Figure 13.6 Formation of a homogeneous solution between CCl_4 and C_6H_{14} upon removal of a barrier separating the two liquids. The solution in (*b*) is more disordered, or random in character, than the separate liquids before solution formation (*a*).

If you let go of a book, it falls to the floor because of gravity. At its initial height it has a potential energy higher than when it is on the floor. Unless it is restrained, the book falls and so loses energy. This fact leads us to the first basic principle identifying spontaneous processes and the direction they take: *Processes in which the energy content of the system decreases tend to occur spontaneously.* Spontaneous processes tend to be exothermic (Section 5.2). Change occurs in the direction that leads to a lower energy content.

However, we can think of processes that do not result in a lower energy—or that may even be endothermic—and still occur spontaneously. For example, NH_4NO_3 readily dissolves in water, even though the solution process is endothermic; $\Delta H_{soln} = 26.4$ kJ/mol. The mixing of CCl_4 and C_6H_{14} provides another simple example. All such processes are characterized by an increase in the disorder, or randomness, of the system. Suppose that we could suddenly remove a barrier that separates 500 mL of CCl_4 from 500 mL of C_6H_{14}, as in Figure 13.6(*a*). Before the barrier is removed, each liquid occupies a volume of 500 mL. We know that we can find all the CCl_4 molecules in the 500 mL to the left of the barrier, and all the C_6H_{14} molecules in the 500 mL to the right. When equilibrium has been established after removal of the barrier, the two liquids together occupy a volume of about 1000 mL.* Formation of a homogeneous solution has resulted in an increased disorder, or randomness, in that the molecules of each substance are now distributed in a volume twice as large as that which they occupied before mixing. This example illustrates our second basic principle: *Processes in which the disorder of the system increases tend to occur spontaneously.*

When molecules of different types are brought together, an increase in disorder occurs spontaneously unless the molecules are restrained by sufficiently strong intermolecular forces or by physical barriers. Thus, because of the strong bonds holding the sodium and chloride ions together, sodium chloride does not spontaneously dissolve in gasoline. Conversely, gases spontaneously expand unless restrained by their containers; in this case intermolecular forces are too weak to restrain the molecules. We shall discuss spontaneous processes again in Chapter 19. At that time, we shall consider the balance between the tendencies toward lower energy and toward increased disorder in greater detail. For the moment, we need to be aware that formation of a solution is always favored by the increase in disorder that accompanies mixing. Consequently, a solution will form unless solute-solute or solvent-solvent interactions are too strong relative to the solute-solvent interactions.

* The slight change in total volume that may occur upon mixing is unimportant for our example.

Figure 13.7 (*a*) Nickel metal and hydrochloric acid. (*b*) Nickel reacts slowly with hydrochloric acid, forming $NiCl_2(aq)$ and $H_2(g)$. (*c*) $NiCl_2(s)$ is obtained when the solution from (*b*) is evaporated to dryness. (© Richard Megna/Fundamental Photographs)

(*a*) (*c*) (*b*)

Solution Formation and Chemical Reactions

In all our discussions of solutions we must be careful to distinguish the *physical* process of solution formation from *chemical* processes that lead to a solution. For example, nickel metal is dissolved on contact with hydrochloric acid solution. The following chemical reaction occurs (see Section 4.6):

$$Ni(s) + 2HCl(aq) \longrightarrow H_2(g) + NiCl_2(aq) \qquad [13.7]$$

In this instance, the chemical form of the substance being dissolved is changed. If the solution is evaporated to dryness, Ni(*s*) is not recovered as such; instead, $NiCl_2(s)$ is recovered (Figure 13.7). In contrast, when NaCl(*s*) is dissolved in water, it can be recovered by evaporation of its solution to dryness. Our focus throughout this chapter is on solutions from which the solute can be recovered unchanged from the solution.

13.3 SATURATED SOLUTIONS AND SOLUBILITY

As a solid solute begins to dissolve in a solvent, the concentration of solute particles in solution increases, so the chances of their colliding with the surface of the solid increase (Figure 13.8). Such a collision may result in the solute particle becoming attached to the solid. This process, which is the opposite of the solution process, is called **crystallization**. Thus, two opposing processes occur in a solution in contact with undissolved solute. This situation is represented in Equation 13.8 by use of a double arrow:

$$\text{Solute} + \text{solvent} \underset{\text{crystallize}}{\overset{\text{dissolve}}{\rightleftharpoons}} \text{solution} \qquad [13.8]$$

When the rates of these opposing processes become equal, no further net increase in the amount of solute in solution occurs. This balance is an example of *dynamic equilibrium*, similar to that discussed in Section 11.5, where the processes of evaporation and condensation were considered.

A solution that is in equilibrium with undissolved solute is said to be **saturated**. Additional solute will not dissolve if added to a saturated solution. The amount of solute needed to form a saturated solution in a given quantity of solvent is known as the **solubility** of that solute. For example, the solubility of NaCl in water at 0°C is 35.7 g per 100 mL of water. This is the maximum amount of NaCl that can be dissolved in water to give a stable, equilibrium solution at that temperature.

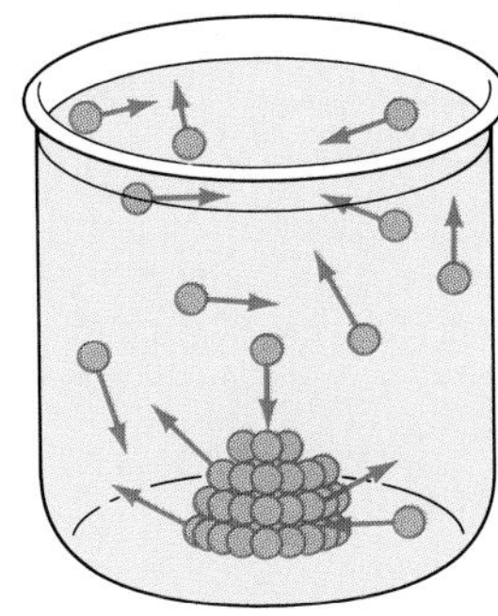

Figure 13.8 Movement of solute particles in solvent containing excess solute. Both dissolution and crystallization occur.

It is, of course, possible to dissolve less solute than that needed to form a saturated solution. Thus, we might choose to form a solution containing only 10.0 g of NaCl per 100 mL of water at 0°C. The solution is then said to be **unsaturated**.

It is also sometimes possible to form solutions that contain a greater amount of solute than that needed to form a saturated solution. Such solutions are said to be **supersaturated**. These solutions can sometimes be prepared by saturating a solution at a high temperature and then carefully cooling it to a temperature at which the solute is less soluble. Supersaturated solutions are unstable, and under proper conditions solute will crystallize from them to give saturated solutions, as shown in Figure 13.9.

Supersaturated solutions of $Na_2S_2O_3$ are used to make instant heat packs to warm skiers' hands. The exothermic crystallization of $Na_2S_2O_3$ from the supersaturated solution is initiated by squeezing out a seed crystal from a small compartment in the pack. The heat pack can be recycled by placing it in boiling water to redissolve the $Na_2S_2O_3$. The pack can be reused several times until the supply of seed crystals runs out.

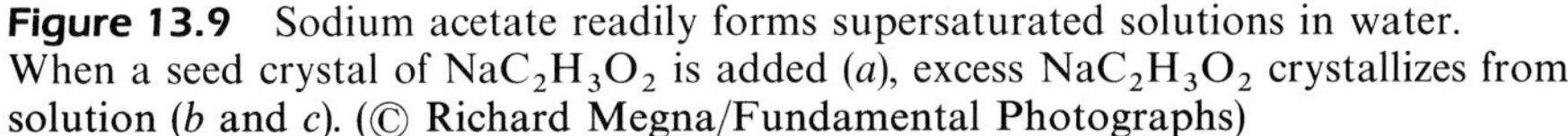

Figure 13.9 Sodium acetate readily forms supersaturated solutions in water. When a seed crystal of $NaC_2H_3O_2$ is added (*a*), excess $NaC_2H_3O_2$ crystallizes from solution (*b* and *c*). (© Richard Megna/Fundamental Photographs)

(*a*)

(*b*)

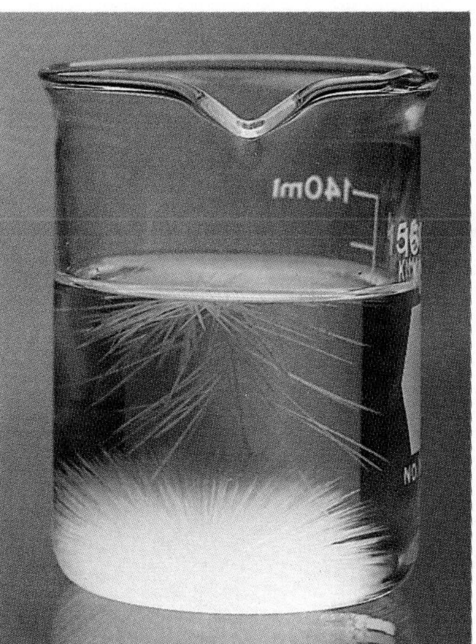

(*c*)

13.4 FACTORS AFFECTING SOLUBILITY

The extent to which one substance dissolves in another depends on the nature of both the solute and the solvent. It also depends on temperature and, at least for gases, on pressure.

Solute-Solvent Interactions

Our discussion of the solution process enables us to understand many observations regarding solubilities. As a simple example, consider the data in Table 13.3 for the solubilities of various simple gases in water. Note that solubility increases with increasing molecular mass. The attractive forces between the gas and solvent molecules are mainly of the London dispersion type, which increase with increasing size and mass of the gas molecules. When a chemical reaction occurs between the gas and solvent, much higher gas solubilities result. We will encounter instances of this in later chapters, but as a simple example, the solubility of Cl_2 in water under the same conditions given in Table 13.3 is 0.102 *M*. This is much higher than would be predicted from the trends in the table, based just on molecular mass. We can infer from this that the dissolving of Cl_2 in water is accompanied by some kind of chemical process. The use of chlorine as a bactericide in municipal water supplies and swimming pools is based on its chemical reaction with water (Section 18.6, Equation 18.17).

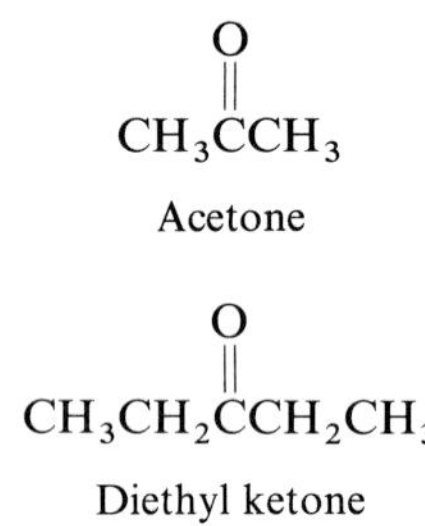

Polar liquids tend to dissolve readily in polar solvents. For example, acetone, a polar molecule whose structure is shown at the left, mixes in all proportions with water. Pairs of liquids that mix in all proportions are said to be **miscible**, and liquids that do not mix are termed **immiscible**. Water and hexane, C_6H_{14}, for example, are immiscible. Diethyl ketone (see structure at left), which is similar to acetone but has a higher molecular mass, dissolves in water to the extent of about 47 g per liter of water at 20°C, but it is not completely miscible.

Hydrogen-bonding interactions between solute and solvent may lead to high solubility. For example, water is completely miscible with ethanol, CH_3CH_2OH. The CH_3CH_2OH molecules are able to form hydrogen bonds with water molecules as well as with each other (see Figure 13.10). Because of this hydrogen-bonding ability, the solute-solute, solvent-solvent, and solute-solvent forces are not appreciably different within a mixture of CH_3CH_2OH

Table 13.3 Solubilities of Several Gases in Water at 20°C, with 1 atm Gas Pressure

Gas	Solubility (*M*)
N_2	6.9×10^{-4}
CO	1.04×10^{-3}
O_2	1.38×10^{-3}
Ar	1.50×10^{-3}
Kr	2.79×10^{-3}

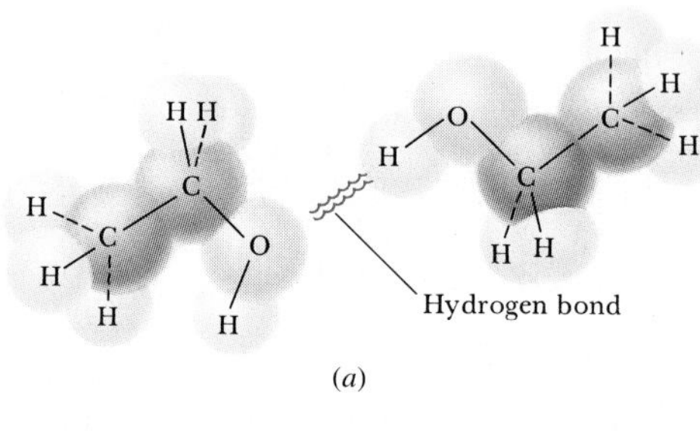

Figure 13.10 Hydrogen-bonding interactions between ethanol molecules (*a*) and between water and ethanol molecules (*b*).

Table 13.4 Solubilities of Some Alcohols in Water

Alcohol	Solubility in H_2O (mol/100 g H_2O at 20°C)[a]
CH_3OH (methanol)	∞
CH_3CH_2OH (ethanol)	∞
$CH_3CH_2CH_2OH$ (propanol)	∞
$CH_3CH_2CH_2CH_2OH$ (butanol)	0.11
$CH_3CH_2CH_2CH_2CH_2OH$ (pentanol)	0.030
$CH_3CH_2CH_2CH_2CH_2CH_2OH$ (hexanol)	0.0058
$CH_3CH_2CH_2CH_2CH_2CH_2CH_2OH$ (heptanol)	0.0008

[a] The infinity symbol indicates that the alcohol is completely miscible in water.

and H_2O. There is no significant change in the environment of the molecules as they are mixed. The increase in disorder accompanying mixing therefore plays a significant role in formation of the solution.

The number of carbon atoms in an alcohol affects its solubility in water, as shown in Table 13.4. As the length of the carbon chain increases, the OH group becomes an ever smaller part of the molecule and the molecule becomes more like a hydrocarbon. The solubility of the alcohol decreases correspondingly. If the number of OH groups along the carbon chain increases, more solute-water hydrogen bonding is possible and solubility generally increases. Glucose, $C_6H_{12}O_6$, has five OH groups on a six-carbon framework, which makes the molecule very soluble in water (83 g dissolve in 100 mL of water at 17.5°C). The glucose molecule is shown in Figure 13.11.

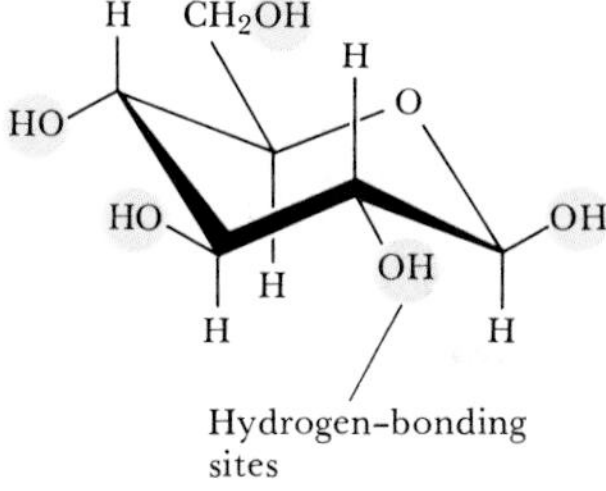

Figure 13.11 Structure of glucose. Colored spheres indicate sites capable of hydrogen bonding with water.

Examination of pairs of substances such as those listed in the preceding paragraphs has led to an important generalization: *Substances with similar intermolecular attractive forces tend to be soluble in one another.* This generalization is often simply stated as *"like dissolves like."* Nonpolar substances are soluble in nonpolar solvents; ionic and polar solutes are soluble in polar solvents. Network solids like diamond and quartz are not soluble in either polar or nonpolar solvents because of the strong bonding forces within the solid.

SAMPLE EXERCISE 13.6

Predict whether each of the following substances is more likely to dissolve in carbon tetrachloride, CCl_4, which is a nonpolar solvent, or in water: C_7H_{16}, $NaHCO_3$, HCl, I_2.

Solution: Both C_7H_{16} and I_2 are nonpolar. We would therefore predict that they would be more soluble in CCl_4 than in H_2O. $NaHCO_3$ is ionic, and HCl is polar covalent. Water would be a better solvent than CCl_4 for these two substances.

PRACTICE EXERCISE

Which of the following is most likely to be miscible with water?

Dioxane: cyclic $\ddot{O}$–CH_2–CH_2–$\ddot{O}$–CH_2–CH_2 CH_3—$\ddot{F}$: $CH_3CH_2CH_2CH_2$—$\ddot{S}H$

Dioxane   Methyl fluoride   Butyl mercaptan

Answer: dioxane

CHEMISTRY AT WORK: Fat- and Water-Soluble Vitamins

Vitamins B and C are water-soluble. Vitamins A, D, E, and K are soluble in nonpolar solvents and in the fatty tissue of the body (which is nonpolar). Because of their water solubility, vitamins B and C are not stored to any appreciable extent in the body, and so foods containing these vitamins should be included in the daily diet. In contrast, the fat-soluble vitamins are stored in sufficient quantities to keep vitamin-deficiency diseases from appearing even after a person has subsisted for a long period on a vitamin-deficient diet. With the ready availability of vitamin supplements, cases of hypervitaminosis, an illness caused by an excessive amount of vitamins, are now being seen by physicians in this country. Because the body can store only the fat-soluble vitamins, true hypervitaminosis has been observed solely for these vitamins.

The different solubility patterns of the water-soluble vitamins and the fat-soluble ones can be rationalized in terms of the structures of the molecules. The chemical structures of vitamin A (retinol) and of vitamin C (ascorbic acid) are shown below. Note that the vitamin A molecule is an alcohol with a very long carbon chain. It is nearly nonpolar, and because the OH group is such a small part of the molecule, the molecule resembles the long-chain alcohols listed in Table 13.4. In contrast, the vitamin C molecule is smaller and has more OH groups that can form hydrogen bonds with water. It is somewhat like glucose, discussed above.

Vitamin A

Vitamin C

Pressure Effects

The solubility of a gas in any solvent is increased as the pressure of the gas over the solvent increases. By contrast, the solubilities of solids and liquids are not appreciably affected by pressure. We can understand the effect of pressure on the solubility of a gas by considering the dynamic equilibrium, illustrated in Figure 13.12. Suppose that we have a gaseous substance distributed between the gas and solution phases. When equilibrium is established, the rate at which gas molecules enter the solution equals the rate at which solute molecules escape from the solution to enter the gas phase. The small arrows in Figure 13.12(*a*) represent the rates of these opposing processes. Now suppose that we exert added

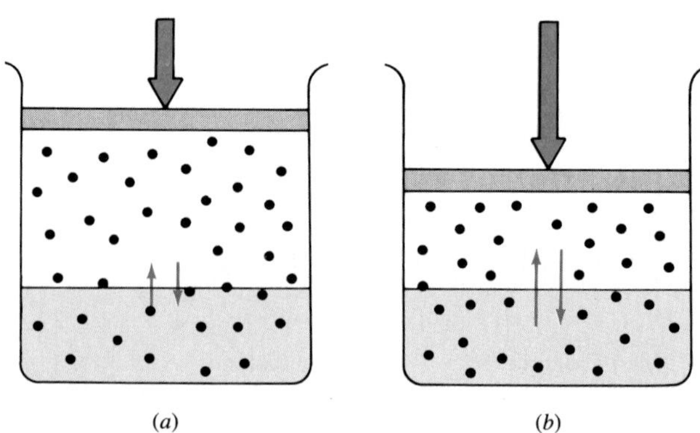

Figure 13.12 Effect of pressure on the solubility of a gas. When the pressure is increased, as in (*b*), the rate at which gas molecules enter the solution increases. The concentration of solute molecules at equilibrium increases in proportion to the pressure.

pressure on the piston and compress the gas above the solution, as shown in Figure 13.12(*b*). If we reduced the volume to half its original value, the pressure of the gas would increase to about twice its original value. The rate at which gas molecules strike the surface to enter the solution phase would therefore increase. Thus, the solubility of the gas in the solution would increase until an equilibrium is again established; that is, solubility increases until the rate at which gas molecules enter the solution equals the rate at which solute molecules escape from the solvent, as indicated by the arrows in Figure 13.12(*b*). Thus, the solubility of the gas should increase in direct proportion to the pressure.

The relationship between pressure and solubility is expressed in terms of a simple equation known as **Henry's law:**

$$C_g = kP_g \qquad [13.9]$$

where C_g is the solubility of the gas in the solution phase, P_g is the partial pressure of the gas over the solution, and k is a proportionality constant known as the *Henry's law constant.* The Henry's law constant is different for each solute-solvent pair. As an example, the solubility of N_2 gas in water at 25°C and 0.78 atm pressure is 5.3×10^{-4} *M*. The Henry's law constant for N_2 in water is $(5.3 \times 10^{-4}$ mol/L)/0.78 atm $= 6.8 \times 10^{-4}$ mol/L-atm. If the partial pressure of N_2 is doubled, Henry's law predicts that the solubility in water will also be doubled to 1.06×10^{-3} *M*.

Bottlers use the effect of pressure on solubility in producing carbonated beverages such as champagne, beer, and many soft drinks. These are bottled under a carbon dioxide pressure slightly greater than 1 atm. When the bottles are opened to the air, the partial pressure of CO_2 above the solution is decreased, and CO_2 bubbles out of the solution.

SAMPLE EXERCISE 13.7

Calculate the concentration of CO_2 in a soft drink that is bottled with a partial pressure of CO_2 of 4.0 atm over the liquid at 25°C. The Henry's law constant for CO_2 in water at this temperature is 3.1×10^{-2} mol/L-atm.

Solution: Henry's law, Equation 13.9, can be applied in a straightforward fashion:

$$\begin{aligned} C_g = kP_g &= (3.1 \times 10^{-2} \text{ mol/L-atm})(4.0 \text{ atm}) \\ &= 0.12 \text{ mol/L} = 0.12\ M \end{aligned}$$

PRACTICE EXERCISE

Calculate the concentration of CO_2 in a soft drink after the bottle is opened and sits at 25°C under a CO_2 partial pressure of 3.0×10^{-4} atm. *Answer:* 9.3×10^{-6} *M*

Temperature Effects

The solubilities of several common gases in water as a function of temperature are graphed in Figure 13.13. These solubilities correspond to a constant pressure of 1 atm of gas over the solution. Note that, in general, solubility decreases with increasing temperature. If a glass of cold tap water is warmed, bubbles of air are seen on the side of the glass. Similarly, a carbonated beverage like soda pop goes "flat" as it is allowed to warm; as the temperature of the solution increases, CO_2 escapes from the solution. The decreased solubility of O_2 in water as temperature increases is one of the effects of the *thermal pollution* of lakes and streams. The effect is particularly serious in deep lakes, because warm

Figure 13.13 Solubilities of several gases in water as a function of temperature. Note that solubilities are in units of millimoles per liter, for a constant pressure of 1 atm in the gas phase.

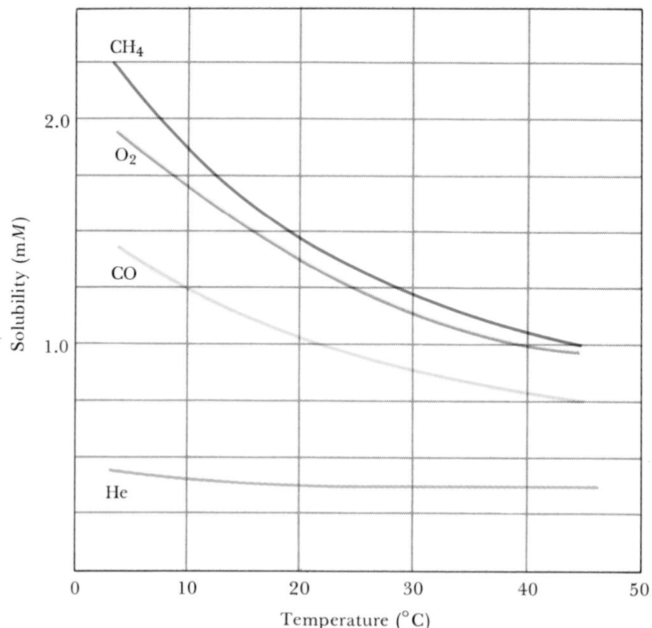

water is less dense than cold water. It therefore tends to remain on top of the cold water, at the surface. This situation impedes the dissolving of oxygen into the deeper layers, thus stifling the respiration of all aquatic life needing oxygen. Fish may suffocate and die in these circumstances.

Figure 13.14 shows the effect of temperature on the solubilities of several ionic substances in water. The relationship here is generally the opposite of that for gases; that is, the solubility of ionic compounds normally increases as the temperature increases. There are a few exceptions to this rule. Note, for example, the curve for $Ce_2(SO_4)_3$.

Figure 13.14 Solubilities of several common ionic solids shown as a function of temperature.

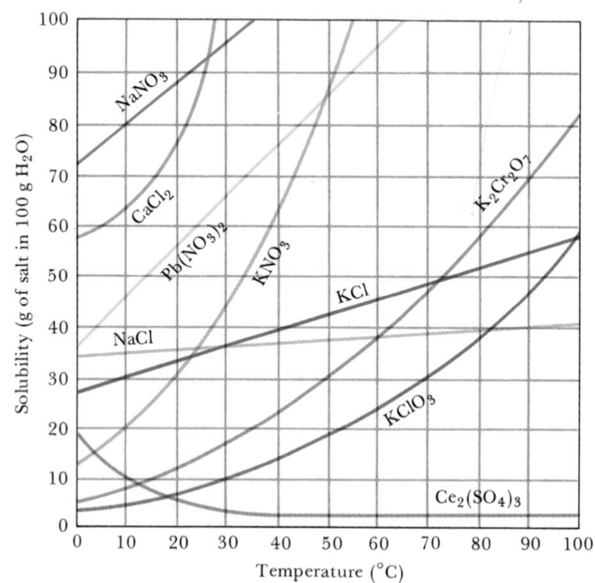

CHEMISTRY AT WORK: Blood Gases and Deep-Sea Diving

Deep-sea divers rely on compressed air for their oxygen supply. According to Henry's law, the solubilities of gases increase with pressure. If a diver is suddenly exposed to atmospheric pressure, where the solubility of gases is less, bubbles form in the bloodstream and in other fluids of the body. These bubbles affect nerve impulses and give rise to the disease known as "the bends," or decompression sickness. Nitrogen is the main problem because it has the highest partial pressure in air and because it can be removed only through the respiratory system. Oxygen is consumed in metabolism. Substitution of helium for nitrogen minimizes this effect because helium has a much lower solubility in biological fluids than does N_2. Jacques Cousteau's divers on *Conshelf III* used a mixture of 98 percent helium and 2 percent oxygen. At the high pressures (10 atm) experienced by the divers, this percentage of oxygen gives an oxygen partial pressure of about 0.2 atm, which is the partial pressure in normal air at 1 atm. If the oxygen partial pressures become too great, the urge to breathe is reduced, CO_2 is not removed from the body, and CO_2 poisoning occurs. At excessive concentrations in the body, carbon dioxide acts as a neurotoxin, interfering with nerve conduction and transmission.

13.5 COLLIGATIVE PROPERTIES

Some physical properties of solutions differ in important ways from those of the pure solvent. For example, pure water freezes at 0°C, but aqueous solutions freeze at lower temperatures. Ethylene glycol is added to the water in radiators of cars as an antifreeze to lower the freezing point of the solution. It also raises the boiling point of the solution above that of pure water, permitting operation of the engine at a higher temperature.

The lowering of the freezing point and the raising of the boiling point are examples of physical properties of solutions that depend on the *concentration* but not the *kind* of solute particles. Such properties are called **colligative properties**. (Colligative means "depending on the collection"; colligative properties depend on the collective effect of the number of solute particles.) In addition to freezing-point lowering and boiling-point elevation, there are two other colligative properties: vapor-pressure reduction and osmotic pressure.

Lowering the Vapor Pressure

The effect of a nonvolatile solute on the vapor pressure of a volatile solvent is illustrated by the simple experiment shown in Figure 13.15. Two beakers are placed side by side in a sealed enclosure. One beaker contains pure water, the

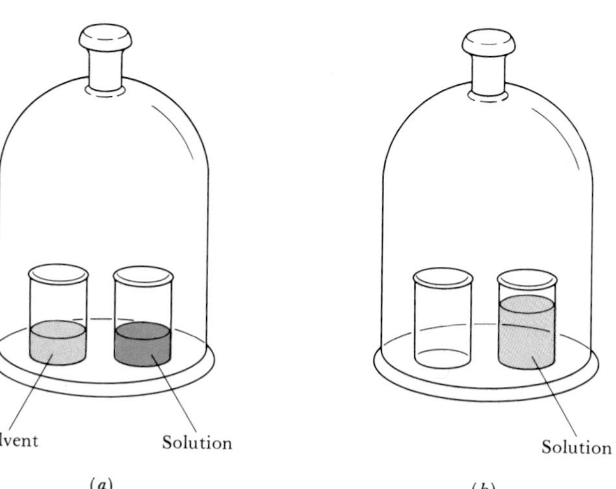

Figure 13.15 Experiment showing that a solution possesses a lower vapor pressure than does the pure solvent.

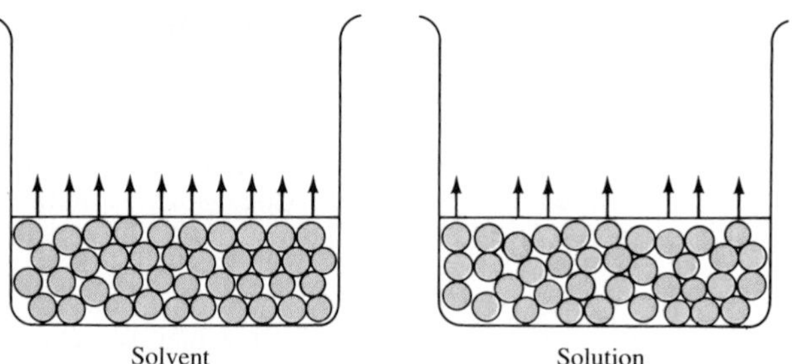

Figure 13.16 A nonvolatile solute reduces the rate of vaporization of the solvent.

other an equal volume of an aqueous solution of sugar. Gradually, the volume of the sugar solution increases, while the volume of the pure water decreases. Eventually, all the water transfers to the sugar solution, as shown in Figure 13.15(*b*). How do we explain this result?

Let's refer back to Section 11.5, particularly Figure 11.15. Recall that the vapor pressure over a liquid is the result of a dynamic equilibrium: The rate at which molecules leave the liquid surface for the gas phase equals the rate at which gas-phase molecules return to the surface of the liquid. A nonvolatile solute added to the liquid reduces the capacity of the solvent molecules to move from the liquid phase to the vapor phase, as shown in Figure 13.16. At the same time, however, there is no change in the rate at which solvent molecules in the gas phase return to the liquid. The shift in equilibrium due to the solute reduces the vapor pressure over the solution. The vapor pressure over the pure solvent is thus higher than that over the solution.

Return to the experiment shown in Figure 13.15. You should now see why the solvent transfers from the beaker holding pure water to the one holding the sugar solution. The vapor pressure necessary to achieve equilibrium with the pure solvent is higher than that required with the solution. Consequently, as the pure solvent seeks to reach equilibrium by forming vapor, the solution seeks to reach equilibrium by removing molecules from the vapor phase. A net movement of solvent molecules from the pure solvent to the solution results. The process continues until no free solvent remains.

The *extent* to which a nonvolatile solute lowers the vapor pressure is proportional to its concentration. Doubling the concentration of solute doubles its effect. In fact, the reduction in vapor pressure is roughly proportional to the total concentration of solute particles, whether they are neutral or charged. For example, 1.0 mol of a nonelectrolyte, like glucose, produces essentially the same reduction in vapor pressure in a given quantity of water as does 0.5 mol of NaCl, a strong electrolyte (Section 4.2). Both solutions have 1.0 mol of particles because 0.5 mol of NaCl dissociates to give 0.5 mol of $Na^+(aq)$ and 0.5 mol of $Cl^-(aq)$.

Raoult's Law

Quantitatively, the vapor pressure of solutions containing nonvolatile solutes is given by **Raoult's law**, Equation 13.10, where P_A is the vapor pressure of the solution, X_A is the mole fraction of the solvent, and P_A° is the vapor pressure of the pure solvent:

$$P_A = X_A P_A^\circ \qquad [13.10]$$

For example, the vapor pressure of water is 17.5 mm Hg at 20°C. Imagine holding the temperature constant while adding glucose, $C_6H_{12}O_6$, to the water so that the resulting solution has $X_{H_2O} = 0.80$ and $X_{C_6H_{12}O_6} = 0.20$. According to Equation 13.10, the vapor pressure of water over the solution will be 14 mm Hg:

$$P_{H_2O} = (0.80)(17.5 \text{ mm Hg}) = 14 \text{ mm Hg}$$

The reduction in vapor pressure, ΔP_A, is $P_A^\circ - P_A$. Using Equation 13.10, we can write this as

$$\begin{aligned} \Delta P_A^\circ &= P_A^\circ - X_A P_A^\circ \\ &= P_A^\circ(1 - X_A) \end{aligned} \quad [13.11]$$

Since $X_A + X_B = 1$, where X_B is the mole fraction of solute particles, $X_B = 1 - X_A$. Thus,

$$\Delta P_A = X_B P_A^\circ \quad [13.12]$$

In our example, $X_{C_6H_{12}O_6} = 0.20$. The vapor pressure will be lowered by 3.5 mm Hg:

$$\Delta P_A = (0.20)(17.5 \text{ mm Hg}) = 3.5 \text{ mm Hg}$$

SAMPLE EXERCISE 13.8

Calculate the reduction in vapor pressure caused by the addition of 100 g of sucrose, $C_{12}H_{22}O_{11}$, to 1000 g of water. The vapor pressure of the pure water at 25°C is 23.8 mm Hg.

Solution

$$X_{C_{12}H_{22}O_{11}} = \frac{\text{moles } C_{12}H_{22}O_{11}}{\text{moles } H_2O + \text{moles } C_{12}H_{22}O_{11}}$$

$$\text{Moles } C_{12}H_{22}O_{11} = (100 \text{ g})\left(\frac{1 \text{ mol } C_{12}H_{22}O_{11}}{342 \text{ g } C_{12}H_{22}O_{11}}\right) = 0.292 \text{ mol}$$

$$\text{Moles } H_2O = (1000 \text{ g})\left(\frac{1 \text{ mol } H_2O}{18.0 \text{ g } H_2O}\right) = 55.5 \text{ mol}$$

$$X_{C_{12}H_{22}O_{11}} = \frac{0.292}{55.5 + 0.292} = \frac{0.292}{55.8}$$

$$\Delta P_A = \left(\frac{0.292}{55.8}\right)(23.8 \text{ mm Hg}) = 0.125 \text{ mm Hg}$$

PRACTICE EXERCISE

The vapor pressure over pure water at 120°C is 1480 mm Hg. If Raoult's law is obeyed, what mole fraction of ethylene glycol must be present in the water to reduce the vapor pressure of water to 1 atm, 760 mm Hg? ***Answer:*** 0.486

Solutions that obey Raoult's law are called **ideal solutions**. Real solutions best approximate ideal behavior when the solute concentration is low and when the solute and solvent are much alike in molecular size and in the strength and type of intermolecular attractions.

A CLOSER LOOK: Ideal Solutions with Two or More Volatile Components

Solutions sometimes have two or more volatile components. For example, gasoline is a complex solution containing several volatile substances. To gain some understanding of such mixtures, consider an ideal solution containing two components, A and B. The partial pressures of A and B vapors above the solution are given by Raoult's law:

$$P_A = X_A P_A^\circ \quad \text{and} \quad P_B = X_B P_B^\circ$$

The total vapor pressure over the solution is the sum of the partial pressures of each volatile component:

$$P_{\text{total}} = P_A + P_B = X_A P_A^\circ + X_B P_B^\circ$$

As an example of such a solution, consider a mixture of benzene, C_6H_6, and toluene, C_7H_8, containing 1 mol of benzene and 2 mol of toluene ($X_{\text{ben}} = 0.33$ and $X_{\text{tol}} = 0.67$). At 20°C, the vapor pressures of the pure substances are

$$\text{Benzene:} \quad P_{\text{ben}}^\circ = 75 \text{ mm Hg}$$

$$\text{Toluene:} \quad P_{\text{tol}}^\circ = 22 \text{ mm Hg}$$

Thus, the partial pressures of benzene and toluene above the solution are

$$P_{\text{ben}} = (0.33)(75 \text{ mm Hg}) = 25 \text{ mm Hg}$$

$$P_{\text{tol}} = (0.67)(22 \text{ mm Hg}) = 15 \text{ mm Hg}$$

The total pressure is

$$P_{\text{total}} = 25 \text{ mm Hg} + 15 \text{ mm Hg} = 40 \text{ mm Hg}$$

These results indicate that the vapor is richer in the more volatile component, benzene. The mole fraction of benzene in the vapor is given by the ratio of its vapor pressure to the total pressure (See Equation 10.17):

$$X_{\text{ben}} \text{ in vapor} = \frac{P_{\text{ben}}}{P_{\text{total}}} = \frac{25 \text{ mm Hg}}{40 \text{ mm Hg}} = 0.63$$

Although benzene comprises only 33 percent of the molecules in the solution, it comprises 63 percent of the molecules in the vapor.

When ideal solutions are in equilibrium with their vapor, the more volatile component of the mixture will be relatively richer in the vapor. This fact forms the basis of the important technique of *distillation*. This technique is used to separate (or partially separate) mixtures containing one or more volatile components. Distillation is the procedure by which a "moonshiner" obtains whiskey using a "still" and by which petrochemical plants achieve the separation of crude petroleum into gasoline, diesel fuel, lubricating oil, and so forth. It is also used routinely on a small scale in the laboratory.. Figure 13.17 shows a simple laboratory distillation apparatus. Imagine that the solution in the distillation flask is a mixture of benzene and toluene. As the mixture is heated to its boiling point, the more volatile benzene will concentrate in the vapor. The vapor is condensed in the cooler part of the apparatus (the condenser) and collected. The liquid may be distilled again to enrich it further in the more volatile component. A specially designed *fractional distillation* apparatus can achieve in a single operation a degree of separation that may require several simple distillations.

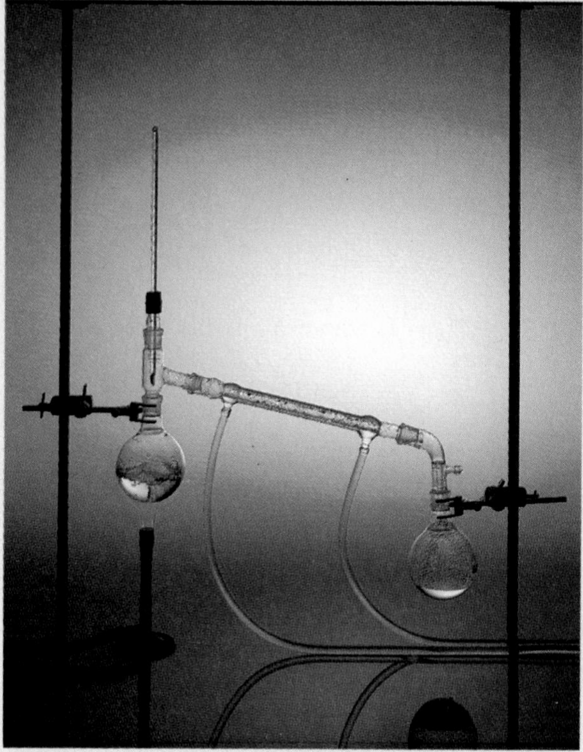

Figure 13.17 Simple laboratory distillation setup, showing the distillation of an aqueous solution. Open flames should be avoided when heating flammable liquids such as benzene and toluene. (© Richard Megna/Fundamental Photographs)

Many solutions are not ideal; that is, they do not obey Raoult's law exactly. The solvent vapor pressure over a solution tends to be greater than predicted by Raoult's law when the intermolecular forces between solvent and solute are weaker than those between solvent and solvent and between solute and solute. Conversely, when the interactions between solute and solvent are

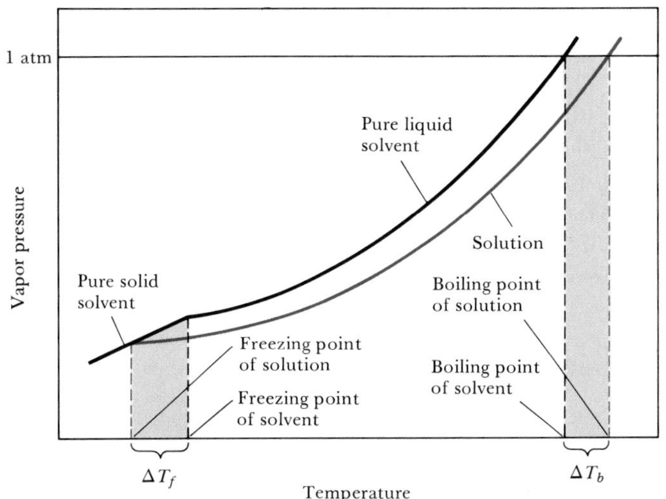

Figure 13.18 Vapor-pressure versus temperature curves of a pure solvent and a solution of a nonvolatile solute, under a constant total pressure of 1 atm. The vapor pressure of the solid solvent is unaffected by the presence of solute if the solid freezes out without containing a significant concentration of solute, as is usually the case.

exceptionally strong, as might be the case when hydrogen bonding or ion-solvent forces exist, the solvent vapor pressure is lower than Raoult's law predicts. Although you should be aware that these departures from an ideal solution do occur, we will ignore them for the remainder of this chapter.

Boiling-Point Elevation

We saw in Section 11.5 that the vapor pressure of a liquid increases with increasing temperature; it boils when its vapor pressure is the same as the external pressure on its surface. Because nonvolatile solutes lower the vapor pressure of the solution, a higher temperature is required to cause the solution to boil. Figure 13.18 shows the variation in vapor pressure of a pure liquid with temperature (the black line), as compared with a solution (the blue line). Because the vapor pressure of the solution is lower than that of the solvent at all temperatures, in accordance with Raoult's law, a higher temperature is required to attain a vapor pressure of 1 atm.

The increase in boiling point, ΔT_b (relative to the boiling point of the pure solvent), is directly proportional to the number of solute particles per mole of solvent molecules. We know that molality expresses the number of moles of solute per 1000 g of solvent, which represents a fixed number of moles of solvent. Thus, ΔT_b is proportional to molality, as shown in the following equation:

$$\Delta T_b = K_b m \qquad [13.13]$$

The magnitude of K_b, which is called the **molal boiling-point-elevation constant**, depends only on the solvent. Some typical values for several common solvents are given in Table 13.5.

Table 13.5 Molal Boiling-Point-Elevation and Freezing-Point-Depression Constants

Solvent	Normal boiling point (°C)	K_b (°C/m)	Normal freezing point (°C)	K_f (°C/m)
Water, H_2O	100.0	0.52	0.0	1.86
Benzene, C_6H_6	80.1	2.53	5.5	5.12
Carbon tetrachloride, CCl_4	76.8	5.02	−22.3	29.8
Ethanol, C_2H_5OH	78.4	1.22	−114.6	1.99
Chloroform, $CHCl_3$	61.2	3.63	−63.5	4.68

For water, K_b is 0.52°C/m; therefore, a 1 m aqueous solution of sucrose or any other aqueous solution that is 1 m in nonvolatile solute particles will boil at a temperature 0.52°C higher than pure water. It is important to notice that the boiling-point elevation is proportional to the number of solute particles present in a given quantity of solution. When NaCl dissolves in water, 2 mol of solute particles—1 mol of Na^+ and 1 mol of Cl^-—are formed for each mole of NaCl that dissolves. Therefore, a 1 m solution of NaCl in water causes a boiling-point elevation twice as large as a 1 m solution of a nonelectrolyte, such as sucrose.

Freezing-Point Depression

The freezing point corresponds to the temperature at which the vapor pressures of the solid and liquid phases are the same. The freezing point of a solution is lowered because the solute is not normally soluble in the solid phase of the solvent. For example, when aqueous solutions freeze, the solid that separates out is almost always pure ice. Thus, the vapor pressure of the solid is unaffected by the presence of solute. However, if the solute is nonvolatile, the vapor pressure of the solution is reduced in proportion to the mole fraction of solute (Equation 13.12). This means that the temperature at which the solution and solid phases will have the same vapor pressure is reduced, as illustrated in Figure 13.18. Note that the point that represents the equilibrium between solid and liquid lies to the left of the corresponding point for pure solvent, at a lower temperature.

Like the boiling-point elevation, the decrease in freezing point, ΔT_f, is directly proportional to the molality of the solute:

$$\Delta T_f = K_f m \qquad [13.14]$$

The values of K_f, the **molal freezing-point-depression constant**, for several common solvents are given in Table 13.5. For water, K_f is 1.86°C/m; therefore, a 0.5 m aqueous solution of NaCl or any aqueous solution that is 1 m in nonvolatile solute particles will freeze 1.86°C lower than pure water. The freezing-point lowering caused by solutes explains the use of antifreeze in cars (Sample Exercise 13.9) and the use of calcium chloride, $CaCl_2$, to melt ice on roads during winter.

SAMPLE EXERCISE 13.9

Calculate the freezing point and the boiling point of a solution of 100 g of ethylene glycol, $C_2H_6O_2$, antifreeze in 900 g of H_2O.

Solution

$$\text{Molality} = \frac{\text{moles } C_2H_6O_2}{\text{kilograms } H_2O}$$

$$= \left(\frac{100 \text{ g } C_2H_6O_2}{900 \text{ g } H_2O}\right)\left(\frac{1 \text{ mol } C_2H_6O_2}{62.0 \text{ g } C_2H_6O_2}\right)\left(\frac{1000 \text{ g } H_2O}{1 \text{ kg } H_2O}\right)$$

$$= 1.79\ m$$

$$\Delta T_f = K_f m = \left(1.86 \frac{°\text{C}}{m}\right)(1.79\ m) = 3.33°\text{C}$$

$$\text{Freezing point} = (\text{normal f.p. of solvent}) - \Delta T_f$$
$$= 0.00°C - 3.33°C = -3.33°C$$

$$\Delta T_b = K_b m = \left(0.52 \frac{°C}{m}\right)(1.79\ m) = 0.93°C$$

$$\text{Boiling point} = (\text{normal b.p. of solvent}) + \Delta T_b$$
$$= 100.00°C + 0.93°C$$
$$= 100.93°C$$

PRACTICE EXERCISE

Calculate the freezing point of a solution containing 600 g of $CHCl_3$ and 42 g of eucalyptol, $C_{10}H_{18}O$, a fragrant substance found in the leaves of eucalyptus trees. ***Answer:*** Freezing point $-65.6°C$

SAMPLE EXERCISE 13.10

List the following aqueous solutions in order of their expected freezing points: 0.050 m $CaCl_2$; 0.15 m NaCl; 0.10 m HCl; 0.050 m $HC_2H_3O_2$; 0.10 m $C_{12}H_{22}O_{11}$.

Solution: First notice that $CaCl_2$, NaCl, and HCl are strong electrolytes, $HC_2H_3O_2$ is a weak electrolyte, and $C_{12}H_{22}O_{11}$ is a nonelectrolyte. The molality of each solution in total particles is as follows:

0.050 m $CaCl_2$	(0.15 m in particles)
0.15 m NaCl	(0.30 m in particles)
0.10 m HCl	(0.20 m in particles)
0.050 m $HC_2H_3O_2$	(between 0.050 and 0.10 m in particles)
0.10 m $C_{12}H_{22}O_{11}$	(0.10 m in particles)

The freezing points are expected to run from 0.15 m NaCl (lowest freezing point) to 0.10 m HCl to 0.050 m $CaCl_2$ to 0.10 m $C_{12}H_{22}O_{11}$ to 0.050 m $HC_2H_3O_2$ (highest freezing point).

PRACTICE EXERCISE

Which of the following solutes will produce the largest total molality of solute particles upon addition to 1 kg of water: 1 mol of $Co(NO_3)_2$, 2 mol of KCl, or 3 mol of C_2H_5OH? ***Answer:*** 2 mol of KCl

Osmosis

Certain materials—including many membranes in biological systems and synthetic substances such as cellophane—are *semipermeable*. That is, when in contact with a solution, they permit the passage of some molecules but not others. They often permit the passage of small solvent molecules such as water but block the passage of larger solute molecules or ions. This semipermeable character is due to a network of tiny pores within the membrane.

Consider a situation in which only solvent molecules are able to pass through a membrane. If such a membrane is placed between two solutions of different concentration, solvent molecules move in both directions through the membrane. However, the concentration of *solvent* is higher in the solution containing less solute than in the more concentrated one. Therefore, the rate of passage of solvent from the less concentrated to the more concentrated solution is greater than the rate in the opposite direction. Thus, there is a net movement of solvent molecules from the less concentrated solution into the more concentrated one. This process is called **osmosis**. The important point to remember is that *the net movement of solvent is always toward the more concentrated solution.*

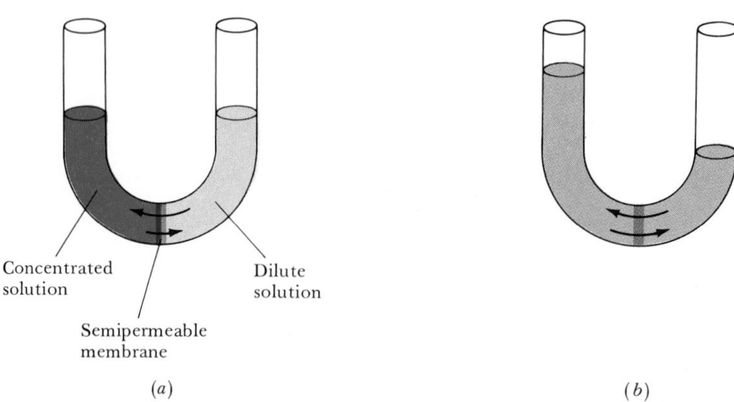

Figure 13.19 Osmosis: (*a*) net movement of solvent from the solution with low solute concentration into the solution with high solute concentration; (*b*) osmosis stops when the column of solution on the left becomes high enough to exert sufficient pressure to stop the osmosis.

Figure 13.19(*a*) shows two solutions separated by a semipermeable membrane. Solvent moves through the membrane from right to left, as if the solutions were driven to attain equal concentrations. As a result, the liquid levels in the two arms become uneven. Eventually, the pressure difference resulting from the uneven heights of the liquid in the two arms becomes so large that the net flow of solvent ceases, as shown in Figure 13.19(*b*). Alternatively, we may apply pressure to the left arm of the apparatus, as shown in Figure 13.20, to halt the net flow of solvent. The pressure required to prevent osmosis is known as the **osmotic pressure**, π, of the solution. The osmotic pressure is found to obey a law similar in form to the ideal-gas law: $\pi V = nRT$, where V is the volume of the solution, R is the ideal-gas constant, and T is the temperature on the Kelvin scale. From this equation, we can write

$$\pi = \left(\frac{n}{V}\right)RT = MRT \qquad [13.15]$$

where M is the molarity of the solution.

If two solutions of identical osmotic pressure are separated by a semipermeable membrane, no osmosis will occur. The two solutions are said to be *isotonic*. If one solution is of lower osmotic pressure, it is described as being *hypotonic* with respect to the more concentrated solution. The more concentrated solution is said to be *hypertonic* with respect to the dilute solution.

Osmosis plays a very important role in living systems. For example, the membranes of red blood cells are semipermeable. Placement of a red blood cell in a solution that is hypertonic relative to the intracellular solution (the solution within the cells) causes water to move out of the cell, as shown in Figure 13.21. This causes the cell to shrivel, a process known as *crenation*. Placement of the cell in a solution that is hypotonic relative to the intracellular fluid causes water to move into the cell. This causes rupturing of the cell, a process known as *hemolysis*. People needing replacement of body fluids or nutrients who cannot be fed orally are administered solutions by intravenous (or IV) infusion, which feeds nutrients directly into the veins. To prevent crenation or hemolysis of red blood cells, the IV solutions must be isotonic with the intracellular fluids of the cells.

Figure 13.20 Applied pressure on the left arm of the apparatus stops net movement of solvent from the right side of the semipermeable membrane. This applied pressure is known as the osmotic pressure of the solution.

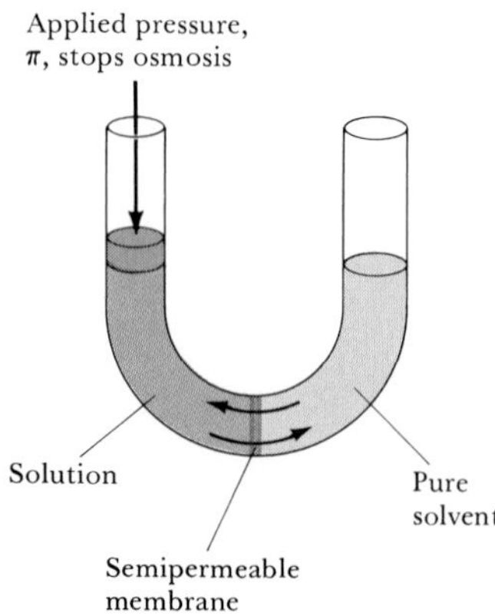

SAMPLE EXERCISE 13.11

The average osmotic pressure of blood is 7.7 atm at 25°C. What concentration of glucose, $C_6H_{12}O_6$, will be isotonic with blood?

Solution

$$\pi = MRT$$

$$M = \frac{\pi}{RT} = \frac{7.7\ \text{atm}}{\left(0.0821\ \dfrac{\text{L-atm}}{\text{K-mol}}\right)(298\ \text{K})} = 0.31\ M$$

In clinical situations, the concentrations of solutions are generally expressed in terms of weight percentages. The weight percentage of a 0.31 *M* solution of glucose is 5.3 percent. The concentration of NaCl that is isotonic with blood is 0.16 *M* because NaCl ionizes to form two particles, Na^+ and Cl^- (a 0.155 *M* solution of NaCl is 0.310 *M* in particles). A 0.16 *M* solution of NaCl is 0.9 percent in NaCl. Such a solution is known as a physiological saline solution.

PRACTICE EXERCISE

What is the osmotic pressure at 20°C of a 0.0020 *M* sucrose, $C_{12}H_{22}O_{11}$, solution? ***Answer:*** 0.048 atm, or 37 mm Hg

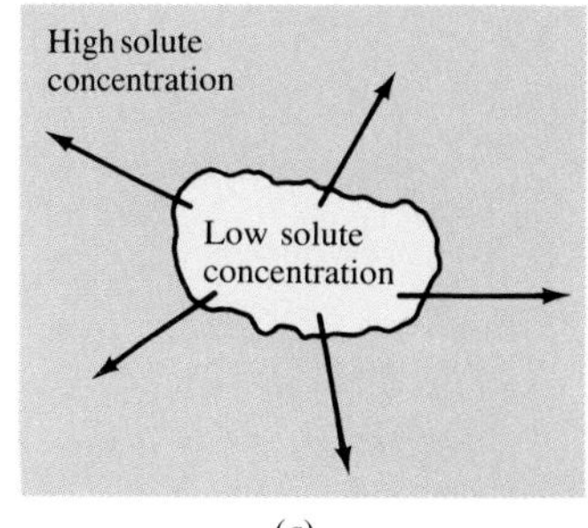

(*a*)

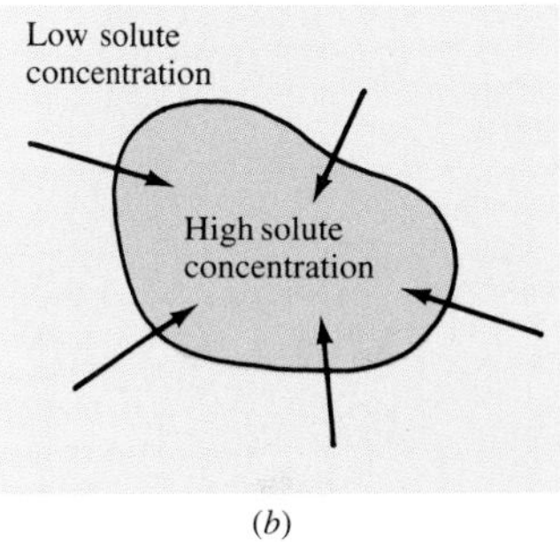

(*b*)

Figure 13.21 Osmosis through the semipermeable membrane of a red blood cell: (*a*) crenation caused by movement of water from the cell; (*b*) hemolysis caused by movement of water into the cell.

There are many interesting examples of osmosis. A cucumber placed in concentrated brine loses water via osmosis and shrivels into a pickle. A carrot that has become limp because of water loss to the atmosphere can be placed in water. Water moves into the carrot through osmosis, making it firm once again. People eating a lot of salty food experience water retention in tissue cells and intercellular spaces because of osmosis. The resultant swelling or puffiness is called *edema.* Movement of water from soil into plant roots and subsequently into the upper portions of the plant is due at least in part to osmosis. The preservation of meat by salting and of fruit by adding sugar protects against bacterial action. Through the process of osmosis, a bacterium on salted meat or candied fruit loses water, shrivels, and dies.

In osmosis, water moves from an area of high water concentration (low solute concentration) into an area of low water concentration (high solute concentration). Such movement of a substance from an area where its concentration is high to an area where it is low is spontaneous. Biological cells transport not only water but also other select materials through their membrane walls. This permits entry of nutrients and allows for disposal of waste materials. In some cases, substances must be moved from an area of low concentration to one of high concentration. This movement is called *active transport.* It is not spontaneous and so requires expenditures of energy by the cell.

Determination of Molar Mass

The colligative properties of solutions provide a useful means of experimentally determining molar mass. Any of the four colligative properties could be used to determine molar mass. The procedures are illustrated in Sample Exercises 13.12 and 13.13.

SAMPLE EXERCISE 13.12

A solution of an unknown nonvolatile nonelectrolyte was prepared by dissolving 0.250 g in 40.0 g of CCl_4. The normal boiling point of the resultant solution was increased by 0.357°C. Calculate the molar mass of the solute.

Solution: Using Equation 13.13, we have

$$\text{Molality} = \frac{\Delta T_b}{K_b} = \frac{0.357°\text{C}}{5.02°\text{C}/m} = 0.0711\ m$$

Thus the solution contains 0.0711 mol of solute per kilogram of solvent. The solution was prepared from 0.250 g of solute and 40.0 g of solvent. The number of grams of solute in a kilogram of solvent is therefore

$$\frac{\text{Grams solute}}{\text{Kilograms CCl}_4} = \left(\frac{0.250\text{ g solute}}{40.0\text{ g CCl}_4}\right)\left(\frac{1000\text{ g CCl}_4}{1\text{ kg CCl}_4}\right) = \frac{6.25\text{ g solute}}{1\text{ kg CCl}_4}$$

Notice that a kilogram of solvent contains 6.25 g, which from the ΔT_b measurement must be 0.0711 *m*. Therefore,

$$0.0711\text{ mol} = 6.25\text{ g}$$

$$1\text{ mol} = \frac{6.25\text{ g}}{0.0711} = 87.9\text{ g}$$

Therefore, $\mathscr{M} = 87.9$ g/mol

PRACTICE EXERCISE

Camphor, $C_{10}H_{16}O$, melts at 179.8°C; it has a particularly large freezing-point depression constant, $K_f = 40°\text{C}/m$. When 0.186 g of an organic substance of unknown molar mass is dissolved in 22.01 g of liquid camphor, the freezing point of the mixture is found to be 176.7°C. What is the approximate molar mass of the solute?
Answer: 108 g/mol

SAMPLE EXERCISE 13.13

The osmotic pressure of an aqueous solution of a certain protein was measured in order to determine its molar mass. The solution contained 3.50 mg of protein dissolved in sufficient water to form 5.00 mL of solution. The osmotic pressure of the solution at 25°C was found to be 1.54 mm Hg. Calculate the molar mass of the protein.

Solution: Using Equation 13.15, we can calculate the molarity of the solution:

$$\text{Molarity} = \frac{\pi}{RT} = \frac{(1.54\text{ mm Hg})\left(\dfrac{1\text{ atm}}{760\text{ mm Hg}}\right)}{\left(0.0821\dfrac{\text{L-atm}}{\text{K-mol}}\right)(298\text{ K})} = 8.28 \times 10^{-5}\,\frac{\text{mol}}{\text{L}}$$

Because the volume of the solution is 5.00 mL $= 5.00 \times 10^{-3}$ L, the number of moles of protein must be

$$\text{Moles} = (8.28 \times 10^{-5}\text{ mol/L})(5.00 \times 10^{-3}\text{ L}) = 4.14 \times 10^{-7}\text{ mol}$$

The molar mass is the number of grams per mole of the substance. The sample has a mass of 3.50 mg $= 3.50 \times 10^{-3}$ g. The molar mass is the number of grams divided by the number of moles:

$$\frac{\text{Grams}}{\text{Moles}} = \frac{3.50 \times 10^{-3}\text{ g}}{4.14 \times 10^{-7}\text{ mol}} = 8.45 \times 10^{3}\text{ g/mol}$$

Because small pressures can be measured easily and accurately, osmotic pressure measurements provide an excellent way to determine the molecular weights of very large molecules.

PRACTICE EXERCISE

A sample of 2.05 g of the plastic polystyrene was dissolved in enough benzene to form 100 mL of solution. The osmotic pressure of this solution was found to be 1.21 kPa at 25°C. Calculate the molar mass of the polystyrene. ***Answer:*** 4.20×10^4 g/mol

A CLOSER LOOK: Colligative Properties of Electrolyte Solutions

We have observed that the colligative properties of solutions depend on the total concentration of solute particles, regardless of whether the particles are ions or molecules. Thus, we would expect a 0.100 *m* solution of NaCl to have a freezing-point depression of $(0.200\ m) \times (1.86°C/m) = 0.372°C$ because it is 0.100 *m* in $Na^+(aq)$ and 0.100 *m* in $Cl^-(aq)$. However, the measured freezing-point depression, 0.348°C, is not quite this large. The situation is similar for other strong electrolytes. For example, a 0.100 *m* solution of KCl freezes at −0.344°C.

The difference between the expected and observed colligative properties for strong electrolytes is due to electrostatic attractions between ions. As the ions move about in solution, ions of opposite charge collide and "stick together" for brief moments. While they are together, they behave as a single particle, called an **ion pair** (Figure 13.22). The number of independent particles is thereby reduced, causing a reduction in the freezing-point depression (as well as in the boiling-point elevation, the vapor-pressure reduction, and the osmotic pressure).

One measure of the extent to which electrolytes dissociate is the *van't Hoff factor, i.* This factor is the ratio of the actual value of a colligative property to the value calculated assuming the substance to be a nonelectrolyte. For example, using the freezing-point depression, we have

$$i = \frac{\Delta T_f(\text{measured})}{\Delta T_f(\text{calculated for nonelectrolyte})} \qquad [13.16]$$

The ideal value of *i* can be determined for a salt by noting the number of ions per formula unit. For example, for NaCl the ideal van't Hoff factor is 2 because the salt consists of one Na^+ and one Cl^- per formula unit; for K_2SO_4, it is 3, the three ions being two K^+ and one SO_4^{2-}. In the absence of any information about the actual value of *i* for a solution, we will use the ideal value in calculations.

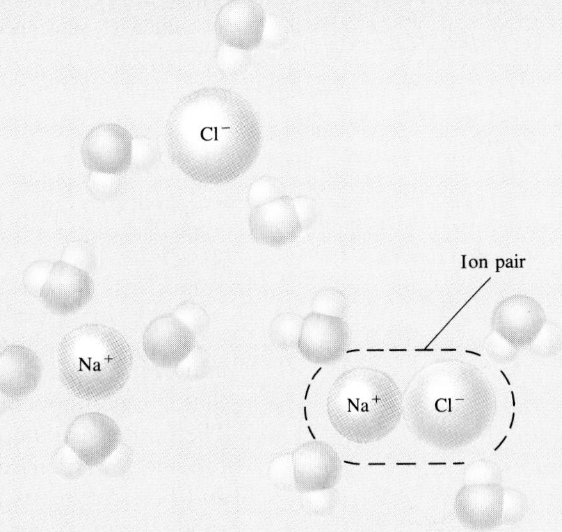

Figure 13.22 A solution of NaCl contains not only separated $Na^+(aq)$ and $Cl^-(aq)$ ions, but ion pairs as well.

Table 13.6 gives the observed van't Hoff factors for several substances at different dilutions. Two trends are evident in these data. First, dilution affects the value of *i* for electrolytes; the more dilute the solution, the more closely *i* approaches the ideal or limiting value. Thus, the extent of ion pairing in electrolyte solutions decreases upon dilution. Second, the lower the charges on the ions, the smaller the departure of *i* from the limiting value, because the extent of ion pairing decreases as the charges of ions decrease. Both trends are consistent with simple electrostatics: The force of interaction between charged particles decreases as their separation increases and as their charges decrease.

Table 13.6 van't Hoff Factors for Several Substances at 25°C

	Concentration			
Compound	**0.100 *m***	**0.0100 *m***	**0.00100 *m***	**Limiting value**
Sucrose	1.00	1.00	1.00	1.00
NaCl	1.87	1.94	1.97	2.00
K_2SO_4	2.32	2.70	2.84	3.00
$MgSO_4$	1.21	1.53	1.82	2.00

13.6 COLLOIDS

When finely divided clay particles are dispersed through water, they do not remain suspended but eventually settle out of the water because of the gravitational pull. The dispersed clay particles are much larger than molecules and consist of many thousands or even millions of atoms. In contrast, the dispersed particles of a solution are of molecular size. Between these extremes is the sit-

Table 13.7 Types of Colloids

Phase of colloid	Dispersing (solventlike) substance	Dispersed (solutelike) substance	Colloid type	Example
Gas	Gas	Gas	—	None (all are solutions)
Gas	Gas	Liquid	Aerosol	Fog
Gas	Gas	Solid	Aerosol	Smoke
Liquid	Liquid	Gas	Foam	Whipped cream
Liquid	Liquid	Liquid	Emulsion	Milk
Liquid	Liquid	Solid	Sol	Paint
Solid	Solid	Gas	Solid foam	Marshmallow
Solid	Solid	Liquid	Solid emulsion	Butter
Solid	Solid	Solid	Solid sol	Ruby glass

uation in which dispersed particles are larger than molecules but not so large that the components of the mixture separate under the influence of gravity. These intermediate types of dispersions or suspensions are called **colloidal dispersions**, or simply **colloids**. Colloids are on the dividing line between solutions and heterogeneous mixtures. Like solutions, colloids can be gases, liquids, or solids. Examples of each are listed in Table 13.7.

The size of the dispersed particle is the property used to classify a mixture as a colloid. Colloid particles range in diameter from approximately 10 to 2000 Å. Solute particles are smaller. The colloid particle may consist of many atoms, ions, or molecules, or it may even be a single giant molecule. For example, the hemoglobin molecule, which carries oxygen in blood, has molecular dimensions of 65 Å × 55 Å × 50 Å and a molecular weight of 64,500 amu.

Although colloid particles may be so small that the dispersion appears uniform even under a microscope, they are large enough to scatter light very effectively. Consequently, most colloids appear cloudy or opaque unless they are very dilute. (Milk is a colloid.) Furthermore, because they scatter light, a light beam can be seen as it passes through a colloidal suspension, as shown in Figure 13.23. This scattering of light by colloidal particles, known as the **Tyndall effect**, makes it possible to see the light beam coming from the projection housing in a smoke-filled theater or the light beam from an automobile on a dusty dirt road.

Figure 13.23 Illustration of the Tyndall effect. The vessel on the left contains a colloidal suspension, that on the right a solution. Note that the path of the beam through the colloidal suspension is clearly seen, because the light is scattered by the colloidal particles. Light is not scattered by the individual solute molecules in the solution. (© Richard Megna/Fundamental Photographs)

Hydrophilic and Hydrophobic Colloids

The most important colloids are those in which the dispersing medium is water. Such colloids are frequently referred to as **hydrophilic** (water-loving) or **hydrophobic** (water-hating). Hydrophilic colloids are most like the solutions that we have previously examined. In the human body, the extremely large molecules that make up such important substances as enzymes and antibodies are kept in suspension by interaction with surrounding water molecules. The molecules fold so that polar, or charged, groups can interact with water molecules at the periphery of the molecules. These hydrophilic groups generally contain oxygen or nitrogen. Some examples are shown in Figure 13.24.

Hydrophobic colloids can be prepared in water only if they are stabilized in some way. Otherwise, their natural lack of affinity for water causes them to separate from the water. Hydrophobic colloids can be stabilized by adsorption of ions on their surface, as shown in Figure 13.25. (*Adsorption* refers to adherence to a surface. It differs from *absorption*, which means passage into the interior, as when water is absorbed by a sponge.) These adsorbed ions can interact with water, thereby stabilizing the colloid. At the same time the mutual repulsion between colloid particles with adsorbed ions of the same charge keeps the particles from colliding and so getting larger.

Hydrophobic colloids can also be stabilized by the presence of other hydrophilic groups on their surfaces. For example, small droplets of oil are hydrophobic. They do not remain suspended in water; instead, they separate, forming an oil slick on the surface of the water. Addition of sodium stearate, whose structure is shown at left, or any similar substance having one end that is hydrophilic (polar, or charged) and one that is hydrophobic (nonpolar), will stabilize a suspension of oil in water, as shown in Figure 13.26. The hydrophobic ends of the stearate ions interact with the oil droplet, and the hydrophilic ends point out toward the water with which they interact.

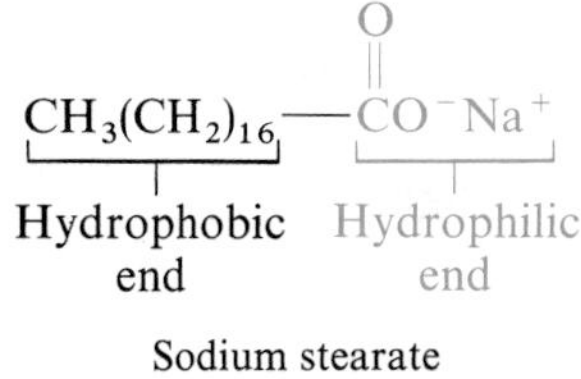

Sodium stearate

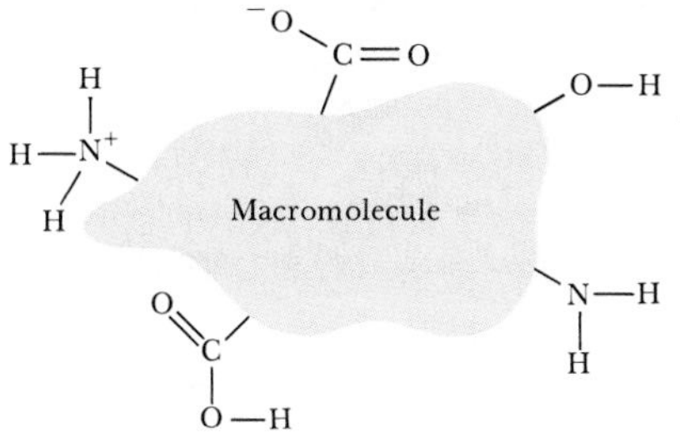

Figure 13.24 Examples of hydrophilic groups at the surface of a giant molecule (macromolecule) that help keep the molecule suspended in water.

Figure 13.25 Schematic representation of the stabilization of a hydrophobic colloid by adsorbed ions.

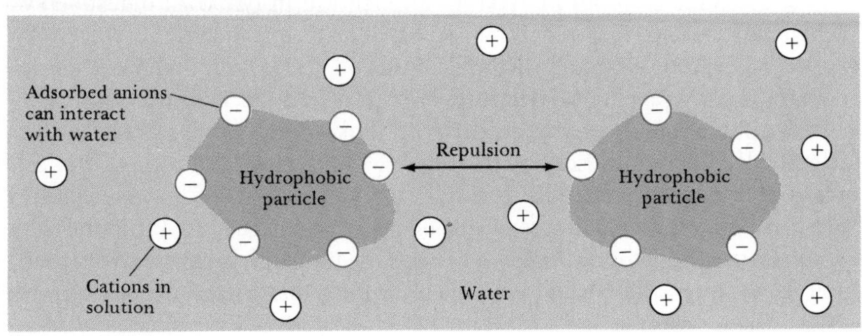

Figure 13.26 Stabilization of an emulsion of oil in water by stearate ions.

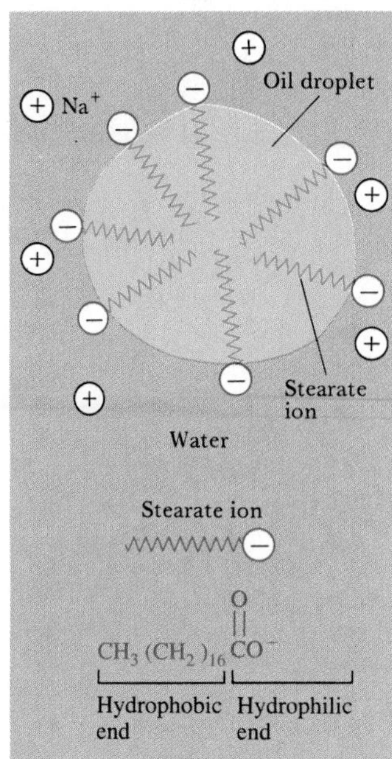

CHEMISTRY AT WORK: Sickle-Cell Anemia

Our blood contains a complex protein called *hemoglobin* that is responsible for carrying oxygen from our lungs to other parts of our body. In the genetic disease known as *sickle-cell anemia*, hemoglobin molecules are abnormal: They have lower solubility, especially in the unoxygenated form. Consequently, as much as 85 percent of the hemoglobin in the red blood cells crystallizes from solution. This distorts the cells into a sickle shape, as shown in Figure 13.27. These clog the capillaries, thus causing gradual deterioration of the vital organs. The disease is hereditary, and if both parents carry the defective genes, it is likely that the child will possess only abnormal hemoglobin. Such children seldom survive more than a few years after birth. The reason for the insolubility of hemoglobin in sickle-cell anemia can be traced to a change in one part of the molecule.

Normal hemoglobin molecules have an amino acid in their makeup that has a side chain protruding from the main body of the molecule:

$$-CH_2-CH_2-\overset{\displaystyle O}{\overset{\|}{C}}-OH$$

Normal

Notice that this side chain terminates in a polar group, which contributes to the solubility of the hemoglobin molecule in water. In the hemoglobin molecules of persons suffering from sickle-cell anemia, the side chain is of a different type:

$$-\underset{\displaystyle CH_3}{\underset{|}{CH}}-CH_3$$

Abnormal

This abnormal group of atoms is nonpolar (hydrophobic), and its presence leads to the aggregation of this defective form of hemoglobin into particles too large to remain suspended in biological fluids.

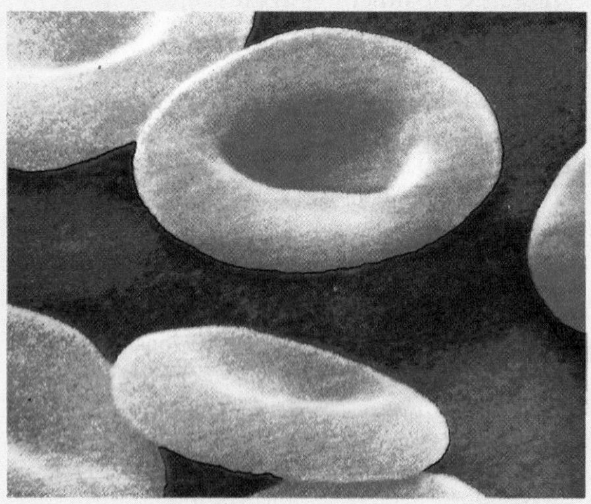

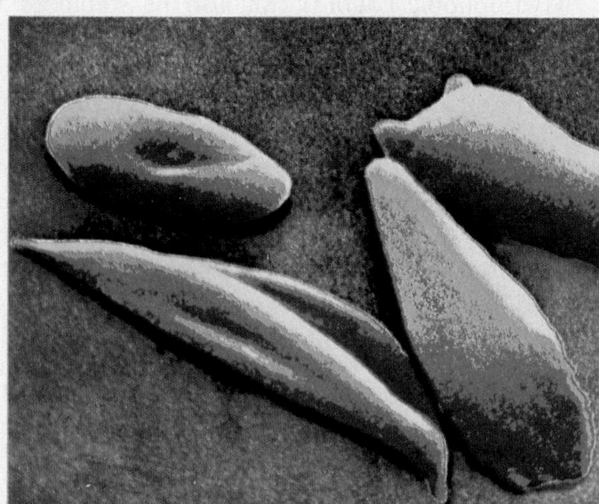

Figure 13.27 Electron micrograph showing normal red blood cells (left) and sickled red blood cells (right). (Bill Longcore/Photo Researchers)

These concepts have an interesting application in our own digestive system. When fats in our diet reach the small intestine, a hormone causes the gallbladder to excrete a fluid called bile. Among the components of bile are compounds that have chemical structures similar to sodium stearate; that is, they have a hydrophilic (polar) end and a hydrophobic (nonpolar) end. These compounds emulsify the fats present in the intestine and thus permit digestion and absorption of fat-soluble vitamins through the intestinal wall. The term *emulsify* means to form an emulsion, that is, a suspension of one liquid in another (Table 13.7). A substance that aids in the formation of an emulsion is called an emulsifying agent. If you read the labels for foods and other materials, you will observe that a variety of chemicals are used as emulsifying agents. These chemicals typically have a hydrophilic end and a hydrophobic end, as in the examples in the text above.

Removal of Colloidal Particles

Colloidal particles frequently must be removed from a dispersing medium, as in the removal of smoke from stacks or butterfat from milk. Because colloidal particles are so small, they cannot be extracted by simple filtration. The colloidal particles must be enlarged, a process called **coagulation**. The resultant, larger particles can then be separated by filtration or merely by allowing them to settle out of the dispersing medium.

Heating or adding an electrolyte to the mixture may bring about coagulation. Heating the colloidal dispersion increases the particle motion and so the number of collisions. The particles increase in size as they stick together after colliding. The addition of electrolytes causes neutralization of the surface charges of the particles, thereby removing the electrostatic repulsions that inhibit their coming together. The effect of electrolytes is seen in the depositing of suspended clay in a river as it mixes with salt water. This results in the formation of river deltas wherever rivers empty into oceans or other salty bodies of water.

Semipermeable membranes can also be used to separate ions from colloidal particles because the ions can pass through the membrane, but the colloid particles cannot. This type of separation is known as **dialysis**. This process is used in the purification of blood in artificial kidney machines. Our kidneys are responsible for removing the waste products of metabolism from blood. In the kidney machine, blood is circulated through a dialyzing tube immersed in a washing solution. That solution is isotonic in ions that must be retained by the blood but is lacking the waste products. Wastes therefore dialyze out of the blood.

FOR REVIEW

SUMMARY

Solutions are homogeneous mixtures of atoms, ions, or molecules. The relative amounts of solute and solvent in a solution can be described qualitatively (dilute or concentrated solutions) or quantitatively (in terms of weight percentage, parts per million, mole fraction, molarity, molality, and normality).

A saturated solution is one in which dissolved and undissolved solute are in dynamic equilibrium. The solubility of a solute is its concentration in a saturated solution. Solubility depends on the relative magnitudes of solute-solute, solute-solvent, and solvent-solvent attractive forces as well as on the change in disorder accompanying mixing. Although the solution process can be either endothermic or exothermic, solutions form only if the solute-solvent interaction is comparable in magnitude to the solute-solute and solvent-solvent interactions. Solubility therefore depends on the identity of the solute and the solvent ("like dissolves like").

For most ionic solutes in water, solubility increases with increasing temperature. For gases, solubility generally decreases with increasing temperature and increases with increasing pressure. As described by Henry's law, the solubility of a gas is proportional to its partial pressure above the solution.

Some physical properties of solutions are colligative; that is, they depend on the concentration of the particles present and not on their chemical identity.

Four colligative properties were discussed: the lowering of vapor pressure, the increase in boiling point, the decrease in freezing point, and osmotic pressure. Osmosis is the passage of solvent through a semipermeable membrane from a solution with a lower solute concentration into one with a higher solute concentration. Osmotic pressure is the pressure that must be applied to a solution to prevent osmosis into a solution from a pure solvent. Colligative properties can be used to determine the molar masses of nonvolatile nonelectrolytes.

True solutions can be differentiated from colloids on the basis of particle size. Colloids play important roles in many chemical and biological systems.

KEY TERMS

weight percentage (Sec. 13.1)
parts per million (Sec. 13.1)
molality (Sec. 13.1)
solvation (Sec. 13.2)
hydration (Sec. 13.2)
crystallization (Sec. 13.3)
saturated solution (Sec. 13.3)
solubility (Sec. 13.3)
unsaturated solution (Sec. 13.3)
supersaturated solution (Sec. 13.3)
miscible (Sec. 13.4)
immiscible (Sec. 13.4)
Henry's law (Sec. 13.4)
colligative properties (Sec. 13.5)
Raoult's law (Sec. 13.5)
ideal solutions (Sec. 13.5)
molal boiling-point-elevation constant (Sec. 13.5)
molal freezing-point-depression constant (Sec. 13.5)
osmosis (Sec. 13.5)
osmotic pressure (Sec. 13.5)
ion pair (Sec. 13.5)
colloidal dispersions (colloids) (Sec. 13.6)
Tyndall effect (Sec. 13.6)
hydrophilic (Sec. 13.6)
hydrophobic (Sec. 13.6)
coagulation (Sec. 13.6)
dialysis (Sec. 13.6)

EXERCISES

Concentrations of Solutions

13.1 **(a)** Calculate the weight percentage of solute in a solution containing 3.25 g of $Ba(NO_3)_2$ in 85.0 g of water. **(b)** How many ppm of gold are present in an ore that contains 5.0 g of gold per ton of ore?

13.2 **(a)** What is the weight percentage of benzene, C_6H_6, in a solution containing 8.75 g of benzene and 25.0 g of carbon tetrachloride, CCl_4? **(b)** Seawater contains 0.412 g of Ca^{2+} per kilogram of water. What is the concentration of Ca^{2+} in ppm?

13.3 Calculate the mole fraction of methyl alcohol, CH_3OH, in the following solutions: **(a)** 6.00 g of CH_3OH in 480 g of H_2O; **(b)** 4.13 g of CH_3OH in 48.6 g of CCl_4.

13.4 Calculate the mole fraction of ethylene glycol, $C_2H_6O_2$, in the following solutions: **(a)** 120 g of $C_2H_6O_2$ dissolved in 120 g of water; **(b)** 120 g of $C_2H_6O_2$ dissolved in 1.20 kg of acetone, C_3H_6O.

13.5 Calculate the molarity of each of the following solutions: **(a)** 1.50 g of KBr in 1.60 L of solution; **(b)** 2.78 g of $Ca(NO_3)_2 \cdot 4H_2O$ in 450 mL of solution; **(c)** 50.0 mL of 0.250 *M* HCl solution diluted to 1.00 L.

13.6 Calculate the molarity of each of the following solutions: **(a)** 3.50 g of NaOH in 0.650 L of solution; **(b)** 2.50 g of $Co(NO_3)_2 \cdot 6H_2O$ in 750 mL of solution; **(c)** 20.0 mL of 6.0 *M* H_2SO_4 diluted to 0.500 L.

13.7 Calculate the molality of each of the following solutions: **(a)** 13.0 g of benzene, C_6H_6, dissolved in 17.0 g of carbon tetrachloride, CCl_4; **(b)** 5.85 g of NaCl dissolved in 0.250 L of water.

13.8 Calculate the molality of each of the following solutions: **(a)** 2.1 g of sulfur, S_8, dissolved in 95.0 g of naphthalene, $C_{10}H_8$; **(b)** 1.50 mol of NaCl dissolved in 15.0 mol water.

13.9 A solution containing 66.0 g of acetone, C_3H_6O, and 46.0 g of H_2O has a density of 0.926 g/mL. Calculate **(a)** the weight percentage; **(b)** the mole fraction; **(c)** the molality; **(d)** the molarity of H_2O in this solution.

13.10 A sulfuric acid solution containing 571.6 g of H_2SO_4 per liter of solution has a density of 1.329 g/cm^3. Calculate **(a)** the weight percentage; **(b)** the mole fraction; **(c)** the molality; **(d)** the molarity of H_2SO_4 in this solution.

13.11 Commercial concentrated nitric acid is 69 percent HNO_3 by weight and has a density of 1.42 g/mL. What is the molarity of this solution?

13.12 Commercial concentrated aqueous ammonia is 28 percent NH_3 by weight and has a density of 0.90 g/mL. What is the molarity of this solution?

13.13 Calculate the number of moles of solute present in each of the following solutions: **(a)** 356 mL of 0.358 M $Ca(NO_3)_2$; **(b)** 4.60×10^2 L of 0.582 M HBr; **(c)** 132 mL of 0.0288 M $Al(NO_3)_3$.

13.14 Calculate the number of moles of solute present in each of the following solutions: **(a)** 60.0 g of an aqueous solution that is 1.25 percent KI by weight; **(b)** 250 g of an aqueous solution that is 0.460 percent NaCl by weight; **(c)** 600 mL of 1.25 M H_2SO_4.

13.15 Describe how you would prepare each of the following aqueous solutions, starting with solid KBr: **(a)** 1.40 L of 1.5×10^{-2} M KBr; **(b)** 250 g of 0.400 m KBr; **(c)** 1.50 L of a solution that is 12.0 percent KBr by weight (the density of the solution is 1.10 g/mL); **(d)** a 0.200 M solution of KBr that contains just enough KBr to precipitate 21.0 g of AgBr from a solution containing 0.480 mol of $AgNO_3$.

13.16 Describe how you would prepare each of the following aqueous solutions: **(a)** 500 mL of 0.200 M Na_2CO_3 solution, starting with solid Na_2CO_3; **(b)** 150 g of a solution that is 1.00 m in $(NH_4)_2SO_4$, starting with the solid solute; **(c)** 1.50 L of a solution that is 20.0 percent $Pb(NO_3)_2$ by weight (the density of the solution is 1.20 g/mL), starting with solid solute; **(d)** a 0.50 M solution of HCl that would just neutralize 5.0 g of $Ba(OH)_2$, starting with 6.0 M HCl.

[13.17] In a certain reaction, Sn^{4+} is reduced to Sn^{2+}. **(a)** What is the normality of a 0.38 M solution of Sn^{4+}? **(b)** What is the molarity of a 0.222 N solution of H_2SO_4 reacted with NaOH solution, in which both of the hydrogens of the H_2SO_4 react?

[13.18] In each of the following reactions, indicate the normality of a 0.1 M solution of the reagent underlined:

(a) $\underline{H_3PO_3(aq)} + 2NaOH(aq) \longrightarrow Na_2HPO_3(aq) + 2H_2O(aq)$

(b) $8HNO_3(aq) + \underline{KMnO_4(aq)} + 5Fe(NO_3)_2(aq) \longrightarrow 5Fe(NO_3)_3(aq) + Mn(NO_3)_2(aq) + KNO_3(aq) + 4H_2O(l)$

The Solution Process; Solubility and Structure

13.19 Explain why solvent-solvent, solute-solute, and solvent-solute interactions are important in determining the extent to which a solute dissolves in a solvent.

13.20 Why are some solution processes exothermic whereas others are endothermic?

13.21 Indicate the type of solute-solvent intermolecular attractive force (Section 11.2) that should be most important in each of the following solutions: **(a)** KCl in water; **(b)** C_6H_6 in CCl_4; **(c)** HF in water; **(d)** acetonitrile, CH_3CN, in C_3H_6O (see Table 11.2).

13.22 Indicate the type of solute-solvent intermolecular force (Section 11.2) that should be most important in each of the following solutions: **(a)** methanol, CH_3OH, in water; **(b)** NaBr in water; **(c)** benzene, C_6H_6, in cyclohexane, C_6H_{12}; **(d)** HCl in acetonitrile, CH_3CN.

13.23 **(a)** Based on the nature of intermolecular forces, explain what the phrase "like dissolves like" means. **(b)** Would naphthalene, $C_{10}H_8$, be more soluble in methanol, CH_3OH, or in benzene, C_6H_6?

13.24 Which member of each of the following pairs is the more soluble in water: **(a)** CH_3CH_3 or CH_3OH; **(b)** CCl_4 or $CaCl_2$? Explain in each case.

13.25 Give an explanation in terms of intermolecular forces for each of the following observations: **(a)** KCl is insoluble in C_6H_6; **(b)** $Br_2(l)$ is more soluble than $I_2(s)$ in CCl_4.

13.26 Give an explanation in terms of intermolecular forces for each of the following observations: **(a)** Methanol, CH_3OH, is not soluble in cyclohexane, C_6H_{12}; **(b)** chloroform, $CHCl_3$, is soluble in water to the extent of only 1 g per 100 g of water, but is miscible in ethanol, C_2H_5OH.

13.27 The enthalpies of solution of hydrated salts are generally more positive than those of anhydrous materials. For example, ΔH of solution for KOH is -57.3 kJ/mol and that for $KOH \cdot H_2O$ is -14.6 kJ/mol. Similarly, ΔH_{soln} for $NaClO_4$ is $+13.8$ kJ/mol and that for $NaClO_4 \cdot H_2O$ is $+22.5$ kJ/mol. Explain this effect in terms of the enthalpy contributions to the solution process depicted in Figure 13.4.

13.28 The enthalpy of solution of KBr in water is about $+19.8$ kJ/mol. The process, then, is endothermic. Nevertheless, the solubility of KBr in water is relatively high. Why does the solution process, although endothermic, proceed?

Effect of Temperature and Pressure on Solubility

13.29 How do the solubilities of most ionic compounds in water change with temperature? How about gases in water?

13.30 How do the solubilities of gases, liquids, and solids in water change with pressure? State Henry's law.

13.31 The Henry's law constant for helium gas in water at 30°C is 3.7×10^{-4} M/atm; that for N_2 at 30°C is 6.0×10^{-4} M/atm. If the two gases are each present at 2.5 atm pressure, calculate the solubility of each gas.

13.32 The partial pressure of O_2 in air at sea level is 0.21 atm. Using the data in Table 13.3, together with Henry's law, calculate the molar concentration of O_2 in the surface water of a lake saturated with air at 20°C.

13.33 In terms of the kinetic-molecular theory, explain why the solubility of gases in liquids generally decreases with increasing temperature.

13.34 Explain how the kinetic-molecular theory accounts for Henry's law. What factor is most important in causing variations in the Henry's law constant among different gases in the same solvent?

13.35 If a saturated solution of KNO_3 at 50°C (Figure 13.14) is cooled to 10°C, how many grams of solute will crystallize for each 100 g of solvent?

13.36 If water is saturated with potassium dichromate, $K_2Cr_2O_7$, at 80°C, and the resultant solution is then cooled to 10°C, how many grams of solute will crystallize for each 100 g of water (Figure 13.14).

Colligative Properties

13.37 What characteristics do colligative properties of solutions share?

13.38 Which of the following properties of solutions are colligative properties? **(a)** density; **(b)** osmotic pressure; **(c)** vapor-pressure lowering; **(d)** freezing-point depression; **(e)** effect of temperature on solubility.

13.39 Calculate **(a)** the amount of $C_2H_6O_2$ that must be added to 1.00 kg of ethanol to reduce its vapor pressure by 9.5 mm Hg at 35°C. The vapor pressure of pure ethanol at this temperature is 100 mm Hg. **(b)** the amount of KBr that must be added to 120 g of water to reduce the vapor pressure by 1.50 mm Hg at 25°C.

13.40 Calculate the vapor pressure of water above a solution prepared by adding **(a)** 10.00 g of lactose, $C_{12}H_{22}O_{11}$, to 82.0 g of water at 338 K; **(b)** 5.00 g of sodium sulfate, Na_2SO_4, to 115 g of water at 338 K; **(c)** 32.5 g of glycerin, $C_3H_8O_3$, to 120 g of water at 338 K. (The vapor pressure of water is given in Appendix B.)

13.41 Using Table 13.5, calculate the freezing and boiling points of each of the following solutions: **(a)** 0.20 *m* glucose in ethanol; **(b)** 4.5 g of $C_{10}H_8$ in 95.5 g of CCl_4; **(c)** 1.8 g of KNO_3 in 50.0 g of water.

13.42 Using Table 13.5, calculate the freezing and boiling points of each of the following solutions: **(a)** 0.45 *m* glycerol in ethanol; **(b)** 25.0 g of menthol, $C_{10}H_{20}O$, in 100 g of benzene; **(c)** 3.50 g of $Ca(NO_3)_2$ in 50.0 g of water.

13.43 List the following aqueous solutions in order of increasing boiling point: **(a)** 0.030 *m* glycerin; **(b)** 0.020 *m* KBr; **(c)** 0.030 *m* benzoic acid, $HC_7H_5O_2$.

13.44 List the following aqueous solutions in order of decreasing freezing point: **(a)** 0.080 *m* glucose; **(b)** 0.060 *m* LiBr; **(c)** 0.030 *m* $Zn(NO_3)_2$.

13.45 Urea, $(NH_2)_2CO$, is the product of protein metabolism in mammals. What is the osmotic pressure of an aqueous solution containing 1.10 g of urea in 100 mL of solution at 20°C?

13.46 Seawater contains 3.4 g of salts for every liter of solution. Assuming that the solute consists entirely of NaCl (over 90 percent is), calculate the osmotic pressure of seawater at 20°C.

13.47 Adrenaline is the hormone that triggers release of extra glucose molecules in times of stress or emergency. A solution of 0.64 g of adrenaline in 36.0 g at CCl_4 causes an elevation of 0.49°C in the boiling point. What is the molar mass of adrenaline?

13.48 Lauryl alcohol is obtained from coconut oil and is used to make detergents. A solution of 5.00 g of lauryl alcohol in 100 g of benzene freezes at 4.1°C. What is the molar mass of this substance?

13.49 Lysozyme is an enzyme that breaks bacterial cell walls. A solution containing 0.150 g of this enzyme in 210 mL of solution has an osmotic pressure of 0.953 mm Hg at 25°C. What is the molar mass of this substance?

13.50 A dilute sugar solution prepared by mixing 16.0 g of an unknown sugar with sufficient water to form 0.200 L of solution is found to have an osmotic pressure of 2.86 atm at 25°C. What is the molar mass of this substance?

13.51 The osmotic pressure of a 0.010 *M* aqueous solution of $CaCl_2$ is found to be 0.674 atm at 25°C. **(a)** Calculate the van't Hoff factor, *i*, for the solution. **(b)** How would you expect the value of *i* to change as the solution becomes more concentrated?

13.52 **(a)** Using data from Table 13.6, calculate the freezing points of 0.100 *m* NaCl and 0.100 *m* $MgSO_4$. **(b)** To what features of these compounds can you ascribe the difference in the values calculated in (a)?

Colloids

13.53 Distinguish between solutions and colloids. Give an example of each.

13.54 Explain how observation of the passage of a beam of light through a liquid can be used to distinguish between a solution and a colloidal dispersion. Account for the different behavior in the two cases.

13.55 Indicate whether each of the following is a hydrophilic or a hydrophobic colloid: **(a)** butterfat in homogenized milk; **(b)** hemoglobin in blood; **(c)** vegetable oil in a salad dressing.

13.56 Explain how each of the following factors operates in determining the stability or instability of a colloidal dispersion: **(a)** particulate mass; **(b)** hydrophobic character; **(c)** charges on colloidal particles.

13.57 What are the usual reasons that colloid particles do not coalesce into larger particles upon collision? How can colloids be coagulated?

13.58 Explain the following observations: **(a)** Soaps (for example, sodium stearate) will stabilize a colloidal dispersion of oil in water; **(b)** milk curdles upon addition of acid.

Additional Exercises

13.59 The Dead Sea contains 58 mol of bromide per 1000 kg of water. What is the concentration of bromide in ppm?

13.60 Acetonitrile, CH_3CN, is a polar organic solvent that dissolves a wide range of solutes, including many salts. The density of a 1.80 *M* acetonitrile solution of LiBr is 0.826 g/cm^3. Calculate the concentration of the solution in **(a)** molality; **(b)** mole fraction of LiBr; **(c)** weight percentage of CH_3CN.

13.61 Copper(II) sulfate, $CuSO_4$, is commonly used to reduce the growth of algae in lakes, ponds, and water reservoirs. An aqueous solution of $CuSO_4$ that is 18.0 percent by weight has a density of 1.208 g/mL. Copper(II) sulfate is commonly available as the pentahydrate, $CuSO_4 \cdot 5H_2O$. What mass of this substance must be used to make up 8.00 L of a solution that is 18.0 percent by weight? What is the molarity of this solution?

13.62 A solution is made by dissolving 2.16 g of benzoic acid, $HC_7H_5O_2$, in 180 mL of CCl_4, density 1.59 g/mL. Another solution is prepared by dissolving the same quantity of benzoic acid in 180 mL of C_2H_5OH, density 0.782 g/mL. **(a)** Calculate the mole fractions and molalities in each case. **(b)** Assuming that the density of the solution is the same as that of pure solvent, calculate the molarity in each case. **(c)** Compare the molarities and molalities of these solutions and comment on their relative magnitudes.

13.63 In terms of the concepts developed in this chapter, indicate why each of the following commercial products is formulated as it is. **(a)** A gas-line antifreeze additive to the gas tank consists almost entirely of methyl alcohol. **(b)** The solvent in a water-repellent spray for treating shoes is methylene chloride, CH_2Cl_2. **(c)** A wax remover for cross-country skis consists of benzene and similar aromatic hydrocarbons (Section 9.5).

13.64 Choose the substance that you would expect to be more soluble in water: **(a)** KBr or CBr_4; **(b)** CH_4 or NH_3; **(c)** Br_2 or HBr.

13.65 Butylated hydroxytoluene (BHT) has the following molecular structure:

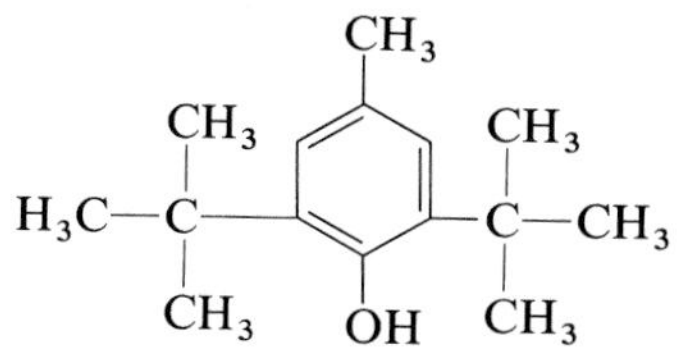

BHT

It is widely used as a preservative in a variety of foods, including dried cereals, cooking oils, and canned goods. The average person in the United States consumes about 2 mg of BHT daily. In terms of its structure, would you expect it to be readily excreted from the body or found stored in body fat? (Incidentally, BHT is not known to have any harmful properties; it is, in fact, a known antiviral agent.)

13.66 Recently concern has grown in many parts of the United States regarding a possible health hazard arising from the presence of the radioactive gas radon (Rn) in well water obtained from aquifers that lie in rock deposits. A sample consisting of various gases contains 3.5×10^{-6} mole fraction of radon. This gas at a total pressure of 36 atm is shaken with water at 30°C. Assume that the solubility of radon in water with 1 atm pressure of the gas over the solution at 30°C is 7.27×10^{-3} *M*. Calculate the molar concentration of radon in the water.

13.67 A "canned heat" product used to warm chafing dishes consists of a homogeneous mixture of ethyl alcohol, C_2H_5OH, and paraffin that has an average formula $C_{24}H_{50}$. What mass of C_2H_5OH should be added to 620 kg of the paraffin in formulating the mixture if the vapor pressure of ethyl alcohol at 35°C over the mixture is to be 8 mm Hg? The vapor pressure of pure ethyl alcohol at 35°C is 100 mm Hg.

13.68 Constant humidity in a closed container can be achieved by placing in the chamber an aqueous solution that has the desired vapor pressure of water. Provide a recipe for preparing 1 kg of a solution of ethylene glycol, $C_2H_6O_2$, in water that has a vapor pressure of 13.0 mm Hg at 24°C. (Assume that ethylene glycol is nonvolatile; see Appendix B for water vapor pressure.)

[13.69] The cooling system of an automobile is filled with a solution formed by mixing equal volumes of water (density = 1.00 g/mL) and ethylene glycol, $C_2H_6O_2$ (density = 1.12 g/mL). Estimate the freezing point and boiling point of the mixture.

13.70 Calculate the freezing point of a 0.100 *m* solution of K_2SO_4 **(a)** ignoring interionic attractions; **(b)** taking interionic attractions into consideration.

[13.71] A 0.0200 *m* solution of hydrofluoric acid freezes at −0.043°C. Calculate the percentage ionization of this weak acid.

[13.72] A lithium salt used in lubricating grease has the formula $LiC_nH_{2n-1}O_2$. The salt is soluble in water to the extent of 0.036 g per 100 g of water at 25°C. The osmotic pressure of this solution is found to be 57.1 mm Hg. Assuming that molality and molarity in such a dilute solution are the same and that the lithium salt is completely dissociated in the solution, determine an appropriate value of *n* in the formula for the salt.

13.73 Pheromones are compounds secreted by the females of many insect species to attract males. One of these compounds contains 80.78 percent C, 13.56 percent H, and 5.66 percent O. A solution of 1.00 g of this substance in 8.50 g of benzene freezes at 3.37°C. What are the molecular weight and molecular formula of the compound?

[13.74] When 23.8 g of $HgCl_2$ is dissolved in 500 g of water, the boiling point of the water is increased by 0.090°C. **(a)** Calculate the van't Hoff factor, *i*, for this solution. **(b)** Would you classify $HgCl_2$ as a strong electrolyte, weak electrolyte, or nonelectrolyte? Explain.

[13.75] Benzene, C_6H_6, and octane, C_8H_{18}, form ideal solutions. At 60°C the vapor pressure of pure benzene is 0.507 atm and that of octane is 0.103 atm. **(a)** What is the vapor pressure at 60°C of a solution composed of 9.80 g of benzene and 40.2 g of octane? **(b)** What is the mole fraction of benzene in the vapor above the solution? (Remember Dalton's law of partial pressures.)

[13.76] A 0.100 *m* solution of $CaCl_2$ freezes at −0.474°C, whereas a 0.100 *m* solution of $NiCl_2$ freezes at −0.538°C. Explain why these temperatures are different.

[13.77] In the bell jar experiment shown in Figure 13.15, 20.0 mL of a 0.050 *M* aqueous solution of a nonvolatile nonelectrolyte is placed in the left beaker, and 20 mL of a 0.030 M aqueous solution of NaCl is placed in the right beaker. What are the volumes in the two beakers when equilibrium is attained?

14 Chemical Kinetics

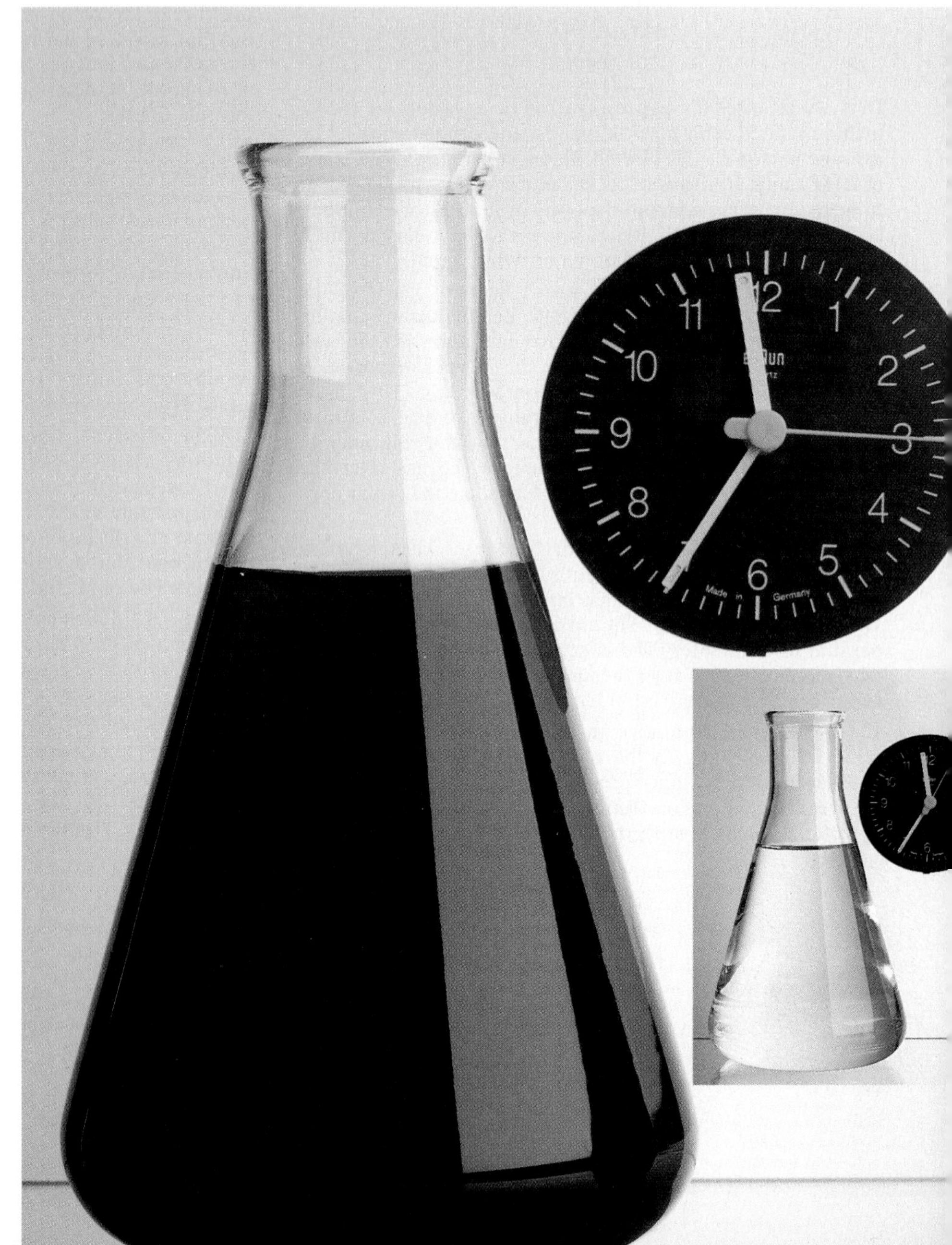

In the iodine "clock" reaction, a dramatic color change indicates how long a reaction takes to proceed from reactants (inset) to products. (© Richard Megna/Fundamental Photographs)

Chemistry is by its very nature concerned with change. Substances with well-defined properties are converted by chemical reactions into other materials with different properties. Chemists want to know which new substances are formed from a given set of starting reactants. However, it is equally important to know how rapidly chemical reactions occur and to understand the factors that control their speeds. For example, what factors are important in determining how rapidly foods spoil? What determines the rate at which steel rusts? How do you design a rapidly setting material for dental fillings? Which biochemical reaction rates determine the contraction and relaxation of smooth muscle in the arteries of the heart? Which factors control the rate at which fuel burns in an auto engine?

The area of chemistry concerned with the speeds, or rates, at which reactions occur is called **chemical kinetics**. In this chapter, we shall learn how to express and determine the rates at which reactions occur. We shall see that the rates of chemical reactions are affected by several factors, most notably:

1. *The concentrations of the reactants:* Most chemical reactions proceed faster if the concentration of one or more of the reactants is increased. For example, the "clock" reaction shown in the chapter-opening photograph proceeds at a specific rate that depends on the initial concentrations of the reactants.
2. *The temperature at which the reaction occurs:* The rates of chemical reactions increase as temperature is increased. It is for this reason that we refrigerate perishable foods such as milk. The bacterial reactions that lead to the spoiling of milk proceed much more rapidly at room temperature than they do at the lower temperatures of a refrigerator.
3. *The presence of a catalyst:* The rates of many reactions can be increased by adding a substance known as a *catalyst*. We will see that a catalyst increases the rate of a reaction without being consumed in the reaction. The physiology of most living species depends crucially on *enzymes*, protein molecules that increase the rates of selected biochemical reactions.
4. *The surface area of solid or liquid reactants or catalysts:* Reactions that involve solids often proceed faster as the surface area of the solid is increased. For example, a medicine in the form of a tablet will dissolve in the stomach and enter the bloodstream more slowly than the same medicine in the form of a fine powder.

We shall consider all the above factors as we proceed through this chapter. We shall also consider what chemical kinetics can teach us about how chemical reactions occur at the molecular level.

CONTENTS

14.1 REACTION RATE

The speed of any event is measured by the change that occurs in a given interval of time. For example, the speed of an automobile is expressed in terms of a change in position in a certain amount of time; the units we use to measure this speed are usually miles per hour (mi/hr). Similarly, the speed of a reaction, or the **reaction rate**, is expressed as the change in concentration of a reactant or product in a certain amount of time. The units of reaction rates are usually molarity per second (*M*/s). As an example, consider the reaction that occurs when butyl chloride, C_4H_9Cl, is placed in water. The resulting reaction produces butyl alcohol, C_4H_9OH, and hydrochloric acid:

$$C_4H_9Cl(l) + H_2O(l) \longrightarrow C_4H_9OH(aq) + HCl(aq) \qquad [14.1]$$

Suppose that we prepare a 0.1000 *M* solution of C_4H_9Cl in water and then measure the concentration of C_4H_9Cl at various times after the solution is prepared. Because C_4H_9Cl is consumed during the reaction, its concentration decreases with time, as shown in the first two columns of Table 14.1. The average rate of the reaction over any time interval is equal to the decrease in the concentration of C_4H_9Cl divided by the time in which that change occurs:

$$\text{Average rate} = \frac{\text{decrease in concentration of } C_4H_9Cl}{\text{length of time interval}} = -\frac{\Delta[C_4H_9Cl]}{\Delta t} \qquad [14.2]$$

As we noted in Section 5.2, the Greek letter delta, Δ, is read "change in"; thus, Δt is the change in time between the beginning and end of the interval:

$$\begin{aligned}\Delta t &= (\text{time at end of interval}) - (\text{time at beginning of interval}) \\ &= (\text{final time}) - (\text{initial time})\end{aligned}$$

The brackets around C_4H_9Cl in Equation 14.2 indicate the concentration of the substance. Thus, $\Delta[C_4H_9Cl]$ is the change in the concentration of C_4H_9Cl during the time interval:

$$\Delta[C_4H_9Cl] = [C_4H_9Cl]_{\text{final time}} - [C_4H_9Cl]_{\text{initial time}}$$

The subscripts on $[C_4H_9Cl]$ indicate when the concentrations were measured.

Table 14.1 Rate Data for Reaction of C_4H_9Cl with Water

Time (s)	$[C_4H_9Cl]$ (*M*)	Average rate (*M*/s)
0	0.1000	
		1.90×10^{-4}
50	0.0905	
		1.70×10^{-4}
100	0.0820	
		1.58×10^{-4}
150	0.0741	
		1.40×10^{-4}
200	0.0671	
		1.22×10^{-4}
300	0.0549	
		1.01×10^{-4}
400	0.0448	
		0.80×10^{-4}
500	0.0368	
		0.56×10^{-4}
800	0.0200	
10,000	0	

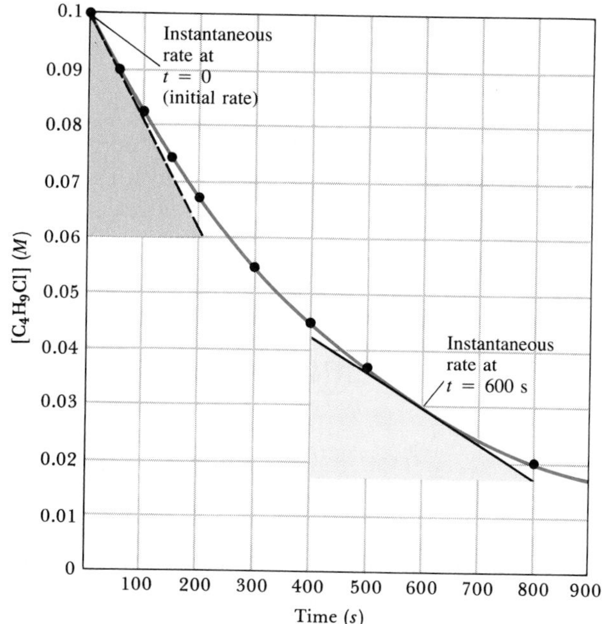

Figure 14.1 Concentration of butyl chloride, C_4H_9Cl, as a function of time. The dots represent the experimental data from the first two columns of Table 14.1; the red line is the smooth curve drawn to connect the data points. The reaction rate at any time is given by the slope of the tangent to the curve at that time. The slope of the tangent is defined as the vertical change divided by the horizontal change in the tangent, that is, $\Delta[C_4H_9Cl]/\Delta t$. Tangents have been drawn that touch the curve at $t = 0$ and $t = 600$ s. The calculation of the slope at $t = 600$ s is performed in the body of the text. The slope at $t = 0$ is calculated in Sample Exercise 14.1.

Because $[C_4H_9Cl]$ decreases as the reaction proceeds, $\Delta[C_4H_9Cl]$ is negative. The negative sign in Equation 14.2 makes the average rate a positive quantity; reaction rates are always positive quantities. We call $-\Delta[C_4H_9Cl]/\Delta t$ the *rate of disappearance* of C_4H_9Cl.

We can use the data in the first two columns of Table 14.1 to calculate the average rate of disappearance of C_4H_9Cl for the various time intervals. For example, from $t = 0$ s to $t = 50$ s, $[C_4H_9Cl]$ decreases from 0.1000 M to 0.0905 M. Therefore, the average rate over this 50-s interval is

$$\text{Average rate} = -\frac{[C_4H_9Cl]_{t=50\text{ s}} - [C_4H_9Cl]_{t=0\text{ s}}}{50\text{ s} - 0\text{ s}}$$

$$= -\frac{(0.0905 - 0.1000)\ M}{(50 - 0)\text{ s}} = 1.90 \times 10^{-4}\ M/\text{s}$$

The calculated average rate for each time interval is shown in the third column of Table 14.1. Notice that the average rate steadily decreases as the reaction proceeds. At some point, the reaction essentially stops; there is no longer any change in concentration with time.

We can also see this decrease in rate if we display the data graphically, as in Figure 14.1. The dots represent the experimental data from the first two columns of Table 14.1. Using this curve we can determine the **instantaneous rate;** this is the rate at a particular time as opposed to the average rate over an interval of time. The instantaneous rate is obtained from the straight-line tangent that touches the curve at the point of interest. We have drawn two such tangents in Figure 14.1, one at $t = 0$ and the other at $t = 600$ s. The slopes of these tangents give the instantaneous rates at these times.* For example, at 600 s we have

* You may wish to review briefly the idea of graphical determination of slopes by referring to Appendix A. If you are familiar with calculus, you may recognize that the average rate approaches the instantaneous rate as the time interval approaches zero. This limit, in the notation of calculus, is represented as $-d[C_4H_9Cl]/dt$.

$$\text{Instantaneous rate} = -\frac{(0.017 - 0.042)\ M}{(800 - 400)\ \text{s}} = 6.2 \times 10^{-5}\ M/\text{s}$$

We will usually refer to the instantaneous rate merely as the rate.

We could equally well have expressed the reaction rate of Equation 14.1 in terms of the changes in the concentration of C_4H_9OH. Because C_4H_9OH is a product of the reaction, its concentration increases with time. The quantity $\Delta[C_4H_9OH]/\Delta t$, which we call the *rate of appearance* of C_4H_9OH, is therefore positive. The stoichiometry of the reaction tells us that, at any given time, the number of moles of C_4H_9OH produced equals the number of moles of C_4H_9Cl consumed. Thus, the rate of appearance of C_4H_9OH equals the rate of disappearance of C_4H_9Cl:

$$\text{Average rate} = -\frac{\Delta[C_4H_9Cl]}{\Delta t} = \frac{\Delta[C_4H_9OH]}{\Delta t}$$

We can always express the rate of a reaction in terms of the disappearance of reactants or the appearance of products. The stoichiometry of the reaction tells us the relationship between these. For example, in the reaction

$$2HI(g) \longrightarrow H_2(g) + I_2(g)$$

we can measure the rate of disappearance of HI or the appearance of either H_2 or I_2. Because 2 mol of HI disappears for each mole of H_2 or I_2 that forms, the rate of disappearance of HI is twice the rate of appearance of H_2 or I_2:

$$-\frac{\Delta[HI]}{\Delta t} = 2\frac{\Delta[H_2]}{\Delta t} = 2\frac{\Delta[I_2]}{\Delta t}$$

SAMPLE EXERCISE 14.1

(a) Using the data in Table 14.1, calculate the average rate of disappearance of C_4H_9Cl over the time interval from 50 to 150 s. **(b)** Using Figure 14.1, estimate the instantaneous rate of disappearance of C_4H_9Cl at $t = 0$ (the initial rate).

Solution: **(a)** From Table 14.1 we have

$$\text{Average rate} = -\frac{\Delta[C_4H_9Cl]}{\Delta t}$$

$$= -\frac{(0.0741 - 0.0905)\ M}{(150 - 50)\ \text{s}} = 1.64 \times 10^{-4}\ M/\text{s}$$

(b) The initial rate is given by the slope of the dashed line in Figure 14.1. The slope of a straight line is given by the change in the vertical axis divided by the corresponding change in the horizontal axis. The straight line falls from $[C_4H_9Cl] = 0.100$ to 0.060 M in the time change from 0 to 200 s. Thus, the instantaneous rate is

$$\text{Rate} = -\frac{(0.060 - 0.100)\ M}{(200 - 0)\ \text{s}} = 2.0 \times 10^{-4}\ M/\text{s}$$

PRACTICE EXERCISE

Using Figure 14.1, estimate the instantaneous rate of disappearance of C_4H_9Cl at $t = 300$ s. ***Answer:*** $1.1 \times 10^{-4}\ M/\text{s}$

14.2 THE DEPENDENCE OF REACTION RATE ON CONCENTRATIONS

The decreasing rate of reaction with passing time that is evident in Figure 14.1 is quite typical of reactions. Reaction rates diminish as the concentrations of reactants diminish. Conversely, rates generally increase when reactant concentrations are increased.

One way of studying the effect of concentration on reaction rate is to determine the way in which the rate at the beginning of a reaction depends on the starting concentrations. To illustrate this approach, consider the following reaction:

$$NH_4^+(aq) + NO_2^-(aq) \longrightarrow N_2(g) + 2H_2O(l)$$

We might study the rate of this reaction by measuring the concentration of NH_4^+ or NO_2^- as a function of time or by measuring the volume of N_2 collected. Because of the 1:1 stoichiometry of the reaction, all of these rates will be equal.

Once we determine the initial reaction rate (the instantaneous rate at $t = 0$) for various starting concentrations of NH_4^+ and NO_2^-, we can tabulate the data as shown in Table 14.2. These data indicate that changing either $[NH_4^+]$ or $[NO_2^-]$ changes the reaction rate. Notice that if we double $[NO_2^-]$ while holding $[NH_4^+]$ constant, the rate doubles (compare experiments 1 and 2). If $[NO_2^-]$ is increased by a factor of 4 (compare experiments 1 and 3), the rate changes by a factor of 4, and so forth. These results indicate that the rate is proportional to $[NO_2^-]$ raised to the first power. When $[NH_4^+]$ is similarly varied while $[NO_2^-]$ is held constant, the rate is affected in the same manner. We conclude that the rate is also directly proportional to the concentration of NH_4^+. We can express the overall concentration dependence in the following way:

$$\text{Rate} = k[NH_4^+][NO_2^-] \qquad [14.3]$$

The proportionality constant, k, in Equation 14.3 is called a **rate constant**. We can evaluate the magnitude of k using the data in Table 14.2. Using the results of experiment 1 and substituting into Equation 14.3, we have

$$5.4 \times 10^{-7}\ M/\text{s} = k(0.0100\ M)(0.200\ M)$$

Table 14.2 Rate Data for the Reaction of Ammonium and Nitrite Ions in Water at 25°C

Experiment number	Initial NO_2^- concentration (M)	Initial NH_4^+ concentration (M)	Observed initial rate (M/s)
1	0.200	0.0100	5.4×10^{-7}
2	0.200	0.0200	10.8×10^{-7}
3	0.200	0.0400	21.5×10^{-7}
4	0.200	0.0600	32.3×10^{-7}
5	0.0202	0.200	10.8×10^{-7}
6	0.0404	0.200	21.6×10^{-7}
7	0.0606	0.200	32.4×10^{-7}
8	0.0808	0.200	43.3×10^{-7}

Solving for k gives

$$k = \frac{5.4 \times 10^{-7}\ M/\text{s}}{(0.0100\ M)(0.200\ M)} = 2.7 \times 10^{-4}\ M^{-1}\text{s}^{-1}$$

You may wish to satisfy yourself that this same value of k is obtained using any of the other experimental results given in Table 14.2. You might also note that given $k = 2.7 \times 10^{-4}\ M^{-1}\text{s}^{-1}$ and using Equation 14.3 we can calculate the rate for any concentration of NH_4^+ and NO_2^-. Suppose that $[NH_4^+] = 0.100\ M$ and $[NO_2^-] = 0.100\ M$; then

$$\text{Rate} = (2.7 \times 10^{-4}\ M^{-1}\text{s}^{-1})(0.100\ M)(0.100\ M) = 2.7 \times 10^{-6}\ M/\text{s}$$

An equation such as Equation 14.3 that relates the rate of a reaction to concentration is called a **rate law**. *The rate law for any chemical reaction must be determined experimentally; it cannot be predicted by merely looking at the chemical equation.* The following are some additional examples of rate laws:

$$2N_2O_5(g) \longrightarrow 4NO_2(g) + O_2(g) \qquad \text{Rate} = k[N_2O_5] \qquad [14.4]$$

$$CHCl_3(g) + Cl_2(g) \longrightarrow CCl_4(g) + HCl(g) \qquad \text{Rate} = k[CHCl_3][Cl_2]^{1/2} \qquad [14.5]$$

$$H_2(g) + I_2(g) \longrightarrow 2HI(g) \qquad \text{Rate} = k[H_2][I_2] \qquad [14.6]$$

The rate laws for most reactions have the general form

$$\text{Rate} = k[\text{reactant 1}]^m[\text{reactant 2}]^n \ldots \qquad [14.7]$$

The exponents m and n in Equation 14.7 are called **reaction orders**, and their sum is the **overall reaction order**. For example, the rate law for the reaction of NH_4^+ with NO_2^- (Equation 14.3) contains the concentration of NH_4^+ raised to the first power. Thus, the reaction order in NH_4^+ is 1; the reaction is *first order* in NH_4^+. It is also first order in NO_2^-. The overall reaction order is $1 + 1 = 2$; we say the reaction is *second order overall.*

In most rate laws, reaction orders are 0, 1, or 2. However, we occasionally encounter rate laws in which the reaction order is fractional (such as Equation 14.5) or even negative.

The units of the rate constant depend on the overall reaction order of the rate law. For example, in a reaction that is second order overall, the units of the rate constant must satisfy

$$\text{Units of rate} = (\text{units of rate constant})(\text{units of concentration})^2$$

Hence, in our usual units of concentration and time:

$$\text{Units of rate constant} = \frac{\text{units of rate}}{(\text{units of concentration})^2} = \frac{M/\text{s}}{M^2} = M^{-1}\text{s}^{-1}.$$

SAMPLE EXERCISE 14.2

(a) What are the overall reaction orders for the reactions described in Equations 14.4 and 14.5. **(b)** What are the usual units of the rate constant for the rate law for Equation 14.4?

Solution: **(a)** The overall reaction order is the sum of the powers to which all the concentrations of reactants are raised in the rate law. The reaction in Equation 14.4 is first order in N_2O_5 and first order overall. The reaction in Equation 14.5 is first order in $CHCl_3$ and one-half order in Cl_2. The overall reaction order is $\frac{3}{2}$. **(b)** For the rate law for Equation 14.4, we have

$$\text{Units of rate} = (\text{units of rate constant})(\text{units of concentration})$$

So

$$\text{Units of rate constant} = \frac{\text{units of rate}}{\text{units of concentration}} = \frac{M/\text{s}}{M} = \text{s}^{-1}.$$

PRACTICE EXERCISE

(a) What is the reaction order of the reactant H_2 in Equation 14.6? **(b)** What are the units of the rate constant for Equation 14.5? ***Answers:*** **(a)** 1; **(b)** $M^{-1/2}\text{s}^{-1}$

Using Initial Rates to Determine Rate Laws

We often determine the rate law for a reaction by the same method we applied to the data in Table 14.2: We observe the effect of changing the initial concentrations of the reactants on the initial rate of the reaction. If a reaction is zero order in a particular reactant, changing its concentration will have no influence on rate (as long as some of the reactant is present). If the reaction is first order in a reactant, changes in the concentration of that substance will produce proportional changes in the rate. Thus, doubling the concentration will double the rate, and so forth. When the rate law is second order in a particular reactant, doubling its concentration increases the rate by a factor of $2^2 = 4$; tripling its concentration causes the rate to increase by a factor of $3^2 = 9$, and so forth.

In working with rate laws, be careful not to confuse the rate constant for a reaction with the reaction rate. The rate of a reaction depends on the concentrations of reactants; the rate constant does not. As we shall see later in this chapter, the rate constant and consequently the reaction rate are affected by temperature and the presence of a catalyst.

SAMPLE EXERCISE 14.3

The initial rate of a reaction $A + B \longrightarrow C$ was measured for several different starting concentrations of A and B, with the results given below:

Experiment number	[A] (M)	[B] (M)	Initial rate (M/s)
1	0.100	0.100	4.0×10^{-5}
2	0.100	0.200	4.0×10^{-5}
3	0.200	0.100	16.0×10^{-5}

Using these data, determine **(a)** the rate law for the reaction; **(b)** the magnitude of the rate constant; **(c)** the rate of the reaction when $[A] = 0.050\ M$ and $[B] = 0.100\ M$.

Solution: **(a)** We may assume that the rate law has the form: Rate $= k[A]^m[B]^n$. Our task is to deduce the values of m and n. Experiments 1 and 2 indicate that the concentration of B has no influence on the reaction rate. The reaction is therefore zero order in B. Experiments 1 and 3 indicate that doubling A increases the rate fourfold. This result indicates that rate is proportional to $[A]^2$; the reaction is second order in A. The rate law is

$$\text{Rate} = k[A]^2[B]^0 = k[A]^2$$

This same conclusion could be reached in a more formal way by taking the ratio of the rates from two experiments:

$$\frac{\text{Rate 1}}{\text{Rate 2}} = \frac{4.0 \times 10^{-5}\ M/\text{s}}{4.0 \times 10^{-5}\ M/\text{s}} = 1$$

Using the rate law, then, we have

$$1 = \frac{\text{rate 1}}{\text{rate 2}} = \frac{\cancel{k[0.100\ M]^m}[0.100\ M]^n}{\cancel{k[0.100\ M]^m}[0.200\ M]^n} = \frac{[0.100]^n}{[0.200]^n} = \left(\frac{1}{2}\right)^n$$

But $(\frac{1}{2})^n$ can equal 1 only if $n = 0$. We can deduce the value of m in a similar fashion:

$$\frac{\text{Rate 1}}{\text{Rate 3}} = \frac{4.0 \times 10^{-5}\ M/\text{s}}{16.0 \times 10^{-5}\ M/\text{s}} = \frac{1}{4}$$

Using the rate law gives us

$$\frac{1}{4} = \frac{\text{rate 1}}{\text{rate 3}} = \frac{\cancel{k}[0.100\ M]^m\cancel{[0.100\ M]^n}}{\cancel{k}[0.200\ M]^m\cancel{[0.100\ M]^n}} = \frac{[0.100]^m}{[0.200]^m} = \left(\frac{1}{2}\right)^m$$

The fact that $(\frac{1}{2})^m = \frac{1}{4}$ indicates that $m = 2$.

(b) Using the rate law and the data from experiment 1, we have

$$k = \frac{\text{rate}}{[\text{A}]^2} = \frac{4.0 \times 10^{-5}\ M/\text{s}}{(0.100\ M)^2} = 4.0 \times 10^{-3}\ M^{-1}\text{s}^{-1}$$

(c) Using the rate law from part (a) and the rate constant from part (b), we have

$$\text{Rate} = k[\text{A}]^2 = (4.0 \times 10^{-3}\ M^{-1}\text{s}^{-1})(0.050\ M)^2 = 1.0 \times 10^{-5}\ M/\text{s}$$

Because [B] is not part of the rate law, it is immaterial to the rate, provided that there is at least some B present to react with A.

PRACTICE EXERCISE

A particular reaction was found to depend on the concentration of the hydrogen ion, $[H^+]$. The initial rates varied as a function of $[H^+]$ as follows:

$[H^+]$ (M)	0.0500	0.100	0.200
Initial rate (M/s)	6.4×10^{-7}	3.2×10^{-7}	1.6×10^{-7}

(a) What is the order of the reaction in $[H^+]$? **(b)** Predict the initial reaction rate when $[H^+] = 0.400\ M$. ***Answers:*** **(a)** -1 (the rate is *inversely* proportional to $[H^+]$); **(b)** $0.8 \times 10^{-7}\ M/\text{s}$

14.3 THE TIME DEPENDENCE OF REACTANT CONCENTRATIONS

The rate law (Equation 14.7) tells us how the rate of a reaction changes as we change reactant concentrations. Rate laws can be converted into equations that tell us what the concentrations of the reactants or products are at any time during the course of a reaction. The mathematics required involves calculus. We don't expect you to be able to perform the calculus operations; however, you should be able to use the resulting equations. We shall apply this conversion to two of the simplest rate laws—those that are first order overall and those that are second order overall.

First-Order Reactions

If the rate of a reaction of the sort A $\longrightarrow$ products is first order in A, that is, a **first-order reaction**, we can write the following rate law:

$$\text{Rate} = -\frac{\Delta[\text{A}]}{\Delta t} = k[\text{A}]$$

Using calculus, this equation can be transformed into an equation that relates the concentration of A at the start of the reaction, $[\text{A}]_0$, to its concentration at any other time t, $[\text{A}]_t$:

$$\ln [\text{A}]_t - \ln [\text{A}]_0 = \ln \frac{[\text{A}]_t}{[\text{A}]_0} = -kt \qquad [14.8]$$

The function "ln" is the natural logarithm (Appendix A-2). In terms of base-10 logarithms, Equation 14.8 can be written

$$\log [\text{A}]_t - \log [\text{A}]_0 = \log \frac{[\text{A}]_t}{[\text{A}]_0} = -\frac{kt}{2.30}$$

The factor 2.30 arises from the conversion of natural logarithms to base-10 logarithms (Appendix A-2).

Recall that the equation for a straight line is

$$y = mx + b$$

where m is the slope and b is the y-intercept of the line (see Appendix A-4). Equation 14.8 can be arranged to give a straight-line equation:

$$\begin{array}{ccccccc} \ln [\text{A}]_t & = & -k & \cdot & t & + & \ln[\text{A}]_0 \\ \updownarrow & & \updownarrow & & \updownarrow & & \updownarrow \\ y & = & m & \cdot & x & + & b \end{array} \qquad [14.9]$$

Thus, for a first-order reaction, a graph of $\ln [\text{A}]_t$ versus time gives a straight line with a slope of $-k$ and a y-intercept of $\ln [\text{A}]_0$.

The conversion of methyl isonitrile to acetonitrile is an example of a first-order reaction:

$$\underset{\text{Methyl isonitrile}}{H_3C{-}N{\equiv}C{:}} \longrightarrow \underset{\text{Acetonitrile}}{H_3C{-}C{\equiv}N{:}} \qquad [14.10]$$

Figure 14.2(*a*) shows how the pressure of methyl isonitrile varies with time as it rearranges in the gas phase at 198.9°C. We can use pressure as a unit of concentration for a gas because, from the ideal-gas law, the number of moles per unit volume is directly proportional to pressure. Figure 14.2(*b*) shows a plot of the natural logarithm of the pressure versus time, a plot that yields a straight line. The slope of this line is $-5.11 \times 10^{-5}\ \text{s}^{-1}$. (You should verify this for yourself, remembering that your result may vary slightly from ours because of inaccuracies associated with reading the graph.) Because the slope of the line equals $-k$, we see that the rate constant for this reaction equals $5.11 \times 10^{-5}\ \text{s}^{-1}$.

For a first-order reaction, Equations 14.8 and 14.9 can be used to determine (1) the concentration of a reactant remaining at any time after the reaction has started, (2) the time required for a given fraction of a sample to react, or (3) the time required for a reactant concentration to reach a certain level.

Figure 14.2 (*a*) Variation in the pressure of methyl isonitrile, CH_3NC, with time at 198.9°C during the reaction $CH_3NC \longrightarrow CH_3CN$. (*b*) A plot of the natural logarithm of the CH_3NC pressure as a function of time.

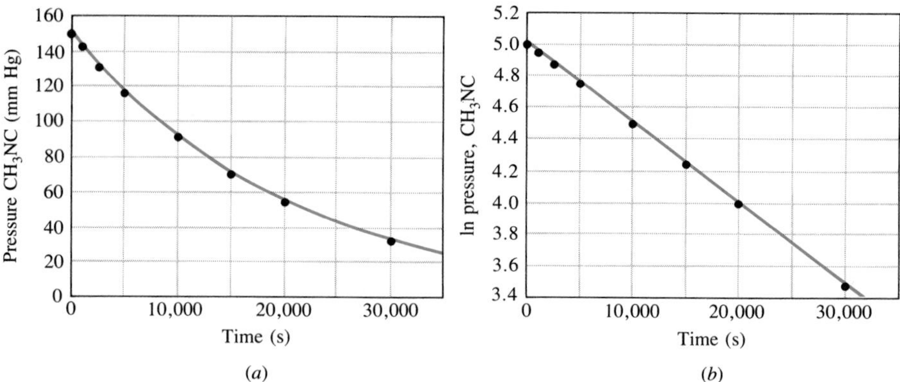

SAMPLE EXERCISE 14.4

The first-order rate constant for the decomposition of a certain insecticide in water at 12°C is 1.45 yr^{-1}. A quantity of this insecticide is washed into a lake in June, leading to a concentration of 5.0×10^{-7} g/cm^3 of water. Assume that the effective temperature of the lake is 12°C. **(a)** What is the concentration of the insecticide in June of the following year? **(b)** How long will it take for the concentration of the insecticide to drop to 3.0×10^{-7} g/cm^3?

Solution: **(a)** Substituting $k = 1.45\ yr^{-1}$, $t = 1.00$ yr, and $[\text{insecticide}]_0 = 5.0 \times 10^{-7}$ g/cm^3 into Equation 14.9 gives

$$\ln[\text{insecticide}]_{t=1\ yr} = -(1.45\ yr^{-1})(1.00\ yr) + \ln(5.0 \times 10^{-7})$$

We use the ln function on a calculator to evaluate the second term on the right, giving

$$\ln[\text{insecticide}]_{t=1\ yr} = -1.45 + (-14.51) = -15.96$$

To obtain $[\text{insecticide}]_{t=1\ yr}$ we use the inverse natural logarithm, or e^x, function on the calculator. Enter -15.96 followed by the e^x key (on some calculators, you may have to enter -15.96, followed by the INV key, followed by the ln key):

$$[\text{insecticide}]_{t=1\ yr} = e^{-15.96} = 1.2 \times 10^{-7}\ g/cm^3$$

Note that the concentration units for $[A]_t$ and $[A]_0$ must be the same.

(b) Again substituting into Equation 14.9, with $[\text{insecticide}]_t = 3.0 \times 10^{-7}$ g/cm^3, gives

$$\ln(3.0 \times 10^{-7}) = -(1.45\ yr^{-1})(t) + \ln(5.0 \times 10^{-7})$$

Solving for t gives

$$t = -[\ln(3.0 \times 10^{-7}) - \ln(5.0 \times 10^{-7})]/1.45\ yr^{-1}$$
$$= -(-15.02 + 14.51)/1.45\ yr^{-1} = 0.35\ yr$$

PRACTICE EXERCISE

The decomposition of N_2O_5 according to the equation

$$2N_2O_5(g) \longrightarrow 4NO_2(g) + O_2(g)$$

has a first-order rate law:

$$\text{Rate} = -\frac{\Delta[N_2O_5]}{\Delta t} = k[N_2O_5]$$

The rate constant k has the value $8.5 \times 10^{-3}\ s^{-1}$. If the initial pressure of N_2O_5 is 240 mm Hg, what will the pressure be after 300 s? ***Answer:*** 18.7 mm Hg

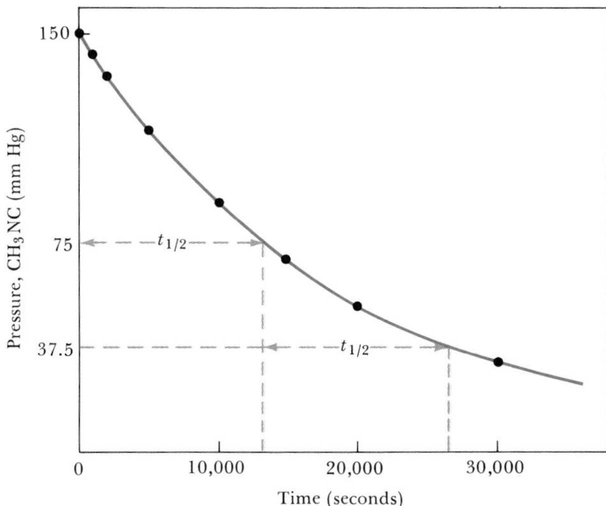

Figure 14.3 Pressure of methyl isonitrile as a function of time. Two successive half-lives of the rearrangement reaction, Equation 14.10, are shown.

Half-life

The **half-life** of a reaction, $t_{1/2}$, is the time required for the concentration of a reactant to drop to one-half of its initial value. That is, $[A]_{t_{1/2}} = \frac{1}{2}[A]_0$. We can determine the half-life of a first-order reaction by substituting $[A]_{t_{1/2}}$ into Equation 14.8:

$$\ln \frac{\frac{1}{2}[A]_0}{[A]_0} = -kt_{1/2}$$

$$\ln \tfrac{1}{2} = -kt_{1/2}$$

$$t_{1/2} = -\frac{\ln \frac{1}{2}}{k} = \frac{0.693}{k} \qquad [14.11]$$

Notice that $t_{1/2}$ is independent of the initial concentration of reactant. Thus, if we measure reactant concentration at *any* time in the course of a first-order reaction, the concentration of reactant will be half of that measured value at a time $0.693/k$ later. The concept of half-life is widely used in describing radioactive decay. This application is discussed in detail in Section 21.4.

The data for the first-order rearrangement of methyl isonitrile at 198.9°C are graphed in Figure 14.3. The first half-life is shown at 13,320 s. At a time 13,320 s later, the isonitrile concentration has decreased to one-half of one half, or one-fourth the original concentration. *In a first-order reaction the concentration of the reactant decreases by factors of $\frac{1}{2}$ in a series of regularly spaced time intervals.*

SAMPLE EXERCISE 14.5

From Figure 14.1 estimate the half-life of the reaction of C_4H_9Cl with water.

Solution: From the figure, we see that the initial value of $[C_4H_9Cl]$ is 0.100 *M*. The half-life for this first-order reaction is the time required for $[C_4H_9Cl]$ to decrease to 0.0500 *M*. This point occurs at approximately 340 s. At the end of the second half-life, which should occur at 680 s, the concentration should have decreased by yet another factor of 2, to 0.025 *M*. Inspection of the graph shows that this is indeed the case.

PRACTICE EXERCISE

Calculate $t_{1/2}$ for the reaction described in Practice Exercise 14.4. ***Answer:*** 81.5 s

Second-Order Reactions

For a reaction that is second order in just one reactant, A, the rate is given by

$$\text{Rate} = k[\text{A}]^2$$

Relying on calculus, this rate law can be used to derive the following equation:

$$\frac{1}{[\text{A}]_t} = kt + \frac{1}{[\text{A}]_0} \quad [14.12]$$

This equation, like Equation 14.9, has the form of a straight line ($y = mx + b$). If the reaction is second order, a plot of $1/[\text{A}]_t$ versus t will yield a straight line with a slope equal to k and a y-intercept equal to $1/[\text{A}]_0$. One way to distinguish between first- and second-order rate laws is to graph both $\ln [\text{A}]_t$ and $1/[\text{A}]_t$ against t. If the $\ln [\text{A}]_t$ plot is linear, the reaction is first order; if the $1/[\text{A}]_t$ plot is linear, the reaction is second order.

Using Equation 14.12, we can show that the half-life of a second-order reaction is

$$t_{1/2} = \frac{1}{k[\text{A}]_0} \quad [14.13]$$

We see that, unlike that of first-order reactions, the half-life of a second-order reaction is *not* independent of the initial concentration of reactant.

SAMPLE EXERCISE 14.6

The following data were obtained for the gas-phase decomposition of nitrogen dioxide at 300°C, $2NO_2(g) \longrightarrow 2NO(g) + O_2(g)$:

Time (s)	$[NO_2]$ (M)
0	0.0100
50	0.0079
100	0.0065
200	0.0048
300	0.0038

Is the reaction first or second order in NO_2?

Solution: To test whether the reaction is first or second order, we can construct plots of $\ln [NO_2]$ and $1/[NO_2]$ against time. In doing so, we will find it useful to prepare the following table from the data given:

Time (s)	$[NO_2]$ (M)	$\ln [NO_2]$	$1/[NO_2]$
0	0.0100	−4.61	100
50	0.0079	−4.84	127
100	0.0065	−5.04	154
200	0.0048	−5.34	208
300	0.0038	−5.57	263

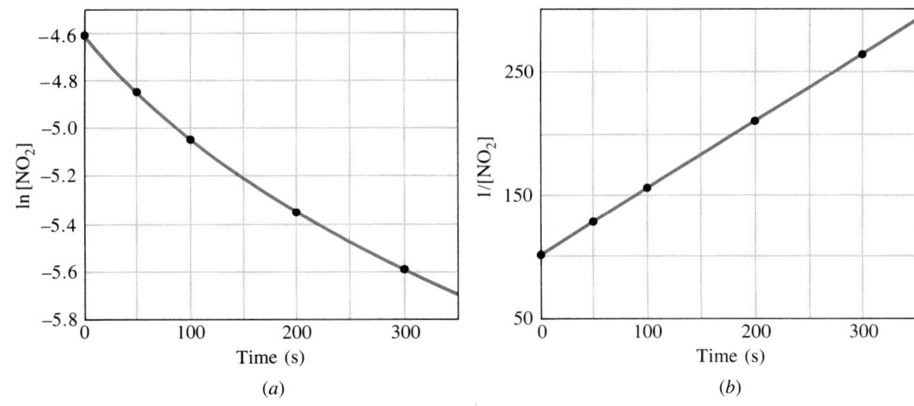

Figure 14.4 Plots of the kinetic data for the reaction $2NO_2(g) \longrightarrow 2NO(g) + O_2(g)$ at 300°C. The plot of $\ln[NO_2]$ versus time (*a*) is not linear; consequently, the reaction is not first order in NO_2. The plot of $1/[NO_2]$ versus time (*b*) is linear; the reaction is second order in NO_2.

As Figure 14.4 shows, only the plot of $1/[NO_2]$ versus time is linear. Thus the reaction obeys a second-order rate law: rate $= k[NO_2]^2$. From the slope of this straight-line graph we have that $k = 0.543\ M^{-1}s^{-1}$.

PRACTICE EXERCISE

What is the half-life measured from time $t = 0$ for the decomposition of NO_2, as represented by the tabular data above? ***Answer:*** 184 s

A reaction may also be second order by having a first-order dependence of the rate on each of two reagents; that is, rate $= k[A][B]$. It is possible to derive an expression for the variation in concentrations of A and B with time. However, we will not consider this and other more complicated rate laws in this text.

14.4 THE TEMPERATURE DEPENDENCE OF REACTION RATES

The rates of most chemical reactions increase as the temperature rises. Examples of this generalization surround us. Plants grow more rapidly in warm weather than in cold. Food cooks faster in boiling water than in merely hot water, and it spoils faster at room temperature than when it is refrigerated. We can literally see the effect of temperature on the rate of a chemical reaction by observing a *chemiluminescent* reaction, a chemical reaction that produces light. Fireflies produce their characteristic glow through a chemiluminescent reaction. Another common example of chemiluminescence is the light produced by Cyalume light sticks. Light sticks contain chemicals in two compartments separated by a divider. When the divider is broken and the chemicals mix, chemiluminescence results. The greater the rate of the reaction, the greater the amount of light produced. Figure 14.5 demonstrates the effect of temperature on the light emitted by Cyalume light sticks. The light stick in cold water produces less light than the one immersed in hot water.

To obtain a clear understanding of how temperature affects reaction rates, we must examine simple reaction systems. As an example, let us consider the reaction about which we have already learned quite a bit, the first-order rearrangement of methyl isonitrile (Equation 14.10). The rate constant for this (and all) reactions is a constant at a given temperature. The value of the rate constant changes as the temperature changes, however. Figure 14.6 shows the experimentally determined rate constant for this reaction as a function of temperature. The rate constant, and hence the rate of the reaction, increases rapidly with temperature in a nonlinear fashion.

Figure 14.5 Temperature affects the rate of the chemiluminescent reaction in Cyalume light sticks. The light stick in hot water (left) glows more brightly than the one in cold water (right); the reaction is faster and produces more light at the higher temperature. (© Richard Megna/ Fundamental Photographs)

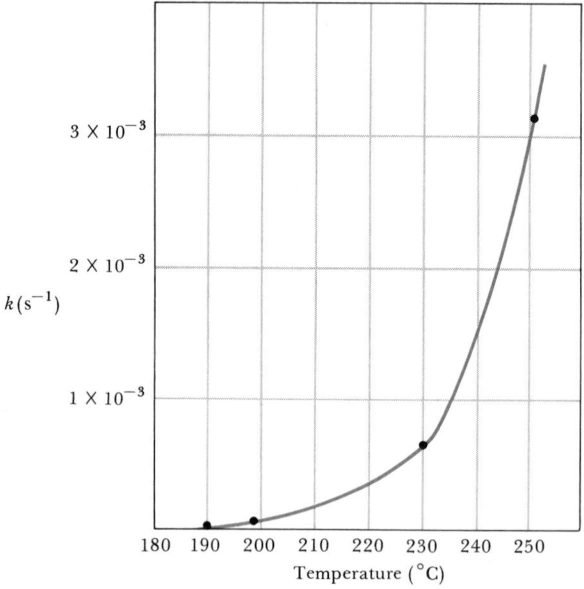

Figure 14.6 Variation in the first-order rate constant for the rearrangement of methyl isonitrile as a function of temperature. (The four points indicated are used in connection with Sample Exercise 14.7.)

As we seek an explanation for this behavior, perhaps the first question to ask is, Why do *any* reactions go slowly? If a reaction is energetically favorable, what keeps it from simply occurring immediately? Why does temperature have such a marked effect on the rates of chemical reactions?

Activation Energy

We know from the kinetic-molecular theory of gases that the average kinetic energy of gas molecules increases as temperature increases (Section 10.8). The fact that the rates of chemical reactions increase with increasing temperature suggests that the kinetic energy of molecules plays a role in determining how quickly the molecules react. In 1888, the Swedish chemist Svante Arrhenius suggested that molecules must possess a certain minimum amount of kinetic energy in order to react. We can envision the kinetic energy as necessary to "propel" the reactants into products. The situation is rather like that shown in Figure 14.7. The boulder is initially at position A. It will be at a lower energy

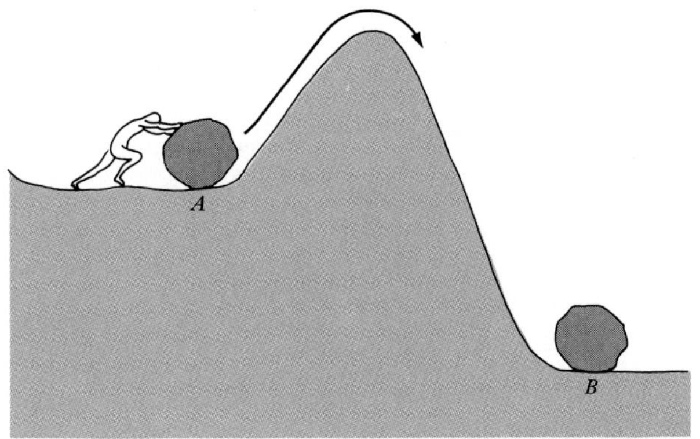

Figure 14.7 Illustration of the potential-energy profile for a boulder. The boulder must be moved over the energy barrier before it can come to rest in the lower-energy location, *B*.

(more stable) at position B. However, in order to get from position A to position B, it must acquire enough energy to get over the hill blocking its passage. In the same way, molecules may require a certain minimum energy to form new bonds during a chemical reaction. For example, in the rearrangement of methyl isonitrile to acetonitrile, we might imagine passing through an intermediate state in which the N≡C portion of the molecule is sitting sideways:

$$H_3C{-}N{\equiv}C\!: \longrightarrow \left[H_3C\cdots\begin{matrix}\ddot{C}\\ |||\\ \ddot{N}\end{matrix}\right] \longrightarrow H_3C{-}C{\equiv}N\!: \quad [14.14]$$

Even though the bonding may be more stable in the product acetonitrile than in the starting compound, energy is required to force the molecule through the relatively unstable intermediate state to the final result. The energy of the molecule as it proceeds along this reaction pathway is shown in Figure 14.8. Arrhenius called the energy barrier between the starting molecule and the highest energy along the reaction pathway the **activation energy**, E_a. The particular arrangement of atoms that has the maximum energy is often called the **activated complex** or **transition state**.

The conversion of $H_3C{-}N{\equiv}C$ to $H_3C{-}C{\equiv}N$ is exothermic; Figure 14.8 therefore shows the product as having a lower energy than the reactant. Notice that the reverse reaction is then endothermic; for that reaction, the activation barrier is equal to the sum of ΔE and E_a for the forward reaction.

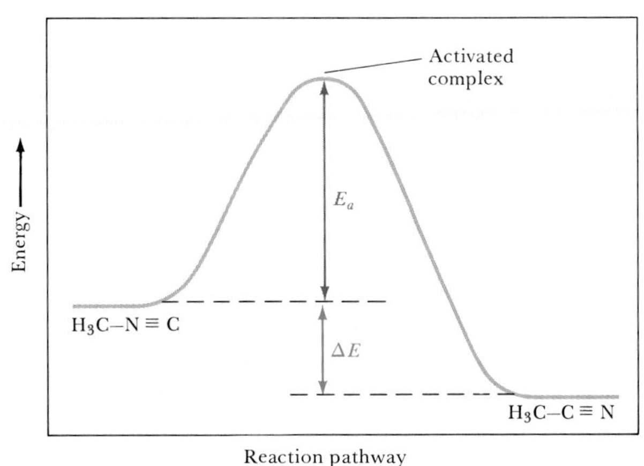

Figure 14.8 Energy profile for the rearrangement of methyl isonitrile. The molecule must surmount the activation-energy barrier before it can form the product, acetonitrile.

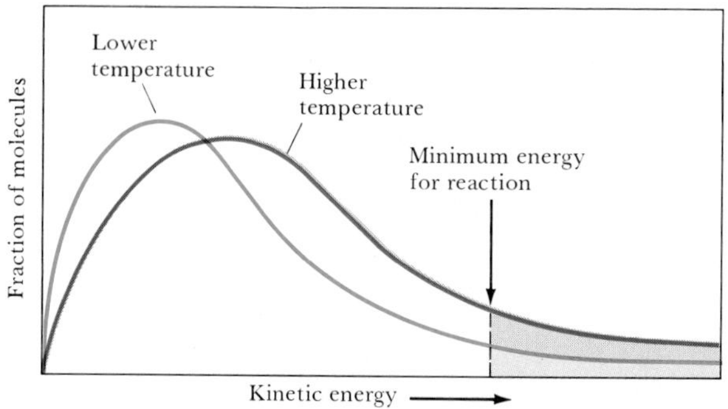

Figure 14.9 Distribution of kinetic energies in a sample of gas molecules at two different temperatures. At the higher temperature a larger number of molecules have higher energies. Thus, a larger fraction at any one instant will have more than the minimum energy required for reaction.

Energy is transferred between molecules through collisions. Thus, within a certain period of time, any particular isonitrile molecule might acquire enough energy to overcome the energy barrier and be converted into acetonitrile. At any given temperature only a small fraction of collisions will occur with sufficient energy to overcome the barrier to reaction. However, as shown in Figure 10.12, the distribution of molecular speeds is more spread out toward higher values when the gas is at a higher temperature. The distribution of kinetic energies of molecules changes in a similar way, as shown in Figure 14.9. This graph shows that at the higher temperature a larger fraction of molecules possess the minimum energy needed for reaction.

In addition to the requirement that the reactant species collide with sufficient energy to begin to rearrange bonds, an orientational requirement exists. The relative orientations of the molecules during their collisions determine whether the atoms are suitably positioned to form new bonds. For most reactions, only a fraction of the collisions that possess enough energy for reaction actually lead to products.

For example, in a mixture of H_2 and I_2 at ordinary temperatures and pressures, each molecule undergoes about 10^{10} collisions per second. If every collision between H_2 and I_2 resulted in the formation of HI, the reaction would be over in much less than a second. Instead, at room temperature the reaction proceeds very slowly. Only about 1 in every 10^{13} collisions occurs with both suitable orientation and sufficient energy to lead to products. If we increase the temperature, the number of collisions increases, as does the fraction that have sufficient energy to react. Consequently, the reaction rate increases. For the reaction between H_2 and I_2, the rate triples with each 10°C rise in temperature.

The Arrhenius Equation

Arrhenius noted that, for most reactions, the increase in rate with increasing temperature is nonlinear, as in the example shown in Figure 14.6. He found that most reaction-rate data obeyed the equation

$$k = Ae^{-E_a/RT} \qquad [14.15]$$

where k is the rate constant. This equation is called the **Arrhenius equation**. The term E_a is the activation energy, R is the gas constant (8.314 J/K-mol), and T is absolute temperature. The term A is constant, or nearly so, as temperature is varied. Called the **frequency factor**, A is related to the frequency of collisions and the probability that the collisions are favorably oriented for reaction. Notice that as the magnitude of E_a increases, k becomes smaller. Thus, reaction rates decrease as the energy barrier increases. Taking the natural log of both sides of Equation 14.15, we have

$$\ln k = -\frac{E_a}{RT} + \ln A \qquad [14.16]$$

Equation 14.16 has the form of a straight line; it predicts that a graph of $\ln k$ versus $1/T$ will be a line with a slope equal to $-E_a/R$ and a y-intercept equal to $\ln A$. Therefore, if we know the rate of a reaction for at least two different temperatures, we can use Equation 14.16 to determine E_a for the reaction. For example, suppose that at two different temperatures T_1 and T_2, a reaction has rate constants k_1 and k_2. At T_1 we have

$$\ln k_1 = -\frac{E_a}{RT_1} + \ln A$$

At T_2,

$$\ln k_2 = -\frac{E_a}{RT_2} + \ln A$$

Subtracting $\ln k_2$ from $\ln k_1$ gives

$$\ln k_1 - \ln k_2 = \left(-\frac{E_a}{RT_1} + \ln A\right) - \left(-\frac{E_a}{RT_2} + \ln A\right)$$

Simplifying this equation and rearranging it gives

$$\ln \frac{k_1}{k_2} = \frac{E_a}{R}\left(\frac{1}{T_2} - \frac{1}{T_1}\right) \qquad [14.17]$$

Equation 14.17 provides a convenient way to calculate the rate constant, k_1, at some temperature, T_1, when we know the activation energy and the rate constant, k_2, at some other temperature, T_2.

SAMPLE EXERCISE 14.7

The following table shows the rate constants for the rearrangement of methyl isonitrile at various temperatures (these are the data that are graphed in Figure 14.6).

Temperature (°C)	k (s^{-1})
189.7	2.52×10^{-5}
198.9	5.25×10^{-5}
230.3	6.30×10^{-4}
251.2	3.16×10^{-3}

(a) From these data calculate the activation energy for the reaction. **(b)** What is the value of the rate constant at 430.0 K?

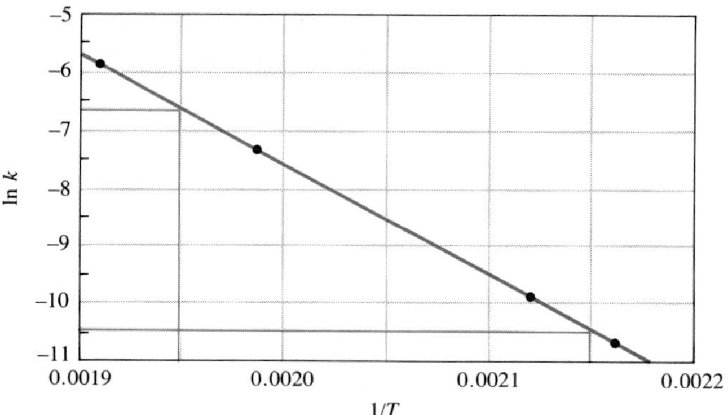

Figure 14.10 The natural logarithm of the rate constant for the rearrangement of methyl isonitrile as a function of $1/T$. The linear relationship is predicted by the Arrhenius equation.

Solution: **(a)** We must first convert the temperatures from °C to K. We then take the inverse of each temperature, $1/T$, and the natural log of each rate constant, ln k. This gives us the following table:

T (K)	$1/T$ (K^{-1})	ln k
462.7	2.161×10^{-3}	−10.59
471.9	2.119×10^{-3}	−9.85
503.3	1.987×10^{-3}	−7.37
524.2	1.908×10^{-3}	−5.76

A graph of ln k versus $1/T$ results in a straight line, as shown in Figure 14.10. The slope of the line is obtained by choosing two well-separated points, as shown, and using the coordinates of each:

$$\text{Slope} = \frac{\Delta y}{\Delta x} = \frac{-6.6 - (-10.4)}{0.00195 - 0.00215} = -1.9 \times 10^4$$

Because logarithms have no units, the numerator in this equation is dimensionless. The denominator has the units of $1/T$, namely, K^{-1}. Thus, the overall units for the slope are K. The slope is equal to $-E_a/R$. We use the value for the molar gas constant R in units of J/K-mol (Table 10.2). We thus obtain

$$\text{Slope} = -\frac{E_a}{R}$$

$$E_a = -(\text{slope})(R) = -(-1.9 \times 10^4 \text{ K})\left(8.31 \frac{\text{J}}{\text{K-mol}}\right)\left(\frac{1 \text{ kJ}}{1000 \text{ J}}\right) = 160 \text{ kJ/mol}$$

Note that we report the activation energy to two significant figures; we are limited by the precision with which we can read the graph in Figure 14.10.

(b) To determine the rate constant, k_1, at 430 K, we can use Equation 14.17 with $E_a = 160$ kJ/mol, $k_2 = 2.52 \times 10^{-5}\ s^{-1}$, $T_2 = 462.7$ K, and $T_1 = 430.0$ K:

$$\ln\left(\frac{k_1}{2.52 \times 10^{-5}\ \text{s}^{-1}}\right) = \left(\frac{160 \text{ kJ/mol}}{8.31 \text{ J/K-mol}}\right)\left(\frac{1}{462.7 \text{ K}} - \frac{1}{430.0 \text{ K}}\right)\left(\frac{1000 \text{ J}}{1 \text{ kJ}}\right) = -3.16$$

Thus,

$$\frac{k_1}{2.52 \times 10^{-5}\ \text{s}^{-1}} = e^{-3.16} = 4.22 \times 10^{-2}$$

$$k_1 = (4.22 \times 10^{-2})(2.52 \times 10^{-5}\ \text{s}^{-1}) = 1.06 \times 10^{-6}\ \text{s}^{-1}$$

Note that the units of k_1 are the same as those of k_2.

PRACTICE EXERCISE

Estimate the rate constant for the rearrangement of methyl isonitrile at 280°C.
Answer: $2.14 \times 10^{-2}\ s^{-1}$

A CLOSER LOOK: Seeing the Transition State

The details of how chemical reactions occur—the events that happen as reactants turn into products—have long fascinated chemists. In the laboratory, you will probably perform experiments that allow you to monitor how long a reaction takes to occur. In so doing, you will observe the result of the passage of reactants through the transition state into products (as in, for example, Figure 14.8), usually on the time scale of several minutes. What happens when the reactants reach the transition state, the energetic "top of the mountain" between reactants and products? The nature of the transition state—how molecules collide, transfer energy, break and make bonds—is the focus of the field of chemistry called *molecular reaction dynamics*. This is an extremely important field of modern chemical research. The 1986 Nobel Prize for Chemistry was awarded jointly to Dudley Herschbach, Yuan Lee, and John Polanyi for their pioneering research in reaction dynamics.

Transition states are very unstable and extremely short-lived; they exist for incomprehensibly short periods of time before falling apart into products or back into reactants. Research in reaction dynamics requires the design of elaborate experiments to take "snapshots" of the transition state, often using pulses of laser light that are extremely short in duration. Recent advances in laser design have allowed chemists to take faster and faster pictures of transition states falling apart into products, on the time scale of femtoseconds (1 fs = 10^{-15} s)!

One of the first examples of chemistry studied on the femtosecond time scale was the dissociation of ICN by light, a so-called photofragmentation reaction:

$$\mathrm{ICN}(g) \xrightarrow{\text{Light}} \underset{\substack{\text{Activated}\\ \text{complex}}}{[\mathrm{I}\cdots\mathrm{CN}]} \longrightarrow \mathrm{I}(g) + \mathrm{CN}(g)$$

The absorption of light by an ICN molecule excites the molecule and ultimately leads to breaking of the I—C bond. In the activated complex that results from the absorption of light, the iodine and carbon atoms are still within bonding distance of one another. In 1987, by using laser pulses less than 100 fs in duration, chemists at the California Institute of Technology determined that it takes only a few hundred femtoseconds for the activated complex to fall apart. Studies such as these hold great promise for providing detailed pictures of how molecules react.

The laser apparatus needed for femtosecond-time-scale studies is extremely complex (Figure 14.11). Excruciating care is required to isolate the equipment from vibrations that could change the distance that the laser light travels. In 1 fs, light travels only 3×10^{-7} m (0.3 μm), so even a minute vibration would throw off the timing of the snapshots.

Figure 14.11 Part of the apparatus at the California Institute of Technology for performing laser experiments on the femtosecond time scale. (Dr. Ahmed H. Zewail, California Institute of Technology)

14.5 REACTION MECHANISMS

A balanced equation for a chemical reaction indicates the substances that are present at the start of the reaction and those produced at the end. However, it provides no information about how the reaction occurs. The process by which a reaction occurs is called the **reaction mechanism**. At the most sophisticated level, a reaction mechanism will describe in great detail the order in which bonds are broken and formed and the changes in relative positions of the atoms in the course of the reaction. We will begin with more rudimentary descriptions of how reactions occur.

Elementary Steps

We have seen (Section 14.4) that reactions take place as a result of collisions among reacting molecules. For example, the collision between methyl isonitrile, CH_3NC, and some other molecule can provide the energy to allow the CH_3NC to rearrange:

$$H_3C{-}N{\equiv}C: \longrightarrow \left[H_3C \cdots \begin{matrix} \ddot{C} \\ ||| \\ \underset{..}{N} \end{matrix} \right] \longrightarrow H_3C{-}C{\equiv}N:$$

Similarly, the reaction of O_3 and NO to form O_2 and NO_2 appears to occur as a result of a single collision involving suitably oriented and sufficiently energetic NO and O_3 molecules:

$$NO(g) + O_3(g) \longrightarrow NO_2(g) + O_2(g) \qquad [14.18]$$

Both of these processes occur in a single event or step and are called **elementary steps** (or elementary processes).

The net change represented by a balanced chemical equation often occurs by a sequence of elementary steps. For example, consider the reaction of NO_2 and CO:

$$NO_2(g) + CO(g) \longrightarrow NO(g) + CO_2(g) \qquad [14.19]$$

Below 225°C, this reaction appears to proceed in two elementary steps. First, two NO_2 molecules collide, and an oxygen atom is transferred from one to the other. The resultant NO_3 then transfers an oxygen atom to CO during a collision between these molecules:

$$NO_2(g) + NO_2(g) \longrightarrow NO_3(g) + NO(g)$$

$$NO_3(g) + CO(g) \longrightarrow NO_2(g) + CO_2(g)$$

The elementary steps in a multistep mechanism must always add to give the chemical equation of the overall process. In the present example, the sum of the elementary steps is

$$2NO_2(g) + NO_3(g) + CO(g) \longrightarrow NO_2(g) + NO_3(g) + NO(g) + CO_2(g)$$

Simplifying this equation by eliminating substances that appear on both sides of the arrow gives the net equation for the process, Equation 14.19. Because NO_3 is neither a reactant nor a product in the overall reaction—it is formed in one elementary step and consumed in the next—it is called an **intermediate**. Multistep mechanisms involve one or more intermediates.

The number of molecules that participate as reactants in an elementary step defines the **molecularity** of the step. If a single molecule is involved, the reaction is said to be **unimolecular**. The rearrangement of methyl isonitrile (Equation 14.14) is a unimolecular process. Elementary steps involving the collision of two reactant molecules are said to be **bimolecular**. The reaction between NO and O_3 (Equation 14.18) is bimolecular. Elementary steps involving the simultaneous collision of three molecules are said to be **termolecular**. Termolecular steps are far less probable than unimolecular or bimolecular processes and are rarely encountered. The chance that four or more molecules will collide simultaneously with any regularity is even more remote; consequently, such collisions are never proposed as part of a reaction mechanism.

SAMPLE EXERCISE 14.8

It has been proposed that the conversion of ozone into O_2 proceeds in two steps:

$$O_3(g) \rightleftharpoons O_2(g) + O(g)$$
$$O_3(g) + O(g) \longrightarrow 2O_2(g)$$

(a) Write the equation for the overall reaction. **(b)** Identify the intermediate, if any. **(c)** Describe the molecularity of each step in the mechanism.

Solution: **(a)** Adding the two elementary steps gives

$$2O_3(g) + O(g) \longrightarrow 3O_2(g) + O(g)$$

Because $O(g)$ appears in equal amounts on both sides of the equation, it can be eliminated to give the net equation for the chemical process:

$$2O_3(g) \longrightarrow 3O_2(g)$$

(b) The intermediate is $O(g)$. It is neither an original reactant nor a final product but is formed in one step and consumed in another.

(c) The first elementary step involves a single reactant and is consequently unimolecular. The second step, which involves two reactant molecules, is bimolecular.

PRACTICE EXERCISE

For the reaction of $Mo(CO)_6$,

$$Mo(CO)_6 + P(CH_3)_3 \longrightarrow Mo(CO)_5P(CH_3)_3 + CO$$

the proposed mechanism is

$$Mo(CO)_6 \longrightarrow Mo(CO)_5 + CO$$
$$Mo(CO)_5 + P(CH_3)_3 \longrightarrow Mo(CO)_5P(CH_3)_3$$

(a) Is the proposed mechanism consistent with the equation for the overall reaction? **(b)** Identify the intermediate or intermediates. *Answers:* **(a)** yes, the two equations add to yield the equation for the reaction; **(b)** $Mo(CO)_5$

Rate Laws of Elementary Steps

In Section 14.2, we stressed that rate laws must be determined experimentally; they cannot, in general, be predicted from the coefficients of balanced chemical reactions. It is not the chemical equation for the overall process that determines the rate law; rather, it is the elementary steps and their relative speeds.

The rate law of any elementary step is based directly on its molecularity. For example, consider the general unimolecular process

$$A \longrightarrow \text{products}$$

As the number of A molecules increases, the number that decompose in a given interval of time will increase proportionally. Thus, the rate of a unimolecular process will be first order:

$$\text{Rate} = k[\text{A}]$$

In the case of bimolecular processes, the rate law will be second order, as in the following examples:

$$\text{A} + \text{B} \longrightarrow \text{products} \qquad \text{Rate} = k[\text{A}][\text{B}]$$
$$\text{A} + \text{A} \longrightarrow \text{products} \qquad \text{Rate} = k[\text{A}]^2$$

The second-order rate law follows from the fact that the rate of collision between A and B molecules is proportional to the concentrations of both A and B.

For termolecular processes (remember, these are extremely rare), the rate law will be third order:

$$\text{A} + \text{B} + \text{C} \longrightarrow \text{products} \qquad \text{Rate} = k[\text{A}][\text{B}][\text{C}]$$

In general, the order for each reactant in an elementary step is equal to the number of reactant molecules reacting in that step. Of course, we cannot tell by looking at a balanced chemical equation whether the reaction involves one or several elementary steps. All we can observe experimentally is the rate law for the overall reaction. One of the great challenges in chemical kinetics is to determine the elementary steps in a reaction that lead to the observed rate law for the overall reaction

SAMPLE EXERCISE 14.9

If the following reaction occurs in a single elementary step, predict the rate law:

$$O_3(g) + NO(g) \longrightarrow NO_2(g) + O_2(g)$$

Solution: The elementary step is bimolecular, involving the collision of one molecule of O_3 with one molecule of NO. The rate law will therefore be first order in each reactant and second order overall:

$$\text{Rate} = k[O_3][NO]$$

Only by performing experiments can we determine whether the rate law is indeed first order in both O_3 and NO and what the value of k is.

PRACTICE EXERCISE

(a) Write the rate law for the following reaction, assuming it involves a single elementary step:

$$2NO(g) + Br_2(g) \longrightarrow 2NOBr(g)$$

(b) Is a single-step mechanism likely for this reaction?

Answer: **(a)** rate $= k[NO]^2[Br_2]$; **(b)** no, because termolecular reactions are very rare

Rate Laws of Multistep Mechanisms

Most chemical reactions occur by a mechanism that involves more than one elementary step. Often, one of the steps is much slower than the others. The

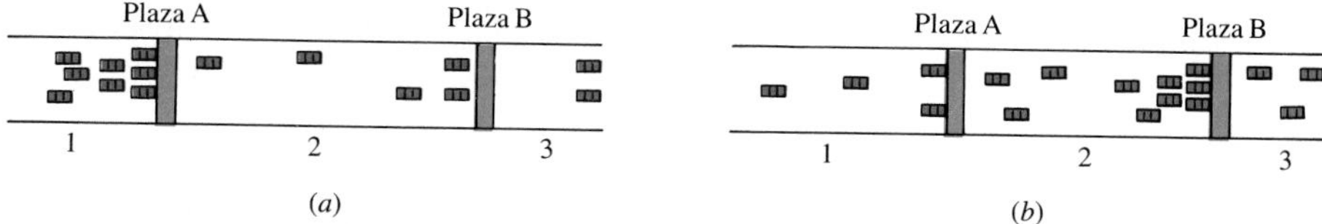

Figure 14.12 The flow of traffic on a toll road is limited by the flow of traffic through the slowest toll plaza. As cars pass from point 1 to point 2, they pass through plaza A; from point 2 to point 3, they pass through plaza B. In (*a*), the rate at which cars can reach point 3 is limited by how quickly they can get through plaza A; getting from point 1 to point 2 is the rate-determining step. In (*b*), getting from point 2 to point 3 is the rate-determining step. Traffic builds up in the intermediate stretch of road between the two toll plazas.

overall rate of a reaction cannot exceed the rate of the slowest elementary step of its mechanism. Because the slow step limits the overall reaction rate, it is called the **rate-determining step**.

In order to understand better the concept of a rate-determining step, it is helpful to think of a toll road with two toll plazas (Figure 14.12). We will measure the rate at which cars exit the toll road. Cars enter the toll road at point 1 and pass through toll plaza A. They then pass an intermediate point 2 before passing through toll plaza B. Upon exiting, they pass point 3. We can therefore envision this trip along the toll road as occurring in two elementary steps:

Step 1:	Point 1	$\longrightarrow$	point 2	(through plaza A)
Step 2:	Point 2	$\longrightarrow$	point 3	(through plaza B)
Overall:	Point 1	$\longrightarrow$	point 3	(through plazas A and B)

Now suppose that several of the gates at toll plaza A are malfunctioning so that traffic backs up behind it [Figure 14.12(*a*)]. The rate at which cars can get to point 3 is limited by the rate at which they can get through the traffic jam at plaza A. Thus, step 1 is the rate-determining step of the journey along the toll road. If, on the other hand, traffic flows quickly through plaza A but gets backed up at plaza B [Figure 14.12(*b*)], there will be a buildup of cars in the intermediate region between the plazas. In this case, step 2 is the rate-determining step: The rate at which cars can travel the toll road is limited by the rate at which they can pass through plaza B.

In the same way, the slowest step in a multistep reaction determines the overall rate. By analogy to Figure 14.12(*a*), the rate of a faster step following the rate-determining step does not affect the overall rate. If the slow step is not the first one, as in Figure 14.12(*b*), the faster preceding steps produce intermediate products that accumulate before being consumed in the slow step.

As an example of a slow first step determining the rate of a reaction, consider the reaction of NO_2 and CO to produce NO and CO_2, mentioned earlier in this section. It is found experimentally that the rate law is second order in NO_2 and zero order in CO: Rate $= k[NO_2]^2$. Can we propose a reaction mechanism that is consistent with this rate law? Consider the following two-step mechanism:*

* The subscript on the rate constant identifies the elementary step involved. Thus, k_1 is the rate constant for step 1, k_2 is the rate constant for step 2, and so forth. A negative subscript refers to the rate constant for the reverse of an elementary step. For example, k_{-1} is the rate constant for the reverse of the first step.

$$\begin{array}{llcll} \text{Step 1:} & NO_2(g) + NO_2(g) & \xrightarrow{k_1} & NO_3(g) + NO(g) & \text{(slow)} \\ \text{Step 2:} & NO_3(g) + CO(g) & \xrightarrow{k_2} & NO_2(g) + CO_2(g) & \text{(fast)} \\ \hline \text{Overall:} & NO_2(g) + CO(g) & \longrightarrow & NO(g) + CO_2(g) & \end{array}$$

Step 2 is much faster than step 1; that is, $k_2 \gg k_1$. Step 1 is rate-determining. Thus, the rate of the overall reaction equals the rate of step 1, and the rate law of the overall reaction equals the rate law of step 1. Step 1 is a bimolecular process that has the rate law

$$\text{Rate} = k_1[NO_2]^2$$

Thus, the rate law predicted by this mechanism agrees with the one observed experimentally.

Could we propose a one-step mechanism for the above reaction? We might suppose that the overall reaction is a single bimolecular elementary process that involves the collision of a molecule of NO_2 with one of CO. However, the rate law predicted by this mechanism would be

$$\text{Rate} = k[NO_2][CO]$$

Because this mechanism predicts a rate law different from that observed experimentally, we can rule it out.

It is not as easy to derive the rate law for a mechanism in which an intermediate is a reactant in the rate-determining step. This situation arises in multistep mechanisms when the first step is *not* rate-determining. Let us consider one example, the gas-phase reaction of nitric oxide, NO, with bromine, Br_2:

$$2NO(g) + Br_2(g) \longrightarrow 2NOBr(g) \qquad [14.20]$$

The experimentally determined rate law for this reaction is second order in NO and first order in Br_2:

$$\text{Rate} = k[NO]^2[Br_2] \qquad [14.21]$$

We seek a reaction mechanism that is consistent with this rate law. One possibility is that the reaction occurs in a single termolecular step:

$$NO(g) + NO(g) + Br_2(g) \longrightarrow 2NOBr(g) \qquad \text{Rate} = k[NO]^2[Br_2] \quad [14.22]$$

This does not seem likely, however, because termolecular processes are so rare. Let us consider an alternative mechanism that does not invoke termolecular steps:

$$\begin{array}{llcll} \text{Step 1:} & NO(g) + Br_2(g) & \underset{k_{-1}}{\overset{k_1}{\rightleftharpoons}} & NOBr_2(g) & \text{(fast)} \\ \text{Step 2:} & NOBr_2(g) + NO(g) & \xrightarrow{k_2} & 2NOBr(g) & \text{(slow)} \end{array}$$

Because the second step is the slow, rate-determining step, the rate of the overall reaction is governed by the rate law for that step:

$$\text{Rate} = k_2[NOBr_2][NO] \qquad [14.23]$$

However, $NOBr_2$ is an intermediate generated in step 1. Intermediates are usually unstable molecules that have a low, unknown concentration. Thus, we have a problem in that our rate law depends on the unknown concentration of an intermediate. Fortunately, with the aid of some assumptions, we can express the concentration of $NOBr_2$ in terms of the concentrations of NO and Br_2. We first assume that $NOBr_2$ is intrinsically unstable. There are two ways for $NOBr_2$ to be consumed after it is formed: Either it can react with NO to form NOBr, or it can fall back apart into NO and Br_2. The first of these possibilities is step 2, a slow process. The second is the reverse of step 1, a unimolecular process:

$$NOBr_2(g) \xrightarrow{k_{-1}} NO(g) + Br_2(g) \qquad [14.24]$$

Because step 2 is slow, we assume that most of the $NOBr_2$ falls apart according to Equation 14.24. Thus, we have both the forward and reverse reactions of step 1 occurring much faster than step 2. We can therefore assume further that the first step achieves an equilibrium in which the rates of the forward and reverse reactions are equal:

$$\underset{\text{Rate of forward reaction}}{k_1[NO][Br_2]} = \underset{\text{Rate of reverse reaction}}{k_{-1}[NOBr_2]}$$

Solving for $[NOBr_2]$, we have

$$[NOBr_2] = \frac{k_1}{k_{-1}}[NO][Br_2]$$

Substituting this relationship into the rate law for the rate-determining step (Equation 14.23), we have

$$\text{Rate} = k_2 \frac{k_1}{k_{-1}}[NO][Br_2][NO] = k[NO]^2[Br_2]$$

This is consistent with the experimental rate law (Equation 14.21). The experimental rate constant k is equal to k_2k_1/k_{-1}. This mechanism, which involves only unimolecular and bimolecular processes, is far more probable than the single termolecular step (Equation 14.22).

In general, *whenever a fast step precedes a slow one, we can solve for the concentration of an intermediate by assuming that an equilibrium is established in the fast step.*

SAMPLE EXERCISE 14.10

Show that the following mechanism for Equation 14.20 also produces a rate law consistent with the experimentally observed one:

Step 1: $NO(g) + NO(g) \underset{k_{-1}}{\overset{k_1}{\rightleftharpoons}} N_2O_2(g)$ (fast, equilibrium)

Step 2: $N_2O_2(g) + Br_2(g) \xrightarrow{k_2} 2NOBr(g)$ (slow)

Solution: The second step is rate-determining, so the overall rate is

$$\text{Rate} = k_2[N_2O_2][Br_2]$$

We solve for the concentration of the intermediate N_2O_2 by assuming that an equilibrium is established in step 1; thus, the rates of the forward and reverse reactions in step 1 are equal:

$$k_1[NO]^2 = k_{-1}[N_2O_2]$$

$$[N_2O_2] = \frac{k_1}{k_{-1}}[NO]^2$$

Substituting this expression into the rate expression gives us

$$\text{Rate} = k_2 \frac{k_1}{k_{-1}}[NO]^2[Br_2] = k[NO]^2[Br_2]$$

Thus, this mechanism also yields a rate law consistent with the experimental one.

PRACTICE EXERCISE

The first step of a mechanism involving the reaction of bromine is

$$Br_2(g) \underset{k_{-1}}{\overset{k_1}{\rightleftharpoons}} 2Br(g) \qquad \text{(fast, equilibrium)}$$

What is the expression relating the concentration of $Br(g)$ to that of $Br_2(g)$?

Answer: $[Br] = \left(\frac{k_1}{k_{-1}}[Br_2]\right)^{1/2}$

14.6 CATALYSIS

A **catalyst** is a substance that changes the speed of a chemical reaction without itself undergoing a permanent chemical change in the process. Catalysts are very common; most reactions occurring in the human body, the atmosphere, the oceans, or in industrial chemical processes are affected by catalysts.

In your laboratory work, you may have carried out the reaction in which oxygen is produced by heating potassium chlorate, $KClO_3$:

$$2KClO_3(s) \xrightarrow{\Delta} 2KCl(s) + 3O_2(g)$$

In the absence of a catalyst, $KClO_3$ does not readily decompose in this manner, even on strong heating. However, mixing black manganese dioxide, MnO_2, with the $KClO_3$ before heating causes the reaction to occur much more readily. The MnO_2 can be recovered largely unchanged from this reaction, so the overall chemical process is clearly still the same. Thus, MnO_2 acts as a catalyst for decomposition of $KClO_3$. As another example, we know that a cube of sugar, when dissolved in water at 37°C, does not undergo oxidation at a significant rate. The sugar can be recovered essentially unchanged from the solution after several days. Yet sugar ingested into the human body at about 37°C is rapidly oxidized and soon ends up mostly as carbon dioxide and water:

$$C_{12}H_{22}O_{11}(aq) + 12O_2(aq) \longrightarrow 12CO_2(aq) + 11H_2O(l)$$

The oxidation of sugar in the biochemical system is greatly speeded up by the presence of one or more catalysts. These biochemical catalysts are **enzymes**, protein molecules that act to catalyze specific biochemical reactions.

Much industrial chemical research is devoted to the search for new and more effective catalysts for reactions of commercial importance. Extensive research efforts also are devoted to finding means of inhibiting or removing certain catalysts that promote undesirable reactions, such as those involved in corrosion of metals, aging, and tooth decay.

A CLOSER LOOK: Evidence of Mechanism

The basic procedure in establishing the mechanism of a chemical reaction is first to determine the rate law experimentally. One or more elementary steps are then postulated that account for that rate law and for other experimental observations. That a mechanism gives a rate law that agrees with the observed one does not prove that that mechanism is correct. Often several mechanisms can be envisioned that give rise to the same rate law. To distinguish between such mechanisms, chemists might search for proposed intermediates or carry out other types of studies. As an example, consider the following reaction:

$$O^+(g) + NO(g) \longrightarrow NO^+(g) + O(g)$$

This reaction occurs in the upper atmosphere. The rate law for this reaction is: Rate $= k[O^+][NO]$, and the reaction is believed to occur in a single elementary step. However, we may ask whether the reaction involves the breaking of the NO bond and the formation of a new bond between N and O^+, or whether it involves merely the transfer of an electron from NO to O^+. These two possibilities are shown in Figure 14.13. The question of which of these mechanisms is operative is answered by performing the reaction using O^+ that has been highly enriched in the rare isotope ^{18}O. Because this isotope is present to the extent of only 0.2 percent in nature, the NO contains only 0.2 percent $N^{18}O$. By measuring the location of ^{18}O in the reaction products, the two mechanisms can be distinguished (see Figure 14.13). The experiment indicates that no ^{18}O is incorporated into NO^+; the experimental results are therefore consistent with the electron-transfer pathway but not the atom-transfer pathway.

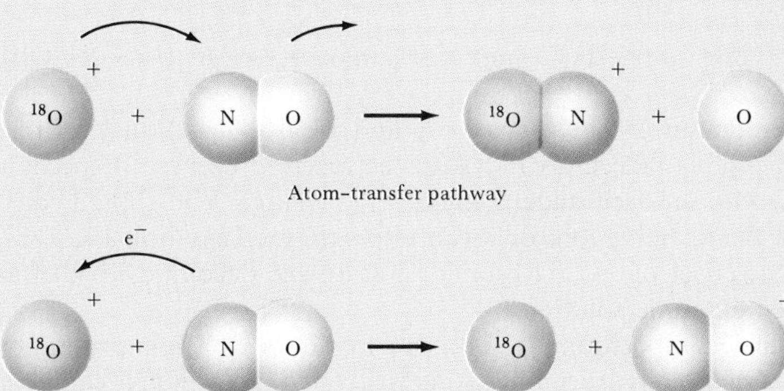

Figure 14.13 Alternative pathways for the reaction $O^+ + NO \longrightarrow O + NO^+$. The results of experiments using ^{18}O labeling show that the reaction proceeds according to the electron-transfer pathway.

Homogeneous Catalysis

A catalyst that is present in the same phase as the components of a chemical reaction is a **homogeneous catalyst**. For example, a homogeneous catalyst for a reaction occurring in solution would itself be dissolved in the solution.

As an example of homogeneous catalysis, we will consider the decomposition of aqueous hydrogen peroxide, $H_2O_2(aq)$, into water and oxygen:

$$2H_2O_2(aq) \longrightarrow 2H_2O(l) + O_2(g) \quad [14.25]$$

In the absence of a catalyst, this reaction occurs at an extremely slow rate. Many different substances are capable of catalyzing the reaction; among these is bromine, Br_2. The bromine reacts with hydrogen peroxide in acidic solution, forming bromide ion and liberating oxygen:

$$Br_2(aq) + H_2O_2(aq) \longrightarrow 2Br^-(aq) + 2H^+(aq) + O_2(g) \quad [14.26]$$

If this were the complete reaction, bromine would not be a catalyst, because it

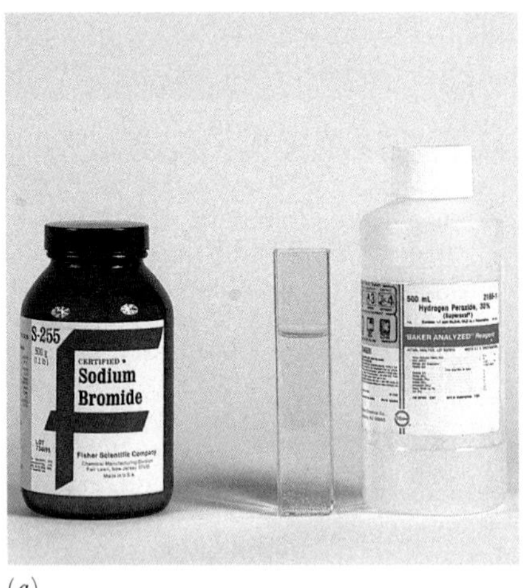

(*a*)

(*b*)

Figure 14.14 Effect of added NaBr on the decomposition of $H_2O_2(aq)$ solution (*a*). Very soon after adding colorless NaBr solution to an H_2O_2 solution, the solution turns orange because Br_2 is formed (Equation 14.27). (*b*) After a time, when the Br_2 concentration has built up, decomposition of H_2O_2 (Equation 14.26) occurs at a rapid pace with evolution of O_2. (Donald Clegg and Roxy Wilson)

undergoes chemical change in the reaction. It happens, however, that hydrogen peroxide reacts with bromide ion in acidic solution to form bromine:

$$2Br^-(aq) + H_2O_2(aq) + 2H^+(aq) \longrightarrow Br_2(l) + 2H_2O(l) \quad [14.27]$$

The overall sum of Equations 14.26 and 14.27 is just Equation 14.25. We see that bromine is indeed a catalyst in the reaction because it speeds the overall reaction without itself undergoing any net change. You should convince yourself that the bromide ion, Br^-, can also catalyze Equation 14.25. Figure 14.14 shows the effect of adding bromide ion on the decomposition of an aqueous hydrogen peroxide solution.

The decomposition of hydrogen peroxide is also an important biochemical process. Because hydrogen peroxide is strongly oxidizing, it can be physiologically harmful. For this reason, the blood and livers of mammals contain an enzyme, *catalase*, that catalyzes the decomposition of hydrogen peroxide into water and oxygen (Figure 14.15).

On the basis of the Arrhenius expression for a chemical reaction, Equation 14.15, the rate constant k is determined by the activation energy E_a and the frequency factor A. A catalyst may affect the rate of reaction by altering the

Figure 14.15 The addition of ground-up beef liver to hydrogen peroxide causes rapid decomposition into water and oxygen. The decomposition is catalyzed by the enzyme *catalase*. Grinding the liver increases its surface area, and hence the number of sites of catalysis. (© Richard Megna/Fundamental Photographs)

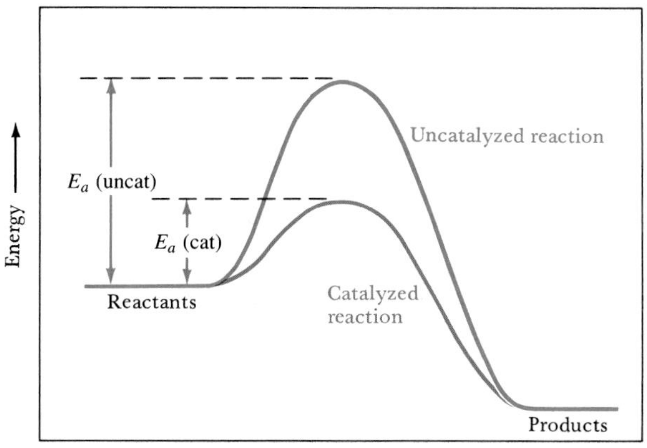

Figure 14.16 Energy profiles for catalyzed and uncatalyzed reactions. The catalyst functions in this example to lower the activation energy for reaction. Notice that the energies of reactants and products are unchanged by the catalyst.

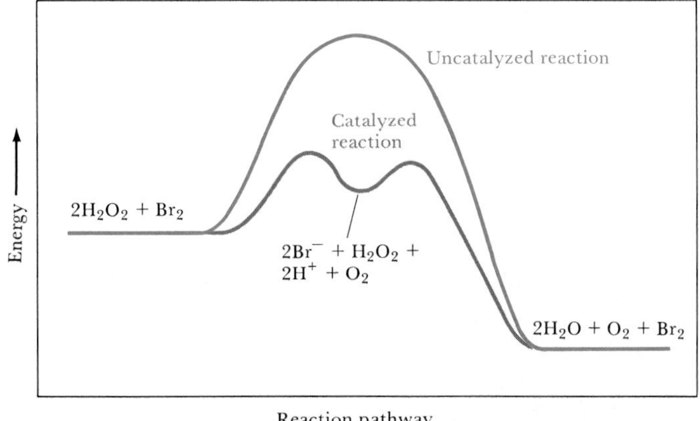

Figure 14.17 Energy profiles for the uncatalyzed decomposition of hydrogen peroxide and for the reaction as catalyzed by Br_2. The catalyzed reaction involves two successive steps, each of which has a lower activation energy than the uncatalyzed reaction.

value for either E_a or A. The most dramatic catalytic effects come from lowering E_a. As a general rule, *a catalyst lowers the overall activation energy for a chemical reaction.* The lowering of E_a by a catalyst is shown schematically in Figure 14.16.

A catalyst usually lowers the overall activation energy for reaction by providing a completely different pathway for reaction. The two examples given above involve a reversible, cyclic reaction of the catalyst with the reactants. For example, in the decomposition of hydrogen peroxide, two successive reactions of H_2O_2, with bromine and then with bromide, take place. Because these two reactions together serve as a catalytic pathway for hydrogen peroxide decomposition, *both* of them must have significantly lower activation energies than the uncatalyzed decomposition, as shown schematically in Figure 14.17.

Heterogeneous Catalysis

A **heterogeneous catalyst** exists in a phase different from the reactant molecules. For example, a reaction between molecules in the gas phase might be catalyzed by a finely divided metal oxide. In the absence of a catalyst, the reaction would occur slowly in the gas phase. However, when the catalyst is present, the reaction occurs more rapidly on the surface of the solid catalyst.

Many industrially important reactions occurring in the gas phase are catalyzed by the surfaces of solids. For example, hydrocarbon molecules are rearranged to form gasoline with the aid of what are called "cracking" catalysts (Section 26.1). Reactions occurring in solution may also be catalyzed by solids. Heterogeneous catalysts are often composed of finely divided metals or metal oxides. Because the catalyzed reaction occurs on the surface, special methods are often used to prepare catalysts so that they have very large surface areas. Some examples of heterogeneous catalysts used in industrially important processes are shown in Figure 14.18

The initial step in heterogeneous catalysis is usually **adsorption** of reactants. As we saw in Chapter 13, *adsorption* refers to the binding of molecules to a surface, whereas *absorption* refers to the uptake of molecules into the interior of another substance. Adsorption occurs because the atoms or ions at the surface of a solid are extremely reactive. Unlike their counterparts in the interior of the substance, they have unfulfilled valence requirements. The unused bonding capability of surface atoms or ions may be used to bond molecules from the gas or solution phase to the surface of the solid. In practice, not all the atoms or ions of the surface are reactive; various impurities may be adsorbed at the surface, and these may occupy many potential reaction sites and block further reaction. The places where reacting molecules may become adsorbed are called **active sites**. The number of active sites per unit amount of catalyst depends on the nature of the catalyst, on its method of preparation, and on its treatment before use.

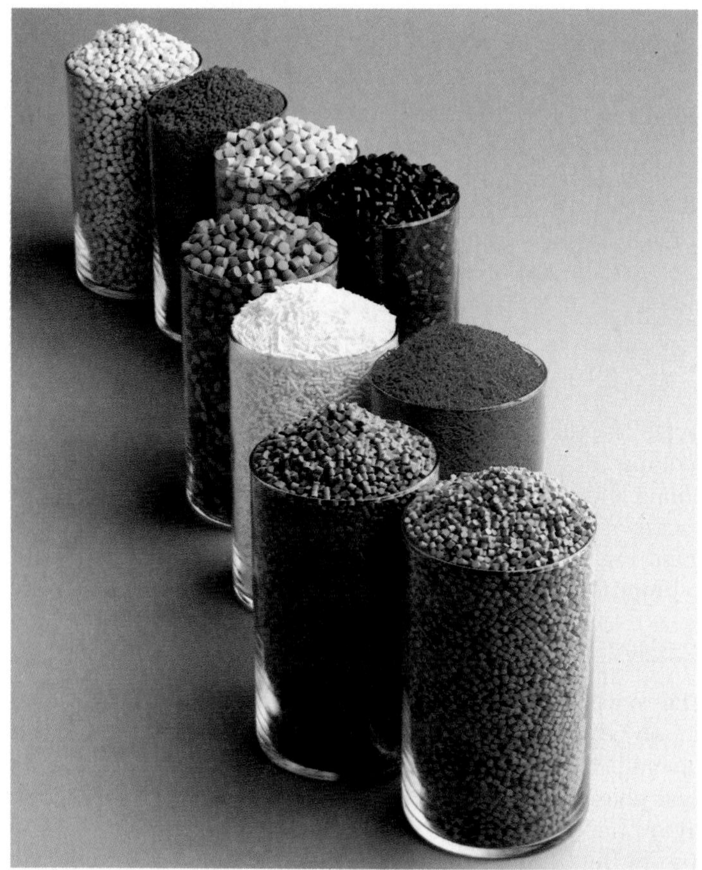

Figure 14.18 Samples of industrially important heterogeneous catalysts. Each catalyst is specially tailored for use in a particular type of reaction. For example, the blue material in the front beaker consists of a cobalt-molybdenum catalyst supported on alumina, Al_2O_3. The catalyst is used for removal of sulfur and nitrogen from crude oil. (Photo courtesy of the Harshaw/Filtrol Partnership)

As an example of heterogeneous catalysis, consider the hydrogenation of ethylene to form ethane:

$$\underset{\text{Ethylene}}{H_2C{=}CH_2} + H_2 \longrightarrow \underset{\text{Ethane}}{H_3C{-}CH_3} \qquad [14.28]$$

In the absence of a catalyst, this reaction occurs slowly. However, in the presence of a very finely divided metal such as nickel, palladium, or platinum, the reaction occurs rather easily at room temperature, under a few hundred atmospheres of hydrogen pressure. The mechanism by which reaction occurs is shown diagrammatically in Figure 14.19. Both ethylene and hydrogen are adsorbed at the metal surface [Figure 14.19(*a*)]. The adsorption of hydrogen results in breaking of the H—H bond and formation of two M—H bonds, where M represents the metal surface [Figure 14.19(*b*)]. The hydrogen atoms are relatively free to move about the surface. When they encounter an adsorbed ethylene, the hydrogen may become bound to the carbon [Figure 14.19(*c*)]. The carbon thus acquires four σ bonds about it, which reduces its tendency to remain adsorbed at the metal. When the other carbon also acquires a hydrogen, the ethane molecule is released from the surface [Figure 14.19(*d*)]. The active site is ready to adsorb another ethylene molecule and thus begin the cycle again.

Figure 14.19 Mechanism for reaction of ethylene with hydrogen on a catalytic surface. (*a*) The hydrogen and ethylene are adsorbed at the metal surface. (*b*) The H—H bond is broken to give adsorbed hydrogen atoms. (*c*) These migrate to the adsorbed ethylene and bond to the carbon atoms. (*d*) As C—H bonds are formed, the adsorption of the molecule to the metal surface is decreased, and ethane is released.

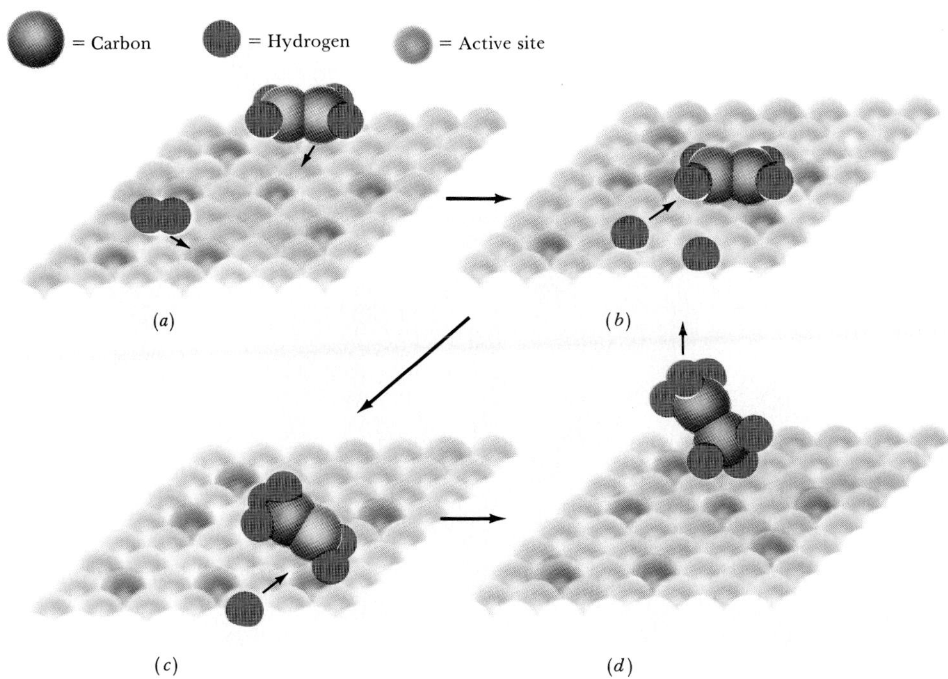

CHEMISTRY AT WORK: Catalytic Converters

Heterogeneous catalysis plays a major role in the fight against urban air pollution. Two components of automobile exhausts that are involved in the formation of photochemical smog are nitrogen oxides and unburned hydrocarbons of various types (Section 18.4). In addition, automobile exhausts may contain considerable quantities of carbon monoxide. Even with the most careful attention to engine design and fuel characteristics, it is not possible under normal driving conditions to reduce the contents of these pollutants to an acceptable level in the exhaust gases coming from the engine. It is therefore necessary somehow to remove them from the exhaust gases before they are vented to the air. This removal is accomplished in the *catalvtic converter*.

The catalytic converter, illustrated in Figure 14.20, must perform two distinct functions: (1) oxidation of CO and unburned hydrocarbons to carbon dioxide and water, and (2) reduction of nitrogen oxides to nitrogen gas:

$$\text{CO, hydrocarbons } (C_xH_y) \xrightarrow{O_2} CO_2 + H_2O$$

$$NO, NO_2 \longrightarrow N_2$$

These two functions require two distinctly different catalysts, so the development of a successful catalyst system is a very difficult challenge. The catalysts must be effective over a wide range of operating temperatures; they must continue to be active in spite of the poisoning action of various gasoline additives emitted along with the exhaust; they must be sufficiently rugged to withstand exhaust gas turbulence and the mechanical shocks of driving under various conditions for thousands of miles.

Catalysts that promote the combustion of CO and hydrocarbons are, in general, the transition-metal oxides and noble metals such as platinum. As an example, a mixture of two different metal oxides—CuO and Cr_2O_3, for example—might be used. These materials are supported on a structure (Figure 14.21) that allows the best possible contact between the flowing exhaust gas and the catalyst surface. Either bead or honeycomb structures made from alumina, Al_2O_3, and impregnated with the catalyst may be employed. Such catalysts operate by first adsorbing oxygen gas, also present in the exhaust gas. This adsorption weakens the O—O bond in O_2, so that oxygen atoms are in effect available for reaction with

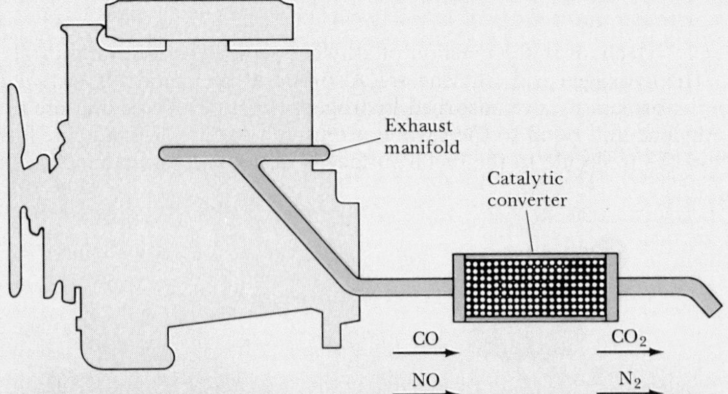

Figure 14.20 Illustration of the arrangement and functions of a catalytic converter.

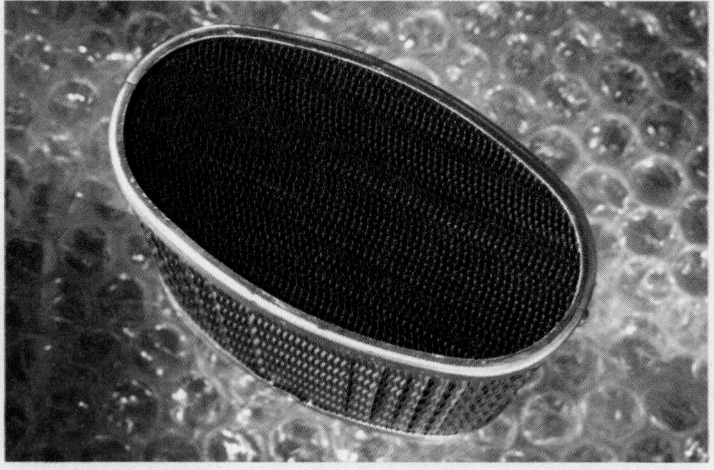

Figure 14.21 A stainless-steel catalytic converter used in automobiles. The zig-zag pattern of the metal foil provides favorable passage of exhaust gases and increases the surface area. Microscopic whiskers of aluminum oxide are grown on the stainless steel to anchor a coating that contains the noble-metal catalysts. The catalysts promote the conversion of noxious exhaust gases into CO_2, H_2O, and N_2. (AC Rochester Division, GMC)

adsorbed CO to form CO_2. Hydrocarbon oxidation probably proceeds somewhat similarly, with the hydrocarbons first being adsorbed by rupture of a C—H bond.

The most effective catalysts for reduction of NO to yield N_2 and O_2 are transition-metal oxides and noble metals, the same kinds of materials that catalyze the oxidation of CO and hydrocarbons. The catalysts that are most effective in one reaction, however, are usually much less effective in the other. It is therefore necessary to have two different catalytic components.

Catalytic converters are remarkably efficient heterogeneous catalysts. The automotive exhaust gases are in contact with the catalyst for only 100 to 400 ms. In this very short time, 96 percent of the hydrocarbons and CO are converted to CO_2 and H_2O. The emission of nitrogen oxides is reduced by 76 percent.

Of course, there are costs as well as benefits associated with the use of catalytic converters. Some of the metals used in the converters are very expensive. Catalytic converters currently account for 35 percent of the platinum and 73 percent of the rhodium used in the United States. Both of these metals are far more expensive than gold. In addition, the catalysts are incompatible with lead-containing antiknock agents that used to be commonly added to gasoline to improve engine performance. Additives such as tetramethyl lead, $Pb(CH_3)_4$, and tetraethyl lead, $Pb(C_2H_5)_4$, "poison" the catalyst; that is, they bind to and block active catalytic sites, decreasing the catalytic activity. Partly because of the severe catalyst poisoning that results from the use of leaded fuels, cars built since 1975 have been engineered for use with unleaded gas.

FOR REVIEW

SUMMARY

In this chapter, we have discussed ways of expressing reaction rates and the factors that influence these rates, namely, concentration, temperature, and catalysts. Rates are expressed as changes in concentration per unit of time; typically, for reactions in solution, the units are M/s. The quantitative relationship between rate and concentration is expressed by the rate law, which often has the following form: Rate = $k[\text{reactant 1}]^m[\text{reactant 2}]^n \ldots$. The constant k in the rate law is called the rate constant; the exponents m, n, and so forth, are called reaction orders. The rate law depends not on the overall reaction but on the mechanism by which it occurs. Thus, rate laws cannot ordinarily be determined from the coefficients in the balanced chemical equation but must be determined by experiment.

In a first-order reaction, the reaction rate is proportional to the concentration of a single reactant raised to the first power: rate = $k[\text{A}]$. In such cases, $\ln [\text{A}]_t = -kt + \ln [\text{A}]_0$, where $[\text{A}]_t$ is the concentration of reactant A at time t, k is the rate constant, and $[\text{A}]_0$ is the initial concentration of reactant A. Thus, a graph of $\ln [\text{A}]_t$ versus time yields a straight line of slope $-k$. First-order reactions are also characterized by having a constant half-life, which is related to the rate constant: $t_{1/2} = 0.693/k$. Reactions more complex than first order yield different expressions for the rate constant and half-life. For example, for second-order reactions, rate = $k[\text{A}]^2$, the following relationship holds: $1/[\text{A}]_t = 1/[\text{A}]_0 + kt$; thus in this case a graph of $1/[\text{A}]_t$ versus time yields a straight line.

Reactions occur as a result of collisions between molecules. Collisions bring atoms together so that new chemical bonds can form. In addition, when molecules collide, part of their kinetic energies may be used to break old bonds. Of course, not all collisions lead to a reaction. In order for a reaction to occur, molecules must collide with sufficient energy and proper orientation. The minimum energy required for a reaction to occur is called the activation energy. When the activation energy is appreciable, only a very small fraction of all collisions between reactants provide the energy required to surmount the activation-energy barrier and form the products of the reaction. By increasing the temperature of a system, we can increase the fraction of molecules whose kinetic energies exceed the activation energy and thereby increase the reaction rate. From the manner in which the rate constant for a reaction varies with temperature, it is possible to determine the activation energy, E_a, for a reaction using the Arrhenius equation: $\ln k = \ln A - E_a/RT$. A graph of $\ln k$ versus $1/T$ yields a straight line whose slope is $-E_a/R$.

From a knowledge of the rate law for a reaction and with other experimental information, a picture of how the reaction proceeds may be formulated.

The overall reaction may occur in a series of elementary steps or in a single elementary step. Elementary steps (or elementary processes) may be unimolecular, bimolecular, or termolecular, depending on whether one, two, or three molecules, respectively, are involved as reactants. The rate law of an elementary step is related directly to its molecularity. Therefore, unimolecular reactions have first-order rate laws, and bimolecular reactions have second-order rate laws. Termolecular reactions, which are very rare, have third-order rate laws.

The sequence of elementary steps by which a reaction proceeds is known as its reaction mechanism. If a reaction proceeds by a multistep mechanism, one of the elementary steps may be much slower than the others and so be the rate-determining step. The rate-determining step governs the rate law for the overall process.

A catalyst speeds a reaction without itself undergoing a net chemical change. It does so by providing a different mechanism for the reaction, one having a lower activation-energy barrier. Catalysts may be either homogeneous—that is, in the same phase with the reactants—or heterogeneous—in a separate phase. Heterogeneous catalysts are particularly important in large-scale industrial chemical processes and in applications such as the catalytic converter of an automobile.

KEY TERMS

chemical kinetics
reaction rate (Sec. 14.1)
instantaneous rate (Sec. 14.1)
rate constant (Sec. 14.2)
rate law (Sec. 14.2)
reaction order (Sec. 14.2)
overall reaction order (Sec. 14.2)
first-order reaction (Sec. 14.2)
half-life (Sec. 14.3)
activation energy (Sec. 14.4)
activated complex (Sec. 14.4)
Arrhenius equation (Sec. 14.4)
frequency factor (Sec. 14.4)
reaction mechanism (Sec. 14.5)
elementary steps (Sec. 14.5)
intermediate (Sec. 14.5)
molecularity (Sec. 14.5)
unimolecular (Sec. 14.5)
bimolecular (Sec. 14.5)
termolecular (Sec. 14.5)
rate-determining step (Sec. 14.5)
catalyst (Sec. 14.6)
enzymes (Sec. 14.6)
homogeneous catalyst (Sec. 14.6)
heterogeneous catalyst (Sec. 14.6)
adsorption (Sec. 14.6)
active sites (Sec. 14.6)

EXERCISES

Reaction Rates

14.1 For each of the following reactions, indicate how the rate of disappearance of each reactant is related to the rate of appearance of each product:
(a) $H_2O_2(g) \longrightarrow H_2(g) + O_2(g)$
(b) $MnO_2(s) + Mn(s) \longrightarrow 2MnO(s)$
(c) $2C_6H_{14}(l) + 13O_2(g) \longrightarrow 12CO(g) + 14H_2O(g)$

14.2 For each of the following reactions, indicate how the rate of disappearance of each reactant is related to the rate of appearance of each product:
(a) $CO(g) + 2H_2(g) \longrightarrow CH_3OH(g)$
(b) $B_2H_6(g) + 3O_2(g) \longrightarrow B_2O_3(s) + 3H_2O(g)$
(c) $MgO(s) + SO_2(g) + \frac{1}{2}O_2(g) \longrightarrow MgSO_4(s)$

14.3 The rate of disappearance of H^+ was measured for the following reaction:

$$CH_3OH(aq) + HCl(aq) \longrightarrow CH_3Cl(aq) + H_2O(l)$$

The following data were collected:

Time (min)	$[H^+]$ (M)
0	1.85
79	1.67
158	1.52
316	1.30
632	1.00

Calculate the average rate of reaction for the time interval between each measurement.

14.4 The rearrangement of methyl isonitrile, CH_3NC, was studied in the gas phase at 215°C, and the following data were obtained:

Time (s)	$[CH_3NC]$ (M)
0	0.0165
2,000	0.0110
5,000	0.00591
8,000	0.00314
12,000	0.00137
15,000	0.00074

Calculate the average rate of reaction for the time interval between each measurement.

14.5 Using the data provided in Exercise 14.3, make a graph of $[H^+]$ versus time. Draw tangents to the curve at $t = 100$ and $t = 500$ min. Determine the rates at these times.

14.6 Using the data provided in Exercise 14.4, make a graph of $[CH_3NC]$ versus time. Draw tangents to the curve at $t = 3500$ and $t = 13{,}500$ s. Determine the rates at these times.

14.7 **(a)** Consider the combustion of $H_2(g)$, $2H_2(g) + O_2(g) \longrightarrow 2H_2O(g)$. If hydrogen is burning at the rate of 4.6 mol/s, what is the rate of consumption of oxygen? What is the rate of formation of water vapor? **(b)** The reaction $2NO(g) + Cl_2(g) \longrightarrow 2NOCl(g)$ is carried out in a closed vessel. If the partial pressure of NO is decreasing at the rate of 30 mm Hg/min, what is the rate of change of the total pressure of the vessel?

14.8 **(a)** Consider the combustion of methane, $CH_4(g) + 2O_2(g) \longrightarrow CO_2(g) + 2H_2O(g)$. If the concentration of CH_4 is decreasing at the rate of 0.40 M/s, what are the rates of change of the concentrations of CO_2 and of H_2O? **(b)** If the rate of increase in NH_3 pressure in a closed reaction vessel from the reaction $N_2(g) + 3H_2(g) \longrightarrow 2NH_3(g)$ is 100 mm Hg/hr, what is the rate of change of the total pressure in the vessel?

Rate Laws

14.9 The decomposition of N_2O_5 in carbon tetrachloride proceeds as follows: $2N_2O_5 \longrightarrow 4NO_2 + O_2$. The rate law for this reaction is first order in N_2O_5. At 45°C, the rate constant is $6.08 \times 10^{-4}\ s^{-1}$. Calculate the rate of reaction when **(a)** $[N_2O_5] = 0.100\ M$; **(b)** $[N_2O_5] = 0.305\ M$.

14.10 Consider the following reaction:

$$2NO(g) + 2H_2(g) \longrightarrow N_2(g) + 2H_2O(g)$$

(a) The rate law for this reaction is first order in H_2 and second order in NO. Write the rate law. **(b)** If the rate constant for this reaction at 1000 K is $6.0 \times 10^4\ M^{-2}s^{-1}$, what is the reaction rate when $[NO] = 0.050\ M$ and $[H_2] = 0.010\ M$? **(c)** What is the reaction rate at 1000 K when the concentration of NO is doubled, to 0.10 M, while the concentration of H_2 is 0.010 M?

14.11 Consider the hypothetical reaction $A + B \longrightarrow$ products. The rate law for the reaction is first order in A and first order in B. When the initial concentrations of A and B are 0.050 M and 0.080 M respectively, the observed initial rate is $1.6 \times 10^{-3}\ M/s$. **(a)** What is the value of the rate constant? **(b)** What are the units of the rate constant? **(c)** What would happen to the rate if the initial concentration of A were tripled?

14.12 The reaction $2NO(g) + O_2(g) \longrightarrow 2NO_2(g)$ is second order in NO and first order in O_2. When $[NO] = 0.030\ M$ and $[O_2] = 0.040\ M$, the observed rate is $7.2 \times 10^{-5}\ M/s$. **(a)** What is the value of the rate constant? **(b)** What are the units of the rate constant? **(c)** What would happen to the rate if the concentration of NO were decreased by a factor of 2?

14.13 Consider the reaction of peroxydisulfate ion, $S_2O_8^{2-}$, with iodide ion, I^-, in aqueous solution:

$$S_2O_8^{2-}(aq) + 3I^-(aq) \longrightarrow 2SO_4^{2-}(aq) + I_3^-(aq)$$

At a particular temperature, the rate of this reaction varies with reactant concentrations in the following manner:

Experiment	$[S_2O_8^{2-}]$ (M)	$[I^-]$ (M)	$\frac{-\Delta[S_2O_8^{2-}]}{\Delta t}$ (M/s)
1	0.038	0.060	1.4×10^{-5}
2	0.076	0.060	2.8×10^{-5}
3	0.076	0.030	1.4×10^{-5}

(a) Write the rate law for the rate of disappearance of $S_2O_8^{2-}$. **(b)** What is the value of the rate constant for the disappearance of $S_2O_8^{2-}$? **(c)** What is the rate of disappearance of $S_2O_8^{2-}$ when $[S_2O_8^{2-}] = 0.025\ M$ and $[I^-] = 0.100\ M$? **(d)** What is the rate of appearance of SO_4^{2-} when $[S_2O_8^{2-}] = 0.025\ M$ and $[I^-] = 0.050\ M$?

14.14 The following data were collected for the gas-phase reaction between nitric oxide and bromine at 273°C:

$$2NO(g) + Br_2(g) \longrightarrow 2NOBr(g)$$

Experiment	[NO] (M)	$[Br_2]$ (M)	Initial rate of appearance of NOBr (M/s)
1	0.10	0.10	12
2	0.10	0.20	24
3	0.20	0.10	48
4	0.30	0.10	108

(a) Determine the rate law. **(b)** Calculate the value of the rate constant for the appearance of NOBr. **(c)** How is the rate of appearance of NOBr related to the rate of disappearance of Br_2? **(d)** What is the rate of appearance of NOBr when $[NO] = 0.15\ M$ and $[Br_2] = 0.25\ M$? **(e)** What is the rate of disappearance of Br_2 when $[NO] = 0.075\ M$ and $[Br_2] = 0.185\ M$?

14.15 The following data were measured for the reaction $BF_3(g) + NH_3(g) \longrightarrow F_3BNH_3(g)$:

Experiment	$[BF_3]$ (M)	$[NH_3]$ (M)	Initial rate (M/s)
1	0.250	0.250	0.2130
2	0.250	0.125	0.1065
3	0.200	0.100	0.0682
4	0.350	0.100	0.1193
5	0.175	0.100	0.0596

(a) What is the rate law for the reaction? **(b)** What is the overall order of the reaction? **(c)** What is the value of the rate constant for the reaction?

14.16 The following data have been measured for the reaction

$$CH_3Cl(g) + H_2O(g) \longrightarrow CH_3OH(g) + HCl(g)$$

Experiment	$[CH_3Cl]$ (M)	$[H_2O]$ (M)	Initial rate (M/s)
1	0.500	0.500	22.700
2	0.750	0.500	34.050
3	0.500	0.750	51.075
4	0.500	0.250	5.675
5	0.750	0.125	2.128

(a) What is the rate law for the reaction? **(b)** What is the overall order of the reaction? **(c)** What is the value of the rate constant for the reaction?

Concentration and Time; Half-lives

14.17 The rate of decomposition of a substance is found to be first order. If $k = 3.6 \times 10^{-3}\ s^{-1}$, what is the half-life of the reaction?

14.18 The decomposition of a substance follows a first-order rate law. If it takes 6.8×10^3 s for the concentration of the substance to decrease to half its initial value, what is the value of the rate constant?

14.19 The reaction

$$SO_2Cl_2(g) \longrightarrow SO_2(g) + Cl_2(g)$$

is first order in SO_2Cl_2. Using the following kinetic data, determine the magnitude of the first-order rate constant:

Time (s)	Pressure, SO_2Cl_2 (atm)
0	1.000
2,500	0.947
5,000	0.895
7,500	0.848
10,000	0.803

14.20 From the following data for the first-order gas-phase rearrangement of CH_3NC at 215°C, calculate the first-order rate constant and half-life for the reaction:

Time (s)	Pressure CH_3NC (mm Hg)
0	502
2,000	335
5,000	180
8,000	95.5
12,000	41.7
15,000	22.4

14.21 Sucrose, $C_{12}H_{22}O_{11}$, which is more commonly known as table sugar, reacts in dilute acid solutions to form two simpler sugars, glucose and fructose. Both of these sugars have the molecular formula $C_6H_{12}O_6$, though they differ in molecular structure. The reaction is

$$C_{12}H_{22}O_{11}(aq) + H_2O(l) \longrightarrow 2C_6H_{12}O_6(aq)$$

The rate of this reaction was studied at 23°C and in 0.5 M HCl; the following data were obtained:

Time (min)	$[C_{12}H_{22}O_{11}]$ (M)
0	0.316
39	0.274
80	0.238
140	0.190
210	0.146

Is the reaction first order or second order with respect to the concentration of sucrose? What is the value of the rate constant?

14.22 The decomposition of N_2O_5 has been studied in carbon tetrachloride solution at 45°C. The following data were obtained:

Time (min)	$[N_2O_5]$ (M)
0	0.90
5	0.75
10	0.63
20	0.44
50	0.15

(a) Is the reaction first order or second order with respect to N_2O_5? **(b)** What is the value of the rate constant?

14.23 The first-order rate constant for the decomposition of N_2O_5 to NO_2 and O_2 at 70°C is $6.82 \times 10^{-3}\ s^{-1}$. Suppose we start with 0.300 mol of $N_2O_5(g)$ in a 500-mL container. **(a)** How many moles of N_2O_5 will remain after 1.5 min? **(b)** How many minutes will it take for the quantity of N_2O_5 to drop to 0.030 mol? **(c)** What is the half-life of N_2O_5 at 70°C?

14.24 The first-order rate constant for the decomposition of a certain antibiotic in water at 20°C is 1.65 yr^{-1}. **(a)** If a 6.0×10^{-3} *M* solution of the antibiotic is stored at 20°C, what will its concentration be after 3 months? After 1 year? **(b)** How long will it take for the concentration of the solution to drop to 1.0×10^{-3} *M*? **(c)** What is the half-life of the antibiotic solution?

14.25 The rate of the gas-phase decomposition of NO_2 to form NO and O_2, $2NO_2(g) \longrightarrow 2NO(g) + O_2(g)$, is second order in NO_2. If the concentration of NO_2 at 383°C varies with time as shown in the table that follows, what is the value of the rate constant?

Time (s)	$[NO_2]$ (*M*)
0.0	0.100
5.0	0.017
10.0	0.0090
15.0	0.0062
20.0	0.0047

14.26 A hypothetical reaction A $\longrightarrow$ products is second order in A. The half-life of a reaction that was initially 1.66 *M* in A is 310 min. What is the value of the rate constant?

14.27 A reaction shows the same half-life regardless of the starting concentration of the reactant. Is it a first-order or a second-order reaction?

14.28 Using Equation 14.12, derive the expression for the half-life of a second-order reaction (Equation 14.13).

Effects of Temperature; Activation Energy

14.29 For the reaction $2N_2O_5(g) \longrightarrow 4NO_2(g) + O_2(g)$, the activation energy, E_a, and overall ΔE are 100 kJ/mol and -23 kJ/mol, respectively. **(a)** Sketch the energy profile for this reaction. **(b)** What is the activation energy for the reverse reaction?

14.30 For the uncatalyzed decomposition of $H_2O_2(aq)$ to form $H_2O(l)$ and $O_2(g)$, the activation energy and overall ΔE are 75.3 kJ/mol and -98.1 kJ/mol, respectively. **(a)** Sketch the energy profile for this reaction. **(b)** What is the activation energy for the reverse reaction?

14.31 Two reactions have identical values for E_a. Does this ensure that they will have the same rate constant if run at the same temperature? Explain.

14.32 Two similar reactions have the same rate constant at 25°C, but at 35°C one of the reactions has a higher rate constant than the other. Account for these observations.

14.33 The rate of the reaction

$$CH_3COOC_2H_5(aq) + OH^-(aq) \longrightarrow CH_3COO^-(aq) + C_2H_5OH(aq)$$

was measured at several temperatures, and the following data were collected:

Temperature (°C)	k ($M^{-1}s^{-1}$)
15	0.0521
25	0.101
35	0.184
45	0.332

Using these data, construct a graph of ln k versus $1/T$. Using your graph, determine the value of E_a.

14.34 The temperature dependence of the rate constant for the reaction

$$CO(g) + NO_2(g) \longrightarrow CO_2(g) + NO(g)$$

is tabulated below. Calculate E_a and A.

Temperature (K)	k ($M^{-1}s^{-1}$)
600	0.028
650	0.22
700	1.3
750	6.0
800	23

14.35 The gas-phase decomposition of HI into H_2 and I_2 is found to have $E_a = 182$ kJ/mol. The rate constant at 700°C is 1.57×10^{-3} $M^{-1}s^{-1}$. What is the value of k at **(a)** 600°C; **(b)** 800°C?

14.36 The rate constant, k, for a reaction is 3.0×10^{-2} s^{-1} at 0°C. Calculate k at 75°C if **(a)** $E_a = 47.8$ kJ/mol; **(b)** $E_a = 125$ kJ/mol.

[14.37] For a particular reaction, raising the temperature from 27°C to 37°C increases the rate by a factor of 2. What is the activation energy for the reaction?

[14.38] The activation energy of a reaction is 38.2 kJ/mol. How many times faster will the reaction occur at 40°C than at 0°C?

Reaction Mechanisms; Catalysis

14.39 Explain the difference between unimolecular, bimolecular, and termolecular elementary processes. Why are termolecular processes very rare?

14.40 Can the rate law for a general reaction A + B $\longrightarrow$ C be predicted without knowing the reaction mechanism? If it is specified that the reaction is an elementary step, what do you know about the rate law of the reaction?

14.41 The following mechanism has been proposed for the reaction of NO with H_2 to form N_2O and H_2O:

$$NO(g) + NO(g) \longrightarrow N_2O_2(g)$$

$$N_2O_2(g) + H_2(g) \longrightarrow N_2O(g) + H_2O(g)$$

(a) Show that the elementary steps of the proposed mechanism add to provide a balanced equation for the reaction.

(b) Write a rate law for each elementary step in the mechanism. (c) Identify any intermediates in the mechanism. (d) The observed rate law is: Rate $= k[NO]^2[H_2]$. If the proposed mechanism is correct, what can we conclude about the relative speeds of the first and second steps?

14.42 The balanced equation for the oxidation of HBr is $4HBr(g) + O_2(g) \longrightarrow 2H_2O(g) + 2Br_2(g)$. The following mechanism has been proposed:

$$HBr + O_2 \longrightarrow HOOBr$$
$$HOOBr + HBr \longrightarrow 2HOBr$$
$$HOBr + HBr \longrightarrow H_2O + Br_2$$

(a) Indicate whether the elementary steps of the proposed mechanism add to give the balanced equation for the reaction. (Hint: You may need to multiply all the coefficients of a given elementary step by some integer before adding.) **(b)** Write the rate law for each elementary step. **(c)** Identify any intermediates in the mechanism. **(d)** The reaction is found to be first order with respect to both HBr and O_2. Neither HOBr nor HOOBr is detected among the products. What can you conclude about the rate-determining step?

14.43 Consider the following reaction: $H_2(g) + 2ICl(g) \longrightarrow 2HCl(g) + I_2(g)$. The rate law for this reaction is first order in both H_2 and ICl: Rate $= k[H_2][ICl]$. Which of the following mechanisms are consistent with the observed rate law?

(a) $2ICl(g) + H_2(g) \longrightarrow 2HCl(g) + I_2(g)$ (termolecular reaction)

(b)
$H_2(g) + ICl(g) \longrightarrow HI(g) + HCl(g)$ (slow)
$HI(g) + ICl(g) \longrightarrow HCl(g) + I_2(g)$ (fast)

(c)
$H_2(g) + ICl(g) \longrightarrow HI(g) + HCl(g)$ (fast)
$HI(g) + ICl(g) \longrightarrow HCl(g) + I_2(g)$ (slow)

(d)
$H_2(g) + ICl(g) \longrightarrow HClI(g) + H(g)$ (slow)
$H(g) + ICl(g) \longrightarrow HCl(g) + I(g)$ (fast)
$HClI(g) \longrightarrow HCl(g) + I(g)$ (fast)
$I(g) + I(g) \longrightarrow I_2(g)$ (fast)

14.44 The decomposition of hydrogen peroxide is catalyzed by iodide ion. The catalyzed reaction is thought to proceed by a two-step mechanism:

$$H_2O_2(aq) + I^-(aq) \longrightarrow H_2O(l) + IO^-(aq) \quad \text{(slow)}$$
$$IO^-(aq) + H_2O_2(aq) \longrightarrow H_2O(l) + O_2(g) + I^-(aq) \quad \text{(fast)}$$

(a) Assuming that the first step of the mechanism is rate-determining, predict the rate law for the overall process. **(b)** Write the chemical equation for the overall process. **(c)** Identify the intermediate, if any, in the mechanism.

14.45 In older texts, catalysts were sometimes defined as substances that speed up a chemical reaction without taking part in the reaction. In what way is this definition misleading? How might the definition be modified to correct it?

14.46 What is the difference between a homogeneous and a heterogeneous catalyst? Why is the activity of a heterogeneous catalyst highly dependent on its method of preparation and prior treatment?

14.47 A reaction in solution is catalyzed by metallic iron. Would you expect a solid chunk of iron metal or an equal mass of finely ground iron filings to be a more effective catalyst? Explain.

14.48 Many metallic catalysts, particularly the precious-metal ones, are often deposited as very thin films on a substance of high surface area per unit mass, such as alumina, Al_2O_3, or silica, SiO_2. Why is this an effective way of utilizing the catalyst material?

14.49 When D_2 is reacted with ethylene, C_2H_4, in the presence of a finely divided catalyst, ethane with two deuteriums, $CH_2D—CH_2D$, is formed. (Deuterium, D, is an isotope of hydrogen of mass 2.) Very little ethane forms in which two deuteriums are bound to one carbon, for example, $CH_3—CHD_2$. Explain why this is so in terms of the sequence of steps involved in hydrogenation.

14.50 Suppose that the hydrogens on ethylene (Equation 14.28) were replaced by bulky organic groups such as $—CH_3$. What effect do you think this would have on the ease with which the compound underwent hydrogenation using a nickel metal catalyst? Explain.

Additional Exercises

14.51 Explain how the units for the rate constant depend on the overall order of the reaction.

14.52 List three factors that can be varied to change the rate of a chemical reaction. Explain, on a molecular level, how each exerts its influence.

14.53 Hydrogen sulfide, H_2S, is a common and troublesome pollutant in industrial wastewaters. One way to remove this substance is to treat the water with chlorine, in which case the following reaction occurs:

$$H_2S(aq) + Cl_2(aq) \longrightarrow S(s) + 2H^+(aq) + 2Cl^-(aq)$$

The rate of this reaction is first order in each reactant. The rate constant for disappearance of H_2S at 28°C is $3.5 \times 10^{-2}\ M^{-1}s^{-1}$. If at a given time the concentration of H_2S is $1.6 \times 10^{-4}\ M$ and that of Cl_2 is 0.070 M, what is the rate of formation of Cl^-?

14.54 Consider the following reaction between mercury(II) chloride and oxalate ion:

$$2HgCl_2(aq) + C_2O_4^{2-}(aq) \longrightarrow 2Cl^-(aq) + 2CO_2(g) + Hg_2Cl_2(s)$$

The initial rate of this reaction was determined for several concentrations of $HgCl_2$ and $C_2O_4^{2-}$, and the following rate data were obtained:

Experiment	$[HgCl_2]$ (M)	$[C_2O_4^{2-}]$ (M)	Rate (M/s)
1	0.105	0.15	1.8×10^{-5}
2	0.105	0.30	7.1×10^{-5}
3	0.052	0.30	3.5×10^{-5}
4	0.052	0.15	8.9×10^{-6}

(a) What is the rate law for this reaction? **(b)** What is the value of the rate constant? **(c)** What is the reaction rate

rate when the concentration of $HgCl_2$ is 0.080 M and that of $C_2O_4^{2-}$ is 0.10 M if the temperature is the same as that used to obtain the data shown above?

14.55 Urea, NH_2CONH_2, is the end product in protein metabolism in animals. The decomposition of urea in 0.1 M HCl occurs according to the reaction

$$NH_2CONH_2(aq) + H^+(aq) + 2H_2O(l) \longrightarrow 2NH_4^+(aq) + HCO_3^-(aq)$$

The reaction is first order in urea. When [urea] = 0.200 M, the rate at 61.05°C is 8.56×10^{-5} M/s. **(a)** What is the value for the rate constant, k? **(b)** What is the concentration of urea in this solution after 5.00×10^3 s if the starting concentration is 0.500 M? **(c)** What is the half-life for this reaction at 61.05°C?

14.56 Cyclopentadiene, C_5H_6, is a hydrocarbon with a low boiling point (40°C). On standing, cyclopentadiene reacts with itself to form dicyclopentadiene, $C_{10}H_{12}$, a reaction that, in organic chemistry, is called a *Diels-Alder* reaction:

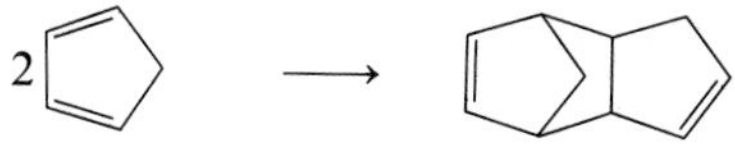

Cyclopentadiene   Dicyclopentadiene

(a) A 0.0400 M solution of C_5H_6 was monitored as a function of time as the reaction $2C_5H_6 \longrightarrow C_{10}H_{12}$ proceeded. The following data were collected:

Time (s)	$[C_5H_6]$ (M)
0	0.0400
50	0.0300
100	0.0240
150	0.0200
200	0.0174

Plot $[C_5H_6]$ versus time, ln $[C_5H_6]$ versus time, and $1/[C_5H_6]$ versus time. What is the order of the reaction? What is the value of the rate constant? **(b)** Dicyclopentadiene can be "cracked" into cyclopentadiene by heating to its boiling point (170°C). The following kinetic data were collected for the cracking reaction, $C_{10}H_{12} \longrightarrow 2C_5H_6$:

Time (s)	$[C_{10}H_{12}]$ (M)
0	0.00200
40	0.00131
80	0.00086
120	0.00056
160	0.00037
200	0.00024

Plot $[C_{10}H_{12}]$ versus time, ln $[C_{10}H_{12}]$ versus time, and $1/[C_{10}H_{12}]$ versus time. What is the order of the reaction? What is the value of the rate constant?

14.57 Sulfuryl chloride, SO_2Cl_2, decomposes in the gas phase into sulfur dioxide, SO_2, and chlorine, Cl_2. The concentration of SO_2Cl_2 is monitored over time. It is found that a plot of ln $[SO_2Cl_2]$ versus time is linear, and that in 240 s the concentration decreases from 0.400 M to 0.280 M. **(a)** What is the rate constant for the reaction $SO_2Cl_2(g) \longrightarrow SO_2(g) + Cl_2(g)$. **(b)** What is the half-life of the reaction?

14.58 A sample of polluted water was oxidized with O_2 at 25°C. The percentage of organic matter in the sample that was oxidized varied with time in the following manner:

Time (days)	1	2	3	4	5	6	7	10	20
Organic matter oxidized (%)	21	37	50	60	68	75	80	90	99

Is the oxidation process first or second order? What is the rate constant for this reaction?

14.59 A possible activated complex in the reaction $H_2(g) + Cl_2(g) \longrightarrow 2HCl(g)$ is the so-called four-center transition state shown below.

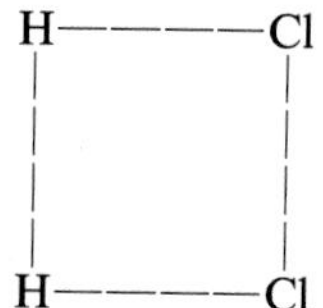

(a) Draw a reaction pathway, analogous to Figure 14.8, showing the positions of the reactants, products, and activated complex. (You should use the data in Appendix C to determine whether the reaction is exothermic or endothermic.) **(b)** For this assumed mechanism, explain why some collisions between H_2 and Cl_2 molecules are more likely than others to lead to products. **(c)** For the reverse reaction, $2HCl(g) \longrightarrow H_2(g) + Cl_2(g)$, there is a four-center arrangement of atoms that cannot lead to products. Explain.

14.60 The first-order rate constant for hydrolysis of a particular organic compound in water varies with temperature as follows:

Temperature (K)	Rate constant (s^{-1})
300	1.0×10^{-5}
320	5.0×10^{-5}
340	2.0×10^{-4}
355	5.0×10^{-4}

From these data calculate the activation energy in units of kJ/mol.

14.61 The activation energy for the reaction

$$2NO_2(g) \longrightarrow 2NO(g) + O_2(g)$$

is 114 kJ/mol. If $k = 0.75\ M^{-1}s^{-1}$ at 600°C, what is the value of k at 500°C?

[14.62] The following mechanism has been proposed for the gas-phase reaction of chloroform, $CHCl_3$, and chlorine:

$$\text{Step 1:} \quad Cl_2(g) \underset{k_{-1}}{\overset{k_1}{\rightleftharpoons}} 2Cl(g) \quad \text{(fast)}$$

$$\text{Step 2:} \quad Cl(g) + CHCl_3(g) \xrightarrow{k_2} HCl(g) + CCl_3(g) \quad \text{(slow)}$$

$$\text{Step 3:} \quad Cl(g) + CCl_3(g) \xrightarrow{k_3} CCl_4(g) \quad \text{(fast)}$$

(a) What is the overall reaction? **(b)** What are the intermediates in the mechanism? **(c)** What is the molecularity of each of the elementary steps? **(d)** What is the rate-determining step? **(e)** What is the rate law predicted by this mechanism? (Hint: The overall reaction order is not an integer.)

[14.63] In hydrocarbon solution, the gold compound $(CH_3)_3AuPH_3$ decomposes into ethane, C_2H_6, and a different gold compound, $(CH_3)AuPH_3$. The following mechanism is proposed for the decomposition of $(CH_3)_3AuPH_3$:

$$\text{Step 1:} \quad (CH_3)_3AuPH_3 \underset{k_{-1}}{\overset{k_1}{\rightleftharpoons}} (CH_3)_3Au + PH_3 \quad \text{(fast)}$$

$$\text{Step 2:} \quad (CH_3)_3Au \xrightarrow{k_2} C_2H_6 + (CH_3)Au \quad \text{(slow)}$$

$$\text{Step 3:} \quad (CH_3)Au + PH_3 \xrightarrow{k_3} (CH_3)AuPH_3 \quad \text{(fast)}$$

(a) What is the overall reaction? **(b)** What are the intermediates in the mechanism? **(c)** What is the molecularity of each of the elementary steps? **(d)** What is the rate-determining step? **(e)** What is the rate law predicted by this mechanism? **(f)** What would be the effect on the reaction rate of adding PH_3 to the solution of $(CH_3)_3AuPH_3$?

[14.64] The gas-phase reaction of chlorine with carbon monoxide to form phosgene, $Cl_2(g) + CO(g) \longrightarrow COCl_2(g)$, obeys the following rate law:

$$\text{Rate} = \frac{\Delta[COCl_2]}{\Delta t} = k[Cl_2]^{3/2}[CO]$$

A mechanism involving the following series of steps is consistent with the rate law:

$$Cl_2 \rightleftharpoons 2Cl$$

$$Cl + CO \rightleftharpoons COCl$$

$$COCl + Cl_2 \rightleftharpoons COCl_2 + Cl$$

Assuming that this mechanism is correct, which of the steps above is the slow, or rate-determining, step? Explain.

14.65 Why is catalysis, rather than an increase in temperature, a better means of increasing the rate in biological systems? What are biological catalysts called?

[14.66] A certain physiologically important first-order reaction has an activation energy equal to 45.0 kJ/mol at normal body temperature (37°C). Without a catalyst, the rate constant for the reaction is $5.0 \times 10^{-4}\ s^{-1}$. To be effective in the human body, where the reaction is catalyzed by an enzyme, the rate constant must be at least $2.0 \times 10^{-2}\ s^{-1}$. If the activation energy is the only factor affected by the presence of the enzyme, by how much must the enzyme lower the activation energy of the reaction to achieve the desired rate?

[14.67] We have seen in this chapter that the hydrogenation of ethylene to form ethane is heterogeneously catalyzed by nickel metal. One particular study found that at low C_2H_4 pressure, the rate of hydrogenation of C_2H_4 was first order with respect to C_2H_4 pressure. However, with all other factors remaining constant, the rate of hydrogenation was found to be zero order with respect to ethylene pressure at high pressures of C_2H_4. **(a)** What is the significance of a zero-order dependence, insofar as the rate of reaction is concerned? **(b)** Suggest an explanation for why the reaction rate has a zero-order dependence on ethylene pressure at high ethylene pressures.

Chemical Equilibrium 15

Chemical equilibrium between the $CoCl_4^{2-}$ ion, which is blue, and the $Co(H_2O)_6^{2+}$ ion, which is pink. Depending on the concentrations of these ions at equilibrium, solutions can appear blue, violet or, pink. (© Richard Megna/ Fundamental Photographs)

CONTENTS

In the laboratory portion of your course, you have had the opportunity to observe a number of simple chemical reactions. For many of these reactions, you may have assumed that after a certain amount of time the reaction "stopped"—colors stopped changing, gases stopped evolving, the temperature no longer changed, and so forth. In many cases, however, chemical reactions come to an apparent halt before the reaction is complete. For example, we can consider the dissociation of colorless N_2O_4 gas into brown NO_2 gas (Figure 15.1):

$$\underset{\text{Colorless}}{N_2O_4(g)} \longrightarrow \underset{\text{Brown}}{2NO_2(g)}$$

When pure, frozen N_2O_4 is warmed above its boiling point (21.2°C), the gas in the reaction flask progressively turns a darker brown as a result of the formation of NO_2. Eventually, the color change stops even though there is still N_2O_4 in the flask. We are left with a mixture of N_2O_4 and NO_2 in which the concentrations of the gases no longer change.

The condition in which the concentrations of all reactants and products cease to change with time is called **chemical equilibrium**. Chemical equilibrium

Figure 15.1 (*a*) Frozen N_2O_4 is nearly colorless. (*b*) As N_2O_4 is warmed above its boiling point, it is partly converted to brown $NO_2(g)$. (*c*) Eventually the color stops changing, even though there is still $N_2O_4(g)$ in the tube. The two gases are in equilibrium. (© Richard Megna/Fundamental Photographs)

(*a*)

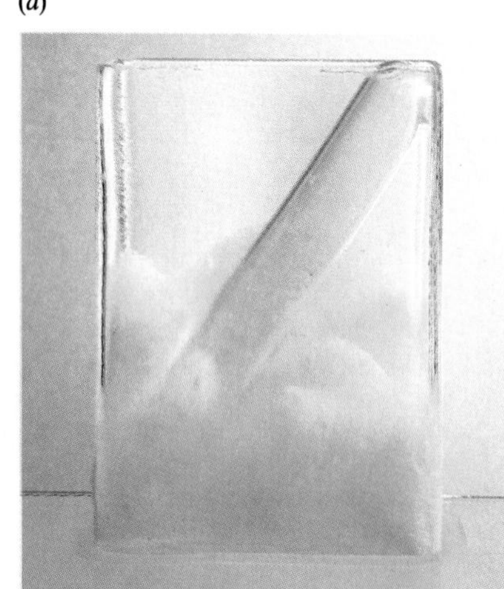

(*b*)

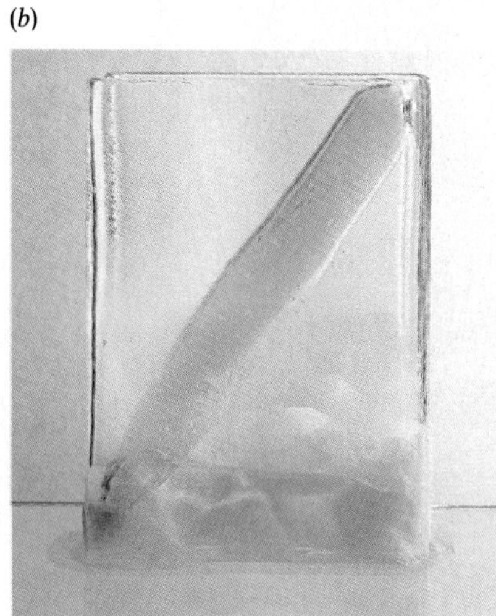

(*c*)

occurs when opposing reactions are proceeding at equal rates: The rate at which products are formed from reactants equals the rate at which reactants are formed from products.

We have already seen several instances of simple equilibria. For example, in a closed container, the vapor above a liquid achieves an equilibrium with the liquid phase (Section 11.2). The rate at which molecules escape from the liquid to the gas phase equals the rate at which molecules in the gas phase strike the surface and become part of the liquid. As another example, in a saturated solution of sodium chloride, solid sodium chloride is in equilibrium with the ions dispersed in water (Section 13.3). The rate at which ions leave the solid surface equals the rate at which other ions are removed from the liquid to become part of the solid.

In this and the next two chapters, we will explore chemical equilibria in some detail. In this chapter we will learn how to express the equilibrium position of a reaction in quantitative terms, and we will study the factors that determine the relative concentrations of reactants and products at equilibrium. We begin by exploring the relationship between the rates of opposing reactions and how this relationship leads to chemical equilibrium.

15.1 THE CONCEPT OF EQUILIBRIUM

At equilibrium, the rate at which products are produced from reactants equals the rate at which reactants are produced from products. We can use some of the concepts we developed in Chapter 14 to illustrate how equilibrium is reached. Let's imagine that we have a simple reaction $A \longrightarrow B$, and that both this reaction and its reverse ($B \longrightarrow A$) are elementary processes. As we learned in Section 14.5, the rates of these unimolecular reactions are

$$\text{Forward reaction:} \quad A \longrightarrow B \qquad \text{Rate} = k_f[A] \qquad [15.1]$$

$$\text{Reverse reaction:} \quad B \longrightarrow A \qquad \text{Rate} = k_r[B] \qquad [15.2]$$

where k_f and k_r are the rate constants for the forward and reverse reactions, respectively.

Now, let us suppose that we start with pure compound A. As A reacts to form compound B, the concentration of A decreases while the concentration of B increases [Figure 15.2(*a*)]. As [A] decreases, the rate of the forward reaction

Figure 15.2 Achieving chemical equilibrium for the reaction $A \rightleftharpoons B$. (*a*) The reaction of pure compound A, with initial concentration $[A]_0$. After a time, the concentrations of A and B do not change. The reason is that (*b*) the rates of the forward reaction ($k_f[A]$) and reverse reaction ($k_r[B]$) become equal.

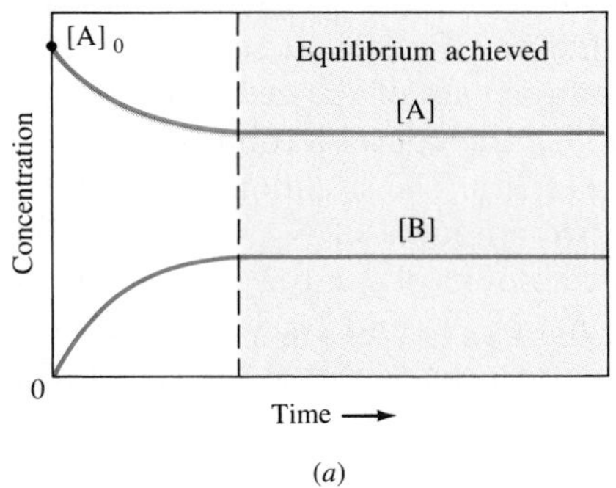

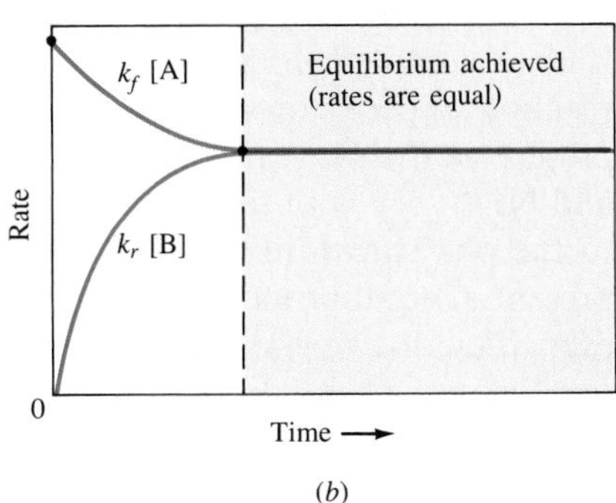

decreases, as shown in Figure 15.2(*b*). Likewise, as [B] increases, the rate of the reverse reaction increases. Eventually, the reaction reaches a point at which the forward and reverse rates are the same [Figure 15.2(*b*)]; compounds A and B are in equilibrium. At equilibrium, therefore,

$$\underset{\text{Forward rate}}{k_f[\text{A}]} = \underset{\text{Reverse rate}}{k_r[\text{B}]}$$

Rearranging this equation gives us

$$\frac{[\text{B}]}{[\text{A}]} = \frac{k_f}{k_r} = \text{a constant} \qquad [15.3]$$

We see that at equilibrium the ratio of the concentrations of A and B equals a constant. (We will consider this constant in Section 15.2.) It makes no difference whether we start with A or B, or even with some mixture of the two. At equilibrium, the ratio of their concentrations equals a definite value.

Once equilibrium is established, the concentrations of A and B do not change. This does *not* mean that A and B stop reacting, however. On the contrary, the equilibrium is *dynamic* (Section 11.5). Compound A is still converted to compound B, and B to A, but *both processes occur at the same rate.* To indicate that the reaction proceeds in both the forward and reverse directions, we use a double arrow:

$$\text{A} \rightleftharpoons \text{B}$$

This example illustrates that opposing reactions naturally lead to an equilibrium situation. In order to examine equilibrium for a real chemical system, we will focus on an extremely important chemical reaction, namely, the synthesis of ammonia from nitrogen and hydrogen:

$$N_2(g) + 3H_2(g) \rightleftharpoons 2NH_3(g) \qquad [15.4]$$

This reaction is the basis for the **Haber process** for synthesizing ammonia.

15.2 THE EQUILIBRIUM CONSTANT

The Haber process consists of putting together N_2 and H_2 in a high-pressure tank at a total pressure of several hundred atmospheres, in the presence of a catalyst, and at a temperature of a few hundred degrees Celsius. Under these conditions, the two gases react to form ammonia. But the reaction does not lead to complete consumption of the N_2 and H_2. Rather, at some point the reaction appears to stop, with all three components of the reaction mixture present at the same time. The manner in which the concentrations of H_2, N_2, and NH_3 vary with time is shown in Figure 15.5(*a*). The situation is analogous to the one shown in Figure 15.2. The relative amounts of N_2, H_2, and NH_3 present at equilibrium do not depend on the amount of catalyst present. However, they do depend on the relative amounts of H_2 and N_2 with which the reaction was begun. Furthermore, if only ammonia is placed into the tank under the usual reaction conditions, at equilibrium there is again a mixture of N_2, H_2, and NH_3. The variations in concentrations as a function of time for this situa-

CHEMISTRY AT WORK: Nitrogen Fixation and the Haber Process

Of all the chemical reactions that humans have learned to carry out and control for their own purposes, the synthesis of ammonia from hydrogen and atmospheric nitrogen is one of the most important. Plant growth requires a substantial store of nitrogen in the soil, in a form usable by plants. The quantity of food required to feed the ever-increasing human population far exceeds the amount that could be produced if we relied solely on naturally available nitrogen in the soil.

The only widely available source of nitrogen is the N_2 present in the atmosphere. The problem thus becomes one of "fixing" atmospheric N_2, that is, converting it to compounds that plants can use. This process is called *nitrogen fixation.*

The N_2 molecule is exceptionally unreactive, in large part because of the strong triple bond between the nitrogen atoms (Section 8.4). For this reason, fixation is not easy to achieve. In nature, the fixation of N_2 is carried out by special nitrogen-fixing bacteria that grow on the roots of certain plants such as clover and alfalfa.

In 1912, the German chemist Fritz Haber (Figure 15.3) developed a process, which still bears his name, for synthesizing ammonia directly from nitrogen and hydrogen (Equation 15.4). The process is sometimes called the *Haber-Bosch process* in honor of Karl Bosch, the engineer who developed the equipment for the industrial production of ammonia. The engineering needed to implement the Haber process requires the use of temperatures and pressures (approximately 500°C and 200 atm) that were difficult to achieve at that time.

The Haber process provides a historically interesting example of the complex impact of chemistry on our lives. At the start of World War I in 1914, Germany was dependent on nitrate deposits in Chile for the nitrogen-containing compounds needed to manufacture explosives. During the war, the Allied blockade of South America cut off this supply. However, by fixing nitrogen from air Germany was able to continue production of explosives (Section 22.5). Experts have estimated that World War I would have ended several years before 1918 had it not been for the Haber process. By thus prolonging the war, the Haber process was indirectly responsible for the deaths of thousands of people.

From these unhappy beginnings as a major factor in international warfare, the Haber process has become the world's principal source of fixed nitrogen. The same process that prolonged World War I has enabled scientists to manufacture fertilizers that have increased crop yields, thereby saving millions of people from starvation. In 1988, 34 billion pounds of ammonia were manufactured in the United States, mostly by the Haber process. The ammonia can be applied directly to the soil as fertilizer (Figure 15.4). It can also be converted into ammonium salts—for example, ammonium sulfate, $(NH_4)_2SO_4$, or ammonium hydrogen phosphate, $(NH_4)_2HPO_4$—that in turn are used as fertilizers.

Figure 15.3 Fritz Haber (1868–1934), the German chemist who developed the process for synthesizing ammonia from nitrogen and hydrogen. (German Information Center)

Haber was a patriotic German who gave enthusiastic support to his nation's war effort. He served as chief of Germany's Chemical Warfare Service during World War I and developed the use of chlorine as a poison-gas weapon. Consequently, the decision to award him the Nobel Prize for Chemistry in 1918 was the subject of considerable controversy and criticism. The ultimate irony, however, came in 1933 when Haber was expelled from Germany because he was Jewish.

Figure 15.4 Liquid ammonia, produced by the Haber process, can be added directly to the soil as a fertilizer. Agricultural use is the largest single application of manufactured NH_3. (Farmland Industries)

Figure 15.5 Variation in gas pressures in formation of the equilibrium $N_2 + 3H_2 \rightleftharpoons 2NH_3$. (*a*) The equilibrium is approached beginning with H_2 and N_2 in the ratio 3:1. (*b*) The equilibrium is approached beginning with NH_3.

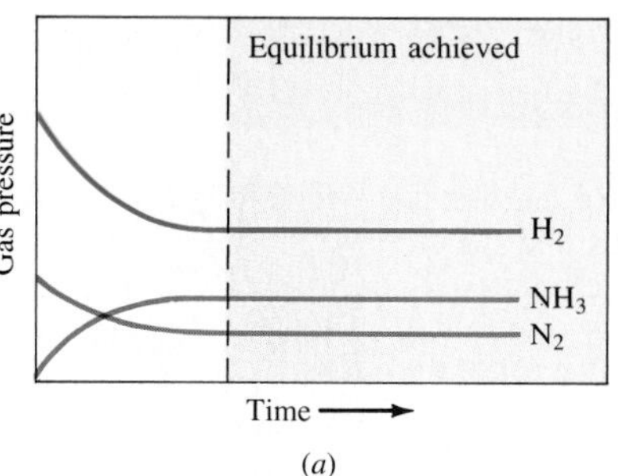

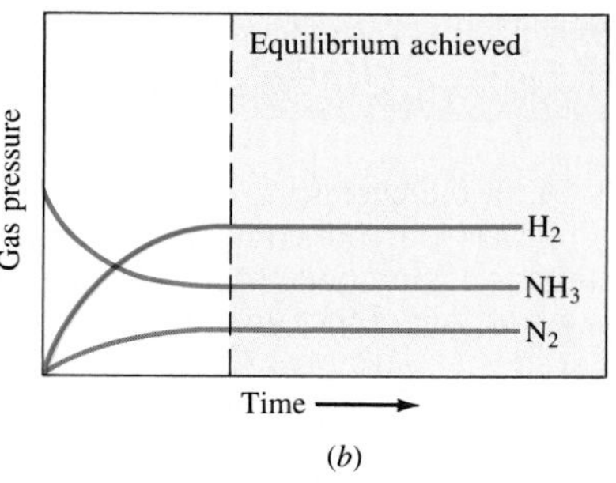

tion are shown in Figure 15.5(*b*). At equilibrium, the relative concentrations of H_2, N_2, and NH_3 are the same, regardless of whether the starting mixture was a 3:1 molar ratio of H_2 and N_2 or pure NH_3. *The equilibrium condition can be reached from either direction.*

Earlier, we saw that when the reaction $A \rightleftharpoons B$ reaches equilibrium, the ratio of the concentrations of A and B has a constant value (Equation 15.3). A similar relationship governs the concentrations of N_2, H_2, and NH_3 at equilibrium. If we were systematically to change the relative amounts of the three gases in the starting mixture and then analyze the gas mixtures at equilibrium, we could determine the relationship among the equilibrium concentrations. Chemists carried out studies of this kind on other chemical systems in the nineteenth century, before Haber's work. In 1864, Cato Maximilian Guldberg (1836–1902) and Peter Waage (1833–1900) proposed their **law of mass action**. This law expresses the relative concentrations of reactants and products at equilibrium in terms of a quantity called the equilibrium constant. Suppose that we have the general reaction

$$jA + kB \rightleftharpoons pR + qS \qquad [15.5]$$

where A, B, R, and S are the chemical species involved, and j, k, p, and q are their coefficients in the balanced chemical equation. According to the law of mass action, the equilibrium condition is expressed by the equation

$$K = \frac{[R]^p[S]^q}{[A]^j[B]^k} \qquad [15.6]$$

where K is a constant, called the **equilibrium constant**, and the square brackets signify the *concentration* of the species within the brackets.

The law of mass action applies only to systems that have attained equilibrium. In general, the equilibrium constant is given by the concentrations of all reaction products multiplied together, each raised to the power of its coefficient in the balanced equation, divided by the concentrations of all reactants multiplied together, each raised to the power of its coefficient in the balanced equation. (Remember that the convention is to write the concentration terms for the *products* in the *numerator* and those for the *reactants* in the *denominator*.) For the reaction $A \rightleftharpoons B$, the equilibrium-constant expression is $K = [B]/[A]$, in accord with Equation 15.3. For the Haber process (Equation 15.4), the equilibrium-constant expression is

$$K = \frac{[NH_3]^2}{[N_2][H_2]^3}$$

Note that once we know the balanced chemical equation for an equilibrium, we can write the equilibrium-constant expression even if we don't know the reaction mechanism. *The equilibrium-constant expression depends only on the stoichiometry of the reaction, not on its mechanism.*

According to the law of mass action, the equilibrium constant is a true constant. Its value at any given temperature does not depend on the initial concentrations of reactants and products. It also does not matter whether other substances are present, as long as they do not react with a reactant or product. The value of the equilibrium constant does, however, vary with temperature.

We can illustrate how the law of mass action was discovered empirically by considering the gas-phase equilibrium between dinitrogen tetroxide and nitrogen dioxide:

$$N_2O_4(g) \rightleftharpoons 2NO_2(g) \qquad [15.7]$$

Figure 15.1 shows this equilibrium being reached after starting with pure N_2O_4. Because NO_2 is a dark-brown gas and N_2O_4 is colorless, the amount of NO_2 in the mixture can be determined by measuring the intensity of the brown color of the gas mixture.

The equilibrium-constant expression for Equation 15.7 is

$$K = \frac{[NO_2]^2}{[N_2O_4]} \qquad [15.8]$$

How can we determine the numerical value for K and verify that it is constant regardless of the starting concentrations of NO_2 and N_2O_4? We could perform experiments in which we start with several sealed tubes containing different concentrations of NO_2 and N_2O_4, as summarized in Table 15.1. The tubes are kept at 100°C until no further change in the color of the gas is noted. We then analyze the mixtures and determine the equilibrium concentrations of NO_2 and N_2O_4, as shown in Table 15.1.

To evaluate the equilibrium constant, K, the equilibrium concentrations are inserted into the equilibrium-constant expression, Equation 15.8. When the concentration unit is molarity, as in the present case, we label the equilibrium constant K_c (where the subscript c stands for concentration).

For example, using the first set of data,

$$[NO_2] = 0.0172\ M \qquad [N_2O_4] = 0.00140\ M$$

$$K_c = \frac{[NO_2]^2}{[N_2O_4]} = \frac{(0.0172)^2}{0.00140} = 0.211$$

Table 15.1 Initial and Equilibrium Concentrations (Molarities) of NO_2 and N_2O_4 in the Gas Phase at 100°C

Experiment	Initial N_2O_4 concentration (M)	Initial NO_2 concentration (M)	Equilibrium N_2O_4 concentration (M)	Equilibrium NO_2 concentration (M)	K_c
1	0.0	0.0200	0.00140	0.0172	0.211
2	0.0	0.0300	0.00280	0.0243	0.211
3	0.0	0.0400	0.00452	0.0310	0.213
4	0.0200	0.0	0.00452	0.0310	0.213

Figure 15.6 The same equilibrium mixture is produced starting with either 0.0400 M NO_2 (experiment 3) or 0.0200 M N_2O_4 (experiment 4).

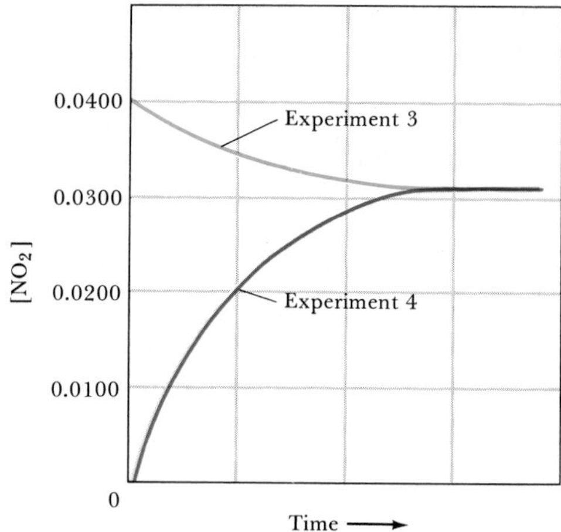

Proceeding in the same way, the values of K_c for the other samples were calculated, as listed in Table 15.1. Note that the value for K_c is essentially constant, even though the initial concentrations vary. Furthermore, the results of experiment 4 show that equilibrium can be attained beginning with N_2O_4 as well as with NO_2. That is, equilibrium can be approached from either direction. Figure 15.6 shows how both experiments 3 and 4 result in the same equilibrium mixture even though one begins with 0.0400 M NO_2 and the other with 0.0200 M N_2O_4.

SAMPLE EXERCISE 15.1

Write the equilibrium-constant expression for each of the following reactions:

(a) $2NO(g) + Cl_2(g) \rightleftharpoons 2NOCl(g)$

(b) $2NOCl(g) \rightleftharpoons 2NO(g) + Cl_2(g)$

Solution: **(a)** As indicated by Equation 15.6, the equilibrium-constant expression has the form of a quotient. The equilibrium concentrations of the products, each raised to a power equal to its coefficient in the balanced equation, appear in the numerator of the quotient. The denominator is similarly obtained using the equilibrium concentrations of the reactants:

$$K = \frac{[NOCl]^2}{[NO]^2[Cl_2]}$$

(b) This reaction is just the reverse of the one given in part (a). Placing products over reactants, we obtain

$$K = \frac{[NO]^2[Cl_2]}{[NOCl]^2}$$

Notice that this expression is just the reciprocal of that given in part (a). It is a general rule that the equilibrium-constant expression for a reaction written in one direction is the reciprocal of the one for the reaction written in the reverse direction. This also means that the numerical value of the equilibrium constant for a reaction written in one direction is the reciprocal of the value of the equilibrium constant for the reaction written in the reverse direction.

PRACTICE EXERCISE

Write the equilibrium constant expression for the reaction $H_2(g) + I_2(g) \rightleftharpoons 2HI(g)$.
Answer: $K = [HI]^2/[H_2][I_2]$

The Magnitude of Equilibrium Constants

Equilibrium constants can be very large or very small. The magnitude of the constant provides us with important information about the equilibrium mixture. For example, consider the reaction of carbon monoxide and chlorine gases at 100°C to form phosgene, $COCl_2$, a highly toxic gas that was used as a poison-gas weapon in World War I. (Phosgene is currently used in the manufacture of certain polymers and insecticides.)

$$CO(g) + Cl_2(g) \rightleftharpoons COCl_2(g) \qquad K_c = \frac{[COCl_2]}{[CO][Cl_2]} = 4.57 \times 10^9$$

In order for the equilibrium constant to be this large, the numerator of the equilibrium-constant expression must be much larger than the denominator. Thus, the equilibrium concentration of $COCl_2$ must be much greater than that of CO or Cl_2; an equilibrium mixture of the three gases is essentially pure $COCl_2$. We say that the equilibrium lies to the right, that is, toward the product side. Likewise, a very small value for the equilibrium constant indicates that an equilibrium mixture contains mostly reactants, as illustrated in Sample Exercise 15.2.

SAMPLE EXERCISE 15.2

The reaction of N_2 with O_2 to form NO might be considered a means of "fixing" nitrogen:

$$N_2(g) + O_2(g) \rightleftharpoons 2NO(g)$$

The value for the equilibrium constant for this reaction at 25°C is $K_c = 1 \times 10^{-30}$ Describe the feasibility of this reaction for nitrogen fixation.

Solution: Because K_c is so small, very little NO will form at 25°C. The equilibrium is said to lie to the left, favoring the reactants. Consequently, this reaction is an extremely poor choice for nitrogen fixation, at least at 25°C.

PRACTICE EXERCISE

(a) Using the value for K_c given in Sample Exercise 15.2, determine the magnitude of the equilibrium constant for decomposition of NO: $2NO(g) \rightleftharpoons N_2(g) + O_2(g)$.

(b) Is it feasible to try to convert undesirable NO in automobile exhausts into N_2 and O_2 using this reaction? ***Answers:*** **(a)** $K = 10^{30}$; **(b)** yes (K is large; it is necessary only to find a way to make the reaction proceed at an acceptable rate).

In general:

$K \gg 1$: Equilibrium lies to the right; products favored.

$K \ll 1$: Equilibrium lies to the left; reactants favored.

One of the first tasks confronting Haber and his co-workers when they approached the problem of ammonia synthesis was the magnitude of the equilibrium constant for the synthesis of NH_3 at various temperatures. If the value of K for Equation 15.4 were very small, then the amount of NH_3 in an equilibrium mixture would be small relative to N_2 and H_2. Clearly, if the equilibrium lies too far to the left, it would not be possible to develop a satisfactory synthesis of ammonia. We discuss some of the methods used to determine the value of equilibrium constants in the next section.

15.3 EVALUATING EQUILIBRIUM CONSTANTS

In their studies on ammonia synthesis, Haber and his co-workers extensively tabulated the equilibrium constant for the reaction at various temperatures. The method they employed is analogous to that described in constructing Table 15.1: They started with various mixtures of N_2, H_2, and NH_3, allowed them to achieve equilibrium at a specific temperature, and measured the concentrations of all three gases at equilibrium. Because the equilibrium concentrations of all products and reactants were known, the equilibrium constant could be calculated directly from the equilibrium-constant expression.

SAMPLE EXERCISE 15.3

In one of their experiments, Haber and co-workers introduced a mixture of hydrogen and nitrogen into a reaction vessel and allowed the system to attain chemical equilibrium at 472°C. The equilibrium mixture of gases was analyzed and found to contain 0.1207 M H_2, 0.0402 M N_2, and 0.00272 M NH_3. From these data, calculate the equilibrium constant, K_c, for

$$N_2(g) + 3H_2(g) \rightleftharpoons 2NH_3(g)$$

Solution

$$K_c = \frac{[NH_3]^2}{[N_2][H_2]^3} = \frac{(0.00272)^2}{(0.0402)(0.1207)^3} = 0.105$$

PRACTICE EXERCISE

Nitryl chloride, NO_2Cl, is in equilibrium in a closed container with NO_2 and Cl_2:

$$2NO_2Cl(g) \rightleftharpoons 2NO_2(g) + Cl_2(g)$$

At equilibrium the concentrations of the substances in the equilibrium are

$$[NO_2Cl] = 0.00106\ M$$
$$[NO_2] = 0.0108\ M$$
$$[Cl_2] = 0.00538\ M$$

From these data, calculate the equilibrium constant, K_c. ***Answer:*** 0.558

We often don't know the equilibrium concentrations of all chemical species in an equilibrium. However, if we know the equilibrium concentration of at least one species, we can generally use the stoichiometry of the reaction to deduce the equilibrium concentrations of the other species in the equation. We will use the following procedure to do this:

1. Tabulate the known initial concentrations of all species involved in the equilibrium.
2. Tabulate the known equilibrium concentrations.
3. For those species for which both the initial and equilibrium concentrations are known, calculate the change in concentration that occurs as the system reaches equilibrium.
4. Use the stoichiometry of the reaction to calculate the changes in concentration for all the other species in the equilibrium.
5. From the initial concentrations and the changes in concentration, calculate the equilibrium concentrations. These are used to evaluate the equilibrium constant.

We illustrate this procedure in Sample Exercise 15.4.

SAMPLE EXERCISE 15.4

A mixture of 5.00×10^{-3} mol of H_2 and 1.00×10^{-2} mol of I_2 is placed in a 5.00-L container at 448°C and allowed to come to equilibrium. Analysis of the equilibrium mixture shows that the concentration of HI is 1.87×10^{-3} *M*. Calculate K_c at 448°C for the reaction

$$H_2(g) + I_2(g) \rightleftharpoons 2HI(g)$$

Solution: We are given the balanced chemical equation for the equilibrium of interest. From this we first set up the equilibrium-constant expression:

$$K_c = \frac{[HI]^2}{[H_2][I_2]}$$

We next build a table listing the initial and equilibrium concentrations of all the species in the equilibrium and the changes in each case in going from initial to final concentrations. The initial concentrations of H_2 and I_2 must be calculated:

$$[H_2]_i = \frac{5.00 \times 10^{-3}\ \text{mol}}{5.00\ \text{L}} = 1.00 \times 10^{-3}\ M$$

$$[I_2]_i = \frac{1.00 \times 10^{-2}\ \text{mol}}{5.00\ \text{L}} = 2.00 \times 10^{-3}\ M$$

The first entries in our table are:

	$H_2(g)$ +	$I_2(g)$ $\rightleftharpoons$	$2HI(g)$
Initial	$1.00 \times 10^{-3}\ M$	$2.00 \times 10^{-3}\ M$	$0\ M$
Change			
Equilibrium			$1.87 \times 10^{-3}\ M$

The equilibrium concentrations of H_2 and I_2 can be calculated from these initial concentrations and the equilibrium concentration of HI. During the course of the reaction the concentration of HI changes from 0 to 1.87×10^{-3} *M*. The balanced equation indicates that 2 mol of HI form from each mole of H_2. Thus, the amount of H_2 consumed is

$$\left(1.87 \times 10^{-3}\frac{\text{mol HI}}{\text{L}}\right)\left(\frac{1\ \text{mol}\ H_2}{2\ \text{mol HI}}\right) = 0.935 \times 10^{-3}\ \text{mol}\ H_2/\text{L}$$

The equilibrium concentration of H_2 is the initial concentration minus that consumed:

$$[H_2] = 1.00 \times 10^{-3}\ M - 0.935 \times 10^{-3}\ M = 0.065 \times 10^{-3}\ M$$

The same line of reasoning gives the equilibrium concentration of I_2:

$$[I_2] = 2.00 \times 10^{-3}\ M - 0.935 \times 10^{-3}\ M = 1.065 \times 10^{-3}\ M$$

The filled-in table now looks like this:

	$H_2(g)$ +	$I_2(g)$ $\rightleftharpoons$	$2HI(g)$
Initial	$1.00 \times 10^{-3}\ M$	$2.00 \times 10^{-3}\ M$	$0\ M$
Change	$-0.935 \times 10^{-3}\ M$	$-0.935 \times 10^{-3}\ M$	$+1.87 \times 10^{-3}\ M$
Equilibrium	$0.065 \times 10^{-3}\ M$	$1.065 \times 10^{-3}\ M$	$1.87 \times 10^{-3}\ M$

Once we have the equilibrium concentrations of each reactant and product, we can calculate the equilibrium constant:

$$K_c = \frac{[HI]^2}{[H_2][I_2]} = \frac{(1.87 \times 10^{-3})^2}{(0.065 \times 10^{-3})(1.065 \times 10^{-3})} = 51$$

PRACTICE EXERCISE

Sulfur trioxide decomposes at high temperature in a sealed container: $2SO_3(g) \rightleftharpoons 2SO_2(g) + O_2(g)$. Initially the vessel is charged at 1000 K with $SO_3(g)$ at a concentration of 6.09×10^{-3} *M*. At equilibrium, the SO_3 concentration is 2.44×10^{-3} *M*. Calculate the value for K_c at 1000 K. ***Answer:*** 4.07×10^{-3}

Concentration Units and Equilibrium Constants

As we have seen, the square brackets around a chemical symbol, as in $[NH_3]$, represent the concentration of that substance. Molarity is the common concentration unit for reactions occurring in solution. For gas-phase reactions, the concentration units used are either molarity or atmospheres of pressure. When the concentration is expressed in molarity, we denote the equilibrium constant as K_c. When the units are atmospheres, we write K_p (where the subscript p stands for pressure). Because the numerical values of K_c and K_p will generally be different, we must take care to indicate which we are using by means of these subscripts.

The ideal-gas equation (Section 10.4) permits us to convert between atmospheres and molarity and therefore to convert between K_p and K_c:

$$PV = nRT$$

$$P = \left(\frac{n}{V}\right)RT = MRT$$

where n/V (the number of moles per liter) is concentration in molarity, M. As a result of the relationship between pressure and molarity, a general expression relating K_p and K_c can be written

$$K_p = K_c(RT)^{\Delta n} \qquad [15.9]$$

The quantity Δn in this equation is the change in the number of moles of gas upon going from reactants to products. It is equal to the number of moles of gaseous products minus the number of moles of gaseous reactants. For example, in the reaction

$$H_2(g) + I_2(g) \rightleftharpoons 2HI(g)$$

there are 2 mol of HI (the coefficient in the balanced equation); there are also 2 mol of gaseous reactants ($1H_2 + 1I_2$). Therefore, $\Delta n = 2 - 2 = 0$, and $K_p = K_c$ for this reaction.

SAMPLE EXERCISE 15.5

Using the value of K_c obtained in Sample Exercise 15.3, calculate K_p for

$$N_2(g) + 3H_2(g) \rightleftharpoons 2NH_3(g)$$

at 472°C.

Solution: There are 2 mol of gaseous products ($2NH_3$) and 4 mol of gaseous reactants ($1N_2 + 3H_2$). Therefore, $\Delta n = 2 - 4 = -2$. (Remember that Δ functions are always based on products minus reactants.) The temperature, T, is $273 + 472 = 745$ K. The value for the ideal-gas constant, R, is 0.0821 L-atm/K-mol. The value of K_c from Sample Exercise 15.3 is 0.105. We therefore have

$$K_p = \frac{P^2_{NH_3}}{P_{N_2}P^3_{H_2}} = K_c(RT)^{\Delta n} = (0.105)(0.0821 \times 745)^{-2} = 2.81 \times 10^{-5}$$

PRACTICE EXERCISE

For the equilibrium $2SO_3(g) \rightleftharpoons 2SO_2(g) + O_2(g)$ at temperature 1000 K, K_c has the value 4.07×10^{-3}. Calculate the value for K_p. ***Answer:*** 0.334

Concentration units can be carried through the calculation of the equilibrium constant to give units for K. For example, for the reaction $N_2O_4(g) \rightleftharpoons 2NO_2(g)$ we have $K = [NO_2]^2/[N_2O_4]$. When concentration is in molarity, the units of the equilibrium constant are $M^2/M = M$; when concentration is in atmospheres, the units are $atm^2/atm = atm$. Attaching units to the equilibrium constant has the advantage of clearly indicating the units in which concentration is expressed. Nevertheless, the more common practice is to write equilibrium constants as dimensionless quantities. We have adopted this practice in this text.

15.4 HETEROGENEOUS EQUILIBRIA

Many equilibria of importance, such as the hydrogen-nitrogen-ammonia system, involve substances all in the same phase. Such equilibria are called **homogeneous equilibria**. On the other hand, the substances in equilibrium may be in different phases, giving rise to **heterogeneous equilibria**. As an example, consider the decomposition of calcium carbonate:

$$CaCO_3(s) \rightleftharpoons CaO(s) + CO_2(g) \qquad [15.10]$$

This system involves a gas in equilibrium with two solids. If we write the equilibrium-constant expression for this process in the usual way, we obtain

$$K = \frac{[CaO][CO_2]}{[CaCO_3]} \qquad [15.11]$$

This example presents us with a problem we have not encountered previously: How do we express the concentration of a solid substance? The concentration of a pure substance, liquid or solid, equals its density divided by its molar mass, $\mathcal{M}$:

$$\frac{\text{Density}}{\mathcal{M}} = \frac{g/cm^3}{g/mol} = \frac{mol}{cm^3}$$

The density of a pure liquid or solid is a constant at any given temperature and changes very little with temperature. Thus, the effective concentration of a pure solid or liquid is a constant, regardless of how much pure solid or liquid is present. We can use this fact to simplify the equilibrium-constant expression. For example, Equation 15.11 simplifies to

$$K = \frac{(\text{constant 1})[CO_2]}{\text{constant 2}}$$

where constant 1 is the concentration of CaO and constant 2 is the concentration of $CaCO_3$. Moving the constants to the left-hand side of the equation, we have

$$K' = K \frac{\text{constant 2}}{\text{constant 1}} = [CO_2] \qquad [15.12]$$

We have in effect included the constant concentrations in the modified equilibrium constant, K'.

Whenever a pure solid or liquid is involved in an equilibrium, we will incorporate its concentration into the equilibrium constant. Thus, we will ignore pure solids and liquids in writing equilibrium-constant expressions, as we did in Equation 15.12.

Equation 15.12 tells us that, at a given temperature, an equilibrium between CaO, $CaCO_3$, and CO_2 will always lead to the same concentration of CO_2 as long as all three components are present. As shown in Figure 15.7, we would have the same pressure of CO_2 regardless of the relative amounts of CaO and $CaCO_3$. Of course, if one of the three components is missing, we cannot have an equilibrium. *Even though they do not appear in the equilibrium-constant expression, pure solids and liquids must be present for an equilibrium to be established.*

SAMPLE EXERCISE 15.6

Each of the mixtures listed below was placed into a closed container and allowed to stand. Which of these mixtures is capable of attaining the equilibrium expressed by Equation 15.10: **(a)** pure $CaCO_3$; **(b)** CaO and a pressure of CO_2 greater than the value of K_p; **(c)** some $CaCO_3$ and a pressure of CO_2 greater than the value of K_p; **(d)** $CaCO_3$ and CaO?

Solution: Equilibrium can be reached in all cases except (c). In **(a)**, $CaCO_3$ simply decomposes, forming CaO(s) and $CO_2(g)$, until the equilibrium pressure of CO_2 is attained. In **(b)**, CO_2 combines with the CaO present until its pressure decreases to the equilibrium value. In **(c)**, equilibrium can't be attained because there is no way in which the CO_2 pressure can decrease so as to attain its equilibrium value. In **(d)**, the situation is essentially the same as in (a); $CaCO_3$ decomposes until equilibrium is attained. The presence of CaO initially makes no difference.

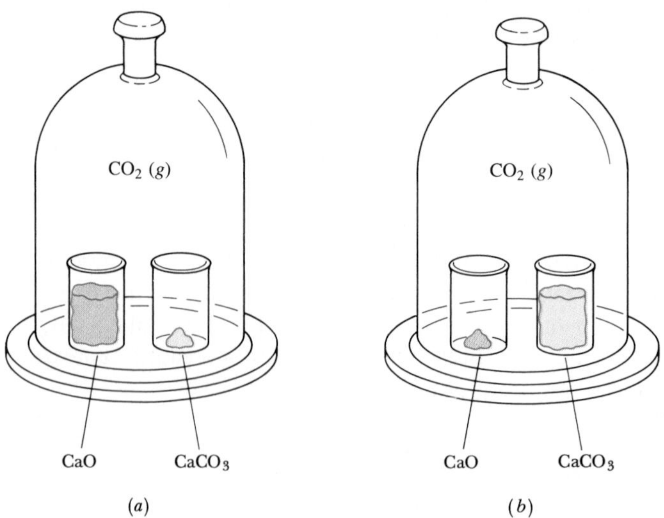

Figure 15.7 The decomposition of $CaCO_3$ is an example of a heterogeneous equilibrium. At the same temperature, the equilibrium pressure of CO_2 is the same in the two bell jars, even though the relative amounts of pure $CaCO_3$ and CaO differ greatly.

PRACTICE EXERCISE

Which of the following substances—$H_2(g)$, $H_2O(g)$, $O_2(g)$—when added to $Fe_3O_4(s)$ in a closed container at high temperature, permits attainment of equilibrium in the reaction $3Fe(s) + H_2O(g) \rightleftharpoons Fe_3O_4(s) + 2H_2(g)$? ***Answer:*** Only $H_2(g)$

SAMPLE EXERCISE 15.7

Write the equilibrium-constant expression for each of the following reactions:
(a) $CO_2(g) + H_2(g) \rightleftharpoons CO(g) + H_2O(l)$
(b) $SnO_2(s) + 2CO(g) \rightleftharpoons Sn(s) + 2CO_2(g)$

Solution: **(a)** The equilibrium-constant expression is

$$K = \frac{[CO]}{[CO_2][H_2]}$$

(Because H_2O is a pure liquid, its concentration does not appear in the equilibrium-constant expression.)

(b) The equilibrium-constant expression is

$$K = \frac{[CO_2]^2}{[CO]^2}$$

(Because SnO_2 and Sn are both pure solids, their concentrations do not appear in the equilibrium-constant expression.)

PRACTICE EXERCISE

Write the equilibrium-constant expression for the reaction $3Fe(s) + 4H_2O(g) \rightleftharpoons Fe_3O_4(s) + 4H_2(g)$. ***Answer:*** $K = [H_2]^4/[H_2O]^4$

15.5 APPLICATIONS OF EQUILIBRIUM CONSTANTS

We have seen that the magnitude of K indicates the extent to which a reaction will proceed. If K is very large, the reaction will tend to proceed far to the right; if K is very small, very little reaction will occur and the equilibrium mixture will contain mainly reactants. The equilibrium constant also allows us (1) to predict the direction in which a reaction mixture will proceed to achieve equilibrium, and (2) to calculate the concentrations of reactants and products once equilibrium has been reached.

Prediction of the Direction of Reaction

Suppose that we place a mixture of 2.00 mol of H_2, 1.00 mol of N_2, and 2.00 mol of NH_3 in a 1-L container at 472°C. Will N_2 and H_2 react to form more NH_3? If we insert the starting concentrations of N_2, H_2, and NH_3 into the equilibrium-constant expression, we have

$$\frac{[NH_3]^2}{[N_2][H_2]^3} = \frac{(2.00)^2}{(1.00)(2.00)^3} = 0.500$$

According to Sample Exercise 15.3, at this temperature $K_c = 0.105$. Therefore, the quotient $[NH_3]^2/[N_2][H_2]^3$ will need to change from 0.500 to 0.105 to move the system toward equilibrium. This change can happen only if $[NH_3]$ decreases and $[N_2]$ and $[H_2]$ increase. Thus, the reaction proceeds toward equilibrium with the formation of N_2 and H_2 from the NH_3; the reaction proceeds from right to left.

Table 15.2 Effect of the Relative Values of Q and K on the Direction of Reaction

Relationship	Direction
$Q > K$	$\longleftarrow$
$Q = K$	Equilibrium
$Q < K$	$\longrightarrow$

When we substitute reactant and product concentrations into the equilibrium-constant expression as we did above, the result is known as the **reaction quotient** and is represented by the letter Q. The reaction quotient will equal the equilibrium constant, K, only if the concentrations are such that the system is at equilibrium: $Q = K$ only at equilibrium. We have seen that when the reaction quotient is larger than K, substances on the right side of the chemical equation will react to form substances on the left; the reaction moves from right to left in approaching equilibrium: If $Q > K$, the reaction moves from right to left. Conversely, if $Q < K$, the reaction will move toward equilibrium with the formation of more products (from left to right). These relationships are summarized in Table 15.2.

SAMPLE EXERCISE 15.8

At 448°C the equilibrium constant, K_c, for the reaction

$$H_2(g) + I_2(g) \rightleftharpoons 2HI(g)$$

is 50.5. Predict the direction in which the reaction will proceed to reach equilibrium at 448°C if we start with 2.0×10^{-2} mol of HI, 1.0×10^{-2} mol of H_2, and 3.0×10^{-2} mol of I_2 in a 2.0-L container.

Solution: The starting concentrations are

$$[HI] = 2.0 \times 10^{-2}\ \text{mol}/2.0\ \text{L} = 1.0 \times 10^{-2}\ M$$

$$[H_2] = 1.0 \times 10^{-2}\ \text{mol}/2.0\ \text{L} = 5.0 \times 10^{-3}\ M$$

$$[I_2] = 3.0 \times 10^{-2}\ \text{mol}/2.0\ \text{L} = 1.5 \times 10^{-2}\ M$$

The reaction quotient is

$$Q = \frac{[HI]^2}{[H_2][I_2]} = \frac{(1.0 \times 10^{-2})^2}{(5.0 \times 10^{-3})(1.5 \times 10^{-2})} = 1.3$$

Because $Q < K_c$, [HI] will need to increase and $[H_2]$ and $[I_2]$ decrease to reach equilibrium; the reaction will proceed from left to right.

PRACTICE EXERCISE

At 1000 K the value of K_c for the reaction $2SO_3(g) \rightleftharpoons 2SO_2(g) + O_2(g)$ is 4.12×10^{-3}. Calculate the value for Q and predict the direction in which the reaction will proceed toward equilibrium if the initial concentrations of reactants are $[SO_3] = 2 \times 10^{-3}\ M$; $[SO_2] = 5 \times 10^{-3}\ M$; $[O_2] = 3 \times 10^{-2}\ M$. ***Answer:*** $Q = 0.2$; the reaction will proceed from right to left, forming SO_3.

Calculation of Equilibrium Concentrations

Chemists frequently need to calculate equilibrium concentrations for a reaction. Our approach in solving problems of this type is similar to the one we used for evaluating equilibrium constants: We tabulate the initial concentrations, the changes in these concentrations, and the final equilibrium concentrations. Usually we end up using the equilibrium-constant expression to derive an equation that must be solved for an unknown quantity, as demonstrated in Sample Exercise 15.9.

SAMPLE EXERCISE 15.9

For the Haber process, $N_2(g) + 3H_2(g) \rightleftharpoons 2NH_3(g)$, $K_p = 1.45 \times 10^{-5}$ at 500°C. In an equilibrium mixture of the three gases at 500°C, the partial pressure of H_2 is 0.928 atm and that of N_2 is 0.432 atm. What is the partial pressure of NH_3 in this equilibrium mixture?

Solution: We are told that the mixture is in equilibrium, so we need not worry about initial concentrations. We tabulate the equilibrium pressures of the gases as follows:

	$N_2(g)$ +	$3H_2(g)$	$\rightleftharpoons$ $2NH_3(g)$
Equilibrium pressure (atm):	0.432	0.928	x

Because we do not know the equilibrium pressure of NH_3, we represent it with a variable, x. At equilibrium, the pressures must satisfy the equilibrium-constant expression:

$$K_p = \frac{P^2_{NH_3}}{P_{N_2}P^3_{H_2}} = \frac{x^2}{(0.432)(0.928)^3} = 1.45 \times 10^{-5}$$

We now rearrange the equation to solve for x:

$$x^2 = (1.45 \times 10^{-5})(0.432)(0.928)^3 = 5.01 \times 10^{-6}$$

$$x = \sqrt{5.01 \times 10^{-6}} = 2.24 \times 10^{-3} \text{ atm} = P_{NH_3}$$

We can always double-check our answer by using it to recalculate the value of the equilibrium constant:

$$K_p = \frac{(2.24 \times 10^{-3})^2}{(0.432)(0.928)^3} = 1.45 \times 10^{-5}$$

PRACTICE EXERCISE

At 500 K, the equilibrium constant K_p for the reaction

$$PCl_5(g) \rightleftharpoons PCl_3(g) + Cl_2(g)$$

has the value 0.497. In an equilibrium mixture at 500 K, the partial pressure of PCl_5 is 0.860 atm and that of PCl_3 is 0.350 atm. What is the partial pressure of Cl_2 in the equilibrium mixture? ***Answer:*** 1.22 atm

In many situations, we will know the value of the equilibrium constant and the initial concentrations of all species. We will then need to solve for the equilibrium concentrations. This usually entails treating as a variable the change in concentration as equilibrium is achieved. The stoichiometry of the reaction gives us the relationship between the changes in the concentrations of all the reactants and products, as illustrated in Sample Exercises 15.10 and 15.11.

SAMPLE EXERCISE 15.10

A 1.000-L flask is filled with 1.000 mol of H_2 and 2.000 mol of I_2 at 448°C. The value of the equilibrium constant, K_c, for the reaction

$$H_2(g) + I_2(g) \rightleftharpoons 2HI(g)$$

at 448°C is 50.5. What are the concentrations of H_2, I_2, and HI in the flask at equilibrium?

Solution: Unlike Sample Exercise 15.4, here we are not given any of the equilibrium concentrations. We are given the initial concentrations of all three gases: $[H_2] = 1.000\ M$, $[I_2] = 2.000\ M$, and $[HI] = 0$. The concentrations of H_2 and I_2 will decrease as equilibrium is established, and that of HI will increase. Let us represent the change in the concentration of H_2 by the variable x. The balanced chemical equation tells us the relationship between the changes in the concentrations of the three gases: For each x moles of H_2 per liter that react, x moles of I_2 per liter are consumed, and $2x$ moles of HI per liter are produced. We can therefore produce the following table:

	$H_2(g)$ +	$I_2(g)$ $\rightleftharpoons$	$2HI(g)$
Initial	1.000 M	2.000 M	0 M
Change	$-x\ M$	$-x\ M$	$+2x\ M$
Equilibrium	$(1.000 - x)\ M$	$(2.000 - x)\ M$	$2x\ M$

We can substitute the equilibrium concentrations into the equilibrium-constant expression and solve for the single unknown, x:

$$K_c = \frac{[HI]^2}{[H_2][I_2]} = \frac{(2x)^2}{(1.000 - x)(2.000 - x)} = 50.5$$

Expanding this expression leads to a quadratic equation in x:

$$4x^2 = 50.5(x^2 - 3.000x + 2.000)$$

$$46.5x^2 - 151.5x + 101.0 = 0$$

Solving the quadratic equation (Appendix A.3) leads to two solutions for x:

$$x = \frac{-(-151.5) \pm \sqrt{(-151.5)^2 - 4(46.5)(101.0)}}{2(46.5)} = 2.323 \text{ or } 0.935$$

The first of these solutions, $x = 2.323$, when substituted into the expressions for the equilibrium concentrations, gives "negative" concentrations of H_2 and I_2. A negative concentration is not chemically meaningful, so we reject this solution. We use the other solution, $x = 0.935$, to find the equilibrium concentrations:

$$[H_2] = 1.0 - x = 0.065\ M$$

$$[I_2] = 2.0 - x = 1.065\ M$$

$$[HI] = 2x = 1.870\ M$$

Finally, we can double-check our solution by putting these numbers into the equilibrium-constant expression:

$$K_c = \frac{[HI]^2}{[H_2][I_2]} = \frac{(1.870)^2}{(0.065)(1.065)} = 50.5$$

Whenever we need to use the quadratic equation to solve an equilibrium problem, one of the solutions will not be meaningful and can be rejected.

PRACTICE EXERCISE

For the equilibrium $PCl_5(g) \rightleftharpoons PCl_3(g) + Cl_2(g)$, the equilibrium constant K_p has the value 0.497 at 500 K. A gas cylinder at 500 K is charged with $PCl_5(g)$ at an initial pressure of 1.66 atm. What are the equilibrium pressures of PCl_5, PCl_3, and Cl_2 at this temperature? ***Answer:*** $P_{PCl_5} = 0.97$ atm; $P_{PCl_3} = P_{Cl_2} = 0.693$ atm

SAMPLE EXERCISE 15.11

As we saw in Sample Exercise 15.3, the equilibrium constant for the Haber process at 472°C is $K_c = 0.105$. A 2.00-L flask is filled with 0.500 mol of NH_3 and is allowed to reach equilibrium at 472°C. What are the equilibrium concentrations of NH_3, N_2, and H_2?

Solution: To achieve equilibrium, NH_3 will be converted into N_2 and H_2. The initial concentration of $NH_3(g)$ is (0.500 mol)/(2.00 L) = 0.250 M. The balanced equation tells us that for every $2x$ moles of NH_3 per liter that decompose, x moles of N_2 per liter and $3x$ moles of H_2 per liter will be produced. We can therefore produce the following table:

$$N_2(g) + 3H_2(g) \rightleftharpoons 2NH_3(g)$$

	$N_2(g)$	$3H_2(g)$	$2NH_3(g)$
Initial	0 M	0 M	0.250 M
Change	$+x$ M	$+3x$ M	$-2x$ M
Equilibrium	x M	$3x$ M	$(0.250 - 2x)$ M

The equilibrium concentrations must satisfy the equilibrium-constant expression:

$$K_c = \frac{[NH_3]^2}{[N_2][H_2]^3} = \frac{(0.250 - 2x)^2}{(x)(3x)^3} = 0.105$$

Expanding $(3x)^3$ and multiplying both sides of the equation by $3^3 = 27$, we have

$$\frac{(0.250 - 2x)^2}{x^4} = (27)(0.105) = 2.835$$

We can solve this equation by taking the square root of both sides, converting it to a quadratic equation:

$$\frac{0.250 - 2x}{x^2} = \sqrt{2.835} = 1.684$$

$$1.684x^2 + 2x - 0.250 = 0$$

Solving the quadratic equation yields $x = 0.114$ or $x = -1.30$. The second of these leads to negative concentrations and is therefore rejected. We are left with the following equilibrium concentrations:

$$[N_2] = x = 0.114\ M$$
$$[H_2] = 3x = 0.342\ M$$
$$[NH_3] = 0.250 - 2x = 0.022\ M$$

PRACTICE EXERCISE

A 1.000-L flask is filled with 0.50 mol of HI at 448°C. Using the equilibrium constant given in Sample Exercise 15.10, determine the concentrations of H_2, I_2, and HI at equilibrium. (Note: You need not use the quadratic equation to solve this problem.)
Answer: $[H_2] = [I_2] = 0.055\ M$; $[HI] = 0.39\ M$

15.6 FACTORS AFFECTING EQUILIBRIUM: LE CHÂTELIER'S PRINCIPLE

In developing his process for making ammonia from N_2 and H_2, Haber sought the factors that might be varied to increase the yield of NH_3. Using the values of the equilibrium constant at various temperatures, he calculated the equilibrium amounts of NH_3 formed under a variety of conditions. The results of some of his calculations are shown in Table 15.3. Notice that the percent of NH_3 present at equilibrium decreases with increasing temperature and increases with increasing pressure. We can understand these effects in terms of a principle first put forward by Henri-Louis Le Châtelier* (1850–1936), a French industrial chemist. **Le Châtelier's principle** can be stated as follows: *If a system at equilibrium is disturbed by a change in temperature, pressure, or the concentration of one of the components, the system will shift its equilibrium position so as to counteract the effect of the disturbance.*

In this section, we will use Le Châtelier's principle to make qualitative

* Pronounced "le-SHOT-lee-ay."

Table 15.3 Effect of Temperature and Total Pressure on the Percentage of Ammonia Present at Equilibrium, Beginning with a 3:1 Molar H_2/N_2 Mixture

	Total pressure (atm)			
Temperature (°C)	200	300	400	500
400	38.7	47.8	54.9	60.6
450	27.4	35.9	42.9	48.8
500	18.9	26.0	32.2	37.8
600	8.8	12.9	16.9	20.8

predictions about the response of a system at equilibrium to various changes in external conditions. We will consider three ways that a chemical equilibrium can be shifted: (1) adding or removing a reactant or product, (2) changing the pressure, and (3) changing the temperature.

Change in Reactant or Product Concentrations

A system at equilibrium is in a dynamic state; the forward and reverse processes are occurring at equal rates, and the system is in a state of balance. An alteration in the conditions of the system may cause the state of balance to be disturbed. If this occurs, the equilibrium shifts until a new state of balance is attained. Le Châtelier's principle states that the shift will be in the direction that minimizes or reduces the effect of the change. Therefore, *if a chemical system is at equilibrium and we add a substance (either a reactant or a product), the reaction will shift so as to reestablish equilibrium by consuming part of the added substance. Conversely, removal of a substance will result in the reaction moving in the direction that forms more of the substance.*

For example, addition of hydrogen to an equilibrium mixture of H_2, N_2, and NH_3 would cause the system to shift so as to reduce the hydrogen pressure toward its original value. This can occur only if the equilibrium is shifted in the direction of forming more NH_3. At the same time, the quantity of N_2 would also be reduced slightly. This situation is illustrated in Figure 15.8. Addition of more N_2 to an equilibrium system would similarly cause a shift in the direction of forming more ammonia. On the other hand, Le Châtelier's principle tells us that if we add NH_3 to the system at equilibrium, the shift will be in such a direction as to reduce the NH_3 concentration toward its original value; that is, some of the added ammonia will decompose to form N_2 and H_2.

We can reach the same conclusions by considering the effect of adding or removing a substance on the reaction quotient (Section 15.5). For example, removal of NH_3 from an equilibrium mixture gives

$$\frac{[NH_3]^2}{[N_2][H_2]^3} = Q < K$$

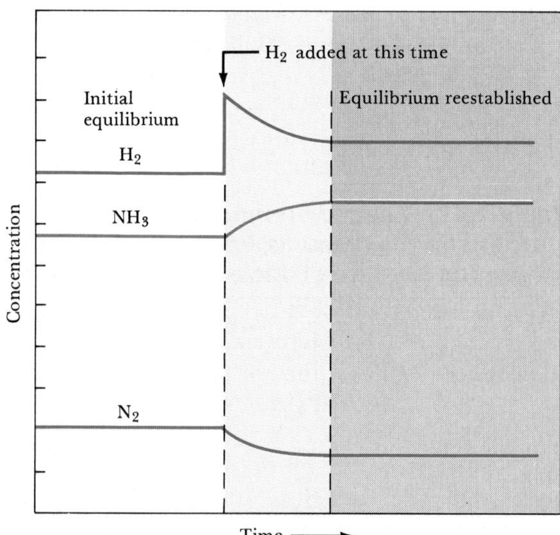

Figure 15.8 When H_2 is added to an equilibrium mixture of N_2, H_2, and NH_3, a portion of the H_2 reacts with N_2 to form NH_3, thereby establishing a new equilibrium position.

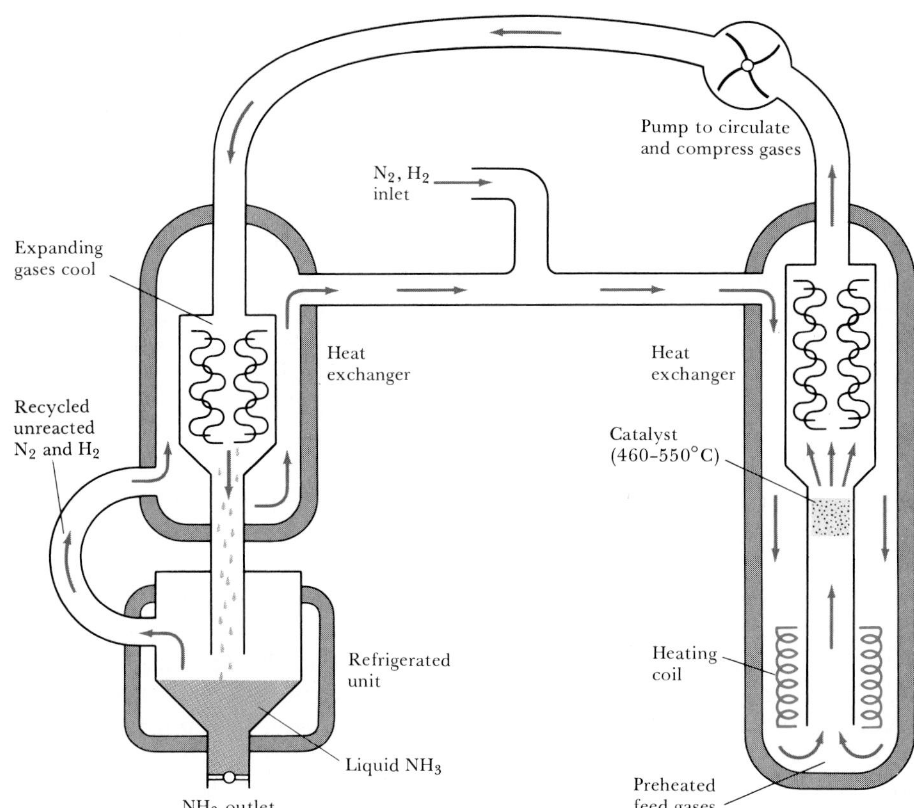

Figure 15.9 Schematic diagram summarizing the industrial production of ammonia. Incoming N_2 and H_2 gases are heated to approximately 500°C and passed over a catalyst. The resultant gas mixture is allowed to expand and cool, causing NH_3 to liquefy. Unreacted N_2 and H_2 gases are recycled.

Because $Q < K$, the reaction shifts from left to right, forming more NH_3 and decreasing $[N_2]$ and $[H_2]$ to restore a new equilibrium that is still governed by K.

If the products of a reaction can be removed continuously, the reacting system can be continuously shifted to form more products. The yield of NH_3 in the Haber process can be increased dramatically by liquefying the NH_3; the liquid NH_3 is removed, and the N_2 and H_2 are recycled to form more NH_3. The general way this is accomplished is shown in Figure 15.9. If a reaction is operated so that equilibrium cannot be achieved because of the escape of products, or if the equilibrium constant is very large, the reaction will proceed essentially to completion. In such instances, the chemical equation for the reaction is usually given with a single arrow: reactants $\longrightarrow$ products.

Effects of Pressure and Volume Changes

If a system is at equilibrium and the total pressure is increased by the application of an external pressure, the system will respond by a shift in equilibrium in the direction that reduces the pressure. If the system is gaseous in whole or part, the equilibrium will shift in the direction that reduces the total number of moles of gas. Conversely, decreasing the pressure by increasing the volume causes a shift in the direction that produces more gas molecules.

For example, let's consider the equilibrium $N_2O_4(g) \rightleftharpoons 2NO_2(g)$, which we saw in Figure 15.1. What happens if the total pressure of an equilibrium mixture is increased by decreasing the volume? According to Le Châtelier's principle, the equilibrium shifts to the side that reduces the total number of moles of gas, which in this case is the reactant side. We therefore expect NO_2 to be converted into N_2O_4 as equilibrium is reestablished. Indeed, we see that the gas mixture gets lighter in color as brown NO_2 is converted into colorless N_2O_4 (Figure 15.10).

For the reaction

$$N_2(g) + 3H_2(g) \rightleftharpoons 2NH_3(g)$$

there are 2 mol of gas on the right side of the chemical equation ($2NH_3$) and 4 mol of gas on the left ($1N_2 + 3H_2$). Consequently, an increase in pressure (decrease in volume) leads to the formation of NH_3; the reaction shifts toward the side with fewer gas molecules. In the case of the reaction

$$H_2(g) + I_2(g) \rightleftharpoons 2HI(g)$$

the number of moles of gaseous products equals the number of moles of gaseous reactants; therefore, changing the pressure will not influence the position of the equilibrium.

Keep in mind that pressure-volume changes do not change the value of K as long as the temperature remains constant. Rather, they change the concentrations of the gaseous substances. In Sample Exercise 15.3, we calculated K_c for an equilibrium mixture at 472°C that contained $[H_2] = 0.1207\ M$, $[N_2] = 0.0402\ M$, and $[NH_3] = 0.00272\ M$. The value of K_c is 0.105. Consider what happens if we suddenly reduce the volume of the system by half. If there were no shift in equilibrium, this volume change would cause the concentrations of all substances to double, giving $[H_2] = 0.2414\ M$, $[N_2] = 0.0804\ M$, and $[NH_3] = 0.00544\ M$. The reaction quotient would then no longer equal the equilibrium constant:

$$Q = \frac{[NH_3]^2}{[N_2][H_2]^3} = \frac{(0.00544)^2}{(0.0804)(0.2414)^3} = 2.62 \times 10^{-2}$$

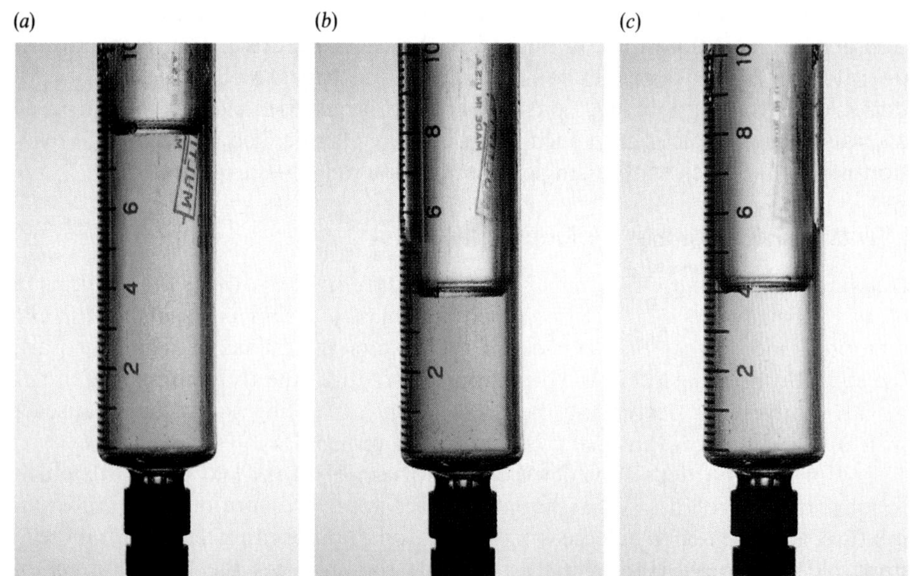

Figure 15.10 (*a*) An equilibrium mixture of brown $NO_2(g)$ and colorless $N_2O_4(g)$. (*b*) When the pressure is increased, the mixture initially appears darker as the concentrations of both gases are increased. (*c*) When equilibrium is reestablished, the color of the gas is lighter than it was initially. In order to decrease the total number of moles of gas, NO_2 is converted into N_2O_4. (© Richard Megna/ Fundamental Photographs)

Because $Q < K_c$, the system is no longer at equilibrium. Equilibrium will be reestablished by increasing $[NH_3]$ and decreasing $[N_2]$ and $[H_2]$ until $Q = K_c = 0.105$. Therefore, the equilibrium shifts to the right as Le Châtelier's principle predicts.

It is possible to change the total pressure of the system without changing its volume. For example, pressure increases if additional amounts of any of the reacting components are added to the system. We have already seen how to deal with a change in concentration of a reactant or product. The total pressure within the reaction vessel might also be increased by addition of a gas that is not involved in the equilibrium. For example, argon might be added to the ammonia equilibrium system. This addition would not alter the partial pressures of any of the reacting components and therefore would not cause a shift in equilibrium.

Effect of Temperature Changes

Changes in concentrations or total pressure cause shifts in equilibrium without changing the value of the equilibrium constant. In contrast, almost every equilibrium constant changes in value as the temperature changes. For example, consider the following equilibrium, which is established when cobalt(II) chloride, $CoCl_2$, is dissolved in hydrochloric acid, HCl(*aq*):

$$\underset{\text{Pale pink}}{Co(H_2O)_6^{2+}(aq)} + 4Cl^-(aq) \rightleftharpoons \underset{\text{Deep blue}}{CoCl_4^{2-}(aq)} + 6H_2O(l) \qquad \Delta H > 0 \quad [15.13]$$

The formation of $CoCl_4^{2-}$ from $Co(H_2O)_6^{2+}$ is an endothermic process. We shall discuss the significance of this enthalpy change shortly. Because $Co(H_2O)_6^{2+}$ is pink and $CoCl_4^{2-}$ is blue, the position of this equilibrium is readily apparent from the color of the solution. Figure 15.11(*a*) shows a room-temperature solution of $CoCl_2$ in HCl(*aq*). Both $Co(H_2O)_6^{2+}$ and $CoCl_4^{2-}$ are present in significant amounts in the solution; the violet color results from the presence of both the pink and blue ions. When the solution is heated [Figure 15.11(*b*)], it becomes intensely blue in color, indicating that the equilibrium has shifted so as to form more $CoCl_4^{2-}$. Cooling the solution, as in Figure 15.11(*c*), leads to a more pink solution, indicating that the equilibrium has shifted to produce more $Co(H_2O)_6^{2+}$. How can we explain the dependence of this equilibrium on temperature?

We can deduce the rules for the temperature dependence of the equilibrium constant by applying Le Châtelier's principle. A simple way to do this is to envision heat as a chemical reagent. In an endothermic reaction, we can treat heat as a reactant, whereas in an exothermic reaction, we can treat heat as a product:

$$\text{Endothermic:} \quad \text{Reactants} + \textit{heat} \rightleftharpoons \text{products}$$

$$\text{Exothermic:} \quad \text{Reactants} \rightleftharpoons \text{products} + \textit{heat}$$

When heat is added to the system, the equilibrium shifts in the direction that absorbs heat. In an endothermic reaction, such as Equation 15.13, heat is

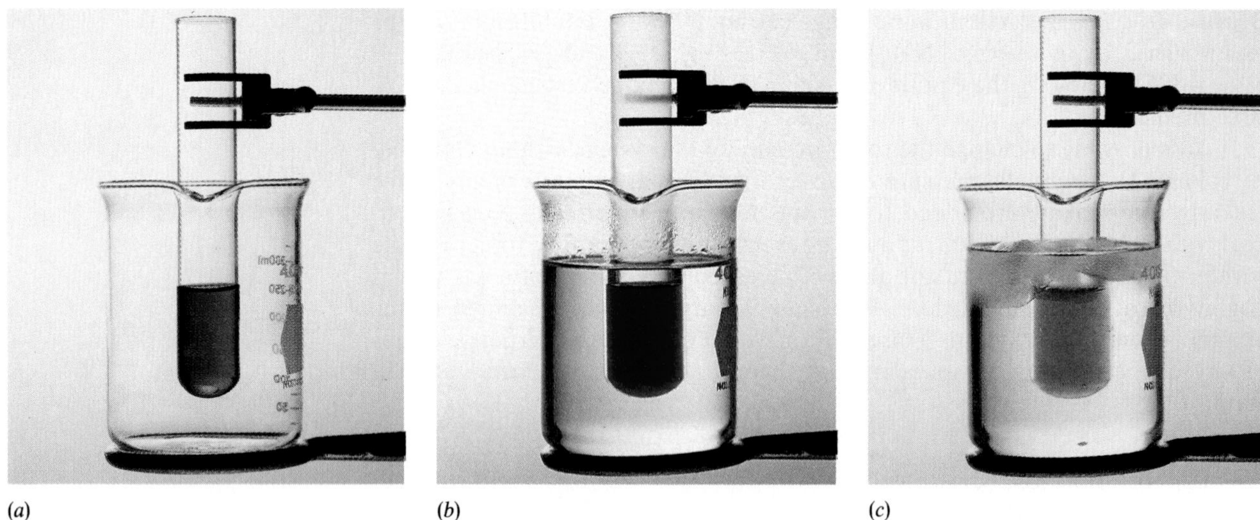

(a) (b) (c)

Figure 15.11 The effect of temperature on the equilibrium $Co(H_2O)_6^{2+}(aq) + 4Cl^-(aq) \rightleftharpoons CoCl_4^{2-}(aq) + 6H_2O(l)$. (*a*) At room temperature, both the pink $Co(H_2O)_6^{2+}$ and blue $CoCl_4^{2-}$ ions are present in significant amounts, giving a violet color to the solution. (*b*) Heating the solution shifts the equilibrium to the right, forming more blue $CoCl_4^{2-}$. (*c*) Cooling the solution shifts the equilibrium to the left, toward pink $Co(H_2O)_6^{2+}$. (© Richard Megna/Fundamental Photographs)

absorbed as reactants are converted to products; thus, the equilibrium shifts to the right, in the direction of products, and K increases. Thus, in the case of Equation 15.13, heating the solution leads to the formation of more $CoCl_4^{2-}$, as observed in Figure 15.11(*b*). In an exothermic reaction, the opposite occurs. Heat is absorbed as products are converted to reactants, so the equilibrium shifts to the left and K decreases. We can summarize these results as follows:

Endothermic: Increasing T results in an increase in K.

Exothermic: Increasing T results in a decrease in K.

Cooling a reaction has the opposite effect of heating it. As we remove heat from the system, the equilibrium shifts to the side that produces heat. Thus, cooling an endothermic reaction shifts the equilibrium to the left, decreasing K. We observed this effect in Figure 15.11(*c*). Cooling an exothermic reaction shifts the equilibrium to the right, increasing K.

SAMPLE EXERCISE 15.12

Consider the following equilibrium:

$$N_2O_4(g) \rightleftharpoons 2NO_2(g) \qquad \Delta H^\circ = 58.0 \text{ kJ}$$

In what direction will the equilibrium shift when each of the following changes is made in a system at equilibrium: **(a)** Add N_2O_4; **(b)** remove NO_2; **(c)** increase the total pressure by adding $N_2(g)$; **(d)** increase the volume; **(e)** decrease the temperature?

Solution: Le Châtelier's principle can be used to determine the effects of each of these changes.

(a) The system will adjust so as to decrease the concentration of the added N_2O_4; the equilibrium consequently shifts to the right, in the direction of products.

(b) The system will adjust to this change by shifting to the side that produces more NO_2; thus, the equilibrium shifts to the right.

(c) Adding N_2 will increase the total pressure of the system, but N_2 is not involved in the reaction. The partial pressures of NO_2 and N_2O_4 are unchanged, and there is no shift in the position of the equilibrium.

(d) The system will shift in the direction that occupies a larger volume (more gas molecules); thus, the equilibrium shifts to the right. (This is the opposite of the effect observed in Figure 15.10.)

(e) The reaction is endothermic; therefore, we can imagine heat as a reagent on the reactant side of the equation. Decreasing the temperature will shift the equilibrium in the direction that produces heat, so the equilibrium shifts to the left, toward the formation of more N_2O_4. Note that only this last change affects the value of the equilibrium constant, K.

PRACTICE EXERCISE

For the reaction

$$PCl_5(g) \rightleftharpoons PCl_3(g) + Cl_2(g) \qquad \Delta H^\circ = 87.9 \text{ kJ}$$

in what direction will the equilibrium shift when **(a)** $Cl_2(g)$ is added; **(b)** the temperature is increased; **(c)** the volume of the reaction system is decreased; **(d)** $PCl_5(g)$ is added?

Answers: **(a)** left: additional $PCl_5(g)$ is formed; **(b)** right: increased dissociation of $PCl_5(g)$ occurs; **(c)** left: more $PCl_5(g)$ is formed; **(d)** right: additional $PCl_3(g)$ and $Cl_2(g)$ are formed.

SAMPLE EXERCISE 15.13

Using the standard heat of formation data in Appendix C, determine the enthalpy change for the reaction

$$N_2(g) + 3H_2(g) \rightleftharpoons 2NH_3(g)$$

From this, determine how the equilibrium constant for the reaction should change with temperature.

Solution: Recall that the standard enthalpy change for a reaction is given by the standard molar enthalpies of formation of the products, each multiplied by its coefficient in the balanced chemical equation, less the same quantities for the reactants. At 25°C, ΔH_f° for $NH_3(g)$ is -46.19 kJ/mol. The ΔH_f° values for $H_2(g)$ and $N_2(g)$ are zero by definition, because the enthalpies of formation of the elements in their normal states at 25°C are defined as zero (Section 5.5). Because 2 mol of NH_3 is formed, the total enthalpy change is

$$2 \text{ mol}(-46.19 \text{ kJ/mol}) - 0 = -92.38 \text{ kJ}$$

The reaction in the forward direction is exothermic. An increase in temperature causes the reaction to shift in the *reverse* direction—in our example, in the direction of less NH_3 and more N_2 and H_2. This is what occurs, as reflected in the values for K_p presented in Table 15.4. Notice that K_p changes very markedly with change in temperature and that it is larger at lower temperatures. This is a matter of great practical importance. To form ammonia at a reasonable rate requires higher temperatures. Yet at higher temperatures, the equilibrium constant is smaller, so the percentage conversion to ammonia is smaller. To compensate for this, higher pressures are needed, because high pressure favors ammonia formation.

Table 15.4 Variation in K_p for the Equilibrium $N_2 + 3H_2 \rightleftharpoons 2NH_3$ as a Function of Temperature

Temperature (°C)	K_p
300	4.34×10^{-3}
400	1.64×10^{-4}
450	4.51×10^{-5}
500	1.45×10^{-5}
550	5.38×10^{-6}
600	2.25×10^{-6}

PRACTICE EXERCISE

Using the thermodynamic data in Appendix C, determine the enthalpy change for the reaction

$$2POCl_3(g) \rightleftharpoons 2PCl_3(g) + O_2(g)$$

From this determine how the equilibrium constant for the reaction should change with temperature. ***Answer:*** $\Delta H^\circ = 620$ kJ; the equilibrium constant for the reaction will increase with increasing temperature.

The Effects of Catalysts

In Section 15.1, we saw that chemical equilibrium is established when the rates of forward and reverse reactions become equal. For the elementary process $A \rightleftharpoons B$, we saw that the equilibrium constant, $K = [B]/[A]$, was equal to the ratio of the rate constants, k_f/k_r (Equation 15.3). The relationship between rate constants and the equilibrium constant is more complicated for reactions that occur by a multistep mechanism. However, remember that even in these cases, the equilibrium-constant expression is given by the law of mass action (Equation 15.6).

In Figure 15.12, we view the equilibrium $A \rightleftharpoons B$ on a reaction pathway similar to that shown in Figure 14.18. As drawn, the energy of B is lower than that of A. Therefore, the activation energy for the forward process, $A \longrightarrow B$, is less than that for the reverse process, $B \longrightarrow A$. Consequently, k_f must be larger than k_r, so K is greater than 1. An equilibrium mixture of A and B will therefore contain a higher concentration of B than A.

What happens if we add a catalyst to the system? As shown in Figure 15.12, the barrier between reactants and products is lowered. The rate at which A is converted to B increases; at the same time, the rate at which B reacts to form A also increases, so we would expect the system to reach equilibrium more quickly. Because the activation energies of both the forward and reverse reactions decrease, the rates of *both* reactions are increased by the presence of a catalyst. In fact, the two rate constants are increased by precisely the same factor, such that their ratio is unchanged. As a result, the equilibrium constant is unchanged by the catalyst. This observation is general for all equilibria, not just the simple one discussed here: *A catalyst changes the rate at which equilibrium is achieved, but it does not change the value of the equilibrium constant.*

The rate at which a reaction approaches equilibrium is an important practical consideration. As an example, let us again consider the synthesis of ammonia from N_2 and H_2. In designing a process for ammonia synthesis, Haber had to deal with a rather serious problem. He wished to synthesize ammonia at the lowest temperature possible, consistent with a reasonable reaction rate. But in the absence of a catalyst, hydrogen and nitrogen do not react with one another at a significant rate either at room temperature or even at much higher temperatures. On the other hand, Haber had to cope with a rapid decrease in

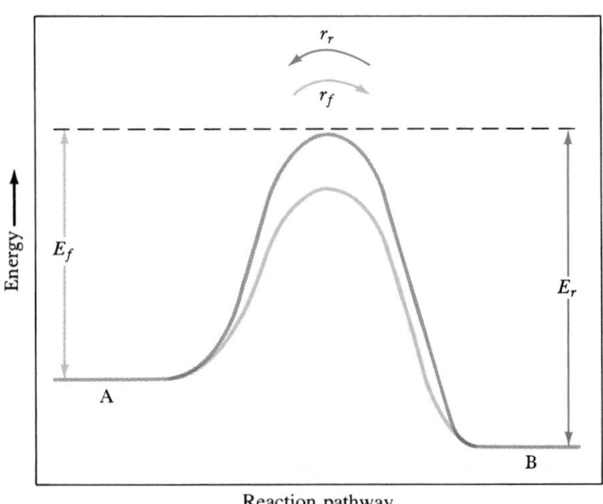

Figure 15.12 Schematic illustration of chemical equilibrium in the reaction $A \rightleftharpoons B$. When equilibrium is attained, the rate of the forward reaction, r_f, equals the rate of the reverse reaction, r_r. The green line refers to the energy profile for a catalyzed reaction in which the activation energy is lowered. The rates of forward and reverse reactions in the catalyzed reaction are increased by the same factor.

CHEMISTRY AT WORK: Controlling Nitrogen Oxide Emissions

The formation of NO from N_2 and O_2 provides another interesting example of the practical importance of changes in equilibrium constant and reaction rate with temperature. The balanced equation and the standard enthalpy change for the reaction are

$$\tfrac{1}{2}N_2(g) + \tfrac{1}{2}O_2(g) \rightleftharpoons NO(g) \qquad \Delta H^\circ = 90.4 \text{ kJ} \quad [15.14]$$

The reaction is endothermic; that is, heat is absorbed when NO is formed from the elements. By applying Le Châtelier's principle, we deduce that an increase in temperature will shift the equilibrium in the direction of more NO. The equilibrium constant, K_c, for formation of 1 mol of NO from the elements at 300 K is only about 10^{-15}. On the other hand, at a much higher temperature, about 2400 K, the equilibrium constant is much larger, about 0.05. The manner in which K_c for reaction Equation 15.14 varies with temperature is shown in Figure 15.13.

This graph helps to explain why NO is a pollution problem. In the cylinder of a modern high-compression auto engine, the temperatures during the fuel-burning part of the cycle may be on the order of 2400 K. Also, there is a fairly large excess of air in the cylinder. These conditions provide an opportunity for the formation of some NO. After the combustion, however, the gases are quickly cooled. As the temperature drops, the equilibrium of Equation 15.14 shifts strongly to the left, that is, in the direction of N_2 and O_2. But the lower temperatures also mean that the rate of the reaction is decreased. The NO formed at high temperatures is essentially "frozen" in that form as the gas cools.

The gases exhausting from the cylinder are still quite hot, perhaps 1200 K. At this temperature, as shown in Figure 15.13, the equilibrium constant for formation of NO is much smaller. However, the rate of conversion of NO to N_2 and O_2 is too slow to permit much loss of NO before the gases are cooled still further. Removing the NO from the exhaust gases depends on finding a catalyst that will work at the temperatures of the exhaust gases and that will cause conversion of NO into something harmless. If a catalyst could be found that would convert the NO back into N_2 and O_2, the equilibrium would be sufficiently favorable. It has not proved possible to find a catalyst capable of withstanding the grueling conditions found in automobile exhaust systems that can catalyze the conversion of NO into N_2 and O_2. Instead, the catalysts used are designed to catalyze the reaction of NO with H_2 or CO.

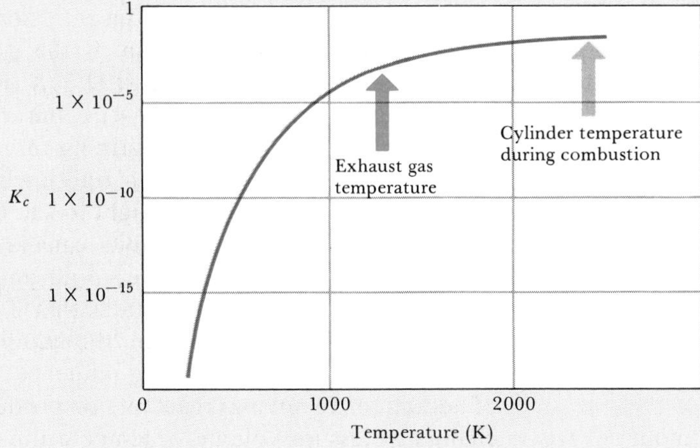

Figure 15.13 Variation in the equilibrium constant for the reaction $\tfrac{1}{2}N_2(g) + \tfrac{1}{2}O_2(g) \rightleftharpoons NO(g)$ as a function of temperature. It is necessary to use a log scale for K_c because the values of K_c vary over such a large range.

equilibrium constant with increasing temperature, as shown in Table 15.4. At temperatures sufficiently high to give a satisfactory reaction rate, the amount of ammonia formed was too small. The solution to this dilemma was to develop a catalyst that would produce a reasonably rapid approach to equilibrium, at a sufficiently low temperature so that the equilibrium constant was still reasonably large. The development of a catalyst thus became the focus of Haber's research efforts.

After trying different substances to see which would be most effective, Haber finally settled on iron mixed with metal oxides. Variants of the original catalyst formulations are still used. These catalysts make it possible to obtain a reasonably rapid approach to equilibrium at temperatures around 400°C to 500°C and with gas pressures of 200 to 600 atm. The high pressures are needed to obtain a satisfactory degree of conversion at equilibrium. You can see from Table 15.3 that if an improved catalyst could be found—one that would lead to sufficiently rapid reaction at temperatures lower than 400°C to 500°C—it would be possible to obtain the same degree of equilibrium conversion at much lower pressures. This would result in great savings in the cost of equipment for ammonia synthesis. In view of the growing need for nitrogen as fertilizer, the fixation of nitrogen is a process of ever-increasing importance.

FOR REVIEW

SUMMARY

If the reactants and products of a reaction are kept in contact, a chemical reaction can achieve a state in which the forward and reverse reactions are occurring at equal rates. This condition is known as chemical equilibrium. A system at equilibrium does not change with time. For such a system we can define an equilibrium constant, K. The equilibrium constant is equal to the product of concentrations of all reaction products, each raised to the power of its coefficient in the balanced equation, divided by the product of concentrations of reactants, each raised to the power of its coefficient in the balanced chemical equation. The equilibrium constant changes with temperature but is not affected by changes in relative concentrations of any reacting substances or by pressure or the presence of catalysts. In heterogeneous equilibria, the concentrations of pure solids or liquids are absent from the equilibrium-constant expression.

If concentrations are expressed in molarity, we label the equilibrium constant K_c; if the concentration units are atmospheres, we use K_p. The constants K_c and K_p are related by the following equation: $K_p = K_c(RT)^{\Delta n}$. A large value for K_c or K_p indicates that the equilibrium mixture contains more products than reactants. A small value for the equilibrium constant means the equilibrium lies toward the reactant side.

The reaction quotient, Q, is found by substituting reactant and product concentrations into the equilibrium-constant expression. If the system is at equilibrium, $Q = K$. However, if $Q \neq K$, nonequilibrium conditions apply; when $Q < K$, the reaction will move toward equilibrium by forming more products (the reaction moves from left to right); when $Q > K$, the reaction will proceed from right to left. Knowledge of the value of K_c or K_p permits calculation of the equilibrium concentrations of reactants and products.

Le Châtelier's principle states that if we disturb a system that is at equilibrium, the equilibrium will shift to minimize the disturbing influence. The effects of adding (or removing) reactants or products and of changing pressure, volume, or temperature can be predicted using this principle. Catalysts affect the speed at which equilibrium is reached but do not affect K.

KEY TERMS

chemical equilibrium
Haber process (Sec. 15.1)
law of mass action (Sec. 15.2)
equilibrium constant (Sec. 15.2)
homogeneous equilibria (Sec. 15.4)
heterogeneous equilibria (Sec. 15.4)
reaction quotient (Sec. 15.5)
Le Châtelier's principle (Sec. 15.6)

EXERCISES

The Concept of Equilibrium; Equilibrium-Constant Expressions

15.1 Explain what is incorrect about the following statements: **(a)** At equilibrium, the reaction stops; no more reactants are transformed into products. **(b)** At equilibrium, there are equal amounts of reactants and products.

15.2 A flask contains a saturated aqueous NaCl solution that is in contact with 10.0 g of undissolved NaCl powder. The flask is stoppered and left undisturbed. A year later it is observed that the system contains a single, large 10.0-g crystal of NaCl in contact with the solution. Explain how this observation can be used to support the notion that equilibrium is a dynamic process.

15.3 Write the equilibrium-constant expression for each of the following reactions. In each case, indicate whether the reaction is homogeneous or heterogeneous.
(a) $2SO_2(g) + O_2(g) \rightleftharpoons 2SO_3(g)$
(b) $H_2(g) + Cl_2(g) \rightleftharpoons 2HCl(g)$
(c) $2H_2S(g) + 3O_2(g) \rightleftharpoons 2H_2O(g) + 2SO_2(g)$
(d) $H_2(g) + FeO(s) \rightleftharpoons H_2O(g) + Fe(s)$
(e) $NH_4NO_2(s) \rightleftharpoons N_2(g) + 2H_2O(g)$

15.4 Write the equilibrium-constant expression for each of the following reactions. In each case, indicate whether the reaction is homogeneous or heterogeneous.
(a) $2CO(g) + O_2(g) \rightleftharpoons 2CO_2(g)$
(b) $N_2O_5(g) \rightleftharpoons NO_2(g) + NO_3(g)$
(c) $Zn(s) + 2HCl(g) \rightleftharpoons ZnCl_2(s) + H_2(g)$
(d) $4NH_3(g) + 3O_2(g) \rightleftharpoons 2N_2(g) + 6H_2O(g)$
(e) $2H_2O_2(l) \rightleftharpoons 2H_2O(l) + O_2(g)$

15.5 For a certain elementary process A $\rightleftharpoons$ B, the forward rate constant is $2.6 \times 10^5\ s^{-1}$, and the reverse rate constant is $4.8 \times 10^3\ s^{-1}$. What is the value of the equilibrium constant for the reaction?

15.6 Consider the reaction A + B $\rightleftharpoons$ C + D. We will assume that both the forward and reverse reactions are elementary processes. Show that the equilibrium constant can be expressed as a ratio of the rate constants for the forward and reverse reactions.

Calculation of Equilibrium Constants

15.7 Gaseous hydrogen bromide is introduced into a flask at 425°C, where it partially decomposes to hydrogen and bromine: $2HBr(g) \rightleftharpoons H_2(g) + Br_2(g)$. At equilibrium, it is found that $[HBr] = 4.30 \times 10^{-1}\ M$, $[H_2] = 2.78 \times 10^{-5}\ M$, and $[Br_2] = 2.78 \times 10^{-5}\ M$. **(a)** What is the value of K_c at this temperature? **(b)** Does the equilibrium favor reactants or products?

15.8 Phosphorus trichloride gas and chlorine gas react to form phosphorus pentachloride gas: $PCl_3(g) + Cl_2(g) \rightleftharpoons PCl_5(g)$. A gas vessel is charged with a mixture of $PCl_3(g)$ and $Cl_2(g)$, which is allowed to equilibrate at 450 K. At equilibrium, the partial pressures of the three gases are $P_{PCl_3} = 0.124$ atm, $P_{Cl_2} = 0.157$ atm, and $P_{PCl_5} = 1.30$ atm. **(a)** What is the value of K_p at this temperature? **(b)** Does the equilibrium favor reactants or products?

15.9 The equilibrium constant for the reaction

$$SO_2(g) + \tfrac{1}{2}O_2(g) \rightleftharpoons SO_3(g)$$

is $K_c = 20.4$ at 700°C.
(a) What is the value of K_c for

$$SO_3(g) \rightleftharpoons SO_2(g) + \tfrac{1}{2}O_2(g)$$

(b) What is the value of K_c for

$$2SO_2(g) + O_2(g) \rightleftharpoons 2SO_3(g)$$

(c) What is the value of K_p for

$$2SO_2(g) + O_2(g) \rightleftharpoons 2SO_3(g)$$

15.10 The equilibrium constant for the reaction

$$2Cl_2(g) + 2H_2O(g) \rightleftharpoons 4HCl(g) + O_2(g)$$

is $K_p = 0.0752$ at 480°C.
(a) What is the value of K_p for the reaction

$$4HCl(g) + O_2(g) \rightleftharpoons 2Cl_2(g) + 2H_2O(g)$$

(b) What is the value of K_p for the reaction

$$Cl_2(g) + H_2O(g) \rightleftharpoons 2HCl(g) + \tfrac{1}{2}O_2(g)$$

(c) What is the value of K_c for the reaction

$$2Cl_2(g) + 2H_2O(g) \rightleftharpoons 4HCl(g) + O_2(g)$$

15.11 At temperatures near 800°C, steam passed over hot coke (a form of carbon obtained from coal) reacts to form CO and H_2:

$$C(s) + H_2O(g) \rightleftharpoons CO(g) + H_2(g)$$

The mixture of gases that results is an important industrial fuel called *water gas*. When equilibrium is achieved at 800°C, $[H_2] = 4.0 \times 10^{-2}\ M$, $[CO] = 4.0 \times 10^{-2}\ M$, and $[H_2O] = 1.0 \times 10^{-2}\ M$. Calculate K_c and K_p at this temperature.

15.12 **(a)** At 700 K, $K_c = 0.11$ for the following reaction:

$$CO_2(g) + H_2(g) \rightleftharpoons CO(g) + H_2O(g)$$

What is the value of K_p? **(b)** At 1000 K, $K_c = 278$ for the reaction

$$2SO_2(g) + O_2(g) \rightleftharpoons 2SO_3(g)$$

What is the value of K_p? **(c)** For the equilibrium

$$C(s) + CO_2(g) \rightleftharpoons 2CO(g)$$

$K_p = 167.5$ at 1000°C. What is the value of K_c?

15.13 A mixture of 0.100 mol of NO, 0.050 mol of H_2, and 0.100 mol of H_2O is placed in a 1.00-L vessel. The following equilibrium is established:

$$2NO(g) + 2H_2(g) \rightleftharpoons N_2(g) + 2H_2O(g)$$

At equilibrium, $[NO] = 0.062\ M$. **(a)** Calculate the equilibrium concentrations of H_2, N_2, and H_2O. **(b)** Calculate K_c.

15.14 A flask is charged with 0.50 atm of $N_2O_4(g)$ and 0.50 atm $NO_2(g)$ at 25°C. The equilibrium reaction is given in Equation 15.7. After reaching equilibrium, the partial pressure of N_2O_4 is 0.60 atm. **(a)** Calculate the equilibrium pressure of NO_2. **(b)** Calculate the value of K_p at 25°C.

15.15 A sample of nitrosyl bromide, NOBr, decomposes according to the following equation:

$$2NOBr(g) \rightleftharpoons 2NO(g) + Br_2(g)$$

An equilibrium mixture in a 5.00-L vessel at 100°C contains 3.22 g of NOBr, 3.08 g of NO, and 4.19 g of Br_2. **(a)** Calculate K_c. **(b)** Calculate K_p. **(c)** What is the total pressure exerted by the mixture of gases?

15.16 A mixture of 1.374 g of H_2 and 70.31 g of Br_2 is heated in a 2.00-L vessel at 700 K. These substances react as follows:

$$H_2(g) + Br_2(g) \rightleftharpoons 2HBr(g)$$

At equilibrium, the vessel is found to contain 0.566 g of H_2. What is the value for K_c at 700 K?

15.17 At 100°C, the equilibrium constant for the reaction $COCl_2(g) \rightleftharpoons CO(g) + Cl_2(g)$ has the value $K_c = 2.19 \times 10^{-10}$. Are the following mixtures of $COCl_2$, CO, and Cl_2 at equilibrium? If not, indicate the direction that the reaction must proceed to achieve equilibrium. **(a)** $[COCl_2] = 5.00 \times 10^{-2}$ *M*, $[CO] = 3.31 \times 10^{-6}$ *M*, $[Cl_2] = 3.31 \times 10^{-6}$ *M*; **(b)** $[COCl_2] = 3.50 \times 10^{-3}$ *M*, $[CO] = 1.11 \times 10^{-5}$ *M*, $[Cl_2] = 3.25 \times 10^{-6}$ *M*; **(c)** $[COCl_2] = 1.45$ *M*, $[CO] = [Cl_2] = 1.56 \times 10^{-6}$ *M*.

15.18 As shown in Table 15.4, K_p for the equilibrium

$$N_2(g) + 3H_2(g) \rightleftharpoons 2NH_3(g)$$

is 4.51×10^{-5} at 450°C. Each of the mixtures listed below may or may not be at equilibrium at 450°C. Indicate in each case whether the mixture is at equilibrium; if it is not at equilibrium, indicate the direction (toward product or toward reactants) in which the mixture must shift to achieve equilibrium: **(a)** 100 atm NH_3, 30 atm N_2, 500 atm H_2; **(b)** 30 atm NH_3, 600 atm H_2, no N_2; **(c)** 26 atm NH_3, 42 atm H_2, 202 atm N_2; **(d)** 100 atm NH_3, 60 atm H_2, 5 atm N_2.

Equilibrium Concentrations

15.19 At 2000°C, the equilibrium constant for the reaction $N_2(g) + O_2(g) \rightleftharpoons 2NO(g)$ is $K_c = 4.1 \times 10^{-4}$. If the equilibrium concentrations of N_2 and NO are 0.24 *M* and 0.011 *M*, respectively, what is the equilibrium concentration of O_2?

15.20 At 25°C, the equilibrium constant for the reaction $2NO(g) + O_2(g) \rightleftharpoons 2NO_2(g)$ is $K_p = 2.23 \times 10^{12}$. In an equilibrium mixture of the three gases, the partial pressures of NO and NO_2 are 1.36×10^{-3} atm and 0.935 atm, respectively. What is the partial pressure of O_2 in the mixture?

15.21 At 1285°C, the equilibrium constant for the reaction $Br_2(g) \rightleftharpoons 2Br(g)$ is $K_c = 1.04 \times 10^{-3}$. A 0.200-L vessel containing an equilibrium mixture of the gases has 0.245 g $Br_2(g)$ in it. What is the mass of Br(*g*) in the vessel?

15.22 For the reaction $H_2(g) + I_2(g) \rightleftharpoons 2HI(g)$, $K_p = 55.3$ at 700 K. In a 2.000-L flask containing an equilibrium mixture of the three gases, there are 0.056 g H_2 and 4.36 g I_2. What is the mass of HI in the flask?

15.23 For the equilibrium

$$C(s) + CO_2(g) \rightleftharpoons 2CO(g)$$

$K_p = 167.5$ at 1000°C. What is the partial pressure of CO_2 that is in equilibrium with CO whose partial pressure is 0.500 atm?

15.24 For the equilibrium

$$2NOBr(g) \rightleftharpoons 2NO(g) + Br_2(g)$$

$K_p = 0.416$ at 373 K. If the pressures of NOBr(*g*) and NO(*g*) are equal, what is the equilibrium pressure of $Br_2(g)$?

15.25 At 21.8°C, the equilibrium constant, K_c, is 1.2×10^{-4} for the following reaction:

$$NH_4HS(s) \rightleftharpoons NH_3(g) + H_2S(g)$$

Calculate the equilibrium concentrations of NH_3 and H_2S if a sample of solid NH_4HS is placed in a closed vessel and allowed to decompose until equilibrium is reached at 21.8°C.

15.26 At 80°C, the equilibrium constant K_p is 1.57 for the following reaction:

$$PH_3BCl_3(s) \rightleftharpoons PH_3(g) + BCl_3(g)$$

(a) Calculate the equilibrium pressures of $PH_3(g)$ and $BCl_3(g)$ if a sample of PH_3BCl_3 is placed in a closed vessel at 80°C and allowed to decompose until equilibrium is attained. **(b)** What is the minimum amount of PH_3BCl_3 that must be placed in a 0.500-L vessel at 80°C if equilibrium is to be attained?

15.27 For the reaction $I_2(g) + Br_2(g) \rightleftharpoons 2IBr(g)$, $K_c = 280$ at 150°C. Suppose that 0.500 mol IBr is placed in a 1.000-L flask and allowed to reach equilibrium at this temperature. What are the equilibrium concentrations of IBr, I_2, and Br_2?

15.28 At 700 K, the reaction $PCl_5(g) \rightleftharpoons PCl_3(g) + Cl_2(g)$ has an equilibrium constant K_p equal to 11.5. A gas vessel is charged with 1.000 atm PCl_5 at 700 K and allowed to reach equilibrium. What are the partial pressures of PCl_5, PCl_3, and Cl_2 at equilibrium?

Le Châtelier's Principle

15.29 Consider the following equilibrium system:

$$C(s) + CO_2(g) \rightleftharpoons 2CO(g) \qquad \Delta H^\circ = 119.8 \text{ kJ}$$

If the reaction is at equilibrium, what would be the effect of **(a)** adding $CO_2(g)$; **(b)** adding C(*s*); **(c)** adding heat; **(d)** increasing the pressure on the system by decreasing the volume; **(e)** adding a catalyst; **(f)** removing CO(*g*)?

15.30 In the reaction

$$6CO_2(g) + 6H_2O(l) \rightleftharpoons C_6H_{12}O_6(s) + 6O_2(g) \qquad \Delta H^\circ = 2816 \text{ kJ}$$

how is the equilibrium yield of $C_6H_{12}O_6$ affected by **(a)** increasing P_{CO_2}; **(b)** increasing temperature; **(c)** removing CO_2; **(d)** increasing the total pressure; **(e)** removing part of the $C_6H_{12}O_6$; **(f)** adding a catalyst?

15.31 How does each of the following changes affect the value of the equilibrium constant for an exothermic reaction: **(a)** removal of a reactant or product; **(b)** increase in the total pressure; **(c)** decrease in the temperature; **(d)** addition of a catalyst?

15.32 Indicate whether each of the following statements is true or false. If the statement is false, correct it so that it is true. **(a)** For a gas-phase reaction in which there is no change in the number of moles of gas in going from reactants to products ($\Delta n = 0$), increasing the total pressure does not markedly affect the equilibrium. **(b)** The value of the equilibrium constant for a reaction is affected only by change in temperature or by the addition of a catalyst. **(c)** The position of equilibrium for an exothermic reaction is shifted to the left by an increase in temperature. **(d)** Increasing the concentration of a reactant in an equilibrium always results in a decrease in the equilibrium concentrations of all other reactants.

Additional Exercises

15.33 Both the forward and reverse reactions of the following equilibrium are believed to be elementary steps:

$$CO(g) + Cl_2(g) \rightleftharpoons COCl(g) + Cl(g)$$

At 25°C, the rate constants for the forward and reverse reactions are $1.4 \times 10^{-28}\ M^{-1}\,s^{-1}$ and $9.3 \times 10^{10}\ M^{-1}\,s^{-1}$, respectively. **(a)** Given the magnitudes of the forward and reverse rate constants, is the reaction likely to be exothermic or endothermic in the forward direction? Explain by using a drawing analogous to Figure 15.12. **(b)** What is the value for the equilibrium constant at 25°C? **(c)** Are reactants or products more plentiful at equilibrium?

15.34 Throughout this chapter we have used gas-phase reactions to illustrate the concept of equilibrium. However, solution reactions could also be used. Write the equilibrium-constant expressions for each of the following aqueous reactions:

(a) $Ag^+(aq) + 2NH_3(aq) \rightleftharpoons Ag(NH_3)_2{}^+(aq)$
(b) $Ag_2CrO_4(s) \rightleftharpoons 2Ag^+(aq) + CrO_4{}^{2-}(aq)$
(c) $HNO_2(aq) \rightleftharpoons H^+(aq) + NO_2{}^-(aq)$
(d) $Zn(s) + Cu^{2+}(aq) \rightleftharpoons Zn^{2+}(aq) + Cu(s)$
(e) $NH_3(aq) + H_2O(l) \rightleftharpoons NH_4{}^+(aq) + OH^-(aq)$

15.35 For the reaction $N_2O_4(g) \rightleftharpoons 2NO_2(g)$, an equilibrium mixture is found to contain 4.27×10^{-2} mol/L of N_2O_4 and 1.41×10^{-2} mol/L of NO_2 at 25°C. What is the value of K_c for this temperature?

15.36 Consider the equilibrium $MgCO_3(s) \rightleftharpoons MgO(s) + CO_2(g)$. **(a)** Write the equilibrium-constant expression for this reaction. **(b)** Why do $MgCO_3$ and MgO not appear in your answer for part (a)? **(c)** If $MgCO_3(s)$ is added to a flask containing pure $CO_2(g)$, will the system achieve equilibrium? What additional information, if any, do you need to answer this question?

15.37 At 500 K, the equilibrium constant for the reaction $2NO(g) + Cl_2(g) \rightleftharpoons 2NOCl(g)$ is $K_p = 52.0$. An equilibrium mixture of the three gases has partial pressures of 0.095 atm and 0.171 atm for NO and Cl_2, respectively. What is the partial pressure of NOCl in the mixture?

15.38 A mixture of 3.0 mol of SO_2, 4.0 mol of NO_2, 1.0 mol of SO_3, and 4.0 mol of NO is placed in a 2.0-L vessel. The following reaction takes place:

$$SO_2(g) + NO_2(g) \rightleftharpoons SO_3(g) + NO(g)$$

When equilibrium is reached at 700°C, the vessel is found to contain 1.0 mol of SO_2. **(a)** Calculate the equilibrium concentrations of SO_2, NO_2, SO_3, and NO. **(b)** Calculate the value of K_c for this reaction at 700°C.

15.39 Consider the hypothetical reaction $A(g) \rightleftharpoons 2B(g)$. A flask is charged with 0.75 atm of pure A, after which it is allowed to reach equilibrium at 0°C. At equilibrium, the partial pressure of A is 0.50 atm. **(a)** What is the total pressure in the flask at equilibrium? **(b)** What is the value of K_p? **(c)** What is the value of K_c?

[15.40] As shown in Table 15.4, at 300°C the equilibrium constant for the reaction $N_2(g) + 3H_2(g) \rightleftharpoons 2NH_3(g)$ is $K_p = 4.34 \times 10^{-3}$. Pure NH_3 is placed in a 1.000-L flask and allowed to reach equilibrium at this temperature. There is 0.753 g NH_3 in the equilibrium mixture. **(a)** What are the masses of N_2 and H_2 in the equilibrium mixture? **(b)** What was the initial mass of ammonia placed in the vessel? **(c)** What is the total pressure in the vessel?

15.41 For the equilibrium

$$PH_3BCl_3(s) \rightleftharpoons PH_3(g) + BCl_3(g)$$

$K_p = 0.052$ at 60°C. **(a)** Calculate K_c. **(b)** Some solid PH_3BCl_3 is added to a closed 0.500-L vessel at 60°C; the vessel is then charged with 0.0216 mol of $BCl_3(g)$. What is the equilibrium concentration of PH_3?

15.42 A mixture of CH_4 and H_2O is passed over a nickel catalyst at 1000 K. The emerging gas is collected in a 5.00-L flask and is found to contain 8.62 g of CO, 2.60 g of H_2, 43.0 g of CH_4, and 48.4 g of H_2O. Assuming that equilibrium has been reached, calculate K_c for the reaction

$$CH_4(g) + H_2O(g) \rightleftharpoons CO(g) + 3H_2(g)$$

15.43 At 1558 K, the equilibrium constant, K_c, for the reaction

$$Br_2(g) \rightleftharpoons 2Br(g)$$

is 1.04×10^{-3}. If a 0.200-L vessel contains 4.53×10^{-2} mol of Br_2 at equilibrium, how many moles of Br are present?

[15.44] Write the equilibrium-constant expression for the equilibrium

$$C(s) + CO_2(g) \rightleftharpoons 2CO(g)$$

The table below shows the relative mole percentages of $CO_2(g)$ and $CO(g)$ at a total pressure of 1 atm for several temperatures. Calculate the value of K_c at each temperature. Is the reaction exothermic or endothermic? Explain.

Temperature (°C)	CO_2 (%)	CO (%)
850	6.23	93.77
950	1.32	98.68
1050	0.37	99.63
1200	0.06	99.94

[15.45] A 0.831-g sample of SO_3 is placed in a 1.00-L container and heated to 1100 K. The SO_3 undergoes decomposition to SO_2 and O_2.

$$2SO_3(g) \rightleftharpoons 2SO_2(g) + O_2(g)$$

At equilibrium, the total pressure in the container is 1.300 atm. Find the values of K_p and K_c for this reaction at 1100 K.

15.46 Calculate K_p for the reaction

$$2SO_2(g) + O_2(g) \rightleftharpoons 2SO_3(g)$$

if at a particular temperature and a total pressure of 112.0 atm the equilibrium mixture consists of 56.6 mole percent SO_2, 10.6 mole percent O_2, and 32.8 mole percent SO_3.

15.47 Nitric oxide, NO, reacts readily with chlorine gas as follows:

$$2NO(g) + Cl_2(g) \rightleftharpoons 2NOCl(g)$$

At 700 K, the equilibrium constant, K_p, for this reaction is 0.26. Predict the behavior of each of the following mixtures at this temperature: **(a)** $P_{NO} = 0.15$ atm, $P_{Cl_2} = 0.31$ atm, and $P_{NOCl} = 0.11$ atm; **(b)** $P_{NO} = 0.12$ atm, $P_{Cl_2} = 0.10$ atm, and $P_{NOCl} = 0.050$ atm; **(c)** $P_{NO} = 0.15$ atm, $P_{Cl_2} = 0.20$ atm, and $P_{NOCl} = 5.10 \times 10^{-3}$ atm.

15.48 Consider the reaction

$$2CO(g) + O_2(g) \rightleftharpoons 2CO_2(g) \qquad \Delta H^\circ = -514.2 \text{ kJ}$$

In which direction will the equilibrium move if **(a)** CO_2 is added; **(b)** CO_2 is removed; **(c)** the volume is increased; **(d)** the pressure is increased; **(e)** the temperature is increased?

15.49 NiO is to be reduced to nickel metal in an industrial process by use of the reaction

$$NiO(s) + CO(g) \rightleftharpoons Ni(s) + CO_2(g)$$

At 1600 K, the equilibrium constant for the reaction is 600. If a CO pressure of 150 mm Hg is to be employed in the furnace, and total pressure never exceeds 760 mm Hg, will reduction occur?

15.50 Consider the reaction

$$CO(g) + 2H_2(g) \rightleftharpoons CH_3OH(l)$$

Using the thermochemical data in Appendix C, determine whether the equilibrium constant for this reaction increases or decreases with increasing temperature. Assuming equal pressures of CO and H_2, how would the extent of conversion of the gas mixture to methanol, CH_3OH, vary with total pressure?

15.51 Calcium carbonate, $CaCO_3$, is also known as limestone. On heating, limestone releases CO_2 according to the reaction $CaCO_3(s) \rightleftharpoons CaO(s) + CO_2(g)$. **(a)** Using data from Appendix C, determine whether the equilibrium constant for this reaction should increase with increasing temperature. **(b)** Will an increase in temperature have an effect on the rate at which equilibrium is reached? Explain. **(c)** The equilibrium constant for this reaction is much less than 1. Why, then, does heating $CaCO_3(s)$ in an open container lead to a complete conversion to $CaO(s)$?

15.52 Nickel carbonyl, $Ni(CO)_4$, is an extremely toxic liquid with a low boiling point. Nickel carbonyl results from the reaction of nickel metal with carbon monoxide. For temperatures above the boiling point (42.2°C) of $Ni(CO)_4$, the reaction is

$$Ni(s) + 4CO(g) \rightleftharpoons Ni(CO)_4(g)$$

Nickel that is more than 99.9 percent pure can be produced by the *carbonyl process*: Impure nickel is reacted with CO at 50°C, which produces $Ni(CO)_4(g)$. The $Ni(CO)_4$ is then heated to 200°C to decompose it back into $Ni(s)$ and $CO(g)$. **(a)** Write the equilibrium-constant expression for the above reaction. **(b)** Given the temperatures used for the steps in the carbonyl process, do you think the above reaction is endothermic or exothermic? **(c)** In the early days of automobiles, nickel-plated exhaust pipes were used. Even though the equilibrium constant for the above reaction is very small at the temperature of automotive exhaust gases, the exhaust pipes quickly corroded. Explain why this occurred.

[15.53] At 700 K, the equilibrium constant for the reaction $CCl_4(g) \rightleftharpoons C(s) + 2Cl_2(g)$ is $K_p = 0.76$. A flask is charged with 2.00 atm of CCl_4, which is then allowed to reach equilibrium at 700 K. **(a)** What is the value of K_c for this reaction at 700 K? **(b)** What fraction of the CCl_4 is converted into C and Cl_2? **(c)** What are the partial pressures of CCl_4 and Cl_2 at equilibrium?

[15.54] At 25°C, the equilibrium constant for the reaction $NH_4SH(s) \rightleftharpoons NH_3(g) + H_2S(g)$ is $K_p = 12.0$. A 10.00-L flask is charged with 1.00 atm of pure $NH_3(g)$ at 25°C. Solid NH_4SH is then added until there is excess unreacted solid remaining. **(a)** Why does no reaction occur until the NH_4SH is added? **(b)** What is the total pressure in the flask at equilibrium? **(c)** What is the minimum mass, in grams, of NH_4SH that must be added in order to achieve equilibrium?

[15.55] The reaction $PCl_3(g) + Cl_2(g) \rightleftharpoons PCl_5(g)$ has $K_p = 0.0870$ at 300°C. A flask is charged with 0.30 atm PCl_3, 0.60 atm Cl_2, and 0.10 atm PCl_5 at this temperature. **(a)** Use the reaction quotient to determine the direction the reaction must proceed in order to reach equilibrium. **(b)** Calculate the equilibrium partial pressures of the gases. **(c)** What effect will increasing the volume of the system have on the mole fraction of PCl_5 in the mixture? **(d)** The reaction is exothermic. What effect will increasing the temperature of the system have on the mole fraction of PCl_5 in the mixture?

[15.56] An equilibrium mixture of H_2, I_2, and HI at 458°C contains 2.24×10^{-2} *M* H_2, 2.24×10^{-2} *M* I_2, and 0.155 *M* HI in a 5.00-L vessel. What are the equilibrium concentrations when equilibrium is reestablished following the addition of 0.100 mol of HI?

[15.57] Consider the hypothetical reaction $A(g) + 2B(g) \rightleftharpoons 2C(g)$, for which $K_c = 0.25$. A 1.00-L reaction vessel is loaded with 1.00 mol of compound C, which is allowed to reach equilibrium. We will let the variable x represent the number of mol/L of compound A present at equilibrium. **(a)** In terms of x, what are the equilibrium concentrations of compounds B and C? **(b)** What limits must be placed on the value of x so that all concentrations are positive? **(c)** By putting the equilibrium concentrations (in terms of x) into the equilibrium-constant expression, derive an equation that can be solved for x. **(d)** The equation from part (c) is a *cubic* equation (one that has the form $ax^3 + bx^2 + cx + d = 0$). In general, cubic equations cannot be solved in closed form. However, you can estimate the solution by plotting the cubic equation in the allowed range for x that you specified in part (b). The point at which the cubic equation crosses the y axis is the solution. **(e)** From the plot in part (d), estimate the equilibrium concentrations of A, B, and C. (You can check the accuracy of your answer by substituting these concentrations into the equilibrium-constant expression.)

15.58 At 1200 K, the approximate temperature of automobile exhaust gases (Figure 15.13), the equilibrium constant for the reaction

$$2CO_2(g) \rightleftharpoons 2CO(g) + O_2(g)$$

is about 1×10^{-13} atm. Assuming that the exhaust gas (total pressure 1 atm) contains 0.2 percent CO by volume, 12 percent CO_2, and 3 percent O_2, is the system at equilibrium with respect to the above reaction? Based on your conclusion, would the CO concentration in the exhaust be lowered or increased by a catalyst that speeds up the reaction above?

15.59 Suppose that you worked at the U.S. Patent Office and a patent application came across your desk claiming that a newly developed catalyst was much superior to the Haber catalyst for ammonia synthesis, because the catalyst led to much greater equilibrium conversion of N_2 and H_2 into NH_3 than the Haber catalyst under the same conditions. What would be your response?

15.60 The hypothetical reaction $A + B \rightleftharpoons C$ occurs in the forward direction in a single step. The energy profile of the reaction is shown in Figure 15.14. **(a)** Is the forward or reverse reaction faster at equilibrium? **(b)** Would you expect the equilibrium to favor reactants or products? **(c)** In general, how would a catalyst affect the energy profile shown? **(d)** How would a catalyst affect the ratio of the rate constants for the forward and reverse reactions? **(e)** How would you expect the equilibrium constant of the reaction to change with increasing temperature?

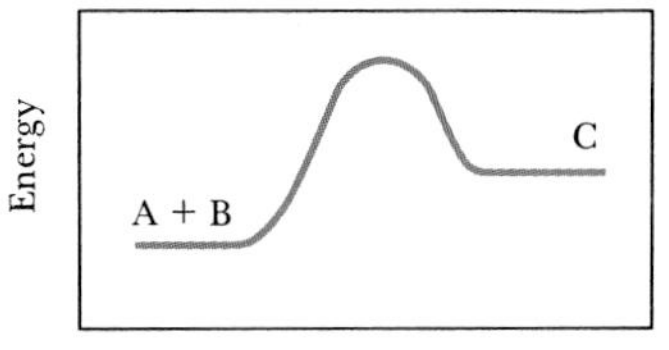

Figure 15.14

[15.61] Consider the equilibrium $A \rightleftharpoons B$ discussed in Equations 15.1 through 15.3. Assume that the only effect of a catalyst on the reaction is to lower the activation energies of the forward and reverse reactions, as shown in Figure 15.8. Using the Arrhenius equation (Section 14.4), prove that the equilibrium constant is the same for the catalyzed reaction as for the uncatalyzed one.

16 Acid-Base Equilibria

Bubbles of CO_2 form as an Alka-Seltzer® tablet dissolves in water. The CO_2 is produced by the reaction of citric acid and sodium bicarbonate in the tablet. (© Paul Silverman/ Fundamental Photographs)

CONTENTS

Acids and bases are important in numerous chemical processes that occur around us, from industrial processes to biological ones, from reactions in the laboratory to those in our environment. The time required for a metal object immersed in water to corrode, the ability of an aquatic environment to support fish and plant life, and the fate of pollutant chemicals washed out of the air by rain are all critically dependent upon the acidity or basicity of the solution.

We have encountered acids and bases many times in earlier discussions. For example, much of Chapter 4 focused on the reactions of these important substances. In this chapter, we will reexamine acids and bases, applying the principles of chemical equilibrium. We will see how to describe the aqueous equilibria quantitatively. We will also examine the relationship between acid or base strength and chemical structure.

16.1 THE NATURE OF ACIDS AND BASES

From the earliest days of experimental chemistry, scientists have realized that certain substances, called acids, possess a sour taste (for example, citric acid in lemon juice) and can dissolve active metals such as iron and zinc. Acids also cause certain vegetable dyes to turn characteristic colors; for example, litmus turns red on contact with acids. Bases likewise possess characteristic properties. They have a bitter taste and feel slippery to the touch (soap is a good example). Whereas acids cause litmus to turn red, bases cause it to turn blue.

Chemists sought to relate the properties of acids and bases to their composition and molecular structure. In the 1880s, the Swedish chemist Svante Arrhenius (1859–1927) suggested that *acids are substances that produce H^+ ions in water solutions and that bases produce OH^- ions.* We use these definitions when we consider the aqueous solutions formed by various solutes. If an aqueous solution contains more $H^+(aq)$ than $OH^-(aq)$, we say that the solution is *acidic*. The $H^+(aq)$ ions produce the characteristic acidic properties of the solution. Conversely, if the solution contains more $OH^-(aq)$ than $H^+(aq)$, we say that the solution is *basic*; the $OH^-(aq)$ ions produce the characteristic basic properties of the solution.

The Proton in Water

If you have had some chemical laboratory experience, you have probably encountered hydrochloric acid. Hydrochloric acid is an aqueous solution of hydrogen chloride gas. The hydrochloric acid normally sold commercially—concentrated hydrochloric acid—is 37 to 38 percent HCl by mass (the solution is 12 *M* in HCl).

A CLOSER LOOK: Nature of the Hydrated Proton

What is it about water that causes a molecule such as HCl to come apart, or dissociate, into H^+ and Cl^- ions? The first and most obvious point to make is that water is polar (Section 9.2). Consequently, the oxygen atom, which bears a partial negative charge, is attracted to the partially positive hydrogen end of the polar HCl molecule. Second, the water molecule has unshared electron pairs on the oxygen atom, which are capable of forming a covalent bond to the hydrogen ion in the following manner:

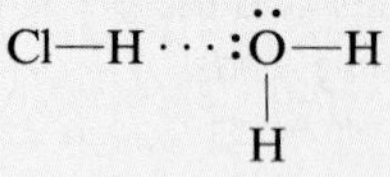

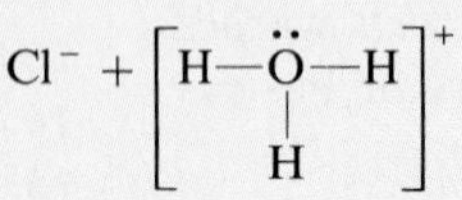 [16.1]

In the reaction between HCl and H_2O shown in Equation 16.4, the proton (an H^+ ion) transfers from HCl to the water molecule, forming the hydronium ion.

The situation in water is actually much more complex than Equation 16.1 suggests. We've learned (Section 11.2) that hydrogen bonds exist throughout liquid water. The existence of this hydrogen-bond network is responsible for many of the special properties of water (for example, its high polarity and high melting and boiling points). Much research has been devoted to learning how H^+ ions fit into the complex structure of liquid water. Experimental studies show that, in part, the H^+ ions must exist as hydronium ions. In fact, it is possible to isolate $H_3O^+Cl^-$, $H_3O^+ClO_4^-$, and other salts in which there is clearly an H_3O^+ ion in the solid lattice. But just as water molecules are strongly hydrogen-bonded to one another, the H_3O^+ ion in solution is hydrogen-bonded to other water molecules. Thus ions such as the two shown in Figure 16.1 may exist and have in fact been found.

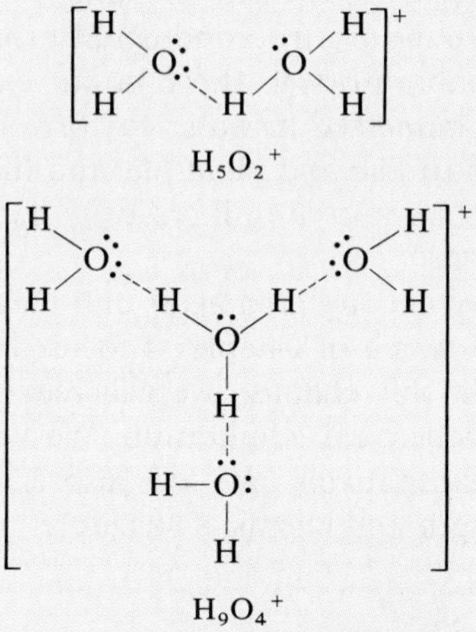

Figure 16.1 Two possible structural forms for the proton in water, in addition to H_3O^+. There is good experimental evidence for the existence of both these species.

Why is HCl so soluble in water? Its solubility in many other solvents, such as benzene, C_6H_6, is very low. Its high solubility in water is due to its chemical reaction with water, which produces hydrated hydrogen and chloride ions:

$$HCl(g) \xrightarrow{H_2O} H^+(aq) + Cl^-(aq)$$

The (*aq*) that follows both H^+ and Cl^- indicates that the ions are separated from one another and hydrated, as illustrated earlier in Figure 13.2. The hydrogen ion fits into the water structure differently than any other ion. Keep in mind that the hydrogen ion is simply the proton, with no surrounding valence electron. The proton interacts strongly with a lone pair of electrons on the oxygen of a water molecule, forming the **hydronium ion**, H_3O^+:

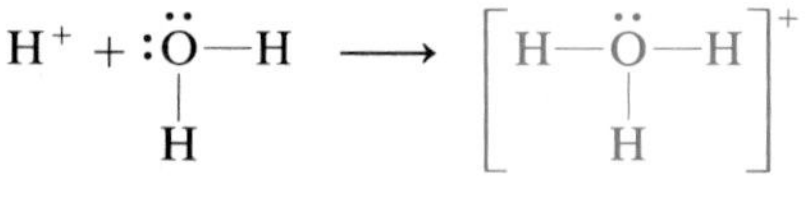 [16.2]

Both $H^+(aq)$ and $H_3O^+(aq)$ are used to represent the hydrated hydrogen ion surrounded by solvent water. We may therefore write the reaction of HCl with water to form hydrochloric acid in either of the following ways:

$$HCl(aq) + H_2O(l) \longrightarrow H_3O^+(aq) + Cl^-(aq) \qquad [16.3]$$

$$HCl(aq) \longrightarrow H^+(aq) + Cl^-(aq) \qquad [16.4]$$

The hydrated proton will ordinarily be represented throughout the remainder of this book as $H^+(aq)$. However, regardless of whether $H^+(aq)$ or $H_3O^+(aq)$ is used to represent the hydrated proton, you should keep in mind that *acidic solutions are formed by a chemical reaction in which an acid transfers a proton* (H^+) *to water.*

The Brønsted-Lowry Concept of Acids and Bases

The nature of the reaction between an acid and water as we have just described it was first appreciated by the Danish chemist Johannes Brønsted (1879–1947) and the English chemist Thomas M. Lowry (1874–1936). Brønsted and Lowry recognized that acid-base behavior could be related to the ability of substances to transfer protons. In 1923 Brønsted and Lowry independently proposed that *acids be defined as substances that are capable of donating a proton, and bases as substances capable of accepting a proton.* In these terms, when HCl dissolves in water (Equation 16.3), it acts as an acid, donating a proton to H_2O; the H_2O simultaneously acts as a base, accepting a proton, as shown in Figure 16.2.

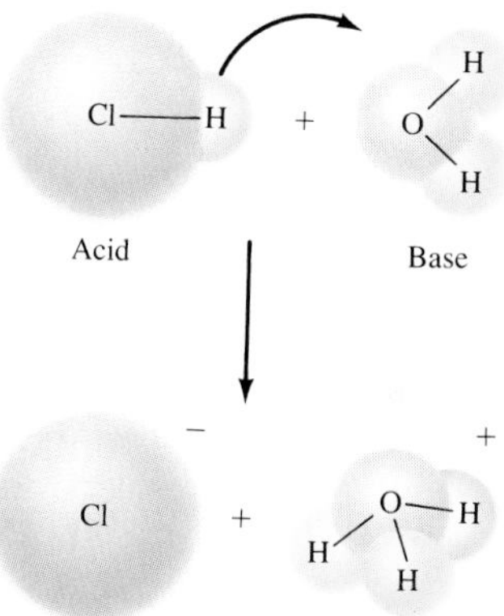

Figure 16.2 When a proton is transferred from HCl to H_2O, HCl acts as the acid and H_2O acts as the base.

Because the emphasis in the Brønsted-Lowry concept is on proton transfer, it is also applicable to reactions that do not occur in aqueous solution. For example, in the reaction between HCl and NH_3, a proton is transferred from the acid HCl to the base NH_3:

$$:\ddot{\underset{\cdot\cdot}{Cl}}-H + :\begin{array}{c}H\\|\\N-H\\|\\H\end{array} \longrightarrow :\ddot{\underset{\cdot\cdot}{Cl}}:^- + \left[\begin{array}{c}H\\|\\H-N-H\\|\\H\end{array}\right]^+ \qquad [16.5]$$

This reaction can occur in the gas phase. The hazy film that forms on the windows of general chemistry laboratories and on glassware in the lab is largely solid NH_4Cl formed by the gas-phase reaction of HCl and NH_3 (Figure 16.3).

To appreciate better the relationship between the Arrhenius definitions and the Brønsted-Lowry definitions of acids and bases, consider an aqueous solution of ammonia, in which the following equilibrium occurs:

$$NH_3(aq) + H_2O(l) \rightleftharpoons NH_4^+(aq) + OH^-(aq) \qquad [16.6]$$

Using the Brønsted-Lowry definitions, we say that H_2O is the acid in the reaction because it donates a H^+ ion. The NH_3 is the base because it accepts the H^+ ion. Using the Arrhenius definition, we say that the resulting solution is basic because it contains $OH^-(aq)$ ions. Thus we may speak of the NH_3 *molecule* as being basic (a proton acceptor) and also of its aqueous *solution* as being basic (having an excess of OH^- ions).

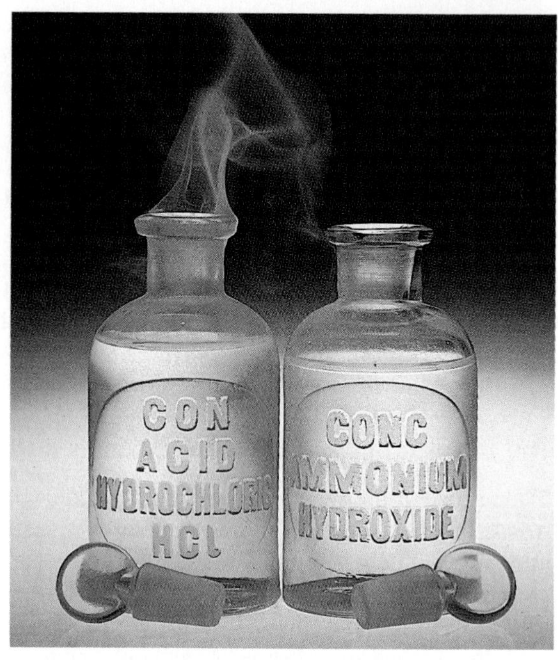

Figure 16.3 The HCl(g) escaping from concentrated hydrochloric acid and the $NH_3(g)$ escaping from aqueous ammonia (here labeled ammonium hydroxide) combine to form a white fog of $NH_4Cl(s)$. (© Richard Megna/Fundamental Photographs)

Sometimes proton donors are called **Brønsted acids**, rather than merely acids, to distinguish between the different definitions of acids. Likewise, proton acceptors may be called **Brønsted bases**.

Conjugate Acid-Base Pairs

In any acid-base equilibrium, both the forward reaction (to the right) and the reverse reaction (to the left) involve proton transfers. For example, consider again the reaction of NH_3 with H_2O:

$$\mathrm{NH_3}(aq) + \mathrm{H_2O}(l) \rightleftharpoons \mathrm{NH_4}^+(aq) + \mathrm{OH}^-(aq)$$

In the forward direction, H_2O donates a proton to NH_3. Therefore, H_2O is the acid, and NH_3 is the base. In the reverse direction, NH_4^+ donates a proton to OH^-. In this case, NH_4^+ is the acid, and OH^- is the base.

An acid and a base such as H_2O and OH^-, which differ only in the presence or absence of a proton, are called a **conjugate acid-base pair**.* Every acid has associated with it a **conjugate acid**, formed from the base by addition of a proton. For example, OH^- is the conjugate base of H_2O. Likewise, every base has associated with it a **conjugate acid**, formed from the base by addition of a proton; NH_4^+ is the conjugate acid of NH_3:

add H^+ ($NH_3 \rightarrow NH_4^+$)

$$\underset{\text{Base}}{\mathrm{NH_3}(aq)} + \underset{\text{Acid}}{\mathrm{H_2O}(l)} \rightleftharpoons \underset{\text{Conjugate acid}}{\mathrm{NH_4}^+(aq)} + \underset{\text{Conjugate base}}{\mathrm{OH}^-(aq)} \tag{16.7}$$

lose H^+ ($H_2O \rightarrow OH^-$)

* The word *conjugate* means "joined together as a pair," or "coupled."

Notice that an acid can be a charged species, such as NH_4^+, or a neutral species, such as H_2O. Similarly, bases can be neutral or charged. Note also that some substances can act as an acid in one situation and a base in another. For example, H_2O is a Brønsted base in its reaction with HCl (Equation 16.3) and a Brønsted acid in its reaction with NH_3 (Equation 16.6).

SAMPLE EXERCISE 16.1

(a) Write the formula for the conjugate base of each of the following: HI; HNO_2; PH_4^+; HSO_4^-. **(b)** Write the formula for the conjugate acid of each of the following: HSO_4^-; H_2O; NH_2^-; F^-.

Solution: **(a)** The conjugate base of each of the species listed is just the species with one proton removed: I^-; NO_2^-; PH_3; SO_4^{2-}.

(b) The conjugate acid of each of the species listed is just the species with one proton added: H_2SO_4; H_3O^+; NH_3; HF.

Notice that HSO_4^- is capable of acting as either an acid or a base.

PRACTICE EXERCISE

Write the formula for the conjugate acid of each of the following: S^{2-}; ClO_3^-; HPO_4^{2-}; CO. *Answers:* HS^-; $HClO_3$; $H_2PO_4^-$; HCO^+

If we arrange acids in order of their ability to donate a proton, we will see an interesting and important fact: The more readily a substance gives up a proton, the less readily its conjugate base accepts a proton. Similarly, the more readily a base accepts a proton, the less readily its conjugate acid gives up a proton. In other words, *the stronger an acid, the weaker its conjugate base; the weaker an acid, the stronger its conjugate base.* Thus, if we know something about the strength of an acid (its ability to donate protons), we also know something about the strength of its conjugate base (its ability to accept protons). For example, we know from earlier discussions that HCl is a strong acid (Section 4.3). Therefore, its conjugate base, Cl^-, must be a very poor proton acceptor.

Figure 16.4 displays some common acids and their conjugate bases. $H^+(aq)$ is the strongest proton donor that can exist at equilibrium in aqueous solution. Thus acids listed above $H^+(aq)$ in Figure 16.4 completely transfer protons to water to form $H^+(aq)$. Likewise, $OH^-(aq)$ is the strongest base that can exist at equilibrium in aqueous solution. Any stronger proton acceptor will completely react with water, removing a proton to form OH^- ions.

SAMPLE EXERCISE 16.2

Hydrocyanic acid, HCN, is ionized in water to a lesser extent than is an HF solution of the same concentration. What is the conjugate base of HCN? Is it a stronger or weaker base than F^-?

Solution: The conjugate base of HCN is CN^-, the ion that remains after a proton has been lost to the solvent. HCN dissociates to a lesser extent than does HF, which means that the tendency of the reverse reaction to occur is greater. In the reverse reaction the conjugate base, CN^-, accepts a proton from the solvent. In other words, CN^- is a stronger base than is F^-.

PRACTICE EXERCISE

The dimethylammonium ion, $(CH_3)_2NH_2^+$, is a weak acid, ionized to a slight degree in water. **(a)** What is the conjugate base of the dimethylammonium ion? **(b)** How does the strength of this base compare with that of Cl^-? *Answers:* **(a)** dimethylamine, $(CH_3)_2NH$; **(b)** dimethylamine is a stronger base than Cl^-.

Figure 16.4 Relative strengths of some common conjugate acid-base pairs, which are listed opposite one another in the two columns.

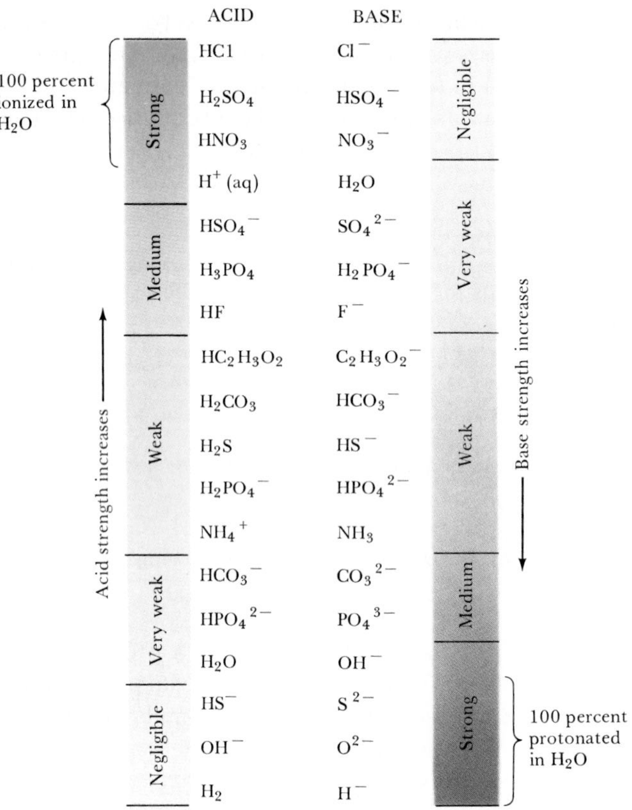

16.2 THE DISSOCIATION OF WATER AND THE pH SCALE

We have seen that H_2O can act as either a proton donor or a proton acceptor, depending on the circumstances. Significantly, H_2O is also capable of acting as a proton donor and proton acceptor toward itself:

$$\mathrm{H{-}\ddot{\underset{|}{\underset{H}{O}}}{:}} + \mathrm{H{-}\ddot{\underset{|}{\underset{H}{O}}}{:}} \rightleftharpoons \left[\mathrm{H{-}\ddot{\underset{|}{\underset{H}{O}}}{-}H}\right]^+ + \mathrm{:\ddot{\underset{..}{O}}{-}H^-} \qquad [16.8]$$

This proton transfer, which we call **autoionization**, amounts to a spontaneous ionization of solvent. The reaction occurs to only a small extent and gives rise to a very small electrical conductivity for pure water. At room temperature, only about one out of every 10^8 molecules is in the ionic form at any instant. No individual molecule remains in the ionic condition for long; the equilibria are extremely rapid. On the average, a proton transfers from one molecule to another in water at a rate of about 1000 times per second.

By expressing the hydrated proton as $H^+(aq)$ rather than $H_3O^+(aq)$, we can rewrite Equation 16.8 as

$$H_2O(l) \rightleftharpoons H^+(aq) + OH^-(aq) \qquad [16.9]$$

The equilibrium expression for this autoionization reaction can be written as

$$K = \frac{[H^+][OH^-]}{[H_2O]}$$

The concentration of water in aqueous solutions is typically very large, about 55 M, and remains essentially constant for dilute solutions. It is therefore customary to exclude the concentration of water from equilibrium-constant expressions for aqueous solutions, just as we exclude the concentrations of pure solids and liquids from equilibrium-constant expressions for heterogeneous reactions (Section 15.4). Thus, we can write the equilibrium-constant expression for the autoionization of water as

$$K[H_2O] = K_w = [H^+][OH^-]$$

The product of two constants, K and $[H_2O]$, defines a new constant, K_w. This important equilibrium constant is called the **ion-product constant** for water. At 25°C, K_w has the value of 1.0×10^{-14}. This is an important equilibrium constant. You should memorize this expression:

$$K_w = [H^+][OH^-] = 1.0 \times 10^{-14} \qquad [16.10]$$

Equation 16.10 is valid for aqueous solutions as well as for pure water. A solution for which $[H^+] = [OH^-]$ is said to be *neutral*. In most solutions, H^+ and OH^- concentrations are not equal. As the concentration of one of these ions increases, the concentration of the other must decrease so that the ion product equals 1.0×10^{-14}. In acidic solutions, $[H^+]$ exceeds $[OH^-]$. In basic solutions, the reverse is true: $[OH^-]$ exceeds $[H^+]$.

SAMPLE EXERCISE 16.3

Calculate the values of $[H^+]$ and $[OH^-]$ in a neutral solution at 25°C.

Solution: By definition, in a neutral solution, $[H^+]$ equals $[OH^-]$. Let us call the concentration of each of these species in neutral solution x. Using Equation 16.10, we have

$$[H^+][OH^-] = (x)(x) = 1.0 \times 10^{-14}$$

$$x^2 = 1.0 \times 10^{-14}$$

$$x = 1.0 \times 10^{-7} = [H^+] = [OH^-]$$

In an acid solution, $[H^+]$ is greater than 1.0×10^{-7} M; in a basic solution it is less than 1.0×10^{-7} M.

PRACTICE EXERCISE

Indicate whether each of the following solutions is neutral, acidic or basic: **(a)** $[H^+] = 2 \times 10^{-5}$ M; **(b)** $[OH^-] = 3 \times 10^{-9}$ M; **(c)** $[OH^-] = 1 \times 10^{-7}$ M.
Answers: **(a)** acidic; **(b)** acidic; **(c)** neutral

SAMPLE EXERCISE 16.4

Calculate the concentration of $H^+(aq)$ in **(a)** a solution in which $[OH^-]$ is 0.010 M; **(b)** a solution in which $[OH^-]$ is 2.0×10^{-9} M.

Solution: In this problem and all that follow, we assume, unless stated otherwise, that the temperature is 25°C.

(a) Using Equation 16.10, we have

$$[H^+][OH^-] = 1.0 \times 10^{-14}$$

$$[H^+] = \frac{1.0 \times 10^{-14}}{[OH^-]} = \frac{1.0 \times 10^{-14}}{0.010}$$

$$= 1.0 \times 10^{-12}\ M$$

This solution is basic because $[OH^-] > [H^+]$

(b) In this instance

$$[H^+] = \frac{1.0 \times 10^{-14}}{[OH^-]} = \frac{1.0 \times 10^{-14}}{2.0 \times 10^{-9}} = 5.0 \times 10^{-6}\ M$$

This solution is acidic because $[H^+] > [OH^-]$.

PRACTICE EXERCISE

Calculate the concentration of $OH^-(aq)$ in a solution in which **(a)** $[H^+] = 2 \times 10^{-6}\ M$; **(b)** $[H^+] = [OH^-]$; **(c)** $[H^+] = 10^2[OH^-]$. ***Answers:*** **(a)** $5 \times 10^{-9}\ M$; **(b)** $1 \times 10^{-7}\ M$; **(c)** $1 \times 10^{-8}\ M$

The pH Scale

Because the concentration of $H^+(aq)$ in an aqueous solution is often quite small, it can be conveniently expressed in terms of **pH**. The pH is defined as the negative logarithm in base 10 of the molar hydrogen-ion concentration:*

$$pH = -\log [H^+] \qquad [16.11]$$

(If you need to review exponential notation and the use of logs, see Appendix A.) As an example of the use of Equation 16.11, let's calculate the pH of a neutral solution at 25°C, that is, one in which $[H^+] = 1.0 \times 10^{-7}\ M$:

$$pH = -\log (1.0 \times 10^{-7}) = -(-7.00) = 7.00$$

Thus, the pH of a neutral solution is 7.00.

Because of the negative sign in Equation 16.11, the *pH decreases as $[H^+]$ increases.* Any acidic solution at 25°C ($[H^+]$ greater than $1.0 \times 10^{-7}\ M$) has a pH below 7.00, whereas a basic solution ($[H^+]$ less than $1.0 \times 10^{-7}\ M$) has a pH above 7.00:

Acidic solutions: pH value < 7.00
Neutral solutions: pH value $= 7.00$
Basic solutions: pH value > 7.00

The pH values characteristic of several familiar solutions are shown in Figure 16.5. Notice that a change in $[H^+]$ by a factor of 10 causes pH to change by 1. Consequently, a solution of pH 6 has 10 times the concentration of $H^+(aq)$ that a solution of pH 7 has.

SAMPLE EXERCISE 16.5

Calculate the pH values for the two solutions described in Sample Exercise 16.4.

Solution: **(a)** In the first instance we found $[H^+]$ to be $1.0 \times 10^{-12}\ M$. The pH of this solution is given by

$$pH = -\log (1.0 \times 10^{-12}) = -(-12.00) = 12.00$$

On most calculators, the calculation is performed with the following keystrokes:

[1] [EE] (or [EXP]) [+/−] [1] [2] [log] [+/−]

* Usually you will see pH defined as $-\log [H^+]$, occasionally as $-\log [H_3O^+]$. As discussed in Section 16.1, the same species is involved in all cases.

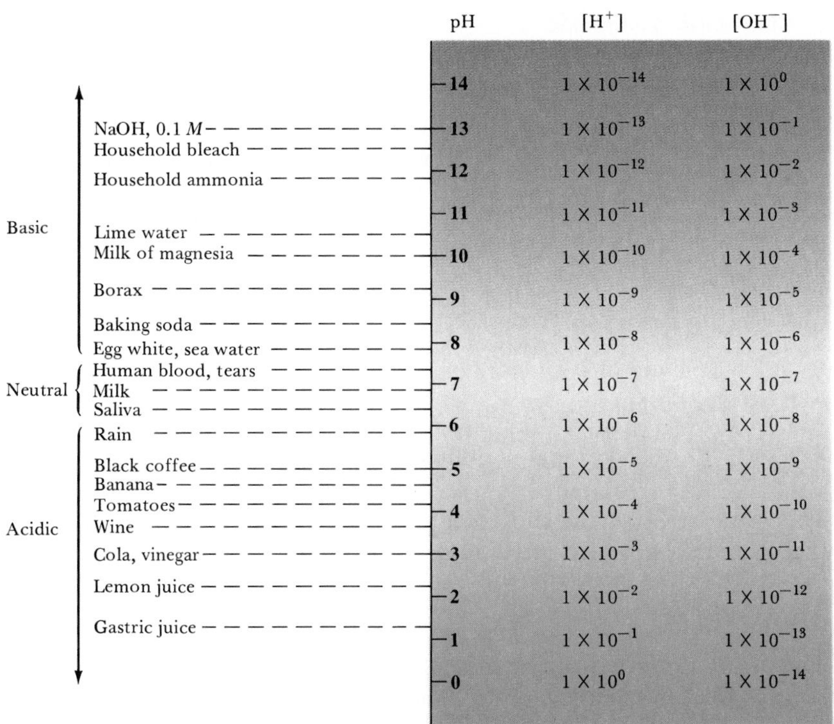

Figure 16.5 Values of pH for some more common solutions. The pH scale in this figure is shown to extend from 0 to 14, because nearly all solutions commonly encountered have pH values in that range. In principle, however, the pH values for strongly acidic solutions can be less than 0, and for strongly basic solutions can be greater than 14.

The rule for using significant figures with logs is that *the number of decimal places in the log equals the number of significant figures in the original number* (see Appendix A). Because **1.0** $\times$ 10^{-12} has two significant figures, the pH has two decimal places, 12.**00**.

(b) The pH of the second solution is given by

$$\text{pH} = -\log(5.0 \times 10^{-6}) = -(-5.30) = 5.30$$

PRACTICE EXERCISE

(a) In a sample of lemon juice, $[H^+]$ is 3.8×10^{-4} *M*. What is the pH? **(b)** A commonly available window-cleaning solution has a $[H^+]$ of 5.3×10^{-9} *M*. What is the pH? ***Answers:*** **(a)** 3.42; **(b)** 8.28

SAMPLE EXERCISE 16.6

A sample of freshly pressed apple juice has a pH of 3.76. Calculate $[H^+]$.

Solution: Because the pH is between 3 and 4, we know immediately that $[H^+]$ will be between 10^{-3} and 10^{-4} *M*. From the equation defining pH, we have

$$\text{pH} = -\log[H^+] = 3.76$$

Thus

$$\log[H^+] = -3.76$$

To find $[H^+]$ we need to determine the *antilog* of -3.76. Many calculators have an antilog function (sometimes labeled INV log), which allows us to perform the calculation with the following key sequence:

3	.	7	6	+/−	INV	log

The result, expressed in standard exponential notation, is 1.7×10^{-4} *M*. Some calculators rely on 10^x or y^x functions to find antilogs: antilog $(-3.76) = 10^{-3.76} = 1.7 \times 10^{-4}$. Consult the user's manual for your calculator to find out how to perform the antilog operation. Notice that the number of significant figures in $[H^+]$ equals the number of decimal places in the pH (2).

PRACTICE EXERCISE

A solution formed by dissolving an antacid tablet has a pH of 9.18. Calculate $[H^+]$.
Answer: $[H^+] = 6.6 \times 10^{-10}\ M$

The negative log is also a convenient way of expressing the magnitudes of other small quantities. We use the convention that the negative log of a quantity is labeled p(quantity). For example, one can express the concentration of OH^- as pOH:

$$pOH = -\log [OH^-]$$

By taking the log of both sides of Equation 16.10 and multiplying through by -1, we can obtain

$$pH + pOH = -\log K_w = 14.00 \qquad [16.12]$$

This expression is often convenient to use. We will see later (Section 16.7) that the pX notation is also useful in dealing with equilibrium constants.

Measuring pH

The pH of a solution can be measured quickly and accurately with a *pH meter* (Figure 16.6). A complete understanding of how these important devices work requires a knowledge of electrochemistry, a subject we take up in Chapter 20. In brief, a pH meter consists of a pair of electrodes connected to a meter capable of measuring small voltages, on the order of millivolts. A voltage, which varies with the pH, is generated when the electrodes are placed in a solution. This voltage is read by the meter, which is calibrated to give pH.

The electrodes used with pH meters come in many shapes and sizes, depending on their intended use. Electrodes have even been developed that are so small that they can be inserted into single living cells in order to monitor the pH of the cell medium. Pocket-sized pH meters are also available for use in environmental studies, in monitoring industrial effluents, and in agricultural work.

Although less precise, *acid-base indicators* are often used to measure pH. An acid-base indicator is a colored substance that can exist in either an acid

Figure 16.6 A digital pH meter. (© Richard Megna/Fundamental Photos)

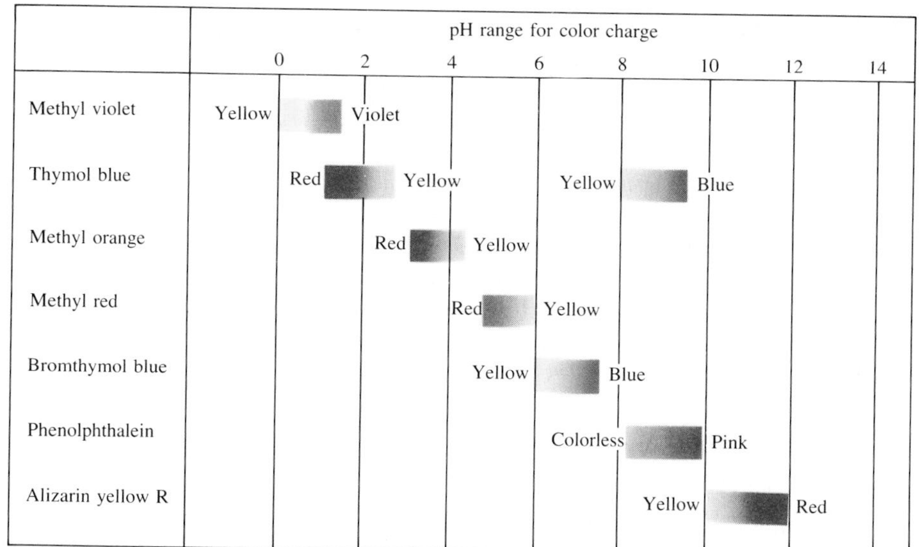

Figure 16.7 The pH ranges for the color changes of some common acid-base indicators. Most indicators have a useful range of about 2 pH units.

or a base form. The two forms are different colors. If you know the pH at which the indicator turns from one form to the other, you can determine whether a solution has a higher or lower pH than this value. For example, litmus, one of the most common indicators, changes color in the vicinity of pH 7. However, the color change is not very sharp. Red litmus indicates a pH of about 5 or lower, and blue litmus indicates a pH of about 8 or higher.

Some of the more common indicators are listed in Figure 16.7. We see from this figure that methyl orange, for example, changes color over the pH interval from 3.1 to 4.4. Below pH 3.1 it is in the acid form, which is red. In the interval between 3.1 and 4.4 it is gradually converted to its basic form, which has a yellow color. By pH 4.4 the conversion is complete, and the solution is yellow. Paper tape that is impregnated with various indicators and comes complete with a comparator color scale is widely used for approximate determinations of pH.

16.3 STRONG ACIDS

In the following sections, we will examine how to determine pH for solutions of acids and bases. The simplest case is that of a strong acid. Strong acids are strong electrolytes (Section 4.3); they react completely with water to form $H^+(aq)$, leaving no undissociated acid molecules in solution [Figure 16.8(*a*)]. In contrast,

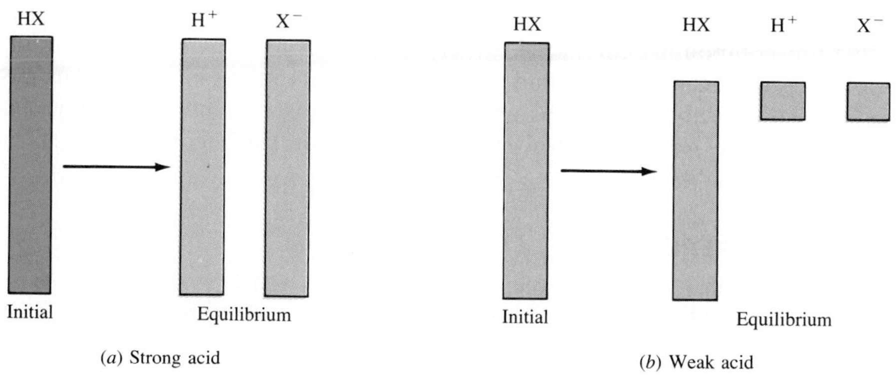

Figure 16.8 The extent of ionization of (*a*) a strong acid and (*b*) a weak acid.

weak acids are weak electrolytes; they ionize only partly. Indeed, undissociated acid molecules are usually the major constituent of aqueous solutions of weak acids [Figure 16.8(*b*)]. As we saw in Section 4.3, the number of strong acids is not very large: The six most common strong acids are HCl, HBr, HI, HNO_3, $HClO_4$, and H_2SO_4.

In an aqueous solution of a strong acid, the acid is normally the only significant source of H^+ ions. (If the concentration of the acid is 10^{-6} *M* or less, $[H^+]$ from the ionization of H_2O needs to be considered; otherwise the concentration of H^+ from H_2O is so small that it can be ignored.) Consequently, $[H^+]$ normally equals the original concentration of the acid. For example, in a 0.10 *M* aqueous solution of HNO_3, $[H^+] = 0.10$ *M* and $[NO_3^-] =$ 0.10 *M*; the concentration of HNO_3 is virtually zero.

SAMPLE EXERCISE 16.7

What is the pH of a 0.010 *M* solution of HCl?

Solution: The key is to recognize HCl as a strong acid. Because it is a strong acid, it is completely ionized, giving $[H^+] = 0.010$ *M* and $\text{pH} = -\log(0.010) = 2.00$.

PRACTICE EXERCISE

If an aqueous solution of HNO_3 has a pH of 2.50, what is the concentration of the acid? ***Answer:*** 3.2×10^{-3} *M*

16.4 WEAK ACIDS

Most substances that are acidic in water are actually weak acids. The extent to which an acid ionizes in an aqueous medium can be expressed by the equilibrium constant for the ionization reaction. In general, we can represent any acid by the symbol HX, where X^- is the formula for the conjugate base that remains when the proton ionizes. The ionization equilibrium is then given by Equation 16.13:

$$HX(aq) \rightleftharpoons H^+(aq) + X^-(aq) \qquad [16.13]$$

The corresponding equilibrium-constant expression is

$$K_a = \frac{[H^+][X^-]}{[HX]} \qquad [16.14]$$

The equilibrium constant is often given the symbol K_a and is called the **acid-dissociation constant**.

Table 16.1 shows the names, structures, and values of K_a for several weak acids. A more complete table is given in Appendix D. Note that many weak acids are compounds composed largely of carbon and hydrogen. Generally speaking, hydrogen atoms bound to carbon are not ionized in an aqueous medium. The ionizable hydrogens are in most instances bound to oxygen. *The smaller the value for K_a, the weaker the acid.* For example, phenol is the weakest acid listed in Table 16.1.

SAMPLE EXERCISE 16.8

A student prepared a 0.10 *M* solution of formic acid, $HCHO_2$, and measured its pH using a pH meter of the type illustrated in Figure 16.6. The pH at 25°C was found to be 2.38. **(a)** Calculate K_a for formic acid at this temperature. **(b)** What percentage of the acid is dissociated in this 0.10 *M* solution?

Table 16.1 Some Weak Acids in Water at 25°C[a]

Acid	Molecular formula	Structural formula[a]	Conjugate base	K_a
Hydrofluoric	HF	H—F	F^-	6.8×10^{-4}
Nitrous	HNO_2	H—O—N=O	NO_2^-	4.5×10^{-4}
Benzoic	$HC_7H_5O_2$	H—O—C(=O)—C₆H₅ (benzene ring)	$C_7H_5O_2^-$	6.5×10^{-5}
Acetic	$HC_2H_3O_2$	H—O—C(=O)—CH₂—H	$C_2H_3O_2^-$	1.8×10^{-5}
Hypochlorous acid	$HClO$	H—O—Cl	ClO^-	3.0×10^{-8}
Hydrocyanic	HCN	H—C≡N	$C{\equiv}N^-$	4.9×10^{-10}
Phenol	HOC_6H_5	H—O—(benzene ring)	$C_6H_5O^-$	1.3×10^{-10}

[a] The proton that ionizes is shown in blue.

Solution: **(a)** The first step in solving any equilibrium problem is to write the equation for the equilibrium reaction. The ionization equilibrium for formic acid can be written as follows:

$$HCHO_2(aq) \rightleftharpoons H^+(aq) + CHO_2^-(aq)$$

The equilibrium-constant expression for this equilibrium is

$$K_a = \frac{[H^+][CHO_2^-]}{[HCHO_2]}$$

From the measured pH we can calculate $[H^+]$:

$$pH = -\log [H^+] = 2.38$$
$$\log [H^+] = -2.38$$
$$[H^+] = 4.2 \times 10^{-3}\ M$$

We can do a little accounting to determine the concentrations of the species involved in the equilibrium. We first imagine that the solution consists of 0.10 M of undissociated $HCHO_2$. We then consider the dissociation of the acid into H^+ and CHO_2^-. For each H^+ produced in solution, one CHO_2^- anion also forms, and there is a loss of one $HCHO_2$ molecule:

	$HCHO_2(aq)$ $\rightleftharpoons$	$H^+(aq)$ +	$CHO_2^-(aq)$
Initial:	$0.10\ M$	0	0
Change:	$-4.2 \times 10^{-3}\ M$	$+4.2 \times 10^{-3}\ M$	$+4.2 \times 10^{-3}\ M$
Equilibrium:	$(0.10 - 4.2 \times 10^{-3})\ M$	$4.2 \times 10^{-3}\ M$	$4.2 \times 10^{-3}\ M$

Notice that the amount of $HCHO_2$ that dissociates is very small in comparison with the initial concentration of the acid. To the number of significant figures we are using, the subtraction yields just 0.10 M.

We can now insert the equilibrium concentrations into the expression for K_a:

$$K_a = \frac{(4.2 \times 10^{-3})(4.2 \times 10^{-3})}{0.10} = 1.8 \times 10^{-4}$$

(b) The percentage of acid that dissociates is given by the concentration of H^+ or CHO_2^- at equilibrium, divided by the initial acid concentration, times 100:

$$\text{Percent dissociation} = \frac{[H^+] \times 100}{[HCHO_2]} = \frac{(4.2 \times 10^{-3}) \times 100}{0.10} = 4.2 \text{ percent}$$

PRACTICE EXERCISE

Niacin, one of the B vitamins, has the following molecular structure:

(structure: pyridine ring with N, attached to C(=O)—O—H)

A 0.020 *M* solution of niacin has a pH of 3.26. What is the acid-dissociation constant, K_a, for the ionizable proton? ***Answer:*** 1.5×10^{-5}

Calculating pH for Solutions of Weak Acids

Using the value for K_a we can calculate the concentration of $H^+(aq)$ in a solution of a weak acid. The procedure is essentially the same as that for other equilibria, outlined in Section 15.5. For example, consider acetic acid, $HC_2H_3O_2$, the substance responsible for the characteristic odor and acidity of vinegar. Let's calculate the concentration of $H^+(aq)$ in a 0.10 *M* solution of acetic acid.

Our first step is to write the ionization equilibrium for acetic acid:

$$HC_2H_3O_2(aq) \rightleftharpoons H^+(aq) + C_2H_3O_2^-(aq) \qquad [16.15]$$

Note from the Lewis structure for acetic acid, shown in Table 16.1, that the hydrogen that ionizes is attached to the oxygen atom. We write this hydrogen separate from the others in the formula to emphasize that only this one hydrogen is readily ionized.

The second step is to write the equilibrium-constant expression and the value for the equilibrium constant, if that is known. From Table 16.1 we have $K_a = 1.8 \times 10^{-5}$. Thus, we can write the following:

$$K_a = \frac{[H^+][C_2H_3O_2^-]}{[HC_2H_3O_2]} = 1.8 \times 10^{-5} \qquad [16.16]$$

As the third step, we need to express the concentrations that make up the equilibrium-constant expression. This can be done with a little accounting, as described in Sample Exercise 16.8. Because we seek to find the equilibrium value for $[H^+]$, let us call this quantity x. The concentration of acetic acid before any of it dissociates is 0.10 *M*. The equation for the equilibrium tells us that for each molecule of $HC_2H_3O_2$ that dissociates, one $H^+(aq)$ and one $C_2H_3O_2^-(aq)$ are formed. Consequently, if x moles per liter of $H^+(aq)$ form at equilibrium, x moles per liter of $C_2H_3O_2^-(aq)$ must also form, and x moles per liter of $HC_2H_3O_2$ must be dissociated. This gives rise to the following equilibrium concentrations:

	$HC_2H_3O_2(aq)$ $\rightleftharpoons$	$H^+(aq)$ +	$C_2H_3O_2^-(aq)$
Initial:	0.10 M	0	0
Change:	$-x\ M$	$+x\ M$	$+x\ M$
Equilibrium:	$(0.10 - x)\ M$	$x\ M$	$x\ M$

As the fourth step of the problem, we need to substitute the equilibrium concentrations into the equilibrium-constant expression. The substitutions give the following equation:

$$K_a = \frac{[H^+][C_2H_3O_2^-]}{[HC_2H_3O_2]} = \frac{(x)(x)}{0.10 - x} = 1.8 \times 10^{-5} \qquad [16.17]$$

Because this equation has only one unknown, it can be solved using algebra. However, the solution is a little tedious because it requires use of the quadratic formula (Appendix A.3). By taking account of what is actually occurring in the solution, we can make things simpler for ourselves. Because the value of K_a is small, we might guess that x will be quite small. (In other words, only a small fraction of the $HC_2H_3O_2$ is actually ionized.) Indeed, if we solve the problem using the quadratic formula, we find that $x = 1.3 \times 10^{-3}\ M$. Now you know that if a small number is subtracted from a much larger one, the result is approximately equal to the larger number. In our example we have

$$0.10 - x = 0.10 - 0.0013 \simeq 0.10$$

We can therefore make the approximation of ignoring x relative to 0.10 in the denominator of Equation 16.17. This leads us to the following simplified expression:

$$K_a = \frac{(x)(x)}{0.10} = 1.8 \times 10^{-5}$$

Solving for x, we have

$$x^2 = (0.10)(1.8 \times 10^{-5}) = 1.8 \times 10^{-6}$$
$$x = \sqrt{1.8 \times 10^{-6}} = 1.3 \times 10^{-3}\ M = [H^+]$$

From the value calculated for x we see that our simplifying approximation is quite reasonable. This type of approximation can be used whenever conditions in a solution are such that only a small fraction of acid ionizes. As a general rule, if the quantity x, which is subtracted from the initial concentration of the acid, is more than about 5 percent of the initial value, it is better to use the quadratic formula. In cases of doubt, assume that the approximation is valid and solve for x in the simplified equation. Compare this approximate value of x with the initial concentration of acid. For example, if the initial concentration of acid is 0.050 M, and x turns out in a given case to be 0.0016 M, then

$$\left(\frac{0.0016\ M}{0.050\ M}\right)(100) = 3.2 \text{ percent}$$

If the result is more than about 5 percent, the problem should be reworked using the quadratic formula.

Finally, we note that the pH of this 0.10 M solution of $HC_2H_3O_2$ is $-\log(1.3 \times 10^{-3}) = 2.89$. By comparison, a 0.10 M solution of a strong acid such as HCl will have a pH of $-\log(0.10) = 1.00$.

SAMPLE EXERCISE 16.9

Calculate the pH of a 0.20 M solution of HCN. (Refer to Table 16.1 or Appendix D for the value of K_a.)

Solution: Proceeding as in the example worked out above, we write both the chemical equation for the dissociation reaction that forms $H^+(aq)$ and the equilibrium-constant (K_a) expression for the reaction:

$$HCN(aq) \rightleftharpoons H^+(aq) + CN^-(aq)$$

$$K_a = \frac{[H^+][CN^-]}{[HCN]} = 4.9 \times 10^{-10}$$

Next we tabulate the concentrations of the species involved in the equilibrium reaction, letting $x = [H^+]$:

	$HCN(aq)$ $\rightleftharpoons$	$H^+(aq)$ +	$CN^-(aq)$
Initial:	0.20 M	0	0
Change:	$-x$ M	$+x$ M	$+x$ M
Equilibrium:	$(0.20 - x)$ M	x M	x M

Substituting the equilibrium concentrations from our table into the equilibrium-constant expression yields

$$K_a = \frac{(x)(x)}{0.20 - x} = 4.9 \times 10^{-10}$$

We next make the simplifying approximation that x, the amount of acid that dissociates, is small in comparison with the initial concentration of acid; that is, $0.20 - x \simeq 0.20$. Thus,

$$\frac{x^2}{0.20} = 4.9 \times 10^{-10}$$

Solving for x, we have

$$x^2 = (0.20)(4.9 \times 10^{-10}) = 0.98 \times 10^{-10}$$

$$x = \sqrt{0.98 \times 10^{-10}} = 9.9 \times 10^{-6}\ M = [H^+]$$

Notice that 9.9×10^{-6} is much smaller than 5 percent of 0.20, the initial HCN concentration. Our simplifying approximation is therefore appropriate.

$$pH = -\log[H^+] = -\log(9.9 \times 10^{-6}) = 5.00$$

PRACTICE EXERCISE

The K_a for niacin (Practice Exercise 16.8) is 1.5×10^{-5}. What is the pH of a 0.010 M solution of niacin? ***Answer:*** 3.41

The result obtained in Sample Exercise 16.9 is typical of the behavior of weak acids; the concentration of $H^+(aq)$ is only a small fraction of the concentration of the acid in solution. Of course, those properties of the acid solution that relate directly to the concentration of $H^+(aq)$, such as electrical conductivity

(*a*)

(*b*)

Figure 16.9 Demonstration of the relative rates of reaction of two acid solutions of the same concentration with an active metal. (*a*) The flask on the left contains 1 *M* $HC_2H_3O_2$; the one on the right contains 1 *M* HCl. Each balloon contains the same amount of magnesium metal. (*b*) When the Mg metal is dumped into the acid, H_2 is formed. The rate of H_2 formation is clearly higher for the 1 *M* HCl solution on the right. (Donald Clegg and Roxy Wilson)

or rate of reaction with an active metal, are much less in evidence for a solution of a weak acid than for a solution of a strong acid. Figure 16.9 illustrates an experiment often carried out in the chemistry laboratory to demonstrate the difference in concentration of $H^+(aq)$ in weak and strong acid solutions of the same concentration. The rate of reaction with the metal is much faster for the solution of a strong acid. Reactions in which the rate depends on $[H^+]$ are common.

Figure 16.10 compares how the electrical conductivities of HCl and HF solutions vary with concentration. The conductivity of the strong acid increases approximately in proportion to its concentration. This is what you would expect. Because all the acid molecules ionize, the concentration of ions in solution is directly proportional to the concentration of acid. The conductivity of the weak acid is much less than that of the strong acid and does not vary linearly with the acid concentration. The nonlinearity of the graph arises from the fact that the percentage of acid ionized varies with the acid concentration. Indeed, the degree of ionization increases as the concentration of the acid decreases, as illustrated in Sample Exercise 16.10.

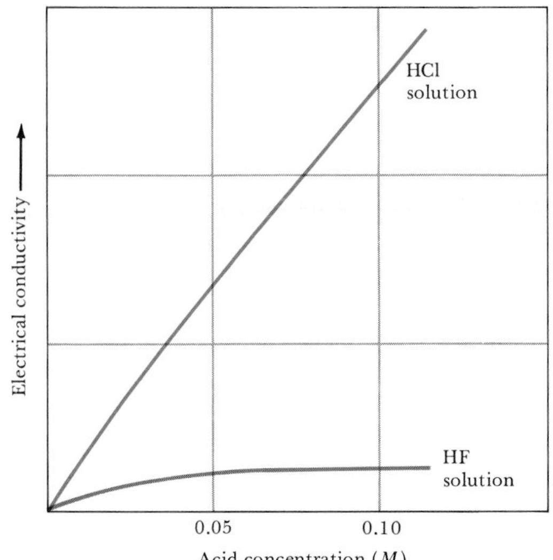

Figure 16.10 Electrical conductivity versus concentration for solutions of HCl, a strong acid, and HF, a weak acid.

SAMPLE EXERCISE 16.10

Calculate the percentage of HF molecules ionized in **(a)** a 0.10 M HF solution; **(b)** a 0.010 M HF solution.

Solution: **(a)** The equilibrium reaction and equilibrium concentrations are as follows:

$$HF(aq) \rightleftharpoons H^+(aq) + F^-(aq)$$

	HF(aq)	H⁺(aq)	F⁻(aq)
Initial:	0.10 M	0	0
Change:	$-x$ M	$+x$ M	$+x$ M
Equilibrium:	$(0.10 - x)$ M	x M	x M

The equilibrium-constant expression is:

$$K_a = \frac{[H^+][F^-]}{[HF]} = \frac{(x)(x)}{0.10 - x} = 6.8 \times 10^{-4}$$

When we try solving this equation using the approximation $0.10 - x = 0.10$ (that is, by neglecting the concentration of acid that ionizes in comparison with the initial concentration), we obtain $x = 8.2 \times 10^{-3}$ M. Because this value is greater than 5 percent of 0.10 M, we should work the problem without the approximation, using the quadratic formula. Rearranging our equation and writing it in standard quadratic form, we have

$$\begin{aligned} x^2 &= (0.10 - x)(6.8 \times 10^{-4}) \\ &= 6.8 \times 10^{-5} - (6.8 \times 10^{-4})x \\ x^2 + (6.8 \times 10^{-4})x - 6.8 \times 10^{-5} &= 0 \end{aligned}$$

Solution of this equation by use of the standard quadratic formula,

$$x = \frac{-b \pm \sqrt{b^2 - 4ac}}{2a}$$

gives

$$\begin{aligned} x &= \frac{-6.8 \times 10^{-4} \pm \sqrt{(6.8 \times 10^{-4})^2 + 4(6.8 \times 10^{-5})}}{2} \\ &= \frac{-6.8 \times 10^{-4} \pm 1.65 \times 10^{-2}}{2} \end{aligned}$$

Of the two solutions, only the one that gives a positive value for x is physically reasonable. Thus,

$$x = 7.9 \times 10^{-3}$$

From our result we can calculate the percent of molecules ionized:

$$\begin{aligned} \text{Percent ionized} &= \left(\frac{\text{concentration ionized}}{\text{original concentration}}\right)(100) \\ &= \left(\frac{7.9 \times 10^{-3}\ M}{0.10\ M}\right)(100) = 7.9\% \end{aligned}$$

(b) Proceeding similarly for the 0.010 M solution, we have

$$\frac{x^2}{0.010 - x} = 6.8 \times 10^{-4}$$

Solving the resultant quadratic expression, we obtain

$$x = [H^+] = [F^-] = 2.3 \times 10^{-3}\ M$$

The percentage of molecules ionized is

$$\left(\frac{0.0023}{0.010}\right)(100) = 23\%$$

Notice that in diluting the solution by a factor of 10, the percentage of molecules ionized increases by a factor of 3. We could have arrived at this conclusion qualitatively by applying Le Châtelier's principle (Section 15.6) to the equilibrium. There are more "particles" or reaction components on the right side of the equation than on the left. Dilution causes the reaction to shift in the direction of the larger number of particles, because this counters the effect of the decreasing concentration of particles.

PRACTICE EXERCISE

Calculate the percentage of niacin molecules ionized in **(a)** the solution in Practice Exercise 16.9; **(b)** a 1.0×10^{-3} M solution of niacin. ***Answers:*** **(a)** 3.9%; **(b)** 11.5%

Polyprotic Acids

Many substances are capable of furnishing more than one proton to water. Substances of this type are called **polyprotic acids**. For example, sulfurous acid, H_2SO_3, can react with water in two successive steps:

$$H_2SO_3(aq) \rightleftharpoons H^+(aq) + HSO_3^-(aq) \qquad K_a = 1.7 \times 10^{-2} \quad [16.18]$$

$$HSO_3^-(aq) \rightleftharpoons H^+(aq) + SO_3^{2-}(aq) \qquad K_a = 6.4 \times 10^{-8} \quad [16.19]$$

The values for K_a in each case show that the both H_2SO_3 and HSO_3^- are weak acids. The smaller value for K_a in the second reaction shows that loss of the second proton occurs much less readily than the first. This trend is intuitively reasonable; on the basis of electrostatic attractions we would expect the positively charged proton to be lost more readily from the neutral H_2SO_3 molecule than from the negatively charged HSO_3^- ion.

The successive acid-dissociation constants of polyprotic acids are sometimes labeled K_{a1}, K_{a2}, and so forth. This notation is often simplified to K_1, K_2, and so forth. For example, the equilibrium constant for the loss of a proton from HSO_3^- (Equation 16.19) can be labeled as K_{a2} or K_2, because this proton is the second one removed from the neutral acid, H_2SO_3. The acid-dissociation constants for a few common polyprotic acids are given in Table 16.2; a more complete list is provided in Appendix D. Notice that the K_a values for successive losses of protons from these acids usually differ by at least a factor of 10^3. The Lewis structures for citric and ascorbic acid are shown in the margin.

Citric acid

Ascorbic acid (vitamin C)

Because K_{a1} is so much larger than subsequent dissociation constants for these polyprotic acids, almost all the $H^+(aq)$ in the solution comes from the first ionization reaction. As long as successive K_a values differ by a factor of 10^3 or more, it is possible to obtain a satisfactory estimate of the pH of polyprotic acid solutions by considering only K_{a1}.

Table 16.2 Acid-Dissociation Constants of Some Common Polyprotic Acids

Name	Formula	K_{a1}	K_{a2}	K_{a3}
Ascorbic	$H_2C_6H_6O_6$	8.0×10^{-5}	1.6×10^{-12}	
Carbonic	H_2CO_3	4.3×10^{-7}	5.6×10^{-11}	
Citric	$H_3C_6H_5O_7$	7.4×10^{-4}	1.7×10^{-5}	4.0×10^{-7}
Oxalic	$H_2C_2O_4$	5.9×10^{-2}	6.4×10^{-5}	
Phosphoric	H_3PO_4	7.5×10^{-3}	6.2×10^{-8}	4.2×10^{-13}
Sulfurous	H_2SO_3	1.7×10^{-2}	6.4×10^{-8}	
Sulfuric	H_2SO_4	Large	1.2×10^{-2}	
Tartaric	$H_2C_4H_4O_6$	1.0×10^{-3}	4.6×10^{-5}	

SAMPLE EXERCISE 16.11

The solubility of CO_2 in pure water at 25°C and 0.1 atm pressure is 0.0037 *M*. The common practice is to assume that all of the dissolved CO_2 is in the form of carbonic acid, H_2CO_3, which is produced by reaction between the CO_2 and H_2O:

$$CO_2(aq) + H_2O(l) \rightleftharpoons H_2CO_3(aq)$$

What is the pH of a 0.0037 *M* solution of H_2CO_3?

Solution: H_2CO_3 is a polyprotic acid; the two acid dissociation constants, K_{a1} and K_{a2} (Table 16.2), differ by more than a factor of 10^3. Consequently, the pH can be determined by considering only K_{a1}, thereby treating the acid as if it were a monoprotic acid. Proceeding as in Sample Exercises 16.9 and 16.10, we can write the equilibrium reaction and equilibrium concentrations as follows:

$$H_2CO_3(aq) \rightleftharpoons H^+(aq) + HCO_3^-(aq)$$

	$H_2CO_3(aq)$	$H^+(aq)$	$HCO_3^-(aq)$
Initial:	$0.0037\ M$	0	0
Change:	$-x\ M$	$+x\ M$	$+x\ M$
Equilibrium:	$(0.0037 - x)\ M$	$x\ M$	$x\ M$

The equilibrium-constant expression is as follows:

$$K_{a1} = \frac{[H^+][HCO_3^-]}{[H_2CO_3]} = \frac{(x)(x)}{0.0037 - x} = 4.3 \times 10^{-7}$$

Because K_{a1} is small, we make the simplifying approximation that x is small so that $0.0037 - x \simeq 0.0037$. Thus

$$\frac{(x)(x)}{0.0037} = 4.3 \times 10^{-7}$$

Solving for x, we have

$$x^2 = (0.0037)(4.3 \times 10^{-7}) = 1.6 \times 10^{-9}$$

$$x = \sqrt{1.6 \times 10^{-9}} = 4.0 \times 10^{-5}\ M = [H^+] = [HCO_3^-]$$

The small value of x indicates that our simplifying assumption was justified. The pH is therefore

$$pH = -\log [H^+] = -\log (4.0 \times 10^{-5}) = 4.40$$

If we are asked to solve for $[CO_3^{2-}]$, we will need to use K_{a2}. Let's illustrate that calculation. Using the values of $[HCO_3^-]$ and $[H^+]$ calculated above, and setting $[CO_3^{2-}] = y$, we have the following initial and equilibrium concentration values:

$$HCO_3^-(aq) \rightleftharpoons H^+(aq) + CO_3^{2-}(aq)$$

	$HCO_3^-(aq)$	$H^+(aq)$	$CO_3^{2-}(aq)$
Initial:	$4.0 \times 10^{-5}\ M$	$4.0 \times 10^{-5}\ M$	0
Change:	$-y\ M$	$+y\ M$	$+y\ M$
Equilibrium:	$(4.0 \times 10^{-5} - y)\ M$	$(4.0 \times 10^{-5} + y)\ M$	$y\ M$

Assuming that y is small compared to 4.0×10^{-5}, we have

$$K_{a2} = \frac{[H^+][CO_3^{2-}]}{[HCO_3^-]} = \frac{(4.0 \times 10^{-5})(y)}{4.0 \times 10^{-5}} = 5.6 \times 10^{-11}$$

$$y = 5.6 \times 10^{-11}\ M = [CO_3^{2-}]$$

The value calculated for y is indeed very small in comparison to 4.0×10^{-5}, showing that our assumption was justified. It also shows that the ionization of HCO_3^- is negligible in comparison to that of H_2CO_3 as far as production of H^+ is concerned. However, it is the *only* source of CO_3^{2-}, which has a very low concentration in the solution.

Our calculations thus tell us that in a solution of carbon dioxide in water most of the CO_2 is in the form of CO_2 or H_2CO_3, a small fraction ionizes to form H^+ and HCO_3^-, and an even smaller fraction ionizes to give CO_3^{2-}.

PRACTICE EXERCISE

Calculate the pH and concentration of oxalate ion, $[C_2O_4^{2-}]$, in a 0.020 M solution of oxalic acid, $H_2C_2O_4$ (see Table 16.2). ***Answers:*** pH = 1.80; $[C_2O_4^{2-}] = 6.4 \times 10^{-5}$ M

16.5 STRONG BASES

The most common soluble strong bases are the hydroxides of group 1A and heavier group 2A metals, such as NaOH, KOH, and $Ca(OH)_2$ (Table 4.1). In solution, the strong bases are normally the only significant source of OH^-. For example, a 0.10 M solution of $Ca(OH)_2$ contains 0.10 M Ca^{2+} and 0.20 M OH^-. To find the pH of a basic solution when we know $[OH^-]$, we can first use Equation 16.10 to calculate $[H^+]$, as shown in Sample Exercise 16.12.

SAMPLE EXERCISE 16.12

What is the pH of a 0.010 M solution of $Ba(OH)_2$?

Solution: Because $Ba(OH)_2$ is a strong base, $[OH^-] = 0.020$ M. Using Equation 16.10, we have

$$[H^+] = \frac{1.00 \times 10^{-14}}{[OH^-]} = \frac{1.00 \times 10^{-14}}{0.020} = 5.0 \times 10^{-13}\ M$$

$$pH = -\log(5.0 \times 10^{-13}) = 12.30$$

PRACTICE EXERCISE

What is the concentration of a NaOH solution whose pH is 12.80?
Answer: 6.3×10^{-2} M

Although all the hydroxides of the alkali metals (family 1A) are strong electrolytes, LiOH, RbOH, and CsOH are not encountered commonly in the laboratory. The hydroxides of all of the alkaline earths (family 2A) except Be are also strong electrolytes. However, they have limited solubilities and consequently are used only when high solubility is not critical. The compound $Mg(OH)_2$ has an especially low solubility (9×10^{-3} g/L of water at 25°C). The least expensive and most common of the alkaline earth hydroxides is $Ca(OH)_2$, with a solubility of 0.97 g/L at 25°C.

Basic solutions are also created when substances react with water to form $OH^-(aq)$. The most common of these is the oxide ion. Ionic metal oxides, especially Na_2O and CaO, are often used in industry when a strong base is needed. Each mole of O^{2-} reacts with water to form 2 mol of OH^-, leaving virtually no O^{2-} remaining in the solution:

$$O^{2-}(aq) + H_2O(l) \longrightarrow 2OH^-(aq) \qquad [16.20]$$

Thus, a solution formed by dissolving 0.010 mol of $Na_2O(s)$ in a liter of water will have $[OH^-] = 0.020$ M and hence a pH of 12.30.

Ionic hydrides and nitrides also react with H_2O to form OH^-:

$$H^-(aq) + H_2O(l) \longrightarrow H_2(g) + OH^-(aq) \quad [16.21]$$

$$N^{3-}(aq) + 3H_2O(l) \longrightarrow NH_3(aq) + 3OH^-(aq) \quad [16.22]$$

Because the anions (O^{2-}, H^-, and N^{3-}) are stronger bases than OH^- (the conjugate base of H_2O), they are able to remove a proton from H_2O.

16.6 WEAK BASES

Many substances behave as weak bases in water. Such substances react with water, removing protons from H_2O, thereby forming the conjugate acid of the base and OH^- ions:

$$\text{Weak base} + H_2O \rightleftharpoons \text{conjugate acid} + OH^- \quad [16.23]$$

The most commonly encountered weak base is ammonia:

$$NH_3(aq) + H_2O(l) \rightleftharpoons NH_4^+(aq) + OH^-(aq) \quad [16.24]$$

The equilibrium-constant expression for this reaction can be written as

$$K = \frac{[NH_4^+][OH^-]}{[NH_3][H_2O]} \quad [16.25]$$

Because the concentration of water is essentially constant even when moderate concentrations of other substances are present, the $[H_2O]$ term is incorporated into the equilibrium constant, giving

$$K[H_2O] = K_b = \frac{[NH_4^+][OH^-]}{[NH_3]} \quad [16.26]$$

The constant K_b is called the **base-dissociation constant**, by analogy with the acid-dissociation constant, K_a, for weak acids. *The constant K_b always refers to the equilibrium in which a base reacts with H_2O to form the conjugate acid and OH^-.* Table 16.3 lists the names, formulas, Lewis structures, equilibrium reactions, and values of K_b for several weak bases in water. Appendix D includes a more extensive list. Notice that these bases contain one or more unshared pairs of electrons. An unshared pair is necessary to form the bond with H^+. Notice also that in the neutral molecules the unshared pairs are on nitrogen atoms and that the other bases are anions of weak acids.

SAMPLE EXERCISE 16.13

Calculate the concentration of OH^- in a 0.15 *M* solution of NH_3.

Solution: We use essentially the same procedure here as used in solving problems involving the dissociation of acids. The first step is to write the ionization reaction and the corresponding equilibrium-constant (K_b) expression:

$$NH_3(aq) + H_2O(l) \rightleftharpoons NH_4^+(aq) + OH^-(aq)$$

$$K_b = \frac{[NH_4^+][OH^-]}{[NH_3]} = 1.8 \times 10^{-5}$$

Table 16.3 Weak Bases and Their Aqueous Solution Equilibria

Base	Lewis structure	Conjugate acid	Equilibrium reaction	K_b
Ammonia (NH_3)	H—N̈—H with H below N	NH_4^+	$NH_3 + H_2O \rightleftharpoons NH_4^+ + OH^-$	1.8×10^{-5}
Pyridine (C_5H_5N)	six-membered ring with N:	$C_5H_5NH^+$	$C_5H_5N + H_2O \rightleftharpoons C_5H_5NH^+ + OH^-$	1.7×10^{-9}
Hydroxylamine (H_2NOH)	H—N̈—ÖH with H below N	H_3NOH^+	$H_2NOH + H_2O \rightleftharpoons H_3NOH^+ + OH^-$	1.1×10^{-8}
Methylamine (NH_2CH_3)	H—N̈—CH₃ with H below N	$NH_3CH_3^+$	$NH_2CH_3 + H_2O \rightleftharpoons NH_3CH_3^+ + OH^-$	4.4×10^{-4}
Nicotine ($C_{10}H_{14}N_2$)	pyridine ring (N:) joined to ring of H₂C, CH₂, CH₂, HC and N̈ bearing CH₃	$HC_{10}H_{14}N_2^+$	$C_{10}H_{14}N_2 + H_2O \rightleftharpoons C_{10}H_{14}N_2H^+ + OH^-$	7×10^{-7} 1.4×10^{-11}
Hydrosulfide ion (HS^-)	[H—S̈:]⁻	H_2S	$HS^- + H_2O \rightleftharpoons H_2S + OH^-$	1.8×10^{-7}
Carbonate ion (CO_3^{2-})	[C bonded to :Ö: (top), :Ö: and O (double bond)]²⁻	HCO_3^-	$CO_3^{2-} + H_2O \rightleftharpoons HCO_3^- + OH^-$	1.8×10^{-4}
Hypochlorite (ClO^-)	[:C̈l—Ö:]⁻	HClO	$ClO^- + H_2O \rightleftharpoons HClO + OH^-$	3.3×10^{-7}

We then tabulate the equilibrium concentrations involved in the equilibrium:

$$NH_3(aq) + H_2O(l) \rightleftharpoons NH_4^+(aq) + OH^-(aq)$$

	$NH_3(aq)$	$H_2O(l)$	$NH_4^+(aq)$	$OH^-(aq)$
Initial:	0.15 *M*	—	0	0
Change:	$-x$ *M*	—	$+x$ *M*	$+x$ *M*
Equilibrium:	$(0.15 - x)$ *M*	—	x *M*	x *M*

(Notice that we ignore the concentration of H_2O, because it is not involved in the equilibrium-constant expression.) Inserting these quantities into the equilibrium-constant expression gives the following:

$$K_b = \frac{[NH_4^+][OH^-]}{[NH_3]} = \frac{(x)(x)}{0.15 - x} = 1.8 \times 10^{-5}$$

Because K_b is small we can neglect the small amount of NH_3 that reacts with water, as compared to the total NH_3 concentration; that is, we can neglect x in comparison to 0.15 *M*. Then we have

$$\frac{x^2}{0.15} = 1.8 \times 10^{-5}$$

$$x^2 = (0.15)(1.8 \times 10^{-5}) = 0.27 \times 10^{-5}$$

$$x = \sqrt{2.7 \times 10^{-6}} = 1.6 \times 10^{-3}\ M = [OH^-]$$

Notice that the value obtained for x is only about 1 percent of the NH_3 concentration, 0.15 *M*. Therefore, our neglect of x in comparison with 0.15 is justified.

PRACTICE EXERCISE

Which of the following compounds should produce the highest pH as a 0.05 *M* solution: pyridine, methylamine, and nitrous acid? ***Answer:*** methylamine

Amines

The weak nitrogen bases listed in Table 16.3 belong to a family known as **amines**. These compounds can be thought of as being formed by replacing one or more of the N—H bonds in NH_3 with N—C bonds. (In hydroxylamine, H_2NOH, one of the N—H bonds of NH_3 has been replaced by an N—OH bond.) Like ammonia, such amines are able to extract a proton from the water molecule by forming an N—H bond. The following equation illustrates this behavior:

$$\mathrm{H{-}\ddot{N}(H){-}CH_3}(aq) + H_2O(l) \rightleftharpoons \left[\mathrm{H{-}N(H)(H){-}CH_3}\right]^+(aq) + OH^-(aq) \quad [16.27]$$

Anions of Weak Acids

A second common class of weak bases is composed of the anions of weak acids. Consider, for example, an aqueous solution of sodium acetate, $NaC_2H_3O_2$. This salt dissolves in water to give Na^+ and $C_2H_3O_2^-$ ions. The Na^+ ion is always a spectator ion in acid-base reactions. However, the $C_2H_3O_2^-$ ion is the conjugate base of a weak acid, acetic acid. Consequently, the $C_2H_3O_2^-$ ion is basic and reacts to a slight extent with water ($K_b = 5.6 \times 10^{-10}$):

$$C_2H_3O_2^-(aq) + H_2O(l) \rightleftharpoons HC_2H_3O_2(aq) + OH^-(aq) \quad [16.28]$$

SAMPLE EXERCISE 16.14

Calculate the pH of a 0.010 *M* solution of sodium hypochlorite, NaClO.

Solution: NaClO is an ionic compound consisting of Na^+ and ClO^- ions. As such it is a strong electrolyte. The hypochlorite ion, ClO^-, supplied by this salt is a weak base. The base dissociation constant for ClO^- is given in Table 16.3: $K_b = 3.3 \times 10^{-7}$. We can write the reaction between ClO^- and water, and the equilibrium concentrations present in this solution, as follows:

	$ClO^-(aq)$	+ $H_2O(l)$ $\rightleftharpoons$	$HClO(aq)$	+ $OH^-(aq)$
Initial:	0.010 *M*	—	0	0
Change:	$-x$ *M*	—	$+x$ *M*	$+x$ *M*
Equilibrium:	$(0.010 - x)$ *M*	—	x *M*	x *M*

CHEMISTRY AT WORK: Amines and Amine Hydrochlorides

Many amines with low molecular weights have unpleasant "fishy" odors. Amines and NH_3 are produced by the anaerobic (absence of O_2) decomposition of dead animal or plant matter. One such amine, $H_2N(CH_2)_5NH_2$, is known as cadaverine.

Many drugs, including quinine, codeine, caffeine, and amphetamine (Benzedrine), are amines. Like other amines, these substances are weak bases; the amine nitrogen is readily protonated by treatment with an acid. The resulting products are called *acid salts.* If we use A as the abbreviation for an amine, the acid salt formed by reaction with hydrochloric acid would be written as AH^+Cl^-. (It is sometimes written as $A \cdot HCl$ and referred to as a hydrochloride.) For example, amphetamine hydrochloride is the acid salt formed by treating amphetamine with HCl:

$$C_6H_5{-}CH_2{-}\underset{\displaystyle CH_3}{\underset{|}{CH}}{-}\ddot{N}H_2(aq) + HCl(aq) \longrightarrow$$

Amphetamine

$$C_6H_5{-}CH_2{-}\underset{\displaystyle CH_3}{\underset{|}{CH}}{-}NH_3{}^+Cl^-(aq)$$

Amphetamine hydrochloride

Such acid salts are less volatile, more stable, and generally more water-soluble than the corresponding neutral amines. Many drugs that are amines are sold and administered as acid salts. Some examples of over-the-counter medications that contain amine hydrochlorides as active ingredients are shown in Figure 16.11.

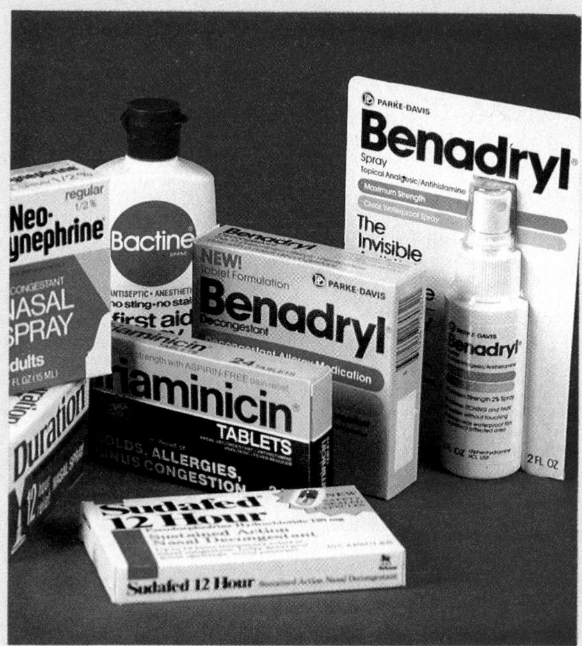

Figure 16.11 Over-the-counter medications in which an amine hydrochloride is a major active ingredient. (Donald Clegg and Roxy Wilson)

Because K_b is small, we anticipate that x will be small, so that $(0.010 - x) \simeq 0.010$. Using this approximation, we have

$$K_b = \frac{[HClO][OH^-]}{[ClO^-]} = \frac{(x)(x)}{0.010} = 3.3 \times 10^{-7}$$

Solving for x yields

$$x^2 = (0.010)(3.3 \times 10^{-7}) = 3.3 \times 10^{-9}$$

$$x = [OH^-] = \sqrt{3.3 \times 10^{-9}} = 5.7 \times 10^{-5}\ M$$

$[H^+]$ can be obtained using the ion-product constant for water:

$$[H^+] = \frac{1.0 \times 10^{-14}}{[OH^-]} = \frac{1.0 \times 10^{-14}}{5.7 \times 10^{-5}} = 1.8 \times 10^{-10}\ M$$

$$pH = -\log[H^+] = -\log(1.8 \times 10^{-10}) = 9.75$$

Thus we see that this solution of NaClO is slightly basic.

PRACTICE EXERCISE

K_b for BrO^- is 5.0×10^{-6}. Calculate the pH of a 0.050 M solution of NaBrO.
Answer: 10.70

16.7 RELATION BETWEEN K_a AND K_b

We've seen in a qualitative way that the stronger acids have the weaker conjugate bases. The fact that this qualitative relationship exists suggests that we might be able to find a quantitative relationship. Let's explore this matter by considering the NH_4^+ and NH_3 conjugate acid-base pair. Each of these species reacts with water:

$$NH_4^+(aq) \rightleftharpoons NH_3(aq) + H^+(aq) \qquad [16.29]$$

$$NH_3(aq) + H_2O(l) \rightleftharpoons NH_4^+(aq) + OH^-(aq) \qquad [16.30]$$

Each of these equilibria is expressed by a characteristic dissociation constant:

$$K_a = \frac{[NH_3][H^+]}{[NH_4^+]} \qquad K_b = \frac{[NH_4^+][OH^-]}{[NH_3]}$$

Now notice something very interesting and important. When Equations 16.29 and 16.30 are added together, the NH_4^+ and NH_3 species cancel, and we are left with just the autoionization of water:

$$\begin{array}{rcl} NH_4^+(aq) & \rightleftharpoons & NH_3(aq) + H^+(aq) \\ NH_3(aq) + H_2O(l) & \rightleftharpoons & NH_4^+(aq) + OH^-(aq) \\ \hline H_2O(l) & \rightleftharpoons & H^+(aq) + OH^-(aq) \end{array}$$

To determine what we should do about the equilibrium constants for the added reactions, we make use of a rule that can be derived from the general principles governing chemical equilibria: *When two reactions are added to give a third reaction, the equilibrium constant for the third reaction is given by the product of the equilibrium constants for the two added reactions.* In general, if

$$\text{reaction 1} + \text{reaction 2} = \text{reaction 3}$$

then

$$K_1 \times K_2 = K_3$$

Applying this to our present example, if we multiply K_a and K_b, we obtain the following result:

$$K_a \times K_b = \left(\frac{[\cancel{NH_3}][H^+]}{[\cancel{NH_4^+}]}\right)\left(\frac{[\cancel{NH_4^+}][OH^-]}{[\cancel{NH_3}]}\right)$$
$$= [H^+][OH^-] = K_w$$

Thus the result of multiplying K_a times K_b is just the ion-product constant, K_w (Equation 16.10). This is, of course, just what we would expect, because addition of Equations 16.29 and 16.30 gave us just the autoionization equilibrium for water, for which the equilibrium constant is K_w.

The relationship we have just found is so important that it should be called to special attention: *The product of the acid-dissociation constant for an acid and the base-dissociation constant for its conjugate base is the ion-product constant for water:*

$$K_a \times K_b = K_w \qquad [16.31]$$

As the strength of an acid increases (larger K_a), the strength of its conjugate base must decrease (smaller K_b), so that the product $K_a \times K_b$ remains equal to 1.0×10^{-14}. This point is illustrated by the data shown in Table 16.4.

Using Equation 16.31 we can calculate K_a for any weak acid if we know K_b for its conjugate base. Similarly, we can calculate K_b for a weak base if we know K_a for its conjugate acid. As a practical consequence, ionization constants are often listed for only one member of a conjugate acid-base pair. For example, Appendix D does not contain K_b values for the anions of weak acids because these can be readily calculated from the tabulated K_a values for their conjugate acids.

If you have the occasion to look up the values for acid- or base-dissociation constants in a chemistry handbook, you may find them expressed as pK_a or pK_b, that is, as $-\log K_a$ or $-\log K_b$ (Section 16.2). Equation 16.31 can be put in terms of pK_a and pK_b by taking the negative log of both sides:

$$pK_a + pK_b = pK_w = 14.00 \qquad [16.32]$$

This form is particularly useful when the tabulated pK value is that for the conjugate acid or base of the substance of interest. Often the dissociation constants for bases are tabulated as pK_a values for the corresponding conjugate acids. An an example, morphine, a nitrogen-containing base, is listed as the protonated cation, with $pK_a = 7.87$. This means that for the reaction

$$C_{17}H_{19}O_3NH^+(aq) \rightleftharpoons C_{17}H_{19}O_3N(aq) + H^+(aq)$$

the equilibrium constant, K_a, has the value $K_a = \text{antilog}\,(-7.87) = 10^{-7.87} = 1.3 \times 10^{-8}$. The reaction

$$C_{17}H_{19}O_3N(aq) + H_2O(l) \rightleftharpoons C_{17}H_{19}O_3NH^+(aq) + OH^-(aq)$$

is described by equilibrium constant K_b. Using Equation 16.32 and $pK_a = 7.87$, we have

$$pK_b = 14.00 - pK_a = 14.00 - 7.87 = 6.13$$

Thus, $K_b = \text{antilog}\,(-6.13) = 10^{-6.13} = 7.4 \times 10^{-7}$.

Table 16.4 Some Conjugate Acid-Base Pairs

Acid	K_a	Base	K_b
HNO_3	(Strong acid)	NO_3^-	(Negligible basicity)
HF	6.8×10^{-4}	F^-	1.5×10^{-11}
$HC_2H_3O_2$	1.8×10^{-5}	$C_2H_3O_2^-$	5.6×10^{-10}
H_2CO_3	4.3×10^{-7}	HCO_3^-	2.3×10^{-8}
NH_4^+	5.6×10^{-10}	NH_3	1.8×10^{-5}
HCO_3^-	5.6×10^{-11}	CO_3^{2-}	1.8×10^{-4}
OH^-	(Negligible acidity)	O^{2-}	(Strong base)

SAMPLE EXERCISE 16.15

Calculate **(a)** the base-dissociation constant, K_b, for the fluoride ion, F^-; **(b)** the acid-dissociation constant, K_a, for the ammonium ion, NH_4^+.

Solution: **(a)** K_b for F^- is not included in Table 16.3 or in Appendix D. However, K_a for its conjugate acid, HF, is given in Table 16.2 and Appendix D as $K_a = 6.8 \times 10^{-4}$. We can therefore use Equation 16.31 to calculate K_b:

$$K_b = \frac{K_w}{K_a} = \frac{1.0 \times 10^{-14}}{6.8 \times 10^{-4}} = 1.5 \times 10^{-11}$$

(b) K_b for NH_3 is listed in Table 16.3 and in Appendix D as $K_b = 1.8 \times 10^{-5}$. Using Equation 16.31, we can calculate K_a for the conjugate acid, NH_4^+:

$$K_a = \frac{K_w}{K_b} = \frac{1.0 \times 10^{-14}}{1.8 \times 10^{-5}} = 5.6 \times 10^{-10}$$

PRACTICE EXERCISE

(a) Which of the following anions has the largest base-dissociation constant: NO_2^-, PO_4^{3-}; N_3^-? **(b)** The base quinoline has the structure

Its conjugate acid is listed in handbooks as having a pK_a of 4.90. What is the base-dissociation constant for quinoline? ***Answers:*** **(a)** PO_4^{3-}; **(b)** 7.9×10^{-10}

16.8 ACID-BASE PROPERTIES OF SALT SOLUTIONS

Even before you began this chapter, you were undoubtedly aware of many substances that are acidic, such as HNO_3, HCl, and H_2SO_4, and others that are basic, such as NaOH and NH_3. However, our recent discussions have indicated that ions can also exhibit acidic or basic properties. For example, we calculated K_a for NH_4^+ and K_b for F^- in Sample Exercise 16.15. Such behavior implies that salt solutions can be acidic or basic. Before proceeding with further discussions of acids and bases, let's summarize some features of salts that should bring their acid and base properties into sharper focus.

We can assume that when salts dissolve in water they are completely ionized; nearly all salts are strong electrolytes. Consequently, the acid-base properties of salt solutions are due to the behavior of their cations and anions. Many ions are able to react with water to generate $H^+(aq)$ or $OH^-(aq)$. This type of reaction is often called **hydrolysis.**

The anions of weak acids, HX, are basic, and consequently they react with water to produce OH^- ions:

$$X^-(aq) + H_2O(l) \rightleftharpoons HX(aq) + OH^-(aq) \qquad [16.33]$$

In contrast, the anions of strong acids, such as the NO_3^- ion, exhibit no significant basicity; these ions do not hydrolyze and consequently do not influence pH.

Anions of polyprotic acids, such as HCO_3^-, that still have ionizable protons are capable of acting as either proton donors or proton acceptors (that is, either acids or bases). Their behavior toward water will be determined by the relative magnitudes of K_a and K_b for the ion, as shown in Sample Exercise 16.16.

SAMPLE EXERCISE 16.16

Predict whether the salt Na_2HPO_4 will form an acidic or basic solution on dissolving in water.

Solution: The two possible reactions that HPO_4^{2-} may undergo on addition to water are

$$HPO_4^{2-}(aq) \rightleftharpoons H^+(aq) + PO_4^{3-}(aq) \qquad [16.34]$$

$$HPO_4^{2-}(aq) + H_2O \rightleftharpoons H_2PO_4^-(aq) + OH^-(aq) \qquad [16.35]$$

Depending on which of these has the larger equilibrium constant, the ion will cause the solution to be acidic or basic. The value of K_a for reaction Equation 16.34, as shown in Table 16.2, is 4.2×10^{-13}. We must calculate the value of K_b for reaction Equation 16.35 from the value of K_a for the conjugate acid formed, $H_2PO_4^-$. We make use of the relationship shown in Equation 16.31.

$$K_a \times K_b = K_w$$

We want to know K_b for the base HPO_4^{2-}, knowing the value of K_a for the conjugate acid $H_2PO_4^-$:

$$K_b(HPO_4^{2-}) \times K_a(H_2PO_4^-) = K_w = 1.0 \times 10^{-14}$$

Because K_a for $H_2PO_4^-$ is 6.2×10^{-8} (Table 16.2), we calculate K_b for HPO_4^{2-} to be 1.6×10^{-7}. This is considerably larger than K_a for HPO_4^{2-}; thus the reaction shown in Equation 16.35 predominates over that in Equation 16.34 and the solution is basic.

PRACTICE EXERCISE

Predict whether the dipotassium salt of citric acid, $K_2HC_6H_5O_7$, will form an acidic or basic solution in water (see Table 16.2 for data). ***Answer:*** acidic

All cations except those of the alkali metals and the heavier alkaline earths (Ca^{2+}, Sr^{2+}, and Ba^{2+}) act as weak acids in water solution. Because the alkali metal and alkaline earth cations do not hydrolyze in water, the presence of any of these ions in solution does not influence pH. It may surprise you that metal ions, such as Al^{3+}, and the transition metal ions form weakly acidic solutions. We can take this observation for now as a point of fact. The reasons for this behavior are discussed in Section 16.10.

Among the cations that produce an acidic solution is, of course, NH_4^+, which is the conjugate acid of the base NH_3. The NH_4^+ ion dissociates in water as follows:

$$NH_4^+(aq) \rightleftharpoons H^+(aq) + NH_3(aq) \qquad [16.36]$$

The pH of a solution of a salt can be qualitatively predicted by considering the cation and anion of which the salt is composed. A convenient way to do this is to consider the relative strengths of the acids and bases from which the salt is derived.*

1. *Salt derived from a strong base and a strong acid:* Examples are NaCl and $Ca(NO_3)_2$, which are derived from NaOH and HCl and from $Ca(OH)_2$ and HNO_3, respectively. Neither cation nor anion hydrolyzes. The solution has a pH of 7.

* These rules apply to what can be called *normal salts.* These salts are ones that contain no ionizable protons on the anion. The pH of an acid salt (such as $NaHCO_3$ or NaH_2PO_4) is affected not only by the hydrolysis of the anion but also by its acid dissociation, as shown in Sample Exercise 16.16.

2. *Salt derived from a strong base and a weak acid:* In this case the anion is a relatively strong conjugate base. Examples are $NaClO$ and $Ba(C_2H_3O_2)_2$. The anion hydrolyzes to produce $OH^-(aq)$ ions. The solution has a pH above 7.
3. *Salt derived from a weak base and a strong acid:* In this case the cation is a relatively strong conjugate acid. Examples are NH_4Cl and $Al(NO_3)_3$. The cation hydrolyzes to produce $H^+(aq)$ ions. The solution has a pH below 7.
4. *Salt derived from a weak base and a weak acid:* Examples are $NH_4C_2H_3O_2$, NH_4CN, and $FeCO_3$. Both cation and anion hydrolyze. The pH of the solution depends upon the extent to which each ion hydrolyzes. The pH of a solution of NH_4CN is greater than 7 because CN^- ($K_b = 2.0 \times 10^{-5}$) is more basic than NH_4^+ ($K_a = 5.6 \times 10^{-10}$) is acidic. Consequently, CN^- hydrolyzes to a greater extent than NH_4^+ does.

Figure 16.12 demonstrates the influence of several salts on pH.

SAMPLE EXERCISE 16.17

List the following solutions in order of increasing pH: (i) 0.1 *M* $Co(ClO_4)_2$; (ii) 0.1 *M* RbCN; (iii) 0.1 *M* $Sr(NO_3)_2$; (iv) 0.1 *M* $KC_2H_3O_2$.

Solution: The most acidic solution will be (i), which has a metal ion that undergoes hydrolysis and an anion derived from a strong acid. Solution (iii) should have pH of about 7, because it is derived from an alkaline earth cation and the anion of a strong acid. Solutions (ii) and (iv) are both derived from an alkali metal ion, which does not undergo hydrolysis, and an anion of a weak acid. Anion hydrolysis should lead to a basic solution in both cases, but solution (ii) will be more strongly basic because CN^- is a stronger base than is $C_2H_3O_2^-$. We see that the order of pH is 0.1 *M* $Co(ClO_4)_2$ < 0.1 *M* $Sr(NO_3)_2$ < 0.1 *M* $KC_2H_3O_2$ < 0.1 *M* RbCN.

Figure 16.12 Depending on the ions involved, salt solutions can be neutral, acidic, or basic. These three solutions contain the acid-base indicator bromthymol blue. (*a*) A NaCl solution is neutral; (*b*) a NH_4Cl solution is acidic; (*c*) a NaClO solution is basic. (© Richard Megna/Fundamental Photographs)

(*a*)

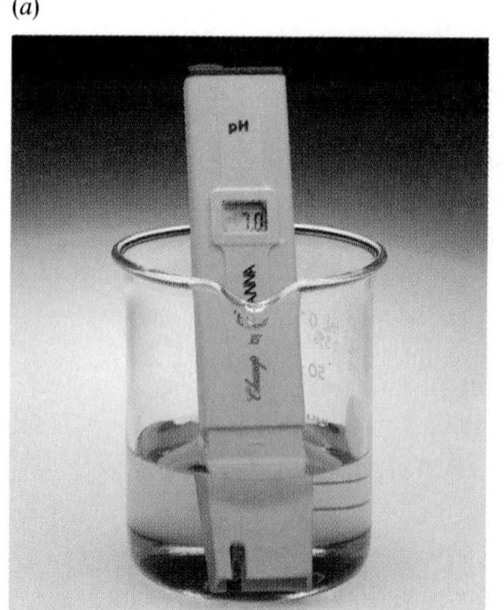

(*b*)

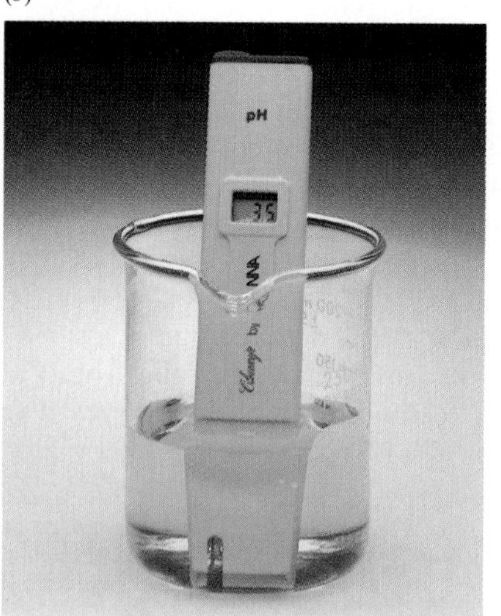

(*c*)

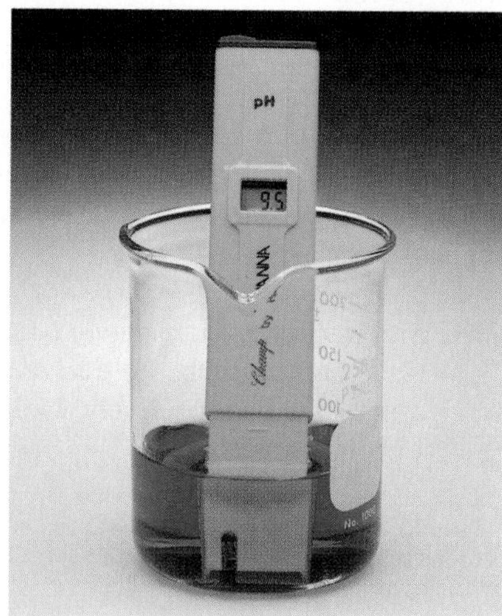

PRACTICE EXERCISE

In each case below, indicate which salt will form the more acidic (or less basic) 0.010 M solution: **(a)** $NaNO_3$, $Fe(NO_3)_3$; **(b)** KBr, KBrO; **(c)** CH_3NH_3Cl, $BaCl_2$; **(d)** NH_4NO_2, NH_4NO_3. *Answers:* **(a)** $Fe(NO_3)_3$; **(b)** KBr; **(c)** CH_3NH_3Cl; **(d)** NH_4NO_3

16.9 ACID-BASE CHARACTER AND CHEMICAL STRUCTURE

From our discussions to this point, we have seen that when a substance is dissolved in water, it may behave as an acid, behave as a base, or exhibit no acid-base properties (neutral). It would be helpful to have further guidelines on how the composition and structure of a molecule relate to its acid or base character.

Effect of Bond Polarity and Bond Strength

Any molecule containing H can potentially act as an acid. However, a substance HX will transfer a proton only if the H—X bond is already polarized in the following way:

$$\overset{+\!\!\longrightarrow}{\text{H—X}}$$

In ionic hydrides such as NaH, the H atom possesses a negative charge and behaves as a proton acceptor (Section 16.5). Nonpolar H—X bonds, such as the H—C bond in CH_4, produce neither acidic nor basic solutions.

There is also a second factor that helps determine whether a molecule containing an H—X bond will donate a proton: the strength of the bond. Very strong bonds are less easily ionized than are weaker ones. This factor is of importance, for example, in the case of the hydrogen halides. The H—F bond is the most polar H—X bond. You therefore might expect that HF would be a very strong acid, if the first rule were all that mattered. However, the energy required to dissociate HF into H and F atoms is much higher than it is for the other hydrogen halides, as shown in Table 8.4. As a result, HF is a weak acid, whereas all the other hydrogen halides are strong acids in water.

The factors we have just considered can be used to relate the acid-base properties of the hydride of an element to its position in the periodic table. In any horizontal row of the table, the most basic hydrides are on the left, the most acidic hydrides on the right. For example, in the second row of the table, NaH is a basic hydride. On addition to water it reacts to form $OH^-(aq)$, as described by Equation 16.21. On the right-hand side of the row, the acidity increases in the order $PH_3 < H_2S < HCl$. This general trend is related to the increasing electronegativity of the element as we move from left to right in any horizontal row. In general, *metal hydrides are either basic or show no pronounced acid-base properties in water*, whereas *nonmetal hydrides range from acidic to showing no pronounced acid-base properties.*

In any vertical row of nonmetallic elements there is a tendency toward increasing acidity with increasing atomic number. For example, among the group 6A elements the acid dissociation constants vary in the order $H_2O < H_2S < H_2Se < H_2Te$. This order arises primarily because the bond strengths steadily decrease in this series as the central atom grows larger and the overlaps of atomic orbitals grow smaller. Figure 16.13 illustrates the application of these general correlations to the nonmetal hydrides of periods 2 and 3.

Figure 16.13 Acid-base properties of the nonmetal hydrides (relative to water) of periods 2 and 3.

	GROUP			
	4A	5A	6A	7A
Period 2	CH_4 No acid or base properties	NH_3 Weak base	H_2O —	HF Weak acid
Period 3	SiH_4 No acid or base properties	PH_3 Weak base	H_2S Weak acid	HCl Strong acid

Oxyacids

Many common acids contain one or more O—H bonds. For example, sulfuric acid contains two such bonds:

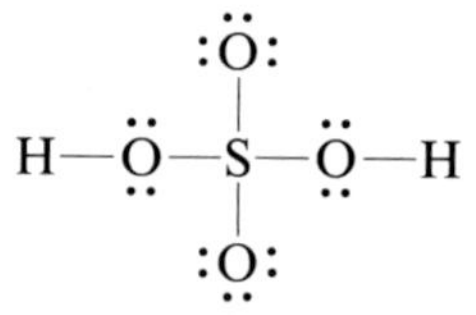

$$\text{H—}\ddot{\text{O}}\text{—S—}\ddot{\text{O}}\text{—H with :}\ddot{\text{O}}\text{: above and below S}$$

Acids in which OH groups and possibly additional oxygen atoms are bound to a central atom are called **oxyacids**. We have seen that the OH group is also present in bases. What factors determine whether an OH group will behave as a base or as an acid? Let's consider an OH group bound to some atom Y, which might in turn have other groups attached to it:

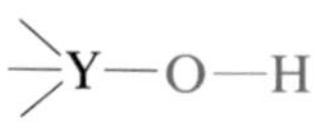

$$\rangle\text{Y—O—H}$$

At one extreme, Y might be a metal, such as Na, K, or Mg. The pair of electrons shared between Y and O is then completely transferred to oxygen, and an ionic compound involving OH^- is formed. Because of the charge that surrounds it, the oxygen of the OH^- ion does not strongly attract to itself the electron pair it shares with hydrogen. That is, the O—H bond in OH^- is not strongly polarized. Therefore, the hydrogen of OH^- has no tendency to transfer to the solvent as $H^+(aq)$. Such compounds therefore behave as bases.

When Y is an element of intermediate electronegativity, around 2.0, the bond to O is more covalent in character, and the substance does not readily lose OH^-. Elements with electronegativities in this range include B, C, P, As, and I (Figure 8.10). Examples of acids of such elements include orthoboric acid, hypoiodous acid, and methanol, the structures of which are shown in Table 16.5.

Such substances might behave as acids in water, depending on the ease with which the proton is lost from oxygen. As a general rule, the more strongly the group Y attracts the electron pair it shares with the oxygen, the more polar the OH bond will be and the more acidic the substance. In the three examples just given, the central atom does not strongly attract the electron pair it shares with oxygen. As a result, the acid-dissociation constants are small, as shown in Table 16.5.

Table 16.5 Acid-Dissociation Constants for Orthoboric Acid, Hypoiodous Acid, and Methanol

Acid	K_a
H—O—B(—O—H)—O—H Orthoboric acid	6.5×10^{-10}
I—O—H Hypoiodous acid	2.3×10^{-11}
CH_3—O—H Methanol	Not measurable

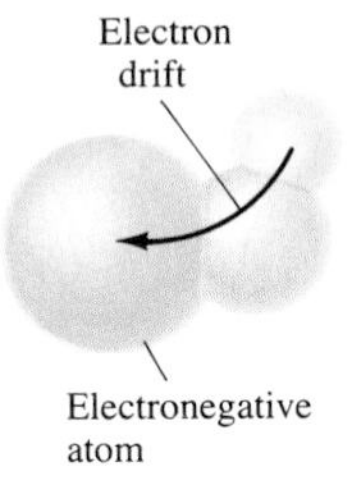

Electron drift

Electronegative atom

Figure 16.14 The greater the electronegativity of the atom attached to an O—H group, the more the electrons in the O—H bond are pulled toward it. This electron drift further polarizes and weakens the O—H bond, resulting in a stronger acid.

As the electronegativity of Y increases, or as groups with greater electron-attracting ability are placed on Y, the acidic properties of the substance increase (Figure 16.14). Two simple rules relate the acid strengths of oxyacids to the electronegativity of Y and to the number of groups attached to Y:

1. *For acids that have the same structure, acid strength generally increases with increasing electronegativity of the central atom, Y.* Examples are shown in Table 16.6.
2. *In a series of acids that have the same central atom, Y, but differ in the number of attached groups, the acid strength increases with increasing oxidation number of the central atom.* For example, in the series of oxyacids of chlorine extending from hypochlorous to perchloric acid, acid strength steadily increases:

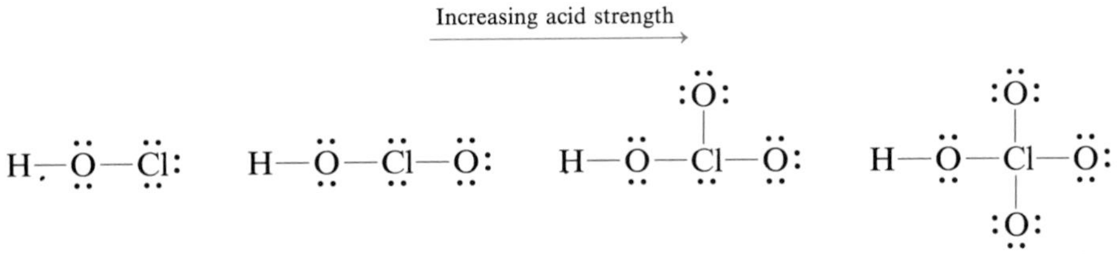

Increasing acid strength →

H—Ö—C̈l:	H—Ö—C̈l—Ö:	H—Ö—Cl(=:Ö:)—Ö:	H—Ö—Cl(:Ö:)(:Ö:)—Ö:	*Acid*
Hypochlorous +1	Chlorous +3	Chloric +5	Perchloric +7	

Chlorine oxidation numbers

Table 16.6 Acid-Dissociation Constants (K_a) of Oxyacids in Comparison with Electronegativity Values (EN) of Atom Y

H—O—Y	K_a	EN of Y	H—O—Y(=O)—O—H	K_{a1}	EN of Y
HClO	3×10^{-8}	3.0	H_2SO_3	1.7×10^{-2}	2.5
HBrO	2×10^{-9}	2.8	H_2SeO_3	3.5×10^{-3}	2.4
HIO	2×10^{-11}	2.5	H_2CO_3	4.3×10^{-7}	2.5
$HOCH_3$	~0	2.5[a]			

[a] This value is the electronegativity for carbon.

In this series the ability of chlorine to withdraw electrons from the OH group—and thus make the O—H bond even more polar—increases as electron-withdrawing oxygen atoms are added to the chlorine. It is interesting to note that this correlation can be stated in another, but equivalent, way: In a series of oxyacids, the acidity increases with the number of nonprotonated oxygens bound to the central atom.

SAMPLE EXERCISE 16.18

Arrange the compounds in each of the following series in order of increasing acid strength: **(a)** AsH_3, HI, NaH, H_2O; **(b)** H_2SeO_3, H_2SeO_4, H_2O.

Solution: **(a)** We have seen that the elements from the left side of the periodic table are the most basic hydrides because the hydrogen in these compounds carries a negative charge. Thus NaH should be the most basic hydride on the list. Because arsenic is a less electronegative element than oxygen, we might expect that AsH_3 would be a weak base toward water. That is also what we would predict by an extension of the trends shown in Figure 16.13. Further, we expect that the hydrides of the halogens, as the most electronegative element in each period, will be acidic relative to water. Finally, we know that HI is one of the strong acids in water. Thus the order of increasing acidity is $NaH < AsH_3 < H_2O < HI$.

(b) The rule for oxyacids is that the acidity increases with increasing oxidation state on the central atom. H_2SeO_4, as the acid representing the highest oxidation state of Se, should be a comparatively strong acid, much like H_2SO_4. Because it is the product of the reaction of a nonmetal oxide with water ($SeO_2 + H_2O \longrightarrow H_2SeO_3$), we expect that H_2SeO_3 will be acidic in water. That is, H_2SeO_3 is a stronger acid than H_2O. However, it should be a weaker acid than H_2SeO_4 because Se is in a lower oxidation state in H_2SeO_3. Thus the order of increasing acidity is $H_2O < H_2SeO_3 < H_2SeO_4$.

PRACTICE EXERCISE

In each of the following pairs, choose the compound that leads to the more acidic (or less basic) solution: **(a)** HBr, HF; **(b)** PH_3, H_2S; **(c)** HNO_2, HNO_3; **(d)** H_2SO_3, H_4SiO_4. ***Answers:*** **(a)** HBr; **(b)** H_2S; **(c)** HNO_3; **(d)** H_2SO_3

Carboxylic Acids

Another large group of acids is illustrated by acetic acid. The structure of acetic acid, whose formula we have usually written as $HC_2H_3O_2$, is as follows:

```
      H  O
      |  ||
   H—C—C—O—H
      |
      H
```

The portion of the structure shown in blue is present in many other acids and is called the *carboxyl group*. Acids containing this group are called **carboxylic acids**. Formic acid and benzoic acid, whose structures are shown below, are further examples. The acidic proton in carboxylic acids is in an O—H bond attached to carbon; the polarity of the O—H bond is due to the presence of the nonprotonated O atom that is also attached to the same carbon.

```
     O
     ||
  H—C—O—H
```

Formic acid

```
         O
         ||
  C6H5—C—O—H
```

Benzoic acid

16.10 LEWIS ACIDS AND BASES

For a substance to be a proton acceptor (that is, a base in the Brønsted-Lowry sense), that substance must possess an unshared pair of electrons for binding the proton. For example, we have seen that NH_3 acts as a proton acceptor. Using Lewis structures we can write the reaction between H^+ and NH_3 as follows:

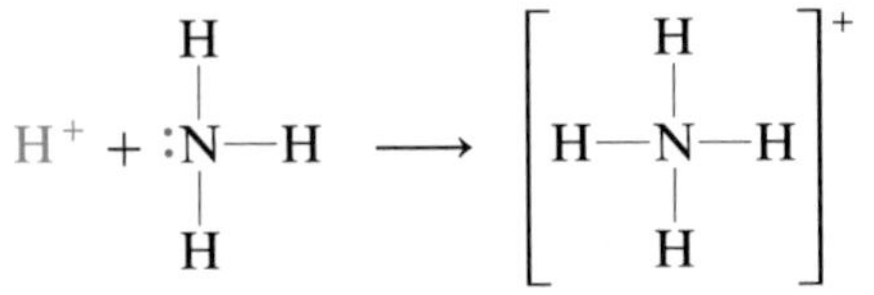

$$H^+ + :\overset{\displaystyle H}{\underset{\displaystyle H}{\overset{|}{\underset{|}{N}}}}-H \longrightarrow \left[H-\overset{\displaystyle H}{\underset{\displaystyle H}{\overset{|}{\underset{|}{N}}}}-H \right]^+ \qquad [16.37]$$

G. N. Lewis was the first to notice this aspect of acid-base reactions. He proposed a definition of acid and base that emphasizes the shared electron pair: A **Lewis acid** is defined as an electron-pair acceptor, and a **Lewis base** as an electron-pair donor.

Every base that we have discussed thus far—whether it be OH^-, H_2O, an amine, or an anion—is an electron-pair donor. Everything that is a base in the Brønsted-Lowry sense (a proton acceptor) is also a base in the Lewis sense (an electron-pair donor). However, in the Lewis theory, a base can donate its electron pair to something other than H^+. The Lewis definition therefore greatly increases the number of species that can be considered acids; H^+ is a Lewis acid, but not the only one. For example, consider the reaction between NH_3 and BF_3. This reaction occurs because BF_3 has a vacant orbital in its valence shell (Section 8.7). It therefore acts as an electron-pair acceptor (a Lewis acid) toward NH_3, which donates the electron pair:

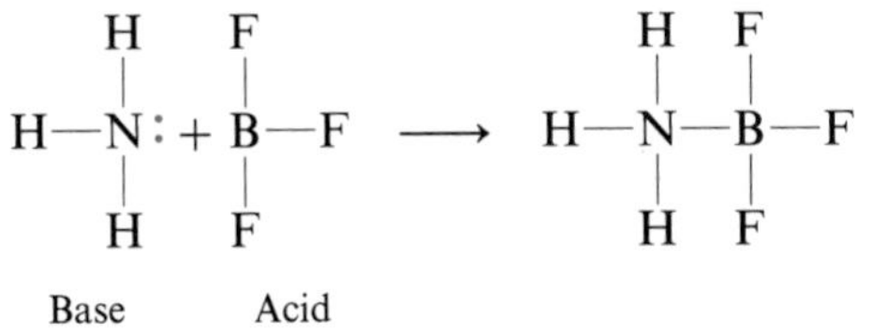

$$\underset{\text{Base}}{H-\overset{\displaystyle H}{\underset{\displaystyle H}{\overset{|}{\underset{|}{N}}}}:} + \underset{\text{Acid}}{\overset{\displaystyle F}{\underset{\displaystyle F}{\overset{|}{\underset{|}{B}}}}-F} \longrightarrow H-\overset{\displaystyle H}{\underset{\displaystyle H}{\overset{|}{\underset{|}{N}}}}-\overset{\displaystyle F}{\underset{\displaystyle F}{\overset{|}{\underset{|}{B}}}}-F \qquad [16.38]$$

Our emphasis throughout this chapter has been on water as the solvent and on the proton as the source of acidic properties. In such cases, we find the Brønsted-Lowry definition of acids and bases to be the most useful. In fact, when we speak of a substance as being acidic or basic, we are usually thinking of aqueous solutions and using these terms in the Arrhenius or Brønsted-Lowry sense. The advantage of the Lewis theory is that it allows us to treat a wider variety of reactions, including those that do not involve proton transfer, as acid-base reactions. To avoid confusion, a substance like BF_3 is rarely called an acid unless it is clear from the context that we are using the term in the sense of the Lewis definition. Instead, substances that function as electron-pair acceptors are referred to explicitly as "Lewis acids."

Lewis acids include molecules that, like BF_3, have an incomplete octet of electrons. In addition, many simple cations can function as Lewis acids. For example, Fe^{3+} interacts strongly with cyanide ions to form the ferricyanide ion, $Fe(CN)_6{}^{3-}$:

$$Fe^{3+} + 6:C{\equiv}N:^- \longrightarrow [Fe(C{\equiv}N:)_6]^{3-}$$

The Fe^{3+} ion has vacant orbitals which accept the electron pairs donated by the CN^- ions; we will learn more in Chapter 25 about just which orbitals are used by the Fe^{3+} ion. That the metal ion is highly charged also contributes to the interaction with CN^- ions.

```
H        :Ö:
  \ ..  ↗ ||
   O: ⤳  C
  /      ||
H        :O:
```

Some compounds with multiple bonds can behave as Lewis acids. For example, the reaction of carbon dioxide with water to form carbonic acid, H_2CO_3, can be pictured as an attack by a water molecule on CO_2, in which the water acts as an electron-pair donor and the CO_2 as an electron-pair acceptor, as shown in the margin. The electron pair of one of the carbon-oxygen double bonds is moved onto the oxygen, leaving a vacant orbital on the carbon that can act as an electron-pair acceptor. We have shown the shift of these electrons with arrows. After forming the initial acid-base product, a proton moves from one oxygen to another, thereby forming carbonic acid:

```
     H⤳:Ö:              H—Ö:
     |   |                   |
  H—O—C      ⟶      H—Ö—C
     ..  ||               ..  ||
        :O:                  :O:
```

A similar kind of Lewis acid-base reaction takes place when any oxide of a nonmetal dissolves in water to form an acidic solution.

Hydrolysis of Metal Ions

The Lewis concept is also helpful in explaining why solutions of many metal ions show acidic properties (Section 16.8). For example, a solution of $Cr(NO_3)_3$ is quite acidic. A solution of $ZnCl_2$ is also acidic, though to a lesser extent. To understand why this is so, we must examine the interaction between a metal ion and water molecules.

Because metal ions are positively charged, they attract the unshared electron pairs of water molecules. It is primarily this interaction, referred to as *hydration*, that causes salts to dissolve in water, as explained in Section 13.2. The strength of attraction increases with the charge of the ion and is strongest for the smallest ions. The ratio of ionic charge to ionic radius provides a good measure of the extent of hydration. This ratio is listed in Table 16.7 for a

Table 16.7 Ionic Charge/Ionic Radius Ratio and Acid Hydrolysis Constant for Metal Ions of Various Charges

Metal ion	Ionic charge / Ionic radius	Acid hydrolysis constant, K_a
Na^+	1.0	Negligible
Li^+	1.5	2×10^{-14}
Ca^{2+}	2.1	2×10^{-13}
Mg^{2+}	3.1	4×10^{-12}
Zn^{2+}	2.7	1×10^{-9}
Cu^{2+}	2.8	1×10^{-8}
Al^{3+}	6.7	1×10^{-5}
Cr^{3+}	4.8	1×10^{-4}
Fe^{3+}	4.7	2×10^{-3}

selection of metal ions. The process of hydration is a Lewis acid-base interaction in which the metal ion acts as a Lewis acid and the water molecules as Lewis bases. When the water molecule interacts with the positively charged metal ion, electron density is drawn from the oxygen, as illustrated in Figure 16.15. This flow of electron density causes the O—H bond to become more polarized; as a result, water molecules bound to the metal ion, M, are more acidic than those in the bulk solvent. The hydrated metal ion thus acts as a source of protons:

$$M(H_2O)_n^{z+} \rightleftharpoons M(H_2O)_{(n-1)}(OH)^{(z-1)+} + H^+(aq) \qquad [16.39]$$

In this equation z is the charge on the metal ion, and n is the number of hydrating water molecules. For the 3+ metal ions listed in Table 16.7, n is 6; for the other ions it is probably closer to 4, although the exact number is difficult to determine. The hydrolysis reaction shown in Equation 16.39 represents the behavior of an acid in just the same way as Equation 16.13, which applies to HX. The "acid" in Equation 16.39 is not just a single molecule, but a collection of molecules. For example it might be $Fe(H_2O)_6^{3+}$, which we usually represent merely as $Fe^{3+}(aq)$. The acid hydrolysis constants (that is, the equilibrium constants for Equation 16.39) are listed in Table 16.7 for several ions whose charge/ionic radius ratios are also given. Notice that there is a general trend toward larger acid hydrolysis constants as the charge/ionic radius ratio increases. The tendency to transfer a proton to water is greatest for the smallest and most highly charged ions, as demonstrated in Figure 16.16. Note that as the charge on the metal ion increases, the solution becomes increasingly more acidic.

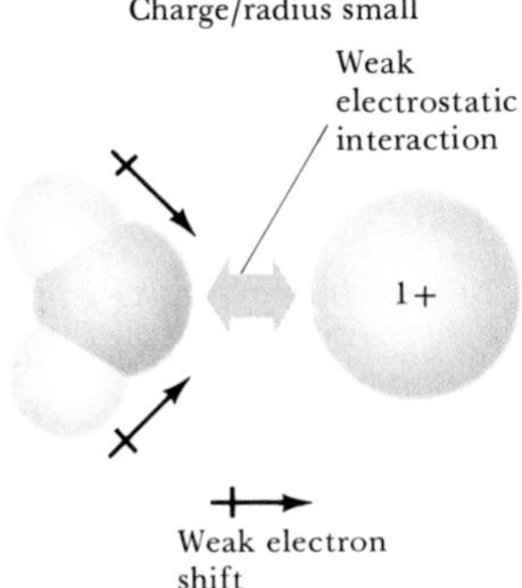

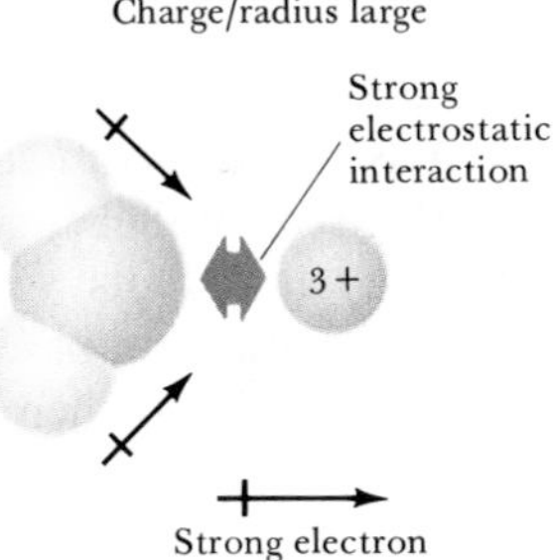

Figure 16.15 Interaction of a water molecule with a cation of 1+ charge or 3+ charge. The interaction is much stronger with the smaller ion of higher charge.

Figure 16.16 The pH values of 1.0 M solutions of a series of nitrate salts, as estimated using acid-base indicators. From left to right, $NaNO_3$, $Ca(NO_3)_2$, $Zn(NO_3)_2$, and $Al(NO_3)_3$. (Donald Clegg and Roxy Wilson)

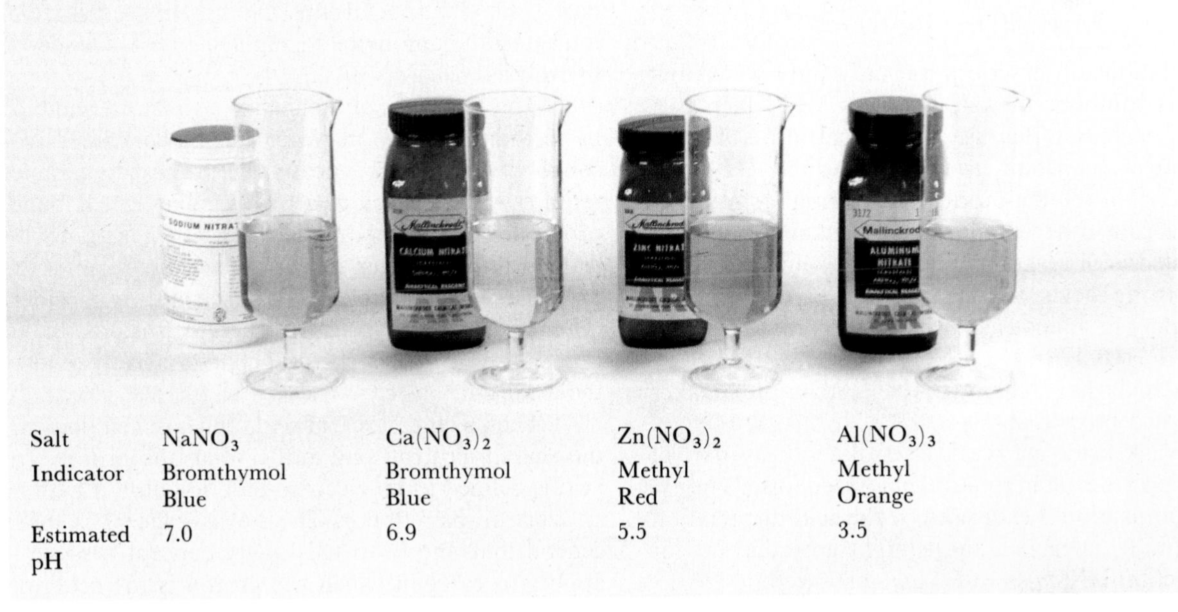

Salt	$NaNO_3$	$Ca(NO_3)_2$	$Zn(NO_3)_2$	$Al(NO_3)_3$
Indicator	Bromthymol Blue	Bromthymol Blue	Methyl Red	Methyl Orange
Estimated pH	7.0	6.9	5.5	3.5

FOR REVIEW

SUMMARY

In this chapter we have considered the general properties of acidic and basic solutions, with emphasis on water as a solvent. We have seen that acidic solutions are those that contain more $H^+(aq)$ than $OH^-(aq)$. The hydrogen ion is strongly bound to water; for this reason the hydronium ion, $H_3O^+(aq)$, is often used to represent the predominant form of H^+ in water. Basic solutions contain more $OH^-(aq)$ than $H^+(aq)$.

The Brønsted-Lowry concept of acids and bases emphasizes the transfer of protons (H^+) from acids to bases. A Brønsted acid is a substance that donates a proton (H^+) to another substance; a Brønsted base is a substance that accepts a proton from another substance. The conjugate base of a Brønsted acid is the species that remains when a proton is removed from the acid. Together, the acid and its conjugate base are called a conjugate acid-base pair. The acid-base strengths of conjugate acid-base pairs are related: The stronger an acid, the weaker its conjugate base; the weaker an acid, the stronger its conjugate base.

Water can function as either a Brønsted acid or a Brønsted base, depending on the circumstances. Water also spontaneously ionizes to a slight degree (autoionization), forming $H^+(aq)$ and $OH^-(aq)$. The extent of ionization is expressed by the ion-product constant for water:

$$K_w = [H^+][OH^-] = 1.0 \times 10^{-14}$$

This relationship describes not only pure water, but aqueous solutions as well. Thus, as $[H^+]$ increases, $[OH^-]$ decreases. Because the concentration of water is effectively constant in dilute solutions, $[H_2O]$ is omitted from the ion-product expression for water as well as from other equilibrium-constant expressions associated with reactions in aqueous solution.

Strong acids are strong electrolytes, ionizing completely in aqueous solution. The common strong acids are HCl, HBr, HI, HNO_3, $HClO_4$, and H_2SO_4. Strong acids have conjugate bases that are weaker proton acceptors than H_2O.

Weak acids are weak electrolytes; only part of the molecules exist in solution in ionized form. The extent of ionization is expressed by the acid-dissociation constant, K_a, which is the equilibrium constant for the reaction $HX(aq) \rightleftharpoons H^+(aq) + X^-(aq)$.

Polyprotic acids, such as H_2SO_3, have more than one ionizable proton. These acids have more than one acid-dissociation constant: K_{a1}, K_{a2}, and so forth. These constants decrease in magnitude in the order $K_{a1} > K_{a2} > K_{a3}$.

Basic solutions are produced by ionic hydroxides such as NaOH and by substances that react with water to increase $[OH^-]$. Strong bases have conjugate acids that are no stronger as proton donors than H_2O. Common strong bases are the hydroxides and oxides of the alkali metals and alkaline earths.

Weak bases include NH_3, amines, and the anions of weak acids. The extent to which a weak base reacts with water to generate OH^- and the conjugate acid of the base is measured by the base-dissociation constant, K_b. This is the equilibrium constant for the reaction $B(aq) + H_2O(l) \rightleftharpoons HB^+(aq) + OH^-(aq)$, where B is the base.

The stronger an acid, the weaker its conjugate base; the weaker an acid, the stronger its conjugate base. This qualitative observation is expressed quantitatively by $K_a \times K_b = K_w$ (where K_a and K_b are dissociation constants for conjugate acid-base pairs).

The acid-base properties of salts can be ascribed to the behavior of their respective cations and anions. The reaction of ions with water with a resultant change in pH is called hydrolysis. The cations of strong bases (the alkali metal ions and alkaline earth metal ions) and the anions of strong acids do not undergo hydrolysis.

The tendency of a substance to show acidic or basic characteristics in water can be correlated reasonably well with chemical structure. Acid character requires the presence of a highly polar H—X bond, promoting loss of hydrogen as H^+ on reaction with water. Basic character, on the other hand, requires the presence of an available pair of electrons. By considering the effects of changes in structure, it is possible to predict how a given structural change is likely to alter the acidity or basicity.

The Lewis concept of acids and bases emphasizes the shared electron pair rather than the proton. An acid is defined as an electron-pair acceptor, a base as an electron-pair donor. The Lewis concept is more general than the Brønsted-Lowry concept because it applies to cases in which the proton is the acid and to others as well.

KEY TERMS

hydronium ion (Sec. 16.1)
Brønsted acid (Sec. 16.1)
Brønsted base (Sec. 16.1)
conjugate acid-base pair (Sec. 16.1)
conjugate base (Sec. 16.1)
conjugate acid (Sec. 16.1)
autoionization (Sec. 16.2)
ion-product constant (Sec. 16.2)
pH (Sec. 16.2)
acid-dissociation constant (Sec. 16.4)
polyprotic acids (Sec. 16.4)
base-dissociation constant (Sec. 16.6)
amines (Sec. 16.6)
hydrolysis (Sec. 16.8)
oxyacids (Sec. 16.9)
Lewis acid (Sec. 16.10)
Lewis base (Sec. 16.10)

EXERCISES

Nature of Acids and Bases

16.1 Although HCl and H_2SO_4 have very different properties as pure substances, their aqueous solutions possess many common properties. List some general properties of these solutions, and explain their common behavior in terms of the species present.

16.2 Although pure NaOH and CaO have very different properties, their aqueous solutions possess many common properties. List some general properties of these solutions, and explain their common behavior in terms of the species present.

16.3 Hydrogen bromide is very soluble in water. Its solutions are good conductors of electricity. Describe the nature of the interaction between HBr and water to account for these observations.

16.4 Compare the dissolution of HCl in water with the dissolution of NaOH in water. In what respects do the processes differ? In what respects are they similar?

16.5 Give the conjugate acid of each of the following Brønsted bases: (a) NH_3; (b) HCO_3^-; (c) CO_3^{2-}; **(d)** OH^-; **(e)** $H_2PO_4^-$.

16.6 Give the conjugate base of each of the following Brønsted acids: **(a)** H_2SO_3; **(b)** $HC_2H_3O_2$; **(c)** HS^-; **(d)** NH_4^+; **(e)** PH_3.

16.7 Identify the acid, the base, the conjugate acid, and the conjugate base in each of the following reactions:

(a) $NH_2^-(aq) + H_2O(l) \longrightarrow NH_3(aq) + OH^-(aq)$

(b) $H_2C_2O_4(aq) + H_2O(l) \rightleftharpoons HC_2O_4^-(aq) + H_3O^+(aq)$

(c) $H^+(aq) + HPO_4^{2-}(aq) \longrightarrow H_2PO_4^-(aq)$

16.8 Identify the acid, the base, the conjugate acid, and the conjugate base in each of the following equilibria:

(a) $HC_2O_4^-(aq) + CO_3^{2-}(aq) \rightleftharpoons C_2O_4^{2-}(aq) + HCO_3^-(aq)$

(b) $PH_4^+(aq) + H_2O(aq) \rightleftharpoons PH_3(aq) + H_3O^+(aq)$

(c) $NH_4^+(aq) + CN^-(aq) \rightleftharpoons NH_3(aq) + HCN(aq)$

16.9 Which of the following would you expect to be the stronger Brønsted base: (a) CN^- or Cl^-; **(b)** NO_3^- or CO_3^{2-}? Briefly explain your choice.

16.10 Which of the following would you expect to be the stronger Brønsted acid: **(a)** NH_3 or H_2O; **(b)** $HClO_4$ or HF? Briefly explain your choice.

Autoionization of Water and the pH Scale

16.11 Indicate whether each solution is acidic, basic, or neutral: **(a)** 0.005 M H^+; **(b)** 8.5×10^{-7} M H^+; **(c)** 0.0003 M OH^-; **(d)** 2.0×10^{-8} M OH^-; **(e)** pH = 5.1; **(f)** pOH = 3.5.

16.12 Indicate whether each solution is acidic, basic, or neutral: **(a)** 2.0×10^{-10} M H^+; **(b)** 1.0×10^{-7} M OH^-; **(c)** 0.0007 M H^+; **(d)** 5.0×10^{-5} M OH^-; **(e)** pH = 8.9; **(f)** pOH = 6.5.

16.13 Calculate the pH of each solution: **(a)** 2.5×10^{-4} M H^+; **(b)** $[H^+] = 0.0073$ M; **(c)** 3.3×10^{-5} M OH^-; **(d)** pOH = 5.88.

16.14 Calculate the pH of each solution: **(a)** 8.9×10^{-4} M H^+; **(b)** $[H^+] = 3.5 \times 10^{-5}$ M; **(c)** 0.0075 M OH^-; **(d)** pOH = 7.50.

16.15 Calculate $[H^+]$ for solutions with the following pH values: (a) 1.25; **(b)** 0.5; **(c)** 11.9.

16.16 Calculate $[H^+]$ for each solution: **(a)** pH = 2.8; **(b)** pOH = 5.53; **(c)** $[OH^-] = 3.65 \times 10^{-4}$ M.

16.17 At normal body temperature, 37°C, $K_w = 2.4 \times 10^{-14}$. Calculate $[H^+]$, $[OH^-]$, pH, and pOH for a neutral aqueous solution at this temperature.

16.18 Deuterium oxide (D_2O, where D = deuterium, the hydrogen-2 nuclide) has an ion-product constant, K_w, of 8.9×10^{-16} at 20°C. Calculate $[D^+]$, $[OD^-]$, pD, and pOD for pure (neutral) D_2O at this temperature.

16.19 By what factor does $[H^+]$ change for a pH change of (a) 2.00 units; **(b)** 0.50 units?

16.20 Can the pH of a solution be zero or negative? If so, give examples to illustrate these values.

Solutions of Acids

16.21 What ions or molecules (besides H_2O) are present in each of the following solutions? Indicate which are the major species: **(a)** 0.10 M $HClO_4$; **(b)** 0.25 M H_2SO_4; **(c)** 0.15 M $HC_2H_3O_2$; **(d)** 1.0 M HBr.

16.22 What ions or molecules (besides H_2O) are present in each of the following solutions? Indicate which are the major species: **(a)** 0.10 *M* HI; **(b)** 1.5 *M* HNO_3; **(c)** 0.20 *M* HCN; **(d)** 0.5 *M* $HC_7H_5O_2$.

16.23 Relying on your knowledge of strong acids and using Table 16.1, indicate which of the following is the strongest and which is the weakest acid: HNO_2; $HClO_4$; HClO; $HC_2H_3O_2$.

16.24 Using Table 16.1 and your knowledge of strong acids, select the strongest and the weakest acid in the following list: HF; HCN; $HC_7H_5O_2$; HBr.

16.25 Calculate the pH of each of the following strong acid solutions: **(a)** 0.025 *M* HNO_3; **(b)** 0.824 g of $HClO_4$ in 0.500 L of solution; **(c)** 5.00 mL of 1.5 *M* HCl diluted to 100 mL; **(d)** a mixture formed by adding 10.0 mL of 0.020 *M* HCl to 30.0 mL of 0.010 *M* HI.

16.26 Calculate the pH of each of the following strong acid solutions: **(a)** 1.8×10^{-4} *M* HBr; **(b)** 1.02 g of HNO_3 in 250 mL of solution; **(c)** 2.00 mL of 0.500 *M* $HClO_4$ diluted to 50.0 mL; **(d)** a solution formed by mixing 10.0 mL of 0.0100 *M* HBr with 20.0 mL of 2.50×10^{-3} *M* HCl.

16.27 Lactic acid, $HC_3H_5O_3$, has one acidic hydrogen. A 0.10 *M* solution of lactic acid has a pH of 2.44. Calculate K_a.

16.28 A 0.050 *M* solution of $KHCrO_4$ has a pH of 3.80. Calculate K_a for $HCrO_4^-$.

16.29 Calculate the concentration of $H^+(aq)$ in each of the following solutions (K_a values are given in Appendix D): **(a)** 0.025 *M* hypochlorous acid, HClO; **(b)** 0.120 *M* hydrazoic acid, HN_3; **(c)** 0.0068 *M* phenol, HOC_6H_5.

16.30 Calculate the pH of each of the following solutions (K_a values are given in Appendix D): **(a)** 0.010 *M* propionic acid; **(b)** 0.035 *M* hypoiodous acid; **(c)** 0.040 *M* benzoic acid.

16.31 Calculate the percent ionization of hydrazoic acid in solutions of each of the following concentrations (K_a given in Appendix D): **(a)** 0.400 *M*; **(b)** 0.100 *M*; **(c)** 0.0400 *M*.

16.32 Calculate the percent ionization of $HCrO_4^-$ in solutions of each of the following concentrations (K_a given in Appendix D): **(a)** 0.250 *M*; **(b)** 0.0800 *M*; **(c)** 0.0200 *M*.

16.33 A 0.200 *M* solution of a weak acid HX is 9.4 percent ionized. Using this information, calculate $[H^+]$, $[X^-]$, [HX], and K_a for HX.

16.34 A 0.100 *M* solution of bromoacetic acid, $CH_2BrCOOH$, is 13.2 percent ionized. Using this information, calculate $[CH_2BrCOO^-]$, $[H^+]$, $[CH_2BrCOOH]$, and K_a for bromoacetic acid.

[16.35] Show that for a weak acid the percent ionization should vary as the inverse square root of the acid concentration.

[16.36] For solutions of a weak acid a graph of pH versus the log of the initial acid concentration should be a straight line. What is the magnitude of the slope of that line?

[16.37] Citric acid is present in citrus fruits. As indicated in Table 16.3, it is a triprotic acid. Calculate the pH of a 0.050 *M* solution of citric acid. Explain any approximations or assumptions that you make in your calculations.

[16.38] Tartaric acid is found in many plants (grapes, for example). It is partly responsible for the dry texture of certain wines. Calculate the pH of a 0.025 *M* solution of tartaric acid, for which the acid dissociation constants are listed in Table 16.3. Explain any approximations or assumptions that you make in your calculation.

Solutions of Bases; K_a-K_b Relationship

16.39 Calculate $[OH^-]$ and pH for each of the following strong base solutions: **(a)** 0.050 *M* KOH; **(b)** 2.33 g of NaOH in 500 mL of solution; **(c)** 10.0 mL of 0.150 *M* $Ca(OH)_2$ diluted to 500 mL; **(d)** a solution formed by mixing 10.0 mL of 0.015 *M* $Ba(OH)_2$ with 30.0 mL of 6.8×10^{-3} *M* NaOH.

16.40 Calculate $[OH^-]$ and pH for **(a)** 3.5×10^{-4} *M* $Sr(OH)_2$; **(b)** 1.50 g of LiOH in 250 mL of solution; **(c)** 1.00 mL of 0.095 *M* NaOH diluted to 2.00 L; **(d)** a solution formed by adding 5.00 mL of 0.0105 *M* KOH to 15.0 mL of 3.5×10^{-3} *M* $Ca(OH)_2$.

16.41 Write the balanced net ionic equation for the reaction of each of the following bases with water. Also write the base-dissociation-constant expression for each substance: **(a)** propylamine, $C_3H_7NH_2$; **(b)** cyanide ion, CN^-; **(c)** formate ion, CHO_2^-.

16.42 Write the balanced net ionic equation for the reaction of each of the following bases with water. Also write the base-dissociation-constant expression for each substance: **(a)** hydrazine, H_2NNH_2; **(b)** methylamine, CH_3NH_2; **(c)** benzoate ion, $C_6H_5CO_2^-$.

16.43 Calculate $[OH^-]$ and pH for each of the following solutions (K_b values in Appendix D): **(a)** 0.050 *M* pyridine; **(b)** 0.020 *M* hydroxylamine; **(c)** 3.0×10^{-3} *M* NH_3.

16.44 Calculate $[OH^-]$ and pH for each of the following solutions (K_a or K_b values in Appendix D): **(a)** 0.020 *M* NaOCl; **(b)** 0.100 *M* methylamine; **(c)** 2.0×10^{-3} *M* dimethylamine, $(CH_3)_2NH$ ($K_b = 5.4 \times 10^{-4}$).

16.45 Using the values of K_a from Appendix D, calculate the base-dissociation constant for each of the following species: **(a)** nitrite ion, NO_2^-; **(b)** azide ion, N_3^-; **(c)** hydrogen phosphate ion, HPO_4^{2-}; **(d)** formate ion, CHO_2^-.

16.46 Using the values of K_b from Appendix D, calculate K_a for each of the following species: **(a)** dimethylammonium ion, $(CH_3)_2NH_2^+$; **(b)** hydrazinium ion, $H_3NNH_2^+$; **(c)** pyridinium ion, $C_5H_5NH^+$; **(d)** hydroxylammonium ion, $HONH_3^+$.

16.47 Calculate $[OH^-]$ and pH for each of the following solutions: **(a)** 0.10 *M* NaCN; **(b)** 0.080 *M* Na_2CO_3; **(c)** a mixture that is 0.10 *M* in $NaNO_2$ and 0.20 *M* in $Ca(NO_2)_2$.

16.48 Calculate $[OH^-]$ and pH for each of the following solutions: **(a)** 0.10 *M* NaF; **(b)** 0.10 *M* Na_2S; **(c)** a mixture that is 0.085 *M* in $NaC_2H_3O_2$ and 0.055 *M* in $Ba(C_2H_3O_2)_2$.

Acid-Base Properties of Salts

16.49 Indicate whether each of the following substances would form an acidic, basic, or neutral solution in water: **(a)** $KC_2H_3O_2$; **(b)** $NaHCO_3$; **(c)** CH_3NH_3Br; **(d)** KNO_2.

16.50 Indicate whether each of the following substances would form an acidic, basic, or neutral solution in water: **(a)** $NaHC_2O_4$; **(b)** CsI; **(c)** $Al(NO_3)_3$; **(d)** NH_4CN.

16.51 Of these salts—KCNO, CH_3NH_3Br, $Ba(C_2H_3O_2)_2$, $Zn(NO_3)_2$, Na_2HPO_4—identify those that **(a)** form an acidic solution; **(b)** form a basic solution. **(c)** Identify the most basic substance in the list.

16.52 Of these salts—Na_2SO_3, $Fe(ClO_4)_3$, CaO, NH_4IO_3—identify those that **(a)** form an acidic solution; **(b)** form a basic solution. **(c)** Identify the most acidic substance in the list.

16.53 Sorbic acid, $HC_6H_7O_2$, is a weak monoprotic acid with $K_a = 1.7 \times 10^{-5}$. Its salt (potassium sorbate) is added to cheese to inhibit the formation of mold. What is the pH of a solution containing 4.93 g of potassium sorbate in 500 mL of solution?

16.54 Trisodium phosphate, Na_3PO_4, is available in hardware stores as TSP and is used as a cleansing agent. The label on a box of TSP warns that the substance is very basic (caustic or alkaline). What is the pH of a solution containing 50.0 g of TSP in a liter of solution?

Acid-Base Character and Chemical Structure

16.55 How does the acid strength of an oxyacid depend on **(a)** the electronegativity of the central atom; **(b)** the number of nonprotonated oxygen atoms in the molecule?

16.56 **(a)** How does the strength of an acid vary with the polarity and strength of the H—X bond? **(b)** How does the acidity of the hydride of an element vary as a function of the electronegativity of the element? How does this relate to the position of the element in the periodic table?

16.57 Based on their compositions and structures, select the member of each of the following pairs that is the stronger acid: **(a)** H_2SO_3 or H_2SeO_3; **(b)** H_3PO_4 or H_3PO_3; **(c)** H_2SO_3 or H_2CO_3.

16.58 Based on their compositions and structures and on conjugate acid-base relationships, select the member of each of the following pairs that is the stronger base: **(a)** BrO^- or BrO_2^-; **(b)** BrO^- or ClO^-; **(c)** HPO_4^{2-} or $H_2PO_4^-$.

16.59 Indicate whether each of the following statements is true or false. For those that are false, correct the statement so that it is true. **(a)** In general, the acidity of hydrides increases from left to right in a given row of the periodic table. **(b)** In a series of acids that have the same central atom, acid strength increases with the number of hydrogen atoms bonded to the central atom. **(c)** Hydrotelluric acid, H_2Te, is a stronger acid than H_2S because Te has a higher electronegativity than S.

16.60 Indicate whether each of the following statements is true or false. For those that are false, correct the statement so that it is true. **(a)** Acid strength in a series H_nX increases with increasing size of X. **(b)** For acids of the same structure but differing electronegativity of the central atom, acid strength decreases with increasing electronegativity of the central atom. **(c)** The strongest acid of all is HF because fluorine is the most electronegative element.

Lewis Acids and Bases

16.61 Prepare a table in which you compare the definitions of acids and bases according to the Lewis, Brønsted, and Arrhenius theories. Which is the most general; that is, which includes the others within its scope? Explain.

16.62 For each of the following descriptive statements, provide an interpretation in terms of the Brønsted-Lowry theory, the Lewis theory, or both, as appropriate. **(a)** Hydrogen bromide, HBr, dissolves in water to form an acidic solution. **(b)** Sodium hydride, NaH, reacts with water to form a basic solution. **(c)** Pyridine reacts with sulfur dioxide in a dry organic solvent to form pyridine-SO_2:

(d) Sulfur dioxide, SO_2, dissolves in water to form an acidic solution.

16.63 Identify the Lewis acid and Lewis base in each of the following reactions:

(a) $Fe(ClO_4)_3(s) + 6H_2O(l) \longrightarrow Fe(H_2O)_6^{3+}(aq) + 3ClO_4^-(aq)$

(b) $CN^-(aq) + H_2O(l) \rightleftharpoons HCN(aq) + OH^-(aq)$

(c) $(CH_3)_3N(g) + BF_3(g) \rightleftharpoons (CH_3)_3NBF_3(s)$

(d) $HIO(lq) + NH_2^-(lq) \rightleftharpoons NH_3(l) + IO^-(lq)$

(lq denotes liquid ammonia as solvent)

16.64 Identify the Lewis acid and Lewis base in each of the following reactions:

(a) $HNO_2(aq) + OH^-(aq) \rightleftharpoons NO_2^-(aq) + H_2O(l)$

(b) $FeBr_3(s) + Br^-(aq) \rightleftharpoons FeBr_4^-(aq)$

(c) $Zn^{2+}(aq) + 4NH_3(aq) \rightleftharpoons Zn(NH_3)_4^{2+}(aq)$

(d) $SO_2(g) + H_2O(l) \rightleftharpoons H_2SO_3(aq)$

16.65 Which member of each of the following pairs would you expect to produce the more acidic solution: **(a)** LiI or CdI_2; **(b)** $Fe(NO_3)_3$ or $Ca(NO_3)_2$; **(c)** $CrCl_2$ or $CrCl_3$?

16.66 Which member of each of the following pairs would you expect to be the stronger Lewis acid: **(a)** BH_4^- or BH_3; **(b)** S_8 or SO_3; **(c)** NaH or HBr; **(d)** Mo^{4+} or Zn^{2+}?

Additional Exercises

16.67 Arrange the following bases in order of their tendency to combine with protons, putting the strongest base first: NH_3, H_2O, OH^-, $C_2H_3O_2^-$, Br^-.

16.68 The dye bromthymol blue, abbreviated HBb, is a weak acid whose ionization can be represented as

$$HBb(aq) \rightleftharpoons H^+(aq) + Bb^-(aq)$$

Which way will this equilibrium shift when NaOH is added? The acid form of the dye is yellow, whereas its conjugate base is blue. What color is the NaOH solution containing this dye?

16.69 Calculate $[H^+]$ for each of the following solutions: **(a)** urine, pH 6.1; **(b)** lemon juice, pH 2.1; **(c)** gastric juice, pH 1.4; **(d)** household ammonia, pH 11.9. Indicate in each case whether the solution is acidic, basic, or neutral.

16.70 What is the pH of each of the following solutions: (a) 0.030 M NaOH; (b) 2.8×10^{-2} M $HClO_4$; (c) 0.0400 M HIO; (d) 0.150 M KCNO; (e) 0.020 M CH_3NH_2; (f) 0.0400 M CH_3NH_3Cl?

16.71 Calculate the number of moles of each of the following substances that must be present in 200 mL of solution to form a solution with pH 3.25: (a) HCl; (b) $HC_7H_5O_2$; (c) HF.

16.72 Predict the effect that each of the following added substances would have on the pH of an aqueous solution of HNO_2; (a) KNO_2; (b) $HClO_4$; (c) NaCN; (d) KOH.

16.73 Although we think of NH_3 as a base, it is capable of donating a proton when a sufficiently strong base is present. (a) Could such a reaction occur in aqueous solution? Explain. (b) Calculate the pH of a solution obtained by adding 0.30 g of $NaNH_2$ to sufficient water to form 0.400 L of solution.

16.74 Hemoglobin plays a part in a series of equilibria involving protonation-deprotonation and oxygenation-deoxygenation. The overall reaction is approximately as follows:

$$HbH^+(aq) + O_2(aq) \rightleftharpoons HbO_2(aq) + H^+(aq)$$

(where Hb stands for hemoglobin, and HbO_2 for oxyhemoglobin). (a) The concentration of O_2 is higher in the lungs and lower in the tissues. What effect does high $[O_2]$ have on the position of this equilibrium? (b) The normal pH of blood is 7.4. What is $[H^+]$ in normal blood? Is the blood acidic, basic, or neutral? (c) If the blood pH is lowered by the presence of large amounts of acidic metabolism products, a condition known as *acidosis* results. What effect does lowering blood pH have on the ability of hemoglobin to transport O_2?

[16.75] Pure sulfuric acid, H_2SO_4, is a colorless liquid that melts at 10°C and boils at 338°C. This pure substance undergoes autoionization to a much larger extent than does water. (a) Write the equilibrium expression for the autoionization of sulfuric acid and identify the acid, base, conjugate acid, and conjugate base. (b) Write the Lewis structures for the conjugate acid and conjugate base formed in the autoionization. (c) Write expressions for the reactions you would expect to occur when a small amount of each of the following substances is added to pure sulfuric acid: H_2O; $HClO_4$ (a stronger acid than H_2SO_4); K_2SO_4.

16.76 Saccharin, a sugar substitute, is a weak acid with $pK_a = 11.68$ at 25°C. It ionizes in aqueous solution as follows:

$$HNC_7H_4SO_3(aq) \rightleftharpoons H^+(aq) + NC_7H_4SO_3^-(aq)$$

What is the pH of a 0.10 M solution of this substance?

16.77 The active ingredient in aspirin is acetylsalicylic acid, $HC_9H_7O_4$, a monoprotic acid with $K_a = 3.3 \times 10^{-4}$. What is the pH of a solution obtained by dissolving two aspirin tablets, each containing 325 mg of acetylsalicylic acid, in 250 mL of water?

[16.78] What are the concentrations of H^+, HSO_4^-, and SO_4^{2-} in a 0.025 M solution of H_2SO_4?

[16.79] What are the concentrations of H^+, $H_2PO_4^-$, HPO_4^{2-}, and PO_4^{3-} in a 0.050 M solution of H_3PO_4?

16.80 Ephedrine, a central nervous system stimulant, is used in nasal sprays as a decongestant. This compound is a weak organic base:

$$C_{10}H_{15}ON(aq) + H_2O(l) \rightleftharpoons C_{10}H_{15}ONH^+(aq) + OH^-(aq)$$

K_b has the value 1.4×10^{-4}. What pH would you expect for a 0.035 M solution of ephedrine, assuming that no other substances are present? What is the value of pK_a for the conjugate acid, ephedrine hydrochloride?

16.81 Morphine, $C_{17}H_{19}NO_3$, is a weak base containing a nitrogen atom, with $pK_b = 6.1$. (a) What is the pH of a 0.050 M solution of morphine? (b) What is the value of pK_a for the conjugate acid, morphine hydrochloride? What is the pH of a 0.050 M solution of this substance?

16.82 Codeine, $C_{18}H_{21}NO_3$, is a weak organic base. A 5.0×10^{-3} M solution of codeine has a pH of 9.95; calculate the value of K_b for this substance.

[16.83] Amino acids contain an amino group, $—NH_2$, located on the carbon atom that also contains a carboxylic acid group, —COOH. Glycine, the simplest amino acid, could exist in water in either form I or II below:

$$\underset{\text{I}}{H_2N—CH_2—\overset{\overset{\displaystyle O}{\|}}{C}—OH} \qquad \underset{\text{II}}{{}^+H_3N—CH_2—\overset{\overset{\displaystyle O}{\|}}{C}—O^-}$$

K_a for the carboxylic acid group of glycine is 4.3×10^{-3}, and K_b for the amino group is 6.0×10^{-5}. (a) What is the pH of a 0.10 M aqueous solution of glycine? (b) In what forms, other than I and II, can glycine exist in solution, depending on pH? (c) What form of glycine would you expect to be present in a solution with pH 10; with pH 2?

[16.84] (a) Using dissociation constants from Appendix D, determine the value for the equilibrium constant for each of the following reactions. (Remember that when reactions are added, the corresponding equilibrium constants are multiplied.)

(i) $HCO_3^-(aq) + OH^-(aq) \rightleftharpoons CO_3^{2-}(aq) + H_2O(l)$
(ii) $NH_4^+(aq) + CO_3^{2-}(aq) \rightleftharpoons NH_3(aq) + HCO_3^-(aq)$

(b) We usually use single arrows for reactions when the forward reaction is appreciable (K much greater than 1) or when products escape from the system so that equilibrium is not achieved. If we follow this convention, which of these equilibria might be written with a single arrow?

[16.85] Many moderately large organic molecules containing basic nitrogen atoms are not very soluble in water as neutral molecules, but they are frequently much more soluble as their acid salts. Assuming that pH in the stomach is 2.5, indicate whether each of the following compounds would be present in the stomach as the neutral base or in the protonated form: nicotine, $K_b = 7 \times 10^{-7}$; caffeine, $K_b = 4 \times 10^{-14}$; strychnine, $K_b = 1 \times 10^{-6}$; quinine, $K_b = 1.1 \times 10^{-6}$.

[16.86] A 1.00 m solution of HF freezes at −1.90°C. Based

on the extent of freezing-point lowering (Section 13.6), calculate the fraction of HF dissociated at this temperature. What is the value of K_a for HF at $-1.90°C$?

[16.87] In an aqueous solution containing only a weak diprotic acid H_2X, the concentration of X^{2-} is numerically equal to K_{a2}. Show why this is so.

[16.88] What is the pH of a 1.0×10^{-9} *M* solution of HBr?

[16.89] The Lewis structure for acetic acid is shown in Table 16.1. Replacement of hydrogen atoms on the carbon by chlorine atoms causes an increase in acidity, as follows:

Acid	Formula	$K_a(25°C)$
Acetic	CH_3COOH	1.8×10^{-5}
Chloroacetic	$CH_2ClCOOH$	1.4×10^{-3}
Dichloroacetic	$CHCl_2COOH$	3.3×10^{-2}
Trichloroacetic	CCl_3COOH	2×10^{-1}

Using Lewis structures as the basis of your discussion, explain the observed trend in acidities in the series. Calculate the pH of a 0.10 *M* solution of each acid.

17 Additional Aspects of Aqueous Equilibria

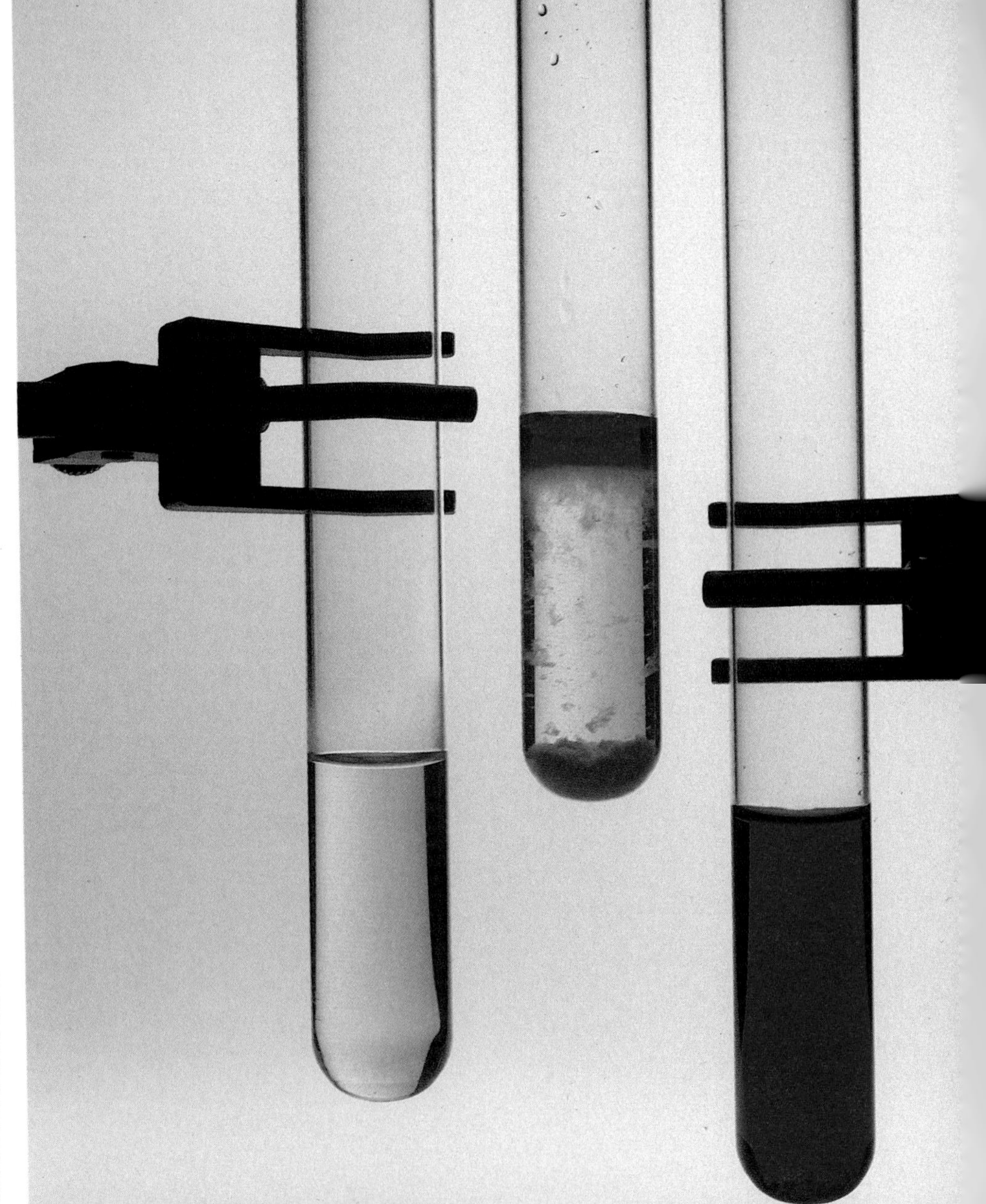

As $NH_3(aq)$ is added to a solution of $Cu^{2+}(aq)$ (*left*), a a pale-blue precipitate of $Cu(OH)_2$ forms (*center*) and then dissolves as additional $NH_3(aq)$ causes the formation of the deep blue $Cu(NH_3)_4{}^{2+}$ ion (*center and right*). (© Paul Silverman/Fundamental Photographs)

CONTENTS

Water is the most common and most important solvent on this planet. In a sense, it is the solvent of life. It is difficult to imagine how living matter in all its complexity could exist with any liquid other than water as solvent. Water occupies its position of importance not only because of its abundance, but also because of its exceptional ability to dissolve a wide variety of substances. Aqueous solutions encountered in nature, such as biological fluids and seawater, contain many solutes. Consequently, many equilibria take place simultaneously in these solutions.

In this chapter, we take a step toward understanding such complex solutions by looking first at further applications of acid-base equilibria. We then broaden our discussion to include two additional types of aqueous equilibria, those involving slightly soluble salts and those forming metal complexes in solution.

17.1 THE COMMON-ION EFFECT

In Chapter 16, we were concerned with determining the equilibrium concentrations of ions in solutions containing a weak acid or a weak base. We now consider solutions that contain not only a weak acid, such as acetic acid, $HC_2H_3O_2$, but also a soluble salt of that acid, such as $NaC_2H_3O_2$. When $NaC_2H_3O_2$ is added to a solution of $HC_2H_3O_2$, the pH of the solution increases ($[H^+]$ is reduced). This result isn't surprising because $C_2H_3O_2^-$ is a weak base; like any other base, it should increase the pH. However, it is instructive to view this effect from the perspective of Le Châtelier's principle (Section 15.6).

Like most salts, $NaC_2H_3O_2$ is a strong electrolyte. Consequently, it ionizes completely in aqueous solution to form Na^+ and $C_2H_3O_2^-$ ions. On the other hand, $HC_2H_3O_2$ is a weak electrolyte that dissociates as follows:

$$HC_2H_3O_2(aq) \rightleftharpoons H^+(aq) + C_2H_3O_2^-(aq) \qquad [17.1]$$

The addition of $C_2H_3O_2^-$, from $NaC_2H_3O_2$, causes this equilibrium, Equation 17.1, to shift to the left, thereby decreasing the equilibrium concentration of $H^+(aq)$:

$$HC_2H_3O_2(aq) \rightleftharpoons H^+(aq) + C_2H_3O_2^-(aq)$$

Addition of $C_2H_3O_2^-$ shifts equilibrium, reducing $[H^+]$

In general, the dissociation of a weak electrolyte is decreased by adding to the solution a strong electrolyte that has an ion in common with the weak electrolyte. This shift in equilibrium position, which occurs when we add an ion that is a component of an equilibrium reaction, is called the **common-ion effect**. Sample Exercises 17.1 and 17.2 illustrate how equilibrium concentrations may be calculated when a solution contains a mixture of a weak electrolyte and a strong electrolyte that share a common ion. You will see that the procedures are similar to those encountered for weak acids and weak bases in Chapter 16.

SAMPLE EXERCISE 17.1

Suppose that we add 8.20 g, or 0.100 mol, of sodium acetate, $NaC_2H_3O_2$, to 1 L of a 0.100 *M* solution of acetic acid, $HC_2H_3O_2$. What is the pH of the resultant solution?

Solution: Our first step in determining the pH of a solution containing a mixture of solutes should always be to identify the major species in solution and consider their acidity or basicity. Because $HC_2H_3O_2$ is a weak electrolyte and $NaC_2H_3O_2$ is a strong electrolyte, the major species in the solution are $HC_2H_3O_2$ (a weak acid), Na^+ (which is neither acidic nor basic), $C_2H_3O_2^-$ (which is the conjugate base of $HC_2H_3O_2$), and H_2O (which is a very weak acid or base).

Our second step is to identify the important equilibrium reaction. Because H_2O is a much weaker acid than $HC_2H_3O_2$, the pH of the solution will be controlled by the dissociation equilibrium of $HC_2H_3O_2$, which involves both $HC_2H_3O_2$ and $C_2H_3O_2^-$:

$$HC_2H_3O_2(aq) \rightleftharpoons H^+(aq) + C_2H_3O_2^-(aq)$$

The Na^+ ion is merely a spectator ion; it has no influence on pH (Section 16.6).

We next tabulate the initial and equilibrium concentrations of each of the species that participates in the equilibrium:

	$HC_2H_3O_2(aq)$ $\rightleftharpoons$	$H^+(aq)$ +	$C_2H_3O_2^-(aq)$
Initial:	0.100 *M*	0	0.100 *M*
Change:	$-x$ *M*	$+x$ *M*	$+x$ *M*
Equilibrium:	$(0.100 - x)$ *M*	x *M*	$(0.100 + x)$ *M*

Notice that the equilibrium concentration of $C_2H_3O_2^-$ (the common ion) is the initial amount coming from the $NaC_2H_3O_2$ (0.100 *M*) plus the amount (x) formed by dissociation of $HC_2H_3O_2$.

The equilibrium-constant expression for the equilibrium is

$$K_a = 1.8 \times 10^{-5} = \frac{[H^+][C_2H_3O_2^-]}{[HC_2H_3O_2]}$$

(The dissociation constant for $HC_2H_3O_2$ is from Appendix D; addition of $NaC_2H_3O_2$ does not change the value of this constant.) Substituting the equilibrium concentrations into the equilibrium-constant equation gives

$$K_a = 1.8 \times 10^{-5} = \frac{x(0.100 + x)}{0.100 - x}$$

If we assume that x is small compared to the original concentrations of $HC_2H_3O_2$ and $C_2H_3O_2^-$ (0.100 *M* each), we can simplify our equation before solving for x:

$$K_a = 1.8 \times 10^{-5} \simeq \frac{x(0.100)}{0.100}$$

$$x = 1.8 \times 10^{-5}\ M = [H^+]$$

The resulting value of x is indeed small relative to 0.100, justifying the approximation made in simplifying the problem.

Finally, we calculate the pH:

$$pH = -\log(1.8 \times 10^{-5}) = 4.74$$

Earlier (Section 16.4) we calculated that in a 0.10 M solution of $HC_2H_3O_2$, $[H^+]$ is 1.3×10^{-3} M, corresponding to a pH value of 2.89. Thus, the addition of $NaC_2H_3O_2$ has substantially increased the solution pH, as expected.

PRACTICE EXERCISE

Calculate the pH of a solution containing 0.060 M formic acid, $HCHO_2$ ($K_a = 1.8 \times 10^{-4}$), and 0.030 M potassium formate, $KCHO_2$. ***Answer:*** 3.44

SAMPLE EXERCISE 17.2

Calculate the fluoride-ion concentration and pH of a solution containing 0.10 mol of HCl and 0.20 mol of HF in 1.0 L of solution.

Solution: Because HCl is a strong acid and HF is a weak acid, the major species in solution are H^+, Cl^-, HF, and the solvent, H_2O. The problem asks for $[F^-]$, which is formed by dissociation of HF. Thus, the important equilibrium is

$$HF(aq) \rightleftharpoons H^+(aq) + F^-(aq)$$

The Cl^- is merely a spectator ion.

Now we can tabulate the initial and equilibrium concentrations of each species involved in this equilibrium:

	$HF(aq)$ $\rightleftharpoons$	$H^+(aq)$ +	$F^-(aq)$
Initial:	0.20 M	0.10 M	0
Change:	$-x$ M	$+x$ M	$+x$ M
Equilibrium:	$(0.20 - x)$ M	$(0.10 + x)$ M	x M

The equilibrium constant, from Appendix D, is 6.8×10^{-4}. Thus,

$$K_a = 6.8 \times 10^{-4} = \frac{[H^+][F^-]}{[HF]} = \frac{(0.10 + x)x}{0.20 - x}$$

If we assume that x is small relative to 0.10 or 0.20 M, this expression simplifies to give

$$\frac{(0.10)x}{0.20} = 6.8 \times 10^{-4}$$

$$x = \frac{0.20}{0.10}(6.8 \times 10^{-4}) = 1.4 \times 10^{-3}\ M = [F^-]$$

This F^- concentration is substantially smaller than it would be in a 0.20 M solution of HF with no added HCl. The common ion, H^+, has repressed the dissociation of HF. The concentration of $H^+(aq)$ is

$$[H^+] = (0.10 + x)\ M \simeq 0.10\ M$$

Thus, pH = 1.00. Notice that $[H^+]$ is for practical purposes due entirely to the HCl; the HF makes a negligible contribution by comparison.

PRACTICE EXERCISE

Calculate the formate ion concentration and pH of a solution that is 0.050 M in formic acid, $HCHO_2$ ($K_a = 1.8 \times 10^{-4}$), and 0.100 M in HNO_3. ***Answer:*** $[CHO_2^-] = 9.0 \times 10^{-5}$; pH = 1.00

Sample Exercises 17.1 and 17.2 both involve weak acids. However, the dissociation of a weak base is also decreased by the addition of a common ion. For example, the addition of NH_4^+ (as from NH_4Cl) causes the base-dissociation equilibrium of NH_3 to shift to the left, decreasing the equilibrium concentration of OH^-:

$$NH_3(aq) + H_2O(l) \rightleftharpoons NH_4^+(aq) + OH^-(aq) \qquad [17.2]$$

Addition of NH_4^+ shifts equilibrium, reducing $[OH^-]$

The equilibrium-constant expression for this reaction contains both the weak base, NH_3, and its conjugate acid, NH_4^+. Interestingly, the acid-dissociation constant for NH_4^+ also contains both these quantities:

$$NH_4^+(aq) \rightleftharpoons NH_3(aq) + H^+(aq) \qquad [17.3]$$

You will see as we move through the chapter that we can calculate the pH of a solution involving this and other conjugate acid-base pairs by working with either the equation for weak-base dissociation or the equation for dissociation of the conjugate acid.

Common Ions Generated by Acid-Base Reactions

The common ion that affects a weak-acid or weak-base equilibrium may be present because it is added as a salt, as in the examples above. However, it might also arise as a result of an acid-base reaction. Indeed, this is a common and important situation. Consider a solution formed by mixing a weak acid, such as $HC_2H_3O_2$, and a strong base, like NaOH. The net ionic equation for the reaction that occurs is

$$HC_2H_3O_2(aq) + OH^-(aq) \longrightarrow C_2H_3O_2^-(aq) + H_2O(l) \qquad [17.4]$$

This reaction proceeds virtually to completion. In general, *reactions between (1) strong acids and strong bases, (2) strong acids and weak bases, and (3) weak acids and strong bases proceed essentially to completion.*

If the number of moles of $HC_2H_3O_2$ exceeds the number of moles of NaOH, the reaction between them produces $C_2H_3O_2^-$ ions, leaving the excess $HC_2H_3O_2$ unreacted. For example, consider a 1-L solution that initially contains 0.20 mol of $HC_2H_3O_2$ and 0.10 mol of NaOH. The reaction (Equation 17.4) consumes 0.10 mol of $HC_2H_3O_2$, producing 0.10 mol of $C_2H_3O_2^-$ and leaving 0.10 mol of $HC_2H_3O_2$ unreacted:

	$HC_2H_3O_2(aq)$ +	$OH^-(aq)$ $\longrightarrow$	$C_2H_3O_2^-(aq)$ +	$H_2O(l)$
Before rxn:	0.20 mol	0.10 mol	0	—
Change:	−0.10 mol	−0.10 mol	+0.10 mol	—
After rxn:	0.10 mol	0	0.10 mol	—

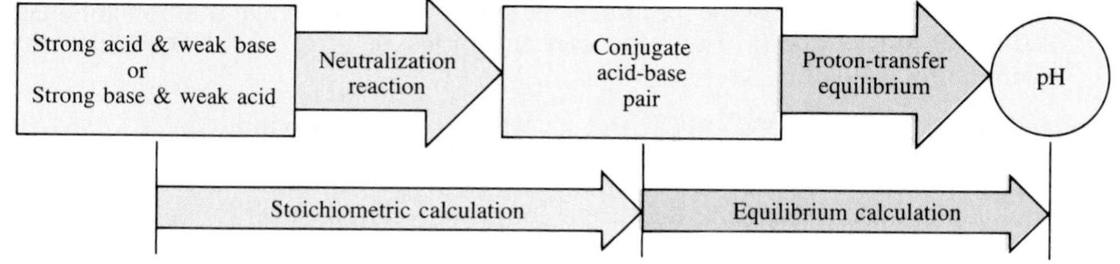

Figure 17.1 Outline of the procedure used to calculate the pH of an acid-base mixture in which the weak acid or weak base is partly neutralized.

The resultant solution, which is 0.10 *M* in $C_2H_3O_2^-$ and 0.10 *M* in $HC_2H_3O_2$, is identical to a solution produced by mixing 0.10 mol of $NaC_2H_3O_2$ and 0.10 mol of $HC_2H_3O_2$ to form a liter of solution. As shown in Sample Exercise 17.1, such a solution has a pH of 4.74.

This example suggests a general approach to determining the pH of any acid-base mixture:

1. If the solution mixture consists only of a conjugate acid-base pair, then you need consider only the proton-transfer equilibrium between these two species, as in Sample Exercise 17.1.
2. If the solution contains a strong acid and a weak base or a strong base and a weak acid, you must first consider the stoichiometry of the acid-base reaction. If a conjugate acid-base pair is present after the neutralization reaction, then you must subsequently consider the proton-transfer equilibrium between them.

The overall approach is diagramed in Figure 17.1.

SAMPLE EXERCISE 17.3

Calculate the pH of a solution produced by mixing 0.60 L of 0.10 *M* NH_4Cl with 0.40 L of 0.10 *M* NaOH.

Solution: Both NH_4Cl and NaOH are strong electrolytes. Before they react, the major species in solution are NH_4^+ (the conjugate acid of NH_3), Cl^- (neither acidic nor basic), Na^+ (neither acidic nor basic), OH^- (strongly basic), and H_2O. The strong base OH^- will abstract a proton from the weakly acidic NH_4^+ ion:

$$NH_4^+(aq) + OH^-(aq) \longrightarrow NH_3(aq) + H_2O(l)$$

Stoichiometric calculations: We first calculate the molar quantities of NH_4^+ and OH^- ions in solution:

$$\text{Moles } NH_4^+ = M_{NH_4Cl} \times V_{NH_4Cl} = \left(0.10\,\frac{\text{mol}}{\text{L}}\right)(0.60\text{ L}) = 0.060\text{ mol}$$

$$\text{Moles } OH^- = M_{NaOH} \times V_{NaOH} = \left(0.10\,\frac{\text{mol}}{\text{L}}\right)(0.40\text{ L}) = 0.040\text{ mol}$$

Assuming a complete reaction between NH_4^+ and OH^-, we calculate the quantities of species remaining after the reaction:

	$NH_4^+(aq)$ +	$OH^-(aq)$ ⟶	$NH_3(aq)$	+ $H_2O(l)$
Before rxn:	0.060 mol	0.040 mol	0	—
Change:	−0.040 mol	−0.040 mol	+0.040 mol	—
After rxn:	0.020 mol	0	0.040 mol	—

The total volume of the solution is the sum of the two original solution volumes: 0.60 L + 0.40 L = 1.00 L. Thus, the concentrations of NH_3 and NH_4^+ after the reaction are

$$[NH_3] = \frac{0.040\ \text{mol}}{1.00\ \text{L}} = 0.040\ M \qquad [NH_4^+] = \frac{0.020\ \text{mol}}{1.00\ \text{L}} = 0.020\ M$$

Equilibrium calculation: Because a conjugate acid-base pair (NH_4^+ and NH_3) is present after the reaction, we next consider their equilibrium to determine the pH of the solution. Viewing the equilibrium from the perspective of the dissociation of the acid, we have:

$$NH_4^+(aq) \rightleftharpoons H^+(aq) + NH_3(aq)$$

$$K_a = \frac{[H^+][NH_3]}{[NH_4^+]}$$

The initial and equilibrium values of the pertinent species are as follows:

$$NH_4^+(aq) \rightleftharpoons H^+(aq) + NH_3(aq)$$

	$NH_4^+(aq)$	$H^+(aq)$	$NH_3(aq)$
Initial:	0.020 M	0	0.040 M
Change:	$-x\ M$	$+x\ M$	$+x$
Equilibrium:	$(0.020 - x)\ M$	$x\ M$	$(0.040 + x)\ M$

K_a for NH_4^+ is related to K_b for NH_3 through the relationship $K_a \times K_b = K_w$. From Appendix D we have $K_b = 1.8 \times 10^{-5}$

$$K_a = \frac{K_w}{K_b} = \frac{1.0 \times 10^{-14}}{1.8 \times 10^{-5}} = 5.6 \times 10^{-10}$$

We can assume that x is small relative to 0.020 and 0.040. Then

$$K_a = 5.6 \times 10^{-10} = \frac{(x)(0.040)}{0.020}$$

$$x = [H^+] = 2.8 \times 10^{-10}\ M$$

$$pH = -\log(2.8 \times 10^{-10}) = 9.55$$

PRACTICE EXERCISE

Calculate the pH of a solution formed by mixing 0.50 L of 0.015 M NaOH solution with 0.50 L of 0.030 M benzoic acid solution. ***Answer:*** 4.19

17.2 BUFFERED SOLUTIONS

Many aqueous solutions resist a change in pH upon addition of small amounts of acid or base. Such solutions are called **buffered solutions** (or merely **buffers**). Human blood, for example, is a complex aqueous medium with a pH buffered at about 7.4. Any significant variation in the pH from this value results in a severe pathological response and, eventually, death. As another example, the chemical behavior of seawater is determined in very important respects by its pH, buffered at about 8.1 to 8.3 near the surface. Addition of a small amount of an acid or base to either blood or seawater does not result in a large change in pH. Buffer solutions find many important applications in the laboratory and in medicine (Figure 17.2).

Composition and Action of Buffered Solutions

Buffers resist changes in pH because they contain an acidic species to neutralize OH^- ions and a basic one to neutralize H^+ ions. It is, of course, necessary

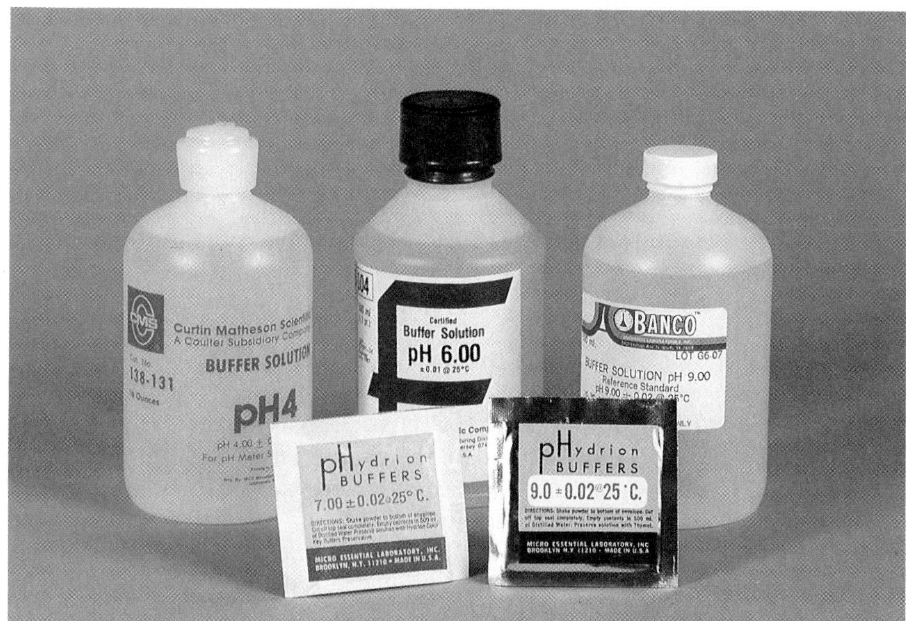

Figure 17.2 Prepackaged buffer solutions and ingredients for forming buffer solutions of predetermined pH. (Donald Clegg and Roxy Wilson)

that these acidic and basic species not consume each other through a neutralization reaction. These requirements are fulfilled by a weak acid-base conjugate pair such as $HC_2H_3O_2$–$C_2H_3O_2^-$ or NH_4^+–NH_3. Thus, buffers are often prepared by mixing a weak acid or a weak base with a salt of that acid or base. For example, the $HC_2H_3O_2$–$C_2H_3O_2^-$ buffer can be prepared by adding $NaC_2H_3O_2$ to a solution of $HC_2H_3O_2$; the NH_4^+–NH_3 buffer can be prepared by adding NH_4Cl to a solution of NH_3. By choosing appropriate components and adjusting their relative concentrations, we can buffer a solution at virtually any pH.

To understand better how a buffer works, consider a buffer composed of a weak acid (HX) and one of its salts (MX, where M^+ could be Na^+, K^+, and so forth). The acid-dissociation equilibrium in this buffered solution involves both the acid and its conjugate base:

$$HX(aq) \rightleftharpoons H^+(aq) + X^-(aq) \qquad [17.5]$$

The corresponding acid-dissociation-constant expression is

$$K_a = \frac{[H^+][X^-]}{[HX]} \qquad [17.6]$$

Solving this expression for $[H^+]$, we have

$$[H^+] = K_a \frac{[HX]}{[X^-]} \qquad [17.7]$$

We see from this expression that $[H^+]$, and thus the pH, is determined by two factors: the value of K_a for the weak-acid component of the buffer, and the ratio of the concentrations of the conjugate acid-base pair, $[HX]/[X^-]$.

If OH^- ions are added to the buffered solution, they react with the acid component of the buffer:

$$OH^-(aq) + HX(aq) \longrightarrow H_2O(l) + X^-(aq) \qquad [17.8]$$

This reaction causes $[HX]$ to decrease and $[X^-]$ to increase. However, as long as the amounts of HX and X^- in the buffer are large compared to the amount of OH^- added, the ratio $[HX]/[X^-]$ doesn't change much, and thus the change in pH is small.

If H^+ ions are added, they react with the base component of the buffer:

$$H^+(aq) + X^-(aq) \longrightarrow HX(aq) \qquad [17.9]$$

In this case, $[HX]$ increases and $[X^-]$ decreases. Once again, as long as the change in the ratio $[HX]/[X^-]$ is small, the change in pH will be small.

Buffered solutions are most effective in buffering against a change in pH in *either* direction when the concentrations of HX and X^- are about the same. Notice that under these conditions $[H^+]$ is approximately equal to K_a. For this reason, we usually try to select a buffer whose acid form has a pK_a close to the desired pH.

Buffer Capacity and pH

Two important characteristics of a buffer are buffering capacity and pH. **Buffering capacity** is the amount of acid or base the buffer can neutralize before the pH begins to change to an appreciable degree. This capacity depends on the amount of acid and base from which the buffer is made. The pH of the buffer depends on K_a for the acid and on the relative concentrations of the acid and base that comprise the buffer. For example, we can see from Equation 17.7 that $[H^+]$ for a 1-L solution that is 1 *M* in $HC_2H_3O_2$ and 1 *M* in $NaC_2H_3O_2$ will be the same as for a 1-L solution that is 0.1 *M* in $HC_2H_3O_2$ and 0.1 *M* in $NaC_2H_3O_2$. However, the first solution has a greater buffering capacity because it contains more $HC_2H_3O_2$ and $C_2H_3O_2^-$. The greater the amounts of the conjugate acid-base pair, the more resistant the ratio of their concentrations, and hence the pH, is to change.

Because conjugate acid-base pairs share a common ion, we can use exactly the same procedures to calculate the pH of a buffer as we used in treating the common-ion effect (see Sample Exercise 17.1). However, an alternate approach is sometimes taken that is based on an equation derived from Equation 17.7. Taking the negative log of both sides of Equation 17.7, we have

$$-\log [H^+] = -\log\left(K_a \frac{[HX]}{[X^-]}\right) = -\log K_a - \log \frac{[HX]}{[X^-]}$$

Because $-\log [H^+] = pH$ and $-\log K_a = pK_a$, we have

$$pH = pK_a - \log \frac{[HX]}{[X^-]} = pK_a + \log \frac{[X^-]}{[HX]} \qquad [17.10]$$

In general,

$$pH = pK_a + \log \frac{[\text{base}]}{[\text{acid}]} \qquad [17.11]$$

This relationship is known as the **Henderson-Hasselbalch equation**. Biologists, biochemists, and others who work frequently with buffers often use this equation to calculate the pH of buffers. What makes the Henderson-Hasselbalch equation particularly convenient is that we can normally neglect the amounts of the acid and base of the buffer that ionize. Therefore, we can use the starting concentrations of the acid and base components of the buffer directly in Equation 17.11.

SAMPLE EXERCISE 17.4

What is the pH of a buffer mixture composed of 0.12 M lactic acid ($HC_3H_5O_3$, $K_a = 1.4 \times 10^{-4}$) and 0.10 M sodium lactate?

Solution: We will first determine the pH using the method described in Section 17.1. The major species in solution are $HC_3H_5O_3$, Na^+, $C_3H_5O_3^-$, and H_2O. The Na^+ ion is neither acidic nor basic; H_2O is a much weaker acid than $HC_3H_5O_3$ and a weaker base than $C_3H_5O_3^-$. Therefore, the pH will be controlled by the acid-dissociation equilibrium of lactic acid shown below. The initial and equilibrium concentrations of the species involved in this equilibrium are

$$HC_3H_5O_3(aq) \rightleftharpoons H^+(aq) + C_3H_5O_3^-(aq)$$

	$HC_3H_5O_3(aq)$	$H^+(aq)$	$C_3H_5O_3^-(aq)$
Initial:	0.12 M	0	0.10 M
Change:	$-x$ M	$+x$ M	$+x$ M
Equilibrium:	$(0.12 - x)$ M	x M	$(0.10 + x)$ M

The equilibrium concentrations are governed by the equilibrium-constant expression:

$$K_a = 1.4 \times 10^{-4} = \frac{[H^+][C_3H_5O_3^-]}{[HC_3H_5O_3]} = \frac{x(0.10 + x)}{0.12 - x}$$

Because of the small K_a and the presence of the common ion, we expect x to be small relative to 0.12 or 0.10 M. Thus, our equilibrium-constant equation can be simplified to give

$$K_a = 1.4 \times 10^{-4} = \frac{x(0.10)}{0.12}$$

Solving for x gives a value that justifies our approximation:

$$x = \left(\frac{0.12}{0.10}\right)(1.4 \times 10^{-4}) = 1.7 \times 10^{-4}\ M$$

$$\text{pH} = -\log(1.7 \times 10^{-4}) = 3.77$$

Alternatively, we could have used the Henderson-Hasselbalch equation to calculate pH directly:

$$\text{pH} = \text{p}K_a + \log\left(\frac{[\text{base}]}{[\text{acid}]}\right) = 3.85 + \log\left(\frac{0.10}{0.12}\right)$$
$$= 3.85 + (-0.08) = 3.77$$

PRACTICE EXERCISE

Calculate the pH of a buffer composed of 0.12 M benzoic acid and 0.20 M sodium benzoate. (Refer to Appendix D.) ***Answer:*** 4.41

SAMPLE EXERCISE 17.5

How many moles of NH_4Cl must be added to 1.0 L of 0.10 M NH_3 to form a buffer whose pH is 9.00?

Solution: The major species in the solution will be NH_4^+, Cl^-, NH_3, and H_2O. The Cl^- ion has negligible basicity (it is the conjugate base of a strong acid), and H_2O is a very weak acid or base. The important acid and base species are therefore NH_4^+ and NH_3. We can consider the equilibrium between this conjugate acid-base pair either from the perspective of the base-dissociation equilibrium of NH_3 (Equation 17.2) or from that of the acid-dissociation equilibrium of NH_4^+ (Equation 17.3). Either approach gives the same results, and we used the acid dissociation in Sample Exercise 17.3. Using the base-dissociation equilibrium this time, we have

$$NH_3(aq) + H_2O(l) \rightleftharpoons NH_4^+(aq) + OH^-(aq)$$

$$K_b = 1.8 \times 10^{-5} = \frac{[NH_4^+][OH^-]}{[NH_3]}$$

(The value of K_b is from Appendix D.) Let's rearrange this equilibrium expression to solve for $[NH_4^+]$:

$$[NH_4^+] = K_b \frac{[NH_3]}{[OH^-]} = (1.8 \times 10^{-5}) \frac{[NH_3]}{[OH^-]}$$

If pH is to be 9.00, then pOH is 5.00, and $[OH^-]$ is 1.0×10^{-5}.

We can assume that $[NH_3]$ is 0.10 M. The amount that reacts with water to form NH_4^+ is very small, especially in the presence of added NH_4^+. Thus,

$$[NH_4^+] = \frac{1.8 \times 10^{-5} \times 0.10\ M}{1.0 \times 10^{-5}} = 0.18\ M$$

Because the amount of NH_4^+ formed from NH_3 is very small, this concentration represents the concentration of NH_4Cl that we must have to achieve a pH of 9.00. Thus, the number of moles of NH_4Cl needed is given by the product of the volume of the solution and its molarity:

$$(1.0\ \text{L})\left(0.18\ \frac{\text{mol}}{\text{L}}\right) = 0.18\ \text{mol}$$

PRACTICE EXERCISE

Calculate the concentration of sodium formate that must be present in a 0.10 M solution of formic acid to produce a pH of 3.80. ***Answer:*** 0.11 M

Addition of Acids or Bases to Buffers

Let us now consider in a more quantitative way the response of a buffer solution to addition of an acid or base. In solving these problems, it is important to recall that the reactions between strong acids and weak bases and those between strong bases and weak acids proceed essentially to completion. Therefore, when a strong acid or strong base is added to a buffered solution, it is best to deal with the stoichiometry of the acid-base neutralization reaction first. After the stoichiometic calculation, we can consider the equilibria between the remaining ions.

SAMPLE EXERCISE 17.6

A liter of solution containing 0.100 mol of $HC_2H_3O_2$ and 0.100 mol of $NaC_2H_3O_2$ forms a buffer solution of pH 4.74. Calculate the pH of this solution **(a)** after 0.020 mol of NaOH is added (neglect any volume changes); **(b)** after 0.020 mol of HCl is added (again, neglect volume changes).

Solution: **(a)** *Stoichiometry Calculation:* The OH^- provided by NaOH reacts with the weak-acid component of the buffer, $HC_2H_3O_2$. The following table summarizes the reaction and the resultant concentrations:

$$HC_2H_3O_2(aq) + OH^-(aq) \longrightarrow H_2O(l) + C_2H_3O_2^-(aq)$$

Before rxn:	0.100 M	0.020 M	—	0.100 M
Change:	−0.020 M	−0.020 M	—	+ 0.020 M
After rxn:	0.080 M	0.0 M	—	0.120 M

Equilibrium Calculation: Because the final solution contains both the weak acid $HC_2H_3O_2$ and its conjugate base, $C_2H_3O_2^-$, we next consider their proton-transfer equilibrium in order to determine the pH of the solution:

$$HC_2H_3O_2(aq) \rightleftharpoons H^+(aq) + C_2H_3O_2^-(aq)$$

Initial:	0.080 M	0	0.120 M
Change:	$-x$ M	$+x$ M	$+x$ M
Equilibrium:	$(0.080 - x)$ M	x M	$(0.120 + x)$ M

$$K_a = \frac{[C_2H_3O_2^-][H^+]}{[HC_2H_3O_2]} = \frac{(0.120 + x)(x)}{0.080 - x} \simeq \frac{(0.120)(x)}{0.080}$$

$$x = [H^+] = \frac{(0.080)K_a}{(0.120)} = 0.67 \times 1.8 \times 10^{-5} = 1.2 \times 10^{-5}\ M$$

$$\text{pH} = -\log(1.2 \times 10^{-5}) = 4.92$$

We could have obtained this answer directly by applying Equation 17.11, but until you are sure you have a thorough understanding of the equilibria of importance in these problems, it is best to set the problem up in full.

(b) *Stoichiometric Calculation:* The H^+ provided by HCl reacts completely with the weak base component of the buffer, $C_2H_3O_2^-$. The following table summarizes the resultant concentrations:

$$C_2H_3O_2^-(aq) + H^+(aq) \longrightarrow HC_2H_3O_2(aq)$$

Before rxn:	0.100 M	0.020 M	0.100 M
Change:	−0.020 M	−0.020 M	+0.020 M
After rxn:	0.080 M	0.0 M	0.120 M

Equilibrium Calculation: To determine the pH of the resultant solution, we must next consider the equilibrium between $HC_2H_3O_2$ and $C_2H_3O_2^-$:

$$HC_2H_3O_2(aq) \rightleftharpoons H^+(aq) + C_2H_3O_2^-(aq)$$

Initial:	0.120 M	0	0.080 M
Change:	$-x$ M	$+x$ M	$+x$ M
Equilibrium:	$(0.120 - x)$ M	x M	$(0.080 + x)$ M

Substituting into the acid-dissociation-equilibrium expression as before, and simplifying, we obtain

$$x = [H^+] = \frac{(0.120)K_a}{0.080} = 1.5 \times 1.8 \times 10^{-5} = 2.7 \times 10^{-5}\ M$$

$$\text{pH} = -\log(2.7 \times 10^{-5}) = 4.57$$

The pH decreases, as expected for addition of a strong acid to a solution. However, the magnitude of the decrease is small, because the solution is a buffer with a capacity to absorb the added acid.

PRACTICE EXERCISE

Consider a liter of a buffered solution that is 0.110 *M* in formic acid and 0.100 *M* in sodium formate. Calculate the pH of the buffer **(a)** before any acid or base is added; **(b)** after the addition of 0.015 mol of HNO_3; **(c)** after the addition of 0.015 mol of KOH. (Assume no volume change.) ***Answer:*** **(a)** 3.70; **(b)** 3.58; **(c)** 3.83

To appreciate more fully the buffer action of the solution of acetic acid and sodium acetate that we've just considered, let's compare its behavior with the action of a solution that is not a buffer. We saw in Sample Exercise 17.1 that the pH of a solution that is 0.100 *M* in acetic acid and 0.100 *M* in sodium acetate is 4.74. A solution of this same pH is obtained by addition of 1.8×10^{-5} mol of HCl to a liter of water. Now suppose that 2.0 mL of 10 *M* HCl solution—that is, 0.020 mol of HCl—is added to a liter of each of these two solutions, as shown in Figure 17.3. As we've seen in Sample Exercise 17.6(b),

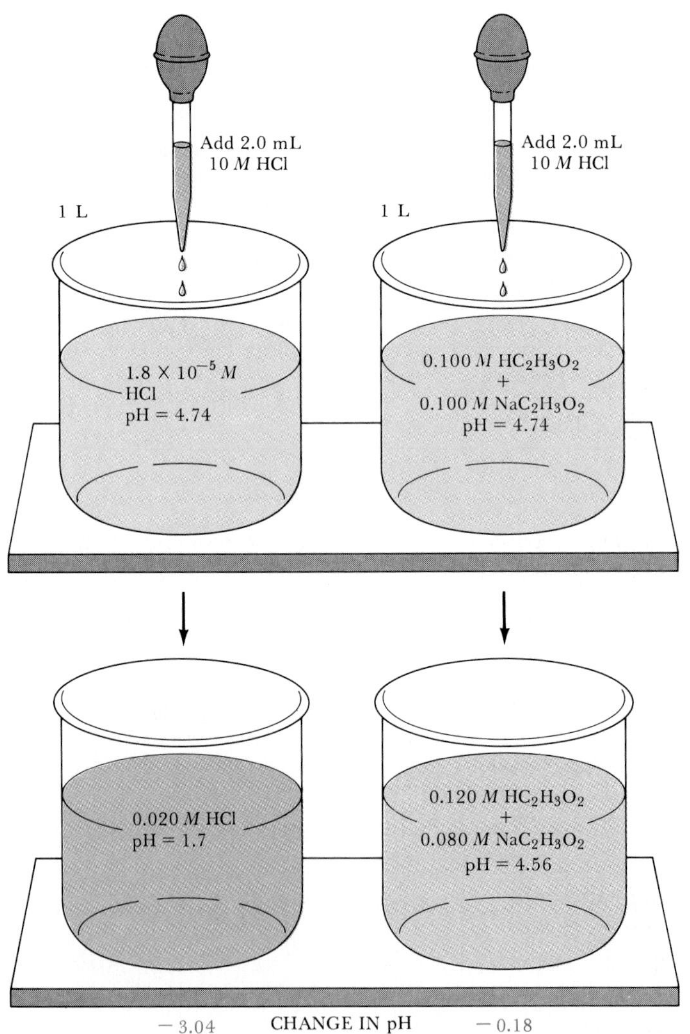

Figure 17.3 The effect of added acid on a buffer solution of pH 4.74 compared with the effect on an HCl solution of pH 4.74.

CHEMISTRY AT WORK: Blood as a Buffered Solution

Blood (Figure 17.4) is an important example of a buffered solution. Human blood is slightly basic with a pH of about 7.39 to 7.45 at 25°C. In a healthy person, the pH never departs more than perhaps 0.2 pH unit from the average value. Whenever pH falls below about 7.4, the condition is called *acidosis;* when pH rises above 7.4, the condition is called *alkalosis.* Death may result if the pH falls below 6.8 or rises above 7.8. Acidosis is the more common tendency because ordinary metabolism produces several acids.

The body uses three primary methods to control blood pH: (1) The blood contains several buffers, including H_2CO_3–HCO_3^- and $H_2PO_4^-$–HPO_4^{2-} pairs, and hemoglobin-containing conjugate acid-base pairs. (2) The kidneys serve to absorb or release $H^+(aq)$. The pH of urine is normally about 5.0 to 7.0. Acidosis is accompanied by increased loss of body fluids as the kidneys work to reduce $H^+(aq)$. (3) The concentration of $H^+(aq)$ is also altered by the rate at which CO_2 is removed from the lungs. The pertinent equilibria are

$$H^+(aq) + HCO_3^-(aq) \rightleftharpoons H_2CO_3(aq) \rightleftharpoons H_2O(l) + CO_2(g)$$

Removal of CO_2 shifts these equilibria to the right, thereby reducing $H^+(aq)$.

Acidosis or alkalosis disrupts the mechanism by which hemoglobin transports oxygen in blood. Hemoglobin (Hb) is involved in a series of equilibria whose overall result is approximately

$$HbH^+(aq) + O_2(aq) \rightleftharpoons HbO_2(aq) + H^+(aq)$$

In acidosis, this equilibrium is shifted to the left, and the ability of hemoglobin to form oxyhemoglobin, HbO_2, is decreased. The lesser amount of O_2 thereby available to cells in the body causes fatigue and headaches; if great enough, it also triggers air hunger (the feeling of being "out of breath" that causes deep breathing).

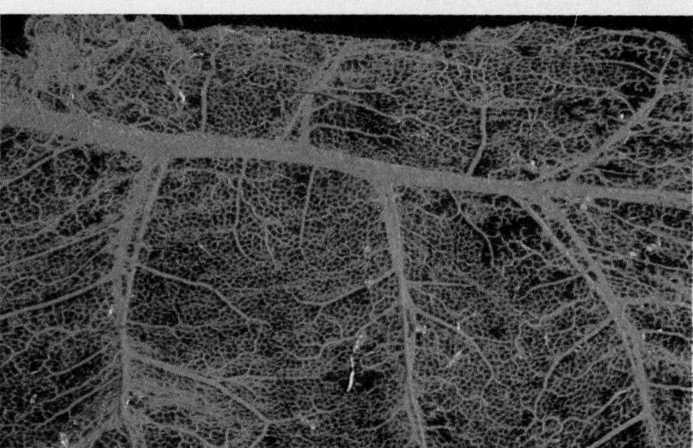

Figure 17.4 Blood vessels; blood is a buffered solution whose pH is maintained at 7.4. (Biophoto Associates, Science Source/Photo Researchers)

Temporary acidosis occurs during strenuous exercise. when energy demands exceed the oxygen available for complete oxidation of glucose to CO_2. In this case, the glucose is converted to an acidic metabolism product, lactic acid, $CH_3CHOHCOOH$. Acidosis also occurs when glucose is unavailable to the cells. This situation can arise, for example, during starvation or as a result of diabetes. In the case of diabetes, glucose is unable to enter the cells because of inadequate insulin, the substance responsible for passage of glucose from the bloodstream to the interior of cells. When glucose is unavailable, the body relies for energy on stored fats, which produce acidic metabolism products.

the pH of the buffer will decrease to 4.57, a change of 0.17 pH unit. In contrast, the pH of the unbuffered solution will decrease to 1.70, a change of 3.04 pH units.

We might have chosen to add 0.020 mol of hydroxide to the solutions illustrated in Figure 17.2. This addition would have caused the pH of the dilute HCl solution to go from 4.74 to 12.3 (you should be able to explain why this is so), whereas the pH of the buffer solution would have increased by 0.18 pH unit. Thus, a buffer solution responds to approximately the same degree, but in opposite direction, to acid or base addition.

17.3 TITRATION CURVES

Many acid-base reactions are used in chemical analyses. For example, the carbonate content in a sample can be determined by a procedure that involves titration with a strong acid, such as HCl. Titrations were described briefly in Section 4.7. In that earlier discussion, we noted that acid-base indicators can be used to signal the equivalence point (that is, the point at which stoichiomet-

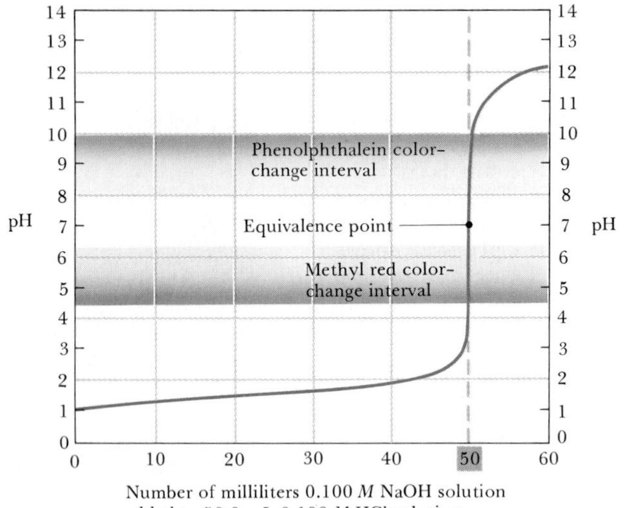

Figure 17.5 The pH curve for titration of a solution of a strong acid with a solution of a strong base, in this case HCl and NaOH.

rically equivalent quantities of acid and base have been brought together). But given the variety of indicators changing colors at different pH values, which indicator is best for a particular titration? This question can be answered by examining a graph of pH changes during a titration. A graph of pH as a function of the volume of added titrant is called a **titration curve**.

Strong Acid-Strong Base Titrations

The titration curve produced when a strong base is added to a strong acid has the general shape shown in Figure 17.5. This curve depicts the pH change that occurs as 0.100 *M* NaOH is added to 50.0 mL of 0.100 *M* HCl. The pH can be calculated at various stages of the titration. The pH of the solution before the addition of any base is determined by the initial concentration of the strong acid: pH = $-\log(0.100) = 1.000$ for 0.100 *M* HCl. Thus, the pH starts out low.

As NaOH is added, the pH increases slowly at first and then rapidly in the vicinity of the equivalence point. The pH of the solution before the equivalence point is determined by the concentration of acid that has not yet been neutralized by base. The pH at the equivalence point is the pH of the resultant salt solution. Because the salt produced by the reaction of a strong acid and a strong base (in this case NaCl) does not hydrolyze (Section 16.8), the equivalence point occurs at pH 7.00. The pH of the solution after the equivalence point is determined by the concentration of the excess base in the solution.

SAMPLE EXERCISE 17.7

Calculate the pH when the following quantities of 0.100 *M* NaOH solution have been added to 50.00 mL of 0.100 *M* HCl solution: **(a)** 49.00 mL; **(b)** 49.90 mL; **(c)** 50.10 mL; **(d)** 51.00 mL.

Solution: **(a)** The number of moles of OH^- in 49.00 mL of 0.100 *M* NaOH is

$$0.04900 \text{ L soln}\left(\frac{0.100 \text{ mol OH}^-}{1 \text{ L soln}}\right) = 4.90 \times 10^{-3} \text{ mol OH}^-$$

The number of moles of H^+ in the original solution is 5.00×10^{-3}. Thus, there remains

$$(5.00 \times 10^{-3}) - (4.90 \times 10^{-3}) = 1.0 \times 10^{-4} \text{ mol } H^+(aq)$$

in 0.0990 L of solution. The concentration of $H^+(aq)$ is thus

$$\frac{1.0 \times 10^{-4} \text{ mol}}{0.0990 \text{ L soln}} = 1.0 \times 10^{-3}\ M$$

The corresponding pH is 3.00.

To answer parts (b), (c), and (d) of the exercise we proceed in the same way. In each case, we calculate the amount of added OH^-, compare that with the amount of H^+ originally present, and determine what amount of H^+ or OH^- exists in excess over the other. The concentration is then calculated from a knowledge of the total volume of the solution. By proceeding in this way we can construct a table of pH versus volume of added NaOH solution, as shown in Table 17.1. By graphing the pH as a function of the volume of added NaOH, we obtain the titration curve shown in Figure 17.5.

PRACTICE EXERCISE

Calculate the pH when the following quantities of 0.100 *M* HCl have been added to 25.00 mL of 0.100 *M* NaOH solution: **(a)** 24.90 mL; **(b)** 25.00 mL; **(c)** 25.10 mL.
Answers: **(a)** 10.30; **(b)** 7.00; **(c)** 3.70

Because the pH change for a strong acid-strong base titration is very large near the equivalence point, the indicator for the titration need not change color precisely at 7.00. Most strong acid-strong base titrations are carried out using phenolphthalein as an indicator because its color change is dramatic (Figure 17.6). From Table 16.1, we see that this indicator changes color in the pH range 8.3 to 10. Thus, a slight excess of NaOH must be present to cause the observed color change. However, such a tiny excess of base is required to bring about the color change that no serious error is introduced. Similarly, methyl red, which changes color in the slightly acid range (Figure 17.7), could also be used. The pH intervals of color change for these two indicators are shown in Table 16.1.

Titration of a solution of a strong base with a solution of a strong acid would yield an entirely analogous curve of pH versus added acid. In this case, however, the pH would be high at the outset of the titration and low at its completion.

Table 17.1 Titration of 50.00 mL of 0.100 *M* HCl Solution with 0.100 *M* NaOH Solution

Volume of HCl	Volume of NaOH	Total volume	Moles H^+	Moles OH^-	Molarity of excess ion	pH
50.00	0.00	50.00	5.00×10^{-3}	0.00	0.100	1.00
50.00	49.00	99.00	5.00×10^{-3}	4.90×10^{-3}	1.00×10^{-3} (H^+)	3.00
50.00	49.90	99.90	5.00×10^{-3}	4.99×10^{-3}	1.00×10^{-4} (H^+)	4.00
50.00	50.10	100.10	5.00×10^{-3}	5.01×10^{-3}	1.00×10^{-4} (OH^-)	10.00
50.00	51.00	101.00	5.00×10^{-3}	5.10×10^{-3}	1.00×10^{-3} (OH^-)	11.00
50.00	60.00	110.00	5.00×10^{-3}	6.00×10^{-3}	9.09×10^{-3} (OH^-)	11.96

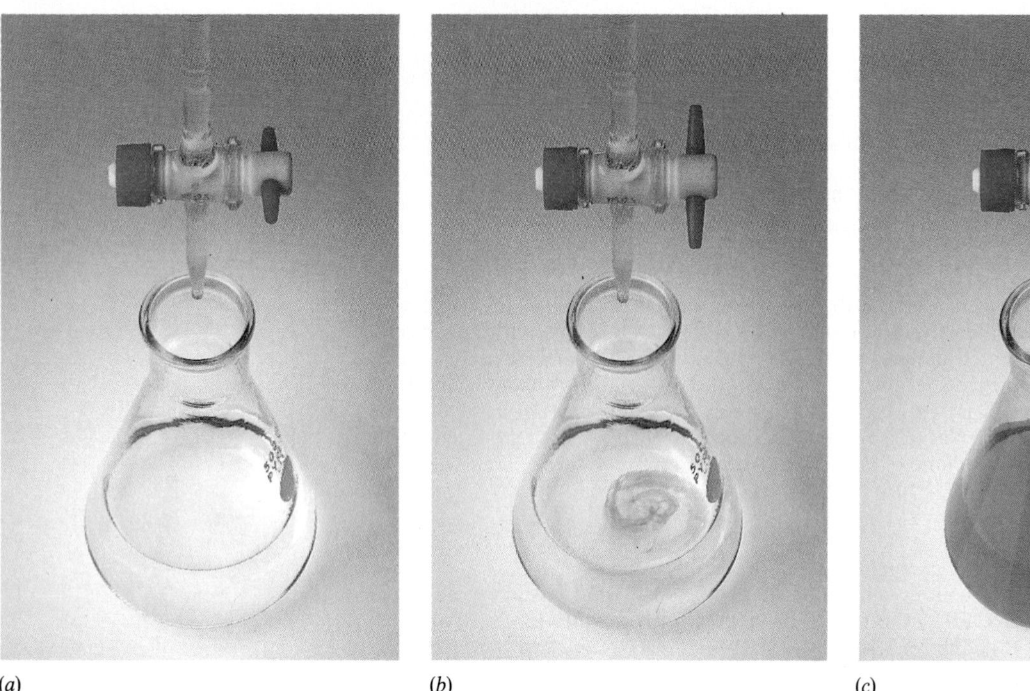

(a) (b) (c)

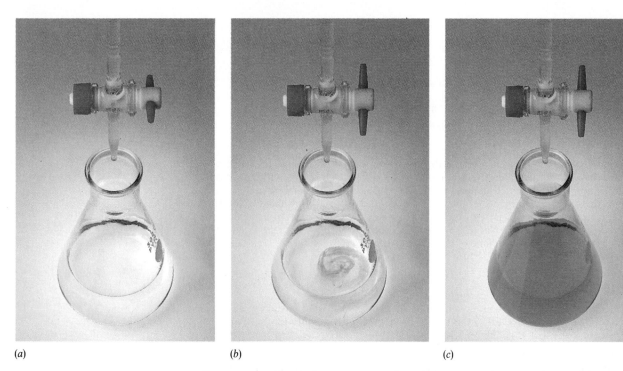

Figure 17.6 Change in appearance of a solution containing phenolphthalein indicator as base is added . Before the endpoint, the solution is colorless (*a*). As the endpoint is approached, a pale pink color forms where the base is added (*b*). At the endpoint, this pale pink color extends throughout the solution after the mixing. As more base is added, the intensity of the pink color increases (*c*). (© Richard Megna/Fundamental Photographs)

Figure 17.7 Change in appearance of a solution containing methyl red indicator in the pH range 4.2 to 6.3. The characteristic acid color is shown in (*a*), the characteristic basic color in (*b*). (Richard Megna/ Fundamental Photographs)

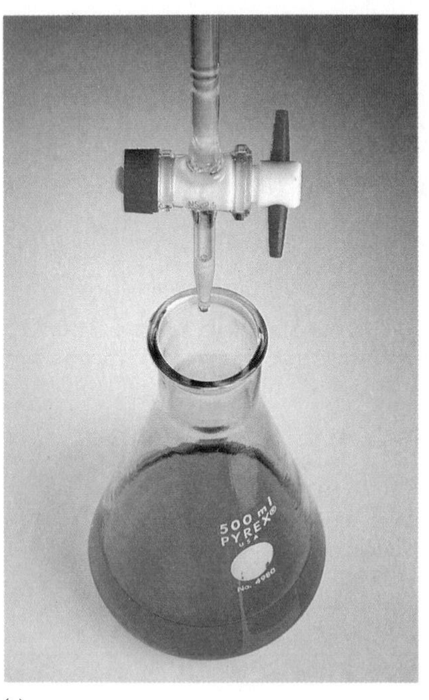

(a)

(b)

Titrations Involving a Weak Acid or Weak Base

Titration of a weak acid by a strong base results in pH curves that look similar to those for strong acid-strong base titrations. Figure 17.8 shows the pH curve produced when 0.100 *M* NaOH is added to 50.0 mL of 0.100 *M* acetic acid. There are three noteworthy differences between this curve and that for a strong acid-strong base titration: (1) The weak-acid solution has a higher initial pH. In Section 16.3, we calculated that the pH of a 0.100 *M* solution of $HC_2H_3O_2$ is 2.89; by contrast, the pH of a 0.100 *M* solution of HCl is 1.00. (2) The pH rises more rapidly in the early part of the titration, but more slowly near the equivalence point. The weaker the acid, the less marked the pH change near the equivalence point. This aspect is illustrated in Figure 17.9, which shows a family of titration curves for weak acids of varying K_a. (3) The pH at the equivalence point is *not* 7.00. Remember that the pH at the equivalence point in any acid-base titration is the pH of the resultant salt solution. In the titration illustrated in Figure 17.8, the solution contains 0.050 *M* $NaC_2H_3O_2$ at the equivalence point. (Remember that the total volume of solution is doubled by addition of the NaOH solution to $HC_2H_3O_2$.) We saw in Chapter 16 how to calculate the pH of a solution of a salt containing a weak base as anion (see, for example, Sample Exercise 16.14). The pH of a 0.050 *M* $NaC_2H_3O_2$ solution is 8.71.

Sample Exercise 17.8 illustrates the calculation of pH at a point along the titration curve of a weak acid with a strong base. The pH before the addition of base is determined by the initial concentration of the weak acid. Calculations of subsequent pH values have two parts. The first part involves simple stoichiometry. We determine how much base has reacted with acid in the neutralization reaction and how much salt forms. From this, we determine the concentration of the salt and the concentration of the excess acid or base (if any) remaining in the solution. The second part of the problem is to calculate the pH of the resultant solution.

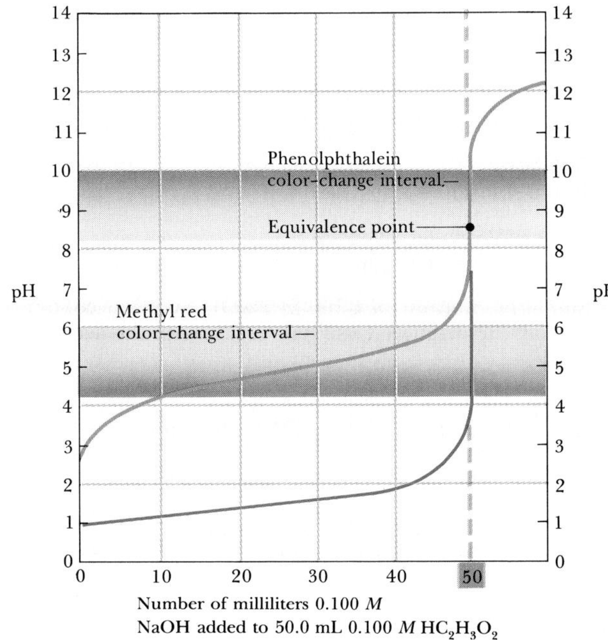

Figure 17.8 The green line shows the variation in pH as 0.100 *M* NaOH solution is added in the titration of 0.100 *M* acetic acid solution. The blue line segment shows the graph of pH versus added base for the titration of 0.100 *M* HCl.

Figure 17.9 Influence of acid strength on the shape of the curve for titration with NaOH. Each curve represents titration of 50.0 mL of 0.10 *M* acid with 0.10 *M* NaOH.

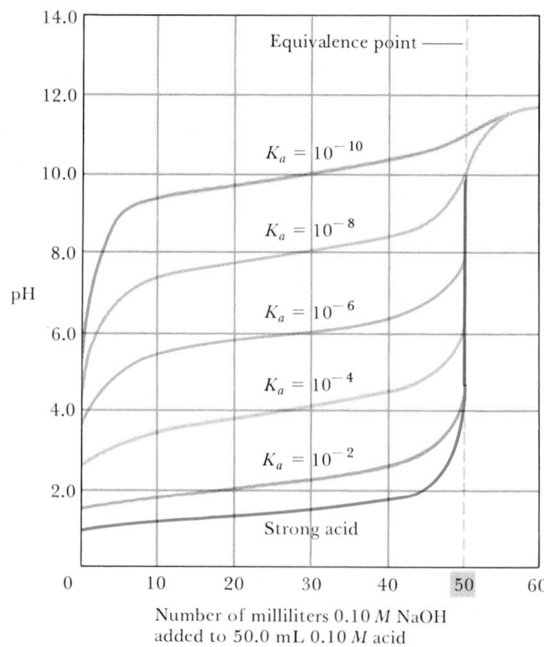

Before the equivalence point is reached, the neutralization reaction results in a solution containing a mixture of the weak acid and its salt. We have encountered the calculation of the pH of these solutions in connection with our discussion of buffers. At the equivalence point, the solution contains only the salt. For the salt of a weak acid and strong base, the solution will be weakly basic because of the hydrolysis of the anion. After the equivalence point, the solution contains a mixture of the salt and the excess strong base; the pH of such solutions is controlled mainly by the concentration of the excess base. The calculation of this pH is the same as that for points after the equivalence point in the titration of a strong acid with a strong base.

SAMPLE EXERCISE 17.8

Calculate the pH in the titration of acetic acid by NaOH after 30.0 mL of 0.100 *M* NaOH has been added to 50.0 mL of 0.100 *M* acetic acid.

Solution: *Stoichiometric Calculation:* The number of moles of $HC_2H_3O_2$ originally in the solution is

$$\left(\frac{0.100 \text{ mol } HC_2H_3O_2}{1 \text{ L soln}}\right)(0.0500 \text{ L soln}) = 5.00 \times 10^{-3} \text{ mol } HC_2H_3O_2$$

Similarly, 30.0 mL of 0.100 *M* NaOH solution contains 3.00×10^{-3} mol of OH^-. During the titration, this OH^- reacts with the acetic acid:

$$OH^-(aq) \quad + \quad HC_2H_3O_2(aq) \quad \longrightarrow \quad C_2H_3O_2^-(aq) \quad + \quad H_2O(l)$$

	$OH^-(aq)$	$HC_2H_3O_2(aq)$	$C_2H_3O_2^-(aq)$	$H_2O(l)$
Before rxn:	3.00×10^{-3} mol	5.00×10^{-3} mol	0	—
Change:	-3.00×10^{-3} mol	-3.00×10^{-3} mol	$+3.00 \times 10^{-3}$ mol	—
After rxn:	0	2.00×10^{-3} mol	3.00×10^{-3} mol	—

The total volume of the solution is 30.0 mL + 50.0 mL = 80.0 mL. The resulting molarities are thus

$$[HC_2H_3O_2] = \frac{2.00 \times 10^{-3} \text{ mol } HC_2H_3O_2}{0.0800 \text{ L}} = 0.0250\ M$$

$$[C_2H_3O_2^-] = \frac{3.00 \times 10^{-3} \text{ mol } C_2H_3O_2^-}{0.0800 \text{ L}} = 0.0375\ M$$

Equilibrium Calculation:
For the weak-acid dissociation equilibrium of $HC_2H_3O_2(aq)$, we have

$$K_a = \frac{[H^+][C_2H_3O_2^-]}{[HC_2H_3O_2]} = 1.8 \times 10^{-5}$$

$$[H^+] = K_a \frac{[HC_2H_3O_2]}{[C_2H_3O_2^-]} = (1.8 \times 10^{-5})\left(\frac{0.0250}{0.0375}\right) = 1.2 \times 10^{-5}\ M$$

$$pH = -\log(1.2 \times 10^{-5}) = 4.92$$

By proceeding in similar fashion, other points on the titration curve shown in Figure 17.8 could be calculated. Incidentally, note that halfway to the equivalence point of the titration, when $[C_2H_3O_2^-]$ equals $[HC_2H_3O_2]$, the pH equals pK_a, which in this case is 4.74.

PRACTICE EXERCISE

(a) Calculate the pH in the titration of ammonia by hydrochloric acid after 15.0 mL of 0.100 M HCl has been added to 25.0 mL of 0.100 M $NH_3(aq)$ solution. **(b)** What is the pH at the equivalence point in this titration? ***Answers:*** **(a)** 9.08; **(b)** 5.28

It is important to note that the pH at the equivalence point in the titration of the weak acid is considerably higher than it is in the titration of the strong acid. Acid-base titrations can be carried out using a pH meter (Section 16.1) to indicate progress of the titration. For occasional titrations, however, it is common to use an indicator to signal the equivalence point. We saw earlier that in titrating 0.100 M HCl with 0.100 M NaOH, either phenolphthalein or methyl red could be used as indicator. Although the pH of color change does not correspond precisely to the equivalence point for either indicator, both were close enough that no significant error would be introduced by using either of them (Figure 17.5). In a titration of acetic acid with NaOH, phenolphthalein is an ideal indicator, because it changes color just at the pH of the equivalence point. However, methyl red is not a good choice. The pH at the midrange of its color change is 5.2. At this pH we are still far short of the equivalence point, as shown in Figure 17.8.

If the acid to be titrated were weaker than acetic acid, the titration curve would look similar to that shown in Figure 17.8. However, the pH at the beginning of the reaction would be higher, and the pH at the equivalence point would also be higher. This means that the pH change at the equivalence point would not be as sharp, and the end point would be more difficult to detect (see Figure 17.9). With very weak acids, for which K_a is less than about 1×10^{-8}, an acid-base titration is not really a good quantitative procedure when using indicators.

Titration of a weak base (for example, 0.10 M NH_3) with a strong acid solution (such as 0.10 M HCl) leads to the titration curve shown in Figure 17.10. In this particular example, the equivalence point occurs at pH 5.3. Thus, methyl red would be an ideal indicator, but phenolphthalein would be a poor choice.

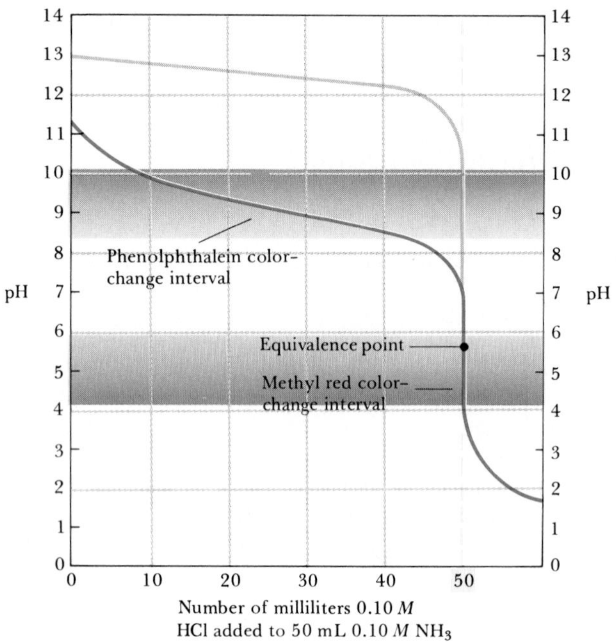

Figure 17.10 The red line shows pH versus volume of added HCl in the titration of 0.10 *M* ammonia with 0.10 *M* HCl. The orange line segment shows the graph of pH versus added acid for the titration of 0.10 *M* NaOH.

Titrations of Polyprotic Acids

In the case of weak acids containing more than one ionizable proton, reaction with OH^- occurs in a series of steps. An important example of the kind of equilibria that occur is carbonic acid, H_2CO_3. We saw in Sample Exercise 16.11 that in an aqueous solution of CO_2 the relative concentrations of the species involved are $[H_2CO_3] \gg [HCO_3^-] \gg [CO_3^{2-}]$. Neutralization of H_2CO_3 proceeds in two stages:

$$H_2CO_3(aq) + OH^-(aq) \longrightarrow H_2O(l) + HCO_3^-(aq) \qquad [17.12]$$

$$HCO_3^-(aq) + OH^-(aq) \longrightarrow H_2O(l) + CO_3^{2-}(aq) \qquad [17.13]$$

Figure 17.11 shows how the relative abundances of H_2CO_3, HCO_3^-, and CO_3^{2-} vary as a function of the solution pH. As OH^- is added to H_2CO_3 (as pH increases), $[H_2CO_3]$ decreases, while $[HCO_3^-]$ increases. By pH 8.5,

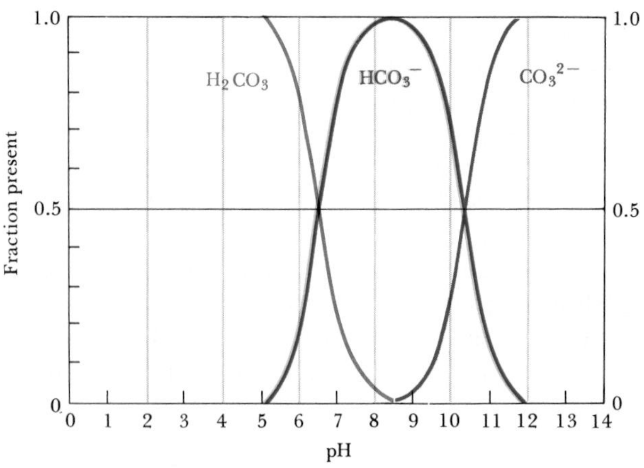

Figure 17.11 Relative abundances of the various species derived from dissolved CO_2 as a function of pH.

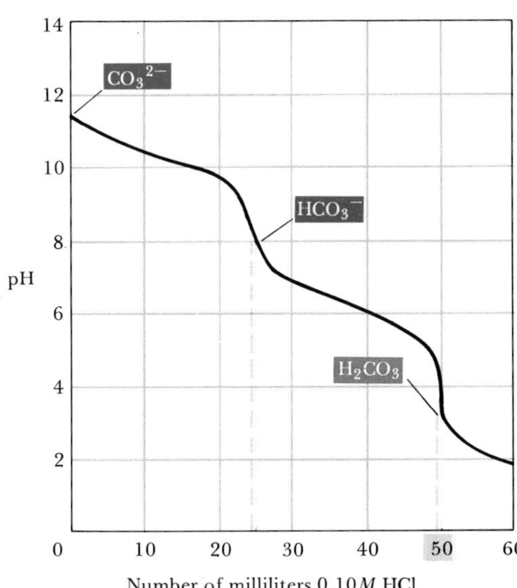

Figure 17.12 Titration curve for the reaction of 25.0 mL of 0.10 *M* Na_2CO_3 with 0.10 *M* HCl.

H_2CO_3 has been nearly completely converted to HCO_3^-. At this stage, Figure 17.11 shows $[H_2CO_3] = 0$, and $[CO_3^{2-}] = 0$; the concentrations of these species are too small to register on the figure. As pH increases beyond 8.5, $[HCO_3^-]$ decreases, while $[CO_3^{2-}]$ increases. Thus, only after the first reaction goes nearly to completion does the second reaction start.

When the neutralization steps of a polyprotic acid or polybasic base are sufficiently separated, the substance exhibits a titration curve with multiple equivalence points. Figure 17.12 shows the titration curve for the H_2CO_3–HCO_3^-–CO_3^{2-} system. Notice that there are two distinct equivalence points along the titration curve.

17.4 SOLUBILITY EQUILIBRIA

The equilibria that we have considered thus far in this chapter have involved acids and bases. Furthermore, they have been homogeneous; that is, all the species involved in an equilibrium have been in the same phase. In this section we will consider the equilibria involved in another important type of solution reaction: the dissolution or precipitation of ionic compounds. These reactions are heterogeneous.

The dissolving and precipitating of compounds are phenomena that occur both within us and around us. For example, the dissolving of tooth enamel in acidic solutions causes tooth decay. The precipitation of certain salts in our kidneys produces kidney stones. The waters of the earth contain salts dissolved as water passes over and through the ground. Precipitation of $CaCO_3$ from groundwater is responsible for the formation of stalactites and stalagmites within limestone caves.

We discussed precipitation reactions briefly in Section 4.5. In that discussion, we considered some general rules for predicting the solubility of common salts in water. These rules give us a qualitative sense of whether a compound will have a low or high solubility in water. By considering solubility equilibria, in contrast, we can make quantitative predictions about the amount of a given compound that will dissolve.

The Solubility-Product Constant, K_{sp}

For an equilibrium to exist between a solid substance and its solution, the solution must be saturated and in contact with undissolved solid. As an example, consider a saturated solution of $BaSO_4$ that is in contact with solid $BaSO_4$. The chemical equation for the relevant equilibrium can be written as

$$BaSO_4(s) \rightleftharpoons Ba^{2+}(aq) + SO_4^{2-}(aq) \quad [17.14]$$

The solid is an ionic compound. Ionic compounds are strong electrolytes. To the extent that they dissolve, we will assume that they dissociate completely in solution, forming hydrated cations and anions.

We observed in Section 15.4 that for heterogeneous reactions the concentration of the solid is constant. It is thus incorporated into the equilibrium constant. The equilibrium-constant expression for the dissolution of $BaSO_4$ can therefore be written as

$$K = [Ba^{2+}][SO_4^{2-}]$$

When concentration is expressed in molarity, the equilibrium constant is called the **solubility-product constant** (or simply the **solubility product**) and is designated as K_{sp}:

$$K_{sp} = [Ba^{2+}][SO_4^{2-}] \quad [17.15]$$

The rules for writing the solubility-product expression are the same as those for the writing of any equilibrium-constant expression: *The solubility product is equal to the product of the concentrations of the ions involved in the equilibrium, each raised to the power of its coefficient in the equilibrium equation.*

SAMPLE EXERCISE 17.9

Write the expression for the solubility-product constant for $Ca_3(PO_4)_2$.

Solution: We first write the equation for the solubility equilibrium:

$$Ca_3(PO_4)_2(s) \rightleftharpoons 3Ca^{2+}(aq) + 2PO_4^{3-}(aq)$$

Following the rule stated above, the power to which the Ca^{2+} concentration is raised is 3 and the power to which the PO_4^{3-} concentration is raised is 2. The resulting expression for K_{sp} is

$$K_{sp} = [Ca^{2+}]^3[PO_4^{3-}]^2$$

In Appendix D, we see that this K_{sp} has a value of 2.0×10^{-29}.

PRACTICE EXERCISE

Give the solubility-product expressions and the values of the solubility-product constants (from Appendix D) for the following: **(a)** cadmium sulfide; **(b)** chromium(III) hydroxide. ***Answers:*** **(a)** $K_{sp} = [Cd^{2+}][S^{2-}] = 8.0 \times 10^{-27}$; **(b)** $K_{sp} = [Cr^{3+}][OH^-]^3 = 6.3 \times 10^{-31}$

Solubility and K_{sp}

It is important to distinguish carefully between solubility and solubility product. The *solubility* of a substance is the quantity that dissolves to form a saturated solution (Section 13.3). Solubility is often expressed as grams of solute per liter

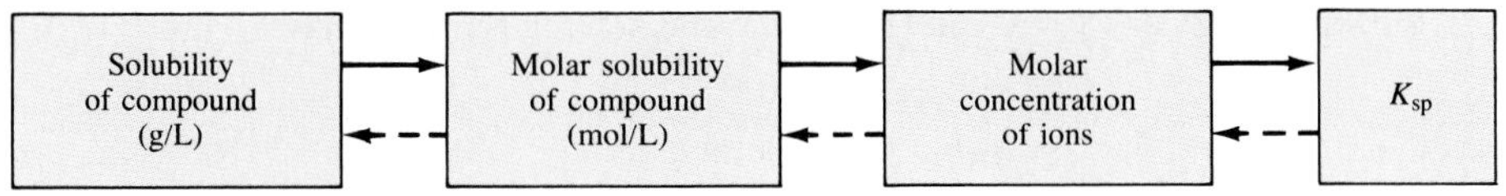

Figure 17.13 Outline of steps involved in interconverting solubility and K_{sp}.

of solution (g/L). The *molar solubility* is the number of moles of the solute that dissolves in forming a liter of a saturated solution of the solute (mol/L). The solubility product (K_{sp}) is the equilibrium constant for the equilibrium between an ionic solid and its saturated solution.

The solubility of a substance changes as conditions such as the concentrations of other solutes change. For example, the solubility of $Mg(OH)_2$ changes as the pH of the solution changes. In contrast, the solubility product has only one value for a given solute at a given temperature.* Appendix D lists the K_{sp} values for a variety of slightly soluble ionic compounds at 25°C.

In studying solubility equilibria, it is important to be able to interconvert solubility and K_{sp}. Sample Exercise 17.10 shows the calculation of K_{sp} from solubility data when the slightly soluble salt is the only solute present. Sample Exercise 17.11 similarly shows the calculation of solubility from K_{sp}. Figure 17.13 summarizes these procedures.

SAMPLE EXERCISE 17.10

Solid barium sulfate is shaken in contact with pure water at 25°C for several days. Each day a sample is withdrawn and analyzed for its barium concentration. After several days, the value of $[Ba^{2+}]$ is constant, indicating that equilibrium has been reached. The concentration of Ba^{2+} is 1.04×10^{-5} *M*. What is K_{sp} for $BaSO_4$?

Solution: Because all of the Ba^{2+} and SO_4^{2-} ions come from $BaSO_4$, there must be one SO_4^{2-} for each Ba^{2+} (Equation 17.14). Consequently, $[Ba^{2+}] = [SO_4^{2-}] = 1.04 \times 10^{-5}$ *M*. Thus, we have

$$K_{sp} = [Ba^{2+}][SO_4^{2-}] = (1.04 \times 10^{-5})(1.04 \times 10^{-5}) = 1.08 \times 10^{-10}$$

PRACTICE EXERCISE

When MgF_2 is shaken with water at 27°C until the solution is saturated, the concentration of Mg^{2+} ions is found to be 1.2×10^{-3} *M*. What is K_{sp} for MgF_2 at this temperature? (Notice that the salt produces unequal numbers of cations and anions when it dissolves.) ***Answer:*** 6.9×10^{-9}

SAMPLE EXERCISE 17.11

The K_{sp} for CaF_2 is 3.9×10^{-11}. What is the solubility of CaF_2 in water in grams per liter?

Solution: The solubility equilibrium involved is

$$CaF_2(s) \rightleftharpoons Ca^{2+}(aq) + 2F^-(aq)$$

For each mole of CaF_2 that dissolves, 1 mol of Ca^{2+} and 2 mol of F^- enter the solution. Thus if we let x equal the molar solubility of CaF_2, the molar concentrations of Ca^{2+} and F^- will be $[Ca^{2+}] = x$ and $[F^-] = 2x$.

The K_{sp} expression for this reaction is

$$K_{sp} = [Ca^{2+}][F^-]^2 = 3.9 \times 10^{-11}$$

* This is strictly true only for very dilute solutions. The values of equilibrium constants are somewhat altered when the total concentration of ionic substances in water is increased. However, we shall ignore these effects, which are taken into consideration only for very accurate work.

Substituting $[Ca^{2+}] = x$ and $[F^-] = 2x$ and solving for x, we have

$$x(2x)^2 = 3.9 \times 10^{-11}$$
$$4x^3 = 3.9 \times 10^{-11}$$
$$x = \sqrt[3]{\frac{3.9 \times 10^{-11}}{4}} = 2.1 \times 10^{-4}\ M$$

Thus, the molar solubility of CaF_2 is 2.1×10^{-4} mol/L. The mass of CaF_2 that dissolves in water to form a liter of solution is

$$\left(\frac{2.1 \times 10^{-4} \text{ mol } CaF_2}{1 \text{ L soln}}\right)\left(\frac{78.1 \text{ g } CaF_2}{1 \text{ mol } CaF_2}\right) = 1.6 \times 10^{-2} \text{ g } CaF_2\text{/L soln}$$

PRACTICE EXERCISE

The K_{sp} for $Cu(N_3)_2$ is 6.3×10^{-10}. What is the solubility of $Cu(N_3)_2$ in water in grams per liter? ***Answer:*** 0.080 g/L

The Common-Ion Effect

In the sample exercises above, we considered the dissolving of ionic compounds in pure water. The presence of other solutes, however, can affect the solubility of a substance. For example, the presence of either $Ca(NO_3)_2$ or NaF will decrease the solubility of CaF_2. We can explain this observation in terms of Le Châtelier's principle. The additional Ca^{2+} or F^- ions shift the solubility equilibrium of CaF_2 to the left, thereby reducing the quantity of CaF_2 that dissolves:

$$CaF_2(s) \rightleftharpoons Ca^{2+}(aq) + 2F^-(aq)$$

Addition of Ca^{2+} or F^- shifts equilibrium, reducing solubility

In general, the solubility of a salt decreases when a solute is added that contains a common cation or anion. This reduction in solubility is another application of the *common-ion effect* (Section 18.1). The procedures used in treating this effect quantitatively are illustrated in Sample Exercise 17.12.

SAMPLE EXERCISE 17.12

Calculate the molar solubility of CaF_2 in a solution containing **(a)** 0.010 M $Ca(NO_3)_2$; **(b)** 0.010 M NaF.

Solution: As in Sample Exercise 17.11, we are dealing with the solubility of CaF_2, so we again use its solubility product:

$$K_{sp} = [Ca^{2+}][F^-]^2 = 3.9 \times 10^{-11}$$

The value of K_{sp} is unchanged by the presence of the additional solutes. In Sample Exercise 17.11, we found that the molar solubility of CaF_2 in pure water is 2.1×10^{-4} mol/L. The molar solubility in the presence of common ions will be less than this.

(a) Let's again let x represent the molar solubility of CaF_2. Then the concentration of Ca^{2+} derived from CaF_2 is x and the concentration of F^- is $2x$. But there is a second source of Ca^{2+}, the 0.010 M contribution from the dissolved $Ca(NO_3)_2$. Thus, we have

$$[Ca^{2+}] = x + 0.010 \qquad \text{and} \qquad [F^-] = 2x$$

A CLOSER LOOK: Limitations of Solubility Products

The concentrations of ions calculated from K_{sp} sometimes deviate appreciably from those found experimentally. In part, these deviations are due to electrostatic interactions between ions in solution. (See the Closer Look box on colligative properties of electrolyte solutions in Section 13.5.) These interactions increase in magnitude as the concentrations of ions in solution increase, and they tend to cause salts to have somewhat higher solubilities than calculated. Chemists have developed procedures for correcting for these "ionic-strength" or "ionic-activity" effects.

Another common source of error in calculating ion concentrations is ignoring other equilibria that occur simultaneously in the solution. For example, we have assumed that *all* the salt that dissolves is present in solution as ions. This assumption is not always valid. Sometimes there is a small concentration of undissociated or partially dissociated salt that is in equilibrium with the dissociated ions in solution. For example, when MgF_2 dissolves, it yields not only Mg^{2+} and F^- ions, but also a very small but not always negligible concentration of MgF^+ ions in solution. Simultaneous solubility and acid-base or complex-ion equilibria are also sometimes overlooked. In particular, both basic anions and cations with high charge-to-size ratios undergo hydrolysis reactions that can measurably increase the solubilities of their salts. For example, if we ignore the hydrolysis of the PO_4^{3-} ion ($K_b = 2.4 \times 10^{-2}$) when we consider the solubility of Li_3PO_4 ($K_{sp} = 3.2 \times 10^{-9}$), we calculate a molar solubility of 3.3×10^{-3} *M*. Hydrolysis of PO_4^{3-} causes the solubility to increase to 3.5×10^{-3} *M*. We consider briefly the effect of acid-base equilibria on solubility in Section 17.5.

Substituting into the solubility-product expression gives

$$K_{sp} = 3.9 \times 10^{-11} = [Ca^{2+}][F^-]^2 = (x + 0.010)(2x)^2$$

This would be a messy problem to solve exactly, but fortunately it is possible to simplify matters greatly. Even without the common-ion effect, the solubility of CaF_2 is very small. We can therefore assume that the 0.010 *M* concentration of Ca^{2+} from $Ca(NO_3)_2$ is very much greater than the small additional concentration resulting from the solubility of CaF_2. That is, x is small compared to 0.010 *M*, and we can therefore assume that $0.010 + x \simeq 0.010$. We then have

$$3.9 \times 10^{-11} = (0.010)(2x)^2$$

$$x^2 = \frac{3.9 \times 10^{-11}}{4(0.010)} = 9.8 \times 10^{-10}$$

$$x = \sqrt{9.8 \times 10^{-10}} = 3.1 \times 10^{-5}\ M$$

The very small value for x validates the simplifying assumption we have made. Our calculation indicates that 3.1×10^{-5} mol of solid CaF_2 dissolves per liter of the 0.010 *M* $Ca(NO_3)_2$ solution.

(b) In this case the common ion is F^-, and we have

$$[Ca^{2+}] = x \quad \text{and} \quad [F^-] = 2x + 0.010$$

Assuming that $2x$ is small compared to 0.010 *M* (that is, $0.010 + 2x \simeq 0.010$), we have

$$3.9 \times 10^{-11} = x(0.010)^2$$

$$x = \frac{3.9 \times 10^{-11}}{(0.010)^2} = 3.9 \times 10^{-7}\ M$$

Thus, 3.9×10^{-7} mol of solid CaF_2 dissolves per liter of 0.010 *M* NaF solution.

PRACTICE EXERCISE

The value of K_{sp} for cerium hydroxide, $Ce(OH)_3$, is 1.5×10^{-20}. What is the molar solubility of $Ce(OH)_3$ in a solution that contains 0.10 *M* NaOH?
Answer: 1.5×10^{-17} *M*

17.5 CRITERIA FOR PRECIPITATION OR DISSOLUTION

Equilibrium can be achieved starting with the substances on either side of the chemical equation. The equilibrium between $BaSO_4(s)$, $Ba^{2+}(aq)$, and $SO_4^{2-}(aq)$ (Equation 17.14) can be achieved starting with solid $BaSO_4$. It can also be achieved starting with solutions of salts containing Ba^{2+} and SO_4^{2-}, say $BaCl_2$ and Na_2SO_4. When these two solutions are mixed, a precipitate of $BaSO_4$ will form if the product of ion concentrations, $Q = [Ba^{2+}][SO_4^{2-}]$, is greater than K_{sp}.

The use of the reaction quotient, Q, to determine the direction a reaction must proceed to reach equilibrium was discussed earlier, in Section 15.5. In the present case, the equilibrium expression contains no denominator and therefore is really not a quotient. Thus, Q is often referred to simply as the *ion product*. The possible relationships between Q and K_{sp} are summarized as follows:

If $Q > K_{sp}$, precipitation occurs until $Q = K_{sp}$.
If $Q = K_{sp}$, equilibrium exists (saturated solution).
If $Q < K_{sp}$, solid dissolves until $Q = K_{sp}$.

SAMPLE EXERCISE 17.13

Will a precipitate form when 0.100 L of 3.0×10^{-3} *M* $Pb(NO_3)_2$ is added to 0.400 L of 5.0×10^{-3} *M* Na_2SO_4?

Solution: The possible metathesis products are $PbSO_4$ and $NaNO_3$. Sodium salts are quite soluble; however, $PbSO_4$ has a K_{sp} of 1.6×10^{-8} (Appendix D). To determine whether the $PbSO_4$ precipitates, we must calculate $Q = [Pb^{2+}][SO_4^{2-}]$ and compare it with K_{sp}.

When the two solutions are mixed, the total volume becomes 0.100 L + 0.400 L = 0.500 L. The number of moles of Pb^{2+} in 0.100 L of 3.0×10^{-3} *M* $Pb(NO_3)_2$ is

$$(0.100 \text{ L})\left(3.0 \times 10^{-3} \frac{\text{mol}}{\text{L}}\right) = 3.0 \times 10^{-4} \text{ mol}$$

The concentration of Pb^{2+} in the 0.500-L mixture is therefore

$$[Pb^{2+}] = \frac{3.0 \times 10^{-4} \text{ mol}}{0.500 \text{ L}} = 6.0 \times 10^{-4}\ M$$

The number of moles of SO_4^{2-} is

$$(0.400 \text{ L})\left(5.0 \times 10^{-3} \frac{\text{mol}}{\text{L}}\right) = 2.0 \times 10^{-3} \text{ mol}$$

Therefore, $[SO_4^{2-}]$ in the 0.500-L mixture is

$$[SO_4^{2-}] = \frac{2.0 \times 10^{-3} \text{ mol}}{0.500 \text{ L}} = 4.0 \times 10^{-3}\ M$$

We then have

$$Q = [Pb^{2+}][SO_4^{2-}] = (6.0 \times 10^{-4})(4.0 \times 10^{-3}) = 2.4 \times 10^{-6}$$

Because $Q > K_{sp}$, precipitation of $PbSO_4$ will occur.

PRACTICE EXERCISE

Assume that 0.10 *M* NaF solution is added slowly, with stirring, to a solution that contains 0.10 *M* $Cr(NO_3)_3$ and 0.10 *M* $Ca(NO_3)_2$. Which salt—CrF_3 or CaF_2—precipitates first? (K_{sp} for CaF_2 is 5.3×10^{-9}; for CrF_3 it is 6.6×10^{-11}).
Answer: CaF_2

SAMPLE EXERCISE 17.14

What concentration of OH^- must be exceeded in a 0.010 *M* solution of $Ni(NO_3)_2$ in order to precipitate $Ni(OH)_2$? (Assume that the added OH^- does not change the concentration of Ni^{2+}.)

Solution: From Appendix D we have $K_{sp} = 1.6 \times 10^{-14}$ for $Ni(OH)_2$. Any OH^- in excess of that in a saturated solution of $Ni(OH)_2$ will cause some $Ni(OH)_2$ to precipitate. For a saturated solution we have

$$K_{sp} = [Ni^{2+}][OH^-]^2 = 1.6 \times 10^{-14}$$

Thus, if $Q = [Ni^{2+}][OH^-]^2 > 1.6 \times 10^{-14}$, precipitation will occur. Letting $[OH^-] = x$ and using $[Ni^{2+}] = 0.010\ M$, we have

$$(0.010)x^2 > 1.6 \times 10^{-14}$$

$$x^2 > \frac{1.6 \times 10^{-14}}{0.010} = 1.6 \times 10^{-12}$$

$$x > \sqrt{1.6 \times 10^{-12}} = 1.3 \times 10^{-6}\ M$$

Thus, $Ni(OH)_2$ will precipitate if $[OH^-]$ exceeds $1.3 \times 10^{-6}\ M$.

This concentration of OH^- corresponds to a solution pH of 8.11 (pH = 14.00 − pOH = 14.00 − 5.89 = 8.11). Thus $Ni(OH)_2$ will precipitate when the solution pH is 8.11 or higher.

PRACTICE EXERCISE

The solubility product for gadolinium hydroxide, $Gd(OH)_3$, is 1.8×10^{-23}. If a solution is 0.010 *M* in Gd^{3+} ion and the pH of the solution is slowly increased, at what pH will $Gd(OH)_3$ begin to precipitate? ***Answer:*** 7.08

Solubility and pH

The solubility of any substance whose anion is basic will be affected to some extent by the pH of the solution. For example, consider $Mg(OH)_2$, for which the solubility equilibrium is

$$Mg(OH)_2(s) \rightleftharpoons Mg^{2+}(aq) + 2OH^-(aq) \qquad [17.16]$$

The value of K_{sp} for $Mg(OH)_2$ is 1.8×10^{-11}. Suppose that solid $Mg(OH)_2$ is equilibrated with a solution buffered at a pH of 9.0. Then pOH is 5.0; that is, $[OH^-] = 1.0 \times 10^{-5}$. Inserting this value for $[OH^-]$ into the solubility-product expression, we have

$$K_{sp} = [Mg^{2+}][OH^-]^2 = 1.8 \times 10^{-11}$$

$$[Mg^{2+}][1.0 \times 10^{-5}]^2 = 1.8 \times 10^{-11}$$

$$[Mg^{2+}] = 0.18\ M$$

Thus, $Mg(OH)_2$ is quite soluble in a buffered, slightly basic medium. If the solution were made more acidic, the solubility of $Mg(OH)_2$ would increase because the OH^- concentration decreases with increasing acidity. The Mg^{2+} concentration would thus increase to maintain the equilibrium condition (Figure 17.14).

The solubility of almost any salt is affected if the solution is made sufficiently acidic or basic. The effects are very noticeable, however, only when one or both ions involved are moderately acidic or basic. The metal hydroxides we've just discussed are good examples of compounds with a strong base, the hydroxide ion. As an additional example, the fluoride ion of CaF_2 is a weak

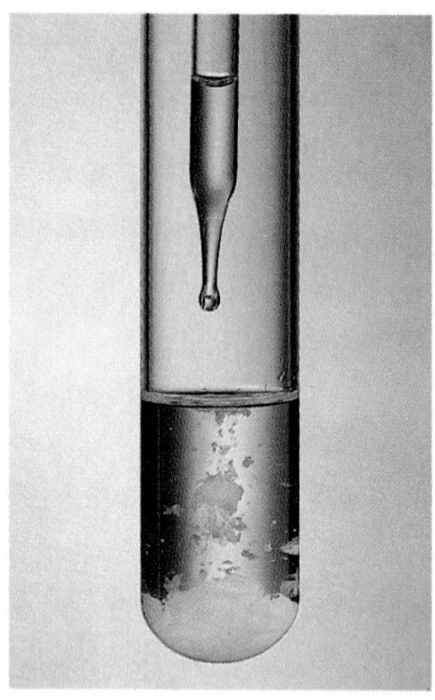

(*a*)

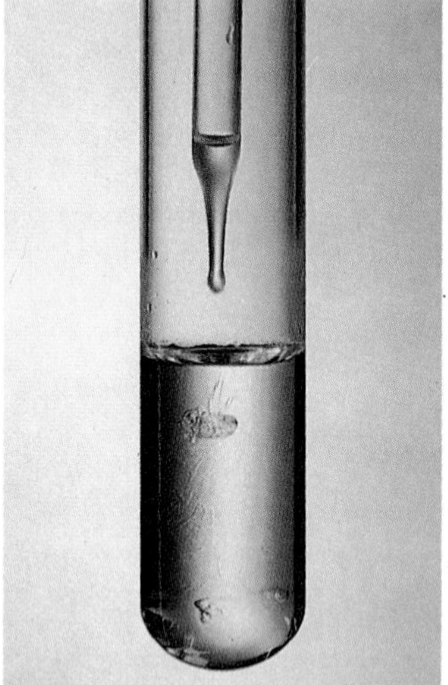

(*b*)

Figure 17.14 (*a*) a precipitate of $Mg(OH)_2$ dissolves (*b*) upon addition of acid. (©) Richard Megna/ Fundamental Photographs

base; it is the conjugate base of the weak acid HF. As a result, CaF_2 is more soluble in acidic solutions than in neutral or basic ones, because of the reaction of F^- with H^+ to form HF. The solution process can be considered as two consecutive reactions:

$$CaF_2(s) \rightleftharpoons Ca^{2+}(aq) + 2F^-(aq) \quad [17.17]$$

$$F^-(aq) + H^+(aq) \rightleftharpoons HF(aq) \quad [17.18]$$

The equation for the overall process is

$$CaF_2(s) + 2H^+(aq) \rightleftharpoons Ca^{2+}(aq) + 2HF(aq) \quad [17.19]$$

Qualitatively, we can understand what occurs in terms of Le Châtelier's principle: The solubility equilibrium is driven to the right because the free F^- concentration is reduced by reaction with H^+. The reduction of $[F^-]$ causes Q to be reduced so that it becomes smaller than K_{sp}. Thus, more CaF_2 dissolves.

These examples illustrate a general rule: *The solubility of slightly soluble salts containing basic anions increases as $[H^+]$ increases (as pH is lowered).* Salts with anions of negligible basicity (the anions of strong acids) are largely unaffected by pH.

SAMPLE EXERCISE 17.15

Which of the following substances will be more soluble in acidic solution than in basic solution: **(a)** $Ni(OH)_2(s)$; **(b)** $CaCO_3(s)$; **(c)** $BaSO_4(s)$; **(d)** $AgCl(s)$?

Solution: **(a)** We can conclude that $Ni(OH)_2(s)$ will be more soluble in acidic solution, because of the basicity of OH^-; the H^+ ion reacts with the OH^- ion, forming water:

$$\begin{array}{rrcl} & Ni(OH)_2(s) & \rightleftharpoons & Ni^{2+}(aq) + 2OH^-(aq) \\ & 2OH^-(aq) + 2H^+(aq) & \rightleftharpoons & 2H_2O(l) \\ \hline \text{Overall:} & Ni(OH)_2(s) + 2H^+(aq) & \rightleftharpoons & Ni^{2+}(aq) + 2H_2O(l) \end{array}$$

(b) Similarly, $CaCO_3(s)$ dissolves in acid solutions because CO_3^{2-} is a basic anion:

$$\begin{array}{rrcl} & CaCO_3(s) & \rightleftharpoons & Ca^{2+}(aq) + CO_3^{2-}(aq) \\ & CO_3^{2-}(aq) + 2H^+(aq) & \rightleftharpoons & H_2CO_3\,(aq) \\ & H_2CO_3(aq) & \longrightarrow & CO_2(g) + H_2O(l) \\ \hline \text{Overall:} & CaCO_3(s) + 2H^+(aq) & \longrightarrow & Ca^{2+}(aq) + CO_2(g) + H_2O(l) \end{array}$$

(c) The solubility of $BaSO_4$ is largely unaffected by changes in solution pH because SO_4^{2-} is a rather weak base and thus has little tendency to combine with a proton. However, $BaSO_4$ is slightly more soluble in strongly acidic solutions.

(d) The solubility of AgCl is unaffected by changes in pH because Cl^- is the anion of a strong acid.

PRACTICE EXERCISE

Write the net ionic equation for the reaction of the following copper(II) compounds with acid: **(a)** $CuCrO_4$; **(b)** $Cu(N_3)_2$.

Answers: **(a)** $CuCrO_4(s) + H^+(aq) \rightleftharpoons Cu^{2+}(aq) + HCrO_4^-(aq)$
(b) $Cu(N_3)_2(s) + 2H^+(aq) \rightleftharpoons Cu^{2+}(aq) + 2HN_3(aq)$

Selective Precipitation of Ions

Ions can be separated from each other on the basis of the solubilities of their salts. Take, for example, a solution containing both Ag^+ and Cu^{2+}. If HCl is

CHEMISTRY AT WORK: Tooth Decay and Fluoridation

Tooth enamel consists mainly of a mineral called hydroxyapatite, $Ca_{10}(PO_4)_6(OH)_2$. It is the hardest substance in the body. Tooth cavities are caused by the dissolving action of acids on tooth enamel:

$$Ca_{10}(PO_4)_6(OH)_2(s) + 8H^+(aq) \longrightarrow 10Ca^{2+}(aq) + 6HPO_4^{2-}(aq) + 2H_2O(l)$$

The resultant Ca^{2+} and HPO_4^{2-} ions diffuse out of the tooth enamel and are washed away by saliva. The acids that attack the hydroxyapatite are formed by the action of specific bacteria on sugars and other carbohydrates present in the plaque adhering to the teeth.

Fluoride ion, present in drinking water, toothpaste, or other sources, can react with hydroxyapatite to form fluoroapatite, $Ca_{10}(PO_4)_6F_2$. This mineral, in which F^- has replaced OH^-, is much more resistant to attack by acids because the fluoride ion is a much weaker Brønsted base than is the hydroxide ion.

Because the fluoride ion is so effective in preventing cavities, it is added to the public water supply in many places to give a concentration of 1 mg/L (that is, 1 ppm). The compound added may be NaF or Na_2SiF_6. The latter compound reacts with water to release fluoride ion by the following reaction:

$$SiF_6^{2-}(aq) + 2H_2O(l) \longrightarrow 6F^-(aq) + 4H^+(aq) + SiO_2(s)$$

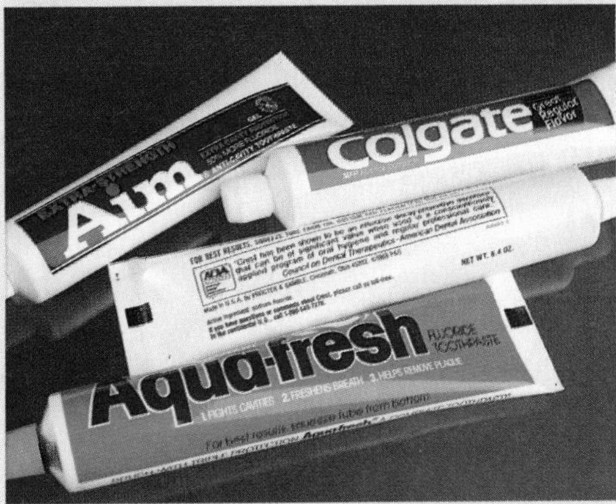

Figure 17.15 Many toothpastes contain fluoride, often as stannous fluoride, SnF_2. (© Robert Mathena/Fundamental Photographs)

About 80 percent of all toothpastes now sold in the United States contain fluoride compounds, usually at the level of 0.1 percent fluoride by weight (Figure 17.15). The most common compounds in toothpastes are stannous fluoride, SnF_2, sodium monofluorophosphate, Na_2PO_3F, and sodium fluoride, NaF.

added to the solution, AgCl ($K_{sp} = 1.8 \times 10^{-10}$) precipitates, while Cu^{2+} remains in solution because $CuCl_2$ is soluble. Separation of ions in an aqueous solution by using a reagent that forms a precipitate with one or a few of the ions is called *selective precipitation.*

Sulfide ion is often used to separate metal ions because the solubilities of sulfide salts span a wide range and $[S^{2-}]$ can be regulated using pH. For example, suppose a solution contains both Cu^{2+} and Zn^{2+}. These ions can be separated by bubbling H_2S gas through a properly acidified solution. Because CuS ($K_{sp} = 6.3 \times 10^{-36}$) is less soluble than ZnS ($K_{sp} = 1.1 \times 10^{-21}$), $[S^{2-}]$ can be adjusted so that CuS precipitates while ZnS does not (Figure 17.16). To understand this separation further, let's consider the equilibria involved.

When H_2S is bubbled through an aqueous solution at 25°C and 1 atm, a saturated solution of H_2S forms that is approximately 0.1 *M* in H_2S. H_2S is a weak diprotic acid:

$$H_2S(aq) \rightleftharpoons H^+(aq) + HS^-(aq) \qquad K_{a1} = 5.7 \times 10^{-8} \qquad [17.21]$$

$$HS^-(aq) \rightleftharpoons H^+(aq) + S^{2-}(aq) \qquad K_{a2} = 1.3 \times 10^{-13} \qquad [17.22]$$

These equations can be combined to give an equation for the overall dissociation of H_2S into S^{2-}:

$$H_2S(aq) \rightleftharpoons 2H^+(aq) + S^{2-}(aq)$$

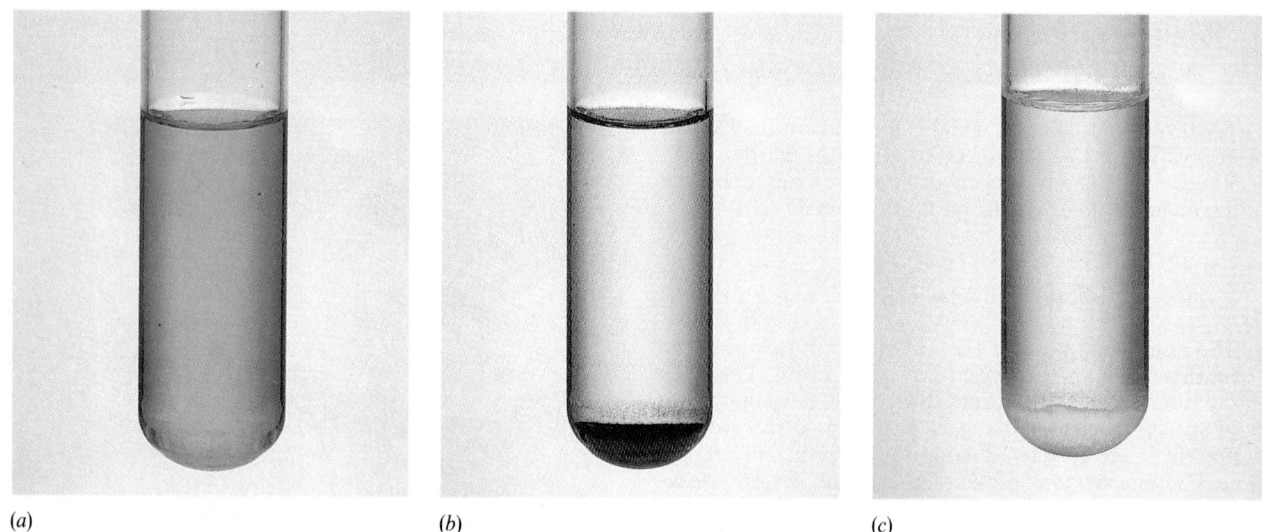

(*a*) (*b*) (*c*)

Figure 17.16 (*a*) Solution containing $Zn^{2+}(aq)$ and $Cu^{2+}(aq)$. (*b*) When H_2S is added to a solution whose pH exceeds 0.6, CuS precipitates. (*c*) After CuS is removed, pH is increased, allowing ZnS to precipitate. (© Richard Megna/ Fundamental Photographs)

The equilibrium-constant expression for this overall dissociation process is

$$K = \frac{[H^+]^2[S^{2-}]}{[H_2S]} = K_{a1} \times K_{a2} = 7.4 \times 10^{-21} \qquad [17.23]$$

Substituting the solubility of H_2S, 0.1 *M*, into this expression gives

$$\frac{[H^+]^2[S^{2-}]}{0.1} = 7.4 \times 10^{-21}$$

$$[H^+]^2[S^{2-}] = (0.1)(7.4 \times 10^{-21}) = 7 \times 10^{-22} \qquad [17.24]$$

Equation 17.24 can be used to calculate the concentration of S^{2-} in saturated solutions of H_2S at various pH values.

Now consider how we can calculate the pH that will allow us to prevent precipitation of ZnS while precipitating CuS from a solution that is 0.10 *M* in Zn^{2+}, 0.10 *M* in Cu^{2+}, and saturated with H_2S. The maximum concentration of S^{2-} that can be present in the solution before precipitation of ZnS occurs can be calculated from the solubility-product expression for this substance:

$$[Zn^{2+}][S^{2-}] = K_{sp} = 1.1 \times 10^{-21}$$

$$[S^{2-}] = \frac{1.1 \times 10^{-21}}{[Zn^{2+}]} = \frac{1.1 \times 10^{-21}}{0.10} = 1.1 \times 10^{-20}\ M$$

The concentration of hydrogen ions necessary to give $[S^{2-}] = 1.1 \times 10^{-20}\ M$ can be calculated using Equation 17.24:

$$[H^+]^2[S^{2-}] = 7 \times 10^{-22}$$

$$[H^+]^2 = \frac{7 \times 10^{-22}}{[S^{2-}]} = \frac{7 \times 10^{-22}}{1.1 \times 10^{-20}} = 6 \times 10^{-2}$$

$$[H^+] = \sqrt{6 \times 10^{-2}} = 0.24\ M$$
$$pH = -\log(2.4 \times 10^{-1}) = 0.6$$

Thus, ZnS will not precipitate if the pH is 0.6 or lower. However, at pH 0.6, where $[S^{2-}] = 1.1 \times 10^{-20}$, the ion product for a 0.10 M Cu^{2+} solution would be

$$Q = [Cu^{2+}][S^{2-}] = (0.10)(1.1 \times 10^{-20}) = 1.1 \times 10^{-21}$$

Because Q exceeds the K_{sp} of CuS (that is, 6.3×10^{-36}), CuS will precipitate under these conditions. Indeed, because $[Cu^{2+}] = 0.10\ M$, $[S^{2-}]$ must be less than $6.3 \times 10^{-35}\ M$ to prevent the precipitation of CuS from this solution. Even under strongly acidic conditions, $[S^{2-}]$ will be high enough to cause precipitation of CuS. Thus, a fairly broad range of sulfide concentration will allow for effective separation of Cu^{2+} (as CuS) from Zn^{2+}.

Effect of Complex Formation on Solubility

A characteristic property of metal ions is their ability to act as Lewis acids, or electron-pair acceptors, toward water molecules, which act as Lewis bases, or electron-pair donors (Section 16.8). Lewis bases other than water can also interact with metal ions, particularly with transition metal ions. Such interactions can have a dramatic effect on the solubility of a metal salt. For example, AgCl, whose $K_{sp} = 1.82 \times 10^{-10}$, will dissolve in the presence of aqueous ammonia because of the interaction between Ag^+ and the Lewis base NH_3, as shown in Figure 17.17. This process can be viewed as the sum of two reactions, the solubility equilibrium of AgCl and the Lewis acid-base interaction between Ag^+ and NH_3:

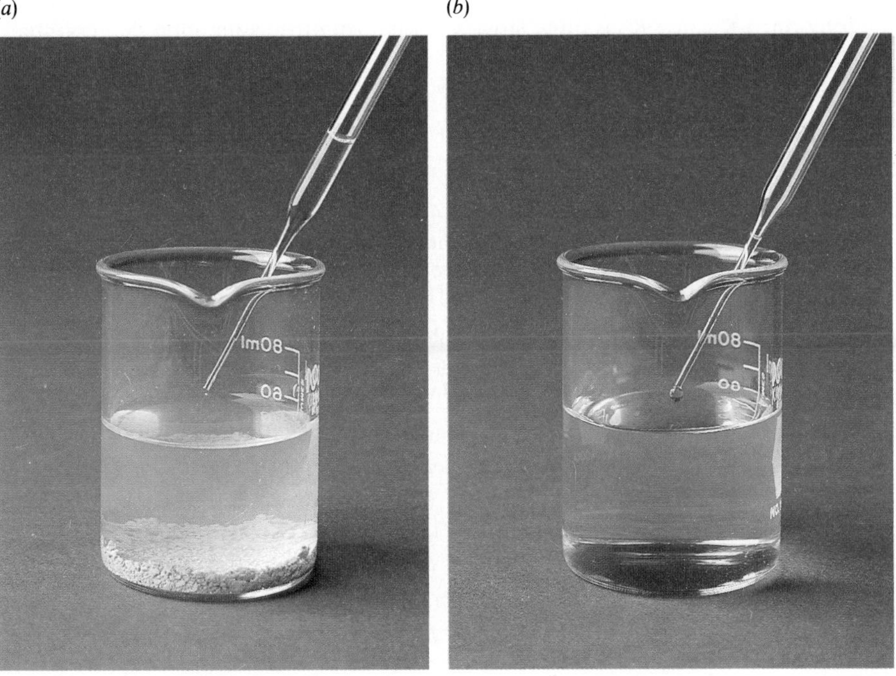

Figure 17.17 (*a*) A precipitate of AgCl dissolves (*b*) upon addition of concentrated $NH_3(aq)$. (Fundamental Photographs)

	$AgCl(s)$	$\rightleftharpoons$	$Ag^+(aq) + Cl^-(aq)$	[17.25]
	$Ag^+(aq) + 2NH_3(aq)$	$\rightleftharpoons$	$Ag(NH_3)_2^+(aq)$	[17.26]
Overall:	$AgCl(s) + 2NH_3(aq)$	$\rightleftharpoons$	$Ag(NH_3)_2^+(aq) + Cl^-(aq)$	[17.27]

The presence of NH_3 drives the top reaction, the solubility equilibrium of AgCl, to the right as $Ag^+(aq)$ is removed to form $Ag(NH_3)_2^+$.

For a Lewis base such as NH_3 to increase the solubility of a metal salt, it must be able to interact more strongly with the metal ion than water does. The NH_3 must displace solvating H_2O molecules (Sections 13.2 and 16.10) in order to form $Ag(NH_3)_2^+$:

$$Ag^+(aq) + 2NH_3(aq) \rightleftharpoons Ag(NH_3)_2^+(aq) \qquad [17.28]$$

An assembly of a metal ion and the Lewis bases bonded to it, such as $Ag(NH_3)_2^+$, is called a **complex ion**. The stability of a complex ion in aqueous solution can be judged by the magnitude of the equilibrium constant for formation of the complex ion from the hydrated metal ion. For example, the equilibrium constant for formation of $Ag(NH_3)_2^+$ (Equation 17.28) is 1.7×10^7:

$$K_f = \frac{[Ag(NH_3)_2^+]}{[Ag^+][NH_3]^2} = 1.7 \times 10^7 \qquad [17.29]$$

Such equilibrium constants are called **formation constants**, K_f. The formation constants for several complex ions are shown in Table 17.2.

SAMPLE EXERCISE 17.16

Calculate the concentration of Ag^+ present in solution at equilibrium when concentrated ammonia is added to a 0.010 M solution of $AgNO_3$ to give an equilibrium concentration of $[NH_3] = 0.20\ M$. Neglect the small volume change that occurs on addition of NH_3.

Solution: Because K_f is quite large, we begin with the assumption that essentially all of the Ag^+ is converted to $Ag(NH_3)_2^+$, in accordance with Equation 17.28. Thus, $[Ag^+]$ will be small at equilibrium. If $[Ag^+]$ was 0.010 M initially, then $[Ag(NH_3)_2^+]$

Table 17.2 Formation Constants for Some Metal Complex Ions in Water at 25°C

Complex ion	K_f	Equilibrium equation
$Ag(NH_3)_2^+$	1.7×10^7	$Ag^+(aq) + 2NH_3(aq) \rightleftharpoons Ag(NH_3)_2^+(aq)$
$Ag(CN)_2^-$	1×10^{21}	$Ag^+(aq) + 2CN^-(aq) \rightleftharpoons Ag(CN)_2^-(aq)$
$Ag(S_2O_3)_2^{3-}$	2.9×10^{13}	$Ag^+(aq) + 2S_2O_3^{2-}(aq) \rightleftharpoons Ag(S_2O_3)_2^{3-}(aq)$
$CdBr_4^{2-}$	5×10^3	$Cd^{2+}(aq) + 4Br^-(aq) \rightleftharpoons CdBr_4^{2-}(aq)$
$Cr(OH)_4^-$	8×10^{29}	$Cr^{3+}(aq) + 4OH^-(aq) \rightleftharpoons Cr(OH)_4^-(aq)$
$Co(SCN)_4^{2-}$	1×10^3	$Co^{2+}(aq) + 4SCN^-(aq) \rightleftharpoons Co(SCN)_4^{2-}(aq)$
$Cu(NH_3)_4^{2+}$	5×10^{12}	$Cu^{2+}(aq) + 4NH_3(aq) \rightleftharpoons Cu(NH_3)_4^{2+}(aq)$
$Cu(CN)_4^{2-}$	1×10^{25}	$Cu^{2+}(aq) + 4CN^-(aq) \rightleftharpoons Cu(CN)_4^{2-}(aq)$
$Ni(NH_3)_6^{2+}$	5.5×10^8	$Ni^{2+}(aq) + 6NH_3(aq) \rightleftharpoons Ni(NH_3)_6^{2+}(aq)$
$Fe(CN)_6^{4-}$	1×10^{35}	$Fe^{2+}(aq) + 6CN^-(aq) \rightleftharpoons Fe(CN)_6^{4-}(aq)$
$Fe(CN)_6^{3-}$	1×10^{42}	$Fe^{3+}(aq) + 6CN^-(aq) \rightleftharpoons Fe(CN)_6^{3-}(aq)$

will be 0.010 *M* following addition of the NH_3. Let the concentration of Ag^+ at equilibrium be x. Then

$$Ag^+(aq) + 2NH_3(aq) \rightleftharpoons Ag(NH_3)_2{}^+(aq)$$
$$x\ M \qquad 0.20\ M \qquad (0.010 - x)\ M$$

Because the concentration of Ag^+ is very small, we can ignore x in comparison with 0.010. Thus, $0.010 - x \simeq 0.010\ M$. Substituting these values into the equilibrium-constant expression, Equation 17.29, we obtain

$$\frac{[Ag(NH_3)_2{}^+]}{[Ag^+][NH_3]^2} = \frac{0.010}{x(0.20)^2} = 1.7 \times 10^7$$

Solving for x, we obtain $x = 1.4 \times 10^{-8}\ M = [Ag^+]$. It is evident that formation of the $Ag(NH_3)_2{}^+$ complex drastically reduces the concentration of free Ag^+ ion in solution.

PRACTICE EXERCISE

Calculate the concentration of free Cr^{3+} ion when 0.010 mol of $Cr(NO_3)_3$ is dissolved in a liter of solution buffered at pH 10.0. ***Answer:*** $[Cr^{3+}] = 1.2 \times 10^{-16}\ M$

The general rule is that metal salts will dissolve in the presence of a suitable Lewis base, such as NH_3, CN^-, or OH^-, if the metal forms a sufficiently stable complex with the base. The ability of metal ions to form complexes is an extremely important aspect of their chemistry. In Chapter 25, we will take a much closer look at complex ions. In that chapter and others, we shall see applications of complex ions to areas such as biochemistry, metallurgy, and photography.

Amphoterism

Many metal hydroxides and oxides that are relatively insoluble in neutral water dissolve in strongly acidic *and* in strongly basic media. Such substances are said to be **amphoteric** (from the Greek word *amphoteros*, which means "both"). Examples include the hydroxides and oxides of Al^{3+}, Cr^{3+}, Zn^{2+}, and Sn^{2+}. (See the Closer Look box in Section 8.10.)

The dissolution of these species in acidic solutions should be anticipated based on the earlier discussions in this section. We have seen that acids promote the dissolving of compounds with basic anions. What makes amphoteric oxides and hydroxides special is that they also dissolve in strongly basic solutions (Figure 17.18). This behavior results from the formation of complex anions containing several (typically four) hydroxides bound to the metal ion:

$$Al(OH)_3(s) + OH^-(aq) \rightleftharpoons Al(OH)_4{}^-(aq) \qquad [17.30]$$

Amphoterism is often interpreted in terms of the behavior of the water molecules that surround the metal ion and that are bonded to it by Lewis acid-base interactions (Section 16.10). For example, $Al^{3+}(aq)$ is more accurately represented as $Al(H_2O)_6{}^{3+}(aq)$; six water molecules are bonded to the Al^{3+} in aqueous solution. As discussed in Section 16.10, this hydrated ion is a weak acid. As a strong base is added, the $Al(H_2O)_6{}^{3+}$ loses protons in a stepwise fashion, eventually forming the neutral and water-insoluble $Al(H_2O)_3(OH)_3$. This substance then dissolves upon removal of an additional proton to form the anion $Al(H_2O)_2(OH)_4{}^-$. The reactions that occur are as follows:

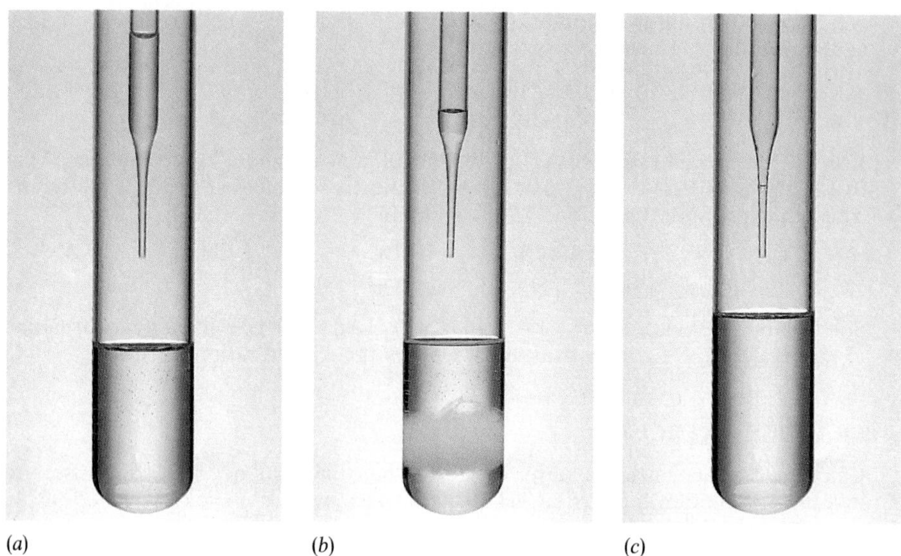

Figure 17.18 As NaOH is added to a solution of Al^{3+} (*a*), a precipitate of $Al(OH)_3$ forms (*b*). As more NaOH is added, the $Al(OH)_3$ dissolves (*c*), demonstrating the amphoterism of the $Al(OH)_3$. (Richard Megna/Fundamental Photographs)

$$Al(H_2O)_6{}^{3+}(aq) + OH^-(aq) \rightleftharpoons Al(H_2O)_5(OH)^{2+}(aq) + H_2O(l)$$
$$Al(H_2O)_5(OH)^{2+}(aq) + OH^-(aq) \rightleftharpoons Al(H_2O)_4(OH)_2{}^+(aq) + H_2O(l)$$
$$Al(H_2O)_4(OH)_2{}^+(aq) + OH^-(aq) \rightleftharpoons Al(H_2O)_3(OH)_3(s) + H_2O(l)$$
$$Al(H_2O)_3(OH)_3(s) + OH^-(aq) \rightleftharpoons Al(H_2O)_2(OH)_4{}^-(aq) + H_2O(l)$$

Further proton removals are possible, but each successive reaction occurs less readily than the one before. As the charge on the ion becomes more negative, it becomes increasingly difficult to remove a positively charged proton. Addition of an acid reverses these reactions. The proton adds in a stepwise fashion to convert the OH^- groups to H_2O, eventually reforming $Al(H_2O)_6{}^{3+}$. The common practice is to simplify the equations for these reactions by excluding the bound H_2O molecules. Thus, we usually write Al^{3+} instead of $Al(H_2O)_6{}^{3+}$, $Al(OH)_3$ instead of $Al(H_2O)_3(OH)_3$, $Al(OH)_4{}^-$ instead of $Al(H_2O)_2(OH)_4{}^-$, and so forth.

The extent to which an insoluble metal hydroxide reacts with either acid or base varies with the particular metal ion involved. Many metal hydroxides—for example, $Ca(OH)_2$, $Fe(OH)_2$, and $Fe(OH)_3$—are capable of dissolving in acidic solution but do not react with excess base. These hydroxides are not amphoteric.

The purification of aluminum ore in the manufacture of aluminum metal provides an interesting application of the property of amphoterism. As we have seen, $Al(OH)_3$ is amphoteric, whereas $Fe(OH)_3$ is not. Aluminum occurs in large quantities as the ore *bauxite*, which is essentially Al_2O_3 with additional water molecules. The ore is contaminated with Fe_2O_3 as an impurity. When bauxite is added to a strongly basic solution, the Al_2O_3 dissolves because the aluminum forms complex ions, such as $Al(OH)_4{}^-$. The Fe_2O_3 impurity, however, is not amphoteric and remains as a solid. The solution is filtered, getting rid of the iron impurity. Aluminum hydroxide is then precipitated by addition of acid. The purified hydroxide receives further treatment and eventually yields aluminum metal.

SAMPLE EXERCISE 17.17

Write the balanced net-ionic equation for the reaction between $Al_2O_3(s)$ and aqueous solutions of NaOH that causes the $Al_2O_3(s)$ to dissolve. (H_2O is also present as a reactant.)

Solution: The NaOH is a strong electrolyte that provides the necessary OH^- ions. The reaction product can be taken to be $Al(OH)_4^-$, an assembly of Al^{3+} and four OH^- ions. The unbalanced equation is

$$Al_2O_3(s) + OH^-(aq) \rightleftharpoons Al(OH)_4^-(aq)$$

Two Al are needed among the products:

$$Al_2O_3(s) + OH^-(aq) \rightleftharpoons 2Al(OH)_4^-(aq)$$

To balance the charge on both sides of the equation requires $2OH^-$. Sufficient H_2O is then added to balance the O and H counts. The balanced equation is

$$Al_2O_3(s) + 2OH^-(aq) + 3H_2O(l) \rightleftharpoons 2Al(OH)_4^-(aq)$$

PRACTICE EXERCISE

Write balanced net-ionic equations for the dissolution of **(a)** $Zn(OH)_2(s)$ and **(b)** $ZnO(s)$ in excess base.

Answers: **(a)** $Zn(OH)_2(s) + 2OH^-(aq) \rightleftharpoons Zn(OH)_4^{2-}(aq)$
(b) $ZnO(s) + H_2O(l) + 2OH^-(aq) \rightleftharpoons Zn(OH)_4^{2-}(aq)$

17.6 QUALITATIVE ANALYSES FOR METALLIC ELEMENTS

In this chapter, we have seen several examples of equilibria involving metal ions in aqueous solution. In this final section, we look very briefly at how solubility equilibria and complex formation can be used to detect the presence of particular metal ions in solution. Before the development of modern analytical instrumentation, it was necessary to analyze mixtures of metals in a sample by so-called "wet" chemical methods. For example, a metallic sample that might contain several metallic elements was dissolved in a concentrated acid solution. This solution was then tested in a systematic way for the presence of various metallic ions.

Qualitative analysis determines only the presence or absence of a particular metal ion. It should be distinguished from **quantitative analysis**, which determines how much of a given substance is present. Wet methods of qualitative analysis have become less important as a means of analysis. However, they are frequently used in general chemistry laboratory programs to illustrate equilibria, teach the properties of common metal ions in solution, and develop laboratory skills. Typically, such analyses proceed in three stages. (1) The ions are separated into broad groups on the basis of solubility properties. (2) The individual ions within each group are then separated by selectively dissolving members in the group. (3) The ions are then identified by means of specific tests.

A scheme in common use divides the common cations into five groups, as shown in Figure 17.19. The order of addition of reagents is important. The most selective separations—that is, those that involve the smallest number of ions—are carried out first. The reactions that are used must proceed so far toward completion that any concentration of cations remaining in the solution is too small to interfere with subsequent tests. Let's take a closer look at each of these five groups of cations, examining briefly the logic used in this qualitative analysis scheme.

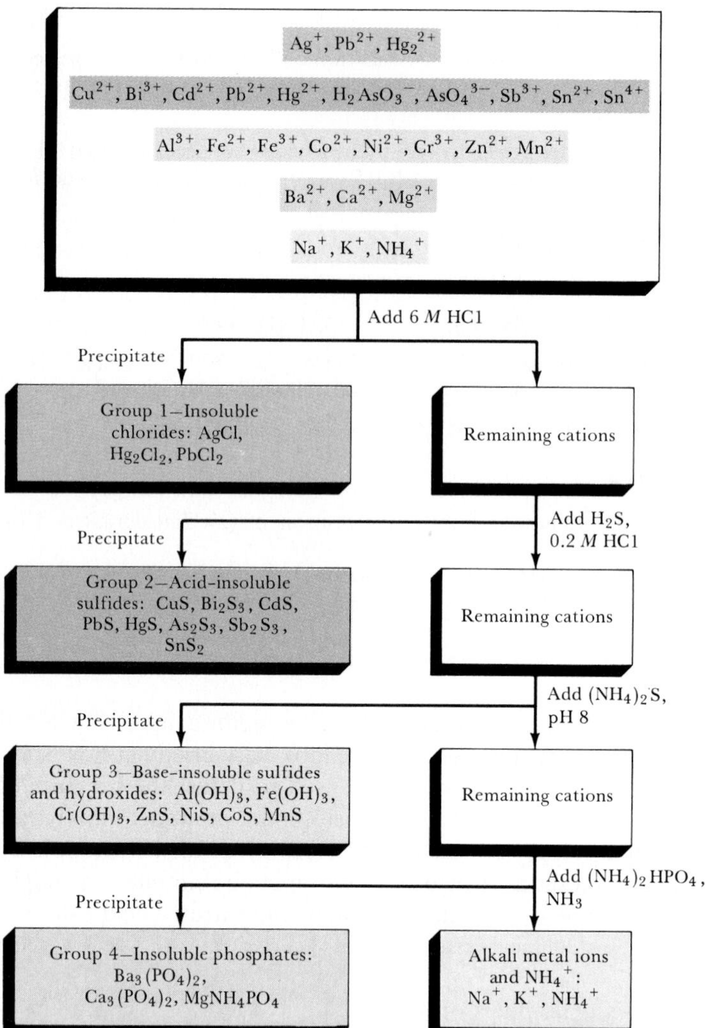

Figure 17.19 Qualitative analysis scheme for separating cations into groups.

1. *Insoluble chlorides:* Of the common metal ions, only Ag^+, Hg_2^{2+}, and Pb^{2+} form insoluble chlorides. Therefore, when dilute HCl is added to a mixture of cations, only AgCl, Hg_2Cl_2, and $PbCl_2$ will precipitate, leaving the other cations in solution. The absence of a precipitate indicates that the starting solution contains no Ag^+, Hg_2^{2+}, or Pb^{2+}.
2. *Acid-insoluble sulfides:* After any insoluble chlorides have been removed, the remaining solution, now acidic, is treated with H_2S. As we saw in Section 17.5, the dissociation of H_2S is repressed in acidic solutions so that the concentration of free S^{2-} is very low. Consequently, only the most insoluble metal sulfides, CuS, Bi_2S_3, CdS, PbS, HgS, As_2S_3, Sb_2S_3, and SnS_2, can precipitate. (Note the very small values of K_{sp} for some of these sulfides in Appendix D.) Those metal ions whose sulfides are somewhat more soluble—for example, ZnS or NiS—remain in solution.
3. *Base-insoluble sulfides:* After the solution is filtered to remove any acid-insoluble sulfides, the remaining solution is made slightly basic, and $(NH_4)_2S$ is added. In basic solutions, the concentration of S^{2-} is higher than in acidic solutions. Thus, the ion products for many of the more

soluble sulfides are caused to exceed their K_{sp} values and precipitation occurs. The metal ions precipitated at this stage are Al^{3+}, Cr^{3+}, Fe^{3+}, Zn^{2+}, Ni^{2+}, Co^{2+}, and Mn^{2+}. (Actually, the Al^{3+}, Fe^{3+}, and Cr^{3+} ions do not form insoluble sulfides; instead they are precipitated as insoluble hydroxides at the same time.)

4. *Insoluble phosphates:* At this point, the solution contains only metal ions from periodic table groups 1A and 2A. Addition of $(NH_4)_2HPO_4$ to a basic solution causes precipitation of the group 2A elements Mg^{2+}, Ca^{2+}, Sn^{2+}, and Ba^{2+}, because these metals form insoluble phosphates.
5. *The alkali metal ions and NH_4^+:* The ions that remain after removal of the insoluble phosphates form a small group. We can test for each ion individually. For example, the flame test is useful to show the presence of K^+ because the flame turns a characteristic violet color if K^+ is present.

Additional separation and testing is necessary to determine which ions are present within each of the groups. As an example, consider the ions of the insoluble chloride group. The precipitate containing the metal chlorides is boiled in water. It happens that $PbCl_2$ is relatively soluble in hot water, whereas AgCl and Hg_2Cl_2 are not. The hot solution is filtered, and a solution of Na_2CrO_4 added to the filtrate. If Pb^{2+} is present, a yellow precipitate of $PbCrO_4$ forms. The test for Ag^+ consists of treating the metal chloride precipitate with dilute ammonia. Only Ag^+ forms an ammonia complex. If AgCl is present in the precipitate, it will dissolve in the ammonia solution:

$$AgCl(s) + 2NH_3(aq) \rightleftharpoons Ag(NH_3)_2^+(aq) + Cl^-(aq) \quad [17.31]$$

After treatment with ammonia, the solution is filtered and the filtrate made acidic by adding nitric acid. The nitric acid removes ammonia from solution by forming NH_4^+, thus releasing Ag^+, which should re-form the AgCl precipitate:

$$Ag(NH_3)_2^+(aq) + Cl^-(aq) + 2H^+(aq) \rightleftharpoons AgCl(s) + 2NH_4^+(aq) \quad [17.32]$$

The analyses for individual ions in the acid-insoluble and base-insoluble sulfides are a bit more complex, but the same general principles are involved. The detailed procedures for carrying out such analyses are given in laboratory manuals.

FOR REVIEW

SUMMARY

In this chapter we've considered several types of important equilibria that occur in aqueous solution. Our primary emphasis has been on acid-base equilibria in solutions containing two or more solutes and on solubility equilibria. We observed that the dissociation of a weak acid or weak base is repressed by the presence of a strong electrolyte that provides an ion common to the equilibrium. This phenomenon is an example of the common-ion effect.

A particularly important type of acid-base mixture is a weak conjugate acid-base pair. Such mixtures function as buffers. Addition of small amounts of additional acid or base to a buffered solution causes only small changes in pH because the buffer reacts with the added acid or base. (Recall that strong acid-strong base, strong acid-weak base, and weak acid-strong base reactions proceed essentially to completion.) Buffer solutions are usually prepared from a weak acid and a salt of that acid or from a weak base and a salt of that base. Two important characteristics of a buffer solution are its buffering capacity and its pH.

The plot of the pH of an acid (or base) as a function of the volume of added base (or acid) is called a titration curve. Titration curves aid in selecting a proper pH indicator for an acid-base titration. The titration curve of a strong acid-strong base titration exhibits a large change in pH in the immediate vicinity of the equivalence point; at the equivalence point for this titration, pH = 7. For strong acid-weak base or weak acid-strong base titrations, the pH change in the vicinity of the equivalence point is not as large. Furthermore, the pH at the equivalence point is not 7 in either or these cases. Rather, it is the pH of the salt solution that results from the neutralization reaction.

The equilibrium between a solid salt and its ions in solution provides an example of heterogeneous equilibrium. The solubility-product constant, K_{sp}, is an equilibrium constant that expresses quantitatively the extent to which the salt dissolves. Addition to the solution of an ion common to a solubility equilibrium causes the solubility of the salt to decrease. This phenomenon is another example of the common-ion effect.

Comparison of the ion product, Q, with the value of K_{sp} can be used to judge whether a precipitate will form when solutions are mixed or whether a slightly soluble salt will dissolve under various conditions. Solubility is affected by the common-ion effect, by pH, and by the presence of certain Lewis bases that react with metal ions to form stable complex ions. Solubility is affected by pH when one or more of the ions in the solubility equilibrium is an acid or base. For example, the solubility of MnS is increased on addition of acid, because S^{2-} is basic.

Amphoteric metal hydroxides are those slightly soluble metal hydroxides that dissolve on addition of either acid or base. The reactions that give rise to the amphoterism are acid-base reactions involving the OH^- or H_2O groups bound to the metal ions. Complex-ion formation in aqueous solution involves the displacement by Lewis bases (such as NH_3 and CN^-) of water molecules attached to the metal ion. The extent to which such complex formation occurs is expressed quantitatively by the formation constant for the complex ion.

The fact that the ions of different metallic elements vary a great deal in the solubilities of their salts, in their acid-base behavior, and in their tendencies to form complexes can be used to separate and detect the presence of metal ions in mixtures. Qualitative analysis determines the presence or absence of a metal ion in a mixture of metal ions in solution. The analysis usually proceeds by separating the ions into groups on the basis of precipitation reactions and then analyzing each group for individual metal ions.

KEY TERMS

common-ion effect (Sec. 17.1)
buffered solution (buffers) (Sec. 17.2)
buffering capacity (Sec. 17.2)
Henderson-Hasselbalch equation (Sec. 17.2)
titration curve (Sec. 17.3)
solubility-product constant (solubility product) (Sec. 17.4)
complex ion (Sec. 17.5)
formation constants (Sec. 17.5)
amphoteric (Sec. 17.5)
qualitative analysis (Sec. 17.6)
quantitative analysis (Sec. 17.6)

EXERCISES

Common-Ion Effect; Buffers

17.1 Describe the effect on pH (increase, decrease, or no change) that results from each of the following additions: **(a)** sodium formate, $NaCHO_2$, to a solution of formic acid, $HCHO_2$, **(b)** ammonium perchlorate, NH_4ClO_4, to a solution of ammonia, NH_3; **(c)** potassium bromide to a solution of potassium nitrite, KNO_2; **(d)** hydrochloric acid, HCl, to a solution of sodium acetate, $NaC_2H_3O_2$.

17.2 Describe the effect on pH (increase, decrease, or no change) that results from each of the following additions: **(a)** ammonia to a solution of HCl; **(b)** ammonium chloride to a solution of HCl; **(c)** sodium cyanide, NaCN, to a solution of HBr; **(d)** pyridinium nitrate, $C_5H_5NHNO_3$, to a solution of pyridine, C_5H_5N.

17.3 Using appropriate equilibrium constants from Appendix D, calculate the pH of the following solutions: **(a)** 0.080 *M* in sodium formate, $NaCHO_2$, and 0.100 *M* in formic acid, $HCHO_2$; **(b)** 0.10 *M* in pyridine, C_5H_5N, and 0.12 *M* in pyridinium chloride, C_5H_5NHCl.

17.4 Using Appendix D, calculate the pH of the following solutions: **(a)** 0.10 *M* in sodium propionate, $NaC_3H_5O_2$, and 0.15 *M* in propionic acid, $HC_3H_5O_2$; **(b)** 0.12 *M* in methylamine, CH_3NH_2, and 0.13 *M* in methylammonium chloride, CH_3NH_3Cl.

17.5 What is the pH of **(a)** a solution prepared by adding 25.5 g of acetic acid, $HC_2H_3O_2$, and 25.0 g of sodium acetate, $NaC_2H_3O_2$, to enough water to form 4.0 L of solution; **(b)** a solution made by mixing 250 mL of 0.22 *M* $HC_2H_3O_2$ and 300 mL of 0.20 *M* $NaC_2H_3O_2$; **(c)** a solution made by adding 0.040 mol of KOH to 825 mL of 0.150 *M* $HC_2H_3O_2$; **(d)** a solution made by adding 0.30 mol NH_3 and 0.50 mol NH_4Cl to enough water to form 2.5 L of solution; **(e)** a solution prepared by adding 0.20 mol of HCl to 2.0 L of 0.35 *M* ammonia?

17.6 Calculate the pH of **(a)** a solution formed by adding 35.0 g of benzoic acid, $HC_7H_5O_2$, and 30.0 g of sodium benzoate, $NaC_7H_5O_2$, to enough water to form 250 mL of solution; **(b)** a solution formed by mixing 30.0 mL of 0.25 *M* $HC_7H_5O_2$ and 20.0 mL of 0.22 *M* $NaC_7H_5O_2$ and diluting to a total volume of 100 mL; **(c)** a solution prepared by adding 1.00 g of NaOH to 1.00 L of 0.105 *M* $HC_7H_5O_2$; **(d)** a solution prepared by adding 0.010 mol of NaOH to 0.100 L of 0.200 *M* NH_3; **(e)** a solution prepared by mixing 32.5 g of $NaC_7H_5O_2$ and 50.0 mL of 2.0 *M* HCl in water and diluting to 500 ml

17.7 A certain organic compound that is used as an indicator for acid-base reactions exists in aqueous solution as equal concentrations of the acid form, HB, and the base form, B^-, at a pH of 7.80. What is pK_a for the acid form of this indicator, HB?

17.8 Dimethylglyoxime, $C_4H_8N_2O_2$, is a weak base with $K_b = 4.0 \times 10^{-4}$. At what pH does a solution of dimethylglyoxime and its conjugate acid contain equal concentrations of the two species?

17.9 Explain why a mixture of HCl and KCl does not function as a buffer, whereas a mixture of $HC_2H_3O_2$ and $NaC_2H_3O_2$ does.

17.10 What factors determine **(a)** the pH, and **(b)** the capacity of a buffer solution?

17.11 **(a)** Write the net ionic equation for the reaction that occurs when a solution of perchloric acid, $HClO_4$, is mixed with a solution of sodium benzoate, $NaC_7H_5O_2$. **(b)** Calculate the equilibrium constant for this reaction. **(c)** Calculate the concentrations of Na^+, ClO_4^-, H^+, $C_7H_5O_2^-$, and $HC_7H_5O_2$ when 0.50 L of 0.20 *M* $HClO_4$ is mixed with 0.50 L of 0.20 *M* $NaC_7H_5O_2$.

17.12 **(a)** Write the net ionic equation for the reaction that occurs when a solution of potassium hydroxide, KOH, is mixed with a solution of dimethylammonium bromide, $(CH_3)_2NH_2Br$. **(b)** Calculate the equilibrium constant for this reaction. **(c)** Calculate the concentrations of K^+, Br^-, H^+, and $(CH_3)_2NH_2^+$ ions and of $(CH_3)_2NH$ when 0.40 L of 0.15 *M* KOH is mixed with 0.40 L of 0.15 *M* $(CH_3)_2NH_2Br$.

17.13 How many moles of sodium hypobromite, NaBrO, should be added to 1.00 L of 0.200 *M* hypobromous acid, HBrO, to form a buffer solution of pH 8.80? Assume that no volume change occurs when the NaBrO is added.

17.14 How many grams of sodium lactate, $NaC_3H_5O_3$, should be added to 1.00 L of 0.0150 *M* lactic acid, $HC_3H_5O_3$, to form a buffer solution with pH 3.90? Assume that no volume change occurs when the $NaC_3H_5O_3$ is added.

17.15 One liter of a buffer solution contains 0.15 mol of acetic acid and 0.10 mol of sodium acetate. **(a)** What is the pH of this buffer? **(b)** What is the pH after addition of 0.010 mol of HNO_3? **(c)** What is the pH after addition of 0.010 mol of NaOH?

17.16 One liter of a buffer solution contains 0.120 mol of benzoic acid, $HC_7H_5O_2$, and 0.105 mol sodium benzoate, $NaC_7H_5O_2$. **(a)** What is the pH of this buffer? **(b)** What is the pH after addition of 0.011 mol of HCl? **(c)** What is the pH after addition of 0.011 mol of NaOH?

17.17 **(a)** What is the ratio of HCO_3^- to H_2CO_3 in blood of pH 7.4? **(b)** What is the ratio of HCO_3^- to H_2CO_3 in an exhausted marathon runner whose blood pH is 7.1?

17.18 A phosphate buffer, consisting of $H_2PO_4^-$ and HPO_4^{2-}, helps control the pH of blood. Many carbonated soft drinks also use this buffer system. What is the pH of a soft drink in which the major buffer ingredients are 6.5 g of NaH_2PO_4 and 8.0 g of Na_2HPO_4 per 355 mL of solution?

Titration Curves

17.19 How does titration of a strong acid with a strong base differ from titration of a weak acid with a strong base with respect to the following points: **(a)** quantity of base required to reach the equivalence point; **(b)** pH at the beginning of the titration; **(c)** pH at the equivalence point; **(d)** pH after addition of a slight excess of base; **(e)** choice of indicator for determining the equivalence point?

17.20 Assume that 30.0 mL of a 0.10 *M* solution of a weak base B that accepts one proton is titrated with a 0.10 *M* solution of the monoprotic strong acid HX. **(a)** How many moles of HX have been added at the equivalence point? **(b)** What is the predominant form of B at the equivalence point? **(c)** What factor determines the pH at the equivalence point? **(d)** Which indicator, phenolphthalein or methyl red, is likely to be the better choice for this titration?

17.21 How many milliliters of 0.048 *M* NaOH are required to reach the equivalence point in titrating each of the following solutions: **(a)** 30.0 mL of 0.038 *M* HBr; **(b)** 28.0 mL of 0.018 *M* H_2SO_4; **(c)** 32.0 mL of 0.034 *M* $HC_2H_3O_2$?

17.22 Four 20.0-mL samples of different HBr solutions were titrated with 0.100 *M* NaOH solution. The volumes of base required to reach the equivalence point in each case were **(a)** 27.5 mL; **(b)** 21.8 mL; **(c)** 48.9 mL; **(d)** 25.5 mL. Calculate concentrations of the four HBr solutions.

17.23 A 20.0-mL sample of 0.200 *M* HBr solution is titrated with 0.200 *M* NaOH solution. Calculate the pH of the solution after the following volumes of base have been added: **(a)** 15.0 mL; **(b)** 19.9 mL; **(c)** 20.0 mL; **(d)** 20.1 mL; **(e)** 35.0 mL.

17.24 A 30.0-mL sample of 0.200 *M* KOH is titrated with 0.150 *M* $HClO_4$ solution. Calculate the pH after the following volumes of acid have been added: **(a)** 30.0 mL; **(b)** 39.5 mL; **(c)** 39.9 mL; **(d)** 40.00 mL; **(e)** 40.10 mL.

17.25 Consider the titration of 30.0 mL of 0.100 *M* propionic acid, $HC_3H_5O_2$. Calculate the pH of this solution upon addition of the following volumes of 0.100 *M* NaOH solution: **(a)** 0; **(b)** 15.0 mL **(c)** 29.5 mL; **(d)** 30.0 mL; **(e)** 30.5 mL; **(f)** 45.0 mL.

17.26 Consider the titration of 40.0 mL of 0.100 *M* NH_3. Calculate the pH of this solution upon addition of the following volumes of 0.100 *M* HCl: **(a)** 0; **(b)** 20.0 mL; **(c)** 39.5 mL; **(d)** 40.0 mL; **(e)** 40.5 mL; **(f)** 50.0 mL.

17.27 Calculate the pH at the equivalence point for titrating 0.200 *M* solutions of each of the following bases with 0.200 *M* HBr: **(a)** sodium hydroxide, NaOH; **(b)** hydroxylamine, NH_2OH; **(c)** aniline, $C_6H_5NH_2$.

17.28 Calculate the pH at the equivalence point in titrating 0.100 *M* solutions of each of the following with 0.080 *M* NaOH: **(a)** hydrobromic acid, HBr; **(b)** lactic acid, $HC_3H_5O_3$; **(c)** sodium hydrogen chromate, $NaHCrO_4$.

Solubility Equilibria

17.29 Write the expression for the solubility-product constant for the solubility equilibrium of each of the following compounds: **(a)** CdS; **(b)** MgC_2O_4; **(c)** CeF_3; **(d)** $Fe_3(AsO_4)_2$.

17.30 Write the expression for the solubility-product constant for the solubility equilibrium of each of the following compounds: **(a)** Cu_2S; **(b)** $Co(OH)_3$; **(c)** $La_2(MoO_4)_3$; **(d)** $Ce(OH)_4$.

17.31 **(a)** If the molar solubility of CaF_2 at 35°C is 1.24×10^{-3} mol/L, what is K_{sp} at this temperature? **(b)** It is found that 1.1×10^{-2} g of SrF_2 dissolves per 100 mL of aqueous solution at 25°C. Calculate the solubility product for SrF_2. **(c)** The K_{sp} of $Ba(IO_3)_2$ at 25°C is 6.0×10^{-10} What is the molar solubility of $Ba(IO_3)_2$?

17.32 **(a)** The molar solubility of $PbBr_2$ at 25°C is 1.0×10^{-2} mol/L. Calculate K_{sp}. **(b)** If 0.0490 g of $AgIO_3$ dissolves per liter of solution, calculate the solubility product. **(c)** Using the appropriate K_{sp} value from Appendix D, calculate the solubility of $Cu(OH)_2$ in grams per liter of solution.

17.33 Using Appendix D, calculate the molar solubility of AgI in **(a)** pure water; **(b)** 5.0×10^{-2} *M* NaI solution; **(c)** 0.10 *M* $AgNO_3$ solution.

17.34 Calculate the molar solubility of Ag_2SO_4 in **(a)** pure water; **(b)** 0.10 *M* Na_2SO_4; **(c)** 0.10 *M* $AgNO_3$.

17.35 Calculate the molar solubility of $Cd(OH)_2$ **(a)** at pH 7.0; **(b)** at pH 9.0.

17.36 Calculate the molar solubility of $Mg(OH)_2$ **(a)** at pH 9.5; **(b)** at pH 11.8.

17.37 A 0.010 *M* solution of NaF is shaken with a mixture of solid CaF_2 and BaF_2. At equilibrium, what is the ratio of the concentration of $Ca^{2+}(aq)$ to that of $Ba^{2+}(aq)$?

17.38 What is the ratio of $[Ca^{2+}]$ to $[Fe^{2+}]$ in a lake in which the water is in equilibrium with deposits of both $CaCO_3$ and $FeCO_3$?

17.39 The solubility-product constant for barium permanganate, $Ba(MnO_4)_2$, is 2.5×10^{-10}. Suppose that solid $Ba(MnO_4)_2$ is in equilibrium with a solution of $KMnO_4$. What concentration of $KMnO_4$ is required to establish a concentration of 2.0×10^{-8} *M* for the Ba^{2+} ion in solution?

17.40 **(a)** The K_{sp} for cerium iodate, $Ce(IO_3)_3$, is 3.2×10^{-10}. Calculate the molar solubility of $Ce(IO_3)_3$ in pure water. **(b)** What concentration of $NaIO_3$ in solution would be necessary to reduce the Ce^{3+} concentration in a saturated solution of $Ce(IO_3)_3$ by a factor of 10 below that calculated in part (a)?

Precipitation; Dissolution

17.41 **(a)** Will $Mn(OH)_2$ precipitate from solution if the pH of a 0.050 *M* solution of $MnCl_2$ is adjusted to 8.0? **(b)** Will Ag_2SO_4 precipitate when 100 mL of 0.010 *M* $AgNO_3$ is mixed with 20 mL of 5.0×10^{-2} *M* H_2SO_4 solution?

17.42 **(a)** Will $Co(OH)_2$ precipitate from solution if the pH of a 0.020 *M* solution of $Co(NO_3)_2$ is adjusted to 8.5? **(b)** Will $AgIO_3$ precipitate when 100 mL of 0.010 *M* $AgNO_3$ is mixed with 10 mL of 0.015 *M* $NaIO_3$? (K_{sp} of $AgIO_3$ is 3.0×10^{-8}.)

17.43 Calculate the minimum pH needed to precipitate $Ni(OH)_2$ so completely that the concentration of Ni^{2+} is less than 1 μg per liter [that is, 1 part per billion (ppb)].

17.44 Suppose that a 50-mL sample of a solution is to be tested for Cl^- ion by addition of one drop (0.2 mL) of 0.10 *M* $AgNO_3$. What is the minimum number of grams of Cl^- that must be present in order for AgCl(*s*) to form?

17.45 Which of the following salts will be substantially more soluble in acid solution than in pure water: **(a)** Ag_2CO_3, **(b)** CeF_3; **(c)** Ag_2SO_4; **(d)** PbI_2; **(e)** $Cd(OH)_2$?

17.46 For each of the following slightly soluble salts, write the net ionic equation for reaction, if any, with acid: **(a)** $Pb(N_3)_2$; **(b)** LaF_3; **(c)** BiI_3; **(d)** AgCN; **(e)** $SrSO_3$.

17.47 A solution contains 2.0×10^{-4} *M* Ag^+ and 1.5×10^{-3} *M* Pb^{2+}. If NaI is added, will AgI ($K_{sp} = 8.3 \times 10^{-17}$) or PbI_2 ($K_{sp} = 1.4 \times 10^{-8}$) precipitate first? Specify the concentration of I^- needed to begin precipitation.

17.48 A solution of Na_2SO_4 is added dropwise to a solution that is 0.010 *M* in Ba^{2+} and 0.010 *M* in Sr^{2+}. **(a)** What concentration of ${SO_4}^{2-}$ is necessary to begin precipitation? (Neglect volume changes; $BaSO_4$: $K_{sp} = 1.1 \times 10^{-10}$; $SrSO_4$: $K_{sp} = 2.8 \times 10^{-7}$.) **(b)** Which cation precipitates first? **(c)** What is the concentration of ${SO_4}^{2-}$ when the second cation begins to precipitate?

17.49 Calculate the solubility of ZnS (in moles per liter) in a 0.10 *M* solution of H_2S if the pH of the solution is 2.40.

17.50 Calculate the solubility of CdS (in moles per liter) in a 0.10 *M* solution of H_2S if the pH of the solution is 1.0.

17.51 Complete and balance each of the following reactions. In each case indicate which species in the reaction can be considered a Lewis base and which a Lewis acid:
(a) $Zn(OH)_2(s) + H^+(aq) \longrightarrow$
(b) $Cd(CN)_2(s) + CN^-(aq) \longrightarrow$
(c) $CrF_3(s) + OH^-(aq) \longrightarrow$

17.52 Complete and balance each of the following reactions. In each case indicate which species in the reaction can be considered a Lewis base and which a Lewis acid:
(a) $Cu^{2+}(aq) + CN^-(aq) \longrightarrow$
(b) $AgCl(s) + S_2O_3^{2-}(aq) \longrightarrow$
(c) $Cu(OH)_2(s) + NH_3(aq) \longrightarrow$

17.53 From the value for K_f listed in Table 17.2 calculate the concentration of Cu^{2+} in 1 L of a solution that contains a total of 1×10^{-3} mol of copper(II) ion and that is 0.10 *M* in NH_3.

17.54 To what final concentration of NH_3 must a solution be adjusted to just dissolve 0.020 mol of NiC_2O_4 in 1 L of solution? (Hint: You can neglect the hydrolysis of $C_2O_4^{2-}$, because the solution will be quite basic.)

17.55 Using the value of K_{sp} for AgCl and K_f for $Ag(CN)_2^-$, calculate the equilibrium constant for the following reaction:

$$AgCl(s) + 2CN^-(aq) \rightleftharpoons Ag(CN)_2^-(aq) + Cl^-(aq)$$

17.56 Using the value of K_{sp} for Ag_2S, K_{a1} and K_{a2} for H_2S, and $K_f = 1.1 \times 10^5$ for $AgCl_2^-$, calculate the equilibrium constant for the following reaction:

$$Ag_2S(s) + 4Cl^-(aq) + 2H^+(aq) \rightleftharpoons 2AgCl_2^-(aq) + H_2S(aq)$$

Qualitative Analysis

17.57 A solution containing an unknown number of metal ions is treated with dilute HCl; no precipitate forms. The pH is adjusted to about 1, and H_2S is bubbled through. Again no precipitate forms. The pH of the solution is then adjusted to about 8. Again H_2S is bubbled through. This time a precipitate forms. The filtrate from this solution is treated wth $(NH_4)_2HPO_4$. No precipitate forms. Which metal ions discussed in Section 17.6 are possibly present? Which are definitely absent within the limits of these tests?

17.58 An unknown solid is entirely soluble in water. On addition of dilute HCl, a precipitate forms. After the precipitate is filtered off, the pH is adjusted to about 1 and H_2S is bubbled in; a precipitate again forms. After filtering off this precipitate, the pH is adjusted to 8 and H_2S is again added; no precipitate forms. No precipitate forms upon addition of $(NH_4)_2HPO_4$. The remaining solution shows a yellow color in a flame test. Based on these observations, which of the following compounds might be present, which are definitely present, and which are definitely absent: CdS, $Pb(NO_3)_2$, HgO, $ZnSO_4$, $Cd(NO_3)_2$, and Na_2SO_4?

17.59 In the course of various qualitative analysis procedures, the following mixtures are encountered: **(a)** Zn^{2+} and Cd^{2+}; **(b)** $Cr(OH)_3$ and $Fe(OH)_3$; **(c)** Mg^{2+} and K^+; **(d)** Ag^+ and Mn^{2+}. Suggest how each mixture might be separated.

17.60 Suggest how the cations in each of the following solution mixtures can be separated: **(a)** Na^+ and Cd^{2+}; **(b)** Cu^{2+} and Mg^{2+}; **(c)** Pb^{2+} and Al^{3+}; **(d)** Ag^+ and Hg^{2+}.

17.61 **(a)** Precipitation of the group 4 cations (Figure 17.19) requires a basic medium. Why is this so? **(b)** What is the most significant difference between the sulfides precipitated in group 2 and those precipitated in group 3? **(c)** Suggest a procedure that would serve to redissolve all of the group 3 cations following their precipitation and separation.

17.62 A student who is in a great hurry to finish his laboratory work decides that his qualitative analysis unknown contains a metal ion from the insoluble phosphate group, group 4 (Figure 17.20). He therefore tests his sample directly with $(NH_4)_2HPO_4$, skipping earlier tests for the metal ions in groups 1 through 3. He observes a precipitate and concludes that a metal ion from group 4 is indeed present. Why is this possibly an erroneous conclusion?

Additional Exercises

17.63 Which of the following solutions has the greatest buffering capacity, and which has the least: **(a)** 0.10 *M* $HC_2H_3O_2$ and 0.10 *M* $NaC_2H_3O_2$, pH 4.74; **(b)** 1.8×10^{-5} *M* HBr, pH 4.74; **(c)** 0.01 *M* $HC_2H_3O_2$ and 0.01 *M* $NaC_2H_3O_2$, pH 4.74? Explain.

17.64 Consider a buffer solution prepared from Na_2CO_3 and $NaHCO_3$. Write chemical equations for the reactions by which this buffer neutralizes H^+ and OH^-.

17.65 A hypothetical weak acid, HA, was combined with NaOH in the following proportions: 0.20 mol of HA, 0.080 mol of NaOH. The mixture was diluted to a total volume of 1 L, and the pH measured. **(a)** If pH = 4.80, what is the pK_a of the acid? **(b)** How many additional moles of NaOH would need to be added to the solution to increase the pH to 5.00?

17.66 Suppose you need to prepare a buffer at pH 10.6. Using Table D.2, select at least two different acid-base pairs that would be appropriate. Describe the composition of your buffers.

[17.67] What is the pH of a solution made by mixing 0.30 mol NaOH, 0.25 mol Na_2HPO_4, and 0.20 mol H_3PO_4 to water and diluting to 1.00 L?

[17.68] Suppose you want to do a physiological experiment that calls for a pH 6.5 buffer. You find that the organism with which you are working is not sensitive to a certain weak acid (H_2X: $K_{a1} = 2 \times 10^{-2}$; $K_{a2} = 5.0 \times 10^{-7}$) and its sodium salts. You have available a 1.0 *M* solution of this acid and a 1.0 *M* solution of NaOH. How much of the NaOH solution should be added to 1.0 L of the acid to give a buffer at pH 6.50? (Ignore any volume change.)

17.69 A 0.30 *M* solution of dimethylamine, $(CH_3)_2NH$, containing an unknown concentration of dimethylammonium chloride, $(CH_3)_2NH_2Cl$, has a pH of 10.40. What is the concentration of dimethylammonium ion in the solution?

17.70 The acid-base indicator bromcresol green is a weak acid. The yellow acid and blue base forms of the indicator are present in equal concentrations in a solution when the pH is 4.68. What is pK_a for bromcresol green?

17.71 When asked as an examination question to give an example of an alkaline buffer solution, a student listed a solution of $NaHCO_3$. If you were the instructor, would you give no credit, partial credit, or full credit for this answer? Explain.

17.72 You have 1.0 M solutions of H_3PO_4 and NaOH. Describe how you would prepare a buffer solution of pH 7.20 with the highest possible buffering capacity from these reagents. Describe the composition of the buffer solution.

17.73 A biochemist needs 750 mL of an acetic acid-sodium acetate buffer with pH 4.50. Solid sodium acetate, $NaC_2H_3O_2$, and glacial acetic acid, $HC_2H_3O_2$, are available. Glacial acetic acid is 99 percent $HC_2H_3O_2$ by weight and has a density of 1.05 g/mL. If the buffer is to be 0.30 M in $HC_2H_3O_2$, how many grams of $NaC_2H_3O_2$ and how many milliliters of glacial acetic acid must be used?

[17.74] Equal quantities of 0.10 M solutions of an acid HA and a base B are mixed. The pH of the resulting solution is 8.8. **(a)** Write the equilibrium equation and equilibrium-constant expression for the reaction between HA and B. **(b)** If K_a for HA is 5.0×10^{-6}, what is the value of the equilibrium constant for the reaction between HA and B? **(c)** What is the value of K_b for B?

17.75 Potassium hydrogen phthalate, often abbreviated KHP, can be obtained in high purity and is used to determine the concentrations of solutions of strong bases. Strong base react with the hydrogen phthalate ion as follows:

$$HP^-(aq) + OH^-(aq) \longrightarrow H_2O(l) + P^{2-}(aq)$$

The molar mass of KHP is 204.2 g/mol and K_a for the HP^- ion is 3.1×10^{-6}. **(a)** If a titration experiment begins with 0.4885 g of KHP and has a final volume of about 100 mL, which indicator from Figure 16.6 would be most appropriate? **(b)** If the titration required 38.55 mL of NaOH solution to reach the endpoint, what is the concentration of the NaOH solution?

17.76 A sample of 0.1355 g of an unknown monoprotic acid was dissolved in 25.0 mL of water and titrated with 0.0950 M NaOH. The acid required 19.3 mL of base to reach the equivalence point. **(a)** What is the molar mass of the acid? **(b)** After 12.0 mL of base had been added in the titration, the pH was found to be 5.10. What is the K_a for the unknown acid?

17.77 If 50.00 mL of 0.100 M Na_2SO_3 is titrated with 0.100 M HCl, calculate **(a)** the pH at the start of the titration; **(b)** the volume of HCl required to reach the first equivalence point. What is the predominant species present? **(c)** what volume of HCl is required to reach the second equivalence point, and what is the pH at the second equivalence point.

17.78 Calculate the pH of a saturated solution of $Ca(OH)_2$ at 25°C.

17.79 The hydrolysis of ions can sometimes appreciably affect the solubility of salts. In which of the following salts would you expect hydrolysis to be the greatest problem? Explain briefly. **(a)** $PbCl_2$; **(b)** $CaCO_3$; **(c)** $Ni(OH)_2$; **(d)** Ag_3PO_4; **(e)** CuS.

17.80 Using K_{sp} values, decide which compound in each of the following pairs has the greater molar solubility: **(a)** $NiCO_3$ or $CdCO_3$; **(b)** $CaCO_3$ or $CaSO_4$ **(c)** CuBr or PbF_2; **(d)** Ag_2S or CoS.

[17.81] The solubility products of $PbSO_4$ and $SrSO_4$ are 1.6×10^{-8} and 2.8×10^{-7}, respectively. What are the values of $[SO_4^{2-}]$, $[Pb^{2+}]$, and $[Sr^{2+}]$ in a solution at equilibrium with both substances.

[17.82] Fluoridation of drinking water is employed in many places to aid prevention of dental caries. Typically the F^- ion concentration is adjusted to about 1 ppb. Some water supplies are also "hard"; that is, they contain certain cations such as Ca^{2+} that interfere with the action of soap. Consider a case where the concentration of Ca^{2+} is 8 ppb. Could a precipitate of CaF_2 form under these conditions? (Make any necessary approximations.)

[17.83] What should be the pH of a buffer solution that will result in a Mg^{2+} concentration of 3.0×10^{-2} M in equilibrium with solid magnesium oxalate?

[17.84] Calculate the solubility of $CuCO_3$ in a strongly buffered solution of pH 7.5.

17.85 What concentration of Pb^{2+} remains in a solution after $PbCl_2$ has been precipitated from a solution that is 0.1 M in Cl^- at 25°C? Will PbS form if the filtrate from the $PbCl_2$ precipitate is saturated with H_2S at pH 1?

[17.86] The value of K_{sp} for $Mg_3(AsO_4)_2$ is 2.1×10^{-20}. The AsO_4^{3-} ion is derived from the weak acid H_3AsO_4 ($pK_{a1} = 2.22$; $pK_{a2} = 6.98$; $pK_{a3} = 11.50$). When asked to calculate the molar solubility of $Mg_3(AsO_4)_2$ in water, a student used the K_{sp} expression and assumed that $[Mg^{2+}] = 1.5[AsO_4^{3-}]$. Why was this a mistake?

[17.87] Cadmium sulfide, CdS, is used commercially as a yellow paint pigment known as cadmium yellow. It is prepared by saturating a slightly acidic solution of Cd^{2+} with H_2S gas. The presence of Fe^{2+}, a common impurity, can result in the precipitate being contaminated with black FeS, thereby ruining the color of the pigment. **(a)** If a solution containing 0.10 M Cd^{2+} and 1×10^{-4} M Fe^{2+} is saturated with H_2S ($[H_2S] = 0.10M$), what pH is needed to keep the Fe^{2+} from precipitating? **(b)** If a $HC_2H_3O_2$–$C_2H_3O_2^-$ buffer is used to control pH, what ratio of $HC_2H_3O_2$ to $C_2H_3O_2^-$ is required?

17.88 The pH of a solution that is 0.020 M in Hg^{2+}, 0.020 M in Ni^{2+}, and 0.020 M in Pb^{2+} is adjusted to 2.0 and then saturated with H_2S gas. Which metal ions, if any, will precipitate?

[17.89] What pH range will permit PbS to precipitate while leaving Mn^{2+} in solution if the solution is 0.010 M in Pb^{2+} and 0.010 M in Mn^{2+} prior to saturation with H_2S?

[17.90] The solubility of manganese oxalate dihydrate, $MnC_2O_4 \cdot 2H_2O$, is 0.035 g per 100 mL of solution. The acid-dissociation constants of oxalic acid are listed in Appendix D. Calculate the pH of a saturated solution of $MnC_2O_4 \cdot 2H_2O$.

[17.91] Calculate the molar solubility of $Fe(OH)_3$ in water. (This problem is not as simple as it might first appear; consider carefully the concentration of OH^-.)

[17.92] The Ag^+ ion forms a complex with the thiosulfate ion:

$$Ag^+(aq) + 2S_2O_3^{2-} \rightleftharpoons Ag(S_2O_3)_2^{3-}$$

Using K_f from Table 17.2, calculate the number of moles of AgBr(*s*) that will dissolve in 1.0 L of 0.050 *M* $Na_2S_2O_3$ solution.

[17.93] The solubility product for $Zn(OH)_2$ is 1.2×10^{-17}. The formation constant for the hydroxo complex, $Zn(OH)_4^{2-}$, is 4.6×10^{17}. What concentration of OH^- is required to dissolve 0.010 mol of $Zn(OH)_2$ in a liter of solution?

18 Chemistry of the Environment

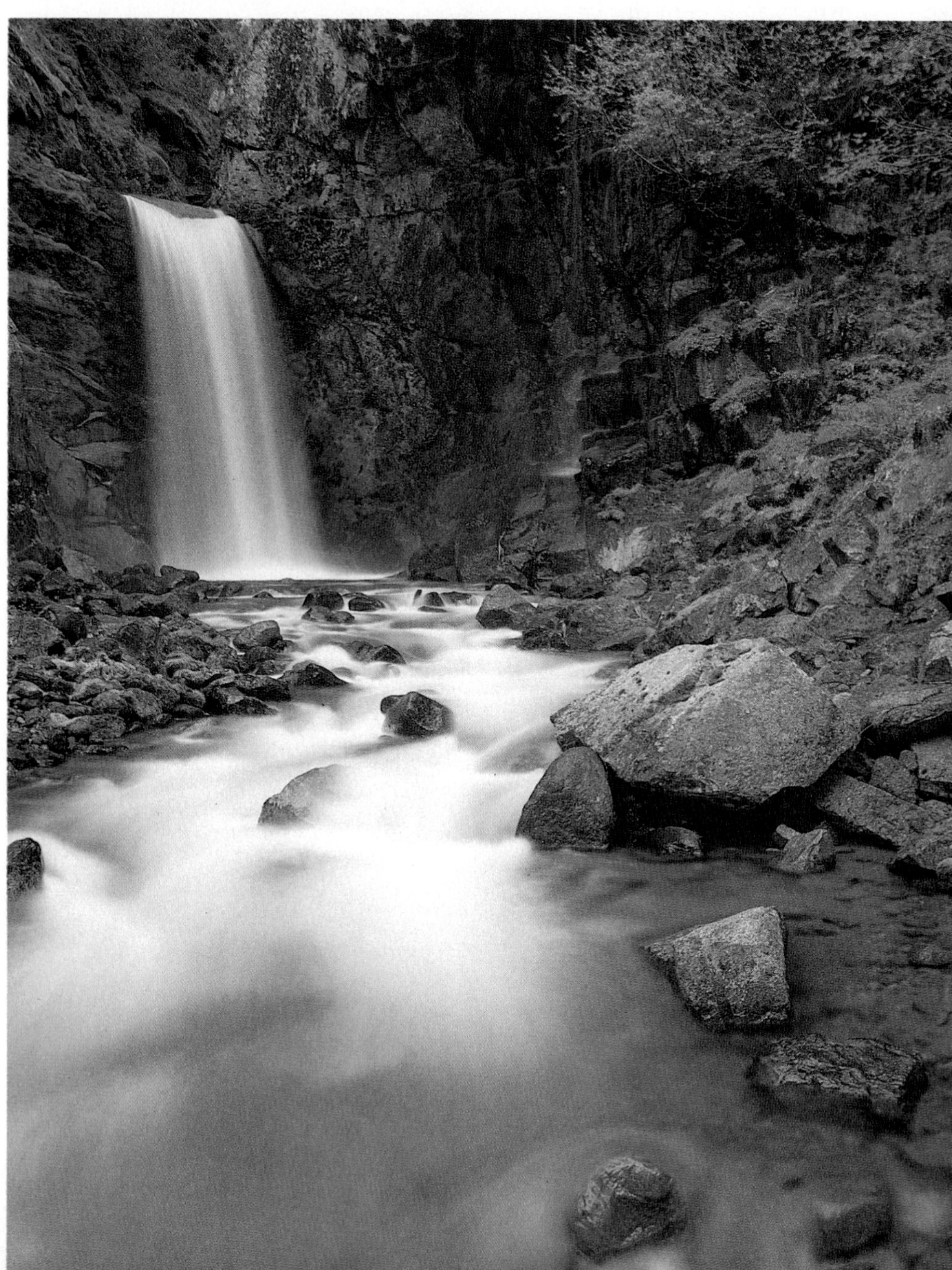

Christine Falls, Mt. Rainier National Park. (Charles A. Mauzy/ Aperture Photobank/Allstock)

In previous chapters, we have dealt for the most part with the principles that govern the chemical and physical behavior of matter. We are now in a position to apply these principles to an understanding of the world in which we live. In this chapter, we consider some aspects of the chemistry of our environment, focusing on the earth's *atmosphere* and on the aqueous environment, which we call the *hydrosphere.*

Both the atmosphere and hydrosphere of our planet make life as we know it possible. Management of this environment so as to maintain and enhance the quality of life is one of the most important concerns of our time. It has become evident that major reforms and much stricter standards are required if we are to preserve the quality of life in the world. As a voting citizen you will be called on to help decide bond issues and referenda that may have an impact on your health as well as on your economic security. Even our simple decisions as consumers require us to weigh the costs versus the benefits of our actions. Unfortunately, the environmental impacts of our decisions are often hidden from view.

The better you understand the chemical principles that underlie environmental issues, the better your chances of forming sound judgments on economic and political matters that affect our environment. Our intent in this chapter is to provide an introduction to the nature of the earth's atmosphere and hydrosphere and to indicate some of the ways in which pollution occurs.

CONTENTS

18.1 EARTH'S ATMOSPHERE

Because most of us have never been very far from the earth's surface, we tend to take for granted the many ways in which the atmosphere determines the environment in which we live. In this section, we will examine some of the important characteristics of our planet's atmosphere.

The temperature of the atmosphere varies in a complex manner as a function of altitude, as shown in Figure 18.1. The atmosphere is divided into four regions based on this temperature profile. Just above the surface, in the **troposphere**, the temperature normally decreases with increasing altitude, reaching a minimum of about 215 K at about 12 km. Nearly all of us live out our entire lives in the troposphere. Howling winds and soft breezes, rain, sunny skies—all that we normally think of as "weather" occurs in this region. Even when we fly in a modern jet aircraft between distant cities, we are still in the troposphere, although we may be near its upper limit, which we call the *tropopause.*

Above the tropopause, the temperature increases with altitude, reaching a maximum of about 275 K at about 50 km. This region is called the **stratosphere**. Beyond the stratosphere are the **mesosphere** and the **thermosphere**. Notice in Figure 18.1 that the temperature extremes that are the boundaries

Figure 18.1 Temperature variations in the atmosphere at altitudes below 110 km.

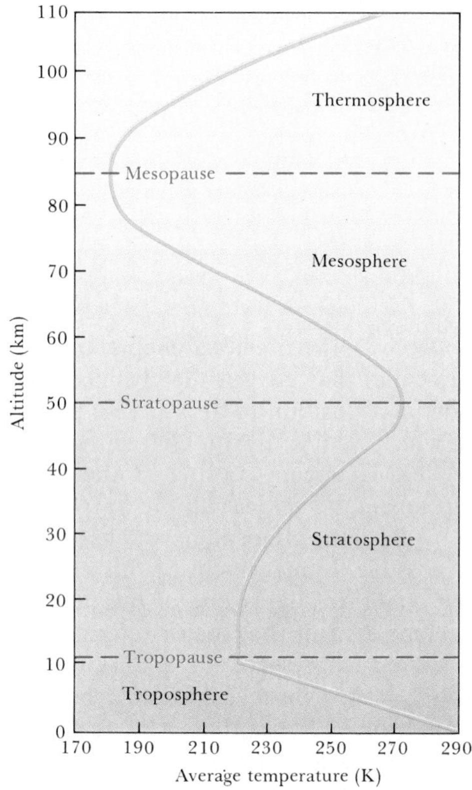

for each region are denoted by the suffix *-pause*. The boundaries are important because mixing of the atmosphere across the boundaries is relatively slow. For example, pollutant gases generated in the troposphere find their way into the stratosphere only very slowly.

In contrast to the temperature changes that occur in the atmosphere, the pressure of the atmosphere decreases in a regular way with increasing elevation, as shown in Figure 18.2. We see that atmospheric pressure drops off much more rapidly at lower elevations than at higher ones. The explanation for this characteristic of the atmosphere lies in its compressibility. As a result of the atmosphere's compressibility, the pressure decreases from an average value of 760 mm Hg at sea level to 2.3×10^{-3} mm Hg at 100 km, to only 1.0×10^{-6} mm Hg at 200 km. The troposphere and stratosphere together account for 99.9 percent of the mass of the atmosphere, with 75 percent of the mass in the troposphere.

Figure 18.2 Variations in atmospheric pressure with altitude.

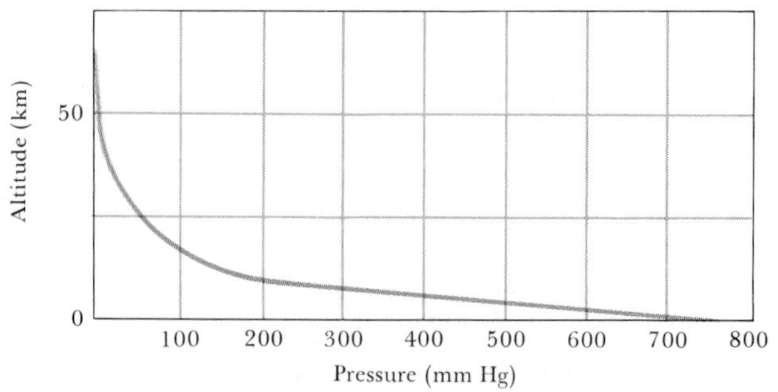

Composition of the Atmosphere

The atmosphere is an extremely complex system. Its temperature and pressure change over a wide range with altitude, as we have just seen. The atmosphere is subjected to bombardment by radiation and energetic particles from the sun and by cosmic radiation from outer space. This barrage of energy has profound chemical effects, especially on the outer reaches of the atmosphere (Figure 18.3). In addition, because of the earth's gravitational field, lighter atoms and molecules tend to rise to the top. As a result of all these factors, the composition of the atmosphere is not uniform.

Table 18.1 shows the composition of dry air near sea level. Although traces of many substances are present, N_2 and O_2 make up about 99 percent of the entire atmosphere. The noble gases and CO_2 make up most of the remainder. Note that the contribution of each component of the atmosphere listed in Table 18.1 is given in terms of its mole fraction. This is simply the total number of moles of a particular component divided by the total number of moles of all the components in the sample (Sections 10.6 and 13.1).

Figure 18.3 The aurora borealis, or northern lights. This luminous display in the northern sky is produced by collisions of high-speed electrons and protons from the sun with air molecules. The charged particles are channeled toward the polar regions by the earth's magnetic field. (Jack Finch/Science Photo Library/Science Source/Photo Researchers)

SAMPLE EXERCISE 18.1

What is the partial pressure of CO_2 in dry air when the total dry air pressure, P_t, is 735 mm Hg?

Solution: The partial pressure of a given component in the atmosphere is given by the product of its mole fraction and the total atmospheric pressure (Section 10.6). Referring to Table 18.1, we see that the mole fraction of CO_2 is 3.30×10^{-4}. Thus, we have

$$P_{CO_2} = X_{CO_2}P_t = (3.3 \times 10^{-4})(735 \text{ mm Hg}) = 0.243 \text{ mm Hg}$$

PRACTICE EXERCISE

What is the partial pressure of argon in dry air when total pressure is 750 mm Hg?
Answer: 7.00 mm Hg

Table 18.1 Composition of Dry Air Near Sea Level

Component[a]	Content (mole fraction)	Molecular weight
Nitrogen	0.78084	28.013
Oxygen	0.20948	31.998
Argon	0.00934	39.948
Carbon dioxide	0.000330	44.0099
Neon	0.00001818	20.183
Helium	0.00000524	4.003
Methane	0.000002	16.043
Krypton	0.00000114	83.80
Hydrogen	0.0000005	2.0159
Nitrous oxide	0.0000005	44.0128
Xenon	0.000000087	131.30

[a] Ozone, sulfur dioxide, nitrogen dioxide, ammonia, and carbon monoxide are present as trace gases in variable amounts.

In speaking of trace constituents of substances, we commonly use *parts per million* (ppm) as the unit of concentration. When applied to substances in solution, parts per million refers to grams of the substance per million grams of solution (Section 13.1). However, in dealing with gases, one part per million refers to one part by *volume* in 1 million volume units of the whole. Because of the properties of gases, volume fraction and mole fraction are the same. Thus, 1 ppm of a trace constituent of the atmosphere amounts to 1 mol of that constituent in 1 million moles of total gas; that is, the concentration in ppm is equal to the mole fraction times 10^6. Note that Table 18.1 lists the mole fraction of CO_2 in the atmosphere as 0.000330. Its concentration in ppm is $0.000330 \times 10^6 = 330$ ppm.

Before we consider the chemical processes that occur in the atmosphere, let's review some of the important chemical properties of the two major components of the atmosphere, N_2 and O_2. We saw in Chapter 8 that the N_2 molecule possesses a triple bond between the nitrogen atoms. This very strong bond is largely responsible for the very low reactivity of N_2, which undergoes reaction only under extreme conditions. The O—O bond energy in O_2 is much lower than that for N_2 (Table 8.4), and O_2 is therefore much more reactive than N_2. Oxygen reacts with many substances to form oxides (Section 7.8). The oxides of nonmetals—for example SO_2—usually form acidic solutions when dissolved in water. The oxides of active metals and other metals in low oxidation states—for example CaO—form basic solutions when dissolved in water.

18.2 THE OUTER REGIONS OF THE ATMOSPHERE

Although the outer portion of the atmosphere, beyond the stratosphere, contains only a small fraction of the atmospheric mass, it plays an important role in determining the conditions of life at the earth's surface. This upper layer forms the outer bastion of defense against the hail of radiation and high-energy particles that continually bombard the planet. As this occurs, the molecules and atoms of the upper atmosphere undergo chemical changes.

Photodissociation

The sun emits radiant energy over a wide range of wavelengths. The shorter-wavelength, higher-energy radiations in the ultraviolet range of the spectrum are sufficiently energetic to cause chemical changes. We have seen in Section 6.2 that electromagnetic radiation can be pictured as a stream of photons. The energy of each photon is given by the relationship $E = h\nu$, where h is Planck's constant and ν is the frequency of the radiation. For a chemical change to occur when radiation falls on the earth's atmosphere, two conditions must be met. First, there must be photons with energy sufficient accomplish whatever chemical process is being considered. Second, molecules must absorb these photons. This requirement means that the energy of the photons is converted into some other form of energy within the molecule.

The rupture of a chemical bond resulting from absorption of a photon by a molecule is called **photodissociation**. One of the most important processes occurring in the upper atmosphere above about 120 km elevation is the photodissociation of the oxygen molecule:

$$O_2(g) + h\nu \longrightarrow 2O(g) \qquad [18.1]$$

The minimum energy required to cause this change is determined by the dissociation energy of O_2, 495 kJ/mol. In Sample Exercise 18.2, we calculate the longest-wavelength photon having sufficient energy to dissociate the O_2 molecule.

SAMPLE EXERCISE 18.2

What is the wavelength of a photon corresponding to a molar bond-dissociation energy of 495 kJ/mol?

Solution: We must first calculate the energy required per molecule and then determine the wavelength of a photon with that energy:

$$\left(495 \times 10^3 \frac{\text{J}}{\text{mol}}\right)\left(\frac{1 \text{ mol}}{6.022 \times 10^{23} \text{ molecules}}\right) = 8.22 \times 10^{-19} \frac{\text{J}}{\text{molecule}}$$

The energy of the photon is given by $E = h\nu$. Rearranging, we have

$$\nu = \frac{E}{h} = \left(\frac{8.22 \times 10^{-19} \text{ J}}{6.625 \times 10^{-34} \text{ J-s}}\right) = 1.24 \times 10^{15} \text{ s}^{-1}$$

Recall from Section 6.1 that the product of frequency and wavelength of radiation equals the velocity of light:

$$\nu\lambda = c = 3.00 \times 10^8 \text{ m/s}$$

Therefore, rearranging this equation, we have

$$\lambda = \frac{c}{\nu} = \left(\frac{3.00 \times 10^8 \text{ m/s}}{1.24 \times 10^{15} \text{ /s}}\right)\left(\frac{10^9 \text{ nm}}{1 \text{ m}}\right) = 242 \text{ nm}$$

PRACTICE EXERCISE

The bond energy in ClO is 268 kJ/mol. What is the longest-wavelength photon that has sufficient energy to cause a Cl—O bond rupture? ***Answer:*** 446 nm

The calculations in Sample Exercise 18.2 tell us that any photon of wavelength *shorter* than 242 nm will have sufficient energy to dissociate the O_2 molecule. (Remember that shorter wavelength means higher energy.)

The second condition that must be met before dissociation actually occurs is that the photon must be absorbed by O_2. Fortunately for us, O_2 absorbs much of the high-energy, short-wavelength radiation from the solar spectrum before it reaches the lower atmosphere. As it does so, atomic oxygen, O, is formed. At higher elevations, the dissociation of O_2 is very extensive. At 400 km, only 1 percent of the oxygen is in the form of O_2; the other 99 percent is in the form of atomic oxygen. At 130 km, O_2 and O are just about equally abundant. Below this elevation, O_2 is more abundant than O.

Because the bond-dissociation energy of N_2 is very high (Table 8.4), only photons of very short wavelength possess sufficient energy to cause dissociation of this molecule. Furthermore, N_2 does not readily absorb photons, even when they do possess sufficient energy. The overall result is that very little atomic nitrogen is formed in the upper atmosphere by dissociation of N_2.

Photoionization

In 1901, Guglielmo Marconi carried out a sensational experiment. He received in St. John's, Newfoundland, a radio signal transmitted from Land's End, England, some 2900 km away. Because radio waves were thought to travel in

Table 18.2 Ionization Processes, Ionization Energies, and Wavelengths Capable of Causing Ionization

Process			Ionization energy (kJ/mol)	λ_{max} (nm)
$N_2 + h\nu$	$\longrightarrow$	$N_2^+ + e^-$	1495	80.1
$O_2 + h\nu$	$\longrightarrow$	$O_2^+ + e^-$	1205	99.3
$O + h\nu$	$\longrightarrow$	$O^+ + e^-$	1313	91.2
$NO + h\nu$	$\longrightarrow$	$NO^+ + e^-$	890	134.5

straight lines, it had been assumed that radio communication over large distances on earth would be impossible. Marconi's successful experiment suggested that the earth's atmosphere in some way substantially affects radio-wave propagation. His discovery led to intensive study of the upper atmosphere. In about 1924, the existence of electrons in the upper atmosphere was established by experimental studies.

For each electron present in the upper atmosphere, there is a corresponding positively charged ion. The electrons in the upper atmosphere result mainly from the **photoionization** of molecules, caused by solar radiation. For photoionization to occur, a photon must be absorbed by the molecule, and this photon must have enough energy to remove the highest-energy electron. Some of the more important ionization processes occurring in the upper atmosphere above about 90 km appear in Table 18.2, together with the ionization energies and λ_{max}, the maximum wavelength of a photon capable of causing ionization. Photons with energies sufficient to cause ionization have wavelengths in the high-energy region of the ultraviolet. These wavelengths are completely filtered out of the radiation reaching earth as a result of their absorption by the upper atmosphere.

18.3 OZONE IN THE UPPER ATMOSPHERE

In contrast to N_2, O_2, and O, which absorb photons with wavelengths shorter than 240 nm, ozone is the key absorber of photons with wavelengths of 240 to 310 nm. Let's consider how ozone forms in the upper atmosphere and how it absorbs photons.

Below 90 km, most of the short-wavelength radiation capable of photoionization has been absorbed. Radiation capable of dissociating the O_2 molecule is sufficiently intense, however, for photodissociation of O_2 (Equation 18.1) to remain important down to 30 km. The chemical processes that occur in the region below about 90 km following photodissociation of O_2 are very different from processes that occur at higher elevations.

In the mesosphere and stratosphere, the concentration of O_2 is much greater than that of atomic oxygen. Therefore, the O atoms that form in the mesosphere and stratosphere undergo frequent collisions with O_2 molecules. These collisions lead to formation of ozone, O_3:

$$O(g) + O_2(g) \longrightarrow O_3^*(g) \qquad [18.2]$$

The asterisk over the O_3 denotes that the ozone molecule contains an excess of energy. Reaction of O with O_2 to form O_3 results in release of 105 kJ/mol. This energy must be gotten rid of by the O_3 molecule in a very short time, or else it will simply fly apart again into O_2 and O. This decomposition is the reverse of the process by which O_3 is formed. The energy-rich O_3 molecules can get rid of the excess energy by colliding with another atom or molecule and transferring some of the excess energy to it. Let us represent the atom or molecule with which O_3 collides as M. (Usually M is N_2 or O_2 because these are the most abundant molecules.) The formation of O_3 and the transfer of excess energy to M are summarized by the following equations:

$$O(g) + O_2(g) \rightleftharpoons O_3^* \quad [18.3]$$

$$O_3^*(g) + M(g) \longrightarrow O_3(g) + M^*(g) \quad [18.4]$$

$$O(g) + O_2(g) + M(g) \longrightarrow O_3(g) + M^*(g) \quad \text{(net)} \quad [18.5]$$

The rate at which O_3 forms depends on the relative rates of the stabilizing collisions between O_3^* and M (Equation 18.4) and the dissociation of O_3^* back to O_2 and O (the reverse process in Equation 18.3). Frequent collisions favor formation of O_3 (Equation 18.4). Because the concentration of molecules is greater at lower altitudes, the frequency of stabilizing collisions is greater. However, at low altitudes most of the radiation energetic enough to dissociate O_2 has been absorbed. Therefore, the highest rate of O_3 formation occurs at about 50-km altitude.

The ozone molecule, once formed, does not last long. Ozone is capable of absorbing solar radiation, which results in its decomposition into O_2 and O. Because only 105 kJ/mol is required for this process, photons of wavelength shorter than 1140 nm are sufficiently energetic to dissociate O_3. The strongest and most important absorptions, however, are of photons from 200 to 310 nm. If it were not for the layer of ozone in the stratosphere, these high-energy photons would penetrate to the earth's surface. Plant and animal life as we know it could not survive in the presence of this high-energy radiation. The "ozone shield" is therefore essential for our continued well being. It should be noted, however, that the ozone molecules that form this essential shield against radiation represent only a tiny fraction of the oxygen atoms present in the stratosphere. This is so because the ozone molecules are continually destroyed even as they are formed.

The photodecomposition of ozone reverses the reaction leading to its formation. We thus have a cyclic process of ozone formation and decomposition, summarized as follows:

$$O_2(g) + h\nu \longrightarrow O(g) + O(g)$$

$$O(g) + O_2(g) + M(g) \longrightarrow O_3(g) + M^*(g) \quad \text{(heat released)}$$

$$O_3(g) + h\nu \longrightarrow O_2(g) + O(g)$$

$$O(g) + O(g) + M(g) \longrightarrow O_2(g) + M^*(g) \quad \text{(heat released)}$$

The first and third processes are photochemical; they use a solar photon to initiate a chemical reaction. The second and fourth processes are exothermic chemical reactions. The net result of all four processes is a cycle in which solar radiant energy is converted into thermal energy. The ozone cycle in the stratosphere is responsible for the temperature rise that reaches its maximum at the stratopause, as illustrated in Figure 18.1.

The scheme described above for the life and death of ozone molecules accounts for some but not all of the known facts about the ozone layer. Many

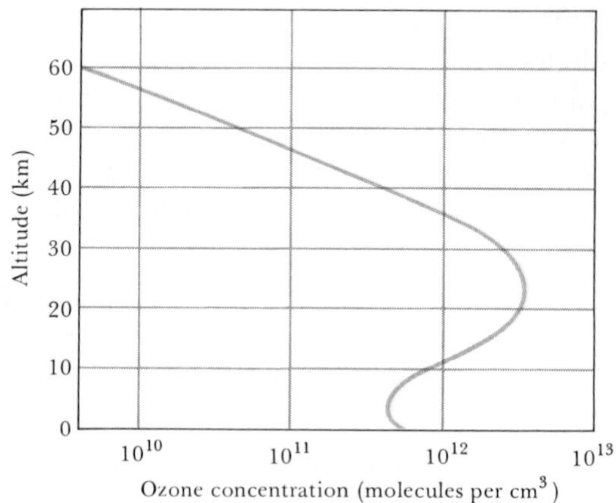

Figure 18.4 Variations in ozone concentration in the atmosphere as a function of altitude.

chemical reactions involving substances other than just oxygen occur. In addition, the effects of turbulence and winds that mix up the stratosphere must be considered. A very complicated picture results. The overall result of ozone formation and removal reactions, coupled with atmospheric turbulence and other factors, is to produce an ozone profile in the upper atmosphere as shown in Figure 18.4.

Depletion of the Ozone Layer

In 1974, F. Sherwood Rowland and Mario Molina of the University of California, Irvine, proposed that chlorine from **chlorofluorocarbons** (CFCs) may deplete the ozone layer. These substances, principally $CFCl_3$ (Freon 11) and CF_2Cl_2 (Freon 12), have been widely used as propellants in spray cans, as refrigerant and air-conditioner gases, and as foaming agents for plastics. They are virtually unreactive in the lower atmosphere. Furthermore, they are relatively insoluble in water and are therefore not removed from the atmosphere by rainfall or by dissolution in the oceans. Unfortunately, the lack of reactivity that makes them commercially useful also allows them to survive in the atmosphere and to diffuse eventually into the stratosphere. It is estimated that several million tons of chlorofluorocarbons are present in the atmosphere.

As the chlorofluorocarbons diffuse into the stratosphere, they become subject to the action of high-energy radiation. Wavelengths in the range of 190 to 225 nm cause the **photolysis**, or light-induced rupture, of a carbon-chlorine bond of the chlorofluorocarbons:

$$CF_xCl_{4-x}(g) + h\nu \longrightarrow CF_xCl_{3-x}(g) + Cl(g) \qquad [18.6]$$

Calculations suggest that chlorine atom formation occurs at the greatest rate at an altitude of about 30 km.

Atomic chlorine is capable of rapid reaction with ozone to form chlorine oxide, ClO, and molecular oxygen, O_2. The ClO can react with atomic oxygen to re-form atomic chlorine:

$$\begin{array}{lcll} Cl(g) + O_3(g) & \longrightarrow & ClO(g) + O_2(g) & [18.7] \\ ClO(g) + O(g) & \longrightarrow & Cl(g) + O_2(g) & [18.8] \\ \hline O_3(g) + O(g) & \longrightarrow & 2O_2(g) \quad \text{net} & [18.9] \end{array}$$

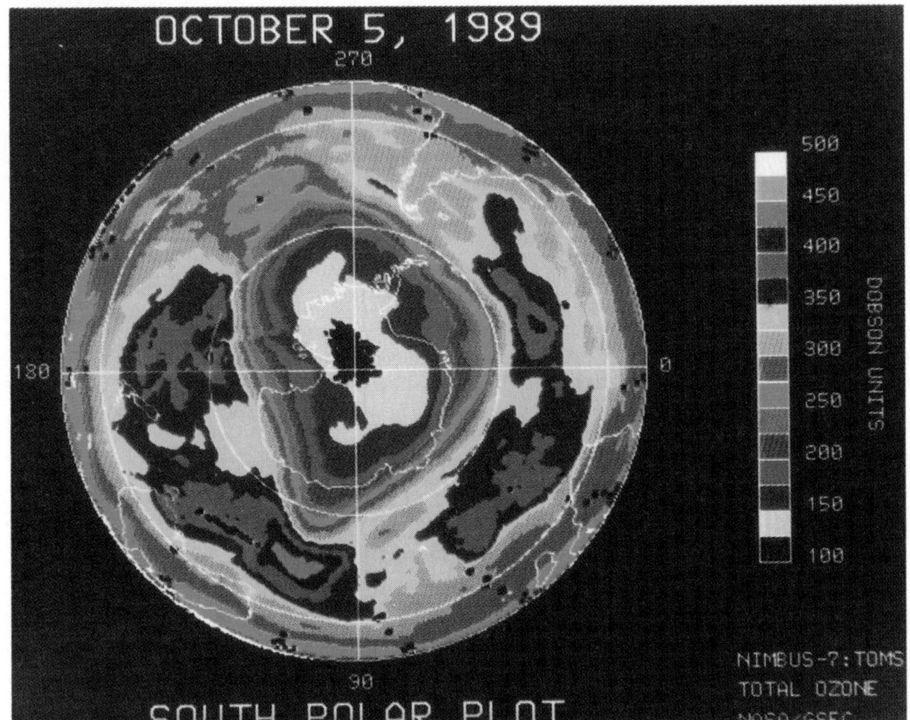

Figure 18.5 Map of the total ozone present in the southern hemisphere, taken in October 1989 from an orbiting satellite. The scale on the right indicates ozone levels; the lower the Dobson unit, the lower the ozone level. The ozone "hole" is the black area in the center of Antarctica. (NASA/Goddard Space Flight Center).

The net result of these reactions is the conversion of ozone into O_2. Because Cl is used in the first step of this mechanism and then re-formed in the second step, it functions as a catalyst. It is estimated that each Cl atom destroys about 100,000 molecules of ozone before it itself is destroyed in other reactions.

Although rates of diffusion of molecules into the stratosphere from the earth's surface are likely to be slow, a loss of ozone believed caused by chlorofluorocarbons has already been observed. Since the late 1970s, researchers have found an annual thinning of the ozone layer over the South Pole, which occurs during the austral (Southern Hemisphere) spring. Ozone levels in October of both 1987 and 1989 (Figure 18.5) dropped to almost 60 percent of the levels in August. Scientists are now finding evidence that the North Pole suffers a similar but less pronounced ozone loss during late winter. There are also preliminary indications of some depletion at lower latitudes.

18.4 CHEMISTRY OF THE TROPOSPHERE

The photodissociation and photoionization reactions that we have discussed in the preceding sections occur in the upper atmosphere. These processes result in nearly complete absorption of all solar radiation of less than about 300-nm wavelength before it reaches the troposphere. Because the major constituents of the troposphere do not interact with radiation longer than 300 nm, the photochemical reactions that occur in the troposphere involve only minor constituents.

Table 18.3 lists some of the important minor constituents of the troposphere and summarizes their major sources and typical concentrations. Many of these substances occur to only a slight extent in the natural environment but exhibit much higher concentrations in certain areas as a result of human activities. Indeed, in some areas the concentrations of these substances have

Table 18.3 Sources and Typical Concentrations of Some Minor Atmospheric Constituents

Minor constituent	Sources	Typical concentrations
Carbon dioxide, CO_2	Decomposition of organic matter; release from the oceans; fossil-fuel combustion	330 ppm throughout troposphere
Carbon monoxide, CO	Decomposition of organic matter; industrial processes; fuel combustion	0.05 ppm in nonpolluted air; 1 to 50 ppm in urban traffic areas
Methane, CH_4	Decomposition of organic matter; natural-gas seepage	1 to 2 ppm throughout troposphere
Nitric oxide, NO	Electrical discharges; internal-combustion engines; combustion of organic matter	0.01 ppm in nonpolluted air; 0.2 ppm in smog atmospheres
Ozone, O_3	Electrical discharges; diffusion from stratosphere; photochemical smog	0 to 0.01 ppm in nonpolluted air; 0.5 ppm in photochemical smog
Sulfur dioxide, SO_2	Volcanic gases; forest fires; bacterial action; fossil-fuel combustion; industrial processes (roasting of ores, and so on)	0 to 0.01 ppm in nonpolluted air; 0.1 to 2 ppm in polluted urban environment

increased considerably during the last 50 years. In this section, we will discuss the most important characteristics of a few of these substances and their chemical role as air pollutants. As we will see, most form as either a direct or indirect result of our widespread use of combustion reactions.

Sulfur Compounds and Acid Rain

Sulfur-containing compounds are present to some extent in the natural, unpolluted atmosphere. They originate in the bacterial decay of organic matter, in volcanic gases, and from other sources listed in Table 18.3. The concentration of sulfur-containing compounds in the atmosphere resulting from natural sources is very small in comparison with the concentrations built up in urban and industrial environments as a result of human activities. Sulfur compounds, chiefly sulfur dioxide, SO_2, are among the most unpleasant and harmful of the common pollutant gases. Table 18.4 shows the concentrations of several pollutant gases in a *typical* urban environment (not one that is particularly affected by smog). According to these data, the level of sulfur dioxide is 0.08 ppm or higher about half the time. This concentration is considerably lower than that of other pollutants, notably carbon monoxide. Nevertheless, sulfur dioxide is regarded as the most serious health hazard among the pollutants shown, especially for people with respiratory difficulties. Studies of the medical case

Table 18.4 Concentrations of Atmospheric Pollutants Exceeded about 50 Percent of the Time in a Typical Urban Atmosphere

Pollutant	Concentration (ppm)
Carbon monoxide	10
Hydrocarbons	3
Sulfur dioxide	0.08
Nitrogen oxides	0.05
Total oxidants (ozone and others)	0.02

histories of large population segments in urban environments have shown clearly that those living in the most heavily polluted parts of cities have higher levels of respiratory disease and shorter life expectancies.

Combustion of coal and oil accounts for about 80 percent of the total SO_2 released in the United States. The extent to which SO_2 emissions are a problem in the burning of coal and oil depends on the level of their sulfur concentration. Oil burned in the power plants at electrical generating stations is the nonvolatile residue that remains after the low-boiling fractions have been distilled off. Some oil, such as that from the Middle East, is relatively low in sulfur, whereas Venezuelan oil is relatively high. Because of concern about SO_2 pollution, low-sulfur oil is in greater demand and is consequently more expensive.

Coals vary considerably in their sulfur content. Much of the coal lying in beds east of the Mississippi is relatively high in sulfur content, up to 6 percent by weight. Much of the coal lying in the western states has a lower sulfur content. (This coal, however, also has a lower heat content per unit weight of coal, so that the difference in sulfur content on the basis of a unit amount of heat produced is not as large as is often assumed.)

Altogether, more than 30 million tons of SO_2 is released into the atmosphere in the United States each year. This material does a great deal of damage to both property and human health. Not all the damage, however, is caused by SO_2 itself; it is likely, in fact, that SO_3, formed by oxidation of SO_2, is the major culprit. Sulfur dioxide may be oxidized to SO_3 by any of several pathways, depending on the particular nature of the atmosphere. Once SO_3 is formed it dissolves in water droplets, forming sulfuric acid, H_2SO_4:

$$SO_3(g) + H_2O(l) \longrightarrow H_2SO_4(aq) \qquad [18.10]$$

The presence of sulfuric acid in rain is largely responsible for the phenomenon of **acid rain**. (Nitrogen oxides, which form nitric acid, also contribute.) About 200 years ago, rain had a pH between 6 and 7.6. Today it is common in many regions for rain to have a pH between 4 and 4.5. In Los Angeles, the pH of fog has actually fallen to 2, about the acidity of lemon juice. Acid rain is now affecting many lakes in northern Europe, the northern United States, and Canada. The acidity has dramatically reduced fish populations and has measurably affected other parts of the ecological network within the lakes and surrounding forests (see Figure 18.6).

The pH of most productive natural waters is between 6.5 and 8.5. At pH levels below 4.0, all vertebrates, most invertebrates, and many microorganisms are destroyed. The lakes that are most susceptible to damage are those with low concentrations of basic ions like HCO_3^- that buffer them against changes in pH. Over 300 lakes in New York State contain no fish, and 140 lakes in Ontario, Canada, are devoid of life. The acid rain that appears to have killed these lakes originates hundreds of kilometers downwind in the Ohio Valley and Great Lakes regions.

Acid rain also corrodes many metals and building materials. For example, marble and limestone, whose major constituent is $CaCO_3$, are readily attacked by acid rain (Figure 18.7). Billions of dollars each year are lost as a result of corrosion due to SO_2 pollution.

Obviously, we all want to reduce the quantity of this noxious gas that is released into the environment. One way to do this is to remove sulfur from

(*a*) (*b*)

Figure 18.6 An area of the Black Forest in Germany photographed recently (*a*) and 10 years previously (*b*). The dramatic change is due largely to the detrimental effects of acid rain. (Régis/Bossu/Sygma)

Figure 18.7 The erosion of this stone railway marker post, erected in the nineteenth century, is the result mainly of the reaction of acidic rainfall with the limestone. (Anne LaBastille/Photo Researchers)

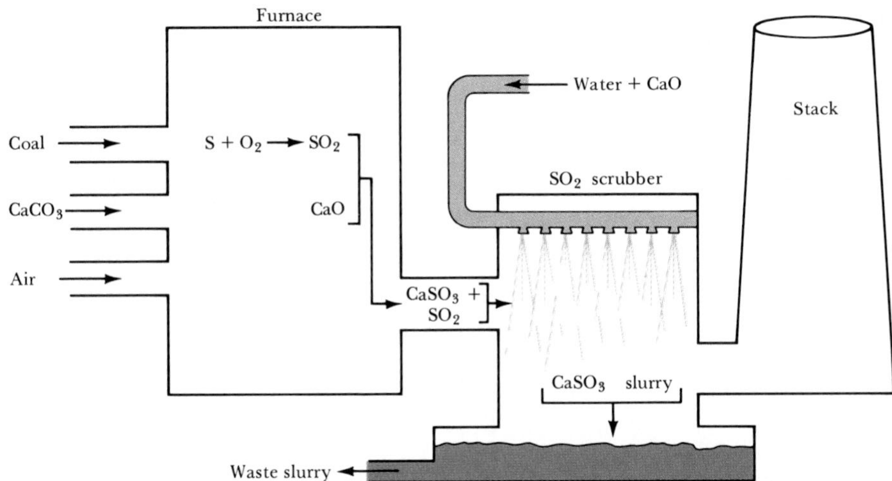

Figure 18.8 Common method for removing SO_2 from combusted fuel. Powdered limestone decomposes into CaO, which reacts with SO_2 to form $CaSO_3$. The $CaSO_3$ and any unreacted SO_2 enter a purification chamber, where a shower of CaO and water converts the remaining SO_2 into $CaSO_3$ and precipitates the $CaSO_3$ into a watery residue called *slurry*.

coal and oil before it is burned. This is presently too difficult and expensive to be technologically feasible. However, several methods have been developed for removing SO_2 from the gases formed when coal and oil are combusted. For example, powdered limestone, $CaCO_3$, may be blown into the combustion chamber. The carbonate (limestone) is decomposed into lime, CaO, and carbon dioxide:

$$CaCO_3(s) \longrightarrow CaO(s) + CO_2(g) \qquad [18.11]$$

The lime then reacts with SO_2 to form calcium sulfite:

$$CaO(s) + SO_2(g) \longrightarrow CaSO_3(s) \qquad [18.12]$$

Only about half the SO_2 is removed by contact with the dry solid. The furnace gas must then be "scrubbed" with an aqueous suspension of lime to remove the $CaSO_3$ and any unreacted SO_2. This process, which is illustrated in Figure 18.8, is difficult to engineer, reduces the heat effectiveness of the fuel, and leaves an enormous solid-waste disposal problem. An electric power plant serving the needs of a population of about 150,000 people would produce about 160,000 tons per year of solid waste if it were equipped with the purification system just described. This is three times the normal fly-ash waste from a plant of this size. Various schemes may be employed to recover elemental sulfur or some other industrially useful chemical from the SO_2, but as yet no process has been found sufficiently attractive from an economic point of view to warrant large-scale development. Pollution by sulfur dioxide will probably remain a major problem for some time.

Carbon Monoxide

Carbon monoxide is formed by the incomplete combustion of carbon-containing materials—for example, fossil fuels. In terms of total mass, CO is the most abundant of all the pollutant gases. The level of CO present in unpolluted air is low, probably on the order of 0.05 to 0.1 ppm. The estimated total amount of CO in the atmosphere is about 5.2×10^{14} g. In the United States alone, however, about 1×10^{14} g of CO is produced each year, about two-thirds of which comes from automobiles.

Figure 18.9 Structure of the heme molecule.

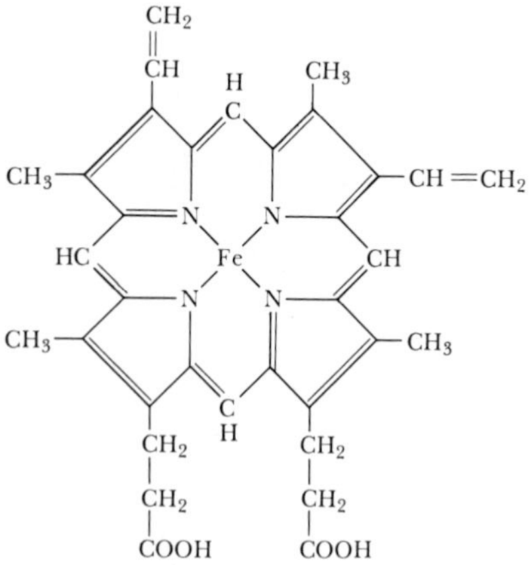

Carbon monoxide is a relatively unreactive molecule and consequently poses no direct threat to vegetation or materials. It does affect humans, however. It has the unusual ability to bind very strongly to **hemoglobin**, the iron-containing protein responsible for oxygen transport in the blood. Hemoglobin consists of four protein chains loosely held together in a cluster. Each chain has within its folds a heme molecule. The structure of heme is shown in Figure 18.9. Note that iron is situated in the center of a plane of four nitrogen atoms. A hemoglobin molecule in the lungs picks up an oxygen molecule, which reacts with the iron atom to form a species called **oxyhemoglobin**. The equilibrium between hemoglobin and oxyhemoglobin is shown in Figure 18.10. As the blood circulates, the oxygen molecule is released in tissues as needed for cell metabolism, that is, for the chemical processes occurring in the cell. (See the Chemistry at Work box on blood as a buffered solution in Section 17.2.)

Carbon monoxide also happens to bind very strongly to the iron in hemoglobin. The complex is called **carboxyhemoglobin** and is represented as COHb. The affinity of human hemoglobin for CO is about 210 times greater than for O_2. As a result, a relatively small quantity of CO can inactivate a substantial fraction of the hemoglobin in the blood for oxygen transport. For example, a person breathing air that contains only 0.1 percent of CO takes in enough CO after a few hours of breathing to convert up to 60 percent of the hemoglobin into COHb, thus reducing the blood's normal oxygen-carrying capacity by 60 percent.

Under normal conditions, a nonsmoker breathing unpolluted air has about 0.3 to 0.5 percent COHb in the bloodstream. This amount arises mainly from the production of small quantities of CO in the course of normal body chemistry and from the small amount of CO present in clean air. Exposure to

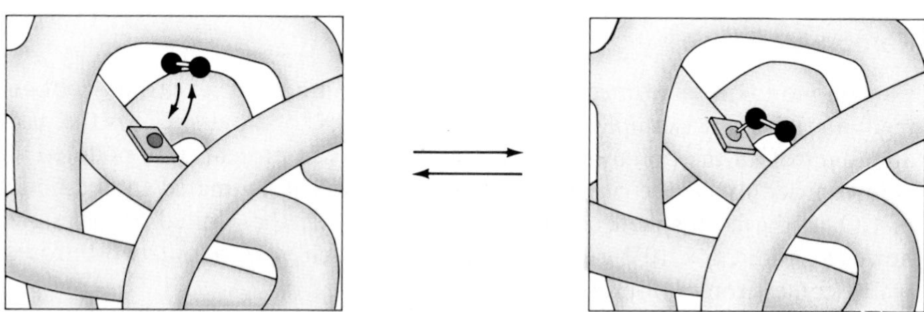

Figure 18.10 Equilibrium between hemoglobin and oxyhemoglobin. The tan regions represent the protein chain. The heme molecule is attached to the protein.

CHEMISTRY AT WORK: Smoking and CO

The CO concentration in city traffic often reaches 50 ppm and may go as high as 140 ppm in traffic jams. The most serious source of carbon monoxide poisoning, however, comes from cigarette smoking. The inhaled smoke from cigarettes contains about 400 ppm of CO (Figure 18.11). The effect of smoking on COHb percentage is evident from the data in Table 18.5. A study of a group of San Francisco dockworkers presented further proof of the dramatic relationship between smoking and COHb percentage. Nonsmokers in the group averaged 1.3 percent COHb; light smokers (less than half a pack per day) averaged 3.0 percent; moderate smokers averaged 4.7 percent; and heavy smokers (two packs or more per day) averaged 6.2 percent COHb.

Widespread evidence indicates that chronic exposure to CO impairs performance on standardized tests. Thus, it is most definitely not a good idea to smoke heavily before and during a test. In addition, motor performance is also impaired by high COHb percentages. For example, evidence shows that drivers responsible for traffic accidents have, on the average, higher than normal percentages of COHb in their blood. As has been mentioned, a chronically high level of COHb means that a certain fraction of the hemoglobin in the blood is not available for oxygen transport. This in turn means that the heart must work that much harder to ensure an adequate supply of oxygen. It is not surprising, therefore, that many medical researchers believe that chronic exposure to CO is a contributing factor in heart disease and in heart attacks.

SURGEON GENERAL'S WARNING: Cigarette Smoke Contains Carbon Monoxide.

Figure 18.11 This warning from a cigarette advertisement reflects the high concentration of carbon monoxide in cigarette smoke.

higher concentrations of CO causes the COHb level to increase. This doesn't happen instantly, but requires several hours. Similarly, when the CO level is suddenly decreased, it requires several hours for the COHb concentration to level off at a lower value. Table 18.5 shows the percentages of COHb in blood that are typical of various groups of people.

Nitrogen Oxides and Photochemical Smog

The atmospheric chemistry of the nitrogen oxides is interesting because these substances are components of smog, a phenomenon with which city dwellers are all too familiar. The term **smog** refers to a particularly unpleasant condition of pollution in certain urban environments that occurs when weather conditions produce a relatively stagnant air mass. The smog made famous by Los Angeles, but now common in many other urban areas as well, is more accurately described as **photochemical smog**, because photochemical processes play a major role in its formation (Figure 18.12).

Table 18.5 Carboxyhemoglobin, COHb Percentages in the Blood of People under Various Conditions

	COHb (%)
Continuous exposure, 10 ppm CO	2.0
Continuous exposure, 30 ppm CO	5.0
Nonsmokers, Chicago (1970)	2.0
Smokers, Chicago (1970)	5.8
Nonsmokers, Milwaukee (1969–1971)	1.1
Smokers, Milwaukee (1969–1971)	5.0
Nonsmokers, clean air	0.3–0.5

Figure 18.12 Smog, now common to many urban areas, is produced largely by the action of sunlight on automobile exhaust gases. (Ted Spiegel/Black Star)

Nitric oxide, NO, forms in small quantities in the cylinders of internal-combustion engines by the direct combination of nitrogen and oxygen:

$$N_2(g) + O_2(g) \rightleftharpoons 2NO(g) \qquad \Delta H = 180.8 \text{ kJ} \qquad [18.13]$$

As noted in the Chemistry at Work box in Section 15.7, the equilibrium constant, K_p, for this reaction increases from about 10^{-15} at 300 K (near room temperature) to about 0.05 at 2400 K (approximately the temperature in the cylinder of an engine during combustion). Thus, the reaction is more favorable at higher temperatures. Before the installation of pollution-control devices, typical emission levels of NO_x were 4 g/mi. (The x is either 1 or 2; both NO and NO_2 are formed, although NO predominates.) Present auto emission standards call for NO_x emission levels of less than 1 g/mi.

In air, NO is rapidly oxidized to nitrogen dioxide, NO_2:

$$2NO(g) + O_2(g) \rightleftharpoons 2NO_2(g) \qquad \Delta H = -113.1 \text{ kJ} \qquad [18.14]$$

The equilibrium constant for this reaction decreases from about 10^{12} at 300 K to about 10^{-5} at 2400 K. The photodissociation of NO_2 initiates the reactions associated with photochemical smog. The dissociation of NO_2 into NO and O requires 304 kJ/mol, which corresponds to a photon wavelength of 393 nm. In sunlight, therefore, NO_2 undergoes dissociation to NO and O:

$$NO_2(g) + h\nu \longrightarrow NO(g) + O(g) \qquad [18.15]$$

The atomic oxygen formed undergoes several possible reactions, one of which forms ozone, as described earlier:

$$O(g) + O_2(g) + M(g) \longrightarrow O_3(g) + M^*(g) \qquad [18.16]$$

Ozone is a key component of photochemical smog. Although it is an essential UV screen in the upper atmosphere, it is an undesirable pollutant in the troposphere. It is extremely reactive and toxic, and breathing air that contains appreciable amounts of ozone can be especially dangerous for asthma sufferers, exercisers, and the elderly. We therefore have two ozone problems: excessive amounts in many urban environments, where it is harmful, and depletion in the stratosphere, where it is vital.

In addition to nitrogen oxides and carbon monoxide, an automobile engine also emits as pollutants unburned *hydrocarbons*. These organic compounds, which are composed entirely of carbon and hydrogen, are the principal components of gasoline (Section 26.1). A typical engine without effective emission controls emits about 10 to 15 g of these compounds per mile. Current standards require that hydrocarbon emissions be less than 0.4 g/mi.

Reactions between unburned hydrocarbons and other pollutants in air create a complex mixture of substances. Table 18.6 shows typical concentrations of trace constituents in photochemical smog. Notice that the smog contains not only NO_x, CO, O_3, and SO_2, but also a variety of organic compounds.

Reduction or elimination of smog requires that the essential ingredients for its formation be removed from automobile exhaust. Catalytic mufflers are designed to reduce drastically the levels of two of the major ingredients of smog: NO_x and hydrocarbons (see the Chemistry at Work box in Section 14.6). However, emission-control systems are notably unsuccessful in poorly maintained automobiles.

Water Vapor, Carbon Dioxide, and Climate

We have seen how the atmosphere makes life as we know it possible on earth by screening out harmful short-wavelength radiation. In addition, the atmosphere is essential in maintaining a reasonably uniform and moderate temperature on the surface of the planet. The two atmospheric components of major importance in maintaining the earth's surface temperature are carbon dioxide and water.

The earth is in overall thermal balance with its surroundings. This means that the planet radiates energy into space at a rate equal to the rate at which it absorbs energy from the sun. The sun has a temperature of about 6000 K.

Table 18.6 Typical Concentrations of Pollutants in Photochemical Smog (Levels Vary Widely from One Situation to Another)

Constituent	Concentration (ppm in air)
NO_x	0.2
NH_3	0.02
CO	40
O_3	0.5
CH_4	2
C_2H_4	0.5
Higher olefins[a]	0.25
C_2H_2 (acetylene)	0.25
Aldehydes	0.6
SO_2	0.2

[a] Higher olefins are compounds that contain a hydrocarbon chain attached to one of the carbons of the C=C bond.

As seen from outer space, the earth is relatively cold, with a temperature of about 254 K. The distribution of wavelengths in the radiation emitted from an object is determined by its temperature. The radiation emitted by relatively cold objects is in the low-energy, or long-wavelength, region of the spectrum. The maximum in the wavelength of radiation from the earth is in the far-infrared region, around 12,000 nm (Figure 6.3). The troposphere, transparent to visible light, is not at all transparent to infrared radiation. Figure 18.13 shows the distribution of radiation from the earth's surface and, on the same scale, the wavelengths absorbed by atmospheric water vapor and carbon dioxide. Clearly, these atmospheric gases absorb much of the outgoing radiation from the earth's surface. It is indeed fortunate for us that they do so; they serve to maintain a livably uniform temperature at the surface by holding in, as it were, the infrared radiation from the surface, which we feel as heat. The influence of water and carbon dioxide on the earth's climate is often called the *greenhouse effect*. (See the Chemistry at Work box in Section 3.2.)

The partial pressure of water vapor in the atmosphere varies greatly from place to place and time to time, but, in general, it is highest near the surface and drops off very sharply with increased elevation. Carbon dioxide, by contrast, is uniformly distributed throughout the atmosphere, at a concentration of about 330 ppm. Because water vapor absorbs infrared radiation so strongly, it plays the major role in maintaining the atmospheric temperature at night, when the surface is emitting radiation into space and not receiving energy from the sun. In very dry desert climates, where the water-vapor concentration is unusually low, it may be extremely hot during the day but very cold at night. In the absence of an extensive layer of water vapor to absorb and then radiate back to the earth part of the infrared radiation, the surface loses this radiation into space and cools off very rapidly.

Carbon dioxide plays a secondary, but very important, role in maintaining the surface temperature. The worldwide combustion of fossil fuels, principally coal and oil, on a prodigious scale in the modern era has materially increased the carbon dioxide level of the atmosphere. From measurements carried out over several decades it is clear that the CO_2 concentration in the atmosphere is steadily increasing. On the basis of present and expected future rates of fossil-fuel use, the atmospheric CO_2 level is expected to increase about 100 percent by 2050. From a knowledge of the infrared-absorbing characteristics of CO_2 and water, and using a theoretical model for the atmosphere, it has been estimated that doubling the concentration of CO_2 from its present level would cause the average surface temperature of the planet to increase by 3°C. Major changes in global climate could result from a temperature change of this or even a smaller magnitude. Because so many factors go into determining climate,

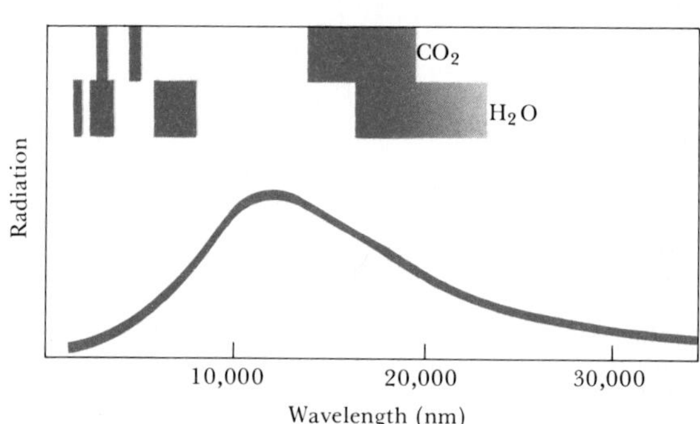

Figure 18.13
Long-wavelength radiation from earth compared with the absorption of infrared radiation by carbon dioxide and water.

we cannot predict with certainty what changes will occur. It is clear, however, that humanity has acquired the potential, by changing the CO_2 concentration in the atmosphere, to substantially alter the climate of the planet.

18.5 THE WORLD OCEAN

Water is the most common liquid on earth. It covers 72 percent of the earth's surface and is essential to life. Our bodies are about 65 percent water by mass. We have learned that water has many exceptional properties. Because of extensive hydrogen bonding (Section 11.2), water has high melting and boiling points and a high heat capacity. Its highly polar character is responsible for its exceptional ability to dissolve a wide range of ionic and polar-covalent substances. Many reactions occur in water, including reactions in which H_2O itself is a reactant. For example, we have seen that H_2O can participate in acid-base reactions as either a proton donor or a proton acceptor (Section 16.1). We will soon see that H_2O can also participate in oxidation-reduction reactions as either a source or a receptor of electrons (Chapter 20). All these properties play a role in our environment.

Seawater

The vast layer of salty water that covers so much of the earth is connected and is generally constant in composition. For this reason, oceanographers (scientists whose major interest is the sea) speak of a world ocean rather than of the separate oceans we learn about in geography books. The world ocean is indeed huge. Its volume is 1.35×10^9 km^3. Almost all the water on earth, 97.2 percent, is in the world ocean. Of the remainder, 2.1 percent is in the form of ice caps and glaciers. All the fresh water—in lakes, rivers, and groundwater—amounts to only 0.6 percent. Most of the remaining 0.1 percent is in brackish (salty) water, such as that in the Great Salt Lake in Utah.

Seawater is often referred to as saline water. The **salinity** of seawater is defined as the mass in grams of dry salts present in 1 kg of seawater. In the world ocean, the salinity varies from 33 to 37, with an average of about 35. To put it another way, seawater contains about 3.5 percent dissolved salts by mass. The list of elements present in seawater is very long. However, most are present only in very low concentrations. Table 18.7 lists the 11 ionic species that are most abundant in seawater.

Table 18.7 Ionic Constituents of Seawater Present in Concentrations Greater than 0.001 g/kg (1 ppm)

Ionic constituent	g/kg seawater	Concentration (M)
Chloride, Cl^-	19.35	0.55
Sodium, Na^+	10.76	0.47
Sulfate, SO_4^{2-}	2.71	0.028
Magnesium, Mg^{2+}	1.29	0.054
Calcium, Ca^{2+}	0.412	0.010
Potassium, K^+	0.40	0.010
Carbon dioxide[a]	0.106	2.3×10^{-3}
Bromide, Br^-	0.067	8.3×10^{-4}
Boric acid, H_3BO_3	0.027	4.3×10^{-4}
Strontium, Sr^{2+}	0.0079	9.1×10^{-5}
Fluoride, F^-	0.001	7×10^{-5}

[a] CO_2 is present in seawater as HCO_3^- and CO_3^{2-}

The sea is so vast that if a substance is present in seawater to the extent of only 1 part per billion (ppb, that is, 1×10^{-6} g per kilogram of water), there is still 5×10^{9} kg of it in the world ocean. Nevertheless, the ocean is not used very much as a source of raw materials, because the cost of extracting the desired substances from the water is too high. Only three substances are recovered from seawater in commercially important amounts: sodium chloride, bromine, and magnesium.

Desalination

Because of its high salt content, seawater is unfit for human consumption and indeed for most of the uses to which we put water. In the United States, the salt content of municipal water supplies is restricted by health codes to no more than about 500 ppm. This amount is much lower than the 3.5 percent dissolved salts present in seawater and the 0.5 percent or so present in brackish water found underground in some regions. The removal of salts from seawater or brackish water to make the water usable is called **desalination**.

Water can be separated from dissolved salts by *distillation* (described in Section 13.5) because water is a volatile substance and the salts are nonvolatile. The principle of distillation is simple enough, but carrying out the process on a large scale presents many problems. For example, as water is distilled from a vessel containing seawater, the salts become more and more concentrated and eventually precipitate out.

Seawater can also be desalinated using **reverse osmosis**. Recall from Section 13.5 that osmosis is the net movement of solvent molecules, but not solute molecules, through a semipermeable membrane. In osmosis, solvent passes from the more dilute solution into the more concentrated one. However, if sufficient external pressure is applied, osmosis can be stopped and, at still higher pressures, reversed. When this occurs, solvent passes from the more concentrated into the more dilute solution. In a modern reverse-osmosis facility, tiny hollow fibers are used as the semipermeable membrane. Water is introduced under pressure into the fibers, and desalinated water is recovered on the outside, as illustrated in Figure 18.14.

The island of Malta, located in the Mediterranean Sea, is formed from limestone, and there is little underground fresh water. Part of the island's water supply is obtained from a reverse-osmosis desalination plant (Figure 18.15) that produces 5.3 million gallons per day of desalinated water. The salt content is reduced from 36,000 ppm of dissolved salts to less than 500 ppm, within the acceptable limits for drinkable water.

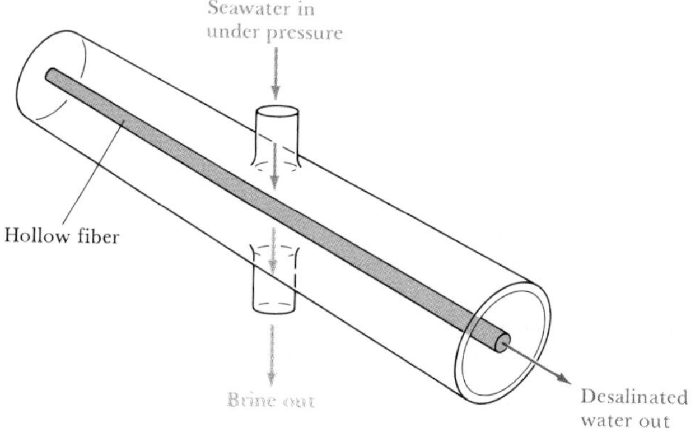

Figure 18.14 Schematic view of a hollow-fiber reverse-osmosis unit. Water is introduced under pressure around the hollow fibers. Desalinated water is recovered from the inside of the fiber. In practice each unit contains more than 3 million fibers, each of which has about the diameter of a human hair. The fibers are bundled together in units such as those shown in Figure 18.15(*b*).

(a)

(b)

Figure 18.15 (*a*) Seawater reverse-osmosis desalination plant at Ghar Lapsi on the island of Malta, in the Mediterranean. (*b*) A permeator room at Ghar Lapsi. Each of the cylinders shown contains several million tiny hollow fibers. When seawater is introduced under pressure into the cylinder, water passes through the fiber wall to the desalinated, or largely salt-free, side. (Courtesy Polymetrics, Inc. and DuPont)

Ocean Pollution

The oceans are exploited by many countries, but they are the responsibility of none. For this reason, they are all the more difficult to safeguard. From the discharge of sewage and industrial wastes, the runoff of agricultural fertilizers, and oil spills, the oceans have been assaulted with a heavy burden of pollutants. In the past, the oceans have demonstrated a remarkable ability to cleanse them-

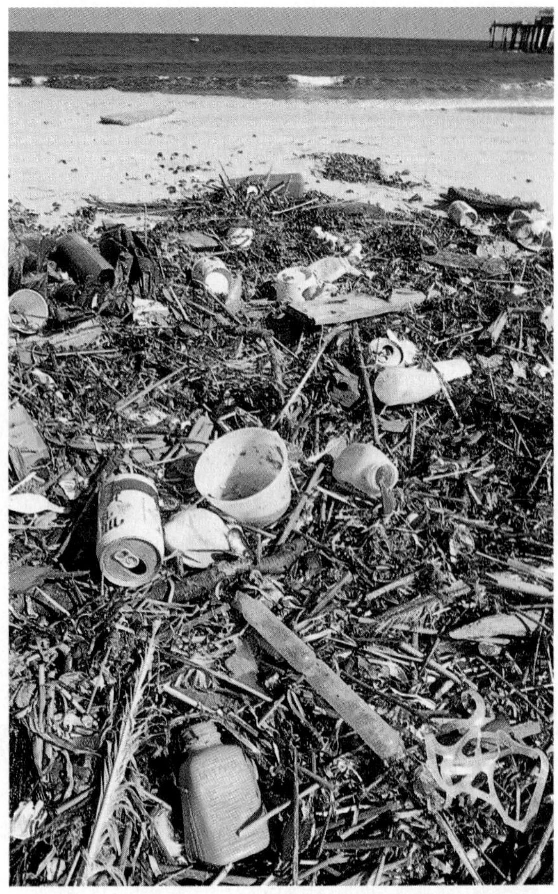

Figure 18.16 Litter along an ocean beach. (John Chiasson/Gamma Liaison)

selves but now they may be perilously close to their capacity to absorb civilization's wastes. The most obvious damage has been to coastal zones, where fouled waters and littered beaches are costing the fishing and resort industries billions of dollars (Figure 18.16). Not all the damage is so readily apparent. The longer we use the oceans as a giant garbage dump, the more difficult and costly the cleanup will be.

Some of the pollution problems of the oceans also affect our lakes and rivers. Principal among these are the problems of oxygen depletion due to the decomposition of oxygen-demanding wastes, the effects of plant nutrients that stimulate algal growth, and the accumulation of toxic substances. We examine some of these pollution problems in the next section.

18.6 FRESH WATER

Water is essential to support life. An adult person needs about 1.5 L/day for drinking. In the United States, our daily use of water per person far exceeds this subsistence level. We use about 7 L per person for cooking and drinking, about 110 L for cleaning (bathing, laundering, and housecleaning), 80 L for flushing toilets, and 80 L for lawns. We use far greater quantities indirectly in agriculture and industry to produce food and other items. For example, about 1×10^5 L of water is used in the manufacture of 1000 kg of steel, about the quantity of steel in an average automobile.

We've seen that the total amount of fresh water on earth is not a very large fraction of the total water present. Indeed, fresh water may well be one of our most precious resources. Fresh water forms by evaporation from the oceans and the land. The water vapor that accumulates in the atmosphere is transported by global atmospheric circulation, eventually returning to earth as rain and snow.

As rain falls and as water runs off the land on its way to the oceans, it dissolves numerous substances. Fresh water contains a variety of cations (mainly Na^+, K^+, Mg^{2+}, Ca^{2+}, and Fe^{2+}), anions (mainly Cl^-, SO_4^{2-}, and HCO_3^-), and dissolved gases (principally O_2, N_2, and CO_2). As we use water, it becomes laden with additional dissolved material, including the wastes of human society. As our population and output of environmental pollutants increase, we find that we must spend ever-increasing amounts of money and resources to guarantee a supply of fresh water.

Dissolved Oxygen and Water Quality

The amount of dissolved O_2 in water is an important indicator of the quality of water. Water fully saturated with air at 1 atm and 20°C contains about 9 ppm of O_2. Oxygen is necessary for fish and much other aquatic life. Cold-water fish require about 5 ppm of dissolved oxygen for survival. Aerobic bacteria consume dissolved oxygen in order to oxidize organic materials and so meet their energy requirements. The organic material that the bacteria are able to oxidize is said to be **biodegradable**. This oxidation occurs by a complex set of chemical reactions, and the organic material disappears gradually.

The presence of an excessive quantity of biodegradable organic materials in water is detrimental because it depletes the water of the oxygen necessary to sustain normal plant and animal life. Indeed, these biodegradable materials are called *oxygen-demanding wastes*. Typical sources of oxygen-demanding wastes include sewage, industrial wastes from food-processing plants and paper mills, and effluent from meat-packing plants.

In the presence of oxygen, the carbon, hydrogen, nitrogen, sulfur, and phosphorus in biodegradable material end up mainly as CO_2, HCO_3^-, H_2O, NO_3^-, SO_4^{2-}, and phosphates. These oxidation reactions sometimes reduce the amount of dissolved oxygen to the point where aerobic bacteria can no longer survive. Anaerobic bacteria then take over the decomposition process, forming CH_4, NH_3, H_2S, PH_3, and other products that contribute to the offensive odors of some polluted waters.

Plant nutrients, particularly nitrogen and phosphorus, contribute to water pollution by stimulating excessive growth of aquatic plants. The most visible results of excessive plant growth are floating algae and murky water. More significantly, however, as plant growth becomes excessive, the amount of dead and decaying plant matter increases rapidly. Decaying plants consume O_2 as they are biodegraded, leading to oxygen depletion in the water. Without sufficient supplies of oxygen, the water in turn cannot sustain any form of animal life. The most important sources of nitrogen and phosphorus compounds in water are domestic sewage (phosphate-containing detergents and nitrogen-containing body wastes), runoff from agricultural land (fertilizers containing both nitrogen and phosphorus), and runoff from livestock areas (animal wastes containing nitrogen).

Treatment of Municipal Water Supplies

The water needed for domestic uses, agriculture, and industrial processes is taken from naturally occurring lakes, rivers, and underground sources or from reservoirs. Much of the water that finds its way into municipal water systems is "used" water; it has already passed through one or more sewage systems or industrial plants. Consequently, this water must be treated before it is distributed to our faucets. Municipal water treatment usually involves five steps: coarse filtration, sedimentation, sand filtration, aeration, and sterilization. Figure 18.17 shows a typical treatment process.

After coarse filtration through a screen, the water is allowed to stand in large settling tanks in which finely divided sand and other minute particles can settle out. To aid in removal of very small particles, the water may first be made slightly basic by adding CaO. Then $Al_2(SO_4)_3$ is added. The aluminum sulfate reacts with OH^- ions to form a spongy, gelatinous precipitate of $Al(OH)_3$ ($K_{sp} = 3.7 \times 10^{-15}$). This precipitate settles slowly, carrying suspended particles down with it, thereby removing nearly all finely divided matter and most bacteria. The water is then filtered through a sand bed. Following filtration, the water may be sprayed into the air to hasten the oxidation of dissolved organic substances.

The final stage of the operation normally involves treating the water with a chemical agent to ensure the destruction of bacteria. Ozone is most effective, but it must be generated at the place where it is used. Chlorine, Cl_2, is therefore more convenient. Chlorine can be shipped in tanks as a liquefied gas and dispensed from the tanks through a metering device directly into the water supply. The amount used depends on the presence of other substances with which the chlorine might react and on the concentrations of bacteria and viruses to be removed. The sterilizing action of chlorine is probably due not to Cl_2 itself but to hypochlorous acid, which forms when chlorine reacts with water:

$$Cl_2(aq) + H_2O(l) \longrightarrow HClO(aq) + H^+(aq) + Cl^-(aq) \qquad [18.17]$$

Figure 18.17 Common steps in treating water for a public water system.

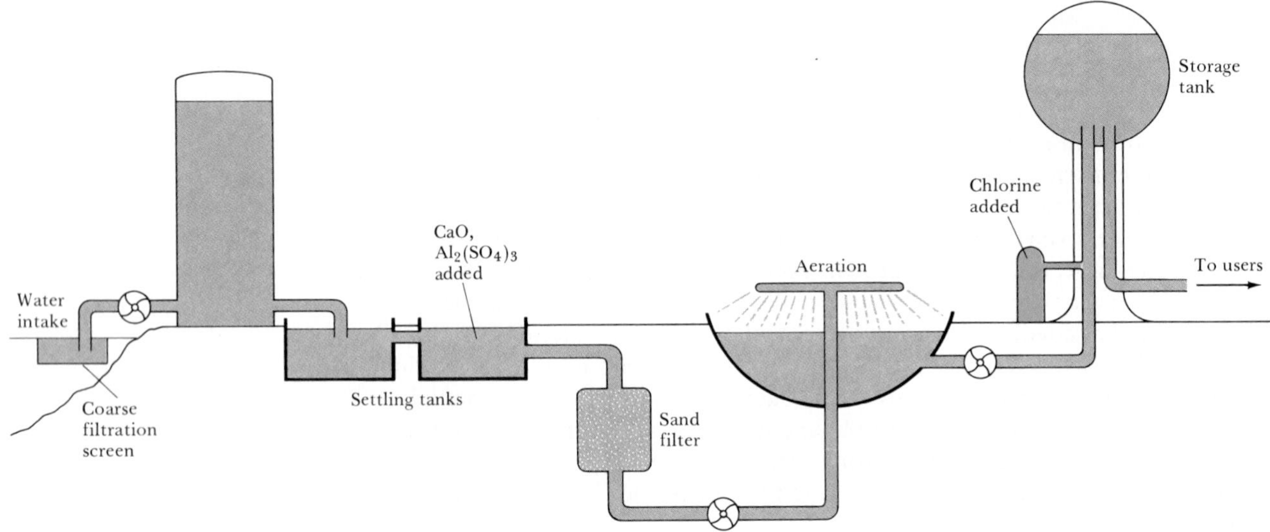

A CLOSER LOOK: Water Softening

Water containing Ca^{2+} and Mg^{2+} is called **hard water.** The presence of these ions makes water unsuitable for some household and industrial uses. For example, these ions react with soaps to form an insoluble soap scum. Although they do not form precipitates with detergents, they adversely affect the performance of such cleaning agents. In addition, mineral deposits may form when water containing these ions is heated. When water containing Ca^{2+} and bicarbonate ions is heated, some carbon dioxide is driven off. As a result, the solution becomes less acidic, and insoluble calcium carbonate forms:

$$Ca^{2+}(aq) + 2HCO_3^-(aq) \xrightarrow{\text{heat}} CaCO_3(s) + CO_2(g) + H_2O(l) \quad [18.18]$$

The solid $CaCO_3$ coats the surface of hot-water systems and the insides of teakettles, thereby reducing heating efficiency. Deposits of scale can be especially serious in boilers in which water is heated under pressure in pipes running through a furnace. Formation of scale reduces the efficiency of heat transfer and may cause the pipes to melt.

Not all municipal water supplies require water softening. In those that do, the water is generally taken from underground sources in which it has had considerable contact with limestone, $CaCO_3$, and other minerals containing Ca^{2+}, Mg^{2+}, and Fe^{2+}. The **lime-soda process** is used for large-scale municipal water-softening operations. The water is treated with "lime," CaO [or "slaked lime," $Ca(OH)_2$], and "soda ash" Na_2CO_3. These chemicals cause precipitation of calcium as $CaCO_3$ ($K_{sp} = 2.8 \times 10^{-9}$) and of magnesium as $Mg(OH)_2$ ($K_{sp} = 1.8 \times 10^{-11}$):

$$Ca^{2+}(aq) + CO_3^{2-}(aq) \longrightarrow CaCO_3(s) \quad [18.19]$$

$$Mg^{2+}(aq) + 2OH^-(aq) \longrightarrow Mg(OH)_2(s) \quad [18.20]$$

The role of Na_2CO_3 is to provide a source of CO_3^{2-}, if needed. If the water already contains a high concentration of bicarbonate ion, calcium can be removed as $CaCO_3$ simply by adding $Ca(OH)_2$:

$$Ca^{2+}(aq) + 2HCO_3^-(aq) + [Ca^{2+}(aq) + 2OH^-(aq)] \longrightarrow 2CaCO_3(s) + 2H_2O(l) \quad [18.21]$$

Lime is used only to the extent that bicarbonate is present: 1 mol of $Ca(OH)_2$ for each 2 mol of HCO_3^-. When bicarbonate is not present, addition of Na_2CO_3 causes removal of Ca^{2+} as $CaCO_3$. The carbonate ion also serves to cause precipitation of $Mg(OH)_2$:

$$Mg^{2+}(aq) + 2CO_3^{2-}(aq) + 2H_2O(l) \longrightarrow 2HCO_3^-(aq) + Mg(OH)_2(s) \quad [18.22]$$

There are two problems with the lime-soda process. First, $CaCO_3$ and $Mg(OH)_2$ may form as very fine particles that do not settle out very well. Second, the solution that results after removal of the precipitates is too strongly basic for subsequent uses. To resolve the first problem, alum, $Al_2(SO_4)_3$, is added to aid the settling of the precipitates. In the basic solution, Al^{3+} from the alum forms a gelatinous precipitate of $Al(OH)_3$, which carries the finely divided $CaCO_3$ and $Mg(OH)_2$ particles with it as it settles out of solution. The basicity of the solution is then decreased by bubbling CO_2 through the water.

Sewage Treatment

Municipal sewage treatment is generally divided into primary, secondary, and tertiary treatments. About 10 percent of sewage handled by public sewers receives no treatment, about 30 percent receives only primary treatment, and about 60 percent is subject to primary and secondary treatment. Tertiary treatment is presently rare, but it is expected to become more common as communities upgrade their treatment facilities to meet federal water-pollution standards.

Primary treatment consists first of screening the incoming sewage to filter out debris and larger suspended solids. The sewage is then passed into settling or sedimentation tanks where suspended solids, called *sludge*, settle out. This type of treatment removes about 60 percent of the suspended solids and 35 percent of dissolved biodegradable materials from the wastewater. If the water receives no secondary treatment, it is then often treated with chlorine before being dumped back into the natural water.

Figure 18.18
Activated-sludge process for aerobic degradation of biodegradable materials in sewage water.

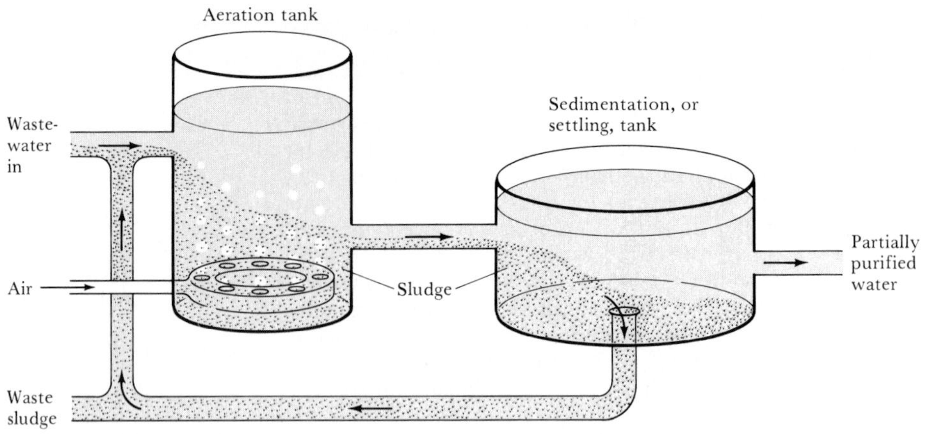

Secondary treatment is based on aerobic decomposition of organic material. The most common type of secondary treatment is known as the activated-sludge method. In this method, the water from primary treatment is passed into an aeration tank, where air is blown through it (see Figure 18.18). This aeration results in rapid growth of aerobic bacteria that feed on the organic wastes. The bacteria form a mass called activated sludge. This sludge settles out in sedimentation tanks, and the water is discharged, often after chlorination. Most of the activated sludge is returned to the aeration tank, where it aids in decomposing the organic wastes in the incoming water. After secondary treatment, about 90 percent of suspended solids and 90 percent of dissolved biodegradable material have been removed.

Many inorganic pollutants—for example, heavy metal ions like Cd^{2+} and Pb^{2+}—are not removed in any appreciable amount by secondary treatment. These pollutants can be removed by tertiary treatment processes, but these processes are generally very costly. Consequently, only a small percentage of wastewater treatment facilities now use any tertiary treatment processes.

FOR REVIEW

SUMMARY

In this chapter we have examined the physical and chemical properties of the earth's atmosphere. The complex temperature variations in the atmosphere give rise to several regions, each with characteristic properties. The lowest of these regions, the troposphere, extends from the surface to about 12 km. Above the troposphere, in order of increasing altitude, are the stratosphere, mesosphere, and thermosphere. In the upper reaches of the atmosphere, only the simplest chemical species can survive the bombardment of highly energetic particles and radiation from the sun. The average molecular weight of the atmosphere at high elevations is lower than at the earth's surface, because the lightest atoms and molecules diffuse upward and because of photodissociation. Absorption of radiation may also lead to photoionization.

Ozone is produced in the mesosphere and stratosphere as a result of the reaction of atomic oxygen with O_2. Ozone is itself decomposed by absorption of a photon or by reaction with an active species such as NO. Human activities could result in addition to the stratosphere of atomic chlorine, which is capable or reacting with ozone in a catalytic cycle to convert ozone to O_2. A marked reduction in the ozone level in the upper atmosphere would have serious adverse consequences, because the ozone layer filters out certain wavelengths of ultraviolet light that are not taken out by any other atmospheric component.

In the troposphere, the chemistry of trace atmospheric components is of major importance. Many of these minor components are pollutants; sulfur dioxide is one of the more noxious and prevalent. It is oxidized in air to form sulfur trioxide, which upon dissolving in water forms sulfuric acid. One method of preventing SO_2 from escaping from industrial operations involves reacting the SO_2 with CaO to form calcium sulfite, $CaSO_3$.

Carbon monoxide is found in high concentrations in the exhaust of automobile engines and in cigarette smoke. This compound is a health hazard because of its ability to form a strong bond with hemoglobin and thus reduce the capacity of blood for oxygen transfer from the lungs.

Photochemical smog is a complex mixture of components in which both nitrogen oxides and ozone play important roles. Smog components are generated mainly in automobile engines, and smog control consists largely of controlling auto emissions.

Carbon dioxide and water vapor are the only major components of the atmosphere that strongly absorb infrared radiation. The level of carbon dioxide in the atmosphere is thus important in determining worldwide climate. As a result of the extensive combustion of fossil fuels (coals, oil, and natural gas), the carbon dioxide level of the atmosphere is steadily increasing.

Seawater contains about 3.5 percent by weight of dissolved salts. Because most of the world's water is in the oceans, humankind may eventually look to the seas for fresh water. Desalination is the removal of dissolved salts from seawater, brine, or brackish water to make it fit for human consumption. Among the means by which desalination may be accomplished are distillation and reverse osmosis.

Fresh water contains many dissolved substances, including dissolved oxygen, which is necessary for fish and other aquatic life. Substances that are decomposed by bacteria are said to be biodegradable. Because the oxidation of biodegradable substances by aerobic bacteria consumes dissolved oxygen, these substances are called oxygen-demanding wastes. The presence of an excess amount of oxygen-demanding wastes in water can deplete the dissolved oxygen sufficiently to kill fish and produce offensive odors. Plant nutrients can contribute to the problem by stimulating the growth of plants that become oxygen-demanding wastes when they die.

The water available from freshwater sources may require treatment before it can be can be used domestically. The several steps usually used in municipal water treatment include coarse filtration, sedimentation, sand filtration, aeration, sterilization, and water softening.

Wastewater treatment is applied to sewage waters or to water that has been used in an industrial operation. Municipal wastewaters are given a primary treatment to remove debris and larger suspended solids. Secondary treatment consists of aeration of sewage sludge to promote the growth of microorganisms that feed on the organic compounds present in sewage. Eventually, clear water is separated from the mass of microorganisms. The water has a lower content of dissolved biodegradable substances than before treatment. The many substances that remain in the water after secondary treatment can be removed only by extensive additional processing, referred to as tertiary treatment.

KEY TERMS

troposphere (Sec. 18.1)
stratosphere (Sec. 18.1)
mesosphere (Sec. 18.1)
thermosphere (Sec. 18.1)
photodissociation (Sec. 18.2)
photoionization (Sec. 18.2)
chlorofluorocarbons (Sec. 18.3)
photolysis (Sec. 18.3)
acid rain (Sec. 18.4)
hemoglobin (Sec. 18.4)
oxyhemoglobin (Sec. 18.4)
carboxyhemoglobin (Sec. 18.4)
smog (Sec. 18.4)
photochemical smog (Sec. 18.4)
salinity (Sec. 18.5)
desalination (Sec. 18.5)
reverse osmosis (Sec. 18.5)
biodegradable (Sec. 18.6)
hard water (Sec. 18.6)
lime-soda process (Sec. 18.6)

EXERCISES

Earth's Atmosphere; The Outer Regions

18.1 Name the regions of the atmosphere, indicate the altitude interval for each region, and describe the variation in temperature in that region.

18.2 Name the boundaries between the regions of the atmosphere, and indicate the temperatures in the boundary regions.

18.3 From the data in Table 18.1, calculate the partial pressures of helium and krypton when the total atmospheric pressure is 735 mm Hg.

18.4 In a particular sample of the atmosphere, the partial pressure of O_2 is found to be 154 mm Hg. Assuming that the sample has the overall composition indicated in Table 15.1, what is the total pressure in the sample?

18.5 The dissociation energy of a carbon-bromine bond is typically about 210 kJ/mol. What is the maximum wavelength of photons that can cause C—Br bond dissociation?

18.6 In CF_3Cl, the C—Cl bond dissociation energy is 339 kJ/mol. In CCl_4, the C—Cl bond dissociation energy is 293 kJ/mol. What is the range of wavelengths of photons that can cause C—Cl bond rupture in one molecule but not in the other?

18.7 What conditions need to be met for radiant energy to cause photodissociation of NO: $NO(g) \longrightarrow N(g) + O(g)$?

18.8 In terms of the energy requirements, explain why photodissociation of oxygen is more important than photoionization of oxygen at altitudes below about 90 km.

Chemistry of the Stratosphere

18.9 Explain why oxygen atoms exist much longer on average at 120-km elevation than at 50-km elevation.

18.10 Explain the means by which ozone, O_3, is formed in the stratosphere. What is the biological significance at the earth's surface of the ozone layer in the stratosphere?

18.11 Using the thermodynamic data in Appendix C, calculate the overall enthalpy change in each step in the following cycle that converts O_3 to O_2:

$$NO(g) + O_3(g) \longrightarrow NO_2(g) + O_2(g)$$
$$NO_2(g) + O(g) \longrightarrow O_2(g) + NO(g)$$

18.12 The standard enthalpies of formation of ClO and ClO_2 are 101 and 102 kJ/mol, respectively. Using these data and the thermodynamic data in Appendix C, calculate the overall enthalpy change for each step in the following catalytic cycle:

$$ClO(g) + O_3(g) \longrightarrow ClO_2(g) + O_2(g)$$
$$ClO_2(g) + O(g) \longrightarrow ClO(g) + O_2(g)$$

On the basis of your results, indicate whether the ClO–ClO_2 pair is at least a possible catalyst for decomposition of ozone in the atmosphere.

Chemistry of the Troposphere

18.13 What are the major health effects of each of the following pollutants: **(a)** CO; **(b)** SO_2; **(c)** O_3?

18.14 Explain how CO, SO_2, and NO_2 can all result from the burning of coal.

18.15 Compare typical concentrations of CO, SO_2, and NO in nonpolluted air (Table 18.3) and urban air (Table 18.4), and indicate in each case at least one possible source of the higher values in Table 18.4.

18.16 For each of the following gases, make a list of known or possible naturally occurring sources: **(a)** CH_4; **(b)** SO_2; **(c)** NO; **(d)** CO.

18.17 In a particular urban environment, the ozone concentration is 0.31 ppm. Assuming a temperature of 16°C and an atmospheric pressure of 745 mm Hg, calculate the partial pressure of ozone and the number of O_3 molecules per cubic meter.

18.18 In a particular urban environment the NO concentration is 0.45 ppm. If the atmospheric pressure is 730 mm Hg and the temperature is 18°C, calculate the partial presssure of NO and the number of NO molecules per cubic meter.

18.19 Show, using chemical equations, how acid rain attacks **(a)** iron; **(b)** limestone.

18.20 Why is NO rather than NO_2 the principal nitrogen oxide formed during combustion reactions?

18.21 Assuming an overall efficiency of about 30 percent, how much calcium carbonate would be required to remove the SO_2 formed in burning a ton of coal containing 2.7 percent sulfur by weight?

18.22 Georgia Power Company's electrical power plant in Taylorsville, Georgia, burned 8,376,726 tons of coal and produced 21,170,999 megawatts of electricity in 1986, a national record. **(a)** Assuming that the coal was 83 percent carbon and 2.5 percent sulfur and that combustion was complete, calculate the number of tons of carbon dioxide and sulfur dioxide produced by the plant during the year. **(b)** If 55 percent of the SO_2 could be removed by reaction with powdered CaO to form $CaSO_3$, how many tons of $CaSO_3$ would be produced?

[18.23] Assume that an equilibrium of N_2, O_2, and NO is achieved at a temperature of 2400 K:

$$N_2(g) + O_2(g) \rightleftharpoons 2NO(g) \qquad K_p = 0.05$$

If the original reaction mixture consists of air at sea level and at 1.0 atm in a 1.0-L vessel, what is the partial pressure of NO at equilibrium? What is the concentration of NO in ppm?

[18.24] A 1.0-L sample of air at 1.0 atm and 900°C containing a mole fraction of 0.78 N_2, 0.20 O_2, and 3500 ppm NO is brought to equilibrium using a heterogeneous catalyst:

$$N_2(g) + O_2(g) \rightleftharpoons 2NO(g) \qquad K_p = 1.0 \times 10^{-5}$$

What are the concentrations of N_2, O_2, and NO at equilibrium?

The World Ocean

18.25 What is the molarity of Na^+ in a solution of NaCl whose salinity is 5 if the solution has a density of 1.0 g/mL?

18.26 Phosphorus is present in seawater to the extent of 0.07 ppm by weight. If the phosphorus is present as phosphate, PO_4^{3-}, calculate the corresponding molar concentration of phosphate.

18.27 Assuming a 10 percent efficiency of recovery, how many liters of seawater must be processed to obtain 10^8 kg of bromine in a commercial production process, assuming the bromide ion concentration listed in Table 18.7?

18.28 A first stage in the recovery of magnesium from seawater is precipitation of $Mg(OH)_2$ by the use of CaO:

$$Mg^{2+}(aq) + CaO(s) + H_2O(l) \longrightarrow Mg(OH)_2(s) + Ca^{2+}(aq)$$

What mass of CaO is needed to precipitate 4.0×10^7 g of $Mg(OH)_2$?

Fresh Water

18.29 Explain why the concentration of dissolved oxygen in fresh water is an important indicator of the quality of the water.

18.30 What forms of decomposition occur when dissolved oxygen levels in water drop below the levels at which aerobic bacteria can survive? Give examples of the decomposition products that form.

18.31 The following organic anion is found in most detergents:

$$H_3C-(CH_2)_9-\overset{H}{\underset{CH_3}{C}}-C_6H_4-SO_3^-$$

Assume that this anion undergoes aerobic decomposition in the following manner:

$$2C_{18}H_{29}O_3S^-(aq) + 51O_2(aq) \longrightarrow 36CO_2(aq) + 28H_2O(l) + 2H^+(aq) + 2SO_4^{2-}(aq)$$

What is the total mass of O_2 required to biodegrade 1.0 g of this substance?

18.32 The average daily mass of O_2 taken up by sewage discharged in the United States is 59 g per person. How many liters of water at 9 ppm O_2 are totally depleted of oxygen in 1 day by a population of 50,000 people?

18.33 Write a balanced chemical equation to describe what occurs when hard water is heated.

18.34 Write a balanced chemical equation to describe how magnesium ion is removed in water treatment by the addition of slaked lime, $Ca(OH)_2$.

18.35 In a particular water supply, the concentration of Ca^{2+} is 2.2×10^{-3} *M*, and the concentration of bicarbonate ion, HCO_3^-, is 1.3×10^{-3} *M*. What weights of $Ca(OH)_2$ and Na_2CO_3 are needed to reduce the level of Ca^{2+} to one-fourth its original level if 1.0×10^7 L of water must be treated?

18.36 How many moles of $Ca(OH)_2$ and of Na_2CO_3 should be added to soften 10^3 L of water in which $[Ca^{2+}] = 5.0 \times 10^{-4}$ *M* and $[HCO_3^-] = 7.0 \times 10^{-4}$ *M*?

18.37 The precipitation of $Al(OH)_3$ ($K_{sp} = 3.7 \times 10^{-15}$) is sometimes used to purify water. **(a)** At what pH will precipitation of $Al(OH)_3$ begin if 2.0 lb of $Al_2(SO_4)_3$ is added to 1000 gal of water? **(b)** Approximately how many pounds of CaO must be added to the water to achieve this pH?

18.38 **(a)** Explain why $Mg(OH)_2$ precipitates when CO_3^{2-} ion is added to a solution containing Mg^{2+}. **(b)** Will $Mg(OH)_2$ precipitate when 5.0 g of Na_2CO_3 is added to 1.00 L of a solution containing 150 ppm of Mg^{2+}?

Additional Exercises

18.39 A friend of yours has seen each of the following terms in newspaper articles and would like an explanation: (a) acid rain; **(b)** greenhouse effect; **(c)** photochemical smog; **(d)** ozone hole. Give a brief explanation of each term and identify one or two of the chemicals associated with each.

18.40 Describe the major factors that lead to a maximum in the ozone concentration at about 22-km elevation.

18.41 The temperature profile in the atmosphere shows a maximum at about 50 km (Figure 18.1). Explain the origin of this maximum.

18.42 Experiments have been performed in which a metal such as sodium or barium has been released into the atmosphere at an altitude of about 120 km. Assuming that the metal is present in the atomic form, what reactions would you expect to occur with the ionic species present? Explain.

18.43 Beginning with the intact chlorofluoromethane, CF_2Cl_2, write equations showing how a catalytic effect for the destruction of ozone may be established in the stratosphere.

18.44 Explain why the stratosphere, which extends from 12 km to about 50 km, contains a smaller total atmospheric mass than the troposphere, which extends from the surface to 12 km.

18.45 Suppose that on another planet the atmosphere consisted of 20 percent Ar, 35 percent CH_4, and 45 percent O_2. What would be the average molecular weight at the surface? What would be the average molecular weight at 200 km, assuming that all the O_2 is photodissociated?

18.46 Explain, using Le Châtelier's principle, why the equilibrium constant for the formation of NO from N_2 and O_2 increases with increasing temperature, whereas the equilibrium constant for the formation of NO_2 from NO and O_2 decreases with increasing temperature.

18.47 If an urban environment contains 10.5 ppm CO at an atmospheric pressure of 745 mm Hg and a temperature of 18°C, how many CO molecules will be present in a room that measures 12 ft × 14 ft × 8 ft?

18.48 A recent study carried out in Canada found that at a particular location far from industrial activity the sulfate in rainfall amounted to 210 mg/m^2 per year. Calculate the total number of moles of sulfate falling in a square mile per year.

18.49 We have noted that the affinity of carbon monoxide for hemoglobin is about 210 times that of O_2. Assume that a person is inhaling air that contains 86 ppm of CO. If all the hemoglobin leaving the lungs carries either oxygen or CO, calculate the fraction in the form of carboxyhemoglobin.

18.50 Most of the earth's atmosphere is transparent to long-wavelength infrared radiation. However, there are two reasonably abundant substances that absorb this radiation. What are these two substances? In what way does the absorption of infrared radiation affect the earth's climate? Explain how increased levels of infrared-absorbing substances in the atmosphere could lead to a higher average surface temperature.

18.51 The Henry's law constant for CO_2 in water at 25°C is 3.1×10^{-2} M/atm. **(a)** What is the solubility of CO_2 in water at this temperature if the solution is in contact with air at normal atmospheric pressure. **(b)** Assume that all of this CO_2 is in the form of H_2CO_3 produced by the reaction between CO_2 and H_2O:

$$CO_2(aq) + H_2O(l) \longrightarrow H_2CO_3(aq)$$

What is the pH of this solution?

18.52 Write balanced chemical equations for each of the following verbal descriptions. **(a)** The nitric oxide molecule undergoes photodissociation in the upper atmosphere. **(b)** The nitric oxide molecule undergoes photoionization in the upper atmosphere. **(c)** Nitric oxide undergoes oxidation by ozone in the stratosphere. **(d)** Nitrogen dioxide dissolves in water to form nitric acid and nitric oxide.

18.53 **(a)** What ions are commonly responsible for the hardness of water? **(b)** What makes these ions objectionable?

18.54 Explain what is meant by the following terms: **(a)** biodegradable; **(b)** aerobic decay; **(c)** anaerobic decomposition.

18.55 Suppose that seawater to be processed in a reverse-osmosis plant has the concentrations of dissolved salts listed in Table 18.7. If the concentration of salts on the pure desalinated water side is taken as zero, what osmotic pressure must be applied before water begins to flow from the seawater through the semipermeable membrane to the desalinated water side? (Refer to Section 13.5.)

18.56 Write balanced chemical equations for each of the following verbal descriptions: **(a)** Hypochlorous acid is formed when chlorine is added to water. **(b)** A sample of water containing Ca^{2+} and bicarbonate ion forms a precipitate when heated.

18.57 How many grams of CaO must be added to hard water containing 65 ppm of Mg^{2+} in order to precipitate $Mg(OH)_2$ from 1000 gal of this water?

18.58 A particular hard water contains 105 ppm HCO_3^-. How many grams of CaO are required to soften 1000 gal of this water, assuming that all the cations in the water are Ca^{2+}?

18.59 The hollow fibers used in the reverse-osmosis desalination plant shown in Figure 18.15 are bundled together in modules, each containing about 3 million fibers of very small diameter. Why is this approach superior to employing a few fibers of much larger diameter?

18.60 Distinguish between primary and secondary sewage treatment. Do these modes of treatment remove the phosphorus that might be present in detergents? Explain.

[18.61] It has recently been pointed out that there may be increased amounts of NO in the troposphere as compared with the past because of massive use of nitrogen-containing compounds in fertilizers. Assuming that NO can eventually diffuse into the stratosphere, what role might it play in affecting the conditions of life on earth? Using the index to this text, look up the chemistry of nitrogen oxides. What chemical pathways might NO in the troposphere follow?

Chemical Thermodynamics 19

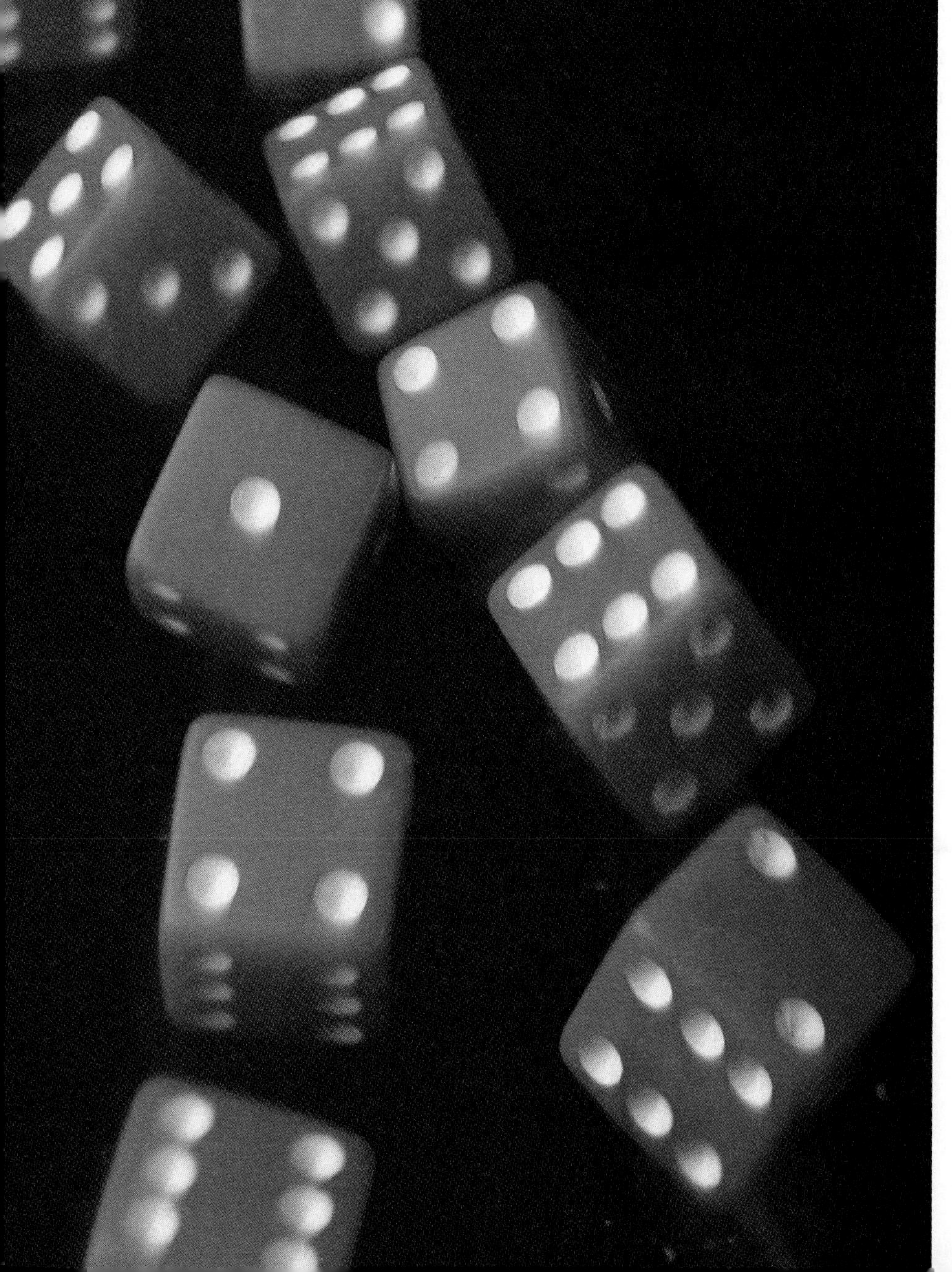

The values shown when dice are rolled are governed by the laws of probability. Random arrangements of the numbers are more probable than well-ordered ones. (© Richard Megna/ Fundamental Photographs)

CONTENTS

In several of the preceding chapters, we have asked two important questions that are at the heart of designing and understanding chemical reactions: How rapidly does the reaction progress? How far toward completion does the reaction proceed? We derive the answer to the first question from a study of the reaction rate, and the answer to the second question from a knowledge of the equilibrium constant. Let's briefly review these concepts.

In Chapter 14, we learned that reaction rates are principally controlled by *energy*, namely the activation energy of the reaction. The lower the activation energy, the faster a reaction proceeds. When we discussed chemical equilibrium in Chapter 15, we defined it from a kinetic point of view: Equilibrium occurs when opposing reactions occur at equal rates. Because reaction rates are intimately tied to energy, it seems logical that chemical equilibrium also depends in some way on energy.

How do we relate chemical equilibrium to the energies of the reactants and products? What kind of energy is important in determining whether a reaction will occur? Questions such as these are the focus of *chemical thermodynamics*, the area of chemistry dealing with energy relationships. We first encountered thermodynamics in Chapter 5, where we discussed the nature of energy, the first law of thermodynamics, the concept of enthalpy, and the enthalpies of reactions. We might be tempted to presume that the enthalpy change of a reaction determines whether the reaction proceeds. After all, an exothermic reaction is "downhill" in enthalpy, so we might expect reactants to be readily converted into products. However, we also know of endothermic processes that proceed readily. For example, ice at room temperature (27°C) melts to liquid water even though the process is endothermic. Ammonium nitrate readily dissolves in water even though ΔH_{soln} is positive (Section 13.2). Clearly, enthalpy alone is not enough to tell us whether a reaction will proceed.

In this chapter, we shall address several additional aspects of chemical thermodynamics that were not discussed in Chapter 5, including the second and third laws of thermodynamics. We shall see that, in addition to enthalpy, the change in the *randomness* or *disorder* of a chemical system during the course of a reaction is extremely important in determining whether equilibrium favors reactants or products (a notion to which we alluded in Section 13.2). Finally, we shall learn how to combine the enthalpy change of a reaction with the change in randomness to define a new type of energy that relates directly to equilibrium.

19.1 SPONTANEOUS PROCESSES

Thermodynamics is based on several fundamental laws that summarize our experience with energy changes. The first law of thermodynamics, which we discussed at length in Chapter 5, states that energy is conserved. By this we mean that energy is neither created nor destroyed in any process, such as the falling of a brick, the melting of an ice cube, or the reacting of chemicals. Energy flows from one part of nature to another, or is converted from one form to another, but the total remains constant. We saw in Chapter 5 that we could express the first law in the form $\Delta E = q + w$, where ΔE is the change in energy of a system, q is the heat absorbed by the system from its surroundings, and w is the work done on the system by its surroundings.

Once we specify a particular process or change, the first law helps us to balance the books, so to speak, on the heat released, work done, and so forth. However, it says nothing about whether the process or change we specify can in fact occur. That question is encompassed in the second law of thermodynamics.

The **second law of thermodynamics** expresses the notion that there is an inherent direction in which any system not at equilibrium moves. For example, if you drop a brick, it falls to the floor. Water placed in a freezer compartment turns into ice. A shiny nail left outdoors eventually rusts. Every one of these processes occurs without outside intervention; such processes are said to be **spontaneous**.

For every spontaneous process, we can imagine a reverse process. For example, we can imagine a brick moving from the floor into your hand, ice cubes melting at $-10°C$, or a rusty iron nail being transformed into a shiny one. It is inconceivable that any of these occurrences is spontaneous. If we saw a film in which these things happened, we would conclude that it was being run backward. Our years of observing nature at work have impressed us with a simple rule: *Processes that are spontaneous in one direction are not spontaneous in the reverse direction.*

Consider a reaction discussed in Chapter 15:

$$N_2(g) + 3H_2(g) \rightleftharpoons 2NH_3(g) \qquad [19.1]$$

When we mix N_2 and H_2 at a particular temperature, say 472°C, the reaction proceeds in the forward direction; this process is spontaneous. However, if we place a mixture of 1.00 mol of N_2, 3.00 mol of H_2, and 1.00 mol of NH_3 in a 1-L container at 472°C, it is not immediately obvious whether the formation of more NH_3 should be spontaneous. Nevertheless, if we know the equilibrium constant, which at this temperature happens to be $K_c = 0.105$, we can predict the direction in which the reaction proceeds to reach equilibrium. In this case

$$Q = \frac{[NH_3]^2}{[N_2][H_2]^3} = \frac{(1.00)^2}{(1.00)(3.00)^3} = 0.0370$$

Because the reaction quotient, Q, is smaller than K_c, the system moves spontaneously toward equilibrium by formation of NH_3 (Section 15.5). The opposite process, conversion of NH_3 into N_2 and H_2, is not spontaneous for this particular reaction mixture at 472°C.

It is important to realize that just because a process is spontaneous does not mean that it will occur at an observable rate. A spontaneous reaction may be very fast, as in the case of an acid-base neutralization, or very slow, as in the case of rusting of iron. Thermodynamics can tell us the *direction* and *extent* of a reaction, but it can say nothing about its *speed*. Reaction rates are the subject of chemical kinetics.

What factors make a process spontaneous? In Chapter 5, we saw that the enthalpy change for a process is an important factor in determining whether the process is favorable. Processes in which the enthalpy of the system decreases (exothermic processes) tend to occur spontaneously. However, we shall see that considering only the enthalpy change of a process is not enough. The spontaneity of a process also depends on how the disorder of the system changes during the process. In the next section, we consider these factors, especially the matter of disorder, in greater detail.

19.2 SPONTANEITY, ENTHALPY, AND ENTROPY

The spontaneous motion of a brick released from your hand is toward the ground. As the brick falls, it loses potential energy. This potential energy is first converted into kinetic energy, the energy of motion of the brick. When the brick hits the floor, its kinetic energy is converted into heat. The overall result of the brick's fall is thus a conversion of the potential energy of the brick into heat in its surroundings. Our experience with other simple mechanical systems is similar: objects fall, clocks run down, stretched rubber bands contract. All of these phenomena can be summarized by saying that such systems seek a resting place of minimum energy.

It is clear that the tendency for a system to achieve the lowest possible energy is one of the driving forces that determine the behavior of molecular systems. For example, just as a brick possesses potential energy because of its position relative to the floor, so also a chemical substance possesses potential energy relative to other substances because of the arrangements of nuclei and electrons. When these arrangements change, energy may be released. For example, the combustion of propane (bottled gas), which is clearly a spontaneous process, is strongly exothermic:

$$C_3H_8(g) + 5O_2(g) \longrightarrow 3CO_2(g) + 4H_2O(l) \qquad \Delta H^\circ = -2202 \text{ kJ} \qquad [19.2]$$

The rearrangements in space of nuclei and electrons in going from propane and oxygen to carbon dioxide and water lead to a lower chemical potential energy, so heat is evolved. Reactions that are exothermic are generally spontaneous. However, it is clear that the tendency toward minimum enthalpy cannot be the only factor that determines spontaneity in molecular processes. It is instructive to consider some spontaneous processes that are not exothermic.

Figure 19.1 Expansion of an ideal gas into an evacuated space. In (*a*), flask *A* holds an ideal gas at 1 atm pressure and flask *B* is evacuated. In (*b*), the stopcock connecting the flasks has been opened. The ideal gas expands to occupy both flasks *A* and *B* at a pressure of 0.5 atm.

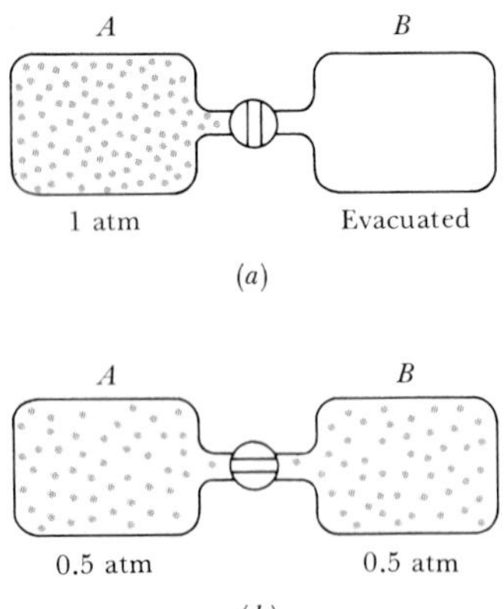

Spontaneity and Entropy Change

We can think of several processes that are spontaneous even though they are not exothermic. For example, consider an ideal gas confined at 1 atm pressure to a 1-L flask, as shown in Figure 19.1. The flask is connected by a closed stopcock to another 1-L flask, which is evacuated, and the two flasks are kept at a constant temperature. Now suppose the stopcock is opened. Is there any doubt about what would happen? We intuitively recognize that the gas would expand into the second flask until the pressure is equally distributed in both flasks, at 0.5 atm. In the course of expanding from the 1-L flask into the larger volume, the ideal gas neither absorbs nor emits heat. Nevertheless, the process is spontaneous. The reverse process, in which the gas that is evenly distributed between the two flasks suddenly moves entirely into one of the flasks, leaving the other vacant, is inconceivable. Yet this process would also involve no emission or absorption of heat. Evidently, some factor other than heat emitted or absorbed is important in making the process of gas expansion spontaneous.

As another example, consider the melting of ice cubes at room temperature. The process

$$H_2O(s) \longrightarrow H_2O(l) \qquad [19.3]$$

is highly spontaneous at 27°C, yet it is an endothermic change. The melting of ice above 0°C thus represents an example of a spontaneous endothermic process.

A similar type of process, discussed in Section 13.2, is the endothermic dissolving of many salts in water. If we add solid potassium chloride, KCl, to a glass of water at room temperature and stir, we can feel the solution growing colder as the salt dissolves. The process is endothermic and yet spontaneous.

The three processes just described have something in common that accounts for their being spontaneous. In each instance, the products of the process are in a more random or disordered state than the reactants. Let's consider each case in turn.

When we have a gas confined to a 1-L volume, as in Figure 19.1(*a*), we can specify the location of each and every gas molecule as being in that liter of space. After the gas has expanded, we can't be sure which gas molecules are at any one instant in the original volume and which are on the other side. We must therefore say that the location of each and every gas molecule is specified as being in the entire 2-L space. In other words, the gas molecules, because they can be anywhere within a 2-L space, are more randomized than when they are confined to a 1-L space.

The molecules of water that make up an ice crystal are held rigidly in place in the ice crystal lattice (Figure 19.2). When the ice melts, the water molecules are free to move about with respect to one another and to tumble around. Thus, in liquid water the individual water molecules are more randomly distributed than in the solid. The highly ordered solid structure is replaced by the highly disordered liquid structure.

A similar situation applies when KCl dissolves in water, although here we must be a little careful not to take too much for granted. In solid KCl, the K^+ and Cl^- are in a highly ordered, crystalline state. When the solid dissolves,

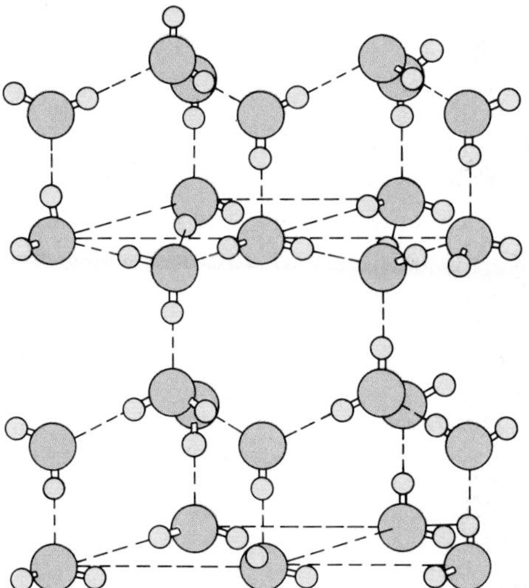

Figure 19.2 Structure of ice.

Figure 19.3 Changes in degree of order in the ions and solvent molecules on dissolving an ionic solid in water. The ions themselves become more randomized, but the water molecules that hydrate the ions become less randomized.

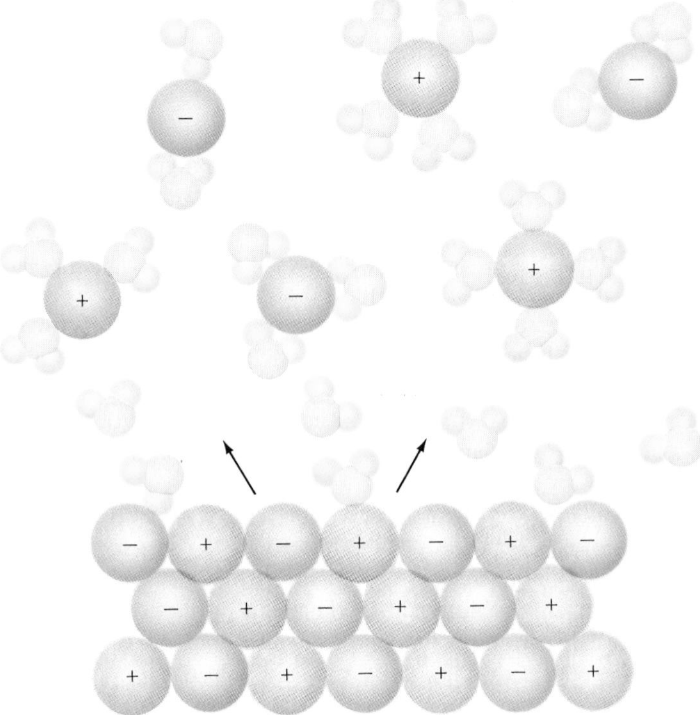

the ions are free to move about in the water. They are obviously in a much more random and disordered state than before. At the same time, though, water molecules are held around the ions as water of hydration (Section 13.2), as shown in Figure 19.3. These water molecules are in a *more* ordered state than before, because they are confined to the immediate environment of the ions. Therefore, the dissolving of a salt involves both ordering and disordering processes. It happens that the disordering processes are dominant, so the overall effect is an increase in disorder upon dissolving a salt in water.

As these examples illustrate, spontaneity is associated with an increase in randomness or disorder of a system. The randomness is expressed by a thermodynamic quantity called **entropy**, given the symbol S. The more disordered a system, the larger its entropy. Like enthalpy, entropy is a state function (Section 5.2). The change in entropy of a system, $\Delta S = S_{\text{final}} - S_{\text{initial}}$, depends only on the initial and final states of the system and not on the particular pathway by which the system changes. Like other thermodynamic quantities, ΔS has two parts, a number giving the magnitude of the change and a sign giving its direction. A positive ΔS indicates an increase in randomness or disorder. A negative ΔS indicates a decrease in randomness or disorder.

SAMPLE EXERCISE 19.1

By considering the relative extents of randomness or disorder in the reactants and products, predict whether ΔS is positive or negative for each of the following processes:

(a) $H_2O(l) \longrightarrow H_2O(g)$
(b) $Ag^+(aq) + Cl^-(aq) \longrightarrow AgCl(s)$
(c) $4Fe(s) + 3O_2(g) \longrightarrow 2Fe_2O_3(s)$

Solution: **(a)** The evaporation of a liquid is accompanied by a large increase in volume. One mole of water (18 g) occupies about 18 mL as a liquid and 22.4 L as a gas at STP. Because the molecules are distributed throughout a much larger volume in the gaseous state than in the liquid state, an increase in disorder accompanies vaporization. Therefore, ΔS is positive.

(b) In this process the ions that are free to move about in the larger volume of the solution form a solid in which the ions are confined to highly ordered positions. Thus, there is a decrease in disorder, and ΔS is negative.

(c) The particles of a solid are much more highly ordered and confined to specific locations than are the molecules of a gas. Because a gas is converted into part of a solid product, disorder decreases, and ΔS is negative.

PRACTICE EXERCISE

Indicate whether each of the following reactions produces an increase or decrease in the entropy of the system:
(a) $CO_2(s) \longrightarrow CO_2(g)$
(b) $CaO(s) + CO_2(g) \longrightarrow CaCO_3(s)$.
Answers: **(a)** increase; **(b)** decrease

The Second Law of Thermodynamics

Our introduction of the concept of entropy allows us to reexamine the second law of thermodynamics and its implications. In the discussion of spontaneity in Section 19.1, we noted that the second law concerns the direction in which processes move; it is associated with the idea that processes that are spontaneous in one direction are not spontaneous in the opposite direction. This idea applies not only to chemical changes but to other processes as well.

We know that heat flows spontaneously from a hot object to a cold one. We also know that to cause heat to flow in the reverse direction—from a cold object to a hotter one or from a system at some temperature to surroundings at a higher temperature—requires an input of energy. For example, electrical energy is required to maintain a refrigerator at a lower temperature than the surrounding kitchen.

There are many ways to state the second law. In chemical contexts it is usually expressed in terms of entropy. To develop such a statement, we must think in terms of an **isolated system**, one that doesn't exchange energy or matter with its surroundings. When a process occurs spontaneously in an isolated system, the system always ends up in a more random state. For example, when a gas expands under the conditions shown in Figure 19.1, there is no exchange of heat, work, or matter with the surroundings; this is an isolated system. The spontaneous expansion corresponds to an increase in entropy.

In the real world, we rarely deal with isolated systems. We are usually concerned with systems that exchange energy with their surroundings in the form of heat or work. The second law of thermodynamics governs the *total* change in entropy for the universe (system plus surroundings). A statement of the second law is: *In any spontaneous process, there is always an increase in the entropy of the universe.* Thus, even if the entropy of a system decreases in a spontaneous process, the *total* change in entropy must be positive.

To examine the implications of the second law, let's consider the oxidation of iron to $Fe_2O_3(s)$:

$$4Fe(s) + 3O_2(g) \longrightarrow 2Fe_2O_3(s) \qquad [19.4]$$

CHEMISTRY AT WORK: Entropy and Life

The second law of thermodynamics has profound implications concerning our existence as humans. We are very complex, highly organized, and well-ordered systems. As a result, our entropy content is much lower than it would be if we were completely decomposed into carbon dioxide, water, and several other simple chemicals. Does this mean that our existence is a violation of the second law? The answer is no, because the thousands of chemical reactions necessary to produce and maintain a human life have caused a very large increase in the entropy of the rest of the universe. Thus, as the second law requires, the overall entropy change during the lifetime of a human, or for that matter any other living system, is positive.

We humans are masters of producing order in the world around us. We build impressive, highly ordered structures and buildings (Figure 19.4). We use tremendous quantities of raw materials to produce highly ordered materials—copper metal from copper ore, silicon for computer chips from sand, paper from wood pulp, and so forth. In so doing, we must expend a great deal of energy to, in essence, "fight" the second law of thermodynamics; for every bit of order we produce in our world, we produce an even greater amount of disorder. Coal and oil are burned to form CO_2 and H_2O. Sulfide ores are roasted (Section 24.3), producing SO_2 that pollutes the atmosphere. Waste products are scattered in various forms throughout the environment. And, of course, we produce huge amounts of heat that have adverse effects on nature and may serve to alter our climate permanently. This *thermal pollution* is one of the prices that, according to the second law, we must pay for demanding a well-ordered world around us.

Modern human society is, in effect, using up its storehouse of energy-rich materials in its headlong rush to create order and exploit technology. We must eventually learn to live within the limits of the energy supply that reaches the earth daily from the sun, for we will soon exhaust the supply of readily available energy of other sorts.

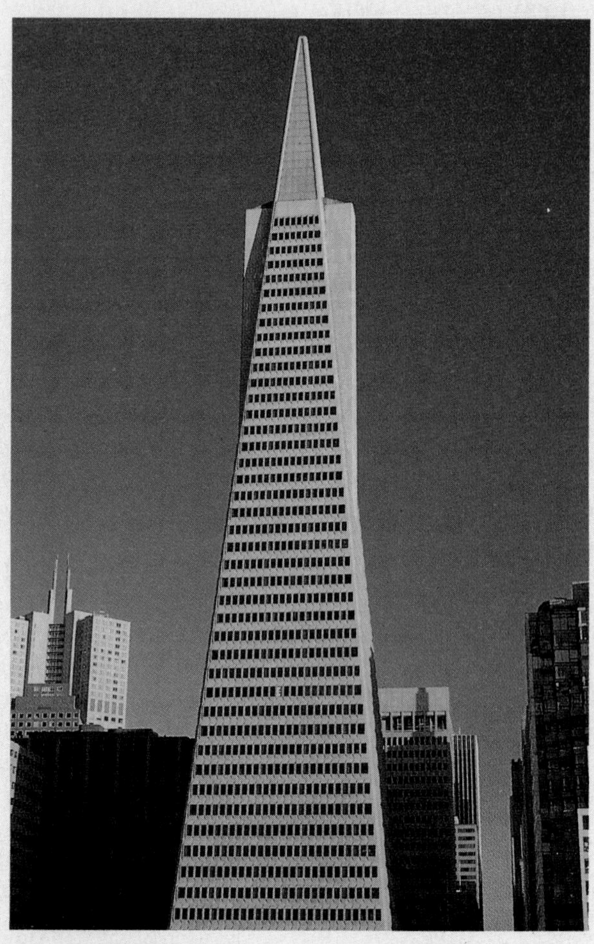

Figure 19.4 In both ancient and modern times, humans have constructed highly ordered structures based on the pyramid. The Great Pyramid of Cheops (left) was built in Egypt around 2600 B.C. The Transamerica Building in San Francisco (right) is a modern skyscraper built in 1972. From the second law of thermodynamics, the introduction of order into these structures requires that an even greater amount of disorder be imparted to the universe. (*a*, © Michael Dalton/Fundamental Photographs; *b*, © Peter Saloutos/The Stock Market)

As discussed in Sample Exercise 19.1, this spontaneous process results in a decrease in the entropy of the system; that is, ΔS for the system is negative. But as the reaction occurs, the surroundings are also affected. Because the reaction is exothermic, heat is evolved, which increases the entropy of the surroundings by increasing the thermal motion of surrounding molecules. The second law tells us that the increase in the entropy of the surroundings *must* be greater than the decrease in the entropy of the system, so that the *total* entropy increases:

$$\Delta S_{\text{universe}} = \Delta S_{\text{system}} + \Delta S_{\text{surroundings}} > 0 \qquad [19.5]$$

No process that produces order (a decrease in entropy) in a system can proceed without producing an even larger disorder (an increase in entropy) in its surroundings. Thus, while the energy of the universe is conserved (the first law), the entropy of the universe continues to increase (the second law).

19.3 A MOLECULAR INTERPRETATION OF ENTROPY

It is useful to develop a qualitative sense of how changes in entropy within a system depend on changes in structure, physical state, and so forth. In Section 19.2, we considered some examples of spontaneous, endothermic processes. We saw, for example, that the increase in volume that occurs when a gas expands results in an increase in the randomness of the system (positive ΔS). In a similar fashion, the distribution of a liquid or solid solute in a solution is accompanied by an increase in entropy. For example, ΔS is positive when KCl dissolves in water. However, the dissolving of a gas, such as CO_2 in H_2O, causes the gas molecules to move in a much smaller volume; consequently, for this process the entropy of the system decreases (negative ΔS). Similarly, a decrease in the number of gaseous particles as the result of a reaction causes a decrease in entropy (negative ΔS). For example, ΔS for the following reaction is negative:

$$2NO(g) + O_2(g) \longrightarrow 2NO_2(g) \qquad [19.6]$$

Entropy changes can also be associated with molecular motions within a substance. A molecule consisting of more than one atom can engage in several types of motion. The entire molecule can move in one direction, as in the movements of gas molecules. We call such movement **translational motion**. The atoms within a molecule may also undergo **vibrational motion**, in which they move periodically toward and away from each other, much as a tuning fork vibrates about its equilibrium shape. Figure 19.5 shows the vibrational motions possible for the water molecule. In addition, molecules may possess **rotational motion**, as though they were spinning like a top. One of the rotational motions of the water molecule is also illustrated in Figure 19.5. These forms of motion are ways that the molecule has of storing energy. As the temperature of a system increases, the amounts of energy stored in these forms of motion increase.

To see what this has to do with entropy, let's imagine that we begin with a pure substance that forms a perfect crystalline lattice at the lowest temperature possible, absolute zero. Under these conditions, none of the kinds of motion that we have been talking about is present. The individual atoms and molecules are as well defined in position and in terms of energy as they can ever be. The **third law of thermodynamics** states that *the entropy of a pure crystalline substance at absolute zero is zero:* $S\ (0\ \text{K}) = 0$. As the temperature is raised, the

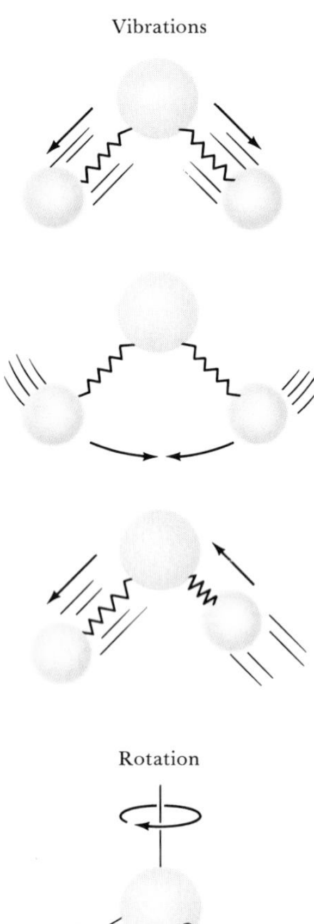

Figure 19.5 Examples of vibrational and rotational motion, illustrated for the water molecule. Vibrational motions involve periodic displacements of the atoms with respect to one another, a phenomenon similar to the vibrations of the arms of a tuning fork. Rotational motions involve the spinning of a molecule about an axis.

units of the solid lattice begin to acquire energy. In a crystalline solid, the molecules or atoms that occupy the lattice sites are constrained to remain essentially in place. Nevertheless, they may store energy in the form of vibrational motion about their lattice positions. Instead of all the molecules necessarily being in the lowest possible energy state, the number of possible energies that the lattice atoms or molecules may have expands. This increase in possible energy states is not unlike the expansion of the gas illustrated in Figure 19.1. The entropy of the gas increases on expansion because the volume through which the gas molecules move is larger. The entropy of the lattice increases with temperature because the number of possible energy states in which the molecules or atoms are distributed is larger.

It is instructive to follow what happens to the entropy of our substance as we continue to heat it. Let's suppose that at some temperature a solid-state phase change occurs, converting the substance from one solid form to another. This means that the arrangement of the lattice units changes in some way, possibly so that the lattice is less regular.* This type of phase change occurs sharply at one temperature, just as do other types of phase changes (for example, the change from a solid to a liquid). When the phase change occurs, entropy changes, because the two lattice arrangements do not have precisely the same degrees of randomness.

Figure 19.6 shows the variation in entropy with temperature for our sample. Note that the change in S with temperature is gradual up to the solid-state phase change and that there is then a sharp increase in S at that temperature. At temperatures above the phase change, the entropy increases with increasing temperature, up to the melting point of the solid.

When the solid melts, the units of the lattice are no longer confined to specific locations relative to other units but are free to move about the entire volume of the substance. This added freedom of motion for the individual molecules adds greatly to the entropy content of the substance. At the temperature of melting, we therefore see a large increase in entropy content. After all the solid has melted, the temperature again increases, and with it the entropy.

* As an example of a solid-state phase change, gray tin converts at 13°C to another solid form called white tin. White tin is stable above the transition temperature, gray tin below it. White tin has a higher entropy than does gray tin.

Figure 19.6 Entropy changes that occur as the temperature of a substance rises from absolute zero.

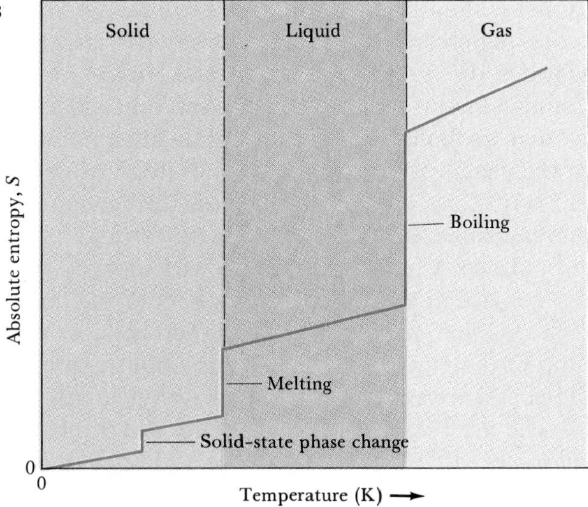

SAMPLE EXERCISE 19.2

Figure 19.6 shows that the entropy of a liquid increases as its temperature increases. What factors are responsible for the increase in entropy?

Solution: The average kinetic energy of the molecules in a liquid increases with temperature. As temperature increases, more molecules at any given instant possess higher energies. This "expansion" in the energies that the molecules possess is measured by the increase in entropy. The increase in S with increasing temperature results from increased energy of motion of all kinds within the liquid.

PRACTICE EXERCISE

Based on Figure 19.6, what general relationship exists among the magnitudes of the entropies of the gas, liquid, and solid phases of a substance?
Answer: $S_{gas} > S_{liquid} > S_{solid}$

At the boiling point of the liquid, another big increase in entropy occurs. The increase in this case results largely from the increased volume in which the molecules may be found. This is intuitively in line with our earlier ideas about entropy, because an increase in volume means an increase in randomness.

As the gas is heated, the entropy increases steadily because more and more energy is being stored in the gas molecules. The distribution of molecular speeds is spread out toward higher values, as illustrated in Figure 10.13. Again, the idea of an expansion in the range of energies in which molecules may be found helps us remember that increased average energy means increased entropy.

SAMPLE EXERCISE 19.3

Choose the substance with greater entropy in each pair and explain your choice: **(a)** 1 mol of $NaCl(s)$ and 1 mol of $HCl(g)$ at 25°C; **(b)** 2 mol of $HCl(g)$ and 1 mol of $HCl(g)$ at 25°C; **(c)** 1 mol of $HCl(g)$ and 1 mol of $Ar(g)$ at 25°C; **(d)** 1 mol of $N_2(s)$ at 24 K and 1 mol of $N_2(g)$ at 298 K.

Solution: **(a)** Gaseous HCl has the higher entropy per mole, because it has acquired a high degree of randomness as a result of being in the gaseous state. **(b)** The sample containing 2 mol of HCl has twice the entropy of the sample containing 1 mol. **(c)** The HCl sample has the higher entropy, because the HCl molecule is capable of storing energy in more ways than is Ar. It may rotate, or the H—Cl distance may change periodically in a vibrational motion. **(d)** The gaseous N_2 sample has the higher entropy, because the entropy increases resulting from melting and then boiling N_2 are included in its total entropy content.

PRACTICE EXERCISE

Choose the substance with the greater entropy in each case: **(a)** 1 mol of $H_2(g)$ at STP or 1 mol of $H_2(g)$ at 100°C and 0.5 atm; **(b)** 1 mol of $H_2O(s)$ at 0°C and 1 mol of $H_2O(l)$ at 25°C; **(c)** 1 mol of $H_2(g)$ at STP and 1 mol of $SO_2(g)$ at STP; **(d)** 1 mol of $N_2O_4(g)$ at STP and 2 mol of $NO_2(g)$ at STP. ***Answers:*** **(a)** 1 mol of $H_2(g)$ at 100°C; **(b)** 1 mol of $H_2O(l)$ at 25°C; **(c)** 1 mol of $SO_2(g)$ at STP; **(d)** 2 mol of $NO_2(g)$ at STP

In general, entropy *increases* are expected to accompany processes in which

Liquids or solutions are formed from solids.

Gases are formed from either solids or liquids.

The number of molecules of gas increases during a chemical reaction.

The temperature of a substance is increased.

A CLOSER LOOK: Entropy, Randomness, and Ludwig Boltzmann

The entropy of a system is related to its randomness or disorder. The fact that we can assign a definite value to the entropy of a system implies that disorder can be quantified in some sense. The quantitative correspondence between entropy and randomness was first advanced by the Austrian physicist Ludwig Boltzmann (1844–1906). Boltzmann reasoned that the randomness of a particular state of a system, and thus its entropy, is related to the number of possible arrangements of molecules in the state.

We can illustrate Boltzmann's idea by using the poker hands shown in Table 19.1. The probability that a poker hand will contain five *specific* cards is the same, regardless of which five cards are specified. Thus, there is an equal probability of dealing either of the specific hands shown in Table 19.1. However, the first hand, a royal flush (the ten through ace of a single suit), strikes us as much more highly ordered than the second hand, a "nothing." The reason for this is clear if we compare the number of arrangements of five cards that correspond to a royal flush to the number corresponding to a nothing. There are only 4 poker hands that are in the "state" of a royal flush; in contrast, there are over 1.3 million nothing hands. The "nothing state" has a higher degree of disorder than the "royal-flush state" because there are so many more arrangements of cards that correspond to the nothing state.

We can use the same reasoning for chemical systems. For example, consider the two states for the gas molecules represented in Figure 19.1. In Figure 19.1(*a*), the atoms are confined to one flask. When the stopcock is opened, as in Figure 19.1(*b*), the volume available to the atoms is doubled. There are more possible arrangements of the atoms in Figure 19.1(*b*) than in Figure 19.1(*a*); thus, the randomness of the former is greater than that of the latter. Likewise, there are more possible arrangements of water molecules in liquid water than there are in ice (Figure 19.2). Hence, liquid water is more disordered than ice.

Boltzmann showed that the entropy of a system equals a constant times the natural logarithm of the number of possible arrangements of atoms or molecules in the system:

$$S = k \ln W \qquad [19.7]$$

In Equation 19.7, W is the number of possible arrangements in the system, and k is a constant known as *Boltzmann's constant*. Boltzmann's constant is the "atomic equivalent" of the gas constant R. It equals the gas constant (usually expressed in joules) divided by Avogadro's number:

$$k = \frac{R}{N} = \frac{(8.31 \text{ J/K-mol})}{(6.02 \times 10^{23} \text{ mol}^{-1})} = 1.38 \times 10^{-23} \text{ J/K}$$

A pure crystalline substance at absolute zero is assumed to have only one arrangement of atoms or molecules, that is, $W = 1$. Thus, Equation 19.7 is consistent with the third law of thermodynamics: If $W = 1$, then $S = k \ln 1 = 0$. At any temperature above absolute zero, the atoms acquire energy, and more arrangements become possible, so $W > 1$ and $S > 0$.

Table 19.1 A Comparison of the Number of Combinations that can Lead to a Royal Flush and a "Nothing" Hand in Poker

Hand	State	Number of hands that lead to this state[a]
10♠ J♠ Q♠ K♠ A♠	Royal flush	4
2♣ 5♦ 6♥ 10♠ J♥	"Nothing"	1,302,540

[a] There are 2,598,960 total possible five-card poker hands.

Boltzmann made many other significant contributions to science, particularly in the area of *statistical mechanics*, which is the derivation of bulk thermodynamic properties for large collections of atoms or molecules by using the laws of probability. For example, the molecular speed distributions shown in Figure 10.13 are derived by using statistical mechanics; such plots are known as *Maxwell-Boltzmann distributions*. Unfortunately, Boltzmann's life had a tragic ending. He strongly believed in the existence of atoms, which, as strange as it seems to us now, was an unpopular viewpoint in physics at the beginning of the twentieth century. In poor health and unable to endure continual intellectual attacks on his beliefs, Boltzmann committed suicide on September 5, 1906. Ironically, it was only a few years later that the work of Thomson, Millikan, and Rutherford led to acceptance of the nuclear atom model (Section 2.2). Although Boltzmann made many contributions to science, the connection between entropy and randomness described in Equation 19.7 is arguably his greatest. This equation is so significant that it is inscribed on his gravestone (Figure 19.7).

Figure 19.7 Ludwig Boltzmann's gravestone in Vienna is inscribed with his famous relationship between the entropy of a state and the number of arrangements available in the state (in Boltzmann's time, "log" was used to represent the natural logarithm). (Institut für Theoretische Physik der Universitat Wien)

SAMPLE EXERCISE 19.4

Predict whether the entropy change of the system in each of the following reactions is positive or negative.

(a) $CaCO_3(s) \longrightarrow CaO(s) + CO_2(g)$
(b) $N_2(g) + 3H_2(g) \longrightarrow 2NH_3(g)$
(c) $N_2(g) + O_2(g) \longrightarrow 2NO(g)$

Solution: **(a)** The entropy change here is positive, because a solid is converted into a solid and a gas. Gaseous substances generally possess more entropy than solids, so whenever the products contain more moles of gas than the reactants, the entropy change is probably positive.

(b) The entropy change in formation of ammonia from nitrogen and hydrogen is negative, because there are fewer moles of gas in the product than in the reactants.

(c) This represents a case in which the entropy change will be small, because the same number of moles of gas is involved in the reactants and in the product. The sign of ΔS is impossible to predict based on our discussions thus far, but we can predict that ΔS will be small.

PRACTICE EXERCISE

Predict whether ΔS is positive or negative in each of the following processes:

(a) $HCl(g) + NH_3(g) \longrightarrow NH_4Cl(s)$
(b) $2SO_2(g) + O_2(g) \longrightarrow 2SO_3(g)$
(c) cooling of nitrogen gas from 20°C to −50°C.

Answers: **(a)** negative; **(b)** negative; **(c)** negative

19.4 CALCULATION OF ENTROPY CHANGES

In Section 5.7, we discussed how calorimetry can be used to measure ΔE and ΔH for chemical reactions. No comparable, easy means exists for measuring the change in entropy during a reaction. However, by using experimental measurements of the variation of heat capacity with temperature, we can determine the absolute entropy, S, for many substances at any temperature. (The thermodynamic theory and the methods used for these measurements are beyond the scope of this text.) These entropies are based on the reference point of zero entropy for perfect crystalline solids (the third law). The entropy values at 298 K and 1 atm pressure are known as *standard entropies* and are denoted $S°$. Standard entropies are expressed in units of joules per mole per Kelvin: J/mol-K. Appendix C gives the values of $S°$ for a variety of substances. You should note that the standard entropies of gases are greater than those of liquids or solids, and that the standard entropy increases with increasing molecular complexity.

The entropy change in a chemical reaction is given by the sum of the entropies of the products less the sum of entropies of reactants. Thus, in the overall reaction

$$aA + bB + \cdots \rightleftharpoons pP + qQ + \cdots \qquad [19.8]$$

the standard entropy change, $\Delta S°$, is given by

$$\Delta S° = [pS°(P) + qS°(Q) + \cdots] - [aS°(A) + bS°(B) + \cdots] \qquad [19.9]$$

In other words, we sum the standard entropies of all the products, multiplying each by the coefficient of the product in the balanced equation, and then subtract the sum of the entropies of the reactants, multiplied by their coefficients.

SAMPLE EXERCISE 19.5

Calculate $\Delta S°$ for the synthesis of ammonia from $N_2(g)$ and $H_2(g)$:

$$N_2(g) + 3H_2(g) \longrightarrow 2NH_3(g)$$

Solution: Using Equation 19.9, we have

$$\Delta S° = 2S°(NH_3) - [S°(N_2) + 3S°(H_2)]$$

Substituting the appropriate $S°$ values from Appendix C yields $\Delta S° =$

$$(2 \text{ mol})\left(192.5 \frac{\text{J}}{\text{mol} - \text{K}}\right) - \left[(1 \text{ mol})\left(191.5 \frac{\text{J}}{\text{mol} - \text{K}}\right) + (3 \text{ mol})\left(130.58 \frac{\text{J}}{\text{mol} - \text{K}}\right)\right]$$
$$= -198.2 \text{ J/K}$$

The value for $\Delta S°$ is negative, as we predicted in Sample Exercise 19.4(b).

PRACTICE EXERCISE

Using the standard entropies in Appendix C, calculate the standard entropy change, $\Delta S°$, for the following reaction at 298 K: $Al_2O_3(s) + 3H_2(g) \longrightarrow 2Al(s) + 3H_2O(g)$.
Answer: 179.9 J/K

19.5 GIBBS FREE ENERGY

We have seen that the spontaneity of a reaction involves two thermodynamic concepts, enthalpy and entropy. Now that we can assign a quantitative value to the entropy change during a reaction, is there a way to use ΔH and ΔS to predict whether a given reaction will be spontaneous? The means for incorporating both ΔH and ΔS into a new quantity that tells us whether a reaction will be spontaneous was first developed by the American mathematician J. Willard Gibbs (1839–1903). Gibbs (Figure 19.8) proposed a new state function, now called the **Gibbs free energy** (or just **free energy**). The Gibbs free energy, G, of a state is defined as

$$G = H - TS \qquad [19.10]$$

where T is the absolute temperature.

For a process occurring at constant temperature, the change in free energy is given by the expression

$$\Delta G = \Delta H - T\,\Delta S \qquad [19.11]$$

A process that is driven spontaneously toward equilibrium both by decreasing energy (negative ΔH) and increasing randomness (positive ΔS) will have a negative ΔG. Indeed, there is a simple relationship between the sign of ΔG for a reaction and the spontaneity of that reaction operated at constant temperature and pressure:

1. If ΔG is negative, the reaction is spontaneous in the forward direction.
2. If ΔG is zero, the reaction is at equilibrium; there is no driving force tending to make the reaction go in either direction.
3. If ΔG is positive, the reaction in the forward direction is nonspontaneous; work must be supplied from the surroundings to make it occur. However, the reverse reaction will be spontaneous.

Figure 19.8 Josiah Willard Gibbs (1839–1903) was the first person to be awarded a Ph.D. in science from an American university (Yale, 1863). From 1871 until his death he held the chair of mathematical physics at Yale. Gibbs developed much of the theoretical foundation that led to the development of chemical thermodynamics. (Culver Pictures)

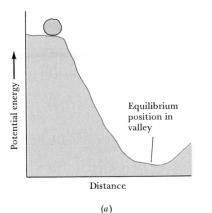

An analogy is often drawn between the free-energy change in a spontaneous reaction and the potential-energy change in a boulder rolling down a hill. Potential energy in a gravitational field "drives" the boulder until it reaches a state of minimum potential energy in the valley [Figure 19.9(a)]. Similarly, the free energy of a chemical system decreases (negative ΔG) until it reaches a minimum value [Figure 19.9(b)]. When this minimum is reached, a state of equilibrium exists. In any spontaneous process at constant temperature and pressure, the free energy always decreases. As shown in Figure 19.9(b), the equilibrium condition can be approached by a spontaneous change from either direction, from the product side or the reactant side.

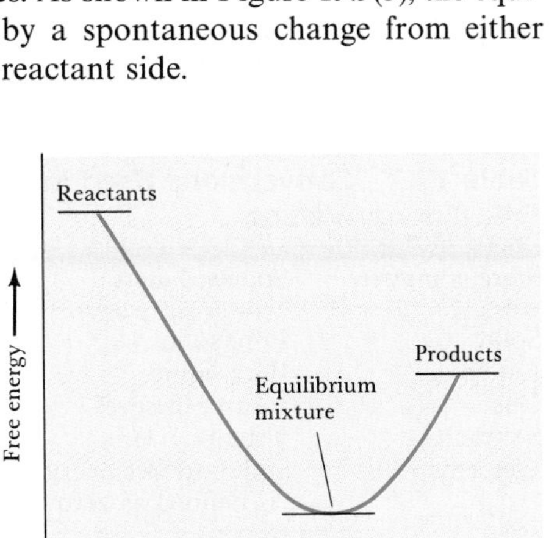

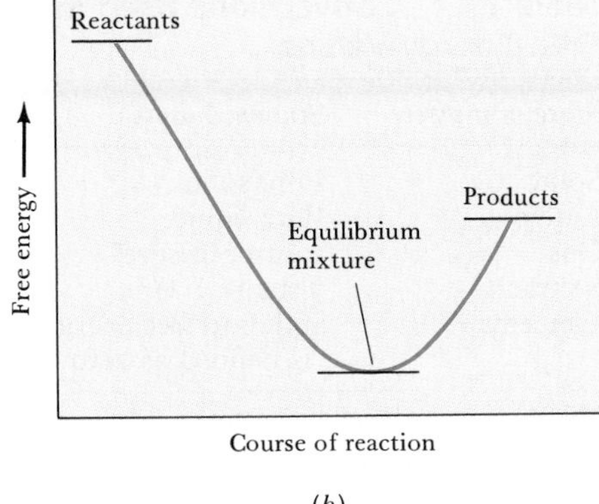

Figure 19.9 Analogy between the potential-energy change of a boulder rolling down a hill (a) and the free-energy change in a spontaneous reaction (b). The equilibrium position in (a) is given by the minimum potential energy available to the system. The equilibrium position in (b) is given by the minimum free energy available to the system.

As an example of these ideas, let's return to the synthesis of ammonia from nitrogen and hydrogen. Imagine that we have a certain number of moles of nitrogen and three times that number of moles of hydrogen in a reaction vessel that permits us to maintain a constant temperature and pressure. We know from our earlier discussions that the formation of ammonia will not be complete; an equilibrium will be reached in which the reaction vessel contains some mixture of N_2, H_2, and NH_3. The free energy of the system decreases (negative ΔG) until this equilibrium is attained. Once the equilibrium has been reached, there is no further spontaneous net formation of NH_3. The equilibrium condition is the one in which the free energy of the system is at a minimum at this temperature and pressure. To form more NH_3 from N_2 and H_2 after equilibrium has been reached requires an increase in free energy (positive ΔG).

It is not necessary that we reach equilibrium beginning only with N_2 and H_2. We could reach the same equilibrium by beginning with an appropriate amount of NH_3. Ammonia held at a constant temperature and pressure will decompose to form N_2 and H_2 until an equilibrium is attained. This process is also spontaneous, involving a decrease in free energy as the system approaches equilibrium. The equilibrium can be reached from either the reactant side or product side, as indicated in a general fashion in Figure 19.9(*b*).

Standard Free-Energy Changes

We have noted that free energy is a state function. This means that it is possible to tabulate **standard free energies of formation** for substances, just as it is possible to tabulate standard enthalpies of formation. It is important to remember that standard values for these functions imply a particular set of conditions, or standard states (Section 5.6). The standard state for gaseous substances is 1 atm pressure. For solid substances, the standard state is the pure solid; for liquids, the pure liquid. For substances in solution, the standard state is normally a concentration of 1 *M*. (In accurate work, it may be necessary to make certain corrections, but we need not worry about these.) The temperature usually chosen for purposes of tabulating data is 25°C. Just as for the standard heats of formation, the free energies of elements in their standard states are arbitrarily set to zero. This arbitrary choice of a reference point has no effect on the quantity in which we are really interested, namely, the *difference* in free energy between reactants and products. The rules about standard states are summarized in Table 19.2. A listing of standard free energies of formation, denoted ΔG_f°, appears in Appendix C.

Table 19.2 Conventions Used in Establishing Standard Free-Energy Values

State of matter	Standard state
Solid	Pure solid
Liquid	Pure liquid
Gas	1 atm pressure[a]
Solution	Usually 1 *M*[b]
Elements	Standard free energy of formation of an element in its normal state is defined as zero

[a] Neglecting nonideal gas behavior.
[b] Neglecting nonideality of solutions.

The standard free energies of formation for substances are useful in calculating the standard free-energy change in a chemical process. For the general reaction

$$aA + bB + \cdots \longrightarrow pP + qQ + \cdots \quad [19.12]$$

the standard free-energy change is

$$\Delta G^\circ = [p\,\Delta G_f^\circ(P) + q\,\Delta G_f^\circ(Q) + \cdots] - [a\,\Delta G_f^\circ(A) + b\,\Delta G_f^\circ(B) + \cdots] \quad [19.13]$$

In this expression, $\Delta G_f^\circ(P)$ represents the standard free energy of formation of product P, and all the other ΔG_f° have similar meanings. Equation 19.13 is therefore analogous to Equation 5.17 for the calculation of ΔH°, and Equation 19.9 for the calculation of ΔS°. Stated verbally, the *standard free-energy change* for a reaction equals the sum of the standard free-energy values per mole of each product, each multiplied by the corresponding coefficient in the balanced equation, less the corresponding sum for the reactants.

What use can be made of this standard free-energy change for a chemical reaction? The quantity ΔG° tells us whether a mixture of reactants and products, each present under standard conditions, would spontaneously react in the forward direction to produce more products (ΔG° negative) or in the reverse direction to form more reactants (ΔG° positive). Because standard free-energy values are readily available for a large number of substances, the standard free-energy change is easy to calculate for many reaction systems of interest.

SAMPLE EXERCISE 19.6

Determine the standard free-energy change for the following reaction at 298 K:

$$N_2(g) + 3H_2(g) \longrightarrow 2NH_3(g)$$

Solution: Using Appendix C, we find that the standard free energies for the three substances of interest are as follows: $N_2(g)$, $\Delta G_f^\circ = 0.0$; $H_2(g)$, $\Delta G_f^\circ = 0.0$; $NH_3(g)$, $\Delta G_f^\circ = -16.66$ kJ/mol.

The standard free-energy change for the reaction of interest is

$$\Delta G^\circ = 2\,\Delta G_f^\circ(NH_3) - [3\,\Delta G_f^\circ(H_2) + \Delta G_f^\circ(N_2)]$$

Inserting numerical quantities we obtain

$$\Delta G^\circ = -33.32 \text{ kJ}$$

The fact that ΔG° is negative tells us that a mixture of H_2, N_2, and NH_3 at 25°C, each present at a pressure of 1 atm, would react spontaneously to form more ammonia. (Remember, however, that this says nothing about the rate at which the reaction occurs.)

PRACTICE EXERCISE

Using the standard free energies of formation tabulated in Appendix C, calculate ΔG° for the following reaction at 298 K: $2CH_3OH(l) + 3O_2(g) \longrightarrow 2CO_2(g) + 4H_2O(g)$.
Answer: -1370.8 kJ

Free Energy and Temperature

It is worthwhile to examine Equation 19.11 closely to see how the change in free energy depends on both the enthalpy and the entropy changes for a given process. Were it not for entropy effects, all exothermic reactions—those in

Table 19.3 Effect of Temperature on Reaction Spontaneity

ΔH	ΔS	ΔG	Reaction characteristics	Example
−	+	Always negative	Reaction is spontaneous at all temperatures; reverse reaction is always nonspontaneous	$2O_3(g) \longrightarrow 3O_2(g)$
+	−	Always positive	Reaction is nonspontaneous at all temperatures; reverse reaction is spontaneous	$3O_2(g) \longrightarrow 2O_3(g)$
−	−	Negative at low temperatures; positive at high temperatures	Reaction is spontaneous at low temperatures but becomes nonspontaneous at high temperatures	$CaO(s) + CO_2(g) \longrightarrow CaCO_3(s)$
+	+	Positive at low temperatures; negative at high temperatures	Reaction is nonspontaneous at low temperatures but becomes spontaneous as temperature is raised	$CaCO_3(s) \longrightarrow CaO(s) + CO_2(g)$

which ΔH is negative—would be spontaneous. The entropy contribution, represented by the quantity $-T\,\Delta S$, may increase or decrease the tendency of the reaction to proceed spontaneously. When ΔS is positive, meaning that the final state is more random or disordered than the initial state, the term $-T\,\Delta S$ makes ΔG less positive (or more negative), increasing the tendency of the reaction to occur spontaneously. When ΔS is negative, however, the term $-T\,\Delta S$ decreases the tendency of the reaction to occur spontaneously.

When ΔH and $-T\,\Delta S$ are of opposite sign, the relative importance of the two terms determines whether ΔG is negative or positive. In these instances, temperature is an important consideration. Both ΔH and ΔS are in principle capable of changing with temperature. However, the only quantity in the equation

$$\Delta G = \Delta H - T\,\Delta S$$

that changes markedly with temperature is $-T\,\Delta S$. Thus, at high temperatures the $-T\,\Delta S$ term becomes relatively more important in determining the sign and magnitude of ΔG.

Various possible situations for the relative signs of ΔH and ΔS are shown in Table 19.3, with examples of each. By applying the concepts we have developed for predicting entropy changes, we may often predict how ΔG will change with change in temperature.

SAMPLE EXERCISE 19.7

(a) Predict the direction in which ΔG° for the equilibrium

$$N_2(g) + 3H_2(g) \rightleftharpoons 2NH_3(g)$$

will change with increase in temperature. **(b)** Calculate ΔG° at 500°C, assuming that ΔH° and ΔS° do not change with temperature.

Solution: **(a)** In Sample Exercise 19.5, we saw that the change in ΔS° for the equilibrium of interest is negative. This means that the term $-T\,\Delta S^\circ$ is positive and grows larger with increasing temperature. The standard free-energy change, ΔG°, is the sum of the negative quantity ΔH° and the positive quantity $-T\,\Delta S^\circ$. Because only the latter grows larger with increasing temperature, ΔG° grows less negative. Thus, the driving force for the reaction becomes smaller with increasing temperature, indicating less conversion to products.

(b) The ΔH_f° and S° values necessary to calculate ΔH° and ΔS° for this reaction can be taken from Appendix C. We have previously performed these calculations in Sample Exercise 15.13 (Section 15.6) and in Sample Exercise 19.5, giving $\Delta H^\circ = -92.38$ kJ and $\Delta S^\circ = -198.2$ J/K. To calculate ΔG° given ΔH° and ΔS°, we use the following relationship:

$$\Delta G^\circ = \Delta H^\circ - T\,\Delta S^\circ$$

(Recall that the superscript $^\circ$ indicates that the process is operated under standard-state conditions.) Assuming that ΔH° and ΔS° do not change with temperature, and using $T = 500 + 273 = 773$ K, we have

$$\Delta G^\circ = -92.38 \text{ kJ} - (773 \text{ K})\left(-198.2 \frac{\text{J}}{\text{K}}\right)\left(\frac{1 \text{ kJ}}{10^3 \text{ J}}\right)$$

$$= -92.38 \text{ kJ} + 153.21 \text{ kJ} = 60.83 \text{ kJ}$$

Notice that we changed $T\,\Delta S^\circ$ from units of joules to kilojoules so it could be added to ΔH°, which is in units of kilojoules.

In Sample Exercise 19.6, we calculated ΔG° for this reaction at 298 K: $\Delta G^\circ_{298} = -33.32$ kJ. Thus, we see that increasing the temperature from 298 K to 773 K changes ΔG° from -33.32 kJ to $+60.83$ kJ. Of course, the result at 773 K is not as accurate as that at 298 K, because ΔH° and ΔS° do change slightly with temperature. Nevertheless, the result should be a reasonable approximation. The positive increase in ΔG° with increasing temperature is in agreement with the qualitative prediction made in part (a) of this exercise. Our result indicates that a mixture of $N_2(g)$, $H_2(g)$, and $NH_3(g)$, each at 1 atm pressure (standard-state conditions), will react spontaneously at 298 K to form more $NH_3(g)$; however, this same reaction is not spontaneous at 773 K. In fact, at 773 K the reaction will proceed in the opposite direction, to form more $N_2(g)$ and $H_2(g)$.

PRACTICE EXERCISE

(a) Using the standard enthalpies of formation and standard entropies in Appendix C, calculate ΔH° and ΔS° at 298 K for the following reaction: $2SO_2(g) + O_2(g) \longrightarrow 2SO_3(g)$. **(b)** Using the values for ΔH° and ΔS° from part (a), estimate ΔG° at 400 K.
Answers: **(a)** $\Delta H^\circ = -196.6$ kJ, $\Delta S^\circ = -189.6$ J/K; **(b)** $\Delta G^\circ = -120.8$ kJ

19.6 FREE ENERGY AND THE EQUILIBRIUM CONSTANT

Although it is useful to have a ready means of determining ΔG° for a reaction from tabulated values, we usually want to know about the direction of spontaneous change for systems that are not at standard conditions. For any chemical process, the general relationship between the free-energy change under standard conditions, ΔG°, and the free-energy change under any other conditions, ΔG, is given by the following expression.

$$\Delta G = \Delta G^\circ + RT \ln Q \qquad [19.14]$$

In this expression, R is the ideal-gas-equation constant, 8.314 J/K-mol; T is the absolute temperature; and Q is the reaction quotient (Section 15.5) that corresponds to the chemical reaction and particular reaction mixture of interest.

SAMPLE EXERCISE 19.8

Calculate ΔG at 298 K for the following reaction if the reaction mixture consists of 1.0 atm N_2, 3.0 atm H_2, and 1.0 atm NH_3:

$$N_2(g) + 3H_2(g) \longrightarrow 2NH_3(g)$$

Solution: For the balanced equation and set of concentrations given, the reaction quotient Q is

$$Q = \frac{P^2_{NH_3}}{P_{N_2}P^3_{H_2}} = \frac{(1.0)^2}{(1.0)(3.0)^3} = 3.7 \times 10^{-2}$$

ΔG°_{298} was calculated in Sample Exercise 19.6: $\Delta G^\circ_{298} = -33.32$ kJ. Therefore, for 1 mol N_2 reacting with 3 mol H_2, we have

$$\begin{aligned}\Delta G &= \Delta G^\circ + RT \ln Q \\ &= (-33.32 \text{ kJ}) + \left(8.314 \frac{\text{J}}{\text{K}}\right)(298 \text{ K})\left(\frac{1 \text{ kJ}}{10^3 \text{ J}}\right) \ln (3.7 \times 10^{-2}) \\ &= -33.32 \text{ kJ} + (-8.17 \text{ kJ}) \\ &= -41.49 \text{ kJ}\end{aligned}$$

The free-energy change becomes more negative, changing from -33.32 kJ to -41.49 kJ, as the pressures of N_2, H_2, and NH_3 are changed from 1.0 atm each (standard-state conditions, ΔG°) to 1.0 atm, 3.0 atm, and 1.0 atm, respectively. The larger negative value for ΔG when the pressure of H_2 is increased from 1.0 atm to 3.0 atm indicates a larger "driving force" to produce NH_3. This result bears out the prediction of Le Châtelier's principle, which indicates that increasing P_{H_2} should shift the reaction more to the product side, thereby forming more NH_3.

PRACTICE EXERCISE

Calculate ΔG at 298 K for the reaction of nitrogen and hydrogen to form ammonia if the reaction mixture consists of 3.0 atm NH_3, 1.0 atm N_2, and 0.50 atm H_2.
Answer: -22.72 kJ

When a system is at equilibrium, ΔG must be zero, and the reaction quotient Q must by definition equal K (Section 15.5). Thus, for a system at equilibrium (when $\Delta G = 0$ and $Q = K$), Equation 19.14 transforms as follows:

$$\Delta G = \Delta G^\circ + RT \ln Q \qquad [19.14]$$

$$0 = \Delta G^\circ + RT \ln K$$

$$\Delta G^\circ = -RT \ln K \qquad [19.15]$$

From Equation 19.15 we can readily see that if ΔG° is negative, ln K must be positive. A positive value for ln K means that $K > 1$. Therefore, the more negative ΔG° is, the larger the equilibrium constant, K. Conversely, if ΔG° is positive, ln K is negative, which means that $K < 1$. To summarize:

ΔG° negative:	$K > 1$
ΔG° zero:	$K = 1$
ΔG° positive:	$K < 1$

It is possible from a knowledge of ΔG° for a reaction to calculate the value for the equilibrium constant, using Equation 19.15. Some care is necessary, however, in the matter of units. When dealing with gases, the concentrations of reactants should be expressed in units of atmospheres. The concentrations of solids are 1 if they are pure solids. The concentrations of liquids are 1 if they are pure liquids; if a mixture of liquids is involved, the concentration of each is expressed as mole fraction. For substances in solution, concentrations in moles per liter are appropriate.

SAMPLE EXERCISE 19.9

From standard free energies of formation, calculate the equilibrium constant for the reaction

$$N_2(g) + 3H_2(g) \rightleftharpoons 2NH_3(g)$$

at 25°C.

Solution: The equilibrium constant for this reaction is written as

$$K_p = \frac{P^2_{NH_3}}{P_{N_2}P^3_{H_2}}$$

where the gas concentrations are expressed in atmospheres pressure. The standard free-energy change for the reaction was determined in Sample Exercise 19.6 to be -33.32 kJ. Inserting this into Equation 19.15, we obtain

$$-33{,}320 \text{ J} = -(8.314 \text{ J/K})(298 \text{ K}) \ln K_p$$

$$\ln K_p = 13.45$$

Thus, K_p is given by

$$K_p = e^{13.45} = 6.9 \times 10^5$$

This is a large equilibrium constant. Compare its magnitude with the equilibrium constants at higher temperature, as listed in Table 15.5. If a catalyst could be found that would permit reasonably rapid reaction of N_2 with H_2 at room temperature, high pressures would not be required to force the equilibrium toward NH_3.

PRACTICE EXERCISE

Using data given in Appendix C, calculate first the standard free-energy change, $\Delta G°$, and then the equilibrium constant at 298 K for the following reaction: $2H_2(g) + O_2(g) \longrightarrow 2H_2O(g)$. ***Answers:*** -457.2 kJ; 1.4×10^{80}

19.7 FREE ENERGY AND WORK

One of the important ways we use chemical reactions is to produce energy for performing work. Naturally, we want to accomplish the work as efficiently and economically as possible. Any process that occurs spontaneously can be utilized to perform work, at least in principle. For example, the burning of gasoline in the cylinders of a car produces the work accomplished in moving the car. How much work is extracted from a particular process depends on how it is carried out. For example, we might burn gasoline in an open container and extract no useful work at all. In an automobile engine, less than 20 percent of the released energy is used to accomplish work. If the gasoline were burned under more favorable conditions, we could extract more work. However, there is a theoretical limit to the amount of work we can obtain from a spontaneous process. In practice, we always obtain less than this maximum possible amount. Nevertheless, it is useful to know the maximum for a process, so we can measure our success in extracting work from it.

Thermodynamics tells us that *the change in free energy for a process, ΔG, equals the maximum useful work that can be done by the system on its surroundings in a spontaneous process occurring at constant temperature and pressure:*

$$w_{max} = \Delta G \qquad [19.16]$$

This relationship explains why ΔG is called the *free* energy. It is the portion of the energy change of a spontaneous reaction that is free to do useful work. The remainder of the energy enters the environment as heat.

CHEMISTRY AT WORK: Driving Nonspontaneous Reactions

Free-energy considerations are very important in thinking about many nonspontaneous reactions that we might wish to carry out for our own purposes or that occur in nature. For example, we might wish to extract a metal from an ore. If we look at a reaction such as

$$Cu_2S(s) \longrightarrow 2Cu(s) + S(s) \qquad \Delta G^\circ = +86.2\ \text{kJ} \quad [19.17]$$

we find that it is highly nonspontaneous. Clearly, then, we cannot hope to obtain copper metal from Cu_2S merely by trying to catalyze the reaction shown in Equation 19.17. Instead, we must "do work" on the reaction in some way in order to force it to occur as we wish. We might do this by coupling the reaction we've written with another reaction so that we arrive at an overall reaction that *is* spontaneous. Consider, for example, the reaction

$$S(s) + O_2(g) \longrightarrow SO_2(g) \qquad \Delta G^\circ = -300.1\ \text{kJ} \quad [19.18]$$

This is a spontaneous reaction. Adding Equations 19.17 and 19.18, we obtain

			$\Delta G^\circ =$
$Cu_2S(s)$	$\longrightarrow$	$2Cu(s) + S(s)$	+86.2 kJ
$S(s) + O_2(g)$	$\longrightarrow$	$SO_2(g)$	−300.1 kJ
$Cu_2S(s) + O_2(g)$	$\longrightarrow$	$2Cu(s) + SO_2(g)$	−213.9 kJ

The free-energy change for the overall reaction is the sum of the free-energy changes of the two reactions. Because the negative free-energy change for the second reaction is much larger than the positive free-energy change for the first, the overall reaction has a large and negative standard free-energy change.

The coupling of two or more reactions to cause a nonspontaneous chemical process to occur is very important in biochemical systems. Many of the reactions that are essential to the maintenance of life do not occur spontaneously within the human body. These necessary reactions are made to occur, however, by coupling them with reactions that are spontaneous and release energy. The energy releases that accompany the metabolism of foodstuffs provide the primary source of necessary free energy. For example, the compound glucose, $C_6H_{12}O_6$, is oxidized in the body, and a substantial amount of energy is released.

$$C_6H_{12}O_6(s) + 6O_2(g) \longrightarrow 6CO_2(g) + 6H_2O(l) \qquad \Delta G^\circ = -2880\ \text{kJ} \quad [19.19]$$

This energy "does work" in the body. However, some means is necessary to couple, or connect, the energy released by glucose oxidation to the reactions that require energy. One means of accomplishing this is shown graphically in Figure 19.10. Adenosine triphosphate (ATP) is a high-energy molecule. When ATP is converted to a lower-energy molecule, adenosine diphosphate (ADP), the energy to drive other chemical reactions becomes available. The energy released in glucose oxidation is used in part to reconvert ADP back to ATP. The ATP-ADP interconversions thus act to store energy and release it to drive needed reactions. The coupling of reactions so that the free energy released in one may be used in another way requires particular enzymes as catalysts. In Chapter 27, which deals with biochemistry, we will examine the energy relationships in living systems in more detail.

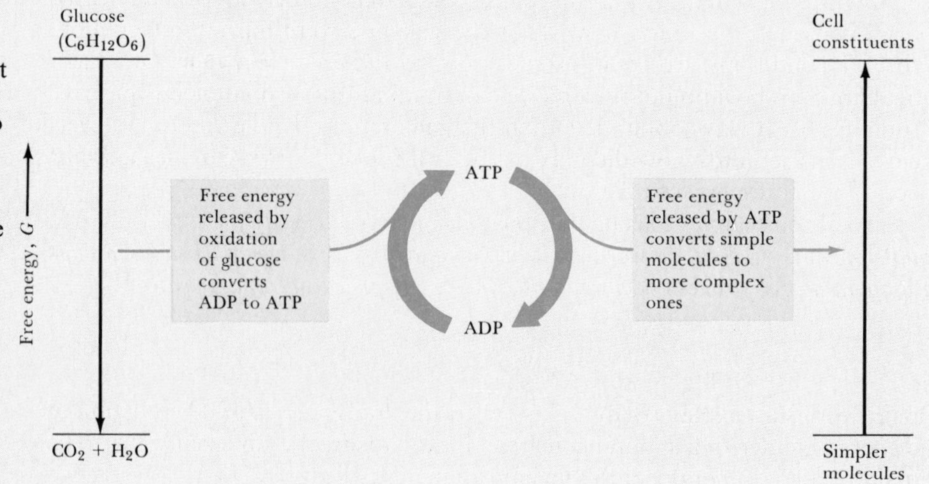

Figure 19.10 Schematic representation of a part of the free-energy changes that occur in cell metabolism. The oxidation of glucose to CO_2 and H_2O produces free energy. This released free energy is used to convert ADP into the more energetic ATP. The ATP is then used, as needed, as an energy source to convert simple molecules into more complex cell constituents. When ATP releases its free energy, it is converted to ADP.

For processes that are not spontaneous ($\Delta G > 0$), the free-energy change is a measure of the *minimum* amount of work that must be done to cause the process to occur. In actual cases, we always need to do more than this theoretical minimum amount because of the inefficiencies in the way the changes occur.

FOR REVIEW

SUMMARY

In this chapter, we have examined the concept of equilibrium from a thermodynamic point of view. The enthalpy change of a system, ΔH, is a measure of the potential energy change in a process. Exothermic processes ($\Delta H < 0$) tend to occur spontaneously. The spontaneous character of a reaction is also determined by the change in randomness or disorder of the system, measured by the entropy, S. Processes that produce an increase in randomness or disorder of the system ($\Delta S > 0$) tend to occur spontaneously.

Entropy changes in a system are associated with an increase in the number of ways the particles of the system can be distributed among possible energy states or spatial arrangements. For example, an increase in volume, in translational energy, or in number of particles leads to an increase in entropy. In any process, the total entropy change is the sum of the entropy change of the system and the surroundings. The second law of thermodynamics tells us that in any spontaneous process the entropy of the universe increases. That is,

$$\Delta S_{\text{system}} + \Delta S_{\text{surroundings}} > 0$$

The standard entropy change in a system, ΔS°, can be calculated from tabulated standard entropy values, S°. Entropies are determined with the aid of the third law of thermodynamics, which states that the entropy of a pure crystalline solid at 0 K is zero.

The Gibbs free energy, G, is a thermodynamic state function that combines the two state functions enthalpy and entropy: $G = H - TS$. For processes that occur at constant temperature, $\Delta G = \Delta H - T\,\Delta S$. The free-energy change for a process occurring at constant temperature and pressure relates directly to reaction spontaneity. For all spontaneous processes, ΔG is negative, whereas a positive value for ΔG indicates a nonspontaneous process (one that is spontaneous in the reverse direction). At equilibrium, ΔG is zero. The free energy also is a measure of the maximum useful work that can be performed by a system in a spontaneous process. In practice, this amount of useful work is never realized because processes in the real world have inherent inefficiencies. If a process is nonspontaneous, the value of ΔG is a measure of the minimum work that must be done on the system to cause the process to occur. In practice, we always need to do more than this minimum.

The standard free-energy change, ΔG°, for any process can be calculated from tabulated standard free energies of formation, ΔG_f°; it can also be calculated from standard enthalpy and entropy changes: $\Delta G^\circ = \Delta H^\circ - T\,\Delta S^\circ$. Temperature changes will change the value of ΔG and can also change its sign.

The free-energy change under nonstandard conditions is related to the standard free-energy change; $\Delta G = \Delta G^\circ + RT \ln Q$. At equilibrium ($\Delta G = 0$, $Q = K$), $\Delta G^\circ = -RT \ln K$. Thus, the standard free-energy change is related to the equilibrium constant. Consequently, we can understand the position of a chemical equilibrium in terms of the ΔH° and $T\,\Delta S^\circ$ functions of which ΔG° is composed.

KEY TERMS

second law of thermodynamics (Sec. 19.1)
spontaneous (Sec. 19.1)
entropy (Sec. 19.2)
isolated system (Sec. 19.2)
translational motion (Sec. 19.3)
vibrational motion (Sec. 19.3)
rotational motion (Sec. 19.3)
third law of thermodynamics (Sec. 19.3)
Gibbs free energy (free energy) (Sec. 19.5)
standard free energy of formation (Sec. 19.5)

EXERCISES

Spontaneity and Entropy

19.1 Which of the following processes are spontaneous and which are nonspontaneous: **(a)** spreading of the fragrance of perfume through a room; **(b)** separation of N_2 and O_2 molecules in air from each other; **(c)** mending a broken clock; **(d)** the reaction of sodium metal with chlorine gas to form sodium chloride; **(e)** the dissolution of $HCl(g)$ in water to form concentrated hydrochloric acid?

19.2 Which of the following processes are spontaneous and which are nonspontaneous: **(a)** the melting of ice cubes at $-5°C$ and 1 atm pressure; **(b)** dissolution of sugar in a cup of hot coffee; **(c)** the reaction of nitrogen atoms to form N_2 molecules at 25°C and 1 atm; **(d)** alignment of iron filings in a magnetic field; **(e)** formation of CH_4 and O_2 molecules from CO_2 and H_2O at room temperature and 1 atm pressure?

19.3 A nineteenth-century chemist, Marcellin Berthelot, suggested that all chemical processes that proceed spontaneously are exothermic. Is this correct? If you think not, offer some counterexamples.

19.4 The freezing of water to ice is an exothermic process. **(a)** In what temperature range is it a spontaneous process? **(b)** In what temperature range is it a nonspontaneous process? **(c)** Why isn't this exothermic process *always* spontaneous?

19.5 When steam condenses to water at 90°C, the entropy of the system decreases. What must be true if the second law of thermodynamics is to be satisfied?

19.6 In a living cell, large molecules are assembled from smaller ones. Is this process consistent with the second law of thermodynamics?

19.7 How does the entropy of the system change when the following processes occur: **(a)** a solid is melted; **(b)** a liquid is vaporized; **(c)** a solid is dissolved in water; **(d)** a gas is liquefied?

19.8 Why is the increase in entropy of the system greater for the vaporization of a substance than for its melting?

19.9 For each of the following pairs, choose the substance with the higher entropy (per mole) at a given temperature: **(a)** $O_2(g)$ at 5 atm or $O_2(g)$ at 0.5 atm; **(b)** $Br_2(l)$ or $Br_2(g)$; **(c)** 1 mol of $N_2(g)$ in 22.4 L or 1 mol of $N_2(g)$ in 2.24 L; **(d)** $CO_2(g)$ or $CO_2(aq)$.

19.10 For each of the following pairs, indicate which substance you would expect to possess the larger standard entropy; **(a)** 1 mol of $Cl_2(g)$ at 273 K, 1 atm pressure, or 1 mol of $Cl_2(g)$ at 373 K, 1 atm pressure; **(b)** 1 mol of $H_2O(g)$ at 100°C, 1 atm pressure, or 1 mol of $H_2O(l)$ at 100°C; **(c)** 1 mol of $O_2(g)$ at 300 K, 30.0 L volume, or 2 mol of $O(g)$ at 300 K, 60.0 L volume; **(d)** 1 mol of $KNO_3(s)$ at 40°C, or 1 mol of $KNO_3(aq)$ at 40°C.

19.11 Using Appendix C, compare the standard entropies at 298 K for the substances in the following pairs; **(a)** $CCl_4(l)$ and $CCl_4(g)$; **(b)** C(diamond) and C(graphite); **(c)** 1 mol of $N_2O_4(g)$ and 2 mol of $NO_2(g)$; **(d)** $KClO_3(s)$ and $KClO_3(aq)$. Explain the origin of the difference in entropy values in each case.

19.12 Using Appendix C, compare the standard entropies at 298 K for the substances in the following pairs: **(a)** $Si(g)$ and $Si(s)$; **(b)** 1 mol of $C_6H_{12}O_6(s)$ and 6 mol C(graphite) plus 6 mol $H_2O(l)$; **(c)** 2 mol $P_2(g)$ and 1 mol $P_4(g)$; **(d)** $NaOH(s)$ and $NaOH(aq)$.

19.13 Predict the sign of ΔS for the system in each of the following processes: **(a)** freezing of 1 mol of $H_2O(l)$; **(b)** evaporation of 1 mol of $Br_2(l)$; **(c)** precipitation of $BaSO_4$ upon mixing $Ba(NO_3)_2(aq)$ and $H_2SO_4(aq)$; **(d)** oxidation of magnesium metal; $2Mg(s) + O_2(g) \longrightarrow 2MgO(s)$.

19.14 For each of the following processes, predict whether the entropy change in the system is positive or negative:

(a) $2C(s) + O_2(g) \longrightarrow 2CO(g)$
(b) $2K(s) + Br_2(l) \longrightarrow 2KBr(s)$
(c) $2MnO_2(s) \longrightarrow 2MnO(s) + O_2(g)$
(d) $O(g) + O_2(g) \longrightarrow O_3(g)$

19.15 Using tabulated $S°$ values from Appendix C, calculate $\Delta S°$ for each of the following reactions:

(a) $2HBr(g) + F_2(g) \longrightarrow 2HF(g) + Br_2(g)$
(b) $2NO(g) + O_2(g) \longrightarrow 2NO_2(g)$
(c) $2CH_3OH(g) + 3O_2(g) \longrightarrow 2CO_2(g) + 4H_2O(g)$
(d) $4FeO(s) + O_2(g) \longrightarrow 2Fe_2O_3(s)$

In each case, account for the sign of $\Delta S°$.

19.16 Using standard entropies tabulated in Appendix C, calculate $\Delta S°$ for each of the following processes:

(a) $2O_3(g) \longrightarrow 3O_2(g)$
(b) $SiCl_4(l) + 2H_2O(l) \longrightarrow SiO_2(s) + 4HCl(g)$
(c) $3C_2H_2(g) \longrightarrow C_6H_6(g)$
(d) $H_2(g) + Cl_2(g) \longrightarrow 2HCl(g)$

Gibbs Free Energy

19.17 **(a)** For a process that occurs at constant temperature, express the change in Gibbs free energy in terms of the changes in the enthalpy and entropy of the system. **(b)** What is the relationship between ΔG for a process and the speed at which it occurs?

19.18 **(a)** What is the significance of $\Delta G = 0$ for any process in a system? **(b)** What is the meaning of the *standard* free-energy change in a process, $\Delta G°$, as contrasted with simply the free-energy change, ΔG?

19.19 Calculate $\Delta G°_{298}$ for

$$H_2O_2(g) \longrightarrow H_2O(g) + \tfrac{1}{2}O_2(g)$$

given that $\Delta H°_{298} = -106$ kJ and $\Delta S°_{298} = +58$ J/K for this process. Would you expect $H_2O_2(g)$ to be very stable at 298 K? Explain briefly.

19.20 Using the data in Appendix C, calculate $\Delta H°$, $\Delta S°$, and $\Delta G°$ for each of the following reactions. In each case, show that $\Delta G° = \Delta H° - T\Delta S°$.

(a) $BaO(s) + CO_2(g) \longrightarrow BaCO_3(s)$
(b) $2KClO_3(s) \longrightarrow 2KCl(s) + 3O_2(g)$
(c) $2CH_3OH(l) + 3O_2(g) \longrightarrow 2CO_2(g) + 4H_2O(g)$
(d) $NOCl(g) + Cl(g) \longrightarrow NO(g) + Cl_2(g)$

19.21 Using the data from Appendix C, calculate ΔG° for each of the following processes. In each case, indicate whether the reaction is spontaneous under standard conditions.
(a) $H_2(g) + Cl_2(g) \longrightarrow 2HCl(g)$
(b) $MgCl_2(s) + H_2O(l) \longrightarrow MgO(s) + 2HCl(g)$
(c) $2NH_3(g) \longrightarrow N_2H_4(g) + H_2(g)$
(d) $2NOCl(g) \longrightarrow 2NO(g) + Cl_2(g)$

19.22 Using the data from Appendix C, calculate the change in Gibbs free energy for each of the following processes. In each case, indicate whether the reaction is spontaneous under standard conditions.
(a) $2SO_2(g) + O_2(g) \longrightarrow 2SO_3(g)$
(b) $NO_2(g) + N_2O(g) \longrightarrow 3NO(g)$
(c) $6Cl_2(g) + 2Fe_2O_3(s) \longrightarrow 4FeCl_3(s) + 3O_2(g)$
(d) $SO_2(g) + 2H_2(g) \longrightarrow S(s) + 2H_2O(g)$

19.23 Classify each of the following reactions as belonging to one of the four possible types summarized in Table 19.3:
(a) $N_2(g) + 3F_2(g) \longrightarrow 2NF_3(g)$
$\Delta H^\circ = -249$ kJ; $\Delta S^\circ = -278$ J/K
(b) $N_2(g) + 3Cl_2(g) \longrightarrow 2NCl_3(g)$
$\Delta H^\circ = 460$ kJ; $\Delta S^\circ = -275$ J/K
(c) $N_2F_4(g) \longrightarrow 2NF_2(g)$
$\Delta H^\circ = 85$ kJ; $\Delta S^\circ = 198$ J/K
(d) $2H_2O(l) \longrightarrow 2H_2(g) + O_2(g)$
$\Delta H^\circ = 572$ kJ; $\Delta S^\circ = 329$ J/K

19.24 From the values given for ΔH° and ΔS°, calculate ΔG° for each of the following reactions at 298 K. If the reaction is not spontaneous under standard conditions at 298 K, at what temperature (if any) would the reaction become spontaneous?
(a) $2PbS(s) + 3O_2(g) \longrightarrow 2PbO(s) + 2SO_2(g)$
$\Delta H^\circ = -844$ kJ; $\Delta S^\circ = -0.165$ kJ/K
(b) $2POCl_3(g) \longrightarrow 2PCl_3(g) + O_2(g)$
$\Delta H^\circ = 572$ kJ; $\Delta S^\circ = 179$ J/K

19.25 A certain reaction is nonspontaneous at 298 K. The entropy change during the reaction is 121 J/K. **(a)** Is the reaction endothermic or exothermic? **(b)** What is the minimum value of ΔH for the reaction?

19.26 A certain reaction is spontaneous at 85°C. The reaction is endothermic by 34 kJ. What is the minimum value of ΔS for the reaction?

19.27 For a certain process, $\Delta H = 178$ kJ, and $\Delta S = 160$ J/K. What is the minimum temperature at which the process will be spontaneous? (You may assume that ΔH and ΔS do not vary with temperature.)

19.28 Reactions in which a substance decomposes by losing CO_2 are called *decarboxylation* reactions. The decarboxylation of acetic acid proceeds as follows:

$$CH_3COOH(l) \longrightarrow CH_4(g) + CO_2(g)$$

By using data from Appendix C, calculate the minimum temperature at which this process will be spontaneous under standard conditions. (You may assume that ΔH° and ΔS° do not vary with temperature.)

19.29 **(a)** Using the data in Appendix C, predict how ΔG° for the following process will change with increasing temperature:

$$P_4(g) \longrightarrow 2P_2(g)$$

(b) Calculate ΔG° at 900 K, assuming that ΔH° and ΔS° do not change with temperature.

19.30 **(a)** Using the data in Appendix C, predict how ΔG° for the following process will change with increasing temperature:

$$SO_3(g) + H_2(g) \longrightarrow SO_2(g) + H_2O(g)$$

(b) Calculate ΔG° at 600 K, assuming that ΔH° and ΔS° do not change with variation in temperature.

Free Energy, Equilibrium, and Work

19.31 Explain qualitatively how ΔG changes for each of the following reactions as the partial pressure of N_2 is increased:
(a) $N_2H_4(g) \longrightarrow N_2(g) + 2H_2(g)$
(b) $N_2(g) + 3F_2(g) \longrightarrow 2NF_3(g)$
(c) $NH_4NO_2(s) \longrightarrow N_2(g) + 2H_2O(g)$

19.32 Indicate whether ΔG increases, decreases, or does not change when the partial pressure of H_2 is increased in each of the following reactions:
(a) $N_2(g) + 3H_2(g) \longrightarrow 2NH_3(g)$
(b) $2HBr(g) \longrightarrow H_2(g) + Br_2(g)$

19.33 Consider the reaction $2NO_2(g) \longrightarrow N_2O_4(g)$. **(a)** Using data from Appendix C, calculate ΔG° at 298 K for this reaction. **(b)** Calculate ΔG at 298 K if the partial pressures of NO_2 and N_2O_4 are 2.00 atm and 0.10 atm, respectively.

19.34 Consider the reaction

$$2CO(g) + O_2(g) \longrightarrow 2CO_2(g).$$

(a) Using data from Appendix C, calculate ΔG° for this reaction at 298 K. **(b)** Calculate ΔG at 298 K if the reaction mixture consists of 6.0-atm CO, 300-atm O_2, and 0.10-atm CO_2.

19.35 By using data from Appendix C, calculate K_p at 298 K for each of the following reactions:
(a) $H_2(g) + Cl_2(g) \rightleftharpoons 2HCl(g)$
(b) $3C_2H_2(g) \rightleftharpoons C_6H_6(g)$
(c) $N_2O(g) + NO_2(g) \rightleftharpoons 3NO(g)$

19.36 Write the equilibrium-constant expression and calculate the magnitude of the equilibrium constant for each of the following reactions at 298 K, using data from Appendix C:
(a) $NaHCO_3(s) \rightleftharpoons NaOH(s) + CO_2(g)$
(b) $2HBr(g) + Cl_2(g) \rightleftharpoons 2HCl(g) + Br_2(g)$
(c) $2SO_2(g) + O_2(g) \rightleftharpoons 2SO_3(g)$

[19.37] Consider the decarboxylation of calcium carbonate:

$$CaCO_3(s) \rightleftharpoons CaO(s) + CO_2(g)$$

By using data from Appendix C, calculate the equilibrium pressure of CO_2 at **(a)** 298 K and **(b)** 800 K.

[19.38] Consider the following reaction:

$$PbCO_3(s) \longrightarrow PbO(s) + CO_2(g)$$

Using data in Appendix C, calculate the equilibrium pressure of CO_2 in the system at **(a)** 25°C and **(b)** 500°C.

19.39 **(a)** Given K_a for benzoic acid at 298 K (Appendix D), calculate ΔG° for the dissociation of this substance in aqueous solution. **(b)** What is the value of ΔG at equilibrium? **(c)** What is the value of ΔG when $[H^+] = 3.0 \times 10^{-3}\ M$, $[C_7H_5O_2^-] = 2.0 \times 10^{-5}\ M$, and $[HC_7H_5O_2] = 0.10\ M$?

19.40 **(a)** Given K_b for ammonia at 298 K (Appendix D), calculate ΔG° for the following reaction:

$$NH_3(aq) + H_2O(l) \longrightarrow NH_4^+(aq) + OH^-(aq)$$

(b) What is the value of ΔG at equilibrium? **(c)** What is the value of ΔG when $[NH_3] = 0.10\ M$, $[NH_4^+] = 0.10\ M$, and $[OH^-] = 0.050\ M$?

19.41 Natural gas consists primarily of methane, CH_4. **(a)** How much heat is produced in burning a mole of CH_4 under standard conditions if reactants and products are brought to 298 K and $H_2O(l)$ is formed? **(b)** What is the maximum amount of useful work that can be accomplished under standard conditions by this system?

19.42 Acetylene gas, $C_2H_2(g)$, is used in welding. **(a)** How much heat is produced in burning a mole of C_2H_2 under standard conditions if both reactants and products are brought to 298 K and $H_2O(l)$ is formed? **(b)** What is the maximum amount of useful work that can be accomplished under standard conditions by this system?

Additional Exercises

19.43 Indicate whether each of the following statements is true or false. If it is false, correct it. **(a)** The feasibility of manufacturing NH_3 from N_2 and H_2 depends entirely on the value of ΔH for the process $N_2(g) + 3H_2(g) \longrightarrow 2NH_3(g)$. **(b)** The reaction of $H_2(g)$ with $Cl_2(g)$ to form $HCl(g)$ is an example of a spontaneous process. **(c)** A spontaneous process is one that occurs rapidly. **(d)** A process that is nonspontaneous in one direction is spontaneous in the opposite direction. **(e)** Spontaneous processes are those that are exothermic and that lead to a higher degree of order in the system.

19.44 For each of the following processes, indicate whether the signs of ΔS and ΔH are expected to be positive, negative, or about zero. **(a)** A solid sublimes. **(b)** The temperature of a solid is lowered by 25°C. **(c)** Ethyl alcohol evaporates from a beaker. **(d)** A diatomic molecule dissociates into atoms. **(e)** A piece of charcoal is combusted to form $CO_2(g)$ and $H_2O(g)$.

19.45 The reaction $2Mg(s) + O_2(g) \longrightarrow 2\,MgO(s)$ has associated with it a negative value for ΔS°, and it is a highly spontaneous process. The second law of thermodynamics states that in any spontaneous process there is always an increase in the entropy of the universe. Is there an inconsistency between the second law and what we know about the oxidation of magnesium? Explain.

19.46 The melting and boiling points of HCl are -115°C and -84°C, respectively. Using this information and the value of S° listed in Appendix C, draw a rough sketch of S° for HCl from absolute zero to 298 K.

19.47 **(a)** Calculate ΔH°, ΔS°, and ΔG° for each of the following reactions at 25°C, using the data in Appendix C:

(i) $2RbCl(s) + 3O_2(g) \longrightarrow 2RbClO_3(s)$
(ii) $2CH_4(g) \longrightarrow C_2H_6(g) + H_2(g)$
(iii) $C_2H_2(g) + 4Cl_2(g) \longrightarrow 2CCl_4(l) + H_2(g)$

(b) Which of these processes are spontaneous at 25°C? **(c)** For each of these processes, predict the manner in which the change in free energy varies with an increase in temperature.

[19.48] **(a)** Listed below are the normal boiling points (Section 11.3) for several substances and their enthalpies of vaporization at those temperatures. From these data calculate ΔS of vaporization for each substance. (Hint: The gaseous and liquid states are in equilibrium: What does this imply about ΔG?)

Substance	**Normal boiling point (°C)**	**Enthalpy of vaporization (kJ/mol)**
Acetone, $(CH_3)_2CO$	56.2	30.3
Benzene, C_6H_6	80.1	30.7
Ammonia, NH_3	-33.4	23.4
Water, H_2O	100.0	40.6

(b) What significance can be attached to the variations in ΔS_{vap} among these compounds? Explain.

19.49 Using the data in Appendix C and given the pressures listed, calculate ΔG for each of the following reactions:

(a) $N_2(g) + 3H_2(g) \longrightarrow 2NH_3(g)$
$P_{N_2} = 8.5$ atm, $P_{H_2} = 2.4$ atm, $P_{NH_3} = 4.7$ atm

(b) $2N_2H_4(g) + 2NO_2(g) \longrightarrow 3N_2(g) + 4H_2O(g)$
$P_{N_2H_4} = P_{NO_2} = 1.0 \times 10^{-3}$ atm, $P_{N_2} = 2.5$ atm, $P_{H_2O} = 1.2$ atm

(c) $N_2H_4(g) \longrightarrow N_2(g) + 2H_2(g)$
$P_{N_2H_4} = 6.0$ atm, $P_{N_2} = 1 \times 10^{-3}$ atm, $P_{H_2} = 2 \times 10^{-4}$ atm

19.50 For the following equilibria, calculate ΔG° at the temperature indicated:

(a) $H_2(g) + I_2(g) \longrightarrow 2HI(g)$ $\quad K_p = 50.2$ at 445°C
(b) $2SO_2(g) + O_2(g) \longrightarrow 2SO_3(g)$ $\quad K_p = 9.1 \times 10^2$ at 800°C

19.51 **(a)** In each of the following reactions, predict the sign of ΔH° and ΔS° and discuss briefly how these factors determine the magnitude of K. **(b)** Based on your general chemical knowledge, predict which of these reactions will have $K > 1$. **(c)** In each case, indicate whether K should increase or decrease with increasing temperature.

(i) $2Mg(s) + O_2(g) \rightleftharpoons 2MgO(s)$
(ii) $2KI(l) \rightleftharpoons 2K(g) + I_2(g)$
(iii) $Na_2(g) \rightleftharpoons 2Na(g)$
(iv) $V_2O_5(s) \rightleftharpoons 2V(s) + \frac{5}{2}O_2(g)$

[19.52] The relationship between the temperature of a reaction, its standard enthalpy change, and the equilibrium constant at that temperature can be expressed in terms of the following linear equation:

$$\ln K = \frac{-\Delta H^\circ}{RT} + \text{constant}$$

(a) Explain how this equation can be used to determine

$\Delta H°$ experimentally from the equilibrium constants at several different temperatures. **(b)** Derive the equation above using relationships given in this chapter. What is the constant equal to?

[19.53] The conversion of natural gas, which is mostly methane, into products that contain two or more carbon atoms, such as ethane, C_2H_6, is a very important industrial chemical process. In principle, methane can be converted into ethane and hydrogen:

$$2CH_4(g) \longrightarrow C_2H_6(g) + H_2(g)$$

In practice, this reaction is carried out in the presence of oxygen:

$$2CH_4(g) + \tfrac{1}{2}O_2(g) \longrightarrow C_2H_6(g) + H_2O(g)$$

(a) Using the data in Appendix C, calculate K_p for the above reactions at 25°C and 500°C. **(b)** Is the difference in $\Delta G°$ for the two reactions primarily enthalpic (due to ΔH) or entropic (due to ΔS)? **(c)** Explain how a comparison of the above reactions is an example of driving a nonspontaneous reaction, as discussed in the Chemistry at Work box in Section 19.7. **(d)** The reaction of CH_4 and O_2 to form C_2H_6 and H_2O must be carried out very carefully to avoid a competing reaction. What is the most likely competing reaction?

[19.54] The potassium ion concentration in blood plasma is about 5.0×10^{-3} *M*, whereas the concentration in muscle-cell fluid is much greater, 0.15 *M*. The plasma and intracellular fluid are separated by the cell membrane, which we assume is permeable only to K^+. **(a)** What is ΔG for the transfer of 1 mol of K^+ from blood plasma to the cellular fluid at body temperature, 37°C? **(b)** What is the minimum amount of work that must be used to transfer this K^+?

[19.55] Acetic acid can be manufactured by reacting methanol with carbon monoxide, an example of a *carbonylation* reaction:

$$CH_3OH(l) + CO(g) \longrightarrow CH_3COOH(l)$$

(a) Calculate the equilibrium constant for the reaction at 25°C. **(b)** At 25°C, a certain equilibrium mixture of methanol, carbon monoxide, and acetic acid has $P_{CO} = 3.0$ atm. Could the liquid portion of the equilibrium mixture be used "as is" in a chemical process that requires 99.99 percent pure acetic acid? **(c)** Industrially, this reaction is run at temperatures above 25°C. Will an increase in temperature produce an increase or decrease in the mole fraction of acetic acid in an equilibrium mixture? Why are elevated temperatures used for the reaction? **(d)** At what temperature will this reaction have an equilibrium constant equal to 1? (You may assume that $\Delta H°$ and $\Delta S°$ are temperature-independent, and you may ignore any phase changes that might occur.)

19.56 The oxidation of glucose, $C_6H_{12}O_6$, in body tissue produces CO_2 and H_2O. In contrast, anaerobic decomposition, which occurs during fermentation, produces ethyl alcohol, C_2H_5OH, and CO_2. **(a)** By using data given in Appendix C, compare the equilibrium constants for the following reactions:

$$C_6H_{12}O_6(s) + 6O_2(g) \rightleftharpoons 6CO_2(g) + 6H_2O(l)$$

$$C_6H_{12}O_6(s) \rightleftharpoons 2C_2H_5OH(l) + 2CO_2(g)$$

(b) Compare the maximum amounts of work that can be obtained from these two processes under standard conditions.

[19.57] Consider the apparatus shown in Figure 19.11, which is much like that shown in Figure 19.1. $N_2(g)$ is initially confined in flask *A*, of 1 L volume, at 2.0 atm pressure. Chamber *B*, of volume 1 L, is evacuated. When the stopcock is opened, N_2 expands to both sides. **(a)** Assuming that N_2 behaves as an ideal gas, what is ΔH for this process? **(b)** Describe the process by which the gas can be returned to its original condition. What change in ΔH occurs in the system in this process? **(c)** What are the changes in ΔG and ΔH of the system for the overall processes of expansion followed by compression to the original condition state? **(d)** In this overall process, is work done by the system on the surroundings or by the surroundings on the system? Explain. **(e)** Describe how this example illustrates the second law of thermodynamics.

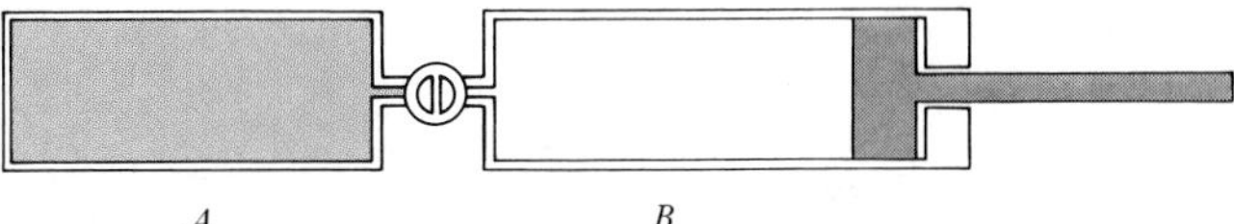

Figure 19.11

[19.58] The reaction

$$SO_2(g) + 2H_2S(g) \rightleftharpoons 3S(s) + 2H_2O(g)$$

is the basis of a suggested method for removal of SO_2 from power-plant stack gases. The standard free energy of each substance is given in Appendix C. **(a)** What is the equilibrium constant for the reaction at 298 K? **(b)** In principle, is this reaction a feasible method of removing SO_2? **(c)** If $P_{SO_2} = P_{H_2S}$ and the vapor pressure of water is 25 mm Hg, calculate the equilibrium SO_2 pressure in the system at 298 K. **(d)** Would you expect the process to be more or less effective at higher temperatures?

[19.59] Cells use the hydrolysis of adenosine triphosphate, ATP, as a source of energy (Figure 19.10). The conversion of ATP to ADP has a standard free-energy change of −30.5 kJ/mol. If all the free energy from the metabolism of glucose,

$$C_6H_{12}O_6(s) + 6O_2(g) \longrightarrow 6CO_2(g) + 6H_2O(l)$$

goes into the conversion of ADP to ATP, how many moles of ATP can be produced for each mole of glucose?

[19.60] Consider the following equilibrium:

$$N_2O_4(g) \rightleftharpoons 2NO_2(g)$$

Thermodynamic data on these gases are given in Appendix C. You may assume that $\Delta H°$ and $\Delta S°$ do not vary with temperature. **(a)** At what temperature will an equilibrium mixture contain equal amounts of the two gases? **(b)** At what temperature will an equilibrium mixture of 1 atm total pressure contain twice as much NO_2 as N_2O_4? **(c)** At what temperature will an equilibrium mixture of 10 atm total pressure contain twice as much NO_2 as N_2O_4? **(d)** Rationalize the results from parts (b) and (c) by using Le Châtelier's principle.

[19.61] The number of ways of arranging n ideal-gas particles in a volume V is proportional to the volume raised to the n power:

$$W \propto V^n$$

(a) By using this relationship and Boltzmann's relationship between entropy and number of arrangements (Equation 19.7), show that the entropy change for the compression or expansion of 1 mol of an ideal gas is given by

$$\Delta S = R \ln\left(\frac{V_2}{V_1}\right)$$

where R is the gas constant and V_1 and V_2 are the initial and final volumes of the gas. **(b)** Calculate the entropy change accompanying the gas expansion shown in Figure 19.1, assuming the two flasks are of equal volume. **(c)** Assuming $\Delta H = 0$ and $T = 298$ K, calculate the change in free energy accompanying the gas expansion shown in Figure 19.1.

Electrochemistry 20

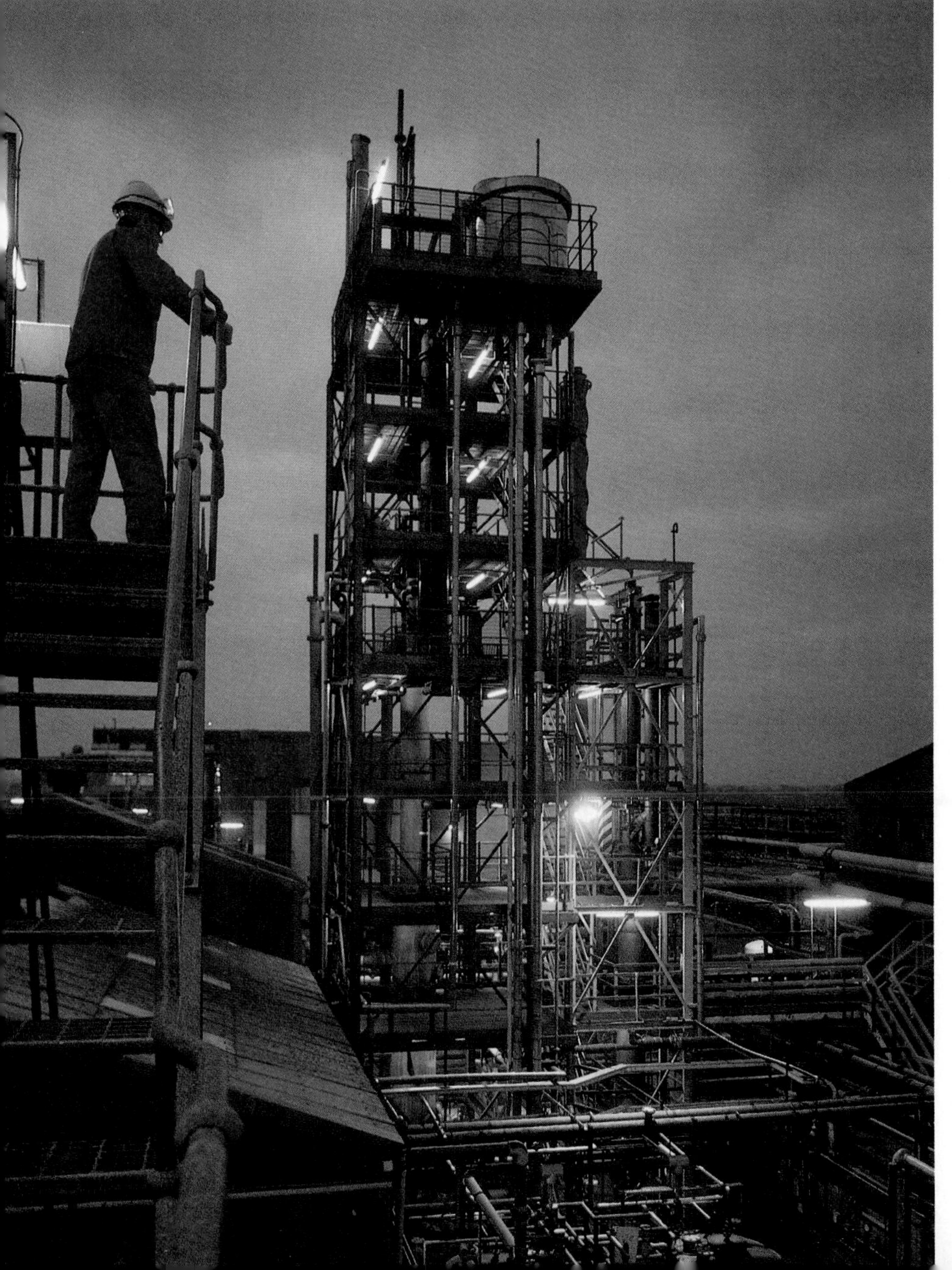

A British chemical plant for the production of chlorine and sodium hydroxide from the electrolysis of brine, concentrated saltwater. (James Holmes, Hays Chemicals/Science Photo Library, Science Source/Photo Researchers)

CONTENTS

The concept of *oxidation state*, which we introduced in Section 8.10, is an extremely useful guide to the changes that occur during chemical reactions. Among the most common and most important chemical reactions are those that involve changes in the oxidation states of atoms. For example, consider the reaction that occurs when zinc metal is added to hydrochloric acid (Figure 20.1):

$$Zn(s) + 2H^{+}(aq) \longrightarrow Zn^{2+}(aq) + H_2(g) \qquad [20.1]$$

During this reaction, the oxidation state of zinc increases from 0 in $Zn(s)$ to $+2$ in $Zn^{2+}(aq)$, whereas that of hydrogen decreases from $+1$ to 0. Chemical reactions in which the oxidation state of one or more substances changes are called **oxidation-reduction reactions** (or *redox* reactions).

As we discussed in Section 4.6, *oxidation* involves the loss of electrons. Conversely, *reduction* involves the gain of electrons. We can view oxidation-reduction reactions as involving the transfer of electrons from the atom that is oxidized to the atom that is reduced. In Equation 20.1, electrons are transferred from zinc atoms (zinc is oxidized) to hydrogen ions (hydrogen is reduced).

The transfer of electrons that occurs in the reaction in Figure 20.1 produces energy in the form of heat; the reaction is thermodynamically "downhill"

(*a*)

(*b*)

Figure 20.1 The addition of zinc metal to hydrochloric acid leads to a spontaneous oxidation-reduction reaction: Zinc metal is oxidized to $Zn^{2+}(aq)$, and $H^{+}(aq)$ is reduced to $H_2(g)$. (© Richard Megna/Fundamental Photographs)

and proceeds spontaneously. The transfer of electrons that occurs during oxidation-reduction reactions can also be used to produce energy in the form of electricity. In other instances, we use electrical energy to make certain nonspontaneous chemical processes occur. **Electrochemistry** is the branch of chemistry that deals with the relationships between electricity and chemical reactions. As we shall see, our discussion of electrochemistry will provide insight into such diverse topics as the construction and operation of batteries, the spontaneity of reactions, electroplating, and the corrosion of metals. We shall begin our study of electrochemistry by learning more about oxidation-reduction reactions.

20.1 OXIDATION-REDUCTION REACTIONS

How do we determine whether a given chemical reaction is an oxidation-reduction reaction? We can do so by keeping track of the oxidation numbers of all the elements involved in the reaction. This procedure tells us which elements (if any) are changing oxidation state. For example, the reaction in Equation 20.1 can be written

$$\underset{(0)}{Zn(s)} + \underset{(+1)}{2H^+(aq)} \longrightarrow \underset{(+2)}{Zn^{2+}(aq)} + \underset{(0)}{H_2(g)} \qquad [20.2]$$

By writing the oxidation number of each element under the equation, we can easily see the oxidation state changes that occur: The oxidation state of Zn changes from 0 to +2, and that of H changes from +1 to 0.

In an oxidation-reduction reaction such as Equation 20.2, a clear transfer of electrons occurs. Zinc must lose electrons as the $Zn(s)$ is converted to $Zn^{2+}(aq)$. Likewise, hydrogen must gain electrons as $H^+(aq)$ is turned into $H_2(g)$. In other reactions, changes occur in the oxidation states, but we can't say that any substance literally gains or loses electrons. For example, consider the combustion of hydrogen gas:

$$\underset{(0)}{2H_2(g)} + \underset{(0)}{O_2(g)} \longrightarrow 2\underset{(+1)}{H_2}\underset{(-2)}{O}(g) \qquad [20.3]$$

Hydrogen has been oxidized from the 0 to the +1 oxidation state, and oxygen has been reduced from the 0 to the −2 oxidation state. Therefore, Equation 20.3 is an oxidation-reduction reaction. However, because water is not an ionic substance, there is not a complete transfer of electrons from hydrogen to oxygen as water is formed. It is important to remember that using oxidation numbers is a convenient form of "bookkeeping." In general, you should not equate the oxidation state of an atom with its actual charge in a chemical compound.

In any oxidation-reduction reaction, both oxidation and reduction must occur. In other words, if one substance is oxidized, then another must be reduced. We can envision the oxidation of one substance as leading to the reduction of another. Therefore, the substance that is oxidized is called a **reducing agent**, or **reductant**. In Equation 20.2, $Zn(s)$ is a reducing agent. Similarly, the reduction of one substance must lead to the oxidation of another. We therefore call the substance that is reduced an **oxidizing agent**, or **oxidant**. In Equation 20.3, $O_2(g)$ is an oxidizing agent.

SAMPLE EXERCISE 20.1

Phosphine, PH_3, a highly toxic gas, burns readily in air to form phosphorus (V) oxide and water vapor. The balanced equation for this reaction is

$$2PH_3(g) + 4O_2(g) \longrightarrow P_2O_5(s) + 3H_2O(g)$$

Identify the elements oxidized and reduced and indicate which substances are the oxidizing and reducing agents.

Solution: We first assign oxidation numbers to all atoms in the reaction

$$\underset{-3\;+1}{2PH_3(g)} + \underset{0}{4O_2(g)} \longrightarrow \underset{+5\;-2}{P_2O_5(s)} + \underset{+1\;-2}{3H_2O(g)}$$

We see that phosphorus undergoes an increase in oxidation number from −3 to +5 and that oxygen undergoes a reduction from 0 to −2. Thus phosphorus is oxidized, and PH_3 is the reducing agent. Oxygen, O_2, is reduced and is therefore the oxidizing agent.

PRACTICE EXERCISE

Identify the oxidizing and reducing agents in the following oxidation-reduction equation:

$$2H_2O(l) + Al(s) + MnO_4^-(aq) \longrightarrow Al(OH)_4^-(aq) + MnO_2(s)$$

Answer: Al is the reducing agent; $MnO_4^-(aq)$ is the oxidizing agent

20.2 BALANCING OXIDATION-REDUCTION REACTIONS

We know that whenever we balance a chemical reaction, we must obey the law of conservation of mass: The amount of each element must be the same on both sides of the equation. When we balance an oxidation-reduction reaction, we have an additional requirement: The gains and losses of electrons must be balanced. In other words, if a substance loses a certain number of electrons during a reaction, then another substance must gain that same number of electrons. In many simple chemical reactions, such as Equation 20.2, this balance of electrons is handled "automatically"; we can balance the equation without explicitly considering the transfer of electrons. However, many redox reactions are more complex than Equation 20.2 and cannot be balanced easily without taking into account the changes in oxidation state that occur. In this section, we shall examine the systematic procedures for balancing oxidation-reduction reactions.

Oxidation Number Method

One simple method for balancing oxidation-reduction equations, called the *oxidation number method*, is to match any increase in oxidation number with a corresponding decrease in oxidation number. As an example, consider the high-temperature process by which aluminum is used to reduce pyrolusite, a manganese ore, to manganese metal. The unbalanced equation for the process is

$$Al(l) + MnO_2(s) \longrightarrow Al_2O_3(s) + Mn(l)$$

We proceed by first assigning oxidation numbers to all the elements in the reaction:

$$Al + MnO_2 \longrightarrow Al_2O_3 + Mn$$
(0) (+4) (−2) (+3) (−2) (0)

We then note the changes in oxidation number in going from reactants to products. The oxidation state of Al increases by 3, and that of Mn decreases by 4:

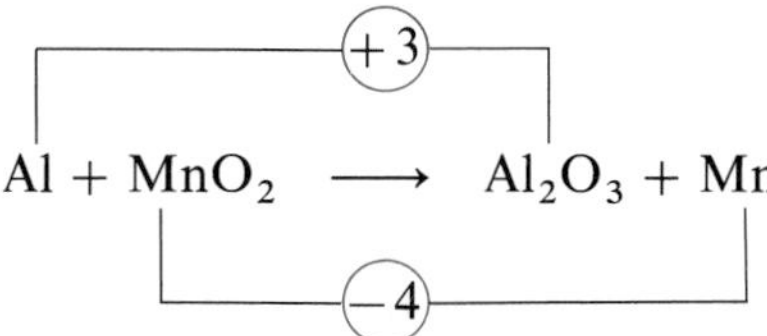

We must now choose coefficients to meet the requirement that the total increase in oxidation state of aluminum equals the total decrease in oxidation state of manganese:

(Coefficient on Al) × (change of 3 in oxidation state per Al)
= (Coefficient on Mn) × (change of 4 in oxidation state per Mn)

To balance these changes, we need 4 Al atoms for every 3 Mn atoms: 4 × 3 = 3 × 4. This gives us

$$4Al + 3MnO_2 \longrightarrow Al_2O_3 + Mn$$

We can now complete the balancing process for each metal by inspection. We need a coefficient of 2 before Al_2O_3 and a coefficient of 3 before Mn:

$$4Al + 3MnO_2 \longrightarrow 2Al_2O_3 + 3Mn$$

Notice that balancing the metal atoms automatically balances the oxygen atoms.

The procedure for balancing an oxidation-reduction equation by the oxidation number method is summarized as follows:

1. Write the unbalanced equation.
2. Assign oxidation numbers and determine which elements undergo changes in oxidation number during the reaction.
3. Choose coefficients that make the total increase in oxidation numbers for the oxidized substances equal to the total decrease in oxidation numbers for the reduced substances.
4. Balance the remaining elements by inspection.

SAMPLE EXERCISE 20.2

Barium chlorite, $Ba(ClO_2)_2$, can be prepared by reacting hydrogen peroxide, H_2O_2, with chlorine dioxide, ClO_2, in an aqueous solution of barium hydroxide, $Ba(OH)_2$. The unbalanced ionic equation is

$$Ba^{2+}(aq) + 2OH^-(aq) + H_2O_2(aq) + ClO_2(aq) \longrightarrow Ba(ClO_2)_2(s) + H_2O(l) + O_2(g)$$

Balance this equation by the oxidation number method.

Solution: We already have the first step in the outline of procedure, the unbalanced chemical equation. The next step is to assign oxidation numbers to all elements:

$$\underset{+2}{Ba^{2+}} + 2\underset{-2}{O}\underset{+1}{H^-} + \underset{+1}{H_2}\underset{-1}{O_2} + \underset{+4}{Cl}\underset{-2}{O_2} \longrightarrow \underset{+2}{Ba}(\underset{+3}{Cl}\underset{-2}{O_2})_2 + \underset{+1}{H_2}\underset{-2}{O} + \underset{0}{O_2}$$

This equation illustrates a situation that is sometimes a source of difficulty for students. A single element, in this case oxygen, appears in more than one place and undergoes an oxidation-state change in one case but not in the other. It is best to focus simply on that portion of the element that undergoes the change in oxidation number; the rest of the element will come into balance later in the process of balancing the equation. Here we see that chlorine undergoes a decrease in oxidation number from +4 to +3 and that each oxygen atom of the hydrogen peroxide undergoes an increase in oxidation number from −1 to 0 (a total of two electrons):

$$Ba^{2+} + 2OH^- + H_2O_2 + ClO_2 \longrightarrow Ba(ClO_2)_2 + H_2O + O_2$$

(+1: $H_2O_2 \rightarrow O_2$; −1: $ClO_2 \rightarrow Ba(ClO_2)_2$)

To balance the changes in oxidation number we need to insert a 2 before the ClO_2. We also need a coefficient of 2 on $H_2O(l)$ to balance hydrogen and oxygen:

$$Ba^{2+}(aq) + 2OH^-(aq) + H_2O_2(aq) + 2ClO_2(aq) \longrightarrow Ba(ClO_2)_2(s) + 2H_2O(l) + O_2(g)$$

We note that now the numbers of oxygen, hydrogen, and barium atoms on both sides are in balance. The overall equation is therefore balanced.

PRACTICE EXERCISE

Iodine pentoxide, which rapidly oxidizes carbon monoxide, can be used as an analytical reagent for determining CO levels in the atmosphere. The unbalanced equation for the reaction is

$$I_2O_5(s) + CO(g) \longrightarrow I_2(s) + CO_2(g)$$

Balance this equation by the oxidation number method.
Answer: $I_2O_5(s) + 5CO(g) \longrightarrow I_2(s) + 5CO_2(g)$

Method of Half-Reactions

Although oxidation and reduction must take place simultaneously, it is often convenient to consider them as separate processes. For example, the oxidation of Sn^{2+} by Fe^{3+}

$$Sn^{2+}(aq) + 2Fe^{3+}(aq) \longrightarrow Sn^{4+}(aq) + 2Fe^{2+}(aq)$$

can be considered to consist of two processes: (1) the oxidation of Sn^{2+} (Equation 20.4) and (2) the reduction of Fe^{3+} (Equation 20.5).

Oxidation: $$Sn^{2+}(aq) \longrightarrow Sn^{4+}(aq) + 2e^- \quad [20.4]$$

Reduction: $$2Fe^{3+}(aq) + 2e^- \longrightarrow 2Fe^{2+}(aq) \quad [20.5]$$

Equations that show either oxidation or reduction alone are called **half-reactions**. As shown in Equations 20.4 and 20.5, the number of electrons lost in an oxidation half-reaction must equal the number of electrons gained in the reduction half-reaction. When this condition is met and each half-reaction is balanced, the two half-reactions can be added to give the overall balanced total oxidation-reduction equation.

The use of half-reactions provides a general method for balancing oxidation-reduction equations. As an example, let's consider the reaction that occurs between permanganate ion, MnO_4^-, and oxalate ion, $C_2O_4^{2-}$, in acidic aqueous solutions. When MnO_4^- is added to an acidified solution of $C_2O_4^{2-}$, the deep purple color of the MnO_4^- ion fades, as illustrated in Figure 20.2. Bubbles of CO_2 form, and the solution takes on the pale pink color of Mn^{2+}. We can therefore write the unbalanced equation as follows:

$$MnO_4^-(aq) + C_2O_4^{2-}(aq) \longrightarrow Mn^{2+}(aq) + CO_2(g) \qquad [20.6]$$

Experiments also show that H^+ is consumed and H_2O produced in the reaction. We shall see that these facts can be deduced in the course of balancing the equation.

To complete and balance Equation 20.6 by the method of half-reactions, we begin with the unbalanced reaction and write two incomplete half-reactions, one involving the oxidant and the other involving the reductant.

$$MnO_4^-(aq) \longrightarrow Mn^{2+}(aq)$$
$$C_2O_4^{2-}(aq) \longrightarrow CO_2(g)$$

The half-reactions are then completed and balanced separately. This means that the number of atoms of each element appearing in the half-reaction must be the same on both sides of the equation. First the atoms undergoing oxidation or reduction are balanced by adding coefficients on one side or the other as necessary. Then the remaining elements are balanced in the same way. If the reaction occurs in acidic aqueous solution, H^+ and H_2O can be added either to reactants or to products to balance hydrogen and oxygen. Similarly, in basic solution the equation can be completed using OH^- and H_2O. These species

(*a*)

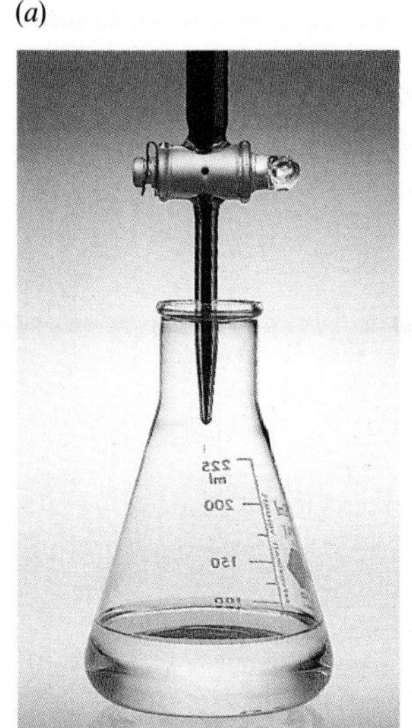

(*b*)

(*c*)

Figure 20.2 (*a*) Reaction between aqueous permanganate solution (deep purple) and an acidic solution of oxalic acid, which contains oxalate ions, $C_2O_4^{2-}$. (*b*) As the reaction proceeds, MnO_4^- is reduced to the $Mn^{2+}(aq)$ ion, which has a pink color too pale to be seen. The reaction is very rapid, and no MnO_4^- persists in solution until all the $C_2O_4^{2-}$ ion has been consumed. (*c*) When the reaction is complete, the purple color of unreacted MnO_4^- is visible in solution. (© Richard Megna/Fundamental Photographs)

are in large supply in the respective solutions, and their formation as products or their use as reactants can easily go undetected experimentally. In the permanganate half-reaction, we already have one manganese atom on each side of the equation. However, we have four oxygens on the left and none on the right side; four H_2O molecules are needed among the products to balance the four oxygen atoms in MnO_4^-:

$$MnO_4^-(aq) \longrightarrow Mn^{2+}(aq) + 4H_2O(l)$$

The eight hydrogen atoms that this introduces among the products can then be balanced by adding $8H^+$ to the reactants:

$$8H^+(aq) + MnO_4^-(aq) \longrightarrow Mn^{2+}(aq) + 4H_2O(l)$$

At this stage there are equal numbers of each type of atom on both sides of the equation, but the charge still needs to be balanced. The total charge of the reactants is $+8 - 1 = +7$, and that of the products is $+2 + 4(0) = +2$. To balance the charge, five electrons are added to the reactant side.*

$$5e^- + 8H^-(aq) + MnO_4^-(aq) \longrightarrow Mn^{2+}(aq) + 4H_2O(l)$$

Proceeding similarly with the oxalate half-reaction, we first realize that mass balance requires the production of two CO_2 molecules for each oxalate ion that reacts:

$$C_2O_4^{2-}(aq) \longrightarrow 2CO_2(g)$$

Charge is balanced by adding two electrons among the products, giving a balanced half-reaction:

$$C_2O_4^{2-}(aq) \longrightarrow 2CO_2(g) + 2e^-$$

Now that we have two balanced half-reactions, we need to multiply each by an appropriate factor so that the number of electrons gained in one half-reaction equals the number of electrons lost in the other. The half-reactions are then added to give the overall balanced equation. In our example, the MnO_4^- half-reaction must be multiplied by 2, and the $C_2O_4^{2-}$ half-reaction must be multiplied by 5 in order that the same number of electrons (ten) appears on both sides of the equation:

$$\begin{array}{rcl} 10e^- + 16H^+(aq) + 2MnO_4^-(aq) & \longrightarrow & 2Mn^{2+}(aq) + 8H_2O(l) \\ 5C_2O_4^{2-}(aq) & \longrightarrow & 10CO_2(g) + 10e^- \\ \hline 16H^+(aq) + 2MnO_4^-(aq) + 5C_2O_4^{2-}(aq) & \longrightarrow & 2Mn^{2+}(aq) + 8H_2O(l) + 10CO_2(g) \end{array}$$

The balanced equation is the sum of the balanced half-reactions. Note that the electrons on the reactant and product sides of the equation cancel each other.

* Although the oxidation numbers of the elements need not be used in balancing a half-reaction by this method, oxidation numbers can be used as a check. In this example MnO_4^- contains manganese in a $+7$ oxidation state. Because manganese changes from a $+7$ to a $+2$ oxidation state, it must gain five electrons, just as we have already concluded.

The steps to balance an oxidation-reduction equation by the method of half-reactions when the reaction occurs in acid solution are summarized as follows:

1. Write the unbalanced equation for the reaction.
2. Divide the overall reaction into two unbalanced half-reactions, one for oxidation, the other for reduction.
3. Balance the elements other than H and O.
4. Balance the O atoms by adding H_2O, and then balance the H atoms by adding H^+.
5. Balance charge on each side of the half-reaction equation by adding e^- to the side with the greater positive charge.
6. Multiply the two half-reactions by coefficients such that the overall electron loss equals the overall electron gain.
7. Add the two half-reactions; simplify where possible by eliminating terms appearing on both sides of the equation.

The equations for reactions that occur in basic solution can be balanced initially as if they occurred in acidic solution. The H^+ ions can then be "neutralized" by adding an equal number of OH^- ions to both sides of the equation. This procedure is shown in Sample Exercise 20.3.

SAMPLE EXERCISE 20.3

Complete and balance the following equations:

(a) $Cr_2O_7^{2-}(aq) + Cl^-(aq) \longrightarrow Cr^{3+}(aq) + Cl_2(g)$ (acidic solution)

(b) $CN^-(aq) + MnO_4^-(aq) \longrightarrow CNO^-(aq) + MnO_2(s)$ (basic solution)

Solution (a) The unbalanced half-reactions are

$$Cr_2O_7^{2-}(aq) \longrightarrow Cr^{3+}(aq)$$
$$Cl^-(aq) \longrightarrow Cl_2(g)$$

The half-reactions are first balanced with respect to the elements. In the first half-reaction the presence of $Cr_2O_7^{2-}$ among the reactants requires two Cr^{3+} among the products. The 7 oxygen atoms in $Cr_2O_7^{2-}$ are balanced by adding $7H_2O$ to the products. The 14 hydrogen atoms in $7H_2O$ are then balanced by adding $14H^+$ among the reactants:

$$14H^+(aq) + Cr_2O_7^{2-}(aq) \longrightarrow 2Cr^{3+}(aq) + 7H_2O(l)$$

Charge is balanced by adding electrons to the left side of the equation so that the total charge is the same on both sides:

$$6e^- + 14H^+(aq) + Cr_2O_7^{2-}(aq) \longrightarrow 2Cr^{3+}(aq) + 7H_2O(l)$$

In the second half-reaction, two Cl^- are required to balance one Cl_2:

$$2Cl^-(aq) \longrightarrow Cl_2(g)$$

We add two electrons to the right side to attain charge balance:

$$2Cl^-(aq) \longrightarrow Cl_2(g) + 2e^-$$

This second half-reaction must be multiplied by 3 to equalize electron loss and gain in the two half-reactions. The two half-reactions are then added to give the balanced equation:

$$14H^+(aq) + Cr_2O_7^{2-}(aq) + 6Cl^-(aq) \longrightarrow 2Cr^{3+}(aq) + 7H_2O(l) + 3Cl_2(g)$$

(b) The incomplete and unbalanced half-reactions are

$$CN^-(aq) \longrightarrow CNO^-(aq)$$
$$MnO_4^-(aq) \longrightarrow MnO_2(s)$$

The equations may be balanced initially as if they took place in acidic solution. The resultant balanced half-reactions are

$$CN^-(aq) + H_2O(l) \longrightarrow CNO^-(aq) + 2H^+(aq) + 2e^-$$
$$3e^- + 4H^+(aq) + MnO_4^-(aq) \longrightarrow MnO_2(s) + 2H_2O(l)$$

Because H^+ cannot exist in any appreciable concentration in basic solution, it is removed from the equations by the addition of an appropriate amount of $OH^-(aq)$. In the CN^- half-reaction, $2OH^-(aq)$ is added to both sides of the equation to neutralize the $2H^+(aq)$. The $2OH^-(aq)$ and $2H^+(aq)$ form $2H_2O(l)$:

$$2OH^-(aq) + H_2O(l) + CN^-(aq) \longrightarrow CNO^-(aq) + 2H_2O(l) + 2e^-$$

The half-reaction can be simplified because H_2O occurs on both sides of the equation. The simplified equation is

$$2OH^-(aq) + CN^-(aq) \longrightarrow CNO^-(aq) + H_2O(l) + 2e^-$$

For the MnO_4^- half-reaction, $4OH^-(aq)$ is added to both sides of the equation:

$$3e^- + 4H_2O(l) + MnO_4^-(aq) \longrightarrow MnO_2(s) + 2H_2O(l) + 4OH^-(aq)$$

Simplifying gives

$$3e^- + 2H_2O(l) + MnO_4^-(aq) \longrightarrow MnO_2(s) + 4OH^-(aq)$$

The top equation is multiplied by 3 and the bottom one by 2 to equalize electron loss and gain in the two half-reactions. The half-reactions are then added:

$$6OH^-(aq) + 3CN^-(aq) \longrightarrow 3CNO^-(aq) + 3H_2O(l) + 6e^-$$
$$6e^- + 4H_2O(l) + 2MnO_4^-(aq) \longrightarrow 2MnO_2(s) + 8OH^-(aq)$$

$$6OH^-(aq) + 3CN^-(aq) + 4H_2O(l) + 2MnO_4^-(aq) \longrightarrow 3CNO^-(aq) + 3H_2O(l) + 2MnO_2(s) + 8OH^-(aq)$$

The overall equation can be simplified because H_2O and OH^- occur on both sides. The simplified equation is

$$3CN^-(aq) + H_2O(l) + 2MnO_4^-(aq) \longrightarrow 3CNO^-(aq) + 2MnO_2(s) + 2OH^-(aq)$$

PRACTICE EXERCISE

Balance the following oxidation-reduction reactions by the method of half-reactions. Reaction **(a)** occurs on acidic solution, reaction **(b)** in basic solution.

(a) $Mn^{2+}(aq) + NaBiO_3(s) \longrightarrow Bi^{3+}(aq) + MnO_4^-(aq)$

(b) $NO_2^-(aq) + Al(s) \longrightarrow NH_3(aq) + Al(OH)_4^-(aq)$

Answers:

(a) $2Mn^{2+}(aq) + 5NaBiO_3(s) + 14H^+(aq) \longrightarrow 2MnO_4^-(aq) + 5Bi^{3+}(aq) + 5Na^+(aq) + 7H_2O(l)$

(b) $NO_2^-(aq) + 2Al(s) + 5H_2O(l) + OH^-(aq) \longrightarrow NH_3(aq) + 2Al(OH)_4^-(aq)$

20.3 VOLTAIC CELLS

In principle, the energy released in any spontaneous redox reaction can be directly harnessed to perform electrical work. This task is accomplished through a **voltaic** (or **galvanic**) **cell**, which is merely a device in which electron transfer is forced to take place through an external pathway rather than directly between reactants.

One such spontaneous reaction occurs when a piece of zinc is placed in contact with a solution containing Cu^{2+}. As the reaction proceeds, the blue color that is characteristic of $Cu^{2+}(aq)$ ions fades, and copper metal begins to

(a)

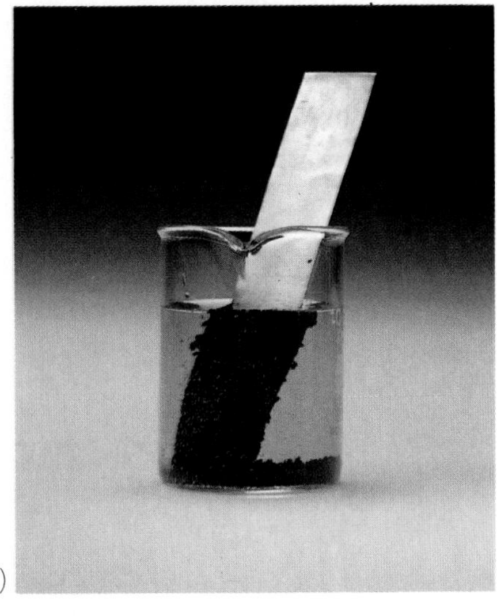

(b)

Figure 20.3 (*a*) A strip of zinc is placed in a solution of copper (II) sulfate. (*b*) Electrons are transferred from the zinc to the Cu^{2+}ion, forming Zn^{2+} ions and copper. Thus, as the reaction proceeds, the zinc dissolves, the blue color due to $Cu^{2+}(aq)$ fades, and copper deposits. (Fundamental Photographs)

deposit on the zinc. At the same time, the zinc begins to dissolve. These transformations are shown in Figure 20.3 and are summarized by Equation 20.7:

$$Zn(s) + Cu^{2+}(aq) \longrightarrow Zn^{2+}(aq) + Cu(s) \qquad [20.7]$$

Figure 20.4 shows a voltaic cell that uses the oxidation-reduction reaction between Zn and Cu^{2+} expressed in Equation 20.7. Although the experimental design shown in Figure 20.4 is more complex than that in Figure 20.3, it is important to recognize that the chemical reaction is the same in both cases. The major difference between the two arrangements is that in Figure 20.4 the zinc metal and $Cu^{2+}(aq)$ are no longer in direct contact. Consequently, reduction of the Cu^{2+} can occur only by a flow of electrons through the wire that connects Zn and Cu (the external circuit).

Figure 20.4 An electrochemical cell based on the reaction shown in Equation 20.7. The compartment on the left contains 1 *M* $CuSO_4$ and a copper electrode. The one on the right contains 1 *M* $ZnSO_4$ and a zinc electrode. The solutions are connected electrically through the bridge, which has a porous glass disc that permits contact of the solutions in the two compartments. The metal electrodes are connected through a digital voltmeter, which reads the standard potential of the cell, 1.10 V. (© Richard Megna/ Fundamental Photographs)

Figure 20.5 Voltaic cell using a salt bridge to complete the electrical circuit.

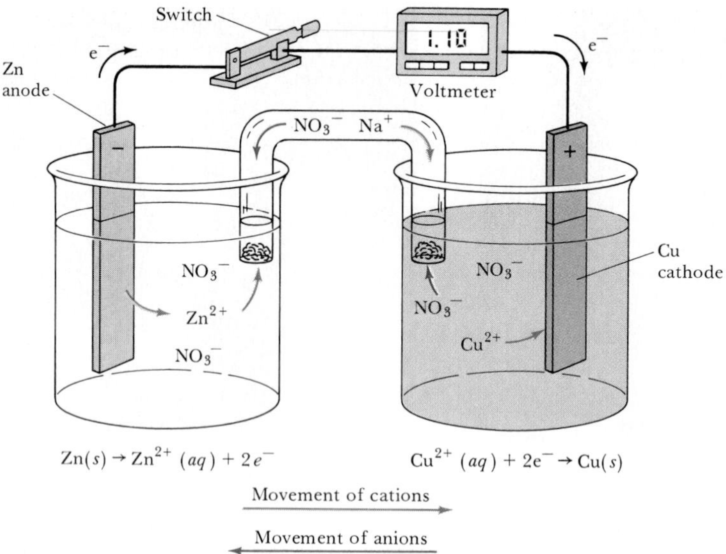

The two solid metals that are connected by the external circuit are called *electrodes.* By definition, the electrode at which oxidation occurs is called the **anode;** the electrode at which reduction occurs is called the **cathode.*** We can think of the voltaic cell as two "half-cells," one corresponding to the oxidation half-reaction and one corresponding to the reduction half-reaction. In our present example, Zn is oxidized and Cu^{2+} is reduced:

$$\text{Anode (oxidation half-reaction):} \quad Zn(s) \longrightarrow Zn^{2+}(aq) + 2e^-$$

$$\text{Cathode (reduction half-reaction):} \quad Cu^{2+}(aq) + 2e^- \longrightarrow Cu(s)$$

Electrons become available as zinc metal is oxidized at the anode. They flow through the external circuit to the cathode, where they are consumed as $Cu^{2+}(aq)$ is reduced. Because $Zn(s)$ is oxidized in the cell, the zinc electrode loses mass, and the concentration of the Zn^{2+} solution increases as the cell operates. Similarly, the Cu electrode gains mass, and the Cu^{2+} solution becomes less concentrated as Cu^{2+} is reduced to $Cu(s)$.

We must be careful about the signs we attach to the electrodes in a voltaic cell. We have seen that electrons are released at the anode as the zinc is oxidized. Electrons are thus flowing out of the anode and into the external circuit, as illustrated in Figure 20.5. Because the electrons are negatively charged, we assign a negative sign to the anode. Conversely, electrons flow into the cathode, where they are consumed in the reduction of copper. A positive sign is thus assigned to the cathode, because it appears to attract the negative electrons.

As the cell pictured in Figure 20.4 operates, oxidation of Zn introduces additional Zn^{2+} ions into the anode compartment. Unless a means is provided to neutralize this positive charge, no further oxidation can take place. Similarly, the reduction of Cu^{2+} at the cathode leaves an excess of negative charge in solution in that compartment. Electrical neutrality is maintained by a migration of ions through the porous glass disc separating the two compartments, or through a *salt bridge*, as illustrated in Figure 20.5. A salt bridge consists of a

* To help remember these definitions, it is useful to note that anode and oxidation both begin with a vowel, and cathode and reduction both begin with a consonant.

U-shaped tube that contains an electrolyte solution, such as $NaNO_3(aq)$, whose ions will not react with other ions in the cell or with the electrode materials. The ends of the U-tube may be loosely plugged with glass wool, or the electrolyte may be incorporated into a gel so that the electrolyte solution does not run out when the U-tube is inverted. As oxidation and reduction proceed at the electrodes, ions from the salt bridge migrate to neutralize charge in the anode and cathode compartments. Anions migrate toward the anode, and cations toward the cathode. In fact, no measurable electron flow will occur through the external circuit unless a means is provided for ions to migrate through the solution from one electrode compartment to another, thereby completing the circuit.

SAMPLE EXERCISE 20.4

The following oxidation-reduction reaction is spontaneous in the direction indicated:

$$Cr_2O_7^{2-}(aq) + 14H^+(aq) + 6I^-(aq) \longrightarrow 2Cr^{3+}(aq) + 3I_2(s) + 7H_2O(l)$$

A solution containing $K_2Cr_2O_7$ and H_2SO_4 is poured into one beaker, and a solution of KI is poured into another. A salt bridge is used to join the beakers. A metallic conductor that will not react with either solution (such as platinum foil) is suspended in each solution, and the two conductors are connected with wires through a voltmeter or some other device to detect an electric current. The resultant voltaic cell generates an electric current. Indicate the reaction occurring at the anode, the reaction at the cathode, the direction of electron and ion migrations, and the signs of the electrodes.

Solution: The half-reactions for the given chemical equation are

$$Cr_2O_7^{2-}(aq) + 14H^+(aq) + 6e^- \longrightarrow 2Cr^{3+}(aq) + 7H_2O(l)$$

$$6I^-(aq) \longrightarrow 3I_2(s) + 6e^-$$

The first half-reaction is the reduction process, which by definition occurs at the cathode. The second half-reaction is an oxidation, which occurs at the anode. The I^- ions are the source of electrons, and the $Cr_2O_7^{2-}$ ions are the receptors. Consequently, the electrons flow through the external circuit from the electrode immersed in the KI solution (the anode) to the electrode immersed in the $K_2Cr_2O_7$–H_2SO_4 solution (the cathode). The electrodes themselves do not react in any way; they merely provide a means of transferring electrons from or to the solutions. The cations move through the solutions toward the cathode, and the anions move toward the anode. The anode is the negative electrode, and the cathode the positive one.

PRACTICE EXERCISE

The two half-reactions in a voltaic cell are

$$Zn(s) \longrightarrow Zn^{2+}(aq) + 2e^-$$

$$ClO_3^-(aq) + 6H^+(aq) + 6e^- \longrightarrow Cl^-(aq) + 3H_2O(l)$$

(a) Indicate which reaction occurs at the anode and which at the cathode. **(b)** Which electrode is consumed in the cell reaction? **(c)** Which electrode is positive?
Answers: **(a)** The first reaction occurs at the anode, the second reaction at the cathode. **(b)** The anode (Zn) is consumed in the cell reaction. **(c)** The cathode is positive.

20.4 CELL EMF

As we have been discussing voltaic cells, you may have wondered why electrons flow spontaneously through the external circuit. In the cell shown in Figure 20.5, for example, what causes the electrons to leave the Zn electrode, pass through the external circuit, and enter the Cu electrode? Like all chemical processes that occur spontaneously, the answer involves energy: *It is energetically*

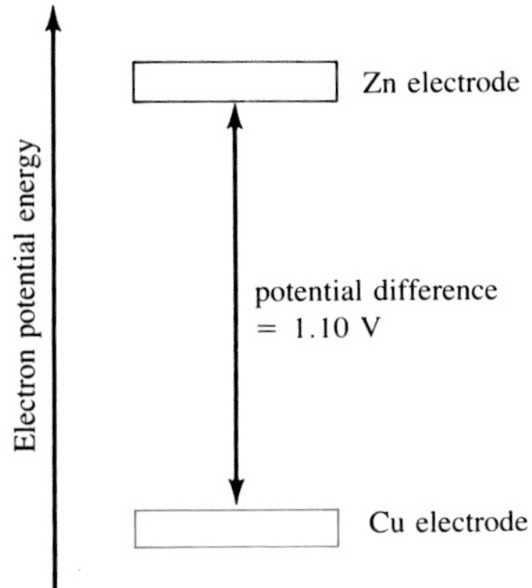

Figure 20.6 A schematic comparison of the potential energy of electrons in the two electrodes in Figure 20.5. The potential energy of electrons is lower in the Cu electrode than in the Zn electrode, so electrons flow through the external circuit in order to become lower in energy. The *potential difference* between the electrodes is 1.10 V.

favorable for electrons to flow from the anode to the cathode of a voltaic cell. In this section, we shall discuss quantitatively the energetic aspects of electrochemical cells.

In each half-cell of a voltaic cell, the electrons have a different potential energy. For example, Figure 20.6 illustrates the difference in potential energy of the electrons in the two electrodes of the voltaic cell in Figure 20.5. The potential energy of electrons is higher in the Zn electrode than in the Cu electrode. If the electrodes are connected by an external circuit, the electrons can lower their potential energy by flowing from the zinc to the copper electrode. The *potential difference* (the difference in potential energy per electrical charge) between two electrodes is measured in units of volts. One volt (V) is the potential difference required to impart 1 J of energy to a charge of 1 C (coulomb):

$$1\ \text{V} = 1\ \frac{\text{J}}{\text{C}}$$

In the Zn/Cu voltaic cell in Figure 20.5, the potential difference is 1.10 V.

We can view the potential difference between two electrodes of a voltaic cell as a "driving force" or "electrical pressure" that pushes electrons through the external circuit. Therefore, we usually call the potential difference between the electrodes of a cell the **electromotive force** (abbreviated **emf**), or **cell potential**. The emf of a cell, denoted E_{cell}, is also measured in volts.

The emf of the Zn/Cu voltaic cell is 1.10 V when operated under standard conditions. You will recall from Section 19.5 that standard conditions include 1 *M* concentrations for reactants and products in solution and 1 atm pressure for those that are gases (Table 19.2). In the present example, the cell would be operated with both $[Cu^{2+}]$ and $[Zn^{2+}]$ at 1 *M*.* Under standard conditions, the emf is called the **standard emf** or the **standard cell potential,** E°_{cell}:

$$\text{Zn}(s) + \text{Cu}^{2+}(aq) \longrightarrow \text{Zn}^{2+}(aq) + \text{Cu}(s) \qquad E^\circ_{\text{cell}} = 1.10\ \text{V}$$

(We often drop the subscript "cell" when it is clear that we are discussing the emf of a cell.) The emf of any voltaic cell depends on the chemical reactions taking place in the cell, the concentrations of reactants and products, and the temperature of the cell, which we shall take to be 25°C unless otherwise noted.

Standard Electrode Potentials

Just as we can think of an overall cell reaction as the sum of two half-reactions, we can think of the emf of a cell as the sum of two half-cell potentials. The half-cell potential due to the loss of electrons at the anode is called the *oxidation potential*, E_{ox}, and that due to electron gain at the cathode is called the *reduction potential*, E_{red}:

$$E_{\text{cell}} = E_{\text{ox}} + E_{\text{red}}$$

* Actually, solutions of 1 *M* are the standard state only if they behave ideally. Salts dissolved in an aqueous medium at concentrations on the order of 1 *M* generally do not behave ideally; corrections need to be made to allow for nonideal behavior. However, in this introduction to electrochemistry we shall not concern ourselves with these corrections.

If all reagents are under standard conditions, the half-cell potentials are called the **standard oxidation potential**, E°_{ox}, and the **standard reduction potential**, E°_{red}; the sum of these equals the standard emf:

$$E^\circ_{cell} = E^\circ_{ox} + E^\circ_{red}$$

We cannot directly measure an isolated oxidation or reduction potential. However, if we arbitrarily assign a standard half-cell potential to a certain half-reaction, we can then determine the standard potentials of other half-reactions relative to that reference. The reference half-reaction is the reduction of H^+ to H_2, which is assigned a standard reduction potential of exactly 0 V:

$$2H^+(1\ M) + 2e^- \longrightarrow H_2(1\ atm) \qquad E^\circ_{cell} = 0\ V$$

An electrode designed to produce this half-reaction is called a **standard hydrogen electrode**, and is one of the electrodes illustrated in Figure 20.7. A standard hydrogen electrode consists of a platinum wire and a piece of platinum foil covered with finely divided platinum that serves as an inert surface for the cathode reaction. The electrode is encased in a glass tube so that hydrogen gas can bubble over the platinum.

Figure 20.7 shows a voltaic cell constructed to use the following redox reaction between Zn and H^+, the reaction shown in Figure 20.1:

$$Zn(s) + 2H^+(aq) \longrightarrow Zn^{2+}(aq) + H_2(g)$$

Oxidation of zinc occurs in the anode compartment, and reduction of H^+ occurs in the cathode compartment, which is a standard hydrogen electrode. This voltaic cell generates a standard emf (E°_{cell}) of 0.76 V. By using the defined standard reduction potential of H^+ ($E^\circ_{red} = 0$), we can calculate the standard oxidation potential of Zn:

$$E^\circ_{cell} = E^\circ_{ox} + E^\circ_{red}$$
$$0.76\ V = E^\circ_{ox} + 0$$

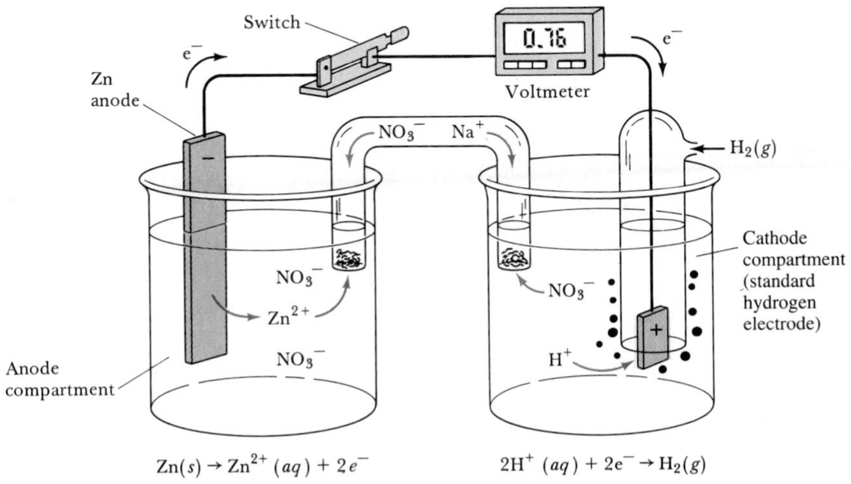

Figure 20.7 Voltaic cell using a standard hydrogen electrode.

Thus, a standard oxidation potential of 0.76 V can be assigned to Zn:

$$Zn(s) \longrightarrow Zn^{2+}(aq) + 2e^- \qquad E^\circ_{ox} = 0.76 \text{ V} \tag{20.8}$$

The standard potentials for other half-reactions can be established from other cell emfs in a similar fashion.

By convention, half-cell potentials are tabulated as standard *reduction* potentials. For the Zn half-reaction, we therefore want to calculate the potential for the reduction of Zn^{2+} to Zn, which is the reverse of Equation 20.8. As with energy-related quantities, such as ΔH and ΔG, reversing a reaction changes the sign of E°. Thus, *the half-cell potential for a reduction is equal in magnitude but opposite in sign to the half-cell potential for the same reaction written as an oxidation*:

$$E^\circ_{red} = -E^\circ_{ox} \tag{20.9}$$

For the Zn half-reaction, we can therefore write

$$Zn^{2+}(aq) + 2e^- \longrightarrow Zn(s) \qquad E^\circ_{red} = -0.76 \text{ V}$$

Table 20.1 lists some standard reduction potentials, also called **standard electrode potentials**. A more complete list of standard electrode potentials is found in Appendix E. These standard electrode potentials can be combined to calculate the standard emfs of a large variety of voltaic cells.

Table 20.1 Standard Electrode Potentials in Water at 25°C

Standard potential (V)	Reduction half-reaction
2.87	$F_2(g) + 2e^- \longrightarrow 2F^-(aq)$
1.51	$MnO_4^-(aq) + 8H^+(aq) + 5e^- \longrightarrow Mn^{2+}(aq) + 4H_2O(l)$
1.36	$Cl_2(g) + 2e^- \longrightarrow 2Cl^-(aq)$
1.33	$Cr_2O_7^{2-}(aq) + 14H^+(aq) + 6e^- \longrightarrow 2Cr^{3+}(aq) + 7H_2O(l)$
1.23	$O_2(g) + 4H^+(aq) + 4e^- \longrightarrow 2H_2O(l)$
1.06	$Br_2(l) + 2e^- \longrightarrow 2Br^-(aq)$
0.96	$NO_3^-(aq) + 4H^+(aq) + 3e^- \longrightarrow NO(g) + 2H_2O(l)$
0.80	$Ag^+(aq) + e^- \longrightarrow Ag(s)$
0.77	$Fe^{3+}(aq) + e^- \longrightarrow Fe^{2+}(aq)$
0.68	$O_2(g) + 2H^+(aq) + 2e^- \longrightarrow H_2O_2(aq)$
0.59	$MnO_4^-(aq) + 2H_2O(l) + 3e^- \longrightarrow MnO_2(s) + 4OH^-(aq)$
0.54	$I_2(s) + 2e^- \longrightarrow 2I^-(aq)$
0.40	$O_2(g) + 2H_2O(l) + 4e^- \longrightarrow 4OH^-(aq)$
0.34	$Cu^{2+}(aq) + 2e^- \longrightarrow Cu(s)$
0	$2H^+(aq) + 2e^- \longrightarrow H_2(g)$
−0.28	$Ni^{2+}(aq) + 2e^- \longrightarrow Ni(s)$
−0.44	$Fe^{2+}(aq) + 2e^- \longrightarrow Fe(s)$
−0.76	$Zn^{2+}(aq) + 2e^- \longrightarrow Zn(s)$
−0.83	$2H_2O(l) + 2e^- \longrightarrow H_2(g) + 2OH^-(aq)$
−1.66	$Al^{3+}(aq) + 3e^- \longrightarrow Al(s)$
−2.71	$Na^+(aq) + e^- \longrightarrow Na(s)$
−3.05	$Li^+(aq) + e^- \longrightarrow Li(s)$

SAMPLE EXERCISE 20.5

Given that E°_{cell} is 1.10 V for the Zn–Cu^{2+} cell shown in Figure 20.5, and that E°_{ox} is 0.76 V for the oxidation of zinc (Equation 20.8), calculate E°_{red} for the reduction of Cu^{2+} to Cu:

$$Cu^{2+}(aq) + 2e^- \longrightarrow Cu(s)$$

Solution:

$$E^\circ_{cell} = E^\circ_{ox} + E^\circ_{red}$$

$$1.10\ V = 0.76\ V + E^\circ_{red}$$

$$E^\circ_{red} = 1.10\ V - 0.76\ V = 0.34\ V$$

PRACTICE EXERCISE

A voltaic cell employs the following half-reactions:

$$In^+(aq) \longrightarrow In^{3+}(aq) + 2e^-$$

$$Br_2(l) + 2e^-(aq) \longrightarrow 2Br^-(aq)$$

The standard cell potential for this cell is 1.46 V. Using the data in Table 20.1, calculate E°_{ox} for the first half-cell reaction. ***Answer:*** +0.40 V

SAMPLE EXERCISE 20.6

Using the standard electrode potentials listed in Table 20.1, calculate the standard emf for the cell described in Sample Exercise 20.4:

$$Cr_2O_7{}^{2-}(aq) + 14H^+(aq) + 6I^-(aq) \longrightarrow 2Cr^{3+}(aq) + 3I_2(s) + 7H_2O(l)$$

Solution: The standard emf of the cell, E°_{cell}, is the sum of the standard oxidation and reduction potentials for the appropriate half-reactions:

$$E^\circ_{cell} = E^\circ_{ox} + E^\circ_{red}$$

The half-reactions and their potentials are as follows:

$$Cr_2O_7{}^{2-}(aq) + 14H^+(aq) + 6e^- \longrightarrow 2Cr^{3+}(aq) + 7H_2O(l) \qquad E^\circ_{red} = 1.33\ V$$

$$6I^-(aq) \longrightarrow 3I_2(s) + 6e^- \qquad E^\circ_{ox} = -0.54\ V$$

$$Cr_2O_7{}^{2-}(aq) + 14H^+(aq) + 6I^-(aq) \longrightarrow 2Cr^{3+}(aq) + 7H_2O(l) + 3I_2(s) \qquad E^\circ_{cell} = 0.79\ V$$

Notice that E°_{ox} for I^- has a sign opposite that of the reduction potential of I_2 listed in Table 20.1. Also notice that even though the iodide half-reaction must be multiplied by 3 in order to obtain a balanced equation for the reaction, the half-cell potential is *not* multiplied by 3. The standard potential is an intensive property; it does not depend on the quantities of reactants and products, but only on their concentrations. Therefore, it does not matter whether there are 6 mol of I^- or 1 mol as long as the concentration of iodide is 1 *M*.

Just as the sum of the half-reactions gives the chemical equation for the overall cell reaction, the sum of the corresponding oxidation and reduction potentials gives the cell emf. In this case, $E^\circ_{cell} = 0.79$ V; a positive emf is characteristic of all voltaic cells.

PRACTICE EXERCISE

Using Table 20.1, calculate the standard emf for a cell that employs the following overall cell reaction:

$$2Al(s) + 3I_2(s) \longrightarrow 2Al^{3+}(aq) + 6I^-(aq)$$

Answer: 2.20 V

Oxidizing and Reducing Agents

The half-cell potential provides us with a tool for quantitatively expressing the ease with which a species is oxidized or reduced. *The more positive the $E°$ value for a half-reaction, the greater the tendency for that reaction to occur as written.* A negative reduction potential indicates that the species is more difficult to reduce than $H^+(aq)$, whereas a negative oxidation potential indicates that the species is more difficult to oxidize than H_2. Examination of the half-reactions in Table 20.1 shows that F_2 is the most easily reduced species and is consequently the strongest oxidizing agent listed:

$$F_2(g) + 2e^- \longrightarrow 2F^-(aq) \qquad E°_{red} = 2.87 \text{ V}$$

Lithium ion, Li^+, is the most difficult to reduce and is therefore the poorest oxidizing agent:

$$Li^+(aq) + e^- \longrightarrow Li(s) \qquad E°_{red} = -3.05 \text{ V}$$

Among the most frequently encountered oxidizing agents are the halogens, oxygen, and oxyanions such as MnO_4^-, $Cr_2O_7^{2-}$, and NO_3^-, whose central atoms have high positive oxidation states. Metal ions in high positive oxidation states—such as Ce^{4+}, which is readily reduced to Ce^{3+}—are also employed as oxidizing agents.

Equation 20.9 tells us that a half-reaction with a very negative $E°_{red}$ will have a very positive $E°_{ox}$. Thus, among the substances listed in Table 20.1, $Li(s)$ is the most easily oxidized and is consequently the strongest reducing agent:

$$Li(s) \longrightarrow Li^+(aq) + e^- \qquad E°_{ox} = 3.05 \text{ V}$$

Fluoride ion, F^-, is the most difficult to oxidize and is therefore the poorest reducing agent:

$$2F^-(aq) \longrightarrow F_2(g) + 2e^- \qquad E°_{ox} = -2.87 \text{ V}$$

Commonly used reducing agents include H_2 and a variety of metals having positive oxidation potentials (such as Zn and Fe). Some metal ions in low oxidation states—such as Sn^{2+}, which is oxidized to Sn^{4+}—also function as reducing agents. Solutions of reducing agents are difficult to store for extended periods because of the ubiquitous presence of O_2, a good oxidizing agent. For example, developer solutions used in photography are mild reducing agents; they have only a limited shelf life because they are readily oxidized by O_2 from the air.

SAMPLE EXERCISE 20.7

Using Table 20.1, determine which of the following species is the strongest oxidizing agent: MnO_4^- (in acid solution), $I_2(s)$, $Zn^{2+}(aq)$.

Solution: The strongest oxidizing agent will be the species that is most readily reduced. Therefore, we should compare reduction potentials. From Table 20.1 we have

$$MnO_4^-(aq) + 8H^+(aq) + 5e^- \longrightarrow Mn^{2+}(aq) + 4H_2O(l) \qquad E°_{red} = 1.51 \text{ V}$$

$$I_2(s) + 2e^- \longrightarrow 2I^-(aq) \qquad E°_{red} = 0.54 \text{ V}$$

$$Zn^{2+}(aq) + 2e^- \longrightarrow Zn(s) \qquad E°_{red} = -0.76 \text{ V}$$

Because the reduction of MnO_4^- has the highest positive potential, MnO_4^- is the strongest oxidizing agent of the three.

PRACTICE EXERCISE

Using Table 20.1, determine which of the following species is the strongest reducing agent: $F^-(aq)$, $Zn(s)$, $I^-(aq)$. ***Answer:*** $Zn(s)$

20.5 SPONTANEITY AND EXTENT OF REDOX REACTIONS

We have observed that voltaic cells use redox reactions that proceed spontaneously. Conversely, any reaction that can occur in a voltaic cell to produce a positive emf must be spontaneous. Consequently, it is possible to decide whether a redox reaction will be spontaneous by using half-cell potentials to calculate the emf associated with it: *A positive emf indicates a spontaneous process, and a negative emf indicates a nonspontaneous one.*

SAMPLE EXERCISE 20.8

Using the standard electrode potentials listed in Table 20.1, determine whether the following reactions are spontaneous under standard conditions:

(a) $Cu(s) + 2H^+(aq) \longrightarrow Cu^{2+}(aq) + H_2(g)$
(b) $Cl_2(g) + 2I^-(aq) \longrightarrow 2Cl^-(aq) + I_2(s)$

Solution: **(a)** We use Table 20.1 to obtain the necessary half-reaction potentials. Because the overall reaction converts Cu to Cu^{2+}, we use the standard oxidation potential for Cu. Because the overall reaction converts H^+ to H_2, we use the reduction potential for H^+. Adding these, we obtain the standard emf for the overall reaction:

$$\begin{array}{rcll} Cu(s) & \longrightarrow & Cu^{2+}(aq) + 2e^- & E^\circ_{ox} = -0.34\ V \\ 2e^- + 2H^+(aq) & \longrightarrow & H_2(g) & E^\circ_{red} = 0\ V \\ \hline Cu(s) + 2H^+(aq) & \longrightarrow & Cu^{2+}(aq) + H_2(g) & E^\circ = -0.34\ V \end{array}$$

Because the standard emf is negative, the reaction is not spontaneous in the direction written. Copper does not react with acids in this fashion. However, the reverse reaction is spontaneous: Cu^{2+} can be reduced by H_2.

(b) We write equations to obtain the standard emf for the overall reaction:

$$\begin{array}{rcll} 2e^- + Cl_2(g) & \longrightarrow & 2Cl^-(aq) & E^\circ_{red} = 1.36\ V \\ 2I^-(aq) & \longrightarrow & I_2(s) + 2e^- & E^\circ_{ox} = -0.54\ V \\ \hline Cl_2(g) + 2I^-(aq) & \longrightarrow & 2Cl^-(aq) + I_2(s) & E^\circ = 0.82\ V \end{array}$$

This reaction is spontaneous and could be used to build a voltaic cell. It is often used as a qualitative test for the presence of I^- in aqueous solution. The solution is treated with a solution of Cl_2. If I^- is present, I_2 forms. If CCl_4 is added, the I_2 dissolves in the CCl_4, imparting a characteristic violet color to the solution (see Figure 23.2).

PRACTICE EXERCISE

Using the standard electrode potentials listed in Appendix E, determine which of the following reactions are spontaneous under standard conditions:
(a) $I_2(s) + 5Cu^{2+}(aq) + 6H_2O(l) \longrightarrow 2IO_3^-(aq) + 5Cu(s) + 12H^+(aq)$
(b) $Hg^{2+}(aq) + 2I^-(aq) \longrightarrow Hg(l) + I_2(s)$
(c) $H_2SO_3(aq) + 2Mn(s) + 4H^+(aq) \longrightarrow S(s) + 2Mn^{2+}(aq) + 3H_2O(l)$
Answer: Reactions **(b)** and **(c)** are spontaneous.

We can use standard electrode potentials to understand the activity series of metals, which we discussed in Section 4.6. Recall the main rule concerning the activity series: Any metal in the series is able to displace the elements below it from their compounds. We can now recognize the origin of this rule on the basis of standard electrode potentials. The activity series is simply the oxidation half-reactions of the metals ordered from highest to lowest E°_{ox} (see Table 4.4). For example, iron lies above silver in the activity series. We therefore expect iron to displace silver according to the net reaction

$$Fe(s) + 2Ag^+(aq) \longrightarrow Fe^{2+}(aq) + 2Ag(s) \qquad [20.10]$$

By using the data in Table 20.1, we can see that this should be a spontaneous reaction:

$$\begin{array}{rcll} Fe(s) & \longrightarrow & Fe^{2+}(aq) + 2e^- & E^\circ_{ox} = 0.44\ V \\ 2Ag^+(aq) + 2e^- & \longrightarrow & 2Ag(s) & E^\circ_{red} = 0.80\ V \\ \hline Fe(s) + 2Ag^+(aq) & \longrightarrow & Fe^{2+}(aq) + 2Ag(s) & E^\circ = 1.24\ V \end{array}$$

The positive value of E° indicates that Equation 20.10 should occur spontaneously.

Figure 20.8 Michael Faraday (1791–1867) was born in England, one of ten children of a poor blacksmith. At the age of 14, he was apprenticed to a bookbinder who gave the young man time to read and even to attend lectures. In 1812, he became an assistant in Humphry Davy's laboratory in the Royal Institution. He eventually succeeded Davy as the most famous and influential scientist in England. During his scientific career he made an amazing number of important discoveries in chemistry and physics. He developed methods for liquefying gases, discovered benzene, and formulated the quantitative relationships between electrical current and the extent of chemical reaction in electrochemical cells that either produce or use electricity. In addition, he worked out the design of the first electric generator and laid the theoretical foundations for the development of our modern theory of electricity. (Culver Pictures)

Emf and Free-Energy Change

We have seen that the free-energy change, ΔG, accompanying a chemical process is a measure of its spontaneity (Chapter 19). Because the cell emf indicates whether a redox reaction is spontaneous, we might expect some relationship to exist between the cell emf and the free-energy change. Indeed, this is the case; the emf, E, and the free-energy change, ΔG, are related by Equation 20.11:

$$\Delta G = -n\mathfrak{F}E \qquad [20.11]$$

In this equation, n is the number of moles of electrons transferred in the reaction and $\mathfrak{F}$ is Faraday's constant, named after Michael Faraday (Figure 20.8). Faraday's constant is the electrical charge on 1 mol of electrons. This quantity of charge is called a **faraday:**

$$1\ \mathfrak{F} = 96{,}500\ \frac{\text{C}}{\text{mol e}^-} = 96{,}500\ \frac{\text{J}}{\text{V-mol e}^-}$$

Notice that, because n and $\mathfrak{F}$ are both positive quantities, a positive value of E in Equation 20.11 leads to a negative value of ΔG. Remember: A positive value of E and a negative value of ΔG both indicate that a reaction is spontaneous.

When both the reactants and products are in their standard states, Equation 20.11 can be modified to relate ΔG° and E°:

$$\Delta G^\circ = -n\mathfrak{F}E^\circ \qquad [20.12]$$

SAMPLE EXERCISE 20.9

Use the standard electrode potentials given in Table 20.1 to calculate the standard free-energy change, ΔG°, for the following reaction:

$$2Br^-(aq) + F_2(g) \longrightarrow Br_2(l) + 2F^-(aq)$$

Solution:

$$\begin{array}{rcll} 2Br^-(aq) & \longrightarrow & Br_2(l) + 2e^- & E^\circ_{ox} = -1.06\ \text{V} \\ F_2(g) + 2e^- & \longrightarrow & 2F^-(aq) & E^\circ_{red} = 2.87\ \text{V} \\ \hline 2Br^-(aq) + F_2(g) & \longrightarrow & Br_2(l) + 2F^-(aq) & E^\circ = 1.81\ \text{V} \end{array}$$

Notice that two electrons are transferred in the reaction, so $n = 2$.

$$\begin{aligned} \Delta G^\circ &= -n\mathfrak{F}E^\circ \\ &= -(2\ \text{mol e}^-)\left(96{,}500\ \frac{\text{J}}{\text{V-mol e}^-}\right)(1.81\ \text{V}) \\ &= -3.49 \times 10^5\ \text{J} = -349\ \text{kJ} \end{aligned}$$

PRACTICE EXERCISE

Use the standard electrode potentials given in Appendix E to calculate the free-energy change for the following reaction:

$$I_2(s) + 5Cu^{2+}(aq) + 6H_2O(l) \longrightarrow 2IO_3^-(aq) + 5Cu(s) + 12H^+(aq)$$

Answer: $+828$ kJ

Emf and Equilibrium Constant

In Section 19.6, we saw that the standard free-energy change, ΔG°, is related to the equilibrium constant, K:

$$\Delta G^\circ = -RT \ln K \qquad [20.13]$$

This relationship indicates that E° should also be related to the equilibrium constant. Substituting Equation 20.13 into Equation 20.12 gives

$$-n\mathfrak{F}E^\circ = -RT \ln K \qquad [20.14]$$

We shall find it convenient to write Equation 20.14 in terms of common, or base-10, logarithms. Using the relationship discussed in Appendix A, we obtain

$$-n\mathfrak{F}E^\circ = -2.30RT \log K$$

$$E^\circ = \frac{2.30RT}{n\mathfrak{F}} \log K \qquad [20.15]$$

When $T = 298$ K, Equation 20.15 can be simplified by substituting the numerical values for R and $\mathfrak{F}$:

$$E^\circ = \frac{2.30(8.314 \text{ J/K-mol})(298 \text{ K})}{n(96{,}500 \text{ J/V-mol})} \log K$$

$$= \frac{0.0591 \text{ V}}{n} \log K \qquad [20.16]$$

Thus the standard emf generated by a cell increases as the equilibrium constant for the cell reaction increases.

SAMPLE EXERCISE 20.10

Using standard electrode potentials listed in Appendix E, calculate the equilibrium constant at 25°C for the reaction

$$O_2(g) + 4H^+(aq) + 4Fe^{2+}(aq) \longrightarrow 4Fe^{3+}(aq) + 2H_2O(l)$$

Solution:

$$\begin{array}{rcll} O_2(g) + 4H^+(aq) + 4e^- & \longrightarrow & 2H_2O(l) & E^\circ_{red} = 1.23 \text{ V} \\ 4Fe^{2+}(aq) & \longrightarrow & 4Fe^{3+}(aq) + 4e^- & E^\circ_{ox} = -0.77 \text{ V} \\ \hline O_2(g) + 4H^+(aq) + 4Fe^{2+}(aq) & \longrightarrow & 4Fe^{3+}(aq) + 2H_2O(l) & E^\circ = 0.46 \text{ V} \end{array}$$

From the half-reactions given above we see that $n = 4$. Using Equation 20.16, we have

$$\log K = \frac{nE^\circ}{0.0591 \text{ V}}$$

$$= \frac{4(0.46 \text{ V})}{0.0591 \text{ V}}$$

$$= 31.1$$

$$K = 1 \times 10^{31}$$

Thus, Fe^{2+} ions are stable in acidic solutions only in the absence of O_2 (unless a suitable reducing agent is present.)

PRACTICE EXERCISE

Using standard electrode potentials (Appendix E), calculate the equilibrium constant at 25°C for the reaction

$$2IO_3^-(aq) + 5Cu(s) + 12H^+(aq) \longrightarrow I_2(s) + 5Cu^{2+}(aq) + 6H_2O(l)$$

Answer: $K = 1 \times 10^{145}$

EMF and Concentration

In practice, voltaic cells are unlikely to be operated under standard-state conditions. However, the emf generated under nonstandard conditions can be calculated from E°, temperature, and the concentrations of reactants and products in the cell. The equation that allows this calculation can be derived from the relationship between ΔG and ΔG°, given in Section 19.6:

$$\Delta G = \Delta G^\circ + RT \ln Q \qquad [20.17]$$

$$\Delta G = \Delta G^\circ + 2.30RT \log Q \qquad [20.18]$$

In Equation 20.18, we have written Equation 20.17 in terms of common logarithms. Substituting $\Delta G = -n\mathfrak{F}E$ (Equation 20.12) into Equation 20.18 gives

$$-n\mathfrak{F}E = -n\mathfrak{F}E^\circ + 2.30RT \log Q$$

Solving this equation for E gives

$$E = E^\circ - \frac{2.30\,RT}{n\mathfrak{F}} \log Q$$

This relationship is known as the **Nernst equation** after Walther Hermann Nernst (1864–1941), a German chemist who established many of the theoretical foundations of electrochemistry. At 298 K, the quantity $2.30RT/\mathfrak{F}$ is equal to 0.0591 V-mol, so the Nernst equation can be written in the following simplified form:

$$E = E^\circ - \frac{0.0591}{n} \log Q \qquad (T = 298\text{ K}) \qquad [20.19]$$

As an example of how Equation 20.19 might be used, consider the following reaction:

$$Zn(s) + Cu^{2+}(aq) \longrightarrow Zn^{2+}(aq) + Cu(s) \qquad E^\circ = 1.10\text{ V}$$

In this case $n = 2$, and at 298 K the Nernst equation gives

$$E = 1.10\text{ V} - \frac{0.0591\text{ V}}{2} \log \frac{[Zn^{2+}]}{[Cu^{2+}]} \qquad [20.20]$$

Recall that Q includes expressions for species in solution but not for solids. Experimentally it is found that the emf generated by a cell is independent of the size or shape of the solid electrodes used. From Equation 20.20, it is evident that the emf of a cell based on this chemical reaction increases as $[Cu^{2+}]$ increases and as $[Zn^{2+}]$ decreases. For example, when $[Cu^{2+}]$ is 5.0 *M* and $[Zn^{2+}]$ is 0.050 *M*, we have

$$E = 1.10 \text{ V} - \frac{0.0591 \text{ V}}{2} \log\left(\frac{0.050}{5.0}\right)$$

$$= 1.10 \text{ V} - \frac{0.0591 \text{ V}}{2}(-2.00) = 1.16 \text{ V}$$

The fact that E (1.16 V) is greater than E° (1.10 V) indicates that, at these concentrations, the driving force for the reaction is greater than under standard conditions. We could have anticipated this result by applying Le Châtelier's principle (Section 15.6). If the concentrations of reactants increase relative to the concentrations of products, the cell reaction becomes more highly spontaneous, and the emf increases. Conversely, if the concentrations of products increase relative to reactants, emf decreases. As a cell operates, reactants are consumed and products form. The decreases in reactant concentrations and increases in product concentrations cause the emf to decrease.

SAMPLE EXERCISE 20.11

Calculate the emf generated by the cell described in Sample Exercise 20.4 when $[Cr_2O_7^{2-}] = 2.0\ M$, $[H^+] = 1.0\ M$, $[I^-] = 1.0\ M$, and $[Cr^{3+}] = 1.0 \times 10^{-5}\ M$:

$$Cr_2O_7^{2-}(aq) + 14H^+(aq) + 6I^-(aq) \longrightarrow 2Cr^{3+}(aq) + 3I_2(s) + 7H_2O(l)$$

Solution: The standard emf for this reaction was calculated in Sample Exercise 20.6: $E^\circ = 0.79$ V. As you will see if you refer back to that exercise, n is 6. The reaction quotient, Q, is

$$Q = \frac{[Cr^{3+}]^2}{[Cr_2O_7^{2-}][H^+]^{14}[I^-]^6} = \frac{(1.0 \times 10^{-5})^2}{(2.0)(1.0)^{14}(1.0)^6} = 5.0 \times 10^{-11}$$

Using Equation 20.19, we have

$$E = 0.79 \text{ V} - \frac{0.0591 \text{ V}}{6} \log(5.0 \times 10^{-11})$$

$$= 0.79 \text{ V} - \frac{0.0591 \text{ V}}{6}(-10.30)$$

$$= 0.79 \text{ V} + 0.10 \text{ V}$$

$$= 0.89 \text{ V}$$

This result is qualitatively what we expect: Because the concentration of $Cr_2O_7^{2-}$ (a reactant) is above 1 *M* and the concentration of Cr^{3+} (a product) is below 1 *M*, the emf is greater than E°.

PRACTICE EXERCISE

Calculate the emf generated by the cell described in the practice exercise accompanying Sample Exercise 20.6 when $[Al^{3+}] = 4.0 \times 10^{-3}\ M$ and $[I^-] = 0.010\ M$.
Answer: $E = 2.36$ V

SAMPLE EXERCISE 20.12

If the measured voltage in a Zn–H^+ cell (such as that shown in Figure 20.7) is 0.45 V at 25°C when $[Zn^{2+}]$ is 1 *M* and P_{H_2} is 1 atm, what is the concentration of H^+?

Solution: The cell reaction is

$$Zn(s) + 2H^+(aq) \longrightarrow Zn^{2+}(aq) + H_2(g)$$

The standard emf is $E° = 0.76$ V. Applying Equation 20.19 with $n = 2$ gives

$$0.45 = 0.76 - \frac{0.0591}{2} \log \frac{[Zn^{2+}]P_{H_2}}{[H^+]^2}$$

$$= 0.76 - \frac{0.0591}{2} \log \frac{1}{[H^+]^2}$$

Because

$$\log \frac{1}{x^2} = -\log x^2 = -2 \log x$$

we can write

$$0.45 = 0.76 - \frac{0.0591}{2}(-2 \log [H^+])$$

Solving for log $[H^+]$ gives us

$$\log [H^+] = \frac{0.45 - 0.76}{0.0591} = -5.25$$

$$[H^+] = 10^{-5.25} = 5.6 \times 10^{-6}\ M$$

This example shows how a voltaic cell whose cell reaction involves H^+ can be used to measure $[H^+]$ or pH. A pH meter (Section 16.2) is merely a specially designed voltaic cell with a voltmeter calibrated to read pH directly.

PRACTICE EXERCISE

What is the pH of the solution in the cathode compartment of the cell pictured in Figure 20.7 when $P_{H_2} = 1$ atm, $[Zn^{2+}]$ in the anode compartment is 0.10 M, and cell emf is 0.542 V? ***Answer:*** pH = 4.19

20.6 COMMERCIAL VOLTAIC CELLS

Voltaic cells are widely used as convenient energy sources whose primary virtue is portability. Although any spontaneous redox reaction can serve as the basis of a voltaic cell, making a commercial cell that utilizes a particular redox reaction can require considerable ingenuity. The salt-bridge cells that we have been discussing provide us with considerable insight into the operation of voltaic cells. However, these cells are generally unsuitable for commercial use because they have high internal resistances. This means that the flow of current within the cell, due to movement of the ions within the cell compartments and the electrolyte bridge, is restricted. Because the flow is restricted, there is resistance. As a result, if we attempt to draw a large current, voltage drops sharply. Furthermore, the cells that we have pictured so far lack the compactness and ruggedness required for portability.

Voltaic cells cannot yet compete with other common energy sources on the basis of cost alone. The cost of electricity from a common flashlight battery is on the order of $80 per kilowatt-hour. By comparison, electrical energy from power plants normally costs the consumer less than 10¢ per kilowatt-hour.

In this section, we shall consider some common batteries (Figure 20.9). A *battery* consists of one or more voltaic cells. When the cells are connected in series (that is, with the positive terminal of one attached to the negative terminal of another), the battery produces an emf that is the sum of the emfs of the individual cells.

Figure 20.9 Batteries are voltaic cells that serve as portable sources of electricity. Batteries vary markedly in size and in the electrochemical reaction used to generate electricity. The large battery in back is a lead-based automobile battery. Those in front are, from left to right, rechargeable nickel-cadmium batteries, alkaline cells, and a conventional zinc-graphite dry cell. (© Richard Megna/ Fundamental Photographs)

Lead Storage Battery

One of the most common batteries is the lead storage battery used in automobiles. A 12-V lead storage battery consists of six cells, each producing 2 V. The anode of each cell is composed of lead; the cathode is composed of lead dioxide, PbO_2, packed on a metal grid. Both electrodes are immersed in sulfuric acid. The electrode reactions that occur during discharge are as follows:

$$\text{Anode:} \quad Pb(s) + SO_4^{2-}(aq) \longrightarrow PbSO_4(s) + 2e^- \qquad E^\circ = +0.356\ \text{V}$$

$$\text{Cathode:} \quad PbO_2(s) + SO_4^{2-}(aq) + 4H^+(aq) + 2e^- \longrightarrow PbSO_4(s) + 2H_2O(l) \qquad E^\circ = +1.685\ \text{V}$$

$$Pb(s) + PbO_2(s) + 4H^+(aq) + 2SO_4^{2-}(aq) \longrightarrow 2PbSO_4(s) + 2H_2O(l) \qquad E^\circ = +2.041\ \text{V} \qquad [20.21]$$

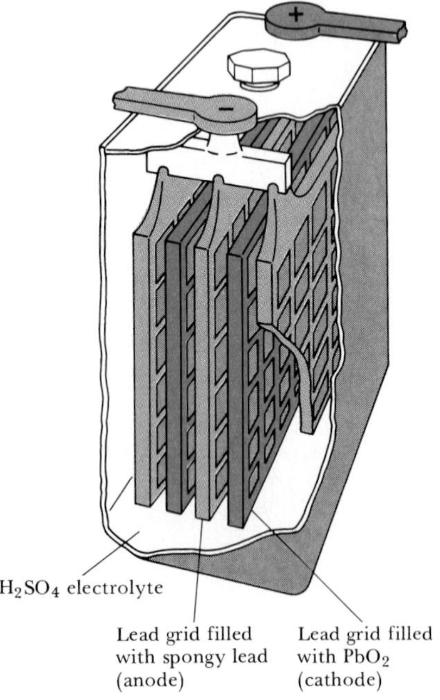

Figure 20.10 Lead storage cell.

The reactants Pb and PbO_2, between which electron transfer occurs, serve as the electrodes. Because the reactants are solids, there is no need to separate the cell into anode and cathode compartments; the Pb and PbO_2 cannot come into direct physical contact unless one electrode plate touches another. To keep the electrodes from touching, wood or glass-fiber spacers are placed between them. To increase the current output, each cell contains a number of anode and cathode plates, as shown in Figure 20.10.

The cell emf of a lead storage battery varies during use, because the concentration of H_2SO_4 varies with the extent of cell discharge. As Equation 20.21 indicates, H_2SO_4 is used up during the discharge of a lead storage battery.

One advantage of a lead storage battery is that it can be recharged. During recharging, an external source of energy is used to reverse the direction of the spontaneous redox reaction, Equation 20.21. Thus, the overall process during charging is as follows:

$$2PbSO_4(s) + 2H_2O(l) \longrightarrow Pb(s) + PbO_2(s) + 4H^+(aq) + 2SO_4^{2-}(aq)$$

The energy necessary for recharging the battery is provided in an automobile by a generator driven by the engine. Recharging is possible because $PbSO_4$ formed during discharge adheres to the electrodes. Thus, as the external source forces electrons from one electrode to another, the $PbSO_4$ is converted to Pb at one electrode and to PbO_2 at the other; these, of course, are the materials of a fully charged cell.

A problem can arise if the battery is charged too rapidly: Water may be decomposed into H_2 and O_2. Besides the explosive potential of H_2–O_2 mixtures, this secondary reaction can shorten the lifetime of the battery. The evolution of these gases can dislodge Pb, PbO_2, or $PbSO_4$ from the plates. These solids can then accumulate as a sludge at the bottom of the battery. In time, they may form a short circuit that renders the cell useless.

The problem of short-circuiting can be substantially reduced by adding calcium (about 0.07 percent by weight) to the lead in forming the electrodes. The presence of the calcium reduces the extent to which water is decomposed during the charging cycle. In newer batteries, electrolysis of water is sufficiently low so that the battery is "sealed"; that is, no provision is made for addition of water and escape of gases, as was necessary in the older designs.

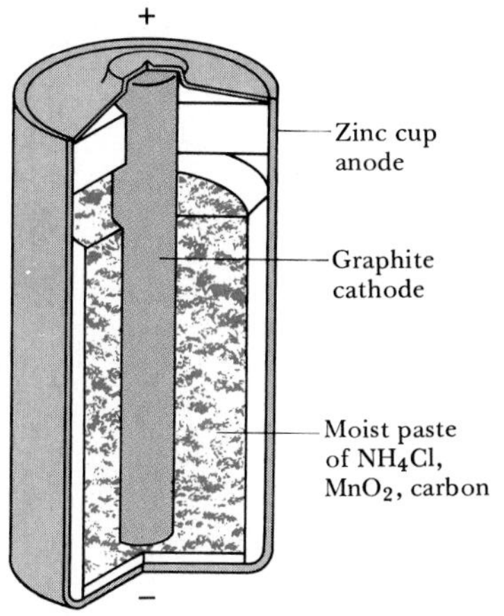

+
Zinc cup anode
Graphite cathode
Moist paste of NH_4Cl, MnO_2, carbon
–

Figure 20.11 Cutaway view of a dry cell.

Dry Cell

The common dry cell is widely used in flashlights, portable radios, and the like. In fact, a dry cell is often called a flashlight battery. It is also known as the Leclanché cell after its inventor, who patented it in 1866. In the acid version, the anode consists of a zinc can in contact with a paste of MnO_2, NH_4Cl, and carbon. An inert cathode, consisting of a graphite rod, is immersed in the center of the paste, as shown in Figure 20.11. The cell has an exterior layer of cardboard or metal to seal it against the atmosphere. The electrode reactions are complex, and the cathode reaction appears to vary with the rate of discharge. The reactions at the electrodes are generally represented as shown below:

$$\text{Anode:} \qquad Zn(s) \longrightarrow Zn^{2+}(aq) + 2e^-$$

$$\text{Cathode:} \qquad 2NH_4^+(aq) + 2MnO_2(s) + 2e^- \longrightarrow Mn_2O_3(s) + 2NH_3(aq) + H_2O(l)$$

Only a fraction of the cathode material, that near the electrode, is electrochemically active because of the limited mobility of the chemicals in the cell.

In the alkaline version of a dry cell, NH_4Cl is replaced by KOH. The anode reaction still involves oxidation of Zn, but the zinc is present as a powder, mixed with the electrolyte in a gel formulation. As in a common dry cell, the cathode reaction involves reduction of MnO_2. Figure 20.12 shows a cutaway view of a miniature alkaline cell of the type used in camera exposure controls, calculators, and some watches. Although more costly than the common dry

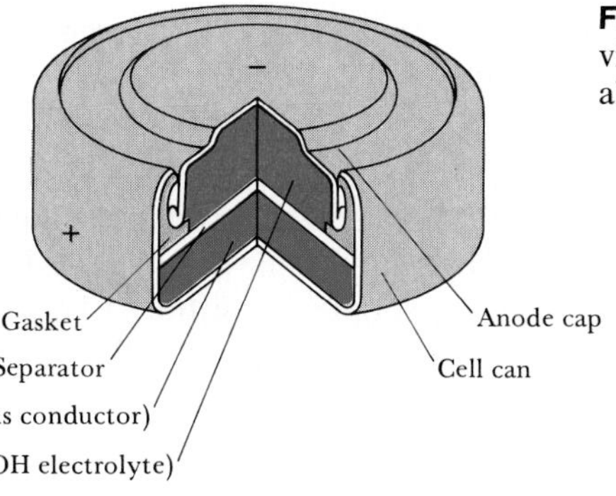

Figure 20.12 Cutaway view of a miniature alkaline Zn–MnO_2 dry cell.

cell, alkaline cells provide improved performance. They maintain usable voltage over a longer fraction of consumption of anode and cathode materials, and they provide up to 50 percent more total energy than a common dry cell of the same size.

Nickel-Cadmium Batteries

Because dry cells are not rechargeable, they have to be replaced frequently. For this reason, a rechargeable cell, the nickel-cadmium battery, has become increasingly popular, especially for use in battery-operated tools and calculators. Cadmium metal acts as the anode, and $NiO_2(s)$, which is reduced to $Ni(OH)_2(s)$, serves as the cathode. The following electrode reactions occur within this cell during discharge:

$$\text{Anode:} \quad Cd(s) + 2OH^-(aq) \longrightarrow Cd(OH)_2(s) + 2e^-$$

$$\text{Cathode:} \quad NiO_2(s) + 2H_2O(l) + 2e^- \longrightarrow Ni(OH)_2(s) + 2OH^-(aq)$$

As in the lead storage cell, the solid reaction products adhere to the electrodes. This permits the reactions to be readily reversed during charging. Because no gases are produced during either charging or discharging, the battery can be sealed.

Fuel Cells

Many substances are used as fuels. The thermal energy released by combustion is often converted to electrical energy. The heat may convert water to steam, which drives a turbine that in turn drives the generator. Typically a maximum of only 40 percent of the energy from combustion is converted to electricity; the remainder is lost as heat. The *direct* production of electricity from fuels by a voltaic cell could, in principle, yield a higher rate of conversion of the chemical energy of the reaction. Voltaic cells that perform this conversion using conventional fuels, such as H_2 and CH_4, are called **fuel cells**.

A great deal of research has gone into attempts to develop practical fuel cells. One of the problems encountered is the high operating temperature of most cells, which not only siphons off energy but also accelerates corrosion of cell parts. A low-temperature cell has been developed that uses H_2, but the present cost of the cell makes it too expensive for large-scale use. However, it has been used in special situations, such as in space vehicles. For example, a H_2–O_2 fuel cell was used as the primary source of electrical energy on the *Apollo* moon flights. The weight of the fuel cell sufficient for 11 days in space was approximately 500 lb. This may be compared to the several tons that would have been required for an engine-generator set.

The electrode reactions in the H_2–O_2 fuel cell are as follows:

$$\begin{array}{lrcl} \text{Anode:} & 2H_2(g) + 4OH^-(aq) & \longrightarrow & 4H_2O(l) + 4e^- \\ \text{Cathode:} & 4e^- + O_2(g) + 2H_2O(l) & \longrightarrow & 4OH^-(aq) \\ \hline & 2H_2(g) + O_2(g) & \longrightarrow & 2H_2O(l) \end{array}$$

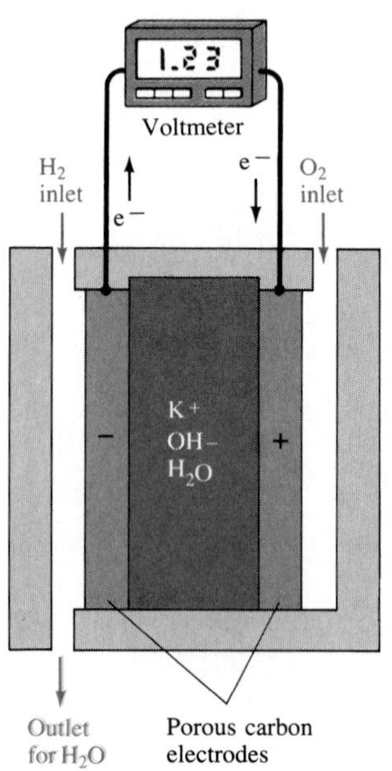

Figure 20.13 Cross section of a H_2–O_2 fuel cell.

The cell is illustrated in Figure 20.13. The electrodes are composed of hollow tubes of porous, compressed carbon impregnated with catalyst; the electrolyte is KOH. Because the reactants are supplied continuously, a fuel cell does not "go dead."

CHEMISTRY AT WORK: Solid-Electrolyte Batteries

All the batteries we have discussed so far have contained an aqueous solution or water-based paste as the electrolyte medium connecting the two half-cells. In recent years, much research effort has been devoted to developing batteries in which the electrolyte, or ion carrier, within the battery is a solid. A solid capable of conducting a current through the motion of ions within it is called a *solid electrolyte* or *fast-ion conductor.*

The existence of solid ionic conductors has been known for a long time. Figure 20.14 shows schematically an experiment performed in 1910 by the German scientist C. Tubandt. It was known that when a pair of electrodes is placed in contact with solid silver iodide, AgI, the solid readily conducts a current. However, it was not known whether the current was carried by electrons moving into and through the solid or by the movement of ions. Tubandt placed Ag electrodes of known mass on each end of a block of AgI. He allowed current to flow for a period of time and measured the total number of faradays of electricity passed. At the end of the experiment, he weighed the two Ag electrodes. One had gained weight; the other had lost an equal amount of weight. The weight changes were consistent with the reduction of Ag^+ at the cathode, adding to the weight of the cathode, and with the oxidation of Ag at the anode, corresponding to the weight loss at this electrode. For these electrode processes to occur, Ag^+ ions produced at the anode must travel through the solid to the cathode to be reduced there. Thus, AgI is an example of a solid electrolyte.

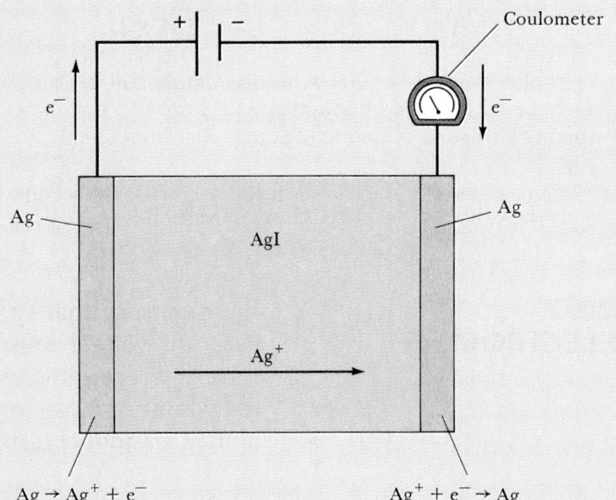

Figure 20.14 Tubandt's classic experiment showing that silver iodide, AgI, conducts electrical current via movement of Ag^+ ions in the solid.

Figure 20.15 shows a schematic diagram of a battery incorporating a solid electrolyte. The anode reaction is the oxidation of lithium:

$$Li(s) \longrightarrow Li^+(s) + e^-$$

The cathode reaction is the reduction of TiS_2:

$$TiS_2(s) + e^- \longrightarrow TiS_2^-(s)$$

The cell potential for these reactions is greater than 2 V. Electrons released at the anode flow through the external circuit to the cathode. The cathode electrode, TiS_2, is a solid electrolyte of a special type, called an *ion-insertion compound.* It is a good electronic conductor as well as an ionic conductor. When TiS_2 is reduced, lithium ions

Figure 20.15 A solid electrolyte battery. Lithium metal is the anode, and TiS_2 is the cathode. Lithium ions are conducted within the battery by movement through a solid polymer electrolyte and through the cathode material, where $LiTiS_2$ is formed.

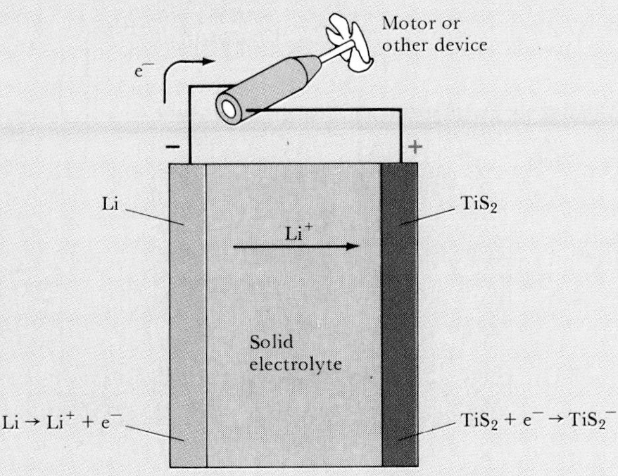

move into the electrode from the solid electrolyte to form $LiTiS_2$:*

$$Li^+(s) + TiS_2(s) + e^- \longrightarrow LiTiS_2(s)$$

The solid electrolyte separating the anode and cathode must be capable of conducting Li^+ ions but not conducting electrons. (Conducting electrons would create an internal short circuit of the external path.) The most promising materials in use today are solid polymers (Chapter 12) that permit the flow of ions through the material.

A battery such as the one outlined in Figure 20.15 could be recharged in the same way as a nickel-cadmium or lead-acid cell. The reactions for recharging the half-cells are the reverse of the above reactions. The big advantage expected for solid electrolyte batteries over other cells is that they will have from two to five times as much energy per unit weight and volume.

* This description is a bit oversimplified chemically, but the principle is correct.

20.7 ELECTROLYSIS

We have seen that spontaneous oxidation-reduction reactions are used as the basis for voltaic cells, electrochemical devices that generate electricity. Conversely, it is possible to use electrical energy to cause nonspontaneous oxidation-reduction reactions to occur. For example, electricity can be used to decompose molten sodium chloride into its component elements:

$$2NaCl(l) \longrightarrow 2Na(l) + Cl_2(g)$$

Such processes, which are driven by an outside source of electrical energy, are called **electrolysis** reactions and take place in **electrolytic cells**.

An electrolytic cell consists of two electrodes in a molten salt or aqueous solution. The cell is driven by a battery or some other source of direct electrical current. The battery acts as an electron pump, pushing electrons into one electrode and pulling them from the other. Withdrawing electrons from an electrode gives it a positive charge, and adding electrons to an electrode makes it negative. In the electrolysis of molten NaCl, shown in Figure 20.16, Na^+ ions pick up electrons at the negative electrode and are thereby reduced. As the Na^+ ions in the vicinity of this electrode are depleted, additional Na^+ ions migrate in. In a related fashion, there is a net movement of Cl^- ions to the positive electrode, where they give up electrons and are thereby oxidized. Just as in voltaic cells, the electrode at which reduction occurs is called the cathode, and the electrode at which oxidation occurs is called the anode.

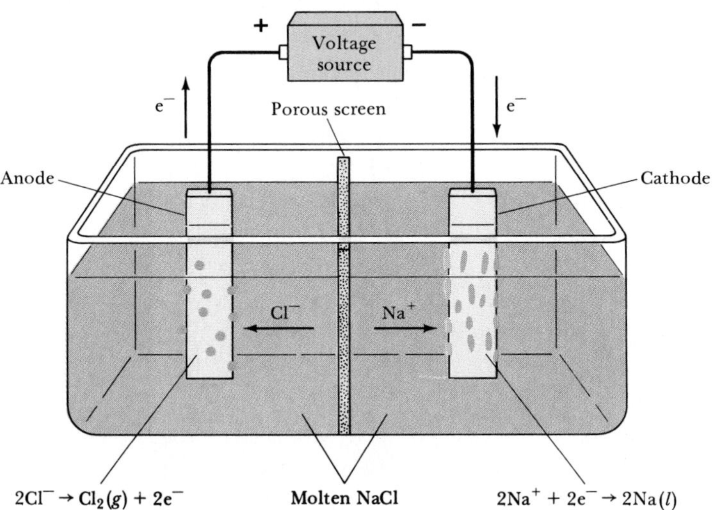

Figure 20.16 Electrolysis of molten sodium chloride.

$$\begin{array}{llcl}
\text{Anode:} & 2Cl^-(l) & \longrightarrow & Cl_2(g) + 2e^- \\
\text{Cathode:} & 2Na^+(l) + 2e^- & \longrightarrow & 2Na(l) \\
\hline
 & 2Na^+(l) + 2Cl^-(l) & \longrightarrow & 2Na(l) + Cl_2(g)
\end{array}$$

Notice that the sign convention for the electrodes in an electrolytic cell is just the opposite of that for a voltaic cell. The cathode in the electrolytic cell is negative because electrons are being forced onto it by the external voltage source. The anode is positive because electrons are being withdrawn by the external source.

Electrolyses of molten salts and molten-salt solutions for the production of active metals such as sodium and aluminum are important industrial processes. We shall have more to say about them in Chapter 24, when we discuss processes for obtaining metals.

Electrolysis of Aqueous Solutions

Sodium cannot be prepared by electrolysis of aqueous solutions of NaCl, because water is more easily reduced than $Na^+(aq)$:

$$2H_2O(l) + 2e^- \longrightarrow H_2(g) + 2OH^-(aq) \qquad E^\circ_{red} = -0.83\ V$$

$$Na^+(aq) + e^- \longrightarrow Na(s) \qquad E^\circ_{red} = -2.71\ V$$

Consequently, H_2 is produced at the cathode.

The possible anode reactions are the oxidation of Cl^- and of H_2O:

$$2Cl^-(aq) \longrightarrow Cl_2(g) + 2e^- \qquad E^\circ_{ox} = -1.36\ V$$

$$2H_2O(l) \longrightarrow 4H^+(aq) + O_2(g) + 4e^- \qquad E^\circ_{ox} = -1.23\ V$$

These standard oxidation potentials are not greatly different, but they do suggest that H_2O should be oxidized more readily than Cl^-. However, the actual voltage required for a reaction is sometimes much greater than the theoretical voltage based on the electrode potentials. The additional voltage required to cause electrolysis is called the *overvoltage*. The overvoltage is believed to be caused by slow reaction rates at the electrodes. Overvoltages for the deposition of metals are low, but those required for the liberation of hydrogen gas or oxygen gas are usually high. In the present case, the overvoltage for O_2 formation is sufficiently high to permit oxidation of Cl^- rather than H_2O. Consequently, electrolysis of aqueous solutions of NaCl, known as brines, produces H_2 and Cl_2 unless the concentration of Cl^- is quite low:

$$\begin{array}{llcl}
\text{Anode:} & 2Cl^-(aq) & \longrightarrow & Cl_2(g) + 2e^- \\
\text{Cathode:} & 2H_2O(l) + 2e^- & \longrightarrow & H_2(g) + 2OH^-(aq) \\
\hline
 & 2Cl^-(aq) + 2H_2O(l) & \longrightarrow & Cl_2(g) + H_2(g) + 2OH^-(aq)
\end{array}$$

The Na^+ ion is merely a spectator ion (Section 4.4) in the electrolysis. This process is used commercially because all of the products (H_2, Cl_2, and NaOH) are commercially important chemicals. The chapter-opening photograph shows a chemical plant that uses this process.

Electrode potentials can be used to determine the minimum emf required for an electrolysis. In the case of the formation of H_2 and Cl_2 from a brine solution under standard conditions, a minimum emf of 2.19 V is required.

$$E^\circ = E^\circ_{ox}(Cl^-) + E^\circ_{red}(H_2O)$$
$$= -1.36 \text{ V} + (-0.83 \text{ V}) = -2.19 \text{ V}$$

The emf calculated above is negative, reminding us that the process is not spontaneous but must be driven by an outside source of energy. Higher voltages than those calculated are invariably needed. One reason is the internal resistance of the cell; another is the overvoltage phenomenon discussed above.

SAMPLE EXERCISE 20.13

Explain why the electrolysis of an aqueous solution of $CuCl_2$ produces Cu(*s*) and $Cl_2(g)$. What is the minimum emf required for this process under standard conditions?

Solution: At the cathode we can envision reduction of either Cu^{2+} or H_2O:

$$Cu^{2+}(aq) + 2e^- \longrightarrow Cu(s) \qquad E^\circ_{red} = 0.34 \text{ V}$$
$$2H_2O(l) + 2e^- \longrightarrow H_2(g) + 2OH^-(aq) \qquad E^\circ_{red} = -0.83 \text{ V}$$

The electrode potentials indicate that reduction of Cu^{2+} occurs more readily. Reduction of H_2O is made even more difficult because of the overvoltage for H_2 formation.

At the anode, we can envision the oxidation of either Cl^- or H_2O. As in the case of NaCl solutions, Cl_2 is usually produced because of the overvoltage for O_2 formation.

The minimum emf required for this electrolysis under standard conditions is 1.02 V:

$$E^\circ = E^\circ_{red}(Cu^{2+}) + E^\circ_{ox}(Cl^-)$$
$$= 0.34 \text{ V} + (-1.36 \text{ V}) = -1.02 \text{ V}$$

PRACTICE EXERCISE

(a) What are the expected products of electrolysis of a 1.0 *M* aqueous HBr solution? **(b)** What is the minimum emf required to produce these products?
Answers: **(a)** at the cathode, $H_2(g)$; at the anode, $Br_2(l)$; **(b)** -1.06 V

Electrolysis with Active Electrodes

In our discussion of the electrolysis of molten NaCl and of NaCl solutions, we considered the electrodes to be inert. Consequently, they did not undergo reaction but merely served as the surface at which oxidation and reduction of solvent or solute occurred. However, the electrodes themselves often participate in the electrolysis process.

When aqueous solutions are electrolyzed using metal electrodes, an electrode will be oxidized if its oxidation potential is more positive than that for water. For example, nickel is oxidized more readily than water:

$$Ni(s) \longrightarrow Ni^{2+}(aq) + 2e^- \qquad E^\circ_{ox} = +0.28 \text{ V}$$
$$2H_2O(l) \longrightarrow 4H^+(aq) + O_2(g) + 4e^- \qquad E^\circ_{ox} = -1.23 \text{ V}$$

If nickel is made the anode in an electrolytic cell, nickel metal is oxidized as the anode reaction. If Ni^{2+} (*aq*) is present in the solution, it is reduced at the cathode in preference to reduction of water. An electrolytic cell of this kind is illustrated in Figure 20.17. As current flows, nickel dissolves from the anode and deposits on the cathode:

Anode: $$Ni(s) \longrightarrow Ni^{2+}(aq) + 2e^-$$
Cathode: $$Ni^{2+}(aq) + 2e^- \longrightarrow Ni(s)$$

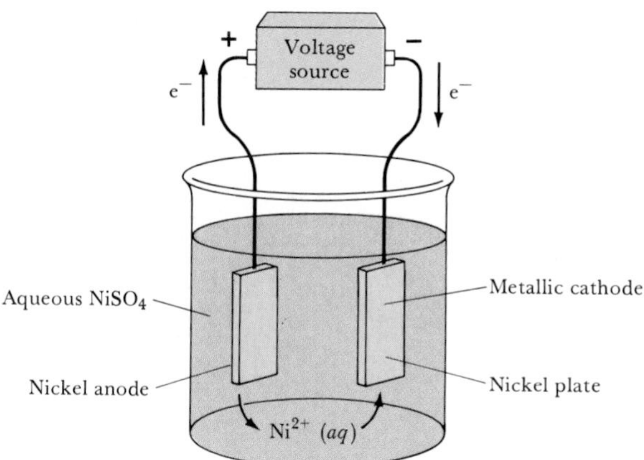

Figure 20.17 Electrolytic cell with an active metal electrode. Nickel dissolves from the anode to form $Ni^{2+}(aq)$. At the cathode, $Ni^{2+}(aq)$ is reduced and forms a nickel "plate."

Electrolytic processes involving active metal electrodes—that is, in which the metal electrodes participate in the cell reaction—have several important applications. We will see in Chapter 24 that electrolysis affords a means of purifying crude metals such as copper, zinc, cobalt, and nickel. An additional important application is in *electroplating*, in which one metal is "plated," or deposited, on another. Electroplating is used to protect objects against corrosion and to improve their appearance. For example, fine dinnerware is silver-plate electrochemically by making the dinnerware utensils the cathode in an electrolytic plating bath (Figure 20.18).

Figure 20.18 Electroplating of silverware. Part (*a*) shows the silverware being withdrawn from the electroplating bath. In part (*b*) we see the polished final product. (Courtesy Oneida Silversmiths)

(*a*)

(*b*)

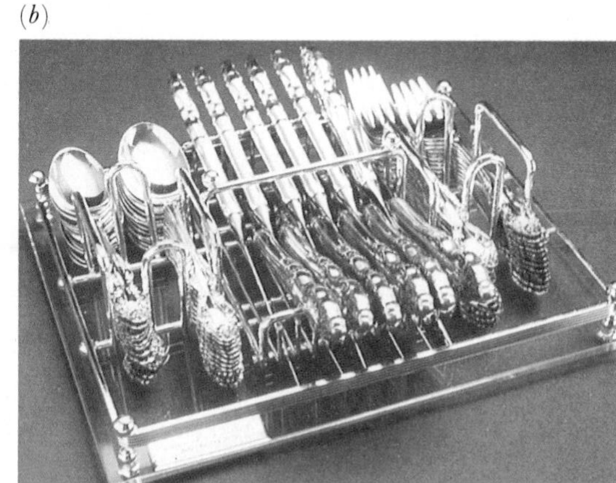

20.8 QUANTITATIVE ASPECTS OF ELECTROLYSIS

The stoichiometry of a half-reaction tells us how many electrons are needed to achieve an electrolytic process. For example, the reduction of Na^+ to Na is a one-electron process:

$$Na^+ + e^- \longrightarrow Na$$

Thus, 1 mol of electrons will plate out 1 mol of Na metal, 2 mol of electrons will plate out 2 mol of Na metal, and so forth. Similarly, 2 mol of electrons are required to produce 1 mol of copper from Cu^{2+}, and 3 mol of electrons are required to produce 1 mol of aluminum from Al^{3+}:

$$Cu^{2+} + 2e^- \longrightarrow Cu$$

$$Al^{3+} + 3e^- \longrightarrow Al$$

For any half-reaction, the amount of a substance that is reduced or oxidized in an electrolytic cell is directly proportional to the number of electrons passed into the cell.

The quantity of charge passing through an electrical circuit, such as that in an electrolytic cell, is generally measured in *coulombs*. As noted in Section 20.5, there are 96,500 C in a faraday:

$$1\ \mathfrak{F} = 96{,}500\ C = \text{charge of 1 mol of electrons}$$

In terms of other, perhaps more familiar electrical units, a coulomb is the quantity of electrical charge passing a point in a circuit in 1 s when the current is 1 ampere (A).* Therefore, the number of coulombs passing through a cell can be obtained by multiplying the amperage and the elapsed time in seconds:

$$\text{Coulombs} = \text{amperes} \times \text{seconds} \qquad [20.22]$$

These ideas are applied in Sample Exercise 20.14. Although the exercise involves electrolytic cells, the same relationships can be applied to voltaic cells.

SAMPLE EXERCISE 20.14

Calculate the mass of aluminum produced in 1.00 hr by the electrolysis of molten $AlCl_3$ if the current is 10.0 A.

Solution: Using Equation 20.22, we can write

$$\text{Coulombs} = (10.0\ A)(1.00\ hr)\left(\frac{3600\ s}{1\ hr}\right)\left(\frac{1\ C}{1\ A\text{-}s}\right)$$

$$= 3.60 \times 10^4\ C$$

The half-reaction for the reduction of Al^{3+} is

$$Al^{3+} + 3e^- \longrightarrow Al$$

The amount of aluminum produced depends on the number of available electrons: 1 mol Al $\simeq$ 3 $\mathfrak{F}$. We can therefore write

$$\text{Grams Al} = (3.60 \times 10^4\ C)\left(\frac{1\ \mathfrak{F}}{96{,}500\ C}\right)\left(\frac{1\ mol\ Al}{3\ \mathfrak{F}}\right)\left(\frac{27.0\ g\ Al}{1\ mol\ Al}\right) = 3.36\ g$$

* Conversely, current is the rate of flow of electricity. An ampere is the current associated with the flow of 1 C past a point each second.

PRACTICE EXERCISE

The half-reaction for formation of magnesium metal upon electrolysis of molten $MgCl_2$ is $Mg^{2+} + 2e^- \rightarrow Mg$. Calculate the mass of magnesium formed upon passage of a current of 60.0 A for a period of 4.00×10^3 s. ***Answer:*** 30.2 g of Mg

Electrical Work

This is a good point at which to consider the relationship between electrochemical processes and work. We have already seen that a positive value for E is associated with a negative value for the free-energy change and thus with a spontaneous process. We also know that for any spontaneous process, ΔG is a measure of the maximum useful work, w_{max}, that can be extracted from the process: $\Delta G = w_{max}$ (Section 19.7). Because $\Delta G = -n\mathfrak{F}E$, the maximum useful electrical work obtainable from a voltaic cell should be simply

$$w_{max} = -n\mathfrak{F}E \qquad [20.23]$$

A word about sign conventions is in order here. Remember that work done *by* the system *on* its surroundings is indicated by a negative sign for w (see Section 5.2). Therefore, a negative value for w_{max} corresponds to a spontaneous process, for which E is positive.

Notice that the maximum work obtainable is proportional to the cell potential E. We can think of E as a kind of pressure, a measure of the driving force for the process. Recall from Section 20.4 that the units of E are J/C; w_{max} is also proportional to the number of coulombs that flow, measured by $n\mathfrak{F}$. When we put these quantities together, we can see how the units cancel to leave us with units of energy:

$$w_{max} = -n \times \mathfrak{F} \times E$$

$$(\text{J}) = (\text{mol}) \times \left(\frac{\text{C}}{\text{mol}}\right) \times \left(\frac{\text{J}}{\text{C}}\right)$$

In an electrolytic cell, we employ an external source of energy to bring about a nonspontaneous electrochemical process. In this case, ΔG for the cell process is positive, and the cell potential is negative. The *minimum* amount of work done on the system to cause the nonspontaneous cell reaction to occur, w_{min}, is given by

$$w_{min} = -n\mathfrak{F}E$$

where E is the calculated cell potential. In practice, we always need to expend more than this minimum amount of work because of inefficiencies in the process. In addition, a higher potential E' may be required to cause the cell reaction to occur, because of overvoltage. In this case, the minimum work required is given by

$$w_{min} = -n\mathfrak{F}E' \qquad [20.24]$$

Electrical work is usually expressed in energy units of watts times time. The *watt* is a unit of electrical power, that is, the rate of energy expenditure:

$$1 \text{ watt (W)} = \frac{1 \text{ J}}{\text{s}}$$

Therefore, a watt-second is a joule. The unit employed by electric utilities is the kilowatt-hour, which works out to be 3.6×10^6 J:

$$1 \text{ kWh} = (1000 \text{ W})(1 \text{ hr})\left(\frac{3600 \text{ s}}{1 \text{ hr}}\right)\left(\frac{1 \text{ J/s}}{1 \text{ W}}\right) = 3.6 \times 10^6 \text{ J} \qquad [20.25]$$

Using these considerations, we can calculate the maximum work obtainable from voltaic cells and the minimum work required to bring about desired electrolysis reactions.

SAMPLE EXERCISE 20.15

Calculate the minimum number of kilowatt-hours of electricity required to produce 1000 kg of aluminum by electrolysis of Al^{3+} if the required emf is 4.50 V.

Solution: We need to employ Equation 20.24 to calculate w_{min} for an applied potential of 4.50 V. First we need to calculate $n\mathfrak{F}$, the number of coulombs required.

$$\text{Coulombs} = (1000 \text{ kg Al})\left(\frac{1000 \text{ g Al}}{1 \text{ kg Al}}\right)\left(\frac{1 \text{ mol Al}}{27.0 \text{ g Al}}\right)\left(\frac{3\ \mathfrak{F}}{1 \text{ mol Al}}\right)\left(\frac{96{,}500 \text{ C}}{1\ \mathfrak{F}}\right)$$
$$= 1.07 \times 10^{10} \text{ C}$$

We can now employ Equation 20.24 to calculate w_{min}. In doing so, we must apply the unit conversion factor of Equation 20.25:

$$\text{Kilowatt-hours} = (1.07 \times 10^{10} \text{ C})(4.50 \text{ V})\left(\frac{1 \text{ J}}{1 \text{ C-V}}\right)\left(\frac{1 \text{ kWh}}{3.6 \times 10^6 \text{ J}}\right)$$
$$= 1.34 \times 10^4 \text{ kWh}$$

(The negative sign on the right in Equation 20.24 is canceled by the negative sign for E' because it is a voltage applied to make the cell reaction occur.)

This quantity of energy does not include the energy used to mine, transport, and process the aluminum ore, and to keep the electrolysis bath molten during electrolysis. A typical electrolytic cell used to reduce aluminum is only 40 percent efficient, 60 percent of the electrical energy being dissipated as heat. It therefore requires on the order of 33 kWh of electricity to produce 1 kg of aluminum. The aluminum industry consumes about 2 percent of the electrical energy generated in the United States. Because this is used mainly for reduction of aluminum, recycling this metal saves large quantities of energy.

PRACTICE EXERCISE

Calculate the minimum number of kilowatt-hours of electricity required to produce 1.00 kg of Mg from electrolysis of molten $MgCl_2$ if the applied emf is 5.00 V.
Answer: 11.0 kWh

SAMPLE EXERCISE 20.16

A 12-V lead storage battery contains 410 g of lead in its anode plates and a stoichiometrically equivalent amount of PbO_2 in the cathodes. **(a)** What is the maximum number of coulombs of electrical charge it can deliver without being recharged? **(b)** For how many hours could the battery deliver a steady current of 1.0 A assuming that the current does not fall during discharge? **(c)** What is the maximum electrical work that the battery can accomplish in kilowatt-hours?

Solution: **(a)** The lead anode undergoes a two-electron oxidation:

$$Pb \longrightarrow Pb^{2+} + 2e^-$$

Consequently, $2\ \mathfrak{F} \simeq 1$ mol Pb. Using this relationship, we have

$$\text{Coulombs} = (410 \text{ g Pb})\left(\frac{1 \text{ mol Pb}}{207 \text{ g Pb}}\right)\left(\frac{2 \mathfrak{F}}{1 \text{ mol Pb}}\right)\left(\frac{96{,}500 \text{ C}}{1 \mathfrak{F}}\right)$$

$$= 3.8 \times 10^5 \text{ C}$$

Although the emf of a cell is independent of the masses of the solid reactants involved in the cell, the total electrical charge the cell can deliver depends on these quantities. The size and surface area further affect the *rate* at which electrical charge can be delivered.

(b) We calculate the number of hours of operation at a current level of 1.0 A by recalling that a coulomb corresponds to a current of 1 A flowing for 1 s:

$$\text{Hours} = (3.8 \times 10^5 \text{ C})\left(\frac{1 \text{ A-s}}{1 \text{ C}}\right)\left(\frac{1 \text{ hr}}{3600 \text{ s}}\right)\left(\frac{1}{1.0 \text{ A}}\right) = 1.1 \times 10^2 \text{ hr}$$

This battery might be described as a 110 A-hr battery.

(c) The maximum work is given by the product $-n\mathfrak{F}E$, Equation 20.23:

$$\text{Kilowatt-hours} = -(3.8 \times 10^5 \text{ C})(12 \text{ V})\left(\frac{1 \text{ J}}{1 \text{ C-V}}\right)\left(\frac{1 \text{ kWh}}{3.6 \times 10^6 \text{ J}}\right)$$

$$= -1.3 \text{ kWh}$$

(The negative sign indicates that the system does work on its surroundings.)

PRACTICE EXERCISE

A "deep discharge" lead-acid battery for marine use is advertised as having an 80-A-hr capacity. What amount of Pb would be oxidized if this battery were discharged so as to consume 80 percent of its capacity? ***Answer:*** 247 g of Pb

20.9 CORROSION

Before we close our discussion of electrochemistry, we should apply some of what we have learned to a very important problem, the **corrosion** of metals. Corrosion reactions are redox reactions in which a metal is attacked by some substance in its environment and converted to an unwanted compound.

All metals except gold and platinum are thermodynamically capable of undergoing oxidation in air at room temperature. When the oxidation process is not inhibited in some way, it can be very destructive. However, oxidation can result in the formation of an insulating, protective oxide layer that prevents further reaction of the underlying metal. On the basis of the standard oxidation potential for aluminum ($E^\circ_{ox} = 1.66$ V), we would expect aluminum to be very readily oxidized. The many aluminum soft-drink and beer cans that litter the environment are ample evidence, however, that aluminum undergoes only very slow chemical corrosion. The exceptional stability of this active metal in air is due to the formation of a thin, protective coat of oxide—a hydrated form of Al_2O_3—on the surface of the metal. The oxide coat is impermeable to the passage of O_2 or H_2O and so protects the underlying metal from further corrosion. Magnesium, which also has a high oxidation potential, is similarly protected. Some metal alloys, such as stainless steel and Nichrome (Section 24.7), also form protective, impervious oxide coats.

Corrosion of Iron

One of the most familiar corrosion processes is the rusting of iron (Figure 20.19). From an economic standpoint, this is a significant process. It is estimated that up to 20 percent of the iron produced annually in this country is used to replace iron objects that have been discarded because of rust damage.

Figure 20.19 Steel pipes used in dredging operations show extensive corrosion after lying on a North Carolina beach for a time. Corrosion of iron is an electrochemical process of great economic importance. The annual cost of metallic corrosion in the U.S. economy is estimated to be \$70 billion. (Jack Dermid/Photo Researchers)

The rusting of iron is known to require oxygen; iron does not rust in water unless O_2 is present. Rusting also requires water; iron does not rust in oil, even if it contains O_2, unless H_2O is also present. Other factors—such as the pH of the solution, the presence of salts, contact with metals more difficult to oxidize than iron, and stress on the iron—can accelerate rusting.

The corrosion of iron is generally believed to be electrochemical in nature. A region on the surface of the iron serves as an anode at which the iron undergoes oxidation:

$$Fe(s) \longrightarrow Fe^{2+}(aq) + 2e^- \qquad E^\circ_{ox} = 0.44 \text{ V}$$

The electrons so produced migrate through the metal to another portion of the surface that serves as the cathode. Here oxygen can be reduced:

$$O_2(g) + 4H^+(aq) + 4e^- \longrightarrow 2H_2O(l) \qquad E^\circ_{red} = 1.23 \text{ V}$$

Notice that H^+ takes part in the reduction of O_2. As the concentration of H^+ is lowered (that is, as pH is increased), the reduction of O_2 becomes less favorable. It is observed that iron in contact with a solution whose pH is above 9 does not corrode. In the course of the corrosion, the Fe^{2+} formed at the anode is further oxidized to Fe^{3+}. The Fe^{3+} forms the hydrated iron(III) oxide known as rust:*

$$4Fe^{2+}(aq) + O_2(g) + 4H_2O(l) + 2xH_2O(l) \longrightarrow 2Fe_2O_3 \cdot xH_2O(s) + 8H^+(aq)$$

Because the cathode is generally the area having the largest supply of O_2, rust often deposits there. If you look closely at a shovel after it has stood outside

* Frequently, metal compounds obtained from aqueous solution have water associated with them. For example, copper(II) sulfate crystallizes from water with 5 mol of water per mole of $CuSO_4$. We represent this formula as $CuSO_4 \cdot 5H_2O$. Such compounds are called hydrates (Section 13.2). Rust is a hydrate of iron(III) oxide with a variable amount of water of hydration. We represent the variable water content by writing the formula as $Fe_2O_3 \cdot xH_2O$.

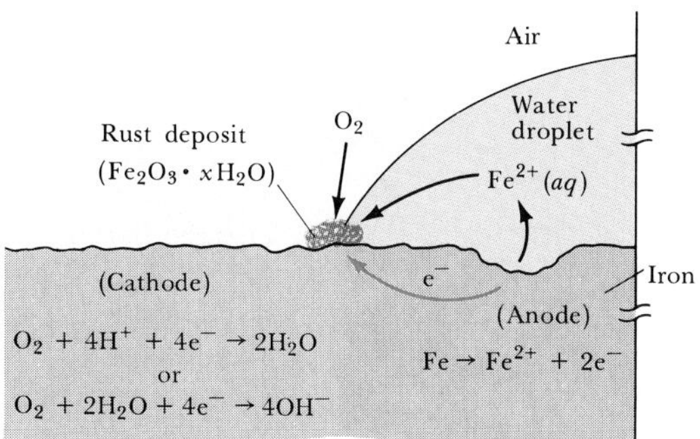

Figure 20.20 Corrosion of iron in contact with water.

in the moist air with wet dirt adhered to its blade, you may notice that pitting has occurred under the dirt but that rust has formed elsewhere, where O_2 is more readily available. The corrosion process is summarized in Figure 20.20.

The enhanced corrosion caused by the presence of salts is usually evident on autos in areas where heavy salting of roads occurs during winter. The effect of salts is readily explained by the voltaic mechanism: The ions of a salt provide the electrolyte necessary for completion of the electrical circuit.

The presence of anodic and cathodic sites on the iron requires two different chemical environments on the surface. These can occur through the presence of impurities or lattice defects (perhaps introduced by strain on the metal). At the sites of such impurities or defects, the atomic-level environment around the iron atom may permit the metal to be either more or less oxidized than at normal lattice sites. Consequently, these sites may serve as either anodes or cathodes. Ultrapure iron, prepared in such a way as to minimize lattice defects, is far less susceptible to corrosion than is ordinary iron.

Iron is often covered with a coat of paint or another metal such as tin, zinc, or chromium to protect its surface against corrosion. Sheet steel used in beverage and food cans can be coated by dipping the sheets in molten tin or by depositing a thin coat (1 to 20 μm) of tin electrochemically. The tin protects the iron only as long as the protective layer remains intact. Once it is broken and the iron exposed to air and water, tin actually promotes the corrosion of the iron by serving as the cathode in the electrochemical corrosion. As shown by the following half-cell potentials, iron is more readily oxidized than tin:

$$Fe(s) \longrightarrow Fe^{2+}(aq) + 2e^- \qquad E^\circ_{ox} = 0.44 \text{ V}$$

$$Sn(s) \longrightarrow Sn^{2+}(aq) + 2e^- \qquad E^\circ_{ox} = 0.14 \text{ V}$$

The iron therefore serves as the anode and is oxidized, as shown in Figure 20.21.

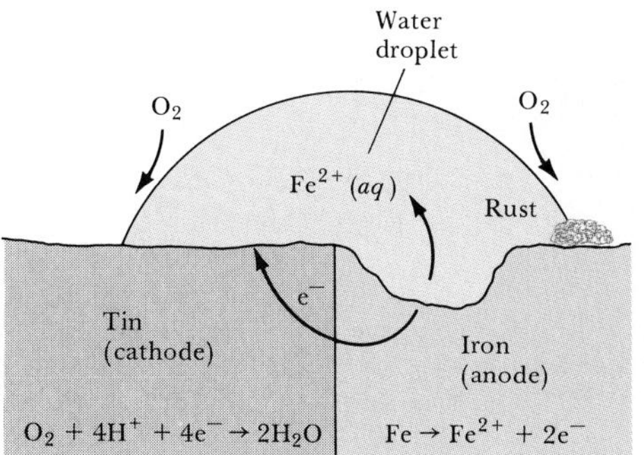

Figure 20.21 Corrosion of iron in contact with tin.

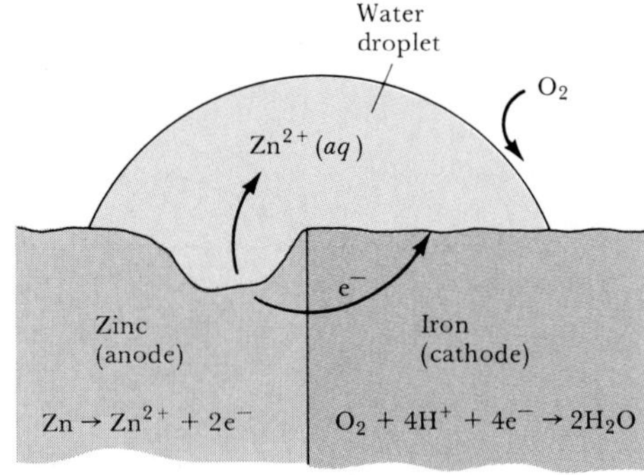

Figure 20.22 Cathodic protection of iron in contact with zinc.

Galvanized iron is produced by coating iron with a thin layer of zinc. The zinc protects the iron against corrosion even after the surface coat is broken. In this case, the iron serves as the cathode in the electrochemical corrosion because zinc is oxidized more easily than iron:

$$Zn(s) \longrightarrow Zn^{2+}(aq) + 2e^- \qquad E^\circ_{ox} = 0.76\text{ V}$$

The zinc therefore serves as the anode and is corroded instead of the iron, as shown in Figure 20.22. Protection of a metal by making it the cathode in an electrochemical cell is known as **cathodic protection**. The metal that oxidizes while protecting the cathode is called the *sacrificial anode*. Underground pipelines are often protected against corrosion by making the pipeline the cathode of a voltaic cell. Pieces of an active metal such as magnesium are buried along the pipeline and connected to it by wire, as shown in Figure 20.23. In moist soil, where corrosion can occur, the active metal serves as the anode, and the pipe experiences cathodic protection.

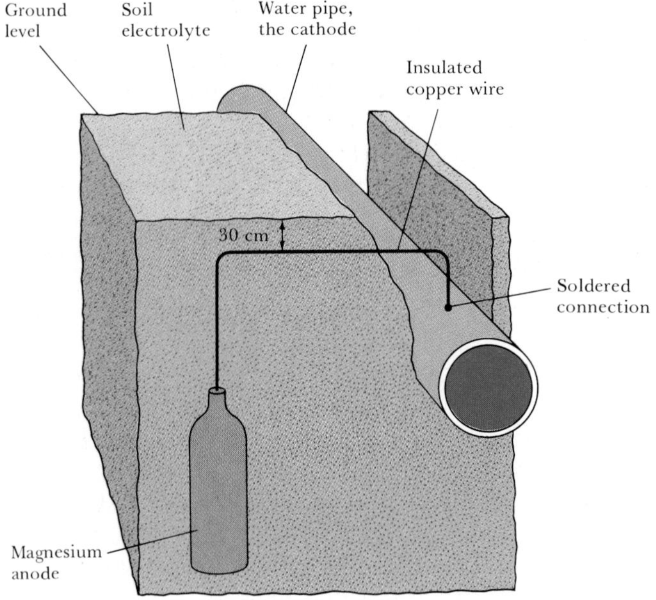

Figure 20.23 Cathodic protection of an iron water pipe. The magnesium anode is surrounded by a mixture of gypsum, sodium sulfate, and clay to promote conductivity of ions. The pipe, in effect, is the cathode of a voltaic cell.

SAMPLE EXERCISE 20.17

Predict the nature of the corrosion that would take place if an iron gutter were nailed to a house using aluminum nails.

Solution: A voltaic cell can be formed at the point of contact of the two metals. The metal that is more easily oxidized will serve as the anode, and the other metal will serve as the cathode. By comparing standard oxidation potentials of Al and Fe, we see that Al will be the anode:

$$Al(s) \longrightarrow Al^{3+}(aq) + 3e^- \qquad E^\circ_{ox} = 1.66 \text{ V}$$

$$Fe(s) \longrightarrow Fe^{2+}(aq) + 2e^- \qquad E^\circ_{ox} = 0.44 \text{ V}$$

The gutter will thus be protected against corrosion in the vicinity of the nail because the iron serves as the cathode. However, the nail will quickly corrode, leaving the gutter on the ground.

What do you think would happen if aluminum siding were nailed to a house using iron nails?

PRACTICE EXERCISE

From examination of Table 20.1, indicate which of the following metals could provide cathodic protection to iron: Al, Cu, Ni, Zn. ***Answer:*** Al, Zn

FOR REVIEW

SUMMARY

Oxidation-reduction reactions are reactions that involve a change in the oxidation state of one or more elements. In every oxidation-reduction reaction, one substance is oxidized, that is, it undergoes an increase in oxidation state. This substance is referred to as a reducing agent or reductant, because it causes the reduction of some other substance. Similarly, a substance that undergoes reduction in oxidation state is referred to as an oxidizing agent, or oxidant, because it causes the oxidation of some other substance.

In one approach to balancing oxidation-reduction reactions, called the oxidation number method, changes in the oxidation numbers of all species are identified and balanced with the overall reaction. The method of half-reactions divides the overall process into two half-reactions, one for oxidation, the other for reduction. Each half-reaction is balanced separately, and the two are brought together with proper coefficients to produce a balance of electrons gained and lost.

Spontaneous oxidation-reduction reactions can be used to generate electricity in voltaic cells. Conversely, electricity can be used to bring about nonspontaneous reactions in electrolytic cells. In either type of cell, the electrode at which oxidation occurs is called the anode, and the electrode at which reduction occurs is called the cathode.

A voltaic cell may be thought to possess a "driving force" that moves the electrons through the external circuit, from anode to cathode. This driving force is called the electromotive force (emf) and is measured in volts. The emf of a cell can be regarded as being composed of two parts: that due to oxidation at the anode and that due to reduction at the cathode ($E_{cell} = E_{ox} + E_{red}$).

Oxidation potentials (E_{ox}) and reduction potentials (E_{red}) can be assigned to half-reactions by defining the standard hydrogen electrode as a reference:

$$2H^+(1\ M) + 2e^- \longrightarrow H_2(1 \text{ atm}) \qquad E^\circ = 0 \text{ V}$$

Standard reduction potentials are referred to as standard electrode potentials and are tabulated for a great variety of reduction half-reactions. The oxidation potential for an oxidation half-reaction will be of the same magnitude as, but opposite in sign to, the electrode potential for the reverse reduction process. The more positive the potential associated with a half-reaction, the greater the tendency for that reaction to occur as written. Electrode potentials can be used to determine the maximum voltages generated by voltaic cells or the minimum voltages required in electrolytic cells. They can also be used to predict whether certain redox reactions are spontaneous (positive E). The emf is related to free-energy changes: $\Delta G = -n\mathfrak{F}E$, where $\mathfrak{F}$ is a unit called the faraday. The faraday is the amount of charge on 1 mol of electrons: $1\ \mathfrak{F} = 96{,}500\ \mathrm{C/mol\ e^-} = 96{,}500\ \mathrm{J/V\text{-}mol\ e^-}$.

The emf of a cell varies in magnitude with temperature and with the concentrations of reactants and products. The Nernst equation relates emf under non-standard conditions to the standard emf:

$$E = E^\circ - \frac{2.30RT}{n\mathfrak{F}} \log Q$$

At equilibrium

$$E = 0 \quad \text{and} \quad E^\circ = \frac{2.30RT}{n\mathfrak{F}} \log K$$

Thus, standard emfs are related to equilibrium constants.

The maximum electrical work that can be obtained from a voltaic cell is the product of the total charge it delivers, $n\mathfrak{F}$, and its emf, E: $w_{max} = -n\mathfrak{F}E$. The principles involved in voltaic cells were illustrated by simple cells utilizing salt bridges. Commercial cells need to be more rugged. Four common batteries were discussed: the lead storage battery, the nickel-cadmium battery, the common dry cell, and the alkaline dry cell. The first two are rechargeable; the latter two are not. Fuel cells, still largely experimental, are voltaic cells that utilize redox reactions involving conventional fuels, such as H_2 and CH_4.

In an electrolytic cell, an external source of electricity is used to "pump" electrons from the anode to the cathode. The current-carrying medium within the cell may be either a molten salt or an aqueous electrolyte solution. The products of electrolysis can generally be predicted by comparing electrode potentials associated with possible oxidation and reduction processes. However, because of the overvoltage phenomenon, some reactions (such as those that generate H_2 or O_2) occur less readily than electrode potentials would suggest.

The extent of chemical reaction in either an electrolytic or voltaic cell can be related to the quantity of electricity that passes through the external circuit.

Our knowledge of electrochemistry allows us to design batteries and to bring about desirable redox reactions, such as those used in electroplating and in the reduction and refining of metals. Electrochemical principles also help us to understand and combat corrosion. Corrosion of a metal such as iron is electrochemical in origin. A metal can be protected against corrosion by putting it in contact with another metal that more readily undergoes oxidation. This process is known as cathodic protection.

KEY TERMS

oxidation-reduction reaction
electrochemistry
reducing agent (reductant) (Sec. 20.1)
oxidizing agent (oxidant) (Sec. 20.1)
half-reaction (Sec. 20.2)
voltaic (galvanic) cell (Sec. 20.3)
anode (Sec. 20.3)
cathode (Sec. 20.3)
electromotive force (emf) (Sec. 20.4)
cell potential (Sec. 20.4)
standard emf (Sec. 20.4)
standard cell potential (E°_{cell}) (Sec. 20.4)
standard oxidation potential (E°_{ox}) (Sec. 20.4)
standard reduction potential (E°_{red}) (Sec. 20.4)
standard hydrogen electrode (Sec. 20.4)
standard electrode potential (Sec. 20.4)
faraday (Sec. 20.5)
Nernst equation (Sec. 20.5)
fuel cell (Sec. 20.6)
electrolysis (Sec. 20.7)
electrolytic cell (Sec. 20.7)
corrosion (Sec. 20.9)
cathodic protection (Sec. 20.9)

EXERCISES

Oxidation-Reduction Reactions

20.1 In each of the following balanced oxidation-reduction equations, identify those elements that undergo changes in oxidation number and indicate the magnitude of the change in each case:

(a) $3H_2S(aq) + 2HNO_3(aq) \longrightarrow 3S(s) + 2NO(g) + 4H_2O(l)$

(b) $5H_2SO_3(aq) + 2MnO_4^-(aq) \longrightarrow 5SO_4^{2-}(aq) + 2Mn^{2+}(aq) + 4H^+(aq) + 3H_2O(l)$

(c) $2CrO_2^-(aq) + 3ClO^-(aq) + 2OH^-(aq) \longrightarrow 2CrO_4^{2-}(aq) + 3Cl^-(aq) + H_2O(l)$

(d) $2Cu^{2+}(aq) + 2H_2O(l) \longrightarrow 2Cu(s) + O_2(g) + 4H^+(aq)$

20.2 In each of the following balanced oxidation-reduction equations, identify those elements that undergo changes in oxidation number and indicate the magnitude of the change in each case:

(a) $2KOH(aq) + Cl_2(aq) \longrightarrow KCl(aq) + KClO(aq) + H_2O(l)$

(b) $BaSO_4(s) + 4C(s) \longrightarrow BaS(s) + 4CO(g)$

(c) $5PbO_2(s) + 2Mn^{2+}(aq) + 5SO_4^{2-}(aq) + 4H^+(aq) \longrightarrow 5PbSO_4(s) + 2MnO_4^-(aq) + 2H_2O(l)$

(d) $CH_4(g) + 2O_2(g) \longrightarrow CO_2(g) + 2H_2O(g)$

20.3 Hydrazine, N_2H_4, and dinitrogen tetroxide, N_2O_4, form a self-igniting mixture that has been used as a rocket propellant. The reaction products are N_2 and H_2O. **(a)** Write a balanced chemical equation for this reaction. **(b)** Which substance serves as reducing agent and which as oxidizing agent?

20.4 Solid lead(II) sulfide reacts at elevated temperatures with oxygen in the air to form lead(II) oxide and sulfur dioxide. **(a)** Write a balanced chemical equation for this reaction. **(b)** Which substances are reductants and which are oxidants?

20.5 Complete and balance each of the following oxidation or reduction half-reactions:

(a) $H_2O_2(aq) \longrightarrow O_2(g)$ (acidic solution)

(b) $NO_3^-(aq) \longrightarrow NO(g)$ (acidic solution)

(c) $SO_3^{2-}(aq) \longrightarrow SO_4^{2-}(aq)$ (basic solution)

(d) $MnO_4^-(aq) \longrightarrow MnO_2(s)$ (basic solution)

20.6 Complete and balance each of the following oxidation or reduction half-reactions:

(a) $H_2O_2(aq) \longrightarrow H_2O(l)$ (acidic solution)

(b) $PbO_2(s) + Cl^- \longrightarrow PbCl_2(s)$ (acidic solution)

(c) $ClO^-(aq) \longrightarrow Cl^-(aq)$ (basic solution)

(d) $Cr(OH)_4^-(aq) \longrightarrow CrO_4^{2-}(aq)$ (basic solution)

20.7 Complete and balance each of the following equations:

(a) $Cr_2O_7^{2-}(aq) + I^-(aq) \longrightarrow Cr^{3+}(aq) + IO_3^-(aq)$ (acidic solution)

(b) $MnO_4^-(aq) + CH_3OH(aq) \longrightarrow Mn^{2+}(aq) + HCO_2H(aq)$ (acidic solution)

(c) $As(s) + ClO_3^-(aq) \longrightarrow H_3AsO_3(aq) + HClO(aq)$ (acidic solution)

(d) $As_2O_3(s) + NO_3^-(aq) \longrightarrow H_3AsO_4(aq) + N_2O_3(aq)$ (acidic solution)

(e) $MnO_4^-(aq) + Br^-(aq) \longrightarrow MnO_2(s) + BrO_3^-(aq)$ (basic solution)

(f) $H_2O_2(aq) + Cl_2O_7(aq) \longrightarrow ClO_2^-(aq) + O_2(g)$ (basic solution)

20.8 Complete and balance each of the following equations:

(a) $Pb(OH)_4^{2-}(aq) + ClO^-(aq) \longrightarrow PbO_2(s) + Cl^-(aq)$ (basic solution)

(b) $Tl_2O_3(s) + NH_2OH(aq) \longrightarrow TlOH(s) + N_2(g)$ (basic solution)

(c) $Cr_2O_7^{2-}(aq) + CH_3OH(aq) \longrightarrow HCO_2H(aq) + Cr^{3+}(aq)$ (acidic solution)

(d) $MnO_4^-(aq) + Cl^-(aq) \longrightarrow Mn^{2+}(aq) + Cl_2(aq)$ (acidic solution)

(e) $H_2O_2(aq) + ClO_2(aq) \longrightarrow ClO_2^-(aq) + O_2(g)$ (basic solution)

(f) $NO_2^-(aq) + Cr_2O_7^{2-}(aq) \longrightarrow Cr^{3+}(aq) + NO_3^-(aq)$ (acidic solution)

Voltaic Cells; Emf; Spontaneity

20.9 What is the role of the salt bridge in the voltaic cell diagrammed in Figure 20.5?

20.10 Using the cell in Figure 20.7 as an example, explain the difference between an inert and an active electrode in a voltaic cell.

20.11 A voltaic cell similar to that shown in Figure 20.5 is constructed. One electrode compartment consists of a cadmium strip placed in a solution of $Cd(NO_3)_2$, and the other has a nickel strip placed in a solution of $NiCl_2$. The overall cell reaction is

$$Cd(s) + Ni^{2+}(aq) \longrightarrow Cd^{2+}(aq) + Ni(s)$$

(a) Write the half-reactions that occur in the two electrode compartments. **(b)** Which electrode is the anode and which is the cathode? **(c)** Indicate the signs of the electrodes. **(d)** Do electrons flow from the cadmium electrode to the nickel electrode, or from nickel to cadmium?

20.12 A voltaic cell similar to that shown in Figure 20.7 is constructed. One electrode compartment has a copper strip placed in a solution of $CuSO_4$, and the other is a standard hydrogen electrode. As the cell operates, the copper strip gains mass. **(a)** Write the half-reactions that occur in the two electrode compartments and the overall cell reaction. **(b)** Which electrode is the anode and which is the cathode? **(c)** Indicate the signs of the electrodes. **(d)** Is the electron flow from the copper electrode to the hydrogen electrode, or the other way around?

20.13 The two half-reactions in a voltaic cell are

$$PtCl_4^{2-}(aq) + 2e^- \longrightarrow Pt(s) + 4Cl^-(aq)$$

$$AgI(s) + e^- \longrightarrow Ag(s) + I^-(aq)$$

(a) By using Appendix E, determine which reaction (or its reverse) occurs at the cathode and which at the anode. (b) What is the standard cell potential?

20.14 A voltaic cell is constructed in which one electrode is a standard hydrogen electrode, as in Figure 20.7. The other electrode compartment consists of a Ag strip in an aqueous solution of $AgNO_3$. **(a)** By referring to Table 20.1, write the half-reactions involved and determine which electrode is the anode and which is the cathode. **(b)** Will the silver strip gain or lose mass as the cell operates? **(c)** What is the standard cell potential?

20.15 Given the following half-reactions and associated standard electrode potentials:

$$AuBr_4^-(aq) + 3e^- \longrightarrow Au(s) + 4Br^-(aq) \qquad E^\circ = -0.858\ V$$

$$Eu^{3+}(aq) + e^- \longrightarrow Eu^{2+}(aq) \qquad E^\circ = -0.43\ V$$

$$IO^-(aq) + H_2O(l) + 2e^- \longrightarrow I^-(aq) + 2OH^-(aq) \qquad E^\circ = 0.49\ V$$

$$Sn^{2+}(aq) + 2e^- \longrightarrow Sn(s) \qquad E^\circ = -0.14\ V$$

(a) Write the cell reaction for the combination of these half-cell reactions that leads to the largest cell emf and calculate the value. **(b)** Write the cell reaction for the combination of half-cell reactions that leads to the smallest cell emf and calculate that value.

20.16 The standard electrode potentials of the following half-reactions are given in Appendix E:

$$Mn^{2+}(aq) + 2e^- \longrightarrow Mn(s)$$
$$Hg^{2+}(aq) + 2e^- \longrightarrow Hg(l)$$
$$Co^{2+}(aq) + 2e^- \longrightarrow Co(s)$$
$$Cu^+(aq) + e^- \longrightarrow Cu(s)$$

(a) Write the cell reaction for the combination of these half-cell reactions that leads to the largest cell emf, and calculate the value. **(b)** Write the cell reaction for the combination of half-cell reactions that leads to the smallest cell emf, and calculate the value.

20.17 A 1 *M* solution of $Cu(NO_3)_2$ is placed in a beaker with a strip of Cu metal. A 1 *M* solution of $SnSO_4$ is placed in a second beaker with a strip of Sn metal. The two beakers are connected by a salt bridge, and the two metal electrodes are linked by wires to a voltmeter. **(a)** Which electrode serves as the anode, and which as the cathode? **(b)** Which electrode gains mass and which loses mass as the cell reaction proceeds? **(c)** Indicate the sign of each electrode. **(d)** What is the emf generated by the cell under standard conditions? **(e)** Write the chemical equation for the overall cell reaction.

20.18 A voltaic cell consists of a strip of lead metal in a solution of $Pb(NO_3)_2$ in one beaker, and in the other beaker a platinum electrode immersed in an NaCl solution, with Cl_2 gas bubbled around the electrode. The two beakers are connected with a salt bridge. **(a)** Which electrode serves as the anode, and which as the cathode? **(b)** Which electrode, if either, gains or loses mass as the cell reaction proceeds? **(c)** Indicate the sign of each electrode. **(d)** What is the emf generated by the cell under standard conditions? **(e)** Write the overall cell reaction.

20.19 A voltaic cell that uses the reaction

$$Tl^{3+}(aq) + 2Cr^{2+}(aq) \longrightarrow Tl^+(aq) + 2Cr^{3+}(aq)$$

has a measured emf of 1.19 V under standard conditions. **(a)** What is E° for the half-cell reaction: $Tl^{3+}(aq) + 2e^- \longrightarrow Tl^+(aq)$? **(b)** What is meant by the expression "under standard conditions"? **(c)** Sketch the voltaic cell, label the anode and cathode, and indicate the direction of electron flow.

20.20 A voltaic cell that uses the reaction

$$PdCl_4^{2-}(aq) + Cd(s) \longrightarrow Pd(s) + 4Cl^-(aq) + Cd^{2+}(aq)$$

has a measured emf under standard conditions of 1.03 V. **(a)** Write the two half-cell reactions. **(b)** From the known value of E° for the half-cell involving Cd (Appendix E), determine the half-cell potential for the reaction involving Pd. **(c)** Sketch the cell, label the anode and cathode, and indicate the direction of electron flow.

20.21 **(a)** Sketch a voltaic cell based on the following half-cell reactions:

$$Ag(CN)_2^-(aq) + e^- \longrightarrow Ag(s) + 2CN^-(aq)$$
$$ClO^-(aq) + H_2O(l) + 2e^- \longrightarrow Cl^-(aq) + 2OH^-(aq)$$

(b) For each half-cell, determine whether it would be more appropriate to use an active or an inert electrode. **(c)** By using data in Appendix E, determine which electrode is the anode and which is the cathode. Label the terminals positive or negative. **(d)** Show the directions of ion and electron movements. **(e)** Write the balanced overall cell reaction. **(f)** Calculate the emf generated by the cell under standard conditions.

20.22 **(a)** Sketch a voltaic cell based on the following half-cell reactions:

$$HSO_4^-(aq) + 3H^+(aq) + 2e^- \longrightarrow H_2SO_3(aq) + H_2O(l)$$
$$Cr^{3+}(aq) + 3e^- \longrightarrow Cr(s)$$

(b) For each half-cell, determine whether it would be more appropriate to use an active or an inert electrode. **(c)** By using data in Appendix E, determine which electrode is the anode and which is the cathode. Label the terminals positive or negative. **(d)** Show the directions of ion and electron movements. **(e)** Write the balanced overall cell reaction. **(f)** Calculate the emf generated by the cell under standard conditions.

20.23 **(a)** Assuming standard conditions, arrange the following species in order of increasing strength as oxidizing agents in acidic solution: $Cr_2O_7^{2-}$, H_2O_2, Cu^{2+}, Cl_2, O_2. **(b)** Arrange the following species in order of increasing strength as reducing agents in acidic solution: Zn, I^-, Sn^{2+}, H_2O_2, Al.

20.24 Based on the data in Appendix E, **(a)** which of the following is the strongest oxidizing agent, and which is the weakest in acidic solution: Ce^{4+}, Br_2, H_2O_2, Zn? **(b)** Which of the following is the strongest reducing agent, and which is the weakest in acidic solution: F^-, Zn, $N_2H_5^+$, I_2, NO?

20.25 Using Table 20.1, suggest one or more agents capable of reducing $Eu^{3+}(aq)$ to $Eu^{2+}(aq)$ under standard conditions ($E^\circ = -0.43$ V).

20.26 Based on Table 20.1, suggest one or more agents capable of oxidizing $RuO_4^{2-}(aq)$ to $RuO_4^-(aq)$ under standard conditions. (E° for the reaction $RuO_4^{2-}(aq) \longrightarrow RuO_4^-(aq) + e^-$ is -0.59 V.)

20.27 Which of the following reactions are spontaneous under standard conditions?
(a) $2Ag(s) + Cl_2(g) \longrightarrow 2AgCl(s)$
(b) $2Cr^{3+}(aq) + 6CO_2(g) + 7H_2O(l) \longrightarrow Cr_2O_7^{2-}(aq) + 3H_2C_2O_4(aq) + 8H^+$
(c) $O_3(g) + H_2O_2(aq) \longrightarrow 2O_2(g) + H_2O(l)$

20.28 Write a balanced equation for each of the following reactions and determine whether the reaction is spontaneous under standard conditions: **(a)** Aqueous iodide ion is oxidized to $I_2(s)$ by $Hg_2^{2+}(aq)$. **(b)** In acidic solution, copper(I) ion is oxidized to copper(II) ion by nitrate ion. **(c)** In basic solution, $Cr(OH)_3(s)$ is oxidized to $CrO_4^{2-}(aq)$ by $ClO^-(aq)$.

Relationships between E°, ΔG°, and K

20.29 Indicate whether each of the following statements is true or false. Correct those that are false. **(a)** For a given set of concentrations, the cell emf is independent of the number of moles of reacting substances present. **(b)** A positive value of E° means that ΔG° will also be positive. **(c)** The units of E° and ΔG° are the same.

20.30 Indicate whether each of the following statements is true or false. Correct those that are false. **(a)** The cell potential E° is directly proportional to the equilibrium constant for the cell reaction. **(b)** The cell emf is independent of the concentrations of solutions species involved in the cell reaction. **(c)** For the cell shown in Figure 20.5, the cell potential is independent of the size of the zinc electrode.

20.31 From the standard potentials, calculate ΔG° for each of the reactions listed in Exercise 20.27.

20.32 Given the following half-cell potentials:

$$Fe^{2+}(aq) \longrightarrow Fe^{3+}(aq) + e^- \qquad E^\circ = -0.771 \text{ V}$$

$$S_2O_6^{2-}(aq) + 4H^+(aq) + 2e^- \longrightarrow 2H_2SO_3(aq) \qquad E^\circ = 0.60 \text{ V}$$

$$N_2O(aq) + 2H^+(aq) + 2e^- \longrightarrow N_2(g) + H_2O(l) \qquad E^\circ = -1.77 \text{ V}$$

$$VO_2^+(aq) + H^+(aq) + e^- \longrightarrow VO^{2+}(aq) + H_2O(l) \qquad E^\circ = 1.00 \text{ V}$$

(a) Write balanced chemical equations for the oxidation of $Fe^{2+}(aq)$ by the other three reagents for which the half-cell reaction is given. **(b)** Calculate ΔG° for each reaction at 298 K.

20.33 Using the standard half-cell potentials listed in Appendix E, calculate the equilibrium constant for each of the following reactions at 298 K:
(a) $Zn(s) + Sn^{2+}(aq) \longrightarrow Zn^{2+}(aq) + Sn(s)$
(b) $Co(s) + 2H^+(aq) \longrightarrow Co^{2+}(aq) + H_2(g)$
(c) $10Br^-(aq) + 2MnO_4^-(aq) + 16H^+(aq) \longrightarrow 2Mn^{2+}(aq) + 8H_2O(l) + 5Br_2(l)$

20.34 Using the standard half-cell potentials listed in Appendix E, calculate the equilibrium constant for each of the following reactions at 298 K:
(a) $2VO_2^+(aq) + 4H^+(aq) + Ni(s) \longrightarrow 2VO^{2+}(aq) + H_2O(l) + Ni^{2+}(aq)$
(b) $3Ce^{4+}(aq) + Bi(s) + H_2O(l) \longrightarrow 3Ce^{3+}(aq) + BiO^+(aq) + 2H^+(aq)$
(c) $N_2H_5^+(aq) + 4Fe(CN)_6^{3-}(aq) \longrightarrow N_2(g) + 5H^+(aq) + 4Fe(CN)_6^{4-}(aq)$

20.35 A cell exhibits a standard emf of 0.35 V at 298 K. What is the value of the equilibrium constant for the cell reaction **(a)** if $n = 1$? **(b)** if $n = 2$? **(c)** if $n = 3$?

20.36 At 298 K, a cell reaction exhibits a standard emf of 0.21 V. The equilibrium constant for the cell reaction is 1.31×10^7. What is the value of n for the cell reaction?

Nernst Equation

20.37 A cell utilizes the following reaction:

$$2Co^{3+}(aq) + Zn(s) \longrightarrow 2Co^{2+}(aq) + Zn^{2+}(aq)$$

What is the effect on cell emf of each of the following changes? **(a)** Cobalt(II) nitrate is dissolved in the cathode compartment. **(b)** Cobalt(III) nitrate is dissolved in the cathode compartment. **(c)** The size of $Zn(s)$ electrode is doubled. **(d)** Additional water is added to the anode compartment.

20.38 What is the effect on the emf of the cell shown in Figure 20.7 of each of the following changes? **(a)** The pressure of the H_2 gas is decreased. **(b)** The area of the anode is doubled. **(c)** Sulfuric acid is added to the cathode compartment. **(d)** Sodium nitrate is added to the anode compartment.

20.39 Calculate the emf of the cell described in Exercise 20.37 under the following conditions: $[Co^{3+}] = 0.25\ M$, $[Co^{2+}] = 0.040\ M$, $[Zn^{2+}] = 0.023\ M$.

20.40 Calculate the emf of a cell that utilizes the reaction

$$Co(s) + I_2(s) \longrightarrow Co^{2+}(aq) + 2I^-(aq)$$

at 298 K when $[Co^{2+}] = 0.016\ M$ and $[I^-] = 0.0060\ M$.

20.41 The cell in Figure 20.7 could be used to provide a measure of the pH in the cathode compartment. Calculate the pH of the cathode compartment solution if the cell emf is measured to be 0.720 V when $[Zn^{2+}] = 0.10\ M$ and $P_{H_2} = 1.0$ atm.

20.42 A voltaic cell is constructed that is based on the following reaction:

$$Sn^{2+}(aq) + Pb(s) \longrightarrow Pb^{2+}(aq) + Sn(s)$$

If the concentration of Sn^{2+} in the cathode compartment is 1.00 M and the cell generates an emf of 0.22 V, what is the concentration of Pb^{2+} in the anode compartment? If the anode compartment contains $[SO_4^{2-}] = 1.00\ M$ in equilibrium with $PbSO_4(s)$, what is the K_{sp} of the $PbSO_4$?

Commercial Voltaic Cells

20.43 If 120 g of zinc is employed in the casing of a particular Leclanché dry cell, and if all of this is consumed in the cell reaction, how many grams of MnO_2 undergo reaction?

20.44 During a period of discharge of a lead-acid battery, 600 g of Pb from the anode is converted into $PbSO_4(s)$. What mass of $PbO_2(s)$ is reduced at the cathode during this same period?

20.45 **(a)** Write the reactions for discharge and charge of a nickel-cadmium rechargeable cell. **(b)** Given the following half-cell potentials, calculate the standard emf of the cell:

$$Cd(OH)_2(s) + 2e^- \longrightarrow Cd(s) + 2OH^-(aq) \qquad E^\circ = -0.76\ V$$

$$NiO_2(s) + 2H_2O(l) + 2e^- \longrightarrow Ni(OH)_2(s) + 2OH^-(aq) \qquad E^\circ = +0.49\ V$$

20.46 Mercuric oxide dry-cell batteries are often used where a high energy density is required, for example, in hearing aids, watches, cameras, and electronic systems. The two half-cell reactions that occur in the battery are as follows:

$$HgO(s) + H_2O(l) + 2e^- \longrightarrow Hg(l) + 2OH^-(aq)$$

$$Zn(s) + 2OH^-(aq) \longrightarrow ZnO(s) + H_2O(l) + 2e^-$$

(a) Write the overall cell reaction. **(b)** The value of E° for the reduction half-reaction written above is 0.098 V. The measured potential for a fresh mercury dry cell is 1.35 V. Assuming that both half-cells are operating under standard conditions, what is the standard potential for the oxidation half-cell reaction? Why is this potential different from that for oxidation of Zn in aqueous acidic medium?

20.47 Suppose that an alkaline dry cell were manufactured using cadmium metal rather than zinc. What effect would this have on the cell emf?

20.48 The most promising experimental batteries that might someday find their way into commercial use employ lithium or sodium as the anodic metal. What advantages might be realized by using these metals rather than zinc, cadmium, lead, or nickel?

Electrolysis

20.49 Why are different products obtained when molten $MgCl_2$ and aqueous $MgCl_2$ are electrolyzed with inert electrodes? Predict the products in each case.

20.50 What are the expected half-reactions at both anode and cathode upon electrolysis of **(a)** molten $CdCl_2$, using inert electrodes; **(b)** aqueous $CdCl_2$ solution, using inert electrodes; **(c)** aqueous $CdCl_2$ solution, using a cadmium metal anode and an iron cathode?

20.51 Sketch a cell for electrolysis of a $CoCl_2$ solution using inert electrodes. Indicate the directions in which ions and electrons move. Give the electrode reactions, and label the anode and cathode, indicating which is positive and which is negative.

20.52 Sketch a cell for the electrolysis of aqueous HCl using copper electrodes. Give the electrode reactions, labeling the anode and cathode. Calculate the minimum applied voltage required to cause electrolysis to occur, assuming standard conditions.

20.53 **(a)** A $Cr^{3+}(aq)$ solution is electrolyzed using a current of 0.850 A. What mass of $Cr(s)$ is plated out after 24 hr? **(b)** What amperage is required to plate out 1.65 g of Cr from a Cr^{3+} solution in a period of 2.50 hr?

20.54 Metallic magnesium can be made by the electrolysis of molten $MgCl_2$. **(a)** What mass of Mg is formed by passing a current of 2.50 A through molten $MgCl_2$ for a period of 550 min? **(b)** How many seconds are required to plate out 2.75 g Mg from molten $MgCl_2$ using 5.00 A current?

20.55 **(a)** In the electrolysis of aqueous NaCl, how many liters of $Cl_2(g)$ (measured at STP) are generated by a current of 5.50 A for a period of 100 min? **(b)** How many moles of $NaOH(aq)$ have formed in the solution in this period?

20.56 If 0.500 L of a 0.600 M $SnSO_4$ solution is electrolyzed for a period of 30.0 min using a current of 4.60 A, and if inert electrodes are used, what is the final concentration of each ion remaining in the solution? (Assume that the volume of the solution does not change.)

Electrical Work

20.57 Indicate whether each of the following statements is true or false. Correct those that are false. **(a)** The theoretical maximum work obtainable from a voltaic cell is exactly equal to the theoretical minimum work required to drive the cell reaction in the opposite direction under the same conditions. **(b)** For a given cell reaction the maximum work obtainable is proportional to the total mass of reactants that form products. **(c)** The minimum work required to cause an electrolytic cell reaction to proceed is often less than the theoretical value, because of overvoltage.

20.58 Provide a simple, direct argument to support the statement that the maximum work obtainable from a given voltaic cell is a state function; that is, it depends only on the initial and final states of the cell. Is this true also of the *actual* work obtainable from the cell? Give examples to illustrate your answer.

20.59 Under standard conditons, what is the maximum electrical work, in joules, that a cell employing the cell reaction

$$Cd(s) + 2H^+(aq) \longrightarrow Cd^{2+}(aq) + H_2(g)$$

can accomplish if 0.780 mol of Cd is consumed?

20.60 Under standard conditions, what is the maximum electrical work, in joules, that a cell employing the cell reaction

$$Sn(s) + I_2(s) \longrightarrow Sn^{2+}(aq) + 2I^-(aq)$$

can accomplish if 0.460 mol of Sn is consumed?

20.61 **(a)** Calculate the mass of magnesium produced at a large facility by electrolysis of molten $MgCl_2$ by a current of 90,000 A flowing for 16 hr. Assume that the cell is 50 percent efficient. **(b)** What is the total energy requirement for this electrolysis if the applied emf is 4.20 V?

20.62 **(a)** Calculate the mass of Li formed by electrolysis of molten LiCl by a current of 6.6×10^4 A flowing for a period of 12 hr. Assume the cell is 85 percent efficient. **(b)** What is the energy requirement for this electrolysis per mole of Li formed if the applied emf is 5.50 V?

Corrosion

20.63 An iron object is plated with a coating of cobalt to protect against corrosion. Does the cobalt protect iron by cathodic protection? Explain.

20.64 When an iron object is plated with tin, does the tin act as a sacrificial anode in protecting against corrosion? Explain.

20.65 *Amines* are compounds related to ammonia. One of their characteristics is their ability to function as Brønsted bases (Section 16.1). Suggest how they act as corrosion inhibitors when added to antifreeze.

20.66 The following quotation is taken from an article dealing with corrosion of electronic materials: "Sulfur dioxide, its acidic oxidation products, and moisture are well established as the principal causes of outdoor corrosion of many metals." Using Ni as an example, explain why the factors cited affect the rate of corrosion. Write chemical equations to illustrate your points. Note that NiO(*s*) is soluble in acidic solution.

Additional Exercises

20.67 Complete and balance each of the following equations:

(a) $Si(s) + HCl(g) \longrightarrow SiHCl_3(g) + H_2(g)$

(b) $NH_3(g) + F_2(g) \longrightarrow NF_3(g) + NH_4F(s)$

(c) $B_2O_3(s) + BrF_3(l) \longrightarrow BF_3(g) + Br_2(l) + O_2(g)$

(d) $Fe^{2+}(aq) + HBrO_3(aq) \longrightarrow Fe^{3+}(aq) + Br_2(l)$ (acidic solution)

(e) $Cu(s) + NO_3^-(aq) \longrightarrow Cu^{2+}(aq) + NO(g)$ (acidic solution)

(f) $IO_3^-(aq) + I^-(aq) \longrightarrow I_3^-(aq)$ (basic solution)

[20.68] A *disproportionation reaction* is an oxidation-reduction reaction in which the same substance is oxidized and reduced. Use the oxidation number method to balance the following disproportionation reactions:

(a) $Fe^{2+}(aq) \longrightarrow Fe(s) + Fe^{3+}(aq)$

(b) $Cl_2(aq) \longrightarrow Cl^-(aq) + ClO^-(aq)$ (basic solution)

(c) $NO_2^-(aq) \longrightarrow NO(g) + NO_3^-(aq)$ (acidic solution)

(d) $MnO_4^{2-}(aq) \longrightarrow MnO_4^-(aq) + MnO_2(s)$ (acidic solution)

[20.69] By using half-reactions listed in Appendix E, balance the following disproportionation reactions (see Exercise 20.68) and determine whether the reaction is spontaneous:

(a) $H_2SO_3(aq) \longrightarrow S(s) + HSO_4^-(aq)$ (acidic solution)

(b) $Br_2(l) \longrightarrow Br^-(aq) + BrO_3^-(aq)$ (acidic solution)

(c) $PbSO_4(s) \longrightarrow Pb(s) + PbO_2(s) + HSO_4^-(aq)$ (acidic solution)

20.70 The following oxidation-reduction reaction is spontaneous in the direction indicated:

$$5Fe^{2+}(aq) + MnO_4^-(aq) + 8H^+(aq) \longrightarrow 5Fe^{3+}(aq) + Mn^{2+}(aq) + 4H_2O(l)$$

A solution containing $KMnO_4$ and H_2SO_4 is poured into one beaker, and a solution of $FeSO_4$ is poured into another. A salt bridge is used to join the beakers. A platinum foil is placed in each solution, and the two solutions are connected by a wire that passes through a voltmeter. **(a)** Indicate the reactions occurring at the anode and at the cathode, the direction of electron movement though the external circuit, the direction of ion migrations through the solutions, and the signs of the electrodes. **(b)** Calculate the emf of the cell under standard conditions.

20.71 A common shorthand way of representing a voltaic cell is to list the reactants and products from left to right in the following form:

anode|anode solution||cathode solution|cathode

A double vertical line represents a salt bridge or porous barrier. A single vertical line represents a change in phase, such as from solid to solution. **(a)** Write the half-reactions and overall cell reaction represented by $Fe|Fe^{2+}||Ag^+|Ag$; sketch the cell. **(b)** Write the half-reactions and overall cell reaction represented by $Zn|Zn^{2+}||H^+|H_2$; sketch the cell. **(c)** Using the notation just described, represent a cell based on the following reaction.

$$ClO_3^-(aq) + Cu(s) + 6H^+(aq) \longrightarrow Cl^-(aq) + Cu^{2+}(aq) + 3H_2O(l)$$

Sketch the cell.

20.72 A voltaic cell is constructed from two half-cells. The first contains a Cd(*s*) electrode immersed in 1 *M* $Cd^{2+}(aq)$ solution. The other contains a Rh(*s*) electrode in 1 *M* $Rh^{3+}(aq)$ solution. The overall cell potential is +1.20 V, and, as the cell operates, the concentration of the $Rh^{3+}(aq)$ solution decreases, and the mass of the Rh electrode increases. **(a)** Write a balanced equation for the overall cell reaction. **(b)** Which electrode is the anode, and which is the cathode? **(c)** What is the standard reduction potential for the reduction of $Rh^{3+}(aq)$ to Rh(*s*)? **(d)** What is the value of $\Delta G°$ for the cell reaction?

20.73 The zinc-silver oxide cell used in hearing aids and electrical watches is based on the following half-reactions:

$$Zn^{2+}(aq) + 2e^- \longrightarrow Zn(s) \qquad E° = -0.763 \text{ V}$$

$$Ag_2O(s) + H_2O(l) + 2e^- \longrightarrow 2Ag(s) + 2OH^-(aq) \qquad E° = 0.344 \text{ V}$$

(a) What substance is oxidized and what substance is reduced in the cell during discharge? **(b)** What is the positive electrode, and what is the negative electrode? **(c)** What emf does this cell generate under standard conditions?

20.74 A voltaic cell consists of Co and a 0.1 *M* solution of

$Co^{2+}(aq)$ in one compartment, and Cu and a 0.1 *M* solution of $Cu^{2+}(aq)$ in the other. **(a)** Sketch the cell and label the anode and cathode. **(b)** What is the emf of the cell? **(c)** What would be the effect on the emf of reducing $[Co^{2+}]$ in the cobalt compartment?

20.75 Predict whether the following reactions will be spontaneous in acidic solution under standard conditions: **(a)** oxidation of Sn to Sn^{2+} by I_2 (to form I^-); **(b)** reduction of Ni^{2+} to Ni by I^- (to form I_2); **(c)** reduction Ce^{4+} to Ce^{3+} by Br^- (to form Br_2); **(d)** reduction of Ag^+ to Ag by H_2O_2.

20.76 Cytochrome, a complicated molecule that we shall represent as $CyFe^{2+}$, reacts with the air we breathe to supply energy required to synthesize adenosine triphosphate, ATP. The body uses ATP as an energy source to drive other reactions (Section 19.7). At pH 7 the following electrode potentials pertain to this oxidation of $CyFe^{2+}$:

$$O_2(g) + 4H^+(aq) + 4e^- \longrightarrow 2H_2O(l) \qquad E = 0.82\text{ V}$$
$$CyFe^{3+}(aq) + e^- \longrightarrow CyFe^{2+}(aq) \qquad E = 0.22\text{ V}$$

(a) What is ΔG for the oxidation of $CyFe^{2+}$ by air? **(b)** If the synthesis of 1 mol of ATP from adenosine diphosphate, ADP, requires a ΔG of 37.7 kJ, how many moles of ATP are synthesized per mole of O_2?

20.77 In a concentration cell, the same reactants are present in both the anode and the cathode compartments, but at different concentrations. Calculate the emf of a cell containing 0.040 *M* Cr^{3+} in one compartment and 1.0 *M* Cr^{3+} in the other if Cr electrodes are used in both. Which is the anode compartment?

[20.78] **(a)** Write the cell reaction for the cell shown in Figure 20.24. **(b)** Write the full Nernst equation expression for this cell and rearrange it so that the pH of the solution is given as a function of the other cell variables. **(c)** If the cell emf is 0.660 V, the pressure of H_2 is 1.0 atm, and $[Cl^-]$ is 1.0×10^{-3} *M*, what is the pH of the solution?

Figure 20.24

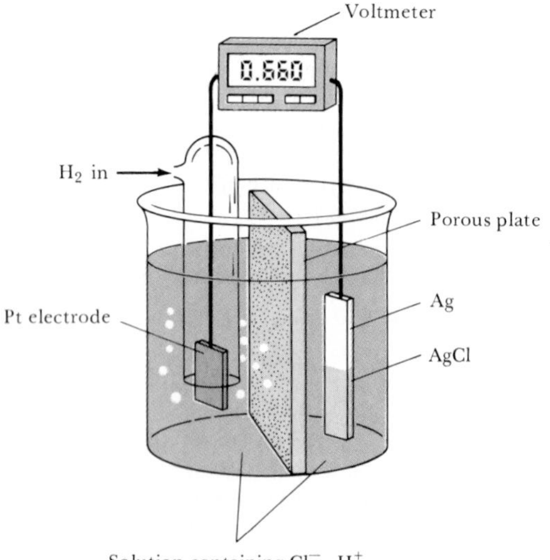

20.79 By using the standard electrode potentials in Appendix E, calculate the equilibrium constant for the disproportionation of copper(I) ion:

$$2Cu^+(aq) \longrightarrow Cu(s) + Cu^{2+}(aq)$$

[20.80] In the presence of $Br^-(aq)$, copper(II) can be reduced to CuBr(*s*):

$$Cu^{2+}(aq) + Br^-(aq) + e^- \longrightarrow CuBr(s) \quad E^\circ = +0.640\text{ V}$$

By using this value of E° together with another electrode potential, calculate K_{sp} for CuBr.

[20.81] The K_{sp} value for PbS(*s*) is 8.0×10^{-28}. By using this value together with an electrode potential from Appendix E, determine the value of the standard electrode potential for the reaction

$$PbS(s) + 2e^- \longrightarrow Pb(s) + S^{2-}(aq)$$

[20.82] Gold exists in two common positive oxidation states, +1 and +3. The standard reduction potentials for these oxidation states are

$$Au^+(aq) + e^- \longrightarrow Au(s) \qquad E^\circ = +1.69\text{ V}$$
$$Au^{3+}(aq) + 3e^- \longrightarrow Au(s) \qquad E^\circ = +1.50\text{ V}$$

(a) Can you use these data to explain why gold does not tarnish in the air? **(b)** Suggest several substances that should be strong enough oxidizing agents to oxidize gold metal. **(c)** Would $Au^+(aq)$ disproportionate (see Exercise 20.68) spontaneously into $Au^{3+}(aq)$ and Au(*s*)? **(d)** Based on your answers to parts (b) and (c), predict the result of reacting gold metal with fluorine gas.

[20.83] A voltaic cell is constructed that uses the following half-cell reactions:

$$Cu^+(aq) + e^- \longrightarrow Cu(s)$$
$$I_2(s) + 2e^- \longrightarrow 2I^-(aq)$$

The cell is operated at 298 K with $[Cu^+] = 2.1$ *M* and $[I^-] = 3.2$ *M*. **(a)** Determine *E* for the cell at these concentrations. **(b)** Which electrode is the anode of the cell? **(c)** Is the answer to part (b) the same as it would be if the cell were operated under standard conditions? **(d)** If $[Cu^+]$ were equal to 1.4 *M*, at what concentration of I^- would the cell have a zero potential?

20.84 **(a)** How many coulombs are required to plate a layer of chromium metal 0.23 mm thick on an auto bumper with a total area of 0.32 m^2 from a solution containing CrO_4^{2-}? The density of chromium metal is 7.20 g/cm^3. **(b)** What current flow is required for this electroplate if the bumper is to be plated in 6.0 s?

20.85 The element indium is to be obtained by electrolysis of a molten halide of the element. Passage of a current of 3.20 A for a period of 40.0 min results in formation of 4.57 g of In. What is the oxidation state of indium in the halide melt?

20.86 An electrolytic cell is set up for the production of aluminum by the Hall process (Chapter 24), which involves the reduction of Al^{3+} to Al. The external source passes a current of 11.2 A through the cell with an emf of 6.0 V. **(a)** How long does it take for the cell to produce 1.0 lb of aluminum metal? **(b)** If the cell has an efficiency of 40 percent, how much electrical power is expended to produce 1.0 lb of aluminum?

20.87 Explain why electrolysis of an aqueous solution of Na_2SO_4 containing litmus develops a blue color at the cathode and a red color at the anode.

20.88 Peroxyborate bleaches, such as those found in Borateem, have replaced older "chlorine" bleaches in many bleaching agents. Sodium peroxyborate, $NaBO_3$, can be prepared by electrolytic oxidation of borax ($Na_2B_4O_7$) solutions:

$$Na_2B_4O_7(aq) + 10NaOH(aq) \longrightarrow 4NaBO_3(aq) + 5H_2O(l) + 8Na^+(aq) + 8e^-$$

How many grams of $NaBO_3$ can be prepared in 24.0 hr if the current is 20.0 A?

20.89 **(a)** Calculate the minimum applied voltage required to cause the following electrolysis reaction to occur, assuming that the anode is platinum and the cathode is nickel:

$$Ni^{2+}(aq) + 2Br^-(aq) \longrightarrow Ni(s) + Br_2(l)$$

(b) In practice, a larger voltage than this calculated minimum is required to produce the electrode reactions. Why is this so?

20.90 **(a)** What is the maximum amount of work that a 6-V golf-cart lead storage battery can accomplish if it is rated at 300 A-h? **(b)** List some of the reasons why this amount of work is never realized.

20.91 The type of lead storage cell used in automobiles does not tolerate deep discharge (in which the cell reaction is allowed to proceed to near completion) very well. Typically, the battery fails after 20 to 30 such cycles. **(a)** Why does deep discharge lead to battery failure? **(b)** How is deep discharge avoided during normal auto use?

20.92 If you were going to apply a small potential to a steel ship resting in the water as a means of inhibiting corrosion, would you apply a negative or a positive charge? Explain.

20.93 Considering the following standard half-cell potentials:

$$Ti(s) \longrightarrow Ti^{2+}(aq) + 2e^- \qquad E^\circ = +1.63 \text{ V}$$

$$Ti(s) + 2H_2O(l) \longrightarrow TiO_2(s) + 4H^+ + 4e^- \qquad E^\circ = +0.86 \text{ V}$$

Why is titanium metal quite corrosion-resistant?

[20.94] Several years ago, a unique proposal was made to raise the *Titanic*. The plan involved placing pontoons within the ship using a surface-controlled submarine-type vessel. The pontoons would contain cathodes and would be filled with hydrogen gas formed by the electrolysis of water. It has been estimated that it would require about 7×10^8 mol of H_2 to provide the buoyancy to lift the ship (*Journal of Chemical Education*, vol. 50, p. 61, 1973). **(a)** How many coulombs of electrical charge would be required? **(b)** What is the minimum voltage required to generate H_2 and O_2 if the pressure on the gases at the depth of the wreckage (2 mi) is 300 atm? **(c)** What is the minimum electrical energy required to raise the *Titanic* by electrolysis? **(d)** What is the minimum cost of the electrical energy required to generate the necessary H_2 if the electricity costs 23¢ per kilowatt-hour?

[20.95] Edison's invention of the light bulb and its public demonstration in December 1879 generated considerable demand for the distribution of electricity to homes. One problem was how to measure the amount of electricity consumed by each household. Edison invented a coulometer (described in the *Journal of Chemical Education*, vol. 49, p. 627, 1972) that could be used with alternating current. Zinc plated out at the cathode of the coulometer. Every month the cathode was removed and weighed to determined the quantity of electricity used. If the cathode increased in mass by 1.62 g and the coulometer drew 0.35 percent of the current entering the home, how many coulombs of electricity were used in that month?

21 Nuclear Chemistry

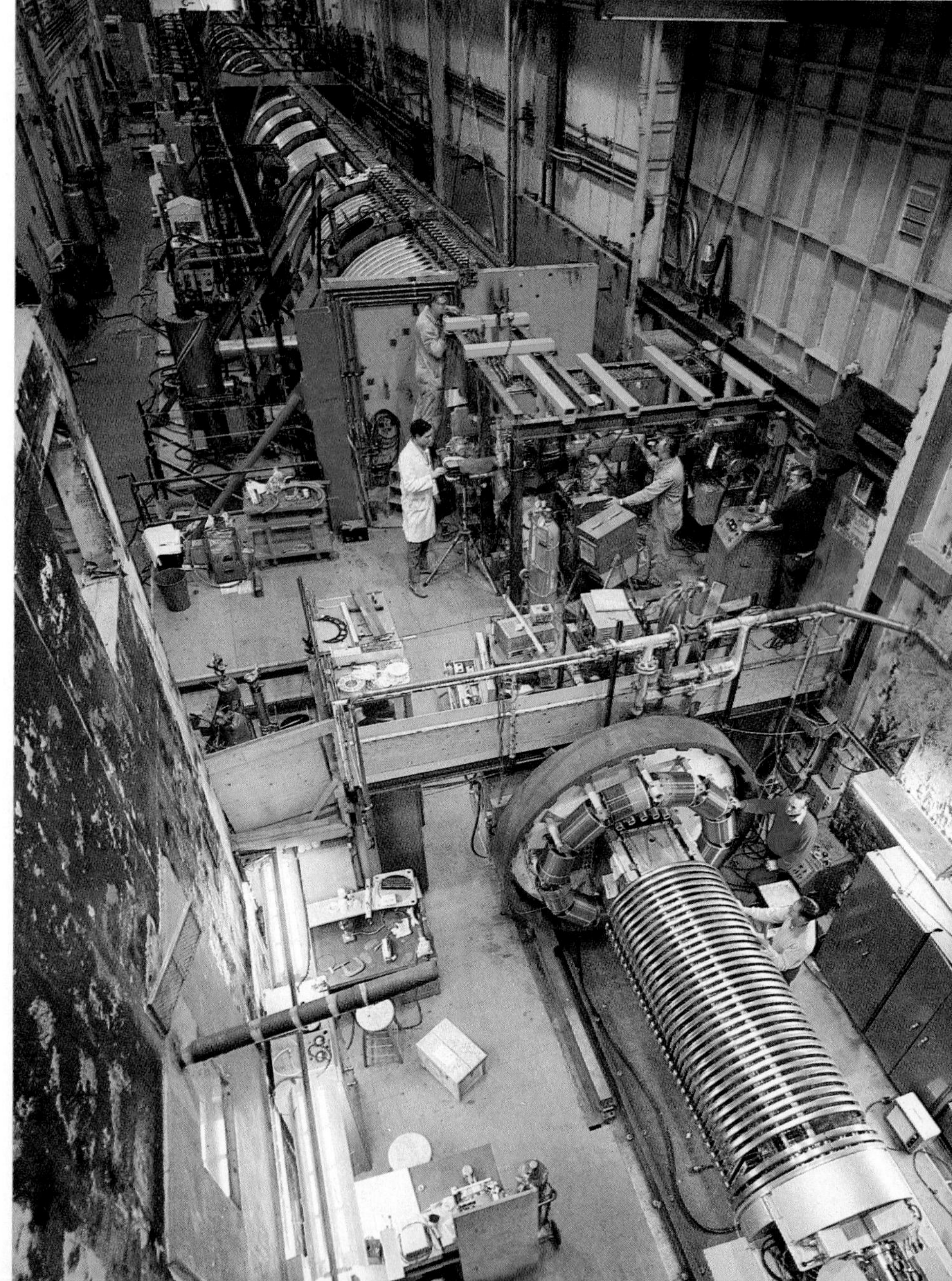

The Superhilac linear particle accelerator at the University of California-Berkeley. The transuranium elements with atomic numbers 102 through 106 were first discovered through nuclear reactions in this accelerator. (Lawrence Berkeley Laboratory, University of California)

As we have progressed through this text, our focus has been on chemical reactions, specifically reactions in which electrons play a dominant role. In this chapter, we shall consider nuclear reactions, changes in matter originating in the nucleus of an atom. Nuclear chemistry has a tremendous impact on our lives, as evidenced by the large number of topics in the news related to nuclear reactions: the accident at Chernobyl (see the Chemistry at Work box in Chapter 2), nuclear weapons treaties, the controversy over disposal of nuclear wastes, radon gas levels in the home, the Strategic Defense Initiative ("Star Wars"), cold fusion as a source of energy, and many others. Nuclear chemistry is used to date important historical artifacts, such as the controversial Shroud of Turin, which some people claimed bears the image of Jesus Christ. Radioactive elements are widely used in medicine as diagnostic tools and as a means of treatment, especially for cancer (Figure 21.1).

Some experts predict that we shall have to depend increasingly on nuclear energy to replace our dwindling supplies of fossil fuels and to meet our rising energy demands. However, the use of nuclear energy is an extremely controversial social and political issue. You should ask yourself how you would feel about having a nuclear power plant in your town. How do you react when you hear the phrase "nuclear waste"? Because the topic of nuclear energy evokes such emotional reactions, it is difficult to sift fact from opinion and make

CONTENTS

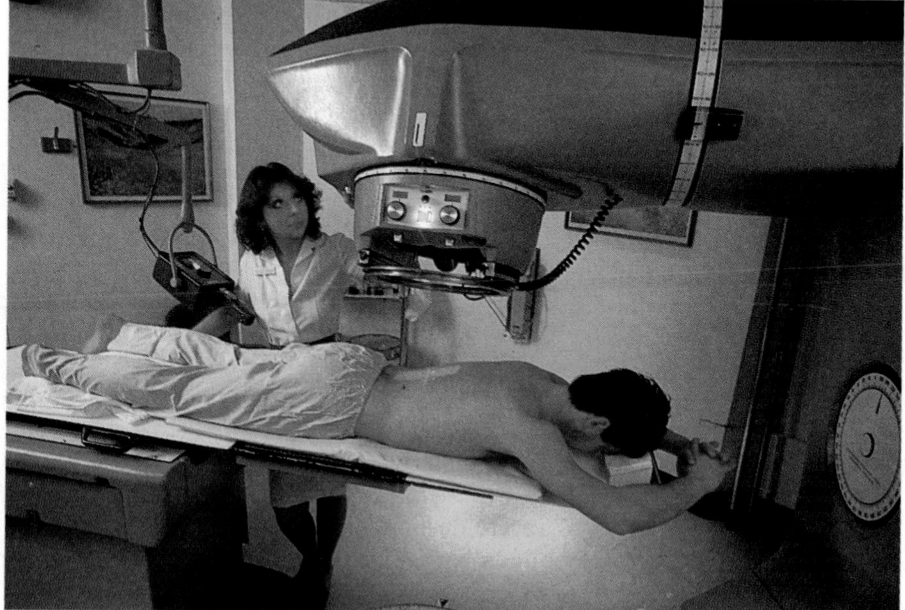

Figure 21.1 Radioactive isotopes, such as ^{60}Co, are used in the treatment of cancer. (Martin Dohan, Science Photo Library, Science Source/Photo Researchers)

rational decisions. It is imperative, therefore, that educated people of our time have some understanding of nuclear reactions and the uses of radioactive substances.

Before we get too deeply involved in our discussions, we should review and develop some ideas introduced in Section 2.3. First, recall that two subatomic particles reside in the nucleus, the *proton* and the *neutron*. We shall refer to these particles as **nucleons**. Recall also that all atoms of a given element have the same number of protons; this number is known as the element's *atomic number*. However, the atoms of a given element can have different numbers of neutrons and therefore different *mass numbers*; the mass number is the total number of nucleons in the nucleus. Atoms with the same atomic number but different mass numbers are known as *isotopes*. The different isotopes of an element are distinguished by citing their mass numbers. For example, the three naturally occurring isotopes of uranium are identified as uranium-233, uranium-235, and uranium-238, where the numerical suffixes represent the mass numbers. These isotopes are also labeled, using chemical symbols, as $^{233}_{92}U$, $^{235}_{92}U$, and $^{238}_{92}U$. The superscript is the mass number; the subscript is the atomic number. Different isotopes have different natural abundances. For example, 99.3 percent of naturally occurring uranium is uranium-238, 0.7 percent is uranium-235, and only a trace is uranium-233. One reason that we must now distinguish between different isotopes is that the nuclear properties of an atom depend on the number of both protons and neutrons in its nucleus. In contrast, we have found that an atom's chemical properties are unaffected by the number of neutrons in the nucleus. Let us now discuss the reactions that a nucleus can undergo.

21.1 NUCLEAR REACTIONS: AN OVERVIEW

A nucleus can undergo a reaction that changes its identity. Some nuclei are unstable and spontaneously emit particles and electromagnetic radiation. Such spontaneous emission from the nucleus of the atom is known as *radioactivity*. The discovery of this phenomenon by Henri Becquerel in 1896 was described in Section 2.2. Those isotopes that are radioactive are known as **radioisotopes**. An example is uranium-238, which spontaneously emits **alpha rays**. These rays consist of streams of helium-4 nuclei known as **alpha particles**. When a uranium-238 nucleus loses an alpha particle, the remaining fragment has an atomic number of 90 and a mass number of 234. It is therefore a thorium-234 nucleus. We can represent this reaction by the following nuclear equation:

$$^{238}_{92}U \longrightarrow ^{234}_{90}Th + ^{4}_{2}He \qquad [21.1]$$

When a nucleus spontaneously decomposes in this way, it is said to have decayed, or undergone *radioactive decay*.

Notice in Equation 21.1 that the sum of the mass numbers is the same on both sides of the equation ($238 = 234 + 4$). Likewise, the sum of the atomic numbers on both sides of the equation is equal ($92 = 90 + 2$). Mass numbers and atomic numbers are similarly balanced in all nuclear equations. The radioactive properties of the nucleus are essentially independent of the state of chemical combination of the atom. Thus, in writing nuclear equations, we are not concerned with the chemical form of the atom in which the nucleus resides. It makes no difference whether we are dealing with the atom in the form of an element or of one of its compounds.

Another way a nucleus can change identity is to be struck by a neutron or by another nucleus. Nuclear reactions that are induced in this way are known as **nuclear transmutations**. Such a transmutation occurs when the chlorine-35 nucleus is struck by a neutron ($^{1}_{0}n$); this collision produces a sulfur-35 nucleus and a proton ($^{1}_{1}p$ or $^{1}_{1}H$). The nuclear equation for this reaction is shown in Equation 21.2:

$$^{35}_{17}Cl + ^{1}_{0}n \longrightarrow ^{35}_{16}S + ^{1}_{1}H \qquad [21.2]$$

Notice again that the sum of mass numbers and of atomic numbers is the same on both sides of the equation. By bombarding nuclei with various particles, it is possible to prepare nuclides* not found in nature. The sulfur-35 produced in Equation 21.2 is an example.

SAMPLE EXERCISE 21.1

Write a balanced nuclear equation for the nuclear transmutation in which an aluminum-27 nucleus is struck by a helium-4 nucleus, producing a phosphorus-30 nucleus and a neutron.

Solution: By referring to a periodic table or a list of elements, we find that aluminum has an atomic number of 13; its chemical symbol is therefore $^{27}_{13}Al$. The atomic number of phosphorus is 15; its chemical symbol is therefore $^{30}_{15}P$. The balanced equation is

$$^{27}_{13}Al + ^{4}_{2}He \longrightarrow ^{30}_{15}P + ^{1}_{0}n$$

PRACTICE EXERCISE

Carbon-11, a radioactive nuclide used in medical imaging, is formed by reaction of a proton with nitrogen-14, yielding carbon-11 and helium-4. Write the nuclear reaction.
Answer: $^{14}_{7}N + ^{1}_{1}H \longrightarrow ^{11}_{6}C + ^{4}_{2}He$

21.2 RADIOACTIVITY

As we were discussing radioactivity in the preceding section, two general questions might have occurred to you: Which nuclei are radioactive, and what types of radiation do they emit? We shall now examine these questions. Let us first consider the types of radiation involved in the phenomenon of radioactivity.

Types of Radioactive Decay

Emission of radiation is one of the ways in which an unstable nucleus is transformed into a more stable one with less energy. The emitted radiation is the carrier of the excess energy. In Section 2.2, we discussed the three most common types of radiation emitted by radioactive substances: alpha (α), beta (β), and gamma (γ) rays. Table 21.1 summarizes some of the important properties of α, β, and γ radiation.

As we noted in Section 21.1, alpha rays consist of streams of helium-4 nuclei known as *alpha particles*. Equation 21.3 gives another example of this type of radioactive decay:

$$^{222}_{86}Rn \longrightarrow ^{218}_{84}Po + ^{4}_{2}He \qquad [21.3]$$

* Recall that the term *nuclide* (Section 2.3) applies to an atom with a specified number of protons and neutrons.

Table 21.1 Summary of the Properties of Alpha, Beta, and Gamma Rays

Property	Type of radiation		
	α	β	γ
Charge	2+	1−	0
Mass	6.64×10^{-24} g	9.11×10^{-28} g	0
Relative penetrating power	1	100	1000
Nature of radiation	$^{4}_{2}$He nuclei	Electrons	High-energy photons

Beta rays consist of streams of electrons. Because the **beta particles** are electrons, they are represented as $_{-1}^{0}e$. The superscript zero indicates the exceedingly small mass of the electron in comparison to the mass of a nucleon. The subscript −1 represents the negative charge of the particle, which is opposite that of the proton. Iodine-131 is an example of an isotope that undergoes decay by beta emission. This reaction is summarized by Equation 21.4:

$$^{131}_{53}I \longrightarrow ^{131}_{54}Xe + _{-1}^{0}e \qquad [21.4]$$

Emission of a beta particle has the effect of converting a neutron within the nucleus into a proton, thereby increasing the atomic number of the nucleus by 1:

$$^{1}_{0}n \longrightarrow ^{1}_{1}p + _{-1}^{0}e \qquad [21.5]$$

However, just because an electron is ejected from the nucleus, we need not think that the nucleus is composed of these particles, any more than we consider a match to be composed of sparks simply because it gives them off when struck. The electron comes into being only when the nucleus is disrupted.

Gamma rays consist of electromagnetic radiation of very short wavelength (that is, high-energy photons). The position of gamma rays in the electromagnetic spectrum is shown in Figure 6.3. Gamma rays can be represented as $^{0}_{0}\gamma$. Such radiation changes neither the atomic number nor the mass number of a nucleus. It almost always accompanies other radioactive emission, because it represents the energy lost when the remaining nucleons reorganize into more stable arrangements. Generally, we shall not show the gamma rays when writing nuclear equations.

Two other types of radioactive decay that occur are positron emission and electron capture. A **positron** is a particle that has the same mass as an electron but an opposite charge.* The positron is represented as $^{0}_{1}e$. Carbon-11 is an example of an isotope that decays by positron emission:

$$^{11}_{6}C \longrightarrow ^{11}_{5}B + ^{0}_{1}e \qquad [21.6]$$

Emission of a positron can be thought of as converting a proton into a neutron, as shown in Equation 21.7. The atomic number of the nucleus is thereby decreased by 1:

* The positron has a very short life because it is annihilated when it collides with an electron, producing gamma rays: $^{0}_{1}e + _{-1}^{0}e \longrightarrow 2\,^{0}_{0}\gamma$.

$$^{1}_{1}p \longrightarrow {}^{1}_{0}n + {}^{0}_{1}e \qquad [21.7]$$

Electron capture is the capture by the nucleus of an inner-shell electron from the electron cloud surrounding the nucleus. Rubidium-81 undergoes decay in this fashion, as shown in Equation 21.8:

$$^{81}_{37}Rb + {}^{0}_{-1}e \text{ (orbital electron)} \longrightarrow {}^{81}_{36}Kr \qquad [21.8]$$

Electron capture has the effect of converting a proton within the nucleus into a neutron, as shown in Equation 21.9

$$^{1}_{1}p + {}^{0}_{-1}e \longrightarrow {}^{1}_{0}n \qquad [21.9]$$

Table 21.2 summarizes the symbols used to represent the various elementary particles in radioactive decay and nuclear transformations.

Table 21.2 Particles Common to Radioactive Decay and Nuclear Transformations

Particle	Symbol
Neutron	$^{1}_{0}n$
Proton	$^{1}_{1}p$ or $^{1}_{1}H$
Electron	$^{0}_{-1}e$
Alpha particle	$^{4}_{2}\alpha$ or $^{4}_{2}He$
Beta particle	$^{0}_{-1}\beta$ or $^{0}_{-1}e$
Positron	$^{0}_{1}e$

SAMPLE EXERCISE 21.2

Write balanced nuclear equations for the following reactions: **(a)** thorium-230 undergoes alpha decay; **(b)** thorium-231 undergoes decay to form protactinium-231.

Solution: **(a)** The information given in the problem can be summarized as

$$^{230}_{90}Th \longrightarrow {}^{4}_{2}He + X$$

The remaining product, X, must be deduced. Because mass numbers must have the same sum on both sides of the equation, we deduce that X has a mass number of 226. Similarly, the atomic number of X must be 88. Element number 88 is radium (refer to the periodic table or list of elements). The equation is therefore as follows:

$$^{230}_{90}Th \longrightarrow {}^{4}_{2}He + {}^{226}_{88}Ra$$

(b) In this case we must determine what type of particle is emitted in the course of the radioactive decay. We can write the following equation:

$$^{231}_{90}Th \longrightarrow {}^{231}_{91}Pa + X$$

The atomic numbers are obtained from a list of elements such as that given on the inside front cover. In order for the mass numbers to balance, X must have a mass number of 0. Its atomic number must be -1. The particle with these characteristics is the beta particle (electron). We therefore write the following:

$$^{231}_{90}Th \longrightarrow {}^{231}_{91}Pa + {}^{0}_{-1}e$$

PRACTICE EXERCISE

Write a balanced nuclear equation for the reaction in which oxygen-15 undergoes positron emission. *Answer:* $^{15}_{8}O \longrightarrow {}^{15}_{7}N + {}^{0}_{1}e$

Nuclear Stability

No single rule allows us to predict whether a particular nucleus is radioactive and how it might decay. However, we can list some empirical observations that are helpful in making predictions:

1. All nuclei with 84 or more protons (atomic number ≥ 84) are unstable. For example, all isotopes of uranium, atomic number 92, are radioactive.
2. Just as certain numbers of electrons (2, 8, 18, 36, 54, and 86) correspond to stable closed-shell electron configurations, certain numbers of nucleons lead to closed shells in nuclei. The protons and neutrons can each achieve a closed shell. Nuclei with 2, 8, 20, 28, 50, or 82 protons or 2, 8, 20, 28,

Table 21.3 The Number of Stable Isotopes with Even and Odd Numbers of Protons and Neutrons

Number of stable isotopes	Protons	Neutrons
157	Even	Even
52	Even	Odd
50	Odd	Even
5	Odd	Odd

50, 82, or 126 neutrons correspond to nuclear closed shells.* Closed-shell nuclei are more stable than those that do not have closed shells. For example, there are three stable nuclei with 18 protons, two with 19, five with 20, and one with 21; there are three stable nuclei with 18 neutrons, none with 19, four with 20, and none with 21. The numbers of nucleons that correspond to nuclear closed shells are called **magic numbers**.

3. Nuclei with even numbers of both protons and neutrons are generally more stable than those with odd numbers of nucleons, as shown in Table 21.3.
4. The stability of a nucleus can be correlated to a certain degree with its neutron-to-proton ratio. All nuclei with two or more protons contain neutrons. Neutrons apparently help to hold protons together within the nucleus. As shown in Figure 21.2, the number of neutrons necessary to create a stable nucleus increases rapidly as the number of protons increases; the neutron-to-proton ratios of stable nuclei increase with increasing atomic number. The area within which all stable nuclei are found is known as the *belt of stability.*

* There are theoretical predictions that a nucleus with 114 protons or 184 neutrons will also be a closed shell. These predictions are being used in the search for new elements beyond the currently known 109.

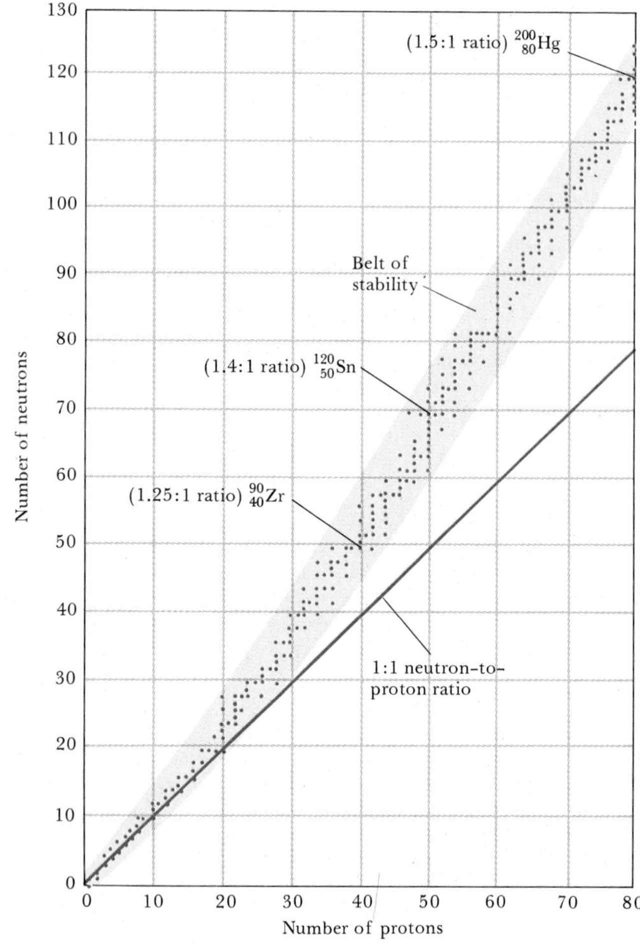

Figure 21.2 Plot of the number of neutrons versus the number of protons in stable nuclei. As the atomic number increases, the neutron-to-proton ratio of the stable nuclei increases. The stable nuclei are located in the shaded area of the graph known as the belt of stability. The majority of radioactive nuclei occur outside this belt.

SAMPLE EXERCISE 21.3

Would you expect the following nuclei to be radioactive: ${}^{4}_{2}He$, ${}^{39}_{20}Ca$, ${}^{210}_{85}At$?

Solution: Helium-4 has a magic number of protons and of neutrons (two each). We would therefore expect ${}^{4}_{2}He$ to be stable.

Calcium-39 has an even number of protons (20) and an odd number of neutrons (19); 20 is one of the magic numbers. Nevertheless, we should suspect that this nuclide is radioactive, because the neutron-to-proton ratio is less than 1. This ratio would place it below the belt of stability.

Astatine-210 is radioactive. Recall that there are no stable nuclei beyond atomic number 83.

PRACTICE EXERCISE

Which of the following nuclides of group 6A elements are likely to be unstable: ${}^{14}_{8}O$, ${}^{32}_{16}S$, ${}^{78}_{34}Se$, ${}^{84}_{34}Se$, ${}^{115}_{52}Te$, ${}^{208}_{84}Po$? ***Answer:*** ${}^{14}_{8}O$, ${}^{84}_{34}Se$, ${}^{115}_{52}Te$, ${}^{208}_{84}Po$

The type of radioactive decay that a particular radioisotope will undergo depends to a large extent on its neutron-to-proton ratio compared to those of nearby nuclei that are within the belt of stability. Consider a nucleus whose high neutron-to-proton ratio places it above the belt of stability. This nucleus can lower its ratio and move toward the belt of stability by emitting a beta particle. Beta emission decreases the number of neutrons and increases the number of protons in a nucleus, as shown in Equation 21.5.

Nuclei that have low neutron-to-proton ratios and thus lie below the belt of stability either emit positrons or undergo electron capture. Both modes of decay decrease the number of protons and increase the number of neutrons in the nucleus, as shown in Equations 21.7 and 21.9. Positron emission is more common than electron capture among the lighter nuclei; however, electron capture becomes increasingly common as nuclear charge increases.

Alpha emission is found primarily among nuclei with an atomic number greater than 83. These nuclei lie beyond the upper right edge of Figure 21.2, outside the belt of stability. Emission of an alpha particle moves the nucleus diagonally toward the belt of stability by decreasing both the number of protons and the number of neutrons by 2. The result of each type of radioactive decay relative to a stable nucleus is shown in Figure 21.3.

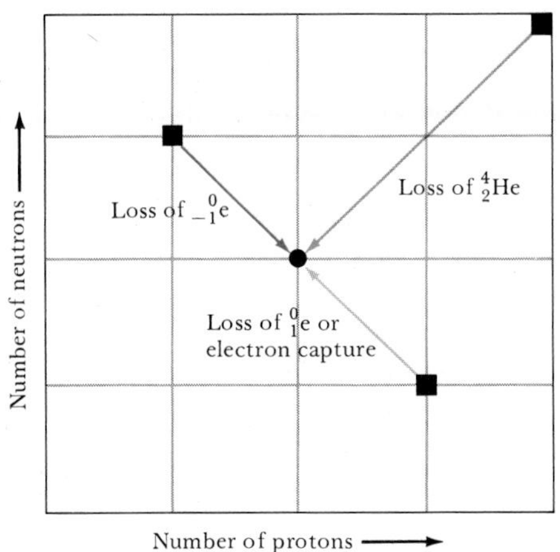

Figure 21.3 Result of alpha emission (${}^{4}_{2}He$), beta emission (${}^{0}_{-1}e$), positron emission (${}^{0}_{1}e$), and electron capture on the number of protons and neutrons in a nucleus. The squares represent unstable nuclei, and the circle represents a stable one. Moving from left to right or from bottom to top, each tick mark represents an additional proton or neutron, respectively. Moving in the reverse direction indicates the loss of a proton or neutron.

SAMPLE EXERCISE 21.4

By referring to Figure 21.2, predict the mode of radioactive decay of the following nuclei: **(a)** $^{20}_{11}Na$; **(b)** $^{97}_{40}Zr$; **(c)** $^{235}_{92}U$.

Solution: To answer this question we use the guidelines given above in the text.

(a) This nucleus has a neutron-to-proton ratio below 1. It therefore lies below the belt of stability. It can gain stability either by positron emission or electron capture. Because the atomic number is small we might predict that the nucleus undergoes positron emission. If we refer to a standard reference such as the *Handbook of Chemistry and Physics*, we find that this prediction is correct. The nuclear reaction is

$$^{20}_{11}Na \longrightarrow {}^{0}_{1}e + {}^{20}_{10}Ne$$

(b) In referring to Figure 21.2, we find that this nucleus has a neutron-to-proton ratio that is too high. We would therefore predict that it undergoes beta decay. Again, the prediction is correct. The nuclear reaction is

$$^{97}_{40}Zr \longrightarrow {}^{0}_{-1}e + {}^{97}_{41}Nb$$

(c) This nucleus lies outside the belt of stability to the upper right. We might therefore predict that it would undergo alpha emission. Again, the prediction is correct. The nuclear equation is

$$^{235}_{92}U \longrightarrow {}^{4}_{2}He + {}^{231}_{90}Th$$

PRACTICE EXERCISE

Predict the mode of decay of the following unstable nuclei: **(a)** ^{15}O; **(b)** ^{139}Xe; **(c)** ^{212}Po. ***Answers:*** **(a)** positron emission; **(b)** beta emission; **(c)** alpha emission

At this point we should note that our guidelines don't always work. For example, thorium-233, $^{233}_{90}Th$, which we might expect to undergo alpha decay, actually undergoes beta decay. Furthermore, a few radioactive nuclei actually

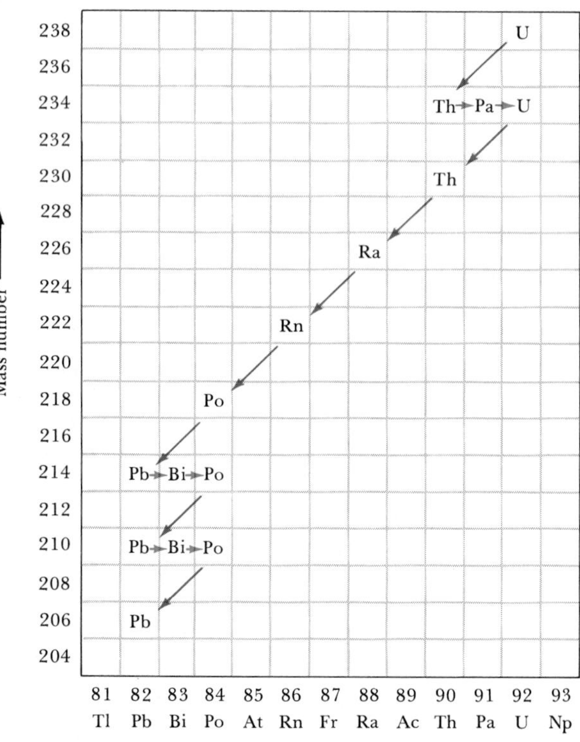

Figure 21.4 Nuclear disintegration series for uranium-238. The $^{238}_{92}U$ nucleus decays to $^{234}_{90}Th$. Subsequent decay processes eventually form the stable $^{206}_{82}Pb$ nucleus. Each of the blue arrows corresponds to the loss of an alpha particle. Each red arrow corresponds to the loss of a beta particle.

lie within the belt of stability. For example, both $^{146}_{60}Nd$ and $^{148}_{60}Nd$ are stable and lie in the belt of stability; however, $^{147}_{60}Nd$, which lies between them, is radioactive.

Radioactive Series

Some nuclei, like uranium-238, cannot gain stability by a single emission. Consequently, a series of successive emissions occurs. As shown in Figure 21.4, uranium-238 decays to thorium-234, which is radioactive and decays to protactinium-234. This nucleus is also unstable and subsequently decays. Such successive reactions continue until a stable nucleus, lead-206, is formed. A series of nuclear reactions that begins with an unstable nucleus and terminates with a stable one is known as a **radioactive series** or a **nuclear disintegration series**. Three such series occur in nature. In addition to the series that begins with uranium-238 and terminates with lead-206, there is one that begins with uranium-235 and ends with lead-207. The third series begins with thorium-232 and ends with lead-208.

21.3 PREPARATION OF NEW NUCLEI

In 1919, Ernest Rutherford performed the first artificial conversion of one nucleus into another. He succeeded in converting nitrogen-14 into oxygen-17, using the high-velocity alpha (α) particles emitted by radium. The reaction is

$$^{14}_{7}N + ^{4}_{2}He \longrightarrow ^{17}_{8}O + ^{1}_{1}H \qquad [21.10]$$

(The symbols $^{1}_{1}H$ and $^{1}_{1}p$ are equivalent; both represent a proton.) This reaction demonstrated that nuclear reactions can be induced by striking nuclei with particles such as alpha particles. Such reactions have permitted synthesis of hundreds of radioisotopes in the laboratory. As noted in Section 21.1, these conversions of one nucleus into another are called nuclear transmutations. Such conversions are commonly represented by listing, in order, the target nucleus, the bombarding particle, the ejected particle, and the product nucleus. Written in this fashion, Equation 21.10 is $^{14}_{7}N(\alpha, p)^{17}_{8}O$. The alpha particle, proton, and neutron are abbreviated as α, p, and n, respectively.

SAMPLE EXERCISE 21.5

Write the balanced nuclear equation for the process summarized as $^{27}_{13}Al(n, \alpha)^{24}_{11}Na$.

Solution: The n is the abbreviation for a neutron, and α represents an alpha particle. The neutron is the bombarding particle, and the alpha particle is a product. Therefore, the nuclear equation is

$$^{27}_{13}Al + ^{1}_{0}n \longrightarrow ^{24}_{11}Na + ^{4}_{2}He$$

PRACTICE EXERCISE

Write the nuclear reaction

$$^{16}_{8}O + ^{1}_{1}H \longrightarrow ^{13}_{7}N + ^{4}_{2}He$$

in a shorthand notation. ***Answer:*** $^{16}_{8}O(p, \alpha)^{13}_{7}N$

Charged particles, such as alpha particles, must be moving very fast in order to overcome the electrostatic repulsion between them and the target nucleus. The higher the nuclear charge on either the projectile or the target,

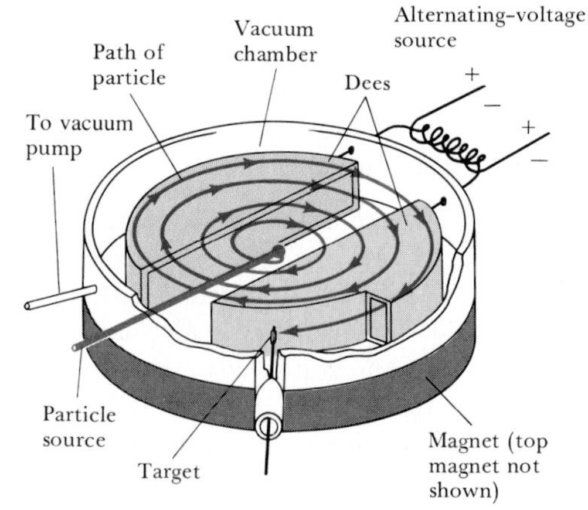

Figure 21.5 Schematic drawing of a cyclotron. Charged particles are accelerated around the ring by the application of alternating voltage to the dees.

the faster the projectile must be moving to bring about a nuclear reaction. Many methods have been devised to accelerate charged particles using strong magnetic and electrostatic fields. These **particle accelerators**, popularly called "atom smashers," bear such names as the *cyclotron* and *synchrotron.* The cyclotron is illustrated in Figure 21.5. The hollow D-shaped electrodes are called "dees." The projectile particles are introduced into a vacuum chamber within the cyclotron. The particles are then accelerated by making the dees alternately positively and negatively charged. Magnets placed above and below the dees keep the particles moving in a spiral path until they are finally deflected out of the cyclotron and emerge to strike a target substance. Particle accelerators have been used mainly to synthesize heavy elements and to investigate the fundamental structure of matter. Figure 21.6 shows an aerial view of Fermilab, the

Figure 21.6 An aerial view of the Fermi National Accelerator Laboratory at Batavia, Illinois. Particles are accelerated to very high energies by circulating them through magnets in the ring, which has a circumference of 6.3 km. (Fermilab Photo Dept.)

National Accelerator Laboratory near Chicago. This facility may one day be supplanted by the even larger and higher-energy Superconducting Supercollider, which is currently proposed for construction in Texas.

Most synthetic isotopes used in quantity in medicine and scientific research are made using neutrons as projectiles. Because neutrons are neutral, they are not repelled by the nucleus; consequently, they do not need to be accelerated, as do the charged particles, in order to cause nuclear reactions (indeed, they cannot be so accelerated). The necessary neutrons are produced by the reactions that occur in nuclear reactors (Section 21.7). Cobalt-60, used in radiation therapy for cancer, is produced by neutron capture. Iron-58 is placed in a nuclear reactor, where it is bombarded by neutrons. The following sequence of reactions takes place:

$$^{58}_{26}\text{Fe} + {}^{1}_{0}\text{n} \longrightarrow {}^{59}_{26}\text{Fe} \qquad [21.11]$$

$$^{59}_{26}\text{Fe} \longrightarrow {}^{59}_{27}\text{Co} + {}^{0}_{-1}\text{e} \qquad [21.12]$$

$$^{59}_{27}\text{Co} + {}^{1}_{0}\text{n} \longrightarrow {}^{60}_{27}\text{Co} \qquad [21.13]$$

Transuranium Elements

Artificial transmutations have been used to produce the elements from atomic number 93 to 109. These are known as the **transuranium elements**, because they occur immediately following uranium in the periodic table. Elements 93 (neptunium) and 94 (plutonium) were first discovered in 1940. They were produced by bombarding uranium-238 with neutrons, as shown in Equations 21.14 and 21.15:

$$^{238}_{92}\text{U} + {}^{1}_{0}\text{n} \longrightarrow {}^{239}_{92}\text{U} \longrightarrow {}^{239}_{93}\text{Np} + {}^{0}_{-1}\text{e} \qquad [21.14]$$

$$^{239}_{93}\text{Np} \longrightarrow {}^{239}_{94}\text{Pu} + {}^{0}_{-1}\text{e} \qquad [21.15]$$

Elements with larger atomic numbers are normally formed in small quantities in particle accelerators. For example, curium-242 is formed when a plutonium-239 target is struck with accelerated alpha particles:

$$^{239}_{94}\text{Pu} + {}^{4}_{2}\text{He} \longrightarrow {}^{242}_{96}\text{Cm} + {}^{1}_{0}\text{n} \qquad [21.16]$$

21.4 HALF-LIFE

Our discussions to this point may have raised some questions in your mind. For example, why are some radioisotopes, such as uranium-238, found in nature, whereas others are not and must be synthesized? The key to answering this question is to realize that different nuclei undergo radioactive decay at different rates. Many radioisotopes decay essentially completely in a matter of seconds; obviously, we do not find such nuclei in nature. In contrast, uranium-238 decays very slowly; therefore despite its instability, we can still observe this isotope in nature. An important characteristic of a radioisotope is its rate of radioactive decay.

Radioactive decay is a first-order process. As shown in Section 14.3, a first-order process has a characteristic **half-life**, which is the time required for half of any given quantity of a substance to react. The rates of decay of nuclei are commonly discussed in terms of their half-lives.

Figure 21.7 Decay of a 10.0-g sample of $^{90}_{38}Sr$ ($t_{1/2}$ = 28.8 yr).

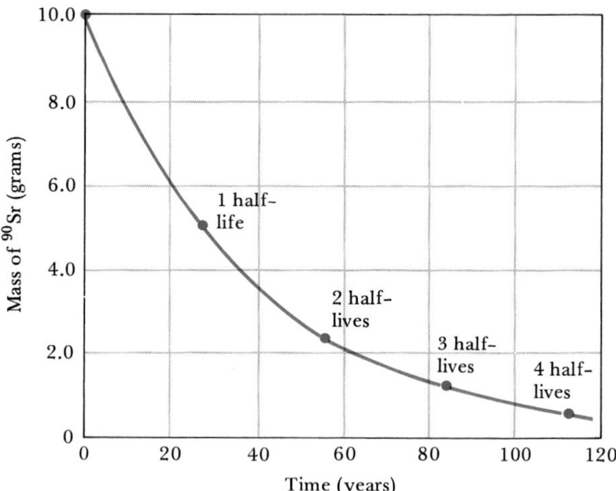

Each isotope has its own characteristic half-life. For example, the half-life of strontium-90 is 29 yr. If we started with 10.0 g of strontium-90, only 5.0 g of that isotope would remain after 29 yr. The other half of the strontium-90 would have been converted to yttrium-90, as shown in Equation 21.17:

$$^{90}_{38}\text{Sr} \longrightarrow {}^{90}_{39}\text{Y} + {}^{\ 0}_{-1}\text{e} \qquad [21.17]$$

After another 29-yr period, half the remaining 5.0 g of strontium-90 would likewise decay. The loss of strontium-90 as a function of time is shown in Figure 21.7.

Half-lives as short as millionths of a second and as long as billions of years have been observed. The half-lives of some radioisotopes are listed in Table 21.4. One important feature of half-lives for nuclear decay is that they are unaffected by external conditions such as temperature, pressure, or state of chemical combination. Therefore, unlike toxic chemicals, radioactive atoms cannot be rendered harmless by chemical reaction or by any other practical treatment. At this point, we can do nothing but allow these nuclei to lose radioactivity at their characteristic rates. In the meantime, of course, we must take maximum precautions to isolate the radioisotopes because of the damage radiation can cause (see Section 21.9).

Table 21.4 Some Radioactive Isotopes and Their Half-Lives and Type of Decay

	Isotope	Half-life (yr)	Type of decay
Natural radioisotopes	$^{238}_{92}U$	4.5×10^9	Alpha
	$^{235}_{92}U$	7.1×10^8	Alpha
	$^{232}_{90}Th$	1.4×10^{10}	Alpha
	$^{40}_{19}K$	1.3×10^9	Beta
	$^{14}_{6}C$	5,730	Beta
Synthetic radioisotopes	$^{239}_{94}Pu$	24,000	Alpha
	$^{137}_{55}Cs$	30	Beta
	$^{90}_{38}Sr$	28.8	Beta
	$^{131}_{53}I$	0.022	Beta

SAMPLE EXERCISE 21.6

The half-life of cobalt-60 is 5.3 yr. How much of a 1.000-mg sample of cobalt-60 is left after a 15.9-yr period?

Solution: A period of 15.9 yr is three half-lives for cobalt-60. At the end of one half-life 0.500 mg of cobalt-60 remains, 0.250 mg at the end of two half-lives, and 0.125 mg at the end of three half-lives.

PRACTICE EXERCISE

Carbon-11, used in medical imaging, has a half-life of 20.4 min. The carbon-11 nuclides are formed and then incorporated into a desired compound. The resulting sample is injected into a patient, and the medical image is obtained. The entire process lasts five half-lives. What percentage of the original carbon-11 activity remains at this time? ***Answer:*** 3.1 percent

Dating

Because the half-life of any particular nuclide is constant, the half-life can serve as a molecular clock to determine the ages of different objects. For example, carbon-14 has been used to determine the age of organic materials (see Figure 21.8). The procedure is based on the formation of carbon-14 by neutron capture in the upper atmosphere:

$$^{14}_{7}\text{N} + ^{1}_{0}\text{n} \longrightarrow ^{14}_{6}\text{C} + ^{1}_{1}\text{H} \qquad [21.18]$$

This reaction provides a small but reasonably constant source of carbon-14. The carbon-14 is radioactive, undergoing beta decay with a half-life of 5730 yr:

$$^{14}_{6}\text{C} \longrightarrow ^{14}_{7}\text{N} + ^{\ 0}_{-1}\text{e} \qquad [21.19]$$

In using radiocarbon dating, we generally assume that the ratio of carbon-14 to carbon-12 in the atmosphere has been constant for at least 50,000 yr. The carbon-14 is incorporated into carbon dioxide, which is in turn incorporated, through photosynthesis, into more complex carbon-containing molecules within plants. When the plants are eaten by animals, the carbon-14 becomes incorporated within them. Because a living plant or animal has a constant intake of carbon compounds, it is able to maintain a ratio of carbon-14

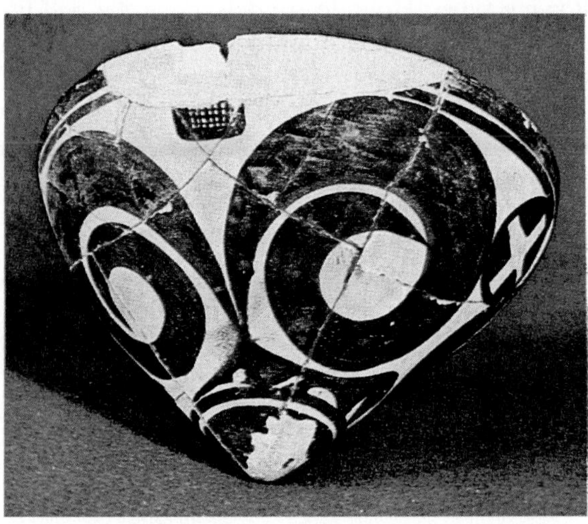

Figure 21.8 A painted bowl from ancient Persia. This object was made over 5000 years ago, as determined by carbon-14 dating. (The Granger Collection)

to carbon-12 that is identical with that of the atmosphere. However, once the organism dies, it no longer ingests carbon compounds to replenish the carbon-14 that is lost through radioactive decay. The ratio of carbon-14 to carbon-12 therefore decreases. By calculating this ratio and contrasting it to that of the atmosphere, we can estimate the age of an object. For example, if the ratio diminishes to half that of the atmosphere, we can conclude that the object is one half-life, or 5730 yr, old. This method cannot be used to date objects older than about 20,000 to 50,000 yr. After this length of time, the radioactivity is too low to be measured accurately.

The radiocarbon-dating technique has been checked by comparing the ages of trees determined by counting their rings and by radiocarbon analysis. As a tree grows, it adds a ring each year. In the old growth, the carbon-14 decays, while the concentration of carbon-12 remains constant. The two dating methods agree to within about 10 percent. Most of the wood used in these tests was from California bristle-cone pines, which reach ages up to 2000 yr. By using trees that died at a known time thousands of years ago, it is possible to make comparisons back to about 5000 B.C.

Other isotopes can be similarly used to date other types of objects. For example, it takes 4.5×10^9 yr for half of a sample of uranium-238 to decay to lead-206. The age of rocks containing uranium can therefore be determined by measuring the ratio of lead-206 to uranium-238. If the lead-206 had somehow become incorporated into the rock by normal chemical processes instead of by radioactive decay, the rock would also contain large amounts of the more abundant isotope lead-208. In the absence of large amounts of this "geonormal" isotope of lead, it is assumed that all of the lead-206 was at one time uranium-238.

The oldest rocks found on the earth are approximately 3×10^9 yr old. This age indicates that the crust of the earth has been solid for at least this length of time. Scientists estimate that it required $1\text{–}1.5 \times 10^9$ yr for the earth to cool and its surface to become solid. This places the age of the earth at 4.0 to 4.5×10^9 (about 4.5 billion) yr.

Calculations Based on Half-Life

So far our discussion has been mainly qualitative. We now consider the topic of half-lives from a more quantitative point of view. This approach enables us to answer questions of the following types: How do we determine the half-life of uranium-238? Similarly, how do we determine quantitatively the age of an object that can be dated by radiometric means?

The rate of radioactive decay of any radioisotope is first order. It can therefore be described by Equation 21.20:

$$\text{Rate} = kN \qquad [21.20]$$

where N is the number of nuclei of a particular radioisotope, and k is the first-order rate constant. This equation can be transformed into Equation 21.21. (Note that the equations that follow are the same as those we encountered in Section 14.3 in dealing with first-order chemical processes.)

$$\ln \frac{N_t}{N_0} = -kt \qquad [21.21]$$

In this equation, t is the time interval of decay, k is the rate constant, N_0 is the initial number of nuclei (at zero time), and N_t is the number remaining after the time interval. The relationship between the rate constant, k, and half-life, $t_{1/2}$, is given by Equation 21.22:

$$k = \frac{0.693}{t_{1/2}} \qquad [21.22]$$

SAMPLE EXERCISE 21.7

A rock contains 0.257 mg of lead-206 for every milligram of uranium-238. The half-life for the decay of uranium-238 to lead-206 is 4.5×10^9 yr. How old is the rock?

Solution: Let us assume that the rock contains 1.000 mg of uranium-238 at present. The amount of uranium-238 in the rock when it was first formed therefore equals 1.000 mg plus the quantity that decayed to lead-206. We obtain the latter quantity by multiplying the present mass of lead-206 by the ratio of the atomic mass of uranium to that of lead, into which it has decayed. The total original $^{238}_{92}U$ was thus

$$\text{Original } ^{238}_{92}\text{U} = 1.000 \text{ mg} + \frac{238}{206}(0.257 \text{ mg})$$
$$= 1.297 \text{ mg}$$

Using Equation 21.22, we can calculate the rate constant for the process from its half-life:

$$k = \frac{0.693}{4.5 \times 10^9 \text{ yr}} = 1.5 \times 10^{-10} \text{ yr}^{-1}$$

Rearranging Equation 21.21 to solve for time, t, and substituting known quantities gives

$$t = -\frac{1}{k} \ln \frac{N_t}{N_0} = -\frac{1}{1.5 \times 10^{-10} \text{ yr}^{-1}} \ln \frac{1.000}{1.297} = 1.7 \times 10^9 \text{ yr}$$

PRACTICE EXERCISE

A wooden object from an archeological site is reduced to carbon. The radioactivity of the sample due to ^{14}C is measured to be 12.4 disintegrations per second. The radioactivity of a carbon sample of equal mass from fresh wood is 19.5 disintegrations per second. The half-life of ^{14}C is 5730 yr. What is the age of the archeological sample? ***Answer:*** 3740 yr

SAMPLE EXERCISE 21.8

If we start with 1.000 g of strontium-90, 0.953 g will remain after 2.00 yr. **(a)** What is the half-life of strontium-90? **(b)** How much strontium-90 will remain after 5.00 yr?

Solution: **(a)** Equation 21.21 is solved for the rate constant, k, and then Equation 21.22 is used to calculate half-life, $t_{1/2}$:

$$k = -\frac{1}{t} \ln \frac{N_t}{N_0} = -\frac{1}{2.00 \text{ yr}} \ln \frac{0.953 \text{ g}}{1.000 \text{ g}}$$
$$= -\frac{1}{2.00 \text{ yr}}(-0.0481) = 0.0241 \text{ yr}^{-1}$$

$$t_{1/2} = \frac{0.693}{k} = \frac{0.693}{0.0241 \text{ yr}^{-1}} = 28.8 \text{ yr}$$

(b) Again using Equation 21.21, with $k = 0.0241\ \text{yr}^{-1}$, we have

$$\ln \frac{N_t}{N_0} = -kt = -(0.0241\ \text{yr}^{-1})(5.00\ \text{yr}) = -0.120$$

Calculation of N_t/N_0 from $\ln(N_t/N_0) = -0.120$ is readily accomplished using the e^x or INV LN function of a calculator:

$$\frac{N_t}{N_0} = e^{-0.120} = 0.887$$

Because $N_0 = 1.000$ g, we have

$$N_t = 0.887N_0 = 0.887(1.000\ \text{g}) = 0.887\ \text{g}$$

PRACTICE EXERCISE

A sample to be used for medical imaging is labeled with ^{18}F, which has a half-life of 110 min. What percentage of the original activity in the sample remains after 300 min? ***Answer:*** 15.1 percent

21.5 DETECTION OF RADIOACTIVITY

A variety of methods have been devised to detect emissions from radioactive substances. Becquerel discovered radioactivity because of the effect of radiation on photographic plates. Photographic plates and film have long been used to detect radioactivity. The radiation affects photographic film in the same way as visible light does. With care, film can be used to give a quantitative measure of activity. The greater the extent of exposure to radiation, the darker the area of the developed negative. People who work with radioactive substances carry film badges to record the extent of their exposure to radiation.

Radioactivity can also be detected and measured using a device known as a **Geiger counter**. The operation of a Geiger counter is based on the ionization of matter caused by radiation (Section 21.9). The ions and electrons produced by the ionizing radiation permit conduction of an electrical current. The basic design of a Geiger counter is shown in Figure 21.9. It consists of a metal tube filled with gas. The cylinder has a "window" made of material that can be penetrated by alpha, beta, or gamma rays. In the center of the tube is a wire. The wire is connected to one terminal of a source of direct current, and the metal cylinder is attached to the other terminal. Current flows between the wire and metal cylinder whenever ions are produced by entering radiation. The current pulse created when radiation enters the tube is amplified; each pulse is counted as a measure of the amount of radiation.

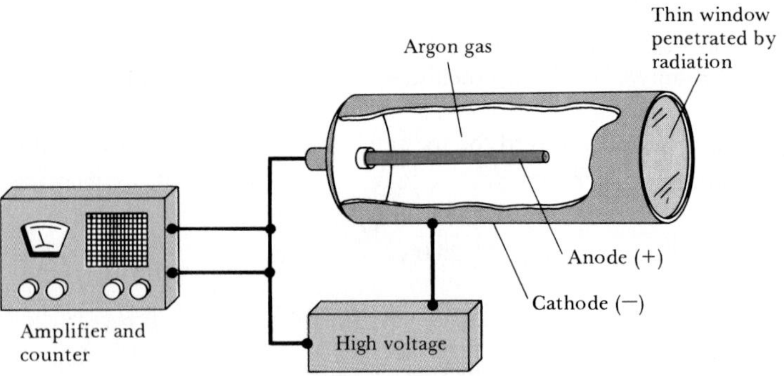

Figure 21.9 Schematic representation of a Geiger counter.

Figure 21.10 The hands and numbers of this luminous watch contain minute amounts of a radium salt. The radioactive decay of radium causes the watch to glow in the dark. (© Richard Megna/Fundamental Photographs)

Certain substances that are electronically excited by radiation can also be used as means for detecting and measuring radiation. Some substances excited by radiation give off light (fluoresce) as electrons return to their lower-energy states. For example, dials of luminous watches used to be painted with a mixture of ZnS and a tiny quantity of $RaSO_4$ (Figure 21.10). The zinc sulfide fluoresces when struck by the radioactive emissions from the radium. An instrument known as a **scintillation counter** can be used to detect and measure fluorescence and thereby the radiation that causes it.

Radiotracers

Because radioisotopes can be detected so readily, they can be used to follow an element through its chemical reactions. For example, the incorporation of carbon atoms from CO_2 into glucose in photosynthesis has been studied using CO_2 containing carbon-14:

$$6CO_2 + 6H_2O \xrightarrow[\text{chlorophyll}]{\text{sunlight}} C_6H_{12}O_6 + 6O_2 \qquad [21.3]$$

The CO_2 is said to be labeled with the carbon-14. Detection devices such as scintillation counters follow the carbon-14 as it moves from the CO_2 through the various intermediate compounds to glucose.

Such use of radioisotopes is possible because all isotopes of an element have essentially identical chemical properties. When a small quantity of a radioisotope is mixed with the naturally occurring stable isotopes of the same element, all of the isotopes go through the same reactions together. The element's path is revealed by the radioactivity of the radioisotope. Because the radioisotope can be used to trace the path of the element, it is called a **radiotracer**.

CHEMISTRY AT WORK: Medical Applications of Radiotracers

Radiotracers have found wide use as a diagnostic tool in medicine. For example, iodine-131 has been used to test the activity of the thyroid gland. This gland is the only important user of iodine in the body. The patient drinks a solution of NaI containing iodine-131. Only a very small amount is used so that the patient does not receive a harmful dose of radioactivity. A Geiger tube placed close to the thyroid, in the neck region, determines the ability of the thyroid to take up the iodine. A normal thyroid will absorb about 12 percent of the iodine within a few hours.

In 1977, Rosalyn S. Yalow received the Nobel Prize in medicine for her pioneering work in developing the technique of radioimmunoassay (RIA). This technique is an extraordinarily sensitive method involving radiotracers for detecting the presence of minute amounts of drugs, hormones, peptides, antibiotics, and a host of other substances. RIA methods can be used, for example, to provide an early indication of pregnancy and to detect substances that signal the early stages of a disease. The tests are carried out on a small sample of tissue, blood, or other fluid taken from the subject.

A new, very promising tool for clinical diagnosis of many diseases is called positron emission transaxial tomography (PET). In this method, compounds that contain radionuclides that decay by positron emission are injected into a patient. These compounds are chosen to enable researchers to monitor blood flow, oxygen and glucose metabolic rates, and other biological functions. Some of the most interesting work involves the study of the brain, which depends on glucose for most of its energy. Changes in how this sugar is metabolized or used by the brain may signal a disease such as cancer, epilepsy, Parkinson's disease, or schizophrenia.

The compound to be detected in the patient must be labeled with a radionuclide that is a positron emitter. The most widely used nuclides are carbon-11 (half-life 20.4 min), fluorine-18 (half-life 110 min), oxygen-15 (half-life 2 min), and nitrogen 13 (half-life 10 min). As an example, glucose can be labeled with ^{11}C. Because the half-lives of positron emitters are so short, the chemist must quickly incorporate the radionuclide into the sugar (or other appropriate) molecule and inject the compound immediately. The patient is placed in an elaborate instrument [Figure 21.11(*a*)] that measures the positron emission and constructs a computer-based image of the organ in which the emitting compound is localized. The nature of this image [Figure 21.11(*b*)] provides clues as to the presence of disease or other abnormality and helps medical researchers understand how a particular disease affects the functioning of the brain.

Figure 21.11 (*a*) An instrument for determining positron emission transaxial tomography (PET). The patient is injected with a solution of radiolabeled compound that quickly moves to the brain. Radioactive nuclei within the compound emit positrons. The PET instrument measures the positron emissions and develops a three-dimensional image of the brain. The level of radioactivity employed in such clinical work is not harmful to the patient. (*b*) PET images of human brains. The red line in the figure at lower right shows the two-dimensional slice through the brain that appears in the images. The brain of a normal person and those of patients suffering from schizophrenia and manic depression show different image patterns, suggesting different patterns of metabolism. (Brookhaven National Laboratory and New York University Medical Center)

(*a*)

(*b*)

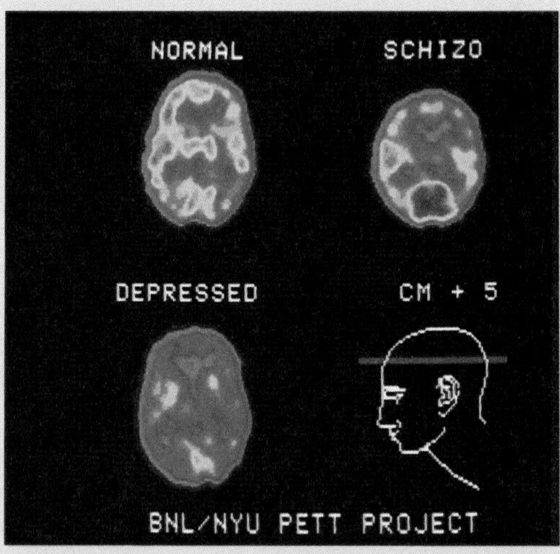

21.6 MASS-ENERGY CONVERSIONS

So far we have said little about the energies associated with nuclear reactions. These energies can be considered with the aid of Einstein's famous equation relating mass and energy:

$$E = mc^2 \quad [21.24]$$

In this equation, E stands for energy, m for mass, and c for the speed of light, 3.00×10^8 m/s. This equation states that the mass and energy of an object are proportional. The greater an object's mass, the greater its energy. Because the proportionality constant in the equation, c^2, is such a large number, even small changes in mass are accompanied by large changes in energy.

The mass changes in chemical reactions are too small to detect easily. For example, the mass change associated with the combustion of a mole of CH_4 is 9.9×10^{-9} g. For this reason, it is possible to speak of the conservation of mass in chemical reactions.

The mass changes and the associated energy changes in nuclear reactions are much greater than in chemical reactions. The energy released through the nuclear fission of only about a pound of uranium (Section 21.7) is equivalent to that released by combustion of 1500 tons of coal.

Nuclear Binding Energies

Scientists discovered in the 1930s that the masses of nuclei are always less than the masses of the individual nucleons of which they are composed. For example, the helium-4 nucleus has a mass of 4.00150 amu. The mass of a proton is 1.00728 amu, and that of a neutron is 1.00867 amu. Consequently, two protons and two neutrons have a total mass of 4.03190 amu:

$$\begin{aligned} \text{Mass of two protons} &= 2(1.00728 \text{ amu}) = 2.01456 \text{ amu} \\ \text{Mass of two neutrons} &= 2(1.00867 \text{ amu}) = \underline{2.01734 \text{ amu}} \\ \text{Total mass} &= 4.03190 \text{ amu} \end{aligned}$$

The individual nucleons weigh 0.03040 amu more than the helium-4 nucleus:

$$\begin{aligned} \text{Mass of two protons and two neutrons} &= 4.03190 \text{ amu} \\ \text{Mass of } {}^4_2\text{He nucleus} &= \underline{4.00150 \text{ amu}} \\ \text{Mass difference} &= 0.03040 \text{ amu} \end{aligned}$$

The mass difference between a nucleus and its constituent nucleons is called the **mass defect**. The origin of the mass defect is readily understood if we consider that energy must be added to a nucleus in order to break it into separated protons and neutrons:

$$\text{Energy} + {}^4_2\text{He} \longrightarrow 2\,{}^1_1\text{p} + 2\,{}^1_0\text{n} \quad [21.25]$$

According to Einstein's mass-energy equivalence relationship (Equation 21.24), the addition of energy to a system must be accompanied by a proportional increase in mass. The mass change, Δm, is defined as the total mass of the products minus the total mass of the reactants. The mass change for the conversion of helium-4 into separated nucleons is $\Delta m = 0.03040$ amu, as shown in the calculations above. The associated energy change, therefore, is readily calculated:

Table 21.5 Mass Differences and Binding Energies for Three Nuclei

Nucleus	Mass of nucleus (amu)	Mass of individual nucleons (amu)	Mass difference (amu)	Binding energy (J)	Binding energy per nucleon (J)
$^{4}_{2}He$	4.00150	4.03190	0.0304	4.54×10^{-12}	1.14×10^{-12}
$^{56}_{26}Fe$	55.92066	56.44938	0.52872	7.90×10^{-11}	1.41×10^{-12}
$^{238}_{92}U$	238.0003	239.9356	1.9353	2.89×10^{-10}	1.22×10^{-12}

$$\Delta E = \Delta(mc^2) = c^2\,\Delta m$$
$$= (3.00 \times 10^8\ \text{m/s})^2(0.0304\ \text{amu})\left(\frac{1.00\ \text{g}}{6.02 \times 10^{23}\ \text{amu}}\right)\left(\frac{1\ \text{kg}}{1000\ \text{g}}\right)$$
$$= 4.54 \times 10^{-12}\ \frac{\text{kg-m}^2}{\text{s}^2} = 4.54 \times 10^{-12}\ \text{J}$$

Decomposition of a mole of helium-4 in this fashion would require a tremendous quantity of energy:

$$\left(6.02 \times 10^{23}\ \frac{\text{nuclei}}{\text{mol}}\right)\left(4.54 \times 10^{-12}\ \frac{\text{J}}{\text{nucleus}}\right) = 2.73 \times 10^{12}\ \text{J/mol}$$

The energy change calculated from the mass defect of a nucleus is called the *binding energy* of the nucleus. It is the energy required to decompose the nucleus into separated protons and neutrons. Therefore, the larger the binding energy, the more stable the nucleus is toward such decomposition. The mass defects and binding energies of three nuclei (helium-4, iron-56, and uranium-238) are compared in Table 21.5. The binding energies per nucleon (that is, the binding energy of each nucleus divided by the total number of nucleons in that nucleus) are also compared in the table. Similar calculations for other nuclei indicate that the binding energy per nucleon increases in magnitude to about 1.4×10^{-12} J for nuclei whose mass numbers are in the vicinity of iron-56. It then decreases slowly to about 1.2×10^{-12} J for very heavy nuclei. This trend is shown in Figure 21.12. These results indicate that heavy nuclei will

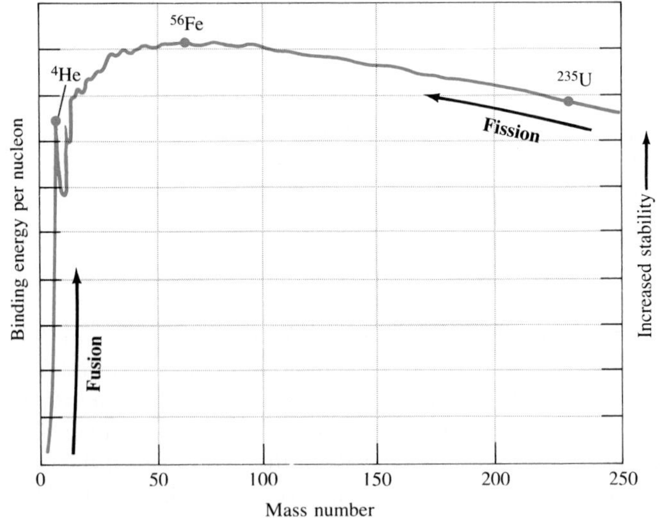

Figure 21.12 The average binding energy per nucleon increases to a maximum at a mass number of 50 to 60 and decreases slowly thereafter. As a result of these trends, fusion of light nuclei and fission of heavy nuclei are exothermic processes.

gain stability, and therefore give off energy, if they are fragmented. This process, known as **fission**, occurs in the atomic bomb and in nuclear power plants. Figure 21.12 also indicates that even greater amounts of energy should be available if very light nuclei are fused together. This **fusion** process takes place in the hydrogen bomb and is the essential energy-producing reaction in the sun. We shall look more closely at fission and fusion in Sections 21.7 and 21.8.

In calculating the mass change in a nuclear reaction, it is usually acceptable to employ the masses of the atoms containing the nuclei of interest, because the number of electrons in the reactants and products is usually the same. Thus, the difference in atomic masses is usually the same as the difference in nuclear masses.

SAMPLE EXERCISE 21.9

How much energy is lost or gained when a mole of cobalt-60 undergoes beta decay: ${}^{60}_{27}Co \longrightarrow {}^{0}_{-1}e + {}^{60}_{28}Ni$? The mass of the ${}^{60}_{27}Co$ atom is 59.9338 amu, and that of a ${}^{60}_{28}Ni$ atom is 59.9308 amu.

Solution: The products of the nuclear reaction are ${}^{60}_{28}Ni^{+}$ (the 27 electrons of cobalt are carried along in the reaction) and an electron (the beta particle). The mass of these products is just the mass of the neutral ${}^{60}_{28}Ni$ atom. The mass change in the reaction, therefore, is given by

$$\begin{aligned}\Delta m &= \text{mass of } {}^{60}_{28}\text{Ni atom} - \text{mass of } {}^{60}_{27}\text{Co atom}\\ &= 59.9308 \text{ amu} - 59.9338 \text{ amu}\\ &= -0.0030 \text{ amu}\end{aligned}$$

Because the mass decreases ($\Delta m < 0$), energy is released when cobalt-60 decays. For a mole of cobalt-60, $\Delta m = -0.0030$ g. The energy produced by the reaction can be calculated from this mass:

$$\begin{aligned}\Delta E &= c^2\,\Delta m\\ &= (3.00 \times 10^8 \text{ m/s})^2(-0.0030 \text{ g})\left(\frac{1 \text{ kg}}{1000 \text{ g}}\right)\\ &= -2.7 \times 10^{11}\,\frac{\text{kg-m}^2}{\text{s}^2} = -2.7 \times 10^{11} \text{ J}\end{aligned}$$

By comparison, it takes only 9×10^5 J to break all the chemical bonds in a mole of water.

PRACTICE EXERCISE

Positron emission from ${}^{11}C$,

$${}^{11}_{6}C \longrightarrow {}^{11}_{5}B + {}^{0}_{1}e$$

occurs with release of 2.87×10^{11} J per mole of ${}^{11}C$. What is the mass change per mole of ${}^{11}C$ in this nuclear reaction? ***Answer:*** 3.19×10^{-3} g

21.7 NUCLEAR FISSION

Our discussion of the energy changes in nuclear reactions (Section 21.6) revealed an important observation: Both the splitting of heavy nuclei (fission) and the union of light nuclei (fusion) are exothermic processes. Commercial nuclear power plants and the most common forms of nuclear weaponry depend on the process of nuclear fission for their operation. The first nuclear fission to be discovered was that of uranium-235. This nucleus, as well as those of uranium-233

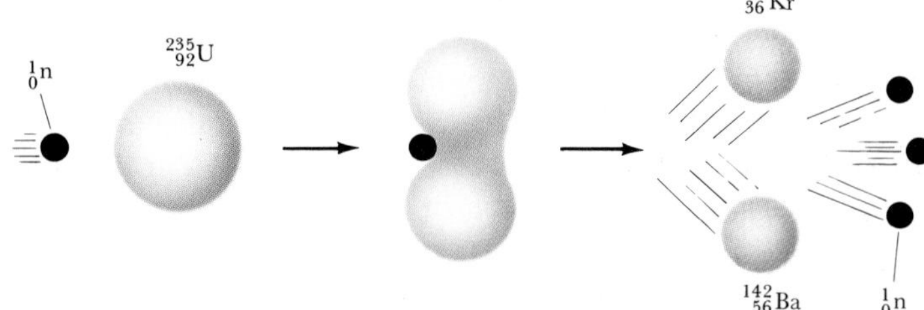

Figure 21.13 Schematic representation of the fission of uranium-235 showing one of its many fission patterns. In this process, 3.5×10^{-11} J of energy is produced per ^{235}U nucleus.

and plutonium-239, undergoes fission when struck by a slow-moving neutrons.* The fission process is illustrated in Figure 21.13. A heavy nucleus can split in many different ways, just as can a piece of glass. Two different ways that the uranium-235 nucleus splits are shown in Equations 21.26 and 21.27:

$$^1_0n + {}^{235}_{92}U \longrightarrow {}^{137}_{52}Te + {}^{97}_{40}Zr + 2^1_0n \qquad [21.26]$$

$$^1_0n + {}^{235}_{92}U \longrightarrow {}^{142}_{56}Ba + {}^{91}_{36}Kr + 3^1_0n \qquad [21.27]$$

Over 200 different isotopes of 35 different elements have been found among the fission products of uranium-235. Most of them are radioactive.

On the average, 2.4 neutrons are produced by every fission of uranium-235. If one fission produces 2 neutrons, these 2 neutrons can cause two fissions. The 4 neutrons thereby released can produce four fissions, and so forth, as shown in Figure 21.14. The number of fissions and the energy released quickly escalate, and, if the process is unchecked, the result is a violent explosion. Reactions that multiply in this fashion are called **chain reactions**.

In order for a fission chain reaction to occur, the sample of fissionable material must have a minimum mass. Otherwise, neutrons escape from the sample before they have the opportunity to strike another nucleus and cause additional fission. The chain stops if enough neutrons are lost. The amount of material is then said to be a **subcritical mass**. The amount of fissionable material large enough to maintain the chain reaction with a constant rate of fission is called the **critical mass**. When a critical mass of material is present, only one neutron from each fission is subsequently effective in producing another fission. The critical mass of uranium-235 is about 1 kg. If more than a critical mass of

* There are other heavy nuclei that can be induced to undergo fission. However, these three are the only ones of practical importance.

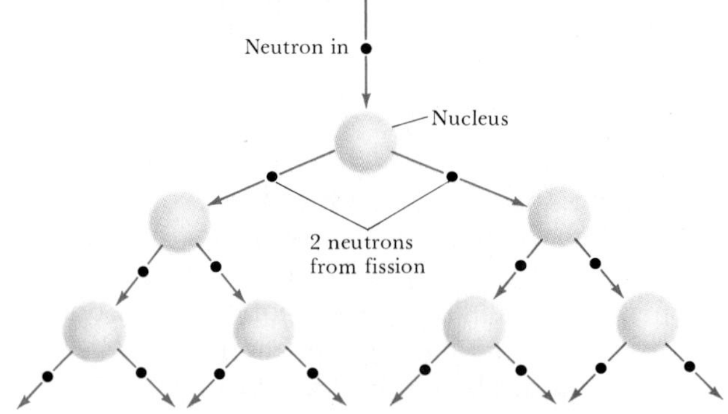

Figure 21.14 Chain fission reaction in which each fission produces two neutrons. The process leads to an accelerating rate of fission, with the number of fissions doubling at each stage.

fissionable material is present, very few neutrons escape, and the chain reaction multiplies the number of fissions, which can lead to a nuclear explosion. A mass in excess of a critical mass is referred to as a **supercritical mass**. The effect of mass on a fission reaction is illustrated in Figure 21.15.

Figure 21.16 shows a schematic diagram of the first atomic bomb used in warfare, the "Little Boy" bomb that was dropped on Hiroshima, Japan, on August 6, 1945. To trigger the fission reaction, two subcritical masses of uranium-235 are slammed together using chemical explosives. The combined masses of the uranium form a supercritical mass, which leads to a rapid, uncontrolled chain reaction and, ultimately, a nuclear explosion (Figure 21.17). The energy released by the bomb dropped on Hiroshima was equivalent to that of 40,000 lb of TNT (it therefore is called a *20-kiloton* bomb). Unfortunately, the basic design of a fission-based atomic bomb is quite simple. The fissionable materials are potentially available to any nation with a nuclear reactor. This simplicity has resulted in the proliferation of atomic weapons. These weapons could come into the hands of terrorist groups or groups intent on using the weapons to extort money from corporations or nations. Such prospects add a frightening aspect to the nuclear age in which we live.

Figure 21.15 The chain reaction in a subcritical mass soon stops because neutrons are lost from the mass without causing fission. As the size of the mass increases, fewer neutrons are able to escape. In a supercritical mass, the chain reaction is able to expand.

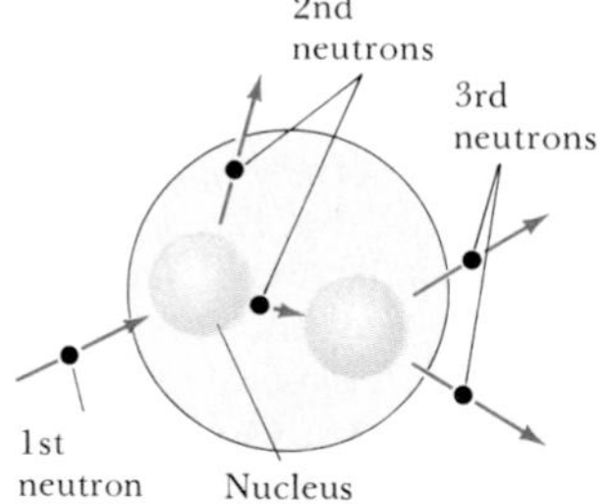

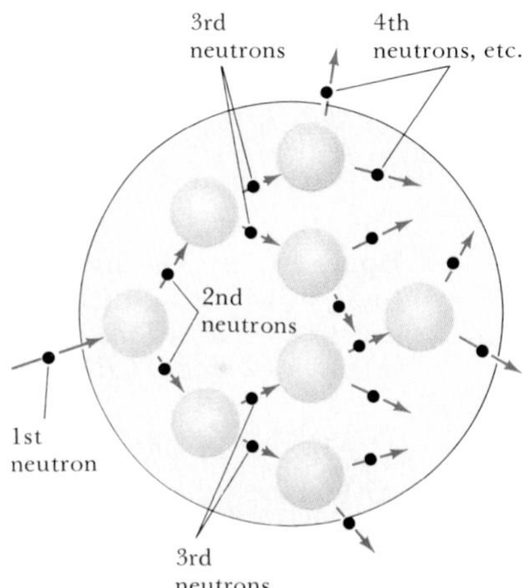

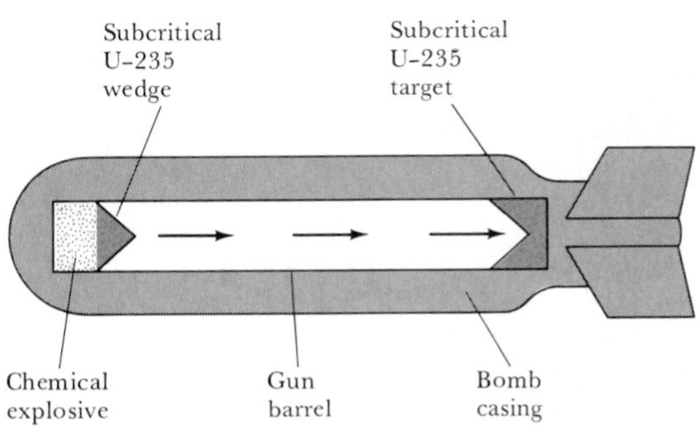

Figure 21.16 One design used in atomic bombs. A conventional explosive is used to bring two subcritical masses together to form a supercritical mass.

Figure 21.17 The explosion of a fission-based nuclear bomb. The characteristic mushroom-shaped cloud is one of the most frightening symbols of our entry into the atomic age. (U.S. Dept. of Navy, Naval Photographic Center)

A HISTORICAL PERSPECTIVE: The Dawning of the Nuclear Age

The fission of uranium-235 was first achieved in the late 1930s by Enrico Fermi and his colleagues in Rome and shortly thereafter by Otto Hahn and his co-workers in Berlin. Both groups were trying to produce transuranium elements. In 1938, Hahn identified barium among his reaction products. He was puzzled by this observation and questioned the identification because the presence of barium was so unexpected. He sent a detailed letter describing his experiments to Lise Meitner, a former co-worker. Meitner had been forced to leave Germany because of the anti-Semitism of the Third Reich and had settled in Sweden. She surmised that Hahn's experiment indicated a new nuclear process was occurring in which the uranium-235 split. She called this process *nuclear fission.* Meitner passed word of this discovery to her nephew, Otto Frisch, a physicist working at Niels Bohr's institute in Copenhagen. He repeated the experiment, verifying Hahn's observations and finding that tremendous energies were involved. In January 1939, Meitner and Frisch published a short article describing this new reaction. In March 1939, Leo Szilard and Walter Zinn at Columbia University discovered that more neutrons are produced than were used in each fission. As we have seen, this allows a chain reaction process to occur. News of these discoveries and an awareness of their potential use in explosive devices spread rapidly within the scientific community. Several scientists finally persuaded Albert Einstein, the most famous physicist of the time, to write a letter to President Roosevelt outlining the implications of these discoveries. Einstein's letter, written in August 1939, outlined the possible military applications of nuclear fission and emphasized the danger that weapons based on fission would pose if they were to be developed by the Nazis. Roosevelt judged it imperative that the United States investigate the possibility of such weapons. Late in 1941, the decision was made to build a bomb based on the fission reaction. An enormous research project, known as the "Manhattan Project," began. On December 2, 1942, the first artificial self-sustaining nuclear fission chain reaction was achieved in an abandoned squash court at the University of Chicago (Figure 21.18). This accomplishment led to the development of the first atomic bomb at Los Alamos National Laboratory in New Mexico (Figure 21.19) in July 1945. In August 1945, the United States dropped atomic bombs on two Japanese cities, Hiroshima and Nagasaki, leading to the end of World War II. The nuclear age had arrived.

Figure 21.18 The first self-sustaining nuclear fission reactor was built in a squash court at the University of Chicago. The painting depicts the scene in which the scientists witnessed the reactor as it became self-sustaining on December 2, 1942. (Argonne National Laboratory)

Figure 21.19 Los Alamos National Laboratory in New Mexico. The first atomic bombs were developed here in 1945. (Los Alamos National Laboratory)

Nuclear Reactors

Nuclear fission produces the energy generated by nuclear power plants. The "fuel" of the nuclear reactor is a fissionable substance, such as uranium-235. Typically, uranium is enriched to about 3 percent uranium-235 and then used in the form of UO_2 pellets. These enriched uranium pellets are encased in zirconium or stainless-steel tubes. Rods composed of materials such as cadmium or boron control the fission process by absorbing neutrons. These *control rods* regulate the flux of neutrons to keep the reaction chain self-sustaining, while preventing the reactor core from overheating.*

The reactor is started up by a neutron-emitting source; it is stopped by inserting the control rods more deeply into the reactor core, the site of the fission (Figure 21.20). The reactor core also contains a *moderator*, which acts to slow down neutrons so that they can be captured more readily by the fuel. A *cooling liquid* circulates through the reactor core to carry off the heat generated by the nuclear fission. The cooling liquid can also serve as the neutron moderator.

The design of a nuclear power plant is basically the same as that of a power plant that burns fossil fuel (except that the burner is replaced by a reactor core). In both instances, steam is used to drive a turbine connected to an electrical generator. The steam must be condensed, so additional cooling water, generally obtained from a large source such as a river or lake, is needed. The

Figure 21.20 Reactor core showing fuel elements, control rods, and cooling fluid.

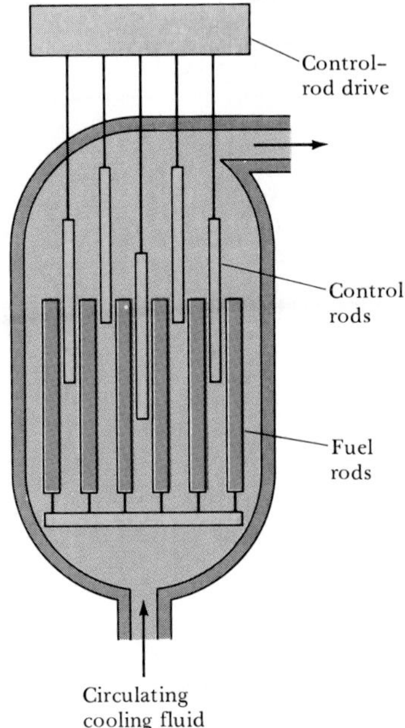

* The reactor core cannot reach supercritical levels and explode with the violence of an atomic bomb because the concentration of uranium-235 is too low. However, if the core overheats, sufficient damage might be done to release materials into the environment.

water is returned to its source at a higher temperature than when it was removed. Power plants are therefore a significant source of thermal pollution. The nuclear power plant design shown in Figure 21.21 is currently the most popular. The primary coolant, which passes through the core, is in a closed system. Other coolants never pass through the reactor core at all. This lessens the chance that radioactive products could escape the core. Additionally, the reactor is surrounded by a concrete shell to shield personnel and nearby residents from radiation.

Fission products accumulate as the reactor operates. These products lessen the efficiency of the reactor by capturing neutrons. The reactor must be stopped periodically so that the nuclear fuel can be replaced or reprocessed. When the fuel rods are removed from the reactor, they are initially very radioactive. It was originally intended that they be stored for several months in pools at the reactor site to allow decay of short-lived radioactive nuclei. They were then to be transported in shielded containers to reprocessing plants, where the fuel would be separated from the fission products. However, reprocessing plants have been plagued with operational difficulties, and there is intense opposition to the transport of nuclear wastes on the nation's highways. Even if the transportation difficulties could be overcome, the high radioactivity of the spent fuel makes reprocessing a hazardous operation. At present, the spent fuel rods are simply being kept in storage at reactor sites.

During reprocessing, the uranium is separated from nuclear waste products and repackaged into new fuel rods. The problem then becomes one of disposal of the leftover fission products. Storage poses a major problem because the fission products are extremely radioactive. It is estimated that 20 half-lives are required for their radioactivity to reach levels acceptable for biological exposure. Based on the 28.8-yr half-life of strontium-90, one of the longer-lived and most dangerous of the products, the wastes must be stored for 600 yr. If plutonium-239 is not removed, storage must be for longer periods; plutonium-239 has a half-life of 24,000 yr. It is advantageous, however, to remove plutonium-239 because it can be used as a fissionable fuel.

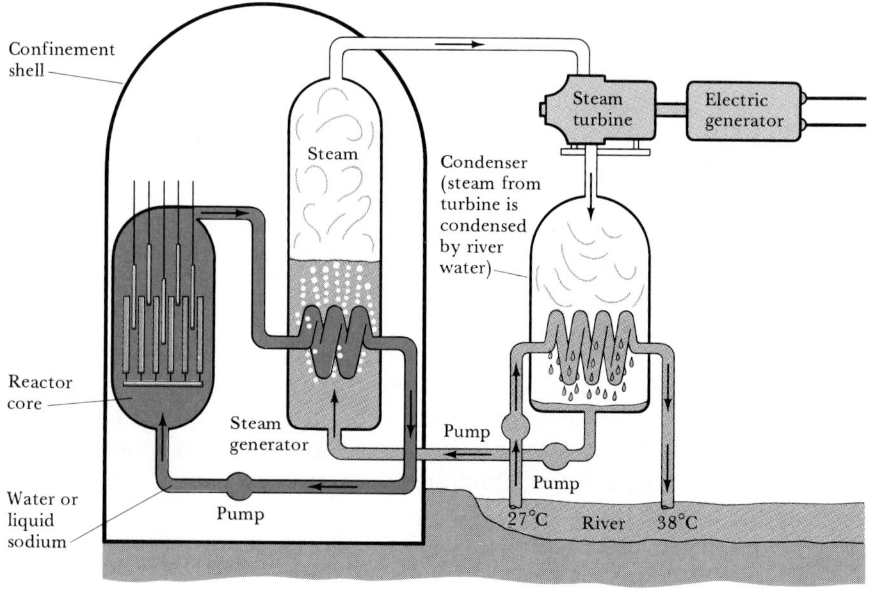

Figure 21.21 Basic design of a nuclear power plant. Heat produced by the reactor core is carried by a cooling fluid such as water or liquid sodium to a steam generator. The steam so produced is used to drive an electrical generator.

A considerable amount of research has been and will continue to be devoted to disposal of radioactive wastes. At present, the most attractive possibilities appear to be formation of a glass, ceramic, or synthetic rock from the wastes, as a means of immobilizing them. The solid materials could then be buried deep underground. Because the radioactivity will persist for a long time, there must be assurances that the solids will not crack from the heat generated by nuclear decay, possibly allowing radioactivity to find its way into underground water supplies.

Breeder Reactors

Uranium-235 is rare, so its supply can readily be exhausted. Methods of generating other fissionable materials are being actively investigated. Fissionable plutonium-239 and uranium-233 can be produced in nuclear reactors from nuclides that are far more abundant than uranium-235. The reactions involved are as follows:

$$^{238}_{92}\text{U} + ^{1}_{0}\text{n} \longrightarrow ^{239}_{92}\text{U} \qquad [21.28]$$

$$^{239}_{92}\text{U} \longrightarrow ^{239}_{93}\text{Np} + ^{0}_{-1}\text{e} \qquad (t_{1/2} = 24 \text{ min}) \qquad [21.29]$$

$$^{239}_{93}\text{Np} \longrightarrow ^{239}_{94}\text{Pu} + ^{0}_{-1}\text{e} \qquad (t_{1/2} = 2.3 \text{ days}) \qquad [21.30]$$

$$^{232}_{90}\text{Th} + ^{1}_{0}\text{n} \longrightarrow ^{233}_{90}\text{Th} \qquad [21.31]$$

$$^{233}_{90}\text{Th} \longrightarrow ^{233}_{91}\text{Pa} + ^{0}_{-1}\text{e} \qquad (t_{1/2} = 22 \text{ min}) \qquad [21.32]$$

$$^{233}_{91}\text{Pa} \longrightarrow ^{233}_{92}\text{U} + ^{0}_{-1}\text{e} \qquad (t_{1/2} = 27 \text{ days}) \qquad [21.33]$$

It is theoretically possible to build a reactor that both produces energy and converts uranium-238 or thorium-232 into fissionable fuel. We can envision a fission of a uranium-235 nucleus producing two neutrons, one causing another fission, the second initiating the change of uranium-238 into plutonium-239. Plutonium-239 is produced in ordinary reactors because of the presence of uranium-238. However, the hope is to be able to produce more fuel than the reactor uses. Reactors able to do this are called **breeder reactors**.

The design of breeder reactors poses many difficult technical problems. In addition, the breeder reactor program has become the subject of intense political debate. Because both breeder reactors and nuclear fuel reprocessing plants produce plutonium-239, any nation that acquired a breeder reactor or reprocessing technology would have the raw materials for atomic weaponry. Fear of nuclear proliferation or the possibility of thefts of nuclear materials by terrorist groups has generated uncertainty as to whether the United States should proceed with breeder reactor development and fuel reprocessing. Breeder reactors are, however, in operation in other countries, notably France and the Soviet Union.

21.8 NUCLEAR FUSION

As shown in Section 21.6, energy is produced when light nuclei are fused to form heavier ones. Reactions of this type are responsible for the energy produced by the sun. Spectroscopic studies indicate that the sun is composed of 73 percent H, 26 percent He, and only 1 percent of all other elements, by mass. Among the several fusion processes that are believed to occur are the following:

$$ {}^{1}_{1}H + {}^{1}_{1}H \longrightarrow {}^{2}_{1}H + {}^{0}_{1}e \qquad [21.34] $$

$$ {}^{1}_{1}H + {}^{2}_{1}H \longrightarrow {}^{3}_{2}He \qquad [21.35] $$

$$ {}^{3}_{2}He + {}^{3}_{2}He \longrightarrow {}^{4}_{2}He + 2{}^{1}_{1}H \qquad [21.36] $$

$$ {}^{3}_{2}He + {}^{1}_{1}H \longrightarrow {}^{4}_{2}He + {}^{0}_{1}e \qquad [21.37] $$

Theories have been proposed for the generation of the other elements through fusion processes.

Fusion is appealing as an energy source because of the availability of light isotopes and because fusion products are generally not radioactive. Fusion is therefore potentially a cleaner process than is fission. Despite this fact, fusion is not presently used to generate energy. The problem is that high energies are needed to overcome the repulsion between nuclei. The required energies are achieved by high temperatures. Fusion reactions are therefore also known as **thermonuclear reactions**. The lowest temperature required for any fusion is that needed to fuse ${}^{2}_{1}H$ and ${}^{3}_{1}H$, shown in Equation 21.38. This reaction requires a temperature of 40,000,000 K:

$$ {}^{2}_{1}H + {}^{3}_{1}H \longrightarrow {}^{4}_{2}He + {}^{1}_{0}n \qquad [21.38] $$

Such high temperatures have been achieved by using an atomic bomb to initiate the fusion process. This is done in the thermonuclear, or hydrogen, bomb. Clearly, this approach is unacceptable for controlled power generation.

Numerous problems must be overcome before fusion becomes a practical energy source. In addition to the high temperatures necessary to initiate the reaction, there is the problem of confining the reaction. No known structural material is able to withstand the enormous temperatures necessary for fusion. Research has centered on the use of an apparatus called a *tokamak*, which uses strong magnetic fields to contain and heat the reaction (Figure 21.22). Tempera-

Figure 21.22 The Tokamak Fusion Test Reactor at Princeton University. A tokamak is essentially a magnetic "bottle" for confining and heating nuclei in an effort to cause them to fuse. (Princeton Plasma Physics Laboratory/Princeton University)

tures of nearly 3,000,000 K have been achieved in a tokamak, but this is not yet hot enough to initiate continuous fusion. Much research has also been directed at the use of powerful lasers to generate the necessary temperatures.

What would happen if controllable nuclear fusion could be achieved at room temperature? Because the fusion isotopes (particularly deuterium, ${}^{2}_{1}H$) are plentiful, the world would have virtually an endless source of energy. Thus, it was no surprise that the announcement of this so-called cold fusion in March 1989 caused worldwide excitement. Two chemists at the University of Utah claimed to achieve room-temperature fusion by the electrolysis (Section 20.7) of deuterium oxide (also called *heavy water*), ${}^{2}_{1}H_2O$. The electrochemical apparatus employed a platinum wire anode and a palladium cathode. Palladium has the property of being able to absorb hydrogen into its metallic structure, with the result being that the hydrogen atoms are packed very close together. The Utah chemists proposed that when the deuterium oxide is reduced at the cathode, the deuterium atoms become so tightly packed in the palladium cathode that they undergo one or both of the following fusion reactions:

$$ {}^{2}_{1}H + {}^{2}_{1}H \longrightarrow {}^{3}_{1}H + {}^{1}_{0}n + {}^{0}_{1}e \qquad [21.39] $$

$$ {}^{2}_{1}H + {}^{2}_{1}H \longrightarrow {}^{4}_{2}He \qquad [21.40] $$

Unfortunately, most scientific evidence seems to show that the claim of achieving cold fusion was in error. It is therefore not yet clear whether fusion at either high or low temperatures will ever be a practical source of energy for humankind.

21.9 BIOLOGICAL EFFECTS OF RADIATION

The increased pace of synthesis and use of radioisotopes has led to increased concern about the effects of radiation on matter, particularly in biological systems. We therefore conclude this chapter by examining the health hazards associated with radioisotopes.

Alpha, beta, and gamma rays (as well as X rays) possess energies far in excess of ordinary bond energies and ionization energies. Consequently, these forms of radiation are able to fragment and ionize molecules, generating unstable, highly reactive particles as they pass through matter. For example, gamma rays are able to ionize water molecules, forming unstable H_2O^+ ions. An H_2O^+ ion can react with another water molecule to form an H_3O^+ ion and a neutral OH molecule:

$$ H_2O^+ + H_2O \longrightarrow H_3O^+ + OH \qquad [21.41] $$

The unstable and highly reactive OH molecule is an example of a **free radical**, a substance with one or more unpaired electrons. In a biological system, such particles can attack a host of other compounds to produce new free radicals, which in turn attack yet other compounds. Thus, the formation of a single free radical can initiate a large number of chemical reactions that are ultimately able to disrupt the normal operations of cells.

The resultant radiation damage to living systems can be classified as either somatic or genetic. **Somatic damage** affects the organism during its own lifetime. **Genetic damage**, as the term implies has a genetic effect; it harms offspring through damage to genes and chromosomes, the body's reproductive material. Genetic effects are more difficult to study than somatic ones because they may not become apparent for several generations. Somatic damage includes "burns," molecular disruptions similar to those produced by high temperatures. It also includes cancer. Cancer is brought about by damage to the growth-regulation mechanism of cells, which causes them to reproduce in an uncontrolled manner. In general, the tissues that show the greatest damage from radiation are those that reproduce at a rapid rate, such as bone marrow, blood-forming tissues, and lymph nodes. Leukemia, which is characterized by excessive growth of white blood cells, is probably the major cancer problem associated with radiation.

The clinical symptoms of acute (short-term) exposure to radiation include a decrease in the number of white blood cells, fatigue, nausea, and diarrhea. Sufficient exposure can result in death from blood disorders, gastrointestinal failure, and damage to the central nervous system. In light of these effects, we must determine whether any levels of exposure to radiation are actually safe. What are the maximum levels of radiation that we should permit from various human activities? Unfortunately, we are hampered in our attempts to set realistic standards by our lack of understanding of the effects of chronic (long-term) exposure to radiation. Most scientists presently believe that the effects of radiation are proportional to exposure, even down to low exposures. This means that *any* amount of radiation causes some finite risk of injury.

Radiation Doses

The SI unit of nuclear radioactivity is called the **becquerel**, after Henri Becquerel. A becquerel is defined as one nuclear disintegration per second. The older and more widely used unit of activity is the **curie** (Ci), after Marie Curie (Section 2.2). A curie is 3.7×10^{10} disintegrations per second, the number of nuclear disintegrations per second from 1 g of radium. For example, a 5.0-mCi sample of cobalt-60 undergoes $(5.0 \times 10^{-3})(3.7 \times 10^{10}) = 1.8 \times 10^{8}$ disintegrations per second.

The damage produced by radiation outside the body depends not only on the nuclear activity but also on the energy and penetrating power of the radiation. Gamma rays are particularly dangerous because they penetrate human tissue very effectively, just as do X rays. Consequently, their damage is not limited to the skin. In contrast, most alpha rays are stopped by skin, and beta rays are able to penetrate only about 1 cm beyond the surface of the skin. Hence, neither is as dangerous as gamma rays unless the radiation source somehow enters the body. Within the body, alpha rays are particularly dangerous because they leave a very dense trail of damaged molecules as they move through matter.

Two units, the rad and rem, are commonly used to measure radiation doses. (A third unit, the roentgen, is essentially the same as the rad.) A **rad** (*r*adiation *a*bsorbed *d*ose) is the amount of radiation that deposits 1×10^{-2} J of energy per kilogram of tissue. A rad of alpha rays can produce more damage than a rad of beta rays. Consequently, the rad is often multiplied by a factor that measures the relative biological damage caused by the radiation. This factor is known as the *relative biological effectiveness* of the radiation, abbreviated **RBE**. The RBE is approximately 1 for beta and gamma rays and 10 for alpha rays. The exact value of the RBE varies with dose rate, total dose, and type of tissue affected. The product of the number of rads and the RBE of the radiation gives the effective dosage in **rems** (*r*oentgen *e*quivalent for *m*an):

$$\text{Number of rems} = (\text{number of rads})(\text{RBE}) \qquad [21.42]$$

Table 21.6 Effects of Short-Time Exposures to Radiation

Dose (rem)	Effect
0 to 25	No detectable clinical effects
25 to 50	Slight, temporary decrease in white blood cell counts
100 to 200	Nausea; marked decrease in white blood cells
500	Death of half the exposed population within 30 days after exposure

The effects of some short-term exposures to radiation appear in Table 21.6. An exposure to 600 rem is fatal to most humans. To put this number in perspective, a typical dental X-ray entails an exposure of about 0.5 mrem. The disaster at Chernobyl led to some truly large radiation exposures. It is estimated that the total radiation released at Chernobyl was 50 to 100 MCi. The service personnel and firefighters at the plant received more than 100 rem of radiation. By comparison, the accident at Three Mile Island in Pennsylvania released 20 Ci of radiation, and no individual was exposed to more than 100 mrem.

Radon

Over the long term, we are all exposed to potentially damaging radiation from both natural sources and human activity. Since the mid-1980s, concern has grown over radiation exposure due to radon, the radioactive noble gas with atomic number 86. As Figure 21.23 indicates, radon exposure accounts for more than half the 360-mrem average annual exposure to radiation. Radon-222 is a decay product in the nuclear disintegration series of uranium-238 (Figure 21.4) and is therefore continually generated as uranium decays.

The U.S. Environmental Protection Agency (EPA) has recommended that ^{222}Rn levels in the home not exceed 4 pCi per liter of air. We are aware today, however, that houses that are built on foundations naturally rich in uranium ore or that are improperly sealed can have levels nearly 1000 times greater than that. Public awareness of the potential health risk of radon has grown rapidly, and the home radon-testing kit has become a common household item in many parts of the country (Figure 1.22).

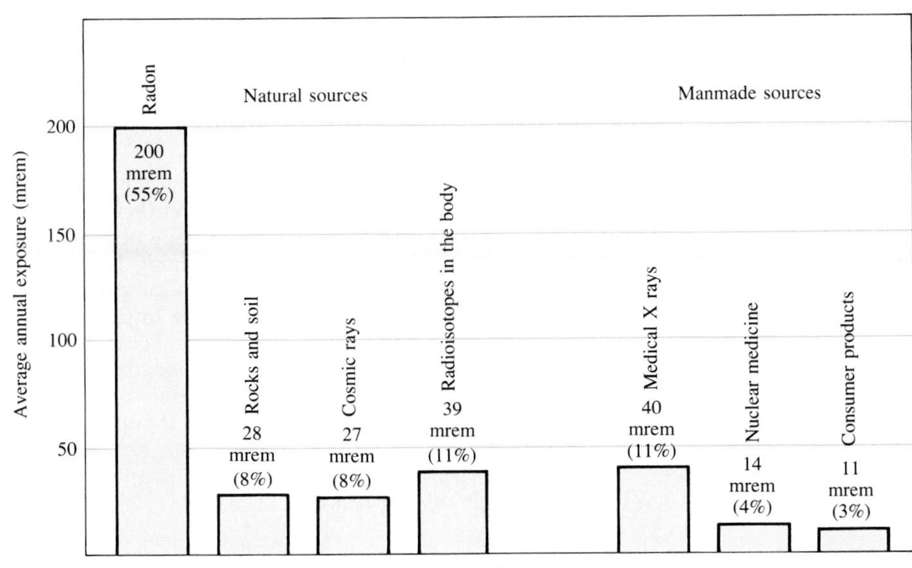

Figure 21.23 A graph of the sources of the average annual exposure of the U. S. population to high-energy radiation. The total average annual exposure is 360 mrem.

CHEMISTRY AT WORK: Radiation Therapy

High-energy radiation poses a health hazard because of the damage it does to cells. Healthy cells are either destroyed or damaged by radiation, leading to physiological disorders. However, radiation can also destroy *unhealthy* cells, including cancerous cells. All cancers are characterized by the runaway growth of abnormal cells. This growth can produce masses of abnormal tissue, called *malignant tumors.* Malignant tumors can be caused by the exposure of healthy cells to high-energy radiation. However, somewhat paradoxically, malignant tumors can be destroyed by exposing them to the same radiation. In fact, cancerous cells are more susceptible to destruction by radiation than are healthy ones, allowing radiation to be used effectively in the treatment of cancer. As early as 1904, physicians attempted to use the radiation emitted by radioactive substances to treat tumors by destroying the mass of unhealthy tissue. The treatment of disease by high-energy radiation is called *radiation therapy.*

Many different radionuclides are currently used in radiation therapy. Some of the more commonly used ones are listed in Table 21.7, along with their half-lives. You will note that most of the half-lives are quite short, meaning that these radioisotopes emit a great deal of radiation in a short period of time (Figure 21.24).

The radiation source used in radiation therapy may be inside or outside the body. In almost all cases, radiation therapy is designed to use the high-energy gamma radiation emitted by radioisotopes. Alpha and beta radiation, which are not as penetrating as gamma radiation, can be blocked by appropriate packaging. For example, ^{192}Ir is often administered as "seeds" consisting of a core of radioactive isotope coated with 0.1 mm of platinum metal. The platinum coating stops the alpha and beta rays, but the gamma rays penetrate it readily. The radioactive seeds can be surgically implanted in a tumor. In other cases, human physiology allows the radioisotope to be ingested. For example, most of the iodine in the human body ends up in the thyroid gland (see the Chemistry at Work box, Section 21.5), a fact that allows thyroid cancer to be treated by using large doses of ^{131}I. Radiation therapy on deep organs, where a surgical implant is impractical, often uses a ^{60}Co "gun" outside the body to shoot a beam of gamma rays at the tumor. Particle accelerators are also used as an external source of high-energy radiation for radiation therapy.

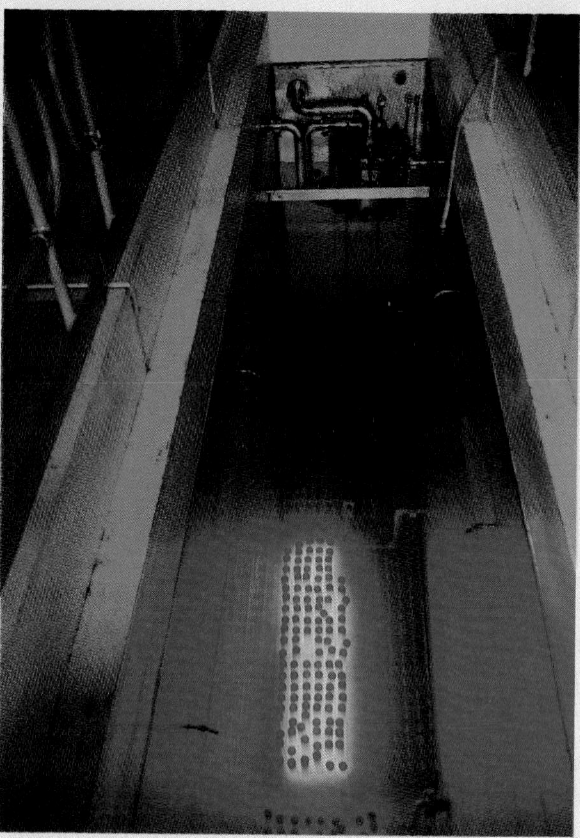

Figure 21.24 A sample of a salt of cesium-137, which is used in radiation therapy. The blue glow is from the radioactivity of the cesium. In Goiânia, Brazil, in 1987, a cylinder containing cesium-137 was left in an abandoned medical clinic. It was discovered by unsuspecting townspeople, who were fascinated by the strange blue glow. The results were tragic: Four people died from radiation exposure, and 249 others were contaminated. (Earl Roberge, Science Source/Photo Researchers)

Table 21.7 Some Radioisotopes Used in Radiation Therapy

Isotope	Half-life	Isotope	Half-life
^{32}P	14.3 days	^{137}Cs	30 yr
^{60}Co	5.26 yr	^{192}Ir	74.2 days
^{90}Sr	28.8 yr	^{198}Au	2.7 days
^{125}I	60.25 days	^{222}Ra	3.82 days
^{131}I	8.06 days	^{226}Rn	1,622 yr

Because gamma radiation is so strongly penetrating, it is nearly impossible to avoid damaging healthy cells during radiation therapy. Most cancer patients undergoing radiation treatments experience unpleasant and dangerous side effects such as fatigue, nausea, hair loss, a weakened immune system, and even death. Thus, in many cases, radiation therapy is used only if other cancer treatments such as *chemotherapy* (the treatment of cancer with powerful drugs) are unsuccessful. Nevertheless, radiation therapy is one of the major weapons we have in the fight against cancer.

It is interesting to consider the interplay between the chemical and nuclear properties of radon that make it a health hazard. Being a noble gas, radon is extremely unreactive and is therefore free to bubble up out of the ground without becoming trapped chemically along the way. It is readily inhaled and exhaled with no direct chemical effects. However, the half-life of ^{222}Rn is a short 3.82 days. It decays through alpha-particle loss into a radioisotope of polonium:

$$^{222}_{86}Rn \longrightarrow ^{218}_{84}Po + ^{4}_{2}He \qquad [21.43]$$

Because radon has such a short half-life and alpha particles have a high RBE, inhaled radon is considered a probable cause of lung cancer. Even worse, however, is the fact that the decay product polonium-218 is an alpha-emitting solid that has even a shorter half-life (3.11 min) than radon-222:

$$^{218}_{84}Po \longrightarrow ^{214}_{82}Pb + ^{4}_{2}He \qquad [21.44]$$

The atoms of polonium-218 can become trapped in the lungs, where they continually bathe the delicate tissue with harmful alpha radiation.

Equations 21.43 and 21.44 give an indication of the complex nuclear reactions that affect our health. As more scientists have become convinced that even low radiation dosage poses some risk, there has been increasing pressure to set tougher standards for radiation exposure. As with many other aspects of human activity, the question at issue is one of risk versus benefit. In order to make intelligent decisions concerning radiation health hazards, we must have a deeper understanding of the risks than we do at present.

FOR REVIEW

SUMMARY

Certain nuclei are radioactive. Most of these nuclei gain stability by emitting alpha particles ($^{4}_{2}He$), beta particles ($^{0}_{-1}e$), and gamma radiation ($^{0}_{0}\gamma$). Some nuclei undergo decay by positron ($^{0}_{1}e$) emission or by electron capture. The neutron-to-proton ratio is one factor determining nuclear stability. The presence of magic numbers of nucleons and an even number of protons and neutrons is also important. Nuclear transmutations can be induced by bombarding nuclei with charged particles using particle accelerators or with neutrons in a nuclear reactor.

Radioisotopes have characteristic rates of decay. These rates are generally expressed in terms of half-lives. The constant half-lives of nuclides permit their use in dating objects. The ease of detection of radioisotopes also permits their use as tracers, to follow elements through their reactions. Three methods of detection are photographic film, scintillation counters, and Geiger counters.

The energy produced in nuclear reactions is accompanied by measurable losses of mass in accordance with Einstein's relationship, $\Delta E = c^2 \Delta m$. The difference in mass between nuclei and the nucleons of which they are composed is known as the mass defect. The mass defect of a nuclide allows calculation of its nuclear binding energy, the energy required to separate the nucleus into individual nucleons. Examination of binding energies per nucleon reveals that energy is produced when heavy nuclei split (fission) and when light nuclei fuse (fusion).

Uranium-235, uranium-233, and plutonium-239 undergo fission when they capture a neutron. The resulting nuclear reaction is a chain reaction. A reaction that maintains a constant rate is said to be critical. If it slows, it is said to be subcritical. In the atomic bomb, subcritical masses are brought together to form a supercritical mass. In nuclear reactors, the fission is controlled to generate a constant power. The reactor core consists of fissionable fuel, control rods, a moderator, and cooling fluid. The nuclear power plant resembles a conventional power plant except that the core replaces the fuel burner. In breeder reactors, more nuclear fuel is produced than is used to generate energy. There is growing concern regarding the safety with which nuclear power plants can be operated. In addition, the reprocessing of spent fuel rods and disposal of highly radioactive nuclear wastes are unsolved problems.

Nuclear fusion requires high temperatures because nuclei must have large kinetic energies to overcome their mutual repulsions. It is not yet possible to generate a controlled fusion process.

The SI unit of radioactivity is the becquerel, defined as one nuclear disintegration per second. Radioactivity is more often measured in curies. One curie corresponds to 3.7×10^{10} disintegrations per second. The amount of energy deposited in biological tissue by radiation is measured in terms of the rad; one rad corresponds to 1×10^{-2} J per kilogram of tissue. The rem is a more useful measure of the biological damage created by the deposited energy. We receive radiation from both naturally occurring and human sources in roughly equal amounts for the general population. The effects of long-term exposure to low levels of radiation are not completely understood, but the most commonly accepted hypothesis is that the extent of biological damage varies in direct proportion to the level of exposure.

KEY TERMS

nucleon
radioisotopes (Sec. 21.1)
alpha rays (Sec. 21.1)
alpha particles (Sec. 21.1)
nuclear transmutations (Sec. 21.1)
beta particles (Sec. 21.2)
gamma rays (Sec. 21.2)
positron (Sec. 21.2)
electron capture (Sec. 21.2)
magic numbers (Sec. 21.2)
radioactive series (Sec. 21.2)
nuclear-disintegration series (Sec. 21.2)
particle accelerators (Sec. 21.3)
transuranium elements (Sec. 21.3)
half-life (Sec. 21.4)
Geiger counter (Sec. 21.5)
scintillation counter (Sec. 21.5)
radiotracer (Sec. 21.5)
mass defect (Sec. 21.6)
fission (Sec. 21.6)
fusion (Sec. 21.6)
chain reactions (Sec. 21.7)
subcritical mass (Sec. 21.7)
critical mass (Sec. 21.7)
supercritical mass (Sec. 21.7)
breeder reactors (Sec. 21.7)
thermonuclear reactions (Sec. 21.8)
free radical (Sec. 21.9)
somatic damage (Sec. 21.9)
genetic damage (Sec. 21.9)
becquerel (Sec. 21.9)
curie (Sec. 21.9)
rad (Sec. 21.9)
RBE (Sec. 21.9)
rems (Sec. 21.9)

EXERCISES

Nuclear Reactions

21.1 Indicate the number of protons and neutrons in each of the following nuclei: **(a)** carbon-13; **(b)** $^{58}_{28}Ni$; **(c)** rubidium-85; **(d)** ^{94}Zr.

21.2 Indicate the number of protons, neutrons, and nucleons in each of the following: **(a)** argon-40; **(b)** ^{103}Rh; **(c)** indium-115; **(d)** $^{26}_{12}Mg$.

21.3 Write balanced nuclear equations for the following transformations: **(a)** Nitrogen-13 undergoes positron emission. **(b)** Bromine-84 undergoes beta decay. **(c)** Tungsten-181 undergoes orbital electron capture. **(d)** Thorium-230 decays to a radium isotope.

21.4 Write balanced equations for each of the following nuclear transformations: **(a)** Iodine-131 undergoes beta decay. **(b)** Gold-194 undergoes positron emission. **(c)** Plutonium-242 emits an alpha particle. **(d)** Polonium-218 is formed by decay of a radon nuclide.

21.5 Complete and balance the following nuclear equations by supplying the missing particle:

(a) $^{32}_{16}S + ^{1}_{0}n \longrightarrow ^{1}_{1}H + ?$

(b) $^{7}_{4}Be + ^{0}_{-1}e$ (orbital electron) $\longrightarrow ?$

(c) $? \longrightarrow ^{187}_{76}Os + ^{0}_{-1}e$

(d) $^{98}_{42}Mo + ^{2}_{1}H \longrightarrow ^{1}_{0}n + ?$

(e) $^{235}_{92}U + ^{1}_{0}n \longrightarrow ^{135}_{54}Xe + ? + 2^{1}_{0}n$

21.6 Complete and balance the following nuclear equations by supplying the missing particle:

(a) $^{252}_{98}Cf + ^{10}_{5}B \longrightarrow 3^{1}_{0}n + ?$

(b) $^{2}_{1}H + ^{3}_{2}He \longrightarrow ^{4}_{2}He + ?$

(c) $^{1}_{1}H + ^{11}_{5}B \longrightarrow 3?$

(d) $^{122}_{53}I \longrightarrow ^{122}_{54}Xe + ?$

(e) $^{59}_{26}Fe \longrightarrow ^{0}_{-1}e + ?$

21.7 The naturally occurring radioactive decay series that begins with $^{235}_{92}U$ stops with formation of the stable $^{207}_{82}Pb$. The decays proceed through a series of alpha particle and beta particle emissions. How many of each type of emission are involved in this series?

21.8 A radioactive decay series that begins with $^{237}_{93}Np$ ends with formation of the stable nuclide $^{209}_{83}Bi$. How many alpha particle emissions and how many beta particle emissions are involved in the sequence of radioactive decays?

21.9 Write balanced equations for the following nuclear reactions: **(a)** $^{238}_{92}U(n, \gamma)^{239}_{92}U$; **(b)** $^{14}_{7}N(p, \alpha)^{11}_{6}C$; **(c)** $^{18}_{8}O(n, \beta)^{19}_{9}F$.

21.10 Write balanced equations for each of the following nuclear reactions: **(a)** $^{14}_{7}N(p, \alpha)^{11}_{6}C$; **(b)** $^{14}_{7}N(\alpha, p)^{17}_{8}O$; **(c)** $^{59}_{26}Fe(\alpha, \beta)^{63}_{29}Cu$.

Nuclear Stability

21.11 Which of the following nuclides would you expect to be radioactive: **(a)** $^{17}_{8}O$; **(b)** $^{176}_{74}W$; **(c)** $^{108}_{50}Sn$; **(d)** $^{92}_{40}Zr$; **(e)** $^{238}_{94}Pu$? Justify your choices.

21.12 In each of the following pairs, which nuclide would you expect to be the more abundant in nature: **(a)** $^{19}_{9}F$ or $^{18}_{9}F$; **(b)** $^{80}_{34}Se$ or $^{81}_{34}Se$; **(c)** $^{56}_{26}Fe$ or $^{57}_{26}Fe$; **(d)** $^{118}_{50}Sn$ or $^{118}_{51}Sb$? Justify your choices.

21.13 Which of the following nuclides is most likely to be a positron emitter: **(a)** $^{53}_{24}Cr$; **(b)** $^{51}_{25}Mn$; **(c)** $^{59}_{26}Fe$? Explain.

21.14 All of the following nuclides are radioactive and undergo either beta or positron emission: **(a)** $^{66}_{32}Ge$; **(b)** $^{105}_{45}Rh$; **(c)** $^{137}_{53}I$; **(d)** $^{133}_{58}Ce$. Indicate which nuclei undergo which type of emission.

21.15 It has been suggested that strontium-90 (derived from nuclear testing) deposited in the hot desert will undergo radioactive decay more rapidly because it will be exposed to much higher average temperatures. Is this a reasonable suggestion?

21.16 Sulfur-35 is radioactive and undergoes beta decay. What differences would you expect in the chemical behavior of atoms containing sulfur-35 as compared with those containing sulfur-32, which is nonradioactive? Explain.

21.17 Indicate whether each of the following nuclides lies within the belt of stability in Figure 21.2: **(a)** $^{108}_{49}In$; **(b)** $^{102}_{47}Ag$; **(c)** $^{17}_{7}N$; **(d)** $^{210}_{86}Rn$. For any that do not, describe a nuclear decay process that would alter the neutron-to-proton ratio in the direction of increased stability.

21.18 Indicate whether each of the following nuclides lies within the belt of stability in Figure 21.2: **(a)** $^{70}_{33}As$; **(b)** $^{34}_{15}P$; **(c)** $^{74}_{32}Ge$; **(d)** $^{248}_{98}Cf$. For any that do not, describe a nuclear decay process that would alter the neutron-to-proton ratio in the direction of increased stability.

21.19 Give three examples of nuclides that have a magic number of both protons and neutrons.

21.20 Explain the following statement: There is an analogy between the stability of the nucleus of ^{208}Pb and the lack of reactivity of argon gas.

Half-life; Dating

21.21 Gallium-68 decays by positron emission, with a half-life of 68.3 min. Write the equation for the nuclear reaction. How much ^{68}Ga remains from a 10.0-mg sample after 683 min?

21.22 The half-life of tritium (hydrogen-3) is 12.3 yr. If 48.0 mg of tritium is released from a nuclear power plant during the course of an accident, what mass of this nuclide will remain after 12.3 yr? After 49.2 yr?

21.23 A sample of the synthetic nuclide curium-243 was prepared. After 1.00 yr, the radioactivity of the sample had declined from 3012 disintegrations per second to 2921 disintegrations per second. What is the half-life of the decay process?

21.24 A sample of a radioactive nuclide exhibits 8540 disintegrations per second. After 350.0 min, the number of disintegrations per second is 1250. What is the half-life of the radionuclide?

21.25 The half-life of iridium-192 is 74.2 days. What mass of ^{192}Ir remains after 1.00 yr from a sample that originally contained 15.0 mg of the nuclide?

21.26 The half-life of ^{239}Pu is 24,000 yr. What fraction of the ^{239}Pu present in nuclear wastes generated today will be present in the year 3000?

21.27 An experiment was designed to determine whether an aquatic plant absorbed iodide ion from water. Iodine-131 ($t_{1/2}$ = 8.1 days) was added as a tracer, in the form of iodide ion, to a tank containing the plants. The initial activity of a 1.00-μL sample of the water was 89 counts per minute. After 32 days the level of activity in a 1.00-μL sample was 5.7 counts per minute. Did the plants absorb iodide from the water?

21.28 A sample of strontium-89 has an initial activity of 4600 counts per minute on a device that measures the level of radioactivity. After exactly 30 days the activity has declined to 3130 counts per minute. What is the half-life for decay of strontium-89?

[21.29] Radium-226, which undergoes alpha decay, has a half-life of 1622 yr. How many alpha particles are emitted in 1.0 min from a 5.0-mg sample of ^{226}Ra?

[21.30] Cobalt-60, which undergoes beta decay, has a half-life of 5.26 yr. How many beta particles are emitted in 10.0 s from a 650-μg sample of ^{60}Co?

21.31 The half-life for the process $^{238}U \longrightarrow {}^{206}Pb$ is 4.5×10^9 yr. A mineral sample contains 50.0 mg of ^{238}U and 14.0 mg of ^{206}Pb. What is the age of the mineral?

21.32 Potassium-40 decays to argon-40 with a half-life of 1.27×10^9 yr. What is the age of a rock in which the weight ratio of ^{40}Ar to ^{40}K is 3.6?

21.33 A wooden artifact from a Chinese temple has a ^{14}C activity of 25.8 counts per minute as compared with an activity of 31.7 counts per minute for a standard of zero age. From the half-life for ^{14}C decay, 5.73×10^3 yr, determine the age of the artifact.

21.34 An ancient wooden object is found to have an activity of 9.6 disintegrations per minute per gram of carbon. By contrast, the carbon in a living tree undergoes 18.4 disintegrations per minute per gram of carbon. Based on the activity of carbon-14 (with a half-life of 5.73×10^3 yr) in the object, calculate its age.

Mass-Energy Relationships

21.35 Calculate the binding energy per nucleon for the following nuclei: **(a)** $^{12}_{6}C$ (atomic mass, 12.00000 amu); **(b)** $^{61}_{28}Ni$ (atomic mass, 60.93106 amu); **(c)** $^{206}_{82}Pb$ (atomic mass, 205.97447 amu).

21.36 Calculate the total binding energy and the binding energy per nucleon for each of the following nuclei: **(a)** $^{64}_{30}Zn$ (atomic mass, 63.92914); **(b)** $^{37}_{17}Cl$ (atomic mass, 36.96590 amu); **(c)** $^{4}_{2}He$ (atomic mass, 4.00260 amu).

21.37 How much energy must be supplied to break a single $^{6}_{3}Li$ nucleus into separated protons and neutrons if the nucleus has a mass of 6.01347 amu? What does this energy correspond to for 1 mol of ^{6}Li nuclei?

21.38 How much energy is absorbed or released when a single $^{19}_{9}F$ nucleus of mass 18.9935 amu is separated into protons and neutrons? What is the amount of energy corresponding to this process for 1 mol of $^{19}_{9}F$ nuclei?

21.39 The solar radiation falling on earth amounts to 1.07×10^{16} kJ/min. What is the mass equivalence of the solar energy falling on earth in a 24-hr period? If the energy released in the reaction

$$^{235}_{92}U + {}^{1}_{0}n \longrightarrow {}^{141}_{56}Ba + {}^{92}_{36}Kr + 3{}^{1}_{0}n$$

(^{235}U atomic mass, 235.0439 amu; ^{141}Ba atomic mass, 140.9140 amu; ^{92}Kr atomic mass, 91.9218 amu)

is taken as typical of that occurring in a nuclear reactor, what mass of uranium-235 is required to equal 0.10 percent of the solar energy that falls on earth in 1 day?

21.40 Based on the following atomic mass values—$^{1}_{1}H$, 1.00782; $^{2}_{1}H$, 2.01410 amu; $^{3}_{1}H$, 3.01605 mu; $^{3}_{2}He$, 3.01603 amu; $^{4}_{2}He$, 4.00260 amu—and the mass of the neutron given in the text, calculate the energy released in each of the following nuclear reactions, all of which are possibilities for a controlled fusion process:

(a) $^{2}_{1}H + {}^{3}_{1}H \longrightarrow {}^{4}_{2}He + {}^{1}_{0}n$

(b) $^{2}_{1}H + {}^{2}_{1}H \longrightarrow {}^{3}_{2}H + {}^{1}_{0}n$

(c) $^{2}_{1}H + {}^{3}_{2}He \longrightarrow {}^{4}_{2}He + {}^{1}_{1}H$

21.41 Which of the following nuclei is likely to have the largest mass defect per nucleon: **(a)** $^{59}_{27}Co$; **(b)** $^{11}_{5}B$; **(c)** $^{118}_{50}Sn$; (d) $^{243}_{96}Cm$? Explain your answer.

21.42 Based on Figure 21.12, explain why we should expect energy to be released in the course of the fission of heavy nuclei that occurs in a nuclear reactor.

Effects and Uses of Radioisotopes

21.43 Complete and balance the nuclear equations for the following fission reactions:

(a) $^{235}_{92}U + {}^{1}_{0}n \longrightarrow {}^{160}_{62}Sm + {}^{72}_{30}Zn + \_\_\_\_ {}^{1}_{0}n$

(b) $^{239}_{94}Pu + {}^{1}_{0}n \longrightarrow {}^{144}_{58}Ce + \_\_\_\_ + 2{}^{1}_{0}n$

21.44 Complete and balance the nuclear equation for the following fission or fusion reactions:

(a) $^{2}_{1}H + {}^{2}_{1}H \longrightarrow {}^{3}_{2}He + \_\_\_\_$

(b) $^{233}_{92}U + {}^{1}_{0}n \longrightarrow {}^{133}_{51}Sb + {}^{98}_{41}Nb + \_\_\_\_ {}^{1}_{0}n$

21.45 Explain how you might use radioactive ^{59}Fe (a beta emitter with $t_{1/2} = 46$ days) to determine the extent to which rabbits are able to convert a particular iron compound in their diet into blood hemoglobin, which contains an iron atom.

21.46 Chlorine-36 is a convenient radiotracer. It is a weak beta emitter, with $t_{1/2} = 3 \times 10^5$ yr. Describe how you would use this radiotracer to carry out each of the following experiments. **(a)** Determine whether trichloroacetic acid, CCl_3COOH, undergoes any ionization of its chlorines as chloride ion in aqueous solution. **(b)** Demonstrate that the equilibrium between dissolved $BaCl_2$ and solid $BaCl_2$ in a saturated solution is a dynamic process. **(c)** Determine the effects of soil pH on the uptake of chloride ion from the soil by soybeans.

21.47 A portion of the sun's energy comes from the reaction

$$4{}^{1}_{1}H \longrightarrow {}^{4}_{2}He + 2{}^{0}_{1}e$$

This reaction requires a temperature of about 10^6 to 10^7 K. Why is such a high temperature required?

21.48 Rutherford was able to carry out the first nuclear transmutation reactions by bombarding nitrogen-14 nuclei with alpha particles. However, in the famous experiment on scattering of alpha particles by gold foil (Section 2.2), a nuclear transmutation reaction did not occur. What is the difference between the two experiments? What would one need to do to carry out a successful nuclear transmutation reaction involving gold nuclei and alpha particles?

Additional Exercises

21.49 Distinguish between the terms in each of the following pairs (a) electron and positron; (b) binding energy and mass defect; (c) curie and rem; **(d)** gamma rays and beta rays.

21.50 Figure 21.4 shows the stepwise decay of uranium-238 to form the stable lead-206 nucleus. Write balanced nuclear equations for each step in this sequence.

21.51 Harmful chemicals are often destroyed by chemical treatment. For example, an acid can be neutralized by a base. Why can't chemical treatment be applied to destroy the fission products produced in a nuclear reactor?

21.52 Cobalt-60 has a half-life of 5.26 yr. The cobalt-60 in a certain radiation therapy unit must be replaced whenever its radioactivity falls to 80 percent of the original sample. If the original sample were purchased in November 1990, when will it be necessary to replace the cobalt-60?

[21.53] The energies of gamma rays are commonly expressed in units of millions of electron volts (MeV). **(a)** By using conversion factors on the back inside cover of this book, determine the energy in kilojoules of a 1.0-MeV gamma ray. **(b)** Cesium-137 emits gamma rays of energy 0.662 MeV. The half-life of ^{137}Cs is 30 yr. Determine the energy in kilojoules emitted in 30 min by a 10.0-mg sample of $^{137}CsCl$. **(c)** Ultraviolet radiation of wavelength 300 nm can cause skinburn. Calculate the ratio of the energy of a 0.662-MeV gamma-ray photon to that of a 300-nm ultraviolet photon. What does your answer tell you about the potential health risk of exposure to gamma rays?

21.54 We have seen that one possible mode of nuclear transformation is orbital electron capture. Which electron in a many-electron atom is most likely to be captured in such a process? How would you expect the likelihood of orbital electron capture to change with atomic number?

21.55 Describe or define the following: **(a)** rem; **(b)** curie; **(c)** moderator; **(d)** breeder reactor; **(e)** critical mass.

21.56 Plutonium-239 emits alpha particles that possess energies of about 5×10^8 kJ/mol. The half-life for the decay is about 24,000 yr. It has been said that the main danger from plutonium is inhalation of plutonium-containing dust. Why is inhalation, rather than mere exposure to plutonium in the environment, the major concern?

21.57 During the past 30 yr, nuclear scientists have synthesized approximately 1600 nuclei not known in nature. Many more might be discovered by using heavy-ion bombardment, which is possible only if high-energy instruments are used to accelerate the ions. Complete and balance the following reactions, which involve heavy-ion bombardments:

(a) $^{6}_{3}Li + ^{63}_{28}Ni \longrightarrow ?$

(b) $^{48}_{20}Ca + ^{248}_{96}Cm \longrightarrow ?$

(c) $^{88}_{38}Sr + ^{84}_{36}Kr \longrightarrow ^{116}_{46}Pd + ?$

(d) $^{48}_{20}Ca + ^{238}_{92}U \longrightarrow ^{70}_{20}Ca + 4^{1}_{0}n + 2?$

21.58 The 13 known nuclides of zinc range from $^{60}_{30}Zn$ to $^{72}_{30}Zn$. The naturally occurring nuclides have mass numbers 64, 66, 67, 68, and 70. What mode or modes of decay would you expect for the least massive radioactive nuclides of zinc? What mode for the most massive nuclides?

21.59 Chlorine has two stable nuclides, ^{35}Cl and ^{37}Cl. In contrast, ^{36}Cl is a radioactive nuclide that decays by beta emission. **(a)** Predict the product of decay of ^{36}Cl. **(b)** Based on the empirical rules about nuclear stability, explain why the nucleus of ^{36}Cl is less stable than either ^{35}Cl or ^{37}Cl.

21.60 The nuclear masses of ^{7}Be, ^{9}Be, and ^{10}Be are 7.0147, 9.0100, and 10.0113 amu, respectively. Which of these nuclei has the largest binding energy per nucleon?

21.61 The sun radiates energy into space at the rate of 3.9×10^{26} J/s. Calculate the rate of mass loss from the sun.

[21.62] The synthetic radioisotope technetium-99, which decays by beta emission, is the most widely used isotope in nuclear medicine. The following data were collected on a sample of ^{99}Tc:

Disintegrations per minute	**Time (hr)**
180	0
130	2.5
104	5.0
77	7.5
59	10.0
46	12.5
24	17.5

Make a graph of these data similar to Figure 21.7 and determine the half-life. (You may wish to make a graph of the natural log of the disintegration rate versus time; a little rearranging of Equation 21.21 will produce an equation for a linear relation between ln N_t and t; from the slope you can obtain k.)

[21.63] According to current regulations, the maximum permissible dose of strontium-90 in the body of an adult is 1 μCi (1×10^{-6} Ci). Using the relationship

$$\text{Rate} = kN$$

calculate the number of atoms of strontium-90 to which this corresponds. To what mass of strontium-90 does this correspond ($t_{1/2}$ for strontium-90 is 27.6 yr)?

[21.64] The half-life for decay of $^{230}_{90}Th$ is 8.0×10^4 yr. Because one cannot take the time to collect data for a graph such as that in Figure 21.7, how might you determine the half-life for such a long-lived isotope?

[21.65] Suppose you had a detection device that could count every decay from a radioactive sample of plutonium-239 ($t_{1/2}$ is 24,000 yr). How many counts per second would you obtain from a sample that contained 0.500 g of plutonium-239? (Hint: Look at Equations 21.21 and 21.22.)

[21.66] Tests on human subjects in Boston in 1965 and 1966, following the era of atomic bomb testing, revealed average quantities of about 2 pCi of plutonium radioactivity in the average person. How many disintegrations per second does this level of activity imply? If each alpha particle deposits 8×10^{-13} J of energy and if the average person weighs 75 kg, calculate the number of rads of radiation dose in 1 yr from such a level of plutonium, and also calculate the number of rems.

21.67 Consider the following statement: A radioisotope with a short half-life is a lesser health hazard than one with a long half-life. Do you agree or disagree with this statement?

[21.68] A 26.00-g sample of water containing tritium, $^{3}_{1}H$, emits 1.50×10^3 beta particles per second. Tritium is a weak beta emitter, with a half-life of 12.26 yr. What fraction of all the hydrogen in the water sample is tritium? (Hint: Use Equations 21.21 and 21.22.)

[21.69] When a positron is annihilated by combination with an electron, two photons of equal energy result. What is the wavelength of these photons? Are they gamma-ray photons?

22 Chemistry of Hydrogen, Oxygen, Nitrogen, and Carbon

Lightning storm. During lightning flashes, N_2 combines with O_2 to form NO. (Gordon Garradd/Science Photo Library, Science Source/Photo Researchers)

CONTENTS

The previous chapters of this book have primarily involved chemical principles, such as rules for bonding, the laws of thermodynamics, the factors influencing reaction rates, and so forth. In the course of explaining these principles, we have described the chemical and physical properties of many substances. However, we have done little to examine systematically the chemical elements and the compounds they form. This aspect of chemistry, often referred to as *descriptive chemistry*, is the subject of the next several chapters.

In this chapter, we examine four important nonmetals: hydrogen, oxygen, nitrogen, and carbon. These nonmetals form many commercially important compounds and are the primary elements in biological systems. In fact, 99 percent of the atoms required by living cells are atoms of these four elements. In Chapter 23, we will examine the remaining nonmetals on a group-by-group basis.

In studying descriptive chemistry, it is important to look for trends and general types of behavior, rather than trying to memorize all the facts presented. The periodic table is, of course, an invaluable tool in this task. Before we begin our examination of particular nonmetals, it is useful to review briefly some general periodic trends.

22.1 PERIODIC TRENDS

Figure 22.1 summarizes the way in which several important properties of elements vary in relation to the periodic chart. One of the most useful features shown is the division of elements into the broad categories of metals and nonmetals. We discussed this division in Section 7.6 and examined the other trends elsewhere in Chapters 7 and 8.

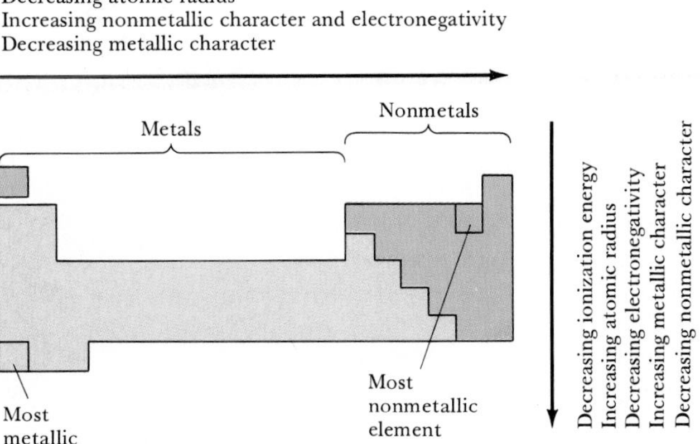

Figure 22.1 Trends in key properties of the elements as a function of position in the periodic table.

As shown in Figure 22.1, electronegativity decreases as we move down a given group and increases from left to right across the table. As a result, nonmetals have higher electronegativities than do metals. Consequently, compounds formed between strongly metallic and strongly nonmetallic elements tend to be ionic (for example, metal fluorides and metal oxides). These substances are solids at room temperature. In contrast, compounds formed between nonmetals are molecular substances. Molecular substances with low molecular weight tend to be gases, liquids, or volatile solids at room temperature.

Among the nonmetals, the chemistry of the first member of each family often differs in several important ways from that of subsequent members. The differences are due in part to the smaller size and greater electronegativity of the first member. In addition, the first member is restricted to forming a maximum of four bonds because it has only the 2*s* and the three 2*p* orbitals for bonding. Subsequent members of the family are able to use *d* orbitals in bonding in addition to *s* and *p* orbitals; they can therefore form more than four bonds. As an example, consider the chlorides of nitrogen and phosphorus, the first two members of group 5A. Nitrogen forms a maximum of three bonds with chlorine, NCl_3. Although phosphorus can form the trichloride compound, PCl_3, it is also able to form five bonds with chlorine, PCl_5. These compounds are shown in Figure 22.2. Because hydrogen has only the 1*s* orbitals available for bonding, it is restricted to forming one bond.

Another difference between the first member of any family and subsequent members of the same family is the greater ability of the former to form π bonds. We can understand this, in part, in terms of atomic size. As atoms increase in size, the sideways overlap of *p* orbitals, which form the strongest type of π bond, becomes less effective. This is shown in Figure 22.3. As an illustration of this effect, consider two differences in the chemistry of carbon and silicon, the first two members of group 4A. Carbon has two crystalline allotropes, diamond and graphite. In diamond there are σ bonds between carbon atoms but no π bonds. In graphite, π bonds result from sideways overlap of *p* orbitals (Section 11.8). Silicon occurs only in the diamondlike crystal form. Silicon does not exhibit a graphitelike structure because of the low stability of π bonds between silicon atoms.

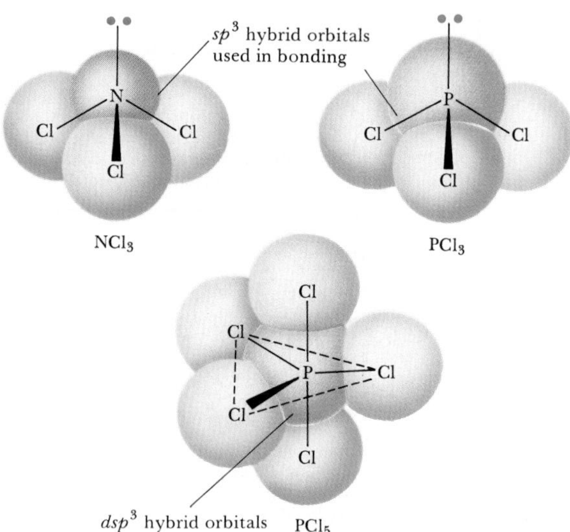

Figure 22.2 Comparison of NCl_3, PCl_3, and PCl_5. Nitrogen is unable to form NCl_5 because it does not have *d* orbitals available for bonding.

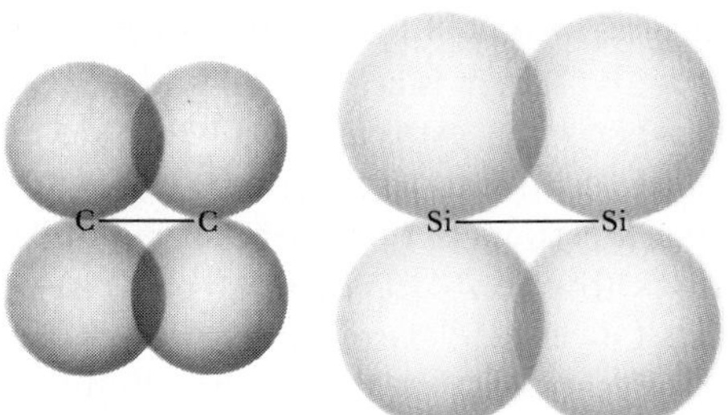

Figure 22.3 Comparison of π-bond formation by sideways overlap of p orbitals between two carbon atoms and between two silicon atoms. The distance between nuclei increases as we move from carbon to silicon. The p orbitals do not overlap as effectively between two silicon atoms because of this greater separation.

We see the same type of difference in the dioxides of these elements (Figure 22.4). In CO_2, carbon forms double bonds to oxygen, thereby achieving its valence by π bonding. In contrast, SiO_2 contains no double bonds. Instead, four oxygens are bonded to each silicon, forming an extended structure.*

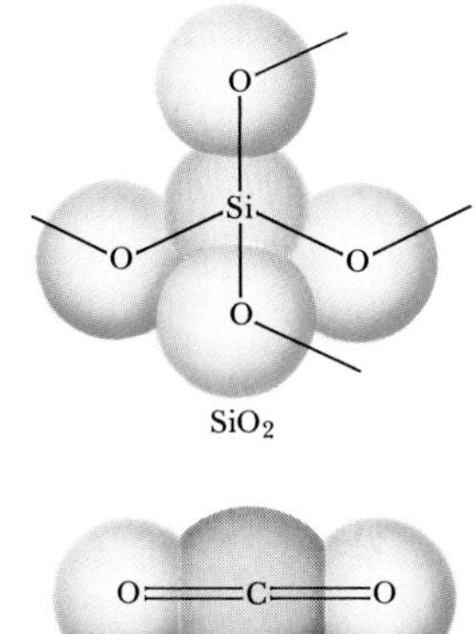

Figure 22.4 Comparison of the structures of CO_2 and SiO_2; CO_2 has double bonds, whereas SiO_2 has only single bonds.

SAMPLE EXERCISE 22.1

Sulfur forms a fluoride compound, SF_6, that has six sulfur-fluorine bonds. In contrast, oxygen is able to form at most two bonds to fluorine, OF_2. **(a)** Rationalize this difference. **(b)** Predict the molecular geometry of each compound and describe the hybridization employed by S and O.

Solution: **(a)** The electron configuration of oxygen is $[He]2s^22p^4$. The next-highest-energy orbitals available for bonding are in the third shell. Thus, oxygen uses only the $2s$ and $2p$ orbitals in bonding. These orbitals can accommodate only eight electrons. In OF_2, oxygen already has eight electrons, six of its own and one from each fluorine atom:

$$F{-}\ddot{\underset{\cdot\cdot}{O}}{-}F$$

Consequently, it is unable to accommodate additional fluorine atoms.

The electron configuration of sulfur is $[Ar]3s^23p^4$. In this case, the $3d$ orbitals are available for bonding. Therefore, sulfur is able to expand its octet to accommodate more than eight electrons:

```
       F
       | F
       |/
    F—S—F
      /|
     F |
       F
```

(b) According to the VSEPR theory (Section 9.1), the four electron pairs around oxygen are disposed at the corners of a tetrahedron. The four electron pairs are accommodated in sp^3 hybrid orbitals (Section 9.4). The geometric arrangement of atoms with two bonded pairs and two nonbonded pairs is nonlinear.

In the case of SF_6, the six electron pairs around sulfur are disposed at the corners of an octahedron. The associated hybridization is sp^3d^2. Because all six electrons are bonded pairs, the geometric arrangement of atoms is also octahedral.

PRACTICE EXERCISE

Consider the following list of elements: Li, K, N, P, Ne. From this list, select the element that **(a)** is most electronegative; **(b)** has the greatest metallic character; **(c)** can bond to more than four surrounding atoms in a molecule; **(d)** forms π bonds most readily. ***Answers:*** **(a)**N; **(b)** K; **(c)** P; **(d)** N

* The formula SiO_2 is consistent with this structure, because each oxygen is shared by two silicon atoms (not shown in Figure 22.4). For bookkeeping purposes, we may therefore count a half of each of the four oxygens that are bound to a given silicon atom as belonging to that silicon. We shall consider silicon-oxygen compounds in some detail in Chapter 23.

22.2 CHEMICAL REACTIONS

Throughout this chapter and subsequent ones, you will find a large number of chemical reactions presented. To remember all of them is an overwhelming task. However, you will probably need to remember some, especially those of a more general nature. In earlier discussions, we have encountered several general categories of reactions: combustion reactions (Section 3.2), metathesis reactions (Section 4.5), single displacement reactions (Section 4.6), Brønsted acid-base (proton-transfer) reactions (Section 16.1), Lewis acid-base reactions (Section 16.10), and redox reactions (Section 20.1). Because O_2 and H_2O are abundant and widespread in our environment, it is particularly important to consider the possible reactions of these substances with other compounds. About one-third of the reactions discussed in this chapter involve either O_2 (oxidation or combustion reactions) or H_2O (especially proton-transfer reactions).

In oxidation reactions with O_2, hydrogen-containing compounds produce H_2O. Carbon-containing ones produce CO_2 (unless the amount of O_2 is insufficient, in which case CO or even C can form). Nitrogen-containing compounds tend to form N_2, although NO can form in special cases. The following reactions illustrate these generalizations:

$$2CH_3OH(l) + 3O_2(g) \longrightarrow 2CO_2(g) + 4H_2O(l) \quad [22.1]$$

$$4CH_3NH_2(g) + 9O_2(g) \longrightarrow 4CO_2(g) + 10H_2O(l) + 2N_2(g) \quad [22.2]$$

The formation of H_2O, CO_2, and N_2 reflects the high thermodynamic stabilities of these substances, which are indicated by the large bond energies for the O—H, C═O, and N≡N bonds that they contain (463, 799, and 941 kJ/mol, respectively). The formation of the stable H_2O and CO_2 molecules also occurs when H_2 and C are used to reduce metal oxides:

$$NiO(s) + H_2(g) \longrightarrow Ni(s) + H_2O(l) \quad [22.3]$$

$$2CuO(s) + C(g) \longrightarrow 2Cu(s) + CO_2(g) \quad [22.4]$$

In dealing with proton-transfer reactions, remember that the weaker a Brønsted acid, the stronger its conjugate base (Section 16.1). For example, H_2, OH^-, NH_3, CH_4, and C_2H_2, which we encounter in this chapter, are exceedingly weak proton donors. In fact, they have *no* tendency to act as acids in water. Thus, species formed from them by removing one or more protons (such as H^-, O^{2-}, NH_2^-, N^{3-}, CH_3^-, C^{4-}, and C_2^{2-}) are extremely strong bases. All react readily with water, removing protons from H_2O to form OH^-. The following reactions are illustrative:

$$CH_3^-(aq) + H_2O(l) \longrightarrow CH_4(g) + OH^-(aq) \quad [22.5]$$

$$N^{3-}(aq) + 3H_2O(l) \longrightarrow NH_3(aq) + 3OH^-(aq) \quad [22.6]$$

Of course, substances that are stronger proton donors than H_2O, such as HCl, H_2SO_4, $HC_2H_3O_2$, and other acids, also react readily with basic anions.

SAMPLE EXERCISE 22.2

Predict the products formed in each of the following reactions and write a balanced chemical equation:

(a) $CH_3NHNH_2(g) + O_2(g) \longrightarrow$

(b) $Mg_3P_2(s) + H_2O(l) \longrightarrow$

(c) $NaCN(s) + HCl(aq) \longrightarrow$

Solution: **(a)** This combustion reaction should produce CO_2, H_2O, and N_2:

$$2CH_3NHNH_2(g) + 5O_2(g) \longrightarrow 2CO_2(g) + 6H_2O(l) + 2N_2(g)$$

(b) The Mg_3P_2 consists of Mg^{2+} and P^{3-} ions. The P^{3-} ion, like N^{3-}, has a strong affinity for protons and reacts with H_2O to form OH^- and PH_3 (PH^{2-}, PH_2^-, and PH_3 are all exceedingly weak proton donors):

$$Mg_3P_2(s) + 6H_2O(l) \longrightarrow 2PH_3(g) + 3Mg(OH)_2(aq)$$

The $Mg(OH)_2$ is of low solubility in water and may precipitate.

(c) The NaCN consists of Na^+ and CN^- ions. The CN^- ion is basic (HCN is a weak acid). Thus CN^- reacts with protons to form its conjugate acid:

$$NaCN(s) + HCl(aq) \longrightarrow HCN(aq) + NaCl(aq)$$

or

$$NaCN(s) + H^+(aq) \longrightarrow HCN(aq) + Na^+(aq)$$

The HCN has limited solubility in water and readily escapes as a gas.

PRACTICE EXERCISE

Write a balanced chemical equation for the reaction of solid sodium hydride with water. ***Answer:*** $NaH(s) + H_2O(l) \longrightarrow NaOH(aq) + H_2(g)$

22.3 HYDROGEN

Formation of elemental hydrogen was first recorded in the sixteenth century by the alchemist Paracelsus (1493–1541), who observed the formation of an "air" (gas) produced by the action of acids on iron. However, the English chemist Henry Cavendish (1731–1810) first isolated pure hydrogen and distinguished it from other gases. Because the element produces water when burned in air, the French chemist Lavoisier gave it the name *hydrogen*, which means "water producer" (Greek: *hydor*, water; *gennao*, to produce).

Hydrogen is the most abundant element in the universe. It is the nuclear fuel consumed by our sun and other stars to produce energy (Section 21.8). Although about 70 percent of the universe is composed of hydrogen, it constitutes only 0.87 percent of the earth's mass. Most of the hydrogen on our planet is found associated with oxygen. Water, which is 11 percent hydrogen by mass, is the most abundant hydrogen compound. Hydrogen is also an important part of petroleum, cellulose, starch, fats, alcohols, acids, and a wide variety of other materials.

Isotopes of Hydrogen

The most common isotope of hydrogen, ${}^{1}_{1}H$, has a nucleus consisting of a single proton. This isotope, sometimes referred to as **protium**,* comprises 99.9844 percent of naturally occurring hydrogen.

Two other isotopes are known: ${}^{2}_{1}H$, whose nucleus contains a proton and a neutron, and ${}^{3}_{1}H$, whose nucleus contains a proton and two neutrons. The ${}^{2}_{1}H$ isotope, called **deuterium**, comprises 0.0156 percent of naturally occurring hydrogen. It is not radioactive. In writing the chemical formulas of compounds containing deuterium, that isotope is often given the symbol D. For example, D_2O, which is called deuterium oxide or *heavy water*, can be obtained by electrolysis of ordinary water. The heavier D_2O undergoes electrolysis at a

* Giving unique names to isotopes is limited to hydrogen. Because of the proportionally large differences in their masses, the isotopes of H show appreciably more differences in their chemical and physical properties than do isotopes of heavier elements.

Table 22.1 Comparison of Properties of H_2O and D_2O

Property	H_2O	D_2O
Melting point (°C)	0.00	3.81
Boiling point (°C)	100.00	101.42
Density at 25°C (g/mL)	0.997	1.104
Heat of fusion (kJ/mol) at m.p.	6.008	6.276
Heat of vaporization (kJ/mol) at b.p.	40.67	41.61
Ion product, (K_w) at 25°C	1.01×10^{-14}	1.95×10^{-15}

slower rate than does the lighter H_2O and therefore becomes concentrated during the electrolysis. Some of the physical properties of H_2O and D_2O are compared in Table 22.1. Notice the small but discernible differences.

The third isotope, ${}^3_1H$, is known as **tritium**. It is radioactive, with a half-life of 12.3 yr:

$$ {}^3_1H \longrightarrow {}^3_2He + {}^{\ 0}_{-1}e \qquad t_{1/2} = 12.3 \text{ yr} \qquad [22.7] $$

It is formed continuously in the upper atmosphere in nuclear reactions induced by cosmic rays; however, because of its short half-life, only trace quantities exist naturally. The isotope can be synthesized in nuclear reactors by neutron bombardment of lithium-6:

$$ {}^6_3Li + {}^1_0n \longrightarrow {}^3_1H + {}^4_2He \qquad [22.8] $$

Each isotope of hydrogen contains a single electron and consequently undergoes identical chemical reactions. However, the heavier deuterium and tritium generally undergo reactions at a somewhat slower rate than protium.

Deuterium and tritium have proved valuable in studying the reactions of compounds containing hydrogen. A compound can be "labeled" by replacing one or more ordinary hydrogen atoms at specific locations within a molecule with deuterium or tritium. By comparing the location of the heavy hydrogen label in the reactants with that in the products, the mechanism of the reaction can often be inferred. As an example of the chemical insight that can be gained by using deuterium, consider what happens when methyl alcohol, CH_3OH, is placed in D_2O. The H atom of the O—H bond exchanges rapidly with the D atoms in D_2O, forming CH_3OD. The H atoms of the CH_3 group do not exchange. This experiment demonstrates the kinetic stability of C—H bonds and reveals the speed at which the O—H bond in the molecule breaks and reforms.

Properties of Hydrogen

Hydrogen is the only element that is not a member of any family in the periodic table. Because of its $1s^1$ electron configuration, it is generally placed above lithium in the periodic table. However, it is definitely not an alkali metal. It forms a positive ion much less readily than any alkali metal; the ionization energy of the hydrogen atom is 1310 kJ/mol, whereas that of lithium is 517 kJ/mol. Furthermore, the hydrogen atom has no electrons below its valence shell; the H^+ ion is just a bare proton. The simple H^+ ion is not known to exist in any compound. Its small size gives it a strong attraction for electrons; either it strips electrons from surrounding matter (forming hydrogen atoms that combine to form H_2), or it shares electron pairs, forming covalent bonds with other atoms. Although we may represent the aquated hydrogen ion as $H^+(aq)$, that proton is bonded to one or more water molecules and is thus often represented as $H_3O^+(aq)$ (Section 16.1).

Hydrogen is also sometimes placed above the halogens in the periodic table because the hydrogen atom can pick up one electron to form the *hydride ion*, H^-, which has a noble-gas electron configuration. However, the electron affinity of hydrogen, $E = -73$ kJ/mol, is not as large as that of any halogen; the electron affinity of fluorine is -332 kJ/mol, and that of iodine is -295 kJ/mol. In general, hydrogen shows no closer resemblance to the halogens than it does to the alkali metals.

In its elemental form, hydrogen exists at room temperature as a colorless, odorless, tasteless gas composed of diatomic molecules, H_2. We can call H_2 dihydrogen, but it is more commonly referred to as molecular hydrogen or merely hydrogen. Because H_2 is nonpolar and has only two electrons, attractive forces between molecules are extremely weak. Consequently, the melting point ($-259°C$) and boiling point ($-253°C$) of H_2 are very low.

The H—H bond-dissociation energy (436 kJ/mol) is high for a single bond (see Table 8.4). By comparison, the Cl—Cl bond-dissociation energy is only 242 kJ/mol. Because H_2 has a strong bond, most reactions of H_2 are slow at room temperature. However, the molecule is readily activated by heating, irradiation, or catalysis. The activation process generally produces hydrogen atoms, which are very reactive. The activation of H_2 by finely divided nickel, palladium, and platinum was considered briefly in our earlier discussions of heterogeneous catalysis (Section 14.6). Once H_2 is activated, it reacts rapidly and exothermically with a wide variety of substances.

Hydrogen forms strong covalent bonds with many elements, including oxygen; the H—O bond-dissociation energy is 464 kJ/mol. The strong H—O bond makes hydrogen an effective reducing agent for many metal oxides. For example, when H_2 is passed over heated CuO, copper is produced:

$$CuO(s) + H_2(g) \longrightarrow Cu(s) + H_2O(g) \qquad [22.9]$$

When H_2 is ignited in air, a vigorous reaction occurs, forming H_2O:

$$2H_2(g) + O_2(g) \longrightarrow 2H_2O(l) \qquad \Delta H = -571.7 \text{ kJ} \qquad [22.10]$$

Air containing as little as 4 percent H_2 (by volume) is potentially explosive. The disastrous burning of the hydrogen-filled airship *Hindenburg* in 1937 (Figure 22.5) dramatically demonstrated the high flammability of H_2.

Preparation of Hydrogen

When a small quantity of H_2 is needed in the laboratory, it is usually obtained by the reaction between an active metal such as zinc and a dilute acid such as HCl or H_2SO_4:

$$Zn(s) + 2H^+(aq) \longrightarrow Zn^{2+}(aq) + H_2(g) \qquad [22.11]$$

Because H_2 has an extremely low solubility in water, it can be collected by displacement of water, as shown in Figure 22.6.

Figure 22.5 Burning of the airship *Hindenburg* while landing at Lakehurst, New Jersey, on May 6, 1937. This picture was taken only 22 seconds after the first explosion occurred. (UPI/Bettmann Newsphotos))

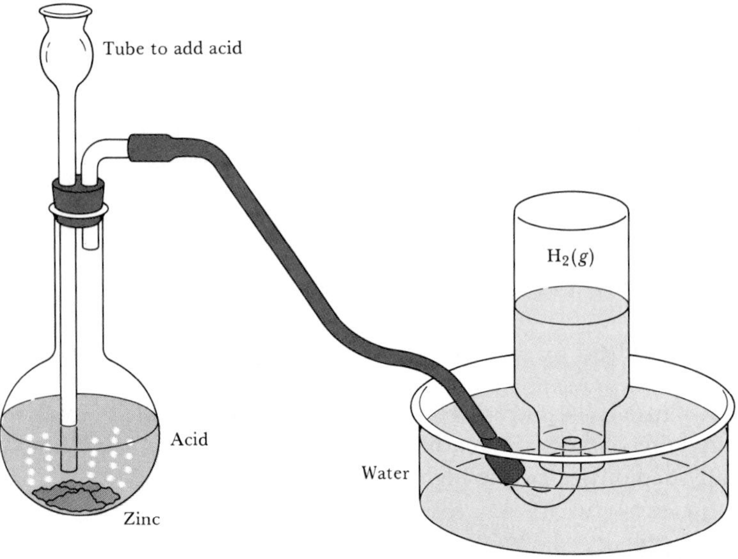

Figure 22.6 Apparatus commonly used in the laboratory for preparation of hydrogen.

When commercial quantities of H_2 are needed, the raw materials are usually hydrocarbons (from either natural gas or petroleum) or water. Hydrocarbons are substances that consist of carbon and hydrogen, such as CH_4 and C_8H_{18}. Much hydrogen is presently obtained in the course of refining petroleum. In the refining process, large hydrocarbons are catalytically broken into smaller molecules with the accompanying production of H_2 as a byproduct.

The method used to produce the largest quantities of H_2 is the reaction of methane, CH_4 (the principal component of natural gas), with steam at 1100°C:

$$CH_4(g) + H_2O(g) \longrightarrow CO(g) + 3H_2(g) \qquad [22.12]$$

$$CO(g) + H_2O(g) \longrightarrow CO_2(g) + H_2(g) \qquad [22.13]$$

When heated to about 1000°C, carbon also reacts with steam to produce a mixture of H_2 and CO gases:

$$C(s) + H_2O(g) \longrightarrow H_2(g) + CO(g) \qquad [22.14]$$

This mixture, known as **water gas**, is used as an industrial fuel.

Simple electrolysis of water consumes too much energy and is consequently too costly to be used commercially to produce H_2. However, H_2 is produced as a byproduct in the electrolysis of brine (NaCl) solutions in the course of Cl_2 and NaOH manufacture:

$$2NaCl(aq) + 2H_2O(l) \xrightarrow{\text{electrolysis}} H_2(g) + Cl_2(g) + 2NaOH(aq) \qquad [22.15]$$

Uses of Hydrogen

Hydrogen is a commercially important substance: About 2×10^8 kg (200,000 tons) is produced annually in the United States. Over two-thirds of the annual production is consumed in the synthesis of ammonia by the Haber process (Section 15.1). Hydrogen is also used to manufacture methanol, CH_3OH. As shown in Equation 22.16, the synthesis involves the catalytic combination of CO and H_2 at high pressure and temperature:

$$CO(g) + 2H_2(g) \xrightarrow[\substack{\text{200–300 atm} \\ \text{catalyst}}]{\text{300—400°C}} CH_3OH(l) \qquad [22.16]$$

Hydrogenation of vegetable oils in the manufacture of margarine and vegetable shortening is another important use. In this process, H_2 is added to carbon-carbon double bonds in the oil. A simple example of a hydrogenation occurs in the conversion of ethylene to ethane:

$$\underset{\text{Ethylene}}{H_2C{=}CH_2}(g) + H_2(g) \xrightarrow{\text{catalyst}} \underset{\text{Ethane}}{H{-}\overset{\displaystyle H}{\underset{\displaystyle H}{C}}{-}\overset{\displaystyle H}{\underset{\displaystyle H}{C}}{-}H}(g) \qquad [22.17]$$

Because organic compounds with double bonds have the ability to add additional hydrogen atoms, they are said to be *unsaturated*. The term *polyunsaturated*, which often appears in food advertisements, refers to molecules that have several (poly) double bonds between carbon atoms. The following molecule is a polyunsaturated hydrocarbon:

$$H_3C(H)C{=}CH{-}CH_2{-}CH_2{-}CH{=}CH_2$$

The polyunsaturated molecules that occur in vegetable oils are much more complex (Section 27.5).

Binary Hydrogen Compounds

Hydrogen reacts with other elements to form compounds of three general types: (1) ionic hydrides, (2) metallic hydrides, and (3) molecular hydrides.

The **ionic hydrides** are formed by the alkali metals and by the heavier alkaline earths (Ca, Sr, and Ba). These active metals are much less electronegative than hydrogen. Consequently, hydrogen acquires electrons from them to form hydride ions, H^-, as shown below:

$$2Li(s) + H_2(g) \longrightarrow 2LiH(s) \qquad [22.18]$$

$$Ca(s) + H_2(g) \longrightarrow CaH_2(s) \qquad [22.19]$$

The resultant ionic hydrides are solids with high melting points (LiH melts at 680°C).

The hydride ion is very basic and reacts readily with compounds having even weakly acidic protons to form H_2. For example, H^- reacts readily with H_2O:

$$H^-(aq) + H_2O(l) \longrightarrow H_2(g) + OH^-(aq) \qquad [22.20]$$

Ionic hydrides can therefore be used as convenient (although expensive) sources of H_2. Calcium hydride, CaH_2, is sold in commercial quantities and is used for inflation of life rafts, weather balloons, and the like, where a simple, compact means of H_2 generation is desired. It is also used to remove H_2O from organic liquids. The reaction of CaH_2 with H_2O is shown in Figure 22.7.

The reaction between H^- and H_2O (Equation 22.20) is not only an acid-base reaction, but a redox reaction as well. The H^- ion can be viewed not only as a good base, but also as a good reducing agent. In fact, hydrides are able to reduce O_2 to H_2O:

$$2NaH(s) + O_2(g) \longrightarrow Na_2O(s) + H_2O(l) \qquad [22.21]$$

Thus, hydrides are normally stored in an environment that is free of both moisture and air.

Figure 22.7 Reaction of CaH_2 with water in the presence of phenolphthalein. The red color of the phenolphthalein is due to the presence of OH^- ions. The gas bubbles are H_2. (Donald Clegg and Roxy Wilson)

CHEMISTRY AT WORK: Using Metal Hydrides

The ready absorption of H_2 by palladium metal has been used to separate H_2 from other gases and in purifying H_2 on an industrial scale. At 300 to 400 K, H_2 dissociates into atomic hydrogen on the Pd surface. The H atoms dissolve in the Pd, and under H_2 pressure they diffuse through, recombining to form H_2 on the opposite surface. Because no other molecules exhibit this property, absolutely pure H_2 results.

Research is in progress investigating metallic hydrides as storage media for hydrogen. In the case of many metals, hydride formation occurs directly upon contact between the metal and H_2. Furthermore, the reaction is often readily reversible so that H_2 can be obtained from the hydride by reducing the pressure of the hydrogen gas above the metal. Metals accommodate an extremely high density of hydrogen, because the hydrogen atoms are closely packed into the interstitial sites between metal atoms. Indeed, the number of hydrogen atoms per unit volume is greater in some metallic hydrides than in liquid H_2. In some hydrides, the metal can accommodate two or three times as many hydrogen atoms as there are metal atoms; that is, the stoichiometry approaches MH_3 (where M is the metal).

The absorption of hydrogen by metals also has detrimental effects. The hydrides tend to be more brittle than the metal. Thus, absorption of hydrogen can weaken and embrittle steel and other structural metals (Figure 22.8)

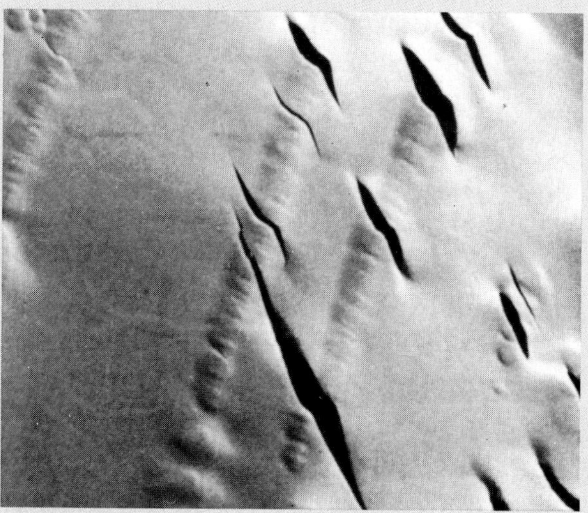

Figure 22.8 Cracking in niobium metal due to hydride formation when the metal is stressed under a 1 percent H_2 atmosphere. The raised lines running nearly vertically are due to formation of NbH at certain planes within the metal. The cracks begin in these regions because NbH is very brittle. (Photo courtesy of Professor Howard Birnbaum, University of Illinois; electron microscope photo, about 200 × magnification. Reprinted with permission from *Acta Metallurgica*, vol. 25, M. L. Grossbeck and H. K. Birnbaum, "Low Temperature Hydrogen Embrittlement of Niobium II—Microscopic Observations," copyright 1977, Pergamon Press Ltd.)

Metallic hydrides are formed when hydrogen reacts with transition metals. These compounds are so named because they retain their metallic conductivity and other metallic properties. In many metallic hydrides the ratio of metal atoms to hydrogen atoms is not a ratio of small whole numbers, nor is it a fixed ratio. The composition can vary within a range, depending on the conditions of synthesis. For example, although TiH_2 can be produced, preparations usually yield substances with about 10 percent less hydrogen than this, $TiH_{1.8}$. These nonstoichiometric metallic hydrides are sometimes called **interstitial hydrides**. They may be considered to be solutions of hydrogen atoms in the metal, with the hydrogen atoms occupying the holes or interstices between metal atoms in the solid lattice. However, this description is an oversimplification; there is evidence for chemical interaction between metal and hydrogen.

The **molecular hydrides**, formed by nonmetals and semimetals, are either gases or liquids under standard conditions. The simple molecular hydrides are listed in Figure 22.9, together with their standard free energies of formation,

Figure 22.9 Standard free energies of formation (kJ/mol) of molecular hydrides.

4A	5A	6A	7A
$CH_4(g)$ −50.8	$NH_3(g)$ −16.7	$H_2O(l)$ −237	$HF(g)$ −271
$SiH_4(g)$ +56.9	$PH_3(g)$ +18.2	$H_2S(g)$ −33.0	$HCl(g)$ −95.3
$GeH_4(g)$ +117	$AsH_3(g)$ +111	$H_2Se(g)$ +71	$HBr(g)$ −53.2
	$SbH_3(g)$ +187	$H_2Te(g)$ +138	$HI(g)$ +1.30

ΔG_f°. In each family, the thermal stability (measured by ΔG_f°) decreases as we move down the family. (Recall that the more stable a compound with respect to its elements under standard conditions, the more negative ΔG_f° is.) We will discuss the molecular hydrides further in the course of examining the other nonmetallic elements.

22.4 OXYGEN

By the middle of the seventeenth century, scientists recognized that air contained a component associated with burning and breathing. That component was not isolated until 1774, however, when Joseph Priestley discovered oxygen (Figure 22.10). Lavoisier subsequently named the element *oxygen*, meaning "acid former."

Oxygen plays an important role in the chemistries of most other elements and is found in combination with other elements in a great variety of compounds. Indeed, oxygen is the most abundant element by mass both in the earth's crust and in the human body. It constitutes 89 percent of water by mass and 20.9 percent of air by volume (23 percent by mass). It also comprises 50 percent by mass of the sand, clay, limestone, and igneous rocks that make up the bulk of the earth's crust.

Properties of Oxygen

Oxygen has two allotropes, O_2 and O_3. When we speak of elemental or molecular oxygen, it is usually understood that we are speaking of dioxygen, O_2, the normal form of the element; O_3 is called **ozone**.

Figure 22.10 Joseph Priestley (1733–1804). Priestley became interested in chemistry at the age of 39, perhaps through his personal acquaintance with Benjamin Franklin. Because Priestley lived next door to a brewery from which he could obtain carbon dioxide, his studies focused on this gas first and were later extended to other gases. Because he was suspected of sympathizing with the American and French Revolutions, his church, home, and laboratory in Birmingham, England, were burned by a mob in 1791. Priestley had to flee in disguise. He eventually emigrated to the United States in 1794, where he lived his remaining years in relative seclusion in Pennsylvania. Although his discovery of "dephlogisticated air" (oxygen) eventually led to the downfall of the phlogiston theory, Priestley stubbornly continued to support this theory even after strong evidence had brought it into serious question. Priestley was a scientific conservative, but he was very liberal in his religious and political views. (The Granger Collection)

Dioxygen exists at room temperature as a colorless, odorless, and tasteless gas. It condenses to a liquid at $-183°C$ and freezes to a solid at $-218°C$. It is only slightly soluble in water, but its presence in water is essential to marine life.

The electron configuration of the oxygen atom is $[He]2s^2 2p^4$. Thus, oxygen can complete its octet of electrons either by picking up two electrons to form the oxide ion, O^{2-}, or by sharing two electrons. In its covalent compounds it tends to form two bonds: either two single bonds, as in H_2O, or a double bond, as in formaldehyde, $H_2C{=}O$. The O_2 molecule itself contains a double bond.

The bond in O_2 is very strong (the bond-dissociation energy is 495 kJ/mol). Oxygen also forms strong bonds with many other elements. Consequently, many oxygen-containing compounds are thermodynamically more stable than O_2. However, in the absence of a catalyst most reactions of O_2 have high activation energies and thus require high temperatures to proceed at a suitable rate. Once a sufficiently exothermic reaction begins, it may accelerate rapidly, producing a reaction of explosive violence.

Preparation of Oxygen

Oxygen can be obtained either from air or from certain oxygen-containing compounds. Nearly all commercial oxygen is obtained by fractional distillation of liquefied air. The normal boiling point of O_2 is $-183°C$, whereas that of N_2, the other principal component of air, is $-196°C$. Thus, when liquefied air is warmed, the N_2 boils off, leaving liquid O_2 contaminated mainly with small amounts of N_2 and Ar.

A common laboratory preparation of O_2 is the thermal decomposition of potassium chlorate, $KClO_3$, with manganese dioxide, MnO_2, added as a catalyst:

$$2KClO_3(s) \xrightarrow{MnO_2} 2KCl(s) + 3O_2(g) \qquad [22.22]$$

Like H_2, O_2 can be collected by displacement of water because of its relatively low solubility.

Uses of Oxygen

Oxygen is one of the most widely used industrial chemicals. In 1988, it ranked behind only sulfuric acid, H_2SO_4, and nitrogen, N_2. About 1.7×10^{10} kg (19 million tons) of O_2 is used annually in the United States. Oxygen can be shipped and stored either as liquid or in steel containers as compressed gas. However, about 70 percent of O_2 output is generated at the site where it is needed.

Oxygen is by far the most widely used oxidizing agent. Over half of the O_2 produced is used in the steel industry, mainly to remove impurities from steel (Section 24.3). It is also used to bleach pulp and paper. (Oxidation of intensely colored compounds often gives colorless products.) In medicine, oxygen eases breathing difficulties. It is also used together with acetylene, C_2H_2, in oxyacetylene welding (Figure 22.11). The reaction between C_2H_2 and O_2 is highly exothermic, producing temperatures in excess of 3000°C:

$$2C_2H_2(g) + 5O_2(g) \longrightarrow 4CO_2(g) + 2H_2O(g) \qquad \Delta H° = -2510 \text{ kJ} \qquad [22.23]$$

Figure 22.11 Welding with an oxyacetylene torch. The heat of combustion of acetylene is exceptionally high, thus giving rise to a very high flame temperature when acetylene is combusted in oxygen. (Lowell Georgia/Photo Researchers)

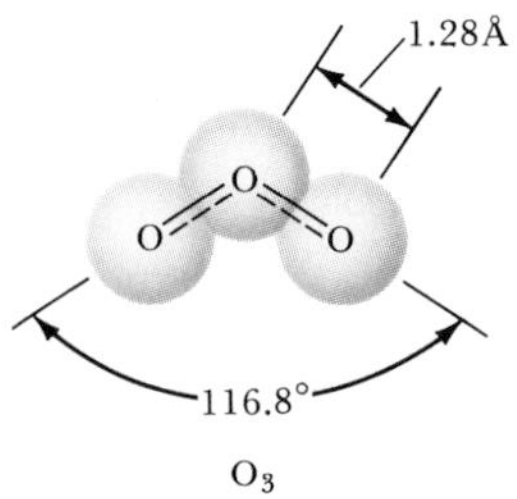

Figure 22.12 Structure of the ozone molecule.

Ozone

Ozone is a pale-blue gas with a sharp, irritating odor. It is poisonous, although no human deaths have been attributed to it. Most people can detect about 0.01 ppm in air. Exposure to 0.1 to 1 ppm produces headaches, burning of the eyes, and irritation to the respiratory passages.

The structure of the O_3 molecule is shown in Figure 22.12. The molecule possesses a π bond that is delocalized over the three oxygen atoms. The molecule dissociates readily, forming reactive oxygen atoms:

$$O_3(g) \longrightarrow O_2(g) + O(g) \qquad \Delta H^\circ = 107 \text{ kJ} \qquad [22.24]$$

Not surprisingly, ozone is a stronger oxidizing agent than dioxygen. One measure of this oxidizing power is the high reduction potential of O_3, compared to that of O_2:

$$O_3(g) + 2H^+(aq) + 2e^- \longrightarrow O_2(g) + H_2O(l) \qquad E^\circ = 2.07 \text{ V} \qquad [22.25]$$

$$O_2(g) + 4H^+(aq) + 4e^- \longrightarrow 2H_2O(l) \qquad E^\circ = 1.23 \text{ V} \qquad [22.26]$$

Ozone forms oxides with many elements under conditions where O_2 will not react; indeed, it oxidizes all of the common metals except gold and platinum.

Ozone can be prepared by passing electricity through dry O_2:

$$3O_2(g) \xrightarrow{\text{electricity}} 2O_3(g) \qquad \Delta H = 285 \text{ kJ} \qquad [22.27]$$

The preparation of O_3 may be accomplished in an apparatus such as that shown in Figure 22.13. The gas cannot be stored for long except at low temperature because it readily decomposes to O_2. The decomposition is catalyzed by certain metals, such as Ag, Pt, and Pd, and by many transition-metal oxides.

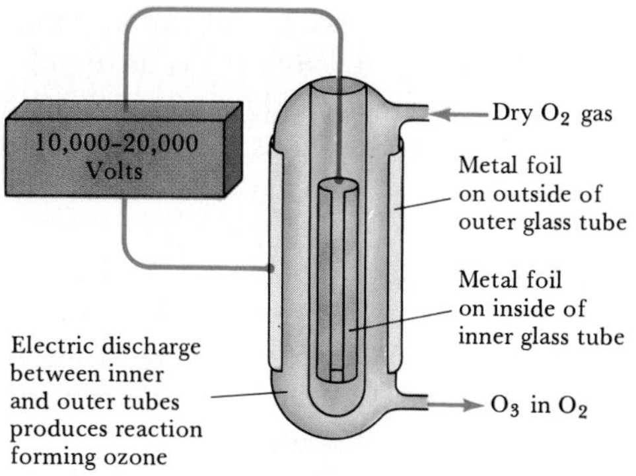

Figure 22.13 Apparatus for producing ozone from O_2.

SAMPLE EXERCISE 22.3

Using ΔG_f° for ozone from Appendix C, calculate the equilibrium constant, K_p, for Equation 22.27 at 298.0 K.

Solution: From Appendix C we have $\Delta G_f^\circ(O_3) = 163.4$ kJ/mol. Thus, for Equation 22.27, $\Delta G^\circ = (2 \text{ mol } O_3)(163.4 \text{ kJ/mol } O_3) = 326.8$ kJ. From Equation 19.15, we have $\Delta G^\circ = -RT \ln K$. Thus,

$$\ln K = \frac{-\Delta G^\circ}{RT} = \frac{-326.8 \times 10^3 \text{ J}}{(8.314 \text{ J/K-mol})(298.0 \text{ K})} = -131.9$$

$$K = 5.2 \times 10^{-58}$$

In spite of the unfavorable equilibrium constant, ozone can be prepared from O_2 as described in the text above. The unfavorable free energy of formation is overcome by energy from the electrical discharge, and O_3 is removed before the reverse reaction can occur, so a nonequilibrium mixture results.

PRACTICE EXERCISE

Using the data in Appendix C, calculate ΔG° and the equilibrium constant, K_p, for Equation 22.24 at 298.0 K. *Answer:* $\Delta G^\circ = 66.7$ kJ; $K_p = 2 \times 10^{-12}$

At the present time, uses of ozone as an industrial chemical are relatively limited. It is sometimes used in treatment of domestic water in place of chlorine. Like Cl_2, it serves to kill bacteria and oxidize organic compounds. The largest use of ozone, however, is in the preparation of pharmaceuticals, synthetic lubricants, and other commercially useful organic compounds, where O_3 is used to sever carbon-carbon double bonds.

Ozone is an important component of the upper atmosphere, where it serves to screen out ultraviolet radiation (Section 18.3). In this way, ozone protects the earth from the effects of these high-energy rays. However, in the lower atmosphere ozone is considered an air pollutant. It is a major constituent of smog. Because of its oxidizing power, it causes damage to living systems and structural materials, especially rubber.

Oxides

The electronegativity of oxygen is second only to that of fluorine. Consequently, oxygen exhibits negative oxidation states in all compounds except those with fluorine, OF_2 and O_2F_2. The -2 oxidation state is by far the most common. As we have seen, compounds in this oxidation state are called *oxides.*

Nonmetals form covalent oxides. Most of these oxides are simple molecules with low melting and boiling points. However, SiO_2 and B_2O_3 have polymeric structures (Sections 23.5 and 23.6). Most nonmetal oxides combine with water to give oxyacids. For example, sulfur dioxide, SO_2, dissolves in water to give sulfurous acid, H_2SO_3:

$$SO_2(g) + H_2O(l) \longrightarrow H_2SO_3(aq) \qquad [22.28]$$

This reaction and that of SO_3 with water to form H_2SO_4 are largely responsible for acid rain (Section 18.4). The analogous reaction of CO_2 with water to form carbonic acid, H_2CO_3, accounts for the acidity of carbonated water.

Oxides that react with water to form acids are called **acidic anhydrides** (anhydride means "without water") or **acidic oxides**. A few nonmetal oxides, especially ones with the nonmetal in a low oxidation state—such as N_2O, NO, and CO—are not acidic anhydrides. These oxides do not react with water.

Most metal oxides are ionic compounds. Those ionic oxides that dissolve in water form hydroxides and are consequently called **basic anhydrides** or **basic oxides**. For example, barium oxide, BaO, dissolves in water to form barium hydroxide, $Ba(OH)_2$:

$$BaO(s) + H_2O(l) \longrightarrow Ba(OH)_2(aq) \qquad [22.29]$$

This reaction is shown in Figure 22.14. Such reactions can be attributed to the high basicity of the O^{2-} ion and consequently its virtually complete hydrolysis in water:

$$O^{2-}(aq) + H_2O(l) \longrightarrow 2OH^-(aq) \qquad [22.30]$$

Even those ionic oxides that are water-insoluble tend to dissolve in acids. Iron(III) oxide, for example, dissolves in acids:

$$Fe_2O_3(s) + 6H^+(aq) \longrightarrow 2Fe^{3+}(aq) + 3H_2O(l) \qquad [22.31]$$

This reaction is used to remove rust, $Fe_2O_3 \cdot nH_2O$, from iron or steel prior to application of a protective coat of zinc or tin.

Oxides that can exhibit both acidic and basic characters are said to be *amphoteric* (Section 17.5). If a metal forms more than one oxide, the basic character of the oxide decreases as the oxidation state of the metal increases:

Compound	Oxidation state of Cr	Nature of oxide
CrO	+2	Basic
Cr_2O_3	+3	Amphoteric
CrO_3	+6	Acidic

Figure 22.14 Barium oxide, BaO, the white solid at the bottom of the container, reacts with water to produce barium hydroxide, $Ba(OH)_2$. The red color of the solution is caused by phenolphthalein and indicates the presence of OH^- ions in solution. (Donald Clegg and Roxy Wilson)

Peroxides and Superoxides

Compounds containing O—O bonds and oxygen in an oxidation state of -1 are called *peroxides*. Oxygen has an oxidation state of $-\frac{1}{2}$ in O_2^-, which is called the *superoxide* ion. The most active metals (Cs, Rb, and K) react with

O_2 to give superoxides (CsO_2, RbO_2, and KO_2). Their active neighbors in the periodic table (Na, Ca, Sr, and Ba) react with O_2, producing peroxides (Na_2O_2, CaO_2, SrO_2, and BaO_2). Less active metals and nonmetals produce normal oxides.

When superoxides dissolve in water, O_2 is produced:

$$2KO_2(s) + 2H_2O(l) \longrightarrow 2K^+(aq) + 2OH^-(aq) + O_2(g) + H_2O_2(aq) \quad [22.32]$$

Because of this reaction, potassium superoxide, KO_2, is used as an oxygen source in masks worn for rescue work. Moisture in the breath causes the compound to decompose to form O_2 and KOH. The KOH so formed serves to remove CO_2 from the exhaled breath:

$$2OH^-(aq) + CO_2(g) \longrightarrow H_2O(l) + CO_3^{2-}(aq) \quad [22.33]$$

Sodium peroxide, Na_2O_2, is used commercially as an oxidizing agent. When dissolved in water it produces hydrogen peroxide:

$$Na_2O_2(s) + 2H_2O(l) \longrightarrow 2Na^+(aq) + 2OH^-(aq) + H_2O_2(aq) \quad [22.34]$$

Hydrogen peroxide, H_2O_2, is the most familiar and commercially important peroxide. The structure of H_2O_2 is shown in Figure 22.15. The peroxide linkage (—O—O—) exists in species other than H_2O_2 and O_2^{2-} ion. For example, the peroxydisulfate ion, $S_2O_8^{2-}$ or $O_3SOOSO_3^{2-}$, contains this linkage.

Pure hydrogen peroxide is a clear, syrupy liquid, density 1.47 g/cm^3 at 0°C. It melts at −0.4°C, and its normal boiling point is 151°C. These properties are characteristic of a highly polar, strongly hydrogen-bonded liquid such as water. Concentrated hydrogen peroxide is a dangerously reactive substance, because the decomposition to form water and oxygen gas is very exothermic:

$$2H_2O_2(l) \longrightarrow 2H_2O(l) + O_2(g) \qquad \Delta H^\circ = -196.0 \text{ kJ} \quad [22.35]$$

The decomposition can occur with explosive violence if highly concentrated hydrogen peroxide comes in contact with substances that can catalyze the reaction. Hydrogen peroxide is marketed as a chemical reagent in aqueous solutions of up to about 30 percent by weight. A solution containing about 3 percent by weight H_2O_2 is commonly used as a mild antiseptic; somewhat more concentrated solutions are employed to bleach fabrics such as cotton, wool, and silk.

The peroxide ion is also important in biochemistry. It is a byproduct of metabolism that results from the reduction of molecular oxygen, O_2. The body disposes of this reactive species with enzymes with such names as peroxidase and catalase. The fizzing that occurs when a dilute H_2O_2 solution is applied to an open wound is due to decomposition of the H_2O_2 into O_2 and H_2O, a reaction catalyzed by the aforementioned enzymes.

Hydrogen peroxide can act as either an oxidizing or a reducing agent. Equations 22.36 and 22.37 show the half-reactions for acid solution.

$$2H^+(aq) + H_2O_2(aq) + 2e^- \longrightarrow 2H_2O(l) \qquad E^\circ = 1.77 \text{ V} \quad [22.36]$$

$$H_2O_2(aq) \longrightarrow O_2(g) + 2H^+(aq) + 2e^- \qquad E^\circ = -0.67 \text{ V} \quad [22.37]$$

Figure 22.15 Structure of the hydrogen peroxide molecule.

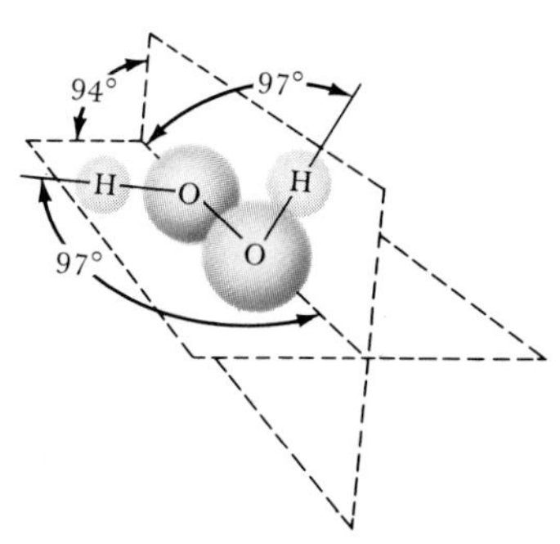

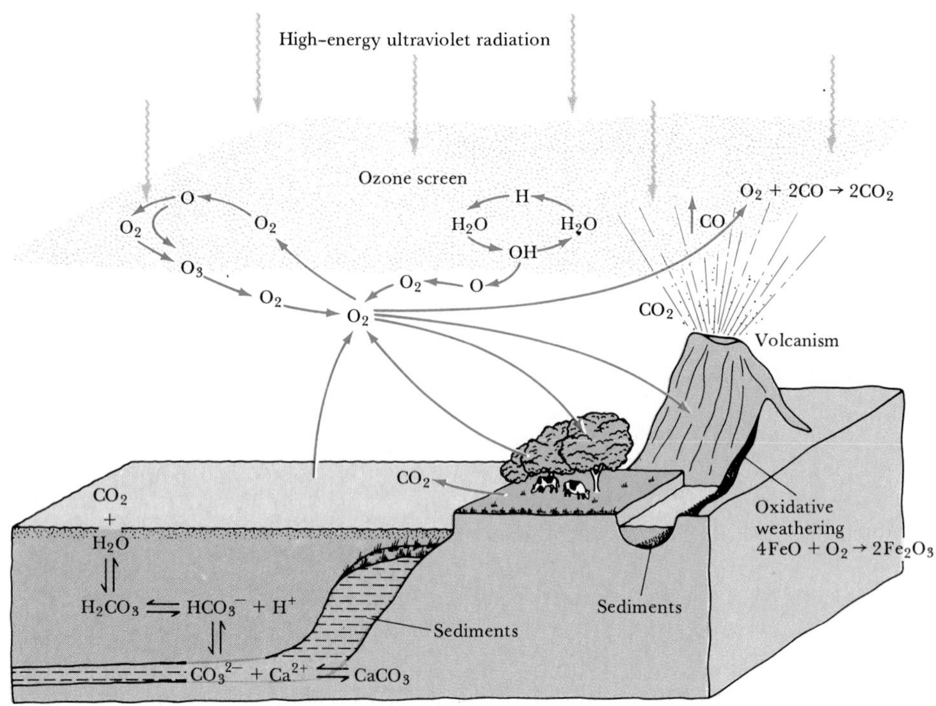

Figure 22.16 Simplified view of the oxygen cycle. The atmosphere, which contains O_2, is one of the primary sources of the element. Some O_2 is produced by radiation-induced dissociation of H_2O in the upper atmosphere. Some O_2 is produced by green plants from H_2O and CO_2 in the course of photosynthesis. Atmospheric CO_2, in turn, results from combustion reactions, animal respiration, and the dissociation of bicarbonate in water. The O_2 is used to produce ozone in the upper atmosphere, in oxidative weathering of rocks, in animal respiration, and in combustion reactions.

The Oxygen Cycle

Oxygen accounts for about one-fourth of the atoms in living matter. The number of oxygen atoms is fixed; as O_2 is removed from air through respiration and other processes, it is replenished. The major nonliving sources of oxygen other than O_2 are CO_2 and H_2O. A simplified picture of the movement of oxygen in our environment is shown in Figure 22.16. This figure points out both how O_2 is removed from the atmosphere and how it is replenished. Oxygen, O_2, is reformed mainly from CO_2 through the process of photosynthesis. Energy is produced when O_2 is converted to CO_2; energy must therefore be supplied to reform O_2 from CO_2. This energy is provided by the sun. Thus, life on earth depends on chemical recycling made possible by solar energy.

22.5 NITROGEN

Nitrogen was discovered in 1772 by the Scottish botanist Daniel Rutherford. He found that when a mouse was enclosed in a sealed jar, the animal quickly consumed the life-sustaining component of air (oxygen) and died. When the "fixed air" (CO_2) in the container was removed, a "noxious air" remained that would not sustain combustion or life. That gas is known to us now as nitrogen.

Nitrogen constitutes 78 percent by volume of the earth's atmosphere, where it occurs as N_2 molecules. Although nitrogen is a key element in living creatures, compounds of nitrogen are not abundant in the earth's crust. The major natural deposits of nitrogen compounds are those of KNO_3 (saltpeter) in India and $NaNO_3$ (Chile saltpeter) in Chile and other desert regions of South America.

Properties of Nitrogen

Nitrogen is a colorless, odorless, and tasteless gas composed of N_2 molecules. Its melting point is $-210°C$, and its normal boiling point is $-196°C$.

The N_2 molecule is very unreactive because of the strong triple bond between nitrogen atoms (the $N \equiv N$ bond-dissociation energy is 941 kJ/mol, nearly twice that for the bond in O_2; see Table 8.4). When substances burn in air they normally react with O_2 but not with N_2. However, when magnesium burns in air, reaction with N_2 also occurs to form magnesium nitride, Mg_3N_2. A similar reaction occurs with lithium:

$$3Mg(s) + N_2(g) \longrightarrow Mg_3N_2(s) \quad [22.38]$$

$$6Li(s) + N_2(g) \longrightarrow 2Li_3N(s) \quad [22.39]$$

The nitride ion is a strong Brønsted base. It reacts with water to form ammonia, NH_3:

$$Mg_3N_2(s) + 6H_2O(l) \longrightarrow 2NH_3(aq) + 3Mg(OH)_2(s) \quad [22.40]$$

The electron configuration of the nitrogen atom is $[He]2s^22p^3$. The element exhibits all formal oxidation states from $+5$ to -3, as shown in Table 22.2. Because nitrogen is the fourth most electronegative element after fluorine, oxygen, and chlorine, it exhibits positive oxidation states only in combination with those three elements.

Figure 22.17 summarizes the standard electrode potentials for interconversion of several common nitrogen species. The potentials in the diagram are large and positive, which indicates that the nitrogen oxides and oxyanions shown are strong oxidizing agents.

Table 22.2 Oxidation States of Nitrogen

Oxidation state	Examples
+5	N_2O_5, HNO_3, NO_3^-
+4	NO_2, N_2O_4
+3	HNO_2, NO_2^-, NF_3
+2	NO
+1	N_2O, $H_2N_2O_2$, $N_2O_2^{2-}$, HNF_2
0	N_2
−1	NH_2OH, H_2NF
−2	N_2H_4
−3	NH_3, NH_4^+, NH_2^-

Preparation and Uses of Nitrogen

Elemental nitrogen is obtained in commercial quantities by fractional distillation of liquid air. About 2.3×10^{10} kg (26 million tons) of N_2 is produced annually in the United States.

Because of its low reactivity, large quantities of N_2 are used as an inert gaseous blanket to exclude O_2 during the processing and packaging of foods, the manufacture of chemicals, the fabrication of metals, and the production of electronic devices. Liquid N_2 is employed as a coolant to freeze foods rapidly.

Figure 22.17 Standard electrode potentials in acid solution for reduction of some common nitrogen-containing compounds in various oxidation states. For example, reduction of NO_3^- to NO_2 in acid solution has a standard electrode potential of 0.79 V (leftmost entry). You should be able to write the complete, balanced half-reaction for this reduction using H^+, H_2O, and e^-, as discussed in Section 20.2.

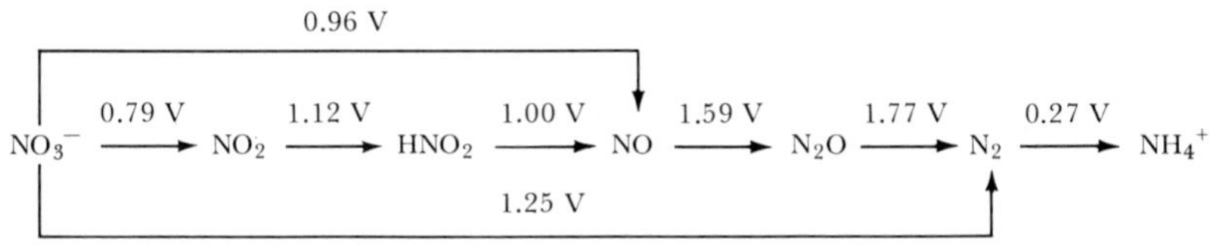

$$NO_3^- \xrightarrow{0.79\text{ V}} NO_2 \xrightarrow{1.12\text{ V}} HNO_2 \xrightarrow{1.00\text{ V}} NO \xrightarrow{1.59\text{ V}} N_2O \xrightarrow{1.77\text{ V}} N_2 \xrightarrow{0.27\text{ V}} NH_4^+$$

$NO_3^- \to NO$: 0.96 V

$NO_3^- \to N_2$: 1.25 V

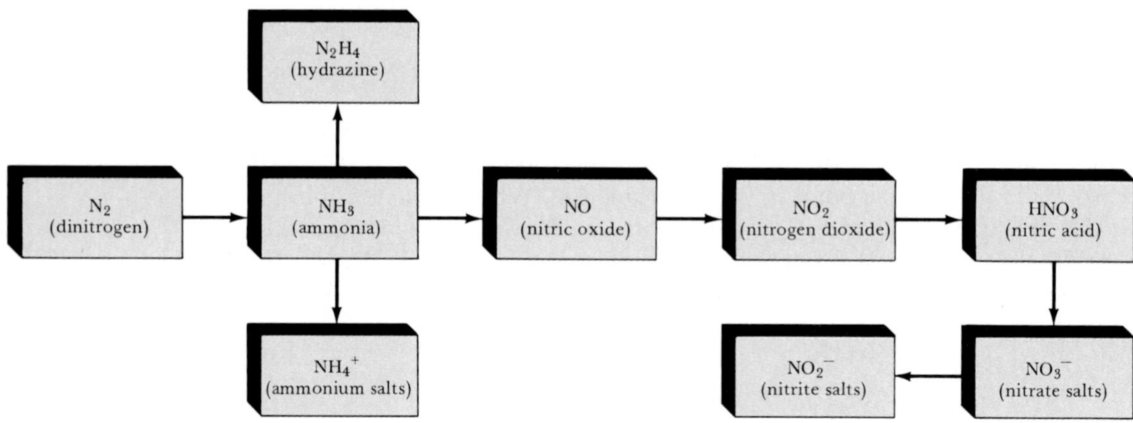

Figure 22.18 Conversion of N_2 into common nitrogen compounds.

The largest use of nitrogen is in the manufacture of nitrogen-containing chemicals. The formation of nitrogen compounds from N_2 is known as *nitrogen fixation.* The demand for fixed nitrogen is high because the element is required in maintaining soil fertility. Although we are immersed in an ocean of air that contains abundant N_2, our supply of food is limited more by the availability of fixed nitrogen than by that of any other plant nutrient. Thus, N_2 is used primarily for the manufacture of nitrogen-containing fertilizers. It is also used to manufacture explosives, plastics, and many important chemicals.

The chain of conversion of N_2 into a variety of useful, simple nitrogen-containing species is given in Figure 22.18. The processes shown are discussed in more detail in later portions of the section.

Hydrogen Compounds of Nitrogen

Ammonia is one of the most important compounds of nitrogen. It is a colorless, toxic gas that has a characteristic, irritating odor. As we have noted in previous discussions (Section 16.4), the NH_3 molecule is basic ($K_b = 1.8 \times 10^{-5}$).

In the laboratory, NH_3 is prepared by the action of NaOH on an ammonium salt. The NH_4^+ ion, which is the conjugate acid of NH_3, loses a proton to OH^-. The resultant NH_3 is volatile and is driven from the solution by mild heating:

$$NH_4Cl(aq) + NaOH(aq) \longrightarrow NH_3(g) + H_2O(l) + NaCl(aq) \qquad [22.41]$$

Commercial production of NH_3 is achieved by the *Haber process* (Section 15.1), in which N_2 and H_2 are catalytically combined at high pressure and high temperature:

$$N_2(g) + 3H_2(g) \longrightarrow 2NH_3(g) \qquad [22.42]$$

Figure 22.19 Lewis structures of hydrazine, N_2H_4, and monomethylhydrazine, CH_3NHNH_2.

```
                             H
                             |
H—N̈—N̈—H        H—N̈—N̈—C—H
  |   |           |   |   |
  H   H           H   H   H

Hydrazine       Monomethyl hydrazine
```

About 1.5×10^{10} kg (17 million tons) of ammonia is produced annually in the United States. About 75 percent is used for fertilizer.

Hydrazine, N_2H_4, bears the same relationship to ammonia that hydrogen peroxide does to water. As shown in Figure 22.19, the hydrazine molecule contains an N—N single bond. Hydrazine is quite poisonous. It can be prepared by the reaction of ammonia with hypochlorite ion, OCl^-, in aqueous solution:

$$2NH_3(aq) + OCl^-(aq) \longrightarrow N_2H_4(aq) + Cl^-(aq) + H_2O(l) \quad [22.43]$$

The reaction is complex, involving several intermediates including chloramine, NH_2Cl. The poisonous NH_2Cl bubbles out of solution when household ammonia and chlorine bleach (which contains OCl^-) are mixed. This reaction is one reason for the frequently cited warning not to mix household cleansing agents.

Pure hydrazine is an oily, colorless liquid that explodes on heating. It is a highly reactive reducing agent. Hydrazine is normally employed in aqueous solution, where it can be handled safely. The substance is weakly basic, and salts of $N_2H_5^+$ can be formed:

$$N_2H_4(aq) + H_2O(l) \rightleftharpoons N_2H_5^+(aq) + OH^-(aq) \quad K_b = 1.3 \times 10^{-6} \quad [22.44]$$

The combustion of hydrazine is highly exothermic:

$$N_2H_4(l) + O_2(g) \longrightarrow N_2(g) + 2H_2O(g) \qquad \Delta H^\circ = -534 \text{ kJ} \quad [22.45]$$

Hydrazine and compounds derived from it, such as monomethylhydrazine (Figure 22.19), are used as rocket fuels.

SAMPLE EXERCISE 22.4

Hydroxylamine, NH_2OH, reduces copper(II) to the free metal in acid solutions. Write a balanced equation for the reaction, assuming that N_2 is the oxidation product.

Solution: The unbalanced and incomplete half-reactions are

$$Cu^{2+}(aq) \longrightarrow Cu(s)$$

$$NH_2OH(aq) \longrightarrow N_2(g)$$

Balancing these equations as described in Section 20.1 gives

$$Cu^{2+}(aq) + 2e^- \longrightarrow Cu(s)$$

$$2NH_2OH(aq) \longrightarrow N_2(g) + 2H_2O(l) + 2H^+(aq) + 2e^-$$

Adding these half-reactions gives the balanced equation:

$$Cu^{2+}(aq) + 2NH_2OH(aq) \longrightarrow Cu(s) + N_2(g) + 2H_2O(l) + 2H^+(aq)$$

PRACTICE EXERCISE

(a) In power plants hydrazine is used to prevent corrosion of the metal parts of steam boilers by the O_2 dissolved in the water. The hydrazine reacts with O_2 in water to give N_2 and H_2O. Write a balanced chemical equation for this reaction. **(b)** Monomethylhydrazine, $N_2H_3CH_3(l)$, is used with the oxidizer dinitrogen tetraoxide, $N_2O_4(l)$, to power the steering rockets of the space shuttle orbiter. The reaction of these two substances produces N_2, CO_2, and H_2O. Write a balanced chemical equation for this reaction.

Answers: **(a)** $N_2H_4(aq) + O_2(aq) \longrightarrow N_2(g) + 2H_2O(l)$

(b) $5N_2O_4(l) + 4N_2H_3CH_3(l) \longrightarrow 9N_2(g) + 4CO_2(g) + 12H_2O(g)$

Oxides and Oxyacids of Nitrogen

Nitrogen forms three common oxides: N_2O (nitrous oxide), NO (nitric oxide), and NO_2 (nitrogen dioxide). It also forms two unstable oxides that we will not discuss, N_2O_3 (dinitrogen trioxide) and N_2O_5 (dinitrogen pentoxide).

(*a*) (*b*) (*c*)

Figure 22.20 (*a*) Nitric oxide, NO, can be prepared by reacting copper with 6 *M* nitric acid. In this photo a jar containing 6 *M* HNO_3 has been inverted over some pieces of copper. Colorless NO, which is only slightly soluble in water, is collected in the jar. The blue color of the solution is due to the presence of Cu^{2+} ions. (*b*) Colorless NO gas, collected as shown on the left. (*c*) When the stopper is removed from the jar of NO, the NO reacts with oxygen in the air to form yellow-brown NO_2. (Donald Clegg and Roxy Wilson)

Nitrous oxide, N_2O, is also known as laughing gas because a person becomes somewhat giddy after inhaling only a small amount of it. This colorless gas was the first substance used as a general anesthetic. It is used as the compressed gas propellant in several aerosols and foams, such as in whipped cream. It can be prepared in the laboratory by carefully heating ammonium nitrate to about 200°C:

$$NH_4NO_3(s) \xrightarrow{\Delta} N_2O(g) + 2H_2O(g) \qquad [22.46]$$

Nitric oxide, NO, is also a colorless gas, but unlike N_2O it is slightly toxic. It can be prepared in the laboratory by reduction of dilute nitric acid, using copper or iron as a reducing agent, as shown in Figure 22.20.

$$Cu(s) + 2NO_3^-(aq) + 8H^+(aq) \longrightarrow 3Cu^{2+}(aq) + 2NO(g) + 4H_2O(l) \qquad [22.47]$$

It is also produced by direct combination of N_2 and O_2 at elevated temperatures. As we saw in Section 18.4, this reaction is a significant source of nitrogen oxide air pollutants, which form during combustion reactions in air. However, the direct combination of N_2 and O_2 is not presently used for commercial production of NO because the yield is low; the equilibrium constant K_c at 2400 K is only 0.05.

The commercial route to NO (and hence to other oxygen-containing compounds of nitrogen) is by means of the catalytic oxidation of NH_3 (Figure 22.21):

$$4NH_3(g) + 5O_2(g) \xrightarrow[850°C]{\text{Pt catalyst}} 4NO(g) + 6H_2O(g) \qquad [22.48]$$

In the absence of the platinum catalyst, NH_3 is converted to N_2 instead of NO:

$$4NH_3(g) + 3O_2(g) \xrightarrow{1000°C} 2N_2(g) + 6H_2O(g) \qquad [22.49]$$

Figure 22.21 The oxidation of ammonia during the production of nitric acid is undertaken on a large industrial scale. The reaction is carried out in the presence of a platinum-rhodium catalyst, which is usually in the form of a woven gauze. Here new catalyst gauzes are being installed in an ammonia oxidation plant. (Johnson-Matthey Metals Limited)

The catalytic conversion of NH_3 to NO is the first step in a three-step process known as the **Ostwald process**, by which NH_3 is converted commercially into nitric acid, HNO_3 (Figure 22.22). Nitric oxide reacts readily with O_2, forming NO_2 when exposed to air (see Figure 22.20):

$$2NO(g) + O_2(g) \longrightarrow 2NO_2(g) \qquad [22.50]$$

When dissolved in water, NO_2 forms nitric acid:

$$3NO_2(g) + H_2O(l) \longrightarrow 2H^+(aq) + 2NO_3^-(aq) + NO(g) \qquad [22.51]$$

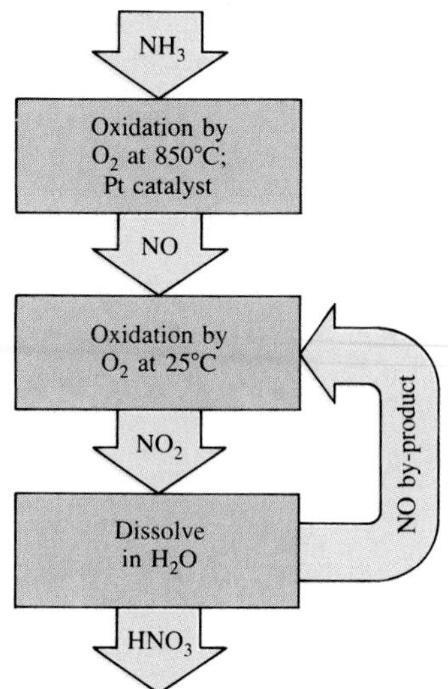

Figure 22.22 The Ostwald process for converting NH_3 to HNO_3.

Note that nitrogen is both oxidized and reduced in this reaction. The NO_2 is said to have undergone **disproportionation**. The reduction product NO can be converted back into NO_2 by exposure to air and thereafter dissolved in water to prepare more HNO_3.

Nitrogen dioxide is a yellow-brown gas. It is poisonous and has a choking odor. Below room temperature two NO_2 molecules combine to form the colorless N_2O_4:

$$2NO_2(g) \longrightarrow N_2O_4(g) \qquad \Delta H^\circ = -58 \text{ kJ} \qquad [22.52]$$

The two common oxyacids of nitrogen are nitric acid, HNO_3, and nitrous acid, HNO_2 (Figure 22.23). *Nitric acid* is a colorless, corrosive liquid. Nitric acid solutions often take on a slightly yellow color (Figure 22.24) as a result of small amounts of NO_2 formed by photochemical decomposition:

$$4HNO_3(aq) \longrightarrow 4NO_2(g) + O_2(g) + 2H_2O(l) \qquad [22.53]$$

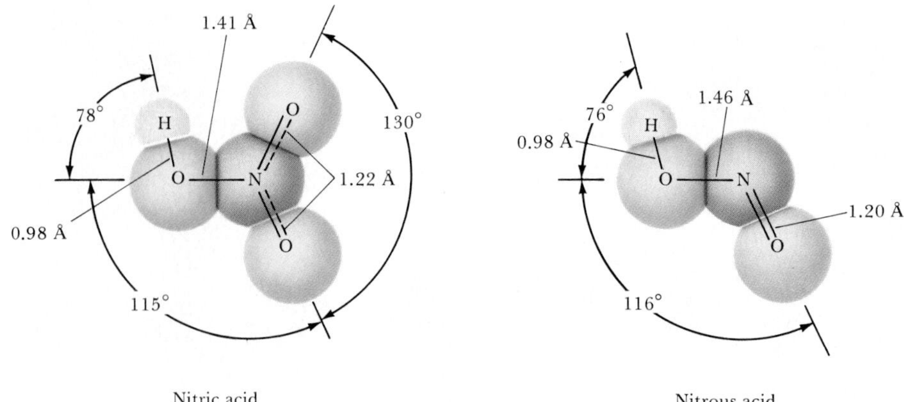

Figure 22.23 Structures of nitric acid and nitrous acid.

Figure 22.24 Colorless nitric acid solution (left) becomes yellow upon standing in sunlight (right). (Kristen Brochmann/Fundamental Photographs)

Nitric acid is a strong acid. It is also a powerful oxidizing agent, as the following electrode potentials indicate:

$$NO_3^-(aq) + 2H^+(aq) + e^- \longrightarrow NO_2(g) + H_2O(l) \qquad E^\circ = 0.79\ V \qquad [22.54]$$

$$NO_3^-(aq) + 4H^+(aq) + 3e^- \longrightarrow NO(g) + 2H_2O(l) \qquad E^\circ = 0.96\ V \qquad [22.55]$$

Concentrated nitric acid will attack and oxidize most metals, except Au, Pt, Rh, and Ir.

About 7.1×10^9 kg (8 million tons) of nitric acid is produced annually in the United States. The largest use of nitric acid is in the manufacture of NH_4NO_3 for fertilizers, which accounts for about 80 percent of the production. The acid is also used in the production of plastics, drugs, and explosives.

The development of the Haber and Ostwald processes in Germany just prior to World War I permitted Germany to make munitions even though naval blockades prevented access to traditional sources of nitrates. Among the explosives made from nitric acid are nitroglycerin, trinitrotoluene (TNT), and nitrocellulose. The reaction of nitric acid with glycerin to form nitroglycerin is shown in Equation 22.56:

$$\begin{array}{c} \text{H} \\ | \\ \text{H—C—OH} \\ | \\ \text{H—C—OH} \\ | \\ \text{H—C—OH} \\ | \\ \text{H} \end{array} + 3HNO_3 \longrightarrow \begin{array}{c} \text{H} \\ | \\ \text{H—CONO}_2 \\ | \\ \text{H—CONO}_2 \\ | \\ \text{H—CONO}_2 \\ | \\ \text{H} \end{array} + 3H_2O \qquad [22.56]$$

When nitroglycerin explodes, the reaction summarized in Equation 22.57 occurs:

$$4C_3H_5N_3O_9(l) \longrightarrow 6N_2(g) + 12CO_2(g) + 10H_2O(g) + O_2(g) \qquad [22.57]$$

All the products of this reaction contain very strong bonds. As a result, the reaction is very exothermic. Furthermore, a considerable amount of gaseous products form from the liquid. The sudden formation of these gases, together with their expansion resulting from the heat generated by the reaction, produces the explosion. (See Chemistry at Work, p. 270.)

Nitrous acid, HNO_2 (Figure 22.23), is considerably less stable than HNO_3 and tends to disproportionate into NO and HNO_3. It is normally made by action of a strong acid such as H_2SO_4 on a cold solution of a nitrite salt such as $NaNO_2$. Nitrous acid is a weak acid ($K_a = 4.5 \times 10^{-4}$).

The Nitrogen Cycle in Nature

Two primary routes for nitrogen fixation exist in nature: Lightning causes the formation of NO from N_2 and O_2 in air, and the root nodules of certain leguminous plants, such as peas, beans, peanuts, and alfalfa, contain nitrogen-fixing bacteria. Both iron and molybdenum are part of the enzyme system responsible for the nitrogen fixation in these root nodules. It is of interest to understand this process and to develop catalysts that, like enzymes, fix nitrogen at ambient pressure and temperature (and hence potentially at an energy savings over the Haber process).

Nitrogen is found in many compounds that are vital to life, including proteins, enzymes, nucleic acids, vitamins, and hormones. Plants employ very simple compounds as starting materials from which such complex, biologically necessary compounds are formed. Plants are able to use several forms of nitrogen, especially NH_3, NH_4^+, and NO_3^-. Liquid ammonia, ammonium nitrate, NH_4NO_3, and urea, $(NH_2)_2CO$, are among the most commonly applied nitrogen fertilizers. Urea is made by the reaction of ammonia and carbon dioxide:

$$2NH_3(aq) + CO_2(aq) \rightleftharpoons \underset{\text{Urea}}{H_2N\overset{\overset{\displaystyle O}{\|}}{C}NH_2(aq)} + H_2O(l) \qquad [22.58]$$

NH_3 is slowly released as the urea reacts with water in the soil.

Animals are unable to synthesize the complex nitrogen compounds they require from the simple substances used by plants. Instead, they rely on more complicated precursors present in foods. Those nitrogen compounds not needed by the animal are excreted as nitrogenous waste. Certain microorganisms are

CHEMISTRY AT WORK: Nitrites in Food

Sodium nitrite is used as a food additive in cured meats such as frankfurters, ham, and cold cuts. The NO_2^- serves two functions. It inhibits growth of bacteria, especially *Clostridium botulinum,* which produces the potentially fatal food poisoning known as *botulism.* It also preserves the red color of the meat and thereby its appetizing appearance. Debate over the continued use of nitrites in cured meat products arises because HNO_2 can react with certain organic compounds to form compounds known as *nitrosoamines.* These reactions occur at high temperatures, as during frying. These are compounds of the type given in Figure 22.25. These organic compounds have been shown to produce cancer in laboratory animals. Although no human cancer has been linked to nitrites in meat, the potential has caused the U.S. Food and Drug Administration to reduce the limits of allowable concentrations of NO_2^- in foods.

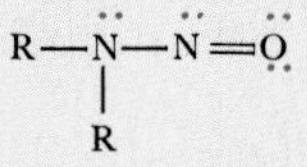

$$R—\ddot{N}—\ddot{N}=\ddot{O}:$$
(with R attached below the first N)

Nitrosoamine

Figure 22.25 General formula for a nitrosoamine (R = groups such as CH_3).

able to convert this waste back into N_2. Nitrogen is recycled in this fashion. As in the case of the cycling of oxygen, energy from the sun is required. A simplified picture of the nitrogen cycle is shown in Figure 22.26.

Large-scale cultivation of nitrogen-fixing legumes and industrial fixation have increased the quantity of fixed nitrogen in the biosphere. Much of this nitrogen ends up as nitrates in the soil. These compounds are highly water-soluble. They are therefore readily washed from the soil into water bodies. Once in a

Figure 22.26 Simplified picture of the nitrogen cycle. The main reservoir of nitrogen is the atmosphere, which contains N_2. This nitrogen is fixed through the action of lightning and leguminous plants. The compounds of nitrogen reside in the soil as NH_3 (and NH_4^+), NO_2^-, and NO_3^-. All are water-soluble and can be washed out of the soil by ground water. These nitrogen compounds are utilized by plants in their growth and are incorporated into animals that eat the plants. Animal waste and dead plants and animals are attacked by certain bacteria that free N_2, which escapes into the atmosphere, thereby completing the cycle.

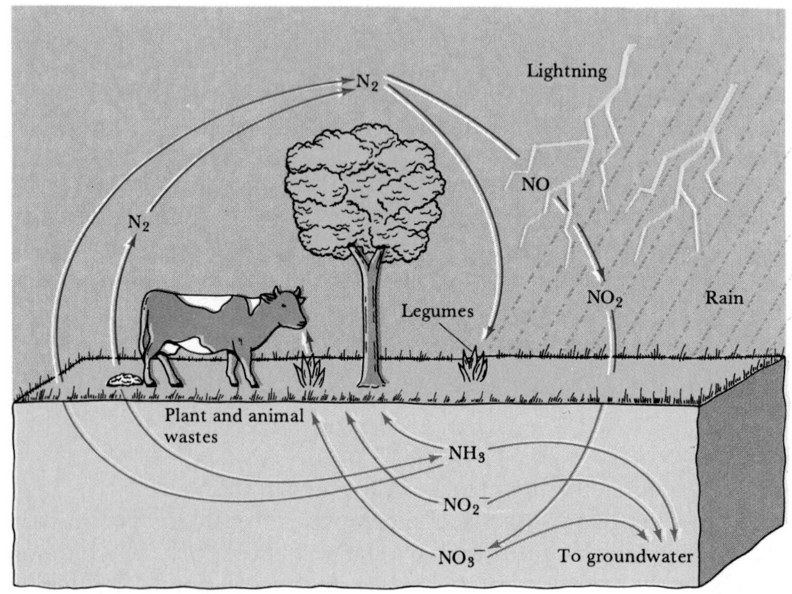

lake, nitrates stimulate plant growth, encouraging, for example, rapid growth of algae. When these plants die, their decay consumes O_2 in the water, thereby killing fish and other oxygen-dependent organisms. The resultant anaerobic environment leads to the foul odors that we associate with highly polluted bodies of water (Section 18.6).

22.6 CARBON

Carbon is not an abundant element; it constitutes only 0.027 percent of the earth's crust. Although some carbon occurs in elemental form as graphite and diamond, most is found in combined form. Over half occurs in carbonate compounds, such as $CaCO_3$. Carbon is also found in coal, petroleum, and natural gas. The importance of the element stems in large part from its occurrence in all living organisms: Life as we know it is based on carbon compounds. About 150 years ago, scientists believed that these life-sustaining compounds could be made only within living systems. They were consequently called *organic compounds*. It is now known that organic compounds can be synthesized in the laboratory from simple inorganic (nonorganic) substances. Although the name *organic chemistry* persists, it is now used to describe the portion of chemistry that focuses on hydrocarbons and compounds derived from them by substituting other atoms for some hydrogen atoms. In this section, we will take a brief look at carbon and its most common inorganic compounds. Organic chemistry is considered in Chapter 26.

Elemental Forms of Carbon

We have seen that carbon exists in two crystalline forms, graphite and diamond (Section 11.8). *Graphite* is a soft, black, slippery solid that has a metallic luster and conducts electricity. Its structure consists of parallel sheets of carbon atoms [Figure 11.40(*b*)] that are held together by London forces. *Diamond* is a clear, hard solid whose structure consists of an interlocking network of carbon atoms [Figure 11.40(*a*)]. Diamond has a more compact structure than graphite ($d = 2.25\ g/cm^3$ for graphite; $d = 3.51\ g/cm^3$ for diamond). At very high pressures and temperatures (on the order of 100,000 atm at 3000°C), graphite converts to diamond, the more compact allotrope. Gem-quality diamonds can be produced (Figure 22.27), but they are costly. However, about 3×10^4 kg (30 tons) of industrial-grade diamonds is synthesized each year for use in cutting, grinding, and polishing tools.

Figure 22.27 Graphite and synthetic diamond prepared from graphite. Most synthetic diamonds lack the size, color, and clarity of natural diamonds and are therefore not used in jewelry. (General Electric Research and Development Center)

CHEMISTRY AT WORK: Carbon Fibers and Composites

The properties of graphite are anisotropic (that is, they differ in different directions through the solid). Along the carbon planes, graphite possesses great strength because of the number and strength of the carbon-carbon bonds in this direction. In contrast, we have seen that the bonds between planes are relatively weak, making graphite weak in that direction.

Fibers of graphite can be prepared in which the carbon planes are aligned to varying extents parallel to the fiber axis. These fibers are also light-weight (density of about 2 g/cm^3) and chemically quite unreactive. The oriented fibers are made by first slowly pyrolyzing (decomposing by action of heat) organic fibers at about 150°C to 300°C. These fibers are then heated to about 2500°C to graphitize them (that is, to convert amorphous carbon to graphite). Stretching the fiber during pyrolysis helps orient the graphite planes parallel to the fiber axis. More amorphous carbon fibers are formed by pyrolysis of organic fibers at lower temperatures, 1200°C to 1400°C. These amorphous materials, commonly called carbon fibers, are the type most commonly used in commercial materials (Figure 22.28).

Composite materials that take advantage of the strength, stability, and low density of carbon fibers are widely used. Composites are combinations of two or more materials. These materials are present as separate phases and are combined to form structures that take advantage of certain desirable properties of each component. In carbon composites, the graphite fibers are often woven into a fabric that is embedded in a matrix that binds them into a solid structure. The fibers transmit loads evenly throughout the matrix. The finished composite is thus stronger than any one of its components.

Epoxy systems are useful matrices because of their excellent adherence, but they are costly and limited to service temperatures below 150°C. More heat-resistant resins are required for many aerospace applications, where carbon composites find wide use. Some parts of high-performance aircraft are now made out of such carbon composites.

Figure 22.28 (*a*) Carbon fibers used to make composite materials. (Courtesy American Cyanamid Co.) (*b*) Extensive use has been made of carbon composites in the construction of the B-2 bomber. (Mark Warmkessel; Department of Defense. U.S. Air Force)

(*a*)

(*b*)

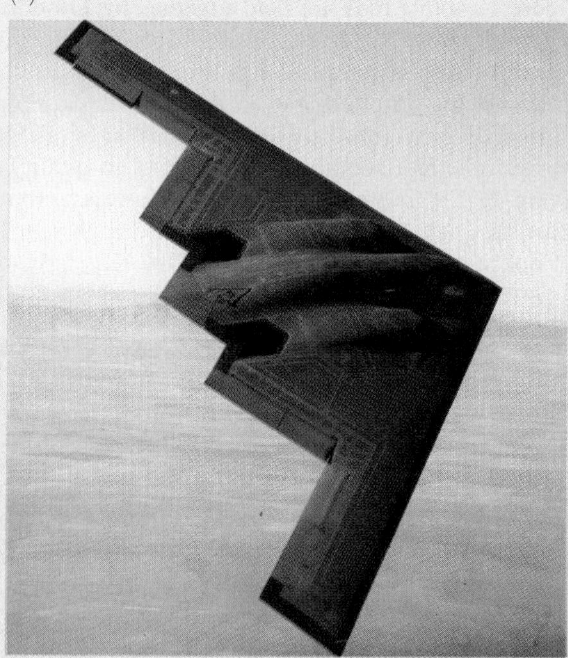

Carbon also exists in three common microcrystalline or amorphous forms of graphite (charcoal, carbon black, and coke). *Carbon black* is formed when hydrocarbons are heated in a very limited supply of oxygen:

$$CH_4(g) + O_2(g) \longrightarrow C(s) + 2H_2O(g) \qquad [22.59]$$

It is used as a pigment in black inks; large amounts are also used in making automobile tires.

Charcoal is formed when wood is heated strongly in the absence of air. Charcoal has a very open structure, giving it an enormous surface area per unit mass. Activated charcoal, a pulverized form whose surface is cleaned by heating with steam, is widely used to adsorb molecules. It is used in filters to remove offensive odors from air and colored or bad-tasting impurities from water.

Coke is an impure form of carbon formed when coal is heated strongly in the absence of air. It is widely used as a reducing agent in metallurgical operations (Chapter 24).

Oxides of Carbon

Carbon forms two principal oxides: carbon monoxide, CO, and carbon dioxide, CO_2. *Carbon monoxide* is formed when carbon or hydrocarbons are burned in a limited supply of oxygen:

$$2C(s) + O_2(g) \longrightarrow 2CO(g) \qquad [22.60]$$

It is a colorless, odorless, and tasteless gas (m.p. = -199°C; b.p. = -192°C). It is toxic because of its ability to bind to hemoglobin and thus interfere with oxygen transport (Section 18.4). Low-level poisoning results in headaches and drowsiness; high-level poisoning can cause death. Carbon monoxide is produced in large quantities by automobile engines, and it is a major air pollutant (Section 18.4).

Carbon monoxide is an unusual carbon compound because it has a lone pair of electrons on carbon: $:C{\equiv}O:$. One might imagine that CO would be unreactive, like the isoelectronic N_2 molecule. Both substances have high bond energies (1072 kJ/mol for $C{\equiv}O$ and 941 kJ/mol for $N{\equiv}N$). However, because of the lower nuclear charge on carbon (compared with either N or O), the lone pair on carbon is not held as strongly as that on N or O. Consequently, CO is better able to function as an electron-pair donor (Lewis base) than is N_2. It is therefore able to form a wide variety of covalent compounds, known as metal carbonyls, with transition metals. An example of such a compound is $Ni(CO)_4$, a volatile, very toxic compound that is formed by simply warming metallic nickel in the presence of CO. The formation of such metal carbonyls is the first step in the transition-metal catalysis of a variety of reactions of CO.

Carbon monoxide has several commercial uses. Because it burns readily, forming CO_2, it is employed as a fuel:

$$2CO(g) + O_2(g) \longrightarrow 2CO_2(g) \qquad \Delta H = -566 \text{ kJ} \qquad [22.61]$$

It is also an important reducing agent, widely used in metallurgical operations to reduce metal oxides. For example, it is the most important reducing agent in the blast furnace reduction of iron(III) oxide:

$$Fe_2O_3(s) + 3CO(g) \longrightarrow 2Fe(s) + 3CO_2(g) \quad [22.62]$$

This reaction is discussed in greater detail in Chapter 24. Carbon monoxide is also used in the preparation of several organic compounds. In Section 22.3 we saw that it can be combined catalytically with H_2 to manufacture methanol, CH_3OH (Equation 22.16).

Carbon dioxide is produced when carbon-containing substances are burned in excess oxygen:

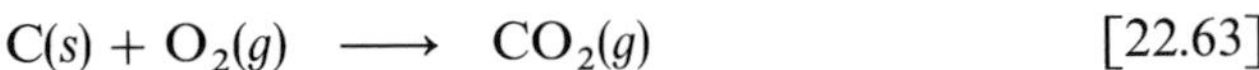

$$C(s) + O_2(g) \longrightarrow CO_2(g) \quad [22.63]$$

$$CH_4(g) + 2O_2(g) \longrightarrow CO_2(g) + 2H_2O(l) \quad [22.64]$$

$$C_2H_5OH(l) + 3O_2(g) \longrightarrow 2CO_2(g) + 3H_2O(l) \quad [22.65]$$

Figure 22.29 Solid $CaCO_3$ reacts with a solution of hydrochloric acid to produce CO_2 gas, seen here as the bubbles in the test tube. (Paul Silverman/ Fundamental Photographs)

It is also produced when many carbonates are heated:

$$CaCO_3(s) \longrightarrow CaO(s) + CO_2(g) \quad [22.66]$$

Large quantities are also obtained as a by-product of the fermentation of sugar during the production of alcohol:

$$\underset{\text{Glucose}}{C_6H_{12}O_6(aq)} \xrightarrow{\text{yeast}} \underset{\text{Ethanol}}{2C_2H_5OH(aq)} + 2CO_2(g) \quad [22.67]$$

In the laboratory, CO_2 is normally produced by the action of acids on carbonates, as shown in Figure 22.29:

$$CO_3{}^{2-}(aq) + 2H^+(aq) \longrightarrow CO_2(g) + H_2O(l) \quad [22.68]$$

Carbon dioxide is a colorless and odorless gas. It is a minor component of the earth's atmosphere but a major contributor to the so-called greenhouse effect (Section 18.4). Although it is not toxic, high concentrations increase respiration rate and can cause suffocation. It is readily liquefied by compression. However, when cooled at atmospheric pressure it condenses as a solid rather than as a liquid. The solid sublimes at atmospheric pressure at $-78°C$. This property makes solid CO_2 valuable as a refrigerant that is always free of the liquid form. Solid CO_2 is known as **dry ice**. About half of the CO_2 consumed annually is used for refrigeration. The other major use is in the production of carbonated beverages. Large quantities are also used to manufacture *washing soda*, $Na_2CO_3 \cdot 10H_2O$, and *baking soda*, $NaHCO_3$. Baking soda is so named because the following reaction occurs in baking:

$$NaHCO_3(s) + H^+(aq) \longrightarrow Na^+(aq) + CO_2(g) + H_2O(l) \quad [22.69]$$

The $H^+(aq)$ is provided by vinegar, sour milk, or hydrolysis of certain salts. The bubbles of CO_2 that form are trapped in the dough, causing it to rise. Washing soda is used to precipitate metal ions that interfere with the cleansing action of soap (Section 13.6).

Carbonic Acid and Carbonates

Carbon dioxide is moderately soluble in H_2O at atmospheric pressure. The resultant solutions are moderately acidic as a result of the formation of carbonic acid, H_2CO_3:

$$CO_2(aq) + H_2O(l) \rightleftharpoons H_2CO_3(aq) \qquad [22.70]$$

Carbonic acid is a weak diprotic acid. Its acidic character causes carbonated beverages to have a sharp, slightly acidic taste.

Although carbonic acid cannot be isolated as a pure compound, two types of salts can be obtained by neutralization of carbonic acid solutions: hydrogen carbonates (bicarbonates) and carbonates. Partial neutralization produces HCO_3^-, and complete neutralization gives CO_3^{2-}.

The HCO_3^- ion is a stronger base than acid ($K_a = 5.6 \times 10^{-11}$; $K_b = 2.3 \times 10^{-8}$). Consequently, aqueous solutions of HCO_3^- are weakly alkaline:

$$HCO_3^-(aq) + H_2O(l) \rightleftharpoons H_2CO_3(aq) + OH^-(aq) \qquad [22.71]$$

The carbonate ion is much more strongly basic ($K_b = 1.8 \times 10^{-4}$):

$$CO_3^{2-}(aq) + H_2O(l) \rightleftharpoons HCO_3^-(aq) + OH^-(aq) \qquad [22.72]$$

Minerals containing the carbonate ion are plentiful.* The principal carbonate minerals are calcite, $CaCO_3$, magnesite, $MgCO_3$, dolomite, $MgCa(CO_3)_2$, and siderite, $FeCO_3$. Calcite is the principal mineral in limestone rock, large deposits of which occur in many parts of the world. It is also the main constituent of marble, chalk, pearls, coral reefs, and the shells of marine animals such as clams and oysters. Although $CaCO_3$ has low solubility in pure water, it dissolves readily in acidic solutions, with evolution of CO_2:

$$CaCO_3(s) + 2H^+(aq) \rightleftharpoons Ca^{2+}(aq) + H_2O(l) + CO_2(g) \qquad [22.73]$$

Because water containing CO_2 is slightly acidic (Equation 22.70), $CaCO_3$ dissolves slowly in this medium:

$$CaCO_3(s) + H_2O(l) + CO_2(g) \longrightarrow Ca^{2+}(aq) + 2HCO_3^-(aq) \qquad [22.74]$$

This reaction occurs when surface waters move underground through limestone deposits. It is the principal way that Ca^{2+} enters groundwater, producing "hard water" (Section 18.6). If the dissolving limestone underlies a comparatively thin layer of earth, sinkholes, such as that shown in Figure 22.30, are produced. If the limestone deposit is deep enough underground, the dissolution of the limestone produces a cave. Two well-known limestone caves are the Mammoth Cave in Kentucky and the Carlsbad Caverns in New Mexico (Figure 22.31).

---

* *Minerals* are solid substances that occur in nature. They are usually known by their common names rather than by their chemical names. What we know as *rock* is merely an aggregate of different kinds of minerals.

Figure 22.30 A sinkhole. Ground on the surface may collapse suddenly after sufficient limestone has dissolved from underground deposits. (M. Timothy O'Keefe/Tom Stack & Associates)

Figure 22.31 Carlsbad Caverns, New Mexico. (Gene Aurens/Shostal Associates/Superstock)

One of the most important reactions of $CaCO_3$ is its decomposition into CaO and CO_2 at elevated temperatures, given earlier in Equation 22.66. Over 1.5×10^{10} kg (16 million tons) of calcium oxide, known as lime or quicklime, is used in the United States each year. Because calcium oxide reacts with water to form $Ca(OH)_2$, it is an important commercial base. It is also important in making mortar, which is a mixture of sand, water, and CaO used in construction to bind bricks, blocks, and rocks together. Calcium oxide reacts with water and CO_2 to form $CaCO_3$, which binds the sand in the mortar:

$$CaO(s) + H_2O(l) \rightleftharpoons Ca^{2+}(aq) + 2OH^-(aq) \quad [22.75]$$

$$Ca^{2+}(aq) + 2OH^-(aq) + CO_2(aq) \longrightarrow CaCO_3(s) + H_2O(l) \quad [22.76]$$

Carbides

The binary compounds of carbon with metals, metalloids, and certain non-metals are called carbides. There are three types: ionic, interstitial, and covalent. The ionic, or saltlike, carbides are formed by the more active metals. The most common ionic carbides contain C_2^{2-} ions, $[:C{\equiv}C:]^{2-}$. This ion hydrolyzes to form acetylene:

$$CaC_2(s) + 2H_2O(l) \longrightarrow Ca(OH)_2(aq) + C_2H_2(g) \qquad [22.77]$$

Carbides containing C_2^{2-} are thus called *acetylides*. The most important ionic carbide is calcium carbide, CaC_2, which is produced industrially by the reduction of CaO with carbon at high temperature:

$$2CaO(s) + 5C(s) \longrightarrow 2CaC_2(s) + CO_2(g) \qquad [22.78]$$

The CaC_2 is used to prepare acetylene, which is used in welding.

Interstitial carbides are formed by many transition metals. The carbon atoms occupy open spaces (interstices) between metal atoms, in a manner analogous to the interstitial hydrides (Section 22.3). An example is tungsten carbide, WC, which is very hard and heat-resistant and consequently is used in making cutting tools.

Covalent carbides are formed by boron and silicon. Silicon carbide, SiC, is known as carborundum. It is made by heating SiO_2 (sand) and carbon to high temperatures:

$$SiO_2(s) + 3C(s) \longrightarrow SiC(s) + 2CO(g) \qquad [22.79]$$

SiC has the same structure as diamond except that silicon atoms alternate with carbon atoms. Like industrial diamond, carborundum is very hard and is thus used as an abrasive and in cutting tools.

Other Inorganic Compounds of Carbon

Hydrogen cyanide, HCN (Figure 22.32), is an extremely toxic gas that has the odor of bitter almonds. In the laboratory it is made by the reaction of a cyanide salt, such as NaCN, with an acid [see Sample Exercise 22.2(c)].

Aqueous solutions of HCN are known as hydrocyanic acid. Neutralization with a base, such as NaOH, produces cyanide salts, such as NaCN. Cyanides find use in the manufacture of several well-known plastics, including nylon and Orlon. The CN^- ion forms very stable complexes with most transition metals (Section 25.1). The toxic action of CN^- is caused by its combination with iron(III) in cytochrome oxidase, a key enzyme that is involved in respiration.

Carbon disulfide, CS_2, is an important industrial solvent for waxes, greases, celluloses, and other nonpolar substances. It is a colorless, volatile liquid (b.p. 46.3°C). The vapor is very poisonous and highly flammable. The compound is formed by direct reaction of carbon and sulfur at high temperature.

Figure 22.32 Structures of hydrogen cyanide and carbon disulfide.

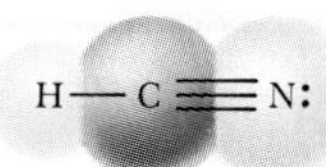

Hydrogen cyanide

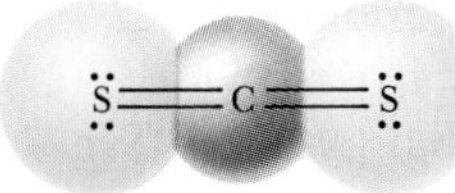

Carbon disulfide

FOR REVIEW

SUMMARY

The periodic table is useful in organizing and remembering the descriptive chemistry of the elements: Among elements of a given family, size increases with increasing atomic number. Correspondingly, electronegativity and ionization energy decrease. Nonmetallic character parallels electronegativity trends. Among the nonmetallic elements, the first member of each family differs dramatically from the other members; it forms a maximum of four bonds to other atoms (that is, it is confined to an octet of valence shell electrons). Also, it exhibits a much greater tendency to form π bonds than do the heavier elements in that family.

Hydrogen has three isotopes: protium (^1_1H), deuterium (^2_1H), and tritium (^3_1H). Hydrogen is not a member of any periodic family, although it is usually placed above lithium. The hydrogen atom can either lose an electron, forming H^+, or gain one, forming H^-, the hydride ion. Because the H—H bond is relatively strong, H_2 is fairly unreactive unless activated by heat, irradiation, or catalysis. Industrially, H_2O or hydrocarbons are used as sources of hydrogen; in the lab H_2 is usually obtained by the action of acids on active metals, such as Zn. The binary compounds of hydrogen are of three general types: ionic hydrides (formed by active metals), metallic hydrides (formed by transition metals), and molecular hydrides (formed by nonmetals).

Oxygen exhibits two allotropes, O_2 and O_3 (ozone). O_2 is separated from air; small amounts can be obtained by heating $KClO_3$. Ozone is obtained by passing an electrical discharge through O_2. Both O_2 and O_3 are good oxidizing agents, but O_3 is the stronger. Oxygen normally exhibits an oxidation state of -2 in compounds (oxides). The soluble oxides of nonmetals generally produce acidic aqueous solutions; they are called acidic anhydrides. In contrast, soluble metal oxides produce basic solutions and are called basic anhydrides. Peroxides contain O—O bonds and oxygen in a -1 oxidation state. In superoxides, which contain the O_2^- ion, oxygen has an oxidation state of $-\frac{1}{2}$.

The primary source of nitrogen is the atmosphere, where it occurs as N_2 molecules. Molecular nitrogen is chemically very stable because of the strong N≡N bond. In its compounds nitrogen exhibits oxidation states ranging from -3 to $+5$. The most important process for converting N_2 into compounds is the Haber process, used to prepare ammonia. Another commercially important process is the Ostwald process, which is the preparation of HNO_3 beginning with the catalytic oxidation of NH_3. Nitrogen has three common oxides: N_2O (nitrous oxide), NO (nitric acid), and NO_2 (nitrogen dioxide). It has two common oxyacids, HNO_2 (nitrous acid) and HNO_3 (nitric acid). Nitric acid is both a strong acid and a good oxidizing agent. Other important nitrogen compounds include hydrazine, N_2H_4, and hydrogen azide, HN_3. Nitrogen compounds are important fertilizers.

Carbon exhibits two allotropes, diamond and graphite. Amorphous forms of graphite include charcoal, carbon black, and coke. Carbon exhibits two common oxides, CO and CO_2. Aqueous solutions of CO_2 produce carbonic acid, H_2CO_3, which is the parent acid of carbonate salts. Binary compounds of carbon are called carbides. Carbides may be ionic, interstitial, or covalent. Calcium carbide, CaC_2, is used to prepare acetylene. Carbon also forms a vast number of organic compounds, discussed in a later chapter.

KEY TERMS

protium (Sec. 22.3)
deuterium (Sec. 22.3)
tritium (Sec. 22.3)
ionic hydrides (Sec. 22.3)
metallic hydrides (Sec. 22.3)
interstitial hydrides (Sec. 22.3)
molecular hydrides (Sec. 22.3)
acidic anhydrides (Sec. 22.4)
acidic oxides (Sec. 22.4)
basic anhydrides (Sec. 22.4)
basic oxides (Sec. 22.4)
Ostwald process (Sec. 22.5)
disproportionation (Sec. 22.5)
charcoal (Sec. 22.6)
coke (Sec. 22.6)
dry ice (Sec. 22.6)

EXERCISES

Periodic Trends and Chemical Reactions

22.1 Identify each of the following elements as a metal, nonmetal, or semimetal: **(a)** antimony; **(b)** strontium; **(c)** cerium; **(d)** selenium; **(e)** rhodium; **(f)** krypton.

22.2 Identify each of the following elements as a metal, nonmetal, or semimetal: **(a)** germanium; **(b)** zirconium; **(c)** xenon; **(d)** indium; **(e)** cesium; **(f)** bromine.

22.3 List two chemical and two physical properties that help us distinguish metals from nonmetals.

22.4 State two important ways in which the first member of each family of nonmetals differs from subsequent members.

22.5 Which element of group 5A would you expect to have the most metallic character? Describe two properties of the elements in the group that should relate to this degree of metallic character.

22.6 Which element of group 3A would you expect to have the most nonmetallic character? Describe two properties of the elements that should demonstrate the greater nonmetallic character of this element compared to the others in the group.

22.7 Consider the following list of elements: Li, K, Cl, C, Ne, Ar. From this list select the element that **(a)** is most electronegative; **(b)** has the greatest metallic character; **(c)** most readily forms a positive ion; **(d)** has the smallest atomic radius; **(e)** forms π bonds most readily.

22.8 Consider the following list of elements: O, Ba, Co, Be, Br, Se. From this list select the element that **(a)** is most electronegative; **(b)** exhibits a maximum oxidation state of $+7$; **(c)** loses an electron most readily; **(d)** forms π bonds most readily; **(e)** is a transition metal.

22.9 In each of the following pairs of substances, select the one that has the more polar bonds: **(a)** NF_3 and IF_3; **(b)** N_2O_3 and B_2O_3; **(c)** BN and BF_3; **(d)** H_2O and H_2Se. Explain your choice in each case.

22.10 In each of the following pairs of substances, select the one that has the more polar bonds: **(a)** CO_2 and SnO_2; **(b)** P_2O_3 and V_2O_3; **(c)** SiO_2 and SO_2; **(d)** CCl_4 and CBr_4. Explain your choice in each case.

22.11 Explain the following observations: **(a)** The highest fluoride compound formed by nitrogen is NF_3, whereas phosphorus readily forms PF_5. **(b)** Although CO is a well-known compound, SiO doesn't exist under ordinary conditions. **(c)** AsH_3 is a stronger reducing agent than NH_3.

22.12 Explain the following observations: **(a)** HNO_3 is a stronger oxidizing agent than H_3PO_4. **(b)** Silicon is able to form a compound with six fluorides, SiF_6, whereas carbon is able to bond to a maximum of four, CF_4. **(c)** There are three compounds formed by carbon and hydrogen that contain two carbon atoms each (C_2H_2, C_2H_4, and C_2H_6), whereas silicon forms only one analogous compound (Si_2H_6).

22.13 Complete and balance the following equations:

(a) $NaNH_2(s) + H_2O(l) \longrightarrow$

(b) $C_3H_7OH(l) + O_2(g) \longrightarrow$

(c) $NiO(s) + C(s) \longrightarrow$

(d) $AlP(s) + H_2O(l) \longrightarrow$

(e) $Na_2S(s) + HCl(aq) \longrightarrow$

22.14 Complete and balance the following equations:

(a) $NaOCH_3(s) + H_2O(l) \longrightarrow$

(b) $Na_2O(s) + HC_2H_3O_2(aq) \longrightarrow$

(c) $WO_3(s) + H_2(g) \longrightarrow$

(d) $NH_2OH(l) + O_2(g) \longrightarrow$

(e) $Al_4C_3(s) + H_2O(l) \longrightarrow$

Hydrogen

22.15 Give the names and chemical symbols for the three isotopes of hydrogen.

22.16 Which isotope of hydrogen is radioactive? Write the nuclear equation for the radioactive decay of this isotope.

22.17 Why is hydrogen placed in either group 1A or 7A of the periodic table?

22.18 Why are the properties of hydrogen different from those of either the group 1A or 7A elements?

22.19 Give a balanced chemical equation for the preparation of H_2 using **(a)** Mg and an acid; **(b)** carbon and steam; **(c)** methane and steam.

22.20 List **(a)** three commercial means of producing H_2; **(b)** three industrial uses of H_2.

22.21 Identify the following hydrides as ionic, metallic, or molecular: **(a)** BaH_2; **(b)** H_2Te; **(c)** $TiH_{1.7}$.

22.22 Identify the following hydrides as ionic, metallic, or molecular: **(a)** B_2H_6; **(b)** RbH; **(c)** $Th_4H_{1.5}$.

22.23 Complete and balance the following equations:

(a) $NaH(s) + H_2O(l) \longrightarrow$

(b) $Fe(s) + H_2SO_4(aq) \longrightarrow$

(c) $H_2(g) + Br_2(g) \longrightarrow$

(d) $Na(l) + H_2(g) \longrightarrow$

(e) $PbO(s) + H_2(g) \longrightarrow$

22.24 Write balanced chemical equations for each of the following reactions (some of these are analogous to reactions shown in the chapter). **(a)** Aluminum metal reacts with acids to form hydrogen gas. **(b)** Steam reacts with magnesium metal to give magnesium oxide and hydrogen. **(c)** Manganese(IV) oxide is reduced to manganese(II) oxide by hydrogen gas. **(d)** Calcium hydride reacts with water to generate hydrogen gas.

22.25 About 1.6×10^{10} kg of ethylene, C_2H_4, is produced annually in the United States by the pyrolysis of ethane, C_2H_6: $C_2H_6(g) \longrightarrow C_2H_4(g) + H_2(g)$. How many kilograms of H_2 are produced as a by-product?

22.26 Hydrogen gas has a higher fuel value than natural gas on a weight basis but not on a volume basis. Thus hydrogen is not competitive with natural gas as a fuel transported long distance through pipelines. Calculate the heat of combustion of H_2 and of CH_4 (the principal component of natural gas) **(a)** per mole of each; **(b)** per gram of each; **(c)** per cubic meter of each at STP. Assume $H_2O(l)$ as a product.

Oxygen

22.27 List three industrial uses of O_2.

22.28 List two industrial uses of O_3.

22.29 Give the structure of ozone. Explain why the O—O bond length in ozone (1.28 Å) is longer than that in O_2(1.21 Å).

22.30 How is O_2 normally prepared in the laboratory?

22.31 Complete and balance the following equations:
(a) $CaO(s) + H_2O(l) \longrightarrow$
(b) $Al_2O_3(s) + H^+(aq) \longrightarrow$
(c) $Na_2O_2(s) + H_2O(l) \longrightarrow$
(d) $N_2O_3(g) + H_2O(l) \longrightarrow$
(e) $KO_2(s) + H_2O(l) \longrightarrow$
(f) $NO(g) + O_3(g) \longrightarrow$

22.30 Write the balanced chemical equations for each of the following reactions. **(a)** When mercury(II) oxide is heated, it decomposes to form O_2 and mercury metal. **(b)** When copper(II) nitrate is heated strongly, it decomposes to form copper(II) oxide, nitrogen dioxide, and oxygen. **(c)** Lead(II) sulfide, PbS(*s*), reacts with ozone to form $PbSO_4(s)$ and $O_2(g)$. **(d)** When heated in air, ZnS(*s*) is converted to ZnO. **(e)** Potassium peroxide reacts with $CO_2(g)$ to give the carbonate ion and O_2. **(f)** Although silver does not react with oxygen at room temperature, it reacts with ozone, forming Ag_2O.

22.33 Predict whether each of the following oxides is acidic, basic, amphoteric, or neutral: **(a)** CO; **(b)** CO_2; **(c)** CaO; **(d)** Al_2O_3.

22.34 Select the more acidic member of each of the following pairs: **(a)** Mn_2O_7 and MnO_2; **(b)** SnO and SnO_2; **(c)** SO_2 and SO_3; **(d)** SiO_2 and SO_2; **(e)** Ga_2O_3 and In_2O_3; **(f)** SO_2 and SeO_2.

22.35 Hydrogen peroxide is capable of oxidizing **(a)** K_2S to S; **(b)** SO_2 to SO_4^{2-}; **(c)** NO_2^- to NO_3^-; **(d)** As_2O_3 to AsO_4^{3-}; **(e)** Fe^{2+} to Fe^{3+}. Write balanced net ionic equations for each of these oxidations.

22.36 Hydrogen peroxide reduces **(a)** MnO_4^- to Mn^{2+}; **(b)** Cl_2 to Cl^-; **(c)** Ce^{4+} to Ce^{3+}; **(d)** O_3 to H_2O. Write balanced net ionic equations for each of these reductions.

Nitrogen

22.37 List three industrial uses for N_2.

22.38 What is meant by the term *nitrogen fixation?* Why is the N_2 molecule so unreactive?

22.39 Write the chemical formula for each of the following compounds and indicate the oxidation state of nitrogen in each: **(a)** nitrous acid; **(b)** hydrazine; **(c)** potassium cyanide: **(d)** sodium nitrate; **(e)** ammonium chloride; **(f)** lithium nitride.

22.40 Write the chemical formula for each of the following compounds and indicate the oxidation state of nitrogen in each: **(a)** sodium nitrite; **(b)** ammonia; **(c)** nitrous oxide; **(d)** sodium cyanide; **(e)** nitric acid; **(f)** nitrogen dioxide.

22.41 Write the Lewis structure for each of the following species and describe its molecular geometry: **(a)** NH_4^+; **(b)** HNO_3; **(c)** N_2O; **(d)** NO_2.

22.42 Write the Lewis structure for each of the following species and describe its molecular geometry: **(a)** HNO_2; **(b)** N_3^-; **(c)** $N_2H_5^+$; **(d)** NO_3^-.

22.43 Complete and balance the following equations:
(a) $Mg_3N_2(s) + H_2O(l) \longrightarrow$
(b) $NO(g) + O_2(g) \longrightarrow$
(c) $NH_3(g) + O_2(g) \xrightarrow{\Delta}$ (no catalyst)
(d) $NaNH_2(s) + H_2O(l) \longrightarrow$

22.44 Complete and balance the following equations:
(a) $N_2O_5(g) + H_2O(l) \longrightarrow$
(b) $Li_3N(s) + H_2O(l) \longrightarrow$
(c) $NH_3(aq) + H^+(aq) \longrightarrow$
(d) $N_2H_4(l) + O_2(g) \longrightarrow$

22.45 Write balanced net ionic equations for each of the following reactions. **(a)** Dilute nitric acid reacts with zinc metal with formation of nitrous oxide. **(b)** Concentrated nitric acid reacts with sulfur with formation of nitrogen dioxide. **(c)** Concentrated nitric acid oxidizes sulfur dioxide with formation of nitric oxide. **(d)** Urea reacts with water to form ammonia and carbon dioxide.

22.46 Write balanced net ionic equations for each of the following reactions. **(a)** Hydrazine is burned in excess fluorine gas, forming NF_3. **(b)** Hydrazine reduces CrO_4^{2-} to $Cr(OH)_4^-$ (hydrazine is oxidized to N_2). **(c)** Hydroxylamine is oxidized to N_2 by Cu^{2+} in aqueous solutions (Cu^{2+} is reduced to copper metal). **(d)** Aqueous azide ion, N_3^-, reacts with Cl_2, forming N_2.

22.47 Write complete balanced half-reactions for **(a)** reduction of nitrate ion to N_2 in acidic solution; **(b)** oxidation of NH_4^+ to N_2 in acidic solution. What is the standard electrode potential in each case? (See Figure 22.18.)

22.48 Write complete balanced half-reactions for **(a)** reduction of nitrate ion to NO in acidic solution; **(b)** oxidation of HNO_2 to NO_2 in acidic solution. What is the standard electrode potential in each case? (See Figure 22.18.)

22.49 It is estimated that 95 percent of the ammonia production in the United States uses H_2 obtained from natural gas (Equations 22.12 and 22.13). If 50 percent of the CH_4 used to manufacture H_2 is burned to maintain proper temperature for ammonia synthesis, how many kilograms of CH_4 are consumed in supplying the 1.5×10^{10} kg of NH_3 produced annually?

22.50 From the thermodynamic data in Appendix C, calculate $\Delta H°$ and $\Delta G°$ at 25°C for the oxidation of NH_3 in the following reactions:

$$4NH_3(g) + 5O_2(g) \longrightarrow 4NO(g) + 6H_2O(g)$$

$$4NH_3(g) + 3O_2(g) \longrightarrow 2N_2(g) + 6H_2O(g)$$

Carbon

22.51 Give three industrial uses of carbon dioxide.

22.52 Give three industrial uses of carbon monoxide.

22.53 Give the chemical formulas for **(a)** hydrocyanic

acid; **(b)** carborundum; **(c)** calcium carbonate; **(d)** calcium acetylide.

22.54 Give the chemical formulas for **(a)** carbonic acid; **(b)** sodium cyanide; **(c)** potassium hydrogen carbonate; **(d)** acetylene.

22.55 Write the Lewis structure of each of the following species: **(a)** CN^-; **(b)** CO; **(c)** C_2^{2-}; **(d)** CS_2; **(e)** CO_2; **(f)** CO_3^{2-}.

22.56 Indicate the geometry and the type of hybrid orbitals used by each carbon atom in the following species: **(a)** $CH_3C{\equiv}CH$; **(b)** NaCN; **(c)** CS_2; **(d)** C_2H_6.

22.57 Complete and balance the following equations:

(a) $ZnCO_3(s) \xrightarrow{\Delta}$
(b) $BaC_2(s) + H_2O(l) \longrightarrow$
(c) $C_2H_4(g) + O_2(g) \longrightarrow$
(d) $CH_3OH(l) + O_2(g) \longrightarrow$
(e) $NaCN(s) + HCl(aq) \longrightarrow$

22.58 Complete and balance the following equations:

(a) $CO_2(g) + OH^-(aq) \longrightarrow$
(b) $NaHCO_3(s) + H^+(aq) \longrightarrow$
(c) $CaO(s) + C(s) \xrightarrow{\Delta}$
(d) $C(s) + H_2O(g) \xrightarrow{\Delta}$
(e) $CuO(s) + CO(g) \longrightarrow$

22.59 Write a balanced chemical equation for each of the following reactions. **(a)** Burning magnesium metal in a carbon dioxide atmosphere reduces the CO_2 to carbon. **(b)** In photosynthesis, solar energy is used to produce glucose, $C_6H_{12}O_6$, and O_2 out of carbon dioxide and water, **(c)** When carbonate salts dissolve in water, they produce basic solutions.

22.60 Write a balanced chemical equation for each of the following reactions. **(a)** Hydrogen cyanide is formed commercially by passing a mixture of methane, ammonia, and air over a catalyst at 800°C. Water is a by-product of the reaction. **(b)** Baking soda reacts with acids to produce carbon dioxide gas. **(c)** When barium carbonate reacts in air with sulfur dioxide, barium sulfate and carbon dioxide form.

22.61 What volume of dry acetylene, measured at 27°C and 720 mm Hg, is formed when 10.0 g of calcium carbide is placed in water?

22.62 A certain carbide of magnesium is reacted with water to form $Mg^{2+}(aq)$ and a volatile hydrocarbon. Hydrolysis of 0.3052 g of the carbide produces 0.1443 g of the hydrocarbon. The hydrocarbon consists of 90.0 percent C and 10.0 percent H. The density of the hydrocarbon gas at 25°C and 742 mm Hg is 1.60 g/L. What is the formula for the carbide, and what is the molecular formula for the hydrocarbon?

Additional Exercises

22.63 The annual production of H_2, N_2, and O_2 in the United States is generally reported in cubic feet, measured at STP. If 2.2×10^8 kg of H_2, 2.4×10^{10} kg of N_2, and 1.7×10^{10} kg of O_2 are produced in a particular year, how many cubic feet are produced of **(a)** H_2; **(b)** N_2; **(c)** O_2?

22.64 **(a)** How many grams of H_2 can be stored in 1.00 kg of the alloy FeTi if the hydride $FeTiH_2$ is formed? **(b)** What volume does this quantity of H_2 occupy at STP?

22.65 Starting with D_2O, suggest a preparation for **(a)** ND_3; **(b)** D_2SO_4; **(c)** NaOD; **(d)** DNO_3; **(e)** C_2D_2; **(f)** DCN.

22.66 A student had three tubes containing colorless gases. One tube held H_2, one O_2, and one CO_2. A glowing splint was inserted into each tube. In tube A, the splint was extinguished; in tube B, a sharp pop was produced; in tube C, the splint burst into flames. Which tube contained which gas?

22.67 Which of the following substances will burn in oxygen: SiH_4; SiO_2; CO; CO_2; Mg; CaO? Why won't some of these substances burn in oxygen?

22.68 Write a balanced chemical equation for the reaction of each of the following compounds with water: **(a)** $SO_2(g)$; **(b)** $Cl_2O(g)$; **(c)** $Na_2O(s)$; **(d)** $BaC_2(s)$; **(e)** $RbO_2(s)$; **(f)** $Mg_3N_2(s)$; **(g)** $Na_2O_2(s)$; **(h)** NaH(*s*).

22.69 What is the anhydride for each of the following acids: **(a)** H_2SO_4; **(b)** $HClO_3$; **(c)** HNO_2; **(d)** H_2CO_3; **(e)** H_3PO_4?

22.70 Write a series of chemical equations that describes how Na_2CO_3 could be prepared starting with only H_2O, NaCl, and $CaCO_3$ as raw materials.

22.71 Hydrogen peroxide can be prepared in the laboratory by air oxidation of barium metal followed by treatment of the resultant product with dilute sulfuric acid. Write balanced chemical equations for the two reactions that take place.

22.72 Both dimethylhydrazine, $(CH_3)_2NNH_2$, and monomethylhydrazine, CH_3HNNH_2, have been used as rocket fuels. When dinitrogen tetraoxide, N_2O_4, is used as the oxidizer, the products are H_2O, CO_2, and N_2. If the thrust of the rocket depends on the volume of the products produced, which of the substituted hydrazines produces a greater thrust per gram total weight of oxidizer plus fuel. [Assume that both fuels generate the same temperature and that $H_2O(g)$ is formed.]

22.73 **(a)** What is the oxidation state of P in PO_4^{3-} and of N in NO_3^-? **(b)** Why doesn't N form a stable NO_4^{3-} ion analogous to P?

22.74 Suggest why each of the following compounds is either unstable or does not exist: **(a)** NCl_5; **(b)** $(CH_3)_2Si{=}O$; **(c)** P_2O; **(d)** H_3.

22.75 Each of the following compounds is used as a nitrogen fertilizer: $(NH_2)_2CO$ (urea), NH_3, $(NH_4)_2SO_4$, and $NaNO_3$. Calculate the percentage of nitrogen by mass in each compound.

22.76 Hydrazine has been employed as a reducing agent for metals. Using standard electrode potentials, predict whether the following metals can be reduced to the metallic state by hydrazine under standard conditions in acidic solution: **(a)** Fe^{2+}; **(b)** Sn^{2+}; **(c)** Cu^{2+}; **(d)** Ag^+; **(e)** Cr^{3+}.

[22.77] Thermodynamic data for $HNO_3(aq)$ at 25°C follow: $\Delta H_f^\circ = -207.3$ kJ/mol; $\Delta G_f^\circ = -111.3$ kJ/mol. Calculate ΔH° and ΔG° for the reaction

$$N_2(g) + H_2O(g) + \tfrac{5}{2}O_2(g) \longrightarrow 2HNO_3(aq)$$

It has been suggested that if a suitable catalyst were present, atmospheric gases could react to make the oceans a dilute nitric acid solution. In thermodynamic terms, is this a possible process?

22.78 The Haber process is carried out at high temperatures, even though the equilibrium constant for the reaction decreases with increasing temperature. Explain why a high temperature is necessary.

22.79 If the lunar lander on the Apollo moon missions used 4.0 tons of dimethylhydrazine, $(CH_3)_2NNH_2$, as fuel, how many tons of N_2O_4 oxidizer were required to react with it? (The reaction produces N_2, CO_2, and H_2O.)

[22.80] The dissolved oxygen present in any highly pressurized, high-temperature steam boiler can be extremely corrosive to its metal parts. Hydrazine, which is completely miscible with water, can be added to remove oxygen by reacting with it to form nitrogen and water. **(a)** Write the balanced chemical equation for the reaction between dissolved hydrazine and oxygen. **(b)** Calculate the enthalpy change accompanying this reaction. **(c)** Oxygen in air dissolves in water to the extent of 9.1 ppm at 20°C and sea level. How many grams of hydrazine are required to react with all of the oxygen in 3.0×10^4 L (the volume of a small swimming pool) under these conditions?

22.81 Rationalize the general trend in Lewis basicities in the following isoelectronic series: $N_2 < CO < CN^-$.

22.82 Complete and balance the following equations:

(a) $Li_3N(s) + H_2O(l) \longrightarrow$
(b) $NH_3(aq) + H_2O(l) \longrightarrow$
(c) $NO_2(g) + H_2O(l) \longrightarrow$
(d) $2NO_2(g) \longrightarrow$
(e) $NH_3(g) + O_2(g) \xrightarrow{\text{catalyst}}$
(f) $CO(g) + O_2(g) \longrightarrow$
(g) $H_2CO_3(aq) \xrightarrow{\Delta}$
(h) $Ni(s) + CO(g) \longrightarrow$
(i) $CS_2(g) + O_2(g) \longrightarrow$
(j) $CaO(s) + SO_2(g) \longrightarrow$
(k) $Na(s) + H_2O(l) \longrightarrow$
(l) $CH_4(g) + H_2O(g) \xrightarrow{\Delta}$
(m) $LiH(s) + H_2O(l) \longrightarrow$
(n) $Fe_2O_3(s) + 3H_2(g) \longrightarrow$

22.83 Complete and balance the following equations:

(a) $MnO_4^-(aq) + H_2O_2(aq) + H^+(aq) \longrightarrow$
(b) $Fe^{2+}(aq) + H_2O_2(aq) \longrightarrow$
(c) $I^-(aq) + H_2O_2(aq) + H^+(aq) \longrightarrow$
(d) $MnO_2(s) + H_2O_2(aq) + H^+(aq) \longrightarrow$
(e) $I^-(aq) + O_3(g) \longrightarrow I_2(s) + O_2(g) + OH^-(aq)$

[22.84] Using this chapter and material elsewhere in the text (use the index), compile a list of physical and chemical properties of the elements oxygen and zinc. How do these properties justify the classification of one as a nonmetal and the other as a metal?

[22.85] The electronegativity of a given element can be regarded as varying with its oxidation state. How would you expect the electronegativity to change as a function of oxidation state? Although manganese (Mn) is completely different from chlorine in properties as an element, the characteristics of MnO_4^- rather closely resemble those of ClO_4^-. Cite at least two instances of this similarity (you may need to look in a handbook; think in terms of acid-base properties, oxidation-reduction behavior, solubility, and so on). Discuss your observations in terms of the oxidation states, electron configurations, and electronegativities of the central atoms.

Chemistry of Other Nonmetallic Elements

23

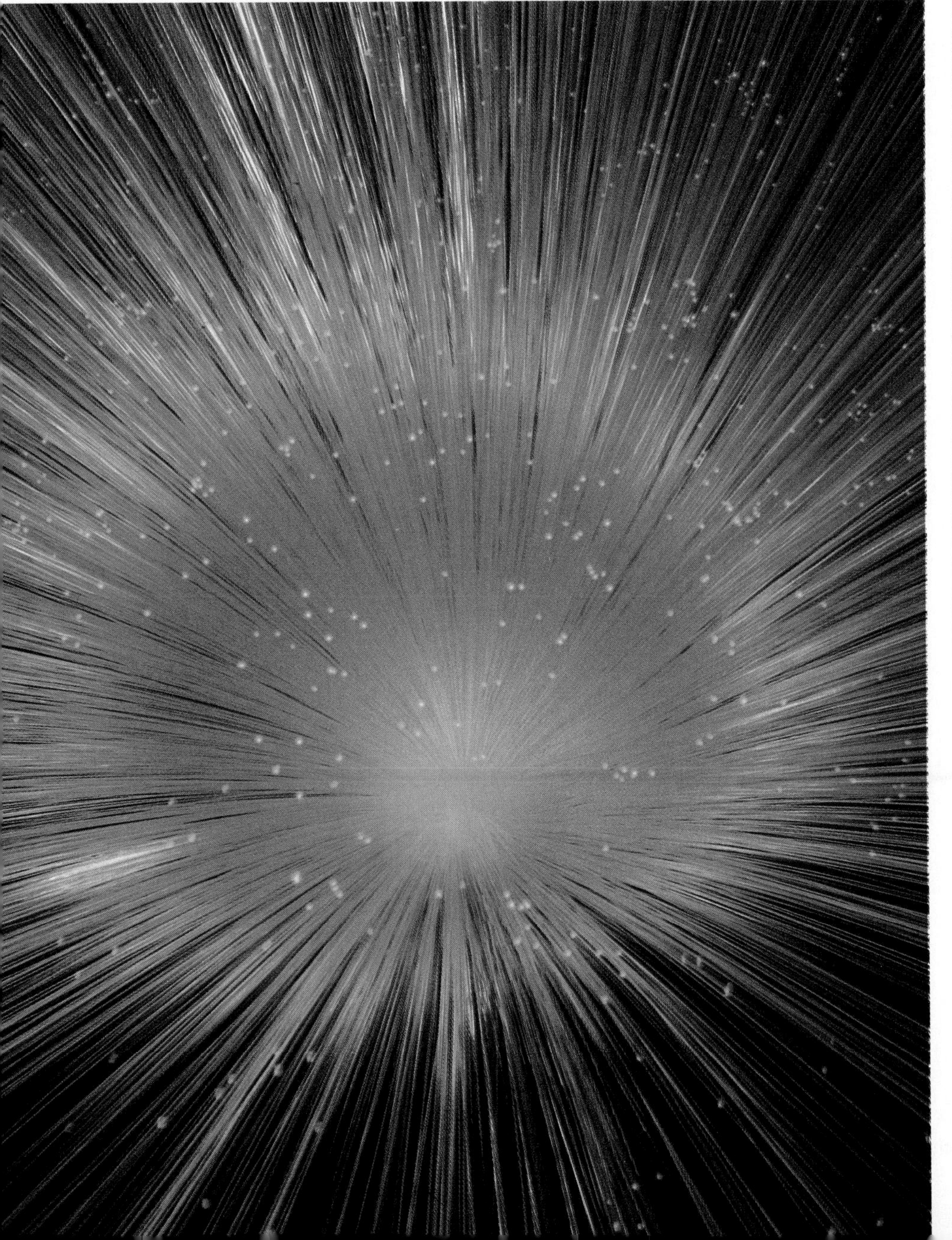

A spray of glass optical fibers seen conducting red light. Each individual fiber has a diameter of only 200 μm. (Simon Fraser/Science Photo Library/Photo Researchers)

CONTENTS

In Chapter 22, we discussed the chemistry of four important nonmetals: hydrogen, oxygen, nitrogen, and carbon. In this chapter, we take a more panoramic view of nonmetals. We will begin with the noble gases and then move, group by group, from right to left through the periodic table. We will, of course, have little to say about those elements discussed in Chapter 22. Of the remaining nonmetals, the most important are the halogens, sulfur, phosphorus, and silicon.

23.1 THE NOBLE-GAS ELEMENTS

8A
2 He
10 Ne
18 Ar
36 Kr
54 Xe
86 Rn

We have mentioned at several points in the text that the elements of group 8A are chemically unreactive. Indeed, most of our references to these elements have been in relation to their physical properties, as when we discussed intermolecular forces in Section 11.2. According to the Lewis theory of chemical bonding, the relative inertness of these elements is due to the formation of a completed octet of valence-shell electrons. The stability of such an arrangement is reflected in the high ionization energies of the group 8A elements (Section 7.4).

The group 8A elements are all gases at room temperature. They are components of the earth's atmosphere, except for radon, which exists only as a short-lived radioisotope. Only argon is relatively abundant (Table 18.1). Neon, argon, krypton, and xenon are recovered from liquid air by distillation. Argon is used as a blanketing atmosphere in electric light bulbs. The gas conducts heat away from the filament but does not react with it. It is also used as a protective atmosphere to prevent oxidation in welding and certain high-temperature metallurgical processes. Neon is used in electric signs; the gas is caused to radiate by passing an electric discharge through the tube. Krypton, xenon, and radon are not commercially important.

Helium is, in many ways, the most important of the noble gases. Liquid helium is used as a coolant to conduct experiments at very low temperatures. Helium boils at 4.2 K under 1 atm pressure, the lowest boiling point of any substance. Because helium has such a low abundance in the atmosphere and boils at such a low temperature, recovery of the gas from the atmosphere would require an immense expenditure of energy. Helium is found in relatively high concentrations in many natural-gas wells. Some of this helium is separated to meet current demands, and a little is kept for later use. Unfortunately, however, most of the helium escapes.

Noble-Gas Compounds

Because the noble gases are exceedingly stable, it is reasonable to expect that they will undergo reaction only under rigorous conditions. Furthermore, we might expect that the heavier noble gases would be most likely to undergo

Table 23.1 Properties of Xenon Compounds

Compound	Oxidation state of Xe	Melting point (°C)	ΔH_f° (kJ/mol)[a]
XeF_2	+2	129	−109(*g*)
XeF_4	+4	117	−218(*g*)
XeF_6	+6	49	−298(*g*)
$XeOF_4$	+6	−41 to −28	+146(*l*)
XeO_3	+6	—[b]	+402(*s*)
XeO_2F_2	+6	31	+145(*s*)
XeO_4	+8	—[c]	—

[a] At 25°C, for the compound in the state indicated.
[b] A solid; decomposes at 40°C.
[c] A solid; decomposes at −40°C.

chemical transformation, because the ionization energies are lower for the heavier elements, as illustrated in Figure 7.5. A lower ionization energy would suggest the possibility of losing an electron in formation of an ionic bond. In addition, because the group 8A elements already contain eight electrons in their valence shell (except, of course, for helium, which contains just two), formation of covalent bonds requires an expanded valence shell. We have seen (Section 8.7) that valence-shell expansion occurs most readily with larger atoms.

The first noble-gas compound was prepared in 1962 by Neil Bartlett while he was on the faculty of the University of British Columbia. His work caused a sensation because it undercut the belief that the noble-gas elements were truly chemically inert. Bartlett initially worked with xenon in combination with fluorine, the element we would expect to be most reactive. Since that time, chemists have prepared several xenon compounds of fluorine and oxygen. Some properties of these substances are listed in Table 23.1. The three fluorides XeF_2, XeF_4, and XeF_6 are made by direct reaction of the elements. By varying the ratio of reactants and altering reaction conditions, one or the other of the three compounds can be obtained. Figure 23.1 shows crystals of XeF_4. The oxygen-containing compounds are formed when the fluorides are reacted with water, as in Equations 23.1 and 23.2.

$$XeF_6(s) + H_2O(l) \longrightarrow XeOF_4(l) + 2HF(g) \qquad [23.1]$$

$$XeF_6(s) + 3H_2O(l) \longrightarrow XeO_3(aq) + 6HF(aq) \qquad [23.2]$$

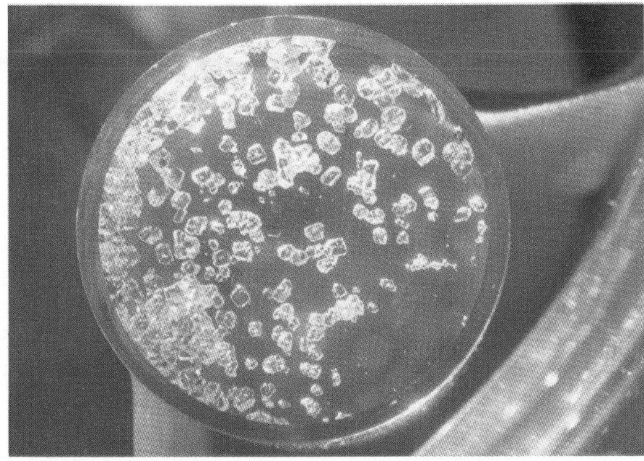

Figure 23.1 Crystals of XeF_4. (Argonne National Laboratory)

SAMPLE EXERCISE 23.1

Predict the structure of XeF_4.

Solution: To predict the structure, we must first write the Lewis structure for the molecule. The total number of valence-shell electrons involved is 36 (8 from xenon and 7 from each of the four fluorines). This leads to the Lewis structure shown in Figure 23.2(*a*). We see that Xe has 12 electrons in its valence shell. We thus expect an octahedral disposition of six electron pairs. Two of these are nonbonded pairs. Because nonbonded pairs have a larger volume requirement than do bonded pairs (Section 9.1), it is reasonable to expect these nonbonded pairs to be opposite one another. The expected structure is square planar, as shown in Figure 23.2(*b*). The experimentally determined structure agrees with this prediction.

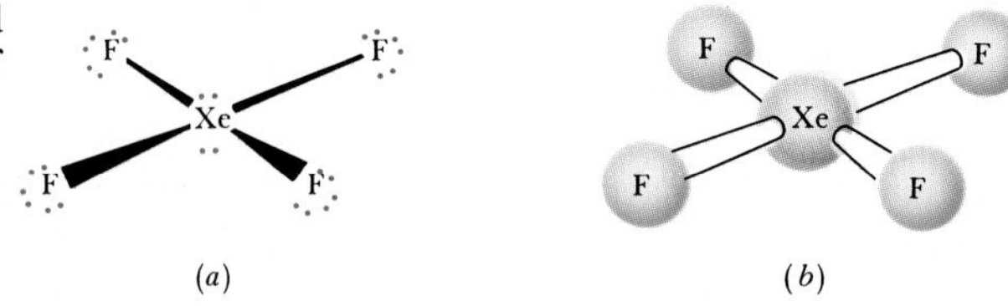

Figure 23.2 Lewis and geometrical structures of XeF_4.

PRACTICE EXERCISE

Describe the electron-pair geometry and the molecular geometry of XeF_2.
Answer: trigonal bipyramidal; linear

The enthalpies of formation of the xenon fluorides are negative, which suggests that these compounds should be reasonably stable, and this is indeed found to be the case. They are, however, powerful fluorinating agents and must be handled in containers that do not readily react to form fluorides. Notice that the enthalpies of formation of the oxyfluorides and oxides of xenon are positive (Table 23.1); these compounds are quite unstable.

As we might expect, formation of compounds by the other noble-gas elements occurs much less readily than in the case of xenon. Only one binary krypton compound, KrF_2, is known with certainty; it decomposes to the elements at $-10°C$.

23.2 THE HALOGENS

9 F
17 Cl
35 Br
53 I
85 At

The elements of group 7A, the halogens, have played an important part in the development of chemistry. Chlorine was first prepared by the Swedish chemist Karl Wilhelm Scheele in 1774, but it was not until 1810 that the English chemist Humphry Davy identified it as an element. Iodine was discovered in 1811, and bromine in 1825. Compounds of fluorine were known for a long time, but not until 1886 did the French chemist Henri Moissan succeed in preparing the very reactive free element.

The general valence-electron configuration of the halogens is ns^2np^5, where n may have values ranging from 2 through 6. The halogens have large electron affinities, and they most often achieve a noble-gas configuration by gaining an electron, which results in a -1 oxidation state. Fluorine, being the most electronegative element, exists in compounds only in the -1 state. The other halogens also exhibit positive oxidation states up to $+7$ by sharing valence electrons with a more electronegative atom such as O. In the positive oxidation states, the halogens tend to be good oxidizing agents (that is, they readily gain electrons).

Table 23.2 Occurrences of the Halogens

Element	Occurrences
Fluorine	Fluorspar, CaF_2; fluorapatite, $Ca_5(PO_4)_3F$; cryolite, Na_3AlF_6 Biologically: teeth, bones
Chlorine	Seawater (0.55 *M*, chiefly as NaCl); salt beds, salt lakes (for example, Dead Sea, Great Salt Lake); NaCl deposits Biologically: gastric juice [as HCl(*aq*)], tissue fluids
Bromine	Seawater (8.3×10^{-4} *M*); underground brines; salt beds, salt lakes Biologically: minor concentrations of bromide along with chloride
Iodine	Seawater (4×10^{-7} *M*); seaweeds; $NaIO_3$ in minor amounts in nitrate deposits; oil-well brines Biologically: in human thyroid gland

Occurrences of the Halogens

Table 23.2 summarizes the occurrences of the halogens in nature. Although fluorine and chlorine are fairly abundant, they are quite differently distributed. This happens because the salts of chlorine are generally soluble, whereas some of those of fluorine are not. Bromine is much less abundant than chlorine or fluorine, and iodine is much rarer still.

Chlorine, bromine, and iodine occur as the halides in seawater and in salt deposits. The concentration of iodine in these sources is generally very small. However, it is concentrated by certain seaweeds. When they are harvested, dried, and burned, iodine can be extracted from the ashes. The element is also extracted commercially from oil-well brines in California. Fluorine occurs in the minerals fluorspar, CaF_2, cryolite, Na_3AlF_6, and fluorapatite $Ca_5(PO_4)_3F$. Only the first of these is an important commercial source of fluorine for the chemical industry.

All isotopes of astatine are radioactive. The longest-lived isotope is astatine-210, which has a half-life of 8.3 hr and decays mainly by electron capture. Astatine was first synthesized by bombarding bismuth-209 with high-energy alpha particles, as shown in Equation 23.3:

$$^{209}_{83}\text{Bi} + ^{4}_{2}\text{He} \longrightarrow ^{211}_{85}\text{At} + 2^{1}_{0}\text{n} \qquad [23.3]$$

A cyclotron must be used to synthesize astatine, which makes it very expensive and limits its application and study.

Properties and Preparation of the Halogens

Some of the properties of the halogens are summarized in Table 23.3. These properties vary in a regular fashion as a function of atomic number. Within each horizontal row of the periodic table, each halogen has a high ionization energy, second only to the noble-gas element adjacent to it. Each halogen has the highest electronegativity of all the elements in its row. Within the halogen family, atomic and ionic radii increase with increasing atomic number. Correspondingly, the ionization energy and electronegativity steadily decrease as we go down the family, from fluorine to iodine.

Under ordinary conditions, the halogens exist as diatomic molecules. The molecules are held together in the solid and liquid states by London dispersion forces (Section 11.2). Because I_2 is the largest and most polarizable of the halogen molecules, it is not surprising that the intermolecular forces between I_2 molecules are the strongest. Thus, I_2 has the highest melting point

Table 23.3 Some Properties of the Halogen Atoms

Property	F	Cl	Br	I
Atomic radius (Å)	0.72	0.99	1.14	1.33
Ionic radius, X^- (Å)	1.33	1.84	1.96	2.20
First ionization energy (kJ/mol)	1681	1256	1143	1009
Electron affinity (kJ/mol)	−332	−349	−325	−295
Electronegativity	4.0	3.2	3.0	2.7
X—X single-bond energy (kJ/mol)	155	242	193	151
Reduction potential: $\frac{1}{2}X_2(aq) + e^- \longrightarrow X^-(aq)$	2.87	1.36	1.07	0.54

and boiling point. At room temperature and 1 atm pressure, I_2 is a solid, Br_2 is a liquid, and Cl_2 and F_2 are gases. Chlorine readily liquifies upon compression at room temperature and is normally stored and handled in liquid form in steel containers.

The comparatively low bond energy in F_2 (155 kJ/mol) accounts in part for the extreme reactivity of elemental fluorine. Because of its high reactivity, F_2 is very difficult to work with. Certain metals, such as copper and nickel, can be used to contain F_2 because their surfaces form a protective coating of metal fluoride. Chlorine and the heavier halogens are also reactive, although less so than fluorine. They combine directly with most elements except the rare gases.

Because of their high electronegativities compared with those of other elements, the halogens tend to gain electrons from other substances and thereby serve as oxidizing agents. The oxidizing ability of the halogens, which is indicated by their reduction potentials, decreases going down the group. As a result, we find that a given halogen is able to oxidize the anions of the halogens below it in the family. For example, Cl_2 will oxidize Br^- and I^-, but not F^-, as seen in Figure 23.3.

Figure 23.3 Aqueous solutions of NaF, NaBr, and NaI (from left to right) to which Cl_2 has been added. Each solution is in contact with carbon tetrachloride, CCl_4, which forms the lower layer in each container. The halogens are more soluble in CCl_4. The F^- ion in the NaF solution (left) does not react with Cl_2, and thus both the aqueous and CCl_4 layers remain colorless. The Br^- ion (center) is oxidized by Cl_2 to form Br_2, producing a yellow color in the water layer and an orange color in the CCl_4 layer. The I^- ion (right) is oxidized to I_2, producing an amber color in the water layer and a violet color in the CCl_4 layer. (Donald Clegg and Roxy Wilson)

SAMPLE EXERCISE 23.2

Write the balanced chemical equation for the reaction, if any, that occurs between **(a)** $I^-(aq)$ and $Br_2(l)$; **(b)** $Cl^-(aq)$ and $I_2(s)$.

Solution: **(a)** Br_2 is able to oxidize (remove electrons) from the anions of the halogens below it in the periodic table. Thus, it will oxidize I^-:

$$2I^-(aq) + Br_2(l) \longrightarrow I_2(s) + 2Br^-(aq)$$

(b) Cl^- is the anion of a halogen above iodine in the periodic table. Thus, I_2 cannot oxidize Cl^-; there is no reaction.

PRACTICE EXERCISE

Write the balanced chemical equation for the reaction that occurs between $Br^-(aq)$ and $Cl_2(aq)$. *Answer:* $2Br^-(aq) + Cl_2(aq) \longrightarrow Br_2(l) + 2Cl^-(aq)$

Notice from Table 23.2 that the reduction potential of F_2 is exceptionally high. Fluorine gas readily oxidizes water according to the reaction

$$F_2(aq) + H_2O(l) \longrightarrow 2HF(aq) + \tfrac{1}{2}O_2(g) \qquad E^\circ = +1.64 \text{ V} \qquad [23.4]$$

Fluorine cannot be prepared by electrolytic oxidation of aqueous solutions of fluoride salts, because water itself is oxidized more readily than F^-. In practice, the element is formed by electrolytic oxidation of a solution of KF in anhydrous HF. The KF reacts with HF to form a salt, $K^+HF_2^-$, which acts as the current carrier in the liquid. (The HF_2^- ion is stable because of the very strong hydrogen bond between the two fluoride ions, as described in Section 11.2.) The overall cell reaction is

$$2KHF_2(l) \longrightarrow H_2(g) + F_2(g) + 2KF(l) \qquad [23.5]$$

Chlorine is produced mainly by electrolysis of either molten or aqueous sodium chloride, as described in Sections 20.5 and 24.5. Both bromine and iodine are obtained commercially from brines containing the halide ions by oxidation with Cl_2.

Uses of the Halogens

Fluorine has become an important industrial chemical. It is used, for example, to prepare fluorocarbons, very stable carbon-fluorine compounds. An example is CF_2Cl_2, known as Freon 12, which is used as a refrigerant and as a propellant for aerosol cans. As we noted in Section 18.3, these substances are suspected of depleting the ozone layer. Fluorocarbons are also used as lubricants and in plastics. Teflon (Figure 23.4) is a polymeric fluorocarbon noted for its high thermal stability and lack of chemical reactivity.

Chlorine is by far the most important halogen commercially. About 1.0×10^{10} kg (11.3 million tons) of Cl_2 is produced in the United States each year. In addition, hydrogen chloride production is about 2.7×10^9 kg (2.9 million tons) annually. About half of this inorganic chlorine finds its way eventually into vinyl chloride, C_2H_3Cl, used in polyvinyl chloride (PVC) plastics manufacture; ethylene dichloride, $C_2H_2Cl_2$, an organic solvent; and other chlorine-containing organic compounds. Much of the remainder of the chlorine is used as a bleach in the paper and textile industries. When Cl_2 dissolves in cold dilute base it disproportionates into Cl^- and the hypochlorite ion, ClO^-:

Figure 23.4 Structure of Teflon, a fluorocarbon polymer.

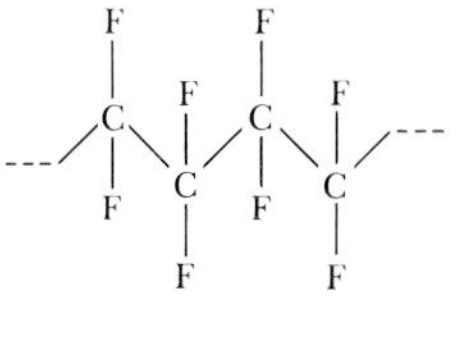

Figure 23.5 Iodized salt contains about 0.02% KI by weight. (© Paul Silverman/ Fundamental Photographs)

$$2OH^-(aq) + Cl_2(aq) \rightleftharpoons Cl^-(aq) + ClO^-(aq) + H_2O(l) \quad [23.6]$$

Sodium hypochlorite, NaClO, is the active ingredient in many liquid bleaches. Chlorine is also used in water treatment to oxidize and thereby destroy bacteria (Section 18.6).

Neither bromine nor iodine is as widely used as fluorine and chlorine. One familiar application of bromine is in the production of silver bromide used in photographic film. A common use of iodine is its addition, as KI, to salt to form iodized table salt, which contains about 0.02 percent potassium iodide by weight. Iodized salt (Figure 23.5) provides the small amount of iodine necessary in our diets; it is essential for the formation of thyroxin, a hormone secreted by the thyroid gland. Lack of iodine in the diet results in an enlarged thyroid gland, a condition called *goiter*.

The Hydrogen Halides

All of the halogens form stable diatomic molecules with hydrogen. These are very important compounds, in part because aqueous solutions of the hydrogen halides other than HF are strongly acidic. Table 23.4 lists some of the more important properties of the hydrogen halides. Notice that the boiling point of HF is abnormally high as compared with those of the other hydrogen halides. The cause of this unusual behavior is the strong hydrogen bonding that exists between HF molecules in the liquid state, as described in Section 11.2.

The hydrogen halides can be formed by direct reaction of the elements. However, the most important means of preparing hydrogen halides is through

Table 23.4 Properties of the Hydrogen Halides

Property	HF	HCl	HBr	HI
Molecular weight	20.01	36.45	80.92	127.91
Melting point (°C)	−83	−115	−89	−51
Boiling point (°C)	19.5	−84.2	−67.1	−35.1
Bond-dissociation energy (kJ/mol)	567	431	366	299
H—X bond length (Å)	0.92	1.27	1.41	1.61
Solubility in H_2O (g/100 g H_2O, 10°C)	∞	78	210	234

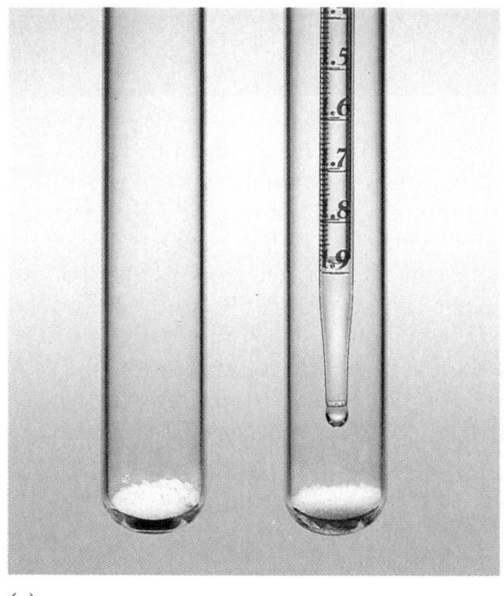

(a)

(b)

Figure 23.6 (a) Sodium iodide in the left test tube, and sodium bromide on the right. Sulfuric acid is in the pipet. (b) Addition of sulfuric acid to the test tubes oxidizes sodium iodide to form the dark-colored iodine on the left. Sodium bromide is oxidized to the yellow-brown bromine on the right. When more concentrated, bromine has a reddish-brown color. (Richard Megna/Fundamental Photographs)

reaction of a salt of the halide with a strong nonvolatile acid. Hydrogen fluoride and hydrogen chloride are prepared in this manner by reaction of a cheap, readily available salt with concentrated sulfuric acid:

$$CaF_2(s) + H_2SO_4(l) \xrightarrow{\Delta} 2HF(g) + CaSO_4(s) \quad [23.7]$$

$$NaCl(s) + H_2SO_4(l) \xrightarrow{\Delta} HCl(g) + NaHSO_4(s) \quad [23.8]$$

Because the hydrogen halide is the only volatile component in the mixture, it can be easily removed. It is usually absorbed in water and marketed as the corresponding acid.

Neither hydrogen bromide nor hydrogen iodide can be prepared by analogous reactions of salts with H_2SO_4, because HBr and HI undergo oxidation by H_2SO_4, as seen in Figure 23.6. The overall reactions are described by Equations 23.9 and 23.10:

$$2NaBr(s) + 2H_2SO_4(l) \longrightarrow Br_2(g) + SO_2(g) + Na_2SO_4(s) + 2H_2O(g) \quad [23.9]$$

$$8NaI(s) + 9H_2SO_4(l) \longrightarrow 8NaHSO_4(s) + H_2S(g) + 4I_2(g) + 4H_2O(g) \quad [23.10]$$

Notice that in the case of the bromide, part of the H_2SO_4 is reduced to SO_2, in which sulfur is in the +4 oxidation state. In the reaction with iodide, sulfur is reduced all the way to H_2S, in which sulfur is in the −2 oxidation state. This difference in products reflects the greater ease of oxidation of the iodide. The difficulties associated with use of H_2SO_4 can be avoided by using a nonvolatile acid that is a poorer oxidizing agent than H_2SO_4; concentrated phosphoric acid, H_3PO_4, serves well.

The hydrogen halides are also formed when certain molecular (covalent) halides are hydrolyzed, as in the following examples (Figure 23.7):

$$PCl_3(l) + 3H_2O(l) \longrightarrow H_3PO_3(l) + 3HCl(g) \quad [23.11]$$

$$SeBr_4(s) + 3H_2O(l) \longrightarrow H_2SeO_3(s) + 4HBr(g) \quad [23.12]$$

Figure 23.7 Phosphorus trichloride is added to water containing methyl orange indicator. As the PCl_3 hydrolyzes to form H_3PO_3 and HCl, the indicator turns red, its color in acid solutions. (© Richard Megna/Fundamental Photographs)

A CLOSER LOOK: The Hydrolysis of Nonmetal Halides

Some nonmetal halides, such as NF_3, CCl_4, and SF_6, are quite unreactive toward water. This lack of reactivity is not due to thermodynamic factors, but rather to the kinetic features of the reaction. For example, ΔG° for the hydrolysis of CCl_4 is highly negative, -377 kJ/mol CCl_4, which implies that the reaction should proceed very nearly to completion (Section 19.6). However, no low-energy reaction mechanism is available; the hydrolysis reaction has a very high activation barrier.

Generally, a molecular halide is unreactive when the central atom is unable to expand further the number of electrons in its valence shell (that is, the central atom has its maximum stable coordination number). For example, carbon can accommodate a maximum of eight electrons in its valence shell. Silicon, on the other hand, can accommodate a greater number of electrons by using *d* orbitals in bonding. The hydrolysis of $SiCl_4$ is believed to occur by attack of the silicon atom by the H_2O molecule which expands the coordination number of Si. This attack is followed by loss of H^+ and Cl^- ions, as shown in Figure 23.8. This reaction pathway is not available to CCl_4.

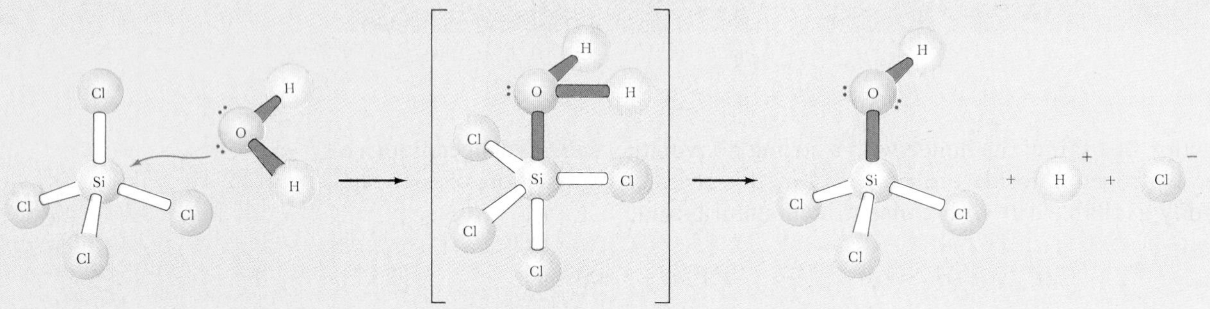

Figure 23.8 Proposed mechanism for the first stage of the reaction between $SiCl_4$ and H_2O. After the H_2O forms an adduct with $SiCl_4$, that adduct loses H^+ and Cl^- ions. The process of H_2O attack is then repeated three additional times, with $Si(OH)_4$ ultimately formed.

SAMPLE EXERCISE 23.3

Write a balanced chemical equation for the formation of hydrogen bromide gas from the reaction of solid sodium bromide with phosphoric acid.

Solution: The chemical formulas of sodium bromide and phosphoric acid are NaBr and H_3PO_4, respectively. Let us assume that only one of the dissociable hydrogens of H_3PO_4 undergoes reaction. (The actual number depends on the reaction conditions.) The balanced equation is then

$$NaBr(s) + H_3PO_4(aq) \longrightarrow NaH_2PO_4(s) + HBr(g)$$

PRACTICE EXERCISE

Write the balanced chemical equation for the preparation of HI from NaI and H_3PO_4. ***Answer:*** $NaI(s) + H_3PO_4(aq) \longrightarrow NaH_2PO_4(s) + HI(g)$

SAMPLE EXERCISE 23.4

Write the balanced chemical equation for the reaction between $SiCl_4$ and H_2O.

Solution: The oxidation state of Si in $SiCl_4$ is $+4$. Because the hydrolysis of nonmetal chlorides is not an oxidation-reduction reaction, the product acid must also have Si in the $+4$ oxidation state. That acid is H_4SiO_4, which we can also write as $Si(OH)_4$. Thus the equation is

$$SiCl_4(l) + 4H_2O(l) \longrightarrow Si(OH)_4(s) + 4HCl(g)$$

Because $Si(OH)_4$ readily undergoes loss of water to form SiO_2, the reaction is also written in the following way:

$$SiCl_4(l) + 2H_2O(l) \longrightarrow SiO_2(s) + 4HCl(g)$$

PRACTICE EXERCISE

What is the oxidation state of P in PCl_5? What is the chemical formula of the oxyacid of P in this oxidation state? Write the balanced chemical equation for the reaction of solid PCl_5 with water.
Answer: $+5$; H_3PO_4; $PCl_5(s) + 4H_2O(l) \longrightarrow H_3PO_4(aq) + 5HCl(aq)$

The hydrogen halides form hydrohalic acid solutions when dissolved in water. These solutions exhibit the characteristic properties of acids, such as reactions with active metals to produce hydrogen gas (Section 4.6). Hydrofluoric acid also reacts readily with silica, SiO_2, and with various silicates to form hexafluorosilicic acid, H_2SiF_6, as in these examples:

$$SiO_2(s) + 6HF(aq) \longrightarrow H_2SiF_6(aq) + 2H_2O(l) \quad [23.13]$$

$$CaSiO_3(s) + 8HF(aq) \longrightarrow H_2SiF_6(aq) + CaF_2(aq) + 3H_2O(l) \quad [23.14]$$

Glass consists mostly of silicate structures (Section 23.5), and these reactions allow HF to etch or frost glass (Figure 23.9). It is also the reason that HF is stored in wax or plastic containers rather than glass.

Interhalogen Compounds

Because the halogens form diatomic molecules in their most stable state at ordinary temperatures and pressures, it is not surprising to discover that diatomic molecules consisting of two different halogen atoms exist. These com-

Figure 23.9 Etched or frosted glass. Designs such as this are produced by first coating the glass with wax. The wax is then removed in the areas to be etched. When treated with hydrofluoric acid, the exposed areas of the glass are attacked, producing the etching effect. The hydrofluoric acid is then washed from the surface and the remaining wax is removed. (Richard Megna/ Fundamental Photographs)

Table 23.5 Properties of Interhalogen Compounds, XX′

Compound	ClF	BrF	BrCl	IF	ICl	IBr
Molecular weight	54.6	98.9	115.4	145.9	162.4	206.8
Melting point (°C)	−156	−33	−66	—	27	41
Boiling point (°C)	−100	20	5	—	98	116
Bond distance (Å)	1.63	1.76	2.14	1.91	2.32	2.49
Dipole moment (D)	0.9	1.3	0.6	—	0.7	1.21
Bond-dissociation energy (kJ/mol)	253	237	218	278	208	175

pounds are the simplest examples of **interhalogens**, that is, compounds formed between two different halogen elements. Some properties of the diatomic interhalogens are listed in Table 23.5. The table is incomplete because certain of the interhalogens are not stable; they undergo decomposition to form the diatomic halogen elements or to form more-complex interhalogen compounds.

The higher interhalogen compounds have formulas of the form XX'_3, XX'_5, or XX'_7, where X is chlorine, bromine, or iodine, and X′ is fluorine. (The one exception is ICl_3, in which X′ is chlorine.) Using the VSEPR model (Section 9.1), we can predict the geometrical structures of these compounds. We can also describe the bonding about the central atom in terms of a hybrid orbital description (Section 9.4).

SAMPLE EXERCISE 23.5

Account for the valence-shell electron distribution and geometrical structure in BrF_3. What hybrid orbital description is most suitable for the central atom in this molecule?

Solution: Bromine has seven valence-shell electrons. When the Br atom is singly bonded to three fluorine atoms, there are three additional electrons from this source. According to the VSEPR model, these ten electrons are disposed as five electron pairs about the central atom at the vertices of a trigonal bipyramid (Table 9.3). Three of the electron pairs are used in bonding to fluorine; the other two are unshared electron pairs. These unshared pairs require a larger space, so they are placed in the equatorial plane of the trigonal bipyramid:

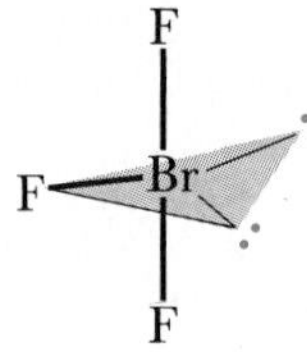

Because the unshared pairs push the bonding pairs back a little, the molecule should have the shape of a bent T. In terms of a hybridization description, we need to employ one of the bromine valence-shell d orbitals (the $4d$) in addition to the $4s$ and three $4p$ orbitals to provide the five atomic orbitals to contain the five electron pairs in the valence shell. Thus the appropriate hybrid orbital description is sp^3d, which results in orbitals directed toward the vertices of a trigonal bipyramid.

PRACTICE EXERCISE

Predict the molecular geometry and hybridization of I orbitals in IF_5.
Answer: square pyramidal; sp^3d^2

The central atom in these higher interhalogen compounds has valence-shell *d* orbitals available for bonding. Thus, the valence shell can be expanded beyond the octet. The central atom in all cases is relatively large compared to the atoms grouped around it. Because fluorine is small and forms very strong bonds, it is ideally suited as the X′ atom. Only when the central atom is very large, as in the case of iodine, can the larger chlorine atom form an interhalogen, as in ICl_3. The importance of size is also seen in the fact that iodine is capable of forming IF_7, but with bromine a maximum of five fluorines can be fitted around the central atom, in BrF_5. Chlorine and fluorine can form the ClF_5 molecule, but only with great difficulty.

The interhalogen compounds are exceedingly reactive. The fluorides attack glass very readily and must be placed in special metal containers. They are very active fluorinating agents, as in these examples:

$$2CoCl_2(s) + 2ClF_3(g) \longrightarrow 2CoF_3(s) + 3Cl_2(g) \quad [23.15]$$

$$Se(s) + 3BrF_5(l) \longrightarrow SeF_6(l) + 3BrF_3(l) \quad [23.16]$$

The *polyhalide ions* are closely related to the interhalogens. Many of these ions are relatively stable as salts of the alkali metal ions—for example, KI_3, $CsIBr_2$, $KICl_4$, and $KBrF_4$. Some of them, notably I_3^-, are also stable in aqueous solution.

Oxyacids and Oxyanions

Table 23.6 summarizes the formulas of the known oxyacids of the halogens and the way they are named.* The oxyacids are rather unstable; they generally decompose (sometimes explosively) when you attempt to isolate them. All of the oxyacids are strong oxidizing agents. The oxyanions, formed on removal of a proton from the oxyacids, are generally more stable than the oxyacids themselves. (Review the nomenclature of the oxyanions, Section 2.6.) Hypochlorite salts are used as bleaches and disinfectants because of the powerful oxidizing capabilities of the hypochlorite ion. Sodium chlorite, which can be isolated as the trihydrate, $NaClO_2 \cdot 3H_2O$, is used as a bleaching agent. Chlorite salts form potentially explosive mixtures with organic materials. Chlorate salts are similarly very reactive. For example, a mixture of potassium chlorate and sulfur may explode when struck. Potassium chlorate is used in making matches and fireworks.

* Fluorine forms one oxyacid, HFO. Because the electronegativity of fluorine is greater than that of oxygen, we must consider fluorine to be in a -1 oxidation state and oxygen to be in the 0 oxidation state.

Table 23.6 The Oxyacids of the Halogens

Oxidation state of halogen	Formula of acid			Name
	Cl	Br	I	
$+1$	$HClO$	$HBrO$	HIO	*Hypo*hal*ous* acid
$+3$	$HClO_2$	—	—	Hal*ous* acid
$+5$	$HClO_3$	$HBrO_3$	HIO_3	Hal*ic* acid
$+7$	$HClO_4$	$HBrO_4$	HIO_4, H_5IO_6	*Per*hal*ic* acid

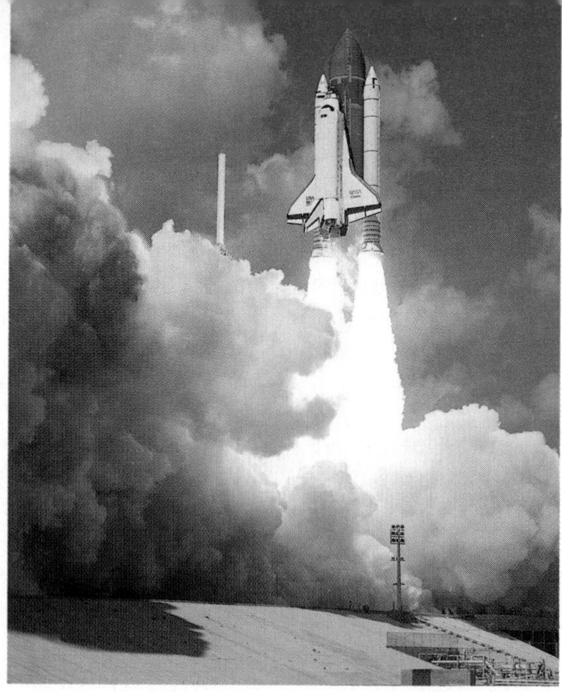

Figure 23.10 Launch of the space shuttle *Atlantis*, October 18, 1989. (NASA, Johnson Space Center)

Perchloric acid and its salts are the most stable of the oxyacids and oxyanions. Dilute perchloric acid solutions are quite safe, and most perchlorate salts are stable except when heated with organic materials. When heated, perchlorates can become vigorous, even violent oxidizers. Therefore, considerable caution should be exercised when handling these substances, and it is crucial to avoid contact between perchlorates and readily oxidized materials such as active metals and combustible organic compounds. The use of ammonium perchlorate as the oxidizer in the solid booster rockets for the space shuttle demonstrates the oxidizing power of perchlorates. The solid propellant contains a mixture of NH_4ClO_4 and powdered aluminum, the reducing agent. Each shuttle launch requires about 6×10^5 kg (700 tons) of NH_4ClO_4 (Figure 23.10).

There are two oxyacids that have iodine in the +7 oxidation state. These periodic acids are HIO_4 (called metaperiodic acid) and H_5IO_6 (called paraperiodic acid). The two forms exist in equilibrium in aqueous solution:

$$H_5IO_6(aq) \rightleftharpoons H^+(aq) + IO_4^-(aq) + 2H_2O(l) \qquad K = 0.015 \qquad [23.17]$$

HIO_4 is a strong acid, and H_5IO_6 is a weak one; the first two acid-dissociation constants for H_5IO_6 are $K_{a1} = 2.8 \times 10^{-2}$ and $K_{a2} = 4.9 \times 10^{-9}$. Crystalline H_5IO_6 is obtained when periodic acid solutions are evaporated at low temperatures. Mild heating of H_5IO_6 under vacuum produces HIO_4. The structure of H_5IO_6 is given in Figure 23.11. The large size of the iodine atom allows it to accommodate six surrounding oxygen atoms. The smaller halogens do not form acids of this type.

Figure 23.11 Paraperiodic acid, H_5IO_6.

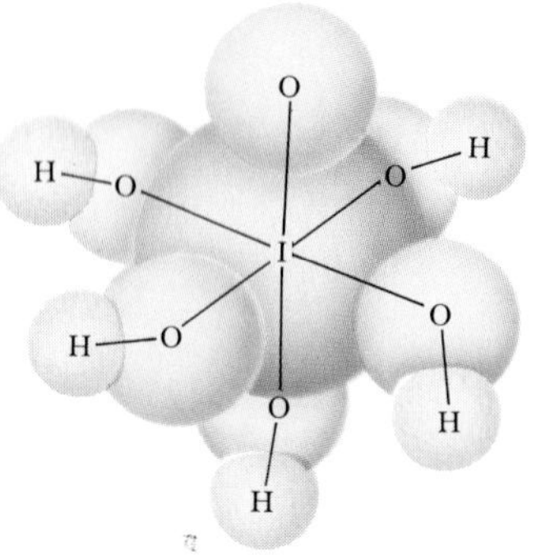

The acid strengths of the oxyacids increase with increasing oxidation state of the central atom; the origin of this trend is discussed in Section 16.9. The stability of the oxyacids and of the corresponding oxyanions toward reduction increases with increasing oxidation state of the central halogen atom. Because the halogens are relatively electronegative elements, compounds in which the halogen has an increasingly positive oxidation number would be expected to be *less* stable. The origins of this unexpected trend are rather complicated and beyond the scope of our survey.

23.3 THE GROUP 6A ELEMENTS

The elements of group 6A are oxygen, sulfur, selenium, tellurium, and polonium. We have already discussed oxygen in Section 22.4. In this section, we will examine the group as a whole and then look at sulfur, selenium, and tellurium, focusing on sulfur. We will not have much to say about polonium, an element produced by radioactive decay of radium. There are no stable isotopes of this element. As a result, it is found only in trace quantities in radium-containing minerals.

General Characteristics of the Group 6A Elements

The group 6A elements possess the general outer-electron configuration ns^2np^4, where n may have values ranging from 2 through 6. These elements thus may attain a noble-gas electron configuration by the addition of two electrons, which results in a -2 oxidation state. Because the group 6A elements are nonmetals, this is a common oxidation state. Except for oxygen, however, the group 6A elements are also commonly found in positive oxidation states up to $+6$, which corresponds to the sharing of all six valence-shell electrons with atoms of a more electronegative element. Sulfur, selenium, and tellurium also differ from oxygen in being able to use d orbitals in bonding. Thus, compounds with expanded valence shells such as SF_6, SeF_6, and TeF_6 occur.

8 O
16 S
34 Se
52 Te
84 Po

Some of the more important properties of the atoms of the group 6A elements are summarized in Table 23.7. The energy of the X—X single bond is estimated from data for the elements, except for oxygen. In this case, because the O—O bond in O_2 is not a single bond (Section 9.6), the estimated O—O bond energy in hydrogen peroxide is employed. The reduction potential listed in the last line of the table refers to the reduction of the element in its standard state to form $H_2X(aq)$ in acidic solution. In most of the properties listed in Table 23.7, we see a regular variation as a function of atomic number. Atomic and ionic radii increase and ionization energies decrease as expected as we move down the family.

The electron affinities listed apply to the process shown in Equation 23.18:

$$X(g) + e^- \longrightarrow X^-(g) \qquad [23.18]$$

This, of course, does not produce the commonly observed stable ion of these elements, X^{2-}, but it is the first step in its formation. It is interesting to note that the addition of an electron to oxygen is less exothermic than the addition of an electron to any other group 6A element. This effect is due to the relatively

Table 23.7 Some Properties of the Atoms of the Group 6A Elements

Property	O	S	Se	Te
Atomic radius (Å)	0.73	1.03	1.40	1.60
X^{2-} ionic radius (Å)	1.40	1.84	1.98	2.21
First ionization energy (kJ/mol)	1314	999	941	869
Electron affinity (kJ/mol)	−141	−201	−195	−186
Electronegativity	3.5	2.5	2.4	2.1
X—X single-bond energy (kJ/mol)	146[a]	266	172	126
Reduction potential to H_2X in acidic solution (V)	1.23	0.14	−0.40	−0.72

[a] Based on O—O bond energy in H_2O_2.

small size of the oxygen atom. Addition of a single extra electron to the smaller atom results in larger electron-electron repulsions, offsetting the gain in stability that results from the closer approach of the electron to the nucleus.

Note that the ease of reduction of the free element to form H_2X varies greatly throughout the series. Whereas oxygen is very readily reduced to the -2 oxidation state, the potential for reduction of tellurium is strongly negative. These observations indicate an increasingly metallic character in the group 6A elements as atomic number increases. The physical properties of the elements are also consistent with an increasing metallic character. At the top of the family we have oxygen, a diatomic molecule, and sulfur, a nonconducting solid that melts at 114°C. Near the bottom we have tellurium, whose stable form has a bright luster, low electrical conductivity, and a melting point of 452°C.

Occurrences and Preparation of Sulfur, Selenium, and Tellurium

Large underground deposits are the principal source of elemental sulfur. The Frasch process, illustrated in Figure 23.12, is used to obtain the element from these deposits. The method is based on the low melting point and low density of sulfur. Superheated water is forced into the deposit, where it melts the sulfur. Compressed air then forces the molten sulfur up a pipe that is concentric with the ones that introduce the hot water and compressed air into the deposit.

Sulfur also occurs widely as sulfide and sulfate minerals. Its presence as a minor component of coal and petroleum poses a major problem. As we saw in Section 18.4, combustion of these "unclean" fuels leads to serious sulfur oxide pollution. Also, operations that use sulfide minerals as sources of metals liberate sulfur oxides. Much effort has been directed at removing this sulfur, and these efforts have increased the availability of sulfur. The sale of this sulfur helps par-

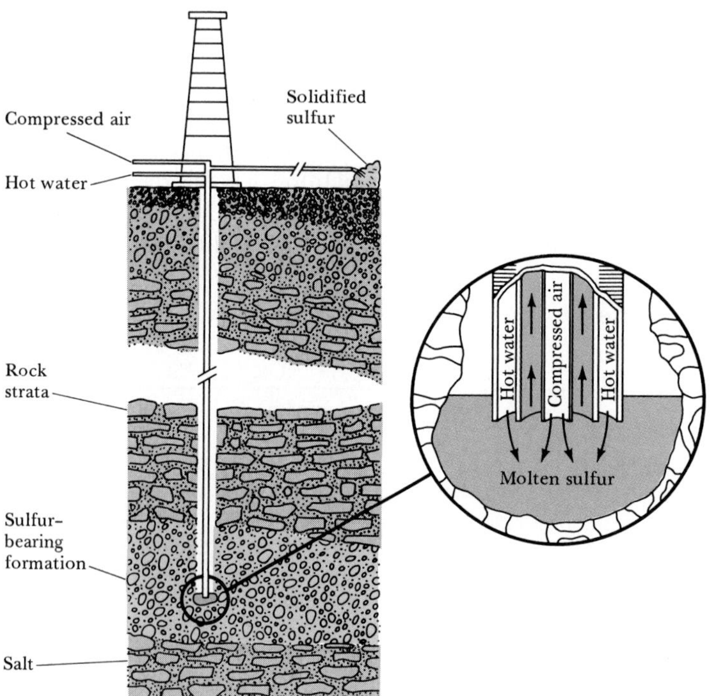

Figure 23.12 Mining of sulfur by the Frasch process. The process is named after Herman Frasch, who invented the process in the early 1890s. The process is particularly useful for recovering sulfur from deposits located under quicksand or water.

tially to offset the costs of the desulfurizing processes and equipment. However, sulfur obtained from sulfur deposits by the Frasch process is about 99.5 percent pure and can be used for most commercial processes without purification. It is therefore relatively cheap. Consequently, sulfur from desulfurizing processes must be sold at prices below its cost in order to compete on the market. Nevertheless, about half the sulfur used in the United States each year is produced by means other than the Frasch process.

Selenium and tellurium occur in rare minerals such as Cu_2Se, PbSe, Ag_2Se, Cu_2Te, PbTe, Ag_2Te, and Au_2Te. They also occur as minor constituents in sulfide ores of copper, iron, nickel, and lead. However, these elements are not very important commercially, so we will not consider the details of how they are obtained.

Properties and Uses of Sulfur, Selenium, and Tellurium

As we normally encounter it, sulfur is yellow, tasteless, and nearly odorless. It is insoluble in water and exists in several allotropic forms. The thermodynamically stable form at room temperature is rhombic sulfur, which consists of puckered S_8 rings, as shown in Figure 23.13. When heated above its melting point, at 113°C, sulfur undergoes a variety of changes. The molten sulfur first contains S_8 molecules and is fluid because the rings readily slip over each other. Further heating of this straw-colored liquid causes rings to break, the fragments joining to form very long molecules that can become entangled. The sulfur consequently becomes highly viscous. This change is marked by a color change to dark reddish brown (Figure 23.14). Further heating breaks the chains, and the viscosity again decreases.

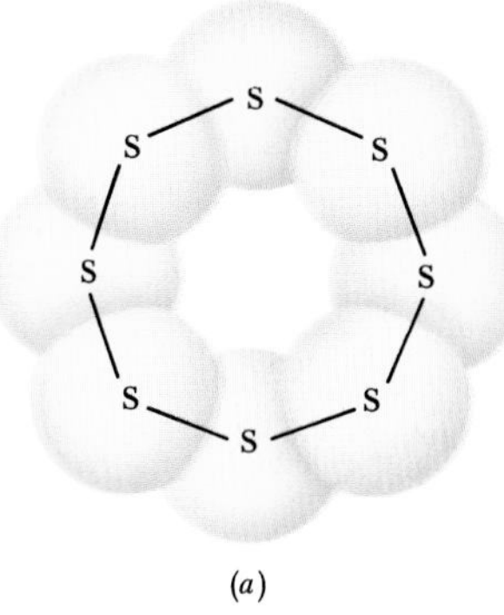

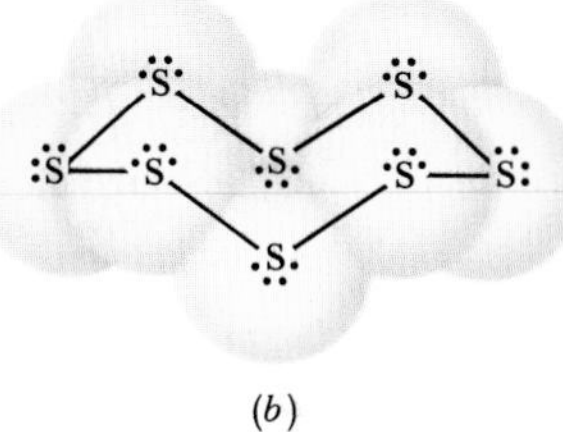

Figure 23.13 Top view (*a*) and side view (*b*) of the sulfur molecule in rhombic sulfur.

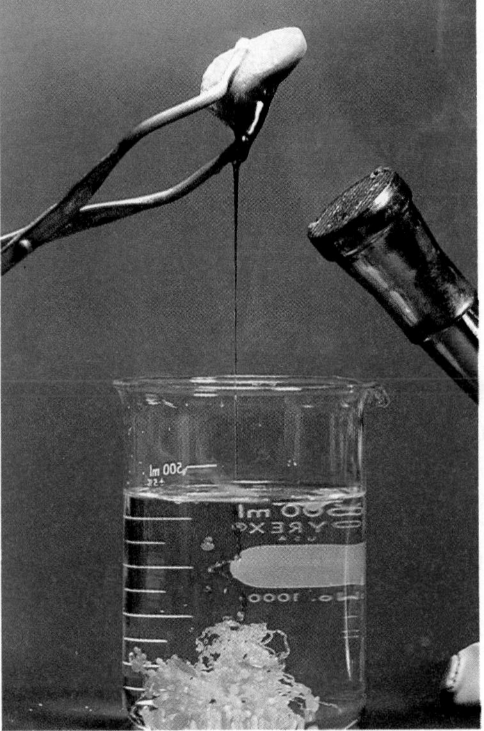

Figure 23.14 When sulfur is heated above its melting point, 113°C, it becomes dark and viscous. Here the liquid is shown falling into cold water, where it again solidifies. (Lawrence Migdale, Science Source/Photo Researchers)

Most of the 1.3×10^{10} kg (14 million tons) of sulfur produced in the United States each year is used in the manufacture of sulfuric acid. Sulfur is also used in vulcanizing rubber, a process that toughens rubber by introducing cross-linking between polymer chains (Section 12.2).

The most stable allotropes of both selenium and tellurium are crystalline substances containing helical chains of atoms, as illustrated in Figure 23.15. Each atom of the chain is close to atoms in adjacent chains, and it appears that some sharing of electron pairs between these atoms occurs, as indicated by the dashed lines in Figure 23.15.

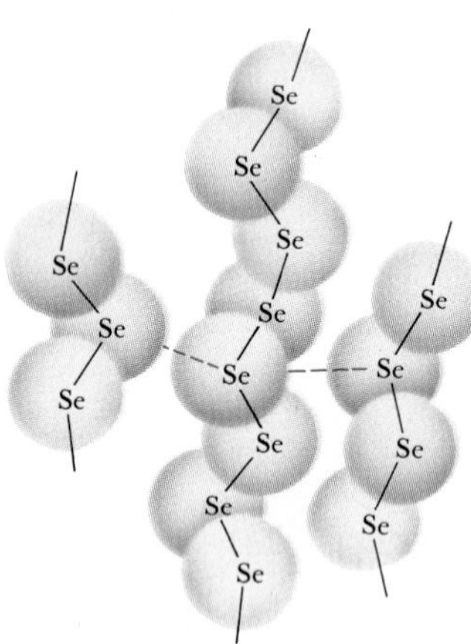

Figure 23.15 Portion of the structure of crystalline selenium. The dashed lines represent weak bonding interactions between atoms in adjacent chains. Tellurium has the same structure.

The electrical conductivity of selenium is very low in the dark but increases greatly upon exposure to light. This property of the element is utilized in photoelectric cells and light meters. Photocopiers also depend on the photoconductivity of selenium. Photocopy machines contain a belt or drum coated with a film of selenium. This drum is electrostatically charged and then exposed to light reflected from the image being photocopied. The electric charge drains from the selenium where the selenium is made conductive by exposure to light. A black powder (the "toner") sticks only to the areas that remain charged. The photocopy is made when the toner is transferred to a sheet of plain paper, which is heated to fuse the toner to the paper.

Sulfur is widely distributed in biological systems. It is present in most proteins, as a component of the amino acids cysteine and methionine (Section 27.2). In contrast, selenium is rare in biological systems. Only recently has the human nutritional requirement for the element been established. Selenium is present in trace quantities in most vegetables, especially spinach. The amount of selenium required for adequate nutrition is very small. In quantities much larger than this small nutritional requirement, the element becomes toxic. Tellurium does not have a known role in human nutrition. Its compounds are poisonous and, if they are volatile, usually have highly offensive odors.

Oxides, Oxyacids, and Oxyanions of Sulfur

Sulfur dioxide was first discovered by Joseph Priestley in 1774, when he heated mercury with concentrated sulfuric acid:

$$Hg(l) + 2H_2SO_4(l) \longrightarrow HgSO_4(s) + SO_2(g) + 2H_2O(l) \quad [23.19]$$

In the laboratory, SO_2 is prepared by the action of aqueous acid on a sulfite salt:

$$2H^+(aq) + SO_3{}^{2-}(aq) \longrightarrow SO_2(g) + H_2O(l) \quad [23.20]$$

Sulfur dioxide is formed when sulfur is combusted in air; it has a choking odor and is poisonous. The gas is particularly toxic to lower organisms, such as fungi, and is consequently used for sterilizing dried fruit. At 1 atm pressure and room temperature, SO_2 dissolves in water to the extent of 45 volumes of gas per volume of water, to produce a solution of about 1.6 *M* concentration. The solution is acidic, and we describe it as $H_2SO_3(aq)$. Actually, there is evidence that much of the sulfur dioxide exists in solution as hydrated SO_2. When the saturated solution is cooled, crystals of the hydrate, $SO_2 \cdot 6H_2O$, can be recovered. It is convenient, however, to assume that all of the SO_2 that dissolves is in the form of the weak acid, H_2SO_3, which ionizes according to Equations 23.21 and 23.22:

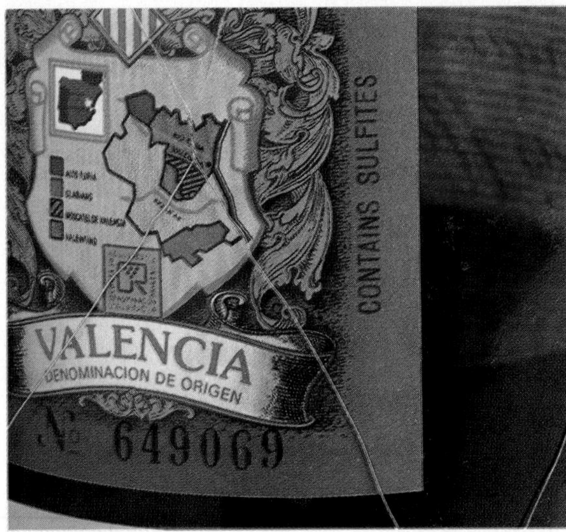

Figure 23.16 The presence of sulfites in foods and beverages is indicated on their labels. (© Richard Megna/Fundamental Photographs)

$$H_2SO_3(aq) \rightleftharpoons H^+(aq) + HSO_3^-(aq) \qquad K_{a1} = 1.7 \times 10^{-2}\ (25°C) \qquad [23.21]$$

$$HSO_3^-(aq) \rightleftharpoons H^+(aq) + SO_3^{2-}(aq) \qquad K_{a2} = 6.4 \times 10^{-8}\ (25°C) \qquad [23.22]$$

Although H_2SO_3 cannot be isolated, salts of SO_3^{2-} (sulfites) and HSO_3^- (hydrogen sulfites or bisulfites) are well known. Small quantities of Na_2SO_3 or $NaHSO_3$ are used as food additives to prevent bacterial spoilage. Because some people are extremely allergic to sulfites, all products with sulfites must now carry a warning disclosing their presence (Figure 23.16).

As we have seen, combustion of sulfur in air produces mainly SO_2, but small amounts of SO_3 are also formed. The reaction produces mainly SO_2 because the activation-energy barrier for further oxidation to SO_3 is very high unless the reaction is catalyzed. Sulfur trioxide is of great commercial importance because it is the anhydride of sulfuric acid. In the manufacture of sulfuric acid, SO_2 is first obtained by burning sulfur. The SO_2 is then oxidized to SO_3 using a catalyst such as V_2O_5 or platinum. The SO_3 is dissolved in H_2SO_4 because it does not dissolve quickly in water. The reaction is shown in Equation 23.23. The $H_2S_2O_7$ formed in this reaction, called pyrosulfuric acid, is then added to water to form H_2SO_4, as shown in Equation 23.24:

$$SO_3(g) + H_2SO_4(l) \longrightarrow H_2S_2O_7(l) \qquad [23.23]$$

$$H_2S_2O_7(l) + H_2O(l) \longrightarrow 2H_2SO_4(l) \qquad [23.24]$$

Commercial sulfuric acid is generally 98 percent H_2SO_4. It is a dense, colorless, oily liquid that boils at 340°C. Sulfuric acid has many useful properties. Most importantly, it is a strong acid, a good dehydrating agent,* and a moderately good oxidizing agent. Its dehydrating ability is demonstrated in Figure 23.17.

* Considerable heat is given off when sulfuric acid is diluted with water. Consequently, dilution must always be done carefully by pouring the acid into water to distribute the heat as uniformly as possible and to avoid spattering of the acid.

Figure 23.17 The reaction between sucrose, $C_{12}H_{22}O_{11}$, and concentrated sulfuric acid. Sucrose is a carbohydrate, containing two H atoms for each O atom. Sulfuric acid, which is an excellent dehydrating agent, removes H_2O from the sucrose to form carbon, the black mass remaining at the end of the reaction. (Kristen Brochman/Fundamental Photographs)

Year after year, the output of sulfuric acid has been the largest of any chemical produced in the United States. About 3.9×10^{10} kg (4.3 million tons) is produced annually in this country. Sulfuric acid is employed in some way in almost all manufacturing. Consequently, its consumption is considered a standard measure of industrial activity. The primary uses for sulfuric acid are shown in Figure 23.18.

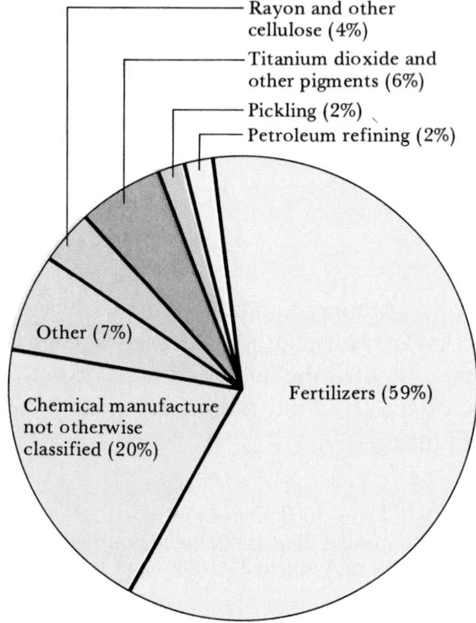

Figure 23.18 Sulfuric acid use in the United States.

Only the first proton in sulfuric acid is completely ionized in aqueous solution. The second proton ionizes only partially:

$$H_2SO_4(aq) \longrightarrow H^+(aq) + HSO_4^-(aq)$$

$$HSO_4^-(aq) \rightleftharpoons H^+(aq) + SO_4^{2-}(aq) \qquad K_a = 1.1 \times 10^{-2}$$

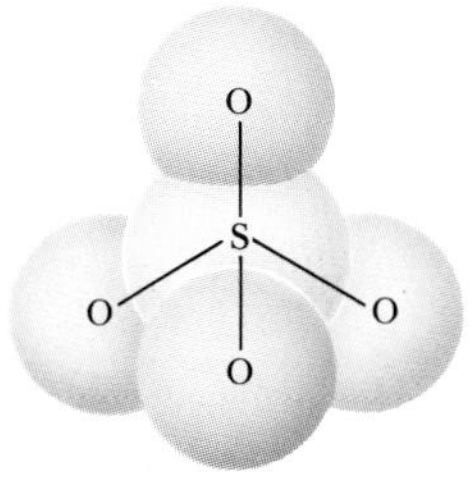

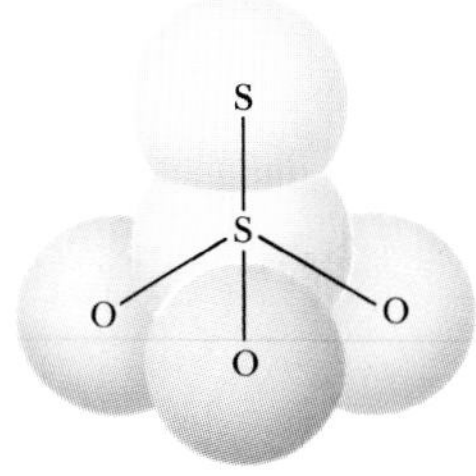

Figure 23.19 Comparison of the structures of the sulfate, SO_4^{2-}, and thiosulfate, $S_2O_3^{2-}$, ions.

Consequently, sulfuric acid forms two series of compounds: sulfates and bisulfates (or hydrogen sulfates). Bisulfate salts are common components of the "dry acids" used for adjusting the pH of swimming pools and hot tubs; they are also components of many toilet bowl cleaners.

Related to the sulfate ion is the thiosulfate ion, $S_2O_3^{2-}$, formed by boiling an alkaline solution of SO_3^{2-} with elemental sulfur:

$$8SO_3^{2-}(aq) + S_8(s) \longrightarrow 8S_2O_3^{2-}(aq) \qquad [23.25]$$

The term *thio* indicates substitution of sulfur for oxygen. The structures of the sulfate and thiosulfate ions are compared in Figure 23.19. When acidified, the thiosulfate ion decomposes to form sulfur and H_2SO_3.

The pentahydrated salt of sodium thiosulfate, $Na_2S_2O_3 \cdot 5H_2O$, known as "hypo," is used in photography. Photographic film consists of a suspension of microcrystals of AgBr in gelatin. When exposed to light, some AgBr decomposes, forming very small grains of silver. When the film is treated with a mild reducing agent (the "developer"), Ag^+ ions in AgBr near the silver grains are reduced, forming an image of black, metallic silver. The film is then treated with sodium thiosulfate solution to remove the remaining, unexposed AgBr. The thiosulfate ion reacts with AgBr to form a soluble silver thiosulfate complex:

$$AgBr(s) + 2S_2O_3^{2-}(aq) \rightleftharpoons Ag(S_2O_3)_2^{3-}(aq) + Br^-(aq) \qquad [23.26]$$

This step in the process is called "fixing." Thiosulfate ion is also used in quantitative analysis as a reducing agent for iodine:

$$2S_2O_3^{2-}(aq) + I_2(s) \longrightarrow 2I^-(aq) + S_4O_6^{2-}(aq) \qquad [23.27]$$

Oxides, Oxyacids, and Oxyanions of Se and Te

Selenium and tellurium form acidic dioxides and trioxides. Selenium dioxide, SeO_2, dissolves in water producing selenous acid, H_2SeO_3, a weak diprotic acid. Tellurium dioxide, TeO_2, is insoluble in water, but tellurite salts such as $NaHTeO_3$ and Na_2TeO_3 form when TeO_2 is dissolved in an aqueous base.

Oxidation of H_2SeO_3 with hydrogen peroxide produces selenic acid, H_2SeO_4, a strong diprotic acid that resembles sulfuric acid:

$$H_2SeO_3(aq) + H_2O_2(aq) \longrightarrow HSeO_4^-(aq) + H_2O(l) + H^+(aq) \qquad [23.28]$$

Telluric acid (also called orthotelluric acid) forms when TeO_2 is treated with various aqueous oxidizing agents. We might expect the formula for this acid to be H_2TeO_4 by analogy with the corresponding acids of sulfur and selenium. However, orthotelluric acid is H_6TeO_6. [The formula can also be written $Te(OH)_6$.] Notice the close similarity to paraperiodic acid, H_5IO_6 (Section 23.2).

The large sizes of Te and I permit them to bond to six surrounding O atoms. Orthotelluric acid is a weak acid, capable of dissociating two protons in aqueous solution (at 25°C, $K_{a1} = 2.4 \times 10^{-8}$, $K_{a2} = 1.0 \times 10^{-11}$).

Sulfides, Selenides, and Tellurides

Sulfur forms compounds by direct combination with many elements. When the element is less electronegative than sulfur, *sulfides*, which contain S^{2-}, form. For example, iron(II) sulfide, FeS, forms by direct combination of iron and sulfur. Many metallic elements are found in the form of sulfide ores, for example, PbS (galena) and HgS (cinnabar). A series of related ores containing the disulfide ion, S_2^{2-} (analogous to the peroxide ion), are known as *pyrites*. Iron pyrite, FeS_2, occurs as golden-yellow cubic crystals (Figure 23.20). Because it has been occasionally mistaken for gold by overeager miners, it is often called "fool's gold."

One of the most important sulfides is hydrogen sulfide, H_2S. This substance is not normally produced by direct union of the elements because it is unstable at elevated temperature and decomposes into the elements. It is normally prepared by action of dilute sulfuric acid on iron(II) sulfide:

$$FeS(s) + 2H^+(aq) \longrightarrow H_2S(aq) + Fe^{2+}(aq) \qquad [23.29]$$

A common laboratory source of H_2S is the reaction of thioacetamide with water:

$$CH_3\overset{\overset{\displaystyle S}{\|}}{C}NH_2(aq) + H^+(aq) + 2H_2O(l) \longrightarrow$$

$$CH_3\overset{\overset{\displaystyle O}{\|}}{C}OH(aq) + NH_4^+(aq) + H_2S(aq) \qquad [23.30]$$

Hydrogen sulfide is often used in the laboratory for qualitative analysis of certain metal ions (Section 17.6).

One of hydrogen sulfide's most readily recognized properties is its odor; H_2S is largely responsible for the offensive odor of rotten eggs. Hydrogen sulfide is actually quite toxic; it has about the same level of toxicity as hydrogen cyanide, the gas that has been used in gas chambers. Fortunately, our noses

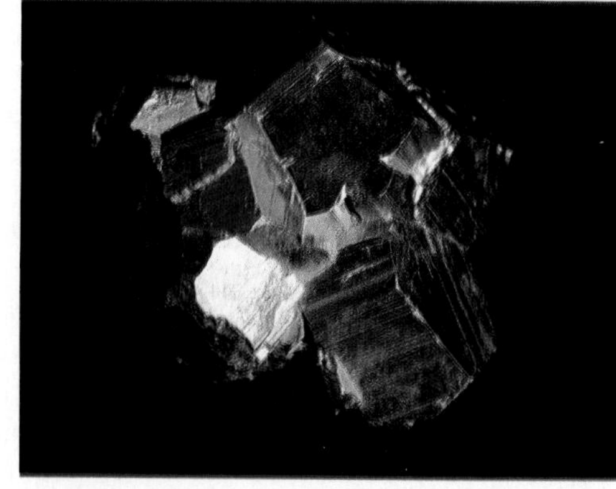

Figure 23.20 Iron pyrite, FeS_2, is also known as fool's gold because its color has fooled people into thinking it was gold. Gold is much more dense and much softer than iron pyrite. (Charles R. Belinky/Photo Researchers)

are able to detect H_2S in extremely low, nontoxic concentrations. Sulfur-containing organic molecules, which are similarly odoriferous, are added to natural gas to give it a detectable odor.

The volatile hydrides H_2Se and H_2Te are similar to H_2S in many respects. Both compounds possess very offensive, lingering odors and are toxic. In aqueous solutions the acid strength increases in the order $H_2S < H_2Se < H_2Te$.

23.4 THE GROUP 5A ELEMENTS

The elements of group 5A are nitrogen, phosphorus, arsenic, antimony, and bismuth. We have already discussed the chemistry of nitrogen (Section 22.5). In our present discussion we will find it convenient to examine the general characteristics of the group, to consider phosphorus, and finally to comment briefly on the heavier elements of the group.

General Characteristics of the Group 5A Elements

7 N
15 P
33 As
51 Sb
83 Bi

The group 5A elements possess the outer-electron configuration ns^2np^3, where n may have values ranging from 2 to 6. A noble-gas configuration results from the addition of three electrons to form the -3 oxidation state. Ionic compounds containing X^{3-} ions are not common, however, except for salts of the more active metals, for example, Na_3N. More commonly, the group 5A element acquires an octet of electrons via covalent bonding. The oxidation number may range from -3 to $+5$, depending on the nature and number of the atoms to which the group 5A element is bonded.

Because of its lower electronegativity, phosphorus is found more frequently in positive oxidation states than is nitrogen. Furthermore, compounds in which phosphorus has the $+5$ oxidation state are not as strongly oxidizing as the corresponding compounds of nitrogen. Conversely, compounds in which phosphorus has a -3 oxidation state are much stronger reducing agents than are corresponding compounds of nitrogen.

Some of the important properties of the atoms of the group 5A elements are listed in Table 23.8. The general pattern that emerges from these data is similar to what we have seen before with other groups; size and metallic character increase as atomic number increases within the group.

The variation in properties among the elements of group 5A is more striking than that seen in groups 6A and 7A. Nitrogen at the one extreme exists as

Table 23.8 Properties of the Atoms of Group 5A Elements

Property	N	P	As	Sb	Bi
Atomic radius (Å)	0.70	1.10	1.20	1.40	1.50
First ionization energy (kJ/mol)	1402	1012	947	834	703
Electron affinity (kJ/mol)	+6.8	−72	−77	−101	−106
Electronegativity	3.0	2.1	2.0	1.9	1.9
X—X single-bond energy (kJ/mol)[a]	163	200	150	120	—
X≡X triple-bond energy (kJ/mol)	941	490	380	295	192

[a] Approximate values only.

a gaseous diatomic molecule; it is clearly nonmetallic in character. At the other extreme, bismuth is a reddish-white, metallic-looking substance that has most of the characteristics of a metal.

The values listed for X—X single-bond energies are not very reliable, because it is difficult to obtain such data from thermochemical experiments. However, there is no doubt about the general trend: a low value for the N—N single bond, an increase at phosphorus, and then a gradual decline to arsenic and antimony. From observations of the group 5A elements in the gas phase (for all except N_2, this requires high temperatures), it is possible to estimate the X≡X triple-bond energy, as listed in Table 23.8. Here we see a trend that is different than that for the X—X single bond. Nitrogen forms a much stronger bond than do the other elements, and there is a steady decline in the triple-bond energy down through the group. These data help us to appreciate why nitrogen alone of the group 5A elements exists as a diatomic molecule in its stable state at 25°C. All the other elements exist in structural forms with single bonds between the atoms.

Occurrence, Isolation, and Properties of Phosphorus

Phosphorus occurs mainly in the form of phosphate minerals. The principal source of phosphorus is phosphate rock, which contains phosphate mainly in the form of $Ca_3(PO_4)_2$. Deposits of phosphate rock occur mostly in Florida, the western United States, North Africa, and parts of the USSR. The element is produced commercially by reduction of phosphate with coke in the presence of SiO_2:

$$2Ca_3(PO_4)_2(s) + 6SiO_2(s) + 10C(s) \xrightarrow{1500^\circ C} P_4(g) + 6CaSiO_3(l) + 10CO(g) \quad [23.31]$$

The phosphorus produced in this fashion is the allotrope known as white phosphorus. This form distills from the reaction mixture as the reaction proceeds.

White phosphorus consists of P_4 tetrahedra, as shown in Figure 23.21. The 60° bond angles in P_4 are unusually small for molecules. There must consequently be much strain in the bonding, a fact that is consistent with the high reactivity of white phosphorus. This allotrope bursts spontaneously into flames if exposed to air. It is a white, waxlike solid that melts at 44.2°C and boils at 280°C. When heated in the absence of air to about 400°C, it is converted to a more stable allotrope known as red phosphorus. This form does not ignite on contact with air. It is also considerably less poisonous than the white form. Both allotropes are shown in Figure 23.22.

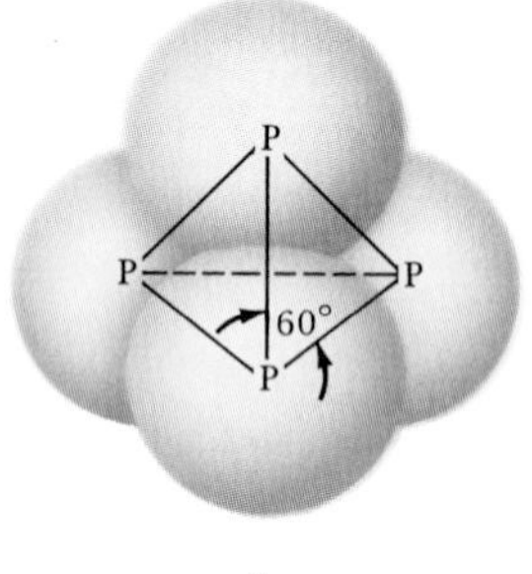

P_4

Figure 23.21 Tetrahedral structure of the P_4 molecule of white phosphorus.

Figure 23.22 White and red allotropes of phosphorus, both stored under water. White phosphorus is very reactive and is normally stored under water to protect it from oxygen. The yellowish cast on the white phosphorus in this photo is due to reactions with air. Red phosphorus is much less reactive than white phosphorus, and it is not necessary to store it under water. (Donald Clegg and Roxy Wilson)

Phosphorus Halides

Phosphorus forms a wide range of compounds with the halogens, the most important of which are the trihalides and pentahalides. Phosphorus trichloride, PCl_3, is commercially the most significant of these compounds and is used to prepare a wide variety of products, including soaps, detergents, plastics, and insecticides. The compounds PCl_3, PBr_3, and PI_3 are made by direct oxidation of elemental phosphorus with the elemental halogen:

$$2P(s) + 3Cl_2(g) \longrightarrow 2PCl_3(l) \qquad [23.32]$$

Because F_2 is such a strong oxidizing agent, however, the reaction of phosphorus with F_2 produces PF_5. Most frequently, PF_3 is prepared from PCl_3 through a halogen-exchange reaction with another fluoride compound, as in the following example:

$$PCl_3(l) + AsF_3(l) \longrightarrow PF_3(g) + AsCl_3(l) \qquad [23.33]$$

The driving force for this reaction is the greater strength of the phosphorus-fluorine bond as compared with the other bond energies (490 kJ/mol for P—F as compared with 406 kJ/mol for As—F, 326 kJ/mol for P—Cl, and 322 kJ/mol for As—Cl).

Both PCl_5 and PBr_5 are normally prepared by reaction of the trihalides with halogens, as in the following example:

$$PCl_3(l) + Cl_2(g) \longrightarrow PCl_5(s) \qquad [23.34]$$

Preparation of PI_5 (first reported only in 1978) is achieved by halogen exchange of PCl_5 with I^- in an appropriate solvent.

In the trihalides, the central phosphorus atom has three bonding electron pairs and one nonbonding pair in its valence shell, giving rise to a pyramidal structure (Section 9.2). In the pentahalides, the phosphorus is surrounded by ten valence-shell electrons. Therefore, we would expect these molecules to have trigonal-bipyramidal structures. In fact, PF_5 has this structure in both the vapor and solid phases. In contrast, PCl_5 has this structure in the vapor, but the solid consists of $[PCl_4]^+$ tetrahedra and $[PCl_6]^-$ octahedra. Solid PBr_5, which dissociates into PBr_3 and Br_2 when vaporized, consists of $[PBr_4]^+$ tetrahedra and Br^- ions.

The phosphorus halides hydrolyze on contact with water. The reactions occur readily, and most of the phosphorus halides fume in air as a result of reaction with water vapor. In the presence of excess water, the products are the corresponding phosphorus oxyacid and hydrogen halide, as in the following examples:

$$PF_3(g) + 3H_2O(l) \longrightarrow H_3PO_3(aq) + 3HF(aq) \qquad [23.35]$$

$$PCl_5(l) + 4H_2O(l) \longrightarrow H_3PO_4(aq) + 5HCl(aq) \qquad [23.36]$$

Oxy Compounds of Phosphorus

Probably the most significant compounds of phosphorus are those in which the element is combined in some way with oxygen. Phosphorus(III) oxide, P_4O_6, is obtained by allowing white phosphorus to oxidize in a limited supply of oxygen. When oxidation takes place in the presence of excess oxygen, phosphorus(V) oxide, P_4O_{10}, forms. This compound is also readily formed by oxidation of P_4O_6. Phosphorus(III) oxide is often called phosphorus trioxide after its empirical formula; similarly, phosphorus(V) oxide is often called phosphorus pentoxide and written as the empirical formula P_2O_5. These two oxides represent the two most common oxidation states for phosphorus, +3 and +5. The structural relationship between P_4O_6 and P_4O_{10} is shown in Figure 23.23. Notice the resemblance these molecules have to the P_4 molecule, Figure 23.19; all three substances have a P_4 core.

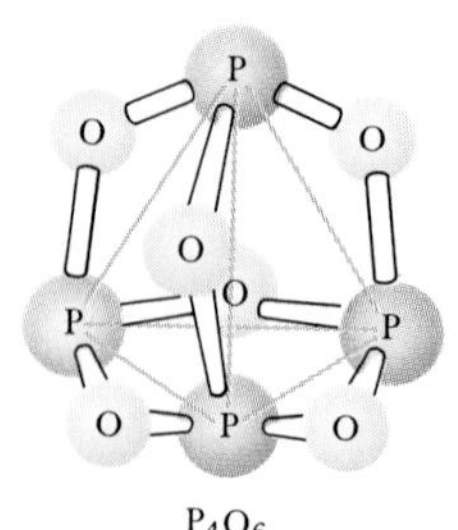

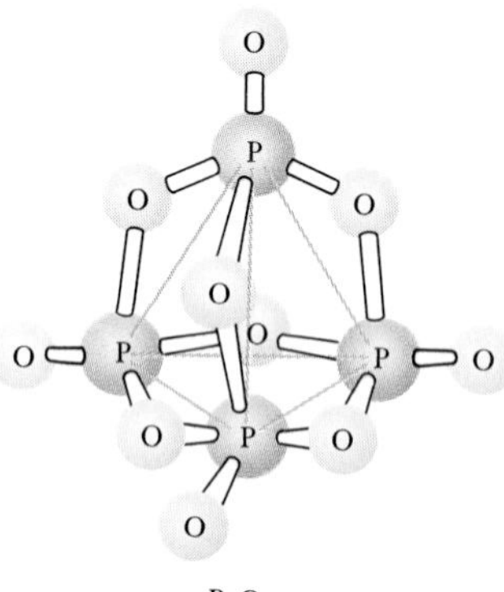

Figure 23.23 Structures of P_4O_6 and P_4O_{10}.

Figure 23.24 Structures of H_3PO_4 and H_3PO_3.

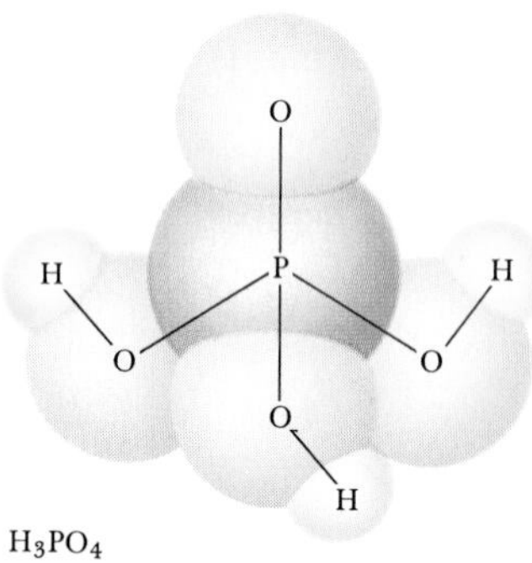

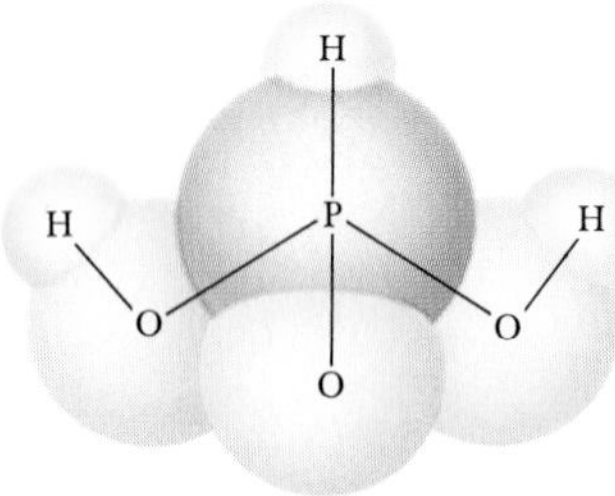

SAMPLE EXERCISE 23.6

The reactive chemicals on the tip of a "strike anywhere" match are usually P_4S_3 and an oxidizing agent such as $KClO_3$. When the match is struck on a rough surface, the heat generated by the friction ignites the P_4S_3, and the oxidizing agent brings about rapid combustion. The products of the combustion are P_4O_{10} and SO_2. Calculate the standard enthalpy for the combustion of P_4S_3 in air, given the following standard enthalpies of formation: P_4S_3 (−154.4 kJ/mol); P_4O_{10} (−2940 kJ/mol); SO_2 (−296.9 kJ/mol).

Solution: The chemical equation for the combustion is

$$P_4S_3(s) + 8O_2(g) \longrightarrow P_4O_{10}(s) + 3SO_2(g)$$

Recalling that the standard enthalpy of formation of any element in its standard states is zero, we have $\Delta H_f^\circ(O_2) = 0$. Thus we can write

$$\begin{aligned}\Delta H^\circ &= \Delta H_f^\circ(P_4O_{10}) + 3\,\Delta H_f^\circ(SO_2) - \Delta H_f^\circ(P_4S_3)\\ &= -2940\text{ kJ} + 3(-296.9)\text{ kJ} - (-154.4\text{ kJ})\\ &= -3616\text{ kJ}\end{aligned}$$

The reaction is strongly exothermic, making it evident why P_4S_3 is used on match tips.

PRACTICE EXERCISE

Write the balanced equation for the reaction of P_4O_{10} with water and calculate ΔH° for this reaction using data from Appendix D.
Answer: $P_4O_{10}(s) + 6H_2O(l) \longrightarrow 4H_3PO_4(aq)$; −498.0 kJ

Phosphorus(V) oxide is the anhydride of phosphoric acid, H_3PO_4, a weak triprotic acid. In fact, P_4O_{10} has a very high affinity for water and is consequently used as a drying agent. Phosphorus(III) oxide is the anhydride of phosphorous acid, H_3PO_3, a weak diprotic acid.* The structures of H_3PO_4 and H_3PO_3 are shown in Figure 23.24. The hydrogen atom that is attached directly to phosphorus in H_3PO_3 is not acidic.

One characteristic of phosphoric and phosphorus acids is their tendency to undergo condensation reactions when heated. A *condensation reaction* is one in which two or more molecules combine to form a larger molecule by eliminating a small molecule, for example, H_2O. The reaction in which two H_3PO_4 molecules are joined by the elimination of one H_2O molecule to form $H_4P_2O_7$ is represented in Equation 23.37.

* Note that the element phosphor*us* (FOS·for·us) has an *-us* suffix, whereas phosphor*ous* (fos·FOR·us) acid has an *-ous* suffix.

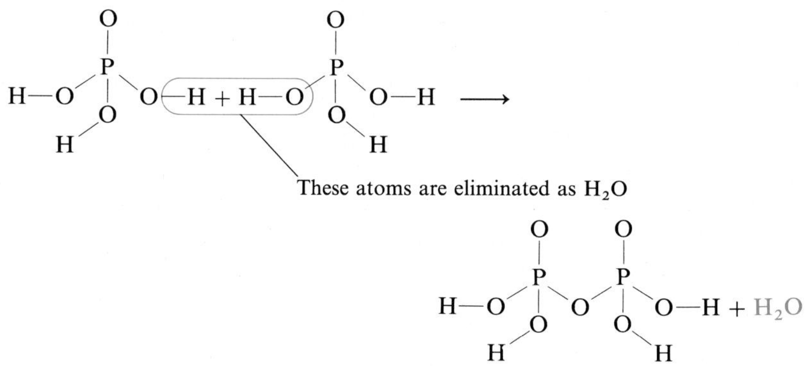

These atoms are eliminated as H_2O

$$+ H_2O \quad [23.37]$$

Further condensation produces phosphates having an empirical formula of HPO_3.

$$nH_3PO_4 \longrightarrow (HPO_3)_n + nH_2O \qquad [23.38]$$

Two phosphates having this empirical formula, one cyclic and the other polymeric, are shown in Figure 23.25. The three acids H_3PO_4, $H_4P_2O_7$, and $(HPO_3)_n$ all contain phosphorus in its $+5$ oxidation state, and all are therefore called phosphoric acids. To differentiate them, the prefixes *ortho-*, *pyro-*, and *meta-* are used: H_3PO_4 is orthophosphoric acid, $H_4P_2O_7$ is pyrophosphoric acid, and HPO_3 is metaphosphoric acid.

Phosphoric acid and its salts find their most important uses in detergents and fertilizers. The phosphates in detergents are often in the form of sodium tripolyphosphate, $Na_5P_3O_{10}$ (see Section 25.2). A typical detergent formulation contains 47 percent phosphate, 16 percent bleaches, perfumes, and abrasives, and 37 percent linear alkylsulfonate (LAS) surfactant (shown as follows):

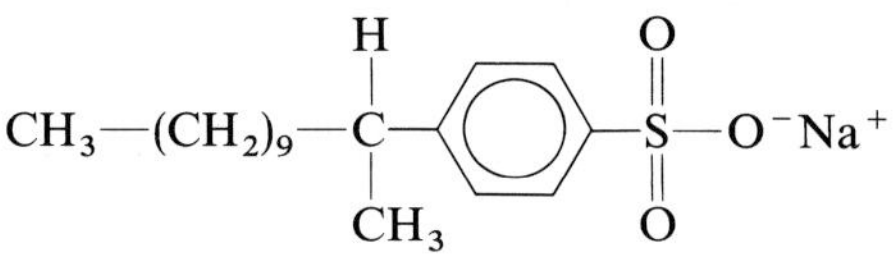

$$CH_3{-}(CH_2)_9{-}CH(CH_3){-}C_6H_4{-}SO_2{-}O^-Na^+$$

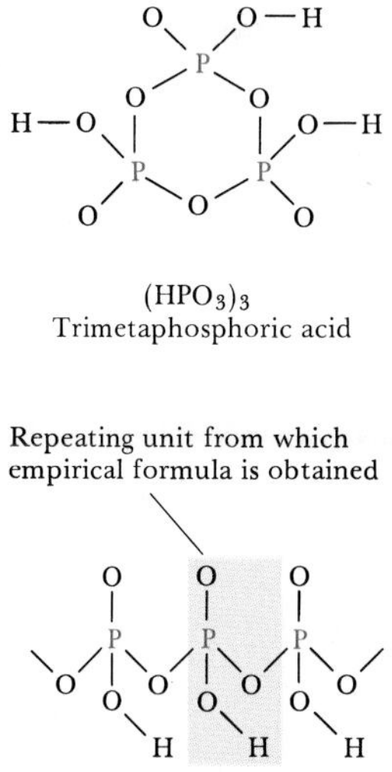

(We have used the notation for the benzene ring described in Section 9.5.) The detergent action of such molecules has been described in Section 13.6. The phosphate ions form bonds with metal ions that contribute to the hardness of water. This keeps the metal ions from interfering with the action of the surfactant. The phosphates also keep the pH above 7 and thus prevent the surfactant molecules from becoming protonated (gaining an H^+ ion).

Most mined phosphate rock is converted to fertilizers. The $Ca_3(PO_4)_2$ in phosphate rock is insoluble ($K_{sp} = 2.0 \times 10^{-29}$). It is converted to a soluble form for use in fertilizers by treating the concentrated phosphate rock with sulfuric or phosphoric acid:

$$Ca_3(PO_4)_2(s) + 4H^+(aq) + 3SO_4^{2-}(aq) \longrightarrow 3CaSO_4(s) + 2H_2PO_4^-(aq) \qquad [23.39]$$

$$Ca_3(PO_4)_2(s) + 4H^+(aq) \longrightarrow 3Ca^{2+}(aq) + 2H_2PO_4^-(aq) \qquad [23.40]$$

Figure 23.25 Structures of trimetaphosphoric acid and polymetaphosphoric acid.

$(HPO_3)_3$
Trimetaphosphoric acid

Repeating unit from which empirical formula is obtained

$(HPO_3)_n$
Polymetaphosphoric acid

The mixture formed when ground phosphate rock is treated with sulfuric acid and then dried and pulverized is known as superphosphate. The $CaSO_4$ formed in this process is of little use in soil except when deficiencies in calcium or sulfur exist. It also dilutes the phosphorus, which is the nutrient of interest. If the phosphate rock is treated with phosphoric acid, the product contains no $CaSO_4$ and has a higher percentage of phosphorus. This product is known as triple superphosphate. Although the solubility of $Ca(H_2PO_4)_2$ allows it to be assimilated by plants, it also allows it to be washed from the soil and into water bodies, thereby contributing to water pollution.

Phosphorus compounds are important in biological systems. The element occurs, for example, in phosphate groups in RNA and DNA, the molecules responsible for control of protein biosynthesis and transmission of genetic information (Section 27.6). It also occurs in adenosine triphosphate (ATP), which stores energy within biological cells:

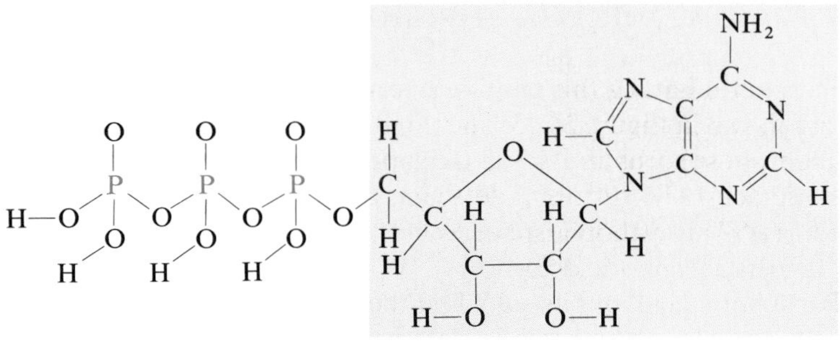

Adenosine

The P—O—P bond of the end phosphate group is broken by hydrolysis with water, forming adenosine diphosphate (ADP). This reaction produces 33 kJ of energy:

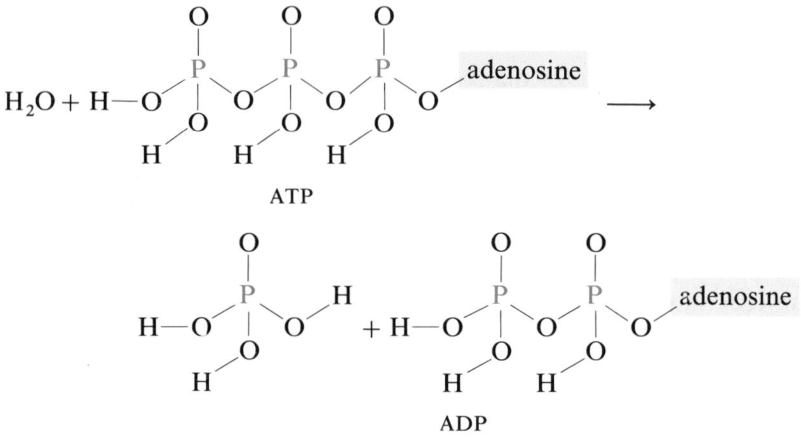

H_2O + ATP (adenosine) $\longrightarrow$ + ADP (adenosine) [23.41]

This energy is used to perform the mechanical work in muscle contraction and in many other biochemical reactions (see Section 19.7 and Figure 19.9).

Arsenic, Antimony, and Bismuth

Arsenic, antimony, and bismuth occur in nature in the form of sulfide minerals, such as As_2S_3, Sb_2S_3, and Bi_2S_3. In addition, the elements are found as minor components in ores of various metals, such as Cu, Pb, Ag, and Hg. Arsenic and antimony exhibit allotropy similar to that of phosphorus. Both elements can be prepared as soft, yellow, nonmetallic solids by quickly cooling the high-temperature vapor of the element. In this allotropic form, analogous to the white allotrope of phosphorus, the elements are present as As_4 or Sb_4 tetrahedra. Heating or the action of light converts the substances into the gray, more metallic forms containing sheets of atoms. In its common form, bismuth has a reddish-white, rather metallic appearance. The element forms alloys with many metallic elements. Alloys of lead, bismuth, and tin are used in the construction of plugs in fire-sprinkler systems. Water is discharged when the fusible metal plug melts.

Arsenic and antimony resemble phosphorus in much of their chemical behavior. For example, the oxy compounds of these two elements are rather similar to those of phosphorus, except that the higher oxidation state is not so easily attained. Thus, the product of burning arsenic in oxygen is As_4O_6, not As_4O_{10}. Arsenic in the higher oxidation state can be obtained by oxidation of As_4O_6 with a strong oxidizing agent, such as nitric acid:

$$As_4O_6(s) + 4NO_3^-(aq) + 6H_2O(l) + 4H^+(aq) \longrightarrow 4H_3AsO_4(aq) + 4HNO_2(aq) \quad [23.42]$$

In terms of the criteria discussed in Section 7.5, bismuth is considered a metal rather than a nonmetal. Bismuth usually appears in the $+3$ oxidation state; there is little tendency to attain the higher $+5$ oxidation state that is so common for phosphorus. The common oxide of bismuth is Bi_2O_3. This substance is insoluble in water or basic solution but is soluble in acidic solution. It is therefore classified as a basic anhydride. As we have seen, the oxides of metals characteristically behave as basic anhydrides.

23.5 THE GROUP 4A ELEMENTS

The elements of group 4A are carbon, silicon, germanium, tin, and lead. The general trend from nonmetallic to metallic as we go down a family is strikingly evident in group 4A. Carbon is strictly nonmetallic; silicon is essentially nonmetallic, although it does exhibit some characteristics of a semimetal, particularly in its electrical and physical properties; germanium is a semimetal; tin and lead are both metallic. The chemistry of carbon was discussed in Section 22.5. In the present discussion we will consider a few general characteristics of group 4A and then look more thoroughly at silicon.

General Characteristics of the Group 4A Elements

Some properties of the group 4A elements are given in Table 23.9. The elements possess the outer-electron configuration ns^2np^2. The electronegativities of the elements are generally low; carbides that formally contain C^{4-} ions are observed only in the case of a very few compounds of carbon with very active metals. Formation of $4+$ ions by electron loss is not observed for any of these elements; the ionization energies are too high. However, the $2+$ state is found

Table 23.9 Some Properties of the Group 4A Elements

Property	C	Si	Ge	Sn	Pb
Atomic radius (Å)	0.77	1.18	1.22	1.41	1.46
First ionization energy (kJ/mol)	1086	786	761	708	715
Electronegativity	2.5	1.8	1.8	1.8	1.9
X—X single-bond energy (kJ/mol)	348	226	188	151	—

6 C
14 Si
32 Ge
50 Sn
82 Pb

in the chemistry of germanium, tin, and lead; it is the principal oxidation state for lead. The vast majority of the compounds of the group 4A elements are covalently bonded. Carbon forms a maximum of four bonds. The other members of the family are able to form higher coordination numbers because of the availability of *d* orbitals for bonding.

Carbon differs from the other group 4A elements in its pronounced ability to form multiple bonds both with itself and with other nonmetals, especially N, O, and S. The origin of this behavior was considered earlier, in Section 22.1.

Table 23.9 shows that the strength of a bond between two atoms of a given element decreases as we go down group 4A. Carbon-carbon bonds are quite strong. As a consequence, carbon has a striking ability to form compounds in which carbon atoms are bonded to each other. This property, called **catenation**, permits the formation of extended chains and rings of carbon atoms and accounts for the large number of organic compounds that exist. Catenation is also exhibited by other elements, especially ones in the vicinity of carbon in the periodic table, such as boron, nitrogen, phosphorus, oxygen, sulfur, silicon, and germanium. However, such self-linkage is far less important in the chemistries of these other elements. For example, the Si—Si bond strength (226 kJ/mol) is far smaller than the Si—O bond strength (368 kJ/mol). As a result, the chemistry of silicon is dominated by the formation of Si—O bonds, and Si—Si bonds play a rather minor role.

Occurrence and Preparation of Silicon

Silicon is the second most abundant element, after oxygen, in the earth's crust. It occurs in SiO_2 and in an enormous variety of silicate minerals. The element is obtained by the reduction of molten silicon dioxide with carbon at high temperature:

$$SiO_2(l) + 2C(s) \longrightarrow Si(l) + 2CO(g) \quad [23.43]$$

Elemental silicon has a diamond-type structure [see Figure 11.40(*a*)]; a graphitelike allotrope is not known, presumably because of the weakness of π bonds between silicon atoms. Crystalline silicon is a gray, metallic-looking solid that melts at 1410°C (Figure 23.26). The element is a semiconductor (Section 24.6) and is thus used in making transistors and solar cells. To be used as a semiconductor, it must be extremely pure. One method of purification is to treat the element with Cl_2 to form $SiCl_4$. The $SiCl_4$ is a volatile liquid that is purified by fractional distillation and then reduced by H_2 to elemental silicon:

$$SiCl_4(g) + 2H_2(g) \longrightarrow Si(s) + 4HCl(g) \quad [23.44]$$

Figure 23.26 Elemental silicon. To prepare electronic devices, silicon (powder, left) is melted, drawn into a single crystal, and purified by zone refining. Wafers of silicon, cut from the crystal, are subsequently treated by a series of elegant techniques to produce various electronic devices. (Courtesy of Texas Instruments)

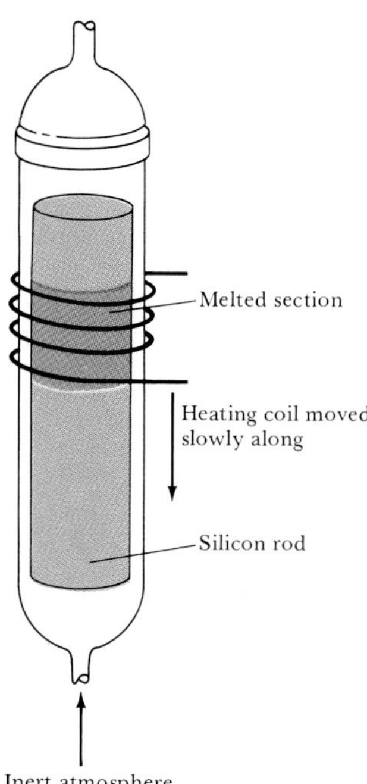

Figure 23.27 Zone-refining apparatus.

The element can be further purified by the process of zone refining. In the zone-refining process, a heated coil is passed slowly along a silicon rod, as shown in Figure 23.27. A narrow band of the element is thereby melted. As the molten area is swept slowly along the length of the rod, the impurities concentrate in the molten region, following it to the end of the rod. The end in which the impurities are collected is cut off and recycled through the purification process starting with the formation of $SiCl_4$. The purified top portion of the rod is retained for manufacture of electronic devices.

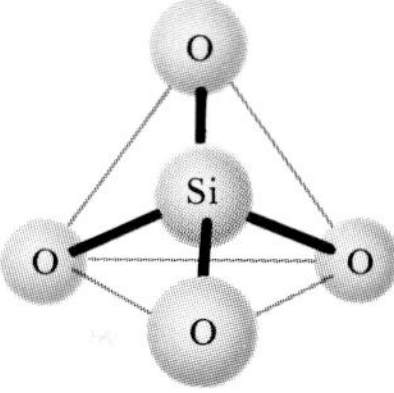

Figure 23.28 Structure of the SiO_4 tetrahedron of the SiO_4^{4-} ion. This ion is found in several minerals, such as zircon, $ZrSiO_4$.

Silicates

It is estimated that over 90 percent of the earth's crust consists of **silicates**, if SiO_2 is included. The basic structural unit of these silicates consists of a silicon atom surrounded in a tetrahedral fashion by four oxygens, as shown in Figure 23.28. Because the silicon has an oxidation state of $+4$ and oxygen -2, a simple ion of this composition has a charge of $+4 + 4(-2) = -4$. The simple SiO_4^{4-} ion, which is known as the orthosilicate ion, is found in very few silicate minerals. Usually, silicate tetrahedra share oxygen atoms to build up more complex structures containing Si—O—Si linkages. When two tetrahedra share a single oxygen, the $Si_2O_7^{6-}$ ion shown in Figure 23.29 results.

It is possible to form a chain of SiO_4 tetrahedra by sharing oxygens at two corners of each tetrahedron, as shown in Figure 23.30(*a*). The simplest formula associated with this single-chain silicate anion is SiO_3^{2-}. This anion occurs, for example, in the mineral enstatite, $MgSiO_3$, which consists of silicate chains (hence the SiO_3^{2-} portion of the mineral's simplest formula) with Mg^{2+} ions between strands to balance charge.

Figure 23.29 Geometric structure of the $Si_2O_7^{6-}$ ion, which is formed by the sharing of an oxygen atom by two silicon atoms. This ion occurs in several minerals, such as hardystonite, $Ca_2Zn(Si_2O_7)$.

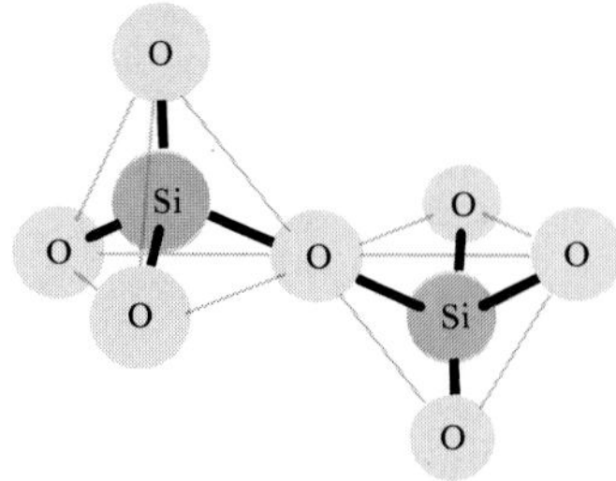

(a) Single-strand chain:

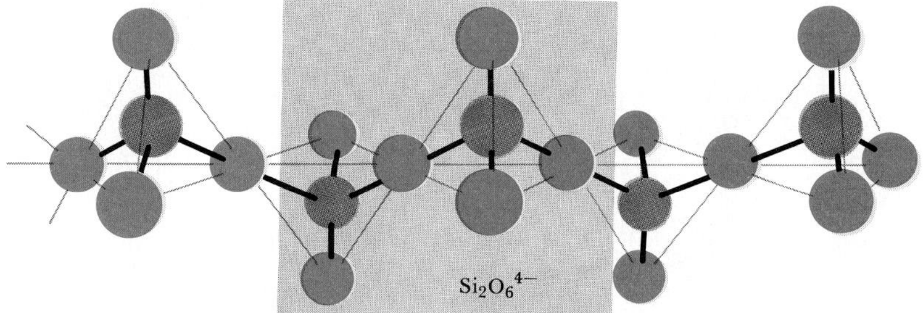

(b) Double-strand chain:

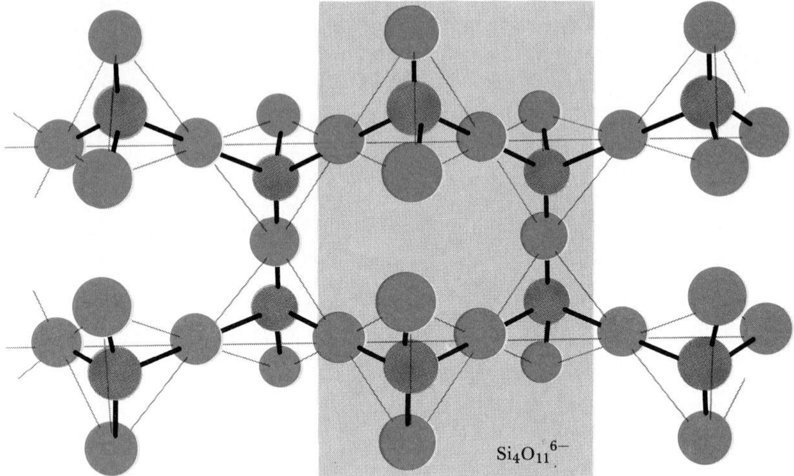

(c) Sheet (or layer) structure:

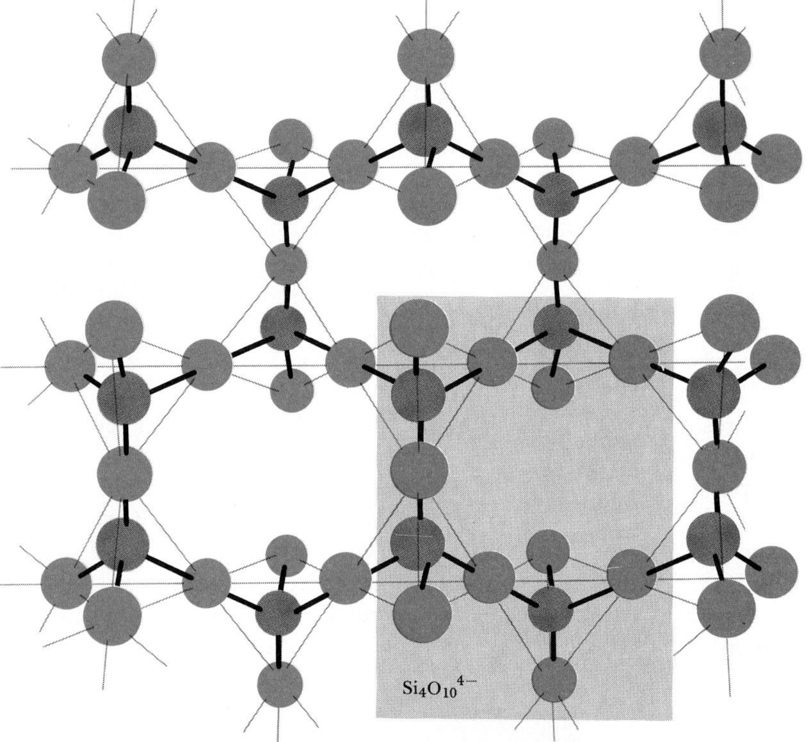

Figure 23.30 Schematic representations of chain and sheet silicates formed by linking together silicate tetrahedra: (*a*) single-strand silicate chain, which has an empirical formula of SiO_3^{2-}, as in the mineral enstatite, $MgSiO_3$; (*b*) double-strand silicate chain, which has an empirical formula of $Si_4O_{11}^{6-}$, as in the mineral tremolite, $Ca_2Mg_5(Si_4O_{11})_2(OH)_2$; (*c*) sheet silicate, which has an empirical formula $Si_2O_5^{2-}$, as in the mineral talc, $Mg_3(Si_2O_5)_2(OH)_2$. Note that in parts (*a*) and (*c*) the repeating unit is double the empirical formula.

A CLOSER LOOK: The Asbestos Minerals

Asbestos is a general term applied to a group of fibrous silicate minerals. These minerals possess chainlike arrangements of the silicate tetrahedra, or sheet structures in which the sheets are formed into rolls. The result is that the minerals have a fibrous character, as shown in Figure 23.31. Asbestos has been widely used as thermal insulation material, especially in high-temperature applications where the high chemical stability of the silicate structure serves well. Until last decade, asbestos was extensively used as a thermal insulator in buildings. However, in recent years there has been a growing awareness that certain forms of asbestos pose a health hazard. Tiny fibers readily penetrate the tissues of the lungs and digestive tract, and they remain there over a long period of time. Eventually, lesions, including cancer, may result.

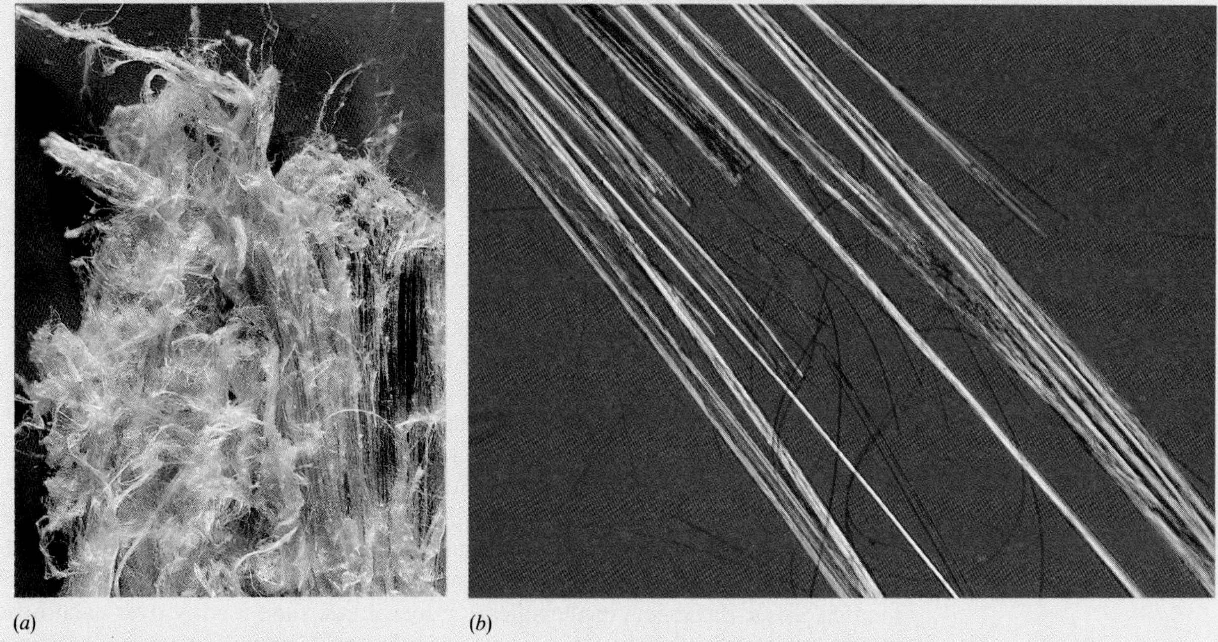

(*a*) (*b*)

Figure 23.31 (*a*) Serpentine asbestos; note the fibrous character of this mineral. (*b*) Fibers of amosite, an asbestos mineral, viewed under a polarizing microscope. (Bureau of Mines, U.S. Department of the Interior; Don Thomson/Science Photo Library/Photo Researchers)

The other structures shown in Figure 23.30 are also found in silicates. For example, it is possible for each tetrahedron to share three corners, giving rise to sheet structures. Such an arrangement results in a simplest formula $Si_2O_5^{2-}$ as in talc, $Mg_3(Si_2O_5)_2(OH)_2$. When all four oxygens of each SiO_4 unit have Si—O—Si linkages, the structure extends in all three dimensions in space, forming quartz, SiO_2. Because the structure is locked together in a three-dimensional array much like diamond, quartz is harder than the strand or sheet-type silicates.

Aluminosilicates

In many silicate minerals, Si^{4+} ions are replaced by Al^{3+} ions within the silicate tetrahedra. This replacement produces **aluminosilicates**. In order to main-

Figure 23.32 Piece of mica, showing its cleavage into thin sheets.

tain charge balance, an extra cation, such as K^+, must accompany each of these substitutions. Muscovite, $KAl_2(AlSi_3O_{10})(OH)_2$,* a mica mineral, is an aluminosilicate.

Replacement of a quarter of the silicon atoms in a sheet silicate, $Si_4O_{10}{}^{4-}$, with aluminum produces the $AlSi_3O_{10}{}^{5-}$ sheets found in muscovite. In both the silicate-sheet and aluminosilicate-sheet minerals, cations are located between sheets to balance the charge. The electrostatic attraction between these cations and the charges in the sheets is greater for the aluminosilicate than for the corresponding silicate, because there are higher charges in the aluminosilicates. Thus, the sheets in the silicate mineral talc slide readily over each other, but the sheets in the aluminosilicate mica do not. Nevertheless, mica does cleave readily into sheets, as illustrated in Figure 23.32.

When aluminum replaces up to half of the silicon atoms in SiO_2, the feldspar minerals result. Orthoclase, $KAlSi_3O_8$, is a common feldspar mineral. The cations that compensate the extra negative charge accompanying this replacement are usually Na^+, K^+, or Ca^{2+} ions. The feldspars are the most abundant rock-forming silicates, comprising about 50 percent of the minerals in the earth's crust.

SAMPLE EXERCISE 23.7

The mineral anorthite is a feldspar mineral formed by replacing half of the silicon atoms in SiO_2 with aluminum and maintaining charge balance with Ca^{2+} ions. What is the simplest formula for this mineral?

Solution: If we write SiO_2 as Si_2O_4, the replacement of half of the Si^{4+} with Al^{3+} produces $AlSiO_4{}^-$. One Ca^{2+} therefore requires two $AlSiO_4{}^-$ units to maintain charge balance, and the empirical formula of the mineral is $CaAl_2Si_2O_8$.

* Aluminum is found in this mineral in two different environments. The first two Al^{3+} ions are located between the aluminosilicate sheets. The aluminum that is shown in parentheses is located within the sheets. It has replaced a silicon and is therefore located in an AlO_4 tetrahedron.

PRACTICE EXERCISE

The mineral albite may be considered to form by replacement of one-fourth of the Si atoms of SiO_2 with Al and maintaining charge balance with Na^+ ions. What is the chemical formula of this mineral? ***Answer:*** $NaAlSi_3O_8$

The **clay minerals** are hydrated aluminosilicates having sheet-type structures. An example is kaolinite, $Al_2Si_2O_5(OH)_4$. These minerals have small particle size and correspondingly large surface areas. They have the ability to adsorb cations on their surfaces. Often the metal ions displace H^+ ions from the OH groups on the surfaces of the clay particle:

$$M^+(aq) + H{-}O{-}clay \rightleftharpoons H^+(aq) + M{-}O{-}clay \quad [23.45]$$

This situation gives rise to pH-dependent equilibria. Notice that the higher the concentration of $H^+(aq)$, the more the equilibrium is shifted to the left. If the soil is basic, the equilibrium lies to the right, and $M^+(aq)$ is not available to plants. Thus, the pH of the soil plays an important role in determining the soil's fertility (that is, its ability to supply plants with essential nutrients).

Glass

Quartz melts at approximately 1600°C, forming a tacky liquid. In the course of melting, many silicon-oxygen bonds are broken. When the liquid is rapidly cooled, silicon-oxygen bonds are re-formed before the atoms are able to arrange themselves in a regular fashion. An amorphous solid, known as quartz glass or silica glass, results (see Figure 11.23). Many different substances can be added to SiO_2 to cause it to melt at a lower temperature. The common **glass** used in windows and bottles is known as soda-lime glass. It contains CaO and Na_2O in addition to SiO_2 from sand. The CaO and Na_2O are produced by heating two inexpensive chemicals: limestone, $CaCO_3$, and soda ash, Na_2CO_3. These carbonates decompose at elevated temperatures:

$$CaCO_3(s) \longrightarrow CaO(s) + CO_2(g) \quad [23.46]$$

$$Na_2CO_3(s) \longrightarrow Na_2O(s) + CO_2(g) \quad [23.47]$$

Other substances can be added to soda-lime glass to produce color or to change the properties of the glass in various ways. For example, addition of CoO produces the deep blue color of "cobalt glass." Replacement of Na_2O by K_2O results in a harder glass that has a higher melting point. Replacement of CaO by PbO results in a denser glass with a higher refractive index. Addition of nonmetal oxides, such as B_2O_3 and P_2O_5, which form network structures related to the silicates, also causes a change in properties of the glass. Addition of B_2O_3 creates a glass with a higher melting point and a greater ability to withstand temperature changes. Such glasses, sold commercially under trade names such as Pyrex and Kimex, are used where resistance to thermal shock is important, for example, in laboratory glassware or coffee makers. Table 23.10 lists the formulations and properties of several representative types of glass.

Photochromic glass contains a dispersion of AgCl or AgBr. The glass darkens when exposed to sunlight because the silver halide decomposes in light to form silver and halogen atoms. The finely divided silver is black. The silver and halogen atoms are kept in close proximity by the glass matrix and reform the clear AgCl or AgBr in the dark.

Table 23.10 Compositions, Properties, and Uses of Various Types of Glass

Type of glass	Composition by weight	Properties and uses
Soda-lime	12% Na_2O, 12% CaO, 76% SiO_2	Window glass, bottles
Aluminosilicate	5% B_2O_3, 10% MgO, 10% CaO, 20% Al_2O_3, 55% SiO_2	High melting—used in cooking ware
Lead alkali	10% Na_2O, 20% PbO, 70% SiO_2	High refractive index—used in lenses, decorative glass
Borosilicate	5% Na_2O, 3% CaO, 16% B_2O_3, 76% SiO_2	Low coefficient of thermal expansion—used in laboratory ware, cooking utensils
Bioglass	24% Na_2O, 24% CaO, 6% P_2O_5, 46% SiO_2	Compatible with bone—used as coating on surgical implants

Silicones

Silicones consist of O—Si—O chains in which the remaining bonding positions on each silicon are occupied by organic groups such as $-CH_3$:

$$\cdots - O - Si(CH_3)_2 - O - Si(CH_3)_2 - O - Si(CH_3)_2 - O - \cdots$$

Depending on the length of the chain and the degree of crosslinking between chains, silicones can be either oils or rubberlike materials. Silicones are nontoxic and have good stability toward heat, light, oxygen, and water. They are used commercially in a wide variety of products, including lubricants, car polishes, sealants, and gaskets. They are also used for waterproofing fabrics. When applied to a fabric, the oxygen atoms form hydrogen bonds with the molecules on the surface of the fabric. The hydrophobic (water-repelling) organic groups of the silicone are then left pointing away from the surface as a barrier.

23.6 BORON

5 B
13 Al
31 Ga
49 In
81 Tl

At this point there is only one additional element to consider in our survey of nonmetallic elements: boron. Boron is the only element of group 3A that can be considered nonmetallic. The element has an extended network structure. The melting point of boron, 2300°C, is intermediate between that of carbon, 3550°C, and that of silicon, 1410°C. The electronic configuration of boron is $[He]2s^22p^1$.

Salts of the borohydride ion, BH_4^-, are widely used as reducing agents. This ion is isoelectronic to CH_4 and NH_4^+. The lower charge of the central atom in BH_4^- means that the hydrogens of the BH_4^- are "hydridic"; that is, they carry a partial negative charge. Thus, it is not surprising that borohydrides are good reducing agents, as illustrated by the reaction in Equation 23.48:

$$3BH_4^-(aq) + 4IO_3^-(aq) \longrightarrow 4I^-(aq) + 3H_2BO_3^-(aq) + 3H_2O(l) \quad [23.48]$$

The only important oxide of boron is boric oxide, B_2O_3. This substance is the anhydride of boric acid, which we may write as H_3BO_3 or $B(OH)_3$. Boric

CHEMISTRY AT WORK: Optical Fibers

Optical fibers (Figure 23.33) represent a relatively new, high-technology application of glassmaking. An optical fiber is a glass thread that conducts light, just as a copper wire conducts electrons. Optical fibers transmit information, using the light waves that pass through the fiber. Optical fibers are capable of carrying much more information for a given cross section of fiber than can be transmitted via the conventional coaxial cable.

The key to long-distance transmission of signals via optical fibers is glass purity. Impurity ions, such as Fe^{2+}, cause absorption of the light waves, with resulting signal loss. Today it is possible to obtain optical fibers that undergo a loss of only about 1 percent of signal over a distance of 1 km. To achieve that level of performance, impurities must be reduced to the level of 1 ppb. This ultrahigh purity is achieved by distilling high-purity liquid glass to separate it from any remaining impurities.

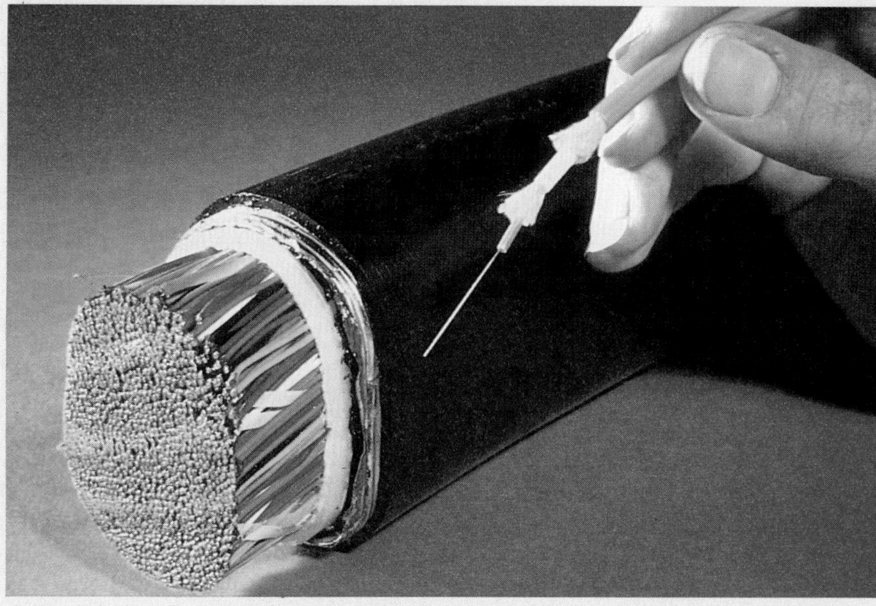

Figure 23.33 The optical fiber on the right has the same information-carrying capacity as the much larger copper cable on the left. (Corning, Inc.)

acid is so weak an acid that solutions of H_3BO_3 are used as an eyewash. Upon warming to 100°C, orthoboric acid loses water by a condensation reaction similar to that described for phosphorus in Section 23.4. The product of the reaction, as shown in Equation 23.49, is metaboric acid, a polymeric substance of formula HBO_2:

$$H_3BO_3(s) \longrightarrow HBO_2(s) + H_2O(g) \qquad [23.49]$$

Heating of metaboric acid results in still more loss of water:

$$4HBO_2(s) \longrightarrow H_2B_4O_7(s) + H_2O(l) \qquad [23.50]$$

$$H_2B_4O_7(s) \longrightarrow 2B_2O_3(s) + H_2O(g) \qquad [23.51]$$

The diprotic acid $H_2B_4O_7$ is called tetraboric acid. The sodium salt, $Na_2B_4O_7 \cdot 10H_2O$, called borax, occurs in dry lake deposits in California and can also be readily prepared from other borate minerals. Solutions of borax are alkaline; the substance is used in various laundry and cleaning products.

FOR REVIEW

SUMMARY

The noble-gas elements exhibit a very limited chemical behavior because of the exceptional stability of their electronic configuration. The xenon fluorides and oxides and KrF_2 are the best established examples of chemical reactivity among these elements.

The halogens occur as diatomic molecules. These elements possess the highest electronegativities of the elements in each row of the periodic table. All except fluorine exhibit oxidation states varying from -1 to $+7$. Fluorine, being the most electronegative element, is restricted to the oxidation states 0 and -1. The tendency to form the -1 oxidation state from the free element (that is, the oxidizing power of the element) decreases with increasing atomic number in the family. The halogens form compounds with one another, called interhalogens. In the higher interhalogens—XX'_n—the element X may be Cl, Br, or I, and X′ is nearly always F; *n* may have the value 3, 5, or 7.

The group 6A elements range from the very abundant and strongly nonmetallic oxygen to the rare and rather metallic tellurium. Sulfur occurs widely in the form of sulfide ores and in elemental sulfur beds. The element has several allotropic forms; the most stable one consists of S_8 rings. The most important compound of the element is sulfuric acid, a strong acid that is a good dehydrating agent and has a high boiling point. Selenium and tellurium are chemically rather similar to sulfur, especially with respect to formation of oxides and oxyanions.

The group 5A elements exhibit a wide range of behavior, from strongly nonmetallic in the case of nitrogen to distinctly metallic in the case of bismuth. Phosphorus occurs in nature in certain phosphate minerals. The element exhibits several allotropes, including one known as white phosphorus, a reactive form consisting of P_4 tetrahedra. Phosphorus forms compounds of formula PX_3 and PX_5 with the halogens. These undergo hydrolysis in water to produce the corresponding oxyacid of phosphorus and HX. Phosphorus forms two oxides, P_4O_6 and P_4O_{10}. Their corresponding acids, phosphorous acid and phosphoric acid, show a strong tendency to undergo condensation reactions when heated. Compounds of phosphorus are important components of fertilizers.

The group 4A elements show great diversity in physical and chemical properties. Carbon excels in being able to form multiple bonds and in undergoing catenation. Silicon is noteworthy as a semiconductor and for its tendency to form Si—O bonds. Silicon is the second most abundant element, and it occurs in a wide variety of silicates. Silicates are composed of SiO_4 tetrahedra, which, by sharing oxygen atoms, are able to link together to form chains, sheets, and three-dimensional arrays. In many minerals, Si^{4+} ions are replaced by Al^{3+} ions, thus forming aluminosilicates. Silicates are important components of glass.

Boron commonly exhibits an oxidation state of $+3$, as in boric oxide, B_2O_3, and boric acid, H_3BO_3. The acid readily undergoes condensation reactions.

KEY TERMS

interhalogens (Sec. 23.2)
catenation (Sec. 23.5)
silicates (Sec. 23.5)
aluminosilicates (Sec. 23.5)
clay minerals (Sec. 23.5)
glass (Sec. 23.5)

EXERCISES

The Noble Gases and Halogens

23.1 Write the chemical formula for each of the following compounds and indicate the oxidation state of the halogen or noble-gas atom in each: (a) iodate ion; **(b)** bromic acid; (c) bromine trifluoride; **(d)** sodium hypochlorite; (e) iodous acid; **(f)** xenon trioxide

23.2 Write the chemical formula for each of the following compounds, and indicate the oxidation state of the halogen or noble-gas atom in each: **(a)** calcium bromide; **(b)** perchloric acid; **(c)** xenon oxytetrafluoride; **(d)** chlorite ion; **(e)** hypobromous acid; **(f)** iodine pentafluoride.

23.3 Name the following compounds: (a) $KClO_3$; **(b)** $Ca(IO_3)_2$; (c) $AlCl_3$; **(d)** $HBrO_3$; (e) H_5IO_6; **(f)** XeF_4.

23.4 Name the following compounds: **(a)** $Fe(ClO_4)_2$; **(b)** $HClO_2$; **(c)** XeF_2; **(d)** IF_5; **(e)** XeO_3; **(f)** HBr (named as an acid).

23.5 Predict the geometrical structures of the following: (a) I_3^-; **(b)** ICl_4^-; (c) ClO_3^-; **(d)** H_5IO_6; (e) XeF_4.

23.6 The interhalogen compound $BrF_3(l)$ reacts with antimony(V) fluoride to form the salt $(BrF_2^+)(SbF_6^-)$. Write the Lewis structure for both the cation and anion in this substance and describe the likely geometrical structure of each.

23.7 What are the major factors responsible for the fact that xenon forms stable compounds with fluorine, whereas argon does not?

23.8 Why were the noble gases the last family of elements to be discovered?

23.9 Explain each of the following observations. **(a)** At room temperature, I_2 is a solid, Br_2 is a liquid, and Cl_2 and F_2 are both gases. **(b)** F_2 cannot be prepared by electrolytic oxidation of aqueous F^- solutions. **(c)** The boiling point of HF is much higher than those of the other hydrogen halides. **(d)** The halogens decrease in oxidizing power in the order $F_2 > Cl_2 > Br_2 > I_2$.

23.10 Explain the following observations: **(a)** For a given oxidation state, the acid strength of the oxyacid in aqueous solution decreases in the order chlorine > bromine > iodine. **(b)** Hydrofluoric acid cannot be stored in glass bottles. **(c)** HI cannot be prepared by treating NaI with sulfuric acid. **(d)** The interhalogen ICl_3 is known, but $BrCl_3$ is not. **(e)** I_2 is more soluble in aqueous solutions of I^- than in pure water.

23.11 List one commercial use of each of the halogens.

23.12 Write a balanced chemical equation for the commercial preparation of each halogen element.

23.13 Write balanced net ionic equations for the reaction of each of the following substances with water: (a) PBr_5; (b) IF_5; (c) $SiBr_4$; **(d)** F_2; **(e)** ClO_2 (chloric acid is a product); **(f)** $HI(g)$.

23.14 Write a balanced chemical equation that describes a suitable means of preparing each of the following substances: **(a)** HF; **(b)** I_2; **(c)** XeF_4; **(d)** $Ca(OCl)Cl$; **(e)** SF_6; **(f)** NaClO.

23.15 Write balanced chemical equations for each of the following reactions (some of which are analogous but not identical to reactions shown in this chapter). **(a)** Bromine forms hypobromite ion on addition to aqueous base. **(b)** Chlorine reacts with an aqueous solution of sodium bromide. **(c)** Bromine reacts with an aqueous solution of hydrogen peroxide, liberating O_2.

23.16 Write balanced chemical equations for each of the following reactions (some of which are analogous but not identical to reactions shown in this chapter). **(a)** Hydrogen bromide is produced upon heating calcium bromide with phosphoric acid. **(b)** Hydrogen bromide is formed upon hydrolysis of aluminum bromide. **(c)** Aqueous hydrogen fluoride reacts with solid calcium carbonate, forming water-insoluble calcium fluoride.

23.17 **(a)** Write the balanced net ionic equation for the reduction of ClO_3^- to Cl_2 by Fe^{2+} in acidic aqueous solution. **(b)** Calculate the standard emf for this reaction.

23.18 Chloride ion is oxidized in aqueous solution to $Cl_2(aq)$ by each of the following reagents: **(a)** $MnO_2(s)$; **(b)** $MnO_4^-(aq)$; **(c)** $Cr_2O_7^{2-}(aq)$. In each case, write a complete, balanced net ionic equation.

[23.19] Using the thermochemical data in Table 23.1 and Appendix C, calculate the average Xe—F bond energies in XeF_2, XeF_4, and XeF_6, respectively. What is the significance of the trend in these quantities?

[23.20] The solubility of Cl_2 in 100 g of water at STP is 310 cm^3. Assume that this quantity of Cl_2 is dissolved and equilibrated as follows:

$$Cl_2(aq) + H_2O(l) \rightleftharpoons Cl^-(aq) + HClO(aq) + H^+(aq)$$

If the equilibrium constant for this reaction is 4.7×10^{-4}, calculate the equilibrium concentration of HClO formed.

Group 6A

23.21 Write the chemical formula for each of the following compounds and indicate the oxidation state of the group 6A element in each: (a) selenous acid; **(b)** potassium hydrogen sulfite; (c) hydrogen telluride; **(d)** carbon disulfide; (e) calcium sulfate; **(f)** sodium thiosulfate.

23.22 Write the chemical formula for each of the following compounds and indicate the oxidation state of the group 6A element in each: **(a)** selenium trioxide; **(b)** orthotelluric acid; **(c)** zinc selenate; **(d)** sulfur tetrafluoride; **(e)** hydrogen sulfide; **(f)** sulfurous acid.

23.23 Name each of the following compounds: **(a)** $K_2S_2O_3$; **(b)** Al_2S_3; **(c)** $NaHSeO_3$; **(d)** SeF_6.

23.24 Name each of the following compounds: **(a)** H_2Se; **(b)** FeS_2; **(c)** $NaHSO_4$; **(d)** Na_2SeO_4.

23.25 Write the Lewis structure for each of the following species and indicate the geometrical structure of each: (a) SeO_4^{2-}; (b) H_6TeO_6; **(c)** $TeO_2(g)$; **(d)** S_2Cl_2; **(e)** chlorosulfonic acid, HSO_3Cl (chlorine is bonded to sulfur).

23.26 The SF_5^- ion is formed when $SF_4(g)$ reacts with fluoride salts containing large cations, such as $CsF(s)$. Draw the Lewis structures for SF_4 and SF_5^- and predict the molecular structure of each.

23.27 Write a balanced chemical equation for each of the following reactions. **(a)** Selenium dioxide dissolves in water. **(b)** Solid zinc sulfide reacts with hydrochloric acid. **(c)** Elemental sulfur reacts with sulfite ion to form thiosulfate. (d) Elemental selenium is heated with sulfuric acid. **(e)** Sulfur trioxide is dissolved in sulfuric acid. (f) Selenous acid is oxidized by hydrogen peroxide.

23.28 Write a balanced chemical equation for each of the following reactions. (You may have to guess at one or more of the reaction products, but you should be able to make a reasonable guess based on your study of this chapter.) **(a)** Selenous acid is reduced by hydrazine in aqueous solution to yield elemental selenium (see Chapter 22 for a discussion of hydrazine). **(b)** Heating orthotelluric acid to temperatures over 200°C yields the acid anhydride. **(c)** Hydrogen selenide can be prepared by reaction of aqueous acid solution on aluminum selenide. **(d)** Sodium thiosulfate is used to remove excess Cl_2 from chlorine-bleached fabrics. The thiosulfate ion forms SO_4^{2-} and elemental sulfur while Cl_2 is reduced to Cl^-.

23.29 An aqueous solution of SO_2 acts as a reducing agent to reduce **(a)** aqueous $KMnO_4$ to $MnSO_4(s)$; (b) acidic aqueous $K_2Cr_2O_7$ to aqueous Cr^{3+}; **(c)** aqueous $Hg_2(NO_3)_2$ to mercury metal. Write balanced chemical equations for these reactions.

23.30 In aqueous solution, hydrogen sulfide reduces **(a)** Fe^{3+} to Fe^{2+}; **(b)** Br_2 to Br^-; **(c)** MnO_4^- to Mn^{2+}; **(d)** HNO_3 to NO_2. In all cases, under appropriate conditions, the product is elemental sulfur. Write a balanced net ionic equation for each reaction.

[23.31] Suggest an explanation for the fact that telluric acid is a weak acid, whereas sulfuric acid and selenic acid are strong acids.

[23.32] SF_4 is very reactive; for example, it reacts readily with water to produce HF and SO_2. By contrast, SF_6 is very stable. It has even been used, with O_2, for X-ray examination of the lungs. **(a)** What features of the molecules make attack of sulfur by Lewis bases much more likely for SF_4 than for SF_6? **(b)** What feature of SF_4 makes attack of sulfur by Lewis acids possible? **(c)** Why is SF_4 subject to oxidation, whereas SF_6 is not?

Group 5A

23.33 Write formulas for the following compounds and indicate the oxidation state of the group 5A element in each: (a) orthophosphoric acid; (b) arsenous acid; **(c)** antimony(III) sulfide; calcium dihydrogen phosphate; (e) potassium phosphide.

23.34 Write formulas for the following compounds and indicate the oxidation state of the group 5A element in each: **(a)** phosphorous acid; **(b)** pyrophosphoric acid; **(c)** antimony trichloride; **(d)** magnesium arsenate; **(e)** diphosphorus pentoxide.

23.35 Name each of the following compounds: **(a)** Na_3P; (b) H_3AsO_4; (c) P_4O_{10}; **(d)** AsF_5.

23.36 Name each of the following compounds: **(a)** K_3As; **(b)** PBr_3; **(c)** Sb_2O_3; **(d)** NaH_2AsO_4.

23.37 Phosphorus pentachloride exists in one form in the solid state as an ionic lattice of PCl_4^+ and PCl_6^- ions. **(a)** Draw the Lewis structures of these ions and predict their geometries. **(b)** What set of hybrid orbitals is employed by phosphorus in each case? **(c)** Why should PCl_5 exist as an ionic substance in the solid state, whereas it is stable as the neutral molecule in the gas phase?

23.38 Sodium trimetaphosphate, $Na_3P_3O_9$, and sodium tetrametaphosphate, $Na_4P_4O_{12}$, are used as water-softening agents. They contain cyclic $P_3O_9^{3-}$ and $P_4O_{12}^{4-}$ ions, respectively. Propose reasonable structures for these ions.

23.39 Account for the following observations (a) H_3PO_3 is a diprotic acid. **(b)** Nitric acid is a strong acid, whereas phosphoric acid is weak. (c) Phosphate rock is not effective as a phosphate fertilizer. **(d)** Phosphorus does not exist at room temperature as diatomic molecules, but nitrogen does. (e) Solutions of Na_3PO_4 are quite basic.

23.40 Account for the following observations. **(a)** Phosphorus forms a pentachloride, but nitrogen does not. **(b)** H_3PO_2 is a monoprotic acid. **(c)** Phosphonium salts, such as PH_4Cl, can be formed under anhydrous conditions, but they can't be made in aqueous solution. **(d)** Whereas PCl_3 hydrolyzes readily in water to form H_3PO_3, $SbCl_3$ hydrolyzes only in part, forming SbOCl. **(e)** White phosphorus is extremely reactive.

23.41 Write a balanced chemical equation for each of the following reactions: (a) preparation of white phosphorus from calcium phosphate; (b) hydrolysis of PCl_3; **(c)** preparation of PCl_3 from P_4; **(d)** reaction of P_4O_{10} with water.

23.42 Write a balanced chemical equation for each of the following reactions: **(a)** Preparation of PF_5 from PCl_5 using AsF_3; **(b)** reaction of As_2O_3 with water; **(c)** dehydration of orthophosphoric acid to form pyrophosphoric acid; **(d)** reaction of elemental arsenic with dilute nitric acid to form NO and arsenous acid, H_3AsO_3.

Group 4A and Boron

23.43 Write the formulas for the following compounds and indicate the oxidation state of the group 4A element or of boron in each: **(a)** silicon dioxide; **(b)** germanium tetrachloride; **(c)** sodium borohydride; **(d)** stannous chloride.

23.44 Write the formulas for the following compounds and indicate the oxidation state of the group 4A element or of boron in each: **(a)** boric acid; **(b)** silicon tetrabromide; **(c)** lead chloride; **(d)** sodium tetraborate decahydrate (borax).

23.45 Covalent silicon-hydrogen compounds are called silanes. The silane Si_2H_6, known as disilane, exists, but no Si_2H_4 and Si_2H_2 compounds are known. In contrast, carbon forms C_2H_6, C_2H_4, and C_2H_2. Explain why silicon doesn't form Si_2H_4 and Si_2H_2.

23.46 Suggest some reasons why carbon is more suitable than silicon as the major structural element in living systems.

23.47 Select the member of group 4A that best fits each of the following descriptions: **(a)** forms the most acidic oxide; **(b)** is most commonly found in the +2 oxidation state; **(c)** is a component of sand.

23.48 Select the member of group 4A that best fits each of the following descriptions: **(a)** catenates to the greatest extent; **(b)** forms the most basic oxide; **(c)** is a metalloid that can form +2 ions.

23.49 Both $GeCl_4$ and $SiCl_4$ fume in moist air because of hydrolysis to GeO_2 and SiO_2. Write balanced equations for these reactions.

23.50 Germanium differs markedly from silicon in that the 2+ halides are fairly stable. These halides can be prepared by the reduction of the tetrahalide with germanium metal.

(a) Write the balanced chemical equation for the formation of $GeCl_2$. (b) Predict the geometrical structure of the gaseous $GeCl_2$ molecule.

23.51 The mineral orthoclase is a feldspar mineral formed by replacing a quarter of the Si^{4+} ions in SiO_2 with Al^{3+} and maintaining charge balance with K^+ ions. What is the empirical formula for this mineral?

23.52 How is the mica mineral $KMg_3(AlSi_3O_{10})(OH)_2$ related structurally to the mica mineral muscovite mentioned in the text?

23.53 What empirical formula and unit charge are associated with each of the following structural types: (a) isolated SiO_4 tetrahedra; (b) a chain structure of SiO_4 tetrahedra joined at corners to adjacent units; (c) a structure consisting of tetrahedra joined at corners to form a six-membered ring of alternating Si and O atoms?

23.54 Propose a reasonable description of the structure of each of the following minerals: (a) albite, $NaAlSi_3O_8$; (b) leucite, $KAlSi_2O_6$; (c) zircon, $ZrSiO_4$; (d) sphene, $CaTiSiO_5$.

Additional Exercises

23.55 Name each of the following compounds: (a) $H_2B_4O_7$; (b) SiC; (c) HPO_3; (d) XeF_2; (e) Na_2S; (f) $KClO_3$.

23.56 Explain each of the following observations. (a) H_2S is a better reducing agent than H_2O. (b) H_2SO_4 is a stronger acid than H_2SeO_4. (c) Astatine is generally not considered in any detail in discussion of halogens. (d) White phosphorus is quite volatile, whereas red phosphorus is not. (e) Xenon hexafluoride is a stable compound, whereas krypton hexafluoride is unknown. (f) Addition of SF_4 to water results in an acidic solution. (g) Silicate-sheet minerals are softer than aluminosilicate-sheet ones. (h) Sulfur occurs naturally as sulfates, but not as sulfites.

23.57 Xenon trioxide disproportionates in strongly alkaline solution to form the thermally stable perxenate ion, XeO_6^{4-}. Predict the geometry of this ion. Describe the bonding in terms of the hybridization of xenon valence-shell orbitals.

23.58 What pressure of gas is formed when 0.654 g of XeO_3 decomposes completely to the free elements at 48°C in a 0.452-L volume?

23.59 Write balanced chemical equations to account for the following observations. (There may not be closely similar reactions shown in the chapter; however, you should be able to make reasonable guesses at the likely products.) (a) When burning sodium metal is immersed in a pure HCl atmosphere, it continues to burn. (b) Bubbling SO_2 gas through liquid bromine that is covered with a layer of water results in formation of a strongly acidic solution; upon distillation, an aqueous HBr solution is collected. The remaining liquid is still strongly acidic. (c) When bromine is added to a basic solution containing potassium hypochlorite, insoluble potassium bromate is formed. (d) When bromic acid is reacted with SO_2, $Br_2(aq)$ is formed. (e) Uranium(VI) fluoride is formed by the action of ClF_3 on uranium(IV) chloride.

23.60 The cyano group behaves in some ways like a halogen. Thus, cyanogen gas, $(CN)_2$, has been called a pseudohalogen, and the cyanide ion, CN^-, a pseudohalide. Cyanogen reacts with an aqueous solution of NaOH in a fashion analogous to Cl_2. (a) Write a balanced chemical equation for this reaction. (b) Write the Lewis structure for $(CN)_2$ and describe its geometrical structure.

23.61 Although the ClO_4^- and IO_4^- ions have been known for a long time, BrO_4^- was not synthesized until 1965. The ion was synthesized by oxidizing the bromate ion with xenon difluoride, producing xenon, hydrofluoric acid, and the perbromate ion. Write the balanced chemical equation for this reaction.

23.62 List the halogens in order of increasing X—X halogen bond energies. Suggest a reason for the low F—F bond energy.

23.63 What structural feature do the molecules P_4, P_4O_6, and P_4O_{10} have in common? What is the common structural feature of all of the acids containing phosphorus(V)?

23.64 Elemental sulfur is capable of reacting under suitable conditions with Fe, F_2, O_2, or H_2. Write balanced chemical equations to describe these reactions. In which reactions is sulfur acting as a reducing agent and in which as an oxidizing agent?

23.65 Sodium sulfide is used in the leather industry to remove hair from hides. It can be synthesized by reducing sodium sulfate with carbon. (Carbon monoxide forms together with the sodium sulfide.) Write a balanced chemical equation for this reaction.

23.66 Sulfur and the group 6A elements below it in the periodic chart exhibit oxidation states ranging from -2 to $+6$. What factors control the lowest and highest oxidation states? Explain.

23.67 Draw the Lewis structures for the following species and predict the relative S—O bond lengths in each: SO_2, SO_3, SO_4^{2-}.

23.68 One method proposed for removal of SO_2 from the flue gases of power plants involves reaction with aqueous H_2S. Elemental sulfur is the product. (a) Write a balanced chemical equation for the reaction. (b) What volume of H_2S at 27°C and 740 mm Hg would be required to remove the SO_2 formed by burning 1.0 ton of coal containing 3.5 percent S by weight? (c) What mass of elemental sulfur is produced? Assume that all reactions are 100 percent efficient.

23.69 Although H_2Se is toxic, no deaths have been attributed to it. One reason is its vile odor, which serves as a sensitive warning of its presence. In addition, H_2Se is readily oxidized by O_2 in air to nontoxic elemental red selenium before harmful amounts of H_2Se can enter the body. (a) Write a balanced chemical equation for this oxidation. (b) Calculate $\Delta G°$ and the equilibrium constant (at 298 K) for the reaction.

23.70 A sulfuric acid plant produces a considerable amount of heat. This heat is used to generate electricity, which helps reduce operating costs. The synthesis of H_2SO_4 consists of three main chemical processes: (1) oxidation of S to SO_2; (2) oxidation of SO_2 to SO_3; (3) the dissolving of SO_3 in H_2SO_4 and its reaction with water to form H_2SO_4. If the third process produces 130 kJ/mol, how much heat is produced in preparing a mole of H_2SO_4 from a mole of S? How much heat is produced in preparing a ton of H_2SO_4?

23.71 Ultrapure germanium, like silicon, is used in semiconductors. Germanium of "ordinary" purity is prepared by the high-temperature reduction of GeO_2 with carbon. The Ge is converted to $GeCl_4$ by treatment with Cl_2 and then purified by distillation; $GeCl_4$ is then hydrolyzed in water to GeO_2 and reduced to the elemental form with H_2. The element is then zone-refined. Write a balanced chemical equation for each of the chemical transformations in the course of forming ultrapure Ge from GeO_2.

23.72 Describe the fundamental structural unit present in all silicate minerals. How is this unit modified in aluminosilicates?

23.73 Draw the Lewis structure for the cyclic $Si_3O_9^{6-}$ ion.

[23.74] The maximum allowable concentration of $H_2S(g)$ in air is 20 mg per kilogram of air (20 ppm by weight). How many grams of FeS would be required to react with hydrochloric acid to produce this concentration in an average room measuring 2.7 m × 4.3 m × 4.3 m?

[23.75] The standard heats of formation of $H_2O(g)$, $H_2S(g)$, $H_2Se(g)$, and $H_2Te(g)$ are -241.8, -20.17, $+29.7$, and 99.6 kJ/mol, respectively. The enthalpies necessary to convert the elements in their standard states to 1 mol of gaseous atoms are 248, 277, 227, and 197 kJ/mol of atoms for O, S, Se, and Te, respectively. The enthalpy for dissociation of H_2 is 436 kJ/mol. Calculate the average H—O, H—S, H—Se, and H—Te bond energies and comment on their trend.

[23.76] **(a)** Calculate the P-to-P distance in both P_4O_6 and P_4O_{10} from the following data: the P—O—P bond angle for P_4O_6 is 127.5°, while that for P_4O_{10} is 124.5°. The P—O distance (to bridging oxygens) is 0.165 nm in P_4O_6 and 0.160 nm in P_4O_{10}. **(b)** Rationalize the relative P-to-P distances in the two compounds.

[23.77] The N—X bond distances in the nitrosyl halides, NOX, are 1.52, 1.98, and 2.14 Å for NOF, NOCl, and NOBr, respectively. Compare the distances with the atomic radii for the halogens (Table 23.3). Is the variation in N—X distance what you expect from these covalent radii? If not, account for the deviations.

[23.78] Considering the chemical formulas and structures of clay minerals, suggest an explanation for why they can be molded and then become hard and brittle when heated.

[23.79] Boron nitride has a graphitelike structure with B—N bond distances of 1.45 Å within sheets and a separation of 3.30 Å between sheets. At high temperatures, the BN assumes a diamondlike form that is harder than diamond. Rationalize the similarity between BN and elemental carbon.

Metals and Metallurgy

24

Molten iron being poured into a basic oxygen furnace. By adding other metals and by blowing $O_2(g)$ through the molten mixture to oxidize impurities, the iron is converted to steel. (Samsung America, Inc.)

CONTENTS

In Chapters 22 and 23, we examined the chemistry of nonmetallic elements. In this chapter, we turn our attention to metals. Metals have played a major role in the development of civilization. Early history is often divided into the Stone Age, the Bronze Age, and the Iron Age, based on the compositions of the tools used in each era. Modern societies rely on a large variety of metals for making tools, machines, and other items. Chemists and other scientists have found uses for even the least abundant metals as they search for materials to meet evolving technological needs. To illustrate this point, Figure 24.1 shows the approximate composition of a high-performance jet engine. Notice that iron, long the dominant metal of technology, is not even present to a significant extent.

In this chapter, we shall consider the chemical forms in which metallic elements occur in nature and the means by which we obtain metals from these sources. We shall also examine the bonding in solids and see how metals and mixtures of metals, called alloys, are employed in modern technology. Finally, we shall look specifically at the properties of transition metals. As we shall see, metals have a varied and interesting chemistry.

24.1 OCCURRENCE AND DISTRIBUTION OF METALS

The portion of our environment that constitutes the solid earth beneath our feet is called the **lithosphere**. The lithosphere provides most of the materials we use to feed, clothe, shelter, support, and entertain ourselves. Although the bulk of the earth is solid, we have access to only a small region near the surface. The

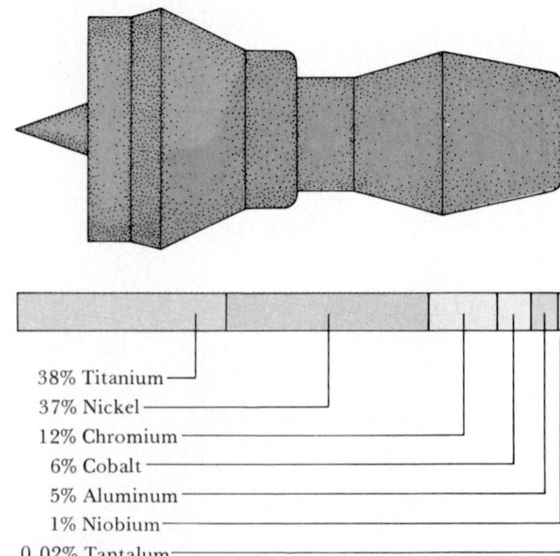

Figure 24.1 Metallic elements employed in construction of a jet engine.

deepest well ever drilled is only about 8 km deep, and the deepest mine extends between 3 and 4 km into the earth. In comparison, the earth has a radius of 6370 km.

Many of the substances most useful to us are not especially abundant in that portion of the lithosphere to which we have ready access. It is interesting to compare the order of abundance of the elements in the lithosphere with the estimated order of their global consumption. The 12 elements listed in Table 24.1 compose 99.5 percent of the lithosphere by mass. The most widely used elements are listed in Table 24.2. The ranking of elements in the two lists shows little correlation. Some elements that are not abundant are widely used (chromium and copper, for example). Consequently, the occurrence and distribution of *concentrated* deposits of some of these elements often play a role in international politics as nations compete for access to these materials.

Deposits that contain metals in economically exploitable quantities are known as **ores**. Usually, the compounds or elements that we desire must be separated from a large quantity of unwanted material and then chemically processed to be rendered useful. By processing large quantities of substances, we have literally changed the surface of our planet (see Figure 24.2). Experts estimate that about 2.3×10^4 kg (25 tons) of materials are extracted from the

Table 24.1 The 12 Most Abundant Elements in the Lithosphere

Element	Percent by weight
Oxygen	50
Silicon	26
Aluminum	7.5
Iron	4.7
Calcium	3.4
Sodium	2.6
Potassium	2.4
Magnesium	1.9
Hydrogen	0.9
Titanium	0.6
Chlorine	0.2
Phosphorus	0.1

Table 24.2 Estimated Annual World Consumption of Elements[a]

Element	Annual consumption (kg)
C	10^{12} to 10^{13}
Na, Fe	10^{11} to 10^{12}
N, O, S, K, Ca	10^{10} to 10^{11}
H, F, Mg, Al, P, Cl, Cr, Mn, Cu, Zn, Ba, Pb	10^{9} to 10^{10}
B, Ti, Ni, Zr, Sn	10^{8} to 10^{9}
Ar, Co, As, Mo, Sb, W, U	10^{7} to 10^{8}
Li, V, Se, Sr, Nb, Ag, Cd, I, rare earths, Au, Hg, Bi	10^{6} to 10^{7}
He, Be, Te, Ta	10^{5} to 10^{6}

[a] Elements in color also appear in Table 24.1.

Figure 24.2 Large open-pit mining operation. (Georg Gerster/Photo Researchers)

lithosphere and processed annually to support each person in our country. Because the richest sources of many substances are becoming exhausted, it will be necessary in the future to process larger volumes and lower-quality raw materials. Consequently, extraction of the compounds and elements we need will cost more in terms of both energy and environmental impact.

Minerals

With the exception of gold and the platinum-group metals (Ru, Rh, Pd, Os, Ir, and Pt), most metallic elements are found in nature in solid inorganic compounds called **minerals**. Table 24.3 lists the principal mineral sources of several common metals, two of which are shown in Figure 24.3. Notice that minerals are identified by common names rather than by chemical names. Names of minerals are usually based on the locations where they were discovered, the

Table 24.3 Principal Mineral Sources of Some Common Metals

Metal	Mineral	Composition
Aluminum	Bauxite	Al_2O_3
Chromium	Chromite	$FeCr_2O_4$
Copper	Chalcocite	Cu_2S
	Chalcopyrite	$CuFeS_2$
	Malachite	$Cu_2CO_3(OH)_2$
Iron	Hematite	Fe_2O_3
	Magnetite	Fe_3O_4
Lead	Galena	PbS
Manganese	Pyrolusite	MnO_2
Mercury	Cinnabar	HgS
Molybdenum	Molybdenite	MoS_2
Tin	Cassiterite	SnO_2
Titanium	Rutile	TiO_2
	Ilmenite	$FeTiO_3$
Zinc	Sphalerite	ZnS

Figure 24.3 Two common minerals: (*a*) cinnabar, (*b*) malachite. Note the different colors and shapes. (George Whiteley/Photo Researchers; Tom McHugh/Photo Researchers)

(*a*)

(*b*)

A CLOSER LOOK: Metals as Strategic Materials

A modern, high-technology society depends critically on the availability of many metallic elements. Dependable access to the mineral sources from which these essential metals are derived is important to the economic and military security of a highly industralized nation. In recent years, this access has become a matter of deep concern in the United States, because this country has no domestic sources of many of the most critical elements. Figure 24.1 shows the extent of our dependence on imports in terms of percentages of annual consumption that are imported. Note that all the elements listed in Figure 24.1 as important components of a modern jet engine are on this list. Furthermore, the most important sources of some of the metals are nations that are not highly developed and have uncertain relations with the United States.

In response to this situation, the United States has developed a strategic minerals policy. Part of this policy establishes stockpiles of the most critical metals, so that if all foreign sources were cut off, the United States could function for an extended time by drawing on the reserves. Intensive efforts to develop new reserves within the United States and to recycle old metal have also been initiated.

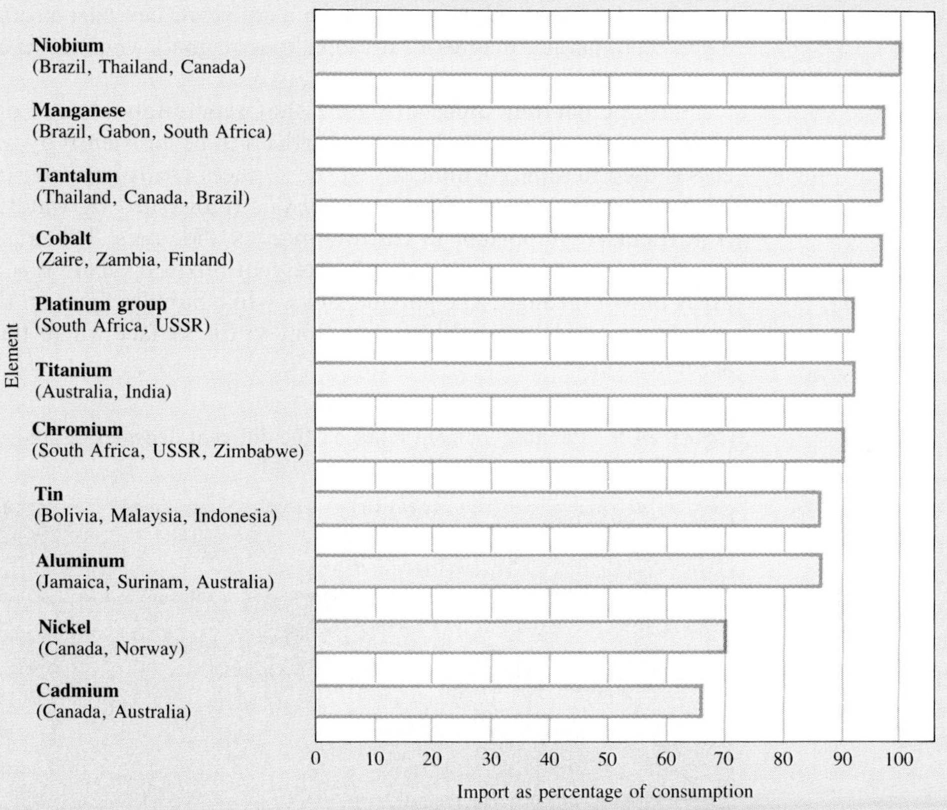

Figure 24.4 Dependence of the United States on imports of strategic metals.

person who discovered them, or some characteristic such as color. For example, the name *malachite* comes from the Greek word *malache*, the name of a type of tree whose leaves are the color of the mineral.

Commercially, the most important sources of metals are oxide, sulfide, and carbonate minerals. Silicate minerals (Section 23.5) are very abundant, but they are generally difficult to concentrate and reduce. Therefore, most silicates are not economical sources of metals.

24.2 METALLURGY

Metallurgy is the science and technology of extracting metals from their natural sources and preparing them for practical use. It usually involves several steps: (1) mining, (2) concentrating the ore or otherwise preparing it for further treatment, (3) reducing the ore to obtain the free metal, (4) refining or purifying the metal, and (5) alloying the metal, if necessary. (An **alloy** is a metallic material that is composed of two or more elements; see Section 24.7.)

Concentration of Ores

After being mined, an ore is usually crushed and ground and then treated to concentrate the desired metal. The concentration stage relies on differences in the properties of the mineral and the undesired material that accompanies it, which is called **gangue** (pronounced "gang"). For example, miners who panned for gold (Figure 24.5) used a pan to rinse away the gangue from the denser gold nuggets. In another example, magnetite, a mineral of iron, can be concentrated by moving finely ground ore on a conveyor belt past a series of magnets. The iron mineral is magnetic (attracted to a magnet), whereas the accompanying gangue is not.

In the **flotation process** (Figure 24.6), separation is based on the different abilities of the mineral and gangue particles to be moistened by water. The process is used to separate minerals whose surfaces are *hydrophobic* (not moistened by water) from gangue that is *hydrophilic* (moistened by water). The process is particularly important in treating many sulfide ores. The crushed ore is fed into a tank containing a water-oil-detergent mixture, where it is agitated while air is blown through. The gangue sinks, while bubbles form on the oil-covered sulfide minerals. The bubbles then float to the surface where the minerals are skimmed off.

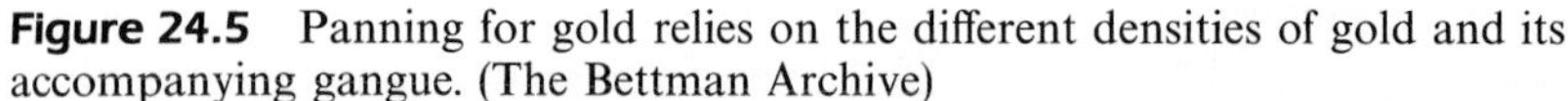

Figure 24.5 Panning for gold relies on the different densities of gold and its accompanying gangue. (The Bettman Archive)

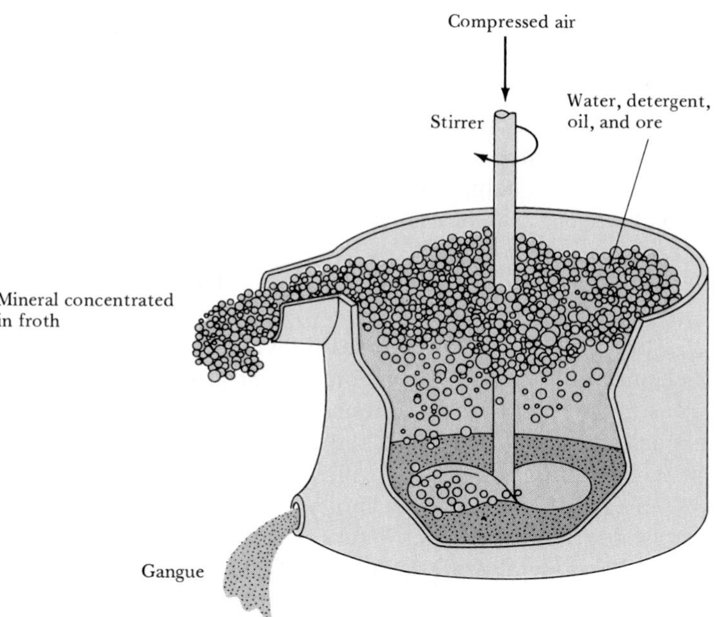

Figure 24.6 Schematic diagram of flotation, by means of which a mineral can be separated from a larger amount of gangue.

24.3 PYROMETALLURGY

A large number of metallurgical processes utilize high temperatures to alter the mineral chemically and ultimately to reduce it to the free metal. The use of heat to alter or reduce the mineral is called **pyrometallurgy**. (*Pyro* means "at high temperature.")

Calcination is the heating of an ore to bring about its decomposition and the elimination of a volatile product. The volatile product could be, for example, CO_2 or H_2O. Carbonates are often calcined to drive off CO_2, forming the metal oxide. For example,

$$PbCO_3(s) \longrightarrow PbO(s) + CO_2(g) \qquad [24.1]$$

Most carbonates decompose reasonably rapidly at temperatures in the range of 400 to 500°C, although $CaCO_3$ requires a temperature of about 1000°C. Most hydrated minerals lose H_2O at temperatures on the order of 100 to 300°C.

Roasting is a thermal treatment that causes chemical reactions between the ore and the furnace atmosphere. Roasting may produce oxidation or reduction and may be accompanied by calcination. An important roasting process is the oxidation of sulfide ores, in which the metal is converted to the oxide, as in these examples:

$$2ZnS(s) + 3O_2(g) \longrightarrow 2ZnO(s) + 2SO_2(g) \qquad [24.2]$$

$$2MoS_2(s) + 7O_2(g) \longrightarrow 2MoO_3(s) + 4SO_2(g) \qquad [24.3]$$

The ore of a less active metal, such as mercury, can be roasted to the free metal:

$$HgS(s) + O_2(g) \longrightarrow Hg(g) + SO_2(g) \qquad [24.4]$$

The free metal can also be produced when a reducing atmosphere is present during the roast. Carbon monoxide, CO, provides such an atmosphere:

$$PbO(s) + CO(g) \longrightarrow Pb(l) + CO_2(g) \qquad [24.5]$$

Roasting is not always a feasible method of reduction. Some metals that are difficult to obtain as the free metal by roasting are best converted to the metal halide, which can then be reduced. To obtain the metal halide, the metal oxide or another compound, such as metal carbide, is roasted in an atmosphere of the halogen, usually chlorine. For example,

$$TiC(s) + 4Cl_2(g) \longrightarrow TiCl_4(g) + CCl_4(g) \qquad [24.6]$$

Smelting is a melting process in which the materials formed in the course of chemical reactions separate into two or more layers. Smelting often involves a roasting stage in the same furnace. Two of the important types of layers formed in smelters are molten metal and slag. The molten metal may consist almost entirely of a single metal, or it may be a solution of two or more metals.

Slag consists mainly of molten silicate minerals, with aluminates, phosphates, fluorides, and other ionic compounds as constituents. A slag is formed when a basic metal oxide such as CaO reacts at high temperatures with molten silica, SiO_2:

$$CaO(l) + SiO_2(l) \longrightarrow CaSiO_3(l) \qquad [24.7]$$

Pyrometallurgical operations may involve not only the concentration and reduction of a mineral, but the refining of the metal as well. **Refining** is the treatment of a crude, relatively impure metal product from a metallurgical process to improve its purity and to define its composition better. Sometimes the goal of the refining process is to obtain the metal itself in pure form. However, the goal may also be to produce a mixture with a well-defined composition, as in the production of steels from crude iron, which we now examine.

The Pyrometallurgy of Iron

The most important pyrometallurgical operation in industry is the reduction of iron. Iron occurs in many different minerals, but the most important sources are the iron oxide minerals, hematite, Fe_2O_3, and magnetite, Fe_3O_4. The reduction of these oxides occurs in a *blast furnace*, such as the one illustrated in Figure 24.7. A blast furnace is essentially a huge chemical reactor capable of continuous operation. The largest furnaces are over 60 m high and 14 m wide. When operating at full capacity, they produce up to 10,000 tons of iron per day. The blast furnace is charged at the top with a mixture of iron ore, coke, and limestone. Coke is coal that has been heated in the absence of air to drive off volatile components. It is about 85 to 90 percent carbon. Coke serves as the fuel, producing heat as it is burned in the lower part of the furnace. It is also the source of the reducing gases CO and H_2. Limestone, $CaCO_3$, serves as the source of basic oxide in slag formation. Air, which enters the blast furnace at the bottom after preheating, is also an important raw material; it is required for combustion of the coke. Production of 1 kg of crude iron called *pig iron*, requires about 2 kg of ore, 1 kg of coke, 0.3 kg of limestone, and 1.5 kg of air.

In the furnace, oxygen reacts with the carbon in the coke to form carbon monoxide:

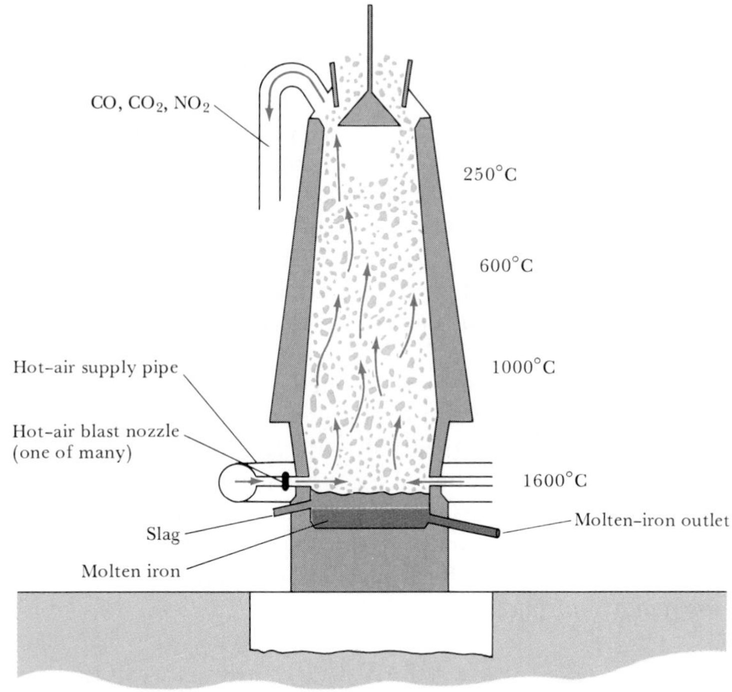

Figure 24.7 Blast furnace used for reduction of iron ore. Notice the approximate temperatures in the various regions of the furnace.

$$2C(s) + O_2(g) \longrightarrow 2CO(g) \qquad \Delta H = -110 \text{ kJ} \qquad [24.8]$$

Water vapor present in the air also reacts with carbon:

$$C(s) + H_2O(g) \longrightarrow CO(g) + H_2(g) \qquad \Delta H = +113 \text{ kJ} \qquad [24.9]$$

Note that the reaction with oxygen is exothermic and provides heat for furnace operation, but that the reaction with water vapor is endothermic. Addition of water to the air thus provides a means of controlling furnace temperature.

In the upper part of the furnace limestone is calcined (Equation 22.67). Here also the iron oxides are reduced by CO and H_2. For example, the important reactions for Fe_3O_4 are

$$Fe_3O_4(s) + 4CO(g) \longrightarrow 3Fe(s) + 4CO_2(g) \qquad \Delta H = -19 \text{ kJ} \qquad [24.10]$$

$$Fe_3O_4(s) + 4H_2(g) \longrightarrow 3Fe(s) + 4H_2O(g) \qquad \Delta H = 149 \text{ kJ} \qquad [24.11]$$

Reduction of other elements present in the ore also occurs in the hottest parts of the furnace, where carbon is the major reducing agent.

Molten iron collects at the base of the furnace, as shown in Figure 24.7. It is overlaid with a layer of molten slag formed by the reaction of CaO with the silica present in the ore, as described by Equation 24.7. The layer of slag over the molten iron helps to protect it from reaction with the incoming air. Periodically, the furnace is tapped to drain off slag and molten iron. The iron produced in the furnace may be cast into solid ingots. However, most is used directly in the manufacture of steel. For this purpose, it is transported, while still liquid, to the steelmaking shop.

Formation of Steel

Steel is an alloy of iron. The production of iron from its ore is a chemical reduction process that results in a crude iron containing many undesired impurities. Iron from a blast furnace typically contains 0.6 to 1.2 percent silicon, about 0.2 percent phosphorus, 0.4 to 2.0 percent manganese, and about 0.03 percent sulfur. In addition, there is considerable dissolved carbon. In the production of steel, these impurity elements are removed by oxidation in a vessel called a *converter*. In modern steel making, the oxidizing agent is either pure O_2 or O_2 diluted with argon. Air cannot be used directly as the source of O_2 because N_2 reacts with the molten iron, forming iron nitride and causing the steel to become brittle.

A cross-sectional view of one converter design appears in Figure 24.8. In this design, O_2, diluted with argon, is blown directly into the molten metal. The oxygen reacts exothermally with carbon, silicon, and impurity metals, reducing the concentrations of these elements in the iron. Carbon and sulfur are expelled as CO and SO_2 gases, respectively. Silicon is oxidized to SiO_2 and adds to whatever slag may have been present initially in the melt. Metal oxides react with the SiO_2 to form silicates. The presence of a basic slag is important for removal of phosphorus:

$$3CaO(l) + P_2O_5(l) \longrightarrow Ca_3(PO_4)_2(l) \qquad [24.12]$$

Nearly all of the O_2 blown into the converter is consumed in the oxidation reactions. By monitoring the O_2 concentration in the gas coming from the converter it is possible to tell when the oxidation is essentially complete. Oxidation of the impurities present in the iron normally requires only about 20 min. At this point, a sample of the molten metal is withdrawn and subjected to analysis for all of the elements whose presence could affect the quality of the steel produced: carbon, manganese, phosphorus, sulfur, silicon, nickel, chromium, copper, and other metallic elements that might be present. Depending on the results, additional oxygen might be blown in or additional ore might be added. When the desired composition is attained, the contents of the con-

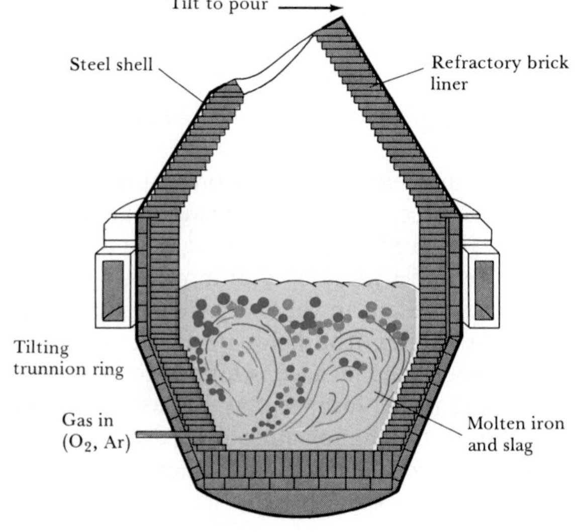

Figure 24.8 Converter for refining of iron. A mixture of oxygen and argon is blown through the molten iron and slag. The heat of oxidation of impurities maintains the mixture in a molten state. When the desired composition is attained, the converter is tilted to pour.

verter are dumped into a large ladle. As the ladle is being filled, alloying elements are added. These added elements produce steels with various kinds of properties. (We will have more to say about alloys in Section 24.7.) The still-molten mixture is then poured into molds for solidification.

24.4 HYDROMETALLURGY

Pyrometallurgical operations require large quantities of energy and are often a source of atmospheric pollution (especially by sulfur dioxide). In the last 100 years, new techniques have been developed in which metals are extracted from their ores by use of aqueous reactions. These processes are called **hydrometallurgy** (*hydro* means "water").

The most important hydrometallurgical process is **leaching**, in which the desired metal-containing compound is selectively dissolved. If the compound is water-soluble, the leaching agent can be merely water. More commonly the agent is an aqueous solution of an acid, a base, or a salt. Often the dissolving process involves formation of a complex ion (see Section 16.10), as in the leaching of gold.

Gold from low-grade ores is often concentrated by the process of cyanidation. The crushed ore is placed on large concrete slabs, and a solution of NaCN is sprayed over it. In the presence of CN^- and air, the gold is oxidized and dissolves, forming the stable $Au(CN)_2^-$ ion:

$$4Au(s) + 8CN^-(aq) + O_2(aq) + 2H_2O(l) \longrightarrow 4Au(CN)_2^-(aq) + 4OH^-(aq) \quad [24.13]$$

After a metal ion is selectively leached from the ore, it is precipitated from solution as the free metal or as an insoluble ionic compound. For example, gold is obtained from its cyanide complex by reduction with zinc powder:

$$2Au(CN)_2^-(aq) + Zn(s) \longrightarrow Zn(CN)_4^{2-}(aq) + 2Au(s) \quad [24.14]$$

The Hydrometallurgy of Aluminum

Among metals, aluminum is second only to iron in commercial use. World production of the metal is about 1.5×10^{10} kg (16 million tons) per year. The most useful ore of aluminum is *bauxite*, in which Al is present as hydrated oxides, $Al_2O_3 \cdot xH_2O$. The value of x varies, depending on the particular mineral present. Large bauxite deposits are found in Guiana, Australia, Jamaica, Mexico, and Brazil. Because bauxite deposits in the United States are very limited, the nation imports most of the ore used in the production of aluminum.

The major impurities found in bauxite are silica, SiO_2, and iron(III) oxide, Fe_2O_3. It is essential to separate alumina, Al_2O_3, from these impurities before the metal is recovered by electrochemical reduction, as described in Section 24.5. The process used to purify bauxite, called the **Bayer process**, is a hydrometallurgical procedure. The ore is first crushed and ground, then digested in a concentrated aqueous NaOH solution, about 30 percent NaOH by weight, at a temperature in the range 150 to 230°C. Sufficient pressure, between 4 and 30 atm, is maintained to prevent boiling. The trihydrate, $Al_2O_3 \cdot 3H_2O$, dissolves more readily than the monohydrate, $Al_2O_3 \cdot H_2O$. The reaction in either case leads to a complex aluminate anion of the form $Al(H_2O)_2(OH)_4^-$ (see Section 17.5). For example,

$$Al_2O_3 \cdot H_2O(s) + 6H_2O(l) + 2OH^- \longrightarrow 2Al(H_2O)_2(OH)_4^-(aq) \quad [24.15]$$

Silica dissolves in the strongly basic medium, then forms insoluble aluminosilicate salts with the aluminate ion. These settle out with a red "mud" consisting mostly of iron(III) oxides, which are not soluble in the strong base. Notice that this procedure works because Al^{3+} is amphoteric and Fe^{3+} is not. After filtration, the solution is diluted to reduce the concentration of hydroxide. In effect, this causes a partial reversal of the reaction shown in Equation 24.15, although the product is not the original mineral but a highly hydrated aluminum hydroxide.

After the aluminum hydroxide precipitate has been filtered, it is calcined in preparation for electroreduction of the ore to the metal. The solution recovered from the filtration must be reconcentrated so that it can be used again. This is accomplished by heating to evaporate water from the solution. The energy requirements of this evaporative stage are high, and it is the most costly part of the process.

24.5 ELECTROMETALLURGY

Electrolysis methods are often used either to reduce a metal compound to obtain the free metal or to refine the metal. Collectively, these processes are referred to as **electrometallurgy**. The principles of electrolysis were discussed in Section 20.7, which you should review if necessary.

Electrometallurgical procedures can be broadly differentiated according to whether they involve electrolysis of a molten salt or of an aqueous solution. Electrolytic methods are very important as a means of obtaining the more active metals, such as sodium, magnesium, and aluminum, in the free state. These metals cannot be reduced in aqueous solution, because water is more easily reduced than the metal ions. The standard potentials for reduction of water under acidic and basic conditions are both more positive than the standard potentials for reduction of such active metals as Na ($E^\circ = -2.71$ V), Mg ($E^\circ = -2.37$ V), and Al ($E^\circ = -1.66$ V):

$$2H^+(aq) + 2e^- \longrightarrow H_2(g) \qquad E^\circ = 0.00 \text{ V} \quad [24.16]$$

$$H_2O(l) + 2e^- \longrightarrow H_2(g) + 2OH^-(aq) \qquad E^\circ = -0.83 \text{ V} \quad [24.17]$$

To form such metals by electrochemical reduction, therefore, we must employ a molten-salt medium in which the metal ion of interest is the most readily reduced species.

Electrometallurgy of Sodium

In the commercial preparation of sodium, molten NaCl is electrolyzed in a specially designed cell called the **Downs cell**, illustrated in Figure 24.9. The electrolyte medium through which current flows is molten NaCl. Calcium chloride, $CaCl_2$, is added to lower the melting point of the cell medium from the normal melting point of NaCl, 804°C, to around 600°C. The Na(l) and $Cl_2(g)$ produced in the electrolysis are kept from coming in contact and reforming NaCl. In addition, the Na must be prevented from contact with oxygen, because the metal would quickly oxidize under the high-temperature conditions of the cell reaction.

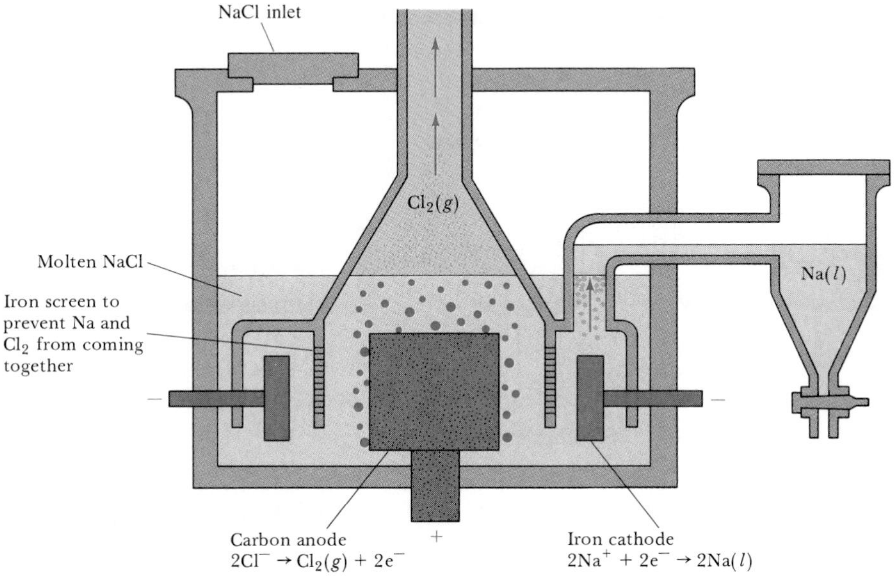

NaCl inlet
$Cl_2(g)$
Molten NaCl
Iron screen to prevent Na and Cl_2 from coming together
$Na(l)$
−
−
Carbon anode
$2Cl^- \rightarrow Cl_2(g) + 2e^-$
+
Iron cathode
$2Na^+ + 2e^- \rightarrow 2Na(l)$

Figure 24.9 Downs cell used in the commercial production of sodium.

Electrometallurgy of Aluminum

In Section 24.4, we discussed the Bayer process, in which bauxite is concentrated to produce a relatively pure hydrous aluminum hydroxide. When this concentrate is calcined at several hundred degrees Celsius, a partially hydrated aluminum oxide, $Al_2O_3 \cdot xH_2O$, called *alumina*, is formed. At temperatures in excess of 1000°C, anhydrous aluminum oxide is formed. Anhydrous aluminum oxide melts at over 2000°C. This is too high to permit its use as a molten medium for electrolytic formation of free aluminum. The electrolytic process used commercially to produce aluminum is known as the **Hall process**, named after its inventor, Charles M. Hall (see the Historical Perspective box). The purified Al_2O_3 is dissolved in molten cryolite, Na_3AlF_6, which has a melting point of 1012°C and is an effective conductor of electric current. A schematic diagram of the electrolysis cell is shown in Figure 24.10. Graphite rods are employed as anodes and are consumed in the electrolysis process. The electrode reactions are given in Equations 24.18 and 24.19.

Graphite anodes
+
−
Al_2O_3 dissolved in molten cryolite
Molten aluminum
Carbon-lined iron

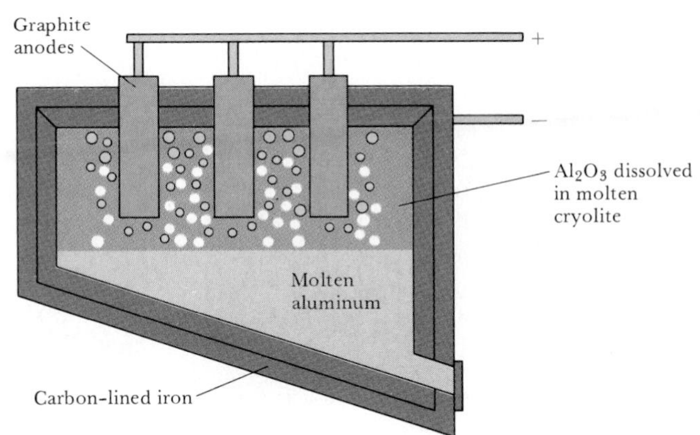

Figure 24.10 Typical Hall-process electrolysis cell used to reduce aluminum. Because molten aluminum is more dense than the molten mixture of Na_3AlF_6 and Al_2O_3, the metal collects at the bottom of the cell.

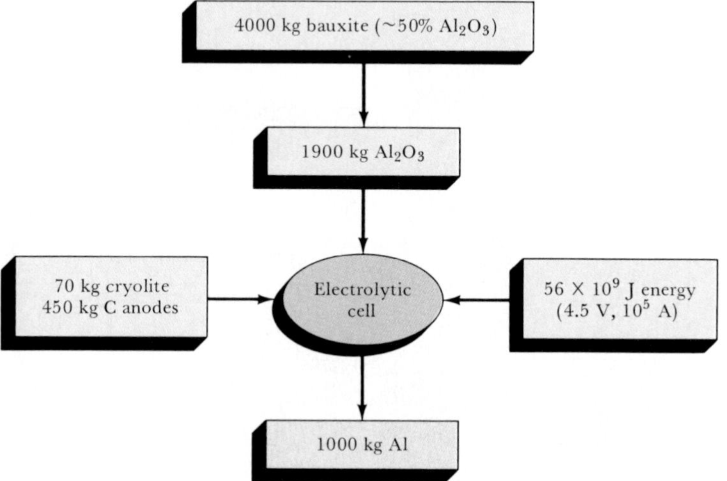

Figure 24.11 The quantities of bauxite, cryolite, graphite, and energy required to produce 1000 kg of aluminum.

$$\text{Anode:} \quad C(s) + 2O^{2-}(l) \longrightarrow CO_2(g) + 4e^- \quad [24.18]$$

$$\text{Cathode:} \quad 3e^- + Al^{3+}(l) \longrightarrow Al(l) \quad [24.19]$$

The amounts of raw materials and energy required to produce 1000 kg of aluminum metal from bauxite by this procedure are summarized in Figure 24.11.

Electrorefining of Copper

Copper is widely used to make electrical wiring and in other applications that utilize its high electrical conductivity. The crude copper obtained by pyrometallurgical methods is not suitable to serve in electrical applications, because impurities greatly reduce the metal's conductivity.

Purification of copper is achieved by electrolysis, as illustrated in Figure 24.12. Large slabs of crude copper serve as the anodes in the cell, and thin sheets of pure copper serve as the cathodes. The electrolyte consists of an acidic solution of $CuSO_4$. Application of a suitable voltage to the electrodes causes oxidation of copper metal at the anode and reduction of Cu^{2+} to form copper metal at the cathode. These reactions occur because copper is oxidized more readily than water:

$$Cu(s) \longrightarrow Cu^{2+}(aq) + 2e^- \qquad E^\circ = -0.34 \text{ V} \quad [24.20]$$

$$2H_2O(l) \longrightarrow 4H^+(aq) + O_2(g) + 4e^- \qquad E^\circ = -1.23 \text{ V} \quad [24.21]$$

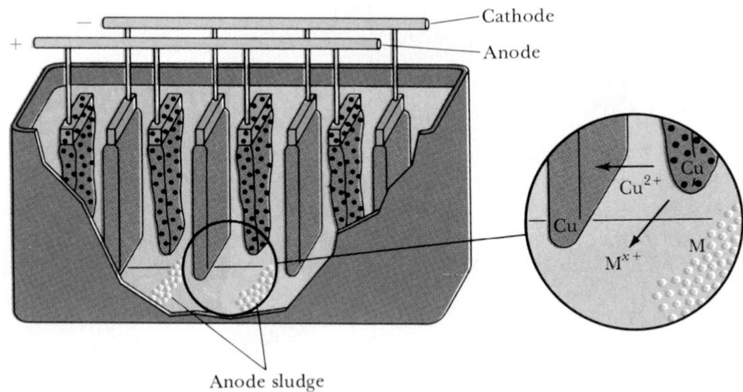

Figure 24.12 Electrolysis cell for refining of copper. Notice that as the anodes dissolve away, the cathodes on which the pure metal is deposited grow in size.

A HISTORICAL PERSPECTIVE: Charles M. Hall

Charles M. Hall (Figure 24.13) began work on the problem of reducing aluminum in about 1885, after he had learned from a professor of the difficulty of reducing ores of very active metals. Prior to the development of his electrolytic process, aluminum was obtained by a chemical reduction using sodium or potassium as the reducing agent. The procedure was very costly; as late as 1852 the cost of aluminum was $545 per pound. During the Paris Exposition in 1855, aluminum was exhibited as a rare metal in spite of the fact that it is the third most abundant element in the earth's crust. Hall, who was 21 years old when he began his research, utilized handmade and borrowed equipment in his studies, and used a woodshed near his home as his laboratory. In about a year's time, he was able to solve the problem of reducing aluminum. His solution consisted of finding an ionic compound that could be melted to form a conducting medium that would dissolve Al_2O_3 but would not interfere in the electrolysis reactions. The relatively rare mineral cryolite, Na_3AlF_6, found in Greenland, met these criteria. Ironically, Paul Héroult, who was the same age as Hall, made the same discovery in France at about the same time. As a result of the research of Hall and Héroult, large-scale production of aluminum became commercially feasible, and aluminum became a common and familiar metal.

Figure 24.13 Charles M. Hall (1863–1914) as a young man. (Courtesy of ALCOA)

The impurities in the copper anode include lead, zinc, nickel, arsenic, selenium, tellurium, and several precious metals including gold and silver. Metallic impurities that are more active than copper are readily oxidized at the anode but do not plate out at the cathode, because their reduction potentials are more negative than that for copper. However, less active metals are not oxidized at the anode. Instead, they collect below the anode as a sludge that is collected and processed to recover the valuable metals. The anode sludges from copper refining cells provide one-fourth of U.S. silver production and about one-eighth of U.S. gold production.

SAMPLE EXERCISE 24.1

Nickel is one of the chief impurities in the crude copper that is subjected to electrorefining. What happens to this nickel in the course of the electrolytic process?

Solution: The standard potential for oxidation of nickel is more positive than that for copper:

$$Ni(s) \longrightarrow Ni^{2+}(aq) + 2e^- \qquad E^\circ = 0.28 \text{ V}$$

$$Cu(s) \longrightarrow Cu^{2+}(aq) + 2e^- \qquad E^\circ = -0.34 \text{ V}$$

Nickel will thus be more readily oxidized than copper at the anode, assuming standard conditions. Of course, we do not have standard conditions in the electrolytic cell. The crude copper anode is nearly all copper, so we can assume that the activity, or effective concentration, of copper in the anode is essentially the same as that for pure metal. However, the nickel is present as a highly dilute impurity in the copper.

Thus the activity, or effective concentration, of nickel in the anode is much lower than it would be if we had a pure nickel anode. Nonetheless, the nickel is preferentially oxidized at the anode. The reduction of Ni^{2+}, which is just the reverse of the oxidation process, occurs less readily than the reduction of Cu^{2+}. The Ni^{2+} thus accumulates in the electrolyte solution, while the Cu^{2+} is reduced at the cathode. After a time it is necessary to recycle the electrolyte solution to remove the accumulated metal ion impurities, such as Ni^{2+}

PRACTICE EXERCISE

Zinc is another common impurity in copper. Using electrode potentials, determine whether zinc will accumulate in the anode sludge or in the electrolytic solution during the electrorefining of copper. ***Answer:*** electrolytic solution

24.6 METALLIC BONDING

In our discussions of metallurgy, we have so far confined ourselves to discussing the methods employed for obtaining metals in pure form. Metallurgy is also concerned with understanding the properties of metals, and with developing useful new materials. As with any branch of science and engineering, our ability to make advances is coupled to our understanding of the fundamental properties of the systems with which we work. At several places in the text, we have referred to the differences between metals and nonmetals with regard to both physical and chemical behavior. Let us now consider the distinctive properties of metals and then relate these properties to a model for metallic bonding.

Physical Properties of Metals

You have no doubt at some time held a length of copper wire or an iron bolt. Perhaps you have had occasion to observe the surface of a freshly cut piece of sodium metal. These substances, although distinct from one another, share certain similarities that enable us to classify them as metallic. A fresh metal surface has a characteristic luster. In addition, metals that we can handle with bare hands have a characteristic cold feeling related to their high heat conductivity. Metals also have high electrical conductivities; electrical current flows easily through metals. Current flow occurs without any displacement of atoms within the metal structure and is due to the flow of electrons within the metal. The heat conductivity of a metal usually parallels its electrical conductivity. For example, silver and copper, which possess the highest electrical conductivities, also possess the highest heat conductivities. This observation suggests that the two types of conductivity have the same origin in metals, which we shall soon discuss.

Most metals are *malleable*, which means that they can be hammered into thin sheets, and *ductile*, which means that they can be drawn into wires (Figure 7.11). These properties indicate that the atoms are capable of slipping with respect to one another. Ionic solids or crystals of most covalent compounds do not exhibit such behavior. These types of solids are typically brittle and fracture easily. Consider, for example, the difference between dropping an ice cube and a block of aluminum metal on a concrete floor.

Metals form solid structures in which the atoms are arranged as close-packed spheres or in a similar packing arrangement. For example, copper possesses a cubic close-packed structure, (Section 11.7) in which each copper atom is in contact with 12 other copper atoms. The number of valence-shell electrons available for bond formation is insufficient for a copper atom to form an electron-pair bond to each of its neighbors. If each atom is to share its bonding electrons with all its neighbors, these electrons must be able to move from one bonding region to another.

Electron-Sea Model for Metallic Bonding

One very simple model that accounts for some of the most important characteristics of metals is the *electron-sea model.* In this model, the metal is pictured as an array of metal cations in a "sea" of electrons, as illustrated in Figure 24.14. The electrons are confined to the metal by electrostatic attractions to the cations, and they are uniformly distributed throughout the structure. However, the electrons are mobile, and no individual electron is confined to any particular metal ion. When a metal wire is connected to the terminals of a battery, electrons flow through the metal toward the positive terminal and into the metal from the battery at the negative terminal. The high heat conductivity of metals is also accounted for by the mobility of the electrons, which permits ready transfer of kinetic energy throughout the solid. The ability of metals to deform (their malleability and ductility) can be explained by the fact that metal ions can move without any specific bonds being broken; the change in the positions of the ions brought about in reshaping the metal is accommodated by a redistribution of electrons.

The electron-sea model accounts in a qualitative way for some properties of metals, such as their conductivity and malleability, but it does not adequately account for others, such as melting points, heats of atomization, and hardness. To understand these properties, we must consider the number of electrons involved in metallic bonding. In the electron-sea model, we expect the strength of bonding to increase as the number of bonding (valence) electrons increases.

If you look back at Figure 7.12 (page 225), you will see that the melting points of transition metals tend to increase with increasing atomic number up to group 6B, near the center of the transition metals, and then to decrease beyond that point. Curves for heat of fusion, hardness, boiling point, and heat of atomization have similar appearances, although the maximum point in each period is often reached in group 8B. These results imply that the strength of metallic bonding first increases with increasing number of electrons, then decreases. This trend can be understood by applying the concepts of molecular-orbital theory to metals.

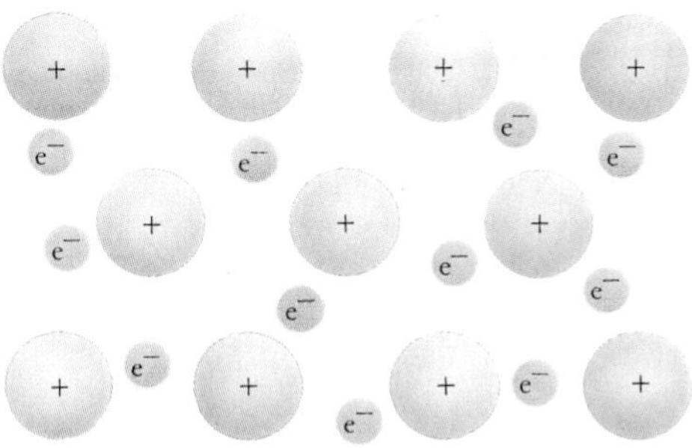

Figure 24.14 Schematic illustration of the electron-sea model of the electronic structure of metals.

Molecular-Orbital Model for Metals

In considering the structures of molecules such as benzene (Section 9.5), we saw that in some cases electrons are delocalized, or distributed, over several atoms. This happens when the atomic orbitals on one atom are able to interact with atomic orbitals on several neighboring atoms. In graphite (Section 11.8) the electrons are delocalized over an entire plane. It is useful to think of the bonding in metals in a similar way. The valence atomic orbitals on one metal atom overlap with those on several nearest neighbors, which in turn overlap with atomic orbitals on still other atoms.

We saw in Section 9.6 that the overlap of atomic orbitals can lead to the formation of molecular orbitals. The number of molecular orbitals is equal to the number of atomic orbitals that overlap. In a metal, the number of atomic orbitals that interact or overlap is very large. Thus, the number of molecular orbitals is also very large. Figure 24.15 shows schematically what happens as increasing numbers of metal atoms come together to form molecular orbitals. As overlap of atomic orbitals occurs, bonding and antibonding molecular-orbital combinations are formed. The energies of these molecular orbitals lie at closely spaced intervals in the energy range between the highest- and lowest-energy orbitals. Consequently, interaction of all the valence atomic orbitals of each metal atom with the orbitals of adjacent metal atoms gives rise to a huge number of very closely spaced molecular orbitals that extend over the entire metal structure. The energy separations between these metal orbitals are so tiny that for all practical purposes we may think of the orbitals as forming a continuous *band* of allowed energy states, referred to as an *energy band*, as shown in Figure 24.15.

The electrons available for metallic bonding do not completely fill the available molecular orbitals; we can think of the energy band as a partially filled container for electrons. The incomplete filling of the energy band gives rise to characteristic metallic properties. The electrons in orbitals near the top of the occupied levels require very little energy input to be "promoted" to still higher energy orbitals, which are unoccupied. Under the influence of any source of excitation, such as an applied electrical potential or an input of thermal energy, electrons move into previously vacant levels and are thus freed to move through the lattice, giving rise to electrical and thermal conductivity.

Trends in properties such as the melting point can readily be explained in terms of the molecular-orbital model. In transition metals, the valence-shell *s*, *p*, and *d* orbitals may overlap. If they do, the resultant energy band can ac-

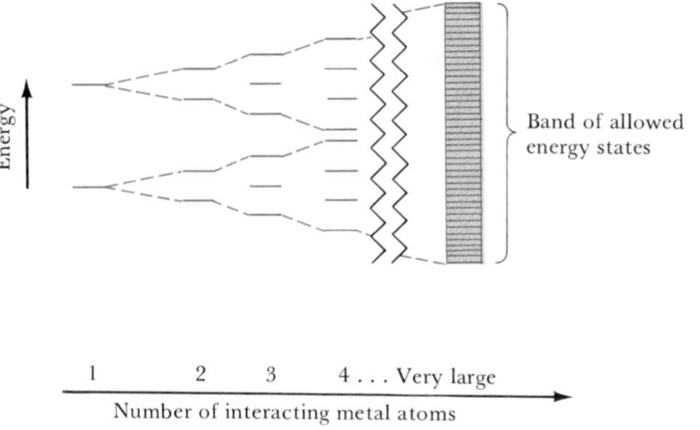

Figure 24.15 Schematic illustration of the interactions of metal atomic orbitals to form the delocalized orbitals of the metal lattice. The two atomic orbitals on each metal atom in this example might represent an *s* orbital and a *d* orbital. Whatever the orbitals, a large number of molecular orbitals with closely spaced energy levels arise. The number of electrons available does not completely fill these orbitals.

A CLOSER LOOK: Insulators and Semiconductors

A solid exhibits metallic character because it has a partially filled energy band, as shown in Figure 24.16(*a*); there are more molecular orbitals in the band than are needed to accommodate all the bonding electrons of the structure. Thus, an excited electron may easily move to the nearby higher orbital. In some solids, however, the electrons completely fill the allowed levels in a band. Such a situation applies, for example, to diamond. When we apply molecular-orbital theory to this solid, we find that the bands of allowed energies are as shown in Figure 24.16(*b*). The carbon 2*s* and 2*p* atomic orbitals combine to form two energy bands, each of which accommodates four electrons per carbon atom. One of these is completely filled with electrons. The other is completely empty. There is a large energy gap between the two bands. Because there is no readily available vacant orbital into which the highest-energy electrons can move under the influence of an applied electrical potential, diamond is not a good electrical conductor. Solids in which the energy bands are either completely filled or completely empty are electrical *insulators*.

Silicon and germanium have electronic structures like diamond. However, the energy gap between the filled and empty bands becomes smaller as we move from carbon (diamond) to silicon to germanium, as seen in Table 24.4. For silicon and germanium, the energy gap is small enough that at ordinary temperatures, a few electrons have sufficient energy to jump from the filled band (called the *valence band*) to the empty band (called the *conduction band*). As a result, there are some empty orbitals in the valence band, permitting electrical conductivity; the electrons in the upper energy band also serve as carriers of electrical current. Therefore, silicon and germanium are *semiconductors*, solids with electrical conductivities between those of metals and those of insulators. Other substances, for example GaAs, also behave as semiconductors.

Figure 24.16 Metallic conductors have partly filled energy bands, as shown in (*a*). Insulators have filled or empty energy bands, as in (*b*).

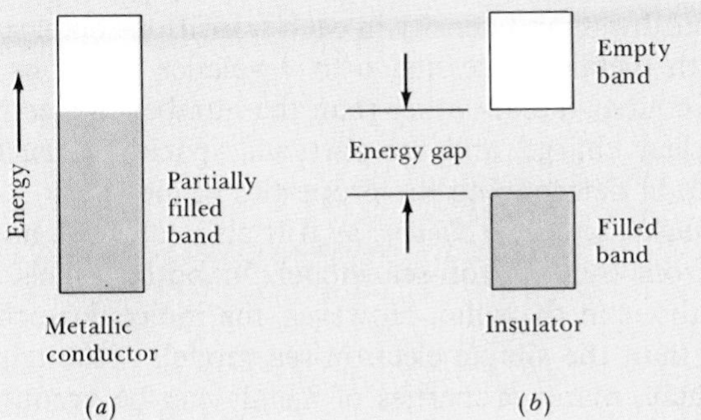

Table 24.4 Energy Gaps in Diamond, Silicon, and Germanium

Element	Energy gap (kJ/mol)
C	502
Si	100
Ge	67

The electrical conductivity of a semiconductor or insulator can be modified by adding small amounts of other substances. This process, called *doping*, causes the solid to have either too few or too many electrons to fill the valence band. Consider what happens to silicon when a small amount of phosphorus or another element from group 5A is added. The phosphorus atoms substitute for silicon atoms at random sites in the structure. Phosphorus, however, possesses five valence-shell electrons per atom, as compared to four for silicon. There is no room for these extra electrons in the valence band. They must therefore occupy the conduction band, as illustrated in Figure 24.17. These higher-energy electrons have access to many vacant orbitals within the energy band they occupy and serve as carriers of electrical current. Silicon doped with phosphorus in this manner is called an *n-type* semiconductor, because this doping introduces extra *n*egative charges (electrons) into the system.

If the silicon is doped instead with a group 3A ele-

Figure 24.17 Effect of doping on the occupancy of the allowed energy levels in silicon. (*a*) Pure silicon. The valence-shell electrons just fill the lower-energy allowed energy levels. (*b*) Doped with phosphorus. Excess electrons occupy the lowest-energy orbitals in the higher-energy band of allowed energies. These electrons are capable of conducting current. (*c*) Silicon doped with gallium. There are not quite enough electrons to fully occupy the orbitals of the lower-energy allowed band. The presence of vacant orbitals in this band permits current flow.

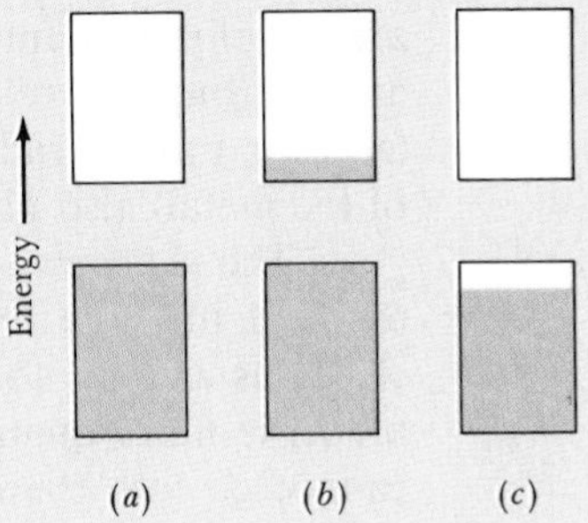

ment, such as gallium, the atoms that substitute for silicon have one electron too few to meet their bonding requirements to the four neighboring silicon atoms. The valence band is thus not completely filled, as illustrated in Figure 24.17(*c*). Under the influence of an applied field, electrons can move from occupied molecular orbitals to the few that are vacant in the valence band. A semiconductor formed by doping silicon with a group 3A element is called a *p-type* semiconductor, because this doping creates electron vacancies that can be thought of as *p*ositive holes in the system. The modern electronics industry is based on integrated circuitry formed from silicon or germanium doped with various elements to create the desired electronic characteristics (Figure 24.18).

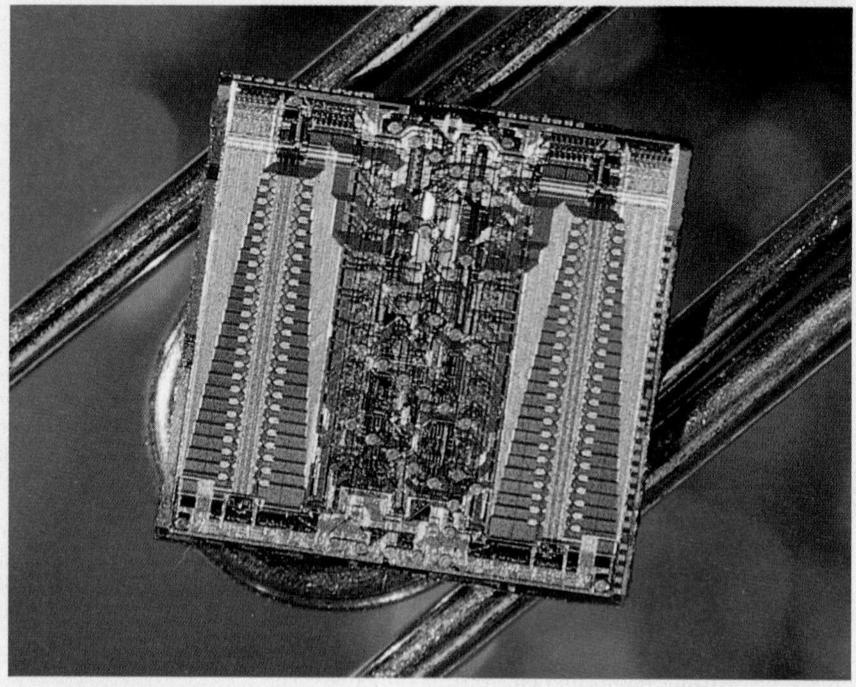

Figure 24.18
Semiconductors permit tremendous miniaturization of electronic devices. (Photo courtesy of IBM)

commodate up to 2 + 6 + 10 = 18 electrons per atom. The lower-energy half of the band, which contributes to the net bonding in the metal, can hold 9 of these electrons. The higher-energy half of the energy band, which is the antibonding portion of the band, can accommodate the other 9 electrons. Thus if the *s*, *p*, and *d* orbitals interact, 9 electrons per atom gives the maximum bond order and thus the highest bond strength. If only the *s* and *d* orbitals overlap, the maximum bond order would correspond to the presence of 6 electrons per atom. This analysis agrees well with what we observe. The highest melting point, heat of atomization, hardness, and density in each transition-metal series are usually encountered with metals possessing 6 to 9 valence electrons per atom (groups 6B to 8B). Of course, factors other than the number of electrons (such as atomic radius, nuclear charge, and the particular packing structure of the metal) also play a role in determining the properties of metals.

The *molecular-orbital model* (or *band theory*, as it is also called) is not so different in some respects from the electron-sea model. In both models, the electrons are free to move about in the solid. However, the molecular-orbital model is more quantitative than the simple electron-sea model. Although we cannot go into the details here, many properties of metals can be accounted for by quantum-mechanical calculations using molecular-orbital theory.

Table 24.5 Some Common Alloys

Primary element	Name of alloy	Composition by weight	Properties	Uses
Bismuth	Wood's metal	50% Bi, 25% Pb, 12.5% Sn, 12.5% Cd	Low melting point (70°C)	Fuse plugs, automatic sprinklers
Copper	Yellow brass	67% Cu, 33% Zn	Ductile, takes polish	Hardware items
Iron	Stainless steel	80.6% Fe, 0.4% C, 18% Cr, 1% Ni	Resists corrosion	Tableware
Lead	Plumber's solder	67% Pb, 33% Sn	Low melting point (275°C)	Soldering joints
Silver	Sterling silver	92.5% Ag, 7.5% Cu	Bright surface	Tableware
	Dental amalgam	70% Ag, 18% Sn, 10% Cu, 2% Hg	Easily worked	Dental fillings

24.7 ALLOYS

An *alloy* is a material that contains more than one element and has the characteristic properties of metals. The alloying of metals is of great importance because it is one of the primary ways of modifying the properties of pure metallic elements. For example, nearly all the common uses of iron involve alloy compositions. As another example, pure gold is too soft to be used in jewelry, whereas alloys of gold and copper are quite hard. Pure gold is termed 24 karat; the common alloy used in jewelry is 14 karat, meaning that it is 58 percent gold ($\frac{14}{24} \times 100$). A gold alloy of this composition has suitable hardness to be used in jewelry. The alloy can be either yellow or white, depending on the elements added. Some further examples of alloys are given in Table 24.5.

Alloys can be classified as solution alloys, heterogeneous alloys, and intermetallic compounds. **Solution alloys** are homogeneous mixtures in which the components are dispersed randomly and uniformly. Atoms of the solute can take positions normally occupied by a solvent atom, thereby forming a *substitutional alloy*, or they can occupy interstitial positions, thereby forming an *interstitial alloy*. These types are diagramed in Figure 24.19.

Substitutional alloys are formed when the two metallic components have similar atomic radii and chemical-bonding characteristics. For example, silver and gold form such an alloy over the entire range of possible compositions. When two metals differ in radii by more than about 15 percent, solubility is more limited. For an interstitial alloy to form, the component present in the interstitial positions between the solvent atoms must have a much smaller covalent radius than the solvent atoms. Typically, an interstitial element is a nonmetal that participates in bonding to neighboring atoms. The presence of the extra bonds provided by the interstitial component causes the metal lattice to become harder, stronger, and less ductile. For example, iron containing up to

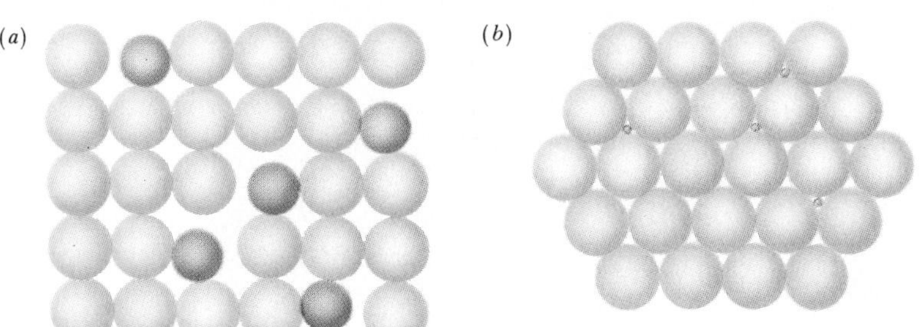

Figure 24.19
(*a*) Substitutional and (*b*) interstitial alloys. The gray spheres are host metal; the purple spheres are the other component of the alloy.

3 percent carbon is much harder than pure iron and has a much higher tensile strength and other desirable physical properties. *Mild steels* contain less than 0.2 percent carbon; they are malleable and ductile and are used to make cables, nails, and chains. *Medium steels* contain 0.2 to 0.6 percent carbon; they are tougher than mild steels and are used to make girders and rails. *High-carbon steel*, used in cutlery, tools, and springs, contains 0.6 to 1.5 percent carbon. In all these cases, other elements may be added to form *alloy steels*. Vanadium and chromium may be added to impart strength and to increase resistance to fatigue and corrosion. For example, a rail steel used in Sweden on lines bearing heavy ore carriers contains 0.7 percent carbon, 1 percent chromium, and 0.1 percent vanadium.

One of the most important iron alloys is stainless steel, which contains 0.4 percent carbon, 18 percent chromium, and 1 percent nickel. The chromium is obtained by carbon reduction of chromite, $FeCr_2O_4$, in an electric furnace. The product of the reduction is *ferrochrome*, $FeCr_2$, which is then added in the appropriate amount to molten iron that comes from the converter to achieve the desired steel composition. Alloys of the type we have been discussing differ from ordinary chemical compounds in that the composition is not fixed. The ratio of elements present may vary over a wide range, imparting a variety of specific physical and chemical properties to the materials.

In **heterogeneous alloys**, the components are not dispersed uniformly. For example, in the form of steel known as pearlite, two distinct phases—essentially pure iron and the compound Fe_3C, known as cementite—are present in alternating layers. In general, the properties of heterogeneous alloys depend not only on the composition but also on the manner in which the solid is formed from the molten mixture. Rapid cooling leads to distinctly different properties than does slow cooling.

Intermetallic Compounds

Intermetallic compounds are homogeneous alloys that have definite properties and composition. For example, copper and aluminum form a compound, $CuAl_2$, known as duraluminum. Intermetallic compounds play many important roles in modern society. The intermetallic compound Ni_3Al is a major component of jet aircraft engines because of its strength and low density. Razor blades are often coated with Cr_3Pt, which adds hardness, allowing the blade to stay sharp longer. The compound Co_5Sm is used in the permanent magnets in lightweight headsets (Figure 24.20) because of its high magnetic strength per unit weight.

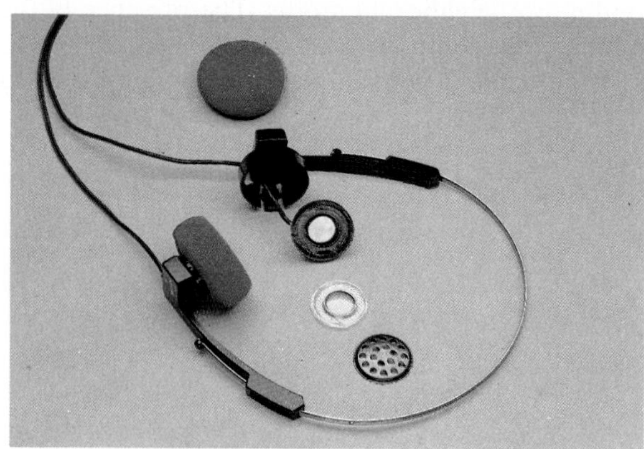

Figure 24.20 Interior of a lightweight audio headset. The assembly can be made small because of the very strong magnetism of the Co_5Sm alloy used. (Fundamental Photographs)

These examples illustrate some unusual ratios of combining elements. Nothing we have yet discussed in this text would lead us to predict such compositions. Among the many fundamental problems that remain unresolved in chemistry is the problem of developing a good theoretical model to predict the stoichiometries of intermetallic compounds.

24.8 TRANSITION METALS

Many of the most important metals of modern society are transition metals. Transition metals, which occupy the *d* block of the periodic table (Figure 24.21), include such familiar elements as chromium, iron, nickel, and copper. They also include less familiar elements that have come to play important roles in modern technology, such as those in the high-performance jet engine pictured earlier in the chapter, Figure 24.1. In this section and those that follow, we consider some of the physical and chemical properties of transition metals.

Physical Properties

Some of the physical properties of the elements of the first transition series are listed in Table 24.6. It is important to distinguish those properties that are characteristic of the bulk element in the liquid or solid phase from those that are characteristic of isolated atoms. Properties such as melting point and density are properties of the solid metal. These properties are related to the extent of metallic bonding that holds the solid together (Section 24.6). Properties of this sort vary over a wide range, but they often exhibit maxima near the middle of each transition series, somewhere in the vicinity of groups 6B to 8B. For example, the heat of atomization, ΔH_{atom}, listed in Table 24.6, reflects the extent of metallic bonding in the solid; in the atomization process [$M(s) \longrightarrow M(g)$], the metal-metal bonds are broken.

Ionization energy and atomic radius are properties of isolated atoms. In contrast to those properties characteristic of the bulk metal, the properties of

Figure 24.21 Transition metals are those elements that occupy the *d* block of the periodic table.

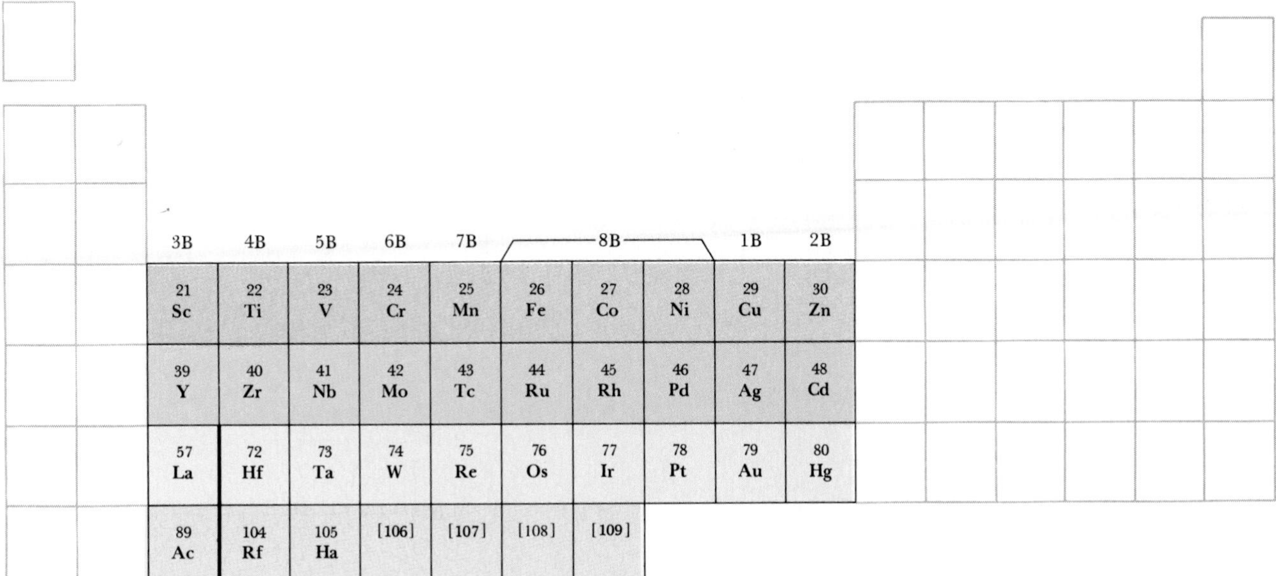

Table 24.6 Properties of the First Transition-Series Element

Group:	3B	4B	5B	6B	7B	8B			1B	2B
Element:	Sc	Ti	V	Cr	Mn	Fe	Co	Ni	Cu	Zn
Electron configuration	$3d^14s^2$	$3d^24s^2$	$3d^34s^2$	$3d^54s^1$	$3d^54s^2$	$3d^64s^2$	$3d^74s^2$	$3d^84s^2$	$3d^{10}4s^1$	$3d^{10}4s^2$
First ionization energy (kJ/mol)	631	658	650	653	717	759	758	737	745	906
Atomic radius (Å)	1.44	1.32	1.22	1.17	1.17	1.16	1.16	1.15	1.17	1.24
Density (g/cm^3)	3.0	4.5	6.1	7.9	7.2	7.9	8.7	8.9	8.9	7.1
ΔH_{atom} (kJ/mol)	377	470	514	398	283	415	428	430	338	130
Melting point (°C)	1538	1660	1917	1857	1244	1537	1494	1455	1084	420

individual atoms show a relatively smooth and rather small variation across each series. For example, the atomic radii of the transition metals are shown in Figure 24.22. Notice that there is a slight decrease in size in moving across the first part of each series. This trend can be understood in terms of the concept of increasing effective nuclear charge, as described in Chapter 6. For example, as we move across the first series, electrons are being added to the 3*d* orbitals. The 3*d* electrons effectively shield the outer 4*s* electrons from the increasing nuclear charge. Thus, the effective nuclear charge experienced by the outer 4*s* electrons increases only slowly, leading to a small decrease in radius across the series. The slight increase in radius at groups 1B and 2B is due to increasing electron-electron repulsions as the 3*d* subshell is completed.

The incomplete screening of the nuclear charge by added electrons produces a very interesting and important effect in the third transition-metal series.

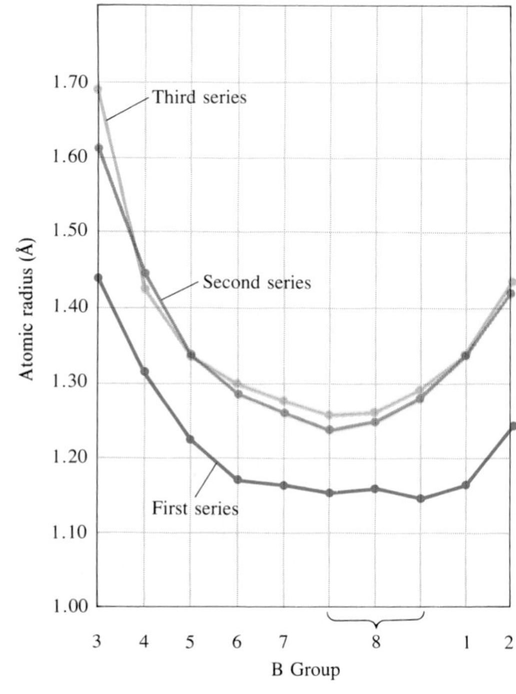

Figure 24.22 Variation in atomic radius of transition elements as a function of the periodic table group number.

In general, atomic size tends to increase going down a family because of the increasing principal quantum number of the outer-shell electrons (Section 7.3). Notice that the radii of the 3B elements follow the expected trend, Sc < Y < La. In group 4B, we would similarly expect the radius to increase in the order Ti < Zr < Hf. However, Hf has virtually the same radius as Zr. This effect has its origin in the lanthanide series, the elements with atomic numbers 58 through 71, which occur between La and Hf. The filling of 4*f* orbitals through the lanthanide elements causes a steady increase in the effective nuclear charge, producing a contraction in size called the **lanthanide contraction**. This contraction just offsets the increase we would expect as we go from the second to the third series. You can see from Figure 24.22 that the second- and third-series transition metals in each group have about the same radii all the way across the series. As a consequence, the second- and third-series metals in a given group have great similarity in their chemical properties. For example, the chemical properties of zirconium and hafnium are remarkably similar. They always occur together in nature, and they are very difficult to separate.

Electron Configurations and Oxidation States

Transition metals owe their location in the periodic table to the filling of the *d* subshells. When these metals are oxidized, however, they lose their outer *s* electrons before they lose electrons from the *d* subshell (Section 8.2). For example, the electron configuration of Fe is $[Ar]3d^64s^2$, whereas that of Fe^{2+} is $[Ar]3d^6$. Formation of Fe^{3+} requires loss of one 3*d* electron, giving $[Ar]3d^5$. Most transition-metal ions contain partially occupied *d* subshells. The existence of these *d* electrons is partially responsible for several characteristics of transition metals:

1. They often exhibit more than one stable oxidation state.
2. Many of their compounds are colored (see Figure 24.23). We shall discuss the origin of these colors in Chapter 25.
3. Transition metals and their compounds exhibit interesting magnetic properties.

Figure 24.23 Salts of transition-metal ions and their solutions. From left to right: Mn^{2+}, Fe^{2+}, Co^{2+}, Ni^{2+}, Cu^{2+}, and Zn^{2+}. (Donald Clegg and Roxy Wilson)

Figure 24.24 summarizes the common nonzero oxidation states for transition metals. The ones shown in grey are those most frequently encountered either in solution or in solid compounds. The ones shown in green are less common, but several well-known examples exist for each of these. For example, manganese is commonly found in solution in the +2 (Mn^{2+}) and +7 (MnO_4^-) oxidation states. In the solid state the +4 oxidation state (as in MnO_2) is common. The +3, +5, and +6 oxidation states are observed as relatively less stable species either in solution or in the solid state. Of course, the figure does not give a complete listing of all the oxidation states observed.

Inspection of Figure 24.24 and further study of transition metals reveal some interesting periodic trends in the occurrences and stabilities of oxidation states.

1. Oxidation states of +2 and +3 are more frequently observed in the first transition series than in the second and third. For example, in group 6B, the +3 oxidation state is common for Cr, occasionally observed for Mo,

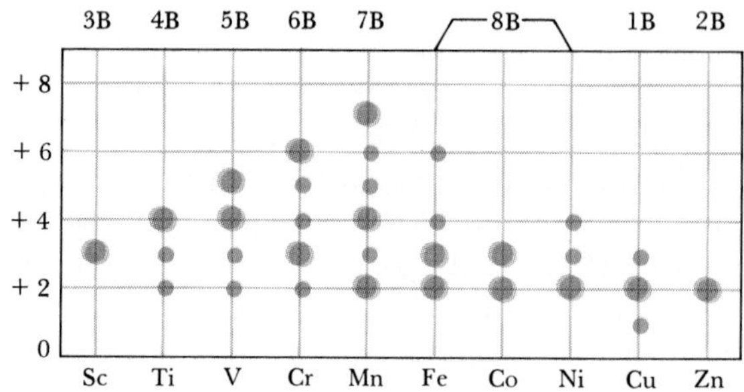

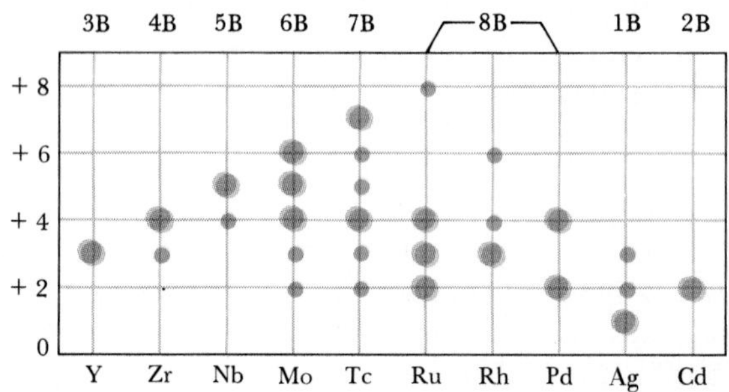

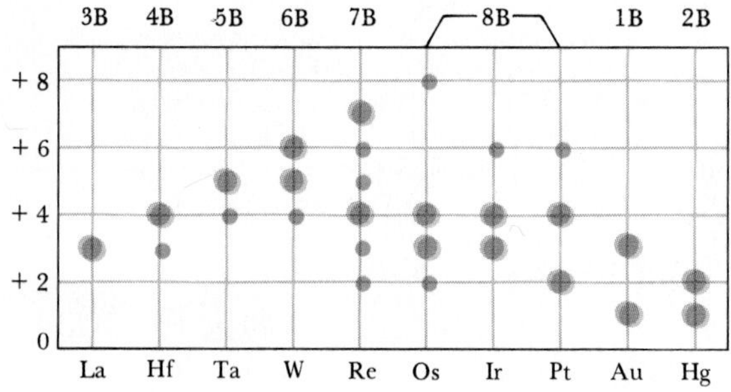

Figure 24.24 Nonzero oxidation states of transition elements. The most common oxidation states are indicated by the larger circles.

and only rarely found for W. Indeed, *as we go down a group of transition metals, the higher oxidation states become increasingly more stable relative to the +2 and +3 states.*

2. From Sc through Mn, the *maximum* oxidation state increases from +3 to +7, equaling in each case the total number of 4*s* plus 3*d* electrons in the atom. In the second and third series, the maximum is +8, an oxidation state found in RuO_4 and OsO_4. As we move to the right beyond Mn, Ru, and Os, the maximum observed oxidation state decreases. In general, the highest oxidation states are found only when the metals are combined with the most electronegative elements: O, F, and possibly Cl.

SAMPLE EXERCISE 24.2

The manganate ion, MnO_4^{2-}, is stable only in strongly basic solution. In acidic solution, it decomposes to form the permanganate ion and solid MnO_2. Write a balanced chemical equation to show this reaction.

Solution: In a disproportionation reaction, the same species is both the oxidizing and the reducing agent. In this case we have manganese going from the +6 to the +7 and +4 oxidation states. Because there is a change in oxidation state of +1 for the oxidation and −2 for the reduction, there must be twice as much MnO_4^- produced as MnO_2. The two half-reactions and overall reaction are:

$$2[MnO_4^{2-}(aq) \longrightarrow MnO_4^-(aq) + e^-]$$
$$MnO_4^{2-}(aq) + 2e^- + 4H^+(aq) \longrightarrow MnO_2(s) + 2H_2O(l)$$
$$3MnO_4^{2-}(aq) + 4H^+(aq) \longrightarrow 2MnO_4^-(aq) + MnO_2(s) + 2H_2O(l)$$

PRACTICE EXERCISE

Which of the following do you expect to be the stronger oxidizing agent, MnO_4^- or ReO_4^-? ***Answer:*** MnO_4^-

24.9 MAGNETISM IN TRANSITION METALS

The magnetic properties of transition metals and their compounds are both interesting and important. Measurements of magnetic properties provide information about chemical bonding. In addition, many important uses are made of magnetic properties in modern technology.

An electron possesses a "spin" that gives it a magnetic moment; that is, it behaves like a tiny magnet (Section 6.8). When all the electrons in an atom or ion are paired, the magnetic moments of the electrons effectively cancel each other, and the substance is *diamagnetic* (Section 9.7). When a diamagnetic substance is placed in a magnetic field, the motions of the electrons are affected in such a way that the substance is very weakly repelled by the magnet. Figure 24.25(*a*) represents a diamagnetic solid, one in which all the electrons in the solid are paired.

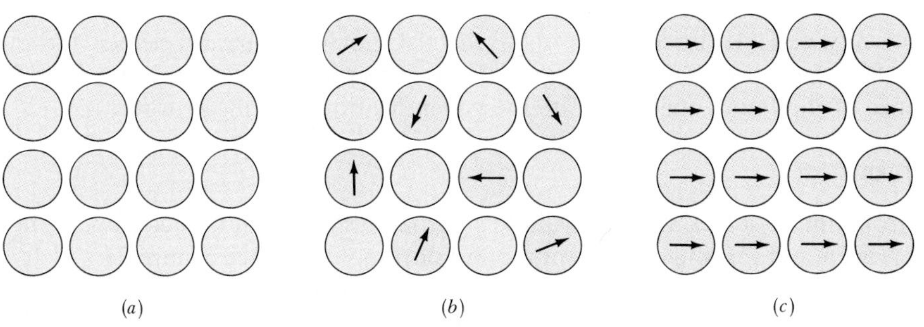

Figure 24.25 Types of magnetic behavior. (*a*) Diamagnetic; no centers with magnetic moments. (*b*) Simple paramagnetic; centers (atoms or ions) with magnetic moments are not aligned unless the substance is in a magnetic field. (*c*) Ferromagnetic; coupled centers aligned in a common direction.

Figure 24.26 Permanent magnets are made from ferromagnetic materials. (© Richard Megna/Fundamental Photographs)

When an atom or ion possesses one or more unpaired electrons, the substance is *paramagnetic* (Section 9.7). In a paramagnetic solid, the unpaired electrons on the atoms or ions of the solid are not influenced by the electrons on adjacent atoms or ions. The magnetic moments on the individual ions are randomly oriented, as shown in Figure 24.25(*b*). When placed in a magnetic field, however, the magnetic moments become aligned roughly parallel to each other, producing a net attractive interaction with the magnet. Thus, a paramagnetic substance is drawn into a magnetic field.

You are probably much more familiar with the magnetic behavior of simple iron magnets (Figure 24.26), a much stronger form of magnetism called **ferromagnetism**. Ferromagnetism arises when the unpaired electrons of the atoms or ions in a solid sense the orientations of electrons of their neighbors. The most stable (lowest-energy) arrangement results when the spins of electrons on adjacent atoms or ions are aligned in the same direction, as shown in Figure 24.25(*c*). When a ferromagnetic solid is placed in a magnetic field, the electrons tend to align strongly along the magnetic field. The attraction for the magnetic field that results may be as much as 1 million times stronger than that for a simple paramagnetic substance. When the external magnetic field is removed, the interactions between the electrons cause the solid as a whole to maintain a magnetic moment. We then refer to it as a *permanent magnet*. The most common examples of ferromagnetic solids are the elements Fe, Co, and Ni. Many alloys exhibit greater ferromagnetism than do the pure metals themselves. Some metal oxides (for example, CrO_2 and Fe_3O_4) are also ferromagnetic. Several ferromagnetic oxides are used in magnetic recording tape.

24.10 CHEMICAL PROPERTIES OF SELECTED TRANSITION METALS

Let's now consider briefly some of the chemistry of four common elements from the first transition series. As you read this material, you should look for evidence of the trends that illustrate the generalizations outlined earlier.

Chromium

Chromium dissolves slowly in dilute hydrochloric or sulfuric acid, liberating hydrogen and forming a blue solution of chromous ion (see Figure 24.27 left);

$$Cr(s) + 2H^+(aq) \longrightarrow Cr^{2+}(aq) + H_2(g) \qquad [24.22]$$

Normally, however, you do not observe this blue color, because the chromium(II) ion is rapidly oxidized in air to the violet-colored chromium(III) ion:

$$4Cr^{2+}(aq) + O_2(g) + 4H^+(aq) \longrightarrow 4Cr^{3+}(aq) + 2H_2O(l) \quad [24.23]$$

When hydrochloric acid solution is used, the solution appears green as a result of the formation of complex ions containing chloride coordinated to chromium, for example, $Cr(H_2O)_4Cl_2^+$ (Figure 24.27, right).

Chromium is frequently encountered in aqueous solution in the +6 oxidation state. In basic solution, the yellow chromate ion, CrO_4^{2-}, is the most stable. In acidic solution, the dichromate ion, $Cr_2O_7^{2-}$, is formed:

$$CrO_4^{2-}(aq) + H^+(aq) \rightleftharpoons HCrO_4^-(aq) \quad [24.24]$$

$$2HCrO_4^-(aq) \rightleftharpoons Cr_2O_7^{2-}(aq) + H_2O(l) \quad [24.25]$$

You might recognize the second of these reactions as a condensation reaction in which water is split out from between two $HCrO_4^-$ ions. Similar reactions occur among the oxyanions of nonmetallic and other metallic elements (Section 23.4). The equilibrium between the dichromate and chromate ions is readily observable, because CrO_4^{2-} is bright yellow and $Cr_2O_7^{2-}$ is deep orange, as seen in Figure 24.28. The dichromate ion in acidic solution is a strong oxidizing agent, as evidenced by its large, positive reduction potential. By contrast, chromate ion in basic solution is not a particularly strong oxidizing agent.

Chromium in trace quantities is an essential element in human nutrition. However, at the concentrations employed in laboratory work the element is highly toxic.

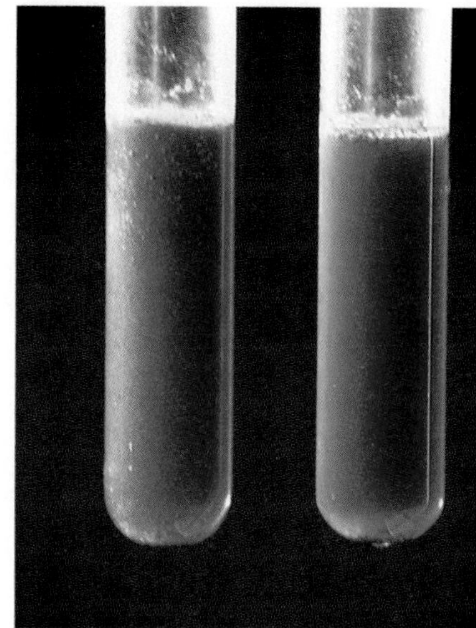

Figure 24.27 The tube on the left shows the reaction of chromium metal with 6 *M* sulfuric acid in the absence of air. If air had not been excluded, the solution would have taken on a violet cast. The tube on the right shows the reaction of chromium metal with 6 *M* hydrochloric acid in the presence of air. (Donald Clegg and Roxy Wilson)

Figure 24.28 Lead chromate, $PbCrO_4$ (on the left), and potassium dichromate, $K_2Cr_2O_7$ (on the right), illustrate the difference in color of the chromate and dichromate ions. (Richard Megna/Fundamental Photographs)

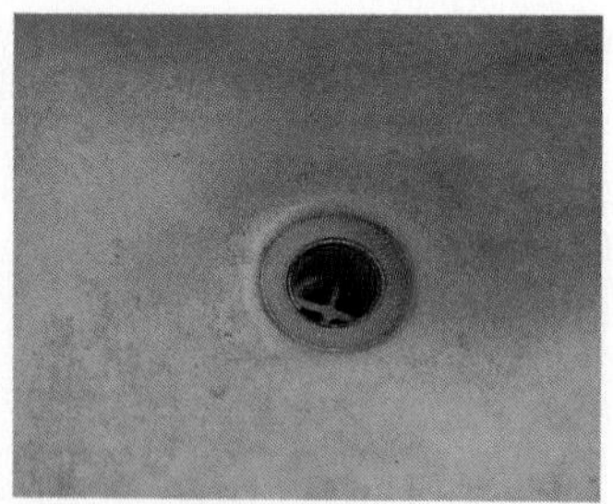

Figure 24.29 Stain in a sink caused by iron in water, which is oxidized to Fe^{3+} and precipitated as Fe_2O_3. (Page Poore)

Iron

We have already discussed the metallurgy of iron in considerable detail in Section 24.3. Let's consider here some of the important aqueous solution chemistry of the metal. Iron exists in aqueous solution in either the +2 (ferrous) or +3 (ferric) oxidation state. It often appears in natural waters because of contact of these waters with deposits of $FeCO_3$ ($K_{sp} = 3.2 \times 10^{-11}$). Dissolved CO_2 in the water can lead to dissolution of the mineral:

$$FeCO_3(s) + CO_2(aq) + H_2O(l) \longrightarrow Fe^{2+}(aq) + 2HCO_3^-(aq) \quad [24.26]$$

The dissolved Fe^{2+}, together with Ca^{2+} and Mg^{2+}, contributes to water hardness (Section 18.6).

The standard reduction potentials in Appendix E tell us much about the kind of chemical behavior we should expect to observe for iron. The potential for reduction from the +2 state to the metal is negative; however, the reduction from the +3 to the +2 state is positive. This tells us that iron should react with nonoxidizing acids such as dilute sulfuric acid or acetic acid to form $Fe^{2+}(aq)$, as indeed occurs. However, in the presence of air, $Fe^{2+}(aq)$ tends to oxidize to $Fe^{3+}(aq)$, as evidenced by the positive standard voltage for Equation 24.27:

$$4Fe^{2+}(aq) + O_2(g) + 4H^+(aq) \longrightarrow 4Fe^{3+}(aq) + 2H_2O(l) \qquad E^\circ = +0.46\text{ V} \quad [24.27]$$

You may have seen instances in which water dripping from a faucet has left a brown stain in a sink (see Figure 24.29). The brown color is due to insoluble iron(III) oxide, formed by oxidation of iron(II) present in the water:

$$4Fe^{2+}(aq) + 8HCO_3^-(aq) + O_2(g) \longrightarrow 2Fe_2O_3(s) + 8CO_2(g) + 4H_2O(l) \quad [24.28]$$

Figure 24.30 Addition of an NaOH solution to an aqueous solution of Fe^{3+} causes $Fe(OH)_3$ to precipitate. (Donald Clegg and Roxy Wilson)

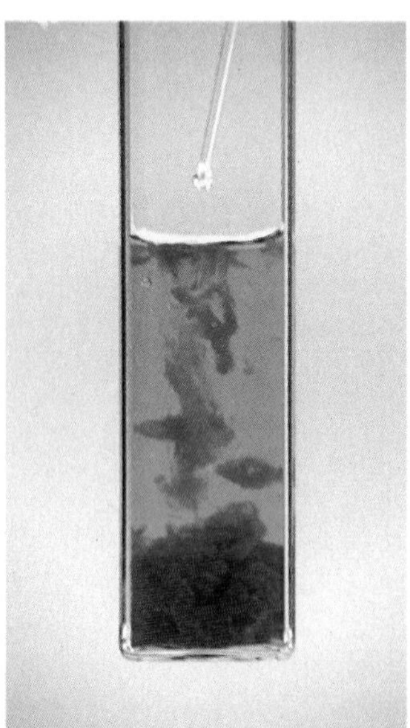

When iron metal reacts with an oxidizing acid such as warm, dilute nitric acid, $Fe^{3+}(aq)$ is formed directly:

$$Fe(s) + NO_3^-(aq) + 4H^+(aq) \longrightarrow Fe^{3+}(aq) + NO(g) + 2H_2O(l) \quad [24.29]$$

In the +3 oxidation state, iron is soluble in acidic solution as the hydrated ion, $Fe(H_2O)_6^{3+}$. However, this ion hydrolyzes readily (Section 16.10):

$$Fe(H_2O)_6^{3+}(aq) \rightleftharpoons Fe(H_2O)_5(OH)^{2+}(aq) + H^+(aq) \quad [24.30]$$

As an acidic solution of iron(III) is made more basic, a gelatinous red-brown precipitate, most accurately described as a hydrous oxide, $Fe_2O_3 \cdot nH_2O$, is formed (see Figure 24.30). In this formulation, n represents an indefinite number of water molecules, depending on the precise conditions of the precipitation. Usually, the precipitate that forms is represented merely as $Fe(OH)_3$. The solubility of $Fe(OH)_3$ is very low ($K_{sp} = 4 \times 10^{-38}$). It dissolves in strongly acidic solution, but not in basic solution. The fact that it does *not* dissolve in basic solution is the basis of the Bayer process, in which aluminum is separated from impurities, primarily iron(III) (Sections 17.5 and 24.4).

Nickel

Figure 24.31 Anhydrous nickel(II) chloride, $NiCl_2$, is yellow; aqueous solutions of this salt are green. (Donald Clegg and Roxy Wilson)

Nickel is not a particularly abundant element. It occurs chiefly in combination with arsenic, antimony, or sulfur. (You will note that both nickel and copper, which we discuss next, are found mainly in combination with sulfur and related nonmetals, whereas the earlier transition elements, such as chromium or iron, are often found as oxide ores.) After nickel is separated from other metallic elements that are usually present, the ore is roasted to NiO, which is then reduced with coke to form the crude metal. Crude nickel is normally refined by electrolysis, using nickel electrodes and an electrolyte containing both nickel sulfate and ammonium sulfate.

About 80 percent of the nickel produced goes into alloys. Among the more important of these are *stainless steels*, which also contains substantial amounts of chromium, and *Nichrome*, composed of 80 percent nickel and 20 percent chromium. Nichrome is used to make heating elements and in other applications that require a corrosion- and heat-resistant metal.

Nickel metal dissolves slowly in dilute mineral acids. Rather surprisingly, it does not dissolve in concentrated nitric acid. Attack of the strongly oxidizing acid on the metal surface results in formation of a protective oxide coat that stops further reaction.

In its compounds, nickel is almost always found in the +2 oxidation state. Anhydrous nickel(II) chloride, $NiCl_2$, is a yellow solid. It dissolves in water to form a green solution that contains the hydrated nickel(II) cation, $Ni(H_2O)_6{}^{2+}$ (see Figure 24.31). The Ni^{2+} ion forms complexes with many different anions and other neutral molecules, such as NH_3; these are discussed in Chapter 25. Among the less soluble compounds of nickel are the carbonate, $NiCO_3$, the oxalate, NiC_2O_4, and the sulfide, NiS.

Copper

In its aqueous-solution chemistry, copper exhibits two oxidation states, +1 (cuprous) and +2 (cupric). Note that in the +1 oxidation state copper possesses a d^{10} electron configuration. Salts of Cu^+ are often water-insoluble and are mostly white in color. In solution, the Cu^+ ion readily disproportionates:

$$2Cu^+(aq) \longrightarrow Cu^{2+}(aq) + Cu(s) \qquad K = 1.2 \times 10^6 \qquad [24.31]$$

Because of this reaction, and because copper(I) is readily oxidized to copper(II) under most solution conditions, the +2 oxidation state is by far the more common.

Many salts of Cu^{2+}, including $Cu(NO_3)_2$, $CuSO_4$, and $CuCl_2$, are water-soluble. Copper sulfate pentahydrate, $CuSO_4 \cdot 5H_2O$, a widely used salt, has four water molecules bound to the copper ion, and a fifth held to the $SO_4{}^{2-}$ ion by hydrogen bonding. The salt is blue (it is called *blue vitriol;* see Figure 24.32). Aqueous solutions of Cu^{2+}, in which the copper ion is coordinated by water molecules, are also blue. Among the insoluble compounds of copper(II) is $Cu(OH)_2$, which is formed when NaOH is added to an aqueous Cu^{2+} solution (see Figure 24.33). This blue compound readily loses water on heating to form black copper(II) oxide:

$$Cu(OH)_2(s) \longrightarrow CuO(s) + H_2O(l) \qquad [24.32]$$

Figure 24.32 Crystals of copper(II) sulfate pentahydrate, $CuSO_4 \cdot 5H_2O$. (Paul Silverman/ Fundamental Photographs)

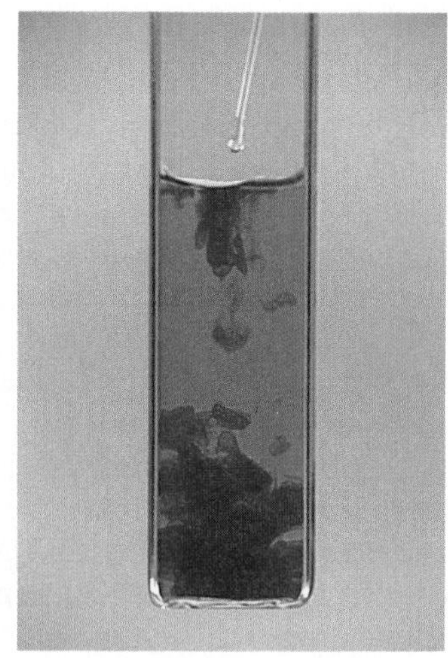

Figure 24.33 Addition of an NaOH solution to an aqueous solution of Cu^{2+} causes $Cu(OH)_2$ to precipitate. (Donald Clegg and Roxy Wilson)

CuS is one of the least soluble copper(II) compounds ($K_{sp} = 6.3 \times 10^{-36}$). This black substance does not dissolve in NaOH, NH_3, or in nonoxidizing acids such as HCl. However, it does dissolve in HNO_3, which oxidizes the sulfide to sulfur:

$$3CuS(s) + 8H^+(aq) + 2NO_3^-(aq) \longrightarrow 3Cu^{2+}(aq) + 3S(s) + 2NO(g) + 4H_2O(l) \quad [24.33]$$

$CuSO_4$ is often added to water to stop algae or fungal growth, and other copper preparations are used to spray or dust plants to protect them from lower organisms and insects. Copper compounds are not generally toxic to human beings, except in massive quantities. Our daily diet normally includes from 2 to 5 mg of copper.

FOR REVIEW

SUMMARY

Metallic elements occur in nature in minerals, solid inorganic substances found in various deposits or ores. Metallurgy is concerned with obtaining metals from these sources and with understanding and modifying the properties of metals. The processes required to obtain the metal from the crude ore (extractive metallurgy) include physical concentration steps, such as flotation.

Pyrometallurgy is the use of heat to bring about chemical reactions that convert an ore from one chemical form to another. One such process is calcination, in which an ore is heated to drive off a volatile substance, as in heating a carbonate ore to drive off CO_2. In roasting, the ore is heated under conditions that bring about reaction with the furnace atmosphere. For example, sulfide ores might be heated to cause oxidation of sulfur to SO_2. In a smelting operation, two or more layers of mutually insoluble materials form in the furnace. One layer consists of molten metal, and the other layer (slag) is composed of molten silicate minerals and other ionic materials such as phosphates.

Iron, the most important metal in modern society, is obtained from its oxide ores by reduction in a blast furnace. The reducing agent is carbon, in the form of coke. Limestone, $CaCO_3$, is added to react with the silicates present in the crude ore to form slag. The raw iron from the blast furnace, called pig iron, is usually taken directly to a converter, in which it is refined to form various kinds of steel. In the converter, the molten iron is reacted with pure oxygen to oxidize impurity elements.

Hydrometallurgy is the use of chemical processes occurring in aqueous solution to effect separation of a mineral from its ore or of one particular element from others. In leaching, an ore is treated with an aqueous reagent to dissolve a component selectively. In the Bayer process, aluminum is selectively dissolved from bauxite by treatment with concentrated NaOH solution.

Electrometallurgy is the use of electrolytic methods for the preparation or purification of a metallic element. Sodium is prepared by electrolysis of molten NaCl in a Downs cell. Aluminum is obtained in the Hall process by electrolysis of Al_2O_3 in molten cryolite, Na_3AlF_6. Copper is purified by electrolysis of aqueous copper sulfate solution using anodes composed of impure copper.

The properties of metals can be accounted for in a general way by the electron-sea model, in which the electrons are free to move throughout the metal structure. In the molecular-orbital model, the valence atomic orbitals of the metal atoms interact to form an energy band, which is not completely filled by valence electrons. The orbitals that comprise the energy band are delocalized over the atoms of the metal, and their energies are closely spaced. Because the energy differences between orbitals in the band are so small, promotion of electrons to higher-energy orbitals requires very little energy, giving rise to high electrical and thermal conductivity, as well as other characteristic metallic properties. In contrast, in an insulator, all the orbitals of the band are completely filled.

Alloys are materials that possess characteristic metallic properties and are composed of more than one element. Usually, one or more metallic elements are major components. Solution alloys are homogeneous alloys in which the components are distributed uniformly throughout. In heterogeneous alloys, the components are not distributed uniformly; instead, two or more distinct phases with characteristic compositions are present. Intermetallic compounds are homogeneous alloys that have definite properties and compositions.

Transition metals are characterized by incomplete filling of the *d* orbitals. The presence of *d* electrons in transition elements leads to the existence of multiple oxidation states. As we proceed through a given series of transition metals (horizontal row), the effective nuclear charge, as seen by the valence electrons, increases slowly. As a result, the later transition elements in a given row tend to adopt lower oxidation states and have slightly smaller ionic radii. In comparing the elements of a given group (vertical column), we find that the first-series metals tend to adopt lower oxidation states. Although the atomic and ionic radii increase in the second series as compared to the first, the elements of the second and third series are similar with respect to these properties. This similarity is due to the lanthanide contraction. Lanthanide elements, atomic numbers 58 through 71, display an increase in effective nuclear charge that compensates for the increase in major quantum number in the third series.

The presence of unpaired electrons in valence orbitals leads to interesting magnetic behavior in transition metals and their compounds. In ferromagnetic substances, the unpaired electron spins on atoms in a solid are affected by those on neighboring atoms. In a magnetic field, the spins become aligned along the direction of the magnetic field. When the magnetic field is removed, this orientation remains, giving the solid a magnetic moment as observed in permanent magnets.

In this chapter, we also briefly considered some chemistry of four of the common transition metals: chromium, iron, nickel, and copper.

KEY TERMS

lithosphere (Sec. 24.1)
ore (Sec. 24.1)
mineral (Sec. 24.1)
metallurgy (Sec. 24.2)
alloy (Sec. 24.2)
gangue (Sec. 24.2)
flotation process (Sec. 24.2)
pyrometallurgy (Sec. 24.3)
calcination (Sec. 24.3)
roasting (Sec. 24.3)
smelting (Sec. 24.3)
slag (Sec. 24.3)
refining (Sec. 24.3)
hydrometallurgy (Sec. 24.4)
leaching (Sec. 24.4)
Bayer process (Sec. 24.4)
electrometallurgy (Sec. 24.5)
Downs cell (Sec. 24.5)
Hall process (Sec. 24.5)
solution alloy (Sec. 24.7)
heterogeneous alloy (Sec. 24.7)
intermetallic compound (Sec. 24.7)
lanthanide contraction (Sec. 24.8)
ferromagnetism (Sec. 24.9)

EXERCISES

Metallurgy

24.1 Two of the most heavily utilized elements listed in Table 24.2 are Al and Fe. Indicate the most important natural sources of these elements and the oxidation state of the metal in each case.

24.2 Malachite is a common mineral of copper. What is the chemical formula for this substance? What is the oxidation state of Cu in malachite?

24.3 Complete and balance each of the following equations:

(a) $ZnCO_3(s) \xrightarrow{\Delta}$
(b) $MnO(s) + CO(g) \longrightarrow$
(c) $Al(OH)_3(s) + OH^-(aq) \longrightarrow$
(d) $TiCl_4(g) + K(l) \longrightarrow$
(e) $CaO(l) + P_2O_5(l) \longrightarrow$

24.4 Complete and balance each of the following equations:

(a) $PbS(s) + O_2(g) \longrightarrow$
(b) $PbCO_3(s) \xrightarrow{\Delta}$
(c) $WO_3(s) + H_2(g) \longrightarrow$
(d) $ZnO(s) + CO(g) \longrightarrow$
(e) $ZrC(s) + Cl_2(g) \longrightarrow$

24.5 Write balanced chemical equations for the reductions of FeO and Fe_2O_3 by H_2 or CO.

24.6 What is the major reducing agent in the reduction of iron ore in a blast furnace? Write a balanced chemical equation for the reduction process.

24.7 A charge of 2.0×10^4 kg of material containing 32 percent Cu_2S and 7 percent FeS is added to a converter and oxidized. What mass of $SO_2(g)$ is formed?

24.8 A charge of 1.2×10^4 kg of concentrate from a partial roast of a chalcopyrite ore, containing 26 percent FeO, is added to a furnace. Then SiO_2 is added to form a slag with FeO. Write the balanced equation for the reaction leading to slag formation, and calculate the mass of SiO_2 required to react with the FeO to form slag.

24.9 What role does each of the following materials play in the chemical processes that occur in the blast furnace: (a) air; (b) limestone; **(c)** coke; **(d)** water? Write balanced chemical equations to illustrate your answers.

24.10 In terms of the concepts discussed in Chapter 13, indicate why the molten metal and slag phases formed in the blast furnace shown in Figure 24.7 are immiscible.

24.11 Describe how electrometallurgy could be employed to purify crude cobalt metal. Describe the compositions of the electrodes and electrolyte, and write out all electrode reactions.

24.12 .The element tin is generally recovered from deposits of the ore cassiterite, SnO_2. The oxide is reduced with carbon, and the crude metal is purified by electrolysis. Write balanced chemical equations for the reduction process and the electrode reactions in the electrolysis, assuming that an acidic solution of $SnSO_4$ is employed as an electrolyte.

24.13 In an electrolytic process nickel sulfide is oxidized in a two-step reaction:

$$Ni_3S_2(s) \longrightarrow Ni^{2+}(aq) + 2NiS(s) + 2e^-$$
$$NiS(s) \longrightarrow Ni^{2+}(aq) + S(s) + 2e^-$$

What mass of Ni^{2+} is produced in solution by passage of a current of 66 A for a period of 8.0 hr, assuming the cell to be 100 percent efficient?

24.14 What mass of copper is deposited in an electrolytic refining cell by a passage of 240 A current for a period of 10 hr, assuming 80 percent current efficiency?

[24.15] Using the data in Appendix C, estimate the free-energy change for each of the following reactions at 1200°C:

(a) $PbO(s) + CO(g) \longrightarrow Pb(s) + CO_2(g)$
(b) $Si(s) + 2MnO(s) \longrightarrow SiO_2(s) + 2Mn(s)$
(c) $FeO(s) + H_2(g) \longrightarrow Fe(s) + H_2O(g)$

[24.16] In terms of thermodynamic functions, particularly free energies of formation, what conditions must be operative for refining iron in the converter shown in Figure 24.8 to be a workable process?

Metals and Alloys

24.17 Sodium is a highly malleable substance, whereas sodium chloride is not. Explain this difference in properties.

24.18 Suppose that you have samples of magnesium, iron, sulfur, high-carbon steel, and calcium sulfate, all formed into solid, pencil-shaped rods. Compare these substances with respect to appearance, flexibility, melting point, and thermal conductivity. Account for the differences you would expect to observe.

24.19 How does the electron-sea model account for the high electrical conductivity of copper metal?

24.20 Discuss briefly the difference in bonding between metals and nonmetals.

24.21 Moving across the fourth period of the periodic table, hardness increases at first, reaching a maximum at Cr; the hardness then decreases as we move from Cr to Zn. Explain this trend.

24.22 The densities of the elements K, Ca, Sc, and Ti are 0.86, 1.5, 3.2, and 4.5 g/cm^3, respectively. What factors are likely to be of major importance in determining this variation?

24.23 Use the band theory to describe the differences among insulators, conductors, and semiconductors.

[24.24] Indicate whether each of the following is likely to be an insulator, a metallic conductor, an *n*-type semiconductor, or a *p*-type semiconductor: **(a)** germanium doped with Ga; **(b)** pure $CuAl_2$; **(c)** pure boron; **(d)** silicon doped with As; **(e)** pure Nb; **(f)** pure MgO.

24.25 Tin exists in two allotropic forms; gray tin has a diamond structure, and white tin has a close-packed structure. Which of these allotropic forms would you expect to be more metallic in character? Explain why the electrical conductivity of white tin is much greater than that of gray tin. Which form would you expect to have the longer Sn—Sn bond distance?

24.26 The interatomic distance in silver metal, 2.88 Å, is much greater than the calculated Ag—Ag single-bond distance, 1.34 Å. Nevertheless, the melting point of silver, 906°C, and its high boiling point, 2164°C, suggest that there is strong bonding between silver atoms in the metal. Describe a model for the structure of silver metal that is consistent with these observed properties.

24.27 Define the term *alloy*. Distinguish among solution alloys, heterogeneous alloys, and intermetallic compounds.

24.28 Distinguish between substitutional and interstitial alloys. What conditions favor formation of substitutional alloys?

Transition Metals

24.29 How would you expect the heat of fusion for vanadium to compare with the heat of vaporization? Explain.

24.30 How would you expect the heat of fusion of titanium to compare with that of vanadium?

24.31 Which of the following properties is most closely related to bonding in the metal, and which is characteristic of individual atoms: **(a)** heat of sublimation; **(b)** boiling point; **(c)** first ionization energy; **(d)** malleability.

24.32 Which of the following properties is most clearly related to the bonding in metals, and which is characteristic of individual atoms: **(a)** heat of fusion; **(b)** hardness; **(c)** atomic radius; **(d)** electrode potential for $M^{2+}(aq) + 2e^- \longrightarrow M(s)$?

24.33 What is the lanthanide contraction? Why does it occur?

24.34 How does the lanthanide contraction affect the properties of transition elements?

24.35 Write the formula for the chloride corresponding to the highest expected oxidation state for each of the following elements: **(a)** Nb; **(b)** W; **(c)** Co; **(d)** Cd; **(e)** Re.

24.36 Write the formula for the oxide corresponding to the highest expected oxidation state for each of the following elements: **(a)** Cr; **(b)** Zn; **(c)** Hf; **(d)** Os; **(e)** Sc.

24.37 What accounts for the fact that chromium exhibits several oxidation states in its compounds, whereas aluminum exhibits only the +3 oxidation state?

24.38 Account for the fact that zinc exhibits only the +2 oxidation state in its compounds, whereas copper exhibits both the +1 and +2 oxidation states.

24.39 Write the expected electron configuration for each of the following ions: (a) Ru^{4+}; (b) Hf^{3+}; **(c)** Ag^{3+}; **(d)** Co^{3+}; **(e)** V^{3+}; **(f)** Pt^{2+}.

24.40 Write the electron configurations for each of the following ions: **(a)** Ti^{2+}; **(b)** Sc^{3+}; **(c)** W^{4+}; **(d)** Ir^{3+}; **(e)** Ni^{2+}; **(f)** Ag^+.

24.41 Which would you expect to be more easily oxidized, Cr^{2+} or Fe^{2+}?

24.42 Which would you expect to be more easily oxidized, Ti^{2+} or Ni^{2+}?

24.43 Which would you expect to be the better oxidizing agent, $CrO_4{}^{2-}$ or $WO_4{}^{2-}$?

24.44 Which would you expect to be harder to reduce, V_2O_5 or Ta_2O_5?

24.45 What are the two most common oxidation states of chromium in aqueous solution?

24.46 Give the chemical formulas and colors of the chromate and dichromate ions. Which of these ions is more stable in acidic solution?

24.47 What are the two most common oxidation states of iron in aqueous solution?

24.48 How does the presence of air affect the relative stabilities of ferrous and ferric ions?

24.49 Write balanced chemical equations for the reaction between iron and **(a)** hydrochloric acid; **(b)** nitric acid.

24.50 Write balanced chemical equations for the reaction of **(a)** $Fe^{3+}(aq)$ and sodium hydroxide solution; **(b)** $FeCO_3(s)$ and hydrochloric acid solution.

24.51 What are the electron configurations of Ni^{2+} and Cu^{2+}? What colors are associated with $Ni^{2+}(aq)$ and $Cu^{2+}(aq)$?

24.52 Write the chemical equation for the disproportionation of Cu^+ in aqueous solution. What color is associated with the Cu^+ ion?

24.53 What, on an atomic level, distinguishes a paramagnetic material from a diamagnetic one? How does each behave in a magnetic field?

24.54 What characteristics of a ferromagnetic material distinguish it from one that is paramagnetic? What type of interaction must occur in the solid to bring about ferromagnetic behavior?

24.55 List three elements that are ferromagnetic.

24.56 Predict whether the following solid substances are diamagnetic, paramagnetic, or ferromagnetic: **(a)** Ni; **(b)** Mg; **(c)** NaCl; **(d)** CoO.

Additional Exercises

24.57 Copper exists in nature in both the free (elemental) and combined states, whereas aluminum is never found in the free state. Explain.

24.58 Write a balanced chemical equation to correspond to each of the following verbal descriptions. **(a)** $NiO(s)$ can be solubilized by leaching with aqueous sulfuric acid. **(b)** After concentration, an ore containing the mineral carrolite, $CuCo_2S_4$, is leached with aqueous sulfuric acid to produce a solution containing copper and cobalt ions. **(c)** Titanium dioxide is treated with chlorine in the presence of carbon as reducing agent to form $TiCl_4$. **(d)** Under oxygen pressure, $ZnS(s)$ reacts at 150°C with aqueous sulfuric acid to form soluble zinc sulfate, with deposition of elemental sulfur.

24.59 Write a balanced chemical equation for each of the following verbal descriptions. **(a)** Vanadium(V) oxide is formed by thermal decomposition of ammonium metavanadate, NH_4VO_3. **(b)** Vanadium(V) oxide is reduced to vanadium(IV) oxide by thermal reduction with gaseous SO_2. **(c)** Vanadium(V) oxide is reduced by magnesium in hydrochloric acid to give a green solution of vanadium(III). **(d)** Niobium(V) chloride reacts with water to yield crystals of niobic acid, $HNbO_3$.

24.60 In the early days of iron mining in the United States, the iron oxide mined in the Lake Superior region was reduced to pig iron in blast furnaces located near the mines. The furnaces used charcoal made from hardwood as the reducing agent (transportation savings prompted this procedure). What mass of pig iron containing 2.5 percent carbon was produced from each kilogram of iron ore, assuming that it is mined as magnetite, Fe_3O_4, of 70 percent purity?

24.61 The reduction of metal oxides is often accomplished using carbon monoxide as a reducing agent. Carbon (coke) and carbon dioxide are usually present, leading to the following reaction:

$$C(s) + CO_2(g) \rightleftharpoons 2CO(g)$$

Calculate the equilibrium constant for this reaction at 289 K and at 2000 K, assuming that the enthalpies and entropies do not depend upon temperature. Consult Chapter 19 as necessary, and use data from Appendix C.

24.62 Magnesium metal is obtained from seawater. First, water is allowed to evaporate in order to concentrate magnesium ions, and then $Ca(OH)_2$ is added to precipitate $Mg(OH)_2$. The $Mg(OH)_2$ is filtered off and dissolved in hydrochloric acid. Evaporating the resultant solution to dryness yields $MgCl_2$, which is melted and electrolyzed. Write balanced chemical equations for these reactions.

24.63 Magnesium is obtained by electrolysis of molten $MgCl_2$. Several cells are connected in parallel by very large copper buses that convey current to the cells. Assuming that the cells are 96 percent efficient in producing the desired products in electrolysis, what mass of Mg is formed by passage of a current of 97,000 A for a period of 24 hr?

24.64 Zinc can be purified by electrolytic refining. A current of 75 A is used at an applied voltage of 3.5 V. Assuming that the electrolytic process is 94 percent efficient, how much electrical energy (kilowatt hours) is required to produce 1.0×10^8 kg of zinc metal?

24.65 The galvanizing of iron sheet can be carried out electrolytically using a bath containing a zinc sulfate solution. The sheet is made the cathode, and a graphite anode is used. Calculate the cost of the electricity required to deposit a 0.30-mm layer of zinc on both sides of an iron sheet 1.0 m wide and 50 m long if the current is 30 A, the voltage is 3.5 V, and the energy efficiency of the process is 90 percent. Assume the cost of electricity is $0.080 per kilowatt hour. The density of zinc is 7.1 g/cm^3.

24.66 The crude copper that is subjected to electrorefining contains selenium and tellurium as impurities. Describe the probable fate of these elements during electrorefining, and relate your answer to the positions of these elements in the periodic table.

[24.67] In the electrolytic purification of nickel, some iron may be present in the crude metal. It is found that as iron accumulates in the electrolyte solution, the current efficiency of the electrolysis cell decreases when the contents of the cathode cell compartment are not isolated from the air. Write reactions involving iron species that could account for the loss in current efficiency. (Hint: Recall that iron can exist in solution in more than one oxidation state.)

[24.68] Write balanced chemical equations that correspond to the steps in the following brief account of the metallurgy of molybdenum: Molybdenum occurs primarily as the sulfide, MoS_2. On boiling with concentrated nitric acid, a white residue of MoO_3 is obtained. This is an acidic oxide; from a solution of hot, excess concentrated ammonia, ammonium molybdate crystallizes on cooling. On heating ammonium molybdate, white MoO_3 is obtained. On heating to 1200°C in hydrogen, a gray powder of metallic molybdenum is obtained.

[24.69] Suppose that a rich deposit of the mineral absolite, $CoO \cdot 2MnO_2 \cdot 4H_2O$, were found. Sketch out a scheme for recovery of cobalt metal from the ore containing this mineral.

24.70 Distinguish between **(a)** a substitutional and an interstitial alloy; **(b)** an intermetallic compound and a heterogeneous alloy; **(c)** a metal and an insulator; **(d)** the bonding in $Na_2(g)$ and in $Na(s)$.

24.71 Pure silicon is a very poor conductor of electricity. Titanium, which also possesses four valence-shell electrons, is a metallic conductor. Explain the difference.

24.72 Treatment of iron surfaces at high temperatures with ammonia leads to "nitriding," in which nitrogen atoms are incorporated into the metal structure. What kind of alloy formation occurs in this case? How would you expect nitriding to alter the properties of the metal?

[24.73] The melting point of neodymium, Nd, is 1019°C. Considering its place in the periodic table and the melting points of other elements as shown in Figure 7.12, what number of electrons would you estimate to be involved in metallic bonding in this element? What do your conclusions suggest about the role of 4*f* electrons in the bonding?

[24.74] The stabilities of the three complexes $Zn(H_2O)_4{}^{2+}$, $Zn(NH_3)_4{}^{2+}$, and $Zn(CN)_4{}^{2-}$ increase from the H_2O to the NH_3 to the CN^- complex. How do you expect the electrode potentials of these three complexes to compare to that of the processes in which $Zn^{2+}(aq)$ is reduced to Zn metal?

24.75 Explain why the heat of atomization of zinc is much smaller than the heat of atomization of vanadium (Table 24.6).

24.76 The statement has been made that the chemistries of zirconium and hafnium are more nearly identical than those for any other two elements of a given group. What accounts for this similarity?

25 Chemistry of Coordination Compounds

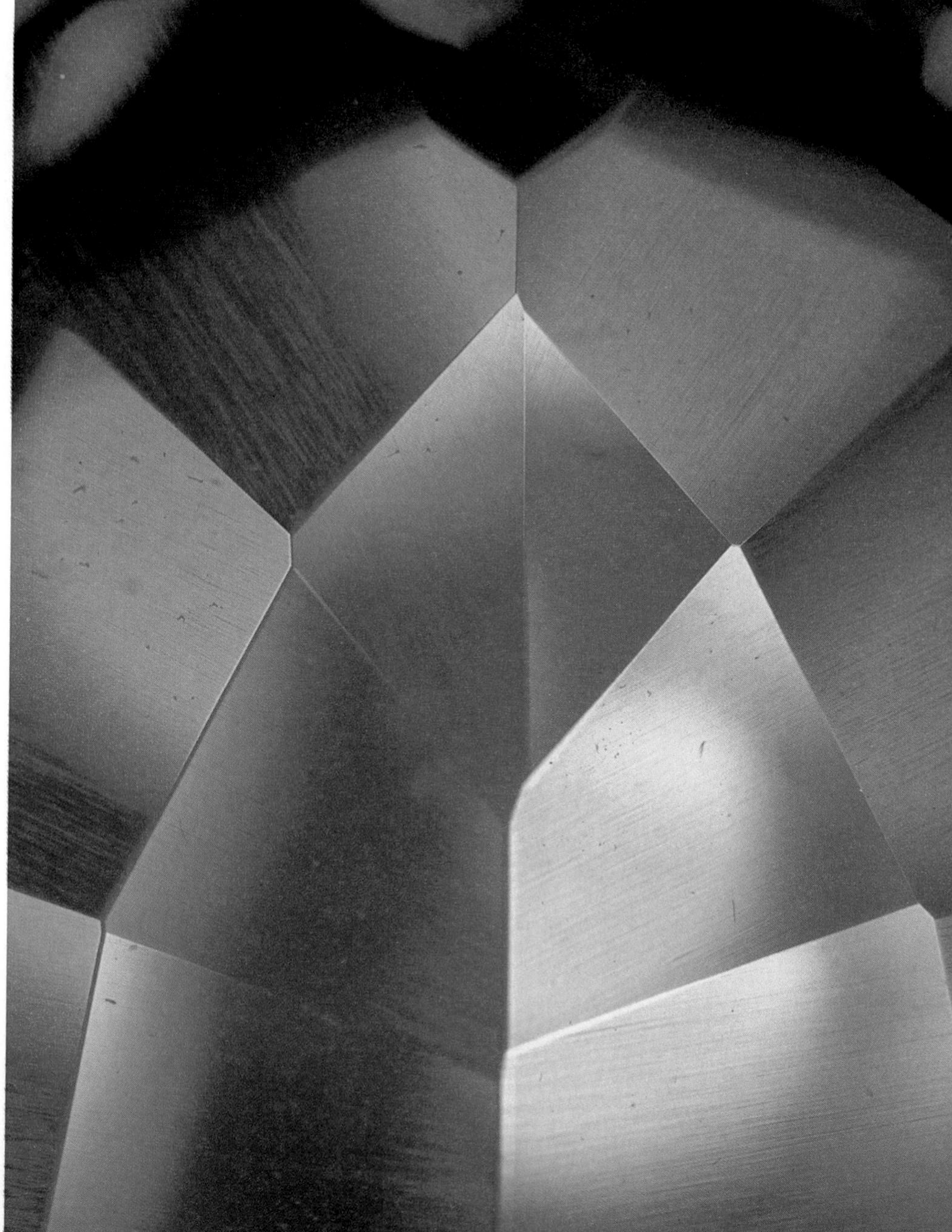

A corundum sapphire. Sapphires are one of several precious stones that consist of Al_2O_3 with small amounts of transition metals that give the stones their colors. Sapphire contains Fe^{2+} and Ti^{4+} ions. (John Walsh, Science Photo Library/Science Source/Photo Researchers)

In earlier chapters, we noted that metallic elements tend to lose electrons in chemical reactions. For this reason, positively charged metal ions play a primary role in the chemical behavior of metals. Of course, metal ions do not exist in isolation. In the first place, they are accompanied by anions whose negative charge balances the positive charge of the metal ions, producing neutral compounds. In addition, metal ions act as Lewis acids (Section 16.10). Anions or molecules with unshared pairs of electrons can act as Lewis bases and bind to the metal center. On several occasions, we have discussed compounds in which a metal ion is surrounded by a group of anions or neutral molecules. Some examples are $[Au(CN)_2]^-$, discussed in connection with metallurgy (Section 24.4); hemoglobin, discussed in connection with the oxygen-carrying capacity of the blood in Section 18.4; and $[Fe(H_2O)_6]^{3+}$ and $[Ag(NH_3)_2]^+$, encountered in our discussion of equilibria in Sections 16.10 and 17.5. Such species are known as **complex ions**, or merely **complexes**. Compounds containing them are called **coordination compounds**.

CONTENTS

25.1 THE STRUCTURE OF COMPLEXES

The molecules or ions that surround a metal ion in a complex are known as *complexing agents* or **ligands** (from the Latin word *ligare*, meaning "to bind"). Ligands are normally either anions or polar molecules. Furthermore, they have at least one unshared pair of valence electrons, as illustrated in the following examples:

$H_2\ddot{O}:$ (:Ö—H with H below) $:NH_3$ (H above, :N—H, H below) $:\ddot{Cl}:^-$ $:C{\equiv}N:^-$

In many cases, we can think of the bonding between a metal ion and its ligands as an electrostatic interaction between the positive cation and the surrounding negative ions or dipoles oriented with their negative ends toward the metal ion. The ability of metal ions to form complexes normally increases as the positive charge of the cation increases and its size decreases. The weakest complexes are formed by the alkali metal ions, such as Na^+ and K^+. Conversely, the 2+ and 3+ ions of the transition elements show a great tendency to form complexes. In fact, many of the transition-metal ions form complexes more readily than their charge and size would suggest. For example, on the basis of size alone we might expect that the Al^{3+} ion (radius = 0.45 Å) would form complexes more readily than the larger Cr^{3+} ion (radius = 0.62 Å). However, with most ligands, Cr^{3+} forms much more stable complexes than does Al^{3+}.

(a)

(b)

Figure 25.1 When an aqueous solution of NH_4SCN is added to an aqueous solution of Fe^{3+}, the intensely colored $[Fe(H_2O)_5SCN]^{2+}$ ion is formed. (Richard Megna/Fundamental Photographs)

Indeed, the bonding in most transition-metal ions cannot be explained entirely on the basis of electrostatic attractions, and we must assume some degree of covalent character in the metal-ligand bond.

Because metal ions have empty valence orbitals, they can act as Lewis acids (electron-pair acceptors). Because ligands have unshared pairs of electrons, they can function as Lewis bases (electron-pair donors). We can picture the bond between metal and ligand as the result of their sharing a pair of electrons that was initially on the ligand:

$$Ag^+(aq) + 2\!:\!\underset{\displaystyle H}{\overset{\displaystyle H}{\underset{|}{\overset{|}{N}}}}\!-\!H(aq) \longrightarrow \left[H\!-\!\underset{\displaystyle H}{\overset{\displaystyle H}{\underset{|}{\overset{|}{N}}}}\!:\!Ag\!:\!\underset{\displaystyle H}{\overset{\displaystyle H}{\underset{|}{\overset{|}{N}}}}\!-\!H \right]^+ (aq) \qquad [25.1]$$

We shall examine the bonding in complexes more closely in Section 25.8.

In forming a complex, the ligands are said to *coordinate* to the metal or to *complex* the metal. The central metal and the ligands bound to it constitute the **coordination sphere**. In writing the chemical formula for a coordination compound, we use square brackets to set off the groups within the coordination sphere from other parts of the compound. For example, the formula $[Cu(NH_3)_4]SO_4$ represents a coordination compound consisting of the $[Cu(NH_3)_4]^{2+}$ complex ion and the $SO_4{}^{2-}$ ion. The four ammonia ligands in the compound are bound directly to the copper(II) ion.

A complex is a distinct chemical species with its own characteristic physical and chemical properties. Thus, a metal complex has different properties than the metal ion and the ligands from which it is formed. For example, complexes may have colors that differ markedly from those of their component metal ions and ligands. Figure 25.1 shows the color change that occurs when aqueous solutions of SCN^- and Fe^{3+} are mixed, forming $[Fe(H_2O)_5SCN]^{2+}$.

Complex formation can also dramatically change other properties of metal ions such as their ease of oxidation or reduction. For example, Ag^+ is readily reduced in water:

$$Ag^+(aq) + e^- \longrightarrow Ag(s) \qquad E^\circ = +0.799\ V \qquad [25.2]$$

In contrast, the $[Ag(CN)_2]^-$ ion is not at all readily reduced, because complexation by CN^- ions stabilizes silver in the $+1$ oxidation state:

$$[Ag(CN)_2]^-(aq) + e^- \longrightarrow Ag(s) + 2CN^-(aq) \qquad E^\circ = -0.31\ V \qquad [25.3]$$

Of course, aquated metal ions are themselves complex ions in which the ligand is water. Thus, $Fe^{3+}(aq)$ consists largely of $[Fe(H_2O)_6]^{3+}$ (Section 16.10). When we speak of complex formation in aqueous solutions, we are actually considering reactions in which ligands such as SCN^- and CN^- replace H_2O molecules in the coordination sphere of the metal ion.

Charges, Coordination Numbers, and Geometries

The charge of a complex is the sum of the charges on the central metal and on its surrounding ligands. In $[Cu(NH_3)_4]SO_4$, we can deduce the charge on the complex if we first recognize SO_4 as being the sulfate ion and therefore having a 2− charge. Because the compound is neutral, the complex ion must have a 2+ charge, $[Cu(NH_3)_4]^{2+}$. We can then use the charge of the complex ion to deduce the oxidation number of copper. Because the NH_3 ligands are neutral, the oxidation number of copper must be +2:

$$\underset{[Cu(NH_3)_4]^{2+}}{+2 + 4(0) = +2}$$

SAMPLE EXERCISE 25.1

What is the oxidation number of the central metal in $[Co(NH_3)_5Cl](NO_3)_2$?

Solution: The NO_3 group is the nitrate anion and has a 1− charge, NO_3^-. The NH_3 ligands are neutral; the Cl is a coordinated chloride ion and therefore has a 1− charge. The sum of all the charges must be zero:

$$\underset{[Co(NH_3)_5Cl](NO_3)_2}{x + 5(0) + (-1) + 2(-1) = 0}$$

The oxidation number of the cobalt, x, must therefore be +3.

PRACTICE EXERCISE

What is the charge of the complex formed by a platinum(IV) metal ion surrounded by three ammonia molecules and three bromide ions? ***Answer:*** 1+

SAMPLE EXERCISE 25.2

Given that a complex ion contains a chromium(III) bound to four water molecules and two chloride ions, write its formula.

Solution: The metal has a +3 oxidation number, water is neutral, and chloride has a 1− charge:

$$\underset{Cr(H_2O)_4Cl_2}{+3 + 4(0) + 2(-1) = +1}$$

Therefore, the charge on the ion is 1+, $[Cr(H_2O)_4Cl_2]^+$.

PRACTICE EXERCISE

Write the formula for the complex ion described in the Practice Exercise accompanying Sample Exercise 25.1. ***Answer:*** $[Pt(NH_3)_3Br_3]^+$

The atom of the ligand bound directly to the metal is called the **donor atom**. For example, nitrogen is the donor atom in the $[Ag(NH_3)_2]^+$ complex shown in Equation 25.1. The number of donor atoms attached to a metal is known as its **coordination number**. In $[Ag(NH_3)_2]^+$, silver has a coordination number of 2; in $[Cr(H_2O)_4Cl_2]^+$, chromium has a coordination number of 6.

Some metal ions exhibit constant coordination numbers. For example, the coordination number of chromium(III) and cobalt(III) is invariably 6, and that of platinum(II) is always 4. However, the coordination numbers of most metal ions vary with the ligand. The most common coordination numbers are 4 and 6.

The coordination number of a metal ion is often influenced by the relative sizes of the metal ion and the surrounding ligands. As the ligand gets larger, fewer can coordinate to the metal. This helps explain why iron(III) is able to

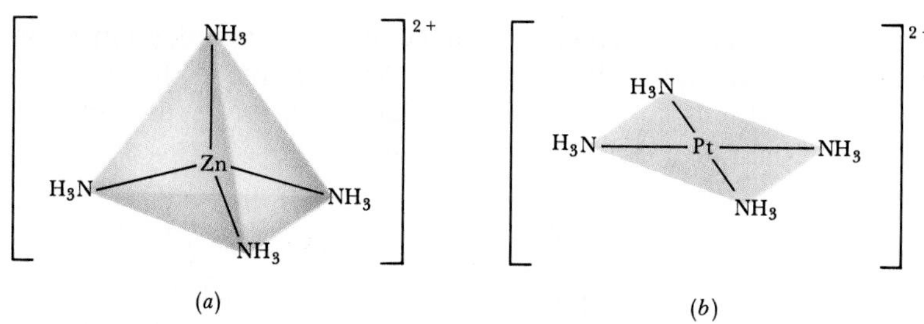

Figure 25.2 Structures of (*a*) $[Zn(NH_3)_4]^{2+}$ and (*b*) $[Pt(NH_3)_4]^{2+}$, illustrating the tetrahedral and square-planar geometries, respectively. These are the two common geometries for complexes in which the metal ion has a coordination number of 4. The shaded surfaces shown in the figure are not bonds; they are included merely to assist in visualizing the shape of the metal complex.

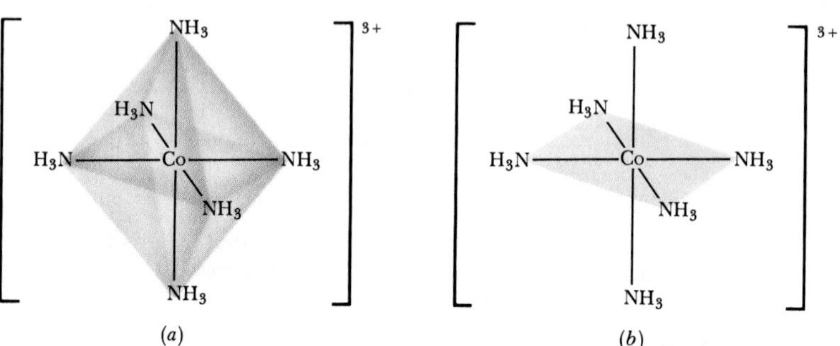

Figure 25.3 Two representations of an octahedral coordination sphere, the common geometric arrangement for complexes in which the metal ion has a coordination number of 6.

coordinate to six fluorides in $[FeF_6]^{3-}$, but to only four chlorides in $[FeCl_4]^-$. Ligands that transfer substantial negative charge to the metal also produce reduced coordination numbers. For example, six neutral ammonia molecules can coordinate to nickel(II), forming $[Ni(NH_3)_6]^{2+}$; however, only four negatively charged cyanide ions coordinate, forming $[Ni(CN)_4]^{2-}$.

Four-coordinate complexes have two common geometries—tetrahedral and square planar—as shown in Figure 25.2. The tetrahedral geometry is the more common of the two and is especially common among nontransition metals. Square-planar geometry is characteristic of transition-metal ions with eight *d* electrons in the valence shell, for example, platinum(II) and gold(III); it is also found in some copper(II) complexes.

The vast majority of 6-coordinate complexes have an octahedral geometry, as shown in Figure 25.3(*a*). The octahedron is often represented as a planar square with ligands above and below the plane, as in Figure 25.3(*b*). Recall, however, that all positions on an octahedron are geometrically equivalent (Section 9.1).

25.2 CHELATES

The ligands that we have discussed so far, such as NH_3 and Cl^-, are called **monodentate ligands** (from the Latin meaning "one-toothed"). These ligands possess a single donor atom and are able to occupy only one site in a coordination sphere. Some ligands have two or more donor atoms that can simultaneously coordinate to a metal ion, thereby occupying two or more coordination sites. They are called **polydentate ligands** ("many-toothed" ligands). Because they appear to grasp the metal between two or more donor atoms, polydentate ligands are also known as **chelating agents** (from the Greek word *chele*, "claw"). One such ligand is *ethylenediamine:*

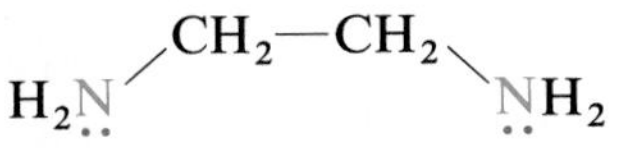

$$H_2\ddot{N}{-}CH_2{-}CH_2{-}\ddot{N}H_2$$

This ligand, which is abbreviated en, has two nitrogen atoms (shown in color) that have unshared pairs of electrons. These donor atoms are sufficiently far apart that the ligand can wrap around a metal ion with the two nitrogen atoms simultaneously complexing to the metal in adjacent positions. The $[Co(en)_3]^{3+}$ ion, which contains three ethylenediamine ligands in the octahedral coordination sphere of cobalt(III), is shown in Figure 25.4. Notice that the ethylenediamine has been written in a shorthand notation as two nitrogen atoms connected by a line. Ethylenediamine is a **bidentate ligand** ("two-toothed" ligand), one that can occupy two coordination sites. The structures of several other bidentate ligands are shown in Figure 25.5.

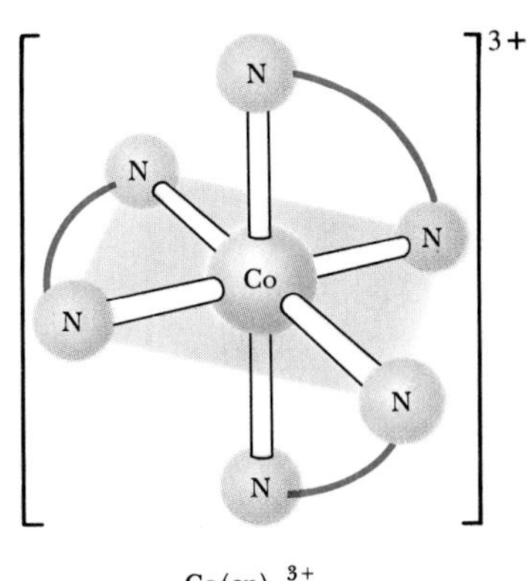

Figure 25.4 $[Co(en)_3]^{3+}$ ion showing how each bidentate ethylenediamine ligand is able to occupy two positions in the coordination sphere.

Oxalate ion ($2-$)

Carbonate ion ($2-$)

Ortho-phenanthroline
(*o*-phen)

Bipyridine
(bipy)

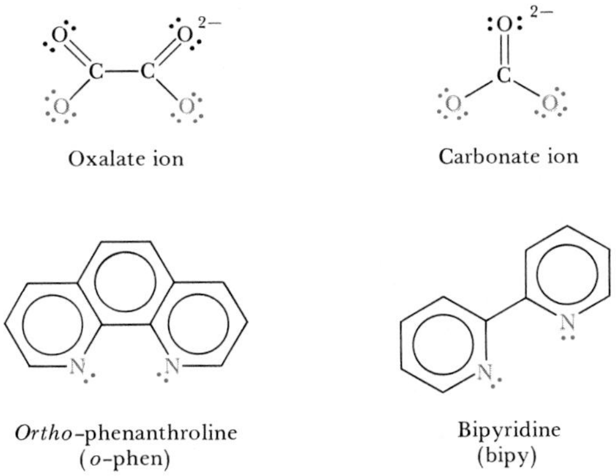

Figure 25.5 Structures of some bidentate ligands. The coordinating atoms are shown in blue.

The ethylenediaminetetraacetate ion is another important polydentate ligand:

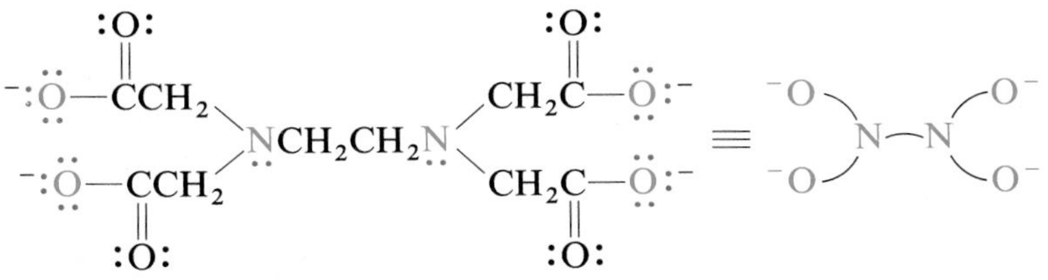

$$\begin{array}{l} ^{-}{:}\ddot{O}{-}C({=}O)CH_2 \\ ^{-}{:}\ddot{O}{-}C({=}O)CH_2 \end{array} \!\!> \ddot{N}CH_2CH_2\ddot{N} <\!\! \begin{array}{l} CH_2C({=}O){-}\ddot{O}{:}^{-} \\ CH_2C({=}O){-}\ddot{O}{:}^{-} \end{array} \equiv \begin{array}{c} {}^{-}O \quad\quad O^{-} \\ N{-}N \\ {}^{-}O \quad\quad O^{-} \end{array}$$

$[EDTA]^{4-}$

This ion, abbreviated $EDTA^{4-}$, has six donor atoms. It can wrap around a metal ion using all six of these donor atoms, as shown in Figure 25.6.

Figure 25.6 The $CoEDTA^{-}$ ion, showing how the ethylenediaminetetraacetate ion is able to wrap around a metal ion, occupying six positions in the coordination sphere.

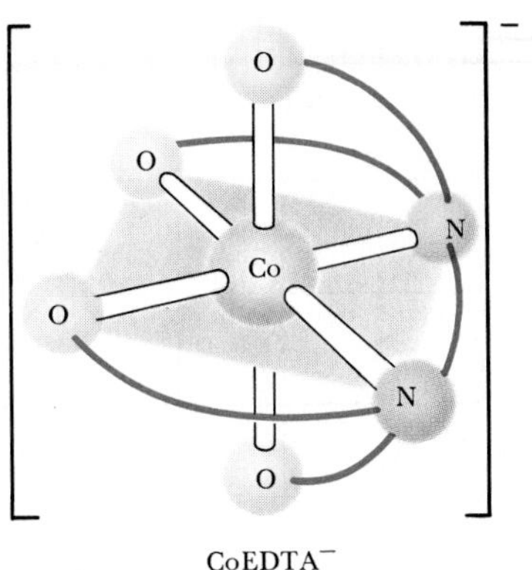

In general, chelating agents form more stable complexes than do related monodentate ligands. This is illustrated by the formation constants for $[Ni(NH_3)_6]^{2+}$ and $[Ni(en)_3]^{2+}$, shown in Equations 25.4 and 25.5.

$$[Ni(H_2O)_6]^{2+}(aq) + 6NH_3(aq) \rightleftharpoons [Ni(NH_3)_6]^{2+}(aq) + 6H_2O(l) \qquad K_f = 4 \times 10^8 \qquad [25.4]$$

$$[Ni(H_2O)_6]^{2+}(aq) + 3en(aq) \rightleftharpoons [Ni(en)_3]^{2+}(aq) + 6H_2O(l) \qquad K_f = 2 \times 10^{18} \qquad [25.5]$$

Although the donor atom is nitrogen in both instances, $[Ni(en)_3]^{2+}$ has a formation constant nearly 10^{10} times larger than that of $[Ni(NH_3)_6]^{2+}$. The generally larger formation constants for polydentate ligands as compared with the corresponding monodentate ligands is known as the **chelate effect**.

Chelating agents are often used to prevent one or more of the customary reactions of a metal ion without actually removing it from solution. For example, a metal ion that interferes with a chemical analysis can often be complexed and its interference thereby removed. In a sense, the chelating agent hides the metal ion. For this reason, scientists sometimes refer to these ligands as *sequestering agents.* Phosphates such as sodium tripolyphosphate, shown below, are used to complex or sequester metal ions in hard water (Section 18.6) so these ions cannot interfere with the action of soap or detergents:

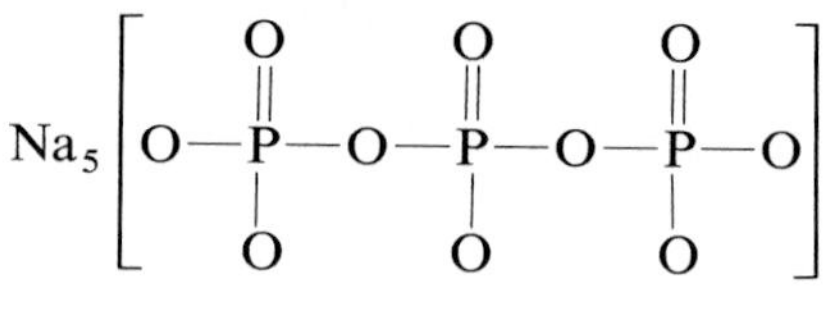

$$\mathrm{Na_5}\left[\mathrm{O{-}\overset{\overset{\displaystyle O}{\|}}{\underset{\underset{\displaystyle O}{|}}{P}}{-}O{-}\overset{\overset{\displaystyle O}{\|}}{\underset{\underset{\displaystyle O}{|}}{P}}{-}O{-}\overset{\overset{\displaystyle O}{\|}}{\underset{\underset{\displaystyle O}{|}}{P}}{-}O}\right]$$

Chelating agents such as EDTA are often used in consumer products, including many prepared foods such as salad dressings and frozen desserts, to complex trace metal ions that catalyze decomposition reactions. Chelating agents are used in medicine to remove metal ions such as Hg^{2+}, Pb^{2+}, and Cd^{2+}, which are detrimental to health. One method of treating lead poisoning is to administer $Na_2[CaEDTA]$. The EDTA chelates the lead, allowing its removal from the body in urine. Chelating agents are also quite common in nature. Mosses and lichens secrete chelating agents to capture metal ions from the rocks they inhabit.

Metals and Chelates in Living Systems

Living systems consist mainly of hydrogen, oxygen, carbon, and nitrogen. Nevertheless, many other elements are known to be essential for life (see Figure 25.7). Nine of the essential elements are transition metals—vanadium, chro-

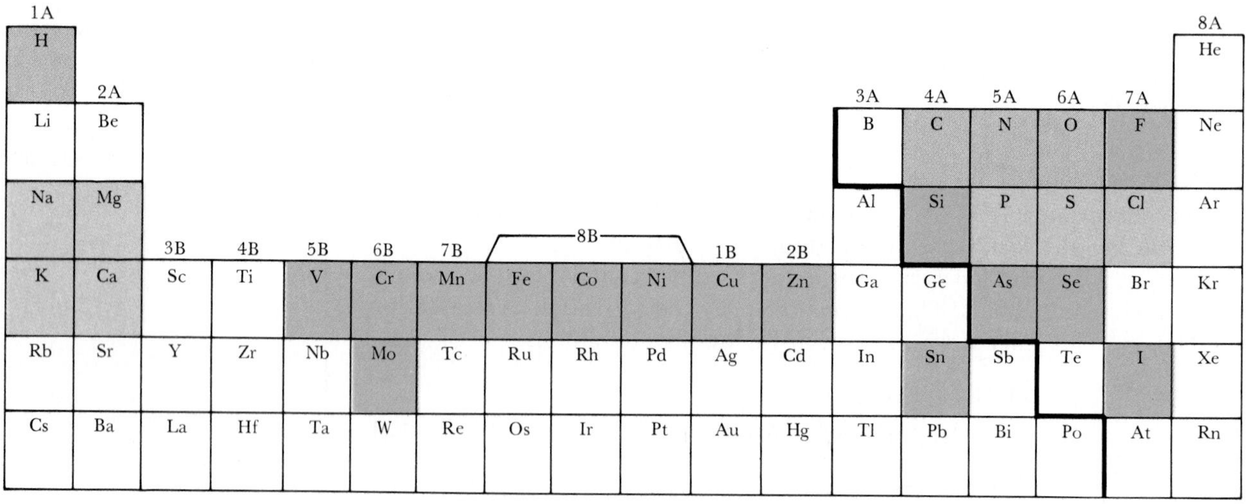

Figure 25.7 The elements that are essential for life are indicated by colors. The red denotes the four most abundant elements in living systems (hydrogen, carbon, nitrogen, and oxygen). The orange indicates the seven next most abundant elements. The green indicates the elements needed in only trace amounts.

1A	2A	3B	4B	5B	6B	7B	8B	8B	8B	1B	2B	3A	4A	5A	6A	7A	8A
H																	He
Li	Be											B	C	N	O	F	Ne
Na	Mg											Al	Si	P	S	Cl	Ar
K	Ca	Sc	Ti	V	Cr	Mn	Fe	Co	Ni	Cu	Zn	Ga	Ge	As	Se	Br	Kr
Rb	Sr	Y	Zr	Nb	Mo	Tc	Ru	Rh	Pd	Ag	Cd	In	Sn	Sb	Te	I	Xe
Cs	Ba	La	Hf	Ta	W	Re	Os	Ir	Pt	Au	Hg	Tl	Pb	Bi	Po	At	Rn

A CLOSER LOOK: The Stability of Chelates

Thermochemical studies of complex formation in aqueous solution show that in nearly all cases *the chelate effect is due to a more favorable entropy change for complex formation involving polydentate ligands.* As an example, let's compare the thermodynamic data at 25°C for formation of two closely related complexes of Cd^{2+}:

$$Cd^{2+}(aq) + 4CH_3NH_2(aq) \rightleftharpoons [Cd(CH_3NH_2)_4]^{2+}(aq)$$

$$\Delta G^\circ = -37.2\text{ kJ};\quad \Delta H^\circ = -57.3\text{ kJ};\quad \Delta S^\circ = -67.3\text{ J/K}$$

$$Cd^{2+}(aq) + 2H_2NCH_2CH_2NH_2(aq) \rightleftharpoons [Cd(H_2NCH_2CH_2NH_2)_2]^{2+}(aq)$$

$$\Delta G^\circ = -60.7\text{ kJ};\quad \Delta H^\circ = -56.5\text{ kJ};\quad \Delta S^\circ = +14.1\text{ J/K}$$

Recall from Section 19.6 that a large negative value for ΔG° corresponds to a large equilibrium constant for the forward reaction. The equilibrium constant for complex formation is thus much greater for the second reaction, with en as ligand. Although the enthalpy changes in the two reactions are nearly the same, the entropy change is much more positive for the second reaction. The more positive ΔS° accounts for the more negative value for ΔG° for the second reaction.

The more positive entropy change associated with reactions involving polydentate ligands is related to some of the ideas regarding entropy discussed in Section 19.3. Ligands bound to the metal ion are constrained to remain with that metal ion; they are not free to move about the solution independently. Therefore, coordination of ligands leads to a reduction in entropy. In the first reaction above, four CH_3NH_2 molecules become bound to the metal ion, and four water molecules are freed to move about the solution. Although it might seem at first that this should lead to a net entropy change of zero, differences in the hydrogen-bonding ability of H_2O as compared with the CH_3NH_2 molecules, and the tighter binding of CH_3NH_2 to the Cd^{2+} ion as compared with water, lead to a net negative entropy change. However, when one ethylenediamine molecule binds to Cd^{2+}, two H_2O molecules are released to move about freely in the solution. There is a net increase in the number of species that move about independently in the solution. The system has become more disordered, and the entropy change is accordingly more positive.

mium, iron, copper, zinc, manganese, cobalt, nickel, and molybdenum. These elements owe their roles in living systems mainly to their ability to form complexes with a variety of donor groups present in biological systems. Many enzymes, the body's catalysts, require metal ions to function. We shall take a closer look at enzymes in Chapter 27.

Although our bodies require only small quantities of metals, deficiencies can lead to serious illness. For example, a deficiency of manganese in the diet can lead to convulsive disorders. Some epilepsy patients have been helped by the addition of manganese to their diets.

Among the most important chelating agents in nature are those derived from the *porphine* molecule, which is shown in Figure 25.8. This molecule can coordinate to a metal using the four nitrogen atoms as donors. Upon coordination to a metal, the two H^+ shown bonded to nitrogen are displaced. Complexes derived from porphine are called **porphyrins**. Different porphyrins contain different metals and have different substituent groups attached to the carbon atoms at the ligand's periphery. Two of the most important porphyrins are heme, which contains iron(II), and chlorophyll, which contains magnesium(II). We discussed heme earlier, in Section 18.4. Hemoglobin contains four heme subunits; one of these subunits is shown in Figure 18.11. The iron is coordinated to the four nitrogen atoms of the porphyrin and also to a nitrogen atom from the protein that composes the bulk of the hemoglobin molecule. The sixth position around the iron is occupied either by O_2 (in oxyhemoglobin, the bright red form) or by water (in deoxyhemoglobin, the purplish-red form). Oxyhemoglobin is shown in Figure 25.9. As noted in Section 18.4, some substances, such as CO, act as poisons because they bind to iron more strongly than O_2 can.

Figure 25.8 Structure of the porphine molecule. This molecule forms a tetradentate ligand with the loss of the two protons bound to nitrogen atoms. Porphine is the basic component of porphyrins, compounds whose complexes play a variety of important roles in nature.

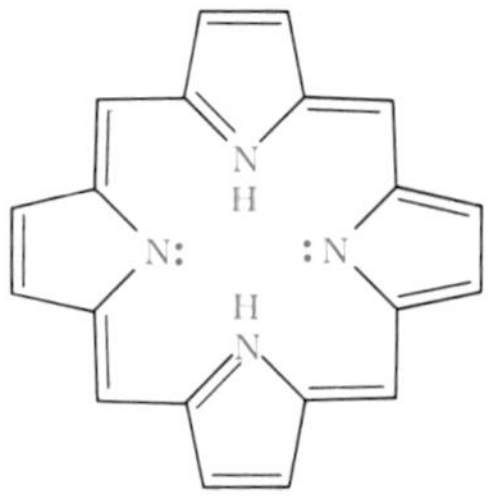

Figure 25.9 Schematic representation of oxyhemoglobin, showing one of the four heme units in the molecule. The iron is bound to four nitrogen atoms of the porphyrin, to a nitrogen from the surrounding protein, and to an O_2 molecule.

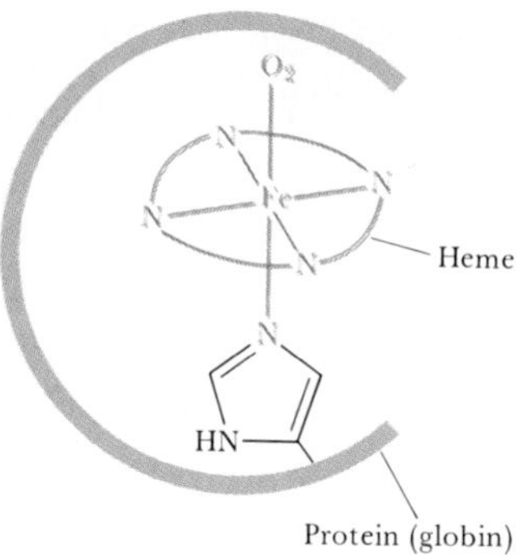

When we have an insufficient quantity of iron in our diet, we suffer from iron-deficiency anemia. The lack of iron leads to a reduction in the amount of hemoglobin; we develop what advertisements have referred to as "iron-poor blood." Without hemoglobin to transport oxygen, our body's cells are unable to produce energy. Therefore, the symptoms of anemia include weakness and drowsiness.

25.3 NOMENCLATURE

When complexes were first discovered and few were known, they were named after the chemist who originally prepared them. A few of these names persist; for example, $NH_4[Cr(NH_3)_2(NCS)_4]$ is known as Reinecke's salt. As the number of known complexes grew, chemists began to give them names based on their color. For example, $[Co(NH_3)_5Cl]Cl_2$, whose formula was then written $CoCl_3 \cdot 5NH_3$, was known as purpureocobaltic chloride after its purple color. Once the structures of complexes were more fully understood, it became possible to name them in a more systematic manner. Let's consider two examples:

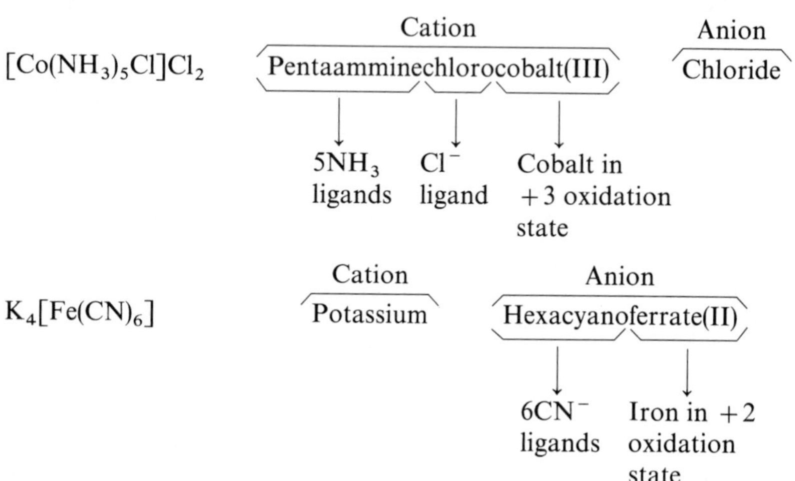

The rules of nomenclature follow on page 912.*

* The rules of nomenclature are approved by the International Union of Pure and Applied Chemistry and are subject to periodic revision. Some rules are clearly more important than others. For example, the order in which ligands are named (rule 2) is not as important as assigning each ligand its correct name.

CHEMISTRY AT WORK: The Battle for Iron in Living Systems

Although iron is the fourth most abundant element in the earth's crust, living systems have difficulty in assimilating enough iron to satisfy their needs. Consequently, iron-deficiency anemia is a common problem in humans. In plants, chlorosis, an iron deficiency that results in yellowing of leaves, is also commonplace. Living systems have difficulty in assimilating iron because of changes that occurred in the earth's atmosphere in the course of geological time. The earliest living systems had a plentiful supply of soluble iron(II) in the oceans. However, when oxygen appeared in the atmosphere, vast deposits of insoluble iron(III) oxide formed. The amount of dissolved iron remaining was too small to support life. Microorganisms adapted to this problem by secreting an iron-binding compound, called a *siderophore*, that forms an extremely stable, water-soluble complex with iron(III). This complex is called *ferrichrome;* its structure is shown in Figure 25.10. The iron-binding strength of the siderophore is so great that it can extract iron from Pyrex glassware, and it readily solubilizes the iron in iron oxides.

The overall charge of ferrichrome is zero, which makes it possible for the complex to pass through the rather hydrophobic membrane walls of cells. When a dilute solution of ferrichrome is added to a cell suspension, iron is found entirely within the cells in an hour. When ferrichrome enters the cell, the iron is removed through an enzyme-catalyzed reaction that reduces the iron to iron(II). Iron in the lower oxidation state is not strongly complexed by the siderophore. Microorganisms thus acquire iron by excreting siderophore into their immediate environment, then taking the ferrichrome complex into the cell. The overall process is illustrated in Figure 25.11.

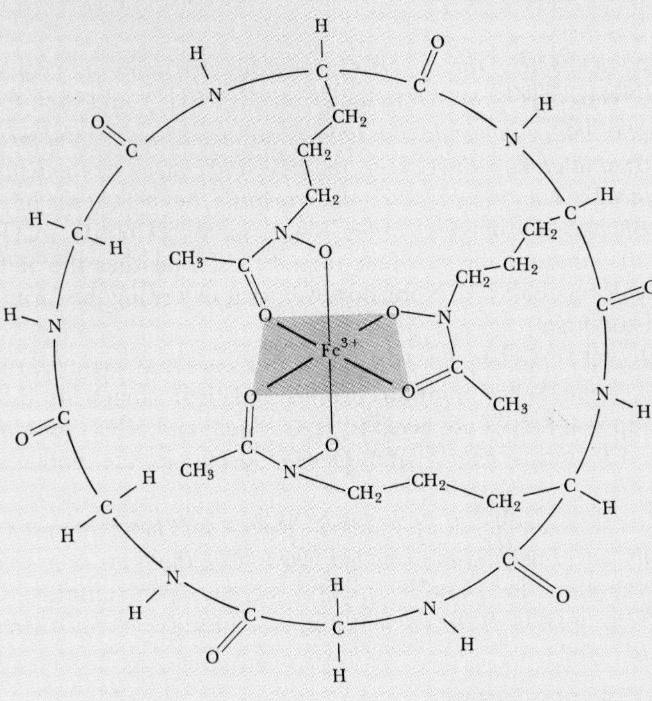

Figure 25.10 The structure of ferrichrome. In this complex an Fe^{3+} ion is coordinated by six oxygen atoms. The complex is very stable; it has a formation constant of about 10^{30}. The overall charge of the complex is zero.

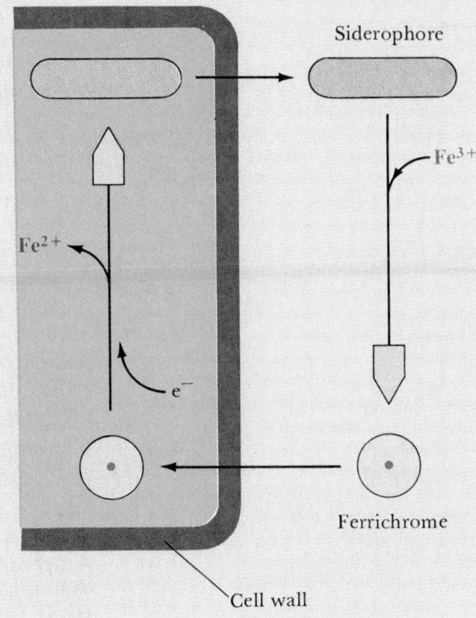

Figure 25.11 The iron-transport system of a bacterial cell. The iron-binding ligand, called the siderophore, is synthesized inside the cell and excreted into the surrounding medium. It reacts with Fe^{3+} ion to form ferrichrome, which is then absorbed by the cell. Inside the cell the ferrichrome is reduced, forming Fe^{2+}, which is not tightly bound by the siderophore. Having released the iron for use in the cell, the siderophore may be recycled back into the medium.

In humans, iron is assimilated from food in the intestine. A protein called *transferrin* binds iron and transports it across the intestinal wall to distribute it to other tissues in the body. The normal adult human carries a total of about 4 g of iron. At any one time, about 3 g, or 75 percent, of the iron is in the blood, mostly in the form of hemoglobin. Most of the remainder is carried by transferrin.

A bacterium that infects the blood requires a source of iron if it is to grow and reproduce. The bacterium excretes siderophore into the blood to compete with transferrin for the iron it holds. The formation constants for iron binding are about the same for transferrin and siderophore. The more iron available to the bacterium, the more rapidly it can reproduce, and thus the more harm it can do. Several years ago, New Zealand clinics regularly gave iron supplements to infants soon after birth. However, the incidence of certain bacterial infections was eight times higher in treated than in untreated infants. Presumably, the presence in the blood of more iron than absolutely necessary makes it easier for bacteria to obtain the iron necessary for growth and reproduction.

In the United States, it is common medical practice to supplement infant formula with iron sometime during the first year of life. This practice is based on the fact that human milk is virtually devoid of iron. Given what is now known about iron metabolism by bacteria, many research workers in nutrition believe that iron supplementation is not generally justified or wise.

For bacteria to continue to multiply in the bloodstream, they must synthesize new supplies of siderophore. It has been discovered that synthesis of siderophore in bacteria slows as the temperature is increased above the normal body temperature of 37°C, and it stops completely at 40°C. This suggests that fever in the presence of an invading microbe is a mechanism used by the body to deprive bacteria of iron.

1. *In naming salts, the name of the cation is given before the name of the anion.* Thus, in $[Co(NH_3)_5Cl]Cl_2$ we name the $[Co(NH_3)_5Cl]^{2+}$ and then Cl^-.
2. *Within a complex ion or molecule, the ligands are named before the metal. Ligands are listed in alphabetical order, regardless of charge on the ligand. Prefixes that give the number of ligands are not considered part of the ligand name in determining alphabetical order.* Thus, in the $[Co(NH_3)_5Cl]^{2+}$ ion, we name the ammonia ligands first, then the chloride, then the metal: pentaamminechlorocobalt(III). Note, however, that in writing the formula the metal is listed first.
3. *Anionic ligands end in the letter o, whereas neutral ones ordinarily bear the name of the molecule.* Some common ligands and their names are listed in Table 25.1. Special names are given to H_2O (aqua) and NH_3 (ammine). For example, the terms *chloro* and *ammine* occur in the name for $[Co(NH_3)_5Cl]Cl_2$.
4. *A Greek prefix (for example, di-, tri-, tetra-, penta-, and hexa-) is used to indicate the number of each kind of ligand when more than one is present.* Therefore, in the name for $[Co(NH_3)_5Cl]^{2+}$ we have pentaammine, indicating five NH_3 ligands. *If the name of the ligand itself contains a Greek*

Table 25.1 Some Common Ligands

Ligand	Ligand name
Azide, N_3^-	Azido
Bromide, Br^-	Bromo
Chloride, Cl^-	Chloro
Cyanide, CN^-	Cyano
Hydroxide, OH^-	Hydroxo
Carbonate, CO_3^{2-}	Carbonato
Oxalate, $C_2O_4^{2-}$	Oxalato
Ammonia, NH_3	Ammine
Ethylenediamine, en	Ethylenediamine
Water, H_2O	Aqua

prefix, such as mono- or di-, the name of the ligand is enclosed in parentheses, and alternate prefixes (bis-, tris-, tetrakis-, pentakis-, and hexakis-) are used. For example, the name for $[Co(en)_3]Cl_3$ is tris(ethylenediamine)cobalt(III) chloride.

5. *If the complex is an anion, its name ends in -ate.* For example, in $K_4[Fe(CN)_6]$ the anion is called the hexacyanoferrate(II) ion. The suffix *-ate* is often added to the Latin stem, as in this example.
6. *The oxidation number of the metal is given in parentheses in Roman numerals following the name of the metal.* For example, the Roman numeral III is used to indicate the +3 oxidation state of cobalt in $[Co(NH_3)_5Cl]^{2+}$.

We apply these rules to the compounds listed below on the left to derive the names given on the right:

$[Ni(C_5H_5N)_6]Br_2$	hexapyridinenickel(II) bromide
$[Co(NH_3)_4(H_2O)CN]Cl_2$	tetraammineaquacyanocobalt(III) chloride
$Na_2[MoOCl_4]$	sodium tetrachlorooxomolybdate(IV)
$Na[Al(OH)_4]$	sodium tetrahydroxoaluminate

In the last example, the oxidation state of the metal is not mentioned in the name because aluminum is always in the +3 oxidation state.

SAMPLE EXERCISE 25.3

Give the name of the following compounds: **(a)** $[Cr(H_2O)_4Cl_2]Cl$; **(b)** $K_4[Ni(CN)_4]$.

Solution: **(a)** We begin with the four water molecules, which are indicated as tetraaqua. Then there are two chloride ions, indicated as dichloro. The oxidation state of Cr is +3:

$$+3 + 4(0) + 2(-1) + (-1) = 0$$
$$[Cr(H_2O)_4Cl_2]Cl$$

Thus, we have chromium(III). Finally, the anion is chloride. Putting these parts together, we have the compound's name: tetraaquadichlorochromium(III) chloride.

(b) The complex has four CN^-, which we indicate as tetracyano. The oxidation state of the nickel is zero:

$$4(+1) + 0 + 4(-1) = 0$$
$$K_4[Ni(CN)_4]$$

Because the complex is an anion, the metal is indicated as nickelate(0). Putting these parts together and naming the cation first, we have: potassium tetracyanonickelate(0).

PRACTICE EXERCISE

Name the following compounds: **(a)** $[Mo(NH_3)_3Br_3]NO_3$; **(b)** $(NH_4)_2[CuBr_4]$.
Answers: **(a)** triamminetribromomolybdenum(IV) nitrate; **(b)** ammonium tetrabromocuprate(II)

SAMPLE EXERCISE 25.4

Write the formula for bis(ethylenediamine)difluorocobalt(III) perchlorate.

Solution: The complex cation contains two fluorides, two ethylenediamines, and a cobalt with a +3 oxidation number. Knowing this, we can determine the charge on the complex:

$$+3 + 2(0) + 2(-1) = +1$$

$$[Co(en)_2F_2]$$

The perchlorate anion has a single negative charge, ClO_4^-. Therefore, only one is needed to balance the charge on the complex cation. The formula is thus $[Co(en)_2F_2]ClO_4$.

PRACTICE EXERCISE

Write the formula for sodium diaquabis(oxalato)ruthenate(III).
Answer: $Na[Ru(H_2O)_2(C_2O_4)_2]$

25.4 ISOMERISM

When two or more compounds have the same compositions but different arrangements of atoms, we call them **isomers**. Isomerism—the existence of isomers—is a characteristic feature of coordination compounds. Although isomers are composed of the same collection of atoms, they differ in one or more physical or chemical properties such as color, solubility, or rate of reaction with some reagent. We will consider two main kinds of isomers: **structural isomers** (which have different bonds) and **stereoisomers** (which have the same bonds but different spatial arrangements of the bonds). Each of these classes also has subclasses, which we now consider (see Figure 25.12).

Structural Isomerism

Many different types of structural isomerism are known in coordination chemistry. The two listed in Figure 25.12 are given as examples. **Linkage isomerism** is a relatively rare but interesting type that arises when a particular ligand is capable of coordinating to a metal in two different ways. For example, the nitrite ion, NO_2^-, can coordinate through either a nitrogen or an oxygen atom, as shown in Figure 25.13. When it coordinates through the nitrogen atom, the NO_2^- ligand is called *nitro;* when it coordinates through the oxygen atom, it is called *nitrito* and is generally written ONO^-. The isomers shown in Figure 25.13 differ in their chemical and physical properties. For example, the N-bonded isomer is yellow, whereas the O-bonded isomer is red. Other ligands capable of coordinating through either of two donor atoms include thiocyanate, SCN^-, whose potential donor atoms are N and S.

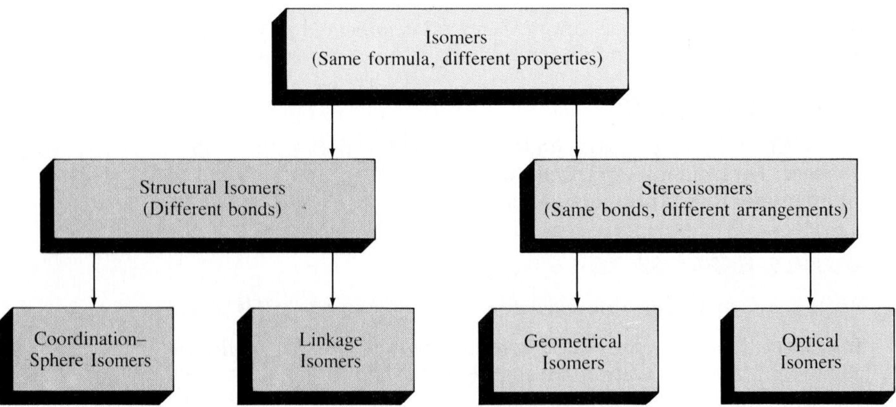

Figure 25.12 Forms of isomerism in coordination compounds.

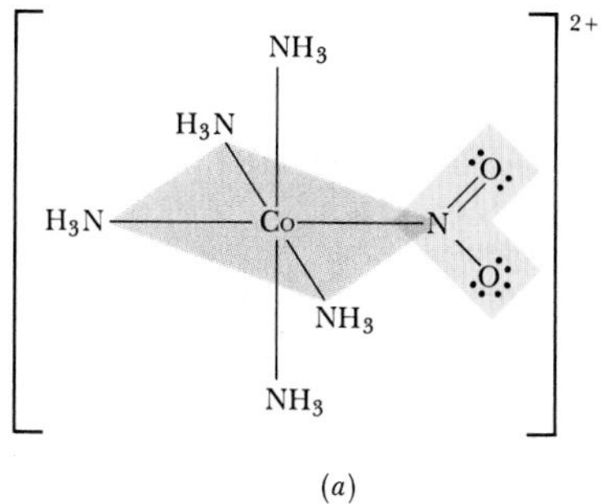

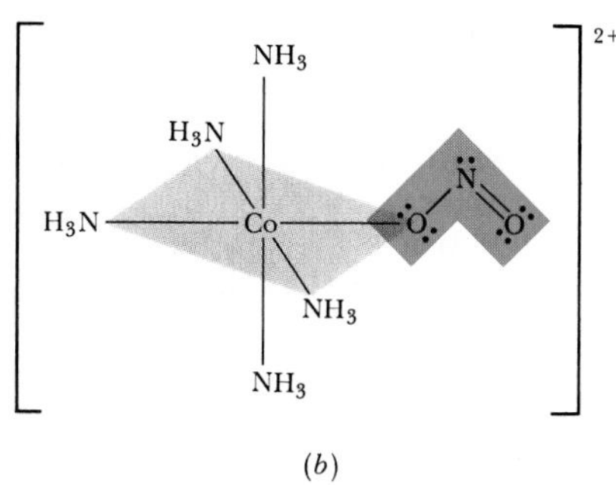

Figure 25.13 (*a*) Yellow N-bound and (*b*) red O-bound isomers of $[Co(NH_3)_5NO_2]^{2+}$.

Coordination-sphere isomers differ in the ligands that are directly bonded to the metal, as opposed to being outside the coordination sphere in the solid lattice. For example, $CrCl_3(H_2O)_6$ exists in three common forms: $[Cr(H_2O)_6]Cl_3$ (a violet compound), $[Cr(H_2O)_5Cl]Cl_2 \cdot H_2O$ (a green compound), and $[Cr(H_2O)_4Cl_2]Cl \cdot 2H_2O$ (also a green compound). In the second and third compounds, the water has been displaced from the coordination sphere by chloride ions and occupies a site in the solid lattice.

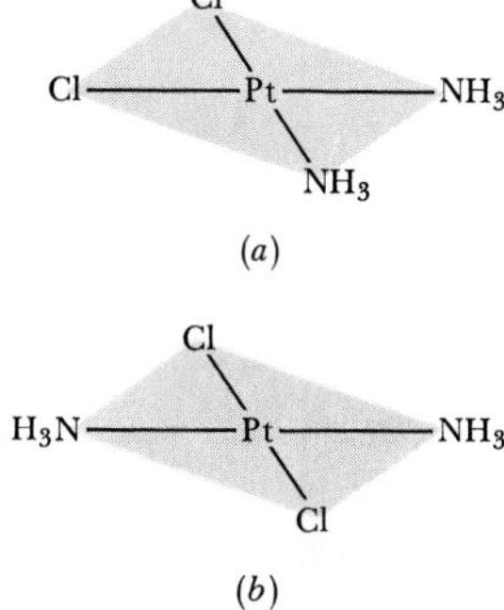

Figure 25.14 (*a*) Cis and (*b*) trans geometric isomers of the square-planar $[Pt(NH_3)_2Cl_2]$.

Stereoisomerism

Stereoisomerism is the most important form of isomerism. Stereoisomers have the same chemical bonds but different spatial arrangements. For example, in $[Pt(NH_3)_2Cl_2]$ the chloro ligands can be either adjacent to or opposite one another, as illustrated in Figure 25.14. This particular form of isomerism, in which the arrangement of the constituent atoms is different though the same bonds are present, is called **geometrical isomerism**. Isomer (*a*), with like groups in adjacent positions, is called the *cis* isomer. Isomer (*b*), with like ligands across from one another, is called the *trans* isomer. The *cis* isomer is used as a chemotherapy agent in treating cancer and carries the name *cisplatin.*

Geometrical isomerism is possible also in octahedral complexes when two or more different ligands are present. The cis and trans isomers of tetraamminedichlorocobalt(III) ion are shown in Figure 25.15. Note that these two isomers have different colors. Their salts also possess different solubilities in water. In general, geometrical isomers possess distinct physical and chemical properties.

Because all of the corners of a tetrahedron are adjacent to one another, cis-trans isomerism is not observed in tetrahedral complexes.

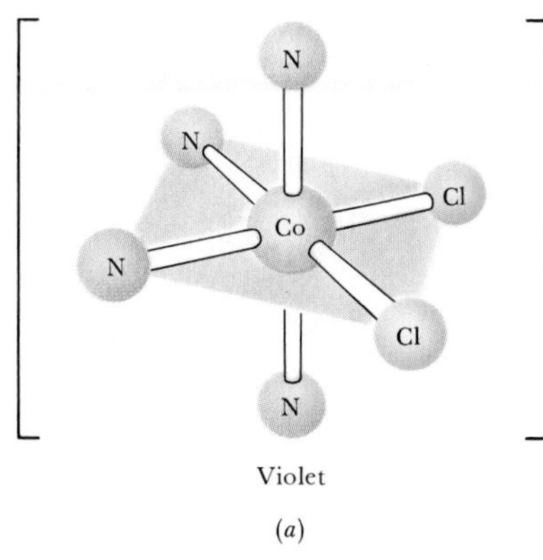

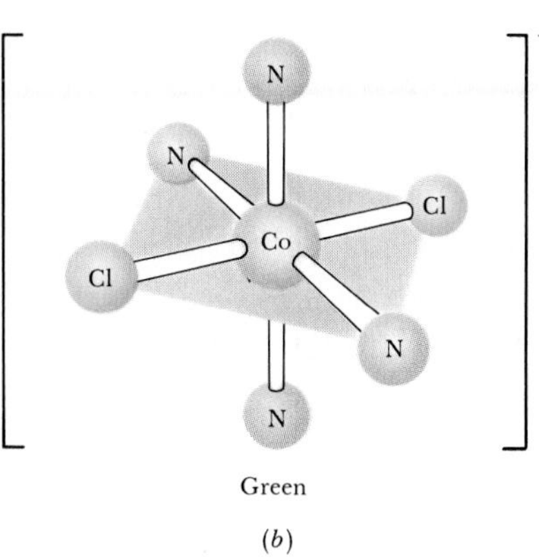

Figure 25.15 (*a*) Cis and (*b*) trans geometric isomers of the octahedral $[Co(NH_3)_4Cl_2]^+$ ion. (The symbol N represents the coordinated NH_3 group.)

SAMPLE EXERCISE 25.5

How many geometric isomers are there for $[Cr(H_2O)_2Br_4]^-$?

Solution: This complex has a coordination number of 6 and therefore presumably has an octahedral geometry. Like $[Co(NH_3)_4Cl_2]^+$ (Figure 25.15), it has four ligands of one type and two of another. Consequently, it possesses two isomers: one with H_2O ligands across the metal from each other (the trans isomer) and one with H_2O ligands adjacent (the cis isomer).

In general, the number of isomers of a complex can be determined by making a series of drawings of the structure with ligands in different locations. It is easy to overestimate the number of geometrical isomers. Sometimes different orientations of a single isomer are incorrectly thought to be different isomers. Therefore, you should keep in mind that if two structures can be rotated so that they are equivalent, they are not isomers of each other. The problem of identifying isomers is compounded by the difficulty we often have in visualizing three-dimensional molecules from their two-dimensional representations. It is easier to determine the number of isomers if we are working with three-dimensional models.

PRACTICE EXERCISE

How many isomers exist for square-planar $[Pt(NH_3)_2ClBr]$? ***Answer:*** two

A second type of stereoisomerism is known as **optical isomerism**. Optical isomers are mirror images that cannot be superimposed on each other. They bear the same resemblance to one another that our left hand bears to our right hand. If you look at your left hand in a mirror, the image is identical to your right hand. Furthermore, your two hands are not superimposable on one another. A good example of a complex that exhibits this type of isomerism is the $[Co(en)_3]^{3+}$ ion. Figure 25.16 shows the two isomers of $[Co(en)_3]^{3+}$ and their mirror-image relationship to each other. Just as there is no way that we can twist or turn our right hand to make it look identical to our left, so also there is no way to rotate one of these isomers to make it identical to the other. If we had models of each that we could handle, perhaps we could more easily satisfy ourselves of this fact. Molecules or ions that have nonsuperimposable

Figure 25.16 Just as our hands are nonsuperimposable mirror images of each other (*a*), so too are optical isomers such as the two optical isomers of $[Co(en)_3]^{3+}$ (*b*).

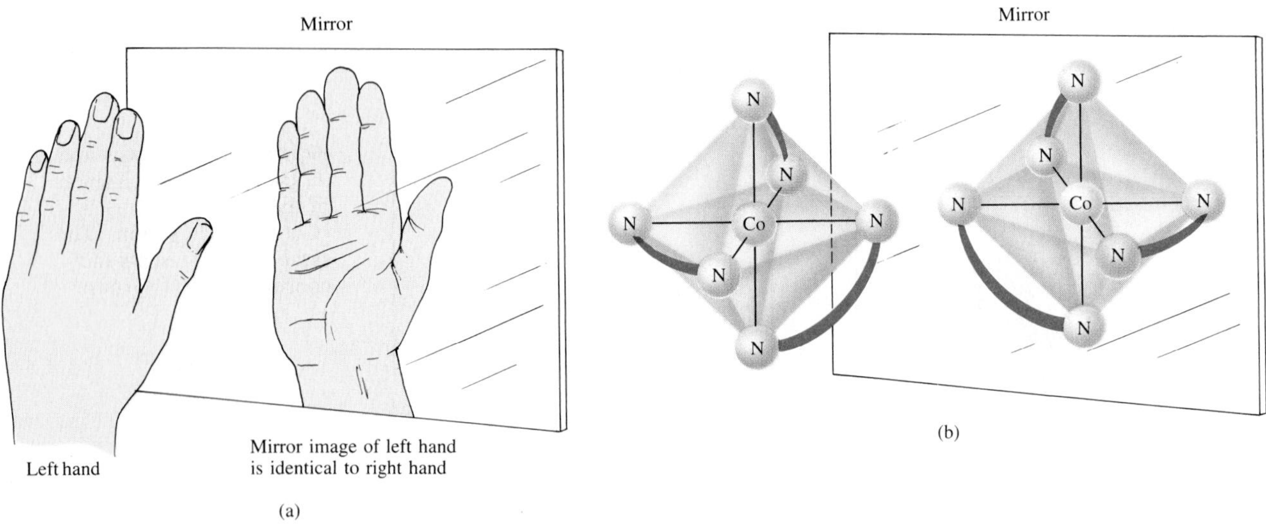

mirror images are said to be **chiral** (pronounced KY-rul). Enzymes are among the most highly chiral molecules known. As noted in Section 25.2, many enzymes contain complexed metal ions.

SAMPLE EXERCISE 25.6

Does either *cis*- or *trans*-$[Co(en)_2Cl_2]^+$ have optical isomers?

Solution: To answer this question you should draw out both the cis and trans isomers of $[Co(en)_2Cl_2]^+$, then their mirror images. Note that the mirror image of the trans isomer is identical to the original. Consequently *trans*-$[Co(en)_2Cl_2]^+$ has no optical isomer. However, the mirror image of *cis*-$[Co(en)_2Cl_2]^+$ is not identical to the original. Consequently, there are optical isomers for this complex.

PRACTICE EXERCISE

Does the square-planar complex ion $[Pt(NH_3)(N_3)ClBr]^-$ have optical isomers?
Answer: no

Most of the physical and chemical properties of optical isomers are identical. The properties of two optical isomers differ only if they are in a chiral environment—that is, one in which there is a sense of right- and left-handedness. For example, in the presence of a chiral enzyme, the reaction of one optical isomer might be catalyzed, whereas the other isomer would remain totally unreacted. Consequently, one optical isomer may produce a specific physiological effect within the body, whereas its mirror image produces a different effect or none at all.

Optical isomers are usually distinguished from each other by their interaction with plane-polarized light. If light is polarized—for example, by passage through a sheet of Polaroid film—the light waves are vibrating in a single plane, as shown in Figure 25.17. If the polarized light is passed through a solution containing one optical isomer, the plane of polarization is rotated either to the right (clockwise) or to the left (counterclockwise). The isomer that rotates the plane of polarization to the right is said to be **dextrorotatory;** it is labeled

Figure 25.17 Effect of an optically active solution on the plane of polarization of plane-polarized light. The unpolarized light is passed through a polarizer. The resultant polarized light thereafter passes through a solution containing a dextrorotatory optical isomer. As a result, the plane of polarization of the light is rotated to the right relative to an observer looking toward the light source.

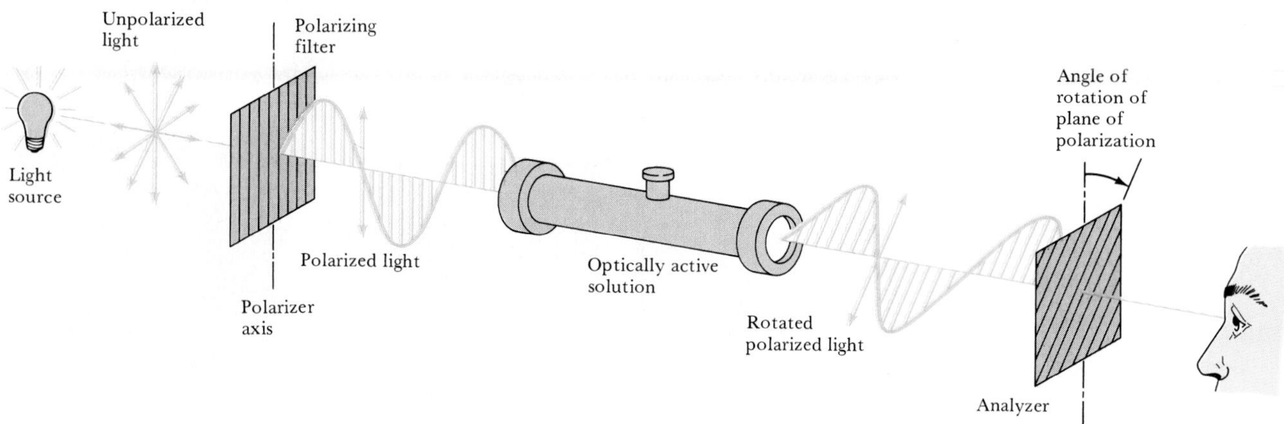

the dextro, or *d*, isomer (Latin *dexter*, "right"). Its mirror image will rotate the plane of polarization to the left; it is said to be **levorotatory** and is labeled the levo, or *l*, isomer (Latin *laevus*, "left"). Because of their effect on plane-polarized light, chiral molecules are said to be **optically active**.

When a substance with optical isomers is prepared in the laboratory, the chemical environment during the synthesis is not usually chiral. Consequently, equal amounts of the two isomers are obtained; the mixture is said to be **racemic**. A racemic mixture will not rotate polarized light because the rotatory effects of the two isomers cancel each other. In order to separate the isomers from the racemic mixture, the isomers must be placed in a chiral environment. For example, one optical isomer of the chiral tartrate anion,* $C_4H_4O_6^{2-}$, can be used to separate a racemic mixture of $[Co(en)_3]Cl_3$. If *d*-tartrate is added to an aqueous solution of $[Co(en)_3]Cl_3$, *d*-$[Co(en)_3](d\text{-}C_4H_4O_6)Cl$ will precipitate, leaving *l*-$[Co(en)_3]^{3+}$ in solution.

25.5 LIGAND-EXCHANGE RATES

Many complexes are prepared by simply mixing solutions of the metal ion with the appropriate ligand. For example, addition of ammonia to an aqueous solution of $CuSO_4$ produces an essentially instantaneous color change as the pale-blue $[Cu(H_2O)_4]^{2+}$ ion is converted to the deep-blue $[Cu(NH_3)_4]^{2+}$ ion (Figure 25.18). When this solution is acidified, the pale-blue color is regenerated at a rapid rate:

$$[Cu(NH_3)_4]^{2+}(aq) + 4H_2O(l) + 4H^+(aq) \longrightarrow [Cu(H_2O)_4]^{2+}(aq) + 4NH_4^+(aq) \quad [25.6]$$

* When sodium ammonium tartrate, $NaNH_4C_4H_4O_6$, is crystallized from solution, the two isomers form separate crystals whose shapes are mirror images of each other. In 1848, Louis Pasteur achieved the first separation of a racemic mixture into optical isomers; using a microscope he picked the "right-handed" crystals of this compound from the "left-handed" ones. For a more detailed discussion of handedness, see the Historical Perspective box in Chapter 1.

Figure 25.18 An aqueous solution of $CuSO_4$ is pale blue because of $Cu(H_2O)_4^{2+}$ (left). When $NH_3(aq)$ is added (middle and right), the deep-blue $Cu(NH_3)_4^{2+}$ ion forms. (© Richard Megna/ Fundamental Photographs)

Reactions in which one ligand replaces another in the coordination sphere of a metal ion are called *ligand-exchange reactions* or *substitution reactions.* Although many ligand-exchange reactions are very rapid, others are slow. For example, $[Co(NH_3)_6]^{3+}$ is more difficult to prepare than $[Cu(NH_3)_4]^{2+}$. However, once it has been formed and placed in an acidic solution, the reaction to form NH_4^+ takes several days. This tells us that the coordinated NH_3 groups are not readily removed from the metal.

Complexes like $[Cu(NH_3)_4]^{2+}$ that undergo rapid ligand exchange are called **labile complexes;** those like $[Co(NH_3)_6]^{3+}$ that undergo slow ligand exchange are called **inert complexes**. The distinction between labile and inert complexes applies to how rapidly equilibrium is attained and not to the position of the equilibrium. For example, although $[Co(NH_3)_6]^{3+}$ is inert in acidified aqueous solutions, the equilibrium constant indicates that the complex is not thermodynamically stable under these conditions:

$$[Co(NH_3)_6]^{3+}(aq) + 6H_2O(l) + 6H^+(aq) \rightleftharpoons [Co(H_2O)_6]^{3+}(aq) + 6NH_4^+(aq) \qquad K_f \simeq 10^{20} \quad [25.7]$$

The kinetic inertness of $[Co(NH_3)_6]^{3+}$ can be attributed to a high activation energy for the reaction.

Cobalt(III) is one of a few metal ions that consistently forms inert complexes; others include chromium(III), platinum(IV), and platinum(II). Complexes of these ions maintain their identity in solution long enough to permit study of their structures and properties. They were therefore among the first complexes studied. Much of our understanding of structure and isomerism comes from studies of these complexes.

25.6 STRUCTURE AND ISOMERISM: A HISTORICAL VIEW

Many early systematic studies of coordination compounds involved ammonia complexes of cobalt(III), chromium(III), platinum(II), and platinum(IV), because these complexes are inert. One of the earliest reports of the preparation of an ammine complex dates from 1798, when a chemist by the name of Tassaert accidentally prepared an ammonia complex of cobalt. The compound was found to have the empirical formula $CoN_6H_{18}Cl_3$. Tassaert wrote the formula as $CoCl_3 \cdot 6NH_3$, suggesting that the compound was analogous to hydrated salts like $CoCl_2 \cdot 6H_2O$.

By 1890, many ammine complexes had been prepared, and a great deal of information about them had been gathered by a number of different investigators. By this time, chemists had begun to wonder how the atoms in these complexes were connected to each other and about the possible effect of these arrangements on the properties of the complexes. Among the observations that any successful theory would have to account for were the electrical conductivity of the complexes in solution and their behavior toward $AgNO_3$. Recall from our previous discussion (Section 4.2) that solutions of ionic substances conduct electrical current. The ease with which the solution conducts current is referred to as the *conductivity*. The conductivity of solutions of ionic substances increases with the total concentration of ions and is also greater for ions of higher charge. By comparing the conductivities of solutions of coordination compounds with those of simple salts, it was possible to determine the number and types of ions

Table 25.2 Properties of Some Ammonia Complexes of Cobalt(III)

Original formulation	Color	Ions per formula unit	Cl^- ions in solution per formula unit	Modern formulation
$CoCl_3 \cdot 6NH_3$	Orange	4	3	$[Co(NH_3)_6]Cl_3$
$CoCl_3 \cdot 5NH_3$	Purple	3	2	$[Co(NH_3)_5Cl]Cl_2$
$CoCl_3 \cdot 4NH_3$	Green	2	1	*trans*-$[Co(NH_3)_4Cl_2]Cl$
$CoCl_3 \cdot 4NH_3$	Violet	2	1	*cis*-$[Co(NH_3)_4Cl_2]Cl$

present in the solution. For example, the molar conductivity of an aqueous solution of $CoCl_3 \cdot 5NH_3$ is about the same as that of $CaCl_2$ and other 1:2 electrolytes. We can thus conclude that an aqueous solution of $CoCl_3 \cdot 5NH_3$ produces three ions, one that carries a 2+ charge and two carrying negative charges. The number of free chloride ions present in the solution was determined by treating solutions with $AgNO_3$. When cold, freshly prepared solutions of $CoCl_3 \cdot 5NH_3$ were treated with $AgNO_3$, 2 mol of AgCl precipitated for each mole of complex; one chloride in the compound did not precipitate. Table 25.2 summarizes these results.

In 1893, Alfred Werner, a 26-year-old Swiss chemist, proposed a theory that successfully explained these facts and became the basis for our subsequent understanding of metal complexes. Werner's first basic postulate was that metals exhibit both primary and secondary valences. We now refer to these as the metal's oxidation state and coordination number, respectively. This postulate had no theoretical basis; it predated Lewis's theory of covalent bonding by 23 yr. However, it allowed Werner to explain many experimental facts. Werner postulated a primary valence of 3 and a secondary valence of 6 for cobalt(III). He therefore wrote the formula for $CoCl_3 \cdot 5NH_3$ as $[Co(NH_3)_5Cl]Cl_2$. The ligands within the brackets satisfied cobalt's secondary valence of 6; the three chlorides satisfied the primary valence of 3. Werner proposed that the chlorides within the coordination sphere of cobalt(III) are bound so tightly that they are unavailable to contribute to the compound's conductivity or to react with $AgNO_3$. Thus, $CoCl_3 \cdot 5NH_3$ consists of a $[Co(NH_3)_5Cl]^{2+}$ ion and two Cl^- ions.

Werner also sought to deduce the arrangement of ligands around the central metal. He postulated that cobalt(III) complexes exhibit an octahedral geometry, and he sought to verify this postulate by comparing the number of observed isomers with the number expected for various geometries. For example, $[Co(NH_3)_4Cl_2]^+$ should exhibit two geometric isomers if it were octahedral. When he first postulated an octahedral geometry for cobalt(III), only one isomer of $[Co(NH_3)_4Cl_2]^+$ was known, the green trans isomer. In 1907, after considerable effort, Werner succeeded in isolating the violet cis isomer. Even before that, however, he had succeeded in isolating other cis and trans isomers of cobalt(III) complexes. The occurrence of two isomers was consistent with his postulate of octahedral geometry. Another result consistent with octahedral geometry was the demonstration that $[Co(en)_3]^{3+}$ and certain other complexes were optically active. In 1913, Werner was awarded the Nobel Prize in chemistry for his outstanding research work in the field of coordination chemistry.

SAMPLE EXERCISE 25.7

Suggest the structure for $CoCl_2 \cdot 6H_2O$.

Solution: By analogy to the ammonia complexes of cobalt(III), we might write the formula for this compound as $[Co(H_2O)_6]Cl_2$. Indeed, experimental evidence indicates that the water molecules are attached to the metal as ligands. Hydrated metal salts generally have water coordinated to the metal. However, water can also be hydrogen-bonded to the anion, particularly to oxyanions. For example, there are four water molecules coordinated to Cu^{2+} and one to SO_4^{2-} in $CuSO_4 \cdot 5H_2O$.

PRACTICE EXERCISE

Suggest the appropriate coordination compound formulation for the compound $PdCl_2 \cdot 3NH_3$. [Pd(II) forms square-planar complexes.] *Answer:* $[Pd(NH_3)_3Cl]Cl$

25.7 COLOR AND MAGNETISM

Werner's theory helps us understand many properties of complexes, including isomerism and conductivity. However, his theory must be extended before we can use it to explain other properties, such as the colors and magnetic properties of transition-metal complexes. Studies of these two properties have played an important role in the development of more modern models for metal-ligand bonding. We have discussed the various types of magnetic behavior in Section 24.9; we also discussed the interaction of radiant energy with matter in Section 6.1. Let's briefly examine the significance of these two properties for transition-metal complexes before we try to develop a model for metal-ligand bonding.

Color

One of the striking features of transition-metal compounds is that they are often colored. We have seen photos of some colored complexes and have noted the colors of others. Table 25.2 lists the colors of several cobalt(III) complexes. From this list you can see that the color of the complex changes as the ligands surrounding the metal ion change. In general, the colors also depend on the particular metal and on its oxidation state. The presence of a partly filled *d* subshell on the metal ion is usually necessary. Metal ions that have completely empty *d* subshells (such as Al^{3+} and Ti^{4+}) or completely filled *d* subshells (such as Zn^{2+}, $3d^{10}$) are colorless.

We consider the relationship between electron configuration and color in Section 25.8. First, however, we need to review our earlier discussion of light and to introduce a few additional concepts. Recall first that visible light consists of electromagnetic radiation whose wavelength, λ, ranges from 400 nm to 700 nm (Figure 6.3). The energy of this radiation is inversely proportional to its wavelength, as discussed earlier, in Section 6.2:

$$E = h\nu = h(c/\lambda) \qquad [25.8]$$

The visible spectrum is shown in Figure 25.19. The colors of the spectrum, indicated by their first letters, spell out what appears to be a man's name: Roy G. Biv.

When a sample absorbs light, what we see is the sum of the remaining colors that strike our eyes. If a sample absorbs all wavelengths of visible light, none reaches our eyes from that sample. Consequently, it appears black. If the

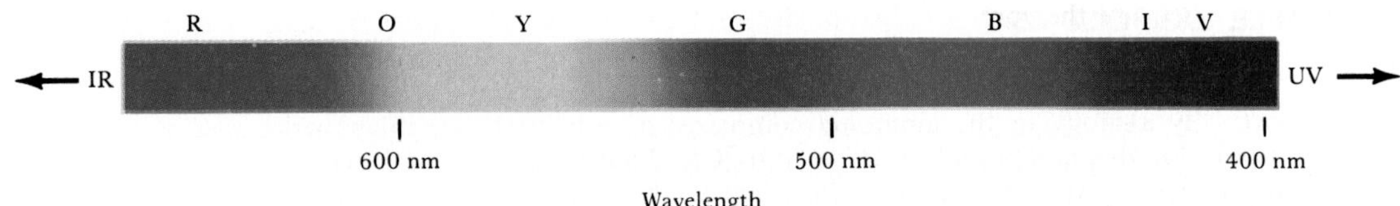

Figure 25.19 Visible spectrum showing the relation between color and wavelength.

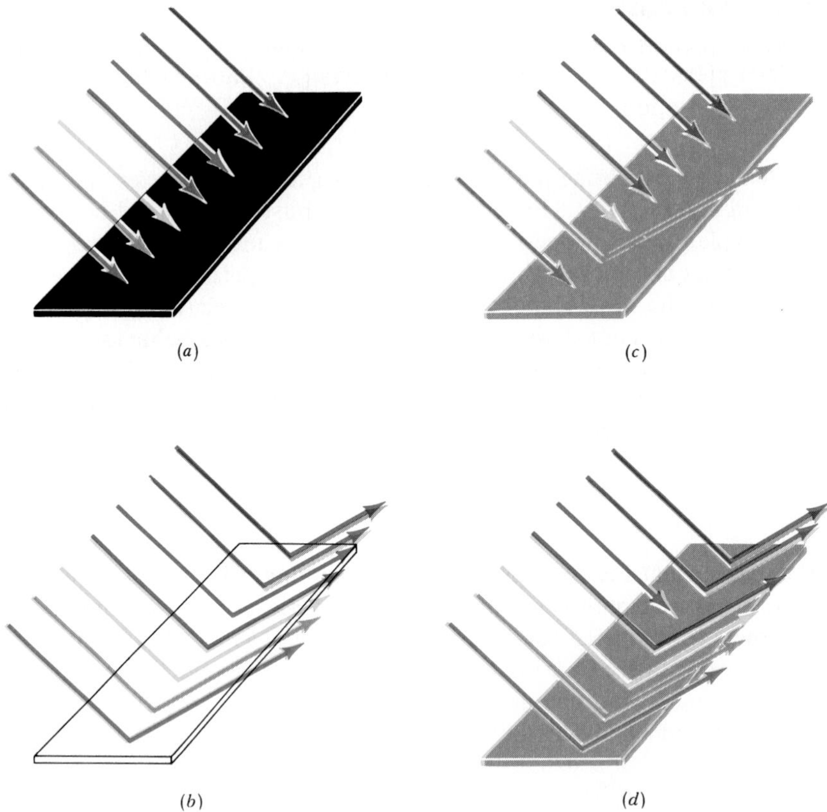

Figure 25.20 (*a*) An object is black if it absorbs all colors of light. (*b*) An object is white if it reflects all colors of light. (*c*) An object is orange if it reflects only this color and absorbs all others. (*d*) An object is also orange if it reflects all colors except blue, the complementary color of orange.

sample absorbs no visible light, it is white or colorless. If it absorbs all but orange, the sample appears orange. Each of these situations is shown in Figure 25.20. That figure shows one further situation; we also perceive an orange color when visible light of all colors except blue strikes our eyes. In a complementary fashion, if the sample absorbed only orange, it would appear blue; blue and orange are said to be **complementary colors**. Complementary colors can be determined with the aid of the artist's color wheel, shown in Figure 25.21. Complementary colors, like orange and blue, are across the wheel from each other.

Figure 25.21 Artist's color wheel, showing the colors that are complementary to one another and the wavelength range of each color.

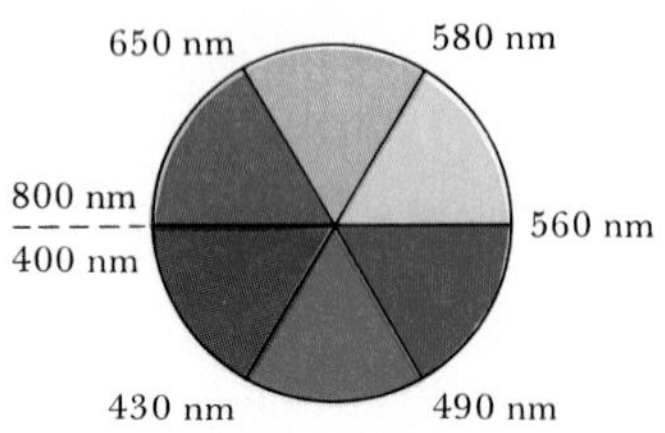

SAMPLE EXERCISE 25.8

The complex ion *trans*-$[Co(NH_3)_4Cl_2]^+$ absorbs light primarily in the red region of the visible spectrum (the most intense absorption is at 680 nm). What is the color of the complex?

Solution: Because the complex absorbs red light, its color will be complementary to red. From Figure 25.21, we see that this is green.

PRACTICE EXERCISE

The $[Cr(H_2O)_6]^{2+}$ ion has an absorption band at about 630 nm. Which of the following colors—sky blue, yellow, green, or deep red—is most likely to describe this ion? ***Answer:*** sky blue

The amount of light absorbed by a sample as a function of wavelength is known as its **absorption spectrum**. The visible absorption spectrum of a transparent sample, such as a solution of *trans*-$[Co(NH_3)_4Cl_2]^+$, can be determined as shown in Figure 25.22. The spectrum of $[Ti(H_2O)_6]^{3+}$, which we shall discuss in the next section, is shown in Figure 25.23. The absorption maximum of $[Ti(H_2O)_6]^{3+}$ is at 510 nm. Because the sample absorbs most strongly in the green and yellow regions of the visible spectrum, it appears purple.

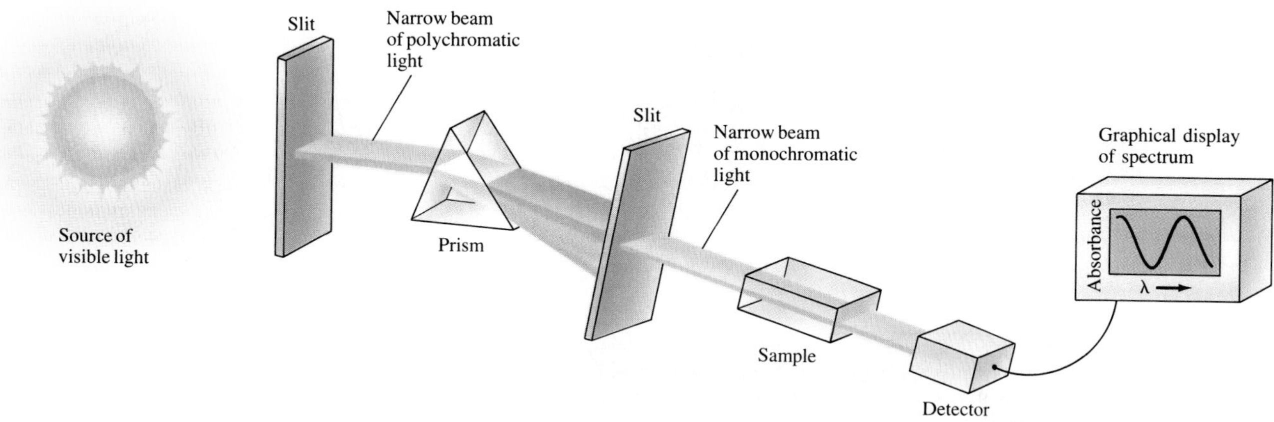

Figure 25.22 Experimental determination of the absorption spectrum of a solution. The prism is rotated so that different wavelengths of light pass through the sample. The detector measures the amount of light reaching it, and this information can be displayed as the absorption at each wavelength.

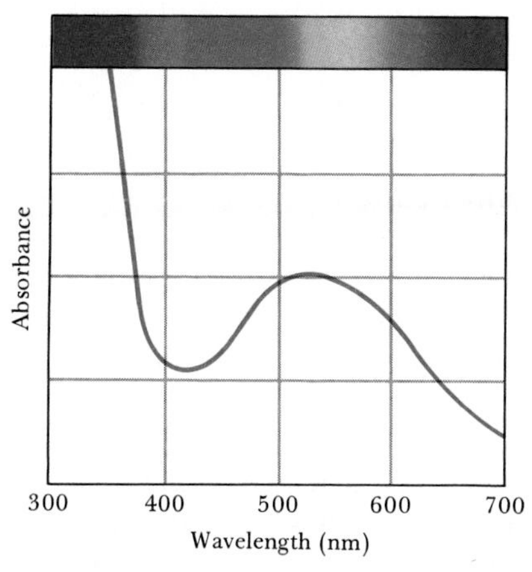

Figure 25.23 Visible absorption spectrum of the $[Ti(H_2O)_6]^{3+}$ ion.

A CLOSER LOOK: Gemstones

Gemstones such as ruby and emerald owe their color to the presence of trace amounts of transition-metal ions. For example, replacement of a fraction of the aluminum in the colorless mineral corundum, Al_2O_3, produces several different gems: chromium forms ruby, manganese forms amethyst, and iron forms topaz. Sapphire, which occurs in a variety of colors but is most often blue, contains titanium and cobalt. Several other gems are produced by replacing a trace of aluminum in the colorless mineral beryl, $Be_3Al_2Si_6O_{18}$, with transition-metal ions. For example, emerald contains chromium, and aquamarine contains iron. See Figure 25.24.

Figure 25.24 A selection of gemstones, some mounted in rings: (*a*) aquamarine, (*b*) emerald, (*c*) amethyst, (*d*) blue sapphire, (*e*) ruby (an example of an uncut mineral sample), and (*f*) citrine quartz. In all these cases the colors are due to transition-metal ions present as minor components in the parent mineral, such as quartz, alumina, or beryl. (Gem Media, a Division of the Gemological Institute of America)

Magnetism

Many transition-metal complexes exhibit simple paramagnetism, as described in Section 24.9. In such compounds, the individual metal ions possess some number of unpaired electrons. It is possible to determine the number of unpaired electrons per metal ion from the degree of paramagnetism. The experiments reveal some interesting comparisons. For example, compounds of the complex ion $[Co(NH_3)_6]^{3+}$ have no unpaired electrons, but compounds of the $[CoF_6]^{3-}$ ion have four per metal ion. Both complexes contain Co(III) with a $3d^6$ electron configuration. Clearly, there is a major difference in the ways in which the electrons are arranged in the metal orbitals in these two cases, even though both are complexes of cobalt(III), with a $3d^6$ electron configuration. Any successful bonding theory must explain this and other related observations. Although the

25.8 CRYSTAL-FIELD THEORY

ability to form complexes is common to all metal ions, the most numerous and interesting complexes are formed by the transition elements. Scientists have long recognized that the magnetic properties and colors of transition-metal complexes are related to the presence of *d* electrons in metal orbitals. In this section, we shall consider a model for bonding in transition-metal complexes, called the **crystal-field theory**, that accounts for the observed properties of these substances.*

We have already noted that the ability of a metal ion to attract ligands such as water around itself can be viewed as a Lewis acid-base interaction (Section 16.10). The base—that is, the ligand—can be considered to donate a pair of electrons into a suitable empty hybrid orbital on the metal, as shown in Figure 25.25. However, we can assume that much of the attractive interaction between the metal ion and the surrounding ligands is due to the electrostatic forces between the positive charge on the metal and negative charges on the ligands. If the ligand is ionic, as in the case of Cl^- or SCN^-, the electrostatic interaction occurs between the positive charge on the metal center and the negative charge on the ligand. When the ligand is neutral, as in the case of H_2O or NH_3, the negative ends of these polar molecules, containing an unshared electron pair, are directed toward the metal. In this case, the attractive interaction is of the ion-dipole type (Section 11.2). In either case, the result is the same; the ligands are attracted strongly toward the metal center. The assembly of metal ion and ligands is lower in energy than the fully separated charges, as illustrated on the left side of Figure 25.26.

In a 6-coordinate octahedral complex, we can envision the ligands approaching along the *x*, *y*, and *z* axes, as shown in Figure 25.27. Using the physical arrangement of ligands and metal ion shown in this figure as our starting point, let us consider what happens to the energies of electrons in the metal *d* orbitals as the ligands approach the metal ion. Keep in mind that the *d* electrons are the outermost electrons of the metal ion. We know that the overall energy of the metal ion plus ligands will be lower (that is, more stable) when the ligands are drawn toward the metal center. At the same time, however, there is a repulsive interaction between the outermost electrons on the metal and the negative charges on the ligands. This interaction is called the *crystal field*. The

* The name *crystal field* arose because the theory was first developed to explain the properties of solid, crystalline materials, such as ruby. The same theoretical model applies to complexes in solution.

Figure 25.25 Representation of the metal-ligand bond in a complex as a Lewis acid-base interaction. The ligand, which acts as a Lewis base, donates charge to the metal via a metal hybrid orbital. The bond that results is strongly polar, with some covalent character. For many purposes it is sufficient to assume that the metal-ligand interaction is entirely electrostatic in character, as is done in the crystal-field model.

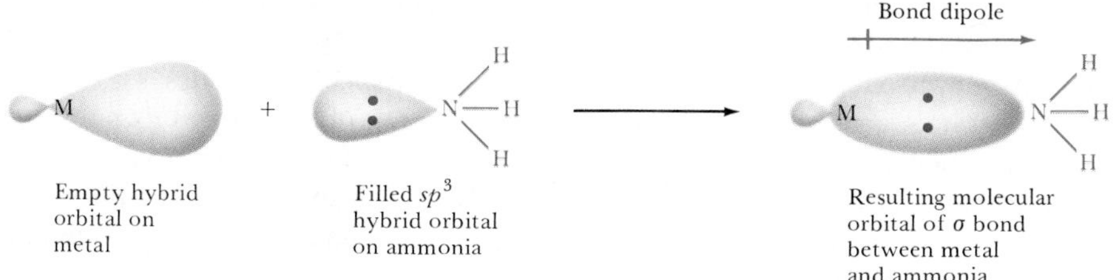

Figure 25.26 In the crystal-field model, the bonding between metal ion and donor atoms is considered to be largely electrostatic. The energy of the metal ion plus coordinated ligands is lower than that of the separated metal ion plus ligands because of the electrostatic attraction. At the same time, the energies of the metal d electrons are increased by the repulsive interaction between these electrons and the electrons of the ligands. These repulsive interactions gives rise to the splitting of the metal d-orbital energies.

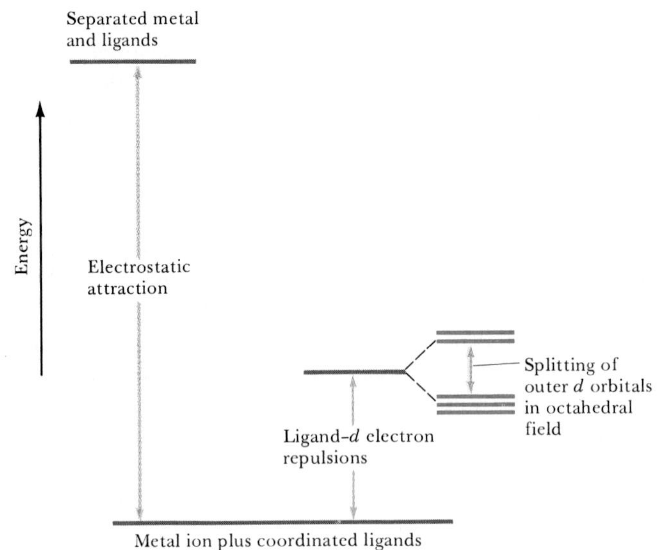

Figure 25.27 Octahedral array of negative charges surrounding a positive charge.

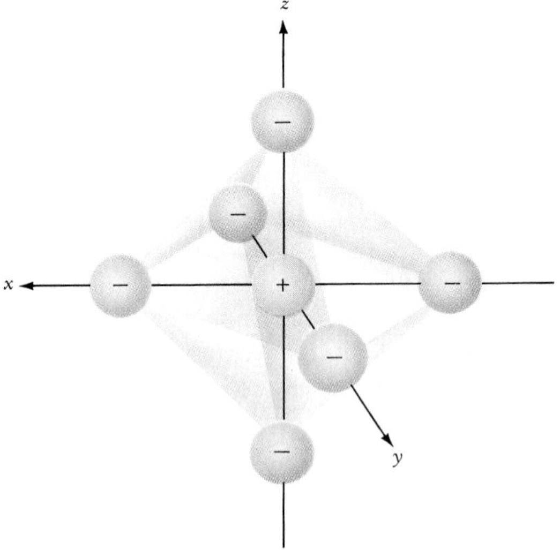

crystal field causes the energies of the d electrons on the metal ion to increase, as shown in Figure 25.26. The d orbitals of the metal ion, however, do not all behave in the same way under the influence of the crystal field. To see why this is so, recall the shapes of the five d orbitals, illustrated in Figure 25.28. In the isolated metal ion, these five orbitals are equivalent in energy. However, as the ligands approach the metal ion, the $d_{x^2-y^2}$ and d_{z^2} orbitals, which are directed *along* the x, y, and z axes, are more strongly repelled by the ligands than the d_{xy}, d_{xz}, and d_{yz} orbitals. These latter are directed *between* the axes along which the ligands approach. Thus, an energy separation, or *splitting*, occurs. The $d_{x^2-y^2}$ and d_{z^2} orbitals are raised in energy, and the d_{xz}, d_{yz}, and d_{xy} orbitals are lowered. This energy splitting is illustrated on the right side of Figure 25.26. In the material that follows, we shall concentrate only on the splitting of the d orbital energies by the crystal field. This process is illustrated in a slightly different form in Figure 25.29.

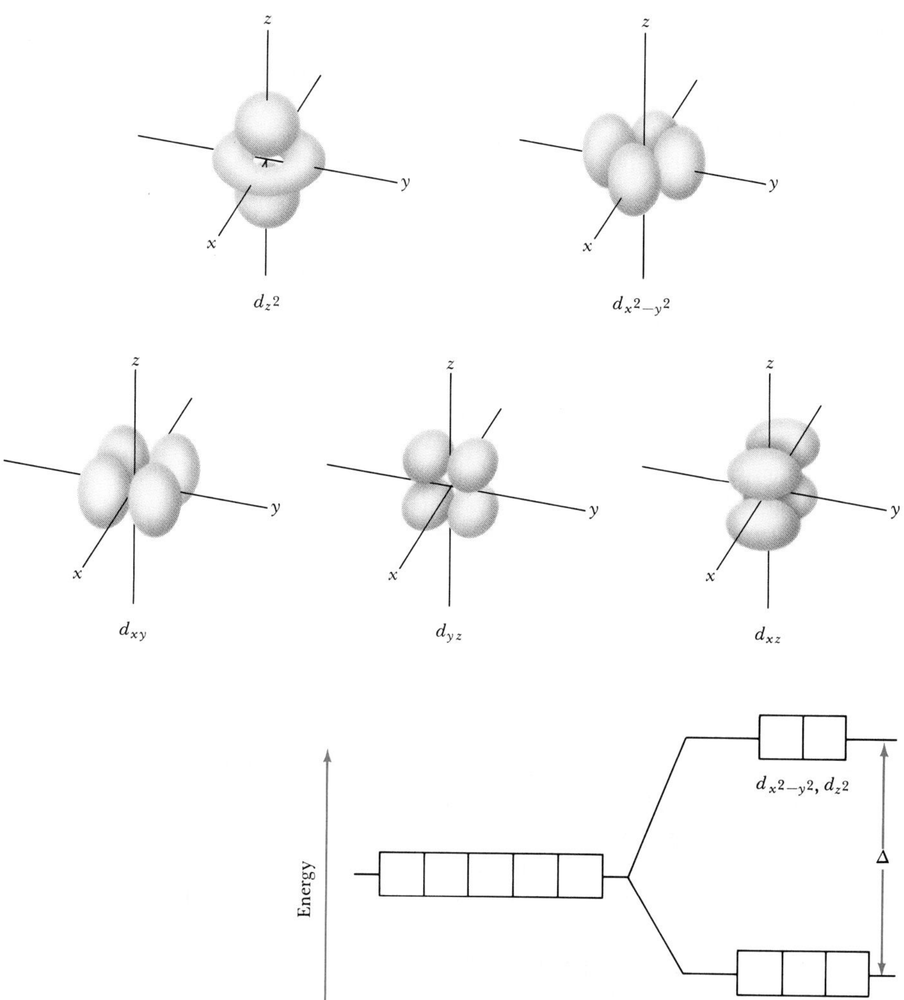

Figure 25.28 Shapes of the five *d* orbitals. Remember that the lobes represent regions in which the electrons occupying an orbital are most likely to be found.

Figure 25.29 Energies of the *d* orbitals in an octahedral crystal field.

Let's examine how the crystal-field model accounts for the observed colors in transition-metal complexes. The energy gap between the *d* orbitals, labeled Δ, is of the same order of magnitude as the energy of a photon of visible light. (The energy gap, Δ, is sometimes referred to as the crystal-field splitting energy.) It is therefore possible for a transition-metal complex to absorb visible light, which thereby excites an electron from the lower-energy *d* orbitals into the higher-energy ones. The $[Ti(H_2O)_6]^{3+}$ ion provides a simple example, because titanium(III) has only one 3*d* electron. As shown in Figure 25.23, $[Ti(H_2O)_6]^{3+}$ has a single absorption peak in the visible region of the spectrum. The maximum absorption is at 510 nm (235 kJ/mol). Light of this wavelength causes the *d* electron to move from the lower-energy set of *d* orbitals into the higher-energy set, as shown in Figure 25.30. The absorption of 510-nm radiation that produces this transition causes substances containing the $Ti(H_2O)_6{}^{3+}$ ion to appear purple.

The magnitude of the energy gap, Δ—and consequently the color of a complex—depend on both the metal and the surrounding ligands. For example, $[Fe(H_2O)_6]^{3+}$ is light violet, $[Cr(H_2O)_6]^{3+}$ is violet, and $[Cr(NH_3)_6]^{3+}$

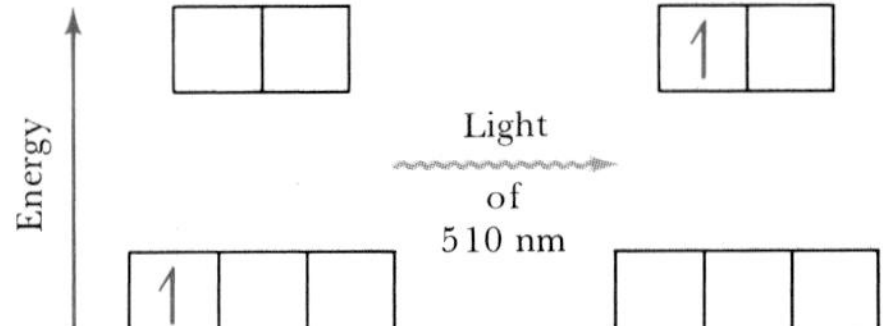

Figure 25.30 The 3*d* electron of $[Ti(H_2O)_6]^{3+}$ is excited from the lower-energy *d* orbitals to the higher-energy ones when irradiated with light of 510-nm wavelength.

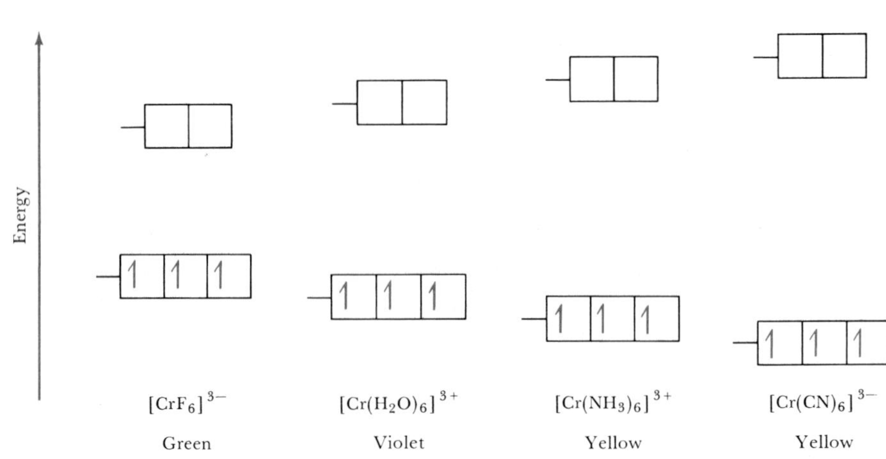

Figure 25.31 Crystal-field splitting in a series of octahedral chromium(III) complexes.

is yellow. Ligands can be arranged in order of their abilities to increase the energy gap, Δ. The following is an abbreviated list of common ligands arranged in order of increasing Δ:

$$Cl^- < F^- < \overrightarrow{H_2O < NH_3 < en < NO_2^-}^{\Delta \text{ increasing}} \text{ (N-bonded)} < CN^-$$

This list is known as the **spectrochemical series.**

Ligands that lie on the low end of the spectrochemical series are termed *weak-field ligands;* those on the high end are termed *strong-field ligands.* Figure 25.31 shows schematically what happens to the crystal-field splitting when the ligand is varied in a series of chromium(III) complexes. (This is a good place to remind you that when a transition metal is ionized, the valence *s* electrons are removed first. Therefore, the outer electron configuration for chromium is $[Ar]3d^54s^1$; that for Cr^{3+} is $[Ar]3d^3$.) Notice that as the field exerted by the six surrounding ligands increases, the splitting of the metal *d* orbitals increases. Because the absorption spectrum is related to this energy separation, these complexes vary in color.

SAMPLE EXERCISE 25.9

Which of the following complexes of Ti^{3+} exhibits the shortest wavelength absorption in the visible spectrum: $[Ti(H_2O)_6]^{3+}$; $[Ti(en)_3]^{3+}$; $[TiCl_6]^{3-}$?

Solution: The wavelength of the absorption is determined by the magnitude of the splitting between the *d* orbital energies in the field of the surrounding ligands. The larger the splitting, the shorter the wavelength of the absorption corresponding to the transition of the electron from the lower- to the higher-energy orbital. The splitting will be largest for ethylenediamine, en, the ligand that is highest in the spectrochemical series. Thus, the complex with the shortest wavelength absorption is $[Ti(en)_3]^{3+}$.

PRACTICE EXERCISE

The absorption spectrum of $[Ti(NCS)_6]^{3-}$ shows a band that lies intermediate in wavelength between those for $[TiCl_6]^{3-}$ and $[TiF_6]^{3-}$. What can we conclude about the place of NCS^- in the spectrochemical series? ***Answer:*** It lies between Cl^- and F^-; that is, $Cl^- < NCS^- < F^-$

Electron Configurations in Octahedral Complexes

The crystal-field model helps us understand the magnetic properties and some important chemical properties of transition-metal ions. From our earlier discussion of electronic structure in atoms (Section 6.9), we expect that electrons will always occupy the lowest-energy vacant orbitals first and that they will occupy a set of degenerate orbitals one at a time with their spins parallel (Hund's rule). Thus, if we have one, two, or three electrons to add to the *d* orbitals in an octahedral complex ion, the electrons will go into the lower-energy set of orbitals, with their spins parallel, as shown in Figure 25.32. When we come to add a fourth electron, a problem arises. If the electron is added to the lower-energy orbital, an energy gain of magnitude Δ is realized, as compared with placing the electron in the higher-energy orbital. However, there is a penalty for doing this, because the electron must now be paired up with the electron already occupying the orbital. The energy required to do this, relative to putting it in another orbital with parallel spin, is called the **spin-pairing energy**. The spin-pairing energy arises from the greater electrostatic repulsion of electrons that share an orbital as compared with two that are in different orbitals.

The ligands that surround the metal ion, and the charge on the metal, often play major roles in determining which of the two electronic arrangements arises. Consider the $[CoF_6]^{3-}$ and $[Co(CN)_6]^{3-}$ ions. In both cases the ligands have a 1− charge. However, the F^- ion, on the low end of the spectrochemical series, is a weak-field ligand. The CN^- ion, on the high end of the spectrochemical series, is a strong-field ligand. It produces a larger energy gap than does the F^- ion. The splittings of the *d*-orbital energies in the complexes are compared in Figure 25.33.

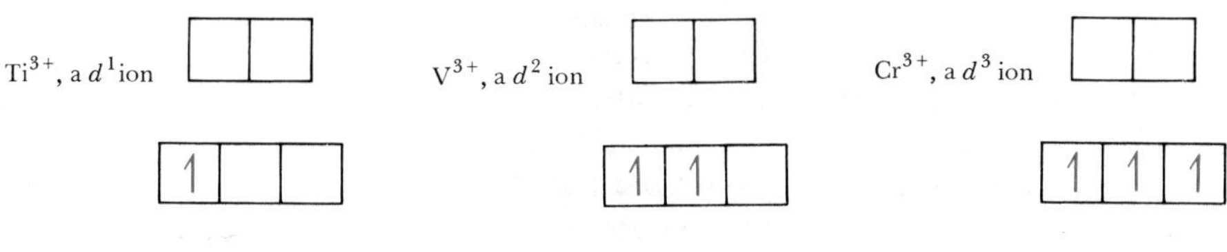

Figure 25.32 Electron configurations associated with one, two, and three electrons in the 3*d* orbitals in octahedral complexes.

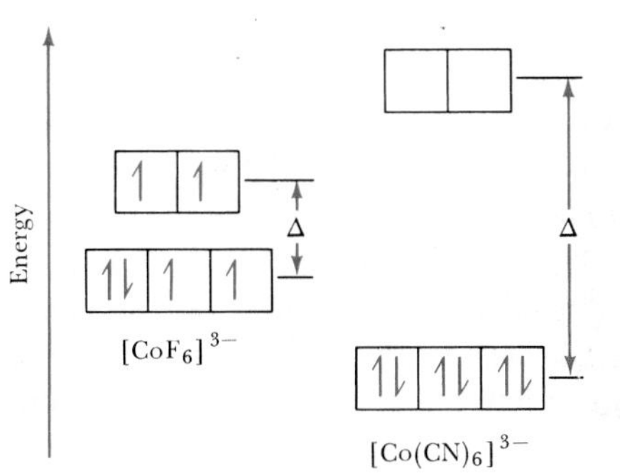

Figure 25.33 Population of *d* orbitals in the high-spin $[CoF_6]^{3-}$ ion (small Δ) and low-spin $[Co(CN)_6]^{3-}$ ion (large Δ).

A count of the electrons in cobalt(III) tells us that we have six electrons to place in the 3*d* orbitals. Let us imagine that we add these electrons one at a time to the *d* orbitals of the CoF_6^{3-} ion. The first three will, of course, go into the lower-energy orbitals with spins parallel. The fourth electron could go into a lower-energy orbital, pairing up with one of those already present. This would result in an energy gain of Δ as compared with putting it in one of the higher-energy orbitals. However, it would cost energy in an amount equal to the spin-pairing energy. Because F^- is a weak-field ligand, Δ is small, and the more stable arrangement is the one in which the electron is placed in the higher-energy orbital. Similarly, the fifth electron we add goes into a higher-energy orbital. With all of the orbitals containing at least one electron, the sixth must be paired up, and it goes into a lower-energy orbital. In the case of the $[Co(CN)_6]^{3-}$ complex, the crystal-field splitting is much larger. The spin-pairing energy is smaller than Δ, so electrons are paired in the lower-energy orbitals, as illustrated in Figure 25.33.

The $[CoF_6]^{3-}$ complex is referred to as a **high-spin complex;** that is, the electrons are arranged so that they remain unpaired as much as possible. On the other hand, the $[Co(CN)_6]^{3-}$ ion is referred to as a **low-spin complex**. These two different electronic arrangements can be readily distinguished by measuring the magnetic properties of the complex, as described earlier. The absorption spectrum also shows characteristic features that indicate the electronic arrangement.

SAMPLE EXERCISE 25.10

Predict the number of unpaired electrons in 6-coordinate high-spin and low-spin complexes of Fe^{3+}.

Solution: The Fe^{3+} ion possesses five 3*d* electrons. In a high-spin complex, these are all unpaired. In a low-spin complex the electrons are confined to the lower-energy set of *d* orbitals, with the result that there is one unpaired electron:

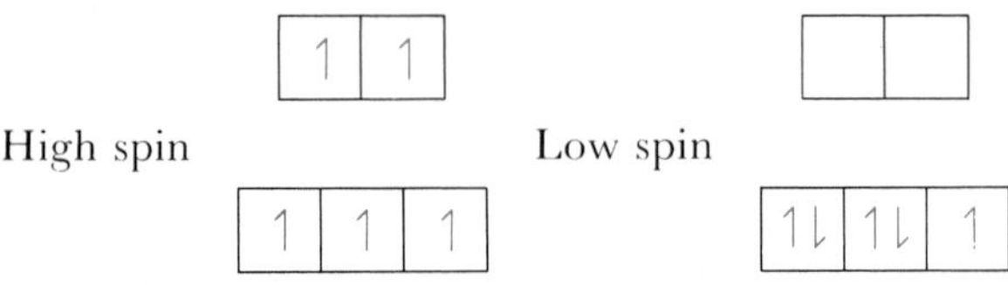

PRACTICE EXERCISE

For which *d* electron configurations does the possibility of a distinction between high-spin and low-spin arrangements exist in octahedral complexes?
Answer: d^4, d^5, d^6, d^7

Tetrahedral and Square-Planar Complexes

Thus far, we have considered the crystal-field model only for complexes of octahedral geometry. When there are only four ligands about the metal, the geometry is tetrahedral, except for the special case of metal ions with a d^8 electron configuration, which we will discuss in a moment. The crystal-field splitting of the metal *d* orbitals in tetrahedral complexes differs from that in octahedral complexes. Four equivalent ligands can interact with a central metal ion most effectively by approaching along the vertices of a tetrahedron. It turns out—and this is not easy to explain in just a few sentences—that the splitting

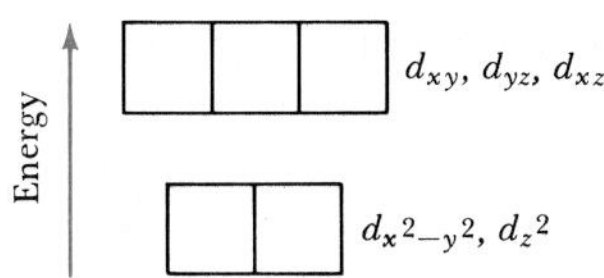

Figure 25.34 Energies of the d orbitals in a tetrahedral crystal field.

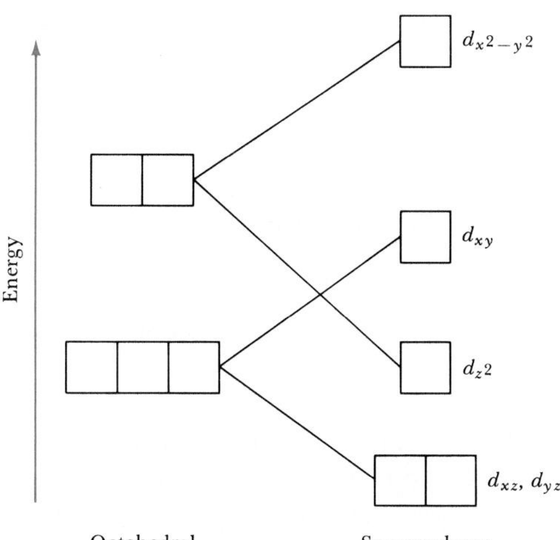

Figure 25.35 Effect on the relative energies of the d orbitals caused by removing the two negative charges from the z axis of an octahedral complex. When the charges are completely removed, the square-planar geometry results.

of the metal d orbitals in a tetrahedral crystal is just the opposite of that for the octahedral case. That is, three of the metal d orbitals are raised in energy, and the other two are lowered, as illustrated in Figure 25.34. Because there are only four ligands instead of six, as in the octahedral case, the crystal-field splitting is much smaller for tetrahedral complexes. Calculations show that for the same metal ion and ligand set, the crystal-field splitting for a tetrahedral complex is only four-ninths as large as for the octahedral complex. For this reason, all tetrahedral complexes are high spin; the crystal field is never large enough to overcome the spin-pairing energies.

Square-planar complexes, in which four ligands are arranged about the metal ion in a plane, represent a common geometric form. You can envision the square-planar complex as formed by removing two ligands from along the vertical z axis of the octahedral complex. As this happens, the four ligands in the plane are drawn in more tightly. The changes that occur in the energy levels of the d orbitals are illustrated in Figure 25.35.

Square-planar complexes are characteristic of metal ions with a d^8 electron configuration. They are nearly always low spin; that is, the eight d electrons are spin-paired to form a diamagnetic complex. Such an electronic arrangement is particularly common among the ions of heavier metals, such as Pd^{2+}, Pt^{2+}, Ir^{+}, and Au^{3+}.

SAMPLE EXERCISE 25.11

Four-coordinate nickel(II) complexes exhibit both square-planar and tetrahedral geometries. The tetrahedral ones, such as $[NiCl_4]^{2-}$, are paramagnetic; the square-planar ones, such as $[Ni(CN)_4]^{2-}$, are diamagnetic. Show how the d electrons of nickel(II) populate the d orbitals in the appropriate crystal-field-splitting diagram in each case.

Solution: Nickel(II) has an electron configuration of [Ar]$3d^8$. The population of the d electrons in the two geometries is given below:

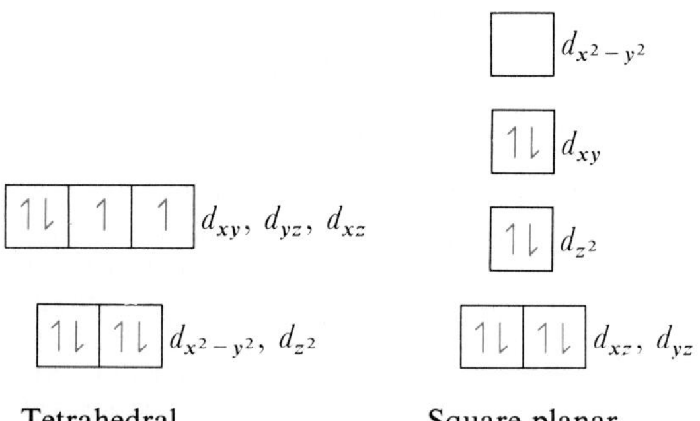

PRACTICE EXERCISE

How many unpaired electrons do you predict for the tetrahedral $[CoCl_4]^{2-}$ ion? ***Answer:*** three

We have seen that the crystal-field model provides a basis for explaining many features of transition-metal complexes. In fact, it can be used to explain many observations in addition to those we have discussed. However, many lines of evidence show that the bonding between transition-metal ions and ligands must have some covalent character. Molecular orbital theory (Sections 9.6 and 9.7) can also be used to describe the bonding in complexes. However, the application of molecular-orbital theory to coordination compounds is beyond the scope of our discussion. The crystal-field model, although not entirely accurate in all details, provides an adequate and useful description.

FOR REVIEW

SUMMARY

Coordination compounds or complexes contain metal ions bonded to several surrounding anions or molecules known as ligands. The metal ion and its ligands comprise the coordination sphere of the complex. The atom of the ligand that bonds to the metal ion is known as the donor atom. The number of donor atoms attached to the metal ion is known as the coordination number of the metal ion. The most common coordination numbers are 4 and 6; the most common coordination geometries are tetrahedral, square planar, and octahedral.

If a ligand has several donor atoms that can coordinate simultaneously to the metal, it is said to be polydentate and is referred to as a chelating agent. Two common examples are ethylenediamine (en), which is potentially bidentate, and ethylenediaminetetraacetate ($EDTA^{4-}$), which has six potential donor atoms. Many biologically important molecules, such as the porphyrins, are complexes of chelating agents.

Isomerism is common among coordination compounds. Structural isomerism involves differences in the bonding arrangements of the ligands. One simple form of structural isomerism, known as linkage isomerism, occurs when a ligand is capable of coordinating to a metal through either of two donor atoms. Coordination-sphere isomerism occurs when two compounds with the same overall formula contain different ligands in the coordination sphere.

Stereoisomerism involves complexes with the same chemical bonding arrangements but with differing spatial arrangements of ligands. The most common forms are geometric and optical isomerism. Geometric isomers differ from one another in the relative locations of donor atoms in the coordination sphere; the most common are cis-trans isomers. Optical isomers differ from one another in that they are nonsuperimposable mirror images of one another. Geometric isomers differ from one another in their chemical and physical properties; however, optical isomers differ only in the presence of a chiral environment. Optical isomers can be distinguished from one another by their interactions with plane-polarized light; solutions of one isomer rotate the plane of polarization to the right (dextrorotatory), and solutions of its mirror image rotate the plane to the left (levorotatory). Chiral molecules are said to be optically active. A 50–50 mixture of two optical isomers does not rotate plane-polarized light and is said to be racemic.

Many of the early studies that served as a basis for our current understanding of complexes focused on complexes of chromium(III), cobalt(III), platinum(II), and platinum(IV). Complexes of these metal ions are inert, in that they undergo ligand exchange at a slow rate. Complexes undergoing rapid exchange are said to be labile.

Studies of the magnetic properties and colors of transition-metal complexes have played an important role in formulation of bonding theories for these compounds. The crystal-field theory successfully accounts for many properties of coordination compounds. In this model, the interaction between metal ion and ligand is viewed as electrostatic. The ligands produce an electric field that causes a splitting in the energies of the metal *d* orbitals. In the spectrochemical series, the ligands are listed in order of their ability to split the *d*-orbital energies in octahedral complexes.

In strong-field complexes, the splitting of *d*-orbital energies is large enough to overcome spin-pairing energies, and *d* electrons preferentially pair up in the lower-energy orbitals. Such complexes are low spin. When the ligands exert a relatively weak crystal field, the electrons occupy the higher-energy *d* orbitals in preference to pairing up in the lower-energy set, and the complexes are high spin.

The crystal-field model is applicable also to tetrahedral and square-planar complexes. However, the ordering of the *d*-orbital energies is different than in octahedral complexes.

KEY TERMS

complex ion
complex
coordination compounds
ligand (Sec. 25.1)
coordination sphere (Sec. 25.1)
donor atom (Sec. 25.1)
coordination number (Sec. 25.1)
monodentate ligand (Sec. 25.2)
polydentate ligand (Sec. 25.2)
chelating agent (Sec. 25.2)
bidentate ligand (Sec. 25.2)
chelate effect (Sec. 25.2)
porphyrin (Sec. 25.2)
isomer (Sec. 25.4)
stereoisomer (Sec. 25.4)
structural isomer (Sec. 25.4)
linkage isomerism (Sec. 25.4)
coordination-sphere isomer (Sec. 25.4)
geometric isomerism (Sec. 25.4)
optical isomerism (Sec. 25.4)
chiral (Sec. 25.4)
dextrorotatory (Sec. 25.4)
levorotatory (Sec. 25.4)
optically active (Sec. 25.4)
racemic (Sec. 25.4)
labile complex (Sec. 25.5)
inert complex (Sec. 25.5)
complementary colors (Sec. 25.7)
absorption spectrum (Sec. 25.7)
crystal-field theory (Sec. 25.8)
spectrochemical series (Sec. 25.8)
spin-pairing energy (Sec. 25.8)
high-spin complex (Sec. 25.8)
low-spin complex (Sec. 25.8)

EXERCISES

Structure and Nomenclature

25.1 Indicate the coordination number about the metal and the oxidation number of the metal in each of the following complexes:

(a) $[Zn(en)_2]Br_2$ **(b)** $[Co(NH_3)_4Cl_2]Cl$
(c) $K[Co(C_2O_4)_2(NH_3)_2]$ **(d)** $K_2[MoOCl_4]$
(e) $K[Au(CN)_2]$.

25.2 Indicate the coordination number about the metal and the oxidation number of the metal in each of the following complexes:
(a) $K_2[V(C_2O_4)_3]$ **(b)** $K_4[Fe(CN)_6]$
(c) $[Pd(NH_3)_2Cl_2]$ **(d)** $[Cr(en)_2F_2]NO_3$
(e) $K_2[HgCl_4]$.

25.3 Sketch the structure of each of the following complexes:
(a) $[AlCl_4]^-$ (b) $[Ag(CN)_2]^-$
(c) $[PtCl_2(en)]$ **(d)** *trans*-$[Cr(NH_3)_4(H_2O)_2]^{3+}$

25.4 Sketch the structure of each of the following complexes:
(a) $[Zn(NH_3)_4]^{2+}$ **(b)** $[Ru(H_2O)Cl_5]^{2-}$
(c) *trans*-$[Co(en)_2(NO_2)_2]^+$ **(d)** *cis*-$[Pt(NH_3)_2H(Br)]$

25.5 Name each of the complexes listed in Exercises 25.3 and 25.4.

25.6 Name each of the following complexes:
(a) $[Cr(NH_3)_3Br_3]$ **(b)** $[Co(en)(NH_3)_2Cl_2]Br$
(c) $K_3[Fe(C_2O_4)_3]$ **(d)** $Cs[Cr(C_2O_4)_2Cl_2]$
(e) $K_3[IrCl_5(S_2O_3)]$ **(f)** $[Pd(en)_2][Cr(NH_3)_2Br_4]_2$

25.7 Using Br^- and NH_3 as ligands, (a) give the formula of a 6-coordinate palladium(IV) complex that would be a nonelectrolyte in aqueous solution; **(b)** give the coordination compound of Pt(II) that has about the same electrical conductivity as RbBr; (c) give an octahedral complex of Fe(III) containing two NH_3 groups.

25.8 Using Cl^- and ethylenediamine, en, as ligands and Rb^+ as cation, **(a)** give the formulas of two 6-coordinate cobalt(III) complexes that have approximately the same electrical conductivity in water as RbCl; **(b)** give the formula of a nickel(II) complex that has about the same electrical conductivity in water as $SrCl_2$; **(c)** give the formula for a tetrahedral complex of Cd^{2+} that contains one en group.

25.9 Write the formula for each of the following compounds, being sure to use brackets to indicate the coordination sphere:
(a) hexaamminechromium(III) nitrate
(b) tetraamminecarbonatocobalt(III) sulfate
(c) dichlorobis(ethylenediamine)platinum(IV) bromide
(d) potassium diaquatetrabromovanadate(III)
(e) bis(ethylenediamine)zinc(II) tetraiodomercurate(II)

25.10 Write the formula for each of the following compounds, being sure to use brackets to indicate the coordination sphere:
(a) pentaamminethiosulfatomanganese(III) sulfate
(b) tris(bipyridyl)ruthenium(II) nitrate
(c) dichlorobis(*ortho*-phenanthroline)iron(III) perchlorate
(d) sodium tetrabromo(ethylenediamine)cobaltate(III)
(e) hexaamminenickel(II) tris(oxalato)chromate(III)

25.11 Polydentate ligands can vary in the number of coordination positions they occupy. In each of the following, indicate the most probable number of coordination positions occupied by the polydentate ligand present:
(a) $[Co(NH_3)_4CO_3]Cl$ **(b)** $[Cr(C_2O_4)_2(H_2O)_2]Br$
(c) $[CrEDTA(H_2O)]^-$ **(d)** $[Zn(en)_2](ClO_4)_2$.

25.12 Indicate the likely coordination number of the metal in each of the following complexes:
(a) $[Zn(en)Cl_2]$ **(b)** $[Co(o\text{-phen})_2Cl_2]Cl$
(c) $K_2[Hg(SCN)_4]$ **(d)** Na[FeEDTA].

Isomerism

25.13 By writing formulas or drawing structures related to any one of the following complexes, illustrate **(a)** geometrical isomerism; **(b)** linkage isomerism; **(c)** optical isomerism; **(d)** coordination-sphere isomerism. The complexes are: $[Co(NH_3)_4Br_2]Cl$; $[Pd(NH_3)_2(ONO)_2]$; and *cis*-$[V(en)_2Cl_2]^+$.

25.14 (a) Draw the two linkage isomers of $[Co(NH_3)_5SCN]^{2+}$; **(b)** Draw the two geometrical isomers of $[Co(NH_3)_3Cl_3]$; **(c)** Two compounds with the formula $Co(NH_3)_5ClBr$ can be prepared. Use structural formulas to show how they differ. What kind of isomerism does this illustrate?

25.15 Draw the four geometrical isomers of $[Co(en)(NH_3)_2BrCl]^+$. Two of these geometrical isomers have optical isomers. Identify them and draw the structures of their optical isomers.

25.16 Sketch all the possible stereoisomers for each of the following complexes:
(a) tetrahedral $[Cd(en)Cl_2]$
(b) square-planar $[Pd(NH_3)Cl(C_2O_4)]^-$
(c) square-planar $[PtCl_2(CN)_2]^{2-}$
(d) octahedral $[Fe(bipy)(en)_2]^+$

25.17 The compound $Co(NH_3)_5(SO_4)Br$ exists in two forms, one red and one violet. Both forms dissociate in solution to form two ions. Solutions of the red compound form a precipitate of AgBr on addition of $AgNO_3$ solution, but no precipitate of $BaSO_4$ on addition of $BaCl_2$ solution. For the violet compound, just the reverse occurs. From this evidence, indicate the structures of the complex ions in each case, and give the correct name of each compound.

25.18 Write the formulas for, and properly name, two possible coordination-sphere isomers with the formula $Cr(H_2O)_4(OH)Br_2$.

25.19 A palladium complex formed from a solution containing bromide ion and pyridine, C_5H_5N (a good electron-pair donor), is found on elemental analysis to contain 37.6 percent bromine, 28.3 percent carbon, 6.60 percent nitrogen, and 2.37 percent hydrogen. The compound is slightly soluble in several organic solvents; its solutions in water or alcohol do not conduct electricity. It is found experimentally to have a zero dipole moment. Write the chemical formula and indicate its probable structure. Name the compound.

25.20 A manganese complex formed from a solution containing potassium bromide and oxalate ion is purified and analyzed. It contains 10.0 percent Mn, 28.6 percent potassium, 8.8 percent carbon, and 29.2 percent bromine. The remainder of the compound is oxygen. An aqueous solution of the complex has about the same electrical conductivity as an equimolar solution of $K_4[Fe(CN)_6]$. Write the formula of the compound, using brackets to denote the manganese and its coordination sphere. Name the compound.

Color; Magnetism; Crystal-Field Theory

25.21 Which of the following compounds would you expect to have color: **(a)** TiO_2; **(b)** Cr_2O_3; **(c)** $[Al(C_2O_4)_3]^{3-}$; **(d)** $[Fe(C_2O_4)_3]^{3-}$; **(e)** $[Zn(NH_3)_4]^{2+}$; **(f)** $[Ni(NH_3)_4]^{2+}$?

25.22 Which of the following hydrated cations would you expect to be colorless: $Sc^{3+}(aq)$, $Ti^{3+}(aq)$, $Pb^{2+}(aq)$, $Hf^{4+}(aq)$, $Co^{2+}(aq)$, $Cu^{+}(aq)$, $V^{2+}(aq)$? Explain briefly.

25.23 What is the observed color of a coordination compound that absorbs radiation of wavelength 580 nm?

25.24 A coordination compound is bright blue in color. Assuming that this color is due to a single absorption band, what is the approximate wavelength of the absorption?

25.25 Give the number of d electrons associated with the central metal in each of the following complexes: (a) $[Co(CN)_5]^{3-}$; **(b)** $[AuCl_4]^{-}$; (c) $[V(NH_3)_3Cl_3]^{-}$; **(d)** $[Ru(en)_3]^{2+}$; **(e)** $[Mn(CN)_6]^{3-}$.

25.26 Give the number of d electrons associated with the central metal ion in each of the following complexes: **(a)** $[Ru(en)_3]Cl_3$; **(b)** $K_2[Cu(CN)_4]$; **(c)** $Na_3[Co(NO_2)_6]$; **(d)** $[MoEDTA]ClO_4$; **(e)** $K_3[ReCl_6]$.

25.27 Explain why the d_{xy}, d_{xz}, and d_{yz} orbitals lie lower in energy than the $d_{x^2-y^2}$ and d_{z^2} orbitals in the presence of an octahedral arrangement of ligands about the central metal ion.

25.28 What properties of the ligand determine the size of the splitting of the d-orbital energies in the presence of an octahedral arrangement of ligands about a central transition-metal ion? Explain.

25.29 Explain why many cyano complexes of divalent transition-metal ions are yellow, whereas many aqua complexes of these ions are blue or green.

25.30 The $[Ni(H_2O)_6]^{2+}$ ion is green, whereas the $[Ni(NH_3)_6]^{2+}$ ion is purple. Predict the predominant color of light absorbed by each ion. Which ion absorbs light with the shorter wavelength? Do your conclusions agree with the spectrochemical series?

25.31 For each of the following metals, write the electronic configuration of the metal atom and the 3+ ion: (a) Rh; (b) Mn; **(c)** Pd. Draw the crystal-field energy-level diagram for the d orbitals of an octahedral complex, and show the placement of the d electrons in each case, assuming a strong-field complex.

25.32 For each of the following metals, write the electronic configuration of the metal atom and the 2+ ion: **(a)** Ni; **(b)** Tc; **(c)** Cr. Draw the crystal-field energy-level diagram for the d electrons of an octahedral complex, and show the placement of the d electrons in each case, assuming a weak-field complex of the 2+ ion.

25.33 Draw the crystal-field energy-level diagram and show the placement of d electrons for each of the following: **(a)** $[V(H_2O)_6]^{2+}$; (b) $[Mn(H_2O)_6]^{3+}$ (high spin); (c) $[Ru(CN)_6]^{3-}$ (one unpaired electron); **(d)** $[IrCl_6]^{3-}$ (low spin); **(e)** $[Cr(en)_3]^{2+}$ (four unpaired electrons); **(f)** $[NiF_6]^{4-}$.

25.34 Draw the crystal-field energy-level diagrams and show the placement of electrons for the following complexes: **(a)** $[ZrCl_6]^{2-}$; **(b)** $[MnF_6]^{3-}$ (a high-spin complex); **(c)** $[Rh(NH_3)_6]^{3+}$ (a low-spin complex); **(d)** $[NiCl_4]^{2-}$ (tetrahedral); **(e)** $[PtBr_4]^{2-}$ (square planar); **(f)** $[Cu(en)_3]^{2+}$.

25.35 The complex $[Mn(NH_3)_6]^{2+}$ contains five unpaired electrons. Sketch the energy-level diagram for the d orbitals and indicate the placement of electrons for this complex ion. Is the ion a high-spin or a low-spin complex?

25.36 Explain why the ion $[Fe(CN)_6]^{3-}$ has one unpaired electron, whereas $[Fe(NCS)_6]^{3-}$ has five unpaired electrons.

Additional Exercises

25.37 Distinguish between the following terms: **(a)** a chelate and a monodentate ligand; **(b)** coordination sphere and coordination number; **(c)** a labile and an inert complex; **(d)** a high-spin and a low-spin complex; **(e)** coordination-sphere isomerism and linkage isomerism; **(f)** optical isomerism and geometrical isomerism.

25.38 Based on the molar conductance values listed below for the series of platinum(IV) complexes, write the formula for each complex so as to show which ligands are in the coordination sphere of the metal.

Complex	Molar conductance (Ω^{-1})[a] of 0.05 M solution
$Pt(NH_3)_6Cl_4$	523
$Pt(NH_3)_4Cl_4$	228
$Pt(NH_3)_3Cl_4$	97
$Pt(NH_3)_2Cl_4$	0
$KPt(NH_3)Cl_5$	108

[a] The ohm (Ω) is a unit of resistance; conductance is the inverse of resistance.

25.39 In Werner's early studies, he observed that when the complex $[Co(NH_3)_4Br_2]Br$ was placed in water, the electrical conductivity of a 0.05 M solution changed from an initial value of 191 Ω^{-1} to a final value of 374 Ω^{-1} over a period of an hour or so. Suggest an explanation of the observed results. Write a balanced chemical equation to describe the reaction.

25.40 A pink solid has the empirical formula $CoCl_3 \cdot 5NH_3 \cdot H_2O$. A solution of this compound is also pink, and 1 mol of the compound gives 3 mol AgCl(s) upon titration with $AgNO_3$. When the pink solid is heated to 110°C, it loses 1 mol of H_2O to give a purple solid whose solution gives 2 mol AgCl(s) when titrated with $AgNO_3$. Propose structures for the two cobalt complexes and name them.

25.41 Draw the geometric structure of each of the following complexes:
(a) $[Pt(en)_2]^{2+}$

(b) *cis*-dichloroethylenediamineoxalatoferrate(III) ion
(c) *trans*-diamminedibromopalladium(II)
(d) *trans*-$[Co(en)_2Br_2]^+$
(e) tetrabromocobaltate(II) ion
(f) *trans*-$[Mo(NCS)_2(en)_2]^+$
(g) pentaamminebromovanadium(II) ion.

25.42 Acetylacetone forms very stable complexes with many metallic ions. It acts as a bidentate ligand, coordinating to the metal at two adjacent positions. Suppose that one of the CH_3 groups of the ligand is replaced by a CF_3 group, as shown:

Trifluoromethyl acetylacetonate (tfac)

$$\left[CF_3-\overset{\displaystyle :O:}{\underset{}{C}}=\overset{H}{C}-\overset{\displaystyle :\ddot{O}:}{C}-CH_3 \right]^-$$

Sketch all possible isomers for the complex with three tfac ligands on cobalt(III). (You can use the symbol ●◡○ to represent the ligand.)

25.43 Many trace metal ions exist in the bloodstream as complexes with amino acids (Figure 27.4) or small peptides. The anion of the amino acid glycine,

$$H_2NCH_2\overset{O}{\overset{\|}{C}}-O^-$$

symbol gly, is capable of acting as a bidentate ligand, coordinating to the metal through nitrogen and $-O^-$ atoms. How many isomers are possible for **(a)** $[Zn(gly)_2]$ (tetrahedral); **(b)** $[Pt(gly)_2]$ (square planar); **(c)** $[Co(gly)_3]$ octahedral). Sketch all possible isomers. Use N◡O to represent the ligand.

25.44 Write balanced chemical equations to represent the following observations. (In some instances, the complex involved has been discussed previously at some point in the text.) **(a)** Solid silver chloride dissolves in an excess of aqueous ammonia. **(b)** The green complex $[Cr(en)_2Cl_2]Cl$, on treatment with water over a long time, converts to a brown-orange complex. Reaction of $AgNO_3$ with 1 L of 1 *M* solution of the product results in precipitation of 3 mol of AgCl. (Write *two* chemical equations.) **(c)** Insoluble zinc hydroxide dissolves in excess aqueous ammonia. **(d)** A pink solution of $Co(NO_3)_2$ turns deep blue on addition of concentrated hydrochloric acid.

25.45 Oxalic acid, $H_2C_2O_4$, can be used to remove rust stains, Fe_2O_3. Write a balanced chemical equation that explains how the oxalic acid works.

25.46 Draw out the *d*-orbital energy-level diagram for each of the following complexes, and indicate the most likely placement of *d* electrons: **(a)** $[Au(CN)_4]^-$; **(b)** $[Ni(C_2O_4)_3]^{4-}$; **(c)** $[PtF_6]$; **(d)** $[Rh(CN)_6]^{3-}$.

25.47 How would you go about determining the number of unpaired electrons in the compound $Na_2[CoCl_4]$? If the compound proved to have three unpaired electrons, how would you account for this in terms of crystal-field theory?

25.48 What changes would you expect in the absorption spectrum of complexes of V(III) as the ligand is varied in the order F^-, NH_3, NO_2^-?

25.49 Solutions containing the $[Co(H_2O)_6]^{2+}$ ion absorb at about 520 nm; those containing the $[CoCl_4]^{2-}$ ion absorb at about 690 nm. What colors do you expect for the solutions? Why is the absorption maximum for the $[CoCl_4]^{2-}$ ion at a longer wavelength than that for the $[Co(H_2O)_6]^{2+}$ ion?

25.50 Indicate whether each of the following statements is true or false. If it is false, modify it to make it true. **(a)** Spin-pairing energy is larger than Δ in low-spin complexes. **(b)** Δ is larger for complexes of Mn^{3+} with a given ligand than for complexes of Mn^{2+}. **(c)** $[NiCl_4]^{2-}$ is more likely to be square planar than is $[Ni(CN)_4]^{2-}$.

25.51 The value of Δ for the $[CrF_6]^{3-}$ complex is 182 kJ/mol. Calculate the expected wavelength of the absorption corresponding to promotion of an electron from the lower-energy to the higher-energy *d*-orbital set in this complex. Should the complex absorb in the visible range? (You may need to review Sample Exercise 6.2; remember to divide by Avogadro's number.)

25.52 In each of the following pairs of complexes, which would you expect to absorb at the longer wavelength: **(a)** $[CoF_6]^{4-}$ or $[Co(CN)_6]^{4-}$; **(b)** $[V(H_2O)_6]^{2+}$ or $[V(H_2O)_6]^{3+}$; **(c)** $[Mn(CN)_6]^{3-}$ or $[MnCl_4]^-$? Explain your reasoning in each case.

[25.53] The red color of ruby is due to the presence of Cr(III) ions at octahedral sites in the close-packed oxide lattice of Al_2O_3. Draw the crystal-field splitting diagram for Cr(III) in this environment. Suppose that the ruby crystal is subjected to high pressure. What do you predict for the variation in the wavelength of absorption of the ruby as a function of pressure? Explain.

[25.54] The d^3 and d^6 electronic configurations are favorable for octahedral coordination, but not for tetrahedral. Explain why this is so in terms of crystal-field theory.

[25.55] Suppose that a transition-metal ion were in a lattice in which it was in contact with just two nearby anions, located on opposite sides of the metal. Diagram the splitting of the metal *d* orbitals that would result from such a crystal field. Assuming a strong field, how many unpaired electrons would you expect for a metal ion with six *d* electrons? (Hint: Consider the linear axis to be the *z* axis.)

[25.56] A Cu electrode is immersed in a solution that is 1.00 *M* in $[Cu(NH_3)_4]^{2+}$ and 1.00 *M* in NH_3. When the cathode is a standard hydrogen electrode, the emf of the cell is found to be +0.08 V. What is the formation constant for $[Cu(NH_3)_4]^{2+}$?

Organic Chemistry

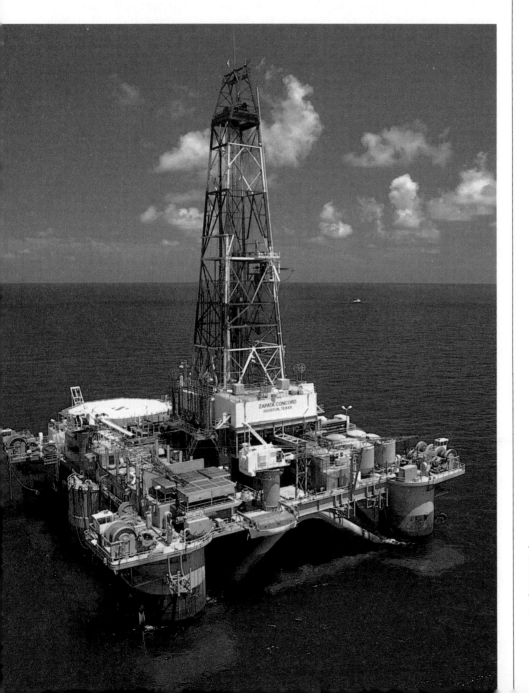

An offshore platform for drilling oil. Hydrocarbons obtained from crude oil are used in the manufacture of a great variety of organic compounds. (Shelly Katz/Black Star)

CONTENTS

Organic chemistry deals primarily with compounds in which carbon is the principal element. Carbon forms a vast number of compounds. Over 7 million carbon compounds are known, and about 90 percent of the new compounds prepared each year contain this element. Thus, the study of carbon compounds has come to constitute a separate branch of chemistry.

The terms *organic chemistry* and *organic compounds* arose in the eighteenth century from the "vitalist theory," which held that organic compounds could be formed only by living organisms. In 1828, Friedrich Wöhler, a German chemist, reacted potassium cyanate, KOCN, with ammonium chloride, NH_4Cl, and, much to his surprise, obtained urea, H_2NCONH_2, a well-known organic substance that had been isolated from the urine of mammals. Today, chemists prepare many organic compounds from inorganic starting materials.

In this chapter, we present a brief view of some of the elementary aspects of organic chemistry. Because the compounds of carbon are so numerous, it is convenient to organize them into families that exhibit structural similarities. The simplest class of organic compounds is the *hydrocarbons*, compounds composed only of carbon and hydrogen. We shall see that organic compounds containing other elements can be considered derivatives of hydrocarbons.

The key structural feature of hydrocarbons, and for that matter of most other organic substances, is the presence of stable carbon-carbon bonds. Carbon is the only element capable of forming stable, extended chains of atoms bonded through single, double, or triple bonds. Hydrocarbons can be divided into four general types, depending on the kinds of carbon-carbon bonds in their molecules. Figure 26.1 shows an example of each of the four types: alkanes, alkenes, alkynes, and aromatic hydrocarbons.

Alkanes are hydrocarbons that contain only single bonds, as in ethane, C_2H_6. Because alkanes contain the largest possible number of hydrogen atoms per carbon atom, they are called *saturated hydrocarbons.* **Alkenes,** also known as olefins, are hydrocarbons with one or more carbon-carbon double bonds, as in ethylene, C_2H_4. **Alkynes** contain at least one carbon-carbon triple bond, as in acetylene, C_2H_2. In **aromatic hydrocarbons**, the carbon atoms are connected in a planar ring structure, joined by both σ and π bonds between carbon atoms. Benzene, C_6H_6, is the best-known example of an aromatic hydrocarbon. Alkenes, alkynes, and aromatic hydrocarbons are called *unsaturated hydrocarbons* because they contain less hydrogen than an alkane having the same number of carbon atoms.

The members of the different series of hydrocarbons exhibit different chemical behaviors, as we shall see shortly. However, they are similar in many ways. Because carbon and hydrogen do not differ greatly in electronegativity (2.5 for

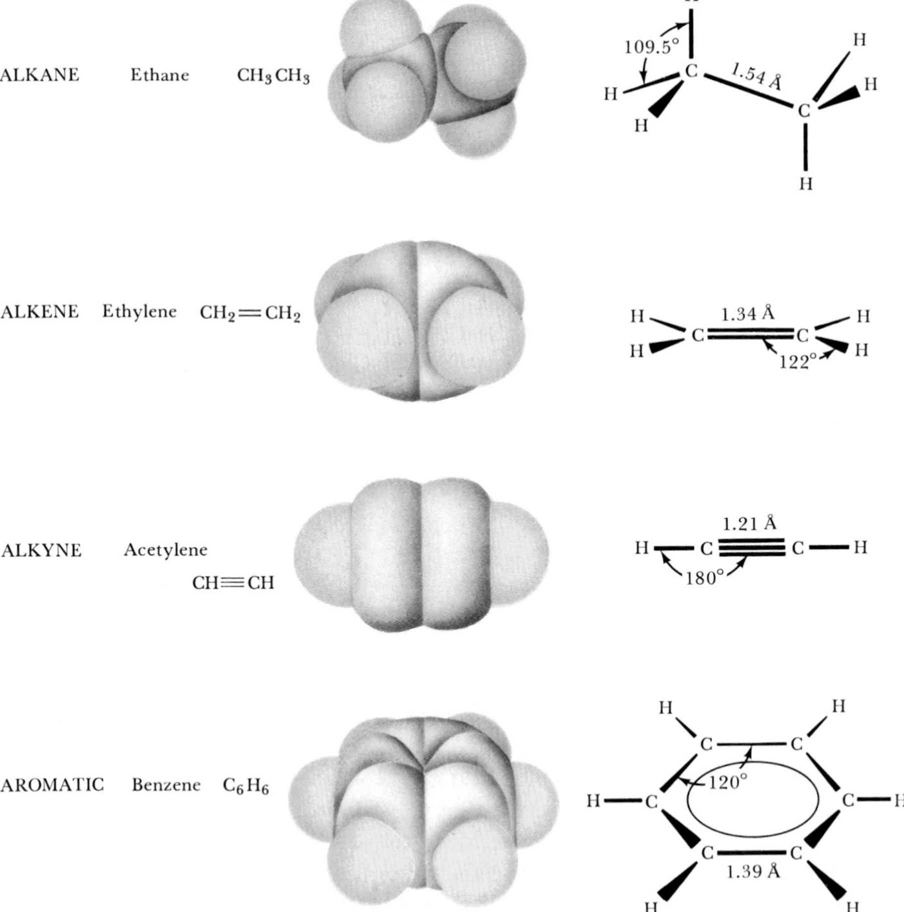

Figure 26.1 Names, geometrical structures, and molecular formulas for examples of each type of hydrocarbon.

carbon, 2.2 for hydrogen), hydrocarbon molecules are relatively nonpolar. Thus, they are almost completely insoluble in water, but they dissolve readily in other nonpolar solvents. Furthermore, they tend to become less volatile with increasing molar mass because of London dispersion forces (Section 11.2).

26.1 ALKANES

Table 26.1 lists several of the simplest alkanes. Many of these substances are familiar because of their widespread use. Methane is a major component of natural gas and is used for home heating and in gas stoves and hot-water heaters. Propane is the major component of bottled, or LP, gas used for home heating and cooking in areas where natural gas is not available. Butane is used in disposable lighters and in fuel cannisters for gas camping stoves and lanterns. Alkanes with from 5 to 12 carbon atoms per molecule are found in gasoline.

The formulas for the alkanes given in Table 26.1 are written in a notation called the *condensed structural formula*. This notation reveals the way in which atoms are bonded to one another but does not require drawing in all the bonds. For example, the Lewis structure and condensed structural formula for butane, C_4H_{10}, are

Table 26.1 First Several Members of the Straight-Chain Alkane Series

Molecular formula	Condensed structural formula	Name	Boiling point (°C)
CH_4	CH_4	Methane	−161
C_2H_6	CH_3CH_3	Ethane	−89
C_3H_8	$CH_3CH_2CH_3$	Propane	−44
C_4H_{10}	$CH_3CH_2CH_2CH_3$	Butane	−0.5
C_5H_{12}	$CH_3CH_2CH_2CH_2CH_3$	Pentane	36
C_6H_{14}	$CH_3CH_2CH_2CH_2CH_2CH_3$	Hexane	68
C_7H_{16}	$CH_3CH_2CH_2CH_2CH_2CH_2CH_3$	Heptane	98
C_8H_{18}	$CH_3CH_2CH_2CH_2CH_2CH_2CH_2CH_3$	Octane	125
C_9H_{20}	$CH_3CH_2CH_2CH_2CH_2CH_2CH_2CH_2CH_3$	Nonane	151
$C_{10}H_{22}$	$CH_3CH_2CH_2CH_2CH_2CH_2CH_2CH_2CH_2CH_3$	Decane	174

```
     H   H   H   H
     |   |   |   |
 H — C — C — C — C — H
     |   |   |   |
     H   H   H   H
```

Lewis structure

$CH_3CH_2CH_2CH_3$

Condensed structural formula

We shall frequently use either Lewis structures or condensed structural formulas to represent organic compounds. You should practice drawing the structural formulas from the condensed ones. Notice that each carbon atom in an alkane has four single bonds, whereas each hydrogen atom forms one single bond.

Each succeeding compound in the series listed in Table 26.1 has an additional CH_2 unit. A series such as that shown in Table 26.1 is known as a **homologous series**. The general formula for all the compounds listed in the table is C_nH_{2n+2}, where n is the number of carbon atoms. One of the characteristics of a homologous series is that all the compounds of the series can be described by the same general formula. We shall see several other examples of homologous series as we proceed.

Structures of Alkanes

The Lewis structures and condensed structural formulas for alkanes do not tell us anything about the three-dimensional structures of these substances. As we would predict from the VSEPR model (Section 9.1), the geometry about each carbon atom in an alkane is tetrahedral; that is, the four groups attached to each carbon are located at the vertices of a tetrahedron. The three-dimensional structures can be represented as shown for methane in Figure 26.2. The bonding may be described as involving sp^3 hybridized orbitals on the carbon, as discussed earlier in Section 9.4.

Rotation about the carbon-carbon single bond is relatively easy. You might imagine grasping the top left methyl group in Figure 26.3, which shows the structure of propane, and twisting it relative to the rest of the structure. Motion of this sort occurs very rapidly in alkanes at room temperature. Consequently, a long-chain alkane is constantly undergoing motions that cause it to change its shape, something like a length of chain that is being shaken.

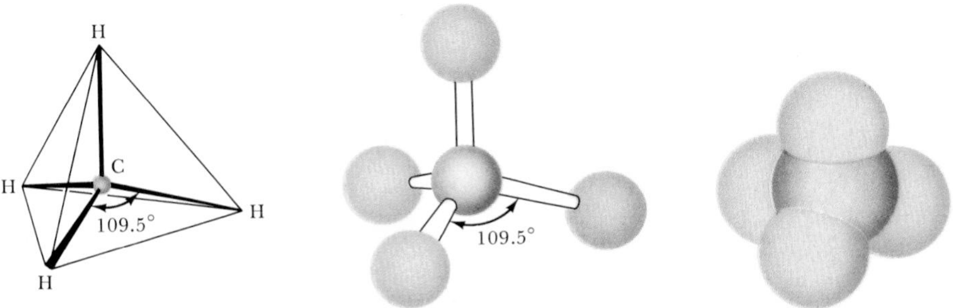

Figure 26.2 Representations of the three-dimensional arrangement of bonds about carbon in methane.

Structural Isomers

The alkanes listed in Table 26.1 are called *straight-chain hydrocarbons* because all the carbon atoms are joined in a continuous chain. Alkanes consisting of four or more carbon atoms can also form branched chains; hydrocarbons with branched chains are called *branched hydrocarbons.* Figure 26.4 shows the condensed formulas and space-filling models for all the possible structures of alkanes containing four and five carbon atoms. Notice that there are two ways that four carbon atoms can be joined to give C_4H_{10}—as a straight chain (left) or a branched chain (right). For alkanes with five carbon atoms, C_5H_{12}, there are three different arrangements.

Compounds with the same molecular formula but with different bonding arrangements and hence different structures are called **structural isomers**. The structural isomers of a given alkane differ slightly from one another in physical properties. Note the melting and boiling points of the isomers of butane and pentane, given in Figure 26.4. The number of possible isomers increases rapidly with the number of carbon atoms in the alkane. For example, there are 18 possible isomers of octane, C_8H_{18}, and 75 possible isomers of decane, $C_{10}H_{22}$.

Nomenclature of Alkanes

The first names given to the structural isomers shown in Figure 26.4 are the so-called common names. The straight-chain isomer is called the normal isomer, abbreviated by the prefix *n*-. The isomer in which one CH_3 group is branched off the major chain is labeled the iso- isomer (for example, isobutane). However, as the number of isomers grows, it becomes impossible to find a suitable prefix to denote each isomer. The need for a systematic means of naming organic compounds was recognized early in the history of organic chemistry. In 1892, an organization called the International Union of Chemistry met in Geneva,

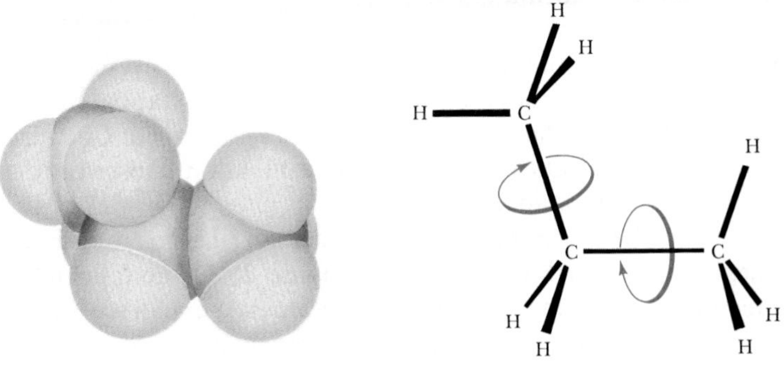

Figure 26.3 Three-dimensional models for propane, C_3H_8, showing rotations about the carbon-carbon single bonds.

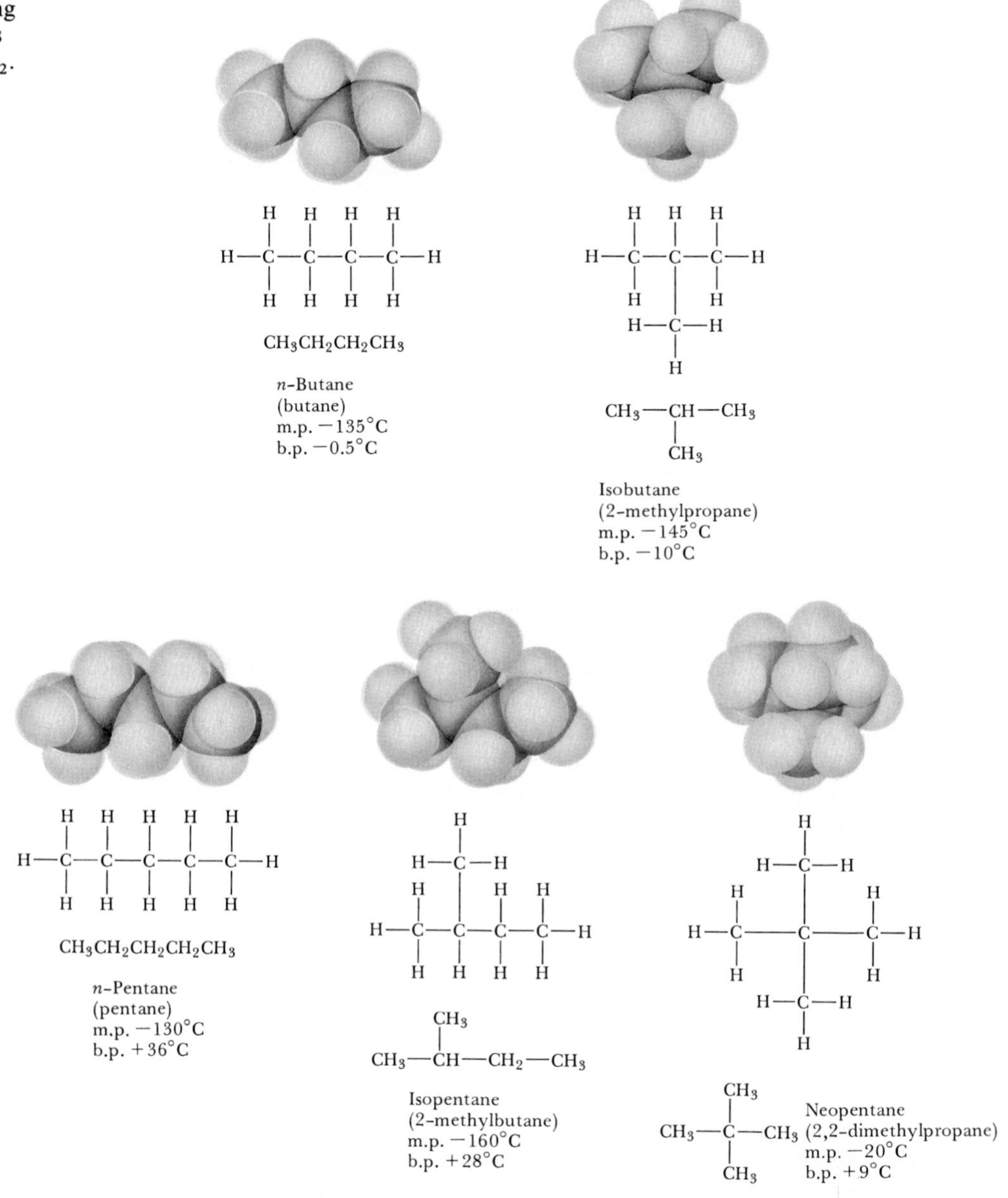

Figure 26.4 Possible structures, names, and melting and boiling points of alkanes of formula C_4H_{10} and C_5H_{12}.

Switzerland, to formulate rules for systematic naming of organic substances. Since that time, the task of keeping the rules for naming compounds up-to-date has fallen to the International Union of Pure and Applied Chemistry (IUPAC). It is interesting to note that through two devastating world wars and major social upheavals, the work of IUPAC has continued. Chemists everywhere, regardless of their nationality or political affiliation, subscribe to a common system for naming compounds.

The IUPAC names for the isomers of butane and pentane are the second ones given for each compound in Figure 26.4. The following rules summarize the procedures used to arrive at these names. We shall see that a similar approach is taken in writing the names for other organic compounds.

Table 26.2 Names and Condensed Structural Formulas for Several Alkyl Groups

Group	Name
$CH_3—$	Methyl
$CH_3CH_2—$	Ethyl
$CH_3CH_2CH_2—$	*n*-Propyl
$CH_3CH_2CH_2CH_2—$	*n*-Butyl
$\begin{array}{l} CH_3 \\ \vert \\ HC— \\ \vert \\ CH_3 \end{array}$	Isopropyl
$\begin{array}{c} \quad\quad CH_3 \\ \quad\quad \vert \\ CH_3—C— \\ \quad\quad \vert \\ \quad\quad CH_3 \end{array}$	*t*-Butyl

1. Each compound is named for the longest continuous chain of carbon atoms present. For example, the longest chain of carbon atoms in isobutane is three (Figure 26.4). Consequently, this compound is named as a derivative of propane, which has three carbon atoms; in the IUPAC system it is called 2-methylpropane.
2. In general, a group that is formed by removing a hydrogen atom from an alkane is called an **alkyl group**. The names for alkyl groups are derived by dropping the *-ane* ending from the name of the parent alkane and adding *-yl*. For example, the methyl group, CH_3, is derived from methane, CH_4; likewise, the ethyl group, C_2H_5, is derived from ethane, C_2H_6. Table 26.2 lists several of the more common alkyl groups.
3. The location of an alkyl group along a carbon-atom chain is indicated by numbering the carbon atoms along the chain. Thus, the name 2-methylpropane indicates the presence of a methyl, CH_3, group on the second carbon atom of a propane (three-carbon) chain. In general, the chain is numbered from the end that gives the lowest numbers for the alkyl positions.
4. If there is more than one substituent group of a certain type along the chain, the number of groups of that type is indicated by a prefix: *di-* (two), *tri-* (three), *tetra-* (four), *penta-* (five), and so forth. Therefore, the IUPAC name for neopentane (Figure 26.4) is 2,2-dimethylpropane. Dimethyl indicates the presence of two methyl groups; the 2,2- prefix indicates that both are on the second carbon atom of the propane chain.

SAMPLE EXERCISE 26.1

Name the following alkane:

$$\begin{array}{l} CH_3—CH—CH_3 \\ \quad\quad\quad \vert \\ CH_3—CH—CH_2 \\ \quad\quad\quad\quad\quad\quad \vert \\ \quad\quad\quad\quad\quad\quad CH_3 \end{array}$$

Solution: To name this compound properly, you must first find the longest continuous chain of carbon atoms. This chain, extending from the upper left CH_3 group to the lower right CH_3 group, is five carbon atoms long:

$$
\begin{array}{l}
{}^{①}CH_3-{}^{②}CH-CH_3 \\
\qquad\quad {}^{③}| \quad {}^{④} \\
CH_3-{}^{}CH-{}^{}CH_2 \\
\qquad\qquad\quad {}^{⑤}| \\
\qquad\qquad\quad CH_3
\end{array}
$$

The compound is thus named as a derivative of pentane. We could number the carbon atoms starting from either end. However, IUPAC rules state that the numbering should be done so that the numbers of those carbons which bear side chains are as low as possible. This means that we should start numbering with the upper carbon. There is a methyl group on carbon 2, and one on carbon 3. The compound is thus called 2,3-dimethylpentane.

PRACTICE EXERCISE

Name the following alkane:

$$
\begin{array}{l}
CH_3-CH_2 \quad CH_3 \\
\qquad\quad | \qquad\quad | \\
CH_3-CH-CH \\
\qquad\qquad\quad\;\; | \\
\qquad\qquad\quad CH_3
\end{array}
$$

Answer: 2,3-dimethylpentane

SAMPLE EXERCISE 26.2

Write the condensed structural formula for 2-methyl-3-ethylpentane.

Solution: The longest continuous chain of carbon atoms in this compound is five. We can therefore begin by writing out a string of five C atoms:

$$C-C-C-C-C$$

We next place a methyl group on the second carbon, and an ethyl group on the middle carbon atom of the chain. Hydrogens are then added to all the other carbon atoms to make the four bonds to each carbon.

$$
\begin{array}{l}
\qquad\quad CH_3 \\
\qquad\quad | \\
CH_3-CH-CH-CH_2-CH_3 \\
\qquad\qquad\quad | \\
\qquad\qquad\quad CH_2CH_3
\end{array}
$$

The condensed structural formula is $CH_3CH(CH_3)CH(C_2H_5)CH_2CH_3$.

PRACTICE EXERCISE

Write the condensed structural formula for 2,3-dimethylhexane.

Answer:

$$
\begin{array}{l}
\qquad\;\; CH_3 \;\; CH_3 \\
\qquad\quad | \qquad | \\
CH_3CH-CHCH_2CH_2CH_3
\end{array}
$$

Cycloalkanes

Alkanes can form not only branched chains, but rings or cycles as well. Alkanes with this form of structure are called **cycloalkanes**. Figure 26.5 illustrates a few examples of cycloalkanes. Cycloalkane structures are sometimes drawn as simple polygons in which each corner of the polygon represents a CH_2 group. This

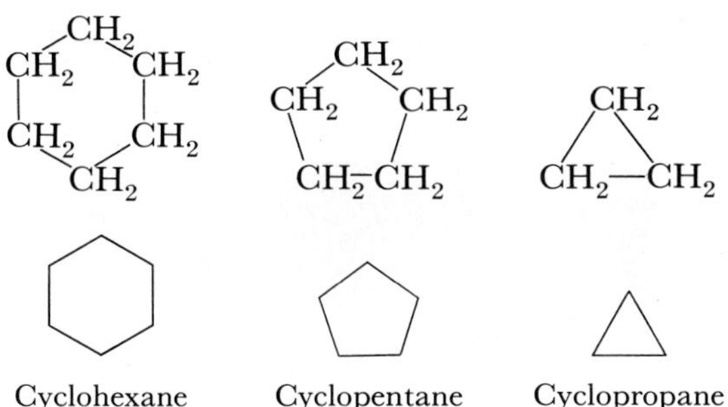

Figure 26.5 Condensed structural formulas for three cycloalkanes.

method of representation is similar to that used for aromatic rings, as illustrated in Section 9.5. In the case of aromatic structures, each corner represents a CH group.

Carbon rings containing fewer than five carbon atoms are strained, because the C—C—C bond angle in the smaller rings must be less than the 109.5° tetrahedral angle. The amount of strain increases as the rings get smaller. In cyclopropane, which has the shape of an equilateral triangle, the angle is only 60°; this molecule is therefore much more reactive than either propane, its straight-chain analog, or cyclohexane, which has no ring strain. Note that the general formula for cycloalkanes is C_nH_{2n}, which differs from that for straight-chain alkanes. Cycloalkanes thus form a separate homologous series.

Reactions of Alkanes

Alkanes are relatively unreactive. For example, at room temperature they do not react with acids, bases, or strong oxidizing agents, and they are not even attacked by boiling nitric acid. One reason for their low chemical reactivity is the strength of the C—C and C—H bonds.

Alkanes are not completely inert, however. One of their most commercially important reactions is *combustion* in air, which is the basis of their use as fuels. For example, the complete combustion of ethane proceeds as follows:

$$2C_2H_6(g) + 7O_2(g) \longrightarrow 4CO_2(g) + 6H_2O(g) \qquad \Delta H = -1888\ \text{kJ}$$

Alkanes also undergo **substitution reactions** with F_2, Cl_2, and Br_2 in which halogen atoms replace one or more hydrogen atoms in the alkane (Figure 26.6). Reactions between F_2 and alkanes are very vigorous, and special precautions must be taken to prevent explosions. Reactions with Cl_2 and Br_2 require heat or light to break Cl—Cl or Br—Br bonds in order to initiate the process. For example, ultraviolet light causes dissociation of the Cl_2 molecule, forming Cl atoms:

$$Cl_2 \longrightarrow 2Cl \qquad [26.1]$$

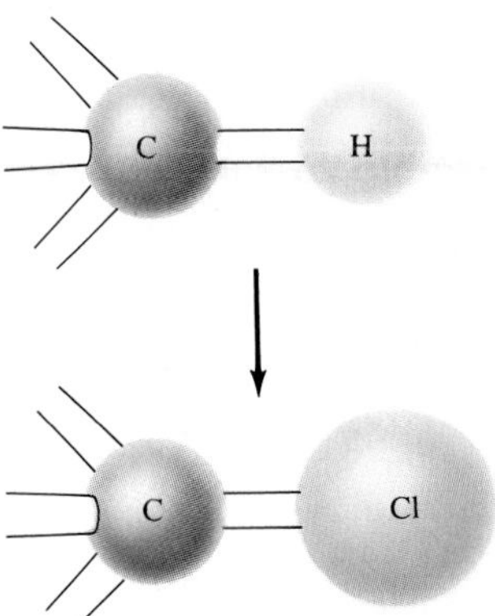

Figure 26.6 In a substitution reaction, one atom replaces another, as Cl replaces H in this example.

A chlorine atom has only seven valence-shell electrons. Species with an odd number of electrons are called **free radicals** (or simply radicals). They are frequently represented in chemical equations by showing a dot next to their chemical formula, representing the unpaired electron. Chlorine atoms are highly

CHEMISTRY AT WORK: Gasoline

Petroleum, or crude oil, is a complex mixture of organic compounds, mainly hydrocarbons, with smaller quantities of other organic compounds containing nitrogen, oxygen, or sulfur. The tremendous demand for petroleum to meet the world's energy needs has led to the tapping of oil wells in such forbidding places as the North Sea and northern Alaska.

The usual first step in the *refining*, or processing, of petroleum is to separate it into fractions on the basis of boiling point. The fractions commonly taken are shown in Table 26.3. Because gasoline is the most commercially important of these fractions, various processes are used to maximize its yield.

Gasoline is a mixture of volatile hydrocarbons containing varying amounts of aromatic hydrocarbons in addition to alkanes. In an automobile engine, a mixture of air and gasoline vapor is compressed by a piston and then ignited by a spark plug. The burning of the gasoline should create a strong, smooth expansion of gas, forcing the piston outward and imparting force along the driveshaft of the engine. If the gas burns too rapidly, the piston receives a single hard slam rather than a strong, smooth push. The result is a "knocking" or "pinging" sound and a reduction in the efficiency with which energy produced by the combustion is converted to power.

The *octane number* of a gasoline is a measure of its resistance to knock. Gasolines with high octane numbers burn more smoothly and are thus more effective fuels (Figure 26.7). The more highly branched alkanes have higher octane numbers than the straight-chain alkanes. The octane number of gasoline is obtained by comparing its knocking characteristics with those of "isooctane" (2,2,4- trimethylpentane) and heptane. Isooctane is assigned an octane number of 100, whereas heptane is assigned 0. Gasoline with the same knocking characteristics as a mixture of 90 percent isooctane and 10 percent heptane is rated as 90 octane.

The gasoline obtained directly from fractionation of petroleum (called *straight-run* gasoline) contains mainly straight-chain hydrocarbons and has an octane number

Figure 26.7 The octane rating of gasoline measures its resistance to knocking when burned in an engine. The octane rating of this gasoline is 87, as shown on the face of the pump. (Marmel Studios/The Stock Market)

Table 26.3 Hydrocarbon Fractions from Petroleum

Fraction	Size range of molecules	Boiling-point range (°C)	Uses
Gas	C_1 to C_5	-160 to 30	Gaseous fuel, production of H_2
Straight-run gasoline	C_5 to C_{12}	30 to 200	Motor fuel
Kerosene, fuel oil	C_{12} to C_{18}	180 to 400	Diesel fuel, furnace fuel, cracking
Lubricants	C_{16} and up	350 and up	Lubricants
Paraffins	C_{20} and up	Low-melting solids	Candles, matches
Asphalt	C_{36}	Gummy residues	Surfacing roads, fuel

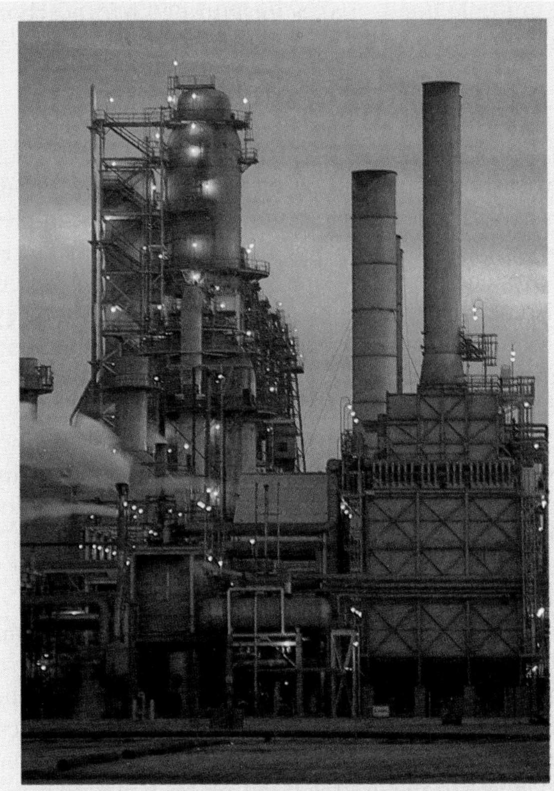

Figure 26.8 Petroleum is separated into fractions by distillation and subjected to catalytic cracking in a refinery, as shown here. (Ken Biggs/Photo Researchers)

around 50. It is therefore subjected to a process called *cracking*, which converts the straight-chain alkanes into more desirable branched-chain ones (Figure 26.8). Cracking is also used to convert some of the less volatile kerosene and fuel-oil fraction into compounds with lower molecular weights that are suitable for use as automobile fuel. In the cracking process, the hydrocarbons are mixed with a catalyst and heated to 400 to 500°C. The catalysts used are naturally occurring clay mineral or synthetic Al_2O_3–SiO_2 mixtures. In addition to forming molecules more suitable for gasoline, cracking results in the formation of hydrocarbons of lower molecular weight, such as ethylene and propene. These substances are used in a variety of processes to form plastics and other chemicals.

The octane rating of gasoline is further improved by adding certain compounds called *antiknock agents*. Until the mid-1970s, the principal antiknock agent was tetraethyllead, $(C_2H_5)_4Pb$. Its use has been drastically curtailed because of the environmental hazards of lead, and because it poisons catalytic converters (see the Chemistry at Work box in Section 14.6). Oxygenated hydrocarbons are now generally used as antiknock agents. One of the most cost-effective is methyl *t*-butyl ether (MTBE):

$$
\begin{array}{c}
\quad CH_3 \\
\quad | \\
CH_3{-}C{-}O{-}CH_3 \\
\quad | \\
\quad CH_3
\end{array}
$$

reactive and attack the alkane, removing a hydrogen atom and forming an alkyl radical. For example, when the alkane is ethane we have

$$Cl\cdot + CH_3CH_3 \longrightarrow HCl + CH_3CH\cdot \qquad [26.2]$$

The ethyl radical, $CH_3CH_2\cdot$, is another reactive species. It combines with molecular chlorine to give ethyl chloride and yet another chlorine atom:

$$CH_3CH_2\cdot + Cl_2 \longrightarrow CH_3CH_2Cl + Cl\cdot \qquad [26.3]$$

The resultant chlorine atom reacts with more ethane, continuing the cycle. Thus, for each quantum of light absorbed by a chlorine molecule, many molecules of ethyl chloride may be formed. This reaction provides an example of a **radical chain** process. A disadvantage of radical chain reactions is that they are not very selective. As the concentration of ethyl chloride builds up in the reaction, chlorine atoms may abstract hydrogen atoms from it, so that eventually dichloroethane and even more highly chlorinated molecules may be formed. Therefore, several products are formed in the reaction, and these must be carefully separated by distillation or some other separation procedure.

26.2 ALKENES AND ALKYNES

The presence of one or more multiple bonds makes unsaturated hydrocarbons significantly different from alkanes both in terms of their structure and their reactivity.

Alkenes

Alkenes are unsaturated hydrocarbons that contain one or more C=C bonds. The general formula for alkenes with one double bond is C_nH_{2n}. The simplest alkene is $CH_2{=}CH_2$, called ethene or ethylene. The next member of the series is $CH_3{-}CH{=}CH_2$, called propene or propylene. For alkenes with four or more carbon atoms, several isomers exist for each molecular formula. For example, there are four isomers of C_4H_8, as shown in Figure 26.9. Notice both their structures and their names.

The names of alkenes are based on the longest continuous chain of carbon atoms that contains the double bond. The name given to the chain is obtained from the name of the corresponding alkane (Table 26.1) by changing the ending from *–ane* to *–ene*. For example, the compound on the left in Figure 26.9 has a three-carbon chain; thus, the parent alkene is considered to be propene.

The location of the double bond along an alkene chain is indicated by a prefix number that designates the number of the carbon atom that is part of the double bond and is nearest an end of the chain. The chain is always numbered from the end that brings us to the double bond sooner and hence gives the smallest number prefix. In propene, the only possible location for the double bond is between the first and second carbons; thus, a prefix indicating its location is unnecessary. For the compound on the left in Figure 26.9, numbering the carbon chain from the end closer to the double bond places a methyl group on the second carbon. Thus, the name of the isomer is 2-methylpropene. For the other compounds in Figure 26.9, the carbon chain contains four carbons, and there are two possible positions for the double bond, either after the first carbon (1-butene) or after the second carbon (2-butene).

If a substance contains two or more double bonds, each is located by a numerical prefix. Furthermore, the ending of the name is altered to identify the number of double bonds: diene (two), triene (three), and so forth. For example, $CH_2{=}CH{-}CH_2{-}CH{=}CH_2$ is 1,4-pentadiene.

Notice that the two isomers on the right in Figure 26.9 differ in the relative locations of their terminal methyl groups. These two compounds are examples of **geometrical isomers**, compounds that have the same molecular formula and the same groups bonded to one another but differ in the spatial arrangement of these groups. In the cis isomer, the two methyl groups are on the same side of the double bond, whereas in the trans isomer, they are on opposite sides. Geometrical isomers possess distinct physical properties and may even differ significantly in their chemical behavior.

Geometrical isomerism in alkenes arises because unlike the C—C bond, the C=C bond is resistant to twisting. Recall that the double bond between

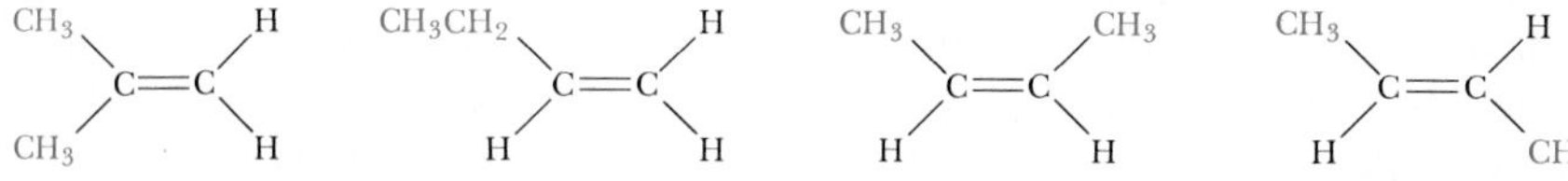

Figure 26.9 Structures, names, and boiling points of alkenes with molecular formula C_4H_8.

2-Methylpropene	1-Butene	*cis*-2-Butene	*trans*-2-Butene
b.p. −7°C	b.p. −6°C	b.p. 4°C	b.p. 1°C

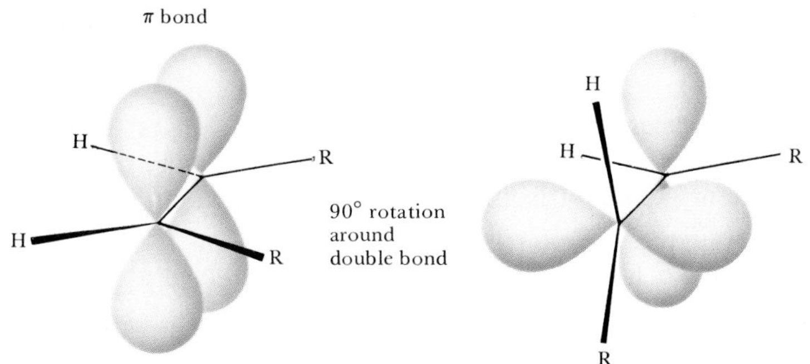

Figure 26.10 Schematic illustration of rotation about a carbon-carbon double bond in an alkene. The overlap of the p orbitals that form the π bond is lost in the rotation. For this reason, rotation about carbon-carbon double bonds does not occur readily.

two carbon atoms consists of a σ and a π part (Sections 9.3 and 9.5). Figure 26.10 shows a cis alkane. The carbon-carbon bond axis and the bonds to the hydrogen atoms and to the alkyl groups (designated R), are all in a plane. The p orbitals whose sideways overlap forms the π bond are perpendicular to the molecular plane. As Figure 26.10 shows, rotation around the carbon-carbon double bond requires the π bond to be broken, a process that requires considerable energy.

SAMPLE EXERCISE 26.3

Name the following compound:

```
                   CH3
                   |
CH3CH2CH2—CH            CH3
                \        /
                 C=C
                /        \
              H            H
```

Solution: Because this compound possesses a double bond, it is an alkene. The longest continuous chain of carbons that contains the double bond is seven in length. The parent compound is therefore considered a heptene. The double bond begins at carbon 2 (numbering from the end closest to the double bond); thus the parent hydrocarbon chain is named 2-heptene. Continuing the numbering along the chain, a methyl group is bound at carbon atom 4. Thus the compound is 4-methyl-2-heptene. Finally, we note that the geometrical configuration at the double bond is cis; that is, the alkyl groups are bonded to the double bond on the same side. For this reason, the full name is 4-methyl-*cis*-2-heptene.

PRACTICE EXERCISE

Draw the structural formula for the compound *trans*-1,3-hexadiene.

Answer:
```
CH2=CH         H
       \      /
        C=C
       /      \
      H        CH2CH3
```

Alkynes

Alkynes are unsaturated hydrocarbons containing one or more C≡C bonds. The general formula for simple alkynes is C_nH_{2n-2}. The simplest alkyne is acetylene, C_2H_2, a highly reactive molecule. When acetylene is burned in a stream of oxygen in an oxyacetylene torch, the flame reaches a very high temperature, about 3200 K (Section 22.4). The oxyacetylene torch is widely used in welding, which requires high temperatures. Alkynes in general are highly reactive molecules. Because of their higher reactivity, they are not as widely distributed in nature as alkenes; however, they are important intermediates in many industrial processes.

Alkynes are named by identifying the longest continuous chain in the molecule containing the triple bond and modifying the ending of the name as listed in Table 26.1 from *-ane* to *-yne*, as shown in Sample Exercise 26.4.

SAMPLE EXERCISE 26.4

Name the following compounds:
(a) $CH_3CH_2CH_2—C{\equiv}C—CH_3$
(b) $CH_3CH_2CH_2\underset{\displaystyle |\atop \displaystyle CH_2CH_2CH_3}{CH}—C{\equiv}CH$

Solution: In (a) the longest chain of carbon atoms is six. There are no side chains. The triple bond begins at carbon 2 (remember, we always arrange the numbering so that the smallest possible number is assigned to the carbon containing the multiple bond). Thus, the name is 2-hexyne.

In (b) the longest continuous chain of carbon atoms is seven; but because this chain does not contain the triple bond we do not count it as derived from heptane. The longest chain containing the triple bond is six, so this compound is named as a derivative of hexyne, 3-propyl-1-hexyne.

PRACTICE EXERCISE

Draw the condensed structural formula for 4-methyl-2-pentyne.

Answer: $CH_3—C{\equiv}C—\underset{\displaystyle |\atop \displaystyle CH_3}{CH}—CH_3$

Addition Reactions of Alkenes and Alkynes

The presence of carbon-carbon double or triple bonds in hydrocarbons markedly increases their chemical reactivity. The most characteristic reactions of alkenes and alkynes are **addition reactions**, in which a reactant is added to the two atoms that form the multiple bond. A simple example is the addition of a halogen such as Br_2 to ethene:

$$H_2C{=}CH_2 + Br_2 \longrightarrow H_2\underset{\displaystyle |\atop \displaystyle Br}{C}—\underset{\displaystyle |\atop \displaystyle Br}{CH_2} \qquad [26.4]$$

The pair of electrons that form the π bonds in ethylene is uncoupled and is used to form two new bonds to the two bromine atoms. The σ bond between the carbon atoms remains.

Addition of H_2 to an alkene converts it to an alkane:

$$CH_3CH{=}CHCH_3 + H_2 \xrightarrow{\text{Ni, 500°C}} CH_3CH_2CH_2CH_3 \qquad [26.5]$$

This reaction, referred to as *hydrogenation*, does not occur readily under ordinary temperature and pressure conditions. One reason for the lack of reactivity of H_2 toward alkenes is the high bond energy of the H_2 bond. To promote the reaction, it is necessary to use a catalyst that assists in rupturing the H—H bond. The most widely used catalysts are finely divided metals on which H_2 is adsorbed (Section 14.6).

The addition of hydrogen halides and other nonsymmetrical molecules to alkenes is more complicated because two products are often possible. For example, addition of HBr to propene might proceed in either of the following ways:

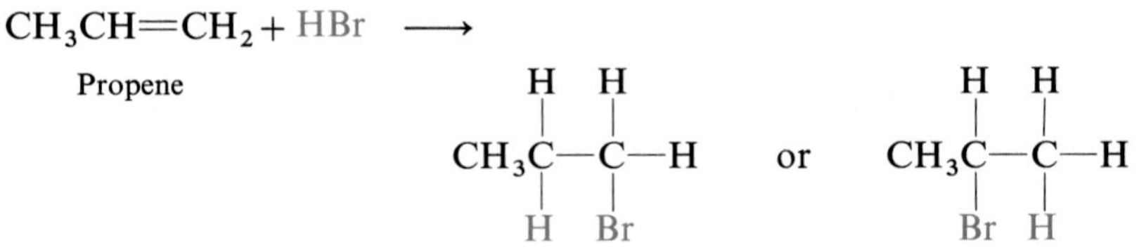

$$CH_3CH{=}CH_2 + HBr \longrightarrow \underset{\text{1-Bromopropane}}{CH_3\overset{H}{\underset{H}{C}}-\overset{H}{\underset{Br}{C}}-H} \quad \text{or} \quad \underset{\text{2-Bromopropane}}{CH_3\overset{H}{\underset{Br}{C}}-\overset{H}{\underset{H}{C}}-H} \qquad [26.6]$$

Propene

In practice, however, only 2-bromopropane forms. *When a hydrogen halide is added to an alkene, the hydrogen atom normally ends up on the carbon atom that already has the most hydrogen atoms.* This observation, known as **Markovnikov's rule**, was first formulated by the Russian chemist V. V. Markovnikov (1838–1904).

With an acid such as H_2SO_4 as catalyst, it is also possible to add H_2O to a double bond. Markovnikov's rule applies here as well. A hydrogen atom adds to the carbon with the most hydrogens, and the OH group adds to the carbon with the fewest:

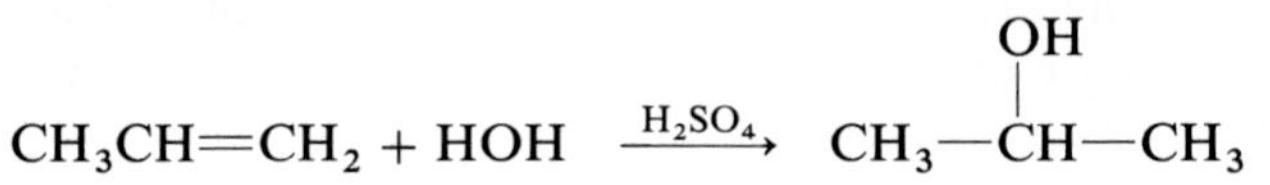

$$CH_3CH{=}CH_2 + HOH \xrightarrow{H_2SO_4} CH_3-\overset{OH}{CH}-CH_3 \qquad [26.7]$$

The addition reactions of alkynes resemble those of alkenes, as shown in the following examples:

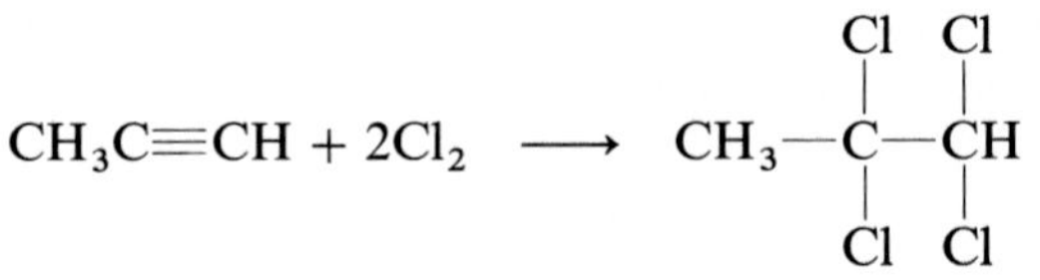

$$CH_3C{\equiv}CH + 2Cl_2 \longrightarrow CH_3-\overset{Cl}{\underset{Cl}{C}}-\overset{Cl}{\underset{Cl}{CH}} \qquad [26.8]$$

$$CH_3CH_2C{\equiv}CH + 2HBr \longrightarrow CH_3CH_2\overset{Br}{\underset{Br}{C}}-CH_3 \qquad [26.9]$$

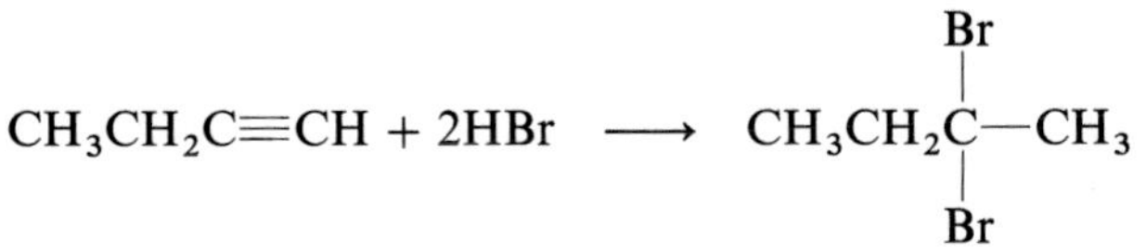

SAMPLE EXERCISE 26.5

Predict the product of the reaction of HCl with the following compound:

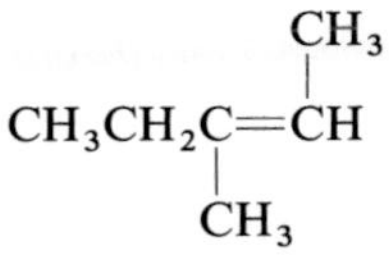

$$CH_3CH_2\underset{CH_3}{C}{=}\overset{CH_3}{CH}$$

Solution: The double-bonded carbon atom on the right has one hydrogen on it; the other has none. According to Markovnikov's rule, the chloride of HCl will end up on the carbon atom on the left:

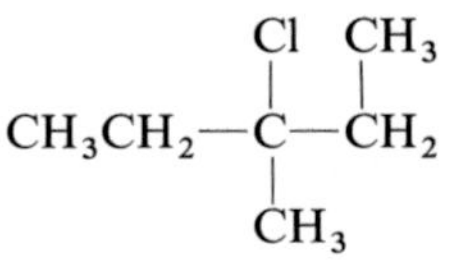

$$CH_3CH_2-\overset{Cl}{\underset{CH_3}{C}}-\overset{CH_3}{CH_2}$$

PRACTICE EXERCISE

The compound

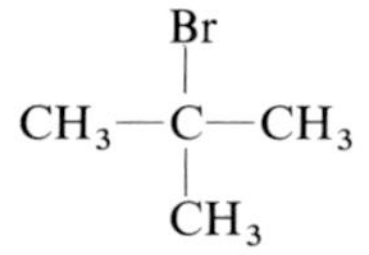

$$CH_3-\overset{\overset{\large Br}{|}}{\underset{\underset{\large CH_3}{|}}{C}}-CH_3$$

was formed by addition of HBr to an alkene. Draw the condensed structural formula for the alkene.

Answer: $CH_3-\underset{\underset{\large CH_3}{|}}{C}=CH_2$

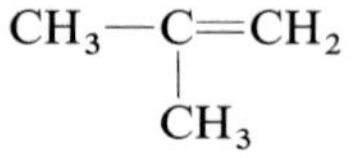

26.3 AROMATIC HYDROCARBONS

Aromatic hydrocarbons are members of a large and important class of hydrocarbons. The simplest member of the series is benzene (see Figure 26.1), with molecular formula C_6H_6. As we have already noted, benzene is a planar, highly symmetrical molecule. The molecular formula for benzene suggests a high degree of unsaturation. You might therefore expect benzene to resemble the unsaturated hydrocarbons and to be highly reactive. In fact, however, benzene is not at all similar to alkenes or alkynes in chemical behavior. The great stability of benzene and the other aromatic hydrocarbons as compared with alkenes and alkynes is due to stabilization of the π electrons through delocalization in the π orbitals (Section 9.5).

There is no widely used systematic nomenclature for naming aromatic rings. Each ring system is given a common name; several aromatic compounds are shown in Figure 26.11. The aromatic rings are represented by hexagons with a circle inscribed inside to denote aromatic character. Each corner represents a carbon atom. Each carbon is bound to three other atoms—either three carbons or two carbons and a hydrogen. The hydrogen atoms are not shown. In naming derivatives of the aromatic hydrocarbons, it is often necessary to indicate the position on the aromatic ring at which some side chain or other group is located. The numbering shown in Figure 26.11 is used for this purpose.

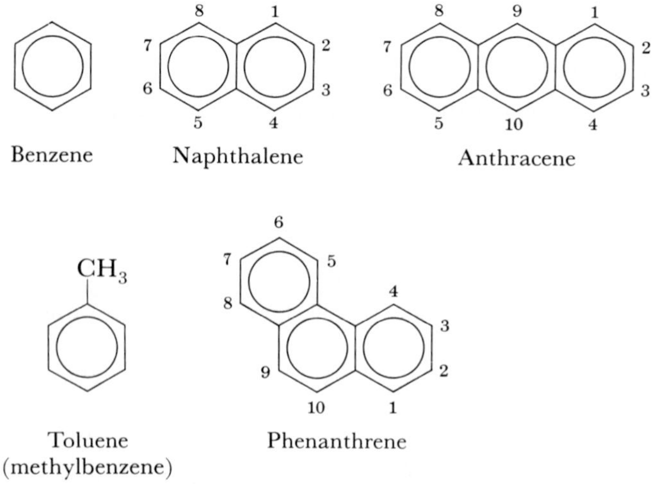

Figure 26.11 Structures, names, and numbering systems of several aromatic compounds.

A CLOSER LOOK: Aromatic Stabilization

We can obtain an estimate of the stabilization of the π electrons in benzene by comparing the energy required to add hydrogen to benzene to form a saturated compound with the energy required to hydrogenate simple alkenes. The hydrogenation of benzene to form cyclohexane can be represented as

$$\text{(benzene)} + 3H_2 \longrightarrow \text{(cyclohexane, s)} \qquad \Delta H^\circ = -208\ \text{kJ/mol} \quad [26.10]$$

(The s in the ring on the right indicates that it is saturated, with CH_2 groups at each corner.) The enthalpy change in this reaction is -208 kJ/mol. The heat of hydrogenation of the cyclic alkene cyclohexene is -120 kJ/mol:

$$\text{Cyclohexene} + H_2 \longrightarrow \text{(cyclohexane, s)} \qquad \Delta H^\circ = -120\ \text{kJ/mol} \quad [26.11]$$

Similarly, the heat released on hydrogenating 1,4-cyclohexadiene is -232 kJ/mol:

$$\text{1,4-Cyclohexadiene} + 2H_2 \longrightarrow \text{(cyclohexane, s)} \qquad \Delta H^\circ = -232\ \text{kJ/mol} \quad [26.12]$$

From these last two reactions, it would appear that the heat of hydrogenating a double bond is about 116 kJ/mol for each bond. There is the equivalent of three double bonds in benzene. Therefore, we might expect that the heat of hydrogenating benzene would be about three times -116, or -348 kJ/mol, if benzene behaved as though it were "cyclohexatriene"; that is, if it behaved as though it had three double bonds in a ring. Instead, the heat released is much less than this, indicating that benzene is more stable than would be expected for three double bonds. The difference of 140 kJ/mol between -348 kJ/mol and the observed heat of hydrogenation, -208 kJ/mol, can be ascribed to stabilization of the π electrons through delocalization in the π orbitals that extend around the ring.

Substitution Reactions of Aromatic Hydrocarbons

Although aromatic hydrocarbons are unsaturated, they do not readily undergo addition reactions. The delocalized π bonding causes aromatic compounds to behave quite differently from alkenes and alkynes. For example, benzene does not add Cl_2 or Br_2 to its double bonds under ordinary conditions. In contrast, aromatic hydrocarbons undergo substitution reactions relatively easily. For example, when benzene is warmed in a mixture of nitric and sulfuric acids, hydrogen is replaced by the nitro group, NO_2:

$$C_6H_6 + HNO_3 \xrightarrow{H_2SO_4} C_6H_5NO_2 + H_2O \qquad [26.13]$$

More vigorous treatment results in substitution of a second nitro group into the molecule:

$$C_6H_5NO_2 + HNO_3 \xrightarrow{H_2SO_4} m\text{-}C_6H_4(NO_2)_2 + H_2O \qquad [26.14]$$

There are three possible isomers of benzene with two nitro groups attached. These three isomers are named *ortho*-, *meta*-, and *para*-dinitrobenzene:

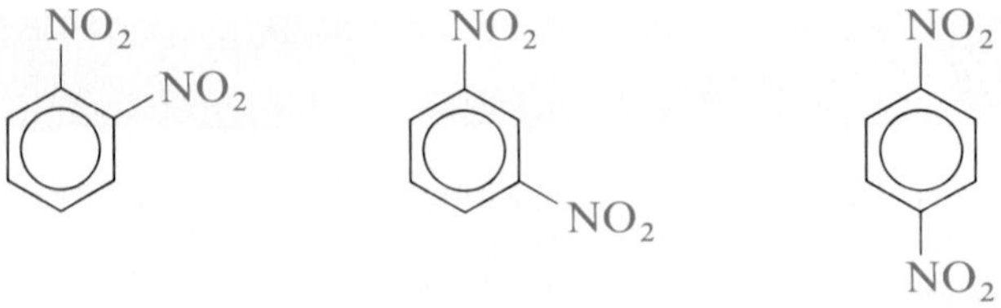

ortho-Dinitrobenzene m.p. 118°C — meta-Dinitrobenzene m.p. 90°C — para-Dinitrobenzene m.p. 174°C

ortho-Dinitrobenzene
m.p. 118°C

meta-Dinitrobenzene
m.p. 90°C

para-Dinitrobenzene
m.p. 174°C

Only the meta isomer is formed in the reaction of nitric acid with nitrobenzene.

Bromination of benzene is carried out using $FeBr_3$ as a catalyst:

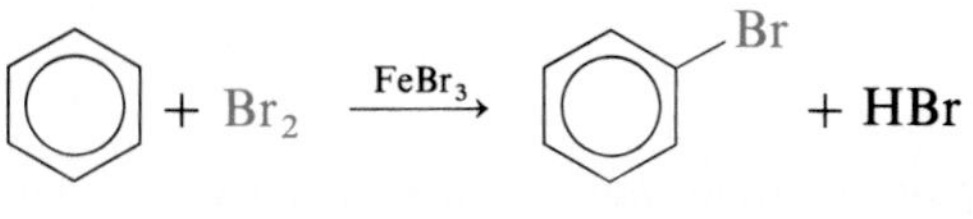

$$C_6H_6 + Br_2 \xrightarrow{FeBr_3} C_6H_5Br + HBr \quad [26.15]$$

In a similar reaction, called the *Friedel-Crafts reaction*, alkyl groups can be substituted onto an aromatic ring by reaction of an alkyl halide with an aromatic compound in the presence of $AlCl_3$ as a catalyst:

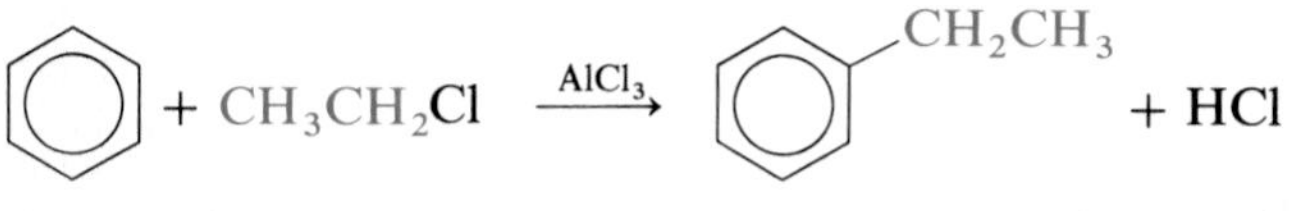

$$C_6H_6 + CH_3CH_2Cl \xrightarrow{AlCl_3} C_6H_5CH_2CH_3 + HCl \quad [26.16]$$

The substitution reactions of aromatic compounds occur through an attack of a positively charged reagent on the ring. Substances such as H_2SO_4, $FeCl_3$, and $AlCl_3$ serve as catalysts by generating the positively charged species. For example, the bromination of benzene proceeds by the following steps:

$$FeBr_3 + Br_2 \rightleftharpoons FeBr_4^- + Br^+ \quad [26.17]$$

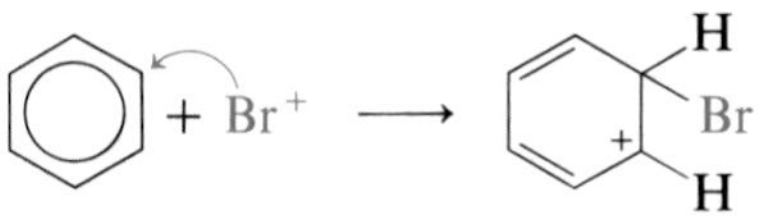

$$C_6H_6 + Br^+ \longrightarrow C_6H_6Br^+ \quad [26.18]$$

$$C_6H_6Br^+ + FeBr_4^- \longrightarrow C_6H_5Br + FeBr_3 + HBr \quad [26.19]$$

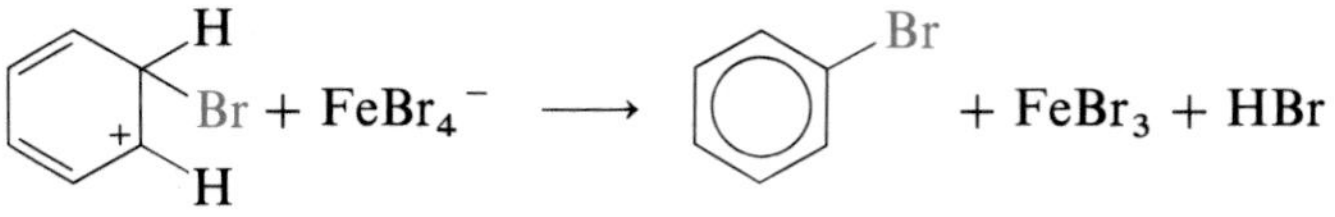

The Br^+ ion has only six electrons in its valence shell. Using a pair of π electrons from the benzene, it forms a single bond to a carbon atom. In doing so, it leaves the adjacent carbon atom with only six electrons in its valence shell. Such a species is not very stable. Loss of a proton from the carbon atom to which the bromine is attached converts the molecule to a more stable one, bromobenzene. The overall result is that bromine has replaced hydrogen on the benzene ring.

26.4 HYDROCARBON DERIVATIVES

Certain groups or arrangements of atoms impart characteristic behavior to an organic molecule. These groups are called **functional groups**. Two of these groups have already been discussed—double and triple carbon-carbon bonds—both of which impart considerable chemical reactivity to a hydrocarbon. The other functional groups contain elements other than carbon and hydrogen, most often oxygen, nitrogen, or a halogen. Compounds containing these elements are generally considered *hydrocarbon derivatives.* In these compounds, one or more hydrogen atoms of a hydrocarbon have been replaced by other atoms or groups of atoms. The compound can then be considered to consist of two parts: a hydrocarbon fragment, such as an alkyl group (often designated R), and one or more functional groups. Because an alkyl group is relatively unreactive, the functional group is generally the chemically active part of the molecule. Because each functional group imparts characteristic properties to organic compounds, we consider them separately.

Alcohols

Alcohols are hydrocarbon derivatives in which one or more hydrogens of a parent hydrocarbon have been replaced by a *hydroxyl* or *alcohol* functional group, OH. Figure 26.12 shows the structural formulas and names of several alcohols. Note that the accepted name for an alcohol ends in *-ol.* The simple alcohols are named by changing the last letter in the name of the corresponding alkane to *-ol*—for example, ethan*e* becomes ethan*ol.* Where necessary, the location of the OH group is designated by an appropriate prefix numeral that indicates the number of the carbon atom bearing the OH group, as shown in the examples in Figure 26.12.

Because the O—H bond in alcohol is polar, alcohols are much more soluble in polar solvents such as water than are hydrocarbons. The OH functional group can participate in hydrogen bonding. As a result, the boiling points of alcohols are much higher than those of their parent alkanes.

Alcohols are classified according to the number of carbon groups bonded to the carbon that contains the OH group. If there is only one other carbon atom, as in ethanol or 1-propanol, the alcohol is termed a *primary alcohol.* If there are two other carbon groups attached, as in 2-propanol, the alcohol is *secondary.* If there are three other carbons, as in 2-methyl-2-propanol, the alcohol is *tertiary.*

Figure 26.12 Structural formulas of several important alcohols.

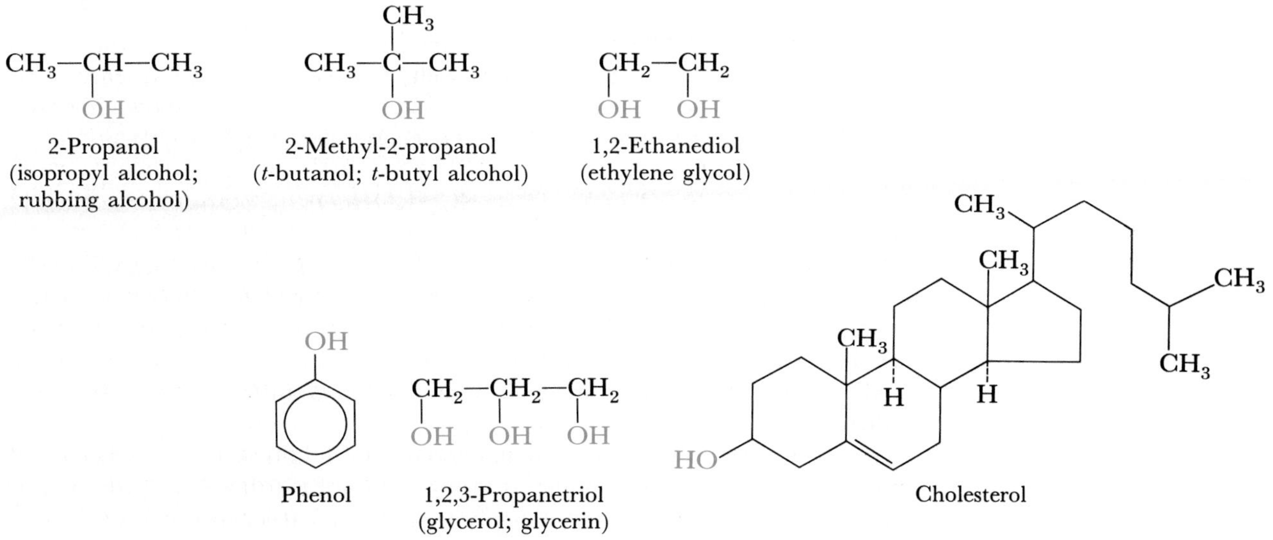

$CH_3-CH(OH)-CH_3$

2-Propanol (isopropyl alcohol; rubbing alcohol)

$CH_3-C(CH_3)(OH)-CH_3$

2-Methyl-2-propanol (*t*-butanol; *t*-butyl alcohol)

$CH_2(OH)-CH_2(OH)$

1,2-Ethanediol (ethylene glycol)

C_6H_5OH

Phenol

$CH_2(OH)-CH(OH)-CH_2(OH)$

1,2,3-Propanetriol (glycerol; glycerin)

Cholesterol

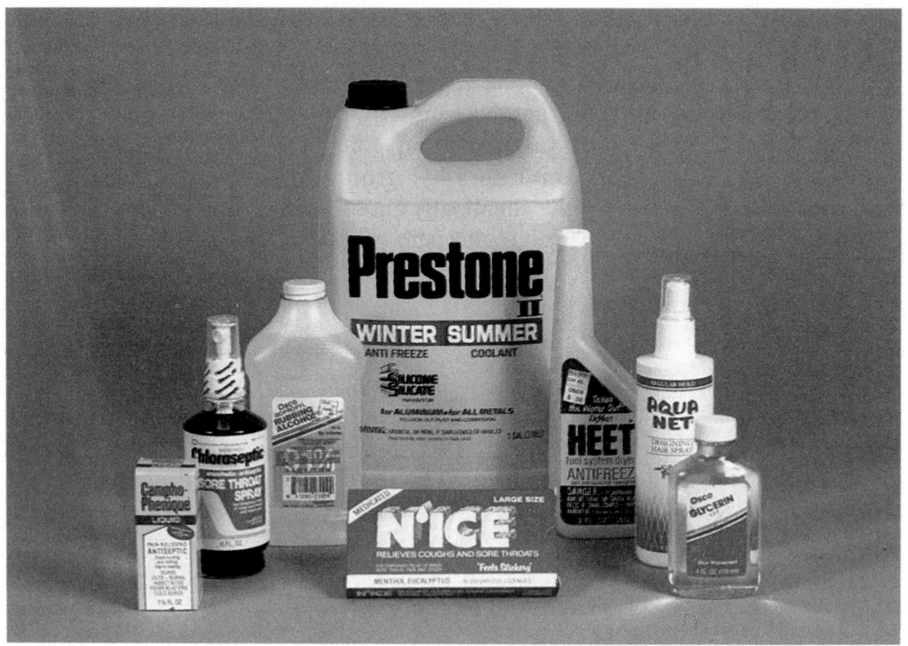

Figure 26.13 Some commercial products that are composed entirely or mainly of alcohols. (Donald Clegg and Roxy Wilson)

Figure 26.13 shows several familiar commercial products that consist entirely or in major part of an organic alcohol. Let's consider how some of the more important alcohols are formed and used.

The simplest alcohol, methanol, has many important industrial uses and is produced on a large scale. Carbon monoxide and hydrogen are heated together under pressure in the presence of a metal oxide catalyst:

$$CO(g) + 2H_2(g) \xrightarrow[400^\circ C]{200\text{–}300 \text{ atm}} CH_3OH(g) \qquad [26.20]$$

Because methanol has a very high octane rating as an automobile fuel, it has received increasing attention in recent years as a gasoline additive and as a fuel in its own right.

Ethanol, C_2H_5OH, is a product of the fermentation of carbohydrates such as sugar and starch. Bacterial cultures such as yeast work in the absence of air to convert carbohydrates into a mixture of ethanol and CO_2, as shown in Equation 26.21. In the process, they derive the energy necessary for growth:

$$C_6H_{12}O_6(aq) \xrightarrow{\text{yeast}} 2C_2H_5OH(aq) + 2CO_2(g) \qquad [26.21]$$

This naturally occurring reaction is carried out under carefully controlled conditions to produce beer, wine, and other beverages in which ethanol is the active ingredient. It is often said that ethanol is the least toxic of the straight-chain alcohols. Although this is true in the strictest sense, the combination of ethanol and automobile driving produces far more human fatalities each year than any other chemical agent.

Many alcohols are formed industrially by addition of the elements of water to an olefin. For example, isopropyl alcohol is formed by hydration of propene using sulfuric acid as a catalyst, as shown earlier in Equation 26.7.

Many polyhydroxyl (more than one OH group) alcohols are known. The simplest of these is 1,2-ethanediol (ethylene glycol), $HOCH_2CH_2OH$. This substance is the major ingredient in automobile antifreeze. Another common polyhydroxyl alcohol is 1,2,3-propanetriol (glycerol), $HOCH_2CH(OH)CH_2OH$. It is a viscous liquid that dissolves readily in water and is widely used as a skin softener in cosmetic preparations. It is also used in foods and candies to keep them moist.

Phenol is the simplest example of a compound with an OH group attached to an aromatic ring. One of the most striking effects of the aromatic group is the greatly increased acidity of the proton. Phenol is about 1 million times more acidic in water than a typical aliphatic alcohol such as ethanol. Even so, it is not a very strong acid ($K_a = 1.3 \times 10^{-10}$). Phenol is used industrially in the making of several kinds of plastics and in the preparation of dyes.

Cholesterol, shown in Figure 26.12, is an example of a biochemically important alcohol. Notice that the OH group forms only a small component of this rather large molecule. As a result, cholesterol is not very soluble in water (0.26 g per 100 mL of H_2O). Cholesterol is a normal component of our bodies. However, when present in excessive amounts it may precipitate from solution. It precipitates in the gallbladder to form crystalline lumps called *gallstones*. It may also precipitate against the walls of veins and arteries and thus contribute to high blood pressure and other cardiovascular problems. The amount of cholesterol in our blood is determined not only by how much cholesterol we eat but also by total dietary intake. There is evidence that excessive caloric intake leads the body to synthesize excessive cholesterol.

Ethers

Compounds in which two hydrocarbon groups are bonded to one oxygen are called **ethers**. Ethers can be formed from two molecules of alcohol by splitting out a molecule of water. The reaction is thus a dehydration process; it is catalyzed by sulfuric acid, which takes up water to remove it from the system:

$$CH_3CH_2{-}OH + H{-}OCH_2CH_3 \xrightarrow{H_2SO_4} CH_3CH_2{-}O{-}CH_2CH_3 + H_2O \quad [26.22]$$

A reaction in which water is split out from two substances is called a *condensation reaction.*

Ethers are used as solvents; both diethyl ether and the cyclic ether tetrahydrofuran are common solvents for organic reactions.

$CH_3CH_2{-}O{-}CH_2CH_3$

Diethyl ether

```
CH2—CH2
|    |
CH2  CH2
  \ /
   O
```

Tetrahydrofuran (THF)

Aldehydes and Ketones

Several classes of organic compounds contain the **carbonyl group**:

$$>C{=}O$$

In **aldehydes**, the carbonyl group has at least one hydrogen atom attached, as in the following examples:

$$\underset{\text{Formaldehyde}}{\mathrm{H\overset{\overset{\displaystyle O}{\|}}{C}H}} \qquad \underset{\text{Acetaldehyde}}{\mathrm{CH_3\overset{\overset{\displaystyle O}{\|}}{C}H}}$$

In **ketones**, the carbonyl group occurs at the interior of a carbon chain and is therefore flanked by carbon atoms:

$$\underset{\text{Acetone}}{\mathrm{CH_3{-}\overset{\overset{\displaystyle O}{\|}}{C}{-}CH_3}} \qquad \underset{\text{Methyl ethyl ketone}}{\mathrm{CH_3{-}\overset{\overset{\displaystyle O}{\|}}{C}{-}CH_2CH_3}}$$

Aldehydes and ketones can be prepared by careful oxidation of alcohols. It is fairly easy to oxidize alcohols. Complete oxidation results in formation of CO_2 and H_2O, as in the burning of methanol:

$$CH_3OH(g) + \tfrac{3}{2}O_2(g) \longrightarrow CO_2(g) + 2H_2O(g) \qquad [26.23]$$

Controlled oxidation to form other organic substances is carried out by using various oxidizing agents such as air, hydrogen peroxide, H_2O_2, ozone, O_3, and potassium dichromate, $K_2Cr_2O_7$. We shall not concern ourselves with balancing most of these oxidation-reduction equations; instead, we shall simply show the source of oxygen as an O in parentheses.

Aldehydes are produced by oxidation of primary alcohols. For example, acetaldehyde is formed by oxidation of ethanol:

$$\underset{\text{Ethanol}}{\mathrm{CH_3CH_2OH}} + (\mathrm{O}) \longrightarrow \underset{\text{Acetaldehyde}}{\mathrm{CH_3\overset{\overset{\displaystyle O}{\|}}{C}H}} + H_2O \qquad [26.24]$$

Ketones are produced by oxidation of secondary alcohols. For example, acetone is formed in large quantities by oxidation of isopropyl alcohol:

$$\mathrm{CH_3{-}\overset{\overset{\displaystyle OH}{|}}{C}H{-}CH_3} + (\mathrm{O}) \longrightarrow \underset{\text{Acetone}}{\mathrm{CH_3{-}\overset{\overset{\displaystyle O}{\|}}{C}{-}CH_3}} + H_2O \qquad [26.25]$$

Ketones are less reactive than aldehydes because ketones do not possess the reactive C—H bond attached to the carbonyl carbon. Ketones are used extensively as solvents. Acetone, which boils at 56°C, is the most widely used ketone. The carbonyl functional group imparts polarity to the solvent. Acetone is completely miscible with water, yet it dissolves a wide range of organic substances. Methyl ethyl ketone, $CH_3COCH_2CH_3$, which boils at 80°C, is also used industrially as a solvent.

Carboxylic Acids

Oxidation of a primary alcohol may yield either an aldehyde or a **carboxylic acid**, depending on reaction conditions. For example, mild oxidation of ethanol produces acetaldehyde (Equation 26.24), which under more vigorous conditions may be further oxidized to acetic acid:

$$CH_3\overset{\overset{\displaystyle O}{\|}}{C}H + (O) \longrightarrow CH_3\overset{\overset{\displaystyle O}{\|}}{C}OH \qquad [26.26]$$

The oxidation of ethanol to acetic acid is responsible for causing wine to turn sour, producing vinegar.

As seen in this example, carboxylic acids contain the following functional group:

$$-\overset{\overset{\displaystyle O}{\|}}{C}-OH$$

Carboxylic acids are generally weak acids (Section 16.9). They are widely distributed in nature, are present in many consumer products (Figure 26.14), and are important in many industrial processes.

Figure 26.15 shows the structural formulas of several substances containing one or more carboxylic acid functional groups. Note that these substances

Figure 26.14 Carboxylic acids and esters are components of many household items: (*a*) Spinach and some cleaners contain oxalic acid; vinegar contains acetic acid; vitamin C is ascorbic acid; citrus fruits contain citric acid; and aspirin is acetylsalicylic acid (also an ester). (*b*) Many sunburn lotions contain benzocaine (an ester); some nail-polish remover is ethyl acetate; vegetable oils, polyester thread, and aspirin are also esters. (© Richard Megna/Fundamental Photographs)

(*a*)

(*b*)

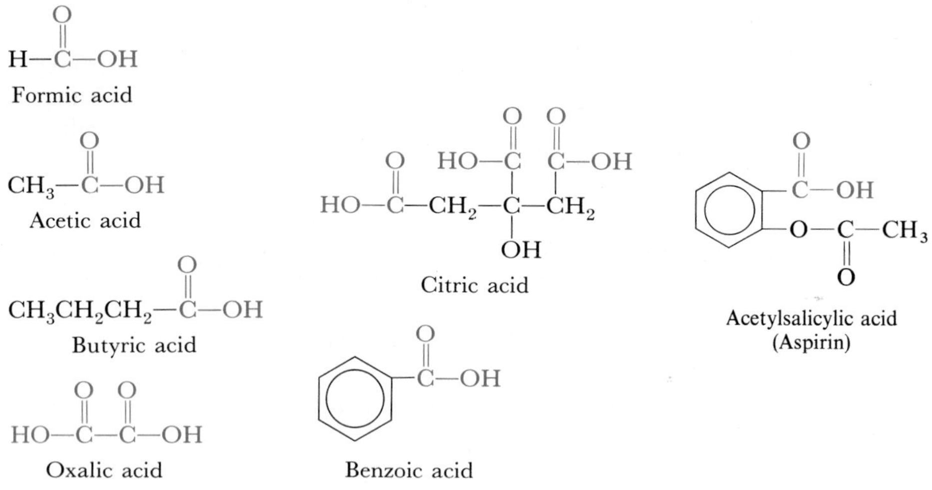

$$\mathrm{H{-}\overset{\overset{\displaystyle O}{\|}}{C}{-}OH}$$

Formic acid

$$\mathrm{CH_3{-}\overset{\overset{\displaystyle O}{\|}}{C}{-}OH}$$

Acetic acid

$$\mathrm{CH_3CH_2CH_2{-}\overset{\overset{\displaystyle O}{\|}}{C}{-}OH}$$

Butyric acid

$$\mathrm{HO{-}\overset{\overset{\displaystyle O}{\|}}{C}{-}\overset{\overset{\displaystyle O}{\|}}{C}{-}OH}$$

Oxalic acid

$$\mathrm{HO{-}\overset{\overset{\displaystyle O}{\|}}{C}{-}CH_2{-}\underset{\underset{\displaystyle OH}{|}}{\overset{\overset{\displaystyle HO{-}\overset{\overset{\displaystyle O}{\|}}{C}}{|}}{C}}{-}\overset{\overset{\displaystyle \overset{\overset{\displaystyle O}{\|}}{C}{-}OH}{|}}{CH_2}}$$

Citric acid

$$\mathrm{C_6H_5{-}\overset{\overset{\displaystyle O}{\|}}{C}{-}OH}$$

Benzoic acid

$$\mathrm{C_6H_4(\overset{\overset{\displaystyle O}{\|}}{C}{-}OH)(O{-}\underset{\underset{\displaystyle O}{\|}}{C}{-}CH_3)}$$

Acetylsalicylic acid
(Aspirin)

Figure 26.15 Structural formulas of several common carboxylic acids.

are not named in a systematic way. The names of many acids are based on their historical origins. For example, formic acid was first prepared by extraction from ants; its name is derived from the Latin word *formica*, meaning "ant." Oxalic acid was first identified as a constituent of members of the plant group *Oxalis*.

Carboxylic acids are important in the manufacture of polymers used to make fibers, films, and paints. Among the compounds having low molecular weights, acetic acid is an industrially important compound. The most important method for manufacture of acetic acid is the reaction of methanol with carbon monoxide in the presence of a rhodium catalyst:

$$\mathrm{CH_3OH + CO \xrightarrow{catalyst} CH_3{-}\overset{\overset{\displaystyle O}{\|}}{C}{-}OH} \qquad [26.27]$$

Notice that this reaction involves, in effect, the insertion of a carbon monoxide molecule between the CH_3 and OH groups. A reaction of this kind is called **carbonylation**.

Esters

Carboxylic acids can be caused to undergo condensation reactions with alcohols to form **esters:**

$$\underset{\text{Acetic acid}}{\mathrm{CH_3{-}\overset{\overset{\displaystyle O}{\|}}{C}{-}OH}} + \underset{\text{Ethanol}}{\mathrm{HO{-}CH_2CH_3}} \longrightarrow$$

$$\underset{\text{Ethyl acetate}}{\mathrm{CH_3{-}\overset{\overset{\displaystyle O}{\|}}{C}{-}O{-}CH_2CH_3}} + \mathrm{H_2O} \qquad [26.28]$$

As seen in this example, esters are compounds in which the H atom of a carboxylic acid is replaced by a hydrocarbon group:

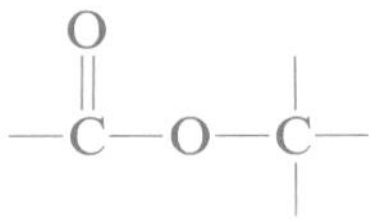

$$-\overset{\overset{\displaystyle O}{\|}}{C}-O-\underset{|}{\overset{|}{C}}-$$

Esters are named by using first the group from which the alcohol is derived and then the group from which the acid is derived. Figure 26.14 shows some common esters.

SAMPLE EXERCISE 26.6

Name the following esters:

(a) 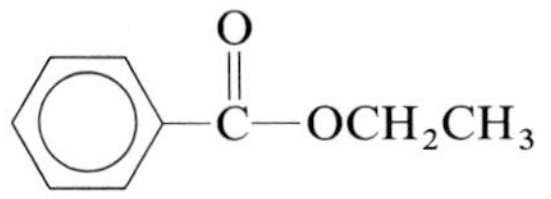 $C_6H_5-\overset{\overset{\displaystyle O}{\|}}{C}-OCH_2CH_3$

(b) 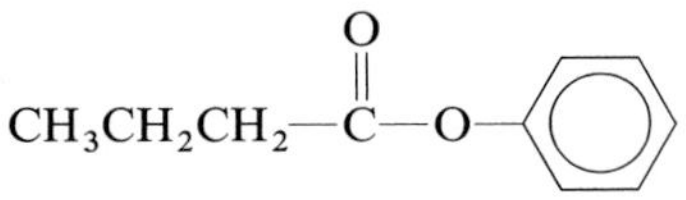 $CH_3CH_2CH_2-\overset{\overset{\displaystyle O}{\|}}{C}-O-C_6H_5$

Solution: In **(a)** the ester is derived from ethanol and benzoic acid. Its name is therefore ethyl benzoate. In **(b)** the ester is derived from phenol and butyric acid. The residue from the phenol, C_6H_5, is called the phenyl group. The ester is therefore named phenyl butyrate.

PRACTICE EXERCISE

Draw the structural formula for the compound propyl propionate.

Answer:

$$CH_3CH_2\overset{\overset{\displaystyle O}{\|}}{C}-O-CH_2CH_2CH_3$$

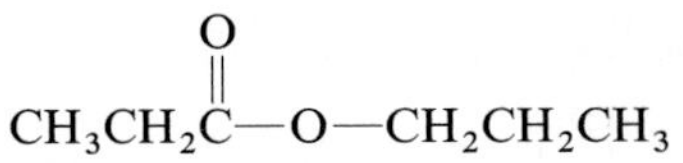

SAMPLE EXERCISE 26.7

Beginning with ethene as the only carbon-containing compound, suggest a series of reactions by which ethyl acetate might be prepared.

Solution: As shown in Equation 26.28, ethyl acetate can be prepared by a condensation reaction between acetic acid and ethanol. The ethanol can be prepared from ethene by H_2SO_4-catalyzed hydration:

$$CH_2{=}CH_2 + H_2O \xrightarrow{H_2SO_4} CH_3CH_2OH$$

This process is analogous to that given in Equation 26.7 for the hydration of propene. A portion of the ethanol thus produced can be oxidized to acetic acid, as shown in Equations 26.24 and 26.25.

PRACTICE EXERCISE

What alcohol and carboxylic acid undergo condensation to form propyl propionate?
Answer: propyl alcohol and propionic acid.

Esters generally have very pleasant odors. They are largely responsible for the pleasant aromas of fruit. For example, one of the esters responsible for the odor of bananas is pentyl acetate, $CH_3COOCH_2CH_2CH_2CH_2CH_3$.

When esters are treated with an acid or a base in aqueous solution, they are hydrolyzed; that is, the molecule is split into its alcohol and acid components:

$$\underset{\text{Methyl propionate}}{CH_3CH_2-\overset{\overset{\displaystyle O}{\|}}{C}-O-CH_3} + Na^+ + OH^- \longrightarrow$$

$$\underset{\text{Sodium propionate}}{CH_3CH_2-\overset{\overset{\displaystyle O}{\|}}{C}-O^-} + Na^+ + \underset{\text{Methanol}}{CH_3OH} \quad [26.29]$$

In this example, the hydrolysis was carried out in basic medium. The products of the reaction are the sodium salt of the carboxylic acid and the alcohol.

The hydrolysis of an ester in the presence of a base is called **saponification**, a term that comes from the Latin word for soap (*sapon*). Naturally occurring esters include fats and oils (Section 27.5). In the soap-making process, an animal fat or a vegetable oil is boiled with a strong base, usually NaOH. The resultant soap consists of a mixture of sodium salts of long-chain carboxylic acids (called fatty acids) which form during the saponification reaction (Figure 26.16).

Amines and Amides

Amines are organic bases (Section 16.6). They have the general formula R_3N, where R may be H or a hydrocarbon group, as in the following examples:

Figure 26.16 Saponification of fats and oils has long been used to make soap. This etching shows soap manufacture in France during the mid-nineteenth century. (The Bettman Archive)

$CH_3CH_2NH_2$ — Ethylamine

$(CH_3)_3N$ — Trimethylamine

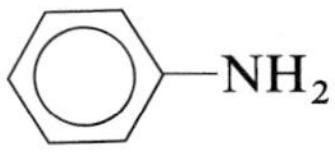

C_6H_5—NH_2 — Aniline

Amines containing a proton can undergo condensation reactions with carboxylic acids to form **amides:**

$$CH_3\overset{\overset{\displaystyle O}{\|}}{C}-OH + H-N(CH_3)_2 \longrightarrow CH_3\overset{\overset{\displaystyle O}{\|}}{C}-N(CH_3)_2 + H_2O \quad [26.30]$$

We may consider the amide functional group to be derived from a carboxylic acid with an NR_2 group replacing the OH of the acid, as in these additional examples:

$CH_3\overset{\overset{\displaystyle O}{\|}}{C}-NH_2$ — Acetamide

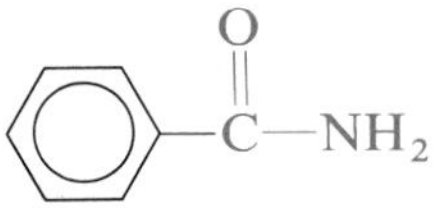

$C_6H_5-\overset{\overset{\displaystyle O}{\|}}{C}-NH_2$ — Benzamide

The amide linkage,

$$R-\overset{\overset{\displaystyle O}{\|}}{C}-\underset{\underset{\displaystyle H}{|}}{N}-R'$$

where R and R′ are organic groups, is the key functional group in the structures of proteins, about which we shall have more to say in Chapter 27.

26.5 ORGANIC POLYMERS

A *polymer* is a material with a high molecular weight, which is formed from simple molecules called *monomers* (Section 12.2). Polymers may be either synthetic or natural in origin. Natural polymers include proteins, starch, and cellulose (Chapter 27). Synthetic polymers include familiar materials such as Teflon, polystyrene, and nylon.

Synthetic polymers are made by taking advantage of the chemical behavior of the functional groups of their monomers. For example, monomers with C=C double bonds, such as tetrafluoroethylene, $F_2C{=}CF_2$, can be caused to react with one another by addition reactions. The polymerization of tetrafluoroethylene produces Teflon:

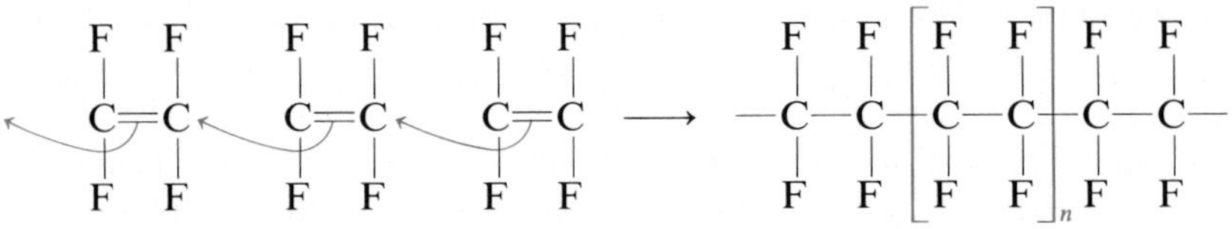

$$F_2C{=}CF_2 \quad F_2C{=}CF_2 \quad F_2C{=}CF_2 \longrightarrow -CF_2-CF_2-\left[CF_2-CF_2\right]_n-CF_2-CF_2- \quad [26.31]$$

Teflon

In an *addition polymerization* reaction, π bonds are broken and the electrons used to form new σ bonds between the monomer units. Many different alkenes have been used to synthesize polymers, as summarized earlier in Table 12.1.

Small molecules can also be joined together to form polymers using condensation reactions *(condensation polymerization)*. The most important condensation polymers are polyamides and polyesters. Polyamides are formed by condensation reactions between carboxylic acid and amine functional groups. The nylons and Kevlar are examples of common polyamides (Section 12.2). Polyesters are formed by condensation reactions between carboxylic acid and alcohol functional groups. The familiar polyester Dacron, from which so much of our clothing is produced, is a condensation polymer of ethylene glycol and *para*-phthalic acid:

$$\underset{\text{Ethylene glycol}}{HOCH_2CH_2OH} + \underset{\textit{para}\text{-phthalic acid}}{HO\overset{O}{\overset{\|}{C}}-C_6H_4-\overset{O}{\overset{\|}{C}}OH} + HOCH_2CH_2OH + HO\overset{O}{\overset{\|}{C}}-C_6H_4-\overset{O}{\overset{\|}{C}}OH\text{---} \xrightarrow{\text{heat}}$$

$$\underset{\text{Dacron}}{\text{---}OCH_2CH_2O\overset{O}{\overset{\|}{C}}-C_6H_4-\overset{O}{\overset{\|}{C}}OCH_2CH_2O\overset{O}{\overset{\|}{C}}-C_6H_4-\overset{O}{\overset{\|}{C}}OCH_2CH_2O\overset{O}{\overset{\|}{C}}-C_6H_4-\overset{O}{\overset{\|}{C}}O\text{---}} \quad [26.32]$$

FOR REVIEW

SUMMARY

Organic compounds are carbon-containing compounds that include both hydrocarbons and derivatives of hydrocarbons. There are four major kinds of hydrocarbons: alkanes, alkenes, alkynes, and aromatic hydrocarbons. Alkanes are composed of only C—H and C—C single bonds. Alkenes contain one or more carbon-carbon double bonds. Alkynes contain one or more carbon-carbon triple bonds. Aromatic hydrocarbons contain cyclic arrangements of carbon atoms bonded through both σ and delocalized π bonds. Alkanes are saturated hydrocarbons; the others are unsaturated. Hydrocarbons may form straight-chain, branched-chain, and cyclic arrangements.

Isomers are substances that possess the same molecular formula but differ in the arrangements of atoms. In structural isomers, the bonding arrangements of the atoms differ. In geometrical isomers, the bonds are the same, but the molecules have different geometries. Geometrical (cis-trans) isomerism is possible in alkenes because of restricted rotation about the C=C double bond.

The naming of hydrocarbons is based on the longest continuous chain of carbon atoms in the structure. The locations of alkyl side groups, which branch off the chain, are specified by numbering along the carbon chain. In naming alkenes and alkynes, the continuous chain must contain the multiple bond, and its location is also specified by a numerical prefix.

Combustion of hydrocarbons is a highly exothermic process. The chief use of hydrocarbons is as sources of heat energy produced by combustion. Alkanes also undergo substitution reactions with halogens, although the reaction is difficult to control and mixtures of products are generally obtained. Alkenes and alkynes readily undergo addition reactions to the carbon-carbon multiple bonds. Markovnikov's rule helps predict the product formed when an unsymmetrical reagent, such as a hydrogen halide or water, is added across a double or triple bond. Addition reactions are difficult to carry out with aromatic hydrocarbons, but substitution reactions are easily accomplished in the presence of acid catalysts.

The chemistry of hydrocarbon derivatives is often dominated by the nature of their functional groups. The functional groups we have considered are summarized here:

$R-O-H$ Alcohol

$R-\overset{\overset{\displaystyle O}{\|}}{C}-H$ Aldehyde

$\rangle C=C \langle$ Alkene

$-C\equiv C-$ Alkyne

$R-\overset{\overset{\displaystyle O}{\|}}{C}-N\langle$ Amide

$R-\overset{\overset{\displaystyle R' \text{ (or H)}}{|}}{N}-R'' \text{ (or H)}$ Amine

$R-\overset{\overset{\displaystyle O}{\|}}{C}-O-H$ Carboxylic acid

$R-\overset{\overset{\displaystyle O}{\|}}{C}-O-R'$ Ester

$R-O-R'$ Ether

$R-\overset{\overset{\displaystyle O}{\|}}{C}-R'$ Ketone

(Remember that R, R′, and R″ represent some hydrocarbon groups—for example, methyl, CH_3, or phenyl, C_6H_5.)

Alcohols are hydrocarbon derivatives containing one or more OH groups. Ethers are related compounds formed by a condensation reaction of two molecules of alcohol, with water splitting out. Oxidation of primary alcohols can lead to formation of aldehydes; further oxidation of the aldehydes produces carboxylic acids. Oxidation of secondary alcohols leads to formation of ketones.

Carboxylic acids can form esters by a condensation reaction with alcohols, or they can form amides by a condensation reaction with amines. Esters undergo hydrolysis (saponification) in the presence of strong bases.

Polymers are large molecules formed by the combination of small, repeating units called monomers. Monomers with C=C double bonds can be caused to undergo addition polymerization. Molecules with suitable functional groups on both ends can be caused to undergo condensation polymerization, forming, for example, polyamides and polyesters.

KEY TERMS

alkanes
alkenes
alkynes
aromatic hydrocarbons
homologous series (Sec. 26.1)
structural isomer (Sec. 26.1)
alkyl group (Sec. 26.1)
cycloalkane (Sec. 26.1)
substitution reaction (Sec. 26.1)
free radical (Sec. 26.1)
radical chain (Sec. 26.1)
geometrical isomers (Sec. 26.2)
addition reaction (Sec. 26.2)
Markovnikov's rule (Sec. 26.2)
functional group (Sec. 26.4)
alcohol (Sec. 26.4)
ether (Sec. 26.4)
carbonyl group (Sec. 26.4)
aldehyde (Sec. 26.4)
ketone (Sec. 26.4)
carboxylic acid (Sec. 26.4)
carbonylation (Sec. 26.4)
ester (Sec. 26.4)
saponification (Sec. 26.4)
amine (Sec. 26.4)
amide (Sec. 26.4)

EXERCISES

Hydrocarbon Structures and Nomenclature

26.1 Give the molecular formula of a hydrocarbon containing five carbon atoms that is **(a)** an alkane; **(b)** a cycloalkane; **(c)** an alkene; **(d)** an alkyne. Which are saturated and which are unsaturated hydrocarbons?

26.2 Give the molecular formula of an alkane, an alkene, an alkyne, and an aromatic hydrocarbon that in each case contains six carbon atoms. Which are saturated and which are unsaturated hydrocarbons?

26.3 Draw the five structural isomers of hexane, C_6H_{14}. Name each compound.

26.4 Write the condensed structural formulas for as many alkanes, alkenes, and alkynes that you can think of that have the molecular formula **(a)** C_5H_8; **(b)** C_5H_{10}.

26.5 Write the Lewis structural formula for each of the following hydrocarbons:
(a) $CH_3CH(CH_3)CH_2CH_3$ **(b)** $CH_3C\equiv CCH_3$
(c) $CH_2=CHCH_3$

26.6 Write the Lewis structural formula for each of the following hydrocarbons:

(a) (cyclopropane ring)$-CH_3$ **(b)** $CH_2{=}CHCH_2CH{=}CH_2$

(c) $(CH_3)_2CHCH_3$.

26.7 What are the characteristic bond angles **(a)** about carbon in an alkane; **(b)** about the carbon-carbon double bond in an alkene; **(c)** about the carbon-carbon triple bond in an alkyne?

26.8 What are the characteristic hybrid orbitals employed by **(a)** carbon in an alkane; **(b)** carbon in a double bond in an alkene; **(c)** carbon in the benzene ring; **(d)** carbon in a triple bond in an alkyne.

26.9 Write the condensed structural formula for each of the following compounds:
(a) 5-methyl-*trans*-2-heptene
(b) 3-chloropropyne
(c) *ortho*-dichlorobenzene
(d) 2,2,4,4-tetramethylpentane
(e) 2-methyl-3-ethylhexane
(f) 2-methyl-6-chloro-3-heptyne
(g) 1,5-dimethylnaphthalene
(h) 1,6-heptadiene

26.10 Write the condensed structural formula for each of the following compounds:
(a) 2,2-dimethylpentane **(b)** 2,3-dimethylhexane
(c) *cis*-2-hexene **(d)** methylcyclopentane
(e) 2-chlorobutane **(f)** 1,2-dibromobenzene
(g) methylcyclobutane **(h)** 4-methyl-2-pentyne

26.11 Name the following compounds:

(a) CH_3CHCH_3 with $CH_2CHCH_2CH_2CH_3$ on the central carbon, and CH_3 on the second carbon of that chain

(b) CH_3CH_2 and H on one carbon, $C{=}C$, H and $CH_2CHCH_2CH_3$ (with CH_3 on CH) on the other carbon

(c) benzene ring with Br at top and Br at bottom

(d) $HC{\equiv}CCH_2CCH_3$ with CH_2CH_3 above and CH_3 below the quaternary C

(e) (cyclobutane ring)$-CH_3$

26.12 Name the following compounds:

(a) $CH_3CHCH_2CHCH_3$ with Br on each CH

(b) $CH_3CH{=}CHCH_2CH{=}CHCH_2CH_3$

(c) $CH_3CHCHCH_2CHCH_3$ with Cl on the third and fifth carbons and a benzene ring on the second carbon

(d) $CH_3{-}CH{-}CH_2Cl$ with a cyclopentane ring on CH

(e) Cl and H on one carbon, $C{=}C$, H and Cl on the other carbon (Cl upper left, H upper right, H lower left, Cl lower right)

26.13 Using butene as an example, distinguish between structural and geometrical isomers.

26.14 Why is structural isomerism possible for alkenes but not for alkanes and alkynes?

26.15 Draw all the structural and geometrical isomers of dichloropropene.

26.16 Indicate whether each of the following molecules is capable of geometrical (cis-trans) isomerism. For those that are, draw the structures of the isomers: **(a)** 2,3-dichlorobutane; **(b)** 2,3-dichloro-2-butene; **(c)** 1,3-dimethylbenzene; **(d)** 4,4-dimethyl-2-pentyne.

26.17 What is the octane number of a mixture of 30 percent *n*-heptane and 70 percent isooctane?

26.18 Describe two ways in which the octane number of a gasoline consisting of alkanes can be increased.

Reactions of Hydrocarbons

26.19 Give an example, in the form of a balanced equation, of each of the following chemical reactions: (a) substitution reaction of an alkane; (b) combustion of an alkene; **(c)** addition reaction of an alkyne; **(d)** substitution reaction of an aromatic hydrocarbon.

26.20 Using condensed structural formulas, write a balanced chemical equation for each of the following reactions: **(a)** hydrogenation of 1-butene; **(b)** addition of H_2O to *cis*-2-butene using H_2SO_4 as a catalyst; **(c)** combustion of cyclobutane; **(d)** chlorination of benzene in the presence of $FeCl_3$.

26.21 Using Markovnikov's rule, predict the product formed when excess HCl is reacted with each of the following compounds:

(a) $CH_3CH{=}CH_2$

(b) CH_3 and CH_3 on one carbon, $C{=}C$, CH_3 and H on the other carbon

(c) $CH_3C{\equiv}CH$

26.22 Using Markovnikov's rule, predict the product of each of the following reactions:
(a) addition of HBr to $CH_3CH_2CH{=}CHBr$
(b) addition of H_2O to

1-methylcyclohexene (cyclohexene ring with CH_3 on the double-bonded carbon)

(c) reaction of excess HBr with $(CH_3)_3CC{\equiv}CH$.

26.23 When cyclopropane is treated with HI, 1-iodopropane is formed. A similar type of reaction does not occur with cyclopentane or cyclohexane. How do you account for the activity of cyclopropane?

26.24 Why do addition reactions occur more readily with alkenes and alkynes than with aromatic hydrocarbons?

26.25 Predict the product or products formed in each of the following reactions:

(a) *cis*-$CH_3CH{=}CHCH_3 + H_2 \xrightarrow{Ni}$

(b) $CH_3CH_2C{\equiv}CH + HI \longrightarrow$

(c) $(CH_3)_2C{=}C(CH_3)(CH_2CH_3) + H_2O \xrightarrow{H_2SO_4}$

(d) $CH_2{=}CHCH_3 + Br_2 \longrightarrow$

26.26 Predict the product or products formed in each of the following reactions:

(a) cyclohexene $+ Br_2 \longrightarrow$

(b) $(CH_3)_2C{=}CHCH_3 + HBr \longrightarrow$

(c) $CH_3CH{=}CHCH_2CH_3 + O_2 \xrightarrow{\Delta}$

(d) benzene $+ CH_3CHClCH_3 \xrightarrow{AlCl_3}$

26.27 Suggest a method of preparing ethylbenzene, starting with benzene and ethylene as the only organic reagents.

26.28 Write a series of reactions leading to *para*-bromoethylbenzene, beginning with benzene and using other reagents as needed.

26.29 The heat of combustion of decahydronaphthalene, $C_{10}H_{18}$, is -6286 kJ/mol. The heat of combustion of naphthalene, $C_{10}H_8$, is -5157 kJ/mol. (In both cases, $CO_2(g)$ and $H_2O(l)$ are the products.) Using these data and data in Appendix C, calculate the heat of hydrogenation of naphthalene. Does this value provide any evidence for aromatic character in naphthalene?

26.30 The molar heat of combustion of cyclopropane is 2089 kJ/mol; that for cyclopentane is 3317 kJ/mol. Calculate the heat of combustion per CH_2 group in the two cases and account for the difference.

Hydrocarbon Derivatives

26.31 Identify the functional groups in each of the following compounds:

(a) $CH_3C({=}O)CH_2CH_3$ **(b)** $CH_3C({=}O)OH$

(c) $HOCH_2CH_2CH_3$ **(d)** $CH_3OC({=}O)CH_2CH_3$

(e) $NH_2C({=}O)CH_3$ **(f)** $CH_3CH_2NH(CH_3)$

26.32 Identify the functional groups in each of the following compounds:

(a) $H_2N{-}C_6H_4{-}C({=}O){-}OH$ (para)

(b) $CH_2{=}CH{-}CH_2OH$ **(c)** $CH_3C({=}O){-}NHCH_3$

(d) $CH_3C({=}O)CH_3$ **(e)** $HC({=}O)CH_2CH_3$

(f) $HC{\equiv}C{-}CH_2OH$

26.33 Give the structural formula for an aldehyde that is an isomer of acetone.

26.34 Give the structural formula for an ether that is an isomer of ethanol.

26.35 Predict the bond angles around each carbon atom in acetone:

$$CH_3C({=}O)CH_3$$

26.36 Identify the carbon atom(s) in the following structure that has each of the following hybridizations: **(a)** sp^3; **(b)** sp; **(c)** sp^2.

$$HC{\equiv}C{-}C({=}O){-}O{-}CH_3$$

26.37 What products, if any, occur when each of the following compounds is mildly oxidized:

(a) $CH_3CH(CH_3)CH_2OH$

(b) $CH_3CH_2CH_2OH$ **(c)** $CH_3C({=}O)CH_3$

26.38 What products, if any, occur when each of the following compounds is mildly oxidized?

(a) $CH_3\overset{\overset{O}{\|}}{C}H$

(b) $HOCH_2CH_2CH_2OH$

(c) 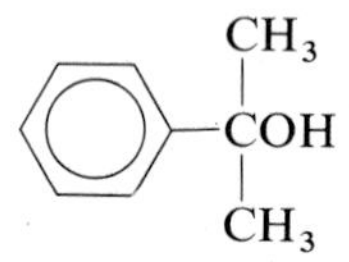 $C_6H_5-\overset{\overset{CH_3}{|}}{\underset{\underset{CH_3}{|}}{C}}OH$

26.39 Draw the structures of the esters formed from **(a)** acetic acid and 2-propanol; **(b)** acetic acid and 1-propanol; **(c)** formic acid and ethanol. Name the compound in each case.

26.40 Draw the structures of the compounds formed by condensation reactions between **(a)** benzoic acid and ethanol; **(b)** propionic acid and methylamine; **(c)** acetic acid and phenol. Name the compound in each case.

26.41 Write a balanced chemical equation for the saponification of ethyl formate.

26.42 Write a balanced chemical equation for the saponification of methyl acetate.

26.43 Write the structural formula for each of the following compounds: **(a)** 2-butanol; (b) 1,2-ethanediol; **(c)** methyl formate; (d) diethyl ketone; (e) diethyl ether.

26.44 Write the structural formula for each of the following compounds: **(a)** 3-chloropropionaldehyde; **(b)** methyl isopropyl ketone; **(c)** *meta*-chlorobenzaldehyde; **(d)** methyl-*cis*-2-butenyl ether; **(e)** *N,N*-diethylpropionamide.

26.45 The IUPAC name for a carboxylic acid is based on the name of the hydrocarbon with the same number of carbon atoms. The ending *-oic* is appended, as in ethanoic acid, which is the IUPAC name for acetic acid,

$$CH_3\overset{\overset{O}{\|}}{C}OH$$

Give the IUPAC name for each of the following acids:

(a) $H\overset{\overset{O}{\|}}{C}OH$

(b) $CH_3CH_2CH_2\overset{\overset{O}{\|}}{C}OH$

(c) $CH_3CH_2\underset{\underset{CH_3}{|}}{C}HCH_2\overset{\overset{O}{\|}}{C}OH$

26.46 Aldehydes and ketones can be named in a systematic way by counting the number of carbon atoms (including the carbonyl carbon) that they contain. The name of the aldehyde or ketone is based on the hydrocarbon with the same number of carbon atoms. The ending *-al*, for aldehyde, or *-one*, for ketone, is added as appropriate. Draw the structural formulas for the following aldehydes or ketones: **(a)** propanal; **(b)** 2-pentanone; **(c)** 3-methyl-2-butanone; **(d)** 2-methylbutanal.

26.47 Write balanced chemical equations, with organic substances given by their condensed formulas, describing the preparation of (a) acetone from 1-propene; (b) propanoic acid (a carboxylic acid with three carbon atoms) from 1-propanol; **(c)** sodium acetate from ethene; **(d)** methyl acetate from methyl alcohol and acetic acid; **(e)** methyl ethyl ketone from 2-butanol.

26.48 Using condensed formulas, write balanced chemical equations for the condensation reactions that occur between the following substances: **(a)** methanol with itself; **(b)** methanol with ethanol; **(c)** acetic acid with 2-propanol; **(d)** formic acid with 1-propanol; **(e)** diethylamine with benzoic acid (see Figure 26.16); **(f)** methylethylamine with acetic acid.

Polymers

26.49 Write the chemical equation for the addition polymerization of propylene.

26.50 If two different monomers are involved in addition polymerization, the resultant polymer is said to be a copolymer of the two monomers. Draw the formula for the copolymer of vinyl chloride, $CH_2{=}CHCl$, and 1,1-dichloroethylene, $CH_2{=}CCl_2$, to form Saran, the polymer used to make food wrapping.

26.51 A rather simple condensation polymer can be made from ethylene glycol, $HOCH_2CH_2OH$, and oxalic acid, $HO-\underset{\underset{O}{\|}}{C}-\underset{\underset{O}{\|}}{C}-OH$. Sketch a portion of the condensation polymer chain obtained from these monomers.

26.52 Draw the portion of the condensation polymer chain formed by the polymerization of 6-aminohexanoic acid,

$$H_2N-(CH_2)_5-\overset{\overset{O}{\|}}{C}-O-H$$

This material, called nylon 6 or Perlon, is used for making strong flexible fibers used in ropes and tire cords.

Additional Exercises

26.53 Draw the structural formulas for two molecules with the formula C_4H_6.

26.54 Unbranched hydrocarbons are often called straight-chain hydrocarbons. Does this mean that the carbon atoms have a linear arrangement? Explain.

26.55 What is the molecular formula for **(a)** an alkane with 20 carbon atoms; **(b)** an alkene with 18 carbon atoms; **(c)** an alkyne with 12 carbon atoms?

26.56 Classify each of the following substances as alkane, alkene, or alkyne (assuming that none are cyclic hydrocarbons): **(a)** C_5H_{12}; **(b)** C_5H_8; **(c)** C_6H_{12}; **(d)** C_8H_{18}.

26.57 Draw the Lewis structures for the cis and trans isomers of 2-pentene. Does cyclopentene exhibit cis-trans isomerism? Explain.

26.58 Give the IUPAC name for each of the following molecules:

(a) $CH_3CH(OH)CH_3$

(b) CH_3OCH_3

(c) $CH_2{=}CHCH_2CH_2OH$

(d) $CH_3C(CH_2CH_3){=}CHCH_2CH_3$

(e) $CH_3C{\equiv}C{-}CH(CH_3)_2$

26.59 Write the formulas for all of the structural isomers of C_3H_8O.

26.60 Explain why *trans*-1,2-dichloroethene has no dipole moment, whereas *cis*-1,2-dichloroethene has a dipole moment.

26.61 Would you expect cyclohexyne to be a stable compound? Explain.

26.62 Describe how you would prepare **(a)** 2-bromo-2-chloropropane from propyne; **(b)** 1,2-dichloroethane from ethene; **(c)** a condensation polymer from $HOCH_2CH_2OH$ and

$$HOCCH_2CH_2COH \text{ (with C=O on each carbonyl carbon)}$$

(d) sodium benzoate from methyl benzoate; **(e)** polypropylene from propene; **(f)** 1-ethylnaphthalene from naphthalene and ethene.

26.63 Aldehydes, ketones, carboxylic acids, esters, and amides all contain the carbonyl group. What distinguishes these functional groups?

26.64 Identify all of the functional groups in each of the following molecules:

(a) $CH_2{=}CH{-}O{-}CH{=}CH_2$ (an anesthetic)

(b) 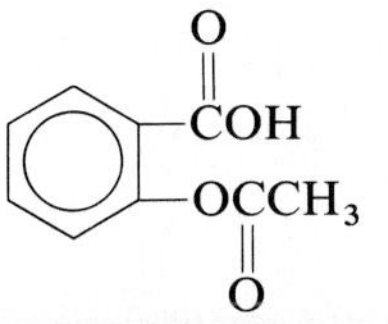 $C_6H_4(COOH)(OCOCH_3)$ (acetylsalicyclic acid, aspirin)

(c) 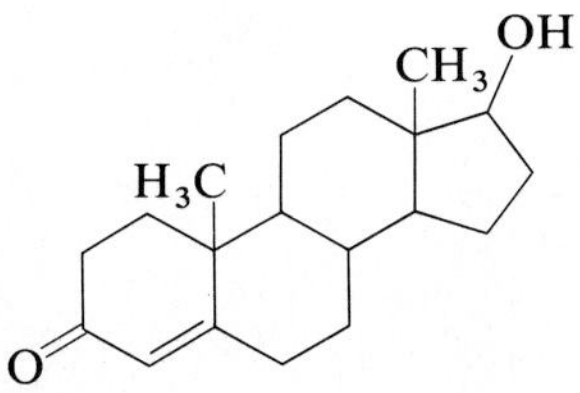 (testosterone, a male sex hormone)

26.65 Write the structural formula for each of the following: **(a)** an ether with the formula C_3H_8O; **(b)** an aldehyde with the formula C_3H_6O; **(c)** a ketone with the formula C_3H_6O; **(d)** a secondary alkyl fluoride with the formula C_3H_7F; **(e)** a primary alcohol with the formula $C_4H_{10}O$; **(f)** a carboxylic acid with the formula $C_3H_6O_2$; **(g)** an ester with the formula $C_3H_6O_2$.

26.66 Give the condensed formulas for the carboxylic acid and the alcohol from which each of the following esters is formed:

(a) 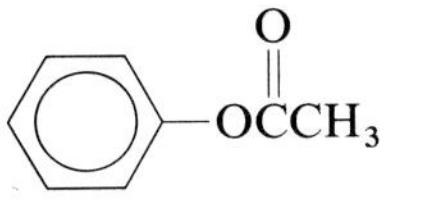 $C_6H_5{-}OCCH_3$ (C=O on the C) **(b)** 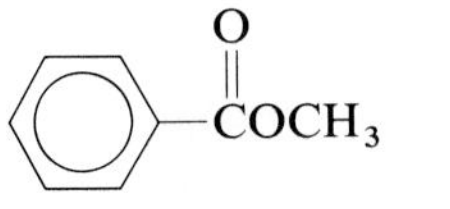 $C_6H_5{-}COCH_3$ (C=O on the C)

26.67 In each of the following pairs, indicate which molecule is the more reactive and give a reason for the greater reactivity: (a) butane and cyclobutane; **(b)** cyclohexane and cyclohexene; (c) benzene and 1-hexene; **(d)** 2-hexyne and 2-hexene.

26.68 A 256-mg sample of an organic compound is combusted producing 512 mg of CO_2 and 209 mg of H_2O. At 127°C, 155 mg of this substance occupies a volume of 0.100 L with a pressure of 1.16 atm. What is the molecular formula of the compound? Describe a possible structure for the compound and suggest some chemical test for the functional group.

26.69 Bromination of butane requires irradiation. Write a series of reaction steps that can account for the bromination of butane under irradiation conditions.

[26.70] Suggest a process that would cause the breakup of a condensation polymer by essentially reversing the condensation reaction.

[26.71] An unknown organic compound is found on analysis to have the empirical formula $C_5H_{12}O$. It is slightly soluble in water. Upon careful oxidation it is converted into a compound of empirical formula $C_5H_{10}O$, which behaves chemically like a ketone. Indicate two or more reasonable structures for the unknown.

[26.72] Consider two hydrocarbons, A and B, whose molecular formulas are both C_4H_8. Both compounds decolorize bromine water. When compound A is treated with water and sulfuric acid, it is converted to $C_4H_{10}O$. Mild oxidation of this product produces C_4H_8O, which does not oxidize further when treated with a stronger oxidizing agent, such as $K_2Cr_2O_7$. Compound B also produces a compound whose formula is $C_4H_{10}O$ when treated with water and sulfuric acid. However, this substance has different properties than the substance obtained from compound A. For example, the $C_4H_{10}O$ obtained from compound B is not oxidized even by a strong oxidizing agent such as $K_2Cr_2O_7$. Identify the two compounds, A and B, and trace all the reactions described that begin with these compounds, writing the names and structural formulas for all the organic compounds involved.

[26.73] An unknown substance is found to contain only carbon and hydrogen. It is a liquid that boils at 49°C at 1 atm pressure. Upon analysis it is found to contain 85.7 percent carbon and 14.3 percent hydrogen. At 100°C and 735 mm Hg, the vapor of this unknown has a density of 2.21 g/L. When it is dissolved in hexane solution and bromine water added, no reaction occurs. Suggest the identity of the unknown compound.

27 Biochemistry

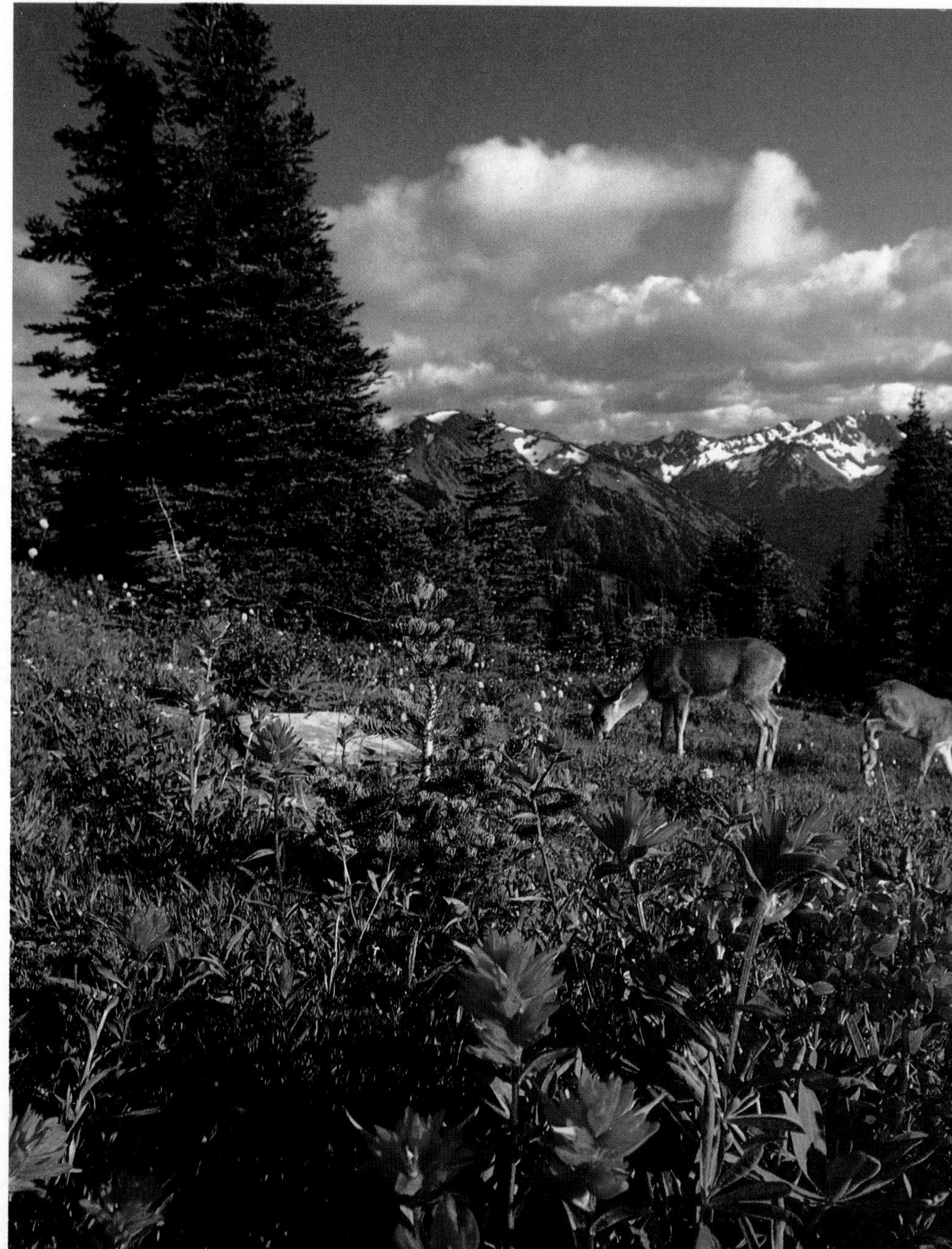

Biochemistry is the study of the chemical processes in living organisms. (Jim Fitzharris, Animals for Advertising Collection/Allstock)

CONTENTS

In earlier chapters of this text, we have considered various aspects of the physical world in which we live. We have discussed the chemistry of the atmosphere and of natural waters (Chapter 18) and of the solid earth (Chapter 24). In this final chapter, we consider the **biosphere**, which is defined as that part of the earth in which living organisms are formed and live out their life cycles. The biosphere is not distinct from the atmosphere, natural waters, or the solid earth; rather, it is an integral part of it in those places where conditions permit life to exist. For living organisms to be sustained, there must be a supply of available energy, because the growth and maintenance of organisms require energy. Second, there must be an adequate supply of water, because organisms are composed largely of water and use it for exchange of materials with their environment.

27.1 ENERGY REQUIREMENTS OF ORGANISMS

Living organisms require energy for their maintenance, growth, and reproduction. The ultimate source of this energy is the sun. However, as living matter proliferated on earth, and as organisms became more and more specialized, many of these organisms developed the capacity for obtaining energy indirectly by using the energy stored in other organisms. For example, our bodies have essentially no capacity for using solar energy directly. We consume animal and plant materials to acquire substances that our bodies can use as energy sources.

There are two distinct reasons why living systems need energy. First, organisms rely on substances readily available in their surroundings for synthesis of compounds needed for their existence. Most of these reactions are endothermic. To make such reactions proceed, energy must be supplied from an outside source. Second, living organisms are highly organized. The complexity of all the substances that make up even the simplest of single-cell organisms and the relationships among all the many chemical processes occurring are truly amazing. In thermodynamic terms, living systems are very low in entropy as compared with the raw materials from which they are formed. The orderliness that characterizes living systems is attained by expenditure of energy.

Recall from the discussion in Section 19.5 that the free-energy change, ΔG, is related to both the enthalpy and the entropy changes that occur in a process:

$$\Delta G = \Delta H - T\,\Delta S$$

If the entropy change in a process that builds up a living organism is negative (in other words, if a more highly ordered state results), the contribution to ΔG is positive. This means that the process becomes less spontaneous. Therefore, both the enthalpy and the entropy changes that result in the formation, main-

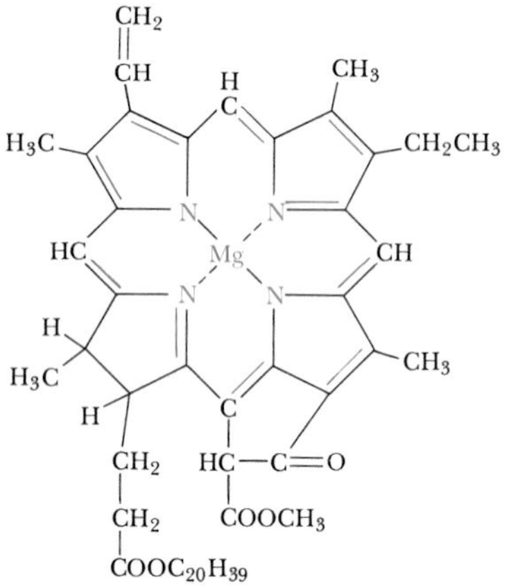

CH_2
CH
H
C
CH_3
H_3C
CH_2CH_3
N
N
HC
Mg
CH
N
N
H
CH_3
H_3C
H
C
CH_2 HC—C=O
CH_2 $COOCH_3$
$COOC_{20}H_{39}$

Figure 27.1 Structure of chlorophyll *a*. All chlorophyll molecules are essentially alike; they differ only in details of the side chains.

tenance, and reproduction of living systems are such that the overall process is nonspontaneous. To overcome the positive values for ΔG associated with the essential processes, living systems must be coupled to some outside source of energy that can be converted into a form useful for driving the biochemical processes. The ultimate source of this needed energy is the sun.

The major means for conversion of solar energy into forms that can be used by living organisms is **photosynthesis**. The photosynthetic reaction that occurs in the leaves of plants is conversion of carbon dioxide and water to carbohydrate, with release of oxygen:

$$6CO_2 + 6H_2O \xrightarrow{48\,h\nu} C_6H_{12}O_6 + 6O_2 \qquad [27.1]$$

Note that formation of a mole of sugar, $C_6H_{12}O_6$, requires the absorption and utilization of 48 mol of photons. This needed radiant energy comes from the visible region of the spectrum (Figure 6.3). The photons are absorbed by photosynthetic pigments in the leaves of plants. The key pigments are the **chlorophylls**. The structure of the most abundant chlorophyll, called chlorophyll *a*, is shown in Figure 27.1.

Chlorophyll is a coordination compound; it contains a Mg^{2+} ion bound to four nitrogen atoms arranged around the metal in a planar array. The nitrogen atoms are part of a porphine-like ring (Figure 25.8). Notice that a series of alternating double bonds in the ring surrounds the metal ion. This system of alternating, or conjugated, double bonds gives rise to the strong absorptions of chlorophyll in the visible region of the spectrum. Figure 27.2 shows the absorption spectrum of chlorophyll as compared with the distribution of solar energy at the earth's surface. Chlorophyll is green in color because it absorbs red light (maximum absorption at 655 nm) and blue light (maximum absorption at 430 nm) and transmits green light.

The solar energy absorbed by chlorophyll is converted by a complex series of steps into chemical energy. This stored energy is then used to drive the reaction in Equation 27.1 to the right, a direction in which it is highly endothermic. Plant photosynthesis is nature's solar-energy-conversion machine; all living systems on earth are dependent on it for continued existence (Figure 27.3).

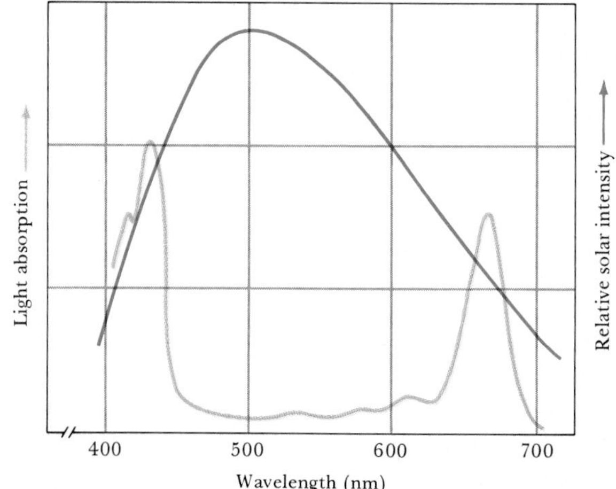

Figure 27.2 Absorption spectrum of chlorophyll (green curve), in comparison with the solar radiation at ground level (red curve).

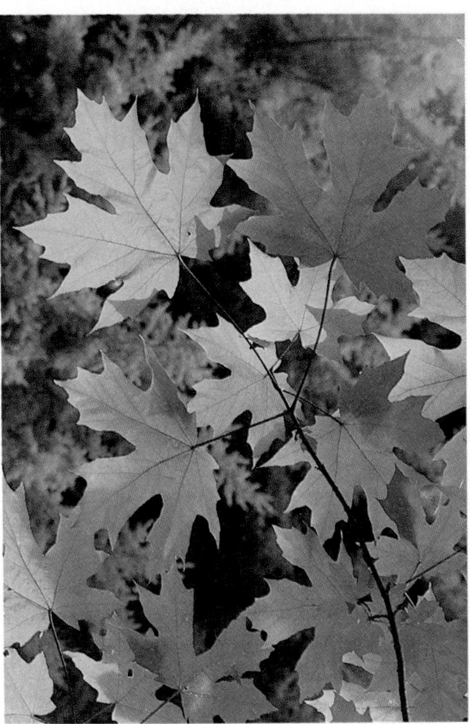

Figure 27.3 The absorption and conversion of solar energy that occurs in the maple leaf provides the energy necessary to drive all the living processes of the tree, including growth. (Calvin Larsen/Photo Researchers)

Let us now turn our attention to the materials from which living organisms are formed. At some level of concentration, in some organism somewhere, almost every element seems to play a role in the biosphere. However, the elements of major importance in terms of their abundances in living systems are carbon, hydrogen, oxygen, nitrogen, phosphorus, and sulfur. Many other elements, including many metallic elements, are present in lesser quantities (refer back to Figure 25.7).

Much of the material present in living systems is in the form of polymers, molecules of high molecular weight (Section 26.5). These polymers in living systems, called **biopolymers**, can be classified into three broad groups: proteins, carbohydrates, and nucleic acids. The proteins and carbohydrates, along with a class of molecules called fats and oils, are the major sources of energy in the food supply of animals. Polymeric carbohydrates are the major construction material that gives form to plant systems; proteins perform a similar role in animal systems. The nucleic acids are the storehouse of information that determines the form of a living system and that regulates its reproduction and development.

27.2 PROTEINS

Proteins are macromolecular substances present in all living cells. About 50 percent of your body's dry weight is protein. Proteins serve as the major structural components in animal tissues; they are a key part of skin, nails, cartilage, and muscles. Other proteins catalyze reactions, transport oxygen, serve as hormones to regulate specific body processes, and perform other tasks. Whatever their function, all proteins are chemically similar, being composed of the same basic building blocks, called **amino acids**.

Amino Acids

The building blocks of all proteins are α-amino acids, which are substances in which the amino group is located on the carbon atom immediately adjacent to the carboxylic acid group. The general formula for an α-amino acid is represented in the following ways:

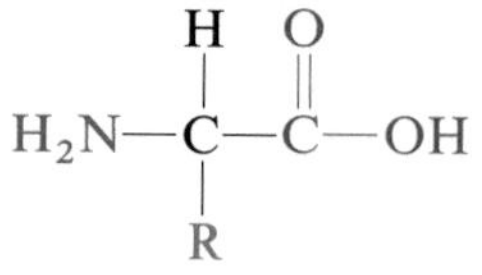

$$H_2N-\underset{\underset{R}{|}}{\overset{\overset{H}{|}}{C}}-\overset{\overset{O}{\|}}{C}-OH \quad \text{or} \quad {}^{+}H_3N-\underset{\underset{R}{|}}{\overset{\overset{H}{|}}{C}}-\overset{\overset{O}{\|}}{C}-O^{-}$$

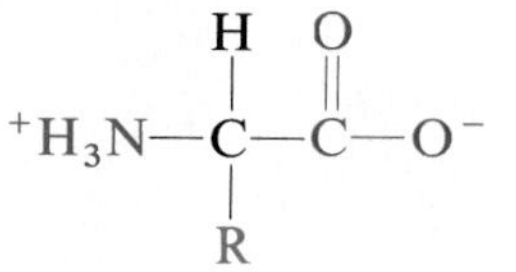

The doubly ionized form predominates in the near-neutral pH of biological systems. This form results from the transfer of a proton from the carboxylic acid group to the basic amine group.

Amino acids differ from one another in the nature of their R groups. Figure 27.4 shows the structural formulas of the 20 amino acids found in most proteins. Our bodies can synthesize 10 of these in sufficient amounts for our needs. The other 10 must be ingested and are called *essential amino acids* because they are necessary components of our diet.

You can see from the structural formulas of amino acids that the α-carbon atom, which bears both the amine and carboxylic acid groups, has four different groups attached to it.* *Any molecule containing a carbon with four different attached groups is chiral.* As we saw in Section 25.4, a **chiral** molecule is capable of existing as isomers that are nonsuperimposable mirror images of each other.

The mirror-image isomers of a chiral substance are called **enantiomers.** The two enantiomers of a mirror-image pair are sometimes labeled *R*- (from the Latin *rectus*, "right") and *S*- (from the Latin *sinister*, "left") to distinguish

* The sole exception is glycine, for which R = H. For this amino acid, there are two H atoms on the α-carbon atom.

Figure 27.4 Condensed structural formulas for the amino acids, with the three-letter abbreviation for each acid.

Amino acids with acidic or basic side chain functional groups

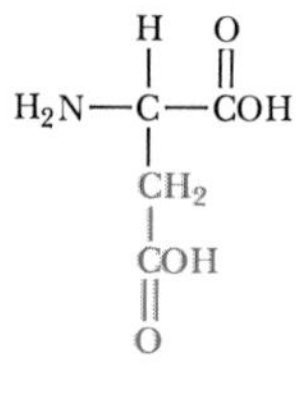

Aspartic acid (asp): $H_2N-CH(COOH)-CH_2-COOH$

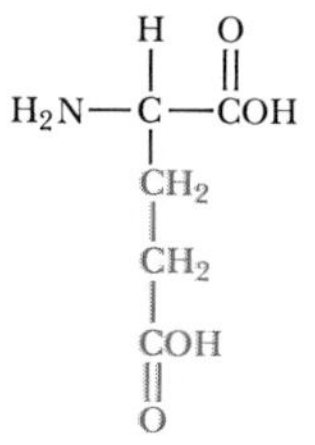

Glutamic acid (glu): $H_2N-CH(COOH)-CH_2-CH_2-COOH$

Tyrosine (tyr): $H_2N-CH(COOH)-CH_2-C_6H_4-OH$

Lysine (lys): $H_2N-CH(COOH)-(CH_2)_4-NH_2$

Arginine (arg): $H_2N-CH(COOH)-(CH_2)_3-HN-C(=NH)-NH_2$

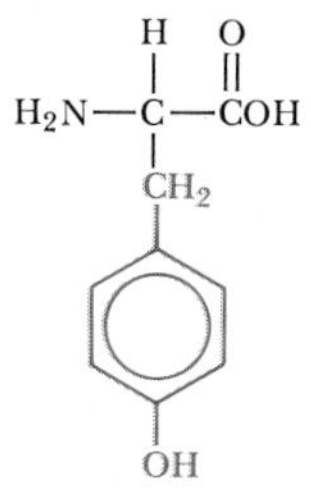

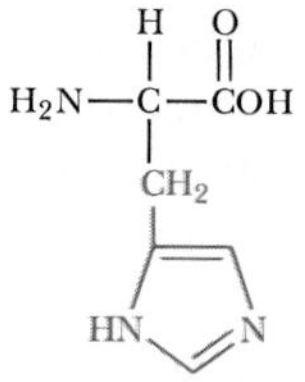

Histidine (his): $H_2N-CH(COOH)-CH_2-$(imidazole ring, HN, N)

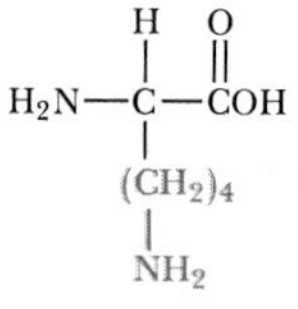

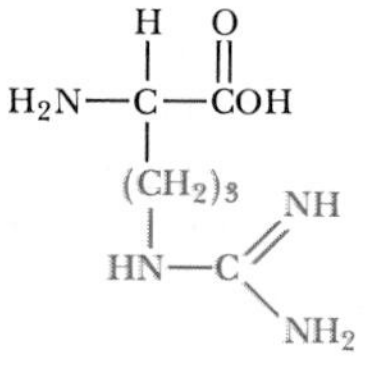

Amino acids with hydrocarbon side chains

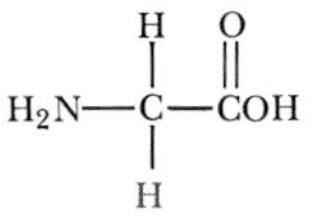

Figure 27.4
(continued)

Glycine (gly) 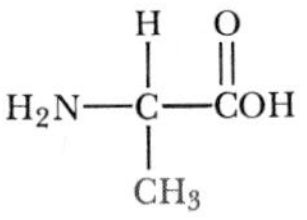 Alanine (ala) 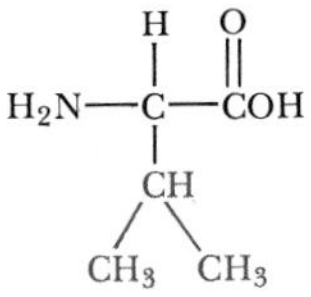 Valine (val) 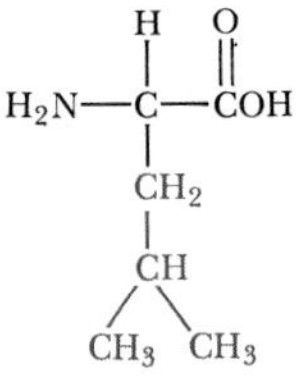 Leucine (leu)

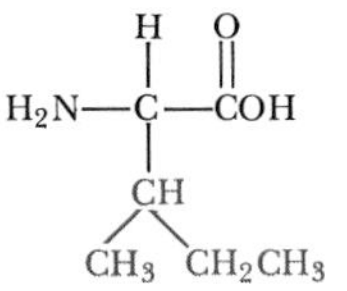

Isoleucine (ile) 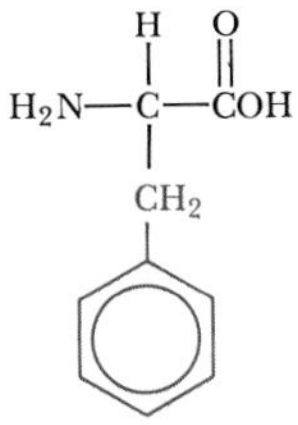 Phenylalanine (phe) 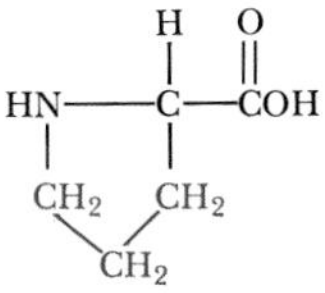 Proline (pro)

Amino acids with polar, neutral side chains

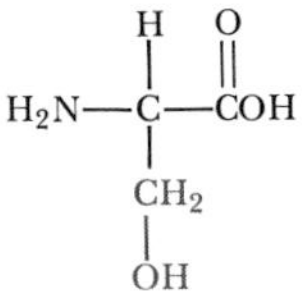

Serine (ser) 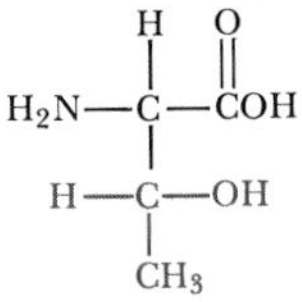 Threonine (thr) 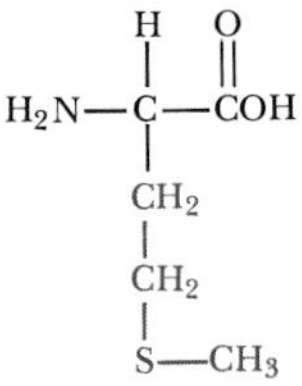 Methionine (met) 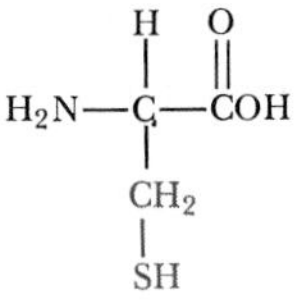 Cysteine (cys)

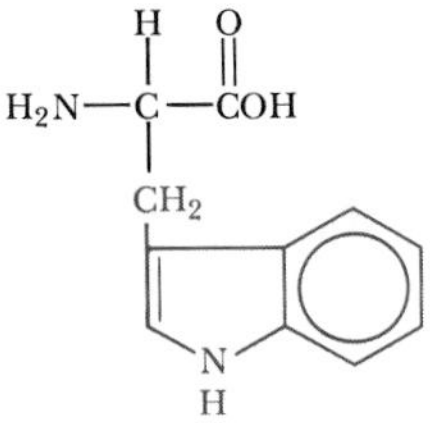

Tryptophan (trp) 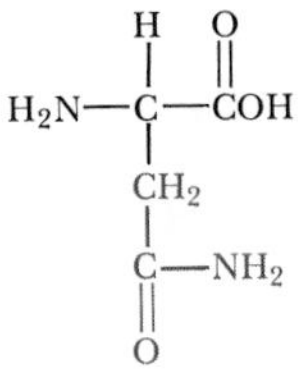 Asparagine (asn) 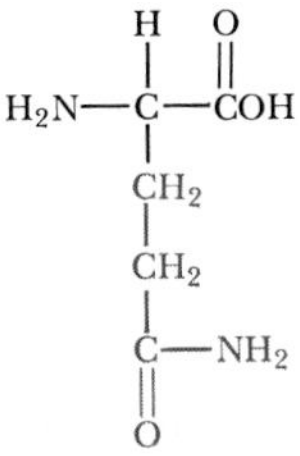 Glutamine (gln)

them. Another widely used notation for distinguishing enantiomers uses the prefix labels D- (from the Latin *dexter*, "right") and L- (from the Latin *laevus*, "left"). The enantiomers of a chiral substance possess the same physical properties, such as solubility, melting point, and so forth. Their chemical behavior toward ordinary chemical reagents is also the same. However, they differ in chemical reactivity toward other chiral molecules. It is a striking fact that all the amino acids in nature are of the *S*-, or L-, configuration at the carbon center (except glycine, which is not chiral). Only amino acids with this specific configuration at the chiral carbon center are biologically effective in forming proteins in most organisms. Figure 27.5 shows the two enantiomers of the amino acid alanine.

Figure 27.5 Alanine is chiral and therefore has two enantiomers, which are nonsuperimposable mirror images of each other.

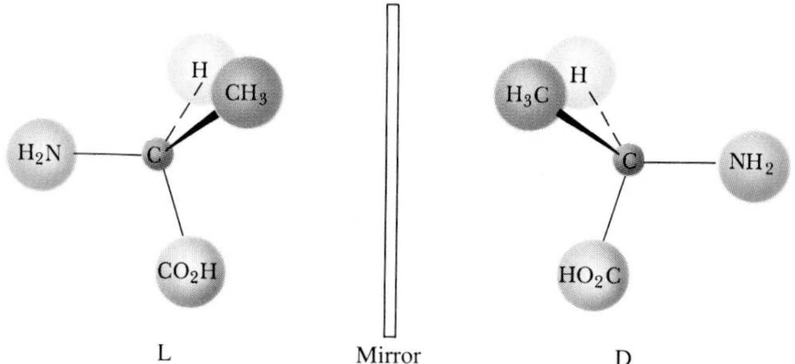

Polypeptides and Proteins

In proteins, amino acids are linked by amide groups (see Section 26.4):

$$\mathrm{R{-}\overset{\overset{\displaystyle O}{\|}}{C}{-}\underset{\underset{\displaystyle H}{|}}{N}{-}R}$$

Each of these amide groups is called a **peptide** bond when it is formed by amino acids. A peptide bond is formed by a condensation reaction between the carboxyl group of one amino acid and the amino group of another amino acid. For example, alanine and glycine can be caused to react to form the dipeptide glycylalanine:

$$\underset{\text{Glycine}}{\mathrm{H_2N{-}\overset{\overset{\displaystyle H}{|}}{\underset{\underset{\displaystyle H}{|}}{C}}{-}\overset{\overset{\displaystyle O}{\|}}{C}{-}OH}} + \underset{\text{Alanine}}{\mathrm{\overset{\overset{\displaystyle H}{|}}{HN}{-}\overset{\overset{\displaystyle H}{|}}{\underset{\underset{\displaystyle CH_3}{|}}{C}}{-}\overset{\overset{\displaystyle O}{\|}}{C}{-}OH}} \longrightarrow$$

$$\underset{\text{Glycylalanine}}{\mathrm{H_2N{-}\overset{\overset{\displaystyle H}{|}}{\underset{\underset{\displaystyle H}{|}}{C}}{-}\overset{\overset{\displaystyle O}{\|}}{C}{-}\underset{\underset{\displaystyle H}{|}}{N}{-}\overset{\overset{\displaystyle H}{|}}{\underset{\underset{\displaystyle CH_3}{|}}{C}}{-}\overset{\overset{\displaystyle O}{\|}}{C}{-}OH}} + H_2O \quad [27.2]$$

Notice that the acid that furnishes the carboxyl group is named first, with a -*yl* ending; then the amino acid furnishing the amino group is named. Based on the three-letter codes for the amino acids from Figure 27.4, glycylalanine can be abbreviated gly-ala. In this notation, it is understood that the amino group is on the left and the carboxyl group on the right, as in the following tetrapeptide:

Amino group ⟶ Carboxyl group

$$\underset{\text{ala}}{\boxed{\mathrm{H_2N{-}\overset{\overset{\displaystyle H}{|}}{\underset{\underset{\displaystyle CH_3}{|}}{C}}{-}\overset{\overset{\displaystyle O}{\|}}{C}}}}{-}\underset{\text{gly}}{\boxed{\mathrm{\underset{\underset{\displaystyle H}{|}}{N}{-}\overset{\overset{\displaystyle H}{|}}{\underset{\underset{\displaystyle H}{|}}{C}}{-}\overset{\overset{\displaystyle O}{\|}}{C}}}}{-}\underset{\text{cys}}{\boxed{\mathrm{\underset{\underset{\displaystyle H}{|}}{N}{-}\overset{\overset{\displaystyle H}{|}}{\underset{\underset{\underset{\displaystyle SH}{|}}{CH_2}}{|}}{C}}{-}\overset{\overset{\displaystyle O}{\|}}{C}}}}{-}\underset{\text{ser}}{\boxed{\mathrm{\underset{\underset{\displaystyle H}{|}}{N}{-}\overset{\overset{\displaystyle H}{|}}{\underset{\underset{\displaystyle CH_2OH}{|}}{C}}{-}\overset{\overset{\displaystyle O}{\|}}{C}{-}OH}}}$$

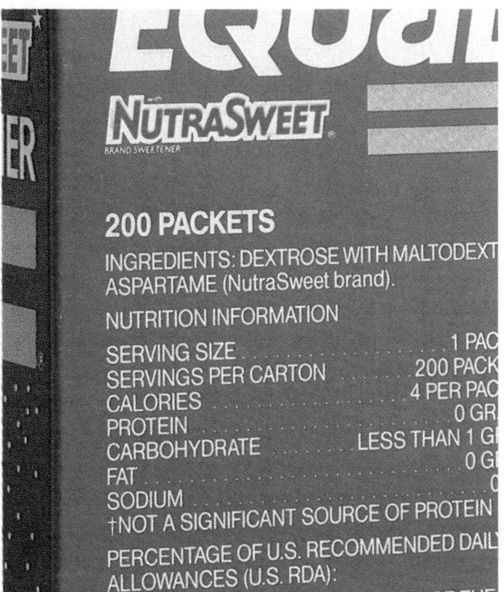

Figure 27.6 The artificial sweetener aspartame is the methyl ester of a dipeptide. (© Richard Megna/Fundamental Photographs)

The artificial sweetener *aspartame* (Figure 27.6) is the methyl ester of the dipeptide of aspartic acid and phenylalanine:

```
       H   O       H   O
       |   ||      |   ||
H2N—C—C—N—C—C—O—CH3
       |       |   |
      CH2      H  CH2
       |           |
       C—OH      C6H5 (benzene ring)
       ||
       O
```

SAMPLE EXERCISE 27.1

Draw the structural formula for histidylglycine (his-gly).

Solution: The name and abbreviation for this dipeptide tell us that the amino group of glycine and the carboxylic acid function of histidine are involved in formation of the peptide bond. The structure is therefore

```
       H   O       H   O
       |   ||      |   ||
H2N—C—C—N—C—C—OH
       |       |   |
      CH2      H   H
       |
   (imidazole ring: HN   N)
```

PRACTICE EXERCISE

Name the dipeptide that has this structure and give its abbreviation:

```
       H   O   H   H   O
       |   ||  |   |   ||
NH2—C—C—N—C—C—OH
       |           |
      CH2OH       CH2
                   |
                HO—C=O
```

Answer: serylaspartic acid; ser-asp

Polypeptides result when a large number of amino acids are linked together by peptide bonds. Proteins are polypeptide molecules with molecular weights ranging from about 6000 to over 50 million amu. *Simple proteins* are composed entirely of amino acids. Other proteins, called *conjugated proteins*, are made up of simple proteins bonded to other kinds of biochemical molecules, for example, carbohydrates. Because 20 different amino acids are present in proteins and because proteins consist of hundreds of amino acids, the number of possible arrangements of amino acids within proteins is immense.

Protein Structure

The arrangement, or sequence, of amino acids along a protein chain is called its **primary structure**. The primary structure gives the protein its unique identity. A change in even one amino acid can alter the biochemical characteristics of the protein. For example, sickle-cell anemia is a genetic disorder resulting from a single misplacement in a protein chain in hemoglobin. The chain that is affected contains 146 amino acids. The first 7 in the chain are valine, histidine, leucine, threonine, proline, glutamic acid, and glutamic acid. In a person suffering from sickle-cell anemia, the sixth amino acid is valine instead of glutamic acid. This one substitution of an amino acid with a hydrocarbon side chain for one that has an acidic functional group in the side chain alters the solubility properties of the hemoglobin, and normal blood flow is impeded (see the Chemistry at Work box in Section 13.6).

Figure 27.7 α-Helix structure for a protein. The symbol R represents any one of the several side chains shown in Figure 27.4.

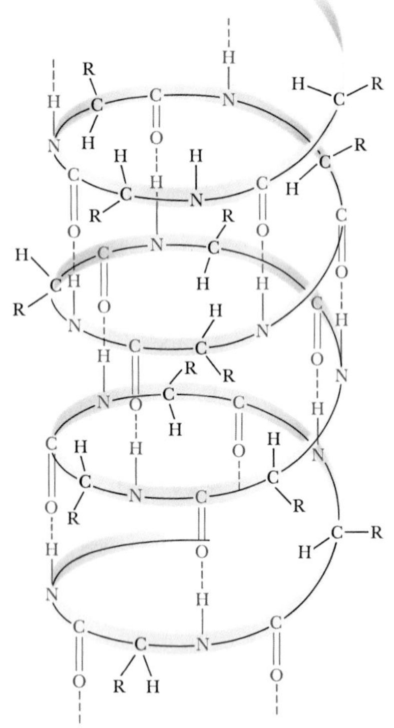

Proteins in living organisms are not simply long, flexible chains with random shapes. Rather, the chains coil or stretch in particular ways. The **secondary structure** of a protein refers to how segments of the protein chain are oriented in a regular pattern.

One of the most important and common secondary structure arrangements is the **α-helix**, first proposed by Linus Pauling and R. B. Corey. The helix arrangement is shown in schematic form in Figure 27.7. Imagine winding a long protein chain in a helical fashion around a long cylinder. The helix is held in position by hydrogen-bond interactions between N—H bonds and the oxygens of nearby carbonyl groups. The pitch of the helix and the diameter of the cylinder must be such that (1) no bond angles are strained and (2) the N—H and C=O functional groups on adjacent turns are in proper position for hydrogen bonding. An arrangement of this kind is possible for some amino acids along the chain, but not for others. Large protein molecules may contain segments of chain that have the α-helical arrangement interspersed with sections in which the chain is in a random coil.

The overall shape of a protein, determined by all the bends, kinks, and sections of rodlike α-helical structure, is called the **tertiary structure**. Figure 27.8 shows the tertiary structure of myoglobin, a protein whose molecular weight is about 18,000 amu and which contains one heme group (Section 25.2). Myoglobin resides in the tissues, where it stores oxygen until it is required for metabolic activities. Notice the helical sections of the protein.

Myoglobin is an example of a *globular protein*, one that folds into a compact, roughly spherical shape. Globular proteins are generally soluble in water and are mobile within cells. They have nonstructural functions, such as combating the invasion of foreign objects, transporting and storing oxygen, and acting as catalysts. In *fibrous proteins*, the long coils align themselves in a more-

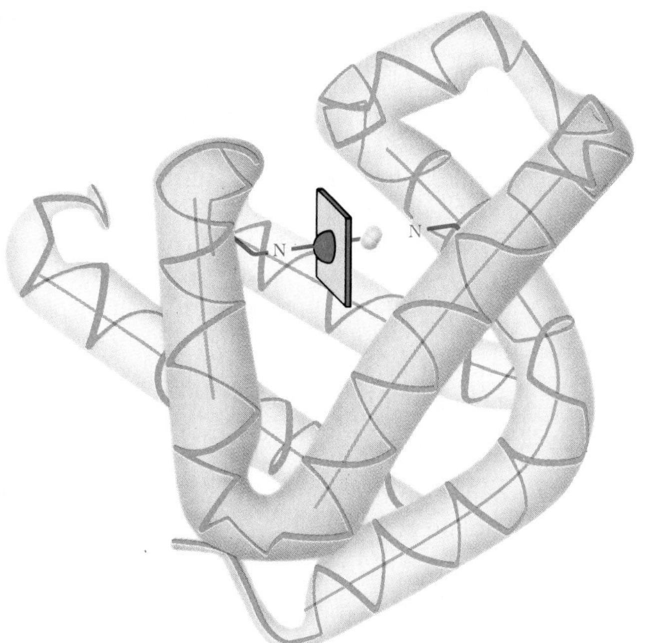

Figure 27.8 The structure of myoglobin, a protein that acts to store oxygen in cells. This protein contains one heme unit, symbolized by the red sphere and surrounding square. The heme unit is bound to the protein through a nitrogen-containing ligand, represented by the red N on the left. In the oxygenated form an O_2 molecule is coordinated to the heme group, as shown. The protein chain is represented by the continuous orange ribbon. The helical sections are denoted by the green lines. The protein wraps around to make a kind of pocket for the heme group.

or-less parallel fashion to form long, water-insoluble fibers. Fibrous proteins provide structural integrity and strength to many kinds of tissue and are the main components of muscle, tendons, and hair.

The tertiary structure of a protein is maintained by many different interactions. Certain foldings of the protein chain lead to lower-energy (more-stable) arrangements than do other folding patterns. For example, a globular protein dissolved in aqueous solution folds in such a way that the nonpolar hydrocarbon portions are tucked within the molecule, away from the polar water molecules. The more polar acidic and basic side chains, however, project into the solution where they can interact with water molecules through ion-dipole, dipole-dipole, or hydrogen-bonding interactions.

27.3 ENZYMES

The human body is characterized by an extremely complex system of interrelated chemical reactions. All these reactions must occur at carefully controlled rates in order to maintain life. A large number of marvelously efficient catalysts known as *enzymes* are necessary for many of these reactions to occur at suitable rates at body temperature (37°C). Most enzymes are globular proteins with molecular weights ranging from about 10,000 to about 1 million amu. An enzyme can consist of just a single protein chain, or it can involve several chains that are loosely held together.

The reaction that a given enzyme catalyzes occurs at a specific site on the protein called the **active site**. The substances that undergo reaction at this site are called **substrates**. Besides the substrate, other substances, called **cofactors** or **coenzymes**, may be required in the enzyme-catalyzed reaction. For example, the enzyme may require the presence of Mg^{2+} or some other metal ion, or it may require the presence and participation of a small organic molecule.

Enzymes possess certain characteristics not common to other types of catalysts. In the first place, they have a special sensitivity to temperature. Experimental studies have shown that the activity of a particular enzyme maximizes at around the normal temperature of the organism in which the enzyme is found. Figure 27.9 illustrates a typical activity-versus-temperature curve. When the temperature is raised above the usual operating temperature of the enzyme, the activity often temporarily increases but then subsequently decreases. The secondary and tertiary structures of the protein chain, on which the activity of the active site depends, are produced by many weak forces that induce the chain to take up a particular arrangement. Heating causes the protein chain to come undone, as it were, and the enzyme loses its activity completely.

A second respect in which enzymes are special is that they often show a sharp change in activity with a change in the acidity or basicity of the solution. This suggests that acid-base reactions are important in the catalyzed reaction. Third, enzymes may be very specific in catalyzing exclusively a particular kind of reaction and no other, or even in catalyzing a particular reaction for only one compound and no other. The degree of specificity of enzymes varies widely. For example, enzymes called carboxypeptidases are rather general in their action.* They catalyze the hydrolysis of polypeptides into amino acids:

$$\text{---}\underset{\underset{\large R'}{|}}{\overset{\overset{\large H}{|}}{C}}-\overset{\overset{\large O}{\|}}{C}-\underset{\underset{\large H}{|}}{N}-\underset{\underset{\large R}{|}}{\overset{\overset{\large H}{|}}{C}}-\overset{\overset{\large O}{\|}}{C}-OH + H_2O \xrightarrow{\text{carboxypeptidase}} \text{---}\underset{\underset{\large R'}{|}}{\overset{\overset{\large H}{|}}{C}}-\overset{\overset{\large O}{\|}}{C}-OH + H_2N-\underset{\underset{\large R}{|}}{\overset{\overset{\large H}{|}}{C}}-\overset{\overset{\large O}{\|}}{C}-OH \quad [27.3]$$

This equation shows only the end member of a peptide chain; the carboxypeptidase attacks only the peptide bond nearest the carboxyl end of the chain. However, it is active regardless of the nature of the particular side chains R and R′.

Another group of related enzymes, called *proteases*, catalyze the hydrolysis of a wide range of peptide bonds, not just those at the end of the chain. It is this ability that accounts for the use of papain, a protease isolated from papaya fruit, in meat tenderizers and contact-lens cleaners (Figure 27.10).

Finally, many enzymes differ from ordinary nonbiochemical catalysts in that they are enormously more efficient. The number of individual, catalyzed-reaction events occurring per second at a particular active site, called the *turnover number*, is generally in the range 10^3 to 10^7 per second. Such large turnover numbers are indicative of reactions with very low activation energies.

Figure 27.9 A typical example of enzyme activity as a function of temperature. At the higher temperature, at which the activity of the enzyme drops to zero, the protein loses its secondary and tertiary structure.

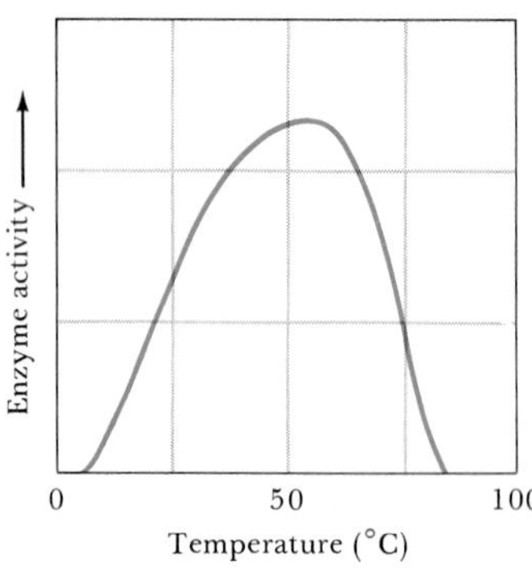

* The names of most enzymes end in *-ase*. The *-ase* ending is attached to the name of the substrate on which the enzyme acts, or to the name of the type of reaction that the enzyme catalyzes. Thus *peptidases* are enzymes that act on peptides; *lipase* is an enzyme that acts on lipids; a *transmethylase* is an enzyme that catalyzes transfer of a methyl group.

Figure 27.10 The active ingredient in meat tenderizers such as the product shown here is usually papain, a protease present in the fruit of the papaya tree. Pineapple also contains a protease, called bromelin. Experienced cooks know that fresh pineapple should not be added to gelatin to make a molded fruit salad. The gelatin never hardens because the protein responsible for the setting action is digested by the bromelin present in the pineapple. (Richard Megna/Fundamental Photographs)

SAMPLE EXERCISE 27.2

Solutions of dilute (about 3 percent) hydrogen peroxide are stable for long periods of time, especially if a small amount of a stabilizer is added. However, when such a solution is poured onto an open cut or wound, the hydrogen peroxide decomposes very rapidly with evolution of oxygen and formation of water:

$$2H_2O_2(aq) \longrightarrow O_2(g) + 2H_2O(l)$$

How can you account for this result?

Solution: Because the reaction occurs so rapidly on contact with the open wound, some chemical species present at the wound must catalyze the decomposition of hydrogen peroxide. That substance is an enzyme called catalase. Catalase is present to ensure that hydrogen peroxide does not accumulate in cells, where it could interfere with many other cell reactions. Catalase is an efficient enzyme, with a very high turnover number (see Figure 14.15).

PRACTICE EXERCISE

In 1926, James Sumner isolated the first pure, crystalline enzyme. This substance catalyzes the hydrolysis of urea to ammonia and carbon dioxide:

$$H_2N-\overset{\overset{\displaystyle O}{\|}}{C}-NH_2(aq) + H_2O(l) \longrightarrow 2NH_3(aq) + CO_2(g)$$

(a) What is the name of this enzyme? **(b)** What type of biological substance is it?
Answers: **(a)** urease; **(b)** protein

Biochemists have been trying for a long time to discover how and why enzymes are so efficient and selective. During the past 20 years, the development of new experimental techniques for studying the structures of molecules and for following the rates of very rapid reactions has led to a much better understanding of how enzymes work.

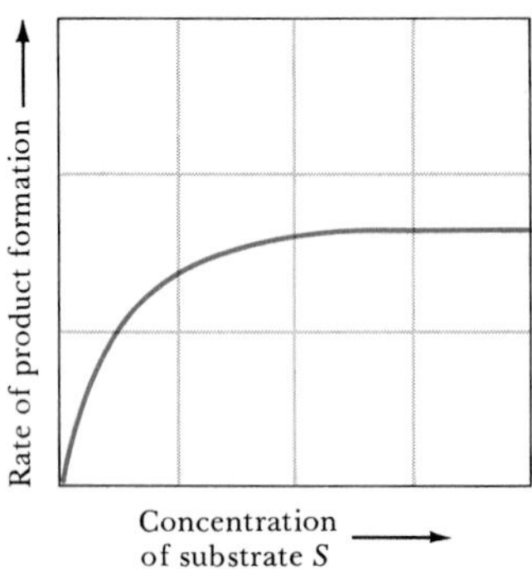

Figure 27.11 Rate of product formation as a function of concentration in a typical enzyme-catalyzed reaction. The rate of product formation is proportional to the substrate concentration in the region of low concentration. At high substrate concentrations, the active sites on the enzyme are all complexed; further addition of substrate does not affect the reaction rate.

The high turnover numbers observed for enzymes suggest that the substrate molecules cannot be very tightly bound to the enzyme; if they were, they might block the active site. Reaction would then be slow, because the active site is not quickly cleared out. Most enzyme systems that have been studied behave as though there were an equilibrium between the substrate (S) and the active site (E); we can write this equilibrium as follows:

$$E + S \underset{k_{-1}}{\overset{k_1}{\rightleftharpoons}} ES \quad [27.4]$$

The symbol ES represents a species in which the substrate is attached in some way to the enzyme. This *enzyme-substrate complex*, as it is called, then reacts to give product (P) and free the active site:

$$ES \xrightarrow{k_2} E + P \quad [27.5]$$

In this picture, it is assumed that the substrate molecules may move off and onto the active site very rapidly compared with the rate at which they undergo reaction to form products P. This means that the equilibrium described by Equation 27.4 is rapidly established between enzyme and substrate. It is also supposed that the equilibrium is such that most of the enzyme sites are not occupied by S when the substrate is present at normal concentrations. Now suppose that we study and graph (Figure 27.11) the rate of the enzyme-catalyzed reaction as a function of increasing concentration of substrate S. When S is present in low concentration, most of the enzyme active sites are not in use. Increasing the concentration of S shifts the equilibrium in Equation 27.4 to the right, so that a larger number of enzyme-substrate complexes are formed. This in turn increases the overall rate of the reaction, because that rate depends on the concentration of ES, [ES]:

$$\text{Rate} = \frac{\Delta[P]}{\Delta t} = k_2[ES] \quad [27.6]$$

However, with still further increases in the concentration of S, a sizable fraction of the active sites are occupied. Adding still more S therefore does not result in the same degree of increase in [ES], and the rate does not increase as much. Eventually, with a still higher concentration of S, *all* the active sites are effectively occupied. Consequently, further increases in S cannot result in a faster reaction. Many enzyme systems have been found to obey the sort of behavior illustrated in Figure 27.11. This strongly indicates that an enzyme-substrate complex is involved in the action of enzymes.

The Lock-and-Key Model

One of the earliest models for the enzyme-substrate interaction was the **lock-and-key model**, illustrated in Figure 27.12. The substrate is pictured as fitting neatly into a special place on the enzyme (the active site), much like a specific key fitting into a lock. The catalyzed reaction occurs while the substrate is bound to the enzyme, and the reaction products then depart. Figure 27.13 shows a computer-drawn structure of an enzyme (red) with its substrate (blue) occupying its active site.

The lock-and-key model for enzyme action goes a long way toward explaining many aspects of enzyme action, but it cannot tell the whole story. As the substrate molecules are drawn into the active site, they are somehow activated so that they are capable of extremely rapid reaction. This activation may result from the withdrawal or donation of electron density at a particular bond by the enzyme. In addition, in the process of fitting into the active site, the molecule may be distorted by the portions of the enzyme protein chain near the active site and thus made more reactive. If coenzymes are involved, the enzyme may promote reaction by binding both the coenzyme and substrate to the same site or to adjacent sites, thus keeping the reactants in close proximity. Enzymes, like all proteins, contain many acidic and basic sites along the chain, and these may be involved in acid-base reactions with the substrate. It is clear that enzyme-coenzyme-substrate interactions can be very complex and are often difficult to unravel.

One of the more interesting corollaries of the lock-and-key model for enzyme action is that certain molecules should be capable of inhibiting the enzyme. Suppose that a certain molecule is capable of fitting the active site on the enzyme but is not reactive for one reason or another. If this molecule is present in the solution along with substrate, it competes with the substrate for binding at the active sites. The substrate is thus prevented from forming the necessary enzyme-substrate complex, and the rate of product formation decreases. Metals that are highly toxic—for example, lead and mercury—probably operate as enzyme inhibitors. Heavy-metal ions are especially strongly bound to sulfur-containing groups on protein side chains. By complexing very strongly to these sites on proteins, they interfere with the normal reactions of enzymes.

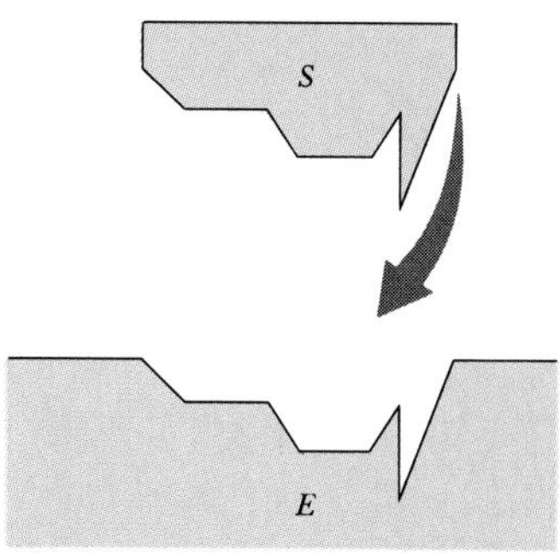

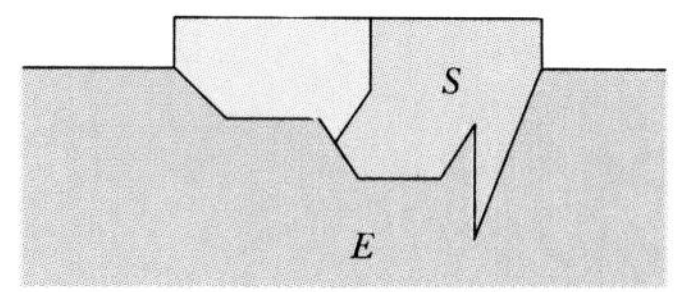

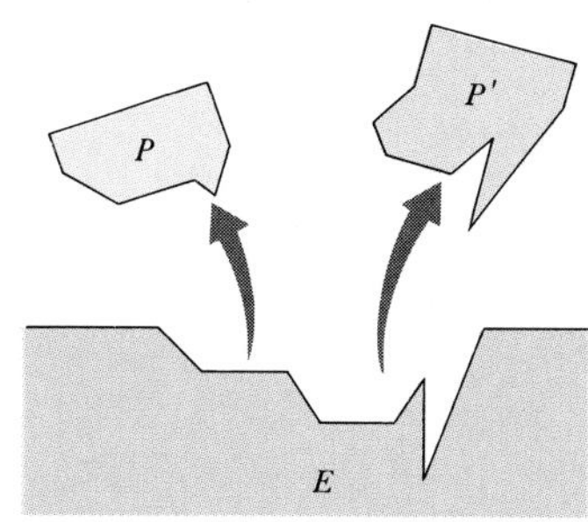

Figure 27.12 Lock-and-key model for formation of the enzyme-substrate complex. Once the substrate (S) is bound to the enzyme (E), one or more products (P) may form and separate from the enzyme.

Figure 27.13 A computer-generated structure of a proteinase enzyme (red), with a substrate (blue) at its active site. The white ribbon shows how the protein chain of the enzyme folds. (From Bott, R., E. Subramanian, and D. R. Davies: "Three-Dimensional Structure of the Complex of the *Rhizopus chinensis* Carboxyl Proteinase and Pepstatin at 2.5-A Resolution." *Biochemistry* (1982), *21*, 6956. Courtesy of Dr. Christine Humblett/Parke-Davis).

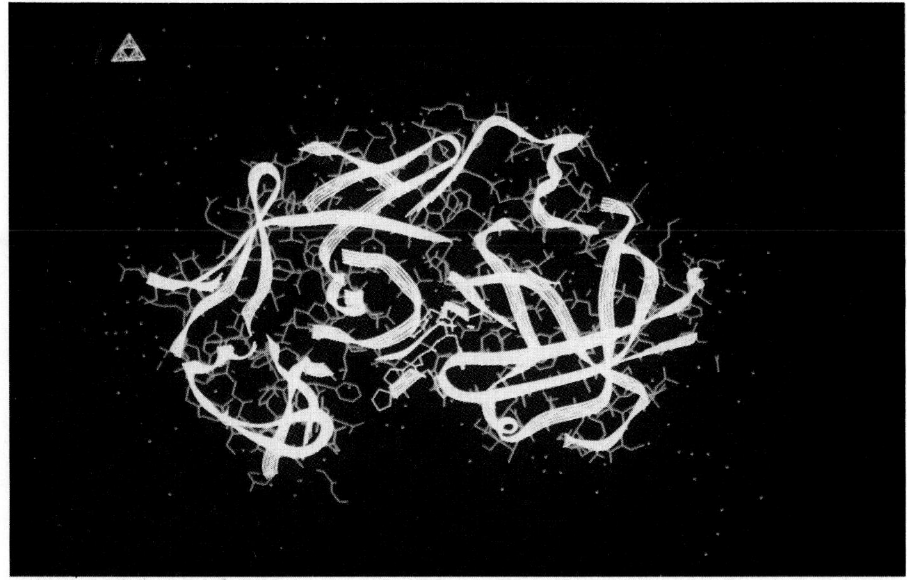

SAMPLE EXERCISE 27.3

Many bacteria employ *para*-aminobenzoic acid (PABA) in a vital enzyme-catalyzed metabolic reaction. Sulfa drugs such as sulfanilamide (*para*-aminobenzenesulfonamide) kill bacteria. Explain this in terms of the concept of the enzyme-substrate complex.

$$H_2N{-}C_6H_4{-}S(=O)_2{-}NH_2 \qquad H_2N{-}C_6H_4{-}C(=O){-}OH$$

Sulfanilamide
(*para*-aminobenzenesulfonamide)

para-aminobenzoic acid (PABA)

Solution: The structural formulas reveal that PABA and sulfanilamide are very similar in shape and in the locations of polar functional groups. The sulfa drug acts as an inhibitor by binding at the active site on the enzyme and blocking access by PABA. Without this needed ingredient, the metabolic processes of the bacteria cease.

PRACTICE EXERCISE

Diisopropylphosphofluoridate (DFP) is a potent nerve poison. It has been found that DFP acts as an inhibitor of enzymatic reactions by binding to the active site. One set of experiments determined that DFP reacts with a serine side chain that is normally involved at the active site. From the structure of the product formed in the reaction of DFP with serine and the structure of DFP itself, suggest the nature of the reaction leading to binding of the inhibitor:

$$[(CH_3)_2HC{-}O]_2P(=O){-}F \qquad {-}HN{-}CH(CH_2{-}O{-}P(=O)[O{-}CH(CH_3)_2]_2){-}C(=O){-}NH{-}$$

DFP

DFP-serine reaction product

Answer: a condensation reaction, leading to formation of HF(*aq*)

27.4 CARBOHYDRATES

Carbohydrates are an important class of naturally occurring substances found in both plant and animal matter. The name **carbohydrate** (hydrate of carbon) comes from the empirical formulas for most substances in this class; they can be written as $C_x(H_2O)_y$. For example, **glucose**, the most abundant carbohydrate, has the molecular formula $C_6H_{12}O_6$, or $C_6(H_2O)_6$. Carbohydrates are not really hydrates of carbon; rather, they are polyhydroxy aldehydes and ketones. For example, glucose is a six-carbon aldehyde sugar, whereas *fructose*, the sugar that occurs widely in fruit, is a six-carbon ketone sugar (Figure 27.14).

Glucose, having both alcohol and aldehyde functional groups and having a reasonably long and flexible backbone, can react with itself to form a six-membered ring structure, as shown in Figure 27.15. Indeed, only a small percentage of the glucose molecules are in the open-chain form in aqueous solution. Although the ring is often drawn in a planar form, the molecules are actually nonplanar, as shown at the bottom of Figure 27.15.

During formation of the ring structure of glucose, the OH group on carbon 1 can be on the same side of the ring as the OH group of carbon 2 (giving the cyclic α form of glucose) or on the opposite side (giving the cyclic β form). Another way to distinguish these two forms is to notice that in the α form the OH group on carbon 1 and the CH_2OH group on carbon 5 point in opposite directions. In the β form, the OH on carbon 1 and the CH_2OH on carbon 5 point in the same direction. Although the difference between the α and β forms might seem small, it has enormous biological consequences. As we shall soon see, this one small change in structure accounts for the vast difference between starch and cellulose.

Fructose can cyclize to form either five- or six-member rings. The five-member ring forms when the OH group on carbon 5 reacts with the carbonyl group on carbon 2:

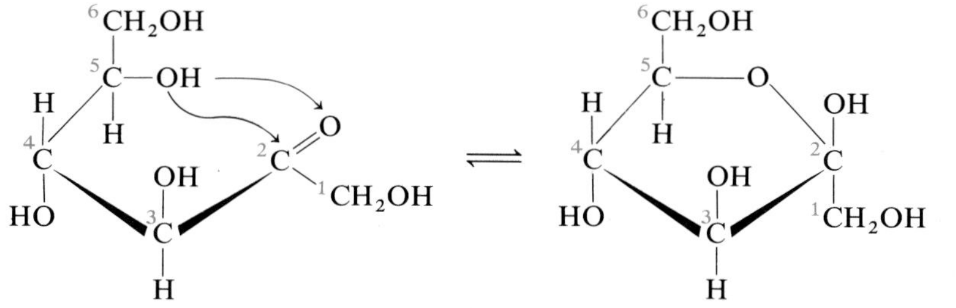

[27.7]

The six-member ring results from the reaction between the OH group on carbon 6 and the carbonyl group on carbon 2.

α-Glucose

β-Glucose

Glucose

Fructose

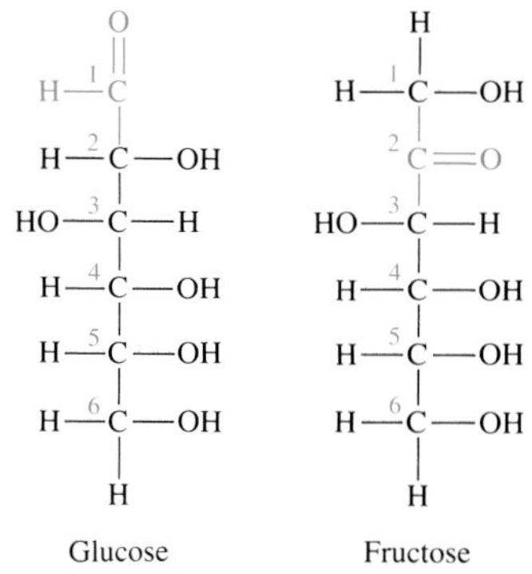

Figure 27.14 Linear structures of glucose and fructose.

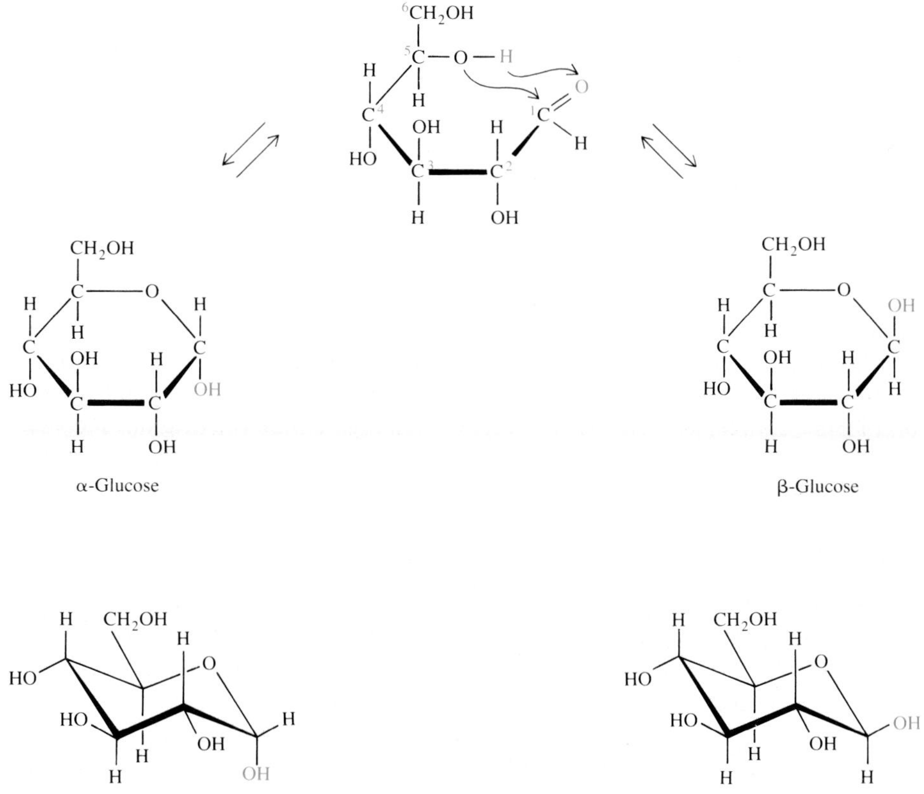

Figure 27.15 Glucose reacts with itself to form two six-member ring structures, designated α and β. Although the structures are often drawn in a planar form (middle), they are actually nonplanar (bottom).

SAMPLE EXERCISE 27.4

How many chiral carbon atoms are in the open-chain form of glucose (Figure 27.14)?

Solution: Notice that the carbon atoms numbered 2, 3, 4, and 5 each have four different groups attached to them, as indicated below:

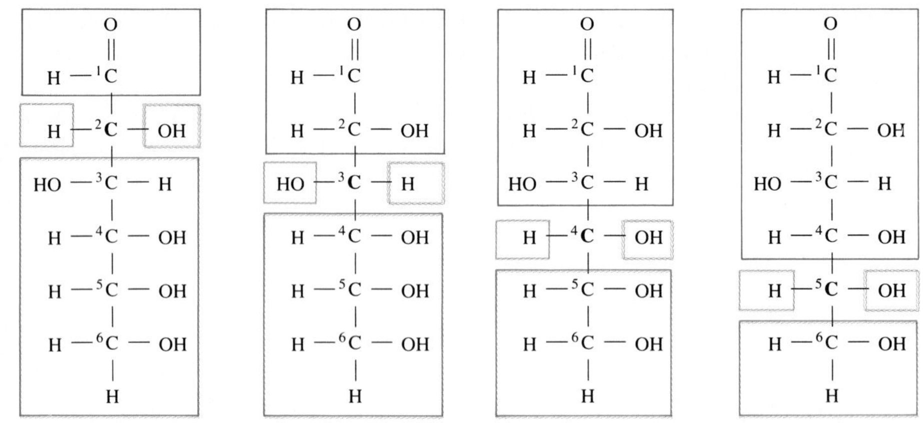

Carbon atoms 1 and 6 have only three different substituents on them. Thus, there are four chiral carbon atoms in the glucose molecule.

PRACTICE EXERCISE

How many chiral carbon atoms are in the open-chain form of fructose (Figure 27.14)?
Answer: three

Disaccharides

Both glucose and fructose are examples of **monosaccharides** (mah-no-SACK-uh-rides), simple sugars that can't be broken into smaller molecules by hydrolysis with aqueous acids. Two monosaccharide units can be linked together by a condensation reaction to form a *disaccharide.* The structures of three common disaccharides—*sucrose* (table sugar), *maltose* (malt sugar), and *lactose* (milk sugar)—are shown in Figure 27.16.

The word *sugar* makes us think of sweetness. All sugars are sweet, but they differ in the degree of sweetness we perceive when we taste them. Sucrose is about six times sweeter than lactose, slightly sweeter than glucose, but only about half as sweet as fructose. Disaccharides can be hydrolyzed; that is, they can be reacted with water in the presence of an acid catalyst to form monosaccharides. When sucrose is hydrolyzed, the mixture of glucose and fructose that forms, called *invert sugar,** is sweeter to the taste than the original sucrose. The sweet syrup present in canned fruits and candies is largely invert sugar formed from hydrolysis of added sucrose.

Polysaccharides

Polysaccharides are made up of several monosaccharide units joined together by a bonding arrangement similar to those shown for disaccharides in Figure 27.16. The most important polysaccharides are starch, glycogen, and cellulose, which are formed from repeating glucose units.

* The term *invert sugar* comes from the fact that rotation of the plane of polarized light by the glucose-fructose mixture is in the opposite direction, or inverted, from that of the sucrose solution.

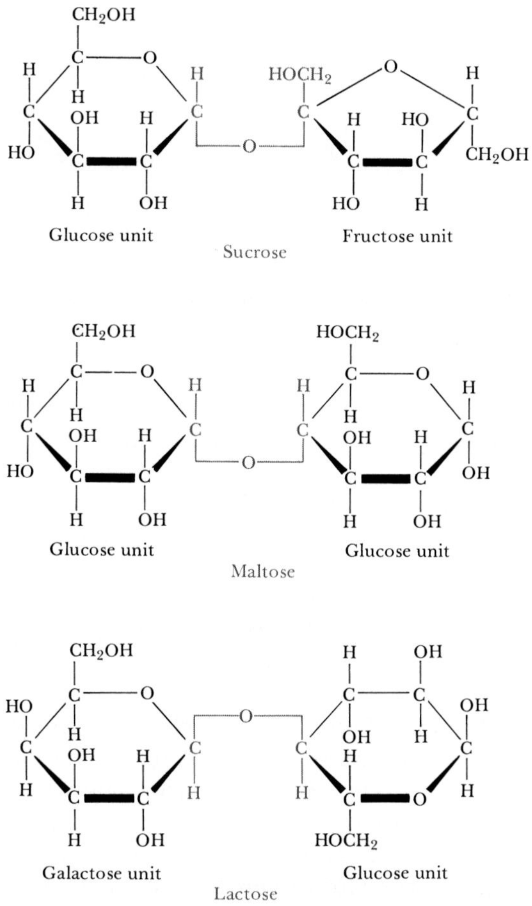

Glucose unit Fructose unit

Sucrose

Glucose unit Glucose unit

Maltose

Galactose unit Glucose unit

Lactose

Figure 27.16 The structures of two disaccharides. Note that in both cases the sugar units are joined by a linkage that is termed the α form.

Starch is not a pure substance. The term refers to a group of polysaccharides found in plants. Starches serve as a major method of food storage in plant seeds and tubers. Corn, potatoes, wheat, and rice all contain substantial amounts of starch. These plant products serve as major sources of needed food energy for humans. Enzymes within the digestive system catalyze the hydrolysis of starch to glucose.

Some starch molecules are unbranched chains, whereas others are branched. Figure 27.17 illustrates an unbranched starch structure. It is particularly important to note that the glucose units that are linked together are in the α form. (Notice that the bridging oxygen atom is opposite the CH_2OH groups.)

Glycogen is a starchlike substance synthesized in the body. Glycogen molecules vary in molecular weight from about 5000 to more than 5 million amu. Glycogen acts as a kind of energy bank in the body. It is concentrated in the

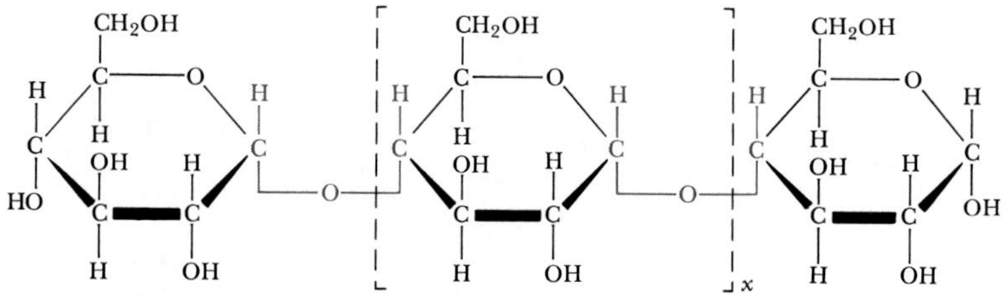

Figure 27.17 Structure of a starch molecule. The molecule consists of many units of the kind enclosed in brackets, joined by linkages of the α form. (You can identify this form by noting that the C—O bonds on the linking carbons are on the opposite side of the ring from the CH_2OH groups.)

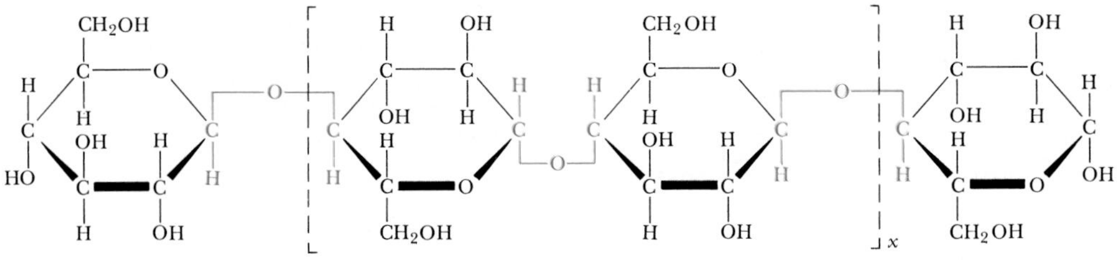

Figure 27.18 Structure of cellulose. Like starch, cellulose is a polymer. The repeating unit is shown between brackets. The linkage in cellulose is of the β form, different from that in starch (see Figure 27.17).

muscles and liver. In muscles, it serves as an immediate source of energy; in the liver, it serves as a storage place for glucose and helps to maintain a constant glucose level in the blood.

Cellulose forms the major structural unit of plants. Wood is about 50 percent cellulose; cotton fibers are almost entirely cellulose. Cellulose consists of a straight chain of glucose units, with molecular weights averaging more than 500,000 amu. The structure of cellulose is shown in Figure 27.18. At first glance, this structure looks very similar to that of starch. However, the differences in the arrangements of bonds that join the glucose units are important. Notice that glucose in cellulose is in the β form.

The distinction between starch and cellulose is made clearer when we examine their structures in a more realistic three-dimensional representation, as shown in Figure 27.19. You can see that the individual glucose units have different relationships to one another in the two structures. Enzymes that readily hydrolyze starches do not hydrolyze cellulose. Thus you might chew up and swallow a pound of cellulose and receive no caloric value from it whatsoever, even though the heat of combustion per unit weight is essentially the same for both cellulose and starch. A pound of starch, on the other hand, would represent a substantial caloric intake. The difference is that the starch is hydrolyzed to glucose, which is eventually oxidized with release of energy. Cellulose, however,

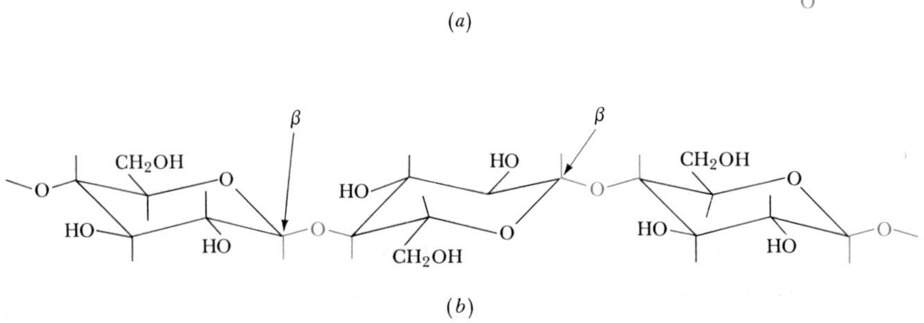

Figure 27.19 Structures of starch (*a*) and cellulose (*b*). These representations show the geometrical arrangements of bonds about each carbon atom. It is easy to see that the glucose rings are oriented differently with respect to one another in the two structures.

is not hydrolyzed by any enzymes present in the body, and so it passes through the body unchanged. Many bacteria contain enzymes, called cellulases, that hydrolyze cellulose. These bacteria are present in the digestive systems of grazing animals, such as cattle, that use cellulose for food.

27.5 FATS AND OILS

Both plant and animal systems need means of storing energy in various chemical forms so that this energy supply can be tapped as needed later. In plant seeds, the stored energy is used to promote rapid growth after germination. In animals that hibernate during cold periods, the stored energy is needed when other sources of food are absent or scarce. One of the most important classes of compounds used for energy storage consists of **fats** and **oils**. These substances are esters of long-chain carboxylic acids with 1,2,3-trihydroxypropane (glycerol) and are called **triglycerides**. Their general structural formula is shown in Figure 27.20. The groups R_1, R_2, and R_3 can be alike or different. Triglycerides are found as fat deposits in animals and as oils in nuts and seeds. Those that are liquid at room temperature are generally referred to as oils; those that are solid are called fats.

Hydrolysis of a fat or oil yields glycerol and the carboxylic acids of the R_1, R_2, and R_3 chains. The acids are commonly called *fatty acids* and usually contain from 12 to 22 carbons. It is an interesting fact that nearly all fatty acids contain an even number of carbon atoms, including the carboxylic acid carbon. Several of the more common fatty acids are listed in Table 27.1. Oils, which are obtained mainly from plant products (corn, peanuts, and soybeans), are formed primarily from unsaturated fatty acids. Animal fats (beef, butter, and pork) contain mainly saturated fatty acids.

Some of the oil extracted from cottonseed, corn, or soybean is used as liquid cooking oil. Other plant-derived oils are hydrogenated to convert the carbon-carbon double bonds in the acid chains to carbon-carbon single bonds. In the process, the oils are converted to solids; they are used to produce oleomargarine, peanut butter, vegetable shortening, and similar food products. As an example, trilinolein, a major constituent of cottonseed oil, is hydrogenated to form tristearin, which is solid at room temperature:

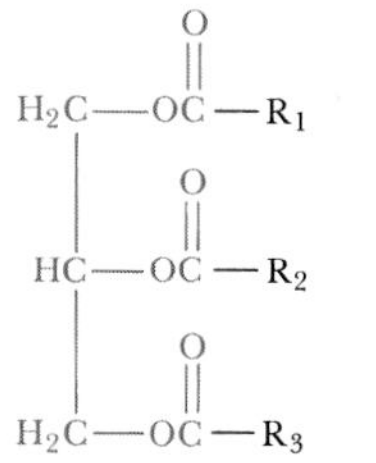

$$\begin{array}{l} H_2C-O\overset{O}{\overset{\|}{C}}-R_1 \\ \;\;| \\ HC-O\overset{O}{\overset{\|}{C}}-R_2 \\ \;\;| \\ H_2C-O\overset{O}{\overset{\|}{C}}-R_3 \end{array}$$

Figure 27.20 Structure of triglycerides. The groups R_1, R_2, and R_3 are hydrocarbon chains of from 3 to 21 carbons. The chains may be entirely saturated or may contain one or more cis olefin groups.

Table 27.1 Structure, Name, and Source of Several Fatty Acids

Formula or structure[a]	Name	Source
$CH_3(CH_2)_{10}COOH$	Lauric acid	Coconuts
$CH_3(CH_2)_{12}COOH$	Myristic acid	Butter
$CH_3(CH_2)_{16}COOH$	Stearic acid	Animal fats
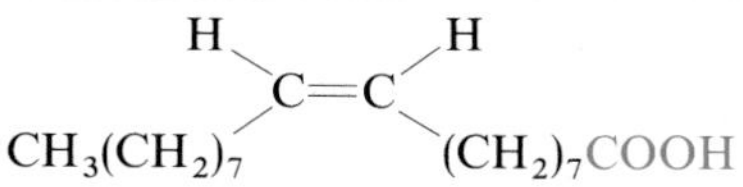 $CH_3(CH_2)_7CH=CH(CH_2)_7COOH$ (cis)	Oleic acid	Corn oil
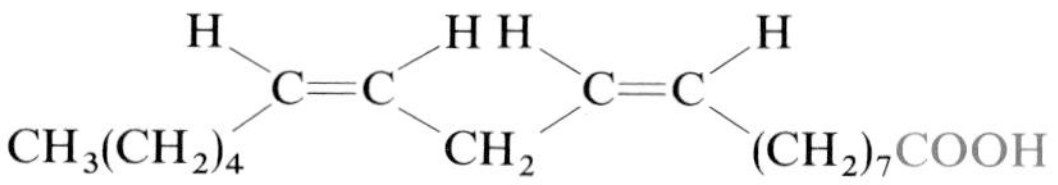 $CH_3(CH_2)_4CH=CHCH_2CH=CH(CH_2)_7COOH$ (cis)	Linoleic acid	Vegetable oils
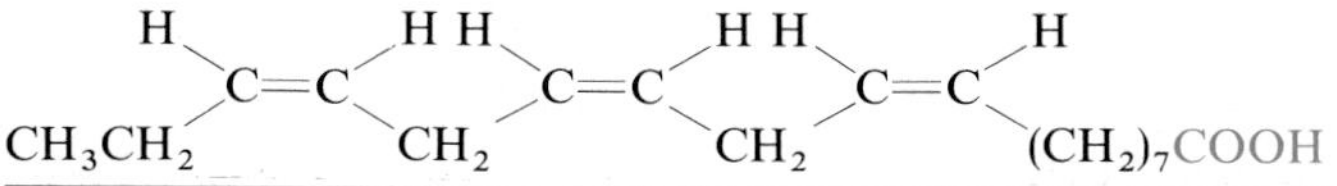 $CH_3CH_2CH=CHCH_2CH=CHCH_2CH=CH(CH_2)_7COOH$ (cis)	Linolenic acid	Linseed oil

[a] Notice that the *cis* configuration is adopted at the double bonds in the unsaturated fatty acids.

$$
\begin{array}{lcl}
\quad\quad\quad\quad\quad\quad\quad\quad\quad\quad\quad\quad\quad\quad\quad\quad\quad\quad\text{O} & & \quad\quad\quad\quad\quad\text{O} \\
CH_3(CH_2)_4{-}CH{=}CH{-}CH_2{-}CH{=}CH{-}(CH_2)_7\overset{\|}{C}OCH_2 & & CH_3(CH_2)_{16}\overset{\|}{C}OCH_2 \\
CH_3(CH_2)_4{-}CH{=}CH{-}CH_2{-}CH{=}CH{-}(CH_2)_7\overset{\overset{O}{\|}}{C}OCH & \xrightarrow[Ni]{6H_2} & CH_3(CH_2)_{16}\overset{\overset{O}{\|}}{C}OCH \\
CH_3(CH_2)_4{-}CH{=}CH{-}CH_2{-}CH{=}CH{-}(CH_2)_7\overset{\overset{O}{\|}}{C}OCH_2 & & CH_3(CH_2)_{16}\overset{\overset{O}{\|}}{C}OCH_2 \\
\text{Trilinolein (liquid)} & & \text{Tristearin (mp 71°C)}
\end{array}
\quad [27.8]
$$

Fats and oils are an important source of energy in our food supply. In the body they undergo hydrolysis to form glycerol and carboxylic acids. The hydrolysis is promoted by enzymes called *lipases:*

$$
\begin{array}{l}
H_2C{-}O\overset{\overset{O}{\|}}{C}{-}R \\
\;| \\
HC{-}O\overset{\overset{O}{\|}}{C}{-}R + 3H_2O \\
\;| \\
H_2C{-}O\overset{\overset{O}{\|}}{C}{-}R
\end{array}
\xrightarrow{\text{lipase}}
\begin{array}{l}
H_2C{-}OH \\
\;| \\
HC{-}OH \\
\;| \\
H_2C{-}OH
\end{array}
+ 3R{-}\overset{\overset{O}{\|}}{C}{-}OH \quad [27.9]
$$

This hydrolysis reaction, which takes place in aqueous solution, is hampered by the fact that fats and oils are essentially insoluble in water. As a result, not much hydrolysis occurs in the stomach. The gallbladder excretes compounds called bile salts into the small intestine to assist in the hydrolysis. The bile salts break up the larger droplets of fats into an emulsion (a suspension of very small droplets) so that hydrolysis can proceed more rapidly.

27.6 NUCLEIC ACIDS

Nucleic acids are a class of biopolymers that are the chemical carriers of an organism's genetic information. **Deoxyribonucleic acids (DNA)** are huge molecules (Figure 27.21) whose molecular weights may range from 6 to 16 million amu. **Ribonucleic acids (RNA)** are smaller molecules, with molecular weights in the range of 20,000 to 40,000 amu. Whereas DNA is found primarily in the nucleus of the cell, RNA is found outside the nucleus in the surrounding fluid called the *cytoplasm.* The DNA stores the genetic information of the cell and controls the production of proteins. The RNA carries the information stored by DNA out of the nucleus of the cell into the cytoplasm, where the information can be used in protein synthesis.

The monomers of nucleic acids, called **nucleotides**, consist of three parts:

1. A phosphoric acid molecule, H_3PO_4
2. A five-carbon sugar
3. A nitrogen-containing organic base.

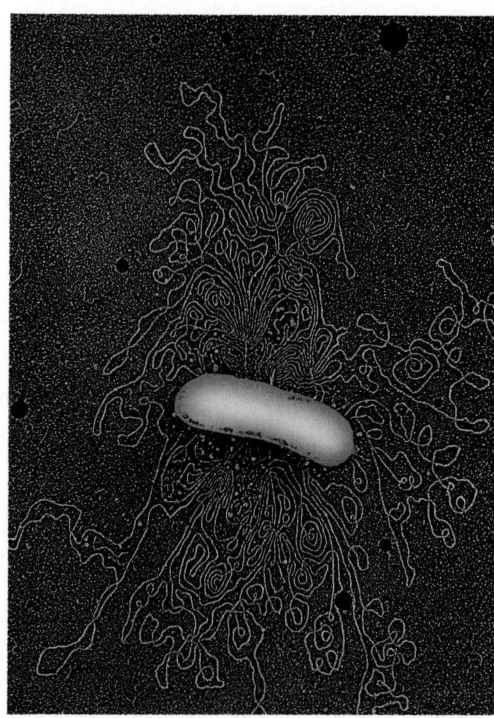

Figure 27.21 The DNA molecule is very large. In this micrograph, a DNA molecule is seen to spill out of the damaged cell wall of a bacterium. (© Dr. Gopal Murti, Science Photo Library, Science Source/Photo Researchers)

The sugar component of RNA is *ribose*, whereas that in DNA is *deoxyribose:*

Ribose   Deoxyribose

Notice that *deoxy*ribose differs from ribose only in having one less oxygen atom (at carbon 2). The following nitrogen bases are the only ones found in DNA and RNA:

Adenine (A)	Guanine (G)	Cytosine (C)	Thymine (T)	Uracil (U)
DNA	DNA	DNA	DNA	RNA
RNA	RNA	RNA		

The base is attached to a ribose or deoxyribose molecule through a bond to the nitrogen atom shown in color. An example of a nucleotide in which the base is adenine and the sugar is deoxyribose is shown in Figure 27.22.

Figure 27.22 Structure of deoxyadenylic acid, a nucleotide formed from phosphoric acid, deoxyribose, and an organic base, adenine.

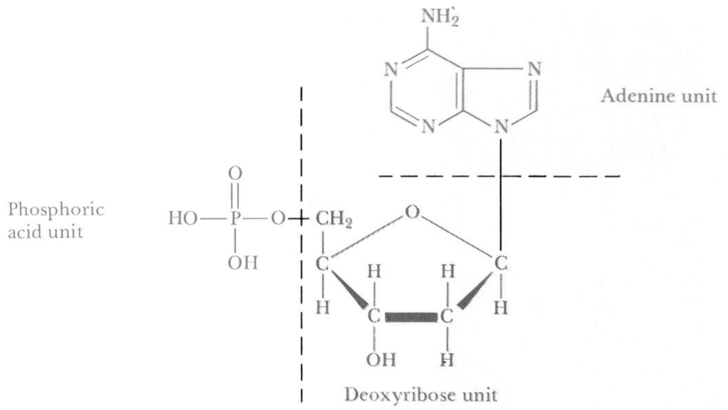

Adenine unit

Phosphoric acid unit

Deoxyribose unit

Nucleic acids are polynucleotides formed by condensation reactions between an OH group of the phosphoric acid on one nucleotide and an OH group of the sugar of another nucleotide. Figure 27.23 shows a portion of the polymeric chain of a DNA molecule.

Figure 27.23 Structure of a polynucleotide. Because the sugar in each nucleotide is deoxyribose, this polynucleotide is of the form found in DNA.

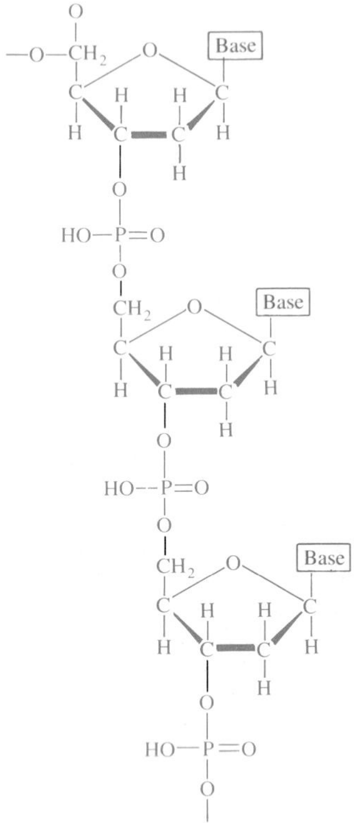

DNA molecules consist of two deoxyribonucleic acid chains or strands that are wound together in the form of a **double helix**, as shown in Figure 27.24. The drawing on the right [Figure 27.24(*b*)] has been simplified to show the essential features of the structure. The sugar and phosphate groups, which form the backbone of each strand, are represented as —S— and —P—, respectively. The bases (represented by the letters T, A, C, and G) are attached to the sugars.

Figure 27.24 (*a*) A computer-generated model of a DNA double helix. The blue and white atoms represent the sugar-phosphate chains that wrap around the outside. Inside the chains are the bases, shown in red and yellow. (*b*) A schematic illustration of the double helix showing the hydrogen-bond interactions between complementary base pairs (green symbols). (Richard Feldman/National Institutes of Health)

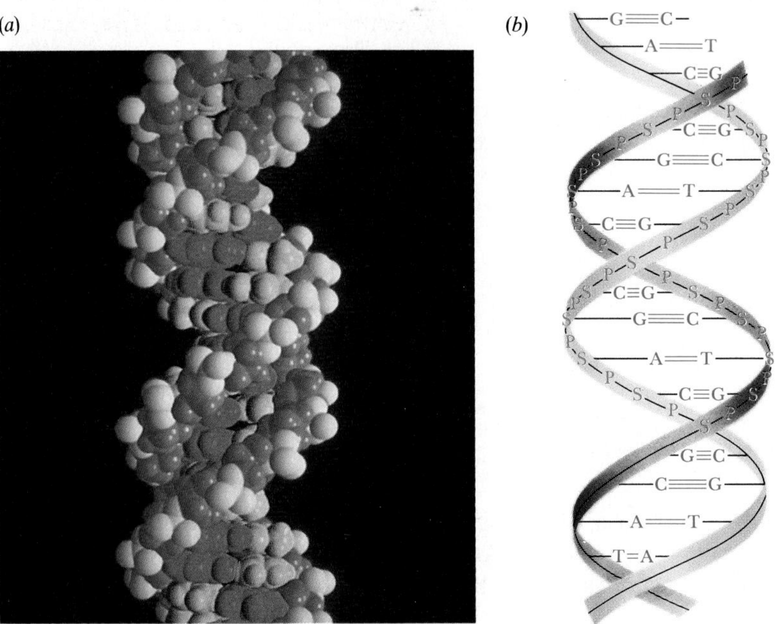

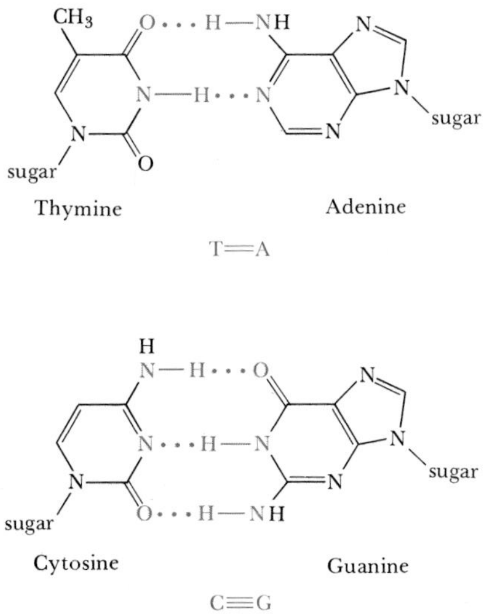

Thymine Adenine

T═A

Cytosine Guanine

C≡G

Figure 27.25 Hydrogen bonding between complementary base pairs. The hydrogen bonds shown here are responsible for formation of the double-stranded helical structure for DNA, as shown in Figure 27.24(*b*).

The two strands are held together by hydrogen bonding between bases on the two strands. As shown in Figure 27.25, the structures of thymine (T) and adenine (A) make them perfect partners for hydrogen bonding. Likewise, cytosine (C) and guanine (G) form ideal hydrogen-bonding partners. In the double-helix structure, each thymine on one strand is always opposite an adenine on the other strand. Likewise, each cytosine is always opposite a thymine. The double-helical structure with complementary bases on the two strands is the key to understanding the functioning of DNA.

Much evidence suggests that the two strands of DNA unwind during cell division and that new complementary strands are constructed on the unraveling strands (Figure 27.26). This process results in two identical double-helix DNA structures, each containing one strand from the original structure and one newly synthesized strand. This replication process allows for the transmission of genetic information when cells divide. The structure of DNA is also the key to understanding protein synthesis, the means by which viruses infect cells, and many other problems of central importance to modern biology. These themes are beyond the scope of this book. However, if you take courses in the area of life sciences, you will learn a good deal about such matters.

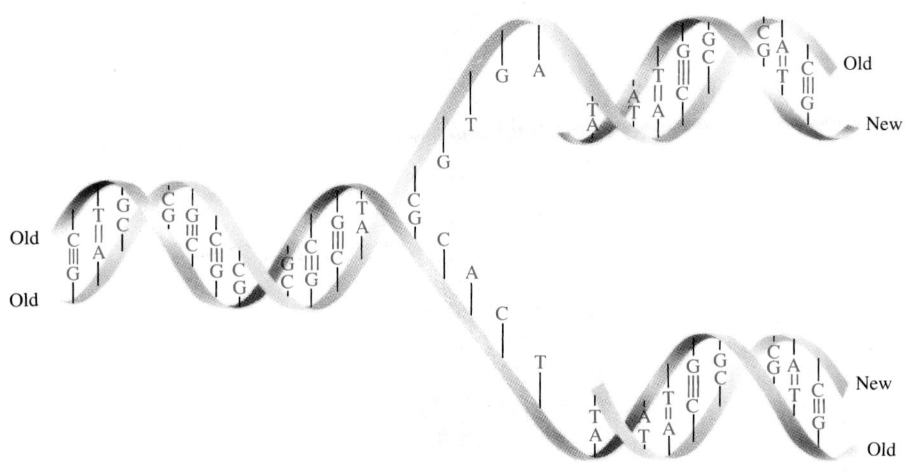

Figure 27.26 A schematic representation of DNA replication. The original DNA double helix partially unwinds, and complementary new nucleotides line up on each strand. When the new nucleotides are joined by condensation reactions, two identical double-helix DNA molecules result.

FOR REVIEW

SUMMARY

In this chapter, we have taken a brief look at some of the major constituents of the biosphere—that part of the physical world in which organisms live out their life cycles. The maintenance of life requires a source of energy as well as appropriate environmental conditions. The ultimate source of the needed energy is the sun. Plants convert solar energy to chemical energy in the process called photosynthesis. Solar energy is absorbed by a plant pigment called chlorophyll and then utilized to form carbohydrate and O_2 from CO_2 and H_2O.

Proteins are polymers of amino acids. They are the major structural materials in animal systems. Enzymes, the catalysts of biochemical reactions, are protein in nature. All naturally occurring proteins are formed from 20 amino acids. Amino acids are chiral substances; that is, they are capable of existing as nonsuperimposable mirror-image isomers called enantiomers. Usually, only one of the enantiomers is found to be biologically active. Protein structure is determined by the sequence of amino acids in the chain, the coiling or stretching of the chain, and the overall shape of the complete molecule. All these aspects of protein structure are important in determining its biological activity.

In this chapter, we also have considered the action of enzymes—their specificity, a model for how they may operate, and the ways in which they are sometimes inhibited.

Carbohydrates, which are formed from polyhydroxy aldehydes and ketones, are the major structural constituent of plants and a source of energy in both plants and animals. The three most important groups of carbohydrates are starch, which is found in plants, glycogen, which is found in mammals, and cellulose, which is also found in plants. All are polysaccharides; that is, they are polymers of a simple sugar, glucose.

Fats and oils, along with proteins and carbohydrates, form the important sources of energy in our food supply. Fats and oils are esters of long-chain acids with glycerol, 1,2,3-trihydroxypropane; they are often called triglycerides. The major sources of these substances are animal fats and the oils of plant seeds, such as corn, peanuts, cottonseeds, and soybeans.

Nucleic acids are biopolymers of high molecular weight that carry the genetic information necessary for cell reproduction. In addition, nucleic acids control cell development through control of protein synthesis. Nucleic acids consist of a polymeric backbone of alternating phosphate and ribose sugar groups, with organic bases attached to the sugar molecules. The DNA polymer is a double-stranded helix held together by hydrogen bonding between matching organic bases situated across from one another on the two strands.

KEY TERMS

biosphere
photosynthesis (Sec. 27.1)
chlorophyll (Sec. 27.1)
biopolymer (Sec. 27.1)
protein (Sec. 27.2)
amino acid (Sec. 27.2)
chiral (Sec. 27.2)
enantiomer (Sec. 27.2)
peptide (Sec. 27.2)
polypeptide (Sec. 27.2)
primary structure (Sec. 27.2)
secondary structure (Sec. 27.2)
α-helix (Sec. 27.2)
tertiary structure (Sec. 27.2)
active site (Sec. 27.3)
substrate (Sec. 27.3)
cofactor (Sec. 27.3)
coenzyme (Sec. 27.3)
lock-and-key model (Sec. 27.3)
carbohydrate (Sec. 27.4)
glucose (Sec. 27.4)
monosaccharide (Sec. 27.4)
polysaccharide (Sec. 27.4)
starch (Sec. 27.4)

glycogen (Sec. 27.4)
cellulose (Sec. 27.4)
fats (Sec. 27.5)
oils (Sec. 27.5)
triglycerides (Sec. 27.5)
nucleic acids (Sec. 27.6)
deoxyribonucleic acid (DNA) (Sec. 27.6)
ribonucleic acid (RNA) (Sec. 27.6)
nucleotide (Sec. 27.6)
double helix (Sec. 27.6)

EXERCISES

Energy Requirements

27.1 Explain the relationship between entropy and the energy needs of organisms.

27.2 Name at least four processes occurring in an organism (such as yourself) that require energy. These may occur at the molecular level, or they may involve the complete organism.

27.3 Given that ΔG°_{298} for oxidation of glucose in solution is -2878 kJ:

$$C_6H_{12}O_6(aq) + 6O_2(g) \longrightarrow 6CO_2(g) + 6H_2O(l) \qquad \Delta G^\circ_{298} = -2878 \text{ kJ}$$

(a) What is ΔG°_{298} for photosynthesis? Is this reaction spontaneous under standard conditions? **(b)** What is ΔG_{298} for photosynthesis if $P_{O_2} = 0.20$ atm, $P_{CO_2} = 3.1 \times 10^{-4}$ atm, and $[C_6H_{12}O_6] = 1.0 \times 10^{-3}$ *M*? (Assume that H_2O can be omitted from the reaction quotient; refer to Section 19.6, if necessary.)

27.4 Assume that 1×10^{14} kg of carbon is photosynthetically fixed as glucose each year on the earth's surface. If the total solar energy falling on the earth's surface is 4×10^{21} kJ/year, and if 2878 kJ is required to produce a mole of glucose, calculate the percentage of the solar energy absorbed annually by photosynthetic processes.

27.5 A tree might convert about 50 g of CO_2 per day into carbohydrate at its greatest rate of photosynthesis. How many grams of oxygen does the tree produce in 1 day at this rate? How many liters of O_2 (at STP) does this correspond to?

27.6 Assuming that half the 48 mol of photons required in Equation 27.1 are of 430-nm wavelength and that half are of 650-nm wavelength, calculate the energy absorption represented by these 48 photons, on a molar basis. The free-energy change for Equation 27.1 is $+2878$ kJ. What is the approximate percentage efficiency of the plant system in converting the incoming radiant energy into glucose, assuming that all incoming photons are employed in the photosynthetic reaction? (See Sample Exercise 6.2.)

Proteins

27.7 **(a)** What is an α-amino acid? **(b)** How do amino acids react to form proteins?

27.8 How do the side chains (R groups) of amino acids affect their behavior?

27.9 Draw the dipeptides formed by a condensation reaction between glycine and valine.

27.10 Write a chemical equation for the formation of aspartylcysteine from the constituent amino acids.

27.11 Draw the structure of the tripeptide trp-gly-ser.

27.12 What amino acids would be obtained by hydrolysis of the following tripeptide?

```
        O       O       O
        ||      ||      ||
H2NCHCNHCHCNHCHCOH
     |        |        |
    CH3      CH2      CH2
              |        |
             OH       OH
```

27.13 In what form would you expect glycine to exist in basic solution? In acidic solution?

27.14 Draw the structural formula for glutamic acid in **(a)** acidic solution; **(b)** basic solution.

27.15 How many different tripeptides can be made from the two amino acids serine and proline? Give the abbreviations for each of these tripeptides using the three-letter codes for amino acids.

27.16 How many different tripeptides can be made from the amino acids glycine, valine, and alanine? Give the abbreviation for each of these tripeptides using the three-letter codes for amino acids.

27.17 Draw the two enantiomeric forms of aspartic acid.

28.18 Draw the two enantiomeric forms of cysteine.

27.19 Describe the primary, secondary, and tertiary structures of proteins.

27.20 Describe the role of hydrogen bonding in determining the α-helix structure of a protein.

Enzymes

27.21 Normally, the rates of enzyme-catalyzed reactions increase linearly with increase in substrate concentration. What does this tell us about the position of equilibrium in Equation 27.4? Explain.

27.22 In terms of the lock-and-key model, what characteristics must an enzyme inhibitor possess?

27.23 One of the many remarkable enzymes in the human body is carbonic anhydrase, which catalyzes the interconversion of carbonic acid with carbon dioxide and water. If it were not for this enzyme, the body could not rid itself rapidly enough of the CO_2 accumulated by cell metabolism. The enzyme has a molecular weight of 30,000 amu, contains one atom of zinc per protein molecule, and catalyzes the dehydration (release to air) of up to 10^7 CO_2 molecules per second. Which components of this description correspond to the terms *enzyme*, *cofactor*, *substrate*, and *turnover number?*

27.24 What general effects would you expect the following changes to have on the speed of an enzyme-catalyzed reaction: **(a)** lowering the reaction temperature from 37°C to 27°C; **(b)** raising the pH from 7.4 to 10.5; **(c)** adding a heavy-metal salt like $Hg(NO_3)_2$; **(d)** raising the temperature from 37°C to 42°C?

Carbohydrates

27.25 What is the difference between α-glucose and β-glucose? Show the condensation of two glucose molecules to form a disaccharide with α linkages; with β linkages.

27.26 Identify each of the following as an α or a β form of a sugar:

(a)
CH_2OH
HO H O H
OH H
H OH
H OH

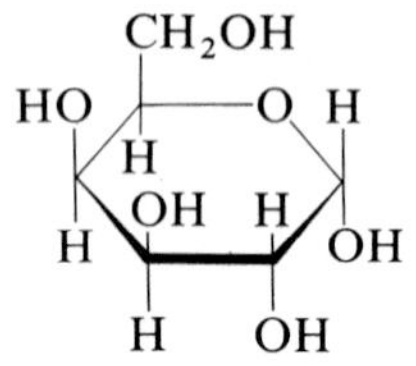

(b)
CH_2OH
O OH
H H
H H
HO OH

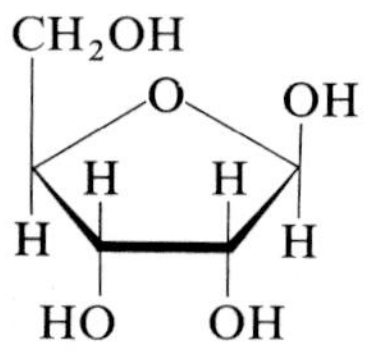

(c)
CH_2OH
H H O OH
OH H
HO H
H OH

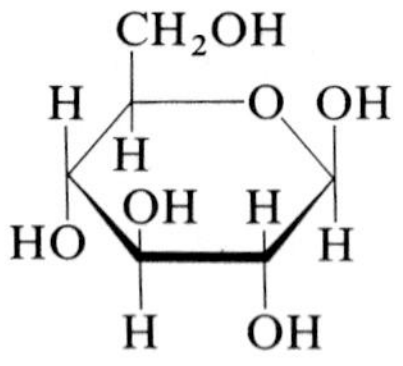

27.27 The structural formula for the linear form of galactose is shown here. Draw the structure of the six-member-ring form of this sugar.

```
      O
      ‖
      CH
      |
   H—C—OH
      |
  HO—C—H
      |
  HO—C—H
      |
   H—C—OH
      |
      CH2OH
```

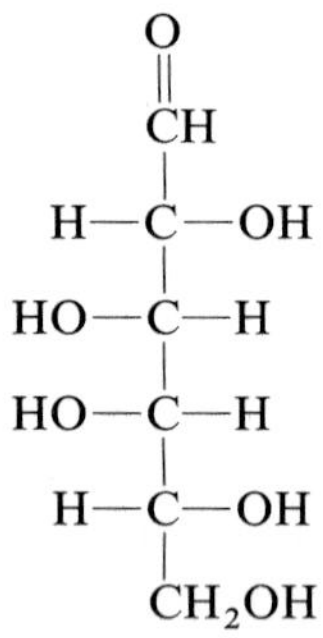

27.28 The structural formula for the linear form of L-sorbose is shown here. Draw the structure of the five-member-ring form of this sugar.

```
      CH2OH
      |
      C=O
      |
  HO—C—H
      |
   H—C—OH
      |
  HO—C—H
      |
      H2COH
```

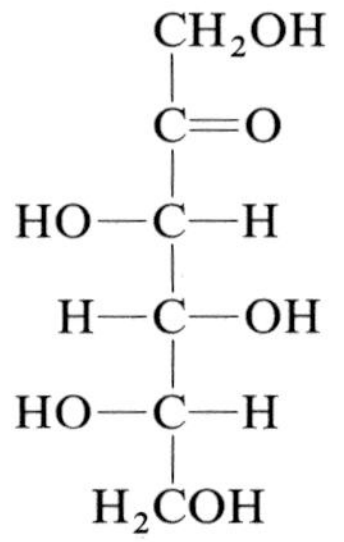

27.29 Which carbon atoms in galactose (see Exercise 27.7) are chiral?

27.30 Which carbon atoms in sorbose (see Exercise 27.28) are chiral?

Fats and Oils

27.31 Draw the structural formula for the triglyceride of oleic acid.

27.32 Write a balanced chemical equation for complete hydrogenation of trilinolein. Assume partial hydrogenation occurs, with uptake of only 4 mol of H_2 per mole of fat. Draw the structural formulas of two possible products.

27.33 The reaction between glycerol and a fatty acid is called *esterification*. Write the chemical equation for the esterification of a mole of glycerol with 3 mol of lauric acid.

27.34 The hydrolysis of fats and oils is catalyzed by bases. Many household cleaners (such as aqueous ammonia solutions and Drāno, which contains NaOH) are basic. Write the chemical equation for the hydrolysis of the triglyceride of myristic acid. (See Table 27.1.)

Nucleic Acids

27.35 Describe a nucleotide. Draw the structural formula for deoxycytidine monophosphate in which cytosine is the organic base.

27.36 A nucleoside consists of an organic base of the kind shown in Section 27.6, bound to ribose or deoxyribose. Draw the structure for thymidine, formed from thymine and deoxyribose.

27.37 Write a balanced chemical equation for the condensation reaction between a mole of deoxyribose and a mole of phosphoric acid.

27.38 An unknown substance undergoes hydrolysis under neutral conditions to yield 1 mol of phosphoric acid and an organic product. The same starting material undergoes hydrolysis under acidic conditions to yield guanine and ribose-monophosphate. Draw the structure of the unknown substance.

27.39 Imagine a single DNA strand containing a section with the following base sequence: A, C, T, C, G, A. What is the base sequence of the complementary strand?

27.40 When samples of double-stranded DNA are analyzed, the quantity of adenine present equals that of thymine. Similarly, the quantity of guanine equals that of cytosine. Explain the significance of these observations.

Additional Exercises

27.41 In a temperate climate, 1.0 m^2 of leaf area absorbs about 2.0×10^4 kJ of energy per day. About 1.2 percent of this energy is used in photosynthesis. **(a)** Calculate the leaf area required to convert 10,000 kJ per day into plant matter. This is the approximate energy requirement of a person doing an average quantity of work per day (Section 4.8). **(b)** In terms of what you know about the composition of plants, explain why more area than this would be required to provide a 10,000-kJ daily diet for a person.

27.42 In each of the following substances, locate the chiral carbon atoms, if any:

(a) 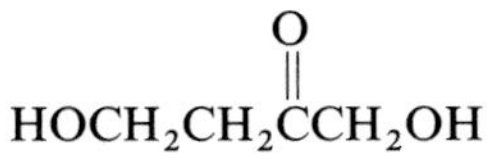 $HOCH_2CH_2\overset{\overset{\Large O}{\|}}{C}CH_2OH$

(b) $HOCH_2\overset{\overset{\Large OH}{|}}{C}H\underset{\underset{\Large O}{\|}}{C}CH_2OH$

(c) $HO\overset{\overset{\Large O}{\|}}{C}\underset{\underset{\Large NH_2}{|}}{C}H\overset{\overset{\Large CH_3}{|}}{C}HC_2H_5$

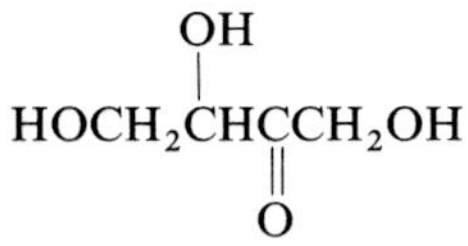

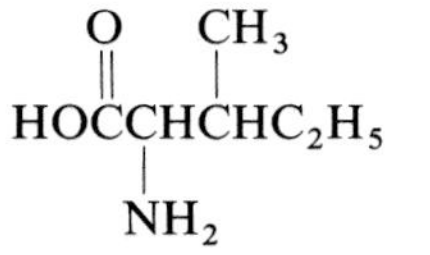

27.43 Predict the products of the hydrolysis of each of the following compounds:

(a) 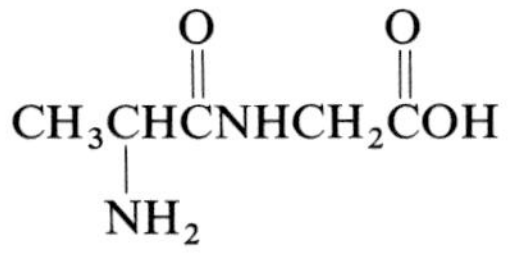 $CH_3\underset{\underset{\Large NH_2}{|}}{C}H\overset{\overset{\Large O}{\|}}{C}NHCH_2\overset{\overset{\Large O}{\|}}{C}OH$

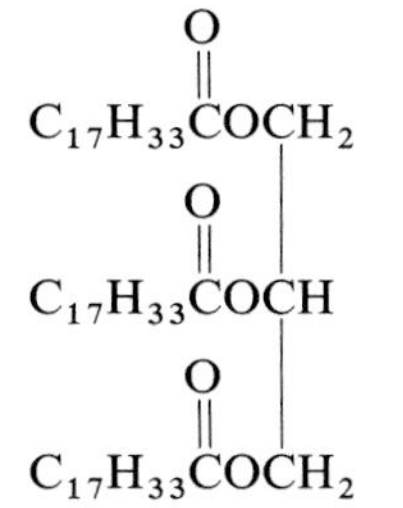

(b)
$C_{17}H_{33}\overset{\overset{\Large O}{\|}}{C}OCH_2$
$C_{17}H_{33}\overset{\overset{\Large O}{\|}}{C}OCH$
$C_{17}H_{33}\overset{\overset{\Large O}{\|}}{C}OCH_2$

(c) [Disaccharide ring structure: two pyranose rings, each with CH_2OH, H, O, H, OH, H, H, OH substituents; left ring with HO, joined by —O— to right ring bearing OH]

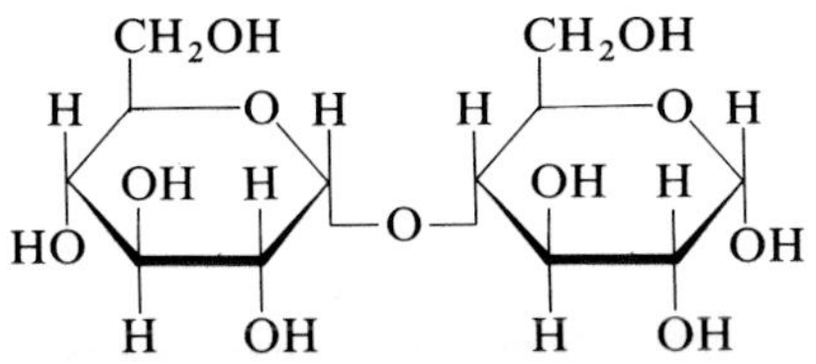

(d) 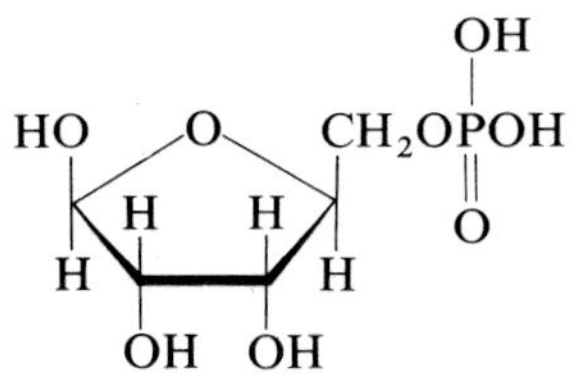 [Furanose ring structure: HO, O, CH_2OPOH (with OH above and =O below P), H, H, H, H, OH, OH]

27.44 Draw the condensed structural formula of each of the following tripeptides: **(a)** val-gly-thr; **(b)** pro-ser-ala.

27.45 Glutathione is a tripeptide found in most living cells. Partial hydrolysis yields cys-gly and glu-cys. What structures are possible for glutathione?

27.46 Phenylketonuria (PKU) is a disease caused by the lack in some individuals of the enzyme phenylalanine hydroxylase. This enzyme catalyzes the conversion of phenylalanine to tyrosine (both amino acids). The disease can lead to severe mental retardation. **(a)** Write the condensed structural formulas for phenylalanine and tyrosine. **(b)** Suggest the origin for the name given to the enzyme.

27.47 The names of most enzymes end in *-ase*. The *-ase* ending is attached to the name of the substrate on which the enzyme acts, or to the type of reaction it catalyzes. Match the following enzyme names and reactions:

1. esterase	**(a)** removal of carboxyl groups from compounds
2. decarboxylase	**(b)** hydrolysis of peptide linkages
3. urease	**(c)** transferral of a methyl group
4. transmethylase	**(d)** formation of ester linkages
5. peptidase	**(e)** hydrolysis of urea

27.48 (a) Describe in qualitative terms how an enzyme works. **(b)** What is meant by *enzyme substrate?* By *enzyme inhibition?*

27.49 The popular flavor enhancer MSG (monosodium glutamate) is the monosodium salt of glutamic acid. **(a)** What is its condensed structural formula? **(b)** Only the L isomer is effective. Is this surprising?

27.50 The standard free energy of formation of glycine(*s*) is -369 kJ/mol, whereas that of glycylglycine(*s*) is -488 kJ/ mol. What is $\Delta G°$ for condensation of glycine to form glycylglycine?

27.51 Give the linear structural formulas for ribose and deoxyribose.

27.52 The enzyme invertase catalyzes the conversion of sucrose, a disaccharide, to invert sugar. When the concentration of invertase is 3×10^{-7} *M* and the concentration of sucrose is 0.01 *M*, invert sugar is formed at the rate of 2×10^{-4} *M*/*s*. When the sucrose concentration is doubled, the rate of formation of invert sugar is doubled also. Assuming that the enzyme-substrate model is operative, is the fraction of enzyme tied up as complex large or small? Explain. Addition of inositol, another sugar, causes a decrease in the rate of formation of invert sugar. Suggest a mechanism by which this occurs.

27.53 Give a specific example of each of the following: **(a)** a disaccharide; **(b)** a sugar present in nucleic acids; **(c)** a sugar present in human blood serum; **(d)** a polysaccharide.

27.54 The standard free energy of formation for aqueous solutions of glucose is -917.2 kJ/mol, whereas that of glycogen is -662.3 kJ/mol of glucose units. Derive a general expression for $\Delta G°$ for the formation of a glycogen molecule that contains *n* units of glucose.

27.55 Define, in your own words, the terms *condensation* and *hydrolysis*. Why are such reactions so important in biochemistry?

27.56 The oil of deep-water fish is rich in omega-3 fatty acids. Some studies have suggested that these fatty acids in a diet lower blood chloresterol levels. The term *omega-3* refers to the location of a carbon-carbon double bond, beginning on the third carbon from the end of the fatty acid. Draw the structure of the omega-3 fatty acid that contains 18 carbon atoms.

27.57 Write a balanced chemical equation for the saponification (base hydrolysis) of tristearin. (See Equation 27.8 for the chemical formula of tristearin.)

27.58 Write a complementary nucleic acid strand for the following strand, using the concept of complementary base pairing: TATGCA.

27.59 The monoanion of adenosine monophosphate (AMP) is an intermediate in phosphate metabolism:

$$\mathrm{A{-}O{-}\overset{\overset{\displaystyle O^-}{|}}{\underset{\underset{\displaystyle O}{\|}}{P}}{-}OH = AMP{-}OH^-}$$

where A = adenosine. If the pK_a for this anion is 7.21, what is the ratio of $[\mathrm{AMP{-}OH^-}]$ to $[\mathrm{AMP{-}O^{2-}}]$ in blood at pH 7.40?

Mathematical Operations A

A.1 EXPONENTIAL NOTATION

The numbers used in chemistry are often either extremely large or extremely small. Such numbers are conveniently expressed in the form

$$N \times 10^n$$

where N is a number between 1 and 10, and n is the exponent. Some examples of this *exponential notation*, which is also called *scientific notation*, follow:

1,200,000 is 1.2×10^6	(read "one point two times ten to the sixth power")
0.000604 is 6.04×10^{-4}	(read "six point zero four times ten to the negative fourth power")

A positive exponent, as in our first example, tells us how many times a number must be multiplied by 10 to give the long form of the number:

$$\begin{aligned} 1.2 \times 10^6 &= 1.2 \times 10 \times 10 \times 10 \times 10 \times 10 \times 10 \quad \text{(six tens)} \\ &= 1{,}200{,}000 \end{aligned}$$

It is also convenient to think of the positive exponent as the number of places the decimal point must be moved to the *left* to obtain a number greater than 1 and less than 10: If we begin with 3450 and move the decimal point three places to left, we end up with 3.45×10^3.

In a related fashion, a negative exponent tells us how many times we must divide a number by 10 to give the long form of the number:

$$6.04 \times 10^{-4} = \frac{6.04}{10 \times 10 \times 10 \times 10} = 0.000604$$

It is convenient to think of the negative exponent as the number of places the decimal point must be moved to the *right* to obtain a number greater than 1 but less than 10: If we begin with 0.0048 and move the decimal point three places to right, we end up with 4.8×10^{-3}.

In the system of exponential notation, with each shift of the decimal point one place to the right, the exponent *decreases* by 1:

$$4.8 \times 10^{-3} = 48 \times 10^{-4}$$

Similarly, with each shift of the decimal point one place to the left, the exponent *increases* by 1:

$$4.8 \times 10^{-3} = 0.48 \times 10^{-2}$$

Most scientific calculators have a key labeled EXP or EE, which is used to enter numbers in exponential notation. To enter the number 5.8×10^3, the key sequence is

[5] [·] [8] [EXP] (or [EE]) [3]

On some calculators the display will show 5.8, then a space, followed by 03, the exponent. On other calculators, a small 10 is shown with an exponent 3.

To enter a negative exponent, use the key labeled +/−. For example, to enter the number 8.6×10^{-5}, the key sequence is

[8] [·] [6] [EXP] [+/−] [5]

When entering a number in exponential notation, do not key in the 10.

In working with exponents, it is important to know that $10^0 = 1$. The following rules are useful for carrying exponents through calculations.

1. **Addition and Subtraction** In order to add or subtract numbers expressed in exponential notation, the powers of 10 must be the same:

$$\begin{aligned}(5.22 \times 10^4) + (3.21 \times 10^2) &= (522 \times 10^2) + (3.21 \times 10^2)\\ &= 525 \times 10^2 \quad \text{(3 significant figures)}\\ &= 5.25 \times 10^4\end{aligned}$$

$$\begin{aligned}(6.25 \times 10^{-2}) - (5.77 \times 10^{-3}) &= (6.25 \times 10^{-2}) - (0.577 \times 10^{-2})\\ &= 5.67 \times 10^{-2} \quad \text{(3 significant figures)}\end{aligned}$$

When you use a calculator to add or subtract, you need not be concerned with having numbers with the same exponents because the calculator automatically takes care of this matter.

2. **Multiplication and Division** When numbers expressed in exponential notation are multiplied, the exponents are added; when numbers expressed in exponential notation are divided, the exponent of the denominator is subtracted from the exponent of the numerator:

$$\begin{aligned}(5.4 \times 10^2)(2.1 \times 10^3) &= (5.4)(2.1) \times 10^{2+3}\\ &= 11 \times 10^5\\ &= 1.1 \times 10^6\end{aligned}$$

$$(1.2 \times 10^5)(3.22 \times 10^{-3}) = (1.2)(3.22) \times 10^{5-3} = 3.9 \times 10^2$$

$$\frac{3.2 \times 10^5}{6.5 \times 10^2} = \frac{3.2}{6.5} \times 10^{5-2} = 0.49 \times 10^3 = 4.9 \times 10^2$$

$$\frac{5.7 \times 10^7}{8.5 \times 10^{-2}} = \frac{5.7}{8.5} \times 10^{7-(-2)} = 0.67 \times 10^9 = 6.7 \times 10^8$$

3. Powers and Roots When numbers expressed in exponential notation are raised to a power, the exponents are multiplied by the power; when the roots of numbers expressed in exponential notation are taken, the exponents are divided by the root:

$$(1.2 \times 10^5)^3 = 1.2^3 \times 10^{5 \times 3}$$
$$= 1.7 \times 10^{15}$$
$$\sqrt[3]{2.5 \times 10^6} = \sqrt[3]{2.5} \times 10^{6/3}$$
$$= 1.3 \times 10^2$$

Scientific calculators usually have keys labeled x^2 and $\sqrt{x}$ for squaring and taking the square root of a number, respectively. To take higher powers or roots, many calculators have y^x and $\sqrt[x]{y}$ (or INV y^x) keys. For example, to perform the operation $\sqrt[3]{7.5 \times 10^{-4}}$ on such a calculator, you would key in 7.5×10^{-4}, press the $\sqrt[x]{y}$ key (or the INV and then the y^x keys), enter the root, 3, and finally press =. The result is 9.1×10^{-2}.

SAMPLE EXERCISE 1

Perform each of the following operations, using your calculator where possible: **(a)** Write the number 0.0054 in standard exponential notation; **(b)** $(5.0 \times 10^{-2}) + (4.7 \times 10^{-3})$; **(c)** $(5.98 \times 10^{12})(2.77 \times 10^{-5})$; **(d)** $\sqrt[4]{1.75 \times 10^{-12}}$.

Solution: **(a)** Because we move the decimal three places to the right to convert 0.0054 to 5.4, the exponent is -3:

$$5.4 \times 10^{-3}$$

Scientific calculators are generally able to convert numbers to exponential notation using one or two key strokes. Consult your instruction manual for how this operation is accomplished on your calculator.

(b) To add these numbers longhand, we must convert them to the same exponent:

$$(5.0 \times 10^{-2}) + (0.47 \times 10^{-2}) = (5.0 + 0.47) \times 10^{-2} = 5.5 \times 10^{-2}$$

(Note that the result has only two significant figures.) To perform this operation on a calculator, we enter the first number, strike the + key, then enter the second number and strike the = key.

(c) Performing this operation longhand, we have

$$(5.98 \times 2.77) \times 10^{12-5} = 16.6 \times 10^7 = 1.66 \times 10^8$$

On a scientific calculator, we enter 5.98×10^{12}, press the × key, enter 2.77×10^{-3}, and press the = key.

(d) To perform this operation on a calculator, we enter the number, press the $\sqrt[x]{y}$ key (or the INV and y^x keys), enter 4, and press the = key. The result is 1.15×10^{-3}.

PRACTICE EXERCISE

Perform the following operations: **(a)** Write 67,000 in exponential notation, showing two significant figures; **(b)** $(3.378 \times 10^{-3}) - (4.97 \times 10^{-5})$; **(c)** $(1.84 \times 10^{15})/(7.45 \times 10^{-2})$; **(d)** $(6.67 \times 10^{-8})^3$. ***Answers:*** **(a)** 6.7×10^4; **(b)** 3.328×10^{-3}; **(c)** 2.47×10^{16}; **(d)** 2.97×10^{-22}

A.2 LOGARITHMS

Common Logarithms

The common, or base-10, logarithm (abbreviated log) of any number is the power to which 10 must be raised to equal the number. For example, the common logarithm of 1000 (written log 1000) is 3 because raising 10 to the third power gives 1000:

$$10^3 = 1000, \text{ therefore } \log 1000 = 3$$

Further examples are

$$\log 10^5 = 5$$
$$\log 1 = 0 \qquad \text{(Remember that } 10^0 = 1\text{)}$$
$$\log 10^{-2} = -2$$

In these examples, the common logarithm can be obtained by inspection. However, it is not possible to obtain the logarithm of a number such as 31.25 by inspection. The logarithm of 31.25 is the number x that satisfies

$$10^x = 31.25$$

Most electronic calculators have a key labeled LOG that can be used to obtain logarithms. For example, we can obtain the value of log 31.25 by entering 31.25 and pressing the LOG key. We obtain the following result:

$$\log 31.25 = 1.4949$$

Notice that 31.25 is greater than 10 (10^1) and less than 100 (10^2). The value for log 31.25 is accordingly between log 10 and log 100, that is, between 1 and 2.

Significant Figures and Common Logarithms

For the common logarithm of a measured quantity, the number of digits after the decimal point equals the number of significant figures in the original number. For example, if 23.5 is a measured quantity (three significant figures), then log 23.5 = 1.371 (three significant figures after the decimal point).

Antilogarithms

The process of determining the number that corresponds to a certain logarithm is known as obtaining an *antilogarithm*. It is the reverse of taking a logarithm. For example, we saw above that log 23.5 = 1.371. This means that the antilogarithm of 1.371 equals 23.5:

$$\log 23.5 = 1.371$$
$$\text{antilog } 1.371 = 23.5$$

The process of taking the antilog of a number is the same as raising 10 to a power equal to that number:

$$\text{antilog } 1.371 = 10^{1.371} = 23.5$$

Many calculators have a key labeled 10^x that allows you to obtain antilogs directly. On others, it will be necessary to press a key labeled INV (for *inverse*), followed by the LOG key.

Natural Logarithms

Logarithms based on the number e are called natural, or base e, logarithms (abbreviated ln). The natural log of a number is the power to which e (which has the value 2.71828 . . .) must be raised to equal the number. For example, the natural log of 10 equals 2.303:

$$e^{2.303} = 10, \text{ therefore } \ln 10 = 2.303$$

Your calculator probably has a key labeled LN that allows you to obtain natural logarithms. For example, to obtain the natural log of 46.8, you enter 46.8 and press the LN key:

$$\ln 46.8 = 3.846$$

The natural antilog of a number is e raised to a power equal to that number. If your calculator can calculate natural logs, it will also be able to calculate natural antilogs. On some calculators, there is a key labeled e^x that allows you to calculate natural antilogs directly; on others, it will be necessary to first press the INV key followed by the LN key. For example, the natural antilog of 1.679 is given by:

$$\text{Natural antilog } 1.679 = e^{1.679} = 5.36$$

The relation between common and natural logarithms is as follows:

$$\ln a = 2.303 \log a$$

Notice that the factor relating the two, 2.303, is the natural log of 10, which we calculated above.

Mathematical Operations Using Logarithms

Because logarithms are exponents, mathematical operations involving logarithms follow the rules for the use of exponents. For example, the product of z^a and z^b (where z is any number) is given by

$$z^a \cdot z^b = z^{(a+b)}$$

Similarly, the logarithm (either common or natural) of a product equals the *sum* of the logs of the individual numbers:

$$\log ab = \log a + \log b \qquad \ln ab = \ln a + \ln b$$

For the log of a quotient:

$$\log (a/b) = \log a - \log b \qquad \ln (a/b) = \ln a - \ln b$$

Using the properties of exponents, we can also derive the rules for the logarithm of a number raised to a certain power:

$$\log a^n = n \log a \qquad \ln a^n = n \ln a$$
$$\log a^{1/n} = (1/n) \log a \qquad \ln a^{1/n} = (1/n) \ln a$$

pH Problems

One of the most frequent uses for common logarithms in general chemistry is in working pH problems. The pH is defined as $-\log [H^+]$, where $[H^+]$ is the hydrogen ion concentration of a solution (Section 16.2). The following sample exercise illustrates this application.

SAMPLE EXERCISE 2

(a) What is the pH of a solution whose hydrogen ion concentration is 0.015 *M*? **(b)** If the pH of a solution is 3.80, what is its hydrogen ion concentration?

Solution: **(a)** We are given the value of $[H^+]$. We use the LOG key of our calculator to calculate the value of $\log [H^+]$. The pH is obtained by changing the sign of the value obtained. (Be sure to change the sign *after* taking the logarithm.):

$$[H^+] = 0.015$$
$$\log [H^+] = -1.82 \qquad \text{(two significant figures)}$$
$$\text{pH} = -(-1.82) = 1.82$$

(b) To obtain the hydrogen ion concentration when given the pH, we must take the antilog of $-$pH:

$$\text{pH} = -\log [H^+] = 3.80$$
$$\log [H^+] = -3.80$$
$$[H^+] = \text{antilog}(-3.80) = 10^{-3.80} = 1.6 \times 10^{-4}\ M$$

A.3 QUADRATIC EQUATIONS

An algebraic equation of the form $ax^2 + bx + c = 0$ is called a *quadratic equation*. The two solutions to such an equation are given by the quadratic formula:

$$x = \frac{-b \pm \sqrt{b^2 - 4ac}}{2a}$$

SAMPLE EXERCISE 3

Find the values of x that satisfy the equation $2x^2 + 4x = 1$.

Solution: To solve the given equation for x, we must first put it in the form

$$ax^2 + bx + c = 0$$

and then use the quadratic formula. If

$$2x^2 + 4x = 1$$

then

$$2x^2 + 4x - 1 = 0$$

Using the quadratic formula, where $a = 2$, $b = 4$, and $c = -1$, we have

$$x = \frac{-4 \pm \sqrt{(4)(4) - 4(2)(-1)}}{2(2)}$$
$$= \frac{-4 \pm \sqrt{16 + 8}}{4} = \frac{-4 \pm \sqrt{24}}{4} = \frac{-4 \pm 4.899}{4}$$

The two solutions are

$$x = \frac{0.899}{4} = 0.225$$

$$x = \frac{-8.899}{4} = -2.225$$

Often in chemical problems the negative solution has no physical meaning, and only the positive answer is used.

A.4 GRAPHS

Often the clearest way to represent the interrelationship between two variables is to graph them. Usually, the variable that is being experimentally varied, called the *independent variable*, is shown along the horizontal axis (x-axis). The variable that responds to the change in the independent variable, called the *dependent variable*, is then shown along the vertical axis (y-axis). For example, consider an experiment in which we vary the temperature of an enclosed gas and measure its pressure. The independent variable is temperature, and the dependent variable is pressure. The data shown in Table 1 could be obtained by means of this experiment. These data are shown graphically in Figure 1. The relationship between temperature and pressure is linear. The equation for any straight-line graph has the form

$$y = mx + b$$

where m is the slope of the line, and b is the intercept with the y-axis. In the case of Figure 1, we could say that the relationship between temperature and pressure takes the form

$$P = mT + b$$

where P is pressure in atm and T is temperature in °C. As shown on Figure 1, the slope is 4.10×10^{-4} atm/°C, and the intercept—the point where the line crosses the y-axis—is 0.112 atm. Therefore, the equation for the line is

$$P = \left(4.10 \times 10^{-4} \frac{\text{atm}}{{}^\circ\text{C}}\right) T + 0.112 \text{ atm}$$

Table 1 Interrelation Between Pressure and Temperature

Temperature (°C)	Pressure (atm)
20.0	0.120
30.0	0.124
40.0	0.128
50.0	0.132

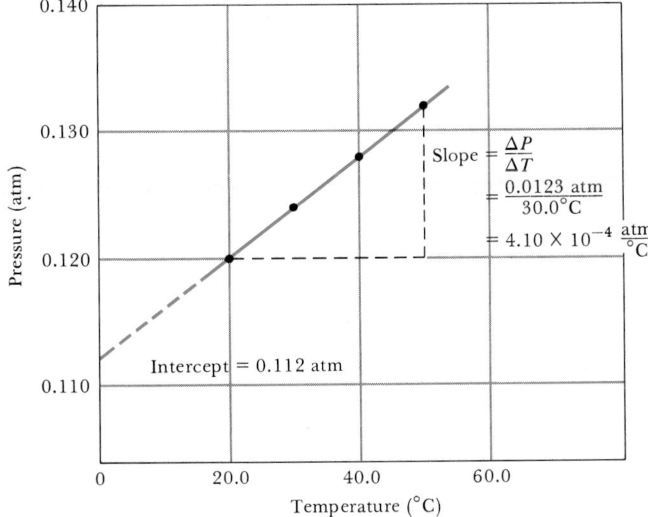

Figure 1

B Properties of Water

Density:	0.99987 g/mL at 0°C 1.00000 g/mL at 4°C 0.99707 g/mL at 25°C 0.95838 g/mL at 100°C
Heat of fusion:	6.008 kJ/mol at 0°C
Heat of vaporization:	44.94 kJ/mol at 0°C 44.02 kJ/mol at 25°C 40.67 kJ/mol at 100°C
Ion-product constant, K_w:	1.14×10^{-15} at 0°C 1.01×10^{-14} at 25°C 5.47×10^{-14} at 50°C
Specific heat:	Ice (−3°C)—2.092 J/g-°C Water at 14.5°C—4.184 J/g-°C Steam (100°C)—1.841 J/g-°C

Vapor pressure (mm Hg):

***T*(°C)**	***P***	***T*(°C)**	***P***	***T*(°C)**	***P***	***T*(°C)**	***P***
0	4.58	21	18.65	35	42.2	92	567.0
5	6.54	22	19.83	40	55.3	94	610.9
10	9.21	23	21.07	45	71.9	96	657.6
12	10.52	24	22.38	50	92.5	98	707.3
14	11.99	25	23.76	55	118.0	100	760.0
16	13.63	26	25.21	60	149.4	102	815.9
17	14.53	27	26.74	65	187.5	104	875.1
18	15.48	28	28.35	70	233.7	106	937.9
19	16.48	29	30.04	80	355.1	108	1004.4
20	17.54	30	31.82	90	525.8	110	1074.6

Thermodynamic Quantities for Selected Substances at 298.15 K (25°C)

Substance	ΔH_f° (kJ/mol)	ΔG_f° (kJ/mol)	S° (J/mol-K)
Aluminum			
$Al(s)$	0	0	28.32
$AlCl_3(s)$	−705.6	−630.0	109.3
$Al_2O_3(s)$	−1669.8	−1576.5	51.00
Barium			
$Ba(s)$	0	0	63.2
$BaCO_3(s)$	−1216.3	−1137.6	112.1
$BaO(s)$	−553.5	−525.1	70.42
Beryllium			
$Be(s)$	0	0	9.44
$BeO(s)$	−608.4	−579.1	13.77
$Be(OH)_2(s)$	−905.8	−817.9	50.21
Bromine			
$Br(g)$	111.8	82.38	174.9
$Br^-(aq)$	−120.9	−102.8	80.71
$Br_2(g)$	30.71	3.14	245.3
$Br_2(l)$	0	0	152.3
$HBr(g)$	−36.23	−53.22	198.49
Calcium			
$Ca(g)$	179.3	145.5	154.8
$Ca(s)$	0	0	41.4
$CaCO_3$ (*s*, calcite)	−1207.1	−1128.76	92.88
$CaCl_2(s)$	−795.8	−748.1	104.6
$CaF_2(s)$	−1219.6	−1167.3	68.87
$CaO(s)$	−635.5	−604.17	39.75
$Ca(OH)_2(s)$	−986.2	−898.5	83.4
$CaSO_4(s)$	−1434.0	−1321.8	106.7
Carbon			
$C(g)$	718.4	672.9	158.0
C(*s*, diamond)	1.88	2.84	2.43
C(*s*, graphite)	0	0	5.69
$CCl_4(g)$	−106.7	−64.0	309.4
$CCl_4(l)$	−139.3	−68.6	214.4
$CF_4(g)$	−679.9	−635.1	262.3
$CH_4(g)$	−74.8	−50.8	186.3
$C_2H_2(g)$	226.7	209.2	200.8
$C_2H_4(g)$	52.30	68.11	219.4
$C_2H_6(g)$	−84.68	−32.89	229.5
$C_3H_8(g)$	−103.85	−23.47	269.9
$C_6H_6(g)$	82.9	129.7	269.2
$C_6H_6(l)$	49.0	124.5	172.8
$CH_3OH(g)$	−201.2	−161.9	237.6
$CH_3OH(l)$	−238.6	−166.23	126.8
$C_2H_5OH(g)$	−235.1	−168.5	282.7
$C_2H_5OH(l)$	−277.7	−174.76	160.7
$C_6H_{12}O_6(s)$	−1273.02	−910.4	212.1
$CO(g)$	−110.5	−137.2	197.9
$CO_2(g)$	−393.5	−394.4	213.6
$HC_2H_3O_2(l)$	−487.0	−392.4	159.8
Cesium			
$Cs(g)$	76.50	49.53	175.6
$Cs(s)$	0	0	85.15
$CsCl(s)$	−442.8	−414.4	101.2
Chlorine			
$Cl(g)$	121.7	105.7	165.2
$Cl^-(aq)$	−167.2	−131.2	56.5
$Cl_2(g)$	0	0	222.96
$HCl(aq)$	−167.2	−131.2	56.5
$HCl(g)$	−92.30	−95.27	186.69
Chromium			
$Cr(g)$	397.5	352.6	174.2
$Cr(s)$	0	0	23.6
$Cr_2O_3(s)$	−1139.7	−1058.1	81.2
Cobalt			
$Co(g)$	439	393	179
$Co(s)$	0	0	28.4

Substance	ΔH_f° (kJ/mol)	ΔG_f° (kJ/mol)	S° (J/mol-K)
Copper			
$Cu(g)$	338.4	298.6	166.3
$Cu(s)$	0	0	33.30
$CuCl_2(s)$	−205.9	−161.7	108.1
$CuO(s)$	−156.1	−128.3	42.59
$Cu_2O(s)$	−170.7	−147.9	92.36
Fluorine			
$F(g)$	80.0	61.9	158.7
$F^-(aq)$	−332.6	−278.8	−13.8
$F_2(g)$	0	0	202.7
$HF(g)$	−268.61	−270.70	173.51
Hydrogen			
$H(g)$	217.94	203.26	114.60
$H^+(aq)$	0	0	0
$H^+(g)$	1536.2	1517.0	108.9
$H_2(g)$	0	0	130.58
Iodine			
$I(g)$	106.60	70.16	180.66
$I^-(aq)$	−55.19	−51.57	111.3
$I_2(g)$	62.25	19.37	260.57
$I_2(s)$	0	0	116.73
$HI(g)$	25.94	1.30	206.3
Iron			
$Fe(g)$	415.5	369.8	180.5
$Fe(s)$	0	0	27.15
$Fe^{2+}(aq)$	−87.86	−84.93	113.4
$Fe^{3+}(aq)$	−47.69	−10.54	293.3
$FeCl_2(s)$	−341.8	−302.3	117.9
$FeCl_3(s)$	−400	−334	142.3
$FeO(s)$	−271.9	−255.2	60.75
$Fe_2O_3(s)$	−822.16	−740.98	89.96
$Fe_3O_4(s)$	−1117.1	−1014.2	146.4
$FeS_2(s)$	−171.5	−160.1	52.92
Lead			
$Pb(s)$	0	0	68.85
$PbBr_2(s)$	−277.4	−260.7	161
$PbCO_3(s)$	−699.1	−625.5	131.0
$Pb(NO_3)_2(aq)$	−421.3	−246.9	303.3
$Pb(NO_3)_2(s)$	−451.9	—	—
$PbO(s)$	−217.3	−187.9	68.70
Lithium			
$Li(g)$	159.3	126.6	138.8
$Li(s)$	0	0	29.09
$Li^+(g)$	685.7	648.5	133.0
$LiCl(s)$	−408.3	−384.0	59.30

Substance	ΔH_f° (kJ/mol)	ΔG_f° (kJ/mol)	S° (J/mol-K)
Magnesium			
$Mg(g)$	147.1	112.5	148.6
$Mg(s)$	0	0	32.51
$MgCl_2(s)$	−641.6	−592.1	89.6
$MgO(s)$	−601.8	−569.6	26.8
$Mg(OH)_2(s)$	−924.7	−833.7	63.24
Manganese			
$Mn(g)$	280.7	238.5	173.6
$Mn(s)$	0	0	32.0
$MnO(s)$	−385.2	−362.9	59.7
$MnO_2(s)$	−519.6	−464.8	53.14
$MnO_4^-(aq)$	−541.4	−447.2	191.2
Mercury			
$Hg(g)$	60.83	31.76	174.89
$Hg(l)$	0	0	77.40
$HgCl_2(s)$	−230.1	−184.0	144.5
$Hg_2Cl_2(s)$	−264.9	−210.5	192.5
Nickel			
$Ni(g)$	429.7	384.5	182.1
$Ni(s)$	0	0	29.9
$NiCl_2(s)$	−305.3	−259.0	97.65
$NiO(s)$	−239.7	−211.7	37.99
Nitrogen			
$N(g)$	472.7	455.5	153.3
$N_2(g)$	0	0	191.50
$NH_3(aq)$	−80.29	−26.50	111.3
$NH_3(g)$	−46.19	−16.66	192.5
$NH_4^+(aq)$	−132.5	−79.31	113.4
$N_2H_4(g)$	95.40	159.4	238.5
$NH_4CN(s)$	0.0	—	—
$NH_4Cl(s)$	−314.4	−203.0	94.6
$NH_4NO_3(s)$	−365.6	−184.0	151
$NO(g)$	90.37	86.71	210.62
$NO_2(g)$	33.84	51.84	240.45
$N_2O(g)$	81.6	103.59	220.0
$N_2O_4(g)$	9.66	98.28	304.3
$NOCl(g)$	52.6	66.3	264
$HNO_3(aq)$	−206.6	−110.5	146
$HNO_3(g)$	−134.3	−73.94	266.4
Oxygen			
$O(g)$	247.5	230.1	161.0
$O_2(g)$	0	0	205.0
$O_3(g)$	142.3	163.4	237.6
$OH^-(aq)$	−230.0	−157.3	−10.7
$H_2O(g)$	−241.8	−228.61	188.7
$H_2O(l)$	−285.85	−236.81	69.96
$H_2O_2(g)$	−136.10	−105.48	232.9
$H_2O_2(l)$	−187.8	−120.4	109.6

Substance	ΔH_f° (kJ/mol)	ΔG_f° (kJ/mol)	S° (J/mol-K)
Phosphorus			
$P(g)$	316.4	280.0	163.2
$P_2(g)$	144.3	103.7	218.1
$P_4(g)$	58.9	24.4	280
$P_4(s, \text{red})$	−17.46	−12.03	22.85
$P_4(s, \text{white})$	0	0	41.08
$PCl_3(g)$	−288.07	−269.6	311.7
$PCl_3(l)$	−319.6	−272.4	217
$PF_5(g)$	−1594.4	−1520.7	300.8
$PH_3(g)$	5.4	13.4	210.2
$P_4O_6(s)$	−1640.1	—	—
$P_4O_{10}(s)$	−2940.1	−2675.2	228.9
$POCl_3(g)$	−542.2	−502.5	325
$POCl_3(l)$	−597.0	−520.9	222
$H_3PO_4(aq)$	−1288.3	−1142.6	158.2
Potassium			
$K(g)$	89.99	61.17	160.2
$K(s)$	0	0	64.67
$KCl(s)$	−435.9	−408.3	82.7
$KClO_3(s)$	−391.2	−289.9	143.0
$KClO_3(aq)$	−349.5	−284.9	265.7
$KNO_3(s)$	−492.70	−393.13	288.1
$K_2O(s)$	−363.2	−322.1	94.14
$KO_2(s)$	−284.5	−240.6	122.5
$K_2O_2(s)$	−495.8	−429.8	113.0
$KOH(s)$	−424.7	−378.9	78.91
$KOH(aq)$	−482.4	−440.5	91.6
Rubidium			
$Rb(g)$	85.8	55.8	170.0
$Rb(s)$	0	0	76.78
$RbCl(s)$	−430.5	−412.0	92
$RbClO_3(s)$	−392.4	−292.0	152
Scandium			
$Sc(g)$	377.8	336.1	174.7
$Sc(s)$	0	0	34.6
Selenium			
$H_2Se(g)$	29.7	15.9	219.0
Silicon			
$Si(g)$	368.2	323.9	167.8
$Si(s)$	0	0	18.7
$SiC(s)$	−73.22	−70.85	16.61
$SiCl_4(l)$	−640.1	−572.8	239.3
$SiO_2(s, \text{quartz})$	−910.9	−856.5	41.84
Silver			
$Ag(s)$	0	0	42.55
$Ag^+(aq)$	105.90	77.11	73.93

Substance	ΔH_f° (kJ/mol)	ΔG_f° (kJ/mol)	S° (J/mol-K)
$AgCl(s)$	−127.0	−109.70	96.11
$Ag_2O(s)$	−31.05	−11.20	121.3
$AgNO_3(s)$	−124.4	−33.41	140.9
Sodium			
$Na(g)$	107.7	77.3	153.7
$Na(s)$	0	0	51.45
$Na^+(aq)$	−240.1	−261.9	59.0
$Na^+(g)$	609.3	574.3	148.0
$NaBr(aq)$	−360.6	−364.7	141
$NaBr(s)$	−361.4	−349.3	86.82
$Na_2CO_3(s)$	−1130.9	−1047.7	136.0
$NaCl(aq)$	−407.1	−393.0	115.5
$NaCl(g)$	−181.4	−201.3	229.8
$NaCl(s)$	−410.9	−384.0	72.33
$NaHCO_3(s)$	−947.7	−851.8	102.1
$NaNO_3(aq)$	−446.2	−372.4	207
$NaNO_3(s)$	−467.9	−367.0	116.5
$NaOH(aq)$	−469.6	−419.2	49.8
$NaOH(s)$	−425.6	−379.5	64.46
Strontium			
$SrO(s)$	−592.0	−561.9	54.9
Sulfur			
$S(s, \text{rhombic})$	0	0	31.88
$SO_2(g)$	−296.9	−300.4	248.5
$SO_3(g)$	−395.2	−370.4	256.2
$SO_4^{2-}(aq)$	−909.3	−744.5	20.1
$SOCl_2(l)$	−245.6	—	—
$H_2S(g)$	−20.17	−33.01	205.6
$H_2SO_4(aq)$	−909.3	−744.5	20.1
$H_2SO_4(l)$	−814.0	−689.9	156.1
Titanium			
$Ti(g)$	468	422	180.3
$Ti(s)$	0	0	30.76
$TiCl_4(g)$	−763.2	−726.8	354.9
$TiCl_4(l)$	−804.2	−728.1	221.9
$TiO_2(s)$	−944.7	−889.4	50.29
Vanadium			
$V(g)$	514.2	453.1	182.2
$V(s)$	0	0	28.9
Zinc			
$Zn(g)$	130.7	95.2	160.9
$Zn(s)$	0	0	41.63
$ZnCl_2(s)$	−415.1	−369.4	111.5
$ZnO(s)$	−348.0	−318.2	43.9

D Aqueous-Equilibrium Constants

D.1 Dissociation Constants for Acids at 25°C

Name	Formula	K_{a1}	K_{a2}	K_{a3}
Acetic	$HC_2H_3O_2$	1.8×10^{-5}		
Arsenic	H_3AsO_4	5.6×10^{-3}	1.0×10^{-7}	3.0×10^{-12}
Arsenous	H_3AsO_3	6×10^{-10}		
Ascorbic	$HC_6H_7O_6$	8.0×10^{-5}	1.6×10^{-12}	
Benzoic	$HC_7H_5O_2$	6.5×10^{-5}		
Boric	H_3BO_3	5.8×10^{-10}		
Carbonic	H_2CO_3	4.3×10^{-7}	5.6×10^{-11}	
Chloroacetic	$HC_2H_2O_2Cl$	1.4×10^{-3}		
Citric	$H_3C_6H_5O_7$	7.4×10^{-4}	1.7×10^{-5}	4.0×10^{-7}
Cyanic	$HCNO$	3.5×10^{-4}		
Formic	$HCHO_2$	1.8×10^{-4}		
Hydroazoic	HN_3	1.9×10^{-5}		
Hydrocyanic	HCN	4.9×10^{-10}		
Hydrofluoric	HF	6.8×10^{-4}		
Hydrogen chromate ion	$HCrO_4^-$	3.0×10^{-7}		
Hydrogen peroxide	H_2O_2	2.4×10^{-12}		
Hydrogen selenate ion	$HSeO_4^-$	2.2×10^{-2}		
Hydrogen sulfide	H_2S	5.7×10^{-8}	1.3×10^{-13}	
Hypobromous	$HBrO$	2×10^{-9}		
Hypochlorous	$HClO$	3.0×10^{-8}		
Hypoiodous	HIO	2×10^{-11}		
Iodic	HIO_3	1.7×10^{-1}		
Lactic	$HC_3H_5O_3$	1.4×10^{-4}		
Malonic	$H_2C_3H_2O_4$	1.5×10^{-3}	2.0×10^{-6}	
Nitrous	HNO_2	4.5×10^{-4}		
Oxalic	$H_2C_2O_4$	5.9×10^{-2}	6.4×10^{-5}	
Paraperiodic	H_5IO_6	2.8×10^{-2}	5.3×10^{-9}	
Phenol	HC_6H_5O	1.3×10^{-10}		
Phosphoric	H_3PO_4	7.5×10^{-3}	6.2×10^{-8}	4.2×10^{-13}
Propionic	$HC_3H_5O_2$	1.3×10^{-5}		
Pyrophosphoric	$H_4P_2O_7$	3.0×10^{-2}	4.4×10^{-3}	
Selenous	H_2SeO_3	2.3×10^{-3}	5.3×10^{-9}	
Sulfuric	H_2SO_4	Strong acid	1.2×10^{-2}	
Sulfurous	H_2SO_3	1.7×10^{-2}	6.4×10^{-8}	
Tartaric	$H_2C_4H_4O_6$	1.0×10^{-3}	4.6×10^{-5}	

D.2 Dissociation Constants for Bases at 25°C

Name	Formula	K_b
Ammonia	NH_3	1.8×10^{-5}
Aniline	$C_6H_5NH_2$	4.3×10^{-10}
Dimethylamine	$(CH_3)_2NH$	5.4×10^{-4}
Ethylamine	$C_2H_5NH_2$	6.4×10^{-4}
Hydrazine	H_2NNH_2	1.3×10^{-6}
Hydroxylamine	$HONH_2$	1.1×10^{-8}
Methylamine	CH_3NH_2	4.4×10^{-4}
Pyridine	C_5H_5N	1.7×10^{-9}
Trimethylamine	$(CH_3)_3N$	6.4×10^{-5}

D.3 Solubility-Product Constants for Compounds at 25°C

Name	Formula	K_{sp}
Barium carbonate	$BaCO_3$	5.1×10^{-9}
Barium chromate	$BaCrO_4$	1.2×10^{-10}
Barium fluoride	BaF_2	1.0×10^{-6}
Barium hydroxide	$Ba(OH)_2$	5×10^{-3}
Barium oxalate	BaC_2O_4	1.6×10^{-7}
Barium phosphate	$Ba_3(PO_4)_2$	3.4×10^{-23}
Barium sulfate	$BaSO_4$	1.1×10^{-10}
Cadmium carbonate	$CdCO_3$	5.2×10^{-12}
Cadmium hydroxide	$Cd(OH)_2$	2.5×10^{-14}
Cadmium sulfide	CdS	8.0×10^{-27}
Calcium carbonate	$CaCO_3$	2.8×10^{-9}
Calcium chromate	$CaCrO_4$	7.1×10^{-4}
Calcium fluoride	CaF_2	3.9×10^{-11}
Calcium hydroxide	$Ca(OH)_2$	5.5×10^{-6}
Calcium phosphate	$Ca_3(PO_4)_2$	2.0×10^{-29}
Calcium sulfate	$CaSO_4$	9.1×10^{-6}
Cerium(III) fluoride	CeF_3	8×10^{-16}
Chromium(III) fluoride	CrF_3	6.6×10^{-11}
Chromium(III) hydroxide	$Cr(OH)_3$	6.3×10^{-31}
Cobalt(II) carbonate	$CoCO_3$	1.4×10^{-13}
Cobalt(II) hydroxide	$Co(OH)_2$	1.6×10^{-15}
Cobalt(III) hydroxide	$Co(OH)_3$	1.6×10^{-44}
α-Cobalt(II) sulfide[a]	CoS	4.0×10^{-21}
Copper(I) bromide	$CuBr$	5.3×10^{-9}
Copper(I) chloride	$CuCl$	1.2×10^{-6}
Copper(I) sulfide	Cu_2S	2.5×10^{-48}
Copper(II) carbonate	$CuCO_3$	1.4×10^{-10}
Copper(II) chromate	$CuCrO_4$	3.6×10^{-6}
Copper(II) hydroxide	$Cu(OH)_2$	2.2×10^{-20}
Copper(II) phosphate	$Cu_3(PO_4)_2$	1.3×10^{-37}
Copper(II) sulfide	CuS	6.3×10^{-36}
Gold(I) chloride	$AuCl$	2.0×10^{-13}
Gold(III) chloride	$AuCl_3$	3.2×10^{-25}
Iron(II) carbonate	$FeCO_3$	3.2×10^{-11}
Iron(II) hydroxide	$Fe(OH)_2$	8.0×10^{-16}
Iron(II) sulfide	FeS	6.3×10^{-18}
Iron(III) hydroxide	$Fe(OH)_3$	4×10^{-38}
Lanthanum fluoride	LaF_3	4×10^{-17}
Lanthanum iodate	$La(IO_3)_3$	6.1×10^{-12}
Lead carbonate	$PbCO_3$	7.4×10^{-14}
Lead chloride	$PbCl_2$	1.6×10^{-5}
Lead chromate	$PbCrO_4$	2.8×10^{-13}
Lead fluoride	PbF_2	2.7×10^{-8}
Lead hydroxide	$Pb(OH)_2$	1.2×10^{-15}
Lead sulfate	$PbSO_4$	1.6×10^{-8}
Lead sulfide	PbS	8.0×10^{-28}
Magnesium hydroxide	$Mg(OH)_2$	1.8×10^{-11}
Magnesium oxalate	MgC_2O_4	8.6×10^{-5}
Manganese carbonate	$MnCO_3$	1.8×10^{-11}
Manganese hydroxide	$Mn(OH)_2$	1.9×10^{-13}
Manganese(II) sulfide	MnS	1.0×10^{-13}
Mercury(I) chloride	Hg_2Cl_2	1.3×10^{-18}
Mercury(I) oxalate	$Hg_2C_2O_4$	2.0×10^{-13}
Mercury(I) sulfide	Hg_2S	1.0×10^{-47}
Mercury(II) hydroxide	$Hg(OH)_2$	3.0×10^{-26}
Mercury(II) sulfide	HgS	4×10^{-53}
Nickel carbonate	$NiCO_3$	6.6×10^{-9}
Nickel hydroxide	$Ni(OH)_2$	1.6×10^{-14}
Nickel oxalate	NiC_2O_4	4×10^{-10}
α-Nickel sulfide[a]	NiS	3.2×10^{-19}
Silver arsenate	Ag_3AsO_4	1.0×10^{-22}
Silver bromide	$AgBr$	5.0×10^{-13}
Silver carbonate	Ag_2CO_3	8.1×10^{-12}
Silver chloride	$AgCl$	1.8×10^{-10}
Silver chromate	Ag_2CrO_4	1.1×10^{-12}
Silver cyanide	$AgCN$	1.2×10^{-16}
Silver iodide	AgI	8.3×10^{-17}
Silver sulfate	Ag_2SO_4	1.4×10^{-5}
Silver sulfide	Ag_2S	6.3×10^{-50}
Strontium carbonate	$SrCO_3$	1.1×10^{-10}
Tin(II) hydroxide	$Sn(OH)_2$	1.4×10^{-28}
Tin(II) sulfide	SnS	1.0×10^{-25}
Zinc carbonate	$ZnCO_3$	1.4×10^{-11}
Zinc hydroxide	$Zn(OH)_2$	1.2×10^{-17}
Zinc oxalate	ZnC_2O_4	2.7×10^{-8}
α-Zinc sulfide[a]	ZnS	1.1×10^{-21}

[a] Some substances exist in more than one crystalline form; the prefix indicates the particular form for which K_{sp} is listed.

E Standard Electrode Potentials at 25°C

Half-reaction	$E°$ (V)
$Ag^+(aq) + e^- \rightleftharpoons Ag(s)$	+0.799
$AgBr(s) + e^- \rightleftharpoons Ag(s) + Br^-(aq)$	+0.095
$AgCl(s) + e^- \rightleftharpoons Ag(s) + Cl^-(aq)$	+0.222
$Ag(CN)_2^-(aq) + e^- \rightleftharpoons Ag(s) + 2CN^-(aq)$	−0.31
$Ag_2CrO_4(s) + 2e^- \rightleftharpoons 2Ag(s) + CrO_4^{2-}(aq)$	+0.446
$AgI(s) + e^- \rightleftharpoons Ag(s) + I^-(aq)$	−0.151
$Ag(S_2O_3)_2^{3-} + e^- \rightleftharpoons Ag(s) + 2S_2O_3^{2-}(aq)$	+0.01
$Al^{3+}(aq) + 3e^- \rightleftharpoons Al(s)$	−1.66
$H_3AsO_4(aq) + 2H^+(aq) + 2e^- \rightleftharpoons H_3AsO_3(aq) + H_2O(l)$	+0.559
$Ba^{2+}(aq) + 2e^- \rightleftharpoons Ba(s)$	−2.90
$BiO^+(aq) + 2H^+(aq) + 3e^- \rightleftharpoons Bi(s) + H_2O(l)$	+0.32
$Br_2(l) + 2e^- \rightleftharpoons 2Br^-(aq)$	+1.065
$BrO_3^-(aq) + 6H^+(aq) + 5e^- \rightleftharpoons \frac{1}{2}Br_2(l) + 3H_2O(l)$	+1.52
$2CO_2(g) + 2H^+(aq) + 2e^- \rightleftharpoons H_2C_2O_4(aq)$	−0.49
$Ca^{2+}(aq) + 2e^- \rightleftharpoons Ca(s)$	−2.87
$Cd^{2+}(aq) + 2e^- \rightleftharpoons Cd(s)$	−0.403
$Ce^{4+}(aq) + e^- \rightleftharpoons Ce^{3+}(aq)$	+1.61
$Cl_2(g) + 2e^- \rightleftharpoons 2Cl^-(aq)$	+1.359
$HClO(aq) + H^+(aq) + e^- \rightleftharpoons \frac{1}{2}Cl_2(g) + H_2O(l)$	+1.63
$ClO^-(aq) + H_2O(l) + 2e^- \rightleftharpoons Cl^-(aq) + 2OH^-(aq)$	+0.89
$ClO_3^-(aq) + 6H^+(aq) + 5e^- \rightleftharpoons \frac{1}{2}Cl_2(g) + 3H_2O(l)$	+1.47
$Co^{2+}(aq) + 2e^- \rightleftharpoons Co(s)$	−0.277
$Co^{3+}(aq) + e^- \rightleftharpoons Co^{2+}(aq)$	+1.842
$Cr^{3+}(aq) + 3e^- \rightleftharpoons Cr(s)$	−0.74
$Cr^{3+}(aq) + e^- \rightleftharpoons Cr^{2+}(aq)$	−0.41
$Cr_2O_7^{2-}(aq) + 14H^+(aq) + 6e^- \rightleftharpoons 2Cr^3(aq) + 7H_2O(l)$	+1.33
$CrO_4^{2-}(aq) + 4H_2O(l) + 3e^- \rightleftharpoons Cr(OH)_3(s) + 5OH^-(aq)$	−0.13
$Cu^{2+}(aq) + 2e^- \rightleftharpoons Cu(s)$	+0.337
$Cu^{2+}(aq) + e^- \rightleftharpoons Cu^+(aq)$	+0.153
$Cu^+(aq) + e^- \rightleftharpoons Cu(s)$	+0.521
$CuI(s) + e^- \rightleftharpoons Cu(s) + I^-(aq)$	−0.185
$F_2(g) + 2e^- \rightleftharpoons 2F^-(aq)$	+2.87
$Fe^{2+}(aq) + 2e^- \rightleftharpoons Fe(s)$	−0.440
$Fe^{3+}(aq) + e^- \rightleftharpoons Fe^{2+}(aq)$	+0.771
$Fe(CN)_6^{3-}(aq) + e^- \rightleftharpoons Fe(CN)_6^{4-}(aq)$	+0.36

Half-reaction	$E°$ (V)
$2H^+(aq) + 2e^- \rightleftharpoons H_2(g)$	0.000
$2H_2O(l) + 2e^- \rightleftharpoons H_2(g) + 2OH^-(aq)$	−0.83
$HO_2^-(aq) + H_2O(l) + 2e^- \rightleftharpoons 3OH^-(aq)$	+0.88
$H_2O_2(aq) + 2H^+(aq) + 2e^- \rightleftharpoons 2H_2O(l)$	+1.776
$Hg_2^{2+}(aq) + 2e^- \rightleftharpoons 2Hg(l)$	+0.789
$2Hg^{2+}(aq) + 2e^- \rightleftharpoons Hg_2^{2+}(aq)$	+0.920
$Hg^{2+}(aq) + 2e^- \rightleftharpoons Hg(l)$	+0.854
$I_2(s) + 2e^- \rightleftharpoons 2I^-(aq)$	+0.536
$IO_3^-(aq) + 6H^+(aq) + 5e^- \rightleftharpoons \frac{1}{2}I_2(s) + 3H_2O(l)$	+1.195
$K^+(aq) + e^- \rightleftharpoons K(s)$	−2.925
$Li^+(aq) + e^- \rightleftharpoons Li(s)$	−3.05
$Mg^{2+}(aq) + 2e^- \rightleftharpoons Mg(s)$	−2.37
$Mn^{2+}(aq) + 2e^- \rightleftharpoons Mn(s)$	−1.18
$MnO_2(s) + 4H^+(aq) + 2e^- \rightleftharpoons Mn^{2+}(aq) + 2H_2O(l)$	+1.23
$MnO_4^-(aq) + 8H^+(aq) + 5e^- \rightleftharpoons Mn^{2+}(aq) + 4H_2O(l)$	+1.51
$MnO_4^-(aq) + 2H_2O(l) + 3e^- \rightleftharpoons MnO_2 + 4OH^-(aq)$	+0.59
$HNO_2(aq) + H^+(aq) + e^- \rightleftharpoons NO(g) + H_2O(l)$	+1.00
$N_2(g) + 4H_2O(l) + 4e^- \rightleftharpoons 4OH^-(aq) + N_2H_4(aq)$	−1.16
$N_2(g) + 5H^+(aq) + 4e^- \rightleftharpoons N_2H_5^+(aq)$	−0.23
$NO_3^-(aq) + 4H^+(aq) + 3e^- \rightleftharpoons NO(g) + 2H_2O(l)$	+0.96
$Na^+(aq) + e^- \rightleftharpoons Na(s)$	−2.71
$Ni^{2+}(aq) + 2e^- \rightleftharpoons Ni(s)$	−0.28
$O_2(g) + 4H^+(aq) + 4e^- \rightleftharpoons 2H_2O(l)$	+1.23
$O_2(g) + 2H_2O(l) + 4e^- \rightleftharpoons 4OH^-(aq)$	+0.40
$O_2(g) + 2H^+(aq) + 2e^- \rightleftharpoons H_2O_2(aq)$	+0.68
$O_3(g) + 2H^+(aq) + 2e^- \rightleftharpoons O_2(g) + H_2O(l)$	+2.07
$Pb^{2+}(aq) + 2e^- \rightleftharpoons Pb(s)$	−0.126
$PbO_2(s) + HSO_4^-(aq) + 3H^+(aq) + 2e^- \rightleftharpoons PbSO_4(s) + 2H_2O(l)$	+1.685
$PbSO_4(s) + H^+(aq) + 2e^- \rightleftharpoons Pb(s) + HSO_4^-(aq)$	−0.356
$PtCl_4^{2-}(aq) + 2e^- \rightleftharpoons Pt(s) + 4Cl^-(aq)$	+0.73
$S(s) + 2H^+(aq) + 2e^- \rightleftharpoons H_2S(g)$	+0.141
$H_2SO_3(aq) + 4H^+(aq) + 4e^- \rightleftharpoons S(s) + 3H_2O(l)$	+0.45
$HSO_4^-(aq) + 3H^+(aq) + 2e^- \rightleftharpoons H_2SO_3(aq) + H_2O(l)$	+0.17
$Sn^{2+}(aq) + 2e^- \rightleftharpoons Sn(s)$	−0.136
$Sn^{4+}(aq) + 2e^- \rightleftharpoons Sn^{2+}(aq)$	+0.154
$VO_2^+(aq) + 2H^+(aq) + e^- \rightleftharpoons VO^{2+}(aq) + H_2O(l)$	+1.00
$Zn^{2+}(aq) + 2e^- \rightleftharpoons Zn(s)$	−0.763

F Selected Nobel Prize Winners in Chemistry and Physics

The following is a list of recipients of Nobel Prizes in Chemistry and Physics whose work is related to material taught in this text. Many of the Nobel laureates are mentioned by name in the text. (*Source:* The Nobel Committee for Physics and Chemistry, The Royal Swedish Academy of Sciences)

Nobel Prize for Chemistry

1901 Professor **Jacobus H. van't Hoff** (Germany), for his discovery of the laws of chemical dynamics and osmotic pressure in solutions.

1903 Professor **Svante Arrhenius** (Sweden), for the services he has rendered to the advancement of chemistry by his electrolytic theory of dissociation.

1904 Sir **William Ramsay** (England), for his discovery of the inert gaseous elements in air and his determination of their place in the periodic system.

1906 Professor **Henri Moissan** (France), for his investigation and isolation of the element fluorine, and for the adoption in the service of science of the electric furnace called after him.

1908 Professor **Ernest Rutherford** (England), for his investigations into the disintegration of the elements and the chemistry of radioactive substances.

1909 Professor **Wilhelm Ostwald** (Germany), for his work on catalysis and for his investigations into the fundamental principles governing chemical equilibria and rates of reactions.

1911 Professor **Marie Curie** (France), for her services to the advancement of chemistry by the discovery of the elements radium and polonium, by the isolation of radium, and the study of the nature and compounds of this remarkable element.

1913 Professor **Alfred Werner** (Switzerland), for his work on the linkage of atoms in molecules by which he has thrown new light on earlier investigations and opened up new fields of research, especially in inorganic chemistry.

1914 Professor **Theodore W. Richards** (United States), for his accurate determinations of the atomic weight of a large number of elements.

1918 Professor **Fritz Haber** (Germany), for the synthesis of ammonia from its elements.

1920 Professor **Walther Nernst** (Germany), in recognition of his work in thermochemistry.

1921 Professor **Frederick Soddy** (England), for his contributions to our knowledge of the chemistry of radioactive substances, and his investigations into the origin and nature of isotopes.

1922 Professor **Francis W. Aston** (English), for his discovery, by means of his mass spectrograph, of isotopes in a large number of nonradioactive elements and for his enunciation of the whole-number rule.

1931 Professor **Carl Bosch** and Dr. **Friedrich Bergius** (Germany), in recognition of their contributions to the invention and development of chemical high-pressure methods.

1932 Dr. **Irving Langmuir** (United States), for his discoveries and investigations in surface chemistry.

1934 Professor **Harold C. Urey** (United States), for his discovery of heavy hydrogen.

1935 Professor **Frederic Joliot** and Dr. **Irène Joliot-Curie** (France), in recognition of their synthesis of new radioactive elements.

1936 Professor **Peter Debye** (Germany), for his contributions to our knowledge of molecular structure through his investigations on dipole moments and on the diffraction of X rays and electrons in gases.

1943 Professor **Georg von Hevesy** (Sweden), for his work on the use of isotopes as tracers in the study of chemical processes.

1944 Professor **Otto Hahn** (Germany), for his discovery of the fission of heavy nuclei.

1949 Professor **William F. Giauque** (United States), for his contributions in the field of chemical thermodynamics, particularly concerning the behavior of substances at extremely low temperatures.

1951 Professor **Edwin M. McMillan** and Professor **Glenn T. Seaborg** (United States), for their discoveries in the chemistry of the transuranium elements.

1954 Professor **Linus Pauling** (United States), for his research into the nature of the chemical bond and its application to the elucidation of the structure of complex substances.

1956 Sir **Cyril N. Hinshelwood** (England) and **Nicolai N. Semenov** (Russia), for their researches into the mechanism of chemical reactions.

1960 Professor **Williard F. Libby** (United States), for his method using carbon-14 for age determination in archeology, geology, geophysics. and other branches of science.

1961 Professor **Melvin Calvin** (United States), for his research on carbon dioxide assimilation in plants.

1966 Professor **Robert S. Mulliken** (United States), for his fundamental work concerning chemical bonds and the electronic structure of molecules by the molecular-orbital method.

1971 Dr. **Gerhard Herzberg** (Canada), for his contributions to the knowledge of electronic structure and geometry of molecules, particularly free radicals.

1986 Professor **Dudley R. Herschbach** (United States), Professor **Yuan T. Lee** (United States), and Professor **John C. Polanyi** (Canada), for their contributions concerning the dynamics of chemical elementary processes.

Nobel Prize for Physics

1901 Professor **Wilhelm C. Röntgen** (Germany), for his discovery of the remarkable rays subsequently named after him (X rays).

1903 Professor **Henri A. Becquerel** (France), for his discovery of spontaneous radioactivity, and Professor **Pierre Curie** and Mme. **Marie Curie** (France), for their joint researches on the radiation phenomena discovered by Professor Henri Becquerel.

1904 Lord **Rayleigh** (England), for his investigations of the densities of the most important gases and for his discovery of argon in connection with these studies.

1905 Professor **Philipp Lenard** (Germany), for his work on cathode rays.

1906 Professor **Joseph J. Thomson** (England), for his theoretical and experimental investigations on the conduction of electricity by gases.

1910 **Johannes D. van der Waals** (The Netherlands), for his work on the equation of state for gases and liquids.

1914 Professor **Max von Laue** (Germany), for his discovery of the diffraction of X rays by crystals.

1915 Professor **William H. Bragg** and **William L. Bragg** (England), for their analysis of crystal structure by means of X rays.

1918 Professor **Max Planck** (Germany), in recognition of the services he rendered to the advancement of physics by his discovery of energy quanta.

1920 **Charles E. Guillaume** (Switzerland), in recognition of the service he has rendered to precise measurements in physics by his discovery of anomalies in nickel-steel alloys.

1921 Professor **Albert Einstein** (Germany), for services to theoretical physics and especially for his discovery of the law of the photoelectric effect.

1922 Professor **Niels Bohr** (Denmark), for his investigations of the structure of atoms and of the radiation emanating from them.

1923 Professor **Robert A. Millikan** (United States), for his work on the elementary charge of electricity and on the photoelectric effect.

1924 Professor **Manne Siegbahn** (Sweden), for his discoveries and researches in the field of X-ray spectroscopy.

1925 Professor **James Franck** and Professor **Gustav Hertz** (Germany), for their discovery of the laws governing the impact of an electron upon an atom.

1929 Prince **Louis V. de Broglie** (France), for his discovery of the wave nature of electrons.

1932 Professor **Werner Heisenberg** (Germany), for the creation of quantum mechanics.

1933 Professor **Erwin Schrödinger** (Germany), and Professor **Paul A. M. Dirac** (England), for the discovery of new productive forms of atomic theory.

1935 Professor **James Chadwick** (England), for his discovery of the neutron.

1937 Dr. **Clinton J. Davisson** (United States), and Professor **George P. Thomson** (England), for their experimental discovery of the diffraction of electrons by crystals.

1938 Professor **Enrico Fermi** (Italy), for his demonstrations of the existence of new radioactive elements produced by neutron irradiation, and for his related discovery of nuclear reactions brought about by slow neutrons.

1939 Professor **Ernest O. Lawrence** (United States), for the invention and development of the cyclotron and for results obtained with it, especially with regard to artificial radioactive elements.

1945 Professor **Wolfgang Pauli** (Switzerland), for the discovery of the Exclusion Principle, also called the Pauli Principle.

1952 Professor **Felix Bloch** and **Edward M. Purcell** (United States), for the development of new methods for nuclear magnetic precision measurements and the discoveries in connection therewith.

1954 Half to Professor **Max Born** (Germany), for his fundamental research in quantum mechanics, especially for his statistical interpretation of the wave function.

1956 Professor **William Shockley**, Professor **John Bardeen**, and Dr. **Walter H. Brattain** (United States), for their investigations on semiconductors and their discovery of the transistor effect.

1972 Professor **John Bardeen**, Professor **Leon N. Cooper**, and Professor **John R. Schrieffer** (United States), for their theory of superconductivity, usually called BCS theory.

1987 Dr. **Georg Bednorz** and Professor **K. Alex Müller** (Switzerland), for the important breakthrough in the discovery of superconductivity in ceramic materials.

Answers to Selected Exercises

Chapter 1

1.1 Behavior when pressure is applied; density; boiling point. **1.3** (a) Solid; (b) liquid; (c) gas; (d) liquid; (e) gas. **1.5** (a) Chemical; (b) physical; (c) physical; (d) chemical; (e) chemical. **1.7** *Physical properties:* silvery white (color); lustrous; melting point = 649°C; boiling point = 1105°C; density at 20°C = 1.738 g/ml; pounded into sheets (malleable); drawn into wires (ductile); good electrical conductor. *Chemical properties:* burns in air to give intense white light; reacts with Cl_2 to produce brittle white solid. **1.9** (a) Pure substance; (b) homogeneous mixture; (c) heterogeneous mixture; (d) pure substance; (e) homogeneous mixture. **1.11** (a) B; (b) Li; (c) Cr; (d) P; (e) K; (f) Ag; (g) W; (h) Sb. **1.13** H_2SO_4, compound; N_2, element; C_2H_4, compound; NH_3, compound; O_2, element; CaO, compound; NaOH, compound; Cl_2, element; H_3PO_4, compound. **1.15** (a) Meters; (b) $(\text{meters})^2$; (c) $(\text{meters})^3$; (d) kilograms; (e) meters/second; (f) kelvins. **1.17** (a) 1×10^{-1}; (b) 1×10^{-2}; (c) 1×10^{-15}; (d) 1×10^{-6}; (e) 1×10^{6}; (f) 1×10^{3}; (g) 1×10^{-9}; (h) 1×10^{-3}; (i) 1×10^{-12}. **1.19** (a) Volume; (b) density; (c) length; (d) volume; (e) area; (f) time; (g) volume. **1.21** (a) 1.2×10^4 kg/m^3; (b) 3.97×10^{-11} m; (c) 1.007×10^{-5} s; (d) $8.\underline{3}645 \times 10^{-2}$ kg; (e) 1.50×10^5 m; (f) 0.320 mol. **1.23** (a) 3.38×10^4 g; (b) 0.64 g/cm^3; it would float; (c) 1.68×10^{-5} cm^3. **1.25** (a) −247.8°C; (b) 38.5°F; (c) 1922 K; (d) −46°C; (e) 627 K; (f) −438°F. **1.27** Exact: (a), (d), (e), (f). **1.29** (a) 4; (b) 1; (c) 4; (d) 2, 3, 4, or 5; (e) 3. **1.31** (a) 1.235×10^7; (b) 2.355; (c) 4.565×10^5; (d) 3.218×10^3; (e) 6.570×10^{-4}; (f) 1.005×10^5. **1.33** (a) 69.040; (b) −476; (c) 1.09×10^4; (d) 3917.0. **1.35** (a) 1.5 m; (b) 9.65×10^3 mL; (c) 2.59×10^5 s; (d) 1.87×10^6 cm^3; (e) 88 km/hr; (f) 10.7 km/L. **1.37** 3.6×10^2 kg. **1.39** 746 g. **1.41** 3.8×10^{10} mL. **1.43** Intensive; (b), (c), and (e). **1.46** 3.515×10^{13} g H_2SO_4. **1.49** 0.953 g/mL. **1.52** (a) 1.91×10^4 cm^3; (b) 0.56 m^3; (c) 7.6×10^3 kg Hg.

Chapter 2

2.1 (a) In nitrogen dioxide, the nitrogen and oxygen atoms have chemically combined to form molecules, the relative number of nitrogen and oxygen atoms is the same in each molecule, and the molecules cannot be broken down by physical means into elemental nitrogen and oxygen. (b) In air, nitrogen and oxygen molecules exist independently; the properties of this mixture are an average of the properties of the components. The mixture can be separated by physical means, such as selective condensation, into the component elements. **2.3** (a) A: 2.37 g fluorine/1 g sulfur; B: 0.59 g fluorine/1 g sulfur; C: 3.55 g fluorine/1 g sulfur. (b) The masses in part (a) are in the ratio of small whole numbers and therefore obey the law of multiple proportions. **2.5** (1) A metal plate exposed to cathode rays acquired a negative charge. (2) Electric and magnetic fields deflected the rays in the same way they would deflect negatively charged particles. **2.7** The total mass of the products of a reaction must equal the total mass of the reactants. Therefore, 11.0 g of carbon dioxide is produced. **2.9** 2.02×10^3 electrons. **2.11** The diffuse positive charge in the "plum pudding" model would not produce a repulsion strong enough to deflect the alpha particles at the large angles observed in the experiment. In Rutherford's nuclear model, the positive charge in the atom is concentrated in the small nucleus. If an alpha particle strikes a gold nucleus directly, it is deflected at a large angle. **2.13** (a) 0.47 nm, 470 pm; (b) 2.1×10^7 Cs atoms. **2.15** (a) ^{51}V—23 p, 28 n, 23 e; (b) ^{119}Sn—50 p, 69 n, 50 e; (c) ^{127}Te—52 p, 75 n, 52 e; (d) ^{165}Ho—67 p, 98 n, 67 e; (e) ^{16}O—8 p, 8 n, 8 e; (f) ^{234}Th—90 p, 144 n, 90 e; (g) ^{112}Cd—48 p, 64 n, 48 e; (h) ^{113}Cd—48 p, 65 n, 48 e. **2.17** $^{235}_{92}U$, $^{238}_{92}U$.

2.19

Symbol	$^{17}_{8}O$	$^{232}_{90}Th$	$^{60}_{27}Co$	$^{100}_{44}Ru$	$^{95}_{42}Mo$
Protons	8	90	27	44	42
Neutrons	9	142	33	56	53
Electrons	8	90	27	44	42
Mass no.	17	232	60	100	95

2.21 (a) S (nonmetal); (b) Ag (metal); (c) Sr (metal); (d) As (semimetal); (e) Xe (nonmetal); (f) Al (metal); (g) Y (metal); (h) I (nonmetal). **2.23** (a) Rn, noble gases, nonmetal; (b) Po, chalcogens, semimetal; (c) Rb, alkali metals, metal; (d) P, pnicogens, nonmetal; (e) Mg, alkaline earth metals, metal; (f) Cl, halogens, nonmetal. **2.25** The molecular formula conveys more information because it indicates *both* the combining ratio of elements in the compound and the exact number of atoms of each element in a molecule of the compound. **2.27** (a) C_2H_3; (b) HO; (c) P_2O_5; (d) CH_2O; (e) NH_2; (f) BH_3. **2.29** (a) Li^+; (b) Br^-; (c) Mg^{2+}; (d) O^{2-}; (e) Al^{3+}; (f) Y^{3+}.
2.31 (a) $Cu(C_2H_3O_2)_2$; (b) NH_4HCO_3; (c) $AlBr_3$; (d) $Ca(ClO_4)_2$; (e) K_2SO_4; (f) $(NH_4)_3PO_4$.
2.33 Molecular: NO_2, BF_3, PF_5, NF_3, Ionic: Li_2O, Sc_2O_3, CsBr, LaP. **2.35** (a) Rubidium iodide; (b) sodium nitrate; (c) silver(I) sulfide; (d) potassium permanganate; (e) mercury(I) chloride; (f) manganese(IV) oxide; (g) calcium hydride; (h) scandium(III) chloride; (i) barium hydroxide; (j) ammonium sulfate.
2.37 (a) $Cu(C_2H_3O_2)_2$; (b) $FeCl_3$; (c) $(NH_4)_2S$; (d) Al_2O_3; (e) Ca_3N_2; (f) $Mg_3(PO_4)_2$. **2.39** (a) Carbon disulfide; (b) SF_6; (c) sulfur trioxide; (d) P_2O_5; (e) carbon tetrabromide; (f) SO_2. **2.41** (a) Phosphoric acid; (b) H_2SO_3; (c) hydrobromic acid; (d) HNO_3; (e) chloric acid; (f) HClO. **2.43** (a) $ZnCO_3$, ZnO, CO_2; (b) HF, SiO_2, SiF_4, H_2O; (c) SO_2, H_2O, H_2SO_3; (d) H_3P (or PH_3); (e) $HClO_4$, Cd, $Cd(ClO_4)_2$; (f) VBr_3.
2.46 Deuterium, ${}^2_1H$; tritium, ${}^3_1H$.

2.48 Symbol	${}^{16}_{8}O$	${}^{185}_{75}Re^{3+}$	${}^{33}_{16}S^{2-}$	${}^{73}_{32}Ge^{2-}$	${}^{195}_{78}Pt^{4+}$
Protons	8	75	16	32	78
Neutrons	8	110	17	41	117
Electrons	8	72	18	34	74
Net charge	0	3+	2−	2−	4+

2.52 (a) ${NH_4}^+$, ${SO_4}^{2-}$, $(NH_4)_2SO_4$, ammonium sulfate; (b) Ca^{2+}, ${HCO_3}^-$, $Ca(HCO_3)_2$, calcium hydrogen carbonate; (c) Al^{3+}, ${HSO_4}^-$, $Al(HSO_4)_3$, aluminum hydrogen sulfate; (d) K^+, ${CrO_4}^{2-}$, K_2CrO_4, potassium chromate; (e) Na^+, ${C_2H_3O_2}^-$, $NaC_2H_3O_2$, sodium acetate; (f) Ag^+, S^{2-}, Ag_2S, silver sulfide; (g) Be^{2+}, Br^-, $BeBr_2$, beryllium bromide; (h) K^+, Te^{2-}, K_2Te, potassium telluride.
2.55 (a) CaS, $Ca(HS)_2$; (b) HBr, $HBrO_3$; (c) AlN, $Al(NO_2)_3$; (d) FeO, Fe_2O_3; (e) NH_3, ${NH_4}^+$; (f) K_2SO_3, $KHSO_3$; (g) Hg_2Cl_2, $HgCl_2$; (h) $HClO_3$, $HClO_4$.

Chapter 3

3.1 (a) Conservation of mass; (b) (*g*), (*l*), (*s*), (*aq*); (c) P_4, four phosphorus atoms bound together into a single molecule; 4P, four separate phosphorus atoms. **3.3** (a) 1, 3, 2; (b) 1, 4, 1, 4; (c) 1, 4, 1, 5; (d) 1, 3, 1, 3; (e) 1, 8, 3, 2; (f) 1, 2, 1, 2; (g) 2, 15, 12, 6; (h) 4, 19, 12, 4, 10.
3.5 (a) $SO_3(g) + H_2O(l) \longrightarrow H_2SO_4(aq)$; (b) $B_2S_3(s) + 6H_2O(l) \longrightarrow 2H_3BO_3(aq) + 3H_2S(g)$; (c) $4PH_3(g) + 8O_2(g) \longrightarrow 6H_2O(g) + P_4O_{10}(s)$; (d) $2Hg(NO_3)_2(s) \longrightarrow 2HgO(s) + 4NO_2(g) + O_2(g)$; (e) $3H_2S(g) + 2Fe(OH)_3(s) \longrightarrow Fe_2S_3(s) + 6H_2O(g)$.
3.7 (a) $C_6H_{12}(l) + 9O_2(g) \longrightarrow 6CO_2(g) + 6H_2O(l)$; (b) $CH_3CO_2C_2H_5(l) + 5O_2(g) \longrightarrow 4CO_2(g) + 4H_2O(l)$; (c) $2Li(s) + 2H_2O(l) \longrightarrow 2LiOH(aq) + H_2(g)$; (d) $PbCO_3(s) \longrightarrow PbO(s) + CO_2(g)$.
3.9 (a) $2H_2O_2(aq) \longrightarrow 2H_2O(l) + O_2(g)$, decomposition; (b) $2Fe(s) + 3Cl_2(g) \longrightarrow 2FeCl_3(s)$, combination; (c) $Na_2CO_3(aq) + Ca(OH)_2(aq) \longrightarrow CaCO_3(s) + 2NaOH(aq)$, double displacement; (d) $Cu(NO_3)_2(aq) + Fe(s) \longrightarrow Cu(s) + Fe(NO_3)_2(aq)$, single displacement.
3.11 (a) $2K(s) + 2NH_3(l) \longrightarrow 2KNH_2(solv) + H_2(g)$; (b) $4CH_3NO_2(g) + 7O_2(g) \longrightarrow 4NO_2(g) + 4CO_2(g) + 6H_2O(l)$; (c) $CH_4(g) + 4F_2(g) \longrightarrow CF_4(g) + 4HF(g)$.
3.13 (a) ${}^{12}_{6}C$. (b) An atomic mass unit is exactly $\frac{1}{12}$ of the mass of one atom of ${}^{12}_{6}C$, 1.66056×10^{-24} g. **3.15** 20.17 amu. **3.17** 207.3 amu. **3.19** 129.0 amu; Berzelius' formula, ZnO_2, was wrong. A ratio of 1 mol O/1 mol Zn in the calculation gives the correct atomic weight. **3.21** (a) 44.01 amu; (b) 92.33 amu; (c) 212.3 amu; (d) 96.09 amu; (e) 58.33 amu; (f) 158.2 amu; (g) 60.09 amu; (h) 392.2 amu. **3.23** (a) 27.29% C, 72.71% O; (b) 29.67% S, 70.33% F; (c) 14.30% N, 4.12% H, 81.58% Br; (d) 52.14% C, 34.73% O, 13.13% H; (e) 16.62% Mg, 1.38% H, 16.42% C, 65.62% O; (f) 34.52% Zn, 14.79% N, 50.69% O.
3.25 (a) A mole is the amount of matter that contains as many objects as the number of atoms in exactly 12 g of ${}^{12}C$. (b) 6.022×10^{23}. (c) The formula weight of a substance in amu has the same numerical value as the molar mass expressed in grams. **3.27** (a) 148.3 g; (b) 13.1 g; (c) 0.0351 mol; (d) 3.04×10^{19} N atoms. **3.29** (a) 0.544 g SO_2; (b) 2.38 g Ar; (c) 4.84×10^{-2} g $C_8H_{10}N_4O_2$.
3.31 (a) 9.03×10^{22} C_2H_2 molecules; (b) 1.71×10^{21} $C_6H_8O_6$ molecules; (c) 1.7×10^{18} H_2O molecules.
3.33 3.28×10^{-8} mol C_2H_3Cl, 1.97×10^{16} molecules C_2H_3Cl. **3.35** An empirical formula gives the relative number and kind of each atom in a compound, but a molecular formula gives the actual number of each kind of atom in the molecule. **3.37** (a) CH_4; (b) Fe_2O_3; (c) NH_2.
3.39 (a) HNO_3; (b) $C_{12}H_{12}N_2O_3$. **3.41** (a) Empirical and molecular formulas are $C_9H_{13}O_3N$; (b) empirical formula: C_5H_7N; molecular formula: $C_{10}H_{14}N_2$. **3.43** CH_2.
3.45 $x = 7$; $MgSO_4 \cdot 7H_2O$. **3.47** Molar mass of fungal laccase is approximately 6.6×10^4 g. **3.49** The mole ratios implicit in the coefficients of a balanced chemical equation are essential for solving stoichiometry problems. If the equation is not balanced, the mole ratios will be incorrect and will lead to erroneous calculated amounts of reactants and products. **3.51** (a) 65.0 mol O_2; (b) 35.8 g O_2.
3.53 (a) 20.0 mol HF; (b) 52.5 g NaF; (c) 3.81 g Na_2SiO_3.
3.55 4.37×10^3 g NH_4ClO_4. **3.57** (a) $Al(OH)_3(s) + 3HCl(aq) \longrightarrow AlCl_3(aq) + 3H_2O(l)$; (b) 7.01 g HCl.
3.59 (a) The limiting reagent is the reactant that regulates the amounts of products produced during a chemical reaction. The excess reagent is the other reactant or reac-

tants. (b) The limiting reagent regulates the amount of products because it is completely used up during the reaction; no more product can be made when one of the reactants is unavailable. **3.61** (a) 3.34 g SiC; (b) 3.00 g C; (c) 1.67 g SiC; (d) SiO_2 is the limiting reactant, and C is present in excess; (e) 1.00 g C remains. **3.63** 1.80 g Na_2S. **3.65** (a) C_6H_6 is the limiting reagent, and the theoretical yield is 60.3 g C_6H_5Br. (b) 94.0% yield.
3.67 (a) $Li_3N(s) + 3H_2O(l) \longrightarrow NH_3(g) + 3LiOH(aq)$; (b) $PBr_5(l) + 4H_2O(l) \longrightarrow H_3PO_4(aq) + 5HBr(aq)$; (c) $2La(NO_3)_3(aq) + 3Ba(OH)_2(aq) \longrightarrow 2La(OH)_3(s) + 3Ba(NO_3)_2(aq)$; (d) $2C_3H_7OH(l) + 9O_2(g) \longrightarrow 6CO_2(g) + 8H_2O(l)$. **3.70** (a) $^{69}_{31}Ga$ – 31 p, 38 n; $^{71}_{31}Ga$ – 31 p, 40 n; (b) 60.3% $^{69}_{31}Ga$ and 39.7% $^{71}_{31}Ga$. **3.72** 4.69×10^{23} Al atoms. **3.74** (a) The atomic weight of X is 138.9 g; (b) X is lanthanum, La, atomic number 57. **3.77** The empirical formula of vanillin is $C_8H_8O_3$. **3.80** 10.2 g $KClO_3$, 20.0 g $KHCO_3$, 13.8 g K_2CO_3, 56.0 g KCl. **3.83** (a) 1.2×10^2 kg $C_7H_6O_3$; (b) 1.5×10^2 kg $C_7H_6O_3$; (c) 2.20×10^5 g $C_9H_8O_4$; (d) 82.4% yield.

Chapter 4

4.1 (a) Calculate the required mass of pure substance: grams of substance = (known molarity × volume of flask in liters × molar mass of substance). (b) Weigh out the required mass of substance, and place it in the volumetric flask. (c) Add some water, and agitate until the substance is dissolved. Add water up to the mark on the neck of the flask; agitate again to ensure a uniform concentration throughout. **4.3** The term 0.50 mol HCl defines an amount of pure HCl. The term 0.50 *M* HCl is a ratio; it indicates that there are 0.50 mol of HCl solute in 1.0 L of solution. **4.5** (a) 0.168 *M* Na_2CrO_4; (b) 0.300 mol HCl; (c) 50.0 mL of 2.00 *M* NaOH. **4.7** (a) 2.98 g KBr; (b) 2.13 g Na_2SO_4; (c) 2.09 g $KBrO_3$; (d) 15.3 g $C_6H_{12}O_6$. **4.9** Weigh 1.802 g $C_6H_{12}O_6$, and add it to a 100-mL volumetric flask; dissolve the solid in a small amount of water; bring the total volume of solution to exactly 100 mL with additional water, and agitate thoroughly to ensure complete mixing. **4.11** Using a pipet, remove 12.5 mL of the 2.00 *M* glucose solution; add this to a 250-mL volumetric flask; add water to bring the total volume to 250 mL, and mix thoroughly. **4.13** Tap water contains enough dissolved electrolytes to complete a circuit between an electrical appliance and our body, producing a shock. **4.15** (a) $HI(g) \longrightarrow H^+(aq) + I^-(aq)$; (b) $K_2SO_4(s) \longrightarrow 2K^+(aq) + SO_4^{2-}(aq)$; (c) $NH_4NO_3(s) \longrightarrow NH_4^+(aq) + NO_3^-(aq)$; (d) $CaCl_2(s) \longrightarrow Ca^{2+}(aq) + 2Cl^-(aq)$. **4.17** (a) HBrO, weak electrolyte; (b) HNO_3, strong electrolyte; (c) KOH, strong electrolyte; (d) $CoSO_4$, strong electrolyte; (e) $C_{12}H_{22}O_{11}$, nonelectrolyte; (f) O_2, nonelectrolyte. **4.19** (a) 0.14 *M* Na^+, 0.14 *M* OH^-, 0.28 *M* total; (b) 0.25 *M* Ca^{2+}, 0.50 *M* Br^-, 0.75 *M* total; (c) 0.25 *M* CH_3OH; (d) 0.13 *M* K^+, 0.13 *M* ClO_3^-, 0.13 *M* Na^+, 0.067 *M* SO_4^{2-}, 0.46 *M* total. **4.21** H_3PO_3. **4.23** The single arrow indicates that HNO_3 is completely dissociated into H^+ and NO_3^- in aqueous solution. The double arrow indicates that HCN is only partially dissociated and exists as a mixture of H^+, CN^-, and undissociated HCN molecules in solution. **4.25** (a) A monoprotic acid has one ionizable (acidic) H, whereas a diprotic acid has two. (b) A strong acid is completely ionized in aqueous solution, whereas only a fraction of weak acid molecules are ionized. (c) An acid is a H^+ donor, and a base is a H^+ acceptor. **4.27** (a) $Fe(OH)_2(s) + 2HClO_3(aq) \longrightarrow Fe(ClO_3)_2(aq) + 2H_2O(l)$; (b) $2HI(aq) + Ca(OH)_2(aq) \longrightarrow CaI_2(aq) + 2H_2O(l)$; (c) $2Al(OH)_3(s) + 3H_2SO_4(aq) \longrightarrow Al_2(SO_4)_3(aq) + 6H_2O(l)$. **4.29** (a) $Pb^{2+}(aq) + SO_4^{2-}(aq) \longrightarrow PbSO_4(s)$; spectators: Na^+, NO_3^-; (b) $Zn(s) + 2H^+(aq) \longrightarrow Zn^{2+}(aq) + H_2(g)$; spectator: Cl^-; (c) $FeO(s) + 2H^+(aq) \longrightarrow H_2O(l) + Fe^{2+}(aq)$; spectator: ClO_4^-. **4.31** The driving force in a metathesis reaction is the formation of a product which removes ions from solution. Driving forces in Exercise 4.30: (a) $H_2O(l)$, nonelectrolyte; (b) $H_2O(l)$, nonelectrolyte; $CO_2(g)$, gas; (c) $Cu(OH)_2(s)$, precipitate. **4.33** (a) $PbCl_2$, insoluble; (b) CsBr, soluble; (c) $Co(OH)_2$, insoluble; (d) $SrSeO_4$, insoluble. **4.35** Ba^{2+}. **4.37** (a) $ZnS(s) + 2H^+(aq) \longrightarrow H_2S(g) + Zn^{2+}(aq)$; (b) $Ba^{2+}(aq) + CO_3^{2-}(aq) \longrightarrow BaCO_3(s)$; (c) $3H^+(aq) + PO_4^{3-}(aq) \rightleftharpoons H_3PO_4(aq)$; (d) $H^+(aq) + OH^-(aq) \longrightarrow H_2O(l)$; (e) $Sr^{2+}(aq) + SO_4^{2-}(aq) \longrightarrow SrSO_4(s)$; (f) $SO_3^{2-}(aq) + 2H^+(aq) \rightleftharpoons H_2SO_3(aq) \rightleftharpoons H_2O(l) + SO_2(g)$; (g) $Pb^{2+}(aq) + H_2S(aq) \longrightarrow PbS(s) + 2H^+(aq)$; (h) $Fe(OH)_3(s) + 3H^+(aq) \longrightarrow Fe^{3+}(aq) + 3H_2O(l)$. **4.39** (a) $Cu(OH)_2(s) + 2HNO_3(aq) \longrightarrow Cu(NO_3)_2(aq) + 2H_2O(l)$; (b) $MnCl_2(aq) + Na_2S(aq) \longrightarrow MnS(s) + 2NaCl(aq)$; (c) $Fe(NO_3)_2(aq) + 2NaOH(aq) \longrightarrow Fe(OH)_2(s) + 2NaNO_3(aq)$. (This reaction should be carried out under an inert N_2 atmosphere to prevent oxidation of Fe(II) to Fe(III).) **4.41** The most easily oxidized metals are near the bottom of the groups on the left side of the chart, especially groups 1A and 2A. The least easily oxidized metals are on the lower right of the transition metals, especially those near the bottom of groups 8B and 1B. **4.43** Molecular equations: (a) $2HCl(aq) + Ni(s) \longrightarrow NiCl_2(aq) + H_2(g)$; (b) $H_2SO_4(aq) + Fe(s) \longrightarrow FeSO_4(aq) + H_2(g)$; (c) $2HBr(aq) + Zn(s) \longrightarrow ZnBr_2(aq) + H_2(g)$; (d) $2HC_2H_3O_2(aq) + Mg(s) \longrightarrow Mg(C_2H_3O_2)_2(aq) + H_2(g)$. **4.45** (a) $2Al(s) + 3NiCl_2(aq) \longrightarrow 2AlCl_3(aq) + 3Ni(s)$; (b) no reaction; (c) $2Cr(s) + 3NiSO_4(aq) \longrightarrow Cr_2(SO_4)_3(aq) + 3Ni(s)$; (d) $Mn(s) + 2HBr(aq) \longrightarrow MnBr_2(aq) + H_2(g)$; (e) $H_2(g) + CuCl_2(aq) \longrightarrow Cu(s) + 2HCl(aq)$; (f) $Ba(s) + 2H_2O(l) \longrightarrow Ba(OH)_2(aq) + H_2(g)$. **4.47** The order of activity is C > A > D > B. (C is most easily oxidized, B is least easily oxidized.) **4.49** HCl is called a nonoxidizing acid because Cl^- cannot oxidize another substance; it is in its most reduced state. HNO_3 is called an oxidizing acid because NO_3^- can oxidize other substances. **4.51** (a) 41.7 mL of 0.105 *M* $HClO_4$; (b) 623 mL of 0.158 *M* HCl; (c) 0.465 *M* $AgNO_3$; (d) 0.235 g KOH. **4.53** 25 g $NaHCO_3$. **4.55** 1.22×10^{-3} *M* $Ca(OH)_2$ solution; the solubility of $Ca(OH)_2$ is 0.0904 g in 100 mL of solution. **4.57** The net charges are not balanced; Just as the number and kinds of atoms must be balanced in a chemical equation, so must the net charges. **4.59** (a) 0.133 *M* NaCl; (b) 1.20 *M* NaOH. **4.62** The two precipitates are AgCl(*s*) and $SrSO_4(s)$; Ni^{2+} and Mn^{2+} must be absent because no precipitate forms on addition of hydroxide ion. **4.65** $H_2C_4H_4O_6(aq) + 2OH^-(aq) \longrightarrow C_4H_4O_6^{2-}(aq) + 2H_2O(l)$; 0.05655 *M* $H_2C_4H_4O_6$ solution. **4.67** 6.86 g $Zn(OH)_2$.

Chapter 5

5.1 A force is any kind of push or pull on an object; energy, the capacity to do work, is used to overcome or counteract the effects of a force. Work equals force times distance: $w = F \times d$. **5.3** (a) $E_k = 88$ J; (b) $E_k = 21$ cal; (c) As the ball collides with the racket, its speed drops to zero and its kinetic energy is converted into the potential energy of a compressed ball on the stretched strings of the racket. As the ball leaves the racket, potential energy is reconverted to kinetic energy. **5.5** (a) Energy can be neither created nor destroyed, but it can be changed in form. (b) The system is the part of the universe whose energy changes are being described. (c) The energy change of a system is equal in magnitude and opposite in sign to that of its surroundings. **5.7** (a) $\Delta E = -450$ J; (b) $\Delta E = +8360$ J; (c) $\Delta E = +154$ J. **5.10** (a) Independent; (b) dependent; (c) dependent. **5.11** At the bench and in nature, many chemical reactions occur open to the atmosphere and thus at atmospheric pressure, which is essentially constant. At constant pressure, the change in enthalpy is equal to the energy transferred as heat, $\Delta H = q$. **5.13** (a) Exothermic; (b) endothermic; (c) exothermic; (d) exothermic; (e) endothermic. **5.15** $\Delta E = +460$ J; $\Delta H = +600$ J. **5.20** (a) -13.1 kJ; (b) -1.14 kJ; (c) $+22.9$ kJ. **5.21** Enthalpy of $H_2O(s) < H_2O(l) < H_2O(g)$. Heat must be added to convert solid to liquid to gas. **5.23** If a reaction can be described as a series of steps, ΔH for the reaction is the sum of the enthalpy changes for each step. As long as we can describe a route where ΔH for each step is known, ΔH for any process can be calculated.
5.26 $\Delta H = +180.8$ kJ. **5.28** $\Delta H = -86$ kJ. **5.29** (a) $Fe(s) + \frac{3}{2}Cl_2(g) \longrightarrow FeCl_3(s)$; $\Delta H_f^\circ = -400$ kJ; (b) $Ba(s) + C(s) + \frac{3}{2}O_2(g) \longrightarrow BaCO_3(s)$; $\Delta H_f^\circ = -1216.3$ kJ; (c) $\frac{1}{2}N_2(g) + \frac{1}{2}O_2(g) + \frac{1}{2}Cl_2(g) \longrightarrow NOCl(g)$; $\Delta H_f^\circ = +52.6$ kJ; (d) $\frac{1}{2}N_2(g) + \frac{3}{2}H_2(g) \longrightarrow NH_3(g)$; $\Delta H_f^\circ = -46.19$ kJ. **5.32** $\Delta H^\circ_{rxn} = -847.6$ kJ. **5.33** ΔH_f° for $NO(g) = +90.4$ kJ. **5.35** (a) $\Delta H^\circ_{rxn} = -38.9$ kJ; (b) $\Delta H^\circ_{rxn} = -36.4$ kJ; (c) $\Delta H^\circ_{rxn} = -1300.0$ kJ; (d) $\Delta H^\circ_{rxn} = -876.9$ kJ. **5.37** ΔH_f° for $CaC_2(s) = -60.6$ kJ. **5.39** (a) 4.184 J/g-°C; (b) 1.46×10^3 J/°C; (c) 355 kJ **5.41** 7.60×10^3 J. **5.43** -44.4 kJ/mol NaOH **5.45** $\Delta H_{comb} = -48.1$ kJ/g C_8H_{18} or -5.49×10^3 kJ/mol C_8H_{18}.
5.47 (a) Heat capacity of the complete calorimeter = 8.67 kJ/°C; (b) heat capacity of the empty calorimeter = 2.39 kJ/°C; (c) $\Delta T = 4.82$°C. **5.49** 88 Cal. **5.51** Fuel value = 11.6×10^2 kJ or 277 Cal.
5.54 (a) $\Delta H_{comb} = -1849.5$ kJ/mol C_3H_4, 4.6162×10^4 kJ/kg C_3H_4; (b) $\Delta H_{comb} = -1926.3$ kJ/mol C_3H_6, 4.5777×10^4 kJ/kg C_3H_6; (c) $\Delta H_{comb} = -2043.9$ kJ/mol C_3H_8, 4.6351×10^4 kJ/kg C_3H_8. These three substances yield nearly identical quantities of heat per unit mass, although propane, C_3H_8, is marginally higher. **5.55** The fuel value of the natural gas is 2.05×10^{14} kJ. The equivalent in anthracite coal is 6.6×10^{12} g or 7.3×10^6 tons of coal. **5.58** $w = -0.23$ kJ. The negative sign for *w* indicates that work is done by the system on the surroundings. **5.62** ΔH_f° for $TiO_2(s) = -939.7$ kJ/mol. **5.64** $\Delta H^\circ_{rxn} = -41.2$ kJ. Substituting $H_2O(l)$, $\Delta H^\circ_{rxn} = +2.85$ kJ. Using $H_2O(l)$ makes the reaction less energetically favorable. **5.67** 1.0×10^4 or 10,000 bricks. **5.68** $\Delta H = -52$ kJ. **5.72** ΔH_f for $NH_3(l) = -50.8$ kJ/mol; ΔH for the combustion of $NH_3(l) = -1247.6$ kJ; ΔH for the combustion of 1 L of $NH_3(l) = 1.5 \times 10^4$ kJ; ΔH for the combustion of 1 L of $CH_3OH(l) = 1.58 \times 10^4$ kJ. In terms of heat produced per unit volume of liquid, CH_3OH is a marginally better fuel than $NH_3(l)$. **5.74** ΔH_{comb} of pure $C_7O_3H_6 = 3.00 \times 10^3$ kJ/mol. ΔH_{comb} of the sample = 1.35×10^3 kJ/138.1 g. The sample yields 45.0% of the expected heat; the mass percent of boric oxide impurity is 55.0%.

Chapter 6

6.1 Wavelength of (e) cosmic rays < (d) ultraviolet light < (b) red (visible) light < (c) infrared radiation < (a) microwave radiation. **6.3** (a) $\lambda = 6.49 \times 10^{-7}$ m; (b) $\nu = 2.86 \times 10^9$ s^{-1}; (c) the radiation in part (a) is visible and that in part (b) is not visible; (d) 5.40×10^{10} m. **6.5** $\nu = 4.87 \times 10^{14}$ s^{-1}; the color is orange. **6.7** In everyday activities, we deal with macroscopic objects such as our bodies or our cars, which gain and lose total amounts of energy much larger than a single quantum, $h\nu$. The gain or loss of the relatively miniscule quantum of energy is unnoticed. **6.9** (a) $E = 5.22 \times 10^{-19}$ J; (b) $E = 2.4 \times 10^{-20}$ J; (c) $\lambda = 2.42 \times 10^{-7}$ m. **6.11** $\lambda = 5.4$ μm, $E = 3.7 \times 10^{-20}$ J; $\lambda = 156$ nm, $E = 1.28 \times 10^{-18}$ J. The longer-wavelength infrared photon is much lower in energy than the ultraviolet photon. **6.13** $E_{photon} = 5.67 \times 10^{-19}$ J; 1.46×10^{22} photons emitted. **6.15** Maximum $\lambda = 242$ nm; this is ultraviolet radiation. **6.17** (a) $E_{min} = 7.23 \times 10^{-19}$ J; (b) $\lambda = 275$ nm; (c) $E_{120} = 1.66 \times 10^{-18}$ J. The difference between this value and E_{min} is the maximum possible kinetic energy of the electron. E_k of the electron = 9.4×10^{-19} J. **6.19** (a) Absorbed; (b) emitted; (c) absorbed. **6.21** (a) $\Delta E = 1.94 \times 10^{-18}$ J absorbed, $\nu = 2.92 \times 10^{15}$ s^{-1}, $\lambda = 1.03 \times 10^{-7}$ m; (b) $\Delta E = 4.58 \times 10^{-19}$ J absorbed, $\nu = 6.91 \times 10^{14}$ s^{-1}, $\lambda = 4.34 \times 10^{-7}$ m; (c) $\Delta E = 1.61 \times 10^{-20}$ J absorbed, $\nu = 2.43 \times 10^{13}$ s^{-1}, $\lambda = 1.24 \times 10^{-5}$ m.
6.23 $\lambda = 434$ nm. This is the "blue" line in the visible por-

tion of the emission spectrum of hydrogen shown in Figure 6.11. **6.25** (a) $\lambda = 2.0 \times 10^{-34}$ m; (b) $\lambda = 4.7 \times 10^{-37}$ m; (c) $\lambda = 6.7 \times 10^{-13}$ m. **6.27** $v = 4.5 \times 10^3$ m/s. **6.29** (a) $n = 5$, $l = 4, 3, 2, 1, 0$; (b) $l = 2$, $m_l = -2, -1, 0, 1, 2$. **6.31** (a) 4, 2, −2; 4, 2, −1; 4, 2, 0; 4, 2, 1; 4, 2, 2; (b) 3, 0, 0; 3, 1, −1; 3, 1, 0; 3, 1, 1; 3, 2, −2; 3, 2, −1; 3, 2, 0; 3, 2, 1; 3, 2, 2. **6.33** (a) Permissible, $2p$; (b) not permissible; (c) permissible, $4d$; (d) not permissible. **6.35** The quantity ψ^2, at a given point in space, is the probability of locating an electron within a small volume element around that point at any given instant.

6.37

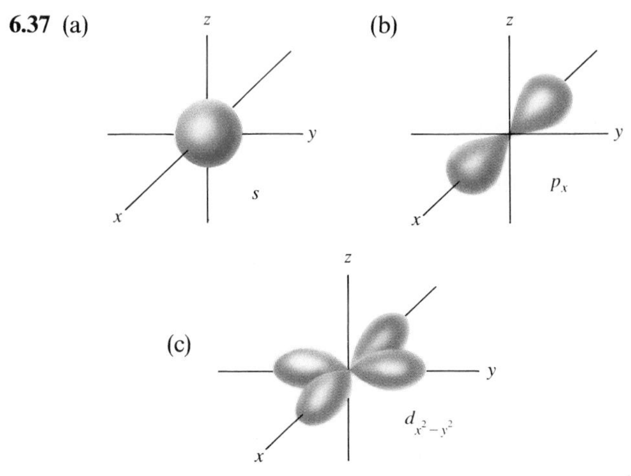

6.39 The $2s$ and $3s$ orbitals have the same overall spherical shape, but the $3s$ orbital has a larger radial extension and one more node than the $2s$ orbital. In the hydrogen atom, $2s$ is lower in energy. **6.41** (a) The principal quantum number, n; (b) both the principal and azimuthal quantum numbers, n and l. **6.43** A $3s$ electron has a greater probability of being close to the nucleus than does a $3d$ electron, so it is less effectively shielded from the nuclear charge by the inner electrons ($1s$, $2s$, $2p$) than is a $3d$ electron. **6.45** The $3s$ electron in Mg experiences a total nuclear charge ($Z = 12$) that is 1 higher than that in Na ($Z = 11$). This added unit of nuclear charge is partially but not totally canceled by the presence of a second $3s$ electron. The overall effect is an increase in the effective nuclear charge for Mg. **6.47** (a) 10; (b) 2; (c) 6; (d) 14. **6.49** Li: $1s^22s^1$; $1s$ electrons: 1, 0, 0, $\frac{1}{2}$; 1, 0, 0, $-\frac{1}{2}$. $2s$ electron: 2, 0, 0, $\frac{1}{2}$ *or* 2, 0, 0, $-\frac{1}{2}$. **6.51** (a) K, $[Ar]4s^1$; (b) Ga, $[Ar]4s^23d^{10}4p^1$; (c) Se, $[Ar]4s^23d^{10}4p^4$; (d) Fe, $[Ar]4s^23d^6$; (e) Sm, $[Xe]6s^24f^6$; (f) Sc, $[Ar]4s^23d^1$.

6.53

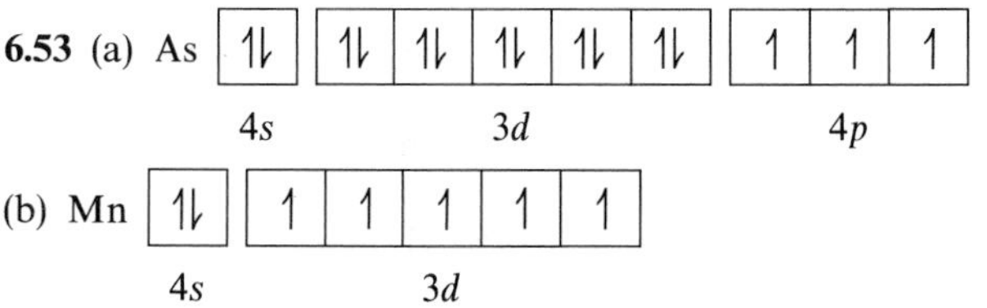

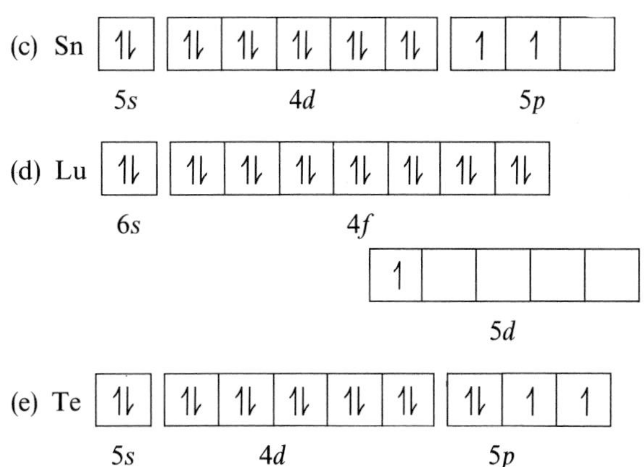

(e) Te ⇅ | ⇅ ⇅ ⇅ ⇅ ⇅ | ⇅ ↑ ↑
5s 4d 5p

(a) 3 unpaired electrons; (b) 5 unpaired electrons; (c) 2 unpaired electrons; (d) 1 unpaired electron; (e) 2 unpaired electrons. **6.55** $3s$, $3p$, $4s$, $3d$, $5s$, $5p$, $4f$. **6.57** (a) O; (b) Cl; (c) K; (d) Cr. **6.59** (a) $\lambda_A = 6.0 \times 10^{-7}$ m, $\lambda_B = 12 \times 10^{-7}$ m; (b) $\nu_A = 5.0 \times 10^{14}\ s^{-1}$, $\nu_B = 2.5 \times 10^{14}\ s^{-1}$; (c) A, visible; B, infrared. **6.62** (a) $\nu = 3.85 \times 10^{14}\ s^{-1}$; (b) $E_{photon} = 2.55 \times 10^{-19}$ J. **6.66** $\Delta H_{rxn} =$ 105.2 kJ/mol O_3; $\Delta H_{rxn} = 1.747 \times 10^{-19}$ J/O_3 atom; maximum $\lambda = 1.137 \times 10^{-6}$ m; radiation with this wavelength is in the infrared portion of the spectrum. **6.68** (a) $\lambda = 9.12 \times 10^{-8}$ m = 91.2 nm; (b) $\lambda = 3.65 \times 10^{-7}$ m = 365 nm; (c) $\lambda = 8.21 \times 10^{-7}$ m = 821 nm. **6.71** (a) $2s$; (b) $4d$; (c) $5p$; (d) $3d$; (e) $4f$. **6.73** In a Cl atom, the $3p$ electrons experience the smallest effective nuclear charge because they are shielded by all the core electrons and to some extent by the $3s$ electrons. The $1s$ electrons experience the greatest effective nuclear charge because they are not shielded by inner electrons. **6.76** (a) Cd: $1s^22s^22p^63s^23p^64s^23d^{10}4p^65s^24d^{10}$; (b) As: $1s^22s^22p^63s^23p^64s^23d^{10}4p^3$; (c) La: $1s^22s^22p^63s^23p^64s^23d^{10}4p^65s^24d^{10}5p^66s^25d^1$; (d) Pd: $1s^22s^22p^63s^23p^64s^23d^{10}4p^65s^24d^8$; (e) S: $1s^22s^22p^63s^23p^4$ **6.79** (a) O, excited; (b) Br, ground; (d) P, excited; (d) In, ground.

Chapter 7

7.1 Lewis proposed that electrons in atoms are arranged in spherical shells around the nucleus. Quantum mechanics calculates that there are certain distances from the nucleus at which there is a high probability of finding an electron. For a particular atom, there are the same number of these distances with high electron density as there are occupied principal quantum levels. Lewis's electron shells correspond to principal quantum levels. **7.4** Rh < Ti < K < P < Mg. **7.6** 0.715 Å. **7.7** F < O < P < Mg < Ca. **7.9** $Sc^+ \longrightarrow Sc^{2+} + 1e^-$; $Sc^{2+} \longrightarrow Sc^{3+} + 1e^-$. **7.11** Much more energy is required to remove a $1s$ core electron close to the nucleus of Li^+ than to remove a $2s$ valence electron further from the nucleus of Be^+. **7.13** (a) Cl; (b) N; (c) Hf; (d)

N (the larger nuclear charge of O is offset by increased electron repulsion); (e) Ge. **7.15** First ionization energies increase slightly going from K to Ar, and atomic sizes decrease. As valence electrons are drawn closer to the nucleus (atom size decreases), more energy is required to completely remove them from the atom (first ionization energy increases). **7.17** The effective nuclear charge in Si^{3+} is greater. According to Table 7.2, I_4 for Si is greater than I_3 for Al; more energy is required to remove a valence electron from Si^{3+} than from Al^{2+}. **7.19** In Cl, the additional electron experiences some nuclear attraction (it is not completely shielded from the nucleus by the other five $3p$ electrons), and it completes the stable octet of valence electrons. In Ar, the stable octet already exists. The added electron occupies a $4s$ orbital and is almost completely shielded from the attraction of the nucleus. This electronic arrangement is not lower in energy than that in Ar atoms. **7.21** Metals are good conductors, and nonmetals are poor ones. Metallic character increases going from right to left across a row and from top to bottom in a column. The order of increasing metallic character and electrical conductivity is S < Si < Ge < Ca. **7.23** (a) Li; (b) Na; (c) Sn; (d) Al. **7.25** Ionic: Na_2O, CaO, Fe_2O_3; molecular: N_2O, CO, P_2O_5, Cl_2O_7. Ionic compounds are formed by combining a metal and a nonmetal; covalent compounds are formed by two or more nonmetals.
7.28 In order of increasing acidity:
$BaO < MgO < Al_2O_3 < SiO_2 < CO_2 < P_2O_5 < SO_3$.
7.29 (a) $Na_2O(s) + H_2O(l) \longrightarrow 2NaOH(aq)$;
(b) $CuO(s) + 2HNO_3(aq) \longrightarrow Cu(NO_3)_2(aq) + H_2O(l)$;
(c) $SO_3(g) + H_2O(l) \longrightarrow H_2SO_4(aq)$;
(d) $SeO_2(s) + 2NaOH(aq) \longrightarrow Na_2SeO_3(aq) + H_2O(l)$.
7.31 (a) Na: $[Ne]3s^1$; Mg: $[Ne]3s^2$. (b) When forming ions, both adopt the stable configuration of Ne, but Na loses 1 electron and Mg loses 2 electrons. (c) The effective nuclear charge of Mg is greater, so its ionization energy is greater. (d) Mg is less reactive because it has a filled subshell and a higher ionization energy. (e) The much lower melting point of Na indicates its greater metallic character. (f) The atomic radius of Mg is smaller because the effective nuclear charge is greater. **7.33** (a) The valence electrons in Sr are less tightly held (greater n value) than those of Be, so Sr is more easily oxidized. (b) Na reacts with O_2 to form Na_2O_2, sodium *per*oxide, instead of the expected metal oxide, Na_2O. (c) The $6s$ valence electrons in Hg experience a much greater effective nuclear charge and are more tightly held than those in Ba. Thus, Hg is much less reactive than Ba.
7.35 (a) $2K(s) + 2H_2O(l) \longrightarrow 2KOH(aq) + H_2(g)$;
(b) $Ba(s) + 2H_2O(l) \longrightarrow Ba(OH)_2(aq) + H_2(g)$;
(c) $6Li(s) + N_2(g) \longrightarrow 2Li_3N(s)$;
(d) $2Mg(s) + O_2(g) \longrightarrow 2MgO(s)$.
7.37 H, $1s^1$; Li, $[He]2s^1$; F, $[He]2s^22p^5$. Like Li, H has only 1 valence electron, and its most common oxidation number is +1. Like F, H needs only 1 additional electron to adopt the stable electron configuration of the nearest noble gas; both H and F can exist in the −1 oxidation state. **7.39** (a) O, $[He]2s^22p^4$; F, $[He]2s^22p^5$. (b) When forming ions, O atoms gain 2 electrons and F atoms gain 1 electron to form the stable electron configuration of Ne. (c) The ionization energy of F is greater because its valence electrons experience a greater Z_{eff}. (d) O_2 is unreactive toward H_2O, whereas F_2 reacts with H_2O to form HF. (e) Both elements react with H_2, forming H_2O and HF, respectively. (f) Electronic repulsions in the small $2p$ orbitals of F counteract its larger Z_{eff}, and the two elements have approximately equal atomic radii. **7.41** Both Xe and Kr were found to react with substances with a strong tendency to remove electrons, such as F_2. Thus, the term *inert* no longer described all the group 8A elements.
7.43 (a) $S_8(s) + 16Li(s) \longrightarrow 8Li_2S(s)$; (b) $2O_3(g) \longrightarrow$
$3O_2(g)$; (c) $2KBr(aq) + 2H_2O(l) \longrightarrow$
$2KOH(aq) + H_2(g) + Br_2(aq)$; (d) $Cl_2(g) + Ca(s) \longrightarrow$
$CaCl_2(s)$. **7.46** (a) Te has more metallic character and is a better electrical conductor. (b) At room temperature, oxygen molecules are diatomic and exist in the gas phase. Sulfur molecules are 8-membered rings and exist in the solid state. (c) Chlorine is generally more reactive than bromine because Cl atoms have a more negative electron affinity than Br atoms. **7.50** (a) Increasing ionization energy:
Si < Se < C < O < F; (b) increasing atomic radius:
F < O < C < Si ≈ Se. **7.53** (a) Zn has a significantly greater Z and Z_{eff} than Ca and thus a smaller atomic radius. (b) In the 2+ ions, both elements have lost their $4s$ electrons; the outermost electrons in Ca are in the $3p$ sublevel, and those in Zn are in the $3d$ sublevel, significantly shielded by core electrons. Thus, the electron clouds in the two ions are more similar in size than in the neutral atoms. **7.55** (a) F and O^- have the same electron configuration; they are isoelectronic. (b) In the first process, an electron is added to a neutral F atom, whereas in the second an electron is added to a negative O^- ion. Also, F has a larger Z_{eff} than does O^-. (c) Following the same trend, $N^{2-}(g)$ will have a more endothermic electron affinity than O^-. **7.57** The electron affinity of the Na^+ ion is −496 kJ/mol, the reverse of the ionization energy of Na. **7.60** Because Xe reacts with F_2 and O_2 has approximately the same ionization energy as Xe, O_2 will probably react with F_2. Possible products would be O_2F_2, analogous to XeF_2, or OF_2.

Chapter 8

8.1 (a) $\cdot\ddot{\text{P}}\cdot$; (b) $\cdot\dot{\text{Ga}}\cdot$; (c) $\cdot\dot{\underset{\cdot}{\text{Si}}}\cdot$; (d) :He.

8.3 (a) $Na\cdot + \cdot H \longrightarrow Na^+ + [:H]^-$.
8.5 (a) $CaCl_2$; (b) YO, Y_2O_3; (c) SrS; (d) Mg_3N_2.
8.7 (a) Ba^{2+}, [Xe], noble-gas configuration; (b) I^-, $[Kr]5s^24d^{10}5p^6$, noble-gas configuration; (c) Fe^{2+}, $[Ar]3d^6$; (d) Zr^{4+}, [Kr], noble gas configuration; (e) Au^{3+}, $[Xe]4f^{14}5d^8$; (f) Se^{2-}, $[Ar]3s^23p^6$, noble gas configuration. **8.10** $Ca(s) \longrightarrow Ca(g)$; $Br_2(l) \longrightarrow 2Br(g)$;
$Ca(g) \longrightarrow Ca^+(g) + 1e^-$; $Ca^+(g) \longrightarrow Ca^{2+}(g) + 1e^-$;
$2Br(g) + 2e^- \longrightarrow 2Br^-(g)$, exothermic;
$Ca^{2+}(g) + 2Br^-(g) \longrightarrow CaBr_2(s)$, exothermic;

$CaBr_2(s) \longrightarrow Ca(s) + Br_2(l)$. **8.12** (a) The charges on the ions in CaS are larger than the charges on the ions in KCl. (b) The ions in LiF are smaller and can approach each other more closely than those in CsBr. (c) Because O^{2-} is smaller than S^{2-} it can approach Mg^{2+} more closely. (d) Because Mg^{2+} is smaller than Ba^{2+} it can approach O^{2-} more closely.
8.13 The lattice energy of RbCl(*s*) is +692 kJ/mol. This value is smaller than the lattice energy for NaCl because Rb^+ has a larger ionic radius than Na^+ and therefore cannot approach Cl^- as closely as Na^+ can.
8.15 $\Delta H^\circ_f = -34$ kJ **8.18** (a) As *Z* stays constant and the number of electrons in the particle increases, the electron-electron repulsions increase, and the size of the particle increases. (b) Going down a family, the increased distance of the electrons from the nucleus and increased shielding by inner electrons outweighs the increase in *Z* and causes the size of particles with like charges to increase. (c) In an isoelectronic series, the electron configuration of each particle is the same, and *Z* changes; the larger the value of *Z*, the smaller the particle. (d) This is also an isoelectronic series [see the explanation for part (c)].
8.19 (a) $Li^+ < K^+ < Rb^+$; (b) $Mg^{2+} < Na^+ < Br^-$; (c) $K^+ < Ar < Cl^- < S^{2-}$; (d) $Ar < Cl < Cl^-$. **8.22** The effective nuclear charge on a 2*p* electron in Na^+ is much greater than that on a 2*p* electron in F^- because the total nuclear charge of Na is greater and electron repulsions are essentially the same. **8.23** (a) Ar; (b) Te; (c) Xe; (d) S.

8.25 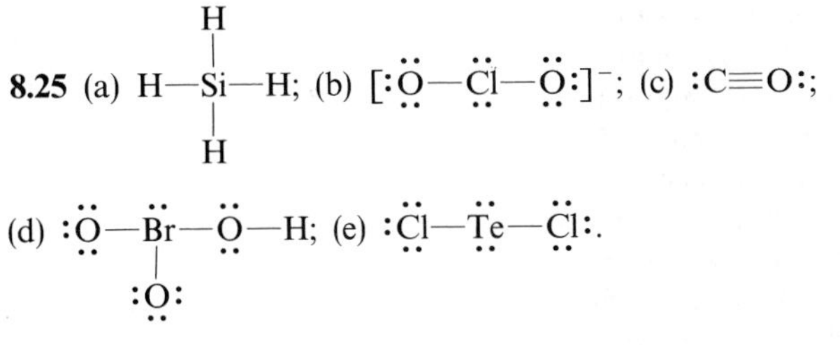

8.27 (a)  the odd electron molecule violates the octet rule. The odd electron is probably on N because it is less electronegative than O.

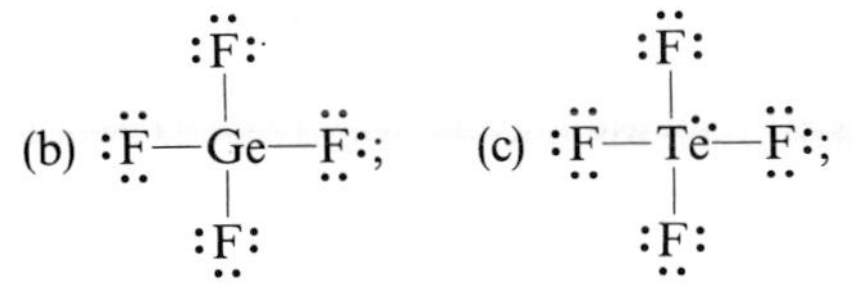
10 electrons around Te.

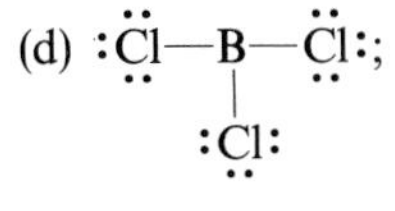
6 electrons around B; (octet structure with positive formal charge on Cl can be drawn.)

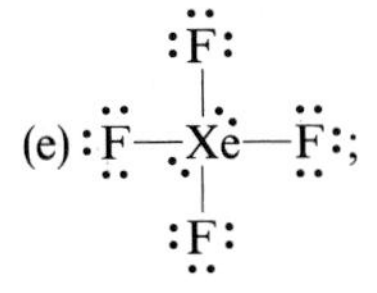
12 electrons around Xe.

8.29 (a) 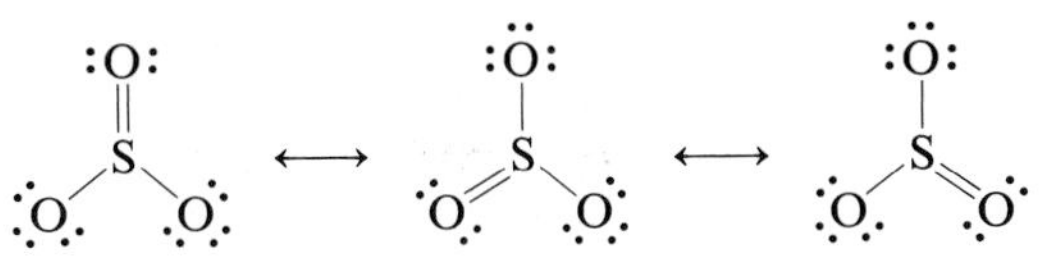

(b) 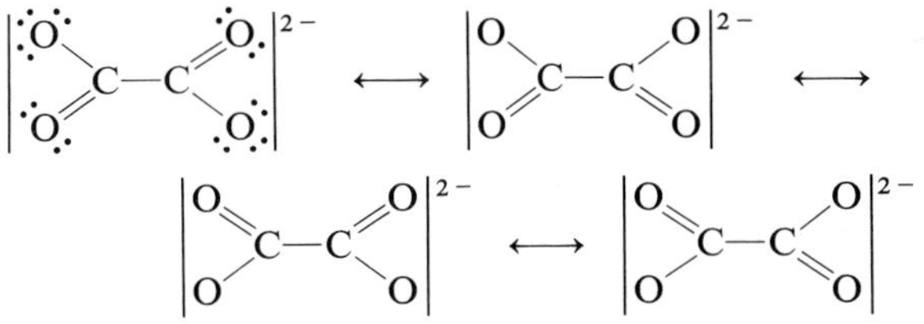

(Lone pairs on terminal O atoms have been omitted on all but the first form.)

(c) 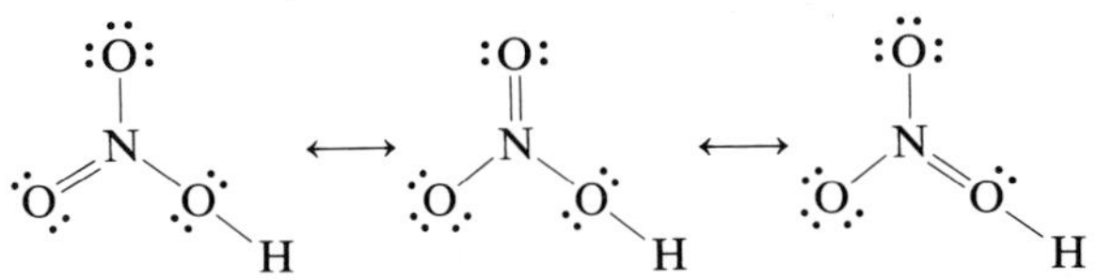

The third form is not a significant contributor because of the nonzero formal charges on all three O atoms.

(d) 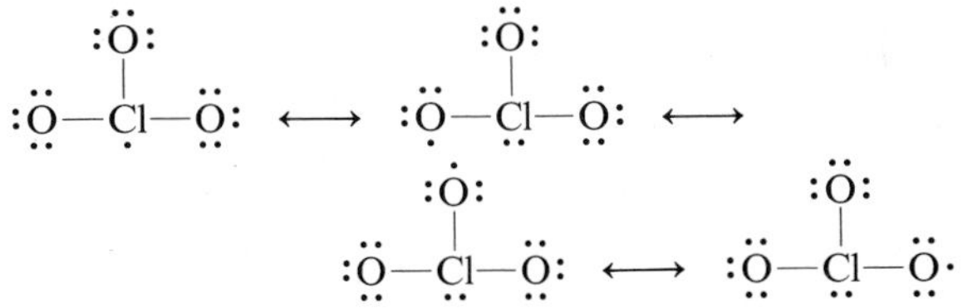

8.31 The N—O bond orders in these four cases (that is, the average number of electron pairs in the N—O bond) are 3.0 for NO^+, 2.0 for NO, 1.5 for NO_2^-, and 1.33 for NO_3^-. The N—O bond length is inversely related to the N—O bond order, so the lengths vary in the order $NO^+ < NO < NO_2^- < NO_3^-$. **8.33** (a) 1+; (b) 1−; (c) 1+ (assuming the odd electron is on N); (d) 0; (e) 3+. **8.35** (a) $\Delta H = -46$ kJ. **8.37** (a) $\Delta H = -213$ kJ. **8.39** The average bond energy for the Ti—Cl bond is 430 kJ/mol.
8.42 (a) The electronegativity of the elements increases going from left to right across a row of the periodic chart. (b) Electronegativity decreases going down a family of the periodic chart. (c) Generally, the greater the ionization energy and the more negative the electron affinity of an element, the greater the electronegativity.
8.43 (a) P < S < O; (b) Mg < Al < Si; (c) S < Br < Cl; (d) Si < C < N. **8.46** (a) C—S < N—O < B—F; (b) Pb—Pb < Pb—C < Pb—Cl; (c) O—F < H—F < Be—F. **8.47** (a) Mn, +7; (b) Fe, +3; (c) Xe, +6; (d) Sn, +4; (e) O, −1; (f) N, +3; (g) U, +6; (h) Cu, +2. **8.49** (a) Br: +7, −1; (b) As: +5, −3; (c) Ba: +2, 0; (d) Cr: +6, 0.
8.51 (a) FeF_3; (b) MoO_3; (c) $AsBr_5$; (d) vanadium(III) oxide; (e) cobalt(III) fluoride; (f) manganese(II) sulfide. **8.53** (a) Xe: 0 to +4, F: 0 to −1; (b) Cu: +2 to +1, I: −1 to 0; (c) N: −3 to +3, Cl: 0 to −1; (d) I: 0 to −1, S: +4 to +6; (e) S: −2 to +4, O: 0 to −2.

8.57 (a) $[Ar]3d^9$; (b) $[Xe]6s^24f^{14}5d^{10}$; (c) $[Ar]3d^6$; (d) [Ar]. **8.61** (a) 2.64×10^3 kJ/mol; (b) 471 kJ/mol. **8.65** (a) SiH_4; (b) SF_2; (c) PCl_3. **8.69** (a) $\Delta H =$ 1547 kJ; (b) $\Delta H = 1394$ kJ; (c) $\Delta H = 1353$ kJ. **8.72** (a) $\Delta H = +40$ kJ; ethanol has the lower enthalpy. (b) $\Delta H = -83$ kJ; acetaldehyde has the lower enthalpy. (c) $\Delta H = +82$ kJ; cyclopentene has the lower enthalpy. (d) $\Delta H = -55$ kJ; acetonitrile has the lower enthalpy. **8.75** On the basis of minimizing formal charge, the structures on the left and in the center contribute equally and the structure on the right is somewhat less important. This argument predicts that the odd electron spends most of its time on a terminal O atom. **8.79** The compound is most likely GeO_2 (m.p. GeO_2 = 1115°C).

Chapter 9

9.1 (a) trigonal planar; (b) tetrahedral; (c) trigonal bipyramidal; (d) octahedral. **9.3** The electron-pair geometry describes the locations of electron pairs as predicted by VSEPR. The molecular geometry indicates the positions of the atoms in the molecule. In H_2O, there are four pairs of electrons around oxygen, so the electron-pair geometry is tetrahedral. Because there are two bonding and two nonbonding pairs, the molecular geometry (the arrangement of the three atoms) is bent. **9.5** (a) Tetrahedral, bent; (b) tetrahedral, tetrahedral; (c) trigonal planar, trigonal planar; (d) tetrahedral, trigonal pyramidal; (e) tetrahedral, trigonal pyramidal; (f) trigonal bipyramidal, T-shaped. **9.7** NF_3, trigonal pyramidal; BF_3, trigonal planar; ClF_3, T-shaped. The molecular geometries or shapes differ because there is a different number of nonbonding pairs of electrons around each central atom. **9.9** Going from NO_2^+ to NO_2 to NO_2^-, the number of nonbonding electrons in the valence shell of N increases from 0 to 1 to 2, and the bond angle decreases correspondingly. **9.11** (a) 1—109°, 2—120°; (b) 3—109°, 4—120°; (c) 5—109°, 6—109°; (d) 7—180°, 8—109°. **9.13** (a) No; (b) yes; (c) no; (d) no; (e) yes; (f) yes. **9.15** In $BeCl_2$, there are two pairs of electrons around Be, the structure is linear, and the bond dipoles cancel. In SCl_2, there are four pairs of electrons around S, and the two nonbonding pairs do *not* cancel with the two S—Cl bonding pairs. **9.17** The middle isomer has a zero net dipole moment. **9.19** 11.7 percent ionic character. **9.21** (a) "Orbital overlap" occurs when valence atomic orbitals on two adjacent atoms share the same region of space. (b) In valence bond theory, orbital overlap allows two bonding electrons to mutually occupy the space between the bonded nuclei. (c) Valence bond theory is a combination of the atomic orbital concept and the Lewis model of electron-pair bonding. **9.23** (a) sp^2, 120° angles; (b) sp^3d, 90°, 120° and 180° angles; (c) sp^3d^2, 90° and 180° angles. **9.25** H—$\ddot{\underset{\cdot\cdot}{S}}$—H; The electron-pair geometry is tetrahedral, and the molecular shape is bent. Each S—H bond is formed by the overlap of an sp^3 hybrid orbital on S with a 1*s* atomic orbital on H. **9.27** (a) sp^3; (b) sp^3d; (c) sp^3; (d) sp; (e) sp^3d^2; (f) sp^2. **9.29** A single valence *p* orbital remains, and one π bond can form. **9.31** ~109° about the leftmost C, sp^3; ~120° about the right-hand C, sp^2. (b) The doubly bonded O can be viewed as sp^2, and the other as sp^3; the nitrogen is sp^3 with ~109° bond angles. (c) 9 σ bonds, 1 π bond. **9.33** There are three equally correct Lewis structures (resonance forms) for nitrate ion. The true model of bonding is taken as an average or composite of the three forms. Each atom in NO_3^- can be viewed as using sp^2 hybrid orbitals for σ bonding, so each has a pure 2*p* orbital available and in the correct orientation to form a delocalized π bond over the entire molecule. **9.35** (a) Bonding MOs are lower in energy than the starting atomic orbitals, and antibonding MOs are higher. (b) The electron density is concentrated between the nuclei in a bonding MO and away from the nuclei in an antibonding MO. **9.37** (a) Bond order is the net number of bonding electron pairs in a molecule $[\frac{1}{2}(\text{bonding } e^- - \text{antibonding } e^-)]$. (b) Paramagnetism is the magnetic behavior of molecules with unpaired electrons. (c) An energy-level diagram shows the interacting atomic orbitals in the left and right columns and the resulting molecular orbitals and their relative energies in the center column. **9.39** (a) BO = 3.0, diamagnetic; (b) BO = 2.0, paramagnetic; (c) BO = 3.0, diamagnetic; (d) BO = 1.5, paramagnetic. **9.41** In order of increasing bond length: $O_2^+ < O_2 < O_2^- < O_2^{2-}$. **9.43** (a) When two atomic orbitals interact, the resulting bonding molecular orbital is always lower in energy because it concentrates electron density between the two bonding nuclei, while the antibonding orbital channels electrons away from the nuclei. (b) The σ^*_{1s} orbital is higher in energy than the H 1*s* atomic orbitals because an electron in this orbital is repelled from the area between the H nuclei and is actually further from a single H nucleus than in the isolated H atom. (c) The Li 2*s* orbitals are larger than the 1*s*, the area of overlap is greater, and the energy separation of the resulting σ_{2s} and σ^*_{2s} orbitals is larger than that between σ_{1s} and σ^*_{1s}. **9.45** (a) Tetrahedral; (b) trigonal pyramidal; (c) linear; (d) trigonal planar; (e) square planar; (f) linear. **9.48** (a) 3 σ, 1 π; (b) 1 σ, 2 π; (c) 2 σ, 1 π; (d) 3 σ, 0 π. **9.50** (a) The bond dipoles in H_2O lie along the O—H bonds with the positive end at H and the negative end at O. The dipole moment vector of the H_2O molecule bisects the H—O—H angle and has a magnitude of 1.85 D with the negative end pointing toward O. (b) The magnitude of the O—H bond dipole is 1.51 D. **9.53** (b) The electron-pair geometry around each N is trigonal planar with bond angles of ~120°; because of the geometric requirements of the π bond, the entire molecule must be planar. (c) Each N atom will use sp^2 hybrid orbitals for σ bonding and pure *p* orbitals for the π bond. (e) The two N—H bonds must be on opposite sides of the molecule to produce a dipole moment of zero. If the two N—H bonds were on one side of the molecule and the two nonbonding pairs on the other side, the molecule would have a net dipole moment. **9.56** (a) $sp^3 \longrightarrow sp^2$; (b) $sp^3d \longrightarrow sp^3$; (c) $sp^3d^2 \longrightarrow sp^3d$.

9.59 From bond dissociation energies, $\Delta H = 5364$ kJ; according to Hess's law, $\Delta H^\circ = 5535$ kJ. The difference in the two results, 171 kJ, is due to the resonance stabilization in benzene. The amount of energy actually required to decompose 1 mol of $C_6H_6(g)$ is greater than the sum of the localized bond energies. **9.62** The Lewis structure of O_2 contains a double bond between O atoms, and the molecular-orbital diagram predicts a bond order of 2.0. However, the MO diagram indicates that O_2 will be paramagnetic, whereas the Lewis structure contains no unpaired electrons. **9.66** Ne_2 has a bond order of zero and is not energetically favored over isolated Ne atoms; it should not exist. Ne_2^+ has a bond order of 0.5 and is slightly lower in energy than isolated Ne atoms, so it will probably exist under special experimental conditions, but be unstable.

Chapter 10

10.1 Most of the volume of a gas is empty space. It is easy to reduce this space by applying pressure. Molecules in liquids and solids are in close contact with one another, and further compression causes strong intermolecular repulsive forces. **10.3** The height of a Hg column is steady when the pressure exerted by the atmosphere is equal to the downward pressure due to the mass of Hg in the column. Pressure is defined as force per unit area. As the diameter of the Hg column changes, the total mass of Hg and thus the force it exerts changes, but so does the area over which the force is exerted. The ratio of force to area remains constant, regardless of the diameter of the Hg column. **10.5** (a) 96.2 kPa; (b) 0.967 atm; (c) 670 torr; (d) 783 mm Hg. **10.7** (a) 32°C, 99.26 kPa; (b) 59°F, 29.5 in Hg. **10.9** 9.1×10^7 Pa. **10.11** (a) $P = 681$ mm Hg; (b) 813 mm Hg. **10.13** (a) If X and Y are *directly proportional*, a change in X will cause a proportional change in Y in the *same* direction. If X and Y are *inversely proportional*, a change in X will cause a proportional change in 1/Y. (b) Volume is inversely proportional to pressure, directly proportional to absolute temperature, and directly proportional to quantity of gas.
10.15 (a) $P = 1.60 \times 10^3$ mm Hg; (b) $V = 3.68$ L.
10.17 An ideal gas exhibits pressure, volume, and temperature relationships described by the equation $PV = nRT$. (b) $PV = nRT$; P in atmospheres, V in liters, n in moles, T in kelvins. **10.19** Solving the ideal-gas equation for V: $V = n(RT/P)$. At constant temperature and pressure, the quantity in parentheses is a constant. Thus, V is proportional to n.
10.21 (a) $P = 9.69$ atm; (b) $n = 0.147$ mol; (c) $V = 125$ L; (d) $T = 19.2$ K. **10.23** 2.42×10^{22} gas molecules.
10.25 (a) $V = 2.59$ L; (b) $V = 0.530$ L.
10.27 (a) 9.83×10^3 g O_2; (b) 6.88×10^3 L at STP.
10.29 (a) $n = 2 \times 10^{-4}$ mol O_2; (b) The roach needs 8×10^{-3} mol O_2 in 48 hr, 100 percent of the O_2 in the jar.
10.31 (a) $d = 1.67$ g/L; (b) $M = 50.0$ g/mol.
10.33 The molecular formula of cyanogen is C_2N_2.
10.35 $M = 89.4$ g/mol. **10.37** (a) The total pressure of a mixture of gases is equal to the sum of the pressures the individual gases would exert if they were present in the same container alone. (b) Partial pressure is the pressure exerted by a single component of a gaseous mixture at the same temperature and volume as the mixture. **10.39** (a) P of $CH_4 = 1.51$ atm; (b) $P_t = 5.45$ atm. **10.41** P of $N_2 = 0.74$ atm; P of $O_2 = 0.41$ atm; P of $CO_2 = 0.16$ atm. **10.43** P of $N_2 = 0.451$ atm; P of $O_2 = 1.71$ atm; $P_t = 2.16$ atm. **10.45** 10.0 L = 0.401 mol H_2; 8.44 g CaH_2. **10.47** $V = 4.25$ L CO_2. **10.49** 0.00470 mol H_2 collected; 0.307 g Zn consumed. **10.51** The kinetic molecular theory assumes that ideal-gas molecules have zero volume and experience no attractive or repulsive forces. Thus, any differences based on molecular formula or size are negligible. **10.53** (a) They have the same number of molecules. (b) SF_6 is more dense. (c) The average kinetic energies are equal. (d) N_2 will effuse faster. **10.55** (a) False; the average kinetic energy per molecule in a collection of gas molecules is the same for all gases at the same temperature. (b) True. (c) False; the molecules in a gas sample at a given temperature exhibit a distribution of kinetic energies. (d) True. **10.57** (a) In order of increasing speed: $SF_6 < HI < Cl_2 < H_2S < CO$. (b) $u_{CO} = 515$ m/s. **10.59** $M = 210$ g/mol. **10.61** (a) Nonideal-gas behavior is observed at very high pressures and low temperatures. (b) The real volumes of gas molecules and attractive intermolecular forces between molecules cause gases to behave nonideally. **10.63** According to the ideal-gas law, the ratio PV/RT should be constant for a given gas sample at all combinations of pressure, volume, and temperature. If this ratio changes with increasing pressure, the gas sample is not behaving ideally. **10.65** O_2 ($a = 1.36$ and $b = 0.0318$) will behave more like an ideal gas than Cl_2 ($a = 6.49$ and $b = 0.0562$) at high pressures. **10.67** (a) $P = 0.918$ atm; $P = 0.896$ atm. **10.69** Water at the top of the Hg column would establish a vapor pressure that would exert additional downward pressure and partially counterbalance the pressure of the atmosphere. The height of the Hg column would be less than the actual atmospheric pressure. **10.71** $P_{gas} = 0.956$ atm. **10.74** $P_{Ar} = 0.9$ mm Hg. **10.77** (a) $n = 0.0425$ mol O_2; (b) 0.388 g C_8H_{18}. **10.80** (a) 2.77 g NH_4NO_3, ignoring the vapor pressure of $H_2O(l)$. (b) The polar N_2O is soluble in H_2O but not in oil. **10.82** 70.1 mol % O_2 in the mixture. **10.85** There are 1.5 mol H_2 per mole of Sc. The equation is $2Sc(s) + 6HCl(aq) \longrightarrow 2ScCl_3(aq) + 3H_2(g)$.

Chapter 11

11.1 Gases are compressible because there are large distances between particles; particles in the solid and liquid states are close together and relatively incompressible. **11.3** Gas-phase molecules have the highest average kinetic energy; liquid-phase molecules have intermediate energy, and solid-phase molecules the least. Lowering the temperature decreases the average kinetic energy of molecules, and the substance changes phases in order of decreasing average kinetic energy. **11.5** (a) Dipole-dipole forces are the attractions between the positive end of a neutral polar molecule and the negative end of another. Example: interactions between acetone, CH_3COCH_3, molecules. (b) Ion-dipole forces are the attractions between one end of the dipole in a polar covalent molecule and an oppositely charged ion. Example: the attraction of Cu^{2+} to the negative end of a H_2O molecule. (c) London dispersion forces are the attractive forces between nonpolar molecules or neutral atoms produced by transient dipoles in the mobile electron clouds of the molecules or atoms. Example: interactions between I_2 molecules. (d) Van der Waals forces are any dipole-dipole or London dispersion forces, excluding hydrogen bonding. Example: interactions between slightly polar molecules such as CH_2Cl_2. (e) A hydrogen bond is the attraction between the H atom in a polar bond and the nonbonded pair of electrons on an adjacent electronegative atom, primarily F, N, or O. Example: interactions between HF molecules. **11.7** (a) Hydrogen bonds; (b) London dispersion forces; (c) ionic bonds; (d) covalent bonds. **11.9** (a) London dispersion; (b) hydrogen bonding; (c) dipole-dipole and London dispersion; (d) ion-dipole. **11.11** (a) Polarizability is the ease with which the electron cloud of an atom or molecule can be distorted. (b) In order of increasing polarizability: $F_2 < Cl_2 < Br_2 < I_2$. In general, the larger the molecule, the less tightly the electron cloud is held. **11.13** (a) O_2; higher molecular weight. (b) SiH_4; higher molecular weight. (c) NaCl; ion-ion forces are stronger than dipole-dipole forces. (d) C_2H_5OH; hydrogen bonding is stronger than dipole-dipole forces. **11.15** Surface tension, high boiling point, high heat capacity per gram. **11.17** Viscosities and surface tensions of liquids both increase as intermolecular forces become stronger. **11.19** (a) The stronger hydrogen bonding interactions between ethanol molecules make the liquid more resistant to flow, more viscous. (b) Polar water molecules have strong cohesive intermolecular interactions relative to the weak adhesive forces between polar water molecules and nonpolar polyethylene molecules in the capillary, so the meniscus is concave downward. **11.21** Endothermic: melting, vaporization, sublimation; exothermic: condensation, freezing, deposition. **11.23** 40.9 kJ released; 142 g CCl_2F_2 evaporated. **11.25** 22.9 kJ. **11.27** The critical temperature is the highest temperature at which a gas can be liquified, regardless of pressure. **11.29** The boiling point is the temperature at which the vapor pressure of the liquid equals the external pressure acting on the liquid; the lower the external pressure, the lower the boiling point. The melting point is the temperature at which the solid and liquid phases are in equilibrium; neither solids nor liquids are significantly affected by pressure because neither is very compressible. **11.31** Because CCl_4 has the lowest molar mass, it experiences the weakest London dispersion forces, has the highest vapor pressure, and is most volatile. **11.33** The temperature of the water in the two pans is the same. **11.35** (a) Approximately 68°C. (b) Approximately 29°C. **11.37** 97°C. **11.39** (a) $H_2O(g)$ will condense to $H_2O(s)$ at approximately 4 mm Hg; at a higher pressure, perhaps 5 atm or so, $H_2O(s)$ will melt to form $H_2O(l)$. (b) $H_2O(s)$ will melt at just above 0°C to form $H_2O(l)$. Just below 70°C, the vapor pressure of $H_2O(l)$ reaches 0.3 atm, and $H_2O(l)$ boils to form $H_2O(g)$.

11.41

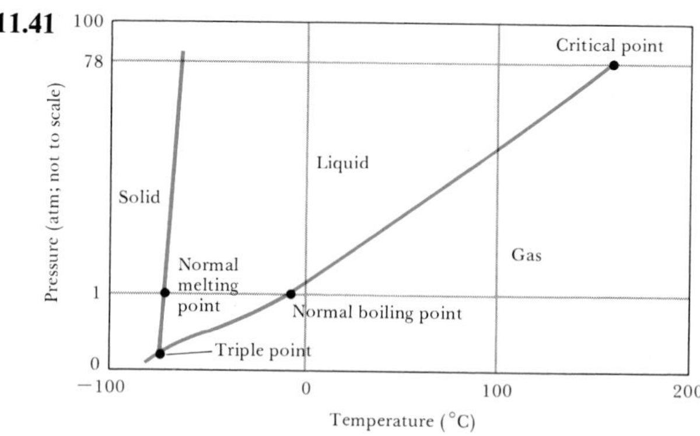

11.43 In a crystalline solid, the component particles are arranged in an ordered repeating pattern. In an amorphous solid, there is no orderly structure. **11.45** The unit cell is the building block of the crystal lattice. When repeated in three dimensions, it produces the crystal lattice. It is a parallelepiped with the characteristic distances and angles, a, b, c, α, β, and γ. Unit cells can be primitive or centered. **11.47** (a) 12; (b) 6; (c) 8. **11.49** (a) 4; (b) 12; (c) $l =$ 351 pm or 3.51 Å; (d) density = 9.02 g/cm^3. **11.51** $l =$ 538.7 pm or 5.387 Å. **11.53** Atomic weight = 55.7 g/mol. **11.55** Constructive interference occurs when two waves are in phase, and destructive interference occurs when two waves are out of phase. When radiation scattered from two different lattice planes in a crystal converges, the two waves can be in phase or out of phase. **11.57** $d = 3.56$ Å. **11.59** (a) Hydrogen bonding, dipole-dipole forces, London dispersion forces; (b) covalent chemical bonds; (c) ionic bonds; (d) metallic bonds. **11.61** Graphite (carbon) and quartz (SiO_2). **11.63** (a) KBr, ion-ion vs. dispersion; (b) SiO_2, (network) covalent vs. dispersion; (c) Se, (network) covalent vs. dispersion; (d) MgF_2, higher charges, stronger electrostatic forces in the solid. **11.65** According to Table 11.6, the solid could be either ionic with low water solubility or network covalent. Because of the extremely high sublimation temperature, it is probably *network covalent*. **11.67** (a) 8; (b) 6; (c) 4. **11.69** $N_2(g)$ can be condensed into a liquid at low temperatures and/or high pressures. **11.71** (a) London dispersion; (b) hydrogen bonding, van der Waals; (c) van der Waals; CH_3OH will have the highest boiling point because hydrogen bonding is the strongest intermolecular force present in the given molecules. **11.74** (a) Viscosity; (b) boiling point; (c) surface tension; (d) heat of vaporization. **11.77** (a) Hydrogen bonding; (b) London dispersion forces; (c) ionic bonding. **11.81** 35.5% relative humidity. **11.85** A plot of log (VP) vs.

$1/T$ is linear; the slope is -3.15×10^3, and $\Delta H_v = 60.3$ kJ/mol. **11.88** $r = 124.8$ pm or 1.248 Å.

Chapter 12

12.2 Reinitzer observed that cholesteryl benzoate has a phase that exhibits properties intermediate between those of the solid and liquid phases. This liquid-crystalline phase is opaque, is defined by sharp transition temperatures, and changes color as the temperature is increased. **12.3** Long, rodlike molecules. **12.5** *Nematic* and *smectic* phases are formed by rodlike molecules aligned with their long axes parallel; smectic phases have some additional ordering. In *cholesteric* phases the molecules are stacked in layers and are often twisted with respect to molecules in adjacent layers. **12.7** The substance must (1) be chemically stable, (2) exist in the liquid-crystalline phase between approximately 0°C and 40°C, (3) be transparent to visible light, (4) not exhibit a color change with changing temperature, and (5) rapidly change orientation in an applied electric field. **12.10** Naturally occurring polymers are: protein materials, DNA, RNA, rubber, silk, starch and forms of cellulose. Synthetic materials include those listed in Table 12.1.

12.12 $n\mathrm{CH_2{=}CH{-}C(Cl){=}CH_2} \longrightarrow \left[\mathrm{{-}CH_2{-}CH{=}C(Cl){-}CH_2{-}}\right]$

12.13 $n\mathrm{HO{-}C_6H_2(CH_3)_2{-}OH} \longrightarrow \left[\mathrm{{-}O{-}C_6H_2(CH_3)_2{-}O{-}C_6H_2(CH_3)_2{-}O{-}}\right]_n$

12.16 The crystallinity of a polymer is the degree of order among the polymer chains. **12.17** (a) An elastomer is a polymer material that recovers its shape when released from a distorting force. (b) A thermoplastic polymer can be shaped and reshaped on application of heat and/or pressure. (c) A thermosetting plastic cannot easily be reshaped once set, because of the presence of chemical bonds that cross-link the polymer chains. **12.20** Engineering ceramics are more resistant to heat than are plastics, are significantly less dense than metals, are resistant to corrosion and wear, and are not easily deformed by stress. **12.22** The C atoms form a face-centered cubic array with Si atoms occupying *alternate* tetrahedral holes in the lattice. Each Si is bound to four C atoms, and each C is bound to four Si atoms in a tetrahedral arrangement, producing an extended three-dimensional network. The extremely high melting point of SiC indicates that it is probably not a purely ionic solid and that the Si—C bonding network has significant covalent character. The extended three-dimensional nature of the structure produces the exceptional hardness, and the covalent character of the bonding network provides the great thermal stability. **12.24** To increase the resistance of a ceramic to mechanical failure, the processed material must be as free of defects as possible. To this end, very pure, small particles of the ceramic are produced and then sintered to form the desired object. Also, ceramic fibers can be imbedded in a host ceramic matrix to toughen the matrix material. **12.25** If we begin with solid metal hydroxide and add the appropriate solvents, we have little control over the particle size of the suspended solid or the uniformity of the sol. These characteristics of the sol are very important in determining the ultimate purity, particle size, and fracture resistance of the ceramic produced. **12.27** A composite is a material in which ceramic fibers have been imbedded in a host ceramic matrix; the composite will be much more fracture-resistant than the pure host ceramic. **12.30** (a) Limited electrical conductivity that changes reproducibly with temperature; (b) hardness and fracture resistance due to fiber reinforcement; (c) structural stability and electrical insulation; (d) heat resistance and low density. **12.32** $YBa_2Cu_3O_7$ is a ceramic and has a much higher superconducting transition temperature, $T_c = 95$ K, than previously known superconducting materials. **12.33** A "thin film" is a very thin coating (0.1 to 300 μm) of a material applied to a substrate. A tantalum carbide thin film on a cutting tool should be heat- and fracture-resistant, adhere well to the metal substrate of the tool, have a uniform thickness, and be of precisely known composition and free of imperfections or dislocations. **12.36** (a) MgO, vacuum deposition or sputtering; (b) Teflon, sputtering; (c) Ti, vacuum deposition or sputtering or chemical vapor deposition (CVD). **12.38** (a) Polymer; (b) ceramic; (c) ceramic; (d) polymer; (e) liquid crystal. **12.42** (a) Thin films of transparent, electrically conducting material on the glass plates of a liquid-crystal display; photo-voltaic devices (solar cells); (b) refractory and protective coatings on optical lenses; (c) to increase the hardness of a metal cutting tool or drill bit.

12.44 (a) $\left[\mathrm{{-}Si(CH_3)_2{-}}\right]_n \xrightarrow{400^\circ\mathrm{C}} \left[\mathrm{{-}SiH(CH_3){-}CH_2{-}}\right]_n$

$\left[\mathrm{{-}SiH(CH_3){-}CH_2{-}}\right]_n \xrightarrow{1200^\circ\mathrm{C}} [\mathrm{SiC}]_n + \mathrm{CH_4}(g) + \mathrm{H_2}(g)$

(b) $2NbBr_5(g) + 5H_2(g) \longrightarrow 2Nb(s) + 10HBr(g)$;
(c) $SiCl_4(l) + C_2H_5OH(l) \longrightarrow Si(OC_2H_5)_4(s) + 4HCl(g)$;

(d) n C_6H_5–CH=CH$_2$ ⟶ [–CH(C_6H_5)–CH$_2$–]$_n$

12.48 These compounds cannot be vaporized without destroying their chemical identities. Under the conditions of vacuum deposition, anions with names ending in *-ate* or *-ite* tend to chemically decompose to form gaseous nonmetal oxides.

Chapter 13

13.1 (a) 3.68% $Ba(NO_3)_2$ by weight; (b) 5.5 ppm Au. **13.3** (a) $X_{CH_3OH} = 6.98 \times 10^{-3}$; (b) $X_{CH_3OH} = 0.290$. **13.5** (a) 0.00788 M KBr; (b) 0.0262 M $Ca(NO_3)_2 \cdot 4H_2O$; (c) 0.0125 M HCl. **13.7** (a) 9.79 m C_6H_6; (b) 0.402 m NaCl. **13.9** (a) 41.1% H_2O by weight; (b) $X_{H_2O} = 0.692$; (c) 38.7 m H_2O; (d) 21.1 M H_2O. **13.11** (a) 16 M HNO_3. **13.13** (a) 0.127 mol $Ca(NO_3)_2$; (b) 2.68×10^2 mol HBr; (c) 3.80×10^3 mol $Al(NO_3)_3$. **13.15** (a) Weigh 2.50 g KBr, dissolve in water, dilute with stirring to 1.40 L. (b) Weigh 11.4 g KBr, dissolve in 238.6 g H_2O to make 250 g of solution. (c) Weigh 198 g KBr, dissolve in 1452 g H_2O to make 1.50 L of solution. (d) Weigh 13.3 g KBr, dissolve in a small amount of H_2O, dilute to 0.560 L. **13.17** (a) 0.76 N Sn^{4+}; (b) 0.111 M H_2SO_4. **13.19** The solubility of a solute in a particular solvent depends on the strength of the "new" solute-solvent interactions relative to the strengths of "old" solute-solute and solvent-solvent interactions. **13.21** (a) Ion-dipole; (b) London dispersion; (c) hydrogen-bonding; (d) dipole-dipole. **13.23** (a) A solute is more likely to dissolve in a solvent if the strengths of the solute-solute, solvent-solvent, and solute-solvent interactions are similar. (b) C_6H_6. **13.25** (a) There are no solute-solvent interactions between ionic KCl and nonpolar covalent C_6H_6 strong enough to compete with the ionic lattice forces in KCl. (b) Because the molar masses of CCl_4 and Br_2 are very similar, the London dispersion forces in Br_2 are more similar to those in CCl_4. **13.27** For ionic solids, the exothermic part of the solution process is ΔH_3, the enthalpy released when solute and solvent particles interact. In hydrated salts, the ions are already interacting with water molecules; therefore, less enthalpy is released during solvation, and the overall enthalpy of solution is more positive. **13.29** The solubility of most ionic solids in water increases with increasing temperature, and the solubility of most gases decreases with increasing temperature. **13.31** 9.2×10^{-4} M He, 1.5×10^{-3} M N_2. **13.33** The solubility of a gas is determined in part by the attractive forces between the gas and solvent molecules. At higher temperatures, the increased kinetic energies of the molecules counteract the effects of these attractive interactions. **13.35** 68 g $KNO_3(s)$ per 100 g $H_2O(l)$. **13.37** Colligative properties depend on concentration (total number of solute particles) but not identity of solute particles. **13.39** (a) 2.29 mol (142 g) $C_2H_6O_2$; (b) 26.7 g KBr. **13.41** (a) $T_f = -115.0°C$, $T_b = 78.6°C$; (b) $T_f = -33.3°C$, $T_b = 78.7°C$; (c) $T_f = -1.3°C$, $T_b = 100.4°C$. **13.43** 0.03 m glycerin < 0.03 m benzoic acid < 0.02 m KBr. **13.45** $\pi = 4.40$ atm. **13.47** $M = 1.8 \times 10^2$ g/mol adrenaline. **13.49** $M = 1.39 \times 10^4$ g/mol lysozyme. **13.51** (a) $i = 2.75$. (b) As the solution becomes more concentrated, the value of i will decrease. **13.53** A solution has very small dispersed solute particles, does not scatter light, and is transparent. A colloid has much larger dispersed particles, scatters light, and is usually cloudy. White vinegar is a solution; milk is a colloid. **13.55** (a) Hydrophobic; (b) hydrophilic; (c) hydrophobic. **13.57** Colloids do not coalesce because of electrostatic repulsions between groups at the surface of the dispersed particles. Hydrophilic colloids can be coagulated by adding electrolytes, and some colloids can be coagulated by heating. **13.59** 4.6×10^3 ppm Br^-. **13.62** (a,b) CCl_4 solution: $X = 9.43 \times 10^{-3}$, 0.0619 m, 0.0978 M; C_2H_5OH solution: $X = 5.77 \times 10^{-3}$, 0.126 m, 0.0967 M. (c) The molarities are very similar; the slight difference is due to the assumption that the densities of the solutions are equal to the densities of the pure solvents. The molalities of the solutions differ by a factor of 2 because the densities of the two solvents are quite different. Equal volumes of the two solvents have different masses, which lead to different calculated molalities. **13.64** (a) KBr; (b) NH_3; (c) HBr. **13.67** 7.4 kg C_2H_5OH. **13.70** (a) $T_f = -0.558°C$; (b) $T_f = -0.432°C$. **13.73** $M = 282$ g/mol; the molecular formula is $C_{19}H_{38}O$. **13.77** The nonelectrolyte beaker contains 18.2 mL, the NaCl(aq) beaker holds 21.8 mL.

Chapter 14

14.1 (a) $-\Delta[H_2O_2]/\Delta t = \Delta[H_2]/\Delta t = \Delta[O_2]/\Delta t$; (b) $-\Delta[MnO_2]/\Delta t = -\Delta[Mn]/\Delta t = \frac{1}{2}\Delta[MnO]/\Delta t$; (c) $-\Delta[C_6H_{14}]/\Delta t = \frac{1}{6}\Delta[CO]/\Delta t = \frac{1}{7}\Delta[H_2O]/\Delta t$; $-\Delta[O_2]/\Delta t = \frac{13}{12}\Delta[CO]/\Delta t = \frac{13}{14}\Delta[H_2O]/\Delta t$.

14.3

Time (min)	Time interval (min)	Concentration (M)	ΔM	Rate (M/min)
0		1.85		
	79		0.18	2.3×10^{-3}
79		1.67		
	79		0.15	1.9×10^{-3}
158		1.52		
	158		0.22	1.4×10^{-3}
316		1.30		
	316		0.30	0.95×10^{-3}
632		1.00		

14.5 From the slopes of the tangents to the graph, the rates are: $t = 100$ min, 1.9×10^{-3} M/min; $t = 500$ min, 9.4×10^{-4} M/min. **14.7** (a) $-\Delta[O_2]/\Delta t = 2.3$ mol/s, $+\Delta[H_2O]/\Delta t = 4.6$ mol/s. (b) $\Delta P_T/\Delta t = -15$ mm Hg/min. **14.9** (a) Rate $= 6.08 \times 10^{-5}$ M/s; (b) rate $= 1.85 \times 10^{-4}$ M/s. **14.11** (a, b) $k = 0.40$ $M^{-1}s^{-1}$; (c) The rate would triple. **14.13** (a) Rate $= k[S_2O_8{}^{2-}][I^-]$; (b) $k = 6.1 \times 10^{-3}$ $M^{-1}s^{-1}$; (c) $-\Delta[S_2O_8{}^{2-}]/\Delta t = 1.5 \times 10^{-5}$ M/s; (d) $+\Delta[SO_4{}^{2-}]/\Delta t = 1.5 \times 10^{-5}$ M/s. **14.15** (a) The reaction is first order in both BF_3 and NH_3. Rate $= k[BF_3][NH_3]$; (b) second order overall; (c) $k = 3.41$ $M^{-1}s^{-1}$. **14.17** (a) $t_{1/2} = 1.9 \times 10^2$ s. **14.19** Plot $\log P_{SO_2Cl_2}$ vs. time, slope $= -9.53 \times 10^{-6}$ s^{-1}; $k = 2.19 \times 10^{-5}$ s^{-1}. **14.21** A plot of $\log[C_{12}H_{22}O_{11}]$ vs. time is linear; the reaction is first order. $k = -2.30$ (slope) $= -2.30\,(-1.60 \times 10^{-3}$ $min^{-1}) = 3.68 \times 10^{-3}$ $min^{-1} = 6.13 \times 10^{-5}$ s^{-1}. **14.23** (a) 0.162 mol N_2O_5; (b) 5.62 min; (c) $t_{1/2} = 102$ s $= 1.69$ min. **14.25** A plot of $1/[NO_2]$ vs. time is a straight line with slope $= k = 10.2$ $M^{-1}s^{-1}$. **14.27** The reaction is first order.

14.29 (a)

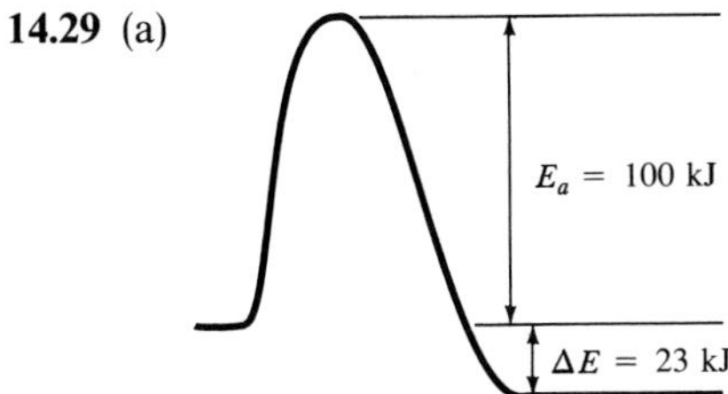

(b) +123 kJ/mol.

14.31 No. The value of A, related to frequency and effectiveness of collisions, is different for each reaction, and k is proportional to A. **14.33** A plot of $\log k$ vs. $1/T$ has a slope of -2.47×10^3; $E_a = -2.3R$(slope) $= 47$ kJ/mol. **14.35** (a) k at 600°C $= 1.19 \times 10^{-4}$ M/s; (b) k at 800°C $= 1.28 \times 10^{-2}$ M/s. **14.37** $E_a = 53.6$ kJ/mol.
14.39 A unimolecular process has one reactant molecule, a bimolecular process has two reactant molecules, and a termolecular process has three reactant molecules. Termolecular processes are rare, because it is highly unlikely that three molecules will collide effectively to form an activated complex. **14.41** (b) First step: $-\Delta[NO]/\Delta t = k[NO]^2$; second step: $-\Delta[H_2]/\Delta t = k[H_2][N_2O_2]$; (c) N_2O_2 is the intermediate. (d) The second step is fast relative to the first.
14.43 Both (b) and (d) are consistent with the observed rate law. **14.45** Catalysts generally *do* take part in the reaction, but there is no net consumption of the catalyst.
14.47 Finely ground iron filings would be a more effective catalyst because they would have a greater surface area and provide a greater number of active sites. **14.49** As illustrated in Figure 14.19, the two C—H bonds that exist on each carbon atom of the ethylene molecule before adsorption are retained in the process in which a single D atom is added to each C (assuming we use D_2 rather than H_2). To put two D atoms on a single carbon, it is necessary that one of the already existing C—H bonds in ethylene be broken while the molecule is adsorbed, so that the H atom moves off as an adsorbed atom and is replaced by a D atom. This requires a larger activation energy than simply adsorbing C_2H_4 and adding one D atom to each carbon.
14.53 $\Delta[Cl^-]/\Delta t = 7.8 \times 10^{-7}$ M/s.
14.57 (a) The reaction is first order; $k = 1.48 \times 10^{-3}$ s^{-1}; (b) $t_{1/2} = 468$ s. **14.60** A plot of $\log k$ vs. $1/T$ is linear with a slope of -3.4×10^3. $E_a = -2.3R$ (slope) $= 65$ kJ/mol.
14.62 (a) $Cl_2(g) + CHCl_3(g) \longrightarrow HCl(g) + CCl_4(g)$; (b) $Cl(g)$, $CCl_3(g)$; (c) step 1—unimolecular, step 2—bimolecular, step 3—bimolecular; (d) step 2 is rate-determining; (e) rate $= k[CHCl_3][Cl_2]^{1/2}$. **14.66** E_a for the catalyzed reaction is 35.5 kJ/mol. The activation energy must be lowered by approximately 10 kJ/mol by the enzyme.

Chapter 15

15.1 (a) At equilibrium, the forward and reverse reactions proceed at equal rates. (b) At equilibrium, the net concentrations of reactants and products are *constant* but not necessarily equal. **15.3** (a) $K = [SO_3]^2/[O_2][SO_2]^2$, homogeneous; (b) $K = [HCl]^2/[H_2][Cl_2]$, homogeneous; (c) $K = [H_2O]^2[SO_2]^2/[H_2S]^2[O_2]^3$, homogeneous; (d) $K = [H_2O]/[H_2]$, heterogeneous; (e) $K = [N_2][H_2O]^2$, heterogeneous; **15.5** $K = 54$.
15.7 (a) $K_c = 4.18 \times 10^{-9}$. (b) Because the value of K_c is much less than 1, the reactants are favored.
15.9 (a) $K_c = 4.9 \times 10^{-2}$; (b) $K_c = 416$; (c) $K_p = 5.21$.
15.11 $K_c = 0.16$, $K_p = 14$. **15.13** (a) $[H_2] = 0.012$ M, $[N_2] = 0.019$ M, $[H_2O] = 0.138$ M; (b) $K_c = 6.5 \times 10^2$.
15.15 (a) $K_c = 6.42 \times 10^{-2}$; (b) $K_p = 1.98$; (c) $P_t = 0.968$ atm. **15.17** (a) $Q = 2.19 \times 10^{-10}$; the mixture is at equilibrium. (b) $Q = 1.03 \times 10^{-8}$; the reaction will proceed to the left. (c) $Q = 1.68 \times 10^{-12}$; the reaction will proceed to the right. **15.19** $[O_2] = 1.2$ M. **15.21** $[Br_2] = 7.67 \times 10^{-3}$ M, $[Br] = 2.82 \times 10^{-3}$ M, 0.0451 g Br.
15.23 $P_{CO_2} = 1.49 \times 10^{-3}$ atm. **15.25** $[H_2S] = [NH_3] = 1.1 \times 10^{-2}$ M. **15.27** $[I_2] = [Br_2] = 0.0267$ M, $[IBr] = 0.447$ M. **15.29** (a) Shift equilibrium to the right; (b) no effect; (c) shift equilibrium to the right; (d) shift equilibrium to the left; (e) no effect; (f) shift equilibrium to the right.
15.31 (a) No effect; (b) no effect; (c) increase equilibrium constant; (d) no effect. **15.34** (a) $K = [Ag(NH_3)_2{}^+]/[Ag^+][NH_3]^2$; (b) $K = [Ag^+]^2[CrO_4{}^{2-}]$; (c) $K = [H^+][NO_2{}^-]/[HNO_2]$; (d) $K = [Zn^{2+}]/[Cu^{2+}]$; (e) $K = [NH_4{}^+][OH^-]/[NH_3]$.
15.37 $P_{NOCl} = 0.28$ atm. **15.41** (a) $K_c = 7.0 \times 10^{-5}$; (b) $[PH_3] = 1.6 \times 10^{-3}$ M. **15.43** 3.07×10^{-3} mol Br.
15.46 $K_p = 0.0282$. **15.48** (a) Left; (b) right; (c) left; (d) right; (e) left. **15.51** (a) $\Delta H^\circ_{rxn} = +178.1$ kJ. Because the reaction is endothermic, the equilibrium constant will *increase* with increasing temperature. (b) The rates of both the forward and reverse reactions will increase, and the rate at which equilibrium is reached will increase. (c) Heating causes the value of K to increase and favors the forma-

tion of products. Also, in an open beaker $CO_2(g)$ escapes to the atmosphere, and more $CaCO_3$ reacts to replace it (Le Châtelier's principle) until no more $CaCO_3$ remains. **15.56** $[H_2] = [I_2] = 2.46 \times 10^{-2}\ M$; $[HI] = 0.170\ M$.

Chapter 16

16.1 Solutions of HCl and H_2SO_4 conduct electricity, taste sour, turn litmus papter red (are acidic), neutralize solutions of bases, and react with active metals to form $H_2(g)$. HCl and H_2SO_4 solutions have these properties in common because both compounds are strong acids. That is, they both dissociate completely in H_2O to form $H^+(aq)$ and an anion. (HSO_4^- is not completely dissociated, but the first dissociation step for H_2SO_4 is complete.) The presence of ions enables the solutions to conduct electricity; the presence of $H^+(aq)$ in excess of $1 \times 10^{-7}\ M$ accounts for all other properties listed. **16.3** In solution, water molecules surround HBr molecules, and the H—Br polar covalent bond is broken to form H^+ and Br^-, each hydrated. These mobile charged particles enable the solution to conduct electricity. **16.5** (a) NH_4^+; (b) H_2CO_3; (c) HCO_3^-. **16.7** (b) Acid: $H_2C_2O_4$; conjugate base: $HC_2O_4^-$; base: H_2O, conjugate acid: H_3O^+. **16.9** (a) CN^-; Cl^- is the conjugate base of the strong acid HCl, which is completely dissociated in water. Cl^- has no tendency to accept $H^+(aq)$. **16.11** (a) Acidic; (b) acidic; (c) basic; (d) acidic; (e) acidic; (f) basic. **16.13** (a) pH = 3.60; (b) pH = 2.14; (c) pH = 9.52; (d) pH = 8.12. **16.15** (a) 0.056 M $H^+(aq)$. **16.17** $[H^+] = [OH^-] = 1.6 \times 10^{-7}\ M$; pH = pOH = 6.81. **16.19** (a) $[H^+]$ changes by a factor of 100. **16.21** (a) H^+, ClO_4^-—both major; (b) H^+, NO_3^-—both major; (c) H^+, $C_2H_3O_2^-$, $HC_2H_3O_2$—$HC_2H_3O_2$ is major (present at a much greater concentration than the other species); (d) H^+, Br^-—both are major. **16.23** Strongest, $HClO_4$; weakest, HClO. **16.25** (a) $[H^+] = 0.025\ M$, pH = 1.60; (b) $[H^+] = 0.0164\ M$, pH = 1.785; (c) $[H^+] = 0.075\ M$, pH = 1.12; (d) $[H^+] = 0.0125\ M$, pH = 1.90. **16.27** $K_a = 1.4 \times 10^{-4}$. **16.29** (a) $[H^+] = 2.7 \times 10^{-5}\ M$; (b) $[H^+] = 1.5 \times 10^{-3}\ M$; (c) $[H^+] = 9.4 \times 10^{-7}\ M$. **16.31** (a) $[H^+] = 2.8 \times 10^{-3}\ M$, 0.70% ionization; (b) $[H^+] = 1.4 \times 10^{-3}\ M$, 1.4% ionization; (c) $[H^+] = 8.7 \times 10^{-4}\ M$, 2.2% ionization. **16.33** $[H^+] = [X^-] = 0.019\ M$, $[HX] = 0.181\ M$, $K_a = 2.0 \times 10^{-3}$. **16.35** $HX(aq) \rightleftharpoons H^+(aq) + X^-(aq)$; $K_a = [H^+][X^-]/[HX]$. Assume that the percent of acid that dissociates (ionizes) is small. Let $[H^+] = [X^-] = y$. $K_a = y^2/[HX]$; $y = K_a^{1/2}[HX]^{1/2}$. Percent ionization = $(y/[HX]) \times 100$. Substituting for y, percent ionization = $100K_a^{1/2}[HX]^{1/2}/[HX]$ or $100K_a^{1/2}/[HX]^{1/2}$. That is, percent ionization varies inversely as the square root of the concentration of HX. **16.37** $[H^+] = 5.7 \times 10^{-3}\ M$, pH = 2.24. The approximation that the first dissociation is less than 5% of the total acid concentration is not valid; the quadratic equation must be solved rigorously. The $[H^+]$ produced from the second and third dissociations is small with respect to that present from the first step; the second and third ionizations can be neglected when calculating the $[H^+]$ and pH. **16.39** (a) $[OH^-] = 0.050\ M$, pH = 12.70; (b) $[OH^-] = 0.116\ M$, pH = 13.066. **16.41** (a) $C_3H_7NH_2(aq) + H_2O(l) \rightleftharpoons C_3H_7NH_3^+(aq) + OH^-(aq)$; $K_b = [C_3H_7NH_3^+][OH^-]/[C_3H_7NH_2]$. (b) $CN^-(aq) + H_2O(l) \rightleftharpoons HCN(aq) + OH^-(aq)$; $K_b = [HCN][OH^-]/[CN^-]$. (c) $CHO_2^-(aq) + H_2O(l) \rightleftharpoons HCHO_2(aq) + OH^-(aq)$; $K_b = [HCHO_2][OH^-]/[CHO_2^-]$. **16.43** (a) $[OH^-] = 9.2 \times 10^{-6}\ M$, pH = 8.96; (b) $[OH^-] = 1.5 \times 10^{-5}\ M$, pH = 9.17; (c) Using the quadratic formula, $[OH^-] = 2.2 \times 10^{-4}\ M$, pH = 10.35. **16.45** (a) $K_b = 2.2 \times 10^{-11}$; (b) $K_b = 5.3 \times 10^{-10}$; (c) $K_b = 1.6 \times 10^{-7}$; (d) $K_b = 5.5 \times 10^{-11}$. **16.47** (a) $[OH^-] = 1.4 \times 10^{-3}\ M$, pH = 11.15; (b) $[OH^-] = 3.8 \times 10^{-3}\ M$, pH = 11.58; (c) $[NO_2^-] = 0.50\ M$, $[OH^-] = 3.3 \times 10^{-6}\ M$, pH = 8.52. **16.49** (a) Basic; (b) basic; (c) acidic; (d) basic. **16.51** (a) Acidic: CH_3NH_3Br, $Zn(NO_3)_2$; (b) basic: KCNO, $Ba(C_2H_3O_2)_2$, Na_2HPO_4; (c) Na_2HPO_4 is the most basic. **16.53** $[OH^-] = 6.6 \times 10^{-6}\ M$, pH = 8.82. **16.55** (a) As the electronegativity of the central atom (X) increases, the strength of the oxyacid increases. (b) As the number of nonprotonated oxygen atoms in the molecule increases, the strength of the oxyacid increases. **16.57** (a) H_2SO_3; (b) H_3PO_4; (c) H_2SO_3. **16.59** (a) True. (b) False. In a series of acids that have the same central atom, acid strength increases with the number of nonprotonated oxygen atoms bonded to the central atom. (c) False. H_2Te is a stronger acid than H_2S because the H—Te bond is longer, weaker, and more easily dissociated than the H—S bond.

16.61

Theory	Acid	Base
Arrhenius	Forms H^+ ions in water	Produces OH^- in water
Brønsted-Lowry	Proton (H^+) donor	Proton acceptor
Lewis	Electron-pair acceptor	Electron-pair donor

The Brønsted-Lowry theory is more general than Arrhenius's definition because it is based on a unified model for the processes responsible for acidic or basic character, and it shows the relationships between these processes. The Lewis theory is more general still because it does not restrict the acidic species to compounds having ionizable hydrogen. Any substance that can be viewed as an electron-pair acceptor is defined as a Lewis acid. **16.63** (a) Acid: $Fe(ClO_4)_3$ or Fe^{3+}; base: H_2O. (b) Acid: H_2O; base: CN^-. (c) Acid: BF_3; base: $(CH_3)_3N$. (d) Acid: HIO; base: NH_2^-. **16.65** (a) CdI_2; (b) $Fe(NO_3)_3$; (c) $CrCl_3$. **16.67** In order of decreasing base strength: $OH^- > NH_3 > C_2H_3O_2^- > H_2O > Br^-$. **16.70** (a) $[OH^-] = 0.030\ M$, pH = 12.48;

(c) $[H^+] = 9 \times 10^{-7}$ M, pH = 6.0; (e) using the quadratic formula, $[OH^-] = 2.8 \times 10^{-3}$ M, pH = 11.44. **16.73** (a) NH_3 cannot be a proton donor in water. If a base strong enough to remove a proton from NH_3 were present in solution, that base would preferentially react with H_2O, because H_2O is a stronger proton donor (acid) than NH_3. There cannot be a stronger base in a solvent than the conjugate base characteristic of that solvent. In water the strongest base possible is OH^-, which cannot remove a proton from NH_3. (b) $[OH^-] = 0.015$ M, pH = 12.18. **16.76** pH = 6.34. **16.78** $[H^+] = 3.2 \times 10^{-2}$ M, $[HSO_4^-] = 1.8 \times 10^{-2}$ M, $[SO_4^{2-}] = 6.8 \times 10^{-3}$ M. **16.80** Using the quadratic formula, $[OH^-] = 2.1 \times 10^{-3}$ M, pH = 11.33, pK_a for ephedrine hydrochloride = 10.15. **16.83** (a) pH = 5.39. (b) In a strongly acidic solution, the $—COO^-$ group would be protonated, so glycine would exist as $^+H_3NCH_2COOH$. In strongly basic solution, the $—NH_3^+$ group would be deprotonated, so glycine would be in the form $H_2NCH_2COO^-$. (c) At pH 10, the ratio of the deprotonated form $H_2NCH_2COO^-$ to form II $^+H_3NCH_2COO^-$ is 1.67 to 1. At pH 2, the ratio of form II to the fully protonated form $^+H_3NCH_2COOH$ is 0.43 to 1. **16.86** There is 2% dissociation; $K_a = 4 \times 10^{-4}$. **16.88** $[H^+]$ from the dissociation of 1.0×10^{-9} M HBr is small compared to $[H^+]$ from the autoionization of H_2O, 1.0×10^{-7} M; pH = 7.00.

Chapter 17

17.1 (b) Decrease; (d) decrease. **17.3** (a) pH = 3.65; (b) pH = 5.15. **17.5** (a) $[HC_2H_3O_2] = 0.11$ M, $[C_2H_3O_2^-] = 0.076$ M, pH = 4.60; (b) $[HC_2H_3O_2] = 0.10$ M, $[C_2H_3O_2^-] = 0.11$ M, pH = 4.79; (c) $[HC_2H_3O_2] = 0.102$ M, $[C_2H_3O_2^-] = 0.048$ M, pH = 4.42; (d) $[NH_4^+] = 0.20$ M, $[NH_3] = 0.12$ M, pH = 9.03; (e) $[NH_4^+] = 0.10$ M, $[NH_3] = 0.25$ M, pH = 9.65. **17.7** $pK_a = 7.80$. **17.9** In a mixture of $HC_2H_3O_2$ and $NaC_2H_3O_2$, $HC_2H_3O_2$ reacts with added base, and $C_2H_3O_2^-$ combines with added acid, leaving $[H^+]$ relatively unchanged. Although HCl and KCl are a conjugate acid/conjugate base pair, Cl^- has no tendency to combine with added acid to form undissociated HCl. Any added acid simply increases $[H^+]$ in an HCl/KCl mixture. **17.11** (a) $H^+(aq) + C_7H_5O_2^-(aq) \longrightarrow HC_7H_5O_2(aq)$; (b) $K = 1.5 \times 10^4$; (c) $[Na^+] = [ClO_4^-] = 0.10$ M, $[H^+] = [C_7H_5O_2^-] = 0.0025$ M, $[HC_7H_5O_2] = 0.098$ M. **17.13** 0.25 mol NaBrO. **17.15** (a) pH = 4.57; (b) $[HC_2H_3O_2] = 0.16$ M, $[C_2H_3O_2^-] = 0.090$ M, pH = 4.49; (c) $[HC_2H_3O_2] = 0.14$ M, $[C_2H_3O_2^-] = 0.11$ M, pH = 4.64. **17.17** (a) $[HCO_3^-]/[H_2CO_3] = 11$; (b) $[HCO_3^-]/[H_2CO_3] = 5.4$. **17.19** (a) The quantity of base required to reach the equivalence point is the same in the two titrations. (c) The pH at the equivalence point is higher in the titration of a weak acid. **17.21** (b) 21.0 mL NaOH solution. **17.23** (b) pH = 3.30; (c) pH = 7.00. **17.25** (a) pH = 2.94; (b) $[HC_3H_5O_2] = [C_3H_5O_2^-] = 0.0333$ M, pH = 4.89; (c) $[HC_3H_5O_2] = 8.4 \times 10^{-4}$ M, $[C_3H_5O_2^-] = 5.0 \times 10^{-2}$ M, pH = 6.66; (d) $[C_3H_5O_2^-] = 5.0 \times 10^{-2}$ M, $[OH^-] = 6.2 \times 10^{-6}$ M, pH = 8.79; (e) $[OH^-] = 8.3 \times 10^{-4}$ M, pH = 10.92; (f) $[OH^-] = 0.020$ M, pH = 12.30. **17.27** (a) pH = 7.00; (b) $[HONH_3^+] = 0.100$ M, pH = 3.52. **17.29** (a) $K_{sp} = [Cd^{2+}][S^{2-}]$; (c) $K_{sp} = [Ce^{3+}][F^-]^3$. **17.31** (a) $K_{sp} = 7.63 \times 10^{-9}$; (b) $K_{sp} = 2.7 \times 10^{-9}$; (c) 5.3×10^{-4} mol $Ba(IO_3)_2$/L. **17.33** (a) 9.1×10^{-9} mol AgI/L; (b) 1.7×10^{-15} mol AgI/L; (c) 8.3×10^{-16} mol AgI/L. **17.35** (a) The solubility is 2.5 mol $Cd(OH)_2$/L. **17.37** $[Ca^{2+}]/[Ba^{2+}] = 3.9 \times 10^{-9}/1.0 \times 10^{-4} = 3.9 \times 10^{-5}$. **17.39** $[KMnO_4] = 0.11$ M. **17.41** (a) $Q < K_{sp}$, no $Mn(OH)_2$ precipitates; (b) $Q < K_{sp}$, no Ag_2SO_4 precipitates. **17.43** $[OH^-] = 9.7 \times 10^{-4}$ M, pH = 10.99. **17.45** More soluble in acid: Ag_2CO_3, CeF_3, $Cd(OH)_2$. Ag_2SO_4 will be slightly more soluble in concentrated acid solutions because of the equilibrium $SO_4^{2-}(aq) + H^+(aq) \rightleftharpoons HSO_4^-(aq)$. **17.47** AgI will precipitate first, at $[I^-] = 4.2 \times 10^{-13}$ M. **17.49** 3×10^{-5} mol ZnS/L. **17.51** (a) $Zn(OH)_2(s) + 2H^+(aq) \longrightarrow Zn^{2+}(aq) + 2H_2O(l)$; Lewis base: OH^-; Lewis acid: H^+; (b) $Cd(CN)_2(s) + 2CN^-(aq) \longrightarrow [Cd(CN)_4]^{2-}(aq)$; Lewis base: CN^-; Lewis acid: Cd^{2+}. **17.53** $[Cu^{2+}] = 2 \times 10^{-12}$ M. **17.55** $K = K_{sp} \times K_f = 2 \times 10^{11}$. **17.57** The first two experiments eliminate group 1 and 2 ions (Figure 17.19). The absence of insoluble carbonate precipitates in the filtrate from the third experiment rules out group 4 ions. The ions that might be in the sample are those from group 3, Al^{3+}, Fe^{2+}, Zn^{2+}, Cr^{3+}, Ni^{2+}, Co^{2+}, or Mn^{2+}, and from group 5, NH_4^+, Na^+, or K^+. **17.59** (a) Make the solution acidic using 0.5 M HCl; saturate with H_2S. CdS will precipitate; ZnS will not. (b) Add excess base; $Fe(OH)_3(s)$ precipitates, but Cr^{3+} forms the soluble complex $[Cr(OH)_4]^-$. **17.61** (b) K_{sp} for those cations in group 3 is much larger, so to exceed K_{sp} a higher $[S^{2-}]$ is required. This is achieved by making the solution more basic. **17.63** The greatest buffer capacity is possessed by solution (a), because the concentrations of the acid and conjugate base available to react with added base or acid are largest. Solution (b) has the smallest buffer capacity, because there is no acid-conjugate base equilibrium present. **17.65** (a) $pK_a = 4.98$; (b) 0.024 mol NaOH. **17.68** 0.39 mol HX^-, 0.61 mol X^{2-}; 1.6 L of 1.0 M NaOH. **17.70** $pK_a = 4.68$. **17.73** $[HC_2H_3O_2] = 0.30$ M, $[C_2H_3O_2^-] = 0.17$ M, 13 mL glacial acetic acid, 10 g $NaC_2H_3O_2$. **17.76** (a) Molar mass = 73.9 g/mol; (b) $[HA] = 0.0186$ M, $[A^-] = 0.0308$ M, $K_a = 1.3 \times 10^{-5}$. **17.79** Hydrolysis will affect the solubility of (b) $CaCO_3$, (d) Ag_3PO_4, and (e) CuS. CuS is most affected, because S^{2-} is the strongest base listed. **17.82** $[Ca^{2+}] = 2 \times 10^{-7}$ M, $[F^-] = 5 \times 10^{-8}$ M; $Q < K_{sp}$, and no CaF_2 will precipitate. **17.85** $[Pb^{2+}]$ in solution is 1.6×10^{-3} M. At pH = 1, $Q = 1 \times 10^{-22}$, and $Q > K_{sp}$. PbS will precipitate from the solution. **17.88** HgS and PbS will precipitate, but NiS will not. **17.91** $[OH^-] = 1 \times 10^{-7}$ M; the solubility is 4×10^{-17} mol $Fe(OH)_3$/L.

Chapter 18

18.1 *Troposphere* (0 to 12 km), temperature varies from 290 K to 220 K; *stratosphere* (12 to 50 km), temperature varies from 220 K to 270 K; *mesosphere* (50 to 85 km), temperature varies from 270 K to 180 K; *thermosphere* (85 km and upward), temperature increases from 180 K beyond 1200 K.
18.3 $P_{He} = 3.85 \times 10^{-3}$ mm Hg; $P_{Kr} = 8.38 \times 10^{-4}$ mm Hg. **18.5** 569 nm. **18.7** The photon must possess sufficient energy to produce rupture of the NO bond. Also, the NO molecule must be able to absorb a photon having at least the minimum required energy. **18.9** For two oxygen atoms to recombine to form O_2, a third body must be present to carry off the excess energy, as described in Section 18.3. This is much more likely to occur at 50 km than at 120 km, because the atmosphere is more dense at the lower elevation. There are many more atoms or molecules per unit volume to dissipate the energy from the reaction, thus reducing the lifetime of $O(g)$ atoms at the lower elevation. **18.11** First step: $\Delta H = -198.9$ kJ; second step: $\Delta H = -190.9$ kJ. **18.13** (a) CO binds with hemoglobin in the blood to block O_2 transport to the cells; people with CO poisoning suffocate from lack of O_2. (b) SO_2 is very corrosive to tissue and contributes to respiratory disease and shorter life expectancy, especially for people with other respiratory problems. It is also a major source of acid rain, which damages forests and wildlife in natural waters. (c) O_3 is extremely reactive and toxic because of its ability to form free radicals upon reaction with organic molecules in the body. The products of its reactions with other atmospheric pollutants cause eye irritation and breathing difficulties. **18.15** CO in unpolluted air is typically 0.05 ppm, about 10 ppm in urban air. A major source is automobile exhaust. SO_2 is less than 0.01 ppm in unpolluted air, about 0.08 ppm in urban air. Major sources are coal and oil-burning power plants. NO is about 0.01 ppm in unpolluted air, about 0.05 ppm in urban air. It comes mainly from auto exhausts. **18.17** $P = 2.0 \times 10^{-4}$ mm Hg; 6.7×10^{18} O_3 molecules/m^3.
18.19 (a) $H_2SO_4(aq) + Fe(s) \longrightarrow FeSO_4(aq) + H_2(g)$; the final product may be red rust, Fe_2O_3, formed by the further oxidation of Fe^{2+} by $O_2(g)$. (b) $H_2SO_4(aq) + CaCO_3(s) \longrightarrow CaSO_4(s) + H_2O(l) + CO_2(g)$.
18.21 2.6×10^5 g $CaCO_3$. **18.23** $P_{NO} = 0.08$ atm $= 8 \times 10^4$ ppm. **18.25** 0.09 M Na^+. **18.27** 1.5×10^{13} L H_2O.
18.29 A high concentration of dissolved O_2 supports biodegradation of organic matter by aerobic bacteria. Lack of O_2 indicates pollution by organic and other materials requiring O_2 for decomposition. **18.31** 2.5 g O_2. **18.33** $Ca^{2+}(aq) + 2HCO_3^-(aq) \longrightarrow CaCO_3(s) + CO_2(g) + H_2O(l)$.
18.35 4.8×10^5 g $Ca(OH)_2$ and 4.2×10^5 g Na_2CO_3.
18.37 (a) $[OH^-] = 1.4 \times 10^{-4}$ M, pH = 10.14; (b) 0.53 mol OH^- needed, 0.033 lb CaO. **18.39** (a) Acid rain is rain with a larger $[H^+]$ and thus a lower pH than normal precipitation. The additional H^+ is produced by the dissolution of sulfur and nitrogen oxides in rain droplets to form sulfuric and nitric acids. **18.42** The ionization energies of the metal atoms are much lower than those of any of the atomic or molecular ions present at 120 km (O_2^+, O^+, NO^+). Thus, any of these ions can react with the metal atoms as illustrated for NO^+: $M + NO^+ \longrightarrow M^+ + NO$.
18.45 The average molecular weight at the surface is 28.0 g/mol; at 200 km it is 19.3 g/mol. **18.48** 5.66×10^3 mol SO_4^{2-}/mi^2 per year. **18.51** (a) 1.0×10^{-5} M; (b) by solving the quadratic formula, $[H^+] = 1.9 \times 10^{-6}$ M, pH = 5.73. **18.53** (a) Ca^{2+}, Mg^{2+}, Fe^{2+}. (b) These ions react with soaps to form insoluble materials. Further, precipitation of carbonate may occur when "hard" water is heated, as described by Equation [18.18]. This precipitation reduces heat transfer and water flow in pipes and may eventually lead to complete blockage. **18.56** (a) $Cl_2(g) + H_2O(l) \longrightarrow HOCl(aq) + H^+(aq) + Cl^-(aq)$;
(b) $Ca^{2+}(aq) + 2HCO_3^-(aq) \xrightarrow{\Delta} CaCO_3(s) + CO_2(g) + H_2O(l)$. **18.58** 397 g HCO_3^-/1000 gal H_2O, 183 g CaO required.

Chapter 19

19.1 Spontaneous: (a), (d), (e); nonspontaneous: (b), (c).
19.3 Berthelot's suggestion is incorrect. Some nonexothermic spontaneous processes are the expansion of certain pressurized gases, the dissolving of one liquid in another, and the dissolving of many salts in water. **19.5** For a process to be spontaneous, $\Delta S_{universe}$ must be positive. When steam condenses at 90°C, there must be an increase in the entropy of the surroundings, and it must be greater than the decrease in the entropy of the system. **19.7** S increases in (a), (b), and (c); S decreases in (d). **19.9** (a) $O_2(g)$ at 0.5 atm; (b) $Br_2(g)$; (c) 1 mol of $N_2(g)$ in 22.4 L; (d) $CO_2(g)$. **19.11** (a) $CCl_4(l)$, 214.4 J/K-mol; $CCl_4(g)$, 309.4 J/K-mol. In general, the gas phase of a substance has a larger S° than the liquid phase becasue of the greater volume and motional freedom of the gas. (b) C(diamond), 2.43 J/K-mol; C(graphite), 5.69 J/K-mol. The internal entropy in graphite is greater because there is translational freedom among the planar sheets of C atoms, but there is very little freedom within the network covalent diamond lattice. (c) 1 mol of $N_2O_4(g)$, 304.3 J/K; 2 mol of $NO_2(g)$, 480.9 J/K; 2 mol of $NO_2(g)$ occupies a larger volume and has greater motional freedom than 1 mol of $N_2O_4(g)$. (d) 1 mol of $KClO_3(s)$, 143.0 J/K; 1 mol of $KClO_3(aq)$, 265.7 J/K. Ions in solution have greater motional freedom than ions in a solid lattice. **19.13** (a) $-\Delta S$; (b) $+\Delta S$; (c) $-\Delta S$; (d) $-\Delta S$. **19.15** (a) $\Delta S^\circ = -7.4$ J/K-mol. The value is near zero because the number of moles and the kinds of molecules are the same on both sides of the equation. The reason for the slightly negative value cannot be determined from the equation alone. (b) $\Delta S^\circ = -145.3$ J/K-mol. The value is negative because the number of moles of gas is lower in the products. (c) $\Delta S^\circ = +91.8$ J/K-mol. The value is positive because there are more moles of gas in the products. (d) $\Delta S^\circ = -268.2$ J/K-mol. The

value is negative because a gaseous reactant is converted to a solid product. **19.17** (a) $\Delta G = \Delta H - T\Delta S$. (b) There is no relationship between ΔG and rate of reaction. **19.19** $\Delta G^\circ = -1.23 \times 10^5$ J. The process is highly spontaneous, so we would not expect that $H_2O_2(g)$ would be stable in the gas phase at 298 K. **19.21** (a) $\Delta G^\circ = -190.5$ kJ, spontaneous; (b) $\Delta G^\circ = +68.8$ kJ, nonspontaneous; (c) $\Delta G^\circ = +192.7$ kJ, nonspontaneous; (d) $\Delta G^\circ = +40.8$ kJ, nonspontaneous. **19.23** (a) The reaction proceeds spontaneously in the forward direction but spontaneously reverses at higher temperatures. (b) The reaction is nonspontaneous in the forward direction at all temperatures. (c) The reaction will proceed spontaneously in the forward direction at high temperatures but is spontaneous in the reverse direction at lower temperatures. (d) The reaction becomes spontaneous in the forward direction at some very high temperature. **19.25** (a) Endothermic; (b) $\Delta H > +36.1$ kJ. **19.27** The process will become spontaneous above 1112 K or 839°C. **19.29** (a) $\Delta H^\circ = +229.7$ kJ; $\Delta S^\circ = +156$ J/K. The reaction is nonspontaneous at low temperatures and will become spontaneous at some very high temperature. (b) ΔG° (900 K) $= +89$ kJ. **19.31** (a) ΔG becomes more positive; (b) ΔG becomes more negative; (c) ΔG becomes more positive. **19.33** (a) $\Delta G^\circ = -5.40$ kJ; (b) $\Delta G = -14.54$ kJ. **19.35** (a) $\Delta G^\circ = -190.5$ kJ, $K_p = 2.4 \times 10^{33}$; (b) $\Delta G^\circ = -497.9$ kJ, $K_p = 1.8 \times 10^{87}$; (c) $\Delta G^\circ = +104.70$ kJ, $K_p = 4.45 \times 10^{-19}$.
19.37 $\Delta H^\circ = +176.1$ kJ, $\Delta S^\circ = +0.1605$ kJ/K; (a) $P_{CO_2} = 3.3 \times 10^{-23}$ atm; (b) $P_{CO_2} = 4.6 \times 10^{-9}$ atm.
19.39 (a) $\Delta G^\circ = +23.9$ kJ. (b) By definition, $\Delta G = 0$ when the system is at equilibrium. (c) $\Delta G = -11.6$ kJ.
19.41 (a) -890.4 kJ of heat produced/mol CH_4 burned; (b) $w_{max} = -817.2$ kJ/mol CH_4. **19.44** (a) $+\Delta H$, $+\Delta S$; (b) $-\Delta H$, $-\Delta S$; (c) $+\Delta H$, $+\Delta S$; (d) $+\Delta H$, $+\Delta S$; (e) $-\Delta H$, $+\Delta S$. **19.47** (a) (i) $\Delta H^\circ = +76.2$ kJ, $\Delta S^\circ = -0.495$ kJ/K, $\Delta G^\circ = +240.0$ kJ; (ii) $\Delta H^\circ = +64.9$ kJ, $\Delta S^\circ = -0.0125$ kJ/K, $\Delta G^\circ = +68.71$ kJ; (iii) $\Delta H^\circ = -505.3$ kJ, $\Delta S^\circ = -0.533$ kJ/K, $\Delta G^\circ = -346.4$ kJ. (b) (i) nonspontaneous; (ii) nonspontaneous; (iii) spontaneous. (c) All three reactions have $-\Delta S^\circ$, so ΔG° will become more positive with increasing temperature. There is little effect for reaction (ii) because the magnitude of ΔS° is so small. **19.50** (a) ΔG° (718 K) $= -23.4$ kJ; (b) ΔG° (1073 K) $= -60.8$ kJ. **19.54** For the aqueous equilibrium K^+(plasma) $\rightleftharpoons$ K^+ (muscle), $K = 1$ and $\Delta G^\circ = 0$. Therefore, $\Delta G = +8.75$ kJ. (b) The minimum amount of work required to transfer 1 mol of K^+ is 8.75 kJ, although in practice more than minimum work is required.
19.59 $\Delta G^\circ = -2876.9$ kJ, 94.3 mol ATP/mol glucose.

Chapter 20

20.1 (a) S, -2 to 0; N, $+5$ to $+2$; (b) S, $+4$ to $+6$; Mn, $+7$ to $+2$; (c) Cr, $+3$ to $+6$; Cl $+1$ to -1; (d) O, -2 to 0, Cu, $+2$ to 0. **20.3** (a) $2N_2H_4(g) + N_2O_4(g) \longrightarrow 3N_2(g) + 4H_2O(l)$; (b) N_2O_4, oxidizing agent; N_2H_4, reducing agent. **20.5** (a) $H_2O_2(aq) \longrightarrow O_2(g) + 2H^+(aq) + 2e^-$; (b) $NO_3^-(aq) + 4H^+(aq) + 3e^- \longrightarrow NO(g) + 2H_2O(l)$; (c) $SO_3^{2-}(aq) + 2OH^-(aq) \longrightarrow SO_4^{2-}(aq) + H_2O(l) + 2e^-$; (d) $MnO_4^-(aq) + 2H_2O(l) + 3e^- \longrightarrow MnO_2(s) + 4OH^-(aq)$. **20.7** (a) $Cr_2O_7^{2-}(aq) + I^-(aq) + 8H^+(aq) \longrightarrow 2Cr^{3+}(aq) + IO_3^-(aq) + 4H_2O(l)$; (b) $4MnO_4^-(aq) + 5CH_3OH(aq) + 12H^+(aq) \longrightarrow 4Mn^{2+}(aq) + 5HCO_2H(aq) + 11H_2O(l)$; (c) $4As(s) + 3ClO_3^-(aq) + 3H^+(aq) + 6H_2O(l) \longrightarrow 4H_3AsO_3(aq) + 3HClO(aq)$; (d) $As_2O_3(s) + 2NO_3^-(aq) + 2H^+(aq) + 2H_2O(l) \longrightarrow 2H_3AsO_4(aq) + N_2O_3(aq)$; (e) $2MnO_4^-(aq) + Br^-(aq) + H_2O(l) \longrightarrow 2MnO_2(s) + BrO_3^-(aq) + 2OH^-(aq)$; (f) $4H_2O_2(aq) + Cl_2O_7(aq) + 2OH^-(aq) \longrightarrow 2ClO_2^- + 4O_2(g) + 5H_2O(l)$.
20.9 The salt bridge provides a mechanism by which ions not directly involved in the redox reaction can migrate into the anode and cathode compartments to maintain charge neutrality of the solutions. **20.11** (a) $Cd(s) \longrightarrow Cd^{2+}(aq) + 2e^-$; $Ni^{2+}(aq) + 2e^- \longrightarrow Ni(s)$. (b) anode, Cd(*s*); cathode, Ni(*s*). (c) Cd (−), Ni (+). (d) Electrons flow away from the Cd (−) electrode and toward the Ni (+) electrode. **20.13** (a) Cathode: $PtCl_4^{2-}(aq) + 2e^- \longrightarrow Pt(s) + 4Cl^-(aq)$; anode: $I^-(aq) + Ag(s) \longrightarrow AgI(s) + e^-$; (b) $E^\circ = +0.88$ V. **20.15** (a) $2Au(s) + 8Br^-(aq) + 3IO^-(aq) + 3H_2O(l) \longrightarrow 2AuBr_4^-(aq) + 3I^-(aq) + 6OH^-(aq)$; $E^\circ = 1.35$ V; (b) $2Eu^{2+}(aq) + Sn^{2+}(aq) \longrightarrow 2Eu^{3+}(aq) + Sn(s)$; $E^\circ = 0.29$ V. **20.17** (a) Anode, Sn(*s*); cathode, Cu(*s*). (b) The copper electrode gains mass as Cu is plated out, and the tin electrode loses mass as Sn is oxidized. (c) Cu (+), Sn (−), (d) $E^\circ = 0.473$ V (e) $Cu^{2+}(aq) + Sn(s) \longrightarrow Cu(s) + Sn^{2+}(aq)$.
20.19 (a) $E^\circ = 0.78$ V. (b) "Under standard conditions" refers to solutions that are 1 *M* in concentration, at 25°C, with all gases at 1 atm pressure. (c) Both electrodes are inert metals, probably Pt(*s*). $Cr^{2+}(aq)/Cr^{3+}(aq)$ are in the anode compartment, $Tl^{+}(aq)/Tl^{3+}(aq)$ in the cathode compartment. Electrons flow from the anode to the cathode.

20.21 (a)–(d)

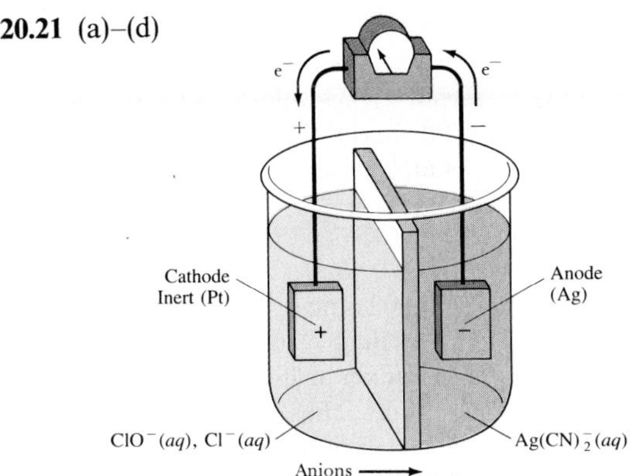

(e) $ClO^-(aq) + 2Ag(s) + 4CN^-(aq) + H_2O(l) \longrightarrow 2Ag(CN)_2^-(aq) + Cl^-(aq) + 2OH^-(aq)$; (f) $E° = 1.20$ V. **20.23** (a) Oxidizing strength: $Cu^{2+}(aq) < O_2(g) < Cr_2O_7^{2-}(aq) < Cl_2(g) < H_2O_2(aq)$; (b) reducing strength: $H_2O_2(aq) < I^-(aq) < Sn^{2+}(aq) < Zn(s) < Al(s)$. **20.25** Any *reduced* species in Table 20.1 for which the oxidation potential is more positive than +0.43 V; these include $Zn(s)$, $H_2(g)$, etc. [$Fe(s)$ is questionable.] **20.27** (a) $E° = +1.137$ V, spontaneous; (b) $E° = -1.82$ V, nonspontaneous; (c) $E° = +1.39$ V, spontaneous. **20.29** (a) True. (b) False. A positive value of $E°$ means that $\Delta G°$ will be negative. (c) False. $\Delta G°$ has units of joules or kJ, and $E°$ is expressed in volts. **20.31** (a) $\Delta G° = -219.4$ kJ; (b) $\Delta G° = +1.05 \times 10^3$ kJ; (c) $\Delta G° = -268$ kJ. **20.33** (a) $E° = +0.627$ V, $K = 2 \times 10^{21}$; (b) $E° = +0.277$ V, $K = 2 \times 10^9$; (c) $E° = +0.44$ V, $K = 1 \times 10^{74}$. **20.35** (a) $K = 8.4 \times 10^5$; (b) $K = 7.0 \times 10^{11}$; (c) $K = 5.8 \times 10^{17}$. **20.37** (a) Decreases E; (b) increases E; (c) no effect; (d) increases E. **20.39** $E° = 2.605$ V, $E = 2.700$ V. **20.41** $E° = 0.763$ V, pH = 1.23. **20.43** 319 g MnO_2. **20.45** Discharge: $Cd(s) + NiO_2(s) + 2H_2O(l) \rightleftharpoons Cd(OH)_2(s) + Ni(OH)_2(s)$; in charging, the reverse reaction occurs. (b) $E° = 1.25$ V. **20.47** $E°$ for the half-reaction $Cd(s) \longrightarrow Cd^{2+}(aq) + 2e^-$ is +0.40 V, smaller than the value of +0.76 V for the corresponding oxidation of Zn. Thus, the overall cell emf will be reduced. **20.50** (a) Anode: $2Cl^-(l) \longrightarrow Cl_2(g) + 2e^-$; cathode: $Cd^{2+}(l) + 2e^- \longrightarrow Cd(l)$; (b) anode: $2Cl^-(aq) \longrightarrow Cl_2(g) + 2e^-$; cathode: $Cd^{2+}(aq) + 2e^- \longrightarrow Cd(s)$; (c) anode: $Cd(s) \longrightarrow Cd^{2+}(aq) + 2e^-$; cathode: $Cd^{2+}(aq) + 2e^- \longrightarrow Cd(s)$ [$Cd(s)$ plates over the iron cathode]. **20.51** Chlorine is oxidized in preference to water because of the large overvoltage for the formation of $O_2(g)$.

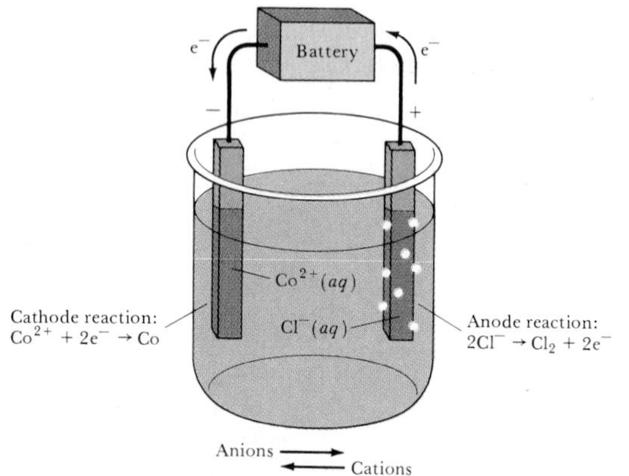

20.53 (a) 13 g $Cr(s)$; (b) 1.02 amps. **20.55** (a) 3.83 L $Cl_2(g)$; (b) 0.342 mol $NaOH(aq)$. **20.57** (a) True. (b) True. (c) False. Overvoltage requires a higher applied potential than calculated theoretically. Thus, more energy than the minimum is expended in producing the cell reaction. **20.59** $w_{max} = -60.7$ kJ [The (−) sign means work is done *by* the cell.] **20.61** (a) 3.2×10^5 g Mg; (b) 6.1×10^3 kWh. **20.63** No. The potential for oxidation of cobalt to Co^{2+}, +0.28 V, is lower than that for oxidation of iron to Fe^{2+}, +0.44 V. To provide cathodic protection, a metal must be more readily oxidized than iron. **20.65** The added amines, acting as Brønsted bases, keep the $[H^+]$ in the antifreeze solution low. This makes the reduction of O_2 less favorable (Section 20.9), and corrosion, therefore slows down. **20.69** (a) $3H_2SO_3(aq) \longrightarrow S(s) + 2HSO_4^-(aq) + H_2O(l) + 2H^+(aq)$, spontaneous; (b) $3Br_2(l) + 3H_2O(l) \longrightarrow 5Br^-(aq) + BrO_3^-(aq) + 6H^+(aq)$, nonspontaneous; (c) $2PbSO_4(s) + 2H_2O(l) \longrightarrow Pb(s) + PbO_2(s) + 2HSO_4^-(aq) + 2H^+(aq)$, nonspontaneous. **20.73** (a) $Zn(s)$ is oxidized to $Zn^{2+}(aq)$; $Ag_2O(s)$ is reduced to $Ag(s)$. (b) Zn is the negative electrode (anode), and Ag is the positive electrode (cathode). (c) $E° = +1.107$ V. **20.75** (a) Spontaneous; (b) nonspontaneous; (c) spontaneous; (d) nonspontaneous. **20.77** $E_{cell} = 0.028$ V; the anode is in the compartment containing the dilute (0.040 *M*) $Cr^{3+}(aq)$ solution. **20.81** $E° = -0.927$ V. **20.84** (a) 10 mol $Cr(s)$ deposited, 5.9×10^6 C required; (b) 9.8×10^5 amps. **20.88** 732 g $NaBO_3$. **20.91** (a) The major reason "deep discharge" causes battery failure is loss of materials from the electrode surfaces. As Pb and PbO_2 are converted to $PbSO_4$ during discharge, the $PbSO_4$ may flake off the electrode and settle to the bottom of the cell, where it is no longer able to take part in the charging reaction. (b) In an auto system, a running engine powers a generator and voltage regulator that drive the charging reaction and keep the battery fully charged.

Chapter 21

21.1 (a) 6 protons, 7 neutrons; (b) 28 protons, 30 neutrons; (c) 37 protons, 48 neutrons; (d) 40 protons, 54 neutrons. **21.3** (a) ${}^{13}_{7}N \longrightarrow {}^{13}_{6}C + {}^{0}_{1}e$; (b) ${}^{84}_{35}Br \longrightarrow {}^{84}_{36}Kr + {}^{0}_{-1}e$; (c) ${}^{181}_{74}W + {}^{0}_{-1}e$ (orbital electron) $\longrightarrow {}^{181}_{73}Ta$; (d) ${}^{230}_{90}Th \longrightarrow {}^{226}_{88}Ra + {}^{4}_{2}He$. **21.5** (a) ${}^{32}_{15}P$; (b) ${}^{7}_{3}Li$; (c) ${}^{187}_{75}Re$; (d) ${}^{99}_{43}Tc$; (e) ${}^{99}_{38}Sr$. **21.7** 7α emissions, 4β emissions. **21.9** (a) ${}^{238}_{92}U + {}^{1}_{0}n \longrightarrow {}^{239}_{92}U + {}^{0}_{0}\gamma$; (b) ${}^{14}_{7}N + {}^{1}_{1}H \longrightarrow {}^{11}_{6}C + {}^{4}_{2}He$; (c) ${}^{18}_{8}O + {}^{1}_{0}n \longrightarrow {}^{19}_{9}F + {}^{0}_{-1}e$. **21.11** Radioactive: (b), low neutron/proton ratio, (c), low neutron/proton ratio, and (e), atomic number greater than 83. Stable: (a) and (d). **21.13** Positron emission (in which a proton is converted to a neutron) is most likely to occur from ^{51}Mn, the nuclide with the lowest neutron/proton ratio. **21.15** The suggestion is not reasonable. The energies of nuclear states are very large relative to ordinary temperatures. Merely changing the temperature by less than 100 K would not significantly affect the behavior of nuclei with regard to nuclear decay rates. **21.17** (a) No (low neutron/proton ratio, should be a positron emitter); (b) no (low neutron/proton ratio, should be a positron emitter or possibly undergo orbital electron capture), (c) no (high neutron/proton ratio, should be a beta emitter); (d) no (high atomic number, should be an alpha emitter). **21.19** $^{4}_{2}He$, $^{16}_{8}O$, $^{40}_{20}Ca$. **21.21** $^{68}_{31}Ga \longrightarrow ^{68}_{30}Zn + ^{0}_{1}e$; 9.78×10^{-6} g ^{68}Ga remain. **21.23** $k = 0.0307\ yr^{-1}$, $t_{1/2} = 22.6$ yr. **21.25** $k = 9.34 \times 10^{-3}\ d^{-1}$; 0.496 mg ^{192}Ir remain. **21.27** Taking only the spontaneous radioactive decay of ^{131}I into account, the activity after 32 days should be 5.7 counts per minute, which is the observed activity. The plants do not absorb iodide. **21.29** 4.0×10^{-12} g ^{226}Ra decays in 1.0 min, 1.1×10^{10} alpha particles emitted. **21.31** $k = 1.5 \times 10^{-10}\ yr^{-1}$; the original rock contained 16.2 mg ^{238}U and is 1.9×10^{9} yr old. **21.33** $k = 1.21 \times 10^{-4}\ yr^{-1}$; 1.70×10^{3} yr. **21.35** (a) $\Delta m = 8.25 \times 10^{-3}$ amu; $\Delta E = 1.23 \times 10^{-12}$ J/nucleon; (b) $\Delta m = 9.41 \times 10^{-3}$ amu; $\Delta E = 1.41 \times 10^{-12}$ J/nucleon; (c) $\Delta m = 8.46 \times 10^{-3}$ amu; $\Delta E = 1.26 \times 10^{-12}$ J/nucleon. **21.37** 5.14×10^{-12} J/nucleus, 3.09×10^{12} J/mol. **21.39** 1.71×10^{5} kg/day; 2.09×10^{8} g ^{235}U. **21.41** (d) ^{59}Co; it has the largest binding energy per nucleon, and binding energy gives rise to mass defect. **21.43** (a) $4\ ^{1}_{0}n$; (b) $^{94}_{36}Kr$. **21.45** The ^{59}Fe is incorporated into the diet component, which in turn is fed to the rabbits. After a time, blood samples are removed from the animals, the red blood cells separated, and the radioactivity of the sample measured. If the iron in the dietary compounds has been incorporated into blood hemoglobin, the blood cell sample should show beta emission. Samples can be taken at various times to determine the rate of iron uptake, rate of loss of the iron from the blood, and so forth. **21.47** The extremely high temperature is required to overcome the electrostatic charge repulsions between the nuclei so that they can come together to react. **21.49** (a) The *electron* and *positron* possess the same mass but opposite electrical charge. (b) *Binding energy* is the energy required to decompose a nucleus into individual nucleons. *Mass defect* is this binding energy expressed as the di erence in mass between the nucleus and the sum of the masses of the component nucleons. (c) A *curie* is a measure of the number of nuclear disintegrations that occur in a sample per second, and a *rem* is a measure of the e ective dose of radiation received by a human. **21.52** The source will last 1.69 yr; it should be replaced in early July 1992. **21.57** (a) $^{6}_{3}Li + ^{63}_{28}Ni \longrightarrow ^{69}_{31}Ga$; (b) $^{48}_{20}Ca + ^{248}_{96}Cm \longrightarrow ^{296}_{116}X$. **21.60** The binding energies/nucleon for ^{9}Be and ^{10}Be are very similar; that for ^{10}Be is slightly higher. **21.63** 3.7×10^{4} disintegrations/s; 4.6×10^{13} ^{90}Sr nuclei or 6.9×10^{-9} g ^{90}Sr. **21.66** 7.4×10^{-2} disintegrations/s; 2.5×10^{-6} rad/yr; 2.5×10^{-5} rem/yr.

Chapter 22

22.1 Metals: Sr, Ce, Rh; nonmetals: Se, Kr; semimetal: Sb. **22.3** Metals have high melting points and large thermal and electrical conductivity, whereas nonmetals have much lower melting points and are poor conductors. Metals react with nonmetals to produce ionic solids that melt at high temperatures; nonmetals react with other nonmetals to form covalent compounds that can be solids, liquids, or gases. Oxides of metals produce basic aqueous solutions, whereas oxides of nonmetals produce acidic aqueous solutions. **22.5** Bismuth should be most metallic. It has a metallic luster and a relatively low ionization energy. Bi_2O_3 is soluble in acid but not in base, characteristic of the basic properties of the oxide of a metal rather than of the oxide of a nonmetal. **22.7** (a) N; (b) K; (c) K in the gas phase (lowest ionization energy), Li in aqueous solution (most positive $E°$ value); (d) N has the smallest *covalent* radius; Ne is difficult to compare because it doesn't form compounds; (e) N. **22.9** (a) IF_3; (b) B_2O_3; (c) BF_3; (d) H_2O. **22.11** (a) N is too small a central atom to fit five fluorine atoms, and it does not have available *d* orbitals, which can help accommodate more than eight electrons. (b) Si does not readily form π bonds, which are necessary to satisfy the octet rule for both atoms in the molecule. (c) As has a lower electronegativity than N; that is, it more readily gives up electrons to an acceptor and is more easily oxidized.
22.13 (a) $NaNH_2(s) + H_2O(l) \longrightarrow NH_3(aq) + Na^+(aq) + OH^-(aq)$; (b) $2C_3H_7OH(l) + 9O_2(g) \longrightarrow 6CO_2(g) + 8H_2O(l)$; (c) $NiO(s) + C(s) \longrightarrow CO(g) + Ni(s)$ or $2NiO(s) + C(s) \longrightarrow CO_2(g) + 2Ni(s)$; (d) $AlP(s) + 3H_2O(l) \longrightarrow PH_3(g) + Al(OH)_3(s)$; (e) $Na_2S(s) + 2HCl(aq) \longrightarrow H_2S(g) + 2Na^+(aq) + 2Cl^-(aq)$.
22.15 $^{1}_{1}H$, protium; $^{2}_{1}H$, deuterium; $^{3}_{1}H$, tritium. **22.17** Like other elements in group 1A, hydrogen has only one valence electron. Like other elements in group 7A, hydrogen needs only one electron to complete its valence shell.
22.19 (a) $Mg(s) + 2H^+(aq) \longrightarrow Mg^{2+}(aq) + H_2(g)$;
(b) $C(s) + H_2O(g) \xrightarrow{1000°C} CO(g) + H_2(g)$;
(c) $CH_4(g) + H_2O(g) \xrightarrow{1100°C} CO(g) + 3H_2(g)$.
22.21 (a) Ionic; (b) molecular; (c) metallic.
22.23 (a) $NaH(s) + H_2O(l) \longrightarrow NaOH(aq) + H_2(g)$; (b) $Fe(s) + H_2SO_4(aq) \longrightarrow Fe^{2+}(aq) + H_2(g) + SO_4^{2-}(aq)$; (c) $H_2(g) + Br_2(g) \longrightarrow 2HBr(g)$; (d) $Na(l) + H_2(g) \longrightarrow 2NaH(s)$; (e) $PbO(s) + H_2(g) \xrightarrow{\Delta} Pb(s) + H_2O(g)$. **22.25** 7.9×10^{8} kg H_2. **22.27** As an oxdizing agent in steel making; to bleach pulp and paper; in oxyacetylene torches; in medicine to assist in breathing.

22.29 $:\ddot{O}{=}\ddot{O}{-}\ddot{O}: \longleftrightarrow :\ddot{O}{-}\ddot{O}{=}\ddot{O}:$

Ozone has two resonance forms; the molecular structure is bent, with an O—O—O bond angle of approximately 120°. The π bond in ozone is delocalized over the entire molecule; neither individual O—O bond is a full double bond, so the observed O—O distance of 1.28 Å is greater than the 1.21 Å distance in O_2, which has a full O—O double bond.
22.31 (a) $CaO(s) + H_2O(l) \longrightarrow Ca^{2+}(aq) + 2OH^-(aq)$; (b) $Al_2O_3(s) + 6H^+(aq) \longrightarrow 2Al^{3+}(aq) + 3H_2O(l)$;

(c) $Na_2O_2(s) + 2H_2O(l) \longrightarrow 2Na^+(aq) + 2OH^-(aq) + H_2O_2(aq)$; (d) $N_2O_3(g) + H_2O(l) \longrightarrow 2HNO_2(aq)$; (e) $2KO_2(s) + 2H_2O(l) \longrightarrow 2K^+(aq) + 2OH^-(aq) + O_2(g) + H_2O_2(aq)$; (f) $NO(g) + O_3(g) \longrightarrow NO_2(g) + O_2(g)$. **22.33** (a) Neutral; (b) Acidic; (c) Basic; (d) Amphoteric. **22.35** Assuming that the reactions occur in basic solution: (a) $H_2O_2(aq) + S^{2-}(aq) \longrightarrow 2OH^-(aq) + S(s)$; (b) $SO_2(g) + 2OH^-(aq) + H_2O_2(aq) \longrightarrow SO_4{}^{2-}(aq) + H_2O(l)$; (c) $NO_2{}^-(aq) + H_2O_2(aq) \longrightarrow NO_3{}^-(aq) + H_2O(l)$; (d) $As_2O_3(s) + 2H_2O_2(aq) + 6OH^-(aq) \longrightarrow 2AsO_4{}^{3-}(aq) + 5H_2O(l)$; (e) This reaction must be occuring in acidic solution, because $Fe(OH)_3$ would form if the solution were basic. $2Fe^{2+}(aq) + H_2O_2(aq) + 2H^+(aq) \longrightarrow 2Fe^{3+}(aq) + 2H_2O(l)$. **22.37** Formation of ammonia, NH_3, and fertilizers; as an inert gas in manufacturing processes; as a coolant in liquid form. **22.39** (a) HNO_2, +3; (b) N_2H_4, −2; (c) KCN, −3; (d) $NaNO_3$, +5; (e) NH_4Cl, −3; (f) Li_3N, −3. **22.41** (a) $\left[\begin{array}{c} H \\ | \\ H{-}N{-}H \\ | \\ H \end{array}\right]^+$; Tetrahedral

(b) $\ddot{O}{=}N{-}\ddot{O}{-}H$ (with $:\ddot{O}:$ on N) $\longleftrightarrow$ $:\ddot{O}{-}N{-}\ddot{O}{-}H$ (with $=\ddot{O}$ on N)

The molecular geometry around nitrogen is trigonal planar, but the hydrogen atom is not required to lie in this plane. (There is a third resonance form, which makes a smaller contribution to the observed structure.) (c) $\ddot{N}{=}N{=}\ddot{O}$ The molecular geometry is linear. (There are two other possible resonance forms involving triple bonds.)

(d) $\ddot{O}{=}\ddot{N}{-}\ddot{O}: \rightleftharpoons :\ddot{O}{-}\ddot{N}{=}\ddot{O}$ The molecule is bent. **22.43** (a) $Mg_3N_2(s) + 6H_2O(l) \longrightarrow 3Mg(OH)_2(s) + 2NH_3(aq)$; (b) $2NO(g) + O_2(g) \longrightarrow 2NO_2(g)$; (c) $4NH_3(g) + 3O_2(g) \longrightarrow 2N_2(g) + 6H_2O(g)$; (d) $NaNH_2(s) + H_2O(l) \longrightarrow Na^+(aq) + OH^-(aq) + NH_3(aq)$. **22.45** (a) $4Zn(s) + 2NO_3{}^-(aq) + 10H^+(aq) \longrightarrow 4Zn^{2+}(aq) + N_2O(g) + 5H_2O(l)$; (b) $4NO_3{}^-(aq) + S(s) + 4H^+(aq) \longrightarrow 4NO_2(g) + SO_2(g) + 2H_2O(l)$ or $6NO_3{}^-(aq) + S(s) + 4H^+(aq) \longrightarrow 6NO_2(g) + SO_4{}^{2-}(aq) + 2H_2O(l)$; (c) $2NO_3{}^-(aq) + 3SO_2(g) + 2H_2O(l) \longrightarrow 2NO(g) + 3SO_4{}^{2-}(aq) + 4H^+(aq)$; (d) $CO(NH_2)_2(aq) + H_2O(l) \longrightarrow 2NH_3(aq) + CO_2(g)$. **22.47** (a) $2NO_3{}^-(aq) + 12H^+(aq) + 10e^- \longrightarrow N_2(g) + 6H_2O(l)$, $E° = +1.25$ V; (b) $2NH_4{}^+(aq) \longrightarrow N_2(g) + 8H^+(aq) + 6e^-$, $E° = -0.27$ V. **22.49** 1.1×10^{10} kg CH_4 is required to produce the NH_3. **22.51** Fire extinguishers; as coolant (dry ice); manufacture of $Na_2CO_3 \cdot 10H_2O$; carbonation of beverages. **22.53** (a) HCN; (b) SiC; (c) $CaCO_3$; (d) CaC_2. **22.55** (a) $[:C{\equiv}N:]^-$; (b) $:C{\equiv}O:$; (c) $[:C{\equiv}C:]^{2-}$; (d) $\ddot{S} = C = \ddot{S}$; (e) $\ddot{O} = C = \ddot{O}$;

(f) $\left[\begin{array}{c} :\ddot{O}: \\ | \\ C \\ \ddot{O}{=}\quad :\ddot{O}: \end{array}\right]^{2-}$; one of three equivalent resonance structures

22.57 (a) $ZnCO_3(s) \xrightarrow{\Delta} ZnO(s) + CO_2(g)$; (b) $BaC_2(s) + 2H_2O(l) \longrightarrow Ba^{2+}(aq) + 2OH^-(aq) + C_2H_2(g)$; (c) $C_2H_4(g) + 3O_2(g) \longrightarrow 2CO_2(g) + 2H_2O(g)$; (d) $2CH_3OH(l) + 3O_2(g) \longrightarrow 2CO_2(g) + 4H_2O(g)$; (e) $NaCN(s) + H^+(aq) \longrightarrow Na^+(aq) + HCN(g)$. **22.59** (a) $2Mg(s) + CO_2(g) \longrightarrow 2MgO(s) + C(s)$; (b) $6CO_2(g) + 6H_2O(l) \xrightarrow{hv} C_6H_{12}O_6(aq) + 6O_2(g)$; (c) $CO_3{}^{2-}(aq) + H_2O(l) \longrightarrow HCO_3{}^-(aq) + OH^-(aq)$. **22.61** 4.06 L $C_2H_2(g)$. **22.63** 1 ft^3 = 28.3 L; at STP, 1 ft^3 = 1.26 mol gas; (a) 7.9×10^{10} ft^3 of H_2; (b) 3.1×10^{11} ft^3 of N_2; (c) 3.5×10^{11} ft^3 of O_2. **22.66** A, CO_2; B, H_2; C, O_2. **22.69** (a) SO_3; (b) Cl_2O_5; (c) N_2O_3; (d) CO_2; (e) P_2O_5. **22.72** $(CH_3)_2NNH_2$: 9 mol gaseous products/244 g reactants = 0.0369 mol gas/1 g reactants; $(CH_3)HNNH_2$: 6.25 mol gaseous products/161 g reactants = 0.0388 mol gas/1 g reactants; $(CH_3)HNNH_2$ has greater thrust. **22.75** $(NH_2)_2CO$ is 46.6% N by mass; NH_3 is 82.3% N by mass; $(NH_4)_2SO_4$ is 21.2% N by mass; $NaNO_3$ is 16.5% N by mass. **22.78** For this reaction, the temperature dependence of the equilibrium constant and the reaction rate are opposing effects. At ambient temperatures, the rate of the reaction is extremely slow, and equilibrium is not established in an economically viable length of time. To establish a reasonable rate, the reaction must be carried out at high temperature. **22.81** N_2, the species with the highest nuclear charge on both atoms, will be the poorest electron-pair donor. The species with the lowest nuclear charge on one center and the lowest total nuclear charge will be the strongest base; this is CN^-.

Chapter 23

23.1 (a) $IO_3{}^-$, +5; (c) BrF_3; Br, +3; F, −1; (e) HIO_2, +3. **23.3** (a) Potassium chlorate; (c) aluminum chloride; (e) paraperiodic acid. **23.5** (a) Linear; (c) trigonal pyramid; (e) square planar. **23.7** Xenon has a lower ionization energy than argon; because the valence electrons are not as strongly attracted to the nucleus, they are more readily promoted to a state in which the atom can form bonds with fluorine. Also, Xe is larger and can more easily accommodate an expanded octet of electrons. **23.9** (a) Van der Waals intermolecular attractive forces increase with increasing number of electrons in the atoms. (b) F_2 reacts with water: $F_2(g) + H_2O(l) \longrightarrow 2HF(g) + O_2(g)$. That is, fluorine is too strong an oxidizing agent to exist in water. (c) HF has extensive hydrogen bonding. (d) Oxidizing power is related to electronegativity. Electronegativity decreases in the order given. **23.11** *Fluorine* is used to prepare fluorocarbons such as Teflon, used as a lubricant. *Chlorine* is used to prepare vinyl chloride, C_2H_3Cl, for plastics, and in bleach containing ClO^-. *Bromine* is used to produce silver bromide, AgBr, used in photographic film. *Iodine* is used in the form of KI as an additive to table salt. **23.13** (a) $PBr_5(l) + 4H_2O(l) \longrightarrow H_3PO_4(aq) + 5H^+(aq) + 5Br^-(aq)$; (b) $IF_5(l) + 3H_2O(l) \longrightarrow H^+(aq) + IO_3^-(aq) + 5HF(aq)$; (c) $SiBr_4(l) + 4H_2O(l) \longrightarrow Si(OH)_4(s) + 4H^+(aq) + 4Br^-(aq)$. **23.15** (a) $Br_2(l) + 2OH^-(aq) \longrightarrow BrO^-(aq) + Br^-(aq) + H_2O(l)$; (b) $Cl_2(g) + 2Br^-(aq) \longrightarrow Br_2(l) + 2Cl^-(aq)$; (c) $Br_2(l) + H_2O_2(aq) \longrightarrow 2Br^-(aq) + O_2(g) + 2H^+(aq)$. **23.17** (a) $2ClO_3^-(aq) + 10Fe^{2+}(aq) + 12H^+(aq) \longrightarrow Cl_2(g) + 10Fe^{3+}(aq) + 6H_2O(l)$; (b) $E° = +0.70$ V. **23.19** The average Xe—F bond enthalpies in the three compounds: XeF_2, 134 kJ; XeF_4, 134 kJ; XeF_6, 130 kJ. They are remarkably constant in the series. **23.21** (a) H_2SeO_3, +4; (c) H_2Te, −2; (e) $CaSO_4$, +6. **23.23** (a) Potassium thiosulfate; (b) aluminum sulfide; (c) sodium hydrogen selenite; (d) selenium hexafluoride.

23.25 (a) $[SeO_4]^{2-}$ Lewis structure (Se bonded to four :Ö: atoms) — Tetrahedral; (b) $Te(OH)_6$ Lewis structure (Te bonded to six O—H groups) — Octahedral

23.27 (d) $Se(s) + 2H_2SO_4(l) \xrightarrow{\Delta} SeO_2(g) + 2SO_2(g) + 2H_2O(g)$; (e) $SO_3(g) + H_2SO_4(l) \longrightarrow H_2S_2O_7(l)$; (f) $H_2SeO_3(aq) + H_2O_2(aq) \longrightarrow 2H^+(aq) + SeO_4^{2-}(aq) + H_2O(l)$. **23.29** (b) $Cr_2O_7^{2-}(aq) + 3H_2SO_3(aq) + 2H^+(aq) \longrightarrow 2Cr^{3+}(aq) + 3SO_4^{2-}(aq) + 4H_2O(l)$. **23.31** Te is less electronegative than Se or S, so it withdraws less charge from the oxygens, making the O—H bonds less polar and less likely to ionize as H^+. Also, Te is larger than Se or S and in aqueous solution is likely to be coordinated by solvent water molecules. This interaction increases the electron density on the central tellurium and further reduces the withdrawal of charge from the O—H bonds in H_2TeO_4. **23.33** (a) H_3PO_4, +5; (b) H_3AsO_3, +3; (e) K_3P, −3. **23.35** (b) Arsenic acid; (c) tetraphosphorus decaoxide. **23.37** (a) PCl_4^+ is tetrahedral, PCl_6^- is octahedral. (b) PCl_4^+, sp^3; PCl_6^-, sp^3d^2. (c) The ionic form is stabilized in the solid state by the lattice energy gained from forming the ions. **23.39** (a) Only two of the hydrogen atoms in H_3PO_3 are bound to oxygen. The third is attached directly to phosphorus and is not readily ionized, because the H—P bond is not very polar. (c) Phosphate rock consists of $Ca_3(PO_4)_2$, which is only slightly soluble in water. The phosphorus is unavailable for plant use. (e) In solution, Na_3PO_4 is completely dissociated into Na^+ and PO_4^{3-}. PO_4^{3-}, the conjugate base of the very weak acid HPO_4^{2-}, has a K_b of 2.4×10^{-2} and produces a considerable amount of OH^- by hydrolysis of H_2O. **23.41** (a) $2Ca_3(PO_4)_2(s) + 6SiO_2(s) + 10C(s) \longrightarrow P_4(g) + 6CaSiO_3(l) + 10CO_2(g)$; (b) $3H_2O(l) + PCl_3(l) \longrightarrow H_3PO_3(aq) + 3H^+(aq) + 3Cl^-(aq)$. **23.43** (a) SiO_2, +4; (b) $GeCl_4$, +4; (c) $NaBH_4$, +3; (d) $SnCl_2$, +2. **23.45** In C_2H_4 and C_2H_2 the carbon atoms are joined by π bonds as well as σ bonds. Silicon does not form Si—Si π bonds of great strength, so Si compounds involving such bonds are unstable relative to other bonding configurations. **23.47** (a) Carbon; (b) lead; (c) silicon. **23.49** $GeCl_4(g) + 2H_2O(l) \longrightarrow GeO_2(s) + 4HCl(g)$; $SiCl_4(g) + 2H_2O(l) \longrightarrow SiO_2(s) + 4HCl(g)$. **23.51** KSi_3AlO_8. **23.53** (a) SiO_4^{4-}; (b) SiO_3^{2-}; (c) SiO_3^{2-}. **23.56** (a) Sulfur is a less electronegative element than oxygen and can be expected to more readily give up its electrons in the course of being oxidized. (c) Astatine is radioactive and not present in nature to a significant extent. (g) In aluminosilicate sheets the overall charge is more negative than in silicates because of replacement of Si^{4+} by Al^{3+}. More cations are present between layers, increasing electrostatic interactions between the cations and the sheets and the rigidity of the structure. **23.58** 0.533 atm. **23.61** $BrO_3^-(aq) + XeF_2(aq) + H_2O(l) \longrightarrow Xe(g) + 2HF(aq) + BrO_4^-(aq)$ **23.64** $8Fe(s) + S_8(s) \longrightarrow 8FeS(s)$; $S_8(s) + 16F_2(g) \longrightarrow 8SF_4(g)$ or $S_8(s) + 24F_2(g) \longrightarrow 8SF_6(g)$; $S_8(s) + 8O_2(g) \longrightarrow 8SO_2(g)$; $S_8(s) + 8H_2(g) \longrightarrow 8H_2S(g)$. **23.66** The valence-shell electron arrangement controls the upper and lower limits of oxidation state. For group 6A the configuration is ns^2np^4. Addition of 2 electrons or loss of 6 electrons produces a closed shell. Thus, the oxidation state limits are −2 to +6. **23.69** (a) $2H_2Se(g) + O_2(g) \longrightarrow 2H_2O(g) + 2Se(s)$; (b) $\Delta G° = -489.0$ kJ; $\Delta G° = -2.303RT \log K$, $K = 5 \times 10^{85}$. **23.71** $GeO_2(s) + C(s) \xrightarrow{\Delta} Ge(l) + CO_2(g)$; $Ge(l) + 2Cl_2(g) \longrightarrow GeCl_4(l)$; $GeCl_4(l) + 2H_2O(l) \longrightarrow GeO_2(s) + 4HCl(g)$; $GeO_2(s) + 2H_2(g) \longrightarrow Ge(s) + 2H_2O(l)$. **23.74** 2.0×10^3 mol air in the room (average molecular weight of air = 29.0 g/mol); 4.0×10^{-2} mol H_2S allowable; 3.5 g FeS required. **23.77** The atomic radii calculated for N in the three compounds are: NOF, 0.80 Å; NOCl, 0.98 Å; NOBr, 0.99 Å. The possible resonance forms are X—N=O ⟷ X=N—O. Because fluorine is a second-row element, we expect that π bond formation is more important than for Cl or Br. The π character of the X—N bond in the second structure suggests that this form will be more important in NOF than NOCl or NOBr and that the N—F bond should be shorter than otherwise expected.

Chapter 24

24.1 Iron: hematite, Fe_2O_3; magnetite, Fe_3O_4; and taconite, mainly Fe_3O_4. Aluminum: bauxite, $Al_2O_3 \cdot xH_2O$; corundum, Al_2O_3; and gibbsite, $Al_2O_3 \cdot 3H_2O$. In ores, iron is present as the +3 ion or as both the +2 and +3 ions, as in magnetite. Aluminum is always present in the +3 oxidation state. **24.3** (a) $ZnCO_3(s) \xrightarrow{\Delta} ZnO(s) + CO_2(g)$; (b) $MnO(s) + CO(g) \longrightarrow Mn(l) + CO_2(g)$; (c) $Al(OH)_3(s) + OH^-(aq) \longrightarrow [Al(OH)_4]^-(aq)$; (d) $TiCl_4(g) + K(l) \longrightarrow Ti(s) + 4KCl(s)$; (e) $3CaO(l) + P_2O_5(l) \longrightarrow Ca_3(PO_4)_2(l)$; **24.5** $FeO(s) + H_2(g) \longrightarrow Fe(l) + H_2O(g)$; $Fe_2O_3(s) +$

$3H_2(g) \longrightarrow 2Fe(l) + 3H_2O(g)$; $FeO(s) + CO(g) \longrightarrow Fe(l) + CO_2(g)$; $Fe_2O_3(s) + 3CO(g) \longrightarrow 2Fe(l) + 3CO_2(g)$. **24.7** 3.6×10^3 kg SO_2 (or, to one significant figure, 4×10^3 kg SO_2). **24.9** (a) Air serves mainly to oxidize coke to CO; this exothermic reaction also provides heat for the furnace: $2C(s) + O_2(g) \longrightarrow 2CO(g)$ $\Delta H = -110$ kJ. (b) Limestone, $CaCO_3$, is the source of basic oxide for slag formation: $CaCO_3(s) \xrightarrow{\Delta} CaO(s) + CO_2(g)$; $CaO(l) + SiO_2(l) \longrightarrow CaSiO_3(l)$. **24.11** To purify electrochemically, use a soluble cobalt salt such as $CoSO_4 \cdot 7H_2O$ as electrolyte. Reduction of H_2O does not occur because of overvoltage. Anode reaction: $Co(s) \longrightarrow Co^{2+}(aq) + 2e^-$; cathode reaction: $Co^{2+}(aq) + 2e^- \longrightarrow Co(s)$. **24.13** 5.8×10^2 g $Ni^{2+}(aq)$. **24.15** (c) $\Delta G = \Delta H - T\Delta S$; $\Delta H° = +30.1$ kJ; $\Delta S° = +24.5$ J/K; $\Delta G° = -6.0$ kJ. **24.17** Sodium is metallic; each atom is bonded to many others. When the metal lattice is distorted, many bonds remain intact. In NaCl the ionic forces are strong, and the ions are arranged in very regular arrays. The ionic forces tend to be broken along certain cleavage planes in the solid, and the substance does not tolerate much distortion before cleaving. **24.19** In the electron-sea model the electrons move about in the metallic lattice while the atoms remain more or less fixed in position. Under the influence of an applied potential the electrons are free to move throughout the structure, giving rise to conductivity. **24.21** According to the molecular-orbital or band theory of metallic bonding, the maximum bond order and highest bond strength occurs in metals with six to eight valence electrons. Assuming that hardness is directly related to the bond strength between metal atoms, Cr, with six valence electrons, should have the highest bond strength and greatest hardness in the fourth row. **24.23** A *conductor* has partially filled energy bands; an *insulator* has energy bands that are either completely filled or completely empty, with a large energy gap between full and empty bands. A *semiconductor* has a filled energy band separated by a small energy gap from an empty band. **24.25** White tin is more metallic in character; it has a higher conductivity and larger Sn—Sn distance (3.02 Å) than gray tin (2.81 Å). **24.27** An *alloy* cotains atoms of more than one element and has the properties of a metal. In a *solution alloy* the components are randomly dispersed. In a *heterogeneous alloy* the components are not evenly dispersed and can be distinguished at a macroscopic level. In an *intermetallic compound* the components have interacted to form a compound substance, as in Cu_3As. **24.29** The heat of fusion of a metal should be much smaller than the heat of vaporization, because fusion does not involve substantial breaking of metalmetal bonds, which hold the atoms together in both the solid and liquid states. **24.31** Isolated atoms: (c); metallic bonding: (a), (b), (d). **24.33** The lanthanide contraction is the name given to the decrease in atomic size due to the buildup in effective nuclear charge as we move through the lanthanides (elements 58 to 71) and beyond them. **24.35** See Figure 24.24 (a) $NbCl_5$; (b) WCl_6. **24.37** Chromium, $[Ar]3d^54s^1$, has 6 valence electrons, some or all of which can be involved in bonding, leading to multiple stable oxidation states. Al, $[Ne]3s^23p^1$, has only three valence electrons, which are all lost or shared during bonding, producing the +3 state exclusively. **24.39** (a) $[Kr]4d^4$; (b) $[Xe]4f^{14}5d^1$. **24.41** Cr^{2+}. **24.43** $CrO_4{}^{2-}$. **24.45** Cr(III) and Cr(VI). **24.47** Fe(II) and Fe(III). **24.49** (a) $Fe(s) + 2HCl(aq) \longrightarrow FeCl_2(aq) + H_2(g)$; (b) $Fe(s) + 4HNO_3(aq) \longrightarrow Fe(NO_3)_3(aq) + NO(g) + 2H_2O(l)$. **24.51** Ni^{2+}: $[Ar]3d^8$, green; Cu^{2+}: $[Ar]3d^9$, blue. **24.53** The unpaired electrons in a paramagnetic material cause it to be weakly attracted into a magnetic field. A diamagnetic material, where all electrons are paired, is very weakly repelled by a magnetic field. **24.55** Fe, Co, Ni. **24.57** Al ($E°_{ox} = +1.66$ V) is a much more active metal than Cu ($E°_{ox} = -0.337$ V), and any of the free element is certain to be oxidized by atmospheric agents such as O_2. **24.60** 493 g pig iron. **24.63** 1.0×10^3 kg Mg **24.65** 3.4×10^2 kWh; $27
24.68 $5MoS_2(s) + 14H^+(aq) + 14NO_3{}^-(aq) \longrightarrow 5MoO_3(s) + 10SO_2(g) + 7N_2(g) + 7H_2O(l)$; $MoO_3(s) + 2NH_3(aq) + H_2O(l) \longrightarrow (NH_4)_2MoO_4(s)$; $(NH_4)_2MoO_4(s) \longrightarrow 2NH_3(g) + H_2O(l) + MoO_3(s)$; $MoO_3(s) + 3H_2(g) \longrightarrow Mo(s) + 3H_2O(g)$.
24.71 In Si, the four valence electrons are localized between the central Si and its four nearest neighbors. In the closest-packed Ti lattice, there are twelve nearest neighbors for each Ti atom and the valence electrons cannot be localized; they are delocalized and free to move throughout the lattice. **24.74** For the process $ZnL_4{}^{2+} + 2e^- \longrightarrow Zn° + 4L$, the reduction potentials ($E°$'s) will be in the order: $Zn(H_2O)_4{}^{2+} > Zn(NH_3)_4{}^{2+} > Zn(CN)_4{}^{2-}$.

Chapter 25

25.1 (a) Coordination number = 4, oxidation number = +2; (c) coordination number = 6, oxidation number = +3.

25.3 (a)

```
       Cl
       |
       Al
      / \ \
    Cl  Cl  Cl
```

Tetrahedral

(b) [:N≡C—Ag—C≡N:]$^-$. Linear
25.5 25.3(a) tetrachloroaluminate(III); (b) dicyanoargentate(I); (c) dichloroethylenediamineplatinum(II); (d) *trans*-tetraamminediaquachromium(III). 25.4(a) tetraamminezinc(II); (b) aquapentachlororuthenate(III); (c) *trans*-bis(ethylenediamine)dinitrocobalt(III); (d) *cis*-diamminebromohydridoplatinum(II). **25.7** (a) $[Pd(NH_3)_2Br_4]$; (b) $K[Pt(NH_3)Br_3]$ or $[Pt(NH_3)_3Br]Br$; (c) $K[Fe(NH_3)_2Br_4]$. **25.9** (a) $[Cr(NH_3)_6](NO_3)_3$; (b) $[Co(NH_3)_4CO_3]_2SO_4$; (d) $K[V(H_2O)_2Br_4]$. **25.11** (a) $CO_3{}^{2-}$ is bidentate; (b) $C_2O_4{}^{2-}$ is bidentate; (c) EDTA is pentadentate; (d) ethylenediamine (en) is bidentate.

25.13 (a)

ONO ONO Pd H_3N NH_3 — *cis*

ONO NH_3 Pd H_3N ONO — *trans*

(b) $[Pd(NH_3)_2(ONO)_2]$, $[Pd(NH_3)_2(NO_2)_2]$

(c) N N N V N Cl Cl $^{+}$ | N N N V Cl N Cl $^{+}$

(d) $[Co(NH_3)_4Br_2]Cl$, $[Co(NH_3)_4BrCl]Br$.

25.15

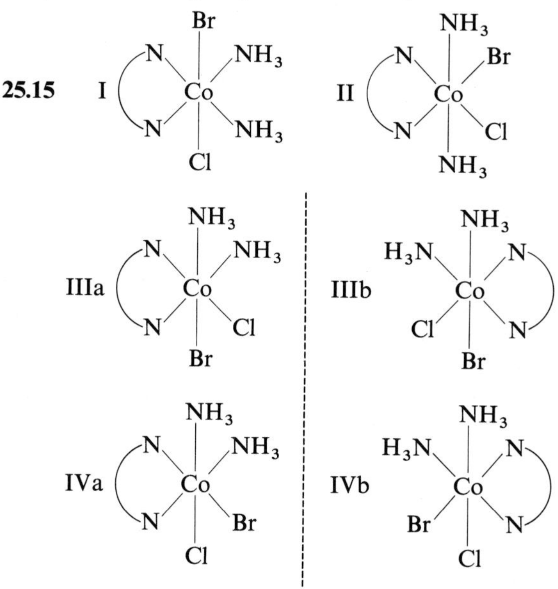

I, II, IIIa, and IVa are geometric isomers; IIIa/IIIb and IVa/IVb are the pairs of optical isomers. **25.17** Cobalt(III) complexes are generally inert. Thus, the ions that form precipitates are outside the coordination sphere. The red complex is $[Co(NH_3)_5SO_4]Br$, pentaaminesulfatocobalt(III) bromide; the violet complex is $[Co(NH_3)_5Br]SO_4$, pentaaminebromocobalt(III) sulfate. **25.19** $[Pd(NC_5H_5)_2Br_2]$; square planar, trans isomer; *trans*-dibromodipyridinepalladium(II). **25.21** Colored compounds: (b), (d), (f). **25.23** Blue to blue-violet. **25.25** (a) Seven; (c) three. **25.27** The $d_{x^2-y^2}$ and d_{z^2} metal orbitals point directly toward the six ligands in an octahedron. Electrons in these metal orbitals experience greater electrostatic repulsion by the negative charge or dipole of the ligands than electrons in the d_{xy}, d_{xz}, and d_{yz} orbitals. Thus, the d_{xy}, d_{xz}, and d_{yz} orbitals are lower in energy. **25.29** A yellow color is due to absorption of light around 400 to 430 nm, a blue color to absorption near 620 nm. The shorter wavelength corresponds to a higher-energy electron transition and larger Δ value. Cyanide is a stronger field ligand, and its complexes are expected to have larger Δ values than aqua complexes.

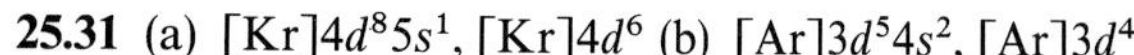

25.31 (a) $[Kr]4d^85s^1$, $[Kr]4d^6$ (b) $[Ar]3d^54s^2$, $[Ar]3d^4$

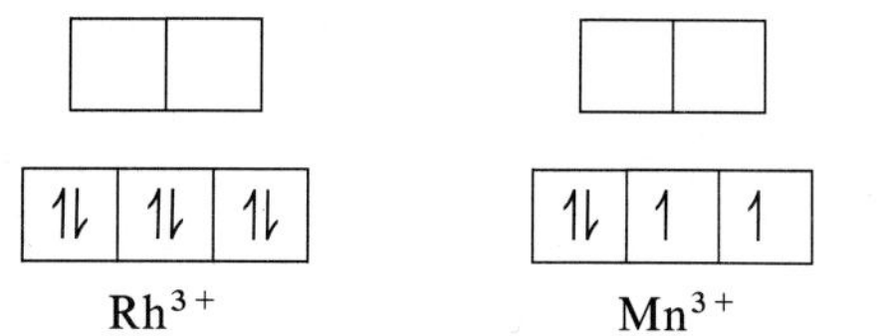

25.33 All complexes in this exercise are 6-coordinate octahedral.

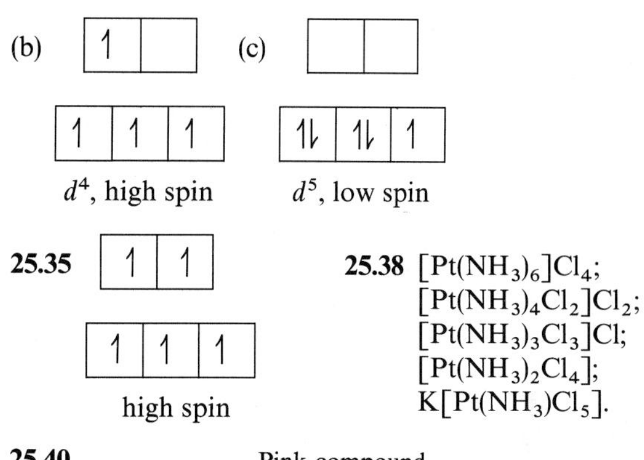

25.38 $[Pt(NH_3)_6]Cl_4$; $[Pt(NH_3)_4Cl_2]Cl_2$; $[Pt(NH_3)_3Cl_3]Cl$; $[Pt(NH_3)_2Cl_4]$; $K[Pt(NH_3)Cl_5]$.

25.40 Pink compound

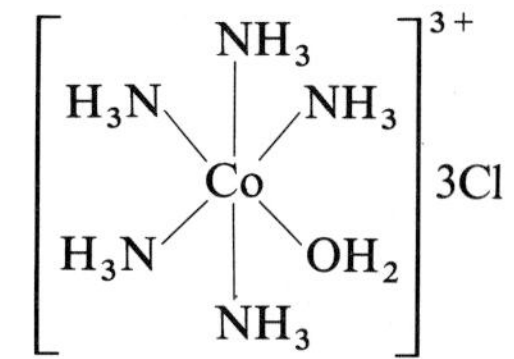

Pentaamminechlorocobalt(III) chloride

Purple compound

$[Co(NH_3)_5Cl]^{3+}$ $2Cl^-$

Pentaammineaquacobalt(III) chloride

25.42 The end of the bidentate ligand containing the CF_3 group is represented by a closed circle, the other end by an open circle.

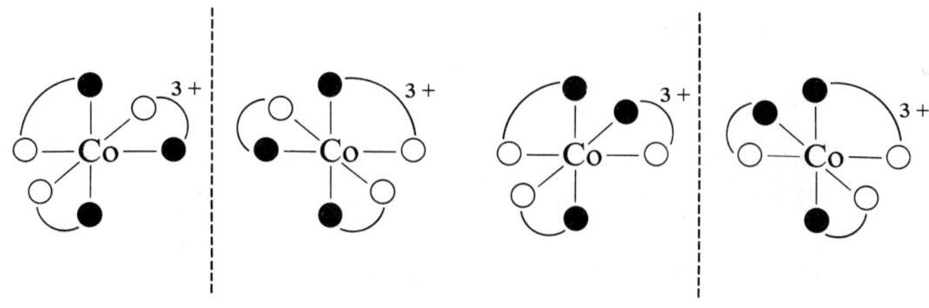

25.45 $Fe_2O_3(s) + 6H_2C_2O_4(aq) \longrightarrow 2[Fe(C_2O_4)_3]^{3-}(aq) + 3H_2O(l) + 6H^+(aq)$. **25.48** The absorption spectra of $[VF_6]^{3-}$, $[V(NH_3)_6]^{3+}$, and $[V(NO_2)_6]^{3-}$ should exhibit a trend toward absorption at shorter wavelength (higher energy). **25.51** The complex will absorb in the visible range, around 660 nm, and appear green. **25.54** For orbital splitting in an octahedral field, the d^3 and d^6 strong field configurations have electrons only in the lower-energy *d* orbitals, but for tetrahedral splitting, they have one or more electrons in the higher-energy *d* orbitals. Thus, tetrahedral arrangements are less favorable.

Chapter 26

26.1 (a) C_5H_{12}; (b) C_5H_{10}; (c) C_5H_{10}; (d) C_5H_8; saturated: (a), (b), unsaturated: (c), (d). **26.3** There are five isomers. Their carbon skeletons are as follows:

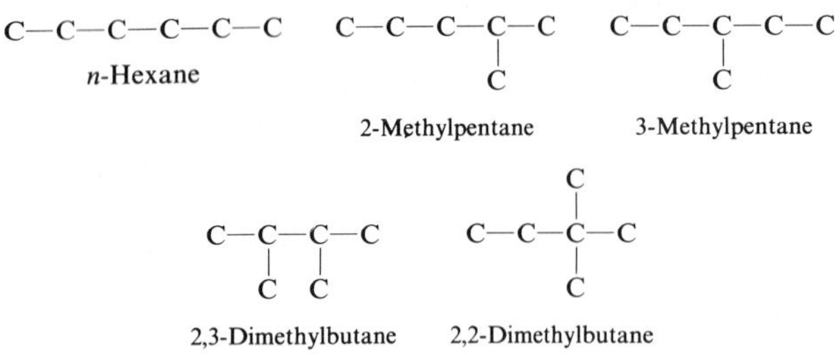
n-Hexane
2-Methylpentane
3-Methylpentane
2,3-Dimethylbutane
2,2-Dimethylbutane

26.5 (a)

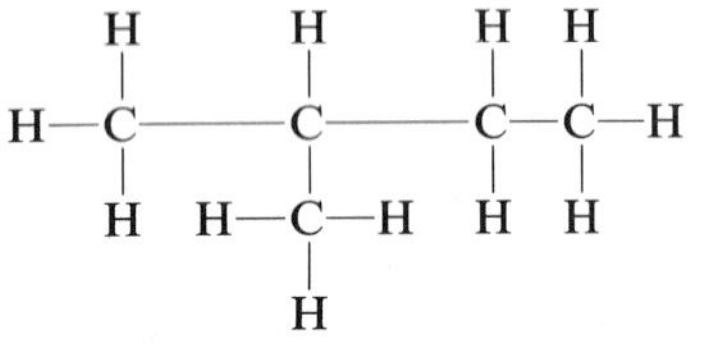

(c)

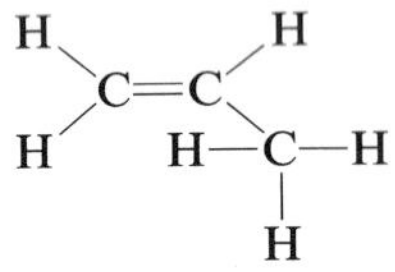

26.7 (a) 109°; (b) 120°; (c) 180°.

26.9 (a)

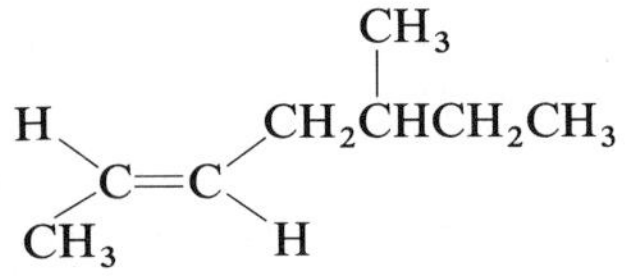

(b) $HC{\equiv}C{-}CH_2Cl$ (c)

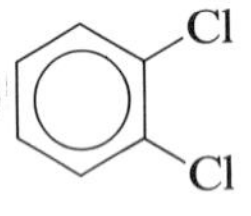

26.11 (b) *cis*-6-methyl-3-octene; (c) *para*-dibromobenzene. **26.13** There are two possible placements for the double bond in butene; these compounds are structural isomers: $CH_2{=}CHCH_2CH_3$ (1-butene) and $CH_3CH{=}CHCH_3$ (2-butene). For 2-butene, there are two arrangements of the carbon skeleton with respect to the double bond; these compounds are geometric isomers:

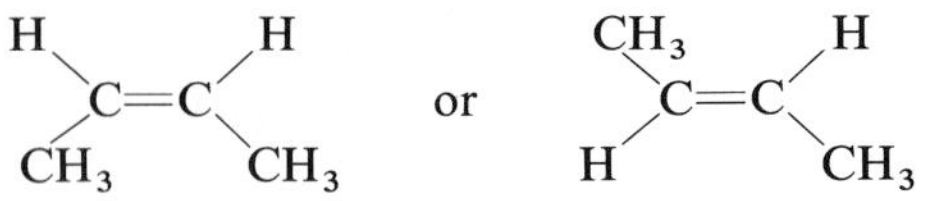

cis-2-Butene *trans*-2-Butene

26.15

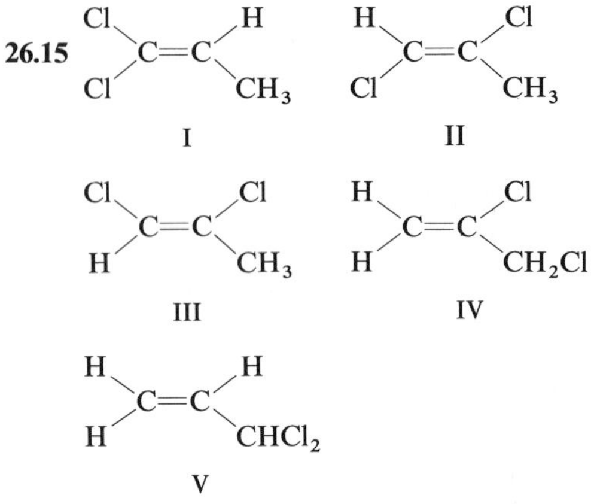
I
II
III
IV
V

II and III are geometric isomers, and the others are structural isomers (different placement of the Cl atoms). **26.17** 70. **26.19** (a) $CH_3CH_2CH_2CH_2CH_3(g) + Cl_2(g) \xrightarrow{hv} CH_3CH_2CH_2CHClCH_3(g) + HCl(g)$; (b) $2CH_3CH{=}CH_2(g) + 9O_2(g) \longrightarrow 6CO_2(g) + 6H_2O(g)$. **26.21** (a)

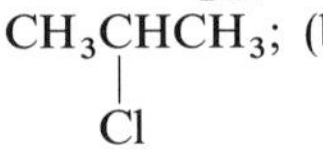

; (b)

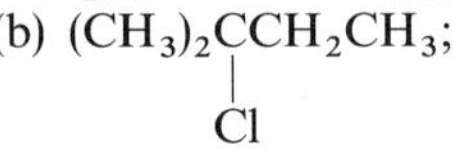

;

(c)

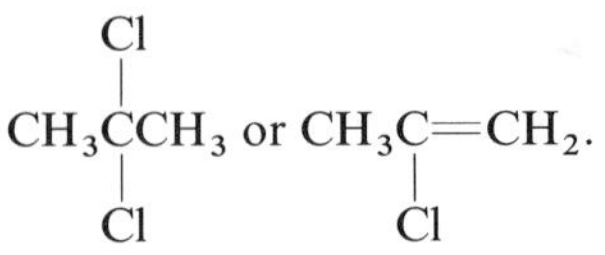

26.23 The 60° C—C—C angles in the cyclopropane ring cause strain that provides a driving force for reactions that result in ring opening. There is no comparable strain in the five- or six-membered rings.

26.25 (b) $CH_3CH_2CI{=}CH_2$ or $CH_3CH_2CI_2CH_3$;

(c)

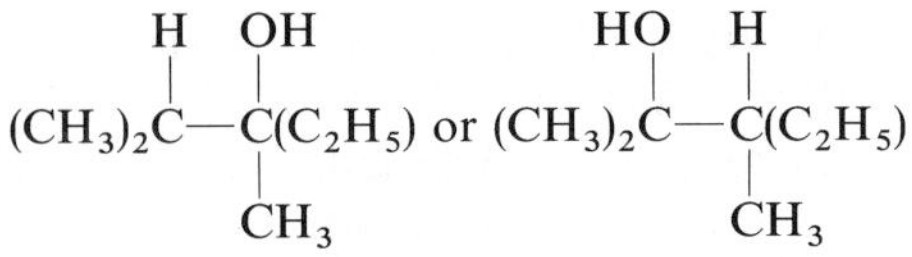

26.27 $C_2H_4(g) + HBr(g) \longrightarrow CH_3CH_2Br(l)$; $C_6H_6(l) + CH_3CH_2Br(l) \xrightarrow{AlCl_3} C_6H_5CH_2CH_3(l) + HBr(g)$. **26.29** Heat of hydrogenation of naphthalene = −300 kJ/mol; for C_2H_4 = −137 kJ/mol. The second value applies to one double bond, so for five double bonds we would expect about −685 kJ. Since hydrogenation of naphthalene produces only −300 kJ, we can conclude that there is some special stability associated with the aromatic system in this molecule. **26.31** (a) Ketone; (b) carboxylic acid; (c) alcohol; (d) ester; (e) amide; (f) amine.

26.33 Propionaldehyde (or propanal): $H-\underset{H}{\overset{H}{C}}-\underset{H}{\overset{H}{C}}-C(=O)H$

26.35 About each methyl (CH_3) carbon, 109°; about the carbonyl carbon, 120°, planar.

26.37 (a) $CH_3\underset{CH_3}{CH}\overset{O}{\overset{\|}{C}}-H$; (b) $CH_3CH_2\overset{O}{\overset{\|}{C}}-H$ or $CH_3CH_2\overset{O}{\overset{\|}{C}}-OH$; (c) no reaction.

26.39 (a) $CH_3\underset{CH_3}{\overset{H}{C}}O-\overset{O}{\overset{\|}{C}}CH_3$ (b) $CH_3CH_2CH_2O-\overset{O}{\overset{\|}{C}}CH_3$

2-propylacetate 1-propylacetate

(c) $CH_3CH_2O-\overset{O}{\overset{\|}{C}}H$

ethylformate

26.41 $H\overset{O}{\overset{\|}{C}}-O-CH_2CH_3 + NaOH \longrightarrow \left[HC\begin{smallmatrix}{=}O\\ \backslash O\end{smallmatrix}\right]^- + Na^+ + CH_3CH_2OH.$

26.43 (b) $HOCH_2CH_2OH$; (d) $CH_3CH_2\overset{O}{\overset{\|}{C}}CH_2CH_3$; (e) $CH_3CH_2OCH_2CH_3$. **26.45** (a) Methanoic acid; (b) butanoic acid; (c) 3-methylpentanoic acid.

26.47 (a) $CH_2{=}CHCH_3 + H_2O \xrightarrow{H_2SO_4} CH_3\underset{OH}{CH}CH_3$;

$CH_3\underset{OH}{CH}CH_3 + (O) \longrightarrow CH_3\underset{O}{\underset{\|}{C}}CH_3 + H_2O$;

(b) $CH_3CH_2CH_2OH + 2(O) \longrightarrow CH_3CH_2\overset{O}{\overset{\|}{C}}-OH + H_2O$

26.49 $nCH_2{=}\underset{CH_3}{CH} \longrightarrow \left[-\underset{H}{\overset{H}{C}}-\underset{CH_3}{\overset{H}{C}}-\right]_n$

26.51 $\left(-O-\overset{O}{\overset{\|}{C}}-\overset{O}{\overset{\|}{C}}-O-CH_2CH_2-O-\overset{O}{\overset{\|}{C}}-\overset{O}{\overset{\|}{C}}-O-CH_2CH_2-O-\right)$

26.53 $HC{\equiv}C-CH_2CH_3$ $CH_3-C{\equiv}C-CH_3$ $CH_2{=}CH-CH{=}CH_2$ $\begin{matrix}HC{=}CH\\ | \quad |\\ H_2C-CH_2\end{matrix}$

26.56 (a) Alkane; (b) alkyne or diene; (c) alkene; (d) alkane.

26.59 $CH_3CH_2\overset{OH}{\overset{|}{C}}H_2$; $CH_3\overset{OH}{\overset{|}{C}}HCH_3$; $CH_3-O-CH_2CH_3$.

26.61 Because of the strain in bond angles about the ring, cyclohexyne would not be stable. The alkyne carbon atoms preferentially have a 180° bond angle, but there are not enough carbon atoms in the ring to make this possible without gross distortions of other bond lengths and angles.

26.64 (a) C—O—C, ether; $-CH{=}CH_2$, alkene. (b) $-\overset{O}{\overset{\|}{C}}-OH$, carboxylic acid; $CH_3\overset{O}{\overset{\|}{C}}-O-$, ester.

26.67 (a) *Cyclobutane*. There is strain in the four-membered ring because the C—C—C angles must be less than the desired 109°. (c) *1-Hexene*. The alkene readily undergoes addition, whereas the aromatic hydrocarbon is extremely stable, due to resonance. **26.70** Hydrolysis of the ester or amide linkages under strong base conditions.

26.73 Cyclopentane.

Chapter 27

27.1 Organisms require energy for maintenance of all living functions. The conversion of external energy to biochemical energy is a spontaneous process, so the accompanying entropy change is positive. The entropy change within the plant may be negative, but the overall entropy of the universe (organism + surroundings) increases.

27.3 (a) $\Delta G^\circ_{298} = 2878$ kJ/mol glucose, nonspontaneous; (b) $\Delta G^\circ_{298} = 2957$ kJ. **27.5** 36 g $O_2(g)$, or 25.4 L.

27.7 (a) An alpha amino acid contains an NH_2 function on the carbon adjacent to the carboxylic acid function. (b) In protein formation, amino acids undergo a condensation reaction between the amino group of one molecule and the carboxylic acid group of another to form the amide linkage.

27.9 Two peptides are possible: $H_2NCH_2CONHCH(CH(CH_3)_2)COOH$ (glycylvaline) and $H_2NCH(CH(CH_3)_2)CONHCH_2COOH$ (valylglycine).

27.11 $H_2N-\underset{CH_2}{\overset{H}{C}}-\overset{O}{\overset{\|}{C}}-\overset{H}{N}-\underset{H}{\overset{H}{C}}-\overset{O}{\overset{\|}{C}}-\overset{H}{N}-\underset{CH_2OH}{\overset{H}{C}}-\overset{O}{\overset{\|}{C}}-OH$ (with the CH_2 attached to an indole ring, N—H)

27.13 In base: $H_2NCH_2CO_2^-$; in acid: $H_3NCH_2COOH^+$. **27.15** Eight: ser-ser-ser, ser-ser-pro, ser-pro-ser, pro-ser-ser, ser-pro-pro, pro-ser-pro, pro-pro-ser, pro-pro-pro. **27.17** The drawing should resemble Figure 27.5, with the $-CH_3$ group of alanine replaced by the $-CH_2COOH$ group of aspartic acid. **27.19** The *primary structure* of a protein refers to the sequence of amino acids in the chain. The *secondary structure* is the configuration (helical, folded, open) of the protein chain. The *tertiary structure* is the overall shape of the protein, determined by the way the segments come together or pack. **27.21** That an increase in substrate concentration produces a roughly proportional increase in the rate indicates that most of the active sites are unoccupied at a particular instant and that the equilibrium in Equation 27.4 lies to the left. **27.23** Enzyme, carbonic anhydrase; cofactor, Zn; substrate, carbonic acid, H_2CO_3; turnover number, 1×10^7 molecules/s.
27.25 When the glucose chain closes to form the hemiacetal (Figure 27.15), α-glucose has the "new" OH group on carbon 1 on the same side of the 6-membered ring as the OH group on the adjacent carbon 2, while the β form has the OH groups on carbon 1 and carbon 2 on opposite sides of the ring.

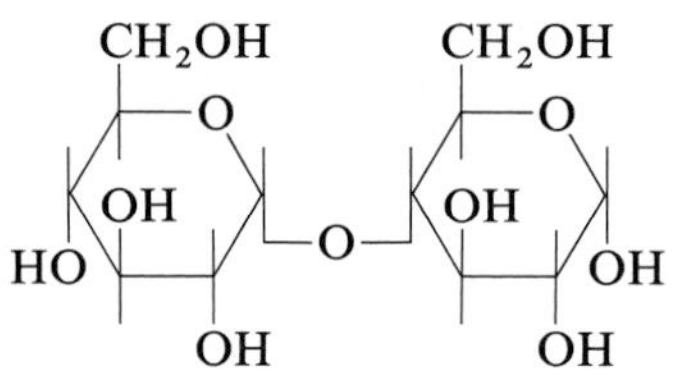

CH₂OH CH₂OH O O OH OH HO O OH OH OH

α-Linkage

CH₂OH CH₂OH O O OH O OH HO OH OH HO

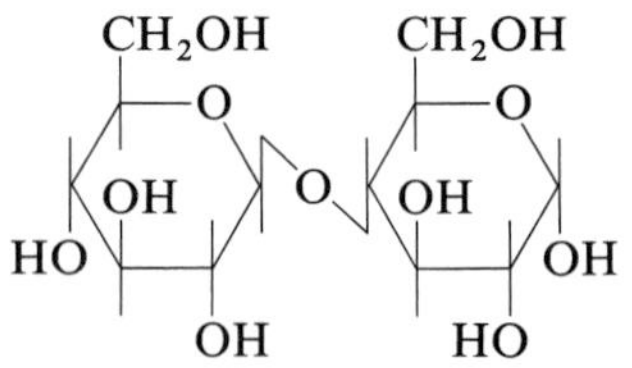

β-Linkage

27.27 Both the β form (shown here) and the α form (OH on C^1 on same side of ring as OH on C^2) are possible.

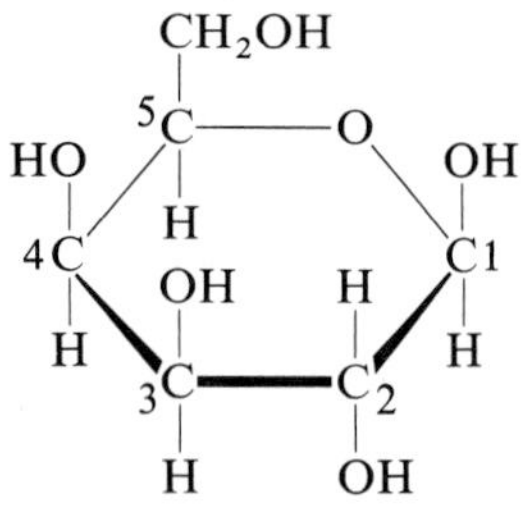

CH₂OH 5C O HO OH H 4C C1 OH H H H 3C C2 H OH

Galactose

27.29 In the 6-membered ring form, carbon atoms 1, 2, 3, 4, and 5 are chiral, because there are four different groups on each.

27.31

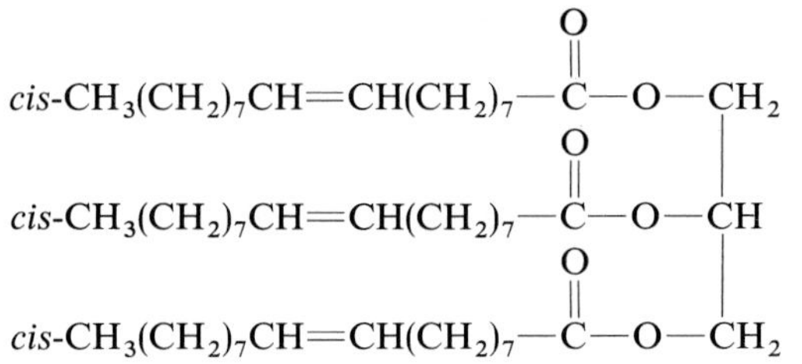

$$\begin{array}{l} cis\text{-}CH_3(CH_2)_7CH{=}CH(CH_2)_7{-}\overset{O}{\overset{\|}{C}}{-}O{-}CH_2 \\ cis\text{-}CH_3(CH_2)_7CH{=}CH(CH_2)_7{-}\overset{O}{\overset{\|}{C}}{-}O{-}CH \\ cis\text{-}CH_3(CH_2)_7CH{=}CH(CH_2)_7{-}\overset{O}{\overset{\|}{C}}{-}O{-}CH_2 \end{array}$$

27.33

$$C_3H_5(OH)_3 + 3C_{11}H_{23}\overset{O}{\overset{\|}{C}}{-}OH \longrightarrow C_3H_5(O\overset{O}{\overset{\|}{C}}C_{11}H_{23})_3 + 3H_2O.$$

27.35 A nucleotide consists of a nitrogen-containing aromatic compound, a sugar in the furanose (5-membered) ring form, and a phosphoric acid molecule. The structure of deoxycytidine monophosphate is:

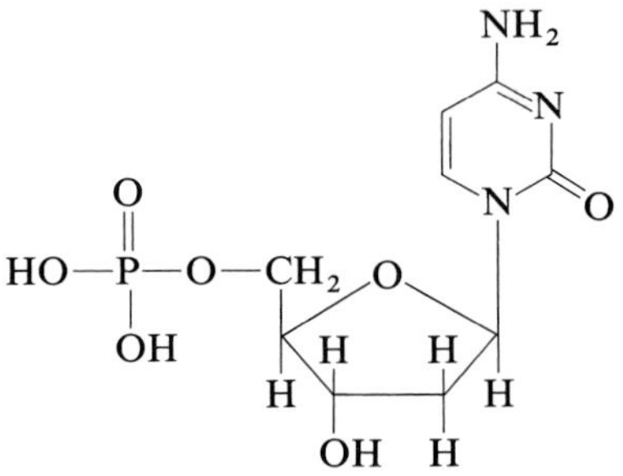

NH₂ N N O O HO—P—O—CH₂ O OH H H H H OH H

27.37 $C_4H_7CH_2OH + H_3PO_4 \longrightarrow C_4H_7CH_2{-}O{-}PO_3H_2 + H_2O.$

27.39

```
—A—C—T—C—G—A—
 |  |  |  |  |  |
 T—G—A—G—C—T—  ⟵ Complementary strand
```

27.41 (a) 42 m^2. (b) This leaf area would result in conversion of about 10,000 kJ per day to plant matter. However, not all this energy is available to the human ingesting the plant. Some energy will be in the form of cellulose, which is not digestible, some goes into root formation, and some is required for various functions within the living plant.

27.43 (a) $CH_3\underset{NH_2}{\underset{|}{C}}H\overset{O}{\overset{\|}{C}}{-}OH + H_2NCH_2\overset{O}{\overset{\|}{C}}{-}OH$;

Alanine Glycine

(b) 3 mol $C_{17}H_{33}\overset{O}{\overset{\|}{C}}{-}OH + C_3H_5(OH)_3$.

(c) 2 mol α-glucose (d) HO, O, CH_2OH [furanose ring: H H / H H; HO OH] + $O-P(OH)_3$

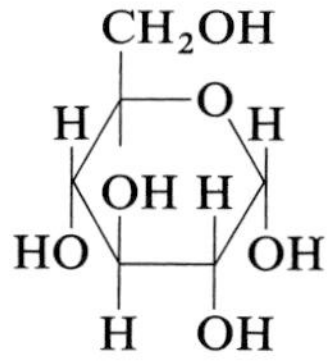

CH_2OH [glucose ring: O, H, H, OH H, HO, OH, H OH]

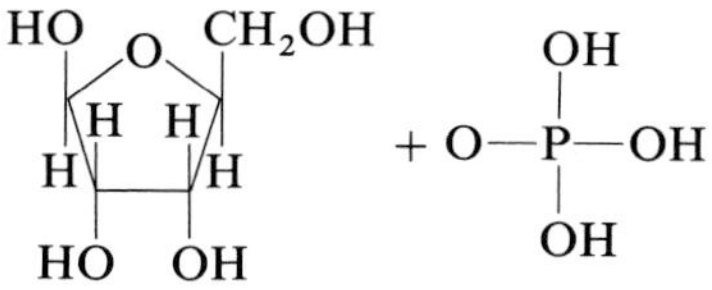

27.46 (a) $H_2N-\underset{CH_2C_6H_5}{\overset{H}{C}}-\overset{O}{\overset{\|}{C}}-OH$ $\quad$ $H_2N-\underset{CH_2C_6H_5OH}{\overset{H}{C}}-\overset{O}{\overset{\|}{C}}-OH$

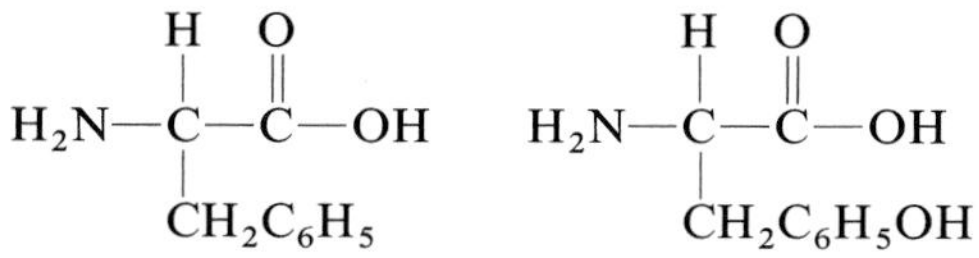

(b) The only difference between the two amino acids is an OH group on the phenyl (C_6H_5) ring. The enzyme catalyzes the placement of this hydroxyl group, so it is called a hydroxylase.

27.49 (a) $NH_2-\underset{CH_2CH_2COOH}{CH}\overset{O}{\overset{\|}{C}}-O^- \quad Na^+$

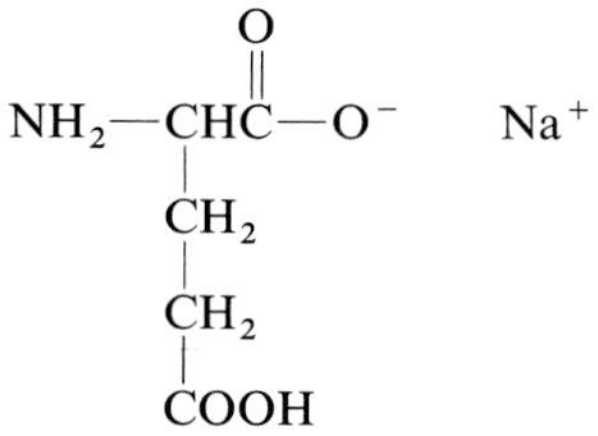

(b) The flavor-enhancing properties of MSG probably relate to the metabolic functions of glutamic acid and its sodium salt. Since the L-isomer is the naturally occurring form of glutamic acid, it is not surprising that only the L-isomer has flavor-enhancing properties. **27.52** That the rate doubles with a doubling of concentration of the sugar tells us that the fraction of enzyme tied up in an enzyme-substrate complex is small. Inositol is behaving like a competitor with sucrose for binding in the active site. **27.55** *Condensation* is the joining of two neutral molecules by a link formed when a small molecule, usually water, is eliminated. *Hydrolysis* is the reverse process: adding water to split up two molecules. These reactions are central to biochemistry because all biopolymers are formed through condensation reactions, and many metabolic processes involve either condensation or hydrolysis.

27.58

—T—A—T—G—C—A—
 | | | | | |
—A—T—A—C—G—T— ⟵ Complementary strand

Glossary

absorption spectrum The amount of light absorbed by a sample as a function of wavelength. (Section 25.7)

accuracy A measure of how closely individual measurements agree with the correct value. (Section 1.4)

acid A substance that is able to donate a H^+ ion (a proton) and hence increases the concentration of $H^+(aq)$ when it dissolves in water. (Section 4.3)

acid-dissociation constant (K_a) An equilibrium constant that expresses the extent to which an acid transfers a proton to solvent water. (Section 16.4)

acidic anhydride (acidic oxide) An oxide that forms an acid when added to water; soluble nonmetal oxides are acidic anhydrides. (Section 22.4)

acidic oxide (acidic anhydride) An oxide that either reacts with a base to form a salt or reacts with water to form an acid. (Section 22.4)

acid rain Rainwater that has become excessively acidic because of absorption of pollutant oxides, notably SO_3, produced by human activities. (Section 18.4)

actinide element Element in which the 5*f* orbital is only partially occupied. (Section 6.9)

activated complex (transition state) The particular arrangement of atoms found at the top of the potential-energy barrier as a reaction proceeds from reactants to products. (Section 14.4)

activation energy (E_a) The minimum energy needed for reaction: the height of the activation barrier to formation of products. (Section 14.4)

active metals Metals in groups 1A and 2A in which electrons occupy only the *s* orbitals of the valence shell. (Section 6.10)

active site Specific site on a heterogeneous catalyst or an enzyme where catalysis occurs. (Sections 14.6 and 27.3)

activity series A list of metals in order of decreasing ease of oxidation. (Section 4.6)

addition polymerization Polymerization that occurs through coupling of monomers with one another, with no other products formed in the reaction. (Section 12.2)

addition reaction A reaction in which a reagent adds to the two carbon atoms of a carbon-carbon multiple bond. (Section 26.2)

adhesive forces Forces that bind a substance to a surface, such as water to glass. (Section 11.3)

adsorption The binding of molecules to a surface. (Section 14.6)

alcohol An organic compound obtained by substituting a hydroxyl group (—OH) for a hydrogen on a hydrocarbon. (Section 26.2)

aldehyde An organic compound in which the carbonyl group has at least one hydrogen atom attached. (Section 26.4)

alkali metals Members of group 1A in the periodic table. (Sections 2.4, 7.7)

alkaline earth metals Members of group 2A in the periodic table. (Sections 2.4 and 7.7)

alkanes Compounds of carbon and hydrogen containing only single carbon-carbon bonds. (Chapter 26: Introduction)

alkenes Hydrocarbons containing one or more carbon-carbon double bonds. (Chapter 26: Introduction)

alkyl group A group that is formed by removing a hydrogen atom from an alkane. (Section 26.1)

alkynes Hydrocarbons containing one or more carbon-carbon triple bonds. (Chapter 26: Introduction)

allotropes Different chemical forms of the same element existing in the same physical state. (Section 7.8)

alloy A substance that has the characteristic properties of a metal and contains more than one element. Often there is one principal metallic component, with other elements present in smaller amounts. Alloys may be homogeneous or heterogeneous in nature. (Section 24.2)

alpha particles Particles that are identical to helium-4 nuclei, consisting of two protons and two neutrons, symbol ${}^4_2He$. (Section 21.1)

alpha rays Streams of alpha particles (helium-4 nuclei). (Section 21.1)

aluminosilicates Compounds that are structurally related to silicates and in which some Si^{4+} ions are replaced by Al^{3+} ions. (Section 23.5)

amide An organic compound that has an NR_2 group attached to a carbonyl. (Section 26.4)

amine A compound that has the general formula R_3N, where R may be H or a hydrocarbon group. (Sections 16.6, 26.4)

amino acid A carboxylic acid that contains an amino ($-NH_2$) group attached to the carbon atom adjacent to the carboxylic acid (CO_2H) functional group. (Section 27.2)

amorphous solid A solid whose molecular arrangement lacks a regular and long-range pattern. (Section 11.7)

amphoteric Capable of dissolving in either an acidic or a basic medium. The solubility in base results from formation of a complex ion via an acid-base reaction. (Section 17.5)

angstrom A common non-SI unit of length that is used to measure atomic dimensions: 1 Å equals 10^{-10} m. (Section 2.3)

anion A negatively charged ion. (Section 2.5)

anode An electrode at which oxidation occurs. (Section 20.3)

antibonding molecular orbital A molecular orbital in which electron density is concentrated outside the region between the two nuclei of bonded atoms. Such orbitals, designated as σ^* or π^*, are less stable (of higher energy) than bonding molecular orbitals. (Section 9.6)

aqueous solution A solution in which water is the solvent. (Chapter 4: Introduction)

aromatic hydrocarbons Hydrocarbon compounds that contain a planar, cyclic arrangement of carbon atoms linked by both σ and delocalized π bonds. (Chapter 26: Introduction)

Arrhenius equation An equation that relates the rate constant for a reaction to the frequency factor, A, the activation energy, E_a, and the temperature, T: $k = Ae^{-E_a/RT}$. In its logarithmic form, it is written: $\ln k = -E_a/RT + \ln A$. (Section 14.4)

atom The smallest representative particle of an element. (Chapter 2: Introduction)

atomic mass unit (amu) A unit based on the value of exactly 12 amu for the isotope of carbon that has six protons and six neutrons in the nucleus. (Section 3.3)

atomic number The number of protons in the nucleus of an atom of an element. (Section 2.3)

atomic radius An estimate of the size of an atom, in which it is assumed to be a spherical object. (Section 7.3)

atomic weight (average atomic mass) The average mass of the atoms of an element in atomic mass units (amu); it is numerically equal to the mass in grams of 1 mol of the element. (Section 3.3)

autoionization The process whereby water spontaneously forms low concentrations of $H^+(aq)$ and $OH^-(aq)$ ions by proton transfer from one water molecule to another. (Section 16.2)

Avogadro's hypothesis A statement that equal volumes of gases at the same temperature and pressure contain equal numbers of molecules. (Section 10.3)

Avogadro's law A statement that the volume of a gas maintained at constant temperature and pressure is directly proportional to the quantity of the gas. (Section 10.3)

Avogadro's number The number of ^{12}C atoms in exactly 12 g of ^{12}C; it equals 6.022×10^{23}. (Section 3.5)

barometer A device for measuring atmospheric pressure. (Section 10.2)

base A substance that is a H^+ acceptor; a base produces an excess of $OH^-(aq)$ ions when it dissolves in water. (Section 4.3)

base-dissociation constant (K_b) An equilibrium constant that expresses the extent to which a base reacts with solvent water, accepting a proton and forming $OH^-(aq)$. (Section 16.6)

basic anhydride (basic oxide) An oxide that forms a base when added to water; soluble metal oxides are basic anhydrides. (Section 22.4)

basic oxide (basic anhydride) An oxide that either reacts with water to form a base or reacts with an acid to form a salt and water. (Section 22.4)

Bayer process A hydrometallurgical procedure for purifying *bauxite* in the recovery of aluminum from bauxite-containing ores. (Section 24.4)

becquerel The SI unit of radioactivity. It corresponds to one nuclear disintegration per second. (Section 21.9)

beta particles Energetic electrons emitted from the nucleus, symbol $_{-1}^{0}e$. (Section 21.2)

bidentate ligand A ligand in which two coordinating atoms are bound to a metal. (Section 25.2)

bimolecular reaction An elementary reaction that involves two molecules. (Section 14.5)

biodegradable Organic material that bacteria are able to oxidize. (Section 18.6)

biopolymer A polymeric molecule of high molecular weight found in living systems. The three major classes of biopolymer are proteins, carbohydrates, and nucleic acids. (Section 27.1)

biosphere The part of the earth in which living organisms can exist and live out their life cycles. (Chapter 27: Introduction)

body-centered cubic cell A cubic unit cell in which the lattice points occur at the corners and at the center. (Section 11.7)

bomb calorimeter A device for measurements of heat evolved in combustion of a substance under constant-volume conditions. (Section 5.7)

bond angles The angles made by the lines joining the nuclei of the atoms in a molecule. (Section 9.1)

bond dipole The contribution to the dipole moment of a molecule by a particular bond. (Section 9.2)

bond-dissociation energy The enthalpy change, ΔH, required to break a particular bond in a mole of gaseous substance; also called **bond energy**. (Section 8.8)

bond energy The enthalpy change, ΔH, required to break a particular bond when the substance is in the gas phase; also called **bond-dissociation energy**. (Section 8.8)

bonding molecular orbital A molecular orbital in which the electron density is concentrated in the internuclear region. The energy of a bonding molecular orbital is lower than the energy of the separate atomic orbitals from which it forms. (Section 9.6)

bonding pair In a Lewis structure, a pair of electrons that is shared by two atoms. (Section 9.1)

bond order The number of bonding electron pairs shared between two atoms, less the number of antibonding electron pairs: bond order $= \frac{1}{2}$ (number of bonding electrons − number of antibonding electrons). (Section 9.6)

bond polarity A measure of the difference in ability of the two atoms in a chemical bond to attract electrons. (Section 8.9)

Born-Haber cycle A thermodynamic cycle based on Hess's law that relates the lattice energy of an ionic substance to its enthalpy of formation and to other measurable quantities. (Section 8.2)

Boyle's law A law stating that at constant temperature, the product of the volume and pressure of a given amount of gas is a constant. (Section 10.3)

breeder reactor A nuclear-fission reactor that produces more fissionable fuel than it consumes. (Section 21.7)

Brønsted acid Any substance that acts as a source of protons. (Section 16.1)

Brønsted base Any substance that acts as a proton acceptor. (Section 16.1)

buffered solution (buffer) A solution that undergoes a limited change in pH upon addition of a small amount of acid or base. (Section 17.2)

buffering capacity The amount of acid or base a buffer can neutralize before the pH begins to change appreciably. (Section 17.2)

calcination The heating of an ore to bring about its decomposition and the elimination of a volatile product. For example, a carbonate ore might be calcined to drive off CO_2. (Section 24.3)

calorie A unit of energy, it is the amount of energy needed to raise the temperature of 1 g of water by 1°C, from 14.5°C to 15.5°C. A related unit is the joule: 1 cal = 4.184 J. (Section 5.1)

calorimeter An apparatus that measures heat flow. (Section 5.7)

calorimetry The experimental measurement of heat effects in chemical and physical processes. (Section 5.7)

capillary action The process by which a liquid rises in a tube because of a combination of adhesion with the walls of the tube and cohesion between liquid particles. (Section 11.3)

carbohydrates A class of substances formed from polyhydroxy aldehydes or ketones. (Section 27.4)

carbonylation A reaction in which the carbonyl functional group is introduced into a molecule. (Section 26.4)

carbonyl group The C=O double bond, a characteristic feature of several organic functional groups, such as ketones and carboxylic acids. (Section 26.4)

carboxyhemoglobin A complex formed between carbon monoxide and hemoglobin, in which CO is bound to the iron atom. (Section 18.4)

carboxylic acid A compound that contains the —COOH functional group. (Section 26.4)

catalyst A substance that acts to change the speed of a chemical reaction without itself undergoing a permanent chemical change in the process. (Section 14.6)

catenation The linking of like atoms to form chains or rings. (Section 23.5)

cathode An electrode at which reduction occurs. (Section 20.3)

cathode rays Streams of electrons that are produced when a high voltage is applied to electrodes in an evacuated tube. (Section 2.2)

cathodic protection A means of protecting a metal against corrosion by making it the cathode in a voltaic cell. This can be achieved by attaching a more easily oxidized metal, which serves as an anode, to the metal to be protected. (Section 20.9)

cation A positively charged ion. (Section 2.5)

cell potential A measure of the driving force, or "electrical pressure," for the completion of an electrochemical reaction; it is measured in volts: 1 V = 1 J/C. Also called **electromotive force**. (Section 20.4)

cellulose A polysaccharide of glucose; it is the major structural element in plant matter. (Section 27.4)

Celsius scale A temperature scale on which water freezes at 0° and boils at 100° at sea level. (Section 1.3)

ceramic A solid inorganic material, either crystalline (oxides, carbides, silicates) or amorphous (glasses). Most ceramics melt at high temperatures. (Section 12.3)

ceramic fiber A ceramic material processed in the form of long, thin fibers with length-to-diameter ratios greater than 100. Such a fiber is often formed by the thermal decomposition of a precursor polymeric material. (Section 12.3)

chain reaction A series of reactions in which one reaction initiates the next. (Section 21.7)

changes of state Transformations of matter from one state to a different one, for example, from a gas to a liquid. (Section 1.1)

charcoal A form of carbon produced when wood is heated strongly in a deficiency of air. (Section 22.6)

Charles's law A law stating that at constant pressure, the volume of a given quantity of gas is proportional to absolute temperature. (Section 10.3)

chelate effect The generally larger formation constants for polydentate ligands as compared with the corresponding monodentate ligands. (Section 25.2)

chelating agent A polydentate ligand that is capable of occupying two or more sites in the coordination sphere. (Section 25.2)

chemical changes Processes in which one or more substances are converted into other substances. (Also called **chemical reactions.**) (Section 1.1)

chemical equation A representation of a chemical reaction using the chemical formulas of the reactants and products; a **balanced chemical equation** contains equal numbers of atoms of each element on both sides of the equation. (Section 3.1)

chemical equilibrium A state of dynamic balance in which the rate of formation of the products of a reaction from the reactants equals the rate of formation of the reactants from the products; at equilibrium, the concentrations of the reactants and products remain constant. (Section 4.2, Chapter 15: Introduction.)

chemical formula A notation that uses atomic symbols with numerical subscripts to convey the relative proportions of atoms of the different elements in a substance. (Section 2.5)

chemical kinetics The area of chemistry concerned with the speeds, or rates, at which chemical reactions occur. (Chapter 14: Introduction)

chemical properties Properties that describe a substance's composition and its reactivity; how the substance reacts, or changes into other substances. (Section 1.1)

chemical reactions Processes in which one or more substances are converted into other substances. (Also called **chemical changes.**) (Section 1.1)

chemical-vapor deposition A method for forming thin films in which a substance is deposited on a surface and then undergoes some form of chemical reaction to form the film. (Section 12.4)

chiral A term describing a molecule or an ion that cannot be superimposed on its mirror image. (Sections 25.4, 27.2)

chlorofluorocarbons Compounds composed entirely of chlorine, fluorine, and carbon. (Section 18.3)

chlorophyll A plant pigment that plays a major role in conversion of solar energy to chemical energy in photosynthesis. (Section 27.1)

cholesteric liquid crystal A crystal formed from flat, disc-shaped molecules that align through a stacking of the molecular discs. (Section 12.1)

clay minerals A class of hydrated aluminosilicates. (Section 23.5)

coagulation A process in which colloidal particles are enlarged; the resultant larger particles can then be separated by filtration or merely by allowing them to settle out of the dispersing medium. (Section 13.6)

coal A naturally occurring solid containing hydrocarbons of high molecular weight as well as compounds containing sulfur, oxygen, and nitrogen. (Section 5.8)

coefficient The number in front of each formula in a balanced chemical equation. (Section 3.1)

coenzyme (cofactor) A substance that is needed along with some enzyme if an enzyme-catalyzed reaction is to occur. (Section 27.3)

cohesive forces Attractive forces between like molecules. (Section 11.3)

coke An impure form of carbon formed when coal is heated strongly in the absence of air. (Section 22.6)

colligative properties Those properties of a solvent (vapor-pressure lowering, freezing-point lowering, boiling-point elevation, osmotic pressure) that depend on the total concentration of solute particles present. (Section 13.5)

colloidal dispersions (colloids) Mixtures containing particles larger than normal solutes but small enough to remain suspended in the dispersing medium. (Section 13.6)

combination reaction A chemical reaction in which two or more substances combine to form a single product. (Section 3.2)

combustion reaction A chemical reaction that proceeds with evolution of heat and usually also a flame; most combustion involves reaction with oxygen, as in the burning of a match. (Section 3.2)

common-ion effect The effect of an ion common to an equilibrium in shifting the equilibrium. For example, added Na_2SO_4 decreases the solubility of the slightly soluble salt $BaSO_4$, or added $NaC_2H_3O_2$ decreases the percent ionization of $HC_2H_3O_2$. (Section 17.1)

complementary colors Colors that, when mixed in proper proportions, appear white or colorless. (Section 25.7)

complete ionic equation An equation written with all soluble strong electrolytes shown as they exist in aqueous solution, that is, as ions. (Section 4.4)

complex ion (complex) An assembly of a metal ion and the Lewis bases bonded to it. (Section 17.5; Chapter 25: Introduction)

composite A complex solid mixture of two or more components. One component is usually present in much greater quantity than the others and acts as the primary host matrix for the other components. (Section 12.3)

compound A substance composed of two or more elements united chemically in definite proportions. (Section 1.2)

concentration The quantity of solute present in a given quantity of solvent or solution. (Section 4.1)

condensation polymerization Polymerization in which molecules are joined together through condensation reactions. (Section 12.2)

condensation reaction A chemical reaction in which a small molecule such as a molecule of water is split out from between two reacting molecules, as for example between an organic acid and an amine function:

$$\underset{\underset{\displaystyle O}{\|}}{-C}-O-H + H-\underset{\underset{\displaystyle H}{|}}{N}- \rightarrow -\underset{\underset{\displaystyle O}{\|}}{C}-\underset{\underset{\displaystyle H}{|}}{N}- + H_2O.$$

(Section 12.3)

conjugate acid A substance formed by addition of a proton to a Brønsted base. (Section 16.1)

conjugate acid-base pair An acid and a base such as H_2O and OH^-, which differ only in the presence or absence of a proton. (Section 16.1)

conjugate base A substance formed by the loss of a proton from a Brønsted acid. (Section 16.1)

continuous spectrum A spectrum that contains radiation distributed over all wavelengths. (Section 6.2)

conversion factor A ratio relating the same quantity in two systems of units; used to convert the units of measurement. (Section 1.5)

coordination compound A compound containing a complex ion. (Chapter 25: Introduction)

coordination number The number of adjacent atoms to which an atom is directly bonded. In a complex, the coordination number of the metal ion is the number of donor atoms to which it is bonded. (Sections 11.7, 25.1)

coordination sphere The metal ion and its surrounding ligands. (Section 25.1)

coordination-sphere isomers Structural isomers of coordination compounds in which the ligands within the coordination sphere differ. (Section 25.4)

core electrons The electrons that are not in the outermost shell of an atom. (Section 6.9)

corrosion The process by which a metal is oxidized by substances in its environment. (Section 20.9)

covalent bond A bond formed between two or more atoms by a sharing of electrons. (Section 8.4)

covalent (network) solids Solids in which the units that make up the three-dimensional network are joined by covalent bonds. (Section 11.8)

critical mass The amount of fissionable material necessary to maintain a chain reaction. (Section 21.7)

critical pressure The pressure at which a gas at its critical temperature is converted to a liquid state. (Section 11.4)

critical temperature The highest temperature at which it is possible to convert the gaseous form of a substance to a liquid. The critical temperature increases with an increase in the magnitude of intermolecular forces. (Section 11.4)

crystal-field theory A theory that accounts for the colors and the magnetic and other properties of transition-metal complexes in terms of the splitting of the energies of metal-ion *d* orbitals by the electrostatic interaction with the ligands. (Section 25.8)

crystal lattice An imaginary network of points on which the repeating unit of the structure of a solid (the contents of the unit cell) may be imagined to be laid down so that the structure of the crystal is obtained. Each point represents an identical environment in the crystal. (Section 11.7)

crystalline solid (crystal) A solid whose internal arrangement of atoms, molecules, or ions shows a regular repetition in any direction through the solid. (Section 11.7)

crystallinity As applied to polymers, this term refers to relatively good ordering of sections of polymer chains in different molecules with respect to one another. (Section 12.2)

crystallization The process in which a dissolved solute comes out of solution and forms a crystalline solid. (Section 13.3)

cubic close packing A close-packing arrangement in which the atoms of the third layer of a solid are not directly over those in the first layer. (Section 11.7)

curie A measure of radioactivity: 1 curie = 3.7×10^{10} nuclear disintegrations per second. (Section 21.9)

cycle A single complete execution of a periodically repeated phenomenon such as a wave form. (Section 6.1)

cycloalkanes Saturated hydrocarbons of general formula C_nH_{2n} in which the carbon atoms form a closed ring. (Section 26.1)

Dalton's law of partial pressures A law stating that the total pressure of a mixture of gases is the sum of the pressures that each gas would exert if it were present alone. (Section 10.6)

decomposition reaction A chemical reaction in which a single compound reacts to give two or more products. (Section 3.2)

degenerate orbitals Orbitals that have the same energy. (Section 6.7)

delocalized electrons Electrons that are spread over a number of atoms in a molecule rather than localized between a pair of atoms. (Section 9.5)

density The ratio of an object's mass to its volume. (Section 1.3)

deoxyribonucleic acid (DNA) A polynucleotide in which the sugar component is deoxyribose. (Section 27.6)

desalination The removal of salts from seawater, brine, or brackish water to make it fit for human consumption. (Section 18.5)

deuterium The isotope of hydrogen whose nucleus contains a proton and a neutron: ${}^2_1H$. (Section 22.3)

dextrorotatory, or merely **dextro** or ***d*** A term used to label a chiral molecule that rotates the plane of polarization of plane-polarized light to the right (clockwise). (Section 25.4)

dialysis The separation of small solute particles from colloid particles by means of a semipermeable membrane. (Section 13.6)

diamagnetism A type of magnetism that causes a substance with no unpaired electrons to be weakly repelled from a magnetic field. (Section 9.7)

diatomic molecule A molecule that contains two atoms. (Section 2.5)

diffusion The spreading of one substance through another. (Section 10.9)

dilution The process of preparing a less concentrated solution from a more concentrated one by adding solvent. (Section 4.1)

dimensional analysis A method of problem solving in which units are carried through all calculations. Dimensional analysis ensures that the final answer of a calculation has the desired units. (Section 1.5)

dipole-dipole force The force that exists between polar molecules. (Section 11.2)

dipole moment A measure of the separation between centers of positive and negative charges in polar molecules. (Section 9.2)

disproportionation A reaction in which a species undergoes simultaneous oxidation and reduction [as in $N_2O_3(g) \rightarrow NO(g) + NO_2(g)$]. (Section 22.5)

donor atom The atom of a ligand that bonds to the metal. (Section 25.1)

double bond A covalent bond involving two electron pairs. (Section 8.4)

double displacement reaction A reaction in which two substances react through an exchange of their component ions: $AX + BY \rightarrow AY + BX$; also called **metathesis reaction**. Precipitation and acid-base neutralization reactions are examples of double displacement reactions. (Section 4.5)

double-helix The structure for DNA that involves the winding of two DNA polynucleotide chains together in a helical arrangement. The two strands of the double helix are complementary in that the organic bases on the two strands are paired for optimal hydrogen-bond interaction. (Section 27.6)

Downs cell A cell used to obtain sodium metal by electrolysis of molten NaCl. (Section 24.5)

dry ice Solid carbon dioxide. (Section 22.6)

dynamic equilibrium A state of balance in which opposing processes occur at the same rate. (Section 11.5)

effective nuclear charge The net positive charge experienced by an electron in a many-electron atom; this charge is not the full nuclear charge because there is some shielding of the nucleus by the other electrons in the atom. (Section 6.7)

effusion The escape of a gas through an orifice or hole. (Section 10.9)

elastomer A material that can undergo a substantial change in shape via stretching, bending, or compression and return to its original shape upon release of the distorting force. (Section 12.2)

electrochemistry The branch of chemistry that deals with the relationships between electricity and chemical reactions. (Chapter 20: Introduction)

electrolysis reaction A reaction in which a nonspontaneous process is brought about by the passage of current under a sufficient external electrical potential. The devices in which electrolysis reactions occur are called **electrolytic cells**. (Section 20.7)

electrolyte A solute that produces ions in solution; an electrolytic solution conducts an electric current. (Section 4.2)

electrolytic cell A device in which a nonspontaneous reaction is caused to occur by passage of current under a sufficient external electrical potential. (Section 20.7)

electromagnetic radiation (radiant energy) A form of energy that has wave characteristics and that propagates through a vacuum at the characteristic speed of 3.00×10^8 m/s. (Section 6.1)

electrometallurgy The use of electrolysis to reduce or refine metals. (Section 24.5)

electromotive force (emf) A measure of the driving force, or "electrical pressure," for the completion of an electrochemical reaction. Electromotive force is measured in volts: 1 V = 1 J/C. Also called the **cell potential**. (Section 20.4)

electron A negatively charged subatomic particle found outside the atomic nucleus; it is a part of all atoms. An electron has a mass 1/1836 times that of a proton. (Section 2.3)

electron affinity The energy change that occurs when an electron is added to a gaseous atom. (Section 7.5)

electron capture A mode of radioactive decay in which an inner-shell orbital electron is captured by the nucleus. (Section 21.2)

electron configuration A particular arrangement of electrons in the orbitals of an atom. (Section 6.9)

electron density The probability of finding an electron at any particular point in an atom; this probability is equal to ψ^2, the square of the wave function. (Section 6.5)

electron-dot symbol (Lewis symbol) The chemical symbol for an element plus a dot for each valence electron. (Section 8.1)

electronegativity A measure of the ability of an atom that is bonded to another atom to attract electrons to itself. (Section 8.9)

electronic charge The negative charge carried by an electron; it has a magnitude of 1.602×10^{-19} C. (Section 2.3)

electronic structure The arrangement of electrons of an atom. (Chapter 6: Introduction)

electron-pair geometry The three-dimensional arrangement of the electron pairs around the atom according to the VSEPR model. (Section 9.1)

electron shell A collection of orbitals that have the same value of n. For example, the orbitals with $n = 3$ (the 3s, 3p, and 3d orbitals) comprise the third shell. (Section 6.5)

electron spin A property of the electron that makes it behave as though it were a tiny magnet. The electron behaves as if it were spinning on its axis; electron spin is quantized. (Section 6.8)

electron spin quantum number (m_s) A quantum number associated with the electron spin; it may have values of $+\frac{1}{2}$ or $-\frac{1}{2}$. (Section 6.8)

element A substance that cannot be separated into simpler substances by chemical means. (Section 1.2)

elementary steps (elementary processes) Processes in a chemical reaction that occur in a single event or step. (Section 14.5)

empirical formula A chemical formula that shows the kinds of atoms and their relative numbers in a substance. (Section 2.5)

enantiomers Two mirror-image molecules of a chiral substance. The enantiomers are nonsuperimposable. (Section 27.2)

endothermic process A process in which a system absorbs heat from its surroundings. (Section 5.2)

energy The ability to do work or to transfer heat. (Section 5.1)

energy-level diagram A diagram that shows the energies of molecular orbitals relative to the atomic orbitals from which they are derived. Also called a **molecular orbital diagram**. (Section 9.6)

engineering ceramic A ceramic material that finds application in replacement of metals, wood, or plastics. (Section 12.3)

enthalpy A quantity defined by the relationship $H = E + PV$; the enthalpy change, ΔH, for a reaction that occurs at constant pressure is the heat evolved or absorbed in the reaction, q_p. (Section 5.3)

entropy A thermodynamic function associated with the number of different, equivalent energy states or spatial arrangements in which a system may be found. It is a thermodynamic state function, which means that once we specify the conditions for a system—that is, the temperature, pressure, and so on—the entropy is defined. (Section 19.2)

enzyme A protein molecule that acts to catalyze specific biochemical reactions. (Section 14.6)

equilibrium constant For a system at equilibrium, the product of the concentrations of all reaction products, each raised to the power of its coefficient in the balanced equation, divided by the product of the concentrations of the reactants, each raised to the power of its coefficient in the reaction. The equilibrium constant is denoted by K. When the concentrations are expressed in mol/L, the constant is denoted K_c; when concentrations are expressed in atm, the constant is denoted K_p. (Section 15.2)

equivalence point The point in a titration at which the added solute reacts completely with the solute present in the solution. (Section 4.7)

ester An organic compound that has an OR group attached to a carbonyl; it is the product of a reaction between a carboxylic acid and an alcohol. (Section 26.4)

ether A compound in which two hydrocarbon groups are bonded to one oxygen. (Section 26.4)

excited state A higher energy state than the ground state. (Section 6.3)

exothermic process A process in which a system releases heat to its surroundings. (Section 5.2)

extensive property A property that depends on the amount of material considered; for example, mass or volume. (Section 1.3)

face-centered cubic cell A cubic unit cell that has lattice points at each corner as well as at the center of each face. (Section 11.7)

family (group) Elements that are in the same column of the periodic table; elements within the same family exhibit similarities in their chemical behavior. (Section 2.4)

faraday A unit of charge that equals the total charge of 1 mol of electrons: 1 $\mathfrak{F}$ = 96,500 C. (Section 20.5)

fats (and oils) Esters of long-chain carboxylic acids and the alcohol 1,2,3-trihydroxypropane (glycerol). These esters are also called **triglycerides**. (Section 27.5)

***f*-block metals** Lanthanide and actinide elements, in which the 4*f* or 5*f* orbitals are partially occupied. (Section 6.10)

ferromagnetism The ability of some substances to become permanently magnetized. (Section 24.9)

fibers Polymer fibers are formed by extrusion (that is, squeezing through a small hole) and then drawn to form a long, threadlike material. (Section 12.2)

first ionization energy (I_1) The amount of energy needed to remove the first electron from a gaseous atom, generating a gaseous 1+ ion. (Section 7.4)

first law of thermodynamics A statement of our experience that energy is conserved in any process. We can express the first law in many ways. One of the more useful expressions is that the change in internal energy, ΔE, of a system in any process is equal to the heat, q, added *to* the system, plus the work, w, done *on* the system by its surroundings: $\Delta E = q + w$. (Section 5.2)

first-order reaction A reaction in which the reaction rate is proportional to the concentration of a single reactant, raised to the first power. (Section 14.2)

fission The splitting of a large nucleus into two smaller ones. (Section 21.6)

flotation process A procedure for concentrating raw, finely ground ore by a process of physical separation of the desired ore from gangue through selective wetting with the use of flotation agents. (Section 24.2)

force A push or a pull. (Section 5.1)

formal charge The number of valence electrons in an isolated atom minus the number of electrons assigned to the atom in the Lewis structure. (Section 8.5)

formation constant For a metal ion complex, the equilibrium constant for formation of the complex from the metal ion and base species present in solution. It is a measure of the tendency of the complex to form. (Section 17.5)

formula weight The mass of the collection of atoms represented by a chemical formula. For example, the formula weight of NO_2 (46.0 amu) is the sum of the masses of one nitrogen atom and two oxygen atoms. (Section 3.3)

fossil fuels Coal, oil, and natural gas, which are presently our major sources of energy. (Section 5.8)

free radical A substance with one or more unpaired electrons. (Sections 21.9, 26.1)

frequency The number of times per second that one complete wavelength passes a given point. (Section 6.1)

frequency factor (A) A term in the Arrhenius equation that is related to the frequency of collision and the probability that the collisions are favorably oriented for reaction. (Section 14.4)

fuel cell A voltaic cell that utilizes the oxidation of a conventional fuel, such as H_2 or CH_4, in the cell reaction. (Section 20.6)

fuel value The energy released when 1 g of a substance is combusted. (Section 5.8)

functional group An atom or group of atoms that imparts characteristic chemical properties to an organic compound. (Section 26.4)

fusion The joining of two light nuclei to form a more massive one. (Section 21.6)

gamma rays Energetic electromagnetic radiation emanating from the nucleus of a radioactive atom. (Section 21.2)

gangue Material of little or no value that accompanies the desired mineral in most raw ores. (Section 24.2)

gas Matter that has no fixed volume or shape; it conforms to the volume and shape of its container. (Section 1.1)

gas constant (R) The constant of proportionality in the ideal-gas equation. (Section 10.4)

Geiger counter A device that can detect and measure radioactivity. (Section 21.5)

gel A semisolid suspension of a material of high molecular mass in a liquid solvent. (Section 12.3)

genetic damage Damage to genes and chromosomes that may have consequences in subsequent generations. (Section 21.9)

geometrical isomers Compounds with the same type and number of atoms and the same chemical bonds but different spatial arrangements of these atoms and bonds. (Sections 25.4, 26.2)

Gibbs free energy A thermodynamic state function that combines enthalpy and entropy, in the form $G = H - TS$. For a change occurring at constant temperature and pressure, the change in free energy is $\Delta G = \Delta H - T\Delta S$. (Section 19.5)

glass An amorphous solid formed by fusion of SiO_2, CaO, and Na_2O. Other oxides may also be used to form glasses with differing characteristics. (Section 23.5)

glucose A polyhydroxy aldehyde, $CH_2OH(CHOH)_4CHO$; it is the most important of the monosaccharides. (Section 27.4)

glycogen The general name given to a group of polysaccharides of glucose that are synthesized in mammals and used to store energy from carbohydrates. (Section 27.4)

Graham's law A law stating that the rate of effusion of a gas is inversely proportional to the square root of its molecular weight. (Section 10.9)

ground state The lowest-energy, or most stable, state. (Section 6.3)

Haber process The catalyst system and conditions of temperature and pressure developed by Fritz Haber and co-workers for the formation of NH_3 from H_2 and N_2. (Section 15.1)

half-life The time required for the concentration of a reactant substance to decrease to half its initial value; the time required for half of a sample of a particular radioisotope to decay. (Sections 14.3, 21.4)

half-reaction An equation for either an oxidation or a reduction that explicitly shows the electrons involved [for example, $Zn^{2+}(aq) + 2e^- \rightarrow Zn(s)$]. (Section 20.2)

Hall process A process used to obtain aluminum by electrolysis of Al_2O_3 dissolved in molten cryolite, Na_3AlF_6. (Section 24.5)

halogens Members of group 7A in the periodic table. (Section 7.8)

hard water Water that contains appreciable concentrations of Ca^{2+} and Mg^{2+}; these ions react with soaps to form an insoluble material. (Section 18.6)

heat The flow of energy from a body at higher temperature to one at lower temperature when they are placed in thermal contact. (Section 5.1)

heat capacity The quantity of heat required to raise the temperature of a sample of matter by 1 °C (or one kelvin). (Section 5.7)

α-helix A protein structure in which the protein is coiled in the form of a helix, with hydrogen bonds between C=O and N—H groups on adjacent turns. (Section 27.2)

heat of formation The heat evolved when a substance is formed from the elements in their standard states. (Section 5.6)

heat of fusion The enthalpy change, ΔH, for melting a solid. (Section 11.4)

heat of vaporization The enthalpy change, ΔH, for vaporizing a liquid. (Section 11.4)

hemoglobin An iron-containing protein responsible for oxygen transport in the blood. (Section 18.4)

Henderson-Hasselbalch equation The relationship between the pH, pK_a, concentration of base, and concentration of acid in an aqueous solution: $\text{pH} = \text{p}K_a + \log \frac{[\text{base}]}{[\text{acid}]}$. (Section 17.2)

Henry's law A law stating that the concentration of a gas in a solution, C_g, is proportional to the pressure of gas over the solution: $C_g = kP_g$. (Section 13.4)

Hess's law The heat evolved in a given process can be expressed as the sum of the heats of several processes that, when added, yield the process of interest. (Section 5.5)

heterogeneous alloy An alloy in which the components are not distributed uniformly; instead, two or more distinct phases with characteristic compositions are present. (Section 24.7)

heterogeneous equilibrium The equilibrium established between substances in two or more different phases, for example, between a gas and a solid or between a solid and a liquid. (Section 15.4)

heterogeneous catalyst A catalyst that is in a different phase from that of the reactant substances. (Section 14.6)

hexagonal close packing A close-packing arrangement in which the atoms of the third layer of a solid lie directly over those in the first layer. (Section 11.7)

high-spin complex A complex whose electrons populate the *d* orbitals to give the maximum number of unpaired electrons. (Section 25.8)

homogeneous catalyst A catalyst that is in the same phase as the reactant substances. (Section 14.6)

homogeneous equilibrium The equilibrium established between reactant and product substances that are all in the same phase. (Section 15.4)

homologous series Compounds that contain common structural elements but differ in the number of atoms making up the molecule. Members of a homologous series have the same general formula. For example, the alkanes have the general formula C_nH_{2n+2}. (Section 26.1)

Hund's rule A rule stating that electrons occupy degenerate orbitals in such a way as to maximize the number of electrons with the same spin. In other words, each orbital will have one electron placed in it before pairing of electrons in orbitals occurs. Note that this rule applies only to orbitals that are *degenerate*, which means that they have the same energy. (Section 6.9)

hybridization The mixing of different types of atomic orbitals to produce a set of equivalent hybrid orbitals. (Section 9.4)

hybrid orbital An orbital that results from the mixing of different kinds of atomic orbitals on the same atom. For example, an sp^3 hybrid that results from the mixing or hybridizing of one *s* orbital and three *p* orbitals. (Section 9.4)

hydration Solvation when the solvent is water. (Section 13.2)

hydride ion An ion formed by the addition of an electron to a hydrogen atom: H^-. (Section 7.7)

hydrogen bonding Bonding that results from intermolecular attractions between molecules containing hydrogen bonded to an electronegative element. The most important examples involve oxygen, nitrogen, or fluorine. (Section 11.2)

hydrolysis A reaction with water. When a cation or anion reacts with water it changes the pH. (Section 16.8)

hydrometallurgy Aqueous chemical processes for recovery of a metal from an ore. (Section 24.4)

hydronium ion (H_3O^+) The predominant form of the proton in aqueous solution. (Section 16.1)

hydrophilic Water-attracting. (Section 13.6)

hydrophobic Water-repelling. (Section 13.6)

ideal gas A hypothetical gas whose pressure, volume, and temperature behavior is completely described by the ideal-gas equation. (Section 10.4)

ideal-gas equation An equation of state for gases that embodies Boyle's law, Charles's law, and Avogadro's hypothesis in the form $PV = nRT$. (Section 10.4)

ideal solution A solution that obeys Raoult's law. (Section 13.5)

immiscible liquids Liquids that do not mix. (Section 13.4)

indicator A substance added to a solution to indicate by a color change the point at which the added solute has just reacted with all the solute present in solution. (Section 4.7)

inert complex A complex that exchanges ligands at a slow rate. (Section 25.5)

instantaneous rate The reaction rate at a particular time as opposed to the average rate over an interval of time. (Section 14.1)

intensive property A property that is independent of the amount of material considered; for example, density. (Section 1.3)

interhalogens Compounds formed between two different halogen elements. Examples include IBr and BrF_3. (Section 23.2)

intermediate A substance formed in one elementary step of a multistep mechanism and consumed in another; it is neither a reactant nor an ultimate product of the overall reaction. (Section 14.5)

intermetallic compound A homogeneous alloy with definite properties and composition. Intermetallic compounds are stoichiometric compounds in the full sense, but their compositions are not readily understood in terms of ordinary chemical bonding theory. (Section 24.7)

intermolecular forces The short-range attractive forces operating between the particles that make up the units of a liquid or solid substance. These same forces also cause gases to liquefy or solidify at low temperatures and high pressures. (Chapter 11: Introduction)

internal energy The total energy possessed by a system. When a system undergoes a change, the change in internal energy, ΔE, is defined as the heat, q, added to the system, plus the work, w, done on the system by its surroundings: $\Delta E = q + w$. (Section 5.2)

interstitial hydrides Nonstoichiometric metallic hydrides, in which the ratio of metal atoms to hydrogen atoms is not a ratio of small whole numbers, nor is it a fixed ratio. (Section 22.3)

ion Electrically charged atom or group of atoms (polyatomic ion); ions can be positively or negatively charged, depending on whether electrons are lost (positive) or gained (negative) by the atoms. (Section 2.5)

ion-dipole force The force that exists between an ion and a neutral polar molecule that possesses a permanent dipole moment. (Section 11.2)

ionic bond A bond formed on the basis of the electrostatic forces that exist between oppositely charged ions. The ions are formed from atoms by transfer of one or more electrons. (Chapter 8: Introduction)

ionic hydrides Compounds formed when hydrogen reacts with alkali metals and also heavier alkaline earths (Ca, Sr, and Ba); these compounds contain the hydride ion, H^-. (Section 22.3)

ionic solids Solids that are composed of ions. (Section 11.8)

ionization Complete removal of an electron from an atom. (Section 6.3)

ionization energy The energy required to remove an electron from a gaseous atom when the atom is in its ground state. (Section 6.3)

ion pair A combination of a single cation and a single anion. (Section 13.5)

ion-product constant For water, K_w is the product of the aquated hydrogen ion and hydroxide ion concentrations: $[H^+][OH^-] = K_w = 1.0 \times 10^{-14}$. (Section 16.2)

isoelectronic series A series of atoms, ions, or molecules having the same number of electrons. (Section 8.3)

isolated system A system that does not exchange energy or matter with its surroundings. (Section 19.2)

isomers Compounds whose molecules have the same overall composition but different structures. (Section 25.4)

isotopes Atoms of the same element containing different numbers of neutrons and therefore having different masses. (Section 2.3)

joule (J) The SI unit of energy, 1 kg-m^2/s^2. A related unit is the calorie: 4.184 J = 1 cal. (Section 5.1)

Kelvin scale The absolute temperature scale; the SI unit for temperature is the kelvin. Zero on the Kelvin scale corresponds to −273.15°C; therefore, K = C° + 273.15. (Section 1.3)

ketone A compound in which the carbonyl group occurs at the interior of a carbon chain and is therefore flanked by carbon atoms. (Section 26.4)

kinetic energy The energy that an object possesses by virtue of its motion. (Section 5.1)

kinetic-molecular theory A set of assumptions about the nature of gases. These assumptions, when translated into mathematical form, yield the ideal-gas equation. (Section 10.8)

labile complex A complex that exchanges ligands at a rapid rate. (Section 25.5)

lanthanide contraction The gradual decrease in atomic and ionic radii with increasing atomic number among the lanthanide elements, atomic numbers 58 through 71. The decrease arises because of a gradual increase in effective nuclear charge through the lanthanide series. (Section 24.8)

lattice energy The energy required to separate completely the ions in an ionic solid. (Section 8.2)

law of conservation of energy A law stating that the energy gained by a system equals that lost by its surroundings. (Section 5.2)

law of constant composition A law that states that the elemental composition of a pure compound is always the same, regardless of its source. (Also called the **law of definite proportions**.) (Section 1.2)

law of definite proportions A law that states that the elemental composition of a pure substance is always the same, regardless of its source. (Also called the **law of constant composition**.) (Section 1.2)

law of mass action The rules according to which the equilibrium constant is expressed in terms of the concentrations of reactants and products, in accordance with the balanced chemical equation for the reaction. (Section 15.2)

law of multiple proportions A law stating that when two elements form more than one compound, the masses of one element that can chemically combine with a given mass of another element are ratios of small whole numbers. (Section 2.1)

leaching The selective dissolution of a desired mineral by passing an aqueous reagent solution through an ore. (Section 24.4)

Le Châtelier's principle A principle stating that when we disturb a system at chemical equilibrium, the relative concentrations of reactants and products will shift so as to undo partially the effects of the disturbance. (Section 15.6)

levorotatory, or merely **levo** or ***l*** A term used to label a chiral molecule that rotates the plane of polarization of plane-polarized light to the left (counterclockwise). (Section 25.4)

Lewis acid An electron-pair acceptor. (Section 16.10)

Lewis base An electron-pair donor. (Section 16.10)

Lewis structure A representation of covalent bonding in a molecule that is drawn using Lewis symbols. Shared electron pairs are shown as lines, and unshared electron pairs are shown as pairs of dots. Only the valence-shell electrons are shown. (Section 8.4)

ligand An ion or molecule that coordinates to a metal atom or to a metal ion to form a complex. (Section 25.1)

lime-soda process A method for removal of Mg^{2+} and Ca^{2+} ions from water to reduce water hardness. The substances added to the water are "lime," CaO [or "slaked lime," $Ca(OH)_2$], and "soda ash," Na_2CO_3, in amounts determined by the concentrations of the offending ions. (Section 18.6)

limiting reactant (reagent) The reactant present in the smallest stoichiometric quantity in a mixture of reactants; the amount of product that can form is limited by the complete consumption of the limiting reactant. (Section 3.6)

line spectrum A spectrum that contains radiation at only certain specific wavelengths. (Section 6.2)

linkage isomers Structural isomers of coordination compounds in which a ligand differs in its mode of attachment to a metal ion. (Section 25.4)

liquid Matter that has a distinct volume but no specific shape. (Section 1.1)

liquid crystal A substance that exhibits one or more partially ordered liquid phases above the melting point of the solid form. By contrast, in nonliquid crystalline substances the liquid phase that forms upon melting is completely unordered, or isotropic. (Section 12.1)

lithosphere That portion of our environment consisting of the solid earth. (Section 24.1)

lock-and-key model A model of enzyme action in which the substrate molecule is pictured as fitting rather specifically into the active site on the enzyme. It is assumed that in being bound to the active site the substrate is somehow activated for reaction. (Section 27.3)

London dispersion forces Intermolecular forces resulting from attractions between induced dipoles. (Section 11.2)

low-spin complex One that has the lowest possible number of unpaired electrons. (Section 25.8)

magic numbers Totals of protons and neutrons in very stable nuclei. (Section 21.2)

manometer A device for measuring the pressure of an enclosed gas. (Section 10.2)

Markovnikov's rule A law stating that when a reagent of general formula XY adds to an alkene, the more negatively charged species, X or Y, ends up on the carbon with the fewer attached hydrogen atoms. For example, on addition of HCl to propene, Cl ends up on the middle carbon atom. (Section 26.2)

mass A measure of material in an object. It measures the resistance of a stationary object to being moved. In SI units, mass is measured in kilograms. (Section 1.3)

mass defect The difference between the mass of a nucleus and the total masses of the individual nucleons that it contains. (Section 21.6)

mass number The sum of the number of protons and neutrons in the nucleus of a particular atom. (Section 2.3)

mass spectrometer An instrument used to measure the precise masses and relative amounts of atomic and molecular ions. (Section 3.4)

mass spectrum A graph of the intensity of the signal from the detector in a mass spectrometer versus the mass of the ion being detected. (Section 3.4)

matter The physical material of the universe; anything that occupies space and has mass. (Section 1.1)

matter waves The term used to describe the wave characteristics of a particle. (Section 6.4)

mean free path The average distance traveled by a gas molecule between collisions. (Section 10.9)

mechanical work The product of the net force, F, and distance, d, through which that force acts: $w = F \times d$. (Section 5.1)

meniscus The curved upper surface of a liquid column. (Section 11.3)

mesosphere The region of the atmosphere directly above the stratosphere. (Section 18.1)

metallic bond Bonding in which the bonding electrons are relatively free to move throughout the three-dimensional structure. (Chapter 8: Introduction)

metallic character The extent to which an element exhibits the physical and chemical properties characteristic of metals, for example, luster, malleability, ductility, and good thermal and electrical conductivity. (Section 7.6)

metallic elements (metals) Elements that are usually solids at room temperature, exhibit high electrical and heat conductivity, and appear lustrous. Most of the elements in the periodic table are metals. (Section 2.4)

metallic hydrides Compounds formed when hydrogen reacts with transition metals; these compounds contain the hydride ion, H^-. (Section 22.3)

metallic solids Solids that are composed of metal atoms. (Section 11.8)

metallurgy The science of extracting metals from their natural sources by a combination of chemical and physical processes. It is also concerned with the properties and structures of metals and alloys. (Section 24.2)

metathesis reaction A reaction in which two substances react through an exchange of their component ions: $AX + BY \rightarrow AY + BX$; also called **double displacement reaction**. Precipitation and acid-base neutralization reactions are examples of metathesis reactions. (Section 4.5)

metric system A system of measurement used in science and in most countries. The meter and the gram are examples of metric units. (Section 1.3)

mineral A solid, inorganic substance occurring in nature, such as calcium carbonate, which occurs as calcite. (Section 24.1)

miscible Liquids that mix in all proportions. (Section 13.4)

mixture A combination of two or more substances in which each substance retains its own chemical identity. (Section 1.1)

molal boiling-point-elevation constant (K_b) A constant characteristic of a particular solvent that gives the change in boiling point as a function of solution molality: $\Delta T_b = K_b m$. (Section 13.5)

molal freezing-point-depression constant (K_f) A constant characteristic of a particular solvent that gives the change in freezing point as a function of solution molality: $\Delta T_f = K_f m$. (Section 13.5)

molality The concentration of a solution expressed as moles of solute per kilogram of solvent; abbreviated *m*. (Section 13.1)

molar heat capacity The heat required to raise the temperature of 1 mol of a substance by 1°C. (Section 5.7)

molarity The concentration of a solution expressed as moles of solute per liter of solution; abbreviated *M*. (Section 4.1)

molar mass The mass of 1 mol of a substance in grams; it is numerically equal to the formula weight in atomic mass units. (Section 3.5)

mole A collection of Avogadro's number (6.022×10^{23}) of objects; for example, a mole of H_2O is 6.022×10^{23} H_2O molecules. (Section 3.5)

molecular equation A chemical equation in which the formula for each substance is written without regard for whether it is an electrolyte or a nonelectrolyte. (Section 4.4)

molecular formula A chemical formula that indicates the actual number of atoms of each element in one molecule of a substance. (Section 2.5)

molecular geometry The arrangement in space of the atoms of a molecule. (Section 9.1)

molecular hydrides Compounds formed when hydrogen reacts with nonmetals and semimetals. (Section 22.3)

molecularity The number of molecules that participate as reactants in an elementary reaction. (Section 14.5)

molecular orbital An allowed state for an electron in a molecule. According to **molecular orbital theory**, a molecular orbital is entirely analogous to an atomic orbital, which is an allowed state for an electron in an atom. A molecular orbital may be classified as σ or π, depending on the disposition of electron density with respect to the internuclear axis. (Section 9.6)

molecular orbital diagram A diagram that shows the energies of molecular orbitals relative to the atomic orbitals from which they are derived. Also called an **energy-level diagram**. (Section 9.6)

molecular orbital theory A theory that accounts for the allowed states for electrons in molecules. (Section 9.6)

molecular solids Solids that are composed of molecules. (Section 11.8)

molecular weight The weight of the collection of atoms represented by the chemical formula for a molecule. (Section 3.3)

molecule A chemical combination of two or more atoms. (Section 2.5)

mole fraction The ratio of the number of moles of one component of a mixture to the total moles of all components; abbreviated X, with a subscript to identify the component. (Section 10.6)

momentum The product of the mass, m, and velocity, v, of a particle. (Section 6.4)

monochromatic Composed of a single wavelength. (Section 6.2)

monodentate ligand A ligand that binds to the metal ion via a single donor atom. It occupies one position in the coordination sphere. (Section 25.2)

monomers Molecules with low molecular weights, which can be joined together (polymerized) to form a polymer. (Section 12.2)

monosaccharide A simple sugar, most commonly containing six carbon atoms. The joining together of monosaccharide units by a condensation reaction results in formation of polysaccharides. (Section 27.4)

multiple bonding Bonding involving two or three electron pairs. (Section 8.4)

natural gas A naturally occurring mixture of gaseous hydrocarbons, compounds composed of hydrogen and carbon. (Section 5.8)

natural rubber The polymer of *cis*-isoprene; it is formed from the resin of the *Hevea brasiliensis* tree. (Section 12.2)

nematic liquid crystal A crystal in which the molecules are aligned in the same general direction, along their long axes, but in which the ends of the molecules are not aligned. (Section 12.1)

Nernst equation An equation that relates the cell emf, E, to the standard emf, E°, and the reaction quotient, Q: $E = E^\circ - 2.30RT/nF \log Q$. (Section 20.5)

net ionic equation A chemical equation for a solution reaction in which soluble strong electrolytes are written as ions and spectator ions are omitted. (Section 4.4)

neutralization reaction A reaction in which an acid and a base react in stoichiometrically equivalent amounts; the neutralization reaction between an acid and a metal hydroxide produces water and a salt. (Section 4.3)

neutron An electrically neutral particle found in the nucleus of an atom; it has approximately the same mass as a proton. (Section 2.3)

noble gases Members of group 8A in the periodic table. (Section 7.8)

nodal surface (node) A locus of points in an atom at which the electron density is zero. For example, the node in a 2*s* orbital (Figure 6.20) is a spherical surface. (Section 6.6)

nomenclature The basic rules for naming substances. (Section 2.6)

nonbonding pair In a Lewis structure, a pair of electrons assigned completely to one atom; also called a *lone pair.* (Section 9.1)

nonelectrolyte A substance that does not ionize in water and consequently gives a nonconducting solution. (Section 4.2)

nonmetallic elements (nonmetals) Elements in the upper-right corner of the periodic table; nonmetals differ from metals in their physical and chemical properties. (Section 2.4)

nonpolar bond A covalent bond in which the electrons are shared equally. (Section 8.9)

normal boiling point The temperature at which a liquid boils when the external pressure is 1 atm (that is, the temperature at which the vapor pressure of the liquid is 1 atm). (Section 11.5)

normal melting point The melting point at 1 atm pressure. (Section 11.6)

nuclear disintegration series A series of nuclear reactions that begins with an unstable nucleus and terminates with a stable one. Also called a **radioactive series.** (Section 21.2)

nuclear transmutation A conversion of one kind of nucleus to another. (Section 21.1)

nucleic acids Polymers of high molecular weights that carry genetic information and control protein synthesis. (Section 27.6)

nucleon A particle found in the nucleus of an atom. (Chapter 21: Introduction)

nucleotide Compounds formed from a molecule of phosphoric acid, a sugar molecule, and an organic nitrogen base. Nucleotides form linear polymers called DNA and RNA, which are involved in protein synthesis and cell reproduction. (Section 27.6)

nucleus The very small, very dense, positively charged portion of an atom; it is composed of protons and neutrons. (Section 2.2)

nuclide An atom of a specific isotope of an element. (Section 2.3)

octet rule A rule stating that bonded atoms tend to possess or share a total of eight valence-shell electrons. (Section 8.1)

oil A naturally occurring combustible liquid composed of hundreds of hydrocarbons and other organic compounds; also known as *petroleum.* (Section 5.8)

optical isomers Stereoisomers in which the two forms of the compound are nonsuperimposable mirror images. (Section 25.4)

optically active Possessing the ability to rotate the plane of polarized light. (Section 25.4)

orbital An allowed energy state of an electron in the quantum mechanical model of the atom; the term *orbital* is also used to describe the spatial distribution of the electron. An orbital is defined by the values of three quantum numbers: n, l, and m_l. (Section 6.5)

orbital diagram A representation of the electrons in which each orbital is represented by a box and each electron by a half-arrow. (Section 6.9)

ore A source of a desired element or mineral, usually accompanied by large quantities of other materials such as sand and clay. (Section 24.1)

osmosis The net movement of solvent through a semipermeable membrane toward the solution with greater solute concentration. (Section 13.5)

osmotic pressure The pressure that must be applied to a solution to stop osmosis from pure solvent into the solution. (Section 13.5)

Ostwald process An industrial process used to make nitric acid from ammonia. The NH_3 is catalytically oxidized by O_2 to form NO; NO in air is oxidized to NO_2; HNO_3 is formed in a disproportionation reaction when NO_2 dissolves in water. (Section 22.5)

overall reaction order The sum of the reaction orders of all the reactants appearing in the rate expression. (Section 14.2)

overlap The extent to which atomic orbitals on different atoms share the same region of space. When the overlap between two orbitals is large, a strong bond may be formed. (Section 9.3)

oxidation A process in which a substance loses one or more electrons. (Section 4.6)

oxidation number (state) A positive or negative whole number assigned to an element in a molecule or ion on the basis of a set of formal rules; to some degree it reflects the positive or negative character of that atom. (Section 8.10)

oxidation-reduction reaction A chemical reaction in which the oxidation state of one or more substances changes. (Chapter 20: Introduction)

oxidizing agent or **oxidant** The substance that is reduced and thereby causes the oxidation of some other substance in an oxidation-reduction reaction. (Section 20.1)

oxyacid A compound in which one or more OH groups, and possibly additional oxygen atoms, are bonded to a central atom. (Section 16.9)

oxyanion A polyatomic ion that contains one or more oxygen atoms. (Section 2.6)

oxyhemoglobin A hemoglobin molecule that has picked up an oxygen molecule. (Section 18.4)

ozone The name given to O_3, an allotrope of oxygen. (Section 7.8)

paramagnetism A property that a substance possesses if it contains one or more unpaired electrons. A paramagnetic substance is drawn into a magnetic field. (Section 9.7)

partial pressure The pressure exerted by a particular gas in a mixture. (Section 10.6)

particle accelerator A device that uses strong magnetic and electrostatic fields to accelerate charged particles. (Section 21.3)

parts per million (ppm) by weight The concentration of a solution in grams of solute per 10^6 (million) grams of solution; equals milligrams of solute per liter of solution for aqueous solutions. (Section 13.1)

pascal The SI unit of pressure: $1\ Pa = 1\ N/m^2$ or $1\ kg/m \cdot s^2$. (Section 10.2)

Pauli exclusion principle A rule stating that no two electrons in an atom may have the same four quantum numbers (n, l, m_l, and m_s). As a consequence of this principle, there can be no more than two electrons in any one atomic orbital. (Section 6.8)

peptide A substance composed of two or more amino acids. (Section 27.2)

percent yield The ratio of the actual (experimental) yield of a product to its theoretical (calculated) yield, multiplied by 100. (Section 3.6)

periodic table The arrangement of elements in order of increasing atomic number, with elements having similar properties placed in vertical columns. (Section 2.4)

pH The negative log in base 10 of the aquated hydrogen-ion concentration: $pH = -\log[H^+]$. (Section 16.2)

phase change The conversion of a substance from one state of matter to another. The phase changes we consider are *melting* and *freezing* (solid $\leftrightarrow$ liquid), *sublimation* and *deposition* (solid $\leftrightarrow$ gas), and *vaporization* and *condensation* (liquid $\leftrightarrow$ gas). (Section 11.4)

phase diagram A graphic representation of the equilibria between the solid, liquid, and gaseous phases of a substance as a function of temperature and pressure. (Section 11.6)

photochemical smog A complex mixture of undesirable substances produced by the action of sunlight on an urban atmosphere polluted with automobile emissions. The major starting ingredients are nitrogen oxides and organic substances, notably olefins and aldehydes. (Section 18.4)

photodissociation The breaking of a molecule into two or more neutral fragments as a result of absorption of light. (Section 18.2)

photoionization The removal of an electron from an atom or molecule by absorption of light. (Section 18.2)

photolysis Light-induced rupture of a chemical bond. (Section 18.3)

photon The smallest increment (a *quantum*) of radiant energy; a photon of light with frequency v has an energy equal to hv. (Section 6.2)

photosynthesis The process that occurs in plant leaves by which light energy is used to convert CO_2 and water to carbohydrates and oxygen. (Section 27.1)

physical changes Changes (such as a phase change) that occur with no change in chemical composition. (Section 1.1)

physical properties Properties that can be measured without changing the composition of a substance, for example, color and freezing point. (Section 1.1)

pi (π) bond A covalent bond in which electron density is concentrated above and below the line joining the bonding atoms. (Section 9.3)

piezoelectric material A crystalline substance that generates an electric potential along its length when subjected to a mechanical stress. Quartz is a well-known piezoelectric material. (Section 12.3)

pi (π) molecular orbital A molecular orbital that concentrates the electron density on opposite sides of a line that passes through the nuclei. (Section 9.7)

plastic A material that can be formed into particular shapes by application of heat and pressure. (Section 12.2)

plasticizers Organic molecules added to a polymer to reduce the intermolecular interactions between polymer chains. The effect of the plasticizer is to make the material more pliable. (Section 12.2)

polar covalent bond A covalent bond in which the electrons are not shared equally. (Section 8.9)

polarizability The ease with which the electron cloud of an atom or a molecule is distorted by an outside influence, thereby inducing a dipole moment. (Section 11.2)

polar molecule A molecule that possesses a nonzero dipole moment. (Section 9.2)

polyatomic ion An electrically charged group of two or more atoms. (Section 2.5)

polydentate ligand A ligand in which two or more donor atoms can coordinate to the same metal ion. (Section 25.2)

polyethylene A polymer formed by the addition polymerization of ethylene, $CH_2{=}CH_2$. (Section 12.2)

polymer A large molecule of high molecular mass, formed by the joining together, or polymerization, of a large number of molecules of low molecular mass. The individual molecules forming the polymer are called *monomers*. (Section 12.2)

polypeptide A polymer of amino acids that has a molecular weight of less than 10,000. (Section 27.2)

polyprotic acid A substance capable of dissociating more than one proton in water; H_2SO_4 is an example. (Section 16.4)

polysaccharide A substance made up of several monosaccharide units joined together. (Section 27.4)

porphyrin A complex derived from the porphine molecule. (Section 25.2)

positron A particle with the same mass as an electron but with a positive charge, symbol ${}^{0}_{1}e$. (Section 21.2)

potential energy The energy that an object possesses as a result of its position with respect to another object or by virtue of its composition. (Section 5.1)

power The rate at which work is done. (Section 5.1)

precipitate An insoluble substance that forms in, and separates from, a solution. (Section 4.5)

precipitation reaction A reaction that occurs between substances in solution in which one of the products is insoluble. (Section 4.5)

precision The closeness of agreement among several measurements of the same quantity; the reproducibility of a measurement. (Section 1.4)

pressure A measure of the force exerted on a unit area. In chemistry, pressure is often expressed in units of atmospheres (atm) or of millimeters of mercury (mm Hg): 760 mm Hg = 1 atm; in SI units, pressure is expressed in pascal (Pa). (Section 10.2)

primary structure The sequence of amino acids along a protein chain. (Section 27.2)

primitive cubic cell A cubic unit cell in which the lattice points are at the corners only. (Section 11.7)

principal quantum number (*n*) The quantum number that relates most directly to the size and energy of an atomic orbital. This number may have any nonnegative integer value. (Section 6.2)

probability density (ψ^2) A value that represents the probability that an electron will be found at a given point in space. (Section 6.5)

product A substance produced in a chemical reaction; it appears to the right of the arrow in a chemical equation. (Section 3.1)

proteins Biopolymers formed from amino acids. (Section 27.2)

protium The most common isotope of hydrogen: ${}^1_1H$. (Section 22.3)

proton A positively charged subatomic particle found in the nucleus of an atom. (Section 2.3)

pure substance Matter that has a fixed composition and distinct properties. (Section 1.1)

pyrometallurgy A process in which heat converts a mineral in an ore from one chemical form to another and eventually to the free metal. (Section 24.3)

qualitative analysis The determination of the presence or absence of a particular substance in a mixture. (Section 17.6)

quantitative analysis The determination of the amount of a given substance that is present in a sample. (Section 17.6)

quantum The smallest increment of radiant energy that may be absorbed or emitted; the magnitude of radiant energy is $h\nu$. (Section 6.2)

quantum theory A theory concerned with the rules that govern the gain or loss of energy from an object. (Chapter 6: Introduction)

quantum (wave) mechanics A modern way of viewing submicroscopic particles such as atoms that describes their wave properties and takes into account the fact that the energy of an object can be changed only by discrete amounts. (Section 6.3)

racemic mixture A mixture of equal amounts of the dextrorotatory and levorotatory forms of a chiral molecule. A racemic mixture will not rotate polarized light. (Section 25.4)

rad A measure of the energy absorbed from radiation by tissue or other biological material; 1 rad = transfer of 1×10^{-2} J of energy per kilogram of material. (Section 21.9)

radical chain A process in which a single highly reactive molecule or atom, such as chlorine, reacts numerous times to produce many molecules of product. (Section 26.1)

radioactive series A series of nuclear reactions that begins with an unstable nucleus and terminates with a stable one. Also called **nuclear disintegration series**. (Section 21.2)

radioactivity The spontaneous disintegration of an unstable atomic nucleus with accompanying emission of radiation. (Section 2.2)

radioisotope An isotope that is radioactive; that is, it is undergoing nuclear changes with emission of radiation. (Section 21.1)

radiotracer A radioisotope that can be used to trace the path of an element. (Section 21.5)

Raoult's law A law stating that the partial pressure of a solvent over a solution, P_A, is given by the vapor pressure of the pure solvent, P°_A, times the mole fraction of solvent in the solution, X_A: $P_A = X_A P^\circ_A$. (Section 13.5)

rare-earth (lanthanide) element Element in which the 4*f* subshell is only partially occupied. (Section 6.9)

rate constant A constant of proportionality between the reaction rate and the concentrations of reactants that appear in the rate law. (Section 14.2)

rate-determining step The slowest elementary step in a reaction mechanism. (Section 14.5)

rate law An equation that relates the reaction rate to the concentrations of reactants (and sometimes of products also). (Section 14.2)

reactant A starting substance in a chemical reaction; it appears to the left of the arrow in a chemical equation. (Section 3.1)

reaction mechanism A detailed picture, or model, of how the reaction occurs; that is, the order in which bonds are broken and formed, and the changes in relative positions of the atoms as the reaction proceeds. (Section 14.5)

reaction order The power to which the concentration of a reactant is raised in a rate law. (Section 14.2)

reaction quotient (*Q*) The value that is obtained when concentrations of reactants and products are inserted into the equilibrium-constant expression. If the concentrations are equilibrium concentrations, $Q = K$; otherwise $Q \neq K$. (Section 15.5)

reaction rate The decrease in concentration of a reactant or the increase in concentration of a product with time. (Section 14.1)

reducing agent or **reductant** The substance that is oxidized and thereby causes the reduction of some other substance in an oxidation-reduction reaction. (Section 20.1)

reduction A process in which a substance gains one or more electrons. (Section 4.6)

refining The process of converting an impure form of a metal into a more usable substance of well-defined composition. For example, crude pig iron from the blast furnace is refined in a converter to produce steels of desired compositions. (Section 24.3)

relative biological effectiveness (RBE) An adjustment factor used to convert rads to rems; it accounts for differences in biological effects of different particles having the same energy. (Section 21.9)

rem A measure of the biological damage caused by radiation; rems = rads × RBE. (Section 21.9)

representative (main-group) element Element in which the *s* and *p* orbitals are partially occupied. (Section 6.10)

resonance forms Individual Lewis structures in cases where two or more Lewis structures are equally good descriptions of a single molecule. The resonance structures in such an instance are "averaged" to give a correct description of the real molecule. (Section 8.6)

reverse osmosis The process by which water molecules move under high pressure through a semipermeable membrane from the more concentrated to the less concentrated solution. (Section 18.5)

ribonucleic acid (RNA) A polynucleotide in which ribose is the sugar component. (Section 27.6)

roasting Thermal treatment of an ore to bring about chemical reactions involving the furnace atmosphere. For example, a sulfide ore might be roasted to form a metal oxide and SO_2. (Section 24.3)

root-mean-square (rms) speed (*u*) The square root of the average of the squared speeds of the gas molecules in a gas sample. (Section 10.8)

rotational motion Movement of a molecule as though it was spinning like a top. (Section 19.3)

salinity A measure of the salt content of seawater, brine, or brackish water. It is equal to the mass in grams of dissolved salts present in 1 kg of seawater. (Section 18.5)

salt An ionic compound formed by replacing one or more H^+ of an acid by other cations. (Section 4.3)

saponification Hydrolysis of an ester in the presence of a base. (Section 26.4)

saturated solution A solution in which undissolved solute and dissolved solute are in equilibrium. (Section 13.3)

Schrödinger's wave equation An equation that relates the wave properties of a particle to its mass, potential energy, kinetic energy, and coordinates. (Section 6.5)

scintillation counter An instrument that is used to detect and measure radiation by the fluorescence it produces in a fluorescing medium. (Section 21.5)

screening effect The effect of inner electrons in decreasing the nuclear charge experienced by outer electrons. (Section 6.7)

secondary structure The manner in which a protein is coiled or stretched. (Section 27.2)

second ionization energy (I_2) The amount of energy needed to remove a second electron from a gaseous 1+ ion, generating a gaseous 2+ ion. The third and higher ionization energies are defined analogously. (Section 7.4)

second law of thermodynamics A statement of our experience that there is a direction to the way events occur in nature. When a process occurs spontaneously in one direction, it is nonspontaneous in the reverse direction. It is possible to state the second law in many different forms, but they all relate back to the same idea about spontaneity. One of the most common statements found in chemical contexts is that in any spontaneous process the entropy of the universe increases. (Section 19.1)

semimetals (metalloids) Elements that lie along the diagonal line separating the metals from the nonmetals in the periodic table; the properties of semimetals are intermediate between those of metals and nonmetals. (Section 2.4)

sigma (σ) bond A covalent bond in which electron density is concentrated along the internuclear axis. (Section 9.3)

sigma (σ) molecular orbital A molecular orbital that centers the electron density about an imaginary line passing through two nuclei. (Section 9.6)

significant figures The digits that indicate the precision with which a measurement is made; all digits of a measured quantity are significant, including the last digit, which is uncertain. (Section 1.4)

silicates Compounds containing silicon and oxygen, structurally based on SiO_4 tetrahedra. (Section 23.5)

single bond A covalent bond involving one electron pair. (Section 8.4)

single displacement reaction Reactions that conform to the general pattern A + BX → AX + B. (Section 4.6)

SI units The preferred metric units for use in science. (Section 1.3)

slag A mixture of molten silicate minerals. Slags may be acidic or basic, according to the acidity or basicity of the oxide added to silica. (Section 24.3)

smectic liquid crystal A crystal in which the molecules are aligned along their long axes and arranged in sheets, with the ends of the molecules aligned. There are several different kinds of smectic phases. (Section 12.1)

smelting A melting process in which the materials formed in the course of the chemical reactions that occur, separate into two or more layers. For example, the layers might be slag and molten metal. (Section 24.3)

smog A particularly unpleasant condition of pollution in certain urban environments that occurs when weather conditions produce a relatively stagnant air mass. (Section 18.4)

sol A suspension of very small solid particles dispersed in a liquid phase. (Section 12.3)

sol-gel process A process in which extremely small particles (0.003 μm to 0.1 μm in diameter) of uniform size are produced in a series of chemical steps followed by controlled heating. (Section 12.3)

solid Matter that has both a definite shape and a definite volume. (Section 1.1)

solubility The amount of a substance that dissolves in a given quantity of solvent at a given temperature to form a saturated solution. (Sections 4.5, 13.3)

solubility-product constant (K_{sp}) An equilibrium constant related to the equilibrium between a solid salt and its ions in solution. It provides a quantitative measure of the solubility of a slightly soluble salt. (Section 17.4)

solute A substance dissolved in a solvent to form a solution; it is normally the component of a solution present in the smaller amount. (Section 4.1)

solution A mixture of substances that has a uniform composition; a homogeneous mixture. (Section 1.1)

solution alloy A homogeneous alloy, with the components distributed uniformly throughout. (Section 24.7)

solvation The clustering of solvent molecules around a solute particle. (Section 13.2)

solvent The dissolving medium of a solution; it is normally the component of a solution present in the greater amount. (Section 4.1)

somatic damage Damage to living systems that affects an organism during its own lifetime. (Section 21.9)

specific heat capacity (specific heat) The heat capacity of 1 g of a substance; the heat required to raise the temperature of 1 g of a substance by 1°C. (Section 5.7)

spectator ions Ions that go through a reaction unchanged and that appear on both sides of the complete ionic equation. (Section 4.4)

spectrochemical series A list of ligands arranged in order of their abilities to split the *d* orbital energies (using the terminology of the crystal-field model). (Section 25.8)

spectrum The distribution among various wavelengths of the radiant energy emitted or absorbed by an object. (Section 6.2)

spin-pairing energy The energy required to pair an electron with another electron occupying an orbital. (Section 25.8)

spontaneous process A process that is capable of proceeding in a given direction, as written or described, without needing to be driven by an outside source of energy. A process may be spontaneous even though it is very slow. For all spontaneous processes, ΔG is negative. (Section 19.1)

sputtering A method for forming thin films in which the material that is to form the film is made the cathode in a high-voltage gaseous discharge of an inert gas. (Section 12.4)

standard atmospheric pressure Defined as 760 mm Hg or, in SI units, 101.325 kPa. (Section 10.2)

standard electrode potential ($E°$) The reduction potential of a half-reaction with all solution species at 1 *M* concentration and all gaseous species at 1 atm pressure. Standard electrode potentials are measured relative to the standard hydrogen electrode, for which $E° = 0$ V. (Section 20.4)

standard emf, also called the **standard cell potential ($E°$)** The emf of a cell when all reagents are at standard conditions at 25°C. (Section 20.4)

standard free energy of formation, ΔG_f° The change in free energy associated with the formation of a substance from its elements under standard conditions. (Section 19.5)

standard heat of formation The change in enthalpy that accompanies the formation of 1 mol of a substance from its elements, with all substances in their standard states. (Section 5.6)

standard hydrogen electrode An electrode based on the half-reaction $2H^+(1M) + 2e^- \rightarrow H_2(1\text{ atm})$. The standard electrode potential of the standard hydrogen electrode is defined as 0 V. (Section 20.4)

standard oxidation potential ($E°_{ox}$) The potential of an oxidation half-reaction under standard conditions, measured relative to the standard hydrogen electrode. (Section 20.4)

standard reduction potential ($E°_{red}$) The potential of a reduction half-reaction under standard conditions, measured relative to the standard hydrogen electrode. Standard reduction potentials are also called **standard electrode potential**. (Section 20.4)

standard solution A solution of known concentration. (Section 4.7)

standard state Pure material in the physical state (gas, liquid, or solid) most stable at the temperature in question (usually 25°C) and under 1 atm pressure. (Section 5.6)

standard temperature and pressure (STP) Defined as 0°C and 1 atm pressure; frequently used as reference conditions for a gas. (Section 10.4)

starch The general name given to a group of polysaccharides that act as energy-storage substances in plants. (Section 27.4)

state function A property of a system that is determined by the state or condition of the system and not by how it got to that state; its value is fixed when temperature, pressure, composition, and physical form are specified; *P*, *V*, *T*, *E*, and *H* are state functions. (Section 5.2)

stereoisomers Compounds possessing the same formula and bonding arrangement but differing in the spatial arrangements of the atoms. (Section 25.4)

stoichiometry The relationships among the quantities of reactants and products involved in chemical reactions. (Chapter 3: Introduction)

stratosphere The region of the atmosphere directly above the troposphere. (Section 18.1)

strong acid An acid that ionizes completely in water. (Section 4.3)

strong base A base that ionizes completely in water. (Section 4.3)

strong electrolyte A substance that is completely ionized in solution, for example, strong acids, strong bases, and most salts. (Section 4.2)

structural formula A formula that shows not only the number and kind of atoms in the molecule but also the arrangement of the atoms. (Section 2.5)

structural isomers Compounds possessing the same formula but differing in the bonding arrangements of the atoms. (Sections 25.4, 26.1)

subcritical mass An amount of fissionable material smaller than the critical mass. (Section 21.7)

subshell One or more orbitals with the same set of quantum numbers n and l. For example, we speak of the $2p$ subshell ($n = 2$, $l = 1$), which is composed of three orbitals ($2p_x$, $2p_y$, and $2p_z$). (Section 6.5)

substitution reaction A reaction in which one atom or functional group bonded to a larger molecule is replaced by another. (Section 26.1)

substrate A substance that undergoes a reaction at the active site in an enzyme. (Section 27.3)

superconducting ceramic A complex metal oxide that undergoes a transition to a superconducting state at a low temperature. (Section 12.3)

superconducting transition temperature (T_c) The temperature below which a substance exhibits superconductivity. (Section 12.3)

superconductivity The "frictionless" flow of electrons that occurs when a substance loses all resistance to the flow of electrical current. (Section 12.3)

superconductor A substance that undergoes a transition at low temperature to a state in which it exhibits zero electrical resistance and in which it excludes all magnetic-field lines from its volume. (Section 12.3)

supercritical mass An amount of fissionable material larger than the critical mass. (Section 21.7)

supersaturated solutions Solutions containing more solute than a saturated solution. (Section 13.3)

surface tension The intermolecular, cohesive attraction that causes a liquid to minimize its surface area. (Section 11.3)

surroundings In thermodynamics, everything that lies outside the system that we study. (Section 5.1)

syngas A mixture of gases, mainly H_2 and CO, that results from heating coal in the presence of steam; the mixture can be used as a fuel or for synthesis of other substances. (Section 5.9)

system In thermodynamics, the portion of the universe that we single out for study: We must be careful to state exactly what the system contains and what transfers of energy it may have with its surroundings. (Section 5.1)

termolecular reaction An elementary reaction that involves three molecules. (Section 14.5)

tertiary structure The overall shape of a large protein, specifically, the manner in which sections of the protein fold back upon themselves or intertwine. (Section 27.2)

theoretical yield The quantity of product that is calculated to form when all of the limiting reagent reacts. (Section 3.6)

thermistor A ceramic material whose electric resistance decreases in a regular and reproducible manner with increasing temperature. (Section 12.3)

thermochemistry The relationship between chemical reactions and energy changes. (Chapter 5: Introduction)

thermodynamics The study of energy and its transformation. (Chapter 5: Introduction)

thermonuclear reaction Another name for fusion reactions, reactions in which two light nuclei are joined to form a more massive one. (Section 21.8)

thermoplastic A polymeric material that can be readily reshaped by application of heat and pressure. (Section 12.2)

thermosetting plastic A plastic that, once formed in a particular mold, is not readily reshaped by application of heat and pressure. (Section 12.2)

thermosphere The region of the atmosphere directly above the mesosphere. (Section 18.1)

thin film A film deposited on an underlying substrate to provide decoration or protection from chemical attack or to enhance a desirable property, such as reflectivity, electrical conductivity, color, or hardness. (Section 12.4)

third law of thermodynamics A law stating that the entropy of a pure, crystalline solid at absolute zero temperature is zero: $S\ (0\ K) = 0$. (Section 19.3)

titration The process of reacting a solution of unknown concentration with one of known concentration (a standard solution). (Section 4.7)

titration curve A graph of pH as a function of added titrant. (Section 17.3)

torr A unit of pressure (1 torr = 1 mm Hg). (Section 10.2)

transition metals Elements in which the d orbitals are partially occupied; also called **transition elements**. (Section 6.9)

transition state (activated complex) The particular arrangement of reactant and product molecules at the point of maximum energy in the rate-determining step of a reaction. (Section 14.4)

translational motion Movement in which an entire molecule moves in a definite direction. (Section 19.3)

transuranium elements Elements that follow uranium in the periodic table. (Section 21.3)

triglycerides Esters of long-chain carboxylic acids and the alcohol 1,2,3-trihydroxypropane (glycerol); they are *fats* and *oils*. (Section 27.5)

triple bond A covalent bond involving three electron pairs. (Section 8.4)

triple point The temperature at which solid, liquid, and gas phases coexist in equilibrium. (Section 11.6)

tritium The isotope of hydrogen whose nucleus contains a proton and two neutrons: ^3_1H. (Section 22.3)

troposphere The region of earth's atmosphere extending from the surface to about 12 km altitude. (Section 18.1)

Tyndall effect The scattering of a beam of visible light by the particles in a colloidal dispersion. (Section 13.6)

uncertainty principle A principle stating there is an inherent uncertainty in the precision with which we can simultaneously specify the position and momentum of a particle. This uncertainty is only significant for extremely small particles, such as electrons. (Section 6.4)

unimolecular reaction An elementary reaction that involves a single molecule. (Section 14.5)

unit cell The smallest portion of a crystal that reproduces the structure of the entire crystal when repeated in different directions in space. It is the repeating unit or "building block" of the crystal lattice. (Section 11.7)

unsaturated solutions Solutions containing less solute than a saturated solution. (Section 13.3)

vacuum deposition A method of forming thin films in which a substance is sublimed at high temperature without decomposition and then deposited on the object to be coated. (Section 12.4)

valence The capacity of an atom for entering into chemical combination with other atoms. Ionic valence is equal to the number of electrons gained or lost in forming the ionic species. Covalence is equal to the number of electrons from an atom that are involved in shared electron-pair bonds with other atoms. (Section 8.1)

valence bond theory A model of chemical bonding in which an electron-pair bond is formed between two atoms by the overlap of orbitals on the two atoms. (Section 9.3)

valence electrons The outermost electrons of an atom; these are the ones the atom uses in bonding. (Sections 6.9, 8.1)

valence orbitals Orbitals that contain the outer-shell electrons of an atom. (Chapter 7: Introduction)

valence-shell electron-pair repulsion (VSEPR) model A model that accounts for the geometric arrangements of shared and unshared electron pairs around a central atom in terms of the repulsions between electron pairs. (Section 9.1)

vapor Gaseous state of any substance that normally exists as a liquid or solid. (Section 10.1)

vapor pressure The pressure exerted by a vapor in equilibrium with its liquid or solid phase. (Section 11.5)

vibrational motion Movement of the atoms within a molecule in which they move periodically toward and away from one another. (Section 19.3)

viscosity A measure of the resistance of fluids to flow. (Section 11.3)

volatile Tending to evaporate readily. (Section 11.5)

voltaic (galvanic) cell A device in which a spontaneous oxidation-reduction reaction occurs with the passage of electrons through an external circuit. (Section 20.3)

vulcanization Reaction of sulfur with natural rubber to form crosslinks between the rubber polymer chains, thus hardening and stabilizing the material. (Section 12.2)

wave function A mathematical description of an allowed energy state (an *orbital*) for an electron in the quantum-mechanical model of the atom; it is usually symbolized by the Greek letter ψ. (Section 6.5)

wavelength The distance between identical points on successive waves. (Section 6.1)

weak acid An acid that only partly ionizes in water. (Section 4.3)

weak base A base that only partly ionizes in water. (Section 4.3)

weak electrolyte A substance that only partly ionizes in solution. (Section 4.2)

weight percentage The number of grams of solute in each 100 g of solution. (Section 13.1)

work The movement of an object against some force. (Section 5.1)

X-ray diffraction The scattering of X rays by the units of a regular crystalline solid. The scattering patterns obtained can be used to deduce the arrangements of particles in the crystal. (Section 11.7)

Index

Note: Pages followed by *f* locate figures; pages followed by *t* locate tables.

A

C

F

G

H

J

K

L

M

O

Q

R

S

T

U

V

W

X

Y

Z

COMMON IONS

Positive ions (Cations)	Negative ions (Anions)
1+	**1−**
Ammonium (NH_4^+)	Acetate ($C_2H_3O_2^-$)
Cesium (Cs^+)	Bromide (Br^-)
Copper (I) or cuprous (Cu^+)	Chlorate (ClO_3^-)
Hydrogen (H^+)	Chloride (Cl^-)
Lithium (Li^+)	Cyanide (CN^-)
Potassium (K^+)	Dihydrogen phosphate ($H_2PO_4^-$)
Silver (Ag^+)	Fluoride (F^-)
Sodium (Na^+)	Hydride (H^-)
	Hydrogen carbonate or bicarbonate (HCO_3^-)
	Hydrogen sulfite or bisulfite (HSO_3^-)
	Hydroxide (OH^-)
	Iodide (I^-)
	Nitrate (NO_3^-)
	Nitrite (NO_2^-)
	Perchlorate (ClO_4^-)
	Permanganate (MnO_4^-)
	Thiocyanate (SCN^-)
2+	**2−**
Barium (Ba^{2+})	Carbonate (CO_3^{2-})
Cadmium (Cd^{2+})	Chromate (CrO_4^{2-})
Calcium (Ca^{2+})	Dichromate ($Cr_2O_7^{2-}$)
Chromium(II) or chromous (Cr^{2+})	Hydrogen phosphate (HPO_4^{2-})
Cobalt(II) or cobaltous (Co^{2+})	Oxide (O^{2-})
Copper(II) or cupric (Cu^{2+})	Peroxide (O_2^{2-})
Iron(II) or ferrous (Fe^{2+})	Sulfate (SO_4^{2-})
Lead(II) or plumbous (Pb^{2+})	Sulfide (S^{2-})
Magnesium (Mg^{2+})	Sulfite (SO_3^{2-})
Manganese(II) or manganous (Mn^{2+})	
Mercury(I) or mercurous (Hg^{2+})	
Mercury(II) or mercuric (Hg^{2+})	
Strontium (Sr^{2+})	
Nickel (Ni^{2+})	
Tin(II) or stannous (Sn^{2+})	
Zinc (Zn^{2+})	
3+	**3−**
Aluminum (Al^{3+})	Arsenate (AsO_4^{3-})
Chromium (III) or chromic (Cr^{3+})	Phosphate (PO_4^{3-})
Iron (III) or ferric (Fe^{3+})	

PHYSICAL AND CHEMICAL CONSTANTS

Atomic mass unit	1 amu $= 1.6605402 \times 10^{-27}$ kg 6.0221367×10^{23} amu = 1 g
Avogadro's number	$N = 6.0221367 \times 10^{23}$/mol
Boltzmann's constant	$k = 1.38066 \times 10^{-23}$ J/K
Electron rest mass	$m_e = 5.485799 \times 10^{-4}$ amu $= 9.1093897 \times 10^{-28}$ g
Electronic charge	$e = 1.60217733 \times 10^{-19}$ C
Faraday's constant	$\mathfrak{F} = Ne = 9.6485309 \times 10^4$ C/mol
Gas constant	$R = Nk = 8.3145$ J/K-mol $= 0.082058$ L-atm/K-mol
Neutron rest mass	$m_n = 1.00866$ amu $= 1.67495 \times 10^{-24}$ g
Pi	$\pi = 3.1415926536$
Planck's constant	$h = 6.6260755 \times 10^{-34}$ J-s
Proton rest mass	$m_p = 1.0072765$ amu $= 1.672623 \times 10^{-24}$ g
Speed of light (in vacuum)	$c = 2.997925 \times 10^8$ m/s